# Study Guide and Solutions Manual

*to Accompany*

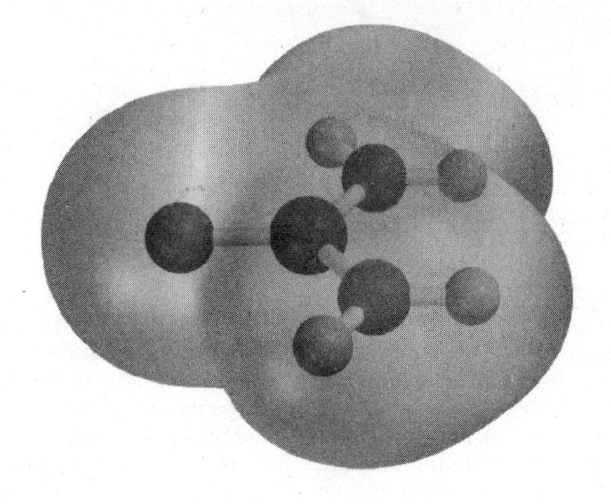

Fourth Edition

# ORGANIC CHEMISTRY

**Robert C. Atkins**
*Department of Chemistry*
*James Madison University*

**Francis A. Carey**
*Department of Chemistry*
*University of Virginia*

Boston Burr Ridge, IL Dubuque, IA Madison, WI New York San Francisco St. Louis
Bangkok Bogotá Caracas Lisbon London Madrid
Mexico City Milan New Delhi Seoul Singapore Sydney Taipei Toronto

*McGraw-Hill Higher Education*
*A Division of The* ***McGraw-Hill*** *Companies*

STUDY GUIDE AND SOLUTIONS MANUAL TO ACCOMPANY ORGANIC CHEMISTRY, FOURTH EDITION



This book is printed on acid-free paper.

1 2 3 4 5 6 7 8 9 0 QPD/QPD 0 9 8 7 6 5 4 3 2 1 0

ISBN 0–07–290510–7

Vice president and editorial director: *Kevin T. Kane*
Publisher: *James M. Smith*
Sponsoring editor: *Kent A. Peterson*
Developmental editor: *Shirley R. Oberbroeckling*
Editorial assistant: *Jennifer Bensink*
Senior marketing manager: *Martin J. Lange*
Senior marketing assistant: *Tami Petsche*
Senior project manager: *Peggy J. Selle*
Senior production supervisor: *Sandra Hahn*
Designer: *K. Wayne Harms*
Interior designer: *Kathy Theis*
Compositor: *Interactive Composition Corporation*
Typeface: *10/12 Times Roman*
Printer: *Quebecor Printing Book Group/Dubuque, IA*

**Library of Congress Cataloging-in-Publication Data**

Carey, Francis A.
Study guide and solutions manual to accompany Organic chemistry, fourth edition / Francis A. Carey, Robert C. Atkins.
p. cm.
Includes index.
ISBN 0–07–290510–7
1. Chemistry, Organic. I. Atkins, Robert C. (Robert Charles), 1944– II. Carey, Francis A.– Organic chemistry. III. Title.

QD251.2. C367 2000
547—dc21 99–047983
CIP

www.mhhe.com

# CONTENTS

Preface v
To the Student vii

CHAPTER 1 CHEMICAL BONDING 1

CHAPTER 2 ALKANES 25

CHAPTER 3 CONFORMATIONS OF ALKANES AND CYCLOALKANES 46

CHAPTER 4 ALCOHOLS AND ALKYL HALIDES 67

CHAPTER 5 STRUCTURE AND PREPARATION OF ALKENES: ELIMINATION REACTIONS 90

CHAPTER 6 REACTIONS OF ALKENES: ADDITION REACTIONS 124

CHAPTER 7 STEREOCHEMISTRY 156

CHAPTER 8 NUCLEOPHILIC SUBSTITUTION 184

CHAPTER 9 ALKYNES 209

CHAPTER 10 CONJUGATION IN ALKADIENES AND ALLYLIC SYSTEMS 230

CHAPTER 11 ARENES AND AROMATICITY 253

CHAPTER 12 REACTIONS OF ARENES: ELECTROPHILIC AROMATIC SUBSTITUTION 279

**CHAPTER 13** SPECTROSCOPY 320

**CHAPTER 14** ORGANOMETALLIC COMPOUNDS 342

**CHAPTER 15** ALCOHOLS, DIOLS, AND THIOLS 364

**CHAPTER 16** ETHERS, EPOXIDES, AND SULFIDES 401

**CHAPTER 17** ALDEHYDES AND KETONES: NUCLEOPHILIC ADDITION TO THE CARBONYL GROUP 426

**CHAPTER 18** ENOLS AND ENOLATES 470

**CHAPTER 19** CARBOXYLIC ACIDS 502

**CHAPTER 20** CARBOXYLIC ACID DERIVATIVES: NUCLEOPHILIC ACYL SUBSTITUTION 536

**CHAPTER 21** ESTER ENOLATES 576

**CHAPTER 22** AMINES 604

**CHAPTER 23** ARYL HALIDES 656

**CHAPTER 24** PHENOLS 676

**CHAPTER 25** CARBOHYDRATES 701

**CHAPTER 26** LIPIDS 731

**CHAPTER 27** AMINO ACIDS, PEPTIDES, AND PROTEINS. NUCLEIC ACIDS 752

**APPENDIX A** ANSWERS TO THE SELF-TESTS 775

**APPENDIX B** TABLES 821

B-1 Bond Dissociation Energies of Some Representative Compounds 821
B-2 Acid Dissociation Constants 822
B-3 Chemical Shifts of Representative Types of Protons 822
B-4 Chemical Shifts of Representative Carbons 823
B-5 Infrared Absorption Frequencies of Some Common Structural Units 823

# PREFACE

It is our hope that in writing this *Study Guide and Solutions Manual* we will make the study of organic chemistry more meaningful and worthwhile. To be effective, a study guide should be more than just an answer book. What we present here was designed with that larger goal in mind.

The *Study Guide and Solutions Manual* contains detailed solutions to all the problems in the text. Learning how to solve a problem is, in our view, more important than merely knowing the correct answer. To that end we have included solutions sufficiently detailed to provide the student with the steps leading to the solution of each problem.

In addition, the Self-Test at the conclusion of each chapter is designed to test the student's mastery of the material. Both fill-in and multiple-choice questions have been included to truly test the student's understanding. Answers to the self-test questions may be found in Appendix A at the back of the book.

The completion of this guide was made possible through the time and talents of numerous people. Our thanks and appreciation also go to the many users of the third edition who provided us with helpful suggestions, comments, and corrections. We also wish to acknowledge the assistance and understanding of Kent Peterson, Terry Stanton, and Peggy Selle of McGraw-Hill. Many thanks also go to Linda Davoli for her skillful copyediting. Last, we thank our wives and families for their understanding of the long hours invested in this work.

**Francis A. Carey**
**Robert C. Atkins**

# TO THE STUDENT

Before beginning the study of organic chemistry, a few words about "how to do it" are in order. You've probably heard that organic chemistry is difficult; there's no denying that. It need not be overwhelming, though, when approached with the right frame of mind and with sustained effort.

First of all you should realize that organic chemistry tends to "build" on itself. That is, once you have learned a reaction or concept, you will find it being used again and again later on. In this way it is quite different from general chemistry, which tends to be much more compartmentalized. In organic chemistry you will continually find previously learned material cropping up and being used to explain and to help you understand new topics. Often, for example, you will see the preparation of one class of compounds using reactions of other classes of compounds studied earlier in the year.

How to keep track of everything? It might be possible to memorize every bit of information presented to you, but you would still lack a fundamental understanding of the subject. It is far better to *generalize* as much as possible.

You will find that the early chapters of the text will emphasize concepts of *reaction theory*. These will be used, as the various classes of organic molecules are presented, to describe *mechanisms* of organic reactions. A relatively few fundamental mechanisms suffice to describe almost every reaction you will encounter. Once learned and understood, these mechanisms provide a valuable means of categorizing the reactions of organic molecules.

There will be numerous facts to learn in the course of the year, however. For example, chemical reagents necessary to carry out specific reactions must be learned. You might find a study aid known as *flash cards* helpful. These take many forms, but one idea is to use $3 \times 5$ index cards. As an example of how the cards might be used, consider the reduction of alkenes (compounds with carbon–carbon double bonds) to alkanes (compounds containing only carbon–carbon single bonds). The front of the card might look like this:

$$\text{Alkenes} \xrightarrow{?} \text{alkanes}$$

The reverse of the card would show the reagents necessary for this reaction:

$H_2$, Pt or Pd catalyst

The card can actually be studied in two ways. You may ask yourself: What reagents will convert alkenes into alkanes? Or, using the back of the card: What chemical reaction is carried out with hydrogen and a platinum or palladium catalyst? This is by no means the only way to use the cards—be creative! Just making up the cards will help you to study.

Although study aids such as flash cards will prove helpful, there is only one way to truly master the subject matter in organic chemistry—*do the problems!* The more you work, the more you will learn. Almost certainly the grade you receive will be a reflection of your ability to solve problems.

Don't just think over the problems, either; write them out as if you were handing them in to be graded. Also, be careful of how you use the *Study Guide.* The solutions contained in this book have been intended to provide explanations to help you understand the problem. Be sure to write out *your* solution to the problem first and only then look it up to see if you have done it correctly.

Students frequently feel that they understand the material but don't do as well as expected on tests. One way to overcome this is to "test" yourself. Each chapter in the *Study Guide* has a self-test at the end. Work the problems in these tests *without* looking up how to solve them in the text. You'll find it is much harder this way, but it is also a closer approximation to what will be expected of you when taking a test in class.

Success in organic chemistry depends on skills in analytical reasoning. Many of the problems you will be asked to solve require you to proceed through a series of logical steps to the correct answer. Most of the individual concepts of organic chemistry are fairly simple; stringing them together in a coherent fashion is where the challenge lies. By doing exercises conscientiously you should see a significant increase in your overall reasoning ability. Enhancement of their analytical powers is just one fringe benefit enjoyed by those students who attack the course rather than simply attend it.

Gaining a mastery of organic chemistry is hard work. We hope that the hints and suggestions outlined here will be helpful to you and that you will find your efforts rewarded with a knowledge and understanding of an important area of science.

**Francis A. Carey**
**Robert C. Atkins**

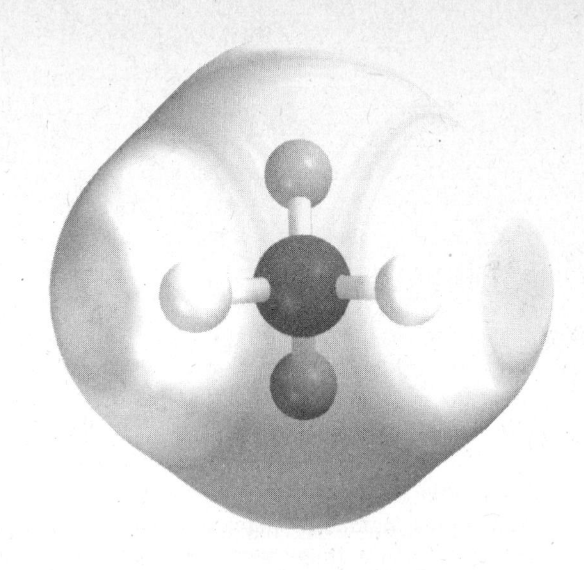

# CHAPTER 1
## CHEMICAL BONDING

### SOLUTIONS TO TEXT PROBLEMS

**1.1** The element carbon has atomic number 6, and so it has a total of six electrons. Two of these electrons are in the 1*s* level. The four electrons in the 2*s* and 2*p* levels (the valence shell) are the valence electrons. Carbon has four valence electrons.

**1.2** Electron configurations of elements are derived by applying the following principles:

(*a*) The number of electrons in a neutral atom is equal to its atomic number Z.
(*b*) The maximum number of electrons in any orbital is 2.
(*c*) Electrons are added to orbitals in order of increasing energy, filling the 1*s* orbital before any electrons occupy the 2*s* level. The 2*s* orbital is filled before any of the 2*p* orbitals, and the 3*s* orbital is filled before any of the 3*p* orbitals.
(*d*) All the 2*p* orbitals ($2p_x$, $2p_y$, $2p_z$) are of equal energy, and each is singly occupied before any is doubly occupied. The same holds for the 3*p* orbitals.

With this as background, the electron configuration of the third-row elements is derived as follows [$2p^6 = 2p_x^{\,2}2p_y^{\,2}2p_z^{\,2}$]:

| | |
|---|---|
| Na (Z = 11) | $1s^22s^22p^63s^1$ |
| Mg (Z = 12) | $1s^22s^22p^63s^2$ |
| Al (Z = 13) | $1s^22s^22p^63s^23p_x^{\,1}$ |
| Si (Z = 14) | $1s^22s^22p^63s^23p_x^{\,1}3p_y^{\,1}$ |
| P (Z = 15) | $1s^22s^22p^63s^23p_x^{\,1}3p_y^{\,1}3p_z^{\,1}$ |
| S (Z = 16) | $1s^22s^22p^63s^23p_x^{\,2}3p_y^{\,1}3p_z^{\,1}$ |
| Cl (Z = 17) | $1s^22s^22p^63s^23p_x^{\,2}3p_y^{\,2}3p_z^{\,1}$ |
| Ar (Z = 18) | $1s^22s^22p^63s^23p_x^{\,2}3p_y^{\,2}3p_z^{\,2}$ |

**1.3** The electron configurations of the designated ions are:

| Ion | Z | Number of Electrons in Ion | Electron Configuration of Ion |
|---|---|---|---|
| (*b*) $He^+$ | 2 | 1 | $1s^1$ |
| (*c*) $H^-$ | 1 | 2 | $1s^2$ |
| (*d*) $O^-$ | 8 | 9 | $1s^22s^22p_x^22p_y^22p_z^1$ |
| (*e*) $F^-$ | 9 | 10 | $1s^22s^22p^6$ |
| (*f*) $Ca^{2+}$ | 20 | 18 | $1s^22s^22p^63s^23p^6$ |

Those with a noble gas configuration are $H^-$, $F^-$, and $Ca^{2+}$.

**1.4** A positively charged ion is formed when an electron is removed from a neutral atom. The equation representing the ionization of carbon and the electron configurations of the neutral atom and the ion is:

$$\underset{1s^22s^22p_x^12p_y^1}{C} \longrightarrow \underset{1s^22s^22p_x^1}{C^+} + e^-$$

A negatively charged carbon is formed when an electron is added to a carbon atom. The additional electron enters the $2p_z$ orbital.

$$\underset{1s^22s^22p_x^12p_y^1}{C} + e^- \longrightarrow \underset{1s^22s^22p_x^1p_y^12p_z^1}{C^-}$$

Neither $C^+$ nor $C^-$ has a noble gas electron configuration.

**1.5** Hydrogen has one valence electron, and fluorine has seven. The covalent bond in hydrogen fluoride arises by sharing the single electron of hydrogen with the unpaired electron of fluorine.

Combine H· and ·F̤̈: to give the Lewis structure for hydrogen fluoride H:F̤̈:

**1.6** We are told that $C_2H_6$ has a carbon–carbon bond.

Thus, we combine two ·Ċ· and six H· to write the Lewis structure of ethane

```
 H H
H:C:C:H
 H H
```

There are a total of 14 valence electrons distributed as shown. Each carbon is surrounded by eight electrons.

**1.7** (*b*) Each carbon contributes four valence electrons, and each fluorine contributes seven. Thus, $C_2F_4$ has 36 valence electrons. The octet rule is satisfied for carbon only if the two carbons are attached by a double bond and there are two fluorines on each carbon. The pattern of connections shown (below left) accounts for 12 electrons. The remaining 24 electrons are divided equally (six each) among the four fluorines. The complete Lewis structure is shown at right below.

```
F       F        :F̤̈       F̤̈:
 \     /           \     /
  C=C               C=C
 /     \           /     \
F       F        :F̤̈       F̤̈:
```

(*c*) Since the problem states that the atoms in $C_3H_3N$ are connected in the order CCCN and all hydrogens are bonded to carbon, the order of attachments can only be as shown (below left) so as to have four bonds to each carbon. Three carbons contribute 12 valence electrons, three hydrogens contribute 3, and nitrogen contributes 5, for a total of 20 valence electrons. The nine bonds indicated in the partial structure account for 18 electrons. Since the octet rule is satisfied for carbon, add the remaining two electrons as an unshared pair on nitrogen (below right).

$$\mathrm{H_2C{=}CH{-}C{\equiv}N} \qquad \mathrm{H_2C{=}CH{-}C{\equiv}N:}$$

**1.8** The degree of positive or negative character at carbon depends on the difference in electronegativity between the carbon and the atoms to which it is attached. From Table 1.2, we find the electronegativity values for the atoms contained in the molecules given in the problem are:

| | |
|---|---|
| Li | 1.0 |
| H | 2.1 |
| **C** | **2.5** |
| Cl | 3.0 |

Thus, carbon is more electronegative than hydrogen and lithium, but less electronegative than chlorine. When bonded to carbon, hydrogen and lithium bear a partial positive charge, and carbon bears a partial negative charge. Conversely, when chlorine is bonded to carbon, it bears a partial negative charge, and carbon becomes partially positive. In this group of compounds, lithium is the least electronegative element, chlorine the most electronegative.

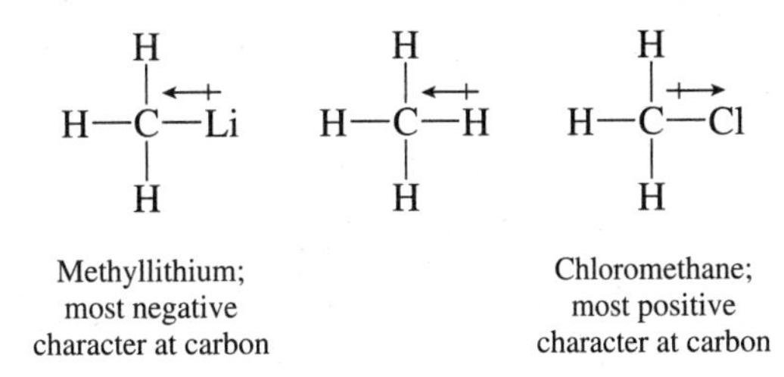

Methyllithium; most negative character at carbon

Chloromethane; most positive character at carbon

**1.9** (*b*) The formal charges in sulfuric acid are calculated as follows:

| | Valence Electrons in Neutral Atom | Electron Count | Formal Charge |
|---|---|---|---|
| Hydrogen: | 1 | $\frac{1}{2}(2) = 1$ | 0 |
| Oxygen (of OH): | 6 | $\frac{1}{2}(4) + 4 = 6$ | 0 |
| Oxygen: | 6 | $\frac{1}{2}(2) + 6 = 7$ | −1 |
| Sulfur: | 6 | $\frac{1}{2}(8) + 0 = 4$ | +2 |

$$\mathrm{H{-}\ddot{O}{-}S^{2+}(\ddot{O}:^{-})_2{-}\ddot{O}{-}H}$$

(*c*) The formal charges in nitrous acid are calculated as follows:

| | Valence Electrons in Neutral Atom | Electron Count | Formal Charge |
|---|---|---|---|
| Hydrogen: | 1 | $\frac{1}{2}(2) = 1$ | 0 |
| Oxygen (of OH): | 6 | $\frac{1}{2}(4) + 4 = 6$ | 0 |
| Oxygen: | 6 | $\frac{1}{2}(4) + 4 = 6$ | 0 |
| Nitrogen: | 5 | $\frac{1}{2}(6) + 2 = 5$ | 0 |

$$\mathrm{H{-}\ddot{O}{-}\ddot{N}{=}\ddot{O}:}$$

**1.10** The electron counts of nitrogen in ammonium ion and boron in borohydride ion are both 4 (one half of 8 electrons in covalent bonds).

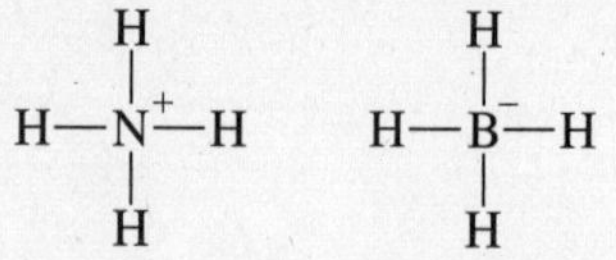

Ammonium ion    Borohydride ion

Since a neutral nitrogen has 5 electrons in its valence shell, an electron count of 4 gives it a formal charge of +1. A neutral boron has 3 valence electrons, and so an electron count of 4 in borohydride ion corresponds to a formal charge of −1.

**1.11** As shown in the text in Table 1.2, nitrogen is more electronegative than hydrogen and will draw the electrons in N—H bonds toward itself. Nitrogen with a formal charge of +1 is even more electronegative than a neutral nitrogen.

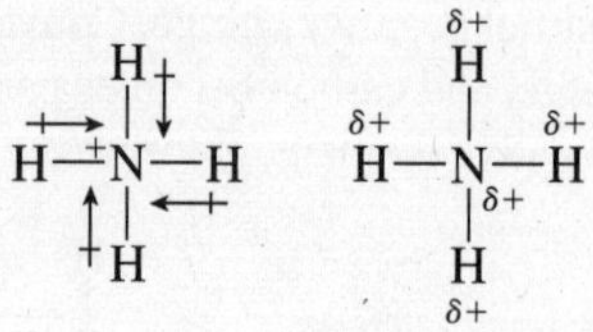

Boron (electronegativity = 2.0) is, on the other hand, slightly less electronegative than hydrogen (electronegativity = 2.1). Boron with a formal charge of −1 is less electronegative than a neutral boron. The electron density in the B—H bonds of $BH_4^-$ is therefore drawn toward hydrogen and away from boron.

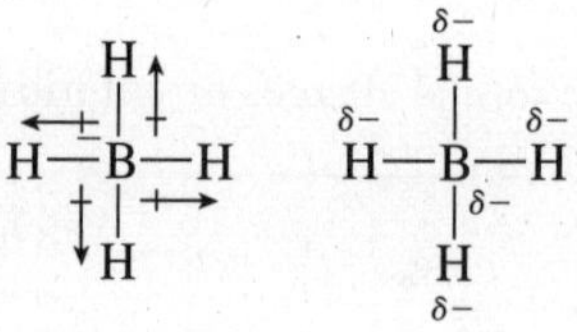

**1.12** (*b*) The compound $(CH_3)_3CH$ has a central carbon to which are attached three $CH_3$ groups and a hydrogen.

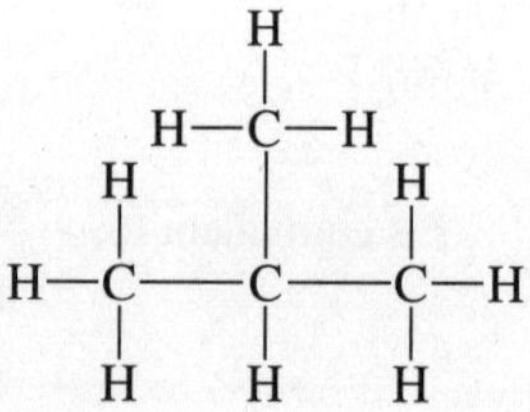

Four carbons and 10 hydrogens contribute 26 valence electrons. The structure shown has 13 covalent bonds, and so all the valence electrons are accounted for. The molecule has no unshared electron pairs.

(*c*) The number of valence electrons in $ClCH_2CH_2Cl$ is 26 (2Cl = 14; 4H = 4; 2C = 8). The constitution at the left below shows seven covalent bonds accounting for 14 electrons. The remaining 12 electrons are divided equally between the two chlorines as unshared electron pairs. The octet rule is satisfied for both carbon and chlorine in the structure at the right below.

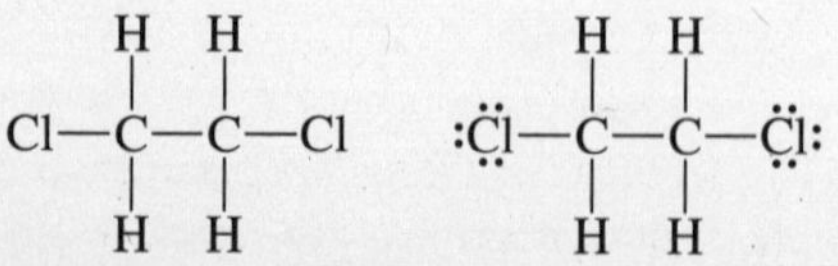

(*d*) This compound has the same molecular formula as the compound in part (*c*), but a different structure. It, too, has 26 valence electrons, and again only chlorine has unshared pairs.

```
     H   H
     |   |
  H—C—C—C̈l:
     |   |
     H  :C̤l:
```

(*e*) The constitution of $CH_3NHCH_2CH_3$ is shown (below left). There are 26 valence electrons, and 24 of them are accounted for by the covalent bonds in the structural formula. The remaining two electrons complete the octet of nitrogen as an unshared pair (below right).

```
   H       H   H              H       H   H
   |       |   |              |       |   |
H—C—N—C—C—H          H—C—N̈—C—C—H
   |   |   |   |              |   |   |   |
   H   H   H   H              H   H   H   H
```

(*f*) Oxygen has two unshared pairs in $(CH_3)_2CHCH{=}O$.

```
              H
              |
           H—C—H
     H        |
     |        |
  H—C——C——C=Ö:
     |     |     |
     H     H     H
```

**1.13** (*b*) This compound has a four-carbon chain to which are appended two other carbons.

```
                          CH3  H
                           |   |
    is equivalent to  CH3—C——C—CH3    which may be      (CH3)2CHCH(CH3)2
                           |   |       rewritten as
                           H  CH3
```

(*c*) The carbon skeleton is the same as that of the compound in part (*b*), but one of the terminal carbons bears an OH group in place of one of its hydrogens.

```
HO                        H
                          |
                      HO—C—H
                          |    H           which may be         CH2OH
    is equivalent to      |    |           rewritten as           |
                    CH3—C——C—CH3                          CH3CHCH(CH3)2
                          |    |
                          H   CH3
```

(*d*) The compound is a six-membered ring that bears a $—C(CH_3)_3$ substituent.

```
                          H   H
                          |   |  H
                        H—C—C/
                       H\ /    \ /H
    is equivalent to     C       C    CH3       which may be      —C(CH3)3
                       H/ \     / \ /           rewritten as
                           C—C     C—CH3
                          /|   |\   |
                         H H   H H  CH3
```

**1.14** The problem specifies that nitrogen and both oxygens of carbamic acid are bonded to carbon and one of the carbon–oxygen bonds is a double bond. Since a neutral carbon is associated with four bonds, a neutral nitrogen three (plus one unshared electron pair), and a neutral oxygen two (plus two unshared electron pairs), this gives the Lewis structure shown.

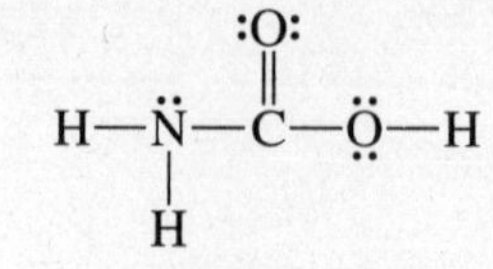

:O:
‖
H—N̈—C—Ö—H
|
H

Carbamic acid

**1.15** (*b*) There are three constitutional isomers of $C_3H_8O$:

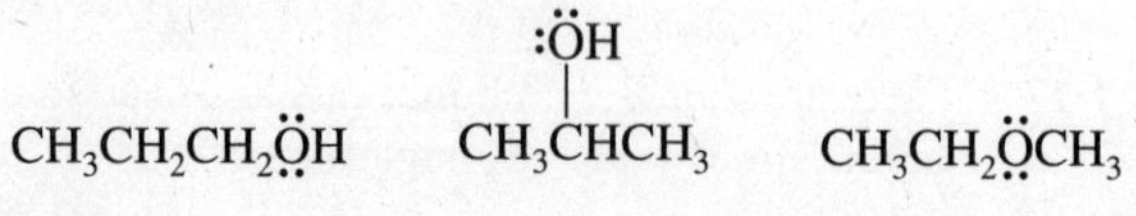

$CH_3CH_2CH_2\ddot{O}H$ $\quad CH_3CH(:\ddot{O}H)CH_3$ $\quad CH_3CH_2\ddot{O}CH_3$

(*c*) Four isomers of $C_4H_{10}O$ have —OH groups:

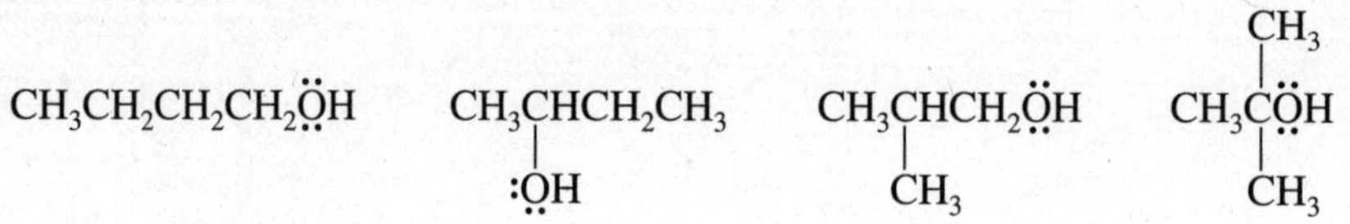

$CH_3CH_2CH_2CH_2\ddot{O}H$ $\quad CH_3CH(:\ddot{O}H)CH_2CH_3$ $\quad CH_3CH(CH_3)CH_2\ddot{O}H$ $\quad CH_3C(CH_3)_2\ddot{O}H$

Three isomers have C—O—C units:

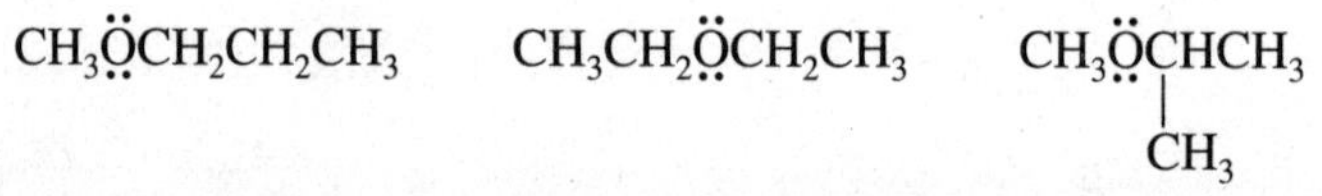

$CH_3\ddot{O}CH_2CH_2CH_3$ $\quad CH_3CH_2\ddot{O}CH_2CH_3$ $\quad CH_3\ddot{O}CH(CH_3)CH_3$

**1.16** (*b*) Move electrons from the negatively charged oxygen, as shown by the curved arrows.

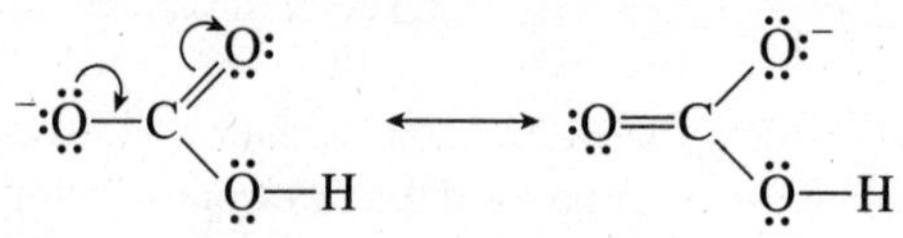

$^{-}$:Ö—C(=O:)—Ö—H ⟷ :O=C(—Ö:$^{-}$)—Ö—H Equivalent to original structure

The resonance interaction shown for bicarbonate ion is more important than an alternative one involving delocalization of lone-pair electrons in the OH group.

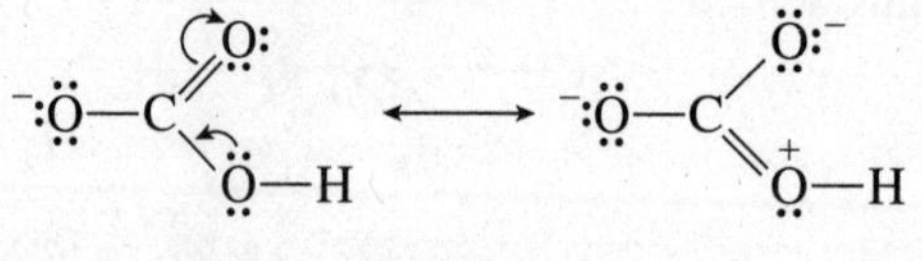

$^{-}$:Ö—C(=O:)—Ö—H ⟷ $^{-}$:Ö—C(—Ö:$^{-}$)=O$^{+}$—H Not equivalent to original structure; not as stable because of charge separation

(*c*) All three oxygens are equivalent in carbonate ion. Either negatively charged oxygen can serve as the donor atom.

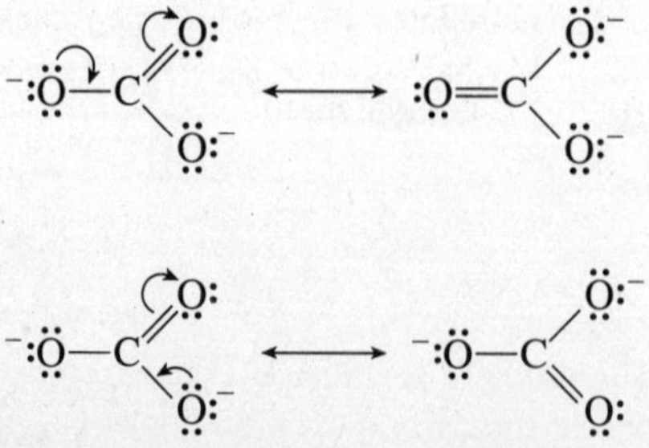

$^{-}$:Ö—C(=O:)—Ö:$^{-}$ ⟷ :O=C(—Ö:$^{-}$)—Ö:$^{-}$

$^{-}$:Ö—C(=O:)—Ö:$^{-}$ ⟷ $^{-}$:Ö—C(—Ö:$^{-}$)=O:

(*d*) Resonance in borate ion is exactly analogous to that in carbonate.

$^-:\ddot{O}-B(-\ddot{O}:^-)(=\ddot{O}:) \longleftrightarrow :\ddot{O}=\bar{B}(-\ddot{O}:^-)(-\ddot{O}:^-)$

and

$^-:\ddot{O}-B(-\ddot{O}:^-)(=\ddot{O}:) \longleftrightarrow {}^-:\ddot{O}-\bar{B}(=\ddot{O}:)(-\ddot{O}:^-)$

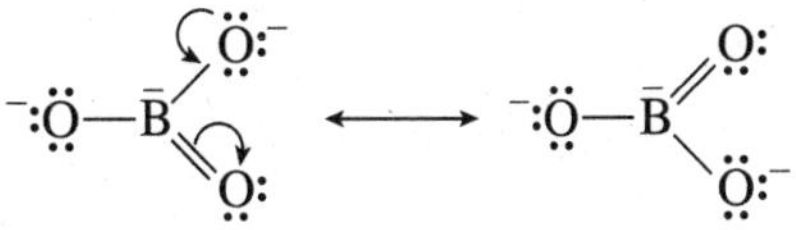

**1.17** There are four B—H bonds in $BH_4^-$. The four electron pairs surround boron in a tetrahedral orientation. The H—B—H angles are 109.5°.

**1.18** (*b*) Nitrogen in ammonium ion is surrounded by 8 electrons in four covalent bonds. These four bonds are directed toward the corners of a tetrahedron.

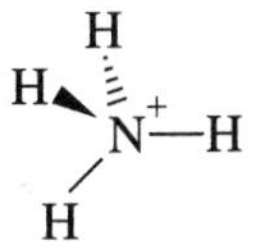

$H_4\overset{+}{N}$ (tetrahedral) Each HNH angle is 109.5°.

(*c*) Double bonds are treated as a single unit when deducing the shape of a molecule using the VSEPR model. Thus azide ion is linear.

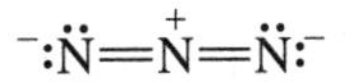

$^-:\ddot{N}=\overset{+}{N}=\ddot{N}:^-$ The NNN angle is 180°.

(*d*) Since the double bond in carbonate ion is treated as if it were a single unit, the three sets of electrons are arranged in a trigonal planar arrangement around carbon.

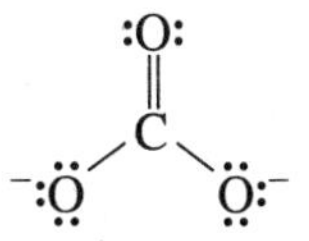

$:O:$ $\|$ $C$ $^-:\ddot{O}$ $\ddot{O}:^-$ The OCO angle is 120°.

**1.19** (*b*) Water is a bent molecule, and so the individual O—H bond dipole moments do not cancel. Water has a dipole moment.

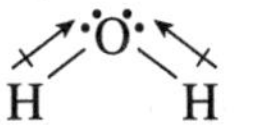

H H H H

Individual OH bond moments in water

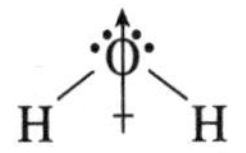

Direction of net dipole moment

(*c*) Methane, $CH_4$, is perfectly tetrahedral, and so the individual (small) C—H bond dipole moments cancel. Methane has no dipole moment.

(*d*) Methyl chloride has a dipole moment.

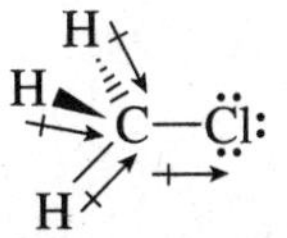

H H C—Cl: H H H C—Cl: H

Directions of bond dipole moments in $CH_3Cl$

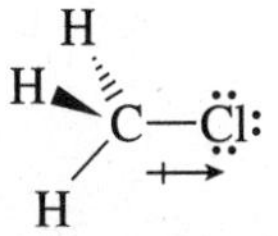

Direction of molecular dipole moment

(*e*) Oxygen is more electronegative than carbon and attracts electrons from it. Formaldehyde has a dipole moment.

Direction of bond dipole moments in formaldehyde

Direction of molecular dipole moment

(*f*) Nitrogen is more electronegative than carbon. Hydrogen cyanide has a dipole moment.

H—C≡N:

Direction of bond dipole moments in HCN

H—C≡N:

Direction of molecular dipole moment

**1.20** The orbital diagram for $sp^3$-hybridized nitrogen is the same as for $sp^3$-hybridized carbon, except nitrogen has one more electron.

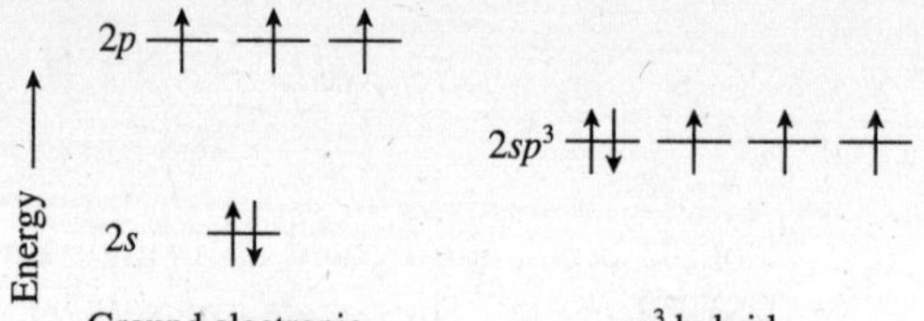

Energy

2*p* ↑ ↑ ↑

2*s* ↑↓

Ground electronic state of nitrogen

(*a*)

$2sp^3$ ↑↓ ↑ ↑ ↑

$sp^3$ hybrid state of nitrogen

(*b*)

The unshared electron pair in ammonia ($:NH_3$) occupies an $sp^3$-hybridized orbital of nitrogen. Each N—H bond corresponds to overlap of a half-filled $sp^3$ hybrid orbital of nitrogen and a 1*s* orbital of hydrogen.

**1.21** Silicon lies below carbon in the periodic table, and it is reasonable to assume that both carbon and silicon are $sp^3$-hybridized in $H_3CSiH_3$. The C—Si bond and all of the C—H and Si—H bonds are $\sigma$ bonds.

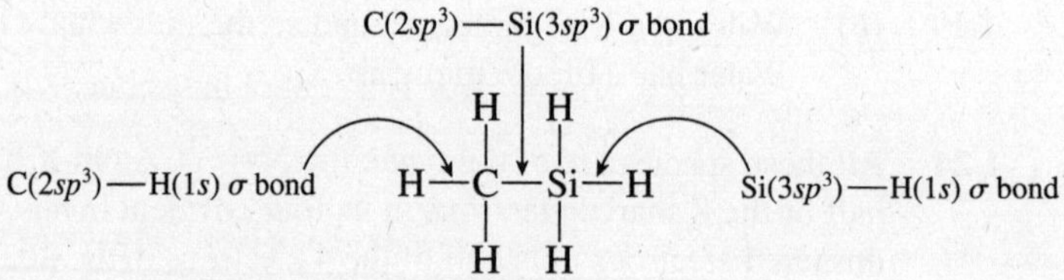

$C(2sp^3)$—$Si(3sp^3)$ $\sigma$ bond

$C(2sp^3)$—$H(1s)$ $\sigma$ bond

$H_3C$—$SiH_3$

$Si(3sp^3)$—$H(1s)$ $\sigma$ bond

The principal quantum number of the carbon orbitals that are hybridized is 2; the principal quantum number for the silicon orbitals is 3.

**1.22** (*b*) Carbon in formaldehyde ($H_2C{=}O$) is directly bonded to three other atoms (two hydrogens and one oxygen). It is $sp^2$-hybridized.

(*c*) Ketene has two carbons in different hybridization states. One is $sp^2$-hybridized; the other is *sp*-hybridized.

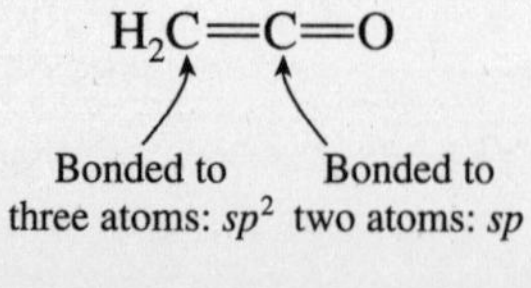

$H_2C{=}C{=}O$

Bonded to three atoms: $sp^2$

Bonded to two atoms: *sp*

(*d*) One of the carbons in propene is $sp^3$-hybridized. The carbons of the double bond are $sp^2$-hybridized.

$$\overset{sp^3}{H_3C}-\overset{sp^2}{CH}=\overset{sp^2}{CH_2}$$

(*e*) The carbons of the $CH_3$ groups in acetone [$(CH_3)_2C{=}O$] are $sp^3$-hybridized. The C=O carbon is $sp^2$-hybridized.

(*f*) The carbons in acrylonitrile are hybridized as shown:

$$\overset{sp^2}{H_2C}=\overset{sp^2}{CH}-\overset{sp}{C}\equiv N$$

**1.23** All these species are characterized by the formula :X≡Y:, and each atom has an electron count of 5.

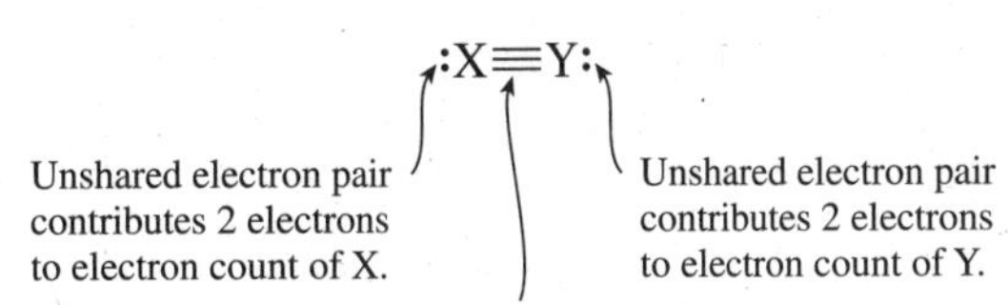
:X≡Y:
Unshared electron pair contributes 2 electrons to electron count of X.
Unshared electron pair contributes 2 electrons to electron count of Y.
Triple bond contributes half of its 6 electrons, or 3 electrons each, to separate electron counts of X and Y.

Electron count X = electron count Y = 2 + 3 = 5

(*a*) :N≡N: A neutral nitrogen atom has 5 valence electrons: therefore, each atom is electrically neutral in molecular nitrogen.

(*b*) :C≡N: Nitrogen, as before, is electrically neutral. A neutral carbon has 4 valence electrons, and so carbon in this species, with an electron count of 5, has a unit negative charge. The species is cyanide anion; its net charge is −1.

(*c*) :C≡C: There are two negatively charged carbon atoms in this species. It is a dianion; its net charge is −2.

(*d*) :N≡O: Here again is a species with a neutral nitrogen atom. Oxygen, with an electron count of 5, has 1 less electron in its valence shell than a neutral oxygen atom. Oxygen has a formal charge of +1; the net charge is +1.

(*e*) :C≡O: Carbon has a formal charge of −1; oxygen has a formal charge of +1. Carbon monoxide is a neutral molecule.

**1.24** All these species are of the type :Ÿ=X=Ÿ:. Atom X has an electron count of 4, corresponding to half of the 8 shared electrons in its four covalent bonds. Each atom Y has an electron count of 6; 4 unshared electrons plus half of the 4 electrons in the double bond of each Y to X.

(*a*) :Ö=C=Ö: Oxygen, with an electron count of 6, and carbon, with an electron count of 4, both correspond to the respective neutral atoms in the number of electrons they "own." Carbon dioxide is a neutral molecule, and neither carbon nor oxygen has a formal charge in this Lewis structure.

(*b*) :N̈=N=N̈: The two terminal nitrogens each have an electron count (6) that is one more than a neutral atom and thus each has a formal charge of −1. The central N has an electron count (4) that is one less than a neutral nitrogen; it has a formal charge of +1. The net charge on the species is (−1 + 1 − 1), or −1.

(*c*) :Ö=N=Ö: As in part (*b*), the central nitrogen has a formal charge of +1. As in part (*a*), each oxygen is electrically neutral. The net charge is +1.

**1.25** (*a, b*) The problem specifies that ionic bonding is present and that the anion is tetrahedral. The cations are the group I metals $Na^+$ and $Li^+$. Both boron and aluminum are group III elements, and thus have a formal charge of $-1$ in the tetrahedral anions $BF_4^-$ and $AlH_4^-$ respectively.

$Na^+$ F—B̄(F)(F)(F) $Li^+$ H—Āl(H)(H)(H)

Sodium tetrafluoroborate Lithium aluminum hydride

(*c*, *d*) Both of the tetrahedral anions have 32 valence electrons. Sulfur contributes 6 valence electrons and phosphorus 5 to the anions. Each oxygen contributes 6 electrons. The double negative charge in sulfate contributes 2 more, and the triple negative charge in phosphate contributes 3 more.

$2K^+$ $^-$:Ö—$S^{2+}$(Ö:$^-$)$_3$ $3Na^+$ $^-$:Ö—$P^+$(Ö:$^-$)$_3$

Potassium sulfate Sodium phosphate

The formal charge on each oxygen in both ions is $-1$. The formal charge on sulfur in sulfate is $+2$; the charge on phosphorus is $+1$. The net charge of sulfate ion is $-2$; the net charge of phosphate ion is $-3$.

**1.26** (*a*) Each hydrogen has a formal charge of 0, as is always the case when hydrogen is covalently bonded to one substituent. Oxygen has an electron count of 5.

H—Ö—H (with H below O) Electron count of oxygen $= 2 + \frac{1}{2}(6) = 5$

Unshared pair Covalently bonded electrons

A neutral oxygen atom has 6 valence electrons; therefore, oxygen in this species has a formal charge of $+1$. The species as a whole has a unit positive charge. It is the hydronium ion, $H_3O^+$.

(*b*) The electron count of carbon is 5; there are 2 electrons in an unshared pair, and 3 electrons are counted as carbon's share of the three covalent bonds to hydrogen.

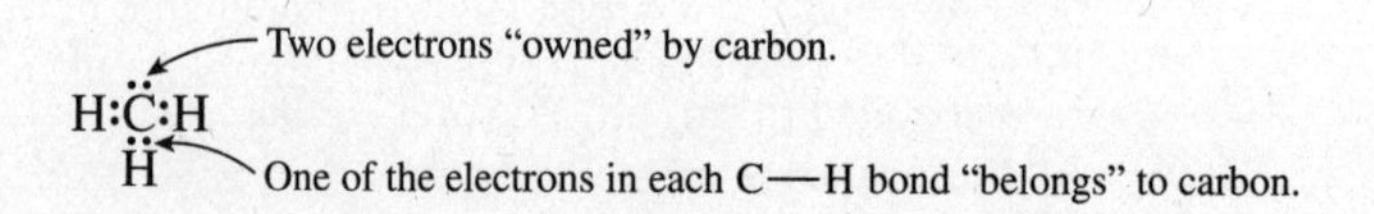


An electron count of 5 is one more than that for a neutral carbon atom. The formal charge on carbon is $-1$, as is the net charge on this species.

(*c*) This species has 1 less electron than that of part (*b*). None of the atoms bears a formal charge. The species is neutral.

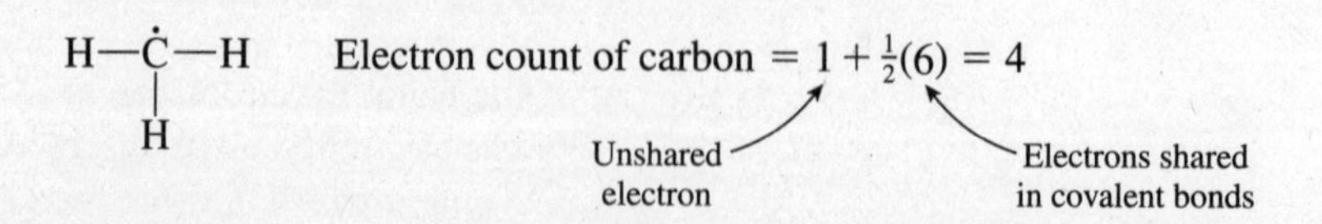


(*d*) The formal charge of carbon in this species is $+1$. Its only electrons are those in its three covalent bonds to hydrogen, and so its electron count is 3. This corresponds to 1 less electron than in a neutral carbon atom, giving it a unit positive charge.

(*e*) In this species the electron count of carbon is 4, or, exactly as in part (*c*), that of a neutral carbon atom. Its formal charge is 0, and the species is neutral.

H—C̈—H

Two unshared electrons contribute 2 to the electron count of carbon.

Half of the 4 electrons in the two covalent bonds contribute 2 to the electron count of carbon.

**1.27** Oxygen is surrounded by a complete octet of electrons in each structure but has a different "electron count" in each one because the proportion of shared to unshared pairs is different.

(*a*) $CH_3\ddot{O}$:  (*b*) $CH_3\ddot{O}CH_3$  (*c*) $CH_3\ddot{O}CH_3$ with $CH_3$ on O

| (*a*) | (*b*) | (*c*) |
|---|---|---|
| Electron count $= 6 + \frac{1}{2}(2) = 7$; formal charge $= -1$ | Electron count $= 4 + \frac{1}{2}(4) = 6$; formal charge $= 0$ | Electron count $= 2 + \frac{1}{2}(6) = 5$; formal charge $= +1$ |

**1.28** (*a*) Each carbon has 4 valence electrons, each hydrogen 1, and chlorine has 7. Hydrogen and chlorine each can form only one bond, and so the only stable structure must have a carbon–carbon bond. Of the 20 valence electrons, 14 are present in the seven covalent bonds and 6 reside in the three unshared electron pairs of chlorine.

H H
H:C:C:C̈l:
H H

or

H H
| |
H—C—C—C̈l:
| |
H H

(*b*) As in part (*a*) the single chlorine as well as all of the hydrogens must be connected to carbon. There are 18 valence electrons in $C_2H_3Cl$, and the framework of five single bonds accounts for only 10 electrons. Six of the remaining 8 are used to complete the octet of chlorine as three unshared pairs, and the last 2 are used to form a carbon–carbon double bond.

H H
H:C::C:C̈l:

or

H H
C=C
H :C̈l:

(*c*) All of the atoms except carbon (H, Br, Cl, and F) are monovalent; therefore, they can only be bonded to carbon. The problem states that all three fluorines are bonded to the same carbon, and so one of the carbons is present as a $CF_3$ group. The other carbon must be present as a CHBrCl group. Connect these groups together to give the structure of halothane.

:F̈ :H
:F̈:C :C:C̈l:
:F̈ :B̈r:

or

F H
| |
F—C—C—Cl
| |
F Br

(Unshared electron pairs omitted for clarity)

(*d*) As in part (*c*) all of the atoms except carbon are monovalent. Since each carbon bears one chlorine, two $ClCF_2$ groups must be bonded together.

:F̈::F̈:
:C̈l:C :C:C̈l:
:F̈::F̈:

or

F F
| |
Cl—C—C—Cl
| |
F F

(Unshared electron pairs omitted for clarity)

**1.29** Place hydrogens on the given atoms so that carbon has four bonds, nitrogen three, and oxygen two. Place unshared electron pairs on nitrogen and oxygen so that nitrogen has an electron count of 5 and oxygen has an electron count of 6. These electron counts satisfy the octet rule when nitrogen has three bonds and oxygen two.

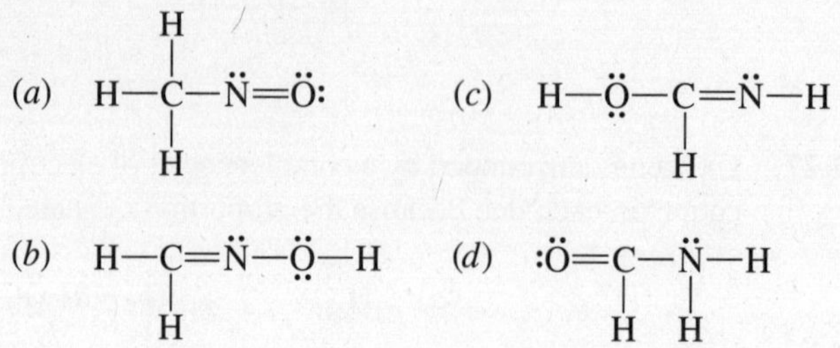

(*a*) H—C(H)(H)—N̈=Ö:  (*c*) H—Ö—C(H)=N̈—H

(*b*) H—C(H)=N̈—Ö—H  (*d*) :Ö=C(H)—N̈(H)—H

**1.30** (*a*) Species A, B, and C have the same molecular formula, the same atomic positions, and the same number of electrons. They differ only in the arrangement of their electrons. They are therefore resonance forms of a single compound.

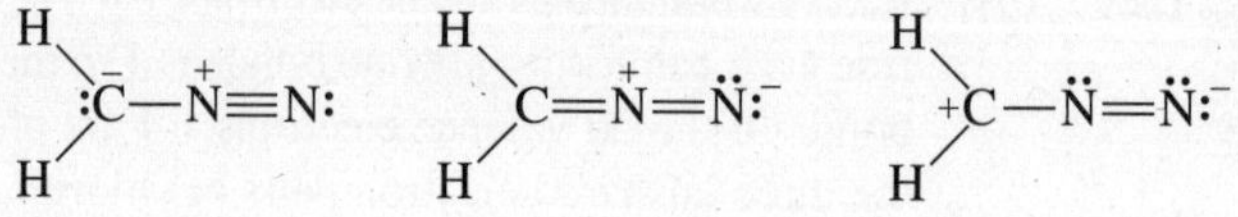

$H_2\overset{-}{C}$:—$\overset{+}{N}$≡N:  $H_2C$=$\overset{+}{N}$=N̈:$^-$  $H_2\overset{+}{C}$—N̈=N̈:$^-$

A B C

(*b*) Structure A has a formal charge of −1 on carbon.
(*c*) Structure C has a formal charge of +1 on carbon.
(*d*) Structures A and B have formal charges of +1 on the internal nitrogen.
(*e*) Structures B and C have a formal charge of −1 on the terminal nitrogen.
(*f*) All resonance forms of a particular species must have the same net charge. In this case, the net charge on A, B, and C is 0.
(*g*) Both A and B have the same number of covalent bonds, but the negative charge is on a more electronegative atom in B (nitrogen) than it is in A (carbon). Structure B is more stable.
(*h*) Structure B is more stable than structure C. Structure B has one more covalent bond, all of its atoms have octets of electrons, and it has a lesser degree of charge separation than C. The carbon in structure C does not have an octet of electrons.
(*i*) The CNN unit is linear in A and B, but bent in C according to VSEPR. This is an example of how VSEPR can fail when comparing resonance structures.

**1.31** The structures given and their calculated formal charges are:

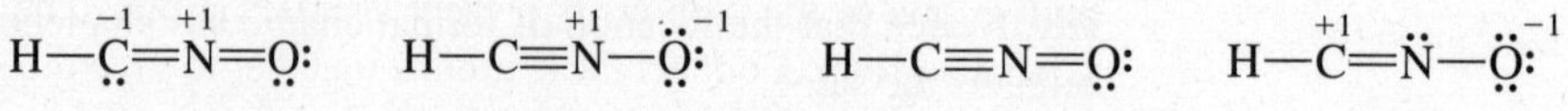

H—$\overset{-1}{C}$=$\overset{+1}{N}$=Ö:  H—C≡$\overset{+1}{N}$—$\overset{-1}{Ö}$:  H—C≡N=Ö:  H—$\overset{+1}{C}$=N̈—$\overset{-1}{Ö}$:

A B C D

(*a*) Structure D contains a positively charged carbon.
(*b*) Structures A and B contain a positively charged nitrogen.
(*c*) None of the structures contain a positively charged oxygen.
(*d*) Structure A contains a negatively charged carbon.
(*e*) None of the structures contain a negatively charged nitrogen.
(*f*) Structures B and D contain a negatively charged oxygen.
(*g*) All the structures are electrically neutral.
(*h*) Structure B is the most stable. All the atoms except hydrogen have octets of electrons, and the negative charge resides on the most electronegative element (oxygen).
(*i*) Structure C is the least stable. Nitrogen has five bonds (10 electrons), which violates the octet rule.

**1.32** (*a*) These two structures are resonance forms since they have the same atomic positions and the same number of electrons.

$^{2-}$:N̤̈—N̈$^{+}$≡N: ⟷ $^{-}$:N̤=N̈$^{+}$=N̤:$^{-}$

16 valence electrons (net charge = −1) | 16 valence electrons (net charge = −1)

(*b*) The two structures have different numbers of electrons and, therefore, can't be resonance forms of each other.

$^{2-}$:N̤̈—N$^{+}$≡N:    :N̈—N$^{2+}$=N̈:$^{-}$

16 valence electrons (net charge −1) | 14 valence electrons (net charge +1)

(*c*) These two structures have different numbers of electrons; they are not resonance forms.

$^{2-}$:N̤̈—N$^{+}$≡N:    $^{2-}$:N̤̈—N̤̈$^{-}$—N̤̈:$^{2-}$

16 valence electrons (net charge = −1) | 20 valence electrons (net charge = −5)

**1.33** Structure C has 10 electrons surrounding nitrogen, but the octet rule limits nitrogen to 8 electrons. Structure C is incorrect.

$CH_2$=N(—$CH_3$)=O̤:    Not a valid Lewis structure!

**1.34** (*a*) The terminal nitrogen has only 6 electrons; therefore, use the unshared pair of the adjacent nitrogen to form another covalent bond.

By moving electrons of the nitrogen lone pair as shown by the arrow   H—C($H_2$)—N̈$^{+}$=N:   a structure that has octets about both nitrogen atoms is obtained.   H—C($H_2$)—N$^{+}$≡N:

In general, move electrons from sites of high electron density toward sites of low electron density. Notice that the location of formal charge has changed, but the net charge on the species remains the same.

(*b*) The dipolar Lewis structure given can be transformed to one that has no charge separation by moving electron pairs as shown:

H—C(—Ö:$^{-}$)=Ö$^{+}$—H ⟷ H—C(=O̤:)—Ö—H

(*c*) Move electrons toward the positive charge. Sharing the lone pair gives an additional covalent bond and avoids the separation of opposite charges.

$^{+}CH_2$—C̈$H_2^{-}$ ⟷ $CH_2$=$CH_2$

(*d*) Octets of electrons at all the carbon atoms can be produced by moving the electrons toward the site of positive charge.

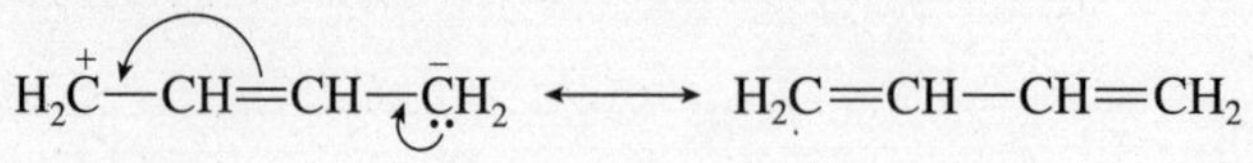

$$H_2\overset{+}{C}-CH=CH-\overset{-}{\ddot{C}}H_2 \longleftrightarrow H_2C=CH-CH=CH_2$$

(*e*) As in part (*d*), move the electron pairs toward the carbon atom that has only 6 electrons.

$$H_2\overset{+}{C}-CH=CH-\ddot{\underset{..}{O}}:^- \longleftrightarrow H_2C=CH-CH=\underset{..}{O}:$$

(*f*) The negative charge can be placed on the most electronegative atom (oxygen) in this molecule by moving electrons as indicated.

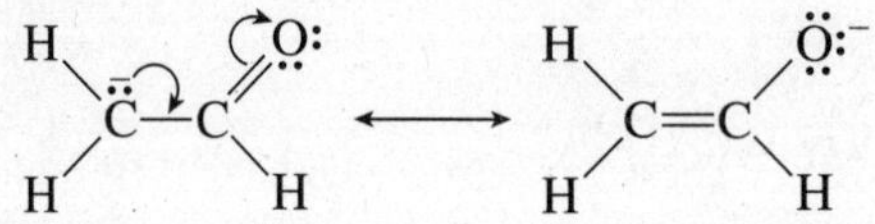

$$H_2\overset{-}{\ddot{C}}-CH=\ddot{O}: \longleftrightarrow H_2C=CH-\ddot{\underset{..}{O}}:^-$$

(*g*) Octets of electrons are present around both carbon and oxygen if an oxygen unshared electron pair is moved toward the positively charged carbon to give an additional covalent bond.

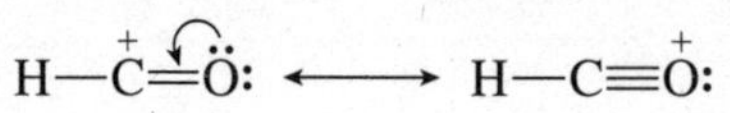

$$H-\overset{+}{C}=\ddot{O}: \longleftrightarrow H-C\equiv\overset{+}{O}:$$

(*h*) This exercise is similar to part (*g*); move electrons from oxygen to carbon so as to produce an additional bond and satisfy the octet rule for both carbon and oxygen.

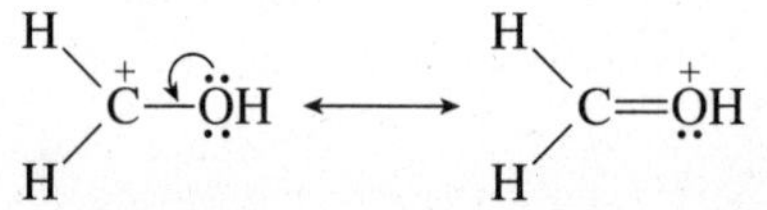

$$H_2\overset{+}{C}-\ddot{\underset{..}{O}}H \longleftrightarrow H_2C=\overset{+}{\underset{..}{O}}H$$

(*i*) By moving electrons from the site of negative charge toward the positive charge, a structure that has no charge separation is generated.

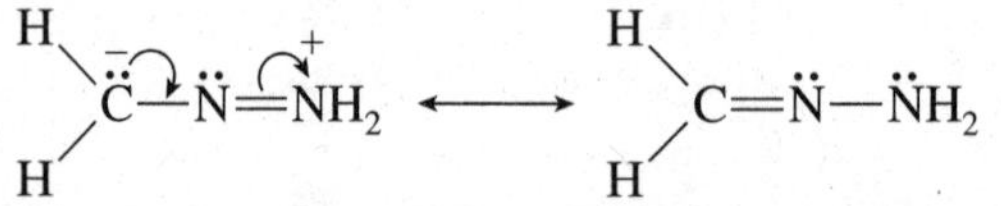

$$H_2\overset{-}{\ddot{C}}-\ddot{N}=\overset{+}{N}H_2 \longleftrightarrow H_2C=\ddot{N}-\ddot{N}H_2$$

**1.35** (*a*) Sulfur is in the same group of the periodic table as oxygen (group VI A) and, like oxygen, has 6 valence electrons. Sulfur dioxide, therefore, has 18 valence electrons. A Lewis structure in which sulfur and both oxygens have complete octets of electrons is:

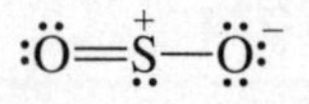

$$:\ddot{O}=\overset{+}{\underset{..}{S}}-\ddot{\underset{..}{O}}:^-$$

(*b*) Move an electron pair from the singly bonded oxygen in part (*a*) to generate a second double bond. The resulting Lewis structure has 10 valence electrons around sulfur. It is a valid Lewis structure because sulfur can expand its valence shell beyond 8 electrons by using its 3*d* orbitals.

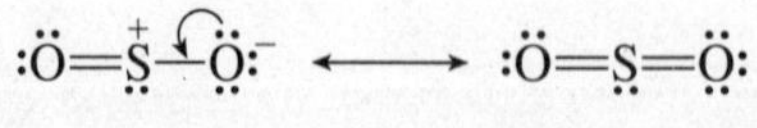

$$:\ddot{O}=\overset{+}{\underset{..}{S}}-\ddot{\underset{..}{O}}:^- \longleftrightarrow :\ddot{O}=\underset{..}{S}=\ddot{O}:$$

**1.36** (*a*) To generate constitutionally isomeric structures having the molecular formula $C_4H_{10}$, you need to consider the various ways in which four carbon atoms can be bonded together. These are

C—C—C—C and C—C(—C)—C

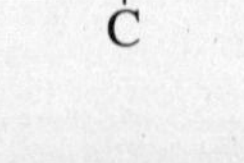

Filling in the appropriate hydrogens gives the correct structures:

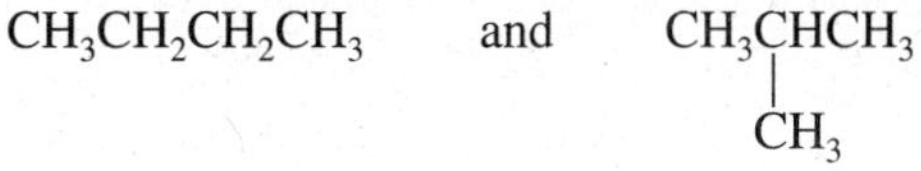

$$CH_3CH_2CH_2CH_3 \quad \text{and} \quad CH_3CH(CH_3)CH_3$$

Continue with the remaining parts of the problem using the general approach outlined for part (*a*).

(*b*) $C_5H_{12}$

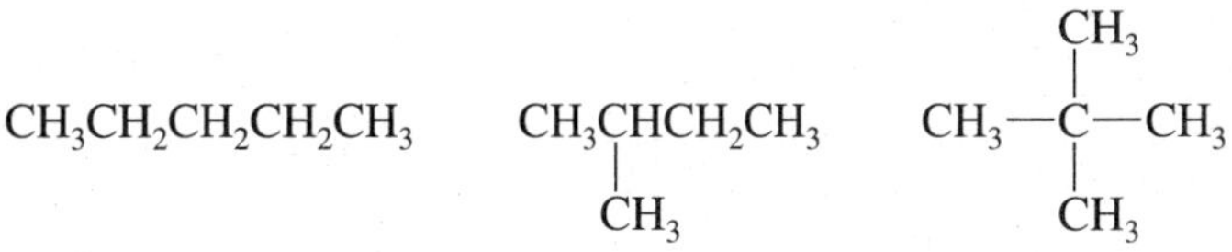

$$CH_3CH_2CH_2CH_2CH_3 \qquad CH_3CH(CH_3)CH_2CH_3 \qquad CH_3—C(CH_3)_2—CH_3$$

(*c*) $C_2H_4Cl_2$

$$CH_3CHCl_2 \quad \text{and} \quad ClCH_2CH_2Cl$$

(*d*) $C_4H_9Br$

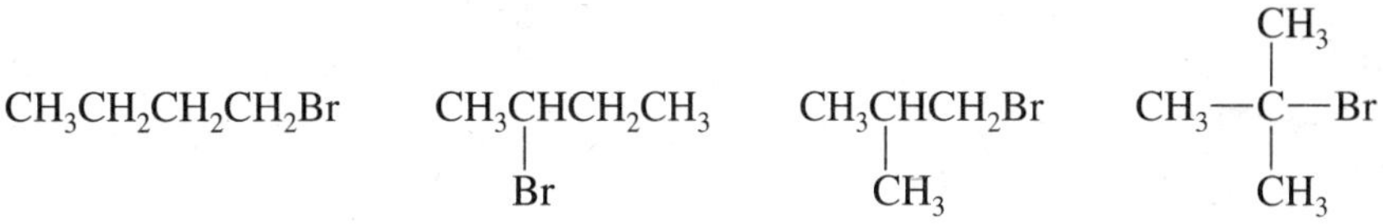

$$CH_3CH_2CH_2CH_2Br \qquad CH_3CH(Br)CH_2CH_3 \qquad CH_3CH(CH_3)CH_2Br \qquad CH_3—C(CH_3)_2—Br$$

(*e*) $C_3H_9N$

$$CH_3CH_2CH_2NH_2 \qquad CH_3CH_2NHCH_3 \qquad CH_3—N(CH_3)_2 \qquad CH_3CH(CH_3)NH_2$$

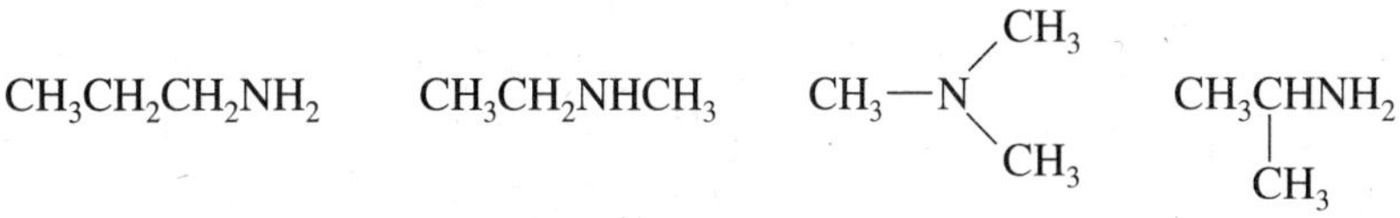

Note that when the three carbons and the nitrogen are arranged in a ring, the molecular formula based on such a structure is $C_3H_7N$, not $C_3H_9N$ as required.

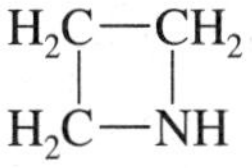

$$\begin{array}{l} H_2C—CH_2 \\ \;|\qquad\;\, | \\ H_2C—NH \end{array}$$

(not an isomer)

**1.37** (*a*) All three carbons must be bonded together, and each one has four bonds; therefore, the molecular formula $C_3H_8$ uniquely corresponds to:

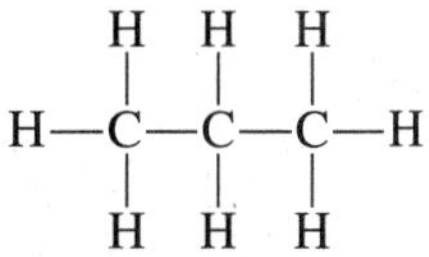

$$\begin{array}{ccccccc} & H & & H & & H & \\ & | & & | & & | & \\ H— & C & — & C & — & C & —H \\ & | & & | & & | & \\ & H & & H & & H & \end{array} \qquad (CH_3CH_2CH_3)$$

(*b*) With two fewer hydrogen atoms than the preceding compound, either $C_3H_6$ must contain a carbon–carbon double bond or its carbons must be arranged in a ring; thus the following structures are constitutional isomers:

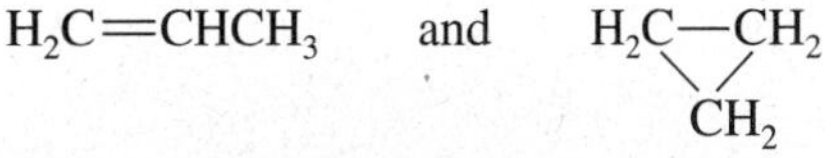

$$H_2C{=}CHCH_3 \quad \text{and} \quad \text{cyclopropane } (H_2C—CH_2, \text{ both bonded to } CH_2)$$

(*c*) The molecular formula $C_3H_4$ is satisfied by the structures

$$H_2C{=}C{=}CH_2 \qquad HC{\equiv}CCH_3 \qquad \underset{\diagdown\ \diagup\atop CH_2}{HC{=}CH}$$

**1.38** (*a*) The only atomic arrangements of $C_3H_6O$ that contain only single bonds must have a ring as part of their structure.

$$\underset{\diagdown\ \diagup\atop CH_2}{H_2C{-}CHOH} \qquad \underset{\diagdown\ \diagup\atop O}{H_2C{-}CHCH_3} \qquad \begin{matrix} H_2C{-}CH_2 \\ | \quad\ \ | \\ O{-}CH_2 \end{matrix}$$

(*b*) Structures corresponding to $C_3H_6O$ are possible in noncyclic compounds if they contain a carbon–carbon or carbon–oxygen double bond.

$$CH_3CH_2\overset{O}{\overset{\|}{C}}H \qquad CH_3\overset{O}{\overset{\|}{C}}CH_3 \qquad CH_3CH{=}CHOH \qquad CH_3OCH{=}CH_2$$

$$CH_3\underset{OH}{\underset{|}{C}}{=}CH_2 \qquad H_2C{=}CHCH_2OH$$

**1.39** The direction of a bond dipole is governed by the electronegativity of the atoms it connects. In each of the parts to this problem, the more electronegative atom is partially negative and the less electronegative atom is partially positive. Electronegativities of the elements are given in Table 1.2 of the text.

(*a*) Chlorine is more electronegative than hydrogen.

$$\overset{+\!\!\longrightarrow}{H{-}Cl}$$

(*b*) Chlorine is more electronegative than iodine.

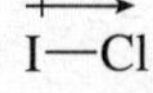

(*c*) Iodine is more electronegative than hydrogen.

$$\overset{+\!\!\longrightarrow}{H{-}I}$$

(*d*) Oxygen is more electronegative than hydrogen.

$$\underset{H \qquad H}{\nearrow O \nwarrow}$$

(*e*) Oxygen is more electronegative than either hydrogen or chlorine.

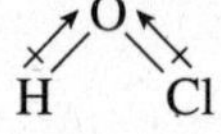

**1.40** The direction of a bond dipole is governed by the electronegativity of the atoms involved. Among the halogens the order of electronegativity is F > Cl > Br > I. Fluorine therefore attracts electrons away from chlorine in FCl, and chlorine attracts electrons away from iodine in ICl.

$$\underset{\mu = 0.9\ D}{\overset{\longleftarrow\!\!+}{F{-}Cl}} \qquad \underset{\mu = 0.7\ D}{\overset{+\!\!\longrightarrow}{I{-}Cl}}$$

Chlorine is the positive end of the dipole in FCl and the negative end in ICl.

**1.41** (*a*) Sodium chloride is ionic; it has a unit positive charge and a unit negative charge separated from each other. Hydrogen chloride has a polarized bond but is a covalent compound. Sodium chloride has a larger dipole moment. The measured values are as shown.

$$\underset{\mu\ 9.4\ D}{Na^+\ Cl^-} \quad \text{is more polar than} \quad \underset{\mu\ 1.1\ D}{H{-}Cl}$$

(*b*) Fluorine is more electronegative than chlorine, and so its bond to hydrogen is more polar, as the measured dipole moments indicate.

H—F is more polar than H—Cl

$\mu$ 1.7 D $\mu$ 1.1 D

(*c*) Boron trifluoride is planar. Its individual B—F bond dipoles cancel. It has no dipole moment.

H—F is more polar than $BF_3$

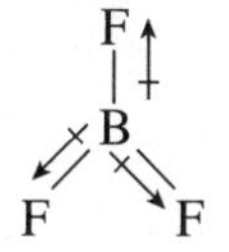

$\mu$ 1.7 D $\mu$ 0 D

(*d*) A carbon–chlorine bond is strongly polar; carbon–hydrogen and carbon–carbon bonds are only weakly polar.

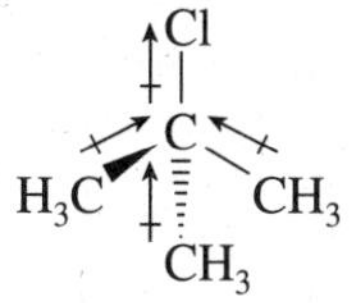

$(CH_3)_3CCl$ is more polar than $(CH_3)_3CH$

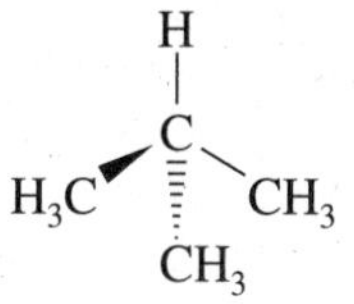

$\mu$ 2.1 D $\mu$ 0.1 D

(*e*) A carbon–fluorine bond in $CCl_3F$ opposes the polarizing effect of the chlorines. The carbon–hydrogen bond in $CHCl_3$ reinforces it. $CHCl_3$ therefore has a larger dipole moment.

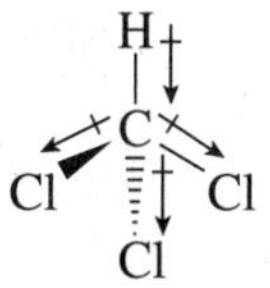

$HCCl_3$ is more polar than $FCCl_3$

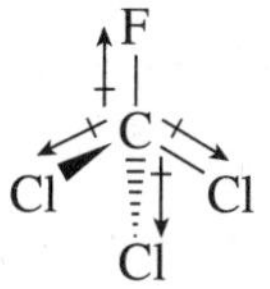

$\mu$ 1.0 D $\mu$ 0.5 D

(*f*) Oxygen is more electronegative than nitrogen; its bonds to carbon and hydrogen are more polar than the corresponding bonds formed by nitrogen.

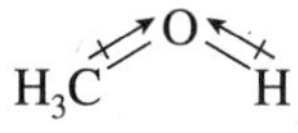

$H_3C$—O—H is more polar than $H_3C$—N($H$)—H

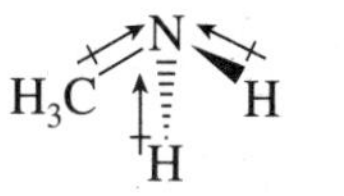

$\mu$ 1.7 D $\mu$ 1.3 D

(*g*) The Lewis structure for $CH_3NO_2$ has a formal charge of +1 on nitrogen, making it more electron-attracting than the uncharged nitrogen of $CH_3NH_2$.

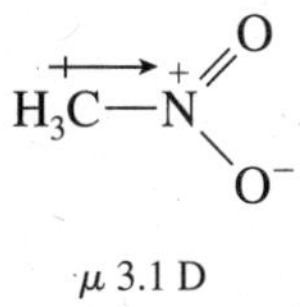

$H_3C$—$\overset{+}{N}$(=O)—$O^-$ is more polar than $H_3C$—$\ddot{N}H_2$

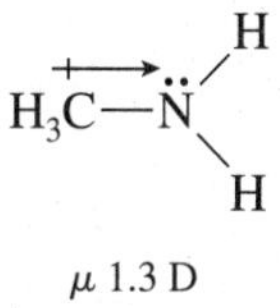

$\mu$ 3.1 D $\mu$ 1.3 D

**1.42** (*a*) There are four electron pairs around carbon in $:\bar{C}H_3$; they are arranged in a tetrahedral fashion. The atoms of this species are in a trigonal pyramidal arrangement.

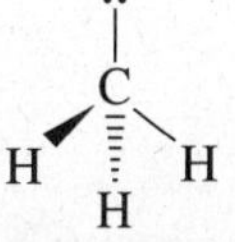

$:\bar{C}H_3$ (C with H, H, H)

(*b*) Only three electron pairs are present in $\overset{+}{C}H_3$, and so it is trigonal planar.

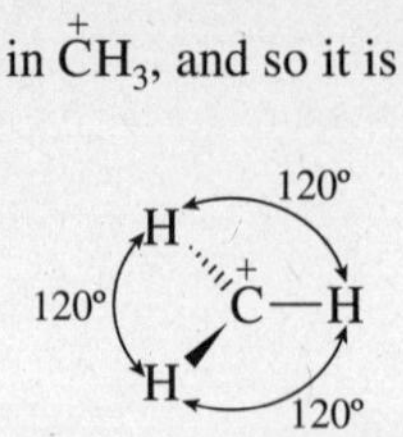

$$\overset{+}{C}H_3 \quad (120^\circ,\ 120^\circ,\ 120^\circ)$$

(*c*) As in part (*b*), there are three electron pairs. When these electron pairs are arranged in a plane, the atoms in $:CH_2$ are not collinear. The atoms of this species are arranged in a bent structure according to VSEPR considerations.

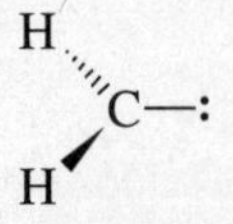

**1.43** The structures, written in a form that indicates hydrogens and unshared electrons, are as shown. Remember: A neutral carbon has four bonds, a neutral nitrogen has three bonds plus one unshared electron pair, and a neutral oxygen has two bonds plus two unshared electron pairs. Halogen substituents have one bond and three unshared electron pairs.

(*a*) 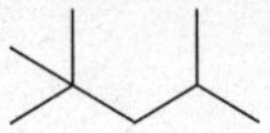 is equivalent to $(CH_3)_3CCH_2CH(CH_3)_2$

(*b*) 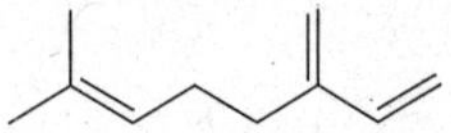 is equivalent to $(CH_3)_2C{=}CHCH_2CH_2\overset{CH_2}{\overset{\|}{C}}CH{=}CH_2$

(*c*) 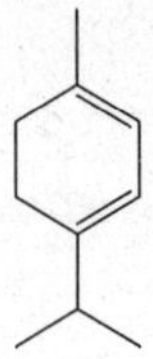 is equivalent to

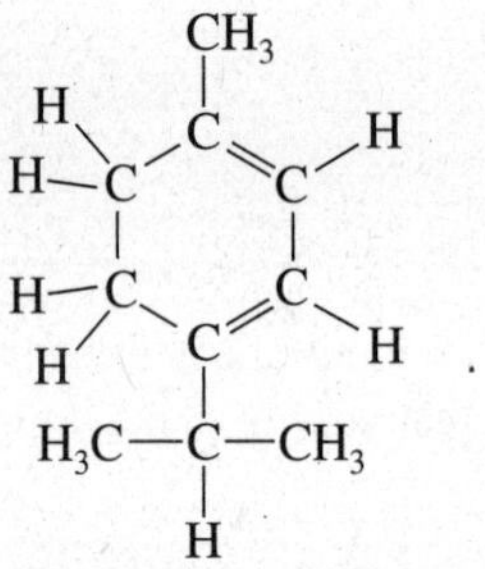

(*d*) 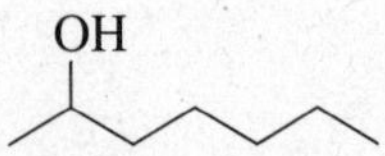 is equivalent to $CH_3\overset{:\ddot{O}H}{\overset{|}{C}}HCH_2CH_2CH_2CH_2CH_3$

(*e*) 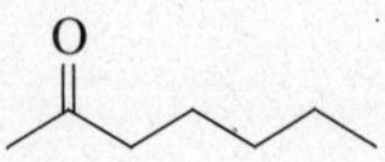 is equivalent to $CH_3\overset{\ddot{O}:}{\overset{\|}{C}}CH_2CH_2CH_2CH_2CH_3$

(*f*) 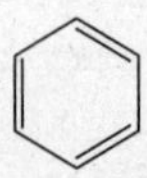 is equivalent to

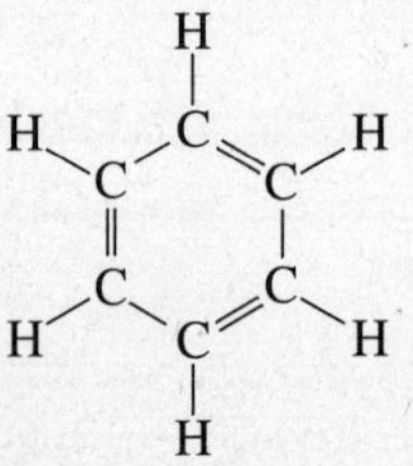

(*g*) 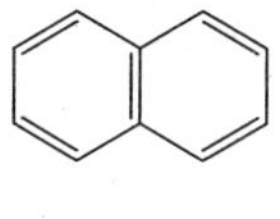 is equivalent to

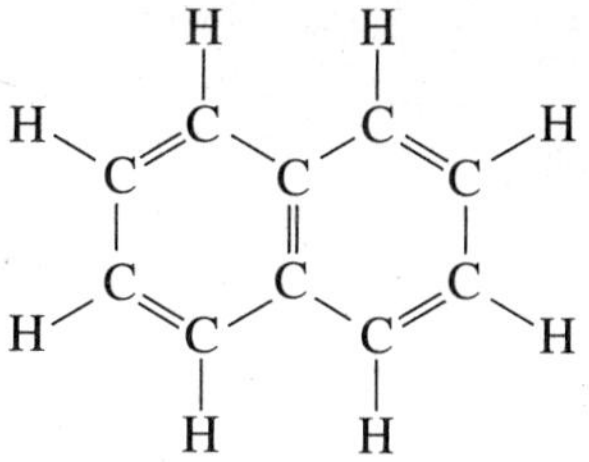

(*h*) 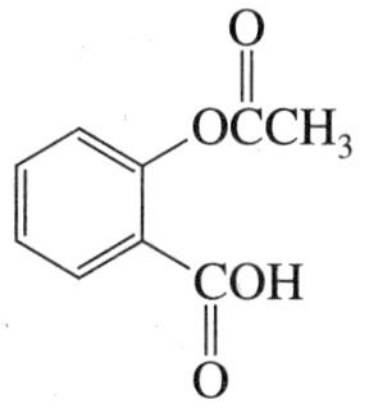 is equivalent to

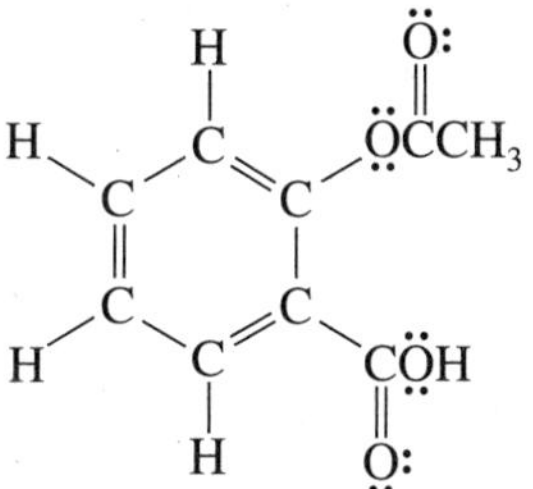

(*i*) 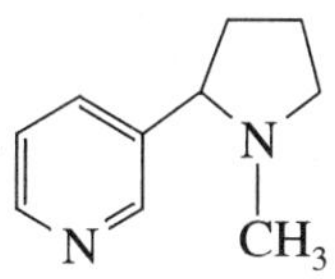 is equivalent to

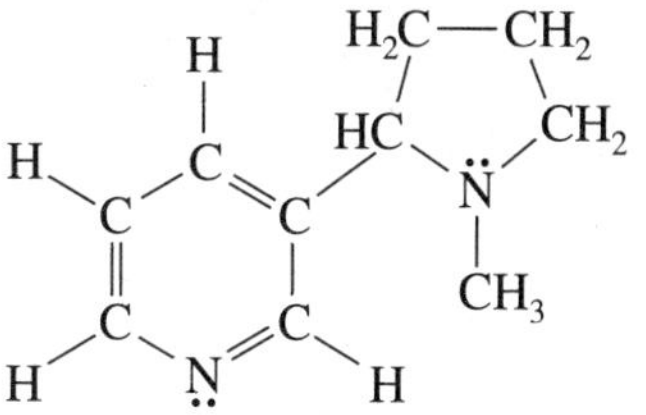

(*j*) 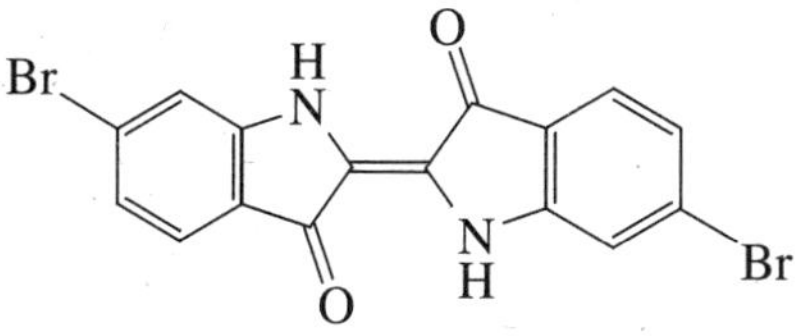 is equivalent to

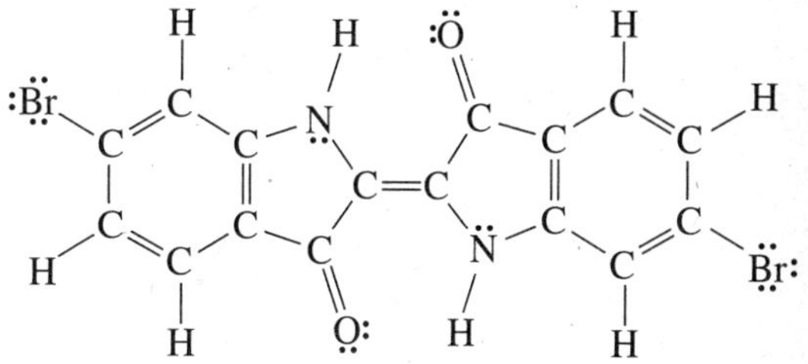

(*k*) 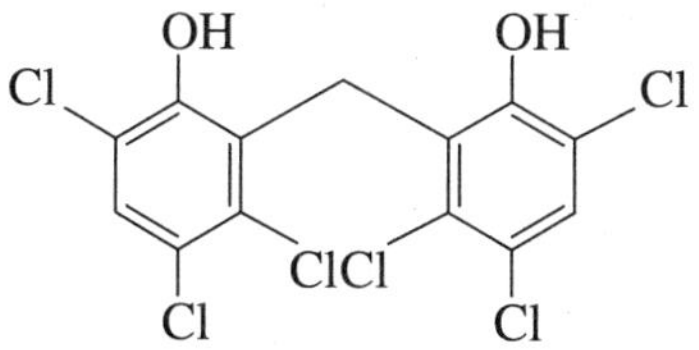 is equivalent to

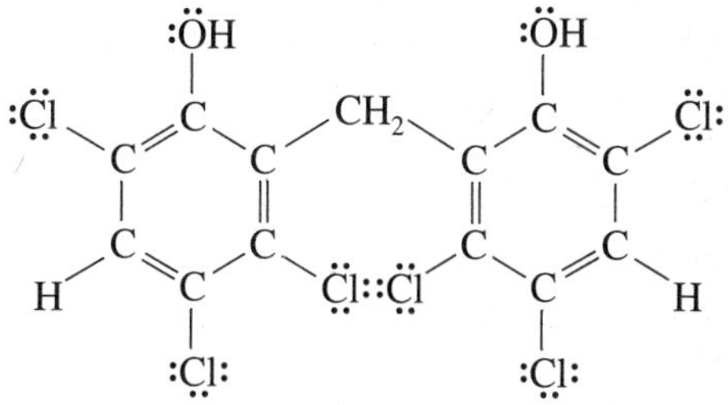

**1.44** (*a*) $C_8H_{18}$
(*b*) $C_{10}H_{16}$
(*c*) $C_{10}H_{16}$
(*d*) $C_7H_{16}O$
(*e*) $C_7H_{14}O$
(*f*) $C_6H_6$
(*g*) $C_{10}H_8$
(*h*) $C_9H_8O_4$
(*i*) $C_{10}H_{14}N_2$
(*j*) $C_{16}H_8Br_2N_2O_2$
(*k*) $C_{13}H_6Cl_6O_2$

Isomers are different compounds that have the same molecular formula. Two of these compounds, (*b*) and (*c*), have the same molecular formula and are isomers of each other.

**1.45** (*a*) Carbon is $sp^3$-hybridized when it is directly bonded to four other atoms. Compounds (*a*) and (*d*) in Problem 1.43 are the only ones in which *all* of the carbons are $sp^3$-hybridized.

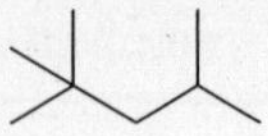

OH

(*a*) 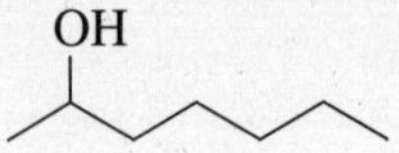 (*d*)

(*b*) Carbon is $sp^2$-hybridized when it is directly bonded to three other atoms. Compounds (*f*), (*g*), and (*j*) in Problem 1.43 have only $sp^2$-hybridized carbons.

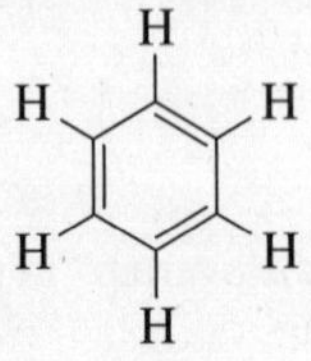

(*f*) 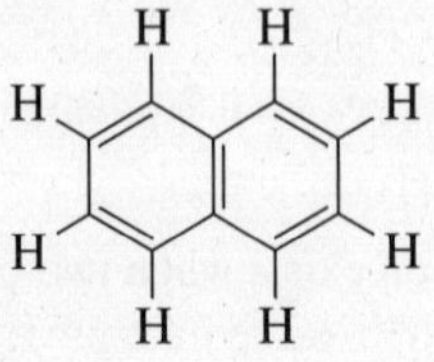 (*g*) 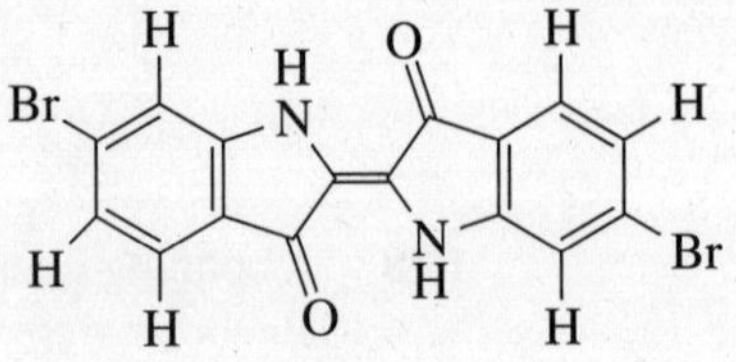 (*j*)

None of the compounds in Problem 1.43 contain an *sp*-hybridized carbon.

**1.46** The problem specifies that the second-row element is $sp^3$-hybridized in each of the compounds. Any unshared electron pairs therefore occupy $sp^3$-hybridized oribitals, and bonded pairs are located in $\sigma$ orbitals.

(*a*) Ammonia

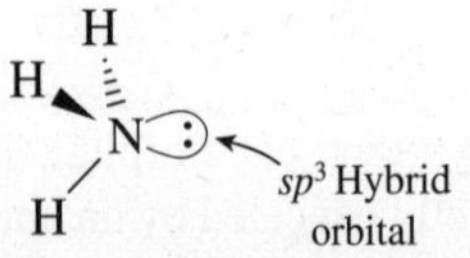

$sp^3$ Hybrid orbital

Three $\sigma$ bonds formed by $sp^3$–$s$ overlap

(*b*) Water

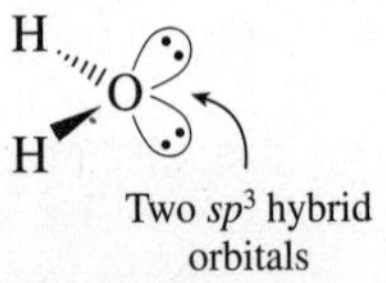

Two $sp^3$ hybrid orbitals

Two $\sigma$ bonds formed by $sp^3$–$s$ overlap

(*c*) Hydrogen fluoride

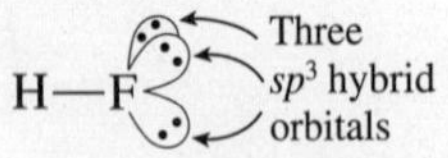

Three $sp^3$ hybrid orbitals

One $\sigma$ bond formed by $sp^3$–$s$ overlap

(*d*) Ammonium ion

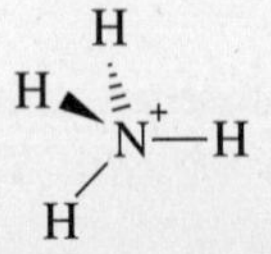

Four $\sigma$ bonds formed by $sp^3$–$s$ overlap

(*e*) Borohydride anion

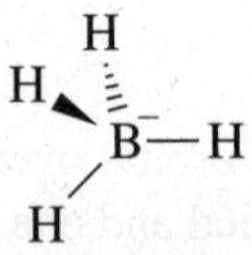

Four $\sigma$ bonds formed by $sp^3$–$s$ overlap

(*f*) Amide anion

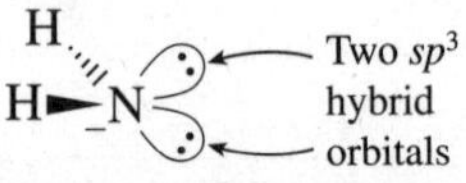

Two $sp^3$ hybrid orbitals

Two $\sigma$ bonds formed by $sp^3$–$s$ overlap

(*g*) Methyl anion

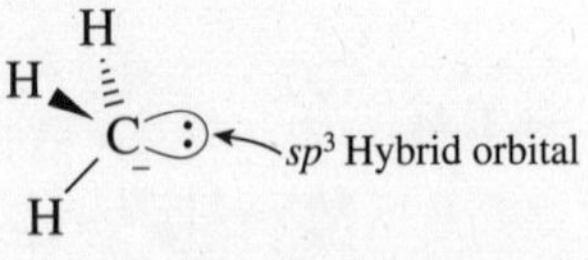

$sp^3$ Hybrid orbital

Three $\sigma$ bonds formed by $sp^3$–$s$ overlap

**1.47** (*a*) The electron configuration of N is $1s^2 2s^2 2p_x^{\ 1} 2p_y^{\ 1} 2p_z^{\ 1}$. If the half-filled $2p_x$, $2p_y$, and $2p_z$ orbitals are involved in bonding to H, then the unshared pair would correspond to the two electrons in the 2*s* orbital.

(*b*) The three *p* orbitals $2p_x$, $2p_y$, and $2p_z$ have their axes at right angles to one another. The H—N—H angles would therefore be 90°.

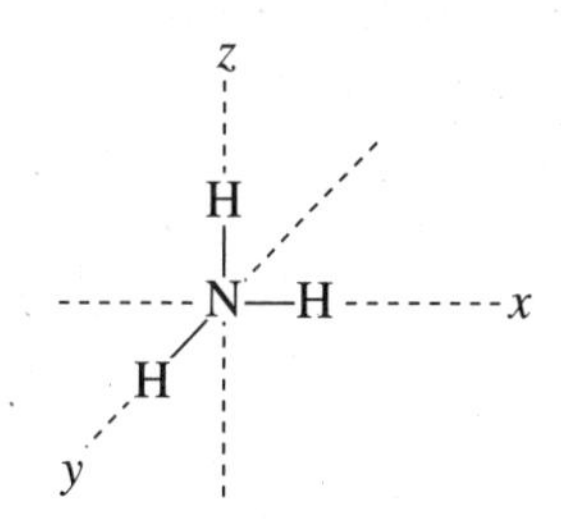


**1.48** A bonding interaction exists when two orbitals overlap "in phase" with each other, that is, when the algebraic signs of their wave functions are the same in the region of overlap. The following orbital is a bonding orbital. It involves overlap of an *s* orbital with the lobe of a *p* orbital of the same sign.

(*c*) (bonding)

On the other hand, the overlap of an *s* orbital with the lobe of a *p* orbital of opposite sign is antibonding.

(*b*) (antibonding)

Overlap in the manner shown next is nonbonding. Both the positive lobe and the negative lobe of the *p* orbital overlap with the spherically symmetrical *s* orbital. The bonding overlap between the *s* orbital and one lobe of the *p* orbital is exactly canceled by an antibonding interaction between the *s* orbital and the lobe of opposite sign.

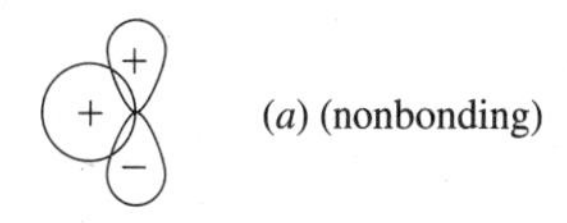

(*a*) (nonbonding)

**1.49–1.55** Solutions to molecular modeling exercises are not provided in this *Study Guide and Solutions Manual.* You should use *Learning By Modeling* for these exercises.

# SELF-TEST

## PART A

**A-1.** Write the electronic configuration for each of the following:
(*a*) Phosphorus (*b*) Sulfide ion in $Na_2S$

**A-2.** Determine the formal charge of each atom and the net charge for each of the following species:

(*a*) :N̈=C=S̈: (*b*) :O≡N—Ö: (*c*) HC(=NH$_2$) with :Ö: attached to C

**A-3.** Write a second Lewis structure that satisfies the octet rule for each of the species in Problem A-2, and determine the formal charge of each atom. Which of the Lewis structures for each species in this and Problem A-2 is more stable?

**A-4.** Write a correct Lewis structure for each of the following. Be sure to show explicitly any unshared pairs of electrons.

(*a*) Methylamine, $CH_3NH_2$

(*b*) Acetaldehyde, $C_2H_4O$ (the atomic order is CCO; all the hydrogens are connected to carbon.)

**A-5.** What is the molecular formula of each of the structures shown? Clearly draw any unshared electron pairs that are present.

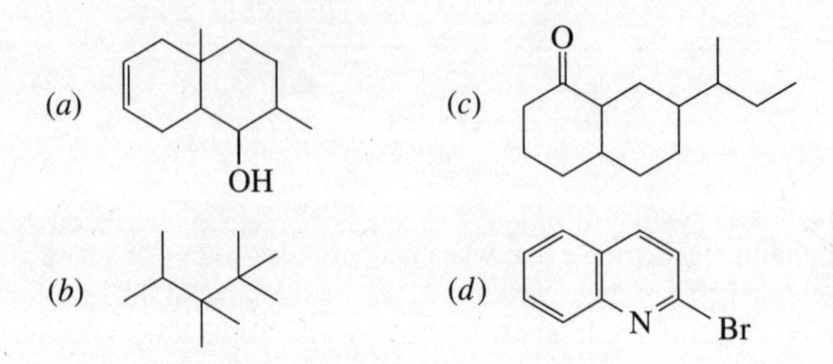


**A-6.** Which compound in Problem A-5 has

(*a*) Only $sp^3$-hybridized carbons

(*b*) Only $sp^2$-hybridized carbons

(*c*) A single $sp^2$-hybridized carbon atom

**A-7.** Account for the fact that all three sulfur–oxygen bonds in $SO_3$ are the same by drawing the appropriate Lewis structure(s).

**A-8.** The cyanate ion contains 16 valence electrons, and its three atoms are arranged in the order OCN. Write the most stable Lewis structure for this species, and assign a formal charge to each atom. What is the net charge of the ion?

**A-9.** Using the VSEPR method,

(*a*) Describe the geometry at each carbon atom and the oxygen atom in the following molecule: $CH_3OCH{=}CHCH_3$.

(*b*) Deduce the shape of $NCl_3$, and draw a three-dimensional representation of the molecule. Is $NCl_3$ polar?

**A-10.** Assign the shape of each of the following as either linear or bent.

(*a*) $CO_2$ (*b*) $NO_2^+$ (*c*) $NO_2^-$

**A-11.** Consider structures A, B, C, and D:

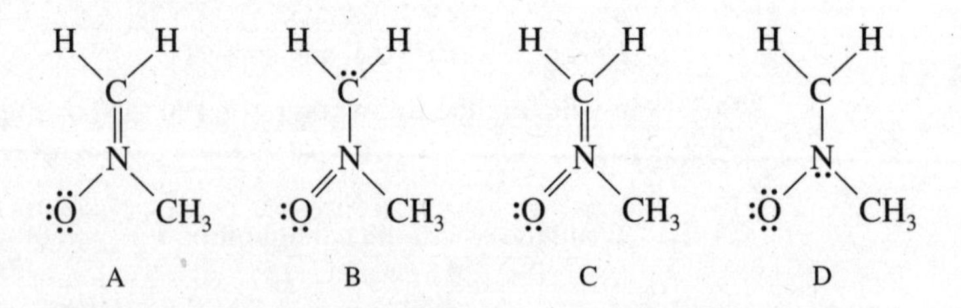


(*a*) Which structure (or structures) contains a positively charged carbon?

(*b*) Which structure (or structures) contains a positively charged nitrogen?

(*c*) Which structure (or structures) contains a positively charged oxygen?

(*d*) Which structure (or structures) contains a negatively charged carbon?

(*e*) Which structure (or structures) contains a negatively charged nitrogen?

(*f*) Which structure (or structures) contains a negatively charged oxygen?

(*g*) Which structure is the most stable?

(*h*) Which structure is the least stable?

**A-12.** Given the following information, write a Lewis structure for urea, $CH_4N_2O$. The oxygen atom and both nitrogen atoms are bonded to carbon, there is a carbon–oxygen double bond, and none of the atoms bears a formal charge. Be sure to include all unshared electron pairs.

**A-13.** How many $\sigma$ and $\pi$ bonds are present in each of the following?

(*a*) $CH_3CH{=}CHCH_3$

(*b*) $HC{\equiv}CCH_2CH_3$

(*c*) O=⟨cyclohexa-2,5-diene ring⟩=O

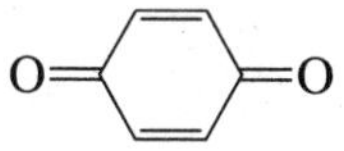

(*d*) cyclopentanone ring bearing C≡N

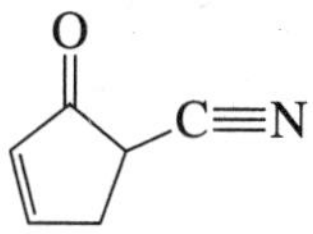

**A-14.** Give the hybridization of each carbon atom in the preceding problem.

## PART B

**B-1.** Which one of the following is most likely to have ionic bonds?

(*a*) HCl (*b*) $Na_2O$ (c) $N_2O$ (*d*) $NCl_3$

**B-2.** Which of the following is *not* an electronic configuration for an atom in its ground state?

(*a*) $1s^22s^22p_x^{\,2}2p_y^{\,1}2p_z^{\,1}$ (c) $1s^22s^22p_x^{\,2}2p_y^{\,2}2p_z^{\,1}$

(*b*) $1s^22s^22p_x^{\,2}2p_y^{\,2}2p_z^{\,0}$ (*d*) $1s^22s^22p_x^{\,2}2p_y^{\,2}2p_z^{\,2}$

**B-3.** The formal charge on phosphorus in $(CH_3)_4P$ is

(*a*) 0 (*b*) −1 (*c*) +1 (*d*) +2

**B-4.** Which of the following is an isomer of compound 1?

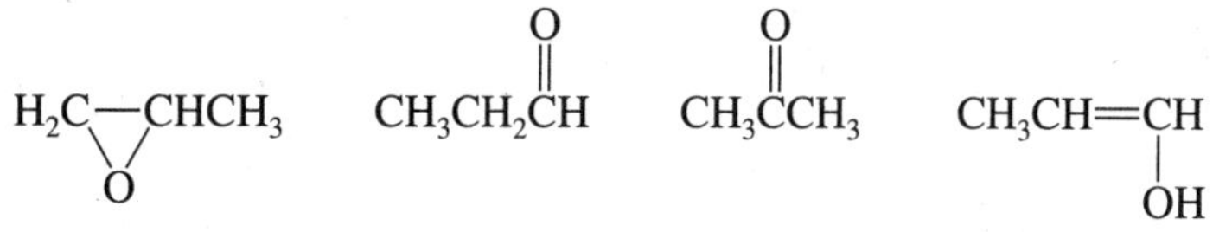

| 1 | 2 | 3 | 4 |
|---|---|---|---|
| $H_2C{-}CHCH_3$ (epoxide, O bridging) | $CH_3CH_2CH{=}O$ | $CH_3C({=}O)CH_3$ | $CH_3CH{=}CH{-}OH$ |

(*a*) 2 (*c*) 2 and 3

(*b*) 4 (*d*) All are isomers.

**B-5.** In which of the following is oxygen the positive end of the bond dipole?

(*a*) O—F (*b*) O—N (*c*) O—S (*d*) O—H

**B-6.** What two structural formulas are resonance forms of one another?

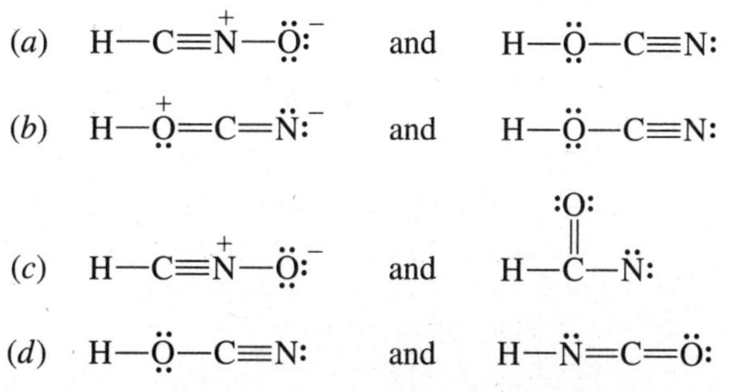

(*a*) $H{-}C{\equiv}\overset{+}{N}{-}\ddot{\underset{..}{O}}:^-$ and $H{-}\ddot{\underset{..}{O}}{-}C{\equiv}N:$

(*b*) $H{-}\overset{+}{\underset{..}{O}}{=}C{=}\ddot{N}:^-$ and $H{-}\ddot{\underset{..}{O}}{-}C{\equiv}N:$

(*c*) $H{-}C{\equiv}\overset{+}{N}{-}\ddot{\underset{..}{O}}:^-$ and $H{-}C({=}\ddot{O}:){-}\ddot{N}:$

(*d*) $H{-}\ddot{\underset{..}{O}}{-}C{\equiv}N:$ and $H{-}\ddot{N}{=}C{=}\ddot{O}:$

**B-7.** The bond identified (with the arrow) in the following structure is best described as:

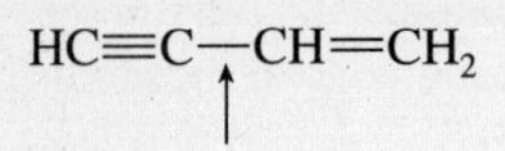


(*a*) $2sp$–$2sp^2$ $\sigma$ (*b*) $2p$–$2p$ $\pi$ (*c*) $2sp^2$–$2sp^3$ $\sigma$ (*d*) $2sp^2$–$2sp^2$ $\sigma$ (*e*) $2p$–$2p$ $\sigma$

**B-8.** The total number of *unshared pairs* of electrons in the molecule

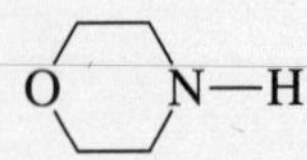


is

(*a*) 0 (*b*) 1 (*c*) 2 (*d*) 3

**B-9.** Which of the following contains a triple bond?

(*a*) $SO_2$ (*b*) HCN (*c*) $C_2H_4$ (*d*) $NH_3$

**B-10.** Which one of the compounds shown is *not* an isomer of the other three?

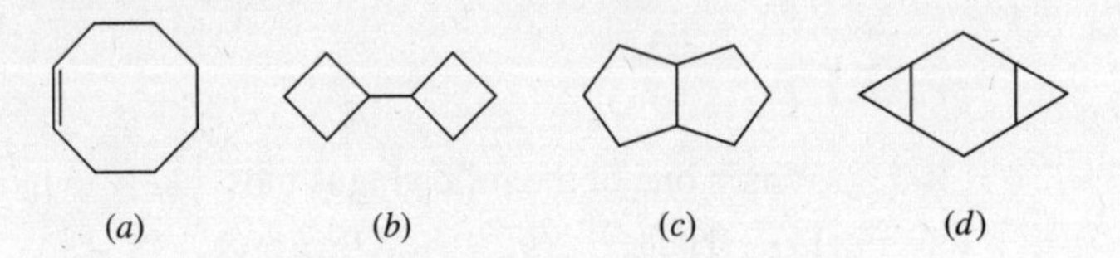


**B-11.** Which one of the following is the most stable Lewis structure? The answer must be correct in terms of bonds, unshared pairs of electrons, and formal charges.

(*a*) :Ö=N=$CH_2$ (*b*) $^-$:Ö̤—N̈=$CH_2$ (*c*) :Ö=N̤̈—C̈$H_2$ (*d*) :Ö=N̤̈=$CH_2$ (*e*) :Ö̤—N̤̈—C̈$H_2$

**B-12.** Repeat the previous question for the following Lewis structures.

(*a*) $^-$:N̤̈—N̤̈—C$^+$$H_2$ (*b*) $^-$:N̤̈—N̈=$CH_2$ (*c*) :N̈=N̈—C̈$^-$$H_2$ (*d*) :N≡N$^+$—$CH_2$ (*e*) $^-$:N̈=N$^+$=$CH_2$

**B-13.** Which of the following molecules would you expect to be *nonpolar*?

1. $CH_2F_2$ 2. $CO_2$ 3. $CF_4$ 4. $CH_3OCH_3$

(*a*) 1 and 2 (*b*) 1 and 3 (*c*) 1 and 4 (*d*) 2 and 3 (*e*) 2, 3, and 4

The remaining two questions refer to the hypothetical compounds:

| A—B—A | A=B̈—A | A=B=A | A—B̈(—A)—A |
|---|---|---|---|
| 1 | 2 | 3 | 4 |

**B-14.** Which substance(s) is (are) linear?

(*a*) 1 only (*b*) 1 and 3 (*c*) 1 and 2 (*d*) 3 only

**B-15.** Assuming A is more electronegative than B, which substance(s) is (are) polar?

(*a*) 1 and 3 (*b*) 2 only (*c*) 4 only (*d*) 2 and 4

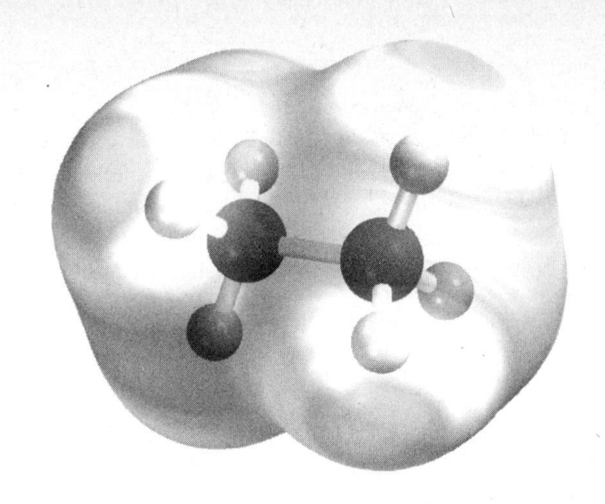

# CHAPTER 2
## ALKANES

### SOLUTIONS TO TEXT PROBLEMS

**2.1** A carbonyl group is C═O. Of the two carbonyl functions in prostaglandin $E_1$ one belongs to the ketone family, the other to the carboxylic acids.

Ketone functional group; Carboxylic acid functional group; O; O; OH; HO; OH

**2.2** An unbranched alkane (*n*-alkane) of 28 carbons has 26 methylene ($CH_2$) groups flanked by a methyl ($CH_3$) group at each end. The condensed formula is $CH_3(CH_2)_{26}CH_3$.

**2.3** The alkane represented by the carbon skeleton formula has 11 carbons. The general formula for an alkane is $C_nH_{2n+2}$, and thus there are 24 hydrogens. The molecular formula is $C_{11}H_{24}$; the condensed structural formula is $CH_3(CH_2)_9CH_3$.

**2.4** In addition to $CH_3(CH_2)_4CH_3$ and $(CH_3)_2CHCH_2CH_2CH_3$, there are three more isomers. One has a five-carbon chain with a one-carbon (methyl) branch:

$CH_3CH_2CH(CH_3)CH_2CH_3$ or

The remaining two isomers have two methyl branches on a four-carbon chain.

$CH_3CH(CH_3)CH(CH_3)CH_3$ or $CH_3CH_2C(CH_3)_2CH_3$ or

**2.5** (*b*) Octacosane is not listed in Table 2.4, but its structure can be deduced from its systematic name. The suffix -cosane pertains to alkanes that contain 20–29 carbons in their longest continuous chain. The prefix octa- means "eight." Octacosane is therefore the unbranched alkane having 28 carbon atoms. It is $CH_3(CH_2)_{26}CH_3$.

(*c*) The alkane has an unbranched chain of 11 carbon atoms and is named **undecane.**

**2.6** The ending -hexadecane reveals that the longest continuous carbon chain has 16 carbon atoms.

1 2 3 4 5 6 7 8 9 10 11 12 13 14 15 16

There are four methyl groups (represented by tetramethyl-), and they are located at carbons 2, 6, 10, and 14.

2,6,10,14-Tetramethylhexadecane
(phytane)

**2.7** (*b*) The systematic name of the unbranched $C_5H_{12}$ isomer is **pentane** (Table 2.4).

$$CH_3CH_2CH_2CH_2CH_3$$

IUPAC name: **pentane**
Common name: *n*-pentane

A second isomer, $(CH_3)_2CHCH_2CH_3$, has four carbons in the longest continuous chain and so is named as a derivative of butane. Since it has a methyl group at C-2, it is **2-methylbutane.**

$$CH_3\underset{\underset{CH_3}{|}}{C}HCH_2CH_3$$

IUPAC name: **2-methylbutane**
Common name: isopentane
methyl group at C-2

The remaining isomer, $(CH_3)_4C$, has three carbons in its longest continuous chain and so is named as a derivative of propane. There are two methyl groups at C-2, and so it is a 2,2-dimethyl derivative of propane.

$$CH_3\overset{\overset{CH_3}{|}}{\underset{\underset{CH_3}{|}}{C}}CH_3$$

IUPAC name: **2,2-dimethylpropane**
Common name: neopentane

(*c*) First write out the structure in more detail, and identify the longest continuous carbon chain.

$$CH_3-\overset{\overset{CH_3}{|}}{\underset{\underset{CH_3}{|}}{C}}-CH_2-\overset{\overset{H}{|}}{\underset{\underset{CH_3}{|}}{C}}-CH_3$$

There are five carbon atoms in the longest chain, and so the compound is named as a derivative of pentane. This five-carbon chain has three methyl substituents attached to it, making it a trimethyl derivative of pentane. Number the chain in the direction that gives the lowest numbers to the substituents at the first point of difference.

$$\overset{1}{CH_3}-\overset{CH_3}{\underset{CH_3}{\overset{2|}{C}}}-\overset{3}{CH_2}-\overset{H}{\underset{CH_3}{\overset{4|}{C}}}-\overset{5}{CH_3} \quad \text{not} \quad \overset{5}{CH_3}\overset{CH_3}{\underset{CH_3}{\overset{4|}{C}}}-\overset{3}{CH_2}-\overset{H}{\underset{CH_3}{\overset{2|}{C}}}-\overset{1}{CH_3}$$

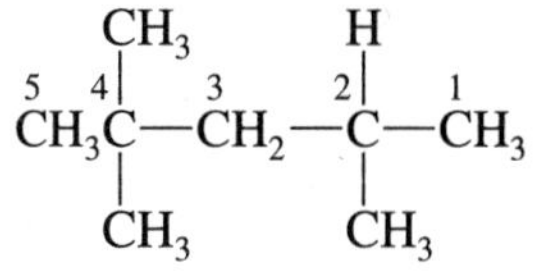

**2,2,4-Trimethylpentane** (correct) 2,4,4-Trimethylpentane (incorrect)

(*d*) The longest continuous chain in $(CH_3)_3CC(CH_3)_3$ contains four carbon atoms.

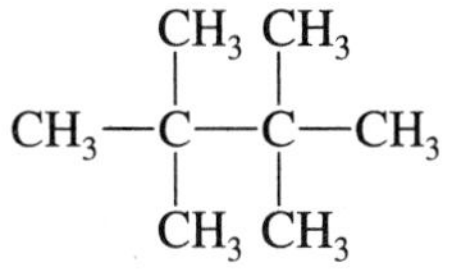

$$CH_3-\overset{CH_3}{\underset{CH_3}{C}}-\overset{CH_3}{\underset{CH_3}{C}}-CH_3$$

The compound is named as a tetramethyl derivative of butane; it is **2,2,3,3-tetramethylbutane.**

**2.8** There are three $C_5H_{11}$ alkyl groups with unbranched carbon chains. One is primary, and two are secondary. The IUPAC name of each group is given beneath the structure. Remember to number the alkyl groups from the point of attachment.

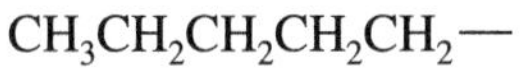

$$CH_3CH_2CH_2CH_2CH_2- \qquad \overset{4}{CH_3}\overset{3}{CH_2}\overset{2}{CH_2}\underset{|}{\overset{1}{C}}HCH_3 \qquad \overset{3}{CH_3}\overset{2}{CH_2}\underset{|}{\overset{1}{C}}HCH_2CH_3$$

Pentyl group (primary) 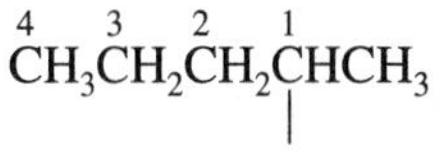 1-Methylbutyl group (secondary)  1-Ethylpropyl group (secondary)

Four alkyl groups are derived from $(CH_3)_2CHCH_2CH_3$. Two are primary, one is secondary, and one is tertiary.

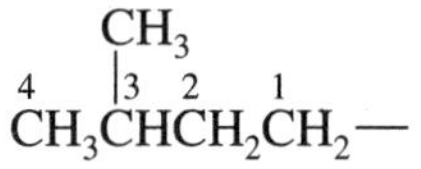

$$\overset{4}{CH_3}\overset{CH_3}{\overset{|3}{C}}H\overset{2}{CH_2}\overset{1}{CH_2}- \qquad -\overset{1}{CH_2}\overset{CH_3}{\overset{|2}{C}}H\overset{3}{CH_2}\overset{4}{CH_3}$$

3-Methylbutyl group (primary) 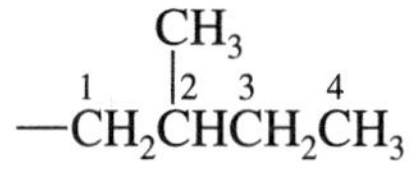 2-Methylbutyl group (primary)

$$CH_3\overset{CH_3}{\underset{|}{\overset{1|}{C}}}\overset{2}{CH_2}\overset{3}{CH_3} \qquad \overset{3}{CH_3}\overset{CH_3}{\overset{|2}{C}}H\underset{|}{\overset{1}{C}}HCH_3$$

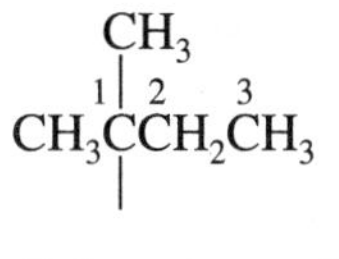

1,1-Dimethylpropyl group (tertiary) 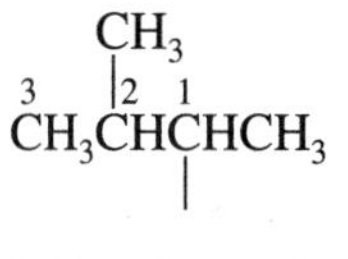 1,2-Dimethylpropyl group (secondary)

**2.9** (*b*) Begin by writing the structure in more detail, showing each of the groups written in parentheses. The compound is named as a derivative of hexane, because it has six carbons in its longest continuous chain.

$$\overset{6}{CH_3}\overset{5}{CH_2}\underset{CH_3CH_2}{\overset{4}{C}}H\overset{3}{CH_2}\underset{CH_3}{\overset{2}{C}}H\overset{1}{CH_3}$$

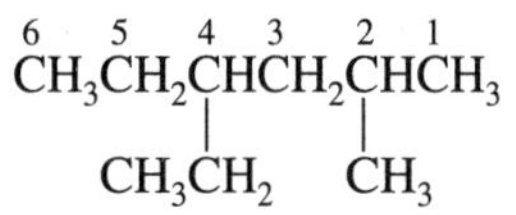

The chain is numbered so as to give the lowest number to the substituent that appears closest to the end of the chain. In this case it is numbered so that the substituents are located at C-2 and C-4 rather than at C-3 and C-5. In alphabetical order the groups are ethyl and methyl; they are listed in alphabetical order in the name. The compound is 4-ethyl-2-methylhexane.

(*c*) The longest continuous chain is shown in the structure; it contains ten carbon atoms. The structure also shows the numbering scheme that gives the lowest number to the substituent at the first point of difference.

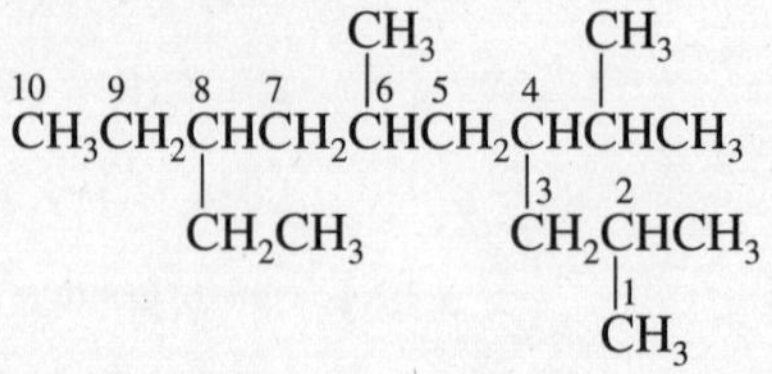

$$\overset{10}{CH_3}\overset{9}{CH_2}\overset{8}{CH}(CH_2CH_3)\overset{7}{CH_2}\overset{6}{CH}(CH_3)\overset{5}{CH_2}\overset{4}{CH}(\overset{3}{CH_2}\overset{2}{CH}(CH_3)\overset{1}{CH_3})CH(CH_3)CH_3$$

In alphabetical order, the substituents are ethyl (at C-8), isopropyl at (C-4), and two methyl groups (at C-2 and C-6). The alkane is 8-ethyl-4-isopropyl-2,6-dimethyldecane. The systematic name for the isopropyl group (1-methylethyl) may also be used, and the name becomes 8-ethyl-2,6-dimethyl-4-(1-methylethyl)decane.

**2.10** (*b*) There are ten carbon atoms in the ring in this cycloalkane, thus it is named as a derivative of cyclodecane.

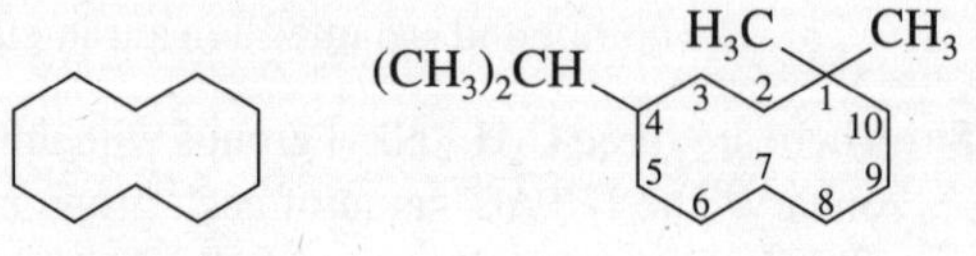

Cyclodecane

$(CH_3)_2CH$ — $H_3C$ $CH_3$ (ring numbered 1–10)

The numbering pattern of the ring is chosen so as to give the lowest number to the substituent at the first point of difference between them. Thus, the carbon bearing two methyl groups is C-1, and the ring is numbered counterclockwise, placing the isopropyl group on C-4 (numbering clockwise would place the isopropyl on C-8). Listing the substituent groups in alphabetical order, the correct name is 4-isopropyl-1,1-dimethylcyclodecane. Alternatively, the systematic name for isopropyl (1-methylethyl) could be used, and the name would become 1,1-dimethyl-4-(1-methylethyl)cyclodecane.

(*c*) When two cycloalkyl groups are attached by a single bond, the compound is named as a cycloalkyl-substituted cycloalkane. This compound is cyclohexylcyclohexane.

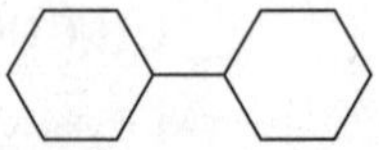

**2.11** The alkane that has the most carbons (nonane) has the highest boiling point (151°C). Among the others, all of which have eight carbons, the unbranched isomer (octane) has the highest boiling point (126°C) and the most branched one (2,2,3,3-tetramethylbutane) the lowest (106°C). The remaining alkane, 2-methylheptane, boils at 116°C.

**2.12** All hydrocarbons burn in air to give carbon dioxide and water. To balance the equation for the combustion of cyclohexane ($C_6H_{12}$), first balance the carbons and the hydrogens on the right side. Then balance the oxygens on the left side.

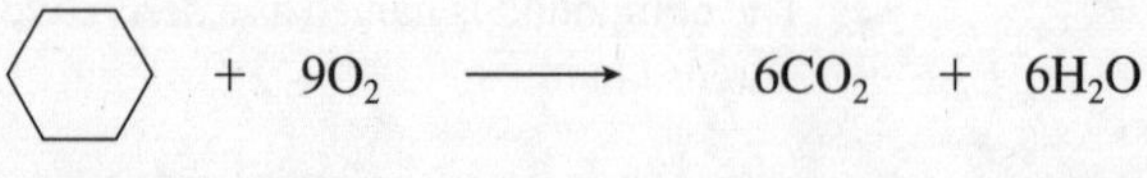

$$\text{Cyclohexane} + 9O_2 \longrightarrow 6CO_2 + 6H_2O$$

Cyclohexane Oxygen Carbon dioxide Water

**2.13** (*b*) Icosane (Table 2.4) is $C_{20}H_{42}$. It has four more methylene ($CH_2$) groups than hexadecane, the last unbranched alkane in Table 2.5. Its calculated heat of combustion is therefore (4 × 653 kJ/mol) higher.

$$\begin{aligned}\text{Heat of combustion of icosane} &= \text{heat of combustion of hexadecane} + 4 \times 653 \text{ kJ/mol}\\ &= 10{,}701 \text{ kJ/mol} + 2612 \text{ kJ/mol}\\ &= 13{,}313 \text{ kJ/mol}\end{aligned}$$

**2.14** Two factors that influence the heats of combustion of alkanes are, in order of decreasing importance, (1) the number of carbon atoms and (2) the extent of chain branching. Pentane, isopentane, and neopentane are all $C_5H_{12}$; hexane is $C_6H_{14}$. Hexane has the largest heat of combustion. Branching leads to a lower heat of combustion; neopentane is the most branched and has the lowest heat of combustion.

| | | |
|---|---|---|
| Hexane | $CH_3(CH_2)_4CH_3$ | Heat of combustion 4163 kJ/mol (995.0 kcal/mol) |
| Pentane | $CH_3CH_2CH_2CH_2CH_3$ | Heat of combustion 3527 kJ/mol (845.3 kcal/mol) |
| Isopentane | $(CH_3)_2CHCH_2CH_3$ | Heat of combustion 3529 kJ/mol (843.4 kcal/mol) |
| Neopentane | $(CH_3)_4C$ | Heat of combustion 3514 kJ/mol (839.9 kcal/mol) |

**2.15** (*b*) In the reaction

$$CH_2{=}CH_2 + Br_2 \longrightarrow BrCH_2CH_2Br$$

carbon becomes bonded to an atom (Br) that is more electronegative than itself. Carbon is *oxidized.*

(*c*) In the reaction

$$6CH_2{=}CH_2 + B_2H_6 \longrightarrow 2(CH_3CH_2)_3B$$

one carbon becomes bonded to hydrogen and is, therefore, *reduced.* The other carbon is also reduced, because it becomes bonded to boron, which is less electronegative than carbon.

**2.16** It is best to approach problems of this type systematically. Since the problem requires all the isomers of $C_7H_{16}$ to be written, begin with the unbranched isomer heptane.

$CH_3CH_2CH_2CH_2CH_2CH_2CH_3$

Heptane

Two isomers have six carbons in their longest continuous chain. One bears a methyl substituent at C-2, the other a methyl substituent at C-3.

$(CH_3)_2CHCH_2CH_2CH_2CH_3$

2-Methylhexane

$CH_3CH_2CH(CH_3)CH_2CH_2CH_3$

3-Methylhexane

Now consider all the isomers that have two methyl groups as substituents on a five-carbon continuous chain.

$(CH_3)_3CCH_2CH_2CH_3$

2,2-Dimethylpentane

$(CH_3CH_2)_2C(CH_3)_2$

3,3-Dimethylpentane

$(CH_3)_2CHCH(CH_3)CH_2CH_3$

2,3-Dimethylpentane

$(CH_3)_2CHCH_2CH(CH_3)_2$

2,4-Dimethylpentane

There is one isomer characterized by an ethyl substituent on a five-carbon chain:

$(CH_3CH_2)_3CH$

3-Ethylpentane

The remaining isomer has three methyl substituents attached to a four-carbon chain.

$(CH_3)_3CCH(CH_3)_2$

2,2,3-Trimethylbutane

**2.17** In the course of doing this problem, you will write and name the 17 alkanes that, in addition to octane, $CH_3(CH_2)_6CH_3$, comprise the 18 constitutional isomers of $C_8H_{18}$.

(*a*) The easiest way to attack this part of the exercise is to draw a bond-line depiction of heptane and add a methyl branch to the various positions.

2-Methylheptane 3-Methylheptane 4-Methylheptane

Other structures bearing a continuous chain of seven carbons would be duplicates of these isomers rather than unique isomers. "5-Methylheptane," for example, is an incorrect name for 3-methylheptane, and "6-methylheptane" is an incorrect name for 2-methylheptane.

(*b*) Six of the isomers named as derivatives of hexane contain two methyl branches on a continuous chain of six carbons.

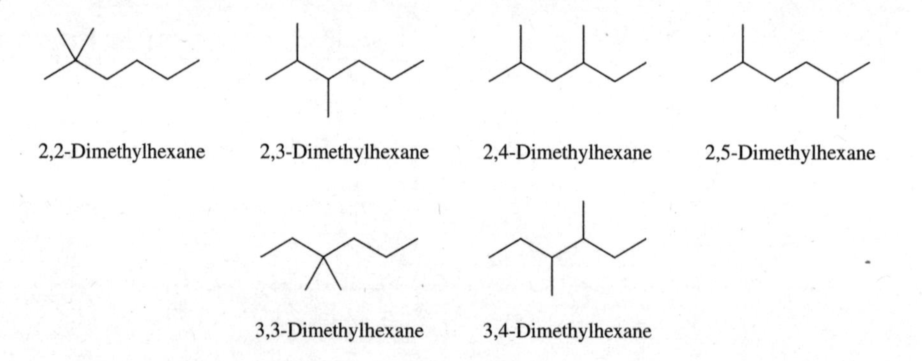

2,2-Dimethylhexane 2,3-Dimethylhexane 2,4-Dimethylhexane 2,5-Dimethylhexane

3,3-Dimethylhexane 3,4-Dimethylhexane

One isomer bears an ethyl substituent:

3-Ethylhexane

(*c*) Four isomers are trimethyl-substituted derivatives of pentane:

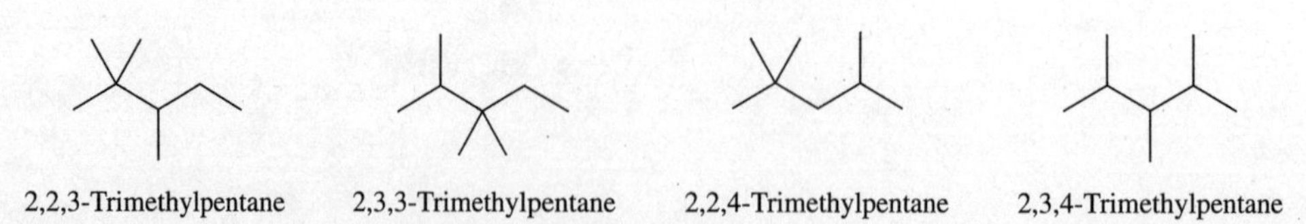

2,2,3-Trimethylpentane 2,3,3-Trimethylpentane 2,2,4-Trimethylpentane 2,3,4-Trimethylpentane

Two bear an ethyl group and a methyl group on a continuous chain of five carbons:

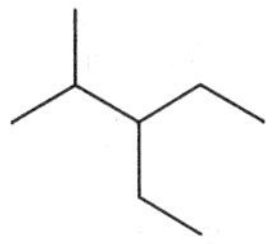

3-Ethyl-2-methylpentane 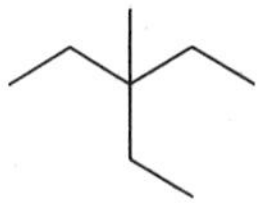    3-Ethyl-3-methylpentane

(*d*) Only one isomer is named as a derivative of butane:

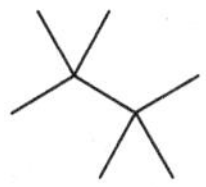

2,2,3,3-Tetramethylbutane

**2.18** (*a*) The longest continuous chain contains nine carbon atoms. Begin the problem by writing and numbering the carbon skeleton of nonane.

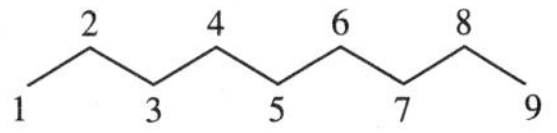

Now add two methyl groups (one to C-2 and the other to C-3) and an isopropyl group (to C-6) to give a structural formula for 6-isopropyl-2,3-dimethylnonane.

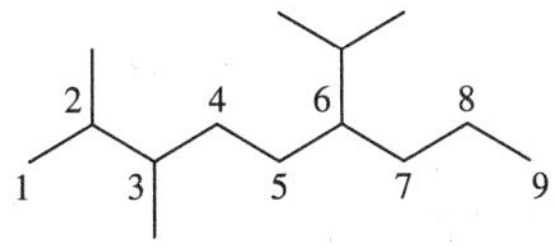

or $CH_3CH(CH_3)CH(CH_3)CH_2CH_2CH(CH(CH_3)_2)CH_2CH_2CH_3$

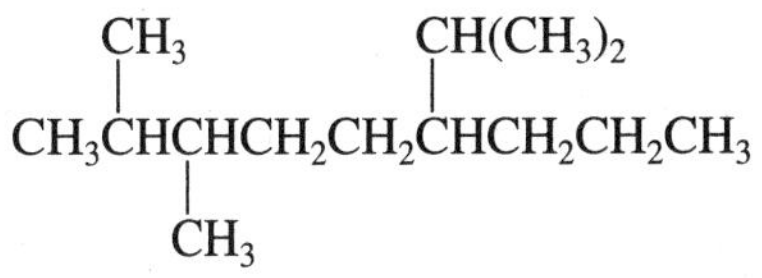

(*b*) To the carbon skeleton of heptane (seven carbons) add a *tert*-butyl group to C-4 and a methyl group to C-3 to give 4-*tert*-butyl-3-methylheptane.

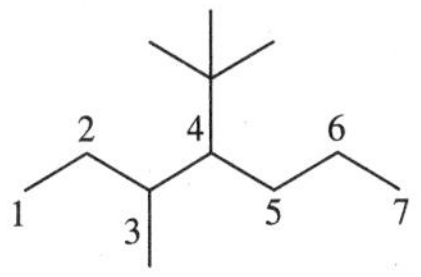

or $CH_3CH_2CH(CH_3)CH(C(CH_3)_3)CH_2CH_2CH_3$

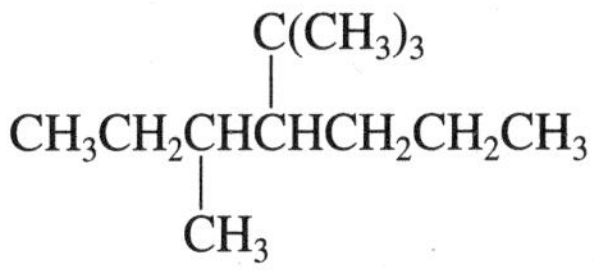

(*c*) An isobutyl group is —$CH_2CH(CH_3)_2$. The structure of 4-isobutyl-1,1-dimethylcyclohexane is as shown.

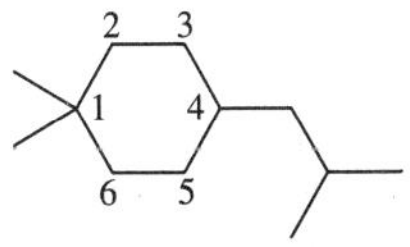

or $(H_3C)_2C_6H_{10}$—$CH_2CH(CH_3)_2$

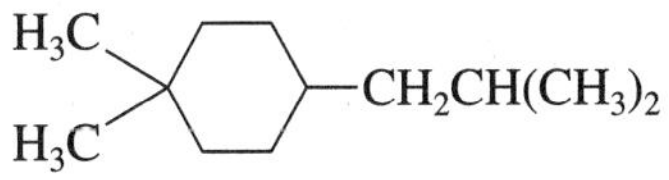

(*d*) A *sec*-butyl group is $CH_3\underset{|}{C}HCH_2CH_3$. *sec*-Butylcycloheptane has a *sec*-butyl group on a seven-membered ring.

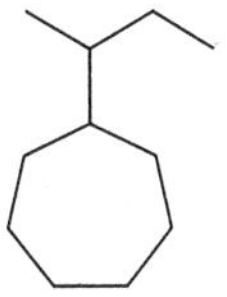

or $CH_3CHCH_2CH_3$ (attached at C-2 to cycloheptane)

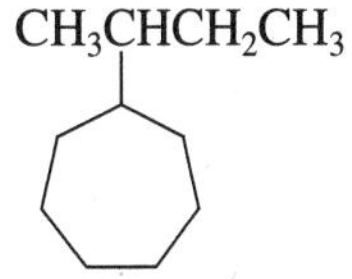

(*e*) A cyclobutyl group is a substituent on a five-membered ring in cyclobutylcyclopentane.

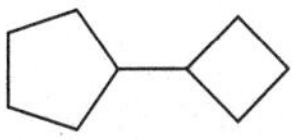

(*f*) Recall that an alkyl group is numbered from the point of attachment. The structure of (2,2-dimethylpropyl)cyclohexane is

$$\text{C}_6\text{H}_{11}\text{—CH}_2\text{—C(CH}_3)_2\text{—CH}_3$$

(*g*) The name "pentacosane" contains no numerical locants or suffixes indicating the presence of alkyl groups. It must therefore be an unbranched alkane. Table 2.4 in the text indicates that the suffix -cosane refers to alkanes with 20–29 carbons. The prefix penta- stands for "five," and so pentacosane must be the unbranched alkane with 25 carbons. Its condensed structural formula is $CH_3(CH_2)_{23}CH_3$.

(*h*) We need to add a 1-methylpentyl group to C-10 of pentacosane. A 1-methylpentyl group is:

$$\overset{1}{\text{—CH}}(\text{CH}_3)\overset{2}{\text{CH}_2}\overset{3}{\text{CH}_2}\overset{4}{\text{CH}_2}\overset{5}{\text{CH}_3}$$

It has five carbons in the longest continuous chain counting from the point of attachment and bears a methyl group at C-1. 10-(1-Methylpentyl)pentacosane is therefore:

$$\text{CH}_3(\text{CH}_2)_8\text{CH}(\text{CH}_2)_{14}\text{CH}_3 \text{ with } \text{CH}_3\text{CHCH}_2\text{CH}_2\text{CH}_2\text{CH}_3 \text{ attached at CH}$$

**2.19** (*a*) This compound is an unbranched alkane with 27 carbons. As noted in part (*g*) of the preceding problem, alkanes with 20–29 carbons have names ending in -cosane. Thus, we add the prefix hepta- ("seven") to -cosane to name the alkane $CH_3(CH_2)_{25}CH_3$ as **heptacosane.**

(*b*) The alkane $(CH_3)_2CHCH_2(CH_2)_{14}CH_3$ has 18 carbons in its longest continuous chain. It is named as a derivative of **octadecane.** There is a single substituent, a methyl group at C-2. The compound is **2-methyloctadecane.**

(*c*) Write the structure out in more detail to reveal that it is **3,3,4-triethylhexane.**

$$(\text{CH}_3\text{CH}_2)_3\text{CCH}(\text{CH}_2\text{CH}_3)_2 \quad \text{is rewritten as} \quad \overset{1}{\text{CH}_3}\overset{2}{\text{CH}_2}\overset{3}{\text{C}}(\text{CH}_2\text{CH}_3)_2\text{—}\overset{4}{\text{CH}}(\text{CH}_2\text{CH}_3)\overset{5}{\text{CH}_2}\overset{6}{\text{CH}_3}$$

(*d*) Each line of a bond-line formula represents a bond between two carbon atoms. Hydrogens are added so that the number of bonds to each carbon atom totals four.

[bond-line formula] is the same as $CH_3CH_2CH(CH_2CH_3)CH_2C(CH_3)_3$

The IUPAC name is **4-ethyl-2,2-dimethylhexane.**

(*e*) [bond-line formula] is the same as $CH_3CH_2CH(CH_3)CH_2CH(CH_3)CH_2CH_3$

The IUPAC name is **3,5-dimethylheptane.**

(*f*)

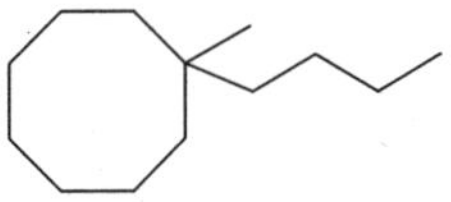

is the same as

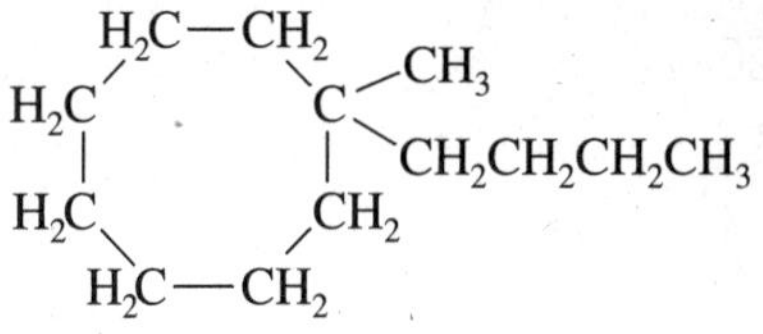

$$\begin{array}{c} H_2C-CH_2 \quad CH_3 \\ H_2C \quad\quad C \\ | \quad\quad | \;\; CH_2CH_2CH_2CH_3 \\ H_2C \quad\quad CH_2 \\ H_2C-CH_2 \end{array}$$

The IUPAC name is **1-butyl-1-methylcyclooctane.**

(*g*) Number the chain in the direction shown to give **3-ethyl-4,5,6-trimethyloctane.** When numbered in the opposite direction, the locants are also 3, 4, 5, and 6. In the case of ties, however, choose the direction that gives the lower number to the substituent that appears first in the name. "Ethyl" precedes "methyl" alphabetically.

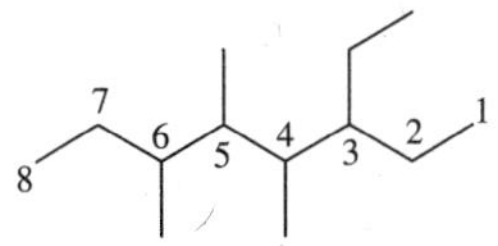

**2.20** (*a*) The alkane contains 13 carbons. Since all alkanes have the molecular formula $C_nH_{2n+2}$, the molecular formula must be $C_{13}H_{28}$.

(*b*) The longest continuous chain is indicated and numbered as shown.

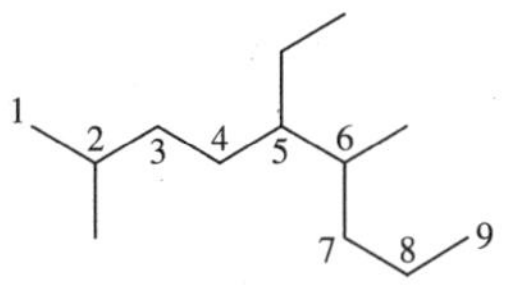

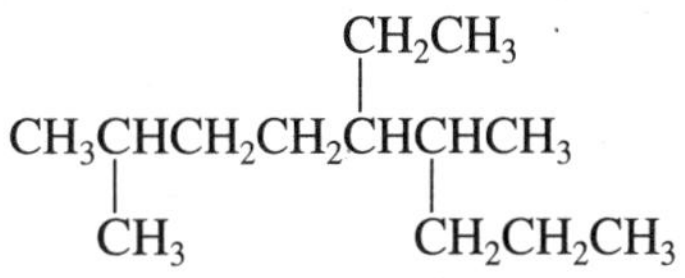

$$\begin{array}{c} CH_2CH_3 \\ CH_3CHCH_2CH_2CHCHCH_3 \\ CH_3 \qquad\qquad CH_2CH_2CH_3 \end{array}$$

In alphabetical order, the substituents are ethyl (at C-5), methyl (at C-2), methyl (at C-6). The IUPAC name is **5-ethyl-2,6-dimethylnonane.**

(*c*) Fill in the hydrogens in the alkane to identify the various kinds of groups present. There are five **methyl** ($CH_3$) groups, five **methylene** ($CH_2$) groups, and three **methine** (CH) groups in the molecule.

(*d*) A primary carbon is attached to one other carbon. There are five primary carbons (the carbons of the five $CH_3$ groups). A secondary carbon is attached to two other carbons, and there are five of these (the carbons of the five $CH_2$ groups). A tertiary carbon is attached to three other carbons, and there are three of these (the carbons of the three methine groups). A quaternary carbon is attached to four other carbons. None of the carbons is a quaternary carbon.

**2.21** (*a*) The group $CH_3(CH_2)_{10}CH_2—$ is an unbranched alkyl group with 12 carbons. It is a **dodecyl group.** The carbon at the point of attachment is directly attached to only one other carbon. It is a primary alkyl group.

(*b*) The longest continuous chain from the point of attachment is six carbons; it is a hexyl group bearing an ethyl substituent at C-3. The group is a **3-ethylhexyl group.** It is a primary alkyl group.

$$\begin{array}{c} 1 \quad 2 \quad 3 \quad 4 \quad 5 \quad 6 \\ —CH_2CH_2CHCH_2CH_2CH_3 \\ CH_2CH_3 \end{array}$$

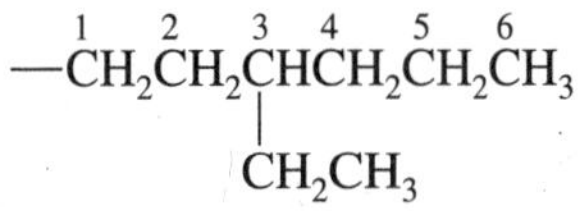

(*c*) By writing the structural formula of this alkyl group in more detail, we see that the longest continuous chain from the point of attachment contains three carbons. It is a **1,1-diethylpropyl** group. Because the carbon at the point of attachment is directly bonded to three other carbons, it is a tertiary alkyl group.

$$—C(CH_2CH_3)_3 \quad \text{is rewritten as} \quad \begin{array}{c} CH_2CH_3 \\ 1 \;\; 2 \;\; 3 \\ —CCH_2CH_3 \\ CH_2CH_3 \end{array}$$

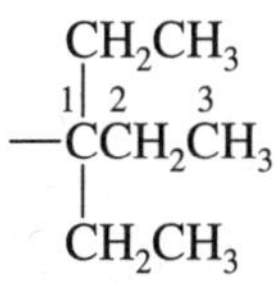

(*d*) This group contains four carbons in its longest continuous chain. It is named as a butyl group with a cyclopropyl substituent at C-1. It is a **1-cyclopropylbutyl** group and is a secondary alkyl group.

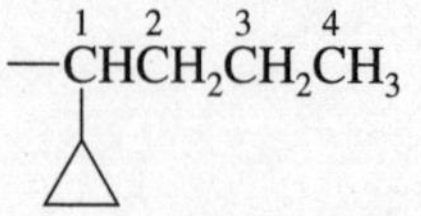

$$\overset{1}{—CH}\overset{2}{CH_2}\overset{3}{CH_2}\overset{4}{CH_3}$$

(*e, f*) A two-carbon group that bears a cyclohexyl substituent is a **cyclohexylethyl** group. Number from the point of attachment when assigning a locant to the cyclohexyl group.

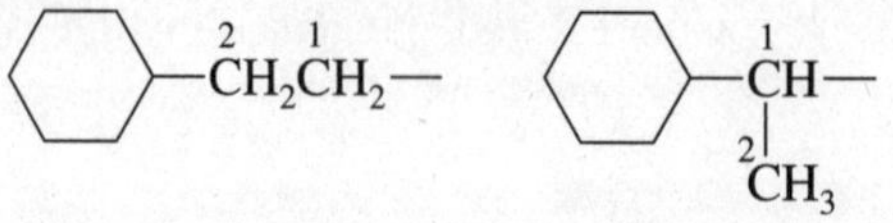

2-Cyclohexylethyl (primary)    1-Cyclohexylethyl (secondary)

**2.22** The IUPAC name for pristane reveals that the longest chain contains 15 carbon atoms (as indicated by -pentadecane). The chain is substituted with four methyl groups at the positions indicated in the name.

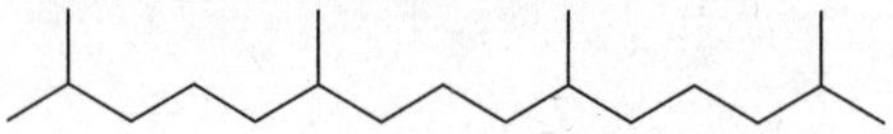

Pristane (2,6,10,14-tetramethylpentadecane)

**2.23** (*a*) An alkane having 100 carbon atoms has 2(100) + 2 = 202 hydrogens. The molecular formula of hectane is $C_{100}H_{202}$ and the condensed structural formula is $CH_3(CH_2)_{98}CH_3$. The 100 carbon atoms are connected by 99 $\sigma$ bonds. The total number of $\sigma$ bonds is 301 (99 C—C bonds + 202 C—H bonds).

(*b*) Unique compounds are formed by methyl substitution at carbons 2 through 50 on the 100-carbon chain (C-51 is identical to C-50, and so on). There are 49 *x*-methylhectanes.

(*c*) Compounds of the type 2,*x*-dimethylhectane can be formed by substitution at carbons 2 through 99. There are 98 of these compounds.

**2.24** Isomers are different compounds that have the same molecular formula. In all these problems the safest approach is to write a structural formula and then count the number of carbons and hydrogens.

(*a*) Among this group of compounds, only butane and isobutane have the same molecular formula; only these two are isomers.

$CH_3CH_2CH_2CH_3$ — Butane, $C_4H_{10}$
Cyclobutane, $C_4H_8$
$CH_3CH(CH_3)CH_3$ — Isobutane, $C_4H_{10}$
$CH_3CH(CH_3)CH_2CH_3$ — 2-Methylbutane, $C_5H_{12}$

(*b*) The two compounds that are isomers, that is, those that have the same molecular formula, are 2,2-dimethylpentane and 2,2,3-trimethylbutane.

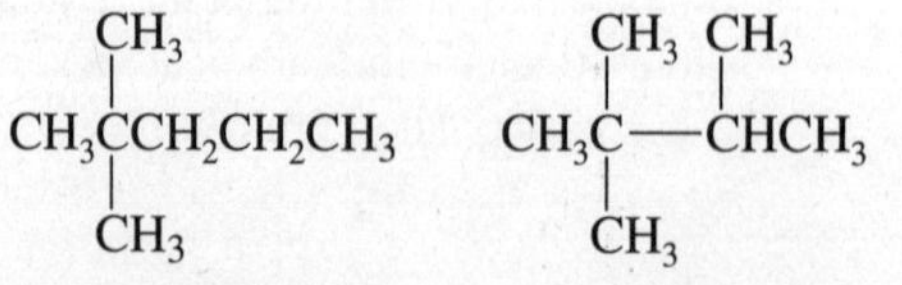

$CH_3C(CH_3)_2CH_2CH_2CH_3$ — 2,2-Dimethylpentane, $C_7H_{16}$
$CH_3C(CH_3)_2—CH(CH_3)CH_3$ — 2,2,3-Trimethylbutane, $C_7H_{16}$

Cyclopentane and neopentane are not isomers of these two compounds, nor are they isomers of each other.

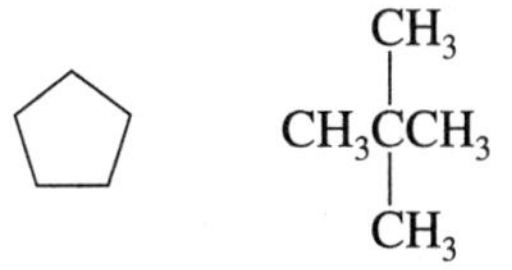

$$CH_3CCH_3$$ (with two $CH_3$ on the central carbon)

Cyclopentane $C_5H_{10}$ Neopentane $C_5H_{12}$

(*c*) The compounds that are isomers are cyclohexane, methylcyclopentane, and 1,1,2-trimethylcyclopropane.

$CH_3$— (cyclopentyl) ; $H_3C$, $CH_3$, $CH_3$ (on cyclopropane)

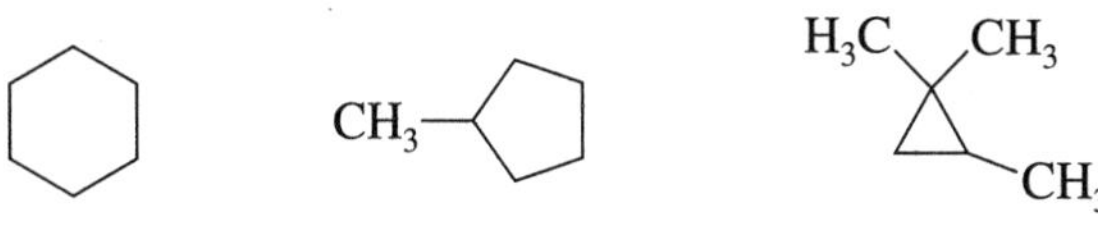

Cyclohexane $C_6H_{12}$ Methylcyclopentane $C_6H_{12}$ 1,1,2-Trimethylcyclopropane $C_6H_{12}$

Hexane, $CH_3CH_2CH_2CH_2CH_2CH_3$, has the molecular formula $C_6H_{14}$; it is not an isomer of the others.

(*d*) The three that are isomers all have the molecular formula $C_5H_{10}$.

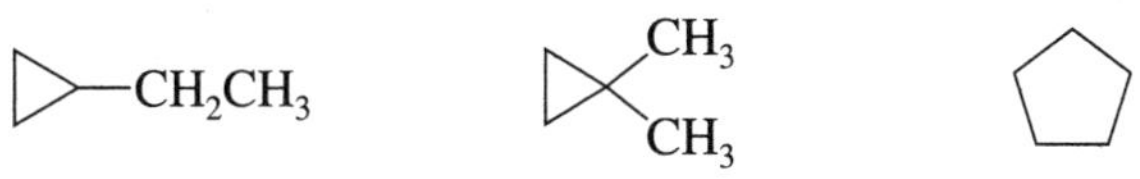

—$CH_2CH_3$ ; $CH_3$, $CH_3$

Ethylcyclopropane $C_5H_{10}$ 1,1-Dimethylcyclopropane $C_5H_{10}$ Cyclopentane $C_5H_{10}$

Propylcyclopropane is not an isomer of the others. Its molecular formula is $C_6H_{12}$.

—$CH_2CH_2CH_3$

(*e*) Only 4-methyltetradecane and pentadecane are isomers. Both have the molecular formula $C_{15}H_{32}$.

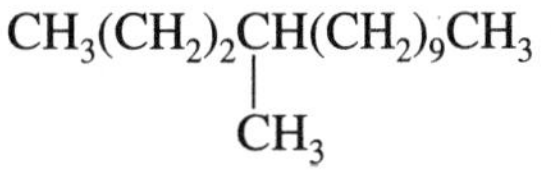

$CH_3(CH_2)_2CH(CH_2)_9CH_3$ (with $CH_3$ on CH) $CH_3(CH_2)_{13}CH_3$

4-Methyltetradecane $C_{15}H_{32}$ Pentadecane $C_{15}H_{32}$

$CH_3CHCHCHCH(CH_2)_4CH_3$ (with $CH_3$ $CH_3$ above and $CH_3$ $CH_3$ below) $CH_3CH_2CH_2CH(CH_2)_5CH_3$ (with cyclobutyl on CH)

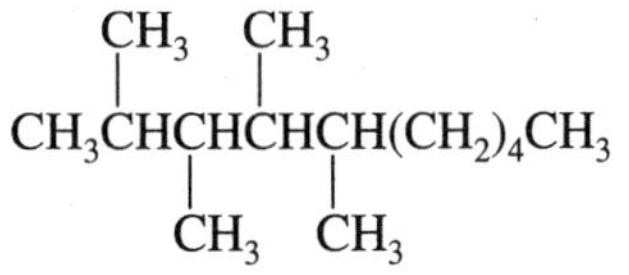

2,3,4,5-Tetramethyldecane $C_{14}H_{30}$ 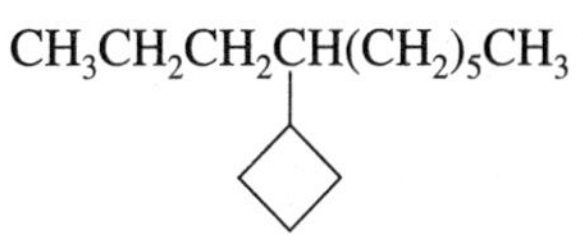 4-Cyclobutyldecane $C_{14}H_{28}$

**2.25** The oxygen and two of the carbons of $C_3H_5ClO$ are part of the structural unit that characterizes epoxides. The problem specifies that a methyl group ($CH_3$) is *not* present; therefore, add the remaining carbon and the chlorine as a $—CH_2Cl$ unit, and fill in the remaining bonds with hydrogen substituents.

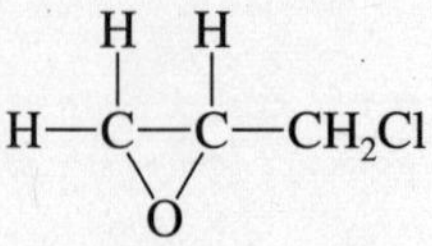

$$\begin{array}{c} \text{H}\quad\text{H} \\ | \quad\; | \\ \text{H—C—C—}CH_2Cl \\ \backslash\;/ \\ \text{O} \end{array}$$

Epichlorohydrin

**2.26** (*a*) Ibuprofen is

$$(CH_3)_2CHCH_2\text{—}C_6H_4\text{—}\underset{\substack{|\\ CH_3}}{CH}\text{—}\overset{\text{O}}{\overset{\|}{C}}OH$$

(*b*) Mandelonitrile is

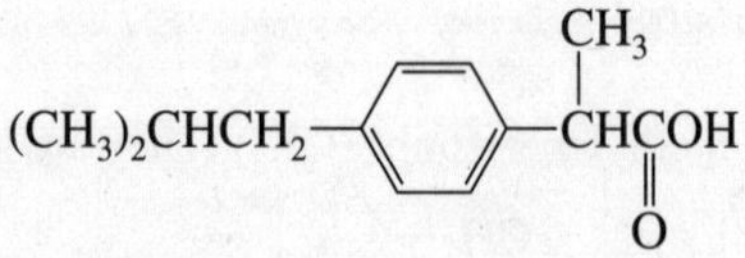

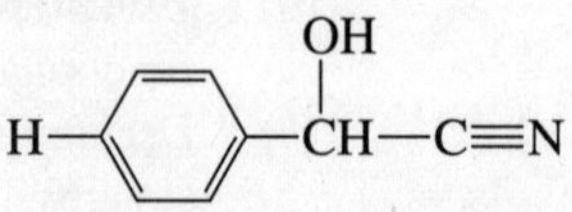

$$\text{H—}C_6H_4\text{—}\underset{}{\overset{\text{OH}}{\overset{|}{\text{CH}}}}\text{—C}\equiv\text{N}$$

**2.27** Isoamyl acetate is

$$\overset{\text{O}}{\overset{\|}{\text{RCOR}'}} \text{ (ester)} \qquad \text{which is} \qquad CH_3\overset{\text{O}}{\overset{\|}{C}}OCH_2CH_2\overset{CH_3}{\overset{|}{C}}HCH_3$$

Methyl 3-Methylbutyl

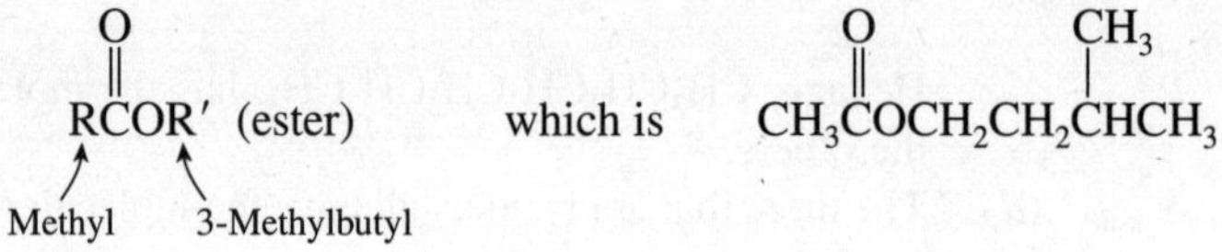

**2.28** Thiols are characterized by the —SH group. *n*-Butyl mercaptan is $CH_3CH_2CH_2CH_2SH$.

**2.29** $\alpha$-Amino acids have the general formula

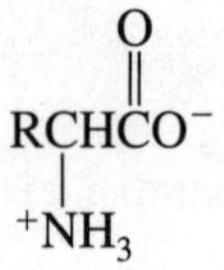

$$\text{R}\underset{\substack{|\\ ^{+}NH_3}}{\text{CH}}\overset{\text{O}}{\overset{\|}{\text{C}}}\text{O}^-$$

The individual amino acids in the problem have the structures shown:

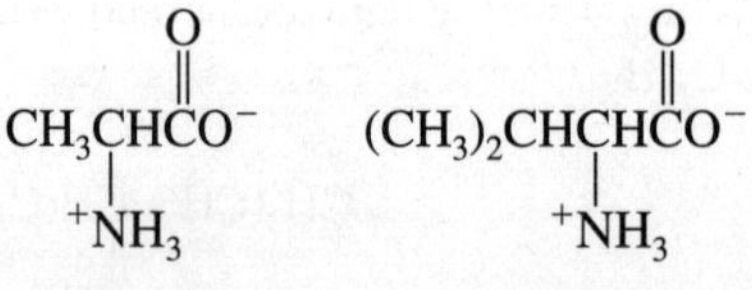

$$CH_3\underset{\substack{|\\ ^{+}NH_3}}{CH}\overset{\text{O}}{\overset{\|}{C}}O^- \qquad (CH_3)_2CH\underset{\substack{|\\ ^{+}NH_3}}{CH}\overset{\text{O}}{\overset{\|}{C}}O^-$$

(*a*) Alanine (*b*) Valine

(*c, d*) An isobutyl group is $(CH_3)_2CHCH_2—$, and a *sec*-butyl group is

$$CH_3\underset{|}{C}HCH_2CH_3$$

The structures of leucine and isoleucine are:

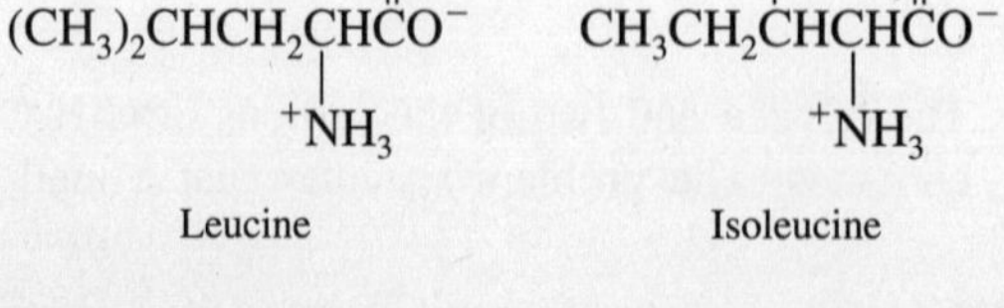

$$(CH_3)_2CHCH_2\underset{\substack{|\\ ^{+}NH_3}}{CH}\overset{\text{O}}{\overset{\|}{C}}O^- \qquad CH_3CH_2\overset{CH_3}{\overset{|}{C}}H\underset{\substack{|\\ ^{+}NH_3}}{CH}\overset{\text{O}}{\overset{\|}{C}}O^-$$

Leucine Isoleucine

(*e–g*) The functional groups that characterize alcohols, thiols, and carboxylic acids are —OH, —SH, and —$CO_2H$, respectively. The structures of serine, cysteine, and aspartic acid are:

$$HOCH_2CH(\overset{+}{N}H_3)C(=O)O^- \qquad HSCH_2CH(\overset{+}{N}H_3)C(=O)O^- \qquad HOC(=O)CH_2CH(\overset{+}{N}H_3)C(=O)O^-$$

Serine — Cysteine — Aspartic acid

**2.30** Uscharidin has the structure shown.

(*a*) There are two alcohol groups, one aldehyde group, one ketone group, and one ester functionality.

(*b*) Uscharidin contains ten methylene groups ($CH_2$). They are indicated in the structure by small squares.

(*c*) The primary carbons in uscharidin are the carbons of the two methyl groups.

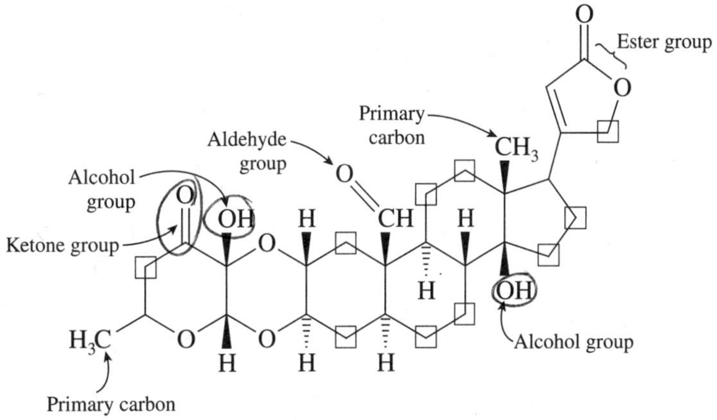

Ester group; Primary carbon; Aldehyde group; Alcohol group; Ketone group; Alcohol group; Primary carbon

**2.31** (*a*) Methylene groups are —$CH_2$—. $ClCH_2CH_2CH_2CH_2Cl$ is therefore the $C_4H_8Cl_2$ isomer in which all the carbons belong to methylene groups.

(*b*) The $C_4H_8Cl_2$ isomers that lack methylene groups are

$$(CH_3)_2CHCHCl_2 \quad \text{and} \quad CH_3CH(Cl)CH(Cl)CH_3$$

**2.32** Since it is an alkane, the sex attractant of the tiger moth has a molecular formula of $C_nH_{2n+2}$. The number of carbons and hydrogens may be calculated from its molecular weight.

$$12n + 1(2n + 2) = 254$$
$$14n = 252$$
$$n = 18$$

The molecular formula of the alkane is $C_{18}H_{38}$. In the problem it is stated that the sex attractant is a 2-methyl-branched alkane. It is therefore 2-methylheptadecane, $(CH_3)_2CHCH_2(CH_2)_{13}CH_3$.

**2.33** When any hydrocarbon is burned in air, the products of combustion are carbon dioxide and water.

(*a*) $CH_3(CH_2)_8CH_3 + \frac{31}{2}O_2 \longrightarrow 10CO_2 + 11H_2O$

Decane ($C_{10}H_{22}$) — Oxygen — Carbon dioxide — Water

(*b*) [cyclodecane ring] + $15O_2$ ⟶ $10CO_2 + 10H_2O$

Cyclodecane ($C_{10}H_{20}$) Oxygen Carbon dioxide Water

(*c*) [methylcyclononane ring]—$CH_3$ + $15O_2$ ⟶ $10CO_2 + 10H_2O$

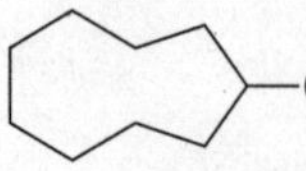

Methylcyclononane ($C_{10}H_{20}$) Oxygen Carbon dioxide Water

(*d*) [cyclopentylcyclopentane] + $\frac{29}{2}O_2$ ⟶ $10CO_2 + 9H_2O$

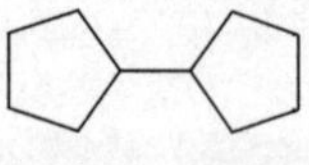

Cyclopentylcyclopentane ($C_{10}H_{18}$) Oxygen Carbon dioxide Water

**2.34** To determine the quantity of heat evolved per unit mass of material, divide the heat of combustion by the molecular weight.

Methane: Heat of combustion = 890 kJ/mol (212.8 kcal/mol)
Molecular weight = 16.0 g/mol
Heat evolved per gram = 55.6 kJ/g (13.3 kcal/g)

Butane: Heat of combustion = 2876 kJ/mol (687.4 kcal/mol)
Molecular weight = 58.0 g/mol
Heat evolved per gram = 49.6 kJ/g (11.8 kcal/g)

When equal masses of methane and butane are compared, methane evolves more heat when it is burned.

Equal volumes of gases contain an equal number of moles, so that when equal volumes of methane and butane are compared, the one with the greater heat of combustion in kilojoules (or kilocalories) per mole gives off more heat. Butane evolves more heat when it is burned than does an equal volume of methane.

**2.35** When comparing heats of combustion of alkanes, two factors are of importance:

1. The heats of combustion of alkanes increase as the number of carbon atoms increases.
2. An unbranched alkane has a greater heat of combustion than a branched isomer.

(*a*) In the group hexane, heptane, and octane, three unbranched alkanes are being compared. Octane ($C_8H_{18}$) has the most carbons and has the greatest heat of combustion. Hexane ($C_6H_{14}$) has the fewest carbons and the lowest heat of combustion. The measured values in this group are as follows:

Hexane: Heat of combustion 4163 kJ/mol (995.0 kcal/mol)
Heptane: Heat of combustion 4817 kJ/mol (1151.3 kcal/mol)
Octane: Heat of combustion 5471 kJ/mol (1307.5 kcal/mol)

(*b*) Isobutane has fewer carbons than either pentane or isopentane and so is the member of the group with the lowest heat of combustion. Isopentane is a 2-methyl-branched isomer of pentane and so has a lower heat of combustion. Pentane has the highest heat of combustion among these compounds.

| | | |
|---|---|---|
| Isobutane | $(CH_3)_3CH$ | Heat of combustion 2868 kJ/mol (685.4 kcal/mol) |
| Isopentane | $(CH_3)_2CHCH_2CH_3$ | Heat of combustion 3529 kJ/mol (843.4 kcal/mol) |
| Pentane | $CH_3CH_2CH_2CH_2CH_3$ | Heat of combustion 3527 kJ/mol (845.3 kcal/mol) |

(*c*) Isopentane and neopentane each have fewer carbons than 2-methylpentane, which therefore has the greatest heat of combustion. Neopentane is more highly branched than isopentane; neopentane has the lowest heat of combustion.

| | | |
|---|---|---|
| Neopentane | $(CH_3)_4C$ | Heat of combustion 3514 kJ/mol (839.9 kcal/mol) |
| Isopentane | $(CH_3)_2CHCH_2CH_3$ | Heat of combustion 3529 kJ/mol (843.4 kcal/mol) |
| 2-Methylpentane | $(CH_3)_2CHCH_2CH_2CH_3$ | Heat of combustion 4157 kJ/mol (993.6 kcal/mol) |

(*d*) Chain branching has a small effect on heat of combustion; the number of carbons has a much larger effect. The alkane with the most carbons in this group is 3,3-dimethylpentane; it has the greatest heat of combustion. Pentane has the fewest carbons in this group and has the smallest heat of combustion.

| | | |
|---|---|---|
| Pentane | $CH_3CH_2CH_2CH_2CH_3$ | Heat of combustion 3527 kJ/mol (845.3 kcal/mol) |
| 3-Methylpentane | $(CH_3CH_2)_2CHCH_3$ | Heat of combustion 4159 kJ/mol (994.1 kcal/mol) |
| 3,3-Dimethylpentane | $(CH_3CH_2)_2C(CH_3)_2$ | Heat of combustion 4804 kJ/mol (1148.3 kcal/mol) |

(*e*) In this series the heat of combustion increases with increasing number of carbons. Ethylcyclopentane has the lowest heat of combustion; ethylcycloheptane has the greatest.

| $CH_3CH_2$–cyclopentyl | $CH_3CH_2$–cyclohexyl | $CH_3CH_2$–cycloheptyl |
|---|---|---|
| Ethylcyclopentane<br>4592 kJ/mol<br>(1097.5 kcal/mol) | Ethylcyclohexane<br>5222 kJ/mol<br>(1248.2 kcal/mol) | Ethylcycloheptane<br>(combustion data not available) |

**2.36** (*a*) The equation for the hydrogenation of ethylene is given by the sum of the following three reactions:

$$\begin{array}{llll}
(1) & H_2(g) + \frac{1}{2}O_2(g) \longrightarrow & H_2O(l) & \Delta H^\circ = -286 \text{ kJ } (-68.4 \text{ kcal}) \\
(2) & H_2C{=}CH_2(g) + 3O_2(g) \longrightarrow & 2CO_2(g) + 2H_2O(l) & \Delta H^\circ = -1410 \text{ kJ } (-337.0 \text{ kcal}) \\
(3) & 3H_2O(l) + 2CO_2(g) \longrightarrow & CH_3CH_3(g) + \frac{7}{2}O_2(g) & \Delta H^\circ = +1560 \text{ kJ } (+372.8 \text{ kcal}) \\
\hline
\text{Sum:} & H_2C{=}CH_2(g) + H_2(g) \longrightarrow & CH_3CH_3(g) & \Delta H^\circ = -136 \text{ kJ } (-32.6 \text{ kcal})
\end{array}$$

Equations (1) and (2) are the combustion of hydrogen and ethylene, respectively, and $\Delta H^\circ$ values for these reactions are given in the statement of the problem. Equation (3) is the reverse of the combustion of ethane, and its value of $\Delta H^\circ$ is the negative of the heat of combustion of ethane.

(*b*) Again we need to collect equations of reactions for which the $\Delta H^\circ$ values are known.

$$\begin{array}{llll}
(1) & H_2(g) + \frac{1}{2}O_2(g) \longrightarrow & H_2O(l) & \Delta H^\circ = -286 \text{ kJ } (-68.4 \text{ kcal}) \\
(2) & HC{\equiv}CH(g) + \frac{5}{2}O_2(g) \longrightarrow & 2CO_2(g) + H_2O(l) & \Delta H^\circ = -1300 \text{ kJ } (-310.7 \text{ kcal}) \\
(3) & 2CO_2(g) + 2H_2O(l) \longrightarrow & CH_2{=}CH_2(g) + 3O_2(g) & \Delta H^\circ = +1410 \text{ kJ } (+337.0 \text{ kcal}) \\
\hline
\text{Sum:} & HC{\equiv}CH(g) + H_2(g) \longrightarrow & CH_2{=}CH_2(g) & \Delta H^\circ = -176 \text{ kJ } (-42.1 \text{ kcal})
\end{array}$$

Equations (1) and (2) are the combustion of hydrogen and acetylene, respectively. Equation (3) is the reverse of the combustion of ethylene, and its value of $\Delta H°$ is the negative of the heat of combustion of ethylene.

The value of $\Delta H°$ for the hydrogenation of acetylene to ethane is equal to the sum of the two reactions just calculated:

$$HC{\equiv}CH(g) + H_2(g) \longrightarrow H_2C{=}CH_2(g) \qquad \Delta H° = -176\text{ kJ } (-42.1\text{ kcal})$$

$$H_2C{=}CH_2(g) + H_2(g) \longrightarrow CH_3CH_3(g) \qquad \Delta H° = -136\text{ kJ } (-32.6\text{ kcal})$$

$$\text{Sum:} \quad HC{\equiv}CH(g) + 2H_2(g) \longrightarrow CH_3CH_3(g) \qquad \Delta H° = -312\text{ kJ } (-74.7\text{ kcal})$$

(*c*) We use the equations for the combustion of ethane, ethylene, and acetylene as shown.

$$(1) \quad 2CH_2{=}CH_2(g) + 6O_2(g) \longrightarrow 4CO_2(g) + 4H_2O(l) \qquad \Delta H° = -2820\text{ kJ } (-674.0\text{ kcal})$$

$$(2) \quad 2CO_2(g) + H_2O(l) \longrightarrow HC{\equiv}CH(g) + \tfrac{5}{2}O_2(g) \qquad \Delta H° = +1300\text{ kJ } (+310.7\text{ kcal})$$

$$(3) \quad 3H_2O(l) + 2CO_2(g) \longrightarrow CH_3CH_3(g) + \tfrac{7}{2}O_2(g) \qquad \Delta H° = +1560\text{ kJ } (+372.8\text{ kcal})$$

$$\text{Sum:} \quad 2CH_2{=}CH_2(g) \longrightarrow CH_3CH_3(g) + HC{\equiv}CH(g) \qquad \Delta H° = +40\text{ kJ } (+9.5\text{ kcal})$$

The value of $\Delta H°$ for reaction (1) is twice that for the combustion of ethylene because 2 mol of ethylene are involved.

**2.37** (*a*) The hydrogen content increases in going from $CH_3C{\equiv}CH$ to $CH_3CH{=}CH_2$. The organic compound $CH_3C{\equiv}CH$ is *reduced.*

(*b*) *Oxidation* occurs because a C—O bond has replaced a C—H bond in going from starting material to product.

H OH —oxidation→ O

(*c*) There are two carbon–oxygen bonds in the starting material and four carbon–oxygen bonds in the products. *Oxidation* occurs.

$$HO{-}CH_2CH_2{-}OH \longrightarrow 2H_2C{=}O$$

Two C—O bonds     Four C—O bonds

(*d*) Although the oxidation state of carbon is unchanged in the process

$$C_6H_5{-}NO_2 \longrightarrow C_6H_5{-}\overset{+}{N}H_3$$

overall, *reduction* of the organic compound has occurred. Its hydrogen content has increased and its oxygen content has decreased.

**2.38** In the reaction

$$2CH_3Cl + Si \longrightarrow (CH_3)_2SiCl_2$$

bonds between carbon and an atom more electronegative than itself (chlorine) are replaced by bonds between carbon and an atom less electronegative than itself (silicon). Carbon is reduced; silicon is oxidized.

**2.39** (*a*) Compound A has the structural unit $C\overset{O}{\overset{\|}{C}}C$; compound A is a ketone.

(*b*) Converting a ketone to an ester increases the oxygen content of carbon and requires an oxidizing agent.

(*c*) Reduction occurs when the hydrogen content increases, as in the conversion of a ketone to an alkane or to an alcohol. Reductions are carried out by using reagents that are reducing agents.

(*d*) Compound A, $CH_3C(=O)C(CH_3)_3$:

$$CH_3\overset{O}{\overset{\|}{C}}C(CH_3)_3 \xrightarrow{\text{oxidation}} CH_3\overset{O}{\overset{\|}{C}}OC(CH_3)_3 \text{ (Ester)}$$

$$CH_3\overset{O}{\overset{\|}{C}}C(CH_3)_3 \xrightarrow{\text{reduction}} CH_3CH_2C(CH_3)_3 \text{ (Alkane)}$$

$$CH_3\overset{O}{\overset{\|}{C}}C(CH_3)_3 \xrightarrow{\text{reduction}} CH_3\underset{OH}{\underset{|}{C}}HC(CH_3)_3 \text{ (Alcohol)}$$

**2.40** Methyl formate is an *ester.*

(*a*) The oxidation numbers of the two carbon atoms in methyl formate and the carbon atoms in the reaction products can be determined by comparison with the entries in text Table 2.6.

$$H\overset{O}{\overset{\|}{C}}OCH_3 \longrightarrow H\overset{O}{\overset{\|}{C}}OH + CH_3OH$$

Oxidation number: +2 −2 → +2, −2

There has been no change in oxidation state in going from reactants to products, and the reaction is neither oxidation nor reduction. The number of carbon–oxygen bonds does not change in this reaction.

(*b*) As in part (*a*), the oxidation states of the carbon atoms in both the reactant and the products do not change in this reaction. The reaction is neither oxidation nor reduction.

$$H\overset{O}{\overset{\|}{C}}OCH_3 \longrightarrow H\overset{O}{\overset{\|}{C}}ONa + CH_3OH$$

Oxidation number: +2 −2 → +2, −2

(*c*) The oxidation number of one carbon of methyl formate has decreased in this reaction.

$$H\overset{O}{\overset{\|}{C}}OCH_3 \longrightarrow 2\,CH_3OH$$

Oxidation number: +2 −2 → −2

This reaction is a reduction and requires a reagent that is a reducing agent.

(*d*) The oxidation number of both carbon atoms of methyl formate has increased. This reaction is an oxidation and requires use of a reagent that is an oxidizing agent.

$$H\overset{O}{\overset{\|}{C}}OCH_3 \longrightarrow 2\,CO_2 + H_2O$$

Oxidation number: +2 −2 → +4

(*e*) Once again the formation of carbon dioxide is an example of an oxidation, and the reaction requires use of an oxidizing agent.

$$\underset{\text{Oxidation number}\quad +2\;-2}{\overset{\overset{\displaystyle O}{\|}}{HCOCH_3}} \longrightarrow \underset{+4}{CO_2} + \underset{-2}{CH_3OH}$$

**2.41** Two atoms appear in their elementary state: Na on the left and $H_2$ on the right. The oxidation state of an atom in its elementary state is 0. Assign an oxidation state of +1 to the hydrogen in the OH group of $CH_3CH_2OH$. H goes from +1 on the left to 0 on the right; it is reduced. Na goes from 0 on the left to +1 on the right; it is oxidized.

$$2CH_3CH_2O\overset{+1}{H} + 2\overset{0}{Na} \longrightarrow 2CH_3CH_2O\overset{+1}{Na} + \overset{0}{H_2}$$

**2.42** Combustion of an organic compound to yield $CO_2$ and $H_2O$ involves oxidation. Heat is given off in each oxidation step. The least oxidized compound ($CH_3CH_2OH$) gives off the most heat. The most oxidized compound $HO_2CCO_2H$ gives off the least. The measured values are:

| | $CH_3CH_2OH$ | $HOCH_2CH_2OH$ | $HO_2CCO_2H$ |
|---|---|---|---|
| kJ/mol | 1371 | 1179 | 252 |
| kcal/mol | 327.6 | 281.9 | 60.2 |

**2.43–2.45** Solutions to molecular modeling exercises are not provided in this *Study Guide and Solutions Manual.* You should use *Learning By Modeling* for these exercises.

# SELF-TEST

## PART A

**A-1.** Write the structure of each of the four-carbon alkyl groups. Give the common name and the systematic name for each.

**A-2.** How many $\sigma$ bonds are present in each of the following?
(*a*) Nonane
(*b*) Cyclononane

**A-3.** Classify each of the following reactions according to whether the organic substrate is oxidized, reduced, or neither.

(*a*) $CH_3CH_3 + Br_2 \xrightarrow{\text{light}} CH_3CH_2Br + HBr$

(*b*) $CH_3CH_2Br + HO^- \longrightarrow CH_3CH_2OH + Br^-$

(*c*) $CH_3CH_2OH \xrightarrow[\text{heat}]{H_2SO_4} H_2C{=}CH_2$

(*d*) $H_2C{=}CH_2 + H_2 \xrightarrow{\text{Pt}} CH_3CH_3$

**A-4.** (*a*) Write a structural formula for 3-isopropyl-2,4-dimethylpentane.
(*b*) How many methyl groups are there in this compound? How many isopropyl groups?

**A-5.** Give the IUPAC name for each of the following substances:

(*a*) 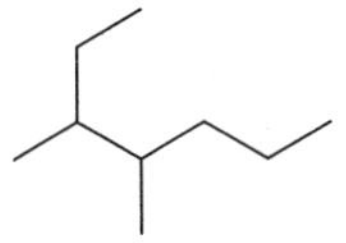 (*b*)

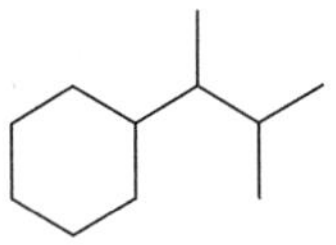

**A-6.** The compounds in each part of the previous question contain ______ primary carbon(s), ______ secondary carbon(s), and ______ tertiary carbon(s).

**A-7.** Give the IUPAC name for each of the following alkyl groups, and classify each one as primary, secondary, or tertiary.

(*a*) $(CH_3)_2CHCH_2CHCH_3$ (with a bond from the CH)

(*b*) $(CH_3CH_2)_3C—$

(*c*) $(CH_3CH_2)_3CCH_2—$

**A-8.** Write a balanced chemical equation for the complete combustion of 2,3-dimethylpentane.

**A-9.** Write structural formulas, and give the names of all the constitutional isomers of $C_5H_{10}$ that contain a ring.

**A-10.** Each of the following names is incorrect. Give the correct name for each compound.
(*a*) 2,3-Diethylhexane
(*b*) (2-Ethylpropyl)cyclohexane
(*c*) 2,3-Dimethyl-3-propylpentane

**A-11.** Which $C_8H_{18}$ isomer
(*a*) Has the highest boiling point?
(*b*) Has the lowest boiling point?
(*c*) Has the greatest number of tertiary carbons?
(*d*) Has only primary and quaternary carbons?

**A-12.** Draw the constitutional isomers of $C_7H_{16}$ that have five carbons in their longest chain, and give an IUPAC name for each of them.

**A-13.** The compound shown is an example of the broad class of organic compounds known as **steroids.** What functional groups does the molecule contain?

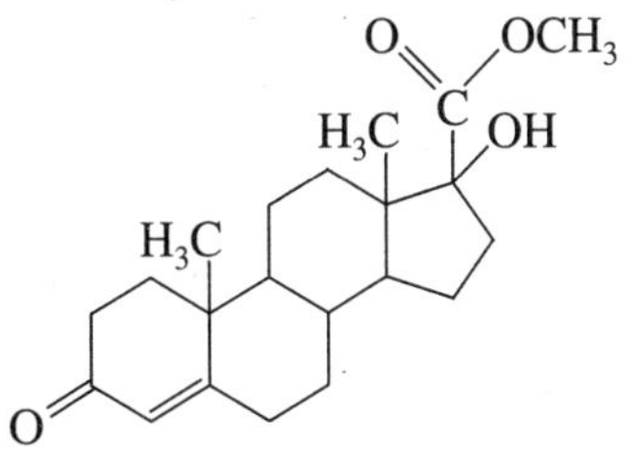

O OCH$_3$
C
H$_3$C OH
H$_3$C
O

**A-14.** Given the following heats of combustion (in kilojoules per mole) for the homologous series of unbranched alkanes: hexane (4163), heptane (4817), octane (5471), nonane (6125), estimate the heat of combustion (in kilojoules per mole) for **pentadecane.**

## PART B

**B-1.** Choose the response that best describes the following compounds:

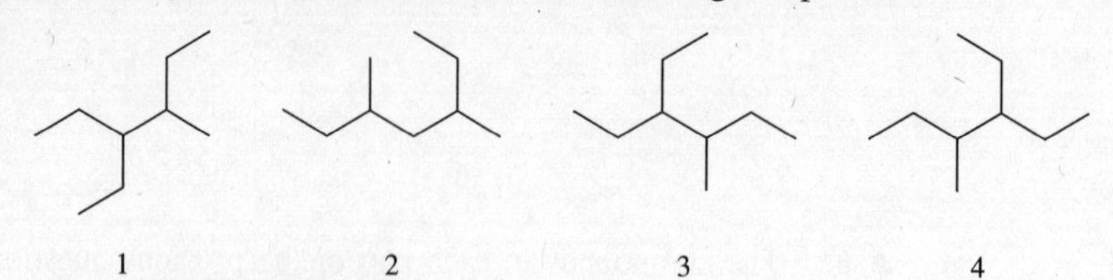


(*a*) 1, 3, and 4 represent the same compound.
(*b*) 1 and 3 are isomers of 2 and 4.
(*c*) 1 and 4 are isomers of 2 and 3.
(*d*) All the structures represent the same compound.

**B-2.** Which of the following is a correct name according to the IUPAC rules?
(*a*) 2-Methylcyclohexane (*c*) 2-Ethyl-2-methylpentane
(*b*) 3,4-Dimethylpentane (*d*) 3-Ethyl-2-methylpentane

**B-3.** Following are the structures of four isomers of hexane. Which of the names given correctly identifies a fifth isomer?

$CH_3CH_2CH_2CH_2CH_2CH_3$ $(CH_3)_3CCH_2CH_3$
$(CH_3)_2CHCH_2CH_2CH_3$ $(CH_3)_2CHCH(CH_3)_2$

(*a*) 2-Methylpentane (*c*) 2-Ethylbutane
(*b*) 2,3-Dimethylbutane (*d*) 3-Methylpentane

**B-4.** Which of the following is cyclohexylcyclohexane?

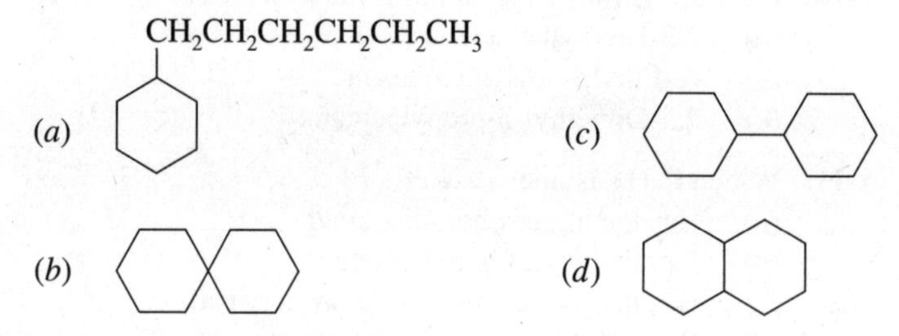


**B-5.** Which of the following structures is a 3-methylbutyl group?
(*a*) $CH_3CH_2CH_2CH_2CH_2—$ (*c*) $(CH_3CH_2)_2CH—$
(*b*) $(CH_3)_2CHCH_2CH_2—$ (*d*) $(CH_3)_3CCH_2—$

**B-6.** Rank the following substances in decreasing order of heats of combustion (most exothermic → least exothermic).

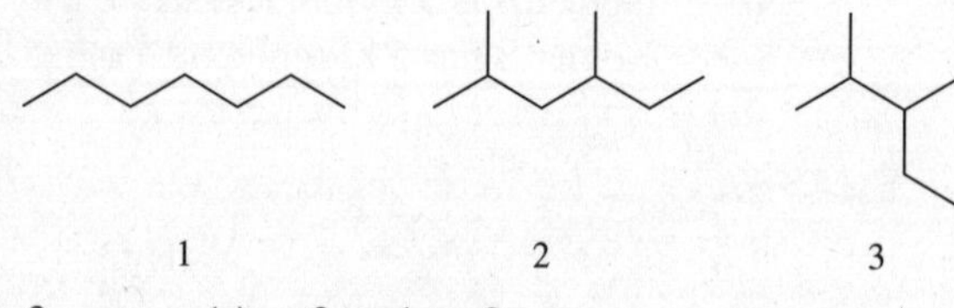


(*a*) 2 > 1 > 3 (*c*) 3 > 1 > 2
(*b*) 2 > 3 > 1 (*d*) 3 > 2 > 1

**B-7.** What is the total number of $\sigma$ bonds present in the molecule shown?

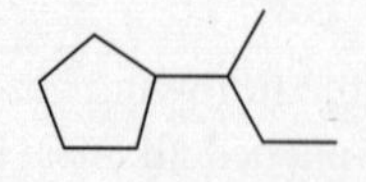

(*a*) 18 (*b*) 26 (*c*) 27 (*d*) 30

**B-8.** Which of the following substances is *not* an isomer of 3-ethyl-2-methylpentane?

(*a*)

(*c*)

(*b*)

(*d*) None of these (all are isomers)

**B-9.** Which alkane has the highest boiling point?
(*a*) Hexane
(*b*) 2,2-Dimethylbutane
(*c*) 2-Methylpentane
(*d*) 2,3-Dimethylbutane
(*e*) 3-Methylpentane

**B-10.** What is the correct IUPAC name of the alkyl group shown?

$$\begin{array}{l} \;\;CH_2CH_3 \\ \;\;| \\ -CHCH_2CH(CH_3)_2 \end{array}$$

(*a*) 1-Ethyl-3-methylbutyl
(*b*) 1-Ethyl-3,3-dimethylpropyl
(*c*) 4-Ethyl-2-methylbutyl
(*d*) 5-Methylhexyl

**B-11.** Which of the following compounds is *not* a constitutional isomer of the others?
(*a*) Methylcyclohexane
(*b*) Cyclopropylcyclobutane
(*c*) Ethylcyclopentane
(*d*) 1,1,2-Trimethylcyclobutane
(*e*) Cycloheptane

**B-12.** The correct IUPAC name for the compound shown is

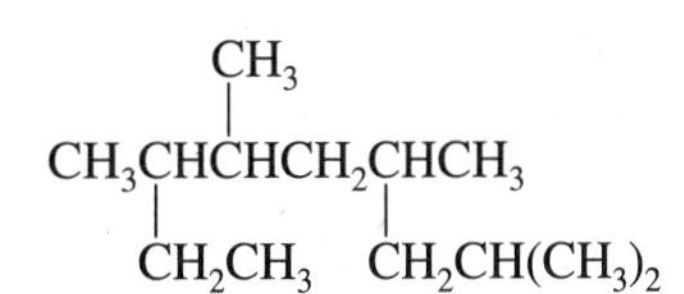


(*a*) 2-Ethyl-5-isobutyl-3-methylhexane
(*b*) 5-*sec*-Butyl-2-ethyl-3-methylhexane
(*c*) 2-Isobutyl-4,5-dimethylheptane
(*d*) 2-Ethyl-3,5,7-trimethyloctane
(*e*) 2,4,6,7-Tetramethylnonane

**B-13.** The heats of combustion of two isomers, A and B, are 4817 kJ/mol and 4812 kJ/mol, respectively. From this information it may be determined that
(*a*) Isomer A is 5 kJ/mol more stable
(*b*) Isomer B is 5 kJ/mol less stable
(*c*) Isomer B has 5 kJ/mol more potential energy
(*d*) Isomer A is 5 kJ/mol less stable

**B-14.** Which of the following reactions requires an *oxidizing agent?*

(*a*) $RCH_2OH \longrightarrow RCH_2Cl$

(*b*) $RCH{=}CH_2 \longrightarrow RCH_2CH_3$

(*c*) $RCH_2Cl \longrightarrow RCH_3$

(*d*) $RCH_2OH \longrightarrow RCH{=}O$

(*e*) None of these

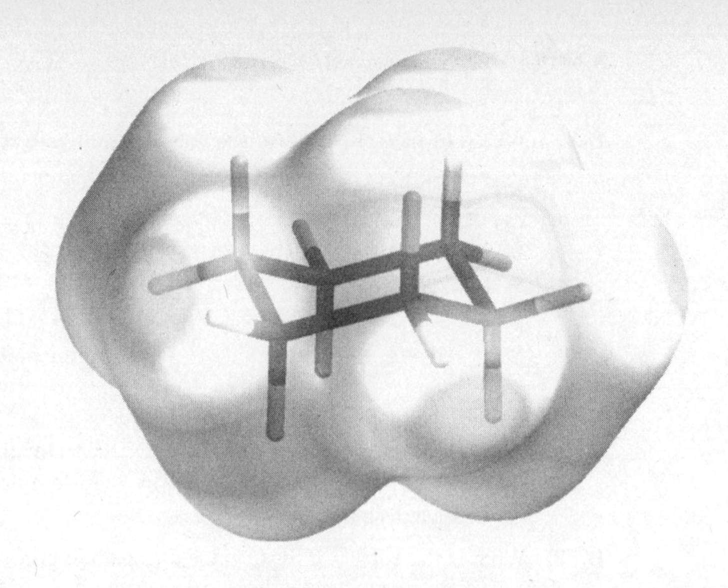

# CHAPTER 3
## CONFORMATIONS OF ALKANES AND CYCLOALKANES

## SOLUTIONS TO TEXT PROBLEMS

**3.1** (*b*) The sawhorse formula contains four carbon atoms in an unbranched chain. The compound is butane, $CH_3CH_2CH_2CH_3$.

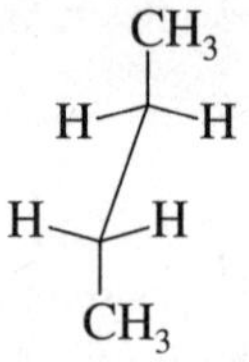

$CH_3$
H—C—H
H—C—H
$CH_3$

(*c*) Rewrite the structure to show its constitution. The compound is $CH_3CH_2CH(CH_3)_2$; it is 2-methylbutane.

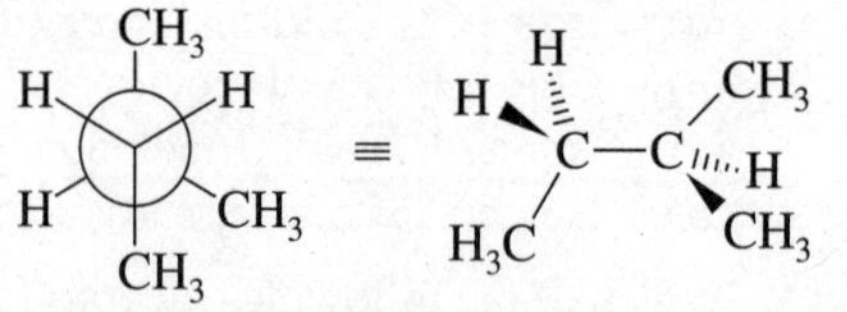

Newman projection: front $CH_3$, H, H; back H, $CH_3$, $CH_3$ ≡ wedge-dash structure: H, H, $H_3C$ on C—C bearing $CH_3$, H, $CH_3$

(*d*) In this structure, we are sighting down the C-3—C-4 bond of a six-carbon chain. It is $CH_3CH_2CH_2\underset{\displaystyle CH_3}{\underset{|}{C}}HCH_2CH_3$, or 3-methylhexane.

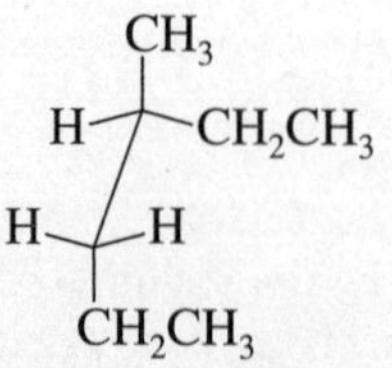

$CH_3$
H—C—$CH_2CH_3$
H—C—H
$CH_2CH_3$

**3.2** Red circles gauche: 60° and 300°. Red circles anti: 180°. Gauche and anti relationships occur only in staggered conformations; therefore, ignore the eclipsed conformations (0°, 120°, 240°, 360°).

**3.3** All the staggered conformations of propane are equivalent to one another, and all its eclipsed conformations are equivalent to one another. The energy diagram resembles that of ethane in that it is a symmetrical one.

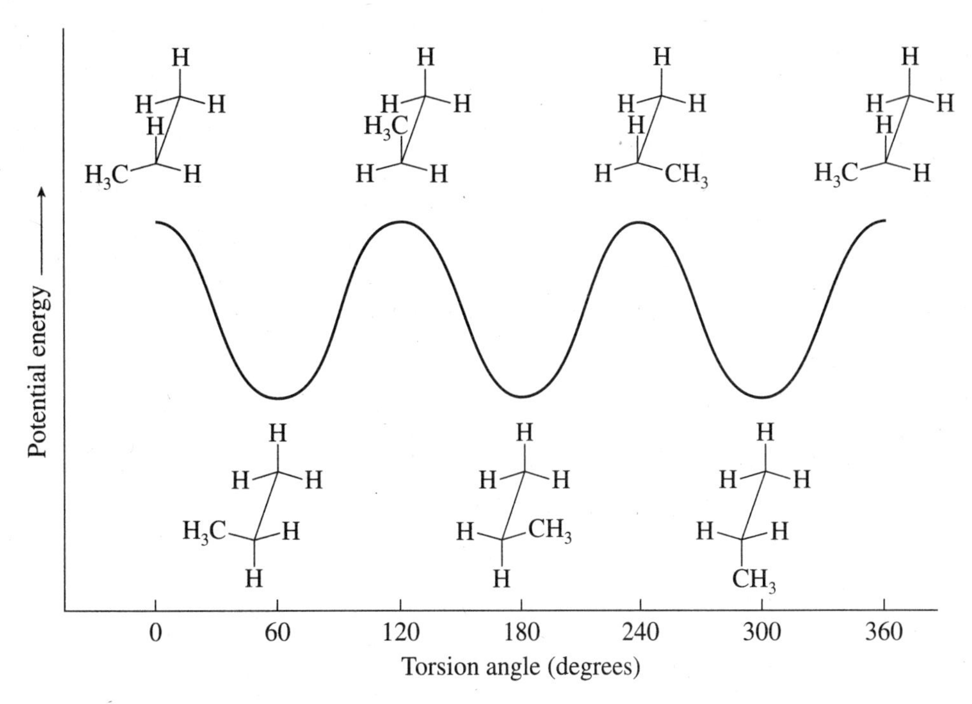


The activation energy for bond rotation in propane is expected to be somewhat higher than that in ethane because of van der Waals strain between the methyl group and a hydrogen in the eclipsed conformation. This strain is, however, less than the van der Waals strain between the methyl groups of butane, which makes the activation energy for bond rotation less for propane than for butane.

**3.4** (*b*) To be gauche, substituents X and A must be related by a 60° torsion angle. If A is axial as specified in the problem, X must therefore be equatorial.

X and A are gauche.

(*c*) For substituent X at C-1 to be anti to C-3, it must be equatorial.

(*d*) When X is axial at C-1, it is gauche to C-3.

**3.5** (*b*) According to the numbering scheme given in the problem, a methyl group is axial when it is "up" at C-1 but is equatorial when it is up at C-4. Since substituents are more stable when they occupy equatorial rather than axial sites, a methyl group that is up at C-1 is less stable than one that is up at C-4.

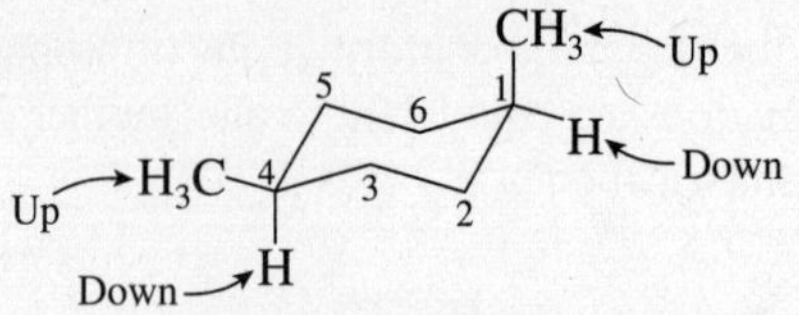

$CH_3$ ← Up
H ← Down
Up → $H_3C$
Down → H

(*c*) An alkyl substituent is more stable in the equatorial position. An equatorial substituent at C-3 is "down."

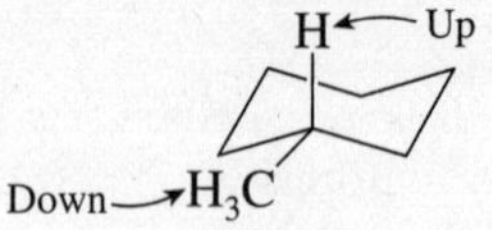

H ← Up
Down → $H_3C$

**3.6** A *tert*-butyl group is much larger than a methyl group and has a greater preference for the equatorial position. The most stable conformation of 1-*tert*-butyl-1-methylcyclohexane has an axial methyl group and an equatorial *tert*-butyl group.

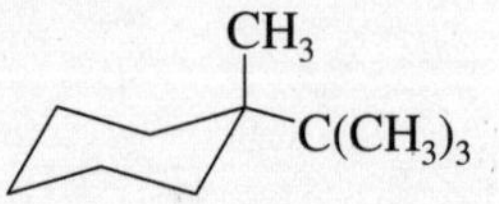

$CH_3$
$C(CH_3)_3$

1-*tert*-Butyl-1-methylcyclohexane

**3.7** Ethylcyclopropane and methylcyclobutane are isomers (both are $C_5H_{10}$). The less stable isomer has the higher heat of combustion. Ethylcyclopropane has more angle strain and is less stable (has higher potential energy) than methylcyclobutane.

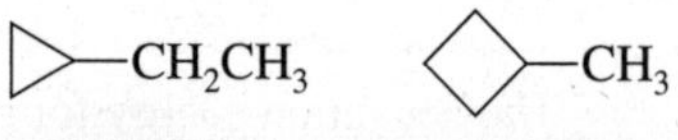

| | $-CH_2CH_3$ (cyclopropane) | $-CH_3$ (cyclobutane) |
|---|---|---|
| | Less stable | More stable |
| Heat of combustion: | 3384 kJ/mol (808.8 kcal/mol) | 3352 kJ/mol (801.2 kcal/mol) |

**3.8** The four constitutional isomers of *cis* and *trans*-1,2-dimethylcyclopropane that do not contain double bonds are

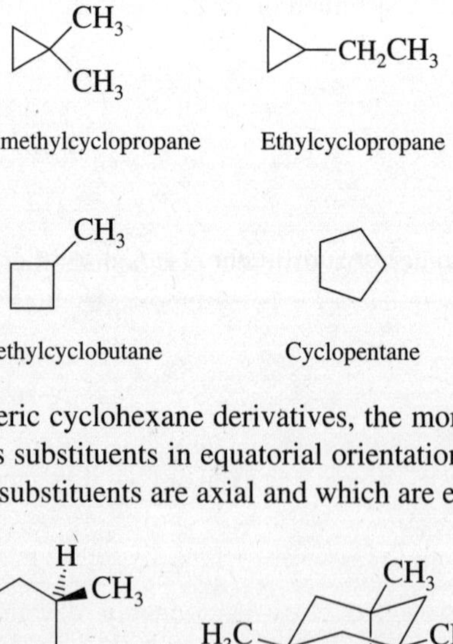

1,1-Dimethylcyclopropane    Ethylcyclopropane

Methylcyclobutane    Cyclopentane

**3.9** When comparing two stereoisomeric cyclohexane derivatives, the more stable stereoisomer is the one with the greater number of its substituents in equatorial orientations. Rewrite the structures as chair conformations to see which substituents are axial and which are equatorial.

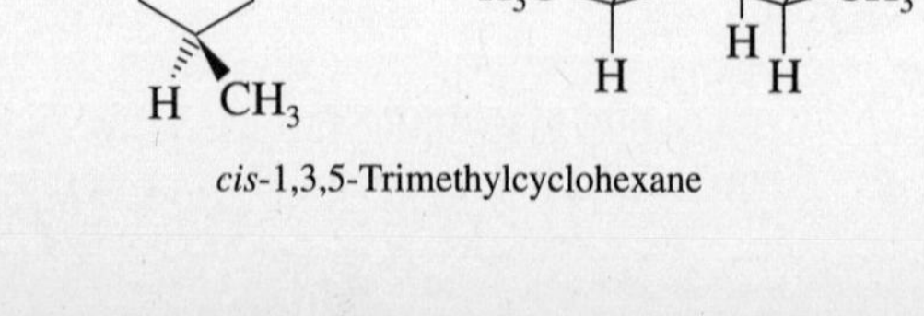

*cis*-1,3,5-Trimethylcyclohexane

All methyl groups are equatorial in *cis*-1,3,5-trimethylcyclohexane. It is more stable than *trans*-1,3,5-trimethylcyclohexane (shown in the following), which has one axial methyl group in its most stable conformation.

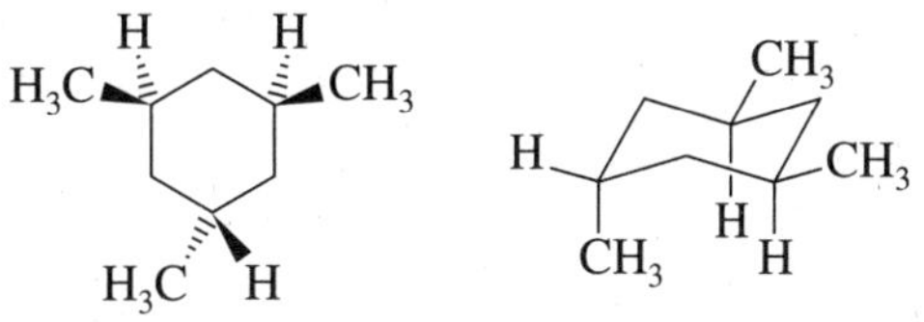

*trans*-1,3,5-Trimethylcyclohexane

**3.10** In each of these problems, a *tert*-butyl group is the larger substituent and will be equatorial in the most stable conformation. Draw a chair conformation of cyclohexane, add an equatorial *tert*-butyl group, and then add the remaining substituent so as to give the required cis or trans relationship to the *tert*-butyl group.

(*b*) Begin by drawing a chair cyclohexane with an equatorial *tert*-butyl group. In *cis*-1-*tert*-butyl-3-methylcyclohexane the C-3 methyl group is equatorial.

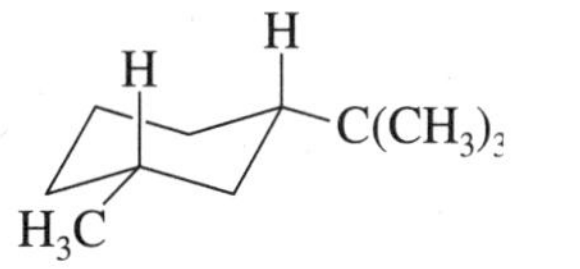

(*c*) In *trans*-1-*tert*-butyl-4-methylcyclohexane both the *tert*-butyl and the C-4 methyl group are equatorial.

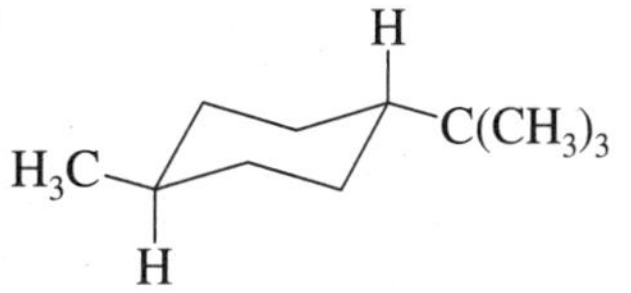

(*d*) Again the *tert*-butyl group is equatorial; however, in *cis*-1-*tert*-butyl-4-methylcyclohexane the methyl group on C-4 is axial.

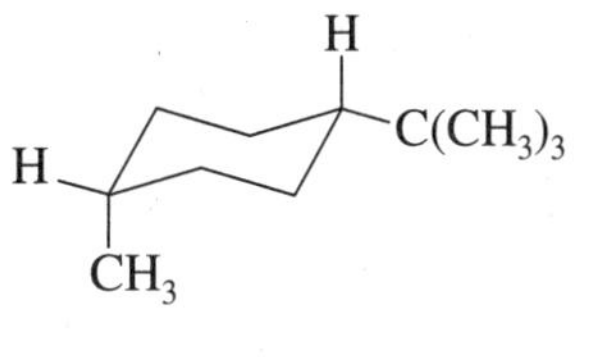

**3.11** Isomers are different compounds that have the same molecular formula. Compare the molecular formulas of the compounds given to the molecular formula of spiropentane.

Spiropentane ($C_5H_8$) $C_5H_8$ $C_6H_{10}$ $C_5H_8$ $C_5H_{10}$

Only the two compounds that have the molecular formula $C_5H_8$ are isomers of spiropentane.

**3.12** Two bond cleavages convert bicyclobutane to a noncyclic species; therefore, bicyclobutane is bicyclic.

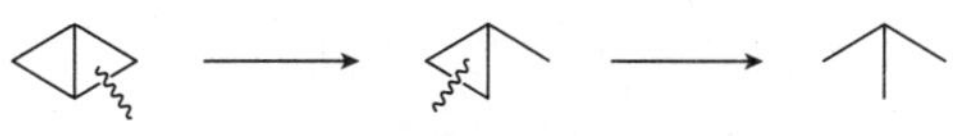

The two bond cleavages shown convert camphene to a noncyclic species; therefore, camphene is bicyclic. (Other pairs of bond cleavages are possible and lead to the same conclusion.)

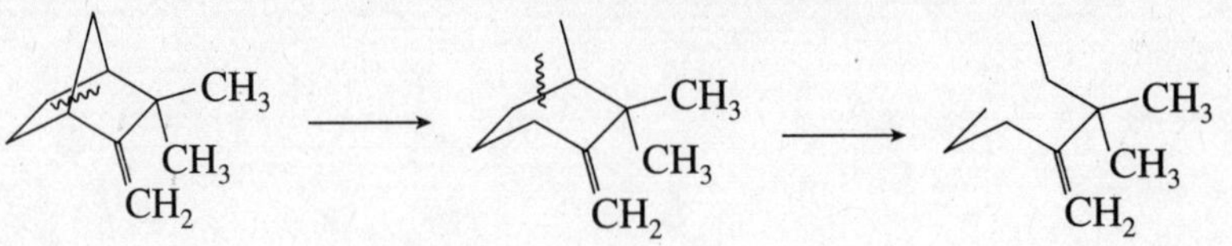

$CH_3$ $CH_3$ $CH_2$ → $CH_3$ $CH_3$ $CH_2$ → $CH_3$ $CH_3$ $CH_2$

**3.13** (*b*) This bicyclic compound contains nine carbon atoms. The name tells us that there is a five-carbon bridge and a two-carbon bridge. The 0 in the name bicyclo[5.2.0]nonane tells us that the third bridge has no atoms in it—the carbons are common to both rings and are directly attached to each other.

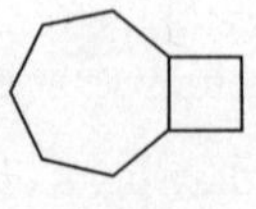

Bicyclo[5.2.0]nonane

(*c*) The three bridges in bicyclo[3.1.1]heptane contain three carbons, one carbon, and one carbon. The structure can be written in a form that shows the actual shape of the molecule or one that simply emphasizes its constitution.

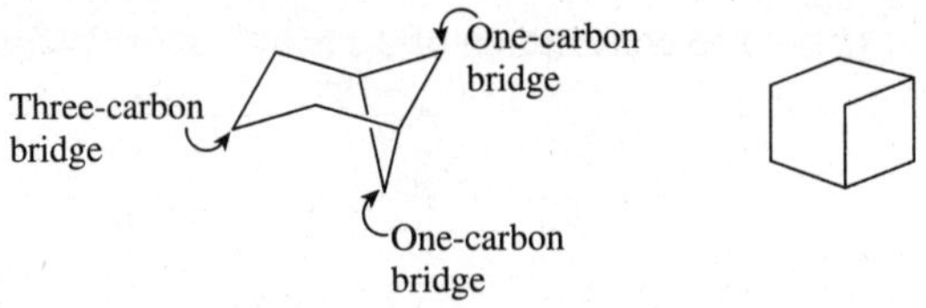

One-carbon bridge
Three-carbon bridge
One-carbon bridge

(*d*) Bicyclo[3.3.0]octane has two five-membered rings that share a common side.

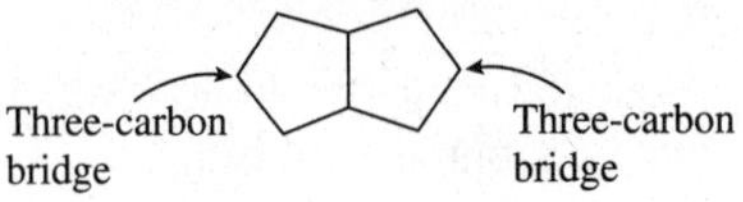

Three-carbon bridge
Three-carbon bridge

**3.14** Since the two conformations are of approximately equal stability when R = H, it is reasonable to expect that the most stable conformation when R = $CH_3$ will have the $CH_3$ group equatorial.

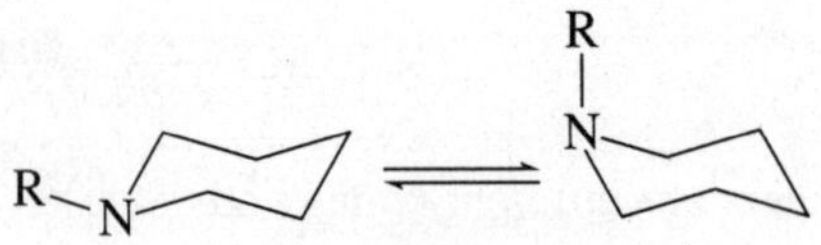

R = H: both conformations similar in energy
R = $CH_3$: most stable conformation has $CH_3$ equatorial

**3.15** (*a*) Recall that a neutral nitrogen atom has three covalent bonds and an unshared electron pair. The three bonds are arranged in a trigonal pyramidal manner around each nitrogen in hydrazine ($H_2NNH_2$).

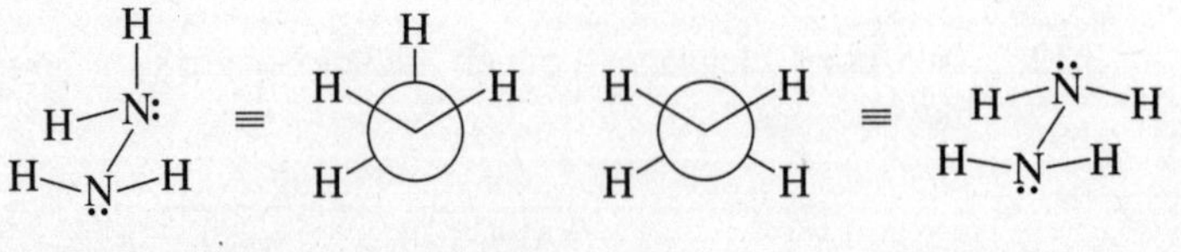

(*b*) The O—H proton may be anti to one N—H proton and gauche to the other (left) or it may be gauche to both (right).

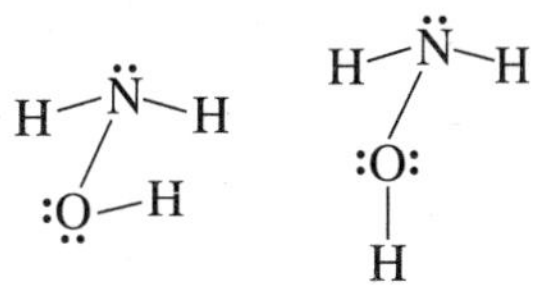

**3.16** Conformation (*a*) is the most stable; all its bonds are staggered. Conformation (*c*) is the least stable; all its bonds are eclipsed.

**3.17** (*a*) First write out the structural formula of 2,2-dimethylbutane in order to identify the substituent groups attached to C-2 and C-3. As shown at left, C-2 bears three methyl groups, and C-3 bears two hydrogens and a methyl group. The most stable conformation is the staggered one shown at right. All other staggered conformations are equivalent to this one.

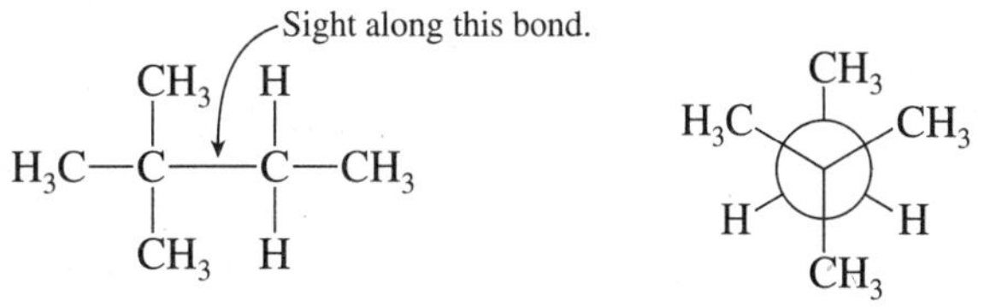

Sight along this bond.

(*b*) The constitution of 2-methylbutane and its two most stable conformations are shown.

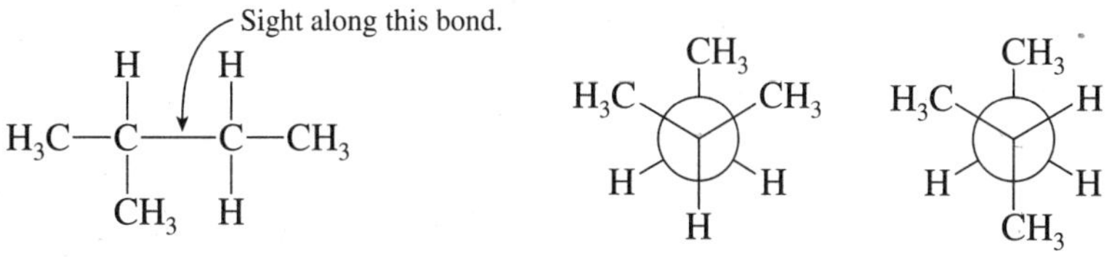

Sight along this bond.

Both conformations are staggered. In one (left), the methyl group at C-3 is gauche to both of the C-2 methyls. In the other (right), the methyl group at C-3 is gauche to one of the C-2 methyls and anti to the other.

(*c*) The hydrogens at C-2 and C-3 may be gauche to one another (left), or they may be anti (right).

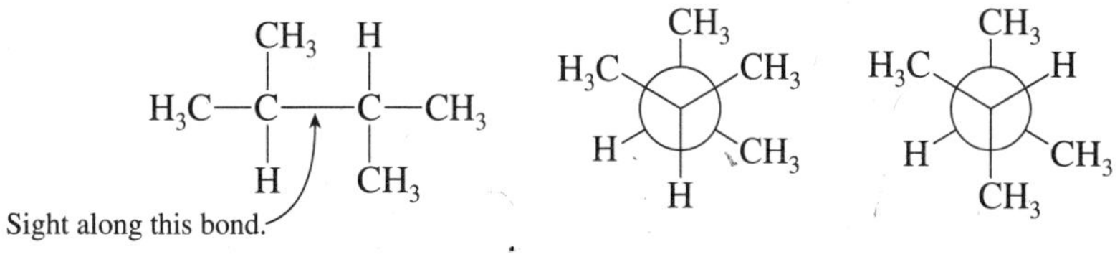

Sight along this bond.

**3.18** The 2-methylbutane conformation with one gauche $CH_3 \cdots CH_3$ and one anti $CH_3 \cdots CH_3$ relationship is more stable than the one with two gauche $CH_3 \cdots CH_3$ relationships. The more stable conformation has less van der Waals strain.

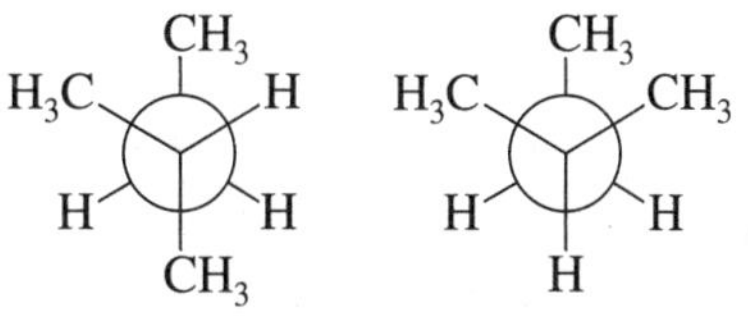

More stable Less stable

**3.19** All the staggered conformations about the C-2—C-3 bond of 2,2-dimethylpropane are equivalent to one another and of equal energy; they represent potential energy minima. All the eclipsed conformations are equivalent and represent potential energy maxima.

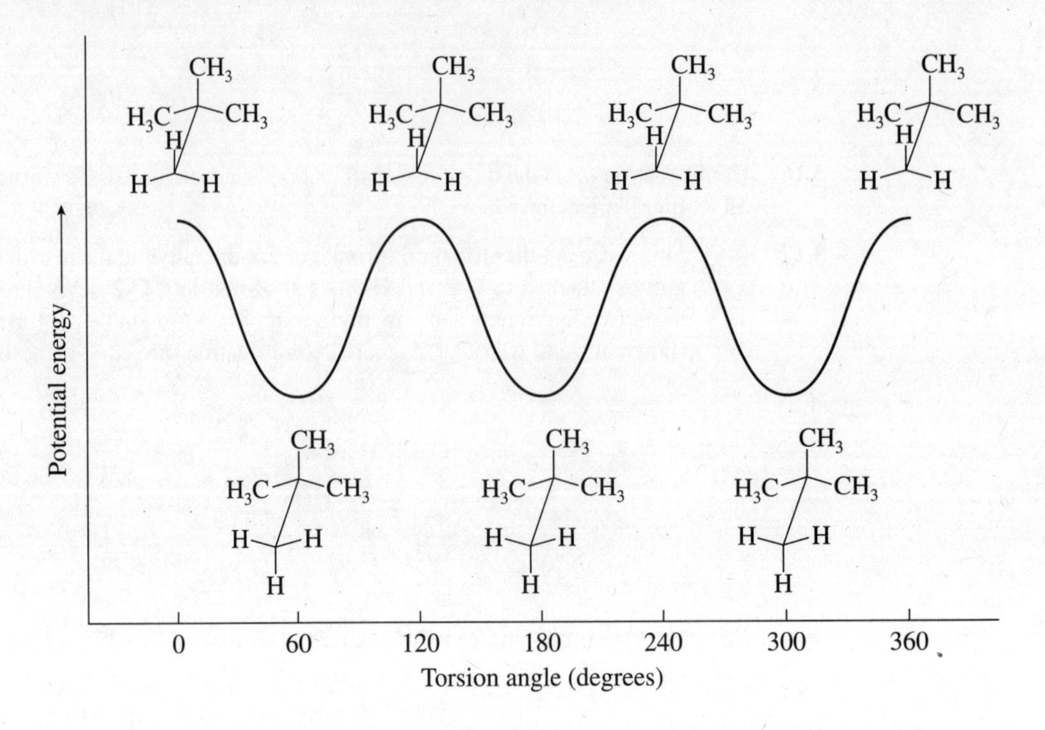


The shape of the potential energy profile for internal rotation in 2,2-dimethylpropane more closely resembles that of ethane than that of butane.

**3.20** The potential energy diagram of 2-methylbutane more closely resembles that of butane than that of propane in that the three staggered forms are not all of the same energy. Similarly, not all of the eclipsed forms are of equal energy.

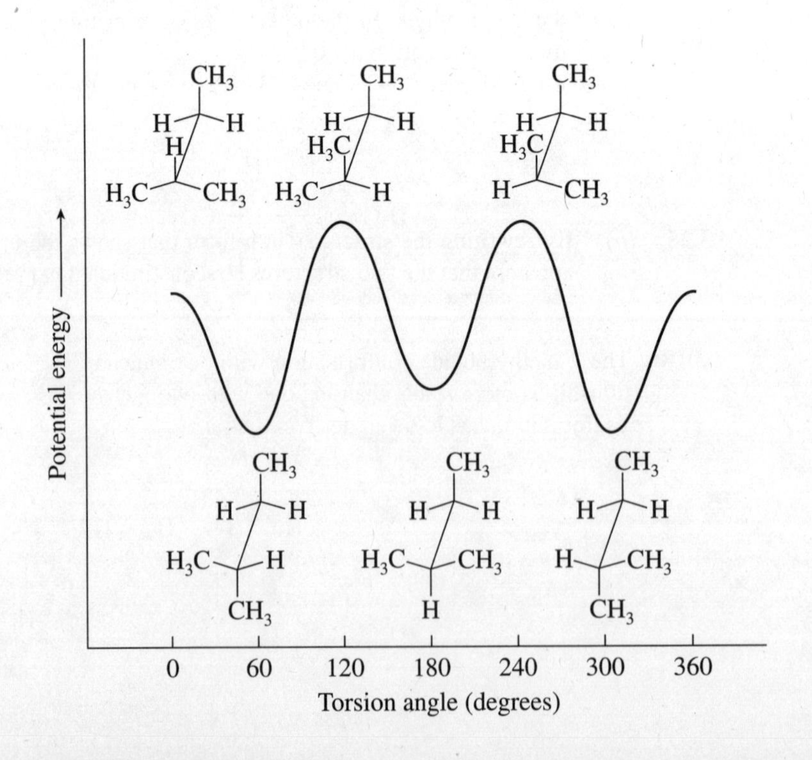

**3.21** Van der Waals strain between the *tert*-butyl groups in 2,2,4,4-tetramethylpentane causes the C-2—C-3—C-4 angle to open to 125–128°.

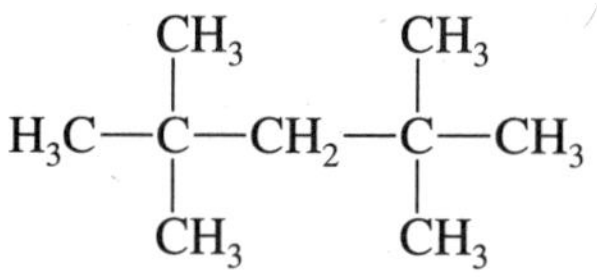

$$H_3C-\underset{CH_3}{\overset{CH_3}{C}}-CH_2-\underset{CH_3}{\overset{CH_3}{C}}-CH_3$$ is equivalent to

This angle is enlarged.

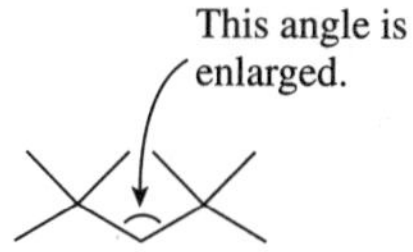

**3.22** The structure shown in the text is not the most stable conformation, because the bonds of the methyl group are eclipsed with those of the ring carbon to which it is attached. The most stable conformation has the bonds of the methyl group and its attached carbon in a staggered relationship.

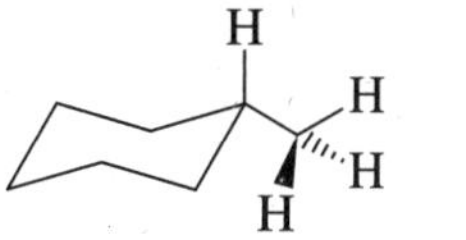

Bonds of methyl group eclipsed with those of attached carbon

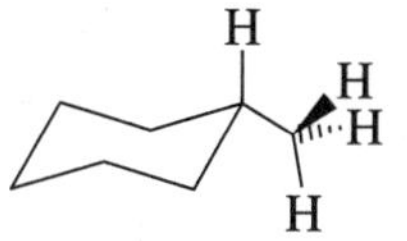

Bonds of methyl group staggered with those of attached carbon

**3.23** Structure A has the hydrogens of its methyl group eclipsed with the ring bonds and is less stable than B. The methyl group in structure B has its bonds and those of its attached ring carbon in a staggered relationship.

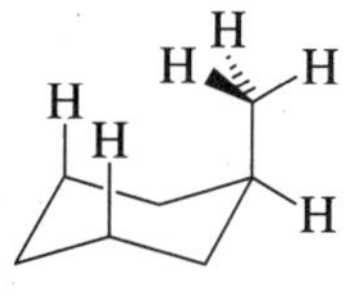

A (less stable) 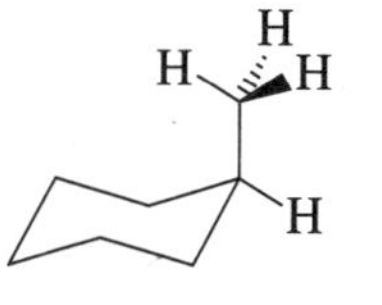 B (more stable)

Furthermore, two of the hydrogens of the methyl group of A are uncomfortably close to two axial hydrogens of the ring.

**3.24** Conformation B is more stable than A. The methyl groups are rather close together in A, resulting in van der Waals strain between them. In B, the methyl groups are farther apart.

Van der Waals strain between cis methyl groups.

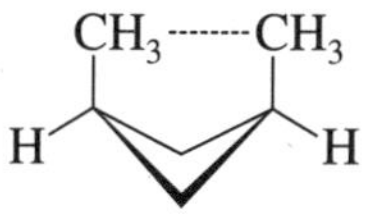

Methyl groups remain cis, but are far apart.

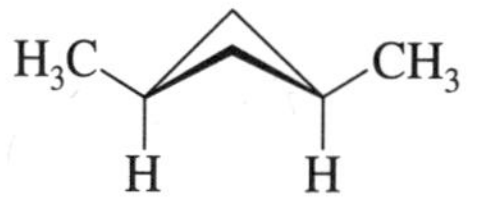

A B

**3.25** (*a*) By rewriting the structures in a form that shows the order of their atomic connections, it is apparent that the two structures are constitutional isomers.

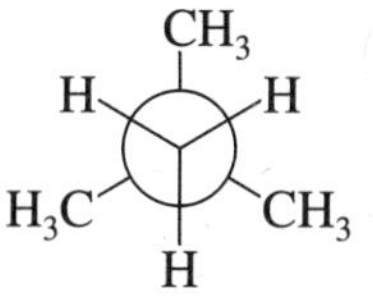

is equivalent to $CH_3\underset{CH_3}{\overset{CH_3}{C}}CH_3$

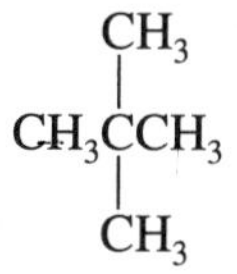

(2,2-Dimethylpropane)

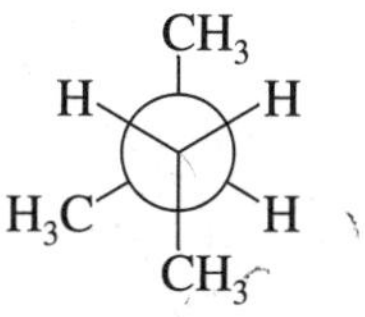

is equivalent to $CH_3CH_2\overset{CH_3}{C}HCH_3$

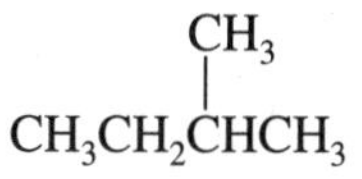

(2-Methylbutane)

(*b*) Both models represent alkanes of molecular formula $C_6H_{14}$. In each one the carbon chain is unbranched. The two models are different conformations of the same compound, $CH_3CH_2CH_2CH_2CH_2CH_3$ (hexane).

(*c*) The two compounds have the same constitution; both are $(CH_3)_2CHCH(CH_3)_2$. The Newman projections represent different staggered conformations of the same molecule: in one the hydrogens are anti to each other, whereas in the other they are gauche.

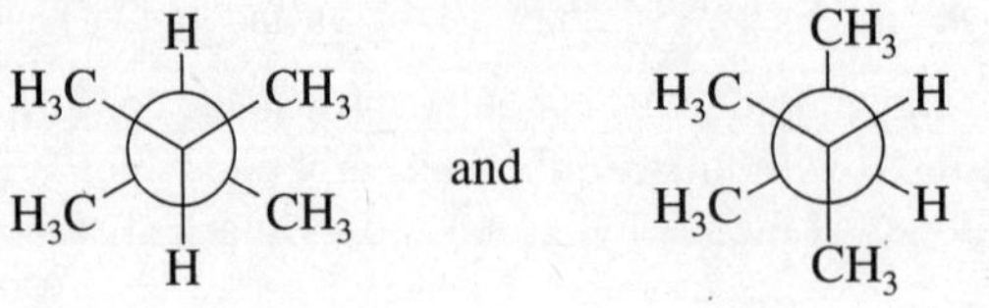

and are different conformations of 2,3-dimethylbutane

Hydrogens at C-2 and C-3 are anti.

Hydrogens at C-2 and C-3 are gauche.

(*d*) The compounds differ in the *order* in which the atoms are connected. They are constitutional isomers. Although the compounds have different stereochemistry (one is cis, the other trans), they are not stereoisomers. Stereoisomers must have the same constitution.

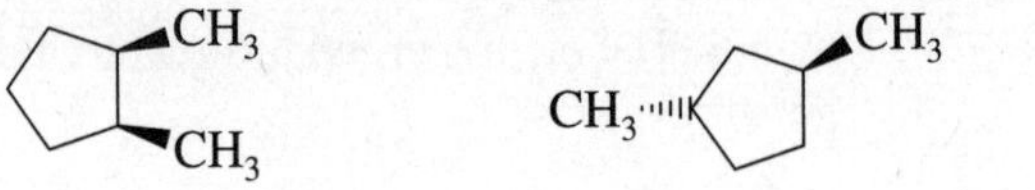

*cis*-1,2-Dimethylcyclopentane

*trans*-1,3-Dimethylcyclopentane

(*e*) Both structures are *cis*-1-ethyl-4-methylcyclohexane (the methyl and ethyl groups are both "up"). In the structure on the left, the methyl is axial and the ethyl equatorial. The orientations are opposite to these in the structure on the right. The two structures are ring-flipped forms of each other—different conformations of the same compound.

(*f*) The methyl and the ethyl groups are cis in the first structure but trans in the second. The two compounds are stereoisomers; they have the same constitution but differ in the arrangement of their atoms in space.

*cis*-1-Ethyl-4-methylcyclohexane (both alkyl groups are up)

*trans*-1-Ethyl-4-methylcyclohexane (ethyl group is down; methyl group is up)

Do not be deceived because the six-membered rings look like ring-flipped forms. Remember, chair–chair interconversion converts all the equatorial bonds to axial and vice versa. Here the ethyl group is equatorial in both structures.

(*g*) The two structures have the same constitution but differ in the arrangement of their atoms in space; they are stereoisomers. They are not different conformations of the same compound, because they are not related by rotation about C—C bonds. In the first structure as shown here the methyl group is trans to the darkened bonds, whereas in the second it is cis to these bonds.

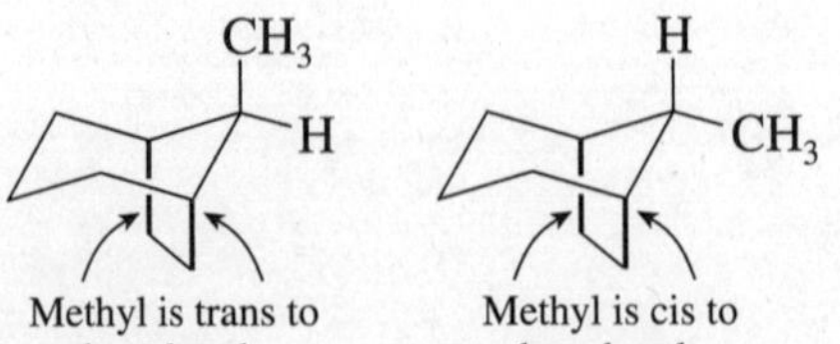

Methyl is trans to these bonds.

Methyl is cis to these bonds.

**3.26** (*a*) Three isomers of $C_5H_8$ contain two rings and have no alkyl substituents:

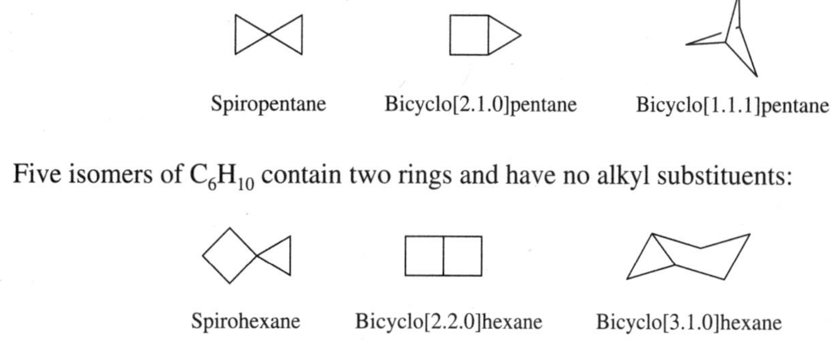

Spiropentane    Bicyclo[2.1.0]pentane    Bicyclo[1.1.1]pentane

(*b*) Five isomers of $C_6H_{10}$ contain two rings and have no alkyl substituents:

Spirohexane    Bicyclo[2.2.0]hexane    Bicyclo[3.1.0]hexane

Bicyclo[2.1.1]hexane    Cyclopropylcyclopropane

**3.27** (*a*) The heat of combustion is highest for the hydrocarbon with the greatest number of carbons. Thus, cyclopropane, even though it is more strained than cyclobutane or cyclopentane, has the lowest heat of combustion.

| | | |
|---|---|---|
| | Cyclopentane | Heat of combustion 3291 kJ/mol (786.6 kcal/mol) |
| | Cyclobutane | Heat of combustion 2721 kJ/mol (650.3 kcal/mol) |
| | Cyclopropane | Heat of combustion 2091 kJ/mol (499.8 kcal/mol) |

A comparison of heats of combustion can only be used to assess relative stability when the compounds are isomers.

(*b*) All these compounds have the molecular formula $C_7H_{14}$. They are isomers, and so the one with the most strain will have the highest heat of combustion.

| | | |
|---|---|---|
| $H_3C$, $CH_3$, $H_3C$, $CH_3$ | 1,1,2,2-Tetramethylcyclopropane (high in angle strain; bonds are eclipsed; van der Waals strain between cis methyl groups) | Heat of combustion 4635 kJ/mol (1107.9 kcal/mol) |
| H, H, $H_3C$, $CH_3$ | *cis*-1,2-Dimethylcyclopentane (low angle strain; some torsional strain; van der Waals strain between cis methyl groups) | Heat of combustion 4590 kJ/mol (1097.1 kcal/mol) |
| $CH_3$ | Methylcyclohexane (minimal angle, torsional, and van der Waals strain) | Heat of combustion 4565 kJ/mol (1091.1 kcal/mol) |

(*c*) These hydrocarbons all have different molecular formulas. Their heats of combustion decrease with decreasing number of carbons, and comparisons of relative stability cannot be made.

| Compound | Heat of combustion |
|---|---|
| 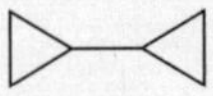 Cyclopropylcyclopropane ($C_6H_{10}$) | Heat of combustion 3886 kJ/mol (928.8 kcal/mol) |
|  Spiropentane ($C_5H_8$) | Heat of combustion 3296 kJ/mol (787.8 kcal/mol) |
| 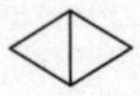 Bicyclo[1.1.0]butane ($C_4H_6$) | Heat of combustion 2648 kJ/mol (633.0 kcal/mol) |

(*d*) Bicyclo[3.3.0]octane and bicyclo[5.1.0]octane are isomers, and their heats of combustion can be compared on the basis of their relative stabilities. The three-membered ring in bicyclo[5.1.0]octane imparts a significant amount of angle strain to this isomer, making it less stable than bicyclo[3.3.0]octane. The third hydrocarbon, bicyclo[4.3.0]nonane, has a greater number of carbons than either of the others and has the largest heat of combustion.

| Compound | Heat of combustion |
|---|---|
| 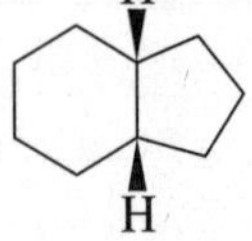 Bicyclo[4.3.0]nonane ($C_9H_{16}$) | Heat of combustion 5652 kJ/mol (1350.9 kcal/mol) |
| 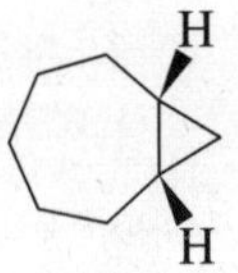 Bicyclo[5.1.0]octane ($C_8H_{14}$) | Heat of combustion 5089 kJ/mol (1216.3 kcal/mol) |
| 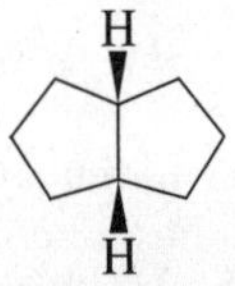 Bicyclo[3.3.0]octane ($C_8H_{14}$) | Heat of combustion 5016 kJ/mol (1198.9 kcal/mol) |

**3.28** (*a*) The structural formula of 2,2,5,5-tetramethylhexane is $(CH_3)_3CCH_2CH_2C(CH_3)_3$. The substituents at C-3 are two hydrogens and a *tert*-butyl group. The substituents at C-4 are the same as those at C-3. The most stable conformation has the large *tert*-butyl groups anti to each other.

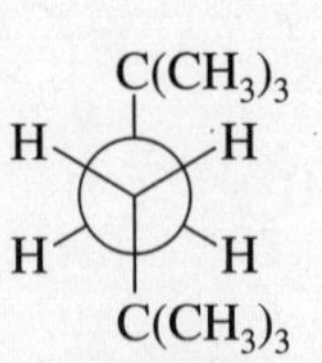

$C(CH_3)_3$
H H
H H
$C(CH_3)_3$

Anti conformation of
2,2,5,5-tetramethylhexane

(*b*) The zigzag conformation of 2,2,5,5-tetramethylhexane is an alternative way of expressing the same conformation implied in the Newman projection of part (*a*). It is more complete, however, in that it also shows the spatial arrangement of the substituents attached to the main chain.

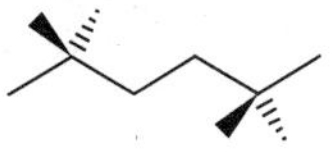

2,2,5,5-Tetramethylhexane

(*c*) An isopropyl group is bulkier than a methyl group, and will have a greater preference for an equatorial orientation in the most stable conformation of *cis*-1-isopropyl-3-methylcyclohexane. Draw a chair conformation of cyclohexane, and place an isopropyl group in an equatorial position.

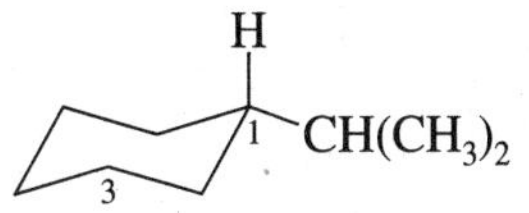

H
1 $CH(CH_3)_2$
3

Notice that the equatorial isopropyl group is down on the carbon atom to which it is attached. Add a methyl group to C-3 so that it is also down.

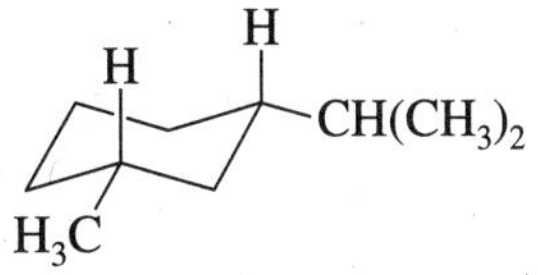

H H $CH(CH_3)_2$
$H_3C$

Both substituents are equatorial in the most stable conformation of *cis*-1-isopropyl-3-methylcyclohexane.

(*d*) One substituent is up and the other is down in the most stable conformation of *trans*-1-isopropyl-3-methylcyclohexane. Begin as in part (*c*) by placing an isopropyl group in an equatorial orientation on a chair conformation of cyclohexane.

H
1 $CH(CH_3)_2$
3

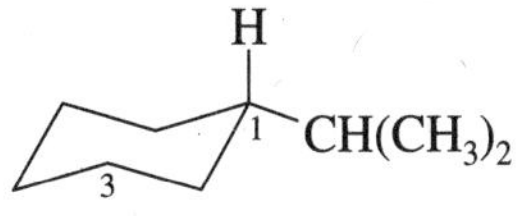

To be trans to the C-1 isopropyl group, the C-3 methyl group must be up.

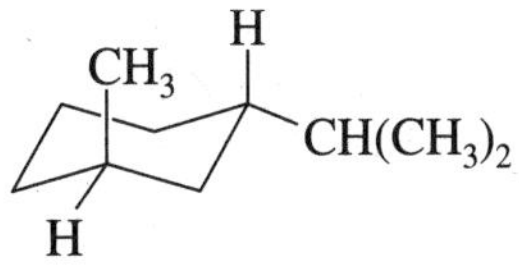

$CH_3$ H $CH(CH_3)_2$
H

The bulkier isopropyl group is equatorial and the methyl group axial in the most stable conformation.

(*e*) To be cis to each other, one substituent must be axial and the other equatorial when they are located at positions 1 and 4 on a cyclohexane ring.

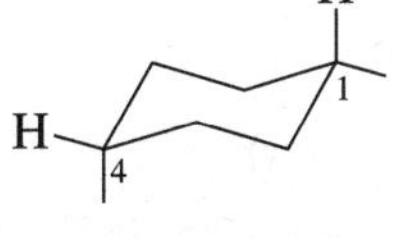

H
1
H 4

Place the larger substituent (the *tert*-butyl group) at the equatorial site and the smaller substituent (the ethyl group) at the axial one.

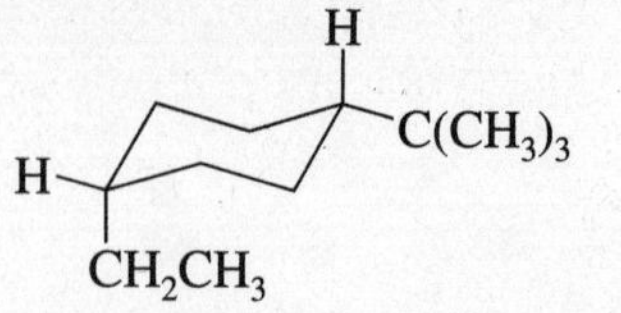

(*f*) First write a chair conformation of cyclohexane, then add two methyl groups at C-1, and draw in the axial and equatorial bonds at C-3 and C-4. Next, add methyl groups to C-3 and C-4 so that they are cis to each other. There are two different ways that this can be accomplished: either the C-3 and C-4 methyl groups are both up or they are both down.

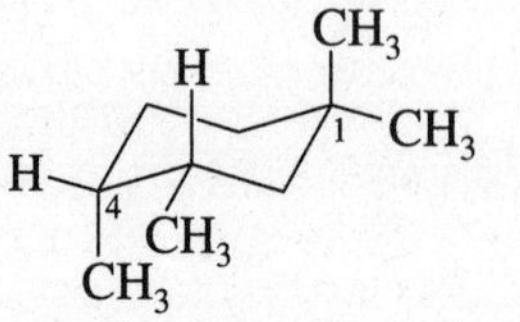

More stable chair conformation: C-3 methyl group is equatorial; no van der Waals strain between axial C-1 methyl group and C-3 methyl

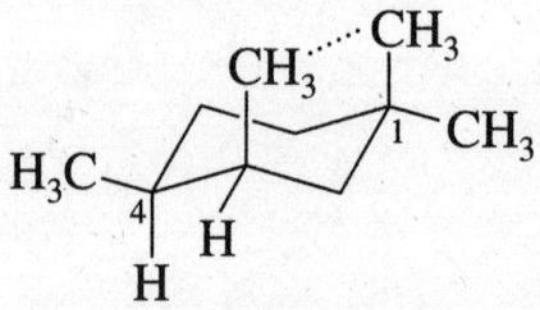

Less stable chair conformation: C-3 methyl group is axial; strong van der Waals strain between axial C-1 and C-3 methyl groups

(*g*) Draw the projection formula as a chair conformation.

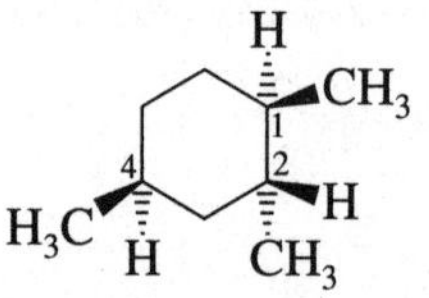

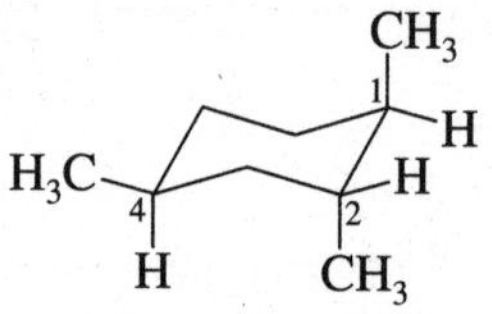

Check to see if this is the most stable conformation by writing its ring-flipped form.

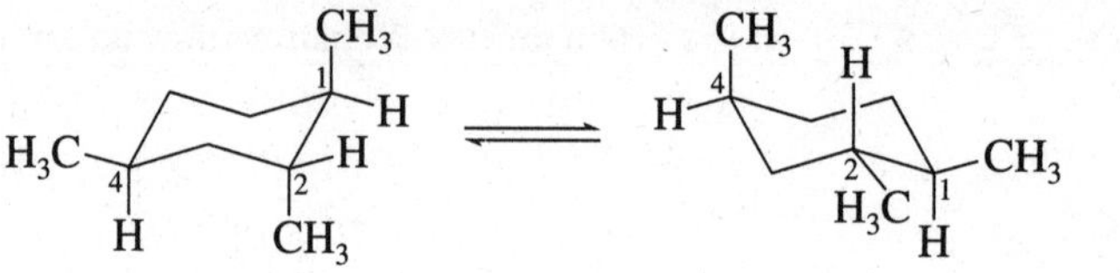

Less stable conformation: two axial methyl groups

More stable conformation: one axial methyl group

The ring-flipped form, with two equatorial methyl groups and one axial methyl group, is more stable than the originally drawn conformation, with two axial ethyl groups and one equatorial methyl group.

**3.29** Begin by writing each of the compounds in its most stable conformation. Compare them by examining their conformations for sources of strain, particularly van der Waals strain arising from groups located too close together in space.

(*a*) Its axial methyl group makes the cis stereoisomer of 1-isopropyl-2-methylcyclohexane less stable than the trans.

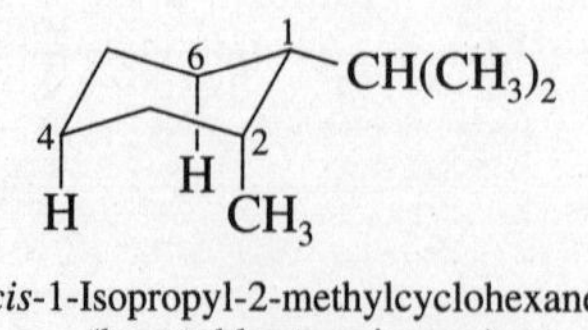

*cis*-1-Isopropyl-2-methylcyclohexane (less stable stereoisomer)

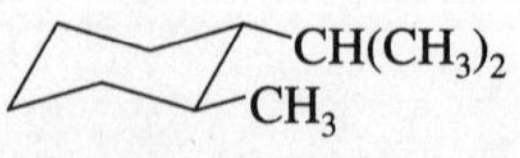

*trans*-1-Isopropyl-2-methylcyclohexane (more stable stereoisomer)

The axial methyl group in the cis stereoisomer is involved in unfavorable repulsions with the C-4 and C-6 axial hydrogens indicated in the drawing.

(*b*) Both groups are equatorial in the cis stereoisomer of 1-isopropyl-3-methylcyclohexane; cis is more stable than trans in 1,3-disubstituted cyclohexanes.

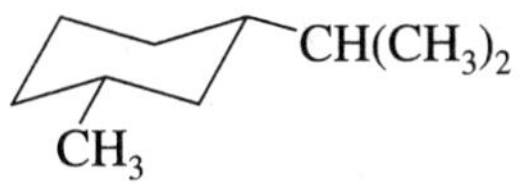

*cis*-1-Isopropyl-3-methylcyclohexane (more stable stereoisomer; both groups are equatorial)

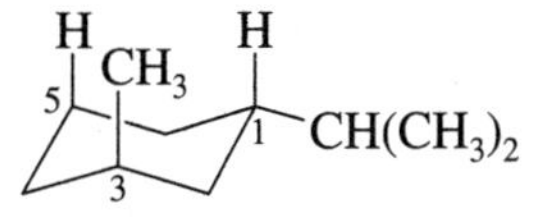

*trans*-1-Isopropyl-3-methylcyclohexane (less stable stereoisomer; methyl group is axial and involved in repulsions with axial hydrogens at C-1 and C-5)

(*c*) The more stable stereoisomer of 1,4-disubstituted cyclohexanes is the trans; both alkyl groups are equatorial in *trans*-1-isopropyl-4-methylcyclohexane.

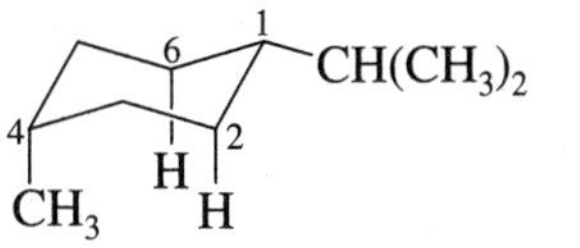

*cis*-1-Isopropyl-4-methylcyclohexane (less stable stereoisomer; methyl group is axial and involved in repulsions with axial hydrogens at C-2 and C-6)

*trans*-1-Isopropyl-4-methylcyclohexane (more stable stereoisomer; both groups are equatorial)

(*d*) The first stereoisomer of 1,2,4-trimethylcyclohexane is the more stable one. All its methyl groups are equatorial in its most stable conformation. The most stable conformation of the second stereoisomer has one axial and two equatorial methyl groups.

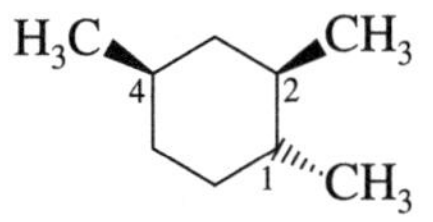

More stable stereoisomer

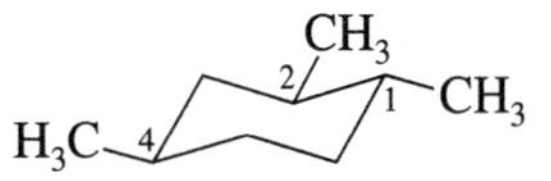

All methyl groups equatorial in most stable conformation

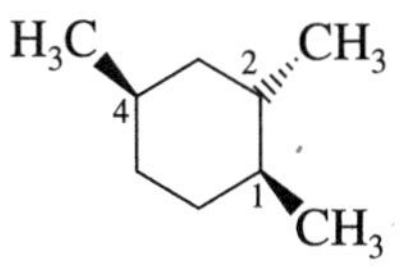

Less stable stereoisomer

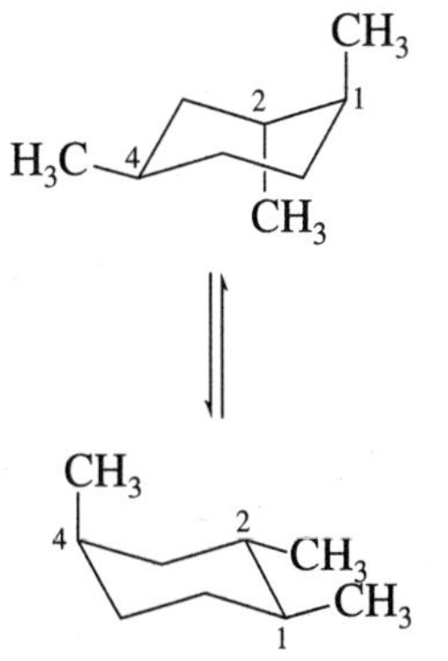

One axial methyl group in most stable conformation

(*e*) The first stereoisomer of 1,2,4-trimethylcyclohexane is the more stable one here, as it was in part (*d*). All its methyl groups are equatorial, but one of the methyl groups is axial in the most stable conformation of the second stereoisomer.

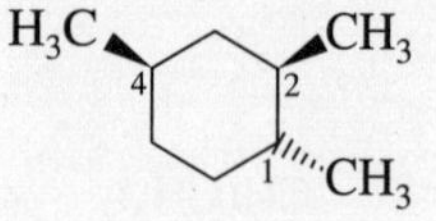

More stable stereoisomer 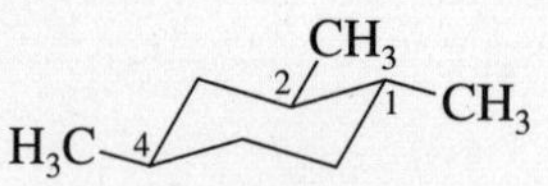 — All methyl groups equatorial in most stable conformation

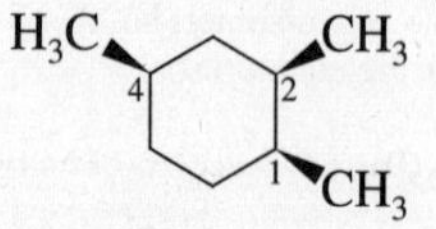

Less stable stereoisomer 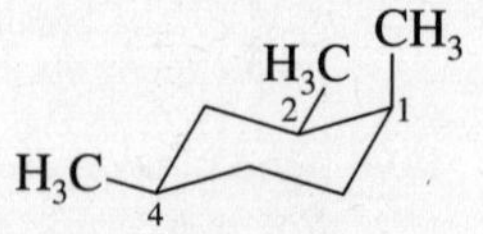 — One axial methyl group in most stable conformation

(*f*) Each stereoisomer of 2,3-dimethylbicyclo[3.2.1]octane has one axial and one equatorial methyl group. The first one, however, has a close contact between its axial methyl group and both methylene groups of the two-carbon bridge. The second stereoisomer has repulsions with only one axial methylene group; it is more stable.

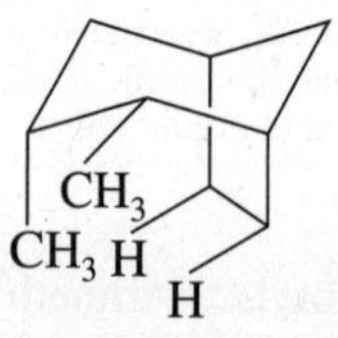

Less stable stereoisomer (more van der Waals strain)

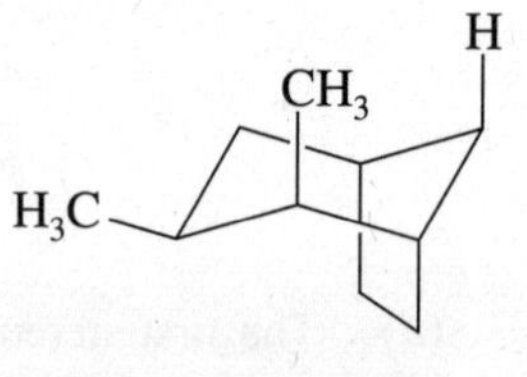

More stable stereoisomer (less van der Waals strain)

**3.30** First write structural formulas showing the relative stereochemistries and the preferred conformations of the two stereoisomers of 1,1,3,5-tetramethylcyclohexane.

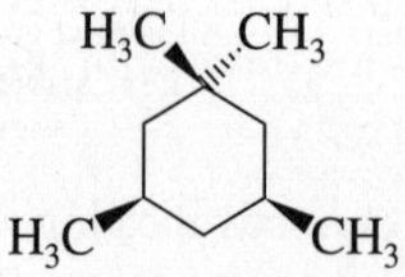

written in its most stable conformation as

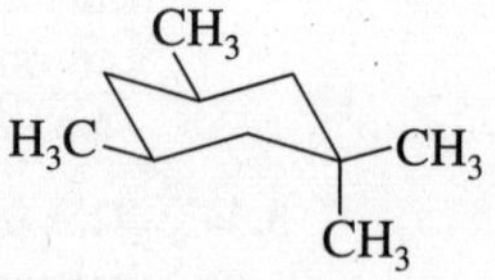

*cis*-1,1,3,5-Tetramethylcyclohexane

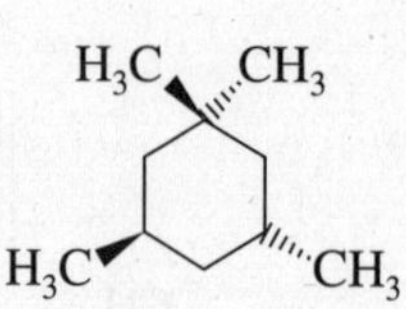

written in its most stable conformation as

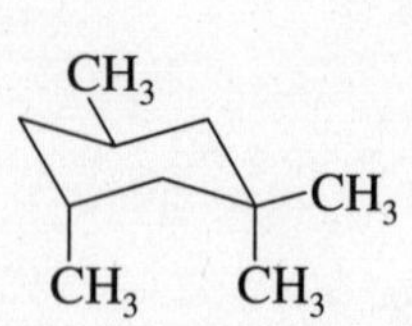

*trans*-1,1,3,5-Tetramethylcyclohexane

The cis stereoisomer is more stable than the trans. It exists in a conformation with only one axial methyl group, while the trans stereoisomer has two axial methyl groups in close contact with each other. The trans stereoisomer is destabilized by van der Waals strain.

**3.31** Both structures have approximately the same degree of angle strain and of torsional strain. Structure B has more van der Waals strain than A because two pairs of hydrogens (shown here) approach each other at distances that are rather close.

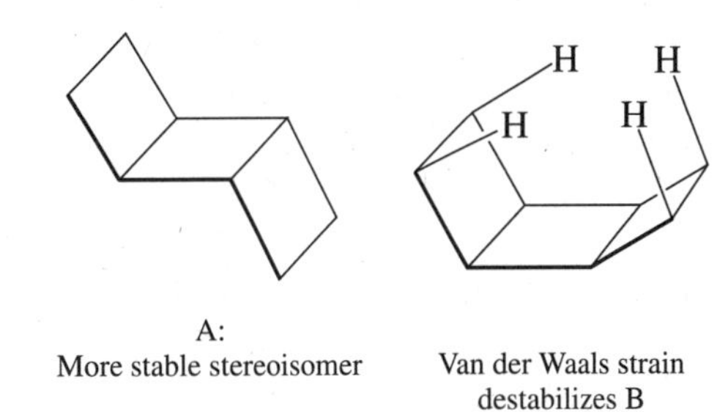


A: More stable stereoisomer

Van der Waals strain destabilizes B

**3.32** Five bond cleavages are required to convert cubane to a noncyclic skeleton; cubane is pentacyclic.

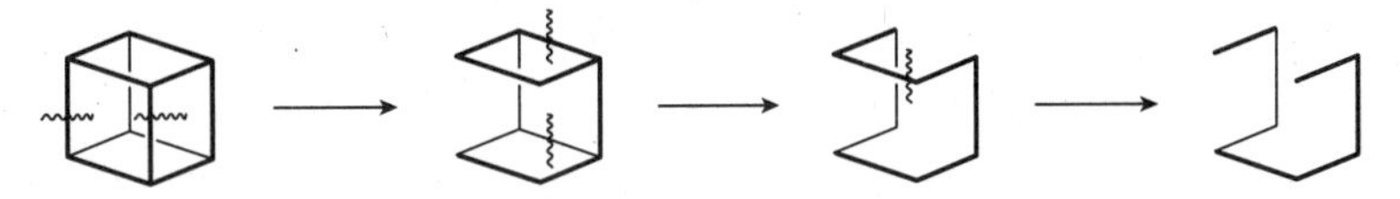

**3.33** Conformational representations of the two different forms of glucose are drawn in the usual way. An oxygen atom is present in the six-membered ring, and we are told in the problem that the ring exists in a chair conformation.

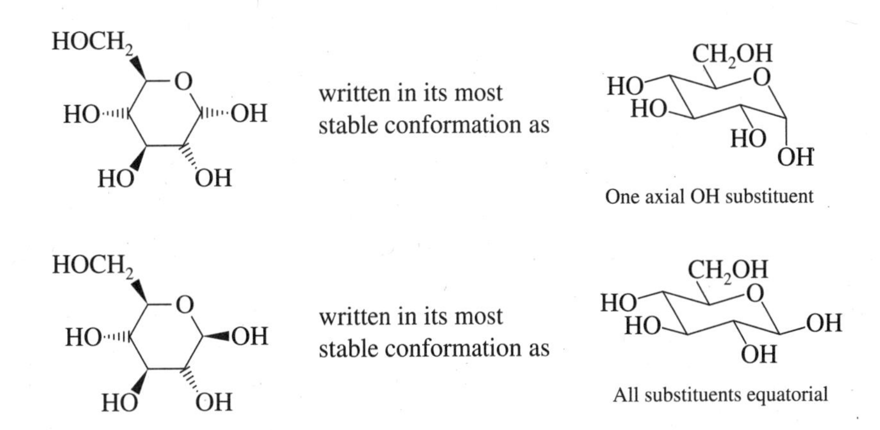


One axial OH substituent

All substituents equatorial

The two structures are not interconvertible by ring flipping; therefore they are not different conformations of the same molecule. Remember, ring flipping transforms all axial substituents to equatorial ones and vice versa. The two structures differ with respect to only one substituent; they are *stereoisomers* of each other.

**3.34** This problem is primarily an exercise in correctly locating equatorial and axial positions in cyclohexane rings that are joined together into a steroid skeleton. Parts (*a*) through (*e*) are concerned with positions 1, 4, 7, 11, and 12 in that order. The following diagram shows the orientation of axial and equatorial bonds at each of those positions.

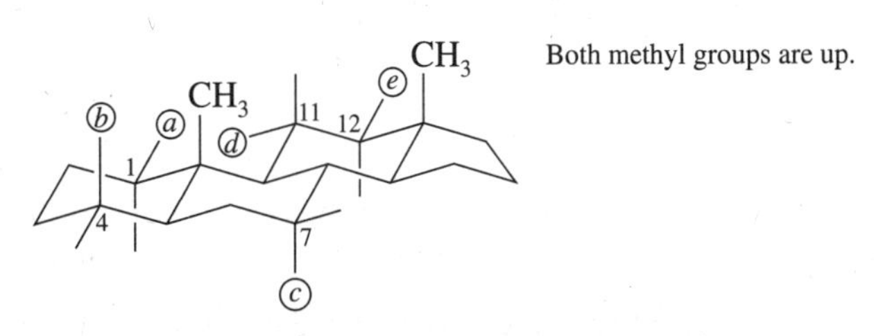


Both methyl groups are up.

(*a*) At C-1 the bond that is cis to the methyl groups is equatorial (up).
(*b*) At C-4 the bond that is cis to the methyl groups is axial (up).

(*c*) At C-7 the bond that is trans to the methyl groups is axial (down).
(*d*) At C-11 the bond that is trans to the methyl groups is equatorial (down).
(*e*) At C-12 the bond that is cis to the methyl groups is equatorial (up).

**3.35** Analyze this problem in exactly the same way as the preceding one by locating the axial and equatorial bonds at each position. It will be seen that the only differences are those at C-1 and C-4.

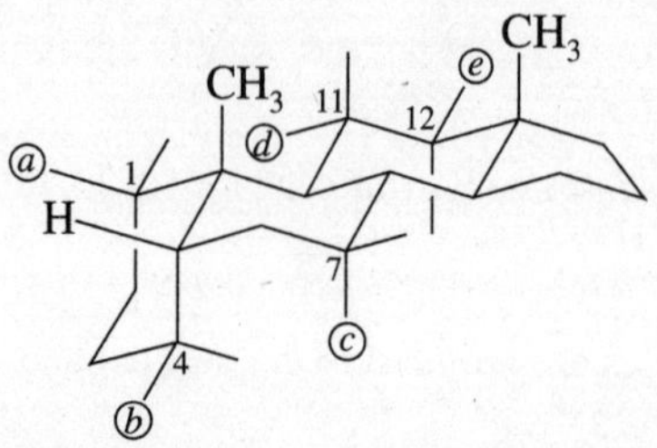

Both methyl groups are up.

(*a*) At C-1 the bond that is cis to the methyl groups is axial (up).
(*b*) At C-4 the bond that is cis to the methyl groups is equatorial (up).
(*c*) At C-7 the bond that is trans to the methyl groups is axial (down).
(*d*) At C-11 the bond that is trans to the methyl groups is equatorial (down).
(*e*) At C-12 the bond that is cis to the methyl groups is equatorial (up).

**3.36** (*a*) The torsion angle between chlorine substituents is 60° in the gauche conformation and 180° in the anti conformation of $ClCH_2CH_2Cl$.

Gauche
(can have a dipole moment)

Anti
(cannot have a dipole moment)

(*b*) All the individual bond dipole moments cancel in the anti conformation of $ClCH_2CH_2Cl$, and this conformation has no dipole moment. Since $ClCH_2CH_2Cl$ has a dipole moment of 1.12 D, it can exist entirely in the gauche conformation or it can be a mixture of anti and gauche conformations, but it cannot exist entirely in the anti conformation. Statement 1 is false.

**3.37–3.40** Solutions to molecular modeling exercises are not provided in this *Study Guide and Solutions Manual.* You should use *Learning By Modeling* for these exercises.

# SELF-TEST

## PART A

**A-1.** Draw Newman projections for both the gauche and the anti conformations of 1-chloropropane, $CH_3CH_2CH_2Cl$. Sight along the C-1, C-2 bond (the chlorine is attached to C-1).

**A-2.** Write Newman projection formulas for
(*a*) The least stable conformation of butane
(*b*) Two different staggered conformations of $CHCl_2CHCl_2$

**A-3.** Give the correct IUPAC name for the compound represented by the following Newman projection.

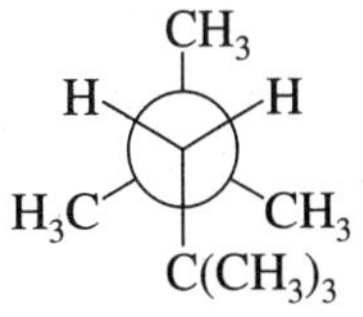

$CH_3$ H H $H_3C$ $CH_3$ $C(CH_3)_3$

**A-4.** Write the structure of the most stable conformation of the *less* stable stereoisomer of 1-*tert*-butyl-3-methylcyclohexane.

**A-5.** Draw the most stable conformation of the following substance:

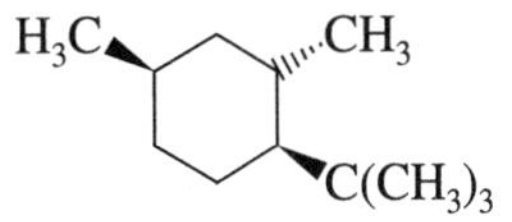

$H_3C$ $CH_3$ $C(CH_3)_3$

Which substituents are axial and which equatorial?

**A-6.** A wedge-and-dash representation of a form of ribose (called $\beta$-D-ribopyranose) is shown here. Draw the most stable chair conformation of this substance.

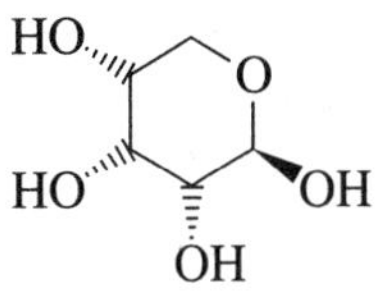

HO O HO OH OH

**A-7.** Consider compounds A, B, C, and D.

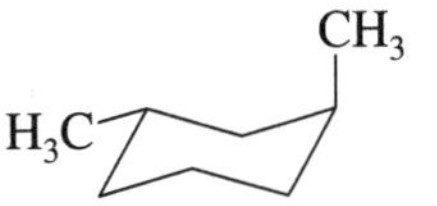

$CH_3$ $H_3C$ A $H_3C$ $CH_3$ B

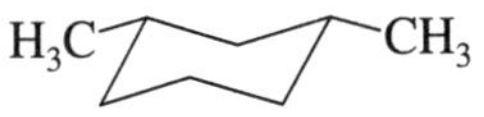

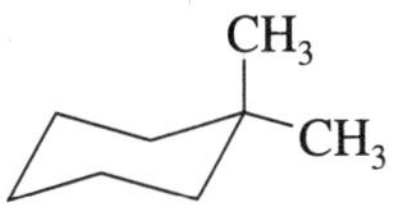

$CH_3$ $CH_3$ C $CH_3$ $H_3C$ $CH_3$ D

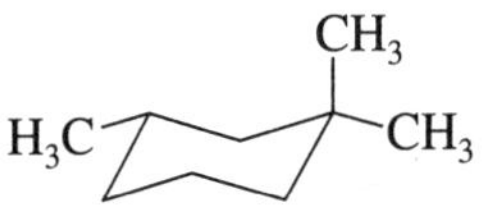

(*a*) Which one is a constitutional isomer of two others?
(*b*) Which two are stereoisomers of one another?
(*c*) Which one has the highest heat of combustion?
(*d*) Which one has the stereochemical descriptor trans in its name?

**A-8.** Draw clear depictions of two nonequivalent chair conformations of *cis*-1-isopropyl-4-methylcyclohexane, and indicate which is more stable.

**A-9.** Which has the lower heat of combustion, *cis*-1-ethyl-3-methylcyclohexane or *cis*-1-ethyl-4-methylcyclohexane?

**A-10.** The hydrocarbon shown is called twistane. Classify twistane as monocyclic, bicyclic, etc. What is the molecular formula of twistane?

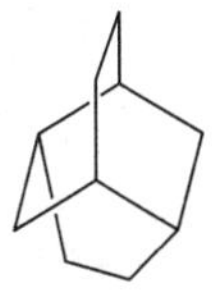

**A-11.** Sketch an approximate potential energy diagram similar to those shown in the text (Figures 3.4 and 3.7) for rotation about a carbon–carbon bond in 2-methylpropane. Does the form of the potential energy curve more closely resemble that of ethane or that of butane?

**A-12.** Draw the structure of the sulfur-containing heterocyclic compound that has a structure analogous to that of tetrahydrofuran.

## PART B

**B-1.** Which of the listed terms best describes the relationship between the methyl groups in the chair conformation of the substance shown?

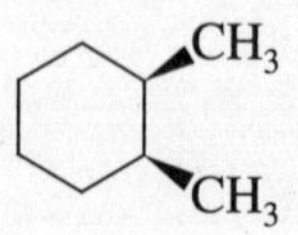

$CH_3$ $CH_3$

(*a*) Eclipsed (*c*) Anti
(*b*) Trans (*d*) Gauche

**B-2.** Rank the following substances in order of decreasing heat of combustion (largest → smallest).

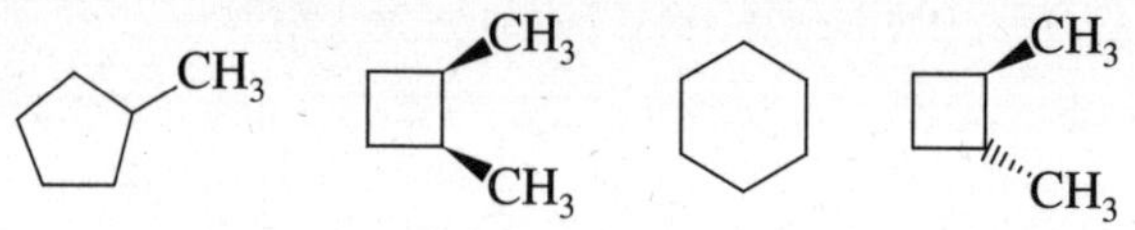

$CH_3$ $CH_3$ $CH_3$ $CH_3$ $CH_3$

1 2 3 4

(*a*) $1 > 2 > 4 > 3$ (*c*) $3 > 4 > 2 > 1$
(*b*) $2 > 4 > 1 > 3$ (*d*) $1 > 3 > 2 > 4$

**B-3.** Which of the following statements best describes the most stable conformation of *trans*-1, 3-dimethylcyclohexane?
(*a*) Both methyl groups are axial.
(*b*) Both methyl groups are equatorial.
(*c*) One methyl group is axial, the other equatorial.
(*d*) The molecule is severely strained and cannot exist.

**B-4.** Compare the stability of the following two compounds:
A: *cis*-1-Ethyl-3-methylcyclohexane
B: *trans*-1-Ethyl-3-methylcyclohexane
(*a*) A is more stable.
(*b*) B is more stable.
(*c*) A and B are of equal stability.
(*d*) No comparison can be made.

**B-5.** What, if anything, can be said about the magnitude of the equilibrium constant *K* for the following process?

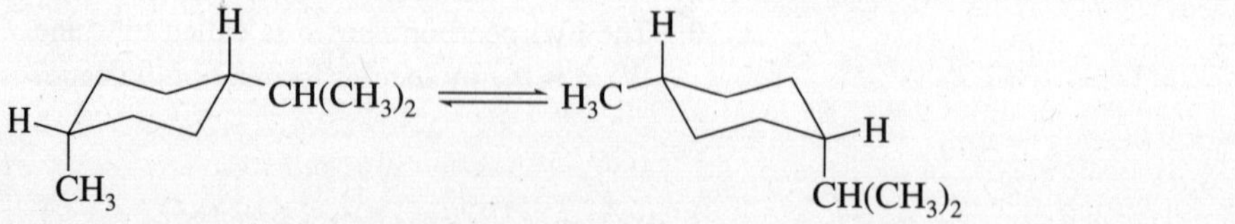

H H
H $CH(CH_3)_2$ ⇌ $H_3C$ H
$CH_3$ $CH(CH_3)_2$

(*a*) $K = 1$ (*c*) $K < 1$
(*b*) $K > 1$ (*d*) No estimate of *K* can be made.

**B-6.** What is the relationship between the two structures shown?

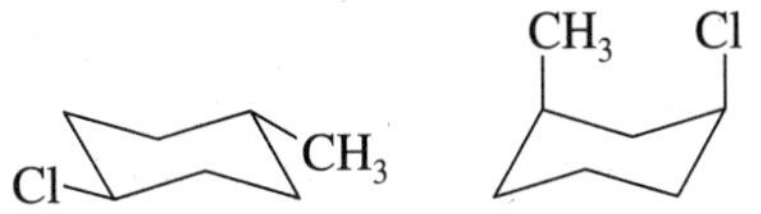

(*a*) Constitutional isomers
(*b*) Stereoisomers
(*c*) Different drawings of the same conformation of the same compound
(*d*) Different conformations of the same compound

**B-7.** The two structures shown here are _______ each other.

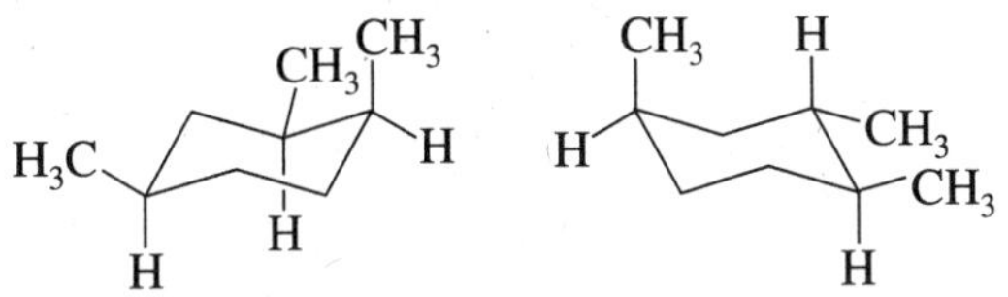

(*a*) identical with
(*b*) conformations of
(*c*) constitutional isomers of
(*d*) stereoisomers of

**B-8.** The most stable conformation of the following compound has

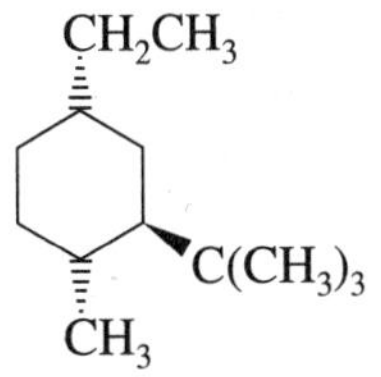

(*a*) An axial methyl group and an axial ethyl group
(*b*) An axial methyl group and an equatorial ethyl group
(*c*) An axial *tert*-buytl group
(*d*) An equatorial methyl group and an equatorial ethyl group
(*e*) An equatorial methyl group and an axial ethyl group

**B-9.** Which of the following statements is *not* true concerning the chair–chair interconversion of *trans*-1,2-diethylcyclohexane?
(*a*) An axial group will be changed into the equatorial position.
(*b*) The energy of repulsions present in the molecule will be changed.
(*c*) Formation of the cis substance will result.
(*d*) One chair conformation is more stable than the other.

**B-10.** The *most stable* conformation of the compound

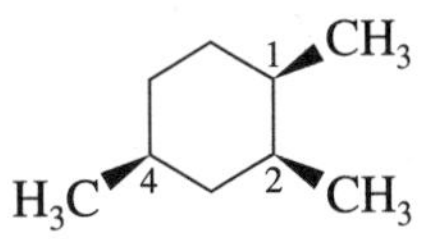

(in which all methyl groups are cis to one another) has:
(*a*) All methyl groups axial
(*b*) All methyl groups equatorial
(*c*) Equatorial methyl groups at C-1 and C-2
(*d*) Equatorial methyl groups at C-1 and C-4
(*e*) Equatorial methyl groups at C-2 and C-4

**B-11.** Which point on the potential energy diagram is represented by the Newman projection shown?

**B-12.** Which of the following statements is true?

(*a*) Van der Waals strain in *cis*-1,2-dimethylcyclopropane is the principal reason for its decreased stability relative to the trans isomer.

(*b*) Cyclohexane gives off more heat per $CH_2$ group on being burned in air than any other cycloalkane.

(*c*) The principal source of strain in the boat conformation of cyclohexane is angle strain.

(*d*) The principal source of strain in the gauche conformation of butane is torsional strain.

**B-13.** Which one of the following has an equatorial methyl group in its most stable conformation?

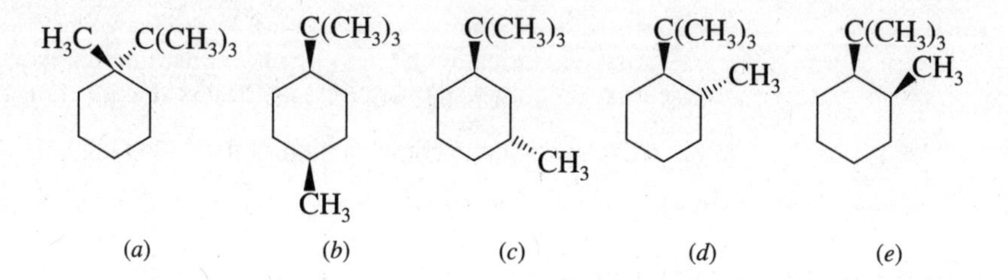


**B-14.** The structure shown is the carbon skeleton of *adamantane,* a symmetrical hydrocarbon having a structure that is a section of the diamond lattice.

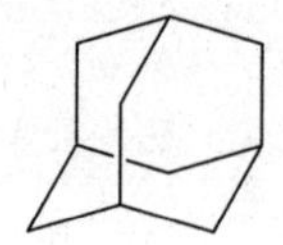

Adamantane is:

(*a*) Bicyclic (*c*) Tetracyclic

(*b*) Tricyclic (*d*) Pentacyclic

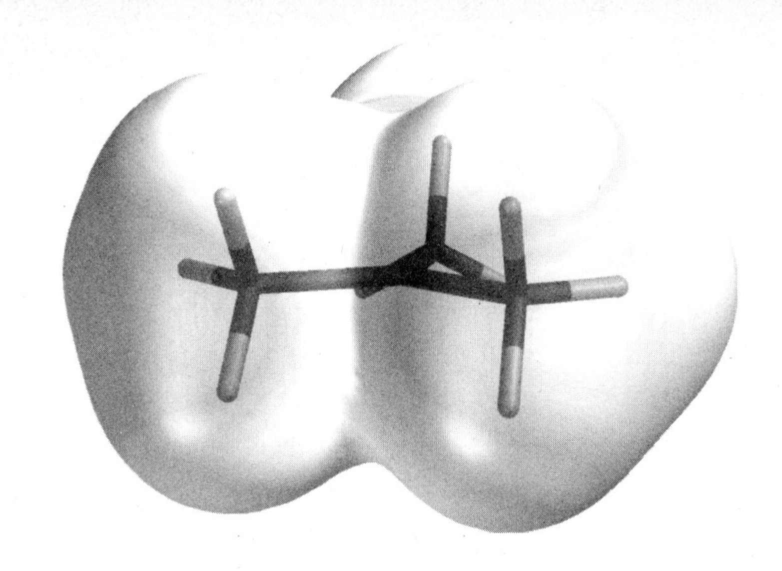

# CHAPTER 4

## ALCOHOLS AND ALKYL HALIDES

## SOLUTIONS TO TEXT PROBLEMS

**4.1** There are four $C_4H_9$ alkyl groups, and so there are four $C_4H_9Cl$ alkyl chlorides. Each may be named by both the functional class and substitutive methods. The functional class name uses the name of the alkyl group followed by the halide as a second word. The substitutive name modifies the name of the corresponding alkane to show the location of the halogen atom.

| | Functional class name | Substitutive name |
|---|---|---|
| $CH_3CH_2CH_2CH_2Cl$ | *n*-Butyl chloride (Butyl chloride) | 1-Chlorobutane |
| $CH_3CHCH_2CH_3$ with Cl on C-2 | *sec*-Butyl chloride (1-Methylpropyl chloride) | 2-Chlorobutane |
| $CH_3CHCH_2Cl$ with $CH_3$ on C-2 | Isobutyl chloride (2-Methylpropyl chloride) | 1-Chloro-2-methylpropane |
| $CH_3CCH_3$ with $CH_3$ and Cl on C-2 | *tert*-Butyl chloride (1,1-Dimethylethyl chloride) | 2-Chloro-2-methylpropane |

**4.2** Alcohols may also be named using both the functional class and substitutive methods, as in the previous problem.

| | Functional class name | Substitutive name |
|---|---|---|
| $CH_3CH_2CH_2CH_2OH$ | *n*-Butyl alcohol (Butyl alcohol) | 1-Butanol |
| 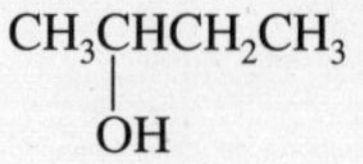 $CH_3CH(OH)CH_2CH_3$ | *sec*-Butyl alcohol (1-Methylpropyl alcohol) | 2-Butanol |
| 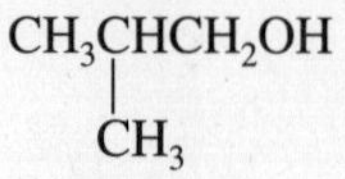 $CH_3CH(CH_3)CH_2OH$ | Isobutyl alcohol (2-Methylpropyl alcohol) | 2-Methyl-1-propanol |
| 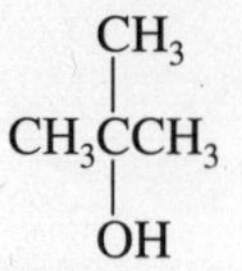 $CH_3C(CH_3)(OH)CH_3$ | *tert*-Butyl alcohol (1,1-Dimethylethyl alcohol) | 2-Methyl-2-propanol |

**4.3** Alcohols are classified as primary, secondary, or tertiary according to the number of carbon substituents attached to the carbon that bears the hydroxyl group.

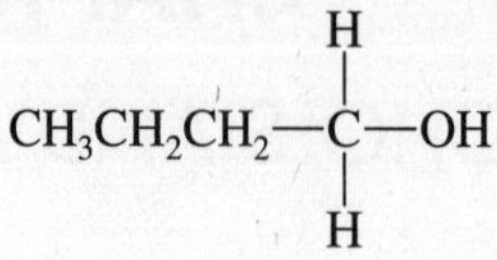

$CH_3CH_2CH_2—CH_2—OH$

Primary alcohol
(one alkyl group bonded to —$CH_2OH$)

$CH_3—CH(OH)—CH_2CH_3$

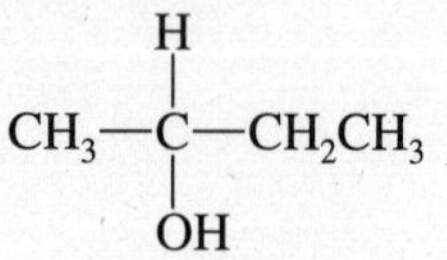

Secondary alcohol
(two alkyl groups bonded to >CHOH)

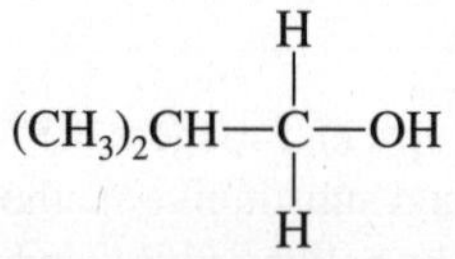

$(CH_3)_2CH—CH_2—OH$

Primary alcohol
(one alkyl group bonded to —$CH_2OH$)

$CH_3—C(CH_3)_2—OH$

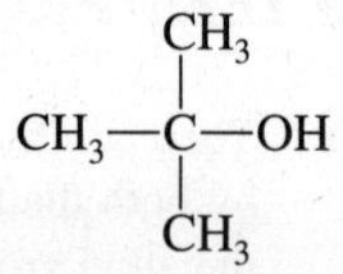

Tertiary alcohol
(three alkyl groups bonded to ≥COH)

**4.4** Dipole moment is the product of charge and distance. Although the electron distribution in the carbon–chlorine bond is more polarized than that in the carbon–bromine bond, this effect is counterbalanced by the longer carbon–bromine bond distance.

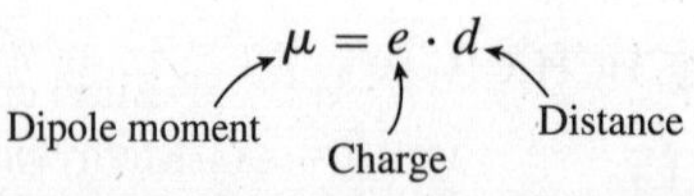

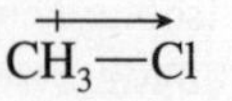

$$\mu = e \cdot d$$

Dipole moment ($\mu$); Charge ($e$); Distance ($d$)

$CH_3—Cl$
Methyl chloride
(greater value of *e*)
$\mu$ 1.9 D

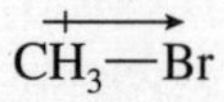

$CH_3—Br$
Methyl bromide
(greater value of *d*)
$\mu$ 1.8 D

**4.5** All the hydrogens in dimethyl ether ($CH_3OCH_3$) are bonded to carbon; therefore, intermolecular hydrogen bonding between dimethyl ether molecules does not take place, and its boiling point is lower than that of ethanol ($CH_3CH_2OH$), where hydrogen bonding involving the —OH group is important.

**4.6** Ammonia is a base and abstracts (accepts) a proton from the acid (proton donor) hydrogen chloride.

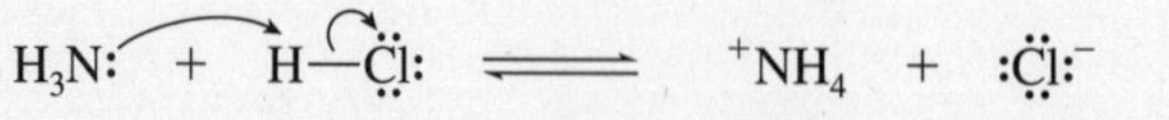

$$H_3N\!: \; + \; H—\ddot{Cl}\!: \;\rightleftharpoons\; {}^{+}NH_4 \; + \; :\ddot{Cl}:^{-}$$

Base; Acid; Conjugate acid; Conjugate base

**4.7** Since the $pK_a$ of HCN is given as 9.1, its $K_a = 10^{-9.1}$. In more conventional notation, $K_a = 8 \times 10^{-10}$. Hydrogen cyanide is a weak acid.

**4.8** Hydrogen cyanide is a weak acid, but it is a stronger acid than water ($pK_a = 15.7$). Since HCN is a stronger acid than water, its conjugate base ($CN^-$) is a weaker base than hydroxide ($HO^-$), which is the conjugate base of water.

**4.9** An unshared electron pair on oxygen abstracts the proton from hydrogen chloride.

$$(CH_3)_3C\ddot{O}H + H{-}\ddot{Cl}: \rightleftharpoons (CH_3)_3C\overset{+}{\ddot{O}}(H){-}H + :\ddot{Cl}:^-$$

Base | Acid | Conjugate acid | Conjugate base

**4.10** In any proton-transfer process, the position of equilibrium favors formation of the weaker acid and the weaker base from the stronger acid and base. Alkyloxonium ions ($ROH_2^+$) have approximately the same acidity as hydronium ion ($H_3O^+$, $pK_a = -1.7$). Thus hydrogen chloride ($pK_a \approx -7$) is the stronger acid. *tert*-Butyl alcohol is the stronger base because it is the conjugate of the weaker acid (*tert*-butyloxonium ion).

$$(CH_3)_3COH + HCl \rightleftharpoons (CH_3)_3C\overset{+}{O}H_2 + Cl^-$$

| Stronger base | ($pK_a \approx -7$) Stronger acid | ($pK_a \approx -1.7$) Weaker acid | Weaker base |
|---|---|---|---|

The equilibrium constant for proton transfer from hydrogen chloride to *tert*-butyl alcohol is much greater than 1.

**4.11** The proton being transferred is partially bonded to the oxygen of *tert*-butyl alcohol and to chloride at the transition state.

$$(CH_3)_3C(H)\overset{\delta+}{\ddot{O}}\text{------}H\text{------}\overset{\delta-}{\ddot{Cl}}:$$

**4.12** (*b*) Hydrogen chloride converts tertiary alcohols to tertiary alkyl chlorides.

$$(CH_3CH_2)_3COH + HCl \longrightarrow (CH_3CH_2)_3CCl + H_2O$$

3-Ethyl-3-pentanol | Hydrogen chloride | 3-Chloro-3-ethylpentane | Water

(*c*) 1-Tetradecanol is a primary alcohol having an unbranched 14-carbon chain. Hydrogen bromide reacts with primary alcohols to give the corresponding primary alkyl bromide.

$$CH_3(CH_2)_{12}CH_2OH + HBr \longrightarrow CH_3(CH_2)_{12}CH_2Br + H_2O$$

1-Tetradecanol | Hydrogen bromide | 1-Bromotetradecane | Water

**4.13** The order of carbocation stability is tertiary > secondary > primary. There is only one $C_5H_{11}^+$ carbocation that is tertiary, and so that is the most stable one.

$$CH_3CH_2\overset{+}{C}(CH_3)_2$$

1,1-Dimethylpropyl cation

**4.14** 1-Butanol is a primary alcohol; 2-butanol is a secondary alcohol. A carbocation intermediate is possible in the reaction of 2-butanol with hydrogen bromide but not in the corresponding reaction of 1-butanol.

The mechanism of the reaction of 1-butanol with hydrogen bromide proceeds by displacement of water by bromide ion from the protonated form of the alcohol (the alkyloxonium ion).

**Protonation of the alcohol:**

$$CH_3CH_2CH_2CH_2\ddot{O}(H): + H{-}\ddot{Br}: \longrightarrow CH_3CH_2CH_2CH_2\ddot{O}^+H_2 + :\ddot{Br}:^-$$

1-Butanol | Hydrogen bromide | Butyloxonium ion | Bromide

**Displacement of water by bromide:**

$$:\ddot{Br}:^- + CH_3CH_2CH_2{-}CH_2{-}\ddot{O}^+H_2 \xrightarrow{\text{slow}} CH_3CH_2CH_2CH_2Br + :\ddot{O}H_2$$

Bromide ion | Butyloxonium ion | 1-Bromobutane | Water

The slow step, displacement of water by bromide from the oxonium ion, is bimolecular. The reaction of 1-butanol with hydrogen bromide follows the $S_N2$ mechanism.

The reaction of 2-butanol with hydrogen bromide involves a carbocation intermediate.

**Protonation of the alcohol:**

$$CH_3CH_2CH(\ddot{O}H)CH_3 + H{-}\ddot{Br}: \longrightarrow CH_3CH_2CH(\ddot{O}^+H_2)CH_3 + :\ddot{Br}:^-$$

2-Butanol | Hydrogen bromide | *sec*-Butyloxonium ion | Bromide ion

**Dissociation of the oxonium ion:**

$$CH_3CH_2CH(\ddot{O}^+H_2)CH_3 \xrightarrow{\text{slow}} CH_3CH_2\overset{+}{C}HCH_3 + H\ddot{O}H$$

*sec*-Butyloxonium ion | *sec*-Butyl cation | Water

**Capture of *sec*-butyl cation by bromide:**

$$:\ddot{Br}:^- + CH_3CH_2\overset{+}{C}HCH_3 \longrightarrow CH_3CH_2CH(Br)CH_3$$

Bromide ion | *sec*-Butyl cation | 2-Bromobutane

The slow step, dissociation of the oxonium ion, is unimolecular. The reaction of 2-butanol with hydrogen bromide follows the $S_N1$ mechanism.

**4.15** The most stable alkyl free radicals are tertiary. The tertiary free radical having the formula $C_5H_{11}$ has the same skeleton as the carbocation in Problem 4.13.

$$CH_3CH_2{-}\dot{C}(CH_3)_2$$

**4.16** (*b*) Writing the equations for carbon–carbon bond cleavage in propane and in 2-methylpropane, we see that a primary ethyl radical is produced by a cleavage of propane whereas a secondary isopropyl radical is produced by cleavage of 2-methylpropane.

$$CH_3CH_2—CH_3 \longrightarrow CH_3\dot{C}H_2 + \cdot CH_3$$

Propane — Ethyl radical — Methyl radical

$$CH_3CH(CH_3)CH_3 \longrightarrow CH_3\dot{C}HCH_3 + \cdot CH_3$$

2-Methylpropane — Isopropyl radical — Methyl radical

A secondary radical is more stable than a primary one, and so carbon–carbon bond cleavage of 2-methylpropane requires less energy than carbon–carbon bond cleavage of propane.

(*c*) Carbon–carbon bond cleavage of 2,2-dimethylpropane gives a tertiary radical.

$$CH_3C(CH_3)_2CH_3 \longrightarrow CH_3—\dot{C}(CH_3)_2 + \cdot CH_3$$

2,2-Dimethylpropane — *tert*-Butyl radical — Methyl radical

As noted in part (*b*), a secondary radical is produced on carbon–carbon bond cleavage of 2-methylpropane. We therefore expect a lower carbon–carbon bond dissociation energy for 2,2-dimethylpropane than for 2-methylpropane, since a tertiary radical is more stable than a secondary one.

**4.17** First write the equation for the overall reaction.

$$CH_3Cl + Cl_2 \longrightarrow CH_2Cl_2 + HCl$$

Chloromethane — Chlorine — Dichloromethane — Hydrogen chloride

The initiation step is dissociation of chlorine to two chlorine atoms.

$$:\ddot{Cl}—\ddot{Cl}: \longrightarrow :\ddot{Cl}\cdot + \cdot\ddot{Cl}:$$

Chlorine — 2 Chlorine atoms

A chlorine atom abstracts a hydrogen atom from chloromethane in the first propagation step.

$$Cl—CH_2—H + \cdot\ddot{Cl}: \longrightarrow Cl—\dot{C}H_2 + H—\ddot{Cl}:$$

Chloromethane — Chlorine atom — Chloromethyl radical — Hydrogen chloride

Chloromethyl radical reacts with $Cl_2$ in the next propagation step.

$$Cl—\dot{C}H_2 + :\ddot{Cl}—\ddot{Cl}: \longrightarrow Cl—CH_2—\ddot{Cl}: + \cdot\ddot{Cl}:$$

Chloromethyl radical — Chlorine — Dichloromethane — Chlorine atom

**4.18** Writing the structural formula for ethyl chloride reveals that there are two nonequivalent sets of hydrogen atoms, in either of which a hydrogen is capable of being replaced by chlorine.

$$\underset{\text{Ethyl chloride}}{CH_3CH_2Cl} \xrightarrow[\text{light or heat}]{Cl_2} \underset{\text{1,1-Dichloroethane}}{CH_3CHCl_2} + \underset{\text{1,2-Dichloroethane}}{ClCH_2CH_2Cl}$$

The two dichlorides are 1,1-dichloroethane and 1,2-dichloroethane.

**4.19** Propane has six primary hydrogens and two secondary. In the chlorination of propane, the relative proportions of hydrogen atom removal are given by the product of the statistical distribution and the relative rate per hydrogen. Given that a secondary hydrogen is abstracted 3.9 times faster than a primary one, we write the expression for the amount of chlorination at the primary relative to that at the secondary position as:

$$\frac{\text{Number of primary hydrogens} \times \text{rate of abstraction of primary hydrogen}}{\text{Number of secondary hydrogens} \times \text{rate of abstraction of a secondary hydrogen}} = \frac{6 \times 1}{2 \times 3.9} = \frac{0.77}{1.00}$$

Thus, the percentage of propyl chloride formed is 0.77/1.77, or 43%, and that of isopropyl chloride is 57%. (The amounts actually observed are propyl 45%, isopropyl 55%.)

**4.20** (*b*) In contrast with free-radical chlorination, alkane bromination is a highly selective process. The major organic product will be the alkyl bromide formed by substitution of a tertiary hydrogen with a bromine.

$$\text{1-Isopropyl-1-methylcyclopentane (cyclopentane bearing } CH_3 \text{ and } CH(CH_3)_2\text{)} \xrightarrow[\text{light}]{Br_2} \text{1-(1-Bromo-1-methylethyl)-1-methylcyclopentane (cyclopentane bearing } CH_3 \text{ and } C(CH_3)_2Br\text{)}$$

Tertiary hydrogen

1-Isopropyl-1-methylcyclopentane — 1-(1-Bromo-1-methylethyl)-1-methylcyclopentane

(*c*) As in part (*b*), bromination results in substitution of a tertiary hydrogen.

$$\underset{\text{2,2,4-Trimethylpentane}}{CH_3C(CH_3)_2CH_2CH(CH_3)CH_3} \xrightarrow[\text{light}]{Br_2} \underset{\text{2-Bromo-2,4,4-trimethylpentane}}{CH_3C(CH_3)_2CH_2C(CH_3)(Br)CH_3}$$

**4.21** (*a*) Cyclobutanol has a hydroxyl group attached to a four-membered ring.

Cyclobutanol

(*b*) *sec*-Butyl alcohol is the functional class name for 2-butanol.

$$\underset{\text{sec-Butyl alcohol}}{CH_3CH(OH)CH_2CH_3}$$

(*c*) The hydroxyl group is at C-3 of an unbranched seven-carbon chain in 3-heptanol.

$$\underset{\text{3-Heptanol}}{CH_3CH_2CH(OH)CH_2CH_2CH_2CH_3}$$

(*d*) A chlorine at C-2 is on the opposite side of the ring from the C-1 hydroxyl group in *trans*-2-chlorocyclopentanol. Note that it is not necessary to assign a number to the carbon that bears the hydroxyl group; naming the compound as a derivative of cyclopentanol automatically requires the hydroxyl group to be located at C-1.

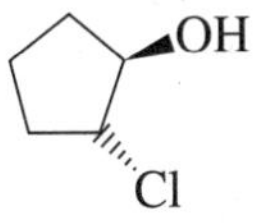

*trans*-2-Chlorocyclopentanol

(*e*) This compound is an alcohol in which the longest continuous chain that incorporates the hydroxyl function has eight carbons. It bears chlorine substituents at C-2 and C-6 and methyl and hydroxyl groups at C-4.

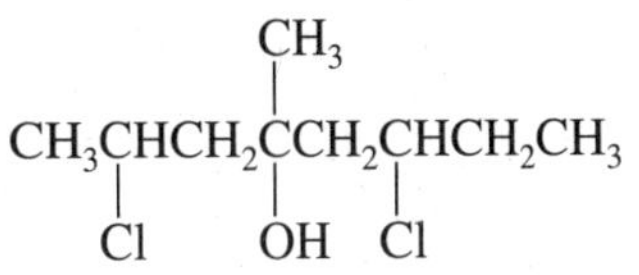

$$CH_3CH(Cl)CH_2C(CH_3)(OH)CH_2CH(Cl)CH_2CH_3$$

2,6-Dichloro-4-methyl-4-octanol

(*f*) The hydroxyl group is at C-1 in *trans*-4-*tert*-butylcyclohexanol; the *tert*-butyl group is at C-4. The structures of the compound can be represented as shown at the left; the structure at the right depicts it in its most stable conformation.

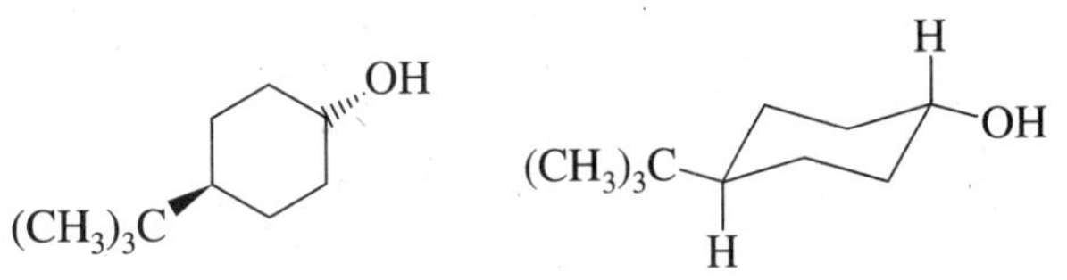

*trans*-4-*tert*-Butylcyclohexanol

(*g*) The cyclopropyl group is on the same carbon as the hydroxyl group in 1-cyclopropylethanol.

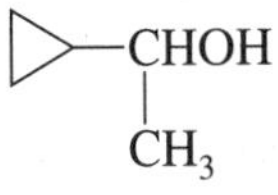

$$\triangleright\!-CH(OH)CH_3$$

1-Cyclopropylethanol

(*h*) The cyclopropyl group and the hydroxyl group are on adjacent carbons in 2-cyclopropylethanol.

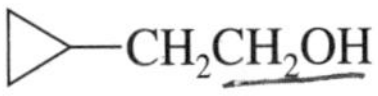

$$\triangleright\!-CH_2CH_2OH$$

2-Cyclopropylethanol

**4.22** (*a*) This compound has a five-carbon chain that bears a methyl substituent and a bromine. The numbering scheme that gives the lower number to the substituent closest to the end of the chain is chosen. Bromine is therefore at C-1, and methyl is a substituent at C-4.

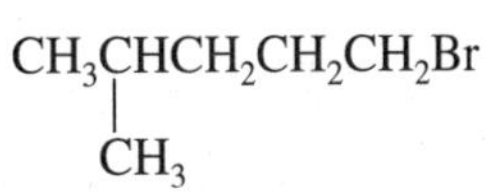

$$CH_3CH(CH_3)CH_2CH_2CH_2Br$$

1-Bromo-4-methylpentane

(*b*) This compound has the same carbon skeleton as the compound in part (*a*) but bears a hydroxyl group in place of the bromine and so is named as a derivative of 1-pentanol.

$$\underset{\displaystyle CH_3}{CH_3\underset{|}{C}HCH_2CH_2CH_2OH}$$

4-Methyl-1-pentanol

(*c*) This molecule is a derivative of ethane and bears three chlorines and one bromine. The name 2-bromo-1,1,1-trichloroethane gives a lower number at the first point of difference than 1-bromo-2,2,2-trichloroethane.

$$Cl_3CCH_2Br$$

2-Bromo-1,1,1-trichloroethane

(*d*) This compound is a constitutional isomer of the preceding one. Regardless of which carbon the numbering begins at, the substitution pattern is 1,1,2,2. Alphabetical ranking of the halogens therefore dictates the direction of numbering. Begin with the carbon that bears bromine.

$$Cl_2CH\underset{\displaystyle Cl}{\underset{|}{C}}HBr$$

1-Bromo-1,2,2-trichloroethane

(*e*) This is a trifluoro derivative of ethanol. The direction of numbering is dictated by the hydroxyl group, which is at C-1 in ethanol.

$$CF_3CH_2OH$$

2,2,2-Trifluoroethanol

(*f*) Here the compound is named as a derivative of cyclohexanol, and so numbering begins at the carbon that bears the hydroxyl group.

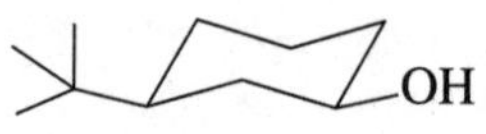

OH

*cis*-3-*tert*-Butylcyclohexanol

(*g*) This alcohol has its hydroxyl group attached to C-2 of a three-carbon continuous chain; it is named as a derivative of 2-propanol.

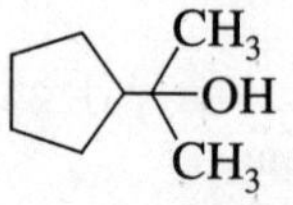

$CH_3$
OH
$CH_3$

2-Cyclopentyl-2-propanol

(*h*) The six carbons that form the longest continuous chain have substituents at C-2, C-3, and C-5 when numbering proceeds in the direction that gives the lowest locants to substituents at the first point of difference. The substituents are cited in alphabetical order.

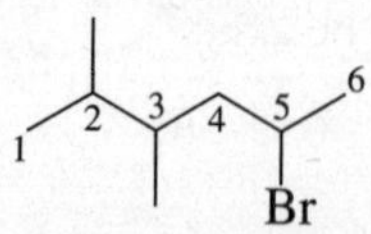

1 2 3 4 5 6
Br

5-Bromo-2,3-dimethylhexane

Had numbering begun in the opposite direction, the locants would be 2,4,5 rather than 2,3,5.

(*i*) Hydroxyl controls the numbering because the compound is named as an alcohol.

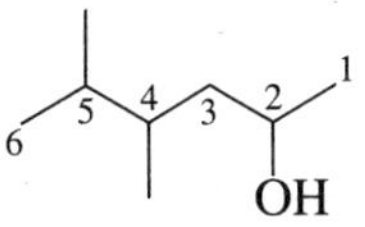

6 5 4 3 2 1 OH

4,5-Dimethyl-2-hexanol

**4.23** Primary alcohols are alcohols in which the hydroxyl group is attached to a carbon atom which has one alkyl substituent and two hydrogens. Four primary alcohols have the molecular formula $C_5H_{12}O$. The functional class name for each compound is given in parentheses.

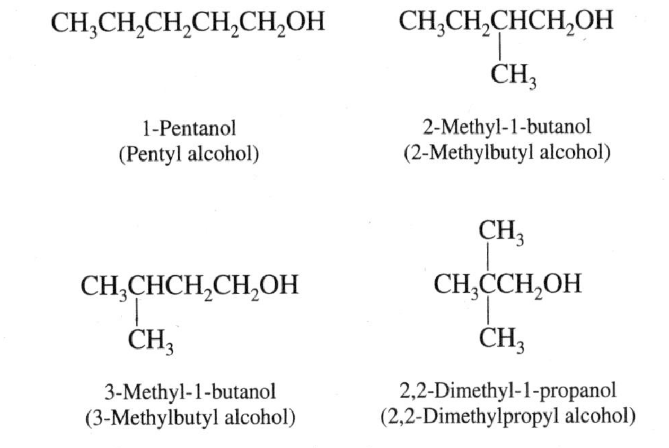


1-Pentanol (Pentyl alcohol)

2-Methyl-1-butanol (2-Methylbutyl alcohol)

3-Methyl-1-butanol (3-Methylbutyl alcohol)

2,2-Dimethyl-1-propanol (2,2-Dimethylpropyl alcohol)

Secondary alcohols are alcohols in which the hydroxyl group is attached to a carbon atom which has two alkyl substituents and one hydrogen. There are three secondary alcohols of molecular formula $C_5H_{12}O$:

$CH_3CHCH_2CH_2CH_3$ (OH)

2-Pentanol (1-Methylbutyl alcohol)

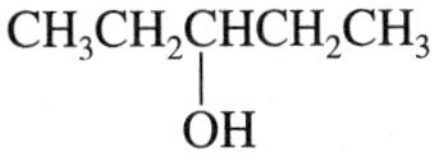

$CH_3CH_2CHCH_2CH_3$ (OH)

3-Pentanol (1-Ethylpropyl alcohol)

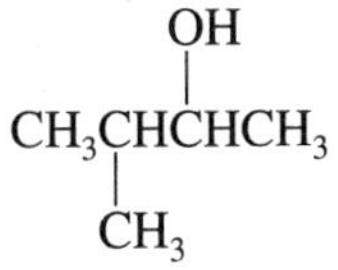

$CH_3CHCHCH_3$ (OH, $CH_3$)

3-Methyl-2-butanol (1,2-Dimethylpropyl alcohol)

Only 2-methyl-2-butanol is a tertiary alcohol (three alkyl substituents on the hydroxyl-bearing carbon):

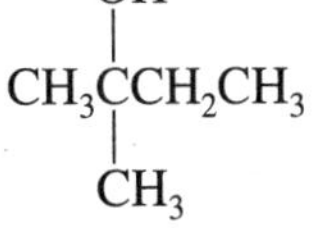

$CH_3CCH_2CH_3$ (OH, $CH_3$)

2-Methyl-2-butanol (1,1-Dimethylpropyl alcohol)

**4.24** The first methylcyclohexanol to be considered is 1-methylcyclohexanol. The preferred chair conformation will have the larger methyl group in an equatorial orientation, whereas the smaller hydroxyl group will be axial.

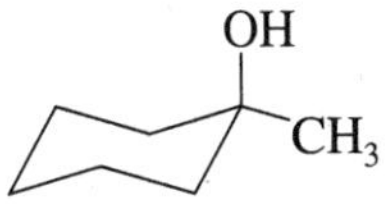

OH $CH_3$

Most stable conformation of 1-methylcyclohexanol

In the other isomers methyl and hydroxyl will be in a 1,2, 1,3, or 1,4 relationship and can be cis or trans in each. We can write the preferred conformation by recognizing that the methyl group will always be equatorial and the hydroxyl either equatorial or axial.

*trans*-2-Methylcyclohexanol *cis*-3-Methylcyclohexanol *trans*-4-Methylcyclohexanol

*cis*-2-Methylcyclohexanol *trans*-3-Methylcyclohexanol *cis*-4-Methylcyclohexanol

**4.25** The assumption is incorrect for the 3-methylcyclohexanols. *cis*-3-Methylcyclohexanol is more stable than *trans*-3-methylcyclohexanol because the methyl group and the hydroxyl group are both equatorial in the cis isomer, whereas one substituent must be axial in the trans.

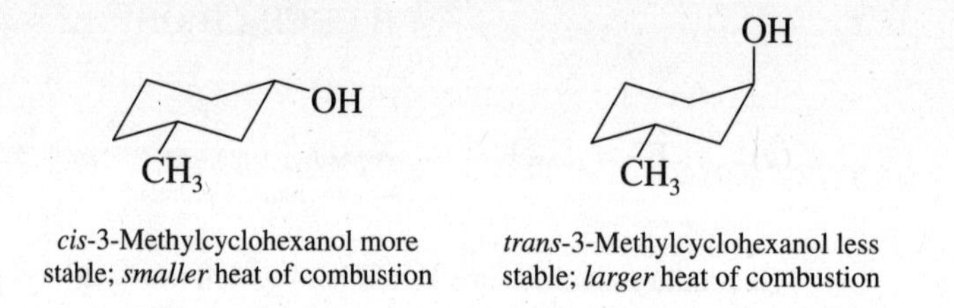


*cis*-3-Methylcyclohexanol more stable; *smaller* heat of combustion

*trans*-3-Methylcyclohexanol less stable; *larger* heat of combustion

**4.26** (*a*) The most stable conformation will be the one with all the substituents equatorial.

The hydroxyl group is trans to the isopropyl group and cis to the methyl group.

(*b*) All three substituents need not always be equatorial; instead, one or two of them may be axial. Since neomenthol is the second most stable stereoisomer, we choose the structure with *one* axial substituent. Furthermore, we choose the structure with the smallest substituent (the hydroxyl group) as the axial one. Neomenthol is shown as follows:

**4.27** In all these reactions the negatively charged atom abstracts a proton from an acid.

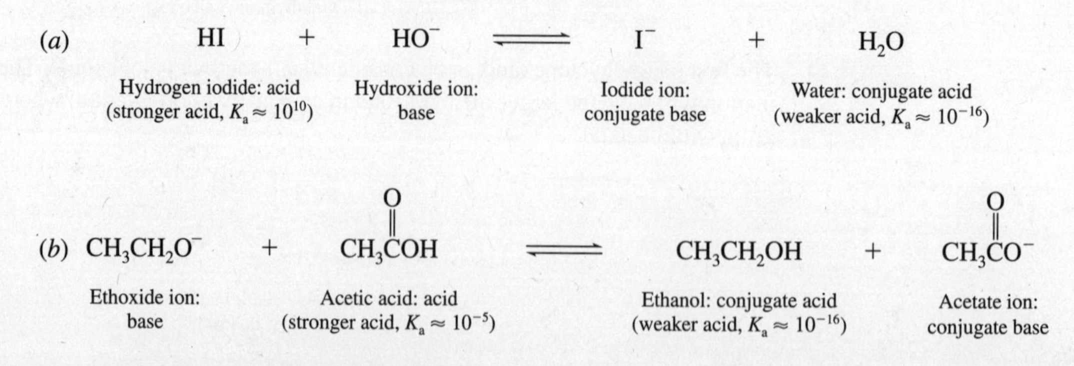

(*c*) $HF + H_2N^- \rightleftharpoons F^- + H_3N$

Hydrogen fluoride: acid (stronger acid, $K_a \approx 10^{-4}$) | Amide ion: base | Fluoride ion: conjugate base | Ammonia: conjugate acid (weaker acid, $K_a \approx 10^{-36}$)

(*d*) $CH_3C(=O)O^- + HCl \rightleftharpoons CH_3C(=O)OH + Cl^-$

Acetate ion: base | Hydrogen chloride: acid (stronger acid, $K_a \approx 10^7$) | Acetic acid: conjugate acid (weaker acid, $K_a \approx 10^{-5}$) | Chloride ion: conjugate base

(*e*) $(CH_3)_3CO^- + H_2O \rightleftharpoons (CH_3)_3COH + HO^-$

*tert*-Butoxide ion: base | Water: acid (stronger acid, $K_a \approx 10^{-16}$) | *tert*-Butyl alcohol: conjugate acid (weaker acid, $K_a \approx 10^{-18}$) | Hydroxide ion: conjugate base

(*f*) $(CH_3)_2CHOH + H_2N^- \rightleftharpoons (CH_3)_2CHO^- + H_3N$

Isopropyl alcohol: acid (stronger acid, $K_a \approx 10^{-17}$) | Amide ion: base | Isopropoxide ion: conjugate base | Ammonia: conjugate acid (weaker acid, $K_a \approx 10^{-36}$)

(*g*) $F^- + H_2SO_4 \rightleftharpoons HF + HSO_4^-$

Fluoride ion: base | Sulfuric acid: acid (stronger acid, $K_a \approx 10^5$) | Hydrogen fluoride: conjugate acid (weaker acid, $K_a \approx 10^{-4}$) | Hydrogen sulfate ion: conjugate base

**4.28** (*a*) The proton-transfer transition state can represent the following reaction, or its reverse:

$$(CH_3)_2CH{-}\ddot{O}{:}^- + H{-}Br \rightleftharpoons (CH_3)_2CH{-}\overset{\delta-}{O}\cdots H \cdots Br^{\delta-} \rightleftharpoons (CH_3)_2CH{-}OH + Br^-$$

Base | Acid (stronger acid) $K_a = 10^9$ | Conjugate acid (weaker acid) $K_a = 10^{-17}$ | Conjugate base

When the reaction proceeds as drawn, the stronger acid (hydrogen bromide) is on the left, the weaker acid (isopropyl alcohol) is on the right, and the equilibrium lies to the right.

(*b*) Hydroxide is a strong base; methyloxonium ion is a strong acid.

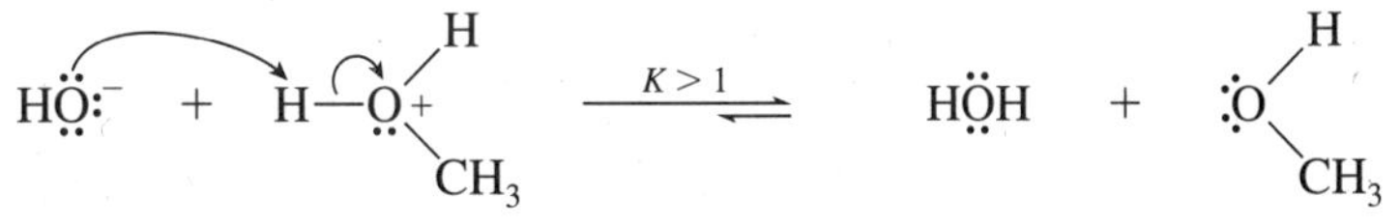

$$H\ddot{O}{:}^- + H{-}\overset{+}{O}(H)CH_3 \xrightarrow{K > 1} H\ddot{O}H + {:}\ddot{O}(H)CH_3$$

Hydroxide ion (base) | Methyloxonium ion (acid) | Water (conjugate acid) | Methanol (conjugate base)

**4.29** (*a*) This problem reviews the relationship between logarithms and exponential numbers. We need to determine $K_a$, given $pK_a$. The equation that relates the two is

$$pK_a = -\log_{10} K_a$$

Therefore

$$K_a = 10^{-pK_a} = 10^{-3.48} = 3.3 \times 10^{-4}$$

(*b*) As described in part (*a*), $K_a = 10^{-pK_a}$, therefore $K_a$ for vitamin C is given by the expression:

$$K_a = 10^{-4.17}$$
$$= 6.7 \times 10^{-5}$$

(*c*) Similarly, $K_a = 1.8 \times 10^{-4}$ for formic acid (p$K_a$ 3.75).
(*d*) $K_a = 6.5 \times 10^{-2}$ for oxalic acid (p$K_a$ 1.19).

In ranking the acids in order of decreasing acidity, remember that the larger the equilibrium constant $K_a$, the stronger the acid; and the lower the p$K_a$ value, the stronger the acid.

| Acid | $K_a$ | p$K_a$ |
|---|---|---|
| Oxalic (strongest) | $6.5 \times 10^{-2}$ | 1.19 |
| Aspirin | $3.3 \times 10^{-4}$ | 3.48 |
| Formic acid | $1.8 \times 10^{-4}$ | 3.75 |
| Vitamin C (weakest) | $6.7 \times 10^{-5}$ | 4.17 |

**4.30** Because the p$K_a$ of $CH_3SH$ (11) is smaller than that of $CH_3OH$ (16), $CH_3SH$ is the stronger acid of the two. Its conjugate base (as in $KSCH_3$) is therefore weaker than the conjugate base of $CH_3OH$ (as in $KOCH_3$).

**4.31** This problem illustrates the reactions of a primary alcohol with the reagents described in the chapter.

(*a*) $$CH_3CH_2CH_2CH_2OH + NaNH_2 \longrightarrow CH_3CH_2CH_2CH_2O^-Na^+ + NH_3$$

Sodium butoxide

(*b*) $$CH_3CH_2CH_2CH_2OH \xrightarrow[\text{heat}]{\text{HBr}} CH_3CH_2CH_2CH_2Br$$

1-Bromobutane

(*c*) $$CH_3CH_2CH_2CH_2OH \xrightarrow[\text{heat}]{\text{NaBr, } H_2SO_4} CH_3CH_2CH_2CH_2Br$$

1-Bromobutane

(*d*) $$CH_3CH_2CH_2CH_2OH \xrightarrow{PBr_3} CH_3CH_2CH_2CH_2Br$$

1-Bromobutane

(*e*) $$CH_3CH_2CH_2CH_2OH \xrightarrow{SOCl_2} CH_3CH_2CH_2CH_2Cl$$

1-Chlorobutane

**4.32** (*a*) This reaction was used to convert the primary alcohol to the corresponding bromide in 60% yield.

$$\text{(cyclohexenyl)}-CH_2CH_2OH \xrightarrow[\text{pyridine}]{PBr_3} \text{(cyclohexenyl)}-CH_2CH_2Br$$

(*b*) Thionyl chloride treatment of this secondary alcohol gave the chloro derivative in 59% yield.

$CH_3$ O ‖ $COCH_2CH_3$ OH $\xrightarrow[\text{pyridine}]{SOCl_2}$ $CH_3$ O ‖ $COCH_2CH_3$ Cl

(c) The starting material is a tertiary alcohol and reacted readily with hydrogen chloride to form the corresponding chloride in 67% yield.

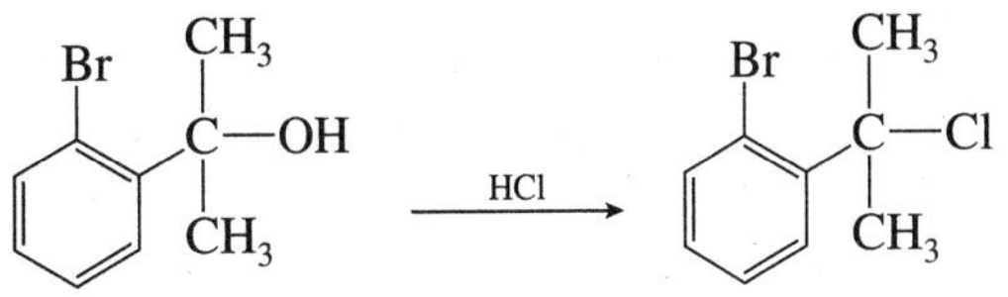

Br, $CH_3$, C—OH, $CH_3$ $\xrightarrow{HCl}$ Br, $CH_3$, C—Cl, $CH_3$

(d) Both primary alcohol functional groups were converted to primary bromides; the yield was 88%.

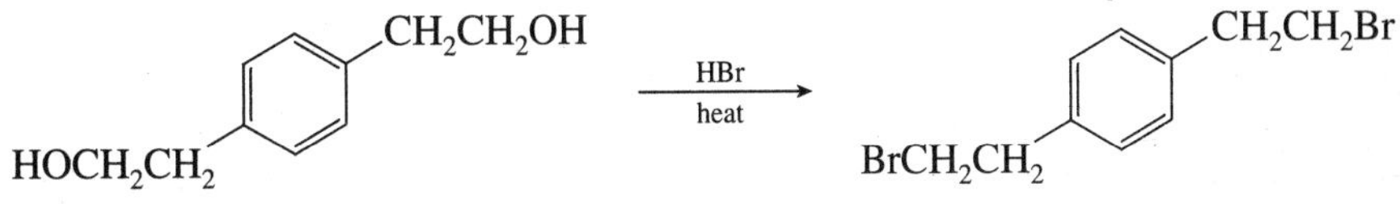

$HOCH_2CH_2$—C6H4—$CH_2CH_2OH$ $\xrightarrow[\text{heat}]{HBr}$ $BrCH_2CH_2$—C6H4—$CH_2CH_2Br$

(e) This molecule is called adamantane. It has six equivalent $CH_2$ groups and four equivalent CH groups. Bromination is selective for tertiary hydrogens, so a hydrogen of one of the CH groups is replaced. The product shown was isolated in 76% yield.

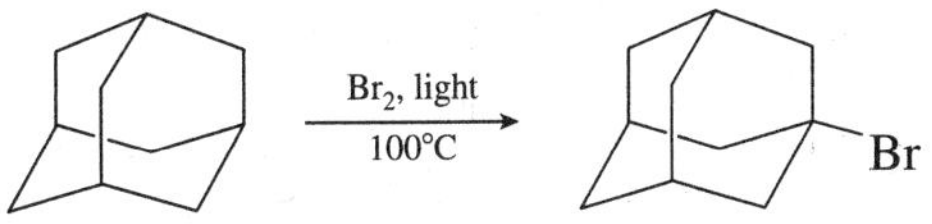

$\xrightarrow[100°C]{Br_2,\ \text{light}}$ Br

$C_{10}H_{15}Br$

**4.33** The order of reactivity of alcohols with hydrogen halides is tertiary > secondary > primary > methyl.

$$ROH + HBr \longrightarrow RBr + H_2O$$

**Reactivity of Alcohols with Hydrogen Bromide:**

| Part | More reactive | Less reactive |
|---|---|---|
| *(a)* | $CH_3CHCH_2CH_3$ (OH on C-2)<br>2-Butanol:<br>secondary | $CH_3CH_2CH_2CH_2OH$<br>1-Butanol:<br>primary |
| *(b)* | $CH_3CHCH_2CH_3$ (OH on C-2)<br>2-Butanol:<br>secondary | $CH_3CH_2CHCH_2OH$ ($CH_3$ on C-2)<br>2-Methyl-1-butanol:<br>primary |
| *(c)* | $(CH_3)_2CCH_2CH_3$ (OH on C)<br>2-Methyl-2-butanol:<br>tertiary | $CH_3CHCH_2CH_3$ (OH on C-2)<br>2-Butanol:<br>secondary |

*(continued)*

| Part | More reactive | Less reactive |
|---|---|---|
| (*d*) | 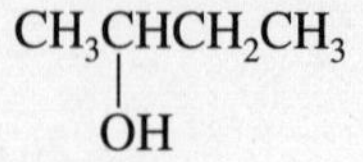 $CH_3CH(OH)CH_2CH_3$<br>2-Butanol | 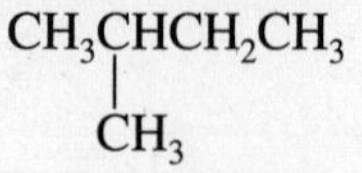 $CH_3CH(CH_3)CH_2CH_3$<br>2-Methylbutane: not an alcohol; does not react with HBr |
| (*e*) | 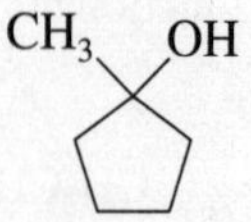 1-Methylcyclopentanol: tertiary | 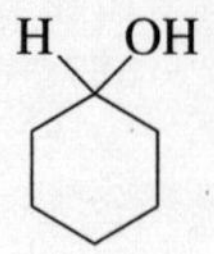 Cyclohexanol: secondary |
| (*f*) | 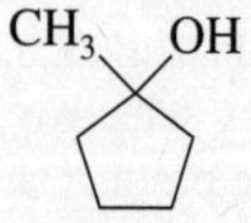 1-Methylcyclopentanol: tertiary | 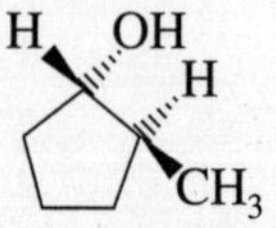 *trans*-2-Methylcyclopentanol: secondary |
| (*g*) | 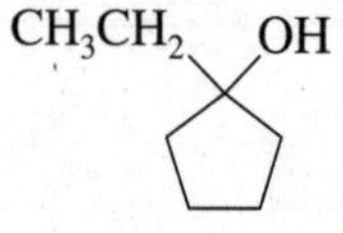 1-Ethylcyclopentanol: tertiary | 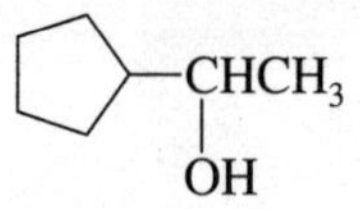 1-Cyclopentylethanol: secondary |

**4.34** The unimolecular step in the reaction of cyclohexanol with hydrogen bromide to give cyclohexyl bromide is the dissociation of the oxonium ion to a carbocation.

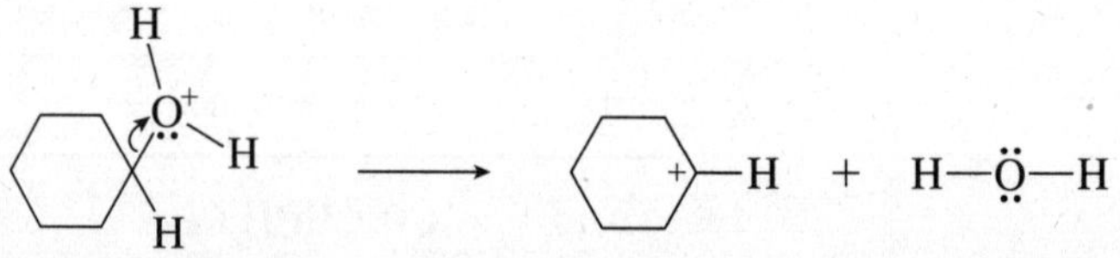

Cyclohexyloxonium ion ⟶ Cyclohexyl cation + Water (H—Ö—H)

**4.35** The nucleophile that attacks the oxonium ion in the reaction of 1-hexanol with hydrogen bromide is bromide ion.

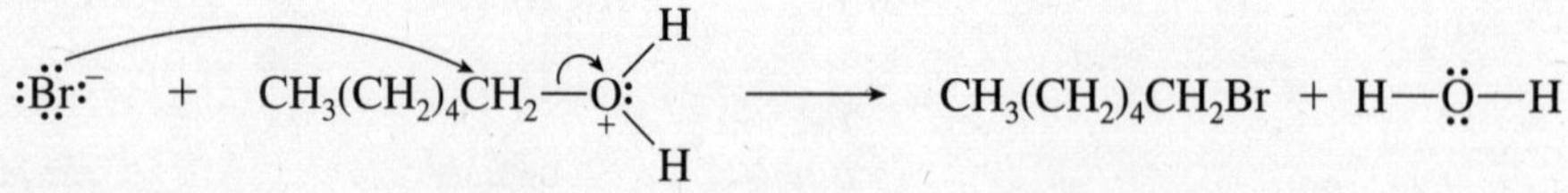

$$:\ddot{Br}:^- + CH_3(CH_2)_4CH_2-\overset{+}{O}H_2 \longrightarrow CH_3(CH_2)_4CH_2Br + H-\ddot{O}-H$$

Bromide ion, Hexyloxonium ion, 1-Bromohexane, Water

**4.36** (*a*) Both the methyl group and the hydroxyl group are equatorial in the most stable conformation of *trans*-4-methylcyclohexanol.

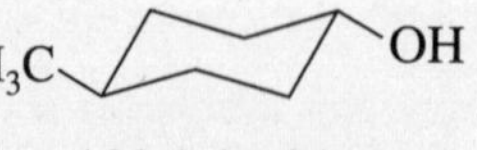

$H_3C$ … OH

*trans*-4-Methylcyclohexanol

(*b*) The positively charged carbon in the carbocation intermediate is $sp^2$-hybridized, and is planar.

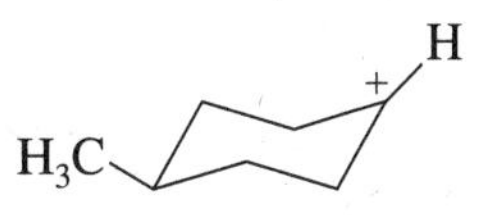


Carbocation intermediate

(*c*) Bromide ion attacks the carbocation from both above and below, giving rise to two stereoisomers, *cis*- and *trans*-1-bromo-4-methylcyclohexane.

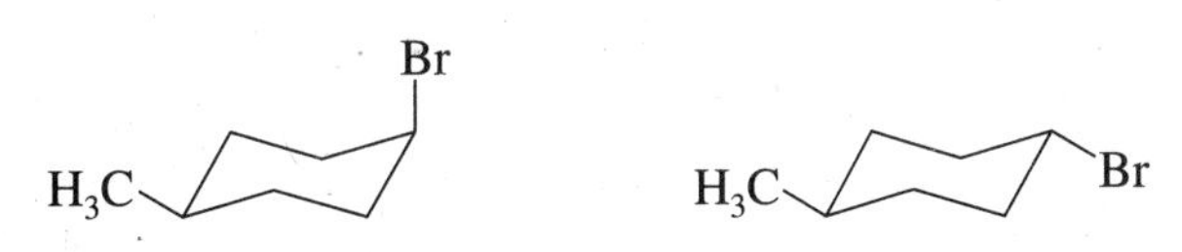


*cis*-1-Bromo-4-methylcyclohexane *trans*-1-Bromo-4-methylcyclohexane

**4.37** Examine the equations to ascertain which bonds are made and which are broken. Then use the bond dissociation energies in Table 4.3 to calculate $\Delta H°$ for each reaction.

(*a*) $(CH_3)_2CH{-}OH + H{-}F \longrightarrow (CH_3)_2CH{-}F + H{-}OH$

| 385 kJ/mol (92 kcal/mol) | 568 kJ/mol (136 kcal/mol) | 439 kJ/mol (105 kcal/mol) | 497 kJ/mol (119 kcal/mol) |
|---|---|---|---|

Bond breaking: 953 kJ/mol (228 kcal/mol) Bond making: 936 kJ/mol (224 kcal/mol)

$$\begin{aligned}\Delta H° &= \text{energy cost of breaking bonds} - \text{energy given off in making bonds}\\ &= 953 \text{ kJ/mol} - 936 \text{ kJ/mol } (228 \text{ kcal/mol} - 224 \text{ kcal/mol})\\ &= +17 \text{ kJ/mol } (+4 \text{ kcal/mol})\end{aligned}$$

The reaction of isopropyl alcohol with hydrogen fluoride is endothermic.

(*b*) $(CH_3)_2CH{-}OH + H{-}Cl \longrightarrow (CH_3)_2CH{-}Cl + H{-}OH$

| 385 kJ/mol (92 kcal/mol) | 431 kJ/mol (103 kcal/mol) | 339 kJ/mol (81 kcal/mol) | 497 kJ/mol (119 kcal/mol) |
|---|---|---|---|

Bond breaking: 816 kJ/mol (195 kcal/mol) Bond making: 836 kJ/mol (200 kcal/mol)

$$\begin{aligned}\Delta H° &= \text{energy cost of breaking bonds} - \text{energy given off in making bonds}\\ &= 816 \text{ kJ/mol} - 836 \text{ kJ/mol } (195 \text{ kcal/mol} - 200 \text{ kcal/mol})\\ &= -20 \text{ kJ/mol } (-5 \text{ kcal/mol})\end{aligned}$$

The reaction of isopropyl alcohol with hydrogen chloride is exothermic.

(*c*) $CH_3CH(H)CH_3 + H{-}Cl \longrightarrow CH_3CH(Cl)CH_3 + H{-}H$

| 397 kJ/mol (95 kcal/mol) | 431 kJ/mol (103 kcal/mol) | 339 kJ/mol (81 kcal/mol) | 435 kJ/mol (104 kcal/mol) |
|---|---|---|---|

Bond breaking: 828 kJ/mol (198 kcal/mol) Bond making: 774 kJ/mol (185 kcal/mol)

$$\begin{aligned}\Delta H° &= \text{energy cost of breaking bonds} - \text{energy given off in making bonds}\\ &= 828 \text{ kJ/mol} - 774 \text{ kJ/mol } (198 \text{ kcal/mol} - 185 \text{ kcal/mol})\\ &= +54 \text{ kJ/mol } (+13 \text{ kcal/mol})\end{aligned}$$

The reaction of propane with hydrogen chloride is endothermic.

**4.38** In the statement of the problem you are told that the starting material is 2,2-dimethylpropane, that the reaction is one of fluorination, meaning that $F_2$ is a reactant, and that the product is $(CF_3)_4C$. You need to complete the equation by realizing that HF is also formed in the fluorination of alkanes. The balanced equation is therefore:

$$(CH_3)_4C + 12F_2 \longrightarrow (CF_3)_4C + 12HF$$

**4.39** The reaction is free-radical chlorination, and substitution occurs at all possible positions that bear a replaceable hydrogen. Write the structure of the starting material, and identify the nonequivalent hydrogens.

$$Cl{-}CF_2{-}CHCl{-}CH_3$$

1,2-Dichloro-1,1-difluoropropane

The problem states that one of the products is 1,2,3-trichloro-1,1-difluoropropane. This compound arises by substitution of one of the methyl hydrogens by chlorine. We are told that the other product is an isomer of 1,2,3-trichloro-1,1-difluoropropane; therefore, it must be formed by replacement of the hydrogen at C-2.

$$Cl{-}CF_2{-}CHCl{-}CH_2Cl \qquad Cl{-}CF_2{-}CCl_2{-}CH_3$$

1,2,3-Trichloro-1,1-difluoropropane    1,2,2-Trichloro-1,1-difluoropropane

**4.40** Free-radical chlorination leads to substitution at each carbon that bears a hydrogen. This problem essentially requires you to recognize structures that possess various numbers of nonequivalent hydrogens. The easiest way to determine the number of constitutional isomers that can be formed by chlorination of a particular compound is to replace one hydrogen with chlorine and assign an IUPAC name to the product. Continue by replacing one hydrogen on each carbon in the compound, and compare names to identify duplicates.

(*a*) 2,2-Dimethylpropane is the $C_5H_{12}$ isomer that gives a single monochloride, since all the hydrogens are equivalent.

$$(CH_3)_4C \xrightarrow[\text{light}]{Cl_2} (CH_3)_3CCH_2Cl$$

2,2-Dimethylpropane → 1-Chloro-2,2-dimethylpropane

(*b*) The $C_5H_{12}$ isomer that has three nonequivalent sets of hydrogens is pentane. It yields three isomeric monochlorides on free-radical chlorination.

$$CH_3CH_2CH_2CH_2CH_3 \xrightarrow{Cl_2} ClCH_2CH_2CH_2CH_2CH_3 + CH_3CHClCH_2CH_2CH_3 + CH_3CH_2CHClCH_2CH_3$$

Pentane → 1-Chloropentane, 2-Chloropentane, 3-Chloropentane

(*c*) 2-Methylbutane forms four different monochlorides.

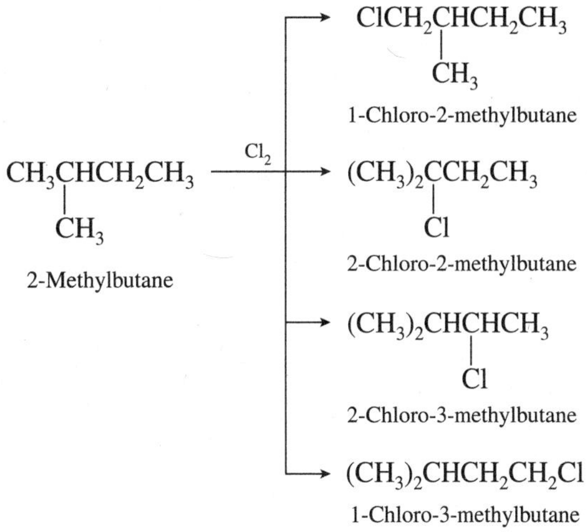

$$CH_3\underset{\displaystyle CH_3}{\underset{|}{C}}HCH_2CH_3 \xrightarrow{Cl_2} \begin{cases} ClCH_2\underset{\displaystyle CH_3}{\underset{|}{C}}HCH_2CH_3 & \text{1-Chloro-2-methylbutane} \\ (CH_3)_2\underset{\displaystyle Cl}{\underset{|}{C}}CH_2CH_3 & \text{2-Chloro-2-methylbutane} \\ (CH_3)_2CH\underset{\displaystyle Cl}{\underset{|}{C}}HCH_3 & \text{2-Chloro-3-methylbutane} \\ (CH_3)_2CHCH_2CH_2Cl & \text{1-Chloro-3-methylbutane} \end{cases}$$

2-Methylbutane

(*d*) For only two dichlorides to be formed, the starting alkane must have a structure that is rather symmetrical; that is, one in which most (or all) of the hydrogens are equivalent. 2,2-Dimethylpropane satisfies this requirement.

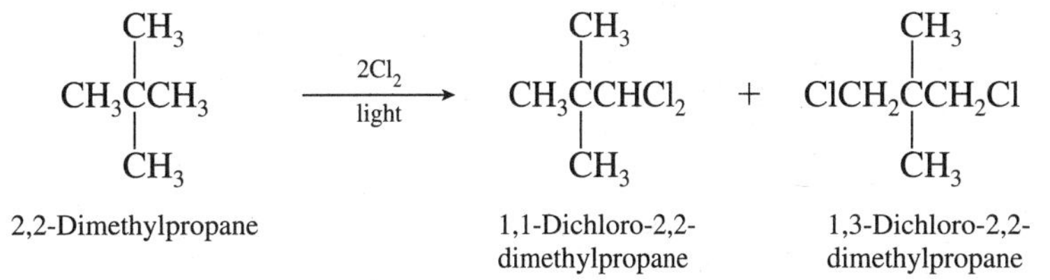

$$CH_3\overset{\displaystyle CH_3}{\overset{|}{\underset{\displaystyle CH_3}{\underset{|}{C}}}}CH_3 \xrightarrow[\text{light}]{2Cl_2} CH_3\overset{\displaystyle CH_3}{\overset{|}{\underset{\displaystyle CH_3}{\underset{|}{C}}}}CHCl_2 + ClCH_2\overset{\displaystyle CH_3}{\overset{|}{\underset{\displaystyle CH_3}{\underset{|}{C}}}}CH_2Cl$$

2,2-Dimethylpropane; 1,1-Dichloro-2,2-dimethylpropane; 1,3-Dichloro-2,2-dimethylpropane

**4.41** (*a*) Heptane has five methylene groups, which on chlorination together contribute 85% of the total monochlorinated product.

$$CH_3(CH_2)_5CH_3 \longrightarrow \underset{15\%}{CH_3(CH_2)_5CH_2Cl} + \underset{85\%}{(\text{2-chloro} + \text{3-chloro} + \text{4-chloro})}$$

Since the problem specifies that attack at each methylene group is equally probable, the five methylene groups each give rise to 85/5, or 17%, of the monochloride product.

Since C-2 and C-6 of heptane are equivalent, we calculate that 2-chloroheptane will constitute 34% of the monochloride fraction. Similarly, C-3 and C-5 are equivalent, and so there should be 34% 3-chloroheptane. The remainder, 17%, is 4-chloroheptane.

These predictions are very close to the observed proportions.

| | Calculated, % | Observed, % |
|---|---|---|
| 2-Chloro | 34 | 35 |
| 3-Chloro | 34 | 34 |
| 4-Chloro | 17 | 16 |

(*b*) There are a total of 20 methylene hydrogens in dodecane, $CH_3(CH_2)_{10}CH_3$. The 19% 2-chlorododecane that is formed arises by substitution of any of the four equivalent methylene hydrogens at C-2 and C-11. The total amount of substitution of methylene hydrogens must therefore be:

$$\frac{20}{4} \times 19\% = 95\%$$

The remaining 5% corresponds to substitution of methyl hydrogens at C-1 and C-12. The proportion of 1-chlorododecane in the monochloride fraction is 5%.

**4.42** (*a*) Two of the monochlorides derived from chlorination of 2,2,4-trimethylpentane are primary chlorides:

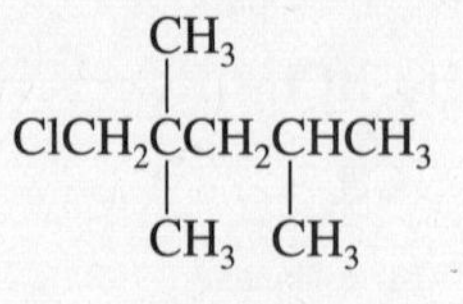

$$\begin{array}{c} CH_3 \\ | \\ ClCH_2CCH_2CHCH_3 \\ |\quad\ \ | \\ CH_3\ CH_3 \end{array} \qquad \begin{array}{c} CH_3 \\ | \\ CH_3CCH_2CHCH_2Cl \\ |\quad\ \ | \\ CH_3\ CH_3 \end{array}$$

1-Chloro-2,2,4-trimethylpentane 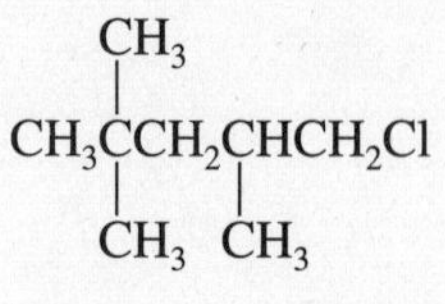    1-Chloro-2,4,4-trimethylpentane

The two remaining isomers are a secondary chloride and a tertiary chloride:

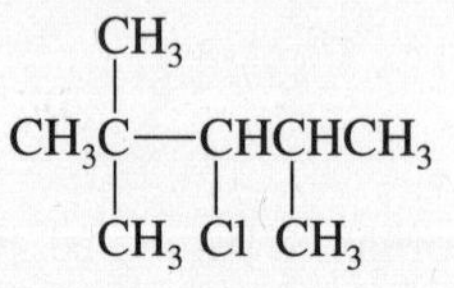

$$\begin{array}{c} CH_3 \\ | \\ CH_3C{-}CHCHCH_3 \\ |\quad\ |\quad | \\ CH_3\ Cl\ CH_3 \end{array} \qquad \begin{array}{c} CH_3\ \ Cl \\ |\quad\ | \\ CH_3CCH_2CCH_3 \\ |\quad\ | \\ CH_3\ CH_3 \end{array}$$

3-Chloro-2,2,4-trimethylpentane 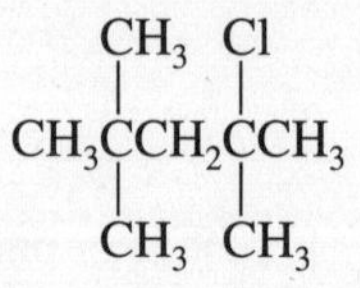    2-Chloro-2,4,4-trimethylpentane

(*b*) Substitution of any one of the nine hydrogens designated as *x* in the structural diagram yields 1-chloro-2,2,4-trimethylpentane. Substitution of any one of the six hydrogens designated as *y* gives 1-chloro-2,4,4-trimethylpentane.

$$\begin{array}{c} x \\ CH_3 \\ | \\ \overset{x}{C}H_3CCH_2CHCH_3^{y} \\ |\quad\ \ | \\ \underset{x}{CH_3}\ \underset{y}{CH_3} \end{array}$$

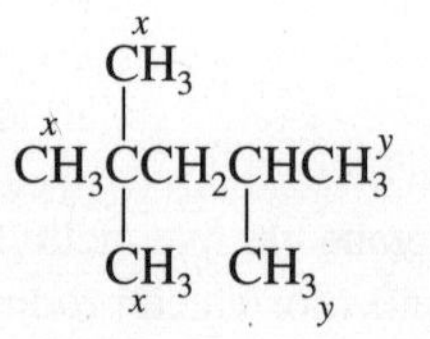

Assuming equal reactivity of a single *x* hydrogen and a single *y* hydrogen, the ratio of the two isomers is then expected to be 9:6. Since together the two primary chlorides total 65% of the monochloride fraction, there will be 39% 1-chloro-2,2,4-trimethylpentane (substitution of *x*) and 26% 1-chloro-2,4,4-trimethylpentane (substitution of *y*).

**4.43** The three monochlorides are shown in the equation

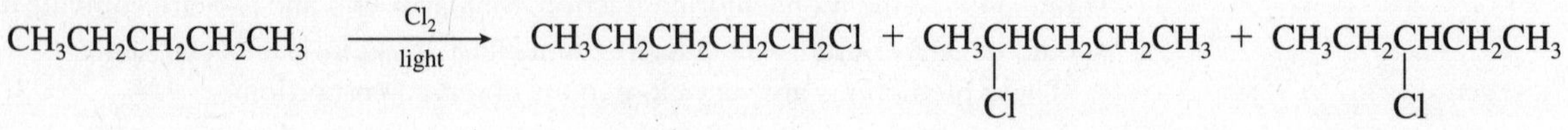

$$CH_3CH_2CH_2CH_2CH_3 \xrightarrow[\text{light}]{Cl_2} CH_3CH_2CH_2CH_2CH_2Cl + \begin{array}{c} CH_3CHCH_2CH_2CH_3 \\ | \\ Cl \end{array} + \begin{array}{c} CH_3CH_2CHCH_2CH_3 \\ | \\ Cl \end{array}$$

Pentane    1-Chloropentane    2-Chloropentane    3-Chloropentane

Pentane has six primary hydrogens (two $CH_3$ groups) and six secondary hydrogens (three $CH_2$ groups). Since a single secondary hydrogen is abstracted three times faster than a single primary hydrogen and there are equal numbers of secondary and primary hydrogens, the product mixture should contain three times as much of the secondary chloride isomers as the primary chloride. The primary chloride 1-chloropentane, therefore, is expected to constitute 25% of the product mixture. The secondary chlorides 2-chloropentane and 3-chloropentane are not formed in equal amounts. Rather, 2-chloropentane may be formed by replacement of a hydrogen at C-2 or at C-4, whereas 3-chloropentane is formed only when a C-3 hydrogen is replaced. The amount of 2-chloropentane is therefore 50%, and that of 3-chloropentane is 25%. We predict the major product to be 2-chloropentane, and the predicted proportion of 50% corresponds closely to the observed 46%.

**4.44** The equation for the reaction is

$$\text{C}_3\text{H}_6 + \text{Cl}_2 \longrightarrow \text{C}_3\text{H}_5\text{Cl} + \text{HCl}$$

Cyclopropane Chlorine Cyclopropyl chloride Hydrogen chloride

The reaction begins with the initiation step in which a chlorine molecule dissociates to two chlorine atoms.

$$:\ddot{\underset{..}{\text{Cl}}}—\ddot{\underset{..}{\text{Cl}}}: \longrightarrow :\ddot{\underset{..}{\text{Cl}}}\cdot + \cdot\ddot{\underset{..}{\text{Cl}}}:$$

Chlorine 2 Chlorine atoms

A chlorine atom abstracts a hydrogen atom from cyclopropane in the first propagation step.

$$\text{C}_3\text{H}_6 + :\ddot{\underset{..}{\text{Cl}}}\cdot \longrightarrow \text{C}_3\text{H}_5\cdot + \text{H}—\ddot{\underset{..}{\text{Cl}}}:$$

Cyclopropane Chlorine atom Cyclopropyl radical Hydrogen chloride

Cyclopropyl radical reacts with $Cl_2$ in the next propagation step.

$$\text{C}_3\text{H}_5\cdot + :\ddot{\underset{..}{\text{Cl}}}—\ddot{\underset{..}{\text{Cl}}}: \longrightarrow \text{C}_3\text{H}_5\ddot{\underset{..}{\text{Cl}}}: + \cdot\ddot{\underset{..}{\text{Cl}}}:$$

Cyclopropyl radical Chlorine Cyclopropyl chloride Chlorine atom

**4.45** (*a*) Acid-catalyzed hydrogen–deuterium exchange takes place by a pair of Brønsted acid–base reactions.

$$\text{R}—\ddot{\underset{..}{\text{O}}}—\text{H} + \text{D}—\overset{+}{\ddot{\text{O}}}\text{D}_2 \rightleftharpoons \text{R}—\underset{..}{\overset{\text{D}}{\overset{|}{\text{O}^+}}}—\text{H} + \text{D}_2\ddot{\text{O}}:$$

Base Acid Conjugate acid Conjugate base

$$\text{R}—\underset{..}{\overset{\text{D}}{\overset{|}{\text{O}^+}}}—\text{H} + :\ddot{\text{O}}\text{D}_2 \rightleftharpoons \text{R}—\ddot{\underset{..}{\text{O}}}—\text{D} + \text{D}—\underset{..}{\overset{\text{H}}{\overset{|}{\text{O}^+}}}—\text{D}$$

Acid Base Conjugate base Conjugate acid

(*b*) Base-catalyzed hydrogen–deuterium exchange occurs by a different pair of Brønsted acid–base equilibria.

$$\text{R}—\ddot{\underset{..}{\text{O}}}—\text{H} + {}^{-}:\ddot{\underset{..}{\text{O}}}\text{D} \rightleftharpoons \text{R}—\ddot{\underset{..}{\text{O}}}:^{-} + \text{D}\ddot{\underset{..}{\text{O}}}\text{H}$$

Acid Base Conjugate base Conjugate acid

$$\text{R}—\ddot{\underset{..}{\text{O}}}:^{-} + \text{D}—\ddot{\underset{..}{\text{O}}}\text{D} \rightleftharpoons \text{R}—\ddot{\underset{..}{\text{O}}}—\text{D} + \text{D}\ddot{\underset{..}{\text{O}}}:^{-}$$

Base Acid Conjugate acid Conjugate base

# SELF-TEST

## PART A

**A-1.** Give the correct substitutive IUPAC name for each of the following compounds:

(*a*) [cyclopentane with wedged $CH_3$ and hashed Br]

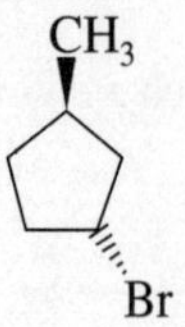

(*b*) $CH_3CH_2CH(CH_2OH)CH_2CH(CH_2CH_3)CH_3$

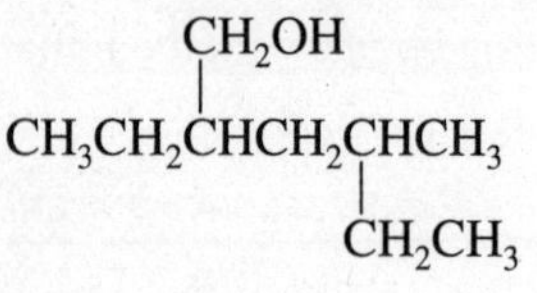

**A-2.** Draw the structures of the following substances:
(*a*) 2-Chloro-1-iodo-2-methylheptane
(*b*) *cis*-3-Isopropylcyclohexanol

**A-3.** Give both a functional class and a substitutive IUPAC name for each of the following compounds:

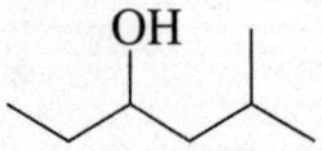

(*a*) [bond-line structure with OH]  (*b*) [bond-line structure with Cl]

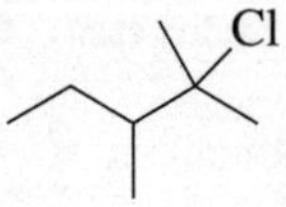

**A-4.** What are the structures of the conjugate acid and the conjugate base of $CH_3OH$?

**A-5.** Supply the missing component for each of the following reactions:

(*a*) $CH_3CH_2CH_2OH \xrightarrow{SOCl_2} ?$

(*b*) $? \xrightarrow{HBr} CH_3CH_2C(Br)(CH_3)_2$

**A-6.** (*a*) Write the products of the acid–base reaction that follows, and identify the stronger acid and base and the conjugate of each. Will the equilibrium lie to the left ($K < 1$) or to the right ($K > 1$)? The approximate p$K_a$ of $NH_3$ is 36; that of $CH_3CH_2OH$ is 16.

$$CH_3CH_2O^- + NH_3 \rightleftharpoons$$

(*b*) Draw a representation of the transition state of the elementary step of the reaction in part (*a*).

**A-7.** (*a*) How many different free radicals can possibly be produced in the reaction between chlorine atoms and 2,4-dimethylpentane?
(*b*) Write their structures.
(*c*) Which is the most stable? Which is the least stable?

**A-8.** Write a balanced chemical equation for the reaction of chlorine with the pentane isomer that gives only one product on monochlorination.

**A-9.** Write the propagation steps for the light-initiated reaction of bromine with methylcyclohexane.

**A-10.** Using the data in Table B-1 of this Study Guide, calculate the heat of reaction ($\Delta H°$) for the light-initiated reaction of bromine ($Br_2$) with 2-methylpropane to give 2-bromo-2-methylpropane and hydrogen bromide.

**A-11.** (*a*) Write out each of the elementary steps in the reaction of *tert*-butyl alcohol with hydrogen bromide. Use curved arrows to show electron movement in each step.
(*b*) Draw the structure of the transition state representing the unimolecular dissociation of the alkyloxonium ion in the preceding reaction.

(*c*) How does the mechanism of the reaction between 1-butanol and hydrogen bromide differ from the reaction in part (*a*)?

**A-12.** (Choose the correct response for each part.) Which species or compound:

(*a*) Reacts faster with sodium bromide and sulfuric acid?

2-methyl-3-pentanol or 3-methyl-3-pentanol

(*b*) Is a stronger base?

$KOC(CH_3)_3$ or $HOC(CH_3)_3$

(*c*) Reacts more vigorously with cyclohexane?

Fluorine or iodine

(*d*) Has an odd number of electrons?

Ethoxide ion or ethyl radical

(*e*) Undergoes bond cleavage in the initiation step in the reaction by which methane is converted to chloromethane?

$CH_4$ or $Cl_2$

## PART B

**B-1.** A certain alcohol has the functional class IUPAC name **1-ethyl-3-methylbutyl alcohol.** What is its substitutive name?

(*a*) 1-Ethyl-3-methyl-1-butanol
(*b*) 2-Methyl-1-hexanol
(*c*) 3-Methyl-1-hexanol
(*d*) 2-Methyl-4-hexanol
(*e*) 5-Methyl-3-hexanol

**B-2.** Rank the following substances in order of increasing boiling point (lowest → highest):

| $CH_3CH_2CH_2CH_2OH$ | $(CH_3)_2CHOCH_3$ | $(CH_3)_3COH$ | $(CH_3)_4C$ |
|---|---|---|---|
| 1 | 2 | 3 | 4 |

(*a*) $1 < 3 < 2 < 4$
(*b*) $2 < 4 < 3 < 1$
(*c*) $4 < 2 < 3 < 1$
(*d*) $2 < 3 < 1 < 4$
(*e*) $4 < 3 < 2 < 1$

**B-3.** Which one of the following reacts with HBr at the fastest rate?

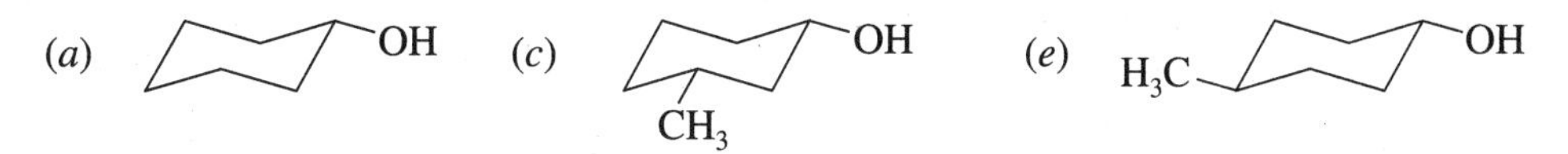


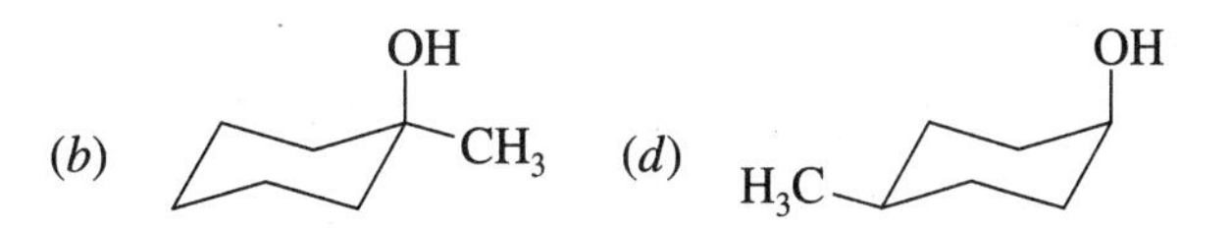


**B-4.** What is the decreasing stability order (most stable → least stable) of the following carbocations?

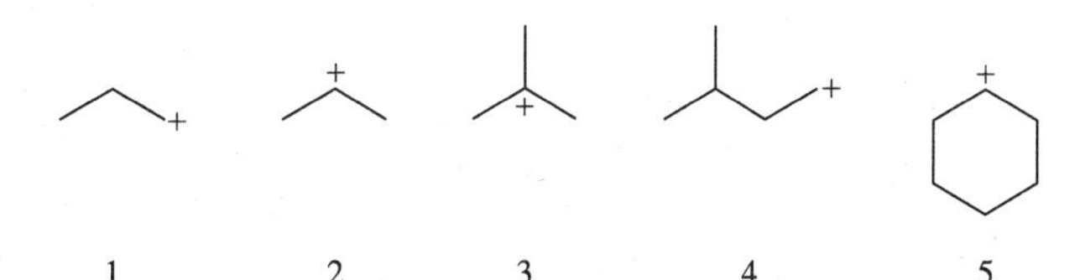


(*a*) $3 > 2 > 1 > 4 > 5$
(*b*) $1 \approx 4 > 2 \approx 5 > 3$
(*c*) $3 > 2 \approx 5 > 1 \approx 4$
(*d*) $3 > 1 \approx 4 > 2 \approx 5$

**B-5.** Rank the bond dissociation energies (BDEs) of the bonds indicated with the arrows from smallest to largest.

(*a*) $1 < 2 < 3$ (*d*) $1 < 3 < 2$
(*b*) $3 < 2 < 1$ (*e*) $3 < 1 < 2$
(*c*) $2 < 3 < 1$

[Structure: cyclohexane ring with H (arrow 1) on one carbon; opposite carbon bearing H (arrow 3) and a $CH_2$–H group (arrow 2)]

**B-6.** What are the chain-propagating steps in the free-radical chlorination of methane?

1. $Cl_2 \longrightarrow 2Cl\cdot$
2. $Cl\cdot + CH_4 \longrightarrow CH_3Cl + H\cdot$
3. $Cl\cdot + CH_4 \longrightarrow \cdot CH_3 + HCl$
4. $H\cdot + Cl_2 \longrightarrow HCl + Cl\cdot$
5. $\cdot CH_3 + Cl_2 \longrightarrow CH_3Cl + Cl\cdot$
6. $\cdot CH_3 + CH_4 \longrightarrow CH_4 + \cdot CH_3$

(*a*) 2, 4 (*b*) 1, 2 (*c*) 3, 5 (*d*) 1, 3, 5
(*e*) A combination different from those listed

**B-7.** Which of the following is *least* able to serve as a nucleophile in a chemical reaction?
(*a*) $Br^-$ (*b*) $OH^-$ (*c*) $NH_3$ (*d*) $CH_3^+$

**B-8.** Thiols are alcohol analogs in which the oxygen has been replaced by sulfur (e.g., $CH_3SH$). Given the fact that the S—H bond is less polar than the O—H bond, which of the following statements comparing thiols and alcohols is correct?
(*a*) Hydrogen bonding forces are weaker in thiols.
(*b*) Hydrogen bonding forces are stronger in thiols.
(*c*) Hydrogen bonding forces would be the same.
(*d*) No comparison can be made without additional information.

**B-9.** Rank the **transition states** that occur during the following reaction steps in order of increasing stability (least → most stable):

1. $CH_3{-}\overset{+}{O}H_2 \longrightarrow CH_3^+ + H_2O$
2. $(CH_3)_3C{-}\overset{+}{O}H_2 \longrightarrow (CH_3)_3C^+ + H_2O$
3. $(CH_3)_2CH{-}\overset{+}{O}H_2 \longrightarrow (CH_3)_2CH^+ + H_2O$

(*a*) $1 < 2 < 3$ (*b*) $2 < 3 < 1$ (*c*) $1 < 3 < 2$ (*d*) $2 < 1 < 3$

**B-10.** Using the data from Appendix B (Table B-1), calculate the heat of reaction $\Delta H°$ for the following:

$$CH_3CH_2\cdot + HBr \longrightarrow CH_3CH_3 + Br\cdot$$

(*a*) +69 kJ/mol (+16.5 kcal/mol)
(*b*) −69 kJ/mol (−16.5 kcal/mol)
(*c*) +44 kJ/mol (+10.5 kcal/mol)
(*d*) −44 kJ/mol (−10.5 kcal/mol)

**B-11.** An alkane with a molecular formula $C_6H_{14}$ reacts with chlorine in the presence of light and heat to give **four** constitutionally isomeric monochlorides of molecular formula $C_6H_{13}Cl$. What is the most reasonable structure for the starting alkane?
(*a*) $CH_3CH_2CH_2CH_2CH_2CH_3$
(*b*) $(CH_3)_2CHCH_2CH_2CH_3$
(*c*) $CH_3CH(CH_2CH_3)_2$
(*d*) $(CH_3)_3CCH_2CH_3$
(*e*) $(CH_3)_2CHCH(CH_3)_2$

**B-12.** The species shown in the box represents ______ of the reaction between isopropyl alcohol and hydrogen bromide.

$$\overset{\delta+}{(CH_3)_2CH}\text{-------}\overset{\delta+}{OH_2}$$

(*a*) the alkyloxonium ion intermediate
(*b*) the transition state of the bimolecular proton transfer step

(*c*) the transition state of the capture of the carbocation by a nucleophile
(*d*) the carbocation intermediate
(*e*) the transition state of the unimolecular dissociation step

For the remaining four questions, consider the following free-radical reaction:

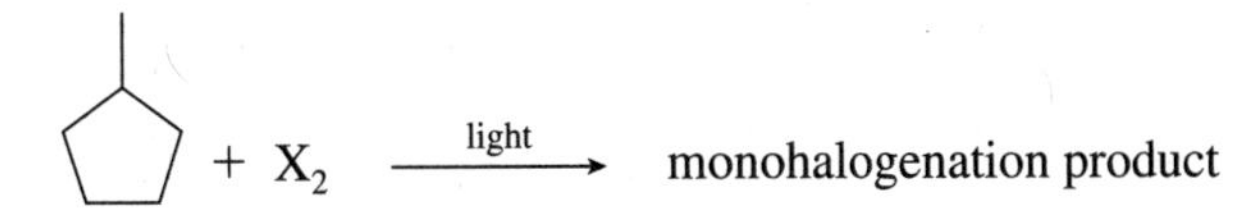


**B-13.** Light is involved in which of the following reaction steps?
(*a*) Initiation only
(*b*) Propagation only
(*c*) Termination only
(*d*) Initiation and propagation

**B-14.** Which of the following statements about the reaction is *not* true?
(*a*) Halogen atoms are consumed in the first propagation step.
(*b*) Halogen atoms are regenerated in the second propagation step.
(*c*) Hydrogen atoms are produced in the first propagation step.
(*d*) Chain termination occurs when two radicals react with each other.

**B-15.** How many monohalogenation products are possible. (Do not consider stereoisomers.)
(*a*) 2 (*b*) 3 (*c*) 4 (*d*) 5

**B-16.** Which halogen ($X_2$) will give the best yield of a single monohalogenation product?
(*a*) $F_2$ (*b*) $Cl_2$ (*c*) $Br_2$ (*d*) $I_2$

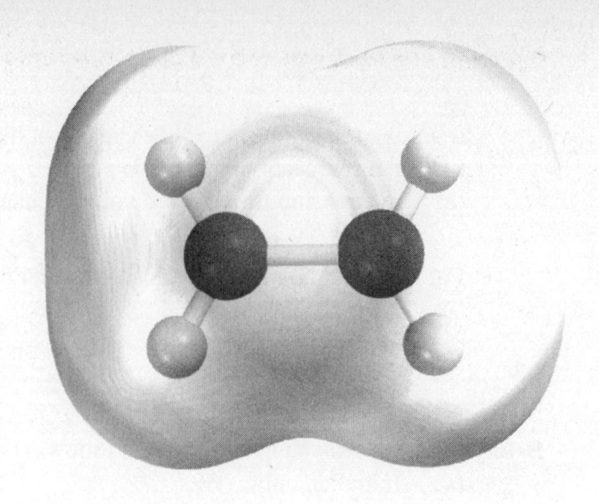

# CHAPTER 5

## STRUCTURE AND PREPARATION OF ALKENES: ELIMINATION REACTIONS

### SOLUTIONS TO TEXT PROBLEMS

**5.1** (*b*) Writing the structure in more detail, we see that the longest continuous chain contains four carbon atoms.

$$\overset{4}{CH_3}-\overset{3}{C}(CH_3)_2-\overset{2}{CH}=\overset{1}{CH_2}$$

The double bond is located at the end of the chain, and so the alkene is named as a derivative of 1-butene. Two methyl groups are substituents at C-3. The correct IUPAC name is 3,3-dimethyl-1-butene.

(*c*) Expanding the structural formula reveals the molecule to be a methyl-substituted derivative of hexene.

$$\overset{1}{CH_3}-\overset{2}{C}(CH_3)=\overset{3}{C}H\overset{4}{C}H_2\overset{5}{C}H_2\overset{6}{C}H_3$$

2-Methyl-2-hexene

(*d*) In compounds containing a double bond and a halogen, the double bond takes precedence in numbering the longest carbon chain.

$$\overset{1}{CH_2}=\overset{2}{C}H\overset{3}{C}H_2\overset{4}{C}H(Cl)\overset{5}{C}H_3$$

4-Chloro-1-pentene

(*e*) When a hydroxyl group is present in a compound containing a double bond, the hydroxyl takes precedence over the double bond in numbering the longest carbon chain.

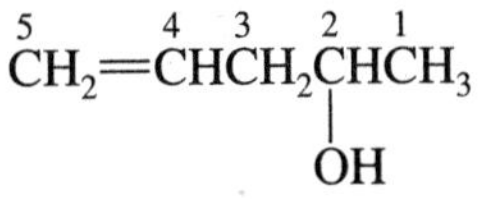

$$\overset{5}{C}H_2{=}\overset{4}{C}H\overset{3}{C}H_2\overset{2}{C}H(OH)\overset{1}{C}H_3$$

4-Penten-2-ol

**5.2** There are three sets of nonequivalent positions on a cyclopentene ring, identified as *a*, *b*, and *c* on the cyclopentene structure shown:

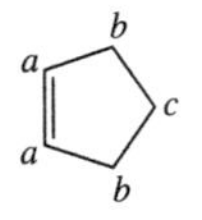

Thus, there are three different monochloro-substituted derivatives of cyclopentene. The carbons that bear the double bond are numbered C-1 and C-2 in each isomer, and the other positions are numbered in sequence in the direction that gives the chlorine-bearing carbon its lower locant.

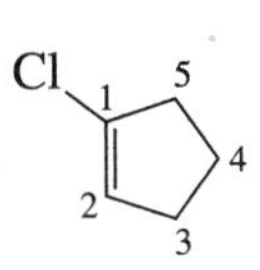

1-Chlorocyclopentene 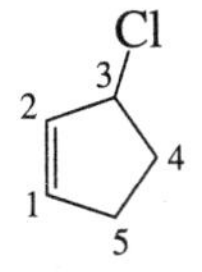 3-Chlorocyclopentene 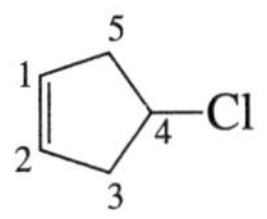 4-Chlorocyclopentene

**5.3** (*b*) The alkene is a derivative of 3-hexene regardless of whether the chain is numbered from left to right or from right to left. Number it in the direction that gives the lower number to the substituent.

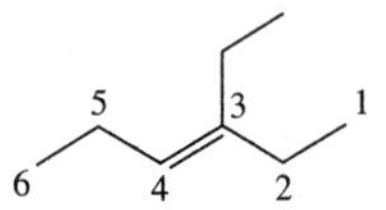

3-Ethyl-3-hexene

(*c*) There are only two $sp^2$-hybridized carbons, the two connected by the double bond. All other carbons (six) are $sp^3$-hybridized.

(*d*) There are three $sp^2$–$sp^3$ $\sigma$ bonds and three $sp^3$–$sp^3$ $\sigma$ bonds.

**5.4** Consider first the $C_5H_{10}$ alkenes that have an unbranched carbon chain:

1-Pentene 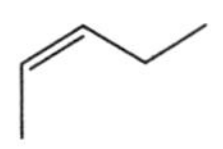 *cis*-2-Pentene 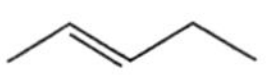 *trans*-2-Pentene

There are three additional isomers. These have a four-carbon chain with a methyl substituent.

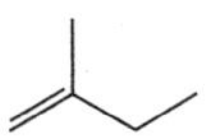

2-Methyl-1-butene 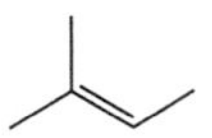 2-Methyl-2-butene 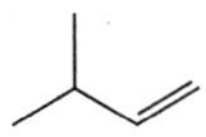 3-Methyl-1-butene

**5.5** First, identify the constitution of 9-tricosene. Referring back to Table 2.4 in Section 2.8 of the text, we see that tricosane is the unbranched alkane containing 23 carbon atoms. 9-Tricosene, therefore, contains an unbranched chain of 23 carbons with a double bond between C-9 and C-10. Since the problem specifies that the pheromone has the cis configuration, the first 8 carbons and the last 13 must be on the same side of the C-9–C-10 double bond.

$$\begin{array}{ccc} CH_3(CH_2)_7 & & (CH_2)_{12}CH_3 \\ & C{=}C & \\ H & & H \end{array}$$

*cis*-9-Tricosene

**5.6** (*b*) One of the carbons of the double bond bears a methyl group and a hydrogen; methyl is of higher rank than hydrogen. The other doubly bonded carbon bears the groups $—CH_2CH_2F$ and $—CH_2CH_2CH_2CH_3$. At the first point of difference between these two, fluorine is of higher atomic number than carbon, and so $—CH_2CH_2F$ is of higher precedence.

$$\begin{array}{ccccc} \text{Higher} & CH_3 & & CH_2CH_2F & \text{Higher} \\ & & C{=}C & & \\ \text{Lower} & H & & CH_2CH_2CH_2CH_3 & \text{Lower} \end{array}$$

Higher ranked substituents are on the same side of the double bond; the alkene has the *Z* configuration.

(*c*) One of the carbons of the double bond bears a methyl group and a hydrogen; as we have seen, methyl is of higher rank. The other doubly bonded carbon bears $—CH_2CH_2OH$ and $—C(CH_3)_3$. Let's analyze these two groups to determine their order of precedence.

| $—CH_2CH_2OH$ | $—C(CH_3)_3$ |
|---|---|
| —C(C,H,H) | —C(C,C,C) |
| Lower priority | Higher priority |

We examine the atoms one by one at the point of attachment before proceeding down the chain. Therefore, $—C(CH_3)_3$ outranks $—CH_2CH_2OH$.

$$\begin{array}{ccccc} \text{Higher} & CH_3 & & CH_2CH_2OH & \text{Lower} \\ & & C{=}C & & \\ \text{Lower} & H & & C(CH_3)_3 & \text{Higher} \end{array}$$

Higher ranked groups are on opposite sides; the configuration of the alkene is *E*.

(*d*) The cyclopropyl ring is attached to the double bond by a carbon that bears the atoms (C, C, H) and is therefore of higher precedence than an ethyl group —C(C, H, H).

$$\begin{array}{ccccc} \text{Higher} & \triangle & & H & \text{Lower} \\ & & C{=}C & & \\ \text{Lower} & CH_3CH_2 & & CH_3 & \text{Higher} \end{array}$$

Higher ranked groups are on opposite sides; the configuration of the alkene is *E*.

**5.7** A trisubstituted alkene has three carbons directly attached to the doubly bonded carbons. There are three trisubstituted $C_6H_{12}$ isomers, two of which are stereoisomers.

$$\begin{array}{ccc} CH_3 & & CH_2CH_3 \\ & C{=}C & \\ CH_3 & & H \end{array} \qquad \begin{array}{ccc} CH_3 & & CH_3 \\ & C{=}C & \\ H & & CH_2CH_3 \end{array} \qquad \begin{array}{ccc} CH_3 & & CH_2CH_3 \\ & C{=}C & \\ H & & CH_3 \end{array}$$

2-Methyl-2-pentene (*E*)-3-Methyl-2-pentene (*Z*)-3-Methyl-2-pentene

**5.8** The most stable $C_6H_{12}$ alkene has a tetrasubstituted double bond:

$(CH_3)_2C{=}C(CH_3)_2$

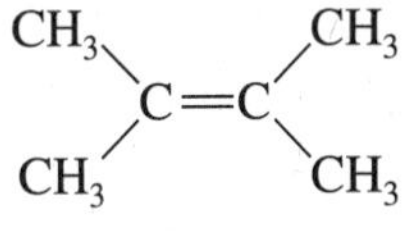

2,3-Dimethyl-2-butene

**5.9** Apply the two general rules for alkene stability to rank these compounds. First, more highly substituted double bonds are more stable than less substituted ones. Second, when two double bonds are similarly constituted, the trans stereoisomer is more stable than the cis. The predicted order of decreasing stability is therefore:

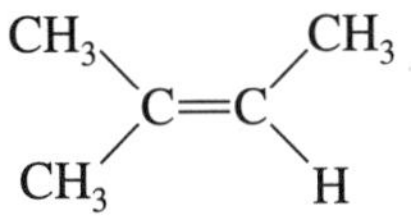

$(CH_3)_2C{=}CHCH_3$ — 2-Methyl-2-butene (trisubstituted): most stable

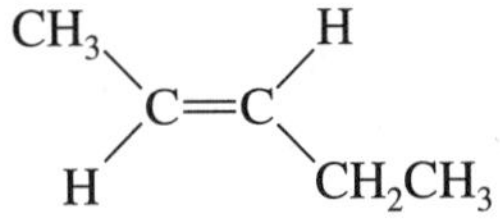

$CH_3CH{=}CHCH_2CH_3$ (trans) — (*E*)-2-Pentene (disubstituted)

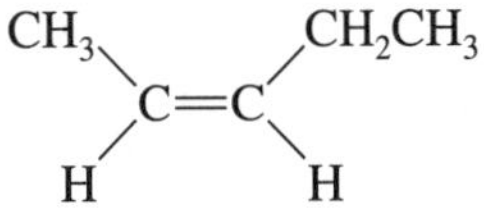

$CH_3CH{=}CHCH_2CH_3$ (cis) — (*Z*)-2-Pentene (disubstituted)

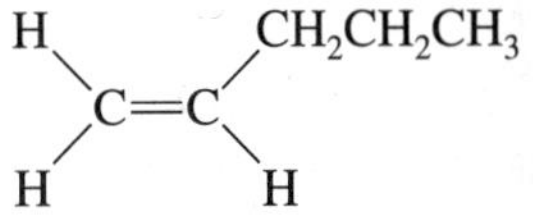

$H_2C{=}CHCH_2CH_2CH_3$ — 1-Pentene (monosubstituted): least stable

**5.10** Begin by writing the structural formula corresponding to the IUPAC name given in the problem. A bond-line depiction is useful here.

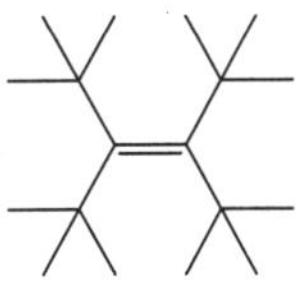

3,4-Di-*tert*-butyl-2,2,5,5-tetramethyl-3-hexene

The alkene is extremely crowded and destabilized by van der Waals strain. Bulky *tert*-butyl groups are cis to one another on each side of the double bond. Highly strained compounds are often quite difficult to synthesize, and this alkene is a good example.

**5.11** Use the zigzag arrangement of bonds in the parent skeleton figure to place *E* and *Z* bonds as appropriate for each part of the problem. From the sample solution to parts (*a*) and (*b*), the ring carbons have the higher priorities. Thus, an *E* double bond will have ring carbons arranged 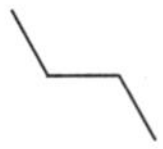 and a *Z* double bond  .

(*c*) (*Z*)-3-Methylcyclodecene

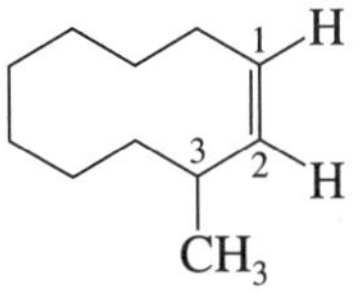

(*e*) (*Z*)-5-Methylcyclodecene

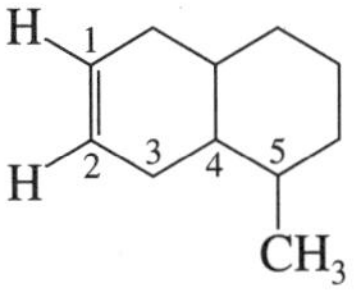

(*d*) (*E*)-3-Methylcyclodecene

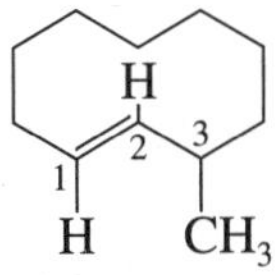

(*f*) (*E*)-5-Methylcyclodecene

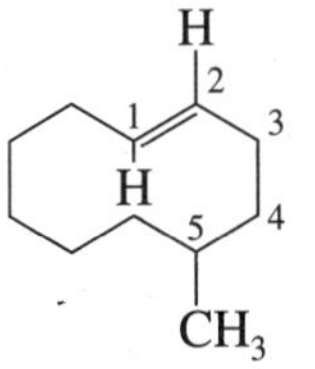

**5.12** Write out the structure of the alcohol, recognizing that the alkene is formed by loss of a hydrogen and a hydroxyl group from adjacent carbons.

(*b, c*) Both 1-propanol and 2-propanol give propene on acid-catalyzed dehydration.

$$\underset{\text{1-Propanol}}{CH_3\overset{\beta}{C}H_2\overset{\alpha}{C}H_2OH} \xrightarrow[\text{heat}]{H^+} \underset{\text{Propene}}{CH_3CH{=}CH_2} \xleftarrow[\text{heat}]{H^+} \underset{\text{2-Propanol}}{\overset{\beta}{C}H_3\overset{\alpha}{C}H(OH)\overset{\beta}{C}H_3}$$

(*d*) Carbon-3 has no hydrogens in 2,3,3-trimethyl-2-butanol. Elimination can involve only the hydroxyl group at C-2 and a hydrogen at C-1.

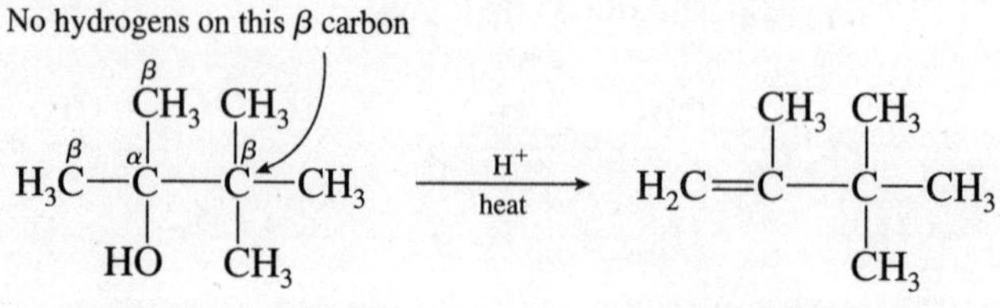

No hydrogens on this $\beta$ carbon

$$\underset{\text{2,3,3-Trimethyl-2-butanol}}{H_3\overset{\beta}{C}{-}\overset{\alpha}{C}(CH_3)(OH){-}\overset{\beta}{C}(CH_3)_2{-}CH_3} \xrightarrow[\text{heat}]{H^+} \underset{\text{2,3,3-Trimethyl-1-butene}}{H_2C{=}C(CH_3){-}C(CH_3)_2{-}CH_3}$$

**5.13** (*b*) Elimination can involve loss of a hydrogen from the methyl group or from C-2 of the ring in 1-methylcyclohexanol.

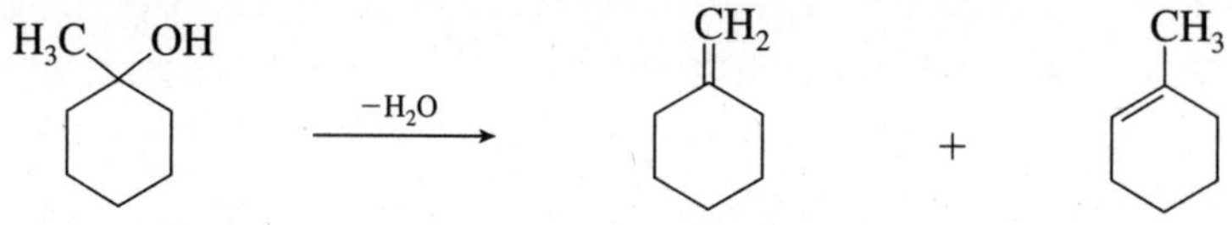

1-Methylcyclohexanol $\xrightarrow{-H_2O}$ Methylenecyclohexane (a disubstituted alkene; minor product) + 1-Methylcyclohexene (a trisubstituted alkene; major product)

According to the Zaitsev rule, the major alkene is the one corresponding to loss of a hydrogen from the alkyl group that has the smaller number of hydrogens. Thus hydrogen is removed from the methylene group in the ring rather than from the methyl group, and 1-methylcyclohexene is formed in greater amounts than methylenecyclohexane.

(*c*) The two alkenes formed are as shown in the equation.

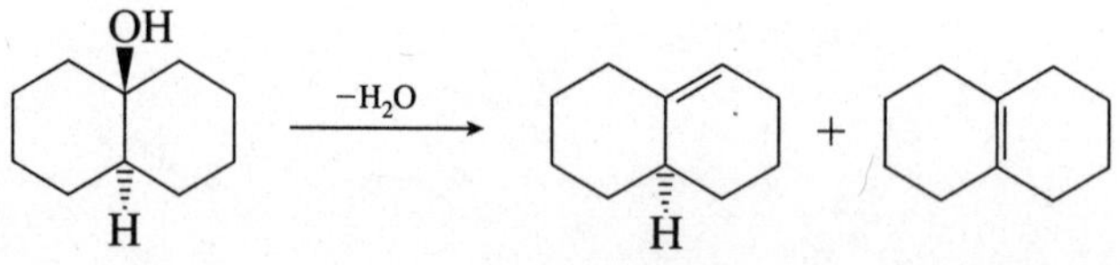

OH-substituted decalin (H) $\xrightarrow{-H_2O}$ Compound has a trisubstituted double bond + Compound has a tetrasubstituted double bond; more stable

The more highly substituted alkene is formed in greater amounts, as predicted by Zaitsev's rule.

**5.14** 2-Pentanol can undergo dehydration in two different directions, giving either 1-pentene or 2-pentene. 2-Pentene is formed as a mixture of the cis and trans stereoisomers.

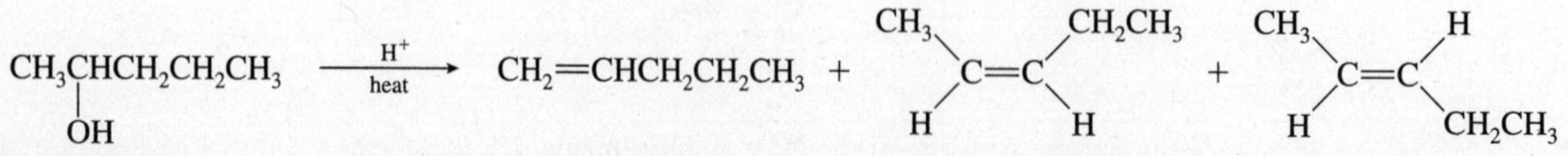

$$\underset{\text{2-Pentanol}}{CH_3CH(OH)CH_2CH_2CH_3} \xrightarrow[\text{heat}]{H^+} \underset{\text{1-Pentene}}{CH_2{=}CHCH_2CH_2CH_3} + \underset{cis\text{-2-Pentene}}{(CH_3)(H)C{=}C(H)(CH_2CH_3)} + \underset{trans\text{-2-Pentene}}{(CH_3)(H)C{=}C(CH_2CH_3)(H)}$$

**5.15** (*b*) The site of positive charge in the carbocation is the carbon atom that bears the hydroxyl group in the starting alcohol.

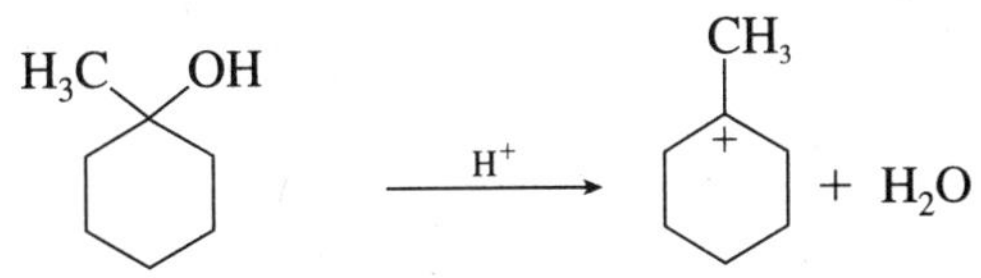

$\xrightarrow{H^+}$ + $H_2O$

1-Methylcyclohexanol

Water may remove a proton from the methyl group, as shown in the following equation:

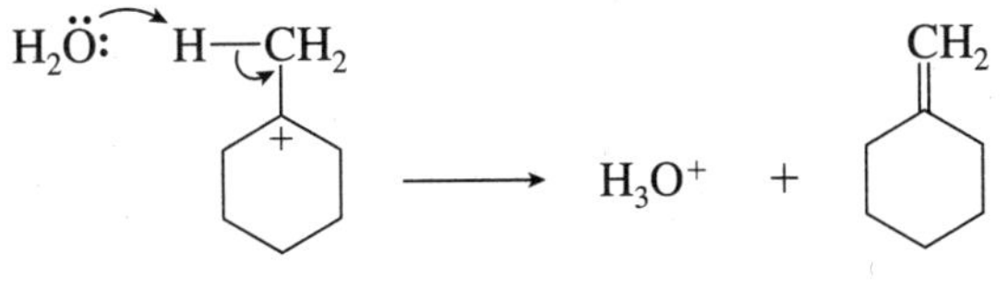

$\longrightarrow$ $H_3O^+$ +

Methylenecyclohexane

Loss of a proton from the ring gives the major product 1-methylcyclohexene.

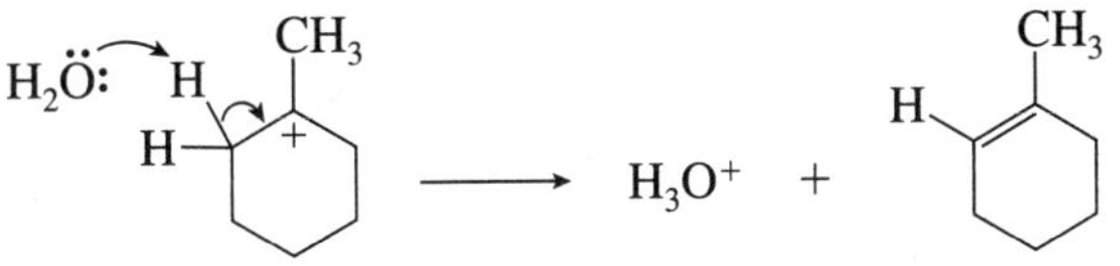

$\longrightarrow$ $H_3O^+$ +

1-Methylcyclohexene

(*c*) Loss of the hydroxyl group under conditions of acid catalysis yields a tertiary carbocation.

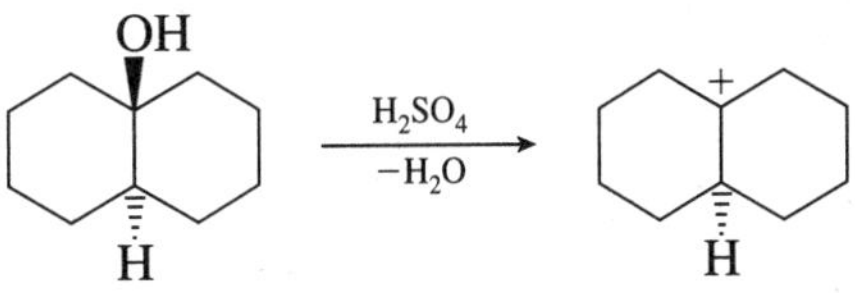

$\xrightarrow[-H_2O]{H_2SO_4}$

Water may remove a proton from an adjacent methylene group to give a trisubstituted alkene.

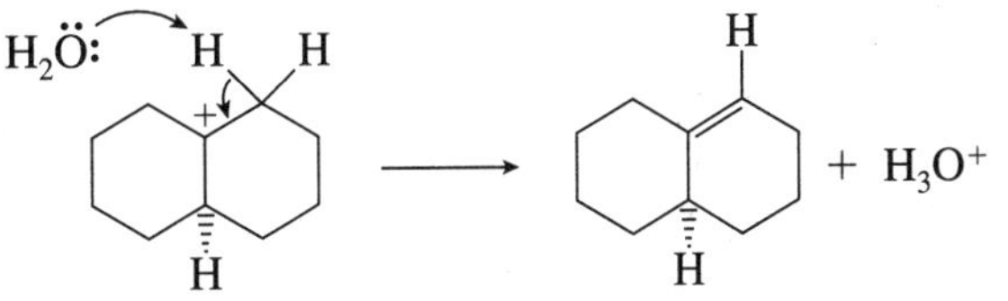

$\longrightarrow$ + $H_3O^+$

Removal of the methine proton gives a tetrasubstituted alkene.

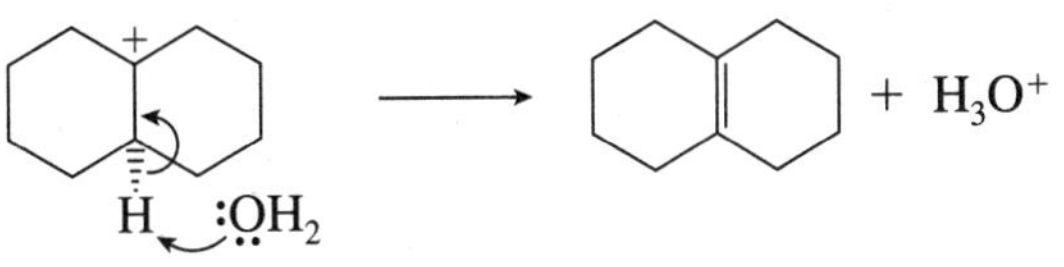

$\longrightarrow$ + $H_3O^+$

**5.16** In writing mechanisms for acid-catalyzed dehydration of alcohols, begin with formation of the carbocation intermediate:

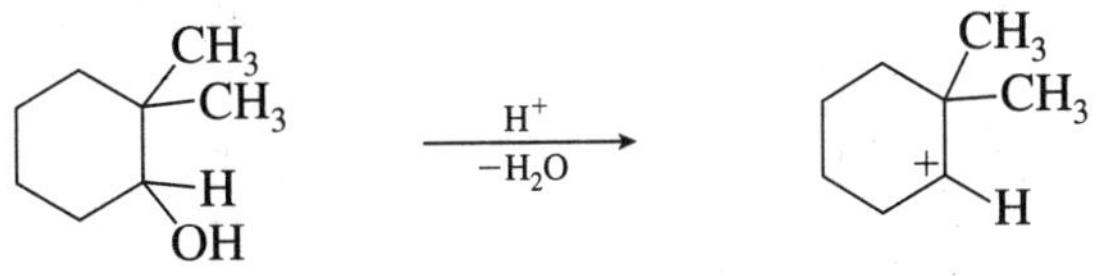

$\xrightarrow[-H_2O]{H^+}$

2,2-Dimethylcyclohexanol

2,2-Dimethylcyclohexyl cation

This secondary carbocation can rearrange to a more stable tertiary carbocation by a methyl group shift.

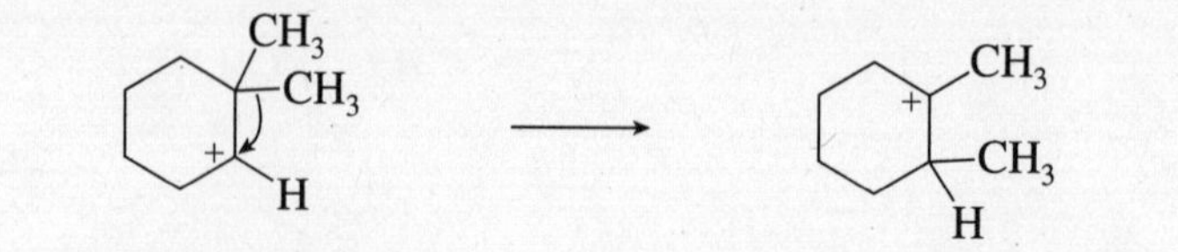

2,2-Dimethylcyclohexyl cation (secondary) ⟶ 1,2-Dimethylcyclohexyl cation (tertiary)

Loss of a proton from the 1,2-dimethylcyclohexyl cation intermediate yields 1,2-dimethylcyclohexene.

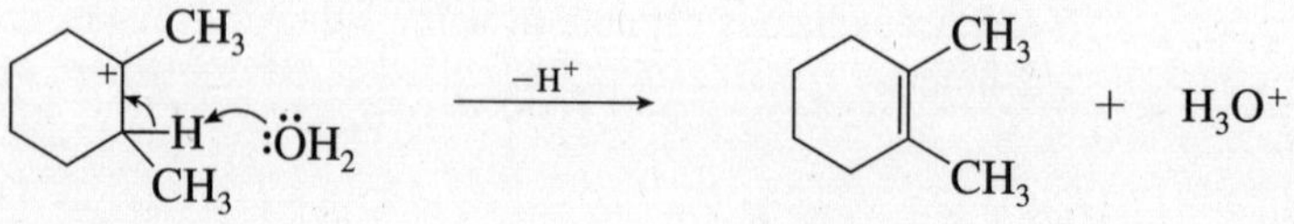

1,2-Dimethylcyclohexyl cation $\xrightarrow{-H^+}$ 1,2-Dimethylcyclohexene + $H_3O^+$

**5.17** (*b*) All the hydrogens of *tert*-butyl chloride are equivalent. Loss of any of these hydrogens along with the chlorine substituent yields 2-methylpropene as the only alkene.

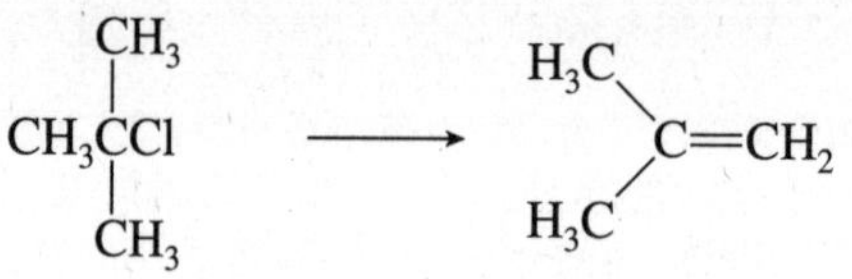

$$(CH_3)_3CCl \longrightarrow (H_3C)_2C{=}CH_2$$

*tert*-Butyl chloride ⟶ 2-Methylpropene

(*c*) All the $\beta$ hydrogens of 3-bromo-3-ethylpentane are equivalent, so that $\beta$-elimination can give only 3-ethyl-2-pentene.

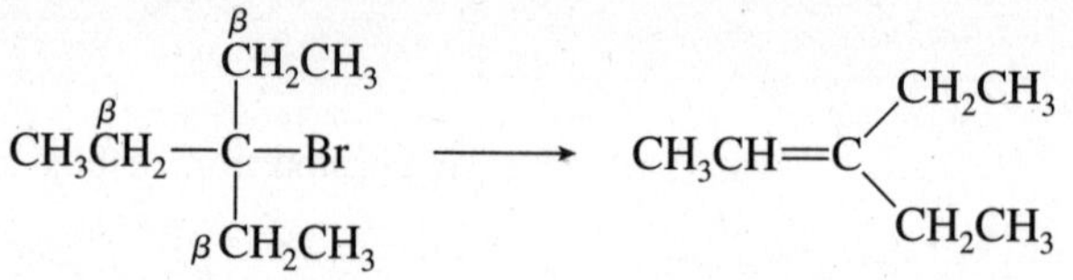

$$CH_3\overset{\beta}{C}H_2{-}C(\overset{\beta}{C}H_2CH_3)_2{-}Br \longrightarrow CH_3CH{=}C(CH_2CH_3)_2$$

3-Bromo-3-ethylpentane ⟶ 3-Ethyl-2-pentene

(*d*) There are two possible modes of $\beta$-elimination from 2-bromo-3-methylbutane. Elimination in one direction provides 3-methyl-1-butene; elimination in the other gives 2-methyl-2-butene.

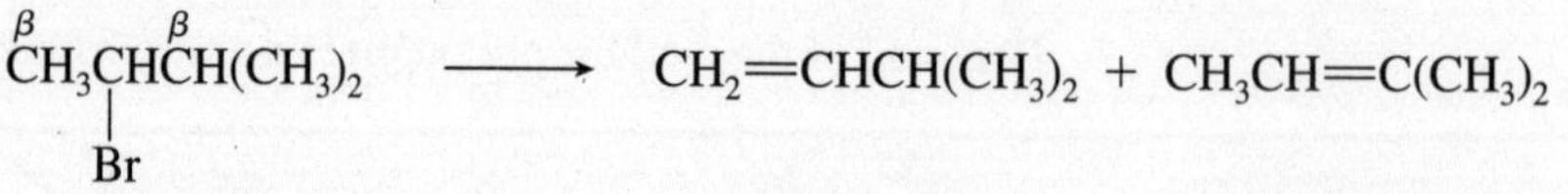

$$\overset{\beta}{C}H_3CH(Br)\overset{\beta}{C}H(CH_3)_2 \longrightarrow CH_2{=}CHCH(CH_3)_2 + CH_3CH{=}C(CH_3)_2$$

2-Bromo-3-methylbutane ⟶ 3-Methyl-1-butene (monosubstituted) + 2-Methyl-2-butene (trisubstituted)

The major product is the more highly substituted alkene, 2-methyl-2-butene. It is the more stable alkene and corresponds to removal of a hydrogen from the carbon that has the fewer hydrogens.

(*e*) Regioselectivity is not an issue here, because 3-methyl-1-butene is the only alkene that can be formed by $\beta$-elimination from 1-bromo-3-methylbutane.

$$BrCH_2\overset{\beta}{C}H_2CH(CH_3)_2 \longrightarrow CH_2{=}CHCH(CH_3)_2$$

1-Bromo-3-methylbutene ⟶ 3-Methyl-1-butene

(*f*) Two alkenes may be formed here. The more highly substituted one is 1-methylcyclohexene, and this is predicted to be the major product in accordance with Zaitsev's rule.

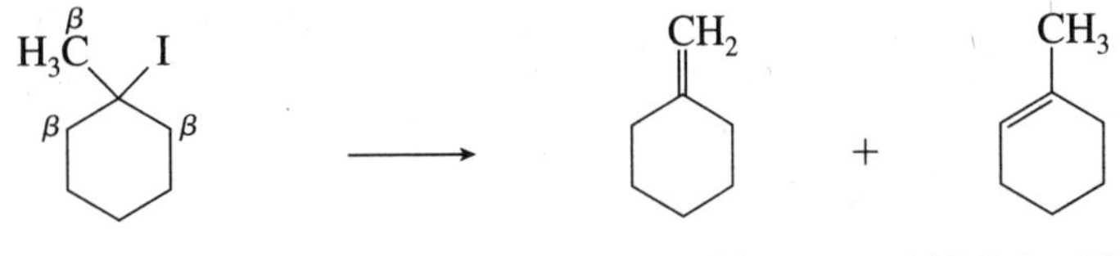

1-Iodo-1-methylcyclohexane → Methylenecyclohexane (disubstituted) + 1-Methylcyclohexene (trisubstituted; major product)

**5.18** Elimination in 2-bromobutane can take place between C-1 and C-2 or between C-2 and C-3. There are three alkenes capable of being formed: 1-butene and the stereoisomers *cis*-2-butene and *trans*-2-butene.

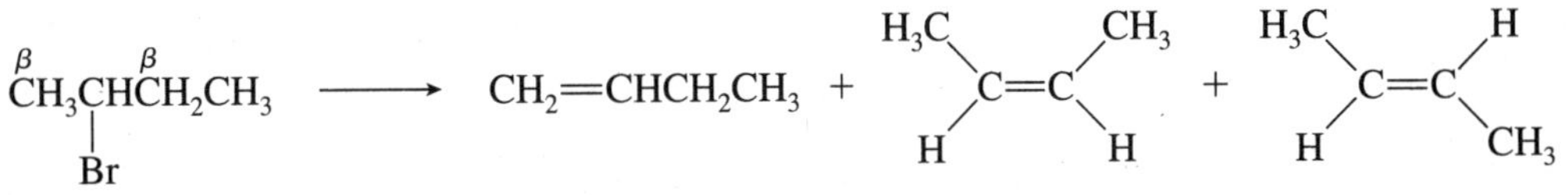

$CH_3CH(Br)CH_2CH_3$ (2-Bromobutane) → $CH_2{=}CHCH_2CH_3$ (1-Butene) + *cis*-2-Butene + *trans*-2-Butene

As predicted by Zaitsev's rule, the most stable alkene predominates. The major product is *trans*-2-butene.

**5.19** An unshared electron pair of the base methoxide ($CH_3O^-$) abstracts a proton from carbon. The pair of electrons in this C—H bond becomes the $\pi$ component of the double bond of the alkene. The pair of electrons in the C—Cl bond becomes an unshared electron pair of chloride ion.

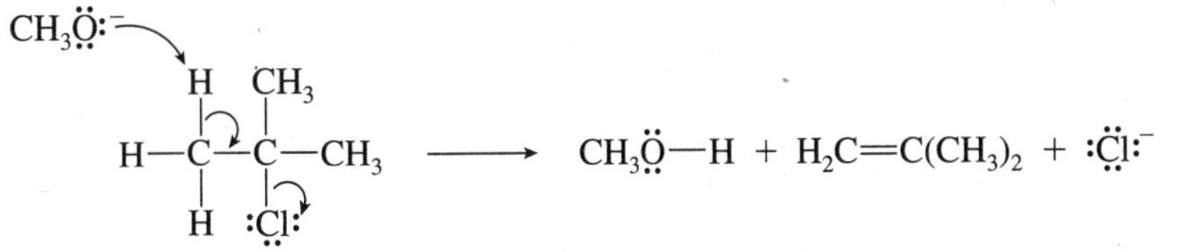

$CH_3\ddot{O}:^- + H_2C(H)\text{—}C(CH_3)(Cl)\text{—}CH_3 \longrightarrow CH_3\ddot{O}\text{—}H + H_2C{=}C(CH_3)_2 + :\ddot{Cl}:^-$

**5.20** The most stable conformation of *cis*-4-*tert*-butylcyclohexyl bromide has the bromine substituent in an axial orientation. The hydrogen that is removed by the base is an axial proton at C-2. This hydrogen and the bromine are anti periplanar to each other in the most stable conformation.

$(CH_3)_3C$ — :Br: — H — $(CH_3)_3C\text{—}\ddot{O}:^-$

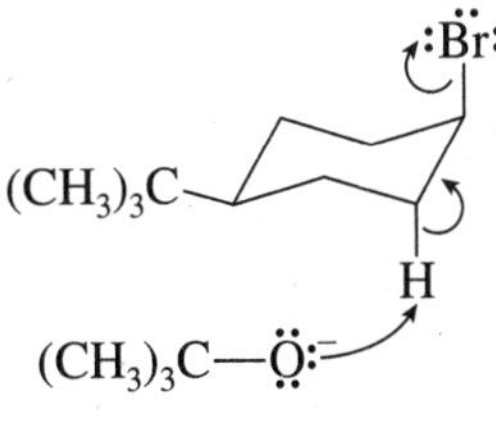

**5.21** (*a*) 1-Heptene is $CH_2{=}CH(CH_2)_4CH_3$.

(*b*) 3-Ethyl-2-pentene is $CH_3CH{=}C(CH_2CH_3)_2$.

(*c*) *cis*-3-Octene is $CH_3CH_2(H)C{=}C(H)CH_2CH_2CH_2CH_3$ (cis)

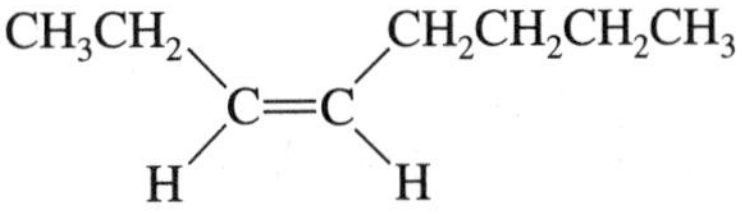

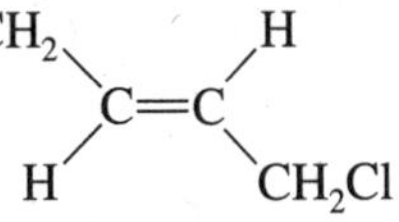

(*d*) *trans*-1,4-Dichloro-2-butene is $ClCH_2(H)C{=}C(H)CH_2Cl$ (trans) (*e*) (*Z*)-3-Methyl-2-hexene is $H_3C(H)C{=}C(CH_2CH_2CH_3)(CH_3)$

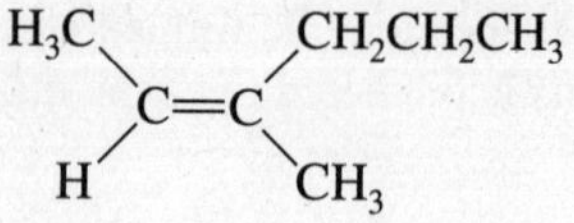

(*f*) (*E*)-3-Chloro-2-hexene is $H_3C(H)C{=}C(CH_2CH_2CH_3)(Cl)$

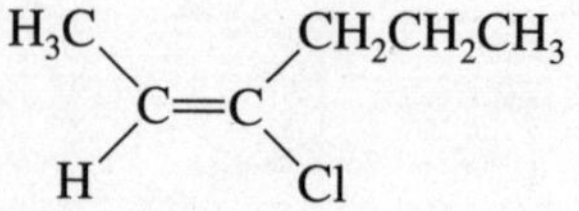

(*g*) 1-Bromo-3-methylcyclohexene is

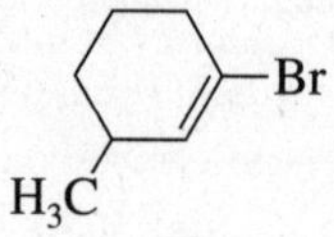

(*h*) 1-Bromo-6-methylcyclohexene is

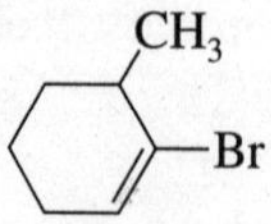

(*i*) 4-Methyl-4-penten-2-ol is $CH_3CH(OH)CH_2C(CH_3){=}CH_2$

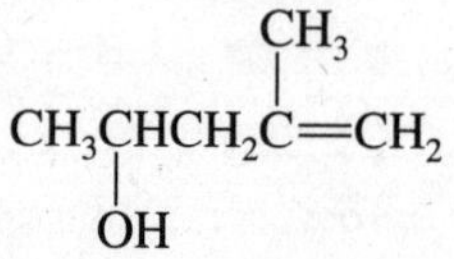

(*j*) Vinylcycloheptane is

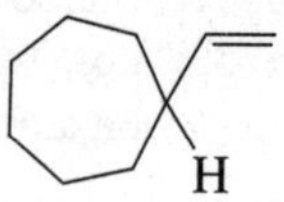

(*k*) An allyl group is $—CH_2CH{=}CH_2$. 1,1-Diallylcyclopropane is

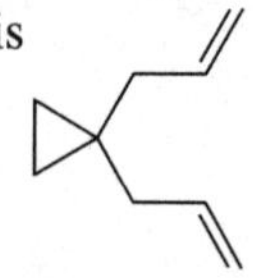

(*l*) An isopropenyl substituent is $—C(CH_3){=}CH_2$. *trans*-1-Isopropenyl-3-methylcyclohexane is therefore

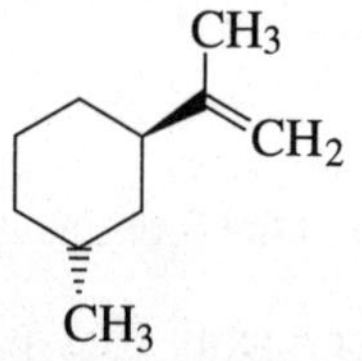

**5.22** Alkenes with tetrasubstituted double bonds have four alkyl groups attached to the doubly bonded carbons. There is only one alkene of molecular formula $C_7H_{14}$ that has a tetrasubstituted double bond, 2,3-dimethyl-2-pentene.

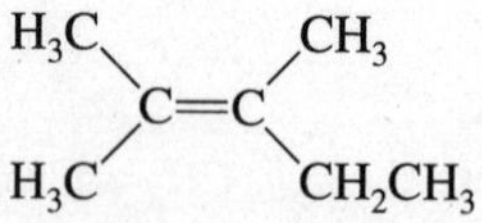

$(H_3C)_2C{=}C(CH_3)(CH_2CH_3)$

2,3-Dimethyl-2-pentene

**5.23** (*a*) The longest chain that includes the double bond in $(CH_3CH_2)_2C{=}CHCH_3$ contains five carbon atoms, and so the parent alkene is a pentene. The numbering scheme that gives the double bond the lowest number is

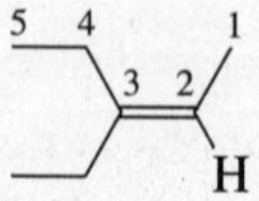

The compound is 3-ethyl-2-pentene.

(*b*) Write out the structure in detail, and identify the longest continuous chain that includes the double bond.

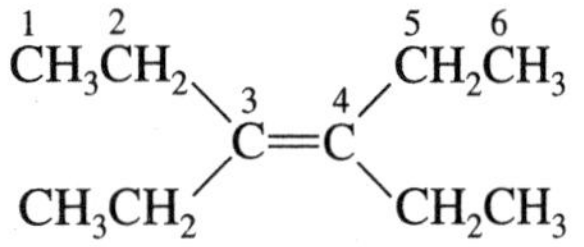

The longest chain contains six carbon atoms, and the double bond is between C-3 and C-4. The compound is named as a derivative of 3-hexene. There are ethyl substituents at C-3 and C-4. The complete name is 3,4-diethyl-3-hexene.

(*c*) Write out the structure completely.

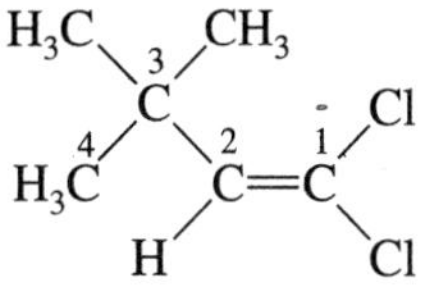

The longest carbon chain contains four carbons. Number the chain so as to give the lowest numbers to the doubly bonded carbons, and list the substituents in alphabetical order. This compound is 1,1-dichloro-3,3-dimethyl-1-butene.

(*d*) The longest chain has five carbon atoms, the double bond is at C-1, and there are two methyl substituents. The compound is 4,4-dimethyl-1-pentene.

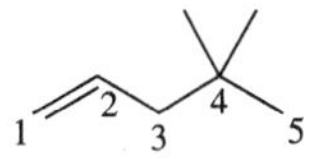

(*e*) We number this trimethylcyclobutene derivative so as to provide the lowest number for the substituent at the first point of difference. We therefore number

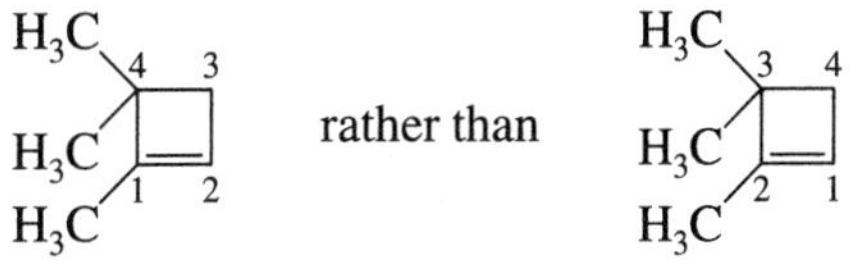

rather than

The correct IUPAC name is 1,4,4-trimethylcyclobutene, not 2,3,3-trimethylcyclobutene.

(*f*) The cyclohexane ring has a 1,2-cis arrangement of vinyl substituents. The compound is *cis*-1,2-divinylcyclohexane.

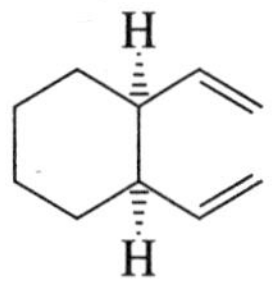

(*g*) Name this compound as a derivative of cyclohexene. It is 1,2-divinylcyclohexene.

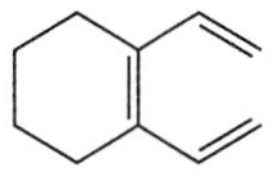

**5.24** (*a*) Go to the end of the name, because this tells you how many carbon atoms are present in the longest chain. In the hydrocarbon name 2,6,10,14-tetramethyl-2-pentadecene, the suffix "2-pentadecene" reveals that the longest continuous chain has 15 carbon atoms and that there is a double bond between C-2 and C-3. The rest of the name provides the information that there are four methyl groups and that they are located at C-2, C-6, C-10, and C-14.

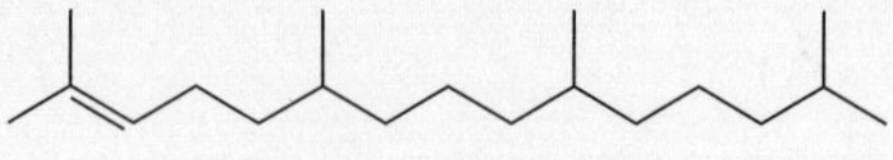

2,6,10,14-Tetramethyl-2-pentadecene

(*b*) An allyl group is $CH_2{=}CHCH_2{-}$. Allyl isothiocyanate is therefore $CH_2{=}CHCH_2N{=}C{=}S$.

**5.25** (*a*) The *E* configuration means that the higher priority groups are on opposite sides of the double bond.

Higher $CH_3CH_2$ (C=C) H Lower

Lower H $CH_2CH_2CH_2CH_2CH_2OH$ Higher

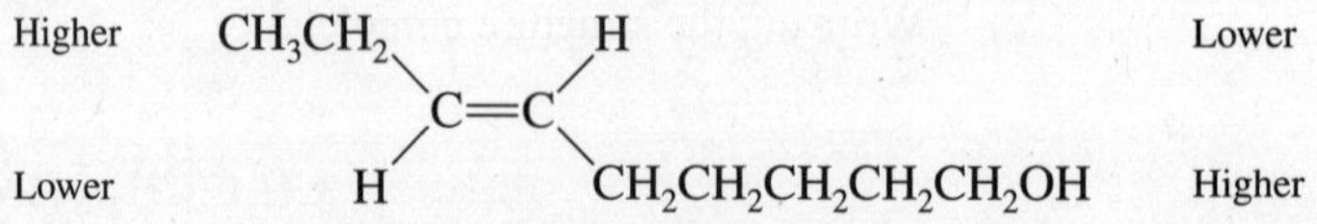

(*E*)-6-Nonen-1-ol

(*b*) Geraniol has two double bonds, but only one of them, the one between C-2 and C-3, is capable of stereochemical variation. Of the groups at C-2, $CH_2OH$ is of higher priority than H. At C-3, $CH_2CH_2$ outranks $CH_3$. Higher priority groups are on opposite sides of the double bond in the *E* isomer; hence geraniol has the structure shown.

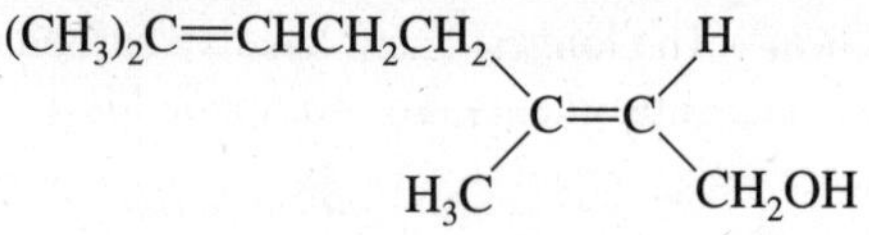

$(CH_3)_2C{=}CHCH_2CH_2$ (C=C) H

$H_3C$ $CH_2OH$

Geraniol

(*c*) Since nerol is a stereoisomer of geraniol, it has the same constitution and differs from geraniol only in having the *Z* configuration of the double bond.

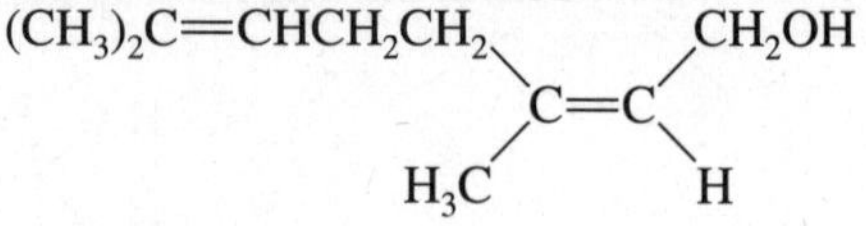

$(CH_3)_2C{=}CHCH_2CH_2$ (C=C) $CH_2OH$

$H_3C$ H

Nerol

(*d*) Beginning at the C-6, C-7 double bond, we see that the propyl group is of higher priority than the methyl group at C-7. Since the C-6, C-7 double bond is *E*, the propyl group must be on the opposite side of the higher priority group at C-6, where the $CH_2$ fragment has a higher priority than hydrogen. We therefore write for the stereochemistry of the C-6, C-7 double bond as:

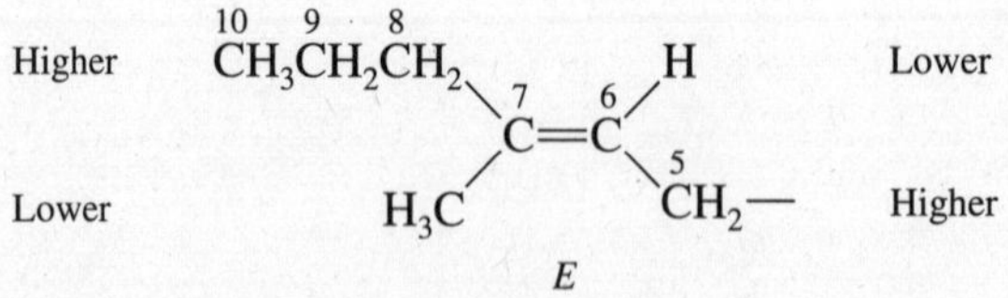

Higher $\overset{10}{C}H_3\overset{9}{C}H_2\overset{8}{C}H_2$ ($\overset{7}{C}{=}\overset{6}{C}$) H Lower

Lower $H_3C$ $\overset{5}{C}H_2{-}$ Higher

*E*

At C-2, $CH_2OH$ is of higher priority than H; and at C-3, $CH_2CH_2C{-}$ is of higher priority than $CH_2CH_3$. The double-bond configuration at C-2 is *Z*. Therefore

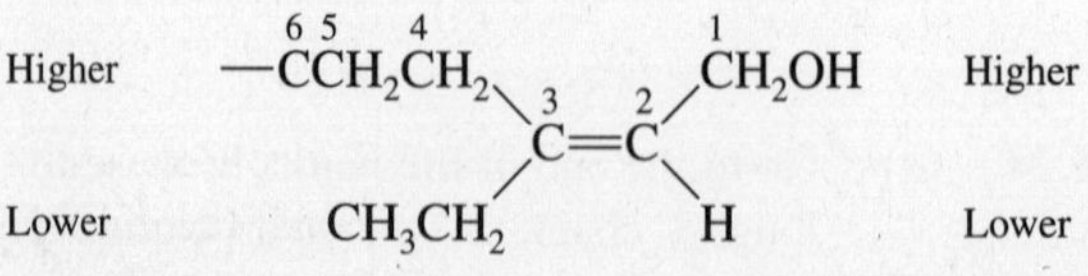

Higher $-\overset{6}{C}\overset{5}{C}H_2\overset{4}{C}H_2$ ($\overset{3}{C}{=}\overset{2}{C}$) $\overset{1}{C}H_2OH$ Higher

Lower $CH_3CH_2$ H Lower

Combining the two partial structures, we obtain for the full structure of the codling moth's sex pheromone

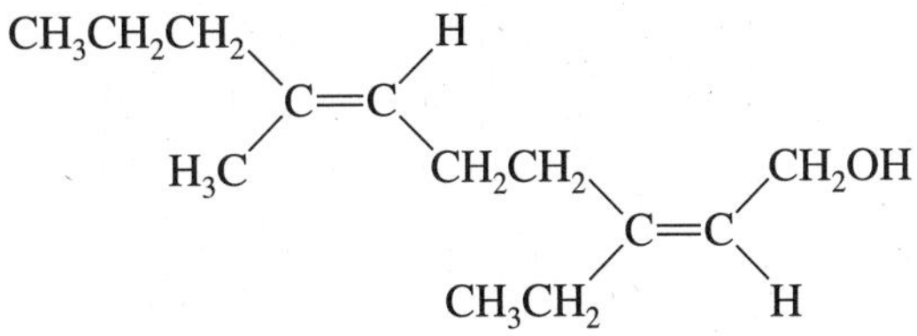

$CH_3CH_2CH_2$, $H_3C$ — C=C — H, $CH_2CH_2$; $CH_2CH_2$, $CH_3CH_2$ — C=C — $CH_2OH$, H

The compound is (2*Z*,6*E*)-3-ethyl-7-methyl-2,6-decadien-1-ol.

(*e*) The sex pheromone of the honeybee is (*E*)-9-oxo-2-decenoic acid, with the structure

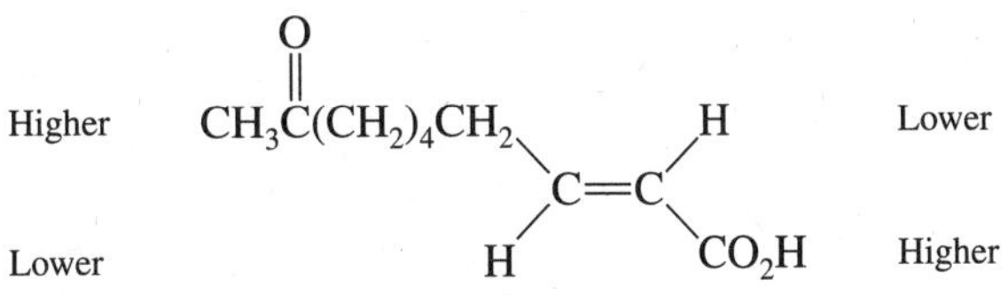

Higher $CH_3C(=O)(CH_2)_4CH_2$, Lower H — C=C — H Lower, $CO_2H$ Higher

(*f*) Looking first at the C-2, C-3 double bond of the cecropia moth's growth hormone

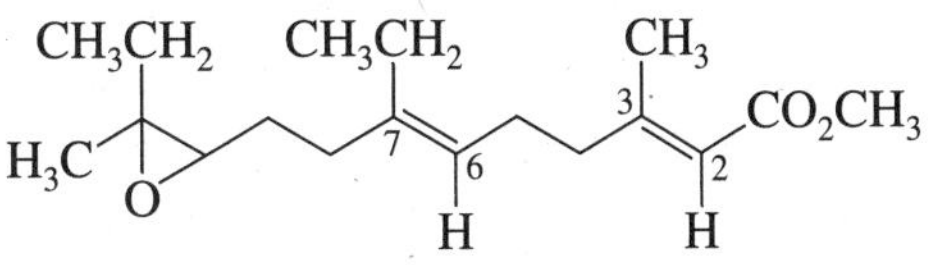

$CH_3CH_2$ $CH_3CH_2$ $CH_3$ $H_3C$ O 7 6 3 2 $CO_2CH_3$ H H

we find that its configuration is *E*, since the higher priority groups are on opposite sides of the double bond.

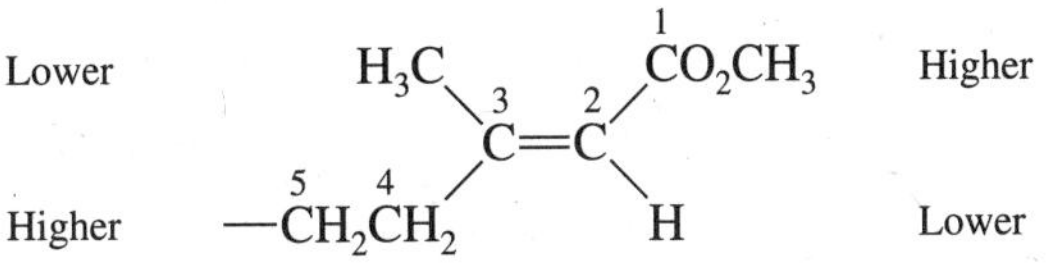

Lower $H_3C$, Higher $-\overset{5}{C}H_2\overset{4}{C}H_2$ — $\overset{3}{C}=\overset{2}{C}$ — $\overset{1}{C}O_2CH_3$ Higher, H Lower

The configuration of the C-6, C-7 double bond is also *E*.

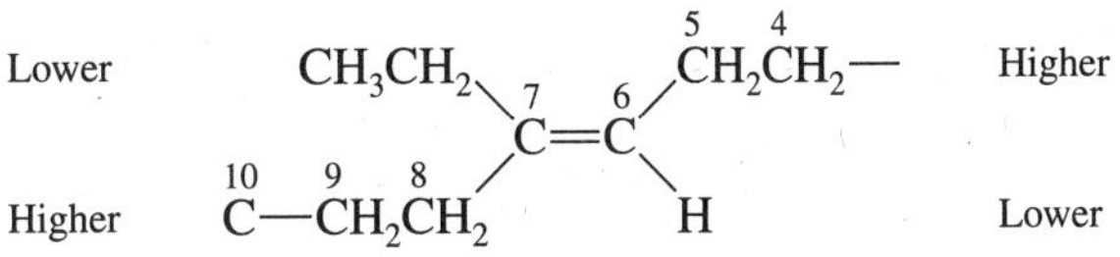

Lower $CH_3CH_2$, Higher $\overset{10}{C}-\overset{9}{C}H_2\overset{8}{C}H_2$ — $\overset{7}{C}=\overset{6}{C}$ — $\overset{5}{C}H_2\overset{4}{C}H_2-$ Higher, H Lower

**5.26** We haven't covered, and won't cover, how to calculate the size of a dipole moment, but we can decide whether a compound has a dipole moment or not. Only compound B has dipole moment. The individual bond dipoles in A, C, and D cancel; therefore, none of these three has a dipole moment.

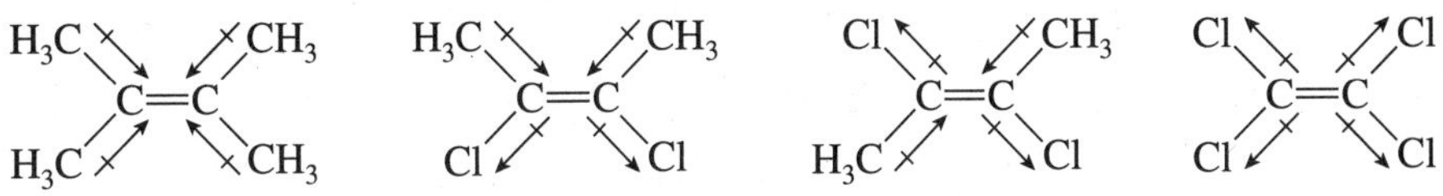

$(H_3C)_2C=C(CH_3)_2$ A ($\mu = 0$ D)

$H_3C(Cl)C=C(CH_3)Cl$ B (has a dipole moment)

$Cl(H_3C)C=C(CH_3)Cl$ C ($\mu = 0$ D)

$Cl_2C=CCl_2$ D ($\mu = 0$ D)

**5.27** The alkenes are listed as follows in order of decreasing heat of combustion:

(*e*) 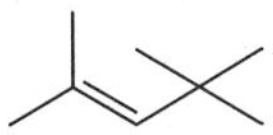 2,4,4-Trimethyl-2-pentene; 5293 kJ/mol (1264.9 kcal/mol). Highest heat of combustion because it is $C_8H_{16}$; all others are $C_7H_{14}$.

(*a*) 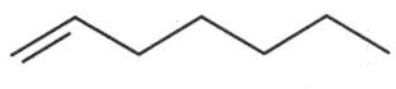 1-Heptene; 4658 kJ/mol (1113.4 kcal/mol). Monosubstituted double bond; therefore least stable $C_7H_{14}$ isomer.

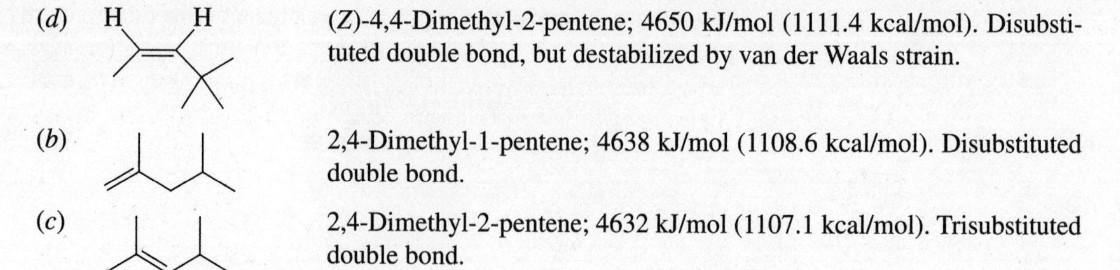

(*d*) (*Z*)-4,4-Dimethyl-2-pentene; 4650 kJ/mol (1111.4 kcal/mol). Disubstituted double bond, but destabilized by van der Waals strain.

(*b*) 2,4-Dimethyl-1-pentene; 4638 kJ/mol (1108.6 kcal/mol). Disubstituted double bond.

(*c*) 2,4-Dimethyl-2-pentene; 4632 kJ/mol (1107.1 kcal/mol). Trisubstituted double bond.

**5.28** (*a*) 1-Methylcyclohexene is more stable; it contains a **trisubstituted** double bond, whereas 3-methylcyclohexene has only a disubstituted double bond.

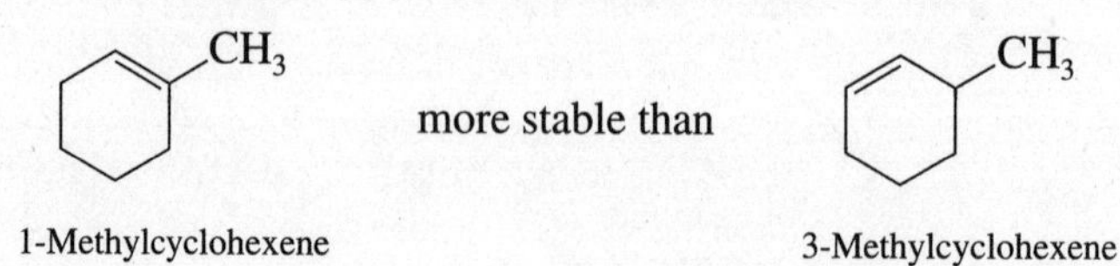


1-Methylcyclohexene 3-Methylcyclohexene

(*b*) Both isopropenyl and allyl are three-carbon alkenyl groups: isopropenyl is $CH_2{=}\underset{|}{C}CH_3$, allyl is $CH_2{=}CHCH_2{-}$.

$CH_2$ / C / $CH_3$ — $CH_2CH{=}CH_2$

Isopropenylcyclopentane Allylcyclopentane

Isopropenylcyclopentane has a disubstituted double bond and so is predicted to be more stable than allylcyclopentane, in which the double bond is monosubstituted.

(*c*) A double bond in a six-membered ring is less strained than a double bond in a four-membered ring; therefore bicyclo[4.2.0]oct-3-ene is more stable.

more stable than

Bicyclo[4.2.0]oct-3-ene Bicyclo[4.2.0]oct-7-ene

(*d*) Cis double bonds are more stable than trans double bonds when the ring is smaller than 11-membered. (*Z*)-Cyclononene has a cis double bond in a 9-membered ring, and is thus more stable than (*E*)-cyclononene.

more stable than

(*Z*)-Cyclononene (*E*)-Cyclononene

(*e*) Trans double bonds are more stable than cis when the ring is large. Here the rings are 18-membered, so that (*E*)-cyclooctadecene is more stable than (*Z*)-cyclooctadecene.

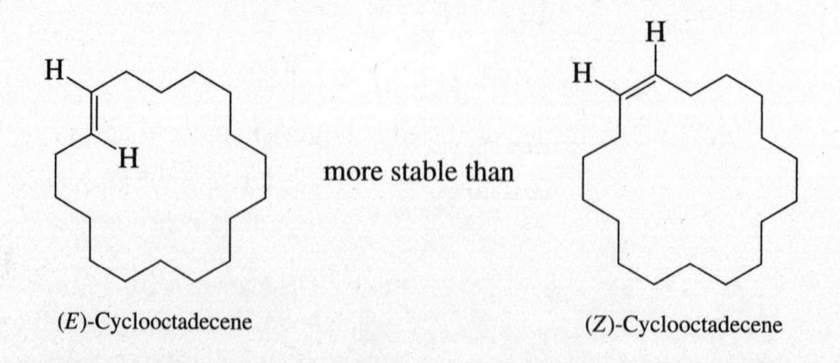


(*E*)-Cyclooctadecene (*Z*)-Cyclooctadecene

**5.29** (*a*) Carbon atoms that are involved in double bonds are $sp^2$-hybridized, with ideal bond angles of 120°. Incorporating an $sp^2$-hybridized carbon into a three-membered ring leads to more angle strain than incorporation of an $sp^3$-hybridized carbon. 1-Methylcyclopropene has two $sp^2$-hybridized carbons in a three-membered ring and so has substantially more angle strain than methylenecyclopropane.

$$\text{(cyclopropene)}\!-\!CH_3 \qquad \text{(cyclopropane)}\!=\!CH_2$$

1-Methylcyclopropene Methylenecyclopropane

The higher degree of substitution at the double bond in 1-methylcyclopropene is not sufficient to offset the increased angle strain, and so 1-methylcyclopropene is less stable than methylenecyclopropane.

(*b*) 3-Methylcyclopropene has a disubstituted double bond and two $sp^2$-hybridized carbons in its three-membered ring. It is the least stable of the isomers.

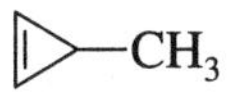

$$\text{(cyclopropene)}\!-\!CH_3$$

3-Methylcyclopropene

**5.30** In all parts of this exercise, write the structure of the alkyl halide in sufficient detail to identify the carbon that bears the halogen and the $\beta$-carbon atoms that bear at least one hydrogen. These are the carbons that become doubly bonded in the alkene product.

(*a*) 1-Bromohexane can give only 1-hexene under conditions of E2 elimination.

$$BrCH_2\overset{\beta}{C}H_2CH_2CH_2CH_2CH_3 \xrightarrow[\text{E2}]{\text{base}} CH_2{=}CHCH_2CH_2CH_2CH_3$$

1-Bromohexane 1-Hexene (only alkene)

(*b*) 2-Bromohexane can give both 1-hexene and 2-hexene on dehydrobromination. The 2-hexene fraction is a mixture of cis and trans stereoisomers.

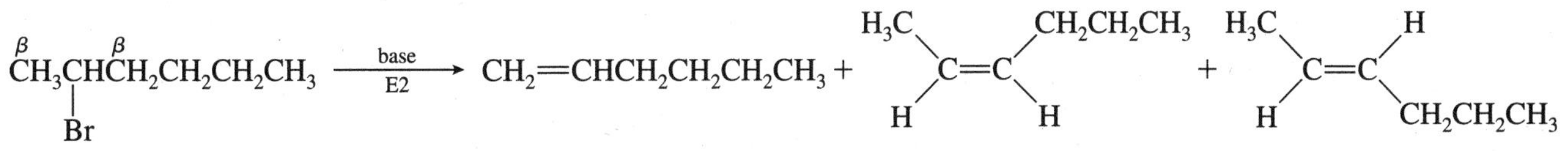

$$\overset{\beta}{C}H_3\underset{|\,Br}{C}H\overset{\beta}{C}H_2CH_2CH_2CH_3 \xrightarrow[\text{E2}]{\text{base}} CH_2{=}CHCH_2CH_2CH_2CH_3 + \underset{H\qquad H}{\overset{H_3C\qquad CH_2CH_2CH_3}{C{=}C}} + \underset{H\qquad CH_2CH_2CH_3}{\overset{H_3C\qquad H}{C{=}C}}$$

2-Bromohexane 1-Hexene *cis*-2-Hexene *trans*-2-Hexene

(*c*) Both a cis–trans pair of 2-hexenes and a cis–trans pair of 3-hexenes are capable of being formed from 3-bromohexane.

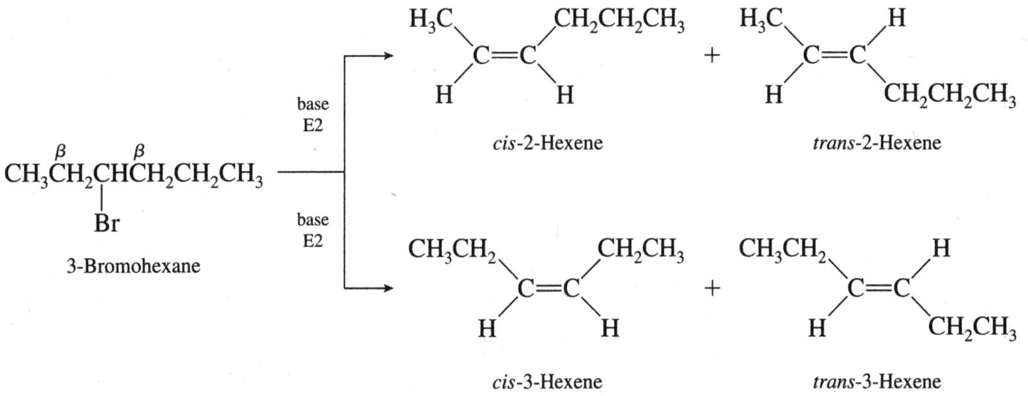

$$CH_3\overset{\beta}{C}H_2\underset{|\,Br}{C}H\overset{\beta}{C}H_2CH_2CH_3 \xrightarrow[\text{E2}]{\text{base}} \underset{H\qquad H}{\overset{H_3C\qquad CH_2CH_2CH_3}{C{=}C}} + \underset{H\qquad CH_2CH_2CH_3}{\overset{H_3C\qquad H}{C{=}C}}$$

3-Bromohexane *cis*-2-Hexene *trans*-2-Hexene

$$CH_3\overset{\beta}{C}H_2\underset{|\,Br}{C}H\overset{\beta}{C}H_2CH_2CH_3 \xrightarrow[\text{E2}]{\text{base}} \underset{H\qquad H}{\overset{CH_3CH_2\qquad CH_2CH_3}{C{=}C}} + \underset{H\qquad CH_2CH_3}{\overset{CH_3CH_2\qquad H}{C{=}C}}$$

*cis*-3-Hexene *trans*-3-Hexene

(*d*) Dehydrobromination of 2-bromo-2-methylpentane can involve one of the hydrogens of either a methyl group (C-1) or a methylene group (C-3).

$$CH_3C(CH_3)(Br)CH_2CH_2CH_3 \xrightarrow[E2]{base} CH_2{=}C(CH_3)CH_2CH_2CH_3 + (CH_3)_2C{=}CHCH_2CH_3$$

2-Bromo-2-methylpentane → 2-Methyl-1-pentene + 2-Methyl-2-pentene

Neither alkene is capable of existing in stereoisomeric forms, and so these two are the only products of E2 elimination.

(*e*) 2-Bromo-3-methylpentane can undergo dehydrohalogenation by loss of a proton from either C-1 or C-3. Loss of a proton from C-1 gives 3-methyl-1-pentene.

$$\text{Base:} \; H{-}CH_2{-}CH(Br)CH(CH_3)CH_2CH_3 \xrightarrow[E2]{base} CH_2{=}CHCH(CH_3)CH_2CH_3$$

2-Bromo-3-methylpentane → 3-Methyl-1-pentene

Loss of a proton from C-3 gives a mixture of (*E*)- and (*Z*)-3-methyl-2-pentene.

$$CH_3CH(Br)CH(CH_3)CH_2CH_3 \xrightarrow[E2]{base} (H_3C)(H)C{=}C(CH_3)(CH_2CH_3) + (H_3C)(H)C{=}C(CH_2CH_3)(CH_3)$$

2-Bromo-3-methylpentane → (*E*)-3-Methyl-2-pentene + (*Z*)-3-Methyl-2-pentene

(*f*) Three alkenes are possible from 3-bromo-2-methylpentane. Loss of the C-2 proton gives 2-methyl-2-pentene.

$$\text{Base:} \; CH_3{-}C(H)(CH_3){-}CH(Br)CH_2CH_3 \xrightarrow{E2} (CH_3)_2C{=}CHCH_2CH_3$$

3-Bromo-2-methylpentane → 2-Methyl-2-pentene

Abstraction of a proton from C-4 can yield either (*E*)- or (*Z*)-4-methyl-2-pentene.

$$(CH_3)_2CHCH(Br){-}CH(H)CH_3 \; \text{:Base} \xrightarrow{E2} ((CH_3)_2CH)(H)C{=}C(H)(CH_3) + ((CH_3)_2CH)(H)C{=}C(CH_3)(H)$$

3-Bromo-2-methylpentane → (*E*)-4-Methyl-2-pentene + (*Z*)-4-Methyl-2-pentene

(*g*) Proton abstraction from the C-3 methyl group of 3-bromo-3-methylpentane yields 2-ethyl-1-butene.

$$\text{Base:} \; CH_3CH_2C(Br)(CH_2{-}H)CH_2CH_3 \xrightarrow{E2} CH_2{=}C(CH_2CH_3)CH_2CH_3$$

3-Bromo-3-methylpentane → 2-Ethyl-1-butene

Stereoisomeric 3-methyl-2-pentenes are formed by proton abstraction from C-2.

$$CH_3—CH(H)—C(CH_3)(Br)CH_2CH_3 \xrightarrow{E2} (E)\text{-}H_3C(H)C{=}C(CH_3)CH_2CH_3 + (Z)\text{-}H_3C(H)C{=}C(CH_2CH_3)CH_3$$

Base: H CH$_3$ Br

3-Bromo-3-methylpentane (*E*)-3-Methyl-2-pentene (*Z*)-3-Methyl-2-pentene

(*h*) Only 3,3-dimethyl-1-butene may be formed under conditions of E2 elimination from 3-bromo-2,2-dimethylbutane.

$$CH_3—C(CH_3)_2—CH(Br)—CH_2—H \xrightarrow{E2} CH_3—C(CH_3)_2—CH{=}CH_2$$

:Base

3-Bromo-2,2-dimethylbutane 3,3-Dimethyl-1-butene

**5.31** (*a*) The reaction that takes place with 1-bromo-3,3-dimethylbutane is an E2 elimination involving loss of the bromine at C-1 and abstraction of the proton at C-2 by the strong base potassium *tert*-butoxide, yielding a single alkene.

$$(CH_3)_3C\ddot{O}:^- \quad (CH_3)_3CCH(H)—CH_2Br \xrightarrow{E2} (CH_3)_3CCH{=}CH_2$$

1-Bromo-3,3-dimethylbutane 3,3-Dimethyl-1-butene

(*b*) Two alkenes are capable of being formed in this $\beta$-elimination reaction.

$$\text{1-Methylcyclopentyl chloride} \xrightarrow[\text{ethanol, 70°C}]{NaOCH_2CH_3} \text{Methylenecyclopentane} + \text{1-Methylcyclopentene}$$

1-Methylcyclopentyl chloride Methylenecyclopentane 1-Methylcyclopentene

The more highly substituted alkene is 1-methylcyclopentene; it is the major product of this reaction. According to Zaitsev's rule, the major alkene is formed by proton removal from the $\beta$ carbon that has the fewest hydrogens.

(*c*) Acid-catalyzed dehydration of 3-methyl-3-pentanol can lead either to 2-ethyl-1-butene or to a mixture of (*E*)- and (*Z*)-3-methyl-2-pentene.

$$CH_3CH_2C(CH_3)(OH)CH_2CH_3 \xrightarrow[80°C]{H_2SO_4} CH_2{=}C(CH_2CH_3)CH_2CH_3 + (E)\text{-}H_3C(H)C{=}C(CH_3)CH_2CH_3 + (Z)\text{-}H_3C(H)C{=}C(CH_2CH_3)CH_3$$

3-Methyl-3-pentanol 2-Ethyl-1-butene (*E*)-3-Methyl-2-pentene (*Z*)-3-Methyl-2-pentene

The major product is a mixture of the trisubstituted alkenes, (*E*)- and (*Z*)-3-methyl-2-pentene. Of these two stereoisomers the *E* isomer is slightly more stable and is expected to predominate.

(*d*) Acid-catalyzed dehydration of 2,3-dimethyl-2-butanol can proceed in either of two directions.

$$\underset{\text{2,3-Dimethyl-2-butanol}}{\overset{\beta}{CH_3}-\underset{OH}{\overset{CH_3}{C}}-\underset{\beta}{\overset{CH_3}{C}}HCH_3} \xrightarrow[\substack{120°C \\ -H_2O}]{H_3PO_4} \underset{\substack{\text{2,3-Dimethyl-1-butene} \\ \text{(disubstituted)}}}{CH_2{=}C(CH_3)CH(CH_3)_2} + \underset{\substack{\text{2,3-Dimethyl-2-butene} \\ \text{(tetrasubstituted)}}}{(CH_3)_2C{=}C(CH_3)_2}$$

The major alkene is the one with the more highly substituted double bond, 2,3-dimethyl-2-butene. Its formation corresponds to Zaitsev's rule in that a proton is lost from the $\beta$ carbon that has the fewest hydrogens.

(*e*) Only a single alkene is capable of being formed on E2 elimination from this alkyl iodide. Stereoisomeric alkenes are not possible, and because all the $\beta$ hydrogens are equivalent, regioisomers cannot be formed either.

$$\underset{\text{3-Iodo-2,4-dimethylpentane}}{CH_3\overset{\beta}{\underset{CH_3}{C}}H\overset{I}{C}H\overset{\beta}{\underset{CH_3}{C}}HCH_3} \xrightarrow[\substack{\text{ethanol} \\ 70°C}]{NaOCH_2CH_3} \underset{\text{2,4-Dimethyl-2-pentene}}{(CH_3)_2C{=}CHCH(CH_3)_2}$$

(*f*) Despite the structural similarity of this alcohol to the alkyl halide in the preceding part of this problem, its dehydration is more complicated. The initially formed carbocation is secondary and can rearrange to a more stable tertiary carbocation by a hydride shift.

$$\underset{\text{2,4-Dimethyl-3-pentanol}}{CH_3\underset{CH_3}{C}H\overset{OH}{C}H\underset{CH_3}{C}HCH_3} \xrightarrow[-H_2O]{H^+} \underset{\substack{\text{Secondary carbocation} \\ \text{(less stable)}}}{CH_3\underset{CH_3}{\overset{H}{C}}-\overset{+}{C}H\underset{CH_3}{C}HCH_3} \xrightarrow[\text{shift}]{\text{hydride}} \underset{\substack{\text{Tertiary carbocation} \\ \text{(more stable)}}}{CH_3\underset{CH_3}{\overset{+}{C}}CH_2\underset{CH_3}{C}HCH_3}$$

The tertiary carbocation, once formed, can give either 2,4-dimethyl-1-pentene or 2,4-dimethyl-2-pentene by loss of a proton.

$$CH_3\underset{CH_3}{\overset{+}{C}}CH_2\underset{CH_3}{C}HCH_3 \longrightarrow \underset{\substack{\text{2,4-Dimethyl-1-pentene} \\ \text{(disubstituted)}}}{H_2C{=}\underset{CH_3}{C}CH_2CH(CH_3)_2} + \underset{\substack{\text{2,4-Dimethyl-2-pentene} \\ \text{(trisubstituted)}}}{(CH_3)_2C{=}CHCH(CH_3)_2}$$

The proton is lost from the methylene group in preference to the methyl group. The major alkene is the more highly substituted one, 2,4-dimethyl-2-pentene.

**5.32** In all parts of this problem you need to reason backward from an alkene to an alkyl bromide of molecular formula $C_7H_{13}Br$ that gives *only* the desired alkene under E2 elimination conditions. Recall that the carbon–carbon double bond is formed by loss of a proton from one of the carbons that becomes doubly bonded and a bromine from the other.

(*a*) Cycloheptene is the only alkene formed by an E2 elimination reaction of cycloheptyl bromide.

$$\underset{\text{Cycloheptyl bromide}}{C_7H_{13}Br} \xrightarrow[E2]{\text{base}} \underset{\text{Cycloheptene}}{C_7H_{12}}$$

(*b*) (Bromomethyl)cyclohexane is the correct answer. It gives methylenecyclohexane as the *only* alkene under E2 conditions.

(Bromomethyl)cyclohexane $\xrightarrow[\text{E2}]{\text{base}}$ Methylenecyclohexane

1-Bromo-1-methylcyclohexane is not correct. It gives a mixture of 1-methylcyclohexene and methylenecyclohexane on elimination.

1-Bromo-1-methylcyclohexane $\xrightarrow[\text{E2}]{\text{base}}$ Methylenecyclohexane + 1-Methylcyclohexene

(*c*) In order for 4-methylcyclohexene to be the only alkene, the starting alkyl bromide must be 1-bromo-4-methylcyclohexane. Either the cis or the trans isomer may be used, although the cis will react more readily, as the more stable conformation (equatorial methyl) has an axial bromine.

*cis*- or *trans*-1-Bromo-4-methylcyclohexane $\xrightarrow[\text{E2}]{\text{base}}$ 4-Methylcyclohexene

1-Bromo-3-methylcyclohexane is incorrect; its dehydrobromination yields a mixture of 3-methylcyclohexene and 4-methylcyclohexene.

1-Bromo-3-methylcyclohexene $\xrightarrow[\text{E2}]{\text{base}}$ 3-Methylcyclohexene + 4-Methylcyclohexene

(*d*) The bromine must be at C-2 in the starting alkyl bromide for a single alkene to be formed on E2 elimination.

2-Bromo-1,1-dimethylcyclopentane $\xrightarrow[\text{E2}]{\text{base}}$ 3,3-Dimethylcyclopentene

If the bromine substituent were at C-3, a mixture of 3,3-dimethyl- and 4,4-dimethylcyclopentene would be formed.

3-Bromo-1,1-dimethylcyclopentane $\xrightarrow[\text{E2}]{\text{base}}$ 3,3-Dimethylcyclopentene + 4,4-Dimethylcyclopentene (*e*) The alkyl bromide must be primary in order for the desired alkene to be the only product of E2 elimination.

$$\text{C}_5\text{H}_9\text{—}CH_2CH_2Br \xrightarrow[\text{E2}]{\text{base}} \text{C}_5\text{H}_9\text{—}CH{=}CH_2$$

2-Cyclopentylethyl bromide → Vinylcyclopentane

If 1-cyclopentylethyl bromide were used, a mixture of regioisomeric alkenes would be formed, with the desired vinylcyclopentane being the minor component of the mixture.

$$\text{C}_5\text{H}_9\text{—}CH(Br)CH_3 \xrightarrow[\text{E2}]{\text{base}} \text{C}_5\text{H}_8{=}CHCH_3 + \text{C}_5\text{H}_9\text{—}CH{=}CH_2$$

1-Cyclopentylethyl bromide → Ethylidenecyclopentane (major product) + Vinylcyclopentane (minor product)

(*f*) Either *cis*- or *trans*-1-bromo-3-isopropylcyclobutane would be appropriate here.

$$\text{1-Br-3-}CH(CH_3)_2\text{-cyclobutane} \xrightarrow[\text{E2}]{\text{base}} \text{3-}CH(CH_3)_2\text{-cyclobutene}$$

*cis*- or *trans*-1-Bromo-3-isopropylcyclobutane → 3-Isopropylcyclobutene

(*g*) The desired alkene is the exclusive product formed on E2 elimination from 1-bromo-1-*tert*-butylcyclopropane.

$$\text{1-Br-1-}C(CH_3)_3\text{-cyclopropane} \xrightarrow[\text{E2}]{\text{base}} \text{cyclopropene—}C(CH_3)_3$$

1-Bromo-1-*tert*-butylcyclopropane → 1-*tert*-Butylcyclopropene

**5.33** (*a*) Both 1-bromopropane and 2-bromopropane yield propene as the exclusive product of E2 elimination.

$$CH_3CH_2CH_2Br \text{ or } CH_3CH(Br)CH_3 \xrightarrow[\text{E2}]{\text{base}} CH_3CH{=}CH_2$$

1-Bromopropane, 2-Bromopropane → Propene

(*b*) Isobutene is formed on dehydrobromination of either *tert*-butyl bromide or isobutyl bromide.

$$(CH_3)_3CBr \text{ or } (CH_3)_2CHCH_2Br \xrightarrow[\text{E2}]{\text{base}} (CH_3)_2C{=}CH_2$$

*tert*-Butyl bromide, Isobutyl bromide → 2-Methylpropene

(*c*) A tetrabromoalkane is required as the starting material to form a tribromoalkene under E2 elimination conditions. Either 1,1,2,2-tetrabromoethane or 1,1,1,2-tetrabromoethane is satisfactory.

$$Br_2CHCHBr_2 \text{ or } BrCH_2CBr_3 \xrightarrow[\text{E2}]{\text{base}} BrCH{=}CBr_2$$

1,1,2,2-Tetrabromoethane, 1,1,1,2-Tetrabromoethane → 1,1,2-Tribromoethene (*d*) The bromine substituent may be at either C-2 or C-3.

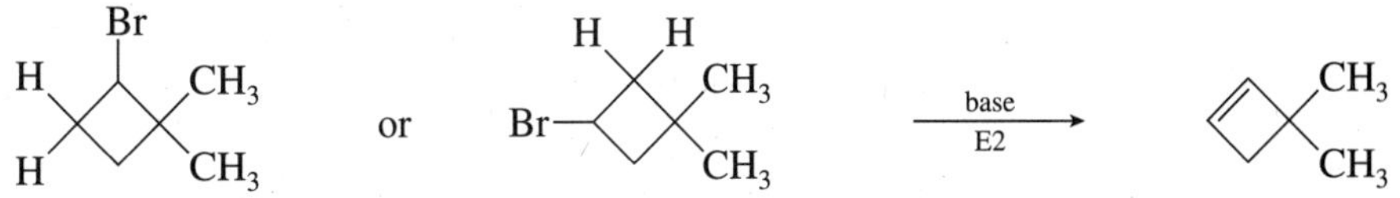

Br; H, H; $CH_3$, $CH_3$ or H, H; Br; $CH_3$, $CH_3$ $\xrightarrow[\text{E2}]{\text{base}}$ $CH_3$, $CH_3$

2-Bromo-1,1-dimethylcyclobutane 3-Bromo-1,1-dimethylcyclobutane 3,3-Dimethylcyclobutene

**5.34** (*a*) The isomeric alkyl bromides having the molecular formula $C_5H_{11}Br$ are:

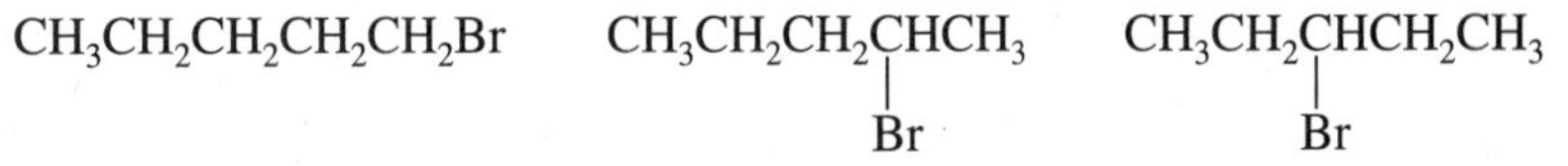

$CH_3CH_2CH_2CH_2CH_2Br$ — 1-Bromopentane

$CH_3CH_2CH_2CHCH_3$ (Br on CH) — 2-Bromopentane

$CH_3CH_2CHCH_2CH_3$ (Br on CH) — 3-Bromopentane

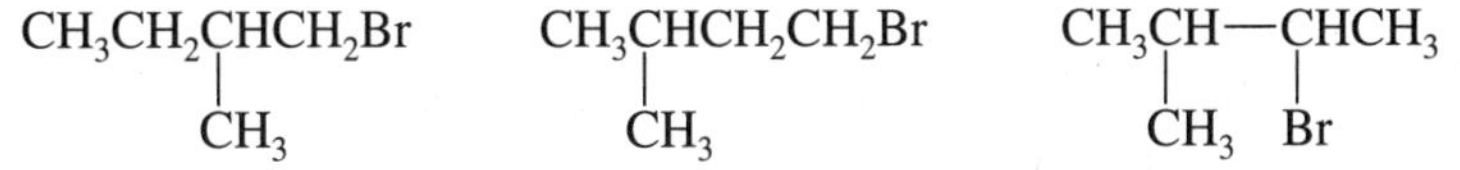

$CH_3CH_2CHCH_2Br$ ($CH_3$ on CH) — 1-Bromo-2-methylbutane

$CH_3CHCH_2CH_2Br$ ($CH_3$ on CH) — 1-Bromo-3-methylbutane

$CH_3CH{-}CHCH_3$ ($CH_3$, Br on CH groups) — 2-Bromo-3-methylbutane

$CH_3CCH_2CH_3$ (Br and $CH_3$ on C) — 2-Bromo-2-methylbutane

$CH_3CCH_2Br$ ($CH_3$ and $CH_3$ on C) — 1-Bromo-2,2-dimethylpropane

(*b*) The order of reactivity toward E1 elimination parallels carbocation stability and is tertiary > secondary > primary. The tertiary bromide 2-bromo-2-methylbutane will undergo E1 elimination at the fastest rate.

(*c*) 1-Bromo-2,2-dimethylpropane has no hydrogens on the $\beta$ carbon and so cannot form an alkene by an E2 process.

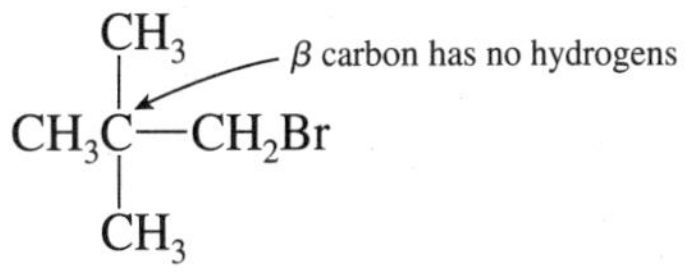

$CH_3C{-}CH_2Br$ ($CH_3$ above and below C) ← $\beta$ carbon has no hydrogens

The only available pathway is E1 with rearrangement.

(*d*) Only the primary bromides will give a single alkene on E2 elimination.

$CH_3CH_2CH_2CH_2CH_2Br \xrightarrow[\text{E2}]{\text{base}} CH_3CH_2CH_2CH{=}CH_2$

1-Bromopentane 1-Pentene

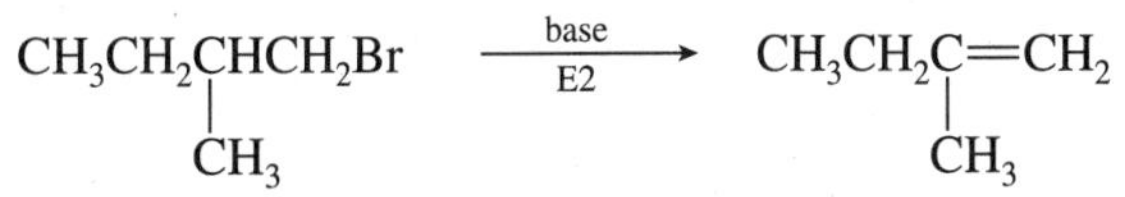

$CH_3CH_2CHCH_2Br$ ($CH_3$ on CH) $\xrightarrow[\text{E2}]{\text{base}}$ $CH_3CH_2C{=}CH_2$ ($CH_3$ on C)

1-Bromo-2-methylbutane 2-Methyl-1-butene

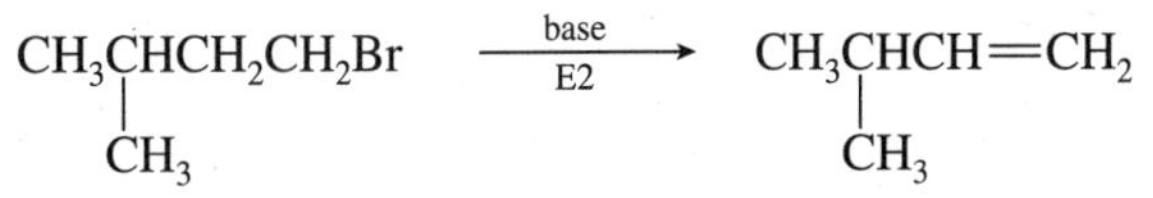

$CH_3CHCH_2CH_2Br$ ($CH_3$ on CH) $\xrightarrow[\text{E2}]{\text{base}}$ $CH_3CHCH{=}CH_2$ ($CH_3$ on CH)

1-Bromo-3-methylbutane 3-Methyl-1-butene

(*e*) Elimination in 3-bromopentane will give the stereoisomers (*E*)- and (*Z*)-2-pentene.

$$CH_3CH_2CHCH_2CH_3\ (\text{Br on C-3}) \xrightarrow[\text{E2}]{\text{base}} (E)\text{-}CH_3CH{=}CHCH_2CH_3 + (Z)\text{-}CH_3CH{=}CHCH_2CH_3$$

3-Bromopentane (*E*)-2-Pentene (*Z*)-2-Pentene

(*f*) Three alkenes can be formed from 2-bromopentane.

$$CH_3CH_2CH_2CHCH_3\ (\text{Br on C-2}) \xrightarrow[\text{E2}]{\text{base}} CH_3CH_2CH_2CH{=}CH_2 + (Z)\text{-}CH_3CH_2CH{=}CHCH_3 + (E)\text{-}CH_3CH_2CH{=}CHCH_3$$

2-Bromopentane 1-Pentene (*Z*)-2-Pentene (*E*)-2-Pentene

**5.35** (*a*) The isomeric $C_5H_{12}O$ alcohols are:

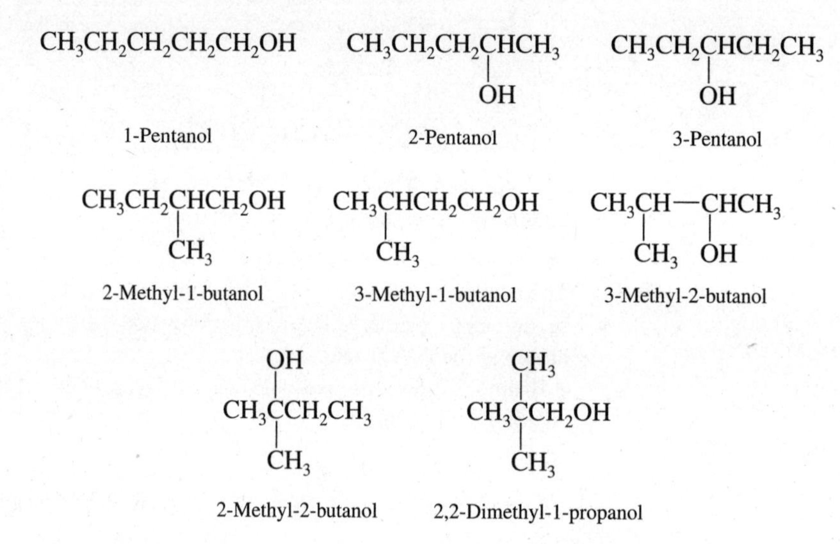


(*b*) The order of reactivity in alcohol dehydration parallels carbocation stability and is tertiary > secondary > primary. The only tertiary alcohol in the group is 2-methyl-2-butanol. It will dehydrate fastest.

(*c*) The most stable $C_5H_{11}$ carbocation is the tertiary carbocation.

$$CH_3\overset{+}{C}(CH_3)CH_2CH_3$$

1,1-Dimethylpropyl cation

(*d*) A proton may be lost from C-1 or C-3:

$$CH_3\overset{+}{C}(CH_3)CH_2CH_3 \longrightarrow H_2C{=}C(CH_3)CH_2CH_3 + CH_3C(CH_3){=}CHCH_3$$

1,1-Dimethylpropyl cation 2-Methyl-1-butene (minor alkene) 2-Methyl-2-butene (major alkene) (*e*) For the 1,1-dimethylpropyl cation to be formed by a process involving a hydride shift, the starting alcohol must have the same carbon skeleton as the 1,1-dimethylpropyl cation.

$$CH_3\overset{+}{C}(CH_3)CH_2CH_3 \quad \text{has same carbon skeleton as} \quad HOCH_2CH(CH_3)CH_2CH_3 \quad \text{and} \quad CH_3CH(CH_3)-CH(OH)CH_3$$

$$HOCH_2CH(CH_3)CH_2CH_3 \xrightarrow{H^+} H\overset{+}{O}(H)-CH_2-CH(CH_3)CH_2CH_3 \longrightarrow CH_3-\overset{+}{C}(CH_3)CH_2CH_3$$

2-Methyl-1-butanol

$$CH_3CH(CH_3)-CH(OH)CH_3 \xrightarrow[-H_2O]{H^+} CH_3-CH(CH_3)-\overset{+}{C}HCH_3 \longrightarrow CH_3-\overset{+}{C}(CH_3)-CH_2CH_3$$

3-Methyl-2-butanol

Although the same carbon skeleton is necessary, it alone is not sufficient; the alcohol must also have its hydroxyl group on the carbon atom adjacent to the carbon that bears the migrating hydrogen. Thus, 3-methyl-1-butanol cannot form a tertiary carbocation by a single hydride shift. It requires two sequential hydride shifts.

$$CH_3CH(CH_3)CH_2CH_2OH \xrightarrow{H^+} CH_3CH(CH_3)CH(H)-CH_2-\overset{+}{O}H_2 \longrightarrow CH_3CH(CH_3)-\overset{+}{C}HCH_3 \longrightarrow CH_3\overset{+}{C}(CH_3)CH_2CH_3$$

3-Methyl-1-butanol

(*f*) 2,2-Dimethyl-1-propanol can yield a tertiary carbocation by a process involving a methyl shift.

$$CH_3C(CH_3)_2-CH_2OH \xrightarrow{H^+} CH_3C(CH_3)_2-CH_2-\overset{+}{O}H_2 \longrightarrow CH_3\overset{+}{C}(CH_3)-CH_2CH_3$$

2,2-Dimethyl-1-propanol

**5.36** (*a*) Heating an alcohol in the presence of an acid catalyst ($KHSO_4$) leads to dehydration with formation of an alkene. In this alcohol, elimination can occur in only one direction to give a mixture of cis and trans alkenes.

$$m\text{-}BrC_6H_4-CH(OH)CH_2CH_3 \xrightarrow[\text{heat}]{KHSO_4} m\text{-}BrC_6H_4-CH{=}CHCH_3$$

Cis–trans mixture

(*b*) Alkyl halides undergo E2 elimination on being heated with potassium *tert*-butoxide.

$$ICH_2CH(OCH_2CH_3)_2 \xrightarrow[\text{(CH}_3)_3\text{COH, heat}]{\text{KOC(CH}_3)_3} H_2C{=}C(OCH_2CH_3)_2$$

(*c*) The exclusive product of this reaction is 1,2-dimethylcyclohexene.

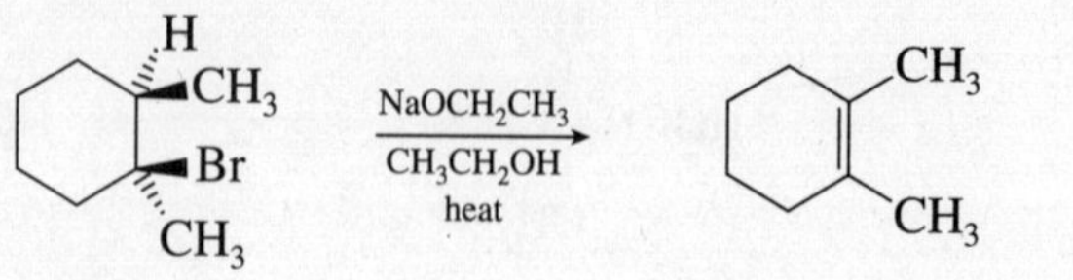

$$\text{1-Bromo-}\textit{trans}\text{-1,2-dimethylcyclohexane} \xrightarrow[\text{CH}_3\text{CH}_2\text{OH, heat}]{\text{NaOCH}_2\text{CH}_3} \text{1,2-Dimethylcyclohexene (100\%)}$$

1-Bromo-*trans*-1,2-dimethylcyclohexane

1,2-Dimethylcyclohexene (100%)

(*d*) Elimination can occur only in one direction, to give the alkene shown.

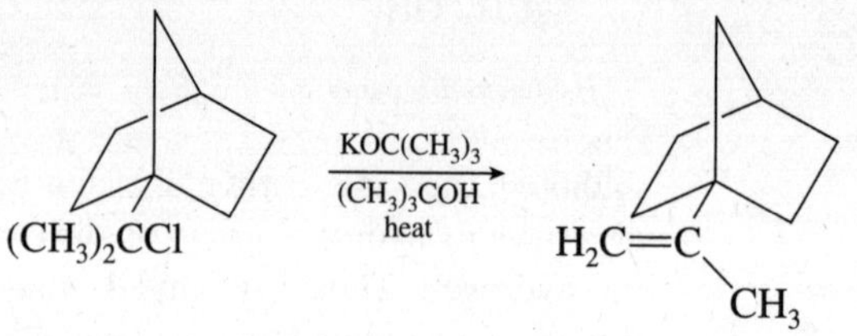

$$(CH_3)_2CCl\text{–(bicyclic)} \xrightarrow[\text{(CH}_3)_3\text{COH, heat}]{\text{KOC(CH}_3)_3} H_2C{=}C(CH_3)\text{–(bicyclic)}$$

(*e*) The reaction is a conventional one of alcohol dehydration and proceeds as written in 76–78% yield.

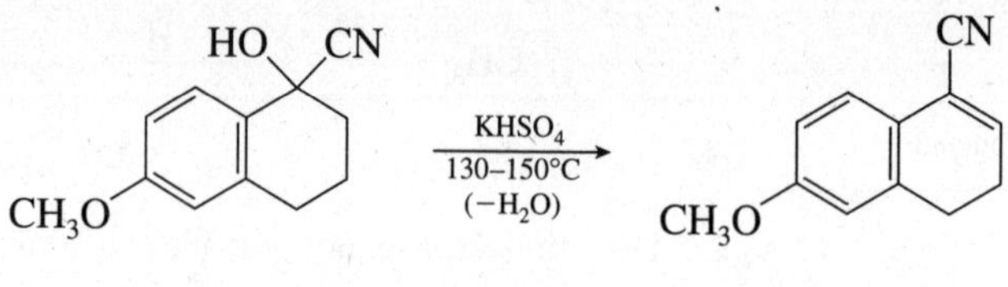

$$\xrightarrow[\text{130–150°C}\ (-H_2O)]{\text{KHSO}_4}$$

(*f*) Dehydration of citric acid occurs, giving aconitic acid.

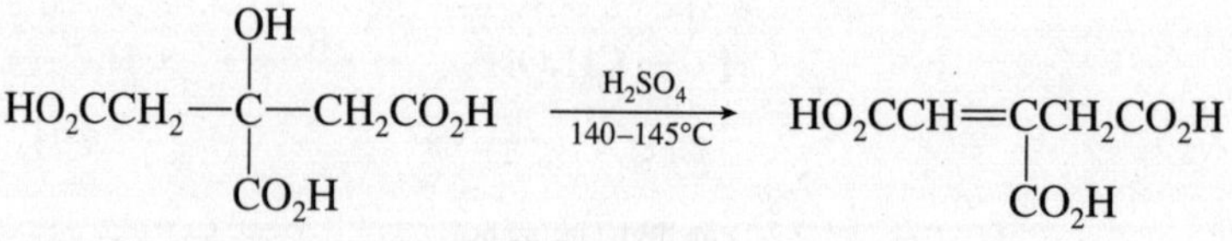

$$HO_2CCH_2{-}\underset{CO_2H}{\overset{OH}{C}}{-}CH_2CO_2H \xrightarrow[\text{140–145°C}]{H_2SO_4} HO_2CCH{=}\underset{CO_2H}{C}CH_2CO_2H$$

Citric acid

Aconitic acid

(*g*) Sequential double dehydrohalogenation gives the diene.

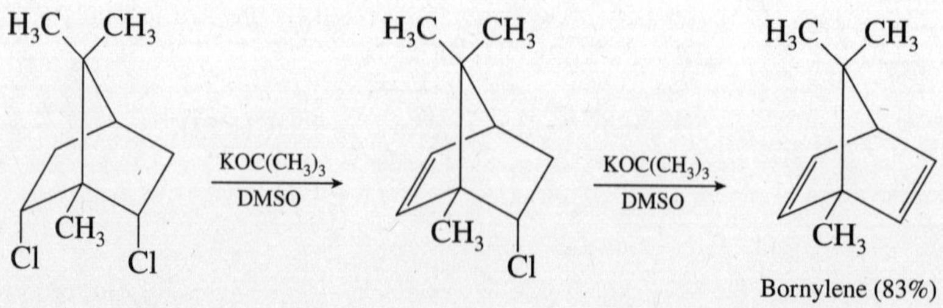

$$\xrightarrow[\text{DMSO}]{\text{KOC(CH}_3)_3} \quad \xrightarrow[\text{DMSO}]{\text{KOC(CH}_3)_3}$$

Bornylene (83%)

(*h*) This example has been reported in the chemical literature, and in spite of the complexity of the starting material, elimination proceeds in the usual way.

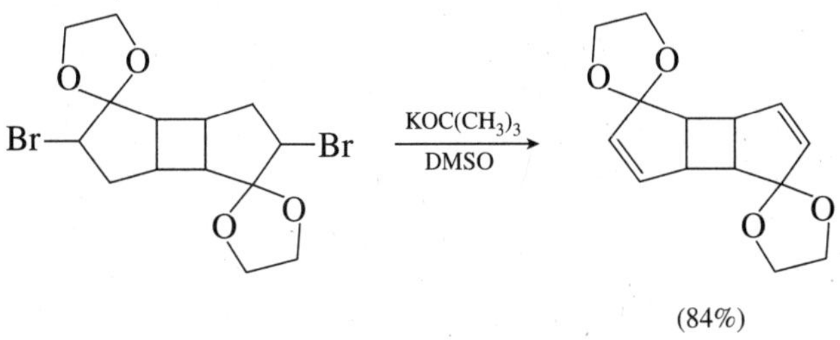

$\xrightarrow[\text{DMSO}]{KOC(CH_3)_3}$

(84%)

(*i*) Again, we have a fairly complicated substrate, but notice that it is well disposed toward E2 elimination of the axial bromine.

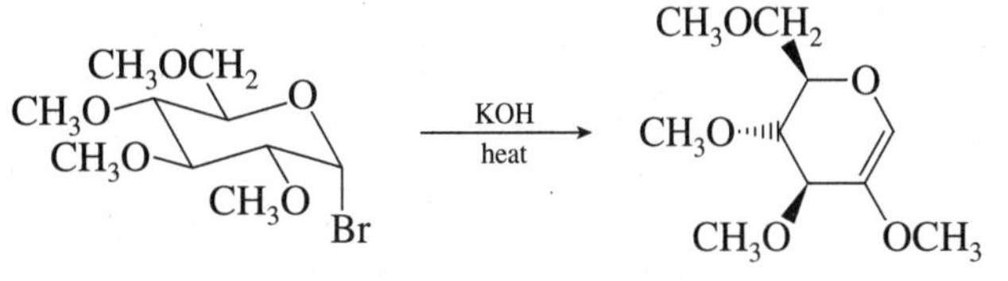

$CH_3OCH_2$, $CH_3O$, $CH_3O$, $CH_3O$, Br $\xrightarrow[\text{heat}]{KOH}$ $CH_3OCH_2$, $CH_3O$, $CH_3O$, $OCH_3$

(*j*) In the most stable conformation of this compound, chlorine occupies an axial site, and so it is ideally situated to undergo an E2 elimination reaction by way of an anti arrangement in the transition state.

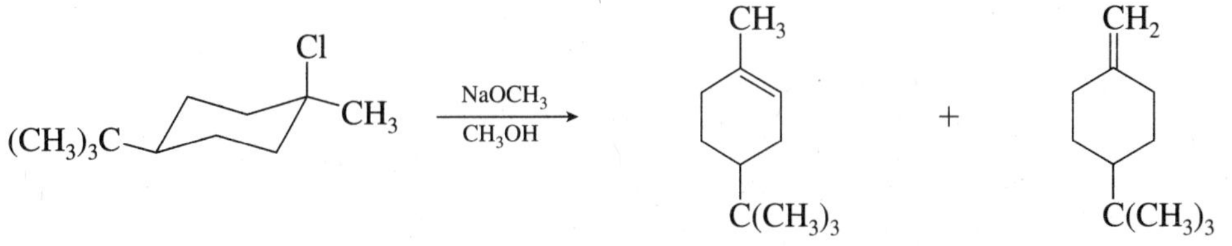

Cl, $CH_3$, $(CH_3)_3C$ $\xrightarrow[CH_3OH]{NaOCH_3}$ $CH_3$, $C(CH_3)_3$ + $CH_2$, $C(CH_3)_3$

4-*tert*-Butyl-1-methylcyclohexene (95%) 4-*tert*-Butyl-(methylene)cyclohexane (5%)

The minor product is the less highly substituted isomer, in which the double bond is exocyclic to the ring.

**5.37** First identify the base as the amide ion ($H_2N^-$) portion of potassium amide ($KNH_2$). Amide ion is a strong base and uses an unshared electron pair to abstract a proton from $\beta$ carbon of the alkyl halide. The pair of electrons in the C—H bond becomes the $\pi$ component of the double bond as the C—Br bond breaks. The electrons in the C—Br bond become an unshared electron pair of bromide ion.

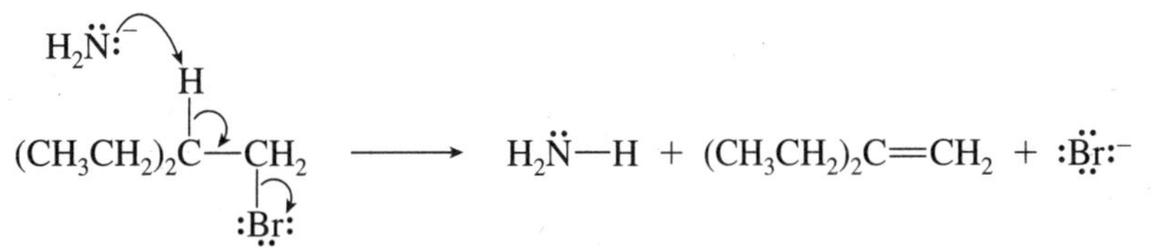

$$H_2\ddot{N}:^- + (CH_3CH_2)_2C(H)—CH_2Br \longrightarrow H_2\ddot{N}—H + (CH_3CH_2)_2C{=}CH_2 + :\ddot{Br}:^-$$

**5.38** The problem states that the reaction is first order in $(CH_3)_3CCl$ (*tert*-butyl chloride) and first order in $NaSCH_2CH_3$ (sodium ethanethiolate). It therefore exhibits the kinetic behavior (overall second order) of a reaction that proceeds by the E2 mechanism. The base that abstracts the proton from carbon is the anion $CH_3CH_2S^-$.

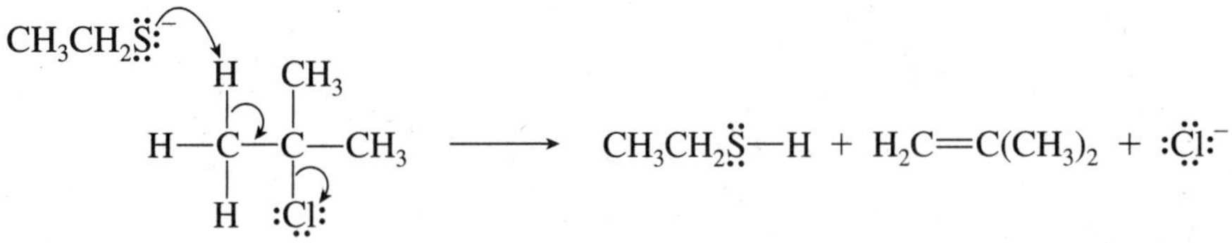

$$CH_3CH_2\ddot{S}:^- + H—CH_2—C(CH_3)_2—Cl \longrightarrow CH_3CH_2\ddot{S}—H + H_2C{=}C(CH_3)_2 + :\ddot{Cl}:^-$$

**5.39** The two starting materials are stereoisomers of each other, and so it is reasonable to begin by examining each one in more stereochemical detail. First, write the most stable conformation of each isomer, keeping in mind that isopropyl is the bulkiest of the three substituents and has the greatest preference for an equatorial orientation.

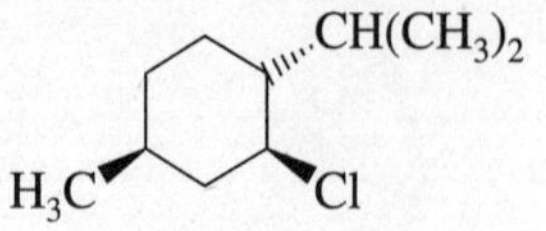

Menthyl chloride

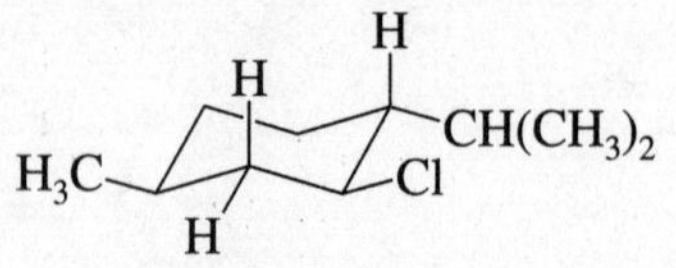

Most stable conformation of menthyl chloride: none of the three $\beta$ protons is anti to chlorine

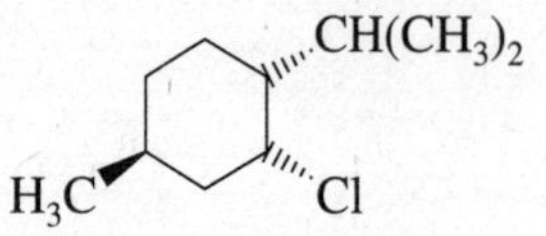

Neomenthyl chloride

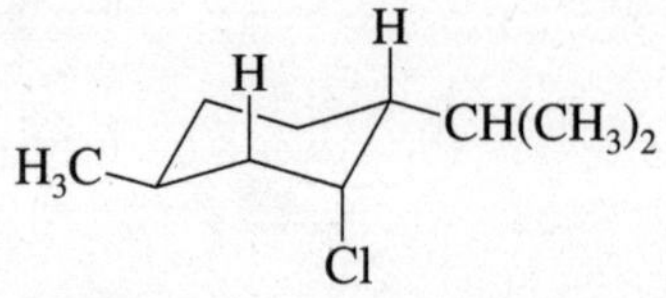

Most stable conformation of neomenthyl chloride: each $\beta$ carbon has a proton that is anti to chlorine

The anti periplanar relationship of halide and proton can be achieved only when the chlorine is axial; this corresponds to the most stable conformation of neomenthyl chloride. Menthyl chloride, on the other hand, must undergo appreciable distortion of its ring to achieve an anti periplanar Cl—C—C—H geometry. Strain increases substantially in going to the transition state for E2 elimination in menthyl chloride but not in neomenthyl chloride. Neomenthyl chloride undergoes E2 elimination at the faster rate.

**5.40** The proton that is removed by the base must be anti to bromine. Thus, the alkyl groups must be gauche to one another in the conformation that leads to *cis*-4-nonene and anti to one another in the one that leads to *trans*-4-nonene.

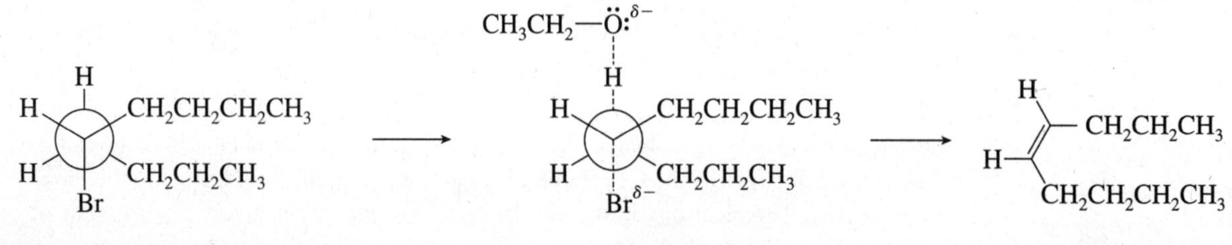

Gauche conformation of 5-bromononane

E2 transition state

*cis*-4-Nonene

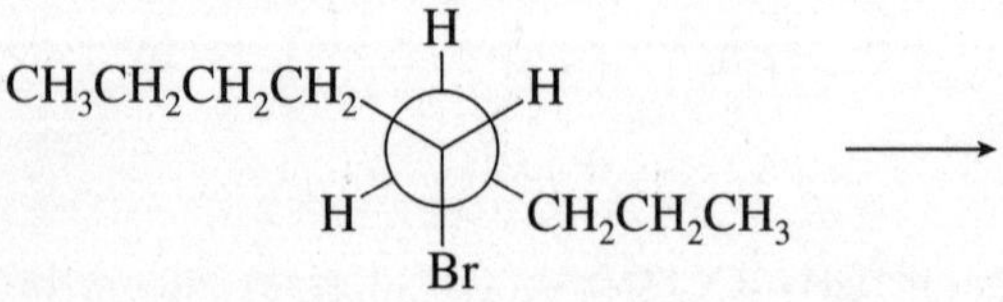

Anti conformation of 5-bromononane

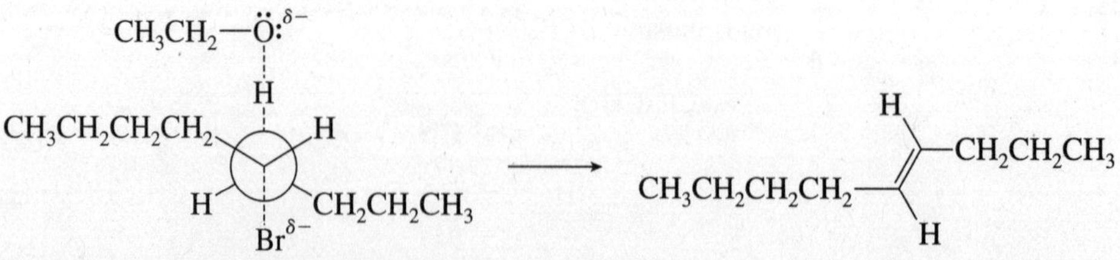

E2 transition state

*trans*-4-Nonene

The alkyl groups move closer together (van der Waals strain increases) as the transition state for formation of *cis*-4-nonene is approached. No comparable increase in strain is involved in going to the transition state for formation of the trans isomer.

**5.41** Begin by writing chemical equations for the processes specified in the problem. First consider rearrangement by way of a hydride shift:

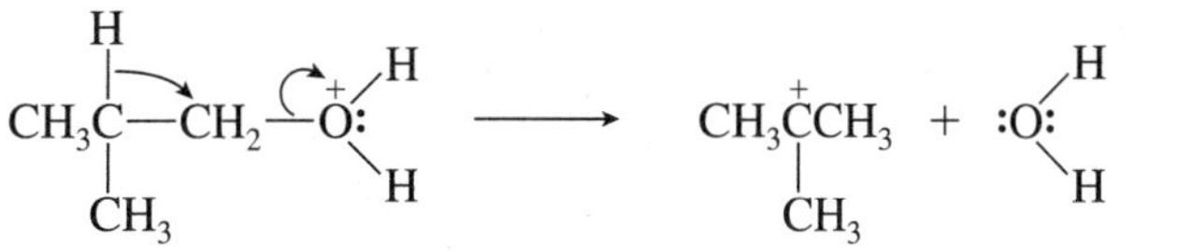

$$CH_3C(H)(CH_3)-CH_2-\overset{+}{O}H_2 \longrightarrow CH_3\overset{+}{C}(CH_3)CH_3 + H_2O$$

Isobutyloxonium ion — Tertiary cation — Water

Rearrangement by way of a methyl group shift is as follows:

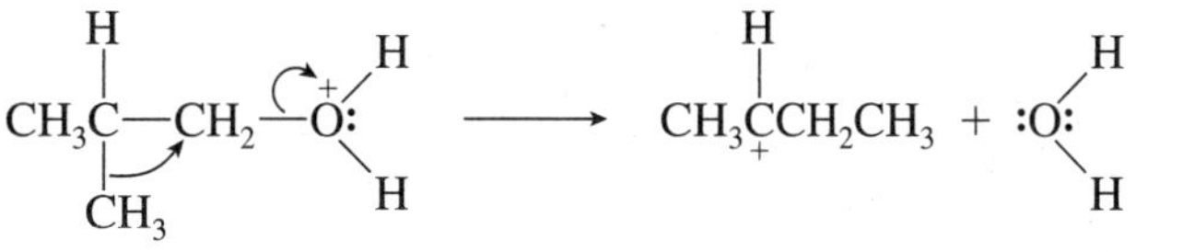

$$CH_3C(H)(CH_3)-CH_2-\overset{+}{O}H_2 \longrightarrow CH_3\overset{+}{C}HCH_2CH_3 + H_2O$$

Isobutyloxonium ion — Secondary cation — Water

A hydride shift gives a tertiary carbocation; a methyl migration gives a secondary carbocation. It is reasonable to expect that rearrangement will occur so as to produce the more stable of these two carbocations because the transition state has carbocation character at the carbon that bears the migrating group. We predict that rearrangement proceeds by a hydride shift rather than a methyl shift, since the group that remains behind in this process stabilizes the carbocation better.

**5.42** Rearrangement proceeds by migration of a hydrogen or an alkyl group from the carbon atom adjacent to the positively charged carbon.

(*a*) A propyl cation is primary and rearranges to an isopropyl cation, which is secondary, by migration of a hydrogen with its pair of electrons.

$$CH_3CH(H)-\overset{+}{C}H_2 \longrightarrow CH_3\overset{+}{C}HCH_3$$

Propyl cation (primary, less stable) — Isopropyl cation (secondary, more stable)

(*b*) A hydride shift transforms the secondary carbocation to a tertiary one.

$$CH_3C(H)(CH_3)-\overset{+}{C}HCH_3 \longrightarrow CH_3\overset{+}{C}(CH_3)CH_2CH_3$$

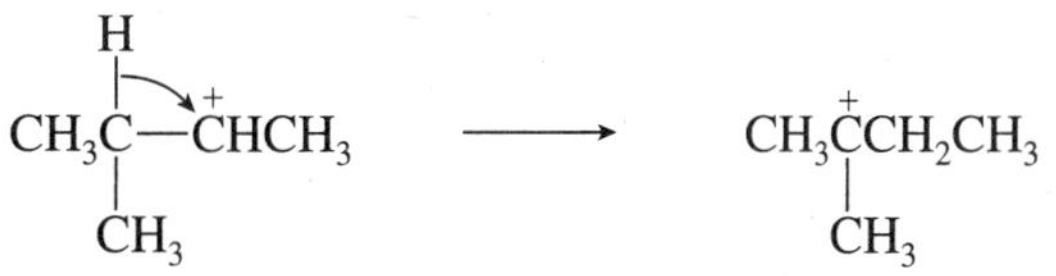

1,2-Dimethylpropyl cation (secondary, less stable) — 1,1-Dimethylpropyl cation (tertiary, more stable)

This hydride shift occurs in preference to methyl migration, which would produce the same secondary carbocation. (Verify this by writing appropriate structural formulas.)

(*c*) Migration of a methyl group converts this secondary carbocation to a tertiary one.

$$CH_3C(CH_3)(CH_3)-\overset{+}{C}HCH_3 \longrightarrow CH_3\overset{+}{C}(CH_3)-CH(CH_3)CH_3$$

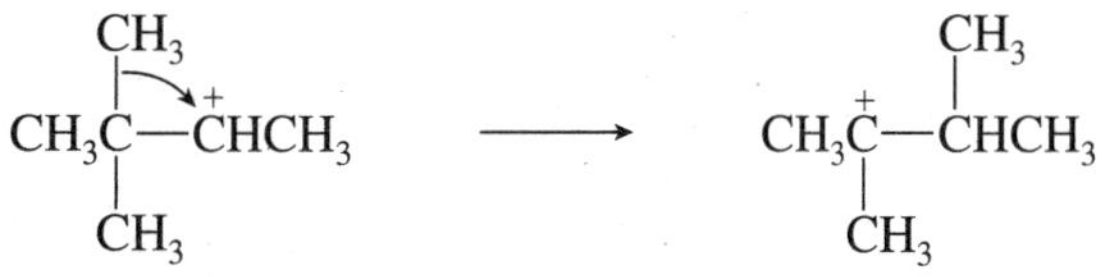

1,2,2-Trimethylpropyl cation (secondary, less stable) — 1,1,2-Trimethylpropyl cation (tertiary, more stable)

(*d*) The group that shifts in this case is the entire ethyl group.

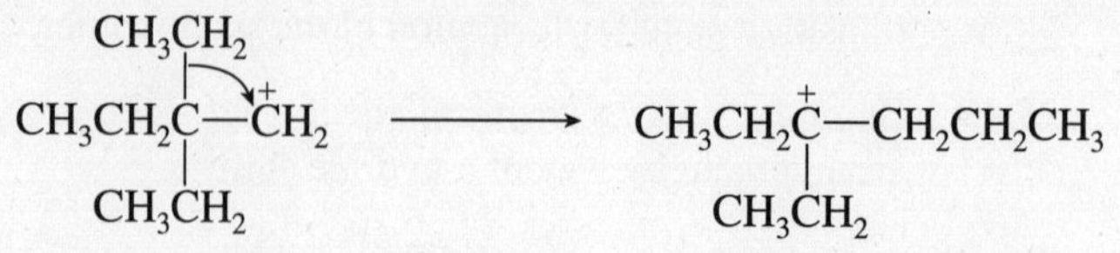


2,2-Diethylbutyl cation (primary, less stable) | 1,1-Diethylbutyl cation (tertiary, more stable)

(*e*) Migration of a hydride from the ring carbon that bears the methyl group produces a tertiary carbocation.

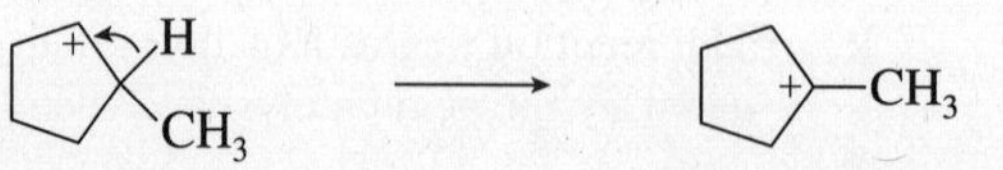


2-Methylcyclopentyl cation (secondary, less stable) | 1-Methylcyclopentyl cation (tertiary, more stable)

**5.43** (*a*) Note that the starting material is an alcohol and that it is treated with an acid. The product is an alkene but its carbon skeleton is different from that of the starting alcohol. The reaction is one of alcohol dehydration accompanied by rearrangement at the carbocation stage. Begin by writing the step in which the alcohol is converted to a carbocation.

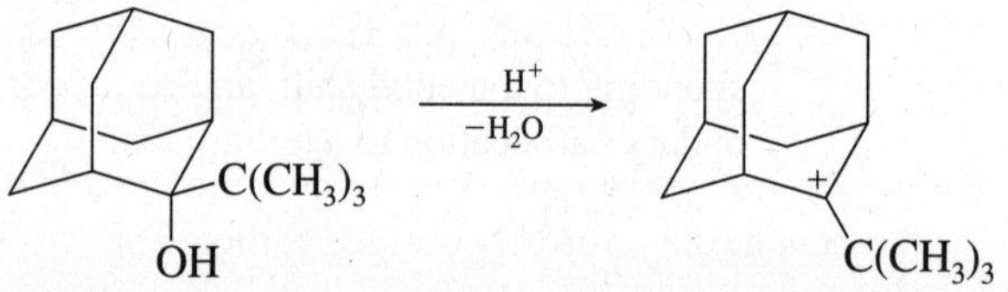


The carbocation is tertiary and relatively stable. Migration of a methyl group from the *tert*-butyl substituent, however, converts it to an isomeric carbocation, which is also tertiary.

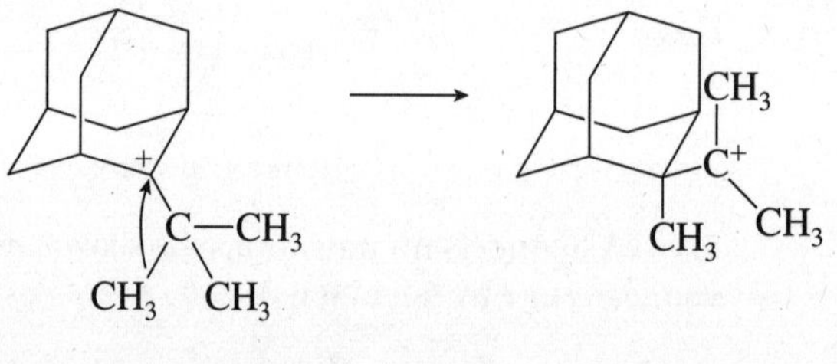


Loss of a proton from this carbocation gives the observed product.

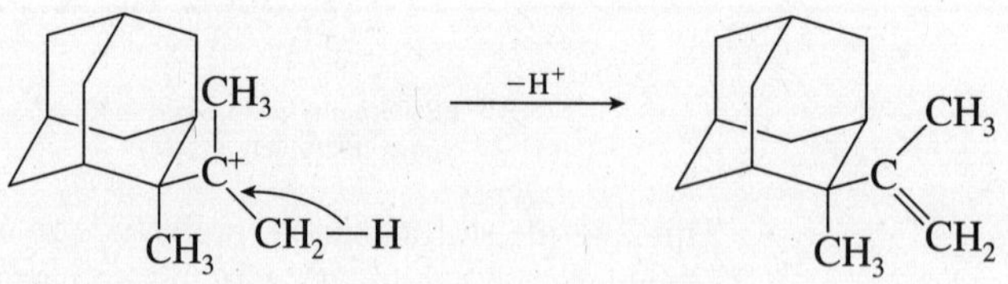


(*b*) Here also we have an alcohol dehydration reaction accompanied by rearrangement. The initially formed carbocation is secondary.

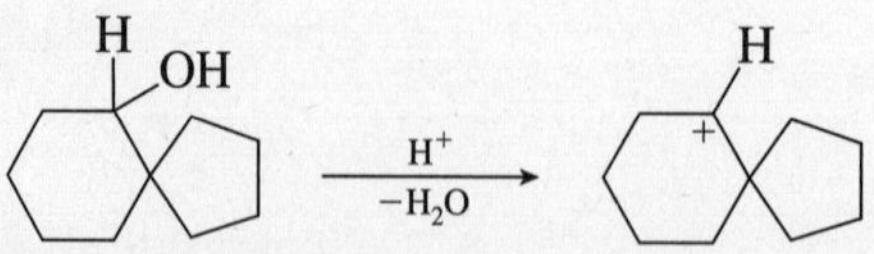

This cation can rearrange to a tertiary carbocation by an alkyl group shift.

H ⟶ H

Loss of a proton from the tertiary carbocation gives the observed alkene.

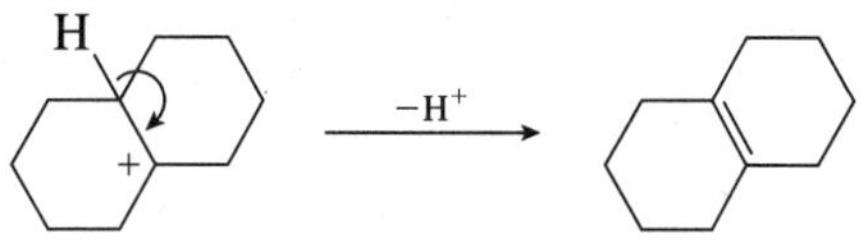

H $\xrightarrow{-H^+}$

(*c*) The reaction begins as a normal alcohol dehydration in which the hydroxyl group is protonated by the acid catalyst and then loses water from the oxonium ion to give a carbocation.

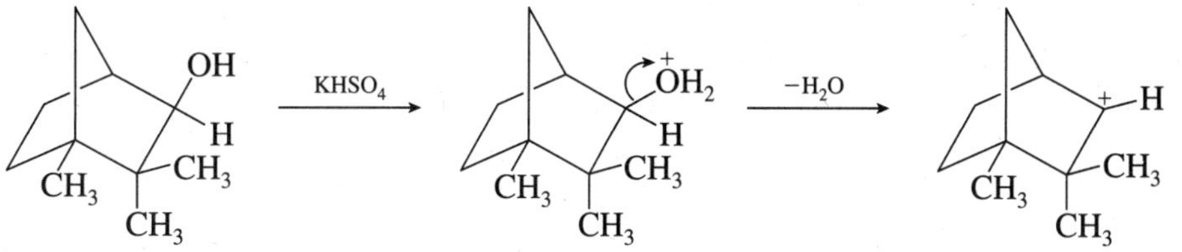

OH, H, $CH_3$, $CH_3$, $CH_3$ $\xrightarrow{KHSO_4}$ $\overset{+}{O}H_2$, H, $CH_3$, $CH_3$, $CH_3$ $\xrightarrow{-H_2O}$ H, $CH_3$, $CH_3$, $CH_3$

4-Methylcamphenilol

Secondary carbocation

We see that the final product, 1-methylsantene, has a rearranged carbon skeleton corresponding to a methyl shift, and so we consider the rearrangement of the initially formed secondary carbocation to a tertiary ion.

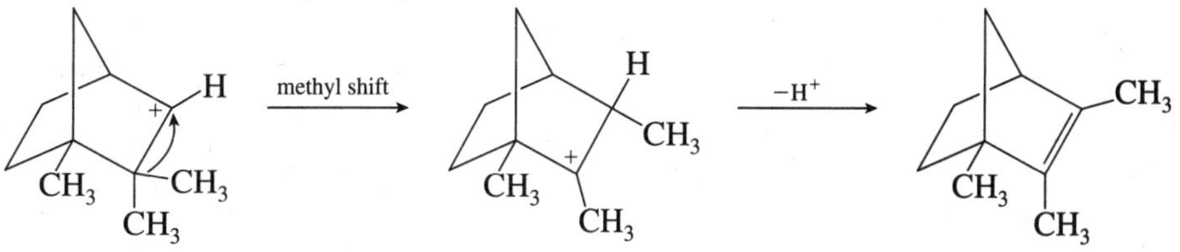

H, $CH_3$, $CH_3$, $CH_3$ $\xrightarrow{\text{methyl shift}}$ H, $CH_3$, $CH_3$, $CH_3$ $\xrightarrow{-H^+}$ $CH_3$, $CH_3$, $CH_3$

Tertiary carbocation

1-Methylsantene

Deprotonation of the tertiary carbocation yields 1-methylsantene.

**5.44** The secondary carbocation can, as we have seen, rearrange by a methyl shift (Problem 5.16). It can also rearrange by migration of one of the ring bonds.

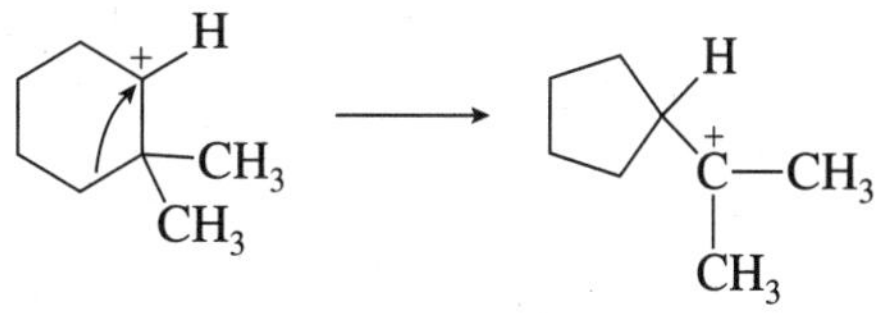

H, $CH_3$, $CH_3$ ⟶ H, $\overset{+}{C}-CH_3$, $CH_3$

Secondary carbocation

Tertiary carbocation

The tertiary carbocation formed by this rearrangement can lose a proton to give the observed byproduct.

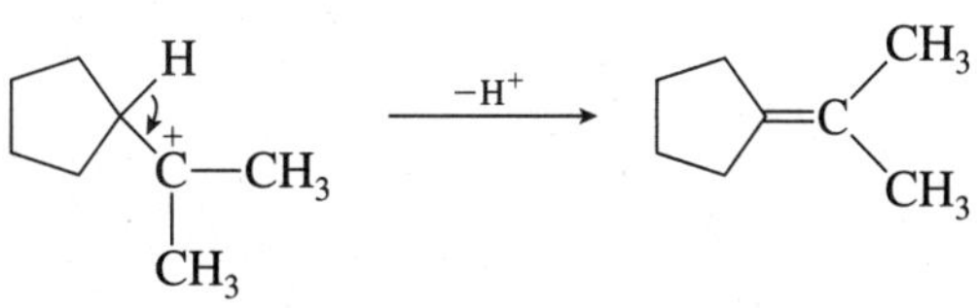

H, $\overset{+}{C}-CH_3$, $CH_3$ $\xrightarrow{-H^+}$ $=C(CH_3)_2$

Isopropylidenecyclopentane

**5.45** Let's do both part (*a*) and part (*b*) together by reasoning mechanistically. The first step in any acid-catalyzed alcohol dehydration is proton transfer to the OH group.

$$CH_3CH_2CH_2CH_2C(CH_3)_2CH_2-\ddot{O}H \xrightarrow{H_2SO_4} CH_3CH_2CH_2CH_2C(CH_3)_2CH_2-\overset{+}{O}H_2$$

But notice that because this alcohol does not have any hydrogens on its $\beta$ carbon, it cannot dehydrate directly. Any alkenes that are formed must arise by rearrangement processes. Consider, for example, migration of either of the two equivalent methyl groups at C-2.

$$CH_3CH_2CH_2CH_2C(CH_3)_2-CH_2-\overset{+}{O}H_2 \longrightarrow CH_3CH_2CH_2CH_2\overset{+}{C}(CH_3)CH_2CH_3 + :OH_2$$

The resulting carbocation can lose a proton in three different directions.

$$CH_3CH_2CH_2CH_2\overset{+}{C}(CH_3)CH_2CH_3 \xrightarrow{-H^+}$$

$$CH_3CH_2CH_2CH{=}C(CH_3)CH_2CH_3 + CH_3CH_2CH_2CH_2C(=CH_2)CH_2CH_3 + CH_3CH_2CH_2CH_2C(CH_3){=}CHCH_3$$

3-Methyl-3-heptene (mixture of *E* and *Z*) | 2-Ethyl-1-hexene | 3-Methyl-2-heptene (mixture of *E* and *Z*)

The alkene mixture shown in the preceding equation constitutes part of the answer to part (*b*). None of the alkenes arising from methyl migration is 2-methyl-2-heptene, the answer to part (*a*), however.

What other group can migrate? The other group attached to the $\beta$ carbon is a butyl group. Consider its migration.

$$CH_3CH_2CH_2CH_2-C(CH_3)_2-CH_2-\overset{+}{O}H_2 \longrightarrow (CH_3)_2\overset{+}{C}CH_2CH_2CH_2CH_2CH_3 + :OH_2$$

Loss of a proton from the carbocation gives the alkene in part (*a*).

$$(CH_3)_2\overset{+}{C}CH_2CH_2CH_2CH_2CH_3 \xrightarrow{-H^+} (CH_3)_2C{=}CHCH_2CH_2CH_2CH_3$$

2-Methyl-2-heptene

A proton can also be lost from one of the methyl groups to give 2-methyl-1-heptene. This is the last alkene constituting the answer to part (*b*).

$$(CH_3)_2\overset{+}{C}CH_2CH_2CH_2CH_2CH_3 \xrightarrow{-H^+} H_2C{=}C(CH_3)CH_2CH_2CH_2CH_2CH_3$$

2-Methyl-1-heptene

**5.46** Only two alkanes have the molecular formula $C_4H_{10}$: butane and isobutane (2-methylpropane)—both of which give two monochlorides on free-radical chlorination. However, dehydrochlorination of one of the monochlorides derived from butane yields a mixture of alkenes.

$$\underset{\text{2-Chlorobutane}}{CH_3\underset{\underset{Cl}{|}}{C}HCH_2CH_3} \xrightarrow[\text{dimethyl sulfoxide}]{KOC(CH_3)_3} \underset{\text{1-Butene}}{H_2C{=}CHCH_2CH_3} + \underset{\text{2-Butene } (cis + trans)}{CH_3CH{=}CHCH_3}$$

Both monochlorides derived from 2-methylpropane yield only 2-methylpropene under conditions of E2 elimination.

$$\underset{tert\text{-Butyl chloride}}{(CH_3)_3CCl} \quad \text{or} \quad \underset{\text{Isobutyl chloride}}{(CH_3)_2CHCH_2Cl} \xrightarrow[\text{dimethyl sulfoxide}]{KOC(CH_3)_3} \underset{\text{2-Methylpropene}}{(CH_3)_2C{=}CH_2}$$

Compound A is therefore 2-methylpropane, the two alkyl chlorides are *tert*-butyl chloride and isobutyl chloride, and alkene B is 2-methylpropene.

**5.47** The key to this problem is the fact that one of the alkyl chlorides of molecular formula $C_6H_{13}Cl$ does not undergo E2 elimination. It must therefore have a structure in which the carbon atom that is $\beta$ to the chlorine bears no hydrogens. This $C_6H_{13}Cl$ isomer is 1-chloro-2,2-dimethylbutane.

$$CH_3CH_2\overset{\overset{CH_3}{|}}{\underset{\underset{CH_3}{|}}{C}}CH_2Cl$$

1-Chloro-2,2-dimethylbutane
(cannot form an alkene)

Identifying this monochloride derivative gives us the carbon skeleton. The starting alkane (compound A) must be 2,2-dimethylbutane. Its free-radical halogenation gives three different monochlorides:

$$\underset{\substack{\text{2,2-Dimethylbutane}\\\text{(compound A)}}}{CH_3CH_2\overset{\overset{CH_3}{|}}{\underset{\underset{CH_3}{|}}{C}}CH_3} \xrightarrow[\text{light}]{Cl_2} \underset{\substack{\text{1-Chloro-2,2-}\\\text{dimethylbutane}}}{CH_3CH_2\overset{\overset{CH_3}{|}}{\underset{\underset{CH_3}{|}}{C}}CH_2Cl} + \underset{\substack{\text{3-Chloro-2,2-}\\\text{dimethylbutane}}}{CH_3\underset{\underset{Cl}{|}}{C}H\overset{\overset{CH_3}{|}}{\underset{\underset{CH_3}{|}}{C}}CH_3} + \underset{\substack{\text{1-Chloro-3,3-}\\\text{dimethylbutane}}}{ClCH_2CH_2\overset{\overset{CH_3}{|}}{\underset{\underset{CH_3}{|}}{C}}CH_3}$$

Both 3-chloro-2,2-dimethylbutane and 1-chloro-3,3-dimethylbutane give only 3,3-dimethyl-1-butene on E2 elimination.

$$\underset{\substack{\text{3-Chloro-2,2-}\\\text{dimethylbutane}}}{CH_3\underset{\underset{Cl}{|}}{C}H\overset{\overset{CH_3}{|}}{\underset{\underset{CH_3}{|}}{C}}CH_3} \quad \text{or} \quad \underset{\substack{\text{1-Chloro-3,3-}\\\text{dimethylbutane}}}{ClCH_2CH_2\overset{\overset{CH_3}{|}}{\underset{\underset{CH_3}{|}}{C}}CH_3} \xrightarrow[(CH_3)_3COH]{KOC(CH_3)_3} \underset{\substack{\text{3,3-Dimethyl-1-butene}\\\text{(alkene B)}}}{CH_2{=}CH\overset{\overset{CH_3}{|}}{\underset{\underset{CH_3}{|}}{C}}CH_3}$$

## SELF-TEST

### PART A

**A-1.** Write the correct IUPAC name for each of the following:

(*a*) $(CH_3)_3CCH{=}C(CH_3)_2$

(*b*) [skeletal structure]

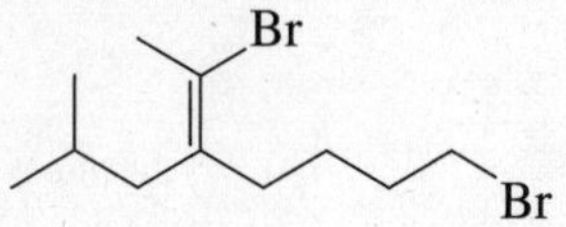

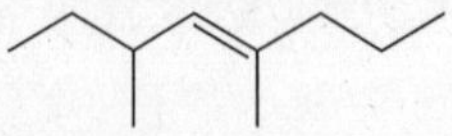

(*c*) [skeletal structure with Br, Br]

(*d*) [skeletal structure with OH]

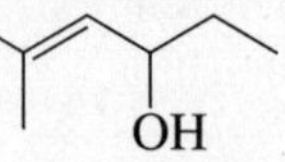

**A-2.** Each of the following is an incorrect name for an alkene. Write the structure and give the correct name for each.

(*a*) 2-Ethyl-3-methyl-2-butene
(*b*) 2-Chloro-5-methyl-5-hexene
(*c*) 2,3-Dimethylcyclohexene
(*d*) 2-Methyl-1-penten-4-ol

**A-3.** (*a*) Write the structures of all the alkenes of molecular formula $C_5H_{10}$.
(*b*) Which isomer is the most stable?
(*c*) Which isomers are the least stable?
(*d*) Which isomers can exist as a pair of stereoisomers?

**A-4.** How many carbon atoms are $sp^2$-hybridized in 2-methyl-2-pentene? How many are $sp^3$-hybridized? How many $\sigma$ bonds are of the $sp^2$–$sp^3$ type?

**A-5.** Write the structure, clearly indicating the stereochemistry, of each of the following:
(*a*) (*Z*)-4-Ethyl-3-methyl-3-heptene
(*b*) (*E*)-1,2-Dichloro-3-methyl-2-hexene
(*c*) (*E*)-3-Methyl-3-penten-1-ol

**A-6.** Write structural formulas for two alkenes of molecular formula $C_7H_{14}$ that are stereoisomers of each other and have a trisubstituted double bond. Give systematic names for each.

**A-7.** Write structural formulas for the reactant or product(s) omitted from each of the following. If more than one product is formed, indicate the major one.

(*a*) $(CH_3)_2C(OH)CH_2CH_2CH_3 \xrightarrow[\text{heat}]{H_2SO_4}$ ?

(*b*) 1-Chloro-1-methylcyclohexane (Cl, $CH_3$) $\xrightarrow[CH_3OH]{NaOCH_3}$ ?

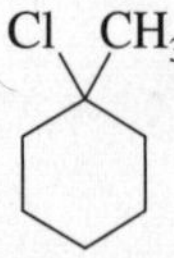

(*c*) ? $\xrightarrow[CH_3CH_2OH]{KOCH_2CH_3}$ [alkene skeletal structure] (only alkene formed)

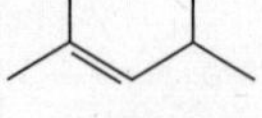

(*d*) $(CH_3)_2C(OH)C(CH_3)_3 \xrightarrow{H_3PO_4}$ ?

**A-8.** Write the structure of the $C_6H_{13}Br$ isomer that is *not* capable of undergoing E2 elimination.

**A-9.** Write a stepwise mechanism for the formation of 2-methyl-2-butene from the dehydration of 2-methyl-2-butanol is sulfuric acid.

**A-10.** Draw the structures of all the alkenes, including stereoisomers, that can be formed from the E2 elimination of 3-bromo-2,3-dimethylpentane with sodium ethoxide ($NaOCH_2CH_3$) in ethanol. Which of these would you expect to be the major product?

**A-11.** Using curved arrows and perspective drawings (of chair cyclohexanes), explain the formation of the indicated product from the following reaction:

$\xrightarrow{NaOCH_3}$

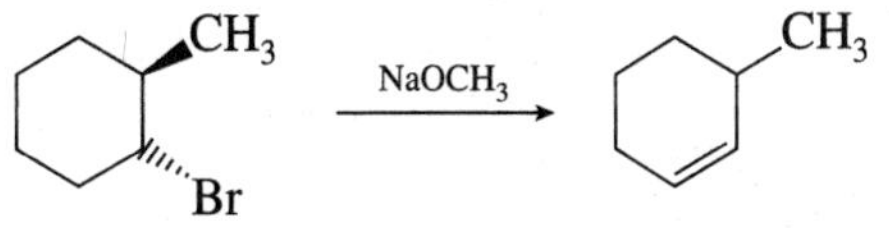

**A-12.** Compare the relative rate of reaction of *cis*- and *trans*-1-chloro-3-isopropylcyclohexane with sodium methoxide in methanol by the E2 mechanism.

**A-13.** Outline a mechanism for the following reaction:

$\xrightarrow[heat]{H_3PO_4}$ + other alkenes

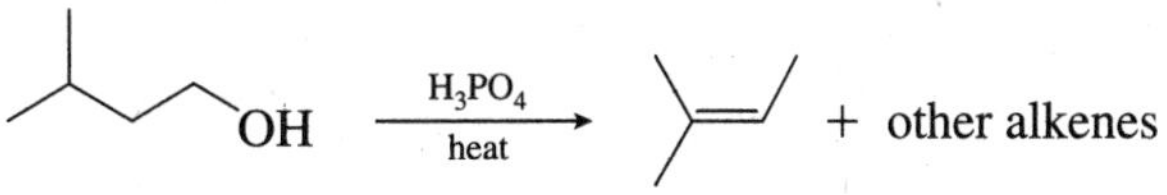

**A-14.** Compound A, on reaction with bromine in the presence of light, gave as the major product compound B ($C_9H_{19}Br$). Reaction of B with sodium ethoxide in ethanol gave 3-ethyl-4,4-dimethyl-2-pentene as the only alkene. Identify compounds A and B.

## PART B

**B-1.** Which one of the alkenes shown below has the *Z* configuration of its double bond?

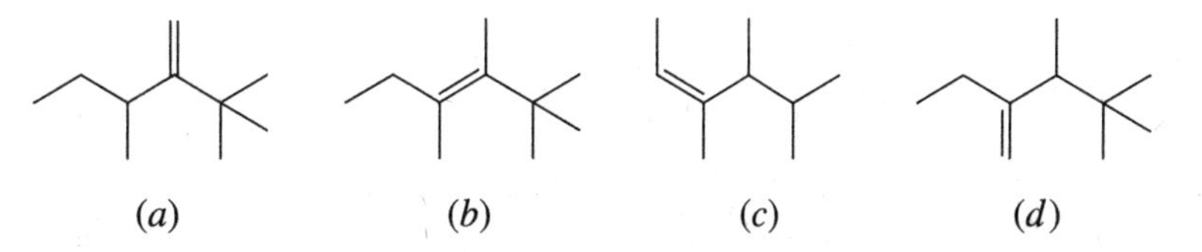

(*a*) (*b*) (*c*) (*d*)

**B-2.** Carbon–carbon double bonds do not undergo rotation as do carbon–carbon single bonds. The reason is that
(*a*) The double bond is much stronger and thus more difficult to rotate
(*b*) Overlap of the $sp^2$ orbitals of the carbon–carbon $\sigma$ bond would be lost
(*c*) Overlap of the *p* orbitals of the carbon–carbon $\pi$ bond would be lost
(*d*) The shorter bond length of the double bond makes it more difficult for the attached groups to pass one another
(*e*) The statement is incorrect—rotation around double bonds does occur.

**B-3.** Rank the following substituent groups in order of decreasing priority according to the Cahn–Ingold–Prelog system:

| $—CH(CH_3)_2$ | $—CH_2Br$ | $—CH_2CH_2Br$ |
|---|---|---|
| 1 | 2 | 3 |

(*a*) 2 > 3 > 1 (*b*) 1 > 3 > 2 (*c*) 3 > 1 > 2 (*d*) 2 > 1 > 3

**B-4.** The heats of combustion for the four $C_6H_{12}$ isomers shown are (not necessarily in order): 955.3, 953.6, 950.6, and 949.7 (all in kilocalories per mole). Which of these values is most likely the heat of combustion of isomer 1?

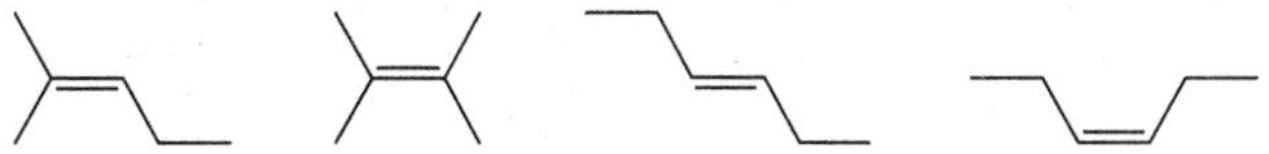

1 2 3 4

(*a*) 955.3 kcal/mol (*c*) 950.6 kcal/mol
(*b*) 953.6 kcal/mol (*d*) 949.7 kcal/mol

**B-5.** Referring to the structures in the previous question, what can be said about isomers 3 and 4?
(*a*) 3 is more stable by 1.7 kcal/mol.
(*b*) 4 is more stable by 1.7 kcal/mol.
(*c*) 3 is more stable by 3.0 kcal/mol.
(*d*) 3 is more stable by 0.9 kcal/mol.

**B-6.** The structure of (*E*)-1-chloro-3-methyl-3-hexene is

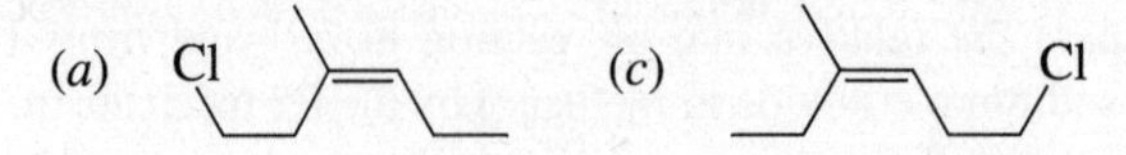


(*b*) Cl (*d*) None of these

**B-7.** In the dehydrohalogenation of 2-bromobutane, which conformation leads to the formation of *cis*-2-butene?

(*a*) CH$_3$, H, CH$_3$, H, Br, H
(*b*) CH$_3$, H, CH$_3$, Br, H, H
(*c*) CH$_3$, H, H, Br, H, CH$_3$
(*d*) CH$_3$, H$_3$C, H, H, Br, H

(e) None of these is the correct conformation.

**B-8.** Which of the following alcohols would be *most* likely to undergo dehydration with rearrangement by a process involving a methyl migration (methyl shift)?

(*a*) OH (*c*) OH (*e*) OH

(*b*) OH (*d*) OH

**B-9.** Rank the following alcohols in order of decreasing reactivity (fastest → slowest) toward dehydration with 85% $H_3PO_4$:

$(CH_3)_2CHCH_2CH_2OH$ (1) $\quad (CH_3)_2C(OH)CH_2CH_3$ (2) $\quad (CH_3)_2CHCH(OH)CH_3$ (3)

(*a*) 2 > 3 > 1 (*b*) 1 > 3 > 2 (*c*) 2 > 1 > 3 (*d*) 1 > 2 > 3

**B-10.** Consider the following reaction:

OH $\xrightarrow[\text{heat}]{H_3PO_4}$ + 

Which response contains all the correct statements about this process and no incorrect ones?

1. Dehydration
2. E2 mechanism
3. Carbon skeleton migration
4. Most stable carbocation forms
5. Single-step reaction

(*a*) 1, 3 (*b*) 1, 2, 3 (*c*) 1, 2, 5 (*d*) 1, 3, 4

**B-11.** Select the formula or statement representing the major product(s) of the following reaction:

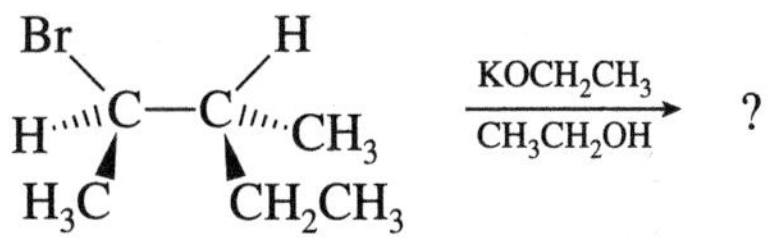

Br, H, $H_3C$ on C — C with H, $CH_3$, $CH_2CH_3$ $\xrightarrow[CH_3CH_2OH]{KOCH_2CH_3}$ ?

(*a*) H, $CH_3$ / $CH_3$, $CH_2CH_3$ (alkene)

(*b*) H, $CH_2CH_3$ / $CH_3$, $CH_3$ (alkene)

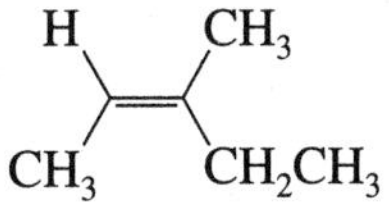

(*c*) $CH_2{=}CHCH(CH_3)CH_2CH_3$

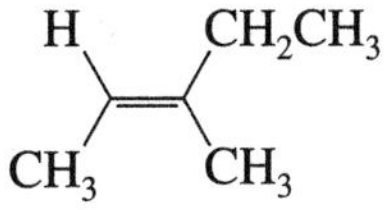

(*d*) Both (*a*) and (*b*) form in approximately equal amounts.

**B-12.** Which one of the following statements concerning E2 reactions of alkyl halides is true?
(*a*) The rate of an E2 reaction depends only on the concentration of the alkyl halide.
(*b*) The rate of an E2 reaction depends only on the concentration of the base.
(*c*) The C—H bond and the C—X (X = halogen) bond are broken in the same step.
(*d*) Alkyl chlorides generally react faster than alkyl bromides.

**B-13.** Which alkyl halide undergoes E2 elimination at the fastest rate?

(*a*) 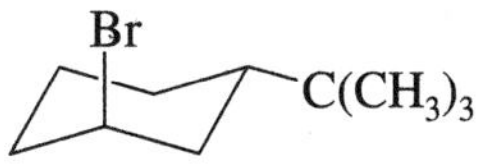 Br (axial) cyclohexane with $C(CH_3)_3$

(*b*) Cl (axial) cyclohexane with $C(CH_3)_3$

(*c*) Br (equatorial) cyclohexane with $C(CH_3)_3$

(*d*) Cl (equatorial) cyclohexane with $C(CH_3)_3$

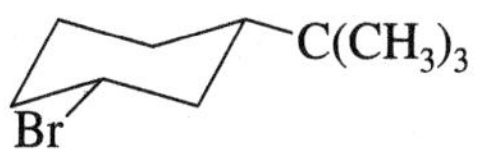

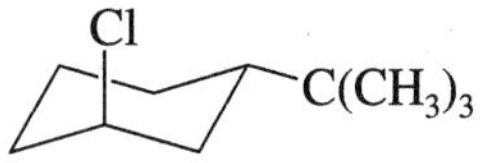

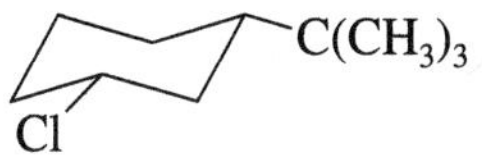

**B-14.** What is the relationship between the pair of compounds shown?

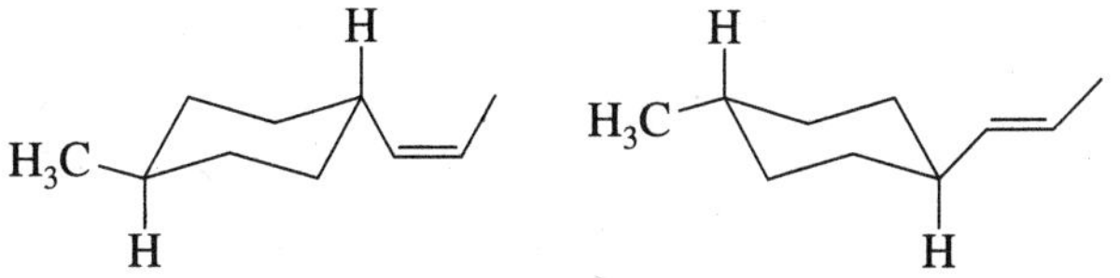

H, $H_3C$, H (cyclohexane with alkenyl group) and H, $H_3C$, H (cyclohexane with alkenyl group)

(*a*) Identical: superimposable without bond rotations
(*b*) Conformations
(*c*) Stereoisomers
(*d*) Constitutional isomers

**B-15.** Which one of the following will give 2-methyl-1-butene as the *only* alkene on treatment with $KOC(CH_3)_3$ in dimethyl sulfoxide?
(*a*) 1-Bromo-2-methylbutane
(*b*) 2-Methyl-1-butanol
(*c*) 2-Bromo-2-methylbutane
(*d*) 2-Methyl-2-butanol

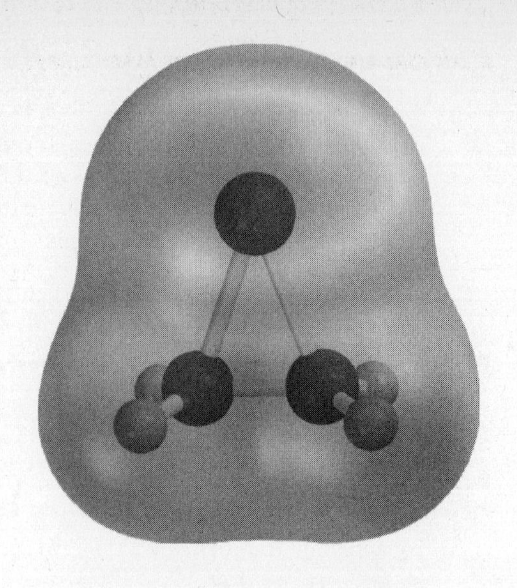

# CHAPTER 6

# REACTIONS OF ALKENES: ADDITION REACTIONS

## SOLUTIONS TO TEXT PROBLEMS

**6.1** Catalytic hydrogenation converts an alkene to an alkane having the same carbon skeleton. Since 2-methylbutane is the product of hydrogenation, all three alkenes must have a four-carbon chain with a one-carbon branch. The three alkenes are therefore:

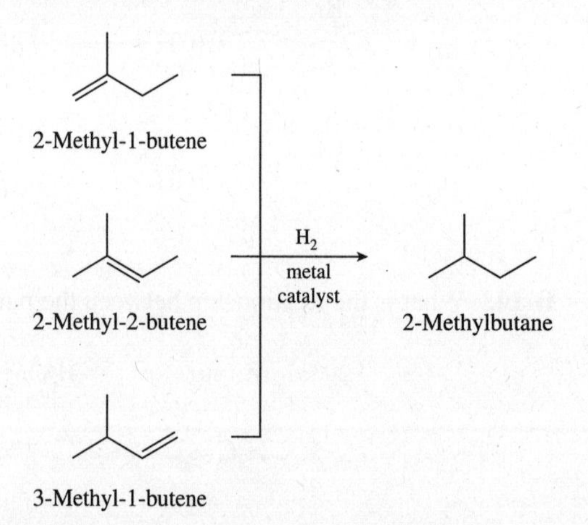


**6.2** The most highly substituted double bond is the most stable and has the smallest heat of hydrogenation.

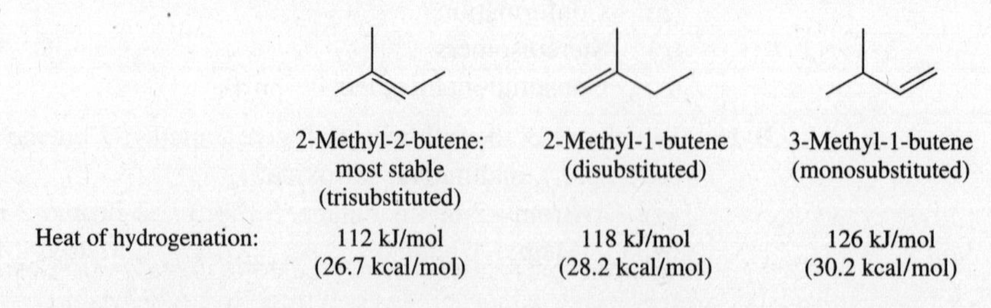

| | 2-Methyl-2-butene: most stable (trisubstituted) | 2-Methyl-1-butene (disubstituted) | 3-Methyl-1-butene (monosubstituted) |
|---|---|---|---|
| Heat of hydrogenation: | 112 kJ/mol (26.7 kcal/mol) | 118 kJ/mol (28.2 kcal/mol) | 126 kJ/mol (30.2 kcal/mol) |

**6.3** (*b*) Begin by writing out the structure of the starting alkene. Identify the doubly bonded carbon that has the greater number of attached hydrogens; this is the one to which the proton of hydrogen chloride adds. Chlorine adds to the carbon atom of the double bond that has the fewer attached hydrogens.

Chlorine adds to this carbon. Hydrogen adds to this carbon.

$$(CH_3CH_2)(H_3C)C{=}CH_2 \xrightarrow{HCl} CH_3CH_2C(CH_3)(Cl)CH_3$$

2-Methyl-1-butene → 2-Chloro-2-methylbutane

By applying Markovnikov's rule, we see that the major product is 2-chloro-2-methylbutane.

(*c*) Regioselectivity of addition is not an issue here, because the two carbons of the double bond are equivalent in *cis*-2-butene. Hydrogen chloride adds to *cis*-2-butene to give 2-chlorobutane.

$$H_3CCH{=}CHCH_3 + HCl \longrightarrow CH_3CH_2CH(Cl)CH_3$$

*cis*-2-Butene + Hydrogen chloride → 2-Chlorobutane

(*d*) One end of the double bond has no attached hydrogens, but the other end has one. In accordance with Markovnikov's rule, the proton of hydrogen chloride adds to the carbon that already has one hydrogen. The product is 1-chloro-1-ethylcyclohexane.

$$CH_3CH{=}C_6H_{10} + HCl \longrightarrow CH_3CH_2(Cl)C_6H_{10}$$

Ethylidenecyclohexane + Hydrogen chloride → 1-Chloro-1-ethylcyclohexane

**6.4** (*b*) A proton is transferred to the terminal carbon atom of 2-methyl-1-butene so as to produce a tertiary carbocation.

$$(CH_3CH_2)(H_3C)C{=}CH_2 + H{-}Cl \longrightarrow CH_3CH_2\overset{+}{C}(CH_3)CH_3 + Cl^-$$

2-Methyl-1-butene + Hydrogen chloride → Tertiary carbocation + Chloride

This is the carbocation that leads to the observed product, 2-chloro-2-methylbutane.

(*c*) A secondary carbocation is an intermediate in the reaction of *cis*-2-butene with hydrogen chloride.

$$H_3CCH{=}CHCH_3 + H{-}Cl \longrightarrow H_3C\overset{+}{C}H{-}CH_2CH_3 + Cl^-$$

*cis*-2-Butene + Hydrogen chloride → Secondary carbocation + Chloride

Capture of this carbocation by chloride gives 2-chlorobutane.

(*d*) A tertiary carbocation is formed by protonation of the double bond.

$$CH_3CH{=}C_6H_{10} + H{-}Cl \longrightarrow CH_3CH_2{-}\overset{+}{C}_6H_{10} + Cl^-$$

Tertiary cation — Chloride

This carbocation is captured by chloride to give the observed product, 1-chloro-1-ethylcyclohexane.

**6.5** The carbocation formed by protonation of the double bond of 3,3-dimethyl-1-butene is secondary. Methyl migration can occur to give a more stable tertiary carbocation.

$$(CH_3)_3CCH{=}CH_2 \xrightarrow{HCl} (CH_3)_3C{-}\overset{+}{C}HCH_3 \xrightarrow{\text{methyl migration}} (CH_3)_2\overset{+}{C}{-}CH(CH_3)CH_3$$

3,3-Dimethyl-1-butene — Secondary carbocation — Tertiary carbocation

Secondary carbocation $\xrightarrow{Cl^-}$ $(CH_3)_3C{-}CHClCH_3$ (3-Chloro-2,2-dimethylbutane)

Tertiary carbocation $\xrightarrow{Cl^-}$ $(CH_3)_2CCl{-}CH(CH_3)CH_3$ (2-Chloro-2,3-dimethylbutane)

The two chlorides are 3-chloro-2,2-dimethylbutane and 2-chloro-2,3-dimethylbutane.

**6.6** The structure of allyl bromide (3-bromo-1-propene) is $CH_2{=}CHCH_2Br$. Its reaction with hydrogen bromide in accordance with Markovnikov's rule proceeds by addition of a proton to the doubly bonded carbon that has the greater number of attached hydrogens.

**Addition according to Markovnikov's rule:**

$$CH_2{=}CHCH_2Br + HBr \longrightarrow CH_3CHBrCH_2Br$$

Allyl bromide — Hydrogen bromide — 1,2-Dibromopropane

Addition of hydrogen bromide opposite to Markovnikov's rule leads to 1,3-dibromopropane.

**Addition contrary to Markovnikov's rule:**

$$CH_2{=}CHCH_2Br + HBr \longrightarrow BrCH_2CH_2CH_2Br$$

Allyl bromide — Hydrogen bromide — 1,3-Dibromopropane

**6.7** (*b*) Hydrogen bromide adds to 2-methyl-1-butene in accordance with Markovnikov's rule when peroxides are absent. The product is 2-bromo-2-methylbutane.

$$(CH_3)(CH_3CH_2)C{=}CH_2 + HBr \longrightarrow CH_3CH_2C(CH_3)(Br)CH_3$$

2-Methyl-1-butene — Hydrogen bromide — 2-Bromo-2-methylbutane

The opposite regioselectivity is observed when peroxides are present. The product is 1-bromo-2-methylbutane.

$$(CH_3)(CH_3CH_2)C{=}CH_2 + HBr \xrightarrow{\text{peroxides}} CH_3CH_2C(CH_3)(H)CH_2Br$$

2-Methyl-1-butene — Hydrogen bromide — 1-Bromo-2-methylbutane

(*c*) Both ends of the double bond in *cis*-2-butene are equivalently substituted, so that the same product (2-bromobutane) is formed by hydrogen bromide addition regardless of whether the reaction is carried out in the presence of peroxides or in their absence.

$$CH_3CH{=}CHCH_3 + HBr \longrightarrow CH_3CH_2CH(Br)CH_3$$

*cis*-2-Butene — Hydrogen bromide — 2-Bromobutane

(*d*) A tertiary bromide is formed on addition of hydrogen bromide to ethylidenecyclohexane in the absence of peroxides.

$$CH_3CH{=}C_6H_{10} + HBr \longrightarrow CH_3CH_2(Br)C_6H_{10}$$

Ethylidenecyclohexane — Hydrogen bromide — 1-Bromo-1-ethylcyclohexane

The regioselectivity of addition is reversed in the presence of peroxides, and the product is (1-bromoethyl)cyclohexane.

$$CH_3CH{=}C_6H_{10} + HBr \xrightarrow{\text{peroxides}} CH_3CH(Br){-}C_6H_{11}$$

Ethylidenecyclohexane — Hydrogen bromide — (1-Bromoethyl)cyclohexane

**6.8** The first step is the addition of sulfuric acid to give cyclohexyl hydrogen sulfate.

$$C_6H_{10} \xrightarrow{H_2SO_4} C_6H_{11}OSO_2OH$$

Cyclohexene — Cyclohexyl hydrogen sulfate

**6.9** The presence of hydroxide ion in the second step is incompatible with the medium in which the reaction is carried out. The reaction as shown in step 1

$$1.\quad (CH_3)_2C{=}CH_2 + H_3O^+ \longrightarrow (CH_3)_3C^+ + H_2O$$

is performed in acidic solution. There are, for all practical purposes, no hydroxide ions in aqueous acid, the strongest base present being water itself. It is quite important to pay attention to the species that are actually present in the reaction medium whenever you formulate a reaction mechanism.

**6.10** The more stable the carbocation, the faster it is formed. The more reactive alkene gives a tertiary carbocation in the rate-determining step.

$$\triangleright\!-\!\underset{}{\overset{CH_3}{\overset{|}{C}}}{=}CH_2 \xrightarrow{H_3O^+} \triangleright\!-\!C^+(CH_3)_2$$

Tertiary carbocation

Protonation of $\triangleright\!-\!CH{=}CHCH_3$ gives a secondary carbocation.

**6.11** The mechanism of electrophilic addition of hydrogen chloride to 2-methylpropene as outlined in text Section 6.6 proceeds through a carbocation intermediate. This mechanism is the reverse of the E1 elimination. The E2 mechanism is concerted—it does not involve an intermediate.

**6.12** (*b*) The carbon–carbon double bond is symmetrically substituted in *cis*-2-butene, and so the regioselectivity of hydroboration–oxidation is not an issue. Hydration of the double bond gives 2-butanol.

$$\text{(H}_3\text{C)(H)C{=}C(H)(CH}_3\text{)} \xrightarrow[\text{2. oxidation}]{\text{1. hydroboration}} CH_3\underset{\displaystyle OH}{\underset{|}{C}}HCH_2CH_3$$

*cis*-2-Butene 2-Butanol

(*c*) Hydroboration–oxidation of alkenes is a method that leads to hydration of the double bond with a regioselectivity opposite to Markovnikov's rule.

$$\square{=}CH_2 \xrightarrow[\text{2. oxidation}]{\text{1. hydroboration}} \square\!\!<^{H}_{CH_2OH}$$

Methylenecyclobutane Cyclobutylmethanol

(*d*) Hydroboration–oxidation of cyclopentene gives cyclopentanol.

$$\text{Cyclopentene} \xrightarrow[\text{2. oxidation}]{\text{1. hydroboration}} \text{Cyclopentyl–OH}$$

Cyclopentene Cyclopentanol

(*e*) When alkenes are converted to alcohols by hydroboration–oxidation, the hydroxyl group is introduced at the less substituted carbon of the double bond.

$$CH_3CH{=}C(CH_2CH_3)_2 \xrightarrow[\text{2. oxidation}]{\text{1. hydroboration}} CH_3\underset{\displaystyle OH}{\underset{|}{C}}HCH(CH_2CH_3)_2$$

3-Ethyl-2-pentene 3-Ethyl-2-pentanol

(*f*) The less substituted carbon of the double bond in 3-ethyl-1-pentene is at the end of the chain. It is this carbon that bears the hydroxyl group in the product of hydroboration–oxidation.

$$H_2C{=}CHCH(CH_2CH_3)_2 \xrightarrow[\text{2. oxidation}]{\text{1. hydroboration}} HOCH_2CH_2CH(CH_2CH_3)_2$$

3-Ethyl-1-pentene → 3-Ethyl-1-pentanol

**6.13** The bottom face of the double bond of $\alpha$-pinene is less hindered than the top face.

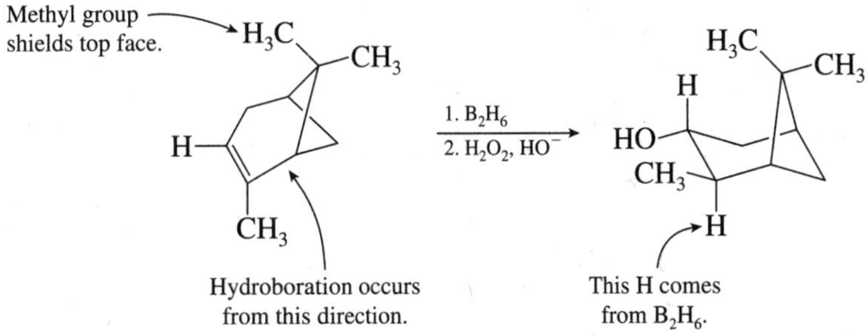

Methyl group shields top face. $H_3C$, $CH_3$, H, $CH_3$ $\xrightarrow[\text{2. } H_2O_2,\ HO^-]{\text{1. } B_2H_6}$ $H_3C$, $CH_3$, H, HO, $CH_3$, H

Hydroboration occurs from this direction.

This H comes from $B_2H_6$.

Syn addition of H and OH takes place and with a regioselectivity opposite to that of Markovnikov's rule.

**6.14** Bromine adds anti to the double bond of 1-bromocyclohexene to give 1,1,2-tribromocyclohexane. The radioactive bromines ($^{82}Br$) are vicinal and trans to each other.

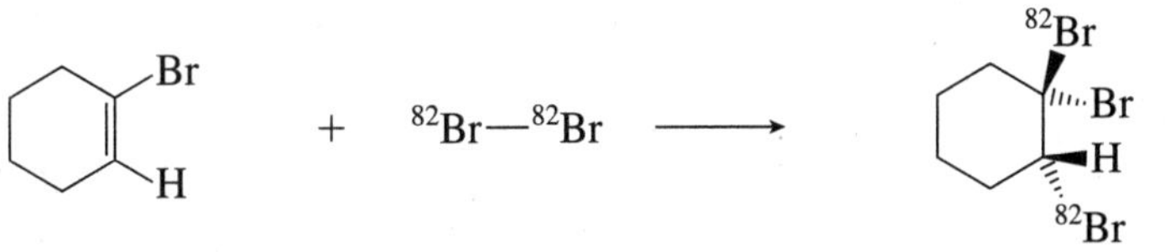

1-Bromocyclohexene + $^{82}Br{-}^{82}Br$ (Bromine) → 1,1,2-Tribromocyclohexane

**6.15** Alkyl substituents on the double bond increase the reactivity of the alkene toward addition of bromine.

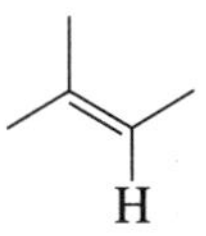

2-Methyl-2-butene (trisubstituted double bond; most reactive)

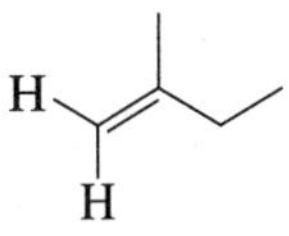

2-Methyl-1-butene (disubstituted double bond)

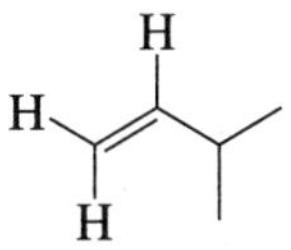

3-Methyl-1-butene (monosubstituted double bond; least reactive)

**6.16** (*b*) Bromine becomes bonded to the less highly substituted carbon of the double bond, the hydroxyl group to the more highly substituted one.

$$(CH_3)_2C{=}CHCH_3 \xrightarrow[H_2O]{Br_2} (CH_3)_2C(OH)CH(Br)CH_3$$

2-Methyl-2-butene → 3-Bromo-2-methyl-2-butanol

(*c*)

$$(CH_3)_2CHCH{=}CH_2 \xrightarrow[H_2O]{Br_2} (CH_3)_2CHCH(OH)CH_2Br$$

3-Methyl-1-butene → 1-Bromo-3-methyl-2-butanol

(*d*) Anti addition occurs.

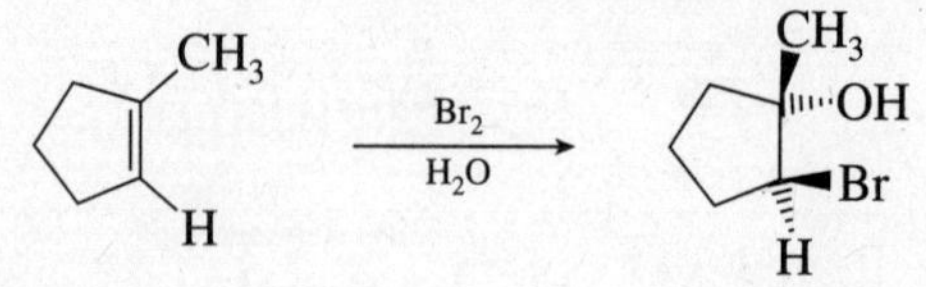

$CH_3$ / H ; $Br_2$, $H_2O$ → $CH_3$, OH, Br, H

1-Methylcyclopentene

*trans*-2-Bromo-1-methylcyclopentanol

**6.17** The structure of disparlure is as shown.

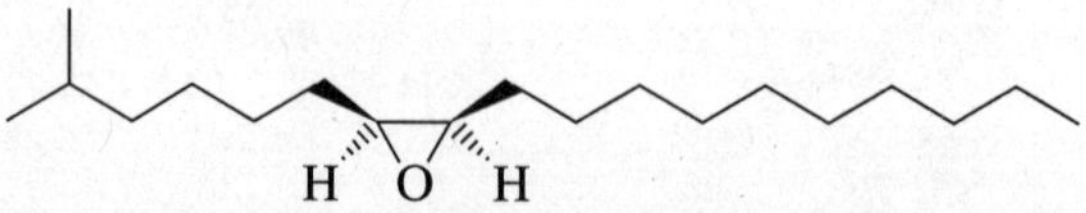

H O H

Its longest continuous chain contains 18 carbon atoms, and so it is named as an epoxy derivative of octadecane. Number the chain in the direction that gives the lowest number to the carbons that bear oxygen. Thus, disparlure is *cis*-2-methyl-7,8-epoxyoctadecane.

**6.18** Disparlure can be prepared by epoxidation of the corresponding alkene. Cis alkenes yield cis epoxides upon epoxidation. *cis*-2-Methyl-7-octadecene is therefore the alkene chosen to prepare disparlure by epoxidation.

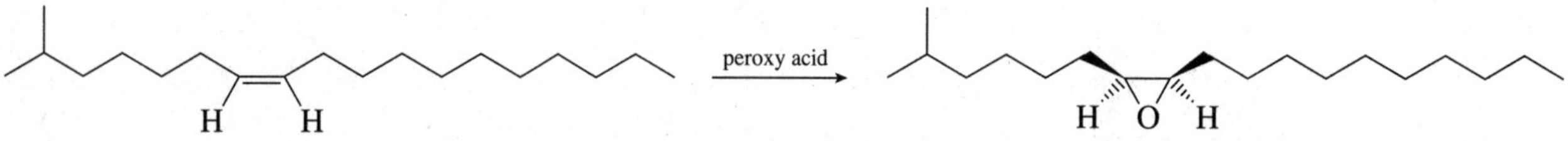

H H —peroxy acid→ H O H

*cis*-2-Methyl-7-octadecene

Disparlure

**6.19** The products of ozonolysis are formaldehyde and 4,4-dimethyl-2-pentanone.

$(H)_2C{=}O$ $\quad$ $O{=}C(CH_3)CH_2C(CH_3)_3$

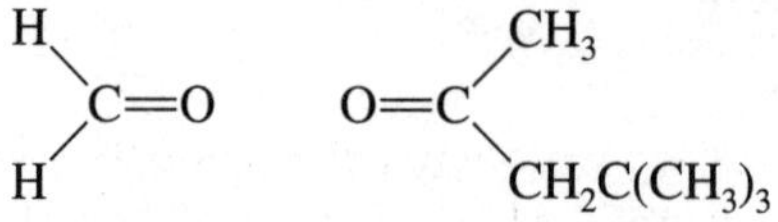

Formaldehyde $\quad$ 4,4-Dimethyl-2-pentanone

The two carbons that were doubly bonded to each other in the alkene become the carbons that are doubly bonded to oxygen in the products of ozonolysis. Therefore, mentally remove the oxygens and connect these two carbons by a double bond to reveal the structure of the starting alkene.

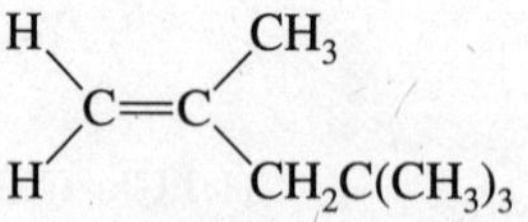

$(H)_2C{=}C(CH_3)CH_2C(CH_3)_3$

2,4,4-Trimethyl-1-pentene

**6.20** From the structural formula of the desired product, we see that it is a **vicinal bromohydrin.** Vicinal bromohydrins are made from alkenes by reaction with bromine in water.

$BrCH_2C(CH_3)_2$ (with OH on the C) is made from $CH_2{=}C(CH_3)_2$

OH

Since the starting material given is *tert*-butyl bromide, a practical synthesis is:

$$(CH_3)_3CBr \xrightarrow[\text{CH}_3\text{CH}_2\text{OH, heat}]{\text{NaOCH}_2\text{CH}_3} (CH_3)_2C{=}CH_2 \xrightarrow[\text{H}_2\text{O}]{\text{Br}_2} (CH_3)_2\underset{\displaystyle OH}{\underset{|}{C}}CH_2Br$$

*tert*-Butyl bromide → 2-Methylpropene → 1-Bromo-2-methyl-2-propanol

**6.21** Catalytic hydrogenation of the double bond converts 2,4,4-trimethyl-1-pentene and 2,4,4-trimethyl-2-pentene to 2,2,4-trimethylpentane.

$$(H)_2C{=}C(CH_3)CH_2C(CH_3)_3 \text{ or } (H_3C)_2C{=}CHC(CH_3)_3 \xrightarrow{\text{H}_2\text{, Pt}} (CH_3)_2CHCH_2C(CH_3)_3$$

2,4,4-Trimethyl-1-pentene; 2,4,4-Trimethyl-2-pentene; 2,2,4-Trimethylpentane

**6.22** This problem illustrates the reactions of alkenes with various reagents and requires application of Markovnikov's rule to the addition of unsymmetrical electrophiles.

(*a*) Addition of hydrogen chloride to 1-pentene will give 2-chloropentane.

$$H_2C{=}CHCH_2CH_2CH_3 + HCl \longrightarrow CH_3\underset{\displaystyle Cl}{\underset{|}{C}}HCH_2CH_2CH_3$$

1-Pentene → 2-Chloropentane

(*b*) Electrophilic addition of hydrogen bromide will give 2-bromopentane.

$$H_2C{=}CHCH_2CH_2CH_3 + HBr \longrightarrow CH_3\underset{\displaystyle Br}{\underset{|}{C}}HCH_2CH_2CH_3$$

2-Bromopentane

(*c*) The presence of peroxides will cause free-radical addition of hydrogen bromide, and regioselective addition opposite to Markovnikov's rule will be observed.

$$H_2C{=}CHCH_2CH_2CH_3 + HBr \xrightarrow{\text{peroxides}} BrCH_2CH_2CH_2CH_2CH_3$$

1-Bromopentane

(*d*) Hydrogen iodide will add according to Markovnikov's rule.

$$H_2C{=}CHCH_2CH_2CH_3 + HI \longrightarrow CH_3\underset{\displaystyle I}{\underset{|}{C}}HCH_2CH_2CH_3$$

2-Iodopentane

(*e*) Dilute sulfuric acid will cause hydration of the double bond with regioselectivity in accord with Markovnikov's rule.

$$H_2C{=}CHCH_2CH_2CH_3 + H_2O \xrightarrow{\text{H}_2\text{SO}_4} CH_3\underset{\displaystyle OH}{\underset{|}{C}}HCH_2CH_2CH_3$$

2-Pentanol

(*f*) Hydroboration–oxidation of an alkene brings about hydration of the double bond opposite to Markovnikov's rule; 1-pentanol will be the product.

$$H_2C{=}CHCH_2CH_2CH_3 \xrightarrow[2.\ H_2O_2,\ HO^-]{1.\ B_2H_6} HOCH_2CH_2CH_2CH_2CH_3$$

1-Pentanol

(*g*) Bromine adds across the double bond to give a vicinal dibromide.

$$H_2C{=}CHCH_2CH_2CH_3 + Br_2 \xrightarrow{CCl_4} BrCH_2\underset{\large Br}{\underset{|}{C}}HCH_2CH_2CH_3$$

1,2-Dibromopentane

(*h*) Vicinal bromohydrins are formed when bromine in water adds to alkenes. Br adds to the less substituted carbon, OH to the more substituted one.

$$H_2C{=}CHCH_2CH_2CH_3 + Br_2 \xrightarrow{H_2O} BrCH_2\underset{\large OH}{\underset{|}{C}}HCH_2CH_2CH_3$$

1-Bromo-2-pentanol

(*i*) Epoxidation of the alkene occurs on treatment with peroxy acids.

$$H_2C{=}CHCH_2CH_2CH_3 + CH_3CO_2OH \longrightarrow H_2C\underset{O}{—}CHCH_2CH_2CH_3 + CH_3CO_2H$$

1,2-Epoxypentane    Acetic acid

(*j*) Ozone reacts with alkenes to give ozonides.

$$H_2C{=}CHCH_2CH_2CH_3 + O_3 \longrightarrow H_2C\overset{O}{\frown}CHCH_2CH_2CH_3 \text{ (with } O{—}O \text{ bridge)}$$

Ozonide

(*k*) When the ozonide in part (*j*) is hydrolyzed in the presence of zinc, formaldehyde and butanal are formed.

$$\text{Ozonide} \xrightarrow[Zn]{H_2O} H\overset{O}{\overset{\|}{C}}H + H\overset{O}{\overset{\|}{C}}CH_2CH_2CH_3$$

Formaldehyde    Butanal

**6.23** When we compare the reactions of 2-methyl-2-butene with the analogous reactions of 1-pentene, we find that the reactions proceed in a similar manner.

(*a*) $$(CH_3)_2C{=}CHCH_3 + HCl \longrightarrow (CH_3)_2\underset{\large Cl}{\underset{|}{C}}CH_2CH_3$$

2-Methyl-2-butene    2-Chloro-2-methylbutane

(*b*) $$(CH_3)_2C{=}CHCH_3 + HBr \longrightarrow (CH_3)_2\underset{\large Br}{\underset{|}{C}}CH_2CH_3$$

2-Bromo-2-methylbutane

(*c*) $$(CH_3)_2C{=}CHCH_3 + HBr \xrightarrow{\text{peroxides}} (CH_3)_2CH\underset{\large Br}{\underset{|}{C}}HCH_3$$

2-Bromo-3-methylbutane

(*d*) $(CH_3)_2C{=}CHCH_3 + HI \longrightarrow (CH_3)_2C(I)CH_2CH_3$

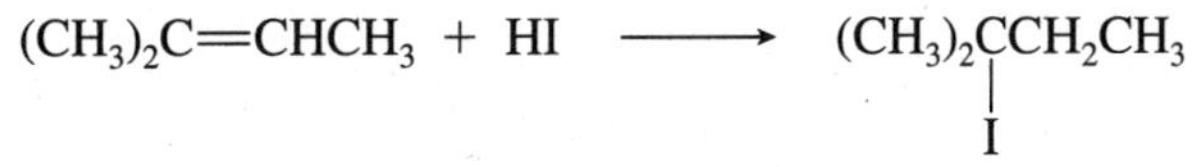

2-Iodo-2-methylbutane

(*e*) $(CH_3)_2C{=}CHCH_3 + H_2O \xrightarrow{H_2SO_4} (CH_3)_2C(OH)CH_2CH_3$

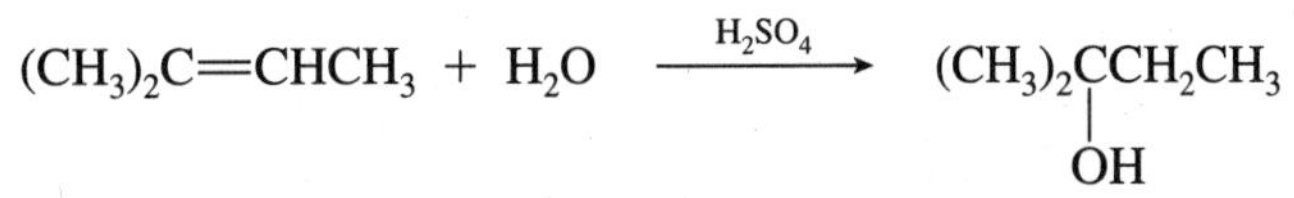

2-Methyl-2-butanol

(*f*) $(CH_3)_2C{=}CHCH_3 \xrightarrow[2.\ H_2O_2,\ HO^-]{1.\ B_2H_6} (CH_3)_2CHCH(OH)CH_3$

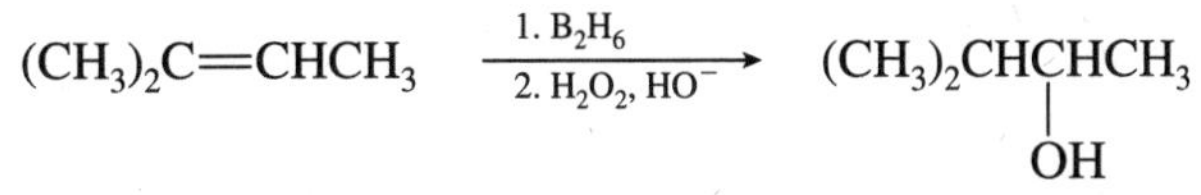

3-Methyl-2-butanol

(*g*) $(CH_3)_2C{=}CHCH_3 + Br_2 \xrightarrow{CCl_4} (CH_3)_2C(Br)CH(Br)CH_3$

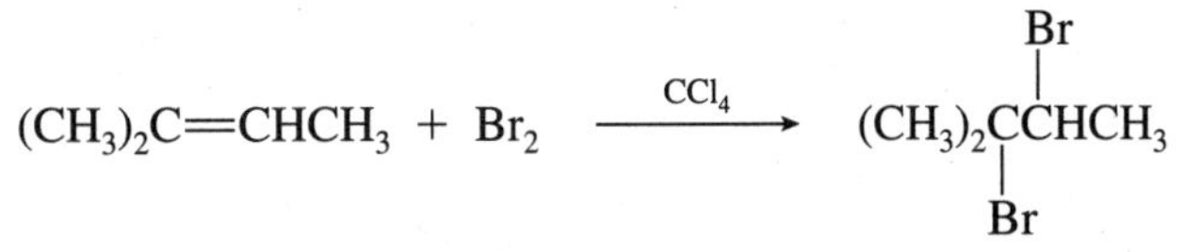

2,3-Dibromo-2-methylbutane

(*h*) $(CH_3)_2C{=}CHCH_3 + Br_2 \xrightarrow{H_2O} (CH_3)_2C(OH)CH(Br)CH_3$

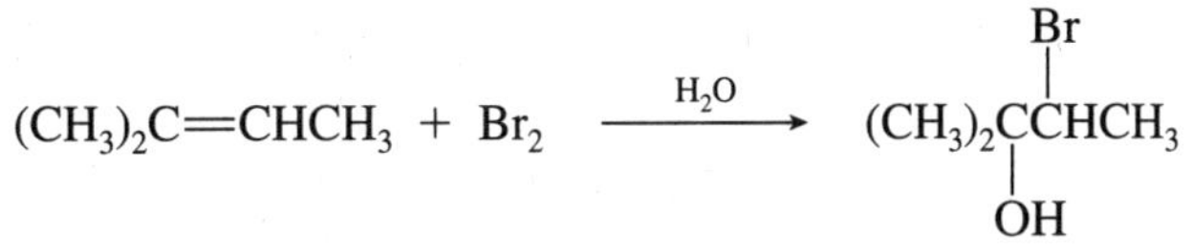

3-Bromo-2-methyl-2-butanol

(*i*) $(CH_3)_2C{=}CHCH_3 + CH_3CO_2OH \longrightarrow (CH_3)_2C\text{—}CHCH_3$ (bridged by O) $+ CH_3CO_2H$

2-Methyl-2,3-epoxybutane

(*j*) $(CH_3)_2C{=}CHCH_3 + O_3 \longrightarrow$ Ozonide (ring: $(H_3C)_2C$–O–$CH(CH_3)$ with O–O bridge)

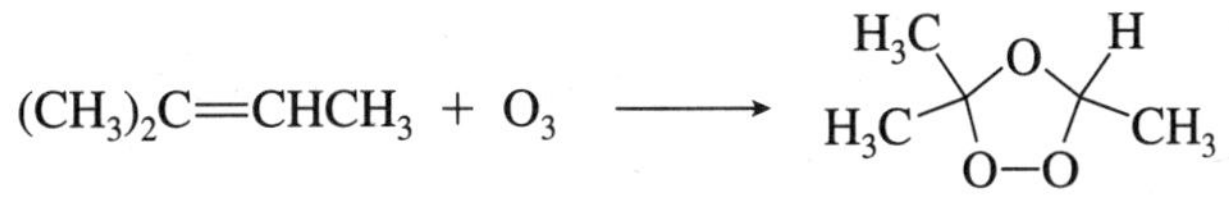

Ozonide

(*k*) Ozonide $\xrightarrow[Zn]{H_2O} CH_3\overset{O}{\overset{\|}{C}}CH_3 + H\overset{O}{\overset{\|}{C}}CH_3$

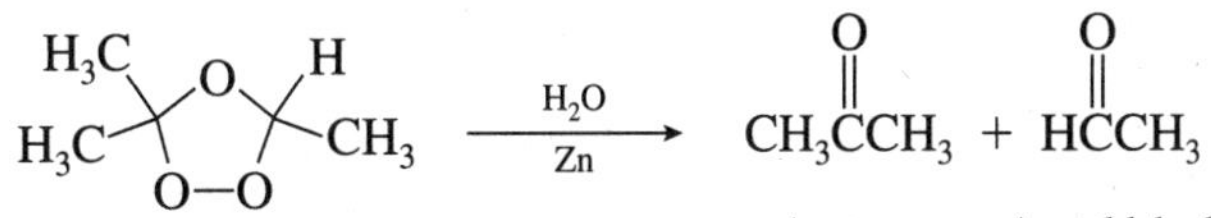

Acetone Acetaldehyde

**6.24** Cycloalkenes undergo the same kinds of reactions as do noncyclic alkenes.

(*a*) 1-Methylcyclohexene $+ HCl \longrightarrow$ 1-Chloro-1-methylcyclohexane

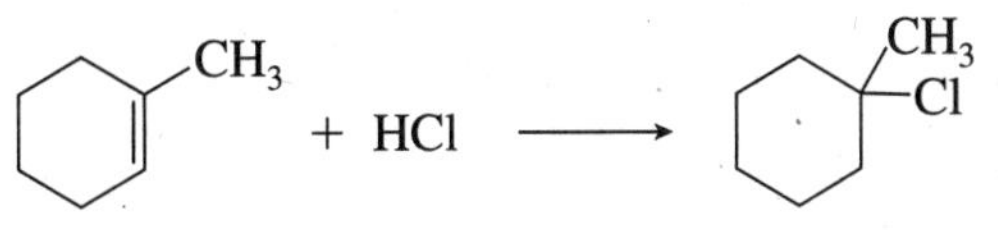

1-Methylcyclohexene 1-Chloro-1-methylcyclohexane

(*b*) 1-Methylcyclohexene $+ HBr \longrightarrow$ 1-Bromo-1-methylcyclohexane

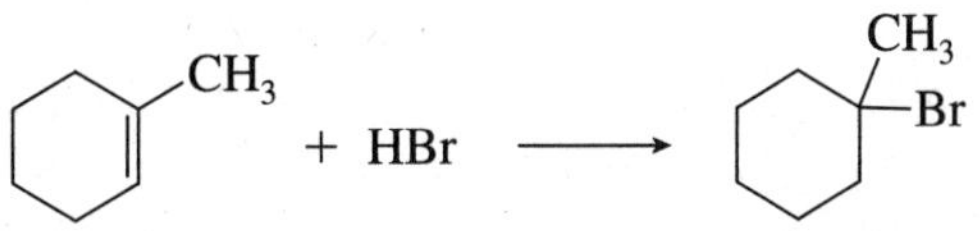

1-Bromo-1-methylcyclohexane

(*c*) $CH_3$ + HBr $\xrightarrow{\text{peroxides}}$ $CH_3$, Br

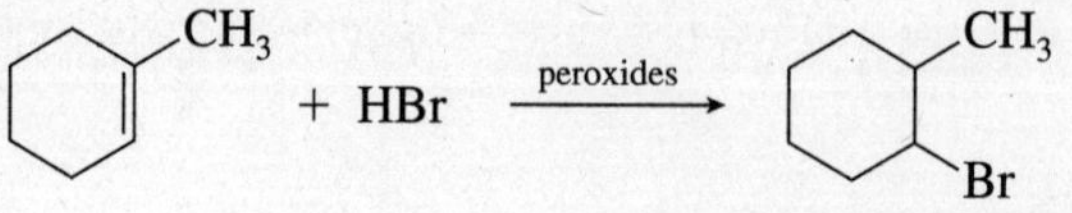

1-Bromo-2-methylcyclohexane
(mixture of cis and trans)

(*d*) $CH_3$ + HI $\longrightarrow$ $CH_3$, I

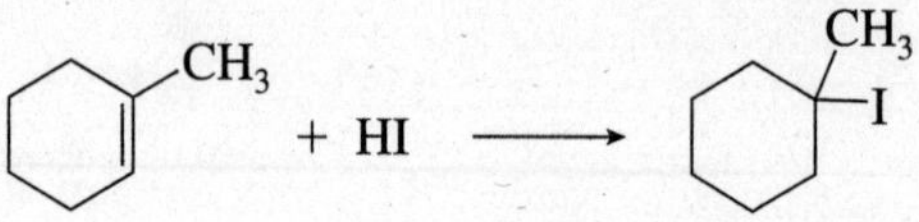

1-Iodo-1-methylcyclohexane

(*e*) $CH_3$ + $H_2O$ $\xrightarrow{H_2SO_4}$ $CH_3$, OH

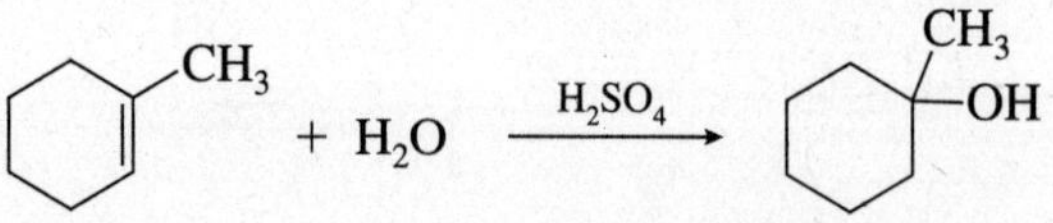

1-Methylcyclohexanol

(*f*) $CH_3$ $\xrightarrow[2.\ H_2O_2,\ HO^-]{1.\ B_2H_6}$ $CH_3$, H, OH

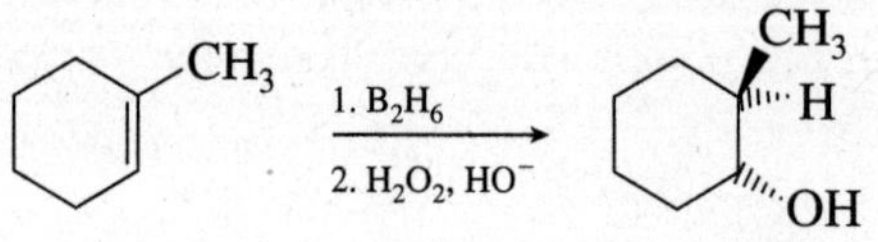

*trans*-2-Methylcyclohexanol

(*g*) $CH_3$ + $Br_2$ $\xrightarrow{CCl_4}$ Br, $CH_3$, Br

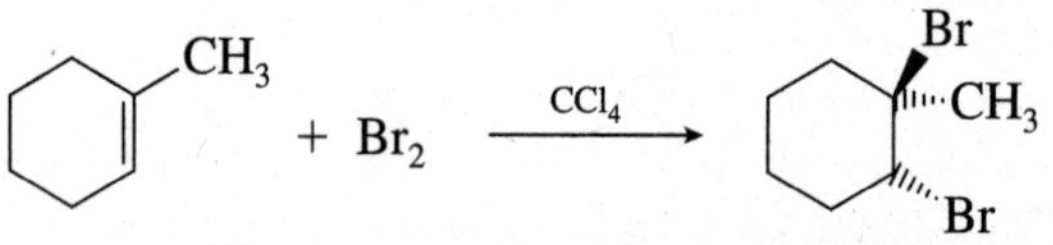

*trans*-1,2-Dibromo-1-
methylcyclohexane

(*h*) $CH_3$ + $Br_2$ $\xrightarrow{H_2O}$ OH, $CH_3$, Br

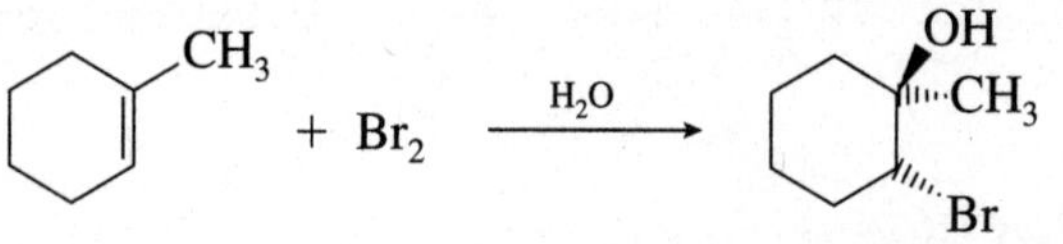

*trans*-2-Bromo-1-
methylcyclohexanol

(*i*) $CH_3$ + $CH_3CO_2OH$ $\longrightarrow$ $CH_3$, O + $CH_3CO_2H$

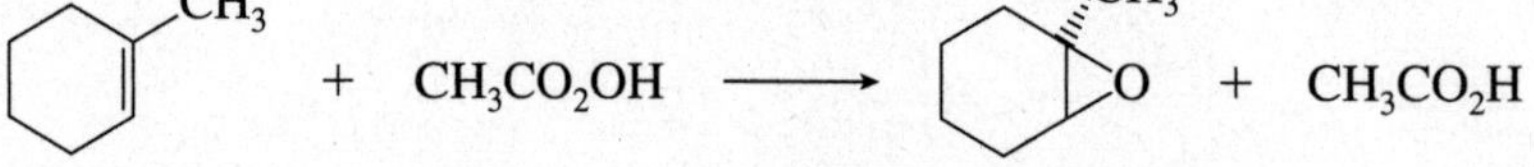

1,2-Epoxy-1-methylcyclohexane

(*j*) $CH_3$ + $O_3$ $\longrightarrow$ $CH_3$, O, O, O

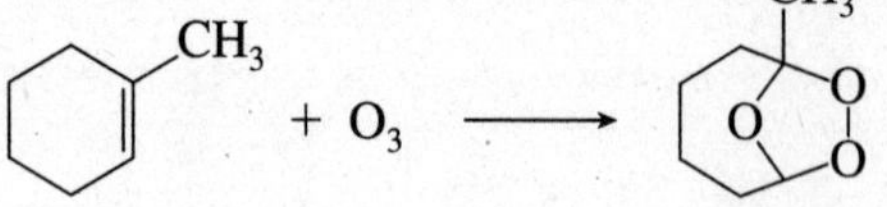

Ozonide

(*k*) $CH_3$, O, O, O $\xrightarrow[Zn]{H_2O}$ $CH_3$, H, O, O $\equiv$ $CH_3\overset{O}{\overset{\|}{C}}CH_2CH_2CH_2CH_2\overset{O}{\overset{\|}{C}}H$

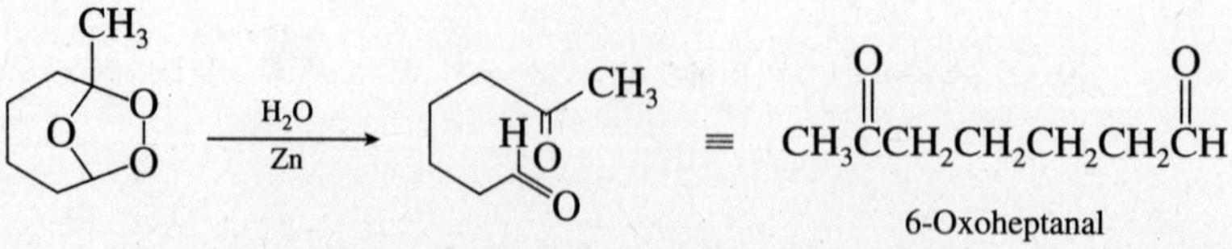

6-Oxoheptanal

**6.25** We need first to write out the structures in more detail to evaluate the substitution patterns at the double bonds.

(*a*) 1-Pentene 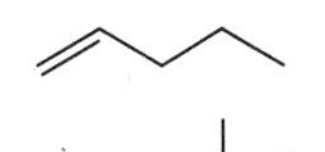 Monosubstituted

(*b*) (*E*)-4,4-Dimethyl-2-pentene trans-Disubstituted

(*c*) (*Z*)-4-Methyl-2-pentene 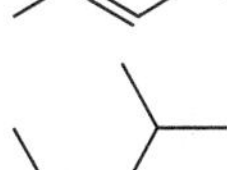 cis-Disubstituted

(*d*) (*Z*)-2,2,5,5-Tetramethyl-3-hexene  Two *tert*-butyl groups cis

(*e*) 2,4-Dimethyl-2-pentene 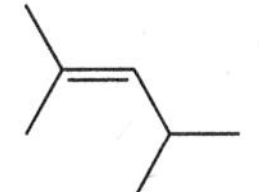 Trisubstituted

Compound *d*, having two cis *tert*-butyl groups, should have the least stable (highest energy) double bond. The remaining alkenes are arranged in order of increasing stability (decreasing heats of hydrogenation) according to the degree of substitution of the double bond: monosubstituted, cis-disubstituted, trans-disubstituted, trisubstituted. The heats of hydrogenation are therefore:

(*d*) 151 kJ/mol (36.2 kcal/mol)

(*a*) 122 kJ/mol (29.3 kcal/mol)

(*c*) 114 kJ/mol (27.3 kcal/mol)

(*b*) 111 kJ/mol (26.5 kcal/mol)

(*e*) 105 kJ/mol (25.1 kcal/mol)

**6.26** In all parts of this exercise we deduce the carbon skeleton on the basis of the alkane formed on hydrogenation of an alkene and then determine what carbon atoms may be connected by a double bond in that skeleton. Problems of this type are best done by using carbon skeleton formulas.

(*a*) Product is 2,2,3,4,4-pentamethylpentane. The only possible alkene precursor is

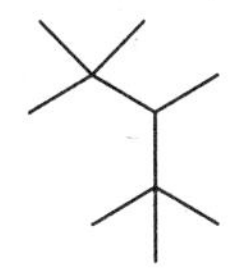

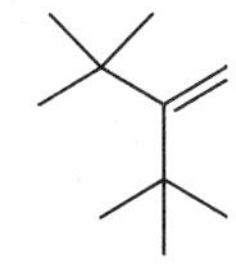

(*b*) Product is 2,3-dimethylbutane. May be formed by hydrogenation of

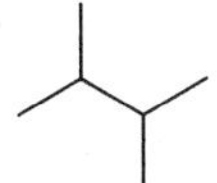

or 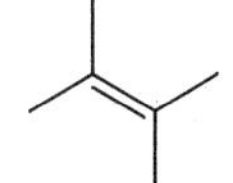

(*c*) Product is methylcyclobutane. May be formed by hydrogenation of

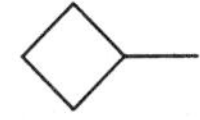

or or

**6.27** Hydrogenation of the alkenes shown will give a mixture of *cis*- and *trans*-1,4-dimethylcyclohexane.

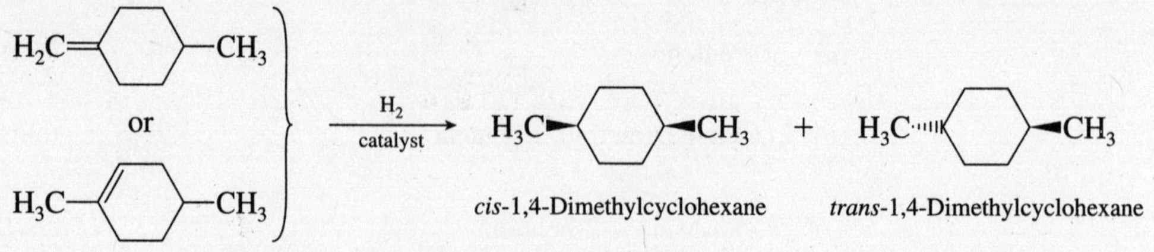

$H_2C$=(cyclohexane)–$CH_3$ or $H_3C$–(cyclohexene)–$CH_3$ $\xrightarrow[\text{catalyst}]{H_2}$ $H_3C$–(cyclohexane)–$CH_3$ + $H_3C$–(cyclohexane)–$CH_3$

*cis*-1,4-Dimethylcyclohexane *trans*-1,4-Dimethylcyclohexane

Only when the methyl groups are cis in the starting alkene will the cis stereoisomer be the sole product following hydrogenation. Hydrogenation of *cis*-3,6-dimethylcyclohexane will yield exclusively *cis*-1,4-dimethylcyclohexane.

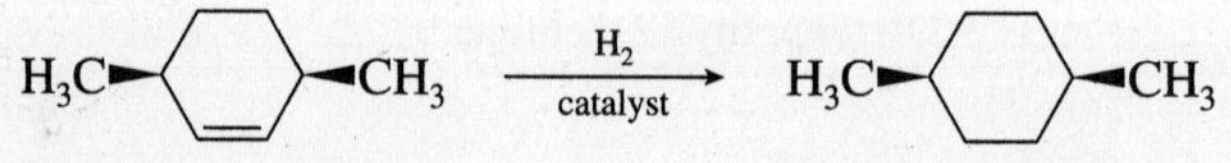

$H_3C$–(cyclohexene)–$CH_3$ $\xrightarrow[\text{catalyst}]{H_2}$ $H_3C$–(cyclohexane)–$CH_3$

*cis*-3,6-Dimethylcyclohexene *cis*-1,4-Dimethylcyclohexane

**6.28** (*a*) The desired transformation is the conversion of an alkene to a vicinal dibromide.

$$CH_3CH{=}C(CH_2CH_3)_2 \xrightarrow[CCl_4]{Br_2} CH_3\underset{Br}{CH}\underset{Br}{C}(CH_2CH_3)_2$$

3-Ethyl-2-pentene 2,3-Dibromo-3-ethylpentane

(*b*) Markovnikov addition of hydrogen chloride is indicated.

$$CH_3CH{=}C(CH_2CH_3)_2 \xrightarrow{HCl} CH_3CH_2\underset{Cl}{C}(CH_2CH_3)_2$$

3-Chloro-3-ethylpentane

(*c*) Free-radical addition of hydrogen bromide opposite to Markovnikov's rule will give the required regiochemistry.

$$CH_3CH{=}C(CH_2CH_3)_2 \xrightarrow[\text{peroxides}]{HBr} CH_3\underset{Br}{CH}CH(CH_2CH_3)_2$$

2-Bromo-3-ethylpentane

(*d*) Acid-catalyzed hydration will occur in accordance with Markovnikov's rule to yield the desired tertiary alcohol.

$$CH_3CH{=}C(CH_2CH_3)_2 \xrightarrow[H_2SO_4]{H_2O} CH_3CH_2\underset{OH}{C}(CH_2CH_3)_2$$

3-Ethyl-3-pentanol

(*e*) Hydroboration–oxidation results in hydration of alkenes with a regioselectivity opposite to that of Markovnikov's rule.

$$CH_3CH{=}C(CH_2CH_3)_2 \xrightarrow[2.\ H_2O_2,\ HO^-]{1.\ B_2H_6} CH_3\underset{OH}{CH}CH(CH_2CH_3)_2$$

3-Ethyl-2-pentanol

(*f*) A peroxy acid will convert an alkene to an epoxide.

$$CH_3CH{=}C(CH_2CH_3)_2 \xrightarrow{CH_3CO_2OH} \text{(epoxide: } CH_3,\ CH_2CH_3,\ H,\ O,\ CH_2CH_3)$$

3-Ethyl-2,3-epoxypentane

(*g*) Hydrogenation of alkenes converts them to alkanes.

$$CH_3CH{=}C(CH_2CH_3)_2 \xrightarrow[Pt]{H_2} CH_3CH_2CH(CH_2CH_3)_2$$

3-Ethylpentane

**6.29** (*a*) Four primary alcohols have the molecular formula $C_5H_{12}O$:

| $CH_3CH_2CH_2CH_2CH_2OH$ | $CH_3CH_2CH(CH_3)CH_2OH$ | $(CH_3)_2CHCH_2CH_2OH$ | $(CH_3)_3CCH_2OH$ |
|---|---|---|---|
| 1-Pentanol | 2-Methyl-1-butanol | 3-Methyl-1-butanol | 2,2-Dimethyl-1-propanol |

2,2-Dimethyl-1-propanol cannot be prepared by hydration of an alkene, because no alkene can have this carbon skeleton.

(*b*) Hydroboration–oxidation of alkenes is the method of choice for converting terminal alkenes to primary alcohols.

$$CH_3CH_2CH_2CH{=}CH_2 \xrightarrow[2.\ H_2O_2,\ HO^-]{1.\ B_2H_6} CH_3CH_2CH_2CH_2CH_2OH$$

1-Pentene → 1-Pentanol

$$CH_3CH_2C(CH_3){=}CH_2 \xrightarrow[2.\ H_2O_2,\ HO^-]{1.\ B_2H_6} CH_3CH_2CH(CH_3)CH_2OH$$

2-Methyl-1-butene → 2-Methyl-1-butanol

$$(CH_3)_2CHCH{=}CH_2 \xrightarrow[2.\ H_2O_2,\ HO^-]{1.\ B_2H_6} (CH_3)_2CHCH_2CH_2OH$$

3-Methyl-1-butene → 3-Methyl-1-butanol

(*c*) The only tertiary alcohol is 2-methyl-2-butanol. It can be made by Markovnikov hydration of 2-methyl-1-butene or of 2-methyl-2-butene.

$$H_2C{=}C(CH_3)CH_2CH_3 \xrightarrow{H_2O,\ H_2SO_4} (CH_3)_2C(OH)CH_2CH_3$$

2-Methyl-1-butene → 2-Methyl-2-butanol

$$(CH_3)_2C{=}CHCH_3 \xrightarrow{H_2O,\ H_2SO_4} (CH_3)_2C(OH)CH_2CH_3$$

2-Methyl-2-butene → 2-Methyl-2-butanol

**6.30** (*a*) Because the double bond is symmetrically substituted, the same addition product is formed under either ionic or free-radical conditions. Peroxides are absent, and so addition takes place by an ionic mechanism to give 3-bromohexane. (It does not matter whether the starting material is *cis*- or *trans*-3-hexene; both give the same product.)

$$CH_3CH_2CH{=}CHCH_2CH_3 + HBr \xrightarrow{\text{no peroxides}} CH_3CH_2CH_2CH(Br)CH_2CH_3$$

3-Hexene | Hydrogen bromide | 3-Bromohexane (observed yield 76%)

(*b*) In the presence of peroxides, hydrogen bromide adds with a regioselectivity opposite to that predicted by Markovnikov's rule. The product is the corresponding primary bromide.

$$(CH_3)_2CHCH_2CH_2CH_2CH{=}CH_2 \xrightarrow[\text{peroxides}]{\text{HBr}} (CH_3)_2CHCH_2CH_2CH_2CH_2CH_2Br$$

6-Methyl-1-heptene | 1-Bromo-6-methylheptane (observed yield 92%)

(*c*) Hydroboration–oxidation of alkenes leads to hydration of the double bond with a regioselectivity contrary to Markovnikov's rule and without rearrangement of the carbon skeleton.

$$H_2C{=}C(C(CH_3)_3)_2 \xrightarrow[\text{2. } H_2O_2,\ HO^-]{\text{1. } B_2H_6} HOCH_2CH(C(CH_3)_3)C(CH_3)_3$$

2-*tert*-Butyl-3,3-dimethyl-1-butene | 2-*tert*-Butyl-3,3-dimethyl-1-butanol (observed yield 65%)

(*d*) Hydroboration–oxidation of alkenes leads to syn hydration of double bonds.

$$\text{1,2-dimethylcyclohexene (}CH_3, CH_3\text{)} \xrightarrow[\text{2. } H_2O_2,\ HO^-]{\text{1. } B_2H_6} \text{cyclohexane with H, }CH_3\text{ and }CH_3\text{, OH (cis)}$$

1,2-Dimethylcyclohexene | *cis*-1,2-Dimethylcyclohexanol (observed yield 82%)

(*e*) Bromine adds across the double bond of alkenes to give vicinal dibromides.

$$H_2C{=}C(CH_3)CH_2CH_2CH_3 + Br_2 \xrightarrow{CHCl_3} BrCH_2C(Br)(CH_3)CH_2CH_2CH_3$$

2-Methyl-1-pentene | 1,2-Dibromo-2-methylpentane (observed yield 60%)

(*f*) In aqueous solution bromine reacts with alkenes to give bromohydrins. Bromine is the electrophile in this reaction and adds to the carbon that has the greater number of attached hydrogens.

$$(CH_3)_2C{=}CHCH_3 + Br_2 \xrightarrow{H_2O} (CH_3)_2C(OH)CH(Br)CH_3$$

2-Methyl-2-butene | Bromine | 3-Bromo-2-methyl-2-butanol (observed yield 77%)

(*g*) An aqueous solution of chlorine will react with 1-methylcyclopentene by an anti addition. Chlorine is the electrophile and adds to the less substituted end of the double bond.

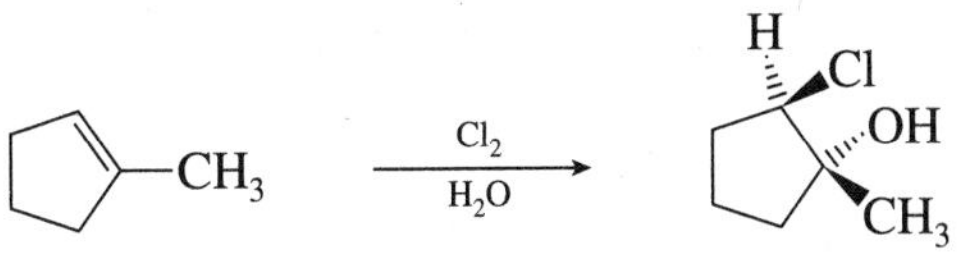

1-Methylcyclopentene $\xrightarrow[H_2O]{Cl_2}$ *trans*-2-Chloro-1-methylcyclopentanol

(*h*) Compounds of the type $RCOOH$ (with C=O) are peroxy acids and react with alkenes to give epoxides.

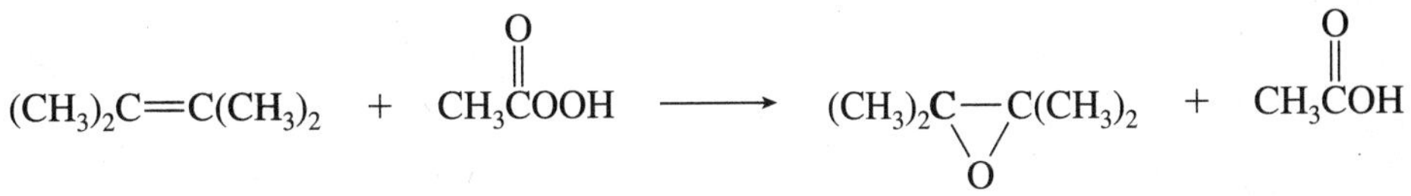

$$(CH_3)_2C{=}C(CH_3)_2 + CH_3\overset{O}{\overset{\|}{C}}OOH \longrightarrow (CH_3)_2C\underset{O}{-}C(CH_3)_2 + CH_3\overset{O}{\overset{\|}{C}}OH$$

2,3-Dimethyl-2-butene; Peroxyacetic acid; 2,3-Dimethyl-2,3-epoxybutane (observed yield 70–80%); Acetic acid

(*i*) The double bond is cleaved by ozonolysis. Each of the doubly bonded carbons becomes doubly bonded to oxygen in the product.

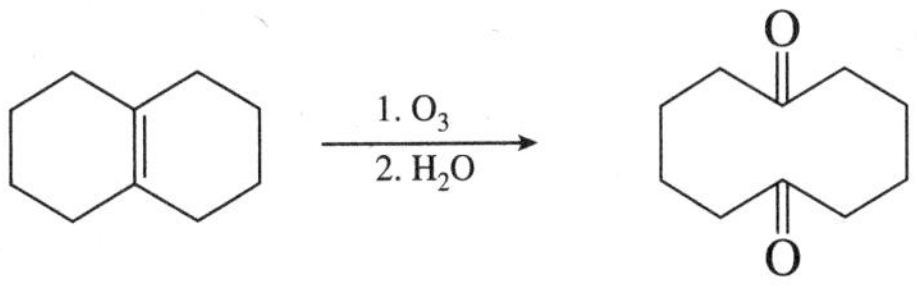

$\xrightarrow[2.\ H_2O]{1.\ O_3}$

Cyclodecan-1,6-dione (observed yield 45%)

**6.31** The product is epoxide B.

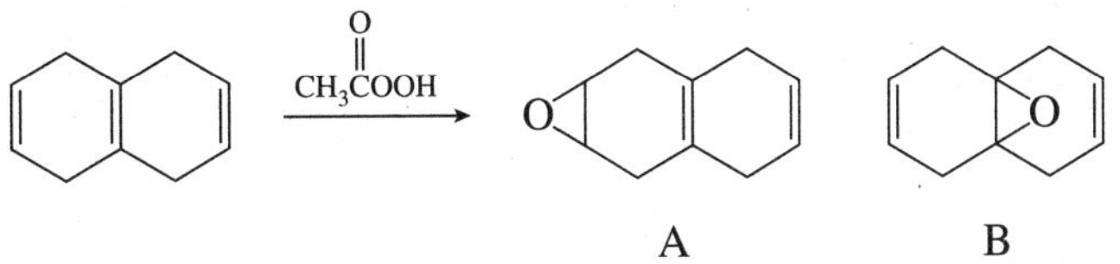

$\xrightarrow{CH_3\overset{O}{\overset{\|}{C}}OOH}$

A

B

Major product; formed faster

Epoxidation is an electrophilic addition; oxygen is transferred to the more electron-rich, more highly substituted double bond. A tetrasubstituted double bond reacts faster than a disubstituted one.

**6.32** (*a*) There is no direct, one-step transformation that moves a hydroxyl group from one carbon to another, and so it is not possible to convert 2-propanol to 1-propanol in a single reaction. Analyze the problem by reasoning backward. 1-Propanol is a primary alcohol. What reactions do we have available for the preparation of primary alcohols? One way is by the hydroboration–oxidation of terminal alkenes.

$$CH_3CH{=}CH_2 \xrightarrow{\text{hydroboration–oxidation}} CH_3CH_2CH_2OH$$

Propene; 1-Propanol

The problem now becomes the preparation of propene from 2-propanol. The simplest way is by acid-catalyzed dehydration.

$$\underset{\text{2-Propanol}}{\underset{\displaystyle |\atop OH}{CH_3CHCH_3}} \xrightarrow{H^+,\text{ heat}} \underset{\text{Propene}}{CH_3CH{=}CH_2}$$

After analyzing the problem in terms of overall strategy, present the synthesis in detail showing the reagents required in each step. Thus, the answer is:

$$\underset{\text{2-Propanol}}{\underset{\displaystyle |\atop OH}{CH_3CHCH_3}} \xrightarrow[\text{heat}]{H_2SO_4} \underset{\text{Propene}}{CH_3CH{=}CH_2} \xrightarrow[\text{2. } H_2O_2,\ HO^-]{\text{1. } B_2H_6} \underset{\text{1-Propanol}}{CH_3CH_2CH_2OH}$$

(*b*) We analyze this synthetic exercise in a manner similar to the preceding one. There is no direct way to move a bromine from C-2 in 2-bromopropane to C-1 in 1-bromopropane. We can, however, prepare 1-bromopropane from propene by free-radical addition of hydrogen bromide in the presence of peroxides.

$$\underset{\text{Propene}}{CH_3CH{=}CH_2} + \underset{\text{Hydrogen bromide}}{HBr} \xrightarrow{\text{peroxides}} \underset{\text{1-Bromopropane}}{CH_3CH_2CH_2Br}$$

We prepare propene from 2-bromopropane by dehydrohalogenation.

$$\underset{\text{2-Bromopropane}}{\underset{\displaystyle |\atop Br}{CH_3CHCH_3}} \xrightarrow{E2} \underset{\text{Propene}}{CH_3CH{=}CH_2}$$

Sodium ethoxide in ethanol is a suitable base-solvent system for this conversion. Sodium methoxide in methanol or potassium *tert*-butoxide in *tert*-butyl alcohol could also be used, as could potassium hydroxide in ethanol.

Combining these two transformations gives the complete synthesis.

$$\underset{\text{2-Bromopropane}}{\underset{\displaystyle |\atop Br}{CH_3CHCH_3}} \xrightarrow[CH_3CH_2OH,\text{ heat}]{NaOCH_2CH_3} \underset{\text{Propene}}{CH_3CH{=}CH_2} \xrightarrow[\text{peroxides}]{HBr} \underset{\text{1-Bromopropane}}{CH_3CH_2CH_2Br}$$

(*c*) Planning your strategy in a forward direction can lead to problems when the conversion of 2-bromopropane to 1,2-dibromopropane is considered. There is a temptation to try to simply add the second bromine by free-radical halogenation.

$$\underset{\text{2-Bromopropane}}{\underset{\displaystyle |\atop Br}{CH_3CHCH_3}} \xrightarrow{Br_2,\text{ light and heat}} \underset{\text{1,2-Dibromopropane}}{\underset{\displaystyle |\atop Br}{CH_3CHCH_2Br}}$$

This is *incorrect!* There is no reason to believe that the second bromine will be introduced exclusively at C-1. In fact, the selectivity rules for bromination tell us that 2,2-dibromopropane is the expected major product.

The best approach is to reason backward. 1,2-Dibromopropane is a vicinal dibromide, and we prepare vicinal dibromides by adding elemental bromine to alkenes.

$$\underset{\text{Propene}}{CH_3CH{=}CH_2} + \underset{\text{Bromine}}{Br_2} \longrightarrow \underset{\text{1,2-Dibromopropane}}{CH_3\underset{\displaystyle |\atop \displaystyle Br}{C}HCH_2Br}$$

As described in part (*b*), we prepare propene from 2-bromopropane by E2 elimination. The correct synthesis is therefore

$$\underset{\text{2-Bromopropane}}{CH_3\underset{\displaystyle |\atop \displaystyle Br}{C}HCH_3} \xrightarrow[CH_3CH_2OH]{NaOCH_2CH_3} \underset{\text{Propene}}{CH_3CH{=}CH_2} \xrightarrow{Br_2} \underset{\text{1,2-Dibromopropane}}{CH_3\underset{\displaystyle |\atop \displaystyle Br}{C}HCH_2Br}$$

(*d*) Do not attempt to reason forward and convert 2-propanol to 1-bromo-2-propanol by free-radical bromination. Reason backward! The desired compound is a vicinal bromohydrin, and vicinal bromohydrins are prepared by adding bromine to alkenes in aqueous solution. The correct solution is

$$\underset{\text{2-Propanol}}{CH_3\underset{\displaystyle |\atop \displaystyle OH}{C}HCH_3} \xrightarrow[\text{heat}]{H_2SO_4} \underset{\text{Propene}}{CH_3CH{=}CH_2} \xrightarrow[H_2O]{Br_2} \underset{\text{1-Bromo-2-propanol}}{CH_3\underset{\displaystyle |\atop \displaystyle OH}{C}HCH_2Br}$$

(*e*) Here we have another problem where reasoning forward can lead to trouble. If we try to conserve the oxygen of 2-propanol so that it becomes the oxygen of 1,2-epoxypropane, we need a reaction in which this oxygen becomes bonded to C-1.

$$\underset{\text{2-Propanol}}{CH_3\underset{\displaystyle |\atop \displaystyle OH}{C}HCH_3} \longrightarrow \underset{\text{1,2-Epoxypropane}}{CH_3CH{-}CH_2 \text{ (bridged by O)}}$$

This will not work as no synthetic method for such a single-step transformation exists!

By reasoning backward, recalling that epoxides are made from alkenes by reaction with peroxy acids, we develop a proper synthesis.

$$\underset{\text{2-Propanol}}{CH_3\underset{\displaystyle |\atop \displaystyle OH}{C}HCH_3} \xrightarrow[\text{heat}]{H_2SO_4} \underset{\text{Propene}}{CH_3CH{=}CH_2} \xrightarrow{CH_3\overset{O}{\overset{\|}{C}}OOH} \underset{\text{1,2-Epoxypropane}}{CH_3CH{-}CH_2 \text{ (bridged by O)}}$$

(*f*) *tert*-Butyl alcohol and isobutyl alcohol have the same carbon skeleton; all that is required is to move the hydroxyl group from C-1 to C-2. As pointed out in part (*a*) of this problem, we cannot do that directly but we can do it in two efficient steps through a synthesis that involves hydration of an alkene.

$$(CH_3)_2CHCH_2OH \xrightarrow[\text{heat}]{H_2SO_4} (CH_3)_2C{=}CH_2 \xrightarrow{H_2O,\ H_2SO_4} (CH_3)_3COH$$

Isobutyl alcohol — 2-Methylpropene — *tert*-Butyl alcohol

Acid-catalyzed hydration of the alkene gives the desired regioselectivity.

(*g*) The strategy of this exercise is similar to that of the preceding one. Convert the starting material to an alkene by an elimination reaction, followed by electrophilic addition to the double bond.

$$(CH_3)_2CHCH_2I \xrightarrow[(CH_3)_3COH,\ \text{heat}]{KOC(CH_3)_3} (CH_3)_2C{=}CH_2 \xrightarrow{HI} (CH_3)_3CI$$

Isobutyl iodide — 2-Methylpropene — *tert*-Butyl iodide

(*h*) This problem is similar to the one in part (*d*) in that it requires the preparation of a halohydrin from an alkyl halide. The strategy is the same. Convert the alkyl halide to an alkene, and then form the halohydrin by treatment with the appropriate halogen in aqueous solution.

$$\text{Cyclohexyl chloride} \xrightarrow[CH_3CH_2OH]{NaOCH_2CH_3} \text{Cyclohexene} \xrightarrow[H_2O]{Cl_2} \textit{trans}\text{-2-Chlorocyclohexanol}$$

Cyclohexyl chloride — Cyclohexene — *trans*-2-Chlorocyclohexanol

(*i*) Halogenation of an alkane is required here. Iodination of alkanes, however, is not a feasible reaction. We can make alkyl iodides from alcohols or from alkenes by treatment with HI. A reasonable synthesis using reactions that have been presented to this point proceeds as shown:

$$\text{Cyclopentane} \xrightarrow[\text{light}]{Cl_2} \text{Cyclopentyl chloride} \xrightarrow[CH_3CH_2OH]{NaOCH_2CH_3} \text{Cyclopentene} \xrightarrow{HI} \text{Cyclopentyl iodide}$$

Cyclopentane — Cyclopentyl chloride — Cyclopentene — Cyclopentyl iodide

(*j*) Dichlorination of cyclopentane under free-radical conditions is not a realistic approach to the introduction of two chlorines in a trans-1,2 relationship without contamination by isomeric dichlorides. Vicinal dichlorides are prepared by electrophilic addition of chlorine to alkenes. The stereochemistry of addition is anti.

$$\text{Cyclopentane} \xrightarrow[\text{light}]{Cl_2} \text{Cyclopentyl chloride} \xrightarrow[CH_3CH_2OH]{NaOCH_2CH_3} \text{Cyclopentene} \xrightarrow{Cl_2} \textit{trans}\text{-1,2-Dichlorocyclopentane}$$

Cyclopentane — Cyclopentyl chloride — Cyclopentene — *trans*-1,2-Dichlorocyclopentane

(*k*) The desired compound contains all five carbon atoms of cyclopentane but is not cyclic. Two aldehyde functions are present. We know that cleavage of carbon–carbon double bonds by ozonolysis leads to two carbonyl groups, which suggests the synthesis shown in the following equation:

$$\text{Cyclopentanol} \xrightarrow[\text{heat}]{H_2SO_4} \text{Cyclopentene} \xrightarrow[2.\ Zn,\ H_2O]{1.\ O_3} \equiv \overset{O}{\overset{\|}{H C}}CH_2CH_2CH_2\overset{O}{\overset{\|}{C}}H$$

Cyclopentanol — Cyclopentene — Pentanedial

**6.33** The two products formed by addition of hydrogen bromide to 1,2-dimethylcyclohexene cannot be regioisomers. Stereoisomers are possible, however.

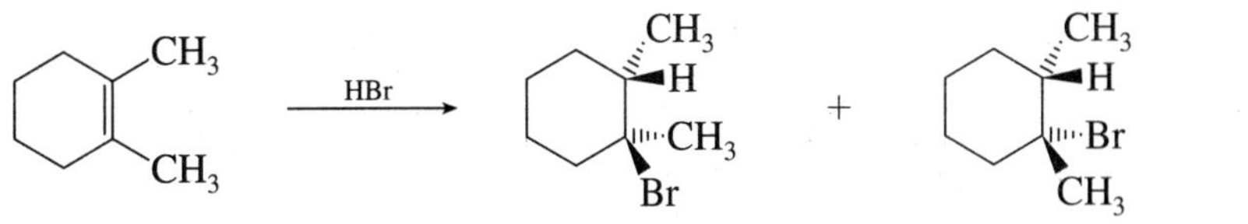

1,2-Dimethylcyclohexene $\xrightarrow{HBr}$ *cis*-1,2-Dimethylcyclohexyl bromide + *trans*-1,2-Dimethylcyclohexyl bromide

The same two products are formed from 1,6-dimethylcyclohexene because addition of hydrogen bromide follows Markovnikov's rule in the absence of peroxides.

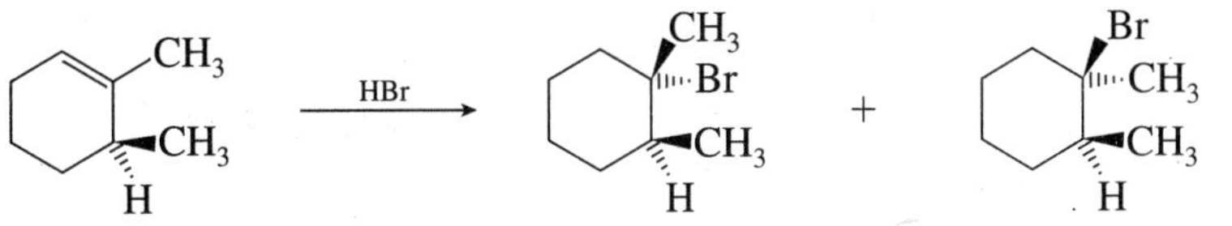

1,6-Dimethylcyclohexene $\xrightarrow{HBr}$ *cis*-1,2-Dimethylcyclohexyl bromide + *trans*-1,2-Dimethylcyclohexyl bromide

**6.34** The problem presents the following experimental observation:

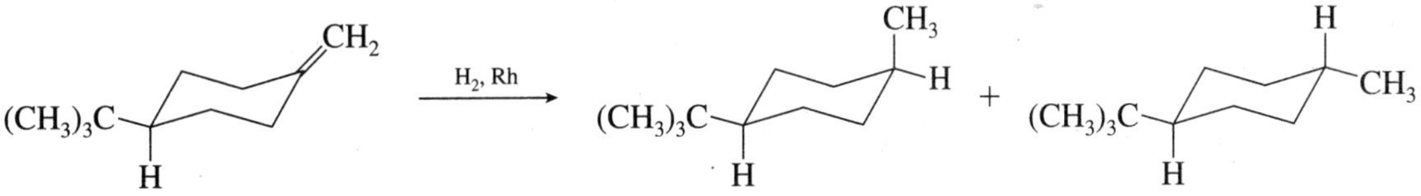

4-*tert*-Butyl(methylene)-cyclohexane $\xrightarrow{H_2,\ Rh}$ *cis*-1-*tert*-Butyl-4-methylcyclohexane (88%) + *trans*-1-*tert*-Butyl-4-methylcyclohexane (12%)

This observation tells us that the predominant mode of hydrogen addition to the double bond is from the equatorial direction. Equatorial addition is the less hindered approach and thus occurs faster.

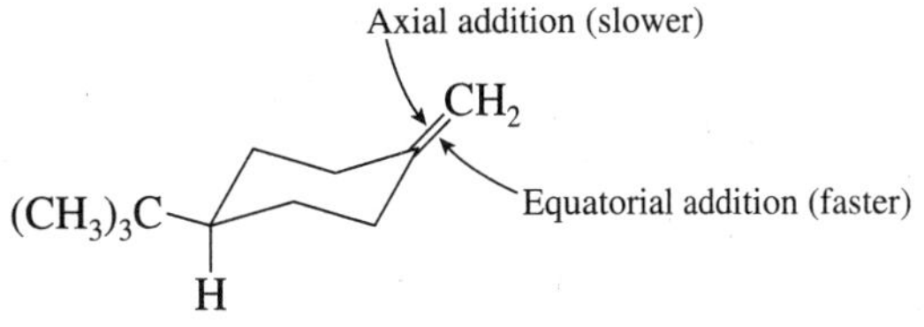

Axial addition (slower)

Equatorial addition (faster)

(*a*) Epoxidation should therefore give the following products:

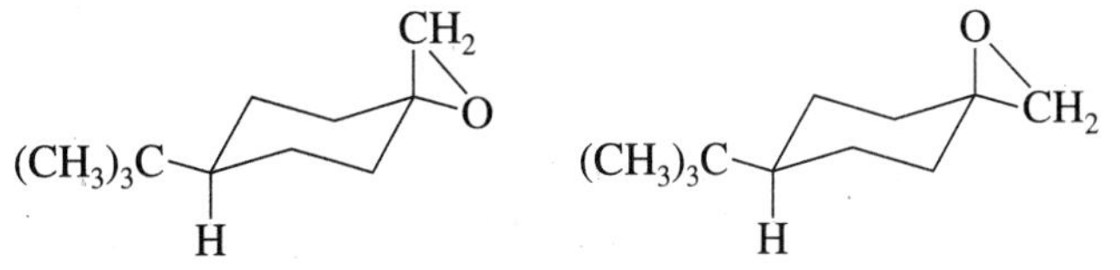

Major product

Minor product

The major product is the stereoisomer that corresponds to transfer of oxygen from the equatorial direction.

(*b*) Hydroboration–oxidation occurs from the equatorial direction.

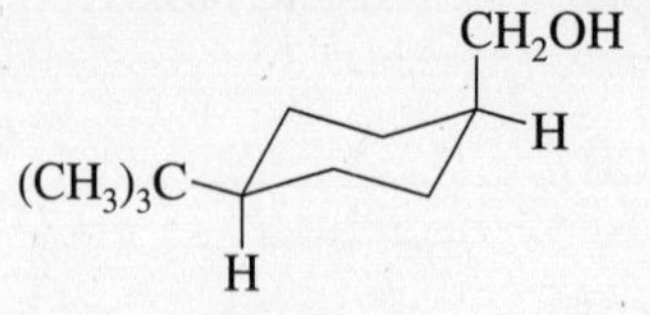

Major product 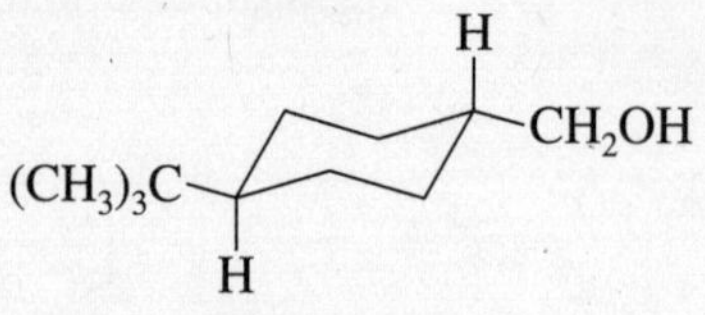    Minor product

**6.35** The methyl group in compound B shields one face of the double bond from the catalyst surface, therefore hydrogen can be transferred only to the bottom face of the double bond. The methyl group in compound A does not interfere with hydrogen transfer to the double bond.

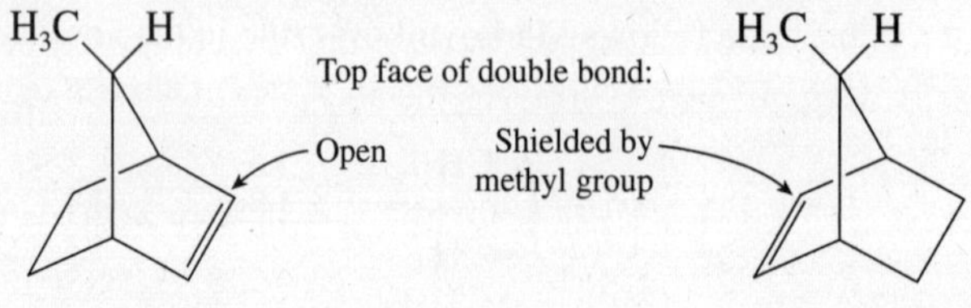

Top face of double bond: Open; Shielded by methyl group

Compound A    Compound B

Thus, hydrogenation of A is faster than that of B because B contains a more sterically hindered double bond.

**6.36** Hydrogen can add to the double bond of 1,4-dimethylcyclopentene either from the same side as the C-4 methyl group or from the opposite side. The two possible products are *cis*- and *trans*-1, 3-dimethylcyclopentane.

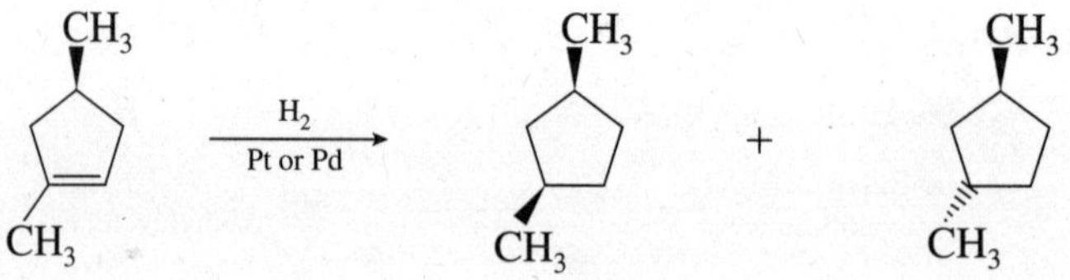

$\xrightarrow[\text{Pt or Pd}]{H_2}$

1,4-Dimethylcyclopentene    *cis*-1,3-Dimethylcyclopentane    *trans*-1,3-Dimethylcyclopentane

Hydrogen transfer occurs to the less hindered face of the double bond, that is, trans to the C-4 methyl group. Thus, the major product is *cis*-1,3-dimethylcyclopentane.

**6.37** Hydrogen can add to either the top face or the bottom face of the double bond. Syn addition to the double bond requires that the methyl groups in the product be cis.

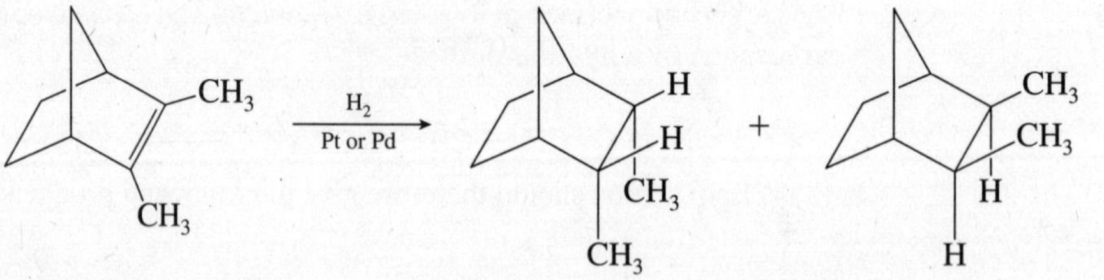

$\xrightarrow[\text{Pt or Pd}]{H_2}$

**6.38** 3-Carene can in theory undergo hydrogenation to give either *cis*-carane or *trans*-carane.

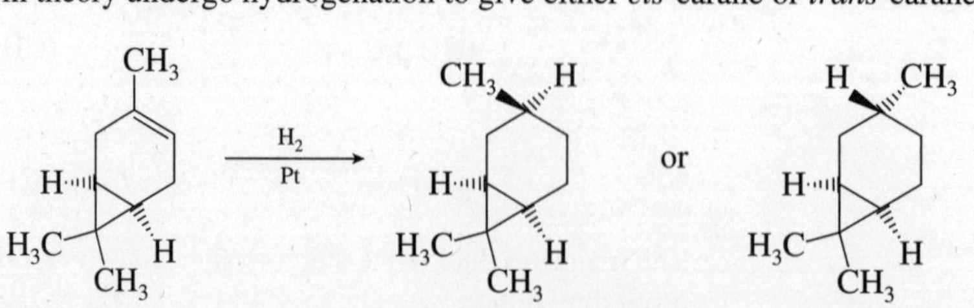

$\xrightarrow[\text{Pt}]{H_2}$    or

*cis*-Carane (98%)    *trans*-Carane

The exclusive product is *cis*-carane, since it corresponds to transfer of hydrogen from the less hindered side.

Methyl groups shield this side of molecule

*cis*-Carane

**6.39** Ethylene and propene react with concentrated sulfuric acid to form alkyl hydrogen sulfates. Addition of water hydrolyzes the alkyl hydrogen sulfates to the corresponding alcohols.

$$H_2C{=}CH_2 \xrightarrow{H_2SO_4} CH_3CH_2OSO_2OH \xrightarrow[\text{heat}]{H_2O} CH_3CH_2OH$$

Ethylene — Ethyl hydrogen sulfate — Ethanol

$$H_2C{=}CHCH_3 \xrightarrow{H_2SO_4} CH_3CH(OSO_2OH)CH_3 \xrightarrow[\text{heat}]{H_2O} CH_3CH(OH)CH_3$$

Propene — Isopropyl hydrogen sulfate — Isopropyl alcohol

Recall that alkyl substituents on the double bond increase the reactivity of alkenes toward electrophilic addition. Propene therefore reacts faster than ethylene with sulfuric acid, and the mixture of alkyl hydrogen sulfates is mainly isopropyl hydrogen sulfate, and the alcohol obtained on hydrolysis is isopropyl alcohol.

**6.40** The first step in the mechanism of acid-catalyzed hydration of alkenes is protonation of the double bond to give a carbocation intermediate.

$$H_3O^+ + CH_2{=}CHCH(CH_3)_2 \longrightarrow H_2O + CH_3\overset{+}{C}HCH(CH_3)_2$$

Hydronium ion — 3-Methyl-1-butene — Water — 1,2-Dimethylpropyl cation (secondary)

The carbocation formed in this step is secondary and capable of rearranging to a more stable tertiary carbocation by a hydride shift.

$$CH_3{-}\overset{+}{C}H{-}C(CH_3)_2(H) \longrightarrow CH_3{-}CH(H){-}\overset{+}{C}(CH_3)_2$$

1,2-Dimethylpropyl cation (secondary) — 1,1-Dimethylpropyl cation (tertiary)

The alcohol that is formed when water reacts with the tertiary carbocation is 2-methyl-2-butanol, not 3-methyl-2-butanol.

$$H_2\ddot{O}: + CH_3CH_2\overset{+}{C}(CH_3)_2 \longrightarrow CH_3CH_2C(CH_3)_2\overset{+}{O}H_2 \xrightarrow{-H^+} CH_3CH_2C(CH_3)_2\ddot{O}H$$

Water — 1,1-Dimethylpropyl cation — 2-Methyl-2-butanol

**6.41** In the presence of sulfuric acid, the carbon–carbon double bond of 2-methyl-1-butene is protonated and a carbocation is formed.

Hydronium ion + 2-Methyl-1-butene → Water + 1,1-Dimethylpropyl cation

$$H_3O^+ + H_2C{=}C(CH_3)CH_2CH_3 \longrightarrow H_2O + H_3C{-}\overset{+}{C}(CH_3)CH_2CH_3$$

This carbocation can then lose a proton from its $CH_2$ group to form 2-methyl-2-butene.

Water + 1,1-Dimethylpropyl cation → Hydronium ion + 2-Methyl-2-butene

$$H_2O + H_3C{-}\overset{+}{C}(CH_3)CH_2CH_3 \longrightarrow H_3O^+ + H_3C{-}C(CH_3){=}CHCH_3$$

**6.42** The first step in the reaction of an alkene with bromine is the formation of a bromonium ion.

$$H_2C{=}CHCH_2CH_2CH_2CH_3 \xrightarrow{Br_2} H_2C{-}CHCH_2CH_2CH_2CH_3 \text{ (bridged by } :\overset{+}{Br}:)$$

This bromonium ion can react with $Br^-$ to form 1,2-dibromohexane, or it can be attacked by methanol.

$$CH_3OH + \text{bromonium ion} \longrightarrow CH_2(Br)CH(\overset{+}{O}HCH_3)CH_2CH_2CH_2CH_3 \xrightarrow{-H^+} CH_2(Br)CH(OCH_3)CH_2CH_2CH_2CH_3$$

1-Bromo-2-methoxyhexane

Attack on the bromonium ion by methanol is analogous to the attack by water in the mechanism of bromohydrin formation.

**6.43** The problem stipulates that a bridged sulfonium ion is an intermediate. Therefore, use the $\pi$ electrons of the double bond to attack one of the sulfur atoms of thiocyanogen and cleave the S—S bond in a manner analogous to cleavage of a Br—Br bond in the reaction of bromine with an alkene.

Cyclooctene + NCS—SCN → bridged sulfonium ion ($\overset{+}{S}$—CN) + $^-$:SCN

The sulfonium ion is then attacked by thiocyanate ($NCS^-$) to give the observed product, which has the trans stereochemistry.

Bridged sulfonium ion + $^-$:S—CN → trans-1,2-bis(thiocyanato)cyclooctane (S—CN groups trans)

**6.44** Alkenes of molecular formula $C_{12}H_{24}$ are **trimers** of 2-methylpropene. The first molecule of 2-methylpropene is protonated to form *tert*-butyl cation, which reacts with a second molecule of 2-methylpropene to give a tertiary carbocation having eight carbons.

$$(CH_3)_3C^+ + CH_2{=}C(CH_3)_2 \longrightarrow (CH_3)_3CCH_2{-}\overset{+}{C}(CH_3)_2$$

*tert*-Butyl cation  2-Methylpropene  1,1,3,3-Tetramethylbutyl cation

This carbocation reacts with a third molecule of 2-methylpropene to give a 12-carbon tertiary carbocation.

$$(CH_3)_3CCH_2\overset{+}{C}(CH_3)_2 + CH_2{=}C(CH_3)_2 \longrightarrow (CH_3)_3CCH_2C(CH_3)_2CH_2{-}\overset{+}{C}(CH_3)_2$$

1,1,3,3-Tetramethylbutyl cation  2-Methylpropene  1,1,3,3,5,5-Hexamethylhexyl cation

The 12-carbon carbocation can lose a proton in either of two directions to give the alkenes shown.

$$(CH_3)_3CCH_2C(CH_3)_2CH_2{-}\overset{+}{C}(CH_3)_2 \xrightarrow{-H^+} (CH_3)_3CCH_2C(CH_3)_2CH_2{-}C(=CH_2)CH_3 + (CH_3)_3CCH_2C(CH_3)_2CH{=}C(CH_3)_2$$

1,1,3,3,5,5-Hexamethylhexyl cation  2,4,4,6,6-Pentamethyl-1-heptene  2,4,4,6,6-Pentamethyl-2-heptene

**6.45** The carbon skeleton is revealed by the hydrogenation experiment. Compounds B and C must have the same carbon skeleton as 3-ethylpentane.

$$\text{3-ethyl-1-pentene} \text{ or } \text{3-ethyl-2-pentene} \xrightarrow[\text{cat}]{H_2} \text{3-ethylpentane}$$

Three alkyl bromides have this carbon skeleton, namely, 1-bromo-3-ethylpentane, 2-bromo-3-ethylpentane, and 3-bromo-3-ethylpentane. Of these three only 2-bromo-3-ethylpentane will give two alkenes on dehydrobromination.

$$\text{1-Bromo-3-ethylpentane} \xrightarrow{E2} \text{3-ethyl-1-pentene} \quad \text{(only product)}$$

1-Bromo-3-ethylpentane

$$\text{3-Bromo-3-ethylpentane} \xrightarrow{E2} \text{3-ethyl-2-pentene} \quad \text{(only product)}$$

3-Bromo-3-ethylpentane

$$\text{2-Bromo-3-ethylpentane} \xrightarrow{E2} \text{3-Ethyl-1-pentene} + \text{3-Ethyl-2-pentene}$$

2-Bromo-3-ethylpentane  3-Ethyl-1-pentene  3-Ethyl-2-pentene

Compound A must therefore be 2-bromo-3-ethylpentane. Dehydrobromination of A will follow Zaitsev's rule, so that the major alkene (compound B) is 3-ethyl-2-pentene and the minor alkene (compound C) is 3-ethyl-1-pentene.

**6.46** The information that compound B gives 2,4-dimethylpentane on catalytic hydrogenation establishes its carbon skeleton.

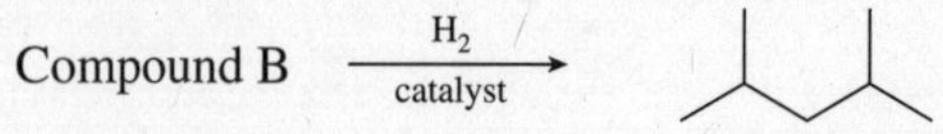

Compound B $\xrightarrow[\text{catalyst}]{H_2}$ 2,4-Dimethylpentane

Compound B is an alkene derived from compound A—an alkyl bromide of molecular formula $C_7H_{15}Br$. We are told that compound A is not a primary alkyl bromide. Compound A can therefore be only:

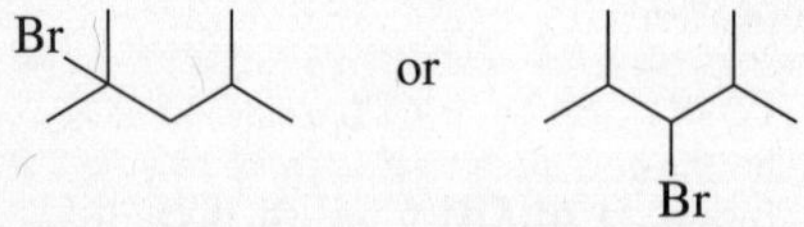

Br ... or ... Br

Since compound A gives a single alkene on being treated with sodium ethoxide in ethanol, it can only be 3-bromo-2,4-dimethylpentane, and compound B must be 2,4-dimethyl-2-pentene.

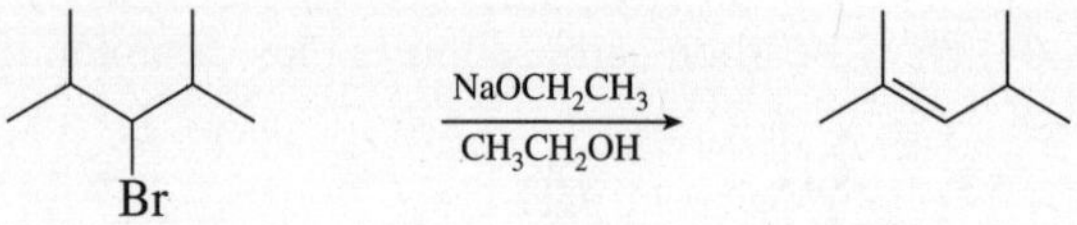

$\xrightarrow[CH_3CH_2OH]{NaOCH_2CH_3}$

3-Bromo-2,4-dimethylpentane (compound A)

2,4-Dimethyl-2-pentene (compound B)

**6.47** Alkene C must have the same carbon skeleton as its hydrogenation product, 2,3,3,4-tetramethylpentane.

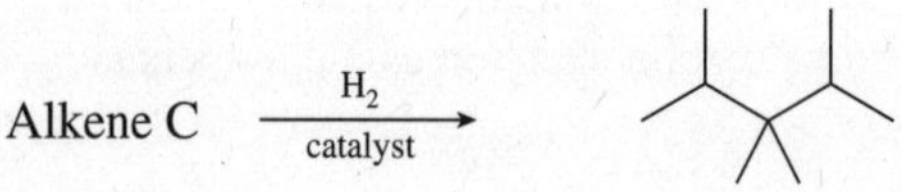

Alkene C $\xrightarrow[\text{catalyst}]{H_2}$ 2,3,3,4-Tetramethylpentane

Alkene C can only therefore be 2,3,3,4-tetramethyl-1-pentene. The two alkyl bromides, compounds A and B, that give this alkene on dehydrobromination have their bromine substituents at C-1 and C-2, respectively.

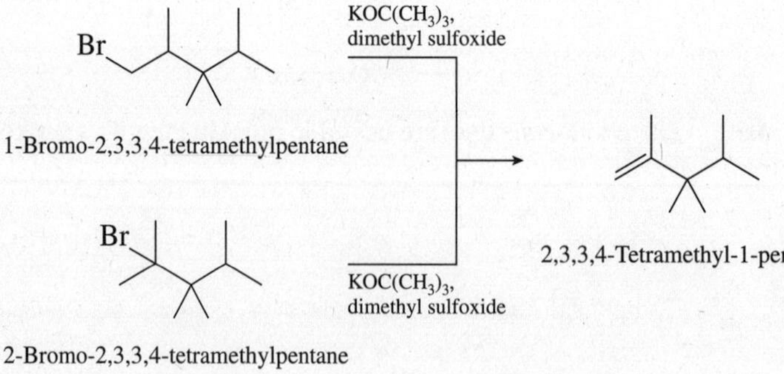

1-Bromo-2,3,3,4-tetramethylpentane $\xrightarrow{KOC(CH_3)_3, \text{ dimethyl sulfoxide}}$

2-Bromo-2,3,3,4-tetramethylpentane $\xrightarrow{KOC(CH_3)_3, \text{ dimethyl sulfoxide}}$

2,3,3,4-Tetramethyl-1-pentene

**6.48** The only alcohol (compound A) that can undergo acid-catalyzed dehydration to alkene B without rearrangement is the one shown in the equation.

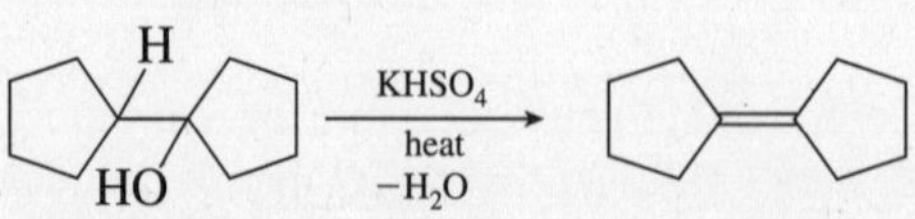

H, HO $\xrightarrow[\text{heat}, -H_2O]{KHSO_4}$

Alcohol A

Alkene B

Dehydration of alcohol A also yields an isomeric alkene under these conditions.

$$\text{Alcohol A} \xrightarrow[\text{heat, } -H_2O]{KHSO_4} \text{Alkene C}$$

Alcohol A     Alkene C

**6.49** Electrophilic addition of hydrogen iodide should occur in accordance with Markovnikov's rule.

$$H_2C{=}CHC(CH_3)_3 \underset{\text{KOH, n-PrOH}}{\overset{HI}{\rightleftharpoons}} CH_3CH(I)C(CH_3)_3$$

3,3-Dimethyl-1-butene     3-Iodo-2,2-dimethylbutane

Treatment of 3-iodo-2,2-dimethylbutane with alcoholic potassium hydroxide should bring about E2 elimination to regenerate the starting alkene. Hence, compound A is 3-iodo-2,2-dimethylbutane.

The carbocation intermediate formed in the addition of hydrogen iodide to the alkene is one which can rearrange by a methyl group migration.

$$H_2C{=}CHC(CH_3)_3 + H^+ \longrightarrow H_3C{-}\overset{+}{C}H{-}C(CH_3)_3 \xrightarrow{\text{methyl migration}} H_3C{-}CH(CH_3){-}\overset{+}{C}(CH_3)_2$$

$$\downarrow I^- \qquad\qquad \downarrow I^-$$

$$CH_3CH(I){-}C(CH_3)_3 \qquad (CH_3)_2CH{-}C(I)(CH_3)_2$$

Compound A     Compound B (2-iodo-2,3-dimethylbutane)

A likely candidate for compound B is therefore the one with a rearranged carbon skeleton, 2-iodo-2,3-dimethylbutane. This is confirmed by the fact that compound B undergoes elimination to give 2,3-dimethyl-2-butene.

$$(CH_3)_2CH{-}C(I)(CH_3)_2 \xrightarrow{E2} (CH_3)_2C{=}C(CH_3)_2$$

Compound B     2,3-Dimethyl-2-butene

**6.50** The ozonolysis data are useful in quickly identifying alkenes A and B.

$$\text{Compound A} \longrightarrow HCH{=}O + (CH_3)_3CC(=O)C(CH_3)_3$$

Compound A is therefore 2-*tert*-butyl-3,3-dimethyl-1-butene.

$$(CH_3)_3CC(=CH_2)C(CH_3)_3$$

Compound A

$$\text{Compound B} \longrightarrow HCH{=}O + CH_3C(=O){-}C(CH_3)_2{-}C(CH_3)_3$$

Compound B is therefore 2,3,3,4,4-pentamethyl-1-pentene.

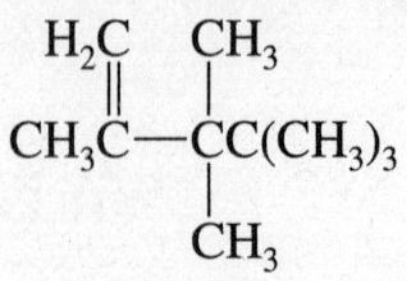

$$CH_3C(=CH_2)\text{—}C(CH_3)_2C(CH_3)_3$$

Compound B

Compound B has a carbon skeleton different from the alcohol that produced it by dehydration. We are therefore led to consider a carbocation rearrangement.

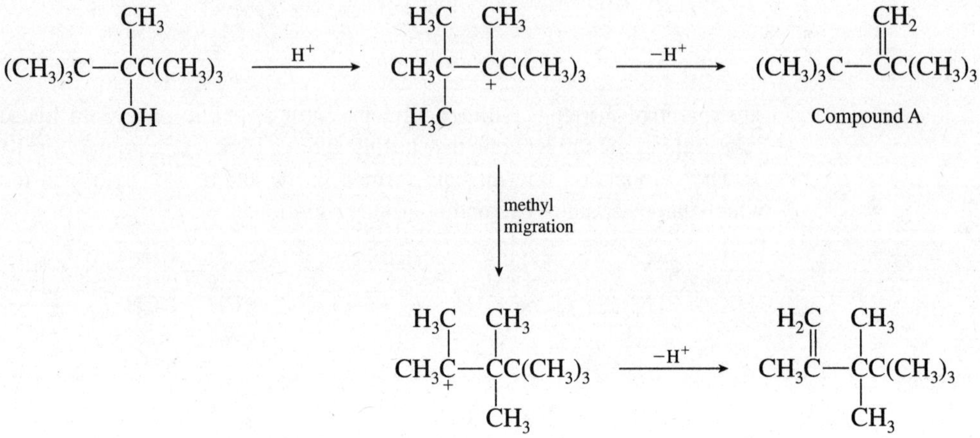

$$(CH_3)_3C\text{—}C(CH_3)(OH)C(CH_3)_3 \xrightarrow{H^+} CH_3C(CH_3)_2\text{—}\overset{+}{C}(CH_3)C(CH_3)_3 \xrightarrow{-H^+} (CH_3)_3C\text{—}C(=CH_2)C(CH_3)_3$$

Compound A

methyl migration

$$CH_3\overset{+}{C}(CH_3)\text{—}C(CH_3)_2C(CH_3)_3 \xrightarrow{-H^+} CH_3C(=CH_2)\text{—}C(CH_3)_2C(CH_3)_3$$

Compound B

**6.51** The important clue to deducing the structures of A and B is the ozonolysis product C. Remembering that the two carbonyl carbons of C must have been joined by a double bond in the precursor B, we write

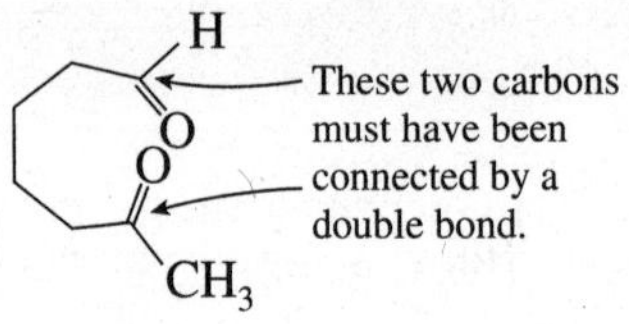

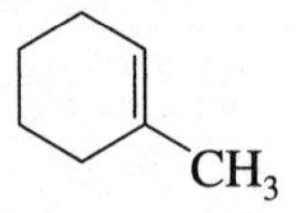

These two carbons must have been connected by a double bond.

Compound C Compound B

The tertiary bromide that gives compound B on dehydrobromination is 1-methylcyclohexyl bromide.

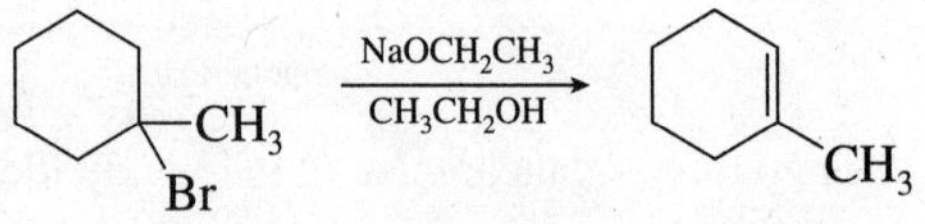

$$\xrightarrow[CH_3CH_2OH]{NaOCH_2CH_3}$$

Compound A Compound B

When tertiary halides are treated with base, they undergo E2 elimination. The regioselectivity of elimination of tertiary halides follows the Zaitsev rule.

**6.52** Since santene and 1,3-diacetylcyclopentane (compound A) contain the same number of carbon atoms, the two carbonyl carbons of the diketone must have been connected by a double bond in santene. The structure of santene must therefore be

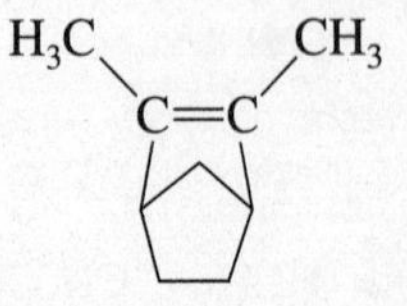

$H_3C$ $CH_3$ C=C more appropriately represented as $CH_3$ $CH_3$

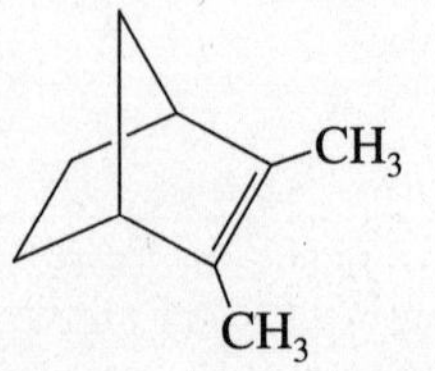

**6.53** (*a*) Compound A contains nine of the ten carbons and 14 of the 16 hydrogens of sabinene. Ozonolysis has led to the separation of one carbon and two hydrogens from the rest of the molecule. The carbon and the two hydrogens must have been lost as formaldehyde, $H_2C{=}O$. This $H_2C$ unit was originally doubly bonded to the carbonyl carbon of compound A. Sabinene must therefore have the structure shown in the equation representing its ozonolysis:

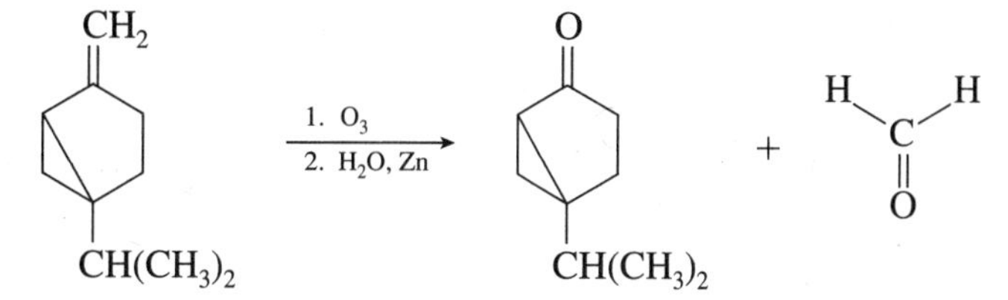

$CH_2$ ... $CH(CH_3)_2$ $\xrightarrow[\text{2. } H_2O, Zn]{\text{1. } O_3}$ O ... $CH(CH_3)_2$ + $H_2C{=}O$

Sabinene — Compound A — Formaldehyde

(*b*) Compound B contains all ten of the carbons and all 16 of the hydrogens of $\Delta^3$-carene. The two carbonyl carbons of compound B must have been linked by a double bond in $\Delta^3$-carene.

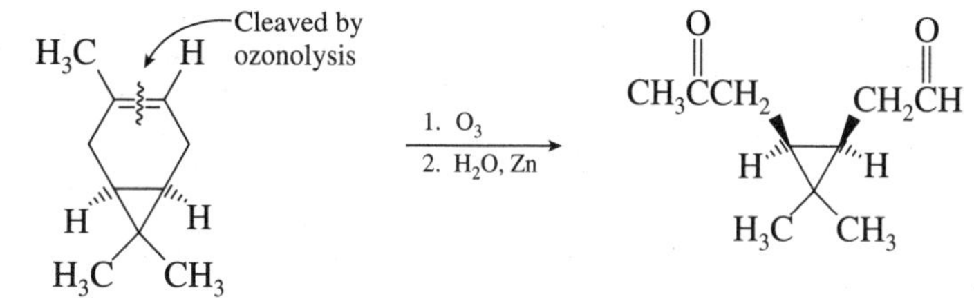

Cleaved by ozonolysis

$\Delta^3$-Carene $\xrightarrow[\text{2. } H_2O, Zn]{\text{1. } O_3}$ Compound B ($CH_3\overset{O}{\overset{\|}{C}}CH_2$, $CH_2\overset{O}{\overset{\|}{C}}H$, H, H, $H_3C$, $CH_3$)

**6.54** The sex attractant of the female housefly consumes one mole of hydrogen on catalytic hydrogenation (the molecular formula changes from $C_{23}H_{46}$ to $C_{23}H_{48}$). Thus, the molecule has one double bond. The position of the double bond is revealed by the ozonolysis data.

$$C_{23}H_{46} \xrightarrow[\text{2. } H_2O, Zn]{\text{1. } O_3} CH_3(CH_2)_7\overset{O}{\overset{\|}{C}}H + CH_3(CH_2)_{12}\overset{O}{\overset{\|}{C}}H$$

An unbranched 9-carbon unit and an unbranched 14-carbon unit make up the carbon skeleton, and these two units must be connected by a double bond. The housefly sex attractant therefore has the constitution:

$$CH_3(CH_2)_7CH{=}CH(CH_2)_{12}CH_3$$

9-Tricosene

The data cited in the problem do not permit the stereochemistry of this natural product to be determined.

**6.55** The hydrogenation data tell us that $C_{19}H_{38}$ contains one double bond and has the same carbon skeleton as 2,6,10,14-tetramethylpentadecane. We locate the double bond at C-2 on the basis of the fact that acetone, $(CH_3)_2C{=}O$, is obtained on ozonolysis. The structures of the natural product and the aldehyde produced on its ozonolysis are as follows:

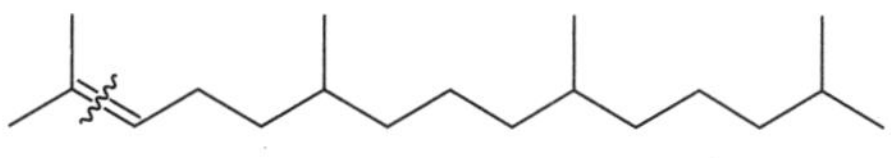

Ozonolysis cleaves molecule here. 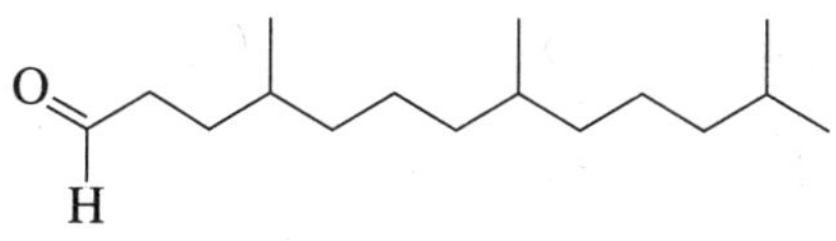 — Aldehyde obtained on ozonolysis

**6.56** Since $HC(=O)CH_2C(=O)H$ (HCCH$_2$CH with O above each terminal carbon) is one of the products of its ozonolysis, the sex attractant of the arctiid moth must contain the unit $=CHCH_2CH=$. This unit must be bonded to an unbranched 12-carbon unit at one end and an unbranched 6-carbon unit at the other in order to give $CH_3(CH_2)_{10}CH=O$ and $CH_3(CH_2)_4CH=O$ on ozonolysis.

$$CH_3(CH_2)_{10}CH \wr\!= CHCH_2CH =\!\wr CH(CH_2)_4CH_3$$

Sex attractant of arctiid moth
(wavy lines show positions of cleavage on ozonolysis)

$$\downarrow \text{1. } O_3 \quad \text{2. } H_2O, Zn$$

$$CH_3(CH_2)_{10}\overset{O}{\overset{\|}{C}}H + H\overset{O}{\overset{\|}{C}}CH_2\overset{O}{\overset{\|}{C}}H + H\overset{O}{\overset{\|}{C}}(CH_2)_4CH_3$$

The stereochemistry of the double bonds cannot be determined on the basis of the available information.

**6.57–6.59** Solutions to molecular modeling exercises are not provided in this *Study Guide and Solutions Manual.* You should use *Learning By Modeling* for these exercises.

# SELF-TEST

## PART A

**A-1.** How many different alkenes will yield 2,3-dimethylpentane on catalytic hydrogenation? Draw their structures, and name them.

**A-2.** Write structural formulas for the reactant, reagents, or product omitted from each of the following:

(*a*) $(CH_3)_2C=CHCH_3 \xrightarrow{H_2SO_4\text{(dilute)}} ?$

(*b*) (cyclohexylidene)$=CH_2 \xrightarrow{?}$ cyclohexyl$-CH_2Br$

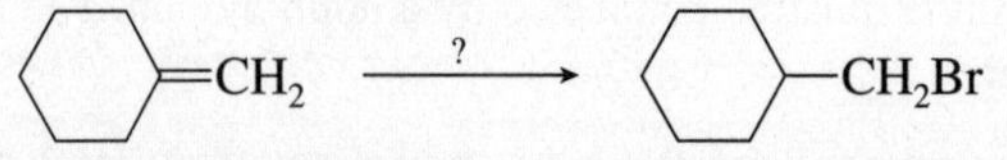

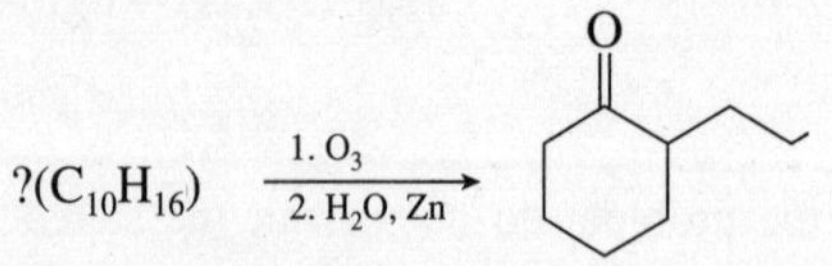

(*c*) $?(C_{10}H_{16}) \xrightarrow[\text{2. } H_2O, Zn]{\text{1. } O_3}$ 2-propylcyclohexanone (cyclohexanone with a propyl group on the carbon adjacent to C=O)

(*d*) 1-methylcyclopentene ($CH_3$) $+ Br_2 \xrightarrow{H_2O} ?$

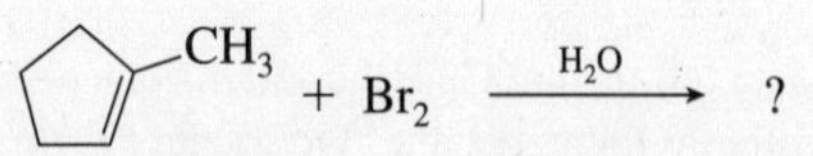

**A-3.** Provide a sequence of reactions to carry out the following conversions. More than one synthetic step is necessary for each. Write the structure of the product of each synthetic step.

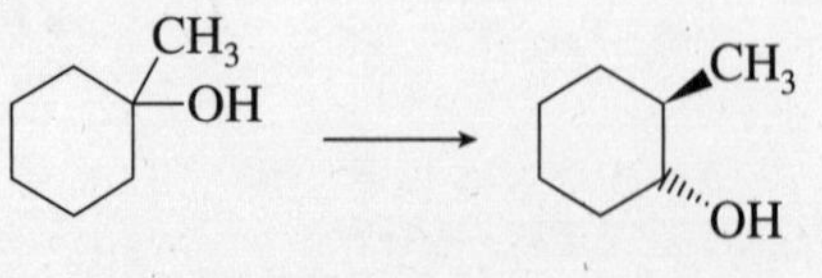

(*a*) 1-methylcyclohexanol ($CH_3$, $-OH$ on same carbon) $\longrightarrow$ trans-2-methylcyclohexanol ($CH_3$ wedge, $OH$ hashed on adjacent carbon) (*b*) $CH_3CH_2CH(Cl)CH(CH_3)_2 \longrightarrow CH_3CH_2\overset{O}{CH-C}(CH_3)_2$ (epoxide)

(*c*) $(CH_3)_3CCH(Br)CH_3 \longrightarrow (CH_3)_3CCH_2CH_2Br$

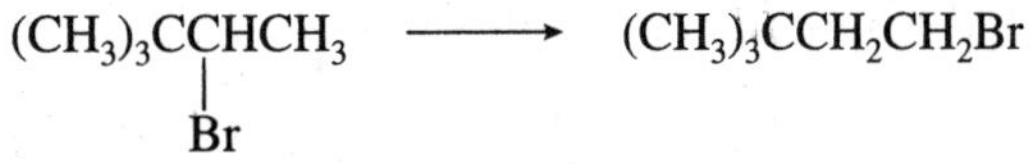

**A-4.** Provide a detailed mechanism describing the elementary steps in the reaction of 1-butene with HBr in the presence of peroxides.

**A-5.** Chlorine reacts with an alkene to give the 2,3-dichlorobutane isomer whose structure is shown. What are the structure and name of the alkene? Outline a mechanism for the reaction.

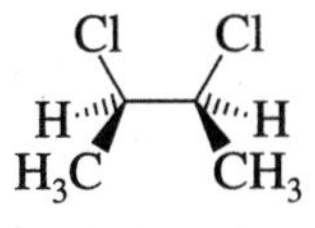

Cl Cl
H H
$H_3C$ $CH_3$

**A-6.** Write a structural formula, including stereochemistry, for the compound formed from *cis*-3-hexene on treatment with peroxyacetic acid.

**A-7.** Give a mechanism describing the elementary steps in the reaction of 2-methyl-1-butene with hydrogen chloride. Use curved arrows to show the flow of electrons.

**A-8.** What two alkenes give 2-chloro-2-methylbutane on reaction with hydrogen chloride?

**A-9.** Give the major organic product formed from the following sequence of reactions.

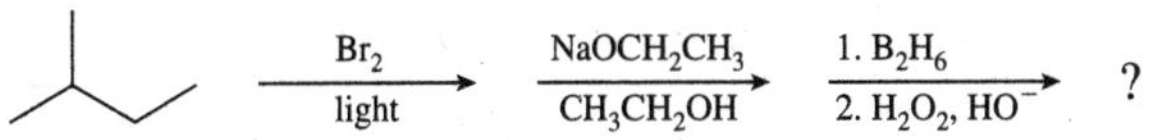

$\xrightarrow[\text{light}]{Br_2} \xrightarrow[CH_3CH_2OH]{NaOCH_2CH_3} \xrightarrow[2.\ H_2O_2,\ HO^-]{1.\ B_2H_6}$ ?

**A-10.** The reaction of 3-methyl-1-butene with hydrogen chloride gives two alkyl halide products; one is a secondary alkyl chloride and the other is tertiary. Write the structures of the products, and provide a mechanism explaining their formation.

**A-11.** A hydrocarbon A ($C_6H_{12}$) undergoes reaction with HBr to yield compound B ($C_6H_{13}Br$). Treatment of B with sodium ethoxide in ethanol yields C, an isomer of A. Reaction of C with ozone followed by treatment with water and zinc gives acetone, $(CH_3)_2C{=}O$, as the only organic product. Provide structures for A, B, and C, and outline the reaction pathway.

## PART B

**B-1.** Rank the following alkenes in order of decreasing heats of hydrogenation (largest first)

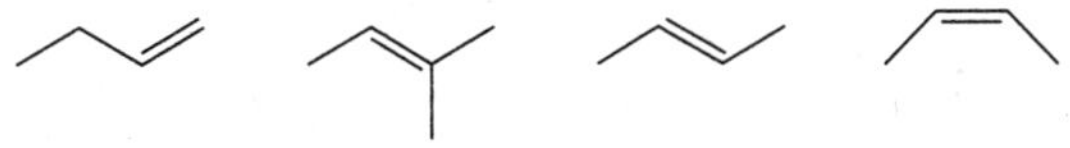

1 2 3 4

(*a*) 2 > 3 > 4 > 1
(*b*) 1 > 3 > 4 > 2
(*c*) 1 > 4 > 3 > 2
(*d*) 2 > 4 > 3 > 1
(*e*) 1 > 2 > 3 > 4

**B-2.** The product from the reaction of 1-pentene with $Cl_2$ in $H_2O$ is named:
(*a*) 1-Chloro-2-pentanol
(*b*) 2-Chloro-2-pentanol
(*c*) 1-Chloro-1-pentanol
(*d*) 2-Chloro-1-pentanol

**B-3.** In the reaction of hydrogen bromide with an alkene (in the absence of peroxides), the first step of the reaction is the _____ to the alkene.
(*a*) Fast addition of an electrophile
(*b*) Fast addition of a nucleophile
(*c*) Slow addition of an electrophile
(*d*) Slow addition of a nucleophile

**B-4.** The major product of the following reaction sequence is

1. $B_2H_6$
2. $H_2O_2$, $HO^-$ → ?

(*a*) OH (*c*) OH (*e*) =O + O= H

(*b*) O (*d*) OH

**B-5.** Which, if any, of the following alcohols *cannot* be prepared from an alkene?

(*a*) OH (*c*) OH

(*b*) OH (*d*) OH

(*e*) None of these—all of the alcohols shown can be prepared from an alkene

**B-6.** Which of the species shown is the most stable form of the intermediate in the electrophilic addition of $Cl_2$ in water to cyclohexene to form a halohydrin? Electron pairs have been omitted for convenience, and their absence should not be considered as part of the problem.

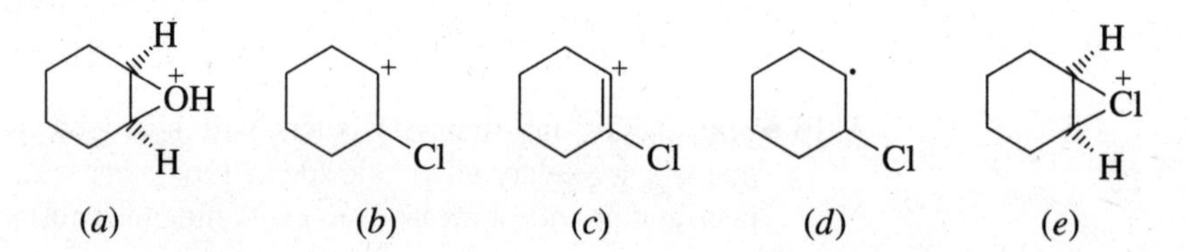


**B-7.** Treatment of 2-methyl-2-butene with HBr in the presence of a peroxide yields
(*a*) A primary alkyl bromide
(*b*) A secondary alkyl bromide
(*c*) A tertiary alkyl bromide
(*d*) A vicinal dibromide

**B-8.** The reaction

$$(CH_3)_2C{=}CH_2 + Br\cdot \longrightarrow (CH_3)_2\dot{C}{-}CH_2Br$$

is an example of a(n) ______ step in a radical chain reaction.
(*a*) Initiation (*c*) Termination
(*b*) Propagation (*d*) Heterolytic cleavage

**B-9.** To which point on the potential energy diagram for the reaction of 2-methylpropene with hydrogen chloride does the figure shown at the right correspond?

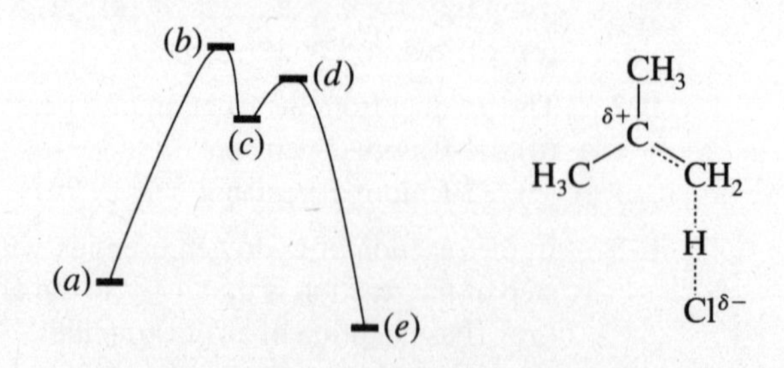

**B-10.** Which of the following most accurately describes the first step in the reaction of hydrogen chloride with 1-butene?

(*a*) 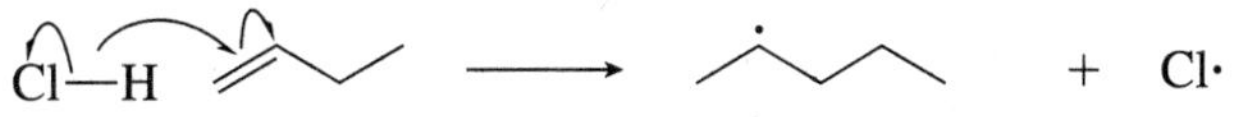

(*b*) 

(*c*) 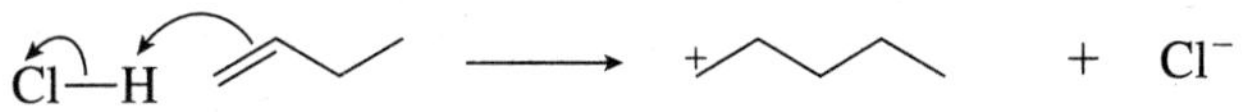

(*d*) 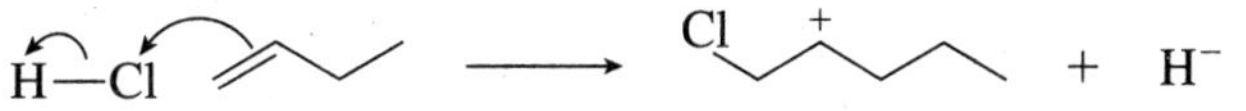

**B-11.** Which of the following best describes the flow of electrons in the acid-catalyzed dimerization of $(CH_3)_2C{=}CH_2$?

(*a*) 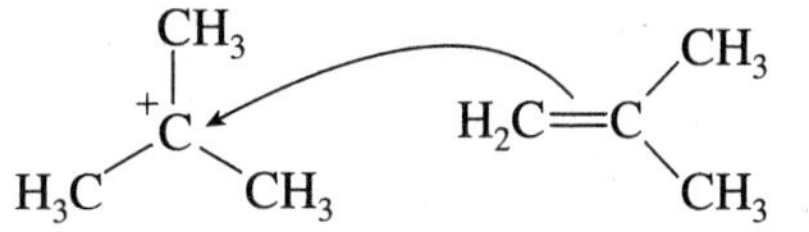

(*c*) 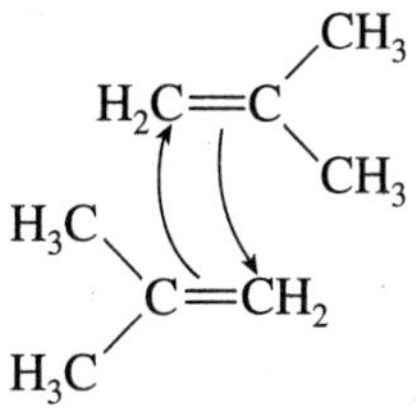

(*b*) 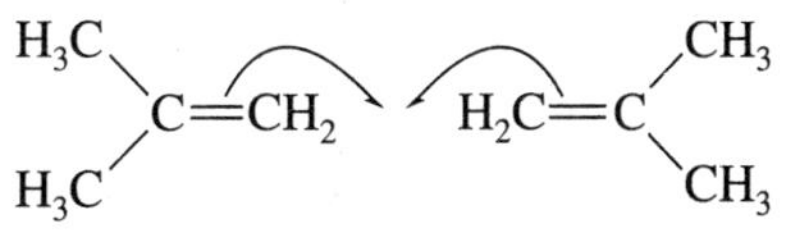

(*d*) 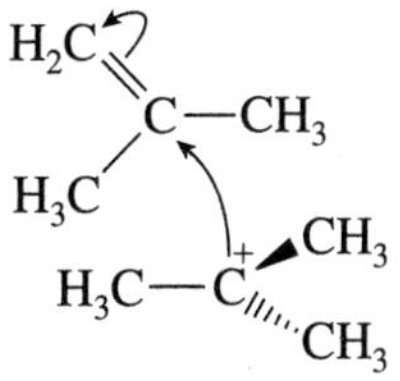

**B-12.** Which one of the following compounds gives acetone $(CH_3)_2C{=}O$ as one of the products of its ozonolysis?

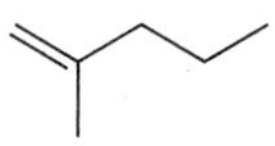
(*a*)

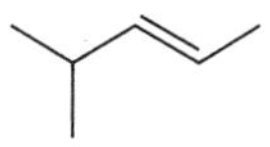
(*b*)

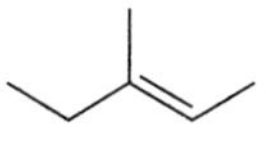
(*c*)

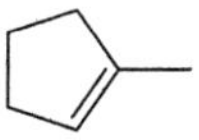
(*d*)

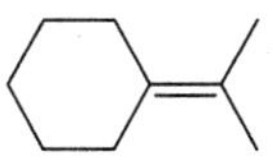
(*e*)

**B-13.** Addition of HCl to 3,3-dimethyl-1-butene yields two products, one of which has a rearranged carbon skeleton. Which of the following cations are intermediates in that reaction?

| | | | |
|---|---|---|---|
| $(CH_3)_3C\overset{+}{C}HCH_2Cl$ | $(CH_3)_3C\overset{+}{C}HCH_3$ | $(CH_3)_2C(Cl)\overset{+}{C}(CH_3)_2$ | $(CH_3)_2\overset{+}{C}CH(CH_3)_2$ |
| 1 | 2 | 3 | 4 |

(*a*) 1, 2 (*b*) 1, 3 (*c*) 1, 4 (*d*) 2, 3 (*e*) 2, 4

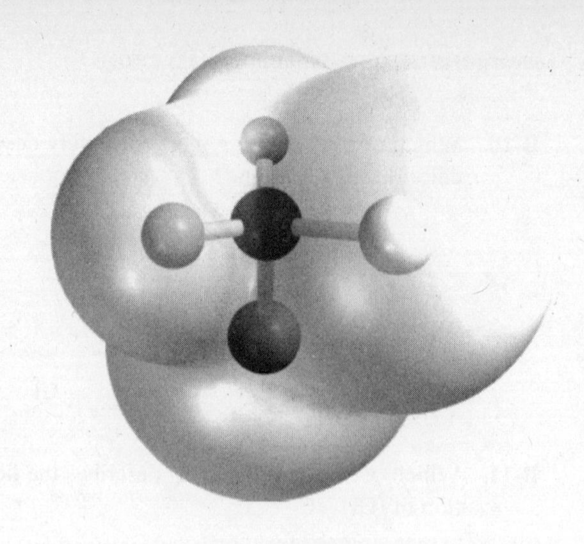

# CHAPTER 7
## STEREOCHEMISTRY

## SOLUTIONS TO TEXT PROBLEMS

**7.1** (*c*) Carbon-2 is a stereogenic center in 1-bromo-2-methylbutane, as it has four different substituents: H, $CH_3$, $CH_3CH_2$, and $BrCH_2$.

$$\begin{array}{c} \text{H} \\ | \\ BrCH_2\text{—}\underset{|}{\overset{}{C}}\text{—}CH_2CH_3 \\ CH_3 \end{array}$$

(*d*) There are no stereogenic centers in 2-bromo-2-methylbutane.

$$\begin{array}{c} \text{Br} \\ | \\ CH_3\text{—}\underset{|}{\overset{}{C}}\text{—}CH_2CH_3 \\ CH_3 \end{array}$$

**7.2** (*b*) Carbon-2 is a stereogenic center in 1,1,2-trimethylcyclobutane.

$CH_3$, H, $H_3C$, $CH_3$ — A stereogenic center; the four substituents to which it is directly bonded [H, $CH_3$, $CH_2$, and $C(CH_3)_2$] are all different from one another.

1,1,3-Trimethylcyclobutane however, has no stereogenic centers.

$H_3C$, $H_3C$, $CH_3$, H — Not a stereogenic center; two of its substituents are the same.

**7.3** (*b*) There are *two* planes of symmetry in (*Z*)-1,2-dichloroethene, of which one is the plane of the molecule and the second bisects the carbon–carbon bond. There is no center of symmetry. The molecule is achiral.

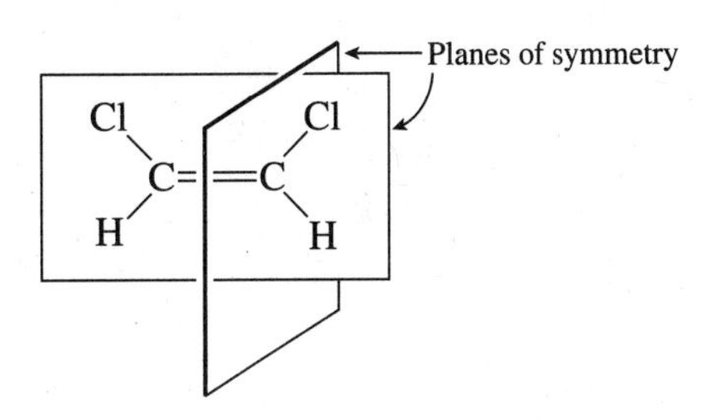


(*c*) There is a plane of symmetry in *cis*-1,2-dichlorocyclopropane that bisects the C-1—C-2 bond and passes through C-3. The molecule is achiral.

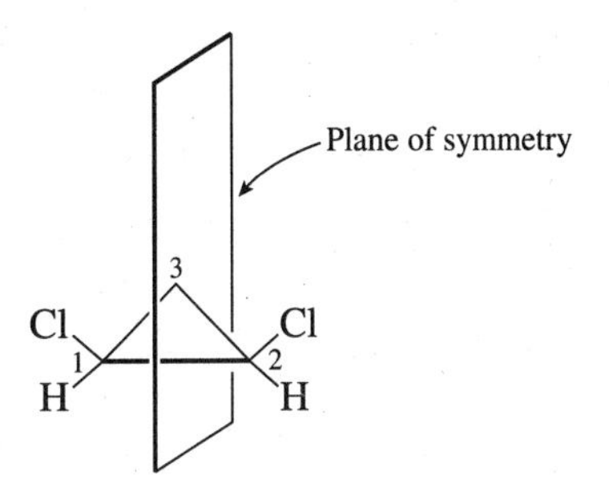


(*d*) *trans*-1,2-Dichlorocyclopropane has neither a plane of symmetry nor a center of symmetry. Its two mirror images cannot be superposed on each other. The molecule is chiral.

Cl, H / H, Cl and H, Cl / Cl, H are nonsuperposable mirror images

**7.4** The equation relating specific rotation $[\alpha]$ to observed rotation $\alpha$ is

$$[\alpha] = \frac{100\alpha}{cl}$$

The concentration $c$ is expressed in grams per 100 mL and the length $l$ of the polarimeter tube in decimeters. Since the problem specifies the concentration as 0.3 g/15 mL and the path length as 10 cm, the specific rotation $[\alpha]$ is:

$$[\alpha] = \frac{100(-0.78^\circ)}{100(0.3\text{ g}/15\text{ mL})(10\text{ cm}/10\text{ cm/dm})}$$
$$= -39^\circ$$

**7.5** From the previous problem, the specific rotation of natural cholesterol is $[\alpha] = -39^\circ$. The mixture of natural $(-)$-cholesterol and synthetic $(+)$-cholesterol specified in this problem has a specific rotation $[\alpha]$ of $-13^\circ$.

$$\text{Optical purity} = \%(-)\text{-cholesterol} - \%\,(+)\text{-cholesterol}$$
$$33.3\% = \%(-)\text{-cholesterol} - [100 - \%\,(-)\text{-cholesterol}]$$
$$133.3\% = 2\,[\%\,(-)\text{-cholesterol}]$$
$$66.7\% = \%\,(-)\text{-cholesterol}$$

The mixture is two thirds natural $(-)$-cholesterol and one third synthetic $(+)$-cholesterol.

**7.6** Draw the molecular model so that it is in the same format as the drawings of (+) and (−)-2-butanol in the text.

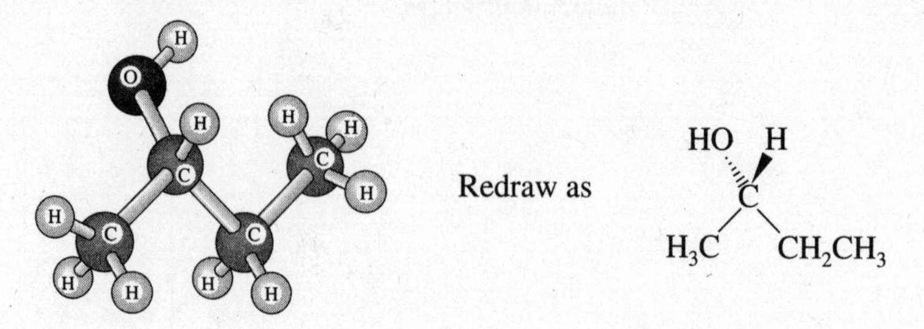


Reorient the molecule so that it can be compared with the drawings of (+) and (−)-2-butanol.

Reorient

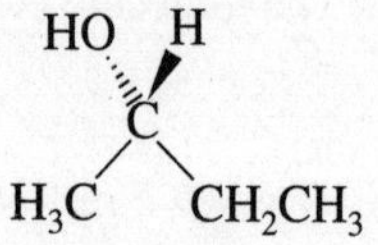

HO H
C
$H_3C$ $CH_2CH_3$

which becomes

$CH_3CH_2$ H
C—OH
$H_3C$

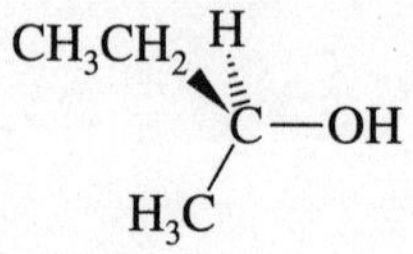

The molecular model when redrawn matches the text's drawing of (+)-2-butanol.

**7.7** (*b*) The solution to this problem is exactly analogous to the sample solution given in the text to part (*a*).

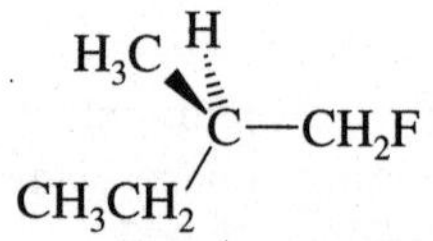

$H_3C$ H
C—$CH_2F$
$CH_3CH_2$

(+)-1-Fluoro-2-methylbutane

**Order of precedence:** $CH_2F > CH_3CH_2 > CH_3 > H$

The lowest ranked substituent (H) at the stereogenic center points away from us in the drawing. The three higher ranked substituents trace a clockwise path from $CH_2F$ to $CH_2CH_3$ to $CH_3$.

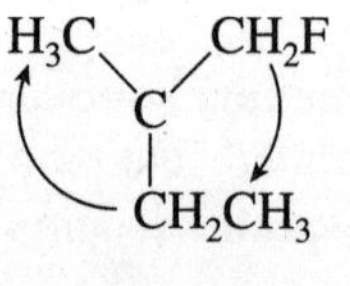

$H_3C$ $CH_2F$
C
$CH_2CH_3$

The absolute configuration is *R*; the compound is (*R*)-(+)-1-fluoro-2-methylbutane.

(*c*) The highest ranked substituent at the stereogenic center of 1-bromo-2-methylbutane is $CH_2Br$, and the lowest ranked substituent is H. Of the remaining two, ethyl outranks methyl.

**Order of precedence:** $CH_2Br > CH_2CH_3 > CH_3 > H$

The lowest ranking substituent (H) is directed toward you in the drawing, and therefore the molecule needs to be reoriented so that H points in the opposite direction.

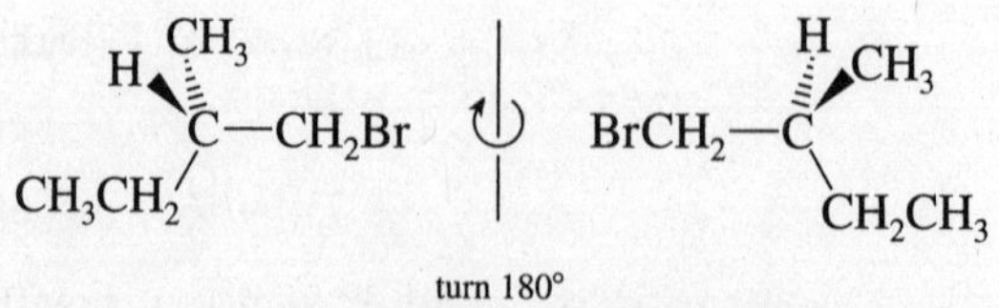

$CH_3$
H
C—$CH_2Br$
$CH_3CH_2$

turn 180°

H $CH_3$
$BrCH_2$—C
$CH_2CH_3$

(+)-1-Bromo-2-methylbutane

The three highest ranking substituents trace a counterclockwise path when the lowest ranked substituent is held away from you.

$BrCH_2$, $CH_3$, C, $CH_2CH_3$

The absolute configuration is *S*, and thus the compound is (*S*)-(+)-1-bromo-2-methylbutane.

(*d*) The highest ranked substituent at the stereogenic center of 3-buten-2-ol is the hydroxyl group, and the lowest ranked substituent is H. Of the remaining two, vinyl outranks methyl.

**Order of precedence:** $HO > CH_2{=}CH > CH_3 > H$

The lowest ranking substituent (H) is directed away from you in the drawing. We see that the order of decreasing precedence appears in a counterclockwise manner.

$H_3C$, H, $C{-}CH_2{=}CH_2$, HO — $H_3C$, $CH{=}CH_2$, C, OH

(+)-3-Buten-2-ol

The absolute configuration is *S*, and the compound is (*S*)-(+)-3-buten-2-ol.

**7.8** (*b*) The stereogenic center is the carbon that bears the methyl group. Its substituents are:

$$-CF_2CH_2 > -CH_2CF_2 > CH_3 > H$$

Highest rank — Lowest rank

When the lowest ranked substituent points away from you, the remaining three must appear in descending order of precedence in a counterclockwise fashion in the *S* enantiomer. (*S*)-1, 1-difluoro-2-methylcyclopropane is therefore

H, $CH_3$, F, F

**7.9** (*b*) The Fischer projection of (*R*)-(+)-1-fluoro-2-methylbutane is analogous to that of the alcohol in part (*a*). The only difference in the two is that fluorine has replaced hydroxyl as a substituent at C-1.

$H_3C$, H, $C{-}CH_2F$, $CH_3CH_2$ is the same as $CH_2F$, $H{-}C{-}CH_3$, $CH_2CH_3$ which becomes the Fischer projection $CH_2F$, $H{-}{+}{-}CH_3$, $CH_2CH_3$

Although other Fischer projections may be drawn by rotating the perspective view in other directions, the one shown is preferred because it has the longest chain of carbon atoms oriented on the vertical axis with the lowest numbered carbon at the top.

(*c*) As in the previous parts of this problem, orient the structural formula of (*S*)-(+)-1-bromo-2-methylbutane so the segment $BrCH_2{-}C{-}CH_2CH_3$ is aligned vertically with the lowest numbered carbon at the top.

H, $CH_3$, $C{-}CH_2Br$, $CH_3CH_2$ is the same as $CH_2Br$, $H_3C{-}C{-}H$, $CH_2CH_3$ which becomes the Fischer projection $CH_2Br$, $H_3C{-}{+}{-}H$, $CH_2CH_3$

(*d*) Here we need to view the molecule from behind the page in order to write the Fischer projection of (*S*)-(+)-3-buten-2-ol.

$CH_3$, H
C—CH=$CH_2$
HO

is the same as

$CH_3$
H—C—OH
CH=$CH_2$

which becomes the Fischer projection

$CH_3$
H—+—OH
CH=$CH_2$

**7.10** In order of decreasing rank, the substituents attached to the stereogenic center in lactic acid are —OH, —$CO_2H$, —$CH_3$, and —H. The Fischer projection given for (+)-lactic acid (*a*) corresponds to the three-dimensional representation (*b*), which can be reoriented as in (*c*). When (*c*) is viewed from the side opposite the lowest ranked substituent (H), the order of decreasing precedence is anticlockwise, as shown in (*d*). (+)-Lactic acid has the *S* configuration.

$CO_2H$
HO—+—H
$CH_3$
(*a*)

$CO_2H$
HO—C—H
$CH_3$
(*b*)

HO $CO_2H$
C—H
$H_3C$
(*c*)

$HO_2C$ OH
$CH_3$
(*d*)

**7.11** The erythro stereoisomers are characterized by Fischer projections in which analogous substituents, in this case OH and $NH_2$, are on the same side when the carbon chain is vertical. There are two erythro stereoisomers that are enantiomers of each other:

$CH_3$
H—+—OH
H—+—$NH_2$
$CH_3$
Erythro

$CH_3$
HO—+—H
$H_2N$—+—H
$CH_3$
Erythro

Analogous substituents are on opposite sides in the threo isomer:

$CH_3$
H—+—OH
$H_2N$—+—H
$CH_3$
Threo

$CH_3$
HO—+—H
H—+—$NH_2$
$CH_3$
Threo

**7.12** There are four stereoisomeric forms of 3-amino-3-butanol:

(2*R*,3*R*) and its enantiomer (2*S*,3*S*)
(2*R*,3*S*) and its enantiomer (2*S*,3*R*)

In the text we are told that the (2*R*,3*R*) stereoisomer is a liquid. Its enantiomer (2*S*,3*S*) has the same physical properties and so must also be a liquid. The text notes that the (2*R*,3*S*) stereoisomer is a solid (mp 49°C). Its enantiomer (2*S*,3*R*) must therefore be the other stereoisomer that is a crystalline solid.

**7.13** Examine the structural formula of each compound for equivalently substituted stereogenic centers. The only one capable of existing in a meso form is 2,4-dibromopentane.

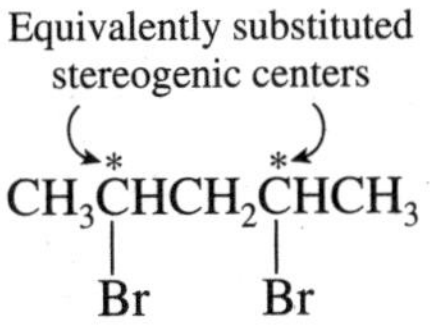

Equivalently substituted stereogenic centers

$$CH_3\overset{*}{C}HCH_2\overset{*}{C}HCH_3 \text{ (Br on C-2 and C-4)}$$

2,4-Dibromopentane

$CH_3$ / H—C—Br / H—C—H (Plane of symmetry) / H—C—Br / $CH_3$

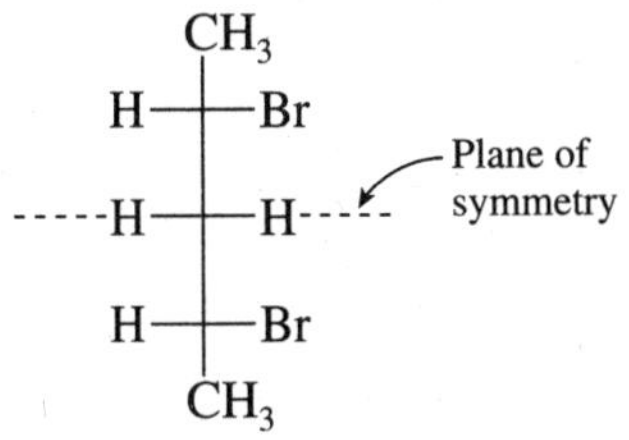

Fischer projection of *meso*-2,4-dibromopentane

None of the other compounds has equivalently substituted stereogenic centers. No meso forms are possible for:

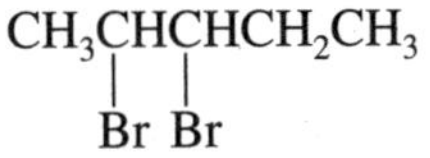

$CH_3CHCHCH_2CH_3$ (Br, Br) — 2,3-Dibromopentane

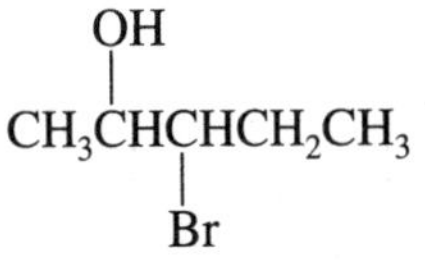

$CH_3CHCHCH_2CH_3$ (OH, Br) — 3-Bromo-2-pentanol

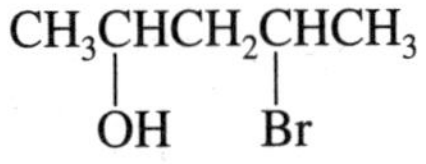

$CH_3CHCH_2CHCH_3$ (OH, Br) — 4-Bromo-2-pentanol

**7.14** There is a plane of symmetry in the cis stereoisomer of 1,3-dimethylcyclohexane, and so it is an achiral substance—it is a meso form.

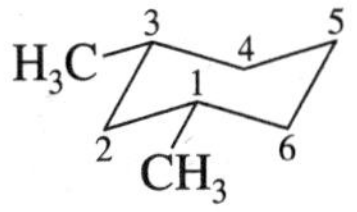

$H_3C$, $CH_3$ (ring positions 1–6)

Plane of symmetry passes through C-2 and C-5 and bisects the ring.

The trans stereoisomer is chiral. It is not a meso form.

**7.15** A molecule with three stereogenic centers has $2^3$, or 8, stereoisomers. The eight combinations of *R* and *S* stereogenic centers are:

| | *Stereogenic center* | | *Stereogenic center* |
|---|---|---|---|
| | **1 2 3** | | **1 2 3** |
| Isomer 1 | *R R R* | Isomer 5 | *S S S* |
| Isomer 2 | *R R S* | Isomer 6 | *S S R* |
| Isomer 3 | *R S R* | Isomer 7 | *S R S* |
| Isomer 4 | *S R R* | Isomer 8 | *R S S* |

**7.16** 2-Hexuloses have three stereogenic centers. They are marked with asterisks in the structural formula.

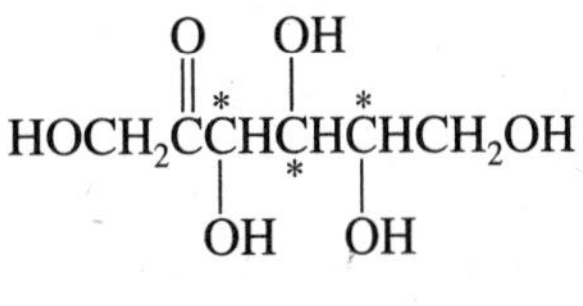

$$HOCH_2\overset{O}{\overset{\|}{C}}\overset{*}{C}H(OH)\overset{*}{C}H(OH)\overset{*}{C}H(OH)CH_2OH$$

No meso forms are possible, and so there are a total of $2^3$, or 8, stereoisomeric 2-hexuloses.

**7.17** Epoxidation of (*Z*)-2-butene gives the meso (achiral) epoxide. Oxygen transfer from the peroxy acid can occur at either face of the double bond, but the product formed is the same because the two mirror-image forms of the epoxide are superposable.

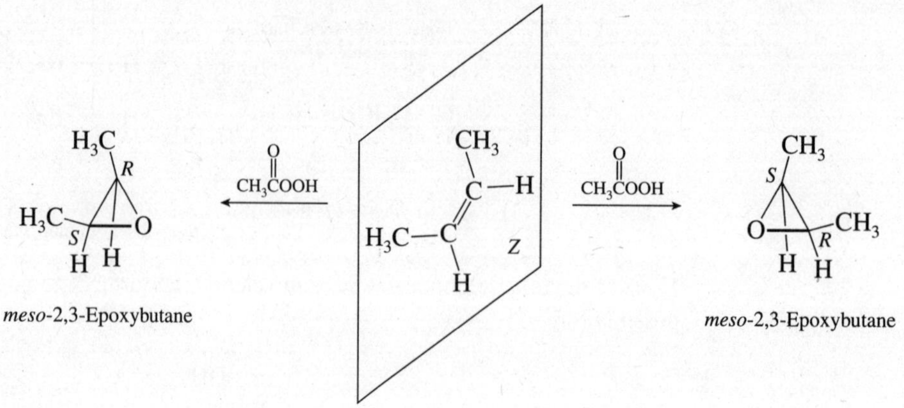


Epoxidation of (*E*)-2-butene gives a racemic mixture of two enantiomeric epoxides.

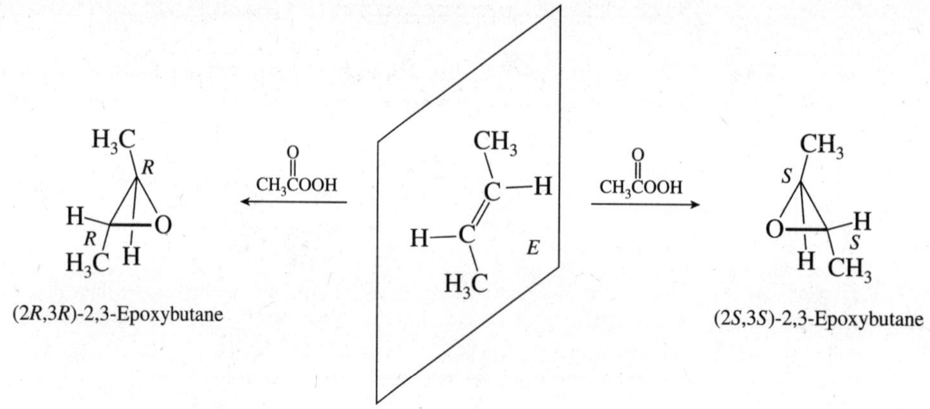


**7.18** The observed product mixture (68% *cis*-1,2-dimethylcyclohexane: 32% *trans*-1,2-dimethylcyclohexane) contains more of the less stable cis stereoisomer than the trans. The relative stabilities of the products therefore play no role in determining the stereoselectivity of this reaction.

**7.19** The tartaric acids incorporate two equivalently substituted stereogenic centers. (+)-Tartaric acid, as noted in the text, is the 2*R*,3*R* stereoisomer. There will be two additional stereoisomers, the enantiomeric (−)-tartaric acid (2*S*,3*S*) and an optically inactive meso form.

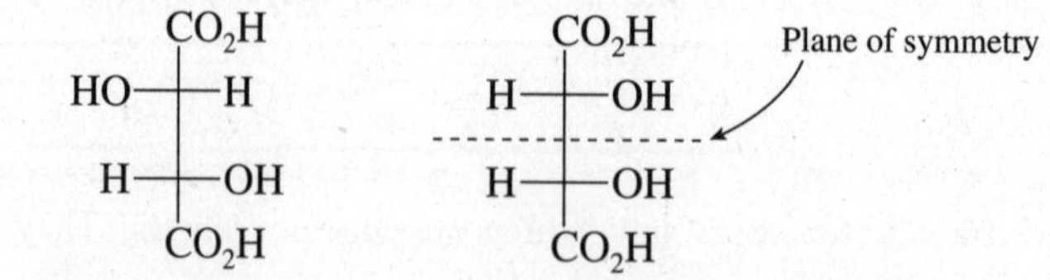


(2*S*,3*S*)-Tartaric acid (optically active) (mp 170°C, $[\alpha]_D - 12°$)

*meso*-Tartaric acid (optically inactive) (mp 140°C)

**7.20** No. Pasteur separated an optically inactive racemic mixture into two optically active enantiomers. A meso form is achiral, is identical to its mirror image, and is incapable of being separated into optically active forms.

**7.21** The more soluble salt must have the opposite configuration at the stereogenic center of 1-phenylethylamine, that is, the *S* configuration. The malic acid used in the resolution is a single enantiomer, *S*. In this particular case the more soluble salt is therefore (*S*)-1-phenylethylammonium (*S*)-malate.

**7.22** In an earlier exercise (Problem 4.23) the structures of all the isomeric $C_5H_{12}O$ alcohols were presented. Those that lack a stereogenic center and thus are *achiral* are

$CH_3CH_2CH_2CH_2CH_2OH$ — 1-Pentanol

$CH_3CHCH_2CH_2OH$ (with $CH_3$ on C-3) — 3-Methyl-1-butanol

$(CH_3)_3CCH_2OH$ — 2,2-Dimethyl-1-propanol

$CH_3CH_2CHCH_2CH_3$ (with OH on C-3) — 3-Pentanol

$CH_3CH_2COH$ (with two $CH_3$ on C-2) — 2-Methyl-2-butanol

The chiral isomers are characterized by carbons that bear four different groups. These are:

$CH_3\overset{*}{C}HCH_2CH_2CH_3$ (with OH on the starred carbon) — 2-Pentanol

$CH_3\overset{*}{C}HCH(CH_3)_2$ (with OH on the starred carbon) — 3-Methyl-2-butanol

$CH_3CH_2\overset{*}{C}HCH_2OH$ (with $CH_3$ on the starred carbon) — 2-Methyl-1-butanol

**7.23** The isomers of trichlorocyclopropane are

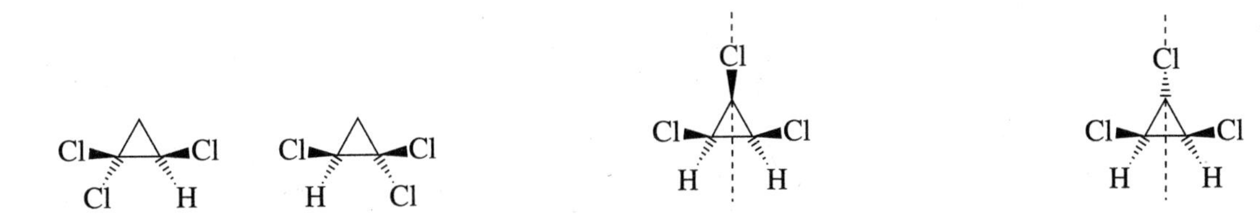

Enantiomeric forms of 1,1,2-trichlorocyclopropane (both chiral)

*cis*-1,2,3-Trichlorocyclopropane (achiral—contains a plane of symmetry)

*trans*-1,2,3-Trichlorocyclopropane (achiral—contains a plane of symmetry)

**7.24** (*a*) Carbon-2 is a stereogenic center in 3-chloro-1,2-propanediol. Carbon-2 has two equivalent substituents in 2-chloro-1,3-propanediol, and is not a stereogenic center.

$ClCH_2\overset{*}{C}HCH_2OH$ (with OH on the starred carbon) — 3-Chloro-1,2-propanediol, Chiral

$HOCH_2CHCH_2OH$ (with Cl on C-2) — 2-Chloro-1,3-propanediol, Achiral

(*b*) The primary bromide is achiral; the secondary bromide contains a stereogenic center and is chiral.

$CH_3CH{=}CHCH_2Br$ — Achiral

$CH_3\overset{*}{C}HCH{=}CH_2$ (with Br on the starred carbon) — Chiral (*c*) Both stereoisomers have two equivalently substituted stereogenic centers, and so we must be alert for the possibility of a meso stereoisomer. The structure at the left is chiral. The one at the right has a plane of symmetry and is the achiral meso stereoisomer.

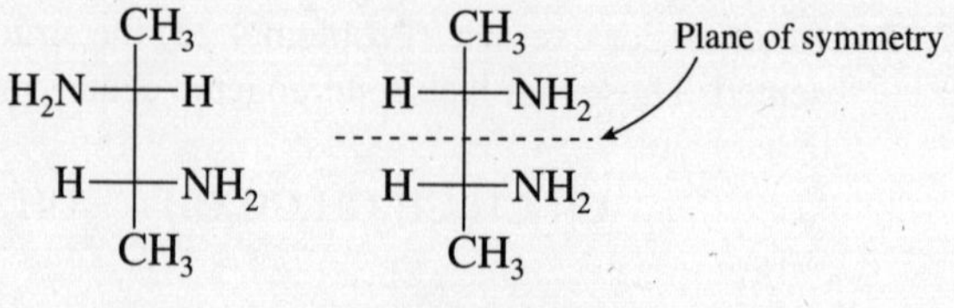

Plane of symmetry

Chiral　　Meso: achiral

(*d*) The first structure is achiral; it has a plane of symmetry.

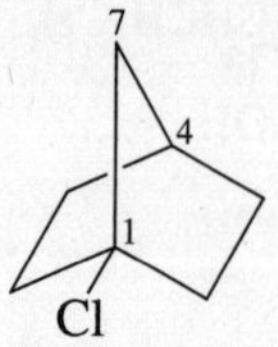

Plane of symmetry passes through C-1, C-4, and C-7.

The second structure cannot be superposed on its mirror image; it is chiral.

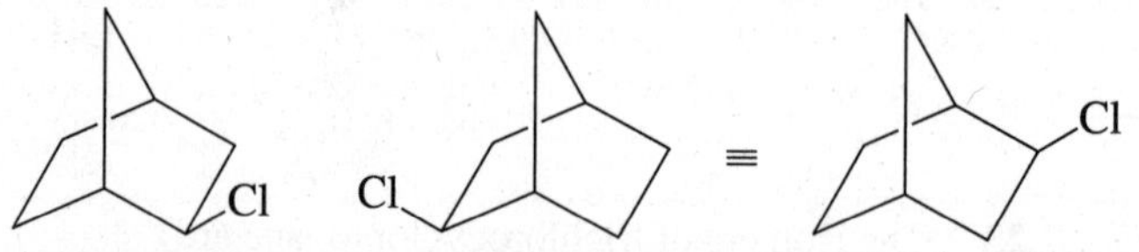

Reference structure　　Mirror image　　Reoriented mirror image

**7.25** There are four stereoisomers of 2,3-pentanediol, represented by the Fischer projections shown. All are chiral.

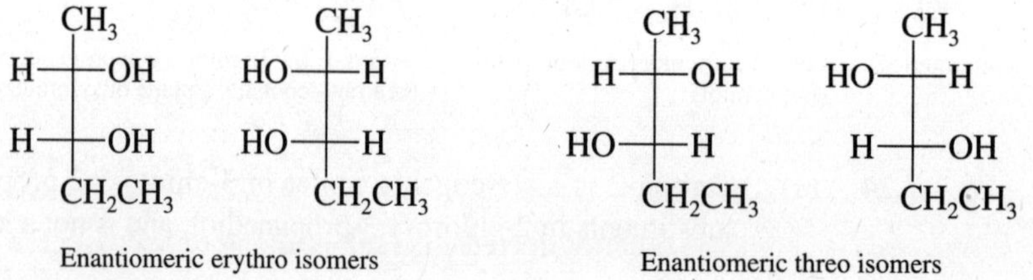

Enantiomeric erythro isomers　　Enantiomeric threo isomers

There are three stereoisomers of 2,4-pentanediol. The meso form is achiral; both threo forms are chiral.

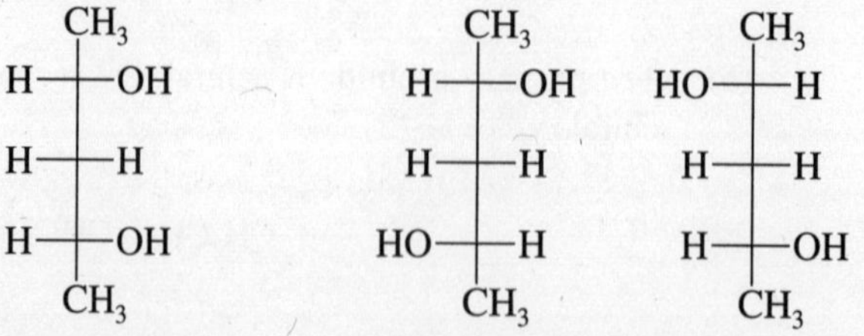

*meso*-2,4-Pentanediol　　Enantiomeric threo isomers

**7.26** Among the atoms attached to the stereogenic center, the order of decreasing precedence is Br > Cl > F > H. When the molecule is viewed with the hydrogen pointing away from us, the order Br → Cl → F appears clockwise in the *R* enantiomer, anticlockwise in the *S* enantiomer.

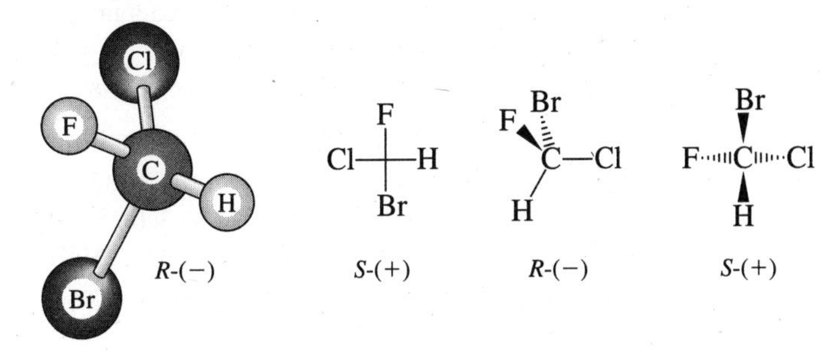


**7.27** (*a*) (−)-2-Octanol has the *R* configuration at C-2. The order of substituent precedence is

$$HO > CH_2CH_2 > CH_3 > H$$

The molecule is oriented so that the lowest ranking substituent is directed away from you and the order of decreasing precedence is clockwise.

HO H

OH, $CH_3$, $CH_2CH_2$

(*b*) In order of decreasing sequence rule precedence, the four substituents at the stereogenic center of monosodium L-glutamate are

$$\overset{+}{N}H_3 > CO_2^- > CH_2 > H$$

$CO_2^-$, $H_3\overset{+}{N}$, H, $CH_2CH_2CO_2^-$ $Na^+$ is the same as $CO_2^-$, $H_3\overset{+}{N}$—C—H, $CH_2CH_2CO_2^-$ $Na^+$

When the molecule is oriented so that the lowest ranking substituent (hydrogen) is directed away from you, the other three substituents are arranged as shown.

$^-O_2C$, $\overset{+}{N}H_3$, $CH_2CH_2CO_2^-$ $Na^+$

The order of decreasing rank is counterclockwise; the absolute configuration is *S*.

**7.28** (*a*) Among the isotopes of hydrogen, T has the highest mass number (3), D next (2), and H lowest (1). Thus, the order of rank at the stereogenic center in the reactant is $CH_3$ > T > D > H. The order of rank in the product is HO > $CH_3$ > T > D.

T, D, C—$CH_3$, H —biological oxidation→ T, D, C—$CH_3$, HO

Orient with lowest ranked substituent away from you.

D, T, $CH_3$ H$_3$C, T, OH

The order of decreasing rank in the reactant is anticlockwise; the configuration is *S*. The order of decreasing rank in the product is clockwise; the configuration is *R*.

(*b*) Retention of configuration means that the three-dimensional arrangement of bonds at the stereogenic center is the same in the reactant and the product. The *R* and *S* descriptors change because the order of precedence changes in going from reactant to product; for example, $CH_3$ is the highest ranked substituent in the reactant, but becomes the second-highest ranked in the product.

**7.29** Two compounds can be stereoisomers only if they have the *same* constitution. Thus, you should compare first the constitution of the two structures and then their stereochemistry. The best way to compare constitutions is to assign a systematic (IUPAC) name to each molecule. Also remember that enantiomers are nonsuperposable mirror images, and diastereomers are stereoisomers that are not enantiomers.

(*a*) The two compounds are constitutional isomers. Their IUPAC names clearly reflect this difference.

$CH_3CH(OH)CH_2Br$ and $CH_3CH(Br)CH_2OH$

1-Bromo-2-propanol　　2-Bromo-1-propanol

(*b*) The two structures have the same constitution. Test them for superposability. To do this we need to place them in comparable orientations.

H, $CH_3$, C—Br, $CH_3CH_2$ is equivalent to $CH_3$ / H—+—Br / $CH_2CH_3$

and

H, Br, C—$CH_3$, $CH_3CH_2$ is equivalent to $CH_3$ / Br—+—H / $CH_2CH_3$

The two are nonsuperposable mirror images of each other. They are enantiomers.

To check this conclusion, work out the absolute configuration of each using the Cahn–Ingold–Prelog system.

H, $CH_3$, C—Br, $CH_3CH_2$　　H, Br, C—$CH_3$, $CH_3CH_2$

(*S*)-2-Bromobutane　　(*R*)-2-Bromobutane

(*c*) Again, place the structures in comparable orientations, and examine them for superposability.

H, $CH_3$, C—Br, $CH_3CH_2$ is equivalent to $CH_3$ / H—+—Br / $CH_2CH_3$

and

$CH_3$, H, C—$CH_2CH_3$, Br is equivalent to $CH_3$ / H—+—Br / $CH_2CH_3$

The two structures represent the same compound, since they are superposable. (As a check, notice that both have the *S* configuration.)

(*d*) If we reorient the first structure,

H, $CH_3$, C—Br, $CH_3CH_2$ becomes $CH_3$ H—+—Br $CH_2CH_3$

which is the enantiomer of

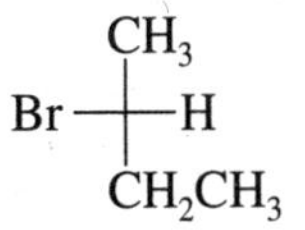

$CH_3$ Br—+—H $CH_2CH_3$

As a check, the first structure is seen to have the *S* configuration, and the second has the *R* configuration.

(*e*) As drawn, the two structures are mirror images of each other; however, they represent an achiral molecule. The two structures are superposable mirror images and are not stereoisomers but identical.

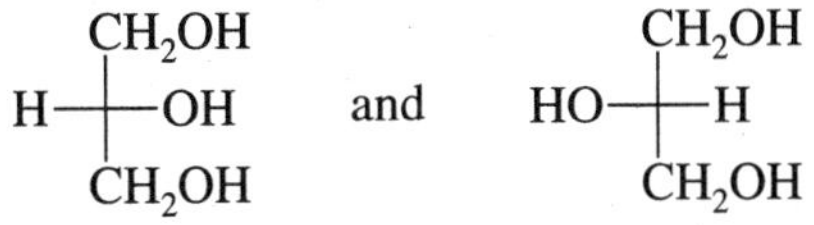

$CH_2OH$ H—+—OH $CH_2OH$ and $CH_2OH$ HO—+—H $CH_2OH$ are both identical

(*f*) The two structures—one cis, the other trans—are stereoisomers that are not mirror images; they are diastereomers.

*trans*-1-Chloro-2-methylcyclopropane *cis*-1-Chloro-2-methylcyclopropane

(*g*) The two structures are enantiomers, since they are nonsuperposable mirror images. Checking their absolute configurations reveals one to be *R*, the other *S*. Both have the *E* configuration at the double bond.

(2*R*,3*E*)-3-Penten-2-ol (2*S*,3*E*)-3-Penten-2-ol

(*h*) These two structures are identical; both have the *E* configuration at the double bond and the *R* configuration at the stereogenic center.

Alternatively, we can show their superposability by rotating the second structure 180° about an axis passing through the doubly bonded carbons.

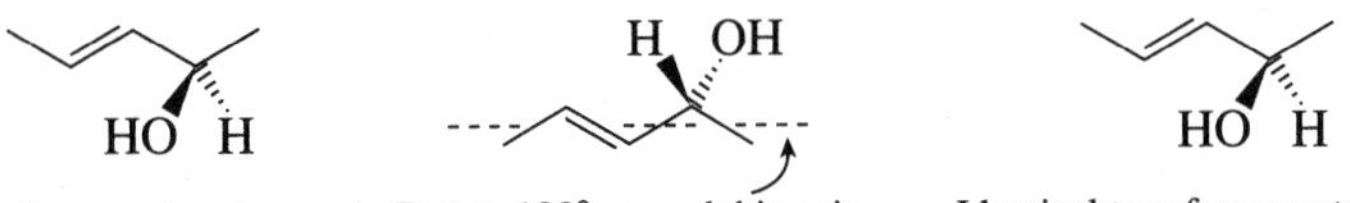

Reference structure Rotate 180° around this axis Identical to reference structure

(*i*) One structure has a cis double bond, the other a trans double bond; therefore, the two are diastereomers. Even though one stereogenic center is *R* and the other is *S*, the two structures are *not* enantiomers. The mirror image of a cis (or *Z*) double bond is cis, and that of a trans (or *E*) double bond is trans.

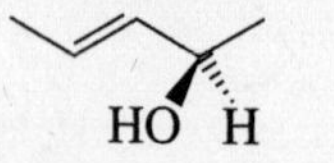

HO H H OH

(2*R*,3*E*)-3-Penten-2-ol 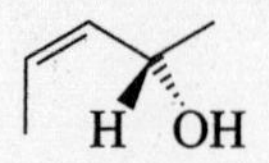 (2*S*,3*Z*)-3-Penten-2-ol

(*j*) Here it will be helpful to reorient the second structure so that it may be more readily compared with the first.

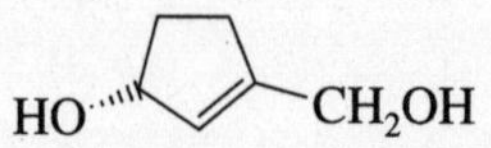

HO $CH_2OH$ and HO $CH_2OH$ which is equivalent to HO $CH_2OH$

Reference structure

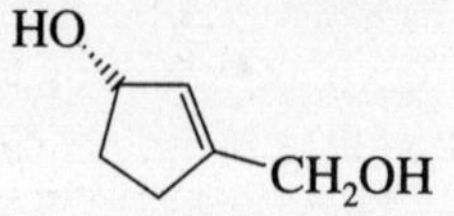

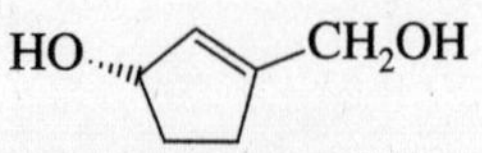

Enantiomer of reference structure

The two compounds are enantiomers.

Examining their absolute configurations confirms the enantiomeric nature of the two compounds.

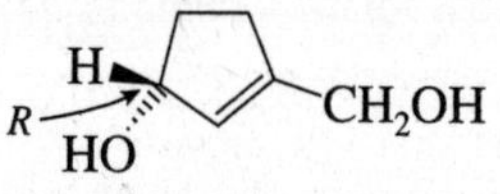

H R HO $CH_2OH$ OH H S $CH_2OH$

(*R*)-3-Hydroxymethyl-2-cyclopenten-1-ol 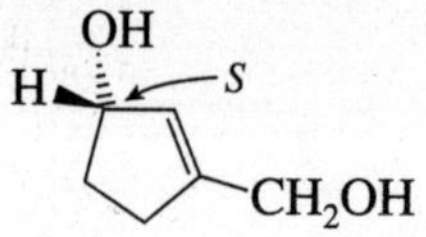 (*S*)-3-Hydroxymethyl-2-cyclopenten-1-ol

(*k*) These two compounds differ in the order in which their atoms are joined together; they are constitutional isomers.

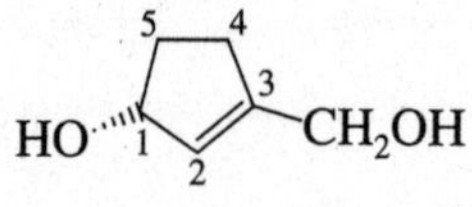

HO 1 2 3 4 5 $CH_2OH$ HO 1 2 3 4 5 $CH_2OH$

3-Hydroxymethyl-2-cyclopenten-1-ol 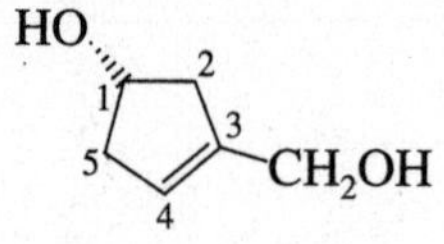 3-Hydroxymethyl-3-cyclopenten-1-ol

(*l*) To better compare these two structures, place them both in the same format.

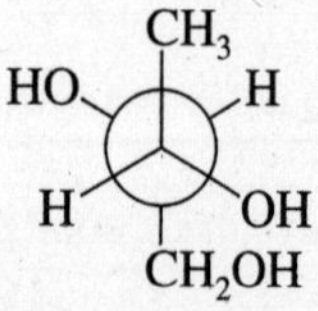

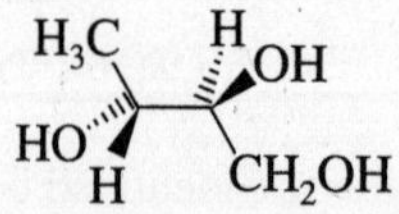

$CH_3$ HO H H OH $CH_2OH$ $H_3C$ H OH HO H $CH_2OH$ which is equivalent to $CH_3$ H OH HO H $CH_2OH$

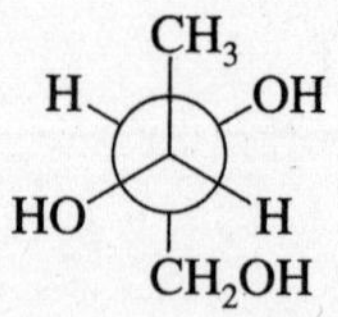

The two are enantiomers.

(*m*) Since *cis*-1,3-dimethylcyclopentane has a plane of symmetry, it is achiral and cannot have an enantiomer. The two structures given in the problem are identical.

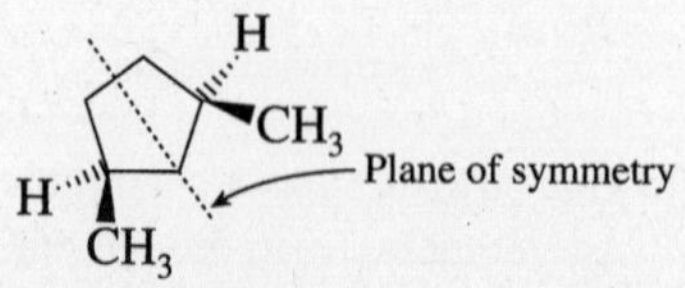

H $CH_3$ H $CH_3$ Plane of symmetry

(*n*) These structures are diastereomers, that is, stereoisomers that are not mirror images. They have the same configuration at C-3 but opposite configurations at C-2.

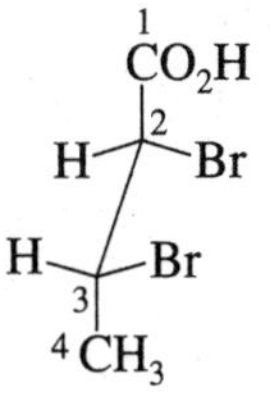

1 $CO_2H$, 2 H—Br, H—Br, 3, 4 $CH_3$ — 2*R*,3*R*

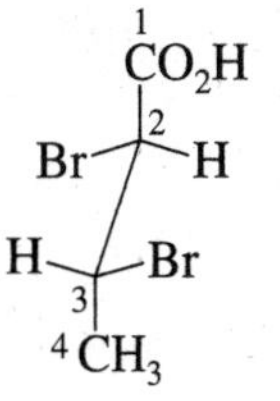

1 $CO_2H$, 2 Br—H, H—Br, 3, 4 $CH_3$ — 2*S*,3*R*

(*o*) To compare these compounds, reorient the first structure so that it may be drawn as a Fischer projection. The first step in the reorientation consists of a 180° rotation about an axis passing through the midpoint of the C-2—C-3 bond.

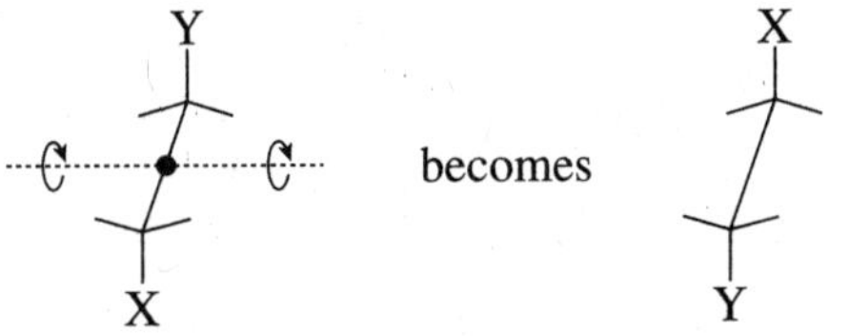

Y ... X becomes X ... Y

Thus

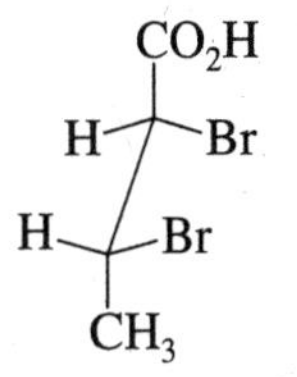

$CO_2H$, H—Br, H—Br, $CH_3$ is $CH_3$, H—Br, H—Br, $CO_2H$

Reference structure

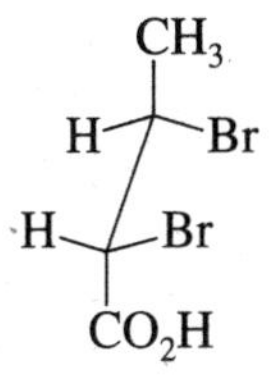

Now rotate the "back" carbon of the reoriented structure to give the necessary alignment for a Fischer projection.

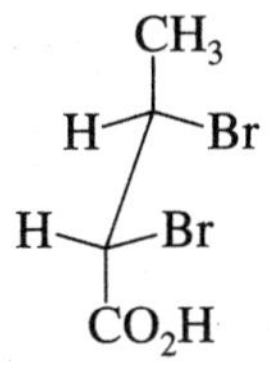

$CH_3$, H—Br, H—Br, $CO_2H$ becomes Br—H, $CH_3$, H—Br, $CO_2H$ which is the same as $CH_3$, Br—H, H—Br, $CO_2H$

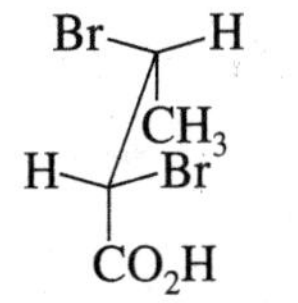

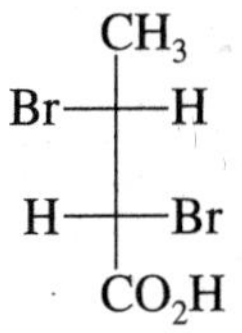

This reveals that the original two structures in the problem are equivalent.

(*p*) These two structures are nonsuperposable mirror images of a molecule with two nonequivalent stereogenic centers; they are enantiomers.

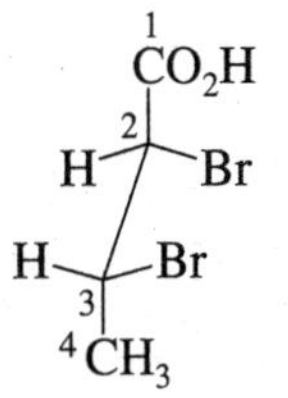

1 $CO_2H$, 2 H—Br, H—Br, 3, 4 $CH_3$ — 2*R*,3*R*

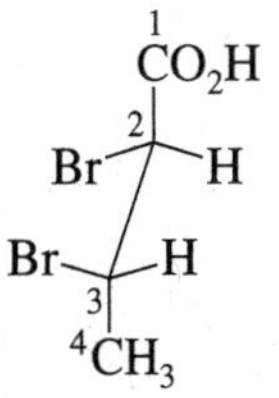

1 $CO_2H$, 2 Br—H, Br—H, 3, 4 $CH_3$ — 2*S*,3*S*

(*q*) The two structures are stereoisomers that are not enantiomers; they are diastereomers.

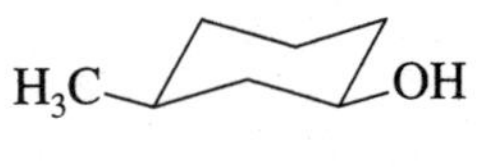

$H_3C$ ... OH — *cis*-3-Methylcyclohexanol

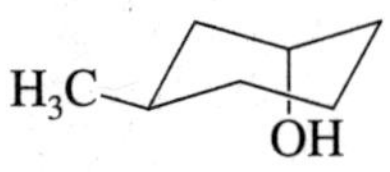

$H_3C$ ... OH — *trans*-3-Methylcyclohexanol

(*r*) These two structures, *cis*- and *trans*-4-*tert*-butylcyclohexyl iodide, are diastereomers.

$(CH_3)_3C$ — I

Trans

$(CH_3)_3C$ — I

Cis

(*s*) The two structures are nonsuperposable mirror images; they are enantiomers.

$CH_3$

Reference structure

$CH_3$ is equivalent to $H_3C$

Enantiomer of reference structure

(*t*) The two structures are identical.

$CH_3$

Reference structure

$H_3C$ is equivalent to $CH_3$

Identical to reference structure

(*u*) As represented, the two structures are mirror images of each other, but because the molecule is achiral (it has a plane of symmetry), the two must be superposable. They represent the same compound.

$H_3C$ H

Reference structure

H $CH_3$ is equivalent to $H_3C$ H

Identical to reference structure

The plane of symmetry passes through C-7 and bisects the C-2—C-3 bond and the C-5—C-6 bond.

(*v*) The structures are stereoisomers but not enantiomers; they are diastereomers. (Both are achiral and so cannot have enantiomers.)

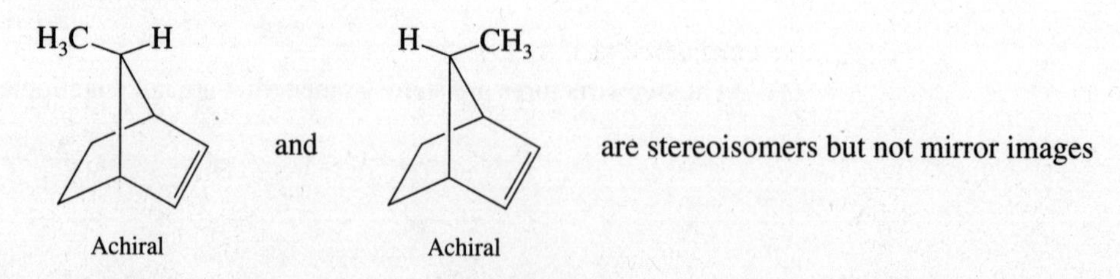

**7.30** Write a structural formula for phytol and count the number of structural units capable of stereochemical variation.

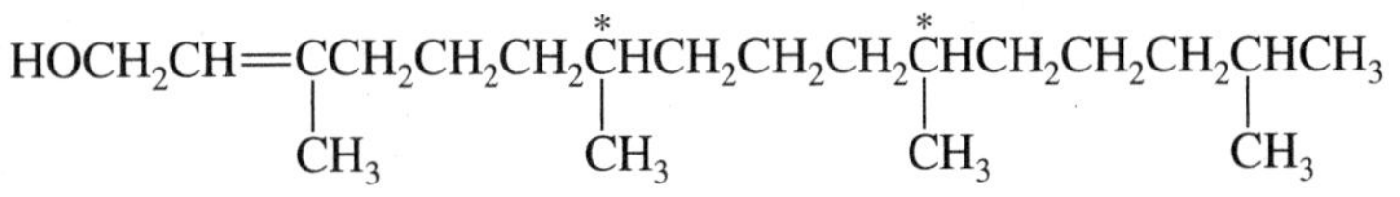

$$\mathrm{HOCH_2CH{=}\underset{\displaystyle CH_3}{C}CH_2CH_2CH_2\underset{\displaystyle CH_3}{\overset{*}{C}}HCH_2CH_2CH_2\underset{\displaystyle CH_3}{\overset{*}{C}}HCH_2CH_2CH_2\underset{\displaystyle CH_3}{C}HCH_3}$$

3,7,11,15-Tetramethyl-2-hexadecen-1-ol

Phytol has two stereogenic centers (C-7 and C-11) and one double bond. The stereogenic centers may be either *R* or *S*, and the double bond may be either *E* or *Z*. Eight stereoisomers are possible.

| | Isomer | | | | | | | |
|---|---|---|---|---|---|---|---|---|
| | 1 | 2 | 3 | 4 | 5 | 6 | 7 | 8 |
| Double bond | *E* | *E* | *E* | *E* | *Z* | *Z* | *Z* | *Z* |
| Carbon-7 | *R* | *S* | *R* | *S* | *R* | *S* | *R* | *S* |
| Carbon-11 | *R* | *S* | *S* | *R* | *R* | *S* | *S* | *R* |

**7.31** (*a*) Muscarine has three stereogenic centers, and so *eight* stereoisomers have this constitution.

(*b*) The three substituents on the ring (at C-2, C-3, and C-5) can be thought of as being either up (U) or down (D) in a perspective drawing. Thus the eight possibilities are:

UUU, UUD, UDU, DUU, UDD, DUD, DDU, DDD

Of these, *six* have one substituent trans to the other two.

(*c*) Muscarine is

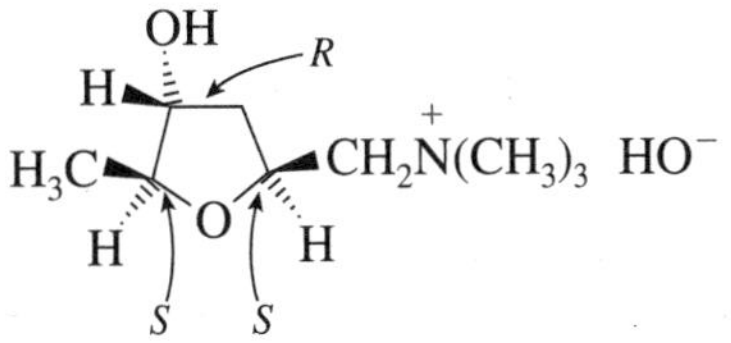

OH, H, *R*, $H_3C$, H, O, H, $\overset{+}{\mathrm{C}}H_2N(CH_3)_3$ $HO^-$, *S*, *S*

**7.32** To write a stereochemically accurate representation of ectocarpene, it is best to begin with the configuration of the stereogenic center, which we are told is *S*.

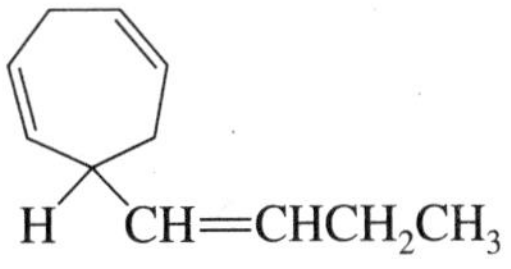

H $CH{=}CHCH_2CH_3$

Clearly, hydrogen is the lowest ranking substituent; among the other three substituents, two are part of the ring and the third is the four-carbon side chain. The priority rankings of these groups are determined by systematically working along the chain.

The substituents

$-CH{=}CHCH_2CH{=}CHCH_2$ (Ring) $\qquad -CH{=}CHCH_2CH_3$ (Side chain) $\qquad -CH_2CH{=}CHCH_2CH{=}CH$ (Ring)

are considered as if they were

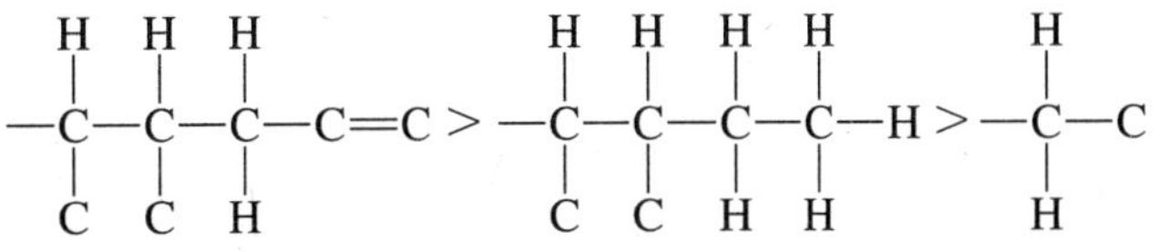

$$-\underset{\displaystyle C}{\overset{\displaystyle H}{C}}-\underset{\displaystyle C}{\overset{\displaystyle H}{C}}-\underset{\displaystyle H}{\overset{\displaystyle H}{C}}-C{=}C > -\underset{\displaystyle C}{\overset{\displaystyle H}{C}}-\underset{\displaystyle C}{\overset{\displaystyle H}{C}}-\underset{\displaystyle H}{\overset{\displaystyle H}{C}}-\underset{\displaystyle H}{\overset{\displaystyle H}{C}}-H > -\underset{\displaystyle H}{\overset{\displaystyle H}{C}}-C$$

(Ring) (Side chain) (Ring)

Orienting the molecule with the hydrogen away from you

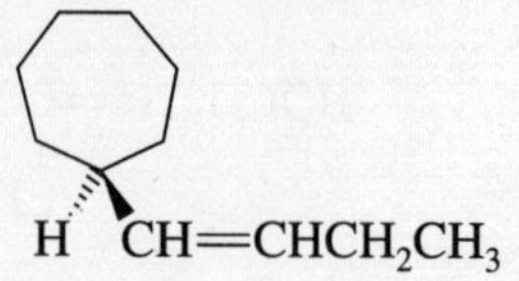

H $CH{=}CHCH_2CH_3$

we place the double bonds in the ring so that the order of decreasing sequence rule precedence is counterclockwise:

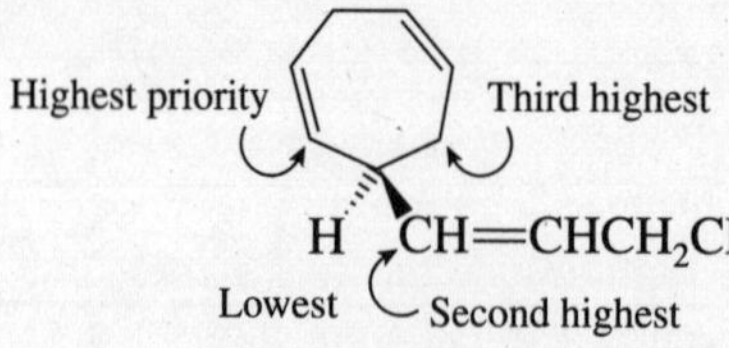

Highest priority Third highest

H $CH{=}CHCH_2CH_3$

Lowest Second highest

Finally, since all the double bonds are cis, the complete structure becomes:

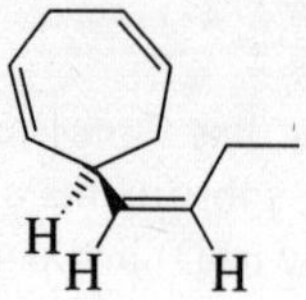

H H H

**7.33** (*a*) Multifidene has two stereogenic centers and three double bonds. Neither the ring double bond nor the double bond of the vinyl substituent can give rise to stereoisomers, but the butenyl side chain can be either *E* or *Z*. Eight ($2^3$) stereoisomers are therefore possible. We can rationalize them as

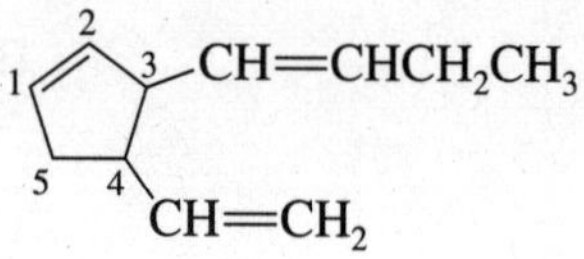

$CH{=}CHCH_2CH_3$ (at C-3), $CH{=}CH_2$ (at C-4); ring positions 1, 2, 3, 4, 5

| Stereoisomer | C-3 | C-4 | Butenyl double bond | |
|---|---|---|---|---|
| 1 | *R* | *R* | *E* | enantiomers |
| 2 | *S* | *S* | *E* | |
| 3 | *R* | *R* | *Z* | enantiomers |
| 4 | *S* | *S* | *Z* | |
| 5 | *R* | *S* | *E* | enantiomers |
| 6 | *S* | *R* | *E* | |
| 7 | *R* | *S* | *Z* | enantiomers |
| 8 | *S* | *R* | *Z* | |

(*b*) Given the information that the alkenyl substituents are cis to each other, the number of stereoisomers is reduced by half. Four stereoisomers are therefore possible.

(*c*) Knowing that the butenyl group has a *Z* double bond reduces the number of possibilities by half. Two stereoisomers are possible.

(*d*) The two stereoisomers are

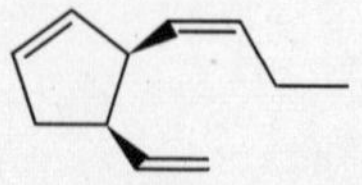

and

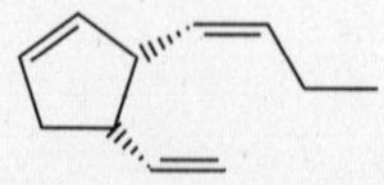

(*e*) These two stereoisomers are enantiomers. They are nonsuperposable mirror images.

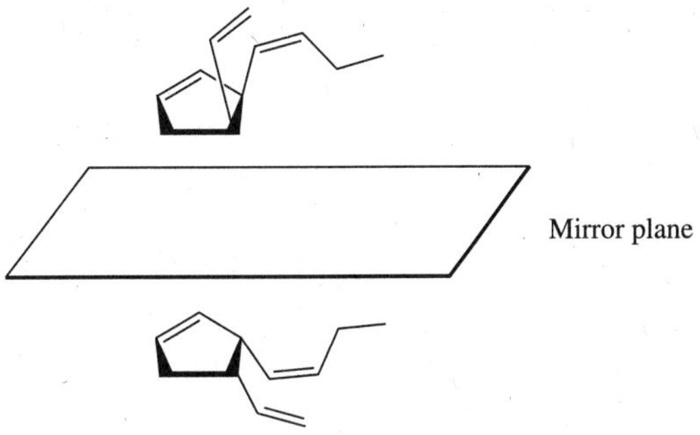

Mirror plane

**7.34** In a substance with more than one stereogenic center, each center is independently specified as *R* or *S*. Streptimidone has two stereogenic centers and two double bonds. Only the internal double bond is capable of stereoisomerism.

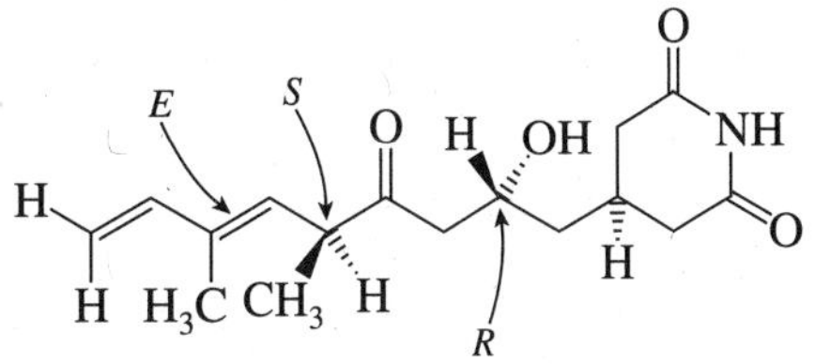

The three stereochemical variables give rise to eight ($2^3$) stereoisomers, of which one is streptimidone and a second is the enantiomer of streptimidone. The remaining six stereoisomers are diastereomers of streptimidone.

**7.35** (*a*) The first step is to set out the constitution of menthol, which we are told is 2-isopropyl-5-methylcyclohexanol.

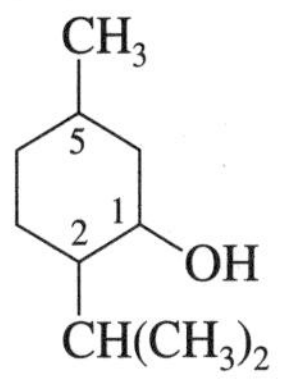

2-Isopropyl-5-methylcyclohexanol

Since the configuration at C-1 is *R* in (−)-menthol, the hydroxyl group must be "up" in our drawing.

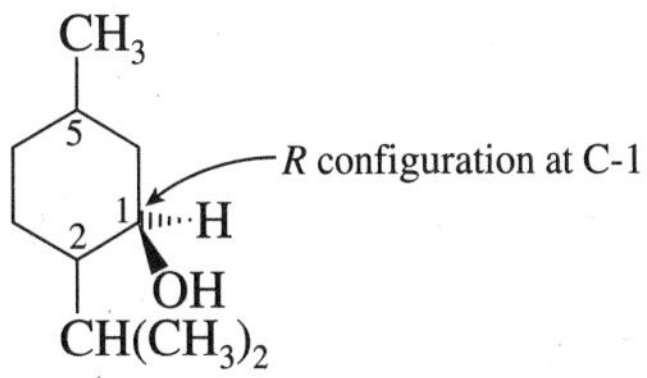

*R* configuration at C-1

Because menthol is the most stable stereoisomer of this constitution, all three of its substituents must be equatorial. We therefore draw the chair form of the preceding structure, which has the hydroxyl group equatorial and up, placing isopropyl and methyl groups so as to preserve the *R* configuration at C-1.

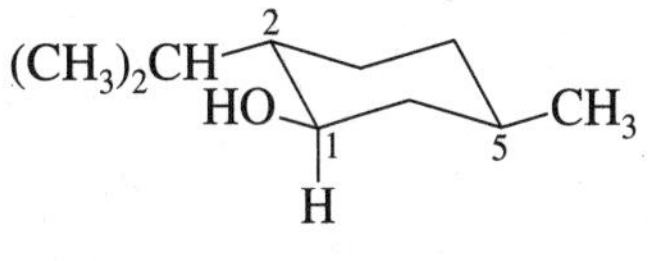

(−)-Menthol

(*b*) To transform the structure of (−)-menthol to that of (+)-isomenthol, the configuration at C-5 must remain the same, whereas those at C-1 and C-2 are inverted.

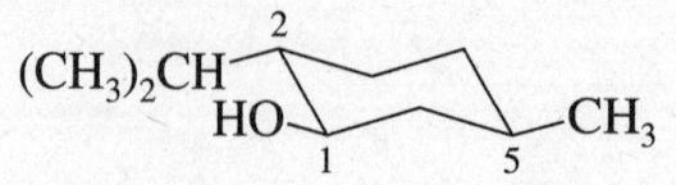

(−)-Menthol 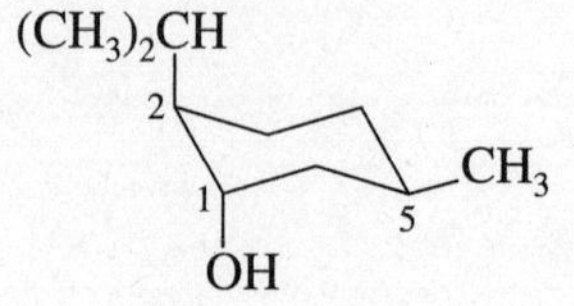 (+)-Isomenthol

(+)-Isomenthol is represented here in its correct configuration, but the conformation with two axial substituents is not the most stable one. The ring-flipped form will be the preferred conformation of (+)-isomenthol:

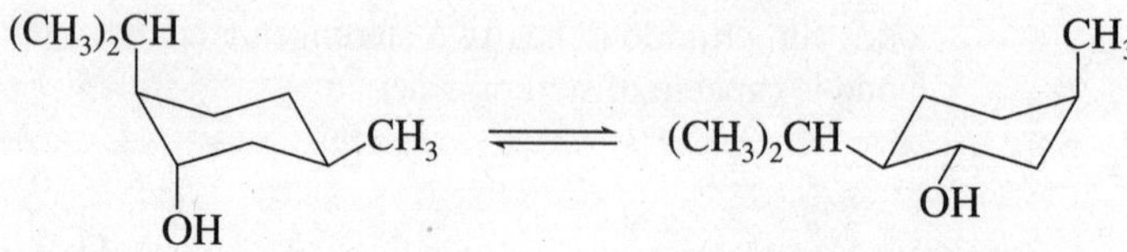

Most stable conformation of (+)-isomenthol

**7.36** Since the only information available about the compound is its optical activity, examine the two structures for chirality, recalling that only chiral substances can be optically active.

The structure with the six-membered ring has a plane of symmetry passing through C-1 and C-4. It is achiral and cannot be optically active.

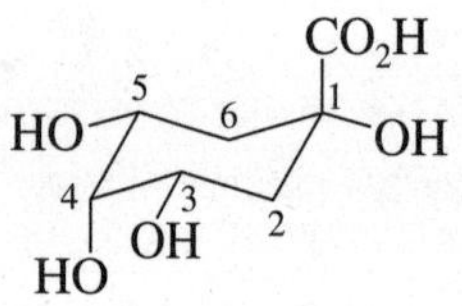

Achiral; $[\alpha]_D$ 0° 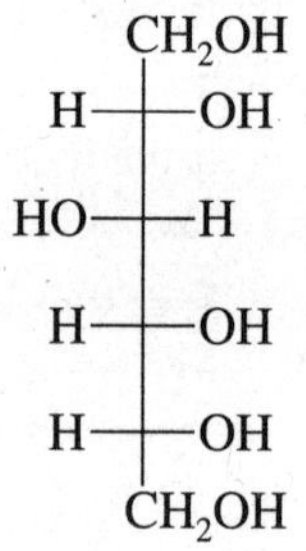 Chiral; can be optically active

The open-chain structure has neither a plane of symmetry nor a center of symmetry; it is not superposable on its mirror image and so is chiral. It can be optically active and is more likely to be the correct choice.

**7.37** Compound B has a center of symmetry, is achiral, and thus cannot be optically active.

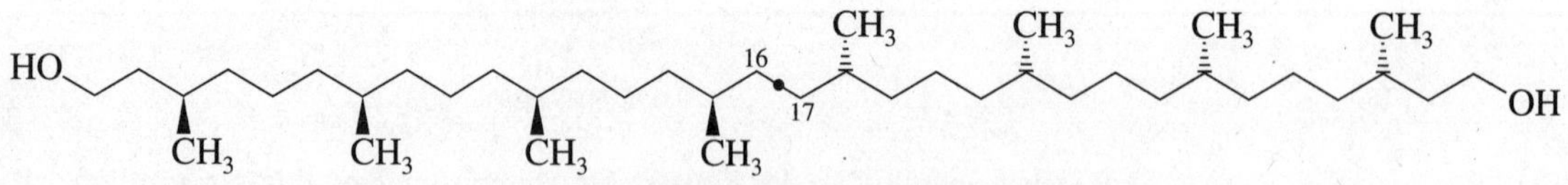

Compound B: not optically active
(center of symmetry is midpoint of C-16—C-17 bond)

The diol in the problem is optically active, and so it must be chiral. Compound A is the naturally occurring diol.

**7.38** (*a*) The equation that relates specific rotation $[\alpha]_D$ to observed rotation $\alpha$ is

$$[\alpha]_D = \frac{100\alpha}{cl}$$

where $c$ is concentration in grams per 100 mL and $l$ is path length in decimeters.

$$[\alpha]_D = \frac{100(-5.20^\circ)}{(2.0 \text{ g}/100 \text{ mL})(2 \text{ dm})}$$

$$= -130^\circ$$

(*b*) The optical purity of the resulting solution is 10/15, or 66.7%, since 10 g of optically pure fructose has been mixed with 5 g of racemic fructose. The specific rotation will therefore be two thirds (10/15) of the specific rotation of optically pure fructose:

$$[\alpha]_D = \tfrac{2}{3}(-130^\circ) = -87^\circ$$

**7.39** (*a*) The reaction of 1-butene with hydrogen iodide is one of electrophilic addition. It follows Markovnikov's rule and yields a racemic mixture of (*R*)- and (*S*)-2-iodobutane.

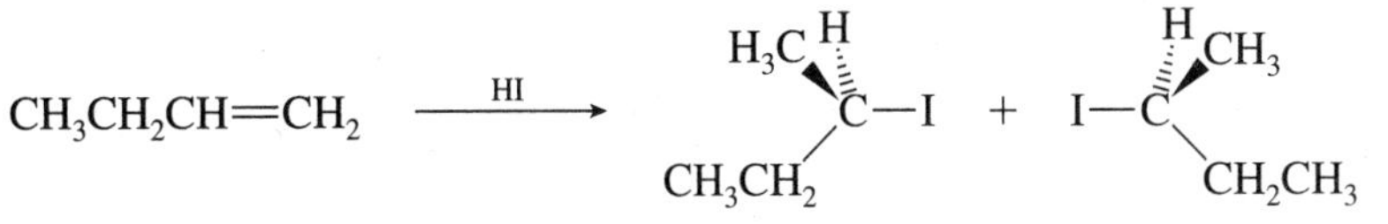

$CH_3CH_2CH{=}CH_2$ $\xrightarrow{HI}$ (*R*)-2-Iodobutane + (*S*)-2-Iodobutane

1-Butene · (*R*)-2-Iodobutane · (*S*)-2-Iodobutane

(*b*) Bromine adds anti to carbon–carbon double bonds to give vicinal dibromides.

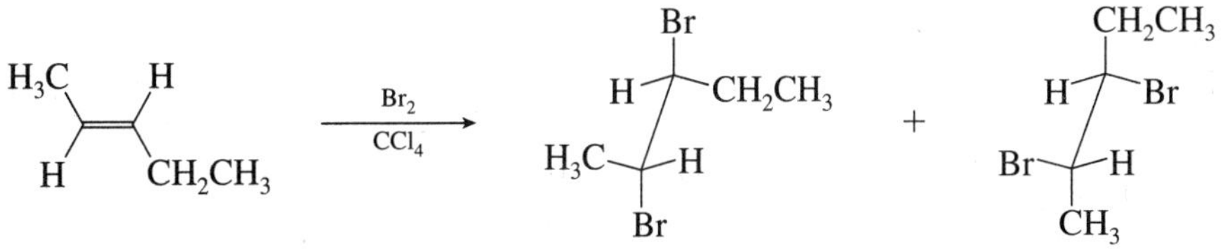

(*E*)-2-Pentene $\xrightarrow[CCl_4]{Br_2}$ (2*R*,3*S*)-2,3-Dibromopentane + (2*S*,3*R*)-2,3-Dibromopentane

The two stereoisomers are enantiomers and are formed in equal amounts.

(*c*) Two enantiomers are formed in equal amounts in this reaction, involving electrophilic addition of bromine to (*Z*)-2-pentene. These two are diastereomeric with those formed in part (*b*).

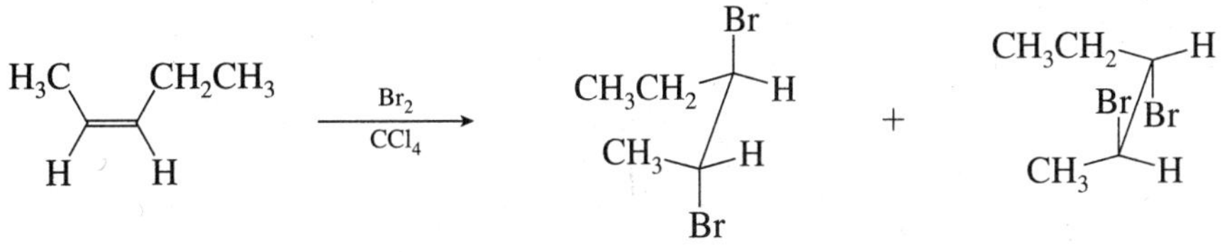

(*Z*)-2-Pentene $\xrightarrow[CCl_4]{Br_2}$ (2*R*,3*R*)-2,3-Dibromopentane + (2*S*,3*S*)-2,3-Dibromopentane

(*d*) Epoxidation of 1-butene yields a racemic epoxide mixture.

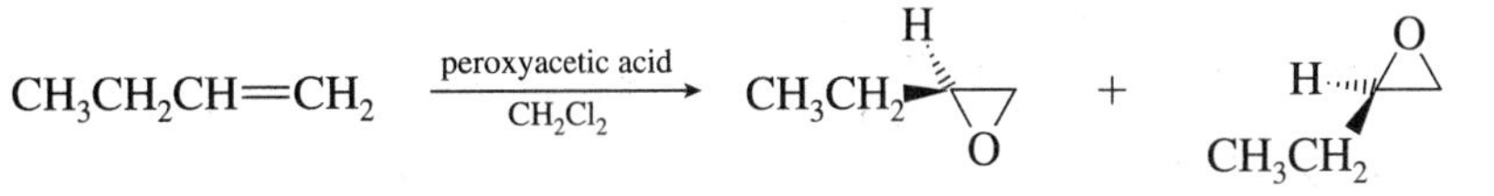

$CH_3CH_2CH{=}CH_2$ $\xrightarrow[CH_2Cl_2]{\text{peroxyacetic acid}}$ (*S*)-1,2-Epoxybutane + (*R*)-1,2-Epoxybutane

(*e*) Two enantiomeric epoxides are formed in equal amounts on epoxidation of (*Z*)-2-pentene.

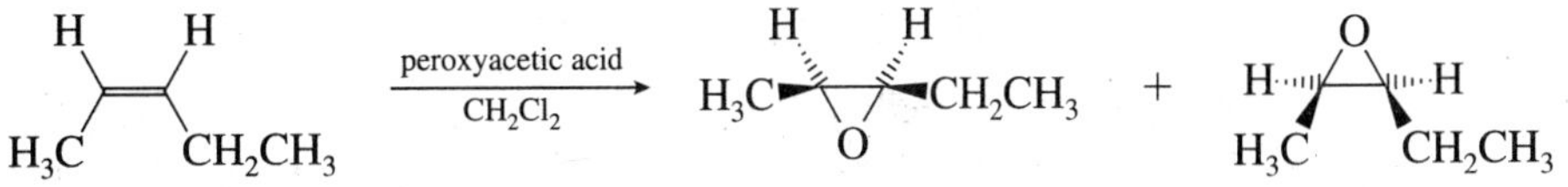

(*Z*)-2-Pentene $\xrightarrow[CH_2Cl_2]{\text{peroxyacetic acid}}$ (2*S*,3*R*)-2,3-Epoxypentane + (2*R*,3*S*)-2,3-Epoxypentane

The reaction is a stereospecific syn addition. The cis alkyl groups in the starting alkene remain cis in the product epoxide.

(*f*) The starting material is achiral, so even though a chiral product is formed, it is a racemic mixture of enantiomers and is optically inactive.

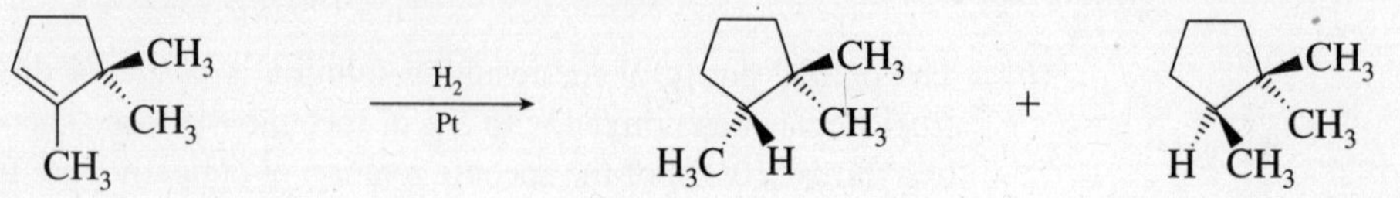
CH3
CH3
CH3
H2
Pt
CH3
CH3
H3C H
+
CH3
CH3
H CH3

1,5,5-Trimethylcyclopentene (*R*)-1,1,2-Trimethylcyclopentane (*S*)-1,1,2-Trimethylcyclopentane

(*g*) Recall that hydroboration–oxidation leads to anti-Markovnikov hydration of the double bond.

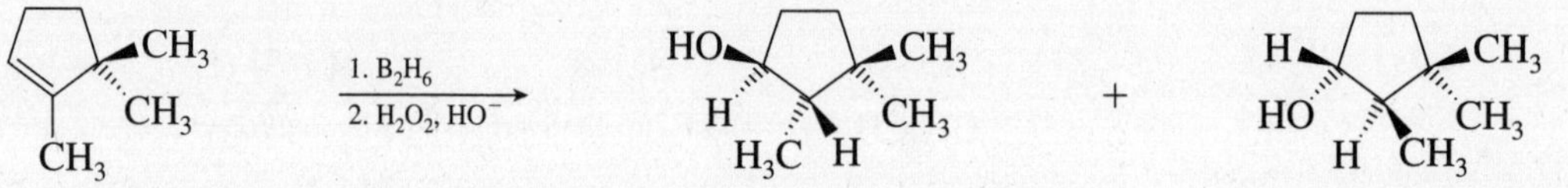
CH3
CH3
CH3
1. B2H6
2. H2O2, HO−
HO
CH3
H
CH3
H3C H
+
H
CH3
HO
CH3
H CH3

1,5,5-Trimethylcyclopentene (1*S*,2*S*)-2,3,3-Trimethylcyclopentanol (1*R*,2*R*)-2,3,3-Trimethylcyclopentanol

The product has two stereogenic centers. It is formed as a racemic mixture of enantiomers.

**7.40** Hydration of the double bond of aconitic acid (shown in the center) can occur in two regiochemically distinct ways:

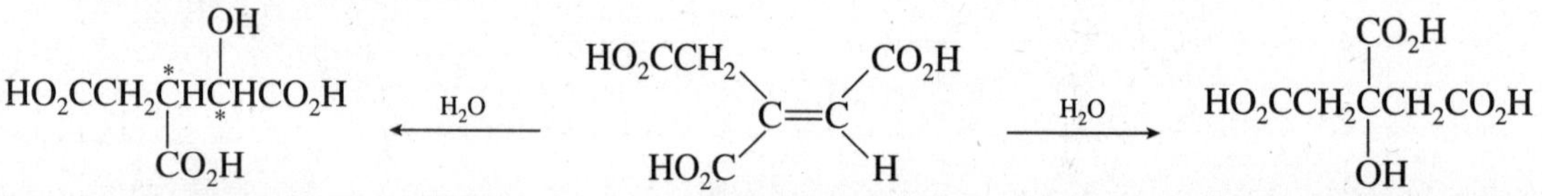
OH
HO2CCH2CHCHCO2H
CO2H
H2O
HO2CCH2
CO2H
C=C
HO2C
H
H2O
CO2H
HO2CCH2CCH2CO2H
OH

Chiral (isocitric acid) Aconitic acid Achiral (citric acid)

One of the hydration products lacks a stereogenic center. It must be citric acid, the achiral, optically inactive isomer. The other one has two different stereogenic centers and must be isocitric acid, the optically active isomer.

**7.41** (*a*) Structures A and B are chiral. Structure C has a plane of symmetry and is an achiral meso form.

(*b*) Ozonolysis of the starting material proceeds with the stereochemistry shown. Compound B is the product of the reaction.

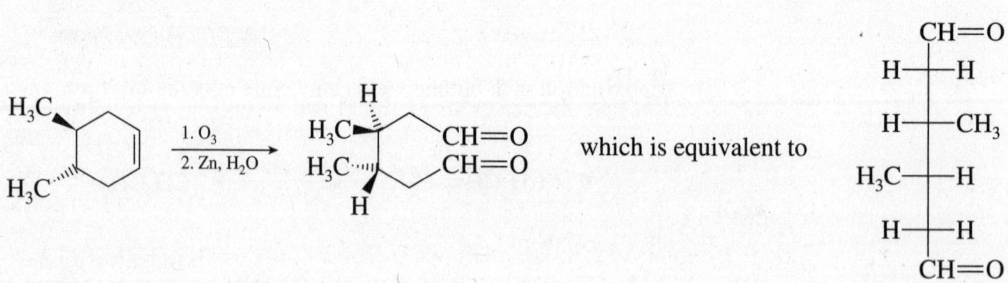
H3C
H3C
1. O3
2. Zn, H2O
H
H3C
CH=O
H3C
CH=O
H
which is equivalent to
CH=O
H H
H CH3
H3C H
H H
CH=O

Compound B

(*c*) If the methyl groups were cis to each other in the cycloalkene, they would be on the same side of the Fischer projection in the product. Compound C would be formed.

**7.42** (*a*) The E2 transition state requires that the bromine and the hydrogen that is lost be antiperiplanar to each other. Examination of the compound given in the problem reveals that loss of bromine and the deuterium will yield *trans*-2-butene, whereas loss of the bromine and the hydrogen on C-3 will yield *cis*-2-butene.

can be redrawn as

rotation around C-2 - C-3 bond

E2 (-DBr) E2 (-HBr)

*trans*-2-Butene *cis*-2-Butene

The *trans*-2-butene that forms does not contain deuterium, but *cis*-2-butene does. 1-Butene also contains deuterium.

$$CH_3CH(Br)CH(D)CH_3 \xrightarrow{E2} CH_2{=}CHCH(D)CH_3$$

1-Butene

(*b*) The starting material in part (*a*) is the erythro isomer. The relative positions of the H and C at C-3 are reversed in the threo isomer. The erythro and threo isomers can be drawn using Fischer projections:

| Erythro | Threo |
|---|---|
| $CH_3$ | $CH_3$ |
| Br—H | Br—H |
| D—H | H—D |
| $CH_3$ | $CH_3$ |

Because the positions of the H and D on C-3 in the threo isomer are opposite that in the erythro, the deuterium content of *cis*- and *trans*-2-butene would be reversed. *trans*-2-Butene obtained from the threo isomer would contain deuterium, and *cis*-2-butene would not. 1-Butene obtained from the threo isomer would also contain deuterium.

**7.43** Bromine adds to the unknown compound, suggesting the presence of a double bond in addition to the five-membered ring. The following are possible structures for the unknown:

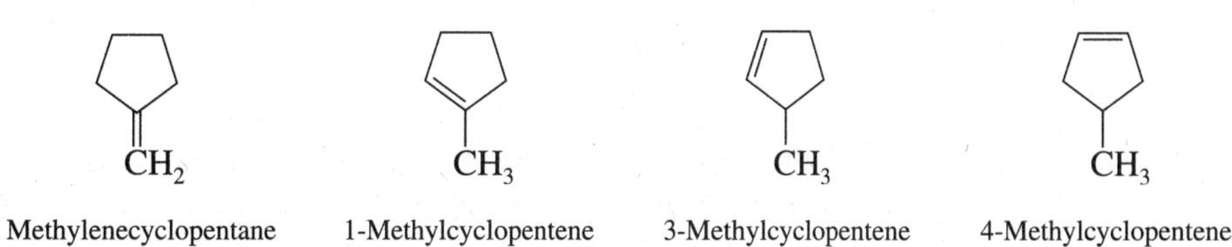


Methylenecyclopentane 1-Methylcyclopentene 3-Methylcyclopentene 4-Methylcyclopentene

Which of these form diastereomeric dibromides on anti addition of bromine?

Methylenecyclopentane $\xrightarrow{Br_2}$ 1-bromo-1-(bromomethyl)cyclopentane (Br, $CH_2Br$) (only product)

1-Methylcyclopentene $\xrightarrow{Br_2}$ (enantiomers)

3-Methylcyclopentene $\xrightarrow{Br_2}$ (diastereomers)

4-Methylcyclopentene $\xrightarrow{Br_2}$ (enantiomers)

We are told in the problem that two diastereomeric bromides were formed, thus the compound must be 3-methylcyclopentene.

**7.44** Dehydration of this tertiary alcohol can yield 2,3-dimethyl-1-pentene or 2,3-dimethyl-2-pentene. Only the terminal alkene in this case is chiral.

$$CH_3CH_2\overset{*}{C}HC(CH_3)_2 \ (H_3C,\ OH) \longrightarrow CH_3CH_2\overset{*}{C}H(CH_3)C(=CH_2)CH_3 + CH_3CH_2C(CH_3)=C(CH_3)_2$$

2,3-Dimethyl-2-pentanol (chiral, optically pure)

2,3-Dimethyl-1-pentene (chiral, optically pure)

2,3-Dimethyl-2-pentene (achiral, optically inactive)

$\downarrow H_2, Pt$     $\downarrow H_2, Pt$

$$CH_3CH_2\overset{*}{C}HCH(CH_3)_2 \ (CH_3) \qquad CH_3CH_2\overset{*}{C}HCH(CH_3)_2 \ (CH_3)$$

2,3-Dimethylpentane (chiral, optically pure)

2,3-Dimethylpentane (chiral, optically inactive)

The 2,3-dimethyl-1-pentene formed in the dehydration reaction must be optically pure because it arises from optically pure alcohol by a reaction that does not involve any of the bonds to the stereogenic center. When optically pure 2,3-dimethyl-1-pentene is hydrogenated, it must yield optically pure 2,3-dimethylpentane—again, no bonds to the stereogenic center are involved in this step.

The 2,3-dimethyl-2-pentene formed in the dehydration reaction is achiral and must yield racemic 2,3-dimethylpentane on hydrogenation.

Because the alkane is 50% optically pure, the alkene fraction must have contained equal amounts of optically pure 2,3-dimethyl-1-pentene and its achiral isomer 2,3-dimethyl-2-pentene.

**7.45** (*a*) Oxygen may be transferred to either the front face or the back face of the double bond when (*R*)-3-buten-2-ol reacts with a peroxy acid. The structure of the minor stereoisomer was given in the problem. The major stereoisomer results from addition to the opposite face of the double bond.

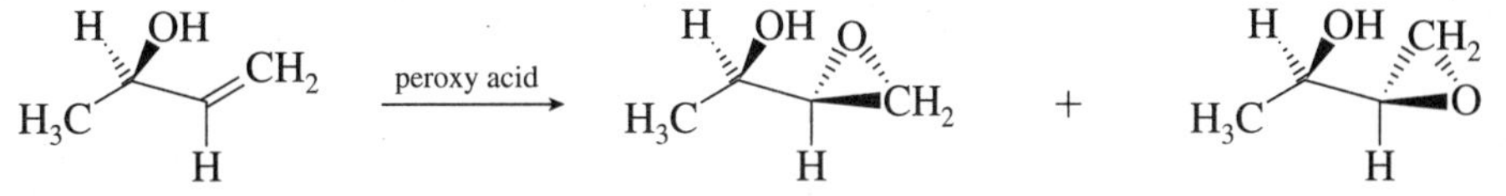

peroxy acid

(*R*)-3-Buten-2-ol — Minor stereoisomer + Major stereoisomer

(*b*) The two epoxides have the same configuration (*R*) at the secondary alcohol carbon, but opposite configurations at the stereogenic center of the epoxide ring. They are diastereomers.

(*c*) In addition to the two diastereomeric epoxides whose structures are shown in the solution to part (*a*), the enantiomers of each will be formed when racemic 3-buten-2-ol is epoxidized. The relative amounts of the four products will be:

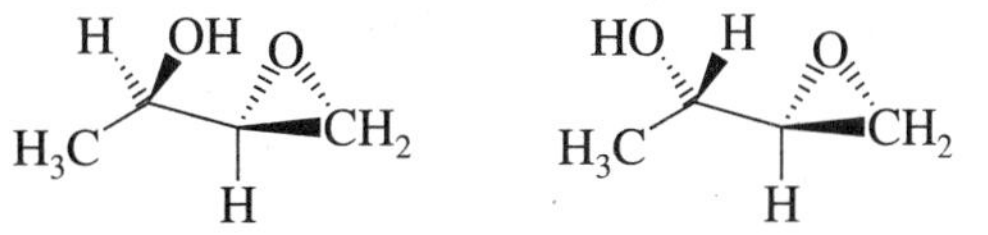

20% 20% — Enantiomeric forms of minor stereoisomer, totaling 40%

30% 30% — Enantiomeric forms of major stereoisomer, totaling 60%

**7.46–7.49** Solutions to molecular modeling exercises are not provided in this *Study Guide and Solutions Manual.* You should use *Learning By Modeling* for these exercises.

# SELF-TEST

## PART A

**A-1.** For each of the following pairs of drawings, identify the molecules as chiral or achiral and tell whether each pair represents molecules that are enantiomers, diastereomers, or identical.

(*a*) ① and 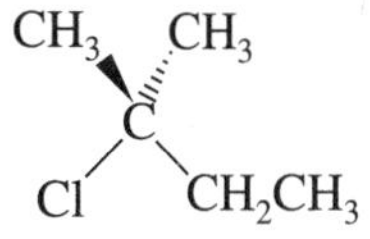 ②

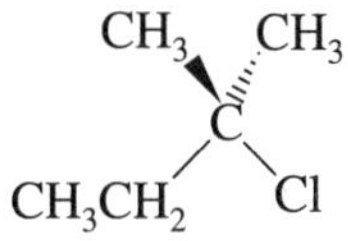

(*b*) ③ and 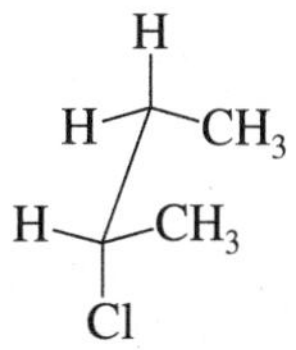 ④

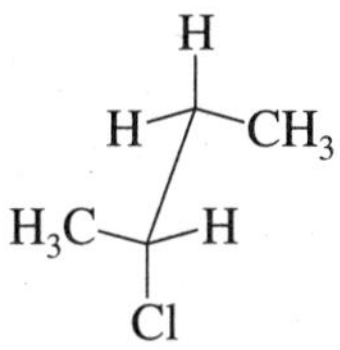

(*c*) 5 and 6

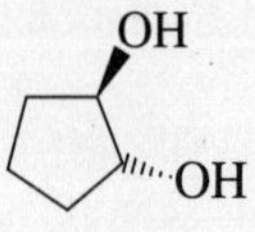

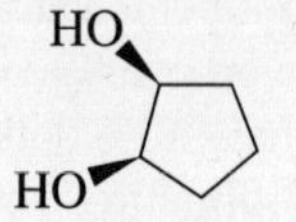

(*d*) 7 and 8

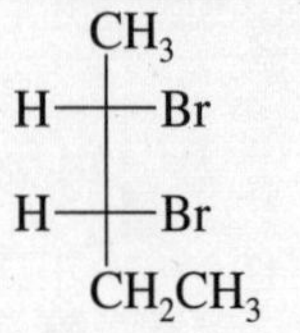

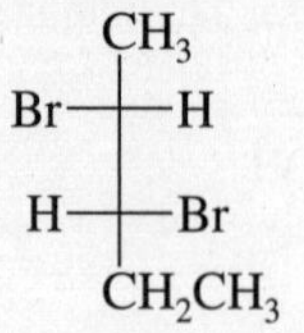

(*e*) 9 and 10

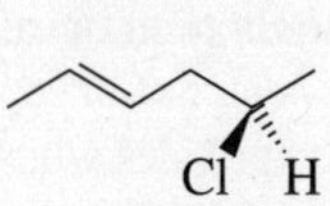

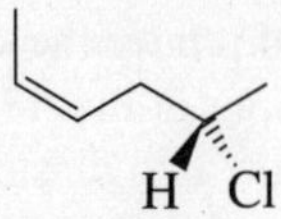

**A-2.** Specify the configuration of each stereogenic carbon in the preceding problem, using the Cahn–Ingold–Prelog *R–S* system.

**A-3.** Predict the number of stereoisomers possible for each of the following constitutions. For which of these will meso forms be possible?

(*a*)

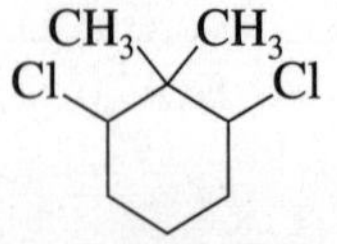

(*c*) $CH_2{=}CHCH(OH)CH{=}CHCH_3$

(*b*) $CH_3CH(Br)CH(Br)CH(Cl)CH_3$

**A-4.** Using the skeletons provided as a guide,

(*a*) Draw a perspective view of (2*R*,3*R*)-3-chloro-2-butanol.

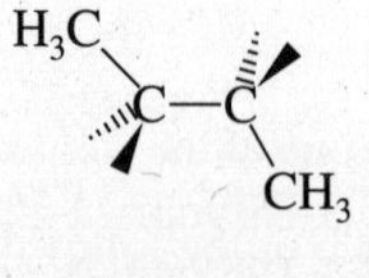

(*b*) Draw a sawhorse diagram of (*R*)-2-bromobutane.

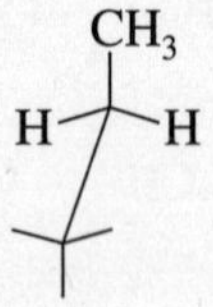

(*c*) Draw Fischer projections of both these compounds.

**A-5.** Draw Fischer projections of each stereoisomer of 2,3-dichlorobutane. Identify each stereogenic center as *R* or *S*. Which stereoisomers are chiral? Which are not? Why?

**A-6.** (*a*) The specific rotation of pure (−)-cholesterol is −39°. What is the specific rotation of a sample of cholesterol containing 10% (+)-cholesterol and 90% (−)-cholesterol.

(*b*) If the rotation of optically pure (*R*)-2-octanol is −10°, what is the percentage of the *S* enantiomer in a sample of 2-octanol that has a rotation of −4°?

**A-7.** Write the organic product(s) expected from each of the following reactions. Show each stereoisomer if more than one forms.
(*a*) 1,5,5-Trimethylcyclopentene and hydrogen bromide
(*b*) (*E*)-2-Butene and chlorine ($Cl_2$)
(*c*) (*Z*)-2-Pentene and peroxyacetic acid

**A-8.** Give the IUPAC name, including stereochemistry, for the following:

(*a*) [Newman projection: front carbon CH$_3$, Cl, H... substituents CH$_3$, Cl, $CH_2Br$, H, Br, H]

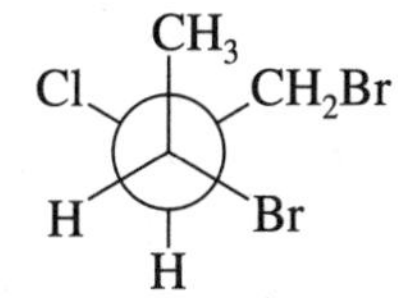

(*b*) [Cyclohexene ring bearing OH and $CH_2CH_3$]

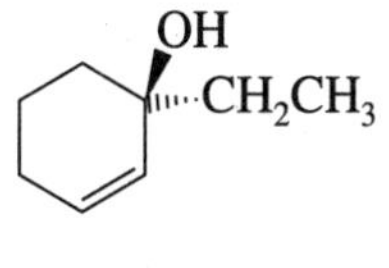

**A-9.** How many stereoisomeric products are obtained from the reaction of (*S*)-3-chloro-1-butene with hydrogen bromide? What is their relationship (enantiomers, diastereomers)?

**A-10.** Write the final product of the following reaction sequence, clearly showing its stereochemistry. Is the product achiral, a meso compound, optically active, or a racemic mixture?

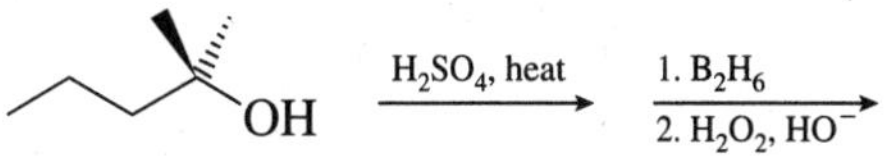

OH $\xrightarrow{H_2SO_4,\ heat}$ $\xrightarrow[2.\ H_2O_2,\ HO^-]{1.\ B_2H_6}$

## PART B

**B-1.** The structure of (*S*)-2-fluorobutane is best represented by

(*a*) $CH_3\overset{F}{\overset{|}{C}}HCH_2CH_3$

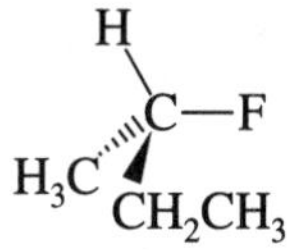

(*c*) H, $H_3C$, C—F, $CH_2CH_3$

(*b*) F, $H_3C$, C—H, $CH_2CH_3$

(*d*) $CH_3$, F—|—H, $CH_2CH_3$

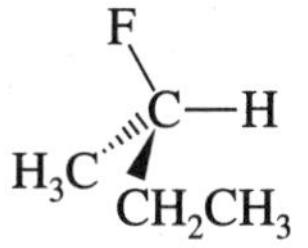

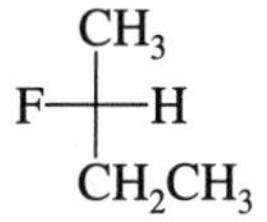

**B-2.** Which one of the following is chiral?
(*a*) 1,1-Dibromo-1-chloropropane
(*b*) 1,1-Dibromo-3-chloropropane
(*c*) 1,3-Dibromo-1-chloropropane
(*d*) 1,3-Dibromo-2-chloropropane

**B-3.** Which of the following compounds are meso forms?

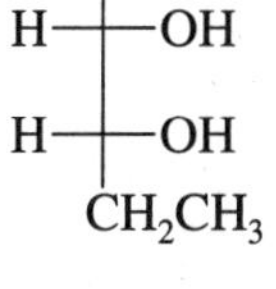

1: $CH_3$; H—OH; H—OH; $CH_2CH_3$

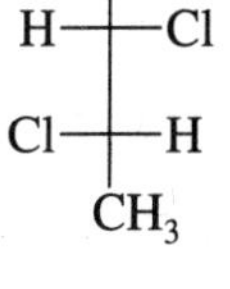

2: $CH_3$; H—Cl; Cl—H; $CH_3$

3: cyclohexane with $CH_3$ and $CH_3$

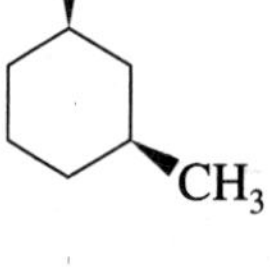

(*a*) 1 only
(*b*) 3 only
(*c*) 1 and 2
(*d*) 2 and 3

**B-4.** The 2,3-dichloropentane whose structure is shown is

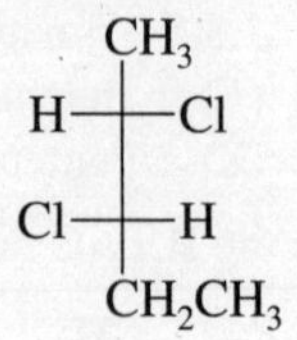

$$\begin{array}{c} CH_3 \\ H-\!\!\!+\!\!\!-Cl \\ Cl-\!\!\!+\!\!\!-H \\ CH_2CH_3 \end{array}$$

(*a*) 2*R*,3*R* (*b*) 2*R*,3*S* (*c*) 2*S*,3*R* (*d*) 2*S*,3*S*

**B-5.** The separation of a racemic mixture into the pure enantiomers is termed
(*a*) Racemization (*c*) Isomerization
(*b*) Resolution (*d*) Equilibration

**B-6.** Order the following groups in order of *R*–*S* ranking (4 is highest):

$-CH(CH_3)_2$ (A) $-CH_2CH_2Br$ (B) $-CH_2Br$ (C) $-C(CH_3)_3$ (D)

| | 4 | 3 | 2 | 1 |
|---|---|---|---|---|
| (*a*) | C | B | D | A |
| (*b*) | A | D | B | C |
| (*c*) | C | D | A | B |
| (*d*) | C | D | B | A |

**B-7.** A meso compound
(*a*) Is an achiral molecule that contains stereogenic centers.
(*b*) Contains a plane of symmetry or a center of symmetry.
(*c*) Is optically inactive.
(*d*) Is characterized by all of these.

**B-8.** The *S* enantiomer of ibuprofen is responsible for its pain-relieving properties. Which one of the structures shown is (*S*)-ibuprofen?

(*a*) [structure: C bearing C(=O)—OH, $CH_3$, H]
(*c*) [structure: C bearing H, $H_3C$, C(=O)OH]
(*b*) [structure: C bearing $CH_3$, H, C(=O)—OH]
(*d*) [structure: C bearing C(=O)—OH, H, $H_3C$]

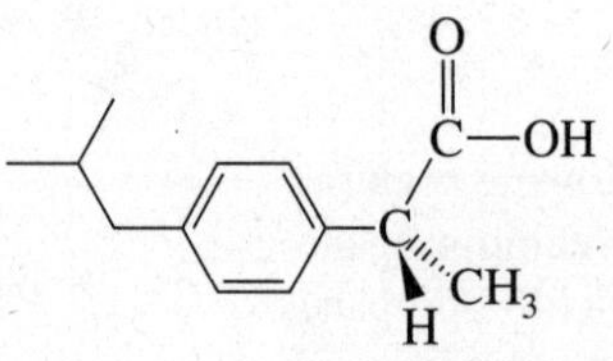

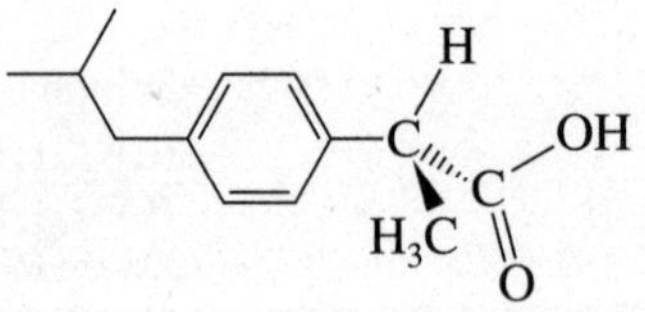

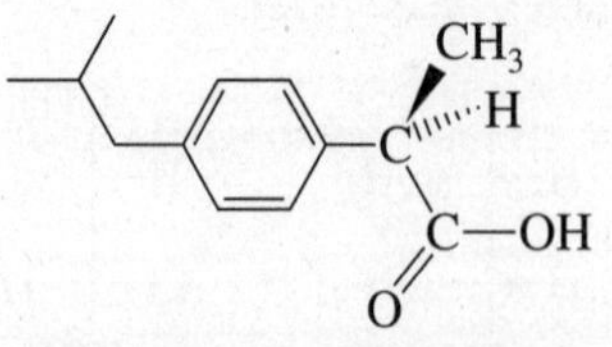

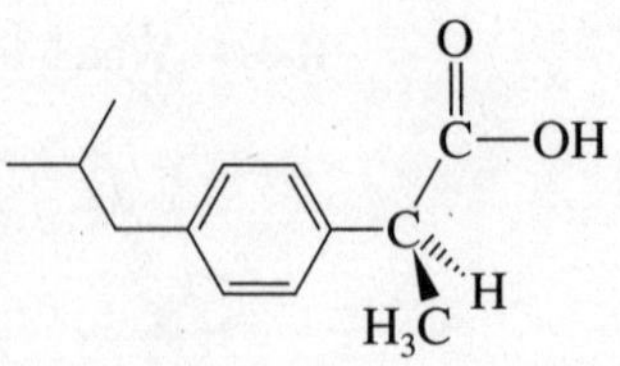

**B-9.** Which one of the following is a diastereomer of (*R*)-4-bromo-*cis*-2-hexene?
(*a*) (*S*)-4-bromo-*cis*-2-hexene
(*b*) (*R*)-4-bromo-*trans*-2-hexene
(*c*) (*R*)-5-bromo-*cis*-2-hexene
(*d*) (*S*)-5-bromo-*trans*-2-hexene
(*e*) (*R*)-5-bromo-*trans*-2-hexene

**B-10.** The reaction sequence

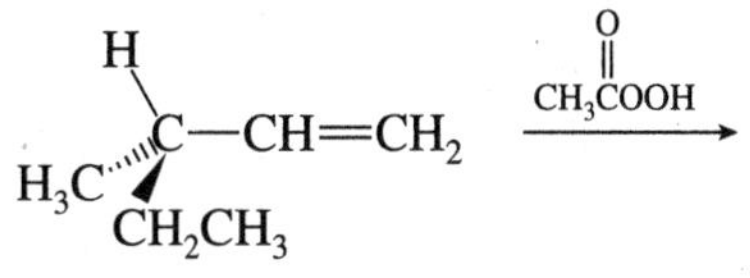

$$H_3C\text{—}C(H)(CH_2CH_3)\text{—}CH{=}CH_2 \xrightarrow{CH_3\overset{O}{\overset{\|}{C}}OOH}$$

will yield:
(*a*) A pair of products that are enantiomers
(*b*) A single product that is optically active
(*c*) A pair of products that are diastereomers
(*d*) A pair of products one of which is meso

**B-11.** Which of the following depict the same stereoisomer?

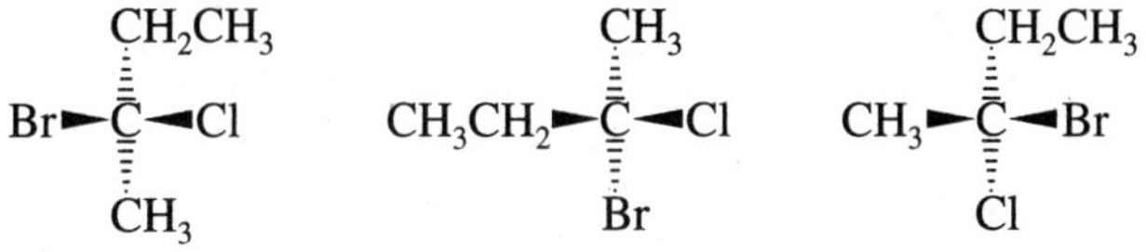

1: top $CH_2CH_3$, left Br, right Cl, bottom $CH_3$
2: top $CH_3$, left $CH_3CH_2$, right Cl, bottom Br
3: top $CH_2CH_3$, left $CH_3$, right Br, bottom Cl

(*a*) 1 and 2 (*b*) 1 and 3 (*c*) 2 and 3 (*d*) 1, 2, and 3

**B-12.** A naturally occurring substance has the constitution shown. How many stereoisomers may have this constitution?

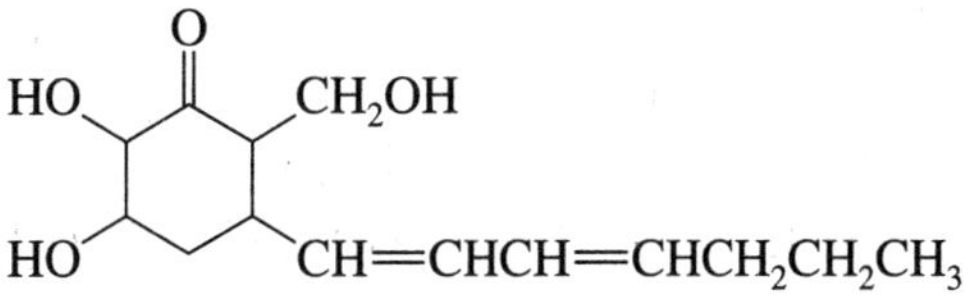

HO, O, $CH_2OH$, HO, $CH{=}CHCH{=}CHCH_2CH_2CH_3$

(*a*) 2 (*b*) 8 (*c*) 16 (*d*) 64 (*e*) 128

**B-13.** Acid-catalyzed hydration of an unknown compound X, $C_6H_{12}$, yielded as the major product a racemic mixture Y, $C_6H_{14}O$. Which (if any) of the following is (are) likely candidate(s) for X?

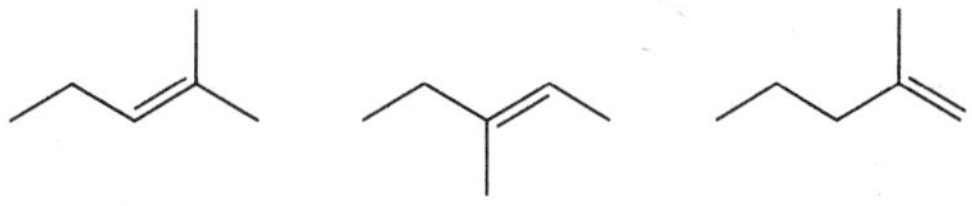

1 2 3

(*a*) 3 only (*c*) 1 and 3 (*e*) None of these
(*b*) 2 only (*d*) 2 and 3

**B-14.** The major product(s) from the reaction of $Br_2$ with (*Z*)-3-hexene is (are)
(*a*) Optically pure
(*b*) A racemic mixture of enantiomers
(*c*) The meso form
(*d*) Both the racemic mixture and the meso form

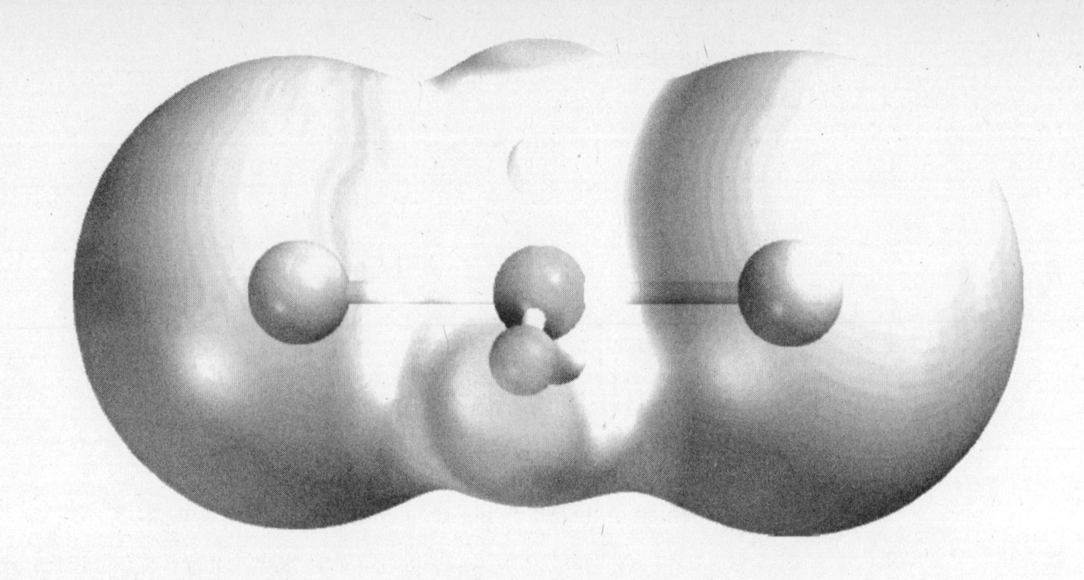

# CHAPTER 8
## NUCLEOPHILIC SUBSTITUTION

## SOLUTIONS TO TEXT PROBLEMS

**8.1** Identify the nucleophilic anion in each reactant. The nucleophilic anion replaces bromine as a substituent on carbon.

(*b*) Potassium ethoxide serves as a source of the nucleophilic anion $CH_3CH_2O^-$.

$$CH_3CH_2\ddot{\underset{..}{O}}:^- + CH_3\ddot{\underset{..}{Br}}: \longrightarrow CH_3CH_2\ddot{\underset{..}{O}}CH_3 + :\ddot{\underset{..}{Br}}:^-$$

Ethoxide ion (nucleophile) | Methyl bromide | Ethyl methyl ether (product) | Bromide ion

(*c*)

$$C_6H_5\text{—}\overset{\overset{\ddot{O}:}{\|}}{C}\text{—}\ddot{\underset{..}{O}}:^- + CH_3\ddot{\underset{..}{Br}}: \longrightarrow C_6H_5\text{—}\overset{\overset{\ddot{O}:}{\|}}{C}\text{—}\ddot{\underset{..}{O}}CH_3 + :\ddot{\underset{..}{Br}}:^-$$

Benzoate ion | Methyl bromide | Methyl benzoate | Bromide ion

(*d*) Lithium azide is a source of the azide ion.

$$:\overset{-}{\underset{..}{N}}=\overset{+}{N}=\overset{-}{\underset{..}{N}}:$$

It reacts with methyl bromide to give methyl azide.

$$:\overset{-}{\underset{..}{N}}=\overset{+}{N}=\overset{-}{\underset{..}{N}}: + CH_3\ddot{\underset{..}{Br}}: \longrightarrow CH_3\underset{..}{N}=\overset{+}{N}=\overset{-}{\underset{..}{N}}: + :\ddot{\underset{..}{Br}}:^-$$

Azide ion (nucleophile) | Methyl bromide | Methyl azide (product) | Bromide ion

(*e*) The nucleophilic anion in KCN is cyanide ($:\overset{-}{C}\equiv N:$). The carbon atom is negatively charged and is normally the site of nucleophilic reactivity.

$$:N\equiv C:^- + CH_3\ddot{\underset{..}{Br}}: \longrightarrow CH_3C\equiv N: + :\ddot{\underset{..}{Br}}:^-$$

Cyanide ion (nucleophile) | Methyl bromide | Methyl cyanide (product) | Bromide ion (*f*) The anion in sodium hydrogen sulfide (NaSH) is $^{-}$:S̈H.

$$HS^{-} + CH_3Br \longrightarrow CH_3SH + Br^{-}$$

Hydrogen sulfide ion — Methyl bromide — Methanethiol — Bromide ion

(*g*) Sodium iodide is a source of the nucleophilic anion iodide ion, $I^{-}$. The reaction of sodium iodide with alkyl bromides is usually carried out in acetone to precipitate the sodium bromide formed.

$$I^{-} + CH_3Br \xrightarrow{\text{acetone}} CH_3I + Br^{-}$$

Iodide ion — Methyl bromide — Methyl iodide — Bromide ion

**8.2** Write out the structure of the starting material. Notice that it contains a primary bromide and a primary chloride. Bromide is a better leaving group than chloride and is the one that is displaced faster by the nucleophilic cyanide ion.

$$ClCH_2CH_2CH_2Br \xrightarrow[\text{ethanol–water}]{\text{NaCN}} ClCH_2CH_2CH_2C{\equiv}N$$

1-Bromo-3-chloropropane — 4-Chlorobutanenitrite

**8.3** No, the two-step sequence is not consistent with the observed behavior for the hydrolysis of methyl bromide. The rate-determining step in the two-step sequence shown is the first step, ionization of methyl bromide to give methyl cation.

1. $CH_3Br \xrightarrow{\text{slow}} CH_3^{+} + Br^{-}$

2. $CH_3^{+} + HO^{-} \xrightarrow{\text{fast}} CH_3OH$

In such a sequence the nucleophile would not participate in the reaction until after the rate-determining step is past, and the reaction rate would depend only on the concentration of methyl bromide and be independent of the concentration of hydroxide ion.

$$\text{Rate} = k[CH_3Br]$$

The predicted kinetic behavior is first order. Second order kinetic behavior is actually observed for methyl bromide hydrolysis, so the proposed mechanism cannot be correct.

**8.4** Inversion of configuration occurs at the stereogenic center. When shown in a Fischer projection, this corresponds to replacing the leaving group on the one side by the nucleophile on the opposite side.

Fischer projection: top $CH_3$; left H, right Br; bottom $CH_2(CH_2)_4CH_3$ $\xrightarrow[S_N2]{\text{NaOH}}$ top $CH_3$; left HO, right H; bottom $CH_2(CH_2)_4CH_3$

(*S*)-(+)-2-Bromooctane — (*R*)-(−)-2-Octanol

**8.5** The example given in the text illustrates inversion of configuration in the $S_N2$ hydrolysis of (*S*)-(+)-2-bromooctane, which yields (*R*)-(−)-2-octanol. The hydrolysis of (*R*)-(−)-2-bromooctane exactly mirrors that of its enantiomer and yields (*S*)-(+)-2-octanol.

Hydrolysis of racemic 2-bromooctane gives racemic 2-octanol. Remember, optically inactive reactants must yield optically inactive products.

**8.6** Sodium iodide in acetone is a reagent that converts alkyl chlorides and bromides into alkyl iodides by an $S_N2$ mechanism. Pick the alkyl halide in each pair that is more reactive toward $S_N2$ displacement.

(*b*) The less crowded alkyl halide reacts faster in an $S_N2$ reaction. 1-Bromopentane is a primary alkyl halide and so is more reactive than 3-bromopentane, which is secondary.

$$BrCH_2CH_2CH_2CH_2CH_3 \qquad \underset{\displaystyle Br}{CH_3CH_2\overset{|}{C}HCH_2CH_3}$$

1-Bromopentane (primary; more reactive in $S_N2$) — 3-Bromopentane (secondary; less reactive in $S_N2$)

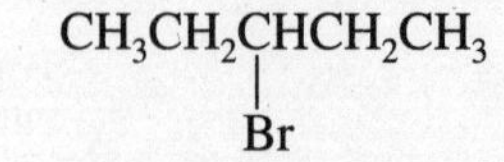

(*c*) Both halides are secondary, but fluoride is a poor leaving group in nucleophilic substitution reactions. Alkyl chlorides are more reactive than alkyl fluorides.

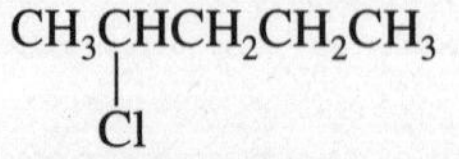

$$\underset{\displaystyle Cl}{CH_3CHCH_2CH_2CH_3} \qquad \underset{\displaystyle F}{CH_3CHCH_2CH_2CH_3}$$

2-Chloropentane (more reactive) — 2-Fluoropentane (less reactive)

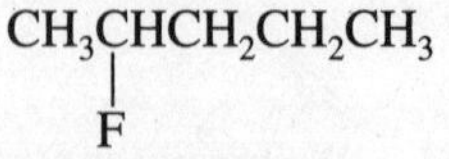

(*d*) A secondary alkyl bromide reacts faster under $S_N2$ conditions than a tertiary one.

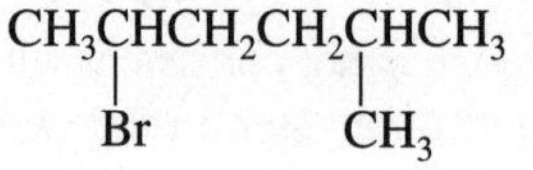

$$CH_3\underset{Br}{C}HCH_2CH_2\underset{CH_3}{C}HCH_3 \qquad CH_3\overset{CH_3}{\underset{Br}{C}}CH_2CH_2CH_2CH_3$$

2-Bromo-5-methylhexane (secondary; more reactive in $S_N2$) — 2-Bromo-2-methylhexane (tertiary; less reactive in $S_N2$)

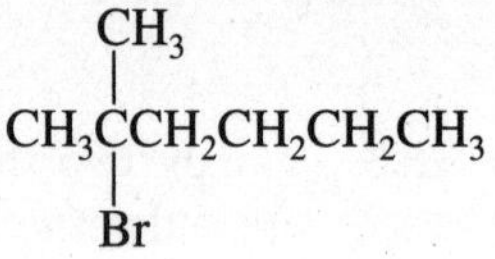

(*e*) The number of carbons does not matter as much as the degree of substitution at the reaction site. The primary alkyl bromide is more reactive than the secondary.

$$BrCH_2(CH_2)_8CH_3 \qquad CH_3\underset{Br}{C}HCH_3$$

1-Bromodecane (primary; more reactive in $S_N2$) — 2-Bromopropane (secondary; less reactive in $S_N2$)

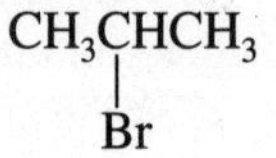

**8.7** Nitrite ion has two potentially nucleophilic sites, oxygen and nitrogen.

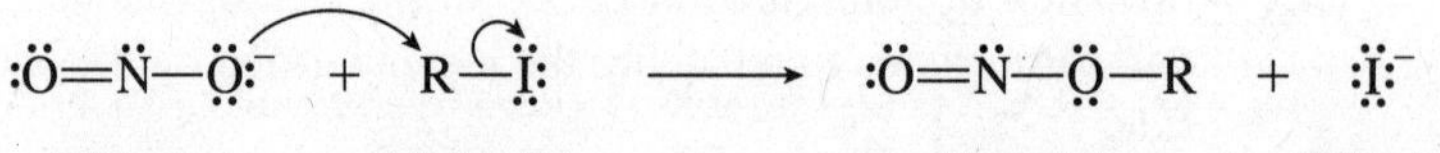

$$:\ddot{O}{=}\ddot{N}{-}\ddot{O}:^- + R{-}\ddot{I}: \longrightarrow :\ddot{O}{=}\ddot{N}{-}\ddot{O}{-}R + :\ddot{I}:^-$$

Nitrite ion — Alkyl iodide — Alkyl nitrite — Iodide ion

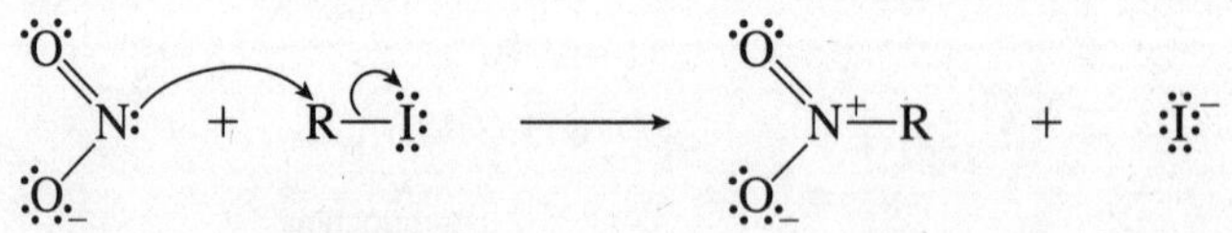

$$O{=}\ddot{N}{-}O^- + R{-}\ddot{I}: \longrightarrow O{=}\overset{+}{N}(R){-}O^- + :\ddot{I}:^-$$

Nitrite ion — Alkyl iodide — Nitroalkane — Iodide ion

Thus, an alkyl iodide can yield either an alkyl nitrite or a nitroalkane depending on whether the oxygen or the nitrogen of nitrite ion attacks carbon. Both do, and the product from 2-iodooctane is a mixture of

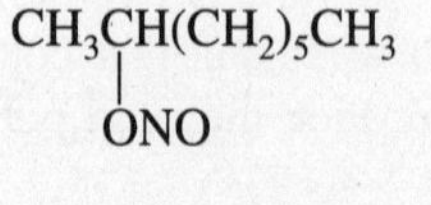

$$CH_3\underset{ONO}{C}H(CH_2)_5CH_3 \quad \text{and} \quad CH_3\underset{NO_2}{C}H(CH_2)_5CH_3$$

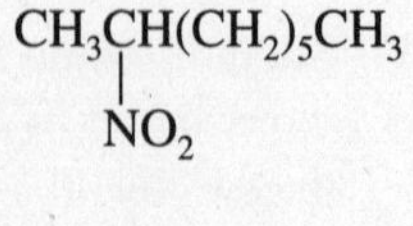

**8.8** Solvolysis of alkyl halides in alcohols yields ethers as the products of reaction.

$$(CH_3)_3CBr + CH_3OH \longrightarrow (CH_3)_3COCH_3 + HBr$$

*tert*-Butyl bromide | Methanol | *tert*-Butyl methyl ether | Hydrogen bromide

The reaction proceeds by an $S_N1$ mechanism.

$$(CH_3)_3C{-}\ddot{\underset{..}{Br}}{:} \xrightarrow{\text{slow}} (CH_3)_3C^+ + :\ddot{\underset{..}{Br}}:^-$$

*tert*-Butyl bromide | *tert*-Butyl cation | Bromide ion

$$(CH_3)_3C^+ + :\ddot{O}(CH_3)H \xrightarrow{\text{fast}} (CH_3)_3C{-}\overset{+}{O}(CH_3)H$$

*tert*-Butyl cation | Methanol | *tert*-Butyloxonium ion

$$(CH_3)_3C{-}\overset{+}{O}(CH_3)H + :\ddot{\underset{..}{Br}}:^- \longrightarrow (CH_3)_3C{-}\ddot{\underset{..}{O}}CH_3 + H{-}\ddot{\underset{..}{Br}}:$$

*tert*-Butyloxonium ion | Bromide ion | *tert*-Butyl methyl ether | Hydrogen bromide

**8.9** The reactivity of an alkyl halide in an $S_N1$ reaction is dictated by the ease with which it ionizes to form a carbocation. Tertiary alkyl halides are the most reactive, methyl halides the least reactive.

(*b*) Cyclopentyl iodide ionizes to form a secondary carbocation, and the carbocation from 1-methylcyclopentyl iodide is tertiary. The tertiary halide is more reactive.

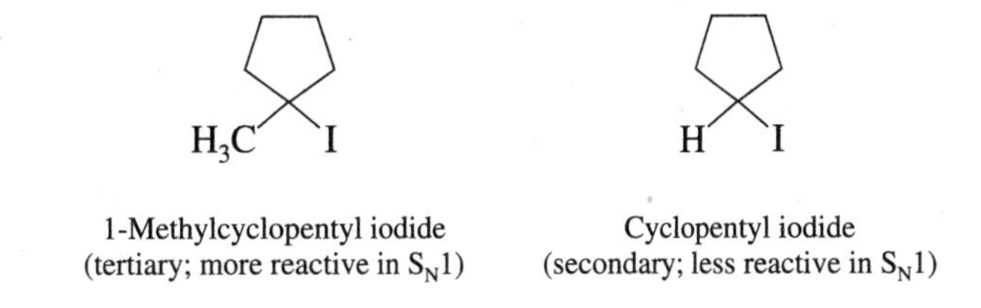


1-Methylcyclopentyl iodide (tertiary; more reactive in $S_N1$) | Cyclopentyl iodide (secondary; less reactive in $S_N1$)

(*c*) Cyclopentyl bromide ionizes to a secondary carbocation. 1-Bromo-2,2-dimethyl-propane is a primary alkyl halide and is therefore less reactive.

H Br | $(CH_3)_3CCH_2Br$

Cyclopentyl bromide (secondary; more reactive in $S_N1$) | 1-Bromo-2,2-dimethylpropane (primary; less reactive in $S_N1$)

(*d*) Iodide is a better leaving group than chloride in both $S_N1$ and $S_N2$ reactions.

$(CH_3)_3CI$ | $(CH_3)_3CCl$

*tert*-Butyl iodide (more reactive) | *tert*-Butyl chloride (less reactive)

**8.10** The alkyl halide is tertiary and so undergoes hydrolysis by an $S_N1$ mechanism. The carbocation can be captured by water at either face. A mixture of the axial and the equatorial alcohols is formed.

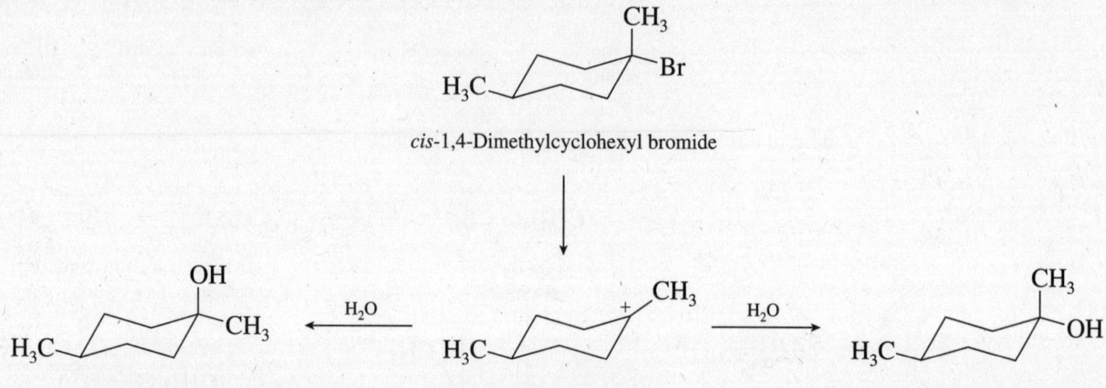

*cis*-1,4-Dimethylcyclohexyl bromide

*trans*-1,4-Dimethylcyclohexanol ⟵($H_2O$)— Carbocation intermediate —($H_2O$)⟶ *cis*-1,4-Dimethylcyclohexanol

The same two substitution products are formed from *trans*-1,4-dimethylcyclohexyl bromide because it undergoes hydrolysis via the same carbocation intermediate.

**8.11** Write chemical equations illustrating each rearrangement process.

**Hydride shift:**

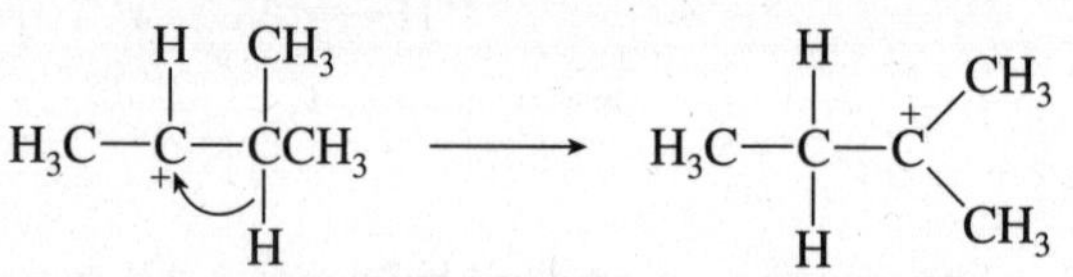

$$H_3C-\overset{+}{C}H-C(H)(CH_3)CH_3 \longrightarrow H_3C-CH_2-\overset{+}{C}(CH_3)_2$$

Tertiary carbocation

**Methyl shift:**

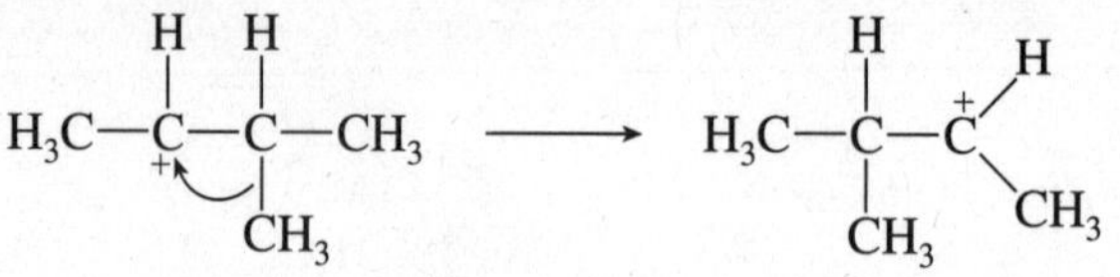

$$H_3C-\overset{+}{C}H-CH(CH_3)-CH_3 \longrightarrow H_3C-CH(CH_3)-\overset{+}{C}H-CH_3$$

Secondary carbocation

Rearrangement by a hydride shift is observed because it converts a secondary carbocation to a more stable tertiary one. A methyl shift gives a secondary carbocation—in this case the same carbocation as the one that existed prior to rearrangement.

**8.12** (*b*) Ethyl bromide is a primary alkyl halide and reacts with the potassium salt of cyclohexanol by substitution.

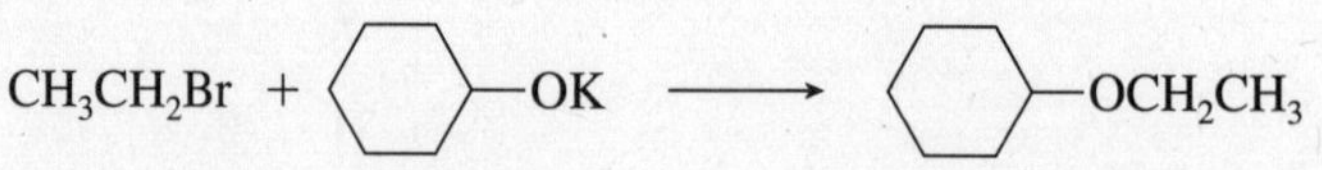

$$CH_3CH_2Br + C_6H_{11}OK \longrightarrow C_6H_{11}OCH_2CH_3$$

Ethyl bromide — Potassium cyclohexanolate — Cyclohexyl ethyl ether

(*c*) No strong base is present in this reaction; the nucleophile is methanol itself, not methoxide. It reacts with *sec*-butyl bromide by substitution, not elimination.

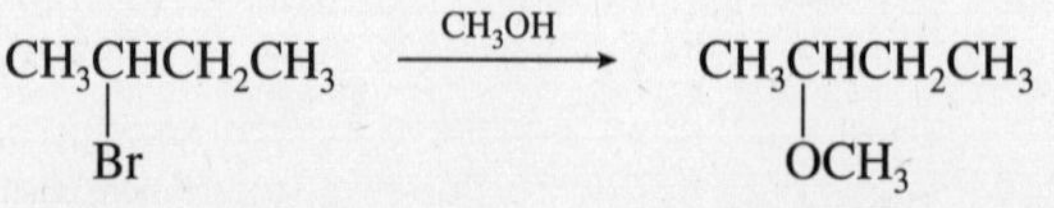

$$CH_3CH(Br)CH_2CH_3 \xrightarrow{CH_3OH} CH_3CH(OCH_3)CH_2CH_3$$

*sec*-Butyl bromide — *sec*-Butyl methyl ether

(*d*) Secondary alkyl halides react with alkoxide bases by E2 elimination.

$$CH_3CHCH_2CH_3 \text{ (with Br on C-2)} \xrightarrow[CH_3OH]{NaOCH_3} CH_3CH{=}CHCH_3 + H_2C{=}CHCH_2CH_3$$

*sec*-Butyl bromide — 2-Butene (major product; mixture of cis and trans) — 1-Butene

**8.13** Alkyl *p*-toluenesulfonates are prepared from alcohols and *p*-toluenesulfonyl chloride.

$$CH_3(CH_2)_{16}CH_2OH + H_3C{-}C_6H_4{-}SO_2Cl \xrightarrow{\text{pyridine}} CH_3(CH_2)_{16}CH_2OSO_2{-}C_6H_4{-}CH_3 + HCl$$

1-Octadecanol — *p*-Toluenesulfonyl chloride — Octadecyl *p*-toluenesulfonate — Hydrogen chloride

**8.14** As in part (*a*), identify the nucleophilic anion in each part. The nucleophile replaces the *p*-toluenesulfonate (tosylate) leaving group by an $S_N2$ process. The tosylate group is abbreviated as OTs.

(*b*)

$$I^- + CH_3(CH_2)_{16}CH_2OTs \longrightarrow CH_3(CH_2)_{16}CH_2I + TsO^-$$

Iodide ion — Octadecyl *p*-toluenesulfonate — Octadecyl iodide — *p*-Toluenesulfonate anion

(*c*)

$$^-C{\equiv}N + CH_3(CH_2)_{16}CH_2OTs \longrightarrow CH_3(CH_2)_{16}CH_2C{\equiv}N + TsO^-$$

Cyanide ion — Octadecyl *p*-toluenesulfonate — Octadecyl cyanide — *p*-Toluenesulfonate anion

(*d*)

$$HS^- + CH_3(CH_2)_{16}CH_2OTs \longrightarrow CH_3(CH_2)_{16}CH_2SH + TsO^-$$

Hydrogen sulfide ion — Octadecyl *p*-toluenesulfonate — 1-Octadecanethiol — *p*-Toluenesulfonate anion

(*e*)

$$CH_3CH_2CH_2CH_2S^- + CH_3(CH_2)_{16}CH_2OTs \longrightarrow CH_3(CH_2)_{16}CH_2SCH_2CH_2CH_2CH_3 + TsO^-$$

Butanethiolate ion — Octadecyl *p*-toluenesulfonate — Butyl octadecyl thioether — *p*-Toluenesulfonate anion

**8.15** The hydrolysis of (*S*)-(+)-1-methylheptyl *p*-toluenesulfonate proceeds with inversion of configuration, giving the *R* enantiomer of 2-octanol.

$$CH_3(CH_2)_5(H)(H_3C)C{-}OTs \xrightarrow{H_2O} HO{-}C(H)(CH_2)_5CH_3(CH_3)$$

(*S*)-(+)-1-Methylheptyl *p*-toluenesulfonate — (*R*)-(−)-2-Octanol

In Section 8.14 of the text we are told that optically pure (*S*)-(+)-1-methylheptyl *p*-toluenesulfonate is prepared from optically pure (*S*)-(+)-2-octanol having a specific rotation $[\alpha]_D^{25}$ +9.9°. The conversion of an alcohol to a *p*-toluenesulfonate proceeds with complete *retention* of configuration. Hydrolysis of this *p*-toluenesulfonate with *inversion* of configuration therefore yields optically pure (*R*)-(−)-2-octanol of $[\alpha]_D^{25}$ −9.9°.

**8.16** Protonation of 3-methyl-2-butanol and dissociation of the alkyloxonium ion gives a secondary carbocation. A hydride shift yields a tertiary, and thus more stable, carbocation. Capture of this carbocation by chloride ion gives the major product, 2-chloro-2-methylbutane.

$$CH_3CH(OH)CH(CH_3)_2 \xrightarrow{HCl} CH_3\overset{+}{C}HCH(CH_3)_2 \xrightarrow[\text{shift}]{\text{hydride}} CH_3CH_2\overset{+}{C}(CH_3)_2$$

3-Methyl-2-butanol; Secondary carbocation; less stable; Tertiary carbocation; more stable

$$CH_3\overset{+}{C}HCH(CH_3)_2 \xrightarrow{Cl^-} CH_3CH(Cl)CH(CH_3)_2$$

2-Chloro-3-methylbutane (trace)

$$CH_3CH_2\overset{+}{C}(CH_3)_2 \xrightarrow{Cl^-} CH_3CH_2C(Cl)(CH_3)_2$$

2-Chloro-2-methylbutane (major product; ≈97%)

**8.17** 1-Bromopropane is a primary alkyl halide, and so it will undergo predominantly $S_N2$ displacement regardless of the basicity of the nucleophile.

(*a*) $CH_3CH_2CH_2Br \xrightarrow[\text{acetone}]{\text{NaI}} CH_3CH_2CH_2I$

1-Bromopropane → 1-Iodopropane

(*b*) $CH_3CH_2CH_2Br \xrightarrow[\text{acetic acid}]{CH_3\overset{O}{\overset{\|}{C}}ONa} CH_3CH_2CH_2O\overset{O}{\overset{\|}{C}}CH_3$

Propyl acetate

(*c*) $CH_3CH_2CH_2Br \xrightarrow[\text{ethanol}]{NaOCH_2CH_3} CH_3CH_2CH_2OCH_2CH_3$

Ethyl propyl ether

(*d*) $CH_3CH_2CH_2Br \xrightarrow[\text{DMSO}]{\text{NaCN}} CH_3CH_2CH_2CN$

Butanenitrite

(*e*) $CH_3CH_2CH_2Br \xrightarrow[\text{ethanol–water}]{NaN_3} CH_3CH_2CH_2N_3$

1-Azidopropane

(*f*) $CH_3CH_2CH_2Br \xrightarrow[\text{ethanol}]{\text{NaSH}} CH_3CH_2CH_2SH$

1-Propanethiol

(*g*) $CH_3CH_2CH_2Br \xrightarrow[\text{ethanol}]{NaSCH_3} CH_3CH_2CH_2SCH_3$

Methyl propyl sulfide

**8.18** Elimination is the major product when secondary halides react with anions as basic as or more basic than hydroxide ion. Alkoxide ions have a basicity comparable with hydroxide ion and react with secondary halides to give predominantly elimination products. Thus ethoxide ion [part (*c*)] will react with 2-bromopropane to give mainly propene.

$$CH_3CH(Br)CH_3 \xrightarrow{NaOCH_2CH_3} CH_3CH{=}CH_2$$

Propene

**8.19** (*a*) The substrate is a primary alkyl bromide and reacts with sodium iodide in acetone to give the corresponding iodide.

$$BrCH_2COCH_2CH_3 \xrightarrow[acetone]{NaI} ICH_2COCH_2CH_3$$

(C=O in each ester: $BrCH_2C(=O)OCH_2CH_3$ → $ICH_2C(=O)OCH_2CH_3$)

Ethyl bromoacetate → Ethyl iodoacetate (89%)

(*b*) Primary alkyl chlorides react with sodium acetate to yield the corresponding acetate esters.

$$O_2N{-}C_6H_4{-}CH_2Cl \xrightarrow[acetic\ acid]{CH_3C(=O)ONa} O_2N{-}C_6H_4{-}CH_2OC(=O)CH_3$$

*p*-Nitrobenzyl chloride → *p*-Nitrobenzyl acetate (78–82%)

(*c*) The only leaving group in the substrate is bromide. Neither of the carbon–oxygen bonds is susceptible to cleavage by nucleophilic attack.

$$CH_3CH_2OCH_2CH_2Br \xrightarrow[ethanol-water]{NaCN} CH_3CH_2OCH_2CH_2CN$$

2-Bromoethyl ethyl ether → 2-Cyanoethyl ethyl ether (52–58%)

(*d*) Hydrolysis of the primary chloride yields the corresponding alcohol.

$$NC{-}C_6H_4{-}CH_2Cl \xrightarrow{H_2O,\ HO^-} NC{-}C_6H_4{-}CH_2OH$$

*p*-Cyanobenzyl chloride → *p*-Cyanobenzyl alcohol (85%)

(*e*) The substrate is a primary chloride.

$$ClCH_2C(=O)OC(CH_3)_3 \xrightarrow[acetone-water]{NaN_3} N_3CH_2C(=O)OC(CH_3)_3$$

*tert*-Butyl chloroacetate → *tert*-Butyl azidoacetate (92%)

(*f*) Primary alkyl tosylates yield iodides on treatment with sodium iodide in acetone.

$$TsOCH_2{-}(2,2\text{-dimethyl-1,3-dioxolane ring}) \xrightarrow[acetone]{NaI} ICH_2{-}(2,2\text{-dimethyl-1,3-dioxolane ring})$$

(2,2-Dimethyl-1,3-dioxolan-4-yl)-methyl *p*-toluenesulfonate → 2,2-Dimethyl-4-(iodomethyl)-1,3-dioxolane (60%)

(*g*) Sulfur displaces bromide from ethyl bromide.

$$(2\text{-furyl}){-}CH_2SNa + CH_3CH_2Br \longrightarrow (2\text{-furyl}){-}CH_2SCH_2CH_3$$

Sodium (2-furyl)-methanethiolate + Ethyl bromide → Ethyl (2-furyl)methyl sulfide (80%)

(*h*) The first reaction is one in which a substituted alcohol is converted to a *p*-toluenesulfonate ester. This is followed by an $S_N2$ displacement with lithium iodide.

$OCH_3$, $CH_3O$, $CH_3O$, $CH_2CH_2CH_2CH_2OH$ $\xrightarrow[\text{pyridine (62\%)}]{\text{TsCl}}$ $OCH_3$, $CH_3O$, $CH_3O$, $CH_2CH_2CH_2CH_2OTs$ $\xrightarrow{\text{LiI, acetone (88\%)}}$ $OCH_3$, $CH_3O$, $CH_3O$, $CH_2CH_2CH_2CH_2I$

4-(2,3,4-Trimethoxyphenyl)-1-butanol

4-(2,3,4-Trimethoxyphenyl)-1-butyl iodide

**8.20** The two products are diastereomers of each other. They are formed by bimolecular nucleophilic substitution ($S_N2$). In each case, a good nucleophile ($C_6H_5S^-$) displaces chloride from a secondary carbon with inversion of configuration.

(*a*) The trans chloride yields a cis substitution product.

Cl–[cyclohexane]–$C(CH_3)_3$ + [benzene]–SNa ⟶ [benzene]–S–[cyclohexane]–$C(CH_3)_3$

*trans*-4-*tert*-Butylcyclohexyl chloride — Sodium benzenethiolate — *cis*-4-*tert*-Butylcyclohexyl phenyl sulfide

(*b*) The cis chloride yields a trans substitution product.

Cl–[cyclohexane]–$C(CH_3)_3$ + NaS–[benzene] ⟶ [benzene]–S–[cyclohexane]–$C(CH_3)_3$

*cis*-4-*tert*-Butylcyclohexyl chloride — Sodium benzenethiolate — *trans*-4-*tert*-Butylcyclohexyl phenyl sulfide

**8.21** The isomers of $C_4H_9Cl$ are:

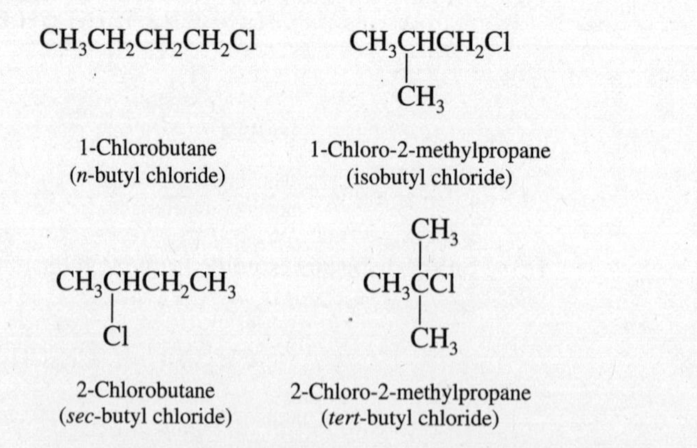


1-Chlorobutane (*n*-butyl chloride)

1-Chloro-2-methylpropane (isobutyl chloride)

2-Chlorobutane (*sec*-butyl chloride)

2-Chloro-2-methylpropane (*tert*-butyl chloride)

The reaction conditions (sodium iodide in acetone) are typical for an $S_N2$ process.

The order of $S_N2$ reactivity is primary > secondary > tertiary, and branching of the chain close to the site of substitution hinders reaction. The unbranched primary halide *n*-butyl chloride will be the most reactive and the tertiary halide *tert*-butyl chloride the least. The order of reactivity will therefore be: 1-chlorobutane > 1-chloro-2-methylpropane > 2-chlorobutane > 2-chloro-2-methylpropane.

**8.22** 1-Chlorohexane is a primary alkyl halide; 2-chlorohexane and 3-chlorohexane are secondary.

$CH_3CH_2CH_2CH_2CH_2CH_2Cl$ — 1-Chlorohexane (primary)

$CH_3CH(Cl)CH_2CH_2CH_2CH_3$ — 2-Chlorohexane (secondary)

$CH_3CH_2CH(Cl)CH_2CH_2CH_3$ — 3-Chlorohexane (secondary)

Primary and secondary alkyl halides react with potassium iodide in acetone by an $S_N2$ mechanism, and the rate depends on steric hindrance to attack on the alkyl halide by the nucleophile.

(*a*) Primary alkyl halides are more reactive than secondary alkyl halides in $S_N2$ reactions. 1-Chlorohexane is the most reactive isomer.

(*b*) Substituents at the carbon adjacent to the one that bears the leaving group slow down the rate of nucleophilic displacement. In 2-chlorohexane the group adjacent to the point of attack is $CH_3$. In 3-chlorohexane the group adjacent to the point of attack is $CH_2CH_3$. 2-Chlorohexane has been observed to be more reactive than 3-chlorohexane by a factor of 2.

**8.23** (*a*) Iodide is a better leaving group than bromide, and so 1-iodobutane should undergo $S_N2$ attack by cyanide faster than 1-bromobutane.

(*b*) The reaction conditions are typical for an $S_N2$ process. The methyl branch in 1-chloro-2-methylbutane sterically hinders attack at C-1. The unbranched isomer, 1-chloropentane, reacts faster.

$CH_3CH_2CH(CH_3)CH_2Cl$ — 1-Chloro-2-methylbutane is more sterically hindered, therefore less reactive.

$CH_3CH_2CH_2CH_2CH_2Cl$ — 1-Chloropentane is less sterically hindered, therefore more reactive.

(*c*) Hexyl chloride is a primary alkyl halide, and cyclohexyl chloride is secondary. Azide ion is a good nucleophile, and so the $S_N2$ reactivity rules apply; primary is more reactive than secondary.

$CH_3CH_2CH_2CH_2CH_2CH_2Cl$ — Hexyl chloride is primary, therefore more reactive in $S_N2$.

Cl (on cyclohexane ring) — Cyclohexyl chloride is secondary, therefore less reactive in $S_N2$.

(*d*) 1-Bromo-2,2-dimethylpropane is too hindered to react with the weakly nucleophilic ethanol by an $S_N2$ reaction, and since it is a primary alkyl halide, it is less reactive in $S_N1$ reactions. *tert*-Butyl bromide will react with ethanol by an $S_N1$ mechanism at a reasonable rate owing to formation of a tertiary carbocation.

$(CH_3)_3CBr$ — *tert*-Butyl bromide; more reactive in $S_N1$ solvolysis

$(CH_3)_3CCH_2Br$ — 1-Bromo-2,2-dimethylpropane; relatively unreactive in nucleophilic substitution reactions

(*e*) Solvolysis of alkyl halides in aqueous formic acid is faster for those that form carbocations readily. The $S_N1$ reactivity order applies here: secondary > primary.

$$\underset{\text{Br}}{CH_3\underset{|}{C}HCH_2CH_3} \qquad (CH_3)_2CHCH_2Br$$

*sec*-Butyl bromide; secondary, therefore more reactive in $S_N1$ — Isobutyl bromide; primary, therefore less reactive in $S_N1$

(*f*) 1-Chlorobutane is a primary alkyl halide and so should react by an $S_N2$ mechanism. Sodium methoxide is more basic than sodium acetate and is a better nucleophile. Reaction will occur faster with sodium methoxide than with sodium acetate.

(*g*) Azide ion is a very good nucleophile, whereas *p*-toluenesulfonate is a very good leaving group but a very poor nucleophile. In an $S_N2$ reaction with 1-chlorobutane, sodium azide will react faster than sodium *p*-toluenesulfonate.

**8.24** There are only two possible products from free-radical chlorination of the starting alkane:

$$(CH_3)_3CCH_2C(CH_3)_3 \xrightarrow[h\nu]{Cl_2} (CH_3)_3CCH_2\underset{CH_2Cl}{C}(CH_3)_2 + (CH_3)_3C\underset{Cl}{C}HC(CH_3)_3$$

2,2,4,4-Tetramethylpentane — 1-Chloro-2,2,4,4-tetramethylpentane (primary) — 3-Chloro-2,2,4,4-tetramethylpentane (secondary)

As revealed by their structural formulas, one isomer is a primary alkyl chloride, the other is secondary. The problem states that the major product (compound A) undergoes $S_N1$ hydrolysis much more slowly than the minor product (compound B). Since secondary halides are much more reactive than primary halides under $S_N1$ conditions, the major (unreactive) product is the primary alkyl halide 1-chloro-2,2,4,4-tetramethylpentane (compound A) and the minor (reactive) product is the secondary alkyl halide 3-chloro-2,2,4,4-tetramethylpentane (compound B).

**8.25** (*a*) The two most stable Lewis structures (resonance forms) of thiocyanate are:

$$^{-}:\ddot{\underset{..}{S}}-C\equiv N: \longleftrightarrow :\ddot{S}=C=\ddot{N}:^{-}$$

(*b*) The two Lewis structures indicate that the negative charge is shared by two atoms: S and N. Thus thiocyanate ion has two potentially nucleophilic sites, and the two possible products are

$$CH_3CH_2CH_2CH_2Br \xrightarrow[DMF]{KSCN} CH_3CH_2CH_2CH_2-\ddot{\underset{..}{S}}-C\equiv N: + CH_3CH_2CH_2CH_2-\ddot{N}=C=\ddot{S}:$$

1-Bromobutane — Butyl thiocyanate — Butyl isothiocyanate

(*c*) Sulfur is more polarizable than nitrogen and is more nucleophilic. The major product is butyl thiocyanate and arises by attack of sulfur of thiocyanate on butyl bromide.

**8.26** Using the unshared electron pair on its nitrogen, triethylamine acts as a nucleophile in an $S_N2$ reaction toward ethyl iodide.

$$(CH_3CH_2)_3N: + \underset{CH_3}{CH_2}-I \longrightarrow (CH_3CH_2)_4N^+ \; I^-$$

Triethylamine — Ethyl iodide — Tetraethylammonium iodide

The product of the reaction is a salt and has the structure shown. The properties given in the problem (soluble in polar solvents, high melting point) are typical of those of an ionic compound.

**8.27** This reaction has been reported in the chemical literature and proceeds as shown (91% yield):

$$CH_3CH_2(H)(H_3C)C{-}CH_2Br \xrightarrow[\text{acetone}]{NaI} CH_3CH_2(H)(H_3C)C{-}CH_2I$$

(*S*)-1-Bromo-2-methylbutane → (*S*)-1-Iodo-2-methylbutane

Notice that the configuration of the product is the *same* as the configuration of the reactant. This is because the stereogenic center is not involved in the reaction. When we say that $S_N2$ reactions proceed with inversion of configuration we refer only to the carbon at which substitution takes place, not a stereogenic center elsewhere in the molecule.

**8.28** (*a*) The starting material incorporates both a primary chloride and a secondary chloride. The nucleophile (iodide) attacks the less hindered primary position.

$$ClCH_2CH_2CH(Cl)CH_2CH_3 \xrightarrow[\text{acetone}]{NaI} ICH_2CH_2CH(Cl)CH_2CH_3$$

1,3-Dichloropentane → 3-Chloro-1-iodopentane ($C_5H_{10}ClI$)

(*b*) Nucleophilic substitution of the first bromide by sulfur occurs in the usual way.

$$^{-}{:}\ddot{S}CH_2CH_2\ddot{S}{:}^{-} + BrCH_2CH_2Br \longrightarrow {}^{-}{:}\ddot{S}CH_2CH_2\ddot{S}CH_2CH_2Br$$

The product of this step cyclizes by way of an intramolecular nucleophilic substitution.

$$^{-}{:}\ddot{S}CH_2CH_2\ddot{S}CH_2CH_2Br \longrightarrow \text{1,4-Dithiane } (C_4H_8S_2)$$

(*c*) The nucleophile is a dianion ($S^{2-}$). Two nucleophilic substitution reactions take place; the second of the two leads to intramolecular cyclization.

$${:}\ddot{S}{:}^{2-} + ClCH_2CH_2CH_2CH_2Cl \longrightarrow {}^{-}{:}\ddot{S}CH_2CH_2CH_2CH_2Cl \longrightarrow \text{Thiolane } (C_4H_8S)$$

**8.29** (*a*) Methyl halides are unhindered and react rapidly by the $S_N2$ mechanism.

(*b*) Sodium ethoxide is a good nucleophile and will react with unhindered primary alkyl halides by the $S_N2$ mechanism.

(*c*) Cyclohexyl bromide is a secondary halide and will react with a strong base (sodium ethoxide) predominantly by the E2 mechanism.

(*d*) The tertiary halide *tert*-butyl bromide will undergo solvolysis by the $S_N1$ mechanism.

(*e*) The presence of the strong base sodium ethoxide will cause the E2 mechanism to predominate.

(*f*) Concerted reactions are those which occur in a single step. The bimolecular mechanisms $S_N2$ and E2 represent concerted processes.

(*g*) In a stereospecific reaction, stereoisomeric reactants yield products that are stereoisomers of each other. Reactions that occur by the $S_N2$ and E2 mechanisms are stereospecific.

(*h*) The unimolecular mechanisms $S_N1$ and E1 involve the formation of carbocation intermediates.

(*i*) Rearrangements are possible when carbocations are intermediates in a reaction. Thus reactions occurring by the $S_N1$ and E1 mechanisms are most likely to have a rearranged carbon skeleton.

(*j*) Iodide is a better leaving group than bromide, and alkyl iodides will react faster than alkyl bromides by any of the four mechanisms $S_N1$, $S_N2$, E1, and E2.

**8.30** (*a*) Cyclopentyl cyanide can be prepared from a cyclopentyl halide by a nucleophilic substitution reaction. The first task, therefore, is to convert cyclopentane to a cyclopentyl halide.

Cyclopentane + $Cl_2$ (Chlorine) $\xrightarrow{\text{light or heat}}$ Cyclopentyl chloride + HCl (Hydrogen chloride)

Cyclopentyl chloride + NaCN (Sodium cyanide) $\xrightarrow[\text{or DMSO}]{\text{ethanol–water}}$ Cyclopentyl cyanide + NaCl (Sodium chloride)

An analogous sequence involving cyclopentyl bromide could be used.

(*b*) Cyclopentene can serve as a precursor to a cyclopentyl halide.

Cyclopentene + HBr (Hydrogen bromide) $\longrightarrow$ Cyclopentyl bromide

Once cyclopentyl bromide has been prepared, it is converted to cyclopentyl cyanide by nucleophilic substitution, as shown in part (*a*).

(*c*) Reaction of cyclopentanol with hydrogen bromide gives cyclopentyl bromide. Then cyclopentyl bromide can be converted to cyclopentyl cyanide, as shown in part (*a*).

Cyclopentanol (cyclopentyl–OH) + HBr $\longrightarrow$ Cyclopentyl bromide (cyclopentyl–Br)

(*d*) Two cyano groups are required here, both of which must be introduced in nucleophilic substitution reactions. The substrate in the key reaction is $BrCH_2CH_2Br$.

$$\underset{\text{1,2-Dibromoethane}}{BrCH_2CH_2Br} + \underset{\text{Sodium cyanide}}{2NaCN} \xrightarrow[\text{DMSO}]{\text{ethanol–water or}} \underset{\text{1,2-Dicyanoethane}}{N{\equiv}CCH_2CH_2C{\equiv}N}$$

1,2-Dibromoethane is prepared from ethylene. The overall synthesis from ethyl alcohol is therefore formulated as shown:

$$\underset{\text{Ethyl alcohol}}{CH_3CH_2OH} \xrightarrow[\text{heat}]{H_2SO_4} \underset{\text{Ethylene}}{CH_2{=}CH_2} \xrightarrow{Br_2} \underset{\text{1,2-Dibromoethane}}{BrCH_2CH_2Br} \xrightarrow{NaCN} \underset{\text{1,2-Dicyanoethane}}{NCCH_2CH_2CN}$$

(*e*) In this synthesis a primary alkyl chloride must be converted to a primary alkyl iodide. This is precisely the kind of transformation for which sodium iodide in acetone is used.

$$\underset{\text{Isobutyl chloride}}{(CH_3)_2CHCH_2Cl} + \underset{\text{Sodium iodide}}{NaI} \xrightarrow{\text{acetone}} \underset{\text{Isobutyl iodide}}{(CH_3)_2CHCH_2I} + \underset{\text{Sodium chloride}}{NaCl}$$

(*f*) First convert *tert*-butyl chloride into an isobutyl halide.

$$(CH_3)_3CCl \xrightarrow{NaOCH_3} (CH_3)_2C{=}CH_2 \xrightarrow[\text{peroxides}]{HBr} (CH_3)_2CHCH_2Br$$

*tert*-Butyl chloride — 2-Methylpropene — Isobutyl bromide

Treating isobutyl bromide with sodium iodide in acetone converts it to isobutyl iodide.

$$(CH_3)_2CHCH_2Br \xrightarrow[\text{acetone}]{NaI} (CH_3)_2CHCH_2I$$

Isobutyl bromide — Isobutyl iodide

A second approach is by way of isobutyl alcohol.

$$(CH_3)_3CCl \xrightarrow{NaOCH_3} (CH_3)_2C{=}CH_2 \xrightarrow[\text{2. } H_2O_2,\ HO^-]{\text{1. } B_2H_6} (CH_3)_2CHCH_2OH$$

*tert*-Butyl chloride — 2-Methylpropene — Isobutyl alcohol

Isobutyl alcohol is then converted to its *p*-toluenesulfonate ester, which reacts with sodium iodide in acetone in a manner analogous to that of isobutyl bromide.

(*g*) First introduce a leaving group into the molecule by converting isopropyl alcohol to an isopropyl halide. Then convert the resulting isopropyl halide to isopropyl azide by a nucleophilic substitution reaction

$$CH_3CH(OH)CH_3 \xrightarrow{HBr} CH_3CH(Br)CH_3 \xrightarrow{NaN_3} CH_3CH(N_3)CH_3$$

Isopropyl alcohol — Isopropyl bromide — Isopropyl azide

(*h*) In this synthesis 1-propanol must be first converted to an isopropyl halide.

$$CH_3CH_2CH_2OH \xrightarrow[\text{heat}]{H_2SO_4} CH_3CH{=}CH_2 \xrightarrow{HBr} CH_3CH(Br)CH_3$$

1-Propanol — Propene — Isopropyl bromide

After an isopropyl halide has been obtained, it can be treated with sodium azide as in part (*g*).

(*i*) First write out the structure of the starting material and of the product so as to determine their relationship in three dimensions.

$$CH_3CH_2C(H)(CH_3)\text{—}OH \qquad N_3\text{—}C(H)(CH_3)CH_2CH_3$$

(*R*)-*sec*-Butyl alcohol — (*S*)-*sec*-Butyl azide

The hydroxyl group must be replaced by azide with inversion of configuration. First, however, a leaving group must be introduced, and it must be introduced in such a way that the configuration at the stereogenic center is not altered. The best way to do this is to convert (*R*)-*sec*-butyl alcohol to its corresponding *p*-toluenesulfonate ester.

$$CH_3CH_2C(H)(CH_3)\text{—}OH \xrightarrow[\text{pyridine}]{p\text{-toluenesulfonyl chloride}} CH_3CH_2C(H)(CH_3)\text{—}OSO_2C_6H_4CH_3$$

(*R*)-*sec*-Butyl alcohol — (*R*)-*sec*-Butyl *p*-toluenesulfonate

Next, convert the *p*-toluenesulfonate to the desired azide by an $S_N2$ reaction.

$$CH_3(H)(CH_3CH_2)C{-}OSO_2{-}C_6H_4{-}CH_3 \xrightarrow{NaN_3} N_3{-}C(H)(CH_3)(CH_2CH_3)$$

(*R*)-*sec*-Butyl *p*-toluenesulfonate (*S*)-*sec*-Butyl azide

(*j*) This problem is carried out in exactly the same way as the preceding one, except that the nucleophile in the second step is $HS^-$.

$$CH_3(H)(CH_3CH_2)C{-}OH \xrightarrow[\text{pyridine}]{p\text{-toluenesulfonyl chloride}} CH_3(H)(CH_3CH_2)C{-}OSO_2{-}C_6H_4{-}CH_3 \xrightarrow{NaSH} HS{-}C(H)(CH_3)(CH_2CH_3)$$

(*R*)-*sec*-Butyl alcohol (*R*)-*sec*-Butyl *p*-toluenesulfonate (*S*)-2-Butanethiol

**8.31** (*a*) The two possible combinations of alkyl bromide and alkoxide ion that might yield *tert*-butyl methyl ether are

**1.** $$CH_3Br + (CH_3)_3CO^- \xrightarrow{\text{fast}} (CH_3)_3COCH_3$$

Methyl bromide *tert*-Butoxide ion *tert*-Butyl methyl ether

**2.** $$CH_3O^- + (CH_3)_3CBr \xrightarrow{\text{slow}} (CH_3)_3COCH_3$$

Methoxide ion *tert*-Butyl bromide *tert*-Butyl methyl ether

We choose the first approach because it is an $S_N2$ reaction on the unhindered substrate, methyl bromide. The second approach requires an $S_N2$ reaction on a hindered tertiary alkyl halide, a very poor choice. Indeed, we would expect that the reaction of methoxide ion with *tert*-butyl bromide could not give any ether at all but would proceed entirely by E2 elimination:

$$CH_3O^- + (CH_3)_3CBr \xrightarrow{\text{fast}} CH_3OH + CH_2{=}C(CH_3)_2$$

Methanol 2-Methylpropene

(*b*) Again, the better alternative is to choose the less hindered alkyl halide to permit substitution to predominate over elimination.

$$C_5H_9{-}OK + CH_3Br \longrightarrow C_5H_9{-}OCH_3$$

Potassium cyclopentoxide Methyl bromide Cyclopentyl methyl ether

An attempt to prepare this compound by the reaction

$$C_5H_9{-}Cl + CH_3ONa \longrightarrow C_5H_9{-}OCH_3$$

Chlorocyclopentane Sodium methoxide Cyclopentyl methyl ether

gave cyclopentyl methyl ether in only 24% yield. Cyclopentene was isolated in 31% yield.

(*c*) A 2,2-dimethylpropyl halide is too sterically hindered to be a good candidate for this synthesis. The only practical method is

$$(CH_3)_3CCH_2OK + CH_3CH_2Br \longrightarrow (CH_3)_3CCH_2OCH_2CH_3$$

Potassium 2,2-dimethylpropoxide Bromoethane Ethyl 2,2-dimethylpropyl ether

**8.32** (*a*) The problem states that the reaction type is nucleophilic substitution. Sodium acetylide is therefore the nucleophile and must be treated with an alkyl halide to give the desired product.

$$Na^+ :\bar{C}{\equiv}CH + CH_3CH_2Br \longrightarrow CH_3CH_2C{\equiv}CH + NaBr$$

Sodium acetylide — Ethyl bromide — 1-Butyne — Sodium bromide

(*b*) The acidity data given in the problem for acetylene tell us that $HC{\equiv}CH$ is a very weak acid ($K_a = 10^{-26}$), so that sodium acetylide must be a very strong base—stronger than hydroxide ion. *Elimination* by the E2 mechanism rather than $S_N2$ substitution is therefore expected to be the principal (probably the exclusive) reaction observed with secondary and tertiary alkyl halides. The substitution reaction will work well with primary alkyl halides but will likely fail for secondary and tertiary ones. Alkynes such as $(CH_3)_2CHC{\equiv}CH$ and $(CH_3)_3CC{\equiv}CH$ could not be prepared by this method.

**8.33** The compound that reacts with *trans*-4-*tert*-butylcyclohexanol is a sulfonyl chloride and converts the alcohol to the corresponding sulfonate.

$(CH_3)_3C$–cyclohexyl–OH + $O_2N$–$C_6H_4$–$SO_2Cl$ $\xrightarrow{\text{pyridine}}$ $(CH_3)_3C$–cyclohexyl–$OSO_2$–$C_6H_4$–$NO_2$

Compound A

Reaction of compound A with lithium bromide in acetone effects displacement of the sulfonate leaving group by bromide with inversion of configuration.

$(CH_3)_3C$–cyclohexyl–$OSO_2$–$C_6H_4$–$NO_2$ $\xrightarrow[\text{acetone}]{\text{LiBr}}$ $(CH_3)_3C$–cyclohexyl–Br

Compound A

*cis*-1-Bromo-4-*tert*-butylcyclohexane
Compound B

**8.34** (*a*) To convert *trans*-2-methylcyclopentanol to *cis*-2-methylcyclopentyl acetate the hydroxyl group must be replaced by acetate with inversion of configuration. Hydroxide is a poor leaving group and so must first be converted to a good leaving group. The best choice is *p*-toluenesulfonate, because this can be prepared by a reaction that alters none of the bonds to the stereogenic center.

$H_3C$, OH (cyclopentane) $\xrightarrow[\text{chloride, pyridine}]{p\text{-toluenesulfonyl}}$ $H_3C$, OTs (cyclopentane)

*trans*-2-Methylcyclopentanol — *trans*-2-Methylcyclopentyl *p*-toluenesulfonate

Treatment of the *p*-toluenesulfonate with potassium acetate in acetic acid will proceed with inversion of configuration to give the desired product.

$H_3C$, OTs (cyclopentane) $\xrightarrow[\text{acetic acid}]{KOCCH_3 (\text{C=O})}$ $H_3C$, $OCCH_3$ (C=O) (cyclopentane)

*trans*-2-Methylcyclopentyl *p*-toluenesulfonate — *cis*-2-Methylcyclopentyl acetate

(*b*) To decide on the best sequence of reactions, we must begin by writing structural formulas to determine what kinds of transformations are required.

1-Methylcyclopentanol $\xrightarrow{?}$ *cis*-2-Methylcyclopentyl acetate

We already know from part (*a*) how to convert *trans*-2-methylcyclopentanol to *cis*-2-methylcyclopentyl acetate. So all that is really necessary is to design a synthesis of *trans*-2-methylcyclopentanol. Therefore,

1-Methylcyclopentanol $\xrightarrow[\text{heat}]{H_2SO_4}$ 1-Methylcyclopentene $\xrightarrow[\text{2. } H_2O_2,\ HO^-]{\text{1. } B_2H_6}$ *trans*-2-Methylcyclopentanol

Hydroboration–oxidation converts 1-methylcyclopentene to the desired alcohol by anti-Markovnikov syn hydration of the double bond. The resulting alcohol is then converted to its *p*-toluenesulfonate ester and treated with acetate ion as in part (*a*) to give *cis*-2-methylcyclopentyl acetate.

**8.35** (*a*) The reaction of an alcohol with a sulfonyl chloride gives a sulfonate ester. The oxygen of the alcohol remains in place and is the atom to which the sulfonyl group becomes attached.

$$\text{H}\!-\!\overset{CH_3}{\underset{CH_2CH_3}{C}}\!-\!\text{OH} \xrightarrow[\text{pyridine}]{CH_3SO_2Cl} \text{H}\!-\!\overset{CH_3}{\underset{CH_2CH_3}{C}}\!-\!OSO_2CH_3$$

(*S*)-(+)-2-Butanol → (*S*)-*sec*-Butyl methanesulfonate

(*b*) Sulfonate is similar to iodide in its leaving-group behavior. The product in part (*a*) is attacked by $NaSCH_2CH_3$ in an $S_N2$ reaction. Inversion of configuration occurs at the stereogenic center.

$$\text{H}\!-\!\overset{CH_3}{\underset{CH_2CH_3}{C}}\!-\!OSO_2CH_3 \xrightarrow{NaSCH_2CH_3} CH_3CH_2S\!-\!\overset{CH_3}{\underset{CH_2CH_3}{C}}\!-\!\text{H}$$

(*S*)-*sec*-Butyl methanesulfonate → (*R*)-(−)-*sec*-Butyl ethyl sulfide

(*c*) In this part of the problem we deduce the stereochemical outcome of the reaction of 2-butanol with $PBr_3$. We know the absolute configuration of (+)-2-butanol (*S*) from the statement of the problem and the configuration of (−)-*sec*-butyl ethyl sulfide (*R*) from part (*b*). We are told that the sulfide formed from (+)-2-butanol via the bromide has a positive rotation. It must therefore have the opposite configuration of the product in part (*b*).

$$\text{H}\!-\!\overset{CH_3}{\underset{CH_2CH_3}{C}}\!-\!\text{OH} \xrightarrow{PBr_3} CH_3\underset{Br}{\underset{|}{C}}HCH_2CH_3 \xrightarrow{NaSCH_2CH_3} \text{H}\!-\!\overset{CH_3}{\underset{CH_2CH_3}{C}}\!-\!SCH_2CH_3$$

(*S*)-(+)-2-Butanol → 2-Bromobutane → (*S*)-(+)-*sec*-Butyl ethyl sulfide

Since the reaction of the bromide with $NaSCH_2CH_3$ proceeds with inversion of configuration at the stereogenic center, and since the final product has the same configuration as the starting alcohol, the conversion of the alcohol to the bromide must proceed with inversion of configuration.

$$\text{(S)-(+)-2-Butanol} \xrightarrow{PBr_3} \text{(R)-(−)-2-Bromobutane}$$

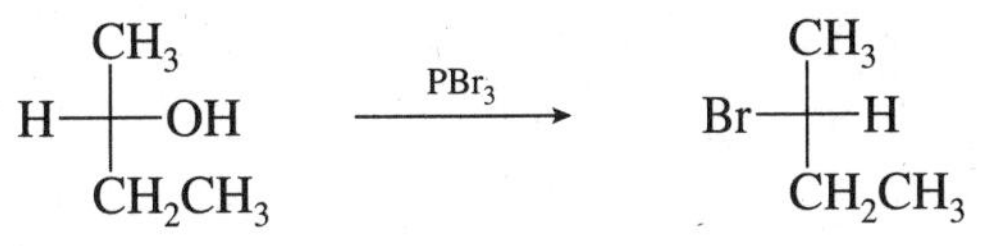

(*S*)-(+)-2-Butanol  (*R*)-(−)-2-Bromobutane

(*d*) The conversion of 2-butanol to *sec*-butyl methanesulfonate does not involve any of the bonds to the stereogenic center, and so it must proceed with 100% retention of configuration. Assuming that the reaction of the methanesulfonate with $NaSCH_2CH_3$ proceeds with 100% inversion of configuration, we conclude that the maximum rotation of *sec*-butyl ethyl sulfide is the value given in the statement of part (*b*), that is, ±25°. Since the sulfide produced in part (*c*) has a rotation of +23°, it is 92% optically pure. It is reasonable to assume that the loss of optical purity occurred in the conversion of the alcohol to the bromide, rather than in the reaction of the bromide with $NaSCH_2CH_3$. If the bromide is 92% optically pure and has a rotation of −38°, optically pure 2-bromobutane therefore has a rotation of 38/0.92, or ±41°.

**8.36** (*a*) If each act of exchange (substitution) occurred with retention of configuration, there would be no observable racemization; $k_{rac} = 0$.

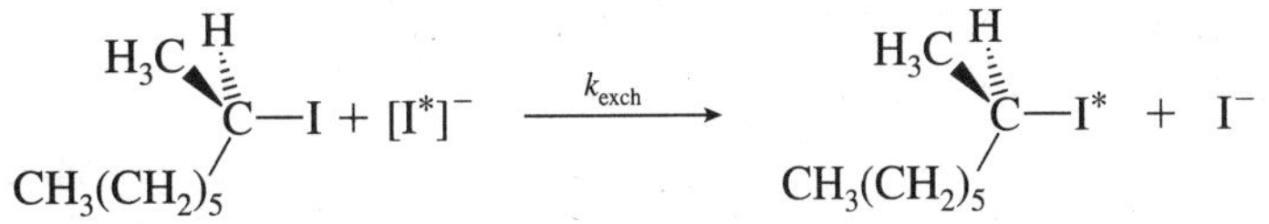

$$\text{(R)-2-iodooctane} + [I^*]^- \xrightarrow{k_{exch}} [\text{(R)-2-iodooctane}]^* + I^-$$

(*R*)-(−)-2-Iodooctane (*indicates radioactive label)  [(*R*)-(−)-2-Iodooctane]*

Therefore $k_{rac}/k_{exch} = 0$.

(*b*) If each act of exchange proceeds with inversion of configuration, (*R*)-(−)-2-iodooctane will be transformed to radioactively labeled (*S*)-(+)-2-iodooctane.

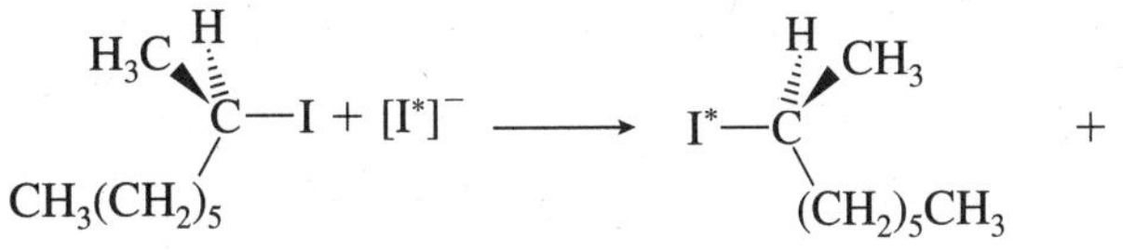

$$\text{(R)-2-iodooctane} + [I^*]^- \longrightarrow [\text{(S)-2-iodooctane}]^* + I^-$$

(*R*)-(−)-2-Iodooctane  [(*S*)-(+)-2-Iodooctane]*

Starting with 100 molecules of (*R*)-(−)-2-iodooctane, the compound will be completely racemized when 50 molecules have become radioactive. Therefore,

$$\frac{k_{rac}}{k_{exch}} = 2$$

(*c*) If radioactivity is incorporated in a stereorandom fashion, then 2-iodooctane will be 50% racemized when 50% of it has reacted. Therefore,

$$\frac{k_{rac}}{k_{exch}} = 1$$

In fact, Hughes found that the rate of racemization was twice the rate of incorporation of radioactive iodide. This experiment provided strong evidence for the belief that bimolecular nucleophilic substitution proceeds stereospecifically with inversion of configuration.

**8.37** (*a*) Tertiary alkyl halides undergo nucleophilic substitution only by way of carbocations: $S_N1$ is the most likely mechanism for solvolysis of the 2-halo-2-methylbutanes.

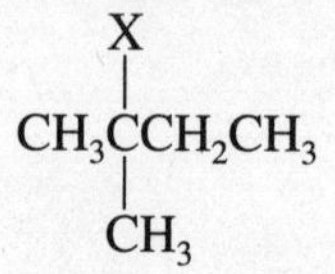

$$CH_3\overset{X}{\underset{CH_3}{C}}CH_2CH_3$$

2-Halo-2-methylbutanes are tertiary alkyl halides.

(*b*) Tertiary alkyl halides can undergo either E1 or E2 elimination. Since no alkoxide base is present, solvolytic elimination most likely occurs by an E1 mechanism.

(*c, d*) Iodides react faster than bromides in substitution and elimination reactions irrespective of whether the mechanism is E1, E2, $S_N1$, or $S_N2$.

(*e*) Solvolysis in aqueous ethanol can give rise to an alcohol or an ether as product, depending on whether the carbocation is captured by water or ethanol.

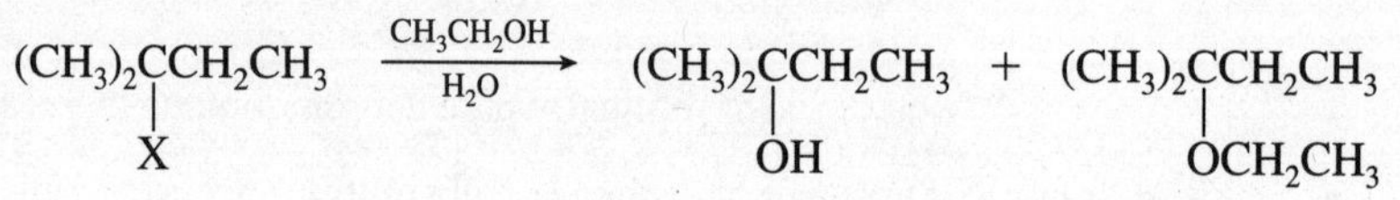

$$(CH_3)_2\underset{X}{C}CH_2CH_3 \xrightarrow[H_2O]{CH_3CH_2OH} (CH_3)_2\underset{OH}{C}CH_2CH_3 + (CH_3)_2\underset{OCH_2CH_3}{C}CH_2CH_3$$

2-Methyl-2-butanol — Ethyl 1,1-dimethylpropyl ether

(*f*) Elimination can yield either of two isomeric alkenes.

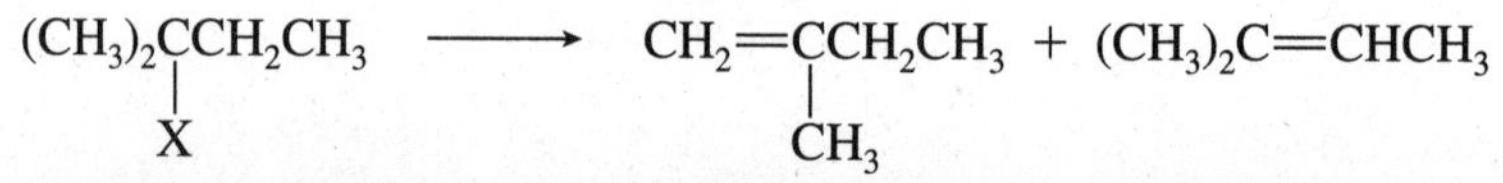

$$(CH_3)_2\underset{X}{C}CH_2CH_3 \longrightarrow CH_2{=}\underset{CH_3}{C}CH_2CH_3 + (CH_3)_2C{=}CHCH_3$$

2-Methyl-1-butene — 2-Methyl-2-butene

Zaitsev's rule predicts that 2-methyl-2-butene should be the major alkene.

(*g*) The product distribution is determined by what happens to the carbocation intermediate. If the carbocation is free of its leaving group, its fate will be the same no matter whether the leaving group is bromide or iodide.

**8.38** Both aspects of this reaction—its slow rate and the formation of a rearranged product—have their origin in the positive character developed at a primary carbon. The alcohol is protonated and the carbon–oxygen bond of the resulting alkyloxonium ion begins to break:

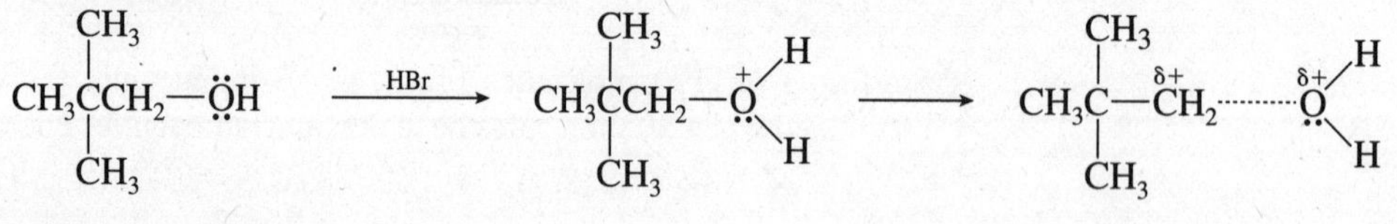

$$CH_3\overset{CH_3}{\underset{CH_3}{C}}CH_2{-}\ddot{O}H \xrightarrow{HBr} CH_3\overset{CH_3}{\underset{CH_3}{C}}CH_2{-}\overset{+}{\ddot{O}}H_2 \longrightarrow CH_3\overset{CH_3}{\underset{CH_3}{C}}{-}\overset{\delta+}{C}H_2\cdots\overset{\delta+}{\ddot{O}}H_2$$

2,2-Dimethyl-1-propanol

As positive character develops at the primary carbon, a methyl group migrates. Rearrangement gives a tertiary carbocation, which is captured by bromide to give the product.

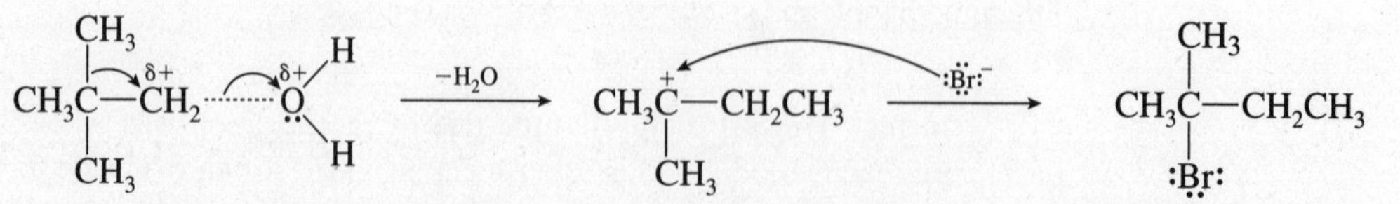

$$CH_3\overset{CH_3}{\underset{CH_3}{C}}{-}\overset{\delta+}{C}H_2\cdots\overset{\delta+}{\ddot{O}}H_2 \xrightarrow{-H_2O} CH_3\overset{+}{\underset{CH_3}{C}}{-}CH_2CH_3 \xrightarrow{:\ddot{Br}:^-} CH_3\overset{CH_3}{\underset{:\ddot{Br}:}{C}}{-}CH_2CH_3$$

2-Bromo-2-methylbutane

**8.39** The substrate is a tertiary alkyl bromide and can undergo $S_N1$ substitution and E1 elimination under these reaction conditions. Elimination in either of two directions to give regioisomeric alkenes can also occur.

$$\underset{\substack{\text{2-Bromo-2-}\\\text{methylbutane}}}{CH_3C(Br)(CH_3)CH_2CH_3} \xrightarrow[CH_3CO_2H]{CH_3CO_2K} \underset{\substack{\text{1,1-Dimethylpropyl}\\\text{acetate}}}{CH_3C(OCCH_3)(CH_3)CH_2CH_3} + \underset{\text{2-Methyl-1-butene}}{CH_2{=}C(CH_3)CH_2CH_3} + \underset{\text{2-Methyl-2-butene}}{(H_3C)_2C{=}C(H)CH_3}$$

(The acetate group OCCH$_3$ has C=O.)

**8.40** Solvolysis of 1,2-dimethylpropyl *p*-toluenesulfonate in acetic acid is expected to give one substitution product and two alkenes.

$$\underset{\substack{\text{1,2-Dimethylpropyl}\\p\text{-toluenesulfonate}}}{(CH_3)_2CHCH(OTs)CH_3} \xrightarrow{CH_3CO_2H} \underset{\substack{\text{1,2-Dimethylpropyl}\\\text{acetate}}}{(CH_3)_2CHCH(OCCH_3)CH_3} + \underset{\text{2-Methyl-2-butene}}{(CH_3)_2C{=}CHCH_3} + \underset{\text{3-Methyl-1-butene}}{(CH_3)_2CHCH{=}CH_2}$$

Since five products are formed, we are led to consider the possibility of carbocation rearrangements in $S_N1$ and E1 solvolysis.

$$\underset{\substack{\text{1,2-Dimethylpropyl}\\\text{cation (secondary)}}}{(CH_3)_2CH\overset{+}{C}HCH_3} \longrightarrow \underset{\substack{\text{1,1-Dimethylpropyl}\\\text{cation (tertiary)}}}{(CH_3)_2\overset{+}{C}CH_2CH_3} \longrightarrow \underset{\substack{\text{1,1-Dimethylpropyl}\\\text{acetate}}}{(CH_3)_2C(OCCH_3)CH_2CH_3} + \underset{\text{2-Methyl-2-butene}}{(CH_3)_2C{=}CHCH_3} + \underset{\text{2-Methyl-1-butene}}{CH_2{=}C(CH_3)CH_2CH_3}$$

Since 2-methyl-2-butene is a product common to both carbocation intermediates, a total of five different products are accounted for. There are two substitution products:

$$\underset{\substack{\text{1,2-Dimethylpropyl}\\\text{acetate}}}{(CH_3)_2CHCH(OCCH_3)CH_3} \qquad \underset{\substack{\text{1,1-Dimethylpropyl}\\\text{acetate}}}{(CH_3)_2C(OCCH_3)CH_2CH_3}$$

and three elimination products:

$$\underset{\text{2-Methyl-2-butene}}{(CH_3)_2C{=}CHCH_3} \qquad \underset{\text{3-Methyl-1-butene}}{(CH_3)_2CHCH{=}CH_2} \qquad \underset{\text{2-Methyl-1-butene}}{CH_2{=}C(CH_3)CH_2CH_3}$$

**8.41** Solution A contains both acetate ion and methanol as nucleophiles. Acetate is more nucleophilic than methanol, and so the major observed reaction is:

$$\underset{\substack{\text{Methyl}\\\text{iodide}}}{CH_3I} + \underset{\text{Acetate}}{CH_3\overset{O}{\overset{\|}{C}}O^-} \xrightarrow{\text{methanol}} \underset{\text{Methyl acetate}}{CH_3O\overset{O}{\overset{\|}{C}}CH_3}$$

Solution B prepared by adding potassium methoxide to acetic acid rapidly undergoes an acid–base reaction:

$$CH_3O^- + CH_3\overset{O}{\overset{\|}{C}}OH \longrightarrow CH_3OH + CH_3\overset{O}{\overset{\|}{C}}O^-$$

Methoxide (stronger base) | Acetic acid (stronger acid) | Methanol (weaker acid) | Acetate (weaker base)

Thus the major base present is not methoxide but acetate. Methyl iodide therefore reacts with acetate anion in solution B to give methyl acetate.

**8.42** Alkyl chlorides arise by the reaction sequence:

$$RCH_2OH + CH_3-C_6H_4-SO_2Cl + C_5H_5N \longrightarrow RCH_2OSO_2-C_6H_4-CH_3 + C_5H_5NH^+\ Cl^-$$

Primary alcohol | p-Toluenesulfonyl chloride | Pyridine | Primary alkyl p-toluenesulfonate | Pyridinium chloride

$$RCH_2OSO_2-C_6H_4-CH_3 + Cl^- \longrightarrow RCH_2Cl + CH_3-C_6H_4-SO_3^-$$

Primary alkyl p-toluenesulfonate | Primary alkyl chloride

The reaction proceeds to form the alkyl *p*-toluenesulfonate as expected, but the chloride anion formed in this step subsequently acts as a nucleophile and displaces *p*-toluenesulfonate from $RCH_2OTs$.

**8.43** Iodide ion is both a better nucleophile than cyanide and a better leaving group than bromide. The two reactions shown are therefore faster than the reaction of cyclopentyl bromide with sodium cyanide alone.

Cyclopentyl bromide $\xrightarrow{NaI}$ Cyclopentyl iodide $\xrightarrow{NaCN}$ Cyclopentyl cyanide

**8.44–8.47** Solutions to molecular modeling exercises are not provided in this *Study Guide and Solutions Manual.* You should use *Learning By Modeling* for these exercises.

# SELF-TEST

## PART A

**A-1.** Write the correct structure of the reactant or product omitted from each of the following. Clearly indicate stereochemistry where it is important.

(*a*) $CH_3CH_2CH_2CH_2Br \xrightarrow[CH_3CH_2OH]{CH_3CH_2ONa}$ ?

(*b*) ? $\xrightarrow{NaCN}$ (cyclopentane bearing CN (hashed) and $CH_3$ (wedged) on adjacent carbons)

(*c*) 1-Chloro-3-methylbutane + sodium iodide $\xrightarrow{\text{acetone}}$ ?

(*d*) 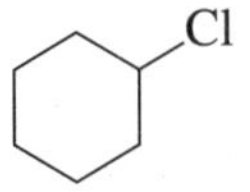 (chloromethyl)cyclohexane $\xrightarrow[CH_3OH]{CH_3ONa}$ ? (major)

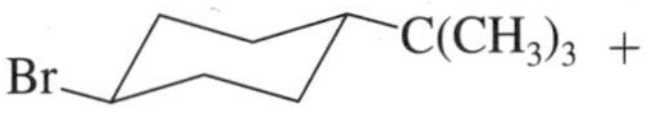

(*e*) Br–(cyclohexane chair)–$C(CH_3)_3$ + $NaN_3$ $\longrightarrow$ ?

(*f*) 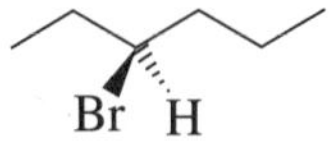 (2-bromohexane, Br wedge, H dash) + $NaSCH_3$ $\longrightarrow$ ?

(g) 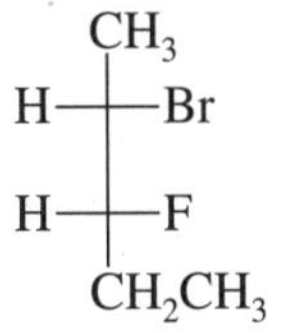 Fischer projection: $CH_3$ (top); H–C–Br; H–C–F; $CH_2CH_3$ (bottom) $\xrightarrow{\text{NaSH}}$ ?

**A-2.** Choose the best pair of reactants to form the following product by an $S_N2$ reaction:

$$(CH_3)_2CHOCH_2CH_2CH_3$$

**A-3.** Outline the chemical steps necessary to convert:

(*a*) 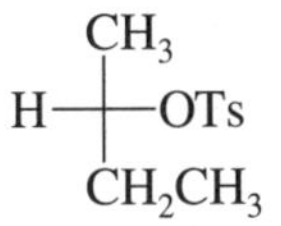 Fischer projection: $CH_3$ (top); H–C–OTs; $CH_2CH_3$ (bottom) to $CH_3$ (top); H–C–CN; $CH_2CH_3$ (bottom)

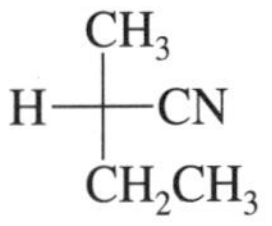

(*b*) (*S*)-2-Pentanol to (*R*)-$CH_3CH(SCH_3)CH_2CH_2CH_3$

**A-4.** Hydrolysis of 3-chloro-2,2-dimethylbutane yields 2,3-dimethyl-2-butanol as the major product. Explain this observation, using structural formulas to outline the mechanism of the reaction.

**A-5.** Identify the class of reaction (e.g., E2), and write the kinetic and chemical equations for:
(*a*) The solvolysis of *tert*-butyl bromide in methanol
(*b*) The reaction of chlorocyclohexane with sodium azide ($NaN_3$).

**A-6.** (*a*) Provide a brief explanation why the halogen exchange reaction shown is an acceptable synthetic method:

$$CH_3CH(Br)CH_3 + NaI \xrightarrow{\text{acetone}} CH_3CH(I)CH_3 + NaBr$$

(*b*) Briefly explain why the reaction of 1-bromobutane with sodium azide occurs faster in dimethyl sulfoxide [$(CH_3)_2S{=}O$] than in water.

**A-7.** Write chemical structures for compounds A through D in the following sequence of reactions. Compounds A and C are alcohols.

$$A \xrightarrow{NaNH_2} B$$

$$C \xrightarrow{\text{HBr, heat}} D$$

$$B + D \longrightarrow CH_3CH_2O\text{–cyclohexyl}$$

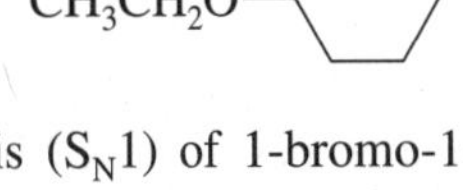

**A-8.** Write a mechanism describing the solvolysis ($S_N1$) of 1-bromo-1-methylcyclohexane in ethanol.

**A-9.** Solvolysis of the compound shown occurs with carbocation rearrangement and yields an alcohol as the major product. Write the structure of this product, and give a mechanism to explain its formation.

$$CH_3CH_2CH_2\underset{Br}{\underset{|}{C}}HCH(CH_3)_2 \xrightarrow{H_2O} ?$$

## PART B

**B-1.** The bimolecular substitution reaction

$$CH_3Br + OH^- \longrightarrow CH_3OH + Br^-$$

is represented by the kinetic equation:

(*a*) Rate = $k[CH_3Br]^2$
(*b*) Rate = $k[CH_3Br][OH^-]$
(*c*) Rate = $k[CH_3Br] + k[OH^-]$
(*d*) Rate = $k/[CH_3Br][OH^-]$

**B-2.** Which compound undergoes nucleophilic substitution with NaCN at the fastest rate?

(*a*) Br (*c*) Br (*e*) Br

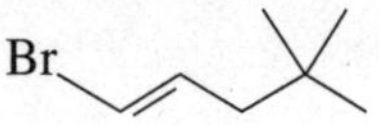

(*b*) Br (*d*) Br

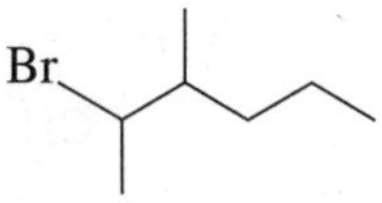

**B-3.** For the reaction

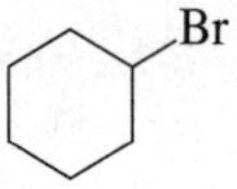

$$\text{(bromocyclohexane)} + CH_3CH_2O^-Na^+ \longrightarrow$$

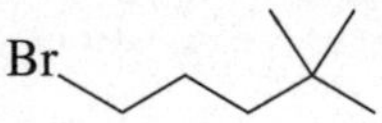

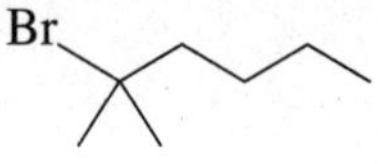

the major product is formed by

(*a*) An $S_N1$ reaction (*c*) An E1 reaction
(*b*) An $S_N2$ reaction (*d*) An E2 reaction

**B-4.** Which of the following statements pertaining to an $S_N2$ reaction are true?

1. The rate of reaction is independent of the concentration of the nucleophile.
2. The nucleophile attacks carbon on the side of the molecule opposite the group being displaced.
3. The reaction proceeds with simultaneous bond formation and bond rupture.
4. Partial racemization of an optically active substrate results.

(*a*) 1, 4 (*b*) 1, 3, 4 (*c*) 2, 3 (*d*) All

**B-5.** Which one of the following alkyl halides would be expected to give the *highest* substitution-to-elimination ratio (most substitution, least elimination) on treatment with sodium ethoxide in ethanol?

(*a*) $(CH_3)_3C$— (cyclohexane) —Br, $CH_2CH_3$ (*c*) $(CH_3)_3C$— (cyclohexane) —$\underset{Br}{CHCH_3}$

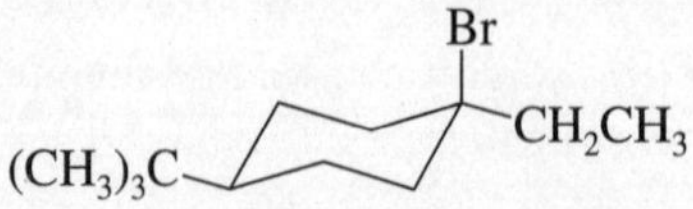

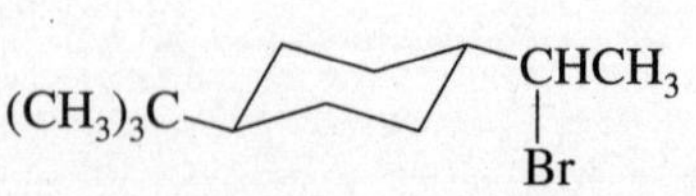

(*b*) $(CH_3)_3C$— (cyclohexane, Br) —$CH_2CH_3$ (*d*) $(CH_3)_3C$— (cyclohexane) —$CH_2CH_2Br$

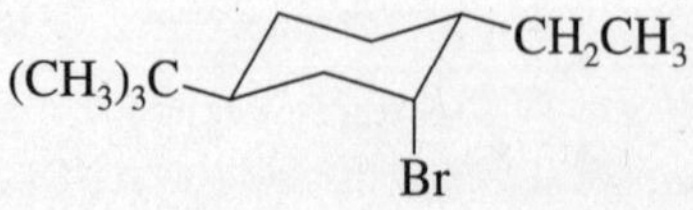

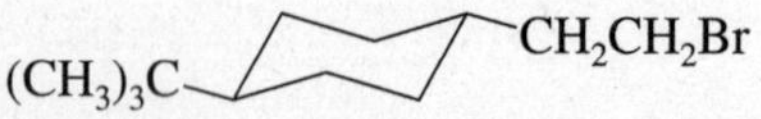

**B-6.** Which of the following phrases are *not* correctly associated with an $S_N1$ reaction?

1. Rearrangement is possible.
2. Rate is affected by solvent polarity.
3. The strength of the nucleophile is important in determining rate.
4. The reactivity series is tertiary > secondary > primary.
5. Proceeds with complete inversion of configuration.

(*a*) 3, 5 (*b*) 5 only (*c*) 2, 3, 5 (*d*) 3 only

**B-7.** Rank the following in order of decreasing rate of solvolysis with aqueous ethanol (fastest → slowest):

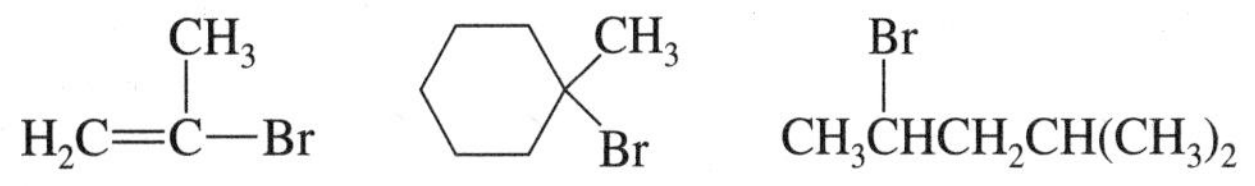

1: $H_2C{=}C(CH_3){-}Br$ 2: 1-bromo-1-methylcyclohexane (cyclohexane ring bearing $CH_3$ and Br on the same carbon) 3: $CH_3CH(Br)CH_2CH(CH_3)_2$

(*a*) 2 > 1 > 3 (*b*) 1 > 2 > 3 (*c*) 2 > 3 > 1 (*d*) 1 > 3 > 2

**B-8.** Rank the following species in order of decreasing nucleophilicity in a polar protic solvent (most → least nucleophilic):

1: $CH_3CH_2CH_2O^-$ 2: $CH_3CH_2CH_2S^-$ 3: $CH_3CH_2C(=O)O^-$

(*a*) 3 > 1 > 2 (*b*) 2 > 3 > 1 (*c*) 1 > 3 > 2 (*d*) 2 > 1 > 3

**B-9.** From each of the following pairs select the compound that will react faster with sodium iodide in acetone.

2-Chloropropane (1) or 2-bromopropane (2)

1-Bromobutane (3) or 2-bromobutane (4)

(*a*) 1, 3 (*b*) 1, 4 (*c*) 2, 3 (*d*) 2, 4

**B-10.** Select the reagent that will yield the greater amount of substitution on reaction with 1-bromobutane.

(*a*) $CH_3CH_2OK$ in dimethyl sulfoxide (DMSO)
(*b*) $(CH_3)_3COK$ in dimethyl sulfoxide (DMSO)
(*c*) Both (*a*) and (*b*) will give comparable amounts of substitution.
(*d*) Neither (*a*) nor (*b*) will give any appreciable amount of substitution.

**B-11.** The reaction of (*R*)-1-chloro-3-methylpentane with sodium iodide in acetone will yield 1-iodo-3-methylpentane that is

(*a*) *R* (*b*) *S* (*c*) A mixture of *R* and *S* (*d*) Meso (*e*) None of these

**B-12.** What is the principal product of the following reaction?

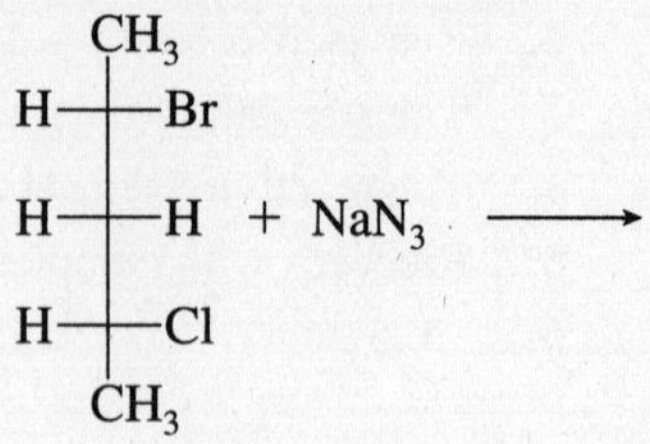

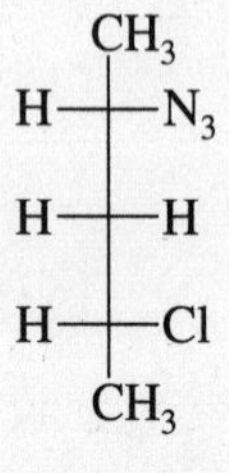

$$\begin{array}{c} CH_3 \\ H-\!\!\!+\!\!\!-Br \\ H-\!\!\!+\!\!\!-H \\ H-\!\!\!+\!\!\!-Cl \\ CH_3 \end{array} + NaN_3 \longrightarrow$$

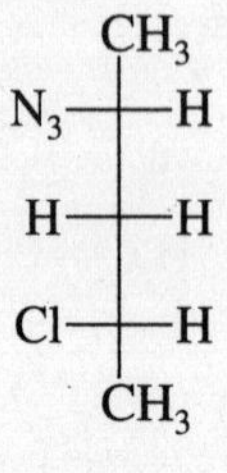

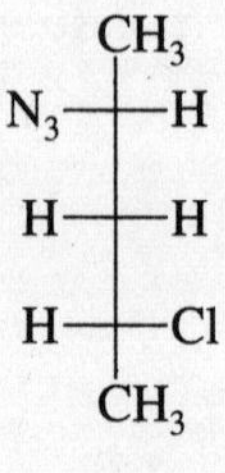

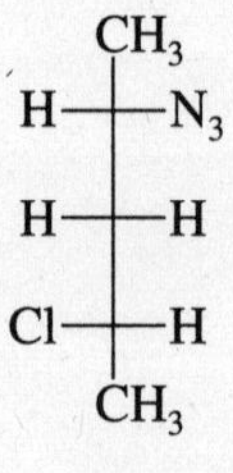

$$\begin{array}{cccc} \begin{array}{c} CH_3 \\ H-\!\!\!+\!\!\!-N_3 \\ H-\!\!\!+\!\!\!-H \\ H-\!\!\!+\!\!\!-Cl \\ CH_3 \end{array} & \begin{array}{c} CH_3 \\ N_3-\!\!\!+\!\!\!-H \\ H-\!\!\!+\!\!\!-H \\ Cl-\!\!\!+\!\!\!-H \\ CH_3 \end{array} & \begin{array}{c} CH_3 \\ N_3-\!\!\!+\!\!\!-H \\ H-\!\!\!+\!\!\!-H \\ H-\!\!\!+\!\!\!-Cl \\ CH_3 \end{array} & \begin{array}{c} CH_3 \\ H-\!\!\!+\!\!\!-N_3 \\ H-\!\!\!+\!\!\!-H \\ Cl-\!\!\!+\!\!\!-H \\ CH_3 \end{array} \\ (a) & (b) & (c) & (d) \end{array}$$

**B-13.** Which of the following statements is true?

(*a*) $CH_3CH_2S^-$ is both a stronger base and more nucleophilic than $CH_3CH_2O^-$.
(*b*) $CH_3CH_2S^-$ is a stronger base but is less nucleophilic than $CH_3CH_2O^-$.
(*c*) $CH_3CH_2S^-$ is a weaker base but is more nucleophilic than $CH_3CH_2O^-$.
(*d*) $CH_3CH_2S^-$ is both a weaker base and less nucleophilic than $CH_3CH_2O^-$.

**B-14.** Which of the following alkyl halides would be most likely to give a rearranged product under $S_N1$ conditions?

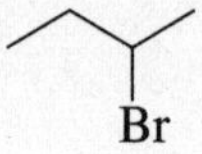

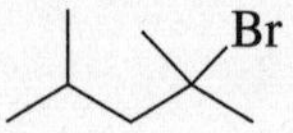

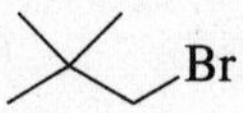

Br (*a*)    Br (*b*)    Br (*c*)    Br (*d*)

(*e*) None of these. Rearrangements only occur under $S_N2$ conditions.

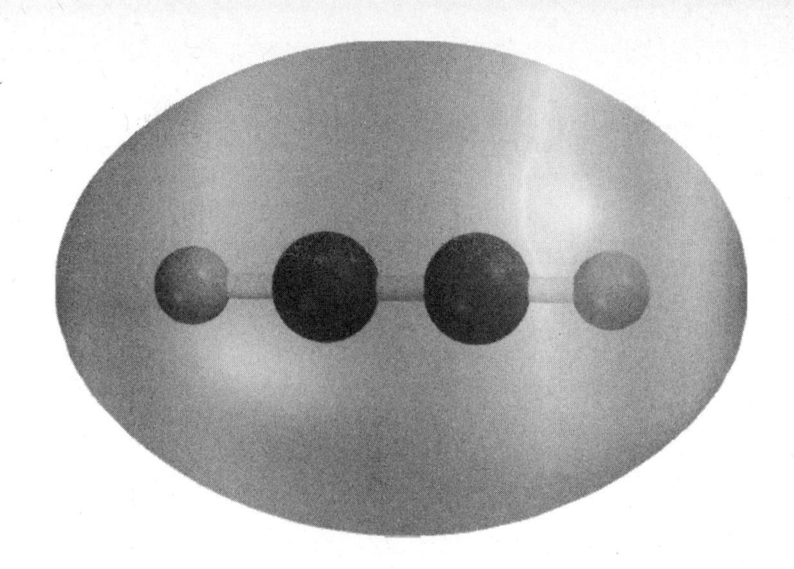

# CHAPTER 9

## ALKYNES

## SOLUTIONS TO TEXT PROBLEMS

**9.1** The reaction is an acid–base process; water is the proton donor. Two separate proton-transfer steps are involved.

$$:\bar{C}\equiv\bar{C}: + H-\ddot{\underset{..}{O}}-H \longrightarrow :\bar{C}\equiv C-H + {}^{-}:\ddot{\underset{..}{O}}-H$$

Carbide ion, Water, Acetylide ion, Hydroxide ion

$$H-\ddot{\underset{..}{O}}-H + :\bar{C}\equiv C-H \longrightarrow H-\ddot{\underset{..}{O}}:^{-} + H-C\equiv C-H$$

Water, Acetylide ion, Hydroxide ion, Acetylene

**9.2** A triple bond may connect C-1 and C-2 or C-2 and C-3 in an unbranched chain of five carbons.

$CH_3CH_2CH_2C\equiv CH$ (1-Pentyne) $\quad CH_3CH_2C\equiv CCH_3$ (2-Pentyne)

One of the $C_5H_8$ isomers has a branched carbon chain.

$CH_3CH(CH_3)C\equiv CH$

3-Methyl-1-butyne

**9.3** The bonds become shorter and stronger in the series as the electronegativity increases.

| | $NH_3$ | $H_2O$ | HF |
|---|---|---|---|
| Electronegativity: | N (3.0) | O (3.5) | F (4.0) |
| Bond distance (pm): | N—H (101) | O—H (95) | F—H (92) |
| Bond dissociation energy (kJ/mol): | N—H (435) | O—H (497) | F—H (568) |
| Bond dissociation energy (kcal/mol): | N—H (104) | O—H (119) | F—H (136) |

**9.4** (*b*) A proton is transferred from acetylene to ethyl anion.

$$HC{\equiv}C{-}H + :\bar{C}H_2CH_3 \rightleftharpoons HC{\equiv}\bar{C}: + CH_3CH_3$$

Acetylene (stronger acid) $K_a$ $10^{-26}$ ($pK_a$ 26); Ethyl anion (stronger base); Acetylide ion (weaker base); Ethane (weaker acid) $K_a \approx 10^{-62}$ ($pK_a \approx 62$)

The position of equilibrium lies to the right. Ethyl anion is a very powerful base and deprotonates acetylene quantitatively.

(*c*) Amide ion is not a strong enough base to remove a proton from ethylene. The equilibrium lies to the left.

$$CH_2{=}CH{-}H + :\ddot{N}H_2^- \rightleftharpoons CH_2{=}\ddot{C}H^- + :NH_3$$

Ethylene (weaker acid) $K_a \approx 10^{-45}$ ($pK_a \approx 45$); Amide ion (weaker base); Vinyl anion (stronger base); Ammonia (stronger acid) $K_a$ $10^{-36}$ ($pK_a$ 36)

(*d*) Alcohols are stronger acids than ammonia; the position of equilibrium lies to the right.

$$CH_3C{\equiv}CCH_2\ddot{O}{-}H + :\ddot{N}H_2^- \rightleftharpoons CH_3C{\equiv}CCH_2\ddot{O}:^- + :NH_3$$

2-Butyn-1-ol (stronger acid) $K_a \approx 10^{-16}$–$10^{-20}$ ($pK_a \approx 16$–$20$); Amide ion (stronger base); 2-Butyn-1-olate anion (weaker base); Ammonia (weaker acid) $K_a 10^{-36}$ ($pK_a$ 36)

**9.5** (*b*) The desired alkyne has a methyl group and a butyl group attached to a —C≡C— unit. Two alkylations of acetylene are therefore required: one with a methyl halide, the other with a butyl halide.

$$\underset{\text{Acetylene}}{HC{\equiv}CH} \xrightarrow[\text{2. } CH_3Br]{\text{1. } NaNH_2,\ NH_3} \underset{\text{Propyne}}{CH_3C{\equiv}CH} \xrightarrow[\text{2. } CH_3CH_2CH_2CH_2Br]{\text{1. } NaNH_2,\ NH_3} \underset{\text{2-Heptyne}}{CH_3C{\equiv}CCH_2CH_2CH_2CH_3}$$

It does not matter whether the methyl group or the butyl group is introduced first; the order of steps shown in this synthetic scheme may be inverted.

(*c*) An ethyl group and a propyl group need to be introduced as substituents on a —C≡C— unit. As in part (*b*), it does not matter which of the two is introduced first.

$$\underset{\text{Acetylene}}{HC{\equiv}CH} \xrightarrow[\text{2. } CH_3CH_2CH_2Br]{\text{1. } NaNH_2,\ NH_3} \underset{\text{1-Pentyne}}{CH_3CH_2CH_2C{\equiv}CH} \xrightarrow[\text{2. } CH_3CH_2Br]{\text{1. } NaNH_2,\ NH_3} \underset{\text{3-Heptyne}}{CH_3CH_2CH_2C{\equiv}CCH_2CH_3}$$

**9.6** Both 1-pentyne and 2-pentyne can be prepared by alkylating acetylene. All the alkylation steps involve nucleophilic substitution of a methyl or primary alkyl halide.

$$\underset{\text{Acetylene}}{HC{\equiv}CH} \xrightarrow[2.\ CH_3CH_2CH_2Br]{1.\ NaNH_2,\ NH_3} \underset{\text{1-Pentyne}}{CH_3CH_2CH_2C{\equiv}CH}$$

$$\underset{\text{Acetylene}}{HC{\equiv}CH} \xrightarrow[2.\ CH_3CH_2Br]{1.\ NaNH_2,\ NH_3} \underset{\text{1-Butyne}}{CH_3CH_2C{\equiv}CH} \xrightarrow[2.\ CH_3Br]{1.\ NaNH_2,\ NH_3} \underset{\text{2-Pentyne}}{CH_3CH_2C{\equiv}CCH_3}$$

A third isomer, 3-methyl-1-butyne, cannot be prepared by alkylation of acetylene, because it requires a secondary alkyl halide as the alkylating agent. The reaction that takes place is elimination, not substitution.

$$\underset{\text{Acetylide ion}}{HC{\equiv}\bar{C}\colon} + \underset{\text{Isopropyl bromide}}{CH_3\underset{\displaystyle Br}{\underset{|}{C}}HCH_3} \xrightarrow{E2} \underset{\text{Acetylene}}{HC{\equiv}CH} + \underset{\text{Propene}}{CH_2{=}CHCH_3}$$

**9.7** Each of the dibromides shown yields 3,3-dimethyl-1-butyne when subjected to double dehydrohalogenation with strong base.

$$\underset{\text{2,2-Dibromo-3,3-dimethylbutane}}{(CH_3)_3C\overset{\displaystyle Br}{\overset{|}{\underset{\displaystyle Br}{\underset{|}{C}}}}CH_3} \text{ or } \underset{\text{1,1-Dibromo-3,3-dimethylbutane}}{(CH_3)_3CCH_2CHBr_2} \text{ or } \underset{\text{1,2-Dibromo-3,3-dimethylbutane}}{(CH_3)_3C\underset{\displaystyle Br}{\underset{|}{C}}HCH_2Br} \xrightarrow[2.\ H_2O]{1.\ 3NaNH_2} \underset{\text{3,3-Dimethyl-1-butyne}}{(CH_3)_3CC{\equiv}CH}$$

**9.8** (*b*) The first task is to convert 1-propanol to propene:

$$\underset{\text{1-Propanol}}{CH_3CH_2CH_2OH} \xrightarrow[\text{heat}]{H_2SO_4} \underset{\text{Propene}}{CH_3CH{=}CH_2}$$

After propene is available, it is converted to 1,2-dibromopropane and then to propyne as described in the sample solution for part (*a*).

(*c*) Treat isopropyl bromide with a base to effect dehydrohalogenation.

$$\underset{\text{Isopropyl bromide}}{(CH_3)_2CHBr} \xrightarrow{NaOCH_2CH_3} \underset{\text{Propene}}{CH_3CH{=}CH_2}$$

Next, convert propene to propyne as in parts (*a*) and (*b*).

(*d*) The starting material contains only two carbon atoms, and so an alkylation step is needed at some point. Propyne arises by alkylation of acetylene, and so the last step in the synthesis is

$$\underset{\text{Acetylene}}{HC{\equiv}CH} \xrightarrow[2.\ CH_3Br]{1.\ NaNH_2,\ NH_3} \underset{\text{Propyne}}{CH_3C{\equiv}CH}$$

The designated starting material, 1,1-dichloroethane, is a geminal dihalide and can be used to prepare acetylene by a double dehydrohalogenation.

$$\underset{\text{1,1-Dichloroethane}}{CH_3CHCl_2} \xrightarrow[2.\ H_2O]{1.\ NaNH_2,\ NH_3} \underset{\text{Acetylene}}{HC{\equiv}CH}$$

(*e*) The first task is to convert ethyl alcohol to acetylene. Once acetylene is prepared it can be alkylated with a methyl halide.

$$\underset{\text{Ethyl alcohol}}{CH_3CH_2OH} \xrightarrow[\text{heat}]{H_2SO_4} \underset{\text{Ethylene}}{H_2C{=}CH_2} \xrightarrow{Br_2} \underset{\text{1,2-Dibromoethane}}{BrCH_2CH_2Br} \xrightarrow[\text{2. } H_2O]{\text{1. } NaNH_2,\ NH_3} \underset{\text{Acetylene}}{HC{\equiv}CH} \xrightarrow[\text{2. } CH_3Br]{\text{1. } NaNH_2,\ NH_3} \underset{\text{Propyne}}{CH_3C{\equiv}CH}$$

**9.9** The first task is to assemble a carbon chain containing eight carbons. Acetylene has two carbon atoms and can be alkylated via its sodium salt to 1-octyne. Hydrogenation over platinum converts 1-octyne to octane.

$$\underset{\text{Acetylene}}{HC{\equiv}CH} \xrightarrow[NH_3]{NaNH_2} \underset{\text{Sodium acetylide}}{HC{\equiv}CNa} \xrightarrow{BrCH_2(CH_2)_4CH_3} \underset{\text{1-Octyne}}{HC{\equiv}CCH_2(CH_2)_4CH_3} \xrightarrow[Pt]{H_2} \underset{\text{Octane}}{CH_3CH_2CH_2(CH_2)_4CH_3}$$

Alternatively, two successive alkylations of acetylene with $CH_3CH_2CH_2Br$ could be carried out to give 4-octyne ($CH_3CH_2CH_2C{\equiv}CCH_2CH_2CH_3$), which could then be hydrogenated to octane.

**9.10** Hydrogenation over Lindlar palladium converts an alkyne to a cis alkene. Oleic acid therefore has the structure indicated in the following equation:

$$\underset{\text{Stearolic acid}}{CH_3(CH_2)_7C{\equiv}C(CH_2)_7CO_2H} \xrightarrow[\text{Lindlar Pd}]{H_2} \underset{\text{Oleic acid}}{\overset{CH_3(CH_2)_7\qquad (CH_2)_7CO_2H}{\underset{H\qquad\quad H}{C{=}C}}}$$

Hydrogenation of alkynes over platinum leads to alkanes.

$$\underset{\text{Stearolic acid}}{CH_3(CH_2)_7C{\equiv}C(CH_2)_7CO_2H} \xrightarrow[Pt]{2H_2} \underset{\text{Stearic acid}}{CH_3(CH_2)_{16}CO_2H}$$

**9.11** Alkynes are converted to trans alkenes on reduction with sodium in liquid ammonia.

$$\underset{\text{Stearolic acid}}{CH_3(CH_2)_7C{\equiv}C(CH_2)_7CO_2H} \xrightarrow[\text{2. } H_3O^+]{\text{1. Na, } NH_3} \underset{\text{Elaidic acid}}{\overset{CH_3(CH_2)_7\qquad H}{\underset{H\qquad (CH_2)_7CO_2H}{C{=}C}}}$$

**9.12** The proper double-bond stereochemistry may be achieved by using 2-heptyne as a reactant in the final step. Lithium–ammonia reduction of 2-heptyne gives the trans alkene; hydrogenation over Lindlar palladium gives the cis isomer. The first task is therefore the alkylation of propyne to 2-heptyne.

$$\underset{\text{Propyne}}{CH_3C{\equiv}CH} \xrightarrow[\text{2. } CH_3CH_2CH_2CH_2Br]{\text{1. } NaNH_2,\ NH_3} \underset{\text{2-Heptyne}}{CH_3C{\equiv}CCH_2CH_2CH_2CH_3}$$

$$\text{2-Heptyne} \xrightarrow{\text{Li, } NH_3} \underset{(E)\text{-2-Heptene}}{\overset{H_3C\qquad H}{\underset{H\qquad CH_2CH_2CH_2CH_3}{C{=}C}}}$$

$$\text{2-Heptyne} \xrightarrow[\text{Lindlar Pd}]{H_2} \underset{(Z)\text{-2-Heptene}}{\overset{H_3C\qquad CH_2CH_2CH_2CH_3}{\underset{H\qquad\quad H}{C{=}C}}}$$

**9.13** (*b*) Addition of hydrogen chloride to vinyl chloride gives the geminal dichloride 1,1-dichloroethane.

$$\underset{\text{Vinyl chloride}}{H_2C{=}CHCl} \xrightarrow{HCl} \underset{\text{1,1-Dichloroethane}}{CH_3CHCl_2}$$

(*c*) Since 1,1-dichloroethane can be prepared by adding 2 mol of hydrogen chloride to acetylene as shown in the sample solution to part (*a*), first convert 1,1-dibromoethane to acetylene by dehydrohalogenation.

$$\underset{\text{1,1-Dibromoethane}}{CH_3CHBr_2} \xrightarrow[\text{2. } H_2O]{\text{1. } NaNH_2,\ NH_3} \underset{\text{Acetylene}}{HC{\equiv}CH} \xrightarrow{2HCl} \underset{\text{1,1-Dichloroethane}}{CH_3CHCl_2}$$

**9.14** The enol arises by addition of water to the triple bond.

$$\underset{\text{2-Butyne}}{CH_3C{\equiv}CCH_3} + H_2O \longrightarrow \underset{\text{2-Buten-2-ol (enol form)}}{\left[CH_3\underset{\substack{|\\OH}}{C}{=}CHCH_3\right]} \longrightarrow \underset{\text{2-Butanone}}{CH_3\overset{\substack{O\\\|}}{C}CH_2CH_3}$$

The mechanism described in the textbook Figure 9.6 is adapted to the case of 2-butyne hydration as shown:

$$\underset{\text{Hydronium ion}}{H_2\overset{+}{O}{-}H} + \underset{\text{2-Buten-2-ol}}{CH_3CH{=}\overset{\substack{:\ddot{O}H\\|}}{C}CH_3} \rightleftharpoons \underset{\text{Water}}{H_2\ddot{O}:} + \underset{\text{Carbocation}}{CH_3CH_2{-}\overset{\substack{:\ddot{O}H\\|}}{\underset{+}{C}}CH_3} \longleftrightarrow CH_3CH_2\overset{\substack{+\ddot{O}H\\\|}}{C}CH_3$$

$$\underset{\text{Carbocation}}{CH_3CH_2{-}\overset{\substack{:\ddot{O}{-}H\\|}}{\underset{+}{C}}CH_3} + \underset{\text{Water}}{:\ddot{O}H_2} \longrightarrow \underset{\text{2-Butanone}}{CH_3CH_2\overset{\substack{:\ddot{O}\\\|}}{C}CH_3} + \underset{\text{Hydronium ion}}{H{-}\overset{+}{O}H_2}$$

**9.15** Hydration of 1-octyne gives 2-octanone according to the equation that immediately precedes this problem in the text. Prepare 1-octyne as described in the solution to Problem 9.9, and then carry out its hydration in the presence of mercury(II) sulfate and sulfuric acid.

Hydration of 4-octyne gives 4-octanone. Prepare 4-octyne as described in the solution to Problem 9.9.

**9.16** Each of the carbons that are part of —$CO_2H$ groups was once part of a —C≡C— unit. The two fragments $CH_3(CH_2)_4CO_2H$ and $HO_2CCH_2CH_2CO_2H$ account for only 10 of the original 16 carbons. The full complement of carbons can be accommodated by assuming that two molecules of $CH_3(CH_2)_4CO_2H$ are formed, along with one molecule of $HO_2CCH_2CH_2CO_2H$. The starting alkyne is therefore deduced from the ozonolysis data to be as shown:

$$CH_3(CH_2)_4C{\equiv}CCH_2CH_2C{\equiv}C(CH_2)_4CH_3$$

$$\swarrow \qquad \downarrow \qquad \searrow$$

$$CH_3(CH_2)_4CO_2H \qquad HO_2CCH_2CH_2CO_2H \qquad HO_2C(CH_2)_4CH_3$$

**9.17** Three isomers have unbranched carbon chains:

$CH_3CH_2CH_2CH_2C\equiv CH$ — 1-Hexyne

$CH_3CH_2CH_2C\equiv CCH_3$ — 2-Hexyne

$CH_3CH_2C\equiv CCH_2CH_3$ — 3-Hexyne

Next consider all the alkynes with a single methyl branch:

$CH_3CH(CH_3)CH_2C\equiv CH$ — 4-Methyl-1-pentyne

$CH_3CH_2CH(CH_3)C\equiv CH$ — 3-Methyl-1-pentyne

$CH_3CH(CH_3)C\equiv CCH_3$ — 4-Methyl-2-pentyne

One isomer has two methyl branches. None is possible with an ethyl branch.

$CH_3C(CH_3)_2C\equiv CH$ — 3,3-Dimethyl-1-butyne

**9.18** (*a*) $\overset{5}{C}H_3\overset{4}{C}H_2\overset{3}{C}H_2\overset{2}{C}\equiv \overset{1}{C}H$ is 1-pentyne

(*b*) $\overset{5}{C}H_3\overset{4}{C}H_2\overset{3}{C}\equiv \overset{2}{C}\overset{1}{C}H_3$ is 2-pentyne

(*c*) $\overset{1}{C}H_3\overset{2}{C}\equiv \overset{3}{C}\overset{4}{C}H(CH_3)\overset{5}{C}H(CH_3)\overset{6}{C}H_3$ is 4,5-dimethyl-2-hexyne

(*d*) (cyclopropyl)$-\overset{5}{C}H_2\overset{4}{C}H_2\overset{3}{C}H_2\overset{2}{C}\equiv \overset{1}{C}H$ is 5-cyclopropyl-1-pentyne

(*e*) $\overset{13}{C}H_2\overset{1}{C}\equiv \overset{2}{C}\overset{3}{C}H_2$ (ring) is cyclotridecyne

(*f*) $CH_3CH_2CH_2CH_2\overset{4}{C}H(\overset{3}{C}\equiv \overset{2}{C}\overset{1}{C}H_3)\overset{5}{C}H_2\overset{6}{C}H_2\overset{7}{C}H_2\overset{8}{C}H_2\overset{9}{C}H_3$ is 4-butyl-2-nonyne

(Parent chain must contain the triple bond.)

(*g*) $\overset{1}{C}H_3\overset{2}{C}(CH_3)_2\overset{3}{C}\equiv \overset{4}{C}\overset{5}{C}(CH_3)_2\overset{6}{C}H_3$ is 2,2,5,5-tetramethyl-3-hexyne

**9.19** (*a*) 1-Octyne is $HC\equiv CCH_2CH_2CH_2CH_2CH_2CH_3$

(*b*) 2-Octyne is $CH_3C\equiv CCH_2CH_2CH_2CH_2CH_3$

(*c*) 3-Octyne is $CH_3CH_2C\equiv CCH_2CH_2CH_2CH_3$

(*d*) 4-Octyne is $CH_3CH_2CH_2C\equiv CCH_2CH_2CH_3$

(*e*) 2,5-Dimethyl-3-hexyne is $CH_3CH(CH_3)C\equiv CCH(CH_3)CH_3$

(*f*) 4-Ethyl-1-hexyne is $CH_3CH_2CH(CH_2CH_3)CH_2C\equiv CH$

(*g*) Ethynylcyclohexane is $C_6H_{11}\text{—}C\equiv CH$ (cyclohexane ring bonded to C≡CH)

(*h*) 3-Ethyl-3-methyl-1-pentyne is $CH_3CH_2\underset{\underset{CH_2CH_3}{|}}{\overset{\overset{CH_3}{|}}{C}}C\equiv CH$

**9.20** Ethynylcyclohexane has the molecular formula $C_8H_{12}$. All the other compounds are $C_8H_{14}$.

**9.21** Only alkynes with the carbon skeletons shown can give 3-ethylhexane on catalytic hydrogenation.

3-Ethyl-1-hexyne or 4-Ethyl-1-hexyne or 4-Ethyl-2-hexyne $\xrightarrow[\text{Pt}]{H_2}$ 3-Ethylhexane

**9.22** The carbon skeleton of the unknown acetylenic amino acid must be the same as that of homoleucine. The structure of homoleucine is such that there is only one possible location for a carbon–carbon triple bond in an acetylenic precursor.

$$HC\equiv C\underset{\underset{CH_3}{|}}{C}HCH_2\underset{\underset{^{+}NH_3}{|}}{C}H\overset{\overset{O}{\|}}{C}O^- \xrightarrow[\text{Pt}]{H_2} CH_3CH_2\underset{\underset{CH_3}{|}}{C}HCH_2\underset{\underset{^{+}NH_3}{|}}{C}H\overset{\overset{O}{\|}}{C}O^-$$

$C_7H_{11}NO_2$ — Homoleucine

**9.23** (*a*)

$$CH_3CH_2CH_2CH_2CH_2CHCl_2 \xrightarrow[\text{2. } H_2O]{\text{1. } NaNH_2,\ NH_3} CH_3CH_2CH_2CH_2C\equiv CH$$

1,1-Dichlorohexane — 1-Hexyne

(*b*)

$$CH_3CH_2CH_2CH_2CH{=}CH_2 \xrightarrow[CCl_4]{Br_2} CH_3CH_2CH_2CH_2\underset{\underset{Br}{|}}{C}HCH_2Br \xrightarrow[\text{2. } H_2O]{\text{1. } NaNH_2,\ NH_3} CH_3CH_2CH_2CH_2C\equiv CH$$

1-Hexene — 1,2-Dibromohexane — 1-Hexyne

(*c*)

$$HC\equiv CH \xrightarrow[NH_3]{NaNH_2} HC\equiv C^-Na^+ \xrightarrow{CH_3CH_2CH_2CH_2Br} CH_3CH_2CH_2CH_2C\equiv CH$$

Acetylene — 1-Hexyne

(*d*)

$$CH_3CH_2CH_2CH_2CH_2CH_2I \xrightarrow[DMSO]{KOC(CH_3)_3} CH_3CH_2CH_2CH_2CH{=}CH_2$$

1-Iodohexane — 1-Hexene

1-Hexene is then converted to 1-hexyne as in part (*b*).

**9.24** (*a*) Working backward from the final product, it can be seen that preparation of 1-butyne will allow the desired carbon skeleton to be constructed.

$$CH_3CH_2C\equiv CCH_2CH_3 \quad \text{prepared from} \quad CH_3CH_2C\equiv C{:}^- + BrCH_2CH_3$$

3-Hexyne

The desired intermediate, 1-butyne, is available by halogenation followed by dehydrohalogenation of 1-butene.

$$CH_3CH_2CH{=}CH_2 + Br_2 \longrightarrow CH_3CH_2\underset{}{\overset{Br}{C}}HCH_2Br \xrightarrow[2.\ H_2O]{1.\ NaNH_2,\ NH_3} CH_3CH_2C{\equiv}CH$$

1-Butene → 1-Butyne

Reaction of the anion of 1-butyne with ethyl bromide completes the synthesis.

$$CH_3CH_2C{\equiv}CH \xrightarrow[NH_3]{NaNH_2} CH_3CH_2C{\equiv}C{:}^-\ Na^+ \xrightarrow{CH_3CH_2Br} CH_3CH_2C{\equiv}CCH_2CH_3$$

1-Butyne → 3-Hexyne

(*b*) Dehydrohalogenation of 1,1-dichlorobutane yields 1-butyne. The synthesis is completed as in part (*a*).

$$CH_3CH_2CH_2CHCl_2 \xrightarrow[2.\ H_2O]{1.\ NaNH_2,\ NH_3} CH_3CH_2C{\equiv}CH$$

1,1-Dichlorobutane → 1-Butyne

(*c*)

$$HC{\equiv}CH \xrightarrow[NH_3]{NaNH_2} HC{\equiv}C{:}^-Na^+ \xrightarrow{CH_3CH_2Br} HC{\equiv}CCH_2CH_3$$

Acetylene → 1-Butyne

1-Butyne is converted to 3-hexyne as in part (*a*).

**9.25** A single dehydrobromination step occurs in the conversion of 1,2-dibromodecane to $C_{10}H_{19}Br$. Bromine may be lost from C-1 to give 2-bromo-1-decene.

$$BrCH_2\underset{Br}{C}H(CH_2)_7CH_3 \xrightarrow[\text{ethanol–water}]{KOH} H_2C{=}\underset{Br}{C}(CH_2)_7CH_3$$

1,2-Dibromodecane → 2-Bromo-1-decene

Loss of bromine from C-2 gives (*E*)- and (*Z*)-1-bromo-1-decene.

$$BrCH_2\underset{Br}{C}H(CH_2)_7CH_3 \xrightarrow[\text{ethanol–water}]{KOH} (H)(Br)C{=}C(H)((CH_2)_7CH_3) + (Br)(H)C{=}C(H)((CH_2)_7CH_3)$$

1,2-Dibromodecane → (*E*)-1-Bromo-1-decene + (*Z*)-1-Bromo-1-decene

**9.26** (*a*)

$$CH_3CH_2CH_2CH_2C{\equiv}CH + 2H_2 \xrightarrow{Pt} CH_3CH_2CH_2CH_2CH_2CH_3$$

1-Hexyne → Hexane

(*b*)

$$CH_3CH_2CH_2CH_2C{\equiv}CH + H_2 \xrightarrow{\text{Lindlar Pd}} CH_3CH_2CH_2CH_2CH{=}CH_2$$

1-Hexyne → 1-Hexene

(*c*)

$$CH_3CH_2CH_2CH_2C{\equiv}CH \xrightarrow[NH_3]{Li} CH_3CH_2CH_2CH_2CH{=}CH_2$$

1-Hexyne → 1-Hexene

(*d*) $CH_3CH_2CH_2CH_2C{\equiv}CH \xrightarrow[NH_3]{NaNH_2} CH_3CH_2CH_2CH_2C{\equiv}C{:}^-Na^+$

1-Hexyne — Sodium 1-hexynide

(*e*)

$CH_3CH_2CH_2CH_2C{\equiv}C{:}^-\ Na^+ + CH_3CH_2CH_2CH_2Br \longrightarrow CH_3CH_2CH_2CH_2C{\equiv}CCH_2CH_2CH_2CH_3$

Sodium 1-hexynide — 1-Bromobutane — 5-Decyne

(*f*)

$CH_3CH_2CH_2CH_2C{\equiv}C{:}^-\ Na^+ + (CH_3)_3CBr \longrightarrow CH_3CH_2CH_2CH_2C{\equiv}CH + (CH_3)_2C{=}CH_2$

Sodium 1-hexynide — *tert*-Butyl bromide — 1-Hexyne — 2-Methylpropene

(*g*) $CH_3CH_2CH_2CH_2C{\equiv}CH \xrightarrow[(1\ mol)]{HCl} CH_3CH_2CH_2CH_2C(Cl){=}CH_2$

1-Hexyne — 2-Chloro-1-hexene

(*h*) $CH_3CH_2CH_2CH_2C{\equiv}CH \xrightarrow[(2\ mol)]{HCl} CH_3CH_2CH_2CH_2CCl_2CH_3$

1-Hexyne — 2,2-Dichlorohexane

(*i*) $CH_3CH_2CH_2CH_2C{\equiv}CH \xrightarrow[(1\ mol)]{Cl_2} (CH_3CH_2CH_2CH_2)(Cl)C{=}C(Cl)(H)$

1-Hexyne — (*E*)-1,2-Dichloro-1-hexene

(*j*) $CH_3CH_2CH_2CH_2C{\equiv}CH \xrightarrow[(2\ mol)]{Cl_2} CH_3CH_2CH_2CH_2CCl_2CHCl_2$

1-Hexyne — 1,1,2,2-Tetrachlorohexane

(*k*) $CH_3CH_2CH_2CH_2C{\equiv}CH \xrightarrow[HgSO_4]{H_2O,\ H_2SO_4} CH_3CH_2CH_2CH_2\overset{O}{\overset{\|}{C}}CH_3$

1-Hexyne — 2-Hexanone

(*l*) $CH_3CH_2CH_2CH_2C{\equiv}CH \xrightarrow[2.\ H_2O]{1.\ O_3} CH_3CH_2CH_2CH_2\overset{O}{\overset{\|}{C}}OH + HO\overset{O}{\overset{\|}{C}}OH$

1-Hexyne — Pentanoic acid — Carbonic acid

**9.27** (*a*) $CH_3CH_2C{\equiv}CCH_2CH_3 + 2H_2 \xrightarrow{Pt} CH_3CH_2CH_2CH_2CH_2CH_3$

3-Hexyne — Hexane

(*b*) $CH_3CH_2C{\equiv}CCH_2CH_3 + H_2 \xrightarrow{Lindlar\ Pd} (CH_3CH_2)(H)C{=}C(H)(CH_2CH_3)$

3-Hexyne — (*Z*)-3-Hexene

(*c*) $CH_3CH_2C{\equiv}CCH_2CH_3 \xrightarrow[NH_3]{Li}$ $CH_3CH_2$(H)$C{=}C$(H)$CH_2CH_3$ (trans)

3-Hexyne → (*E*)-3-Hexene

(*d*) $CH_3CH_2C{\equiv}CCH_2CH_3 \xrightarrow[(1\ mol)]{HCl}$ $CH_3CH_2$(Cl)$C{=}C$(H)$CH_2CH_3$

3-Hexyne → (*Z*)-3-Chloro-3-hexene

(*e*) $CH_3CH_2C{\equiv}CCH_2CH_3 \xrightarrow[(2\ mol)]{HCl} CH_3CH_2CCl_2CH_2CH_2CH_3$

3-Hexyne → 3,3-Dichlorohexane

(*f*) $CH_3CH_2C{\equiv}CCH_2CH_3 \xrightarrow[(1\ mol)]{Cl_2}$ $CH_3CH_2$(Cl)$C{=}C$(Cl)$CH_2CH_3$

3-Hexyne → (*E*)-3,4-Dichloro-3-hexene

(*g*) $CH_3CH_2C{\equiv}CCH_2CH_3 \xrightarrow[(2\ mol)]{Cl_2} CH_3CH_2CCl_2{-}CCl_2CH_2CH_3$

3-Hexyne → 3,3,4,4-Tetrachlorohexane

(*h*) $CH_3CH_2C{\equiv}CCH_2CH_3 \xrightarrow[HgSO_4]{H_2O,\ H_2SO_4} CH_3CH_2\overset{O}{\overset{\|}{C}}CH_2CH_2CH_3$

3-Hexyne → 3-Hexanone

(*i*) $CH_3CH_2C{\equiv}CCH_2CH_3 \xrightarrow[2.\ H_2O]{1.\ O_3} 2CH_3CH_2\overset{O}{\overset{\|}{C}}OH$

3-Hexyne → Propanoic acid

**9.28** The two carbons of the triple bond are similarly but not identically substituted in 2-heptyne, $CH_3C{\equiv}CCH_2CH_2CH_2CH_3$. Two regioisomeric enols are formed, each of which gives a different ketone.

$$CH_3C{\equiv}CCH_2CH_2CH_2CH_3 \xrightarrow[HgSO_4]{H_2O,\ H_2SO_4} CH_3\overset{OH}{\overset{|}{C}}{=}CHCH_2CH_2CH_2CH_3 + CH_3CH{=}\overset{OH}{\overset{|}{C}}CH_2CH_2CH_2CH_3$$

2-Heptyne → 2-Hepten-2-ol + 2-Hepten-3-ol

$$\downarrow \qquad\qquad \downarrow$$

$$CH_3\overset{O}{\overset{\|}{C}}CH_2CH_2CH_2CH_2CH_3 \qquad CH_3CH_2\overset{O}{\overset{\|}{C}}CH_2CH_2CH_2CH_3$$

2-Heptanone 3-Heptanone

**9.29** The alkane formed by hydrogenation of (*S*)-3-methyl-1-pentyne is achiral; it cannot be optically active.

$$CH_3CH_2(H)(H_3C)C{-}C{\equiv}CH \xrightarrow[Pt]{H_2} CH_3CH_2CH(CH_3)CH_2CH_3$$

(*S*)-3-Methyl-1-pentyne

3-Methylpentane (does not have a stereogenic center; optically inactive)

The product of hydrogenation of (*S*)-4-methyl-1-hexyne is optically active because a stereogenic center is present in the starting material and is carried through to the product.

$$CH_3CH_2(H)(H_3C)C{-}CH_2C{\equiv}CH \xrightarrow[Pt]{H_2} CH_3CH_2(H)(H_3C)CCH_2CH_2CH_3$$

(*S*)-4-Methyl-1-hexyne

(*S*)-3-Methylhexane

Both (*S*)-3-methyl-1-pentyne and (*S*)-4-methyl-1-hexyne yield optically active products when their triple bonds are reduced to double bonds.

**9.30** (*a*) The dihaloalkane contains both a primary alkyl chloride and a primary alkyl iodide functional group. Iodide is a better leaving group than chloride and is the one replaced by acetylide.

$$NaC{\equiv}CH + ClCH_2CH_2CH_2CH_2CH_2CH_2I \longrightarrow ClCH_2CH_2CH_2CH_2CH_2CH_2C{\equiv}CH$$

Sodium acetylide — 1-Chloro-6-iodohexane — 8-Chloro-1-octyne

(*b*) Both vicinal dibromide functions are converted to alkyne units on treatment with excess sodium amide.

$$BrCH_2CH(Br)CH_2CH_2CH(Br)CH_2Br \xrightarrow[2.\ H_2O]{1.\ \text{excess } NaNH_2,\ NH_3} HC{\equiv}CCH_2CH_2C{\equiv}CH$$

1,2,5,6-Tetrabromohexane — 1,5-Hexadiyne

(*c*) The starting material is a geminal dichloride. Potassium *tert*-butoxide in dimethyl sulfoxide is a sufficiently strong base to convert it to an alkyne.

$$\text{cyclopropyl}{-}CCl_2CH_3 \xrightarrow[\text{heat}]{KOC(CH_3)_3,\ DMSO} \text{cyclopropyl}{-}C{\equiv}CH$$

1,1-Dichloro-1-cyclopropylethane — Ethynylcyclopropane

(*d*) Alkyl *p*-toluenesulfonates react similarly to alkyl halides in nucleophilic substitution reactions. The alkynide nucleophile displaces the *p*-toluenesulfonate leaving group from ethyl *p*-toluenesulfonate.

$$C_6H_5C{\equiv}C:^- + CH_3CH_2{-}OS(=O)_2C_6H_4CH_3 \longrightarrow C_6H_5C{\equiv}CCH_2CH_3$$

Phenylacetylide ion — Ethyl *p*-toluenesulfonate — 1-Phenyl-1-butyne (*e*) Both carbons of a —C≡C— unit are converted to carboxyl groups (—$CO_2H$) on ozonolysis.

$$\text{Cyclodecyne} \xrightarrow[\text{2. } H_2O]{\text{1. } O_3} HO\overset{O}{\overset{\|}{C}}(CH_2)_8\overset{O}{\overset{\|}{C}}OH$$

Cyclodecyne — Decanedioic acid

(*f*) Ozonolysis cleaves the carbon–carbon triple bond.

$$\text{1-Ethynylcyclohexanol} \xrightarrow[\text{2. } H_2O]{\text{1. } O_3} \text{1-Hydroxycyclohexanecarboxylic acid} + HO\overset{O}{\overset{\|}{C}}OH$$

1-Ethynylcyclohexanol — 1-Hydroxycyclohexane-carboxylic acid — Carbonic acid

(*g*) Hydration of a terminal carbon–carbon triple bond converts it to a —$\overset{O}{\overset{\|}{C}}CH_3$ group.

$$CH_3\underset{CH_3}{\underset{|}{C}}HCH_2\overset{OH}{\overset{|}{\underset{CH_3}{\underset{|}{C}}}}C\equiv CH \xrightarrow[HgO]{H_2O,\ H_2SO_4} CH_3\underset{CH_3}{\underset{|}{C}}HCH_2\overset{HO}{\overset{|}{\underset{CH_3}{\underset{|}{C}}}}-\overset{O}{\overset{\|}{C}}CH_3$$

3,5-Dimethyl-1-hexyn-3-ol — 3-Hydroxy-3,5-dimethyl-2-hexanone

(*h*) Sodium-in-ammonia reduction of an alkyne yields a trans alkene. The stereochemistry of a double bond that is already present in the molecule is not altered during the process.

$$CH_3CH_2CH_2CH_2(H)C{=}C(H)CH_2(CH_2)_7C\equiv CCH_2CH_2OH$$

(*Z*)-13-Octadecen-3-yn-1-ol

$$\downarrow \text{1. Na, } NH_3;\ \text{2. } H_2O$$

$$CH_3CH_2CH_2CH_2(H)C{=}C(H)CH_2(CH_2)_7(H)C{=}C(H)CH_2CH_2OH$$

(3*E*,13*Z*)-3,13-Octadecadien-1-ol

(*i*) The primary chloride leaving group is displaced by the alkynide nucleophile.

$$\text{THP}{-}O(CH_2)_8Cl + NaC\equiv CCH_2CH_2CH_2CH_3 \longrightarrow \text{THP}{-}O(CH_2)_8C\equiv CCH_2CH_2CH_2CH_3$$

8-Chlorooctyl tetrahydropyranyl ether — Sodium 1-hexynide — 9-Tetradecyn-1-yl tetrahydropyranyl ether (*j*) Hydrogenation of the triple bond over the Lindlar catalyst converts the compound to a cis alkene.

$$\text{(tetrahydropyranyl)}-O(CH_2)_8C{\equiv}CCH_2CH_2CH_2CH_3 \xrightarrow[\text{Lindlar Pd}]{H_2} \text{(tetrahydropyranyl)}-O(CH_2)_8CH{=}CHCH_2CH_2CH_2CH_3 \ (cis)$$

9-Tetradecyn-1-yl tetrahydropyranyl ether → (*Z*)-9-Tetradecen-1-yl tetrahydropyranyl ether

**9.31** Ketones such as 2-heptanone may be readily prepared by hydration of terminal alkynes. Thus, if we had 1-heptyne, it could be converted to 2-heptanone.

$$HC{\equiv}C(CH_2)_4CH_3 \xrightarrow[H_2SO_4,\ HgSO_4]{H_2O} CH_3\overset{O}{\overset{\|}{C}}(CH_2)_4CH_3$$

1-Heptyne → 2-Heptanone

Acetylene, as we have seen in earlier problems, can be converted to 1-heptyne by alkylation.

$$HC{\equiv}CH \xrightarrow[NH_3]{NaNH_2} HC{\equiv}C{:}^-\ Na^+$$

$$HC{\equiv}C{:}^-\ Na^+ + CH_3CH_2CH_2CH_2CH_2Br \longrightarrow HC{\equiv}C(CH_2)_4CH_3$$

**9.32** Apply the technique of reasoning backward to gain a clue to how to attack this synthesis problem. A reasonable final step is the formation of the *Z* double bond by hydrogenation of an alkyne over Lindlar palladium.

$$CH_3(CH_2)_7C{\equiv}C(CH_2)_{12}CH_3 \xrightarrow[\text{Lindlar Pd}]{H_2} \underset{H}{\overset{CH_3(CH_2)_7}{}}C{=}C\underset{H}{\overset{(CH_2)_{12}CH_3}{}}$$

9-Tricosyne → (*Z*)-9-Tricosene

The necessary alkyne 9-tricosyne can be prepared by a double alkylation of acetylene.

$$HC{\equiv}CH \xrightarrow[2.\ CH_3(CH_2)_7Br]{1.\ NaNH_2,\ NH_3} CH_3(CH_2)_7C{\equiv}CH \xrightarrow[2.\ CH_3(CH_2)_{12}Br]{1.\ NaNH_2,\ NH_3} CH_3(CH_2)_7C{\equiv}C(CH_2)_{12}CH_3$$

Acetylene → 1-Decyne → 9-Tricosyne

It does not matter which alkyl group is introduced first.

The alkyl halides are prepared from the corresponding alcohols.

$$CH_3(CH_2)_7OH \xrightarrow[\text{or } PBr_3]{HBr} CH_3(CH_2)_7Br$$

1-Octanol → 1-Bromooctane

$$CH_3(CH_2)_{12}OH \xrightarrow[\text{or } PBr_3]{HBr} CH_3(CH_2)_{12}Br$$

1-Tridecanol → 1-Bromotridecane

**9.33** (*a*) 2,2-Dibromopropane is prepared by addition of hydrogen bromide to propyne.

$$CH_3C{\equiv}CH + 2HBr \longrightarrow CH_3C(Br)_2CH_3$$

Propyne — Hydrogen bromide — 2,2-Dibromopropane

The designated starting material, 1,1-dibromopropane, is converted to propyne by a double dehydrohalogenation.

$$CH_3CH_2CHBr_2 \xrightarrow[2.\ H_2O]{1.\ NaNH_2,\ NH_3} CH_3C{\equiv}CH$$

1,1-Dibromopropane — Propyne

(*b*) As in part (*a*), first convert the designated starting material to propyne, and then add hydrogen bromide.

$$CH_3CH(Br)CH_2Br \xrightarrow[2.\ H_2O]{1.\ NaNH_2,\ NH_3} CH_3C{\equiv}CH \xrightarrow{2HBr} CH_3C(Br)_2CH_3$$

1,2-Dibromopropane — Propyne — 2,2-Dibromopropane

(*c*) Instead of trying to introduce two additional chlorines into 1,2-dichloropropane by free-radical substitution (a mixture of products would result), convert the vicinal dichloride to propyne, and then add two moles of $Cl_2$.

$$CH_3CH(Cl)CH_2Cl \xrightarrow[2.\ H_2O]{1.\ NaNH_2,\ NH_3} CH_3C{\equiv}CH \xrightarrow{2Cl_2} CH_3C(Cl)_2CHCl_2$$

1,2-Dichloropropane — Propyne — 1,1,2,2-Tetrachloropropane

(*d*) The required carbon skeleton can be constructed by alkylating acetylene with ethyl bromide.

$$HC{\equiv}CH \xrightarrow[NH_3]{NaNH_2} HC{\equiv}C{:}^-\ Na^+ \xrightarrow{CH_3CH_2Br} HC{\equiv}CCH_2CH_3$$

Acetylene — Sodium acetylide — 1- Butyne

Addition of 2 mol of hydrogen iodide to 1-butyne gives 2,2-diiodobutane.

$$HC{\equiv}CCH_2CH_3 + 2HI \longrightarrow CH_3C(I)_2CH_2CH_3$$

1-Butyne — Hydrogen iodide — 2,2-Diiodobutane

(*e*) The six-carbon chain is available by alkylation of acetylene with 1-bromobutane.

$$HC{\equiv}CH \xrightarrow[2.\ CH_3CH_2CH_2CH_2Br]{1.\ NaNH_2,\ NH_3} HC{\equiv}CCH_2CH_2CH_2CH_3$$

Acetylene — 1-Hexyne

The alkylating agent, 1-bromobutane, is prepared from 1-butene by free-radical (anti-Markovnikov) addition of hydrogen bromide.

$$\underset{\text{1-Butene}}{CH_3CH_2CH{=}CH_2} + \underset{\text{Hydrogen bromide}}{HBr} \xrightarrow{\text{peroxides}} \underset{\text{1-Bromobutane}}{CH_3CH_2CH_2CH_2Br}$$

Once 1-hexyne is prepared, it can be converted to 1-hexene by hydrogenation over Lindlar palladium or by sodium–ammonia reduction.

$$\underset{\text{1-Hexyne}}{CH_3CH_2CH_2CH_2C{\equiv}CH} \xrightarrow[\text{or Na, }NH_3]{H_2\text{, Lindlar Pd}} \underset{\text{1-Hexene}}{CH_3CH_2CH_2CH_2CH{=}CH_2}$$

(*f*) Dialkylation of acetylene with 1-bromobutane, prepared in part (*f*), gives the necessary ten-carbon chain.

$$\underset{\text{Acetylene}}{HC{\equiv}CH} \xrightarrow[2.\ CH_3CH_2CH_2CH_2Br]{1.\ NaNH_2,\ NH_3} \underset{\text{1-Hexyne}}{CH_3CH_2CH_2CH_2C{\equiv}CH} \xrightarrow[2.\ CH_3CH_2CH_2CH_2Br]{1.\ NaNH_2,\ NH_3} \underset{\text{5-Decyne}}{CH_3CH_2CH_2CH_2C{\equiv}CCH_2CH_2CH_2CH_3}$$

Hydrogenation of 5-decyne yields decane.

$$\underset{\text{5-Decyne}}{CH_3(CH_2)_3C{\equiv}C(CH_2)_3CH_3} \xrightarrow[Pt]{2H_2} \underset{\text{Decane}}{CH_3(CH_2)_3CH_2CH_2(CH_2)_3CH_3}$$

(*g*) A standard method for converting alkenes to alkynes is to add $Br_2$ and then carry out a double dehydrohalogenation.

$$\underset{\text{Cyclopentadecene}}{\text{cyclo-}[CH{=}CH{-}(CH_3)_{13}]} \xrightarrow{Br_2} \underset{\text{1,2-Dibromocyclopentadecane}}{\text{cyclo-}[CH(Br){-}CH(Br){-}(CH_2)_{13}]} \xrightarrow[NH_3]{NaNH_2} \underset{\text{Cyclopentadecyne}}{\text{cyclo-}[C{\equiv}C{-}(CH_2)_{13}]}$$

(*h*) Alkylation of the triple bond gives the required carbon skeleton.

$$\underset{\text{1-Ethynylcyclohexene}}{C_6H_9{-}C{\equiv}CH} \xrightarrow[2.\ CH_3Br]{1.\ NaNH_2,\ NH_3} \underset{\text{1-(1-Propynyl)cyclohexene}}{C_6H_9{-}C{\equiv}CCH_3}$$

Hydrogenation over the Lindlar catalyst converts the carbon–carbon triple bond to a cis double bond.

$$\underset{\text{1-(1-Propynyl)cyclohexene}}{C_6H_9{-}C{\equiv}CCH_3} \xrightarrow[\text{Lindlar Pd}]{H_2} \underset{(Z)\text{-1-(1-Propenyl)cyclohexene}}{C_6H_9{-}C(H){=}C(H){-}CH_3}$$

(*i*) The stereochemistry of *meso*-2,3-dibromobutane is most easily seen with a Fischer projection:

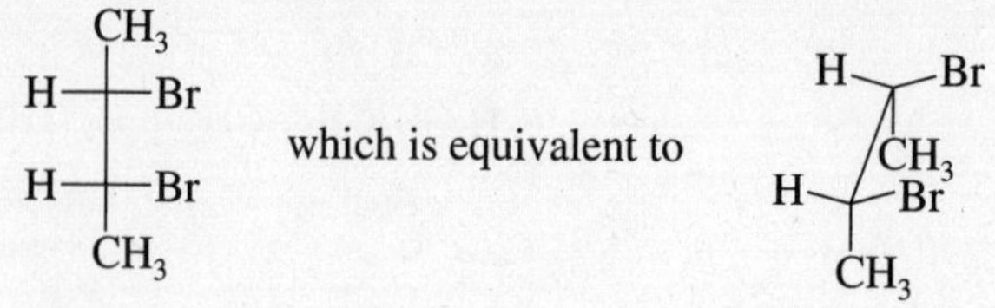

$$\begin{array}{c} CH_3 \\ H-\!\!\!+\!\!\!-Br \\ H-\!\!\!+\!\!\!-Br \\ CH_3 \end{array} \quad \text{which is equivalent to} \quad \text{(sawhorse: H, Br / CH}_3\text{; H, Br / CH}_3\text{)}$$

Recalling that the addition of $Br_2$ to alkenes occurs with anti stereochemistry, rotate the sawhorse diagram so that the bromines are anti to each other:

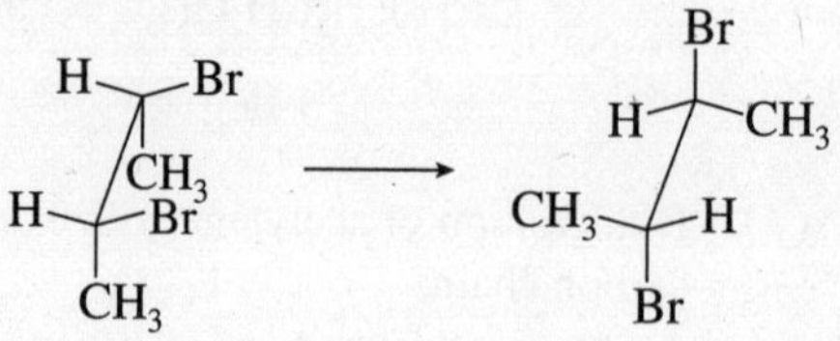

$$\text{(sawhorse: H, Br, CH}_3\text{ / H, Br, CH}_3\text{)} \longrightarrow \text{(sawhorse: Br, H, CH}_3\text{ / CH}_3\text{, H, Br)}$$

Thus, the starting alkene must be *trans*-2-butene. *trans*-2-Butene is available from 2-butyne by metal-ammonia reduction:

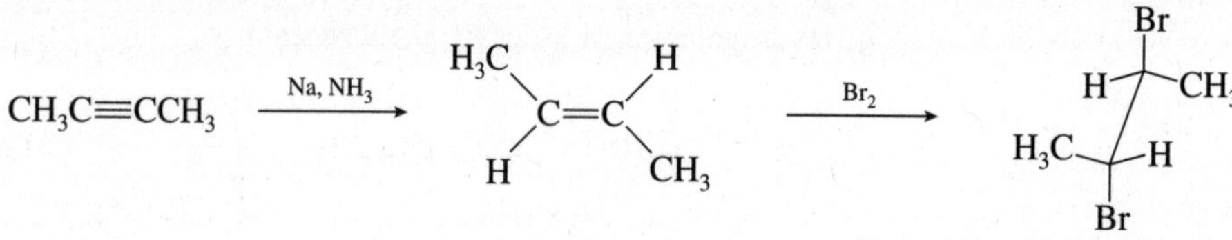

$$CH_3C{\equiv}CCH_3 \xrightarrow{Na,\ NH_3} \underset{\text{trans-2-Butene}}{(H_3C)(H)C{=}C(H)(CH_3)} \xrightarrow{Br_2} \text{(sawhorse: Br, H, CH}_3\text{ / H}_3\text{C, H, Br)}$$

2-Butyne     *trans*-2-Butene     *meso*-2,3-Dibromobutane

**9.34** Attack this problem by first planning a synthesis of 4-methyl-2-pentyne from any starting material in a single step. Two different alkyne alkylations suggest themselves:

$$\underset{\text{4-Methyl-2-pentyne}}{CH_3C{\equiv}CCH(CH_3)_2} \quad \begin{cases} (a) & \text{from } CH_3C{\equiv}C{:}^- \text{ and } BrCH(CH_3)_2 \\ (b) & \text{from } CH_3I \text{ and } {}^-{:}C{\equiv}CCH(CH_3)_2 \end{cases}$$

Isopropyl bromide is a secondary alkyl halide and cannot be used to alkylate $CH_3C{\equiv}C{:}^-$ according to reaction (*a*). A reasonable last step is therefore the alkylation of $(CH_3)_2CHC{\equiv}CH$ via reaction of its anion with methyl iodide.

The next question that arises from this analysis is the origin of $(CH_3)_2CHC{\equiv}CH$. One of the available starting materials is 1,1-dichloro-3-methylbutane. It can be converted to $(CH_3)_2CHC{\equiv}CH$ by a double dehydrohalogenation. The complete synthesis is therefore:

$$\underset{\text{1,1-Dichloro-3-methylbutane}}{(CH_3)_2CHCH_2CHCl_2} \xrightarrow[2.\ H_2O]{1.\ NaNH_2,\ NH_3} \underset{\text{3-Methyl-1-butyne}}{(CH_3)_2CHC{\equiv}CH} \xrightarrow[2.\ CH_3I]{1.\ NaNH_2} \underset{\text{4-Methyl-2-pentyne}}{(CH_3)_2CHC{\equiv}CCH_3}$$

**9.35** The reaction that produces compound A is reasonably straightforward. Compound A is 14-bromo-1-tetradecyne.

$$\underset{\text{Sodium acetylide}}{NaC{\equiv}CH} + \underset{\text{1,12-Dibromododecane}}{Br(CH_2)_{12}Br} \longrightarrow \underset{\text{Compound A }(C_{14}H_{25}Br)}{Br(CH_2)_{12}C{\equiv}CH}$$

Treatment of compound A with sodium amide converts it to compound B. Compound B on ozonolysis gives a diacid that retains all the carbon atoms of B. Compound B must therefore be a cyclic alkyne, formed by an intramolecular alkylation.

$$Br(CH_2)_{12}C{\equiv}CH \xrightarrow{NaNH_2} \left[ \begin{array}{l} (CH_2)_{11}C{\equiv}C{:}^{-}\,Na^{+} \\ H_2C \\ \quad Br \end{array} \right] \longrightarrow \text{cyclo-}[C{\equiv}C(CH_2)_{12}] \xrightarrow[2.\ H_2O]{1.\ O_3} HO\overset{O}{\overset{\|}{C}}(CH_2)_{12}\overset{O}{\overset{\|}{C}}OH$$

Compound A    Compound B

Compound B is cyclotetradecyne.

Hydrogenation of compound B over Lindlar palladium yields *cis*-cyclotetradecene (compound C).

$$\text{cyclo-}[C{\equiv}C(CH_2)_{12}] \xrightarrow[\text{Lindlar Pd}]{H_2} \text{cis-cyclotetradecene}$$

Compound C ($C_{14}H_{26}$)

Hydrogenation over platinum gives cyclotetradecane (compound D).

$$\text{cyclo-}[C{\equiv}C(CH_2)_{12}] \xrightarrow[Pt]{H_2} \text{cyclotetradecane}$$

Compound D ($C_{14}H_{28}$)

Sodium–ammonia reduction of compound B yields *trans*-cyclotetradecene.

$$\text{cyclo-}[C{\equiv}C(CH_2)_{12}] \xrightarrow{Na,\ NH_3} \text{trans-cyclotetradecene}$$

Compound E ($C_{14}H_{26}$)

The cis and trans isomers of cyclotetradecene are both converted to $O{=}CH(CH_2)_{12}CH{=}O$ on ozonolysis, whereas cyclotetradecane does not react with ozone.

**9.36–9.37** Solutions to molecular modeling exercises are not provided in this *Study Guide and Solutions Manual.* You should use *Learning By Modeling* for these exercises.

# SELF-TEST

## PART A

**A-1.** Provide the IUPAC names for the following:

(*a*) $CH_3C{\equiv}C\underset{\underset{CH_3}{|}}{C}HCH(CH_3)_2$

(b) $\underset{\displaystyle CH_2CH_2CH_3}{CH_3CH_2CH_2\overset{\displaystyle CH_2CH_3}{CH}CHC{\equiv}CH}$

(c)

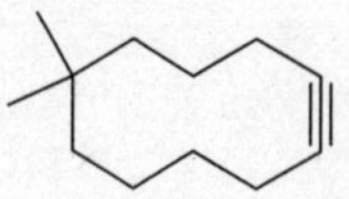

**A-2.** Give the structure of the reactant, reagent, or product omitted from each of the following reactions.

(a) $CH_3CH_2CH_2C{\equiv}CH \xrightarrow{HCl\ (1\ mol)}$ ?

(b) $CH_3CH_2CH_2C{\equiv}CH \xrightarrow{HCl\ (2\ mol)}$ ?

(c) $CH_3CH_2CH_2C{\equiv}CH \xrightarrow{?} CH_3CH_2CH_2\overset{O}{\overset{\|}{C}}CH_3$

(d) $CH_3C{\equiv}CCH_3 \xrightarrow[\text{Lindlar Pd}]{H_2}$ ?

(e) ? $\xrightarrow[2.\ CH_3CH_2Br]{1.\ NaNH_2} (CH_3)_2CHC{\equiv}CCH_2CH_3$

(f) $CH_3C{\equiv}CCH_2CH_3 \xrightarrow{?}$ (E)-2-pentene

(g) $CH_3C{\equiv}CCH_2CH_3 \xrightarrow{Cl_2\ (1\ mol)}$ ?

(h) $CH_3CH_2CH_2\underset{\displaystyle CH_3}{\underset{|}{C}}HC{\equiv}CCH_2CH_3 \xrightarrow[2.\ H_2O]{1.\ O_3}$ ?

**A-3.** Which one of the following two reactions is effective in the synthesis of 4-methyl-2-hexyne? Why is the other not effective?

1. $CH_3CH_2\overset{\displaystyle Br}{\overset{|}{C}}HCH_3 + CH_3C{\equiv}CNa \longrightarrow$

2. $CH_3CH_2\overset{\displaystyle CH_3}{\overset{|}{C}}HC{\equiv}CNa + CH_3I \longrightarrow$

**A-4.** Outline a series of steps, using any necessary organic and inorganic reagents, for the preparation of:

(a) 1-Butyne from ethyl bromide as the source of all carbon atoms
(b) 3-Hexyne from 1-butyne
(c) 3-Hexyne from 1-butene
(d) $CH_3\overset{O}{\overset{\|}{C}}CH_2CH(CH_3)_2$ from acetylene

**A-5.** Treatment of propyne in successive steps with sodium amide, 1-bromobutane, and sodium in liquid ammonia yields as the final product ______.

**A-6.** Give the structures of compounds A through D in the following series of equations.

$$A \xrightarrow{NaNH_2,\ NH_3} B$$
$$C \xrightarrow{HBr,\ heat} D$$
$$B + D \longrightarrow CH_3CH_2CH_2C{\equiv}CC(CH_3)_3$$

**A-7.** What are the structures of compounds E and F in the following sequence of reactions?

$$\text{Compound E} \xrightarrow[\text{2. } CH_3CH_2Br]{\text{1. } NaNH_2,\ NH_3} \text{Compound F} \xrightarrow[HgSO_4]{H_2O,\ H_2SO_4} CH_3CH_2\overset{\overset{\displaystyle O}{\|}}{C}CH_2CH_2CH_3$$

**A-8.** Give the reagents that would be suitable for carrying out the following transformation. Two or more reaction steps are necessary.

$$\text{(cyclopentyl)}{-}C{\equiv}CH \longrightarrow \text{(cyclopentyl)}{-}CH_2CH_2OH$$

## PART B

**B-1.** The IUPAC name for the compound shown is

$$CH_3\underset{}{\overset{\overset{\displaystyle CH_2CH_3}{|}}{C}}HCH_2C{\equiv}CCH(CH_3)_2$$

(*a*) 2,6-Dimethyl-3-octyne
(*b*) 6-Ethyl-2-methyl-3-heptyne
(*c*) 2-Ethylpropyl isopropyl acetylene
(*d*) 2-Ethyl-6-methyl-4-heptyne

**B-2.** Which of the following statements best explains the greater acidity of terminal alkynes ($RC{\equiv}CH$) compared with monosubstituted alkenes ($RCH{=}CH_2$)?

(*a*) The *sp*-hybridized carbons of the alkyne are less electronegative than the $sp^2$ carbons of the alkene.
(*b*) The two $\pi$ bonds of the alkyne are better able to stabilize the negative charge of the anion by resonance.
(*c*) The *sp*-hybridized carbons of the alkyne are more electronegative than the $sp^2$ carbons of the alkene.
(*d*) The question is incorrect—alkenes are more acidic than alkynes.

**B-3.** Referring to the following equilibrium (R = alkyl group)

$$RCH_2CH_3 + RC{\equiv}C\!:^- \rightleftharpoons RCH_2\ddot{C}H_2^- + RC{\equiv}C{-}H$$

(*a*) $K < 1$; the equilibrium would lie to the left.
(*b*) $K > 1$; the equilibrium would lie to the right.
(*c*) $K = 1$; equal amounts of all species would be present.
(*d*) Not enough information is given; the structure of R must be known.

**B-4.** Which of the following is an effective way to prepare 1-pentyne?

(*a*) 1-Pentene $\xrightarrow[\text{2. } NaNH_2\text{, heat}]{\text{1. } Cl_2}$

(*b*) Acetylene $\xrightarrow[\text{2. } CH_3CH_2CH_2Br]{\text{1. } NaNH_2}$

(*c*) 1,1-Dichloropentane $\xrightarrow[\text{2. } H_2O]{\text{1. } NaNH_2,\ NH_3}$

(*d*) All these are effective.

**B-5.** Which alkyne yields butanoic acid ($CH_3CH_2CH_2CO_2H$) as the only organic product on treatment with ozone followed by hydrolysis?

(*a*) 1-Butyne (*c*) 1-Pentyne
(*b*) 4-Octyne (*d*) 2-Hexyne

**B-6.** Which of the following produces a significant amount of acetylide ion on reaction with acetylene?

(*a*) Conjugate base of $CH_3OH$ ($pK_a 16$)
(*b*) Conjugate base of $H_2$ ($pK_a 35$)
(*c*) Conjugate base of $H_2O$ ($pK_a 16$)
(*d*) Both (*a*) and (*c*).

**B-7.** Which of the following is the product of the reaction of 1-hexyne with 1 mol of $Br_2$?

(*a*) Br Br

(*b*) Br Br

(*c*) Br Br

(*d*) Br Br

(*e*) Br Br

**B-8.** Choose the sequence of steps that describes the best synthesis of 1-butene from ethanol.

(*a*) (1) $NaC{\equiv}CH$; (2) $H_2$, Lindlar Pd
(*b*) (1) $NaC{\equiv}CH$; (2) Na, $NH_3$
(*c*) (1) HBr, heat; (2) $NaC{\equiv}CH$; (3) $H_2$, Lindlar Pd
(*d*) (1) HBr, heat; (2) $KOC(CH_3)_3$, DMSO; (3) $NaC{\equiv}CH$; (4) $H_2$, Lindlar Pd

**B-9.** What is (are) the major product(s) of the following reaction?

$$(CH_3)_3CBr + HC{\equiv}C\!:^- Na^+ \longrightarrow ?$$

(*a*) $(CH_3)_3CC{\equiv}CH$

(*b*) $H_2C{=}C(CH_3)CH_3 + HC{\equiv}CH$

(*c*) $H_3C$, $H_3C$, $CH_3$

(*d*) $HC{\equiv}CCH_2CH(CH_3)_2$

**B-10.** Which would be the best sequence of reactions to use to prepare *cis*-3-nonene from 1-butyne?

(*a*) (1) $NaNH_2$ in $NH_3$; (2) 1-bromopentane; (3) $H_2$, Lindlar Pd
(*b*) (1) $NaNH_2$ in $NH_3$; (2) 1-bromopentane; (3) Na, $NH_3$
(*c*) (1) $H_2$, Lindlar Pd; (2) $NaNH_2$ in $NH_3$; (3) 1-bromopentane
(*d*) (1) Na, $NH_3$; (2) $NaNH_2$ in $NH_3$; (3) 1-bromopentane

**B-11.** Which one of the following is the intermediate in the preparation of a ketone by hydration of an alkyne in the presence of sulfuric acid and mercury(II) sulfate?

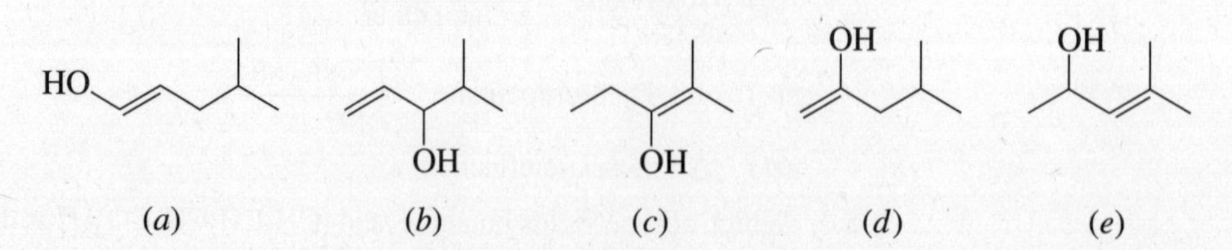

**B-12.** Which combination is best for preparing the compound shown in the box?

$$\mathrm{(H)(CH_3)(CH_3CH_2)C{-}CH_2CH_2CH_2C{\equiv}CH}$$

(*a*) $\mathrm{(H_3C)(H)(CH_3CH_2)C{-}CH_2CH_2CH_2Br \xrightarrow{NaC\equiv CH}}$

(*b*) $\mathrm{(H)(CH_3)(CH_3CH_2)C{-}CH_2CH_2CH_2Br \xrightarrow{NaC\equiv CH}}$

(*c*) $\mathrm{(H_3C)(H)(CH_3CH_2)C{-}Br \xrightarrow[2.\ BrCH_2CH_2CH_2C\equiv CH]{1.\ NaNH_2,\ NH_3}}$

(*d*) $\mathrm{(H)(CH_3)(CH_3CH_2)C{-}Br \xrightarrow[2.\ BrCH_2CH_2CH_2C\equiv CH]{1.\ NaNH_2,\ NH_3}}$

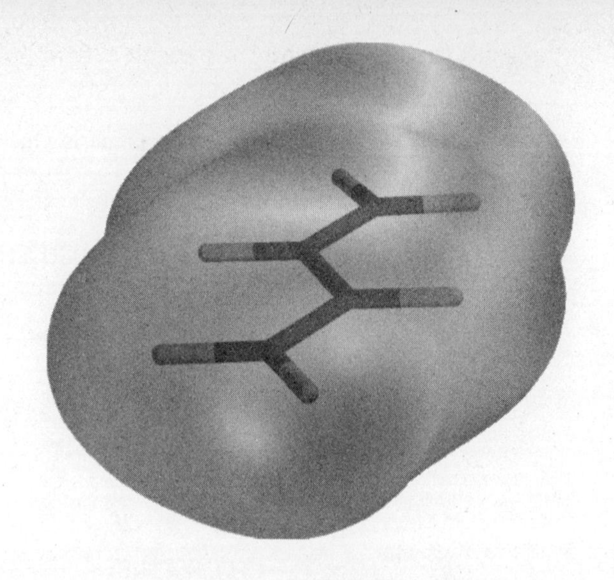

# CHAPTER 10

## CONJUGATION IN ALKADIENES AND ALLYLIC SYSTEMS

## SOLUTIONS TO TEXT PROBLEMS

**10.1** As noted in the sample solution to part (*a*), a pair of electrons is moved from the double bond toward the positively charged carbon.

(*b*) $H_2C{=}\overset{+}{C}$... $H_2C{=}C(CH_3){-}\overset{+}{C}H_2 \longleftrightarrow H_2\overset{+}{C}{-}C(CH_3){=}CH_2$

(*c*) (cyclohexyl ring with positive charge)$=C(CH_3)_2 \longleftrightarrow$ (cyclohexenyl)$-\overset{+}{C}(CH_3)_2$

**10.2** For two isomeric halides to yield the same carbocation on ionization, they must have the same carbon skeleton. They may have their leaving group at a different location, but the carbocations must become equivalent by allylic resonance.

3-Bromo-1-methylcyclohexene → [allylic carbocation resonance forms] ← 3-Chloro-3-methylcyclohexene

4-Bromo-1-methylcyclohexene → Not an allylic carbocation

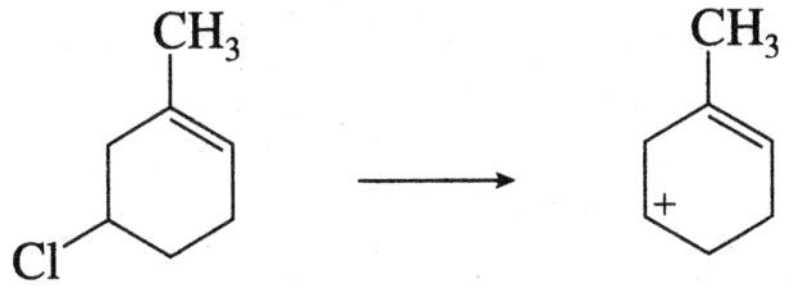

$CH_3$ / Cl → $CH_3$ (+) Not an allylic carbocation

5-Chloro-1-methylcyclohexene

$CH_3$ / Br → $CH_3$ (+) Not an allylic carbocation

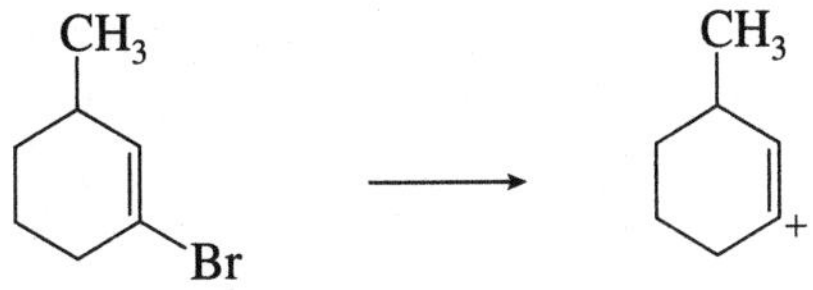

1-Bromo-3-methylcyclohexene

**10.3** The allylic hydrogens are the ones shown in the structural formulas.

(*b*) H, H, $CH_3$, H, H

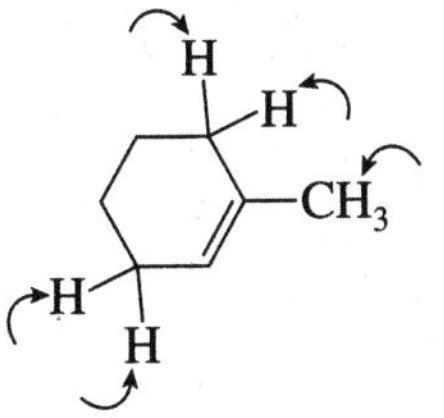

1-Methylcyclohexene

(*c*) $CH_3$

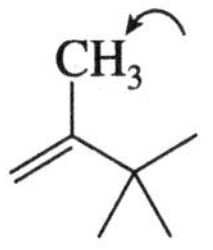

2,3,3-Trimethyl-1-butene

(*d*) H H

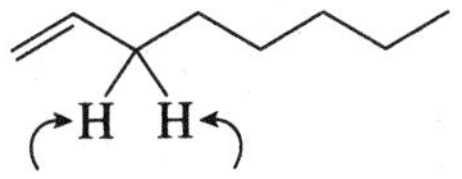

1-Octene

**10.4** The statement of the problem specifies that in allylic brominations using *N*-bromosuccinimide the active reagent is $Br_2$. Thus, the equation for the overall reaction is

| Cyclohexene | + | Bromine ($Br_2$) | ⟶ | 3-Bromocyclohexene | + | Hydrogen bromide (HBr) |
|---|---|---|---|---|---|---|

The propagation steps are analogous to those of other free-radical brominations. An allylic hydrogen is removed by a bromine atom in the first step.

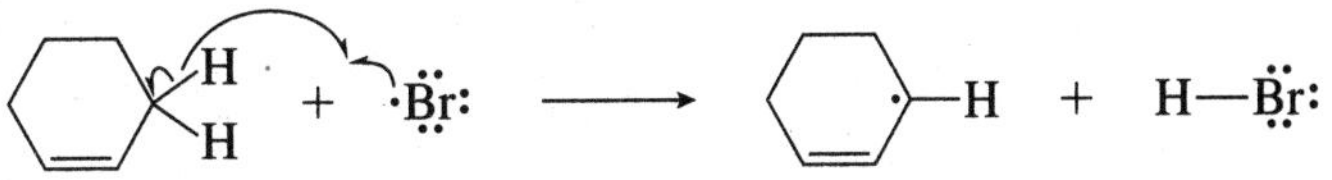

| Cyclohexene | + | Bromine atom (·$\ddot{B}r$:) | ⟶ | 2-Cyclohexenyl radical | + | Hydrogen bromide (H—$\ddot{B}r$:) |
|---|---|---|---|---|---|---|

The allylic radical formed in the first step abstracts a bromine atom from $Br_2$ in the second propagation step.

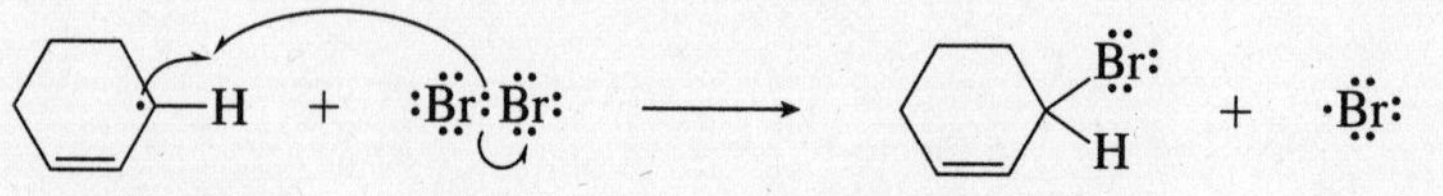

2-Cyclohexenyl radical + Bromine ⟶ 3-Bromocyclohexene + Bromine atom

**10.5** Write both resonance forms of the allylic radicals produced by hydrogen atom abstraction from the alkene.

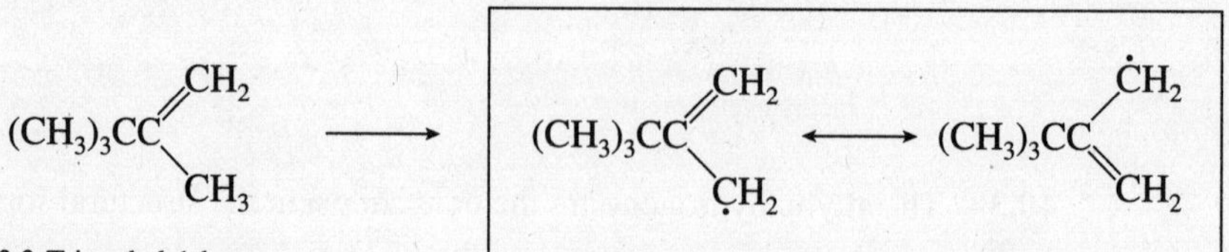

$$(CH_3)_3CC(=CH_2)CH_3 \longrightarrow (CH_3)_3CC(=CH_2)\dot{C}H_2 \longleftrightarrow (CH_3)_3CC(\dot{C}H_2)=CH_2$$

2,3,3-Trimethyl-1-butene

Both resonance forms are equivalent, and so 2,3,3-trimethyl-1-butene gives a single bromide on treatment with *N*-bromosuccinimide (NBS).

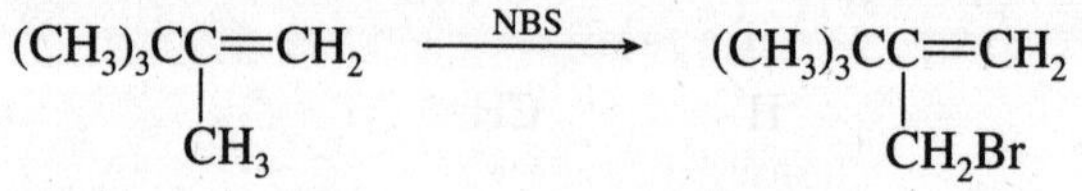

$$(CH_3)_3CC(CH_3)=CH_2 \xrightarrow{NBS} (CH_3)_3CC(CH_2Br)=CH_2$$

2,3,3-Trimethyl-1-butene ⟶ 2-(Bromomethyl)-3,3-dimethyl-1-butene

Hydrogen atom abstraction from 1-octene gives a radical in which the unpaired electron is delocalized between two nonequivalent positions.

$$CH_2{=}CHCH_2(CH_2)_4CH_3 \longrightarrow CH_2{=}CH\dot{C}H(CH_2)_4CH_3 \longleftrightarrow \dot{C}H_2CH{=}CH(CH_2)_4CH_3$$

1-Octene

Allylic bromination of 1-octene gives a mixture of products

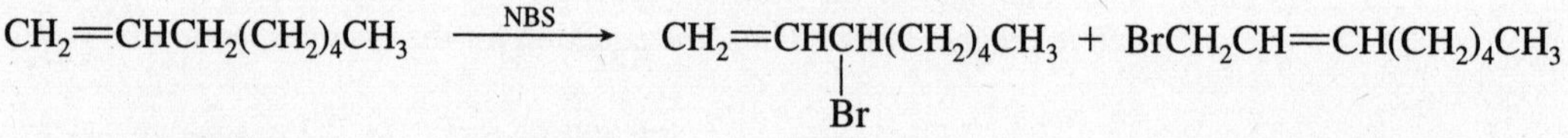

$$CH_2{=}CHCH_2(CH_2)_4CH_3 \xrightarrow{NBS} CH_2{=}CHCH(Br)(CH_2)_4CH_3 + BrCH_2CH{=}CH(CH_2)_4CH_3$$

1-Octene ⟶ 3-Bromo-1-octene + 1-Bromo-2-octene (cis and trans)

**10.6** (*b*) All the double bonds in humulene are isolated, because they are separated from each other by one or more $sp^3$ carbon atoms.

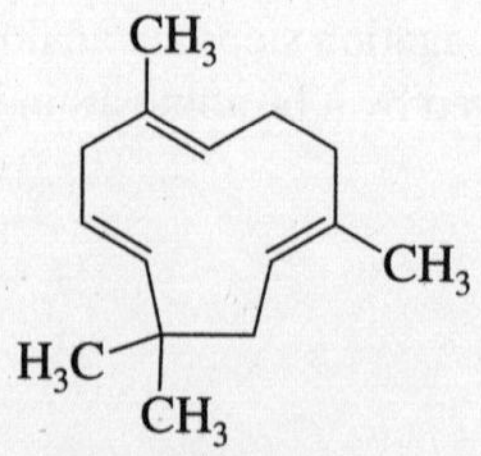

Humulene

(*c*) The C-1 and C-3 double bonds of cembrene are conjugated with each other.

Cembrene

The double bonds at C-6 and C-10 are isolated from each other and from the conjugated diene system.

(*d*) The sex attractant of the dried-bean beetle has a cumulated diene system involving C-4, C-5, and C-6. This allenic system is conjugated with the C-2 double bond.

$$CH_3(CH_2)_6CH_2\overset{6}{C}H{=}\overset{5}{C}{=}\overset{4}{C}H\overset{3}{C}H{=}\overset{2}{C}H\overset{1}{C}O_2CH_3$$

**10.7** The more stable the isomer, the lower its heat of combustion. The conjugated diene is the most stable and has the lowest heat of combustion. The cumulated diene is the least stable and has the highest heat of combustion.

| (*E*)-1,3-Pentadiene | $H_2C{=}CHCH_2CH{=}CH_2$ | $H_2C{=}C{=}CHCH_2CH_3$ |
|---|---|---|
| (*E*)-1,3-Pentadiene | 1,4-Pentadiene | 1,2-Pentadiene |
| Most stable | | Least stable |
| 3186 kJ/mol | 3217 kJ/mol | 3251 kJ/mol |
| (761.6 kcal/mol) | (768.9 kcal/mol) | (777.1 kcal/mol) |

**10.8** Compare the mirror-image forms of each compound for superposability. For 2-methyl-2,3-pentadiene,

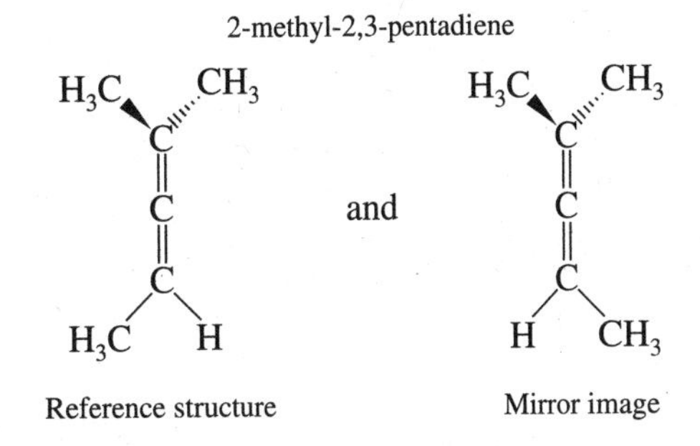


Rotation of the mirror image 180° around an axis passing through the three carbons of the C=C=C unit demonstrates that the reference structure and its mirror image are superposable.

Rotate 180°

Mirror image

Reoriented mirror image

2-Methyl-2,3-pentadiene is an achiral allene.

Comparison of the mirror-image forms of 2-chloro-2,3-pentadiene reveals that they are not superposable. 2-Chloro-2,3-pentadiene is a chiral allene.

2-Chloro-2,3-pentadiene

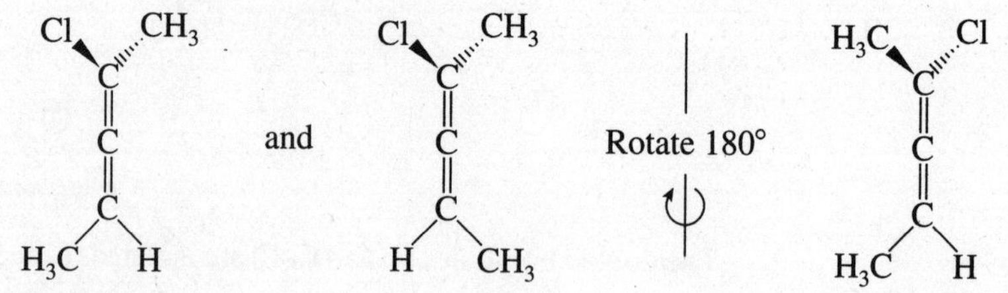

Cl, $CH_3$ / C=C=C / $H_3C$, H — Reference structure

and

Cl, $CH_3$ / C=C=C / H, $CH_3$ — Mirror image

Rotate 180°

$H_3C$, Cl / C=C=C / $H_3C$, H — Reoriented mirror image

**10.9** Both starting materials undergo β-elimination to give a conjugated diene system. Two minor products result, both of which have isolated double bonds.

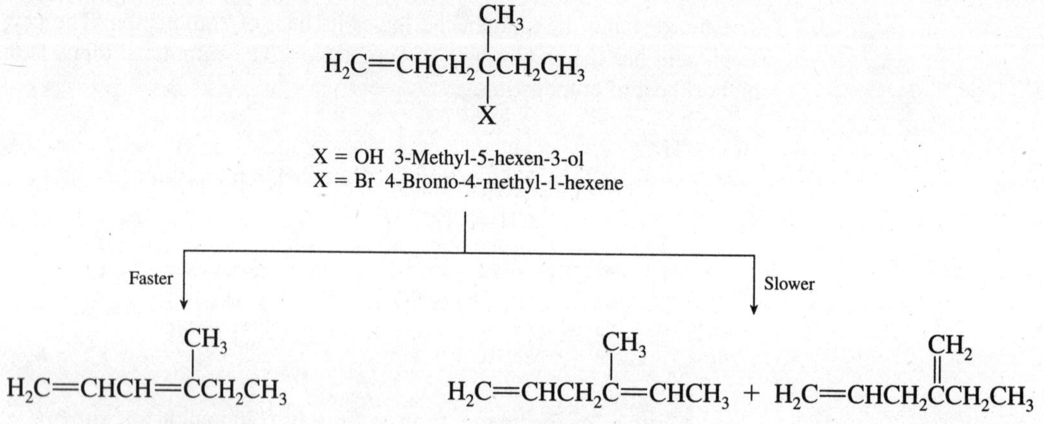

$$H_2C{=}CHCH_2C(CH_3)(X)CH_2CH_3$$

X = OH 3-Methyl-5-hexen-3-ol
X = Br 4-Bromo-4-methyl-1-hexene

Faster → $H_2C{=}CHCH{=}C(CH_3)CH_2CH_3$
4-Methyl-1,3-hexadiene (mixture of *E* and *Z* isomers; major product)

Slower → $H_2C{=}CHCH_2C(CH_3){=}CHCH_3$ + $H_2C{=}CHCH_2C({=}CH_2)CH_2CH_3$
4-Methyl-1,4-hexadiene (mixture of *E* and *Z* isomers; minor product)
2-Ethyl-1,4-pentadiene (minor product)

**10.10** The best approach is to work through this reaction mechanistically. Addition of hydrogen halides always proceeds by protonation of one of the terminal carbons of the diene system. Protonation of C-1 gives an allylic cation for which the most stable resonance form is a tertiary carbocation. Protonation of C-4 would give a less stable allylic carbocation for which the most stable resonance form is a secondary carbocation.

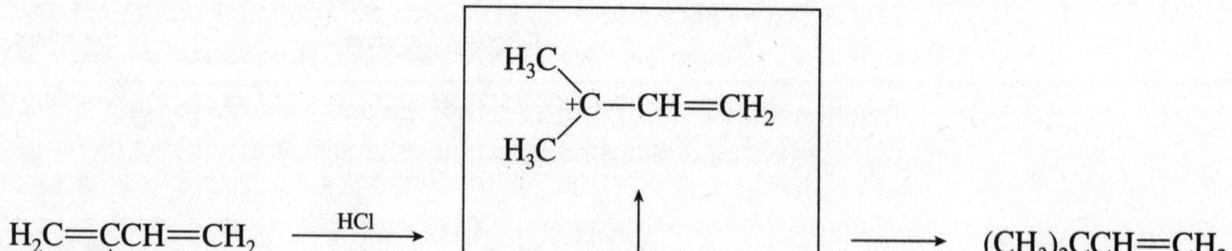

$$H_2C{=}C(CH_3)CH{=}CH_2 \xrightarrow{HCl} \left[(H_3C)_2\overset{+}{C}{-}CH{=}CH_2 \longleftrightarrow (H_3C)_2C{=}CH{-}\overset{+}{C}H_2\right] \longrightarrow (CH_3)_2C(Cl)CH{=}CH_2$$

2-Methyl-1,3-butadiene → 3-Chloro-3-methyl-1-butene (major product)

Under kinetically controlled conditions the carbocation is captured at the carbon that bears the greatest share of positive charge, and the product is the tertiary chloride.

**10.11** The two double bonds of 2-methyl-1,3-butadiene are not equivalent, and so two different products of direct addition are possible, along with one conjugate addition product.

$$H_2C{=}C(CH_3)CH{=}CH_2 \xrightarrow{Br_2} BrCH_2C(Br)(CH_3)CH{=}CH_2 + H_2C{=}C(CH_3)CH(Br)CH_2Br + BrCH_2C(CH_3){=}CHCH_2Br$$

2-Methyl-1,3-butadiene; 3,4-Dibromo-3-methyl-1-butene (direct addition); 3,4-Dibromo-2-methyl-1-butene (direct addition); 1,4-Dibromo-2-methyl-2-butene (conjugate addition)

**10.12** The molecular formula of the product, $C_{10}H_9ClO_2$, is that of a 1:1 Diels–Alder adduct between 2-chloro-1,3-butadiene and benzoquinone.

2-Chloro-1,3-butadiene + Benzoquinone ⟶ $C_{10}H_9ClO_2$

**10.13** "Unravel" the Diels–Alder adduct as described in the sample solution to part (*a*).

(*b*) Diels–Alder adduct (with two C≡N groups) is prepared from Diene + Dienophile (cyano groups are cis)

(*c*) Diels–Alder adduct (with $CH_3$) is prepared from Diene ($CH_3$) + Dienophile

**10.14** Two stereoisomeric Diels–Alder adducts are possible from the reaction of 1,3-cyclopentadiene and methyl acrylate. In one stereoisomer the $CO_2CH_3$ group is syn to the HC═CH bridge, and is called the *endo* isomer. In the other stereoisomer the $CO_2CH_3$ group is anti to the HC═CH bridge and is called the *exo* isomer.

1,3-Cyclopentadiene + $H_2C{=}CHCOCH_3$ (Methyl acrylate) ⟶ Endo isomer (75%) + Exo isomer (25%)

(Stereoisomeric forms of methyl bicyclo[2.2.1]hept-5-ene-2-carboxylate)

**10.15** An electrophile is by definition an *electron-seeker.* When an electrophile attacks ethylene, it interacts with the $\pi$ orbital because this is the orbital that contains electrons. The $\pi^*$ orbital of ethylene is unoccupied.

**10.16** Analyze the reaction of two butadiene molecules by the Woodward–Hoffmann rules by examining the symmetry properties of the highest occupied molecular orbital (HOMO) of one diene and the lowest unoccupied molecular orbital (LUMO) of the other.

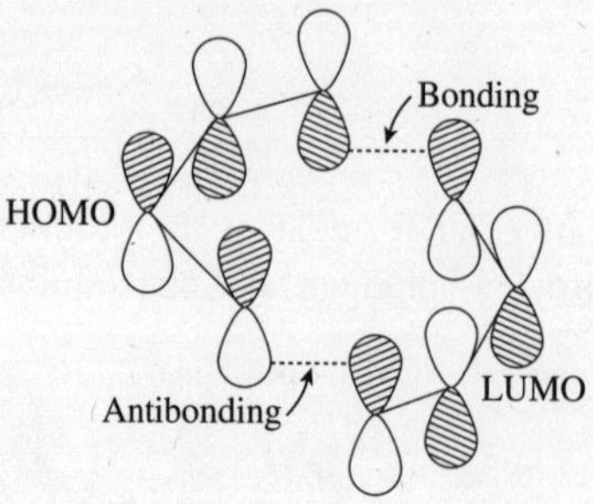

Bonding

HOMO

LUMO

Antibonding

This reaction is forbidden by the Woodward–Hoffmann rules. Both interactions involving the ends of the dienes need to be bonding for concerted cycloaddition to take place. Here, one is bonding and the other is antibonding.

**10.17** Dienes and trienes are named according to the IUPAC convention by replacing the *-ane* ending of the alkane with *-adiene* or *-atriene* and locating the positions of the double bonds by number. The stereoisomers are identified as *E* or *Z* according to the rules established in Chapter 5.

(*a*) 3,4-Octadiene: $CH_3CH_2CH{=}C{=}CHCH_2CH_2CH_3$

(*b*) (*E*,*E*)-3,5-Octadiene: $CH_3CH_2$, H, C=C, H, C=C, H, H, $CH_2CH_3$

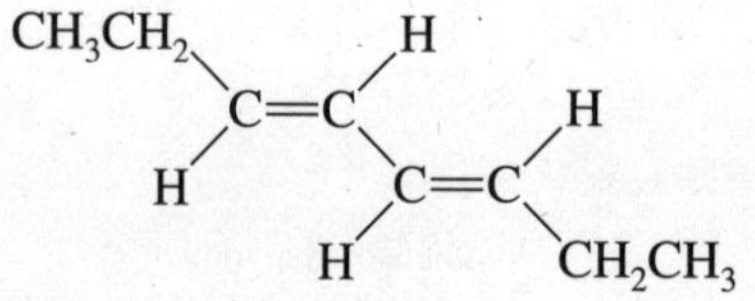

(*c*) (*Z*,*Z*)-1,3-Cyclooctadiene:

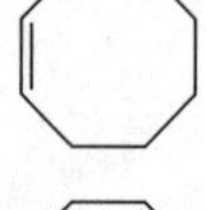

(*d*) (*Z*,*Z*)-1,4-Cyclooctadiene:

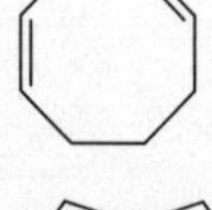

(*e*) (*E*,*E*)-1,5-Cyclooctadiene:

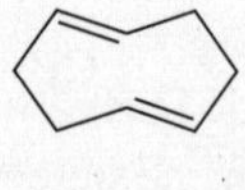

(*f*) (2*E*,4*Z*,6*E*)-2,4,6-Octatriene: $H_3C$, H H, $CH_3$, H, H, H, H

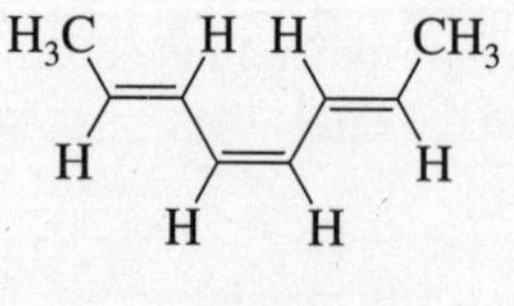

(*g*) 5-Allyl-1,3-cyclopentadiene: 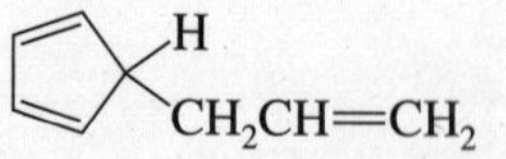 H, $CH_2CH{=}CH_2$

(*h*) *trans*-1,2-Divinylcyclopropane: $H_2C{=}CH$, H, H, $CH{=}CH_2$

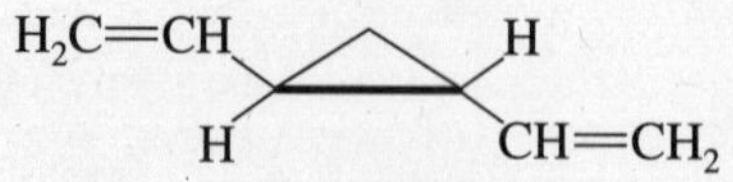

(*i*) 2,4-Dimethyl-1,3-pentadiene: $H_2C{=}C(CH_3)CH{=}C(CH_3)CH_3$

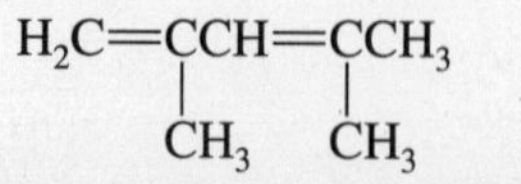

**10.18** (*a*) $H_2C{=}CH(CH_2)_5CH{=}CH_2$

1,8-Nonadiene

(*b*) $(CH_3)_2C{=}C(CH_3)C(CH_3){=}C(CH_3)_2$

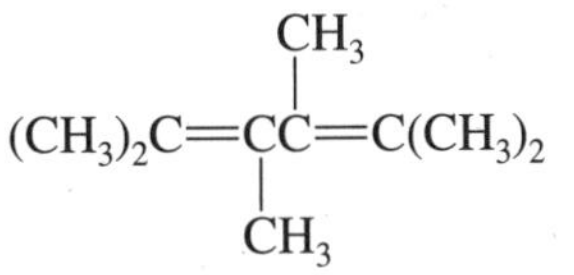

2,3,4,5-Tetramethyl-2,4-hexadiene

(*c*) $CH_2{=}CH{-}CH(CH{=}CH_2){-}CH{=}CH_2$

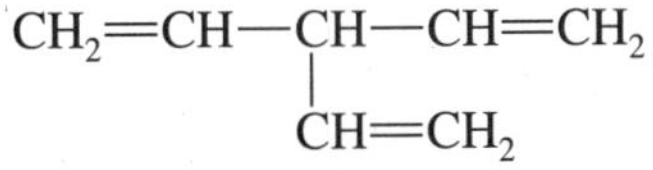

3-Vinyl-1,4-pentadiene

(*d*) H, $CH_3$, $CH_2$

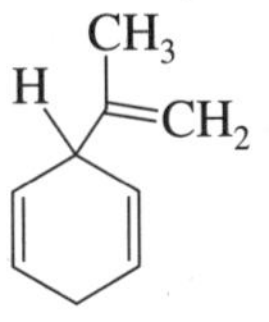

3-Isopropenyl-1,4-cyclohexadiene

(*e*) H H Cl H H Cl H H

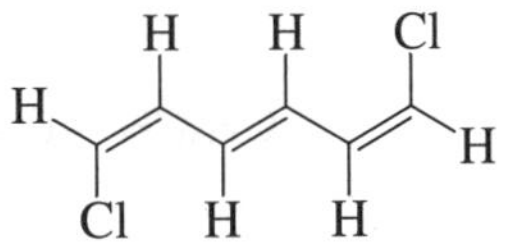

(1*Z*,3*E*,5*Z*)-1,6-Dichloro-1,3,5-hexatriene

(*f*) $H_2C{=}C{=}CHCH{=}CHCH_3$

1,2,4-Hexatriene

(*g*)

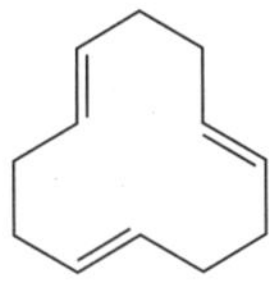

(1*E*,5*E*,9*E*)-1,5,9-Cyclododecatriene

(*h*) $H_3C$, $CH{=}CH_2$, $C{=}C$, $CH_3CH_2$, $CH_2CH_3$

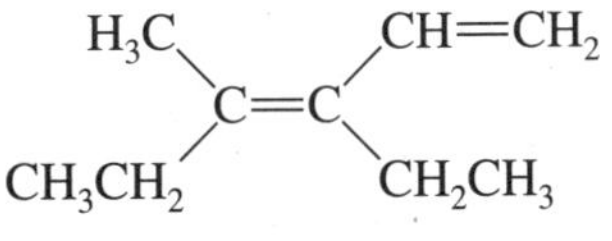

(*E*)-3-Ethyl-4-methyl-1,3-hexadiene

**10.19** (*a*) Since the product is 2,3-dimethylbutane we know that the carbon skeleton of the starting material must be

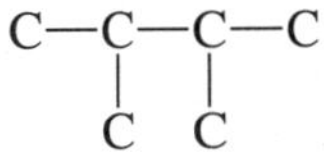

$$C{-}C(C){-}C(C){-}C$$

Since 2,3-dimethylbutane is $C_6H_{14}$ and the starting material is $C_6H_{10}$, *two* molecules of $H_2$ must have been taken up and the starting material must have two double bonds. The starting material can only be 2,3-dimethyl-1,3-butadiene.

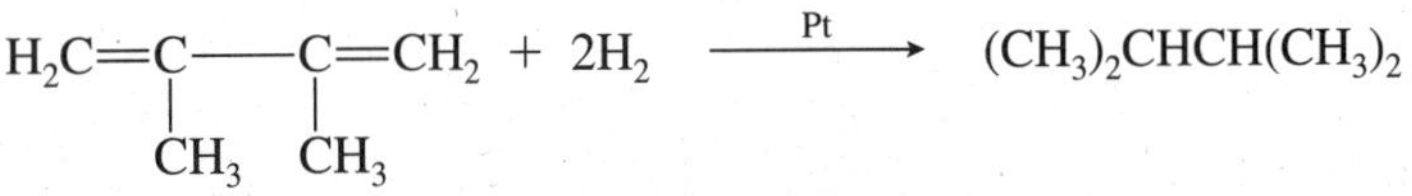

$$H_2C{=}C(CH_3){-}C(CH_3){=}CH_2 + 2H_2 \xrightarrow{Pt} (CH_3)_2CHCH(CH_3)_2$$

(*b*) Write the carbon skeleton corresponding to 2,2,6,6-tetramethylheptane.

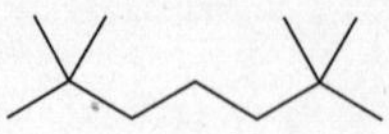

Compounds of molecular formula $C_{11}H_{20}$ have two double bonds or one triple bond. The only compounds with the proper carbon skeleton are the alkyne and the allene shown.

$(CH_3)_3CC{\equiv}CCH_2C(CH_3)_3$ — 2,2,6,6-Tetramethyl-3-heptyne

$(CH_3)_3CCH{=}C{=}CHC(CH_3)_3$ — 2,2,6,6-Tetramethyl-3,4-heptadiene

**10.20** The dienes that give 2,4-dimethylpentane on catalytic hydrogenation must have the same carbon skeleton as that alkane.

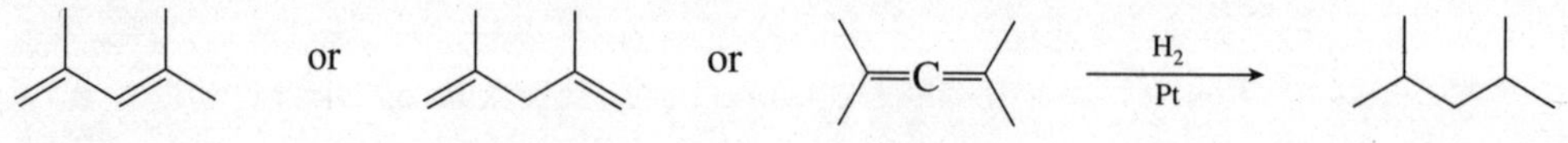

(*a*) 2,4-Dimethyl-1,3-pentadiene, conjugated diene; or (*b*) 2,4-Dimethyl-1,4-pentadiene, isolated diene; or (*c*) 2,4-Dimethyl-2,3-pentadiene, cumulated diene $\xrightarrow[\text{Pt}]{H_2}$ 2,4-Dimethylpentane

**10.21** The important piece of information that allows us to complete the structure properly is that the ant repellent is an *allenic* substance. The allenic unit cannot be incorporated into the ring, because the three carbons must be collinear. The only possible constitution is therefore

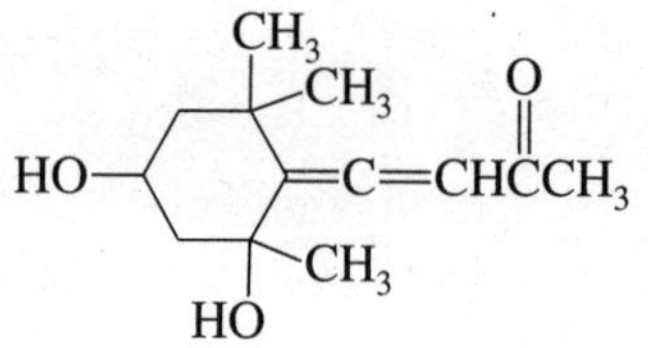

HO–(cyclohexane ring bearing CH₃, CH₃, HO, CH₃ substituents)$=C{=}CHCCH_3$ (with C=O)

**10.22** (*a*) Allylic halogenation of propene with *N*-bromosuccinimide gives allyl bromide.

$$\underset{\text{Propene}}{H_2C{=}CHCH_3} \xrightarrow[\text{CCl}_4,\ \text{heat}]{N\text{-bromosuccinimide}} \underset{\text{Allyl bromide}}{H_2C{=}CHCH_2Br}$$

(*b*) Electrophilic addition of bromine to the double bond of propene gives 1,2-dibromopropane.

$$\underset{\text{Propene}}{H_2C{=}CHCH_3} \xrightarrow{Br_2} \underset{\text{1,2-Dibromopropane}}{BrCH_2CH(Br)CH_3}$$

(*c*) 1,3-Dibromopropane is made from allyl bromide from part (*a*) by free-radical addition of hydrogen bromide.

$$\underset{\text{Allyl bromide}}{H_2C{=}CHCH_2Br} \xrightarrow[\text{peroxides}]{HBr} \underset{\text{1,3-Dibromopropane}}{BrCH_2CH_2CH_2Br}$$

(*d*) Addition of hydrogen chloride to allyl bromide proceeds in accordance with Markovnikov's rule.

$$H_2C{=}CHCH_2Br \xrightarrow{HCl} CH_3\underset{\displaystyle Cl}{\underset{|}{C}}HCH_2Br$$

Allyl bromide → 1-Bromo-2-chloropropane

(*e*) Addition of bromine to allyl bromide gives 1,2,3-tribromopropane.

$$H_2C{=}CHCH_2Br \xrightarrow{Br_2} BrCH_2\underset{\displaystyle Br}{\underset{|}{C}}HCH_2Br$$

Allyl bromide → 1,2,3-Tribromopropane

(*f*) Nucleophilic substitution by hydroxide on allyl bromide gives allyl alcohol.

$$H_2C{=}CHCH_2Br \xrightarrow{NaOH} H_2C{=}CHCH_2OH$$

Allyl bromide → Allyl alcohol

(*g*) Alkylation of sodium acetylide using allyl bromide gives the desired 1-penten-4-yne.

$$H_2C{=}CHCH_2Br \xrightarrow{NaC{\equiv}CH} H_2C{=}CHCH_2C{\equiv}CH$$

Allyl bromide → 1-Penten-4-yne

(*h*) Sodium–ammonia reduction of 1-penten-4-yne reduces the triple bond but leaves the double bond intact. Hydrogenation over Lindlar palladium could also be used.

$$H_2C{=}CHCH_2C{\equiv}CH \xrightarrow[\text{or } H_2\text{, Lindlar Pd}]{Na,\ NH_3} H_2C{=}CHCH_2CH{=}CH_2$$

1-Penten-4-yne → 1,4-Pentadiene

**10.23** (*a*) The desired allylic alcohol can be prepared by hydrolysis of an allylic halide. Cyclopentene can be converted to an allylic bromide by free-radical bromination with *N*-bromosuccinimide (NBS).

$$\text{Cyclopentene} \xrightarrow[\text{heat, peroxides}]{NBS} \text{3-Bromocyclopentene} \xrightarrow[Na_2CO_3]{H_2O} \text{2-Cyclopenten-1-ol}$$

Cyclopentene　3-Bromocyclopentene　2-Cyclopenten-1-ol

(*b*) Reaction of the allylic bromide from part (*a*) with sodium iodide in acetone converts it to the corresponding iodide.

$$\text{3-Bromocyclopentene} \xrightarrow[\text{acetone}]{NaI} \text{3-Iodocyclopentene}$$

3-Bromocyclopentene　3-Iodocyclopentene

(*c*) Nucleophilic substitution by cyanide converts the allylic bromide to 3-cyanocyclopentene.

$$\text{3-Bromocyclopentene} \xrightarrow{\text{NaCN}} \text{3-Cyanocyclopentene}$$

3-Bromocyclopentene 3-Cyanocyclopentene

(*d*) Reaction of the allylic bromide with a strong base will yield cyclopentadiene by an E2 elimination.

$$\text{3-Bromocyclopentene} \xrightarrow[\text{CH}_3\text{CH}_2\text{OH, heat}]{\text{NaOCH}_2\text{CH}_3} \text{1,3-Cyclopentadiene}$$

3-Bromocyclopentene 1,3-Cyclopentadiene

(*e*) Cyclopentadiene formed in part (*d*) is needed in order to form the required Diels–Alder adduct.

$$\text{1,3-Cyclopentadiene} \xrightarrow{\text{CH}_3\text{OCC}{\equiv}\text{CCOCH}_3\ (\text{C=O at both})} \text{bicyclo[2.2.1]heptadiene with two } \overset{\text{O}}{\overset{\|}{\text{C}}}\text{OCH}_3$$

1,3-Cyclopentadiene Dimethyl bicyclo[2.2.1]heptadiene-2,3-dicarboxylate

**10.24** The starting material in all cases is 2,3-dimethyl-1,3-butadiene.

$$\text{H}_2\text{C}{=}\text{C}(\text{CH}_3){-}\text{C}(\text{CH}_3){=}\text{CH}_2$$

2,3-Dimethyl-1,3-butadiene

(*a*) Hydrogenation of both double bonds will occur to yield 2,3-dimethylbutane.

$$\text{H}_2\text{C}{=}\text{C}(\text{CH}_3){-}\text{C}(\text{CH}_3){=}\text{CH}_2 \xrightarrow[\text{Pt}]{\text{H}_2} (\text{CH}_3)_2\text{CHCH}(\text{CH}_3)_2$$

(*b*) Direct addition of 1 mol of hydrogen chloride will give the product of Markovnikov addition to one of the double bonds, 3-chloro-2,3-dimethyl-1-butene.

$$\text{H}_2\text{C}{=}\text{C}(\text{CH}_3){-}\text{C}(\text{CH}_3){=}\text{CH}_2 \xrightarrow{\text{HCl}} (\text{CH}_3)_2\text{C}(\text{Cl}){-}\text{C}(\text{CH}_3){=}\text{CH}_2$$

(*c*) Conjugate addition will lead to double bond migration and produce 1-chloro-2,3-dimethyl-2-butene.

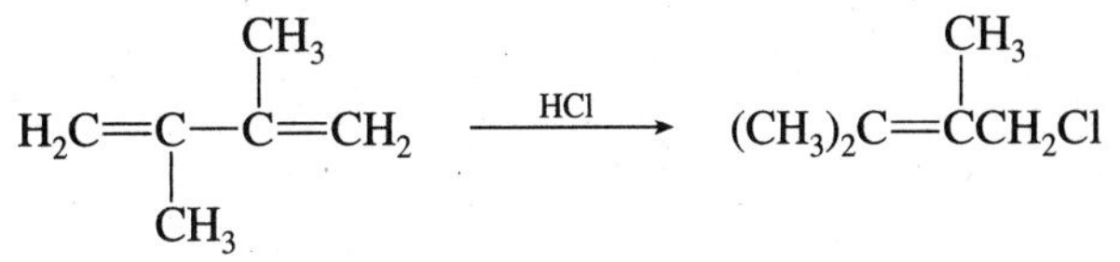

(*d*) The direct addition product is 3,4-dibromo-2,3-dimethyl-1-butene.

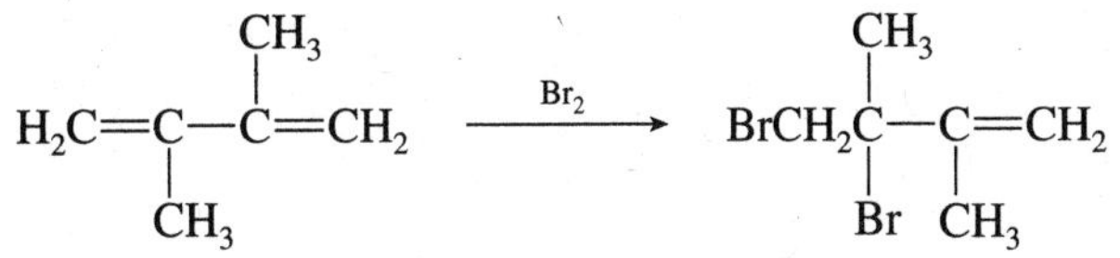

(*e*) The conjugate addition product will be 1,4-dibromo-2,3-dimethyl-2-butene.

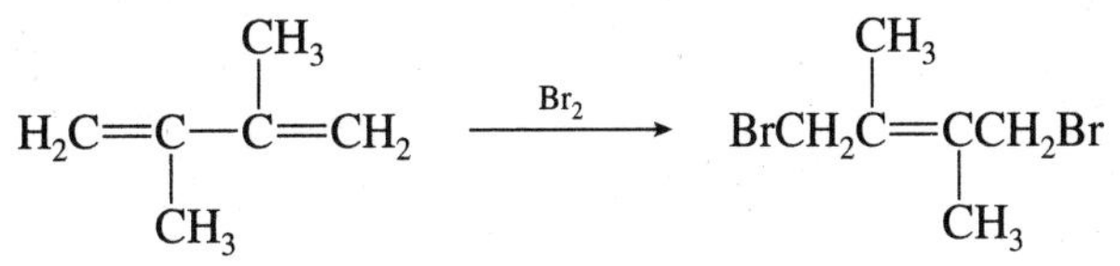

(*f*) Bromination of both double bonds will lead to 1,2,3,4-tetrabromo-2,3-dimethylbutane irrespective of whether the first addition step occurs by direct or conjugate addition.

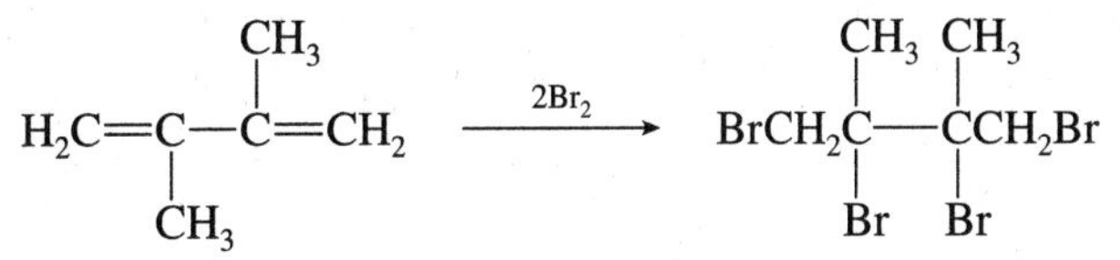

(*g*) The reaction of a diene with maleic anhydride is a Diels–Alder reaction.

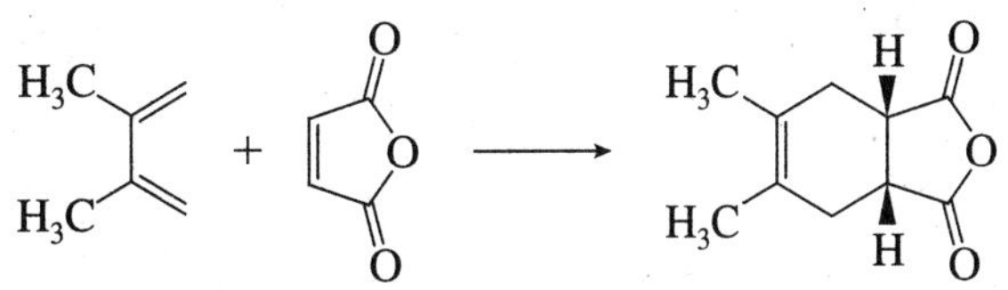

**10.25** The starting material in all cases is 1,3-cyclohexadiene.

(*a*) Cyclohexane will be the product of hydrogenation of 1,3-cyclohexadiene:

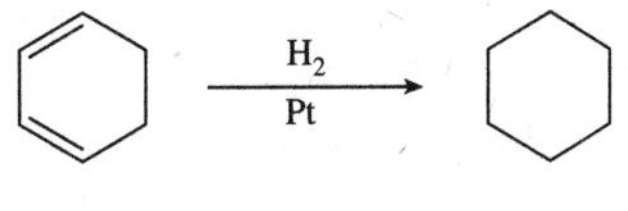

(*b*) Direct addition will occur according to Markovnikov's rule to give 3-chlorocyclohexene

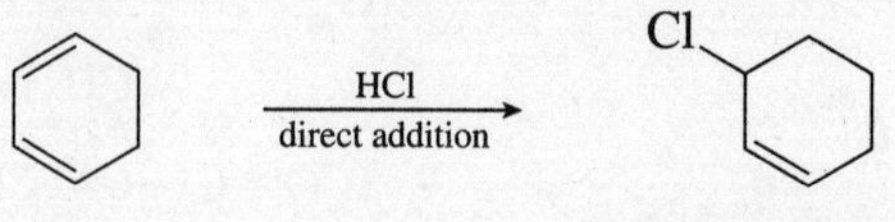

HCl, direct addition

3-Chlorocyclohexene

(*c*) The product of conjugate addition is 3-chlorocyclohexene also. Direct addition and conjugate addition of hydrogen chloride to 1,3-cyclohexadiene give the same product.

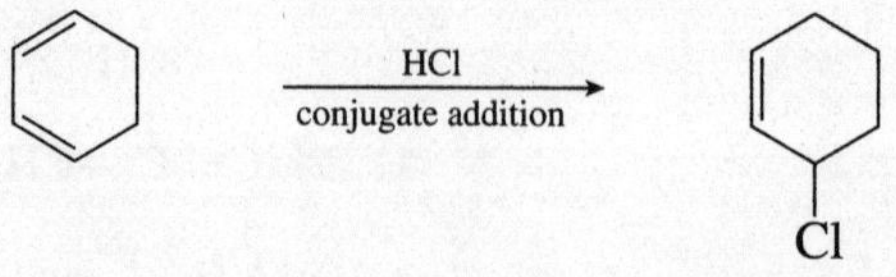

HCl, conjugate addition

3-Chlorocyclohexene

(*d*) Bromine can add directly to one of the double bonds to give 3,4-dibromocyclohexene:

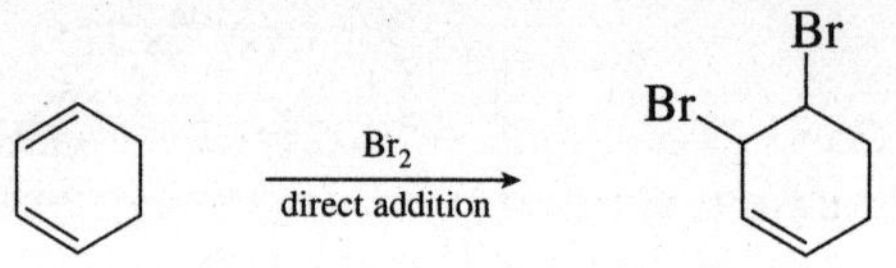

$Br_2$, direct addition

3,4-Dibromocyclohexene

(*e*) Conjugate addition of bromine will give 3,6-dibromocyclohexene:

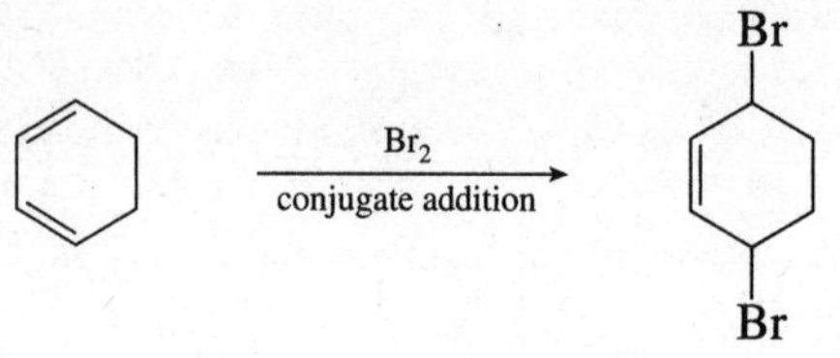

$Br_2$, conjugate addition

3,6-Dibromocyclohexene

(*f*) Addition of 2 moles of bromine will yield 1,2,3,4-tetrabromocyclohexane.

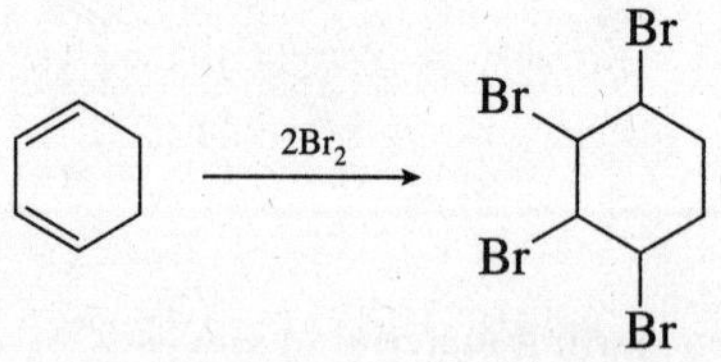

$2Br_2$

(*g*) The constitution of the Diels–Alder adduct of 1,3-cyclohexadiene and maleic anhydride will have a bicyclo [2.2.2]octyl carbon skeleton.

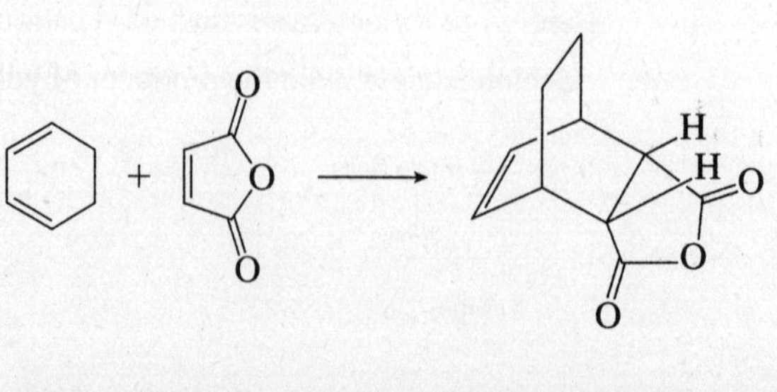

**10.26** Bond formation takes place at the end of the diene system to give a bridged bicyclic ring system.

$$\text{1,3-Cyclohexadiene} + CH_3O_2C{-}C{\equiv}C{-}CO_2CH_3 \longrightarrow \text{Dimethyl bicyclo[2.2.2]octa-2,5-diene-2,3-dicarboxylate}$$

1,3-Cyclohexadiene — Dimethyl acetylenedicarboxylate — Dimethyl bicyclo[2.2.2]octa-2,5-diene-2,3-dicarboxylate

**10.27** The two Diels–Alder adducts formed in the reaction of 1,3-pentadiene with acrolein arise by the two alignments shown:

1,3-pentadiene ($CH_3$) + acrolein (CHO) ⟶ 3-Methylcyclohexene-4-carboxaldehyde

and

1,3-pentadiene ($CH_3$) + acrolein (CHO) ⟶ 3-Methylcyclohexene-5-carboxaldehyde

**10.28** Compound B arises by way of a Diels–Alder reaction between compound A and dimethyl acetylenedicarboxylate. Compound A must therefore have a conjugated diene system.

$$\text{Compound A } (H_2C{=}, {=}CH_2, {=}CH_2) + CH_3O_2CC{\equiv}CCO_2CH_3 \longrightarrow \text{Compound B } (H_2C{=}, CO_2CH_3, CO_2CH_3)$$

Compound A — Compound B

**10.29** The reaction is a nucleophilic substitution in which the nucleophile ($C_6H_5S^-$) becomes attached to the carbon that bore the chloride leaving group. Allylic rearrangement is not observed; therefore, it is reasonable to conclude that an allylic carbocation is *not* involved. The mechanism is $S_N2$.

$$CH_3CH{=}CHCH_2Cl + C_6H_5SNa \xrightarrow{\text{ethanol}} CH_3CH{=}CHCH_2S{-}C_6H_5$$

1-Chloro-2-butene — Sodium benzenethiolate — 2-Butenyl phenyl sulfide

**10.30** (*a*) Solvolysis of $(CH_3)_2C{=}CHCH_2Cl$ in ethanol proceeds by an $S_N1$ mechanism and involves a carbocation intermediate.

$$(CH_3)_2C{=}CHCH_2Cl \xrightarrow{-Cl^-} \boxed{(CH_3)_2C{=}CH{-}\overset{+}{C}H_2 \longleftrightarrow (CH_3)_2\overset{+}{C}{-}CH{=}CH_2}$$

1-Chloro-3-methyl-2-butene

This carbocation has some of the character of a tertiary carbocation. It is more stable and is therefore formed faster than allyl cation, $CH_2{=}CH\overset{+}{C}H_2$.

(*b*) An allylic carbocation is formed from the alcohol in the presence of an acid catalyst.

$$CH_2{=}CHCH(OH)CH_3 \xrightarrow[H_2O]{H_2SO_4} CH_2{=}CHCH(\overset{+}{O}H_2)CH_3 \longrightarrow CH_2{=}CH\overset{+}{C}HCH_3 + H_2O$$

3-Buten-2-ol

This carbocation is a delocalized one and can be captured at either end of the allylic system by water acting as a nucleophile.

$$\left[ CH_2{=}CH\overset{+}{C}HCH_3 \longleftrightarrow \overset{+}{C}H_2CH{=}CHCH_3 \right] \xrightarrow{H_2O} \begin{cases} CH_2{=}CHCH(\overset{+}{O}H_2)CH_3 \xrightarrow{-H^+} CH_2{=}CHCH(OH)CH_3 \quad \text{3-Buten-2-ol} \\ H_2\overset{+}{O}{-}CH_2CH{=}CHCH_3 \xrightarrow{-H^+} HOCH_2CH{=}CHCH_3 \quad \text{2-Buten-1-ol} \end{cases}$$

(*c*) Hydrogen bromide converts the alcohol to an allylic carbocation. Bromide ion captures this carbocation at either end of the delocalized allylic system.

$$\underset{\text{2-Buten-1-ol}}{CH_3CH{=}CHCH_2OH} \xrightarrow{HBr} CH_3CH{=}CHCH_2{-}\overset{+}{O}H_2 \longrightarrow CH_3CH{=}CH\overset{+}{C}H_2$$

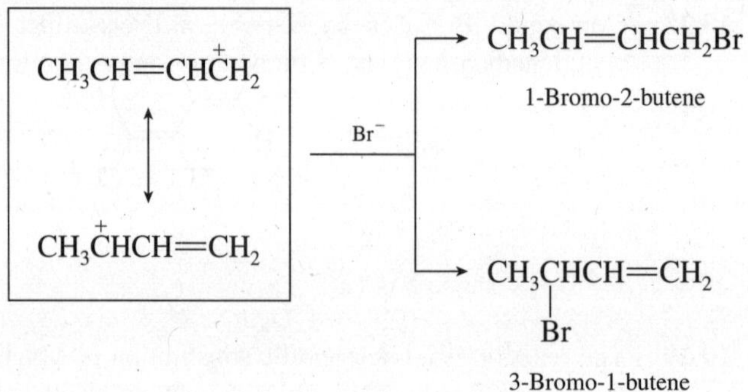

(*d*) The same delocalized carbocation is formed from 3-buten-2-ol as from 2-buten-1-ol.

$$\underset{\text{3-Buten-2-ol}}{CH_3CH(OH)CH{=}CH_2} \xrightarrow{HBr} CH_3\overset{+}{C}HCH{=}CH_2 \longleftrightarrow CH_3CH{=}CH\overset{+}{C}H_2$$

Since this carbocation is the same as the one formed in part (*c*), it gives the same mixture of products when it reacts with bromide.

(*e*) We are told that the major product is 1-bromo-2-butene, not 3-bromo-1-butene.

$$\underset{\substack{\text{1-Bromo-2-butene} \\ \text{(major)}}}{CH_3CH{=}CHCH_2Br} \qquad \underset{\substack{\text{3-Bromo-1-butene} \\ \text{(minor)}}}{CH_3CH(Br)CH{=}CH_2}$$

The major product is the more stable one. It is a primary rather than a secondary halide and contains a more substituted double bond. The reaction is therefore governed by thermodynamic (equilibrium) control.

**10.31** Since both products of reaction of hydrogen chloride with vinylacetylene are chloro-substituted dienes, the first step in addition must involve the triple bond. The carbocation produced is an allylic vinyl cation for which two Lewis structures may be written. Capture of this cation gives the products of 1,2 and 1,4 addition. The 1,2 addition product is more stable because of its conjugated system. The observations of the experiment tell us that the 1,4 addition product is formed faster, although we could not have predicted that.

$$\underset{\text{Vinylacetylene}}{HC\equiv C{-}CH{=}CH_2} \xrightarrow{HCl} \boxed{H_2C{=}\overset{+}{C}{-}CH{=}CH_2 \longleftrightarrow H_2C{=}C{=}CH{-}\overset{+}{C}H_2}$$

$$\downarrow$$

$$\underset{\substack{\text{2-Chloro-1,3-butadiene}\\ \text{(1,2 addition)}}}{H_2C{=}\underset{\substack{|\\ Cl}}{C}{-}CH{=}CH_2} + \underset{\substack{\text{4-Chloro-1,2-butadiene}\\ \text{(1,4 addition)}}}{H_2C{=}C{=}CH{-}CH_2Cl}$$

**10.32** (*a*) The two equilibria are:

**For (*E*)-1,3-pentadiene:**

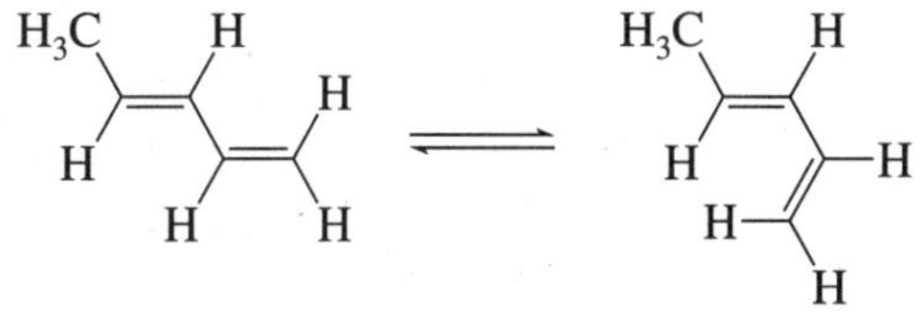
H₃C
H
H
H
H
H
H₃C
H
H
H
H
H

*s*-trans *s*-cis

**For (*Z*)-1,3-pentadiene:**

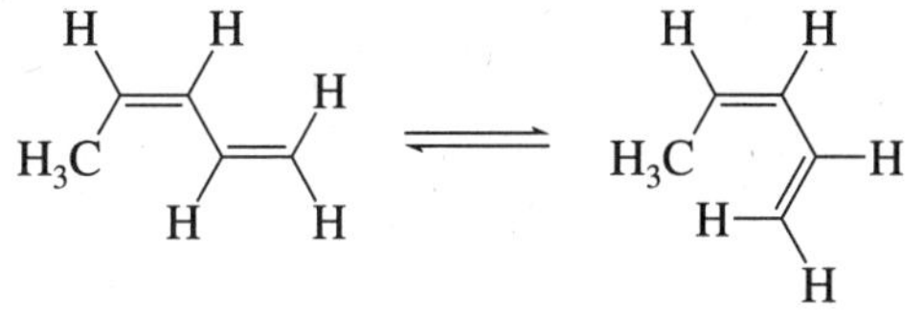
H
H
H
H₃C
H
H
H
H
H
H₃C
H
H

*s*-trans *s*-cis

(*b*) The *s*-cis conformation of (*Z*)-1,3-pentadiene is destabilized by van der Waals strain involving the methyl group.

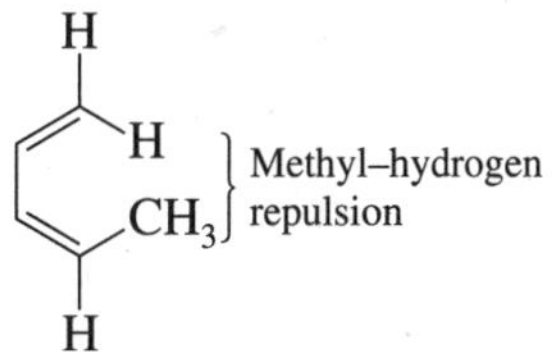
H
H
CH₃
Methyl–hydrogen repulsion
H

*s*-cis conformation of (*Z*)-1,3-pentadiene

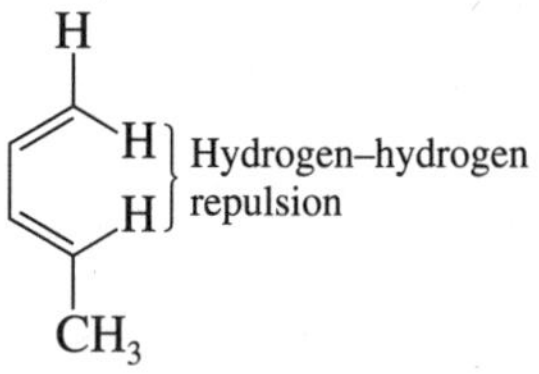
H
H
H
Hydrogen–hydrogen repulsion
CH₃

*s*-cis conformation of (*E*)-1,3-pentadiene

The equilibrium favors the *s*-trans conformation of (*Z*)-1,3-pentadiene more than it does that of the *E* isomer because the *s*-cis conformation of the *Z* isomer has more van der Waals strain.

**10.33** Compare the mirror-image forms of each compound for superposability.

(*a*) 2-Methyl-2,3-hexadiene

$H_3C$, $CH_3$ / C=C=C / $CH_3CH_2$, H — Reference structure

and

$H_3C$, $CH_3$ / C=C=C / H, $CH_2CH_3$ — Mirror image

Rotation of the mirror image 180° around an axis passing through the three carbons of the C=C=C unit demonstrates that the reference structure and its mirror image are superposable.

$H_3C$, $CH_3$ / C=C=C / H, $CH_2CH_3$ — Mirror image

Rotate 180°

$H_3C$, $CH_3$ / C=C=C / $CH_3CH_2$, H — Reoriented mirror image

2-Methyl-2,3-hexadiene is an achiral allene.

(*b*) The two mirror-image forms of 4-methyl-2,3-hexadiene are as shown:

$H_3C$, H / C=C=C / $H_3C$, $CH_2CH_3$ — Reference structure

$H_3C$, H / C=C=C / $CH_3CH_2$, $CH_3$ — Mirror image

Rotate 180°

H, $CH_3$ / C=C=C / $H_3C$, $CH_2CH_3$ — Reoriented mirror image

The two structures cannot be superposed. 4-Methyl-2,3-hexadiene is chiral. Rotation of either representation 180° around an axis that passes through the three carbons of the C=C=C unit leads to superposition of the groups at the "bottom" carbon but not at the "top."

(*c*) 2,4-Dimethyl-2,3-pentadiene is achiral. Its two mirror-image forms are superposable.

$H_3C$, $CH_3$ / C=C=C / $H_3C$, $CH_3$ — Reference structure

$H_3C$, $CH_3$ / C=C=C / $H_3C$, $CH_3$ — Mirror image

The molecule has two planes of symmetry defined by the three carbons of each $CH_3CCH_3$ unit.

**10.34** (*a*) Carbons 2 and 3 of 1,2,3-butatriene are *sp*-hybridized, and the bonding is an extended version of that seen in allene. Allene is nonplanar; its two $CH_2$ units must be in perpendicular planes in order to maximize overlap with the two mutually perpendicular *p* orbitals at C-2. With one more *sp*-hybridized carbon, 1,2,3-butatriene has an "extra turn" in its carbon chain, making the molecule **planar.**

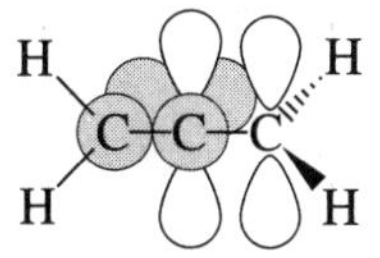

Nonplanar geometry of allene

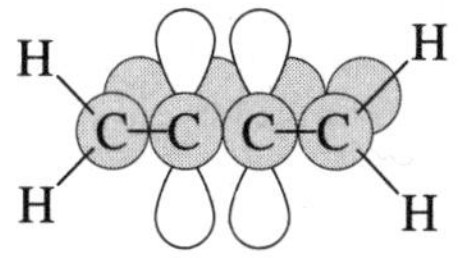

All atoms of 1,2,3-butatriene lie in same plane.

(*b*) The planar geometry of the cumulated triene system leads to the situation where cis and trans stereoisomers are possible for 2,3,4-hexatriene ($CH_3CH{=}C{=}C{=}CHCH_3$). Cis–trans stereoisomers are diastereomers of each other.

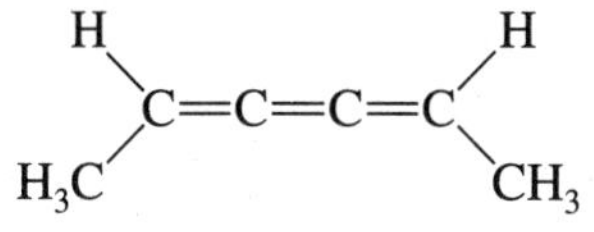

*cis*-2,3,4-Hexatriene

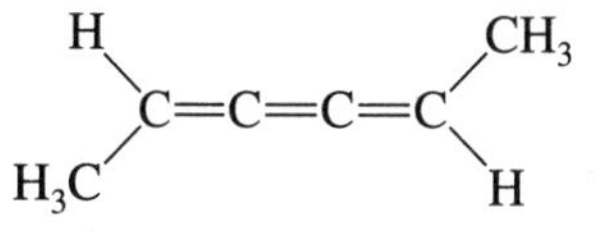

*trans*-2,3,4-Hexatriene

**10.35** Reaction (*a*) is an electrophilic addition of bromine to an alkene; the appropriate reagent is **bromine in carbon tetrachloride.**

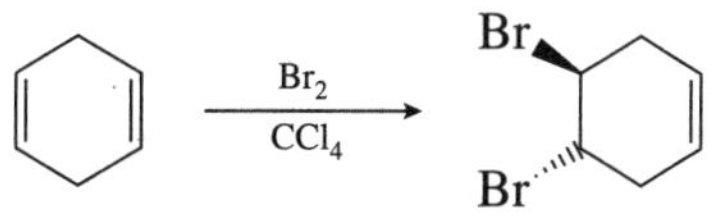

$Br_2$ / $CCl_4$

(74%)

Reaction (*b*) is an epoxidation of an alkene, for which almost any peroxy acid could be used. **Peroxybenzoic acid** was actually used.

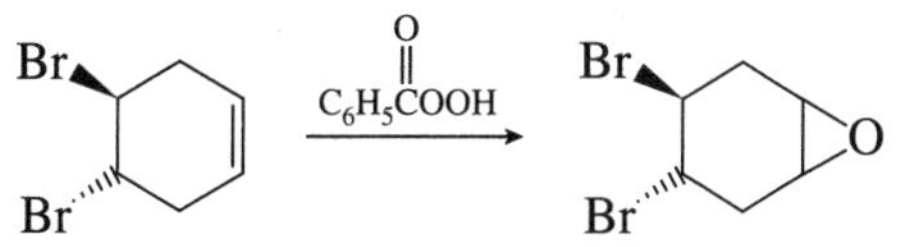

$C_6H_5COOH$ (peroxy acid)

(69%)

Reaction (*c*) is an elimination reaction of a vicinal dibromide to give a conjugated diene and requires E2 conditions. **Sodium methoxide in methanol** was used.

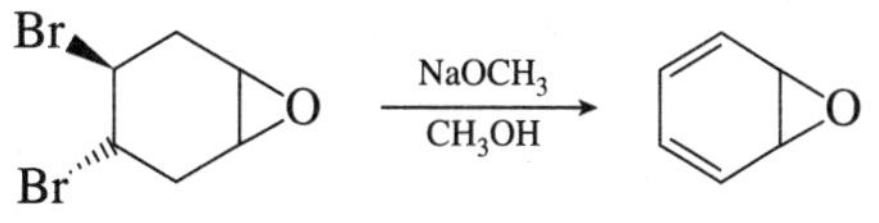

$NaOCH_3$ / $CH_3OH$

(80%)

Reaction (*d*) is a Diels–Alder reaction in which the dienophile is **maleic anhydride.** The dienophile adds from the side opposite that of the epoxide ring.

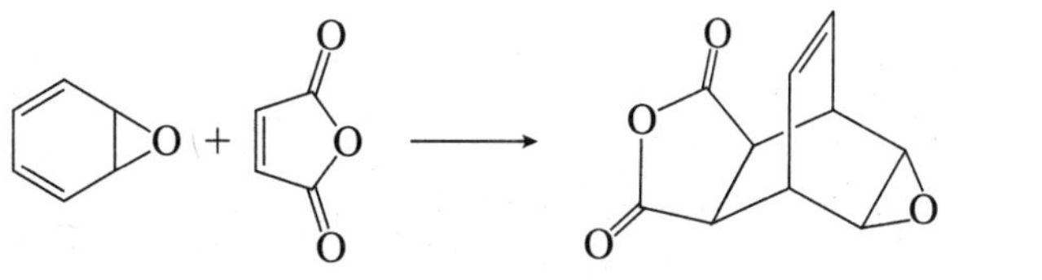

**10.36** To predict the constitution of the Diels–Alder adducts, we can ignore the substituents and simply remember that the fundamental process is

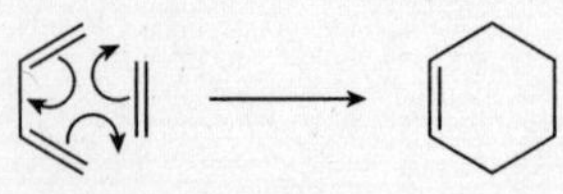

(a) $(CH_3)_3SiO$-diene-$OCH_3$ + $CH_3O_2C{-}C{\equiv}C{-}CO_2CH_3$ ⟶ product with $OCH_3$, $CO_2CH_3$, $CO_2CH_3$, $(CH_3)_3SiO$

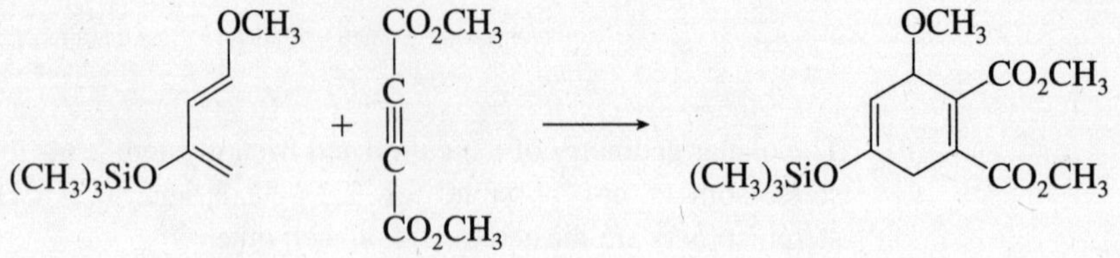

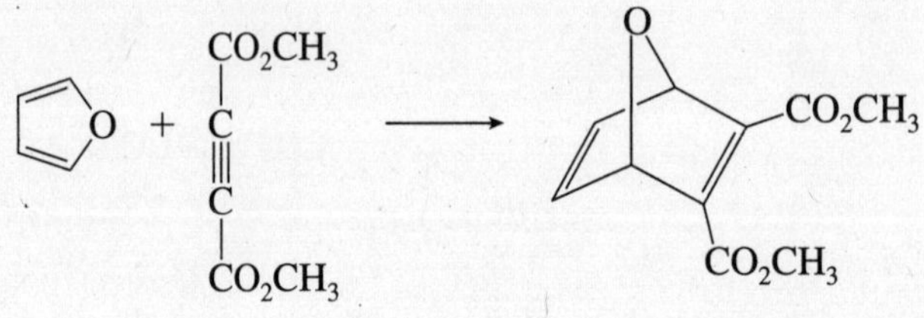

(b) Furan + $CH_3O_2C{-}C{\equiv}C{-}CO_2CH_3$ ⟶ O-bridged adduct with $CO_2CH_3$, $CO_2CH_3$

(c) Cyclopentadiene$-CH_2OCH_3$ + $CH_2{=}CHNO_2$ ⟶ bicyclic adduct with $CH_3OCH_2$, H, $NO_2$

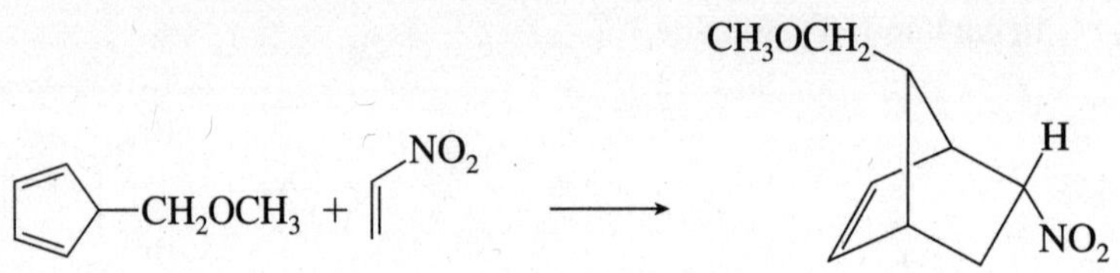

**10.37** The carbon skeleton of dicyclopentadiene must be the same as that of its hydrogenation product, and dicyclopentadiene must contain two double bonds, since 2 mol of hydrogen are consumed in its hydrogenation ($C_{10}H_{12} \longrightarrow C_{10}H_{16}$).

The molecular formula of dicyclopentadiene ($C_{10}H_{12}$) is twice that of 1,3-cyclopentadiene ($C_5H_6$), and its carbon skeleton suggests that 1,3-cyclopentadiene is undergoing a Diels–Alder reaction with itself. Therefore:

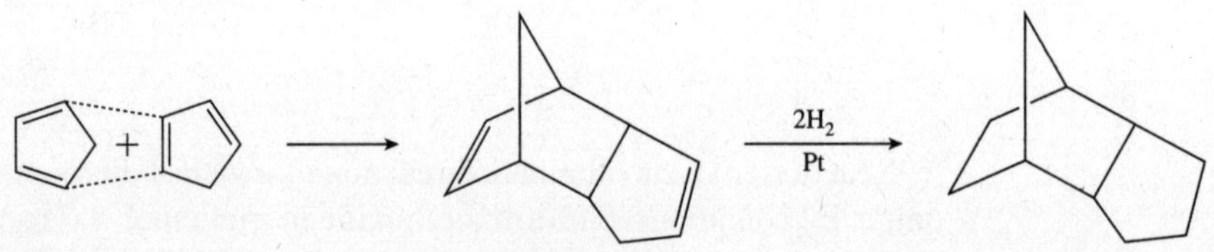

$C_5H_6$ + $C_5H_6$ ⟶ Dicyclopentadiene $C_{10}H_{12}$ $\xrightarrow[\text{Pt}]{2H_2}$ $C_{10}H_{16}$

One molecule of 1,3-cyclopentadiene acts as the diene, and the other acts as the dienophile in this Diels–Alder reaction.

**10.38** (*a*) Since allyl cation is positively charged, examine the process in which electrons "flow" from the HOMO of ethylene to the LUMO of allyl cation.

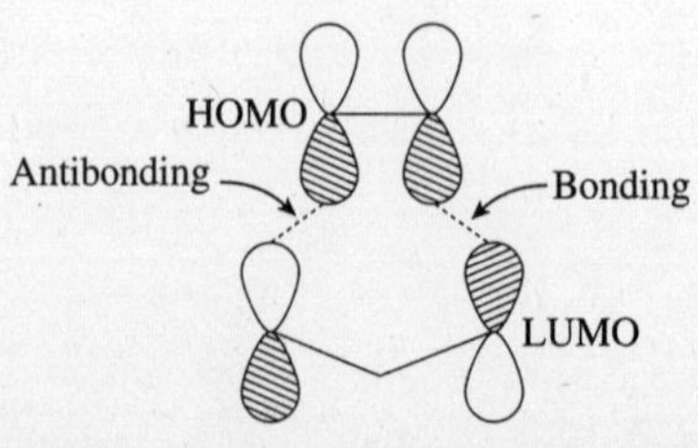

HOMO
Antibonding
Bonding
LUMO

This reaction is forbidden. The symmetries of the orbitals are such that one interaction is bonding and the other is antibonding.

The same answer is obtained if the HOMO of allyl cation and the LUMO of ethylene are examined.

(*b*) In this part of the exercise we consider the LUMO of allyl cation and the HOMO of 1,3-butadiene.

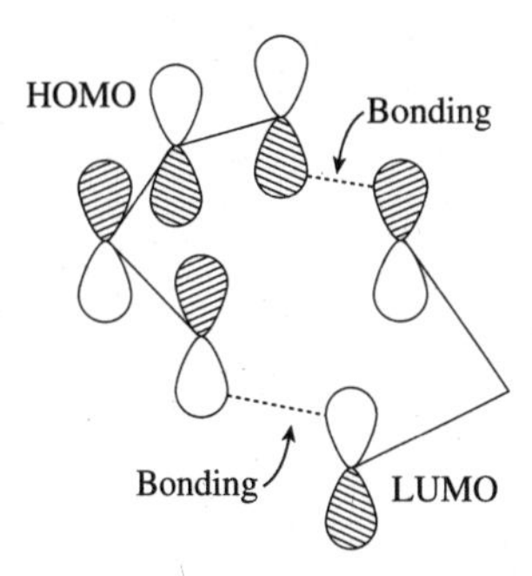


This reaction is allowed by the Woodward–Hoffmann rules. Both interactions are bonding. The same prediction would be arrived at if the HOMO of allyl cation and LUMO of 1,3-butadiene were the orbitals considered.

**10.39** Since oxygen has two unpaired electrons, it can abstract a hydrogen atom from the allylic position of cyclohexene to give a free-radical intermediate.

+ ·Ö—Ö· ⟶ + HÖ—Ö·

The cyclohexenyl radical is resonance-stabilized. It reacts further via the following two propagation steps:

+ ·Ö—Ö· ⟶

+ ⟶ +

**10.40–10.41** Solutions to molecular modeling exercises are not provided in this *Study Guide and Solutions Manual.* You should use *Learning By Modeling* for these exercises.

# SELF-TEST

## PART A

**A-1.** Give the structures of all the constitutionally isomeric alkadienes of molecular formula $C_5H_8$. Indicate which are conjugated and which are allenes.

**A-2.** Provide the IUPAC name for each of the conjugated dienes of the previous problem, *including stereoisomers.*

**A-3.** Hydrolysis of 3-bromo-3-methylcyclohexene yields two isomeric alcohols. Draw their structures and the structure of the intermediate that leads to their formation.

**A-4.** Give the chemical structure of the reactant, reagent, or product omitted from each of the following:

(*a*) $CH_3CH{=}CHCH{=}CHCH_3 \xrightarrow{Br_2}$ ? (two products)

(*b*) $CH_2{=}CHCH{=}CH_2 \xrightarrow{HCl\ (1\ mol)}$ ? (two products)

(*c*) ? $\xrightarrow{\text{Diels–Alder}}$ [structure]

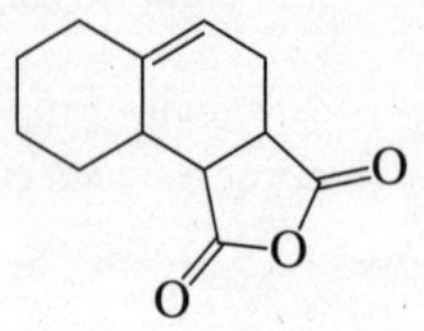

(*d*) [structure] $\xrightarrow{?}$ [structure with Br] + [structure with Br]

(*e*) [cyclopentadiene] + $CH_3O\overset{O}{\overset{\|}{C}}C{\equiv}C\overset{O}{\overset{\|}{C}}OCH_3 \longrightarrow$ ?

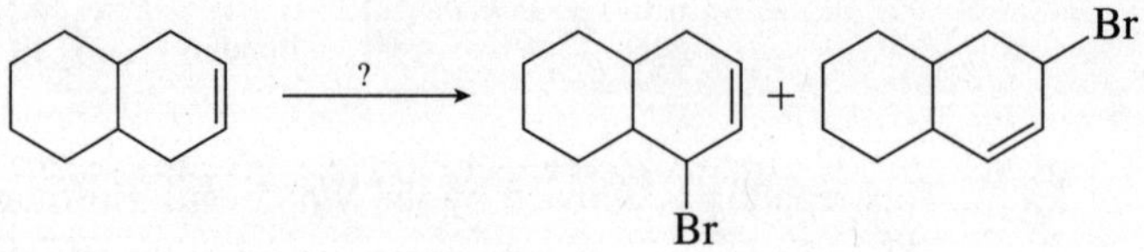

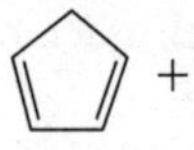

**A-5.** One of the isomeric conjugated dienes having the formula $C_6H_8$ is not able to react with a dienophile in a Diels–Alder reaction. Draw the structure of this compound.

**A-6.** Draw the structure of the carbocation formed on ionization of the compound shown. A constitutional isomer of this compound gives the same carbocation; draw its structure.

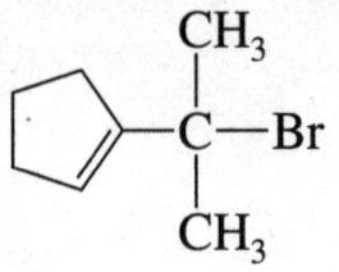

[structure: cyclopentenyl–C($CH_3$)$_2$–Br]

**A-7.** Give the structures of compounds A and B in the following reaction scheme.

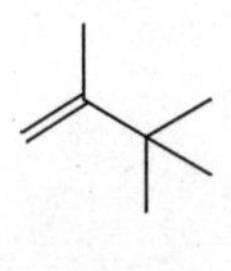

[structure] $\xrightarrow[CCl_4,\ heat]{NBS}$ Compound A $\xrightarrow[acetone]{NaI}$ Compound B

**A-8.** Give the reagents necessary to carry out the following conversion. Note that more than one reaction step is necessary.

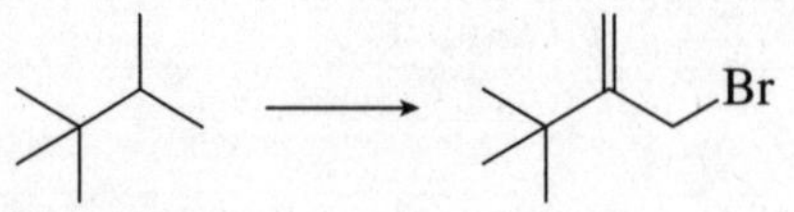

[structure] $\longrightarrow$ [structure with Br]

## PART B

**B-1.** 2,3-Pentadiene, $CH_3CH{=}C{=}CHCH_3$, is

(*a*) A planar substance
(*b*) An allene
(*c*) A conjugated diene
(*d*) A substance capable of cis-trans isomerism

**B-2.** Rank the following carbocations in order of increasing stability (least → most):

$$\underset{1}{CH_3\overset{+}{C}HCH_3} \qquad \underset{2}{CH_3\overset{+}{C}HCH{=}CHCH_3} \qquad \underset{3}{(CH_3)_3C\overset{+}{C}H_2}$$

(*a*) 1 < 2 < 3 (*c*) 3 < 1 < 2
(*b*) 2 < 3 < 1 (*d*) 2 < 1 < 3

**B-3.** Hydrogenation of cyclohexene releases 120 kJ/mol (28.6 kcal/mol) of heat. Which of the following most likely represents the observed heat of hydrogenation of 1,3-cyclohexadiene?

(*a*) 232 kJ/mol (55.4 kcal/mol)
(*b*) 240 kJ/mol (57.2 kcal/mol)
(*c*) 247 kJ/mol (59.0 kcal/mol)
(*d*) 120 kJ/mol (28.6 kcal/mol)

**B-4.** Which of the following compounds give the *same* carbocation on ionization?

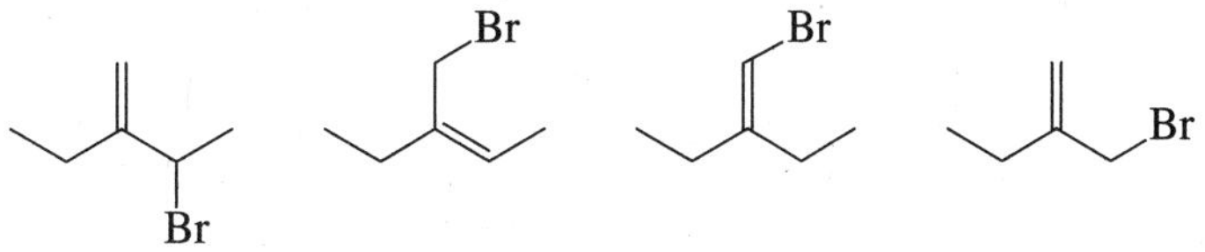

Br Br Br Br

1 2 3 4

(*a*) 1 and 3 (*c*) 1 and 2
(*b*) 2 and 4 (*d*) 1 and 4

**B-5.** For the following reactions the major products are shown:

$$CH_2{=}CH{-}CH{=}CH_2 \xrightarrow[0°C]{HBr} CH_2{=}CH\underset{Br}{\underset{|}{C}}HCH_3 \xrightarrow{+25°C} \underset{Br}{\underset{|}{C}}H_2CH{=}CHCH_3$$

These provide an example of $\underset{1}{\underline{\qquad}}$ control at low temperature and $\underset{2}{\underline{\qquad}}$ control at higher temperature.

| | 1 | 2 |
|---|---|---|
| (*a*) | kinetic | thermodynamic |
| (*b*) | thermodynamic | kinetic |
| (*c*) | kinetic | kinetic |
| (*d*) | thermodynamic | thermodynamic |

**B-6.** Which of the following C—H bonds would have the smallest bond dissociation energy?

(*a*) $CH_3\overset{H}{\overset{|\leftarrow}{C}}HCH_3$ (*c*) $CH_3CH_2\overset{H}{\overset{|\leftarrow}{C}}H_2$

(*b*) $CH_3CH{=}\overset{H}{\overset{|\leftarrow}{C}}H$ (*d*) $H_2C{=}CH{-}\overset{H}{\overset{|\leftarrow}{C}}HCH_3$

**B-7.** Which of the following compounds would undergo solvolysis ($S_N1$) most rapidly in aqueous ethanol?

(*a*) $CH_3$ Br (*c*) $CH_3$ Br

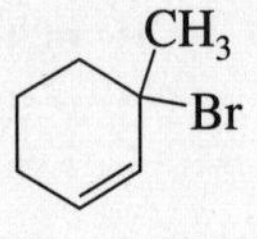

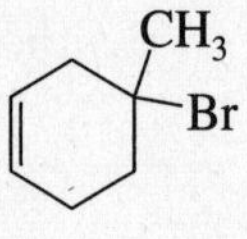

(*b*) $CH_3$ Br (*d*) Br $CH_3$

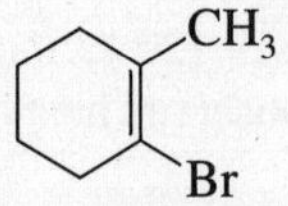

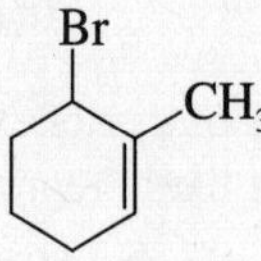

**B-8.** What is the product of 1,4-addition in the reaction shown?

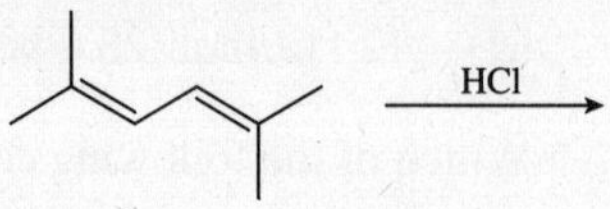

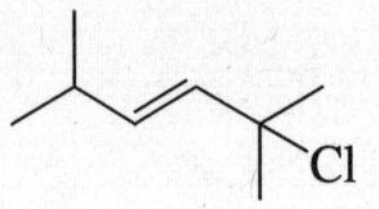

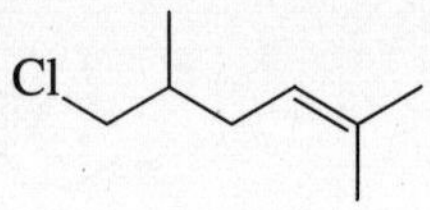

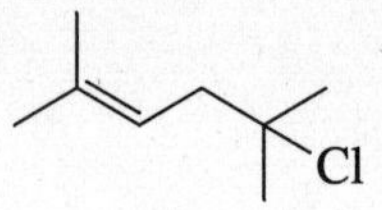

HCl

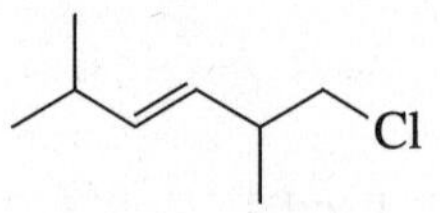

(*a*) Cl (*c*) Cl (*e*) Cl

(*b*) Cl (*d*) Cl

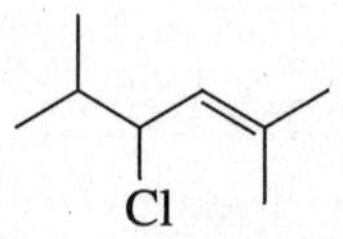

**B-9.** Which of the following compounds will undergo hydrolysis ($S_N1$) to give a mixture of two alcohols that are constitutional isomers?

(*a*) Cl (*c*) Cl

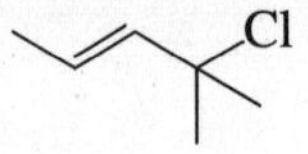

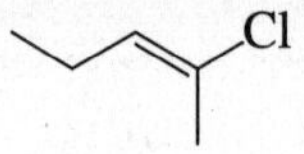

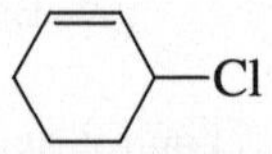

(*b*) Cl (*d*) Cl

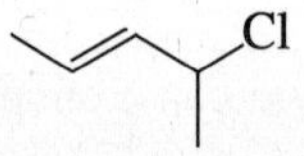

**B-10.** What hydrocarbon reacts with the compound shown (on heating) to give the indicated product?

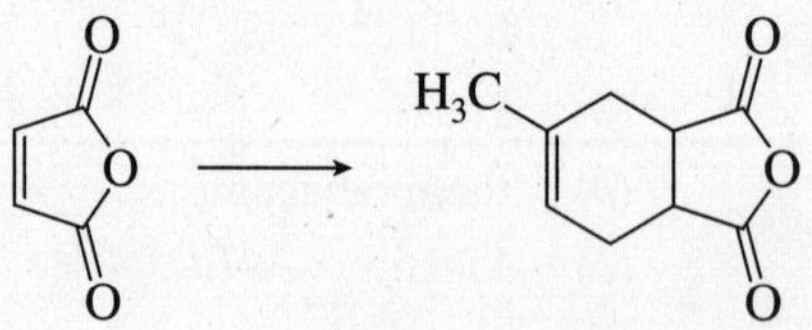

O O O → $H_3C$ O O O

(*a*) 2-Methyl-1-butene
(*b*) 2-Methyl-2-butene
(*c*) 3-Methyl-1-butyne
(*d*) 2-Methyl-1,3-butadiene
(*e*) 1,3-Pentadiene

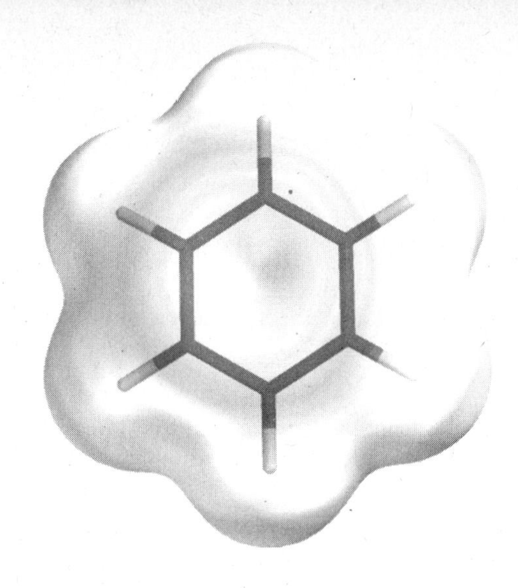

# CHAPTER 11

## ARENES AND AROMATICITY

## SOLUTIONS TO TEXT PROBLEMS

**11.1** Toluene is $C_6H_5CH_3$; it has a methyl group attached to a benzene ring.

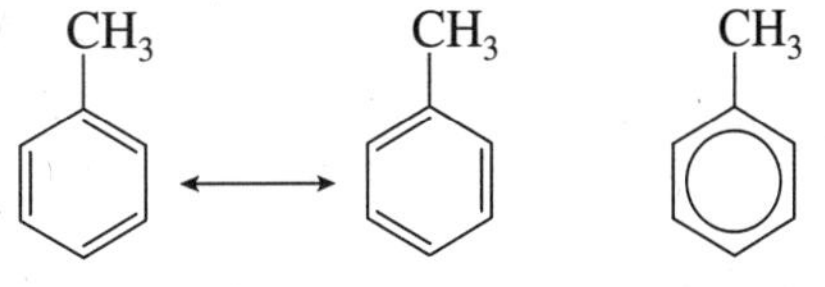

$CH_3$ $CH_3$ $CH_3$

Kekulé forms of toluene — Robinson symbol for toluene

Benzoic acid has a —$CO_2H$ substituent on the benzene ring.

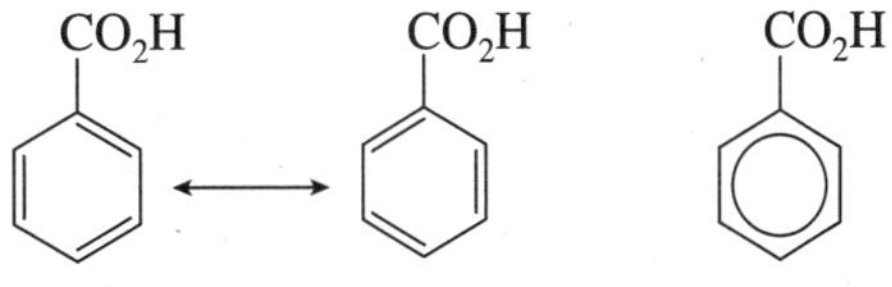

$CO_2H$ $CO_2H$ $CO_2H$

Kekulé forms of benzoic acid — Robinson symbol for benzoic acid

**11.2** Given

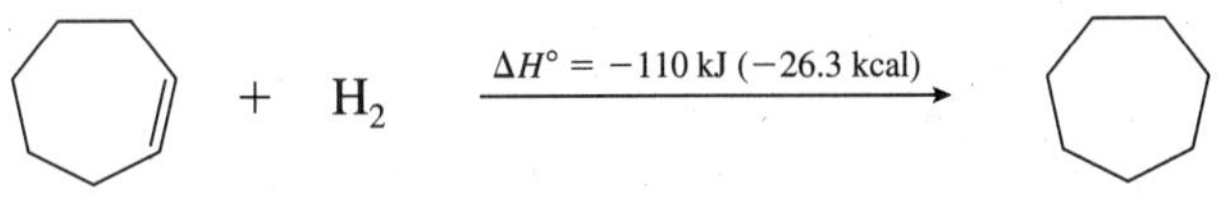

Cycloheptene + $H_2$ $\xrightarrow{\Delta H° = -110 \text{ kJ } (-26.3 \text{ kcal})}$ Cycloheptane

and assuming that there is no resonance stabilization in 1,3,5-cycloheptatriene, we predict that its heat of hydrogenation will be three times that of cycloheptene or 330 kJ/mol (78.9 kcal/mol).

The measured heat of hydrogenation is

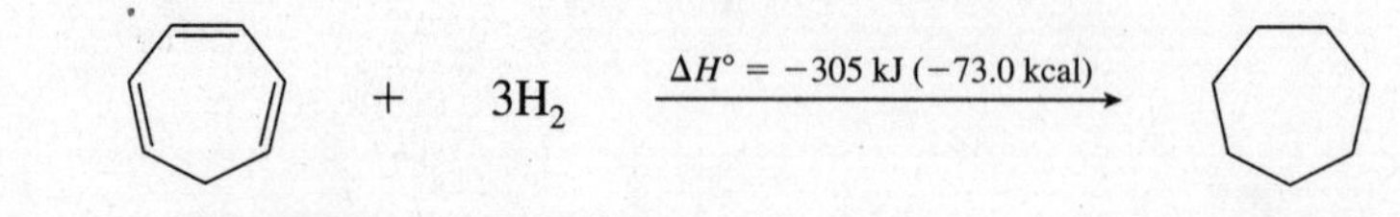

$$\text{1,3,5-Cycloheptatriene} + 3H_2 \xrightarrow{\Delta H° = -305\ \text{kJ}\ (-73.0\ \text{kcal})} \text{Cycloheptane}$$

1,3,5-Cycloheptatriene

Cycloheptane

Therefore

Resonance energy = 330 kJ/mol (predicted for no delocalization) − 305 kJ/mol (observed)

= 25 kJ/mol (5.9 kcal/mol)

The value given in the text for the resonance energy of benzene (152 kJ/mol) is six times larger than this. 1,3,5-Cycloheptatriene is *not* aromatic.

**11.3** (*b*) The parent compound is styrene, $C_6H_5CH{=}CH_2$. The desired compound has a chlorine in the meta position.

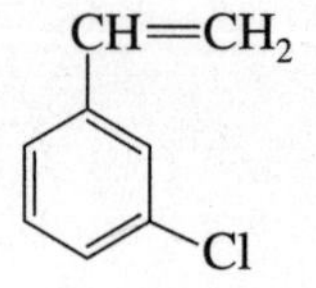

*m*-Chlorostyrene

(*c*) The parent compound is aniline, $C_6H_5NH_2$. *p*-Nitroaniline is therefore

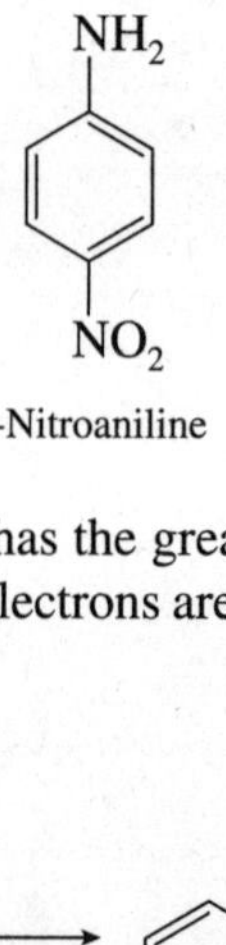

*p*-Nitroaniline

**11.4** The most stable resonance form is the one that has the greatest number of rings that correspond to Kekulé formulations of benzene. For chrysene, electrons are moved in pairs from the structure given to generate a more stable one:

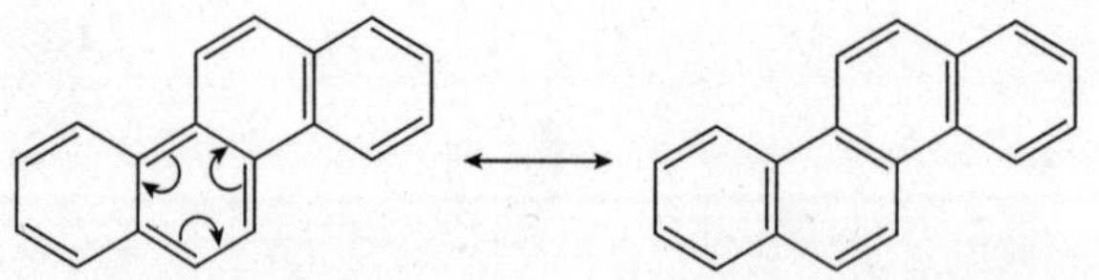

Less stable: two rings have benzene bonding pattern.

More stable: four rings have benzene bonding pattern.

**11.5** Birch reductions of monosubstituted arenes yield 1,4-cyclohexadiene derivatives in which the alkyl group is a substituent on the double bond. With *p*-xylene, both methyl groups are double-bond substituents in the product.

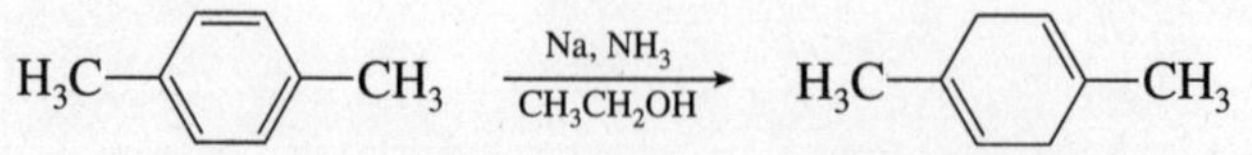

$$\text{H}_3\text{C}\text{–}\text{C}_6\text{H}_4\text{–}\text{CH}_3 \xrightarrow[\text{CH}_3\text{CH}_2\text{OH}]{\text{Na, NH}_3} \text{H}_3\text{C}\text{–}\text{C}_6\text{H}_6\text{–}\text{CH}_3$$

*p*-Xylene

1,4-Dimethyl-1,4-cyclohexadiene

**11.6** (*b*) Only the benzylic hydrogen is replaced by bromine in the reaction of 4-methyl-3-nitroanisole with *N*-bromosuccinimide.

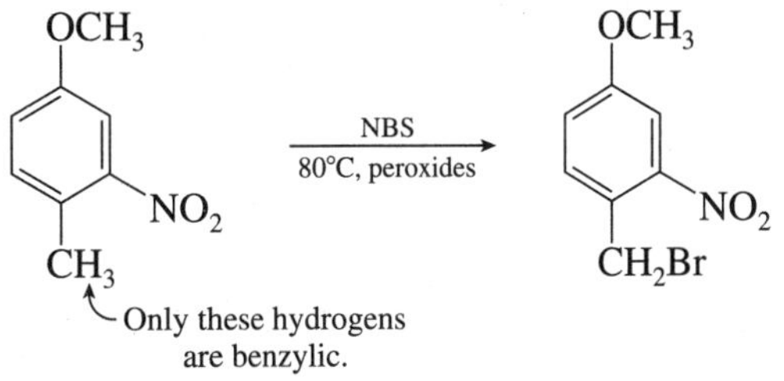

$OCH_3$ $NO_2$ $CH_3$ $\xrightarrow[\text{80°C, peroxides}]{\text{NBS}}$ $OCH_3$ $NO_2$ $CH_2Br$

Only these hydrogens are benzylic.

**11.7** The molecular formula of the product is $C_{12}H_{14}O_4$. Since it contains four oxygens, the product must have two —$CO_2H$ groups. None of the hydrogens of a *tert*-butyl substituent on a benzene ring is benzylic, and so this group is inert to oxidation. Only the benzylic methyl groups of 4-*tert*-butyl-1,2-dimethylbenzene are susceptible to oxidation; therefore, the product is 4-*tert*-butylbenzene-1,2-dicarboxylic acid.

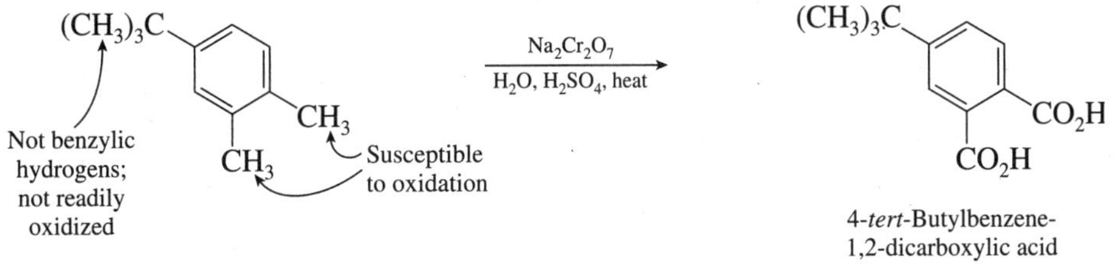

$(CH_3)_3C$ $CH_3$ $CH_3$ $\xrightarrow[\text{H}_2\text{O, H}_2\text{SO}_4\text{, heat}]{\text{Na}_2\text{Cr}_2\text{O}_7}$ $(CH_3)_3C$ $CO_2H$ $CO_2H$

Not benzylic hydrogens; not readily oxidized

Susceptible to oxidation

4-*tert*-Butylbenzene-1,2-dicarboxylic acid

**11.8** Each of these reactions involves nucleophilic substitution of the $S_N2$ type at the benzylic position of benzyl bromide.

(*b*) $(CH_3)_3C\ddot{O}:^-$ $CH_2$—Br $\longrightarrow$ $CH_2OC(CH_3)_3$

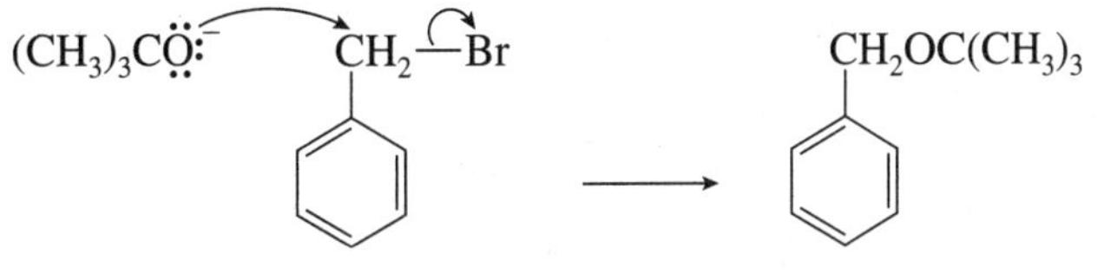

*tert*-Butoxide ion    Benzyl bromide    Benzyl *tert*-butyl ether

(*c*) $^-:\ddot{N}=\overset{+}{N}=\ddot{N}:^-$ $CH_2$—Br $\longrightarrow$ $CH_2\ddot{N}=\overset{+}{N}=\ddot{N}:^-$

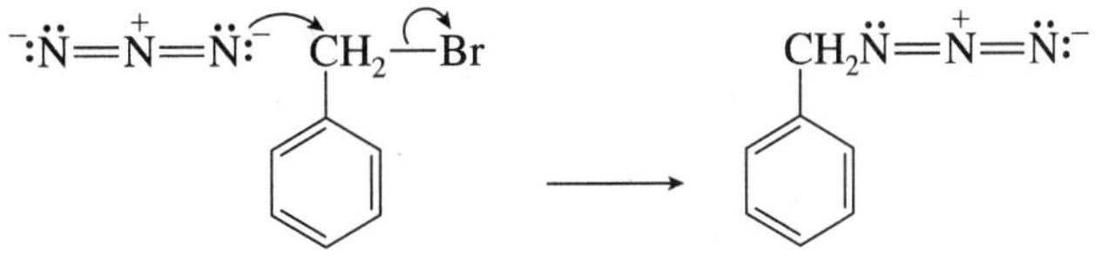

Azide ion    Benzyl bromide    Benzyl azide

(*d*) $H\ddot{S}:^-$ $CH_2$—Br $\longrightarrow$ $CH_2SH$

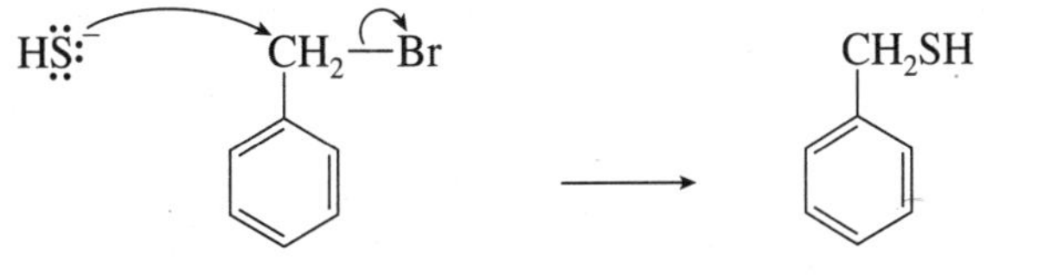

Hydrogen sulfide ion    Benzyl bromide    Phenylmethanethiol

(*e*) $:\ddot{I}:^-$ $CH_2$—Br $\longrightarrow$ $CH_2I$

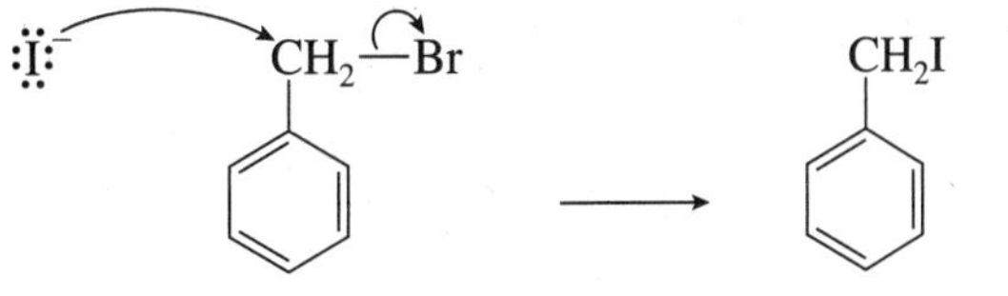

Iodide ion    Benzyl bromide    Benzyl iodide

**11.9** The dihydronaphthalene in which the double bond is conjugated with the aromatic ring is more stable; thus 1,2-dihydronaphthalene has a lower heat of hydrogenation than 1,4-dihydronaphthalene.

1,2-Dihydronaphthalene
Heat of hydrogenation
101 kJ/mol (24.1 kcal/mol)

1,4-Dihydronaphthalene
Heat of hydrogenation
113 kJ/mol (27.1 kcal/mol)

**11.10** (*b*) The regioselectivity of alcohol formation by hydroboration–oxidation is opposite that predicted by Markovnikov's rule.

$C_6H_5C(CH_3){=}CH_2 \xrightarrow[2.\ H_2O_2,\ HO^-]{1.\ B_2H_6} C_6H_5CH(CH_3)CH_2OH$

2-Phenylpropene → 2-Phenyl-1-propanol (92%)

(*c*) Bromine adds to alkenes in aqueous solution to give bromohydrins. A water molecule acts as a nucleophile, attacking the bromonium ion at the carbon that can bear most of the positive charge, which in this case is the benzylic carbon.

$C_6H_5CH{=}CH_2 \xrightarrow[H_2O]{Br_2} C_6H_5CH(OH)CH_2Br$

Styrene → 2-Bromo-1-phenylethanol (82%)

(*d*) Peroxy acids convert alkenes to epoxides.

$C_6H_5CH{=}CH_2 + C_6H_5COOH(\text{peroxy}) \longrightarrow C_6H_5\overset{O}{CH{-}CH_2} + C_6H_5COH(=O)$

Styrene + Peroxybenzoic acid → Epoxystyrene (69–75%) + Benzoic acid

**11.11** Styrene contains a benzene ring and will be appreciably stabilized by resonance, which makes it lower in energy than cyclooctatetraene.

$C_6H_5$—CH=CH$_2$

Structure contains an aromatic ring.

Styrene: heat of combustion 4393 kJ/mol (1050 kcal/mol)

Cyclooctatetraene (not aromatic): heat of combustion 4543 kJ/mol (1086 kcal/mol)

**11.12** The dimerization of cyclobutadiene is a Diels–Alder reaction in which one molecule of cyclobutadiene acts as a diene and the other as a dienophile.

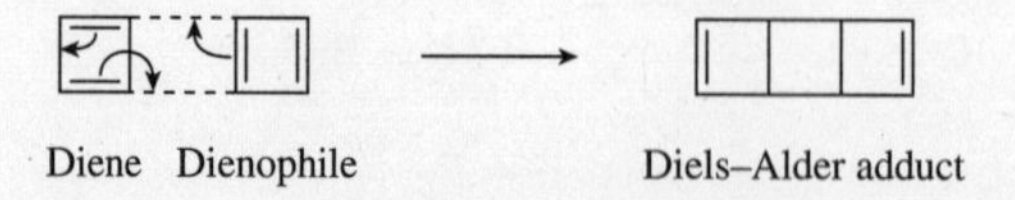

Diene Dienophile Diels–Alder adduct

**11.13** (*b*) Since twelve 2*p* orbitals contribute to the cyclic conjugated system of [12]-annulene, there will be 12$\pi$ molecular orbitals. These MOs are arranged so that one is of highest energy, one is of lowest energy, and the remaining ten are found in pairs between the highest and lowest energy orbitals. There are $12\pi$ electrons, and so the lowest 5 orbitals are each doubly occupied, whereas each of the next 2 orbitals—orbitals of equal energy—is singly occupied.

Antibonding orbitals (5)

Nonbonding orbitals (2)

Bonding orbitals (5)

**11.14** One way to evaluate the relationship between heats of combustion and structure for compounds that are not isomers is to divide the heat of combustion by the number of carbons so that heats of combustion are compared on a "per carbon" basis.

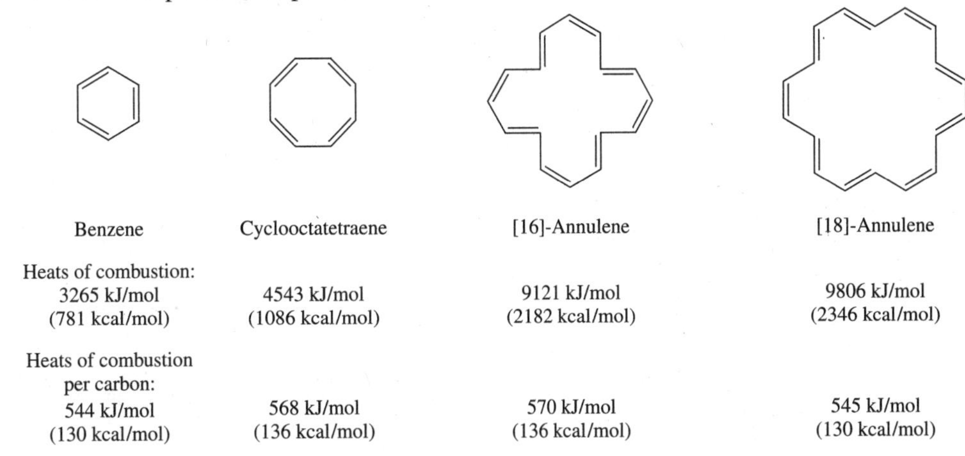

| | Benzene | Cyclooctatetraene | [16]-Annulene | [18]-Annulene |
|---|---|---|---|---|
| Heats of combustion: | 3265 kJ/mol (781 kcal/mol) | 4543 kJ/mol (1086 kcal/mol) | 9121 kJ/mol (2182 kcal/mol) | 9806 kJ/mol (2346 kcal/mol) |
| Heats of combustion per carbon: | 544 kJ/mol (130 kcal/mol) | 568 kJ/mol (136 kcal/mol) | 570 kJ/mol (136 kcal/mol) | 545 kJ/mol (130 kcal/mol) |

As the data indicate (within experimental error), the heats of combustion *per carbon* of the two aromatic hydrocarbons, benzene and [18]-annulene, are equal. Similarly, the heats of combustion per carbon of the two nonaromatic hydrocarbons, cyclooctatetraene and [16]-annulene, are equal. The two aromatic hydrocarbons have heats of combustion per carbon that are less than those of the nonaromatic hydrocarbons. On a per carbon basis, the aromatic hydrocarbons have lower potential energy (are more stable) than the nonaromatic hydrocarbons.

**11.15** The seven resonance forms for tropylium cation (cycloheptatrienyl cation) may be generated by moving $\pi$ electrons in pairs toward the positive charge. The resonance forms are simply a succession of allylic carbocations.

**11.16** Resonance structures are generated for cyclopentadienide anion by moving the unshared electron pair from the carbon to which it is attached to a position where it becomes a shared electron pair in a $\pi$ bond.

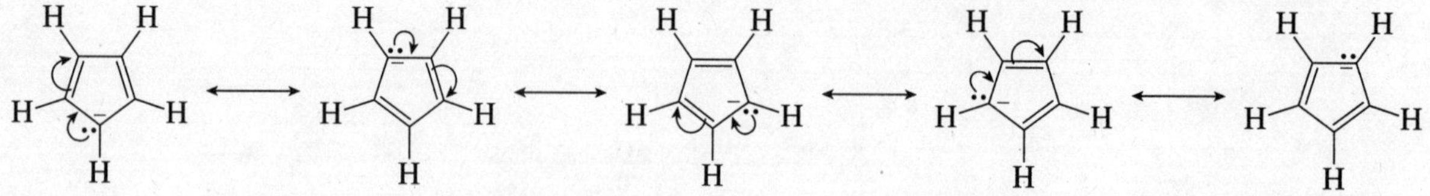

**11.17** The process is an acid–base reaction in which cyclopentadiene transfers a proton to amide ion (the base) to give the aromatic cyclopentadienide anion. The sodium ion ($Na^+$) has been omitted from the equation.

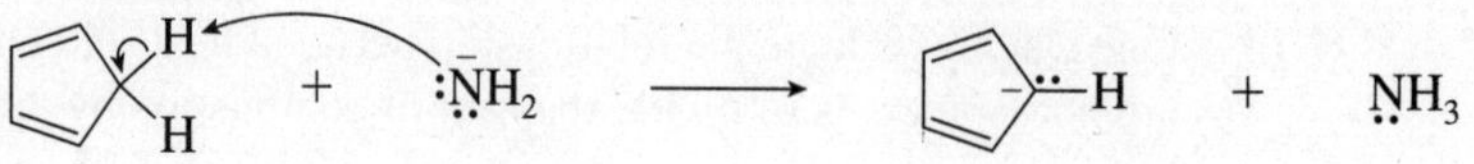

1,3-Cyclopentadiene + Amide ion ⟶ Cyclopentadienide anion + Ammonia

**11.18** (*b*) Cyclononatetraenide anion has 10 $\pi$ electrons; it is aromatic. The 10 $\pi$ electrons are most easily seen by writing a Lewis structure for the anion: there are 2 $\pi$ electrons for each of four double bonds, and the negatively charged carbon contributes 2.

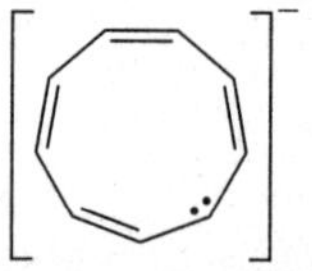

**11.19** Indole is more stable than isoindole. Although the bonding patterns in both five-membered rings are the same, the six-membered ring in indole has a pattern of bonds identical to benzene and so is highly stabilized. The six-membered ring in isoindole is not of the benzene type.

Six-membered ring corresponds to benzene.

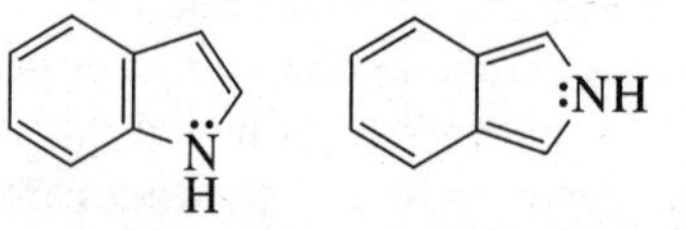

Six-membered ring does not have same pattern of bonds as benzene.

Indole
more stable

Isoindole
less stable

**11.20** The prefix *benz-* in benzimidazole (structure given in text) signifies that a benzene ring is fused to an imidazole ring. By analogy, benzoxazole has a benzene ring fused to oxazole.

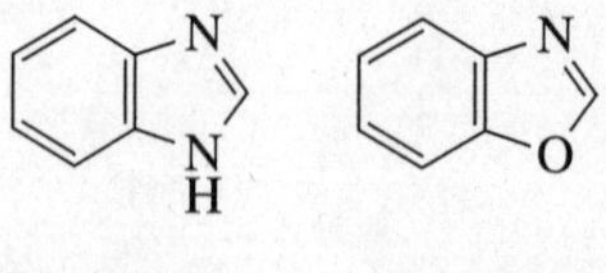

Benzimidazole Benzoxazole

Similarly, benzothiazole has a benzene ring fused to thiazole.

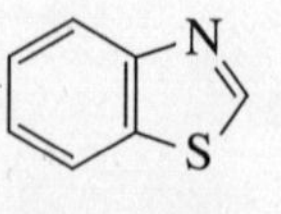

Benzothiazole

**11.21** Write structural formulas for the species formed when a proton is transferred to either of the two nitrogens of imidazole.

**Protonation of N-1:**

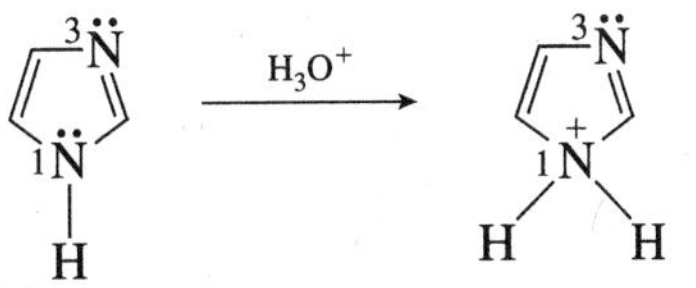

$\xrightarrow{H_3O^+}$

The species formed on protonation of N-1 is not aromatic. The electron pair of N-1 that contributes to the aromatic 6 $\pi$-electron system of imidazole is no longer available for this purpose because it is used to form a covalent bond to the proton in the conjugate acid.

**Protonation of N-3:**

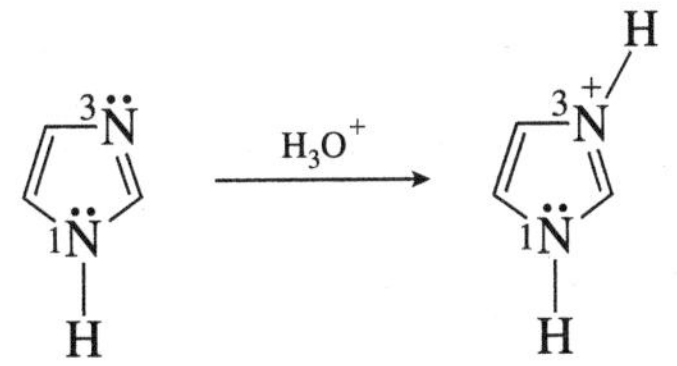

$\xrightarrow{H_3O^+}$

The species formed on protonation of N-3 is aromatic. Electron delocalization represented by the resonance forms shown allows the 6 $\pi$-electron aromatic system of imidazole to be retained in its conjugate acid. The positive charge is shared equally by both nitrogens.

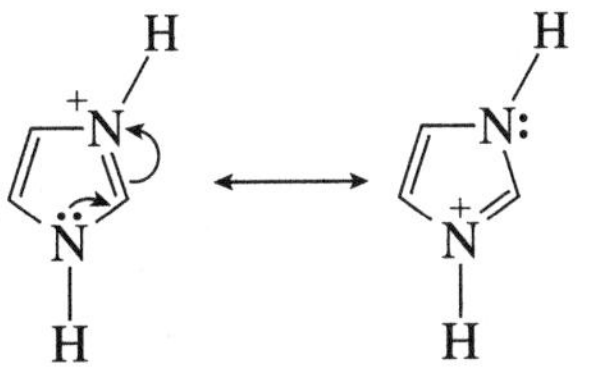

**11.22** Since the problem requires that the benzene ring be monosubstituted, all that needs to be examined are the various isomeric forms of the $C_4H_9$ substituent.

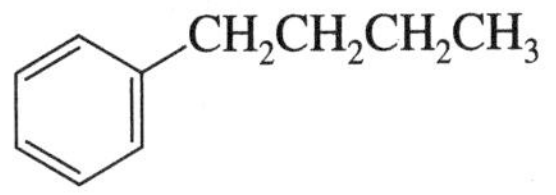

$C_6H_5CH_2CH_2CH_2CH_3$ — Butylbenzene (1-phenylbutane)

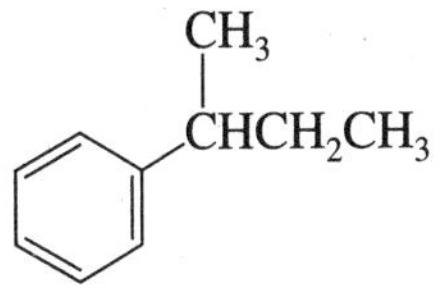

$C_6H_5CH(CH_3)CH_2CH_3$ — *sec*-Butylbenzene (2-phenylbutane)

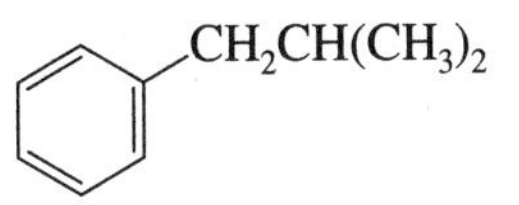

$C_6H_5CH_2CH(CH_3)_2$ — Isobutylbenzene (2-methyl-1-phenylpropane)

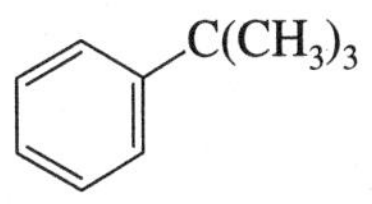

$C_6H_5C(CH_3)_3$ — *tert*-Butylbenzene (2-methyl-2-phenylpropane)

These are the four constitutional isomers. *sec*-Butylbenzene is chiral and so exists in enantiomeric *R* and *S* forms.

**11.23** (*a*) An allyl substituent is $—CH_2CH{=}CH_2$.

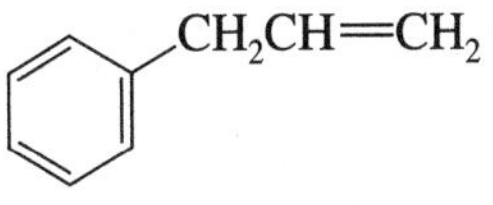

$C_6H_5CH_2CH{=}CH_2$

Allylbenzene

(*b*) The constitution of 1-phenyl-1-butene is $C_6H_5CH{=}CHCH_2CH_3$. The *E* stereoisomer is

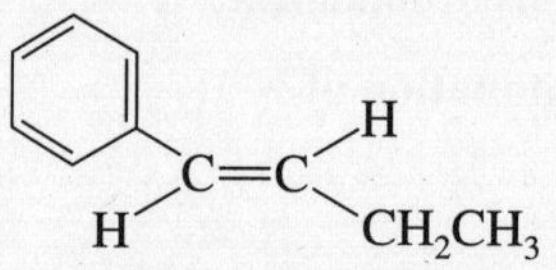

(*E*)-1-Phenyl-1-butene

The two higher ranked substituents, phenyl and ethyl, are on opposite sides of the double bond.

(*c*) The constitution of 2-phenyl-2-butene is $CH_3C(C_6H_5){=}CHCH_3$. The *Z* stereoisomer is

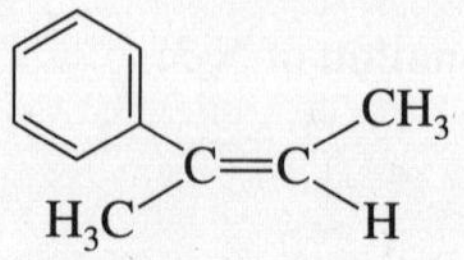

(*Z*)-2-Phenyl-2-butene

The two higher ranked substituents, phenyl and methyl, are on the same side of the double bond.

(*d*) 1-Phenylethanol is chiral and has the constitution $CH_3CH(OH)C_6H_5$. Among the substituents attached to the stereogenic center, the order of decreasing precedence is

$$HO > C_6H_5 > CH_3 > H$$

In the *R* enantiomer the three highest ranked substituents must appear in a clockwise sense in proceeding from higher ranked to next lower ranked when the lowest ranked substituent is directed away from you.

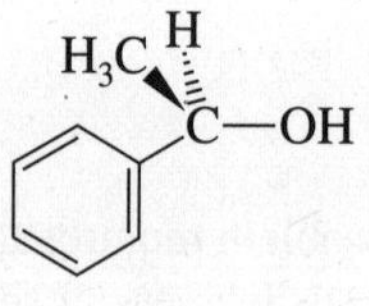

(*R*)-1-Phenylethanol

(*e*) A benzyl group is $C_6H_5CH_2$—. Benzyl alcohol is therefore $C_6H_5CH_2OH$ and *o*-chlorobenzyl alcohol is

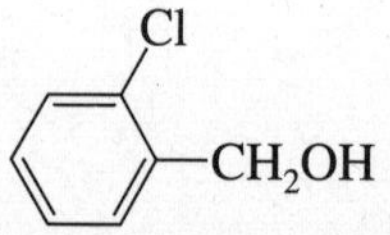

(*f*) In *p*-chlorophenol the benzene ring bears a chlorine and a hydroxyl substituent in a 1,4-substitution pattern.

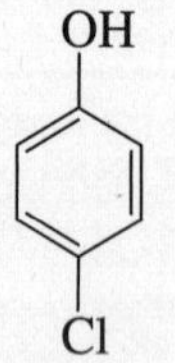

*p*-Chlorophenol

(*g*) Benzenecarboxylic acid is an alternative IUPAC name for benzoic acid.

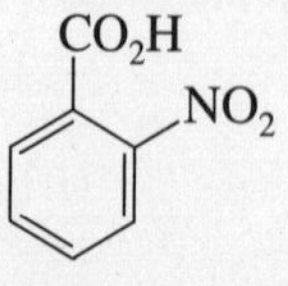

2-Nitrobenzenecarboxylic acid

(*h*) Two isopropyl groups are in a 1,4 relationship in *p*-diisopropylbenzene.

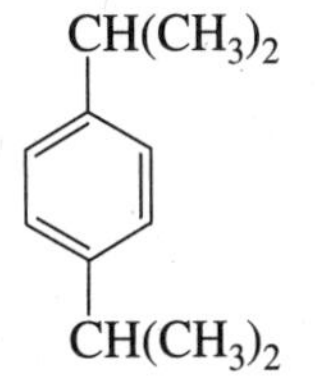

$CH(CH_3)_2$

$CH(CH_3)_2$

*p*-Diisopropylbenzene

(*i*) Aniline is $C_6H_5NH_2$. Therefore

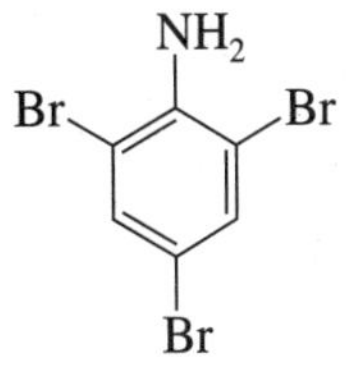

$NH_2$ Br Br Br

2,4,6-Tribromoaniline

(*j*) Acetophenone (from text Table 11.1) is $C_6H_5\overset{O}{\overset{\|}{C}}CH_3$. Therefore

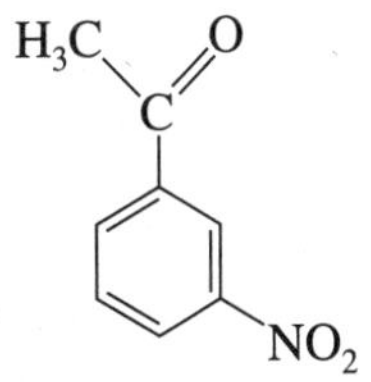

$H_3C$ O C $NO_2$

*m*-Nitroacetophenone

(*k*) Styrene is $C_6H_5CH{=}CH_2$ and numbering of the ring begins at the carbon that bears the side chain.

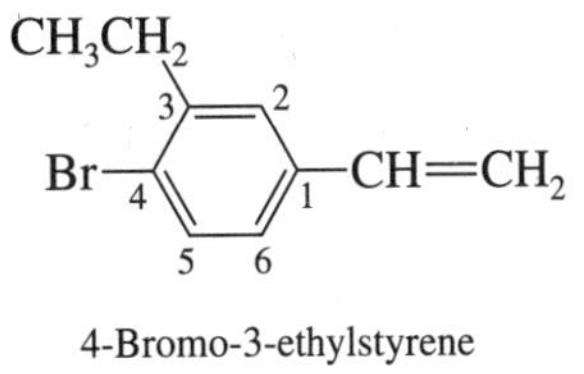

$CH_3CH_2$ Br 4 3 2 1 5 6 $CH{=}CH_2$

4-Bromo-3-ethylstyrene

**11.24** (*a*) Anisole is the name for $C_6H_5OCH_3$, and allyl is an acceptable name for the group $H_2C{=}CHCH_2—$. Number the ring beginning with the carbon that bears the methoxy group.

(*b*) Phenol is the name for $C_6H_5OH$. The ring is numbered beginning at the carbon that bears the hydroxyl group, and the substituents are listed in alphabetical order.

(*c*) Aniline is the name given to $C_6H_5NH_2$. This compound is named as a dimethyl derivative of aniline. Number the ring sequentially beginning with the carbon that bears the amino group.

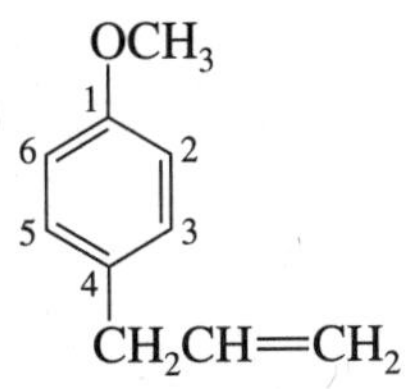

$OCH_3$ 1 2 3 4 5 6 $CH_2CH{=}CH_2$

Estragole
4-Allylanisole

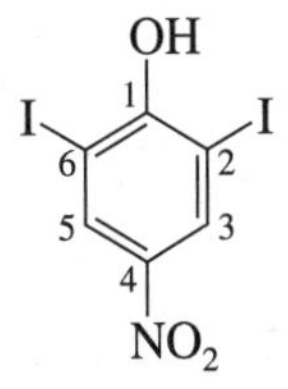

OH I I 1 2 3 4 5 6 $NO_2$

Diosphenol
2,6-Diiodo-4-nitrophenol

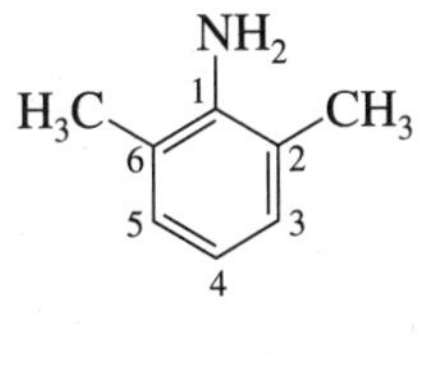

$NH_2$ $H_3C$ $CH_3$ 1 2 3 4 5 6

*m*-Xylidine
2,6-Dimethylaniline

**11.25** (*a*) There are three isomeric nitrotoluenes, because the nitro group can be ortho, meta, or para to the methyl group.

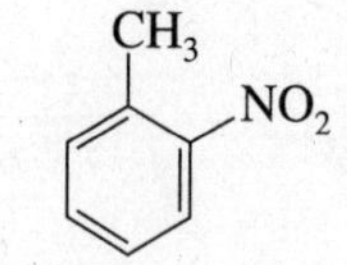

*o*-Nitrotoluene (2-nitrotoluene) 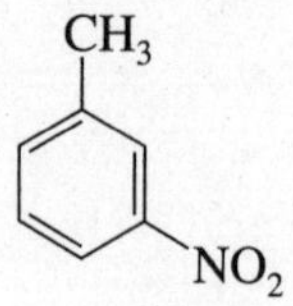 *m*-Nitrotoluene (3-nitrotoluene) 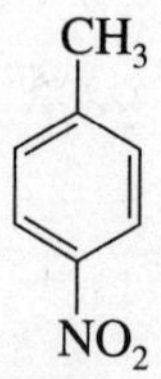 *p*-Nitrotoluene (4-nitrotoluene)

(*b*) Benzoic acid is $C_6H_5CO_2H$. In the isomeric dichlorobenzoic acids, two of the ring hydrogens of benzoic acid have been replaced by chlorines. The isomeric dichlorobenzoic acids are

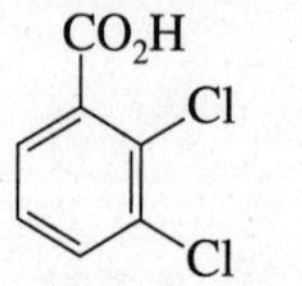

2,3-Dichlorobenzoic acid 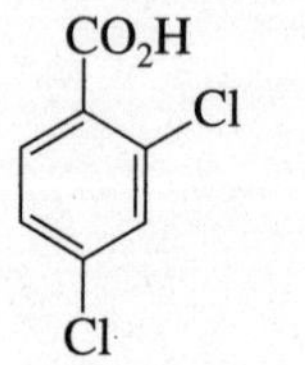 2,4-Dichlorobenzoic acid 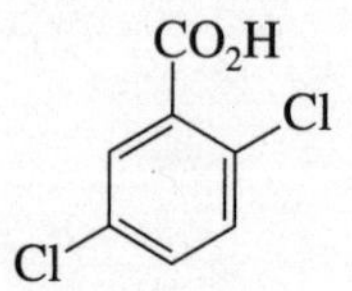 2,5-Dichlorobenzoic acid

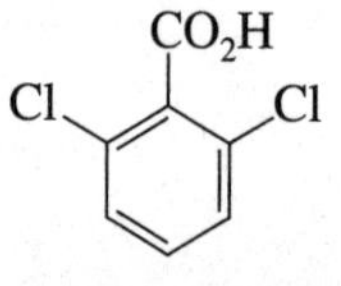

2,6-Dichlorobenzoic acid 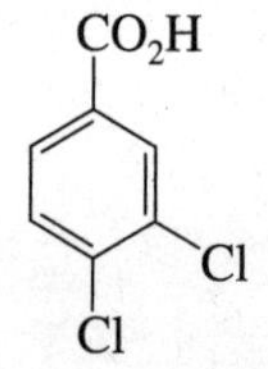 3,4-Dichlorobenzoic acid 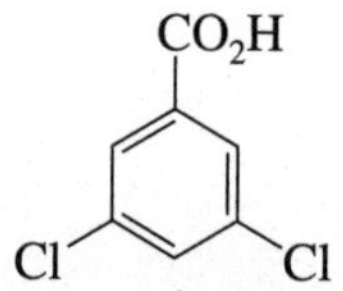 3,5-Dichlorobenzoic acid

The prefixes *o*-, *m*-, and *p*- may not be used in trisubstituted arenes; numerical prefixes are used. Note also that **benzenecarboxylic** may be used in place of **benzoic.**

(*c*) In the various tribromophenols, we are dealing with tetrasubstitution on a benzene ring. Again, *o*-, *m*-, and *p*- are not valid prefixes. The hydroxyl group is assigned position 1 because the base name is phenol.

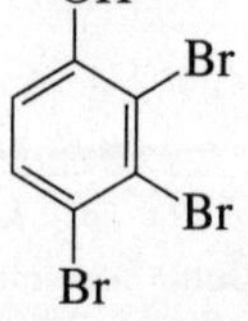

2,3,4-Tribromophenol 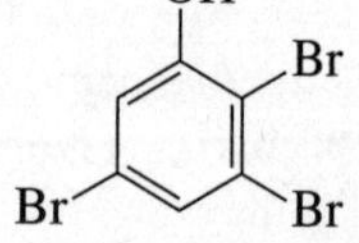 2,3,5-Tribromophenol 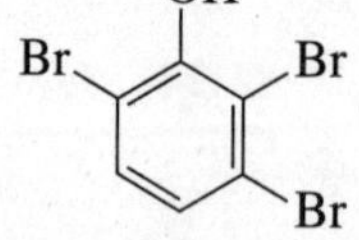 2,3,6-Tribromophenol

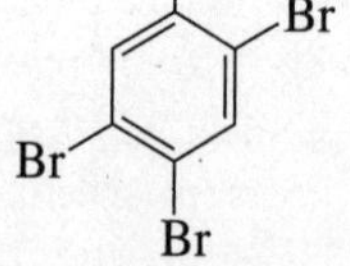

2,4,5-Tribromophenol 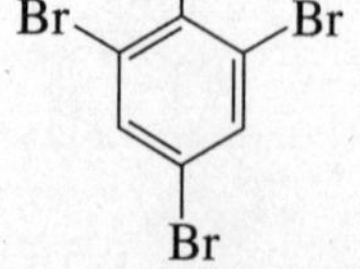 2,4,6-Tribromophenol 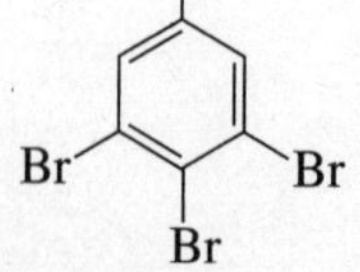 3,4,5-Tribromophenol

(*d*) There are only three tetrafluorobenzenes. The two hydrogens may be ortho, meta, or para to each other.

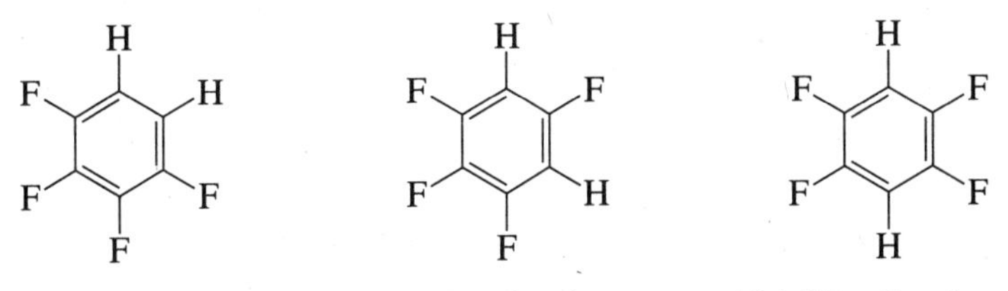


(*e*) There are only two naphthalenecarboxylic acids.

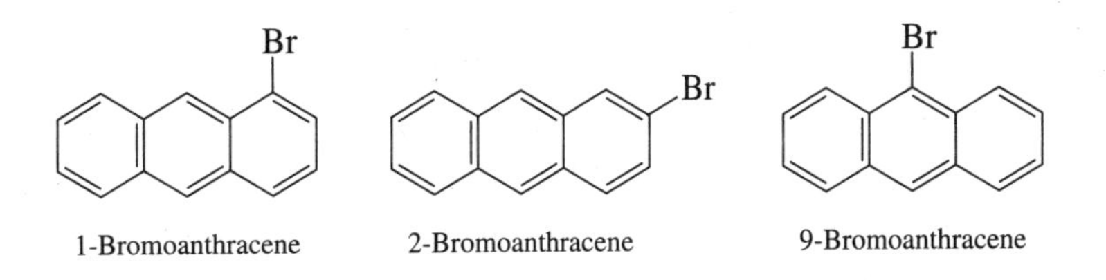


(*f*) There are three isomeric bromoanthracenes. All other positions are equivalent to one of these.



**11.26** There are three isomeric trimethylbenzenes:

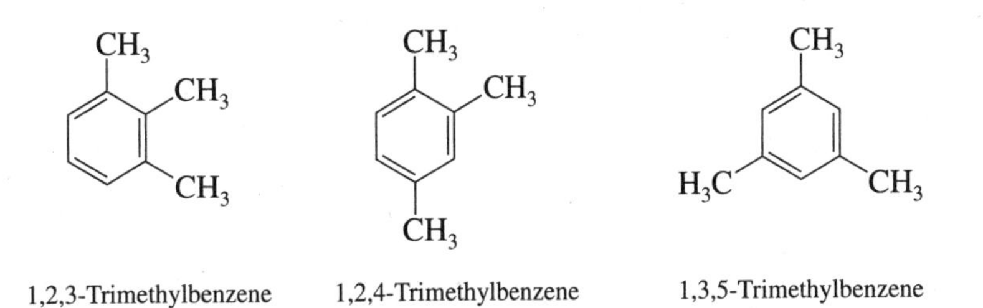


Their relative stabilities are determined by steric effects. Mesitylene (the 1,3,5-trisubstituted isomer) is the most stable because none of its methyl groups are ortho to any other methyl group. Ortho substituents on a benzene ring, depending on their size, experience van der Waals strain in the same way that cis substituents on a carbon–carbon double bond do. Because the carbon–carbon bond length in benzene is somewhat longer than in an alkene, these effects are smaller in magnitude, however. The 1,2,4-substitution pattern has one methyl–methyl repulsion between ortho substituents. The least stable isomer is the 1,2,3-trimethyl derivative, because it is the most crowded. The energy differences between isomers are relatively small, heats of combustion being 5198, 5195, and 5193 kJ/mol (1242.4, 1241.6, and 1241.2 kcal/mol) for the 1,2,3, 1,2,4, and 1,3,5 isomers, respectively.

**11.27** *p*-Dichlorobenzene has a center of symmetry. Each of its individual bond moments is balanced by an identical bond dipole oriented opposite to it. *p*-Dichlorobenzene has no dipole moment. *o*-Dichlorobenzene has the largest dipole moment.

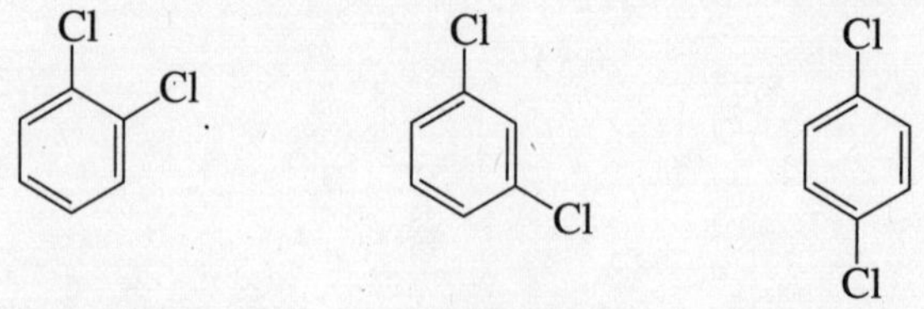

Cl Cl Cl Cl Cl Cl

*o*-Dichlorobenzene $\mu = 2.27$ D — *m*-Dichlorobenzene $\mu = 1.48$ D — *p*-Dichlorobenzene $\mu = 0$ D

**11.28** The shortest carbon–carbon bond in styrene is the double bond of the vinyl substituent; its length is much the same as the double-bond length of any other alkene. The carbon–carbon bond lengths of the ring are intermediate between single- and double-bond lengths. The longest carbon–carbon bond is the $sp^2$ to $sp^2$ single bond connecting the vinyl group to the benzene ring.

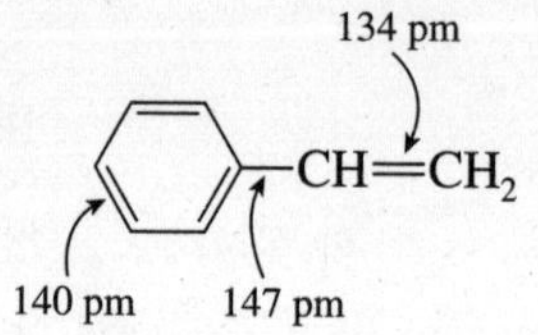

134 pm

$-CH{=}CH_2$

140 pm 147 pm

**11.29** Move $\pi$ electron pairs as shown so that both six-membered rings have an arrangement of bonds that corresponds to benzene.

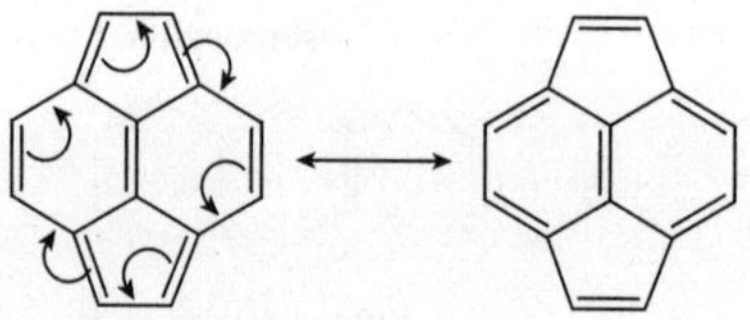

Less stable — More stable

**11.30** (*a*) In the structure shown for naphthalene, one ring but not the other corresponds to a Kekulé form of benzene. We say that one ring is **benzenoid,** and the other is not.

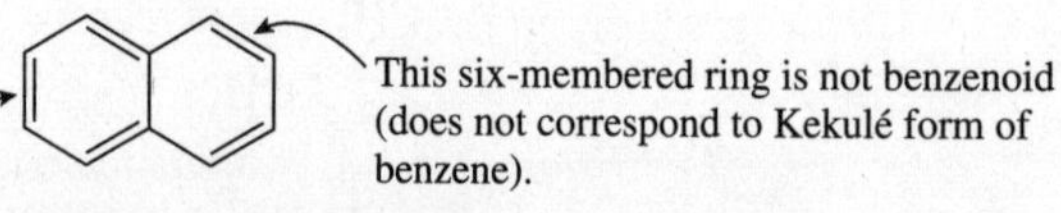

This six-membered ring is not benzenoid (does not correspond to Kekulé form of benzene).

This six-membered ring is benzenoid (corresponds to a Kekulé form of benzene).

By rewriting the benzenoid ring in its alternative Kekulé form, *both* rings become benzenoid.

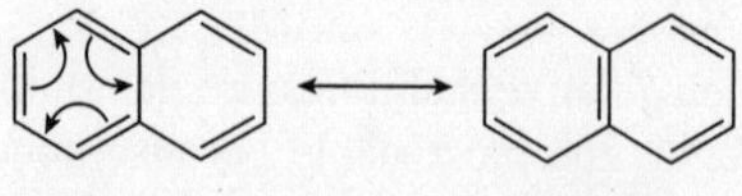

Both rings are benzenoid.

(*b*) Here a cyclobutadiene ring is fused to benzene. By writing the alternative resonance form of cyclobutadiene, the six-membered ring becomes benzenoid.

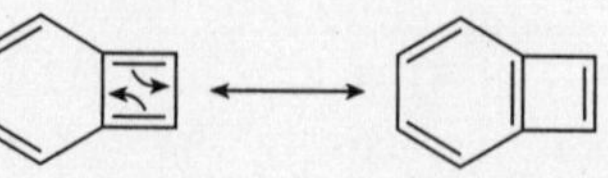

(*c*) The structure portrayed for phenanthrene contains two terminal benzenoid rings and a nonbenzenoid central ring. All three rings may be represented in benzenoid forms by converting one of the terminal six-membered rings to its alternative Kekulé form as shown:

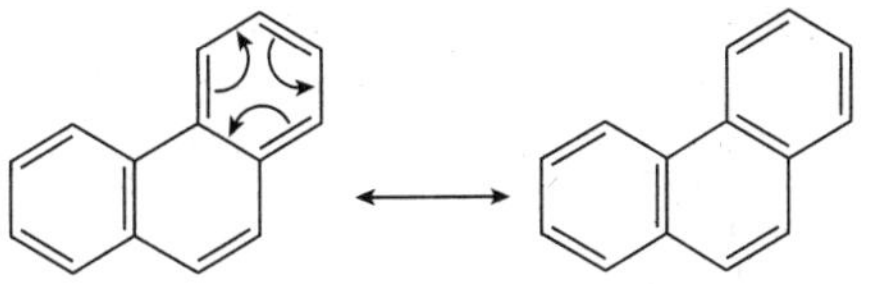

Central ring not benzenoid ⟷ All three rings benzenoid

(*d*) Neither of the six-membered rings is benzenoid in the structure shown. By writing the cyclooctatetraene portion of the molecule in its alternative representation, the two six-membered rings become benzenoid.

Six-membered rings are not benzenoid. ⟷ Six-membered rings are benzenoid.

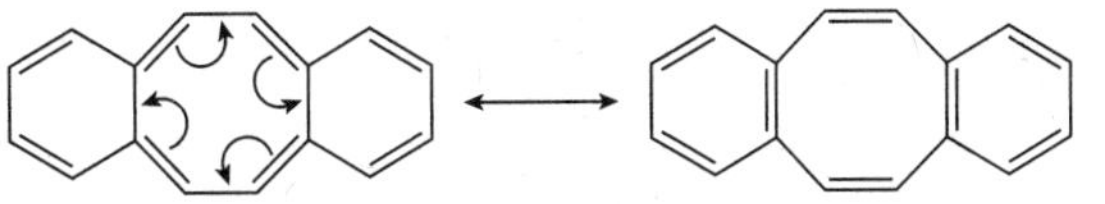

**11.31** (*a*) Hydrogenation of isopropylbenzene converts the benzene ring to a cyclohexane unit.

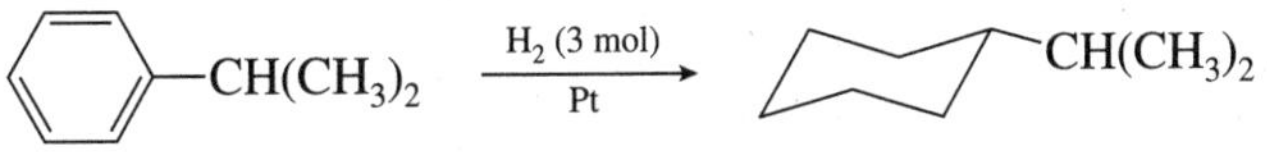

$$C_6H_5CH(CH_3)_2 \xrightarrow[\text{Pt}]{H_2\text{ (3 mol)}} C_6H_{11}CH(CH_3)_2$$

Isopropylbenzene → Isopropylcyclohexane

(*b*) Sodium and ethanol in liquid ammonia is the combination of reagents that brings about Birch reduction of benzene rings. The 1,4-cyclohexadiene that is formed has its isopropyl group as a substituent on one of the double bonds.

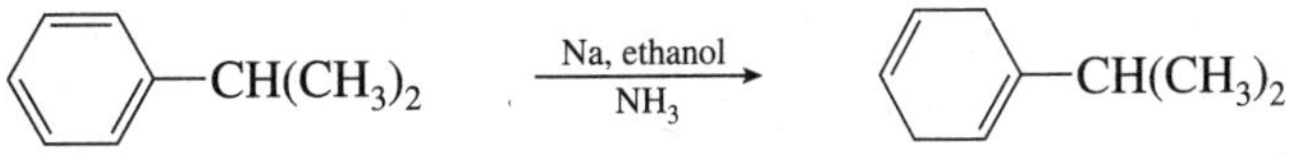

$$C_6H_5CH(CH_3)_2 \xrightarrow[NH_3]{\text{Na, ethanol}} \text{1-Isopropyl-1,4-cyclohexadiene}$$

Isopropylbenzene → 1-Isopropyl-1,4-cyclohexadiene

(*c*) Oxidation of the isopropyl side chain occurs. The benzene ring remains intact.

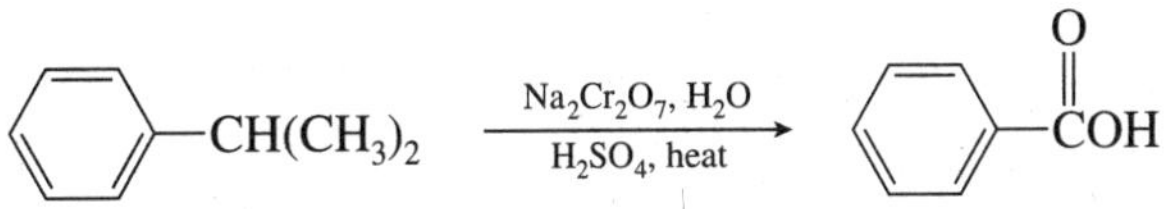

$$C_6H_5CH(CH_3)_2 \xrightarrow[H_2SO_4\text{, heat}]{Na_2Cr_2O_7\text{, }H_2O} C_6H_5COH \text{ (C=O)}$$

Isopropylbenzene → Benzoic acid

(*d*) *N*-Bromosuccinimide is a reagent effective for the substitution of a benzylic hydrogen.

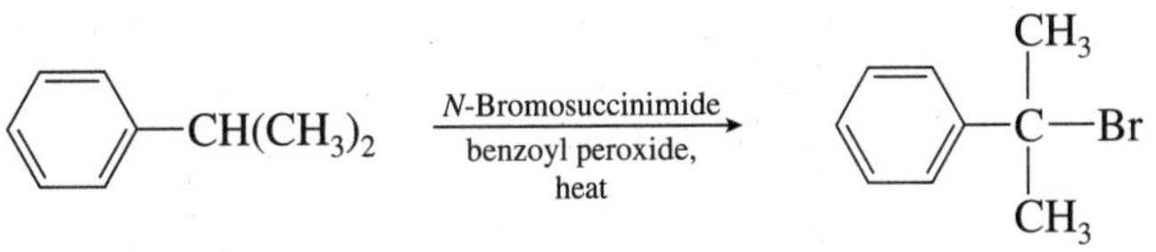

$$C_6H_5CH(CH_3)_2 \xrightarrow[\text{benzoyl peroxide, heat}]{N\text{-Bromosuccinimide}} C_6H_5C(CH_3)_2Br$$

Isopropylbenzene → 2-Bromo-2-phenylpropane

(*e*) The tertiary bromide undergoes E2 elimination to give a carbon–carbon double bond.

$$C_6H_5C(CH_3)_2Br \xrightarrow[CH_3CH_2OH]{NaOCH_2CH_3} C_6H_5C(CH_3)=CH_2$$

2-Bromo-2-phenylpropane → 2-Phenylpropene

**11.32** All the specific reactions in this problem have been reported in the chemical literature with results as indicated.

(*a*) Hydroboration–oxidation of alkenes leads to syn anti-Markovnikov hydration of the double bond.

$$\text{1-Phenylcyclobutene} \xrightarrow[2.\ H_2O_2,\ HO^-]{1.\ B_2H_6} \text{trans-2-Phenylcyclobutanol (82\%)}$$

1-Phenylcyclobutene → *trans*-2-Phenyl-cyclobutanol (82%)

(*b*) The compound contains a substituted benzene ring and an alkene-like double bond. When hydrogenation of this compound was carried out, the alkene-like double bond was hydrogenated cleanly.

$$\text{1-Ethylindene} + H_2 \xrightarrow{Pt} \text{1-Ethylindan (80\%)}$$

1-Ethylindene → 1-Ethylindan (80%)

(*c*) Free-radical chlorination will lead to substitution of benzylic hydrogens. The starting material contains four benzylic hydrogens, all of which may eventually be replaced.

$$(C_6H_5)_2CH\text{—}C_6H_4\text{—}CH_3 \xrightarrow[CCl_4,\ light]{excess\ Cl_2} (C_6H_5)_2CCl\text{—}C_6H_4\text{—}CCl_3$$

(65%)

(*d*) Epoxidation of alkenes is stereospecific.

$$C_6H_5CH=CHC_6H_5 \xrightarrow[acetic\ acid]{CH_3COOH\ (peroxy)} \text{trans-1,2-Diphenylepoxyethane}$$

(*E*)-1,2-Diphenylethene → *trans*-1,2-Diphenylepoxyethane (78–83%)

(*e*) The reaction is one of acid-catalyzed alcohol dehydration.

$$\text{cis-4-Methyl-1-phenylcyclohexanol} \xrightarrow[acetic\ acid]{H_2SO_4} \text{4-Methyl-1-phenylcyclohexene (81\%)}$$

*cis*-4-Methyl-1-phenylcyclohexanol → 4-Methyl-1-phenylcyclohexene (81%)

(*f*) This reaction illustrates identical reactivity at two equivalent sites in a molecule. Both alcohol functions are tertiary and benzylic and undergo acid-catalyzed dehydration readily.

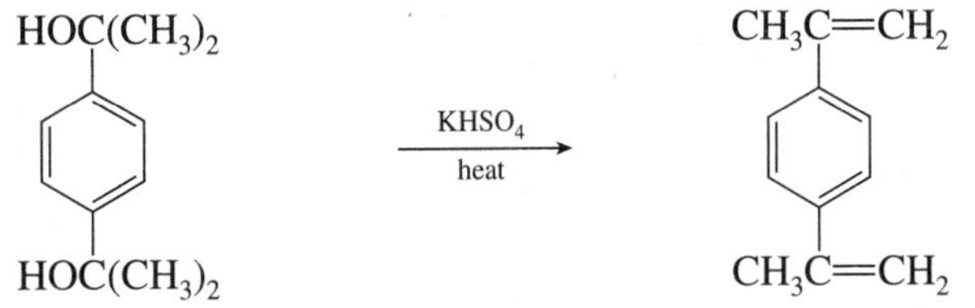

$$\text{HOC(CH}_3)_2\text{-C}_6\text{H}_4\text{-C(CH}_3)_2\text{OH} \xrightarrow[\text{heat}]{\text{KHSO}_4} \text{CH}_2\text{=C(CH}_3)\text{-C}_6\text{H}_4\text{-C(CH}_3)\text{=CH}_2$$

1,4-Di-(1-hydroxy-1-methylethyl) benzene — 1,4-Diisopropenylbenzene (68%)

(*g*) The compound shown is DDT (standing for the nonsystematic name **dichlorodiphenyltrichloroethane**). It undergoes $\beta$-elimination to form an alkene.

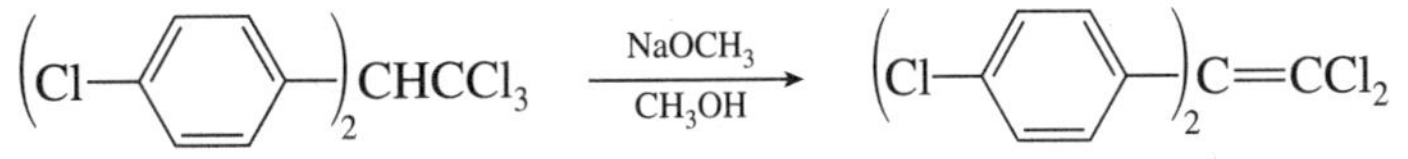

$$(\text{Cl-C}_6\text{H}_4)_2\text{CHCCl}_3 \xrightarrow[\text{CH}_3\text{OH}]{\text{NaOCH}_3} (\text{Cl-C}_6\text{H}_4)_2\text{C=CCl}_2$$

(100%)

(*h*) Alkyl side chains on naphthalene undergo reactions analogous to those of alkyl groups on benzene.

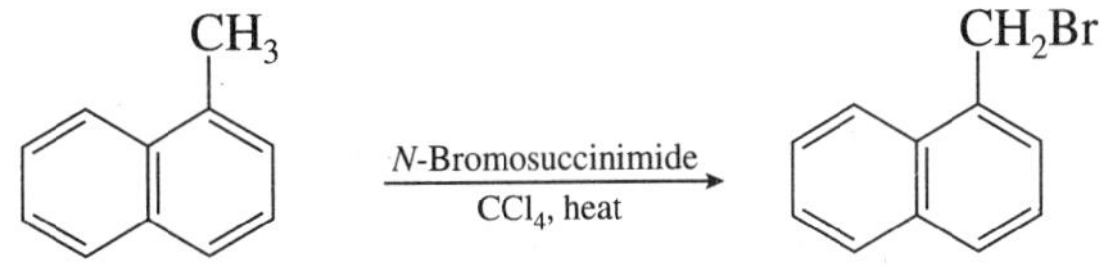

$$\text{1-CH}_3\text{-C}_{10}\text{H}_7 \xrightarrow[\text{CCl}_4\text{, heat}]{N\text{-Bromosuccinimide}} \text{1-BrCH}_2\text{-C}_{10}\text{H}_7$$

1- Methylnaphthalene — 1-(Bromomethyl) naphthalene (46%)

(*i*) Potassium carbonate is a weak base. Hydrolysis of the primary benzylic halide converts it to an alcohol.

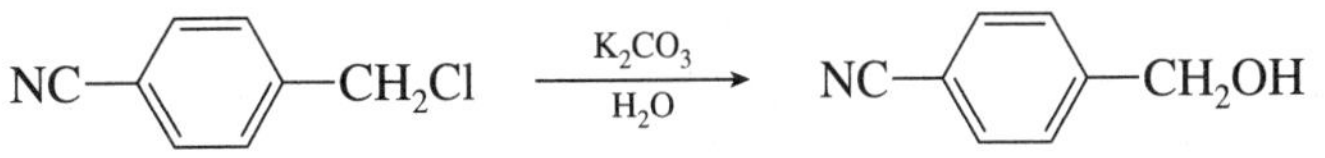

$$\text{NC-C}_6\text{H}_4\text{-CH}_2\text{Cl} \xrightarrow[\text{H}_2\text{O}]{\text{K}_2\text{CO}_3} \text{NC-C}_6\text{H}_4\text{-CH}_2\text{OH}$$

*p*-Cyanobenzyl chloride — *p*-Cyanobenzyl alcohol (85%)

**11.33** Only benzylic (or allylic) hydrogens are replaced by *N*-bromosuccinimide. Among the four bromines in 3,4,5-tribromobenzyl bromide, three are substituents on the ring and are not capable of being introduced by benzylic bromination. The starting material must therefore have these three bromines already in place.

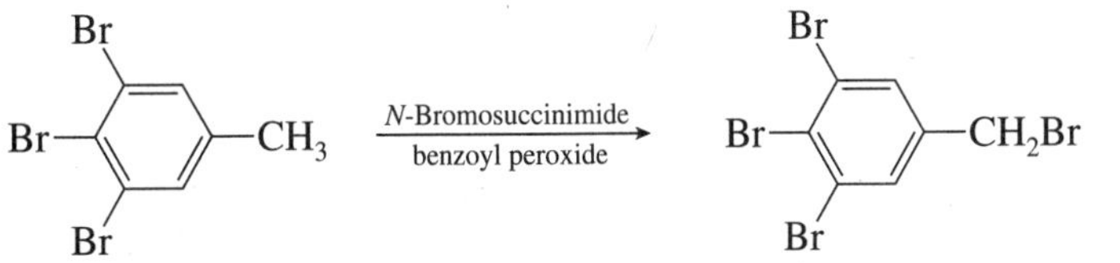

$$\text{3,4,5-Br}_3\text{C}_6\text{H}_2\text{-CH}_3 \xrightarrow[\text{benzoyl peroxide}]{N\text{-Bromosuccinimide}} \text{3,4,5-Br}_3\text{C}_6\text{H}_2\text{-CH}_2\text{Br}$$

3,4,5-Tribromotoluene Compound A — 3,4,5-Tribromobenzyl bromide

**11.34** 2,3,5-Trimethoxybenzoic acid has the structure shown. The three methoxy groups occupy the same positions in this oxidation product that they did in the unknown compound. The carboxylic acid function must have arisen by oxidation of the —$CH_2CH{=}C(CH_3)_2$ side chain. Therefore

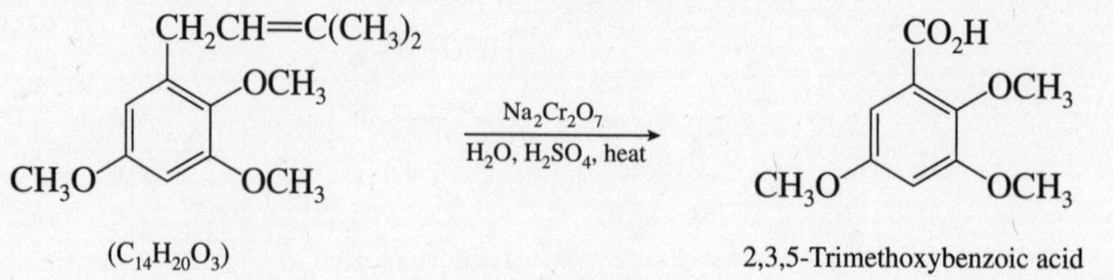

$(C_{14}H_{20}O_3)$ $\xrightarrow[H_2O,\ H_2SO_4,\ heat]{Na_2Cr_2O_7}$ 2,3,5-Trimethoxybenzoic acid

**11.35** Hydroboration–oxidation leads to stereospecific syn addition of H and OH across a carbon–carbon double bond. The regiochemistry of addition is opposite to that predicted by Markovnikov's rule. Hydroboration–oxidation of the *E* alkene gives alcohol A.

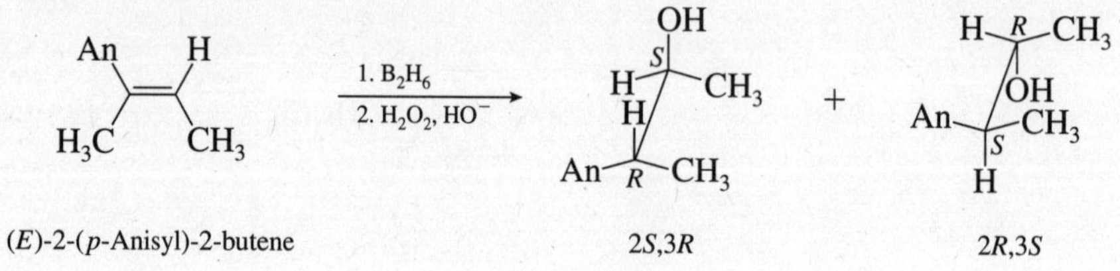

(*E*)-2-(*p*-Anisyl)-2-butene $\xrightarrow[2.\ H_2O_2,\ HO^-]{1.\ B_2H_6}$ 2*S*,3*R* + 2*R*,3*S*

An = $CH_3O$–$C_6H_4$–

Alcohol A is a racemic mixture of the 2*S*,3*R* and 2*R*,3*S* enantiomers of 3-(*p*-anisyl)-2-butanol. Hydroboration–oxidation of the Z alkene gives alcohol B.

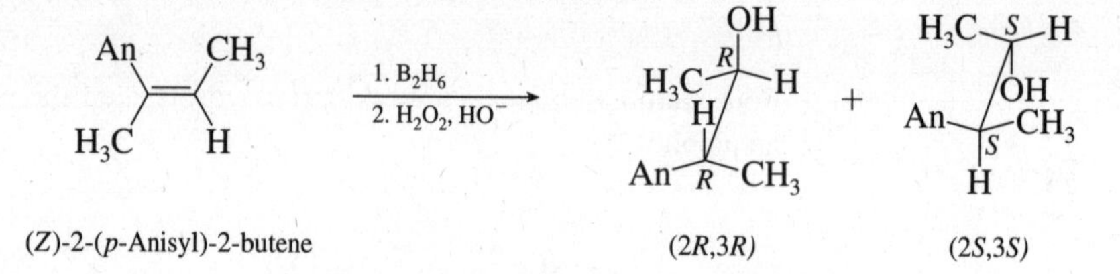

(*Z*)-2-(*p*-Anisyl)-2-butene $\xrightarrow[2.\ H_2O_2,\ HO^-]{1.\ B_2H_6}$ (2*R*,3*R*) + (2*S*,3*S*)

Alcohol B is a racemic mixture of the 2*R*,3*R* and 2*S*,3*S* enantiomers of 3-(*p*-anisyl)-2-butanol. Alcohols A and B are stereoisomers that are not enantiomers; they are diastereomers.

**11.36** Dehydrohalogenation of alkyl halides is stereospecific, requiring an anti arrangement between the hydrogen being lost and the leaving group in the transition state. (*Z*)-1,2-Diphenylpropene must therefore be formed from the diastereomer shown.

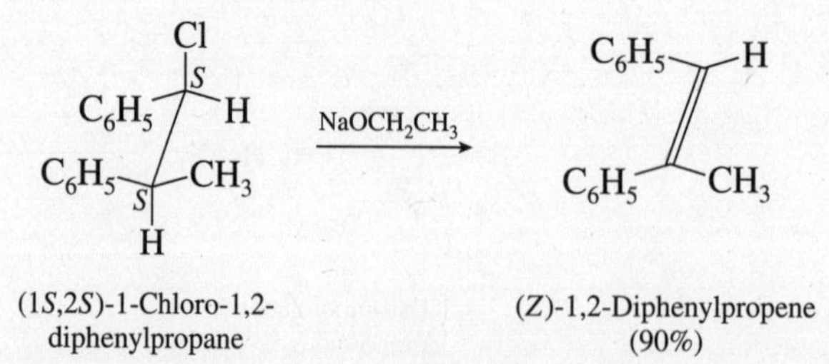

(1*S*,2*S*)-1-Chloro-1,2-diphenylpropane $\xrightarrow{NaOCH_2CH_3}$ (*Z*)-1,2-Diphenylpropene (90%)

The mirror-image chloride, 1*R*,2*R*, will also give the *Z* alkene. In fact, the reaction was carried out on a racemic mixture of the 1*R*,2*R* and 1*S*,2*S* stereoisomers.

The $E$ isomer is formed from either the 1$R$,2$S$ or the 1$S$,2$R$ chloride (or from a racemic mixture of the two).

$$\text{(1}R\text{,2}S\text{)-}C_6H_5CH(Cl)CH(CH_3)C_6H_5 \xrightarrow{NaOCH_2CH_3} (E)\text{-}C_6H_5C(CH_3){=}CHC_6H_5$$

(1$R$,2$S$)-1-Chloro-1,2-diphenylpropane

($E$)-1,2-Diphenylpropene (87%)

**11.37** ($a$) The conversion of ethylbenzene to 1-phenylethyl bromide is a benzylic bromination. It can be achieved by using either bromine or $N$-bromosuccinimide (NBS).

$$C_6H_5CH_2CH_3 \xrightarrow[\text{NBS, heat}]{Br_2\text{, light or}} C_6H_5\underset{\displaystyle Br}{\underset{|}{C}}HCH_3$$

Ethylbenzene → 1-Phenylethyl bromide

($b$) The conversion of 1-phenylethyl bromide to 1,2-dibromo-1-phenylethane

$$C_6H_5\underset{\displaystyle Br}{\underset{|}{C}}HCH_3 \longrightarrow C_6H_5\underset{\displaystyle Br}{\underset{|}{C}}HCH_2Br$$

cannot be achieved cleanly in a single step. We must reason backward from the target molecule, that is, determine how to make 1,2-dibromo-1-phenylethane in one step from any starting material. Vicinal dibromides are customarily prepared by addition of bromine to alkenes. This suggests that 1,2-dibromo-1-phenylethane can be prepared by the reaction

$$C_6H_5CH{=}CH_2 + Br_2 \longrightarrow C_6H_5\underset{\displaystyle Br}{\underset{|}{C}}HCH_2Br$$

Styrene → 1,2-Dibromo-1-phenylethane

The necessary alkene, styrene, is available by dehydrohalogenation of the given starting material, 1-phenylethyl bromide.

$$C_6H_5\underset{\displaystyle Br}{\underset{|}{C}}HCH_3 \xrightarrow[CH_3CH_2OH]{NaOCH_2CH_3} C_6H_5CH{=}CH_2$$

1-Phenylethyl bromide → Styrene

Thus, by reasoning backward from the target molecule, the synthetic scheme becomes apparent.

$$C_6H_5\underset{\displaystyle Br}{\underset{|}{C}}HCH_3 \xrightarrow[CH_3CH_2OH]{NaOCH_2CH_3} C_6H_5CH{=}CH_2 \xrightarrow{Br_2} C_6H_5\underset{\displaystyle Br}{\underset{|}{C}}HCH_2Br$$

1-Phenylethyl bromide → Styrene → 1,2-Dibromo-1-phenylethane

($c$) The conversion of styrene to phenylacetylene cannot be carried out in a single step. As was pointed out in Chapter 9, however, a standard sequence for converting terminal alkenes to alkynes consists of bromine addition followed by a double dehydrohalogenation in strong base.

$$\underset{\text{Styrene}}{C_6H_5CH{=}CH_2} \xrightarrow{Br_2} \underset{\text{1,2-Dibromo-1-phenylethane}}{C_6H_5\underset{|\atop Br}{C}HCH_2Br} \xrightarrow[NH_3]{NaNH_2} \underset{\text{Phenylacetylene}}{C_6H_5C{\equiv}CH}$$

(*d*) The conversion of phenylacetylene to butylbenzene requires both a carbon–carbon bond formation step and a hydrogenation step. The acetylene function is essential for carbon–carbon bond formation by alkylation. The correct sequence is therefore:

$$\underset{\text{Phenylacetylene}}{C_6H_5C{\equiv}CH} \xrightarrow[NH_3]{NaNH_2} C_6H_5C{\equiv}C{:}^-\ Na^+$$

$$C_6H_5C{\equiv}C{:}^-\ Na^+ + CH_3CH_2Br \longrightarrow C_6H_5C{\equiv}CCH_2CH_3$$

$$C_6H_5C{\equiv}CCH_2CH_3 \xrightarrow[Pt]{H_2} \underset{\text{Butylbenzene}}{C_6H_5CH_2CH_2CH_2CH_3}$$

(*e*) The transformation corresponds to alkylation of acetylene, and so the alcohol must first be converted to a species with a good leaving group such as its halide derivative.

$$\underset{\text{2-Phenylethanol}}{C_6H_5CH_2CH_2OH} \xrightarrow{PBr_3} \underset{\text{2-Phenylethyl bromide}}{C_6H_5CH_2CH_2Br}$$

$$\underset{\text{2-Phenylethyl bromide}}{C_6H_5CH_2CH_2Br} + \underset{\text{Sodium acetylide}}{NaC{\equiv}CH} \longrightarrow \underset{\text{4-Phenyl-1-butyne}}{C_6H_5CH_2CH_2C{\equiv}CH}$$

(*f*) The target compound is a bromohydrin. Bromohydrins are formed by addition of bromine and water to alkenes.

$$\underset{\text{2-Phenylethyl bromide}}{C_6H_5CH_2CH_2Br} \xrightarrow[(CH_3)_3COH]{KOC(CH_3)_3} \underset{\text{Styrene}}{C_6H_5CH{=}CH_2} \xrightarrow[H_2O]{Br_2} \underset{\text{2-Bromo-1-phenylethanol}}{C_6H_5\underset{|\atop OH}{C}HCH_2Br}$$

**11.38** The stability of free radicals is reflected in their ease of formation. Toluene, which forms a benzyl radical, reacts with bromine 64,000 times faster than does ethane, which forms a primary alkyl radical. Ethylbenzene, which forms a secondary benzylic radical, reacts 1 million times faster than ethane.

$$\underset{\text{Ethylbenzene (most reactive)}}{C_6H_5CH_2CH_3} + Br\cdot \longrightarrow \underset{\text{Secondary benzylic radical}}{C_6H_5\dot{C}HCH_3} + HBr$$

$$\underset{\text{Toluene}}{C_6H_5CH_3} + Br\cdot \longrightarrow \underset{\text{Primary benzylic radical}}{C_6H_5\dot{C}H_2} + HBr$$

$$\underset{\text{Ethane (least reactive)}}{CH_3CH_3} + Br\cdot \longrightarrow \underset{\text{Primary radical}}{CH_3\dot{C}H_2} + HBr$$

**11.39** A good way to develop alternative resonance structures for carbocations is to move electron pairs toward sites of positive charge.

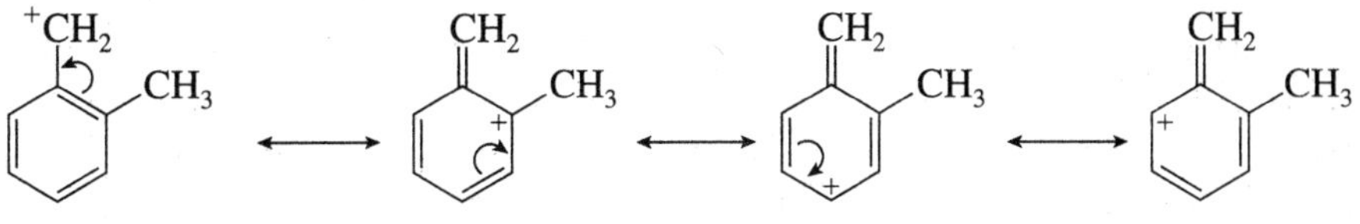

$^+CH_2$ $CH_3$ ⟷ $CH_2$ $CH_3$ ⟷ $CH_2$ $CH_3$ ⟷ $CH_2$ $CH_3$

*o*-Methylbenzyl cation — Tertiary carbocation

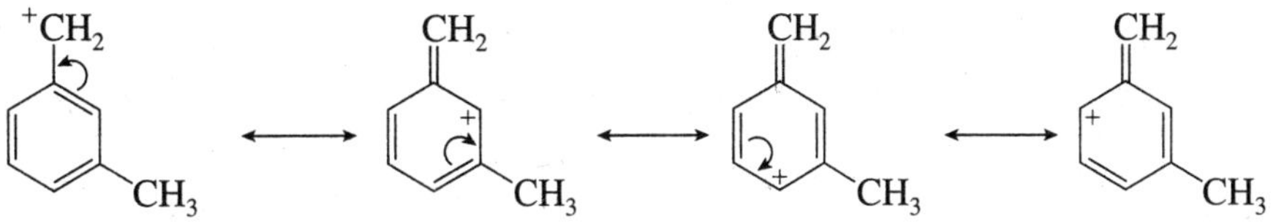

$^+CH_2$ $CH_3$ ⟷ $CH_2$ $CH_3$ ⟷ $CH_2$ $CH_3$ ⟷ $CH_2$ $CH_3$

*m*-Methylbenzyl cation

Only one of the Lewis structures shown is a tertiary carbocation. *o*-Methylbenzyl cation has tertiary carbocation character; *m*-methylbenzyl cation does not.

**11.40** The resonance structures for the cyclopentadienide anions formed by loss of a proton from 1-methyl-1,3-cyclopentadiene and 5-methyl-1,3-cyclopentadiene are equivalent.

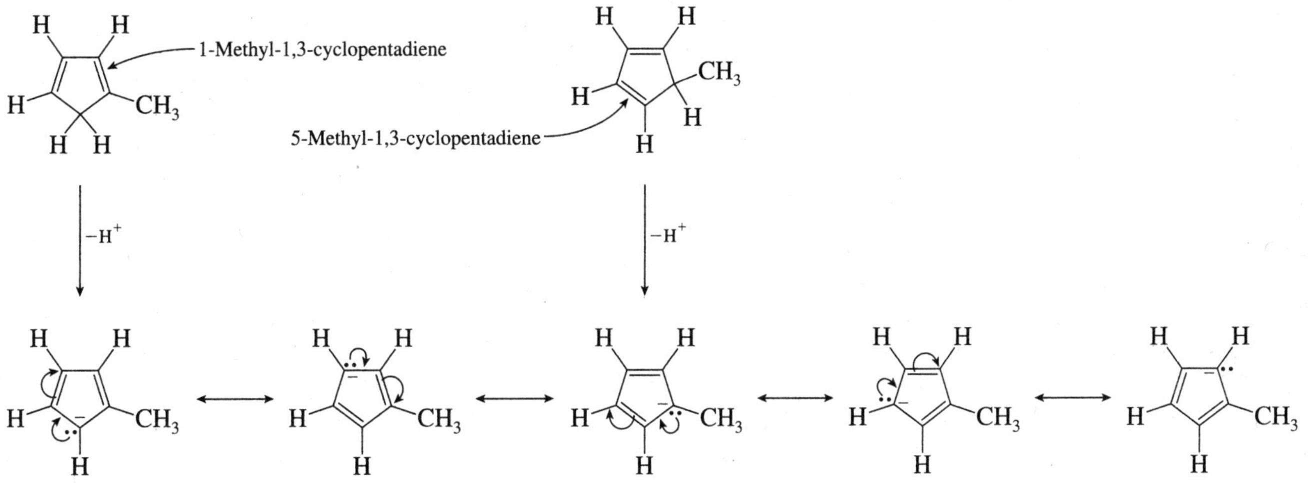

1-Methyl-1,3-cyclopentadiene

5-Methyl-1,3-cyclopentadiene

$-H^+$ $-H^+$

**11.41** Cyclooctatetraene is not aromatic. 1,2,3,4-Tetramethylcyclooctatetraene and 1,2,3,8-tetramethylcyclooctatetraene are constitutional isomers.

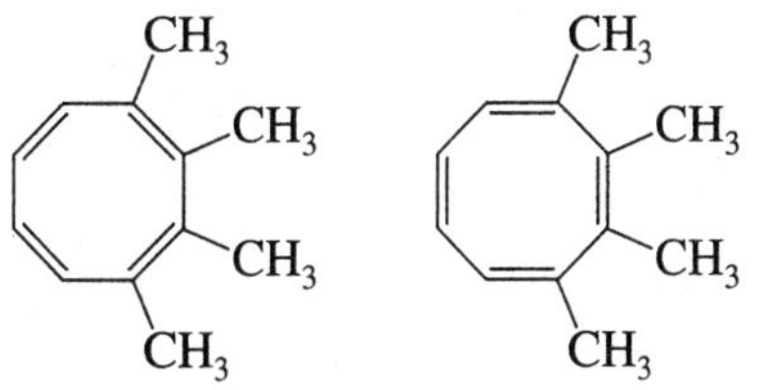

1,2,3,4-Tetramethylcyclooctatetraene

1,2,3,8-Tetramethylcyclooctatetraene

Leo A. Paquette at Ohio State University synthesized each of these compounds independently of the other and showed them to be stable enough to be stored separately without interconversion.

**11.42** Cyclooctatetraene has eight $\pi$ electrons and thus does not satisfy the $(4n + 2)$ $\pi$ electron requirement of the Hückel rule.

Cyclooctatetraene.
Each double bond contributes
2 $\pi$ electrons to give a total of 8.

All of the exercises in this problem involve counting the number of $\pi$ electrons in the various species derived from cyclooctatetraene and determining whether they satisfy the $(4n + 2)$ $\pi$ electron rule.

(*a*) Adding 1 $\pi$ electron gives a species ($C_8H_8^-$) with 9 $\pi$ electrons. $4n + 2$, where $n$ is a whole number, can never equal 9. This species is therefore **not aromatic.**

(*b*) Adding 2 $\pi$ electrons gives a species ($C_8H_8^{2-}$) with 10 $\pi$ electrons. $4n + 2 = 10$ when $n = 2$. The species $C_8H_8^{2-}$ **is aromatic.**

(*c*) Removing 1 $\pi$ electron gives a species ($C_8H_8^+$) with 7 $\pi$ electrons. $4n + 2$ cannot equal 7. The species $C_8H_8^+$ **is not aromatic.**

(*d*) Removing 2 $\pi$ electrons gives a species ($C_8H_8^{2+}$) with 6 $\pi$ electrons. $4n + 2 = 6$ when $n = 1$. The species $C_8H_8^{2+}$ **is aromatic.** (It has the same number of $\pi$ electrons as benzene.)

**11.43** (*a, b*) Cyclononatetraene does not have a continuous conjugated system of $\pi$ electrons. Conjugation is incomplete because it is interrupted by a $CH_2$ group. Thus (*a*) adding one more $\pi$ electron or (*b*) two more $\pi$ electrons will **not** give an aromatic system.

*sp*³ carbon in ring

(*c*) Removing a proton from the $CH_2$ group permits complete conjugation. The species produced has 10 $\pi$ electrons and is aromatic, since $4n + 2 = 10$ when $n = 2$.

$-H^+$

2 $\pi$ electrons for each double bond
+
2 $\pi$ electrons for unshared pair
= 10 $\pi$ electrons

(*d*) Removing a proton from one of the $sp^2$-hybridized carbons of the ring does not produce complete conjugation; the $CH_2$ group remains present to interrupt cyclic conjugation. The anion formed is **not** aromatic.

**11.44** (*a*) Cycloundecapentaene is **not aromatic.** Its $\pi$ system is not conjugated; it is interrupted by an $sp^3$-hybridized carbon.

*sp*³-hybridized carbon;
not a completely conjugated
monocyclic $\pi$ system

(*b*) Cycloundecapentaenyl radical is **not aromatic.** Its $\pi$ system is completely conjugated and monocyclic but contains 11 $\pi$ electrons—a number not equal to $(4n + 2)$ where $n$ is an integer.

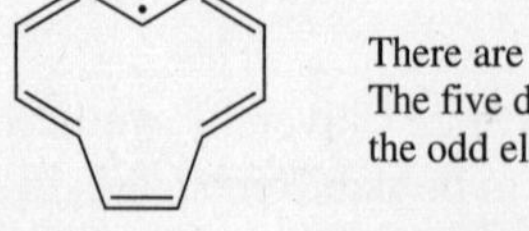

There are 11 electrons in the conjugated $\pi$ system.
The five double bonds contribute 10 $\pi$ electrons;
the odd electron of the radical is the eleventh.

(*c*) Cycloundecapentaenyl cation is **aromatic.** It includes a completely conjugated $\pi$ system which contains 10 $\pi$ electrons (10 equals $4n + 2$ where $n = 2$).

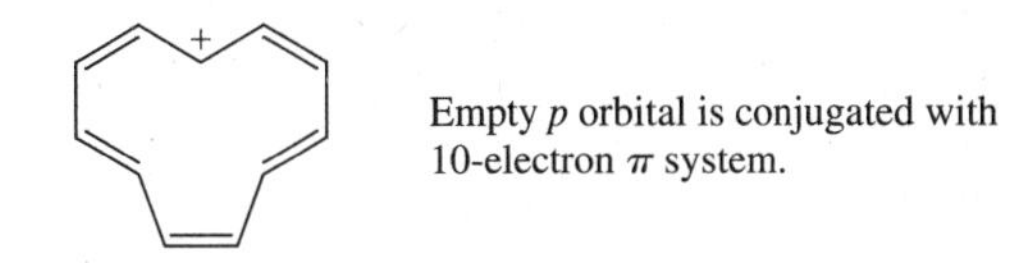


(*d*) Cycloundecapentadienide anion is **not aromatic.** It contains 12 $\pi$ electrons and thus does not satisfy the $(4n + 2)$ rule.

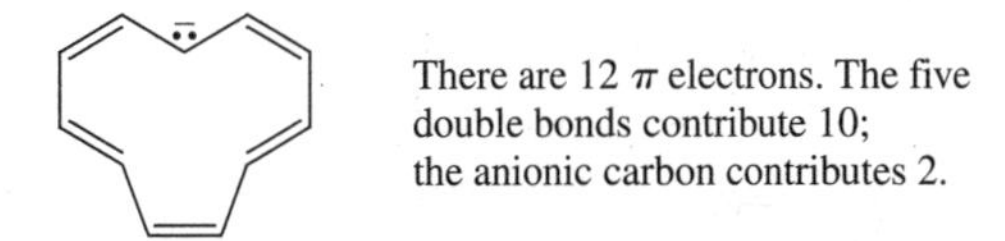

There are 12 π electrons. The five
double bonds contribute 10;
the anionic carbon contributes 2.


**11.45** (*a*) The more stable dipolar resonance structure is A because it has an aromatic cyclopentadienide anion bonded to an aromatic cyclopropenyl cation. In structure B neither ring is aromatic.

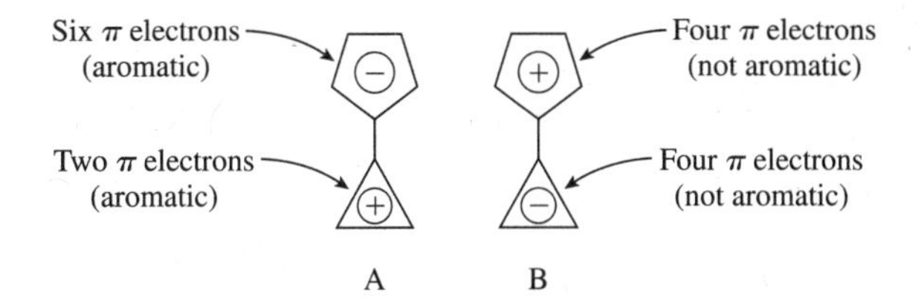


(*b*) Structure D can be stabilized by resonance involving the dipolar form.

Six $\pi$ electrons (aromatic)

Six $\pi$ electrons (aromatic)

D

Comparable stabilization is not possible in structure C because neither a cyclopropenyl system nor a cycloheptatrienyl system is aromatic in its anionic form. Both are aromatic as cations.

Eight $\pi$ electrons (not aromatic)

Two $\pi$ electrons (aromatic)

Six $\pi$ electrons (aromatic)

Four $\pi$ electrons (not aromatic)

C

**11.46** (*a*) This molecule, called **oxepin,** is **not aromatic.** The three double bonds each contribute 2 $\pi$ electrons, and an oxygen atom contributes 2 $\pi$ electrons to the conjugated system, giving a total of 8 $\pi$ electrons. Only one of the two unshared pairs on oxygen can contribute to the $\pi$ system; the other unshared pair is in an $sp^2$-hybridized orbital and cannot interact with it.

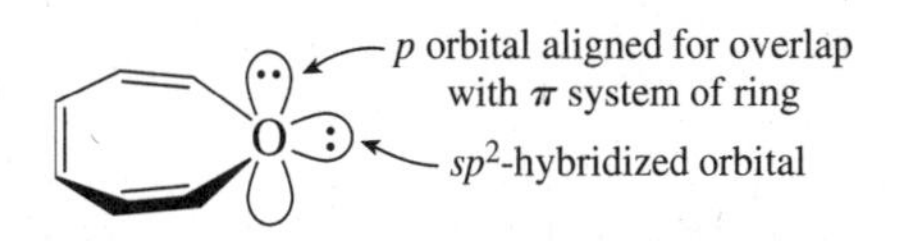

(*b*) This compound, called **azonine,** has 10 electrons in a completely conjugated planar monocyclic $\pi$ system and therefore satisfies Hückel's rule for $(4n + 2)$ $\pi$ electrons where $n = 2$. There are 8 $\pi$ electrons from the conjugated tetraene and 2 electrons contributed by the nitrogen unshared pair.

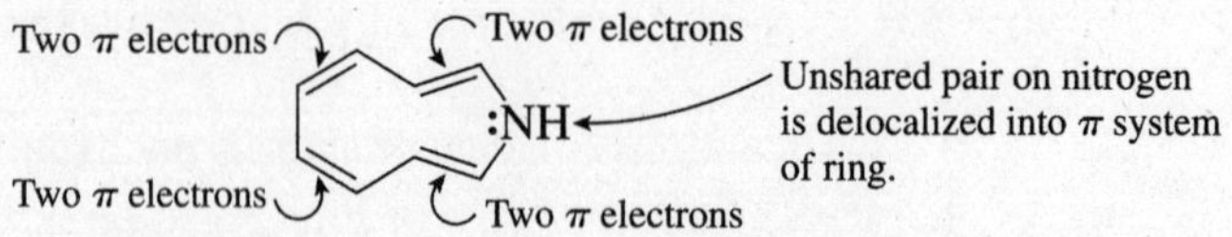

Two $\pi$ electrons
Two $\pi$ electrons
:NH
Unshared pair on nitrogen is delocalized into $\pi$ system of ring.
Two $\pi$ electrons
Two $\pi$ electrons

(*c*) Borazole, sometimes called **inorganic benzene,** is **aromatic.** Six $\pi$ electrons are contributed by the unshared pairs of the three nitrogen atoms. Each boron contributes a *p* orbital to maintain the conjugated system but no electrons.

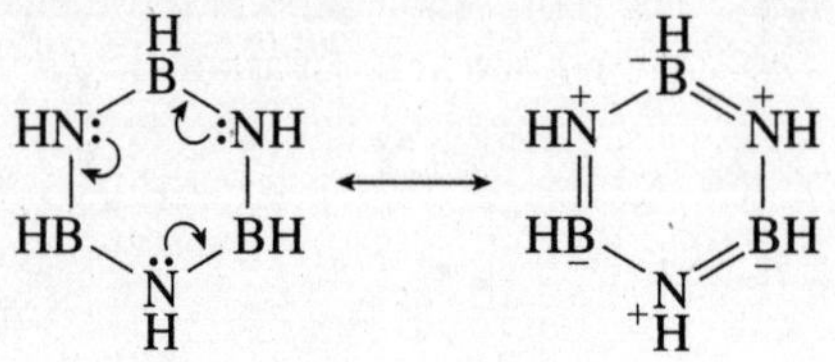

(*d*) This compound has 8 $\pi$ electrons and is **not aromatic.**

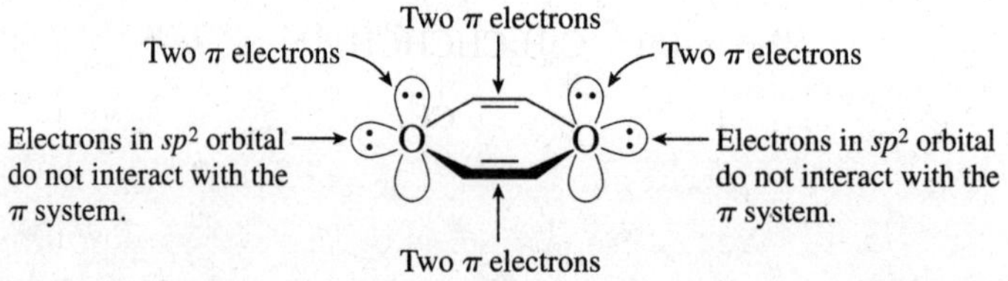

Two $\pi$ electrons
Two $\pi$ electrons
Two $\pi$ electrons
Electrons in $sp^2$ orbital do not interact with the $\pi$ system.
Electrons in $sp^2$ orbital do not interact with the $\pi$ system.
Two $\pi$ electrons

**11.47** The structure and numbering system for pyridine are given in Section 11.21, where we are also told that pyridine is aromatic. Oxidation of 3-methylpyridine is analogous to oxidation of toluene. The methyl side chain is oxidized to a carboxylic acid.

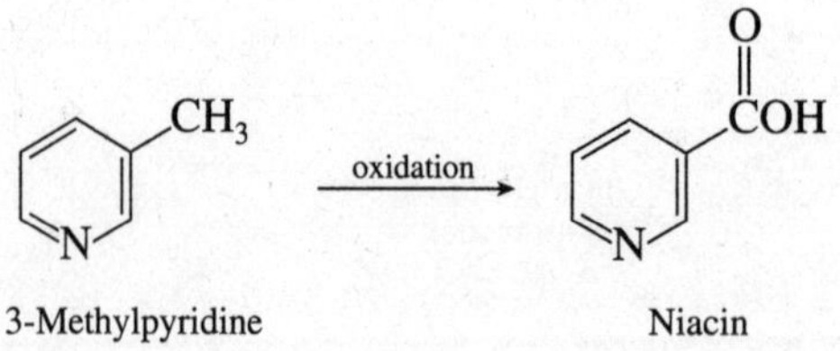

$$\text{3-Methylpyridine} \xrightarrow{\text{oxidation}} \text{Niacin}$$

3-Methylpyridine　　　　Niacin

**11.48** The structure and numbering system for quinoline are given in Section 11.21. **Nitroxoline** has the structural formula:

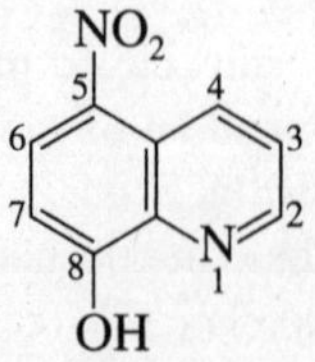

5-Nitro-8-hydroxyquinoline

**11.49** We are told that the ring system of **acridine** ($C_{13}H_9N$) is analogous to that of anthracene (i.e., tricyclic and linearly fused). Furthermore, the two most stable resonance forms are equivalent to each other.

The nitrogen atom must therefore be in the central ring, and the structure of acridine is

The two resonance forms would not be equivalent if the nitrogen were present in one of the terminal rings. Can you see why?

**11.50** Solutions to molecular modeling exercises are not provided in this *Study Guide and Solutions Manual.* You should use *Learning By Modeling* for these exercises.

# SELF-TEST

## PART A

**A-1.** Give an acceptable IUPAC name for each of the following:

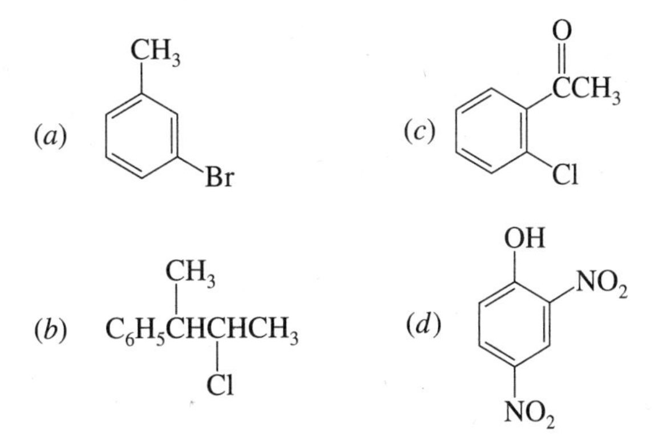


**A-2.** Draw the structure of each of the following:
(*a*) 3,5-Dichlorobenzoic acid
(*b*) *p*-Nitroanisole
(*c*) 2,4-Dimethylaniline
(*d*) *m*-Bromobenzyl chloride

**A-3.** Write a positive (+) or negative (−) charge at the appropriate position so that each of the following structures contains the proper number of $\pi$ electrons to permit it to be considered an aromatic ion. For purposes of this problem ignore strain effects that might destabilize the molecule.

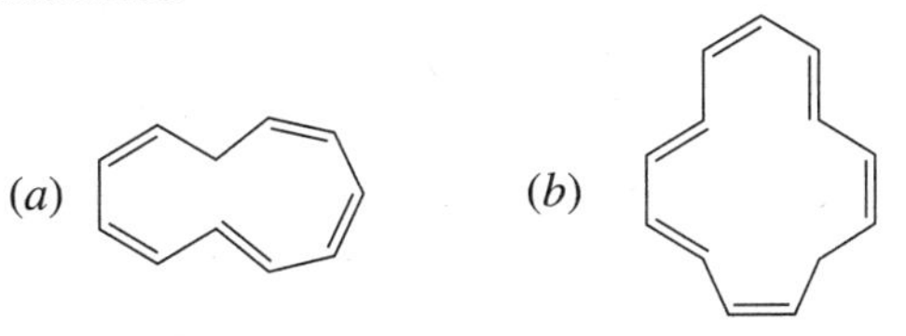


**A-4.** For each of the following, determine how many $\pi$ electrons are counted toward satisfying Hückel's rule. Assuming the molecule can adopt a planar conformation, is it aromatic?

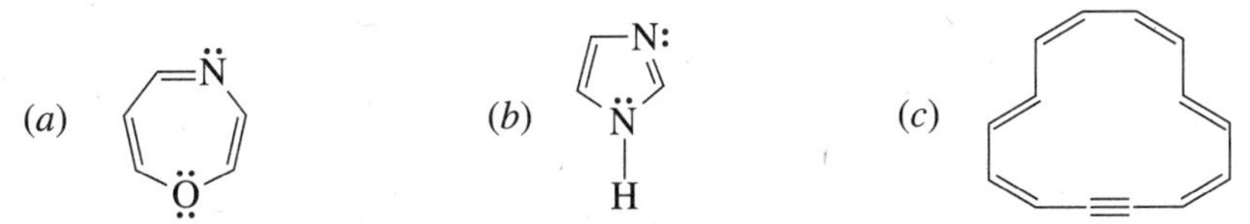


**A-5.** Azulene, shown in the following structure, is highly polar. Draw a dipolar resonance structure to explain this fact.

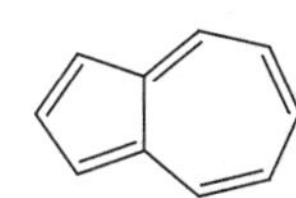

Azulene

**A-6.** Give the reactant, reagent, or product omitted from each of the following:

(*a*) [tetralin] $\xrightarrow[\text{peroxides, heat}]{\text{NBS}}$ ?

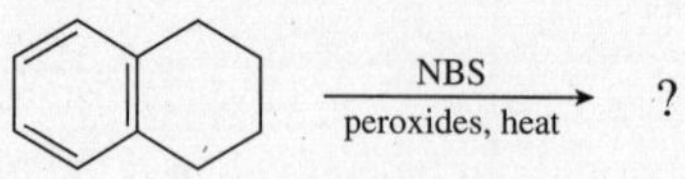

(*b*) ? $\xrightarrow[CH_3OH]{NaOCH_3}$ $C_6H_5CH_2OCH_3$

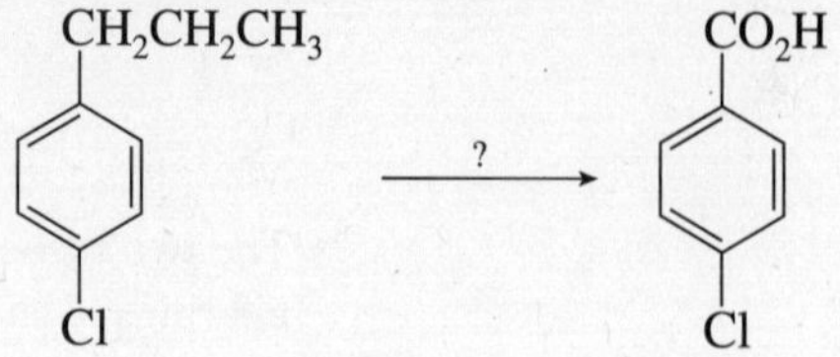

(*c*) [indene] $\xrightarrow[H^+]{H_2O}$ ?

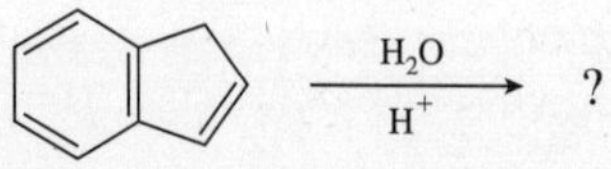

(*d*) [p-chloropropylbenzene: $CH_2CH_2CH_3$, Cl] $\xrightarrow{?}$ [p-chlorobenzoic acid: $CO_2H$, Cl]

(*e*) [1,2-dihydronaphthalene] $\xrightarrow{CH_3COOH \text{ (peroxy, } CH_3\overset{O}{\overset{\|}{C}}OOH)}$ ?

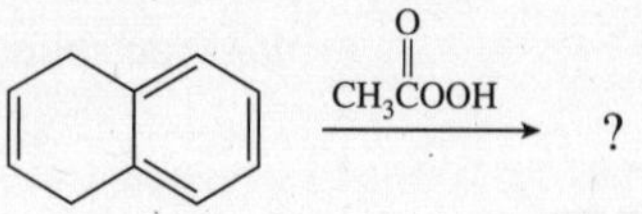

(*f*) $C_6H_5CH{=}CHCH_3$ $\xrightarrow[H_2O]{Br_2}$ ?

**A-7.** Provide two methods for the synthesis of 1-bromo-1-phenylpropane from an aromatic hydrocarbon.

**A-8.** Write the structures of the resonance forms that contribute to the stabilization of the intermediate in the reaction of styrene ($C_6H_5CH{=}CH_2$) with hydrogen bromide in the absence of peroxides.

**A-9.** Write one or more resonance structures that represent the delocalization of the following carbocation.

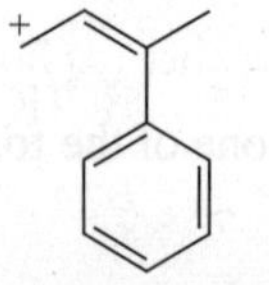

[structure: allylic cation, + on terminal carbon, phenyl-substituted]

**A-10.** An unknown compound, $C_{12}H_{18}$ reacts with sodium dichromate ($Na_2Cr_2O_7$) in warm aqueous sulfuric acid to give *p-tert*-butylbenzoic acid. What is the structure of the unknown?

## PART B

**B-1.** The number of possible dichloronitrobenzene isomers is
(*a*) 3
(*b*) 4
(*c*) 6
(*d*) 8

**B-2.** Which of the following statements is correct concerning the class of reactions to be expected for benzene and cyclooctatetraene?
(*a*) Both substances undergo addition reactions.
(*b*) Both substances undergo substitution reactions.
(*c*) Benzene undergoes substitution; cyclooctatetraene undergoes addition.
(*d*) Benzene undergoes addition; cyclooctatetraene undergoes substitution.

**B-3.** Which, if any, of the following structures represents an aromatic species?

(*a*) [cyclopentadienyl anion, H] (*b*) [cyclopentadienyl radical, H] (*c*) [cyclopentadienyl cation, H] (*d*) None of these is aromatic.

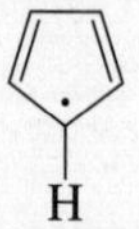

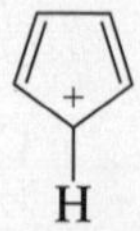

**B-4.** Which of the following compounds has a double bond that is conjugated with the $\pi$ system of the benzene ring?
(*a*) *p*-Benzyltoluene
(*b*) 2-Phenyl-1-decene
(*c*) 3-Phenylcyclohexene
(*d*) 3-Phenyl-1,4-pentadiene

**B-5.** Rank the following compounds in order of increasing rate of solvolysis ($S_N1$) in aqueous acetone (slowest → fastest):

1: $(CH_3)_2CHCH_2CH_2Br$

2: $(CH_3)_2CHCH(Br)CH_3$

3: $C_6H_5CH(Br)CH(CH_3)_2$

(*a*) 1 < 2 < 3 (*b*) 2 < 1 < 3 (*c*) 3 < 2 < 1 (*d*) 1 < 3 < 2

**B-6.** When comparing the hydrogenation of benzene with that of a hypothetical 1,3,5-cyclohexatriene, benzene _______ than the cyclohexatriene.

(*a*) Absorbs 152 kJ/mol (36 kcal/mol) more heat
(*b*) Absorbs 152 kJ/mol (36 kcal/mol) less heat
(*c*) Gives off 152 kJ/mol (36 kcal/mol) more heat
(*d*) Gives off 152 kJ/mol (36 kcal/mol) less heat

**B-7.** The reaction

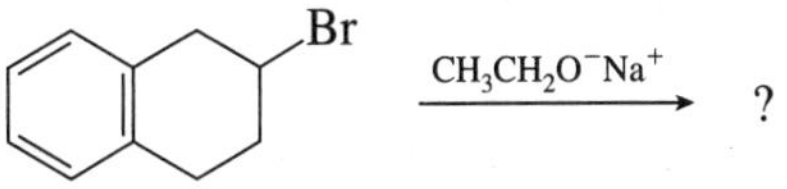

Br (tetralin bromide) $\xrightarrow{CH_3CH_2O^-Na^+}$ ?

gives as the major elimination product

(*a*) 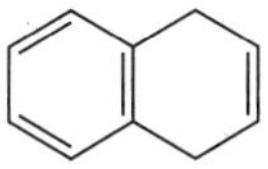 (*b*) 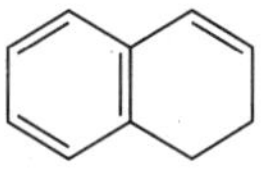 (*c*) *Equal* amounts of (*a*) and (*b*)

(*d*) Neither (*a*) nor (*b*)

**B-8.** Which one of the following is best classified as a **heterocyclic aromatic** compound?

(*a*) :N—H (*c*) N: (*e*) $\ddot{N}H_2$

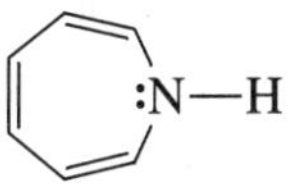

(*b*) N: (*d*) N:

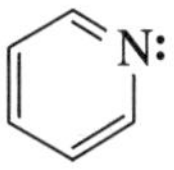

**B-9.** Which of the following has the smallest heat of combustion?

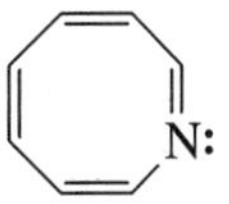

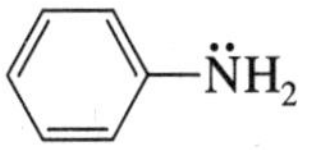

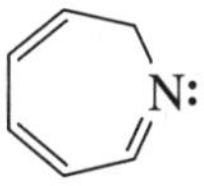

(*a*) 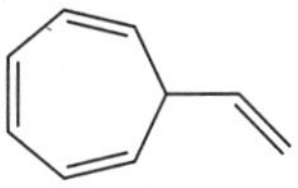 (*c*)

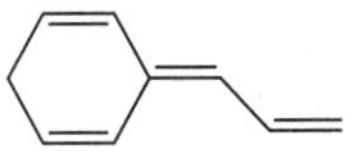

(*b*) 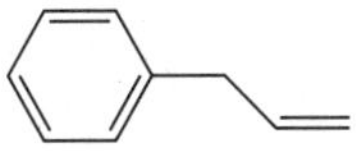 (*d*)

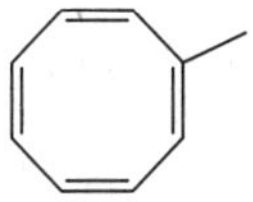

(*e*) The compounds are all isomers; the heats of combustion would be the same.

**B-10.** Which one of the following alcohols undergoes dehydration at the *fastest* rate on being heated with sulfuric acid? (The potential for rearrangement does not affect the rate.)

(*a*) $C_6H_5$—$CH_2CH_2CH_2CH_2OH$

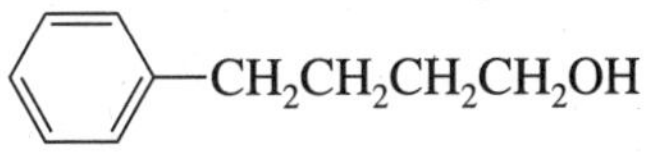

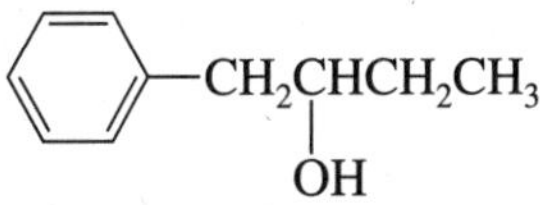

(*c*) $C_6H_5$—$CH_2CH(OH)CH_2CH_3$

(*b*) $C_6H_5$—$CH_2CH_2CH(OH)CH_3$

(*d*) $C_6H_5$—$CH(OH)CH_2CH_2CH_3$

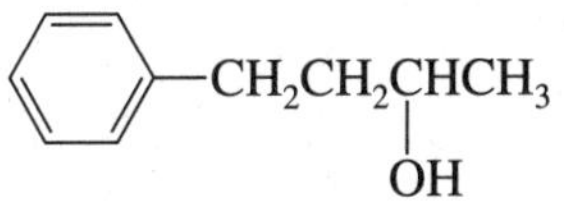

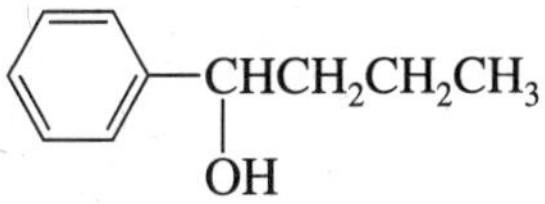

**B-11.** Ethylbenzene is treated with the reagents listed, in the order shown.

1. NBS, peroxides, heat
2. $CH_3CH_2O^-$
3. $B_2H_6$
4. $H_2O_2$, $HO^-$

The structure of the final product is:

(*a*) $C_6H_5CH_2CH_2OH$  (*d*) $C_6H_5\overset{OH}{\overset{|}{C}}HCH_2OH$

(*b*) $C_6H_5\overset{Br}{\overset{|}{C}}HCH_2OH$  (*e*) $C_6H_5\overset{OH}{\overset{|}{C}}HCH_2Br$

(*c*) $C_6H_5\overset{OH}{\overset{|}{C}}HCH_3$

**B-12.** Which of the following hydrogens is most easily abstracted (removed) on reaction with bromine atoms, Br·?

$CH_3CH_2$—⟨benzene ring⟩—$CH_3$

(*a*) (*b*) (*c*) (*d*)

**B-13.** All the hydrocarbons shown are very weak acids. One, however, is far more acidic than the others. Which one is the strongest acid?

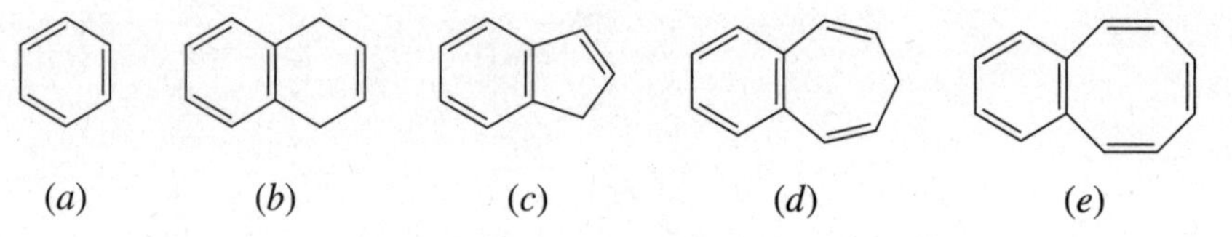


**B-14.** The compound shown is planar, and all the carbon–carbon bond lengths are the same. What (if anything) can you deduce about the bonding of boron from these observations?

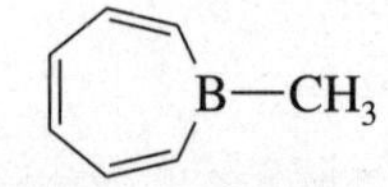


(*a*) The boron is $sp^2$-hybridized, and the *p* orbital contains an unshared pair of electrons.
(*b*) The boron is $sp^3$-hybridized, and a hybrid orbital contains an unshared pair of electrons.
(*c*) The boron is $sp^3$-hybridized, and a hybrid orbital is vacant.
(*d*) The boron is $sp^2$-hybridized, and the *p* orbital is vacant.
(*e*) Nothing about the bonding of boron can be deduced from these observations.

**B-15.** How many benzylic hydrogens are present in the hydrocarbon shown?

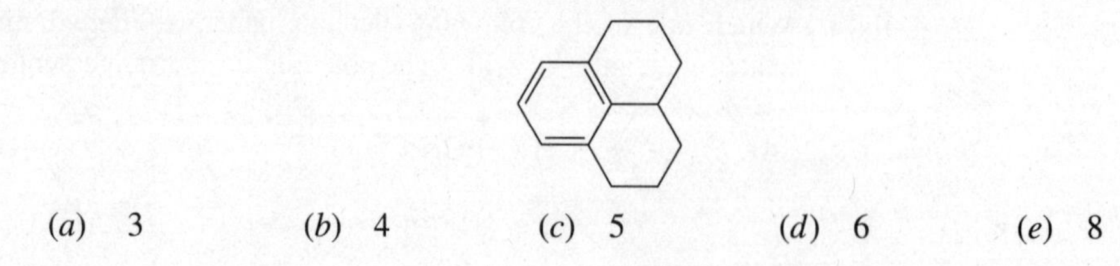

(*a*) 3  (*b*) 4  (*c*) 5  (*d*) 6  (*e*) 8

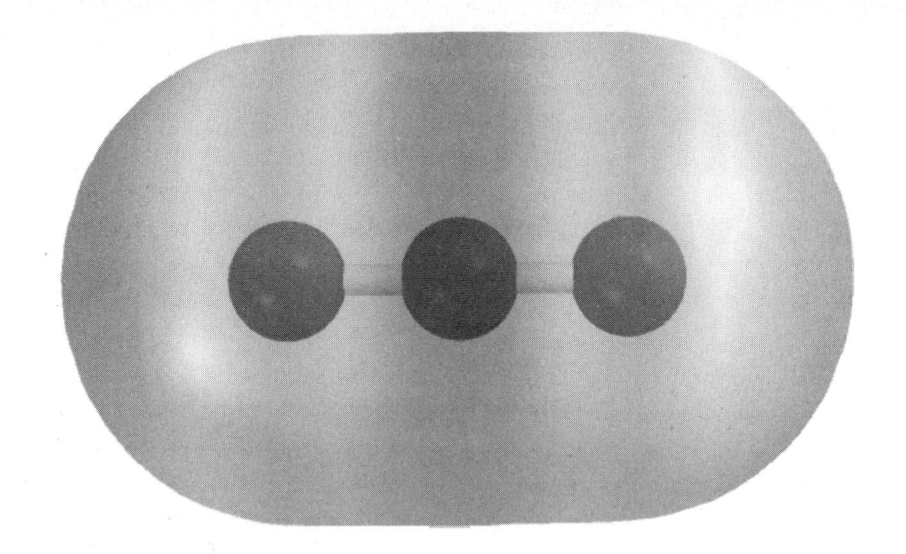

# CHAPTER 12

## REACTIONS OF ARENES: ELECTROPHILIC AROMATIC SUBSTITUTION

## SOLUTIONS TO TEXT PROBLEMS

**12.1** The three most stable resonance structures for cyclohexadienyl cation are

The positive charge is shared equally by the three carbons indicated. Thus the two carbons ortho to the $sp^3$-hybridized carbon and the one para to it each bear one third of a positive charge (+0.33). None of the other carbons is charged. The resonance picture and the simple MO treatment agree with respect to the distribution of charge in cyclohexadienyl cation.

**12.2** Electrophilic aromatic substitution leads to replacement of one of the hydrogens directly attached to the ring by the electrophile. All four of the ring hydrogens of *p*-xylene are equivalent; so it does not matter which one is replaced by the nitro group.

$$\xrightarrow[H_2SO_4]{HNO_3}$$

*p*-Xylene → 1,4-Dimethyl-2-nitrobenzene

**12.3** The aromatic ring of 1,2,4,5-tetramethylbenzene has two equivalent hydrogen substituents. Sulfonation of the ring leads to replacement of one of them by —$SO_3H$.

$SO_3$, $H_2SO_4$

1,2,4,5-Tetramethylbenzene → 2,3,5,6-Tetramethylbenzenesulfonic acid

**12.4** The major product is isopropylbenzene.

Benzene + $CH_3CH_2CH_2Cl$ (1-Chloropropane) $\xrightarrow{AlCl_3}$ Propylbenzene (20% yield) + Isopropylbenzene (40% yield)

Aluminum chloride coordinates with 1-chloropropane to give a Lewis acid/Lewis base complex, which can be attacked by benzene to yield propylbenzene or can undergo an intramolecular hydride shift to produce isopropyl cation. Isopropylbenzene arises by reaction of isopropyl cation with benzene.

$$CH_3—CH(H)—CH_2—\overset{+}{Cl}—\overset{-}{Al}Cl_3 \xrightarrow[\text{migration}]{\text{hydride}} CH_3—\overset{+}{C}H—CH_3 + AlCl_4^-$$

Isopropyl cation

**12.5** The species that attacks the benzene ring is cyclohexyl cation, formed by protonation of cyclohexene.

Cyclohexene + Sulfuric acid ($H—\ddot{O}—SO_2OH$) → Cyclohexyl cation + Hydrogen sulfate ion ($^-:\ddot{O}—SO_2OH$)

The mechanism for the reaction of cyclohexyl cation with benzene is analogous to the general mechanism for electrophilic aromatic substitution.

Benzene + Cyclohexyl cation → Cyclohexadienyl cation intermediate $\xrightarrow{-H^+}$ Cyclohexylbenzene

**12.6** The preparation of cyclohexylbenzene from cyclohexene and benzene was described in text Section 12.6. Cyclohexylbenzene is converted to 1-phenylcyclohexene by benzylic bromination, followed by dehydrohalogenation.

Benzene + Cyclohexene $\xrightarrow{H_2SO_4}$ Cyclohexylbenzene $\xrightarrow{\text{N-Bromosuccinimide (NBS), benzoyl peroxide, heat}}$ 1-Bromo-1-phenylcyclohexane $\xrightarrow{NaOCH_2CH_3}$ 1-Phenylcyclohexene

**12.7** Treatment of 1,3,5-trimethoxybenzene with an acyl chloride and aluminum chloride brings about Friedel–Crafts acylation at one of the three equivalent positions available on the ring.

$$\text{1,3,5-}(CH_3O)_3C_6H_3 + (CH_3)_2CHCH_2\overset{O}{\overset{\|}{C}}Cl \xrightarrow{AlCl_3} (CH_3O)_3C_6H_2\overset{O}{\overset{\|}{C}}CH_2CH(CH_3)_2$$

1,3,5-Trimethoxybenzene   3-Methylbutanoyl chloride   Isobutyl 1,3,5-trimethoxyphenyl ketone

**12.8** Because the anhydride is cyclic, its structural units are not incorporated into a ketone and a carboxylic acid as two separate product molecules. Rather, they become part of a four-carbon unit attached to benzene by a ketone carbonyl. The acyl substituent terminates in a carboxylic acid functional group.

$$C_6H_6 + \text{succinic anhydride} \xrightarrow{AlCl_3} C_6H_5\overset{O}{\overset{\|}{C}}CH_2CH_2\overset{O}{\overset{\|}{C}}OH$$

Benzene   Succinic anhydride   4-Oxo-4-phenylbutanoic acid

**12.9** (*b*) A Friedel–Crafts alkylation of benzene using 1-chloro-2,2-dimethylpropane would not be a satisfactory method to prepare neopentylbenzene because of the likelihood of a carbocation rearrangement. The best way to prepare this compound is by Friedel–Crafts acylation followed by Clemmensen reduction.

$$(CH_3)_3C\overset{O}{\overset{\|}{C}}Cl + C_6H_6 \xrightarrow{AlCl_3} (CH_3)_3C\overset{O}{\overset{\|}{C}}C_6H_5 \xrightarrow{Zn(Hg),\ HCl} (CH_3)_3CCH_2C_6H_5$$

2,2-Dimethylpropanoyl chloride   Benzene   2,2-Dimethyl-1-phenyl-1-propanone   Neopentylbenzene

**12.10** (*b*) Partial rate factors for nitration of toluene and *tert*-butylbenzene, relative to a single position of benzene, are as shown:

Toluene ($CH_3$): ortho 42, 42; meta 2.5, 2.5; para 58

*tert*-Butylbenzene ($C(CH_3)_3$): ortho 4.5, 4.5; meta 3, 3; para 75

The sum of these partial rate factors is 147 for toluene, 90 for *tert*-butylbenzene. Toluene is 147/90, or 1.7, times more reactive than *tert*-butylbenzene.

(*c*) The product distribution for nitration of *tert*-butylbenzene is determined from the partial rate factors.

$$\text{Ortho:}\quad \frac{2(4.5)}{90} = 10\%$$

$$\text{Meta:}\quad \frac{2(3)}{90} = 6.7\%$$

$$\text{Para:}\quad \frac{75}{90} = 83.3\%$$

**12.11** The compounds shown all undergo electrophilic aromatic substitution more slowly than benzene. Therefore, —$CH_2Cl$, —$CHCl_2$, and —$CCl_3$ are *deactivating* substituents.

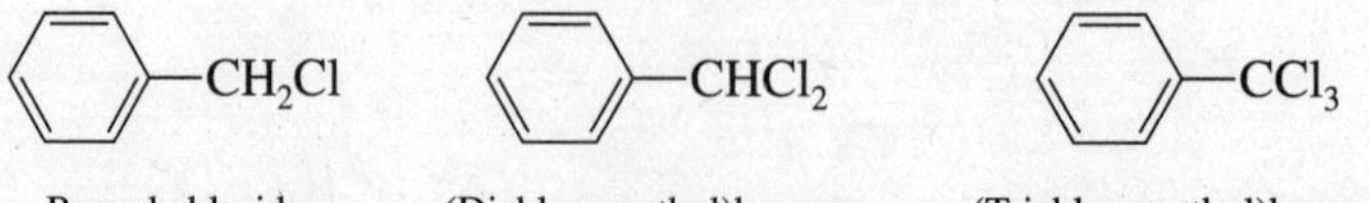

$C_6H_5$—$CH_2Cl$ Benzyl chloride | $C_6H_5$—$CHCl_2$ (Dichloromethyl)benzene | $C_6H_5$—$CCl_3$ (Trichloromethyl)benzene

The electron-withdrawing power of these substituents, and their tendency to direct incoming electrophiles meta to themselves, will increase with the number of chlorines each contains. Thus, the substituent that gives 4% meta nitration (96% ortho + para) contains the fewest chlorine atoms (—$CH_2Cl$), and the one that gives 64% meta nitration contains the most (—$CCl_3$).

| —$CH_2Cl$ | —$CHCl_2$ | —$CCl_3$ |
|---|---|---|
| Deactivating, ortho, para-directing | Deactivating, ortho, para-directing | Deactivating, meta-directing |

**12.12** (*b*) Attack by bromine at the position meta to the amino group gives a cyclohexadienyl cation intermediate in which delocalization of the nitrogen lone pair cannot participate in dispersal of the positive charge.

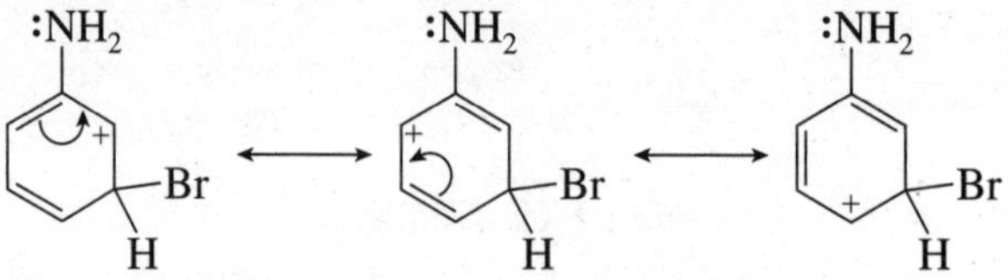

:$NH_2$ :$NH_2$ :$NH_2$

Br ⟷ Br ⟷ Br

H H H

(*c*) Attack at the position para to the amino group yields a cyclohexadienyl cation intermediate that is stabilized by delocalization of the electron pair of the amino group.

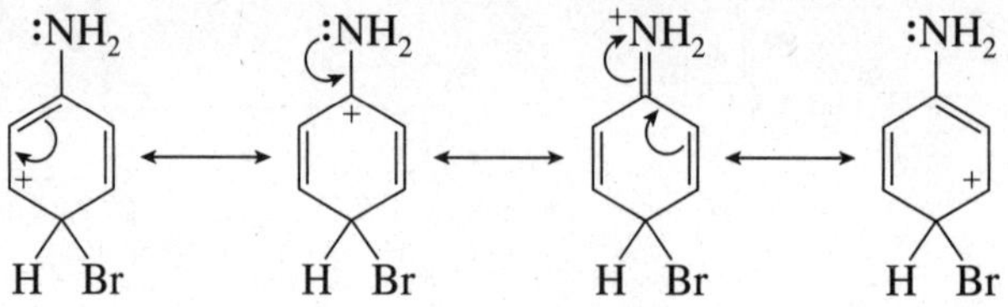

:$NH_2$ ⟷ :$NH_2$ ⟷ $\overset{+}{N}H_2$ ⟷ :$NH_2$

H Br H Br H Br H Br

**12.13** Electrophilic aromatic substitution in biphenyl is best understood by considering one ring as the functional group and the other as a substituent. An aryl substituent is ortho, para-directing. Nitration of biphenyl gives a mixture of *o*-nitrobiphenyl and *p*-nitrobiphenyl.

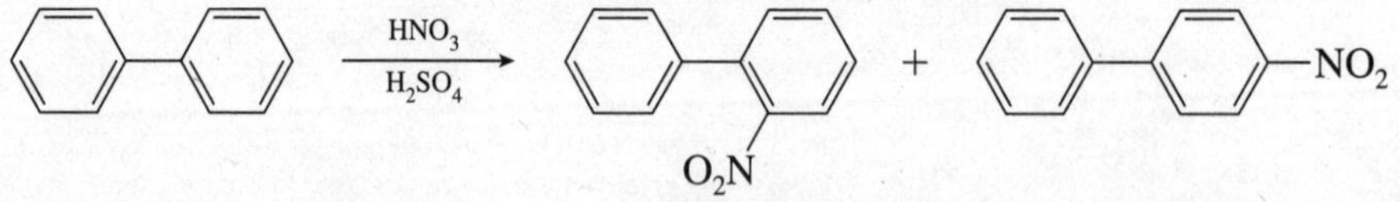

Biphenyl $\xrightarrow[H_2SO_4]{HNO_3}$ *o*-Nitrobiphenyl (37%) ($O_2N$) + *p*-Nitrobiphenyl (63%) (—$NO_2$)

**12.14** (*b*) The carbonyl group attached directly to the ring is a signal that the substituent is a meta-directing group. Nitration of methyl benzoate yields methyl *m*-nitrobenzoate.

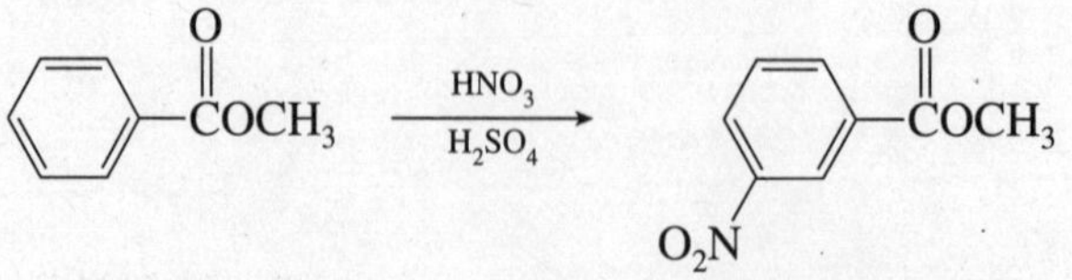

Methyl benzoate ($C_6H_5\overset{O}{\overset{\|}{C}}OCH_3$) $\xrightarrow[H_2SO_4]{HNO_3}$ Methyl *m*-nitrobenzoate ($O_2N$, $\overset{O}{\overset{\|}{C}}OCH_3$) (isolated in 81–85% yield)

(*c*) The acyl group in 1-phenyl-1-propanone is meta-directing; the carbonyl is attached directly to the ring. The product is 1-(*m*-nitrophenyl)-1-propanone.

$C_6H_5CCH_2CH_3$ (O above C) $\xrightarrow{\text{nitration}}$ $m$-$O_2NC_6H_4CCH_2CH_3$

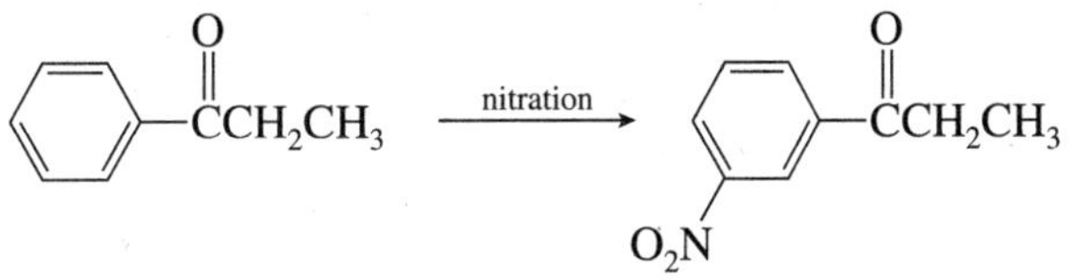

1-Phenyl-1-propanone

1-(*m*-Nitrophenyl)-1-propanone (isolated in 60% yield)

**12.15** Writing the structures out in more detail reveals that the substituent $-\overset{+}{N}(CH_3)_3$ lacks the unshared electron pair of $-\ddot{N}(CH_3)_2$ .

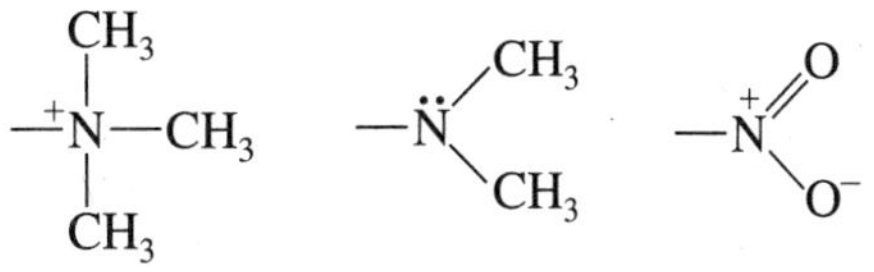

$-\overset{+}{N}(CH_3)_3$ $\quad$ $-\ddot{N}(CH_3)_2$ $\quad$ $-\overset{+}{N}(=O)O^-$

This unshared pair is responsible for the powerful activating effect of an $-\ddot{N}(CH_3)_2$ group. On the other hand, the nitrogen in $-\overset{+}{N}(CH_3)_3$ is positively charged and in that respect resembles the nitrogen of a nitro group. We expect the substituent $-\overset{+}{N}(CH_3)_3$ to be deactivating and meta-directing.

**12.16** The reaction is a Friedel–Crafts alkylation in which 4-chlorobenzyl chloride serves as the carbocation source and chlorobenzene is the aromatic substrate. Alkylation occurs at the positions ortho and para to the chlorine substituent of chlorobenzene.

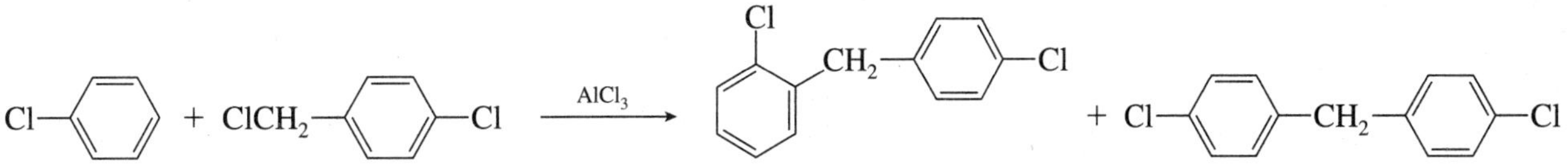

$ClC_6H_5 + ClCH_2C_6H_4Cl \xrightarrow{AlCl_3}$ $o$-$ClC_6H_4CH_2C_6H_4Cl$ + $p$-$ClC_6H_4CH_2C_6H_4Cl$

Chlorobenzene

4-Chlorobenzyl chloride

1-Chloro-2-(4′-chlorobenzyl)-benzene

1-Chloro-4-(4′-chlorobenzyl)-benzene

**12.17** (*b*) Halogen substituents are ortho, para-directing, and the disposition in *m*-dichlorobenzene is such that their effects reinforce each other. The major product is 2,4-dichloro-1-nitrobenzene. Substitution at the position between the two chlorines is slow because it is a sterically hindered position.

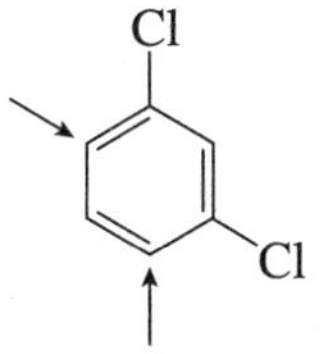

Most reactive positions in electrophilic aromatic substitution of *m*-dichlorobenzene

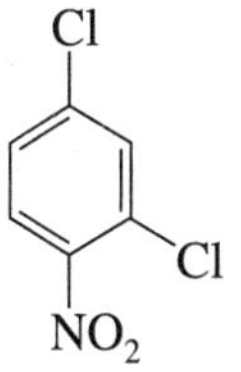

2,4-Dichloro-1-nitrobenzene (major product of nitration)

(*c*) Nitro groups are meta-directing. Both nitro groups of *m*-dinitrobenzene direct an incoming substituent to the same position in an electrophilic aromatic substitution reaction. Nitration of *m*-nitrobenzene yields 1,3,5-trinitrobenzene.

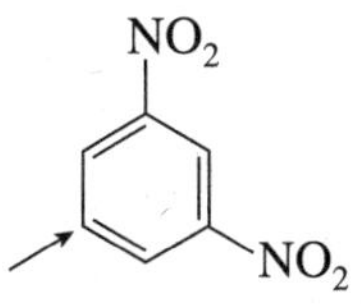

Both nitro groups of *m*-dinitrobenzene direct electrophile to same position.

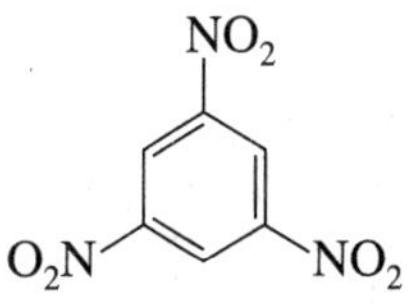

1,3,5-Trinitrobenzene (principal product of nitration of *m*-dinitrobenzene)

(*d*) A methoxy group is ortho, para-directing, and a carbonyl group is meta-directing. The open positions of the ring that are activated by the methoxy group in *p*-methoxyacetophenone are also those that are meta to the carbonyl, so the directing effects of the two substituents reinforce each other. Nitration of *p*-methoxyacetophenone yields 4-methoxy-3-nitroacetophenone.

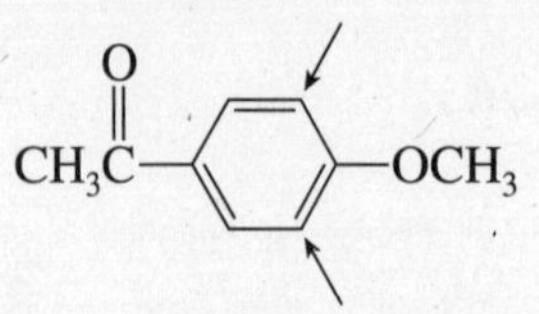

Positions ortho to the methoxy group are meta to the carbonyl.

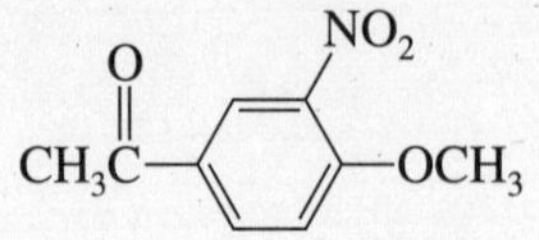

4-Methoxy-3-nitroacetophenone

(*e*) The methoxy group of *p*-methylanisole activates the positions that are ortho to it; the methyl activates those ortho to itself. Methoxy is a more powerful activating substituent than methyl, so nitration occurs ortho to the methoxy group.

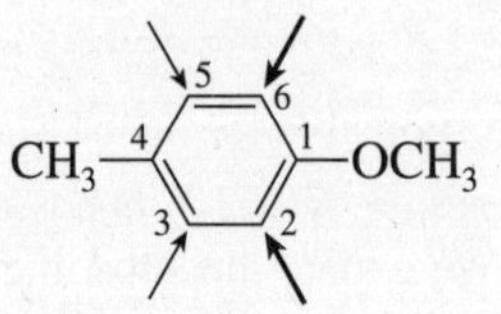

Methyl activates C-3 and C-5; methoxy activates C-2 and C-6.

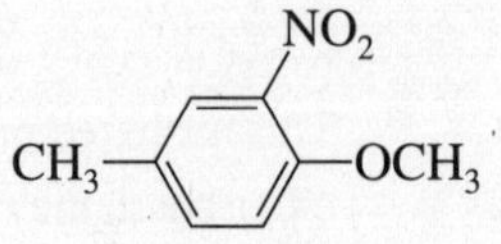

4-Methyl-2-nitroanisole (principal product of nitration)

(*f*) All the substituents in 2,6-dibromoanisole are ortho, para-directing, and their effects are felt at different positions. The methoxy group, however, is a far more powerful activating substituent than bromine, so it controls the regioselectivity of nitration.

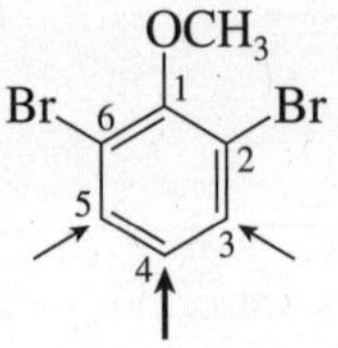

Methoxy directs toward C-4; bromines direct toward C-3 and C-5.

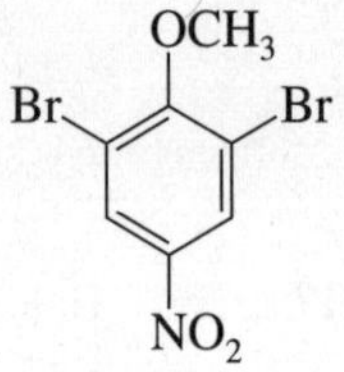

2,6-Dibromo-4-nitroanisole (principal product of nitration)

**12.18** The product that is obtained when benzene is subjected to bromination and nitration depends on the order in which the reactions are carried out. A nitro group is meta-directing, and so if it is introduced prior to the bromination step, *m*-bromonitrobenzene is obtained.

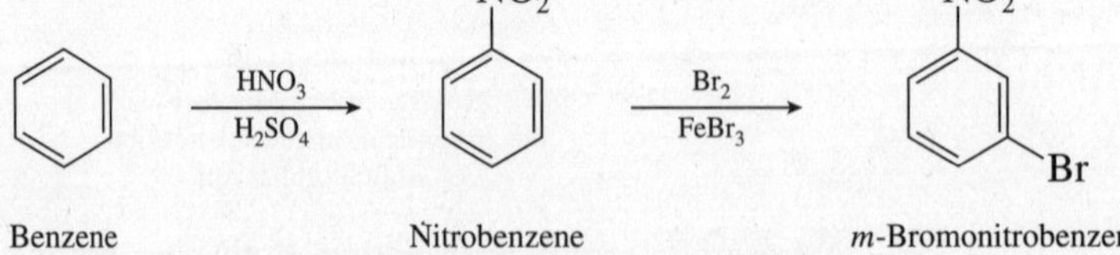

Benzene Nitrobenzene *m*-Bromonitrobenzene

Bromine is an ortho, para-directing group. If it is introduced first, nitration of the resulting bromobenzene yields a mixture of *o*-bromonitrobenzene and *p*-bromonitrobenzene.

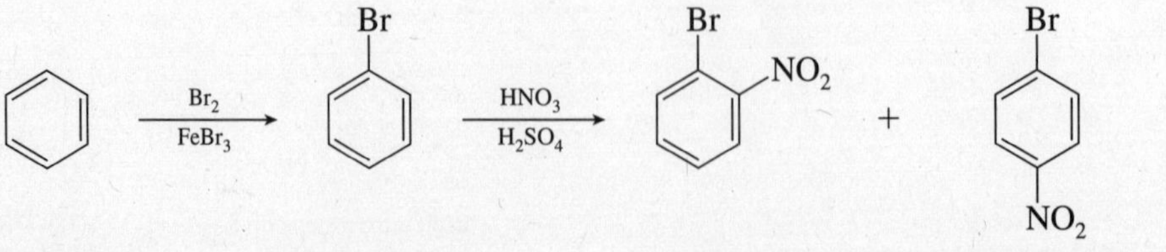

Benzene Bromobenzene *o*-Bromonitrobenzene *p*-Bromonitrobenzene

**12.19** A straightforward approach to the synthesis of *m*-nitrobenzoic acid involves preparation of benzoic acid by oxidation of toluene, followed by nitration. The carboxyl group of benzoic acid is meta-directing. Nitration of toluene prior to oxidation would lead to a mixture of ortho and para products.

$CH_3$ — $Na_2Cr_2O_7$ / $H_2O$, $H_2SO_4$, heat → $CO_2H$ — $HNO_3$ / $H_2SO_4$ → $CO_2H$, $NO_2$

Toluene — Benzoic acid — *m*-Nitrobenzoic acid

**12.20** The text points out that C-1 of naphthalene is more reactive than C-2 toward electrophilic aromatic substitution. Thus, of the two possible products of sulfonation, naphthalene-1-sulfonic acid should be formed faster and should be the major product under conditions of kinetic control. Since the problem states that the product under conditions of thermodynamic control is the other isomer, naphthalene-2-sulfonic acid is the major product at elevated temperature.

H, 1, 2 — $H_2SO_4$ → H, $SO_3H$ + H, $SO_3H$

Naphthalene — Naphthalene-1-sulfonic acid major product at 0°C; formed faster — Naphthalene-2-sulfonic acid major product at 160°C; more stable

Naphthalene-2-sulfonic acid is the more stable isomer for steric reasons. The hydrogen at C-8 (the one shown in the equation) crowds the $—SO_3H$ group in naphthalene-1-sulfonic acid.

**12.21** The text states that electrophilic aromatic substitution in furan, thiophene, and pyrrole occurs at C-2. The sulfonation of thiophene gives thiophene-2-sulfonic acid.

S — $H_2SO_4$ → S, $SO_3H$

Thiophene — Thiophene-2-sulfonic acid

**12.22** (*a*) Nitration of benzene is the archetypical electrophilic aromatic substitution reaction.

$HNO_3$ / $H_2SO_4$ → $NO_2$

Benzene — Nitrobenzene

(*b*) Nitrobenzene is much less reactive than benzene toward electrophilic aromatic substitution. The nitro group on the ring is a meta director.

$NO_2$ — $HNO_3$ / $H_2SO_4$ → $NO_2$, $NO_2$

Nitrobenzene — *m*-Dinitrobenzene

(*c*) Toluene is more reactive than benzene in electrophilic aromatic substitution. A methyl substituent is an ortho, para director.

$CH_3$ $\xrightarrow[FeBr_3]{Br_2}$ $CH_3$, Br + $CH_3$, Br

Toluene — *o*-Bromotoluene — *p*-Bromotoluene

(*d*) Trifluoromethyl is deactivating and meta-directing.

$CF_3$ $\xrightarrow[FeBr_3]{Br_2}$ $CF_3$, Br

(Trifluoromethyl)-benzene — *m*-Bromo(trifluoromethyl)-benzene

(*e*) Anisole is ortho, para-directing, strongly activated toward electrophilic aromatic substitution, and readily sulfonated in sulfuric acid.

$OCH_3$ $\xrightarrow{H_2SO_4}$ $OCH_3$, $SO_3H$ + $OCH_3$, $SO_3H$

Anisole — *o*-Methoxybenzene-sulfonic acid — *p*-Methoxybenzene-sulfonic acid

Sulfur trioxide could be added to the sulfuric acid to facilitate reaction. The para isomer is the predominant product.

(*f*) Acetanilide is quite similar to anisole in its behavior toward electrophilic aromatic substitution.

$HN\ddot{}C(=O)CH_3$ $\xrightarrow{H_2SO_4}$ $HNC(=O)CH_3$, $SO_3H$ + $HNC(=O)CH_3$, $SO_3H$

Acetanilide — *o*-Acetamidobenzene-sulfonic acid — *p*-Acetamidobenzene-sulfonic acid

(*g*) Bromobenzene is less reactive than benzene. A bromine substituent is ortho, para-directing.

Br $\xrightarrow[FeCl_3]{Cl_2}$ Br, Cl + Br, Cl

Bromobenzene — *o*-Bromochloro-benzene — *p*-Bromochloro-benzene

(*h*) Anisole is a reactive substrate toward Friedel–Crafts alkylation and yields a mixture of *o*- and *p*-benzylated products when treated with benzyl chloride and aluminum chloride.

$OCH_3$ $CH_2Cl$ $\xrightarrow{AlCl_3}$ $OCH_3$ $CH_2C_6H_5$ + $OCH_3$ $CH_2C_6H_5$

Anisole + Benzyl chloride → *o*-Benzylanisole + *p*-Benzylanisole

(*i*) Benzene will undergo acylation with benzoyl chloride and aluminum chloride.

$\overset{O}{\|}CCl$ $\xrightarrow{AlCl_3}$ $\overset{O}{\|}C$

Benzene + Benzoyl chloride → Benzophenone

(*j*) A benzoyl substituent is meta-directing and deactivating.

$\overset{O}{\|}C$ $\xrightarrow[H_2SO_4]{HNO_3}$ $\overset{O}{\|}C$ $NO_2$

Benzophenone → *m*-Nitrobenzophenone

(*k*) Clemmensen reduction conditions involve treating a ketone with zinc amalgam and concentrated hydrochloric acid.

$\overset{O}{\|}C$ $\xrightarrow[HCl]{Zn(Hg)}$ $CH_2$

Benzophenone → Diphenylmethane

(*l*) Wolff–Kishner reduction utilizes hydrazine, a base, and a high-boiling alcohol solvent to reduce ketone functions to methylene groups.

$\overset{O}{\|}C$ $\xrightarrow[\text{KOH, diethylene glycol}]{H_2NNH_2}$ $CH_2$

Benzophenone → Diphenylmethane

**12.23** (*a*) There are three principal resonance forms of the cyclohexadienyl cation intermediate formed by attack of bromine on *p*-xylene.

$CH_3$ H Br + $CH_3$ $\longleftrightarrow$ $CH_3$ H Br + $CH_3$ $\longleftrightarrow$ $CH_3$ H Br + $CH_3$

Any one of these resonance forms is a satisfactory answer to the question. Because of its tertiary carbocation character, this carbocation is more stable than the corresponding intermediate formed from benzene.

(*b*) Chlorination of *m*-xylene will give predominantly 4-chloro-1,3-dimethylbenzene.

$CH_3$ $CH_3$ $\xrightarrow{Cl_2}$ $CH_3$ $CH_3$ Cl via $CH_3$ $CH_3$ H Cl

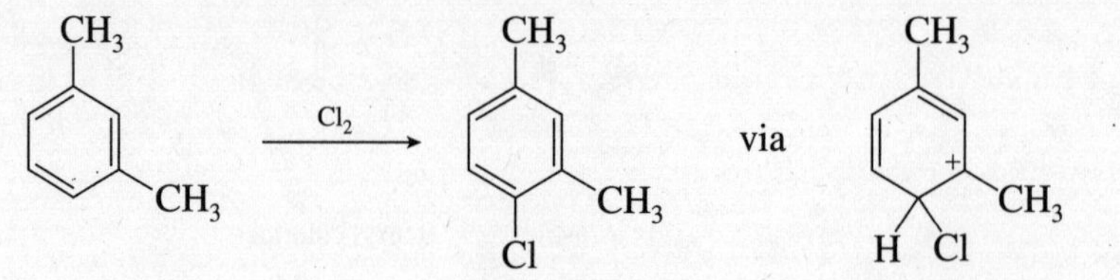

*m*-Xylene 4-Chloro-1,3-dimethylbenzene More stable cyclohexadienyl cation

The intermediate shown (or any of its resonance forms) is more stable for steric reasons than

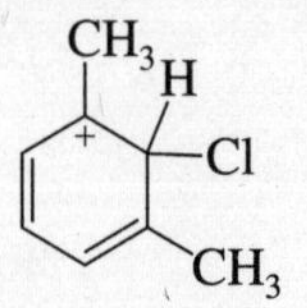

$CH_3$ H Cl $CH_3$

Less stable cyclohexadienyl cation

The cyclohexadienyl cation intermediate leading to 4-chloro-1,3-dimethylbenzene is more stable and is formed faster than the intermediate leading to chlorobenzene because of its tertiary carbocation character.

$CH_3$ $CH_3$ H Cl more stable than H H Cl

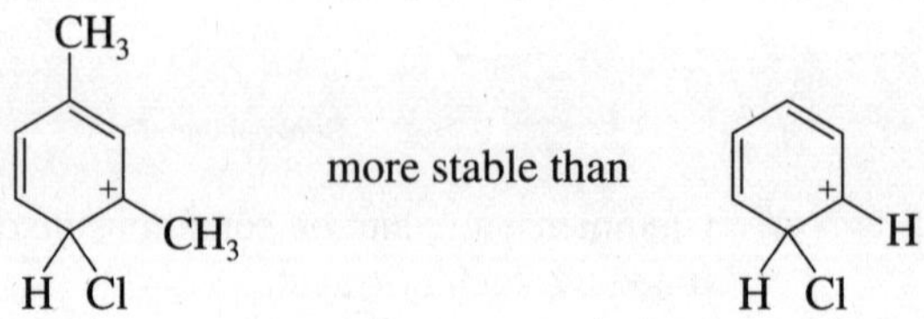

(*c*) The most stable carbocation intermediate formed during nitration of acetophenone is the one corresponding to meta attack.

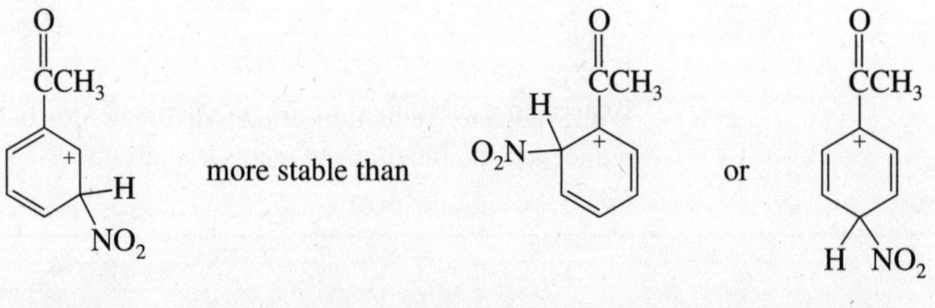

O ‖ $CCH_3$ H $NO_2$ more stable than O ‖ $CCH_3$ H $O_2N$ or O ‖ $CCH_3$ H $NO_2$

An acyl group is electron-withdrawing and destabilizes a carbocation to which it is attached. The most stable carbocation intermediate in the nitration of acetophenone is less stable and is formed more slowly than is the corresponding carbocation formed during nitration of benzene.

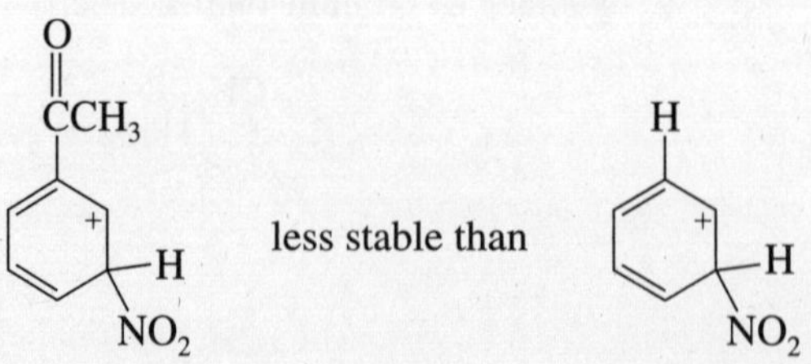

O ‖ $CCH_3$ H $NO_2$ less stable than H H $NO_2$

(*d*) The methoxy group in anisole is strongly activating and ortho, para-directing. For steric reasons and because of inductive electron withdrawal by oxygen, the intermediate leading to para substitution is the most stable.

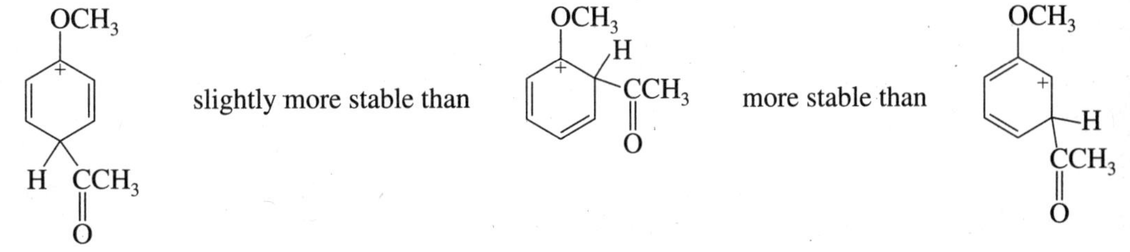

$OCH_3$ / H $\underset{\|}{C}CH_3$ / O  slightly more stable than  $OCH_3$ / H / $\underset{\|}{C}CH_3$ / O  more stable than  $OCH_3$ / H / $\underset{\|}{C}CH_3$ / O

Of the various resonance forms for the most stable intermediate, the most stable one has eight electrons around each oxygen and carbon atom.

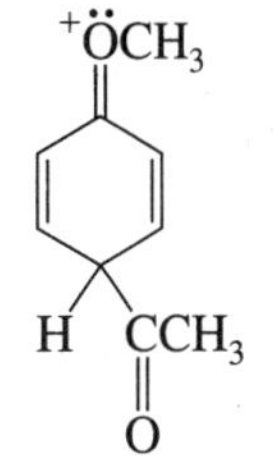

$^{+}\ddot{O}CH_3$ / H $\underset{\|}{C}CH_3$ / O

Most stable resonance form

This intermediate is much more stable than the corresponding intermediate from acylation of benzene.

(*e*) An isopropyl group is an activating substituent and is ortho, para-directing. Attack at the ortho position is sterically hindered. The most stable intermediate is

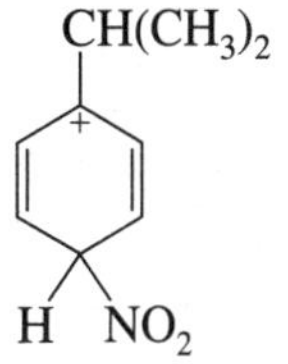

$CH(CH_3)_2$ / H $NO_2$

or any of its resonance forms. Because of its tertiary carbocation character, this cation is more stable than the corresponding cyclohexadienyl cation intermediate from benzene.

(*f*) A nitro substituent is deactivating and meta-directing. The most stable cyclohexadienyl cation formed in the bromination of nitrobenzene is

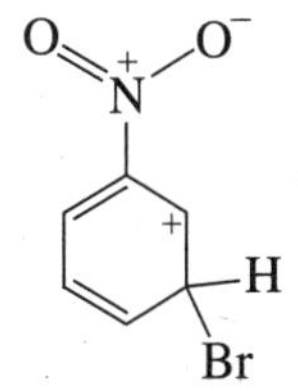

O $\overset{+}{N}$ $O^-$ / H / Br

This ion is less stable than the cyclohexadienyl cation formed during bromination of benzene.

(*g*) Sulfonation of furan takes place at C-2. The cationic intermediate is more stable than the cyclohexadienyl cation formed from benzene because it is stabilized by electron release from oxygen.

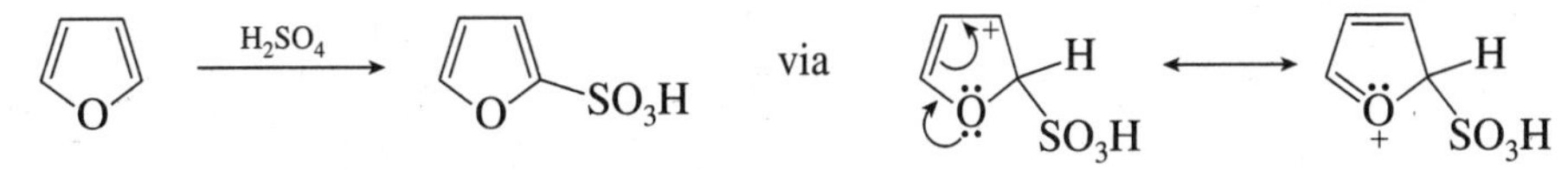

Furan $\xrightarrow{H_2SO_4}$ Furan-2-sulfonic acid ($SO_3H$) via [H, $SO_3H$, $\ddot{O}$] $\longleftrightarrow$ [H, $SO_3H$, $\overset{+}{\ddot{O}}$]

Furan    Furan-2-sulfonic acid

(*h*) Pyridine reacts with electrophiles at C-3. It is less reactive than benzene, and the carbocation intermediate is less stable than the corresponding intermediate formed from benzene.

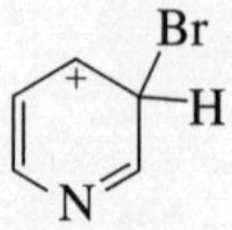

Br H N +

**12.24** (*a*) Toluene is more reactive than chlorobenzene in electrophilic aromatic substitution reactions because a methyl substituent is activating but a halogen substituent is deactivating. Both are ortho, para-directing, however. Nitration of toluene is faster than nitration of chlorobenzene.

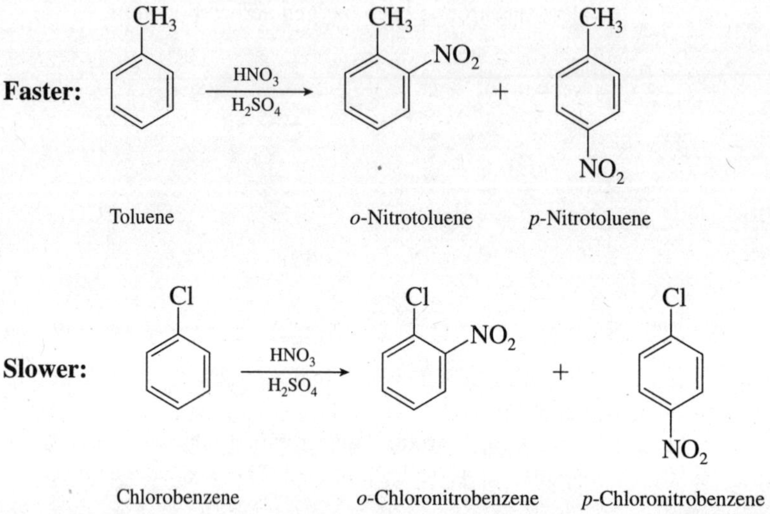

**Faster:** Toluene $\xrightarrow[H_2SO_4]{HNO_3}$ *o*-Nitrotoluene + *p*-Nitrotoluene

**Slower:** Chlorobenzene $\xrightarrow[H_2SO_4]{HNO_3}$ *o*-Chloronitrobenzene + *p*-Chloronitrobenzene

(*b*) A fluorine substituent is not nearly as strongly deactivating as a trifluoromethyl group. The reaction that takes place is Friedel–Crafts alkylation of fluorobenzene.

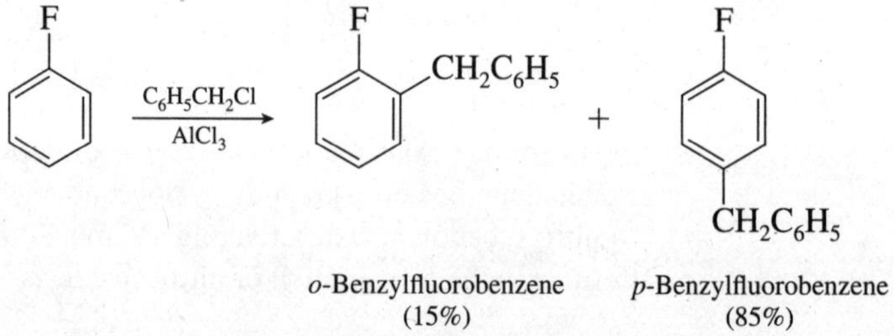

Fluorobenzene $\xrightarrow[AlCl_3]{C_6H_5CH_2Cl}$ *o*-Benzylfluorobenzene (15%) + *p*-Benzylfluorobenzene (85%)

Strongly deactivated aromatic compounds do not undergo Friedel–Crafts reactions.

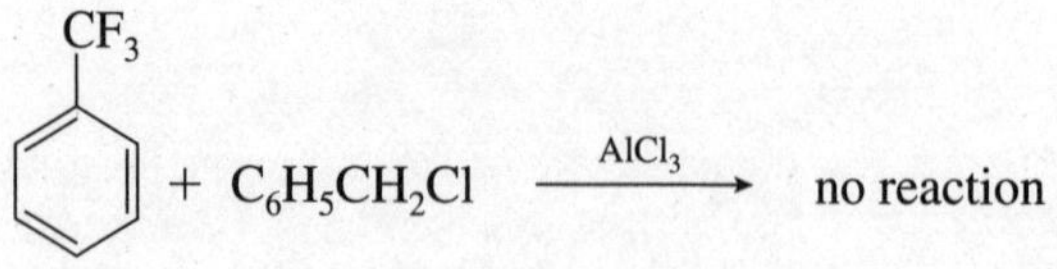

$C_6H_5CF_3 + C_6H_5CH_2Cl \xrightarrow{AlCl_3}$ no reaction

(*c*) A carbonyl group directly bonded to a benzene ring strongly **deactivates** it toward electrophilic aromatic substitution. Methyl benzoate is much less reactive than benzene.

$C_6H_5\overset{O}{\overset{\|}{C}}OCH_3 \longleftrightarrow C_6H_5\overset{O^-}{\overset{|}{C^+}}OCH_3$

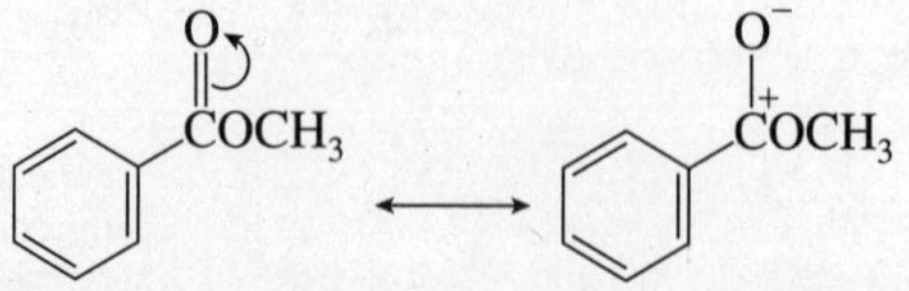

An oxygen substituent directly attached to the ring strongly **activates** it toward electrophilic aromatic substitution. Phenyl acetate is much more reactive than benzene or methyl benzoate.

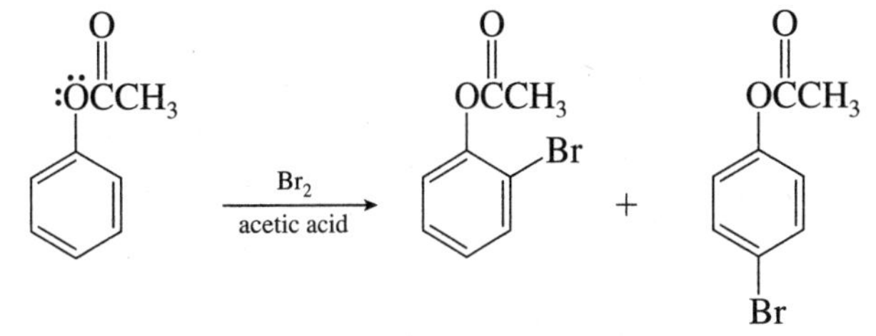

$$\text{Phenyl acetate} \xrightarrow[\text{acetic acid}]{Br_2} \textit{o}\text{-Bromophenyl acetate} + \textit{p}\text{-Bromophenyl acetate}$$

Phenyl acetate; *o*-Bromophenyl acetate; *p*-Bromophenyl acetate

Bromination of methyl benzoate requires more vigorous conditions; catalysis by iron(III) bromide is required for bromination of deactivated aromatic rings.

(*d*) Acetanilide is strongly activated toward electrophilic aromatic substitution and reacts faster than nitrobenzene, which is strongly deactivated.

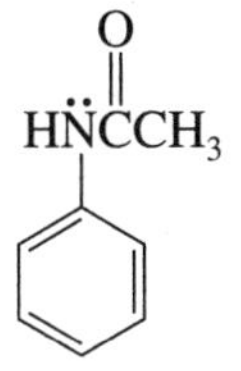

Acetanilide (Lone pair on nitrogen can stabilize cyclohexadienyl cation intermediate.)

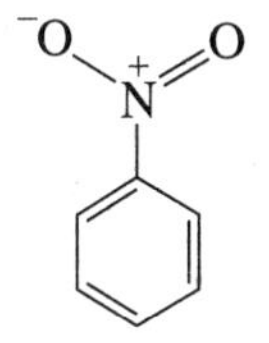

Nitrobenzene (Nitrogen is positively charged and is electron-withdrawing.)

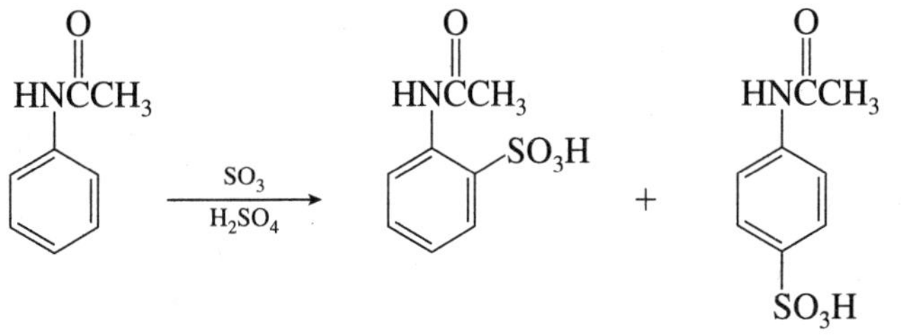

$$\text{Acetanilide} \xrightarrow[H_2SO_4]{SO_3} \textit{o}\text{-Acetamidobenzenesulfonic acid} + \textit{p}\text{-Acetamidobenzenesulfonic acid}$$

Acetanilide; *o*-Acetamidobenzenesulfonic acid; *p*-Acetamidobenzenesulfonic acid

(*e*) Both substrates are of the type

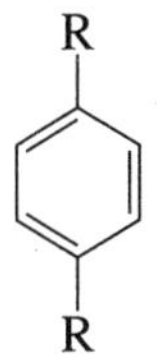

$R = \text{alkyl}$

and are activated toward Friedel–Crafts acylation. Since electronic effects are comparable, we look to differences in steric factors and conclude that reaction will be faster for $R = CH_3$ than for $R = (CH_3)_3C—$.

$$\textit{p}\text{-Xylene} + CH_3CCl \xrightarrow{AlCl_3} \text{2,5-Dimethylacetophenone}$$

*p*-Xylene; Acetyl chloride; 2,5-Dimethylacetophenone

(*f*) A phenyl substituent is activating and ortho, para-directing. Biphenyl will undergo chlorination readily.

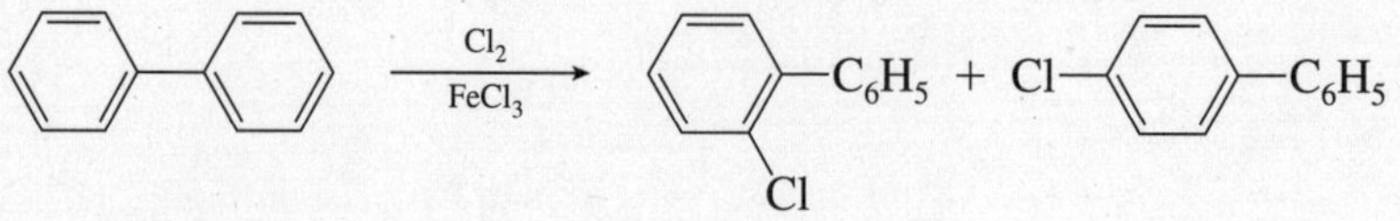

$Cl_2$, $FeCl_3$

Biphenyl → *o*-Chlorobiphenyl ($C_6H_5$, Cl) + *p*-Chlorobiphenyl ($Cl{-}C_6H_4{-}C_6H_5$)

Each benzene ring of benzophenone is deactivated by the carbonyl group.

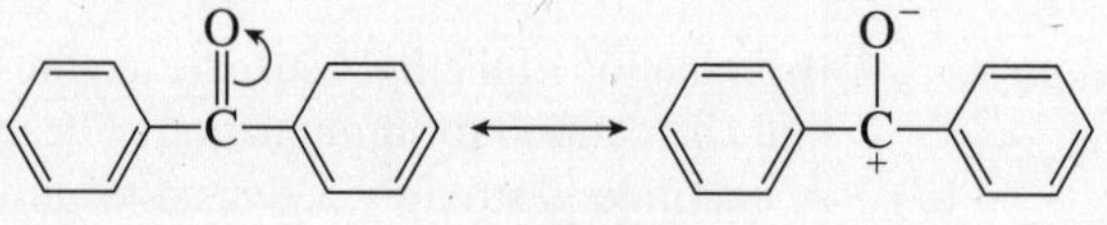

Benzophenone

Benzophenone is much less reactive than biphenyl in electrophilic aromatic substitution reactions.

**12.25** Reactivity toward electrophilic aromatic substitution increases with increasing number of electron-releasing substituents. Benzene, with no methyl substituents, is the least reactive, followed by toluene, with one methyl group. 1,3,5-Trimethylbenzene, with three methyl substituents, is the most reactive.

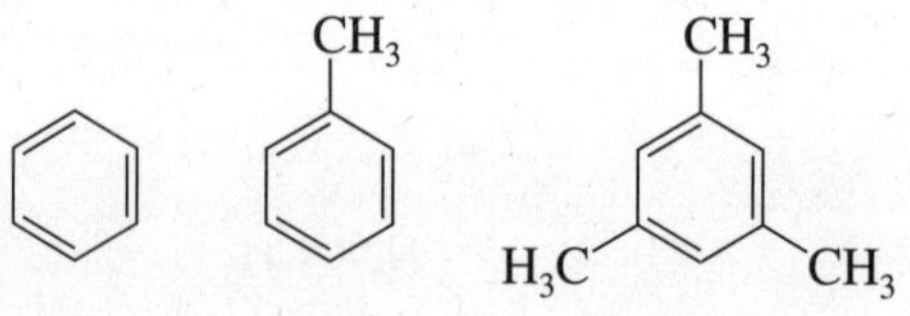

| | Benzene | Toluene | 1,3,5-Trimethylbenzene |
|---|---|---|---|
| Relative reactivity: | 1 | 60 | $2 \times 10^7$ |

*o*-Xylene and *m*-xylene are intermediate in reactivity between toluene and 1,3,5-trimethylbenzene. Of the two, *m*-xylene is more reactive than *o*-xylene because the activating effects of the two methyl groups reinforce each other.

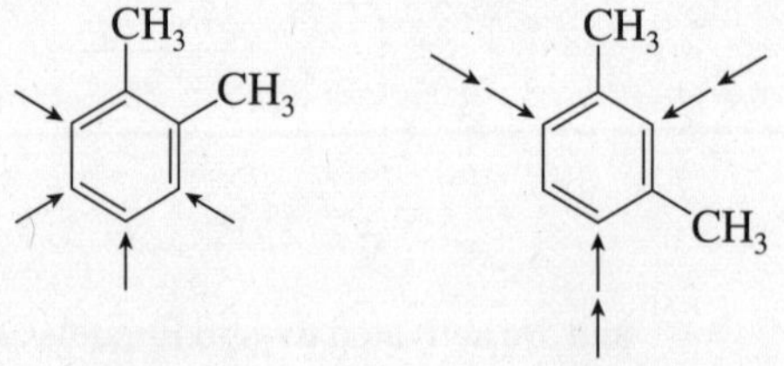

| | *o*-Xylene (all positions somewhat activated) | *m*-Xylene (activating effects reinforce each other) |
|---|---|---|
| Relative reactivity: | $5 \times 10^2$ | $5 \times 10^4$ |

**12.26** (*a*) Chlorine is ortho, para-directing, carboxyl is meta-directing. The positions that are ortho to the chlorine are meta to the carboxyl, so that both substituents direct an incoming electrophile to the same position. Introduction of the second nitro group at the remaining position that is ortho to the chlorine puts it meta to the carboxyl and meta to the first nitro group.

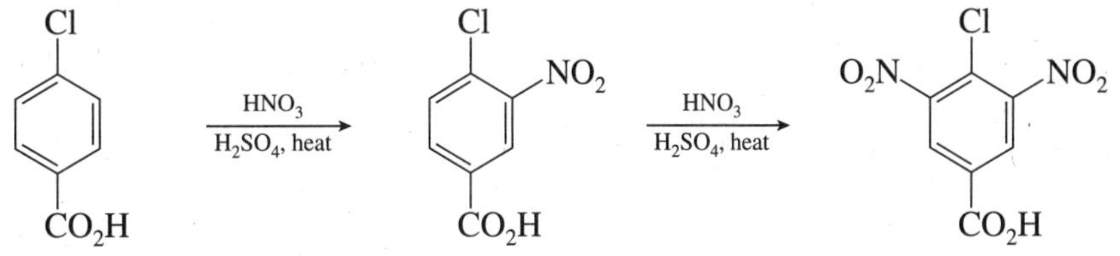

$\xrightarrow[H_2SO_4,\ heat]{HNO_3}$ $\xrightarrow[H_2SO_4,\ heat]{HNO_3}$

*p*-Chlorobenzoic acid → 4-Chloro-3,5-dinitrobenzoic acid (90%)

(*b*) An amino group is one of the strongest activating substituents. The para and both ortho positions are readily substituted in aniline. When aniline is treated with excess bromine, 2,4,6-tribromoaniline is formed in quantitative yield.

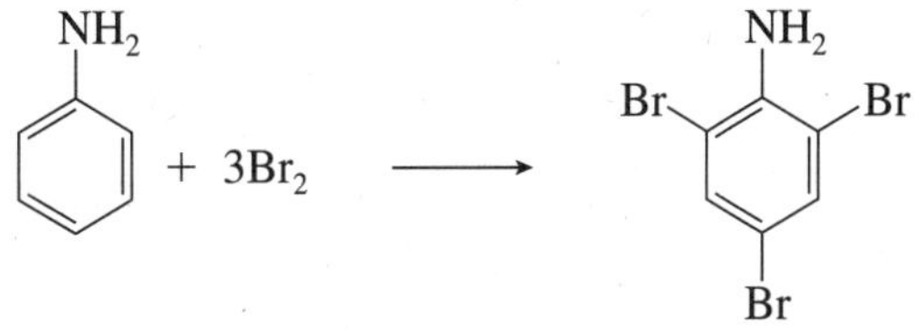

Aniline $+ 3Br_2 \longrightarrow$ 2,4,6-Tribromoaniline (100%)

(*c*) The positions ortho and para to the amino group in *o*-aminoacetophenone are the ones most activated toward electrophilic aromatic substitution.

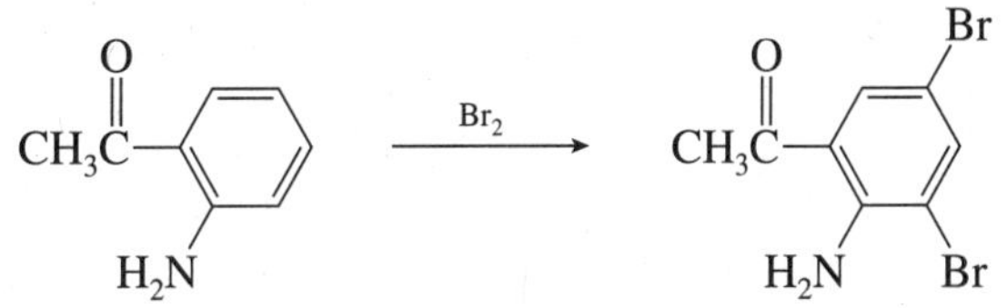

*o*-Aminoacetophenone $\xrightarrow{Br_2}$ 2-Amino-3,5-dibromoacetophenone (65%)

(*d*) The carboxyl group in benzoic acid is meta-directing, and so nitration gives *m*-nitrobenzoic acid. The second nitration step introduces a nitro group meta to both the carboxyl group and the first nitro group.

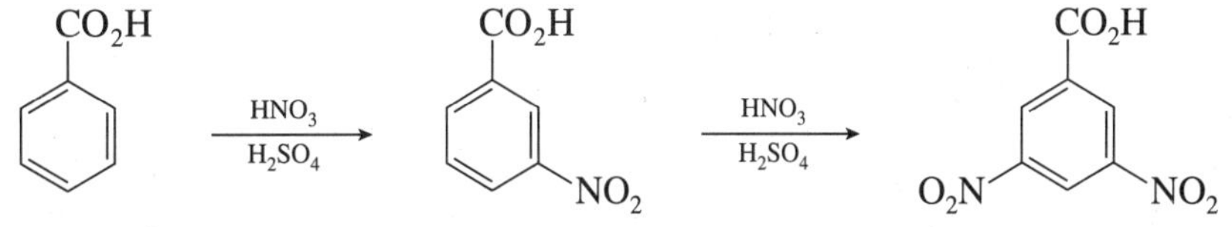

Benzoic acid $\xrightarrow[H_2SO_4]{HNO_3}$ *m*-Nitrobenzoic acid $\xrightarrow[H_2SO_4]{HNO_3}$ 3,5-Dinitrobenzoic acid (54–58%)

(*e*) Both bromine substituents are introduced ortho to the strongly activating hydroxyl group in *p*-nitrophenol.

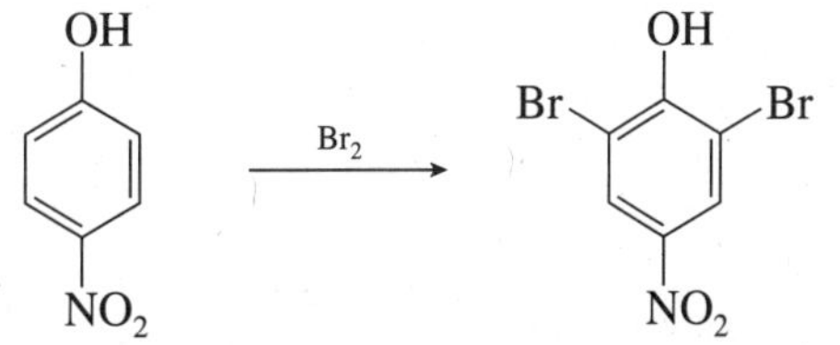

*p*-Nitrophenol $\xrightarrow{Br_2}$ 2,6-Dibromo-4-nitrophenol (96–98%)

(*f*) Friedel–Crafts alkylation occurs when biphenyl is treated with *tert*-butyl chloride and iron (III) chloride (a Lewis acid catalyst); the product of monosubstitution is *p-tert*-butylbiphenyl. All the positions of the ring that bears the *tert*-butyl group are sterically hindered, so the second alkylation step introduces a *tert*-butyl group at the para position of the second ring.

$C(CH_3)_3$ $C(CH_3)_3$

$\xrightarrow[FeCl_3]{(CH_3)_3CCl}$ $\xrightarrow[FeCl_3]{(CH_3)_3CCl}$

$C(CH_3)_3$

Biphenyl 4,4′-Di-*tert*-butylbiphenyl (70%)

(*g*) Disulfonation of phenol occurs at positions ortho and para to the hydroxyl group. The ortho, para product predominates over the ortho, ortho one.

OH OH

$\xrightarrow{H_2SO_4}$ $SO_3H$

$SO_3H$

Phenol 2-Hydroxy-1,5-benzenedisulfonic acid

**12.27** When carrying out each of the following syntheses, evaluate how the structure of the product differs from that of benzene or toluene; that is, determine which groups have been substituted on the benzene ring or altered in some way. The sequence of reaction steps when multiple substitution is desired is important; recall that some groups direct ortho, para and others meta.

(*a*) Isopropylbenzene may be prepared by a Friedel–Crafts alkylation of benzene with isopropyl chloride (or bromide, or iodide).

$$+ \ (CH_3)_2CHCl \xrightarrow{AlCl_3} C_6H_5{-}CH(CH_3)_2$$

Benzene Isopropyl chloride Isopropylbenzene

It would not be appropriate to use propyl chloride and trust that a rearrangement would lead to isopropylbenzene, because a mixture of propylbenzene and isopropylbenzene would be obtained.

Isopropylbenzene may also be prepared by alkylation of benzene with propene in the presence of sulfuric acid.

$$+ \ CH_3CH{=}CH_2 \xrightarrow{H_2SO_4} C_6H_5{-}CH(CH_3)_2$$

Benzene Propene Isopropylbenzene

(*b*) Since the isopropyl and sulfonic acid groups are para to each other, the first group introduced on the ring must be the ortho, para director, that is, the isopropyl group. We may therefore use the product of part (*a*), isopropylbenzene, in this synthesis. An isopropyl group is a fairly bulky ortho, para director, and so sulfonation of isopropylbenzene gives mainly *p*-isopropylbenzenesulfonic acid.

$$(CH_3)_2CH{-}C_6H_5 \xrightarrow[H_2SO_4]{SO_3} (CH_3)_2CH{-}C_6H_4{-}SO_3H$$

Isopropylbenzene → *p*-Isopropylbenzenesulfonic acid

A sulfonic acid group is meta-directing, so that the order of steps must be alkylation followed by sulfonation rather than the reverse.

(*c*) Free-radical halogenation of isopropylbenzene occurs with high regioselectivity at the benzylic position. *N*-Bromosuccinimide (NBS) is a good reagent to use for benzylic bromination reactions.

$$C_6H_5{-}CH(CH_3)_2 \xrightarrow[\text{or NBS}]{Br_2,\ \text{light}} C_6H_5{-}CBr(CH_3){-}CH_3$$

Isopropylbenzene → 2-Bromo-2-phenylpropane

(*d*) Toluene is an obvious starting material for the preparation of 4-*tert*-butyl-2-nitrotoluene. Two possibilities, both involving nitration and alkylation of toluene, present themselves; the problem to be addressed is in what order to carry out the two steps. Friedel–Crafts alkylation must precede nitration.

$$\text{Toluene} \xrightarrow[AlCl_3]{(CH_3)_3CCl} p\text{-}tert\text{-Butyltoluene} \xrightarrow[H_2SO_4]{HNO_3} 4\text{-}tert\text{-Butyl-2-nitrotoluene}$$

Toluene; *p*-*tert*-Butyltoluene; 4-*tert*-Butyl-2-nitrotoluene

Introduction of the nitro group as the first step is an unsatisfactory approach since Friedel–Crafts reactions cannot be carried out on nitro-substituted aromatic compounds.

(*e*) Two electrophilic aromatic substitution reactions need to be performed: chlorination and Friedel–Crafts acylation. The order in which the reactions are carried out is important; chlorine is an ortho, para director, and the acetyl group is a meta director. Since the groups are meta in the desired compound, introduce the acetyl group first.

$$\text{Benzene} \xrightarrow[AlCl_3]{CH_3CCl\ (=O)} C_6H_5CCH_3\ (=O) \xrightarrow[AlCl_3]{Cl_2} m\text{-}ClC_6H_4CCH_3\ (=O)$$

Benzene; Acetophenone; *m*-Chloroacetophenone

(*f*) Reverse the order of steps in part (*e*) to prepare *p*-chloroacetophenone.

$$\text{Benzene} \xrightarrow[FeCl_3]{Cl_2} Cl{-}C_6H_5 \xrightarrow[AlCl_3]{CH_3CCl\ (=O)} Cl{-}C_6H_4{-}CCH_3\ (=O)$$

Benzene; Chlorobenzene; *p*-Chloroacetophenone

Friedel–Crafts reactions can be carried out on halobenzenes but not on arenes that are more strongly deactivated.

(*g*) Here again the problem involves two successive electrophilic aromatic substitution reactions, in this case using toluene as the initial substrate. The proper sequence is Friedel–Crafts acylation first, followed by bromination of the ring.

Toluene $\xrightarrow[AlCl_3]{CH_3CCl\ (O)}$ *p*-Methylacetophenone $\xrightarrow[AlBr_3]{Br_2}$ 3-Bromo-4-methylacetophenone

If the sequence of steps had been reversed, with halogenation preceding acylation, the first intermediate would be *o*-bromotoluene, Friedel–Crafts acylation of which would give a complex mixture of products because both groups are ortho, para-directing. On the other hand, the orienting effects of the two groups in *p*-methylacetophenone reinforce each other, so that its bromination is highly regioselective and in the desired direction.

(*h*) Recalling that alkyl groups attached to the benzene ring by $CH_2$ may be prepared by reduction of the appropriate ketone, we may reduce 3-bromo-4-methylacetophenone, as prepared in part (*g*), by the Clemmensen on Wolff–Kishner procedure to give 2-bromo-4-ethyltoluene.

3-Bromo-4-methylacetophenone $\xrightarrow[\text{or } H_2NNH_2,\ KOH\text{, diethylene glycol, heat}]{Zn(Hg),\ HCl}$ 2-Bromo-4-ethyltoluene

(*i*) This is a relatively straightforward synthetic problem. Bromine is an ortho, para-directing substituent; nitro is meta-directing. Nitrate first, and then brominate to give 1-bromo-3-nitrobenzene.

Benzene $\xrightarrow[H_2SO_4]{HNO_3}$ Nitrobenzene $\xrightarrow[AlBr_3]{Br_2}$ 1-Bromo-3-nitrobenzene

(*j*) Take advantage of the ortho, para-directing properties of bromine to prepare 1-bromo-2,4-dinitrobenzene. Brominate first, and then nitrate under conditions that lead to disubstitution. The nitro groups are introduced at positions ortho and para to the bromine and meta to each other.

Benzene $\xrightarrow{Br_2,\ FeBr_3}$ Bromobenzene $\xrightarrow[H_2SO_4\text{, heat}]{HNO_3}$ 1-Bromo-2,4-dinitrobenzene

(*k*) Although bromo and nitro substituents are readily introduced by electrophilic aromatic substitution, the only methods we have available so far to prepare carboxylic acids is by oxidation of alkyl side chains. Thus, use toluene as a starting material, planning to convert the methyl group to a carboxyl group by oxidation. Nitrate next; nitro and carboxyl are both meta-directing groups, so that the bromination in the last step occurs with the proper regioselectivity.

Toluene $\xrightarrow[H_2O,\ H_2SO_4,\ heat]{Na_2Cr_2O_7}$ Benzoic acid $\xrightarrow[H_2SO_4]{HNO_3}$ 3-Nitrobenzoic acid $\xrightarrow[FeBr_3]{Br_2}$ 3-Bromo-5-nitrobenzoic acid

If bromination is performed prior to nitration, the bromine substituent will direct an incoming electrophile to positions ortho and para to itself, giving the wrong orientation of substituents in the product.

(*l*) Again toluene is a suitable starting material, with its methyl group serving as the source of the carboxyl substituent. The orientation of the substituents in the final product requires that the methyl group be retained until the final step.

Toluene $\xrightarrow[H_2SO_4]{HNO_3}$ *p*-Nitrotoluene $\xrightarrow[FeBr_3]{Br_2}$ 2-Bromo-4-nitrotoluene $\xrightarrow[H_2O,\ H_2SO_4,\ heat]{Na_2Cr_2O_7}$ 2-Bromo-4-nitrobenzoic acid

Nitration must precede bromination, as in the previous part, in order to prevent formation of an undesired mixture of isomers.

(*m*) Friedel–Crafts alkylation of benzene with benzyl chloride (or benzyl bromide) is a satisfactory route to diphenylmethane.

Benzene + $C_6H_5CH_2Cl$ (Benzyl chloride) $\xrightarrow{AlCl_3}$ $C_6H_5CH_2C_6H_5$ (Diphenylmethane)

Benzyl chloride is prepared by free-radical chlorination of toluene.

$C_6H_5CH_3$ (Toluene) $\xrightarrow[light\ or\ heat]{Cl_2}$ $C_6H_5CH_2Cl$ (Benzyl chloride)

Alternatively, benzene could have been subjected to Friedel–Crafts acylation with benzoyl chloride to give benzophenone. Clemmensen or Wolff–Kishner reduction of benzophenone would then furnish diphenylmethane.

(*n*) 1-Phenyloctane cannot be prepared efficiently by direct alkylation of benzene, because of the probability that rearrangement will occur. Indeed, a mixture of 1-phenyloctane and 2-phenyloctane is formed under the usual Friedel–Crafts conditions, along with 3-phenyloctane.

$$C_6H_6 + CH_3(CH_2)_6CH_2Br \xrightarrow{AlBr_3} C_6H_5CH_2(CH_2)_6CH_3 + C_6H_5CH(CH_3)(CH_2)_5CH_3 + C_6H_5CH(CH_2CH_3)(CH_2)_4CH_3$$

Benzene; 1-Bromooctane; 1-Phenyloctane (40%); 2-Phenyloctane (30%); 3-Phenyloctane (30%)

A method that permits the synthesis of 1-phenyloctane free of isomeric compounds is acylation followed by reduction.

$$\underset{\text{Benzene}}{C_6H_6} + \underset{\text{Octanoyl chloride}}{CH_3(CH_2)_6\overset{O}{\overset{\|}{C}}Cl} \xrightarrow{AlCl_3} \underset{\text{1-Phenyl-1-octanone}}{C_6H_5\overset{O}{\overset{\|}{C}}(CH_2)_6CH_3} \xrightarrow[HCl]{Zn(Hg)} \underset{\text{1-Phenyloctane}}{C_6H_5CH_2(CH_2)_6CH_3}$$

Alternatively, Wolff–Kishner conditions (hydrazine, potassium hydroxide, diethylene glycol) could be used in the reduction step.

(*o*) Direct alkenylation of benzene under Friedel–Crafts reaction conditions does not take place, and so 1-phenyl-1-octene cannot be prepared by the reaction

$$\underset{\text{Benzene}}{C_6H_6} + \underset{\text{1-Chloro-1-octene}}{ClCH{=}CH(CH_2)_5CH_3} \xrightarrow{AlCl_3} \underset{\text{1-Phenyl-1-octene}}{C_6H_5CH{=}CH(CH_2)_5CH_3}$$

**No!** Reaction effective only with *alkyl* halides, not 1-haloalkenes.

Having already prepared 1-phenyloctane in part (*n*), however, we can functionalize the benzylic position by bromination and then carry out a dehydrohalogenation to obtain the target compound.

$$\underset{\text{1-Phenyloctane}}{C_6H_5CH_2(CH_2)_6CH_3} \xrightarrow[\text{or NBS}]{Br_2,\ \text{light}} \underset{\text{1-Bromo-1-phenyloctane}}{C_6H_5\underset{Br}{\underset{|}{C}}H(CH_2)_6CH_3} \xrightarrow[CH_3OH]{KOCH_3} \underset{\text{1-Phenyl-1-octene}}{C_6H_5CH{=}CH(CH_2)_5CH_3}$$

(*p*) 1-Phenyl-1-octyne cannot be prepared in one step from benzene; 1-haloalkynes are unsuitable reactants for a Friedel–Crafts process. In Chapter 9, however, we learned that alkynes may be prepared from the corresponding alkene:

$$RC{\equiv}CR \quad \text{obtained from} \quad \underset{Br}{\underset{|}{RCH}}{-}\underset{Br}{\underset{|}{CHR}} \quad \text{obtained from} \quad RCH{=}CHR$$

Using the alkene prepared in part (*o*),

$$\underset{\text{1-Phenyl-1-octene}}{C_6H_5CH{=}CH(CH_2)_5CH_3} \xrightarrow{Br_2} \underset{\text{1,2-Dibromo-1-phenyloctane}}{C_6H_5\underset{Br}{\underset{|}{C}}H\underset{Br}{\underset{|}{C}}H(CH_2)_5CH_3} \xrightarrow[NH_3]{NaNH_2} \underset{\text{1-Phenyl-1-octyne}}{C_6H_5C{\equiv}C(CH_2)_5CH_3}$$

(*q*) Nonconjugated cyclohexadienes are prepared by Birch reduction of arenes. Thus the last step in the synthesis of 1,4-di-*tert*-butyl-1,4-cyclohexadiene is the Birch reduction of 1,4-di-*tert*-butylbenzene.

$$\text{Benzene} \xrightarrow[AlCl_3]{(CH_3)_3CCl} \textit{tert}\text{-Butylbenzene } (C(CH_3)_3) \xrightarrow[AlCl_3]{(CH_3)_3CCl} p\text{-Di-}\textit{tert}\text{-butylbenzene } (C(CH_3)_3,\ C(CH_3)_3) \xrightarrow[\text{ethanol}]{Na,\ NH_3} \text{1,4-Di-}\textit{tert}\text{-butyl-1,4-cyclohexadiene } (C(CH_3)_3,\ C(CH_3)_3)$$

**12.28** (*a*) Methoxy is an ortho, para-directing substituent. All that is required to prepare *p*-methoxybenzenesulfonic acid is to sulfonate anisole.

Anisole ($OCH_3$) $\xrightarrow{H_2SO_4}$ *p*-Methoxybenzenesulfonic acid ($OCH_3$, $SO_3H$)

(*b*) In reactions involving disubstitution of anisole, the better strategy is to introduce the para substituent first. The methoxy group is ortho, para-directing, but para substitution predominates.

Anisole $\xrightarrow[H_2SO_4]{HNO_3}$ *p*-Nitroanisole $\xrightarrow[FeBr_3]{Br_2}$ 2-Bromo-4-nitroanisole

(*c*) Reversing the order of the steps used in part (*b*) yields 4-bromo-2-nitroanisole.

Anisole $\xrightarrow[FeBr_3]{Br_2}$ *p*-Bromoanisole $\xrightarrow[H_2SO_4]{HNO_3}$ 4-Bromo-2-nitroanisole

(*d*) Direct introduction of a vinyl substituent onto an aromatic ring is not a feasible reaction. *p*-Methoxystyrene must be prepared in an indirect way by adding an ethyl side chain and then taking advantage of the reactivity of the benzylic position by bromination (e.g., with *N*-bromosuccinimide) and dehydrohalogenation.

Anisole $\xrightarrow[AlCl_3]{CH_3CH_2Cl}$ *p*-Ethylanisole ($CH_2CH_3$) $\xrightarrow[\text{peroxides, heat}]{NBS}$ *p*-(1-Bromoethyl)anisole ($BrCHCH_3$) $\xrightarrow{NaOCH_3}$ *p*-Methoxystyrene ($CH{=}CH_2$)

**12.29** (*a*) Methyl is an ortho, para-directing substituent, and toluene yields mainly *o*-nitrotoluene and *p*-nitrotoluene on mononitration. Some *m*-nitrotoluene is also formed.

Toluene ($CH_3$) $\xrightarrow[40°C]{HNO_3}$ *o*-Nitrotoluene + *m*-Nitrotoluene + *p*-Nitrotoluene (*b*) There are six isomeric dinitrotoluenes:

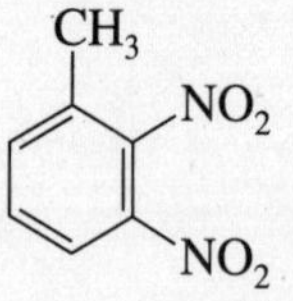

2,3-Dinitrotoluene 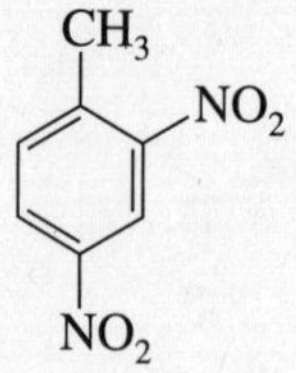 2,4-Dinitrotoluene 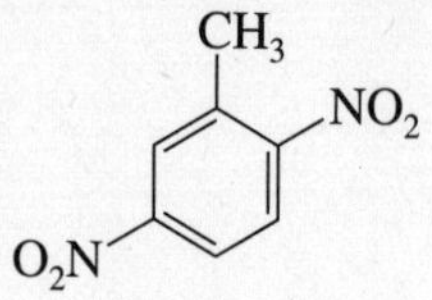 2,5-Dinitrotoluene 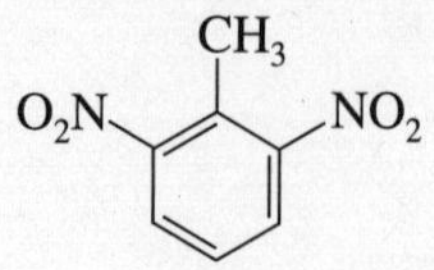 2,6-Dinitrotoluene

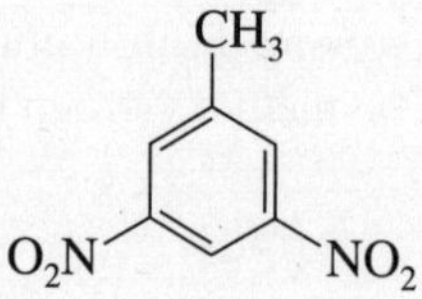

3,5-Dinitrotoluene 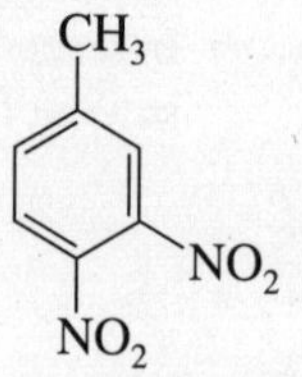 3,4-Dinitrotoluene

The least likely product is 3,5-dinitrotoluene because neither of its nitro groups is ortho or para to the methyl group.

(*c*) There are six trinitrotoluene isomers:

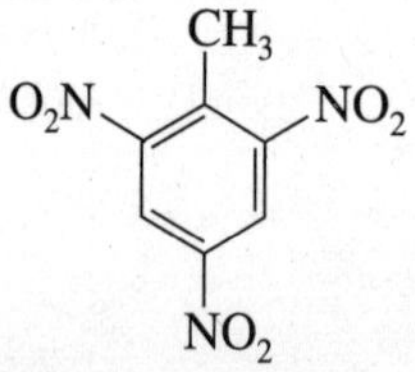

2,4,6-Trinitrotoluene 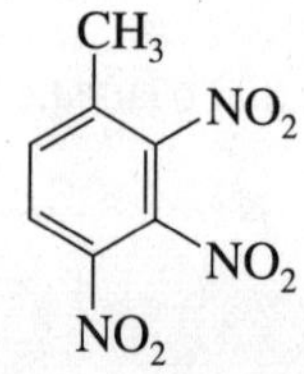 2,3,4-Trinitrotoluene 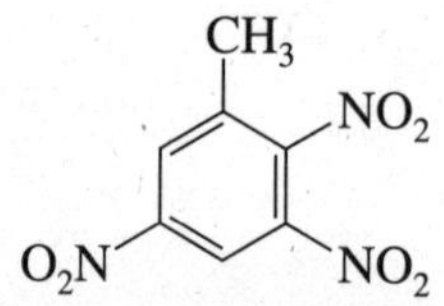 2,3,5-Trinitrotoluene

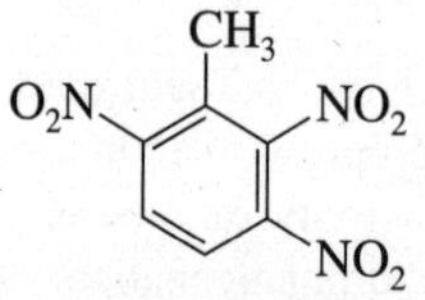

2,3,6-Trinitrotoluene 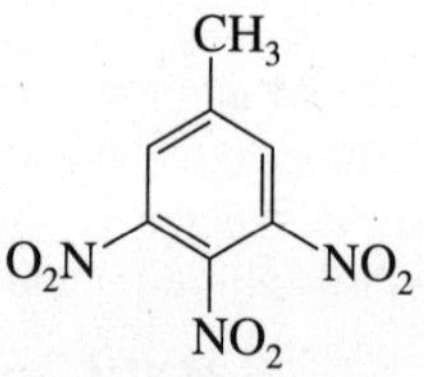 3,4,5-Trinitrotoluene 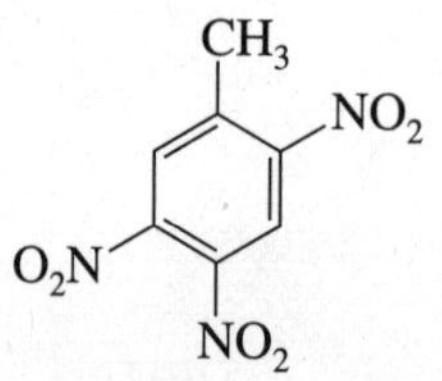 2,4,5-Trinitrotoluene

The most likely major product is 2,4,6-trinitrotoluene because all the positions activated by the methyl group are substituted. This is, in fact, the compound commonly known as TNT.

**12.30 From *o*-xylene:**

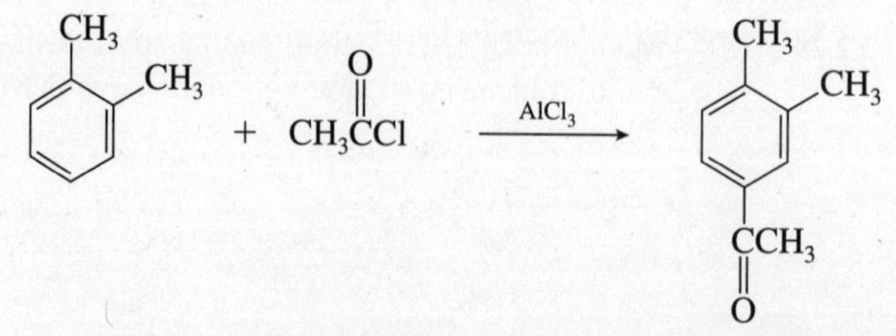

$$\text{o-Xylene} + CH_3\overset{O}{\overset{\|}{C}}Cl \xrightarrow{AlCl_3} \text{3,4-Dimethylacetophenone (94\%)}$$

*o*-Xylene Acetyl chloride 3,4-Dimethyl-acetophenone (94%)

**From *m*-xylene:**

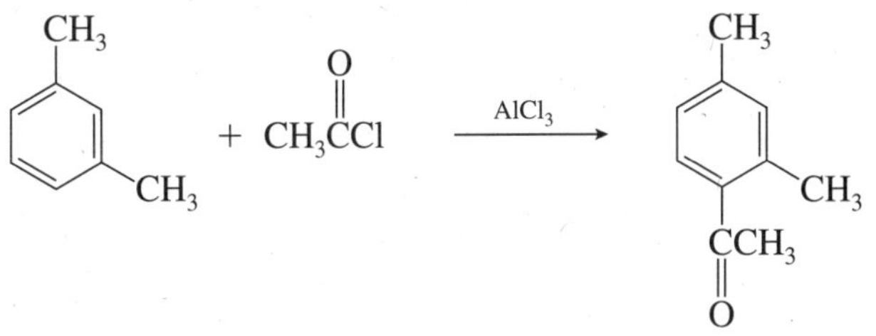

*m*-Xylene + $CH_3\overset{O}{\overset{\|}{C}}Cl$ $\xrightarrow{AlCl_3}$ 2,4-Dimethylacetophenone (86%)

**From *p*-xylene:**

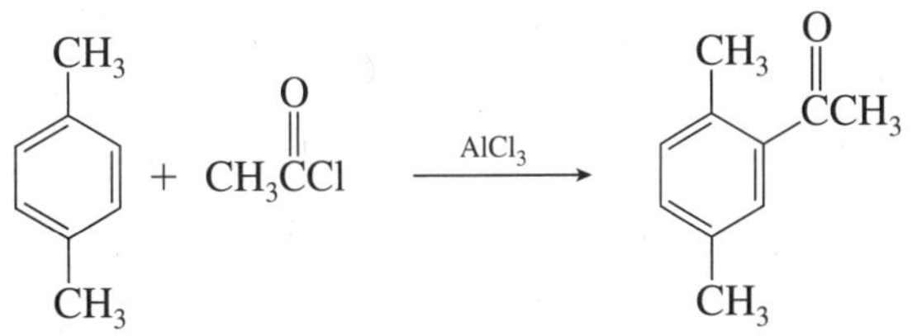

*p*-Xylene + $CH_3\overset{O}{\overset{\|}{C}}Cl$ $\xrightarrow{AlCl_3}$ 2,5-Dimethylacetophenone (99%)

**12.31** The ring that bears the nitrogen in benzanilide is activated toward electrophilic aromatic substitution. The ring that bears the C═O is strongly deactivated.

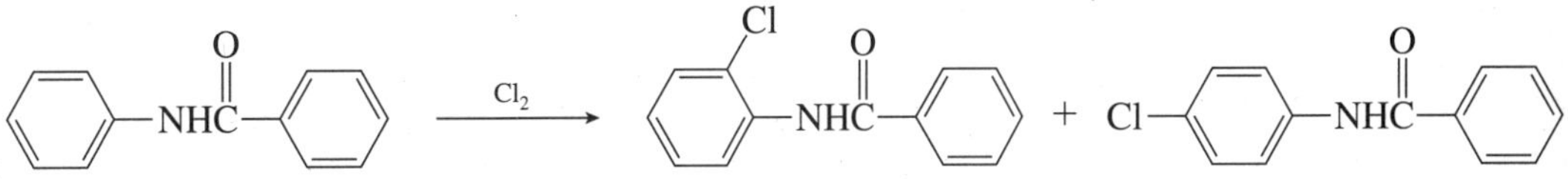

Benzanilide $\xrightarrow{Cl_2}$ *N*-(*o*-Chlorophenyl)benzamide + *N*-(*p*-Chlorophenyl)benzamide

**12.32** (*a*) Nitration of the ring takes place para to the ortho, para-directing chlorine substituent; this position is also meta to the meta-directing carboxyl groups.

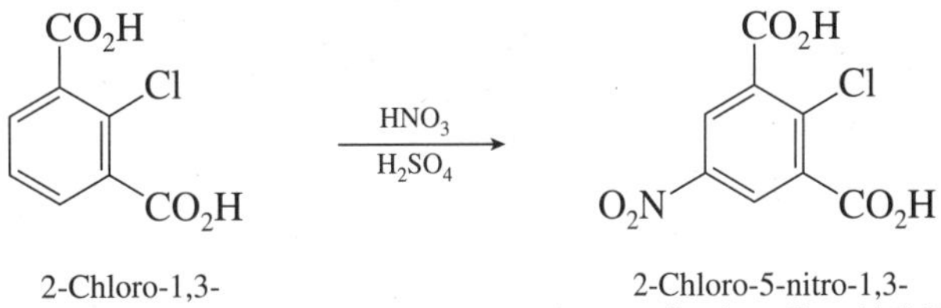

2-Chloro-1,3-benzenedicarboxylic acid $\xrightarrow[H_2SO_4]{HNO_3}$ 2-Chloro-5-nitro-1,3-benzenedicarboxylic acid (86%)

(*b*) Bromination of the ring occurs at the only available position activated by the amino group, a powerful activating substituent and an ortho, para director. This position is meta to the meta-directing trifluoromethyl group and to the meta-directing nitro group.

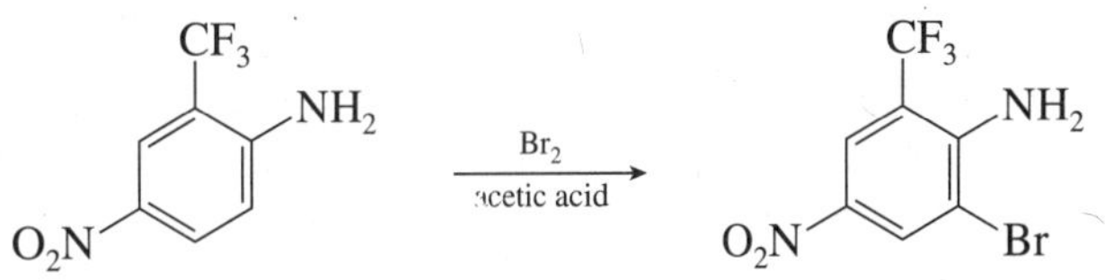

4-Nitro-2-(trifluoromethyl)aniline $\xrightarrow[\text{acetic acid}]{Br_2}$ 2-Bromo-4-nitro-6-(trifluoromethyl)aniline (81%)

(*c*) This may be approached as a problem in which there are two aromatic rings. One of them bears two activating substituents and so is more reactive than the other, which bears only one activating substituent. Of the two activating substituents (—OH and $C_6H_5$—), the hydroxyl substituent is the more powerful and controls the regioselectivity of substitution.

$$p\text{-Phenylphenol} \xrightarrow[CHCl_3]{Br_2} \text{2-Bromo-4-phenylphenol}$$

*p*-Phenylphenol → 2-Bromo-4-phenylphenol

(*d*) Both substituents are activating, nitration occurring readily even in the absence of sulfuric acid; both are ortho, para-directing and comparable in activating power. The position at which substitution takes place is therefore

$C(CH_3)_3$

Not here; too hindered

Not here; too hindered

$CH(CH_3)_2$

Ortho to isopropyl,
para to *tert*-butyl

$$\text{1-}tert\text{-Butyl-3-isopropylbenzene} \xrightarrow[\text{acetic acid}]{HNO_3} \text{4-}tert\text{-Butyl-2-isopropyl-1-nitrobenzene (78\%)}$$

1-*tert*-Butyl-3-isopropylbenzene → 4-*tert*-Butyl-2-isopropyl-1-nitrobenzene (78%)

(*e*) Protonation of 1-octene yields a secondary carbocation, which attacks benzene.

$$C_6H_6 + CH_2{=}CH(CH_2)_5CH_3 \xrightarrow{H_2SO_4} C_6H_5CH(CH_3)(CH_2)_5CH_3$$

Benzene + 1-Octene → 2-Phenyloctane (84%)

(*f*) The reaction that occurs with arenes and acid anhydrides in the presence of aluminum chloride is Friedel–Crafts acylation. The methoxy group is the more powerful activating substituent, so acylation occurs para to it.

$$o\text{-}FC_6H_4OCH_3 + CH_3\overset{O}{\overset{\|}{C}}O\overset{O}{\overset{\|}{C}}CH_3 \xrightarrow{AlCl_3} CH_3\overset{O}{\overset{\|}{C}}\text{—}C_6H_3(F)\text{—}OCH_3 + CH_3CO_2H$$

*o*-Fluoroanisole + Acetic anhydride → 3-Fluoro-4-methoxyacetophenone (70–80%)

(*g*) The isopropyl group is ortho, para-directing, and the nitro group is meta-directing. In this case their orientation effects reinforce each other. Electrophilic aromatic substitution takes place ortho to isopropyl and meta to nitro.

$CH(CH_3)_2$ / $NO_2$ $\xrightarrow[H_2SO_4]{HNO_3}$ $CH(CH_3)_2$ / $NO_2$ / $NO_2$

*p*-Nitroisopropylbenzene — 2,4-Dinitroisopropylbenzene (96%)

(*h*) In the presence of an acid catalyst ($H_2SO_4$), 2-methylpropene is converted to *tert*-butyl cation, which then attacks the aromatic ring ortho to the strongly activating methoxy group.

$$(CH_3)_2C{=}CH_2 + H^+ \longrightarrow (CH_3)_3C^+$$

$OCH_3$ / $CH_3$ + $(CH_3)_3C^+$ $\longrightarrow$ $OCH_3$ / $C(CH_3)_3$ / $CH_3$

In this particular example, 2-*tert*-butyl-4-methylanisole was isolated in 98% yield.

(*i*) There are two things to consider in this problem: (1) In which ring does bromination occur, and (2) what is the orientation of substitution in that ring? All the substituents are activating groups, so substitution will take place in the ring that bears the greater number of substituents. Orientation is governed by the most powerful activating substituent, the hydroxyl group. Both positions ortho to the hydroxyl group are already substituted, so that bromination takes place para to it. The product shown was isolated from the bromination reaction in 100% yield.

$H_3C$, $OH$, $CH_2$, $CH_3$ $\xrightarrow[CHCl_3]{Br_2}$ $H_3C$, $OH$, $CH_2$, $CH_3$, Br

3-Benzyl-2,6-dimethylphenol — 3-Benzyl-4-bromo-2,6-dimethylphenol (100%)

(*j*) Wolff–Kishner reduction converts benzophenone to diphenylmethane.

O, C $\xrightarrow[\text{triethylene glycol, heat}]{H_2NNH_2,\ KOH}$ $CH_2$

Benzophenone — Diphenylmethane (83%)

(*k*) Fluorine is an ortho, para-directing substituent. It undergoes Friedel–Crafts alkylation on being treated with benzyl chloride and aluminum chloride to give a mixture of *o*-fluorodiphenylmethane and *p*-fluorodiphenylmethane.

F

+ $C_6H_5CH_2Cl$ $\xrightarrow{AlCl_3}$ $C_6H_5CH_2$ F + F $C_6H_5CH_2$

Fluorobenzene Benzyl chloride *o*-Fluorodiphenylmethane (15%) *p*-Fluorodiphenylmethane (85%)

(*l*) The $-\ddot{N}HCCH_3$ (with C=O) substituent is a more powerful activator than the ethyl group. It directs Friedel–Crafts acylation primarily to the position para to itself.

$CH_3\overset{O}{\overset{\|}{C}}NH$, $CH_2CH_3$ + $CH_3\overset{O}{\overset{\|}{C}}Cl$ $\xrightarrow{AlCl_3}$ $CH_3\overset{O}{\overset{\|}{C}}NH$, $CH_2CH_3$, $\overset{O}{\overset{\|}{C}}CH_3$ + HCl

*o*-Ethylacetanilide Acetyl chloride 4-Acetamido-3-ethylacetophenone (57%)

(*m*) Clemmensen reduction converts the carbonyl group to a $CH_2$ unit.

$CH_3$, $\overset{O}{\overset{\|}{C}}CH_3$, $H_3C$, $CH_3$ $\xrightarrow[HCl]{Zn(Hg)}$ $CH_3$, $CH_2CH_3$, $H_3C$, $CH_3$

2,4,6-Trimethylacetophenone 2-Ethyl-1,3,5-trimethylbenzene (74%)

(*n*) Bromination occurs at C-5 on thiophene-3-carboxylic acid. Reaction does not occur at C-2 since substitution at this position would place a carbocation adjacent to the electron-withdrawing carboxyl group.

$CO_2H$, 5, 2, S $\xrightarrow[\text{acetic acid}]{Br_2}$ $CO_2H$, Br, S

Thiophene-3-carboxylic acid 2-Bromo-thiophene-4-carboxylic acid (69%)

**12.33** In a Friedel–Crafts acylation reaction an acyl chloride or acid anhydride reacts with an arene to yield an aryl ketone.

$$ArH + R\overset{O}{\overset{\|}{C}}Cl \xrightarrow{AlCl_3} Ar\overset{O}{\overset{\|}{C}}R$$

or

$$ArH + RCOCR \xrightarrow{AlCl_3} ArCR + RCOH$$

(with C=O on each carbonyl carbon: RC(=O)OC(=O)R, ArC(=O)R, RC(=O)OH)

The ketone carbonyl is bonded directly to the ring. In each of these problems, therefore, you should identify the bond between the aromatic ring and the carbonyl group and realize that it arises as shown in this general reaction.

(*a*) The compound is derived from benzene and $C_6H_5CH_2\overset{O}{\overset{\|}{C}}Cl$. The observed yield in this reaction is 82%.

$C_6H_5\overset{O}{\overset{\|}{C}}CH_2C_6H_5$ arises from $C_6H_5$—H and $Cl\overset{O}{\overset{\|}{C}}CH_2C_6H_5$

(*b*) The presence of the $Ar\overset{O}{\overset{\|}{C}}CH_2CH_2CO_2H$ unit suggests an acylation reaction using succinic anhydride.

2,5-Dimethylphenyl—$\overset{O}{\overset{\|}{C}}CH_2CH_2CO_2H$ (ring bearing $H_3C$— and —$CH_3$ para to each other) arises from $H_3C$—$C_6H_4$—$CH_3$ and succinic anhydride

Succinic anhydride

In practice, this reaction has been carried out in 55% yield.

(*c*) Two methods seem possible here but only one actually works. The only effective combination is

$$O_2N{-}C_6H_4{-}\overset{O}{\overset{\|}{C}}Cl + C_6H_6 \xrightarrow{AlCl_3} O_2N{-}C_6H_4{-}\overset{O}{\overset{\|}{C}}{-}C_6H_5$$

*p*-Nitrobenzoyl chloride — Benzene — *p*-Nitrobenzophenone (87%)

The alternative combination

$$O_2N{-}C_6H_5 + Cl\overset{O}{\overset{\|}{C}}{-}C_6H_5 \xrightarrow{AlCl_3} \text{no reaction}$$

fails because it requires a Friedel–Crafts reaction on a strongly deactivated aromatic ring (nitrobenzene).

(*d*) Here also two methods seem possible, but only one is successful in practice. The valid synthesis is

$$3,5\text{-}(H_3C)_2C_6H_3{-}\overset{O}{\overset{\|}{C}}Cl + C_6H_6 \xrightarrow{AlCl_3} 3,5\text{-}(H_3C)_2C_6H_3{-}\overset{O}{\overset{\|}{C}}{-}C_6H_5$$

3,5-Dimethylbenzoyl chloride — Benzene — 3,5-Dimethylbenzophenone (89%)

The alternative combination will not give 3,5-dimethylbenzophenone, because of the ortho, para-directing properties of the methyl substituents in *m*-xylene. The product will be 2,4-dimethylbenzophenone.

*m*-Xylene + Benzoyl chloride $\xrightarrow{AlCl_3}$ 2,4-Dimethylbenzophenone

(*e*) The combination that follows is not effective, because it involves a Friedel–Crafts reaction on a deactivated aromatic ring.

*p*-Methylbenzoyl chloride + Benzoic acid $\xrightarrow{AlCl_3}$ no reaction

The following combination, utilizing toluene, therefore seems appropriate:

$H_3C$–$C_6H_5$ + ClC(=O)–$C_6H_4$–$HO_2C$ $\xrightarrow{AlCl_3}$ $H_3C$–$C_6H_4$–C(=O)–$C_6H_4$–$HO_2C$

The actual sequence used a cyclic anhydride, phthalic anhydride, in a reaction analogous to that seen in part (*b*).

Toluene + Phthalic anhydride $\xrightarrow{AlCl_3}$ *o*-(4-Methylbenzoyl)benzoic acid (96%)

**12.34** (*a*) The problem to be confronted here is that two meta-directing groups are para to each other in the product. However, by recognizing that the carboxylic acid function can be prepared by oxidation of the isopropyl group

$CH(CH_3)_2$–$C_6H_4$–$SO_3H$ $\xrightarrow{\text{oxidize}}$ $CO_2H$–$C_6H_4$–$SO_3H$

we have a reasonable last step in the synthesis. The key intermediate has its sulfonic acid group para to the ortho, para-directing isopropyl group, which suggests the following approach:

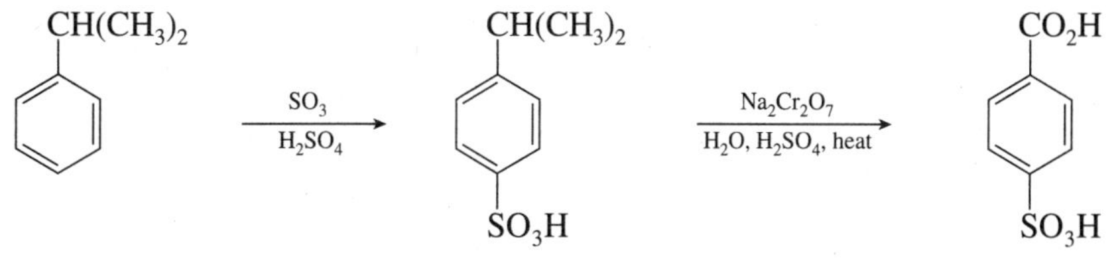

$$\text{Isopropylbenzene} \xrightarrow[H_2SO_4]{SO_3} p\text{-Isopropylbenzenesulfonic acid} \xrightarrow[H_2O,\ H_2SO_4,\ \text{heat}]{Na_2Cr_2O_7} p\text{-Carboxylbenzenesulfonic acid}$$

$CH(CH_3)_2$ ... $SO_3H$ ... $CO_2H$

Isopropylbenzene — *p*-Isopropylbenzene-sulfonic acid — *p*-Carboxylbenzene-sulfonic acid

(*b*) In this problem two methyl groups must be oxidized to carboxylic acid functions, and a *tert*-butyl group must be introduced, most likely by a Friedel–Crafts reaction. Since Friedel–Crafts alkylations cannot be performed on deactivated aromatic rings, oxidation must *follow,* not precede, alkylation. The following reaction sequence therefore seems appropriate:

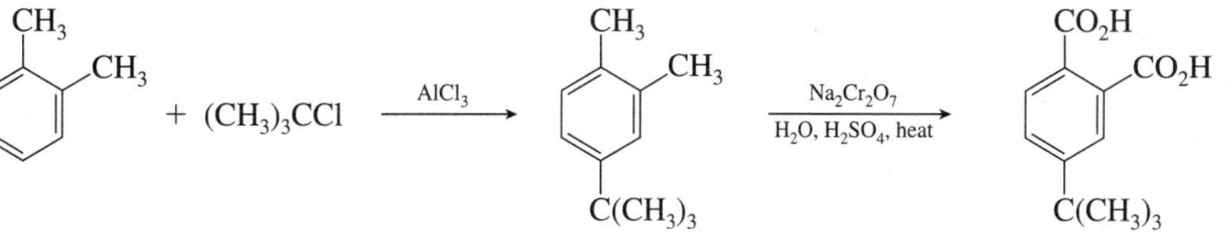

$$o\text{-Xylene} + (CH_3)_3CCl \xrightarrow{AlCl_3} 4\text{-}tert\text{-Butyl-1,2-dimethylbenzene} \xrightarrow[H_2O,\ H_2SO_4,\ \text{heat}]{Na_2Cr_2O_7} 4\text{-}tert\text{-Butylbenzene-1,2-dicarboxylic acid}$$

*o*-Xylene — 4-*tert*-Butyl-1,2-dimethylbenzene — 4-*tert*-Butylbenzene-1,2-dicarboxylic acid

In practice, zinc chloride was used as the Lewis acid to catalyze the Friedel–Crafts reaction (64% yield). Oxidation of the methyl groups occurs preferentially because the *tert*-butyl group has no benzylic hydrogens.

(*c*) The carbonyl group is directly attached to the naphthalene unit in the starting material. Reduce it in the first step so that a Friedel–Crafts acylation can be accomplished on the naphthalene ring. An aromatic ring that bears a strongly electron-withdrawing group such as C=O does not undergo Friedel–Crafts reactions.

$$\xrightarrow{Zn(Hg),\ HCl} \qquad \xrightarrow[AlCl_3]{CH_3CCl\ (\text{C=O})} \qquad CH_3C(=O)\text{–}$$

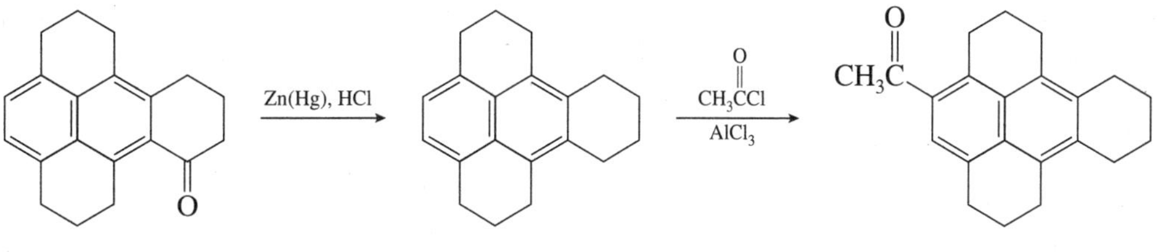

(*d*) *m*-Dimethoxybenzene is a strongly activated aromatic compound and so will undergo electrophilic aromatic substitution readily. The ring position between the two methoxy groups is sterically hindered and less reactive than the other activated positions.

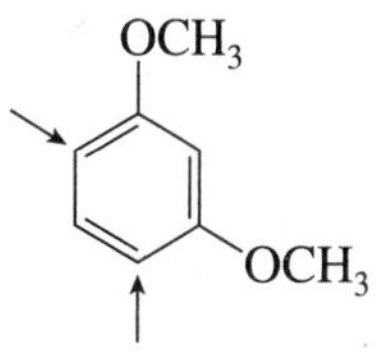

$OCH_3$, $OCH_3$

Arrows indicate equivalent ring positions strongly activated by methoxy groups.

Because Friedel–Crafts reactions may not be performed on deactivated aromatic rings, the *tert*-butyl group must be introduced before the nitro group. The correct sequence is therefore

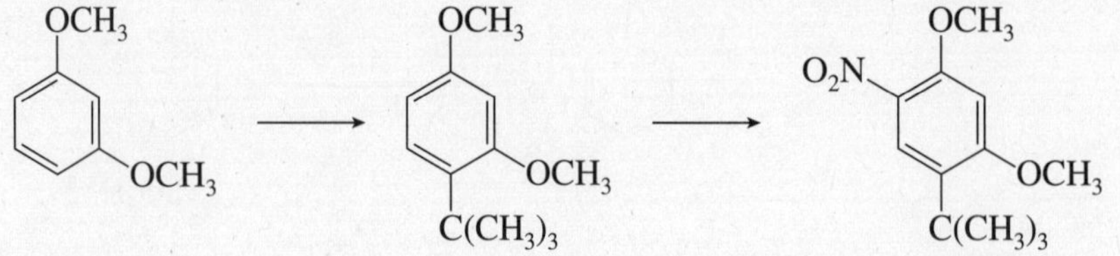

$OCH_3$ / $OCH_3$ → $OCH_3$ / $OCH_3$ / $C(CH_3)_3$ → $O_2N$ / $OCH_3$ / $OCH_3$ / $C(CH_3)_3$

This is essentially the procedure actually followed. Alkylation was effected, however, not with *tert*-butyl chloride and aluminum chloride but with 2-methylpropene and phosphoric acid.

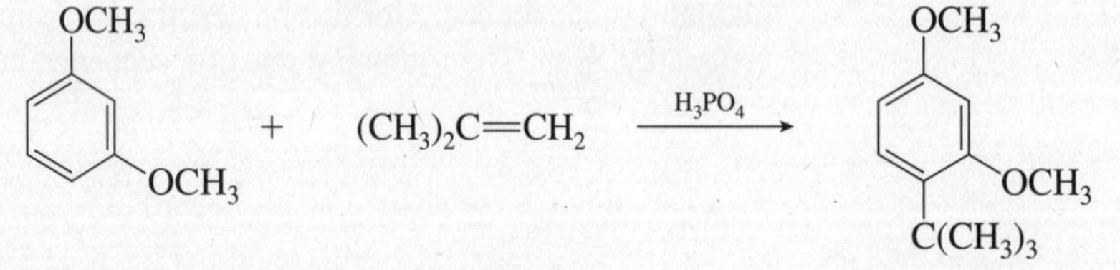

$$\text{m-Dimethoxybenzene} + (CH_3)_2C{=}CH_2 \xrightarrow{H_3PO_4} \text{1-tert-Butyl-2,4-dimethoxybenzene (75\%)}$$

*m*-Dimethoxybenzene 2-Methylpropene 1-*tert*-Butyl-2,4-dimethoxybenzene (75%)

Nitration was carried out in the usual way. the orientation of nitration is controlled by the more powerfully activating methoxy groups rather than by the weakly activating *tert*-butyl.

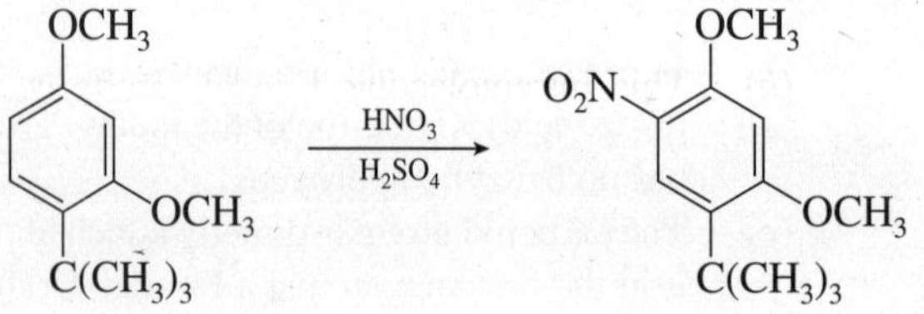

$$\text{1-tert-Butyl-2,4-dimethoxybenzene} \xrightarrow[H_2SO_4]{HNO_3} \text{1-tert-Butyl-2,4-dimethoxy-5-nitrobenzene}$$

1-*tert*-Butyl-2,4-dimethoxy-5-nitrobenzene

**12.35** The first step is a Friedel–Crafts acylation reaction. The use of a cyclic anhydride introduces both the acyl and carboxyl groups into the molecule.

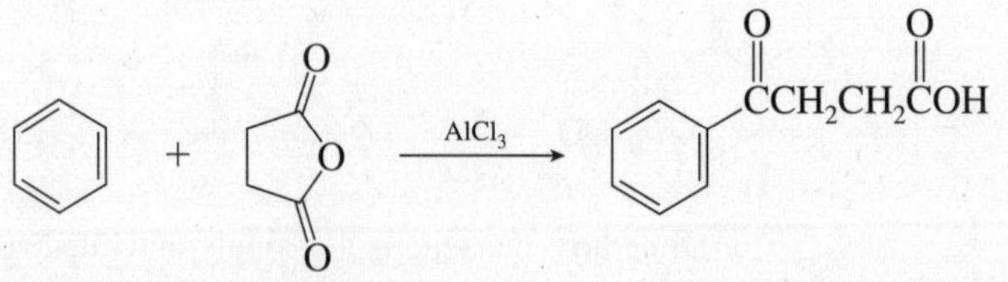

$$\text{Benzene} + \text{Succinic anhydride} \xrightarrow{AlCl_3} C_6H_5CCH_2CH_2COH \text{ (both C=O)}$$

Benzene Succinic anhydride 4-Oxo-4-phenylbutanoic acid

The second step is a reduction of the ketone carbonyl to a methylene group. A Clemmensen reduction is normally used for this step.

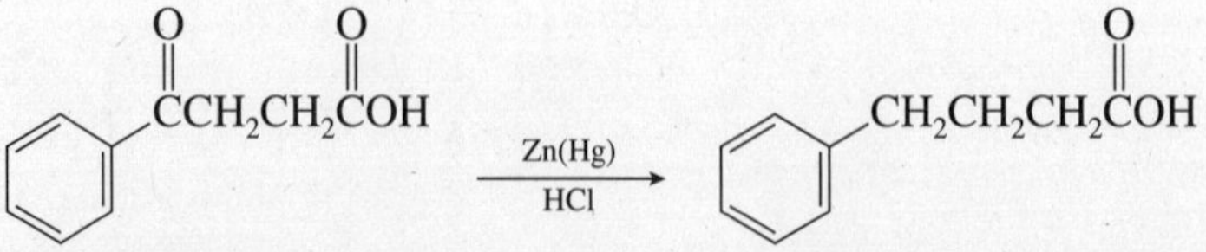

$$C_6H_5\overset{O}{\overset{\|}{C}}CH_2CH_2\overset{O}{\overset{\|}{C}}OH \xrightarrow[HCl]{Zn(Hg)} C_6H_5CH_2CH_2CH_2\overset{O}{\overset{\|}{C}}OH$$

4-Oxo-4-phenylbutanoic acid 4-Phenylbutanoic acid

The cyclization phase of the process is an intramolecular Friedel–Crafts acylation reaction. It requires conversion of the carboxylic acid to the acyl chloride (thionyl chloride is a suitable reagent) followed by treatment with aluminum chloride.

$C_6H_5CH_2CH_2CH_2COOH$ (4-Phenylbutanoic acid) $\xrightarrow{SOCl_2}$ $C_6H_5CH_2CH_2CH_2COCl$ (4-Phenylbutanoyl chloride) $\xrightarrow{AlCl_3}$ 1-tetralone

**12.36** Intramolecular Friedel–Crafts acylation reactions that produce five-membered or six-membered rings occur readily. Cyclization must take place at the position ortho to the reacting side chain.

$C_6H_5CH_2CH_2COCl \xrightarrow{AlCl_3}$ 1-indanone and $C_6H_5CH_2CH_2CH_2COCl \xrightarrow{AlCl_3}$ 1-tetralone

(*a*) A five-membered cyclic ketone is formed here.

$(CH_3)_3C-C_6H_4-C(CH_3)_2CH_2COCl \xrightarrow{AlCl_3}$ $(CH_3)_3C$-substituted 3,3-dimethylindanone (46%)

(*b*) This intramolecular Friedel–Crafts acylation takes place to form a six-membered cyclic ketone in excellent yield.

2-phenylcyclohexyl-$CH_2CCl$ (C=O) $\xrightarrow{AlCl_3}$ tricyclic ketone 93%

(*c*) In this case two aromatic rings are available for attack in the acylation reaction. The more reactive ring is the one that bears the two activating methoxy groups, and cyclization occurs on it.

$(CH_3O)_2C_6H_3-CH_2CHCH_2-C_6H_5$ (with $CCl$=O on CH)

faster → dimethoxy indanone bearing $CH_2-C_6H_5$: Only product (78% yield)

slower → $(CH_3O)_2C_6H_3-CH_2$-indanone: (Not observed)

**12.37** (*a*) To determine the total rate of chlorination of biphenyl relative to that of benzene, we add up the partial rate factors for all the positions in each substrate and compare them.

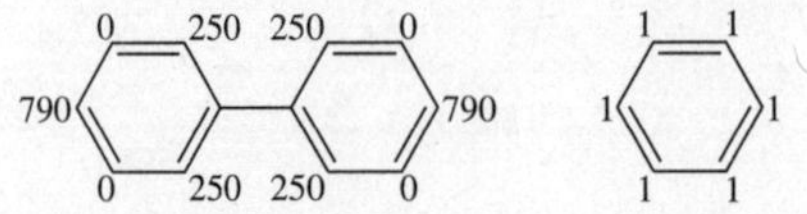

Biphenyl (sum = 2580)  Benzene (sum = 6)

$$\text{Relative rate of chlorination: } \frac{\text{Biphenyl}}{\text{Benzene}} = \frac{2580}{6} = \frac{430}{1}$$

(*b*) The relative rate of attack at the para position compared with the ortho positions is given by the ratio of their partial rate factors.

$$\frac{\text{Para}}{\text{Ortho}} = \frac{1580}{1000} = \frac{1.58}{1}$$

Therefore, 15.8 g of *p*-chlorobiphenyl is formed for every 10 g of *o*-chlorobiphenyl.

**12.38** The problem stipulates that the reactivity of various positions in *o*-bromotoluene can be estimated by multiplying the partial rate factors for the corresponding positions in toluene and bromobenzene. Therefore, given the partial rate factors:

$CH_3$ 4.5 4.5 4.8 4.8 750 and Very small Br 0.0003 0.084 Very small 0.0003

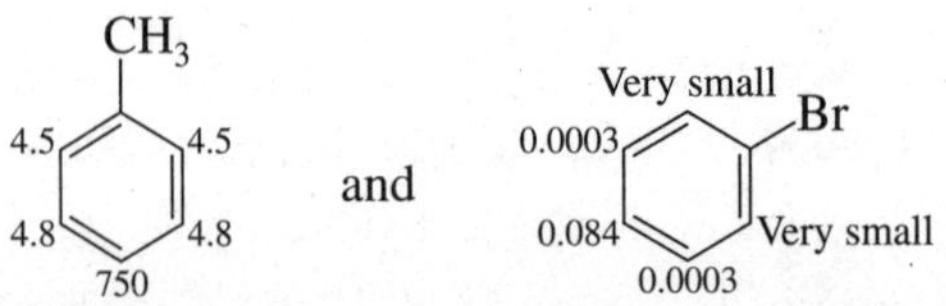

the two are multiplied together to give the combined effects of the two substituents at the various ring positions.

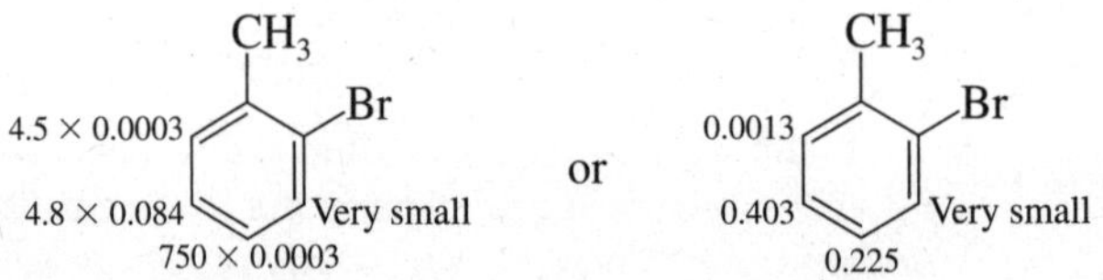

$CH_3$ Br 4.5 × 0.0003 4.8 × 0.084 Very small 750 × 0.0003 or $CH_3$ Br 0.0013 0.403 Very small 0.225

The most reactive position is the one that is para to bromine. The predicted product is therefore 4-bromo-3-methylacetophenone. Indeed, this is what is observed experimentally.

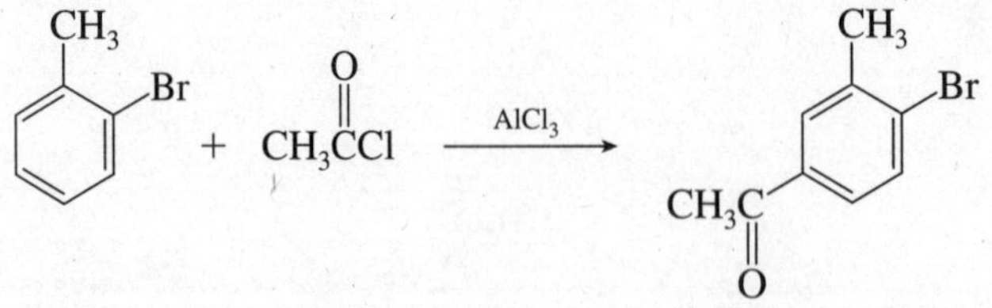

$CH_3$ Br + $CH_3\overset{O}{\overset{\|}{C}}Cl$ $\xrightarrow{AlCl_3}$ $CH_3$ Br $CH_3\underset{O}{\underset{\|}{C}}$

*o*-Bromotoluene  Acetyl chloride  4-Bromo-3-methyl-acetophenone

This was first considered to be "anomalous" behavior on the part of *o*-bromotoluene, but, as can be seen, it is consistent with the individual directing properties of the two substituents.

**12.39** The isomerization is triggered by protonation of the aromatic ring, an electrophilic attack by HCl catalyzed by $AlCl_3$.

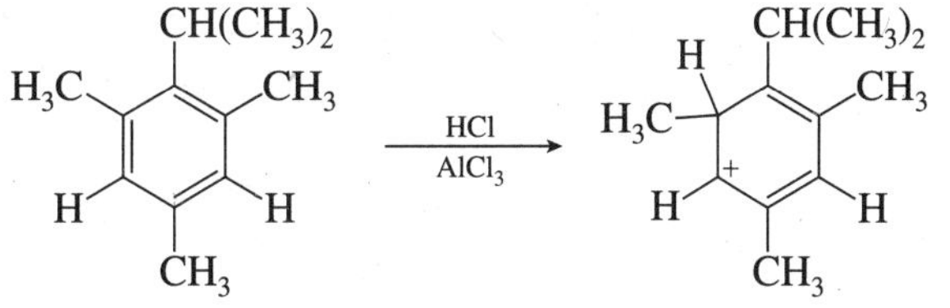

CH(CH3)2
H3C
CH3
H
H
CH3
HCl
AlCl3
H
CH(CH3)2
H3C
CH3
+
H
H
CH3


2-Isopropyl-1,3,5-trimethylbenzene

The carbocation then rearranges by a methyl shift, and the rearranged cyclohexadienyl cation loses a proton to form the isomeric product

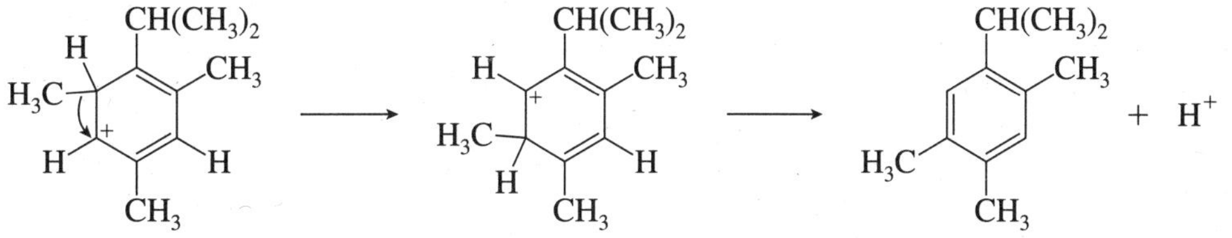

CH(CH3)2
H
H3C
CH3
+
H
H
CH3
CH(CH3)2
H
CH3
+
H3C
H
H
CH3
CH(CH3)2
CH3
H3C
CH3
+ H+


1-Isopropyl-2,4,5-trimethylbenzene

The driving force for rearrangement is relief of steric strain between the isopropyl group and one of its adjacent methyl groups. Isomerization is acid-catalyzed. Protonation of the ring generates the necessary carbocation intermediate and rearomatization occurs by loss of a proton.

**12.40** The relation of compound A to the starting material is

$$\underset{(C_{12}H_{15}ClO)}{C_6H_5(CH_2)_5\overset{O}{\overset{\|}{C}}Cl} \longrightarrow \underset{(C_{12}H_{14}O)}{A} + HCl$$

The starting acyl chloride has lost the elements of HCl in the formation of A. Because A forms benzene-1,2-dicarboxylic acid on oxidation, it must have two carbon substituents ortho to each other.

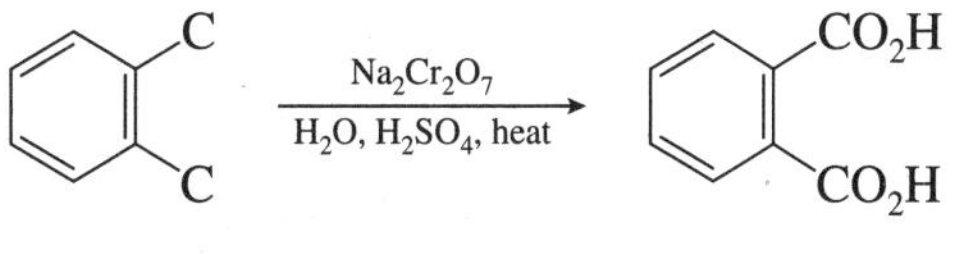

C
C
Na2Cr2O7
H2O, H2SO4, heat
CO2H
CO2H


These facts suggest the following process:

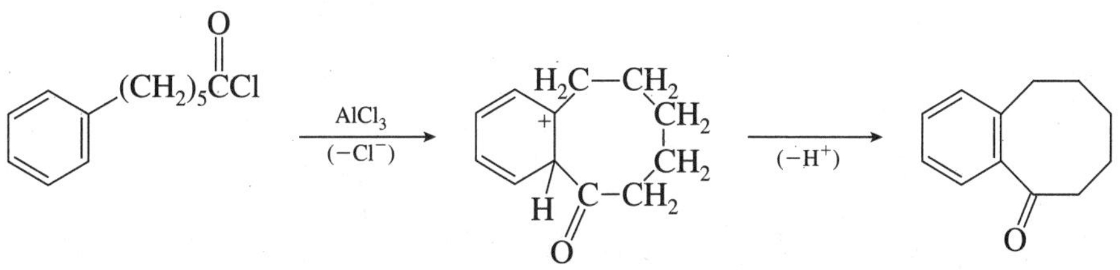

O
(CH2)5CCl
AlCl3
(−Cl−)
H2C—CH2
CH2
CH2
+
H
C—CH2
O
(−H+)
O


A

The reaction leading to compound A is an intramolecular Friedel–Crafts acylation. Since cyclization to form an eight-membered ring is difficult, it must be carried out in dilute solution to minimize competition with intermolecular acylation.

**12.41** Although hexamethylbenzene has no positions available at which ordinary electrophilic aromatic *substitution* might occur, electrophilic *attack* on the ring can still take place to form a cyclohexadienyl cation.

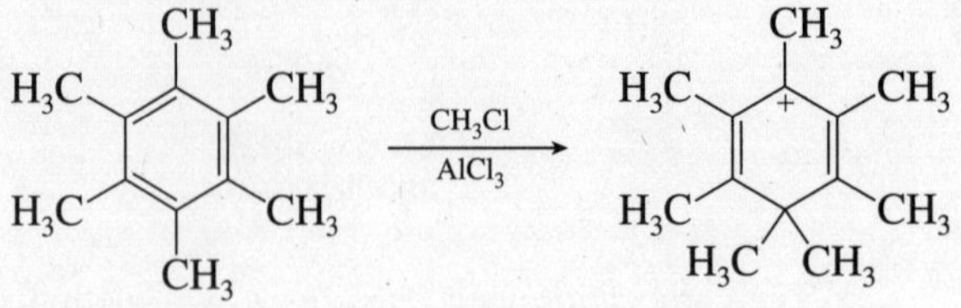

$CH_3$, $H_3C$, $CH_3$, $H_3C$, $CH_3$, $CH_3$ $\xrightarrow[AlCl_3]{CH_3Cl}$ $CH_3$, $H_3C$, $CH_3$, +, $H_3C$, $CH_3$, $H_3C$ $CH_3$

Compound A is the tetrachloroaluminate ($AlCl_4^-$) salt of the carbocation shown. It undergoes deprotonation on being treated with aqueous sodium bicarbonate.

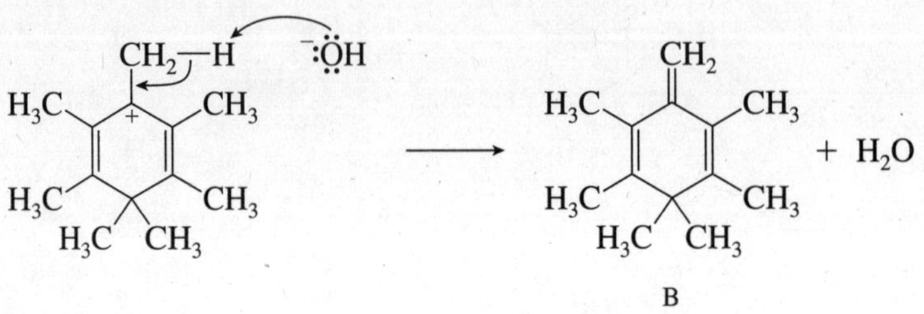

$CH_2$—H $^-$:ÖH, $H_3C$, $CH_3$, +, $H_3C$, $CH_3$, $H_3C$ $CH_3$ $\longrightarrow$ $CH_2$, $H_3C$, $CH_3$, $H_3C$, $CH_3$, $H_3C$ $CH_3$ + $H_2O$

B

**12.42** By examining the structure of the target molecule, compound C, we see that the bond indicated in the following structure joins two fragments that are related to the given starting materials A and B:

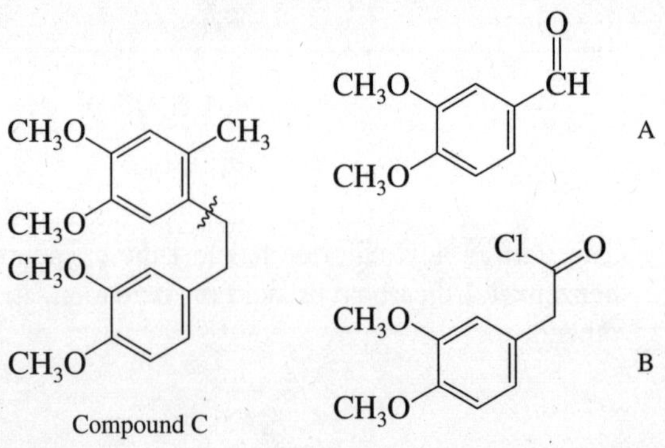

$CH_3O$, $CH_3$, $CH_3O$, $CH_3O$, $CH_3O$ — Compound C

$CH_3O$, $CH_3O$, O, CH — A

Cl, O, $CH_3O$, $CH_3O$ — B

The bond connecting the two fragments can be made by a Friedel–Crafts acylation-reduction sequence using the acyl chloride B.

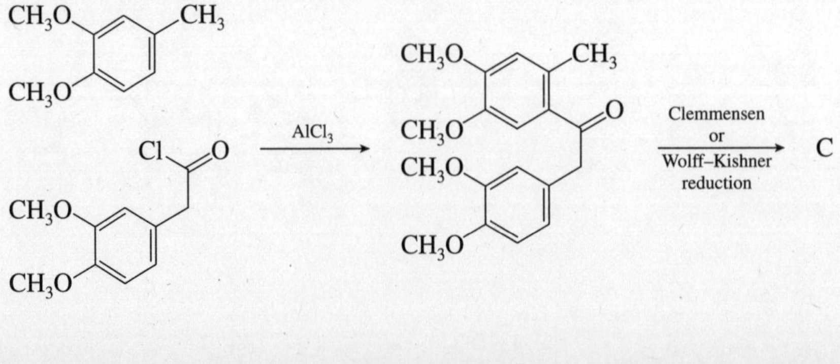

$CH_3O$, $CH_3$, $CH_3O$ + Cl, O, $CH_3O$, $CH_3O$ $\xrightarrow{AlCl_3}$ $CH_3O$, $CH_3$, $CH_3O$, O, $CH_3O$, $CH_3O$ $\xrightarrow[\text{Wolff–Kishner reduction}]{\text{Clemmensen or}}$ C

The orientation is right; attack is para to one of the methoxy groups and ortho to the methyl. The substrate for the Friedel–Crafts acylation reaction, 3,4-dimethoxytoluene, is prepared from compound A by a Clemmensen or Wolff–Kishner reduction. Compound A cannot be acylated directly because it bears a strongly deactivating —CH(=O) substituent.

**12.43** In the presence of aqueous sulfuric acid, the side-chain double bond of styrene undergoes protonation to form a benzylic carbocation.

$$C_6H_5CH{=}CH_2 + H^+ \longrightarrow C_6H_5\overset{+}{C}HCH_3$$

Styrene — 1-Phenylethyl cation

This carbocation then reacts with a molecule of styrene in the manner we have seen earlier (Chapter 6) for alkene dimerization.

$$C_6H_5\overset{+}{C}HCH_3 + C_6H_5CH{=}CH_2 \longrightarrow C_6H_5\overset{+}{C}HCH_2\underset{CH_3}{\underset{|}{C}}HC_6H_5$$

The carbocation produced in this step can lose a proton to form 1,3-diphenyl-1-butene

$$C_6H_5\overset{+}{C}HCH_2\underset{CH_3}{\underset{|}{C}}HC_6H_5 \longrightarrow C_6H_5CH{=}CH\underset{CH_3}{\underset{|}{C}}HC_6H_5 + H^+$$

1,3-Diphenyl-1-butene

or it can undergo a cyclization reaction in what amounts to an intramolecular Friedel–Crafts alkylation

$$C_6H_4[CH(CH_3)CH_2\overset{+}{C}HC_6H_5] \longrightarrow \text{1-methyl-3-phenylindan} + H^+$$

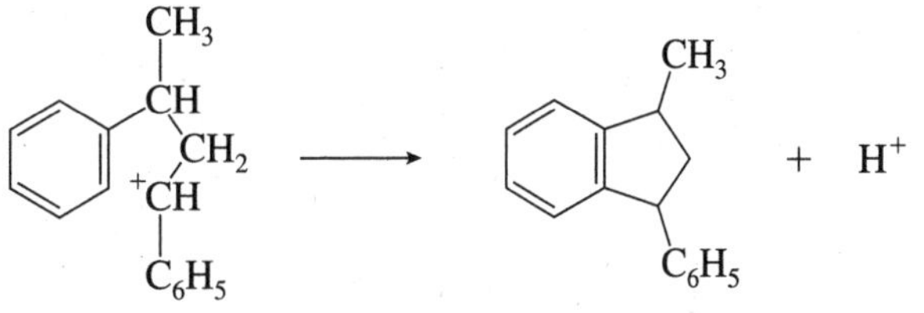

1-Methyl-3-phenylindan

**12.44** The alcohol is tertiary and benzylic. In the presence of sulfuric acid a carbocation is formed.

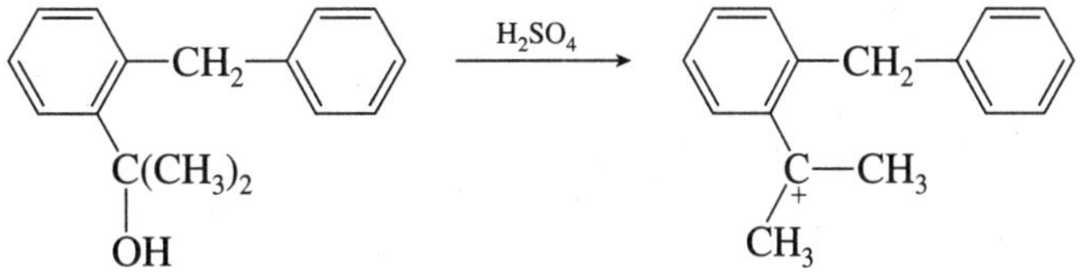

$$o\text{-}(C_6H_5CH_2)C_6H_4C(CH_3)_2OH \xrightarrow{H_2SO_4} o\text{-}(C_6H_5CH_2)C_6H_4\overset{+}{C}(CH_3)_2$$

An intramolecular Friedel–Crafts alkylation reaction follows, in which the carbocation attacks the adjacent aromatic ring.

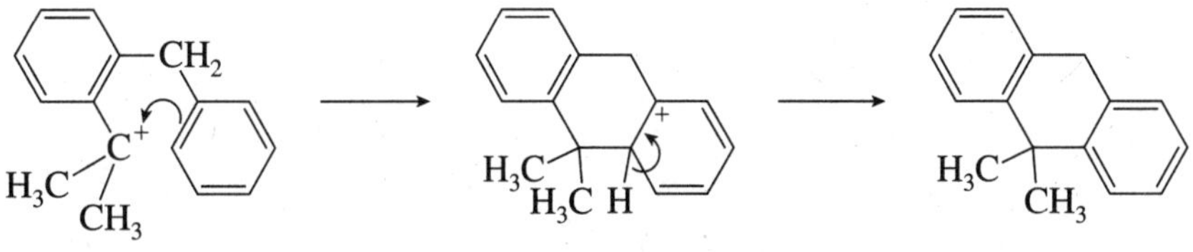

$$\longrightarrow \text{(cyclized cation, } H_3C, H_3C, H) \longrightarrow \text{9,9-dimethyl-9,10-dihydroanthracene } (H_3C, CH_3)$$

$C_{16}H_{16}$

# SELF-TEST

## PART A

**A-1.** Write the three most stable resonance contributors to the cyclohexadienyl cation found in the ortho bromination of toluene.

**A-2.** Give the major product(s) for each of the following reactions. Indicate whether the reaction proceeds faster or slower than the corresponding reaction of benzene.

(*a*) Nitrobenzene ($C_6H_5NO_2$) $\xrightarrow[H_2SO_4]{HNO_3}$ ?

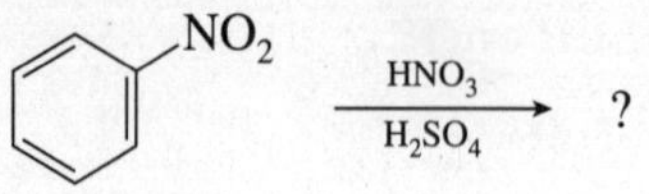

(*b*) Ethylbenzene ($C_6H_5CH_2CH_3$) $\xrightarrow[FeBr_3]{Br_2}$ ?

(*c*) Chlorobenzene ($C_6H_5Cl$) $\xrightarrow[H_2SO_4]{SO_3}$ ?

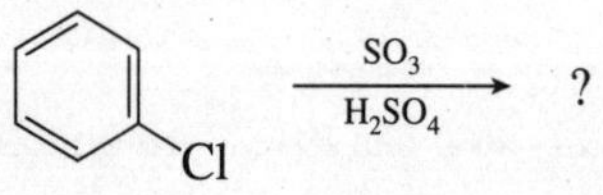

**A-3.** Write the formula of the electrophilic reagent species present in each reaction of the preceding problem.

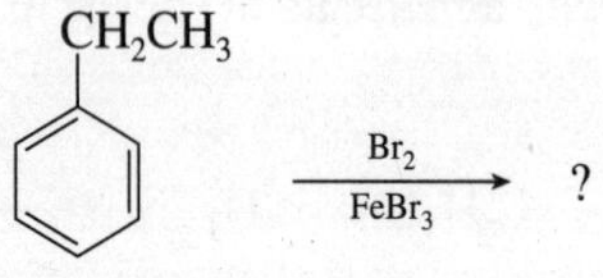

**A-4.** Provide the reactant, reagent, or product omitted from each of the following:

(*a*) ? $\xrightarrow[HCl]{Zn(Hg)}$ 1-$CH_2CH(CH_3)_2$-4-$C(CH_3)_3$-benzene

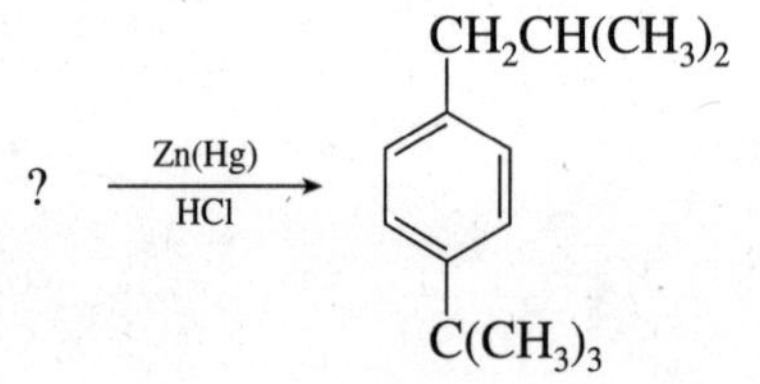

(*b*) $C_6H_5CH_2CH_2CH_2C(Cl)(CH_3)_2$ $\xrightarrow{AlCl_3}$ ?

(*c*) $C_6H_5OCH_3$ $\xrightarrow{?}$ o-$CH_3OC_6H_4CC_6H_5$ (C=O) + p-$CH_3OC_6H_4C(=O)C_6H_5$

(*d*) $C_6H_5C(=O)CH_2CH_3$ $\xrightarrow[H_2SO_4]{HNO_3}$ ?

(*e*) Thiophene $\xrightarrow[FeCl_3]{Cl_2}$ ?

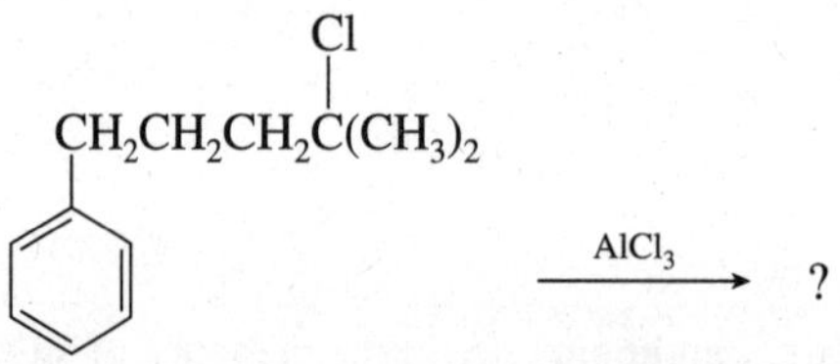

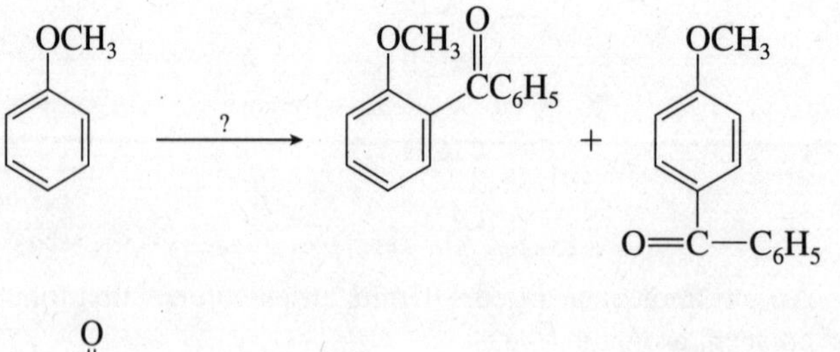

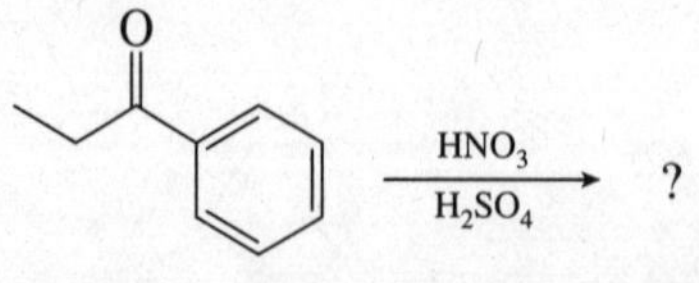

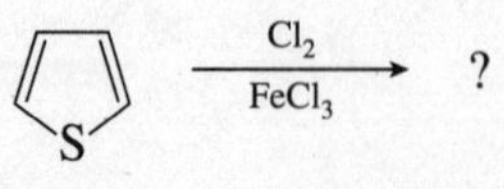

**A-5.** Draw the structure(s) of the major product(s) formed by reaction of each of the following compounds with $Cl_2$ and $FeCl_3$. If two products are formed in significant amounts, draw them both.

(*a*) $NO_2$ / $CH_3$ (para-substituted benzene)

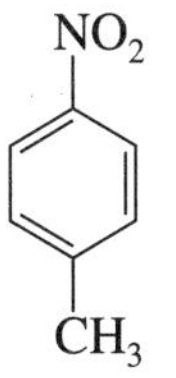

(*c*) O (1-indanone)

(*b*) CN / $OCH_3$ (meta-substituted benzene)

(*d*) $CH_3CH_2$—C₆H₄—OH

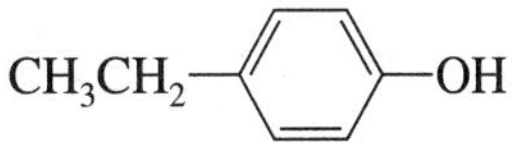

**A-6.** Provide the necessary reagents for each of the following transformations. More than one step may be necessary.

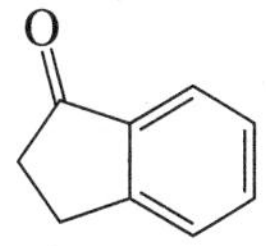

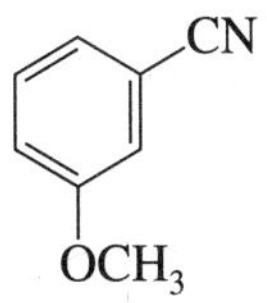

(*a*) C₆H₅—$CH(CH_3)_2$ $\xrightarrow{?}$ $HO_3S$—C₆H₄—$CO_2H$

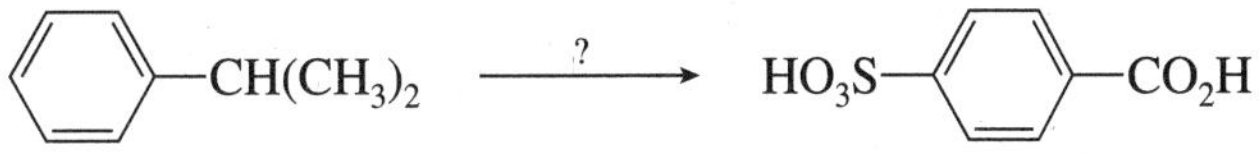

(*b*) C₆H₆ $\xrightarrow{?}$ Cl—C₆H₄—$CH_2CH_2C_6H_5$ (meta)

(*c*) C₆H₅—CH=O $\xrightarrow{?}$ $CH_3C(=O)$—C₆H₄—$CH_3$

(*d*) C₆H₆ $\xrightarrow{?}$ Br—C₆H₄—$SO_3H$ (meta)

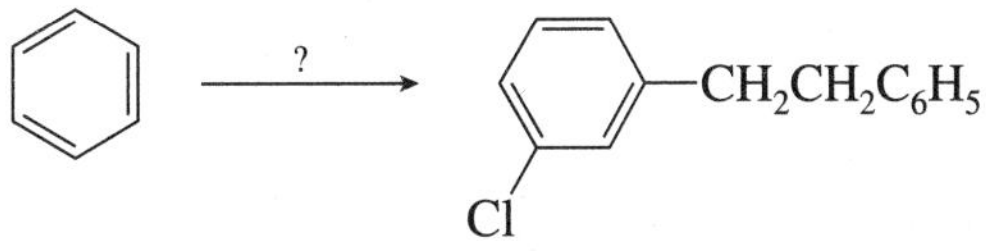

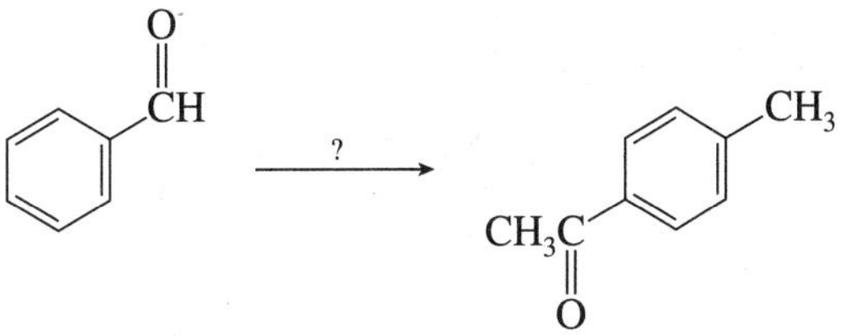

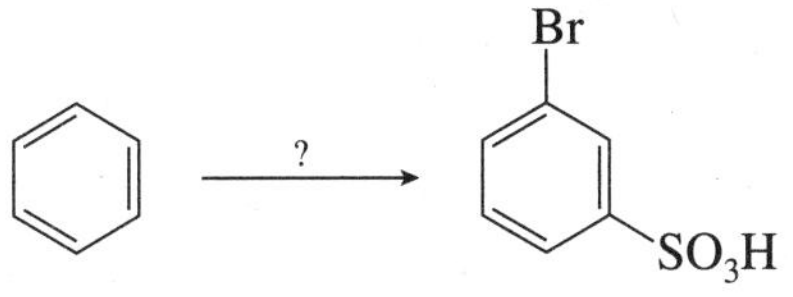

(*e*) C₆H₆ $\xrightarrow{?}$ $(CH_3)_2CH$—C₆H₄—$NO_2$

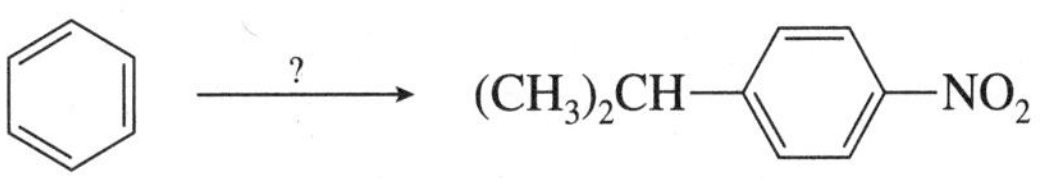

**A-7.** Outline a reasonable synthesis of each of the following from either benzene or toluene and any necessary organic or inorganic reagents.

(a) 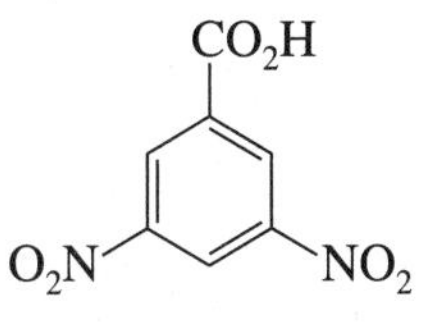 $CO_2H$, $O_2N$, $NO_2$ (3,5-dinitrobenzoic acid)

(b) 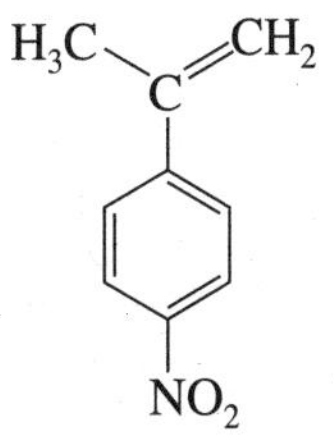 $H_3C$—C(=$CH_2$)—C₆H₄—$NO_2$

(c) $CH_2CH(CH_3)_2$, Cl, $CH_3$

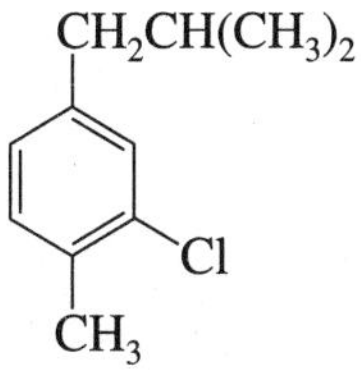

**A-8.** Outline a reasonable synthesis of the compound shown using anisole ($C_6H_5OCH_3$) and any necessary inorganic reagents.

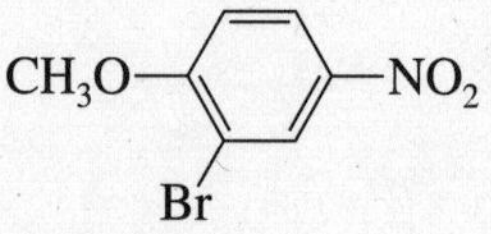

$CH_3O$—(ring)—$NO_2$, with Br on the ring

## PART B

**B-1.** Consider the following statements concerning the effect of a trifluoromethyl group, —$CF_3$, on an electrophilic aromatic substitution.

1. The $CF_3$ group will activate the ring.
2. The $CF_3$ group will deactivate the ring.
3. The $CF_3$ group will be a meta director.
4. The $CF_3$ group will be an ortho, para director.

Which of these statements are correct?

(*a*) 1, 3 (*b*) 1, 4 (*c*) 2, 3 (*d*) 2, 4

**B-2.** Which of the following resonance structures is **not** a contributor to the cyclohexadienyl cation intermediate in the nitration of benzene?

(*a*) 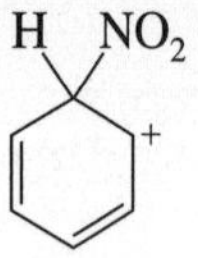 H $NO_2$ + (*c*) H $NO_2$ +

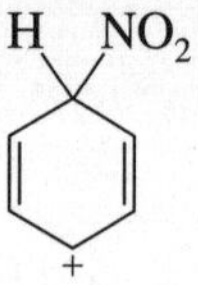

(*b*) $NO_2$ H H— + (*d*) None of these (all are contributors)

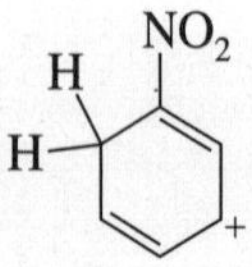

**B-3.** All the following groups are activating ortho, para directors when attached to a benzene ring *except*

(*a*) —$OCH_3$ (*c*) —Cl

(*b*) —$NH\overset{O}{\overset{\|}{C}}CH_3$ (*d*) —$N(CH_3)_2$

**B-4.** Rank the following in terms of increasing reactivity toward nitration with $HNO_3$, $H_2SO_4$ (least → most):

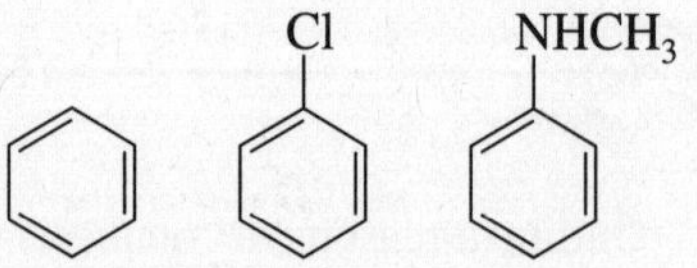

Cl NHCH$_3$

1 2 3

(*a*) 1 < 2 < 3 (*c*) 3 < 1 < 2
(*b*) 2 < 1 < 3 (*d*) 3 < 2 < 1

**B-5.** For the reaction

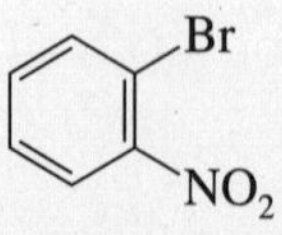

? ⟶ Br, $NO_2$

the best reactants are:

(*a*) $C_6H_5Br$ + $HNO_3$, $H_2SO_4$

(*b*) $C_6H_5NO_2$ + $Br_2$, $FeBr_3$

(*c*) $C_6H_5Br + H_2SO_4$, heat

(*d*) $C_6H_5NO_2$ + HBr

**B-6.** For the reaction

? ⟶ (3-chlorophenyl)–$\overset{O}{\overset{\|}{C}}C_6H_5$

the best reactants are

(*a*) $C_6H_5Cl + C_6H_5\overset{O}{\overset{\|}{C}}Cl$, $AlCl_3$

(*b*) $C_6H_5\overset{O}{\overset{\|}{C}}C_6H_5 + Cl_2$, $FeCl_3$

(*c*) $C_6H_5CH_2C_6H_5 + Cl_2$, $FeCl_3$, followed by oxidation with chromic acid

(*d*) None of these yields the desired product.

**B-7.** The reaction

$H_3C$–$C_6H_4$–$C_6H_4$–Cl $\xrightarrow{HNO_3/H_2SO_4}$

gives as the major product:

(*a*) $H_3C$–$C_6H_4$–$C_6H_3(NO_2)$–Cl

(*b*) $H_3C$–$C_6H_4$–$C_6H_3(NO_2)$–Cl

(*c*) $H_3C$–$C_6H_3(NO_2)$–$C_6H_4$–Cl

(*d*) $H_3C$–$C_6H_3(NO_2)$–$C_6H_4$–Cl

**B-8.** Which one of the following compounds undergoes bromination of its aromatic ring (electrophilic aromatic substitution) at the **fastest** rate?

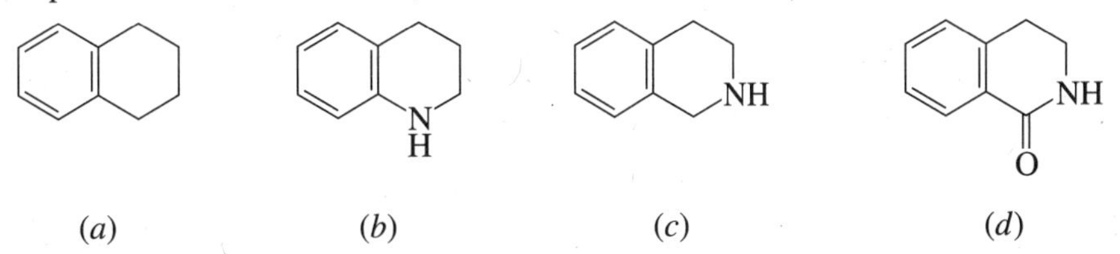


**B-9.** Which one of the following is the **most stable**?

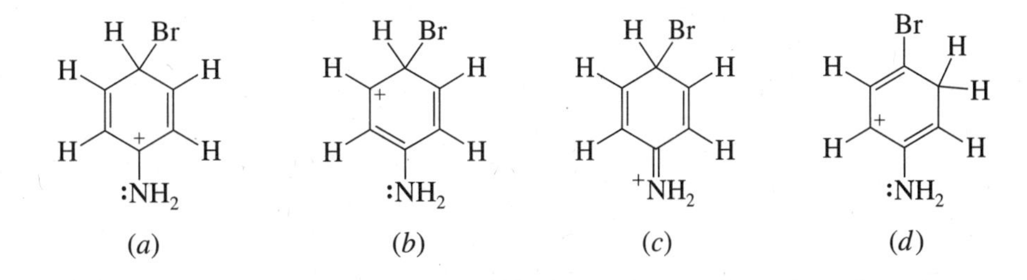

**B-10.** The major product of the reaction

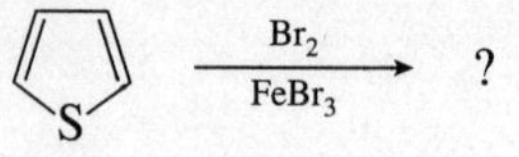

Thiophene $\xrightarrow[FeBr_3]{Br_2}$ ?

is

(*a*) 2-Bromothiophene (Br on carbon adjacent to S)    (*b*) 3-Bromothiophene (Br on carbon not adjacent to S)

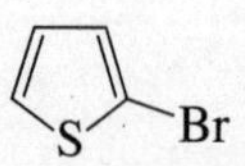

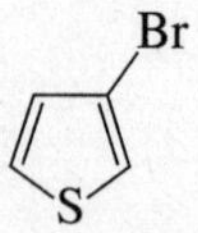

(*c*) An **equal** mixture of compound (*a*) and (*b*) would form.
(*d*) None of these; substitution would not occur.

**B-11.** What is the product of the following reaction?

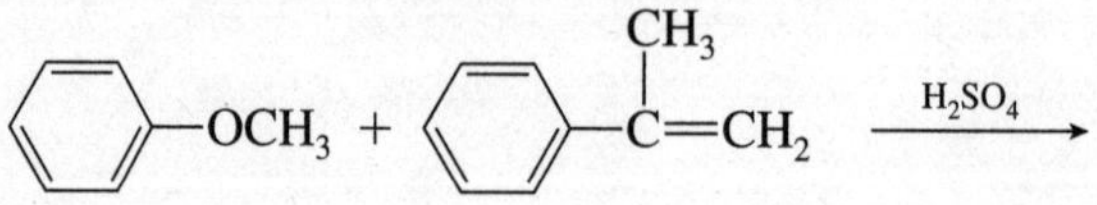

$$C_6H_5{-}OCH_3 + C_6H_5{-}C(CH_3){=}CH_2 \xrightarrow{H_2SO_4}$$

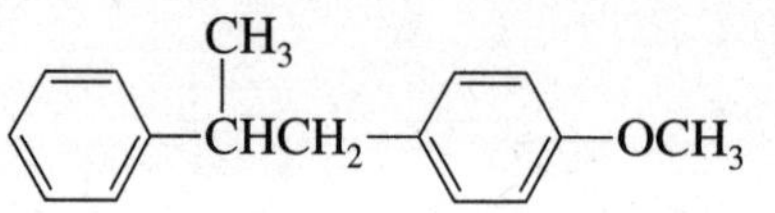

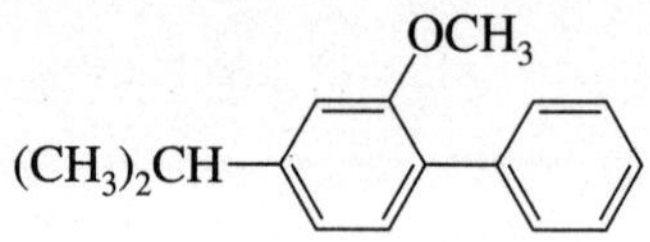

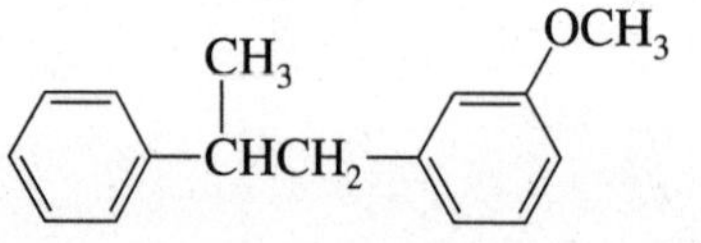

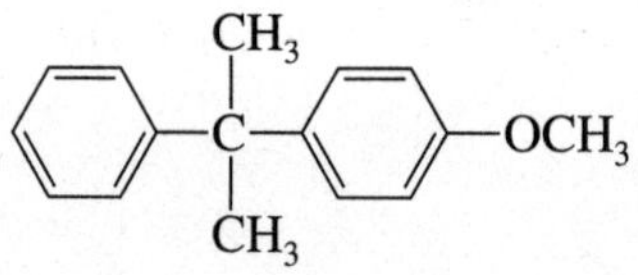

(*a*) $C_6H_5{-}CH(CH_3)CH_2{-}C_6H_4{-}OCH_3$ (para)

(*d*) $(CH_3)_2CH{-}C_6H_3(OCH_3){-}C_6H_5$

(*b*) $C_6H_5{-}CH(CH_3)CH_2{-}C_6H_4{-}OCH_3$ (ortho)

(*e*) $C_6H_5{-}C(CH_3)_2{-}C_6H_4{-}OCH_3$ (para)

(*c*) $(CH_3)_2CH{-}C_6H_4{-}C_6H_4{-}OCH_3$

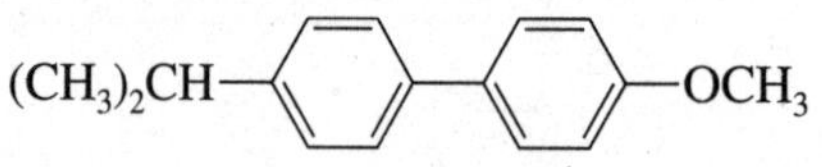

**B-12.** Partial rate factors are shown for nitration of a particular aromatic compound. Based on these data, the most reasonable choice for substituent X is:

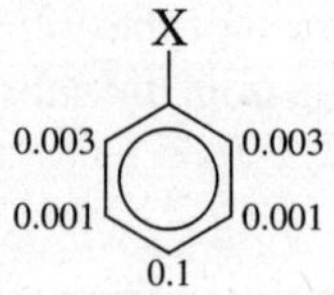

X on benzene ring; ortho: 0.003, 0.003; meta: 0.001, 0.001; para: 0.1

(*a*) $-N(CH_3)_2$    (*c*) $-Br$    (*e*) $-CH{=}O$
(*b*) $-SO_3H$    (*d*) $-CH(CH_3)_2$

**B-13.** Which reactants combine to give the species shown at the right as a reactive intermediate?

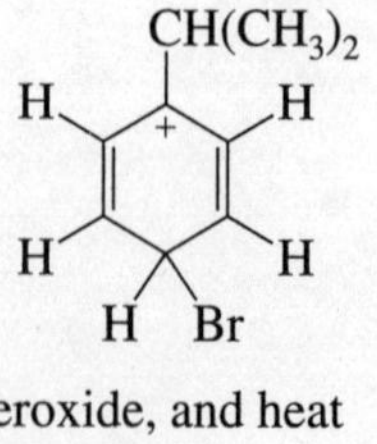

(*a*) Benzene, isopropyl bromide, and HBr
(*b*) Bromobenzene, isopropyl chloride, and $AlCl_3$
(*c*) Isopropylbenzene, $Br_2$, and $FeBr_3$
(*d*) Isopropylbenzene, $Br_2$, light, and heat
(*e*) Isopropylbenzene, *N*-bromosuccinimide, benzoyl peroxide, and heat

Structure: cyclohexadienyl cation with $CH(CH_3)_2$ at top, H's on ring carbons, and H, Br on the sp³ carbon para to the isopropyl group.

**B-14.** Which sequence of steps describes the best synthesis of the compound shown?

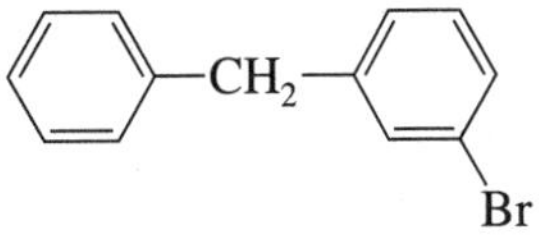

$C_6H_5$–$CH_2$–$C_6H_4$–Br (meta)

(*a*) Benzene $\xrightarrow[AlCl_3]{C_6H_5CH_2Cl}$ $\xrightarrow[FeBr_3]{Br_2}$

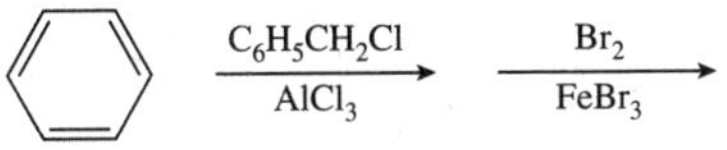

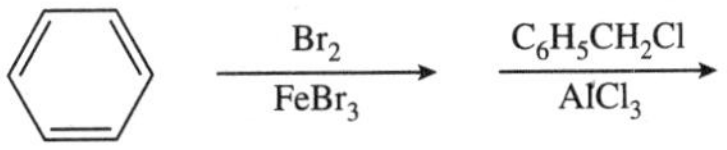

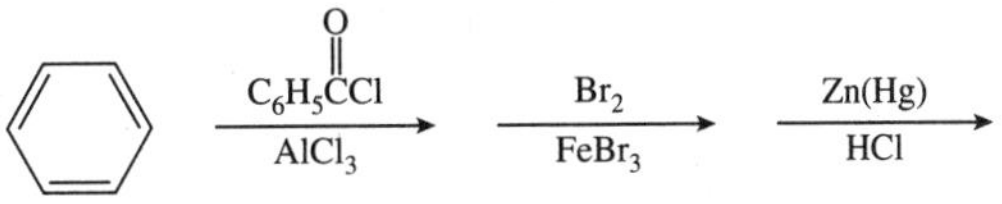

(*b*) Benzene $\xrightarrow[FeBr_3]{Br_2}$ $\xrightarrow[AlCl_3]{C_6H_5CH_2Cl}$

(*c*) Benzene $\xrightarrow[AlCl_3]{C_6H_5\overset{O}{\overset{\|}{C}}Cl}$ $\xrightarrow[FeBr_3]{Br_2}$ $\xrightarrow[HCl]{Zn(Hg)}$

(*d*) Benzene $\xrightarrow[FeBr_3]{Br_2}$ $\xrightarrow[AlCl_3]{C_6H_5\overset{O}{\overset{\|}{C}}Cl}$ $\xrightarrow[HCl]{Zn(Hg)}$

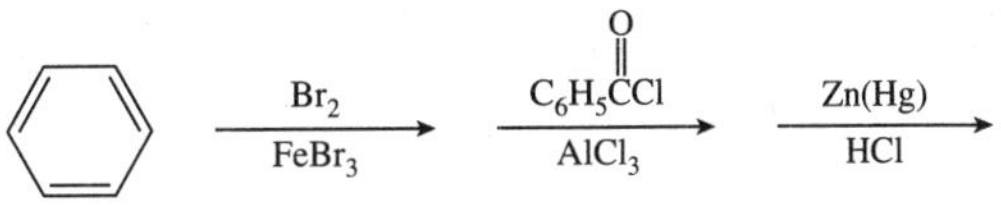

**B-15.** Which one of the following is the best synthesis of 2-chloro-4-nitrobenzoic acid?

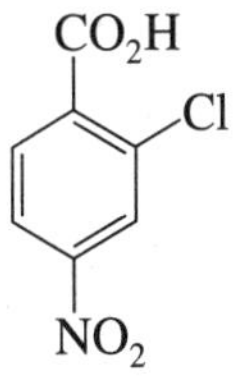

$CO_2H$, Cl, $NO_2$

2-Chloro-4-nitrobenzoic acid

(*a*) 1. Heat benzoic acid with $HNO_3$, $H_2SO_4$
2. $Cl_2$, $FeCl_3$, heat

(*b*) 1. Treat toluene with $HNO_3$, $H_2SO_4$
2. $K_2Cr_2O_7$, $H_2O$, $H_2SO_4$, heat
3. $Cl_2$, $FeCl_3$, heat

(*c*) 1. Treat toluene with $HNO_3$, $H_2SO_4$
2. $Cl_2$, $FeCl_3$, heat
3. $K_2Cr_2O_7$, $H_2O$, $H_2SO_4$, heat

(*d*) 1. Treat nitrobenzene with $Cl_2$, $FeCl_3$, heat
2. $CH_3Cl$, $AlCl_3$
3. $K_2Cr_2O_7$, $H_2O$, $H_2SO_4$, heat

(*e*) 1. Treat chlorobenzene with $HNO_3$, $H_2SO_4$
2. $CH_3Cl$, $AlCl_3$
3. $K_2Cr_2O_7$, $H_2O$, $H_2SO_4$, heat

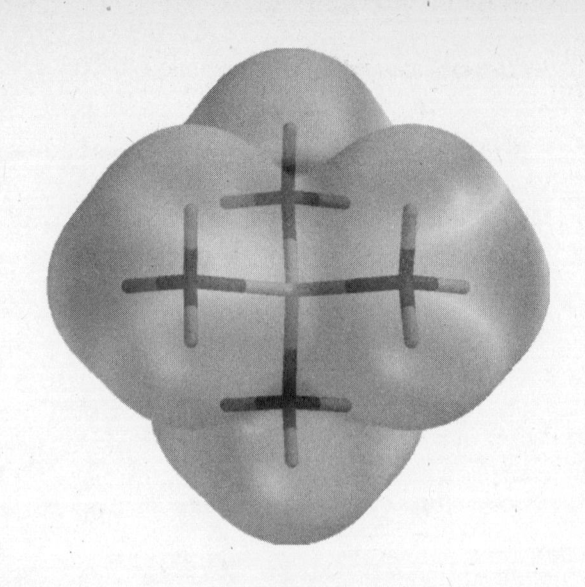

# CHAPTER 13
## SPECTROSCOPY

## SOLUTIONS TO TEXT PROBLEMS

**13.1** The field strength of an NMR spectrometer magnet and the frequency of electromagnetic radiation used to observe an NMR spectrum are directly proportional. Thus, the ratio 4.7 T/200 MHz is the same as 1.41 T/60 MHz. The magnetic field strength of a 60-MHz NMR spectrometer is 1.41 T.

**13.2** The ratio of $^1$H and $^{13}$C resonance frequencies remains constant. When the $^1$H frequency is 200 MHz, $^{13}$C NMR spectra are recorded at 50.4 MHz. Thus, when the $^1$H frequency is 100 MHz, $^{13}$C NMR spectra will be observed at 25.2 MHz.

**13.3** (*a*) Chemical shifts reported in parts per million (ppm) are independent of the field strength of the NMR spectrometer. Thus, to compare the $^1$H NMR signal of bromoform ($CHBr_3$) recorded at 300 MHz with that of chloroform ($CHCl_3$) recorded at 200 MHz as given in the text, the chemical shift of bromoform must be converted from hertz to parts per million. The chemical shift for the proton in bromoform is

$$\delta = \frac{2065 \text{ Hz}}{300 \text{ MHz}} = 6.88 \text{ ppm}$$

(*b*) The chemical shift of the proton in bromoform ($\delta$ 6.88 ppm) is less than that of chloroform ($\delta$ 7.28 ppm). The proton signal of bromoform is farther upfield and thus is **more shielded** than the proton in chloroform.

**13.4** In both chloroform ($CHCl_3$) and 1,1,1-trichloroethane ($CH_3CCl_3$) three chlorines are present. In $CH_3CCl_3$, however, the protons are one carbon removed from the chlorines, and thus the deshielding effect of the halogens will be less. The $^1$H NMR signal of $CH_3CCl_3$ appears 4.6 ppm **upfield** from the proton signal of chloroform. The chemical shift of the protons in $CH_3CCl_3$ is $\delta$ 2.6 ppm.

**13.5** 1,4-Dimethylbenzene has two types of protons: those attached directly to the benzene ring and those of the methyl groups. Aryl protons are significantly less shielded than alkyl protons. As shown in text Table 13.1 they are expected to give signals in the chemical shift range $\delta$ 6.5–8.5 ppm. Thus, the signal at $\delta$ 7.0 ppm is due to the protons of the benzene ring. The signal at $\delta$ 2.2 ppm is due to the methyl protons.

$H_3C$–$C_6H_4$–$CH_3$ (2.2 ppm → $CH_3$; 7.0 ppm → ring H)

**13.6** (*b*) Four nonequivalent sets of protons are bonded to carbon in 1-butanol as well as a fifth distinct type of proton, the one bonded to oxygen. There should be five signals in the $^1$H NMR spectrum of 1-butanol.

$CH_3CH_2CH_2CH_2OH$

Five different proton environments in 1-butanol; five signals

(*c*) Apply the "proton replacement" test to butane.

| $CH_3CH_2CH_2CH_3$ | $ClCH_2CH_2CH_2CH_3$ | $CH_3CHClCH_2CH_3$ |
|---|---|---|
| Butane | 1-Chlorobutane | 2-Chlorobutane |

| $CH_3CH_2CHClCH_3$ | $CH_3CH_2CH_2CH_2Cl$ |
|---|---|
| 2-Chlorobutane | 1-Chlorobutane |

Butane has **two** different types of protons; it will exhibit **two** signals in its $^1$H NMR spectrum.

(*d*) Like butane, 1,4-dibromobutane has two different types of protons. This can be illustrated by using a chlorine atom as a test group.

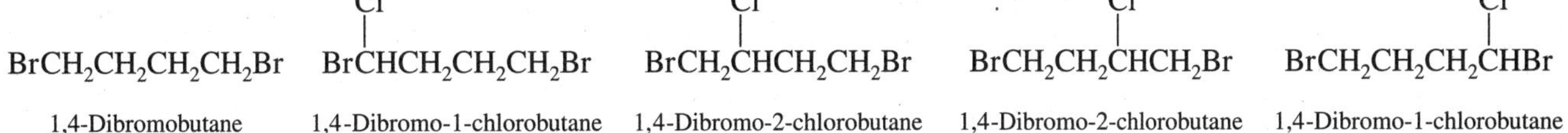

| $BrCH_2CH_2CH_2CH_2Br$ | $BrCHClCH_2CH_2CH_2Br$ | $BrCH_2CHClCH_2CH_2Br$ | $BrCH_2CH_2CHClCH_2Br$ | $BrCH_2CH_2CH_2CHClBr$ |
|---|---|---|---|---|
| 1,4-Dibromobutane | 1,4-Dibromo-1-chlorobutane | 1,4-Dibromo-2-chlorobutane | 1,4-Dibromo-2-chlorobutane | 1,4-Dibromo-1-chlorobutane |

The $^1$H NMR spectrum of 1,4-dibromobutane is expected to consist of two signals.

(*e*) All the carbons in 2,2-dibromobutane are different from each other, and so protons attached to one carbon are not equivalent to the protons attached to any of the other carbons. This compound should have **three** signals in its $^1$H NMR spectrum.

$CH_3CBr_2CH_2CH_3$

2,2-Dibromobutane has three nonequivalent sets of protons.

(*f*) All the protons in 2,2,3,3-tetrabromobutane are equivalent. Its $^1$H NMR spectrum will consist of one signal.

$CH_3CBr_2—CBr_2CH_3$

2,2,3,3-Tetrabromobutane

(*g*) There are **four** nonequivalent sets of protons in 1,1,4-tribromobutane. It will exhibit four signals in its $^1$H NMR spectrum.

$$\mathrm{BrC(Br)(H)CH_2CH_2CH_2Br}$$

1,1,4-Tribromobutane

(*h*) The seven protons of 1,1,1-tribromobutane belong to three nonequivalent sets, and hence the $^1$H NMR spectrum will consist of three signals.

$$\mathrm{Br_3CCH_2CH_2CH_3}$$

1,1,1-Tribromobutane

**13.7** (*b*) Apply the replacement test to each of the protons of 1,1-dibromoethene.

$$\mathrm{Br_2C{=}CH_2} \qquad \mathrm{Br_2C{=}CHCl} \qquad \mathrm{Br_2C{=}CHCl}$$

1,1-Dibromoethene — 1,1-Dibromo-2-chloroethene — 1,1-Dibromo-2-chloroethene

Replacement of one proton by a test group (Cl) gives exactly the same compound as replacement of the other. The two protons of 1,1-dibromoethene are equivalent, and there is only one signal in the $^1$H NMR spectrum of this compound.

(*c*) The replacement test reveals that both protons of *cis*-1,2-dibromoethene are equivalent.

$$\mathrm{BrHC{=}CHBr} \qquad \mathrm{BrClC{=}CHBr} \qquad \mathrm{BrHC{=}CClBr}$$

*cis*-1,2-Dibromoethene — (*Z*)-1,2-Dibromo-1-chloroethene — (*Z*)-1,2-Dibromo-1-chloroethene

Because both protons are equivalent, the $^1$H NMR spectrum of *cis*-1,2-dibromoethene consists of one signal.

(*d*) Both protons of *trans*-1,2-dibromoethene are equivalent; each is cis to a bromine substituent.

$$\mathrm{BrHC{=}CHBr}$$

*trans*-1,2-Dibromoethene
(one signal in the $^1$H NMR spectrum)

(*e*) **Four** nonequivalent sets of protons occur in allyl bromide.

$$\mathrm{H_2C{=}CHCH_2Br}$$

Allyl bromide (four signals in the $^1$H NMR spectrum)

(*f*) The protons of a single methyl group are equivalent to one another, but all three methyl groups of 2-methyl-2-butene are nonequivalent. The vinyl proton is unique.

$$(H_3C)_2C=CH(CH_3)$$

2-Methyl-2-butene (four signals in the $^1$H NMR spectrum)

**13.8** (*b*) The three methyl protons of 1,1,1-trichloroethane ($Cl_3CCH_3$) are equivalent. They have the same chemical shift and do not split each other's signals. The $^1$H NMR spectrum of $Cl_3CCH_3$ consists of a single sharp peak.

(*c*) Separate signals will be seen for the methylene ($CH_2$) protons and for the methine (CH) proton of 1,1,2-trichloroethane.

$$Cl_2C(H)—CH_2Cl$$

Triplet (H) Doublet ($CH_2$)

1,1,2-Trichloroethane

The methine proton splits the signal for the methylene protons into a doublet. The two methylene protons split the methine proton's signal into a triplet.

(*d*) Examine the structure of 1,2,2-trichloropropane.

$$ClCH_2CCl_2CH_3$$

1,2,2-Trichloropropane

The $^1$H NMR spectrum exhibits a signal for the two equivalent methylene protons and one for the three equivalent methyl protons. Both these signals are sharp singlets. The protons of the methyl group and the methylene group are separated by more than three bonds and do not split each other's signals.

(*e*) The methine proton of 1,1,1,2-tetrachloropropane splits the signal of the methyl protons into a doublet; its signal is split into a quartet by the three methyl protons.

$$Cl_3CC(Cl)(H)—CH_3$$

$CH_3$: Doublet; H: Quartet

1,1,1,2-Tetrachloropropane

**13.9** (*b*) The ethyl group appears as a triplet–quartet pattern and the methyl group as a singlet.

$$CH_3CH_2OCH_3$$

Triplet ($CH_3$) Quartet ($CH_2$) Singlet ($OCH_3$); not vicinal to any other protons in molecule

(*c*) The two ethyl groups of diethyl ether are equivalent to each other. The two methyl groups appear as one triplet and the two methylene groups as one quartet.

$CH_3CH_2OCH_2CH_3$

Triplet Quartet Quartet Triplet

(*d*) The two ethyl groups of *p*-diethylbenzene are equivalent to each other and give rise to a single triplet–quartet pattern.

$CH_3CH_2$–$C_6H_4$–$CH_2CH_3$

Three signals:
$CH_3$ triplet;
$CH_2$ quartet;
aromatic H singlet

All four protons of the aromatic ring are equivalent, have the same chemical shift, and do not split either each other's signals or any of the signals of the ethyl group.

(*e*) Four nonequivalent sets of protons occur in this compound:

$ClCH_2CH_2OCH_2CH_3$

Triplet Triplet Quartet Triplet

Vicinal protons in the $ClCH_2CH_2O$ group split one another's signals, as do those in the $CH_3CH_2O$ group.

**13.10** Both $H_b$ and $H_c$ in *m*-nitrostyrene appear as doublets of doublets. $H_b$ is coupled to $H_a$ by a coupling constant of 12 Hz and to $H_c$ by a coupling constant of 2 Hz. $H_c$ is coupled to $H_a$ by a coupling constant of 16 Hz and to $H_b$ by a coupling constant of 2 Hz.

$H_b$ — 12 Hz — 2 Hz, 2 Hz
$H_c$ — 16 Hz — 2 Hz, 2 Hz

(diagrams not to scale)

**13.11** (*b*) The signal of the proton at C-2 is split into a quartet by the methyl protons, and each line of this quartet is split into a doublet by the aldehyde proton. It appears as a doublet of quartets. (*Note:* It does not matter whether the splitting pattern is described as a doublet of quartets or a quartet of doublets. There is no substantive difference in the two descriptions.)

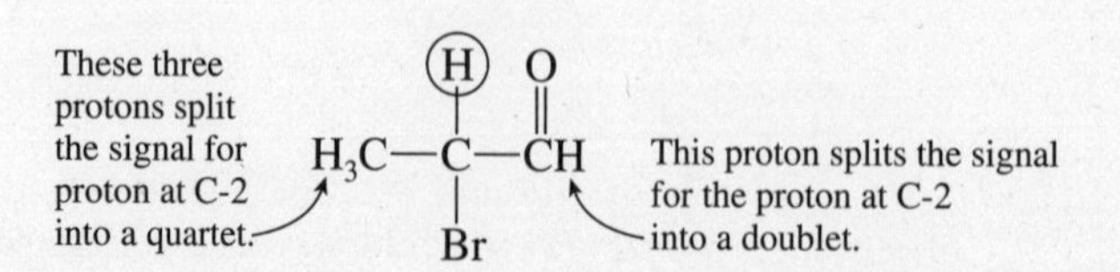
These three protons split the signal for proton at C-2 into a quartet.
H
O
H3C—C—CH
Br
This proton splits the signal for the proton at C-2 into a doublet.

**13.12** (*b*) The two methyl carbons of the isopropyl group are equivalent.

Four different types of carbons occur in the aromatic ring and two different types are present in the isopropyl group. The $^{13}C$ NMR spectrum of isopropylbenzene contains **six** signals.

(*c*) The methyl substituent at C-2 is different from those at C-1 and C-3:

The four nonequivalent ring carbons and the two different types of methyl carbons give rise to a $^{13}C$ NMR spectrum that contains **six** signals.

(*d*) The three methyl carbons of 1,2,4-trimethylbenzene are different from one another:

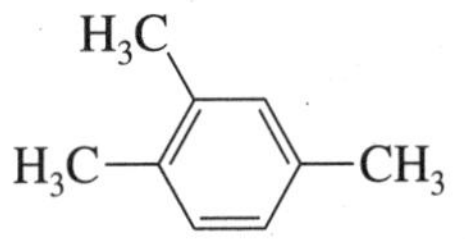

Also, all the ring carbons are different from each other. The nine different carbons give rise to **nine** separate signals.

(*e*) All three methyl carbons of 1,3,5-trimethylbenzene are equivalent.

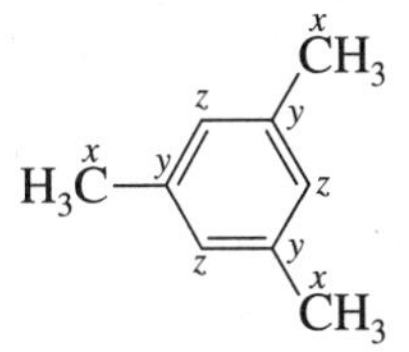

Because of its high symmetry 1,3,5-trimethylbenzene has only **three** signals in its $^{13}C$ NMR spectrum.

**13.13** $sp^3$-Hybridized carbons are more shielded than $sp^2$-hybridized ones. Carbon $x$ is the most shielded, and has a chemical shift of $\delta$ 20 ppm. The oxygen of the $OCH_3$ group decreased the shielding of carbon $z$; its chemical shift is $\delta$ 55 ppm. The least shielded is carbon $y$ with a chemical shift of $\delta$ 157 ppm.

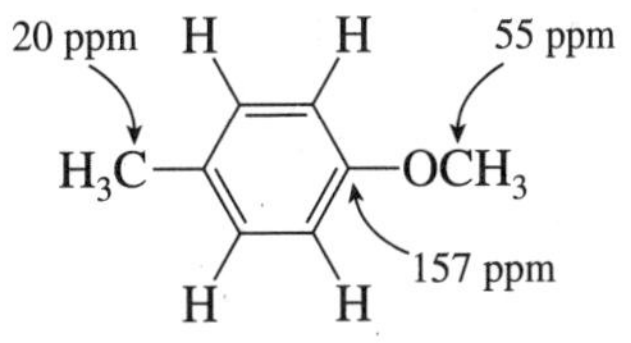

**13.14** The $^{13}C$ NMR spectrum in Figure 13.22 shows nine signals and is the spectrum of 1,2,4-trimethylbenzene from part (*d*) of Problem 13.12. Six of the signals, in the range $\delta$ 127–138 ppm, are due to the six nonequivalent carbons of the benzene ring. The three signals near $\delta$ 20 ppm are due to the three nonequivalent methyl groups.

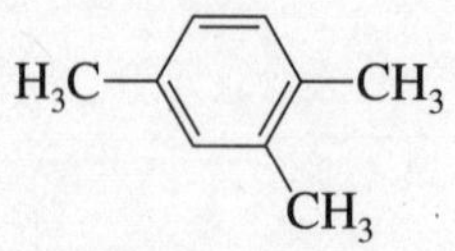

1,2,4-Trimethylbenzene

**13.15** The infrared spectrum of Figure 13.31 has no absorption in the 1600–1800-cm$^{-1}$ region, and so the unknown compound cannot contain a carbonyl (C═O) group. It cannot therefore be acetophenone or benzoic acid.

The broad, intense absorption at 3300 cm$^{-1}$ is attributable to a hydroxyl group. Although both phenol and benzyl alcohol are possibilities, the peaks at 2800–2900 cm$^{-1}$ reveal the presence of hydrogen bonded to $sp^3$-hybridized carbon. All carbons are $sp^2$-hybridized in phenol. The infrared spectrum is that of **benzyl alcohol.**

**13.16** The energy of electromagnetic radiation is inversely proportional to its wavelength. Since excitation of an electron for the $\pi \rightarrow \pi^*$ transition of ethylene occurs at a shorter wavelength ($\lambda_{max}$ = 170 nm) than that of *cis, trans*-1,3-cyclooctadiene ($\lambda_{max}$= 230 nm), the HOMO–LUMO energy difference in ethylene is **greater.**

**13.17** Conjugation shifts $\lambda_{max}$ to longer wavelengths in alkenes. The conjugated diene 2-methyl-1,3-butadiene has the longest wavelength absorption, $\lambda_{max}$ = 222 nm. The isolated diene 1,4-pentadiene and the simple alkene cyclopentene both absorb below 200 nm.

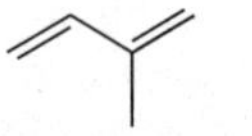

2-Methyl-1,3-butadiene
($\lambda_{max}$ = 222 nm)

**13.18** (*b*) The distribution of molecular-ion peaks in *o*-dichlorobenzene is identical to that in the para isomer. As the sample solution to part (*a*) in the text describes, peaks at $m/z$ 146, 148, and 150 are present for the molecular ion.

(*c*) The two isotopes of bromine are $^{79}$Br and $^{81}$Br. When both bromines of *p*-dibromobenzene are $^{79}$Br, the molecular ion appears at $m/z$ 234. When one is $^{79}$Br and the other is $^{81}$Br, $m/z$ for the molecular ion is 236. When both bromines are $^{81}$Br, $m/z$ for the molecular ion is 238.

(*d*) The combinations of $^{35}$Cl, $^{37}$Cl, $^{79}$Br, and $^{81}$Br in *p*-bromochlorobenzene and the values of $m/z$ for the corresponding molecular ion are as shown.

| | |
|---|---|
| ($^{35}$Cl, $^{79}$Br) | $m/z$ = 190 |
| ($^{37}$Cl, $^{79}$Br) or ($^{35}$Cl, $^{81}$Br) | $m/z$ = 192 |
| ($^{37}$Cl, $^{81}$Br) | $m/z$ = 194 |

**13.19** The base peak in the mass spectrum of alkylbenzenes corresponds to carbon–carbon bond cleavage at the benzylic carbon.

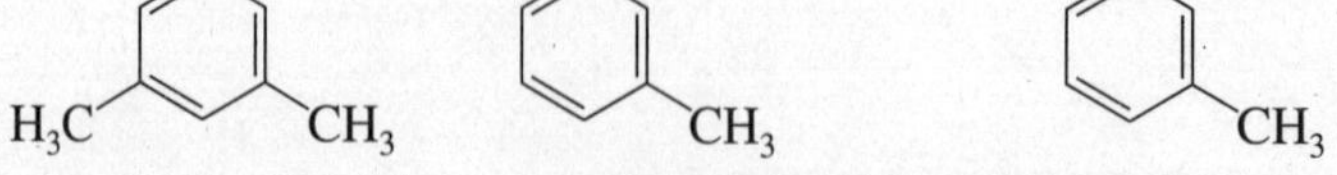

Base peak: $C_9H_{11}^+$
$m/z$ 119

Base peak: $C_8H_9^+$
$m/z$ 105

Base peak: $C_9H_{11}^+$
$m/z$ 119

**13.20** (*b*) The index of hydrogen deficiency is given by the following formula:

$$\text{Index of hydrogen deficiency} = \tfrac{1}{2}(C_nH_{2n+2} - C_nH_x)$$

The compound given contains eight carbons ($C_8H_8$); therefore,

$$\text{Index of hydrogen deficiency} = \tfrac{1}{2}(C_8H_{18} - C_8H_8) = 5$$

The problem specifies that the compound consumes 2 mol of hydrogen, and so it contains two double bonds (or one triple bond). Since the index of hydrogen deficiency is equal to 5, there must be three rings.

(*c*) Chlorine substituents are equivalent to hydrogens when calculating the index of hydrogen deficiency. Therefore, consider $C_8H_8Cl_2$ as equivalent to $C_8H_{10}$. Thus, the index of hydrogen deficiency of this compound is 4.

$$\text{Index of hydrogen deficiency} = \tfrac{1}{2}(C_8H_{18} - C_8H_{10}) = 4$$

Since the compound consumes 2 mol of hydrogen on catalytic hydrogenation, it must therefore contain two rings.

(*d*) Oxygen atoms are ignored when calculating the index of hydrogen deficiency. Thus, $C_8H_8O$ is treated as if it were $C_8H_8$.

$$\text{Index of hydrogen deficiency} = \tfrac{1}{2}(C_8H_{18} - C_8H_8) = 5$$

Since the problem specifies that 2 mol of hydrogen is consumed on catalytic hydrogenation, this compound contains three rings.

(*e*) Ignoring the oxygen atoms in $C_8H_{10}O_2$, we treat this compound as if it were $C_8H_{10}$.

$$\text{Index of hydrogen deficiency} = \tfrac{1}{2}(C_8H_{18} - C_8H_{10}) = 4$$

Because 2 mol of hydrogen is consumed on catalytic hydrogenation, there must be two rings.

(*f*) Ignore the oxygen, and treat the chlorine as if it were hydrogen. Thus, $C_8H_9ClO$ is treated as if it were $C_8H_{10}$. Its index of hydrogen deficiency is 4, and it contains two rings.

**13.21** Since each compound exhibits only a single peak in its $^1H$ NMR spectrum, all the hydrogens are equivalent in each one. Structures are assigned on the basis of their molecular formulas and chemical shifts.

(*a*) This compound has the molecular formula $C_8H_{18}$ and so must be an alkane. The 18 hydrogens are contributed by six equivalent methyl groups.

$$(CH_3)_3CC(CH_3)_3$$

2,2,3,3-Tetramethylbutane
($\delta$ 0.9 ppm)

(*b*) A hydrocarbon with the molecular formula $C_5H_{10}$ has an index of hydrogen deficiency of 1 and so is either a cycloalkane or an alkene. Since all ten hydrogens are equivalent, this compound must be cyclopentane.

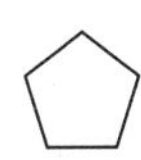

Cyclopentane
($\delta$ 1.5 ppm)

(*c*) The chemical shift of the eight equivalent hydrogens in $C_8H_8$ is $\delta$ 5.8 ppm, which is consistent with protons attached to a carbon–carbon double bond.

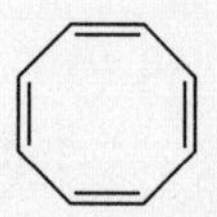

1,3,5,7-Cyclooctatetraene
($\delta$ 5.8 ppm)

(*d*) The compound $C_4H_9Br$ has no rings or double bonds. The nine hydrogens belong to three equivalent methyl groups.

$(CH_3)_3CBr$

*tert*-Butyl bromide ($\delta$ 1.8 ppm)

(*e*) The dichloride has no rings or double bonds (index of hydrogen deficiency = 0). The four equivalent hydrogens are present as two —$CH_2Cl$ groups.

$ClCH_2CH_2Cl$

1,2-Dichloroethane ($\delta$ 3.7 ppm)

(*f*) All three hydrogens in $C_2H_3Cl_3$ must be part of the same methyl group in order to be equivalent.

$CH_3CCl_3$

1,1,1-Trichloroethane ($\delta$ 2.7 ppm)

(*g*) This compound has no rings or double bonds. To have eight equivalent hydrogens it must have four equivalent methylene groups.

$CH_2Cl$
$ClCH_2CCH_2Cl$
$CH_2Cl$

1,3-Dichloro-2,2-di(chloromethyl)propane
($\delta$ 3.7 ppm)

(*h*) A compound with a molecular formula of $C_{12}H_{18}$ has an index of hydrogen deficiency of 4. A likely candidate for a compound with 18 equivalent hydrogens is one with six equivalent $CH_3$ groups. Thus, 6 of the 12 carbons belong to $CH_3$ groups, and the other 6 have no hydrogens. The compound is hexamethylbenzene.

$CH_3$
$H_3C$ $CH_3$
$H_3C$ $CH_3$
$CH_3$

A chemical shift of $\delta$ 2.2 ppm is consistent with the fact that all of the protons are benzylic hydrogens.

(*i*) The molecular formula of $C_3H_6Br_2$ tells us that the compound has no double bonds and no rings. All six hydrogens are equivalent, indicating two equivalent methyl groups. The compound is 2,2-dibromopropane, $(CH_3)_2CBr_2$.

**13.22** In each of the parts to this problem, nonequivalent protons must *not* be bonded to adjacent carbons, because we are told that the two signals in each case are singlets.

(*a*) Each signal corresponds to four protons, and so each must result from two equivalent $CH_2$ groups. The four $CH_2$ groups account for four of the carbons of $C_6H_8$, leaving two carbons that bear no hydrogens. A molecular formula of $C_6H_8$ corresponds to an index of hydrogen deficiency of 3. A compound consistent with these requirements is

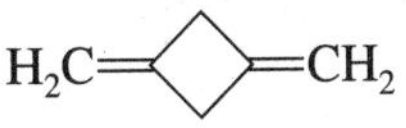

$$H_2C{=}\langle\text{cyclobutane}\rangle{=}CH_2$$

The signal at $\delta$ 5.6 ppm is consistent with that expected for the four vinylic protons. The signal at $\delta$ 2.7 ppm corresponds to that for the allylic protons of the ring.

(*b*) The compound has a molecular formula of $C_5H_{11}Br$ and therefore has no double bonds or rings. A 9-proton singlet at $\delta$ 1.1 ppm indicates three equivalent methyl groups, and a 2-proton singlet at $\delta$ 3.3 ppm indicates a $CH_2Br$ group. The correct structure is $(CH_3)_3CCH_2Br$.

(*c*) This compound ($C_6H_{12}O$) has three equivalent $CH_3$ groups, along with a fourth $CH_3$ group that is somewhat less shielded. Its molecular formula indicates that it can have either one double bond or one ring. This compound is $(CH_3)_3C\overset{O}{\overset{\|}{C}}CH_3$.

(*d*) A molecular formula of $C_6H_{10}O_2$ corresponds to an index of hydrogen deficiency of 2. The signal at $\delta$ 2.2 ppm (6H) is likely due to two equivalent $CH_3$ groups, and the one at $\delta$ 2.7 ppm (4H) to two equivalent $CH_2$ groups. The compound is $CH_3\overset{O}{\overset{\|}{C}}CH_2CH_2\overset{O}{\overset{\|}{C}}CH_3$.

**13.23** (*a*) A 5-proton signal at $\delta$ 7.1 ppm indicates a monosubstituted aromatic ring. With an index of hydrogen deficiency of 4, $C_8H_{10}$ contains this monosubstituted aromatic ring and no other rings or multiple bonds. The triplet–quartet pattern at high field suggests an ethyl group.

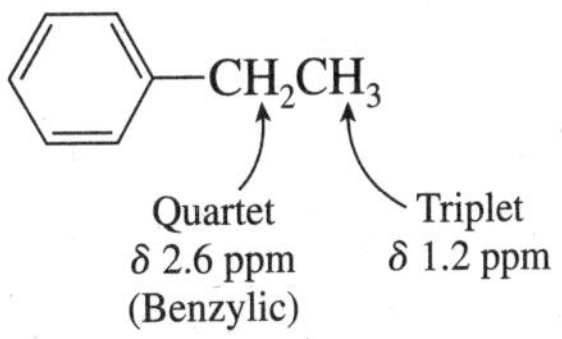

$$C_6H_5{-}CH_2CH_3$$

$CH_2$: Quartet, $\delta$ 2.6 ppm (Benzylic); $CH_3$: Triplet, $\delta$ 1.2 ppm

Ethylbenzene

(*b*) The index of hydrogen deficiency of 4 and the 5-proton multiplet at $\delta$ 7.0 to 7.5 ppm are accommodated by a monosubstituted aromatic ring. The remaining four carbons and nine hydrogens are most reasonably a *tert*-butyl group, since all nine hydrogens are equivalent.

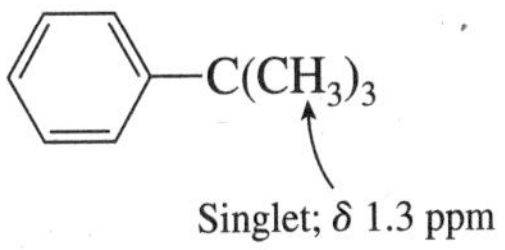

$$C_6H_5{-}C(CH_3)_3$$

$CH_3$: Singlet; $\delta$ 1.3 ppm

*tert*-Butylbenzene

(*c*) Its molecular formula requires that $C_6H_{14}$ be an alkane. The doublet–septet pattern is consistent with an isopropyl group, and the total number of protons requires that two of these groups be present.

$$(CH_3)_2CHCH(CH_3)_2$$

$CH_3$: Doublet, $\delta$ 0.8 ppm; $CH$: Septet, $\delta$ 1.4 ppm

2,3-Dimethylbutane

Note that the methine (CH) protons do not split each other, because they are equivalent and have the same chemical shift.

(*d*) The molecular formula $C_6H_{12}$ requires the presence of one double bond or ring. A peak at $\delta$ 5.1 ppm is consistent with —C═CH, and so the compound is a noncyclic alkene. The vinyl proton gives a triplet signal, and so the group $C{=}CHCH_2$ is present. The $^1$H NMR spectrum shows the presence of the following structural units:

C═C(H)(CH$_2$) — H: 5.1 ppm (Triplet); CH$_2$: 2.0 ppm (Allylic)

$CH_2CH_3$ — CH$_3$: 0.9 ppm (Triplet)

C═C(CH$_3$)(CH$_3$) — 1.6 ppm (Singlet; allylic); 1.7 ppm (Singlet; allylic)

Putting all these fragments together yields a unique structure.

$(H_3C)_2C{=}CHCH_2CH_3$ — H$_3$C: Singlet; H$_3$C: Singlet; H: Triplet; CH$_2$: Pentet; CH$_3$: Triplet

2-Methyl-2-pentene

(*e*) The compound $C_4H_6Cl_4$ contains no double bonds or rings. There are no high-field peaks ($\delta$ 0.5 to 1.5 ppm), and so there are no methyl groups. At least one chlorine substituent must therefore be at each end of the chain. The most likely structure has the four chlorines divided into two groups of two.

$Cl_2CHCH_2CH_2CHCl_2$

$\delta$ 4.6 ppm (Triplet) — CH; $\delta$ 3.9 ppm (Doublet) — CH$_2$

1,1,4,4-Tetrachlorobutane

(*f*) The molecular formula $C_4H_6Cl_2$ indicates the presence of one double bond or ring. A signal at $\delta$ 5.7 ppm is consistent with a proton attached to a doubly bonded carbon. The following structural units are present:

C═C(H)(CH$_2$) — H: 5.7 ppm (Triplet)

$C{=}CCH_3$ — 2.2 ppm (Allylic)

For the methyl group to appear as a singlet and the methylene group to appear as a doublet, the chlorine substituents must be distributed as shown:

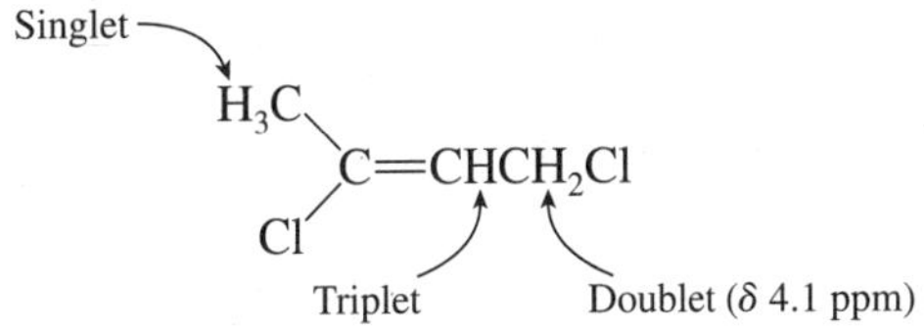

Singlet → $H_3C$
$\begin{matrix} H_3C \\ \quad \diagdown \\ \quad C{=}CHCH_2Cl \\ \quad \diagup \\ Cl \end{matrix}$

Triplet — Doublet (δ 4.1 ppm)

1,3-Dichloro-2-butene

The stereochemistry of the double bond (*E* or *Z*) is not revealed by the $^1H$ NMR spectrum.

(*g*) A molecular formula of $C_3H_7ClO$ is consistent with the absence of rings and multiple bonds (index of hydrogen deficiency = 0). None of the signals is equivalent to three protons, and so no methyl groups are present. Three methylene groups occur, all of which are different from each other. The compound is therefore:

$ClCH_2CH_2CH_2OH$

δ 3.7 or 3.8 ppm (Triplet) — δ 2.0 ppm (Pentet) — δ 3.7 or 3.8 ppm (Triplet) — δ 2.8 ppm (Singlet)

(*h*) The compound has a molecular formula of $C_{14}H_{14}$ and an index of hydrogen deficiency of 8. With a 10-proton signal at δ 7.1 ppm, a logical conclusion is that there are two monosubstituted benzene rings. The other four protons belong to two equivalent methylene groups.

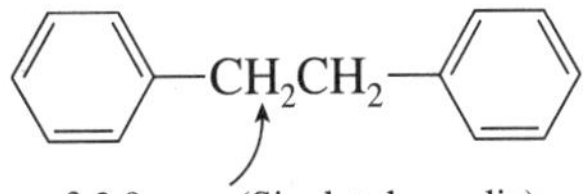

$C_6H_5$—$CH_2CH_2$—$C_6H_5$

δ 2.9 ppm (Singlet; benzylic)

1,2-Diphenylethane

**13.24** The compounds of molecular formula $C_4H_9Cl$ are the isomeric chlorides: butyl, isobutyl, *sec*-butyl, and *tert*-butyl chloride.

(*a*) All nine methyl protons of *tert*-butyl chloride $(CH_3)_3CCl$ are equivalent; its $^1H$ NMR spectrum has only one peak.

(*b*) A doublet at δ 3.4 ppm indicates a —$CH_2Cl$ group attached to a carbon that bears a single proton.

$(CH_3)_2CHCH_2Cl$

δ 3.4 ppm (Doublet)

Isobutyl chloride

(*c*) A triplet at δ 3.5 ppm means that a methylene group is attached to the carbon that bears the chlorine.

$CH_3CH_2CH_2CH_2Cl$

δ 3.5 ppm (Triplet)

Butyl chloride

(*d*) This compound has two nonequivalent methyl groups.

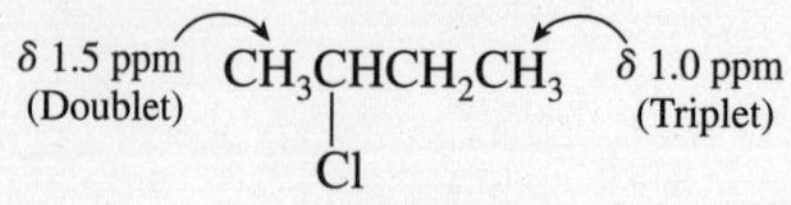

$\delta$ 1.5 ppm (Doublet) $CH_3CHCH_2CH_3$ $\delta$ 1.0 ppm (Triplet)
with Cl on C-2

*sec*-Butyl chloride

**13.25** Compounds with the molecular formula $C_3H_5Br$ have either one ring or one double bond.

(*a*) The two peaks at $\delta$ 5.4 and 5.6 ppm have chemical shifts consistent with the assumption that each peak is due to a vinyl proton (C═CH). The remaining three protons belong to an allylic methyl group ($\delta$ 2.3 ppm).

The compound cannot be $CH_3CH$═CHBr, because the methyl signal would be split into a doublet. Isomer A can only be

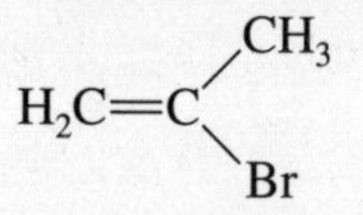

$H_2C$═$C(CH_3)Br$

2-Bromo-1-propene

(*b*) Two of the carbons of isomer B have chemical shifts characteristic of $sp^2$-hybridized carbon. One of these bears two protons ($\delta$ 118.8 ppm); the other bears one proton ($\delta$ 134.2 ppm). The remaining carbon is $sp^3$-hybridized and bears two hydrogens. Isomer B is allyl bromide.

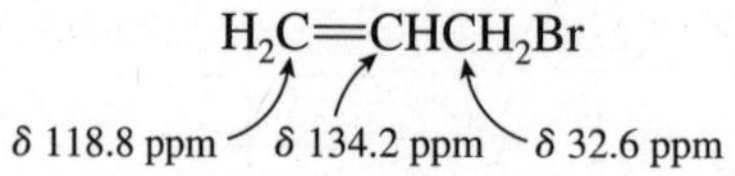

$H_2C$═$CHCH_2Br$

$\delta$ 118.8 ppm $\delta$ 134.2 ppm $\delta$ 32.6 ppm

Allyl bromide

(*c*) All the carbons are $sp^3$-hybridized in this isomer. Two of the carbons belong to equivalent $CH_2$ groups, and the other bears only one hydrogen. Isomer C is cyclopropyl bromide.

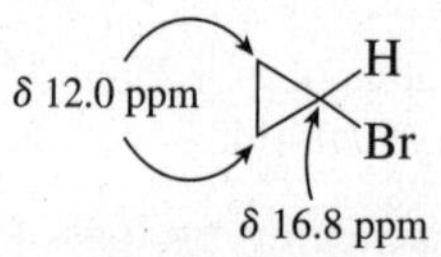

$\delta$ 12.0 ppm H Br $\delta$ 16.8 ppm

Cyclopropyl bromide

**13.26** All these compounds have the molecular formula $C_4H_{10}O$. They have neither multiple bonds nor rings.

(*a*) Two equivalent $CH_3$ groups occur at $\delta$ 18.9 ppm. One carbon bears a single hydrogen. The least shielded carbon, presumably the one bonded to oxygen, has two hydrogen substituents.

Putting all the information together reveals this compound to be isobutyl alcohol.

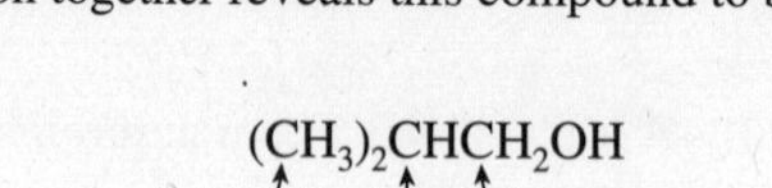

$(CH_3)_2CHCH_2OH$

$\delta$ 18.9 ppm $\delta$ 30.8 ppm $\delta$ 69.4 ppm

Isobutyl alcohol

(*b*) This compound has four distinct peaks, and so none of the four carbons is equivalent to any of the others. The signal for the least shielded carbon represents CH, and so the oxygen is attached to a secondary carbon. Only one carbon appears at low field; the compound is an alcohol, not an ether. Therefore;

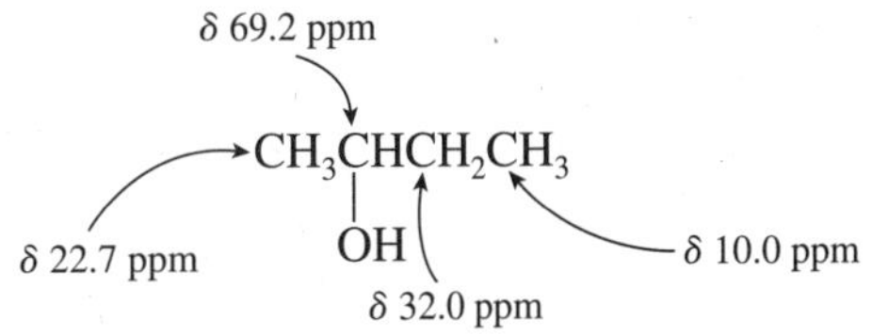

δ 69.2 ppm

$CH_3CHCH_2CH_3$ with OH on C-2

δ 22.7 ppm   δ 32.0 ppm   δ 10.0 ppm

*sec*-Butyl alcohol

(*c*) Signals for three equivalent $CH_3$ carbons indicate that this isomer is *tert*-butyl alcohol. This assignment is reinforced by the observation that the least shielded carbon has no hydrogens attached to it.

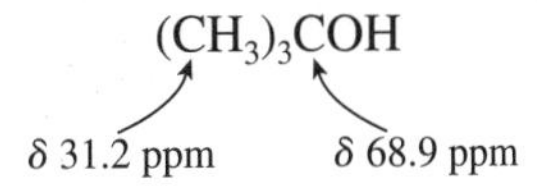

$(CH_3)_3COH$

δ 31.2 ppm   δ 68.9 ppm

*tert*-Butyl alcohol

**13.27** The molecular formula of $C_6H_{14}$ for each of these isomers requires that all of them be alkanes.

(*a*) This compound contains only $CH_3$ and CH carbons.

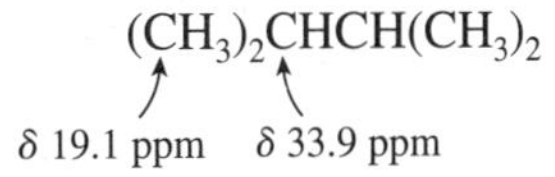

$(CH_3)_2CHCH(CH_3)_2$

δ 19.1 ppm   δ 33.9 ppm

2,3-Dimethylbutane

(*b*) This isomer has no CH carbons, and two different kinds of $CH_2$ groups.

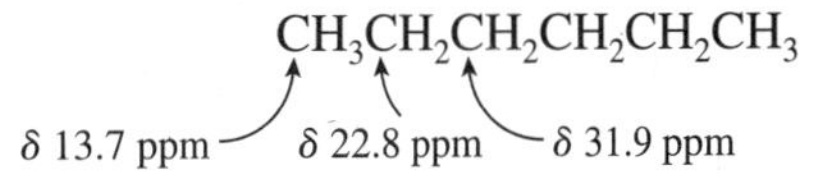

$CH_3CH_2CH_2CH_2CH_2CH_3$

δ 13.7 ppm   δ 22.8 ppm   δ 31.9 ppm

Hexane

(*c*) $CH_3$, $CH_2$, and CH carbons are all present in this isomer. There are two different kinds of $CH_3$ groups.

δ 29.1 ppm   δ 36.4 ppm

$CH_3CH_2CHCH_2CH_3$ with $CH_3$ on C-3

δ 11.1 ppm   δ 18.4 ppm

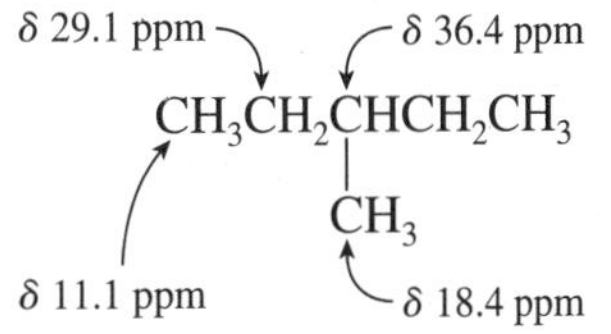

3-Methylpentane

(*d*) This isomer contains a quaternary carbon in addition to a $CH_2$ group and two different kinds of $CH_3$ groups.

δ 28.7 ppm

$H_3C—C(CH_3)_2—CH_2CH_3$

δ 8.5 ppm

δ 30.2 ppm   δ 36.5 ppm

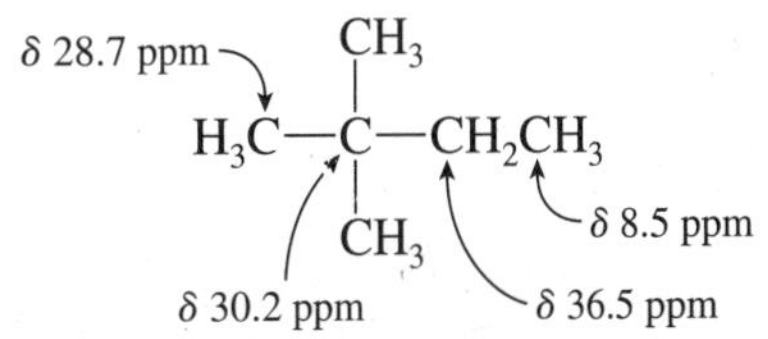

2,2-Dimethylbutane

(*e*) This isomer contains two different kinds of $CH_3$ groups, two different kinds of $CH_2$ groups, and a CH group.

$\delta$ 27.6 ppm; $\delta$ 20.5 ppm

$CH_3CHCH_2CH_2CH_3$ (with $CH_3$ on C-2)

$\delta$ 14.0 ppm; $\delta$ 22.4 ppm; $\delta$ 41.6 ppm

2-Methylpentane

**13.28** The index of hydrogen deficiency of the compound $C_4H_6$ is 2. It can have two double bonds, two rings, one ring and one double bond, or one triple bond.

The chemical shift data indicate that two carbons are $sp^3$-hybridized and two are $sp^2$. The most reasonable structure that is consistent with $^{13}C$ NMR data is cyclobutene.

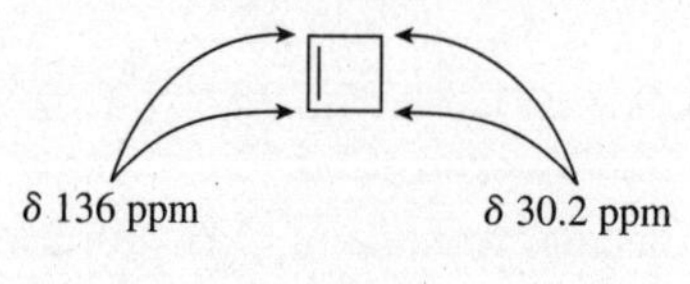


Cyclobutene

The compound cannot be 1- or 2-methylcyclopropene. Neither of the carbon signals represents a methyl group.

**13.29** Each of the carbons in the compound gives its $^{13}C$ NMR signal at relatively low field; it is likely that each one bears an electron-withdrawing substituent. The compound is

$\delta$ 72.0 ppm

$ClCH_2CHCH_2OH$ (with OH on C-2)

$\delta$ 46.8 ppm; $\delta$ 63.5 ppm

3-Chloro-1,2-propanediol

The isomeric compound 2-chloro-1,3-propanediol

$HOCH_2CHCH_2OH$ (with Cl on C-2)

cannot be correct. The C-1 and C-3 positions are equivalent; the $^{13}C$ NMR spectrum of this compound exhibits only two peaks, not three.

**13.30** (*a*) All the hydrogens are equivalent in *p*-dichlorobenzene; therefore it has the simplest $^1H$ NMR spectrum of the three compounds chlorobenzene, *o*-dichlorobenzene, and *p*-dichlorobenzene.

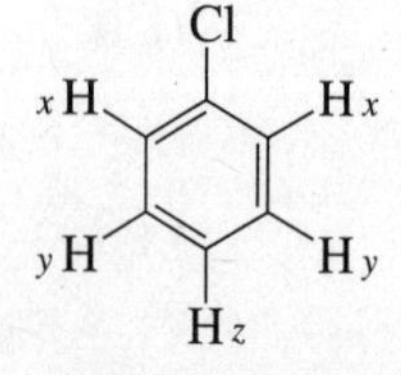


Chlorobenzene
(three different kinds of protons)

Cl; xH; Cl; yH; Hx; Hy

*o*-Dichlorobenzene
(two different kinds of protons)

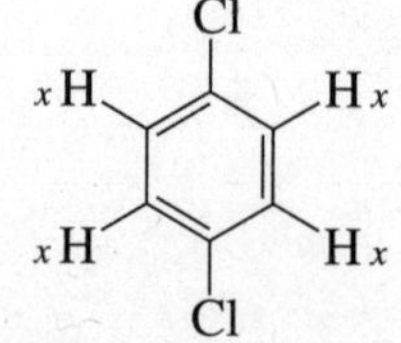


*p*-Dichlorobenzene
(all protons are equivalent)

(*b–d*) In addition to giving the simplest $^1$H NMR spectrum, *p*-dichlorobenzene gives the simplest $^{13}$C NMR spectrum. It has two peaks in its $^{13}$C NMR spectrum, chlorobenzene has four, and *o*-dichlorobenzene has three.

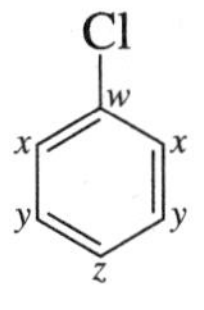

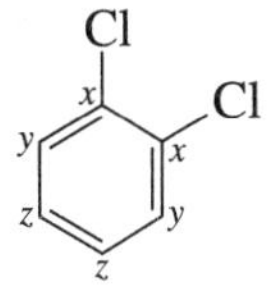

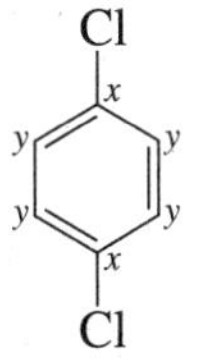

Chlorobenzene (four different kinds of carbon)

*o*-Dichlorobenzene (three different kinds of carbon)

*p*-Dichlorobenzene (two different kinds of carbon)

**13.31** Compounds A and B ($C_{10}H_{14}$) have an index of hydrogen deficiency of 4. Both have peaks in the $\delta$ 130–140-ppm range of their $^{13}$C NMR spectra, so that the index of hydrogen deficiency can be accommodated by a benzene ring.

The $^{13}$C NMR spectrum of compound A shows only a single peak in the upfield region, at $\delta$ 20 ppm. Thus, the four remaining carbons, after accounting for the benzene ring, are four equivalent methyl groups. The benzene ring is symmetrically substituted as there are only two signals in the aromatic region at $\delta$ 132 and 135 ppm. Compound A is 1,2,4,5-tetramethylbenzene.

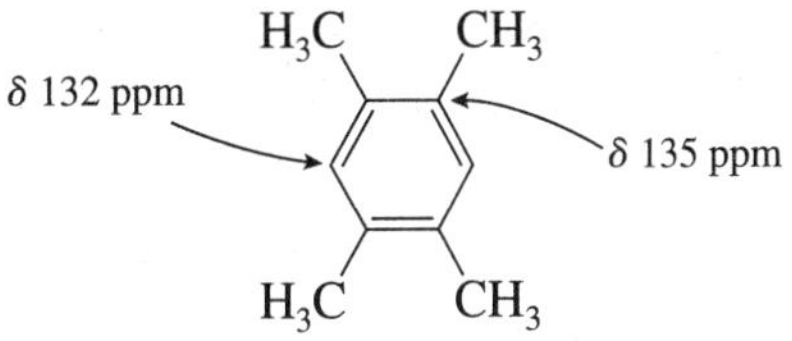

$\delta$ 132 ppm — $\delta$ 135 ppm

1,2,4,5-Tetramethylbenzene
(Compound A)

In compound B the four methyl groups are divided into two pairs. Three different carbons occur in the benzene ring, as noted by the appearance of three signals in the aromatic region ($\delta$ 128–135 ppm). Compound B is 1,2,3,4-tetramethylbenzene.

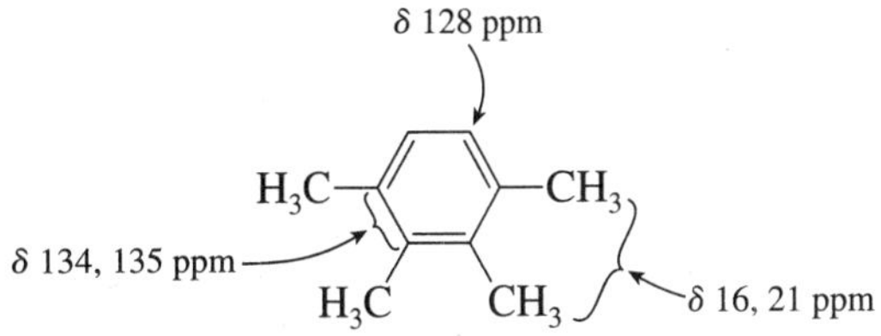

$\delta$ 128 ppm — $\delta$ 134, 135 ppm — $\delta$ 16, 21 ppm

1,2,3,4-Tetramethylbenzene
(Compound B)

**13.32** Since the compound has a 5-proton signal at $\delta$ 7.4 ppm and an index of hydrogen deficiency of 4, we conclude that six of its eight carbons belong to a monosubstituted benzene ring. The infrared spectrum exhibits absorption at 3300 cm$^{-1}$, indicating the presence of a hydroxyl group. The compound is an alcohol. A 3-proton doublet at $\delta$ 1.6 ppm, along with a 1-proton quartet at $\delta$ 4.9 ppm, suggests the presence of a $CH_3CH$ unit.

The compound is 1-phenylethanol.

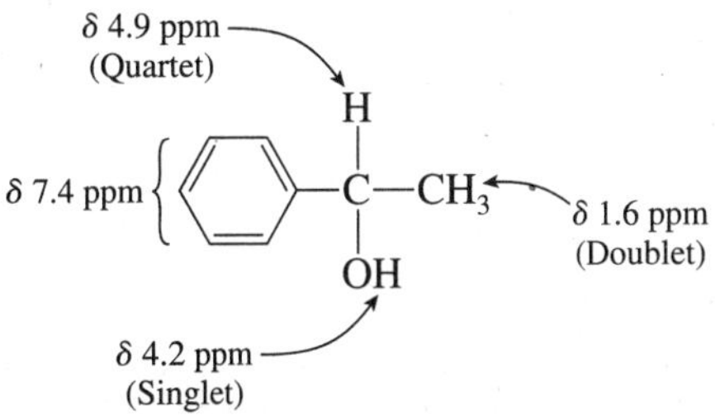

$\delta$ 4.9 ppm (Quartet) — $\delta$ 7.4 ppm — $\delta$ 1.6 ppm (Doublet) — $\delta$ 4.2 ppm (Singlet)

1-Phenylethanol

**13.33** The peak at highest $m/z$ in the mass spectrum of the compound is $m/z = 134$; this is likely to correspond to the molecular ion. Among the possible molecular formulas, $C_{10}H_{14}$ correlates best with the information from the $^1H$ NMR spectrum. What is evident is that there is a signal due to aromatic protons, as well as a triplet–quartet pattern of an ethyl group. A molecular formula of $C_{10}H_{14}$ suggests a benzene ring that bears two ethyl groups. Because the signal for the aryl protons is so sharp, they are probably equivalent. The compound is *p*-diethylbenzene.

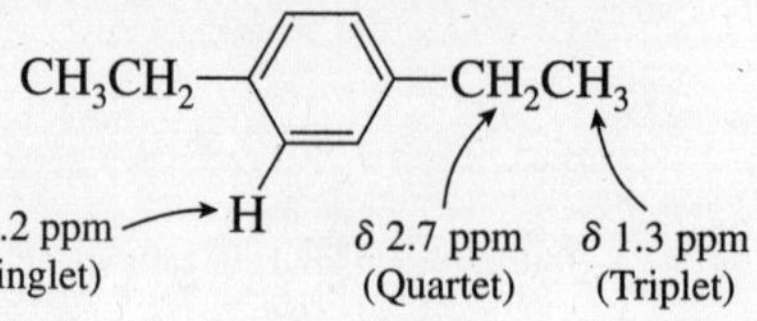

$CH_3CH_2$—$C_6H_4$—$CH_2CH_3$

δ 7.2 ppm (Singlet) → H; δ 2.7 ppm (Quartet); δ 1.3 ppm (Triplet)

*p*-Diethylbenzene

**13.34** There is a prominent peak in the infrared spectrum of the compound at 1725 $cm^{-1}$, characteristic of C═O stretching vibrations.

The $^1H$ NMR spectrum shows only two sets of signals, a triplet at δ 1.1 ppm and a quartet at δ 2.4 ppm. The compound contains a $CH_3CH_2$ group as its only protons.

Its $^{13}C$ NMR spectrum has three peaks, one of which is at very low field. The signal at δ 211 ppm is in the region characteristic of carbons of C═O groups.

If one assumes that the compound contains only carbon, hydrogen, and one oxygen atom and that the peak at highest $m/z$ in its mass spectrum ($m/z$ 86) corresponds to the molecular ion, then the compound has the molecular formula $C_5H_{10}O$.

All the information points to the conclusion that the compound has the structure shown.

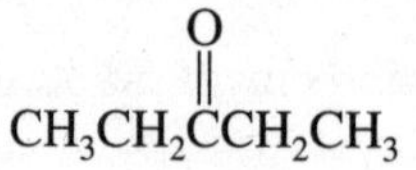

$CH_3CH_2C(=O)CH_2CH_3$

3-Pentanone

**13.35** [18]-Annulene has *two* different kinds of protons; the 12 protons on the outside periphery of the ring are different from the 6 on the inside.

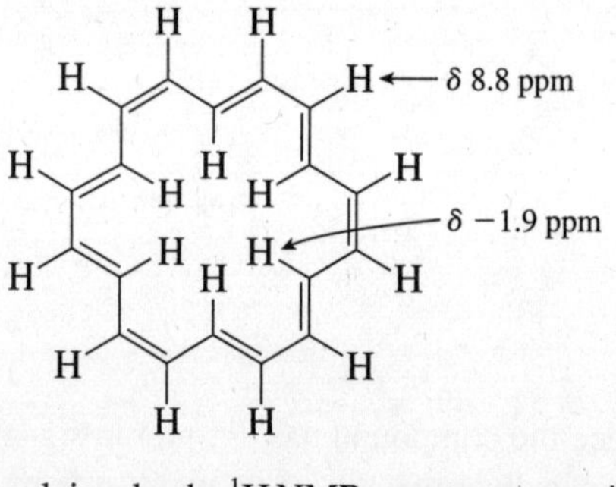

H ← δ 8.8 ppm

H ← δ −1.9 ppm

These different environments explain why the $^1H$ NMR spectrum contains two peaks in a 2:1 ratio. The less intense signal, that for the interior protons, is more shielded than the signal for the outside protons. This results from the magnetic field induced by the circulating π electrons of this aromatic ring, which reinforces the applied field in the region of the outside protons but opposes it in the interior of the ring.

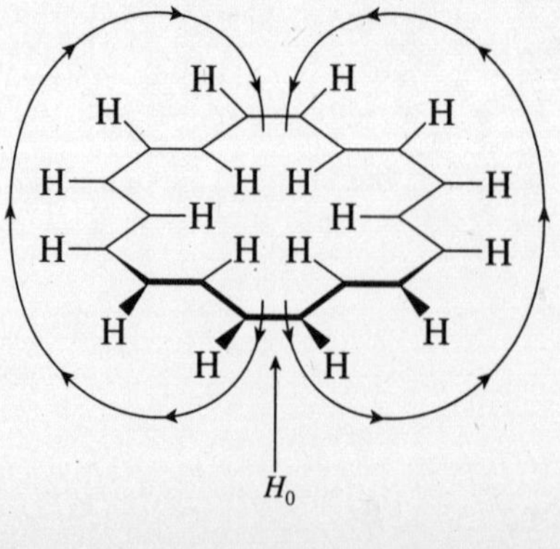

$H_0$

Protons inside the ring are shielded by the induced field to a significant extent—so much so that their signal appears at $\delta$ −1.9 ppm.

**13.36** (*a*) The nuclear spin of $^{19}F$ is $\pm\frac{1}{2}$, that is, the same as that of a proton. The splitting rules for $^{19}F$–$^{1}H$ couplings are the same as those for $^{1}H$–$^{1}H$. Thus, the single fluorine atom of $CH_3F$ splits the signal for the protons of the methyl group into a **doublet.**

(*b*) The set of three equivalent protons of $CH_3F$ splits the signal for fluorine into a **quartet.**

(*c*) The proton signal in $CH_3F$ is a doublet centered at $\delta$ 4.3 ppm. The separation between the two halves of this doublet is 45 Hz, which is equivalent to 0.225 ppm at 200 MHz (200 Hz = 1 ppm). Thus, one line of the doublet appears at $\delta$ (4.3 + 0.225) ppm and the other at $\delta$ (4.3 − 0.225) ppm.

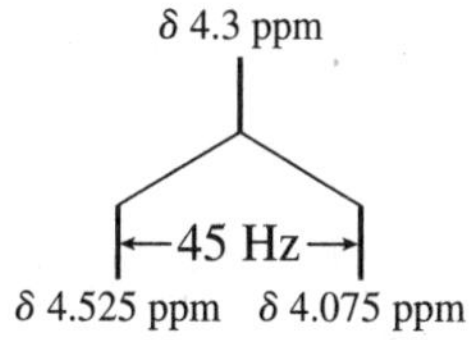

$\delta$ 4.3 ppm

←45 Hz→

$\delta$ 4.525 ppm $\delta$ 4.075 ppm

**13.37–13.38** Solutions to molecular modeling exercises are not provided in this *Study Guide and Solutions Manual.* You should use *Learning By Modeling* for these exercises.

**13.39** Because $^{31}P$ has a spin of $\pm\frac{1}{2}$, it is capable of splitting the $^{1}H$ NMR signal of protons in the same molecule. The problem stipulates that the methyl protons are coupled through three bonds to phosphorus in trimethyl phosphite.

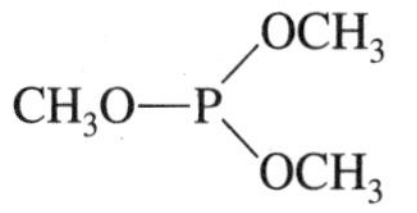

$CH_3O$—P($OCH_3$)($OCH_3$)

(*a*) The reciprocity of splitting requires that the protons split the $^{31}P$ signal of phosphorus. There are 9 equivalent protons, and so the $^{31}P$ signal is split into ten peaks.

(*b*) Each peak in the $^{31}P$ multiplet is separated from the next by a value equal to the $^{1}H$–$^{31}P$ coupling constant of 12 Hz. There are nine such intervals in a ten-line multiplet, and so the separation is 108 Hz between the highest and lowest field peaks in the multiplet.

**13.40** The trans and cis isomers of 1-bromo-4-*tert*-butylcyclohexane can be taken as models to estimate the chemical shift of the proton of the CHBr group when it is axial and equatorial, respectively, in the two chair conformations of bromocyclohexane. An axial proton is more shielded ($\delta$ 3.81 ppm for *trans*-1-bromo-4-*tert*-butylcyclohexane) than an equatorial one ($\delta$ 4.62 ppm for *cis*-1-bromo-4-*tert*-butylcyclohexane).

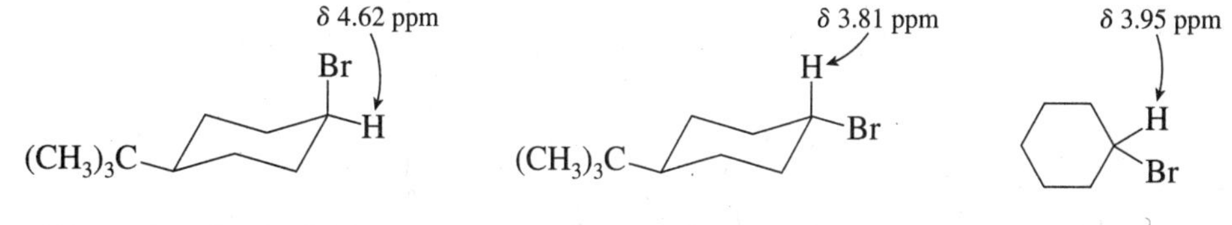

$\delta$ 4.62 ppm Br H $(CH_3)_3C$

*cis*-1-Bromo-4-*tert*-butylcyclohexane; less shielded

$\delta$ 3.81 ppm H Br $(CH_3)_3C$

*trans*-1-Bromo-4-*tert*-butylcyclohexane; more shielded

$\delta$ 3.95 ppm H Br

Bromocyclohexane

The difference in chemical shift between these stereoisomers is 0.81 ppm. The corresponding proton in bromocyclohexane is 0.67 ppm more shielded than in the equatorial proton in *cis*-1-bromo-4-*tert*-butylcyclohexane. The proportion of bromocyclohexane that has an axial hydrogen is therefore 0.67/0.81, or 83%. For bromocyclohexane, 83% of the molecules have an equatorial bromine, and 17% have an axial bromine.

**13.41** The two staggered conformations of 1,2-dichloroethane are the anti and the gauche:

Anti conformation has center of symmetry.

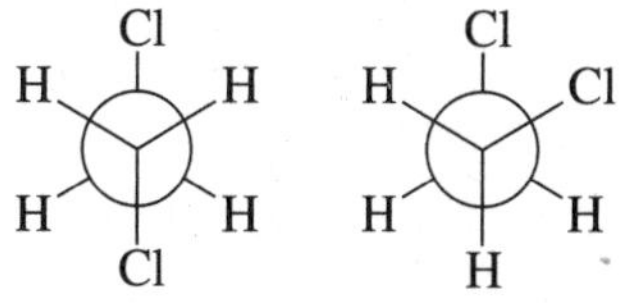

Cl H H H H Cl

Anti

Cl H Cl H H H

Gauche

The species present at low temperature (crystalline 1,2-dichloroethane) has a center of symmetry and is therefore the anti conformation. Liquid 1,2-dichloroethane is a mixture of the anti and the gauche conformations.

**13.42** (*a*) Energy is proportional to frequency and inversely proportional to wavelength. The longer the wavelength, the lower the energy. Microwave photons have a wavelength in the range of $10^{-2}$ m, which is longer than that of infrared photons (on the order of $10^{-5}$ m). Thus, microwave radiation is lower in energy than infrared radiation, and the separation between rotational energy levels (measured by microwave) is less than the separation between vibrational energy levels (measured by infrared).

(*b*) Absorption of a photon occurs only when its energy matches the energy difference between two adjacent energy levels in a molecule. Microwave photons have energies that match the differences between the rotational energy levels of water. They are not sufficiently high in energy to excite a water molecule to a higher vibrational or electronic energy state.

**13.43** A shift in the UV-Vis spectrum of acetone from 279 nm in hexane to 262 nm in water is a shift to shorter wavelength on going from a less polar solvent to a more polar one. This means that the energy difference between the starting electronic state (the **ground** state, $n$) and the excited electronic state ($\pi^*$) is *greater* in water than in hexane. Hexane as a solvent does not interact appreciably with either the ground or the excited state of acetone. Water is polar and solvates the ground state of acetone, lowering its energy. Because the energy gap between the ground state and the excited state increases, it must mean that the ground state is more solvated than the excited state and therefore more polar than the excited state.

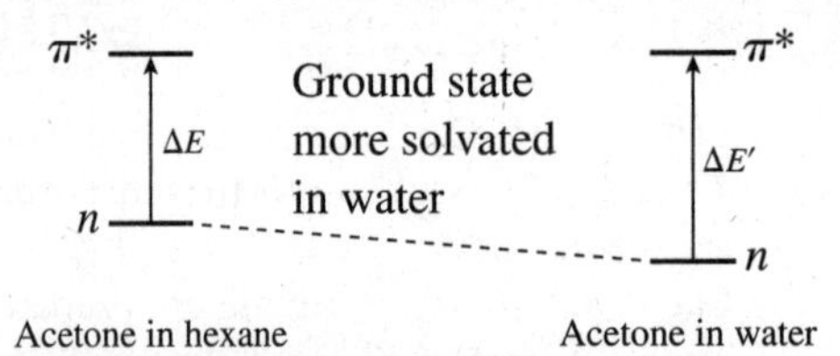


**13.44** The dipole moment of carbon dioxide is zero and does not change during the symmetric stretching vibration. The symmetric stretch is not "infrared-active." The antisymmetric stretch generates a dipole moment in carbon dioxide and is infrared-active.

←O=C=O→      →O=C=O→

Symmetric stretch: no change in dipole moment

Antisymmetric stretch: dipole moment present as a result of unequal C—O bond distances

**13.45** Solutions to molecular modeling exercises are not provided in this *Study Guide and Solutions Manual.* You should use *Learning By Modeling* for these exercises.

# SELF-TEST

## PART A

**A-1.** Complete the following table relating to $^1$H NMR spectra by supplying the missing data for entries 1 through 4.

| | Spectrometer frequency | Chemical shift | |
|---|---|---|---|
| | | ppm | Hz |
| (*a*) | 60 MHz | ____ (1) | 366 |
| (*b*) | 300 MHz | 4.35 | ____ (2) |
| (*c*) | ____ (3) | 3.50 | 700 |
| (*d*) | 100 MHz | ____ (4) | of TMS |

**A-2.** Indicate the number of signals to be expected and the multiplicity of each in the $^1H$ NMR spectrum of each of the following substances:

(*a*) $BrCH_2CH_2CH_2Br$

(*b*) $CH_3CH_2CCl_2CH_2CH_3$ (the central C bears two Cl atoms, one above and one below)

(*c*) $CH_3OCH_2C(=O)OCH_3$

**A-3.** Two isomeric compounds having the molecular formula $C_6H_{12}O_2$ both gave $^1H$ NMR spectra consisting of only two singlets. Given the chemical shifts and integrations shown, identify both compounds.

Compound A: $\delta$ 1.45 ppm (9H), $\delta$ 1.95 ppm (3H)

Compound B: $\delta$ 1.20 ppm (9H), $\delta$ 3.70 ppm (3H)

**A-4.** Identify each of the following compounds on the basis of the IR and $^1H$ NMR information provided

(*a*) $C_{10}H_{12}O$:
IR: 1710 $cm^{-1}$
NMR: $\delta$ 1.0 ppm (triplet, 3H)
$\delta$ 2.4 ppm (quartet, 2H)
$\delta$ 3.6 ppm (singlet, 2H)
$\delta$ 7.2 ppm (singlet, 5H)

(*b*) $C_6H_{14}O_2$:
IR: 3400 $cm^{-1}$
NMR: $\delta$ 1.2 ppm (singlet, 12H)
$\delta$ 2.0 ppm (broad singlet, 2H)

(*c*) $C_{10}H_{16}O_6$:
IR: 1740 $cm^{-1}$
NMR: $\delta$ 1.3 ppm (triplet, 9H)
$\delta$ 4.2 ppm (quartet, 6H)
$\delta$ 4.4 ppm (singlet, 1H)

(*d*) $C_4H_7NO$:
IR: 2240 $cm^{-1}$
3400 $cm^{-1}$ (broad)
NMR: $\delta$ 1.65 ppm (singlet, 6H)
$\delta$ 3.7 ppm (singlet, 1H)

**A-5.** Predict the number of signals and their approximate chemical shifts in the $^{13}C$ NMR spectrum of the compound shown.

$C_6H_5C(=O)CH_2CH_3$ (benzene ring bonded to the carbonyl carbon)

**A-6.** How many signals will appear in the $^{13}C$ NMR spectrum of each of the three $C_5H_{12}$ isomers?

**A-7.** The $^{13}C$ NMR spectrum of an alkane of molecular formula $C_6H_{14}$ exhibits two signals at $\delta$ 23 ppm (4C) and 37 ppm (2C). What is the structure of this alkane?

## PART B

The following three problems refer to the $^1H$ NMR spectrum of $CH_3CH_2OCH_2OCH_2CH_3$.

**B-1.** How many signals are expected?
(*a*) 12 (*b*) 5 (*c*) 4 (*d*) 3

**B-2.** The signal farthest downfield (relative to TMS) will be a
(*a*) Singlet
(*b*) Triplet
(*c*) Doublet
(*d*) Quartet

**B-3.** The signal farthest upfield (closest to TMS) will be a
(*a*) Singlet (*c*) Doublet
(*b*) Triplet (*d*) Quartet

**B-4.** The relationship between magnetic field strength and the energy difference between nuclear spin states is
(*a*) They are independent of each other.
(*b*) They are directly proportional.
(*c*) They are inversely proportional.
(*d*) The relationship varies from molecule to molecule.

**B-5.** An infrared spectrum exhibits a broad band in the 3000–3500-$cm^{-1}$ region and a strong peak at 1710 $cm^{-1}$. Which of the following substances best fits the data?

(*a*) $C_6H_5CH_2CH_2OH$ (*c*) $C_6H_5CH_2\overset{O}{\overset{\|}{C}}CH_3$

(*b*) $C_6H_5CH_2\overset{O}{\overset{\|}{C}}OH$ (*d*) $C_6H_5CH_2\overset{O}{\overset{\|}{C}}OCH_3$

**B-6.** Considering the $^1$H NMR spectrum of the following substance, which set of protons appears farthest downfield relative to TMS?

*a* {$C_6H_5$}–$\underbrace{OCH_2}_{b}\underbrace{C(CH_3)_3}_{c}$

**B-7.** Which of the following substances does *not* give a $^1$H NMR spectrum consisting of only two peaks?

(*a*) $H_3C-C(CH_3)_2-OCH_3$ (*c*) $H_3C-C_6H_4-CH_3$ (para)

(*b*) $H_3C-C(Br)(CH_3)-C(Br)(CH_3)-CH_3$ (*d*) None of these (all satisfy the spectrum)

**B-8.** The multiplicity of the *a* protons in the $^1$H NMR spectrum of the following substance is

$\underset{a}{(CH_3)_2}\overset{OH}{\overset{|}{C}}\underset{b}{CH_2}Cl$

(*a*) Singlet (*b*) Doublet (*c*) Triplet (*d*) Quartet

**B-9.** An unknown compound $C_4H_8O$ gave a strong infrared absorption at 1710 $cm^{-1}$. The $^{13}$C NMR spectrum exhibited four peaks at δ 9, 29, 37, and 209 ppm. The $^1$H NMR spectrum had three signals at δ 1.1 (triplet), 2.1 (singlet), and 2.3 (quartet) ppm. Which, if any, of the following compounds is the unknown?

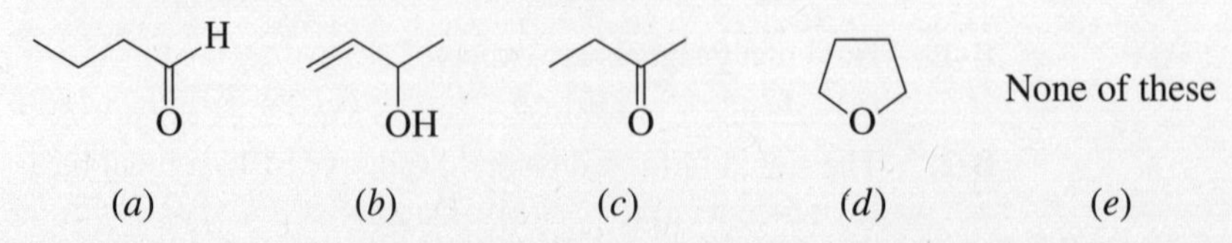

**B-10.** How many signals are expected in the $^{13}$C NMR spectrum of the following substance?

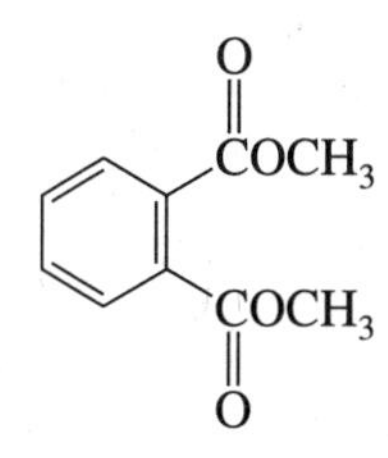


(*a*) 5 (*b*) 6 (*c*) 8 (*d*) 10

**B-11.** Which one of the following has the *greatest* number of signals in its $^{13}$C NMR spectrum? (The spectrum is run under conditions in which splitting due to $^{13}$C–$^{1}$H coupling is not observed.)

(*a*) Hexane (*c*) 1-Hexene (*e*) 1,5-Hexadiene
(*b*) 2-Methylpentane (*d*) *cis*-3-Hexene

**B-12.** Which of the following $C_9H_{12}$ isomers has the *fewest* signals in its $^{13}$C NMR spectrum?

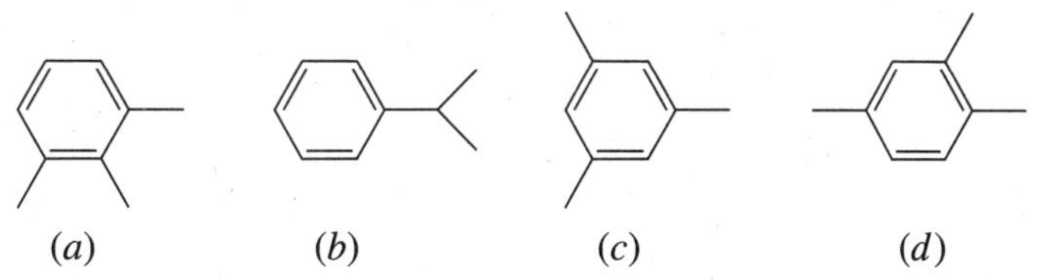

(*a*) (*b*) (*c*) (*d*)

**B-13.** Which of the following compounds would best fit a $^{13}$C NMR spectrum having peaks at $\delta$ 16, 21, 32, 36, 115, and 140 ppm?

(*a*) (*c*)

(*b*) 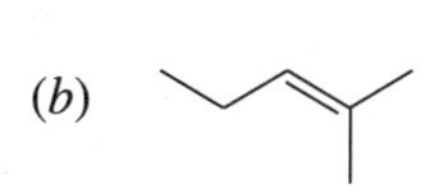 (*d*)

**B-14.** Which of the following compounds would have the *fewest* peaks in its $^{13}$C NMR spectrum?

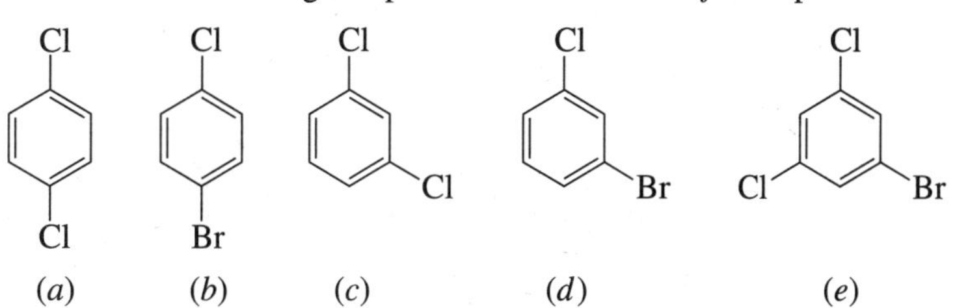


(*a*) (*b*) (*c*) (*d*) (*e*)

**B-15.** Which of the compounds in the previous problem would have the *most* peaks in its $^{13}$C NMR spectrum?

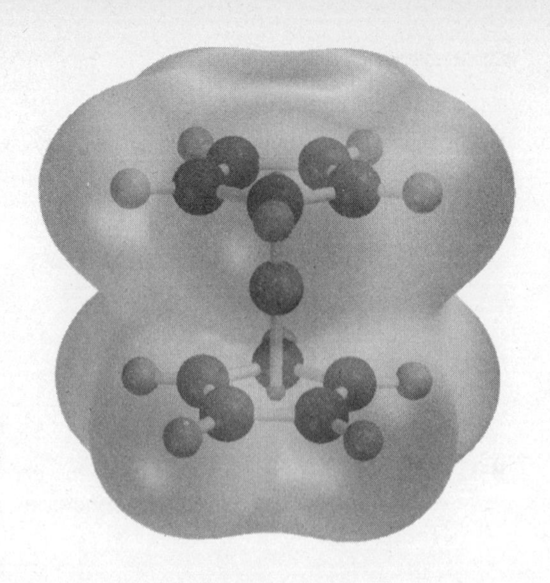

# CHAPTER 14

# ORGANOMETALLIC COMPOUNDS

## SOLUTIONS TO TEXT PROBLEMS

**14.1** (*b*) Magnesium bears a cyclohexyl substituent and a chlorine. Chlorine is named as an anion. The compound is cyclohexylmagnesium chloride.

**14.2** (*b*) The alkyl bromide precursor to *sec*-butyllithium must be *sec*-butyl bromide.

$$\underset{\text{Br}}{CH_3\overset{|}{C}HCH_2CH_3} + 2Li \longrightarrow \underset{\text{Li}}{CH_3\overset{|}{C}HCH_2CH_3} + LiBr$$

2-Bromobutane (*sec*-butyl bromide) — 1-Methylpropyllithium (*sec*-butyllithium)

**14.3** (*b*) Allyl chloride is converted to allylmagnesium chloride on reaction with magnesium.

$$H_2C{=}CHCH_2Cl \xrightarrow[\text{diethyl ether}]{\text{Mg}} H_2C{=}CHCH_2MgCl$$

Allyl chloride — Allylmagnesium chloride

(*c*) The carbon–iodine bond of iodocyclobutane is replaced by a carbon–magnesium bond in the Grignard reagent.

$$\text{C}_4\text{H}_7\text{I} \xrightarrow[\text{diethyl ether}]{\text{Mg}} \text{C}_4\text{H}_7\text{MgI}$$

Iodocyclobutane — Cyclobutylmagnesium iodide

(*d*) Bromine is attached to an $sp^2$-hybridized carbon in 1-bromocyclohexene. The product of its reaction with magnesium has a carbon–magnesium bond in place of the carbon–bromine bond.

$$\text{1-Bromocyclohexene} \xrightarrow[\text{diethyl ether}]{\text{Mg}} \text{1-Cyclohexenylmagnesium bromide}$$

**14.4** (*b*) 1-Hexanol will protonate butyllithium because its hydroxyl group is a proton donor only slightly less acidic than water. This proton-transfer reaction could be used to prepare lithium 1-hexanolate.

$$\underset{\text{1-Hexanol}}{CH_3CH_2CH_2CH_2CH_2CH_2OH} + \underset{\text{Butyllithium}}{CH_3CH_2CH_2CH_2Li} \longrightarrow \underset{\text{Butane}}{CH_3CH_2CH_2CH_3} + \underset{\text{Lithium 1-hexanolate}}{CH_3CH_2CH_2CH_2CH_2CH_2OLi}$$

(*c*) The proton donor here is benzenethiol.

$$\underset{\text{Benzenethiol}}{C_6H_5SH} + \underset{\text{Butyllithium}}{CH_3CH_2CH_2CH_2Li} \longrightarrow \underset{\text{Butane}}{CH_3CH_2CH_2CH_3} + \underset{\text{Lithium benzenethiolate}}{C_6H_5SLi}$$

**14.5** (*b*) Propylmagnesium bromide reacts with benzaldehyde by addition to the carbonyl group.

$$CH_3CH_2CH_2\text{—}MgBr + C_6H_5CH{=}O \xrightarrow{\text{diethyl ether}} C_6H_5CH(CH_2CH_2CH_3)\text{—}OMgBr \xrightarrow{H_3O^+} \underset{\text{1-Phenyl-1-butanol}}{C_6H_5CH(OH)CH_2CH_2CH_3}$$

(*c*) Tertiary alcohols result from the reaction of Grignard reagents and ketones.

$$CH_3CH_2CH_2MgBr + \text{cyclohexanone} \xrightarrow[\text{2. } H_3O^+]{\text{1. diethyl ether}} \underset{\text{1-Propylcyclohexanol}}{\text{1-}(CH_2CH_2CH_3)\text{-1-}(OH)\text{cyclohexane}}$$

(*d*) The starting material is a ketone and so reacts with a Grignard reagent to give a tertiary alcohol.

$$\underset{\text{Propylmagnesium bromide + 2-butanone}}{CH_3CH_2CH_2\text{—}MgBr + (H_3C)(CH_3CH_2)C{=}O} \xrightarrow{\text{diethyl ether}} H_3C\text{—}C(CH_2CH_2CH_3)(CH_2CH_3)\text{—}OMgBr \xrightarrow{H_3O^+} \underset{\text{3-Methyl-3-hexanol}}{CH_3CH_2CH_2C(CH_3)(OH)CH_2CH_3}$$

**14.6** Ethyl anion reacts as a Brønsted base to remove a proton from the alkyne. The proton at C-1 is removed because it is the most acidic, having a p$K_a$ of approximately 25.

$$CH_3\ddot{C}H_2^- + H{-}C{\equiv}CCH_2CH_2CH_2CH_3 \longrightarrow CH_3CH_3 + :\bar{C}{\equiv}CCH_2CH_2CH_2CH_3$$

Ethyl anion — 1-Hexyne — Ethane — Conjugate base of 1-hexyne

**14.7** (*b*) The target alcohol is tertiary and so is prepared by addition of a Grignard reagent to a ketone. The retrosynthetic transformations are:

$$^-{:}CH_3 + C_6H_5C(=O){-}CH_3 \Longleftarrow CH_3{-}C(OH)(C_6H_5){-}CH_3 \Longrightarrow C_6H_5^{:-} + CH_3C(=O)CH_3$$

Because two of the alkyl groups on the hydroxyl-bearing carbon are the same (methyl), only two, not three,different ketones are possible starting materials:

$$CH_3MgI + C_6H_5\overset{O}{\overset{\|}{C}}CH_3 \xrightarrow[2.\ H_3O^+]{1.\ \text{diethyl ether}} C_6H_5C(OH)(CH_3)CH_3$$

Methylmagnesium iodide — Acetophenone — 2-Phenyl-2-propanol

$$C_6H_5MgBr + CH_3\overset{O}{\overset{\|}{C}}CH_3 \xrightarrow[2.\ H_3O^+]{1.\ \text{diethyl ether}} C_6H_5C(OH)(CH_3)CH_3$$

Phenylmagnesium bromide — Acetone — 2-Phenyl-2-propanol

**14.8** (*b*) Recall that the two identical groups bonded to the hydroxyl-bearing carbon of the alcohol arose from the Grignard reagent. That leads to the following retrosynthetic analysis:

$$(C_6H_5)_2C(OH){-}C_3H_5 \Longrightarrow C_3H_5{-}\overset{O}{\overset{\|}{C}}OR + 2C_6H_5MgX$$

Thus, the two phenyl substituents arise by addition of a phenyl Grignard reagent to an ester of cyclopropanecarboxylic acid.

$$2C_6H_5MgBr + C_3H_5{-}\overset{O}{\overset{\|}{C}}OCH_3 \xrightarrow[2.\ H_3O^+]{1.\ \text{diethyl ether}} (C_6H_5)_2C(OH){-}C_3H_5 + CH_3OH$$

Phenylmagnesium bromide — Methyl cyclopropanecarboxylate — Cyclopropyldiphenylmethanol — Methanol

**14.9** (*b*) Of the three methyl groups of 1,3,3-trimethylcyclopentene, only the one connected to the double bond can be attached by way of an organocuprate reagent. Attachment of either of the other methyls would involve a tertiary carbon, a process that does not occur very efficiently.

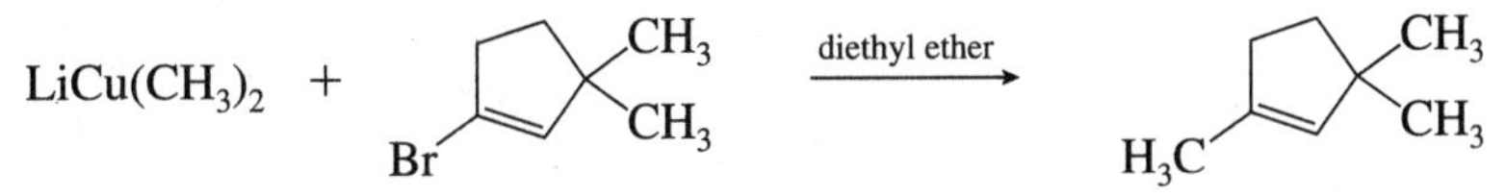

$LiCu(CH_3)_2$ + 1-Bromo-3,3-dimethylcyclopentene —diethyl ether→ 1,3,3-Trimethylcyclopentene

Lithium dimethylcuprate    1-Bromo-3,3-dimethylcyclopentene    1,3,3-Trimethylcyclopentene

**14.10** (*b*) Methylenecyclobutane is the appropriate precursor to the spirohexane shown.

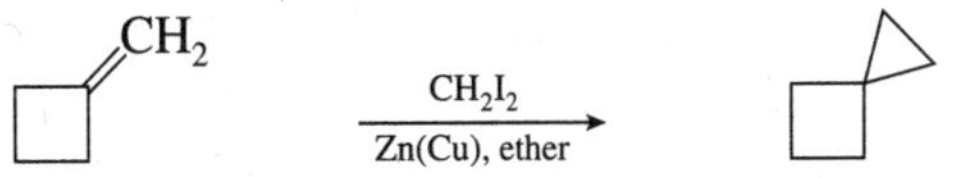

Methylenecyclobutane —$CH_2I_2$, Zn(Cu), ether→ Spiro[3.2]hexane (22%)

**14.11** Syn addition of dibromocarbene to *cis*-2-butene yields a cyclopropane derivative in which the methyl groups are cis.

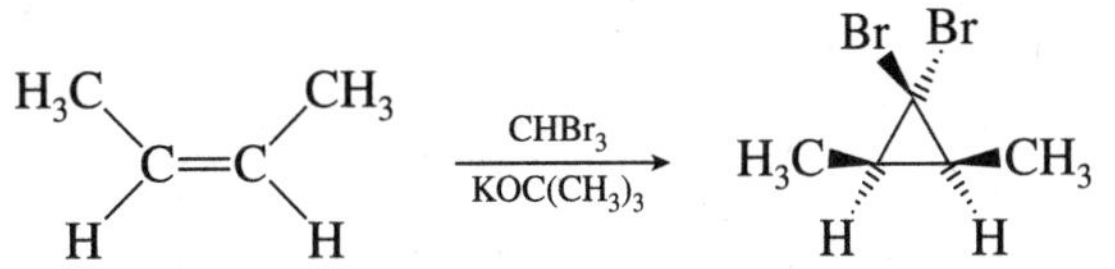

*cis*-2-Butene —$CHBr_3$, $KOC(CH_3)_3$→ *cis*-1,1-Dibromo-2,3-dimethylcyclopropane

Conversely, the methyl groups in the cyclopropane derivative of *trans*-2-butene are trans to one another.

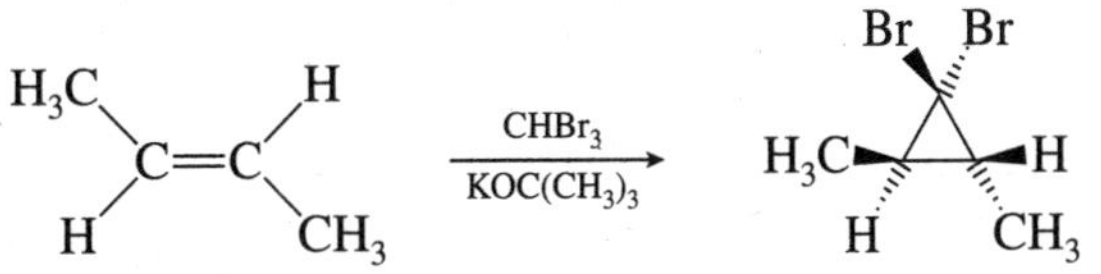

*trans*-2-Butene —$CHBr_3$, $KOC(CH_3)_3$→ *trans*-1,1-Dibromo-2,3-dimethylcyclopropane

**14.12** Iron has an atomic number of 26 and an electron configuration of $[Ar]4s^23d^6$. Thus, it has 8 valence electrons and requires 10 more to satisfy the 18-electron rule. Five CO ligands, each providing two electrons, are therefore needed. The compound is $Fe(CO)_5$.

**14.13** (*a*) Cyclopentyllithium is

(cyclopentane ring bearing H and Li)

It has a carbon–lithium bond. It satisfies the requirement for classification as an organometallic compound.

(*b*) Ethoxymagnesium chloride does not have a carbon–metal bond. It is not an organometallic compound.

$$CH_3CH_2OMgCl \quad \text{or} \quad CH_3CH_2O^- \; Mg^{2+} \; Cl^-$$

(*c*) 2-Phenylethylmagnesium iodide is an example of a Grignard reagent. It is an organometallic compound.

$C_6H_5$—$CH_2CH_2MgI$

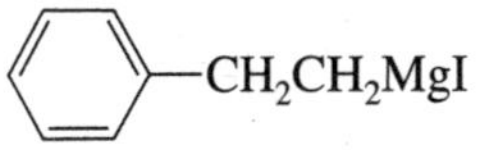

(*d*) Lithium divinylcuprate has two vinyl groups bonded to copper. It is an organometallic compound.

$$Li^+(H_2C{=}CH{-}\bar{C}u{-}CH{=}CH_2)$$

(*e*) Sodium carbonate, $Na_2CO_3$ can be represented by the Lewis structure.

$$Na^+\ {}^{-}{:}\ddot{O}{-}\overset{\overset{\ddot{O}:}{\|}}{C}{-}\ddot{O}{:}^{-}\ Na^+$$

There is no carbon–metal bond, and sodium carbonate is not an organometallic compound.

(*f*) Benzylpotassium is represented as

$$C_6H_5{-}CH_2K \quad \text{or} \quad C_6H_5{-}\ddot{C}H_2\ K^+$$

It has a carbon–potassium bond and thus is an organometallic compound.

**14.14** The two alkyl groups attached to aluminum in $[(CH_3)_2CHCH_2]_2AlH$ are isobutyl groups. The hydrogen bonded to aluminum is named in a separate word as hydride. Thus, "dibal" is a shortened form of the systematic name **diisobutylaluminum hydride.**

**14.15** (*a*) Grignard reagents such as pentylmagnesium iodide are prepared by reaction of magnesium with the corresponding alkyl halide.

$$\underset{\text{1-Iodopentane}}{CH_3CH_2CH_2CH_2CH_2I} + Mg \xrightarrow{\text{diethyl ether}} \underset{\text{Pentylmagnesium iodide}}{CH_3CH_2CH_2CH_2CH_2MgI}$$

(*b*) Acetylenic Grignard reagents are normally prepared by reaction of a terminal alkyne with a readily available Grignard reagent such as an ethylmagnesium halide. The reaction that takes place is an acid–base reaction in which the terminal alkyne acts as a proton donor.

$$\underset{\text{1-Butyne}}{CH_3CH_2C{\equiv}CH} + \underset{\text{Ethylmagnesium iodide}}{CH_3CH_2MgI} \xrightarrow{\text{diethyl ether}} \underset{\text{1-Butynylmagnesium iodide}}{CH_3CH_2C{\equiv}CMgI} + \underset{\text{Ethane}}{CH_3CH_3}$$

(*c*) Alkyllithiums are formed by reaction of lithium with an alkyl halide.

$$\underset{\substack{\text{1-Halopentane}\\(X = Cl, Br, or I)}}{CH_3CH_2CH_2CH_2CH_2X} + 2Li \longrightarrow \underset{\text{Pentyllithium}}{CH_3CH_2CH_2CH_2CH_2Li} + LiX$$

(*d*) Lithium dialkylcuprates arise by the reaction of an alkyllithium with a Cu(I) salt.

$$\underset{\text{Pentyllithium, from part (c)}}{2CH_3CH_2CH_2CH_2CH_2Li} + \underset{(X = Cl, Br, or I)}{CuX} \longrightarrow \underset{\text{Lithium dipentylcuprate}}{LiCu(CH_2CH_2CH_2CH_2CH_3)_2} + LiX$$

**14.16** The polarity of a covalent bond increases with an increase in the electronegativity difference between the connected atoms. Carbon has an electronegativity of 2.5 (Table 14.1). Metals are less electronegative than carbon. When comparing two metals, the less electronegative one therefore has the more polar bond to carbon.

(*a*) Table 14.1 gives the electronegativity of lithium as 1.0, whereas that for aluminum is 1.5. The carbon–lithium bond in $CH_3CH_2Li$ is more polar than the carbon–aluminum bond in $(CH_3CH_2)_3Al$.

(*b*) The electronegativity of magnesium (1.2) is less than that of zinc (1.6). $(CH_3)_2Mg$ therefore has a more polar carbon–metal bond than $(CH_3)_2Zn$.

(*c*) In this part of the problem two Grignard reagents are compared. Magnesium is the metal in both cases. The difference is the hybridization state of carbon. The *sp*-hybridized carbon in $HC\equiv CMgBr$ is more electronegative than the $sp^3$-hybridized carbon in $CH_3CH_2MgBr$, and $HC\equiv CMgBr$ has a more polar carbon–magnesium bond.

**14.17** (*a*)

$$\underset{\text{1-Bromopropane}}{CH_3CH_2CH_2Br} + 2Li \xrightarrow{\text{diethyl ether}} \underset{\text{Propyllithium}}{CH_3CH_2CH_2Li} + LiBr$$

(*b*)

$$\underset{\text{1-Bromopropane}}{CH_3CH_2CH_2Br} + Mg \xrightarrow{\text{diethyl ether}} \underset{\text{Propylmagnesium bromide}}{CH_3CH_2CH_2MgBr}$$

(*c*)

$$\underset{\text{2-Iodopropane}}{CH_3\underset{|}{\underset{I}{C}}HCH_3} + 2Li \xrightarrow{\text{diethyl ether}} \underset{\text{Isopropyllithium}}{CH_3\underset{Li}{C}HCH_3} + LiI$$

(*d*)

$$\underset{\text{2-Iodopropane}}{CH_3\underset{I}{C}HCH_3} + Mg \xrightarrow{\text{diethyl ether}} \underset{\text{Isopropylmagnesium iodide}}{CH_3\underset{MgI}{C}HCH_3}$$

(*e*)

$$\underset{\text{Propyllithium}}{2CH_3CH_2CH_2Li} + CuI \longrightarrow \underset{\text{Lithium dipropylcuprate}}{(CH_3CH_2CH_2)_2CuLi}$$

(*f*)

$$\underset{\text{Lithium dipropylcuprate}}{(CH_3CH_2CH_2)_2CuLi} + \underset{\text{1-Bromobutane}}{CH_3CH_2CH_2CH_2Br} \longrightarrow \underset{\text{Heptane}}{CH_3CH_2CH_2CH_2CH_2CH_2CH_3}$$

(*g*)

$$\underset{\text{Lithium dipropylcuprate}}{(CH_3CH_2CH_2)_2CuLi} + \underset{\text{Iodobenzene}}{C_6H_5I} \longrightarrow \underset{\text{Propylbenzene}}{C_6H_5CH_2CH_2CH_3}$$

(*h*)

$$\underset{\text{Propylmagnesium bromide}}{CH_3CH_2CH_2MgBr} \xrightarrow[\text{DCl}]{D_2O} \underset{\text{1-Deuteriopropane}}{CH_3CH_2CH_2D}$$

(*i*)

$$\underset{\text{Isopropyllithium}}{CH_3\underset{Li}{C}HCH_3} \xrightarrow[\text{DCl}]{D_2O} \underset{\text{2-Deuteriopropane}}{CH_3\underset{D}{C}HCH_3}$$

(*j*)

$$\underset{\text{Propyllithium}}{CH_3CH_2CH_2Li} + H\overset{O}{\overset{\|}{C}}H \xrightarrow[\text{2. } H_3O^+]{\text{1. diethyl ether}} \underset{\text{1-Butanol}}{CH_3CH_2CH_2CH_2OH}$$

(*k*) $CH_3CH_2CH_2MgBr$ + $C_6H_5CH{=}O$ $\xrightarrow[2.\ H_3O^+]{1.\ \text{diethyl ether}}$ $C_6H_5CH(OH)CH_2CH_2CH_3$

Propylmagnesium bromide — Benzaldehyde — 1-Phenyl-1-butanol

(*l*) $CH_3CH(Li)CH_3$ + cycloheptanone $\xrightarrow[2.\ H_3O^+]{1.\ \text{diethyl ether}}$ 1-$(CH_3)_2CH$, 1-OH cycloheptane

Isopropyllithium — Cycloheptanone — 1-Isopropylcycloheptanol

(*m*) $CH_3CH(MgI)CH_3$ + $CH_3\overset{O}{\overset{\|}{C}}CH_2CH_3$ $\xrightarrow[2.\ H_3O^+]{1.\ \text{diethyl ether}}$ $CH_3CH(CH_3){-}C(OH)(CH_3)CH_2CH_3$

Isopropylmagnesium iodide — 2-Butanone — 2,3-Dimethyl-3-pentanol

(*n*) $2CH_3CH_2CH_2MgBr$ + $C_6H_5\overset{O}{\overset{\|}{C}}OCH_3$ $\xrightarrow[2.\ H_3O^+]{1.\ \text{diethyl ether}}$ $C_6H_5C(OH)(CH_2CH_2CH_3)_2$ + $CH_3OH$

Propylmagnesium bromide — Methyl benzoate — 4-Phenyl-4-heptanol — Methanol

(*o*) $H_2C{=}CH(CH_2)_5CH_3$ $\xrightarrow[\text{Zn(Cu), diethyl ether}]{CH_2I_2}$ $H_2C{-}CH(CH_2)_5CH_3$ (ring closed by $CH_2$)

1-Octene — 1-Cyclopropylhexane

(*p*) $H_3C(H)C{=}C(H)(CH_2)_6CH_3$ $\xrightarrow[\text{Zn(Cu), diethyl ether}]{CH_2I_2}$ cyclopropane with $H_3C$, H, H, $(CH_2)_6CH_3$

(*E*)-2-Decene — *trans*-1-Heptyl-2-methylcyclopropane

(*q*) $CH_3CH_2(H)C{=}C(H)(CH_2)_5CH_3$ $\xrightarrow[\text{Zn(Cu), diethyl ether}]{CH_2I_2}$ cyclopropane with $CH_3CH_2$, $(CH_2)_5CH_3$, H, H

(*Z*)-3-Decene — *cis*-1-Ethyl-2-hexylcyclopropane

(*r*) $H_2C{=}CHCH_2CH_2CH_3$ $\xrightarrow[KOC(CH_3)_3]{CHBr_3}$ cyclopropane with Br, Br, H, $CH_2CH_2CH_3$, H, H

1-Pentene — 1,1-Dibromo-2-propylcyclopropane

**14.18** In the solutions to this problem, the Grignard reagent butylmagnesium bromide is used. In each case the use of butyllithium would be equally satisfactory.

(*a*) 1-Pentanol is a primary alcohol having one more carbon atom than 1-bromobutane. Retrosynthetic analysis suggests the reaction of a Grignard reagent with formaldehyde.

$$CH_3CH_2CH_2CH_2 \wr CH_2OH \Rightarrow CH_3CH_2CH_2CH_2MgX + H_2C{=}O$$

1-Pentanol; Butylmagnesium halide; Formaldehyde

An appropriate synthetic scheme is

$$CH_3CH_2CH_2CH_2Br \xrightarrow[\text{diethyl ether}]{Mg} CH_3CH_2CH_2CH_2MgBr \xrightarrow[2.\ H_3O^+]{1.\ HCH(=O)} CH_3CH_2CH_2CH_2CH_2OH$$

1-Bromobutane; Butylmagnesium bromide; 1-Pentanol

(*b*) 2-Hexanol is a secondary alcohol having two more carbon atoms than 1-bromobutane. As revealed by retrosynthetic analysis, it may be prepared by reaction of ethanal (acetaldehyde) with butylmagnesium bromide.

$$CH_3CH_2CH_2CH_2 \wr CH(OH)CH_3 \Rightarrow CH_3CH_2CH_2CH_2MgX + CH_3CH(=O)$$

2-Hexanol; Butylmagnesium halide; Ethanal (acetaldehyde)

The correct reaction sequence is

$$CH_3CH_2CH_2CH_2Br \xrightarrow[\text{diethyl ether}]{Mg} CH_3CH_2CH_2CH_2MgBr \xrightarrow[2.\ H_3O^+]{1.\ CH_3CH(=O)} CH_3CH_2CH_2CH_2CH(OH)CH_3$$

1-Bromobutane; Butylmagnesium bromide; 2-Hexanol

(*c*) 1-Phenyl-1-pentanol is a secondary alcohol. Disconnection suggests that it can be prepared from butylmagnesium bromide and an aldehyde; benzaldehyde is the appropriate aldehyde.

$$CH_3CH_2CH_2CH_2 \wr CH(OH){-}C_6H_5 \Rightarrow CH_3CH_2CH_2CH_2MgX + C_6H_5CH(=O)$$

1-Phenyl-1-pentanol; Butylmagnesium halide; Benzaldehyde

$$CH_3CH_2CH_2CH_2MgBr + C_6H_5CH(=O) \xrightarrow[2.\ H_3O^+]{1.\ \text{ether}} C_6H_5{-}CH(OH)CH_2CH_2CH_2CH_3$$

Butylmagnesium bromide; Benzaldehyde; 1-Phenyl-1-pentanol

(*d*) The target molecule 3-methyl-3-heptanol has the structure

$$CH_3CH_2CH_2CH_2{-}C(CH_3)(OH){-}CH_2CH_3$$

By retrosynthetically disconnecting the butyl group from the carbon that bears the hydroxyl substituent, we see that the appropriate starting ketone is 2-butanone.

$$CH_3CH_2CH_2CH_2 \text{⫶} C(CH_3)(OH)-CH_2CH_3 \Longrightarrow CH_3CH_2CH_2CH_2MgX + CH_3C(=O)CH_2CH_3$$

Butylmagnesium halide — 2-Butanone

Therefore

$$CH_3CH_2CH_2CH_2MgBr + CH_3\overset{O}{\overset{\|}{C}}CH_2CH_3 \xrightarrow[2.\ H_3O^+]{1.\ \text{diethyl ether}} CH_3CH_2CH_2CH_2C(CH_3)(OH)CH_2CH_3$$

Butylmagnesium bromide — 2-Butanone — 3-Methyl-3-heptanol

(*e*) 1-Butylcyclobutanol is a tertiary alcohol. The appropriate ketone is cyclobutanone.

$$CH_3CH_2CH_2CH_2MgBr + \text{cyclobutanone} \xrightarrow[2.\ H_3O^+]{1.\ \text{diethyl ether}} \text{1-butylcyclobutanol (OH, } CH_2CH_2CH_2CH_3\text{)}$$

Butylmagnesium bromide — Cyclobutanone — 1-Butylcyclobutanol

**14.19** In each part of this problem in which there is a change in the carbon skeleton, disconnect the phenyl group of the product to reveal the aldehyde or ketone precursor that reacts with the Grignard reagent derived from bromobenzene. Recall that reaction of a Grignard reagent with formaldehyde ($H_2C{=}O$) yields a primary alcohol, reaction with an aldehyde (other than formaldehyde) yields a secondary alcohol, and reaction with a ketone yields a tertiary alcohol.

(*a*) Conversion of bromobenzene to benzyl alcohol requires formation of the corresponding Grignard reagent and its reaction with formaldehyde. Retrosynthetically, this can be seen as

$$C_6H_5 \text{⫶} CH_2OH \Longrightarrow C_6H_5-MgX + H_2C{=}O$$

Therefore,

$$C_6H_5Br \xrightarrow[\text{diethyl ether}]{Mg} C_6H_5MgBr \xrightarrow[2.\ H_3O^+]{1.\ H\overset{O}{\overset{\|}{C}}H} C_6H_5CH_2OH$$

Bromobenzene — Phenylmagnesium bromide — Benzyl alcohol

(*b*) The product is a secondary alcohol and is formed by reaction of phenylmagnesium bromide with hexanal.

$$C_6H_5 \text{⫶} CH(OH)(CH_2)_4CH_3 \Longrightarrow C_6H_5-MgX + H\overset{O}{\overset{\|}{C}}(CH_2)_4CH_3$$

1-Phenyl-1-hexanol — Phenylmagnesium halide — Hexanal

$$C_6H_5MgBr + CH_3CH_2CH_2CH_2CH_2\overset{O}{\overset{\|}{C}}H \xrightarrow[2.\ H_3O^+]{1.\ \text{diethyl ether}} C_6H_5CH(OH)CH_2CH_2CH_2CH_2CH_3$$

Phenylmagnesium bromide — Hexanal — 1-Phenyl-1-hexanol

(*c*) The desired product is a secondary alkyl **bromide.** A reasonable synthesis would be to first prepare the analogous secondary alcohol by reaction of phenylmagnesium bromide with benzaldehyde, followed by a conversion of the alcohol to the bromide. Retrosynthetically this can be seen as

$$C_6H_5CH(Br)—C_6H_5 \Rightarrow C_6H_5CH(OH)—C_6H_5 \Rightarrow C_6H_5MgX + C_6H_5CH{=}O$$

$$C_6H_5MgBr + C_6H_5CH{=}O \xrightarrow[\text{2. } H_3O^+]{\text{1. diethyl ether}} C_6H_5CH(OH)C_6H_5 \xrightarrow{\text{HBr or } PBr_3} C_6H_5CH(Br)C_6H_5$$

Phenylmagnesium bromide — Benzaldehyde — Diphenylmethanol — Bromodiphenylmethane

(*d*) The target molecule is a tertiary alcohol, which requires that phenylmagnesium bromide react with a ketone. By mentally disconnecting the phenyl group from the carbon that bears the hydroxyl group, we see that the appropriate ketone is 4-heptanone.

$$CH_3CH_2CH_2C(OH)(C_6H_5)CH_2CH_2CH_3 \Rightarrow CH_3CH_2CH_2C(=O)CH_2CH_2CH_3 + C_6H_5^-$$

4-Phenyl-4-heptanol — 4-Heptanone

The synthesis is therefore

$$C_6H_5MgBr + CH_3CH_2CH_2C(=O)CH_2CH_2CH_3 \xrightarrow[\text{2. } H_3O^+]{\text{1. diethyl ether}} CH_3CH_2CH_2C(OH)(C_6H_5)CH_2CH_2CH_3$$

Phenylmagnesium bromide — 4-Heptanone — 4-Phenyl-4-heptanol

(*e*) Reaction of phenylmagnesium bromide with cyclooctanone will give the desired tertiary alcohol.

$$C_6H_5MgBr + \text{cyclooctanone} \xrightarrow[\text{2. } H_3O^+]{\text{1. diethyl ether}} \text{1-phenylcyclooctanol}$$

Phenylmagnesium bromide — Cyclooctanone — 1-Phenylcyclooctanol

(*f*) The 1-phenylcyclooctanol prepared in part (*e*) of this problem can be subjected to acid-catalyzed dehydration to give 1-phenylcyclooctene. Hydroboration–oxidation of 1-phenylcyclooctene gives *trans*-2-phenylcyclooctanol.

$$\text{1-phenylcyclooctanol} \xrightarrow{H_2SO_4,\ \text{heat}} \text{1-phenylcyclooctene} \xrightarrow[\text{2. } H_2O_2,\ HO^-]{\text{1. } B_2H_6} \textit{trans}\text{-2-phenylcyclooctanol}$$

1-Phenylcyclooctanol — 1-Phenylcyclooctene — *trans*-2-Phenylcyclooctanol

**14.20** In these problems the principles of retrosynthetic analysis are applied. The alkyl groups attached to the carbon that bears the hydroxyl group are mentally disconnected to reveal the Grignard reagent and carbonyl compound.

(*a*)

$$CH_3CH_2 \overset{(1)}{\wr} CH(OH) \overset{(2)}{\wr} CH_2CH(CH_3)_2$$

5-Methyl-3-hexanol

(1) ⇒ $CH_3CH_2MgX$ + $HC(=O)CH_2CH(CH_3)_2$

Ethylmagnesium halide + 3-Methylbutanal

(2) ⇒ $CH_3CH_2CH(=O)$ + $XMgCH_2CH(CH_3)_2$

Propanal + Isobutylmagnesium halide

(*b*)

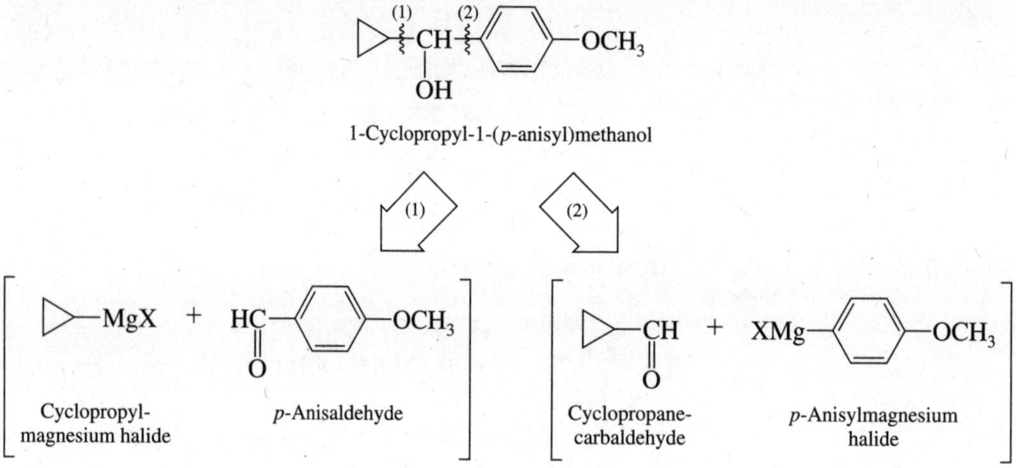

$$\triangleright \overset{(1)}{\wr} CH(OH) \overset{(2)}{\wr} C_6H_4\text{-}OCH_3$$

1-Cyclopropyl-1-(*p*-anisyl)methanol

(1) ⇒ $\triangleright$—MgX + $HC(=O)\text{-}C_6H_4\text{-}OCH_3$

Cyclopropylmagnesium halide + *p*-Anisaldehyde

(2) ⇒ $\triangleright$—$CH(=O)$ + $XMg\text{-}C_6H_4\text{-}OCH_3$

Cyclopropanecarbaldehyde + *p*-Anisylmagnesium halide

(*c*)

$$(CH_3)_3C \wr CH_2OH \Rightarrow (CH_3)_3CMgX + HCH(=O)$$

2,2-Dimethyl-1-propanol ⇒ *tert*-Butylmagnesium halide + Formaldehyde

(*d*)

$$(CH_3)_2C{=}CHCH_2CH_2 \overset{(1)}{\wr} CH(OH) \overset{(2)}{\wr} CH_3$$

6-Methyl-5-hepten-2-ol

(1) ⇒ $(CH_3)_2C{=}CHCH_2CH_2MgX$ + $HC(=O)CH_3$

4-Methyl-3-hexen-1-ylmagnesium halide + Ethanal

(2) ⇒ $(CH_3)_2C{=}CHCH_2CH_2CH(=O)$ + $XMgCH_3$

5-Methyl-4-hexenal + Methylmagnesium halide

(*e*)

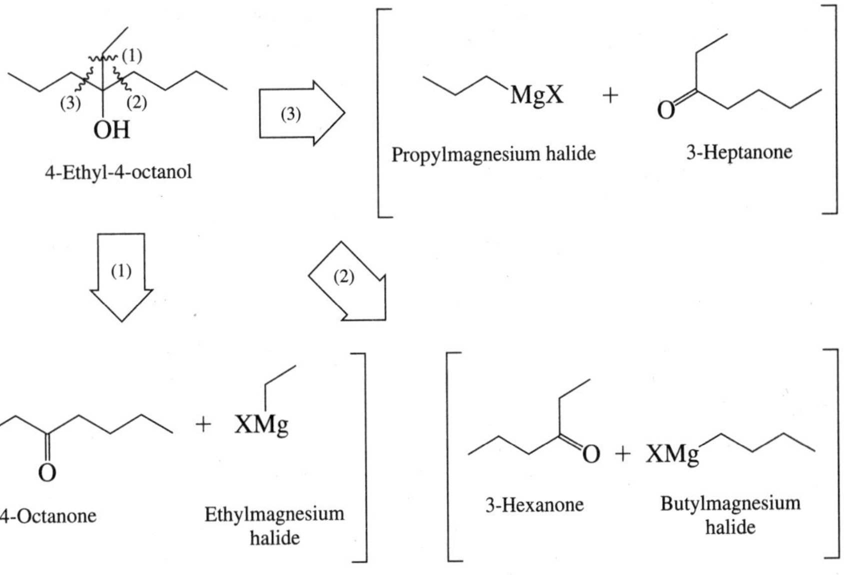

4-Ethyl-4-octanol

(1) ⇒ [4-Octanone + Ethylmagnesium halide]

(2) ⇒ [3-Hexanone + Butylmagnesium halide]

(3) ⇒ [Propylmagnesium halide + 3-Heptanone]

**14.21** (*a*) Meparfynol is a tertiary alcohol and so can be prepared by addition of a carbanionic species to a ketone. Use the same reasoning that applies to the synthesis of alcohols from Grignard reagents. On mentally disconnecting one of the bonds to the carbon bearing the hydroxyl group

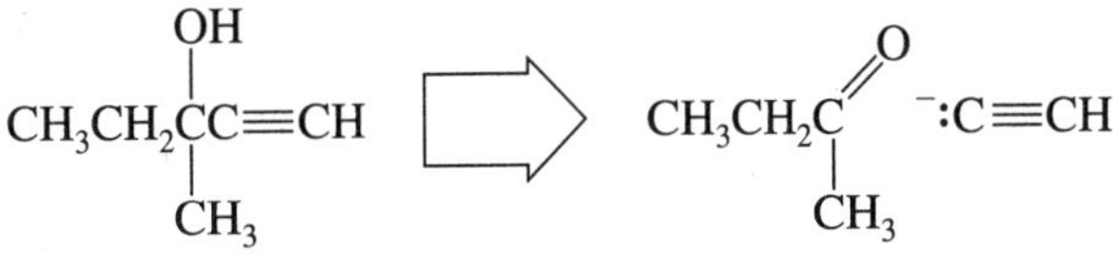

$$CH_3CH_2C(OH)(CH_3)C{\equiv}CH \Longrightarrow CH_3CH_2C(=O)CH_3 \quad {}^{-}{:}C{\equiv}CH$$

we see that the addition of acetylide ion to 2-butanone will provide the target molecule.

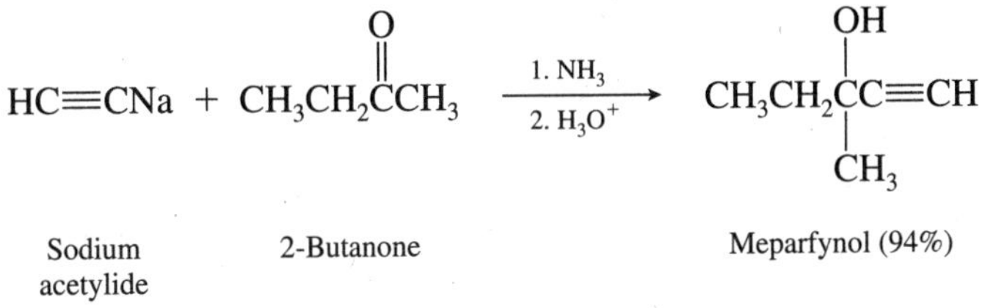

$$\underset{\text{Sodium acetylide}}{HC{\equiv}CNa} + \underset{\text{2-Butanone}}{CH_3CH_2\overset{O}{\overset{\|}{C}}CH_3} \xrightarrow[\text{2. } H_3O^+]{\text{1. } NH_3} \underset{\text{Meparfynol (94\%)}}{CH_3CH_2C(OH)(CH_3)C{\equiv}CH}$$

The alternative, reaction of a Grignard reagent with an alkynyl ketone, is not acceptable in this case. The acidic terminal alkyne C—H would transfer a proton to the Grignard reagent.

(*b*) Diphepanol is a tertiary alcohol and so may be prepared by reaction of a Grignard or organolithium reagent with a ketone. Retrosynthetically, two possibilities seem reasonable:

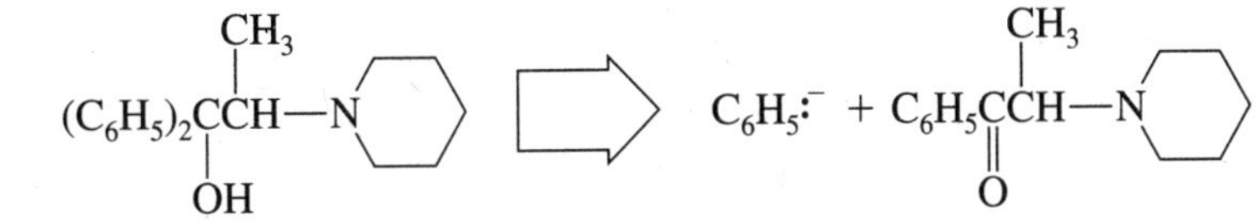

$$(C_6H_5)_2C(OH)CH(CH_3)\text{—N(piperidinyl)} \Longrightarrow C_6H_5{:}^{-} + C_6H_5\overset{O}{\overset{\|}{C}}CH(CH_3)\text{—N(piperidinyl)}$$

and

$$(C_6H_5)_2C(OH)CH(CH_3)\text{—N(piperidinyl)} \Longrightarrow (C_6H_5)_2C{=}O + {}^{-}{:}CH(CH_3)\text{—N(piperidinyl)}$$

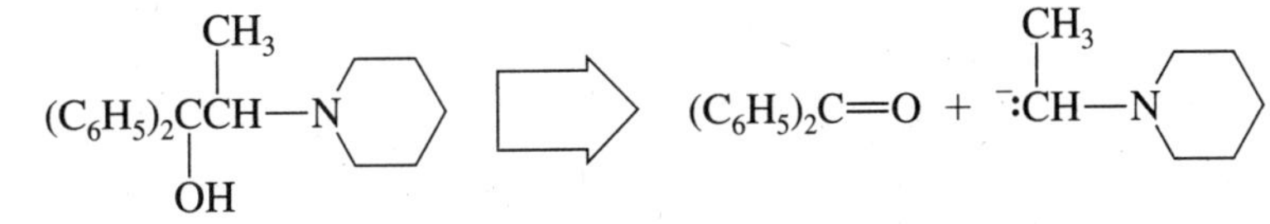

In principle either strategy is acceptable; in practice the one involving phenylmagnesium bromide is used.

$$C_6H_5MgBr + C_6H_5\overset{O}{\overset{\|}{C}}\underset{CH_3}{CH}\text{—N(piperidine)} \xrightarrow[2.\ H_3O^+]{1.\ \text{diethyl ether}} (C_6H_5)_2\underset{OH}{C}\underset{CH_3}{CH}\text{—N(piperidine)}$$

Phenylmagnesium bromide

Diphepanol

(*c*) A reasonable last step in the synthesis of mestranol is the addition of sodium acetylide to the ketone shown.

Shields top face

$$\text{ketone } (H_3C,\ O;\ CH_3O) \xrightarrow[2.\ H_3O^+]{1.\ NaC{\equiv}CH,\ NH_3} \text{Mestranol } (H_3C,\ OH,\ C{\equiv}CH;\ CH_3O)$$

Mestranol

Acetylide anion adds to the carbonyl from the less sterically hindered side. The methyl group shields the top face of the carbonyl, and so acetylide adds from the bottom.

**14.22** (*a*) Sodium acetylide adds to ketones to give tertiary alcohols.

$$C_6H_5\overset{O}{\overset{\|}{C}}C_6H_5 + NaC{\equiv}CH \xrightarrow[2.\ H_3O^+]{1.\ \text{liquid ammonia}} (C_6H_5)_2\overset{OH}{C}\text{—}C{\equiv}CH$$

Benzophenone

1,1-Diphenyl-2-propyn-1-ol (50%)

(*b*) The substrate is a ketone, which reacts with ethyllithium to yield a tertiary alcohol.

$$\text{2-Adamantanone} + CH_3CH_2Li \xrightarrow[2.\ H_3O^+]{1.\ \text{diethyl ether}} \text{2-adamantyl}(OH)CH_2CH_3$$

2-Adamantanone

2-Ethyl-2-adamantanol (83%)

(*c*) The first step is conversion of bromocyclopentene to the corresponding Grignard reagent, which then reacts with formaldehyde to give a primary alcohol.

$$\text{C}_5\text{H}_7\text{Br} \xrightarrow[\text{THF}]{Mg} \text{C}_5\text{H}_7\text{MgBr} \xrightarrow[2.\ H_3O^+]{1.\ H\overset{O}{\overset{\|}{C}}H} \text{C}_5\text{H}_7CH_2OH$$

1-Bromocyclopentene

1-Cyclopentenylmagnesium bromide

1-Cyclopentenylmethanol (53%)

(*d*) The reaction is one in which an alkene is converted to a cyclopropane through use of the Simmons–Smith reagent, iodomethylzinc iodide.

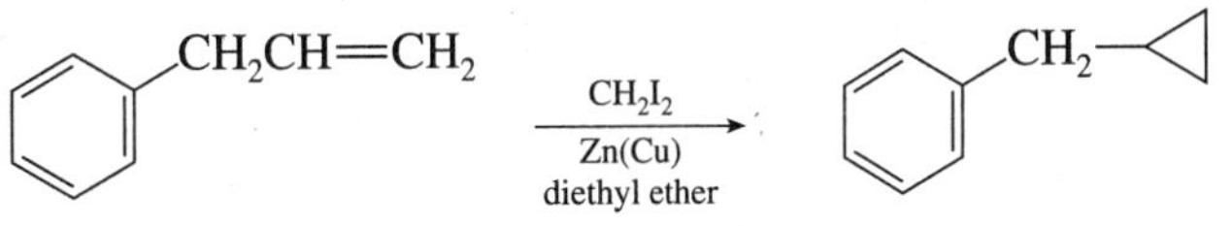

$C_6H_5CH_2CH{=}CH_2 \xrightarrow[\text{Zn(Cu), diethyl ether}]{CH_2I_2} C_6H_5CH_2\text{-}C_3H_5$

Allylbenzene → Benzylcyclopropane (64%)

(*e*) Methylene transfer using the Simmons–Smith reagent is stereospecific. The trans arrangement of substituents in the alkene is carried over to the cyclopropane product.

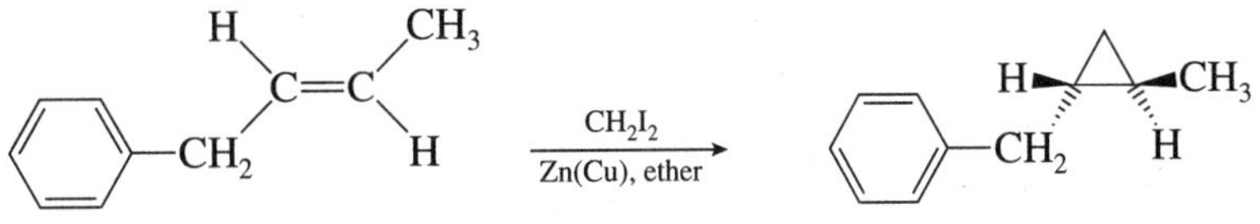

$\xrightarrow[\text{Zn(Cu), ether}]{CH_2I_2}$

(*E*)-1-Phenyl-2-butene → *trans*-1-Benzyl-2-methylcyclopropane (50%)

(*f*) Lithium dimethylcuprate transfers a methyl group, which substitutes for iodine on the iodoalkene. Even halogens on $sp^2$-hybridized carbon are reactive in substitution reactions with lithium dialkylcuprates.

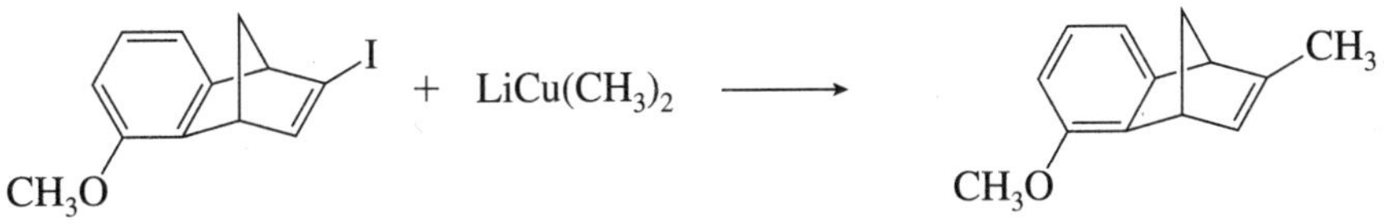

2-Iodo-8-methoxybenzonorbornadiene + $LiCu(CH_3)_2$ → 8-Methoxy-2-methylbenzonorbornadiene (73%)

(*g*) The starting material is a *p*-toluenesulfonate ester. *p*-Toluenesulfonates are similar to alkyl halides in their reactivity. Substitution occurs; a butyl group from lithium dibutylcuprate replaces *p*-toluenesulfonate.

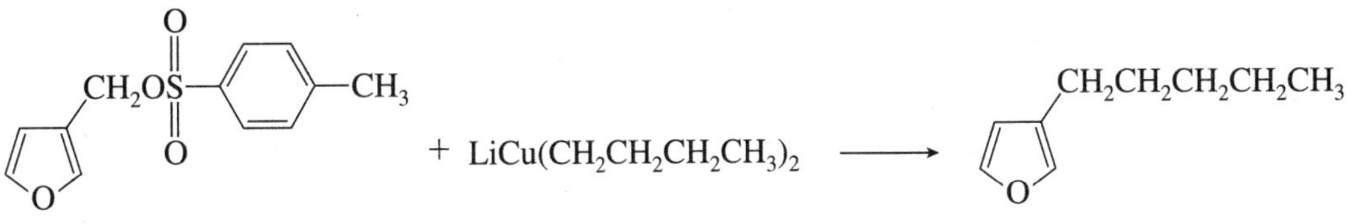

(3-Furyl)methyl *p*-toluenesulfonate + Lithium dibutylcuprate $LiCu(CH_2CH_2CH_2CH_3)_2$ → 3-Pentylfuran ($CH_2CH_2CH_2CH_2CH_3$)

**14.23** Phenylmagnesium bromide reacts with 4-*tert*-butylcyclohexanone as shown.

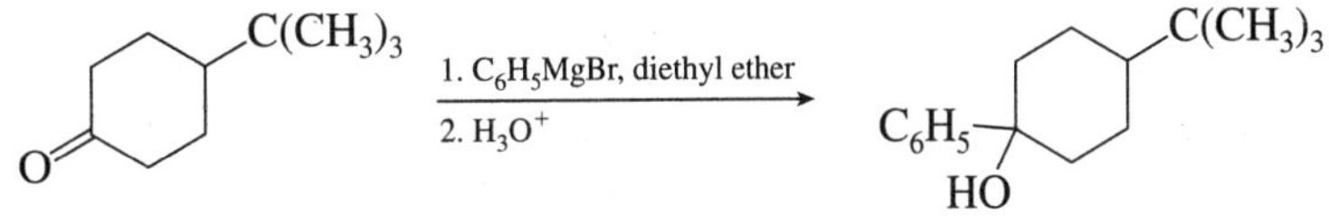

$\xrightarrow[\text{2. } H_3O^+]{\text{1. } C_6H_5MgBr\text{, diethyl ether}}$

4-*tert*-Butylcyclohexanone → 4-*tert*-Butyl-1-phenylcyclohexanol

The phenyl substituent can be introduced either cis or trans to the *tert*-butyl group. The two alcohols are therefore stereoisomers (diastereomers).

Dehydration of either alcohol yields 4-*tert*-butyl-1-phenylcyclohexene.

$$\text{C}_6\text{H}_5,\text{OH-cyclohexane-C(CH}_3)_3 \quad \text{or} \quad \text{HO},\text{C}_6\text{H}_5\text{-cyclohexane-C(CH}_3)_3 \xrightarrow{\text{H}^+,\ \text{heat}} \text{C}_6\text{H}_5\text{-cyclohexene-C(CH}_3)_3$$

4-*tert*-Butyl-1-phenylcyclohexene

**14.24** (*a*) By working through the sequence of reactions that occur when ethyl formate reacts with a Grignard reagent, we can see that this combination leads to **secondary alcohols.**

$$\underset{\text{Grignard reagent}}{\text{RMgX}} + \underset{\text{Ethyl formate}}{\text{HCOCH}_2\text{CH}_3}\ (\text{C=O}) \longrightarrow \underset{\text{Aldehyde}}{\text{RCH}}\ (\text{C=O}) + \text{CH}_3\text{CH}_2\text{OMgX} \xrightarrow[\text{2. H}_3\text{O}^+]{\text{1. RMgX, diethyl ether}} \underset{\text{Secondary alcohol}}{\text{RCHR}}\ (\text{OH})$$

This is simply because the substituent on the carbonyl carbon of the ester, in this case a hydrogen, is carried through and becomes a substituent on the hydroxyl-bearing carbon of the alcohol.

(*b*) Diethyl carbonate has the potential to react with 3 moles of a Grignard reagent.

$$\underset{\text{Grignard reagent}}{\text{RMgX}} + \underset{\text{Diethyl carbonate}}{\text{CH}_3\text{CH}_2\text{OCOCH}_2\text{CH}_3}\ (\text{C=O}) \longrightarrow \underset{\text{Ester}}{\text{RCOCH}_2\text{CH}_3}\ (\text{C=O}) + \text{CH}_3\text{CH}_2\text{OMgX}$$

$$\downarrow \text{RMgX}$$

$$\underset{\text{Tertiary alcohol}}{\text{R}_3\text{COH}} \xleftarrow[\text{2. H}_3\text{O}^+]{\text{1. RMgX}} \underset{\text{Ketone}}{\text{RCR}}\ (\text{C=O}) + \text{CH}_3\text{CH}_2\text{OMgX}$$

The tertiary alcohols that are formed by the reaction of diethyl carbonate with Grignard reagents have three identical R groups attached to the carbon that bears the hydroxyl substituent.

**14.25** If we use the 2-bromobutane given, along with the information that the reaction occurs with net inversion of configuration, the stereochemical course of the reaction may be written as

$$\text{CH}_3\text{CH}_2(\text{CH}_3)\text{C(H)Br} \xrightarrow{\text{LiCu(C}_6\text{H}_5)_2} \text{CH}_3(\text{CH}_3\text{CH}_2)\text{C(H)C}_6\text{H}_5$$

The phenyl group becomes bonded to carbon from the opposite side of the leaving group.

Applying the Cahn–Ingold–Prelog notational system described in Section 7.6 to the product, the order of decreasing precedence is

$$\text{C}_6\text{H}_5 > \text{CH}_3\text{CH}_2 > \text{CH}_3 > \text{H}$$

Orienting the molecule so that the lowest ranked substituent (H) is away from us, we see that the order of decreasing precedence is clockwise.

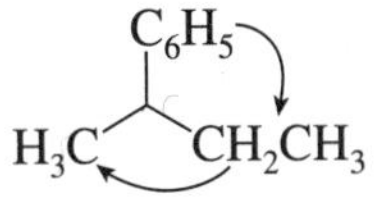

$C_6H_5$ $H_3C$ $CH_2CH_3$

The absolute configuration is *R*.

**14.26** The substrates are secondary alkyl *p*-toluenesulfonates, and so we expect elimination to compete with substitution. Compound B is formed in both reactions and has the molecular formula of 4-*tert*-butylcyclohexene. Because the two *p*-toluenesulfonates are diastereomers, it is likely that compounds A and C, especially since they have the same molecular formula, are also diastereomers. Assuming that the substitution reactions proceed with inversion of configuration, we conclude that the products are as shown.

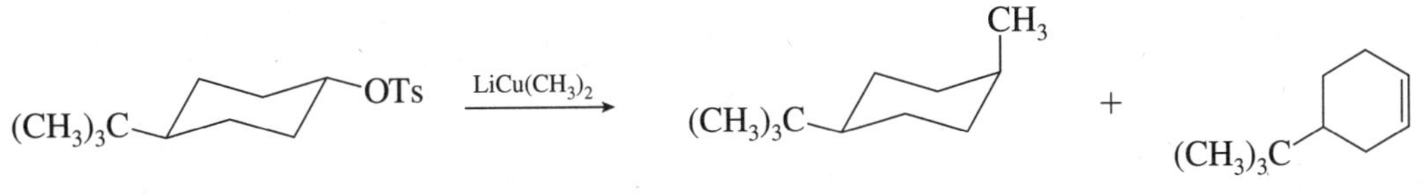

$(CH_3)_3C$–(cyclohexyl)–OTs $\xrightarrow{LiCu(CH_3)_2}$ $(CH_3)_3C$–(cyclohexyl)–$CH_3$ + $(CH_3)_3C$–(cyclohexenyl)

*trans*-4-*tert*-Butylcyclohexyl *p*-toluenesulfonate

*cis*-1-*tert*-Butyl-4-methylcyclohexane (compound A, $C_{11}H_{22}$)

4-*tert*-Butylcyclohexene (compound B, $C_{10}H_{18}$)

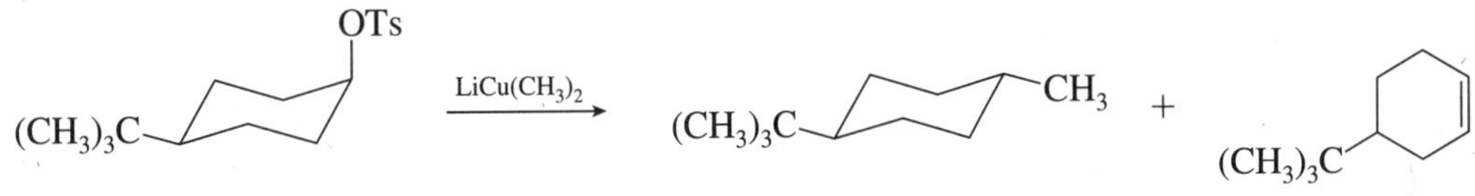

$(CH_3)_3C$–(cyclohexyl)–OTs $\xrightarrow{LiCu(CH_3)_2}$ $(CH_3)_3C$–(cyclohexyl)–$CH_3$ + $(CH_3)_3C$–(cyclohexenyl)

*cis*-4-*tert*-Butylcyclohexyl *p*-toluenesulfonate

*trans*-1-*tert*-Butyl-4-methylcyclohexane (compound C, $C_{11}H_{22}$)

Compound B

Inversion of configuration is borne out by the fact given in the problem that compound C is more stable than compound A. Both substituents are equatorial in C; the methyl group is axial in A.

**14.27** We are told in the statement of the problem that the first step is conversion of the alcohol to the corresponding *p*-toluenesulfonate. This step is carried out as follows:

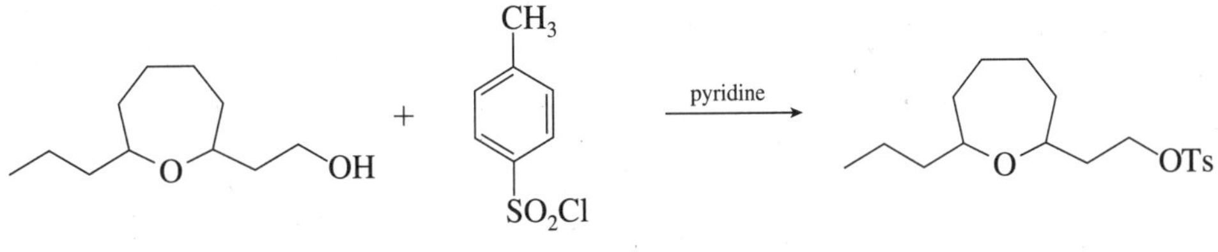

(oxepane)–OH + $CH_3$–$C_6H_4$–$SO_2Cl$ $\xrightarrow{\text{pyridine}}$ (oxepane)–OTs

3,8-Epoxy-1-undecanol

*p*-Toluenesulfonyl chloride (TsCl)

3,8-Epoxyundecyl *p*-toluenesulfonate

Alkyl *p*-toluenesulfonates react with lithium dialkylcuprates in the same way that alkyl halides do. Treatment of the preceding *p*-toluenesulfonate with lithium dibutylcuprate gives the desired compound.

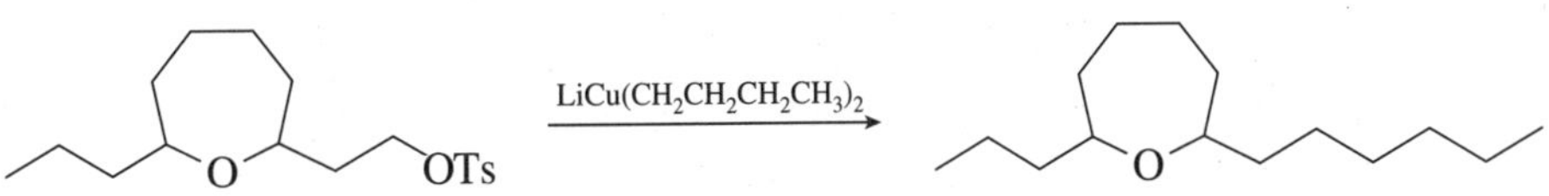

(oxepane)–OTs $\xrightarrow{LiCu(CH_2CH_2CH_2CH_3)_2}$ (oxepane)

3,8-Epoxyundecyl *p*-toluenesulfonate

4,9-Epoxypentadecane

As actually performed, a 91% yield of the desired product was obtained in the reaction of the *p*-toluenesulfonate with lithium dibutylcuprate.

**14.28** (*a*) The desired 1-deuteriobutane can be obtained by reaction of $D_2O$ with butyllithium or butylmagnesium bromide.

$$\underset{\text{Butyllithium}}{CH_3CH_2CH_2CH_2Li} + \underset{\text{Deuterium oxide}}{D_2O} \longrightarrow \underset{\text{1-Deuteriobutane}}{CH_3CH_2CH_2CH_2D}$$

or

$$\underset{\text{Butylmagnesium bromide}}{CH_3CH_2CH_2CH_2MgBr} + D_2O \longrightarrow \underset{\text{1-Deuteriobutane}}{CH_3CH_2CH_2CH_2D}$$

Preparation of the organometallic compounds requires an alkyl bromide, which is synthesized from the corresponding alcohol.

$$\underset{\text{1-Butanol}}{CH_3CH_2CH_2CH_2OH} \xrightarrow[\text{or HBr}]{PBr_3} \underset{\text{1-Bromobutane}}{CH_3CH_2CH_2CH_2Br}$$

$$\underset{\text{Butyllithium}}{CH_3CH_2CH_2CH_2Li} \xleftarrow[\text{ether}]{Li} \underset{\text{1-Bromobutane}}{CH_3CH_2CH_2CH_2Br} \xrightarrow[\text{ether}]{Mg} \underset{\text{Butylmagnesium bromide}}{CH_3CH_2CH_2CH_2MgBr}$$

(*b*) In a sequence identical to that of part (*a*) in design but using 2-butanol as the starting material, 2-deuteriobutane may be prepared.

$$\underset{\text{2-Butanol}}{CH_3\underset{OH}{\underset{|}{C}}HCH_2CH_3} \xrightarrow{PBr_3} \underset{\text{2-Bromobutane}}{CH_3\underset{Br}{\underset{|}{C}}HCH_2CH_3} \xrightarrow[\text{ether}]{Mg} \underset{\textit{sec}\text{-Butylmagnesium bromide}}{CH_3\underset{MgBr}{\underset{|}{C}}HCH_2CH_3} \xrightarrow{D_2O} \underset{\text{2-Deuteriobutane}}{CH_3\underset{D}{\underset{|}{C}}HCH_2CH_3}$$

An analogous procedure involving *sec*-butyllithium in place of the Grignard reagent can be used.

**14.29** All the protons in benzene are equivalent. In diphenylmethane and in triphenylmethane, protons are attached either to the $sp^2$-hybridized carbons of the ring or to the $sp^3$-hybridized carbon between the rings. The large difference in acidity between diphenylmethane and benzene suggests that it is not a ring proton that is lost on ionization in diphenylmethane but rather a proton from the methylene group.

$$\underset{\text{Diphenylmethane}}{(C_6H_5)CH_2} \rightleftharpoons (C_6H_5)_2\ddot{C}H + H^+$$

The anion produced is stabilized by resonance. It is a **benzylic** carbanion.

$$C_6H_5\text{–}\ddot{C}H^-\text{–}C_6H_5 \longleftrightarrow \text{(ring carbanion)}=CH\text{–}C_6H_5 \longleftrightarrow \text{etc.}$$

Both rings are involved in delocalizing the negative charge. The anion from triphenylmethane is stabilized by resonance involving all three rings.

etc.

Delocalization of the negative charge by resonance is not possible in the anion of benzene. The pair of unshared electrons in phenyl anion is in an $sp^2$ hybrid orbital that does not interact with the $\pi$ system.

Not delocalized into $\pi$ system

**14.30** The titanium-containing compound is a metallocene. (It has cyclopentadienyl rings as ligands.) With an atomic number of 22, titanium has an electron configuration of $[Ar]4s^23d^2$. As the following accounting shows, this titanium complex is 2 electrons short of satisfying the 18-electron rule.

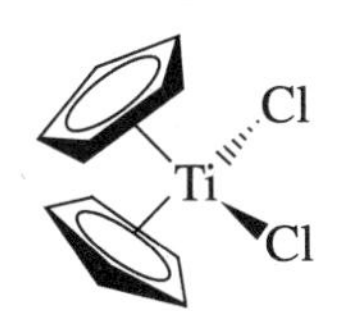


Ti: 4 electrons
Two cyclopentadienyl rings: 10 electrons
Two chlorine atoms: 2 electrons
Total: 16 electrons

1,3-Butadiene(tricarbonyl)iron satisfies the 18-electron rule. The electron configuration of iron is $[Ar]4s^23d^6$.

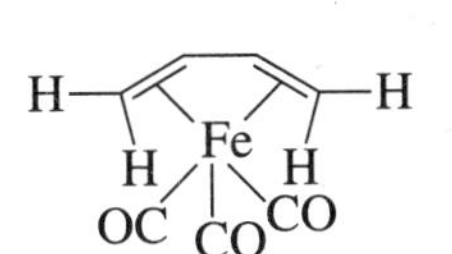


Fe: 8 electrons
1,3-Butadiene ligand: 4 electrons
Three CO ligands: 6 electrons
Total: 18 electrons

**14.31** Using 1-decene as an example, we can see from the following schematic that the growing polymer will incorporate a $C_8$ side chain at every point where 1-decene replaces ethylene.

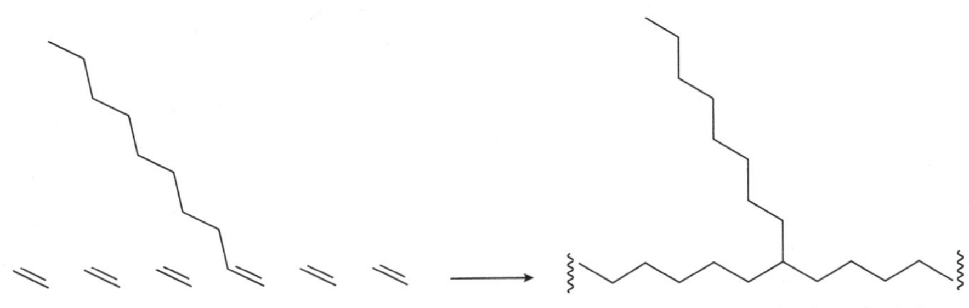


**14.32–14.36** Solutions to molecular modeling exercises are not provided in this *Study Guide and Solutions Manual.* You should use *Learning By Modeling* for these exercises.

# SELF-TEST

## PART A

**A-1.** Give a method for the preparation of each of the following organometallic compounds, using appropriate starting materials:
(*a*) Cyclohexyllithium
(*b*) *tert*-Butylmagnesium bromide
(*c*) Lithium dibenzylcuprate

**A-2.** Give the structure of the product obtained by each of the following reaction schemes:

(*a*) $CH_3CO_2CH_2CH_3 \xrightarrow[2.\ H_3O^+]{1.\ 2C_6H_5MgBr}$ ?

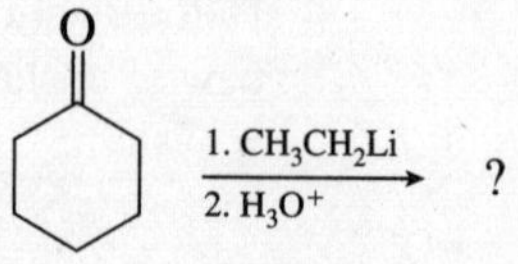

(*b*) $(CH_3)_2CHCH_2Li \xrightarrow{D_2O}$ ?

(*c*) $H_3C-C_6H_4-Br$ (para) $\xrightarrow[3.\ H_3O^+]{1.\ Mg;\ 2.\ H_2C{=}O}$ ?

(*d*) Cyclohexanone $\xrightarrow[2.\ H_3O^+]{1.\ CH_3CH_2Li}$ ?

(*e*) 2-Butene + $CHBr_3$ $\xrightarrow[(CH_3)_3COH]{KOC(CH_3)_3}$ ?

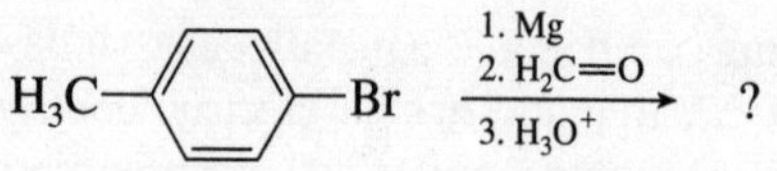

**A-3.** Give two combinations of an organometallic reagent and a carbonyl compound that may be used for the preparation of each of the following:

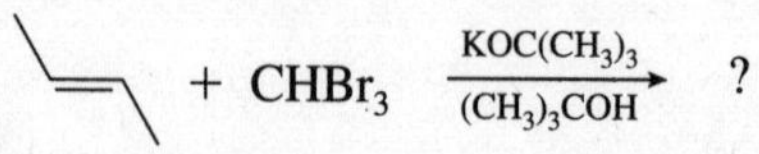

(*a*) $C_6H_5CH(OH)C(CH_3)_3$ (*b*) $CH_3CH_2CH_2CH(OH)CH_2CH_2CH_2CH_3$

**A-4.** Gives the structure of the organometallic reagent necessary to carry out each of the following:

(*a*) 1-Methyl-2-iodocyclohexene $\xrightarrow{?}$ 1-methyl-2-propylcyclohexene ($CH_3$, $CH_2CH_2CH_3$)

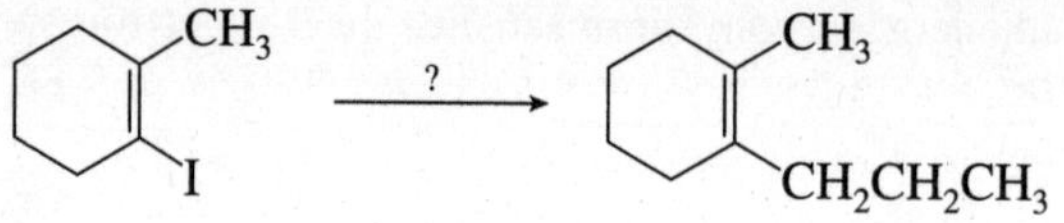

(*b*) $C_6H_5CH_2CO_2CH_3 \xrightarrow[2.\ H_3O^+]{1.\ ?}$ $(CH_3)_2CHC(OH)(CH_2C_6H_5)CH(CH_3)_2$

(*c*) 1,2-Dimethylcyclohexene $\xrightarrow{?}$ 1,6-dimethylbicyclo[4.1.0]heptane ($CH_3$, $CH_3$)

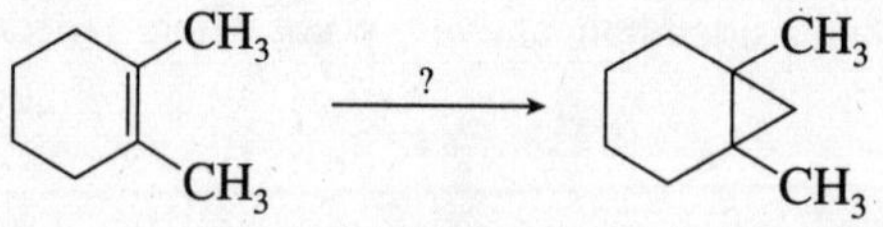

**A-5.** Compounds A through F are some common organic solvents. Which ones would be suitable for use in the preparation of a Grignard reagent? For those that are not suitable, give a brief reason why.

A: $CH_3CH_2CH_2CH_2OCH_2CH_2CH_2CH_3$
B: $CH_3OCH_2CH_2OCH_3$
C: $HOCH_2CH_2OH$
D: $CH_3COCH_2CH_3$ (C=O)
E: tetrahydrofuran (five-membered ring with O)
F: $CH_3COH$ (C=O)

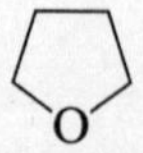

**A-6.** Show by a series of chemical equations how you would prepare octane from 1-butanol as the source of all its carbon atoms.

**A-7.** Synthesis of the following alcohol is possible by three schemes using Grignard reagents. Give the reagents necessary to carry out each of them.

$$(CH_3)_2CH\overset{\overset{OH}{|}}{C}(CH_3)_2$$

**A-8.** Using ethylbenzene and any other necessary organic or inorganic reagents, outline a synthesis of 3-phenyl-2-butanol.

**A-9.** Give the structure of the final product of each of the following sequences of reactions.

(*a*) Benzene $\xrightarrow[FeBr_3]{Br_2}$ $\xrightarrow{Mg}$ $\xrightarrow{CH_3\overset{O}{\overset{\|}{C}}CH_2CH_3}$ $\xrightarrow{H_3O^+}$ ?

(*b*) 1-Butene $\xrightarrow{HCl}$ $\xrightarrow{Mg}$ $\xrightarrow{CH_3\overset{O}{\overset{\|}{C}}H}$ $\xrightarrow{H_3O^+}$ ?

(*c*) $CH_3C{\equiv}CH$ $\xrightarrow{NaNH_2}$ $\xrightarrow{\text{cyclopentanone}}$ $\xrightarrow{H_3O^+}$ ?

## PART B

**B-1.** Which (if any) of the following would *not* be classified as an organometallic substance?
(*a*) Triethylaluminum
(*b*) Ethylmagnesium iodide
(*c*) Potassium *tert*-butoxide
(*d*) None of these (all are organometallic compounds)

**B-2.** Rank the following species in order of increasing polarity of the carbon–metal bond (least → most polar):

| $CH_3CH_2MgCl$ | $CH_3CH_2Na$ | $(CH_3CH_2)_3Al$ |
|---|---|---|
| 1 | 2 | 3 |

(*a*) $3 < 1 < 2$ (*b*) $2 < 1 < 3$ (*c*) $1 < 3 < 2$ (*d*) $2 < 3 < 1$

**B-3.** Which sequence of reagents would carry out the following conversion?

$$CH_3CH_2\underset{\underset{OH}{|}}{C}HCH_3 \xrightarrow{?} CH_3CH_2\underset{\underset{D}{|}}{C}HCH_3$$

(*a*) $H_2SO_4$, heat; then $B_2D_6$; then $H_2O_2$, $HO^-$
(*b*) $H_2SO_4$, heat; then $D_2$, Pt
(*c*) $CH_3MgBr$; then $D_2O$
(*d*) HBr; then Mg; then $D_2O$

**B-4.** Arrange the following intermediates in order of decreasing basicity (strongest → weakest):

| $H_2C{=}CHNa$ | $CH_3CH_2Na$ | $CH_3CH_2ONa$ | $HC{\equiv}CNa$ |
|---|---|---|---|
| 1 | 2 | 3 | 4 |

(*a*) $2 > 1 > 4 > 3$ (*c*) $3 > 4 > 1 > 2$
(*b*) $4 > 1 > 2 > 3$ (*d*) $3 > 2 > 4 > 1$

**B-5.** Which, if any, of the following pairs of reagents could be used to prepare 2-phenyl-2-butanol?

$$CH_3CH_2\overset{\overset{OH}{|}}{\underset{\underset{CH_3}{|}}{C}}C_6H_5$$

2-Phenyl-2-butanol

(*a*) $CH_3CH_2MgBr + C_6H_5CH_2\overset{O}{\overset{\|}{C}}CH_3$

(*b*) $CH_3CH_2MgBr + C_6H_5CH_2\overset{O}{\overset{\|}{C}}H$

(*c*) $CH_3MgI + C_6H_5CH_2\overset{O}{\overset{\|}{C}}CH_3$

(*d*) $C_6H_5MgCl + CH_3\overset{O}{\overset{\|}{C}}CH_2CH_2CH_3$

(*e*) None of these combinations would be effective.

**B-6.** Which of the following reagents would be effective for the following reaction sequence?

$$C_6H_5C{\equiv}CH \xrightarrow[3.\ H_3O^+]{1.\ ?\quad 2.\ H_2C{=}O} C_6H_5C{\equiv}CCH_2OH$$

(*a*) Sodium ethoxide
(*b*) Magnesium in diethyl ether
(*c*) Butyllithium
(*d*) Potassium hydroxide

**B-7.** What is the product of the following reaction?

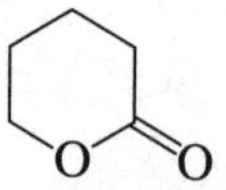

(δ-valerolactone: six-membered ring with O and C=O) $+ 2CH_3MgBr \xrightarrow[2.\ H_3O^+]{1.\ \text{diethyl ether}}$

(*a*) $HO\underset{\underset{CH_3}{|}}{C}HCH_2CH_2CH_2\underset{\underset{CH_3}{|}}{C}HOH$

(*b*) $HOCH_2CH_2CH_2CH_2\overset{\overset{CH_3}{|}}{\underset{\underset{CH_3}{|}}{C}}OH$

(*c*) $CH_3OCH_2CH_2CH_2CH_2\underset{\underset{OH}{|}}{C}HCH_3$

(*d*) $HOCH_2CH_2CH_2CH_2\underset{\underset{CH_3}{|}}{C}HOCH_3$

**B-8.** Which of the following combinations of reagents will yield a chiral product after hydrolysis in aqueous acid?

(*a*) $CH_3CH_2\overset{O}{\overset{\|}{C}}H$ + $CH_3MgBr$

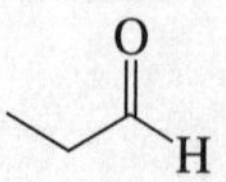

(*b*) $CH_3CH_2\overset{O}{\overset{\|}{C}}CH_3$ + $CH_3MgBr$

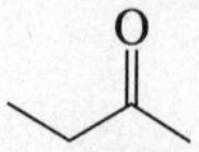

(*c*) $CH_3CH_2\overset{O}{\overset{\|}{C}}OCH_3 + 2CH_3MgBr$

(*d*) Both (*a*) and (*c*)

**B-9.** Which sequence of steps describes the best synthesis of 2-phenylpropene?

(*a*) Benzene + 2-chloropropene, $AlCl_3$
(*b*) Benzene + propene, $H_2SO_4$
(*c*) 1. Benzaldehyde ($C_6H_5CH{=}O$) + $CH_3CH_2MgBr$, diethyl ether
2. $H_3O^+$
3. $H_2SO_4$, heat
(*d*) 1. Bromobenzene + Mg, diethyl ether
2. Propanal ($CH_3CH_2CH{=}O$)
3. $H_3O^+$
4. $H_2SO_4$, heat
(*e*) 1. Bromobenzene + Mg, diethyl ether
2. Acetone [$(CH_3)_2C{=}O$]
3. $H_3O^+$
4. $H_2SO_4$, heat

**B-10.** What sequence of steps represents the best synthesis of 4-heptanol ($(CH_3CH_2CH_2)_2CHOH$)?

(*a*) $CH_3CH_2CH_2MgBr$ (2 mol) + formaldehyde ($CH_2{=}O$) in diethyl ether followed by $H_3O^+$
(*b*) $CH_3CH_2CH_2MgBr$ + butanal ($CH_3CH_2CH_2CH{=}O$) in diethyl ether followed by $H_3O^+$
(*c*) $CH_3CH_2CH_2CH_2MgBr$ + acetone [$(CH_3)_2C{=}O$] in diethyl ether followed by $H_3O^+$
(*d*) $(CH_3CH_2CH_2)_2CHMgBr$ + formaldehyde ($CH_2{=}O$) in diethyl ether followed by $H_3O^+$
(*e*) $CH_3CH_2CH_2MgBr$ + ethyl acetate ($CH_3\overset{O}{\overset{\|}{C}}OCH_2CH_3$) in diethyl ether followed by $H_3O^+$

**B-11.** All of the following compounds react with ethylmagnesium bromide. Alcohols are formed from four of the compounds. Which one does *not* give an alcohol?

(*a*) $C_6H_5\overset{O}{\overset{\|}{C}}H$
(*b*) $C_6H_5\overset{O}{\overset{\|}{C}}OH$
(*c*) $C_6H_5\overset{O}{\overset{\|}{C}}OCH_3$
(*d*) $C_6H_5\overset{O}{\overset{\|}{C}}CH_3$
(*e*) $C_6H_5CH_2O\overset{O}{\overset{\|}{C}}CH_3$

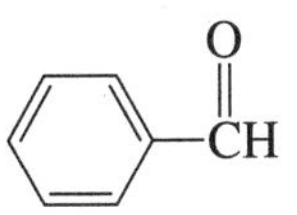

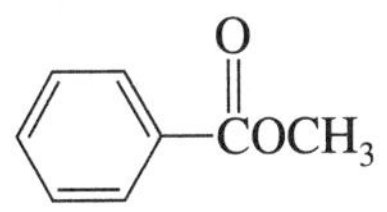

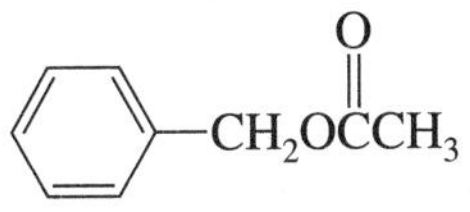

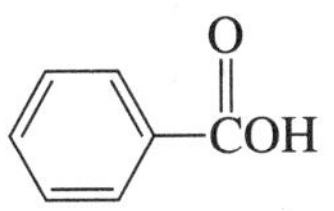

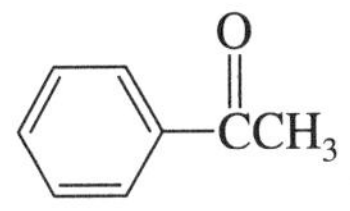

**B-12.** Give the major product of the following reaction:

$$(E)\text{-2-pentene} \xrightarrow{CH_2I_2,\ Zn(Cu)} ?$$

(*a*) *cis*-1-Ethyl-2-methylcyclopropane
(*b*) *trans*-1-Ethyl-2-methylcyclopropane
(*c*) 1-Ethyl-1-methylcyclopropane
(*d*) An equimolar mixture of products (*a*) and (*b*)

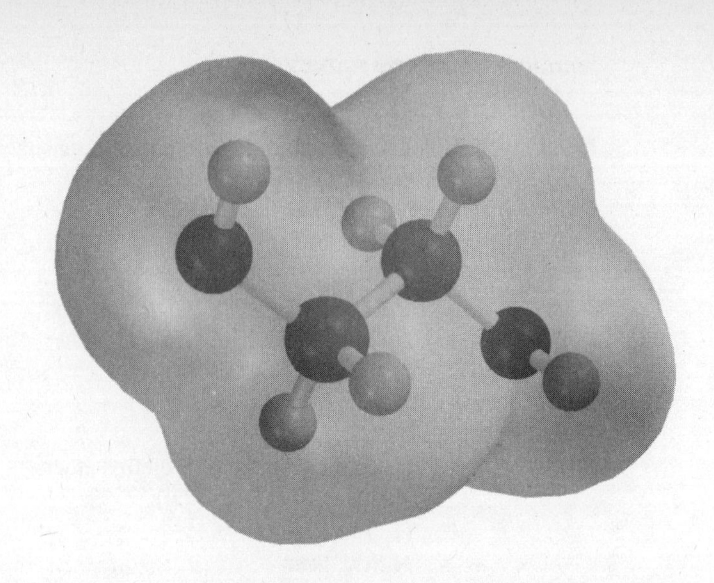

# CHAPTER 15

## ALCOHOLS, DIOLS, AND THIOLS

## SOLUTIONS TO TEXT PROBLEMS

**15.1** The two primary alcohols, 1-butanol and 2-methyl-1-propanol, can be prepared by hydrogenation of the corresponding aldehydes.

$$\underset{\text{Butanal}}{CH_3CH_2CH_2\overset{\overset{O}{\|}}{C}H} \xrightarrow{H_2,\ Ni} \underset{\text{1-Butanol}}{CH_3CH_2CH_2CH_2OH}$$

$$\underset{\text{2-Methylpropanal}}{(CH_3)_2CH\overset{\overset{O}{\|}}{C}H} \xrightarrow{H_2,\ Ni} \underset{\text{2-Methyl-1-propanol}}{(CH_3)_2CHCH_2OH}$$

The secondary alcohol 2-butanol arises by hydrogenation of a ketone.

$$\underset{\text{2-Butanone}}{CH_3\overset{\overset{O}{\|}}{C}CH_2CH_3} \xrightarrow{H_2,\ Ni} \underset{\text{2-Butanol}}{CH_3\underset{\underset{OH}{|}}{C}HCH_2CH_3}$$

Tertiary alcohols such as 2-methyl-2-propanol, $(CH_3)_3COH$, cannot be prepared by hydrogenation of a carbonyl compound.

**15.2** (*b*) A deuterium atom is transferred from $NaBD_4$ to the carbonyl group of acetone.

$$CH_3\underset{\underset{CH_3}{|}}{C}{=}O \ (D{-}\bar{B}D_3) \longrightarrow CH_3\underset{\underset{CH_3}{|}}{\overset{\overset{D}{|}}{C}}{-}O\bar{B}D_3 \xrightarrow{3(CH_3)_2C{=}O} \left(CH_3\underset{\underset{CH_3}{|}}{\overset{\overset{D}{|}}{C}}O{-}\right)_4\bar{B}$$

On reaction with $CH_3OD$, deuterium is transferred from the alcohol to the oxygen of $[(CH_3)_2CDO]_4\bar{B}$.

$$CH_3\underset{CH_3}{\overset{D}{C}}—O—\bar{B}[OCD(CH_3)_2]_3 \quad D—OCH_3 \longrightarrow CH_3\underset{CH_3}{\overset{D}{C}}OD + \underset{OCH_3}{\bar{B}}[OCD(CH_3)_2]_3$$

**Overall:**

$$(CH_3)_2C{=}O \xrightarrow[CH_3OD]{NaBD_4} (CH_3)_2\overset{D}{C}OD$$

Acetone → 2-Propanol-2-d-*O*-d

(*c*) In this case $NaBD_4$ serves as a deuterium donor to carbon, and $CD_3OH$ is a proton (not deuterium) donor to oxygen.

$$C_6H_5\overset{O}{\overset{\|}{C}}H \xrightarrow[CD_3OH]{NaBD_4} C_6H_5\overset{D}{C}HOH$$

Benzaldehyde → Benzyl alcohol-1-d

(*d*) Lithium aluminum deuteride is a deuterium donor to the carbonyl carbon of formaldehyde.

$$D—\bar{A}lD_3 \quad H\underset{H}{C}{=}O \longrightarrow H\underset{H}{\overset{D}{C}}—O\bar{A}lD_3 \xrightarrow{3H\overset{O}{\overset{\|}{C}}H} (DCH_2O)_4\bar{A}l$$

On hydrolysis with $D_2O$, the oxygen–aluminum bond is cleaved and $DCH_2OD$ is formed.

$$\bar{A}l(OCH_2D)_4 \xrightarrow{4D_2O} 4DCH_2OD + \bar{A}l(OD)_4$$

Methanol-d-*O*-d

**15.3** The acyl portion of the ester gives a primary alcohol on reduction. The alkyl group bonded to oxygen may be primary, secondary, or tertiary and gives the corresponding alcohol.

$$CH_3CH_2\overset{O}{\overset{\|}{C}}OCH(CH_3)_2 \xrightarrow[2.\ H_2O]{1.\ LiAlH_4} CH_3CH_2CH_2OH + HOCH(CH_3)_2$$

Isopropyl propanoate → 1-Propanol + 2-Propanol

**15.4** (*b*) Reaction with ethylene oxide results in the addition of a $—CH_2CH_2OH$ unit to the Grignard reagent. Cyclohexylmagnesium bromide (or chloride) is the appropriate reagent.

$$C_6H_{11}MgBr + H_2C\underset{O}{—}CH_2 \xrightarrow[2.\ H_3O^+]{1.\ diethyl\ ether} C_6H_{11}CH_2CH_2OH$$

Cyclohexylmagnesium bromide + Ethylene oxide → 2-Cyclohexylethanol

**15.5** Lithium aluminum hydride is the appropriate reagent for reducing carboxylic acids or esters to alcohols.

$$HO\overset{O}{\overset{\|}{C}}CH_2\underset{CH_3}{C}HCH_2\overset{O}{\overset{\|}{C}}OH \xrightarrow[2.\ H_2O]{1.\ LiAlH_4} HOCH_2CH_2\underset{CH_3}{C}HCH_2CH_2OH$$

3-Methyl-1,5-pentanedioic acid → 3-Methyl-1,5-pentanediol

Any alkyl group may be attached to the oxygen of the ester function. In the following example, it is a methyl group.

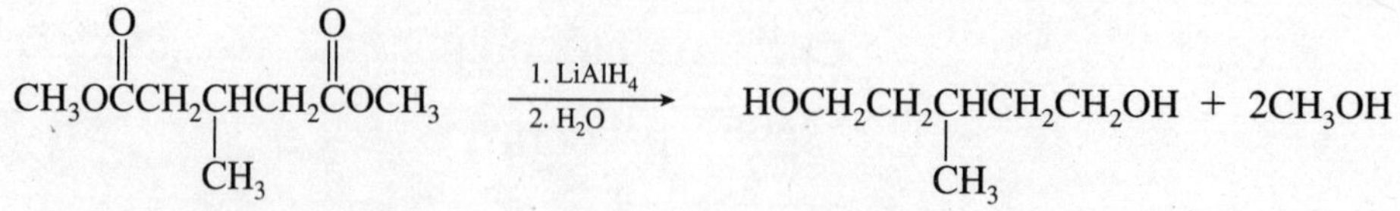

$$CH_3O\overset{O}{\overset{\|}{C}}CH_2\underset{CH_3}{\underset{|}{C}H}CH_2\overset{O}{\overset{\|}{C}}OCH_3 \xrightarrow[2.\ H_2O]{1.\ LiAlH_4} HOCH_2CH_2\underset{CH_3}{\underset{|}{C}H}CH_2CH_2OH + 2CH_3OH$$

Dimethyl 3-methyl-1,5-pentanedioate — 3-Methyl-1,5-pentanediol — Methanol

**15.6** Hydroxylation of alkenes using osmium tetraoxide is a syn addition of hydroxyl groups to the double bond. *cis*-2-Butene yields the meso diol.

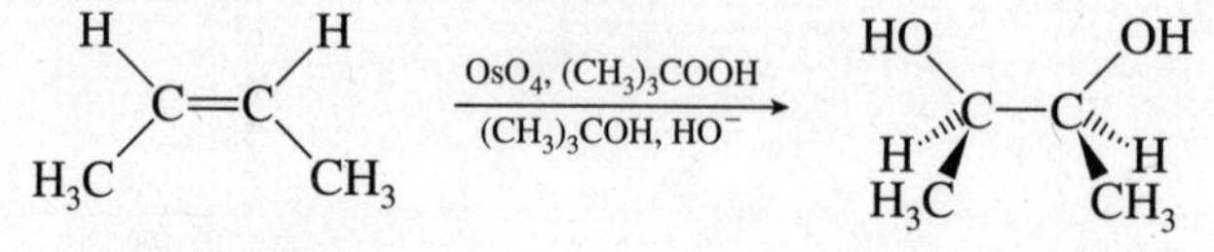

$$\text{cis-2-Butene} \xrightarrow[(CH_3)_3COH,\ HO^-]{OsO_4,\ (CH_3)_3COOH} \text{meso-2,3-Butanediol}$$

*cis*-2-Butene — *meso*-2,3-Butanediol

*trans*-2-Butene yields a racemic mixture of the two enantiomeric forms of the chiral diol.

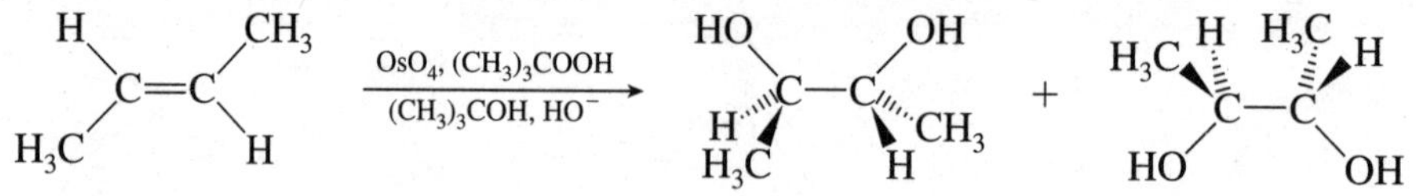

$$\text{trans-2-Butene} \xrightarrow[(CH_3)_3COH,\ HO^-]{OsO_4,\ (CH_3)_3COOH} (2R,3R)\text{-2,3-Butanediol} + (2S,3S)\text{-2,3-Butanediol}$$

*trans*-2-Butene — (2*R*,3*R*)-2,3-Butanediol — (2*S*,3*S*)-2,3-Butanediol

The Fischer projection formulas of the three stereoisomers are

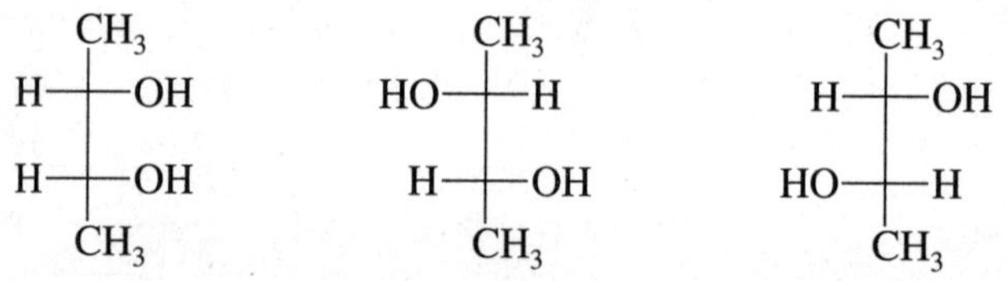

| *meso*-2,3-Butanediol | (2*R*,3*R*)-2,3-Butanediol | (2*S*,3*S*)-2,3-Butanediol |
|---|---|---|
| $CH_3$ | $CH_3$ | $CH_3$ |
| H—C—OH | HO—C—H | H—C—OH |
| H—C—OH | H—C—OH | HO—C—H |
| $CH_3$ | $CH_3$ | $CH_3$ |

**15.7** The first step is proton transfer to 1,5-pentanediol to form the corresponding alkyloxonium ion.

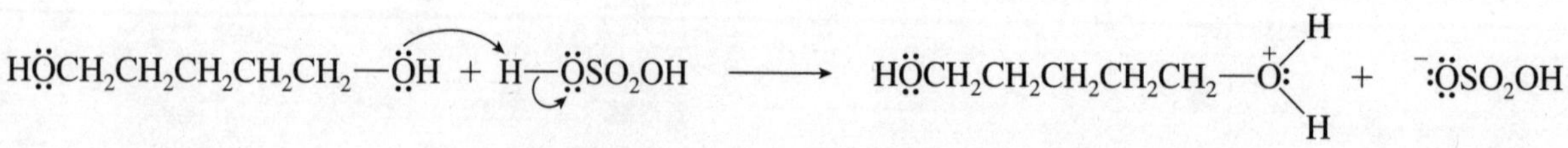

$$H\ddot{O}CH_2CH_2CH_2CH_2CH_2{-}\ddot{O}H + H{-}\ddot{O}SO_2OH \longrightarrow H\ddot{O}CH_2CH_2CH_2CH_2CH_2{-}\overset{+}{O}H_2 + {}^-{:}\ddot{O}SO_2OH$$

1,5-Pentanediol — Sulfuric acid — Conjugate acid of 1,5-pentanediol — Hydrogen sulfate

Rewriting the alkyloxonium ion gives

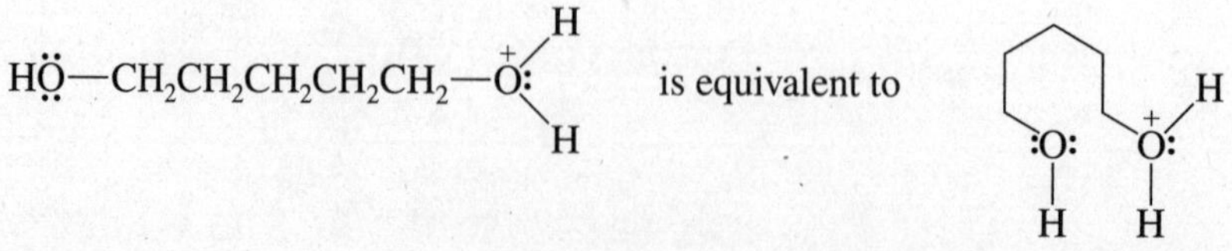

$$H\ddot{O}{-}CH_2CH_2CH_2CH_2CH_2{-}\overset{+}{O}H_2$$ is equivalent to a ring-drawn form with :O:–H and $\overset{+}{O}$:–H ends.

The oxonium ion undergoes cyclization by intramolecular nucleophilic attack of its alcohol function on the carbon that bears the leaving group.

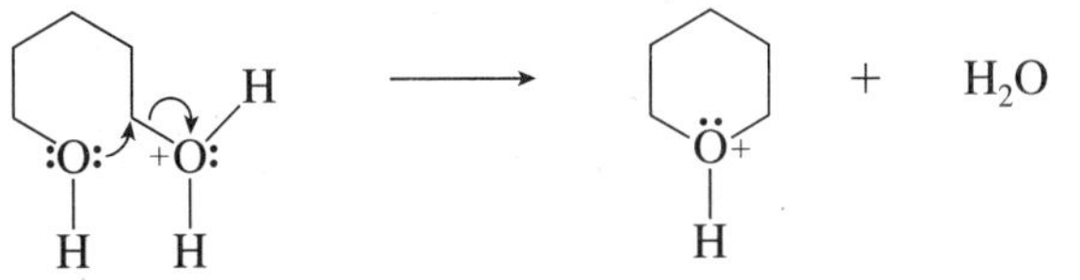

Conjugate acid of 1,5-pentanediol → Conjugate acid of oxane + $H_2O$ (Water)

Loss of a proton gives oxane.

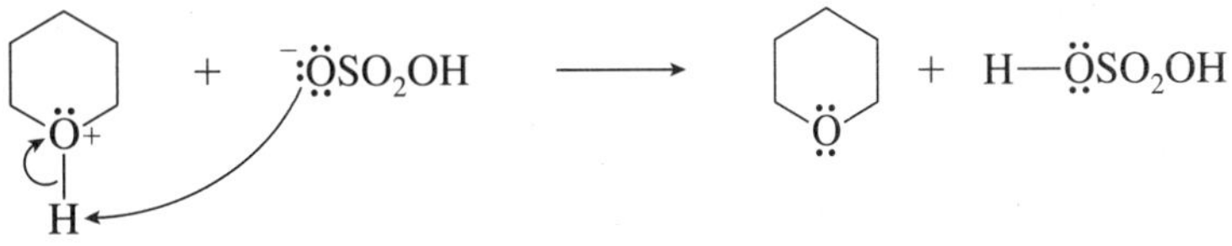

Conjugate acid of oxane + $^{-}:\ddot{O}SO_2OH$ (Hydrogen sulfate) → Oxane + $H-\ddot{O}SO_2OH$ (Sulfuric acid)

**15.8** (*b*) The relationship of the molecular formula of the ester ($C_{10}H_{10}O_4$) to that of the starting dicarboxylic acid ($C_8H_6O_4$) indicates that the diacid reacted with 2 moles of methanol to form a diester.

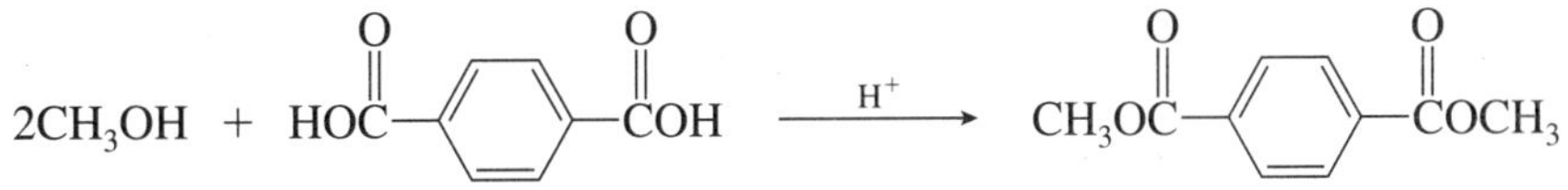

$$2CH_3OH + HOC(=O)-C_6H_4-C(=O)OH \xrightarrow{H^+} CH_3OC(=O)-C_6H_4-C(=O)OCH_3$$

Methanol — 1,4-Benzenedicarboxylic acid — Dimethyl 1,4-benzenedicarboxylate

**15.9** While neither *cis*- nor *trans*-4-*tert*-butylcyclohexanol is a chiral molecule, the stereochemical course of their reactions with acetic anhydride becomes evident when the relative stereochemistry of the ester function is examined for each case. The cis alcohol yields the cis acetate.

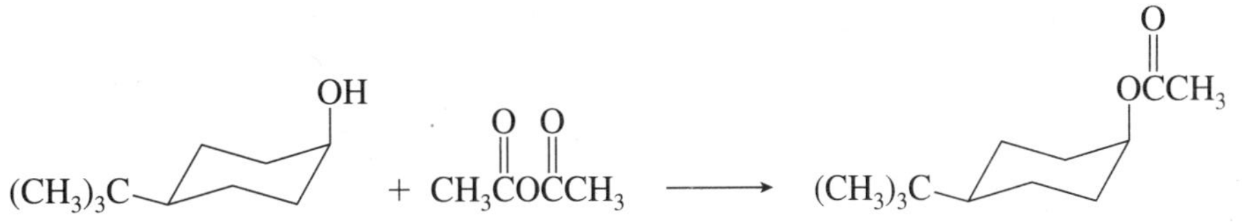

*cis*-4-*tert*-Butylcyclohexanol + $CH_3COCCH_3$ (Acetic anhydride) → *cis*-4-*tert*-Butylcyclohexyl acetate

The trans alcohol yields the trans acetate.

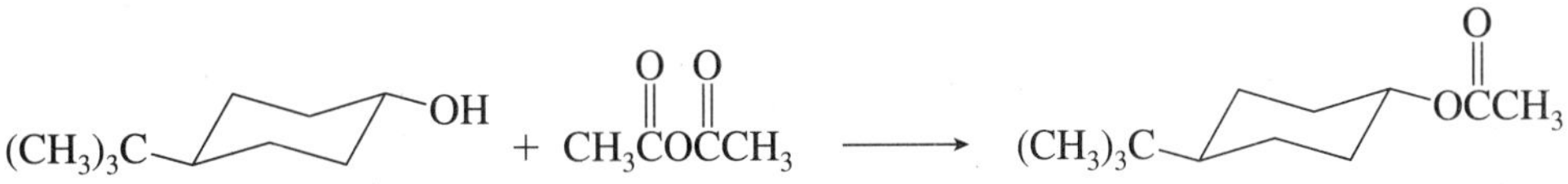

*trans*-4-*tert*-Butylcyclohexanol + $CH_3COCCH_3$ (Acetic anhydride) → *trans*-4-*tert*-Butylcyclohexyl acetate

**15.10** Glycerol has three hydroxyl groups, each of which is converted to a nitrate ester function in nitroglycerin.

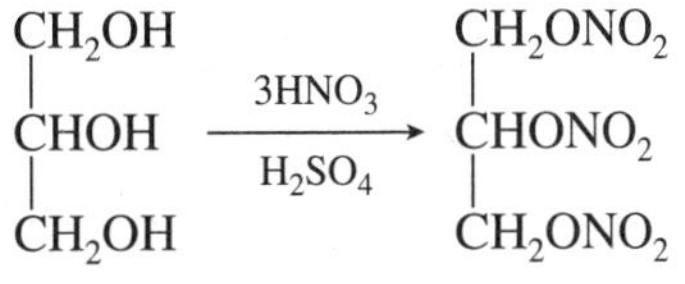

$$CH_2OH-CHOH-CH_2OH \xrightarrow[H_2SO_4]{3HNO_3} CH_2ONO_2-CHONO_2-CH_2ONO_2$$

Glycerol — Nitroglycerin

**15.11** (*b*) The substrate is a secondary alcohol and so gives a ketone on oxidation with sodium dichromate. 2-Octanone has been prepared in 92–96% yield under these reaction conditions.

$$CH_3CH(OH)(CH_2)_5CH_3 \xrightarrow[H_2SO_4,\ H_2O]{Na_2Cr_2O_7} CH_3C(=O)(CH_2)_5CH_3$$

2-Octanol → 2-Octanone

(*c*) The alcohol is primary, and so oxidation can produce either an aldehyde or a carboxylic acid, depending on the reaction conditions. Here the oxidation is carried out under anhydrous conditions using pyridinium chlorochromate (PCC), and the product is the corresponding aldehyde.

$$CH_3CH_2CH_2CH_2CH_2CH_2CH_2OH \xrightarrow[CH_2Cl_2]{PCC} CH_3CH_2CH_2CH_2CH_2CH_2CH{=}O$$

1-Heptanol → Heptanal

**15.12** (*b*) Biological oxidation of $CH_3CD_2OH$ leads to loss of one of the C-1 deuterium atoms to $NAD^+$. The dihydropyridine ring of the reduced form of the coenzyme will bear a single deuterium.

$$CH_3CD_2OH + NAD^+ \xrightarrow{\text{alcohol dehydrogenase}} CH_3CD{=}O + NADD + H^+$$

1,1-Dideuterio-ethanol, $NAD^+$ → 1-Deuterio-ethanal, NADD

(*c*) The deuterium atom of $CH_3CH_2OD$ is lost as $D^+$. The reduced form of the coenzyme contains no deuterium.

$$CH_3CH_2OD + NAD^+ \xrightarrow{\text{alcohol dehydrogenase}} CH_3CH{=}O + NADH + D^+$$

Ethanol-*O*-d, $NAD^+$ → Ethanal, NADH

**15.13** (*b*) Oxidation of the carbon–oxygen bonds to carbonyl groups accompanies their cleavage.

$$(CH_3)_2CHCH_2CH(OH){-}CH(OH)CH_2C_6H_5 \xrightarrow{HIO_4} (CH_3)_2CHCH_2CH{=}O + O{=}CHCH_2C_6H_5$$

1-Phenyl-5-methyl-2,3-hexanediol → 3-Methylbutanal + 2-Phenylethanal

(*c*) The $CH_2OH$ group is cleaved from the ring as formaldehyde to leave cyclopentanone.

$$\text{1-(hydroxymethyl)cyclopentanol} \xrightarrow{HIO_4} \text{cyclopentanone} + HCH{=}O$$

1-(Hydroxymethyl)-cyclopentanol → Cyclopentanone + Formaldehyde

**15.14** Thiols may be prepared from the corresponding alkyl halide by reaction with thiourea followed by treatment of the isothiouronium salt with base.

$$\underset{\text{Alkyl bromide}}{RBr} + \underset{\text{Thiourea}}{(H_2N)_2C{=}S} \longrightarrow \text{Isothiouronium salt (not isolated)} \xrightarrow{NaOH} \underset{\text{Thiol}}{RSH}$$

Thus, an acceptable synthesis of 1-hexanethiol from 1-hexanol would be

$$\underset{\text{1-Hexanol}}{CH_3(CH_2)_4CH_2OH} \xrightarrow[\text{HBr, heat}]{PBr_3} \underset{\text{1-Bromohexane}}{CH_3(CH_2)_4CH_2Br} \xrightarrow[\text{2. NaOH}]{\text{1. }(H_2N)_2C=S} \underset{\text{1-Hexanethiol}}{CH_3(CH_2)_4CH_2SH}$$

**15.15** The three main components of "essence of skunk" are

$$\underset{\text{3-Methyl-1-butanethiol}}{CH_3\overset{CH_3}{\overset{|}{C}}HCH_2CH_2SH}$$

*trans*-2-Butene-1-thiol: $H_3C(H)C{=}C(H)CH_2SH$ (H₃C and CH₂SH trans)

*cis*-2-Butene-1-thiol: $H_3C(H)C{=}C(H)CH_2SH$ (H₃C and CH₂SH cis)

**15.16** The molecular weight of 2-methyl-2-butanol is 88. A peak in its mass spectrum at $m/z$ 70 corresponds to loss of water from the molecular ion. The peaks at $m/z$ 73 and $m/z$ 59 represent stable cations corresponding to the cleavages shown in the equation.

$$CH_3-\overset{+\cdot\ddot{O}H}{\overset{|}{\underset{CH_3}{\underset{|}{C}}}}-CH_2CH_3$$

$$\cdot CH_3 + \underset{m/z\ 73}{CH_3\overset{+\ddot{O}H}{\overset{\|}{C}}CH_2CH_3} \qquad\qquad \underset{m/z\ 59}{CH_3\overset{+\ddot{O}H}{\overset{\|}{C}}CH_3} + \cdot CH_2CH_3$$

**15.17** (*a*) The appropriate alkene for the preparation of 1-butanol by a hydroboration–oxidation sequence is 1-butene. Remember, hydroboration–oxidation leads to hydration of alkenes with a regioselectivity opposite to that seen in acid-catalyzed hydration.

$$\underset{\text{1-Butene}}{CH_3CH_2CH{=}CH_2} \xrightarrow[\text{2. }H_2O_2,\ HO^-]{\text{1. }B_2H_6} \underset{\text{1-Butanol}}{CH_3CH_2CH_2CH_2OH}$$

(*b*) 1-Butanol can be prepared by reaction of a Grignard reagent with formaldehyde.

$$CH_3CH_2CH_2CH_2OH \Longrightarrow CH_3CH_2\overset{-}{\ddot{C}}H_2 + H\overset{O}{\overset{\|}{C}}H$$

An appropriate Grignard reagent is propylmagnesium bromide.

$$\underset{\text{1-Bromopropane}}{CH_3CH_2CH_2Br} \xrightarrow[\text{diethyl ether}]{Mg} \underset{\text{Propylmagnesium bromide}}{CH_3CH_2CH_2MgBr}$$

$$CH_3CH_2CH_2MgBr + H\overset{O}{\overset{\|}{C}}H \xrightarrow[\text{2. }H_3O^+]{\text{1. diethyl ether}} \underset{\text{1-Butanol}}{CH_3CH_2CH_2CH_2OH}$$

(*c*) Alternatively, 1-butanol may be prepared by the reaction of a Grignard reagent with ethylene oxide.

$$CH_3CH_2CH_2CH_2OH \Longrightarrow CH_3\ddot{C}H_2 + H_2C\text{—}CH_2 \text{ (bridged by O)}$$

In this case, ethylmagnesium bromide would be used.

$$\underset{\text{Ethyl bromide}}{CH_3CH_2Br} \xrightarrow[\text{diethyl ether}]{\text{Mg}} \underset{\text{Ethylmagnesium bromide}}{CH_3CH_2MgBr}$$

$$CH_3CH_2MgBr + \underset{\text{Ethylene oxide}}{H_2C\text{—}CH_2 \text{ (bridged by O)}} \xrightarrow[\text{2. } H_3O^+]{\text{1. diethyl ether}} \underset{\text{1-Butanol}}{CH_3CH_2CH_2CH_2OH}$$

(*d*) Primary alcohols may be prepared by reduction of the carboxylic acid having the same number of carbons. Among the reagents we have discussed, the only one that is effective in the reduction of carboxylic acids is lithium aluminum hydride. The four-carbon carboxylic acid butanoic acid is the proper substrate.

$$\underset{\text{Butanoic acid}}{CH_3CH_2CH_2\overset{O}{\overset{\|}{C}}OH} \xrightarrow[\text{2. } H_2O]{\text{1. } LiAlH_4\text{, diethyl ether}} \underset{\text{1-Butanol}}{CH_3CH_2CH_2CH_2OH}$$

(*e*) Reduction of esters can be accomplished using lithium aluminum hydride. The correct methyl ester is methyl butanoate.

$$\underset{\text{Methyl butanoate}}{CH_3CH_2CH_2\overset{O}{\overset{\|}{C}}OCH_3} \xrightarrow[\text{2. } H_2O]{\text{1. } LiAlH_4} \underset{\text{1-Butanol}}{CH_3CH_2CH_2CH_2OH} + \underset{\text{Methanol}}{CH_3OH}$$

(*f*) A butyl ester such as butyl acetate may be reduced with lithium aluminum hydride to prepare 1-butanol.

$$\underset{\text{Butyl acetate}}{CH_3\overset{O}{\overset{\|}{C}}OCH_2CH_2CH_2CH_3} \xrightarrow[\text{2. } H_2O]{\text{1. } LiAlH_4} \underset{\text{1-Butanol}}{CH_3CH_2CH_2CH_2OH} + \underset{\text{Ethanol}}{CH_3CH_2OH}$$

(*g*) Because 1-butanol is a primary alcohol having four carbons, butanal must be the aldehyde that is hydrogenated. Suitable catalysts are nickel, palladium, platinum, and ruthenium.

$$\underset{\text{Butanal}}{CH_3CH_2CH_2\overset{O}{\overset{\|}{C}}H} \xrightarrow{H_2\text{, Pt}} \underset{\text{1-Butanol}}{CH_3CH_2CH_2CH_2OH}$$

(*h*) Sodium borohydride reduces aldehydes and ketones efficiently. It does not reduce carboxylic acids, and its reaction with esters is too slow to be of synthetic value.

$$\underset{\text{Butanal}}{CH_3CH_2CH_2\overset{O}{\overset{\|}{C}}H} \xrightarrow[\text{water, ethanol, or methanol}]{NaBH_4} \underset{\text{1-Butanol}}{CH_3CH_2CH_2CH_2OH}$$

**15.18** (*a*) Both (*Z*)- and (*E*)-2-butene yield 2-butanol on hydroboration–oxidation.

$$\underset{(Z)\text{- or }(E)\text{-2-butene}}{CH_3CH{=}CHCH_3} \xrightarrow[\text{2. } H_2O_2,\ HO^-]{\text{1. } B_2H_6} \underset{\text{2-Butanol}}{CH_3\underset{\displaystyle OH}{\underset{|}{C}}HCH_2CH_3}$$

(*b*) Disconnection of one of the bonds to the carbon that bears the hydroxyl group reveals a feasible route using a Grignard reagent and propanal.

Disconnect this bond.

$$H_3C \nmid \underset{\displaystyle OH}{\underset{|}{C}}HCH_2CH_3 \Longrightarrow {}^-{:}CH_3 + \underset{\text{Propanal}}{H\overset{\displaystyle O}{\overset{\|}{C}}CH_2CH_3}$$

The synthetic sequence is

$$\underset{\text{Methyl bromide}}{CH_3Br} \xrightarrow[\text{diethyl ether}]{Mg} \underset{\text{Methylmagnesium bromide}}{CH_3MgBr} \xrightarrow[\text{2. } H_3O^+]{\text{1. } CH_3CH_2\overset{O}{\overset{\|}{C}}H} \underset{\text{2-Butanol}}{CH_3\underset{\displaystyle OH}{\underset{|}{C}}HCH_2CH_3}$$

(*c*) Another disconnection is related to a synthetic route using a Grignard reagent and acetaldehyde.

Disconnect this bond.

$$CH_3\underset{\displaystyle OH}{\underset{|}{C}}H \nmid CH_2CH_3 \Longrightarrow \underset{\text{Acetaldehyde}}{CH_3\overset{\displaystyle O}{\overset{\|}{C}}H} + CH_3\ddot{C}H_2$$

$$\underset{\text{Ethyl bromide}}{CH_3CH_2Br} \xrightarrow[\text{diethyl ether}]{Mg} \underset{\text{Ethylmagnesium bromide}}{CH_3CH_2MgBr} \xrightarrow[\text{2. } H_3O^+]{\text{1. } CH_3\overset{O}{\overset{\|}{C}}H} \underset{\text{2-Butanol}}{CH_3CH_2\underset{\displaystyle OH}{\underset{|}{C}}HCH_3}$$

(*d–f*) Because 2-butanol is a secondary alcohol, it can be prepared by reduction of a ketone having the same carbon skeleton, in this case 2-butanone. All three reducing agents indicated in the equations are satisfactory.

$$\underset{\text{2-Butanone}}{CH_3\overset{\displaystyle O}{\overset{\|}{C}}CH_2CH_3} \xrightarrow[\text{(or Pt, Ni, Ru)}]{H_2,\ Pd} \underset{\text{2-Butanol}}{CH_3\underset{\displaystyle OH}{\underset{|}{C}}HCH_2CH_3}$$

$$\underset{\text{2-Butanone}}{CH_3\overset{\displaystyle O}{\overset{\|}{C}}CH_2CH_3} \xrightarrow[CH_3OH]{NaBH_4} \underset{\text{2-Butanol}}{CH_3\underset{\displaystyle OH}{\underset{|}{C}}HCH_2CH_3}$$

$$\underset{\text{2-Butanone}}{CH_3\overset{\displaystyle O}{\overset{\|}{C}}CH_2CH_3} \xrightarrow[\text{2. } H_2O]{\text{1. } LiAlH_4} \underset{\text{2-Butanol}}{CH_3\underset{\displaystyle OH}{\underset{|}{C}}HCH_2CH_3}$$

**15.19** (*a*) All the carbon–carbon disconnections are equivalent.

$$H_3C-\overset{CH_3}{\underset{CH_3}{C}}-OH \Longrightarrow {}^{-}{:}CH_3 + CH_3\overset{O}{\overset{\|}{C}}CH_3$$

Acetone

The synthesis via a Grignard reagent and acetone is

$$\underset{\text{Methyl bromide}}{CH_3Br} \xrightarrow[\text{diethyl ether}]{Mg} \underset{\text{Methylmagnesium bromide}}{CH_3MgBr} \xrightarrow[2.\ H_3O^+]{1.\ CH_3\overset{O}{\overset{\|}{C}}CH_3} \underset{\textit{tert}\text{-Butyl alcohol}}{(CH_3)_3COH}$$

(*b*) An alternative route to *tert*-butyl alcohol is addition of a Grignard reagent to an ester. Esters react with *2 moles* of Grignard reagent. Thus, *tert*-butyl alcohol may be formed by reacting methyl acetate with 2 moles of methylmagnesium iodide. Methyl alcohol is formed as a by-product of the reaction.

$$\underset{\text{Methylmagnesium iodide}}{2CH_3MgI} + \underset{\text{Methyl acetate}}{CH_3\overset{O}{\overset{\|}{C}}OCH_3} \xrightarrow[2.\ H_3O^+]{1.\ \text{diethyl ether}} \underset{\textit{tert}\text{-Butyl alcohol}}{CH_3-\overset{CH_3}{\underset{CH_3}{C}}-OH} + \underset{\text{Methyl alcohol}}{CH_3OH}$$

**15.20** (*a*) All of the primary alcohols having the molecular formula $C_5H_{12}O$ may be prepared by reduction of aldehydes. The appropriate equations are

$$\underset{\text{Pentanal}}{CH_3CH_2CH_2CH_2\overset{O}{\overset{\|}{C}}H} \xrightarrow[2.\ H_2O]{1.\ LiAlH_4,\ \text{diethyl ether}} \underset{\text{1-Pentanol}}{CH_3CH_2CH_2CH_2CH_2OH}$$

$$\underset{\text{2-Methylbutanal}}{CH_3CH_2\underset{CH_3}{C}H\overset{O}{\overset{\|}{C}}H} \xrightarrow[2.\ H_2O]{1.\ LiAlH_4,\ \text{diethyl ether}} \underset{\text{2-Methyl-1-butanol}}{CH_3CH_2\underset{CH_3}{C}HCH_2OH}$$

$$\underset{\text{3-Methylbutanal}}{(CH_3)_2CHCH_2\overset{O}{\overset{\|}{C}}H} \xrightarrow[2.\ H_2O]{1.\ LiAlH_4,\ \text{diethyl ether}} \underset{\text{3-Methyl-1-butanol}}{(CH_3)_2CHCH_2CH_2OH}$$

$$\underset{\text{2,2-Dimethylpropanal}}{(CH_3)_3C\overset{O}{\overset{\|}{C}}H} \xrightarrow[2.\ H_2O]{1.\ LiAlH_4,\ \text{diethyl ether}} \underset{\text{2,2-Dimethyl-1-propanol}}{(CH_3)_3CCH_2OH}$$

(*b*) The secondary alcohols having the molecular formula $C_5H_{12}O$ may be prepared by reduction of ketones.

$$\underset{\text{2-Pentanone}}{CH_3CH_2CH_2\overset{O}{\overset{\|}{C}}CH_3} \xrightarrow[\text{2. } H_2O]{\text{1. LiAlH}_4\text{, diethyl ether}} \underset{\text{2-Pentanol}}{CH_3CH_2CH_2\overset{OH}{\overset{|}{C}}HCH_3}$$

$$\underset{\text{3-Pentanone}}{CH_3CH_2\overset{O}{\overset{\|}{C}}CH_2CH_3} \xrightarrow[\text{2. } H_2O]{\text{1. LiAlH}_4\text{, diethyl ether}} \underset{\text{3-Pentanol}}{CH_3CH_2\overset{OH}{\overset{|}{C}}HCH_2CH_3}$$

$$\underset{\text{3-Methyl-2-butanone}}{(CH_3)_2CH\overset{O}{\overset{\|}{C}}CH_3} \xrightarrow[\text{2. } H_2O]{\text{1. LiAlH}_4\text{, diethyl ether}} \underset{\text{3-Methyl-2-butanol}}{(CH_3)_2CH\overset{OH}{\overset{|}{C}}HCH_3}$$

(*c*) As with the reduction of aldehydes in part (*a*), reduction of carboxylic acids yields primary alcohols. For example, 1-pentanol may be prepared by reduction of pentanoic acid.

$$\underset{\text{Pentanoic acid}}{CH_3CH_2CH_2CH_2\overset{O}{\overset{\|}{C}}OH} \xrightarrow[\text{2. } H_2O]{\text{1. LiAlH}_4\text{, diethyl ether}} \underset{\text{1-Pentanol}}{CH_3CH_2CH_2CH_2CH_2OH}$$

The remaining primary alcohols, 2-methyl-1-butanol, 3-methyl-1-butanol, and 2,2-dimethyl-1-propanol, may be prepared in the same way.

(*d*) As with carboxylic acids, esters may be reduced using lithium aluminum hydride to give primary alcohols. For example, 2,2-dimethyl-1-propanol may be prepared by reduction of methyl 2,2-dimethylpropanoate.

$$\underset{\text{Methyl 2,2-dimethylpropanoate}}{(CH_3)_3C\overset{O}{\overset{\|}{C}}OCH_3} \xrightarrow[\text{2. } H_2O]{\text{1. LiAlH}_4\text{, diethyl ether}} \underset{\text{2,2-Dimethyl-1-propanol}}{(CH_3)_3CCH_2OH}$$

**15.21** (*a*) The suggested synthesis

$$\underset{\text{Butane}}{CH_3CH_2CH_2CH_3} \xrightarrow[\text{light or heat}]{Br_2} \underset{\text{1-Bromobutane}}{CH_3CH_2CH_2CH_2Br} \xrightarrow{KOH} \underset{\text{1-Butanol}}{CH_3CH_2CH_2CH_2OH}$$

is a poor one because bromination of butane yields a mixture of 1-bromobutane and 2-bromobutane, 2-bromobutane being the major product.

$$\underset{\text{Butane}}{CH_3CH_2CH_2CH_3} \xrightarrow[\text{light or heat}]{Br_2} \underset{\substack{\text{1-Bromobutane}\\ \text{(minor product)}}}{CH_3CH_2CH_2CH_2Br} + \underset{\substack{\text{2-Bromobutane}\\ \text{(major product)}}}{CH_3\underset{Br}{\underset{|}{C}}HCH_2CH_3}$$

(*b*) The suggested synthesis

$$(CH_3)_3CH \xrightarrow[\text{light or heat}]{Br_2} (CH_3)_3CBr \xrightarrow{KOH} (CH_3)_3COH$$

2-Methylpropane — 2-Bromo-2-methylpropane — 2-Methyl-2-propanol

will fail because the reaction of 2-bromo-2-methylpropane with potassium hydroxide will proceed by elimination rather than by substitution. The first step in the process, selective bromination of 2-methylpropane to 2-bromo-2-methylpropane, is satisfactory because bromination is selective for substitution of tertiary hydrogens in the presence of secondary and primary ones.

(*c*) Benzyl alcohol, unlike 1-butanol and 2-methyl-2-propanol, can be prepared effectively by this method.

$$C_6H_5CH_3 \xrightarrow[\text{light or heat}]{Br_2} C_6H_5CH_2Br \xrightarrow{KOH} C_6H_5CH_2OH$$

Toluene — Benzyl bromide — Benzyl alcohol

Free-radical bromination of toluene is selective for the benzylic position. Benzyl bromide cannot undergo elimination, and so nucleophilic substitution of bromide by hydroxide will work well.

(*d*) The desired transformation

$$C_6H_5CH_2CH_3 \xrightarrow[\text{light or heat}]{Br_2} C_6H_5CH(Br)CH_3 \xrightarrow{KOH} C_6H_5CH(OH)CH_3$$

Ethylbenzene — 1-Bromo-1-phenylethane — 1-Phenylethanol

fails because it produces more than one enantiomer. The reactant ethylbenzene is achiral and although its bromination will be highly regioselective for the benzylic position, the product will be a racemic mixture of (*R*) and (*S*)-1-bromo-1-phenylethane. The alcohol produced by hydrolysis will also be racemic. Furthermore, the hydrolysis step will give mostly styrene by an E2 elimination, rather than 1-phenylethanol by nucleophilic substitution.

**15.22** Glucose contains five hydroxyl groups and an aldehyde functional group. Its hydrogenation will not affect the hydroxyl groups but will reduce the aldehyde to a primary alcohol.

$$\text{Glucose} \xrightarrow[\text{Ni, 140°C}]{H_2\ (120\ \text{atm})} \text{Sorbitol}$$

Glucose — Sorbitol

**15.23** (*a*) 1-Phenylethanol is a secondary alcohol and so can be prepared by the reaction of a Grignard reagent with an aldehyde. One combination is phenylmagnesium bromide and ethanal (acetaldehyde).

$$C_6H_5CH(OH)CH_3 \Rightarrow C_6H_5MgBr + HC(=O)CH_3$$

1-Phenylethanol — Phenylmagnesium bromide — Ethanal (acetaldehyde)

Grignard reagents—phenylmagnesium bromide in this case—are always prepared by reaction of magnesium metal and the corresponding halide. Starting with bromobenzene, a suitable synthesis is described by the sequence

$$\underset{\text{Bromobenzene}}{C_6H_5Br} \xrightarrow[\text{diethyl ether}]{Mg} \underset{\text{Phenylmagnesium bromide}}{C_6H_5MgBr} \xrightarrow[2.\ H_3O^+]{1.\ CH_3CH{=}O} \underset{\text{1-Phenylethanol}}{C_6H_5CH(OH)CH_3}$$

(*b*) An alternative disconnection of 1-phenylethanol reveals a second route using benzaldehyde and a methyl Grignard reagent.

$$\underset{\text{1-Phenylethanol}}{C_6H_5CH(OH)CH_3} \Longrightarrow \underset{\text{Benzaldehyde}}{C_6H_5CH{=}O} + \underset{\text{Methylmagnesium iodide}}{CH_3MgI}$$

Equations representing this approach are

$$\underset{\text{Iodomethane}}{CH_3I} \xrightarrow[\text{diethyl ether}]{Mg} \underset{\text{Methylmagnesium iodide}}{CH_3MgI} \xrightarrow[2.\ H_3O^+]{1.\ C_6H_5CH{=}O} \underset{\text{1-Phenylethanol}}{C_6H_5CH(OH)CH_3}$$

(*c*) Aldehydes are, in general, obtainable by oxidation of the corresponding primary alcohol. By recognizing that benzaldehyde can be obtained by oxidation of benzyl alcohol with PCC, we write

$$\underset{\text{Benzyl alcohol}}{C_6H_5CH_2OH} \xrightarrow[CH_2Cl_2]{PCC} \underset{\text{Benzaldehyde}}{C_6H_5CH{=}O} \xrightarrow[2.\ H_3O^+]{1.\ CH_3MgI,\ \text{diethyl ether}} \underset{\text{1-Phenylethanol}}{C_6H_5CH(OH)CH_3}$$

(*d*) The conversion of acetophenone to 1-phenylethanol is a reduction.

$$\underset{\text{Acetophenone}}{C_6H_5C({=}O)CH_3} \xrightarrow{\text{reducing agent}} \underset{\text{1-Phenylethanol}}{C_6H_5CH(OH)CH_3}$$

Any of a number of reducing agents could be used. These include

**1.** $NaBH_4$, $CH_3OH$

**2.** $LiAlH_4$ in diethyl ether, then $H_2O$

**3.** $H_2$ and a Pt, Pd, Ni, or Ru catalyst

(*e*) Benzene can be employed as the ultimate starting material in a synthesis of 1-phenylethanol. Friedel–Crafts acylation of benzene gives acetophenone, which can then be reduced as in part (*d*).

$$\text{C}_6\text{H}_6 + \text{CH}_3\overset{\overset{\text{O}}{\|}}{\text{C}}\text{Cl} \xrightarrow{\text{AlCl}_3} \text{C}_6\text{H}_5\overset{\overset{\text{O}}{\|}}{\text{C}}\text{CH}_3$$

Benzene Acetyl chloride Acetophenone

Acetic anhydride ($CH_3\overset{O}{\overset{\|}{C}}O\overset{O}{\overset{\|}{C}}CH_3$) can be used in place of acetyl chloride.

**15.24** 2-Phenylethanol is an ingredient in many perfumes, to which it imparts a rose-like fragrance. Numerous methods have been employed for its synthesis.

(*a*) As a primary alcohol having two more carbon atoms than bromobenzene, it can be formed by reaction of a Grignard reagent, phenylmagnesium bromide, with ethylene oxide.

$$\text{C}_6\text{H}_5\text{CH}_2\text{CH}_2\text{OH} \Longrightarrow \text{C}_6\text{H}_5\text{MgBr} + \text{H}_2\text{C}\overset{\text{O}}{\text{—}}\text{CH}_2$$

The desired reaction sequence is therefore

$$\text{C}_6\text{H}_5\text{Br} \xrightarrow[\text{diethyl ether}]{\text{Mg}} \text{C}_6\text{H}_5\text{MgBr} \xrightarrow[\text{2. H}_3\text{O}^+]{\text{1. H}_2\text{C—CH}_2\text{ (O)}} \text{C}_6\text{H}_5\text{CH}_2\text{CH}_2\text{OH}$$

Bromobenzene Phenylmagnesium bromide 2-Phenylethanol

(*b*) Hydration of sytrene with a regioselectivity contrary to that of Markovnikov's rule is required. This is accomplished readily by hydroboration–oxidation.

$$\text{C}_6\text{H}_5\text{CH}{=}\text{CH}_2 \xrightarrow[\text{2. H}_2\text{O}_2\text{, HO}^-]{\text{1. B}_2\text{H}_6\text{, diglyme}} \text{C}_6\text{H}_5\text{CH}_2\text{CH}_2\text{OH}$$

Styrene 2-Phenylethanol

(*c*) Reduction of aldehydes yields primary alcohols.

$$\text{C}_6\text{H}_5\text{CH}_2\overset{\overset{\text{O}}{\|}}{\text{C}}\text{H} \xrightarrow{\text{reducing agent}} \text{C}_6\text{H}_5\text{CH}_2\text{CH}_2\text{OH}$$

2-Phenylethanal 2-Phenylethanol

Among the reducing agents that could be (and have been) used are

**1.** $NaBH_4$, $CH_3OH$
**2.** $LiAlH_4$ in diethyl ether, then $H_2O$
**3.** $H_2$ and a Pt, Pd, Ni, or Ru catalyst

(*d*) Esters are readily reduced to primary alcohols with lithium aluminum hydride.

$$\text{C}_6\text{H}_5\text{CH}_2\overset{\overset{\text{O}}{\|}}{\text{C}}\text{OCH}_2\text{CH}_3 \xrightarrow[\text{2. H}_2\text{O}]{\text{1. LiAlH}_4\text{, diethyl ether}} \text{C}_6\text{H}_5\text{CH}_2\text{CH}_2\text{OH}$$

Ethyl 2-phenylethanoate 2-Phenylethanol

(*e*) The only reagent that is suitable for the direct reduction of carboxylic acids to primary alcohols is lithium aluminum hydride.

$$C_6H_5CH_2\overset{O}{\overset{\|}{C}}OH \xrightarrow[2.\ H_2O]{1.\ LiAlH_4,\ \text{diethyl ether}} C_6H_5CH_2CH_2OH$$

2-Phenylethanoic acid → 2-Phenylethanol

Alternatively, the carboxylic acid could be esterified with ethanol and the resulting ethyl 2-phenylethanoate reduced.

$$C_6H_5CH_2\overset{O}{\overset{\|}{C}}OH + CH_3CH_2OH \xrightarrow{H^+} C_6H_5CH_2\overset{O}{\overset{\|}{C}}OCH_2CH_3 \xrightarrow{\text{reduce as in part } (d)} C_6H_5CH_2CH_2OH$$

2-Phenylethanoic acid + Ethanol → Ethyl 2-phenylethanoate → 2-Phenylethanol

**15.25** (*a*) Thiols are made from alkyl halides by reaction with thiourea, followed by hydrolysis of the isothiouronium salt in base. The first step must therefore be a conversion of the alcohol to an alkyl bromide.

$$CH_3CH_2CH_2CH_2OH \xrightarrow[\text{or } PBr_3]{HBr} CH_3CH_2CH_2CH_2Br \xrightarrow[2.\ NaOH]{1.\ (H_2N)_2C{=}S} CH_3CH_2CH_2CH_2SH$$

1-Butanol → 1-Bromobutane → 1-Butanethiol

(*b*) To obtain 1-hexanol from alcohols having four carbons or fewer, a two-carbon chain extension must be carried out. This suggests reaction of a Grignard reagent with ethylene oxide. The retrosynthetic path for this approach is

$$CH_3CH_2CH_2CH_2 \nmid CH_2CH_2OH \Rightarrow CH_3CH_2CH_2CH_2MgBr + H_2C\overset{O}{—}CH_2$$

The reaction sequence therefore becomes

$$CH_3CH_2CH_2CH_2Br \xrightarrow[\text{diethyl ether}]{Mg} CH_3CH_2CH_2CH_2MgBr \xrightarrow[2.\ H_3O^+]{1.\ H_2C\underset{O}{—}CH_2} CH_3CH_2CH_2CH_2CH_2CH_2OH$$

1-Bromobutane from part (*a*) → Butylmagnesium bromide → 1-Hexanol

Given the constraints of the problem, we prepare ethylene oxide by the sequence

$$CH_3CH_2OH \xrightarrow[\text{heat}]{H_2SO_4} H_2C{=}CH_2 \xrightarrow{CH_3\overset{O}{\overset{\|}{C}}OOH} H_2C\underset{O}{—}CH_2$$

Ethanol → Ethylene

(*c*) The target molecule 2-hexanol may be mentally disconnected as shown to a four-carbon unit and a two-carbon unit.

$$CH_3\underset{OH}{\underset{|}{CH}} \nmid CH_2CH_2CH_2CH_3 \Rightarrow CH_3\overset{O}{\overset{\|}{C}}H + {}^-{:}CH_2CH_2CH_2CH_3$$

The alternative disconnection to $^{-}{:}CH_3$ and $HCCH_2CH_2CH_2CH_3$ (with C=O) reveals a plausible approach to 2-hexanol but is inconsistent with the requirement of the problem that limits starting materials to four carbons or fewer. The five-carbon aldehyde would have to be prepared first, making for a lengthy overall synthetic scheme.

An appropriate synthesis based on alcohols as starting materials is

$$\underset{\text{Ethanol}}{CH_3CH_2OH} \xrightarrow[CH_2Cl]{PCC} \underset{\text{Ethanal}}{CH_3\overset{O}{\overset{\|}{C}}H}$$

$$\underset{\substack{\text{Butylmagnesium bromide}\\ \text{from part }(b)}}{CH_3CH_2CH_2CH_2MgBr} + \underset{\text{Ethanal}}{CH_3\overset{O}{\overset{\|}{C}}H} \xrightarrow[2.\ H_3O^+]{1.\ \text{diethyl ether}} \underset{\text{2-Hexanol}}{CH_3\underset{OH}{\underset{|}{C}}HCH_2CH_2CH_2CH_3}$$

(*d*) Hexanal may be obtained from 1-hexanol [prepared in part (*b*)] by oxidation in dichloromethane using pyridinium chlorochromate (PCC) or pyridinium dichromate (PDC).

$$\underset{\text{1-Hexanol from part }(b)}{CH_3(CH_2)_4CH_2OH} \xrightarrow[CH_2Cl_2]{\text{PCC or PDC}} \underset{\text{Hexanal}}{CH_3(CH_2)_4\overset{O}{\overset{\|}{C}}H}$$

(*e*) Oxidation of 2-hexanol from part (*c*) yields 2-hexanone.

$$\underset{\text{2-Hexanol}}{CH_3\underset{OH}{\underset{|}{C}}HCH_2CH_2CH_2CH_3} \xrightarrow[H_2SO_4,\ H_2O]{Na_2Cr_2O_7} \underset{\text{2-Hexanone}}{CH_3\overset{O}{\overset{\|}{C}}CH_2CH_2CH_2CH_3}$$

PCC or PDC can also be used for this transformation.

(*f*) Oxidation of 1-hexanol with chromic acid (sodium or potassium dichromate in aqueous sulfuric acid) yields hexanoic acid. Use of PDC or PCC in dichloromethane is not acceptable because those reagents yield aldehydes on reaction with primary alcohols.

$$\underset{\text{1-Hexanol from part }(b)}{CH_3(CH_2)_4CH_2OH} \xrightarrow[H_2SO_4,\ H_2O]{K_2Cr_2O_7} \underset{\text{Hexanoic acid}}{CH_3(CH_2)_4CO_2H}$$

(*g*) Fischer esterification of hexanoic acid with ethanol produces ethyl hexanoate.

$$\underset{\substack{\text{Hexanoic acid}\\ \text{from part }(f)}}{CH_3(CH_2)_4CO_2H} + \underset{\text{Ethanol}}{CH_3CH_2OH} \xrightarrow{H^+} \underset{\text{Ethyl hexanoate}}{CH_3(CH_2)_4\overset{O}{\overset{\|}{C}}OCH_2CH_3}$$

(*h*) Vicinal diols are normally prepared by hydroxylation of alkenes with osmium tetraoxide and *tert*-butyl hydroperoxide.

$$\underset{\text{2-Methylpropene}}{(CH_3)_2C{=}CH_2} \xrightarrow[\substack{(CH_3)_3COOH,\ HO^-\\ (CH_3)_3COH}]{OsO_4} \underset{\text{2-Methyl-1,2-propanediol}}{(CH_3)_2\underset{OH}{\underset{|}{C}}CH_2OH}$$

The required alkene is available by dehydration of 2-methyl-2-propanol.

$$(CH_3)_3COH \xrightarrow[\text{heat}]{H_3PO_4} (CH_3)_2C{=}CH_2$$

2-Methyl-2-propanol → 2-Methylpropene

(*i*) The desired aldehyde can be prepared by oxidation of the corresponding primary alcohol with PCC or PDC.

$$(CH_3)_3CCH_2OH \xrightarrow[CH_2Cl_2]{\text{PCC or PDC}} (CH_3)_3C\overset{O}{\overset{\|}{C}}H$$

2,2-Dimethyl-1-propanol → 2,2-Dimethylpropanal

The necessary alcohol is available through reaction of a *tert*-butyl Grignard reagent with formaldehyde, as shown by the disconnection

$$(CH_3)_3CCH_2OH \Rightarrow (CH_3)_3CMgCl + H_2C{=}O$$

$$CH_3OH \xrightarrow[CH_2Cl_2]{\text{PCC or PDC}} H\overset{O}{\overset{\|}{C}}H$$

Methanol → Formaldehyde

$$(CH_3)_3COH \xrightarrow{HCl} (CH_3)_3CCl \xrightarrow[\text{diethyl ether}]{Mg} (CH_3)_3CMgCl$$

2-Methyl-2-propanol (*tert*-butyl alcohol) → 2-Chloro-2-methylpropane (*tert*-butyl chloride) → 1,1-Dimethylethylmagnesium chloride (*tert*-butylmagnesium chloride)

$$(CH_3)_3CMgCl \xrightarrow[2.\ H_3O^+]{1.\ H_2C{=}O,\ \text{diethyl ether}} (CH_3)_3CCH_2OH$$

1,1-Dimethylethylmagnesium chloride (*tert*-butylmagnesium chloride) → 2,2-Dimethyl-1-propanol

**15.26** (*a*) The simplest route to this primary chloride from benzene is through the corresponding alcohol. The first step is the two-carbon chain extension used in Problem 15.24*a*.

$$\text{Benzene} \xrightarrow[FeBr_3]{Br_2} \text{Bromobenzene} \xrightarrow[2.\ H_2C{-}CH_2\ (\text{ethylene oxide});\ 3.\ H_3O^+]{1.\ Mg,\ \text{diethyl ether}} C_6H_5CH_2CH_2OH\ (\text{2-Phenylethanol})$$

$$C_6H_5CH_2CH_2OH \xrightarrow{SOCl_2} C_6H_5CH_2CH_2Cl$$

2-Phenylethanol → 1-Chloro-2-phenylethane

The preparation of ethylene oxide is shown in Problem 15.25*b*.

(*b*) A Friedel–Crafts acylation is the best approach to the target ketone.

$$C_6H_6 + (CH_3)_2CHCCl(=O) \xrightarrow{AlCl_3} C_6H_5CCH(CH_3)_2(=O)$$

Benzene; 2-Methylpropanoyl chloride; 2-Methyl-1-phenyl-1-propanone

Because carboxylic acid chlorides are prepared from the corresponding acids, we write

$$(CH_3)_2CHCH_2OH \xrightarrow[H_2SO_4,\ heat]{K_2Cr_2O_7} (CH_3)_2CHCOH(=O) \xrightarrow{SOCl_2} (CH_3)_2CHCCl(=O)$$

2-Methyl-1-propanol; 2-Methylpropanoic acid; 2-Methylpropanoyl chloride

(*c*) Wolff–Kishner or Clemmensen reduction of the ketone just prepared in part (*b*) affords isobutylbenzene.

$$C_6H_5CCH(CH_3)_2(=O) \xrightarrow[or\ Zn(Hg),\ HCl]{H_2NNH_2,\ HO^-,\ triethylene\ glycol,\ heat} C_6H_5CH_2CH(CH_3)_2$$

2-Methyl-1-phenyl-1-propanone; Isobutylbenzene

A less direct approach requires three steps:

$$C_6H_5CCH(CH_3)_2(=O) \xrightarrow[CH_3OH]{NaBH_4} C_6H_5CH(OH)CH(CH_3)_2 \xrightarrow[heat]{H_2SO_4} C_6H_5CH{=}C(CH_3)_2 \xrightarrow[Pt]{H_2} C_6H_5CH_2CH(CH_3)_2$$

2-Methyl-1-phenyl-1-propanone; 2-Methyl-1-phenyl-1-propanol; 2-Methyl-1-phenylpropene; Isobutylbenzene

**15.27** (*a*) Because 1-phenylcyclopentanol is a tertiary alcohol, a likely synthesis would involve reaction of a ketone and a Grignard reagent. Thus, a reasonable last step is treatment of cyclopentanone with phenylmagnesium bromide.

$$\text{Cyclopentanone} \xrightarrow[2.\ H_3O^+]{1.\ C_6H_5MgBr,\ diethyl\ ether} \text{1-Phenylcyclopentanol}$$

Cyclopentanone; 1-Phenylcyclopentanol

Cyclopentanone is prepared by oxidation of cyclopentanol. Any one of a number of oxidizing agents would be suitable. These include PDC or PCC in $CH_2Cl_2$ or chromic acid ($H_2CrO_4$) generated from $Na_2Cr_2O_7$ in aqueous sulfuric acid.

$$\text{Cyclopentanol} \xrightarrow{oxidize} \text{Cyclopentanone}$$

Cyclopentanol; Cyclopentanone

(*b*) Acid-catalyzed dehydration of 1-phenylcyclopentanol gives 1-phenylcyclopentene.

$$\text{1-Phenylcyclopentanol} \xrightarrow[or\ H_3PO_4,\ heat]{H_2SO_4,\ heat} \text{1-Phenylcyclopentene}$$

1-Phenylcyclopentanol; 1-Phenylcyclopentene

(*c*) Hydroboration–oxidation of 1-phenylcyclopentene gives *trans*-2-phenylcyclopentanol. The elements of water (H and OH) are added across the double bond opposite to Markovnikov's rule and syn to each other.

1-Phenylcyclopentene $\xrightarrow[\text{2. } H_2O_2,\ HO^-]{\text{1. } B_2H_6,\ \text{diglyme}}$ *trans*-2-Phenylcyclopentanol

(*d*) Oxidation of *trans*-2-phenylcyclopentanol converts this secondary alcohol to the desired ketone. Any of the Cr(VI)-derived oxidizing agents mentioned in part (*a*) for oxidation of cyclopentanol to cyclopentanone is satisfactory.

*trans*-2-Phenylcyclopentanol $\xrightarrow{\text{Cr(VI) oxidation}}$ 2-Phenylcyclopentanone

(*e*) The standard procedure for preparing *cis*-1,2-diols is by hydroxylation of alkenes with osmium tetraoxide.

1-Phenylcyclopentene $\xrightarrow[(CH_3)_3COH,\ HO^-]{OsO_4,\ (CH_3)_3COOH}$ 1-Phenyl-*cis*-1,2-cyclopentanediol

(*f*) The desired compound is available either by ozonolysis of 1-phenylcyclopentene:

1-Phenylcyclopentene $\xrightarrow[\text{2. } H_2O,\ Zn]{\text{1. } O_3}$ $C_6H_5\overset{O}{\overset{\|}{C}}CH_2CH_2CH_2\overset{O}{\overset{\|}{C}}H$

5-Oxo-1-phenyl-1-pentanone

or by periodic acid cleavage of the diol in part (*e*):

1-Phenyl-*cis*-1,2-cyclopentanediol $\xrightarrow{HIO_4}$ $C_6H_5\overset{O}{\overset{\|}{C}}CH_2CH_2CH_2\overset{O}{\overset{\|}{C}}H$

5-Oxo-1-phenyl-1-pentanone

(*g*) Reduction of both carbonyl groups in the product of part (*f*) gives the desired diol.

$$C_6H_5\overset{O}{\overset{\|}{C}}CH_2CH_2CH_2\overset{O}{\overset{\|}{C}}H \xrightarrow[\text{or 1. } LiAlH_4,\ \text{diethyl ether; 2. } H_2O]{H_2,\ \text{Pt (or Pd, Ni, Ru) or } NaBH_4,\ H_2O} C_6H_5\underset{OH}{\underset{|}{C}}HCH_2CH_2CH_2CH_2OH$$

5-Oxo-1-phenyl-1-pentanone → 1-Phenyl-1,5-pentanediol

**15.28** (*a, b*) Primary alcohols react in two different ways on being heated with acid catalysts: they can condense to form dialkyl ethers or undergo dehydration to yield alkenes. Ether formation is favored at lower temperature, and alkene formation is favored at higher temperature.

$$\underset{\text{1-Propanol}}{2CH_3CH_2CH_2OH} \xrightarrow[140°C]{H_2SO_4} \underset{\text{Dipropyl ether}}{CH_3CH_2CH_2OCH_2CH_2CH_3} + \underset{\text{Water}}{H_2O}$$

$$\underset{\text{1-Propanol}}{CH_3CH_2CH_2OH} \xrightarrow[200°C]{H_2SO_4} \underset{\text{Propene}}{CH_3CH{=}CH_2} + \underset{\text{Water}}{H_2O}$$

(*c*) Nitrate esters are formed by the reaction of alcohols with nitric acid in the presence of a sulfuric acid catalyst.

$$\underset{\text{1-Propanol}}{CH_3CH_2CH_2OH} + \underset{\text{Nitric acid}}{HONO_2} \xrightarrow{H_2SO_4(cat)} \underset{\text{Propyl nitrate}}{CH_3CH_2CH_2ONO_2} + \underset{\text{Water}}{H_2O}$$

(*d*) Pyridinium chlorochromate (PCC) oxidizes primary alcohols to aldehydes.

$$\underset{\text{1-Propanol}}{CH_3CH_2CH_2OH} \xrightarrow[CH_2Cl_2]{PCC} \underset{\text{Propanal}}{CH_3CH_2\overset{\overset{O}{\|}}{C}H}$$

(*e*) Potassium dichromate in aqueous sulfuric acid oxidizes primary alcohols to carboxylic acids.

$$\underset{\text{1-Propanol}}{CH_3CH_2CH_2OH} \xrightarrow[H_2SO_4,\ H_2O,\ heat]{K_2Cr_2O_7} \underset{\text{Propanoic acid}}{CH_3CH_2\overset{\overset{O}{\|}}{C}OH}$$

(*f*) Amide ion, a strong base, abstracts a proton from 1-propanol to form ammonia and 1-propanolate ion. This is an acid–base reaction.

$$\underset{\text{1-Propanol}}{CH_3CH_2CH_2OH} + \underset{\text{Sodium amide}}{NaNH_2} \longrightarrow \underset{\text{Sodium 1-propanolate}}{CH_3CH_2CH_2ONa} + \underset{\text{Ammonia}}{NH_3}$$

(*g*) With acetic acid and in the presence of an acid catalyst, 1-propanol is converted to its acetate ester.

$$\underset{\text{1-Propanol}}{CH_3CH_2CH_2OH} + \underset{\text{Acetic acid}}{CH_3\overset{\overset{O}{\|}}{C}OH} \xrightleftharpoons{HCl} \underset{\text{Propyl acetate}}{CH_3\overset{\overset{O}{\|}}{C}OCH_2CH_2CH_3} + \underset{\text{Water}}{H_2O}$$

This is an equilibrium process that slightly favors products.

(*h*) Alcohols react with *p*-toluenesulfonyl chloride to give *p*-toluenesulfonate esters.

$$\underset{\text{1-Propanol}}{CH_3CH_2CH_2OH} + \underset{p\text{-Toluenesulfonyl chloride}}{CH_3{-}C_6H_4{-}SO_2Cl} \xrightarrow{pyridine} \underset{\text{Propyl } p\text{-toluenesulfonate}}{CH_3CH_2CH_2O\overset{\overset{O}{\|}}{\underset{\underset{O}{\|}}{S}}{-}C_6H_4{-}CH_3} + HCl$$

(*i*) Acyl chlorides convert alcohols to esters.

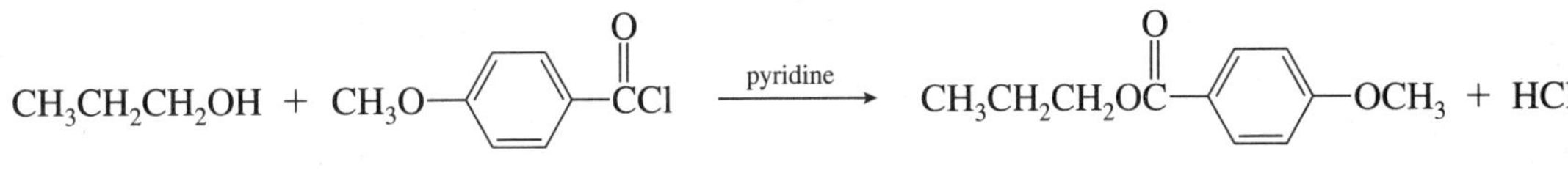

$$CH_3CH_2CH_2OH + CH_3O\text{—}C_6H_4\text{—}\overset{O}{\overset{\|}{C}}Cl \xrightarrow{\text{pyridine}} CH_3CH_2CH_2O\overset{O}{\overset{\|}{C}}\text{—}C_6H_4\text{—}OCH_3 + HCl$$

1-Propanol    *p*-Methoxybenzoyl chloride    Propyl *p*-methoxybenzoate

(*j*) The reagent is benzoic anhydride. Carboxylic acid anhydrides react with alcohols to give esters.

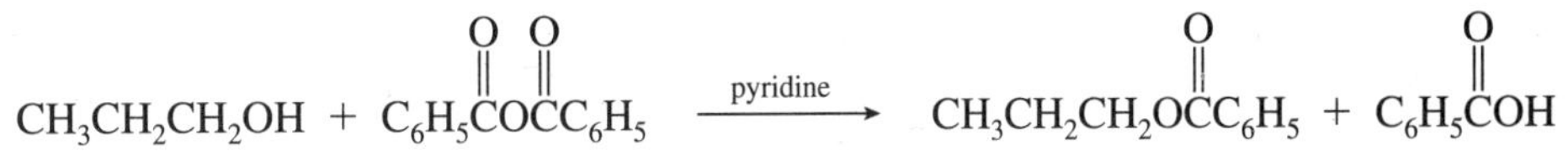

$$CH_3CH_2CH_2OH + C_6H_5\overset{O}{\overset{\|}{C}}O\overset{O}{\overset{\|}{C}}C_6H_5 \xrightarrow{\text{pyridine}} CH_3CH_2CH_2O\overset{O}{\overset{\|}{C}}C_6H_5 + C_6H_5\overset{O}{\overset{\|}{C}}OH$$

1-Propanol    Benzoic anhydride    Propyl benzoate    Benzoic acid

(*k*) The reagent is succinic anhydride, a cyclic anhydride. Esterification occurs, but in this case the resulting ester and carboxylic acid functions remain part of the same molecule.

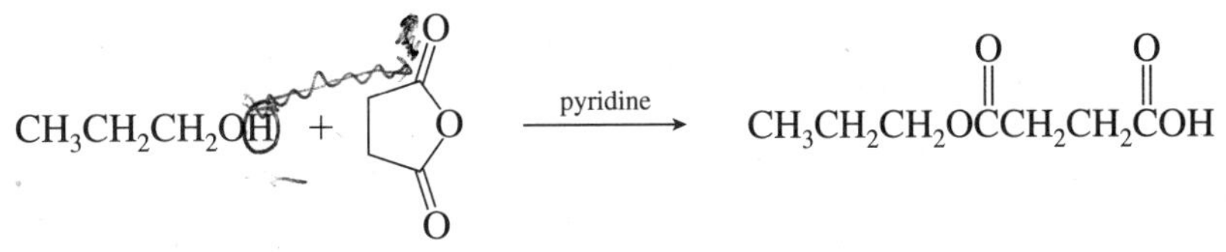

$$CH_3CH_2CH_2OH + \text{(succinic anhydride)} \xrightarrow{\text{pyridine}} CH_3CH_2CH_2O\overset{O}{\overset{\|}{C}}CH_2CH_2\overset{O}{\overset{\|}{C}}OH$$

1-Propanol    Succinic anhydride    Hydrogen propyl succinate

**15.29** (*a*) On being heated in the presence of sulfuric acid, tertiary alcohols undergo elimination.

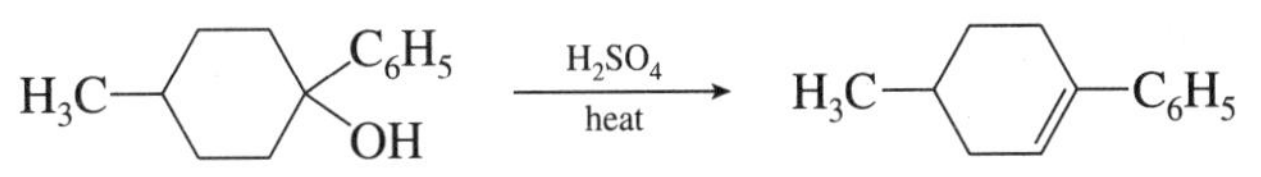

$$H_3C\text{—(cyclohexane)—}C_6H_5, OH \xrightarrow[\text{heat}]{H_2SO_4} H_3C\text{—(cyclohexene)—}C_6H_5$$

4-Methyl-1-phenylcyclohexanol    4-Methyl-1-phenylcyclohexene (81%)

(*b*) The combination of reagents specified converts alkenes to vicinal diols.

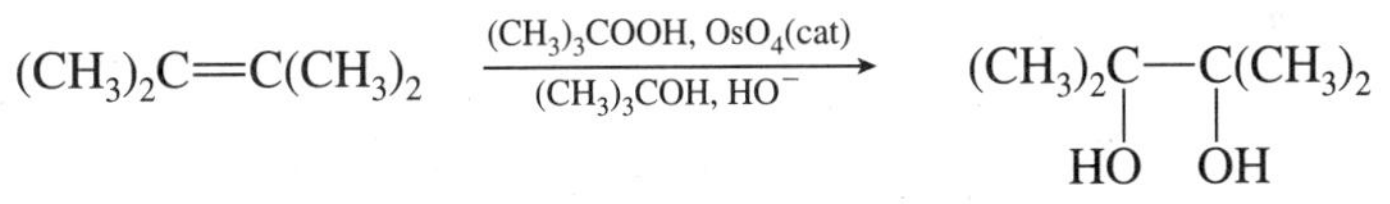

$$(CH_3)_2C{=}C(CH_3)_2 \xrightarrow[(CH_3)_3COH,\ HO^-]{(CH_3)_3COOH,\ OsO_4(\text{cat})} (CH_3)_2\underset{HO}{\underset{|}{C}}\text{—}\underset{OH}{\underset{|}{C}}(CH_3)_2$$

2,3-Dimethyl-2-butene    2,3-Dimethyl-2,3-butanediol (72%)

(*c*) Hydroboration–oxidation of the double bond takes place with a regioselectivity that is opposite to Markovnikov's rule. The elements of water are added in a stereospecific syn fashion.

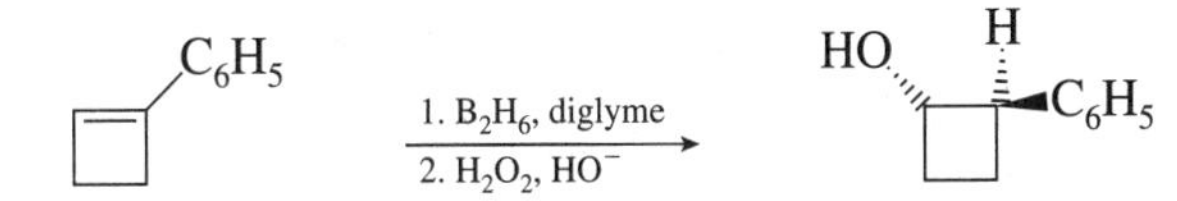

$$\text{1-Phenylcyclobutene} \xrightarrow[2.\ H_2O_2,\ HO^-]{1.\ B_2H_6,\ \text{diglyme}} \text{trans-2-Phenylcyclobutanol}$$

1-Phenylcyclobutene    *trans*-2-Phenylcyclobutanol (82%)

(*d*) Lithium aluminum hydride reduces carboxylic acids to primary alcohols, but does not reduce carbon–carbon double bonds.

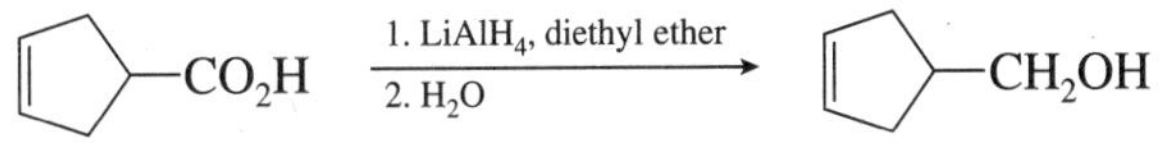

$$\text{(cyclopentenyl)—}CO_2H \xrightarrow[2.\ H_2O]{1.\ LiAlH_4,\ \text{diethyl ether}} \text{(cyclopentenyl)—}CH_2OH$$

Cyclopentene-4-carboxylic acid    (3-Cyclopentenyl)-methanol

(*e*) Chromic acid oxidizes the secondary alcohol to the corresponding ketone but does not affect the triple bond.

$$CH_3CH(OH)C{\equiv}C(CH_2)_3CH_3 \xrightarrow[\text{acetone}]{H_2CrO_4,\ H_2SO_4,\ H_2O} CH_3\overset{O}{\overset{\|}{C}}C{\equiv}C(CH_2)_3CH_3$$

3-Octyn-2-ol → 3-Octyn-2-one (80%)

(*f*) Lithium aluminum hydride reduces carbonyl groups efficiently but does not normally react with double bonds.

$$CH_3\overset{O}{\overset{\|}{C}}CH_2CH{=}CHCH_2\overset{O}{\overset{\|}{C}}CH_3 \xrightarrow[2.\ H_2O]{1.\ LiAlH_4,\ \text{diethyl ether}} CH_3CH(OH)CH_2CH{=}CHCH_2CH(OH)CH_3$$

4-Octen-2,7-dione → 4-Octen-2,7-diol (75%)

(*g*) Alcohols react with acyl chlorides to yield esters. The O—H bond is broken in this reaction; the C—O bond of the alcohol remains intact on ester formation.

*trans*-3-Methylcyclohexanol + 3,5-Dinitrobenzoyl chloride $\xrightarrow{\text{pyridine}}$ *trans*-3-Methylcyclohexyl-3,5-dinitrobenzoate (74%)

(*h*) Carboxylic acid anhydrides react with alcohols to give esters. Here, too, the spatial orientation of the C—O bond remains intact.

*exo*-Bicyclo[2.2.1]heptan-2-ol + $CH_3\overset{O}{\overset{\|}{C}}O\overset{O}{\overset{\|}{C}}CH_3$ (Acetic anhydride) ⟶ *exo*-Bicyclo[2.2.1]hept-2-yl acetate (90%) + $CH_3\overset{O}{\overset{\|}{C}}OH$ (Acetic acid)

(*i*) The substrate is a carboxylic acid and undergoes Fischer esterification with methanol.

4-Chloro-3,5-dinitrobenzoic acid $\xrightarrow[H_2SO_4]{CH_3OH}$ Methyl 4-chloro-3,5-dinitrobenzoate (96%)

(*j*) Both ester functions are cleaved by reduction with lithium aluminum hydride. The product is a diol.

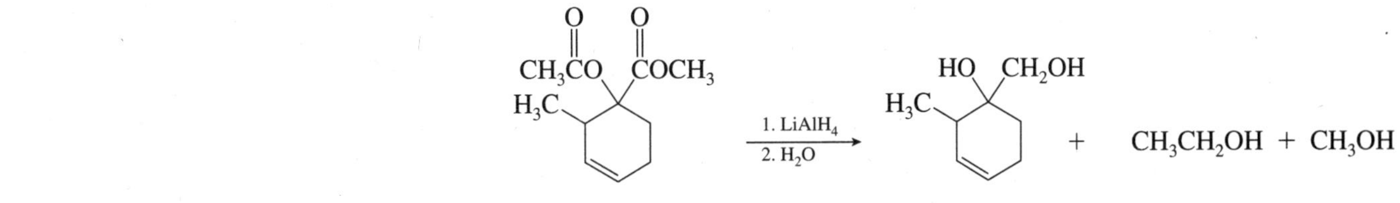

$$\xrightarrow[\text{2. } H_2O]{\text{1. LiAlH}_4} \quad + \quad CH_3CH_2OH + CH_3OH$$

(96%)

(*k*) Treatment of the diol obtained in part (*j*) with periodic acid brings about its cleavage to two carbonyl compounds.

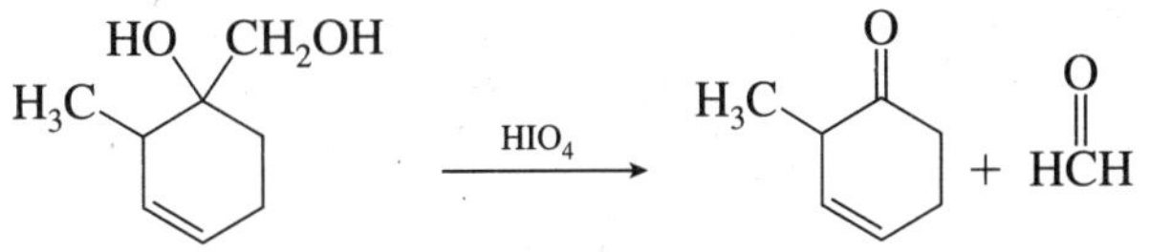

$$\xrightarrow{HIO_4} \quad + \quad HCH(=O)$$

(74%)

**15.30** Only the hydroxyl groups on C-1 and C-4 can be involved, since only these two can lead to a five-membered cyclic ether.

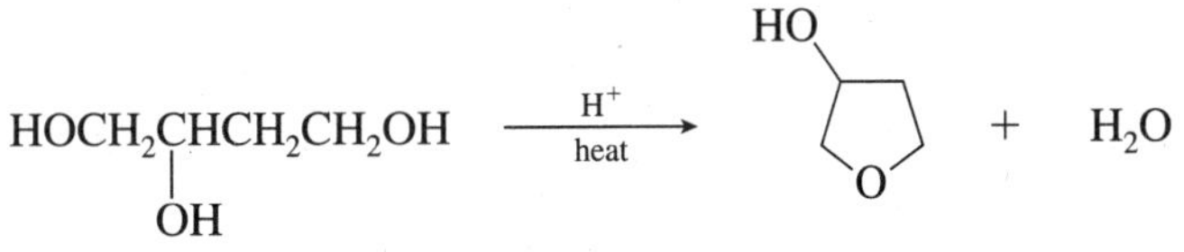

$$HOCH_2CHCH_2CH_2OH \text{ (with OH on C-2)} \xrightarrow[\text{heat}]{H^+} \quad + \quad H_2O$$

1,2,4-Butanetriol — 3-Hydroxyoxolane ($C_4H_8O_2$)

Any other combination of hydroxyl groups would lead to a strained three-membered or four-membered ring and is unfavorable under conditions of acid catalysis.

**15.31** Hydroxylation of alkenes with osmium tetraoxide is a syn addition. A racemic mixture of the 2*R*,3*S* and 2*S*,3*R* stereoisomers is formed from *cis*-2-pentene.

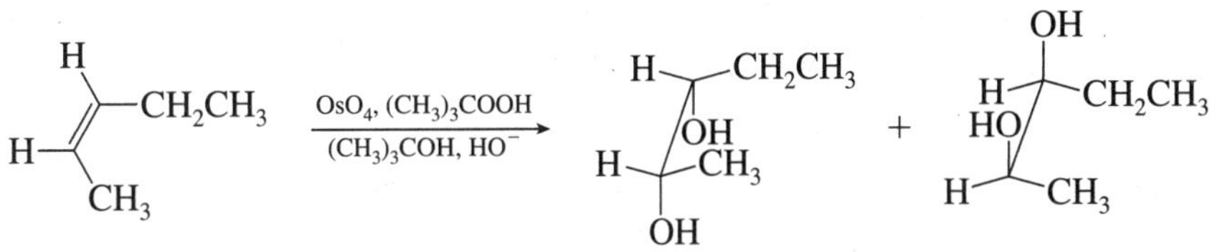

$$\xrightarrow[(CH_3)_3COH,\ HO^-]{OsO_4,\ (CH_3)_3COOH}$$

*cis*-2-Pentene — 2*S*,3*R*-2,3-Pentanediol + 2*R*,3*S*-2,3-Pentanediol

*trans*-2-Pentene gives a racemic mixture of the 2*R*,3*R* and 2*S*,3*S* stereoisomers.

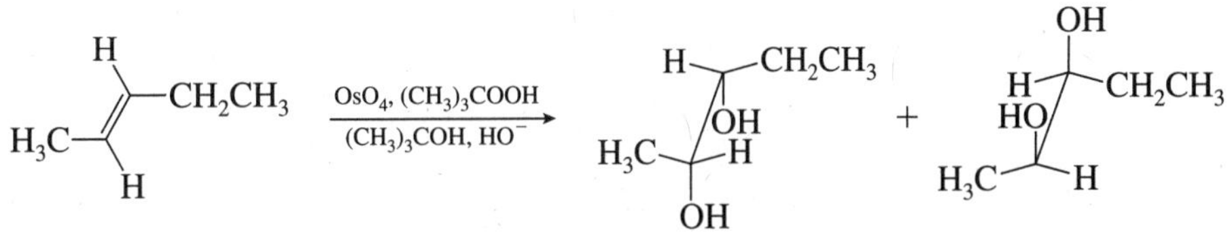

$$\xrightarrow[(CH_3)_3COH,\ HO^-]{OsO_4,\ (CH_3)_3COOH}$$

*trans*-2-Pentene — 2*R*,3*R*-2,3-Pentanediol + 2*S*,3*S*-2,3-Pentanediol

**15.32** (*a*) The task of converting a ketone to an alkene requires first the reduction of the ketone to an alcohol and then dehydration. In practice the two-step transformation has been carried out in 54% yield by treating the ketone with sodium borohydride and then heating the resulting alcohol with *p*-toluenesulfonic acid.

$NaBH_4$, $CH_3OH$ → OH; $H^+$, heat →

Of course, sodium borohydride may be replaced by other suitable reducing agents, and *p*-toluenesulfonic acid is not the only acid that could be used in the dehydration step.

(*b*) This problem and the next one illustrate the value of reasoning backward. The desired product, cyclohexanol, can be prepared cleanly from cyclohexanone.

OH ⇒ O ⇒ OH, $CH_2OH$

Once cyclohexanone is recognized to be a key intermediate, the synthetic pathway becomes apparent—what is needed is a method to convert the indicated starting material to cyclohexanone. The reagent ideally suited to this task is periodic acid. The synthetic sequence to be followed is therefore

OH, $CH_2OH$ —$HIO_4$→ O —$NaBH_4$, $CH_3OH$→ OH

1-(Hydroxymethyl)-cyclohexanol — Cyclohexanone — Cyclohexanol

(*c*) No direct method allows a second hydroxyl group to be introduced at C-2 of 1-phenylcyclohexanol in a single step. We recognize the product as a vicinal diol and recall that such compounds are available by hydroxylation of alkenes.

OH, $C_6H_5$, OH ⇒ $C_6H_5$ ⇒ OH, $C_6H_5$

This tells us that we must first dehydrate the tertiary alcohol, then hydroxylate the resulting alkene.

OH, $C_6H_5$ —$H_2SO_4$, heat→ $C_6H_5$ —$(CH_3)_3COOH$, $(CH_3)_3COH$, $OsO_4$, $HO^-$→ OH, $C_6H_5$, OH

1-Phenylcyclohexanol — 1-Phenylcyclohexene — 1-Phenyl-*cis*-1,2-cyclohexanediol

The syn stereoselectivity of the hydroxylation step ensures that the product will have its hydroxyl groups cis, as the problem requires.

**15.33** Because the target molecule is an eight-carbon secondary alcohol and the problem restricts our choices of starting materials to alcohols of five carbons or fewer, we are led to consider building up the carbon chain by a Grignard reaction.

$$CH_3CH_2CH(OH)\text{-}\!\!\nmid\!\!\text{-}CH(CH_3)CH_2CH_2CH_3 \Longrightarrow CH_3CH_2CH{=}O + {}^{-}\!:CH(CH_3)CH_2CH_2CH_3$$

4-Methyl-3-heptanol

The disconnection shown leads to a three-carbon aldehyde and a five-carbon Grignard reagent. Starting with the corresponding alcohols, the following synthetic scheme seems reasonable.

First, propanal is prepared.

$$\underset{\text{1-Propanol}}{CH_3CH_2CH_2OH} \xrightarrow[CH_2Cl_2]{\text{PCC or PDC}} \underset{\text{Propanal}}{CH_3CH_2CH{=}O}$$

After converting 2-pentanol to its bromo derivative, a solution of the Grignard reagent is prepared.

$$\underset{\text{2-Pentanol}}{CH_3CH(OH)CH_2CH_2CH_3} \xrightarrow{PBr_3} \underset{\text{2-Bromopentane}}{CH_3CH(Br)CH_2CH_2CH_3} \xrightarrow[\text{diethyl ether}]{Mg} \underset{\text{1-Methylbutylmagnesium bromide}}{CH_3CH(MgBr)CH_2CH_2CH_3}$$

Reaction of the Grignard reagent with the aldehyde yields the desired 4-methyl-3-heptanol.

$$\underset{\text{1-Methylbutylmagnesium bromide}}{CH_3CH(MgBr)CH_2CH_2CH_3} + \underset{\text{Propanal}}{CH_3CH_2CH{=}O} \xrightarrow[2.\ H_3O^+]{1.\ \text{diethyl ether}} \underset{\text{4-Methyl-3-heptanol}}{CH_3CH(CH_2CH_2CH_3)CH(OH)CH_2CH_3}$$

**15.34** Our target molecule is void of functionality and so requires us to focus attention on the carbon skeleton. Notice that it can be considered to arise from three ethyl groups.

$$CH_3CH_2 \,\vdots\, CH(CH_3) \,\vdots\, CH_2CH_3$$

3-Methylpentane

Considering the problem retrosynthetically, we can see that a key intermediate having the carbon skeleton of the desired product is 3-methyl-3-pentanol. This becomes apparent from the fact that alkanes may be prepared from alkenes, which in turn are available from alcohols. The desired alcohol may be prepared from reaction of an acetate ester with a Grignard reagent, ethylmagnesium bromide.

$$CH_3CH_2CH(CH_3)CH_2CH_3 \Longrightarrow CH_3CH_2C(CH_3)(OH)CH_2CH_3 \Longrightarrow CH_3C({=}O)OR + 2CH_3CH_2MgBr$$

The carbon skeleton can be assembled in one step by the reaction of ethylmagnesium bromide and ethyl acetate.

$$2CH_3CH_2MgBr + CH_3\overset{O}{\overset{\|}{C}}OCH_2CH_3 \xrightarrow[2.\ H_3O^+]{1.\ \text{diethyl ether}} CH_3\underset{CH_2CH_3}{\overset{OH}{C}}CH_2CH_3$$

Ethylmagnesium bromide — Ethyl acetate — 3-Methyl-3-pentanol

The resulting tertiary alcohol is converted to the desired hydrocarbon by acid-catalyzed dehydration and catalytic hydrogenation of the resulting mixture of alkenes.

$$CH_3\underset{CH_2CH_3}{\overset{OH}{C}}CH_2CH_3 \xrightarrow{H^+} CH_3\underset{CH_2CH_3}{C}{=}CHCH_3 + H_2C{=}C(CH_2CH_3)_2$$

3-Methyl-3-pentanol — 3-Methyl-2-pentene (cis + trans) — 2-Ethyl-1-butene

$$\xrightarrow{H_2,\ Ni} CH_3CH(CH_2CH_3)_2$$

3-Methylpentane

Because the problem requires that ethanol be the ultimate starting material, we need to show the preparation of the ethylmagnesium bromide and ethyl acetate used in constructing the carbon skeleton.

$$CH_3CH_2OH \xrightarrow{PBr_3} CH_3CH_2Br \xrightarrow[\text{diethyl ether}]{Mg} CH_3CH_2MgBr$$

Ethanol — Ethyl bromide — Ethylmagnesium bromide

$$CH_3CH_2OH \xrightarrow[H_2SO_4,\ H_2O,\ \text{heat}]{K_2Cr_2O_7} CH_3\overset{O}{\overset{\|}{C}}OH$$

Ethanol — Acetic acid

$$CH_3\overset{O}{\overset{\|}{C}}OH + CH_3CH_2OH \xrightarrow{H^+} CH_3\overset{O}{\overset{\|}{C}}OCH_2CH_3$$

Acetic acid — Ethyl acetate

**15.35** (*a*) Retrosynthetically, we can see that the cis carbon–carbon double bond is available by hydrogenation of the corresponding alkyne over the Lindlar catalyst.

$$CH_3CH_2CH{=}CHCH_2CH_2OH \Longrightarrow CH_3CH_2C{\equiv}CCH_2CH_2OH$$

The $—CH_2CH_2OH$ unit can be appended to an alkynide anion by reaction with ethylene oxide.

$$CH_3CH_2C{\equiv}CCH_2CH_2OH \Longrightarrow CH_3CH_2C{\equiv}C{:}^- + H_2C{-}CH_2 \text{ (epoxide, O)}$$

The alkynide anion is derived from 1-butyne by alkylation of acetylene. This analysis suggests the following synthetic sequence:

$$\underset{\text{Acetylene}}{HC{\equiv}CH} \xrightarrow[2.\ CH_3CH_2Br]{1.\ NaNH_2,\ NH_3} \underset{\text{1-Butyne}}{CH_3CH_2C{\equiv}CH} \xrightarrow[2.\ H_2C{-}CH_2\ (\text{O})]{1.\ NaNH_2,\ NH_3} \underset{\text{3-Hexyn-1-ol}}{CH_3CH_2C{\equiv}CCH_2CH_2OH}$$

$$\xrightarrow{\text{Lindlar Pd},\ H_2} \underset{cis\text{-3-Hexen-1-ol}}{(CH_3CH_2)HC{=}CH(CH_2CH_2OH)}$$

(*b*) The compound cited is the aldehyde derived by oxidation of the primary alcohol in part (*a*). Oxidize the alcohol with PDC or PCC in $CH_2Cl_2$.

$$\underset{cis\text{-3-Hexen-1-ol}}{(CH_3CH_2)HC{=}CH(CH_2CH_2OH)} \xrightarrow[\text{in } CH_2Cl_2]{\text{PDC or PCC}} \underset{cis\text{-3-Hexenal}}{(CH_3CH_2)HC{=}CH(CH_2\overset{O}{\overset{\|}{C}}H)}$$

**15.36** Even though we are given the structure of the starting material, it is still better to reason backward from the target molecule rather than forward from the starting material.

The desired product contains a cyano (—CN) group. The only method we have seen so far for introducing such a function into a molecule is by nucleophilic substitution. The last step in the synthesis must therefore be

$$m\text{-}CH_3OC_6H_4CH_2X + CN^- \longrightarrow m\text{-}CH_3OC_6H_4CH_2CN + X^-$$

This step should work very well, since the substrate is a primary benzylic halide, cannot undergo elimination, and is very reactive in $S_N2$ reactions.

The primary benzylic halide can be prepared from the corresponding alcohol by any of a number of methods.

$$m\text{-}CH_3OC_6H_4CH_2OH \longrightarrow m\text{-}CH_3OC_6H_4CH_2X$$

Suitable reagents include HBr, $PBr_3$, or $SOCl_2$.

Now we only need to prepare the primary alcohol from the given starting aldehyde, which is accomplished by reduction.

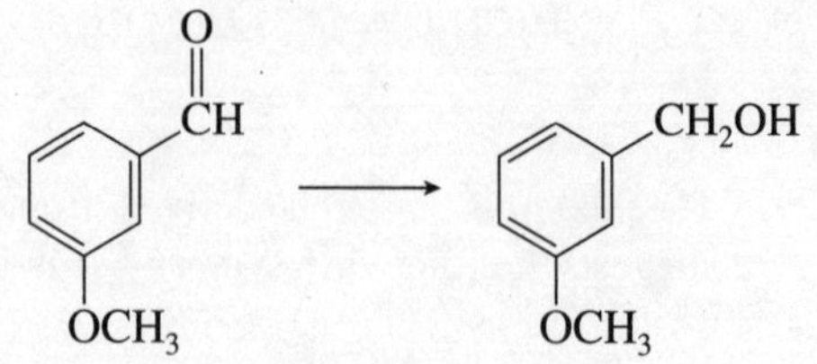

Reduction can be achieved by catalytic hydrogenation, with lithium aluminum hydride, or with sodium borohydride.

The actual sequence of reactions as carried out is as shown.

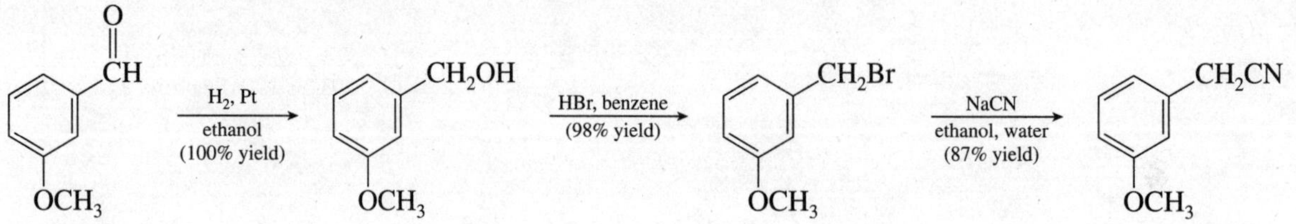

$H_2$, Pt, ethanol (100% yield); HBr, benzene (98% yield); NaCN, ethanol, water (87% yield)

*m*-Methoxybenzaldehyde → *m*-Methoxybenzyl alcohol → *m*-Methoxybenzyl bromide → *m*-Methoxybenzyl cyanide

Another three-step synthesis, which is reasonable but does not involve an alcohol as an intermediate, is

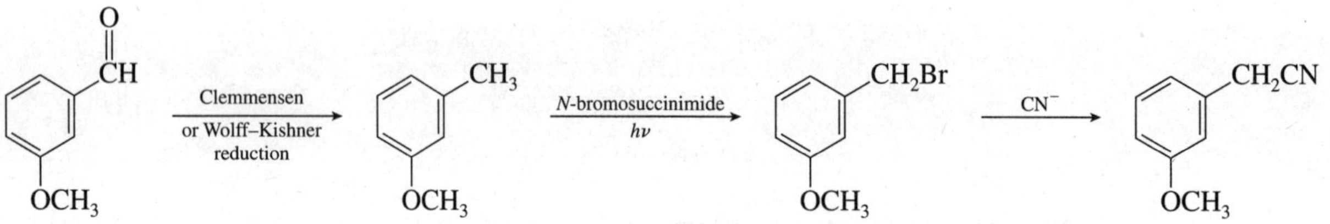

Clemmensen or Wolff–Kishner reduction; *N*-bromosuccinimide, *hν*; $CN^-$

*m*-Methoxybenzaldehyde → *m*-Methoxytoluene → *m*-Methoxybenzyl bromide → *m*-Methoxybenzyl cyanide

**15.37** (*a*) Addition of hydrogen chloride to cyclopentadiene takes place by way of the most stable carbocation. In this case it is an allylic carbocation.

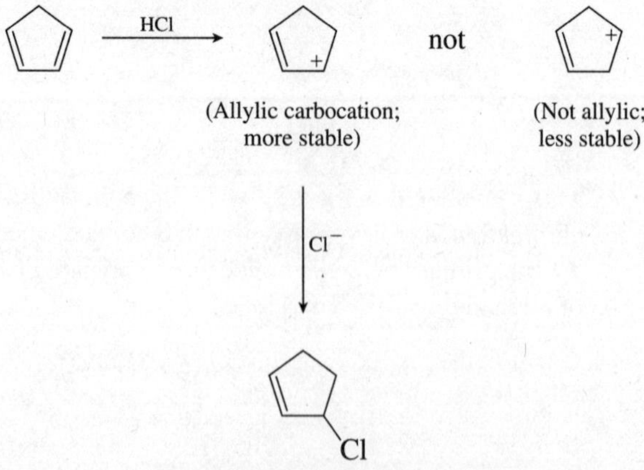

HCl → (Allylic carbocation; more stable) not (Not allylic; less stable)

$Cl^-$

3-Chlorocyclopentene (80–90%) (Compound A)

Hydrolysis of 3-chlorocyclopentene gives the corresponding alcohol. Sodium bicarbonate in water is a weakly basic solvolysis medium.

$$\text{Compound A} \xrightarrow[H_2O]{NaHCO_3} \text{2-Cyclopenten-1-ol (88\%) (compound B)}$$

Cl — Compound A

OH — 2-Cyclopenten-1-ol (88%) (compound B)

Oxidation of compound B (a secondary alcohol) gives the ketone 2-cyclopenten-1-one.

$$\text{Compound B} \xrightarrow[H_2SO_4,\ H_2O]{Na_2Cr_2O_7} \text{2-Cyclopenten-1-one (60–68\%) (compound C)}$$

OH — Compound B

O — 2-Cyclopenten-1-one (60–68%) (compound C)

(*b*) Thionyl chloride converts alcohols to alkyl chlorides.

$$H_2C{=}CHCH_2CH_2\underset{\displaystyle OH}{\underset{|}{C}}HCH_3 \xrightarrow[\text{pyridine}]{SOCl_2} H_2C{=}CHCH_2CH_2\underset{\displaystyle Cl}{\underset{|}{C}}HCH_3$$

5-Hexen-2-ol — 5-Chloro-1-hexene (compound D)

Ozonolysis cleaves the carbon–carbon double bond.

$$H_2C{=}CHCH_2CH_2\underset{\displaystyle Cl}{\underset{|}{C}}HCH_3 \xrightarrow[\text{2. reductive workup}]{1.\ O_3} H\overset{\displaystyle O}{\overset{||}{C}}CH_2CH_2\underset{\displaystyle Cl}{\underset{|}{C}}HCH_3 + H\overset{\displaystyle O}{\overset{||}{C}}H$$

Compound D — 4-Chloropentanal (compound E) — Formaldehyde

Reduction of compound E yields the corresponding alcohol.

$$H\overset{\displaystyle O}{\overset{||}{C}}CH_2CH_2\underset{\displaystyle Cl}{\underset{|}{C}}HCH_3 \xrightarrow{NaBH_4} HOCH_2CH_2CH_2\underset{\displaystyle Cl}{\underset{|}{C}}HCH_3$$

4-Chloropentanal — 4-Chloro-1-pentanol (compound F)

(*c*) *N*-Bromosuccinimide is a reagent designed to accomplish benzylic bromination.

$$\text{1-Bromo-2-methylnaphthalene} \xrightarrow[\text{benzoyl peroxide, heat}]{NBS} \text{1-Bromo-2-(bromomethyl)naphthalene (compound G)}$$

$CH_3$, Br — 1-Bromo-2-methylnaphthalene

$CH_2Br$, Br — 1-Bromo-2-(bromomethyl)naphthalene (compound G)

Hydrolysis of the benzylic bromide gives the corresponding benzylic alcohol. The bromine that is directly attached to the naphthalene ring does not react under these conditions.

$$\text{1-Bromo-2-(bromomethyl)naphthalene } (C_{10}H_6(Br)CH_2Br) \xrightarrow[\text{heat}]{H_2O,\ CaCO_3} \text{(1-Bromo-2-naphthyl)methanol } (C_{10}H_6(Br)CH_2OH)$$

1-Bromo-2-(bromomethyl)naphthalene — (1-Bromo-2-naphthyl)methanol (compound H)

Oxidation of the primary alcohol with PCC gives the aldehyde.

$$C_{10}H_6(Br)CH_2OH \xrightarrow[CH_2Cl_2]{PCC} C_{10}H_6(Br)CH{=}O$$

(1-Bromo-2-naphthyl)methanol — 1-Bromonaphthalene-2-carboxaldehyde (compound I)

**15.38** The alcohol is tertiary and benzylic and yields a relatively stable carbocation.

$$C_6H_5C(CH_3)(OH)CH_2CH_3 \xrightarrow{H_2SO_4} C_6H_5\overset{+}{C}(CH_3)CH_2CH_3$$

2-Phenyl-2-butanol — 1-Methyl-1-phenylpropyl cation

The alcohol is chiral, but the carbocation is not. Thus, irrespective of which enantiomer of 2-phenyl-2-butanol is used, the same carbocation is formed. The carbocation reacts with ethanol to give an optically inactive mixture containing equal quantities of enantiomers (racemic).

$$C_6H_5\overset{+}{C}(CH_3)CH_2CH_3 + CH_3CH_2OH \longrightarrow C_6H_5C(CH_3)(OCH_2CH_3)CH_2CH_3 + H^+$$

1-Methyl-1-phenylpropyl cation — Ethanol — 2-Ethoxy-2-phenylbutane (50% *R*, 50% *S*)

**15.39** The difference between the two ethers is that 1-*O*-benzylglycerol contains a vicinal diol function, but 2-*O*-benzylglycerol does not. Periodic acid will react with 1-*O*-benzylglycerol but not with 2-*O*-benzylglycerol.

$$C_6H_5CH_2OCH_2CH(OH)CH_2OH \xrightarrow{HIO_4} C_6H_5CH_2OCH_2CH{=}O + HCH{=}O$$

1-*O*-Benzylglycerol — 2-Benzyloxyethanal — Formaldehyde

$$HOCH_2CH(OCH_2C_6H_5)CH_2OH \xrightarrow{HIO_4} \text{no reaction}$$

2-*O*-Benzylglycerol

**15.40** The formation of an alkanethiol by reaction of an alkyl halide or alkyl *p*-toluenesulfonate with thiourea occurs with inversion of configuration in the step in which the carbon–sulfur bond is formed. Thus, the formation of (*R*)-2-butanethiol requires (*S*)-*sec*-butyl *p*-toluenesulfonate, which then reacts with thiourea by an $S_N2$ pathway. The *p*-toluenesulfonate is formed from the corresponding alcohol by a reaction that does not involve any of the bonds to the stereogenic center. Therefore, begin with (*S*)-2-butanol.

$CH_3CH_2(H)(CH_3)C—OH$ →(*p*-toluenesulfonyl chloride; retention of configuration)→ $CH_3CH_2(H)(CH_3)C—OTs$ →(1. $(H_2N)_2C{=}S$, inversion of configuration; 2. NaOH)→ $HS—C(H)(CH_2CH_3)(CH_3)$

(*S*)-2-Butanol — (*S*)-*sec*-Butyl *p*-toluenesulfonate — (*R*)-2-Butanethiol

**15.41** (*a*) Cysteine contains an —SH group and is a thiol. Oxidation of thiols gives rise to disulfides.

$$2RSH \xrightarrow{\text{oxidize}} RSSR$$

Thiol — Disulfide

Biological oxidation of cysteine gives the disulfide cystine.

$$2HSCH_2CH(\overset{+}{N}H_3)CO_2^- \xrightarrow{\text{oxidize}} {}^-O_2CCH(\overset{+}{N}H_3)CH_2S—SCH_2CH(\overset{+}{N}H_3)CO_2^-$$

Cysteine — Cystine

(*b*) Oxidation of a thiol yields a series of acids, including a sulfinic acid and a sulfonic acid.

$$RSH \longrightarrow R\overset{+}{S}(O^-)—OH \longrightarrow R\overset{2+}{S}(O^-)_2—OH$$

Thiol — Sulfinic acid — Sulfonic acid

Biological oxidation of cysteine can yield, in addition to the disulfide cystine, cysteine sulfinic acid and the sulfonic acid cysteic acid.

$$HSCH_2CH(\overset{+}{N}H_3)CO_2^- \xrightarrow{\text{oxidize}} HO—\overset{+}{S}(O^-)CH_2CH(\overset{+}{N}H_3)CO_2^- \xrightarrow{\text{oxidize}} HO—\overset{2+}{S}(O^-)_2CH_2CH(\overset{+}{N}H_3)CO_2^-$$

Cysteine — Cysteine sulfinic acid ($C_3H_7NO_4S$) — Cysteic acid ($C_3H_7NO_5S$)

**15.42** The ratio of carbon to hydrogen in the molecular formula is $C_nH_{2n+2}$ ($C_8H_{18}O_2$), and so the compound has no double bonds or rings. The compound cannot be a vicinal diol, because it does not react with periodic acid.

The NMR spectrum is rather simple as all peaks are singlets. The 12-proton singlet at $\delta$ 1.2 ppm must correspond to four equivalent methyl groups and the four-proton singlet at $\delta$ 1.6 ppm to two equivalent methylene groups. No nonequivalent protons can be vicinal, because no splitting is observed. The two-proton singlet at $\delta$ 2.0 ppm is due to the hydroxyl protons of the diol.

The compound is 2,5-dimethyl-2,5-hexanediol.

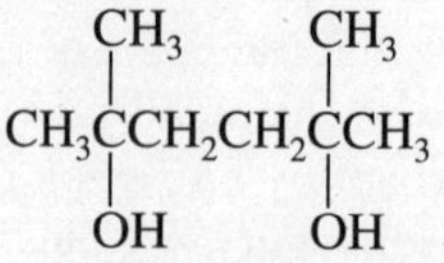

```
    CH3      CH3
    |        |
  CH3CCH2CH2CCH3
    |        |
    OH       OH
```

**15.43** The molecular formula of compound A ($C_8H_{10}O$) corresponds to an index of hydrogen deficiency of 4. The 4 hydrogen signal at $\delta$ 7.2 ppm in the $^1H$ NMR spectrum suggests these unsaturations are due to a disubstituted benzene ring. That the ring is para-substituted is supported by the symmetry of the signal; it is a pair of doublets, not a quartet.

The broad signal (1H) at $\delta$ 2.1 ppm undergoes rapid exchange with $D_2O$, indicating it is the proton of the hydroxyl group of an alcohol. As the remaining signals are singlets, with areas of 2H and 3H, respectively, compound A can be identified as 4-methylbenzyl alcohol.

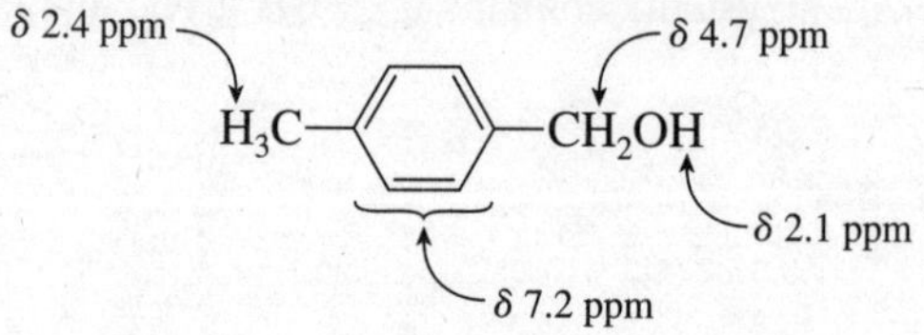

$H_3C$—$C_6H_4$—$CH_2OH$ ($CH_3$: $\delta$ 2.4 ppm; ring H: $\delta$ 7.2 ppm; $CH_2$: $\delta$ 4.7 ppm; OH: $\delta$ 2.1 ppm)

**15.44** (*a*) This compound has only two different types of carbons. One type of carbon comes at low field and is most likely a carbon bonded to oxygen and three other equivalent carbons. The spectrum leads to the conclusion that this compound is *tert*-butyl alcohol.

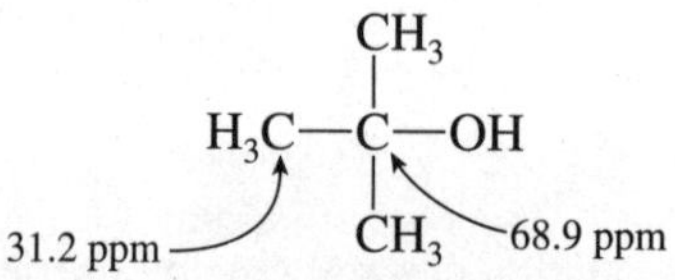

```
         CH3
         |
    H3C—C—OH
         |
         CH3
```
($CH_3$: 31.2 ppm; C–OH: 68.9 ppm)

(*b*) Four different types of carbons occur in this compound. The only $C_4H_{10}O$ isomers that have four nonequivalent carbons are $CH_3CH_2CH_2CH_2OH$, $CH_3CH(OH)CH_2CH_3$, and $CH_3OCH_2CH_2CH_3$.

The lowest field signal, the one at 69.2 ppm from the carbon that bears the oxygen substituent, is a methine (CH). The compound is therefore 2-butanol.

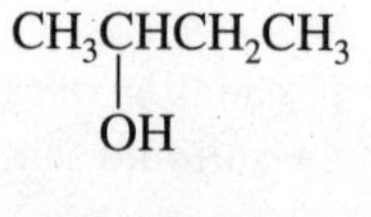

```
  CH3CHCH2CH3
      |
      OH
```

(*c*) This compound has two equivalent $CH_3$ groups, as indicated by the signal at 18.9 ppm. Its lowest field carbon is a $CH_2$, and so the group —$CH_2O$ must be present. The compound is 2-methyl-1-propanol.

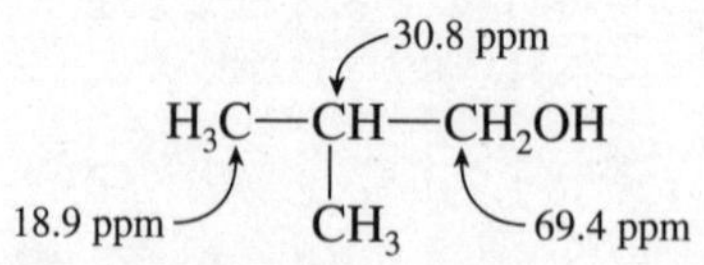

```
    H3C—CH—CH2OH
        |
        CH3
```
($CH_3$: 18.9 ppm; CH: 30.8 ppm; $CH_2OH$: 69.4 ppm)

**15.45** The compound has only three carbons, none of which is a $CH_3$ group. Two of the carbon signals arise from $CH_2$ groups; the other corresponds to a CH group. The only structure consistent with the observed data is that of 3-chloro-1,2-propanediol.

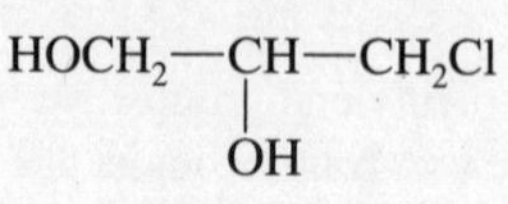

```
  HOCH2—CH—CH2Cl
        |
        OH
```

The structure $HOCH_2CHCH_2OH$ (with Cl on the central carbon) cannot be correct. It would exhibit only two peaks in its $^{13}C$ NMR spectrum, because the two terminal carbons are equivalent to each other.

**15.46** The observation of a peak at $m/z$ 31 in the mass spectrum of the compound suggests the presence of a primary alcohol. This fragment is most likely $H_2C{=}\overset{+}{O}H$. On the basis of this fact and the appearance of four different carbons in the $^{13}C$ NMR spectrum, the compound is 2-ethyl-1-butanol.

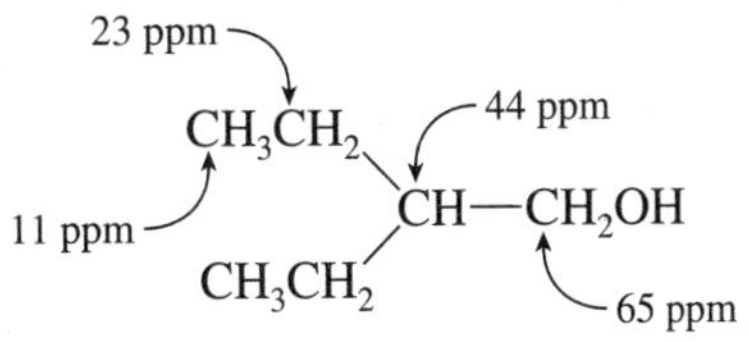

$$(CH_3CH_2)_2CH{-}CH_2OH$$

23 ppm ($CH_2$); 11 ppm ($CH_3$); 44 ppm (CH); 65 ppm ($CH_2OH$)

**15.47–15.49** Solutions to molecular modeling exercises are not provided in this *Study Guide and Solutions Manual.* You should use *Learning By Modeling* for these exercises.

# SELF-TEST

## PART A

**A-1.** For each of the following reactions give the structure of the missing reactant or reagent.

(*a*) ? $\xrightarrow[\text{2. } H_2O]{\text{1. } LiAlH_4}$ cyclohexanol (H, OH on ring carbon)

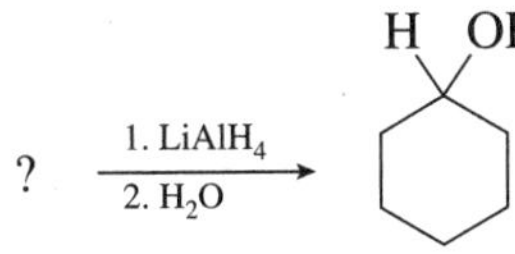

(*b*) ? + $2CH_3CH_2MgBr$ $\xrightarrow[\text{2. } H_3O^+]{\text{1. diethyl ether}}$ $C_6H_5C(OH)(CH_2CH_3)_2$ + $CH_3CH_2OH$

(*c*) $C_6H_5CH_2C(CH_3){=}CH_2$ $\xrightarrow{?}$ $C_6H_5CH_2CH(CH_3)CH_2OH$

(*d*) 1-methylcyclohexene ($CH_3$) $\xrightarrow{?}$ 1-methylcyclohexane-1,2-diol ($CH_3$, OH, OH)

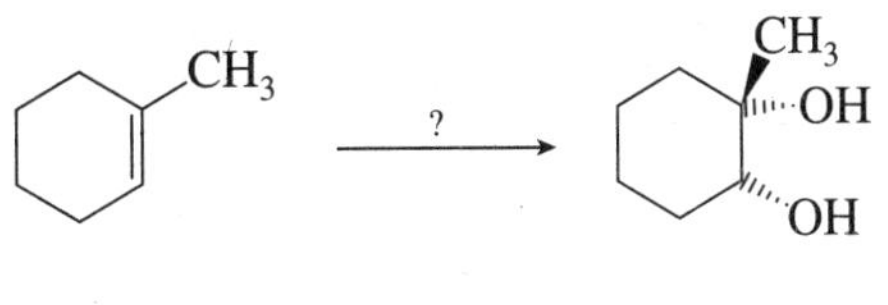

(*e*) $C_6H_5CH_2Br$ $\xrightarrow[\text{2. NaOH}]{\text{1. ?}}$ $C_6H_5CH_2SH$

**A-2.** For the following reactions of 2-phenylethanol, $C_6H_5CH_2CH_2OH$, give the correct reagent or product(s) omitted from the equation.

(*a*) $C_6H_5CH_2CH_2OH$ $\xrightarrow[CH_2Cl_2]{PCC}$ ?

(*b*) $C_6H_5CH_2CH_2OH$ $\xrightarrow{?}$ $CH_3CO_2CH_2CH_2{-}C_6H_5$

(*c*) $C_6H_5CH_2CH_2OH$ (2 mol) $\xrightarrow[\text{heat}]{H^+}$ $H_2O$ + ?

(*d*) $C_6H_5CH_2CH_2OH$ $\xrightarrow{?}$ $C_6H_5CH_2CO_2H$

**A-3.** Write the structure of the major organic product formed in the reaction of 2-propanol with each of the following reagents:

(*a*) Sodium amide ($NaNH_2$)
(*b*) Potassium dichromate ($K_2Cr_2O_7$) in aqueous sulfuric acid, heat
(*c*) PDC in dichloromethane
(*d*) Acetic acid ($CH_3\overset{O}{\overset{\|}{C}}OH$) in the presence of dissolved hydrogen chloride
(*e*) $H_3C$—$C_6H_4$—$SO_2Cl$ in the presence of pyridine
(*f*) $CH_3CH_2$—$C_6H_4$—$\overset{O}{\overset{\|}{C}}Cl$ in the presence of pyridine
(*g*) $CH_3\overset{O}{\overset{\|}{C}}O\overset{O}{\overset{\|}{C}}CH_3$ in the presence of pyridine

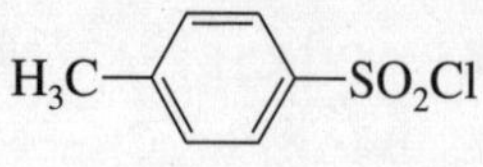

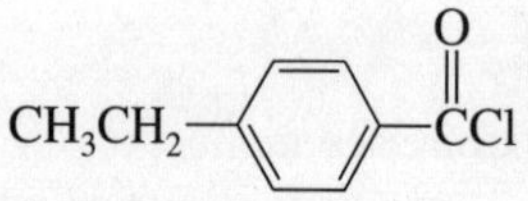

**A-4.** Outline two synthetic schemes for the preparation of 3-methyl-1-butanol using different Grignard reagents.

**A-5.** Give the structure of the reactant, reagent, or product omitted from each of the following. Show stereochemistry where important.

(*a*) [cyclohexane-1,2-diol: OH, H; OH, $CH_3$] $\xrightarrow{HIO_4}$ ?

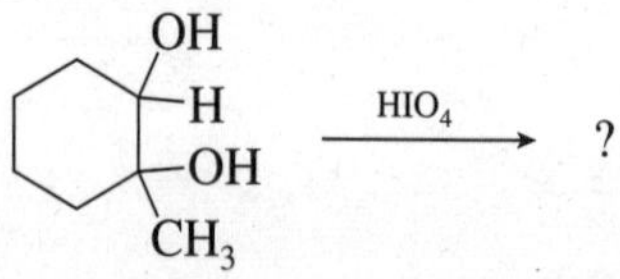

(*b*) ? (a diol) $\xrightarrow[\text{heat}]{H^+}$ [2-methyltetrahydropyran: O, $CH_3$]

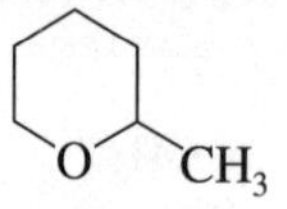

(*c*) ? $\xrightarrow[(CH_3)_3COH,\ HO^-]{OsO_4,\ (CH_3)_3COOH}$ 2,3-butanediol (chiral diastereomer)

**A-6.** Give the reagents necessary to carry out each of the following transformations:

(*a*) Conversion of benzyl alcohol ($C_6H_5CH_2OH$) to benzaldehyde ($C_6H_5CH{=}O$)
(*b*) Conversion of benzyl alcohol to benzoic acid ($C_6H_5CO_2H$)
(*c*) Conversion of $H_2C{=}CHCH_2CH_2CO_2H$ to $H_2C{=}CHCH_2CH_2CH_2OH$
(*d*) Conversion of cyclohexene to *cis*-1,2-cyclohexanediol

**A-7.** Provide structures for compounds A to C in the following reaction scheme:

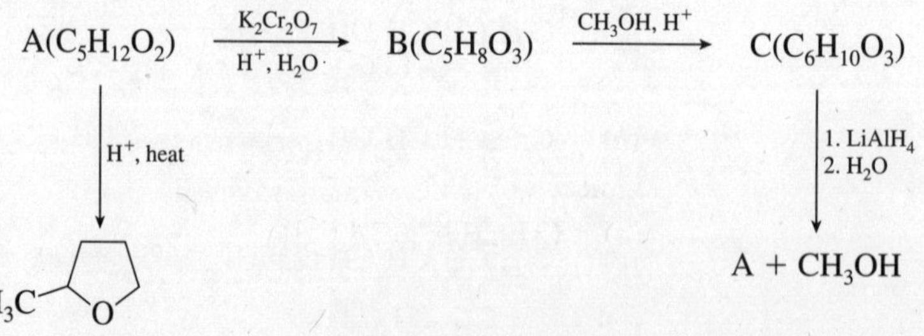

$A(C_5H_{12}O_2) \xrightarrow[H^+,\ H_2O]{K_2Cr_2O_7} B(C_5H_8O_3) \xrightarrow{CH_3OH,\ H^+} C(C_6H_{10}O_3)$

A $\xrightarrow{H^+,\ heat}$ [2-methyltetrahydrofuran: $H_3C$, O]

C $\xrightarrow{1.\ LiAlH_4;\ 2.\ H_2O}$ A + $CH_3OH$

**A-8.** Using any necessary organic or inorganic reagents, outline a scheme for each of the following conversions.

(*a*) $(CH_3)_2C{=}CHCH_3 \xrightarrow{?} (CH_3)_2CHCCH_3$ (with C=O on the second C from the right: $(CH_3)_2CHC(=O)CH_3$)

(*b*) cyclopropyl–$\overset{O}{\overset{\|}{C}}H \xrightarrow{?}$ cyclopropyl–$\overset{O}{\overset{\|}{C}}CH_2CH_3$

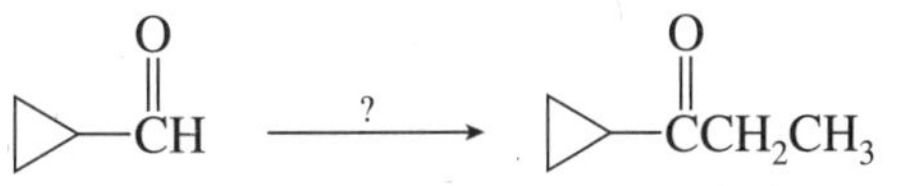

(*c*) $C_6H_5CH_3 \xrightarrow{?} C_6H_5CH_2CH_2CO_2CH_2CH_3$

## PART B

**B-1.** Ethanethiol ($CH_3CH_2SH$) is a gas at room temperature, but ethanol is a liquid. The reason for this is

(*a*) The C—S—H bonds in ethanethiol are linear.
(*b*) The C—O—H bonds in ethanol are linear.
(*c*) Ethanol has a lower molecular weight.
(*d*) Ethanethiol has a higher boiling point.
(*e*) Ethanethiol is less polar.

**B-2.** Which of the following would yield a secondary alcohol after the indicated reaction, followed by hydrolysis if necessary?

(*a*) $LiAlH_4$ + a ketone
(*b*) $CH_3CH_2MgBr$ + an aldehyde
(*c*) 2-Butene + aqueous $H_2SO_4$
(*d*) All of these

**B-3.** What is the major product of the following reaction?

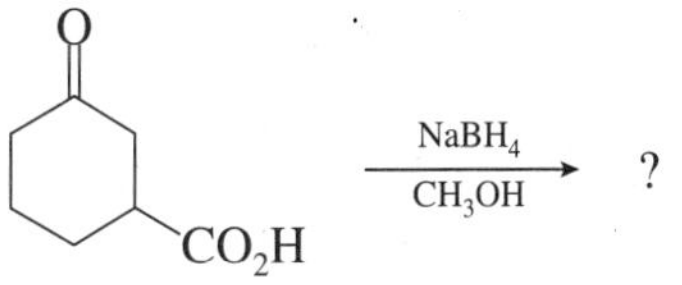

3-oxocyclohexane with $CO_2H$ substituent $\xrightarrow[CH_3OH]{NaBH_4}$ ?

(*a*) cyclohexane with OH and $CH_2OH$ (1,3)

(*b*) cyclohexanone with $CH_2OH$ (3-position)

(*c*) cyclohexane with OH and $CO_2H$ (1,3)

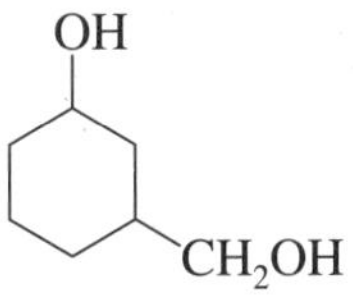

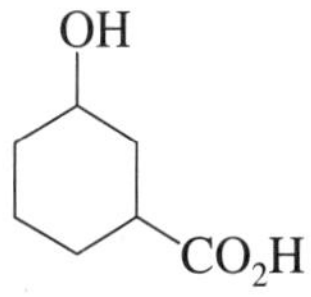

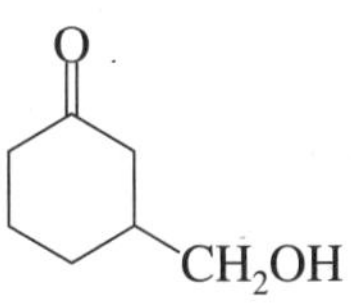

(*d*) cyclohexane with OH and $CO_2CH_3$ (1,3)

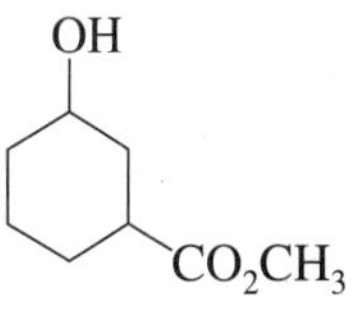

**B-4.** Which of the esters shown, after reduction with $LiAlH_4$ and aqueous workup, will yield two molecules of only a single alcohol?

(*a*) $CH_3CH_2CO_2CH_2CH_3$
(*b*) $C_6H_5CO_2C_6H_5$
(*c*) $C_6H_5CO_2CH_2C_6H_5$
(*d*) None of these

**B-5.** For the following reaction, select the statement that best describes the situation.

$$RCH_2OH + PCC\ [C_5H_5NH^+ClCrO_3^-] \longrightarrow$$

(*a*) The alcohol is oxidized to an acid, and the Cr(VI) is reduced.
(*b*) The alcohol is oxidized to an aldehyde, and the Cr(VI) is reduced.
(*c*) The alcohol is reduced to an aldehyde, and the Cr(III) is oxidized.
(*d*) The alcohol is oxidized to a ketone, and the Cr(VI) is reduced.

**B-6.** What is the product from the following esterification?

$$C_6H_5CH_2CO_2H + CH_3CH_2\text{—}^{18}OH \xrightarrow[\text{heat}]{H^+} ?$$

(*a*) $C_6H_5CH_2\overset{^{18}O}{\overset{\|}{C}}OCH_2CH_3$ (*c*) $C_6H_5CH_2\overset{^{18}O}{\overset{\|}{C}}\text{—}^{18}OCH_2CH_3$

(*b*) $C_6H_5CH_2\overset{O}{\overset{\|}{C}}\text{—}^{18}OCH_2CH_3$ (*d*) $CH_3CH_2\overset{^{18}O}{\overset{\|}{C}}OCH_2C_6H_5$

**B-7.** The following substance acts as a coenzyme in which of the following biological reactions?

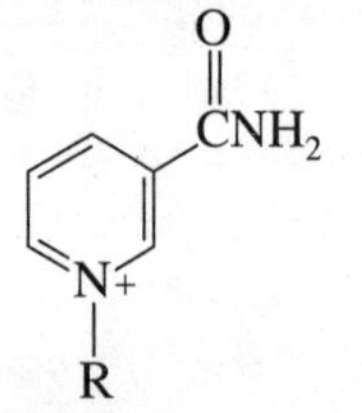

(Structure: pyridinium ring with $\overset{O}{\overset{\|}{C}}NH_2$ substituent, $N^+$–R)

(R = adenine dinucleotide)

(*a*) Alcohol oxidation (*c*) Aldehyde reduction
(*b*) Ketone reduction (*d*) None of these

**B-8.** Which of the following alcohols gives the best yield of dialkyl ether on being heated with a trace of sulfuric acid?
(*a*) 1-Pentanol (*c*) Cyclopentanol
(*b*) 2-Pentanol (*d*) 2-Methyl-2-butanol

**B-9.** What is the major organic product of the following sequence of reactions?

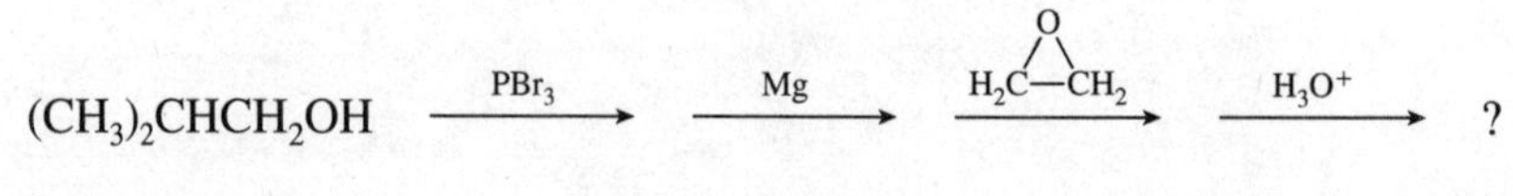

$$(CH_3)_2CHCH_2OH \xrightarrow{PBr_3} \xrightarrow{Mg} \xrightarrow{H_2C\overset{O}{—}CH_2} \xrightarrow{H_3O^+} ?$$

(*a*) $(CH_3)CH\overset{OH}{\overset{|}{C}}HCH_2CH_3$ (*c*) $(CH_3)_2CHCH_2CH_2OH$

(*b*) $(CH_3)_2CHCH_2\overset{OH}{\overset{|}{C}}HCH_3$ (*d*) $(CH_3)_2CHCH_2CH_2CH_2OH$

**B-10.** What is the product of the following reaction?

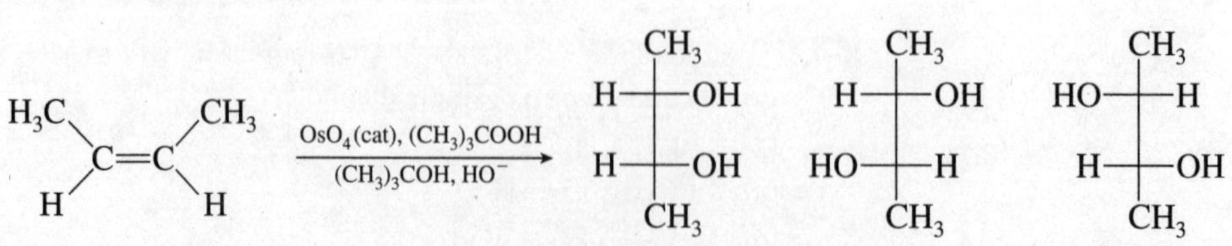

(cis-2-butene: $H_3C$, $CH_3$ on same side; H, H) $\xrightarrow[(CH_3)_3COH,\ HO^-]{OsO_4(\text{cat}),\ (CH_3)_3COOH}$

Fischer projections (top $CH_3$, bottom $CH_3$):

1: H—C—OH / H—C—OH

2: H—C—OH / HO—C—H

3: HO—C—H / H—C—OH

(*a*) Only 1
(*b*) Only 2
(*c*) Only 3
(*d*) A 1:1 mixture of 2 and 3.
(*e*) A 1:1:1 mixture of 1, 2, and 3.

**B-11.** Which reaction is the best method for preparing (*R*)-2-butanol?

(*a*) $CH_3CH_2\overset{O}{\overset{\|}{C}}CH_3 \xrightarrow[\text{2. } H_2O]{\text{1. } LiAlH_4\text{, diethyl ether}}$

(*b*) $(H_3C)(H)(CH_3CH_2)C{-}O\overset{O}{\overset{\|}{C}}CH_3 \xrightarrow[\text{2. } H_2O]{\text{1. } LiAlH_4\text{, diethyl ether}}$

(*c*) $CH_3CH_2\overset{O}{\overset{\|}{C}}H \xrightarrow[\text{2. } H_3O^+]{\text{1. } CH_3MgBr\text{, diethyl ether}}$

(*d*) $CH_3\overset{O}{\overset{\|}{C}}H \xrightarrow[\text{2. } H_3O^+]{\text{1. } CH_3CH_2Li\text{, diethyl ether}}$

(*e*) None of these would be effective.

**B-12.** An organic compound B is formed by the reaction of ethylmagnesium iodide ($CH_3CH_2MgI$) with a substance A, followed by treatment with dilute aqueous acid. Compound B does *not* react with PCC or PDC in dichloromethane. Which of the following is a possible candidate for A?

(*a*) $CH_3\overset{O}{\overset{\|}{C}}H$
(*b*) $H_2C{=}O$
(*c*) $H_2C{-}CH_2$ (epoxide, O bridging)
(*d*) $CH_3CH_2\overset{O}{\overset{\|}{C}}CH_3$
(*e*) None of these

**B-13.** Which alcohol of molecular formula $C_5H_{12}O$ has the fewest signals in its $^{13}C$ NMR spectrum?

(*a*) 1-Pentanol
(*b*) 2-Pentanol
(*c*) 2-Methyl-2-butanol
(*d*) 3-Methyl-2-butanol
(*e*) 2,2-Dimethyl-1-propanol

**B-14.** Which of the following reagents would carry out the following transformation? (D = $^2H$, the mass-2 isotope of hydrogen)

$C_6H_5\overset{O}{\overset{\|}{C}}CH_3 \xrightarrow{?} C_6H_5C(OH)(D)CH_3$

(*a*) $NaBD_4$ in $CH_3OH$
(*b*) $NaBD_4$ in $CH_3OD$
(*c*) $LiAlH_4$, then $D_2O$
(*d*) $LiAlD_4$, then $D_2O$
(*e*) $NaBH_4$ in $CH_3OD$

**B-15.** Which sequence of steps describes the best synthesis of 2-methyl-3-pentanone?

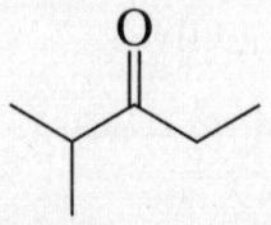

2-Methyl-3-pentanone

(*a*) 1. 1-Propanol + $(CH_3)_2CHMgBr$, diethyl ether
2. $H_3O^+$
3. PDC, $CH_2Cl_2$

(*b*) 1. 1-Propanol + $Na_2Cr_2O_7$, $H_2SO_4$, $H_2O$, heat
2. $SOCl_2$
3. $(CH_3)_2CHCl$, $AlCl_3$

(*c*) 1. 1-Propanol + PCC, $CH_2Cl_2$
2. $(CH_3)_2CHLi$, diethyl ether
3. $H_3O^+$
4. $Na_2Cr_2O_7$, $H_2SO_4$, $H_2O$, heat

(*d*) 1. 2-Propanol + $Na_2Cr_2O_7$, $H_2SO_4$, $H_2O$, heat
2. $CH_3CH_2CH_2Li$, diethyl ether
3. $H_3O^+$
4. PCC, $CH_2Cl_2$

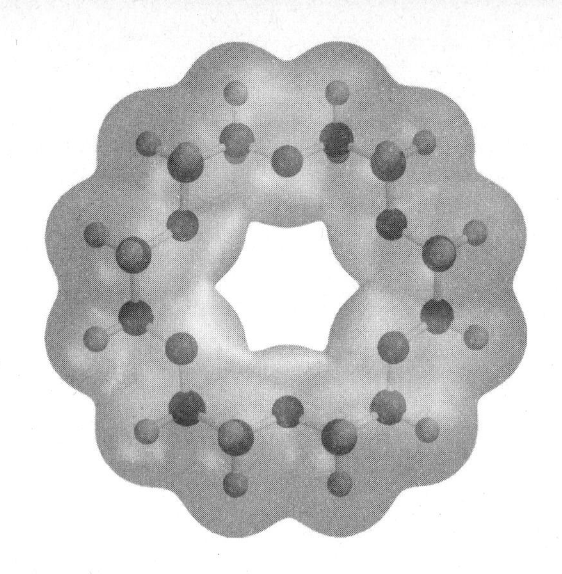

# CHAPTER 16

## ETHERS, EPOXIDES, AND SULFIDES

## SOLUTIONS TO TEXT PROBLEMS

**16.1** (*b*) Oxirane is the IUPAC name for ethylene oxide. A chloromethyl group ($ClCH_2$—) is attached to position 2 of the ring in 2-(chloromethyl)oxirane.

$$\underset{\text{Oxirane}}{\overset{3}{H_2C}—\overset{2}{C}H_2 \text{ (bridged by O)}} \qquad \underset{\text{2-(Chloromethyl)oxirane}}{H_2C—CHCH_2Cl \text{ (bridged by O)}}$$

This compound is more commonly known as **epichlorohydrin.**

(*c*) Epoxides may be named by adding the prefix *epoxy* to the IUPAC name of a parent compound, specifying by number both atoms to which the oxygen is attached.

$$\underset{\text{1-Butene}}{CH_3CH_2CH{=}CH_2} \qquad \underset{\text{3,4-Epoxy-1-butene}}{H_2C—CHCH{=}CH_2 \text{ (bridged by O)}}$$

**16.2** 1,2-Epoxybutane and tetrahydrofuran both have the molecular formula $C_4H_8O$—that is, they are constitutional isomers—and so it is appropriate to compare their heats of combustion directly. Angle strain from the three-membered ring of 1,2-epoxybutane causes it to have more internal energy than tetrahydrofuran, and its combustion is more exothermic.

$$H_2C—CHCH_2CH_3 \text{ (bridged by O)}$$

1,2-Epoxybutane;
heat of combustion 2546 kJ/mol
(609.1 kcal/mol)

Tetrahydrofuran;
heat of combustion 2499 kJ/mol
(597.8 kcal/mol)

**16.3** An ether can function only as a proton acceptor in a hydrogen bond, but an alcohol can be either a proton acceptor or a donor. The only hydrogen bond possible between an ether and an alcohol is therefore the one shown:

$$R_2\ddot{O}: \cdots\cdots H{-}\ddot{O}{-}R$$

Ether Alcohol

**16.4** The compound is 1,4-dioxane; it has a six-membered ring and two oxygens separated by $CH_2$—$CH_2$ units.

1,4-dioxane
("6-crown-2")

**16.5** Protonation of the carbon–carbon double bond leads to the more stable carbocation.

$$(CH_3)_2C{=}CH_2 + H^+ \longrightarrow (CH_3)_2\overset{+}{C}{-}CH_3$$

2-Methylpropene → *tert*-Butyl cation

Methanol acts as a nucleophile to capture *tert*-butyl cation.

$$(CH_3)_2\overset{+}{C}{-}CH_3 + :\ddot{O}(CH_3)H \longrightarrow (CH_3)_3C{-}\overset{+}{\ddot{O}}CH_3 \text{ (with H on O)}$$

Deprotonation of the alkyloxonium ion leads to formation of *tert*-butyl methyl ether.

$$(CH_3)_3C{-}\overset{+}{\ddot{O}}CH_3\text{(H)} + :\ddot{O}(CH_3)H \longrightarrow (CH_3)_3C\ddot{O}CH_3 + H_2\overset{+}{\ddot{O}}CH_3$$

*tert*-Butyl methyl ether

**16.6** Both alkyl groups in benzyl ethyl ether are primary, thus either may come from the alkyl halide in a Williamson ether synthesis. The two routes to benzyl ethyl ether are

$$C_6H_5CH_2ONa + CH_3CH_2Br \longrightarrow C_6H_5CH_2OCH_2CH_3 + NaBr$$

Sodium benzyloxide, Bromoethane → Benzyl ethyl ether, Sodium bromide

$$C_6H_5CH_2Br + CH_3CH_2ONa \longrightarrow C_6H_5CH_2OCH_2CH_3 + NaBr$$

Benzyl bromide, Sodium ethoxide → Benzyl ethyl ether, Sodium bromide

**16.7** (*b*) A primary carbon and a secondary carbon are attached to the ether oxygen. The secondary carbon can only be derived from the alkoxide, because secondary alkyl halides cannot be used in the preparation of ethers by the Williamson method. The only effective method uses an allyl halide and sodium isopropoxide.

$$(CH_3)_2CHONa + H_2C{=}CHCH_2Br \longrightarrow H_2C{=}CHCH_2OCH(CH_3)_2 + NaBr$$

Sodium isopropoxide, Allyl bromide → Allyl isopropyl ether, Sodium bromide

Elimination will be the major reaction of an isopropyl halide with an alkoxide base.

(*c*) Here the ether is a mixed primary–tertiary one. The best combination is the one that uses the primary alkyl halide.

$$(CH_3)_3COK + C_6H_5CH_2Br \longrightarrow (CH_3)_3COCH_2C_6H_5 + KBr$$

Potassium *tert*-butoxide; Benzyl bromide; Benzyl *tert*-butyl ether; Potassium bromide

The reaction between $(CH_3)_3CBr$ and $C_6H_5CH_2O^-$ is elimination, not substitution.

**16.8**

$$CH_3CH_2OCH_2CH_3 + 6O_2 \longrightarrow 4CO_2 + 5H_2O$$

Diethyl ether; Oxygen; Carbon dioxide; Water

**16.9** (*b*) If benzyl bromide is the only organic product from reaction of a dialkyl ether with hydrogen bromide, then both alkyl groups attached to oxygen must be benzyl.

$$C_6H_5CH_2OCH_2C_6H_5 \xrightarrow[\text{heat}]{\text{HBr}} 2C_6H_5CH_2Br + H_2O$$

Dibenzyl ether; Benzyl bromide; Water

(*c*) Since *1 mole of a dihalide,* rather than 2 moles of a monohalide, is produced per mole of ether, the ether must be cyclic.

$$\text{Tetrahydropyran} \xrightarrow[\text{heat}]{\text{2HBr}} BrCH_2CH_2CH_2CH_2CH_2Br + H_2O$$

Tetrahydropyran; 1,5-Dibromopentane; Water

**16.10** As outlined in text Figure 16.4, the first step is protonation of the ether oxygen to give a dialkyloxonium ion.

$$\text{Tetrahydrofuran} + H{-}\ddot{I}: \rightleftharpoons \text{Dialkyloxonium ion } (:\overset{+}{O}{-}H) + :\ddot{I}:^-$$

Tetrahydrofuran; Hydrogen iodide; Dialkyloxonium ion; Iodide ion

In the second step, nucleophilic attack of the halide ion on carbon of the oxonium ion gives 4-iodo-1-butanol.

$$:\ddot{I}:^- + \text{Dialkyloxonium ion} \longrightarrow ICH_2CH_2CH_2CH_2\ddot{O}H$$

Iodide ion; Dialkyloxonium ion; 4-Iodo-1-butanol

The remaining two steps of the mechanism correspond to those in which an alcohol is converted to an alkyl halide, as discussed in Chapter 4.

$$ICH_2CH_2CH_2CH_2\ddot{O}H + H{-}\ddot{I}: \rightleftharpoons ICH_2CH_2CH_2CH_2\overset{+}{O}H_2 + :\ddot{I}:^-$$

4-Iodo-1-butanol; Hydrogen iodide

$$ICH_2CH_2CH_2CH_2\overset{+}{O}H_2 + :\ddot{I}:^- \longrightarrow ICH_2CH_2CH_2CH_2I + H_2\ddot{O}:$$

1,4-Diiodobutane; Water

**16.11** The cis epoxide is achiral. It is a meso form containing a plane of symmetry. The trans isomer is chiral; its two mirror-image representations are not superposable.

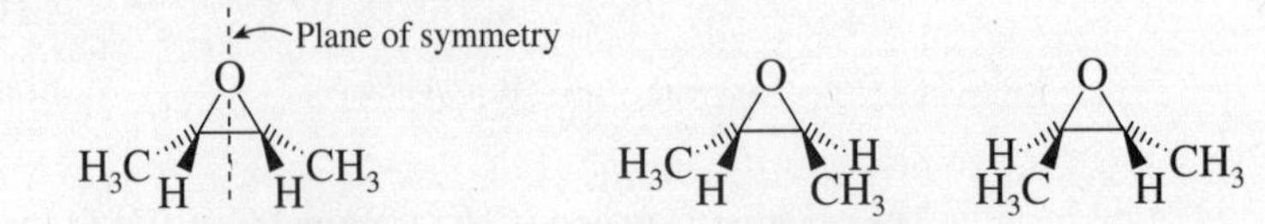

Plane of symmetry

*cis*-2,3-Epoxybutane (Plane of symmetry passes through oxygen and midpoint of carbon–carbon bond.)

Nonsuperposable mirror-image (enantiomeric) forms of *trans*-2,3-epoxybutane

Neither the cis nor the trans epoxide is optically active when formed from the alkene. The cis epoxide is achiral; it cannot be optically active. The trans epoxide is capable of optical activity but is formed as a racemic mixture because achiral starting materials are used.

**16.12** (*b*) Azide ion $[:\ddot{N}{=}N{=}\ddot{N}:]^-$ is a good nucleophile, reacting readily with ethylene oxide to yield 2-azidoethanol.

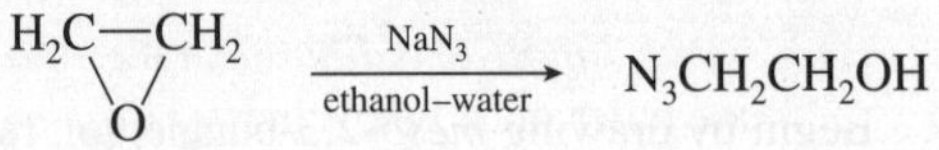

$$H_2C\text{—}CH_2 \text{ (epoxide)} \xrightarrow[\text{ethanol–water}]{NaN_3} N_3CH_2CH_2OH$$

Ethylene oxide → 2-Azidoethanol

(*c*) Ethylene oxide is hydrolyzed to ethylene glycol in the presence of aqueous base.

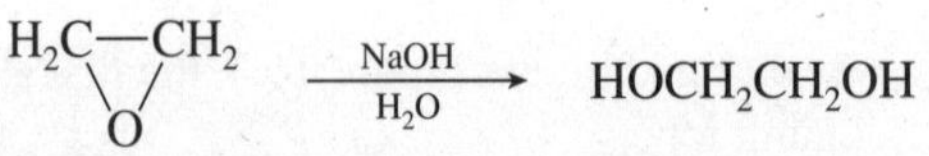

$$H_2C\text{—}CH_2 \text{ (epoxide)} \xrightarrow[H_2O]{NaOH} HOCH_2CH_2OH$$

Ethylene oxide → Ethylene glycol

(*d*) Phenyllithium reacts with ethylene oxide in a manner similar to that of a Grignard reagent.

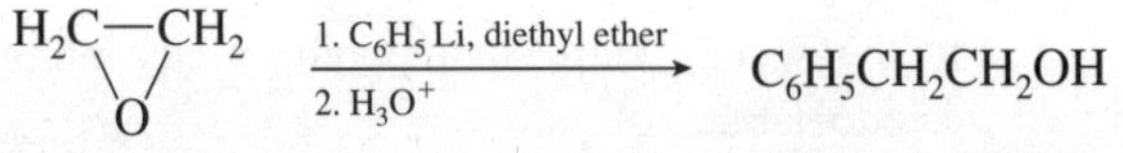

$$H_2C\text{—}CH_2 \text{ (epoxide)} \xrightarrow[2.\ H_3O^+]{1.\ C_6H_5Li,\ \text{diethyl ether}} C_6H_5CH_2CH_2OH$$

Ethylene oxide → 2-Phenylethanol

(*e*) The nucleophilic species here is the acetylenic anion $CH_3CH_2C{\equiv}C:^-$, which attacks a carbon atom of ethylene oxide to give 3-hexyn-1-ol.

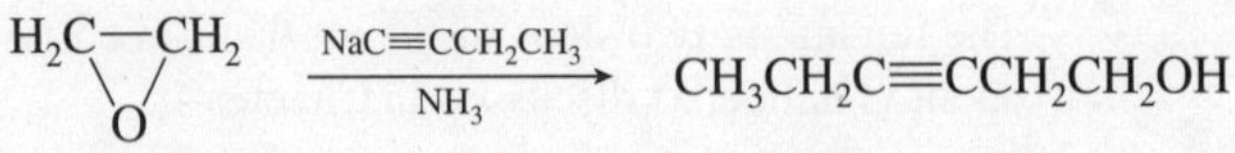

$$H_2C\text{—}CH_2 \text{ (epoxide)} \xrightarrow[NH_3]{NaC{\equiv}CCH_2CH_3} CH_3CH_2C{\equiv}CCH_2CH_2OH$$

Ethylene oxide → 3-Hexyn-1-ol (48%)

**16.13** Nucleophilic attack at C-2 of the starting epoxide will be faster than attack at C-1, because C-1 is more sterically hindered. Compound A, corresponding to attack at C-1, is not as likely as compound B. Compound B not only arises by methoxide ion attack at C-2 but also satisfies the stereochemical requirement that epoxide ring opening take place with inversion of configuration at the site of substitution. Compound B is correct. Compound C, although it is formed by methoxide substitution at the less crowded carbon of the epoxide, is wrong stereochemically. It requires substitution with retention of configuration, which is not the normal mode of epoxide ring opening.

**16.14** Acid-catalyzed nucleophilic ring opening proceeds by attack of methanol at the more substituted carbon of the protonated epoxide. Inversion of configuration is observed at the site of attack. The correct product is compound A.

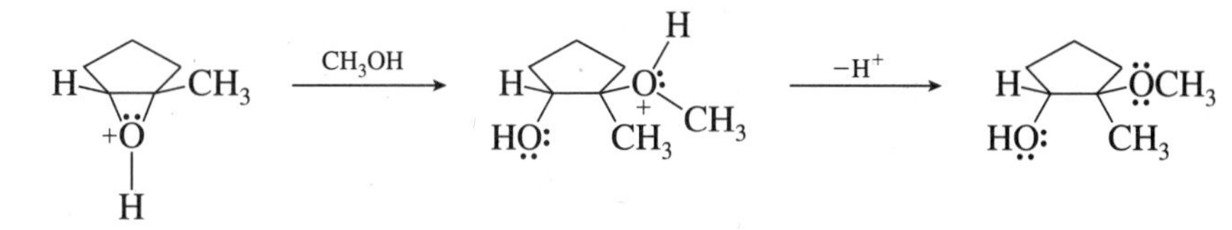

$\xrightarrow{CH_3OH}$ $\xrightarrow{-H^+}$

Protonated form of 1-methyl-1,2-epoxycyclopentane

Compound A

The nucleophilic ring openings in both this problem and Problem 16.13 occur by inversion of configuration. Attack under basic conditions by methoxide ion, however, occurs at the *less* hindered carbon of the epoxide ring, whereas attack by methanol under acid-catalyzed conditions occurs at the *more* substituted carbon.

**16.15** Begin by drawing *meso*-2,3-butanediol, recalling that a meso form is achiral. The eclipsed conformation has a plane of symmetry.

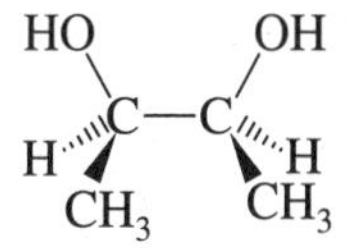

*meso*-2,3-Butanediol

Epoxidation followed by acid-catalyzed hydrolysis results in anti addition of hydroxyl groups to the double bond. *trans*-2-Butene is the required starting material.

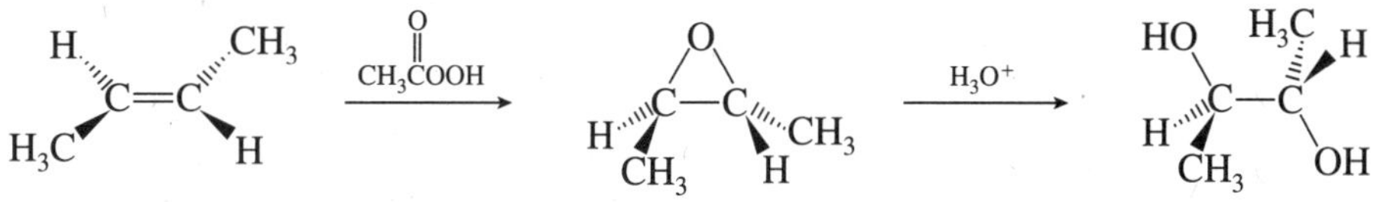

$\xrightarrow{CH_3COOH}$ (peroxyacetic acid) $\xrightarrow{H_3O^+}$

*trans*-2-Butene — *trans*-2,3-Epoxybutane — *meso*-2,3-Butanediol

Osmium tetraoxide hydroxylation is a method of achieving syn hydroxylation. The necessary starting material is *cis*-2-butene.

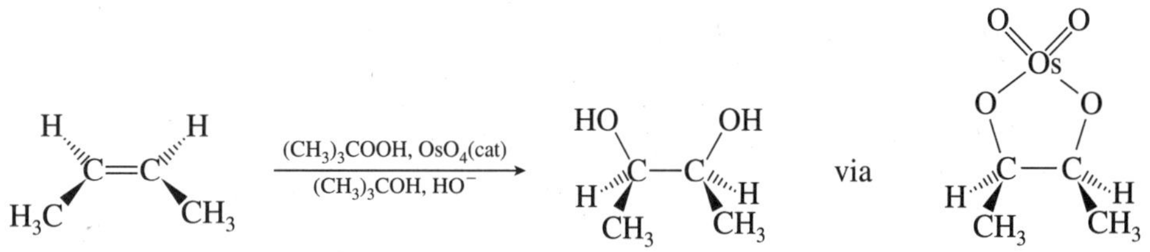

$\xrightarrow[(CH_3)_3COH,\ HO^-]{(CH_3)_3COOH,\ OsO_4(cat)}$ via

*cis*-2-Butene — *meso*-2,3-Butanediol

**16.16** Reaction of (*R*)-2-octanol with *p*-toluenesulfonyl chloride yields a *p*-toluenesulfonate ester (tosylate) having the same configuration; the stereogenic center is not involved in this step. Reaction of the tosylate with a nucleophile proceeds by inversion of configuration in an $S_N2$ process. The product has the *S* configuration.

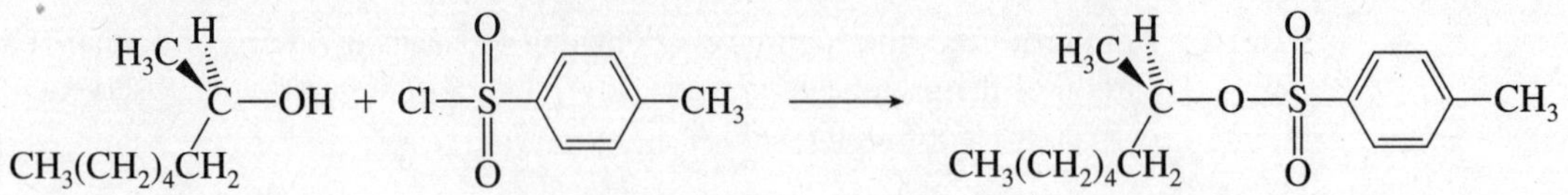

$$\text{(R)-2-Octanol} + p\text{-Toluenesulfonyl chloride } (Cl{-}SO_2{-}C_6H_4{-}CH_3) \longrightarrow \text{(R)-1-Methylheptyl tosylate}$$

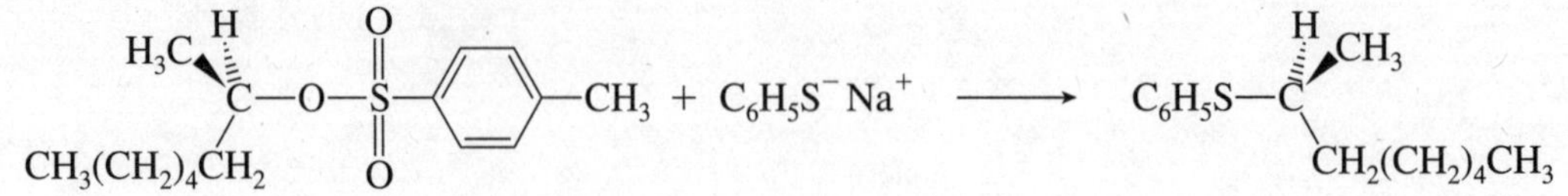

(*R*)-2-Octanol — *p*-Toluenesulfonyl chloride — (*R*)-1-Methylheptyl tosylate

$$\text{(R)-1-Methylheptyl tosylate} + C_6H_5S^-\,Na^+ \longrightarrow C_6H_5S{-}CH(CH_3)CH_2(CH_2)_4CH_3$$

(*R*)-1-Methylheptyl tosylate — Sodium benzenethiolate — (*S*)-1-Methylheptyl phenyl sulfide

**16.17** Phenyl vinyl sulfoxide lacks a plane of symmetry and is chiral. Phenyl vinyl sulfone is achiral; a plane of symmetry passes through the phenyl and vinyl groups and the central sulfur atom.

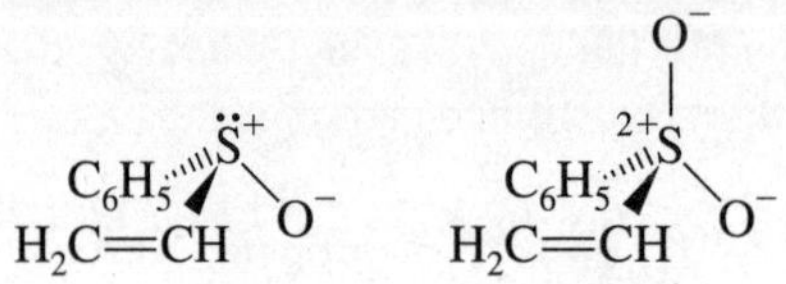

Phenyl vinyl sulfoxide (chiral) — Phenyl vinyl sulfone (achiral)

**16.18** As shown in the text, dodecyldimethylsulfonium iodide may be prepared by reaction of dodecyl methyl sulfide with methyl iodide. An alternative method is the reaction of dodecyl iodide with dimethyl sulfide.

$$(CH_3)_2S + CH_3(CH_2)_{10}CH_2I \longrightarrow CH_3(CH_2)_{10}CH_2\overset{+}{S}(CH_3)_2\ I^-$$

Dimethyl sulfide — Dodecyl iodide — Dodecyldimethylsulfonium iodide

The reaction of a sulfide with an alkyl halide is an $S_N2$ process. The faster reaction will be the one that uses the less sterically hindered alkyl halide. The method presented in the text will proceed faster.

**16.19** The molecular ion from *sec*-butyl ethyl ether can also fragment by cleavage of a carbon–carbon bond in its ethyl group to give an oxygen-stabilized cation of $m/z$ 87.

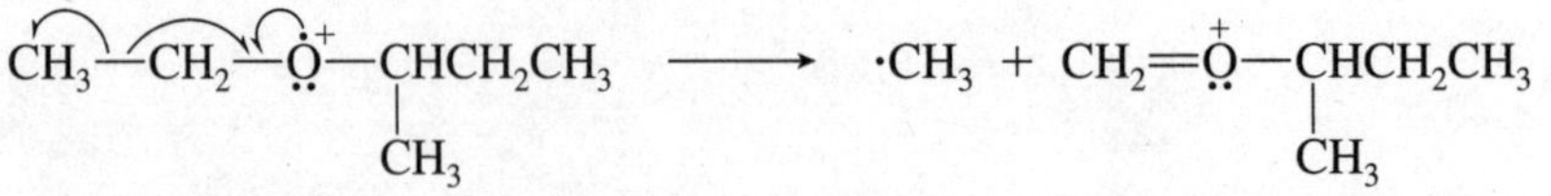

$$CH_3{-}CH_2{-}\overset{+}{\ddot{O}}{-}CH(CH_3)CH_2CH_3 \longrightarrow \cdot CH_3 + CH_2{=}\overset{+}{\ddot{O}}{-}CH(CH_3)CH_2CH_3$$

$m/z$ 87

**16.20** All the constitutionally isomeric ethers of molecular formula $C_5H_{12}O$ belong to one of two general groups: $CH_3OC_4H_9$ and $CH_3CH_2OC_3H_7$. Thus, we have

$CH_3OCH_2CH_2CH_2CH_3$ — Butyl methyl ether

$CH_3OCH(CH_3)CH_2CH_3$ — *sec*-Butyl methyl ether

$CH_3OCH_2CH(CH_3)_2$ — Isobutyl methyl ether

$CH_3OC(CH_3)_3$ — *tert*-Butyl methyl ether

$CH_3CH_2OCH_2CH_2CH_3$ — Ethyl propyl ether

and

$CH_3CH_2OCH(CH_3)_2$ — Ethyl isopropyl ether

These ethers could also have been named as "alkoxyalkanes." Thus, *sec*-butyl methyl ether would become 2-methoxybutane.

**16.21** Isoflurane and enflurane are both halogenated derivatives of ethyl methyl ether.

**Isoflurane:**

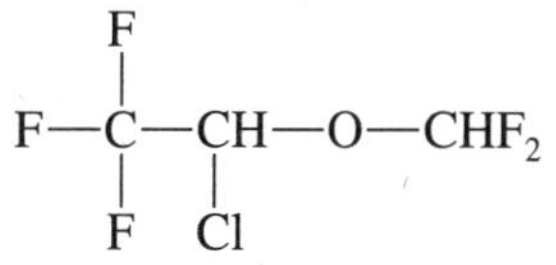

$$\mathrm{F{-}\overset{\displaystyle F}{\underset{\displaystyle F}{C}}{-}\underset{\displaystyle Cl}{CH}{-}O{-}CHF_2}$$

1-Chloro-2,2,2-trifluoroethyl difluoromethyl ether

**Enflurane:**

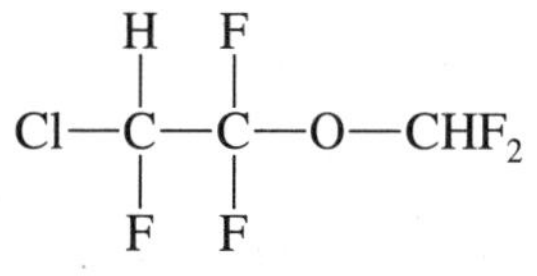

$$\mathrm{Cl{-}\overset{\displaystyle H}{\underset{\displaystyle F}{C}}{-}\overset{\displaystyle F}{\underset{\displaystyle F}{C}}{-}O{-}CHF_2}$$

2-Chloro-1,1,2-trifluoroethyl difluoromethyl ether

**16.22** (*a*) The parent compound is cyclopropane. It has a three-membered epoxide function, and thus a reasonable name is epoxycyclopropane. Numbers locating positions of attachment (as in "1,2-epoxycyclopropane") are not necessary, because no other structures (1,3 or 2,3) are possible here.

O

Epoxycyclopropane

(*b*) The longest continuous carbon chain has seven carbons, and so the compound is named as a derivative of heptane. The epoxy function bridges C-2 and C-4. Therefore

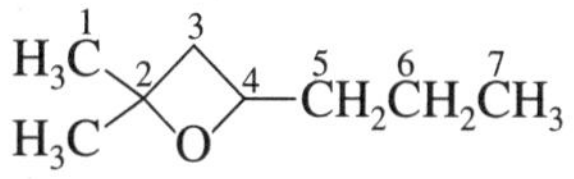

$H_3C$ 1 2 3 4 O $CH_2CH_2CH_3$ 5 6 7; $H_3C$

is 2-methyl-2,4-epoxyheptane.

(*c*) The oxygen atom bridges the C-1 and C-4 atoms of a cyclohexane ring.

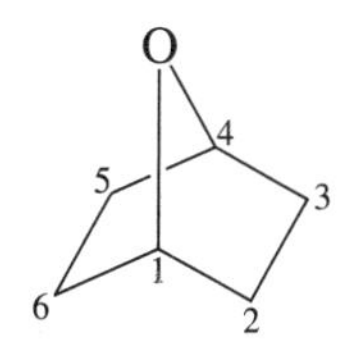


1,4-Epoxycyclohexane

(*d*) Eight carbon atoms are continuously linked and bridged by an oxygen. We name the compound as an epoxy derivative of cyclooctane.

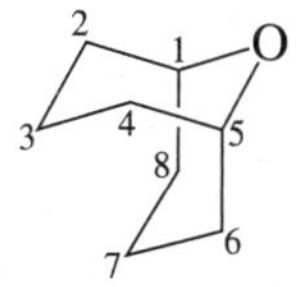

O 1 2 3 4 5 6 7 8

1,5-Epoxycyclooctane

**16.23** (*a*) There are three methyl-substituted thianes, two of which are chiral.

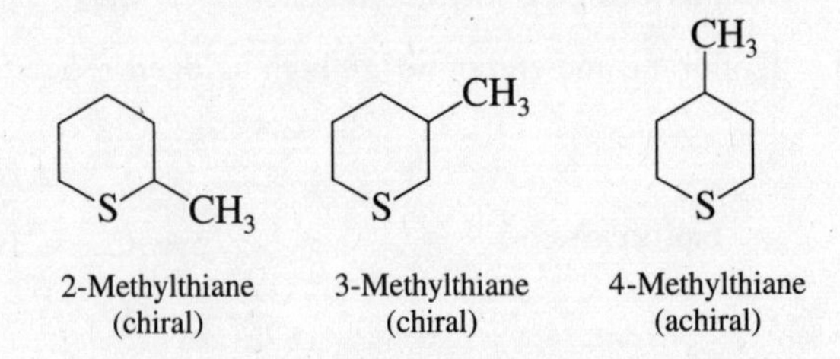

2-Methylthiane (chiral) 3-Methylthiane (chiral) 4-Methylthiane (achiral)

(*b*) The locants in the name indicate the positions of the sulfur atoms in 1,4-dithiane and 1,3,5-trithiane.

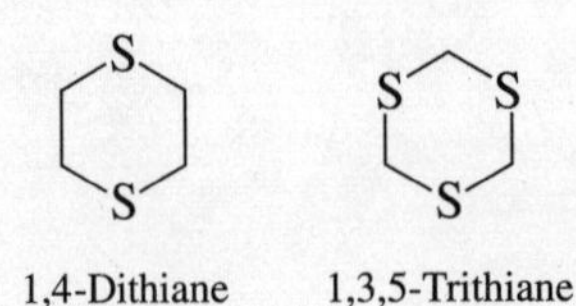

1,4-Dithiane 1,3,5-Trithiane

(*c*) Disulfides possess two adjacent sulfur atoms. 1,2-Dithiane is a disulfide.

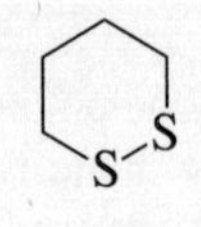

1,2-Dithiane

(*d*) Two chair conformations of the sulfoxide derived from thiane are possible; the oxygen atom may be either equatorial or axial.

**16.24** Intramolecular hydrogen bonding between the hydroxyl group and the ring oxygens is possible when the hydroxyl group is axial but not when it is equatorial.

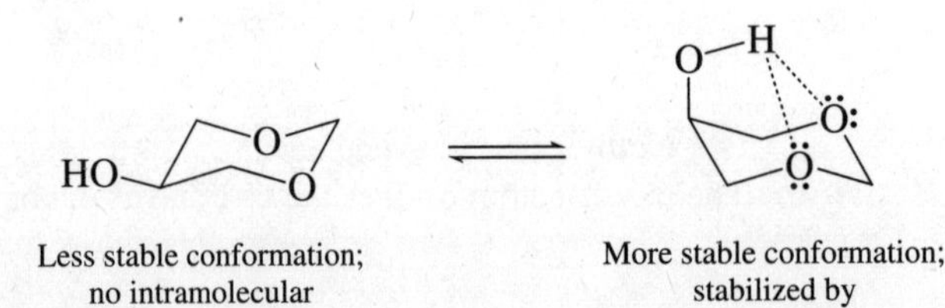

Less stable conformation; no intramolecular hydrogen bonding

More stable conformation; stabilized by hydrogen bonding

**16.25** The ethers that are to be prepared are

$$CH_3OCH_2CH_2CH_3 \qquad CH_3OCH(CH_3)_2 \qquad \text{and} \qquad CH_3CH_2OCH_2CH_3$$

Methyl propyl ether Isopropyl methyl ether Diethyl ether

First examine the preparation of each ether by the Williamson method. Methyl propyl ether can be prepared in two ways:

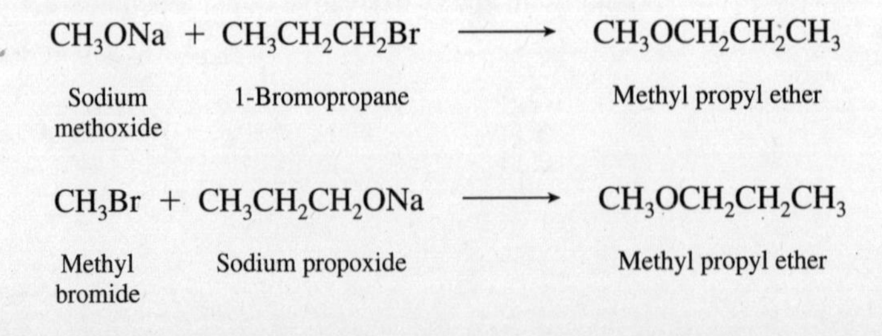

$$CH_3ONa + CH_3CH_2CH_2Br \longrightarrow CH_3OCH_2CH_2CH_3$$

Sodium methoxide 1-Bromopropane Methyl propyl ether

$$CH_3Br + CH_3CH_2CH_2ONa \longrightarrow CH_3OCH_2CH_2CH_3$$

Methyl bromide Sodium propoxide Methyl propyl ether

Either combination is satisfactory. The necessary reagents are prepared as shown.

$$CH_3OH \xrightarrow{Na} CH_3ONa$$

Methanol → Sodium methoxide

$$CH_3CH_2CH_2OH \xrightarrow[\text{(or HBr)}]{PBr_3} CH_3CH_2CH_2Br$$

1-Propanol → 1-Bromopropane

$$CH_3OH \xrightarrow[\text{(or HBr)}]{PBr_3} CH_3Br$$

Methanol → Methyl bromide

$$CH_3CH_2CH_2OH \xrightarrow{Na} CH_3CH_2CH_2ONa$$

1-Propanol → Sodium propoxide

Isopropyl methyl ether is best prepared by the reaction

$$CH_3Br + (CH_3)_2CHONa \longrightarrow CH_3OCH(CH_3)_2$$

Methyl bromide + Sodium isopropoxide → Isopropyl methyl ether

The reaction of sodium methoxide with isopropyl bromide will proceed mainly by elimination. Methyl bromide is prepared as shown previously; sodium isopropoxide can be prepared by adding sodium to isopropyl alcohol.

Diethyl ether may be prepared as outlined:

$$CH_3CH_2OH \xrightarrow{Na} CH_3CH_2ONa$$

Ethanol → Sodium ethoxide

$$CH_3CH_2OH \xrightarrow[\text{(or HBr)}]{PBr_3} CH_3CH_2Br$$

Ethanol → Ethyl bromide

$$CH_3CH_2ONa + CH_3CH_2Br \longrightarrow CH_3CH_2OCH_2CH_3 + NaBr$$

Sodium ethoxide + Ethyl bromide → Diethyl ether + Sodium bromide

**16.26** (*a*) Secondary alkyl halides react with alkoxide bases by E2 elimination as the major pathway. The Williamson ether synthesis is not a useful reaction with secondary alkyl halides.

$$CH_3CH_2CH(ONa)CH_3 + C_6H_{11}Br \longrightarrow CH_3CH_2CH(OH)CH_3 + C_6H_{10} + NaBr$$

Sodium 2-butanolate + Bromocyclohexane → 2-Butanol + Cyclohexene + Sodium bromide

(*b*) Sodium alkoxide acts as a nucleophile toward iodoethane to yield an alkyl ethyl ether.

$$(CH_3CH_2)(CH_3)(H)C{-}O^- \ Na^+ + CH_3CH_2I \longrightarrow (CH_3CH_2)(CH_3)(H)C{-}OCH_2CH_3$$

(*R*)-2-Ethoxybutane

The ether product has the same absolute configuration as the starting alkoxide because no bonds to the stereogenic center are made or broken in the reaction.

(*c*) Vicinal halohydrins are converted to epoxides on being treated with base.

$CH_3CH_2CH(OH)CH_2Br \xrightarrow{NaOH} CH_3CH_2CH(O^-)—CH_2—Br \longrightarrow CH_3CH_2CH\text{-}CH_2$ (epoxide, O bridging)

1-Bromo-2-butanol → 1,2-Epoxybutane

(*d*) The reactants, an alkene plus a peroxy acid, are customary ones for epoxide preparation. The reaction is a stereospecific syn addition of oxygen to the double bond.

(*Z*)-1-Phenylpropene + Peroxybenzoic acid ($C_6H_5COOH$, O above C) → *cis*-2-Methyl-3-phenyloxirane + Benzoic acid ($C_6H_5COH$, O above C)

(*e*) Azide ion is a good nucleophile and attacks the epoxide function. Substitution occurs at carbon with inversion of configuration. The product is *trans*-2-azidocyclohexanol.

1,2-Epoxycyclohexane $\xrightarrow[\text{dioxane–water}]{NaN_3}$ *trans*-2-Azidocyclohexanol (61%)

(*f*) Ammonia is a nucleophile capable of reacting with epoxides. It attacks the less hindered carbon of the epoxide function.

2-(*o*-Bromophenyl)-2-methyloxirane $\xrightarrow[\text{methanol}]{NH_3}$ 1-Amino-2-(*o*-bromophenyl)-2-propanol

Aryl halides do not react with nucleophiles under these conditions, and so the bromine substituent on the ring is unaffected.

(*g*) Methoxide ion attacks the less substituted carbon of the epoxide ring with inversion of configuration.

$^-:\ddot{O}CH_3$ + 1-Benzyl-1,2-epoxycyclohexane → 1-Benzyl-*trans*-2-methoxycyclohexanol (98%)

(*h*) Under acidic conditions, substitution is favored at the carbon that can better support a positive charge. Aryl substituents stabilize carbocations, making the benzylic position the one that is attacked in an aryl substituted epoxide.

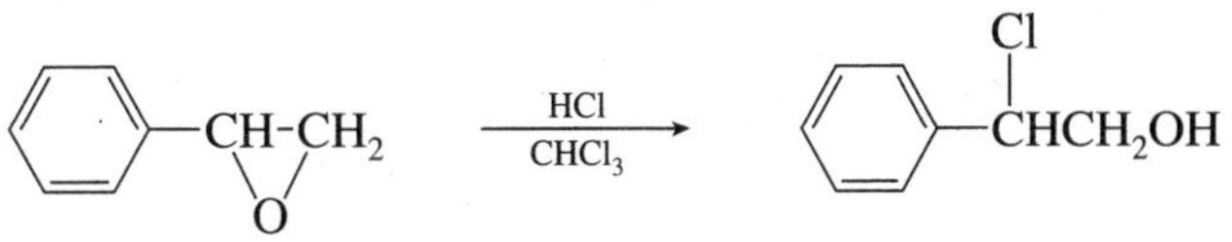

$$C_6H_5\text{-}CH\text{-}CH_2 \text{ (epoxide, O bridge)} \xrightarrow[CHCl_3]{HCl} C_6H_5\text{-}CH(Cl)CH_2OH$$

2-Phenyloxirane → 2-Chloro-2-phenylethanol (71%)

(*i*) Tosylate esters undergo substitution with nucleophiles such as sodium butanethiolate.

$$CH_3(CH_2)_{16}CH_2OTs + CH_3CH_2CH_2CH_2SNa \longrightarrow CH_3CH_2CH_2CH_2SCH_2(CH_2)_{16}CH_3$$

Octadecyl tosylate + Sodium butanethiolate → Butyl octadecyl sulfide

(*j*) Nucleophilic substitution proceeds with inversion of configuration.

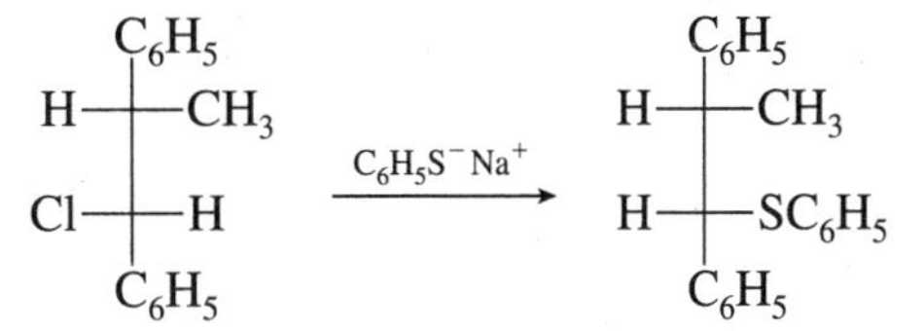

Fischer projection: $C_6H_5$ (top); H—C—$CH_3$; Cl—C—H; $C_6H_5$ (bottom) $\xrightarrow{C_6H_5S^-\,Na^+}$ $C_6H_5$ (top); H—C—$CH_3$; H—C—$SC_6H_5$; $C_6H_5$ (bottom)

**16.27** Oxidation of 4-*tert*-butylthiane yields two sulfoxides that are diastereomers of each other.

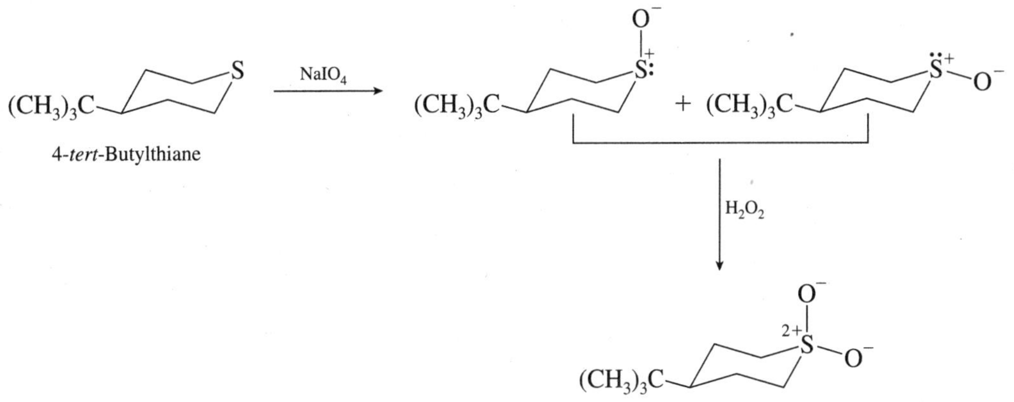

$(CH_3)_3C$-thiane (4-*tert*-Butylthiane) $\xrightarrow{NaIO_4}$ $(CH_3)_3C$-thiane-S$^+$–O$^-$ (axial O) + $(CH_3)_3C$-thiane-S$^+$–O$^-$ (equatorial O)

Both $\xrightarrow{H_2O_2}$ $(CH_3)_3C$-thiane-S$^{2+}$(O$^-$)(O$^-$)

Oxidation of both stereoisomeric sulfoxides yields the same sulfone.

**16.28** Protonation of oxygen to form an alkyloxonium ion is followed by loss of water. The resulting carbocation has a plane of symmetry and is achiral. Capture of the carbocation by methanol yields both enantiomers of 2-methoxy-2-phenylbutane. The product is racemic.

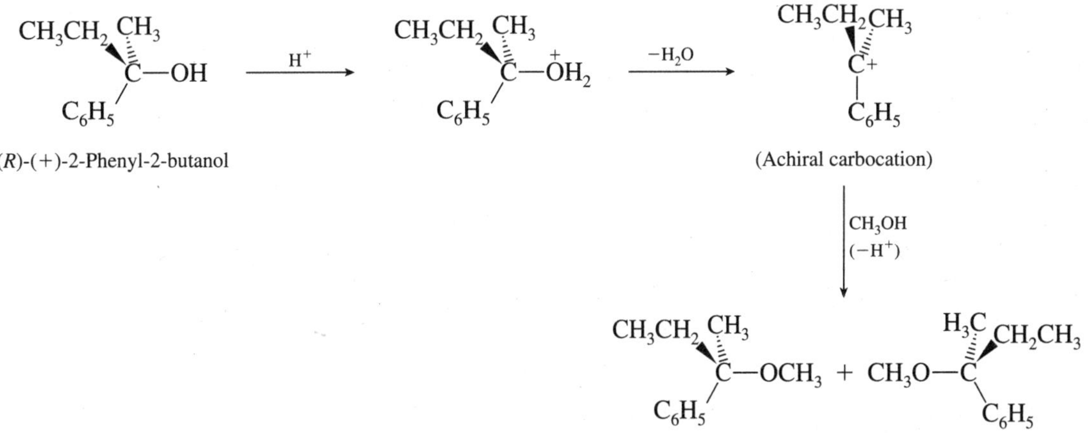

$(CH_3CH_2)(CH_3)(C_6H_5)C\text{—}OH \xrightarrow{H^+} (CH_3CH_2)(CH_3)(C_6H_5)C\text{—}\overset{+}{O}H_2 \xrightarrow{-H_2O} (CH_3CH_2)(CH_3)(C_6H_5)C^+$

(*R*)-(+)-2-Phenyl-2-butanol → (Achiral carbocation)

$\xrightarrow[(-H^+)]{CH_3OH}$ $(CH_3CH_2)(CH_3)(C_6H_5)C\text{—}OCH_3 + CH_3O\text{—}C(CH_3)(CH_2CH_3)(C_6H_5)$

2-Methoxy-2-phenylbutane (racemic)

**16.29** The proper approach to this problem is to first write the equations in full stereochemical detail.

(*a*)

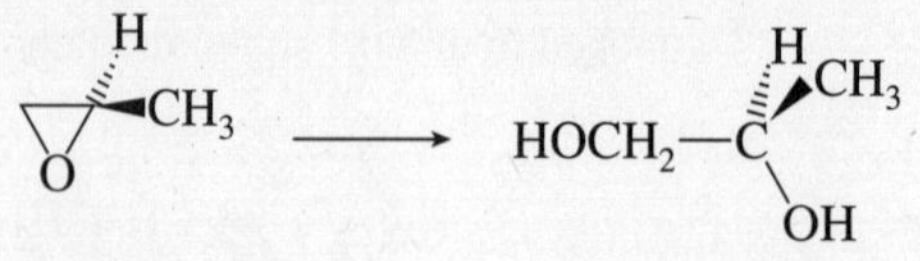

$$\text{(R)-1,2-Epoxypropane} \longrightarrow HOCH_2-C(H)(CH_3)(OH)\ \text{(R)-1,2-Propanediol}$$

(*R*)-1,2-Epoxypropane (*R*)-1,2-Propanediol

It now becomes clear that the arrangement of groups around the stereogenic center remains unchanged in going from starting materials to products. Therefore, choose conditions such that the nucleophile attacks the $CH_2$ group of the epoxide rather than the stereogenic center. Base-catalyzed hydrolysis is required; aqueous sodium hydroxide is appropriate.

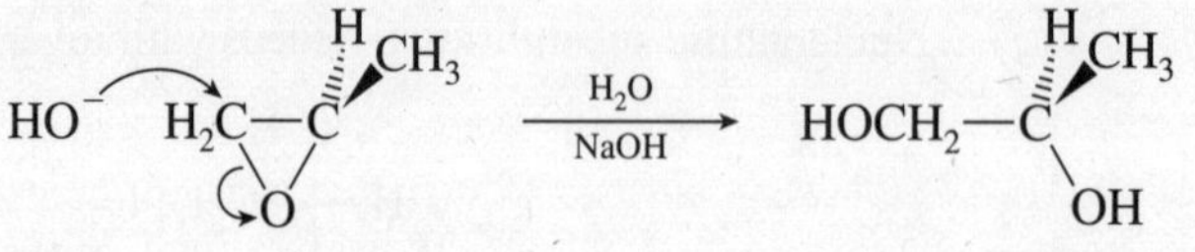

$$HO^- \quad H_2C-C(H)(CH_3)\text{ (epoxide O)} \xrightarrow[NaOH]{H_2O} HOCH_2-C(H)(CH_3)(OH)$$

The nucleophile (hydroxide ion) attacks the less hindered carbon of the epoxide ring.

(*b*)

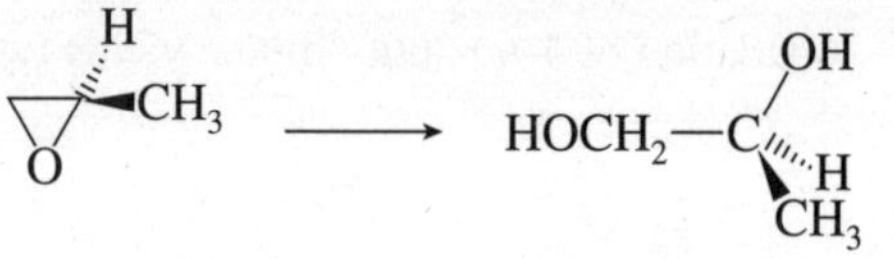

$$\text{epoxide} \longrightarrow HOCH_2-C(OH)(H)(CH_3)$$

(*S*)-1,2-Propanediol

Inversion of configuration at the stereogenic center is required. The nucleophile must therefore attack the stereogenic center, and acid-catalyzed hydrolysis should be chosen. Dilute sulfuric acid would be satisfactory.

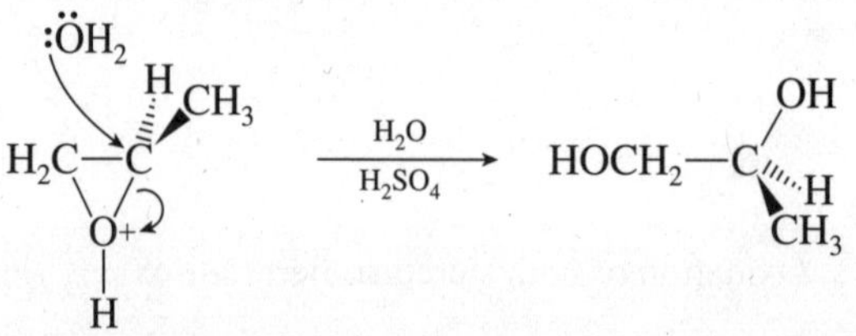

$$:\ddot{O}H_2 \quad H_2C-C(H)(CH_3)\text{ (ring }O^+-H) \xrightarrow[H_2SO_4]{H_2O} HOCH_2-C(OH)(H)(CH_3)$$

The nucleophile (a water molecule) attacks that carbon atom of the ring that can better support a positive charge. Carbocation character develops at the transition state and is better supported by the carbon atom that is more highly substituted.

**16.30** The key intermediate in the preparation of bis(2-chloroethyl) ether from ethylene is 2-chloroethanol, formed from ethylene by reaction with chlorine in water. Heating 2-chloroethanol in acid gives the desired ether.

$$H_2C{=}CH_2 \xrightarrow{Cl_2,\ H_2O} ClCH_2CH_2OH \xrightarrow{H^+,\ heat} ClCH_2CH_2OCH_2CH_2Cl$$

Ethylene 2-Chloroethanol Bis(2-chloroethyl) ether

**16.31** (*a*) There is a temptation to try to do this transformation in a single step by using a reducing agent to convert the carbonyl to a methylene group. No reagent is available that reduces esters in this way! The Clemmensen and Wolff–Kishner reduction methods are suitable only for aldehydes and ketones. The best way to approach this problem is by reasoning backward. The desired product is an ether. Ethers can be prepared by the Williamson ether synthesis involving an alkyl halide and an alkoxide ion.

$$C_6H_5CH_2OCH_3 \Rightarrow C_6H_5CH_2X + CH_3\ddot{\underset{..}{O}}:^-$$

or

$$C_6H_5CH_2OCH_3 \Rightarrow C_6H_5CH_2\ddot{\underset{..}{O}}:^- + CH_3X$$

Both the alkyl halide and the alkoxide ion are prepared from alcohols. The problem then becomes one of preparing the appropriate alcohol (or alcohols) from the starting ester. This is readily done using lithium aluminum hydride.

$$\underset{\text{Methyl benzoate}}{C_6H_5\overset{O}{\overset{\|}{C}}OCH_3} \xrightarrow[2.\ H_2O]{1.\ LiAlH_4} \underset{\text{Benzyl alcohol}}{C_6H_5CH_2OH} + \underset{\text{Methanol}}{CH_3OH}$$

Then

$$\underset{\text{Methanol}}{CH_3OH} \xrightarrow{Na} \underset{\text{Sodium methoxide}}{CH_3ONa}$$

$$\underset{\text{Benzyl alcohol}}{C_6H_5CH_2OH} \xrightarrow[\text{or } PBr_3]{HBr} \underset{\text{Benzyl bromide}}{C_6H_5CH_2Br}$$

and

$$\underset{\text{Benzyl bromide}}{C_6H_5CH_2Br} + \underset{\text{Sodium methoxide}}{NaOCH_3} \longrightarrow \underset{\text{Benzyl methyl ether}}{C_6H_5CH_2OCH_3} + NaBr$$

The following sequence is also appropriate once methanol and benzyl alcohol are obtained by reduction of methyl benzoate:

$$\underset{\text{Benzyl alcohol}}{C_6H_5CH_2OH} \xrightarrow{Na} \underset{\text{Sodium benzyloxide}}{C_6H_5CH_2ONa}$$

$$\underset{\text{Methanol}}{CH_3OH} \xrightarrow[\text{or } HBr]{PBr_3} \underset{\text{Bromomethane}}{CH_3Br}$$

and

$$\underset{\text{Sodium benzyloxide}}{C_6H_5CH_2ONa} + \underset{\text{Bromomethane}}{CH_3Br} \longrightarrow \underset{\text{Benzyl methyl ether}}{C_6H_5CH_2OCH_3} + NaBr$$

(*b*) All the methods that we have so far discussed for the preparation of epoxides are based on alkenes as starting materials. This leads us to consider the partial retrosynthesis shown.

Target molecule (1-phenylcyclohexene oxide, $C_6H_5$) ⟹ Key intermediate (1-phenylcyclohexene, $C_6H_5$) ⟹ 1-phenylcyclohexanol (OH, $C_6H_5$)

The key intermediate, 1-phenylcyclohexene, is both a proper precursor to the desired epoxide and readily available from the given starting materials. A reasonable synthesis is

$$\text{1-phenylcyclohexanol} \xrightarrow[\text{heat}]{H^+} \text{1-phenylcyclohexene} \xrightarrow{CH_3\overset{O}{\overset{\|}{C}}OOH} \text{1-phenylcyclohexene oxide}$$

Preparation of the required tertiary alcohol, 1-phenylcyclohexanol, completes the synthesis.

$$\text{Cyclohexanol} \xrightarrow[H_2SO_4,\ H_2O]{K_2Cr_2O_7} \text{Cyclohexanone}$$

$$\underset{\text{Bromobenzene}}{C_6H_5Br} \xrightarrow[\text{2. cyclohexanone}\quad \text{3. } H_3O^+]{\text{1. Mg, diethyl ether}} \text{1-Phenylcyclohexanol}$$

(*c*) The necessary carbon skeleton can be assembled through the reaction of a Grignard reagent with 1,2-epoxypropane.

$$C_6H_5CH_2\underset{}{\overset{OH}{\overset{|}{C}}}HCH_3 \Longrightarrow C_6H_5{:}^- + H_2\overset{O}{C{-}}CHCH_3$$

The reaction sequence is therefore

$$C_6H_5MgBr + H_2C\overset{O}{-}CHCH_3 \longrightarrow C_6H_5CH_2\underset{OH}{\underset{|}{C}}HCH_3$$

Phenylmagnesium bromide (from bromobenzene and magnesium) — 1,2-Epoxypropane — 1-Phenyl-2-propanol

The epoxide required in the first step, 1,2-epoxypropane, is prepared as follows from isopropyl alcohol:

$$CH_3\underset{OH}{\underset{|}{C}}HCH_3 \xrightarrow[\text{heat}]{H_2SO_4} CH_3CH{=}CH_2 \xrightarrow{CH_3\overset{O}{\overset{\|}{C}}OOH} CH_3CH\overset{O}{-}CH_2$$

2-Propanol (isopropyl alcohol) — Propene — 1,2-Epoxypropane

(*d*) Because the target molecule is an ether, it ultimately derives from two alcohols.

$$C_6H_5CH_2CH_2CH_2OCH_2CH_3 \Rightarrow C_6H_5CH_2CH_2CH_2OH + CH_3CH_2OH$$

Our first task is to assemble 3-phenyl-1-propanol from the designated starting material benzyl alcohol. This requires formation of a primary alcohol with the original carbon chain extended by two carbons. The standard method for this transformation involves reaction of a Grignard reagent with ethylene oxide.

$$\underset{\text{Benzyl alcohol}}{C_6H_5CH_2OH} \xrightarrow[\text{or HBr}]{PBr_3} \underset{\text{Benzyl bromide}}{C_6H_5CH_2Br} \xrightarrow[\text{2. } H_2C\text{—}CH_2 \text{ (epoxide, O)} \quad \text{3. } H_3O^+]{\text{1. Mg, diethyl ether}} \underset{\text{3-Phenyl-1-propanol}}{C_6H_5CH_2CH_2CH_2OH}$$

After 3-phenyl-1-propanol has been prepared, its conversion to the corresponding ethyl ether can be accomplished in either of two ways:

$$\underset{\text{3-Phenyl-1-propanol}}{C_6H_5CH_2CH_2CH_2OH} \xrightarrow[\text{or HBr}]{PBr_3} \underset{\text{1-Bromo-3-phenylpropane}}{C_6H_5CH_2CH_2CH_2Br} \xrightarrow[\text{ethanol}]{NaOCH_2CH_3} \underset{\text{Ethyl 3-phenylpropyl ether (1-ethoxy-3-phenylpropane)}}{C_6H_5CH_2CH_2CH_2OCH_2CH_3}$$

or alternatively

$$\underset{\text{3-Phenyl-1-propanol}}{C_6H_5CH_2CH_2CH_2OH} \xrightarrow[\text{2. } CH_3CH_2Br]{\text{1. Na}} \underset{\text{Ethyl 3-phenylpropyl ether}}{C_6H_5CH_2CH_2CH_2OCH_2CH_3}$$

The reagents in each step are prepared from ethanol.

$$\underset{\text{Ethanol}}{CH_3CH_2OH} \xrightarrow{Na} \underset{\text{Sodium ethoxide}}{CH_3CH_2ONa}$$

$$\underset{\text{Ethanol}}{CH_3CH_2OH} \xrightarrow[\text{or HBr}]{PBr_3} \underset{\text{Ethyl bromide}}{CH_3CH_2Br}$$

(*e*) The target epoxide can be prepared in a single step from the corresponding alkene.

$$\text{Bicyclo[2.2.2]oct-2-ene} \xrightarrow{CH_3\overset{O}{\overset{\|}{C}}OOH} \text{2,3-Epoxybicyclo[2.2.2]octane}$$

Bicyclo[2.2.2]oct-2-ene 2,3-Epoxybicyclo[2.2.2]octane

Disconnections show that this alkene is available through a Diels–Alder reaction.

$$\text{Bicyclo[2.2.2]oct-2-ene} \Rightarrow \text{1,3-cyclohexadiene} + H_2C{=}CH_2$$

The reaction of 1,3-cyclohexadiene with ethylene gives the desired substance.

1,3-Cyclohexadiene + $H_2C{=}CH_2$ ⟶ Bicyclo[2.2.2]oct-2-ene

1,3-Cyclohexadiene is one of the given starting materials. Ethylene is prepared from ethanol.

$$\underset{\text{Ethanol}}{CH_3CH_2OH} \xrightarrow[\text{heat}]{H_2SO_4} \underset{\text{Ethylene}}{H_2C{=}CH_2}$$

(*f*) Retrosynthetic analysis reveals that the desired target molecule may be prepared by reaction of an epoxide with an ethanethiolate ion.

$$C_6H_5CH(OH)CH_2SCH_2CH_3 \Rightarrow C_6H_5\overset{O}{CH{-}CH_2} + {}^{-}SCH_2CH_3$$

Styrene oxide may be prepared by reaction of styrene with peroxyacetic acid.

$$\underset{\text{Styrene}}{C_6H_5CH{=}CH_2} + \underset{\text{Peroxyacetic acid}}{CH_3\overset{O}{\overset{\|}{C}}OOH} \longrightarrow \underset{\text{Styrene oxide}}{C_6H_5\overset{O}{CH{-}CH_2}} + \underset{\text{Acetic acid}}{CH_3CO_2H}$$

The necessary thiolate anion is prepared from ethanol by way of the corresponding thiol.

$$\underset{\text{Ethanol}}{CH_3CH_2OH} \xrightarrow[\text{3. NaOH}]{\text{1. HBr; 2. }(H_2N)_2C{=}S} \underset{\text{Ethanethiol}}{CH_3CH_2SH} \xrightarrow{\text{NaOH}} \underset{\text{Sodium ethanethiolate}}{CH_3CH_2SNa}$$

Reaction of styrene oxide with sodium ethanethiolate completes the synthesis.

$$\underset{\text{Styrene oxide}}{C_6H_5\overset{O}{CH{-}CH_2}} + \underset{\text{Sodium ethanethiolate}}{CH_3CH_2SNa} \xrightarrow{CH_3CH_2OH} C_6H_5\overset{OH}{\overset{|}{C}}HCH_2SCH_2CH_3$$

**16.32** (*a*) A reasonable mechanism is one that parallels the usual one for acid-catalyzed ether formation from alcohols, modified to accommodate these particular starting materials and products. Begin with protonation of one of the oxygen atoms of ethylene glycol.

$$\underset{\text{Ethylene glycol}}{HOCH_2CH_2OH} + H_2SO_4 \rightleftharpoons HOCH_2CH_2{-}\overset{+}{\ddot{O}}(H){-}H + HSO_4^-$$

The protonated alcohol then reacts in the usual way with another molecule of alcohol to give an ether. (This ether is known as **diethylene glycol.**)

$$HOCH_2CH_2{-}\overset{+}{\ddot{O}}(H){-}H + HOCH_2CH_2\ddot{O}H \xrightarrow{-H_2O} HOCH_2CH_2{-}\overset{+}{\ddot{O}}(CH_2CH_2OH){-}H \xrightarrow{-H^+} \underset{\text{Diethylene glycol}}{HOCH_2CH_2OCH_2CH_2OH}$$

Diethylene glycol then undergoes intramolecular ether formation to yield 1,4-dioxane.

$$HOCH_2CH_2OCH_2CH_2OH + H_2SO_4 \rightleftharpoons HOCH_2CH_2OCH_2CH_2{-}\overset{+}{O}(H){-}H + HSO_4^-$$

$$O(CH_2CH_2{-}\overset{+}{O}H_2)(CH_2CH_2OH) \xrightarrow{-H_2O} O(CH_2CH_2)_2\overset{+}{O}H \xrightarrow{-H^+} \text{1,4-dioxane}$$

(*b*) The substrate is a primary alkyl halide and reacts with aqueous sodium hydroxide by nucleophilic substitution.

$$\underset{\text{Bis(2-chloroethyl) ether}}{ClCH_2CH_2OCH_2CH_2Cl} + HO^- \xrightarrow{H_2O} HOCH_2CH_2OCH_2CH_2Cl + Cl^-$$

The product of this reaction now has an alcohol function and a primary chloride built into the same molecule. It contains the requisite functionality to undergo an intramolecular Williamson reaction.

$$HOCH_2CH_2OCH_2CH_2Cl + HO^- \rightleftharpoons {}^-OCH_2CH_2OCH_2CH_2Cl + H_2O$$

$$H_2C(O)CH_2 \ldots H_2C{-}\ddot{O}{:}^- \; CH_2{-}Cl \longrightarrow \text{1,4-Dioxane} + Cl^-$$

1,4-Dioxane

**16.33** (*a*) The first step is a standard Grignard synthesis of a primary alcohol using formaldehyde. Compound A is 3-buten-1-ol.

$$\underset{\text{Allyl bromide}}{H_2C{=}CHCH_2Br} \xrightarrow[\text{2. } H_2C{=}O \text{ 3. } H_3O^+]{\text{1. Mg}} \underset{\text{3-Buten-1-ol (compound A)}}{H_2C{=}CHCH_2CH_2OH}$$

Addition of bromine to the carbon–carbon double bond of 3-buten-1-ol takes place readily to yield the vicinal dibromide.

$$\underset{\text{3-Buten-1-ol}}{H_2C{=}CHCH_2CH_2OH} \xrightarrow{Br_2} \underset{\text{3,4-Dibromo-1-butanol (compound B)}}{BrCH_2CH(Br)CH_2CH_2OH}$$

When compound B is treated with potassium hydroxide, it loses the elements of HBr to give compound C. Because further treatment of compound C with potassium hydroxide converts it to D by a second dehydrobromination, a reasonable candidate for C is 3-bromotetrahydrofuran.

$$\underset{\text{3,4-Dibromo-1-butanol (compound B)}}{BrCH_2CH(Br)CH_2CH_2OH} \xrightarrow[25°C]{KOH} \text{3-Bromotetrahydrofuran (compound C)} \xrightarrow[\text{heat}]{KOH} \text{Compound D}$$

3,4-Dibromo-1-butanol (compound B) — 3-Bromotetrahydrofuran (compound C) — Compound D

Ring closure occurs by an intramolecular Williamson reaction.

$$BrCH_2CHCH_2CH_2OH \text{ (Br on C-2)} \xrightleftharpoons{KOH} \text{Br—}CH_2\text{—CH(Br)—}CH_2\text{—}CH_2\text{—}\ddot{O}^- \longrightarrow \text{3-bromotetrahydrofuran}$$

Compound B — Compound C

Dehydrohalogenation of compound C converts it to the final product, D.

The alternative series of events, in which double-bond formation proceeds ring closure, is unlikely, because it requires nucleophilic attack by the alkoxide on a vinyl bromide.

$$BrCH_2CHCH_2CH_2OH \text{ (Br on C-2)} \xrightarrow{KOH} BrCH{=}CHCH_2CH_2OH \xrightleftharpoons{KOH} BrCH{=}CHCH_2CH_2O^-$$

(Cyclization of this intermediate does not occur.)

(*b*) Lithium aluminum hydride reduces the carboxylic acid to the corresponding primary alcohol, compound E. Treatment of the vicinal chlorohydrin with base results in formation of an epoxide, compound F.

$$\text{Cl—C(}CO_2H\text{)(H)(}CH_3\text{)} \xrightarrow[2.\ H_2O]{1.\ LiAlH_4} \text{Cl—C(}CH_2OH\text{)(H)(}CH_3\text{)} \equiv H_3C(H)C(Cl)\text{—}CH_2OH \xrightarrow[H_2O]{KOH} \text{H}(H_3C)C\text{—}CH_2 \text{ (epoxide)}$$

(*S*)-2-Chloro-1-propanol (compound E) — (*R*)-1,2-Epoxypropane (compound F)

As actually carried out, the first step proceeded in 56–58% yield, the second step in 65–70% yield.

(*c*) Treatment of the vicinal chlorohydrin with base results in ring closure to form an epoxide (compound G). Recall that attack occurs on the side opposite that of the carbon–chlorine bond. Compound G undergoes ring opening on reaction with sodium methanethiolate to give compound H.

$$\text{(2}R\text{,3}S\text{)-3-Chloro-2-butanol} \xrightarrow{NaOH} trans\text{-2,3-Epoxybutane}$$

(2*R*,3*S*)-3-Chloro-2-butanol — *trans*-2,3-Epoxybutane (compound G)

$$\text{Compound G} \xrightarrow{NaSCH_3} \text{Compound H (HO(H)(}H_3C\text{)C—C(}H_3C\text{)(H)}SCH_3\text{)}$$

Compound G — Compound H

(*d*) Because it gives an epoxide on treatment with a peroxy acid, compound I must be an alkene; more specifically, it is 1,2-dimethylcyclopentene.

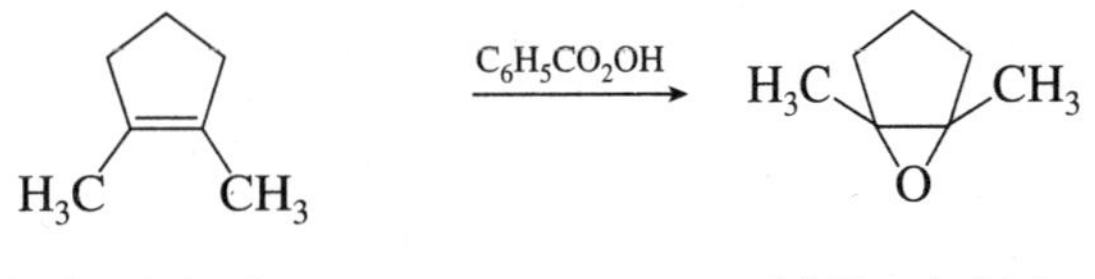
$C_6H_5CO_2OH$
$H_3C$
$CH_3$
O

1,2-Dimethylcyclopentene (compound I)

1,2-Dimethyl-1,2-epoxycyclopentane (compound K)

Compounds J and L have the same molecular formula, $C_7H_{14}O_2$, but J is a liquid and L is a crystalline solid. Their molecular formulas correspond to the addition of two OH groups to compound I. Osmium tetraoxide brings about syn hydroxylation of an alkene; therefore compound J must be the cis diol.

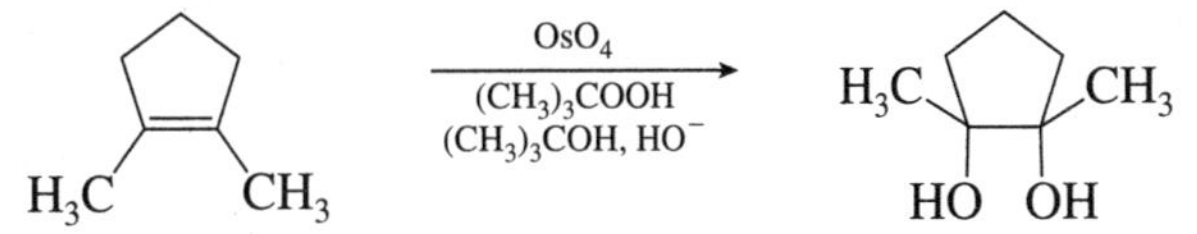
$OsO_4$
$(CH_3)_3COOH$
$(CH_3)_3COH$, $HO^-$
$H_3C$
$CH_3$
HO OH

1,2-Dimethylcyclopentene (compound I)

*cis*-1,2-Dimethylcyclopentane-1,2-diol (compound J)

Acid-catalyzed hydrolysis of an epoxide yields a trans diol (compound L):

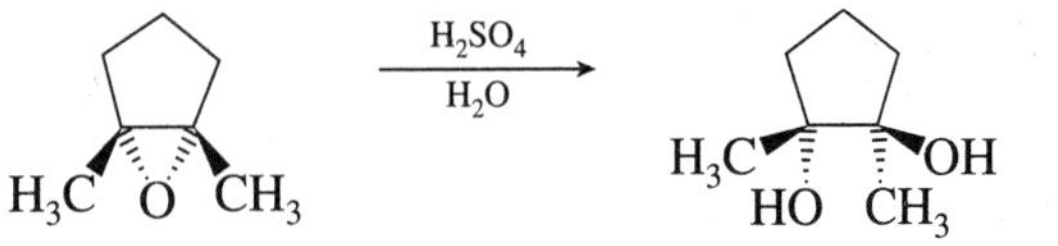
$H_2SO_4$
$H_2O$
$H_3C$ O $CH_3$
$H_3C$ OH
HO $CH_3$

*trans*-1,2-Dimethylcyclopentane-1,2-diol (compound L)

**16.34** Cineole contains no double or triple bonds and therefore must be bicyclic, on the basis of its molecular formula ($C_{10}H_{18}O$, index of hydrogen deficiency = 2). When cineole reacts with hydrogen chloride, one of the rings is broken and water is formed.

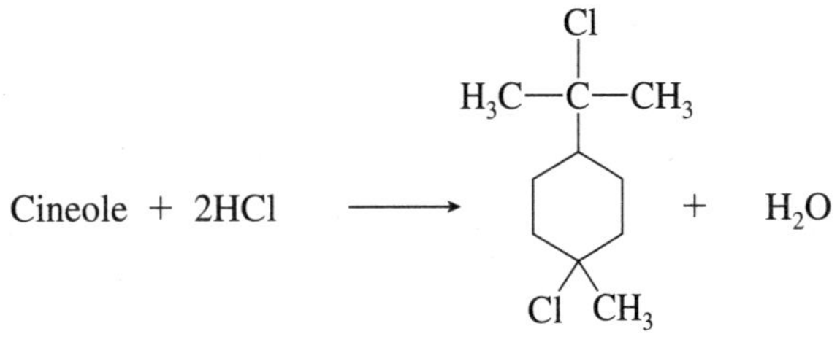
Cineole + 2HCl
Cl
$H_3C$—C—$CH_3$
+ $H_2O$
Cl $CH_3$

($C_{10}H_{18}O$) ($C_{10}H_{18}Cl_2$)

The reaction that takes place is hydrogen halide-promoted ether cleavage. In such a reaction with excess hydrogen halide, the C—O—C unit is cleaved and two carbon–halogen bonds are formed. This suggests that cineole is a cyclic ether because the product contains both newly formed carbon–halogen bonds. A reasonable structure consistent with these facts is

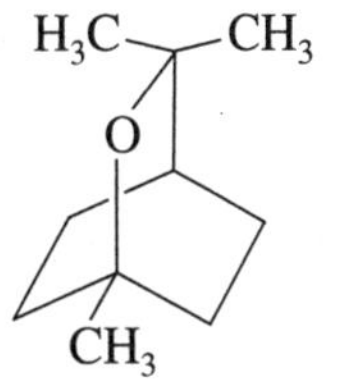
$H_3C$
$CH_3$
O
$CH_3$

Cineole

**16.35** Recall that *p*-toluenesulfonate (tosylate) is a good leaving group in nucleophilic substitution reactions. The nucleophile that displaces tosylate from carbon is the alkoxide ion derived from the hydroxyl group within the molecule. The product is a cyclic ether, and the nature of the union of the two rings is that they are spirocyclic.

:ÖH, $CH_2CH_2CH_2OTs$, $C_6H_5$ —base→ :Ö:⁻, $CH_2$, $CH_2$, $H_2C$–OTs, $C_6H_5$ —intramolecular nucleophilic substitution→ :Ö, $C_6H_5$

$C_{15}H_{20}O$

**16.36** (*a*) Because all the peaks in the $^1H$ NMR spectrum of this ether are singlets, none of the protons can be vicinal to any other nonequivalent proton. The only $C_5H_{12}O$ ether that satisfies this requirement is *tert*-butyl methyl ether.

$H_3C—O—C(CH_3)_3$

Singlet at δ 3.2 ppm (for $H_3C$); Singlet at δ 1.2 ppm (for $C(CH_3)_3$)

(*b*) A doublet–septet pattern is characteristic of an isopropyl group. Two isomeric $C_5H_{12}O$ ethers contain an isopropyl group: ethyl isopropyl ether and isobutyl methyl ether.

| $(CH_3)_2CHOCH_2CH_3$ | $(CH_3)_2CHCH_2OCH_3$ |
| --- | --- |
| Ethyl isopropyl ether | Isobutyl methyl ether |

The signal of the methine proton in isobutyl methyl ether will be split into more than a septet, however, because in addition to being split by two methyl groups, it is coupled to the two protons in the methylene group. Thus, isobutyl methyl ether does not have the correct splitting pattern to be the answer. The correct answer is ethyl isopropyl ether.

(*c*) The low-field signals are due to the protons on the carbon atoms of the C—O—C linkage. Because one gives a doublet, it must be vicinal to only one other proton. We can therefore specify the partial structure:

C—O—C(H)—C(H)(C)—C

Low-field doublet (H on the C bonded to O)

This partial structure contains all the carbon atoms in the molecule. Fill in the remaining valences with hydrogen atoms to reveal isobutyl methyl ether as the correct choice.

$H_3C—O—CH_2—CH(CH_3)_2$

Low-field singlet ($H_3C$); Low-field doublet ($CH_2$)

(*d*) Here again, signals at low field arise from protons on the carbons of the C—O—C unit. One of these signals is a quartet and so corresponds to a proton on a carbon bearing a methyl group.

$H_3C—C(H)—O—C$

Quartet (H)

The other carbon of the C—O—C unit has a hydrogen whose signal is split into a triplet. This hydrogen must therefore be attached to a carbon that bears a methylene group.

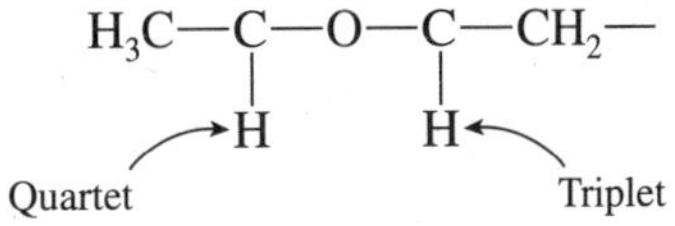

$$H_3C—\underset{\underset{H}{|}}{C}—O—\underset{\underset{H}{|}}{C}—CH_2—$$

Quartet (left H) Triplet (right H)

These data permit us to complete the structure by adding an additional carbon and the requisite number of hydrogens in such a way that the signals of the protons attached to the carbons of the ether linkage are not split further. The correct structure is ethyl propyl ether.

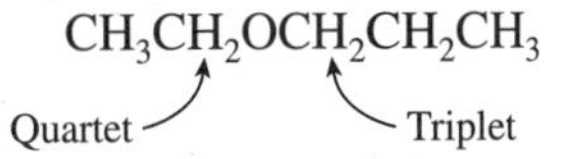

$$CH_3CH_2OCH_2CH_2CH_3$$

Quartet Triplet

**16.37** A good way to address this problem is to consider the dibromide derived by treatment of compound A with hydrogen bromide. The presence of an NMR signal equivalent to four protons in the aromatic region at $\delta$ 7.3 ppm indicates that this dibromide contains a disubstituted aromatic ring. The four remaining protons appear as a sharp singlet at $\delta$ 4.7 ppm and are most reasonably contained in two equivalent methylene groups of the type $ArCH_2Br$. Because the dibromide contains all the carbons and hydrogens of the starting material and is derived from it by treatment with hydrogen bromide, it is likely that compound A is a cyclic ether in which a $CH_2OCH_2$ unit spans two of the carbons of a benzene ring. This can occur only when the positions involved are ortho to each other. Therefore

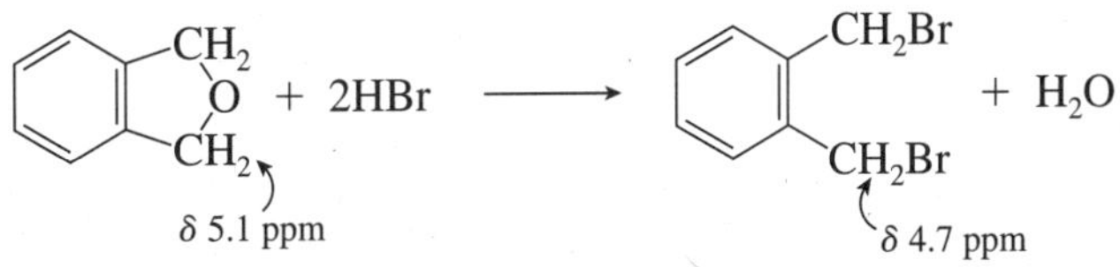

$$\text{(benzene ring fused to } CH_2—O—CH_2\text{)} + 2HBr \longrightarrow \text{(benzene ring with two } CH_2Br\text{ ortho)} + H_2O$$

$\delta$ 5.1 ppm $\delta$ 4.7 ppm

Compound A

**16.38** The molecular formula of a compound ($C_{10}H_{13}BrO$) indicates an index of hydrogen deficiency of 4. One of the products obtained on treatment of the compound with HBr is benzyl bromide ($C_6H_5CH_2Br$), which accounts for seven of its ten carbons and all the double bonds and rings. Thus, the compound is a benzyl ether having the formula $C_6H_5CH_2OC_3H_6Br$. The $^1H$ NMR spectrum includes a five-proton signal at $\delta$ 7.4 ppm for a monosubstituted benzene ring and a two-proton singlet at $\delta$ 4.6 ppm for the benzylic protons. This singlet appears at low field because the benzylic protons are bonded to oxygen.

$$C_6H_5CH_2OC_3H_6Br \xrightarrow[\text{heat}]{\text{HBr}} C_6H_5CH_2Br + C_3H_6Br_2$$

The 6 remaining protons appear as two overlapping 2-proton triplets at $\delta$ 3.6 and 3.7 ppm, along with a 2-proton pentet at $\delta$ 2.2 ppm, consistent with the unit $—OCH_2CH_2CH_2Br$. The compound is $C_6H_5CH_2OCH_2CH_2CH_2Br$.

**16.39** The high index of hydrogen deficiency (5) of the unknown compound $C_9H_{10}O$ and the presence of six signals in the 120–140-ppm region of the $^{13}C$ NMR spectrum suggests the presence of an aromatic ring. The problem states that the compound is a cyclic ether, thus the oxygen atom is contained in a second ring fused to the benzene ring. As oxidation yields 1,2-benzenedicarboxylic acid, the second ring must be attached to the benzene ring by carbon atoms.

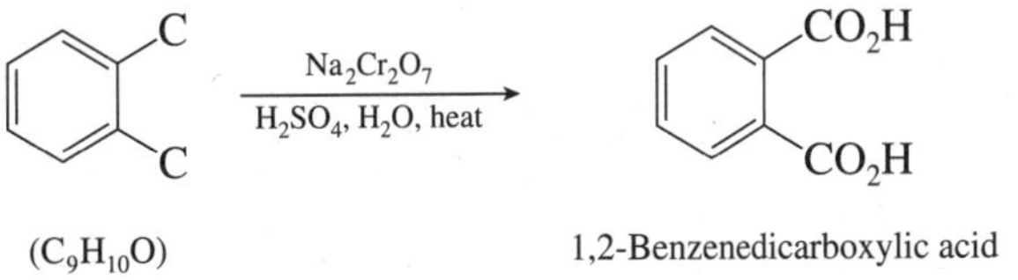

$$\text{(benzene ring with two ortho C)} \xrightarrow[H_2SO_4,\ H_2O,\ \text{heat}]{Na_2Cr_2O_7} \text{(benzene ring with two ortho } CO_2H\text{)}$$

($C_9H_{10}O$) 1,2-Benzenedicarboxylic acid

Two structures are possible with this information; however, only one of them is consistent with the presence of three $CH_2$ groups in the $^{13}C$ NMR spectrum. The compound is

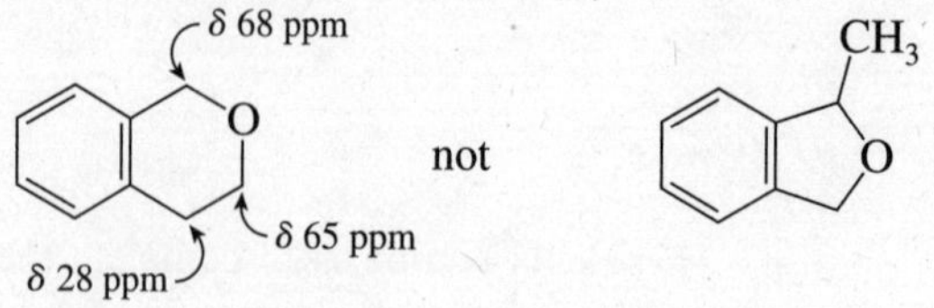

δ 68 ppm

not

δ 65 ppm

δ 28 ppm

$CH_3$

**16.40–16.45** Solutions to molecular modeling exercises are not provided in this *Study Guide and Solutions Manual.* You should use *Learning By Modeling* for these exercises.

## SELF-TEST

### PART A

**A-1.** Write the structures of all the isomeric ethers of molecular formula $C_4H_{10}O$, and give the correct name for each.

**A-2.** Give the structure of the product obtained from each of the following reactions. Show stereochemistry where it is important.

(*a*) $CH_3CH_2$–C(H)($CH_3$)–OH $\xrightarrow[2.\ CH_3I]{1.\ Na}$ ?

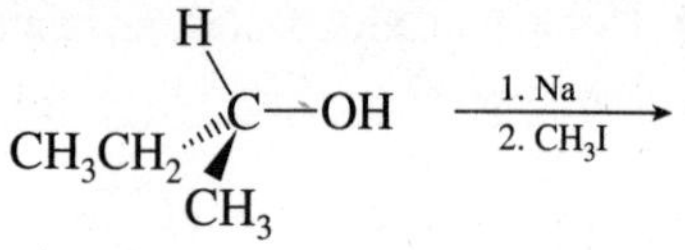

(*d*) dimethylcyclohexene oxide ($CH_3$, $CH_3$, O) $\xrightarrow[CH_3CH_2OH]{CH_3CH_2SNa}$ ?

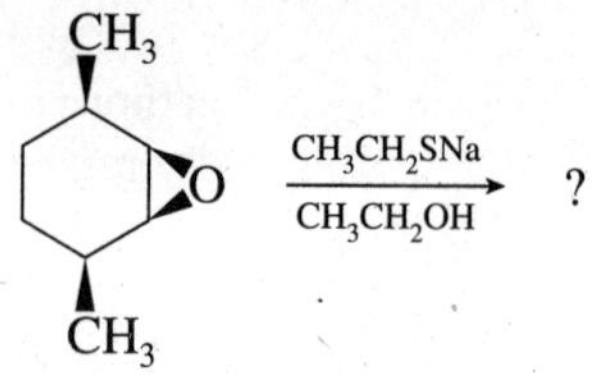

(*b*) (*Z*)-2-butene $\xrightarrow[2.\ H_3O^+]{1.\ CH_3\overset{O}{\overset{\|}{C}}OOH}$ ?

(*e*) $C_6H_5SNa$ $\xrightarrow{CH_3CH_2Br}$ ?

(c) $C_6H_5CH\text{-}CH_2$ (epoxide, O) $\xrightarrow{HI}$ ?

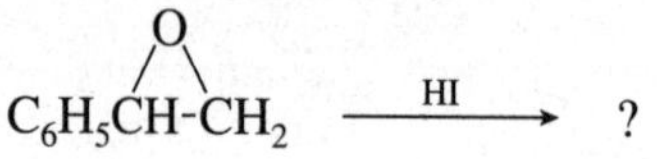

(*f*) Product of part (*e*) $\xrightarrow{NaIO_4}$ ?

**A-3.** Outline a scheme for the preparation of cyclohexyl ethyl ether using the Williamson method.

**A-4.** Outline a synthesis of 2-ethoxyethanol, $CH_3CH_2OCH_2CH_2OH$, using ethanol as the source of all the carbon atoms.

**A-5.** Provide the reagents necessary to complete each of the following conversions. In each case give the structure of the intermediate product.

(*a*) cyclopentane with HO and Br $\longrightarrow$ ? $\longrightarrow$ cyclopentane with HO and $SCH_3$

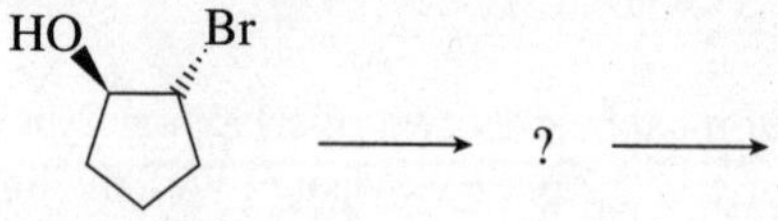

(*b*) $C_6H_5C(CH_3)_2OH$ ($CH_3$, COH, $CH_3$) $\longrightarrow$ ? $\longrightarrow$ $C_6H_5$–C($CH_3$)(–O–$CH_2$) ($CH_2$, C–O, $CH_3$)

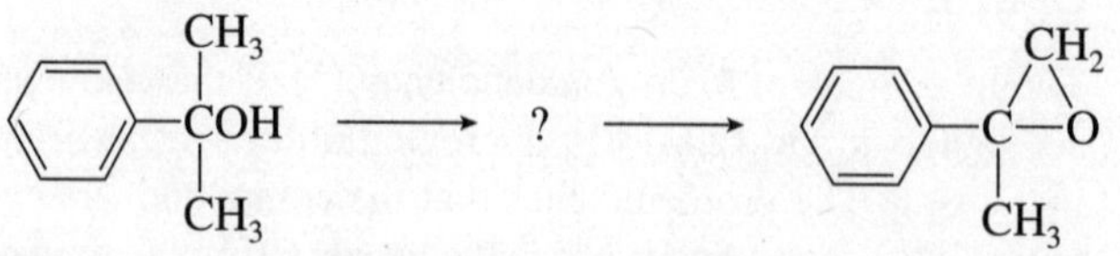

**A-6.** Provide structures for compounds A and B in the following reaction scheme:

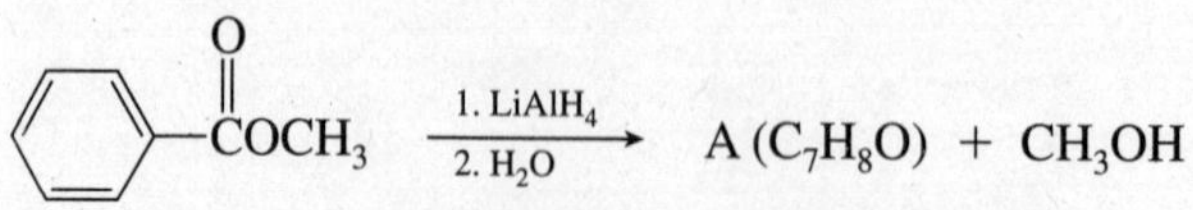

$C_6H_5\overset{O}{\overset{\|}{C}}OCH_3$ $\xrightarrow[2.\ H_2O]{1.\ LiAlH_4}$ A ($C_7H_8O$) + $CH_3OH$

A $\xrightarrow[2.\ CH_3CH_2I]{1.\ Na}$ B ($C_9H_{12}O$)

**A-7.** Using any necessary organic or inorganic reagents, provide the steps to carry out the following synthetic conversion:

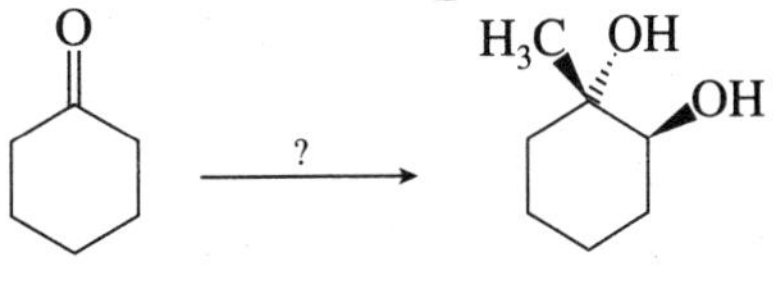

Cyclohexanone $\xrightarrow{?}$ 1-methylcyclohexane-1,2-diol ($H_3C$, OH, OH)

**A-8.** Give the final product, including stereochemistry, of the following reaction sequence:

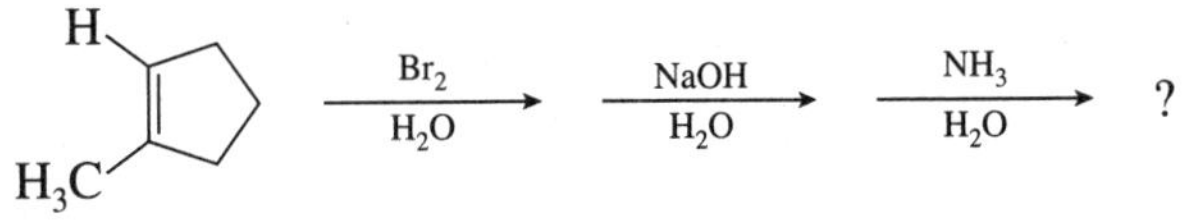

1-methylcyclopentene (H, $H_3C$) $\xrightarrow[H_2O]{Br_2}$ $\xrightarrow[H_2O]{NaOH}$ $\xrightarrow[H_2O]{NH_3}$ ?

## PART B

**B-1.** An acceptable IUPAC name of the compound shown is

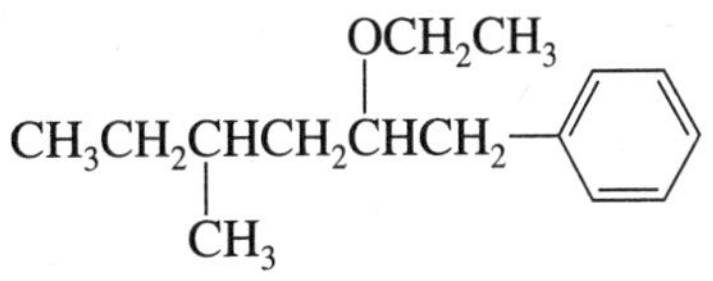

$CH_3CH_2CH(CH_3)CH_2CH(OCH_2CH_3)CH_2—C_6H_5$

(*a*) 1-Benzyl-3-methylpentyl ethyl ether
(*b*) Ethyl 3-methyl-1-methylphenyl-2-hexyl ether
(*c*) Ethyl 4-methyl-1-phenyl-2-hexyl ether
(*d*) 5-Ethoxy-3-methyl-6-phenylhexane

**B-2.** The most effective pair of reagents for the preparation of *tert*-butyl ethyl ether is
(*a*) Potassium *tert*-butoxide and ethyl bromide
(*b*) Potassium *tert*-butoxide and ethanol
(*c*) Sodium ethoxide and *tert*-butyl bromide
(*d*) *tert*-Butyl alcohol and ethyl bromide

**B-3.** The best choice of reactant(s) for the following conversion is

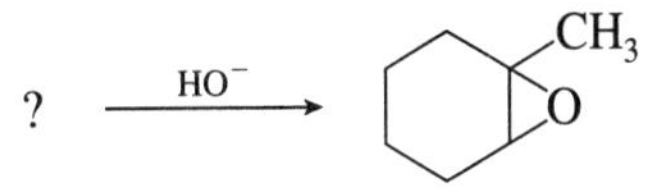

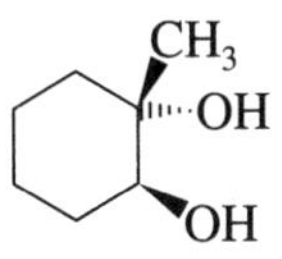

? $\xrightarrow{HO^-}$ 1-methylcyclohexene oxide ($CH_3$, O)

(*a*) $CH_3$, OH, OH
(*b*) OH, $CH_3$, OH
(*c*) OH, $CH_3$, Br
(*d*) OH, $CH_3$, Br

**B-4.** For which of the following ethers would the $^1H$ NMR spectrum consist of only two singlets?

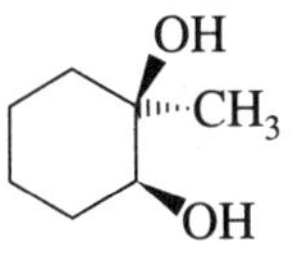

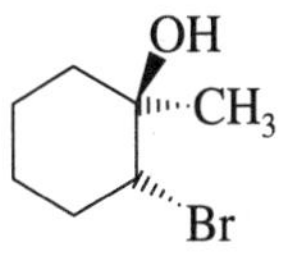

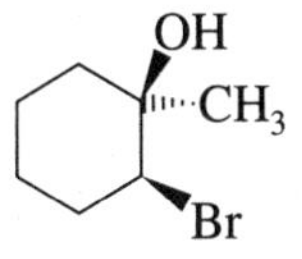

(*a*) $CH_3OC(CH_3)_2CH_3$
(*b*) $CH_3OCH_2CH_2OCH_3$
(*c*) 2,2-dimethyloxirane (O, $CH_3$, $CH_3$)
(*d*) All these

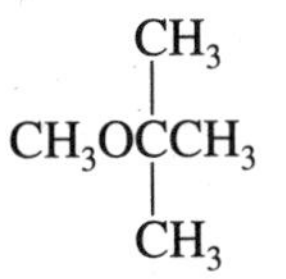

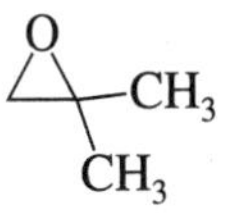

**B-5.** Heating a particular ether with HBr yielded a single organic product. Which of the following conclusions may be reached?

(*a*) The reactant was a methyl ether.
(*b*) The reactant was a symmetric ether.
(*c*) The reactant was a cyclic ether.
(*d*) Both (*b*) and (*c*) are correct.

**B-6.** Treating anisole ($C_6H_5OCH_3$) with the following reagents will give, as the major product,

1. $(CH_3)_3CCl$, $AlCl_3$; 2. $Cl_2$, $FeCl_3$; 3. HBr, heat

(*a*) OH, Br, $C(CH_3)_3$ (benzene ring: OH at C-1, Br at C-2, $C(CH_3)_3$ at C-4)
(*c*) $OCH_3$, Br, $C(CH_3)_3$ (benzene ring: $OCH_3$ at C-1, Br at C-2, $C(CH_3)_3$ at C-4)
(*e*) OH, Cl, $C(CH_3)_3$ (benzene ring: OH at C-1, Cl at C-2, $C(CH_3)_3$ at C-4)
(*b*) Br, Br, $C(CH_3)_3$ (benzene ring: Br at C-1, Br at C-2, $C(CH_3)_3$ at C-4)
(*d*) Br, Cl, $C(CH_3)_3$ (benzene ring: Br at C-1, Cl at C-2, $C(CH_3)_3$ at C-4)

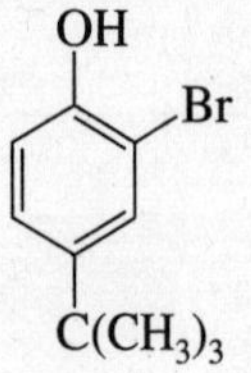

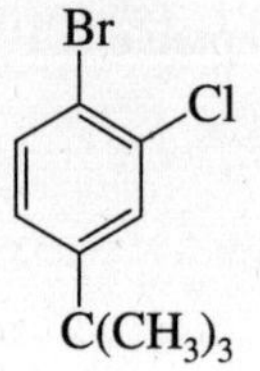

**B-7.** What is the product of the following reaction?

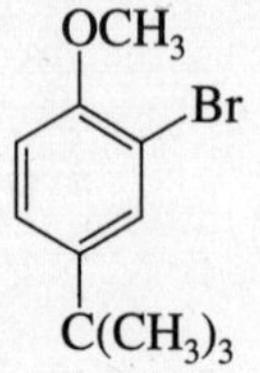

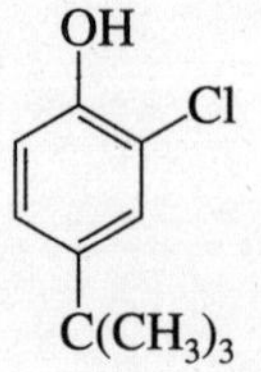

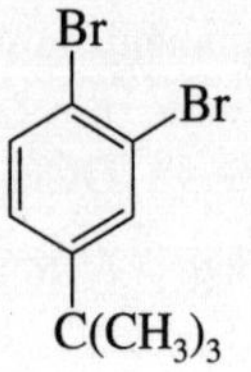

$$(CH_3)_3C\text{-epoxide (O)} + NaSCH_3 \xrightarrow{\text{ethanol}} ?$$

(*a*) $CH_3SCH_2CHC(CH_3)_3$ with OH on CH
(*c*) $CH_3SCH_2CHC(CH_3)_3$ with $OCH_2CH_3$ on CH
(*b*) $(CH_3)_3CCHCH_2OH$ with $SCH_3$ on CH
(*d*) $(CH_3)_3CCH_2CHSCH_3$ with OH on CH

**B-8.** Identify product Z in the following reaction sequence:

$$HOCH_2{-}C{\equiv}C{-}CH_2CH_2OH \xrightarrow[\text{Lindlar Pd}]{H_2} X \xrightarrow[\text{heat}]{H_2SO_4} Y \xrightarrow{CH_3\overset{O}{\overset{\|}{C}}OOH} Z$$

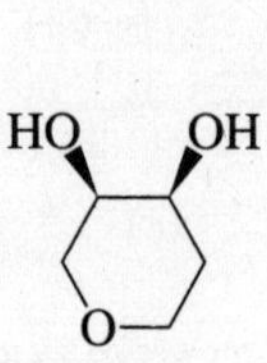

(*a*) HO, OH (tetrahydropyran ring with two wedged OH groups, ring O)
(*b*) $O\overset{O}{\overset{\|}{C}}CH_3$ (tetrahydropyran ring bearing an acetate group, ring O)
(*c*) O (epoxide fused to tetrahydropyran ring, ring O)
(*d*) $CH_3\overset{O}{\overset{\|}{C}}OCH_2$, H / C=C / $CH_2CH_2O\overset{O}{\overset{\|}{C}}CH_3$, H (cis)
(*e*) $CH_3\overset{O}{\overset{\|}{C}}OCH_2$, H / C=C / H, $CH_2CH_2O\overset{O}{\overset{\|}{C}}CH_3$ (trans)

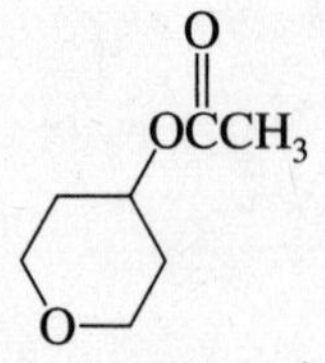

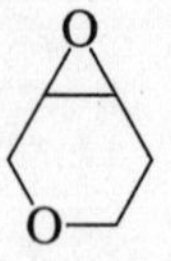

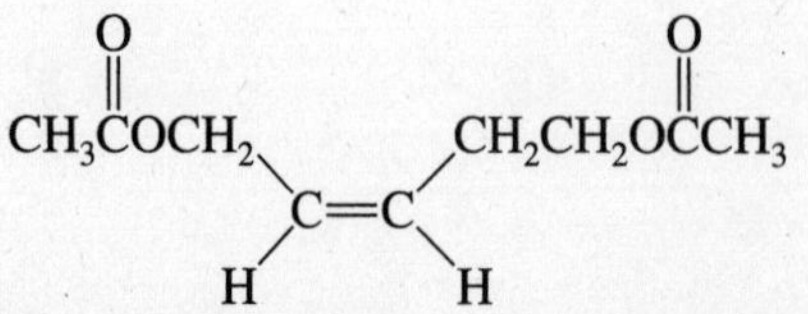

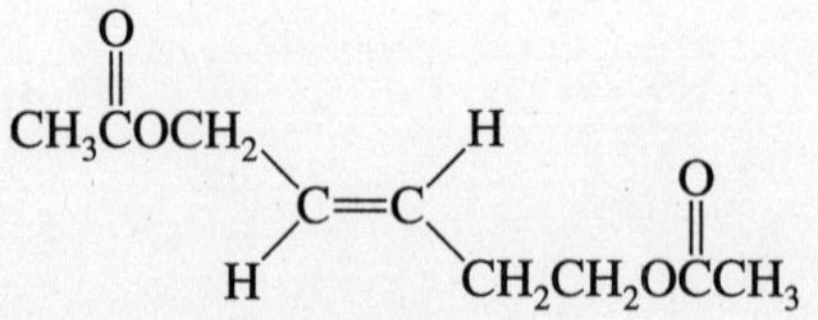

**B-9.** The major product of the following sequence is

(2-bromotoluene) $\xrightarrow[\text{light}]{Br_2}$ $\xrightarrow{CH_3CH_2OK}$ ?

(*a*) 2-$CH_3$, 1-$OCH_2CH_3$ benzene

(*b*) 2-$CH_2Br$, 1-$OCH_2CH_3$ benzene

(*c*) 2-$OCH_2CH_3$, 1-Br benzene

(*d*) 2-$CH_2OCH_2CH_3$, 1-Br benzene

(*e*) 2-$CH_2OCH_2CH_3$, 1-$OCH_2CH_3$ benzene

**B-10.** Which of the following best represents the rate-determining transition state for the reaction shown?

$C_6H_5$–ONa + $CH_3Br$ ⟶

(*a*) $\overset{\delta-}{Br}$·······$\overset{}{C}\delta+$ (with three H)

(*b*) $\overset{\delta+}{Na}$·······$\overset{\delta-}{Br}$·······$C\delta+$ (with three H)

(*c*) $\delta+C$·······$\overset{\delta-}{O}$–$C_6H_5$ (C with three H)

(*d*) $\overset{\delta-}{Br}$·······C·······$\overset{\delta-}{O}$–$C_6H_5$ (C with three H)

(*e*) C (with three H) bonded ······ to $\overset{\delta-}{Br}$ and ······ to $\overset{\delta-}{O}$–$C_6H_5$

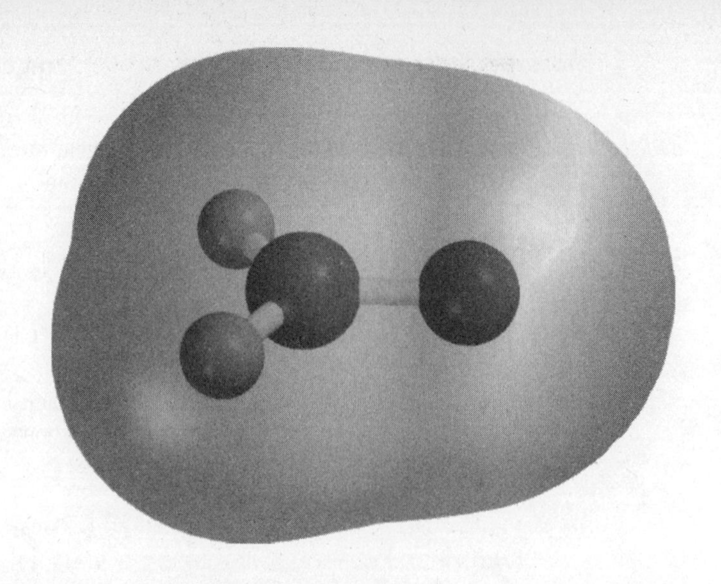

# CHAPTER 17

## ALDEHYDES AND KETONES: NUCLEOPHILIC ADDITION TO THE CARBONYL GROUP

## SOLUTIONS TO TEXT PROBLEMS

**17.1** (*b*) The longest continuous chain in glutaraldehyde has five carbons and terminates in aldehyde functions at both ends. **Pentanedial** is an acceptable IUPAC name for this compound.

$$\overset{\text{O}}{\underset{1}{\overset{\|}{\text{HC}}}}\overset{2}{\text{C}}\text{H}_2\overset{3}{\text{C}}\text{H}_2\overset{4}{\text{C}}\text{H}_2\overset{\text{O}}{\overset{\|}{\underset{5}{\text{C}}}}\text{H}$$

Pentanedial (glutaraldehyde)

(*c*) The three-carbon parent chain has a double bond between C-2 and C-3 and a phenyl substituent at C-3.

$$\text{C}_6\text{H}_5\overset{3}{\text{C}}\text{H}{=}\overset{2}{\text{C}}\text{H}\overset{\text{O}}{\overset{\|}{\underset{1}{\text{C}}}}\text{H}$$

3-Phenyl-2-propenal
(cinnamaldehyde)

(*d*) Vanillin can be named as a derivative of benzaldehyde. Remember to cite the remaining substituents in alphabetical order.

$$\text{HO}-\text{C}_6\text{H}_3(\text{OCH}_3)-\overset{\text{O}}{\overset{\|}{\text{C}}}\text{H}$$

4-Hydroxy-3-methoxybenzaldehyde
(vanillin)

**17.2** (*b*) First write the structure from the name given. Ethyl isopropyl ketone has an ethyl group and an isopropyl group bonded to a carbonyl group.

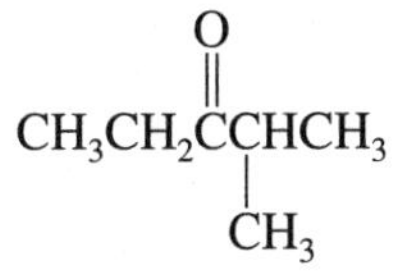

$$CH_3CH_2\overset{\overset{O}{\|}}{C}\underset{\underset{CH_3}{|}}{C}HCH_3$$

Ethyl isopropyl ketone may be alternatively named 2-methyl-3-pentanone. Its longest continuous chain has five carbons. The carbonyl carbon is C-3 irrespective of the direction in which the chain is numbered, and so we choose the direction that gives the lower number to the position that bears the methyl group.

(*c*) Methyl 2,2-dimethylpropyl ketone has a methyl group and a 2,2-dimethylpropyl group bonded to a carbonyl group.

$$CH_3\overset{\overset{O}{\|}}{C}CH_2\overset{\overset{CH_3}{|}}{\underset{\underset{CH_3}{|}}{C}}CH_3$$

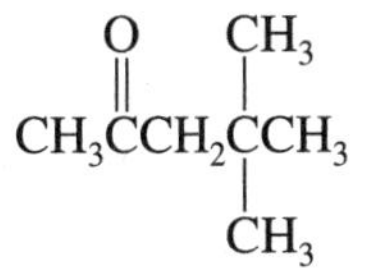

The longest continuous chain has five carbons, and the carbonyl carbon is C-2. Thus, methyl 2,2-dimethylpropyl ketone may also be named 4,4-dimethyl-2-pentanone.

(*d*) The structure corresponding to allyl methyl ketone is

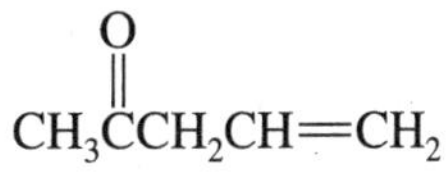

$$CH_3\overset{\overset{O}{\|}}{C}CH_2CH{=}CH_2$$

Because the carbonyl group is given the lowest possible number in the chain, the substitutive name is 4-penten-2-one *not* 1-penten-4-one.

**17.3** No. Lithium aluminum hydride is the only reagent we have discussed that is capable of reducing carboxylic acids (Section 15.3).

**17.4** The target molecule, 2-butanone, contains four carbon atoms. The problem states that all of the carbons originate in acetic acid, which has two carbon atoms. This suggests the following disconnections:

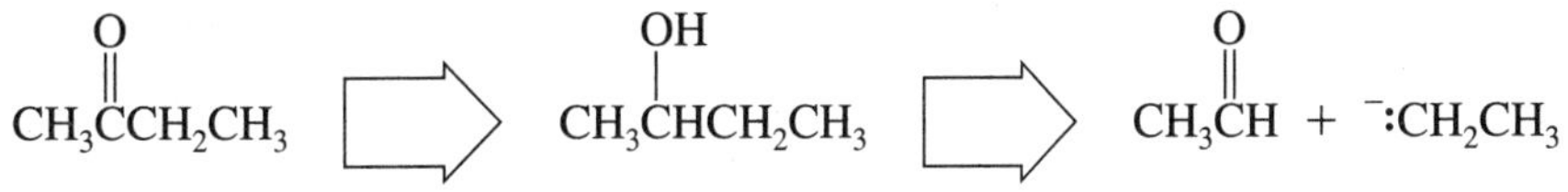

$$\underset{\text{2-Butanone}}{CH_3\overset{\overset{O}{\|}}{C}CH_2CH_3} \Longrightarrow CH_3\overset{\overset{OH}{|}}{C}HCH_2CH_3 \Longrightarrow CH_3\overset{\overset{O}{\|}}{C}H + {}^{-}\!:CH_2CH_3$$

The necessary aldehyde (acetaldehyde) is prepared from acetic acid by reduction followed by oxidation in an anhydrous medium.

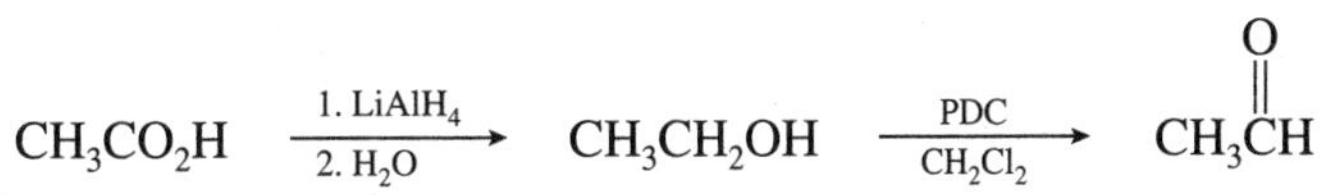

$$\underset{\text{Acetic acid}}{CH_3CO_2H} \xrightarrow[\text{2. } H_2O]{\text{1. } LiAlH_4} \underset{\text{Ethanol}}{CH_3CH_2OH} \xrightarrow[CH_2Cl_2]{\text{PDC}} \underset{\text{Acetaldehyde}}{CH_3\overset{\overset{O}{\|}}{C}H}$$

Ethylmagnesium bromide may be obtained from acetic acid by the following sequence:

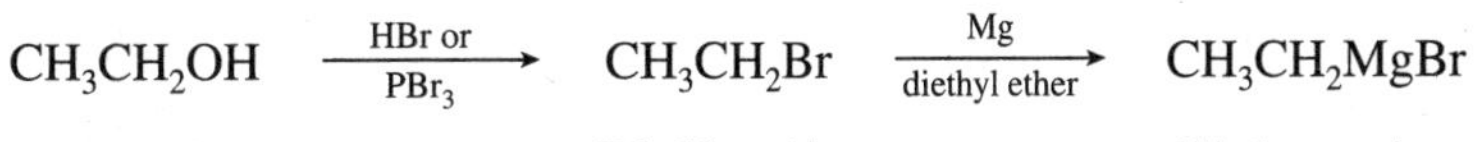

$$\underset{\text{Ethanol (Prepared as previously)}}{CH_3CH_2OH} \xrightarrow[PBr_3]{\text{HBr or}} \underset{\text{Ethyl bromide}}{CH_3CH_2Br} \xrightarrow[\text{diethyl ether}]{\text{Mg}} \underset{\text{Ethylmagnesium bromide}}{CH_3CH_2MgBr}$$

The preparation of 2-butanone is completed as follows:

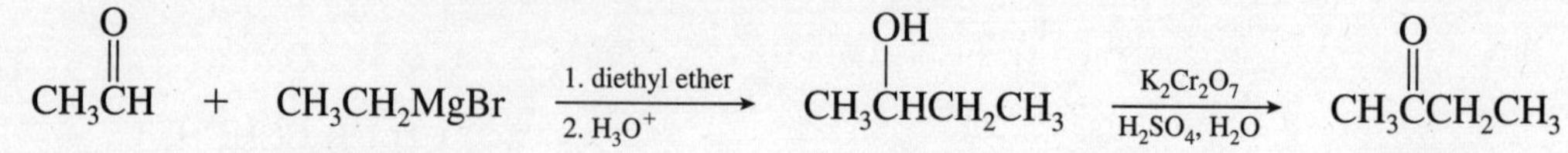

$$\underset{\text{Acetaldehyde}}{CH_3\overset{O}{\overset{\|}{C}}H} + \underset{\text{Ethylmagnesium bromide}}{CH_3CH_2MgBr} \xrightarrow[\text{2. } H_3O^+]{\text{1. diethyl ether}} \underset{\text{2-Butanol}}{CH_3\overset{OH}{\overset{|}{C}}HCH_2CH_3} \xrightarrow[H_2SO_4,\ H_2O]{K_2Cr_2O_7} \underset{\text{2-Butanone}}{CH_3\overset{O}{\overset{\|}{C}}CH_2CH_3}$$

**17.5** Chloral is trichloroethanal, $CCl_3\overset{O}{\overset{\|}{C}}H$. Chloral hydrate is the addition product of chloral and water.

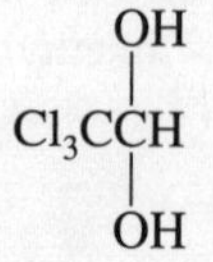

$$\underset{\text{Chloral hydrate}}{Cl_3C\underset{OH}{\underset{|}{\overset{OH}{\overset{|}{C}}}}H}$$

**17.6** Methacrylonitrile is formed by the dehydration of acetone cyanohydrin, and thus has the structure shown.

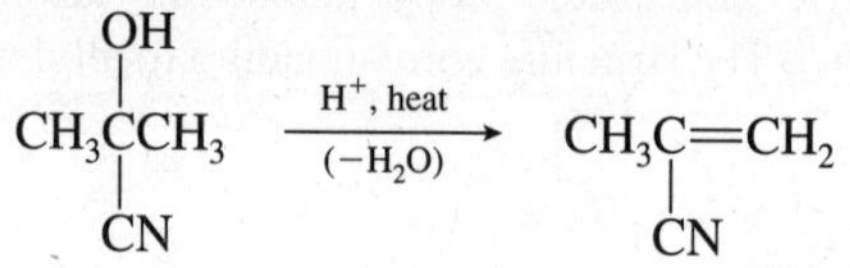

$$\underset{\text{Acetone cyanohydrin}}{CH_3\underset{CN}{\underset{|}{\overset{OH}{\overset{|}{C}}}}CH_3} \xrightarrow[(-H_2O)]{H^+,\ \text{heat}} \underset{\text{Methacrylonitrile}}{CH_3\underset{CN}{\underset{|}{C}}{=}CH_2}$$

**17.7** The overall reaction is

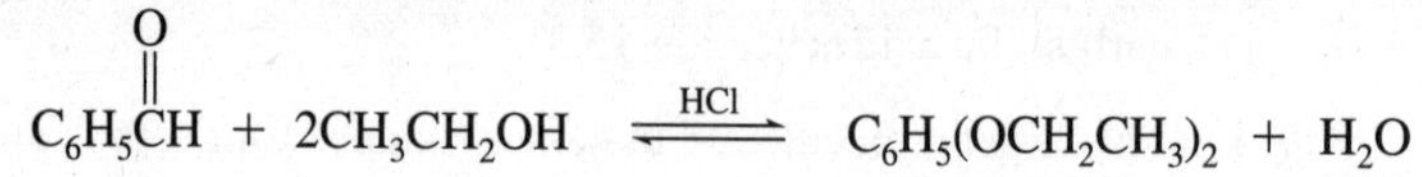

$$\underset{\text{Benzaldehyde}}{C_6H_5\overset{O}{\overset{\|}{C}}H} + \underset{\text{Ethanol}}{2CH_3CH_2OH} \underset{}{\overset{HCl}{\rightleftharpoons}} \underset{\text{Benzaldehyde diethyl acetal}}{C_6H_5(OCH_2CH_3)_2} + \underset{\text{Water}}{H_2O}$$

HCl is a strong acid and, when dissolved in ethanol, transfers a proton to ethanol to give ethyloxonium ion. Thus, we can represent the acid catalyst as the conjugate acid of ethanol.

The first three steps correspond to acid-catalyzed addition of ethanol to the carbonyl group to yield a hemiacetal.

**Step 1:**

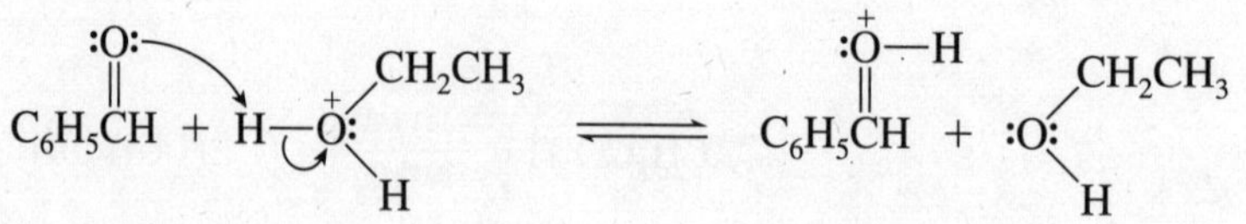

$$C_6H_5\overset{:O:}{\overset{\|}{C}}H + H{-}\overset{+}{O}(CH_2CH_3)(H) \rightleftharpoons C_6H_5\overset{:\overset{+}{O}-H}{\overset{\|}{C}}H + :\ddot{O}(CH_2CH_3)(H)$$

**Step 2:**

$$C_6H_5\overset{\overset{+}{\ddot{O}}-H}{\overset{\|}{C}}H + :\ddot{O}(CH_2CH_3)(H) \rightleftharpoons C_6H_5\overset{:\ddot{O}H}{\overset{|}{C}}H{-}\overset{+}{O}(CH_2CH_3)(H)$$

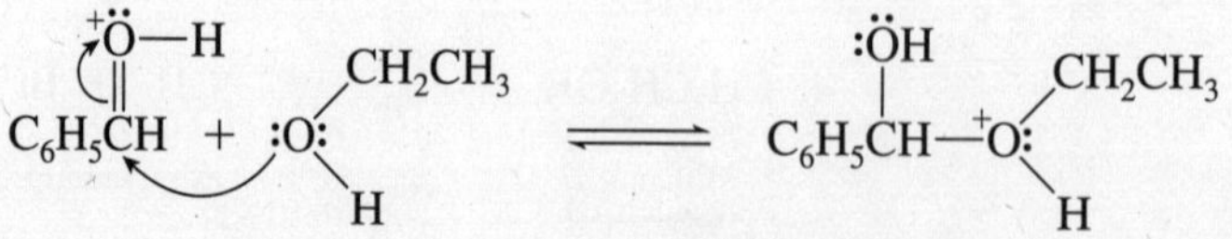

**Step 3:**

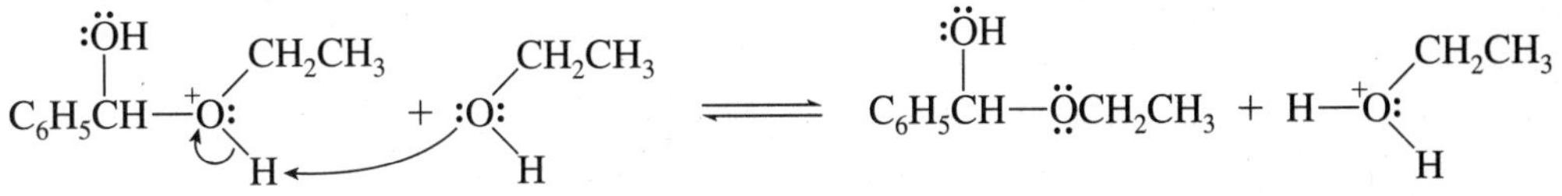

$$C_6H_5CH(OH){-}\overset{+}{O}(H)CH_2CH_3 + :\ddot{O}(H)CH_2CH_3 \rightleftharpoons C_6H_5CH(OH){-}\ddot{O}CH_2CH_3 + H{-}\overset{+}{O}(H)CH_2CH_3$$

Hemiacetal

Formation of the hemiacetal is followed by loss of water to give a carbocation.

**Step 4:**

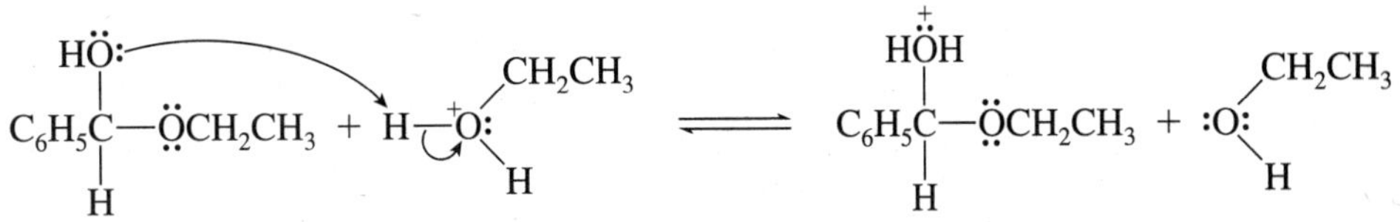

$$C_6H_5CH(\ddot{O}H){-}\ddot{O}CH_2CH_3 + H{-}\overset{+}{O}(H)CH_2CH_3 \rightleftharpoons C_6H_5CH(\overset{+}{O}H_2){-}\ddot{O}CH_2CH_3 + :\ddot{O}(H)CH_2CH_3$$

**Step 5:**

$$C_6H_5CH(\overset{+}{O}H_2){-}\ddot{O}CH_2CH_3 \rightleftharpoons C_6H_5\overset{+}{C}H{-}\ddot{O}CH_2CH_3 + H{-}\ddot{O}{-}H$$

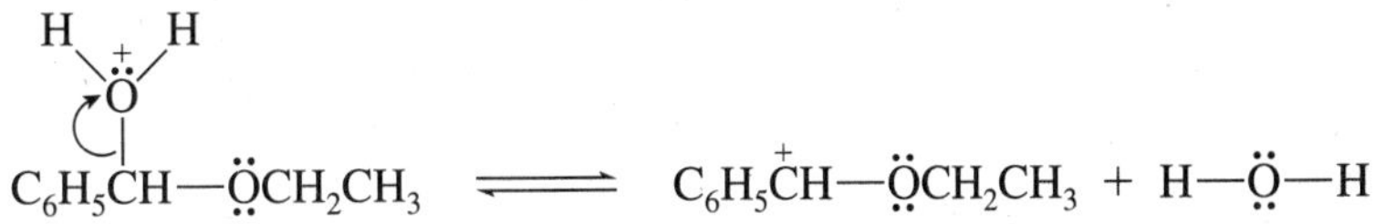

The next two steps describe the capture of the carbocation by ethanol to give the acetal:

**Step 6:**

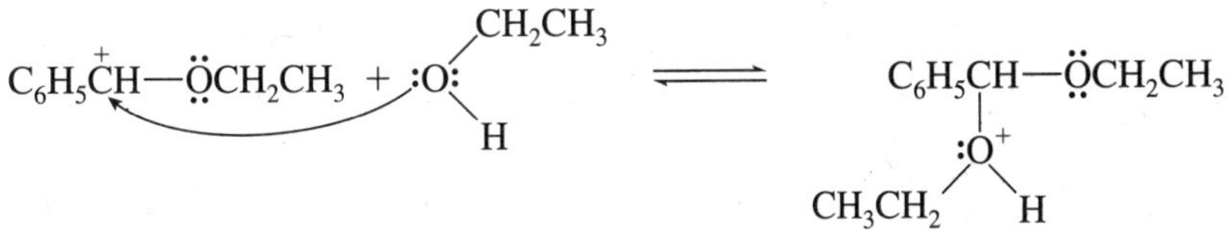

$$C_6H_5\overset{+}{C}H{-}\ddot{O}CH_2CH_3 + :\ddot{O}(H)CH_2CH_3 \rightleftharpoons C_6H_5CH(\overset{+}{O}(H)CH_2CH_3){-}\ddot{O}CH_2CH_3$$

**Step 7:**

$$C_6H_5CH(\overset{+}{O}(H)CH_2CH_3){-}\ddot{O}CH_2CH_3 + :\ddot{O}(H)CH_2CH_3 \rightleftharpoons C_6H_5CH(\ddot{O}CH_2CH_3){-}\ddot{O}CH_2CH_3 + H{-}\overset{+}{O}(H)CH_2CH_3$$

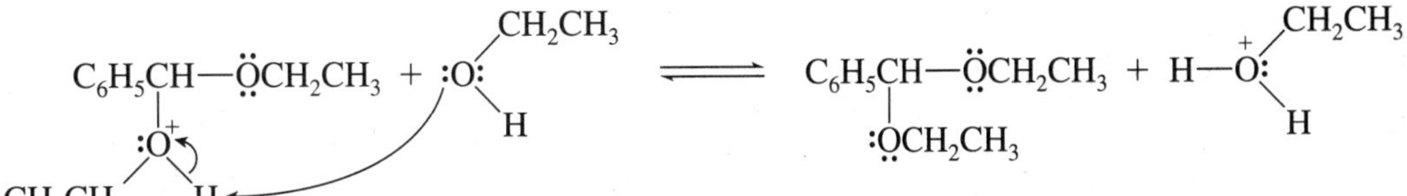

Acetal

**17.8** (*b*) 1,3-Propanediol forms acetals that contain a six-membered 1,3-dioxane ring.

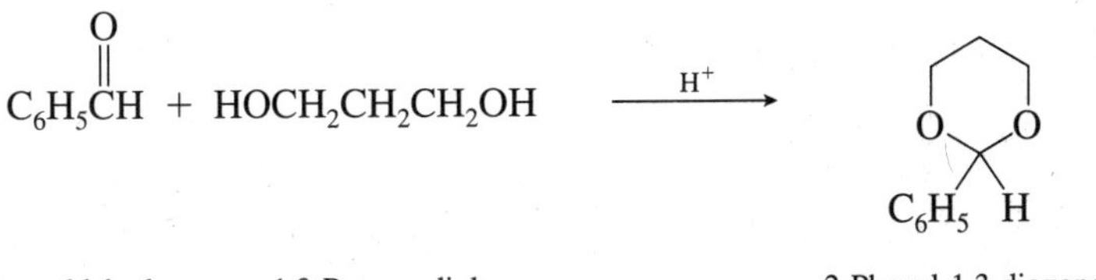

$$C_6H_5CH{=}O + HOCH_2CH_2CH_2OH \xrightarrow{H^+} \text{2-phenyl-1,3-dioxane } (C_6H_5,\ H \text{ on C-2})$$

Benzaldehyde 1,3-Propanediol 2-Phenyl-1,3-dioxane

(*c*) The cyclic acetal derived from isobutyl methyl ketone and ethylene glycol bears an isobutyl group and a methyl group at C-2 of a 1,3-dioxolane ring.

$$(CH_3)_2CHCH_2\overset{O}{\overset{\|}{C}}CH_3 + HOCH_2CH_2OH \xrightarrow{H^+} \text{2-Isobutyl-2-methyl-1,3-dioxolane}$$

Isobutyl methyl ketone — Ethylene glycol — 2-Isobutyl-2-methyl-1,3-dioxolane

(*d*) Because the starting diol is 2,2-dimethyl-1,3-propanediol, the cyclic acetal is six-membered and bears two methyl substituents at C-5 in addition to isobutyl and methyl groups at C-2.

$$(CH_3)_2CHCH_2\overset{O}{\overset{\|}{C}}CH_3 + HOCH_2C(CH_3)_2CH_2OH \xrightarrow{H^+} \text{2-Isobutyl-2,5,5-trimethyl-1,3-dioxane}$$

Isobutyl methyl ketone — 2,2-Dimethyl-1,3-propanediol — 2-Isobutyl-2,5,5-trimethyl-1,3-dioxane

**17.9** The overall reaction is

$$C_6H_5CH(OCH_2CH_3)_2 + H_2O \underset{}{\overset{HCl}{\rightleftharpoons}} C_6H_5\overset{O}{\overset{\|}{C}}H + 2CH_3CH_2OH$$

Benzaldehyde diethyl acetal — Water — Benzaldehyde — Ethanol

The mechanism of acetal hydrolysis is the reverse of acetal formation. The first four steps convert the acetal to the hemiacetal.

**Step 1:**

$$C_6H_5CH(\ddot{O}CH_2CH_3)-\ddot{O}CH_2CH_3 + H-\overset{+}{O}(H)CH_2CH_3 \rightleftharpoons C_6H_5CH(\overset{+}{O}(H)CH_2CH_3)-\ddot{O}CH_2CH_3 + :\ddot{O}(H)CH_2CH_3$$

**Step 2:**

$$C_6H_5CH(\overset{+}{O}(H)CH_2CH_3)-\ddot{O}CH_2CH_3 \rightleftharpoons C_6H_5\overset{+}{C}H-\ddot{O}CH_2CH_3 + :\ddot{O}(H)CH_2CH_3$$

**Step 3:**

$$C_6H_5\overset{+}{C}H-\ddot{O}CH_2CH_3 + :\ddot{O}H_2 \rightleftharpoons C_6H_5CH(\overset{+}{O}H_2)-\ddot{O}CH_2CH_3$$

**Step 4:**

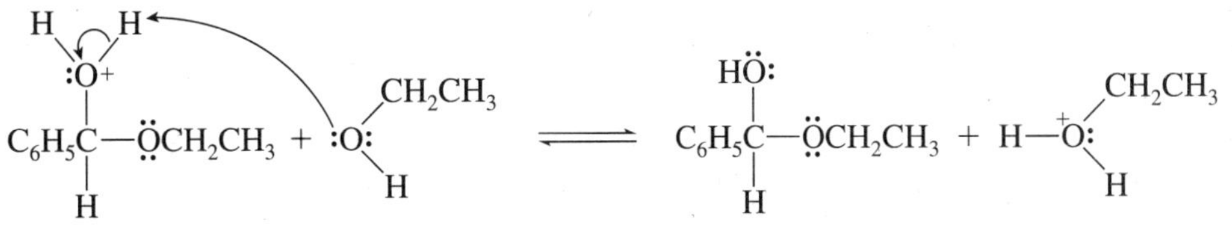

$$C_6H_5CH(\overset{+}{O}H_2)-\ddot{O}CH_2CH_3 + :\ddot{O}(H)CH_2CH_3 \rightleftharpoons C_6H_5CH(\ddot{O}H)-\ddot{O}CH_2CH_3 + H-\overset{+}{\ddot{O}}(H)CH_2CH_3$$

Hemiacetal

**Step 5:**

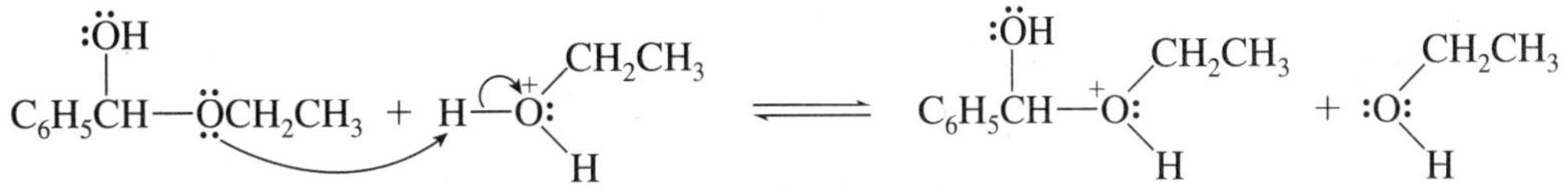

$$C_6H_5CH(:\ddot{O}H)-\ddot{O}CH_2CH_3 + H-\overset{+}{\ddot{O}}(H)CH_2CH_3 \rightleftharpoons C_6H_5CH(:\ddot{O}H)-\overset{+}{\ddot{O}}(H)CH_2CH_3 + :\ddot{O}(H)CH_2CH_3$$

**Step 6:**

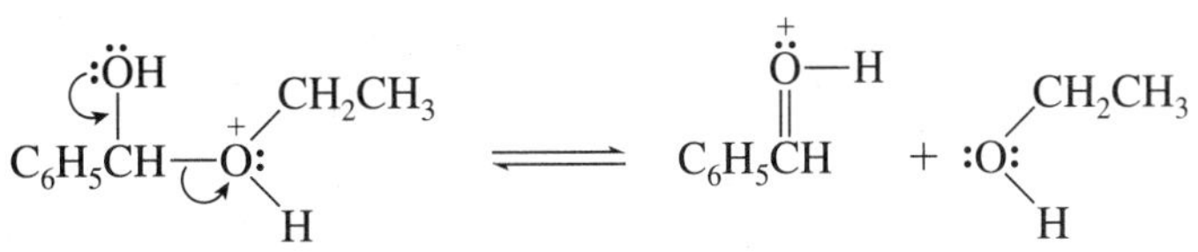

$$C_6H_5CH(:\ddot{O}H)-\overset{+}{\ddot{O}}(H)CH_2CH_3 \rightleftharpoons C_6H_5CH=\overset{+}{\ddot{O}}-H + :\ddot{O}(H)CH_2CH_3$$

**Step 7:**

$$C_6H_5CH=\overset{+}{O}-H + :\ddot{O}(H)CH_2CH_3 \rightleftharpoons C_6H_5CH=\ddot{O}: + H-\overset{+}{\ddot{O}}(H)CH_2CH_3$$

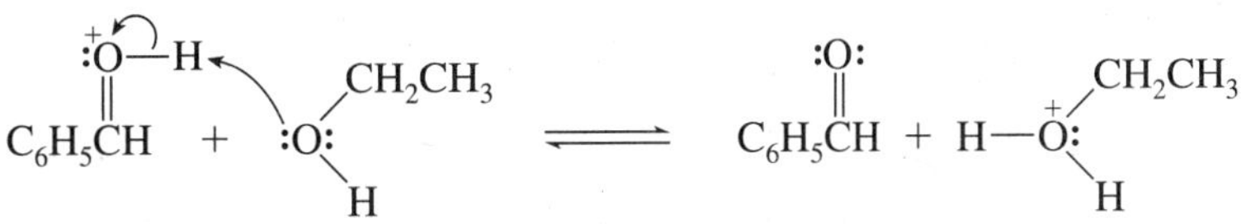

**17.10** The conversion requires reduction; however, the conditions necessary ($LiAlH_4$) would also reduce the ketone carbonyl. The ketone functionality is therefore protected as the cyclic acetal.

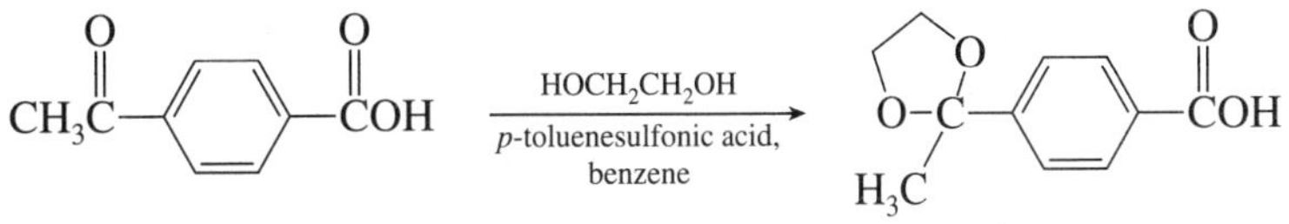

$$CH_3C(=O)-C_6H_4-COH(=O) \xrightarrow[\textit{p}\text{-toluenesulfonic acid, benzene}]{HOCH_2CH_2OH} \text{(cyclic acetal, } H_3C\text{)}-C_6H_4-COH(=O)$$

4-Acetylbenzoic acid

Reduction of the carboxylic acid may now be carried out.

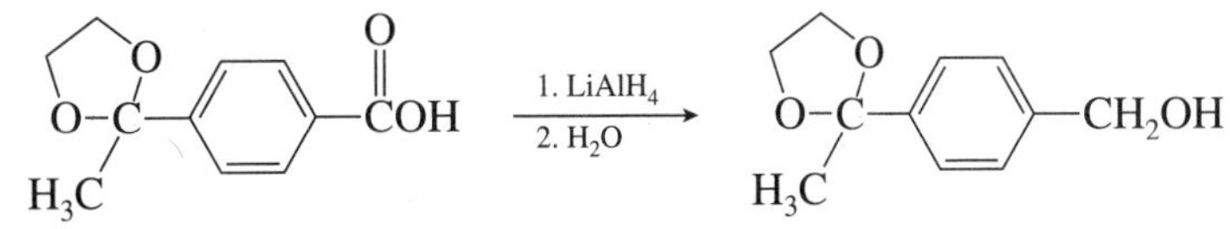

$$\text{(cyclic acetal, } H_3C\text{)}-C_6H_4-COH(=O) \xrightarrow[2.\ H_2O]{1.\ LiAlH_4} \text{(cyclic acetal, } H_3C\text{)}-C_6H_4-CH_2OH$$

Hydrolysis to remove the protecting group completes the synthesis.

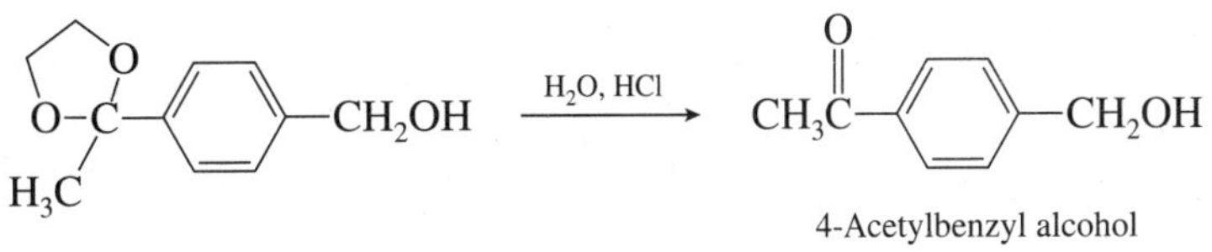

$$\text{(cyclic acetal, } H_3C\text{)}-C_6H_4-CH_2OH \xrightarrow{H_2O,\ HCl} CH_3C(=O)-C_6H_4-CH_2OH$$

4-Acetylbenzyl alcohol

**17.11** (*b*) Nucleophilic addition of butylamine to benzaldehyde gives the carbinolamine.

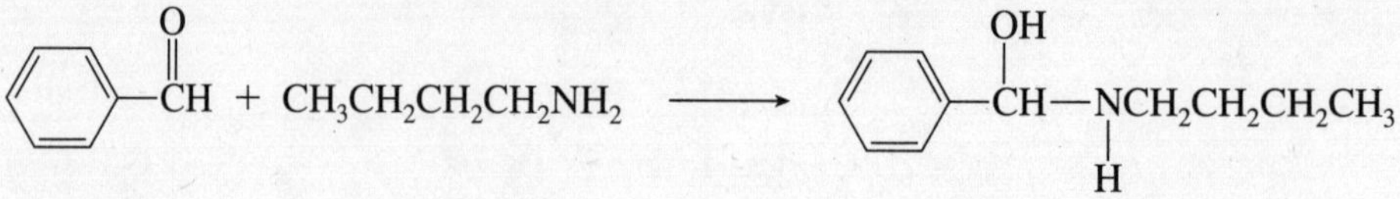

$$C_6H_5CH{=}O + CH_3CH_2CH_2CH_2NH_2 \longrightarrow C_6H_5CH(OH)-NHCH_2CH_2CH_2CH_3$$

Benzaldehyde — Butylamine — Carbinolamine intermediate

Dehydration of the carbinolamine produces the imine.

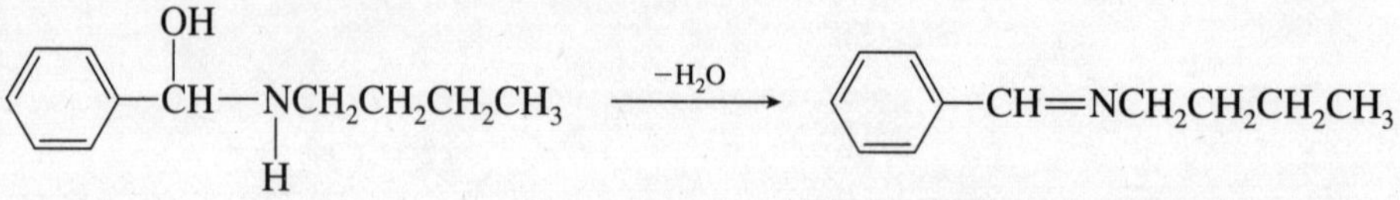

$$C_6H_5CH(OH)-NHCH_2CH_2CH_2CH_3 \xrightarrow{-H_2O} C_6H_5CH{=}NCH_2CH_2CH_2CH_3$$

*N*-Benzylidenebutylamine

(*c*) Cyclohexanone and *tert*-butylamine react according to the equation

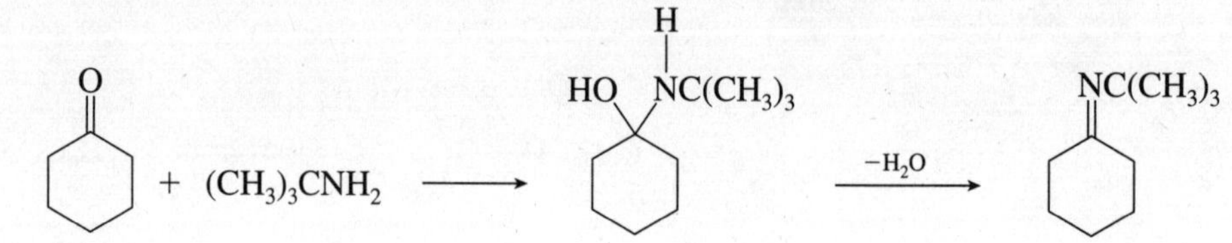

$$\text{Cyclohexanone} + (CH_3)_3CNH_2 \longrightarrow \text{1-HO-1-}[NHC(CH_3)_3]\text{cyclohexane} \xrightarrow{-H_2O} \text{cyclohexylidene}{=}NC(CH_3)_3$$

Cyclohexanone — *tert*-Butylamine — Carbinolamine intermediate — *N*-Cyclohexylidene-*tert*-butylamine

(*d*)

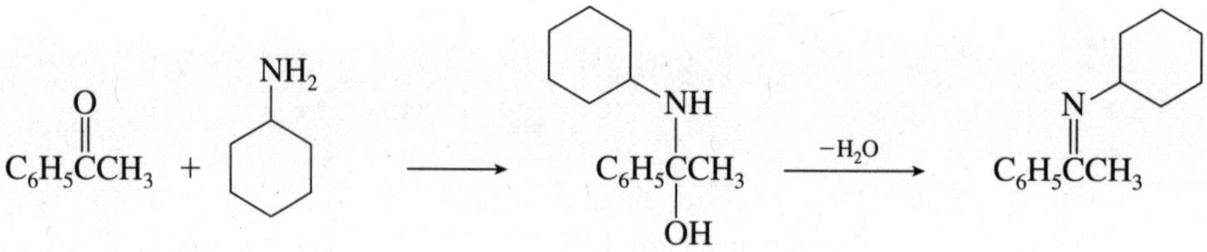

$$C_6H_5\overset{O}{\overset{\|}{C}}CH_3 + C_6H_{11}NH_2 \longrightarrow C_6H_5C(OH)(NHC_6H_{11})CH_3 \xrightarrow{-H_2O} C_6H_5C({=}NC_6H_{11})CH_3$$

Acetophenone — Cyclohexylamine — Carbinolamine intermediate — *N*-(1-Phenylethylidene)-cyclohexylamine

**17.12** (*b*) Pyrrolidine, a secondary amine, adds to 3-pentanone to give a carbinolamine.

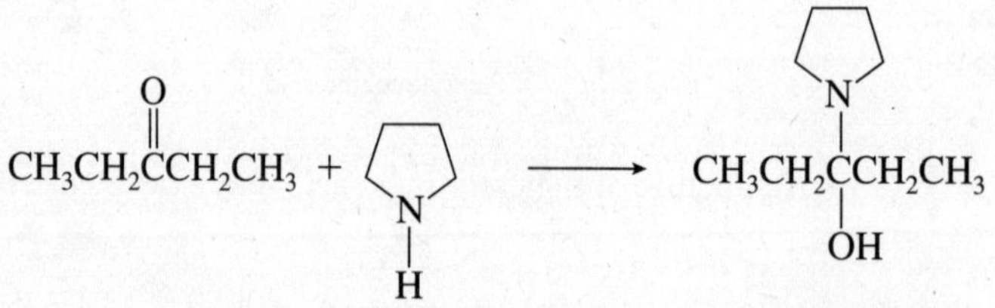

$$CH_3CH_2\overset{O}{\overset{\|}{C}}CH_2CH_3 + C_4H_8NH \longrightarrow CH_3CH_2C(OH)(NC_4H_8)CH_2CH_3$$

3-Pentanone — Pyrrolidine — Carbinolamine intermediate

Dehydration produces the enamine.

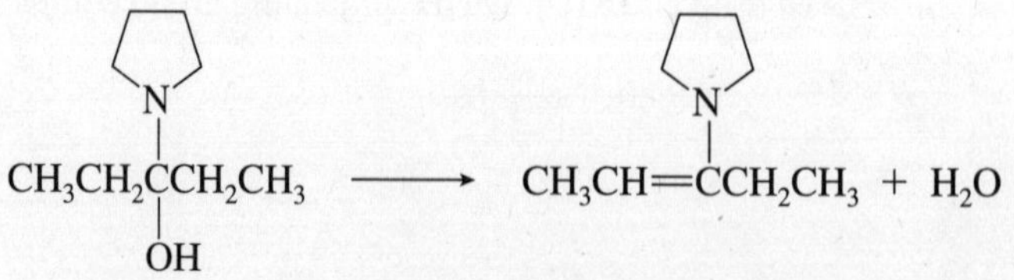

$$CH_3CH_2C(OH)(NC_4H_8)CH_2CH_3 \longrightarrow CH_3CH{=}C(NC_4H_8)CH_2CH_3 + H_2O$$

Carbinolamine intermediate — 3-Pyrrolidino-2-pentene

(*c*)

$$C_6H_5\overset{O}{\overset{\|}{C}}CH_3 + \text{piperidine (C}_5H_{10}NH) \longrightarrow C_6H_5\underset{OH}{\underset{|}{\overset{N(C_5H_{10})}{\overset{|}{C}}}}CH_3 \xrightarrow{-H_2O} C_6H_5\overset{N(C_5H_{10})}{\overset{|}{C}}{=}CH_2$$

Acetophenone | Piperidine | Carbinolamine intermediate | 1-Piperidino-1-phenylethene

**17.13** (*b*) Here we see an example of the Wittig reaction applied to diene synthesis by use of an ylide containing a carbon–carbon double bond.

$$CH_3CH_2CH_2\overset{O}{\overset{\|}{C}}H + (C_6H_5)_3\overset{+}{P}{-}\ddot{C}HCH{=}CH_2 \longrightarrow CH_3CH_2CH_2CH{=}CHCH{=}CH_2 + (C_6H_5)_3\overset{+}{P}{-}O^-$$

Butanal | Allylidenetriphenylphosphorane | 1,3-Heptadiene (52%) | Triphenylphosphine oxide

(*c*) Methylene transfer from methylenetriphenylphosphorane is one of the most commonly used Wittig reactions.

$$C_6H_{11}{-}\overset{O}{\overset{\|}{C}}CH_3 + (C_6H_5)_3\overset{+}{P}{-}\overset{-}{\ddot{C}}H_2 \longrightarrow C_6H_{11}{-}\overset{CH_2}{\overset{\|}{C}}CH_3 + (C_6H_5)_3\overset{+}{P}{-}O^-$$

Cyclohexyl methyl ketone | Methylenetriphenylphosphorane | 2-Cyclohexylpropene (66%) | Triphenylphosphine oxide

**17.14** A second resonance structure can be written for a phosphorus ylide with a double bond between phosphorus and carbon. As a third-row element, phosphorus can have more than 8 electrons in its valence shell.

$$(C_6H_5)_3\overset{+}{P}{-}\overset{-}{\ddot{C}}H_2 \longleftrightarrow (C_6H_5)_3P{=}CH_2$$

Methylenetriphenylphosphorane

**17.15** (*b*) Two Wittig reaction routes lead to 1-pentene. One is represented retrosynthetically by the disconnection

$$CH_3CH_2CH_2CH{=}CH_2 \Longrightarrow CH_3CH_2CH_2\overset{O}{\overset{\|}{C}}H + (C_6H_5)_3\overset{+}{P}{-}\overset{-}{\ddot{C}}H_2$$

1-Pentene | Butanal | Methylenetriphenylphosphorane

The other route is

$$CH_3CH_2CH_2CH{=}CH_2 \Longrightarrow CH_3CH_2CH_2\overset{-}{\ddot{C}}H{-}\overset{+}{P}(C_6H_5)_3 + H\overset{O}{\overset{\|}{C}}H$$

1-Pentene | Butylidenetriphenylphosphorane | Formaldehyde

**17.16** Ylides are prepared by the reaction of an alkyl halide with triphenylphosphine, followed by treatment with strong base. 2-Bromobutane is the alkyl halide needed in this case.

$$(C_6H_5)_3P + CH_3\underset{\displaystyle}{\overset{\displaystyle Br}{\overset{|}{C}}}HCH_2CH_3 \longrightarrow (C_6H_5)_3\overset{+}{P}-\underset{\underset{\displaystyle CH_3}{|}}{C}HCH_2CH_3\ Br^-$$

Triphenylphosphine; 2-Bromobutane; (1-Methylpropyl)triphenylphosphonium bromide

$$(C_6H_5)_3\overset{+}{P}-\underset{\underset{\displaystyle CH_3}{|}}{C}HCH_2CH_3\ Br^- + NaCH_2\overset{\overset{\displaystyle O}{\|}}{S}CH_3 \longrightarrow (C_6H_5)_3\overset{+}{P}-\underset{\underset{\displaystyle CH_3}{|}}{\ddot{C}}CH_2CH_3$$

(1-Methylpropyl)triphenylphosphonium bromide; Sodiomethyl methyl sulfoxide; Ylide

**17.17** The overall reaction is

$$C_6H_{11}\overset{\overset{\displaystyle O}{\|}}{C}CH_3 + C_6H_5\overset{\overset{\displaystyle O}{\|}}{C}OOH \longrightarrow C_6H_{11}O\overset{\overset{\displaystyle O}{\|}}{C}CH_3 + C_6H_5\overset{\overset{\displaystyle O}{\|}}{C}OH$$

Cyclohexyl methyl ketone; Peroxybenzoic acid; Cyclohexyl acetate; Benzoic acid

In the first step, the peroxy acid adds to the carbonyl group of the ketone to form a peroxy monoester of a *gem*-diol.

$$C_6H_{11}\overset{\overset{\displaystyle O}{\|}}{C}CH_3 + C_6H_5\overset{\overset{\displaystyle O}{\|}}{C}OOH \longrightarrow C_6H_{11}\underset{\underset{\displaystyle OO\overset{\|}{C}C_6H_5\ (C{=}O)}{|}}{\overset{\overset{\displaystyle OH}{|}}{C}}CH_3$$

Peroxy monoester

The intermediate then undergoes rearrangement. Alkyl group migration occurs at the same time as cleavage of the O—O bond of the peroxy ester. In general, the more substituted group migrates.

$$C_6H_{11}C(CH_3)(\ddot{O}H)-O-O\overset{\overset{\displaystyle O}{\|}}{C}C_6H_5 \longrightarrow H_3C-\overset{\overset{\displaystyle O}{\|}}{C}-O-C_6H_{11} + HO\overset{\overset{\displaystyle O}{\|}}{C}C_6H_5$$

**17.18** The formation of a carboxylic acid from Baeyer–Villiger oxidation of an aldehyde requires hydrogen migration.

$$m\text{-}O_2NC_6H_4\overset{\overset{\displaystyle O}{\|}}{C}H \xrightarrow{C_6H_5\overset{O}{\overset{\|}{C}}OOH} m\text{-}O_2NC_6H_4\underset{\underset{\displaystyle O-O\overset{\|}{C}C_6H_5}{|}}{\overset{\overset{\displaystyle :\ddot{O}H}{|}}{C}}-H \longrightarrow m\text{-}O_2NC_6H_4\overset{\overset{\displaystyle O}{\|}}{C}OH$$

*m*-Nitrobenzaldehyde; *m*-Nitrobenzoic acid

**17.19** (*a*) First consider all the isomeric aldehydes of molecular formula $C_5H_{10}O$.

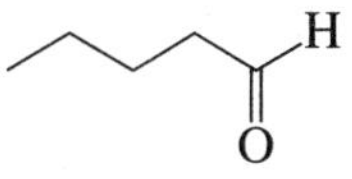

Pentanal 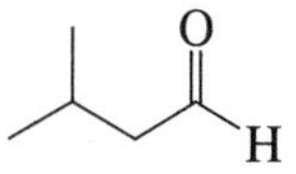 3-Methylbutanal

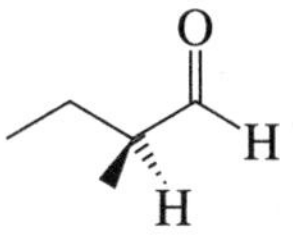

(*S*)-2-Methylbutanal 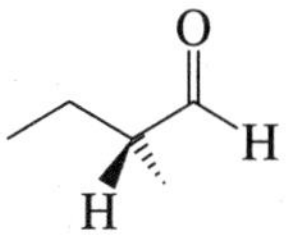 (*R*)-2-Methylbutanal 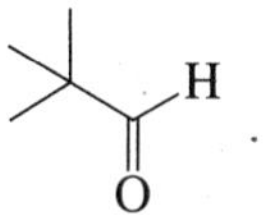 2,2-Dimethylpropanal

There are three isomeric ketones:

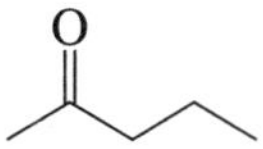

2-Pentanone 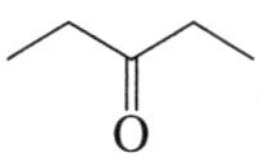 3-Pentanone 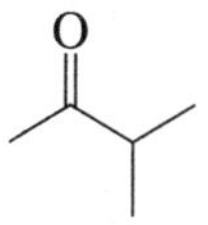 3-Methyl-2-butanone

(*b*) Reduction of an aldehyde to a primary alcohol does not introduce a stereogenic center into the molecule. The only aldehydes that yield chiral alcohols on reduction are therefore those that already contain a stereogenic center.

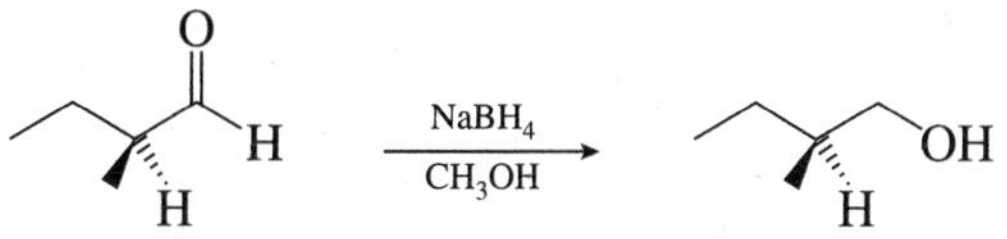

(*S*)-2-Methylbutanal $\xrightarrow[CH_3OH]{NaBH_4}$ (*S*)-2-Methyl-1-butanol

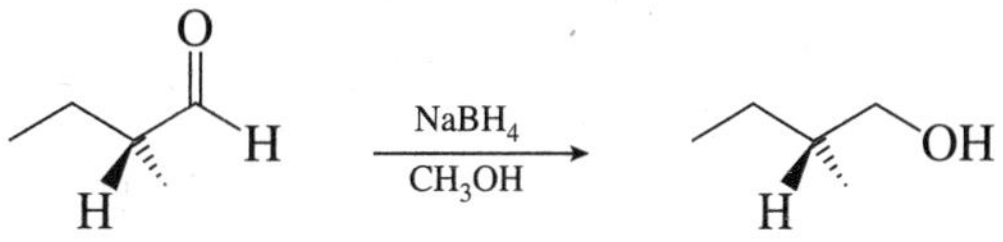

(*R*)-2-Methylbutanal $\xrightarrow[CH_3OH]{NaBH_4}$ (*R*)-2-Methyl-1-butanol

Among the ketones, 2-pentanone and 3-methyl-butanone are reduced to chiral alcohols.

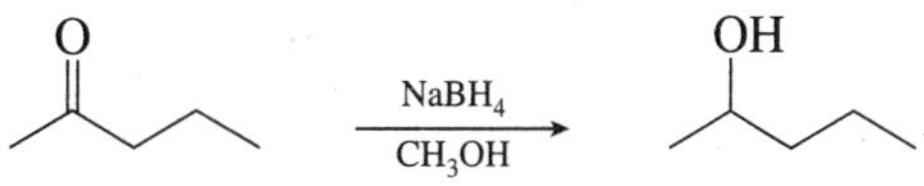

2-Pentanone $\xrightarrow[CH_3OH]{NaBH_4}$ 2-Pentanol (chiral but racemic)

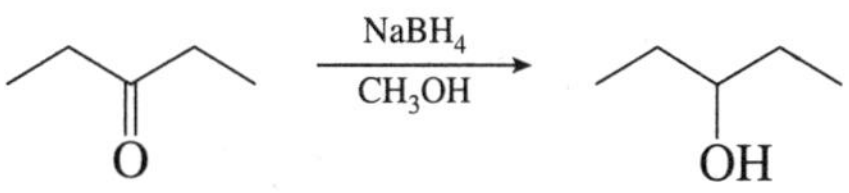

3-Pentanone $\xrightarrow[CH_3OH]{NaBH_4}$ 3-Pentanol (achiral)

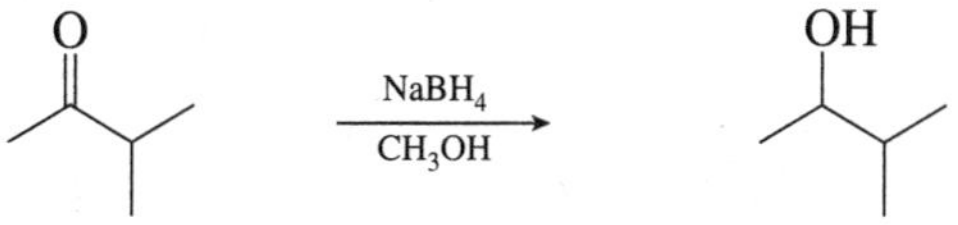

3-Methyl-2-butanone $\xrightarrow[CH_3OH]{NaBH_4}$ 3-Methyl-2-butanol (chiral but racemic)

(*c*) All the aldehydes yield chiral alcohols on reaction with methylmagnesium iodide. Thus,

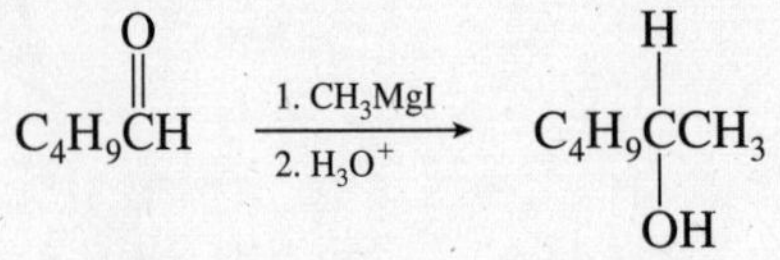

$$C_4H_9\overset{\overset{\displaystyle O}{\|}}{C}H \xrightarrow[2.\ H_3O^+]{1.\ CH_3MgI} C_4H_9\underset{\underset{\displaystyle OH}{|}}{\overset{\overset{\displaystyle H}{|}}{C}}CH_3$$

A stereogenic center is introduced in each case. None of the ketones yield chiral alcohols.

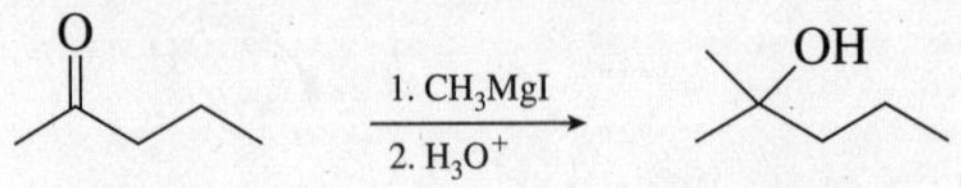

2-Pentanone $\xrightarrow[2.\ H_3O^+]{1.\ CH_3MgI}$ 2-Methyl-2-pentanol (achiral)

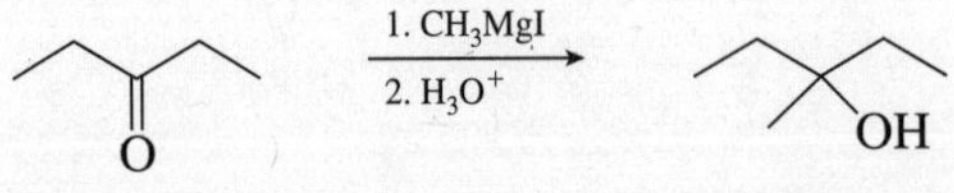

3-Pentanone $\xrightarrow[2.\ H_3O^+]{1.\ CH_3MgI}$ 3-Methyl-3-pentanol (achiral)

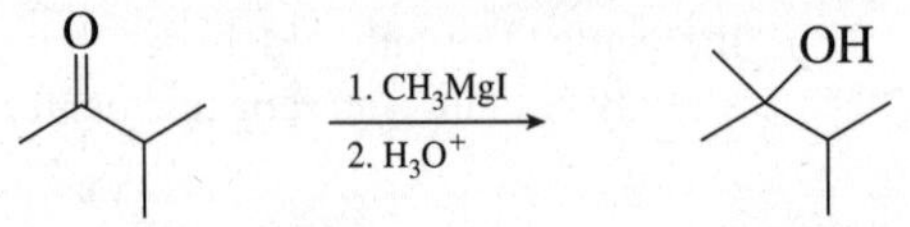

3-Methyl-2-butanone $\xrightarrow[2.\ H_3O^+]{1.\ CH_3MgI}$ 2,3-Dimethyl-2-butanol (achiral)

**17.20** (*a*) Chloral is the trichloro derivative of ethanal (acetaldehyde).

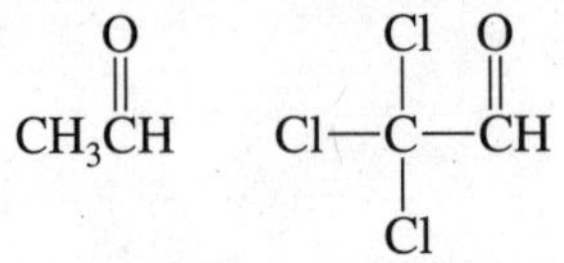

$$CH_3\overset{\overset{\displaystyle O}{\|}}{C}H \qquad Cl-\underset{\underset{\displaystyle Cl}{|}}{\overset{\overset{\displaystyle Cl}{|}}{C}}-\overset{\overset{\displaystyle O}{\|}}{C}H$$

Ethanal — Trichloroethanal (chloral)

(*b*) Pivaldehyde has two methyl groups attached to C-2 of propanal.

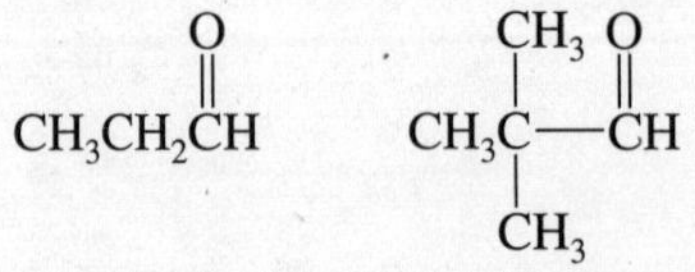

$$CH_3CH_2\overset{\overset{\displaystyle O}{\|}}{C}H \qquad CH_3\underset{\underset{\displaystyle CH_3}{|}}{\overset{\overset{\displaystyle CH_3}{|}}{C}}-\overset{\overset{\displaystyle O}{\|}}{C}H$$

Propanal — 2,2-Dimethylpropanal (pivaldehyde)

(*c*) Acrolein has a double bond between C-2 and C-3 of a three-carbon aldehyde.

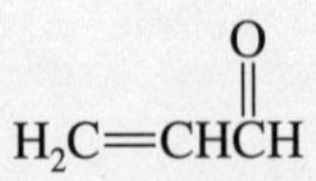

$$H_2C{=}CH\overset{\overset{\displaystyle O}{\|}}{C}H$$

2-Propenal (acrolein)

(*d*) Crotonaldehyde has a trans double bond between C-2 and C-3 of a four-carbon aldehyde.

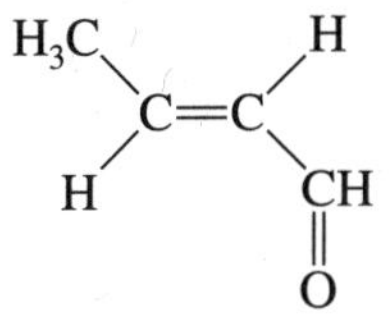

$H_3C$, H, C=C, H, CH=O

(*E*)-2-Butenal
(crotonaldehyde)

(*e*) Citral has two double bonds: one between C-2 and C-3 and the other between C-6 and C-7. The one at C-2 has the *E* configuration. There are methyl substituents at C-3 and C-7.

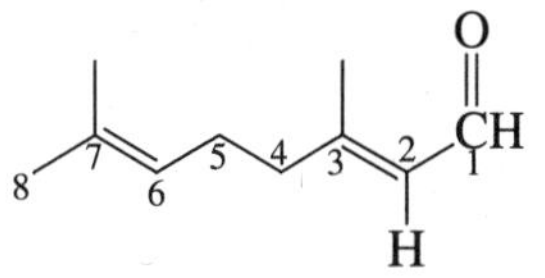

(*E*)-3,7-Dimethyl-2,6-octadienal
(citral)

(*f*) Diacetone alcohol is

$\equiv CH_3CCH_2C(CH_3)_2$ (with O on the C=O carbon and OH on the C(CH$_3$)$_2$ carbon)

4-Hydroxy-4-methyl-
2-pentanone

(*g*) The parent ketone is 2-cyclohexenone.

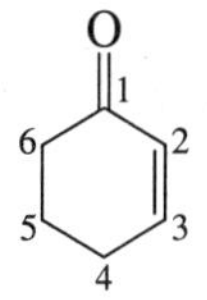

2-Cyclohexenone

Carvone has an isopropenyl group at C-5 and a methyl group at C-2.

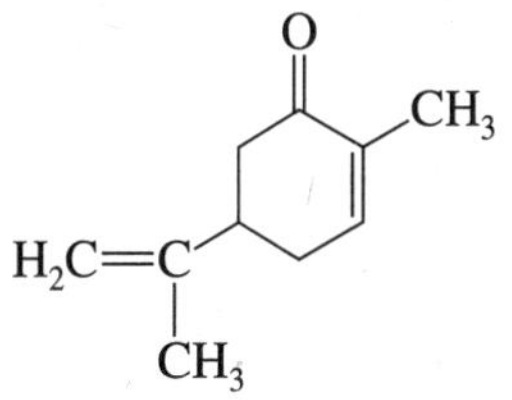

5-Isopropenyl-2-methyl-2-
cyclohexenone (carvone)

(*h*) Biacetyl is 2,3-butanedione. It has a four-carbon chain that incorporates ketone carbonyls as C-2 and C-3.

$CH_3CCCH_3$ (each central C bearing =O)

2,3-Butanedione
(biacetyl)

**17.21** (*a*) Lithium aluminum hydride reduces aldehydes to primary alcohols.

$$\underset{\text{Propanal}}{CH_3CH_2\overset{O}{\overset{\|}{C}}H} \xrightarrow[\text{2. } H_2O]{\text{1. } LiAlH_4} \underset{\text{1-Propanol}}{CH_3CH_2CH_2OH}$$

(*b*) Sodium borohydride reduces aldehydes to primary alcohols.

$$\underset{\text{Propanal}}{CH_3CH_2\overset{O}{\overset{\|}{C}}H} \xrightarrow[CH_3OH]{NaBH_4} \underset{\text{1-Propanol}}{CH_3CH_2CH_2OH}$$

(*c*) Aldehydes can be reduced to primary alcohols by catalytic hydrogenation.

$$\underset{\text{Propanal}}{CH_3CH_2\overset{O}{\overset{\|}{C}}H} \xrightarrow[Ni]{H_2} \underset{\text{1-Propanol}}{CH_3CH_2CH_2OH}$$

(*d*) Aldehydes react with Grignard reagents to form secondary alcohols.

$$\underset{\text{Propanal}}{CH_3CH_2\overset{O}{\overset{\|}{C}}H} \xrightarrow[\text{2. } H_3O^+]{\text{1. } CH_3MgI,\text{ diethyl ether}} \underset{\text{2-Butanol}}{CH_3CH_2\overset{OH}{\overset{|}{C}}HCH_3}$$

(*e*) Sodium acetylide adds to the carbonyl group of propanal to give an acetylenic alcohol.

$$\underset{\text{Propanal}}{CH_3CH_2\overset{O}{\overset{\|}{C}}H} \xrightarrow[\text{2. } H_3O^+]{\text{1. } HC{\equiv}CNa,\text{ liquid ammonia}} \underset{\text{1-Pentyn-3-ol}}{CH_3CH_2\overset{OH}{\overset{|}{C}}HC{\equiv}CH}$$

(*f*) Alkyl- or aryllithium reagents react with aldehydes in much the same way that Grignard reagents do.

$$\underset{\text{Propanal}}{CH_3CH_2\overset{O}{\overset{\|}{C}}H} \xrightarrow[\text{2. } H_3O^+]{\text{1. } C_6H_5Li,\text{ diethyl ether}} \underset{\text{1-Phenyl-1-propanol}}{CH_3CH_2\underset{OH}{\underset{|}{C}}HC_6H_5}$$

(*g*) Aldehydes are converted to acetals on reaction with alcohols in the presence of an acid catalyst.

$$\underset{\text{Propanal}}{CH_3CH_2\overset{O}{\overset{\|}{C}}H} + \underset{\text{Methanol}}{2CH_3OH} \xrightarrow{HCl} \underset{\text{Propanal dimethyl acetal}}{CH_3CH_2CH(OCH_3)_2}$$

(*h*) Cyclic acetal formation occurs when aldehydes react with ethylene glycol.

$$CH_3CH_2\overset{O}{\overset{\|}{C}}H + HOCH_2CH_2OH \xrightarrow[\text{benzene}]{p\text{-toluenesulfonic acid}} \text{2-ethyl-1,3-dioxolane (ring O–CH}_2\text{–CH}_2\text{–O; C bearing H and } CH_2CH_3)$$

Propanal — Ethylene glycol — 2-Ethyl-1,3-dioxolane

(*i*) Aldehydes react with primary amines to yield imines.

$$CH_3CH_2\overset{O}{\overset{\|}{C}}H + C_6H_5NH_2 \xrightarrow{-H_2O} CH_3CH_2CH{=}NC_6H_5$$

Propanal — Aniline — *N*-Propylideneaniline

(*j*) Secondary amines combine with aldehydes to yield enamines.

$$CH_3CH_2\overset{O}{\overset{\|}{C}}H + (CH_3)_2NH \xrightarrow[\text{benzene}]{p\text{-toluenesulfonic acid}} CH_3CH{=}\underset{}{\overset{N(CH_3)_2}{\overset{|}{C}}}H$$

Propanal — Dimethylamine — 1-(Dimethylamino)propene

(*k*) Oximes are formed on reaction of hydroxylamine with aldehydes.

$$CH_3CH_2\overset{O}{\overset{\|}{C}}H \xrightarrow{H_2NOH} CH_3CH_2CH{=}NOH$$

Propanal — Propanal oxime

(*l*) Hydrazine reacts with aldehydes to form hydrazones.

$$CH_3CH_2\overset{O}{\overset{\|}{C}}H \xrightarrow{H_2NNH_2} CH_3CH_2CH{=}NNH_2$$

Propanal — Propanal hydrazone

(*m*) Hydrazone formation is the first step in the Wolff–Kishner reduction (Section 12.8).

$$CH_3CH_2CH{=}NNH_2 \xrightarrow[\text{triethylene glycol, heat}]{\text{NaOH}} CH_3CH_2CH_3 + N_2$$

Propanal hydrazone — Propane

(*n*) The reaction of an aldehyde with *p*-nitrophenylhydrazine is analogous to that with hydrazine.

$$CH_3CH_2\overset{O}{\overset{\|}{C}}H + O_2N\text{–}C_6H_4\text{–}NHNH_2 \longrightarrow CH_3CH_2CH{=}NNH\text{–}C_6H_4\text{–}NO_2 + H_2O$$

Propanal — *p*-Nitrophenylhydrazine — Propanal *p*-nitrophenylhydrazone (*o*) Semicarbazide converts aldehydes to the corresponding semicarbazone.

$$CH_3CH_2\overset{O}{\overset{\|}{C}}H + H_2NNH\overset{O}{\overset{\|}{C}}NH_2 \longrightarrow CH_3CH_2CH{=}NNH\overset{O}{\overset{\|}{C}}NH_2 + H_2O$$

Propanal Semicarbazide Propanal semicarbazone

(*p*) Phosphorus ylides convert aldehydes to alkenes by a Wittig reaction.

$$CH_3CH_2\overset{O}{\overset{\|}{C}}H + (C_6H_5)_3\overset{+}{P}{-}\overset{-}{\ddot{C}}HCH_3 \longrightarrow CH_3CH_2CH{=}CHCH_3 + (C_6H_5)_3\overset{+}{P}{-}\overset{-}{O}$$

Propanal Ethylidenetriphenylphosphorane 2-Pentene Triphenylphosphine oxide

(*q*) Acidification of solutions of sodium cyanide generates HCN, which reacts with aldehydes to form cyanohydrins.

$$CH_3CH_2\overset{O}{\overset{\|}{C}}H + HCN \longrightarrow CH_3CH_2\overset{OH}{\overset{|}{C}}HCN$$

Propanal Hydrogen cyanide Propanal cyanohydrin

(*r*) Chromic acid oxidizes aldehydes to carboxylic acids.

$$CH_3CH_2\overset{O}{\overset{\|}{C}}H \xrightarrow{H_2CrO_4} CH_3CH_2CO_2H$$

Propanal Propanoic acid

**17.22** (*a*) Lithium aluminum hydride reduces ketones to secondary alcohols.

$$\text{Cyclopentanone} \xrightarrow[2.\ H_2O]{1.\ LiAlH_4} \text{Cyclopentanol (H, OH)}$$

Cyclopentanone Cyclopentanol

(*b*) Sodium borohydride converts ketones to secondary alcohols.

$$\text{Cyclopentanone} \xrightarrow[CH_3OH]{NaBH_4} \text{Cyclopentanol (H, OH)}$$

Cyclopentanone Cyclopentanol

(*c*) Catalytic hydrogenation of ketones yields secondary alcohols.

$$\text{Cyclopentanone} \xrightarrow[Ni]{H_2} \text{Cyclopentanol (H, OH)}$$

Cyclopentanone Cyclopentanol

(*d*) Grignard reagents react with ketones to form tertiary alcohols.

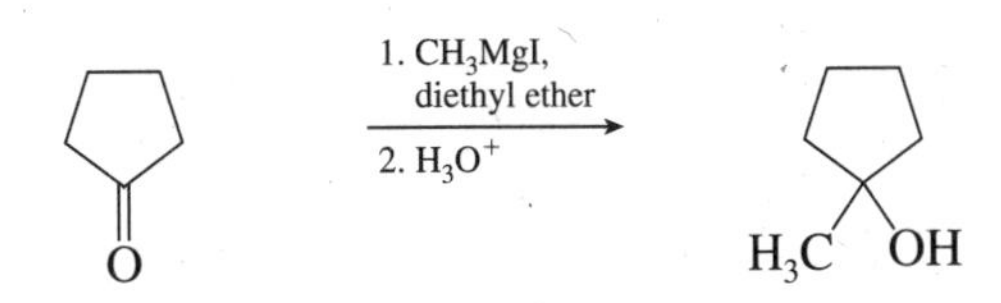


Cyclopentanone 1-Methylcyclopentanol

(*e*) Addition of sodium acetylide to cyclopentanone yields a tertiary acetylenic alcohol.

1. $HC{\equiv}CNa$, liquid ammonia; 2. $H_3O^+$

O → $HC{\equiv}C$ OH

Cyclopentanone 1-Ethynylcyclopentanol

(*f*) Phenyllithium adds to the carbonyl group of cyclopentanone to yield 1-phenylcyclopentanol.

1. $C_6H_5Li$, diethyl ether; 2. $H_3O^+$

O → $C_6H_5$ OH

Cyclopentanone 1-Phenylcyclopentanol

(*g*) The equilibrium constant for acetal formation from ketones is generally unfavorable.

O + $2CH_3OH$ →(HCl) $CH_3O$ $OCH_3$ $K < 1$

Cyclopentanone Methanol Cyclopentanone dimethyl acetal

(*h*) Cyclic acetal formation is favored even for ketones.

O + $HOCH_2CH_2OH$ →(*p*-toluenesulfonic acid, benzene) O O

Cyclopentanone Ethylene glycol 1,4-Dioxaspiro[4.4]nonane

(*i*) Ketones react with primary amines to form imines.

O + $C_6H_5NH_2$ → $NC_6H_5$

Cyclopentanone Aniline *N*-Cyclopentylideneaniline

(*j*) Dimethylamine reacts with cyclopentanone to yield an enamine.

O + $(CH_3)_2NH$ →(*p*-toluenesulfonic acid, benzene) $N(CH_3)_2$

Cyclopentanone Dimethylamine 1-(Dimethylamino)-cyclopentene

(*k*) An oxime is formed when cyclopentanone is treated with hydroxylamine.

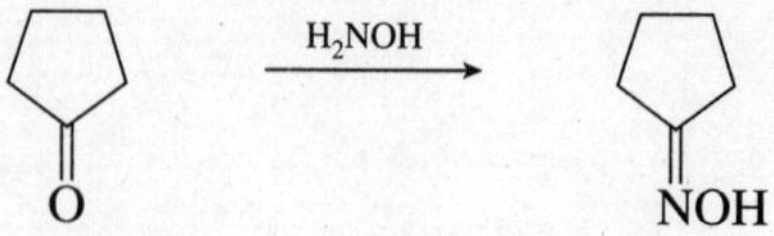

$H_2NOH$

O

NOH

Cyclopentanone

Cyclopentanone oxime

(*l*) Hydrazine reacts with cyclopentanone to form a hydrazone.

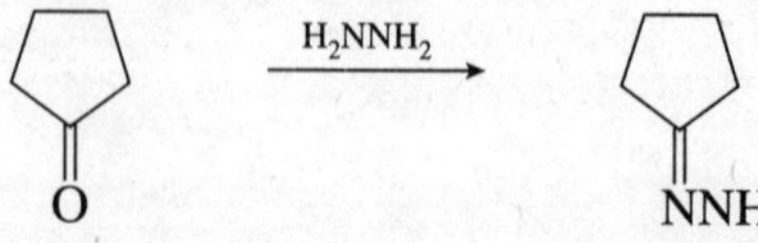

$H_2NNH_2$

O

$NNH_2$

Cyclopentanone

Cyclopentanone hydrazone

(*m*) Heating a hydrazone in base with a high-boiling alcohol as solvent converts it to an alkane.

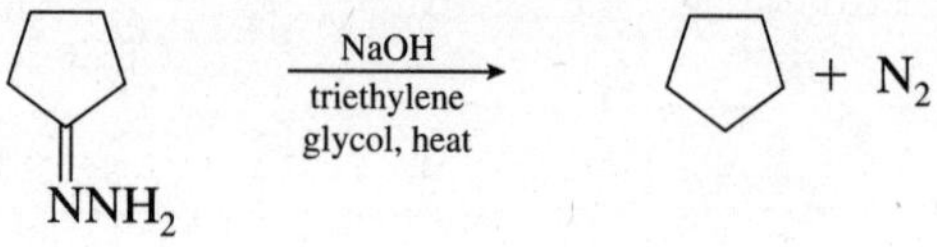

NaOH triethylene glycol, heat

$NNH_2$

$+ N_2$

Cyclopentanone hydrazone

Cyclopentane

(*n*) A *p*-nitrophenylhydrazone is formed.

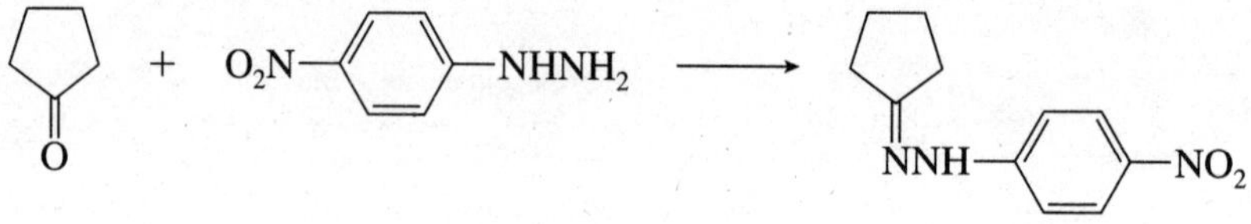

$+ O_2N$—$C_6H_4$—$NHNH_2$ ⟶

O

NNH—$C_6H_4$—$NO_2$

Cyclopentanone

*p*-Nitrophenylhydrazine

Cyclopentanone *p*-nitrophenylhydrazone

(*o*) Cyclopentanone is converted to a semicarbazone on reaction with semicarbazide.

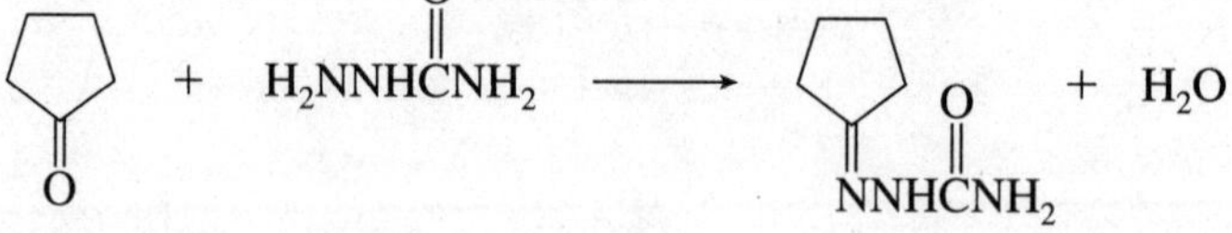

$+ H_2NNHCNH_2$ (C=O) ⟶ $+ H_2O$

O

$NNHCNH_2$ (C=O)

Cyclopentanone

Semicarbazide

Cyclopentanone semicarbazone

(*p*) A Wittig reaction takes place, forming ethylidenecyclopentane.

$+ (C_6H_5)_3\overset{+}{P}—\overset{-}{\ddot{C}}HCH_3$ ⟶ $+ (C_6H_5)_3\overset{+}{P}—\overset{-}{O}$

O

$CHCH_3$

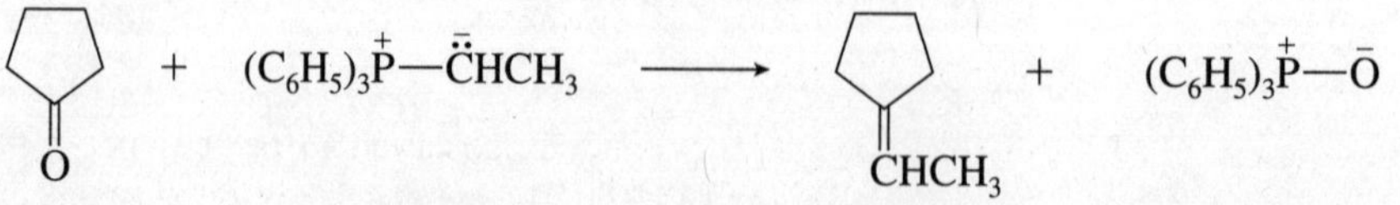

Cyclopentanone

Ethylidenetriphenylphosphorane

Ethylidenecyclopentane

Triphenylphosphine oxide

(*q*) Cyanohydrin formation takes place.

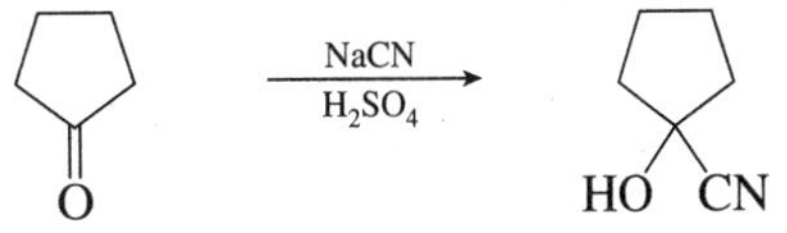

NaCN, $H_2SO_4$

Cyclopentanone → Cyclopentanone cyanohydrin

(*r*) Cyclopentanone is not oxidized readily with chromic acid.

**17.23** (*a*) The first step in analyzing this problem is to write the structure of the starting ketone in stereochemical detail.

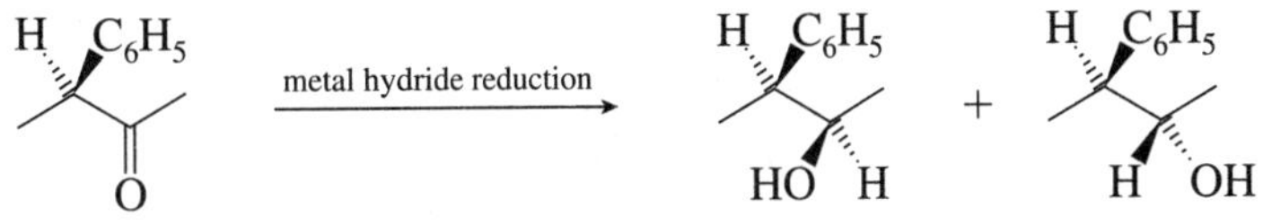

metal hydride reduction

(*S*)-3-Phenyl-2-butanone → (2*R*,3*S*)-3-Phenyl-2-butanol + (2*S*,3*S*)-3-Phenyl-2-butanol

Reduction of the ketone introduces a new stereogenic center, which may have either the *R* or the *S* configuration; the configuration of the original stereogenic center is unaffected. In practice the 2*R*,3*S* diastereomer is observed to form in greater amounts than the 2*S*,3*S* (ratio 2.5:1 for $LiAlH_4$ reduction).

(*b*) Reduction of the ketone can yield either *cis*- or *trans*-4-*tert*-butylcyclohexanol.

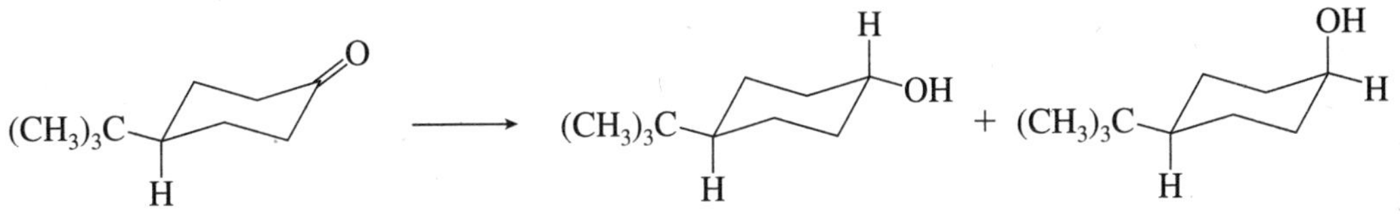

4-*tert*-Butylcyclohexanone → *trans*-4-*tert*-Butylcyclohexanol + *cis*-4-*tert*-Butylcyclohexanol

It has been observed that the major product obtained on reduction with either lithium aluminum hydride or sodium borohydride is the trans alcohol (trans/cis ≈9:1).

(*c*) The two reduction products are the exo and endo alcohols.

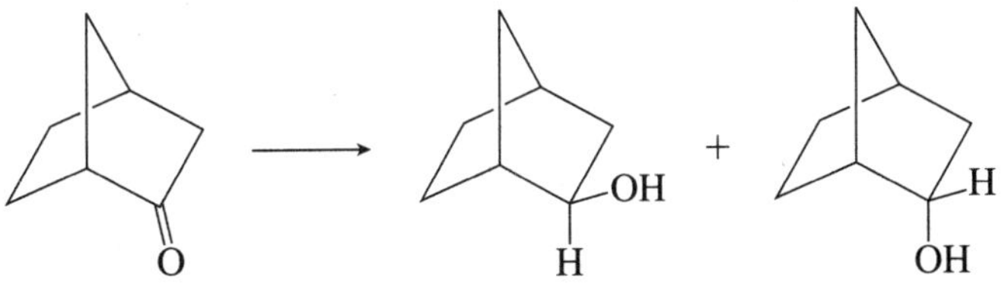

Bicyclo[2.2.1]heptan-2-one → *exo*-Bicyclo[2.2.1]heptan-2-ol + *endo*-Bicyclo[2.2.1]heptan-2-ol

The major product is observed to be the endo alcohol (endo/exo 9:1) for reduction with $NaBH_4$ or $LiAlH_4$. The stereoselectivity observed in this reaction is due to decreased steric hindrance to attack of the hydride reagent from the exo face of the molecule, giving rise to the endo alcohol.

(*d*) The hydroxyl group may be on the same side as the double bond or on the opposite side.

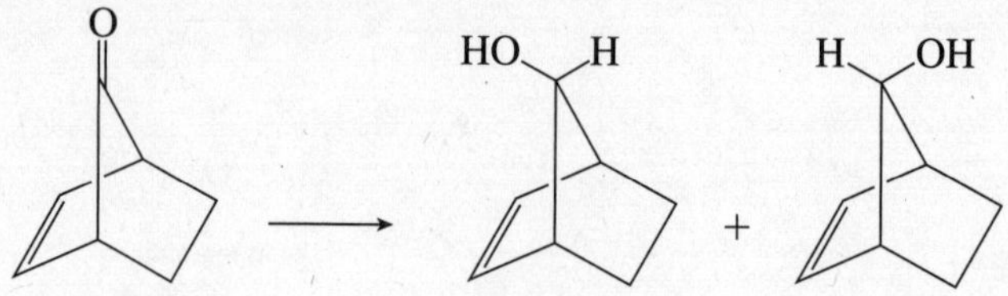

Bicyclo[2.2.1]-hept-2-en-7-one → *syn*-Bicyclo[2.2.1]-hept-2-en-7-ol + *anti*-Bicyclo[2.2.1]-hept-2-en-7-ol

The anti alcohol is observed to be formed in greater amounts (85:15) on reduction of the ketone with $LiAlH_4$. Steric factors governing attack of the hydride reagent again explain the major product observed.

**17.24** (*a*) Aldehydes undergo nucleophilic addition faster than ketones. Steric crowding in the rate-determining step of the ketone reaction raises the energy of the transition state, giving rise to a slower rate of reaction. Thus benzaldehyde is reduced by sodium borohydride more rapidly than is acetophenone. The measured relative rates are

$$k_{rel} = \frac{C_6H_5\overset{O}{\overset{\|}{C}}H}{C_6H_5\overset{O}{\overset{\|}{C}}CH_3} = 440$$

(*b*) The presence of an electronegative substituent on the $\alpha$-carbon atom causes a dramatic increase in $K_{hydr}$. Trichloroethanal (chloral) is almost completely converted to its geminal diol (chloral hydrate) in aqueous solution.

$$Cl_3\overset{O}{\overset{\|}{C}}H + H_2O \longrightarrow Cl_3C(OH)_2H$$

Trichloroethanal (chloral) → 2,2,2-Trichloro-1,1-ethanediol (chloral hydrate)

Electron-withdrawing groups such as $Cl_3C$ destabilize carbonyl groups to which they are attached and make the energy change favoring the products of nucleophilic addition more favorable.

$$K_{rel} = \frac{Cl_3\overset{O}{\overset{\|}{C}}H}{CH_3\overset{O}{\overset{\|}{C}}H} \approx 20{,}000$$

(*c*) Recall that the equilibrium constants for nucleophilic addition to carbonyl groups are governed by a combination of electronic effects and steric effects. Electronically there is little difference between acetone and 3,3-dimethyl-2-butanone, but sterically there is a significant difference. The cyanohydrin products are more crowded than the starting ketones, and so the bulkier the alkyl groups that are attached to the carbonyl, the more strained and less stable will be the cyanohydrin.

$$\underset{\text{Ketone}}{(R)(H_3C)C{=}O} + \underset{\text{Hydrogen cyanide}}{HCN} \rightleftharpoons \underset{\text{Cyanohydrin [less strained for R = CH}_3\text{ than for R = C(CH}_3)_3]}{CH_3C(OH)(R)CN}$$

$$K_{rel} = \frac{CH_3\overset{O}{\overset{\|}{C}}CH_3}{CH_3\overset{O}{\overset{\|}{C}}C(CH_3)_3} = 40$$

(*d*) Steric effects influence the rate of nucleophilic addition to these two ketones. Carbon is on its way from tricoordinate to tetracoordinate at the transition state, and alkyl groups are forced closer together than they are in the ketone.

$$\underset{sp^2}{(R)(H_3C)C{=}O} \xrightarrow{NaBH_4} \underset{\text{Transition state}}{[(R)(H_3C)(H\cdots\overset{\delta-}{B}H_3)C\cdots\overset{\delta-}{O}]} \longrightarrow \underset{sp^3}{(R)(H_3C)(H)C{-}O\bar{B}H_3}$$

The transition state is of lower energy when R is smaller. Acetone (for which R is methyl) is reduced faster than 3,3-dimethyl-2-butanone (where R is *tert*-butyl).

$$k_{rel} = \frac{CH_3\overset{O}{\overset{\|}{C}}CH_3}{CH_3\overset{O}{\overset{\|}{C}}C(CH_3)_3} = 12$$

(*e*) In this problem we examine the rate of hydrolysis of acetals to the corresponding ketone or aldehyde. The rate-determining step is carbocation formation.

$$R_2C(\overset{+}{O}HCH_2CH_3)(OCH_2CH_3) \longrightarrow R_2\overset{+}{C}{-}OCH_2CH_3 + CH_3CH_2OH$$

Hybridization at carbon changes from $sp^3$ to $sp^2$; crowding at this carbon is relieved as the carbocation is formed. The more crowded acetal (R = $CH_3$) forms a carbocation faster than the less crowded one (R = H). Another factor of even greater importance is the extent of stabilization of the carbocation intermediate; the more stable carbocation (R = $CH_3$) is formed faster than the less stable one (R = H).

$$k_{rel} = \frac{(CH_3)_2C(OCH_2CH_3)_2}{CH_2(OCH_2CH_3)_2} = 1.8 \times 10^7$$

**17.25** (*a*) The reaction as written is the reverse of cyanohydrin formation, and the principles that govern equilibria in nucleophilic addition to carbonyl groups apply in reverse order to the dissociation of cyanohydrins to aldehydes and ketones. Cyanohydrins of ketones dissociate more at equilibrium than do cyanohydrins of aldehydes. More strain due to crowding is relieved when a ketone cyanohydrin dissociates and a more stabilized carbonyl group is formed. The equilibrium constant $K_{diss}$ is larger for

$$CH_3C(OH)(CN)CH_3 \xrightleftharpoons{K_{diss}} CH_3\overset{O}{\overset{\|}{C}}CH_3 + HCN$$

Acetone cyanohydrin — Acetone — Hydrogen cyanide

than it is for

$$CH_3CH_2CH(OH)CN \xrightleftharpoons{K_{diss}} CH_3CH_2\overset{O}{\overset{\|}{C}}H + HCN$$

Propanal cyanohydrin — Propanal — Hydrogen cyanide

(*b*) Cyanohydrins of ketones have a more favorable equilibrium constant for dissociation than do cyanohydrins of aldehydes. Crowding is relieved to a greater extent when a ketone cyanohydrin dissociates and a more stable carbonyl group is formed. The measured dissociation constants are

$$C_6H_5CH(OH)CN \rightleftharpoons C_6H_5\overset{O}{\overset{\|}{C}}H + HCN \qquad K = 4.7 \times 10^{-3}$$

Benzaldehyde cyanohydrin — Benzaldehyde

$$C_6H_5C(OH)(CH_3)CN \rightleftharpoons C_6H_5\overset{O}{\overset{\|}{C}}CH_3 + HCN \qquad K = 1.3$$

Acetophenone cyanohydrin — Acetophenone

**17.26** (*a*) The reaction of an aldehyde with 1,3-propanediol in the presence of *p*-toluenesulfonic acid forms a cyclic acetal.

2-Bromo-3,4,5-trimethoxybenzaldehyde + $HOCH_2CH_2CH_2OH$ (1,3-Propanediol) $\xrightarrow[\text{benzene, heat}]{p\text{-toluenesulfonic acid}}$ 2-(2′-Bromo-3′,4′,5′-trimethoxyphenyl)-1,3-dioxane (81%)

(*b*) The reagent $CH_3ONH_2$ is called *O*-methylhydroxylamine, and it reacts with aldehydes in a manner similar to hydroxylamine.

4-Hydroxy-2-methoxybenzaldehyde + $CH_3ONH_2$ (*O*-Methylhydroxylamine) ⟶ 4-Hydroxy-2-methoxybenzaldehyde *O*-methyloxime ($CH{=}NOCH_3$)

(*c*) Propanal reacts with 1,1-dimethylhydrazine to yield the corresponding hydrazone.

$$\underset{\text{Propanal}}{CH_3CH_2\overset{O}{\overset{\|}{C}}H} + \underset{\text{1,1-Dimethylhydrazine}}{(CH_3)_2NNH_2} \longrightarrow \underset{\text{Propanal dimethylhydrazone}}{CH_3CH_2CH{=}NN(CH_3)_2}$$

(*d*) Acid-catalyzed hydrolysis of the acetal gives the aldehyde in 87% yield.

$$H_3C\text{—}C_6H_4\text{—}\overset{CH_3}{\overset{|}{C}}HCH_2CH_2\text{—(1,3-dioxolan-2-yl)} \xrightarrow[\text{heat}]{H_2O,\ HCl} \underset{\text{4-(}p\text{-Methylphenyl)pentanal}}{H_3C\text{—}C_6H_4\text{—}\overset{CH_3}{\overset{|}{C}}HCH_2CH_2\overset{O}{\overset{\|}{C}}H}$$

(*e*) Hydrogen cyanide adds to carbonyl groups to form cyanohydrins.

$$\underset{\text{Acetophenone}}{C_6H_5\overset{O}{\overset{\|}{C}}CH_3} \xrightarrow[HCl]{NaCN} \underset{\text{Acetophenone cyanohydrin}}{C_6H_5\overset{OH}{\overset{|}{\underset{CH_3}{\underset{|}{C}}}}CN}$$

(*f*) The reagent is a secondary amine known as **morpholine.** Secondary amines react with ketones to give enamines.

$$\underset{\text{Acetophenone}}{C_6H_5\overset{O}{\overset{\|}{C}}CH_3} + \underset{\text{Morpholine}}{HN(CH_2CH_2)_2O} \xrightarrow{H^+} \underset{\text{Carbinolamine intermediate}}{C_6H_5\overset{OH}{\overset{|}{C}}(N(CH_2CH_2)_2O)CH_3} \xrightarrow{H^+(-H_2O)} \underset{\text{1-Morpholinostyrene (57–64\%)}}{C_6H_5C(N(CH_2CH_2)_2O){=}CH_2}$$

(*g*) Migration of the alkyl group in a Baeyer–Villiger oxidation occurs with retention of configuration.

$$\underset{(R)\text{-3-Methyl-3-phenyl-2-pentanone}}{CH_3CH_2\text{—}C(CH_3)(C_6H_5)\text{—}\overset{O}{\overset{\|}{C}}CH_3} + \underset{\text{Peroxybenzoic acid}}{C_6H_5\overset{O}{\overset{\|}{C}}OOH} \longrightarrow \underset{(R)\text{-1-Methyl-1-phenylpropyl acetate}}{CH_3CH_2\text{—}C(CH_3)(C_6H_5)\text{—}O\overset{O}{\overset{\|}{C}}CH_3}$$

**17.27** Wolff–Kishner reduction converts a carbonyl group (C=O) to a methylene group ($CH_2$).

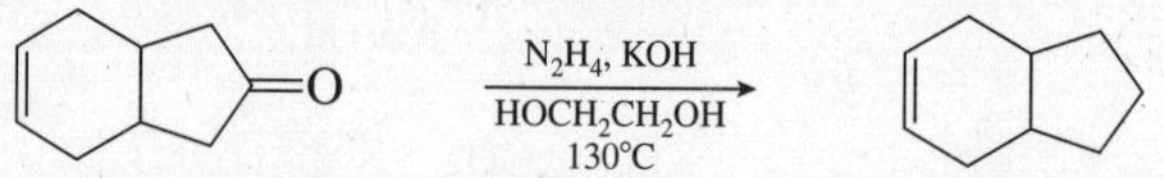

$N_2H_4$, KOH; $HOCH_2CH_2OH$, 130°C

Bicyclo[4.3.0]non-3-en-8-one → Bicyclo[4.3.0]non-3-ene (compound A, 90%)

Treatment of the alkene with *m*-chloroperoxybenzoic acid produces an epoxide, compound B.

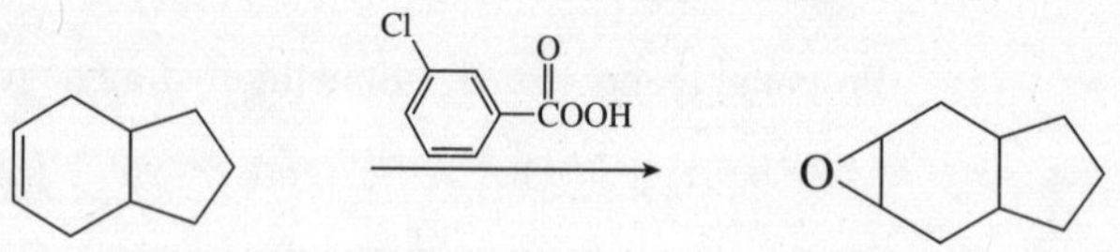

Bicyclo[4.3.0]non-3-ene → 3,4-Epoxybicyclo[4.3.0]nonane (compound B, 92%)

Epoxides undergo reduction with lithium aluminum hydride to form alcohols (Section 16.12).

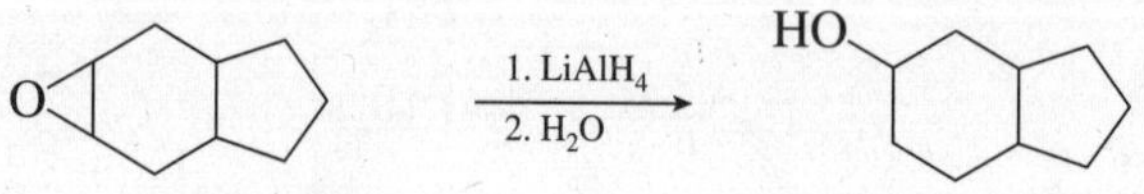

1. $LiAlH_4$ 2. $H_2O$

3,4-Epoxybicyclo[4.3.0]nonane → Bicyclo[4.3.0]nonan-3-ol (compound C, 90%)

Chromic acid oxidizes the alcohol to a ketone.

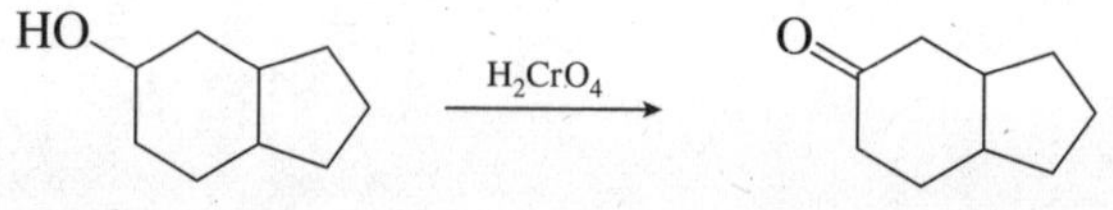

$H_2CrO_4$

Bicyclo[4.3.0]nonan-3-ol → Bicyclo[4.3.0]nonan-3-one (compound D, 75%)

**17.28** Hydration of formaldehyde by $H_2{}^{17}O$ produces a *gem*-diol in which the labeled and unlabeled hydroxyl groups are equivalent. When this *gem*-diol reverts to formaldehyde, loss of either of the hydroxyl groups is equally likely and leads to eventual replacement of the mass-16 isotope of oxygen by $^{17}O$.

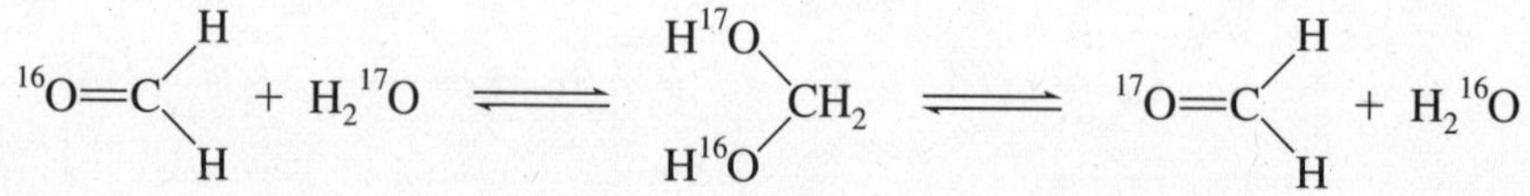

$$H_2C{=}{}^{16}O + H_2{}^{17}O \rightleftharpoons (H^{17}O)(H^{16}O)CH_2 \rightleftharpoons H_2C{=}{}^{17}O + H_2{}^{16}O$$

This reaction has been monitored by $^{17}O$ NMR spectroscopy; $^{17}O$ gives an NMR signal, but $^{16}O$ does not.

**17.29** First write out the chemical equation for the reaction that takes place. Vicinal diols (1,2-diols) react with aldehydes to give cyclic acetals.

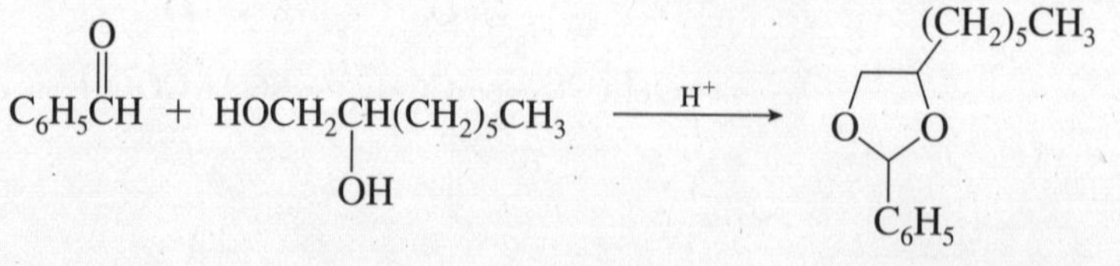

$$C_6H_5CHO + HOCH_2CH(OH)(CH_2)_5CH_3 \xrightarrow{H^+}$$

Benzaldehyde + 1,2-Octanediol → 4-Hexyl-2-phenyl-1,3-dioxolane

Notice that the phenyl and hexyl substituents may be either cis or trans to each other. The two products are the cis and trans stereoisomers.

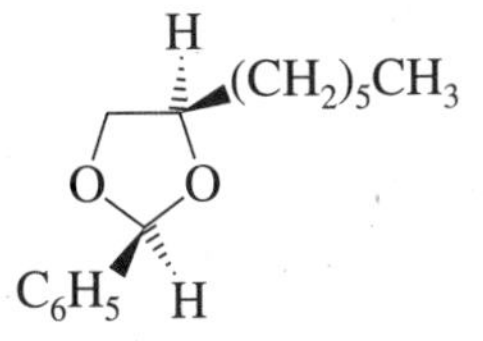

*cis*-4-Hexyl-2-phenyl-1,3-dioxolane

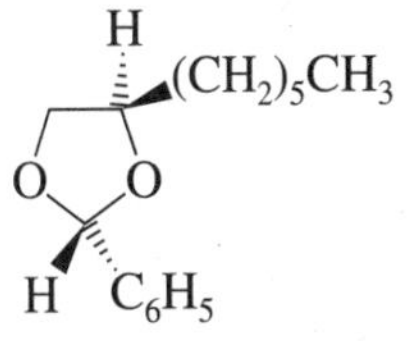

*trans*-4-Hexyl-2-phenyl-1,3-dioxolane

**17.30** Cyclic hemiacetals are formed by intramolecular nucleophilic addition of a hydroxyl group to a carbonyl.

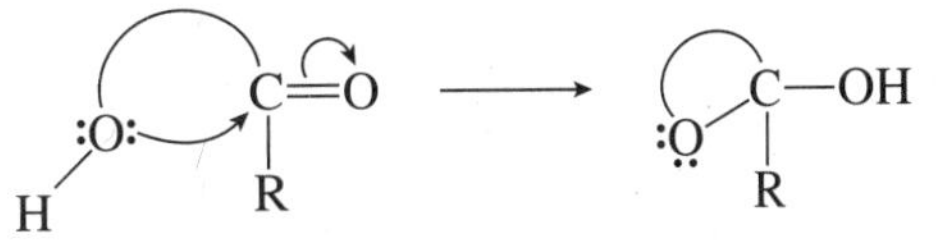

Cyclic hemiacetal

The ring oxygen is derived from the hydroxyl group; the carbonyl oxygen becomes the hydroxyl oxygen of the hemiacetal.

(*a*) This compound is the cyclic hemiacetal of 5-hydroxypentanal.

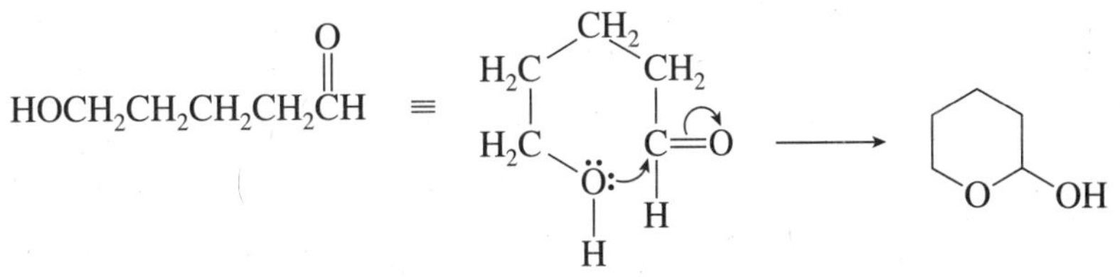

$$HOCH_2CH_2CH_2CH_2\overset{O}{\overset{\|}{C}}H$$

Indeed, 5-hydroxypentanal seems to exist entirely as the cyclic hemiacetal. Its infrared spectrum lacks absorption in the carbonyl region.

(*b*) The carbon connected to two oxygens is the one that is derived from the carbonyl group. Using retrosynthetic symbolism, disconnect the ring oxygen from this carbon.

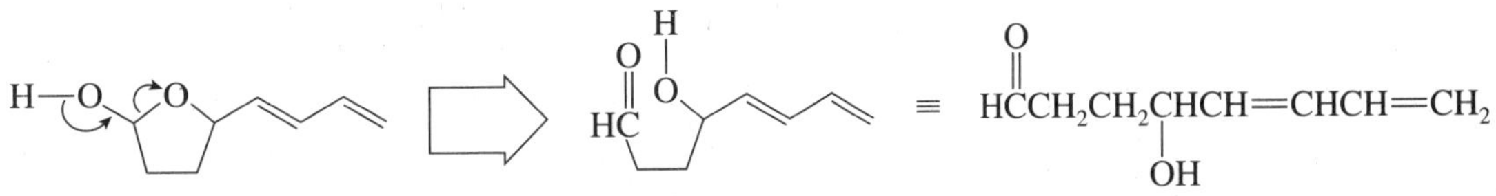

$$\equiv H\overset{O}{\overset{\|}{C}}CH_2CH_2\underset{OH}{\underset{|}{C}}HCH{=}CHCH{=}CH_2$$

4-Hydroxy-5,7-octadienal

The next two compounds are cyclic acetals. The original carbonyl group is identifiable as the one that bears two oxygen substituents, which originate as hydroxyl oxygens of a diol.

(*c*)

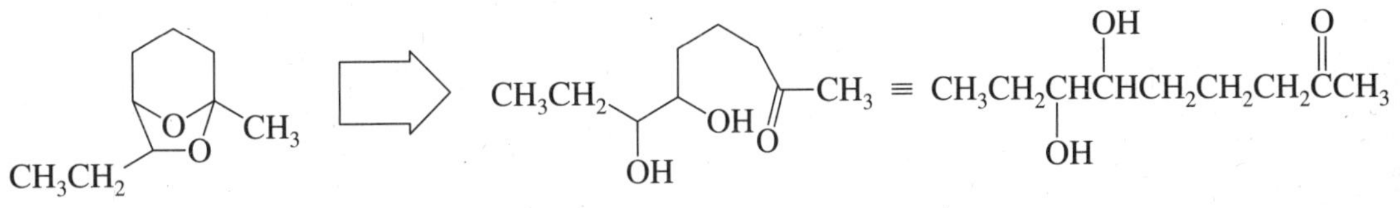

Brevicomin

$$\equiv CH_3CH_2\overset{OH}{\overset{|}{C}}H\underset{OH}{\underset{|}{C}}HCH_2CH_2CH_2\overset{O}{\overset{\|}{C}}CH_3$$

6,7-Dihydroxy-2-nonanone

(*d*)

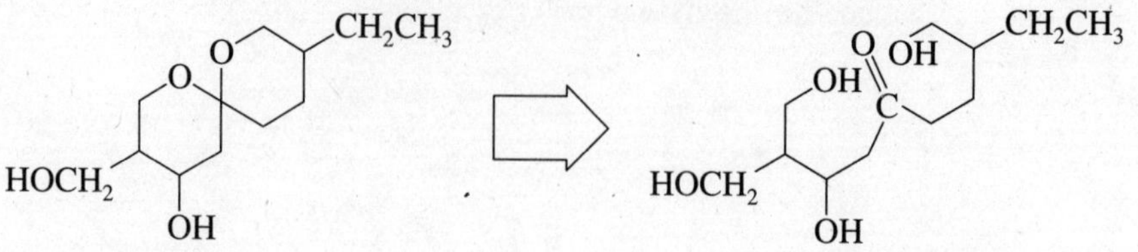

Talaromycin A ⟹ 2,8-Di(hydroxymethyl)-1,3-dihydroxy-5-decanone

**17.31** (*a*) The *Z* stereoisomer of $CH_3CH{=}NCH_3$ has its higher ranked substituents on the same side of the double bond,

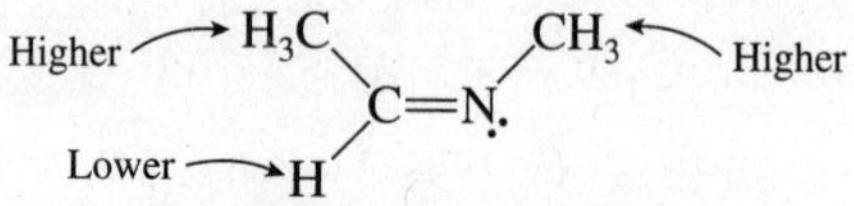

(*Z*)-*N*-Ethylidenemethylamine

The lone pair of nitrogen is lower in rank than any other substituent.

(*b*) Higher ranked groups are on opposite sides of the carbon–nitrogen double bond in the *E* oxime of acetaldehyde.

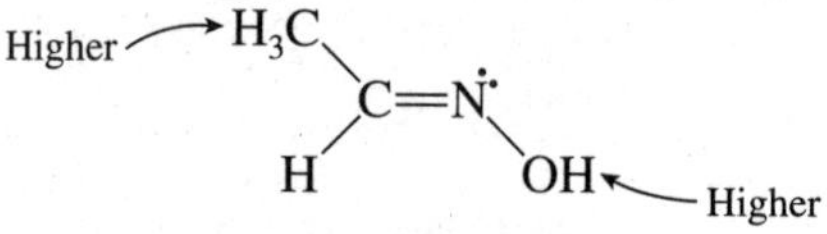

(*E*)-Acetaldehyde oxime

(*c*) (*Z*)-2-Butanone hydrazone is

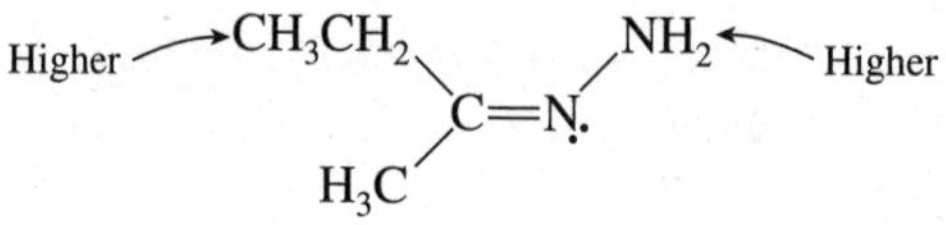

(*d*) (*E*)-Acetophenone semicarbazone is

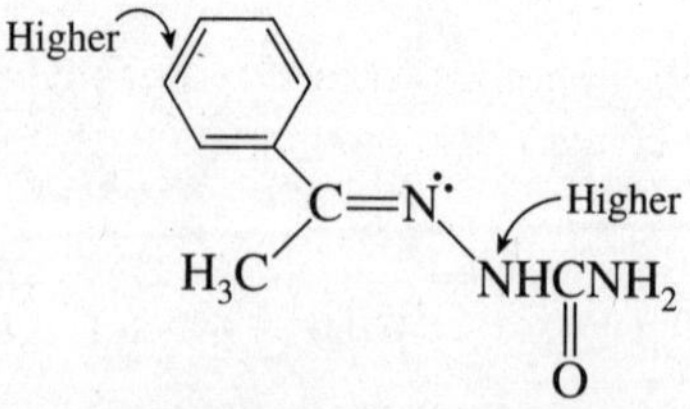

**17.32** Cyclopentanone reacts with peroxybenzoic acid to form a peroxy monoester. The alkyl group that migrates is the ring itself, leading to formation of a six-membered lactone.

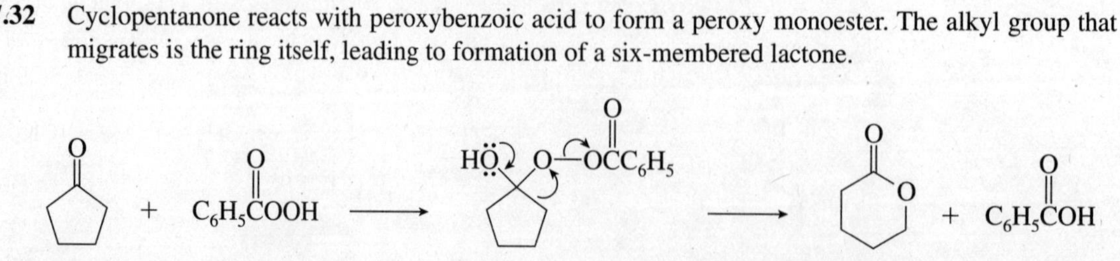

Cyclopentanone + $C_6H_5COOH$ (Peroxybenzoic acid) ⟶ [peroxy monoester] ⟶ 5-Pentanolide + $C_6H_5COH$ (Benzoic acid)

**17.33** (*a*) The bacterial enzyme cyclohexanone monooxygenase was described in Section 17.16 as able to catalyze a biological Baeyer–Villiger reaction. Compound A is 4-methylcyclohexanone.

$$\text{4-Methylcyclohexanone (Compound A)} \xrightarrow[\text{and coenzymes}]{O_2,\ \text{cyclohexanone monoxygenase,}} \text{Compound B}$$

(*b*) The product of Baeyer–Villiger oxidation of 4-methylcyclohexanone with peroxyacetic acid would be the racemic cyclic ester (lactone), not the single enantiomer shown in part (*a*) from the enzyme-catalyzed oxidation.

**17.34** (*a*) Nucleophilic ring opening of the epoxide occurs by attack of methoxide at the less hindered carbon.

$$(CH_3)_3C(Cl)\overset{O}{C{-}CH_2} + {}^-{:}\ddot{O}CH_3 \longrightarrow (CH_3)_3C{-}\underset{Cl}{\overset{:\ddot{O}:^-}{C}}{-}CH_2OCH_3$$

The anion formed in this step loses a chloride ion to form the carbon–oxygen double bond of the product.

$$(CH_3)_3C{-}\underset{Cl}{\overset{:\ddot{O}:^-}{C}}{-}CH_2OCH_3 \longrightarrow (CH_3)_3C\overset{O}{\overset{\|}{C}}CH_2OCH_3 + Cl^-$$

(*b*) Nucleophilic addition of methoxide ion to the aldehyde carbonyl generates an oxyanion, which can close to an epoxide by an intramolecular nucleophilic substitution reaction.

$$(CH_3)_3C\underset{Cl}{C}H\overset{O}{\overset{\|}{C}}H + {}^-{:}\ddot{O}CH_3 \longrightarrow (CH_3)_3\underset{Cl}{C}CH{-}\overset{{}^-:\ddot{O}:}{C}HOCH_3 \longrightarrow (CH_3)_3C\overset{O}{CH{-}C}HOCH_3 + Cl^-$$

The epoxide formed in this process then undergoes nucleophilic ring opening on attack by a second methoxide ion.

$$(CH_3)_3C\overset{O}{CH{-}C}HOCH_3 + {}^-{:}\ddot{O}CH_3 \longrightarrow (CH_3)_3C\overset{:\ddot{O}:^-}{C}H{-}\underset{OCH_3}{C}HOCH_3 \xrightarrow{CH_3OH} (CH_3)_3C\overset{OH}{C}H\underset{OCH_3}{C}HOCH_3$$

**17.35** Amygdalin is a derivative of the cyanohydrin formed from benzaldehyde; thus the structure (without stereochemistry) is

$$C_6H_5\overset{O-R}{\underset{CN}{C}}H$$

R = H, benzaldehyde cyanohydrin

The order of decreasing sequence rule precedence is HO > CN > $C_6H_5$ > H. The groups are arranged in a clockwise orientation in order of decreasing precedence in the *R* enantiomer.

$$C_6H_5-C(H)(OH)(CN)$$

(*R*)-Benzaldehyde cyanohydrin

**17.36** (*a*) The target molecule is the diethyl acetal of acetaldehyde (ethanal).

$$CH_3CH(OCH_2CH_3)_2 \Rightarrow CH_3\overset{O}{\overset{\|}{C}}H \text{ and } CH_3CH_2OH$$

Acetaldehyde diethyl acetal

Acetaldehyde may be prepared by oxidation of ethanol.

$$CH_3CH_2OH \xrightarrow[CH_2Cl_2]{PCC} CH_3\overset{O}{\overset{\|}{C}}H$$

Ethanol → Acetaldehyde

Reaction with ethanol in the presence of hydrogen chloride yields the desired acetal.

$$CH_3\overset{O}{\overset{\|}{C}}H + 2CH_3CH_2OH \xrightarrow{HCl} CH_3CH(OCH_2CH_3)_2$$

Acetaldehyde + Ethanol → Acetaldehyde diethyl acetal

(*b*) In this case the target molecule is a cyclic acetal of acetaldehyde.

$$\text{2-Methyl-1,3-dioxolane} \Rightarrow CH_3\overset{O}{\overset{\|}{C}}H \text{ and } HOCH_2CH_2OH$$

2-Methyl-1,3-dioxolane

Acetaldehyde has been prepared in part (*a*). Recalling that vicinal diols are available from the hydroxylation of alkenes, 1,2-ethanediol may be prepared by the sequence

$$CH_3CH_2OH \xrightarrow[heat]{H_2SO_4} H_2C{=}CH_2 \xrightarrow[(CH_3)_3COH,\ HO^-]{OsO_4,\ (CH_3)_3COOH} HOCH_2CH_2OH$$

Ethanol → Ethylene → 1,2-Ethanediol

Hydrolysis of ethylene oxide is also reasonable.

$$H_2C{=}CH_2 \xrightarrow{CH_3\overset{O}{\overset{\|}{C}}OOH} H_2C\underset{O}{-}CH_2 \xrightarrow[HO^-]{H_2O} HOCH_2CH_2OH$$

Ethylene → Ethylene oxide → 1,2-Ethanediol

Reaction of acetaldehyde with 1,2-ethanediol yields the cyclic acetal.

$$\underset{\text{Acetaldehyde}}{CH_3\overset{O}{\overset{\|}{C}}H} + \underset{\text{1,2-Ethanediol}}{HOCH_2CH_2OH} \xrightarrow{H^+} \underset{\text{2-Methyl-1,3-dioxolane}}{\text{(cyclic acetal: O–C(H)(CH}_3\text{)–O ring)}}$$

(*c*) The target molecule is, in this case, the cyclic acetal of 1,2-ethanediol and formaldehyde.

$$\underset{\text{1,3-Dioxolane}}{\text{(ring)}} \Longrightarrow H\overset{O}{\overset{\|}{C}}H \quad \text{and} \quad HOCH_2CH_2OH$$

The preparation of 1,2-ethanediol was described in part (*b*). One method of preparing formaldehyde is by ozonolysis of ethylene.

$$\underset{\text{Ethanol}}{CH_3CH_2OH} \xrightarrow[\text{heat}]{H_2SO_4} \underset{\text{Ethylene}}{H_2C{=}CH_2} \xrightarrow[\text{2. } H_2O,\ Zn]{\text{1. } O_3} \underset{\text{Formaldehyde}}{2H\overset{O}{\overset{\|}{C}}H}$$

Another method is periodate cleavage of 1,2-ethanediol.

$$\underset{\text{1,2-Ethanediol}}{HOCH_2CH_2OH} \xrightarrow{HIO_4} \underset{\text{Formaldehyde}}{2H\overset{O}{\overset{\|}{C}}H}$$

Cyclic acetal formation is then carried out in the usual way.

$$\underset{\text{Formaldehyde}}{H\overset{O}{\overset{\|}{C}}H} + \underset{\text{1,2-Ethanediol}}{HOCH_2CH_2OH} \xrightarrow{H^+} \underset{\text{1,3-Dioxolane}}{\text{(ring)}}$$

(*d*) Acetylenic alcohols are best prepared from carbonyl compounds and acetylide anions.

$$\underset{\text{3-Butyn-2-ol}}{CH_3\underset{OH}{\underset{|}{C}}HC{\equiv}CH} \Longrightarrow CH_3\overset{O}{\overset{\|}{C}}H + {}^{-}\!:C{\equiv}CH$$

Acetaldehyde is available as in part (*a*). Alkynes such as acetylene are available from the corresponding alkene by bromination followed by double dehydrobromination. Using ethylene, prepared in part (*b*), the sequence becomes

$$\underset{\text{Ethylene}}{H_2C{=}CH_2} \xrightarrow{Br_2} \underset{\text{1,2-Dibromoethane}}{BrCH_2CH_2Br} \xrightarrow[NH_3]{NaNH_2} \underset{\text{Acetylene}}{HC{\equiv}CH}$$

Then

$$\underset{\text{Acetylene}}{HC{\equiv}CH} \xrightarrow{NaNH_2} \underset{\text{Sodium acetylide}}{HC{\equiv}CNa} \xrightarrow[\text{2. } H_3O^+]{\text{1. } CH_3\overset{O}{\overset{\|}{C}}H} \underset{\text{3-Butyn-2-ol}}{HC{\equiv}C\underset{OH}{\underset{|}{C}}HCH_3}$$

(*e*) The target aldehyde may be prepared from the corresponding alcohol.

$$\underset{\text{3-Butynal}}{H\overset{O}{\overset{\|}{C}}CH_2C{\equiv}CH} \Longrightarrow \underset{\text{3-Butyn-1-ol}}{HOCH_2CH_2C{\equiv}CH}$$

The best route to this alcohol is through reaction of an acetylide ion with ethylene oxide.

$$\underset{\substack{\text{Sodium acetylide}\\ \text{[prepared in part } (d)]}}{HC{\equiv}CNa} + \underset{\substack{\text{Ethylene oxide}\\ \text{[prepared in part } (b)]}}{H_2C\overset{O}{—}CH_2} \xrightarrow[\text{2. } H_3O^+]{\text{1. diethyl ether}} \underset{\text{3-Butyn-1-ol}}{HC{\equiv}CCH_2CH_2OH}$$

Oxidation with PCC or PDC is appropriate for the final step.

$$\underset{\text{3-Butyn-1-ol}}{HC{\equiv}CCH_2CH_2OH} \xrightarrow[CH_2Cl_2]{\text{PCC or PDC}} \underset{\text{3-Butynal}}{HC{\equiv}CCH_2\overset{O}{\overset{\|}{C}}H}$$

(*f*) The target molecule has four carbon atoms, suggesting a route involving reaction of an ethyl Grignard reagent with ethylene oxide.

$$CH_3CH_2CH_2CH_2OH \Longrightarrow CH_3\ddot{C}H_2^{-} + H_2C\overset{O}{—}CH_2$$

Ethylmagnesium bromide is prepared in the usual way.

$$\underset{\text{Ethanol}}{CH_3CH_2OH} \xrightarrow[PBr_3]{\text{HBr or}} \underset{\text{Bromoethane}}{CH_3CH_2Br} \xrightarrow[\text{diethyl ether}]{Mg} \underset{\text{Ethylmagnesium bromide}}{CH_3CH_2MgBr}$$

Reaction of the Grignard reagent with ethylene oxide, prepared in part (*b*), completes the synthesis.

$$\underset{\text{Ethylmagnesium bromide}}{CH_3CH_2MgBr} + \underset{\text{Ethylene oxide}}{H_2C\overset{O}{—}CH_2} \xrightarrow[\text{2. } H_3O^+]{\text{1. diethyl ether}} \underset{\text{1-Butanol}}{CH_3CH_2CH_2CH_2OH}$$

**17.37** (*a*) Friedel–Crafts acylation of benzene with benzoyl chloride is a direct route to benzophenone.

$$C_6H_5COCl + C_6H_6 \xrightarrow{AlCl_3} C_6H_5COC_6H_5$$

Benzoyl chloride Benzene Benzophenone

(*b*) On analyzing the overall transformation retrosynthetically, we see that the target molecule may be prepared by a Grignard synthesis followed by oxidation of the alcohol formed.

$$C_6H_5\overset{O}{\overset{\|}{C}}C_6H_5 \Rightarrow C_6H_5\overset{OH}{\overset{|}{C}}HC_6H_5 \Rightarrow C_6H_5\overset{O}{\overset{\|}{C}}H + C_6H_5MgBr$$

In the desired synthesis, benzyl alcohol must first be oxidized to benzaldehyde.

$$C_6H_5CH_2OH \xrightarrow[CH_2Cl_2]{PDC} C_6H_5\overset{O}{\overset{\|}{C}}H$$

Benzyl alcohol Benzaldehyde

Reaction of benzaldehyde with the Grignard reagent of bromobenzene followed by oxidation of the resulting secondary alcohol gives benzophenone.

$$C_6H_5\overset{O}{\overset{\|}{C}}H + C_6H_5MgBr \xrightarrow[2.\ H_3O^+]{1.\ \text{diethyl ether}} C_6H_5\underset{OH}{\underset{|}{C}}HC_6H_5 \xrightarrow[CH_2Cl_2]{PDC} C_6H_5\overset{O}{\overset{\|}{C}}C_6H_5$$

Benzaldehyde Phenylmagnesium bromide Diphenylmethanol Benzophenone

(*c*) Hydrolysis of bromodiphenylmethane yields the corresponding alcohol, which can be oxidized to benzophenone as in part (*b*).

$$C_6H_5\underset{Br}{\underset{|}{C}}HC_6H_5 \xrightarrow{H_2O} C_6H_5\underset{OH}{\underset{|}{C}}HC_6H_5 \xrightarrow{\text{oxidize}} C_6H_5\overset{O}{\overset{\|}{C}}C_6H_5$$

Bromodiphenylmethane Diphenylmethanol Benzophenone

(*d*) The starting material is the dimethyl acetal of benzophenone. All that is required is acid-catalyzed hydrolysis.

$$C_6H_5\overset{OCH_3}{\underset{OCH_3}{C}}C_6H_5 + 2H_2O \xrightarrow{H^+} C_6H_5\overset{O}{\overset{\|}{C}}C_6H_5 + 2CH_3OH$$

Dimethoxydiphenylmethane Water Benzophenone Methanol

(*e*) Oxidative cleavage of the alkene yields benzophenone. Ozonolysis may be used.

$$(C_6H_5)_2C{=}C(C_6H_5)_2 \xrightarrow[\text{2. } H_2O,\ Zn]{\text{1. } O_3} 2(C_6H_5)_2C{=}O$$

1,1,2,2-Tetraphenylethene Benzophenone

**17.38** The two alcohols given as starting materials contain all the carbon atoms of the desired product.

$$CH_3(CH_2)_8CH{=}CHCH_2CH{=}CHCH_2CH{=}CHCH{=}CH_2 \Longrightarrow$$

$$CH_3(CH_2)_8CH{=}CHCH_2CH{=}CHCH_2CH_2OH \quad \text{and} \quad HOCH_2CH{=}CH_2$$

3,6-Hexadecadien-1-ol Allyl alcohol

What is needed is to attach the two groups together so that the two primary alcohol carbons become doubly bonded to each other. This can be accomplished by using a Wittig reaction as the key step.

$$CH_3(CH_2)_8CH{=}CHCH_2CH{=}CHCH_2CH_2OH \xrightarrow[CH_2Cl_2]{PCC} CH_3(CH_2)_8CH{=}CHCH_2CH{=}CHCH_2\overset{O}{\overset{\|}{C}}H$$

3,6-Hexadecadien-1-ol 3,6-Hexadecadienal

$$H_2C{=}CHCH_2OH \xrightarrow{PBr_3} H_2C{=}CHCH_2Br \xrightarrow{(C_6H_5)_3P} (C_6H_5)_3\overset{+}{P}CH_2CH{=}CH_2\ Br^-$$

Allyl alcohol Allyl bromide Allyltriphenylphosphonium bromide

$$\xrightarrow{CH_3CH_2CH_2CH_2Li,\ THF} (C_6H_5)_3\overset{+}{P}{-}\overset{..}{\overset{-}{C}}HCH{=}CH_2$$

Allylidenetriphenylphosphorane

$$CH_3(CH_2)_8CH{=}CHCH_2CH{=}CHCH_2\overset{O}{\overset{\|}{C}}H + (C_6H_5)_3\overset{+}{P}{-}\overset{..}{\overset{-}{C}}HCH{=}CH_2 \longrightarrow$$

3,6-Hexadecadienal Allylidenetriphenylphosphorane

$$CH_3(CH_2)_8CH{=}CHCH_2CH{=}CHCH_2CH{=}CHCH{=}CH_2$$

1,3,6,9-Nonadecatetraene

Alternatively, allyl alcohol could be oxidized to $CH_2{=}CHCHO$ for subsequent reaction with the ylide derived from $CH_3(CH_2)_8CH{=}CHCH_2CH{=}CHCH_2CH_2OH$ via its bromide and triphenylphosphonium salt.

**17.39** The expected course of the reaction would be hydrolysis of the acetal to the corresponding aldehyde.

$$C_6H_5\underset{OH}{\underset{|}{C}}HCH(OCH_3)_2 \xrightarrow[HCl]{H_2O} C_6H_5\underset{OH}{\underset{|}{C}}H\overset{O}{\overset{\|}{C}}H + 2CH_3OH$$

Compound A (mandelaldehyde dimethyl acetal) Mandelaldehyde Methanol

The molecular formula of the observed product (compound B, $C_{16}H_{16}O_4$) is exactly twice that of mandelaldehyde. This suggests that it might be a dimer of mandelaldehyde resulting from hemiacetal formation between the hydroxyl group of one mandelaldehyde molecule and the carbonyl group of another.

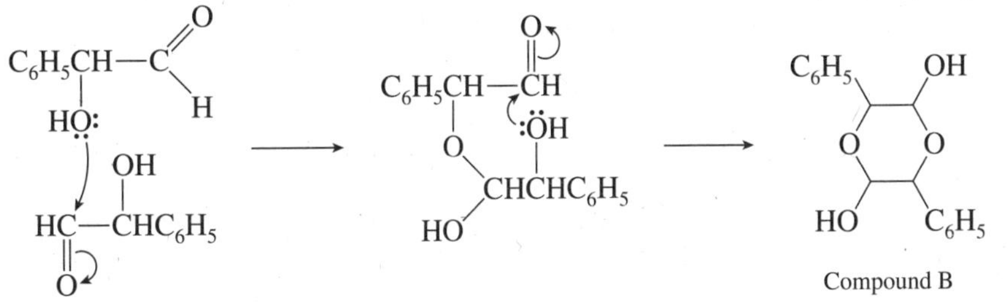

Compound B

Because compound B lacks carbonyl absorption in its infrared spectrum, the cyclic structure is indicated.

**17.40** (*a*) Recalling that alkanes may be prepared by hydrogenation of the appropriate alkene, a synthesis of the desired product becomes apparent. What is needed is to convert —C═O into —C═$CH_2$; a Wittig reaction is appropriate.

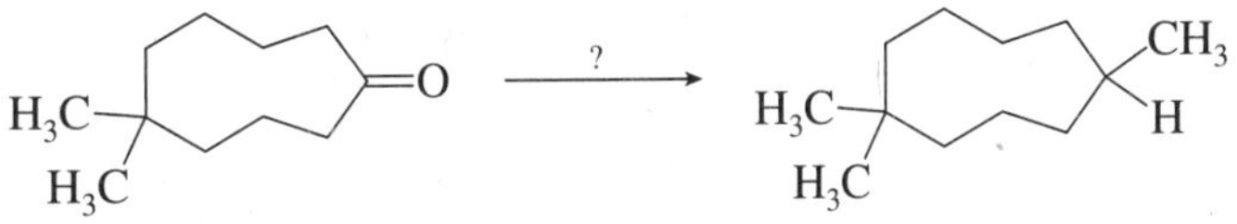

5,5-Dimethylcyclononanone → 1,1,5-Trimethylcyclononane

The two-step procedure that was followed used a Wittig reaction to form the carbon–carbon bond, then catalytic hydrogenation of the resulting alkene.

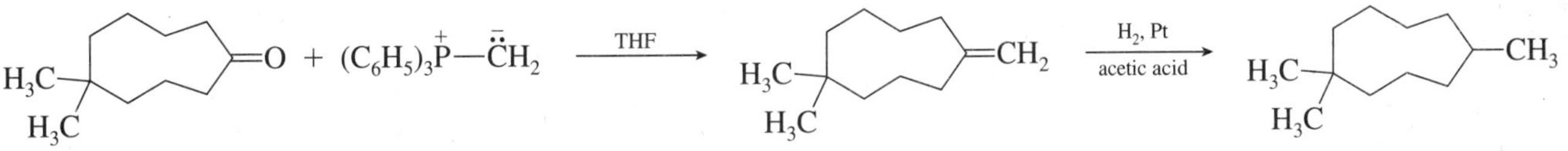

5,5-Dimethylcyclononanone + $(C_6H_5)_3\overset{+}{P}—\overset{-}{C}H_2$ $\xrightarrow{THF}$ 5,5-Dimethyl-1-methylenecyclononane (59%) $\xrightarrow{H_2, Pt, \text{acetic acid}}$ 1,1,5-Trimethylcyclononane (73%)

(*b*) In putting together the carbon skeleton of the target molecule, a methyl group has to be added to the original carbonyl carbon.

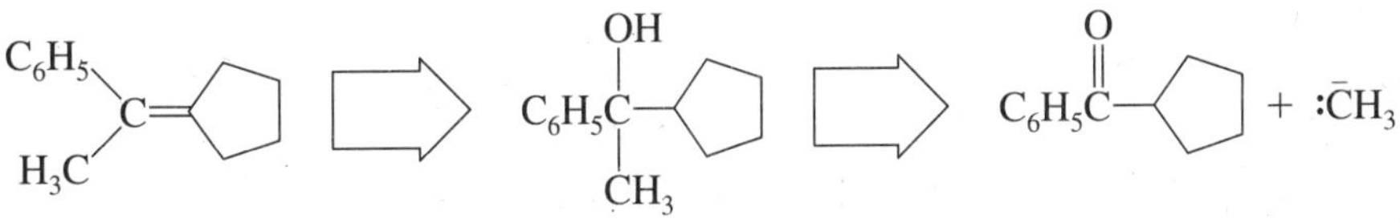

The logical way to do this is by way of a Grignard reagent.

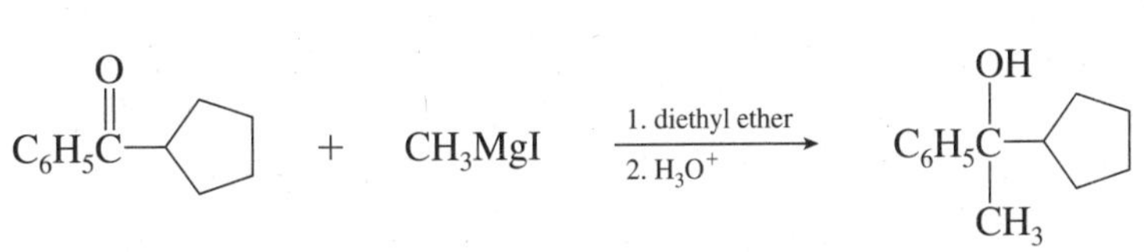

Cyclopentyl phenyl ketone + $CH_3MgI$ $\xrightarrow{\text{1. diethyl ether; 2. } H_3O^+}$ 1-Cyclopentyl-1-phenylethanol

Acid-catalyzed dehydration yields the more highly substituted alkene, the desired product, in accordance with the Zaitsev rule.

1-Cyclopentyl-1-phenylethanol $\xrightarrow[\text{heat}]{H_3PO_4}$ (1-Phenylethylidene)cyclopentane

(*c*) Analyzing the transformation retrosynthetically, keeping in mind the starting materials stated in the problem, we see that the carbon skeleton may be constructed in a straightforward manner.

$o$-$CH_3C_6H_4C(=CH_2)(CH_2)_3CH{=}CH_2 \Rightarrow o$-$CH_3C_6H_4C(=O)(CH_2)_3CH{=}CH_2 + (C_6H_5)_3\overset{+}{P}{-}\ddot{C}H_2$

$o$-$CH_3C_6H_4CH(OH)(CH_2)_3CH{=}CH_2 \Rightarrow o$-$CH_3C_6H_4MgBr + HC(=O)(CH_2)_3CH{=}CH_2$

Proceeding with the synthesis in the forward direction, reaction between the Grignard reagent of *o*-bromotoluene and 5-hexenal produces most of the desired carbon skeleton.

$o$-Methylphenylmagnesium bromide + $H_2C{=}CHCH_2CH_2CH_2CH{=}O$ (5-Hexenal) $\xrightarrow[\text{2. } H_3O^+]{\text{1. diethyl ether}}$ 1-(*o*-Methylphenyl)-5-hexen-1-ol

Oxidation of the resulting alcohol to the ketone followed by a Wittig reaction leads to the final product.

1-(*o*-Methylphenyl)-5-hexen-1-ol $\xrightarrow[CH_2Cl_2]{PCC}$ 1-(*o*-Methylphenyl)-5-hexen-1-one $\xrightarrow{(C_6H_5)_3\overset{+}{P}{-}\ddot{C}H_2}$ 2-(*o*-Methylphenyl)-1,6-heptadiene

Acid-catalyzed dehydration of the corresponding tertiary alcohol would *not* be suitable, because the major elimination product would have the more highly substituted double bond.

2-(*o*-Methylphenyl)-6-hepten-2-ol $\xrightarrow[\text{heat}]{H^+}$ 6-(*o*-Methylphenyl)-1,5-heptadiene ($o$-$CH_3C_6H_4C(CH_3){=}CHCH_2CH_2CH{=}CH_2$)

(*d*) Remember that terminal acetylenes can serve as sources of methyl ketones by hydration.

$$CH_3\overset{O}{\overset{\|}{C}}CH_2CH_2\overset{O}{\overset{\|}{C}}(CH_2)_5CH_3 \Longrightarrow HC{\equiv}CCH_2CH_2\overset{O}{\overset{\|}{C}}(CH_2)_5CH_3$$

This gives us a clue as to how to proceed, since the acetylenic ketone may be prepared from the starting acetylenic alcohol.

$$HC{\equiv}CCH_2CH_2\overset{O}{\overset{\|}{C}}(CH_2)_5CH_3 \Longrightarrow HC{\equiv}CCH_2CH_2\overset{O}{\overset{\|}{C}}H + CH_3(CH_2)_4\ddot{C}H_2 \Longrightarrow HC{\equiv}CCH_2CH_2CH_2OH$$

The first synthetic step is oxidation of the primary alcohol to the aldehyde and construction of the carbon skeleton by a Grignard reaction.

$$\underset{\text{4-Pentyn-1-ol}}{HC{\equiv}CCH_2CH_2CH_2OH} \xrightarrow[CH_2Cl_2]{PDC} \underset{\text{4-Pentynal}}{HC{\equiv}CCH_2CH_2\overset{O}{\overset{\|}{C}}H} \xrightarrow[2.\ H_3O^+]{1.\ CH_3(CH_2)_5MgBr} \underset{\text{1-Undecyn-5-ol}}{HC{\equiv}CCH_2CH_2\underset{OH}{\underset{|}{C}}H(CH_2)_5CH_3}$$

Oxidation of the secondary alcohol to a ketone and hydration of the terminal triple bond complete the synthesis.

$$\underset{\text{1-Undecyn-5-ol}}{HC{\equiv}CCH_2CH_2\underset{OH}{\underset{|}{C}}H(CH_2)_5CH_3} \xrightarrow[CH_2Cl_2]{PDC} \underset{\text{1-Undecyn-5-one}}{HC{\equiv}CCH_2CH_2\overset{O}{\overset{\|}{C}}(CH_2)_5CH_3} \xrightarrow[HgSO_4]{H_2O,\ H_2SO_4} \underset{\text{2,5-Undecanedione}}{CH_3\overset{O}{\overset{\|}{C}}CH_2CH_2\overset{O}{\overset{\|}{C}}(CH_2)_5CH_3}$$

(*e*) The desired product is a benzylic ether. To prepare it, the aldehyde must first be reduced to the corresponding primary alcohol. Sodium borohydride was used in the preparation described in the literature, but lithium aluminum hydride or catalytic hydrogenation would also be possible. Once the alcohol is prepared, it can be converted to its alkoxide ion and this alkoxide ion treated with methyl iodide.

$$\text{2-Cl-3-}H_3C\text{-}C_6H_3\text{-}\overset{O}{\overset{\|}{C}}H \xrightarrow[H_2O]{NaBH_4} \text{2-Cl-3-}CH_3\text{-}C_6H_3\text{-}CH_2OH \xrightarrow[2.\ CH_3I]{1.\ NaH} \text{2-Cl-3-}CH_3\text{-}C_6H_3\text{-}CH_2OCH_3$$

Alternatively, the alcohol could be treated with hydrogen bromide or with phosphorus tribromide to give the benzylic bromide and the bromide then allowed to react with sodium methoxide.

$$\text{2-Cl-3-}CH_3\text{-}C_6H_3\text{-}CH_2OH \xrightarrow[\text{or } PBr_3]{HBr} \text{2-Cl-3-}CH_3\text{-}C_6H_3\text{-}CH_2Br \xrightarrow[CH_3OH]{NaOCH_3} \text{2-Cl-3-}CH_3\text{-}C_6H_3\text{-}CH_2OCH_3$$

**17.41** Step 1 of the synthesis is formation of a cyclic acetal protecting group; the necessary reagents are ethylene glycol ($HOCH_2CH_2OH$) and *p*-toluenesulfonic acid, with heating in benzene. In step 2 the ester function is reduced to a primary alcohol. Lithium aluminum hydride ($LiAlH_4$) is the reagent of choice. Oxidation with PCC in $CH_2Cl_2$ converts the primary alcohol to an aldehyde in step 3.

Wolff–Kishner reduction ($N_2H_4$, KOH, ethylene glycol, heat) converts the aldehyde group to a methyl group in step 4. The synthesis is completed in step 5 by hydrolysis ($H_3O^+$) of the acetal-protecting group.

O, O, $\overset{O}{\|}$COCH$_3$ —①→ O, O, $\overset{O}{\|}$COCH$_3$ —②→ O, O, $CH_2OH$

↓③

O, $CH_3$ ←⑤— O, O, $CH_3$ ←④— O, O, $\overset{O}{\|}$CH

**17.42** We need to assess the extent of resonance donation to the carbonyl group by the $\pi$ electrons of the aromatic rings. Such resonance for benzaldehyde may be written as

O, CH ⟷ $O^-$, +, =CH ⟷ $O^-$, +, =CH ⟷ $O^-$, =CH, +

Electron-releasing groups such as methoxy at positions ortho and para to the aldehyde function increase the "single-bond character" of the aldehyde by stabilizing the dipolar resonance forms and increasing their contribution to the overall electron distribution in the molecule. Electron-withdrawing groups such as nitro decrease this single-bond character. The aldehyde with the lowest carbonyl stretching frequency is 2,4,6-trimethoxybenzaldehyde; the one with the highest is 2,4,6-trinitrobenzaldehyde. The measured values are

CHO, $CH_3O$, $OCH_3$, $OCH_3$ — CHO — CHO, $O_2N$, $NO_2$, $NO_2$

2,4,6-Trimethoxybenzaldehyde ($1665\ cm^{-1}$) Benzaldehyde ($1700\ cm^{-1}$) 2,4,6-Trinitrobenzaldehyde ($1715\ cm^{-1}$)

**17.43** The signal in the $^1H$ NMR spectrum at $\delta$ 9.7 ppm tells us that the compound is an aldehyde rather than a ketone. The 2H signal at $\delta$ 2.4 ppm indicates that the group adjacent to the carbonyl is a $CH_2$ group. The remaining signals support the assignment of the compound as butanal.

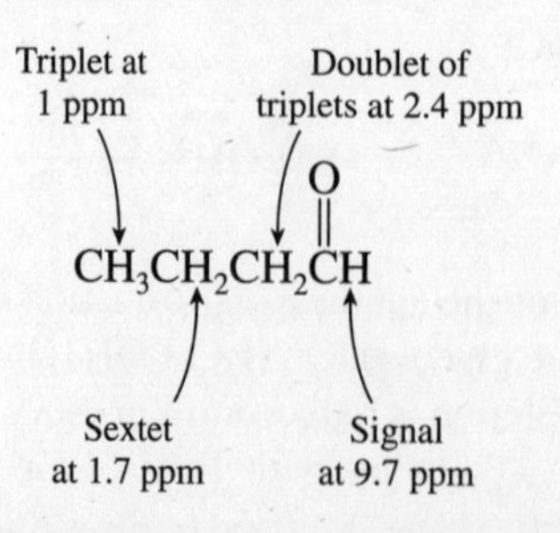

**17.44** A carbonyl group is evident from the strong infrared absorption at 1710 $cm^{-1}$. Since all the $^1H$ NMR signals are singlets, there are no nonequivalent hydrogens in a vicinal or "three-bond" relationship. The three-proton signal at $\delta$ 2.1 ppm, and the 2-proton signal at $\delta$ 2.3 ppm can be understood as arising from a $CH_2\overset{O}{\overset{\|}{C}}CH_3$ unit. The intense 9-proton singlet at $\delta$ 1.0 ppm is due to the three equivalent methyl groups of a $(CH_3)_3C$ unit. The compound is 4,4-dimethyl-2-pentanone.

$$CH_3\overset{O}{\overset{\|}{C}}CH_2C(CH_3)_3$$

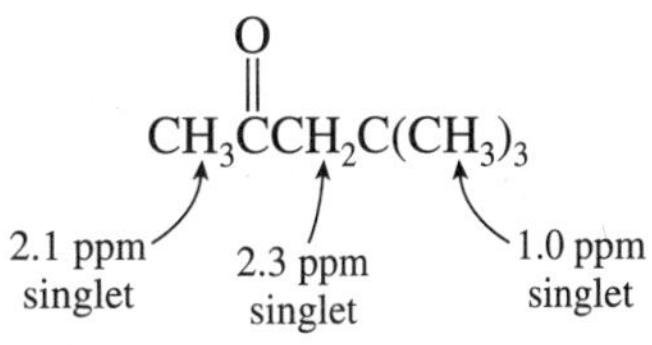

2.1 ppm singlet; 2.3 ppm singlet; 1.0 ppm singlet

4,4-Dimethyl-2-pentanone

**17.45** The molecular formula of compounds A and B ($C_6H_{10}O_2$) indicates an index of hydrogen deficiency of 2. Because we are told the compounds are diketones, the two carbonyl groups account for all the unsaturations.

The $^1H$ NMR spectrum of compound A has only two peaks, both singlets, at $\delta$ 2.2 and 2.8 ppm. Their intensity ratio (6:4) is consistent with two equivalent methyl groups and two equivalent methylene groups. The chemical shifts are appropriate for

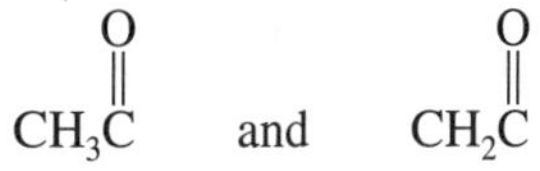

$$CH_3\overset{O}{\overset{\|}{C}} \quad \text{and} \quad CH_2\overset{O}{\overset{\|}{C}}$$

The simplicity of the spectrum can be understood if we are dealing with a symmetric diketone.

The correct structure is

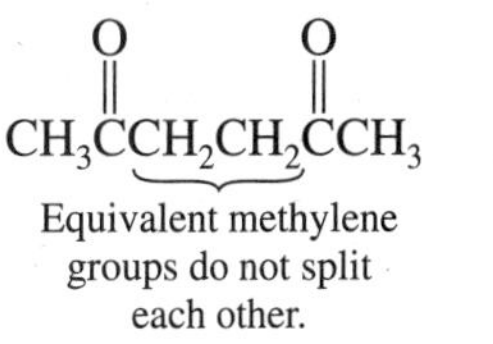

$$CH_3\overset{O}{\overset{\|}{C}}\underbrace{CH_2CH_2}\overset{O}{\overset{\|}{C}}CH_3$$

Equivalent methylene groups do not split each other.

2,5-Hexanedione (compound A)

Compound B is an isomer of compound A. The triplet–quartet pattern in the $^1H$ NMR spectrum is consistent with an ethyl group and, because the triplet is equivalent to 6 protons and the quartet to 4, it is likely that two equivalent ethyl groups are present. The two ethyl groups account for four carbons, and because the problem stipulates that the molecule is a diketone, all the carbons are accounted for. The only $C_6H_{10}O_2$ diketone with two equivalent ethyl groups is 3,4-hexanedione.

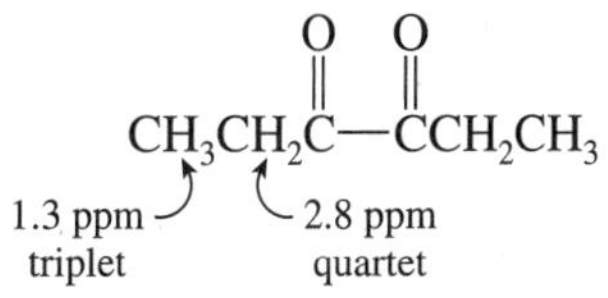

$$CH_3CH_2\overset{O}{\overset{\|}{C}}-\overset{O}{\overset{\|}{C}}CH_2CH_3$$

1.3 ppm triplet; 2.8 ppm quartet

3,4-Hexanedione (compound B)

**17.46** From its molecular formula ($C_{11}H_{14}O$), the compound has a total of five double bonds and rings. The presence of signals in the region $\delta$ 7 to 8 ppm suggests an aromatic ring is present, accounting for four of the elements of unsaturation. The presence of a strong peak at 1700 $cm^{-1}$ in the infrared spectrum indicates the presence of a carbonyl group, accounting for the remaining element of unsaturation. The highest field peak in the NMR spectrum is a 3-proton triplet, corresponding to the methyl group of a $CH_3CH_2$ unit. The 2-proton signal at $\delta$ 3.0 ppm corresponds to a $CH_2$ unit adjacent to the carbonyl group and, because it is a triplet, suggests the grouping $CH_2CH_2C{=}O$. The compound is butyl phenyl ketone (1-phenyl-1-pentanone).

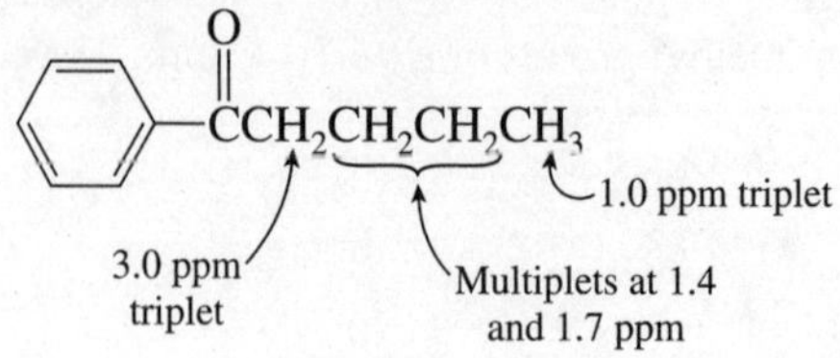

$C_6H_5C(=O)CH_2CH_2CH_2CH_3$

3.0 ppm triplet; Multiplets at 1.4 and 1.7 ppm; 1.0 ppm triplet

Butyl phenyl ketone

**17.47** With a molecular formula of $C_7H_{14}O$, the compound has an index of hydrogen deficiency of 1. We are told that it is a ketone, so it has no rings or double bonds other than the one belonging to its C=O group. The peak at 211 ppm in the $^{13}C$ NMR spectrum corresponds to the carbonyl carbon. Only three other signals occur in the spectrum, and so there are only three types of carbons other than the carbonyl carbon. This suggests that the compound is the symmetrical ketone 4-heptanone.

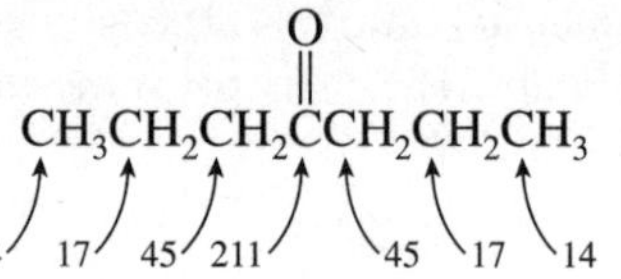

$CH_3CH_2CH_2C(=O)CH_2CH_2CH_3$

14 17 45 211 45 17 14

4-Heptanone
(all chemical shifts in ppm)

**17.48** Compounds A and B are isomers and have an index of hydrogen deficiency of 5. Signals in the region 125–140 ppm in their $^{13}C$ NMR spectra suggest an aromatic ring, and a peak at 200 ppm indicates a carbonyl group. An aromatic ring contributes one ring and three double bonds, and a carbonyl group contributes one double bond, and so the index of hydrogen deficiency of 5 is satisfied by a benzene ring and a carbonyl group. The carbonyl group is attached directly to the benzene ring, as evidenced by the presence of a peak at $m/z$ 105 in the mass spectra of compounds A and B.

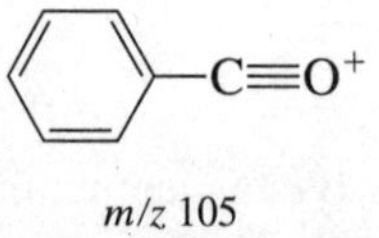

$C_6H_5-C{\equiv}O^+$

$m/z$ 105

Each $^{13}C$ NMR spectrum shows four aromatic signals, and so the rings are monosubstituted.

Compound A has three unique carbons in addition to $C_6H_5C{=}O$ and so must be 1-phenyl-1-butanone. Compound B has only two additional signals and so must be 2-methyl-1-phenyl-1-propanone.

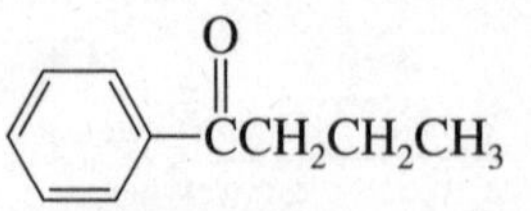

$C_6H_5C(=O)CH_2CH_2CH_3$ $\qquad$ $C_6H_5C(=O)CH(CH_3)_2$

Compound A 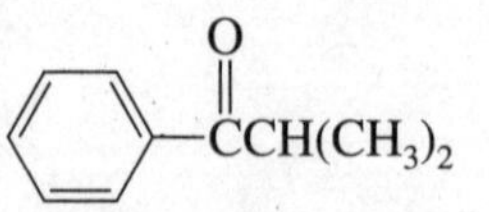 $\qquad$ Compound B

**17.49–17.50** Solutions to molecular modeling exercises are not provided in this *Study Guide and Solutions Manual.* You should use *Learning By Modeling* for these exercises.

# SELF-TEST

## PART A

**A-1.** Give the correct IUPAC name for each of the following:

(*a*) $CH_3CH_2CH(CH_3)CH(CH_3)CH_2CH{=}O$

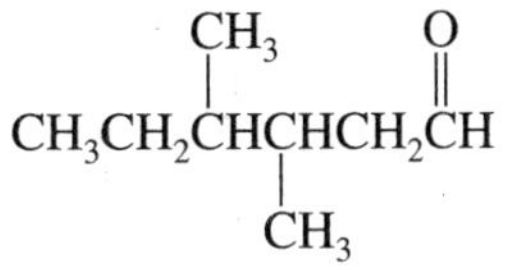

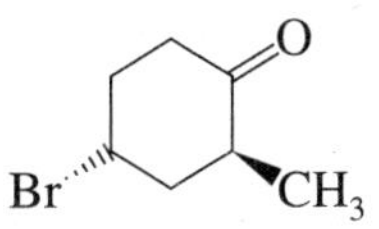

(*b*) $(CH_3)_3CC(=O)CH_2CH(CH_3)_2$

(*c*) Br, O, $CH_3$ (cyclohexanone ring structure)

(*d*) O (line structure)

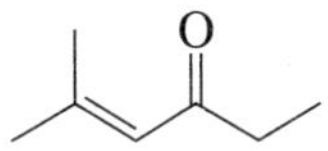

**A-2.** Write the structural formulas for

(*a*) (*E*)-3-Hexen-2-one
(*b*) 3-Cyclopropyl-2,4-pentanedione
(*c*) 3-Ethyl-4-phenylpentanal

**A-3.** For each of the following reactions supply the structure of the missing reactant, reagent, or product:

(*a*) Cyclohexanone$=O$ + HCN $\xrightarrow{CN^-}$ ?

(*b*) $C_6H_5CH(=O)$ + ? $\longrightarrow$ $C_6H_5CH{=}NOH$

(*c*) $(CH_3)_2CHCH$(1,3-dioxane ring) $\xrightarrow{H_2O,\ H^+}$ ? (two products)

(*d*) Cyclopentanone$=O$ + ? $\longrightarrow$ cyclopentylidene$=CHCH_2CH_3$

(*e*) 2-Methylcyclohexanone ($=O$, $CH_3$) + $C_6H_5NHNH_2$ $\longrightarrow$ ?

(*f*) $CH_3CH_2CH_2CH(=O)$ + $2CH_3CH_2OH$ $\xrightarrow{HCl}$ ?

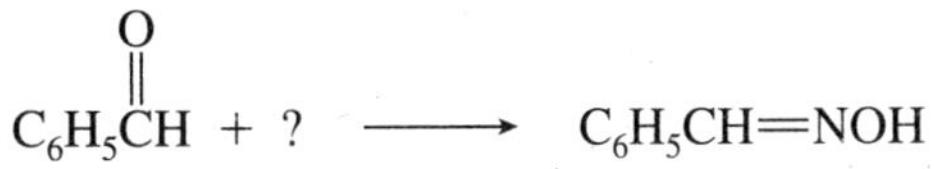

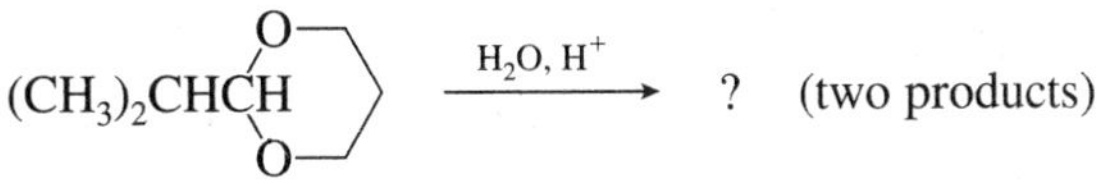

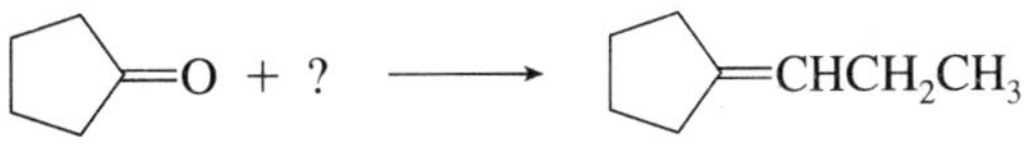

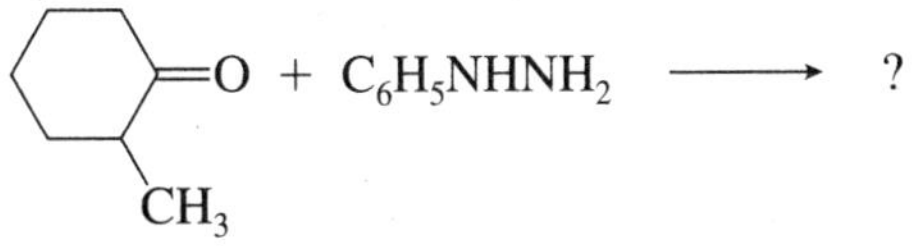

(*g*) ? + ? $\longrightarrow$ $C_6H_5C(N(CH_3)_2){=}CHCH_3$

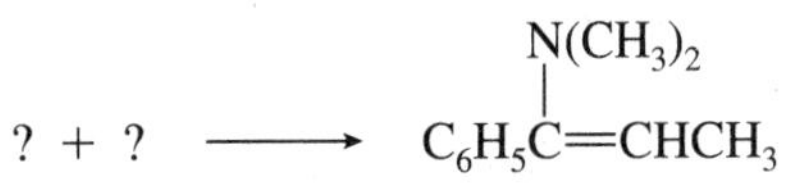

(*h*) $CH_3CH_2CH_2CH{=}O$ $\xrightarrow[H^+,\ H_2O]{Na_2Cr_2O_7}$ ?

**A-4.** Write the structures of the products, compounds A through E, of the reaction steps shown.

(*a*) $(C_6H_5)_3P + (CH_3)_2CHCH_2Br \longrightarrow A$

$A + CH_3CH_2CH_2CH_2Li \longrightarrow B + C_4H_{10}$

$B + \text{benzaldehyde} \longrightarrow C + (C_6H_5)_3\overset{+}{P}{-}O^-$

(*b*) $C_6H_5CH_2\overset{OH}{\overset{|}{C}}HCH_3 \xrightarrow[CH_2Cl_2]{PCC} D$

$D + CH_3\overset{O}{\overset{\|}{C}}OOH \longrightarrow E + CH_3\overset{O}{\overset{\|}{C}}OH$

**A-5.** Give the reagents necessary to convert cyclohexanone into each of the following compounds. More than one step may be necessary.

(*a*) $CH_3$ (*c*) O O

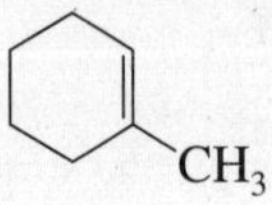

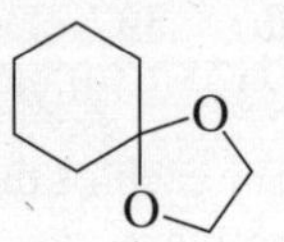

(*b*) $CH_2$ (*d*) O O

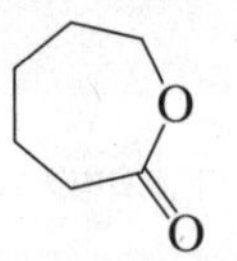

**A-6.** (*a*) What two organic compounds react together (in the presence of an acid catalyst) to give the compound shown, plus a molecule of water?

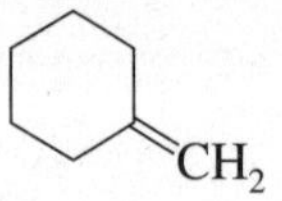

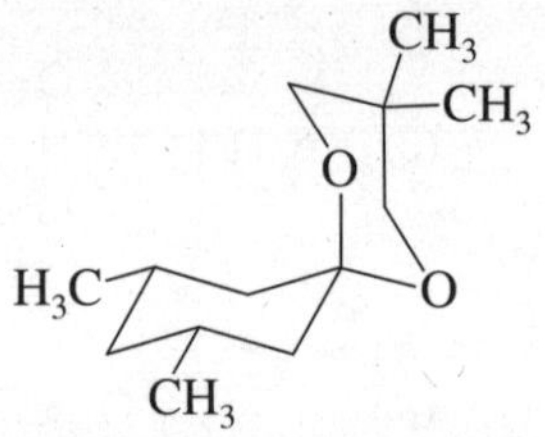

$CH_3$ $CH_3$ O $H_3C$ O $CH_3$

(*b*) Draw the structure of the open-chain form of the following cyclic acetal:

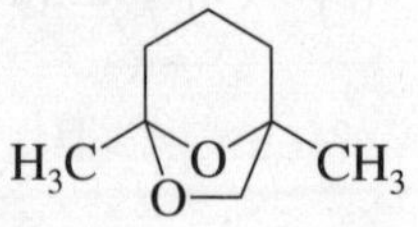

$H_3C$ O O $CH_3$

**A-7.** Outline reaction schemes to carry out each of the following interconversions, using any necessary organic or inorganic reagents.

(*a*) $(CH_3)_2C{=}O$ to $(CH_3)_2\overset{O}{C{-}C}HCH_3$

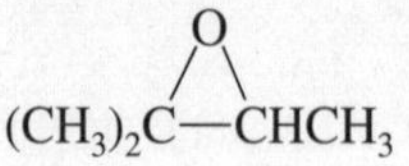

(*b*) O $CH_3$ OH to O $CH_3$ HO $CH_3$

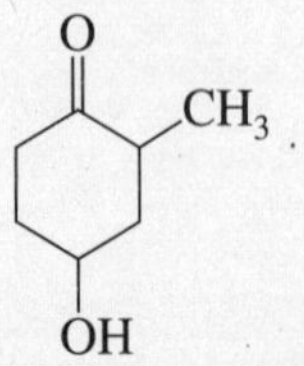

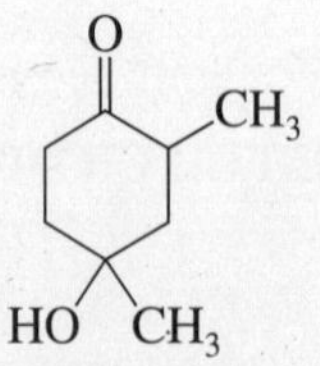

(*c*) [3-oxocyclohexyl $CO_2H$] ⟶ [3-oxocyclohexyl $CH_2Br$]

**A-8.** Write a stepwise mechanism for the formation of $CH_3CH(OCH_3)_2$ from acetaldehyde and methanol under conditions of acid catalysis.

**A-9.** Suggest a structure for an unknown compound, $C_9H_{10}O$, that exhibits a strong infrared absorption at 1710 $cm^{-1}$ and has a $^1H$ NMR spectrum that consists of three singlets at $\delta$ 2.1 ppm (3H), 3.7 ppm (2H), and 7.2 ppm (5H).

## PART B

**B-1.** Which of the compounds shown is (are) correctly named as pentane derivatives, either as pentanals or pentanones?

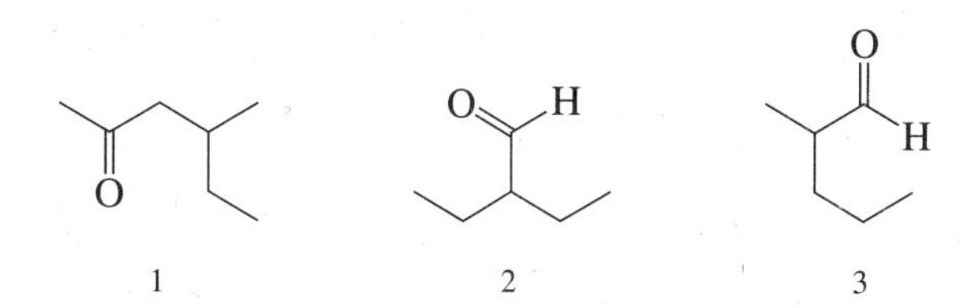


(*a*) 1 only (*b*) 2 only (*c*) 3 only (*d*) 1 and 3 (*e*) None of them

**B-2.** The compound shown is best classified as a(an)

$$(CH_3)_3CCH_2CH{=}NCH_3$$

(*a*) Carbinolamine
(*b*) Enamine
(*c*) Hydrazone
(*d*) Imine
(*e*) Oxime

**B-3.** When a nucleophile encounters a ketone, the site of attack is
(*a*) The carbon atom of the carbonyl
(*b*) The oxygen atom of the carbonyl
(*c*) Both the carbon and oxygen atoms, with equal probability
(*d*) No attack occurs—ketones do not react with nucleophiles.

**B-4.** What reagent and/or reaction conditions would you choose to bring about the following conversion?

[2-phenyl-1,3-dioxolane] ⟶ [$C_6H_5$]—$CH{=}O$ + $HOCH_2CH_2OH$

(*a*) 1. $LiAlH_4$, 2. $H_2O$
(*b*) $H_2O$, NaOH, heat
(*c*) $H_2O$, $H_2SO_4$, heat
(*d*) PCC, $CH_2Cl_2$

**B-5.** Rank the following in order of increasing value of the equilibrium constant for hydration, $K_{hyd}$ (smallest value first).

[cyclobutanecarbaldehyde] (1) [cyclopropanone] (2) $(CH_3)_3CC(=O)C(CH_3)_3$ (3)

(*a*) $1 < 2 < 3$ (*b*) $3 < 1 < 2$ (*c*) $2 < 1 < 3$ (*d*) $2 < 3 < 1$

**B-6.** The structure

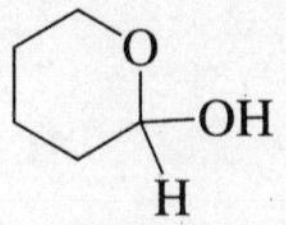

[Structure: six-membered ring containing O, with ring carbon bearing OH and H]

would be best classified as a(n)

(*a*) Acetal
(*b*) Hemiacetal
(*c*) Hydrate
(*d*) Cyanohydrin

**B-7.** Which of the following pairs of reactants is most effective in forming an enamine?

(*a*) $CH_3CH_2\overset{O}{\overset{\|}{C}}H$ + $[(CH_3)_2CH]_2NH$

(*b*) $(CH_3)_3C\overset{O}{\overset{\|}{C}}H$ + $(CH_3)_2NH$

(*c*) cyclohexanone + $(CH_3)_3CNH_2$

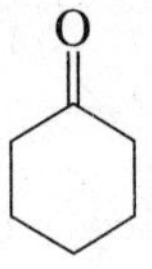

(*d*) None of these forms an enamine.

**B-8.** Which of the following species is an ylide?

(*a*) $(C_6H_5)_3\overset{+}{P}CH_2CH_3\ Br^-$

(*b*) $(C_6H_5)_3\overset{+}{P}\overset{\cdot\cdot}{\underset{-}{C}}HCH_3$

(*c*) $(C_6H_5)_3P{-}CHCH_3$ / $O{-}CH_2$ (four-membered ring: P bonded to O, CH bonded to $CH_2$)

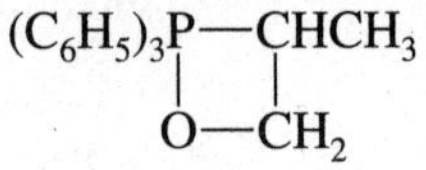

(*d*) None of these

**B-9.** Which pair of the following compounds could serve as the reagents X and Y in the following reaction sequence?

$$X \xrightarrow{(C_6H_5)_3P} \xrightarrow{CH_3CH_2CH_2CH_2Li} \xrightarrow{Y} (CH_3)_2CHCH{=}C(CH_3)(CH_2CH_3)$$

| $(CH_3)_2CHCH_2Br$ | $(CH_3)_2CHBr$ | $CH_3CH_2\overset{CH_3}{\overset{\vert}{C}}HBr$ | $(CH_3)_2CH\overset{O}{\overset{\|}{C}}H$ | $CH_3\overset{O}{\overset{\|}{C}}CH_3$ |
|---|---|---|---|---|
| 1 | 2 | 3 | 4 | 5 |

| | X | Y |
|---|---|---|
| (*a*) | 1 | 5 |
| (*b*) | 1 | 4 |
| (*c*) | 2 | 4 |
| (*d*) | 2 | 5 |
| (*e*) | 3 | 4 |

**B-10.** The final product of the following sequence of reactions is.

$$(CH_3O)_2CHCH_2CH_2CH_2Br \xrightarrow{Mg} \xrightarrow{H_2C{=}O} \xrightarrow[\text{heat}]{H_3O^+} ?$$

(*a*) $CH_3OC(=O)CH_2CH_2CH_2CH_2OH$

(*b*) $CH_3C(=O)CH_2CH_2CH_2CH_2OH$

(*c*) $HC(=O)CH_2CH_2CH_2CH_2OH$

(*d*) $HC(=O)CH_2CH_2CH_2CH(=O)$

(*e*) $(CH_3O)_2CHCH_2CH_2CH_2CH(=O)$

**B-11.** Which of the following sets of reagents, used in the order shown, would successfully accomplish the conversion shown?

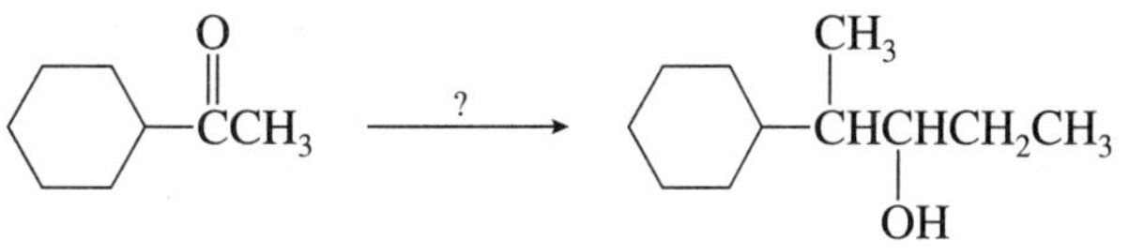

$$C_6H_{11}{-}C(=O)CH_3 \xrightarrow{?} C_6H_{11}{-}CH(OH)CH(CH_3)CH_2CH_3$$

(*a*) $CH_3CH_2CH_2MgBr$; $H_3O^+$; PCC, $CH_2Cl_2$
(*b*) $CH_3CH_2CH_2MgBr$; $H_3O^+$; $H_2SO_4$, heat; PCC, $CH_2Cl_2$
(*c*) $(C_6H_5)_3\overset{+}{P}{-}\ddot{C}HCH_2CH_3$; $B_2H_6$; $H_2O_2$, $HO^-$
(*d*) $(C_6H_5)_3\overset{+}{P}{-}\ddot{C}HCH_2CH_3$; $H_2SO_4$, $H_2O$

**B-12.** Which of the following species is the conjugate acid of the hemiacetal formed by reaction of benzaldehyde with methanol containing a trace of acid?

(*a*) $C_6H_5CH(OH)(OCH_3)$

(*b*) $C_6H_5CH{=}\overset{+}{O}CH_3$

(*c*) $C_6H_5CH(OH)(\overset{+}{O}HCH_3)$

(*d*) $C_6H_5CH(OH)(\overset{+}{O}H_2)$

(*e*) $C_6H_5CH{=}\overset{+}{O}H$

**B-13.** Which sequence represents the best synthesis of hexanal?

$$CH_3CH_2CH_2CH_2CH_2CH{=}O$$

Hexanal

(*a*)
1. $CH_3CH_2CH_2CH_2Br + NaC{\equiv}CH$
2. $H_2O$, $H_2SO_4$, $HgSO_4$

(*b*)
1. $CH_3CH_2CH_2CH_2\overset{O}{\overset{\|}{C}}CH_3$
2. $CH_3\overset{O}{\overset{\|}{C}}OOH$
3. $LiAlH_4$
4. $H_2O$
5. PCC, $CH_2Cl_2$

(*c*)
1. $CH_3CH_2CH_2CH{=}CH_2 + CH_3\overset{O}{\overset{\|}{C}}OOH$
2. $CH_3MgBr$, diethyl ether
3. $H_3O^+$
4. PCC, $CH_2Cl_2$

(*d*)
1. $CH_3CH_2CH_2CH_2MgBr$ + $H_2C{-}CH_2$ (epoxide, bridged by O)
2. $H_3O^+$
3. PCC, $CH_2Cl_2$

**B-14.** The amino ketone shown undergoes a spontaneous cyclization on standing. What is the product of this intramolecular reaction?

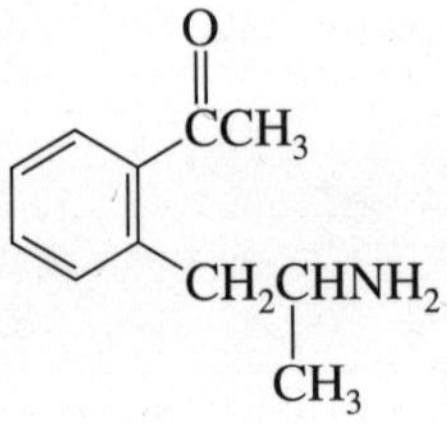

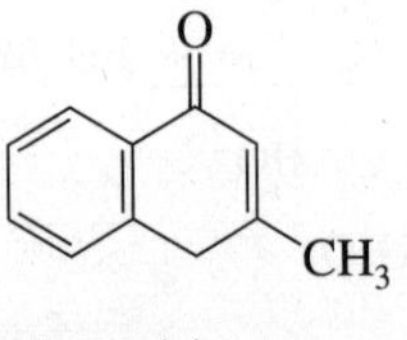

$CH_3$, $\overset{O}{\overset{\|}{C}}CH_3$, $CH_2CHNH_2$ (with $CH_3$ on CH) — ortho-substituted benzene

(*a*) O, $CH_3$

(*b*) $CH_3$, $NH_2$

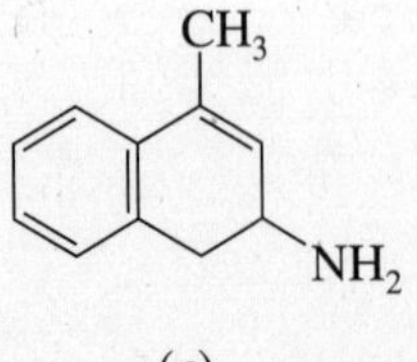

(*c*) $CH_3$, $NH_2$

(*d*) $CH_3$, N, $CH_3$

(*e*) $CH_3$, N, $CH_3$

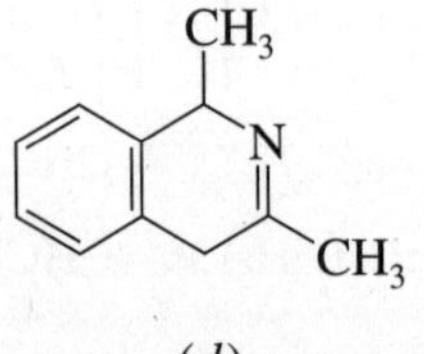

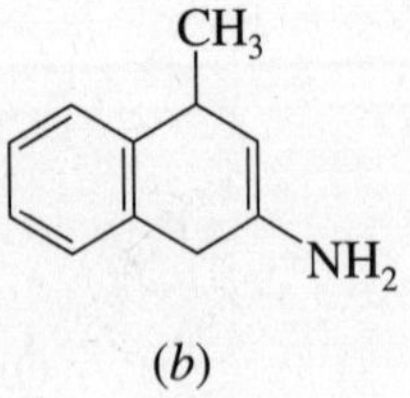

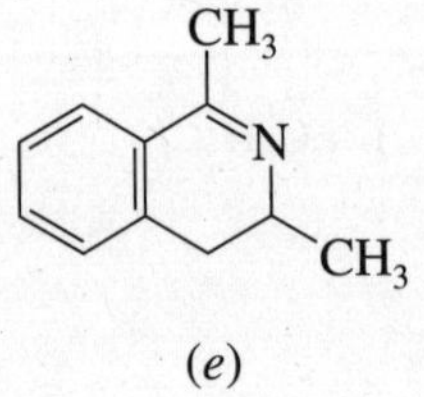

**B-15.** Which of the following compounds would have a $^1H$ NMR spectrum consisting of three singlets?

(*a*) $CH_3C(CH_3)_2CH_2C(=O)CH_3$

(*b*) $CH_3CH_2CH_2C(=O)CH_2CH_2CH_3$

(*c*) $HC(=O)CH_2CH_2CH_2C(=O)H$

(*d*) $C_6H_5$–$CH_2CH_2C(=O)H$

**B-16.** Which of the following compounds would have the fewest number of signals in its $^{13}C$ NMR spectrum?

(*a*) $CH_3C(CH_3)_2CH_2C(=O)CH_3$

(*b*) $CH_3CH_2CH_2C(=O)CH_2CH_2CH_3$

(*c*) $CH_3CH(CH_3)C(=O)CH(CH_3)CH_3$

(*d*) $CH_3CH(CH_3)C(=O)CH_2CH_3$

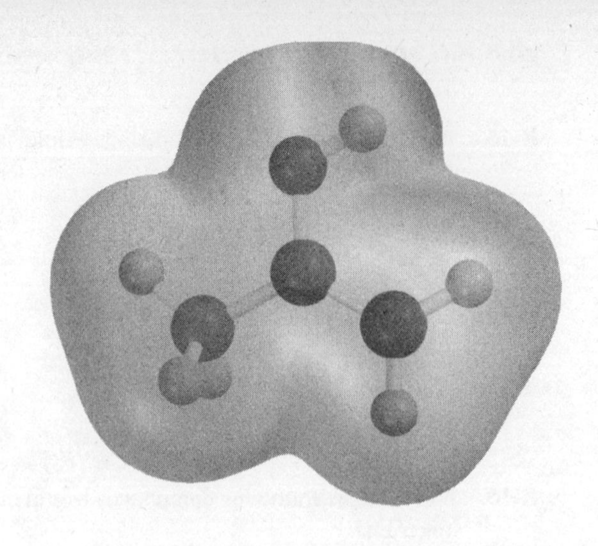

# CHAPTER 18

## ENOLS AND ENOLATES

## SOLUTIONS TO TEXT PROBLEMS

**18.1** (*b*) There are no $\alpha$-hydrogen atoms in 2,2-dimethylpropanal, because the $\alpha$-carbon atom bears three methyl groups.

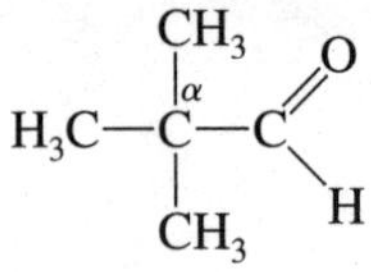

$$H_3C-\overset{\overset{CH_3}{|\alpha}}{\underset{\underset{CH_3}{|}}{C}}-C(=O)H$$

2,2-Dimethylpropanal

(*c*) All three protons of the methyl group, as well as the two benzylic protons, are $\alpha$ hydrogens.

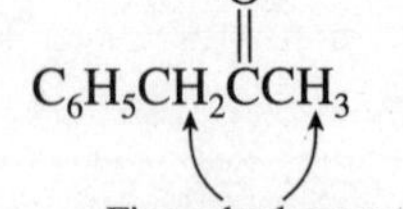

$$C_6H_5CH_2\overset{O}{\overset{\|}{C}}CH_3$$

Five $\alpha$ hydrogens

Benzyl methyl ketone

(*d*) Cyclohexanone has four equivalent $\alpha$ hydrogens.

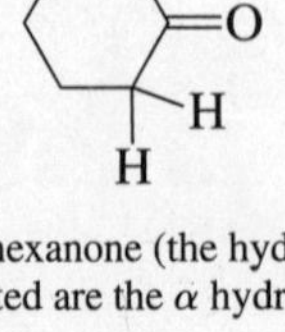

Cyclohexanone (the hydrogens indicated are the $\alpha$ hydrogens)

**18.2** As shown in the general equation and the examples, halogen substitution is specific for the $\alpha$-carbon atom. The ketone 2-butanone has two nonequivalent $\alpha$ carbons, and so substitution is possible at both positions. Both 1-chloro-2-butanone and 3-chloro-2-butanone are formed in the reaction.

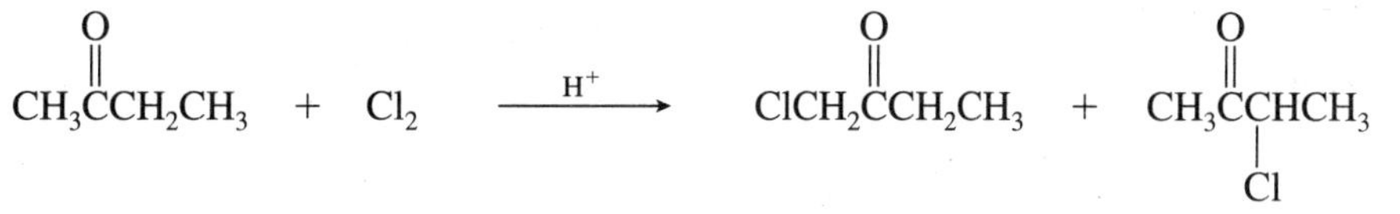

$$\underset{\text{2-Butanone}}{CH_3\overset{O}{\overset{\|}{C}}CH_2CH_3} + \underset{\text{Chlorine}}{Cl_2} \xrightarrow{H^+} \underset{\text{1-Chloro-2-butanone}}{ClCH_2\overset{O}{\overset{\|}{C}}CH_2CH_3} + \underset{\text{3-Chloro-2-butanone}}{CH_3\overset{O}{\overset{\|}{C}}\underset{Cl}{\underset{|}{C}}HCH_3}$$

**18.3** The carbon–carbon double bond of the enol always involves the original carbonyl carbon and the $\alpha$-carbon atom. 2-Butanone can form two different enols, each of which yields a different $\alpha$-chloro ketone.

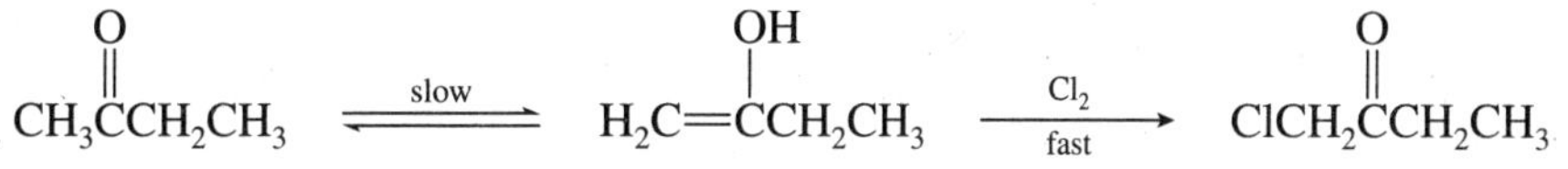

$$\underset{\text{2-Butanone}}{CH_3\overset{O}{\overset{\|}{C}}CH_2CH_3} \overset{\text{slow}}{\rightleftharpoons} \underset{\text{1-Buten-2-ol (enol)}}{H_2C{=}\overset{OH}{\overset{|}{C}}CH_2CH_3} \xrightarrow[\text{fast}]{Cl_2} \underset{\text{1-Chloro-2-butanone}}{ClCH_2\overset{O}{\overset{\|}{C}}CH_2CH_3}$$

$$\underset{\text{2-Butanone}}{CH_3\overset{O}{\overset{\|}{C}}CH_2CH_3} \overset{\text{slow}}{\rightleftharpoons} \underset{\text{2-Buten-2-ol (enol)}}{CH_3\overset{OH}{\overset{|}{C}}{=}CHCH_3} \xrightarrow[\text{fast}]{Cl_2} \underset{\text{3-Chloro-2-butanone}}{CH_3\overset{O}{\overset{\|}{C}}\underset{Cl}{\underset{|}{C}}HCH_3}$$

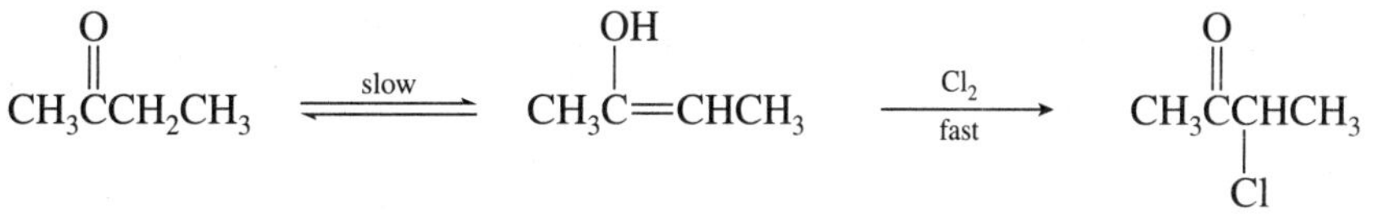

**18.4** Chlorine attacks the carbon–carbon double bond of each enol.

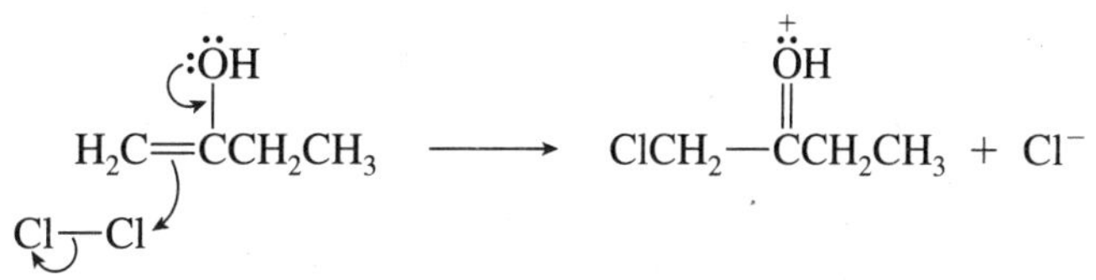

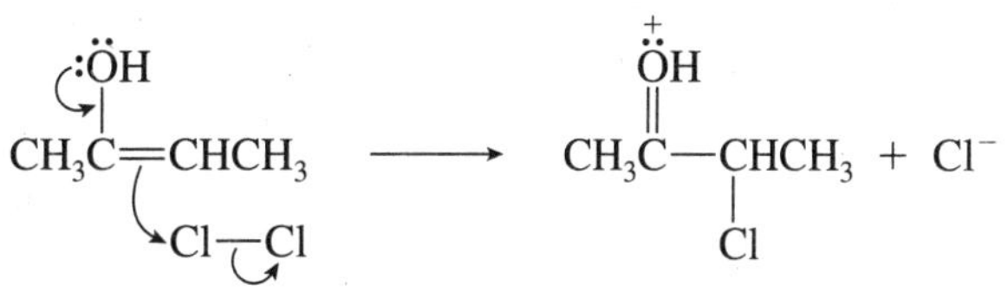

$$H_2C{=}\overset{:\ddot{O}H}{\overset{|}{C}}CH_2CH_3 \;\; Cl{-}Cl \longrightarrow ClCH_2{-}\overset{\overset{+}{\ddot{O}}H}{\overset{\|}{C}}CH_2CH_3 + Cl^-$$

$$CH_3\overset{:\ddot{O}H}{\overset{|}{C}}{=}CHCH_3 \;\; Cl{-}Cl \longrightarrow CH_3\overset{\overset{+}{\ddot{O}}H}{\overset{\|}{C}}{-}\underset{Cl}{\underset{|}{C}}HCH_3 + Cl^-$$

**18.5** (*b*) Acetophenone can enolize only in the direction of the methyl group.

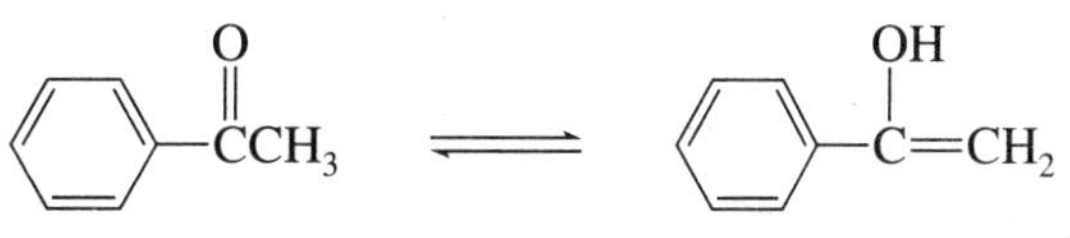

$$\underset{\text{Acetophenone}}{C_6H_5\overset{O}{\overset{\|}{C}}CH_3} \rightleftharpoons \underset{\text{Enol form of acetophenone}}{C_6H_5\overset{OH}{\overset{|}{C}}{=}CH_2}$$

(*c*) Enolization of 2-methylcyclohexanone can take place in two different directions.

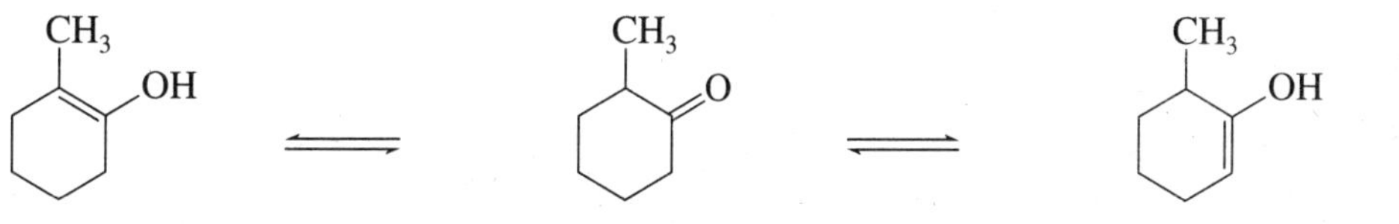

2-Methylcyclohex-1-enol (enol form) ⇌ 2-Methylcyclohexanone ⇌ 6-Methylcyclohex-1-enol (enol form)

**18.6** (*b*) Enolization of the central methylene group can involve either of the two carbonyl groups.

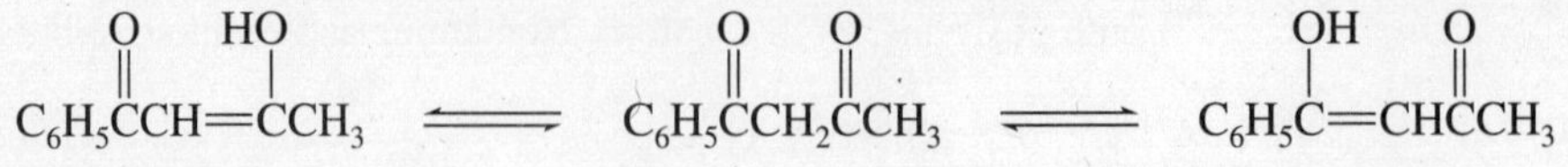

$$C_6H_5\overset{O}{\overset{\|}{C}}CH{=}\overset{HO}{\overset{|}{C}}CH_3 \rightleftharpoons C_6H_5\overset{O}{\overset{\|}{C}}CH_2\overset{O}{\overset{\|}{C}}CH_3 \rightleftharpoons C_6H_5\overset{OH}{\overset{|}{C}}{=}CH\overset{O}{\overset{\|}{C}}CH_3$$

Enol form — 1-Phenyl-1,3-butanedione — Enol form

**18.7** (*b*) Removal of a proton from 1-phenyl-1,3-butanedione occurs on the methylene group between the carbonyls.

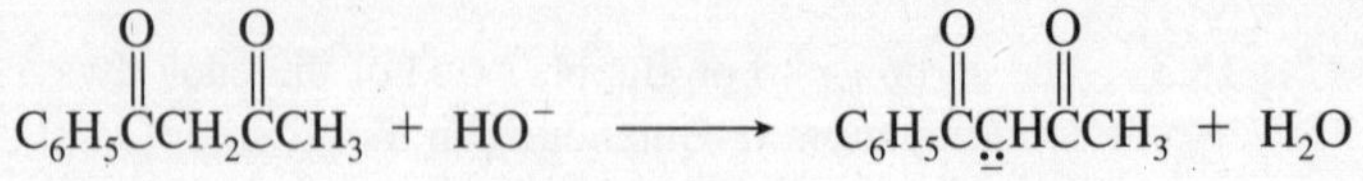

$$C_6H_5\overset{O}{\overset{\|}{C}}CH_2\overset{O}{\overset{\|}{C}}CH_3 + HO^- \longrightarrow C_6H_5\overset{O}{\overset{\|}{C}}\underset{..}{C}H\overset{O}{\overset{\|}{C}}CH_3 + H_2O$$

The three most stable resonance forms of this anion are

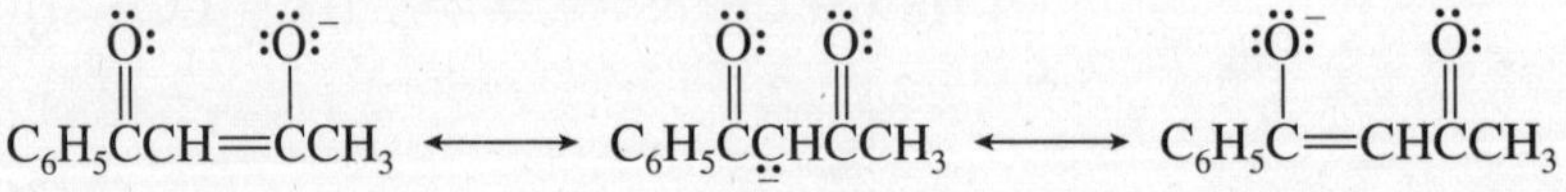

$$C_6H_5\overset{\ddot{O}:}{\overset{\|}{C}}CH{=}\overset{:\ddot{O}:^-}{\overset{|}{C}}CH_3 \longleftrightarrow C_6H_5\overset{\ddot{O}:}{\overset{\|}{C}}\underset{..}{C}H\overset{\ddot{O}:}{\overset{\|}{C}}CH_3 \longleftrightarrow C_6H_5\overset{:\ddot{O}:^-}{\overset{|}{C}}{=}CH\overset{\ddot{O}:}{\overset{\|}{C}}CH_3$$

(*c*) Deprotonation at C-2 of this $\beta$-dicarbonyl compound yields the carbanion shown.

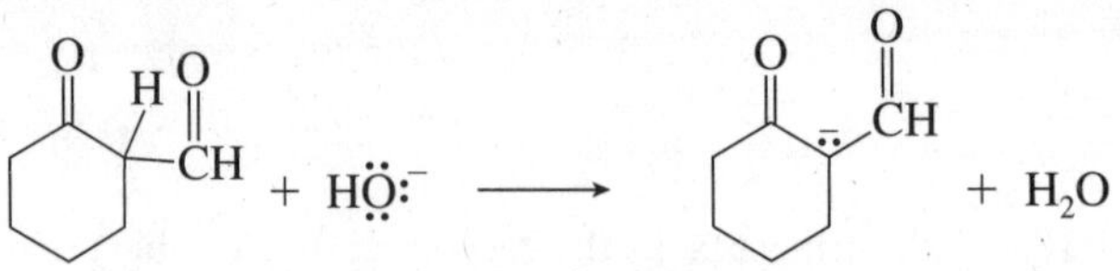

[Structures: 2-oxocyclohexanecarbaldehyde (H on C-2, CHO group) + $H\ddot{O}:^-$ $\longrightarrow$ carbanion at C-2 + $H_2O$]

The three most stable resonance forms of the anion are:

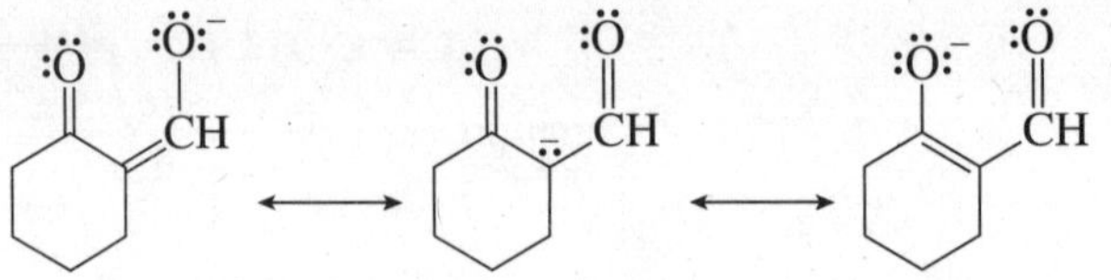

[Structures: enolate with $:\ddot{O}:^-$ on CH (C=C in ring) $\longleftrightarrow$ carbanion at C-2 $\longleftrightarrow$ enolate with ring $:\ddot{O}:^-$ and C=C in ring; labels :Ö, CH]

**18.8** Each of the five $\alpha$ hydrogens has been replaced by deuterium by base-catalyzed enolization. Only the $OCH_3$ hydrogens and the hydrogens on the aromatic ring are observed in the $^1H$ NMR spectrum at $\delta$ 3.9 ppm and $\delta$ 6.7–6.9 ppm, respectively.

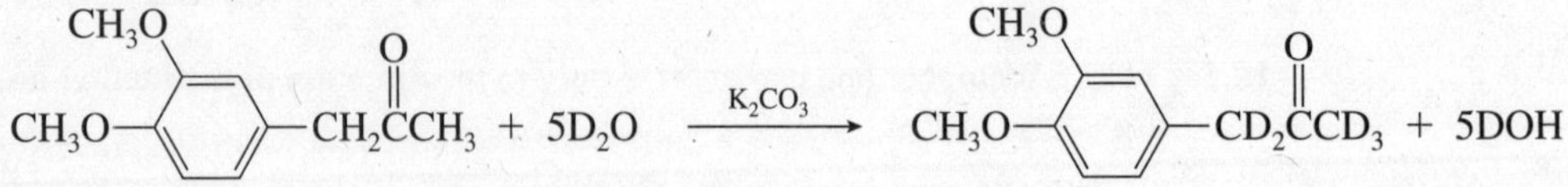

$$(CH_3O)_2C_6H_3{-}CH_2\overset{O}{\overset{\|}{C}}CH_3 + 5D_2O \xrightarrow{K_2CO_3} (CH_3O)_2C_6H_3{-}CD_2\overset{O}{\overset{\|}{C}}CD_3 + 5DOH$$

**18.9** $\alpha$-Chlorination of (*R*)-*sec*-butyl phenyl ketone in acetic acid proceeds via the enol. The enol is achiral and yields equal amounts of (*R*)- and (*S*)-2-chloro-2-methyl-1-phenyl-1-butanone. The product is chiral. It is formed as a racemic mixture, however, and this mixture is not optically active.

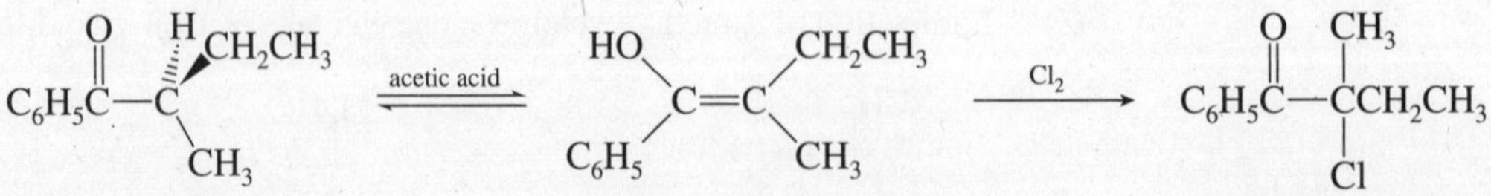

$$C_6H_5\overset{O}{\overset{\|}{C}}{-}C(H)(CH_2CH_3)(CH_3) \xrightleftharpoons{\text{acetic acid}} (HO)(C_6H_5)C{=}C(CH_2CH_3)(CH_3) \xrightarrow{Cl_2} C_6H_5\overset{O}{\overset{\|}{C}}{-}\underset{Cl}{\overset{CH_3}{C}}CH_2CH_3$$

(*R*)-*sec*-Butyl phenyl ketone — Enol — 2-Chloro-2-methyl-1-phenyl-1-butanone (50% *R*; 50% *S*)

**18.10** (*b*) Approaching this problem mechanistically in the same way as part (*a*), write the structure of the enolate ion from 2-methylbutanal.

$$CH_3CH_2CH(CH_3)CH{=}O + HO^- \rightleftharpoons CH_3CH_2\ddot{C}^-(CH_3)CH{=}O \longleftrightarrow CH_3CH_2C(CH_3){=}CHO^-$$

2-Methylbutanal — Enolate of 2-methylbutanal

This enolate adds to the carbonyl group of the aldehyde.

$$CH_3CH_2CH(CH_3)CH{=}O + {:}\bar{C}(CH_3)(CH_2CH_3)HC{=}O \longrightarrow CH_3CH_2CH(CH_3)CH(O^-){-}C(CH_3)(CH_2CH_3)HC{=}O$$

2-Methylbutanal — Enolate of 2-methylbutanal

A proton transfer from solvent yields the product of aldol addition.

$$CH_3CH_2CH(CH_3)CH(O^-){-}C(CH_3)(CH_2CH_3)HC{=}O + H_2O \longrightarrow CH_3CH_2CH(CH_3)CH(OH){-}C(CH_3)(CH_2CH_3)HC{=}O + HO^-$$

2-Ethyl-3-hydroxy-2,4-dimethylhexanal

(*c*) The aldol addition product of 3-methylbutanal can be identified through the same mechanistic approach.

$$(CH_3)_2CHCH_2CH{=}O + HO^- \rightleftharpoons (CH_3)_2CH\ddot{C}^-HCH{=}O + H_2O$$

3-Methylbutanal — Enolate of 3-methylbutanal

$$(CH_3)_2CHCH_2CH{=}O + {:}\bar{C}H(HC{=}O)CH(CH_3)_2 \longrightarrow (CH_3)_2CHCH_2CH(O^-){-}CH(HC{=}O)CH(CH_3)_2$$

3-Methylbutanal — Enolate of 3-methylbutanal

$$\xrightarrow{H_2O} (CH_3)_2CHCH_2CH(OH){-}CH(HC{=}O)CH(CH_3)_2 + OH^-$$

3-Hydroxy-2-isopropyl-5-methylhexanal

**18.11** Dehydration of the aldol addition product involves loss of a proton from the $\alpha$-carbon atom and hydroxide from the $\beta$-carbon atom.

$$R_2C(OH)\text{—}CH(H)CH{=}O \xrightarrow{\text{heat}} R_2C{=}CHCH{=}O + H_2O + HO^-$$

(with $^-$:ÖH: abstracting H)

(*b*) The product of aldol addition of 2-methylbutanal has no $\alpha$ hydrogens. It cannot dehydrate to an aldol condensation product.

$$2CH_3CH_2CH(CH_3)CH{=}O \xrightleftharpoons{HO^-} CH_3CH_2CH(CH_3)CH(HO)\text{—}\overset{\alpha}{C}(CH_3)(HC{=}O)CH_2CH_3$$

2-Methylbutanal — (No protons on $\alpha$-carbon atom)

(*c*) Aldol condensation is possible with 3-methylbutanal.

$$2(CH_3)_2CHCH_2CH{=}O \xrightleftharpoons{HO^-} (CH_3)_2CHCH_2CH(HO)CH(HC{=}O)CH(CH_3)_2 \xrightarrow{-H_2O} (CH_3)_2CHCH_2CH{=}C(HC{=}O)CH(CH_3)_2$$

3-Methylbutanal — Aldol addition product — 2-Isopropyl-5-methyl-2-hexenal

**18.12** The carbon skeleton of 2-ethyl-1-hexanol is the same as that of the aldol condensation product derived from butanal. Hydrogenation of this compound under conditions in which both the carbon–carbon double bond and the carbonyl group are reduced gives 2-ethyl-1-hexanol.

$$CH_3CH_2CH_2CH{=}O \xrightarrow[\text{heat}]{NaOH,\ H_2O} CH_3CH_2CH_2CH{=}C(CH_2CH_3)CH{=}O \xrightarrow{H_2,\ Ni} CH_3CH_2CH_2CH_2CH(CH_2CH_3)CH_2OH$$

Butanal — 2-Ethyl-2-hexenal — 2-Ethyl-1-hexanol

**18.13** (*b*) The only enolate that can be formed from *tert*-butyl methyl ketone arises by proton abstraction from the methyl group.

$$(CH_3)_3C\overset{O}{\overset{\|}{C}}CH_3 + HO^- \rightleftharpoons (CH_3)_3C\overset{O}{\overset{\|}{C}}\ddot{C}H_2^-$$

*tert*-Butyl methyl ketone — Enolate of *tert*-butyl methyl ketone

This enolate adds to the carbonyl group of benzaldehyde to give the mixed aldol addition product, which then dehydrates under the reaction conditions.

$$C_6H_5CH{=}O \;+\; :\bar{C}H_2C(=O)C(CH_3)_3 \longrightarrow C_6H_5CH(O^-)CH_2C(=O)C(CH_3)_3 \xrightarrow{H_2O} C_6H_5CH(OH)CH_2C(=O)C(CH_3)_3$$

Benzaldehyde; Enolate of *tert*-butyl methyl ketone; Product of mixed aldol addition

$$\xrightarrow{-H_2O} C_6H_5CH{=}CHC(=O)C(CH_3)_3$$

4,4-Dimethyl-1-phenyl-1-penten-3-one (product of mixed aldol condensation)

(*c*) The enolate of cyclohexanone adds to benzaldehyde. Dehydration of the mixed aldol addition product takes place under the reaction conditions to give the following mixed aldol condensation product.

$$\text{Cyclohexanone} + C_6H_5CH{=}O \xrightarrow{HO^-} \text{(2-}{=}CHC_6H_5\text{)cyclohexanone} + H_2O$$

Cyclohexanone; Benzaldehyde; Benzylidenecyclohexanone

**18.14** Mesityl oxide is an $\alpha,\beta$-unsaturated ketone. Traces of acids or bases can catalyze its isomerization so that some of the less stable $\beta,\gamma$-unsaturated isomer is present.

$$(H_3C)_2C{=}CHC(=O)CH_3 \rightleftharpoons H_2C{=}C(CH_3)-CH_2C(=O)CH_3$$

Mesityl oxide; 4-methyl-3-penten-2-one (more stable); 4-Methyl-4-penten-2-one (less stable)

**18.15** The relationship between the molecular formula of acrolein ($C_3H_4O$) and the product ($C_3H_5N_3O$) corresponds to the addition of $HN_3$ to acrolein. Because propanal ($CH_3CH_2CH{=}O$) does not react under these conditions, the carbon-carbon, not the carbon-oxygen, double bond of acrolein is the reactive site. Conjugate addition is the reaction that occurs.

$$H_2C{=}CHC(=O)H \xrightarrow[\text{acetic acid}]{NaN_3} N_3CH_2CH_2C(=O)H$$

Acrolein; 3-Azidopropanal

**18.16** The enolate of dibenzyl ketone adds to methyl vinyl ketone in the conjugate addition step.

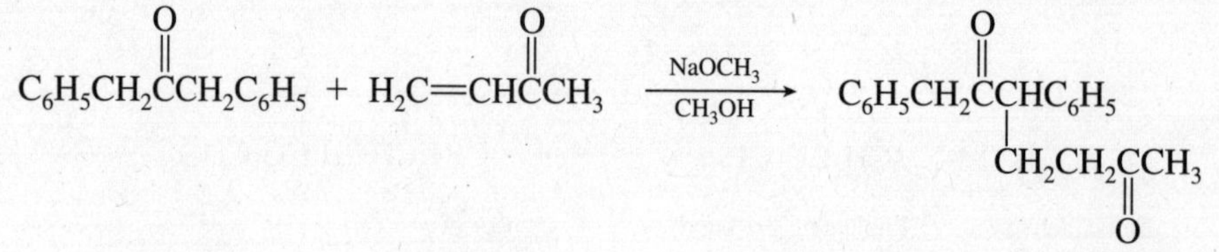

$$C_6H_5CH_2\overset{O}{\overset{\|}{C}}CH_2C_6H_5 + H_2C{=}CH\overset{O}{\overset{\|}{C}}CH_3 \xrightarrow[CH_3OH]{NaOCH_3} C_6H_5CH_2\overset{O}{\overset{\|}{C}}\underset{\underset{\underset{O}{\|}}{CH_2CH_2CCH_3}}{CH}C_6H_5$$

Dibenzyl ketone | Methyl vinyl ketone | 1,3-Diphenyl-2,6-heptanedione

via

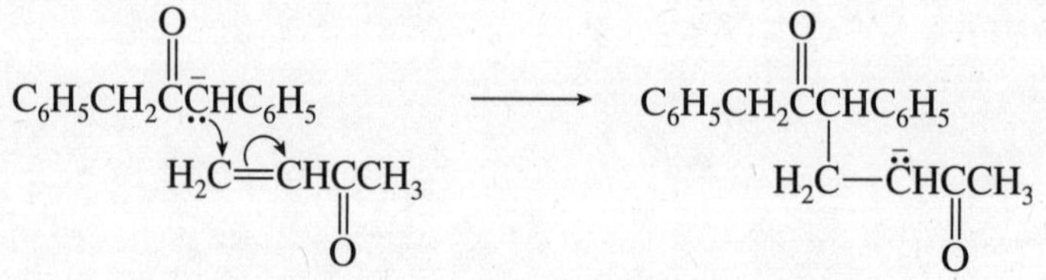

$$C_6H_5CH_2\overset{O}{\overset{\|}{C}}\ddot{\bar{C}}HC_6H_5 + H_2C{=}CH\overset{O}{\overset{\|}{C}}CH_3 \longrightarrow C_6H_5CH_2\overset{O}{\overset{\|}{C}}CHC_6H_5 \text{ with } H_2C{-}\ddot{\bar{C}}H\overset{O}{\overset{\|}{C}}CH_3$$

The intramolecular aldol condensation that gives the observed product is

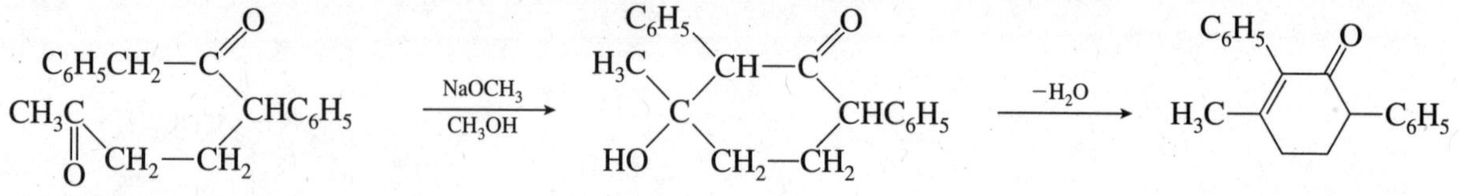

$$\text{1,3-Diphenyl-2,6-heptanedione} \xrightarrow[CH_3OH]{NaOCH_3} \text{(cyclic aldol: HO, }H_3C\text{, }C_6H_5\text{)} \xrightarrow{-H_2O} \text{3-Methyl-2,6-diphenyl-2-cyclohexen-1-one}$$

1,3-Diphenyl-2,6-heptanedione | 3-Methyl-2,6-diphenyl-2-cyclohexen-1-one

**18.17** A second solution to the synthesis of 4-methyl-2-octanone by conjugate addition of a lithium dialkylcuprate reagent to an $\alpha,\beta$-unsaturated ketone is revealed by the disconnection shown:

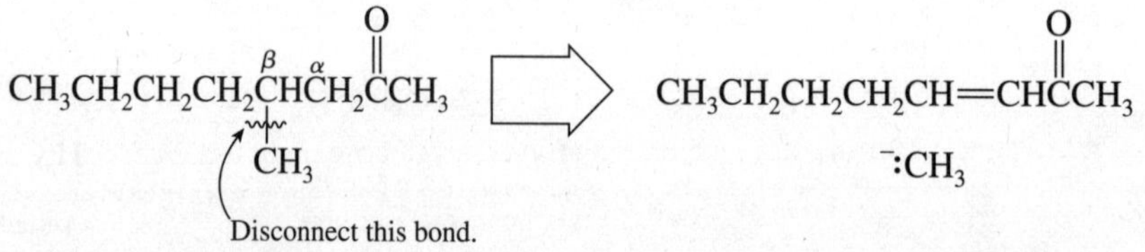

$$CH_3CH_2CH_2CH_2\overset{\beta}{C}H(CH_3)\overset{\alpha}{C}H_2\overset{O}{\overset{\|}{C}}CH_3 \Longrightarrow CH_3CH_2CH_2CH_2CH{=}CH\overset{O}{\overset{\|}{C}}CH_3 + {}^-{:}CH_3$$

Disconnect this bond.

According to this disconnection, the methyl group is derived from lithium dimethylcuprate.

$$CH_3CH_2CH_2CH_2CH{=}CH\overset{O}{\overset{\|}{C}}CH_3 + LiCu(CH_3)_2 \longrightarrow CH_3CH_2CH_2CH_2\underset{CH_3}{\underset{|}{C}H}CH_2\overset{O}{\overset{\|}{C}}CH_3$$

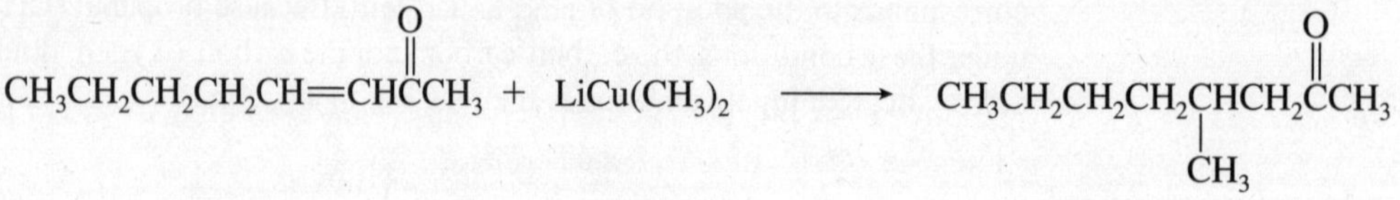

3-Octen-2-one | Lithium dimethylcuprate | 4-Methyl-2-octanone

**18.18** (*a*) In addition to the double bond of the carbonyl group, there must be a double bond elsewhere in the molecule in order to satisfy the molecular formula $C_4H_6O$ (the problem states that the compounds are noncyclic). There are a total of five isomers:

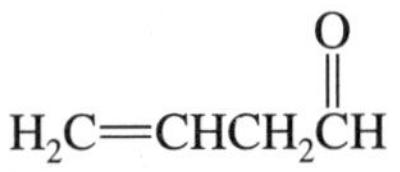

$H_2C{=}CHCH_2CH{=}O$ — 3-Butenal

$(E)$-$CH_3CH{=}CHCH{=}O$ (H and CHO trans... $H_3C$ and CHO on opposite sides) — $(E)$-2-Butenal

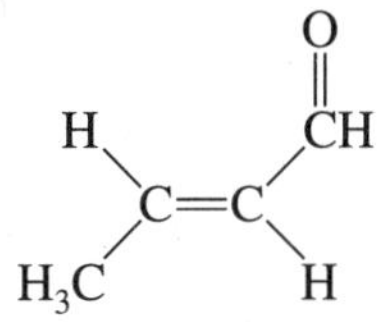

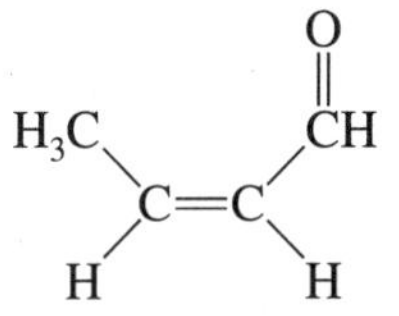

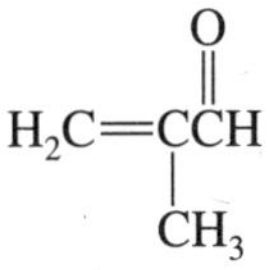

$(Z)$-$CH_3CH{=}CHCH{=}O$ — $(Z)$-2-Butenal

$H_2C{=}C(CH_3)CH{=}O$ — 2-Methylpropenal

$H_2C{=}CHC(=O)CH_3$ — 3-Buten-2-one (methyl vinyl ketone)

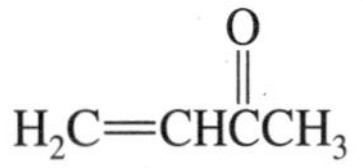

(*b*) The *E* and *Z* isomers of 2-butenal are stereoisomers.
(*c*) None of the $C_4H_6O$ aldehydes and ketones is chiral.
(*d*) The $\alpha,\beta$-unsaturated aldehydes are (*E*)- and (*Z*)-$CH_3CH{=}CHCHO$; and $H_2C{=}C(CH_3)CHO$.

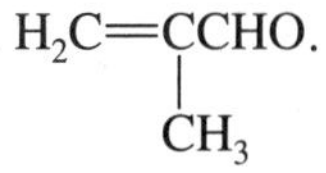

There is one $\alpha,\beta$-unsaturated ketone in the group: $H_2C{=}CHC(=O)CH_3$.
(*e*) The *E* and *Z* isomers of 2-butenal are formed by the aldol condensation of acetaldehyde.

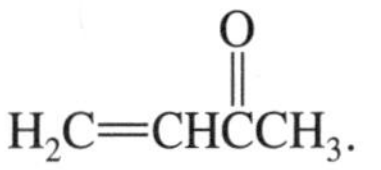

**18.19** The main flavor component of the hazelnut has the structure shown.

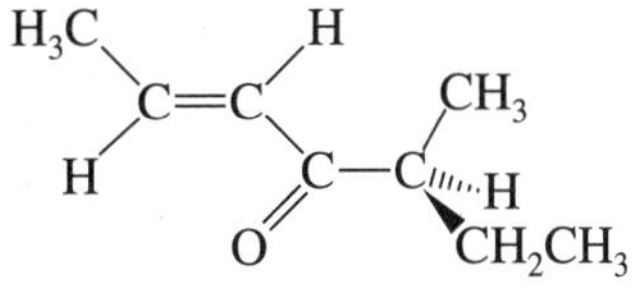

$(E)$-$CH_3CH{=}CHC(=O)CH(CH_3)CH_2CH_3$

(2*E*,5*S*)-5-Methyl-2-hepten-4-one

**18.20** The characteristic reaction of an alcohol on being heated with $KHSO_4$ is acid-catalyzed dehydration. Secondary alcohols dehydrate faster than primary alcohols, and so a reasonable first step is

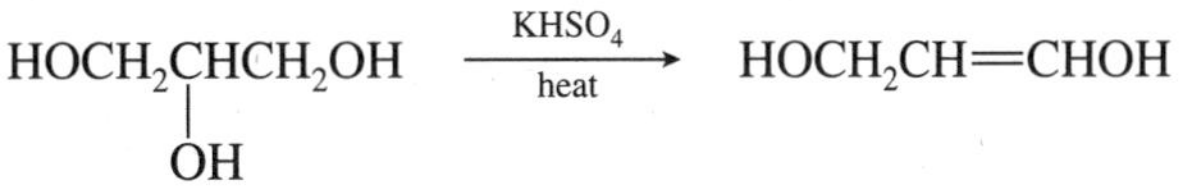

$$\underset{\text{1,2,3-Propanetriol}}{HOCH_2CH(OH)CH_2OH} \xrightarrow[\text{heat}]{KHSO_4} \underset{\text{Propene-1,3-diol}}{HOCH_2CH{=}CHOH}$$

The product of this dehydration is an enol, which tautomerizes to an aldehyde. The aldehyde then undergoes dehydration to form acrolein.

$$\underset{\text{Propene-1,3-diol}}{HOCH_2CH{=}CHOH} \longrightarrow \underset{\text{3-Hydroxypropanal}}{HOCH_2CH_2CH{=}O} \xrightarrow[\text{heat}\ (-H_2O)]{KHSO_4} \underset{\text{Acrolein}}{H_2C{=}CHCH{=}O}$$

**18.21** (*a*) 2-Methylpropanal has the greater enol content.

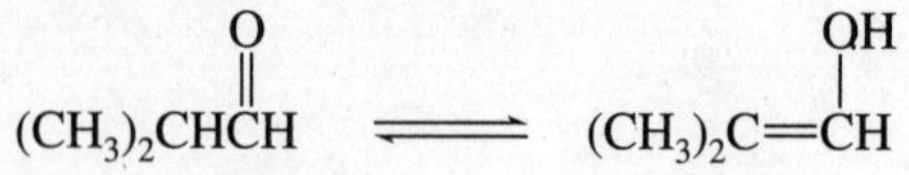

$$(CH_3)_2CH\overset{\overset{\displaystyle O}{\|}}{C}H \rightleftharpoons (CH_3)_2C{=}\overset{\overset{\displaystyle OH}{|}}{C}H$$

2-Methylpropanal — Enol form

Although the enol content of 2-methylpropanal is quite small, the compound is nevertheless capable of enolization, whereas the other compound, 2,2-dimethylpropanal, cannot enolize—it has no $\alpha$ hydrogens.

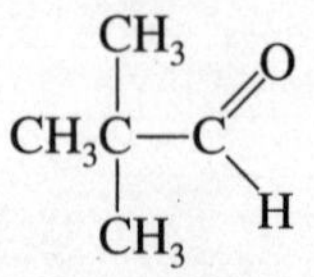

$$CH_3\overset{\overset{\displaystyle CH_3}{|}}{\underset{\underset{\displaystyle CH_3}{|}}{C}}{-}CH{=}O$$

(Enolization is impossible.)

(*b*) Benzophenone has no $\alpha$ hydrogens; it cannot form an enol.

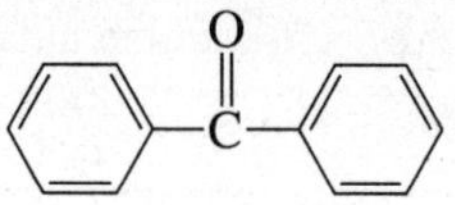

$$C_6H_5\overset{\overset{\displaystyle O}{\|}}{C}C_6H_5$$

(Enolization is impossible.)

Dibenzyl ketone enolizes slightly to form a small amount of enol.

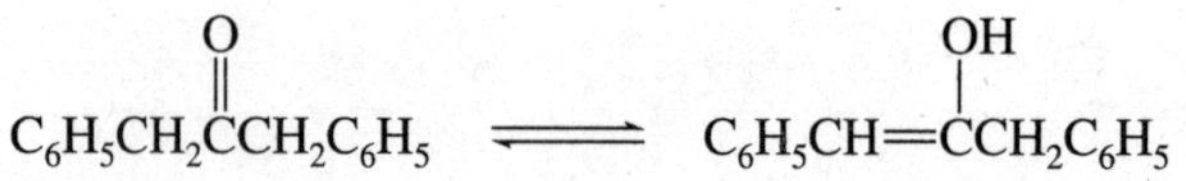

$$C_6H_5CH_2\overset{\overset{\displaystyle O}{\|}}{C}CH_2C_6H_5 \rightleftharpoons C_6H_5CH{=}\overset{\overset{\displaystyle OH}{|}}{C}CH_2C_6H_5$$

Dibenzyl ketone — Enol form

(*c*) Here we are comparing a simple ketone, dibenzyl ketone, with a $\beta$-diketone. The $\beta$-diketone enolizes to a much greater extent than the simple ketone because its enol form is stabilized by conjugation of the double bond with the remaining carbonyl group and by intramolecular hydrogen bonding.

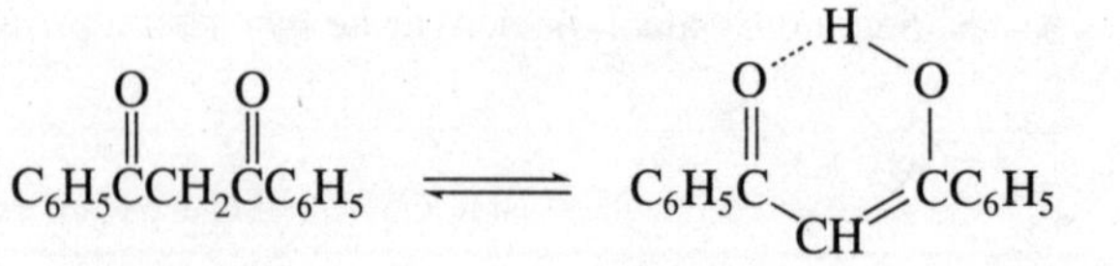

$$C_6H_5\overset{\overset{\displaystyle O}{\|}}{C}CH_2\overset{\overset{\displaystyle O}{\|}}{C}C_6H_5 \rightleftharpoons C_6H_5\overset{\overset{\displaystyle O\cdots H}{\|}}{C}{-}CH{=}\overset{\overset{\displaystyle O}{|}}{C}C_6H_5$$

1,3-Diphenyl-1,3-propanedione — Enol form

(*d*) The enol content of cyclohexanone is quite small, whereas the enol form of 2,4-cyclohexadienone is the aromatic compound phenol, and therefore enolization is essentially complete.

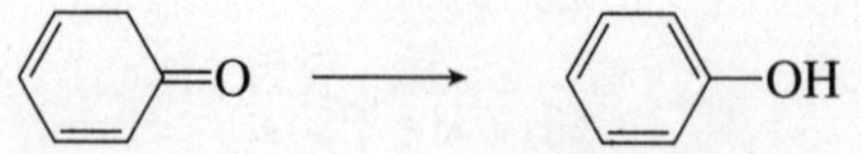

$$\text{2,4-cyclohexadienone} \longrightarrow C_6H_5OH$$

Keto form — Enol form (aromatic; much more stable)

(*e*) A small amount of enol is in equilibrium with cyclopentanone.

Cyclopentanone ⇌ Enol form

Cyclopentadienone does not form a stable enol. Enolization would lead to a highly strained allene-type compound.

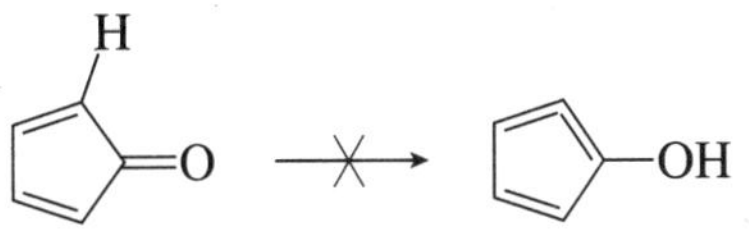

(Not stable; highly strained)

(*f*) The $\beta$-diketone is more extensively enolized.

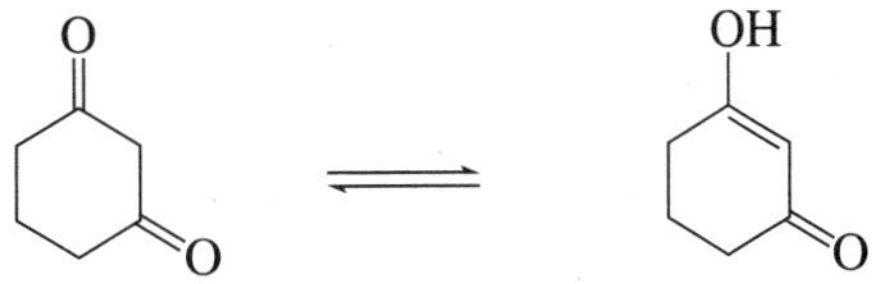

1,3-Cyclohexanedione ⇌ Enol form (double bond conjugated with carbonyl group)

The double bond of the enol form of 1,4-cyclohexanedione is not conjugated with the carbonyl group. Its enol content is expected to be similar to that of cyclohexanone.

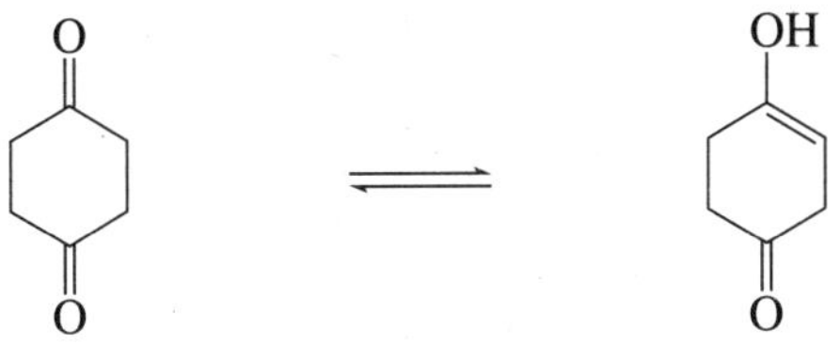

1,4-Cyclohexanedione ⇌ Enol form (not particularly stable; double bond and carbonyl group not conjugated)

**18.22** (*a*) Chlorination of 3-phenylpropanal under conditions of acid catalysis occurs via the enol form and yields the $\alpha$-chloro derivative.

$$C_6H_5CH_2CH_2CH{=}O + Cl_2 \xrightarrow{\text{acetic acid}} C_6H_5CH_2CH(Cl)CH{=}O + HCl$$

3-Phenylpropanal → 2-Chloro-3-phenylpropanal

(*b*) Aldehydes undergo aldol addition on treatment with base.

$$2C_6H_5CH_2CH_2CH{=}O \xrightarrow[\text{ethanol, 10°C}]{\text{NaOH}} C_6H_5CH_2CH_2CH(OH)CH(HC{=}O)CH_2C_6H_5$$

3-Phenylpropanal → 2-Benzyl-3-hydroxy-5-phenylpentanal

(*c*) Dehydration of the aldol addition product occurs when the reaction is carried out at elevated temperature.

$$2C_6H_5CH_2CH_2\overset{O}{\overset{\|}{C}}H \xrightarrow[\text{ethanol, 70°C}]{\text{NaOH}} C_6H_5CH_2CH_2CH{=}\overset{HC=O}{\overset{|}{C}}CH_2C_6H_5$$

3-Phenylpropanal → 2-Benzyl-5-phenyl-2-pentenal

(*d*) Lithium aluminum hydride reduces the aldehyde function to the corresponding primary alcohol.

$$C_6H_5CH_2CH_2CH{=}\overset{HC=O}{\overset{|}{C}}CH_2C_6H_5 \xrightarrow[\text{2. } H_2O]{\text{1. } LiAlH_4} C_6H_5CH_2CH_2CH{=}\overset{CH_2OH}{\overset{|}{C}}CH_2C_6H_5$$

2-Benzyl-5-phenyl-2-pentenal → 2-Benzyl-5-phenyl-2-penten-1-ol

(*e*) A characteristic reaction of $\alpha,\beta$-unsaturated carbonyl compounds is their tendency to undergo conjugate addition on treatment with weakly basic nucleophiles.

$$C_6H_5CH_2CH_2CH{=}\overset{HC=O}{\overset{|}{C}}CH_2C_6H_5 \xrightarrow[H^+]{\text{NaCN}} C_6H_5CH_2CH_2\underset{CN}{\underset{|}{C}H}\overset{HC=O}{\overset{|}{C}}HCH_2C_6H_5$$

2-Benzyl-5-phenyl-2-pentenal → 2-Benzyl-3-cyano-5-phenylpentanal

**18.23** (*a*) Ketones undergo $\alpha$ halogenation by way of their enol form.

$$\text{1-(o-Chlorophenyl)-1-propanone (}o\text{-}ClC_6H_4\overset{O}{\overset{\|}{C}}CH_2CH_3) \xrightarrow[CH_2Cl_2]{Cl_2} o\text{-}ClC_6H_4\overset{O}{\overset{\|}{C}}\underset{Cl}{\underset{|}{C}H}CH_3$$

1-(*o*-Chlorophenyl)-1-propanone → 2-Chloro-1-(*o*-chlorophenyl)-1-propanone

(*b*) The combination of $C_6H_5CH_2SH$ and NaOH yields $C_6H_5CH_2S^-$ (as its sodium salt), which is a weakly basic nucleophile and adds to $\alpha,\beta$-unsaturated ketones by conjugate addition.

$$\text{2-Isopropylidene-5-methylcyclohexanone} \xrightarrow[NaOH,\ H_2O]{C_6H_5CH_2SH} \text{2-(1-Benzylthio-1-methylethyl)-5-methylcyclohexanone}$$

Reactant: $H_3C$–cyclohexanone with $=C(CH_3)_2$; product: $H_3C$–cyclohexanone with H and $C(CH_3)_2SCH_2C_6H_5$

2-Isopropylidene-5-methylcyclohexanone → 2-(1-Benzylthio-1-methylethyl)-5-methylcyclohexanone (89–90%)

(*c*) Bromination occurs at the carbon atom that is $\alpha$ to the carbonyl group.

$$\text{2,2-Diphenylcyclopentanone} \xrightarrow[\text{diethyl ether}]{Br_2} \text{2-Bromo-5,5-diphenylcyclopentanone}$$

2,2-Diphenylcyclopentanone → 2-Bromo-5,5-diphenylcyclopentanone (76%)

(*d*) The reaction is a mixed aldol condensation. The enolate of 2,2-diphenylcyclohexanone reacts with *p*-chlorobenzaldehyde. Elimination of the aldol addition product occurs readily to yield the $\alpha,\beta$-unsaturated ketone as the isolated product.

*p*-Chlorobenzaldehyde + 2,2-Diphenylcyclohexanone $\xrightarrow[\text{ethanol}]{\text{KOH}}$ (Not isolated) $\xrightarrow{-H_2O}$ 2-(*p*-Chlorobenzylidene)-6,6-diphenylcyclohexanone (84%)

(*e*) The aldehyde given as the starting material is called **furfural** and is based on a furan unit as an aromatic ring. Furfural cannot form an enolate. It reacts with the enolate of acetone in a manner much as benzaldehyde would.

Furfural + $CH_3CCH_3$ (Acetone) $\xrightarrow[\text{water}]{\text{NaOH}}$ $CHCH_2CCH_3$ with OH (Not isolated) $\xrightarrow{-H_2O}$ $CH{=}CHCCH_3$ 4-Furyl-3-buten-2-one (60–66%)

(*f*) Lithium dialkylcuprates transfer an alkyl group to the $\beta$-carbon atom of $\alpha,\beta$-unsaturated ketones.

2,4,4-Trimethyl-2-cyclohexenone + $LiCu(CH_3)_2$ $\xrightarrow[\text{2. } H_2O]{\text{1. diethyl ether}}$ 2,3,4,4-Tetramethylcyclohexanone

A mixture of stereoisomers was obtained in 67% yield in this reaction.

(*g*) Two nonequivalent $\alpha$-carbon atoms occur in the starting ketone. Although enolate formation is possible at either position, only reaction at the methylene carbon leads to an intermediate that can undergo dehydration.

$\xrightarrow[C_6H_5CHO]{HO^-}$ $CHC_6H_5$ with OH $\xrightarrow{-H_2O}$ $CHC_6H_5$

Observed product (75% yield)

Reaction at the other $\alpha$ position gives an intermediate that cannot dehydrate.

$$\text{(decalone)} \underset{}{\overset{HO^-,\ C_6H_5CHO}{\rightleftharpoons}} \text{(aldol adduct with } C_6H_5CH(OH)\text{ at ring-fusion carbon)}$$

(Cannot dehydrate; reverts to starting materials)

(*h*) $\beta$-Diketones readily undergo alkylation by primary halides at the most acidic position, on the carbon between the carbonyls.

$$\text{1,3-Cyclohexanedione} + H_2C{=}CHCH_2Br \xrightarrow{KOH} \text{2-}(CH_2CH{=}CH_2)\text{-1,3-cyclohexanedione}$$

1,3-Cyclohexanedione; Allyl bromide; 2-Allyl-1,3-cyclohexanedione (75%)

**18.24** (*a*) Conversion of 3-pentanone to 2-bromo-3-pentanone is best accomplished by acid-catalyzed bromination via the enol. Bromine in acetic acid is the customary reagent for this transformation.

$$CH_3CH_2\overset{O}{\overset{\|}{C}}CH_2CH_3 \xrightarrow[\text{acetic acid}]{Br_2} CH_3\underset{Br}{\underset{|}{C}}H\overset{O}{\overset{\|}{C}}CH_2CH_3$$

3-Pentanone; 2-Bromo-3-pentanone

(*b*) Once 2-bromo-3-pentanone has been prepared, its dehydrohalogenation by base converts it to the desired $\alpha,\beta$-unsaturated ketone 1-penten-3-one.

$$CH_3\underset{Br}{\underset{|}{C}}H\overset{O}{\overset{\|}{C}}CH_2CH_3 \xrightarrow{KOC(CH_3)_3} H_2C{=}CH\overset{O}{\overset{\|}{C}}CH_2CH_3$$

2-Bromo-3-pentanone; 1-Penten-3-one

Potassium *tert*-butoxide is a good base for bringing about elimination reactions of secondary alkyl halides; suitable solvents include *tert*-butyl alcohol and dimethyl sulfoxide.

(*c*) Reduction of the carbonyl group of 1-penten-3-one converts it to the desired alcohol.

$$H_2C{=}CH\overset{O}{\overset{\|}{C}}CH_2CH_3 \xrightarrow[\text{2. } H_2O]{\text{1. } LiAlH_4\text{, diethyl ether}} H_2C{=}CH\underset{OH}{\underset{|}{C}}HCH_2CH_3$$

1-Penten-3-one; 1-Penten-3-ol

Catalytic hydrogenation would not be suitable for this reaction because reduction of the double bond would accompany carbonyl reduction.

(*d*) Conversion of 3-pentanone to 3-hexanone requires addition of a methyl group to the $\beta$-carbon atom.

$$CH_3CH_2CH_2\overset{O}{\overset{\|}{C}}CH_2CH_3 \Rightarrow {}^{-}{:}CH_3 + H_2\overset{\beta}{C}{=}CH\overset{O}{\overset{\|}{C}}CH_2CH_3$$

The best way to add an alkyl group to the $\beta$ carbon of a ketone is via conjugate addition of a dialkylcuprate reagent to an $\alpha,\beta$-unsaturated ketone.

$$H_2C{=}CH\overset{O}{\overset{\|}{C}}CH_2CH_3 \xrightarrow[2.\ H_2O]{1.\ LiCu(CH_3)_2} CH_3CH_2CH_2\overset{O}{\overset{\|}{C}}CH_2CH_3$$

1-Penten-3-one [prepared as described in part (*b*)] → 3-Hexanone

(*e*) The compound to be prepared is the mixed aldol condensation product of 3-pentanone and benzaldehyde.

$$\underset{\underset{C_6H_5CH}{\|}}{CH_3C}\overset{O}{\overset{\|}{C}}CH_2CH_3 \Rightarrow CH_3\overset{..}{\bar{C}}H\overset{O}{\overset{\|}{C}}CH_2CH_3 + C_6H_5\underset{\underset{O}{\|}}{C}H$$

2-Methyl-1-phenyl-1-penten-3-one

The desired reaction sequence is

$$CH_3CH_2\overset{O}{\overset{\|}{C}}CH_2CH_3 \xrightarrow{HO^-} CH_3\overset{..}{\bar{C}}H\overset{O}{\overset{\|}{C}}CH_2CH_3 \xrightarrow{C_6H_5\overset{O}{\overset{\|}{C}}H} \underset{\underset{C_6H_5CHOH}{|}}{CH_3CH}\overset{O}{\overset{\|}{C}}CH_2CH_3 \xrightarrow{-H_2O} \underset{\underset{C_6H_5CH}{\|}}{CH_3C}\overset{O}{\overset{\|}{C}}CH_2CH_3$$

3-Pentanone; Enolate of 3-pentanone; Aldol addition product (not isolated; dehydration occurs under conditions of its formation); 2-Methyl-1-phenyl-1-penten-3-one

**18.25** (*a*) The first step is an $\alpha$ halogenation of a ketone. This is customarily accomplished under conditions of acid catalysis.

$$(CH_3)_3C\overset{O}{\overset{\|}{C}}CH_3 \xrightarrow[H^+]{Br_2} (CH_3)_3C\overset{O}{\overset{\|}{C}}CH_2Br$$

3,3-Dimethyl-2-butanone → 1-Bromo-3,3-dimethyl-2-butanone (58%)

In the second step the carbonyl group of the $\alpha$-bromo ketone is reduced to a secondary alcohol. As actually carried out, sodium borohydride in water was used to achieve this transformation.

$$(CH_3)_3C\overset{O}{\overset{\|}{C}}CH_2Br \xrightarrow[H_2O]{NaBH_4} (CH_3)_3C\overset{OH}{\overset{|}{C}}HCH_2Br$$

1-Bromo-3,3-dimethyl-2-butanone → 1-Bromo-3,3-dimethyl-2-butanol (54%)

The third step is conversion of a vicinal bromohydrin to an epoxide in aqueous base.

$$(CH_3)_3CCH(OH)CH_2Br \xrightarrow[H_2O]{KOH} (CH_3)_3C\overset{O}{C}H{-}CH_2$$

1-Bromo-3,3-dimethyl-2-butanol → 2-*tert*-Butyloxirane (68%)

(*b*) The overall yield is the product of the yields of the individual steps.

$$\text{Yield} = 100(0.58 \times 0.54 \times 0.68)$$
$$= 21\%$$

**18.26** The product is a sulfide (thioether). Retrosynthetic analysis reveals a pathway that begins with benzene and acetic anhydride.

$$C_6H_5\overset{O}{\overset{\|}{C}}CH_2SCH_2CH_2CH_3 \Longrightarrow C_6H_5\overset{O}{\overset{\|}{C}}CH_2Br + {}^{-}{:}\ddot{S}CH_2CHCH_2CH_3$$

$$C_6H_5\overset{O}{\overset{\|}{C}}CH_3 \Longrightarrow CH_3\overset{O}{\overset{\|}{C}}O\overset{O}{\overset{\|}{C}}CH_3 + C_6H_6$$

The desired synthesis can be accomplished with the following series of reactions:

$$C_6H_6 + CH_3\overset{O}{\overset{\|}{C}}O\overset{O}{\overset{\|}{C}}CH_3 \xrightarrow{AlCl_3} C_6H_5\overset{O}{\overset{\|}{C}}CH_3 \xrightarrow[H^+]{Br_2} C_6H_5\overset{O}{\overset{\|}{C}}CH_2Br$$

Benzene, Acetic anhydride → Acetophenone → Bromomethyl phenyl ketone

The synthesis is completed by reacting bromomethyl phenyl ketone with 1-propanethiolate anion.

$$CH_3CH_2CH_2SH \xrightarrow{KOH} CH_3CH_2CH_2\ddot{S}{:}^{-} \xrightarrow{C_6H_5\overset{O}{\overset{\|}{C}}CH_2Br} C_6H_5\overset{O}{\overset{\|}{C}}CH_2SCH_2CH_2CH_3$$

1-Propanethiol → 1-Propanethiolate → Phenyl (1-thiopropyl)-methyl ketone

**18.27** All these problems begin in the same way, with exchange of all the $\alpha$ protons for deuterium (Section 18.8).

$$\text{Cyclopentanone (2,2,5,5-}H_4) \xrightarrow[D_2O]{NaOD} \text{Cyclopentanone-2,2,5,5-}d_4$$

Cyclopentanone → Cyclopentanone-2,2,5,5-$d_4$

Once the tetradeuterated cyclopentanone has been prepared, functional group transformations are employed to convert it to the desired products.

(*a*) Reduction of the carbonyl group can be achieved by using any of the customary reagents.

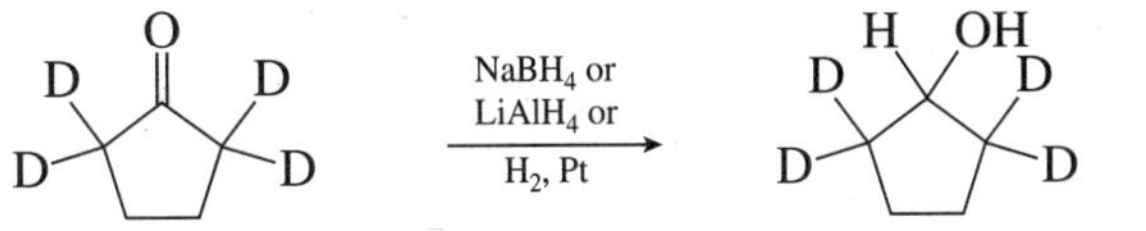

Cyclopentanone-2,2,5,5-$d_4$ $\xrightarrow[H_2,\ Pt]{NaBH_4\ \text{or}\ LiAlH_4\ \text{or}}$ Cyclopentanol-2,2,5,5-$d_4$

(*b*) Acid-catalyzed dehydration of the alcohol prepared in part (*a*) yields the desired alkene.

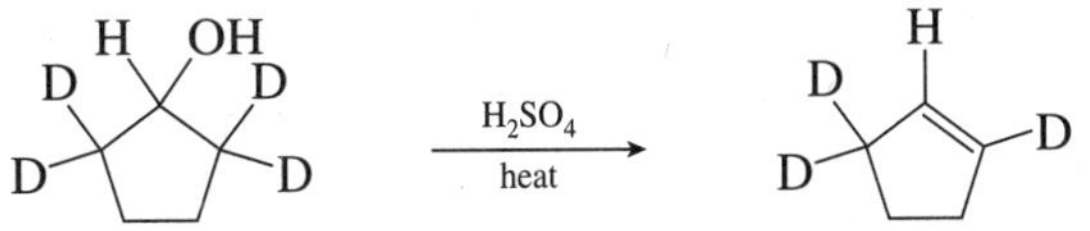

Cyclopentanol-2,2,5,5-$d_4$ $\xrightarrow[heat]{H_2SO_4}$ Cyclopentene-1,3,3-$d_3$

(*c*) Catalytic hydrogenation of the alkene in part (*b*) yields cyclopentane-1,1,3-$d_3$.

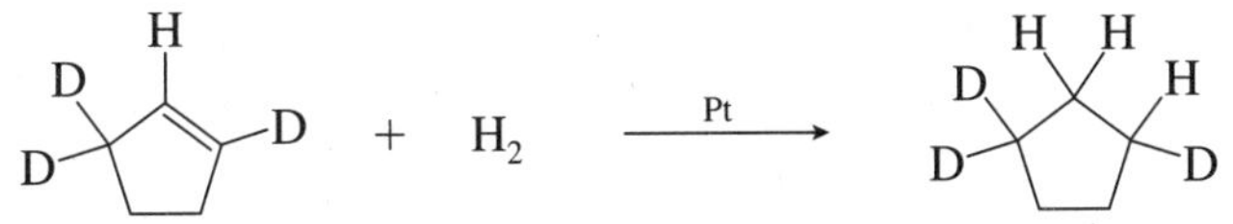

Cyclopentene-1,3,3-$d_3$ + $H_2$ (Hydrogen) $\xrightarrow{Pt}$ Cyclopentane-1,1,3-$d_3$

(*d*) Carbonyl reduction of the tetradeuterated ketone under Wolff–Kishner conditions furnishes the desired product.

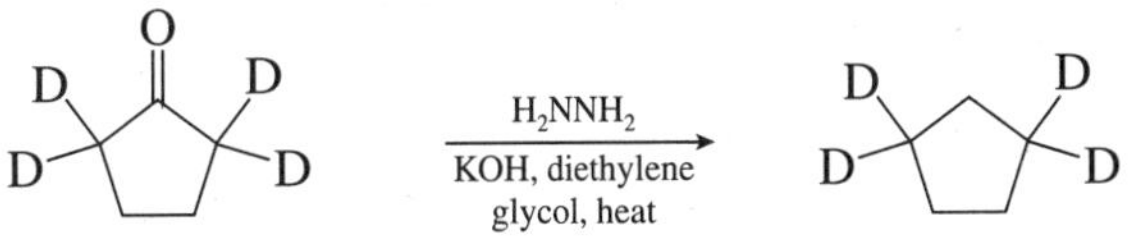

Cyclopentanone-2,2,5,5-$d_4$ $\xrightarrow[\text{KOH, diethylene glycol, heat}]{H_2NNH_2}$ Cyclopentane-1,1,3,3-$d_4$

Alternatively, Clemmensen reduction conditions (Zn, HCl) could be used.

**18.28** (*a*) Hydroformylation converts alkenes to aldehydes having one more carbon atom by reaction with carbon monoxide and hydrogen in the presence of a cobalt octacarbonyl catalyst.

$$CH_3CH{=}CH_2 + CO + H_2 \xrightarrow{Co_2(CO)_8} CH_3CH_2CH_2\overset{O}{\overset{\|}{C}}H$$

Propene, Carbon monoxide, Hydrogen → Butanal

(*b*) Aldol condensation of acetaldehyde to 2-butenal, followed by catalytic hydrogenation of the carbon–carbon double bond, gives butanal.

$$2CH_3\overset{O}{\overset{\|}{C}}H \xrightarrow[heat]{NaOH} CH_3CH{=}CH\overset{O}{\overset{\|}{C}}H \xrightarrow[Ni]{H_2} CH_3CH_2CH_2\overset{O}{\overset{\|}{C}}H$$

Acetaldehyde → 2-Butenal → Butanal

**18.29** (*a*) The first conversion is the $\alpha$ halogenation of an aldehyde. As described in Section 18.2, this particular conversion has been achieved in 80% yield simply by treatment with bromine in chloroform.

Cyclohexanecarbaldehyde $\xrightarrow[CHCl_3]{Br_2}$ 1-Bromocyclohexanecarbaldehyde

Cyclohexane-carbaldehyde

1-Bromocyclohexane-carbaldehyde

Dehydrohalogenation of this compound can be accomplished under E2 conditions by treatment with base. Sodium methoxide in methanol would be appropriate, for example, although almost any alkoxide could be employed to dehydrohalogenate this tertiary bromide.

1-Bromocyclohexanecarbaldehyde $\xrightarrow[CH_3OH]{NaOCH_3}$ Cyclohexene-1-carbaldehyde

1-Bromocyclohexane-carbaldehyde

Cyclohexene-1-carbaldehyde

As the reaction was actually carried out, the bromide was heated with the weak base *N,N*-diethylaniline to effect dehydrobromination in 71% yield.

(*b*) Cleavage of vicinal diols to carbonyl compounds can be achieved by using periodic acid ($HIO_4$) (Section 15.12).

*trans*-1,2-Cyclohexanediol $\xrightarrow{HIO_4}$ 1,6-Hexanedial

*trans*-1,2-Cyclohexanediol

1,6-Hexanedial

The conversion of this dialdehyde to cyclopentene-1-carbaldehyde is an intramolecular aldol condensation and is achieved by treatment with potassium hydroxide.

1,6-Hexanedial $\overset{HO^-}{\rightleftharpoons}$ enolate $\longrightarrow$ 2-hydroxycyclopentanecarbaldehyde $\xrightarrow{-H_2O}$ Cyclopentene-1-carbaldehyde

Cyclopentene-1-carbaldehyde

As the reaction was actually carried out, cyclopentene-1-carbaldehyde was obtained in 58% yield from *trans*-1,2-cyclohexanediol by this method.

(*c*) The first transformation requires an oxidative cleavage of a carbon–carbon double bond. Ozonolysis followed by hydrolysis in the presence of zinc is indicated.

4-Isopropyl-1-methylcyclohexene $\xrightarrow[2.\ H_2O,\ Zn]{1.\ O_3}$ 3-Isopropyl-6-oxoheptanal

4-Isopropyl-1-methylcyclohexene

3-Isopropyl-6-oxoheptanal

Cyclization of the resulting keto aldehyde is an intramolecular aldol condensation. Base is required.

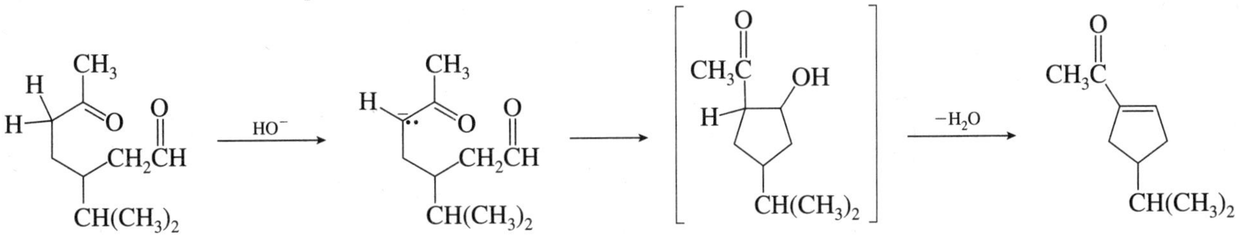

$\xrightarrow{HO^-}$ $\longrightarrow$ $\xrightarrow{-H_2O}$

$CH_3$, $CH(CH_3)_2$, $CH_2CH$, $CH_3C$, OH

(*d*) The first step in this synthesis is the hydration of the alkene function to an alcohol. Notice that this hydration must take place with a regioselectivity opposite to that of Markovnikov's rule and therefore requires a hydroboration–oxidation sequence.

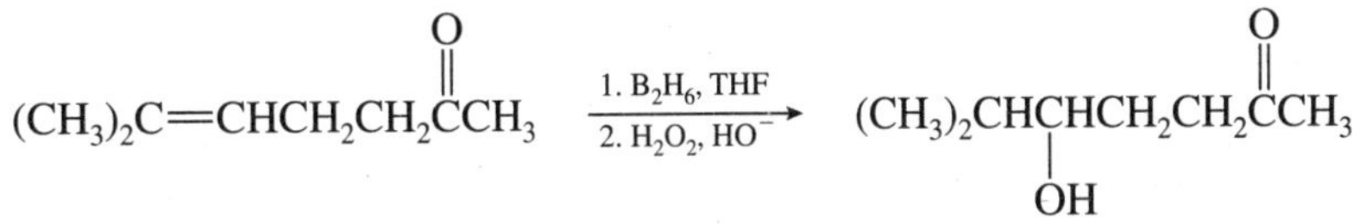

$$(CH_3)_2C{=}CHCH_2CH_2\overset{O}{\overset{\|}{C}}CH_3 \xrightarrow[2.\ H_2O_2,\ HO^-]{1.\ B_2H_6,\ THF} (CH_3)_2CH\underset{OH}{\underset{|}{C}}HCH_2CH_2\overset{O}{\overset{\|}{C}}CH_3$$

6-Methyl-5-hepten-2-one — 5-Hydroxy-6-methyl-2-heptanone

Conversion of the secondary alcohol function to a carbonyl group can be achieved with any of a number of oxidizing agents.

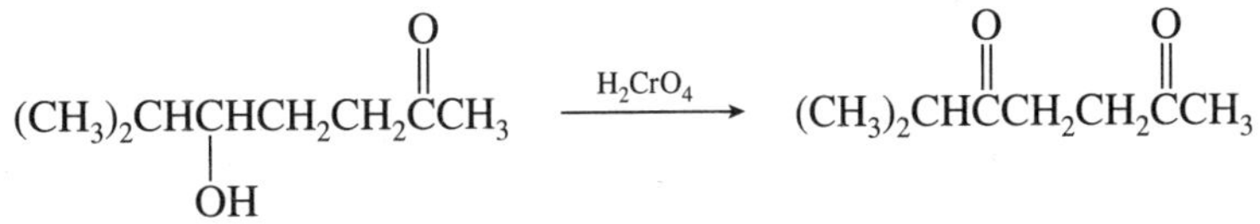

$$(CH_3)_2CH\underset{OH}{\underset{|}{C}}HCH_2CH_2\overset{O}{\overset{\|}{C}}CH_3 \xrightarrow{H_2CrO_4} (CH_3)_2CH\overset{O}{\overset{\|}{C}}CH_2CH_2\overset{O}{\overset{\|}{C}}CH_3$$

5-Hydroxy-6-methyl-2-heptanone — 6-Methyl-2,5-heptanedione

Cyclization of the dione to the final product is a base-catalyzed intramolecular aldol condensation and was accomplished in 71% yield by treatment of the dione with a 2% solution of sodium hydroxide in aqueous ethanol.

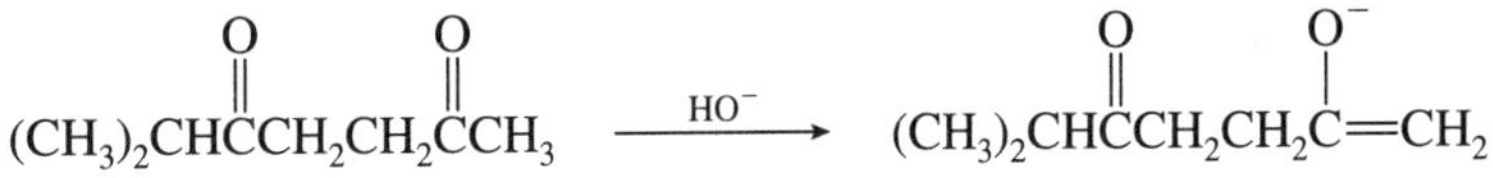

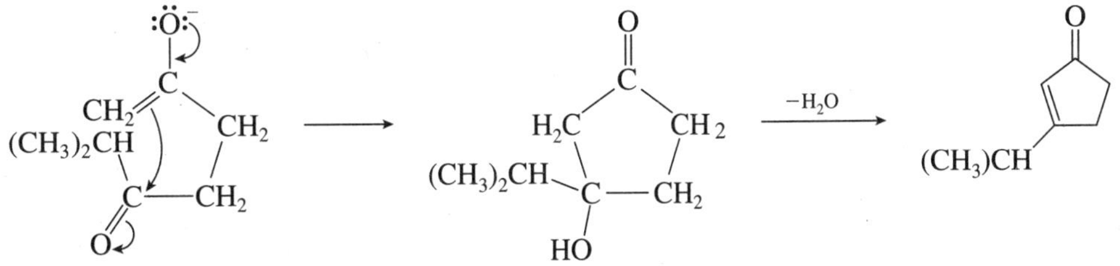

$$(CH_3)_2CH\overset{O}{\overset{\|}{C}}CH_2CH_2\overset{O}{\overset{\|}{C}}CH_3 \xrightarrow{HO^-} (CH_3)_2CH\overset{O}{\overset{\|}{C}}CH_2CH_2\overset{O^-}{\overset{|}{C}}{=}CH_2$$

$\longrightarrow$ $\xrightarrow{-H_2O}$

$(CH_3)_2CH$, $CH_2$, $H_2C$, HO, $(CH_3)CH$

**18.30** Intramolecular aldol condensations occur best when a five- or six-membered ring is formed. Carbon–carbon bond formation therefore involves the aldehyde and the methyl group attached to the ketone carbonyl.

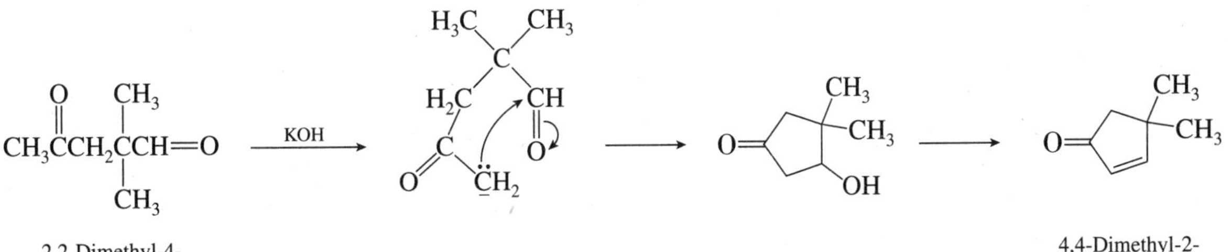

$$CH_3\overset{O}{\overset{\|}{C}}CH_2\overset{CH_3}{\underset{CH_3}{C}}CH{=}O \xrightarrow{KOH} \longrightarrow \longrightarrow$$

$H_3C$, $CH_3$, $H_2C$, CH, $CH_2$, OH

2,2-Dimethyl-4-oxopentanal — 4,4-Dimethyl-2-cyclopentenone (63%)

**18.31** (*a*) By realizing that the primary alcohol function of the target molecule can be introduced by reduction of an aldehyde, it can be seen that the required carbon skeleton is the same as that of the aldol addition product of 2-methylpropanal.

$$(CH_3)_2CHCH(OH)C(CH_3)_2CH_2OH \Longrightarrow (CH_3)_2CHCH(OH)\text{-}\!\!\!\not\;C(CH_3)_2CHO \Longrightarrow 2(CH_3)_2CHCHO$$

The synthetic sequence is

$$(CH_3)_2CHCHO \xrightarrow[\text{ethanol}]{\text{NaOH}} (CH_3)_2CHCH(OH)C(CH_3)_2CHO \xrightarrow[CH_3OH]{NaBH_4} (CH_3)_2CHCH(OH)C(CH_3)_2CH_2OH$$

2-Methylpropanal → 3-Hydroxy-2,2,4-trimethylpentanal → 2,2,4-Trimethyl-1,3-pentanediol

The starting aldehyde is prepared by oxidation of 2-methyl-1-propanol.

$$(CH_3)_2CHCH_2OH \xrightarrow[CH_2Cl_2]{PCC} (CH_3)_2CHCHO$$

2-Methyl-1-propanol → 2-Methylpropanal

(*b*) Retrosynthetic analysis of the desired product shows that the carbon skeleton can be constructed by a mixed aldol condensation between benzaldehyde and propanal.

$$C_6H_5CH{=}C(CH_3)CH_2OH \Longrightarrow C_6H_5CH{=}C(CH_3)CHO \Longrightarrow C_6H_5CHO + CH_3CH_2CHO$$

The reaction scheme therefore becomes

$$C_6H_5CHO + CH_3CH_2CHO \xrightarrow{HO^-} C_6H_5CH{=}C(CH_3)CHO$$

Benzaldehyde + Propanal → 2-Methyl-3-phenyl-2-propenal

Reduction of the aldehyde to the corresponding primary alcohol gives the desired compound.

$$C_6H_5CH{=}C(CH_3)CHO \xrightarrow[\text{or } NaBH_4,\ CH_3OH]{LiAlH_4,\ \text{then } H_2O} C_6H_5CH{=}C(CH_3)CH_2OH$$

2-Methyl-3-phenyl-2-propenal → 2-Methyl-3-phenyl-2-propen-1-ol

The starting materials for the mixed aldol condensation—benzaldehyde and propanal—are prepared by oxidation of benzyl alcohol and 1-propanol, respectively.

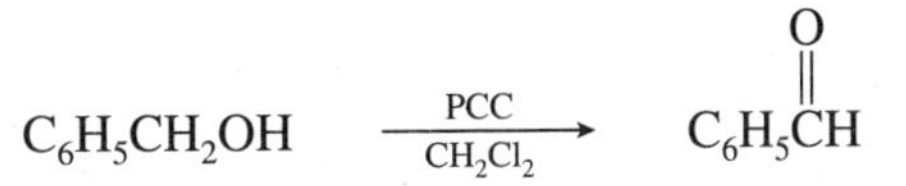

$$C_6H_5CH_2OH \xrightarrow[CH_2Cl_2]{PCC} C_6H_5\overset{O}{\overset{\|}{C}}H$$

Benzyl alcohol → Benzaldehyde

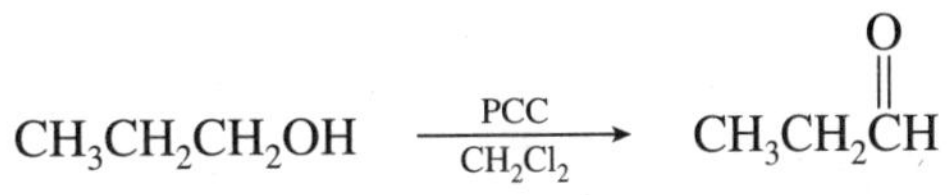

$$CH_3CH_2CH_2OH \xrightarrow[CH_2Cl_2]{PCC} CH_3CH_2\overset{O}{\overset{\|}{C}}H$$

1-Propanol → Propanal

(*c*) The cyclohexene ring in this case can be assembled by a Diels–Alder reaction.

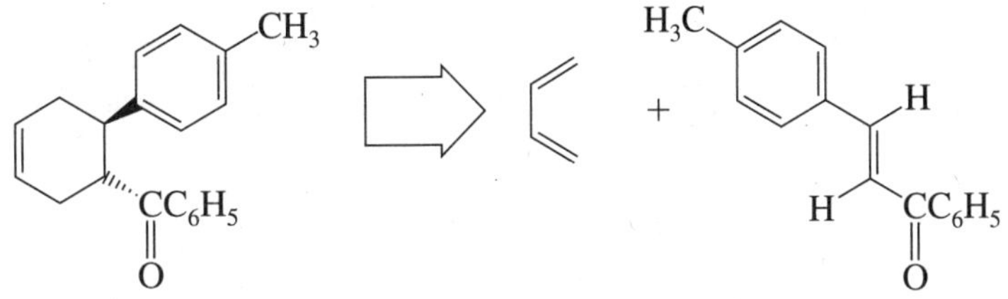

1,3-Butadiene is one of the given starting materials; the $\alpha,\beta$-unsaturated ketone is the mixed aldol condensation product of 4-methylbenzaldehyde and acetophenone.

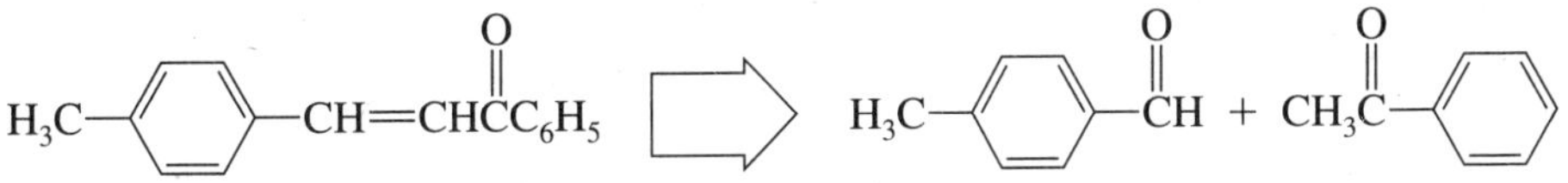

$$H_3C-C_6H_4-CH{=}CH\overset{O}{\overset{\|}{C}}C_6H_5 \Rightarrow H_3C-C_6H_4-\overset{O}{\overset{\|}{C}}H + CH_3\overset{O}{\overset{\|}{C}}-C_6H_5$$

The complete synthetic sequence is

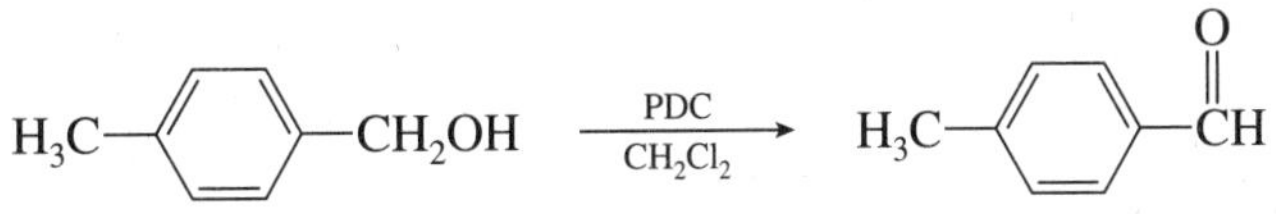

$$H_3C-C_6H_4-CH_2OH \xrightarrow[CH_2Cl_2]{PDC} H_3C-C_6H_4-\overset{O}{\overset{\|}{C}}H$$

4-Methylbenzyl alcohol → 4-Methylbenzaldehyde

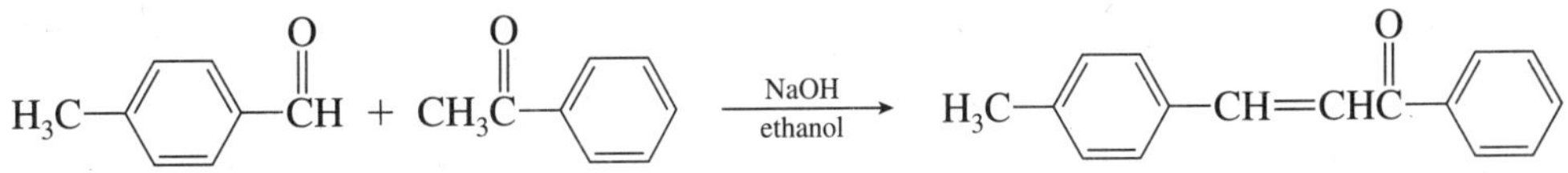

$$H_3C-C_6H_4-\overset{O}{\overset{\|}{C}}H + CH_3\overset{O}{\overset{\|}{C}}-C_6H_5 \xrightarrow[ethanol]{NaOH} H_3C-C_6H_4-CH{=}CH\overset{O}{\overset{\|}{C}}-C_6H_5$$

4-Methylbenzaldehyde, Acetophenone

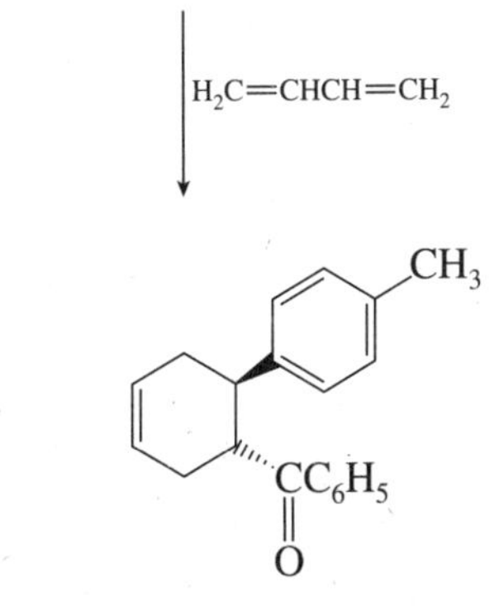

$$\xrightarrow{H_2C{=}CHCH{=}CH_2}$$

*trans*-4-Benzoyl-5-(4-methylphenyl)cyclohexene

$\alpha,\beta$-Unsaturated ketones are good dienophiles in Diels–Alder reactions.

**18.32** It is the carbon atom flanked by two carbonyl groups that is involved in the enolization of terreic acid.

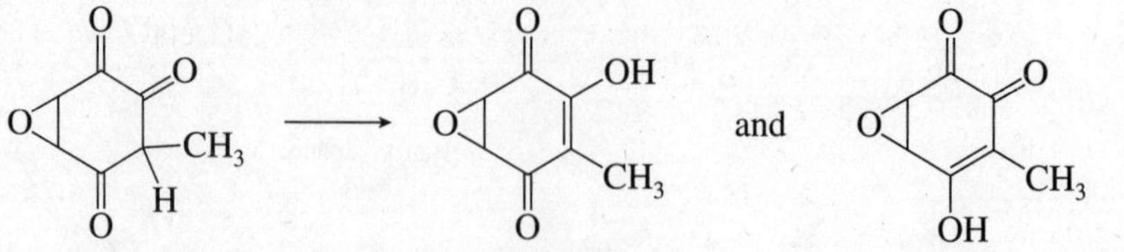

Terreic acid → Enol A and Enol B

Of these two structures, enol A, with its double bond conjugated to two carbonyl groups, is more stable than enol B, in which the double bond is conjugated to only one carbonyl.

**18.33** (*a*) Recall that aldehydes and ketones are in equilibrium with their *hydrates* in aqueous solution (Section 17.6). Thus, the principal substance present when $(C_6H_5)_2CHCH{=}O$ is dissolved in aqueous acid is $(C_6H_5)_2CHCH(OH)_2$ (81%).

(*b*) The problem states that the major species present in aqueous base is *not* $(C_6H_5)_2CHCH{=}O$, its enol, or its hydrate. The most reasonable species is the **enolate ion:**

$$(C_6H_5)_2CHCH{=}O \xrightarrow{-H^+} (C_6H_5)_2\ddot{C}^{-}{-}CH{=}\ddot{O}: \longleftrightarrow (C_6H_5)_2C{=}CH{-}\ddot{O}:^{-}$$

**18.34** (*a*) At first glance this transformation seems to be an internal oxidation–reduction reaction. An aldehyde function is reduced to a primary alcohol, and a secondary alcohol is oxidized to a ketone.

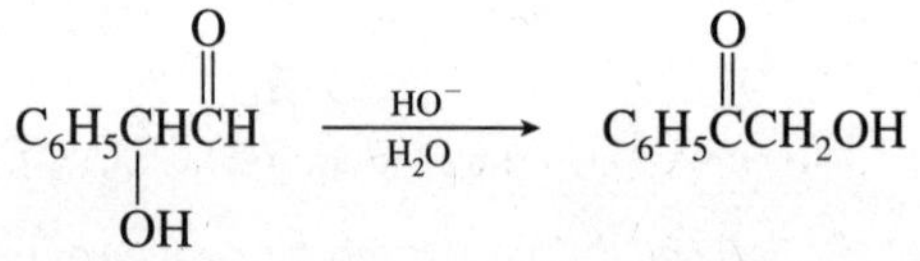

$$\underset{\text{Compound A}}{C_6H_5CH(OH)CH{=}O} \xrightarrow[H_2O]{HO^-} \underset{\text{Compound B}}{C_6H_5C({=}O)CH_2OH}$$

Once one realizes that enolization can occur, however, a simpler explanation, involving only proton-transfer reactions, emerges.

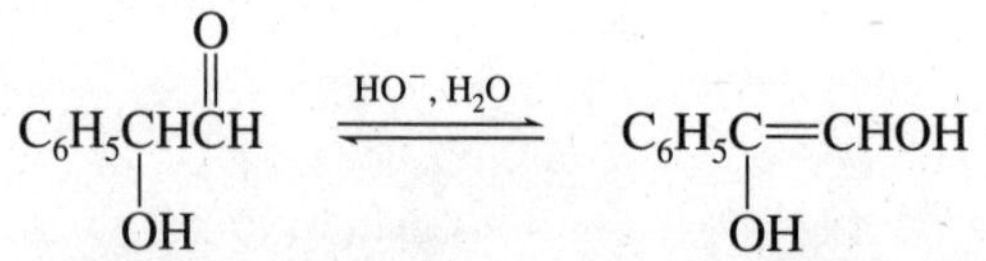

$$\underset{\text{Compound A}}{C_6H_5CH(OH)CH{=}O} \xrightleftharpoons{HO^-,\ H_2O} \underset{\text{Enol form of compound A}}{C_6H_5C(OH){=}CHOH}$$

The enol form of compound A is an enediol; it is at the same time the enol form of compound B. The enediol can revert to compound A or to compound B.

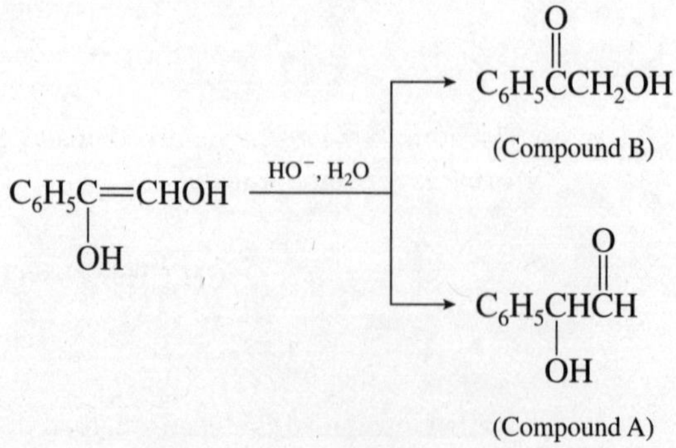

$$C_6H_5C(OH){=}CHOH \xrightarrow{HO^-,\ H_2O} \underset{\text{(Compound B)}}{C_6H_5C({=}O)CH_2OH} \quad\text{or}\quad \underset{\text{(Compound A)}}{C_6H_5CH(OH)CH{=}O}$$

At equilibrium, compound B predominates because it is more stable than A. A ketone carbonyl is more stabilized than an aldehyde, and the carbonyl in B is conjugated with the benzene ring.

(*b*) The isolated product is the double hemiacetal formed between two molecules of compound A.

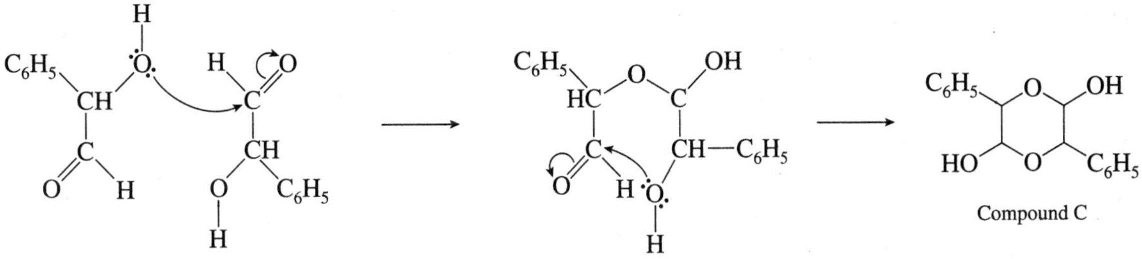

Compound C

**18.35** (*a*) The only stereogenic center in piperitone is adjacent to a carbonyl group. Base-catalyzed enolization causes this carbon to lose its stereochemical integrity.

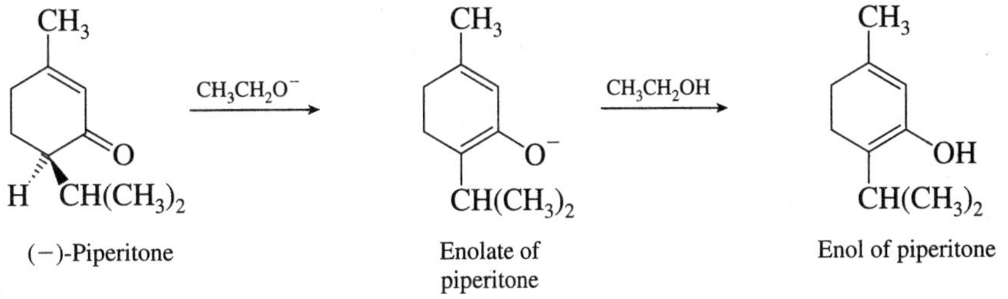

$\xrightarrow{CH_3CH_2O^-}$ $\xrightarrow{CH_3CH_2OH}$

(−)-Piperitone | Enolate of piperitone | Enol of piperitone

Both the enolate and enol of piperitone are achiral and can revert only to a racemic mixture of piperitones.

(*b*) The enol formed from menthone can revert to either menthone or isomenthone.

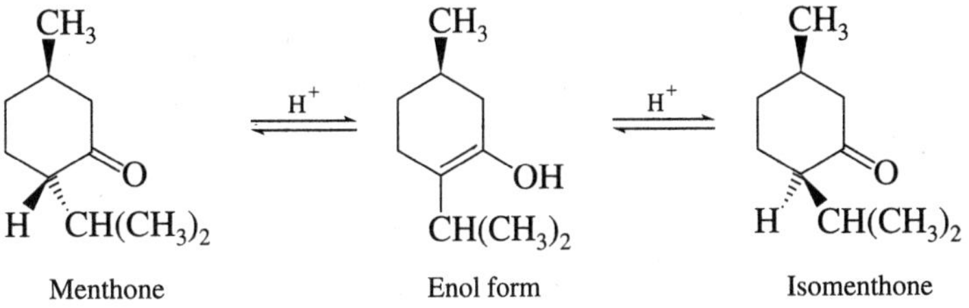

$\overset{H^+}{\rightleftharpoons}$ $\overset{H^+}{\rightleftharpoons}$

Menthone | Enol form | Isomenthone

Only the stereochemistry at the $\alpha$-carbon atom is affected by enolization. The other stereogenic center in menthone (the one bearing the methyl group) is not affected.

**18.36** In all parts of this problem the bonding change that takes place is described by the general equation

$$HX{-}N{=}Z \rightleftharpoons X{=}N{-}ZH$$

(*a*) The compound given is nitrosoethane. Nitrosoalkanes are less stable than their oxime isomers formed by proton transfer.

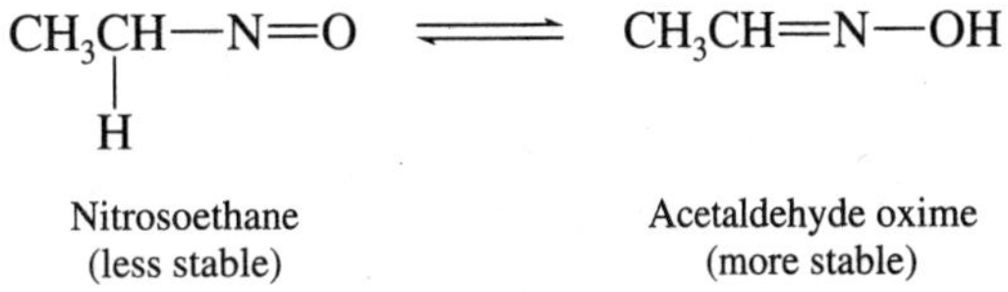

$$CH_3CH_2{-}N{=}O \rightleftharpoons CH_3CH{=}N{-}OH$$

Nitrosoethane (less stable) | Acetaldehyde oxime (more stable)

(*b*) You may recognize this compound as an enamine. It is slightly different, however, from the enamines we discussed earlier (Section 17.11) in that nitrogen bears a hydrogen substituent. Stable enamines are compounds of the type

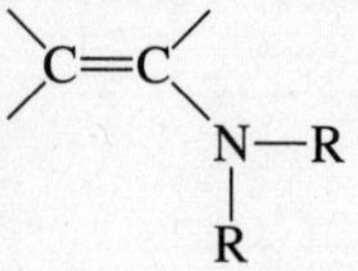

$$\mathrm{>C{=}C<}\ \ \mathrm{N{-}R\ (R)}$$

where neither R group is hydrogen; both R's must be alkyl or aryl. Enamines that bear a hydrogen substituent are converted to imines in a proton-transfer equilibrium.

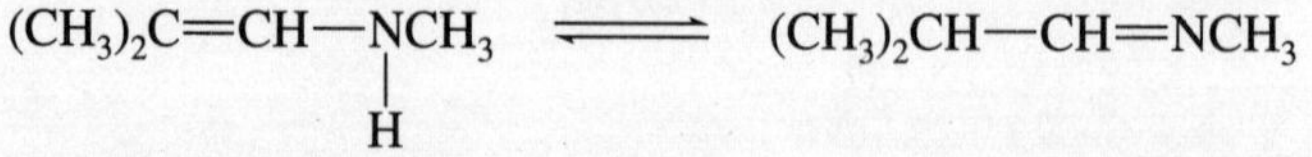

$$(CH_3)_2C{=}CH{-}N(H)CH_3 \rightleftharpoons (CH_3)_2CH{-}CH{=}NCH_3$$

Enamine (less stable) Imine (more stable)

(*c*) The compound given is known as a **nitronic acid;** its more stable tautomeric form is a nitroalkane.

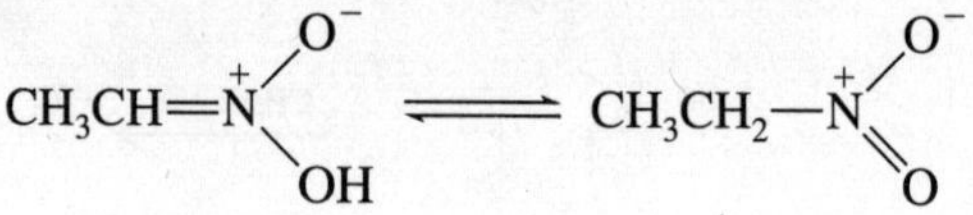

$$CH_3CH{=}\overset{+}{N}(O^-)OH \rightleftharpoons CH_3CH_2{-}\overset{+}{N}(O^-){=}O$$

Nitronic acid Nitroalkane

(*d*) The six-membered ring is aromatic in the tautomeric form derived from the compound given.

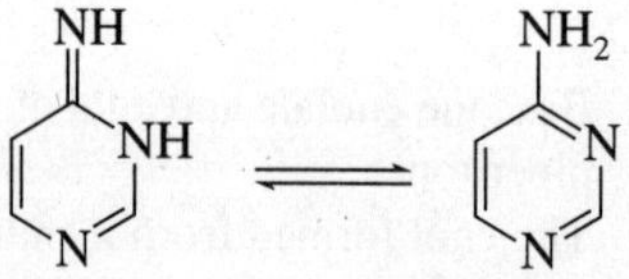

NH, NH ⇌ NH₂, N, N

(*e*) This compound is called **isourea.** Urea has a carbon–oxygen double bond and is more stable.

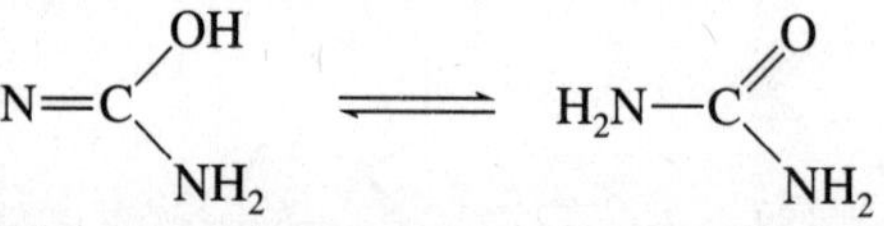

$$HN{=}C(OH)NH_2 \rightleftharpoons H_2N{-}C({=}O)NH_2$$

Isourea (less stable) Urea (more stable)

**18.37** (*a*) This reaction is an intramolecular alkylation of a ketone. Although alkylation of a ketone with a separate alkyl halide molecule is usually difficult, **intramolecular** alkylation reactions can be carried out effectively. The enolate formed by proton abstraction from the $\alpha$-carbon atom carries out a nucleophilic attack on the carbon that bears the leaving group.

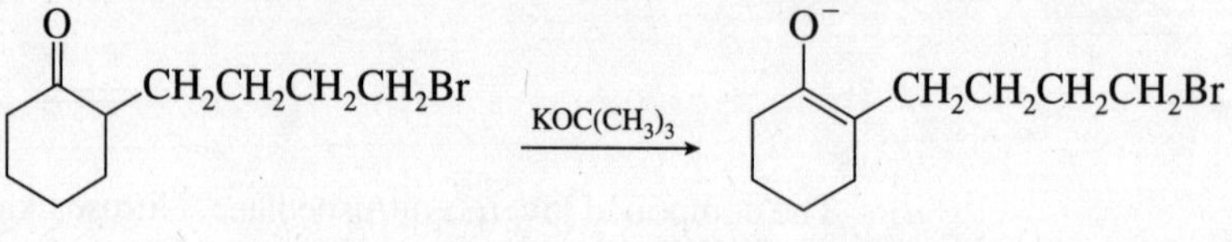

O, $CH_2CH_2CH_2CH_2Br$ $\xrightarrow{KOC(CH_3)_3}$ $O^-$, $CH_2CH_2CH_2CH_2Br$

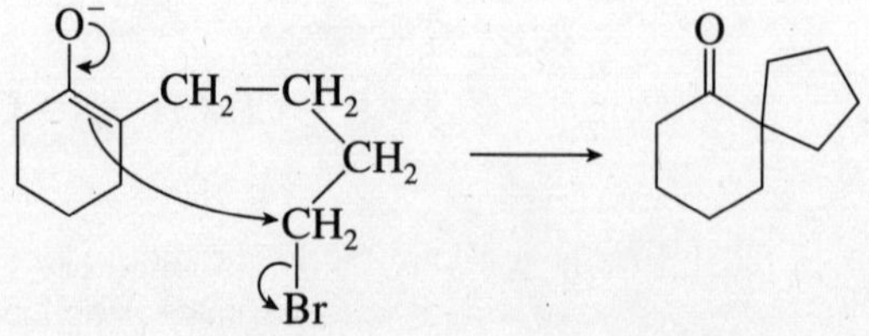

$O^-$, $CH_2{-}CH_2$, $CH_2$, $CH_2$, Br ⟶ O

(*b*) The starting material, known as **citral,** is converted to the two products by a reversal of an aldol condensation. The first step is conjugate addition of hydroxide.

$$(CH_3)_2C{=}CHCH_2CH_2\overset{CH_3}{\underset{}{C}}{=}CH{-}\overset{O}{\overset{\|}{C}}H \;+\; {}^{-}{:}\ddot{O}H \longrightarrow (CH_3)_2C{=}CHCH_2CH_2\overset{CH_3}{\underset{OH}{C}}{-}CH{=}\overset{O^-}{C}H$$

$$\xrightarrow{H_2O} (CH_3)_2C{=}CHCH_2CH_2\overset{CH_3}{\underset{OH}{C}}{-}CH_2\overset{O}{\overset{\|}{C}}H$$

The product of this conjugate addition is a $\beta$-hydroxy ketone. It undergoes base-catalyzed cleavage to the observed products.

$$(CH_3)_2C{=}CHCH_2CH_2\overset{CH_3}{\underset{OH}{C}}{-}CH_2\overset{O}{\overset{\|}{C}}H \underset{}{\overset{HO^-}{\rightleftharpoons}} (CH_3)_2C{=}CHCH_2CH_2\overset{CH_3}{\underset{{:}\ddot{O}{:}^-}{C}}{-}CH_2{-}\overset{O}{\overset{\|}{C}}H$$

$$\longrightarrow H_2C{=}\overset{O^-}{C}H + (CH_3)_2C{=}CHCH_2CH_2\underset{O}{\underset{\|}{C}}CH_3$$

$$H_2C{=}\overset{O^-}{C}H \xrightarrow{H_2O} CH_3\overset{O}{\overset{\|}{C}}H$$

(*c*) The product is formed by an intramolecular aldol condensation.

$$H\overset{O}{\overset{\|}{C}}CH_2CH_2\underset{CH_3}{CH}\overset{O}{\overset{\|}{C}}CH_3 + HO^- \longrightarrow H\overset{O}{\overset{\|}{C}}CH_2CH_2\underset{CH_3}{CH}\overset{O}{\overset{\|}{C}}\ddot{C}H_2 + H_2O \longrightarrow$$

(cyclization of enolate onto aldehyde carbonyl, followed by protonation with $H_2O$ to give 3-hydroxy-2-methylcyclohexanone, then dehydration with $HO^-$ $(-H_2O)$ to give 2-methylcyclohex-2-enone)

(*d*) In this problem stereochemical isomerization involving a proton attached to the $\alpha$-carbon atom of a ketone takes place. Enolization of the ketone yields an intermediate in which the stereochemical integrity of the $\alpha$ carbon is lost. Reversion to ketone eventually leads to the formation of the more stable stereoisomer at equilibrium.

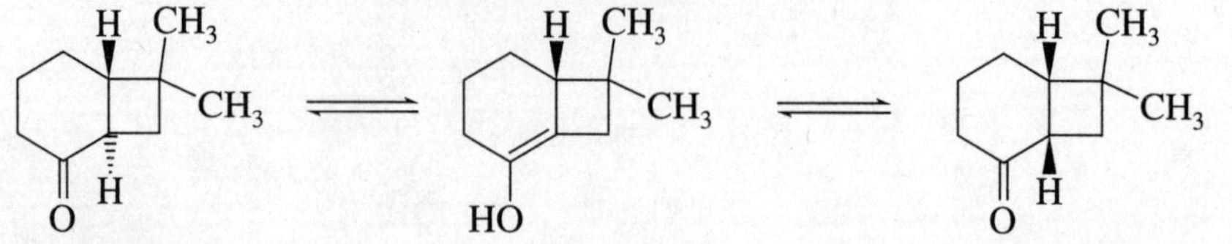

Less stable ketone; starting material

Enol

More stable ketone; preferred at equilibrium

The rate of enolization is increased by heating or by base catalysis. The cis ring fusion in the product is more stable than the trans because there are not enough atoms in the six-membered ring to span *trans*-1,2 positions in the four-membered ring without excessive strain.

(*e*) Working backward from the product, we can see that the transformation involves two aldol condensations: one intermolecular and the other intramolecular.

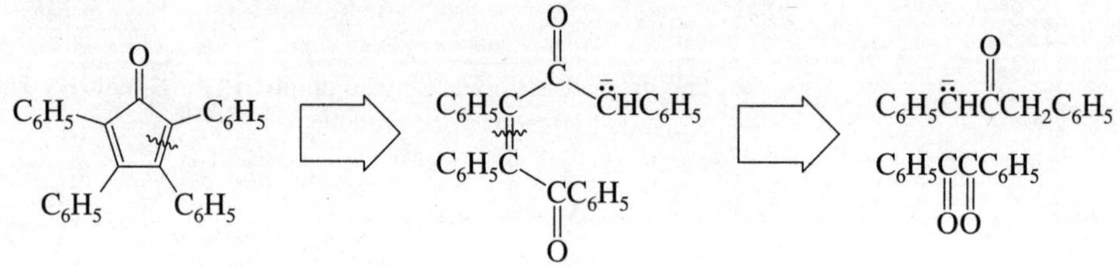

The first reaction is a mixed aldol condensation between the enolate of dibenzyl ketone and one of the carbonyl groups of the dione.

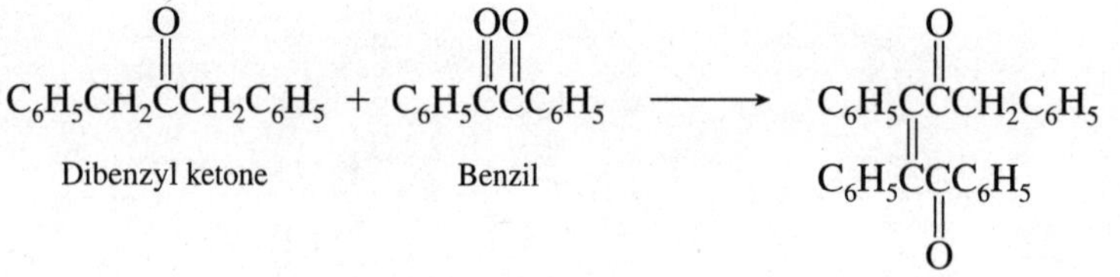

$$C_6H_5CH_2\overset{O}{\overset{\|}{C}}CH_2C_6H_5 + C_6H_5\overset{O}{\overset{\|}{C}}\overset{O}{\overset{\|}{C}}C_6H_5 \longrightarrow$$

Dibenzyl ketone

Benzil

This is followed by an intramolecular aldol condensation.

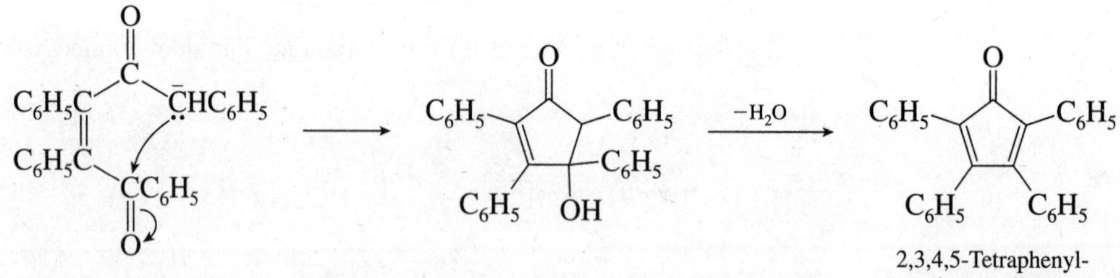

$-H_2O$

2,3,4,5-Tetraphenylcyclopentadienone

(*f*) This is a fairly difficult problem because it is not obvious at the outset which of the two possible enolates of benzyl ethyl ketone undergoes conjugate addition to the $\alpha,\beta$-unsaturated ketone. A good idea here is to work backward from the final product—in effect, do a retrosynthetic analysis. The first step is to recognize that the enone arises by dehydration of a $\beta$-hydroxy ketone.

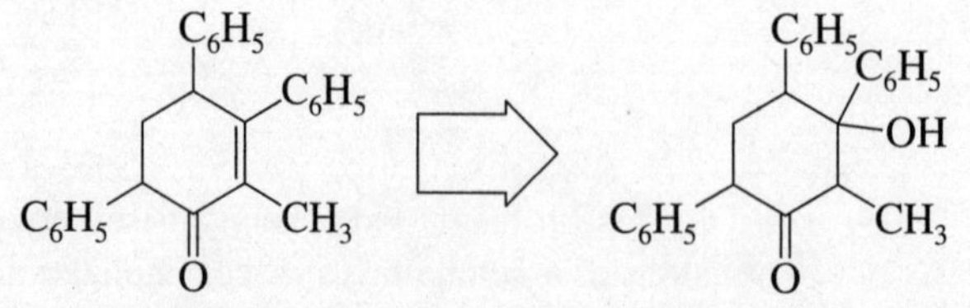

Now, mentally disconnect the bond between the $\alpha$-carbon atom and the carbon that bears the hydroxyl group to reveal the intermediate that undergoes intramolecular aldol condensation.

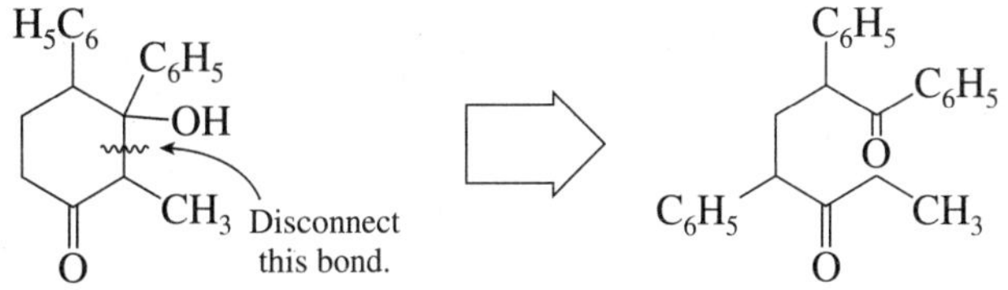

Disconnect this bond.

The $\beta$-hydroxy ketone is the intermediate formed in the intramolecular aldol addition step, and the diketone that leads to it is the intermediate that is formed in the conjugate addition step. The relationship of the starting materials to the intermediates and product is now more evident.

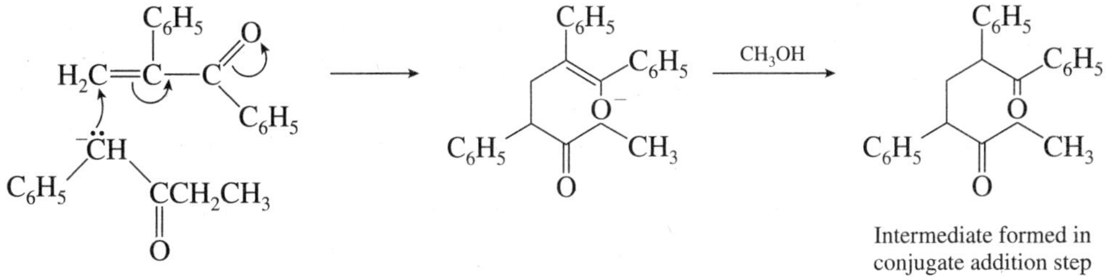

$CH_3OH$

Intermediate formed in conjugate addition step

**18.38** (*a*) The reduced C=O stretching frequency of $\alpha,\beta$-unsaturated ketones is consistent with an enhanced degree of single bond character as compared with simple dialkyl ketones.

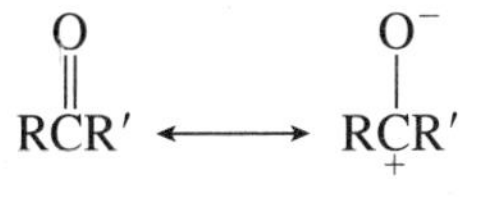

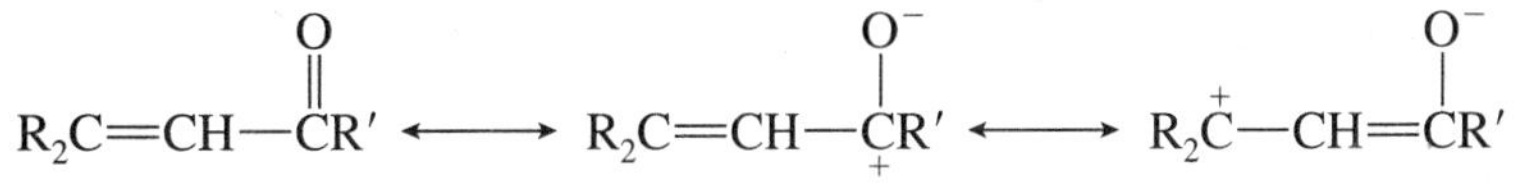

$$RC(=O)R' \longleftrightarrow R\overset{+}{C}(O^-)R'$$

$$R_2C{=}CH{-}C(=O)R' \longleftrightarrow R_2C{=}CH{-}\overset{+}{C}(O^-)R' \longleftrightarrow R_2\overset{+}{C}{-}CH{=}C(O^-)R'$$

Resonance is more important in $\alpha,\beta$-unsaturated ketones. Conjugation of the carbonyl group with the carbon–carbon double bond increases opportunities for electron delocalization.

(*b*) Even more single-bond character is indicated in the carbonyl group of cyclopropenone than in that of typical $\alpha,\beta$-unsaturated ketones. The dipolar resonance form contributes substantially to the electron distribution because of the aromatic character of the three-membered ring. Recall that cyclopropenyl cation satisfies the $4n + 2$ rule for aromaticity (text Section 11.20).

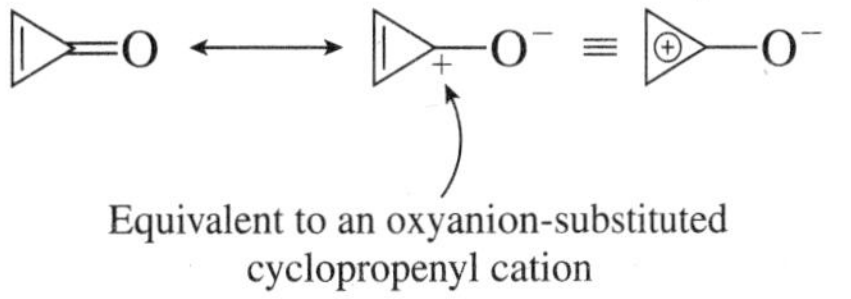

Equivalent to an oxyanion-substituted cyclopropenyl cation

(*c*) The dipolar resonance form is a more important contributor to the electron distribution in diphenylcyclopropenone than in benzophenone.

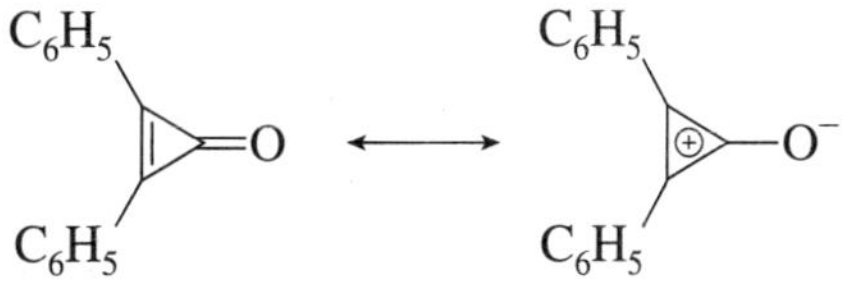

is more pronounced than

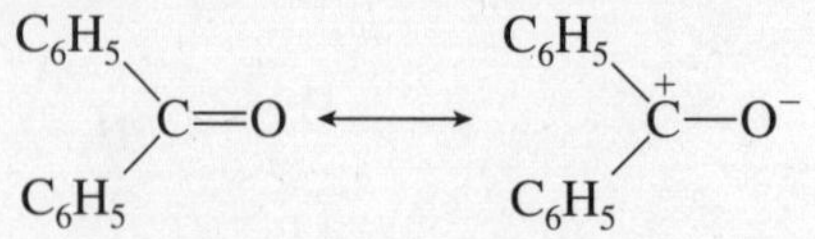

$$(C_6H_5)_2C{=}O \longleftrightarrow (C_6H_5)_2\overset{+}{C}{-}O^-$$

The dipolar resonance form of diphenylcyclopropenone has aromatic character. Its stability leads to increased charge separation and a larger dipole moment.

(*d*) Decreased electron density at the $\beta$ carbon atom of an $\alpha,\beta$-unsaturated ketone is responsible for its decreased shielding. The decreased electron density arises from the polarization of its $\pi$ electrons as represented by a significant contribution of the dipolar resonance form.

$$H_2C{=}CH{-}\overset{O}{\overset{\|}{C}}R \longleftrightarrow H_2\overset{+}{C}{-}CH{=}\overset{O^-}{\overset{|}{C}}R$$

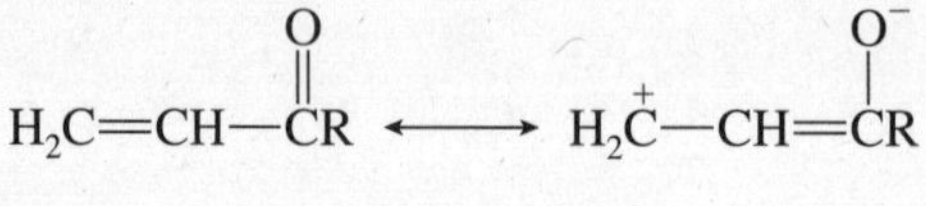

**18.39** Bromination can occur at either of the two $\alpha$-carbon atoms.

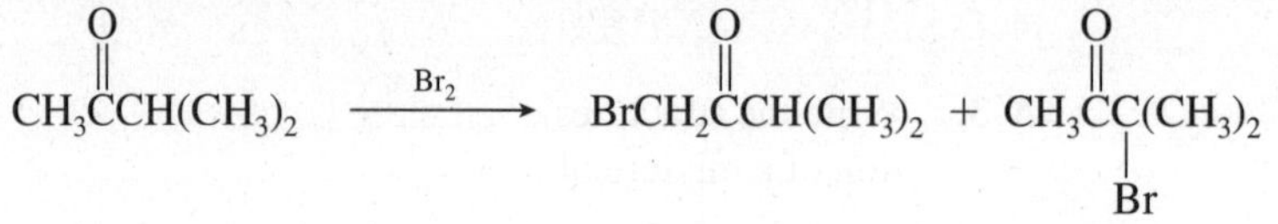

$$CH_3\overset{O}{\overset{\|}{C}}CH(CH_3)_2 \xrightarrow{Br_2} BrCH_2\overset{O}{\overset{\|}{C}}CH(CH_3)_2 + CH_3\overset{O}{\overset{\|}{C}}\underset{Br}{\underset{|}{C}}(CH_3)_2$$

3-Methyl-2-butanone — 1-Bromo-3-methyl-2-butanone — 3-Bromo-3-methyl-2-butanone

The $^1H$ NMR spectrum of the major product, compound A, is consistent with the structure of 1-bromo-3-methyl-2-butanone. The minor product B is identified as 3-bromo-3-methyl-2-butanone on the basis of its NMR spectrum.

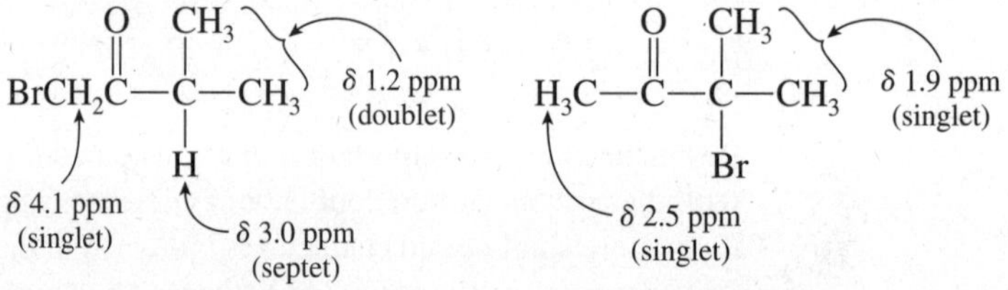

Compound A: $BrCH_2\overset{O}{\overset{\|}{C}}{-}\underset{H}{\underset{|}{\overset{CH_3}{\overset{|}{C}}}}{-}CH_3$; $BrCH_2$: $\delta$ 4.1 ppm (singlet); H: $\delta$ 3.0 ppm (septet); $CH_3$: $\delta$ 1.2 ppm (doublet)

Compound B: $H_3C{-}\overset{O}{\overset{\|}{C}}{-}\underset{Br}{\underset{|}{\overset{CH_3}{\overset{|}{C}}}}{-}CH_3$; $H_3C$: $\delta$ 2.5 ppm (singlet); $CH_3$: $\delta$ 1.9 ppm (singlet)

**18.40** Three dibromination products are possible from $\alpha$ halogenation of 2-butanone.

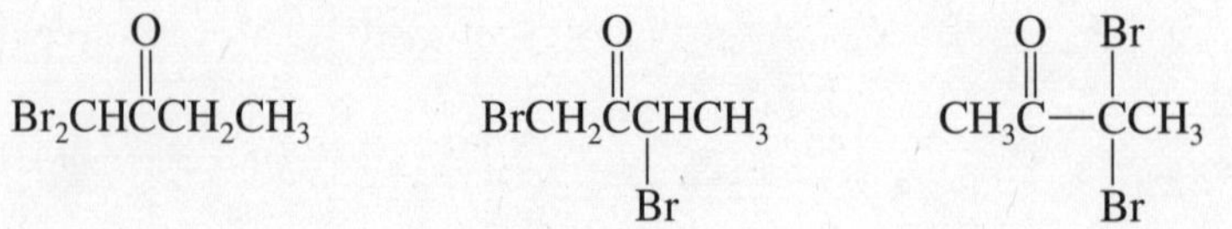

$$Br_2CH\overset{O}{\overset{\|}{C}}CH_2CH_3 \qquad BrCH_2\overset{O}{\overset{\|}{C}}\underset{Br}{\underset{|}{C}}HCH_3 \qquad CH_3\overset{O}{\overset{\|}{C}}{-}\underset{Br}{\underset{|}{\overset{Br}{\overset{|}{C}}}}CH_3$$

1,1-Dibromo-2-butanone — 1,3-Dibromo-2-butanone — 3,3-Dibromo-2-butanone

The product is **1,3-dibromo-2-butanone,** on the basis of its observed $^1H$ NMR spectrum, which showed two signals at low field. One is a two-proton singlet at $\delta$ 4.6 ppm assignable to $CH_2Br$ and the other a one-proton quartet at $\delta$ 5.2 ppm assignable to CHBr.

**18.41** Solutions to molecular modeling exercises are not provided in this *Study Guide and Solutions Manual.* You should use *Learning By Modeling* for this exercise.

# SELF-TEST

## PART A

**A-1.** Write the correct structure(s) for each of the following:

(*a*) The two enol forms of 2-butanone

(*b*) The enolate ion derived from reaction of 1,3-cyclohexanedione with sodium methoxide

(*c*) The carbonyl form of the following enol

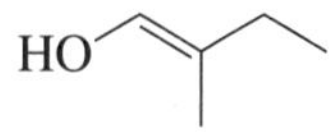

HO

**A-2.** Give the correct structures for compounds A and B in the following reaction schemes:

(*a*) $2C_6H_5CH_2\overset{O}{\overset{\|}{C}}H \xrightarrow[\text{2. heat }(-H_2O)]{\text{1. HO}^-} \text{A}$

(*b*) $CH_3CH_2CH{=}CH\overset{O}{\overset{\|}{C}}CH_2CH_3 + LiCu(CH_2CH_3)_2 \xrightarrow[\text{2. }H_2O]{\text{1. ether}} \text{B}$

**A-3.** Write the structures of all the possible aldol addition products that may be obtained by reaction of a mixture of propanal and 2-methylpropanal with base.

$CH_3CH_2\overset{O}{\overset{\|}{C}}H$ Propanal

$CH_3\overset{CH_3}{\overset{|}{C}}HCH{=}O$ 2-Methylpropanal

**A-4.** Using any necessary organic or inorganic reagents, outline a synthesis of 1,3-butanediol from ethanol as the only source of carbons.

**A-5.** Outline a series of reaction steps that will allow the preparation of compound B from 1,3-cyclopentanedione, compound A.

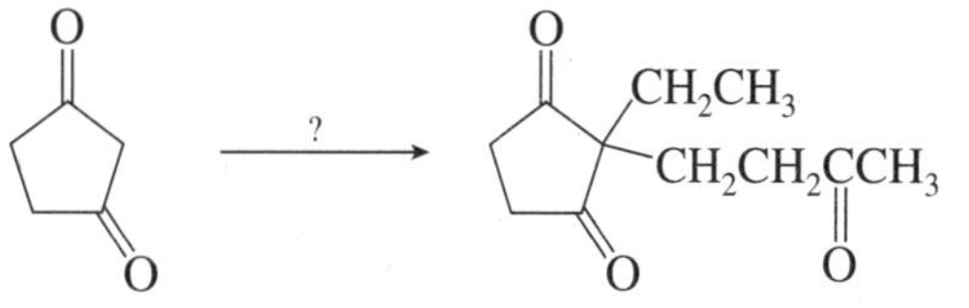

A $\xrightarrow{?}$ B

$CH_2CH_3$

$CH_2CH_2\overset{O}{\overset{\|}{C}}CH_3$

**A-6.** Give the structure of the product formed in each of the following reactions:

(*a*) $CH_3CH_2CH_2CH_2\overset{O}{\overset{\|}{C}}H \xrightarrow[\text{acetic acid}]{Br_2}$

(*b*) $2CH_3CH_2CH_2CH_2\overset{O}{\overset{\|}{C}}H \xrightarrow[5°C]{\text{NaOH}}$

(*c*) 2,2-dimethylcyclopentanone ($CH_3$, $CH_3$) + $C_6H_5\overset{O}{\overset{\|}{C}}H \xrightarrow[(-H_2O)]{\text{NaOH, heat}}$

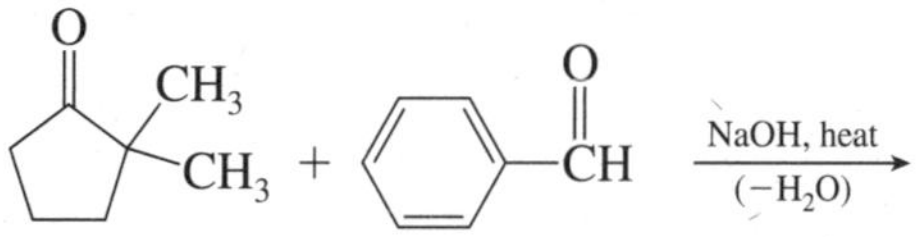

(*d*) $C_6H_5$—CH=CH$\overset{O}{\overset{\|}{C}}$—$C_6H_5$ $\xrightarrow[\text{ethanol}]{\text{NaSCH}_3}$

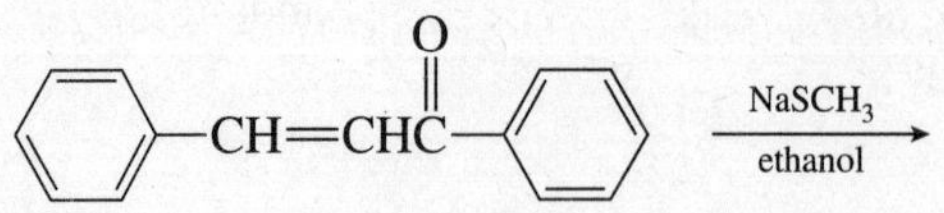

**A-7.** Write out the mechanism, using curved arrows to show electron movement, of the following aldol addition reaction.

$$H\overset{O}{\overset{\|}{C}}H + CH_3CH_2\overset{O}{\overset{\|}{C}}H \xrightarrow[5°C]{NaOH,\ H_2O}$$

**A-8.** Identify the two starting materials needed to make the following compound by a mixed aldol condensation.

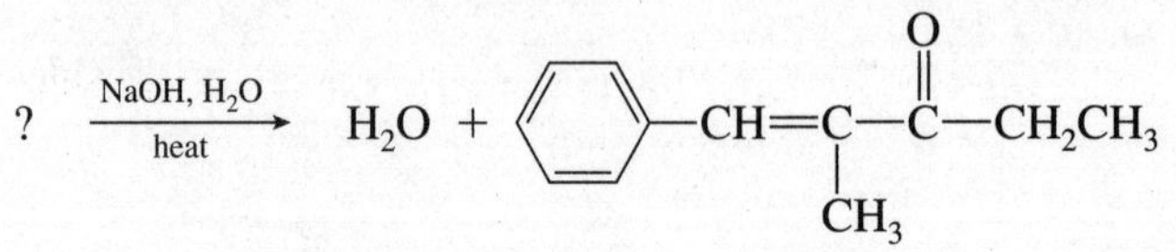

$$? \xrightarrow[heat]{NaOH,\ H_2O} H_2O + C_6H_5-CH{=}\underset{\underset{CH_3}{|}}{C}-\overset{O}{\overset{\|}{C}}-CH_2CH_3$$

## PART B

**B-1.** When enolate A is compared with enolate B

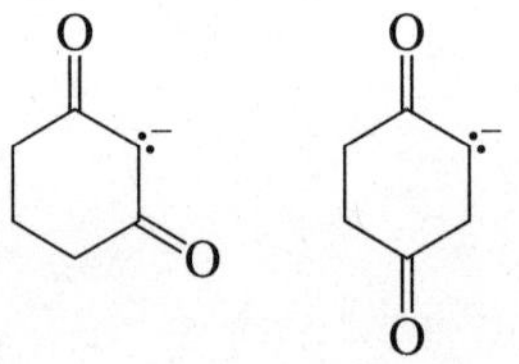

A B

which of the following statements is true?

(*a*) A is more stable than B.
(*b*) B is more stable than A.
(*c*) A and B have the same stability.
(*d*) No comparison of stability can be made.

**B-2.** Which structure is the most stable?

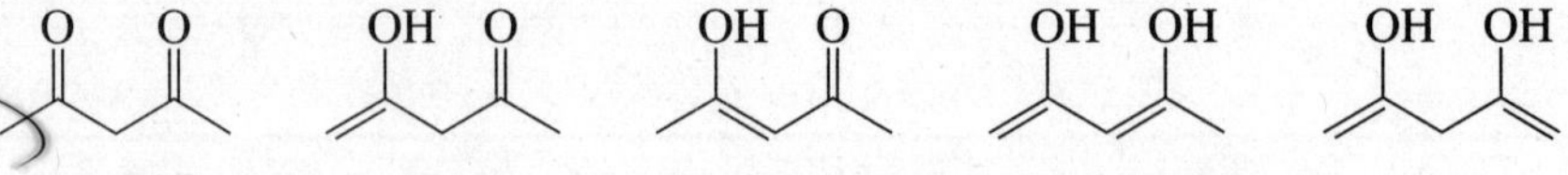

O O (*a*)   OH O (*b*)   OH O (*c*)   OH OH (*d*)   OH OH (*e*)

**B-3.** Which one of the following molecules contains deuterium ($^2H$ = D) after reaction with NaOD in $D_2O$?

(*a*) $C_6H_5\overset{O}{\overset{\|}{C}}H$   (*c*) $C_6H_5\overset{O}{\overset{\|}{C}}C(CH_3)_3$

(*b*) $C_6H_5CH_2\overset{O}{\overset{\|}{C}}H$   (*d*) $(CH_3)_3C\overset{O}{\overset{\|}{C}}C(CH_3)_3$

**B-4.** Which of the following RX compounds is (are) the best alkylating agent(s) in the reaction shown?

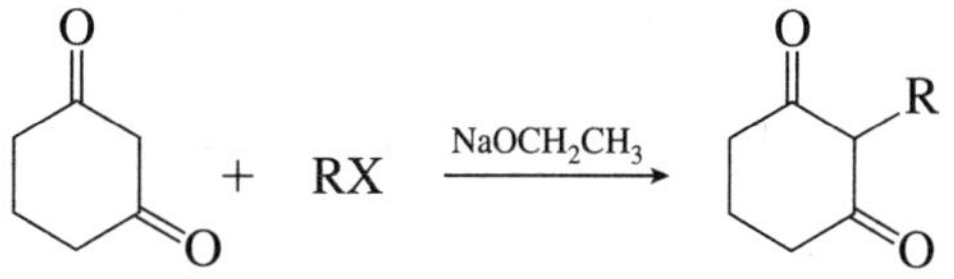

$$\text{1,3-cyclohexanedione} + RX \xrightarrow{NaOCH_2CH_3} \text{2-R-1,3-cyclohexanedione}$$

| $(CH_3)_3CBr$ | $(CH_3)_2CHCH_2OH$ | $C_6H_5Br$ | $(CH_3)_2CHCH_2Br$ |
|---|---|---|---|
| 1 | 2 | 3 | 4 |

(*a*) 1 and 4
(*b*) 4 only
(*c*) 2 and 4
(*d*) 1, 3, and 4

**B-5.** Which of the following pairs of aldehydes gives a single product in a mixed aldol condensation?

(*a*) $C_6H_5CH_2\overset{O}{\overset{\|}{C}}H + C_6H_5\overset{O}{\overset{\|}{C}}H$

(*b*) $C_6H_5\overset{O}{\overset{\|}{C}}H + (CH_3)_3C\overset{O}{\overset{\|}{C}}H$

(*c*) $C_6H_5\overset{O}{\overset{\|}{C}}H + H_2C{=}O$

(*d*) $CH_3\overset{O}{\overset{\|}{C}}H + (CH_3)_2CH\overset{O}{\overset{\|}{C}}H$

**B-6.** What is the principal product of the following reaction?

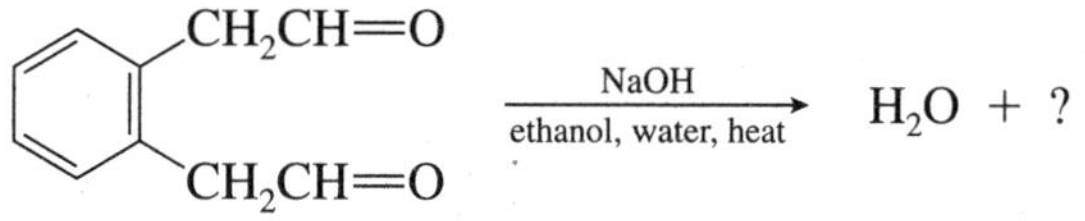

$$o\text{-}C_6H_4(CH_2CH{=}O)_2 \xrightarrow[\text{ethanol, water, heat}]{NaOH} H_2O + ?$$

(*a*) benzocyclobutene bearing two CH=O groups

(*b*) naphthalene bearing two OH groups

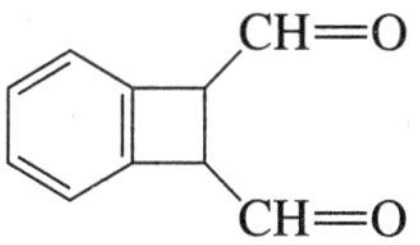

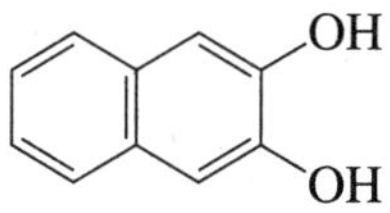

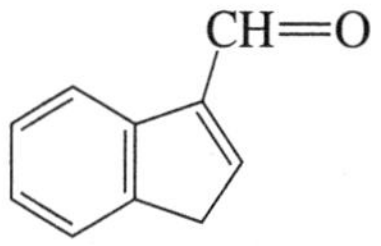

(*c*) indene bearing CH=O

(*d*) tetralin-2,3-dione (ring with two C=O)

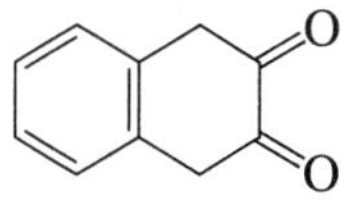

**B-7.** Which of the following forms an enol to the greatest extent?

(*a*) $CH_3CH_2\overset{O}{\overset{\|}{C}}H$

(*b*) $CH_3\overset{O}{\overset{\|}{C}}CH_2CH_2\overset{O}{\overset{\|}{C}}CH_3$

(*c*) $CH_3\overset{O}{\overset{\|}{C}}CH_2\overset{O}{\overset{\|}{C}}H$

(*d*) $CH_3\overset{O}{\overset{\|}{C}}\overset{O}{\overset{\|}{C}}CH_2CH_3$

**B-8.** Which of the following species is (are) *not* intermediates in the aldol condensation of acetaldehyde (ethanal) in aqueous base?

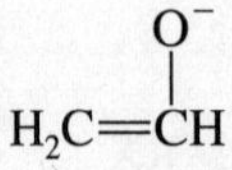

1: $H_2C{=}CH{-}O^-$ 2: $CH_3CH(O^-)CH_2CH{=}O$ 3: $CH_3CH{=}\overset{+}{O}H$ 4: $CH_3CH(O^-)CH{=}CH{-}O^-$

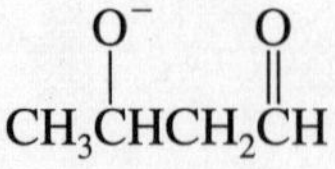

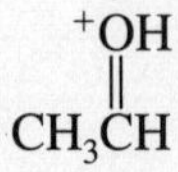

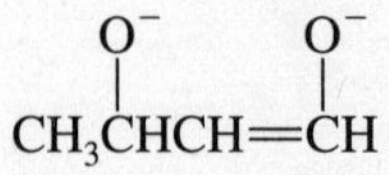

(*a*) 1 and 2
(*b*) 3 only
(*c*) 4 only
(*d*) 2 and 3
(*e*) 3 and 4

**B-9.** The compound shown in the box undergoes racemization on reaction with aqueous acid. Which of the following structures best represents the intermediate responsible for this process?

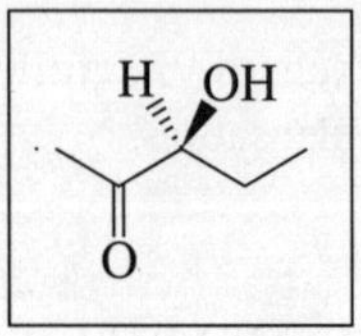

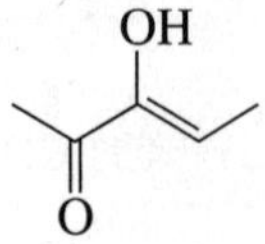

H OH
O

OH (*a*) O; H OH (*b*) OH; OH (*c*) OH; H OH (*d*) HO OH; O (*e*) O

**B-10.** Which one of the following compounds is the best candidate for being prepared by an efficient mixed aldol addition reaction?

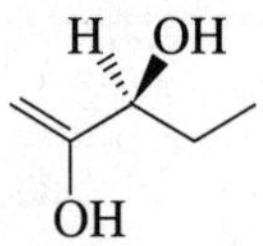

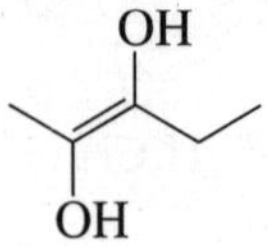

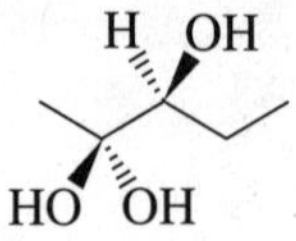

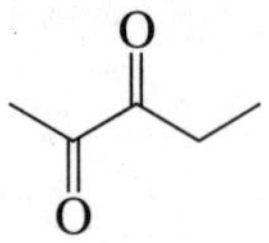

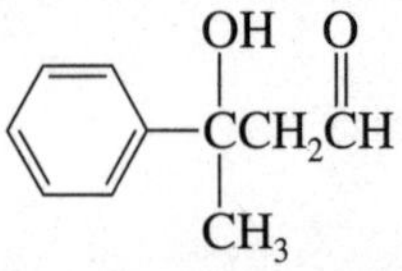

(*a*) $C_6H_5C(OH)(CH_3)CH_2CH{=}O$
(*b*) $C_6H_5C(=O)CH(CH_3)CH_2OH$
(*c*) $C_6H_5CH_2C(=O)CH(CH_3)C(OH)(CH_3)_2$
(*d*) $C_6H_5C(=O)CH_2CH_2CH{=}O$
(*e*) $C_6H_5C(=O)CH_2C(=O)CH_3$

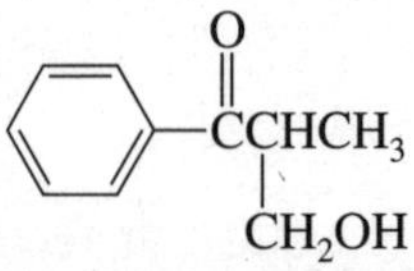

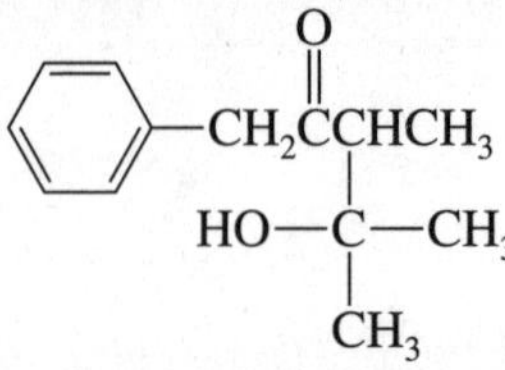

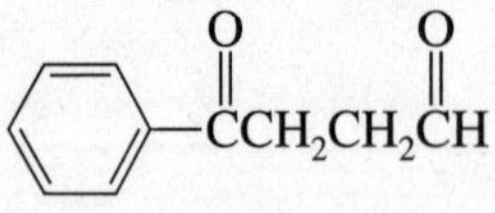

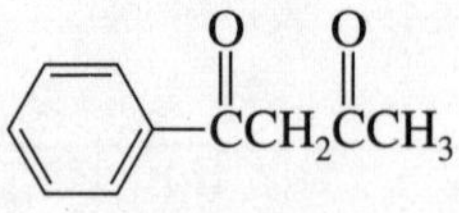

**B-11.** Which one of the following undergoes 1,4-addition with $CH_3SK$ (in ethanol)?

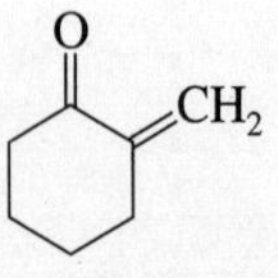

(*a*) O, $CH_2$ (*b*) O, $CH_3$ (*c*) O, O (*d*) O, O=CH, $CH_3$

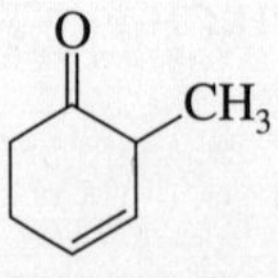

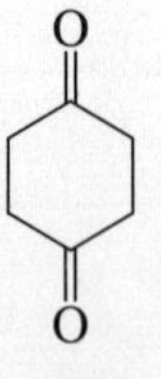

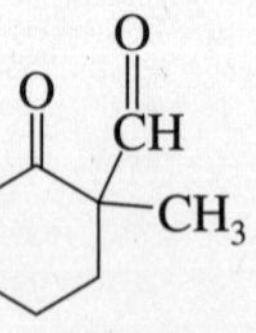

**B-12.** Benzalacetone is the mixed aldol condensation product formed between benzaldehyde ($C_6H_5CH{=}O$) and acetone [$(CH_3)_2C{=}O$]. What is its structure?

(*a*) $C_6H_5CH{=}CH\overset{\overset{\displaystyle O}{\|}}{C}CH_3$

(*b*) $C_6H_5\overset{\overset{\displaystyle O}{\|}}{C}CH{=}CHCH_3$

(*c*) $C_6H_5CH{=}C(CH_3)_2$

(*d*) $C_6H_5CH_2\overset{\overset{\displaystyle O}{\|}}{C}CH{=}CH_2$

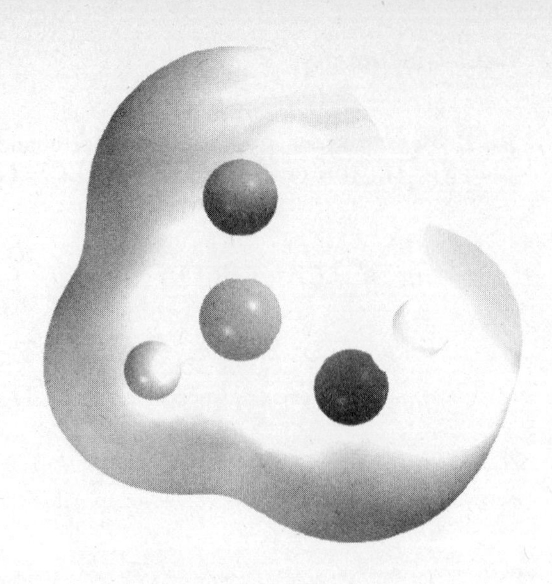

# CHAPTER 19

## CARBOXYLIC ACIDS

## SOLUTIONS TO TEXT PROBLEMS

**19.1** (*b*) The four carbon atoms of crotonic acid form a continuous chain. Because there is a double bond between C-2 and C-3, crotonic acid is one of the stereoisomers of 2-butenoic acid. The stereochemistry of the double bond is *E*.

$$\begin{array}{ccc} H_3C & & H \\ & C{=}C & \\ H & & CO_2H \end{array}$$

(*E*)-2-Butenoic acid
(crotonic acid)

(*c*) Oxalic acid is a dicarboxylic acid that contains two carbons. It is **ethanedioic acid.**

$$HO_2CCO_2H$$

Ethanedioic acid
(oxalic acid)

(*d*) The name given to $C_6H_5CO_2H$ is benzoic acid. Because it has a methyl group at the para position, the compound shown is ***p*-methylbenzoic acid,** or **4-methylbenzoic acid.**

$$H_3C{-}C_6H_4{-}CO_2H$$

*p*-Methylbenzoic acid or
4-methylbenzoic acid
(*p*-toluic acid)

**19.2** Ionization of peroxy acids such as peroxyacetic acid yields an anion that cannot be stabilized by resonance in the same way that acetate can.

$$CH_3C(=\ddot{O}:)\ddot{O}-\ddot{O}:^-$$

Delocalization of negative charge into carbonyl group is not possible in peroxyacetate ion.

**19.3** Recall from Chapter 4 (text Section 4.6) that an acid–base equilibrium favors formation of the weaker acid and base. Also remember that the weaker acid forms the stronger conjugate base, and vice versa.

(*b*) The acid–base reaction between acetic acid and *tert*-butoxide ion is represented by the equation

$$CH_3CO_2H + (CH_3)_3CO^- \rightleftharpoons CH_3CO_2^- + (CH_3)_3COH$$

| Acetic acid (stronger acid) | *tert*-Butoxide (stronger base) | Acetate ion (weaker base) | *tert*-Butyl alcohol (weaker acid) |
|---|---|---|---|

Alcohols are weaker acids than carboxylic acids; the equilibrium lies to the right.

(*c*) Bromide ion is the conjugate base of hydrogen bromide, a strong acid.

$$CH_3CO_2H + Br^- \rightleftharpoons CH_3CO_2^- + HBr$$

| Acetic acid (weaker acid) | Bromide ion (weaker base) | Acetate ion (stronger base) | Hydrogen bromide (stronger acid) |
|---|---|---|---|

In this case, the position of equilibrium favors the starting materials, because acetic acid is a weaker acid than hydrogen bromide.

(*d*) Acetylide ion is a rather strong base, and acetylene, with a $K_a$ of $10^{-26}$, is a much weaker acid than acetic acid. The position of equilibrium favors the formation of products.

$$CH_3CO_2H + HC{\equiv}C:^- \rightleftharpoons CH_3CO_2^- + HC{\equiv}CH$$

| Acetic acid (stronger acid) | Acetylide ion (stronger base) | Acetate ion (weaker base) | Acetylene (weaker acid) |
|---|---|---|---|

(*e*) Nitrate ion is a very weak base; it is the conjugate base of the strong acid nitric acid. The position of equilibrium lies to the left.

$$CH_3CO_2H + NO_3^- \rightleftharpoons CH_3CO_2^- + HNO_3$$

| Acetic acid (weaker acid) | Nitrate ion (weaker base) | Acetate ion (stronger base) | Nitric acid (stronger acid) |
|---|---|---|---|

(*f*) Amide ion is a very strong base; it is the conjugate base of ammonia, $pK_a = 36$. The position of equilibrium lies to the right.

$$CH_3CO_2H + H_2N^- \rightleftharpoons CH_2CO_2^- + NH_3$$

| Acetic acid (stronger acid) | Amide ion (stronger base) | Acetate ion (weaker base) | Ammonia (weaker acid) |
|---|---|---|---|

**19.4** (*b*) Propanoic acid is similar to acetic acid in its acidity. A hydroxyl group at C-2 is electron-withdrawing and stabilizes the carboxylate ion of lactic acid by a combination of inductive and field effects.

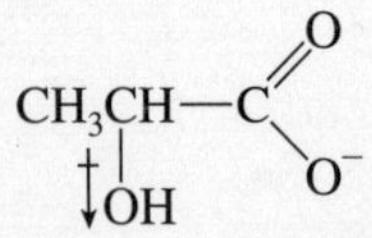

$$CH_3\underset{\downarrow OH}{CH}-C(=O)O^-$$

Hydroxyl group stabilizes negative charge by attracting electrons.

Lactic acid is more acidic than propanoic acid. The measured ionization constants are

| $CH_3\underset{OH}{CH}CO_2H$ | $CH_3CH_2CO_2H$ |
|---|---|
| Lactic acid | Propanoic acid |
| $K_a$ $1.4 \times 10^{-4}$ ($pK_a$ 3.8) | $K_a$ $1.3 \times 10^{-5}$ ($pK_a$ 4.9) |

(*c*) A carbonyl group is more strongly electron-withdrawing than a carbon–carbon double bond. Pyruvic acid is a stronger acid than acrylic acid.

| $CH_3\overset{O}{\overset{\|}{C}}CO_2H$ | $H_2C{=}CHCO_2H$ |
|---|---|
| Pyruvic acid | Acrylic acid |
| $K_a$ $5.1 \times 10^{-4}$ ($pK_a$ 3.3) | $K_a$ $5.5 \times 10^{-5}$ ($pK_a$ 4.3) |

(*d*) Viewing the two compounds as substituted derivatives of acetic acid, $RCH_2CO_2H$, we judge $CH_3S(=O)_2$ to be strongly electron-withdrawing and acid-strengthening, whereas an ethyl group has only a small effect.

| $CH_3S(=O)_2CH_2CO_2H$ | $CH_3CH_2CH_2CO_2H$ |
|---|---|
| Methanesulfonylacetic acid | Butanoic acid |
| $K_a$ $4.3 \times 10^{-3}$ ($pK_a$ 2.4) | $K_a$ $1.5 \times 10^{-5}$ ($pK_a$ 4.7) |

**19.5** The compound can only be a carboxylic acid; no other class containing only carbon, hydrogen, and oxygen is more acidic. A reasonable choice is $HC{\equiv}CCO_2H$; C-2 is *sp*-hybridized and therefore rather electron-withdrawing and acid-strengthening. This is borne out by its measured ionization constant $K_a$, which is $1.4 \times 10^{-2}$ ($pK_a$ 1.8).

**19.6** For carbonic acid, the "true $K_1$" is given by

$$\text{True } K_1 = \frac{[H^+][HCO_3^-]}{[H_2CO_3]}$$

The "observed $K$" is given by the expression

$$4.3 \times 10^{-7} = \frac{[H^+][HCO_3^-]}{[CO_2]}$$

which can be rearranged to

$$[H^+][HCO_3^-] = (4.3 \times 10^{-7})[CO_2]$$

and therefore

$$\text{True } K_1 = \frac{(4.3 \times 10^{-7})[CO_2]}{[H_2CO_3]}$$

$$= \frac{(4.3 \times 10^{-7})(99.7)}{0.3}$$

$$= 1.4 \times 10^{-4}$$

Thus, when corrected for the small degree to which carbon dioxide is hydrated, it can be seen that carbonic acid is actually a stronger acid than acetic acid. Carboxylic acids dissolve in sodium bicarbonate solution because the equilibrium that leads to carbon dioxide formation is favorable, not because carboxylic acids are stronger acids than carbonic acid.

**19.7** (*b*) 2-Chloroethanol has been converted to 3-hydroxypropanoic acid by way of the corresponding nitrile.

$$\underset{\text{2-Chloroethanol}}{HOCH_2CH_2Cl} \xrightarrow[H_2O]{NaCN} \underset{\text{2-Cyanoethanol}}{HOCH_2CH_2CN} \xrightarrow[\text{heat}]{H_3O^+} \underset{\text{3-Hydroxypropanoic acid}}{HOCH_2CH_2CO_2H}$$

The presence of the hydroxyl group in 2-chloroethanol precludes the preparation of a Grignard reagent from this material, and so any attempt at the preparation of 3-hydroxypropanoic acid via the Grignard reagent of 2-chloroethanol is certain to fail.

(*c*) Grignard reagents can be prepared from tertiary halides and react in the expected manner with carbon dioxide. The procedure shown is entirely satisfactory.

$$\underset{\textit{tert}\text{-Butyl chloride}}{(CH_3)_3CCl} \xrightarrow[\text{diethyl ether}]{Mg} \underset{\textit{tert}\text{-Butylmagnesium chloride}}{(CH_3)_3CMgCl} \xrightarrow[2.\ H_3O^+]{1.\ CO_2} \underset{\text{2,2-Dimethylpropanoic acid (61–70\%)}}{(CH_3)_3CCO_2H}$$

Preparation by way of the nitrile will not be feasible. Rather than react with sodium cyanide by substitution, *tert*-butyl chloride will undergo elimination exclusively. The $S_N2$ reaction with cyanide ion is limited to primary and secondary alkyl halides.

**19.8** Incorporation of $^{18}O$ into benzoic acid proceeds by a mechanism analogous to that of esterification. The nucleophile that adds to the protonated form of benzoic acid is $^{18}O$-enriched water (the $^{18}O$ atom is represented by the shaded letter ⊘ in the following equations).

$$\underset{\text{Benzoic acid}}{C_6H_5C(=\ddot{O}:)\ddot{O}H} + H^+ \rightleftharpoons C_6H_5C(=\overset{+}{\ddot{O}}H)\ddot{O}H \overset{H_2⊘:}{\rightleftharpoons} C_6H_5C(:\ddot{O}H)(\ddot{O}H)(\overset{+}{⊘}H_2) \rightleftharpoons \underset{\text{Tetrahedral intermediate}}{C_6H_5C(:\ddot{O}H)(\ddot{O}H)(:⊘H)}$$

The three hydroxyl groups of the tetrahedral intermediate are equivalent except that one of them is labeled with $^{18}O$. Any one of these three hydroxyl groups may be lost in the dehydration step; when the hydroxyl group that is lost is unlabeled, an $^{18}O$ label is retained in the benzoic acid.

$$\underset{\text{Tetrahedral intermediate}}{C_6H_5C(OH)(OH)(⊘H)} + H^+ \rightleftharpoons C_6H_5C(=\overset{+}{O}H)(⊘H) + H_2O \overset{-H^+}{\rightleftharpoons} \underset{\text{}^{18}\text{O-enriched benzoic acid}}{C_6H_5C(=O)(⊘H) + C_6H_5C(=⊘)(OH)}$$

**19.9** (*b*) The 16-membered ring of 15-pentadecanolide is formed from 15-hydroxypentadecanoic acid.

Disconnect this bond.

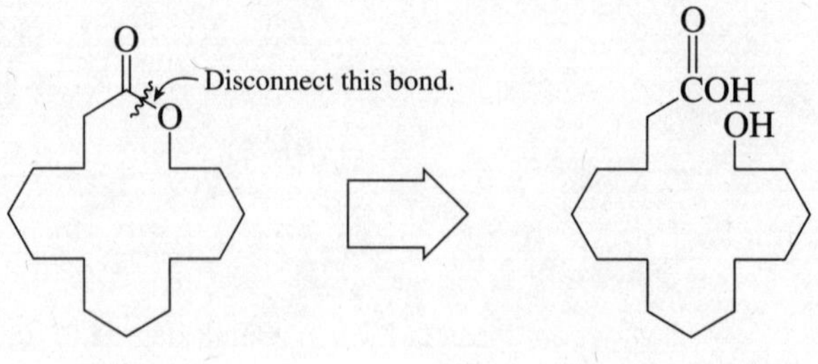

15-Pentadecanolide 15-Hydroxypentadecanoic acid

(*c*) Vernolepin has two lactone rings, which can be related to two hydroxy acid combinations.

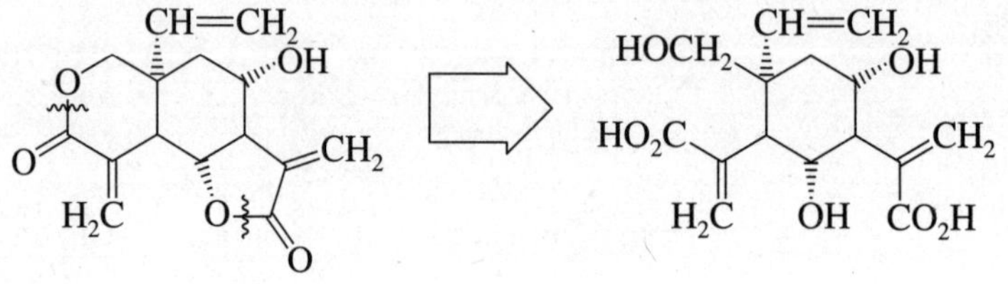

Be sure to keep the relative stereochemistry unchanged. Remember, the carbon–oxygen bond of an alcohol remains intact when the alcohol reacts with a carboxylic acid to give an ester.

**19.10** Alkyl chlorides and bromides undergo nucleophilic substitution when treated with sodium iodide in acetone (Section 8.1). A reasonable approach is to brominate octadecanoic acid at its $\alpha$-carbon atom, then replace the bromine substituent with iodine by nucleophilic substitution.

$$CH_3(CH_2)_{15}CH_2CO_2H \xrightarrow{Br_2,\ PCl_3} CH_3(CH_2)_{15}\underset{\displaystyle Br}{\underset{|}{C}}HCO_2H \xrightarrow[\text{acetone}]{NaI} CH_3(CH_2)_{15}\underset{\displaystyle I}{\underset{|}{C}}HCO_2H$$

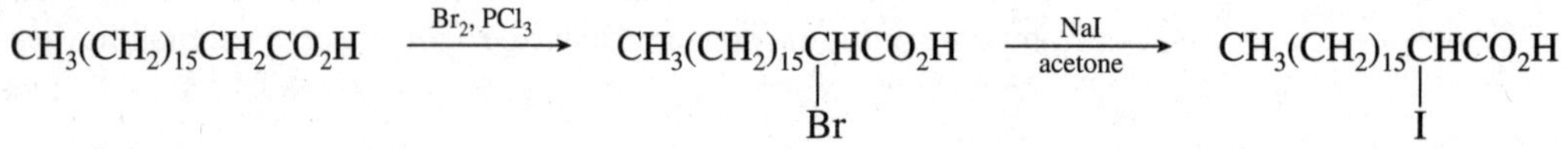

Octadecanoic acid 2-Bromooctadecanoic acid 2-Iodooctadecanoic acid

**19.11** (*b*) The starting material is a derivative of malonic acid. It undergoes efficient thermal decarboxylation in the manner shown.

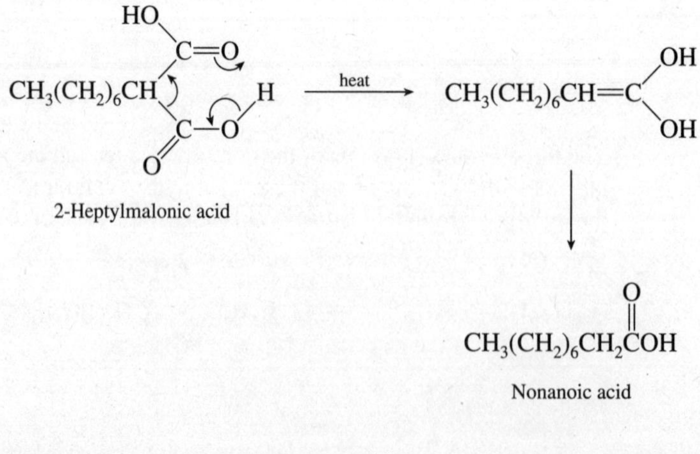

$$CH_3(CH_2)_6CH(CO_2H)_2 \xrightarrow{\text{heat}} CH_3(CH_2)_6CH{=}C(OH)_2 + CO_2$$

2-Heptylmalonic acid Carbon dioxide

$$\downarrow$$

$$CH_3(CH_2)_6CH_2\overset{O}{\overset{\|}{C}}OH$$

Nonanoic acid

(*c*) The phenyl and methyl substituents attached to C-2 of malonic acid play no role in the decarboxylation process.

$$C_6H_5C(CH_3)(CO_2H)_2 \xrightarrow{\text{heat}} C_6H_5(H_3C)C{=}C(OH)_2 + CO_2 \;(\text{Carbon dioxide}) \longrightarrow C_6H_5CH(CH_3)COH\,(=O)$$

2-Phenylpropanoic acid

**19.12** (*b*) The thermal decarboxylation of $\beta$-keto acids resembles that of substituted malonic acids. The structure of 2,2-dimethylacetoacetic acid and the equation representing its decarboxylation were given in the text. The overall process involves the bonding changes shown.

$$CH_3C(=O)C(CH_3)_2CO_2H \xrightarrow{-CO_2} CH_3C(OH){=}C(CH_3)CH_3 \longrightarrow CH_3\overset{O}{\overset{\|}{C}}CH(CH_3)_2$$

2,2-Dimethylacetoacetic acid; Enol form of 3-methyl-2-butanone; 3-Methyl-2-butanone

**19.13** (*a*) Lactic acid (2-hydroxypropanoic acid) is a three-carbon carboxylic acid that bears a hydroxyl group at C-2.

$$\overset{3}{C}H_3\overset{2}{C}H(OH)\overset{1}{C}O_2H$$

2-Hydroxypropanoic acid

(*b*) The parent name **ethanoic acid** tells us that the chain that includes the carboxylic acid function contains only two carbons. A hydroxyl group and a phenyl substituent are present at C-2.

$$C_6H_5CH(OH)CO_2H$$

2-Hydroxy-2-phenylethanoic acid (mandelic acid)

(*c*) The parent alkane is **tetradecane,** which has an unbranched chain of 14 carbons. The terminal methyl group is transformed to a carboxyl function in tetradecanoic acid.

$$CH_3(CH_2)_{12}\overset{O}{\overset{\|}{C}}OH$$

Tetradecanoic acid (myristic acid)

(*d*) Undecane is the unbranched alkane with 11 carbon atoms, undecanoic acid is the corresponding carboxylic acid, and **undecenoic acid** is an 11-carbon carboxylic acid that contains a double bond. Because the carbon chain is numbered beginning with the carboxyl group, 10-undecenoic acid has its double bond at the opposite end of the chain from the carboxyl group.

$$H_2C{=}CH(CH_2)_8CO_2H$$

10-Undecenoic acid
(undecylenic acid)

(*e*) Mevalonic acid has a five-carbon chain with hydroxyl groups at C-3 and C-5, along with a methyl group at C-3.

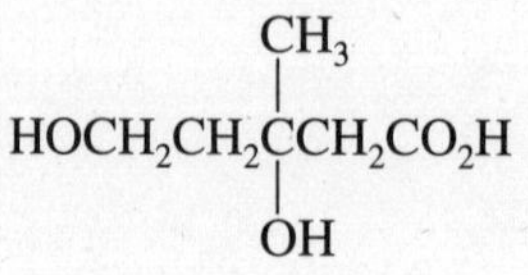

$$HOCH_2CH_2\overset{\displaystyle CH_3}{\underset{\displaystyle OH}{C}}CH_2CO_2H$$

3,5-Dihydroxy-3-methylpentanoic acid
(mevalonic acid)

(*f*) The constitution represented by the systematic name 2-methyl-2-butenoic acid gives rise to two stereoisomers.

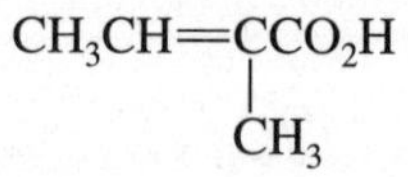

$$CH_3CH{=}\underset{\displaystyle CH_3}{C}CO_2H$$

2-Methyl-2-butenoic acid

Tiglic acid is the *E* isomer, and the *Z* isomer is known as **angelic acid.** The higher ranked substituents, methyl and carboxyl, are placed on opposite sides of the double bond in tiglic acid and on the same side in angelic acid.

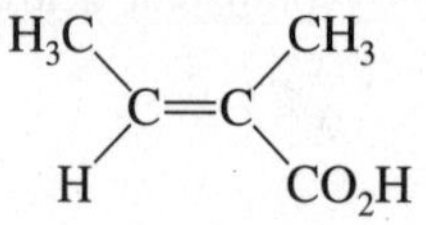

$$(H_3C)(H)C{=}C(CH_3)(CO_2H) \qquad (H_3C)(H)C{=}C(CO_2H)(CH_3)$$

(*E*)-2-Methyl-2-butenoic acid (tiglic acid)

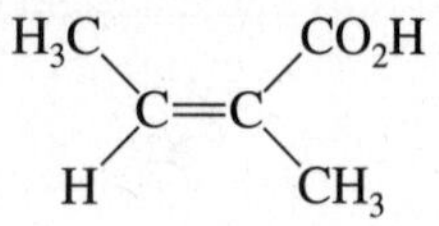

(*Z*)-2-Methyl-2-butenoic acid (angelic acid)

(*g*) Butanedioic acid is a four-carbon chain in which both terminal carbons are carboxylic acid groups. Malic acid has a hydroxyl group at C-2.

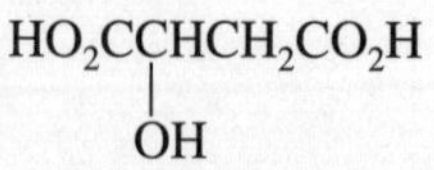

$$HO_2C\underset{\displaystyle OH}{C}HCH_2CO_2H$$

2-Hydroxybutanedioic acid
(malic acid)

(*h*) Each of the carbon atoms of propane bears a carboxyl group as a substituent in 1,2,3-propanetricarboxylic acid. In citric acid C-2 also bears a hydroxyl group.

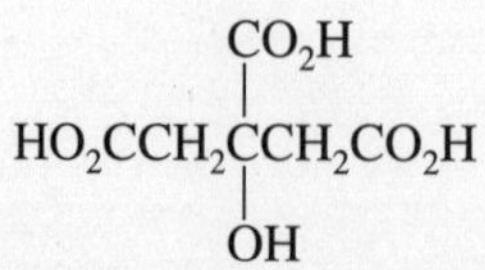

$$HO_2CCH_2\overset{\displaystyle CO_2H}{\underset{\displaystyle OH}{C}}CH_2CO_2H$$

2-Hydroxy-1,2,3-propanetricarboxylic acid
(citric acid)

(*i*) There is an aryl substituent at C-2 of propanoic acid in ibuprofen. This aryl substituent is a benzene ring bearing an isobutyl group at the para position.

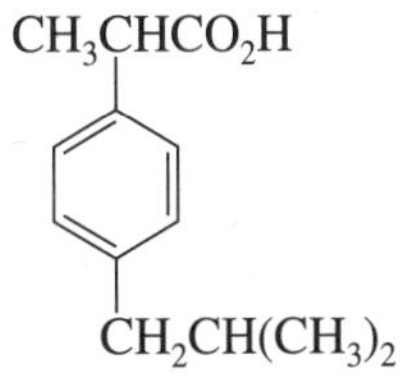

$CH_3CHCO_2H$ (benzene ring, para) $CH_2CH(CH_3)_2$

2-(*p*-Isobutylphenyl)-propanoic acid

(*j*) Benzenecarboxylic acid is the systematic name for benzoic acid. **Salicylic acid** is a derivative of benzoic acid bearing a hydroxyl group at the position ortho to the carboxyl.

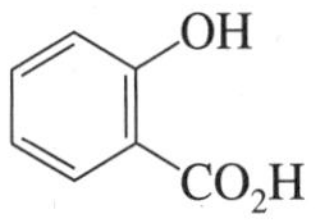

OH, $CO_2H$

*o*-Hydroxybenzenecarboxylic acid (salicylic acid)

**19.14** (*a*) The carboxylic acid contains a linear chain of eight carbon atoms. The parent alkane is **octane,** and so the systematic name of $CH_3(CH_2)_6CO_2H$ is **octanoic acid.**

(*b*) The compound shown is the potassium salt of octanoic acid. It is **potassium octanoate.**

(*c*) The presence of a double bond in $CH_2{=}CH(CH_2)_5CO_2H$ is indicated by the ending *-enoic acid.* Numbering of the chain begins with the carboxylic acid, and so the double bond is between C-7 and C-8. The compound is **7-octenoic acid.**

(*d*) Stereochemistry is systematically described by the *E–Z* notation. Here, the double bond between C-6 and C-7 in octenoic acid has the *Z* configuration; the higher ranked substituents are on the same side.

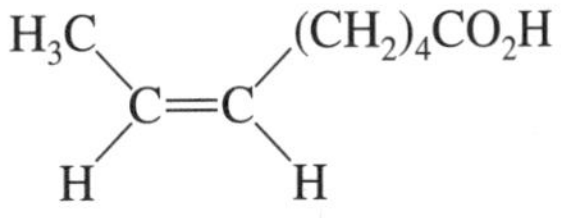

$H_3C$, $(CH_2)_4CO_2H$, C=C, H, H

(Z)-6-Octenoic acid

(*e*) A dicarboxylic acid is named as a **dioic acid.** The carboxyl functions are the terminal carbons of an eight-carbon chain; $HO_2C(CH_2)_6CO_2H$ is **octanedioic acid.** It is not necessary to identify the carboxylic acid locations by number because they can only be at the ends of the chain when the *-dioic acid* name is used.

(*f*) Pick the longest continuous chain that includes both carboxyl groups and name the compound as a *-dioic acid.* This chain contains only three carbons and bears a pentyl group as a substituent at C-2. It is not necessary to specify the position of the pentyl group, because it can only be attached to C-2.

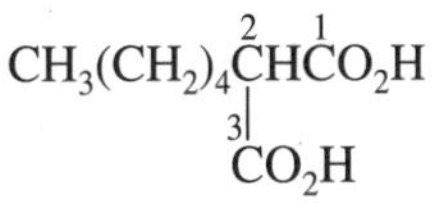

$CH_3(CH_2)_4\overset{2}{C}H\overset{1}{C}O_2H$, $\overset{3}{C}O_2H$

Pentylpropanedioic acid

Malonic acid is an acceptable synonym for propanedioic acid; this compound may also be named **pentylmalonic acid.**

(*g*) A carboxylic acid function is attached as a substituent on a seven-membered ring. The compound is **cycloheptanecarboxylic acid.**

$CO_2H$

(*h*) The aromatic ring is named as a substituent attached to the eight-carbon carboxylic acid. Numbering of the chain begins with the carboxyl group.

$C_6H_5CH(CH_2CH_3)(CH_2)_4CO_2H$

6-Phenyloctanoic acid

**19.15** (*a*) Carboxylic acids are the most acidic class of organic compounds containing only the elements C, H, and O. The order of decreasing acidity is

| | | $K_a$ | $pK_a$ |
|---|---|---|---|
| Acetic acid | $CH_3CO_2H$ | $1.8 \times 10^{-5}$ | 4.7 |
| Ethanol | $CH_3CH_2OH$ | $10^{-16}$ | 16 |
| Ethane | $CH_3CH_3$ | $\approx 10^{-46}$ | $\approx 46$ |

(*b*) Here again, the carboxylic acid is the strongest acid and the hydrocarbon the weakest:

| | | $K_a$ | $pK_a$ |
|---|---|---|---|
| Benzoic acid | $C_6H_5CO_2H$ | $6.7 \times 10^{-5}$ | 4.2 |
| Benzyl alcohol | $C_6H_5CH_2OH$ | $10^{-16}$–$10^{-18}$ | 16–18 |
| Benzene | $C_6H_6$ | $\approx 10^{-43}$ | $\approx 43$ |

(*c*) Propanedioic acid is a stronger acid than propanoic acid because the electron-withdrawing effect of one carboxyl group enhances the ionization of the other. Propanedial is a 1,3-dicarbonyl compound that yields a stabilized enolate; it is more acidic than 1,3-propanediol.

| | | $K_a$ | $pK_a$ |
|---|---|---|---|
| Propanedioic acid | $HO_2CCH_2CO_2H$ | $1.4 \times 10^{-3}$ | 2.9 |
| Propanoic acid | $CH_3CH_2CO_2H$ | $1.3 \times 10^{-5}$ | 4.9 |
| Propanedial | $O{=}CHCH_2CH{=}O$ | $\approx 10^{-9}$ | $\approx 9$ |
| 1,3-Propanediol | $HOCH_2CH_2CH_2OH$ | $\approx 10^{-16}$ | $\approx 16$ |

(*d*) Trifluoromethanesulfonic acid is by far the strongest acid in the group. It is structurally related to sulfuric acid, but its three fluorine substituents make it much stronger. Fluorine substituents increase the acidity of carboxylic acids and alcohols relative to their nonfluorinated analogs, but not enough to make fluorinated alcohols as acidic as carboxylic acids.

| | | $K_a$ | p$K_a$ |
|---|---|---|---|
| Trifluoromethanesulfonic acid | $CF_3SO_2OH$ | $10^6$ | −6 |
| Trifluoroacetic acid | $CF_3CO_2H$ | $5.9 \times 10^{-1}$ | 0.2 |
| Acetic acid | $CH_3CO_2H$ | $1.8 \times 10^{-5}$ | 4.7 |
| 2,2,2-Trifluoroethanol | $CF_3CH_2OH$ | $4.2 \times 10^{-13}$ | 12.4 |
| Ethanol | $CH_3CH_2OH$ | $\approx 10^{-16}$ | ≈16 |

(*e*) The order of decreasing acidity is carboxylic acid > $\beta$-diketone > ketone > hydrocarbon.

| | | $K_a$ | p$K_a$ |
|---|---|---|---|
| Cyclopentanecarboxylic acid | (cyclopentane)–$CO_2H$ | $1 \times 10^{-5}$ | 5.0 |
| 2,4-Pentanedione | $CH_3\overset{O}{\overset{\parallel}{C}}CH_2\overset{O}{\overset{\parallel}{C}}CH_3$ | $10^{-9}$ | 9 |
| Cyclopentanone | (cyclopentane)=O | $10^{-20}$ | 20 |
| Cyclopentene | (cyclopentene) | $10^{-45}$ | 45 |

**19.16** (*a*) A trifluoromethyl group is strongly electron-withdrawing and acid-strengthening. Its ability to attract electrons from the carboxylate ion decreases as its distance down the chain increases. 3,3,3-Trifluoropropanoic acid is a stronger acid than 4,4,4-trifluorobutanoic acid.

| $CF_3CH_2CO_2H$ | $CF_3CH_2CH_2CO_2H$ |
|---|---|
| 3,3,3-Trifluoropropanoic acid | 4,4,4-Trifluorobutanoic acid |
| $K_a$ $9.6 \times 10^{-4}$ (p$K_a$ 3.0) | $K_a$ $6.9 \times 10^{-5}$ (p$K_a$ 4.2) |

(*b*) The carbon that bears the carboxyl group in 2-butynoic acid is *sp*-hybridized and is, therefore, more electron-withdrawing than the $sp^3$-hybridized $\alpha$ carbon of butanoic acid. The anion of 2-butynoic acid is therefore stabilized better than the anion of butanoic acid, and 2-butynoic acid is a stronger acid.

| $CH_3C{\equiv}CCO_2H$ | $CH_3CH_2CH_2CO_2H$ |
|---|---|
| 2-Butynoic acid | Butanoic acid |
| $K_a$ $2.5 \times 10^{-3}$ (p$K_a$ 2.6) | $K_a$ $1.5 \times 10^{-5}$ (p$K_a$ 4.8) |

(*c*) Cyclohexanecarboxylic acid is a typical aliphatic carboxylic acid and is expected to be similar to acetic acid in acidity. The greater electronegativity of the $sp^2$-hybridized carbon attached to the carboxyl group in benzoic acid stabilizes benzoate anion better than the corresponding $sp^3$-hybridized carbon stabilizes cyclohexanecarboxylate. Benzoic acid is a stronger acid.

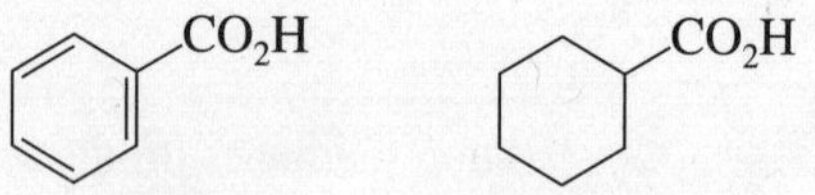

Benzoic acid
$K_a$ 6.7 × $10^{-5}$
(p$K_a$ 4.2)

Cyclohexanecarboxylic acid
$K_a$ 1.2 × $10^{-5}$
(p$K_a$ 4.9)

(*d*) Its five fluorine substituents make the pentafluorophenyl group more electron-withdrawing than an unsubstituted phenyl group. Thus, pentafluorobenzoic acid is a stronger acid than benzoic acid.

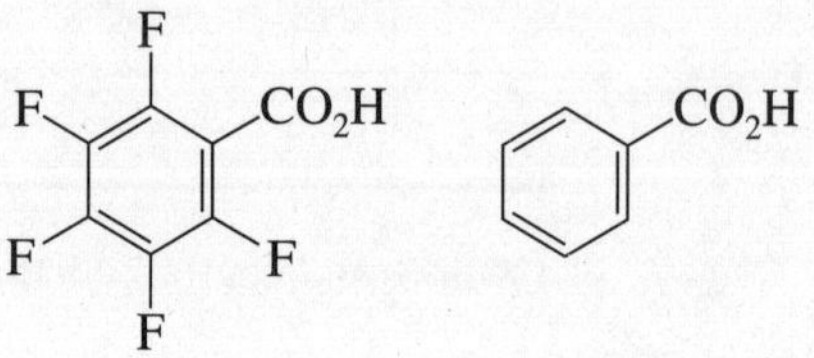

Pentafluorobenzoic acid
$K_a$ 4.1 × $10^{-4}$
(p$K_a$ 3.4)

Benzoic acid
$K_a$ 6.7 × $10^{-5}$
(p$K_a$ 4.2)

(*e*) The pentafluorophenyl substituent is electron-withdrawing and increases the acidity of a carboxyl group to which it is attached. Its electron-withdrawing effect decreases with distance. Pentafluorobenzoic acid is a stronger acid than *p*-(pentafluorophenyl)benzoic acid.

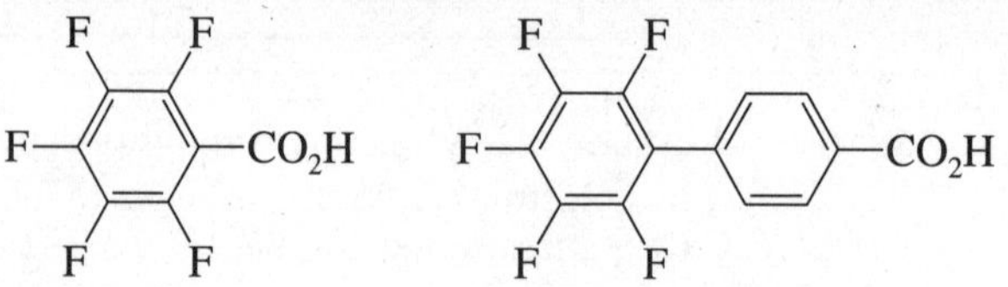

Pentafluorobenzoic acid
$K_a$ 4.1 × $10^{-4}$
(p$K_a$ 3.4)

*p*-(Pentafluorophenyl)benzoic acid
($K_a$ not measured in water; comparable with benzoic acid in acidity)

(*f*) The oxygen of the ring exercises an acidifying effect on the carboxyl group. This effect is largest when the oxygen is attached directly to the carbon that bears the carboxyl group. Furan-2-carboxylic acid is thus a stronger acid than furan-3-carboxylic acid.

Furan-2-carboxylic acid
$K_a$ 6.9 × $10^{-4}$
(p$K_a$ 3.2)

Furan-3-carboxylic acid
$K_a$ 1.1 × $10^{-4}$
(p$K_a$ 3.9)

(*g*) Furan-2-carboxylic acid has an oxygen attached to the carbon that bears the carboxyl group, whereas pyrrole-2-carboxylic acid has a nitrogen in that position. Oxygen is more electronegative than nitrogen and so stabilizes the carboxylate anion better. Furan-2-carboxylic acid is a stronger acid than pyrrole-2-carboxylic acid.

| Furan-2-carboxylic acid | Pyrrole-3-carboxylic acid |
| --- | --- |
| $K_a$ $6.9 \times 10^{-4}$ | $K_a$ $3.5 \times 10^{-5}$ |
| ($pK_a$ 3.2) | ($pK_a$ 4.4) |

**19.17** (*a*) The conversion of 1-butanol to butanoic acid is simply the oxidation of a primary alcohol to a carboxylic acid. Chromic acid is a suitable oxidizing agent.

$$\underset{\text{1-Butanol}}{CH_3CH_2CH_2CH_2OH} \xrightarrow{H_2CrO_4} \underset{\text{Butanoic acid}}{CH_3CH_2CH_2CO_2H}$$

(*b*) Aldehydes may be oxidized to carboxylic acids by any of the oxidizing agents that convert primary alcohols to carboxylic acids.

$$\underset{\text{Butanal}}{CH_3CH_2CH_2\overset{O}{\overset{\|}{C}}H} \xrightarrow[H_2SO_4,\ H_2O]{K_2Cr_2O_7} \underset{\text{Butanoic acid}}{CH_3CH_2CH_2\overset{O}{\overset{\|}{C}}OH}$$

(*c*) The starting material has the same number of carbon atoms as does butanoic acid, and so all that is required is a series of functional group transformations. Carboxylic acids may be obtained by oxidation of the corresponding primary alcohol. The alcohol is available from the designated starting material, 1-butene.

$$CH_3CH_2CH_2CO_2H \Longrightarrow CH_3CH_2CH_2CH_2OH \Longrightarrow CH_3CH_2CH{=}CH_2$$

Hydroboration–oxidation of 1-butene yields 1-butanol, which can then be oxidized to butanoic acid as in part (*a*).

$$\underset{\text{1-Butene}}{CH_3CH_2CH{=}CH_3} \xrightarrow[2.\ H_2O_2,\ HO^-]{1.\ B_2H_6} \underset{\text{1-Butanol}}{CH_3CH_2CH_2CH_2OH} \xrightarrow{H_2CrO_4} \underset{\text{Butanoic acid}}{CH_3CH_2CH_2CO_2H}$$

(*d*) Converting 1-propanol to butanoic acid requires the carbon chain to be extended by one atom. Both methods for achieving this conversion, carboxylation of a Grignard reagent and formation and hydrolysis of a nitrile, begin with alkyl halides. Alkyl halides in turn are prepared from alcohols.

$$CH_3CH_2CH_2CO_2H \Longrightarrow CH_3CH_2CH_2MgBr \text{ or } CH_3CH_2CH_2CN$$

$$\Downarrow$$

$$CH_3CH_2CH_2OH \Longleftarrow CH_3CH_2CH_2Br$$

Either of the two following procedures is satisfactory:

$$\underset{\text{1-Propanol}}{CH_3CH_2CH_2OH} \xrightarrow[\text{or HBr}]{PBr_3} \underset{\text{1-Bromopropane}}{CH_3CH_2CH_2Br} \xrightarrow[\text{diethyl ether}]{Mg} CH_3CH_2CH_2MgBr \xrightarrow[2.\ H_3O^+]{1.\ CO_2} \underset{\text{Butanoic acid}}{CH_3CH_2CH_2CO_2H}$$

$$\underset{\text{1-Propanol}}{CH_3CH_2CH_2OH} \xrightarrow[\text{or HBr}]{PBr_3} \underset{\text{1-Bromopropane}}{CH_3CH_2CH_2Br} \xrightarrow[\text{DMSO}]{KCN} \underset{\text{Butanenitrile}}{CH_3CH_2CH_2CN} \xrightarrow[\text{heat}]{H_2O,\ HCl} \underset{\text{Butanoic acid}}{CH_3CH_2CH_2CO_2H}$$

(*e*) Dehydration of 2-propanol to propene followed by free-radical addition of hydrogen bromide affords 1-bromopropane.

$$\underset{\text{2-Propanol}}{CH_3\underset{OH}{\underset{|}{C}}HCH_3} \xrightarrow[\text{heat}]{H_2SO_4} \underset{\text{Propene}}{CH_3CH{=}CH_2} \xrightarrow[\text{peroxides}]{HBr} \underset{\text{1-Bromopropane}}{CH_3CH_2CH_2Br}$$

Once 1-bromopropane has been prepared it is converted to butanoic acid as in part (*d*).

(*f*) The carbon skeleton of butanoic acid may be assembled by an aldol condensation of acetaldehyde.

$$CH_3CH_2CH_2CO_2H \Longrightarrow CH_3CH{\not=}CH\overset{O}{\overset{\|}{C}}H \Longrightarrow 2CH_3\overset{O}{\overset{\|}{C}}H$$

$$\underset{\text{Acetaldehyde}}{2CH_3\overset{O}{\overset{\|}{C}}H} \xrightarrow[\text{heat}]{\text{KOH, ethanol}} \underset{\text{2-Butenal}}{CH_3CH{=}CH\overset{O}{\overset{\|}{C}}H}$$

Oxidation of the aldehyde followed by hydrogenation of the double bond yields butanoic acid.

$$\underset{\text{2-Butenal}}{CH_3CH{=}CH\overset{O}{\overset{\|}{C}}H} \xrightarrow{H_2CrO_4} \underset{\text{2-Butenoic acid}}{CH_3CH{=}CHCO_2H} \xrightarrow[Pt]{H_2} \underset{\text{Butanoic acid}}{CH_3CH_2CH_2CO_2H}$$

(*g*) Ethylmalonic acid belongs to the class of substituted malonic acids that undergo ready thermal decarboxylation. Decarboxylation yields butanoic acid.

$$\underset{\text{Ethylmalonic acid}}{CH_3CH_2\underset{CO_2H}{\underset{|}{C}}HCO_2H} \xrightarrow{\text{heat}} \underset{\text{Butanoic acid}}{CH_3CH_2CH_2CO_2H} + \underset{\text{Carbon dioxide}}{CO_2}$$

**19.18** (*a*) The Friedel–Crafts alkylation of benzene by methyl chloride can be used to prepare $^{14}C$-labeled toluene ($C^* = {}^{14}C$). Once prepared, toluene could be oxidized to benzoic acid.

$$\underset{\text{Benzene}}{C_6H_6} + \underset{\text{Methyl chloride}}{\overset{*}{C}H_3Cl} \xrightarrow{AlCl_3} \underset{\text{Toluene}}{C_6H_5\overset{*}{C}H_3} \xrightarrow[H_2O,\ \text{heat}]{K_2Cr_2O_7,\ H_2SO_4} \underset{\text{Benzoic acid}}{C_6H_5\overset{*}{C}O_2H}$$

(*b*) Formaldehyde can serve as a one-carbon source if it is attacked by the Grignard reagent derived from bromobenzene.

$$\text{C}_6\text{H}_6 \xrightarrow[\text{FeBr}_3]{\text{Br}_2} \text{C}_6\text{H}_5\text{Br} \xrightarrow[\text{diethyl ether}]{\text{Mg}} \text{C}_6\text{H}_5\text{MgBr} \xrightarrow[\text{2. H}_3\text{O}^+]{\text{1. HC}^*\text{(=O)H}} \text{C}_6\text{H}_5\overset{*}{\text{C}}\text{H}_2\text{OH}$$

Benzene Bromobenzene Phenylmagnesium bromide Benzyl alcohol

This sequence yields $^{14}$C-labeled benzyl alcohol, which can be oxidized to $^{14}$C-labeled benzoic acid.

$$\text{C}_6\text{H}_5\overset{*}{\text{C}}\text{H}_2\text{OH} \xrightarrow[\text{H}_2\text{O}]{\text{K}_2\text{Cr}_2\text{O}_7,\ \text{H}_2\text{SO}_4} \text{C}_6\text{H}_5\overset{*}{\text{C}}\text{O}_2\text{H}$$

Benzyl alcohol Benzoic acid

(*c*) A direct route to $^{14}$C-labeled benzoic acid utilizes a Grignard synthesis employing $^{14}$C-labeled carbon dioxide.

$$\text{C}_6\text{H}_6 \xrightarrow[\text{FeBr}_3]{\text{Br}_2} \text{C}_6\text{H}_5\text{Br} \xrightarrow[\text{diethyl ether}]{\text{Mg}} \text{C}_6\text{H}_5\text{MgBr} \xrightarrow[\text{2. H}_3\text{O}^+]{\text{1. }\overset{*}{\text{C}}\text{O}_2} \text{C}_6\text{H}_5\overset{*}{\text{C}}\text{O}_2\text{H}$$

Benzene Bromobenzene Phenylmagnesium bromide Benzoic acid

**19.19** (*a*) An acid–base reaction takes place when pentanoic acid is combined with sodium hydroxide.

$$\text{CH}_3\text{CH}_2\text{CH}_2\text{CH}_2\text{CO}_2\text{H} + \text{NaOH} \longrightarrow \text{CH}_3\text{CH}_2\text{CH}_2\text{CH}_2\text{CO}_2\text{Na} + \text{H}_2\text{O}$$

Pentanoic acid Sodium hydroxide Sodium pentanoate Water

(*b*) Carboxylic acids react with sodium bicarbonate to give carbonic acid, which dissociates to carbon dioxide and water, so that the actual reaction that takes place is

$$\text{CH}_3\text{CH}_2\text{CH}_2\text{CH}_2\text{CO}_2\text{H} + \text{NaHCO}_3 \longrightarrow \text{CH}_3\text{CH}_2\text{CH}_2\text{CH}_2\text{CO}_2\text{Na} + \text{CO}_2 + \text{H}_2\text{O}$$

Pentanoic acid Sodium bicarbonate Sodium pentanoate Carbon dioxide Water

(*c*) Thionyl chloride is a reagent that converts carboxylic acids to the corresponding acyl chlorides.

$$\text{CH}_3\text{CH}_2\text{CH}_2\text{CH}_2\text{CO}_2\text{H} + \text{SOCl}_2 \longrightarrow \text{CH}_3\text{CH}_2\text{CH}_2\text{CH}_2\overset{\text{O}}{\overset{\|}{\text{C}}}\text{Cl} + \text{SO}_2 + \text{HCl}$$

Pentanoic acid Thionyl chloride Pentanoyl chloride Sulfur dioxide Hydrogen chloride

(*d*) Phosphorus tribromide is used to convert carboxylic acids to their acyl bromides.

$$3\text{CH}_3\text{CH}_2\text{CH}_2\text{CH}_2\text{CO}_2\text{H} + \text{PBr}_3 \longrightarrow 3\text{CH}_3\text{CH}_2\text{CH}_2\text{CH}_2\overset{\text{O}}{\overset{\|}{\text{C}}}\text{Br} + \text{H}_3\text{PO}_3$$

Pentanoic acid Phosphorus tribromide Pentanoyl bromide Phosphorus acid (*e*) Carboxylic acids react with alcohols in the presence of acid catalysts to give esters.

$$\mathrm{CH_3CH_2CH_2CH_2CO_2H + C_6H_5CH_2OH \xrightleftharpoons{H_2SO_4} CH_3CH_2CH_2CH_2\overset{\overset{\displaystyle O}{\|}}{C}OCH_2C_6H_5 + H_2O}$$

Pentanoic acid, Benzyl alcohol, Benzyl pentanoate, Water

(*f*) Chlorine is introduced at the $\alpha$-carbon atom of a carboxylic acid. The reaction is catalyzed by a small amount of phosphorus or a phosphorus trihalide and is called the Hell–Volhard–Zelinsky reaction.

$$\mathrm{CH_3CH_2CH_2CH_2CO_2H + Cl_2 \xrightarrow[(catalyst)]{PBr_3} CH_3CH_2CH_2\underset{\underset{\displaystyle Cl}{|}}{C}HCO_2H + HCl}$$

Pentanoic acid, Chlorine, 2-Chloropentanoic acid, Hydrogen chloride

The $\alpha$-halo substituent is derived from the halogen used, not from the phosphorus trihalide.

(*g*) In the case, bromine is introduced at the $\alpha$ carbon.

$$\mathrm{CH_3CH_2CH_2CH_2CO_2H + Br_2 \xrightarrow[(catalyst)]{PCl_3} CH_3CH_2CH_2\underset{\underset{\displaystyle Br}{|}}{C}HCO_2H + HBr}$$

Pentanoic acid, Bromine, 2-Bromopentanoic acid, Hydrogen bromide

(*h*) $\alpha$-Halo carboxylic acids are reactive substrates in nucleophilic substitution. Iodide acts as a nucleophile to displace bromide from 2-bromopentanoic acid.

$$\mathrm{CH_3CH_2CH_2\underset{\underset{\displaystyle Br}{|}}{C}HCO_2H + NaI \xrightarrow{acetone} CH_3CH_2CH_2\underset{\underset{\displaystyle I}{|}}{C}HCO_2H + NaBr}$$

2-Bromopentanoic acid, Sodium iodide, 2-Iodopentanoic acid, Sodium bromide

(*i*) Aqueous ammonia converts $\alpha$-halo acids to $\alpha$-amino acids.

$$\mathrm{CH_3CH_2CH_2\underset{\underset{\displaystyle Br}{|}}{C}HCO_2H + 2NH_3 \longrightarrow CH_3CH_2CH_2\underset{\underset{\displaystyle NH_2}{|}}{C}HCO_2H + NH_4Br}$$

2-Bromopentanoic acid, Ammonia, 2-Aminopentanoic acid, Ammonium bromide

(*j*) Lithium aluminum hydride is a powerful reducing agent and reduces carboxylic acids to primary alcohols.

$$\mathrm{CH_3CH_2CH_2CH_2CO_2H \xrightarrow[2.\ H_2O]{1.\ LiAlH_4} CH_3CH_2CH_2CH_2CH_2OH}$$

Pentanoic acid, 1-Pentanol

(*k*) Phenylmagnesium bromide acts as a base to abstract the carboxylic acid proton.

$$\mathrm{CH_3CH_2CH_2CH_2CO_2H + C_6H_5MgBr \longrightarrow CH_3CH_2CH_2CH_2CO_2MgBr + C_6H_6}$$

Pentanoic acid, Phenylmagnesium bromide, Bromomagnesium pentanoate, Benzene

Grignard reagents are not compatible with carboxylic acids; proton transfer converts the Grignard reagent to the corresponding hydrocarbon.

**19.20** (*a*) Conversion of butanoic acid to 1-butanol is a reduction and requires lithium aluminum hydride as the reducing agent.

$$\underset{\text{Butanoic acid}}{CH_3CH_2CH_2\overset{O}{\overset{\|}{C}}OH} \xrightarrow[2.\ H_2O]{1.\ LiAlH_4} \underset{\text{1-Butanol}}{CH_3CH_2CH_2CH_2OH}$$

(*b*) Carboxylic acids cannot be reduced directly to aldehydes. The following two-step procedure may be used:

$$\underset{\text{Butanoic acid}}{CH_3CH_2CH_2\overset{O}{\overset{\|}{C}}OH} \xrightarrow[2.\ H_2O]{1.\ LiAlH_4} \underset{\text{1-Butanol}}{CH_3CH_2CH_2CH_2OH} \xrightarrow[CH_2Cl_2]{PCC} \underset{\text{Butanal}}{CH_3CH_2CH_2\overset{O}{\overset{\|}{C}}H}$$

(*c*) Remember that alkyl halides are usually prepared from alcohols. 1-Butanol is therefore needed in order to prepare 1-chlorobutane.

$$\underset{\text{1-Butanol [from part (a)]}}{CH_3CH_2CH_2CH_2OH} \xrightarrow{SOCl_2} \underset{\text{1-Chlorobutane}}{CH_3CH_2CH_2CH_2Cl}$$

(*d*) Carboxylic acids are converted to their corresponding acyl chlorides with thionyl chloride.

$$\underset{\text{Butanoic acid}}{CH_3CH_2CH_2\overset{O}{\overset{\|}{C}}OH} \xrightarrow{SOCl_2} \underset{\text{Butanoyl chloride}}{CH_3CH_2CH_2\overset{O}{\overset{\|}{C}}Cl}$$

(*e*) Aromatic ketones are frequently prepared by Friedel–Crafts acylation of the appropriate acyl chloride and benzene. Butanoyl chloride, prepared in part (*d*), can be used to acylate benzene in a Friedel–Crafts reaction.

$$\underset{\text{Benzene}}{C_6H_6} + \underset{\text{Butanoyl chloride}}{CH_3CH_2CH_2\overset{O}{\overset{\|}{C}}Cl} \xrightarrow{AlCl_3} \underset{\text{Phenyl propyl ketone}}{C_6H_5\overset{O}{\overset{\|}{C}}CH_2CH_2CH_3}$$

(*f*) The preparation of 4-octanone using compounds derived from butanoic acid may be seen by using disconnections in a retrosynthetic analysis.

$$CH_3CH_2CH_2\overset{O}{\overset{\|}{C}}CH_2CH_2CH_2CH_3 \Longrightarrow CH_3CH_2CH_2\overset{OH}{\overset{|}{C}}H \wr CH_2CH_2CH_2CH_3$$

$$\Downarrow$$

$$CH_3CH_2CH_2\overset{O}{\overset{\|}{C}}H + CH_3CH_2CH_2CH_2MgBr$$

The reaction scheme which may be used is

$$\underset{\substack{\text{1-Chlorobutane}\\ \text{[from part (c)]}}}{CH_3CH_2CH_2CH_2Cl} \xrightarrow{Mg} \underset{\text{Butylmagnesium chloride}}{CH_3CH_2CH_2CH_2MgCl}$$

$$\underset{\text{Butylmagnesium chloride}}{CH_3CH_2CH_2CH_2MgCl} + \underset{\text{Butanal [from part (b)]}}{CH_3CH_2CH_2\overset{O}{\overset{\|}{C}}H} \xrightarrow[\text{2. } H_3O^+]{\text{1. diethyl ether}} \underset{\text{4-Octanol}}{CH_3CH_2CH_2\underset{OH}{\underset{|}{C}}HCH_2CH_2CH_2CH_3}$$

$$\xrightarrow{H_2CrO_4} \underset{\text{4-Octanone}}{CH_3CH_2CH_2\overset{O}{\overset{\|}{C}}CH_2CH_2CH_2CH_3}$$

(*g*) Carboxylic acids are halogenated at their $\alpha$-carbon atom by the Hell–Volhard–Zelinsky reaction.

$$\underset{\text{Butanoic acid}}{CH_3CH_2CH_2CO_2H} \xrightarrow[Br_2]{P} \underset{\text{2-Bromobutanoic acid}}{CH_3CH_2\underset{Br}{\underset{|}{C}}HCO_2H}$$

A catalytic amount of $PCl_3$ may be used in place of phosphorus in the reaction.

(*h*) Dehydrohalogenation of 2-bromobutanoic acid gives 2-butenoic acid.

$$\underset{\text{2-Bromobutanoic acid}}{CH_3CH_2\underset{Br}{\underset{|}{C}}HCO_2H} \xrightarrow[\text{DMSO; 2. } H^+]{\text{1. } KOC(CH_3)_3} \underset{\text{2-Butenoic acid}}{CH_3CH{=}CHCO_2H}$$

**19.21** (*a*) The compound to be prepared is **glycine,** an $\alpha$-amino acid. The amino functional group can be introduced by a nucleophilic substitution reaction on an $\alpha$-halo acid, which is available by way of the Hell–Volhard–Zelinsky reaction.

$$\underset{\text{Acetic acid}}{CH_3CO_2H} \xrightarrow[P]{Br_2} \underset{\text{Bromoacetic acid}}{BrCH_2CO_2H} \xrightarrow{NH_3,\ H_2O} \underset{\substack{\text{Aminoacetic acid}\\ \text{(glycine)}}}{H_2NCH_2CO_2H}$$

(*b*) Phenoxyacetic acid is used as a fungicide. It can be prepared by a nucleophilic substitution using sodium phenoxide and bromoacetic acid.

$$\underset{\text{Bromoacetic acid}}{BrCH_2CO_2H} \xrightarrow[\text{2. } H^+]{\text{1. } C_6H_5ONa} \underset{\text{Phenoxyacetic acid}}{C_6H_5OCH_2CO_2H} + NaCl$$

(*c*) Cyanide ion is a good nucleophile and will displace bromide from bromoacetic acid.

$$BrCH_2CO_2H \xrightarrow[Na_2CO_3,\ H_2O]{NaCN} NCCH_2CO_2Na \xrightarrow{H^+} NCCH_2CO_2H$$

Bromoacetic acid [from part (*a*)] — Sodium cyanoacetate — Cyanoacetic acid

(*d*) Cyanoacetic acid, prepared as in part (*c*), serves as a convenient precursor to malonic acid. Hydrolysis of the nitrile substituent converts it to a carboxyl group.

$$NCCH_2CO_2H \xrightarrow[heat]{H_2O,\ H^+} HO_2CCH_2CO_2H$$

Cyanoacetic acid — Malonic acid

(*e*) Iodoacetic acid is not prepared directly from acetic acid but is derived by nucleophilic substitution of iodide in bromoacetic acid.

$$BrCH_2CO_2H \xrightarrow[acetone]{NaI} ICH_2CO_2H$$

Bromoacetic acid — Iodoacetic acid

(*f*) Two transformations need to be accomplished, $\alpha$ bromination and esterification. The correct sequence is bromination followed by esterification.

$$CH_3CO_2H \xrightarrow[P]{Br_2} BrCH_2CO_2H \xrightarrow[H^+]{CH_3CH_2OH} BrCH_2CO_2CH_2CH_3$$

Acetic acid — Bromoacetic acid — Ethyl bromoacetate

Reversing the order of steps is not appropriate. It must be the carboxylic acid that is subjected to halogenation because the Hell–Volhard–Zelinsky reaction is a reaction of carboxylic acids, not esters.

(*g*) The compound shown is an ylide. It can be prepared from ethyl bromoacetate as shown

$$BrCH_2CO_2CH_2CH_3 + (C_6H_5)_3P \longrightarrow (C_6H_5)_3\overset{+}{P}CH_2CO_2CH_2CH_3\ Br^-$$

Ethyl bromoacetate — Triphenylphosphine

$$\xrightarrow{NaOCH_2CH_3} (C_6H_5)_3\overset{+}{P}—\overset{\cdot\cdot\ -}{C}HCO_2CH_2CH_3$$

Ylide

The first step is a nucleophilic substitution of bromide by triphenylphosphine. Treatment of the derived triphenylphosphonium salt with base removes the relatively acidic $\alpha$ proton, forming the ylide. (For a review of ylide formation, refer to Section 17.12.)

(*h*) Reaction of the ylide formed in part (*g*) with benzaldehyde gives the desired alkene by a Wittig reaction.

$$(C_6H_5)_3\overset{+}{P}—\overset{\cdot\cdot\ -}{C}HCO_2CH_2CH_3 + C_6H_5\overset{O}{\overset{\|}{C}}H \longrightarrow C_6H_5CH{=}CHCO_2CH_2CH_3 + (C_6H_5)_3\overset{+}{P}—\overset{-}{O}$$

Ylide from part (*g*) — Benzaldehyde — Ethyl cinnamate — Triphenylphosphine oxide

**19.22** (*a*) Carboxylic acids are converted to ethyl esters when they are allowed to stand in ethanol in the presence of an acid catalyst.

$$(E)\text{-}CH_3CH{=}C(CH_3)CO_2H + CH_3CH_2OH \xrightarrow{H_2SO_4} (E)\text{-}CH_3CH{=}C(CH_3)COCH_2CH_3\ (C{=}O) + H_2O$$

(*E*)-2-Methyl-2-butenoic acid; Ethanol; Ethyl (*E*)-2-methyl-2-butenoate (74–80%); Water

(*b*) Lithium aluminum hydride, $LiAlH_4$, reduces carboxylic acids to primary alcohols. When $LiAlD_4$ is used, deuterium is transferred to the carbonyl carbon.

$$\text{cyclopropyl}{-}COH\ (C{=}O) \xrightarrow[\text{2. } H_2O]{\text{1. } LiAlD_4\text{, diethyl ether}} \text{cyclopropyl}{-}CD_2OH$$

Cyclopropanecarboxylic acid; 1-Cyclopropyl-1,1-dideuteriomethanol (75%)

Notice that deuterium is bonded only to carbon. The hydroxyl proton is derived from water, not from the reducing agent.

(*c*) In the presence of a catalytic amount of phosphorus, bromine reacts with carboxylic acids to yield the corresponding $\alpha$-bromo derivative.

$$\text{cyclohexyl}{-}CO_2H \xrightarrow[P]{Br_2} \text{1-Br-cyclohexyl}{-}CO_2H$$

Cyclohexanecarboxylic acid; 1-Bromocyclohexanecarboxylic acid (96%)

(*d*) Alkyl fluorides are not readily converted to Grignard reagents, and so it is the bromine substituent that is attacked by magnesium.

$$m\text{-}BrC_6H_4CF_3 \xrightarrow[\text{diethyl ether}]{Mg} m\text{-}BrMgC_6H_4CF_3 \xrightarrow[\text{2. } H_3O^+]{\text{1. } CO_2} m\text{-}HO_2CC_6H_4CF_3$$

*m*-Bromo(trifluoromethyl)benzene; *m*-(Trifluoromethyl)benzoic acid

(*e*) Cyano substituents are hydrolyzed to carboxyl groups in the presence of acid catalysts.

$$m\text{-}ClC_6H_4CH_2CN \xrightarrow[H_2SO_4\text{, heat}]{H_2O\text{, acetic acid}} m\text{-}ClC_6H_4CH_2CO_2H$$

*m*-Chlorobenzyl cyanide; *m*-Chlorophenylacetic acid (61%)

(*f*) The carboxylic acid function plays no part in this reaction; free-radical addition of hydrogen bromide to the carbon–carbon double bond occurs.

$$\underset{\text{10-Undecenoic acid}}{H_2C{=}CH(CH_2)_8CO_2H} \xrightarrow[\text{benzoyl peroxide}]{\text{HBr}} \underset{\substack{\text{11-Bromoundecanoic acid}\\(66\text{–}70\%)}}{BrCH_2CH_2(CH_2)_8CO_2H}$$

Recall that hydrogen bromide adds to alkenes in the presence of peroxides with a regioselectivity opposite to that of Markovnikov's rule.

**19.23** (*a*) The desired product and the starting material have the same carbon skeleton, and so all that is required is a series of functional group transformations. Recall that, as seen in Problem 19.17, a carboxylic acid may be prepared by oxidation of the corresponding primary alcohol. The needed alcohol is available from the appropriate alkene.

$$\underset{\substack{\textit{tert}\text{-Butyl}\\\text{alcohol}}}{(CH_3)_3COH} \xrightarrow[\text{heat}]{H^+} \underset{\text{2-Methylpropene}}{(CH_3)_2C{=}CH_2} \xrightarrow[\text{2. } H_2O_2,\ HO^-]{\text{1. } B_2H_6} \underset{\text{2-Methyl-1-propanol}}{(CH_3)_2CHCH_2OH} \xrightarrow[H_2SO_4,\ H_2O]{K_2Cr_2O_7} \underset{\substack{\text{2-Methylpropanoic}\\\text{acid}}}{(CH_3)_2CHCO_2H}$$

(*b*) The target molecule contains one more carbon than the starting material, and so a carbon–carbon bond-forming step is indicated. Two approaches are reasonable; one proceeds by way of nitrile formation and hydrolysis, the other by carboxylation of a Grignard reagent. In either case the key intermediate is 1-bromo-2-methylpropane.

$$\underset{\text{3-Methylbutanoic acid}}{(CH_3)_2CHCH_2CO_2H} \Longrightarrow \underset{\substack{\text{1-Bromo-2-}\\\text{methylpropane}}}{(CH_3)_2CHCH_2Br}$$

The desired alkyl bromide may be prepared by free-radical addition of hydrogen bromide to 2-methylpropene.

$$\underset{\substack{\textit{tert}\text{-Butyl}\\\text{alcohol}}}{(CH_3)_3COH} \xrightarrow[\text{heat}]{H^+} \underset{\text{2-Methylpropene}}{(CH_3)_2C{=}CH_2} \xrightarrow[\text{peroxides}]{\text{HBr}} \underset{\substack{\text{1-Bromo-2-}\\\text{methylpropane}}}{(CH_3)_2CHCH_2Br}$$

Another route to the alkyl bromide utilizes the alcohol prepared in part (*a*).

$$\underset{\text{2-Methyl-1-propanol}}{(CH_3)_2CHCH_2OH} \xrightarrow[\text{or HBr}]{PBr_3} \underset{\substack{\text{1-Bromo-2-}\\\text{methylpropane}}}{(CH_3)_2CHCH_2Br}$$

Conversion of the alkyl bromide to the desired acid is then carried out as follows:

$$\underset{\text{1-Bromo-2-methylpropane}}{(CH_3)_2CHCH_2Br} \xrightarrow{\text{KCN}} (CH_3)_2CHCH_2CN \xrightarrow[\text{heat}]{H_2O,\ H^+} \underset{\text{3-Methylbutanoic acid}}{(CH_3)_2CHCH_2CO_2H}$$

$$\underset{\text{1-Bromo-2-methylpropane}}{(CH_3)_2CHCH_2Br} \xrightarrow[\text{diethyl ether}]{\text{Mg}} (CH_3)_2CHCH_2MgBr \xrightarrow[\text{2. } H_3O^+]{\text{1. } CO_2} \underset{\text{3-Methylbutanoic acid}}{(CH_3)_2CHCH_2CO_2H}$$

(*c*) Examining the target molecule reveals that it contains two more carbon atoms than the indicated starting material, suggesting use of ethylene oxide in a two-carbon chain-extension process.

$$(CH_3)_3CCH_2CO_2H \Longrightarrow (CH_3)_3C\text{-}\!\!\!\wr\,CH_2CH_2OH \Longrightarrow (CH_3)_3CMgX + H_2C\overset{O}{\text{—}}CH_2$$

This suggests the following sequence of steps:

$$\underset{\textit{tert}\text{-Butyl alcohol}}{(CH_3)_3COH} \xrightarrow{HBr} \underset{\text{2-Bromo-2-methylpropane}}{(CH_3)_3CBr} \xrightarrow{Mg} \underset{\textit{tert}\text{-Butylmagnesium bromide}}{(CH_3)_3CMgBr} \xrightarrow[2.\ H_3O^+]{1.\ H_2C\overset{O}{\text{—}}CH_2} \underset{\text{3,3-Dimethyl-1-butanol}}{(CH_3)_3CCH_2CH_2OH}$$

$$\xrightarrow{K_2Cr_2O_7,\ H^+,\ H_2O} \underset{\text{3,3-Dimethylbutanoic acid}}{(CH_3)_3CCH_2CO_2H}$$

(*d*) This synthesis requires extending a carbon chain by two carbon atoms. One way to form dicarboxylic acids is by hydrolysis of dinitriles.

$$HO_2C(CH_2)_5CO_2H \Longrightarrow NC(CH_2)_5CN \Longrightarrow Br(CH_2)_5Br$$

This suggests the following sequence of steps:

$$\underset{\text{Pentanedioic acid}}{HO_2C(CH_2)_3CO_2H} \xrightarrow[2.\ H_2O]{1.\ LiAlH_4} \underset{\text{1,5-Pentanediol}}{HOCH_2(CH_2)_3CH_2OH} \xrightarrow[PBr_3]{HBr\ or} \underset{\text{1,5-Dibromopentane}}{BrCH_2(CH_2)_3CH_2Br}$$

$$\xrightarrow{KCN} \underset{\text{1,5-Dicyanopentane}}{NCCH_2(CH_2)_3CH_2CN}$$

$$\underset{\text{Heptanedioic acid}}{HO_2CCH_2(CH_2)_3CH_2CO_2H \equiv HO_2C(CH_2)_5CO_2H} \xleftarrow[heat]{H_2O,\ H^+} NCCH_2(CH_2)_3CH_2CN$$

(*e*) The desired alcohol cannot be prepared directly from the nitrile. It is available, however, by lithium aluminum hydride reduction of the carboxylic acid obtained by hydrolysis of the nitrile.

$$\underset{\text{3-Phenylbutanenitrile}}{CH_3\underset{C_6H_5}{\underset{|}{C}}HCH_2CN} \xrightarrow[heat]{H_2O,\ H^+} \underset{\text{3-Phenylbutanoic acid}}{CH_3\underset{C_6H_5}{\underset{|}{C}}HCH_2\overset{O}{\overset{\|}{C}}OH} \xrightarrow[2.\ H_2O]{1.\ LiAlH_4} \underset{\text{3-Phenyl-1-butanol}}{CH_3\underset{C_6H_5}{\underset{|}{C}}HCH_2CH_2OH}$$

(*f*) In spite of the structural similarity between the starting material and the desired product, a one-step transformation cannot be achieved.

Cyclopentyl bromide $\not\rightarrow$ 1-Bromocyclopentanecarboxylic acid

Instead, recall that $\alpha$-bromo acids are prepared from carboxylic acids by the Hell–Volhard–Zelinsky reaction:

Cyclopentanecarboxylic acid $\xrightarrow[P]{Br_2}$ 1-Bromocyclopentanecarboxylic acid

The problem now simplifies to one of preparing cyclopentanecarboxylic acid from cyclopentyl bromide. Two routes are possible:

Cyclopentyl bromide $\xrightarrow{KCN}$ Cyclopentyl cyanide $\xrightarrow[heat]{H_2O, H^+}$ Cyclopentanecarboxylic acid

Cyclopentyl bromide $\xrightarrow[\text{diethyl ether}]{Mg}$ Cyclopentylmagnesium bromide $\xrightarrow[2.\ H_3O^+]{1.\ CO_2}$ Cyclopentanecarboxylic acid

The Grignard route is better; it is a "one-pot" transformation. Converting the secondary bromide to a nitrile will be accompanied by elimination, and the procedure requires two separate operations.

(*g*) In this case the halogen substituent is present at the $\beta$ carbon rather than the $\alpha$ carbon atom of the carboxylic acid. The starting material, a $\beta$-chloro unsaturated acid, can lead to the desired carbon skeleton by a Diels–Alder reaction.

1,3-Butadiene + (*E*)-3-Chloropropenoic acid $\longrightarrow$ *trans*-2-Chloro-4-cyclohexenecarboxylic acid

The required trans stereochemistry is a consequence of the stereospecificity of the Diels–Alder reaction.

Hydrogenation of the double bond of the Diels–Alder adduct gives the required product.

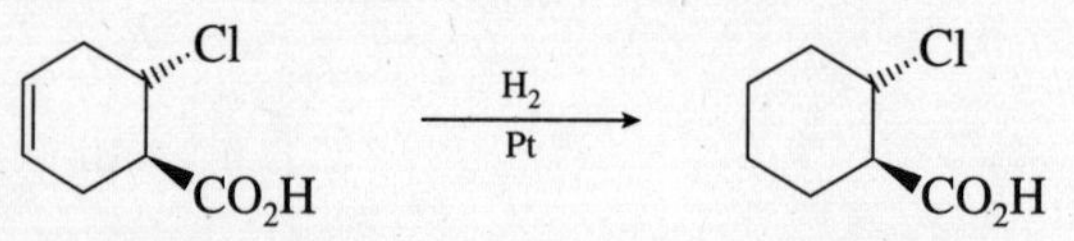

$\xrightarrow[\text{Pt}]{H_2}$

*trans*-2-Chloro-4-cyclohexenecarboxylic acid — *trans*-2-Chlorocyclohexanecarboxylic acid

(*h*) The target molecule is related to the starting material by the retrosynthesis

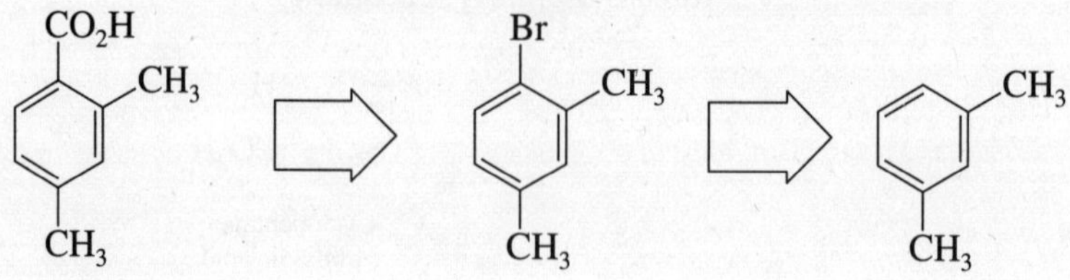

2,4-Dimethylbenzoic acid ⇒ (1-bromo-2,4-dimethylbenzene) ⇒ *m*-Xylene

The necessary bromine substituent can be introduced by electrophilic substitution in the activated aromatic ring of *m*-xylene.

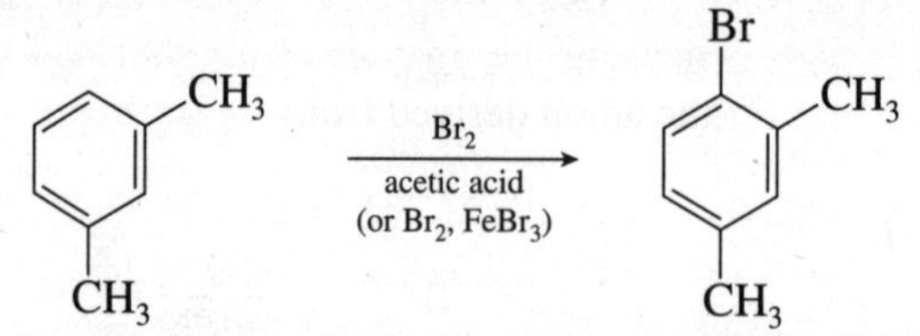

$\xrightarrow[\text{acetic acid (or } Br_2,\ FeBr_3)]{Br_2}$

*m*-Xylene — 1-Bromo-2,4-dimethylbenzene

The aryl bromide cannot be converted to a carboxylic acid by way of the corresponding nitrile, because aryl bromides are not reactive toward nucleophilic substitution. The Grignard route is necessary.

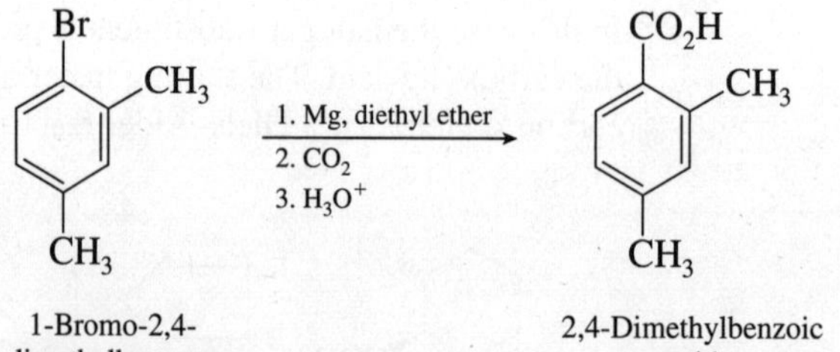

1. Mg, diethyl ether
2. $CO_2$
3. $H_3O^+$

1-Bromo-2,4-dimethylbenzene — 2,4-Dimethylbenzoic acid

(*i*) The relationship of the target molecule to the starting material

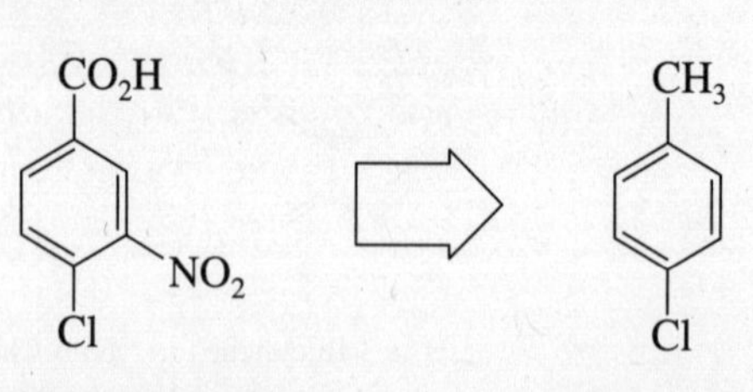

4-Chloro-3-nitrobenzoic acid ⇒ *p*-Chlorotoluene

requires that there be two synthetic operations: oxidation of the methyl group and nitration of the ring. The orientation of the nitro group requires that nitration must follow oxidation of the

methyl group of the starting material

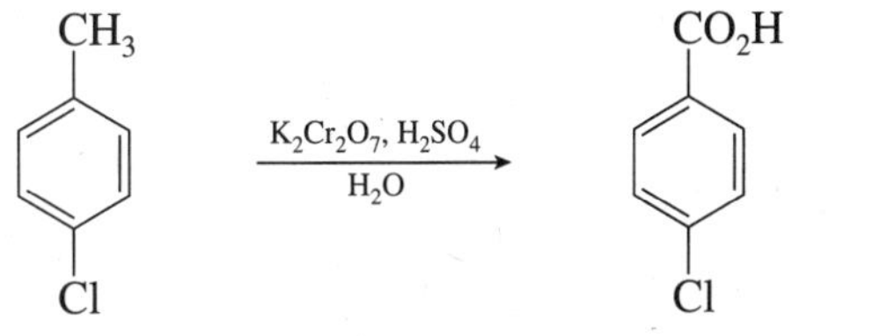

$$p\text{-Chlorotoluene } (CH_3 \text{ para to } Cl) \xrightarrow[H_2O]{K_2Cr_2O_7,\ H_2SO_4} p\text{-Chlorobenzoic acid } (CO_2H \text{ para to } Cl)$$

p-Chlorotoluene — p-Chlorobenzoic acid

Nitration of *p*-chlorobenzoic acid gives the desired product, because the directing effects of the chlorine (ortho, para) and the carboxyl (meta) groups reinforce each other.

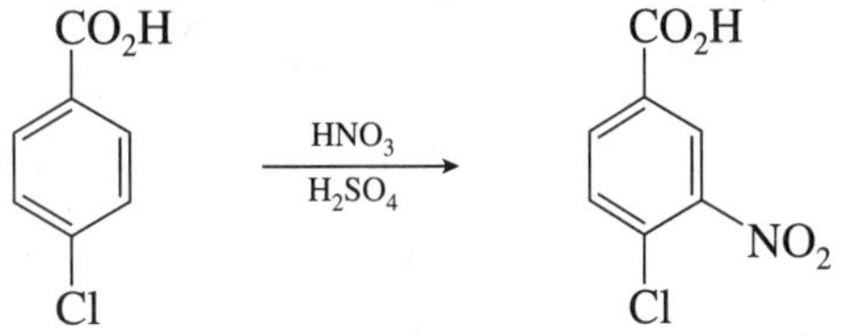

$$p\text{-Chlorobenzoic acid} \xrightarrow[H_2SO_4]{HNO_3} \text{4-Chloro-3-nitrobenzoic acid}$$

p-Chlorobenzoic acid — 4-Chloro-3-nitrobenzoic acid

(*j*) The desired synthetic route becomes apparent when it is recognized that the *Z* alkene stereoisomer may be obtained from an alkyne, which, in turn, is available by carboxylation of the anion derived from the starting material.

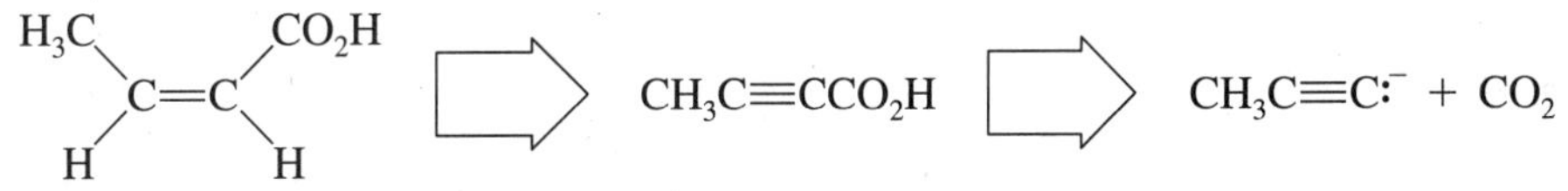

$$(Z)\text{-}H_3C(H)C{=}C(H)CO_2H \Rightarrow CH_3C{\equiv}CCO_2H \Rightarrow CH_3C{\equiv}C{:}^- + CO_2$$

The desired reaction sequence is

$$\underset{\text{Propyne}}{CH_3C{\equiv}CH} \xrightarrow[NH_3]{NaNH_2} \underset{\text{Propynylsodium}}{CH_3C{\equiv}CNa} \xrightarrow[2.\ H_3O^+]{1.\ CO_2} \underset{\text{2-Butynoic acid}}{CH_3C{\equiv}CCO_2H}$$

Hydrogenation of the carbon–carbon triple bond of 2-butynoic acid over the Lindlar catalyst converts this compound to the *Z* isomer of 2-butenoic acid.

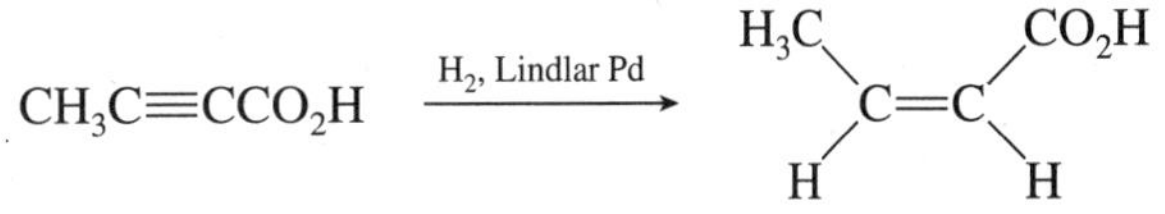

$$\underset{\text{2-Butynoic acid}}{CH_3C{\equiv}CCO_2H} \xrightarrow{H_2,\ \text{Lindlar Pd}} \underset{(Z)\text{-2-Butenoic acid}}{H_3C(H)C{=}C(H)CO_2H}$$

**19.24** (*a*) Only the cis stereoisomer of 4-hydroxycyclohexanecarboxylic acid is capable of forming a lactone, as can be seen in the following drawings or with a molecular model. The most stable conformation of the starting hydroxy acid is a chair conformation; however, in the lactone, the cyclohexane ring adopts a boat conformation.

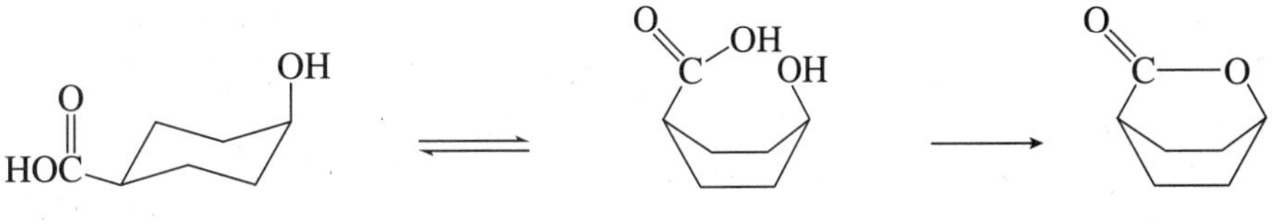

*cis*-4-Hydroxycyclohexane carboxylic acid (chair conformation) ⇌ *cis*-4-Hydroxycyclohexane carboxylic acid (boat conformation) ⟶ Lactone

(*b*) As in part (*a*), lactone formation is possible only when the hydroxyl and carboxyl groups are cis.

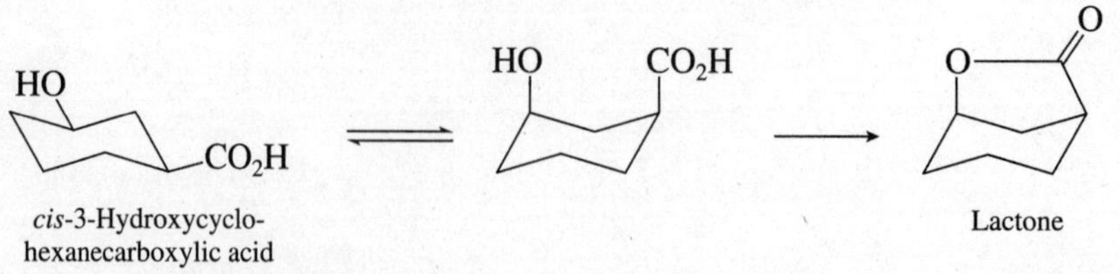

HO, $CO_2H$ ⇌ HO, $CO_2H$ ⟶ O, O

*cis*-3-Hydroxycyclohexanecarboxylic acid Lactone

Although the most stable conformation of *cis*-3-hydroxycyclohexanecarboxylic acid has both substituents equatorial and is unable to close to a lactone, the diaxial orientation is accessible and is capable of lactone formation.

Neither conformation of *trans*-3-hydroxycyclohexanecarboxylic acid has the substituents close enough to each other to form an unstrained lactone.

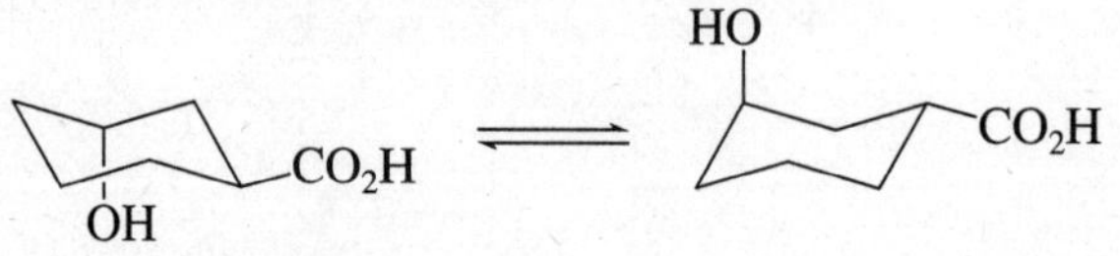

$CO_2H$, OH ⇌ HO, $CO_2H$

*trans*-3-Hydroxycyclohexanecarboxylic acid: lactone formation impossible

**19.25** (*a*) The most stable conformation of formic acid is the one that has both hydrogens anti.

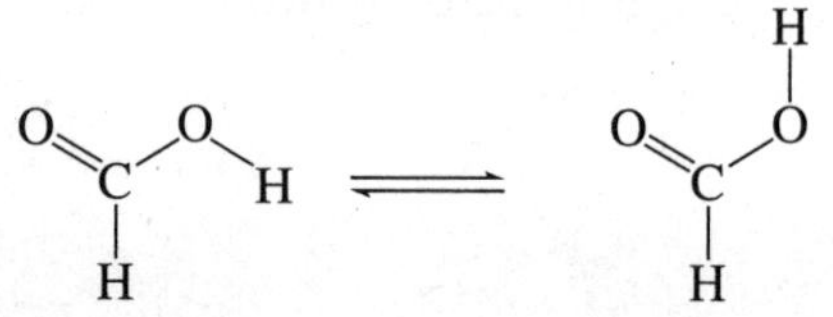

Syn: less stable conformation of formic acid ⇌ Anti: more stable conformation of formic acid

A plausible explanation is that the syn conformation is destabilized by lone-pair repulsions.

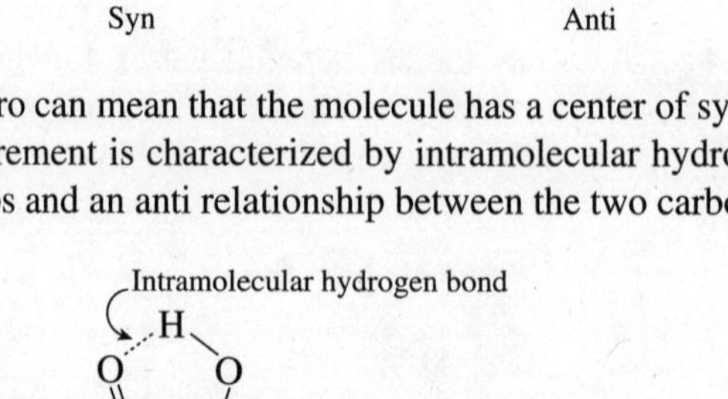

Repulsion between lone pairs

Syn ⇌ Anti

(*b*) A dipole moment of zero can mean that the molecule has a center of symmetry. One structure that satisfies this requirement is characterized by intramolecular hydrogen bonding between the two carboxyl groups and an anti relationship between the two carbonyls.

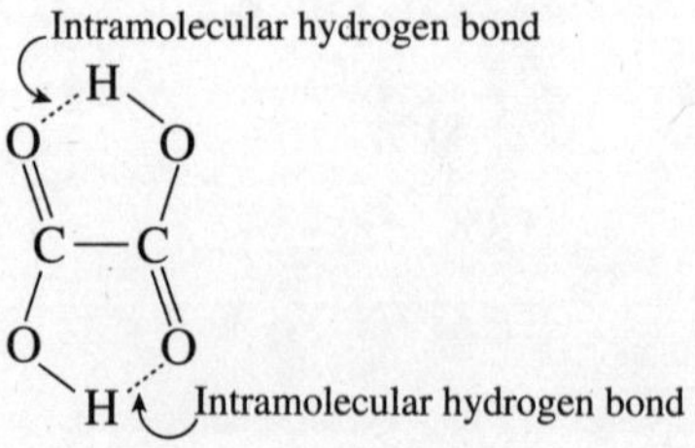

Intramolecular hydrogen bond

Intramolecular hydrogen bond

Another possibility is the following structure; it also has a center of symmetry and an anti relationship between the two carbonyls.

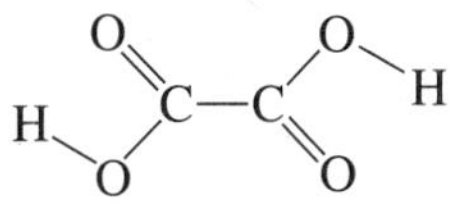

Other centrosymmetric structures can be drawn; these have the two hydrogen atoms out of the plane of the carboxyl groups, however, and are less likely to occur, in view of the known planarity of carboxyl groups. Structures in which the carbonyl groups are syn to each other do not have a center of symmetry.

(*c*) The anion formed on dissociation of *o*-hydroxybenzoic acid can be stabilized by an intramolecular hydrogen bond.

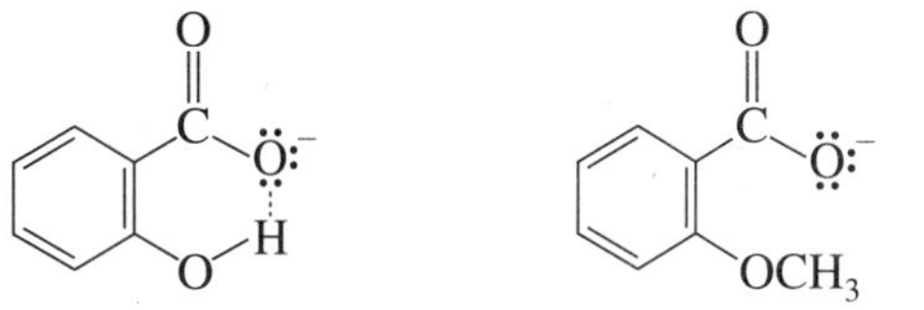

*o*-Hydroxybenzoate ion (stabilized by hydrogen bonding)

*o*-Methoxybenzoate ion (hydrogen bonding is not possible)

(*d*) Ascorbic acid is relatively acidic because ionization of its enolic hydroxyl at C-3 gives an anion that is stabilized by resonance in much the same way as a carboxylate ion; the negative charge is shared by two oxygens.

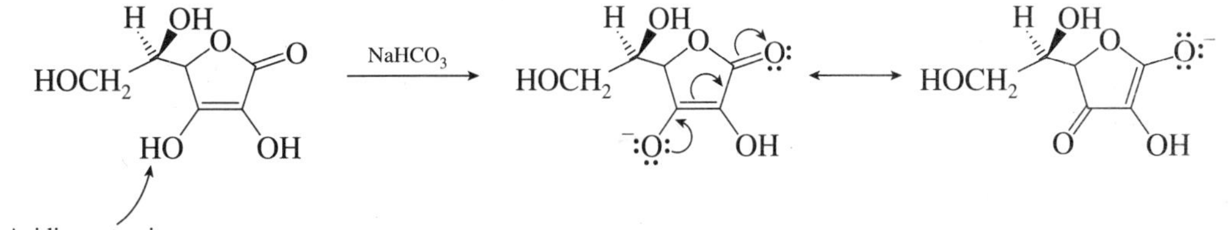

$NaHCO_3$

Acidic proton in ascorbic acid

**19.26** Dicarboxylic acids in which both carboxyl groups are attached to the same carbon undergo ready thermal decarboxylation to produce the enol form of an acid.

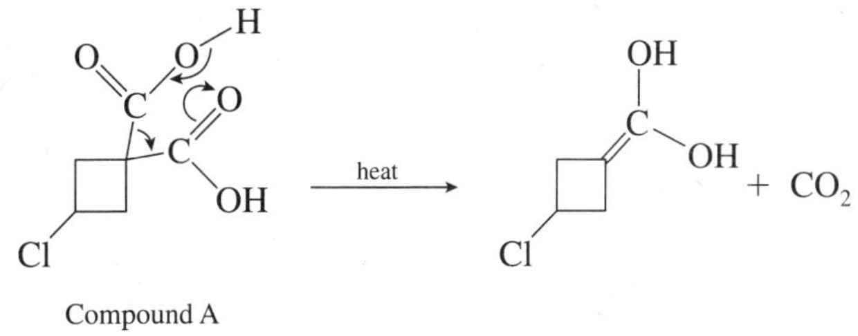

heat

$+ CO_2$

Compound A

This enol yields a mixture of *cis*- and *trans*-3-chlorocyclobutanecarboxylic acid. The two products are stereoisomers.

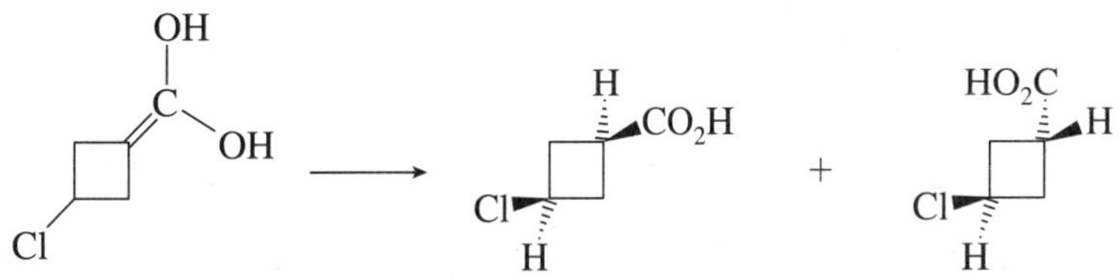

*cis*-3-Chlorocyclobutanecarboxylic acid

*trans*-3-Chlorocyclobutanecarboxylic acid

**19.27** Examination of the molecular formula $C_{14}H_{26}O_2$ reveals that the compound has an index of hydrogen deficiency of 2. Because we are told that the compound is a carboxylic acid, one of these elements of unsaturation must be a carbon–oxygen double bond. The other must be a carbon–carbon double bond because the compound undergoes cleavage on ozonolysis. Examining the products of ozonolysis serves to locate the position of the double bond.

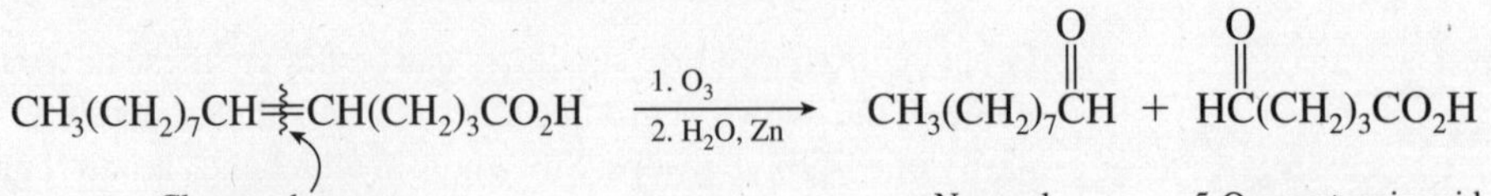

$$CH_3(CH_2)_7CH{=}CH(CH_2)_3CO_2H \xrightarrow[2.\ H_2O,\ Zn]{1.\ O_3} CH_3(CH_2)_7\overset{O}{\overset{\|}{C}}H + H\overset{O}{\overset{\|}{C}}(CH_2)_3CO_2H$$

Cleavage by ozone occurs here. — Nonanal — 5-Oxopentanoic acid

The starting acid must be 5-tetradecenoic acid. The stereochemistry of the double bond is not revealed by these experiments.

**19.28** Hydrogenation of the starting material is expected to result in reduction of the ketone carbonyl while leaving the carboxyl group unaffected. Because the isolated product lacks a carboxyl group, however, that group must react in some way. The most reasonable reaction is intramolecular esterification to form a $\gamma$-lactone.

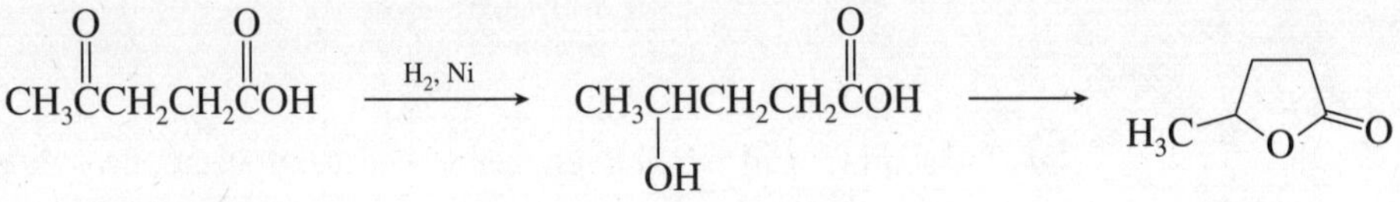

$$CH_3\overset{O}{\overset{\|}{C}}CH_2CH_2\overset{O}{\overset{\|}{C}}OH \xrightarrow{H_2,\ Ni} CH_3\underset{OH}{\underset{|}{C}H}CH_2CH_2\overset{O}{\overset{\|}{C}}OH \longrightarrow \text{(γ-lactone ring, } H_3C\text{, =O)}$$

Levulinic acid — 4-Hydroxypentanoic acid (not isolated) — 4-Pentanolide ($C_5H_8O_2$)

**19.29** Compound A is a cyclic acetal and undergoes hydrolysis in aqueous acid to produce acetaldehyde, along with a dihydroxy carboxylic acid.

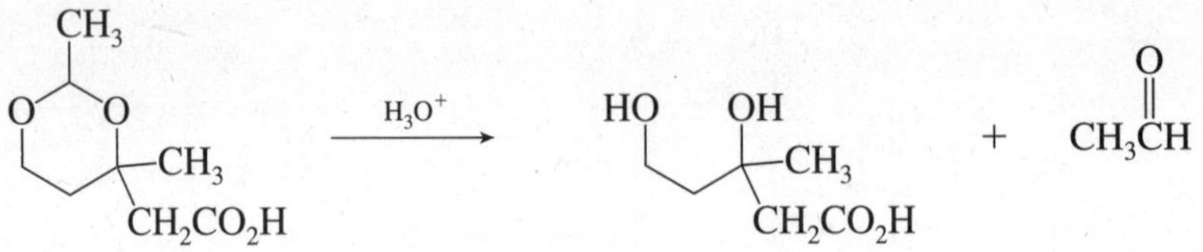

Compound A ($CH_3$, O, O, $CH_3$, $CH_2CO_2H$) $\xrightarrow{H_3O^+}$ HO, OH, $CH_3$, $CH_2CO_2H$ + $CH_3\overset{O}{\overset{\|}{C}}H$

Compound A — 3,5-Dihydroxy-3-methylpentanoic acid — Acetaldehyde

The dihydroxy acid that is formed in this step cyclizes to the $\delta$-lactone mevalonolactone.

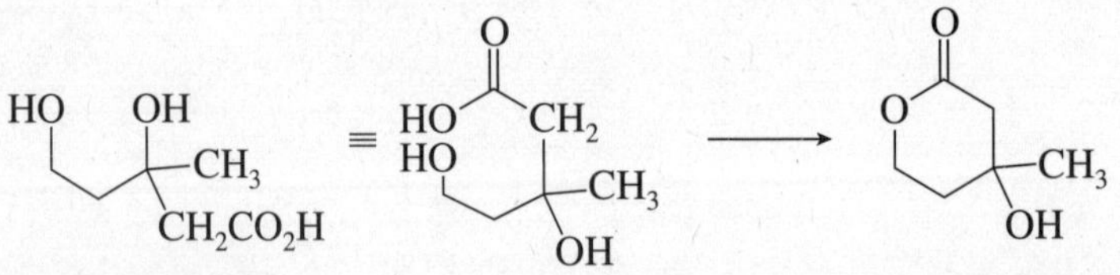

HO, OH, $CH_3$, $CH_2CO_2H$ $\equiv$ HO(C=O)$CH_2$, HO, $CH_3$, OH $\longrightarrow$ (O, C=O, $CH_3$, OH)

3,5-Dihydroxy-3-methylpentanoic acid — Mevalonolactone

**19.30** Compound A is a $\delta$-lactone. To determine its precursor, disconnect the ester linkage to a hydroxy acid.

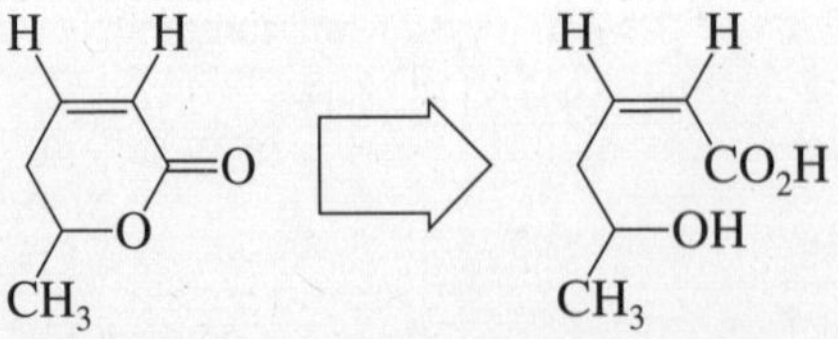

(H, H, O, =O, $CH_3$) $\Longrightarrow$ (H, H, $CO_2H$, OH, $CH_3$)

Compound A

The precursor has the same carbon skeleton as the designated starting material. All that is necessary is to hydrogenate the double bond of the alkynoic acid to the cis alkene. This can be done by using the Lindlar catalyst. Cyclization of the hydroxy acid to the lactone is spontaneous.

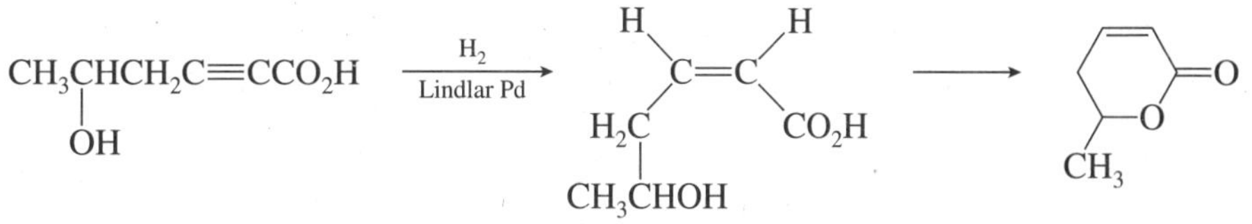

$$\underset{\text{5-Hydroxy-2-hexynoic acid}}{CH_3CH(OH)CH_2C{\equiv}CCO_2H} \xrightarrow[\text{Lindlar Pd}]{H_2} \underset{\text{(Not isolated)}}{(H)(CH_3CHOHCH_2)C{=}C(H)(CO_2H)} \longrightarrow \text{Compound A}$$

5-Hydroxy-2-hexynoic acid (Not isolated) Compound A

**19.31** Hydration of the double bond can occur in two different directions:

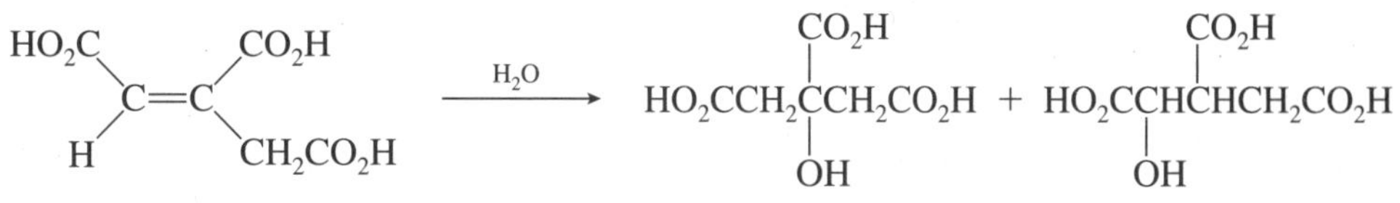

$$(HO_2C)(H)C{=}C(CO_2H)(CH_2CO_2H) \xrightarrow{H_2O} HO_2CCH_2C(CO_2H)(OH)CH_2CO_2H + HO_2CCH(OH)CH(CO_2H)CH_2CO_2H$$

(*a*) The achiral isomer is citric acid.

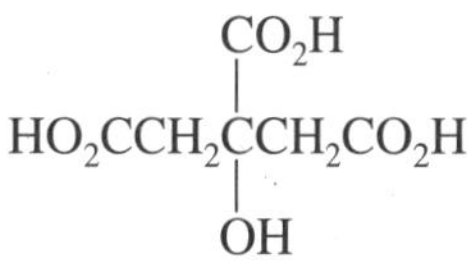

$$HO_2CCH_2C(CO_2H)(OH)CH_2CO_2H$$

Citric acid has no stereogenic centers.

(*b*) The other isomer, isocitric acid, has two stereogenic centers (marked with an asterisk*). Isocitric acid has the constitution

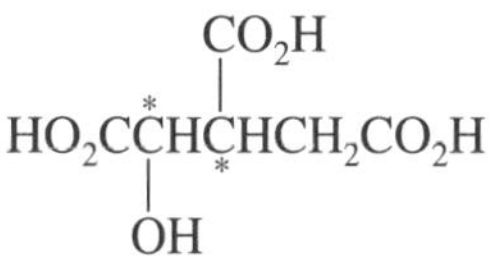

$$HO_2C\overset{*}{C}H(OH)\overset{*}{C}H(CO_2H)CH_2CO_2H$$

Isocitric acid

With two stereogenic centers, there are $2^2$, or four, stereoisomers represented by this constitution. The one that is actually formed in this enzyme-catalyzed reaction is the 2*R*,3*S* isomer.

**19.32** Carboxylic acid protons give signals in the range δ 10–12 ppm. A signal in this region suggests the presence of a carboxyl group but tells little about its environment. Thus, in assigning structures to compounds A, B, and C, the most useful data are the chemical shifts of the protons other than the carboxyl protons. Compare the three structures:

$$\underset{\text{Formic acid}}{HCOH\ (C{=}O)} \qquad \underset{\text{Maleic acid}}{(H)(HO_2C)C{=}C(H)(CO_2H)} \qquad \underset{\text{Malonic acid}}{HO_2CCH_2CO_2H}$$

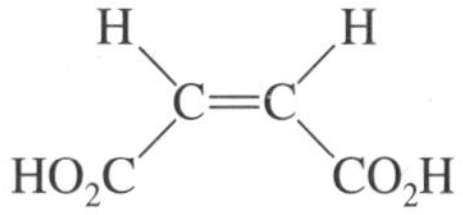

The proton that is diagnostic of structure in formic acid is bonded to a carbonyl group; it is an aldehyde proton. Typical chemical shifts of aldehyde protons are 8–10 ppm, and therefore formic acid is compound C.

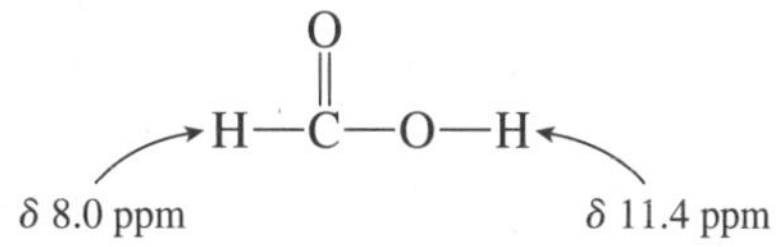

$$\underset{\delta\ 8.0\ \text{ppm}}{H}{-}\overset{O}{\overset{\|}{C}}{-}O{-}\underset{\delta\ 11.4\ \text{ppm}}{H}$$

Compound C

The critical signal in maleic acid is that of the vinyl protons, which normally is found in the range $\delta$ 5–7 ppm. Maleic acid is compound B.

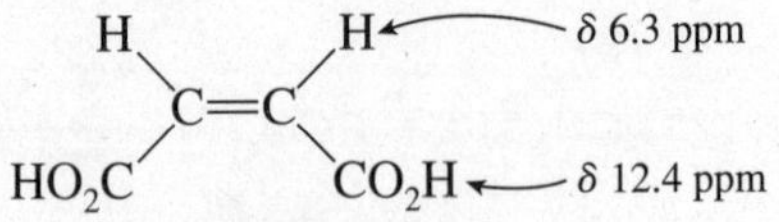

$$\ce{(HO_2C)(H)C=C(H)(CO_2H)}$$

H ← $\delta$ 6.3 ppm; $CO_2H$ ← $\delta$ 12.4 ppm

Compound B

Compound A is malonic acid. Here we have a methylene group bearing two carbonyl substituents. These methylene protons are more shielded than the aldehyde proton of formic acid or the vinyl protons of maleic acid.

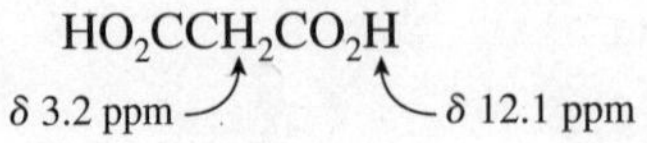

$$HO_2CCH_2CO_2H$$

$\delta$ 3.2 ppm ($CH_2$); $\delta$ 12.1 ppm ($CO_2H$)

Compound A

**19.33** Compounds A and B both exhibit $^1$H NMR absorptions in the region $\delta$ 11–12 ppm characteristic of carboxylic acids. The formula $C_4H_8O_3$ suggests an index of hydrogen deficiency of 1, accounted for by the carbonyl of the carboxyl group. Compound A has the triplet–quartet splitting indicative of an ethyl group, and compound B has two triplets, suggesting $—CH_2CH_2—$.

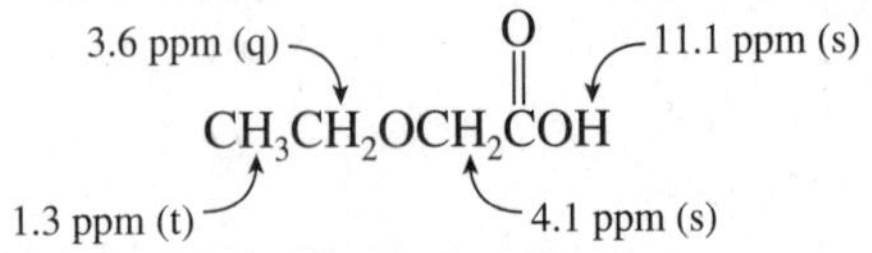

$$CH_3CH_2OCH_2\overset{O}{\overset{\|}{C}}OH$$

3.6 ppm (q); 11.1 ppm (s); 1.3 ppm (t); 4.1 ppm (s)

Compound A

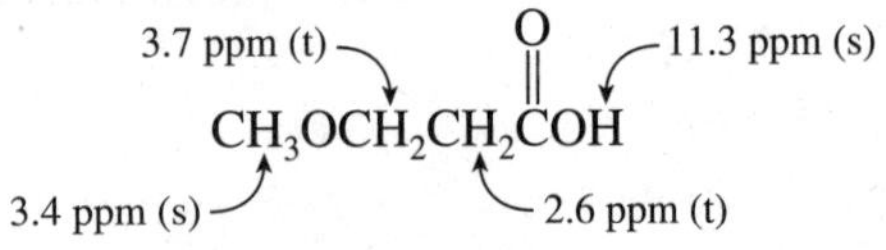

$$CH_3OCH_2CH_2\overset{O}{\overset{\|}{C}}OH$$

3.7 ppm (t); 11.3 ppm (s); 3.4 ppm (s); 2.6 ppm (t)

Compound B

**19.34** (*a*) The formula of compound A ($C_3H_5ClO_2$) has an index of hydrogen deficiency of 1—the carboxyl group. Only two structures are possible:

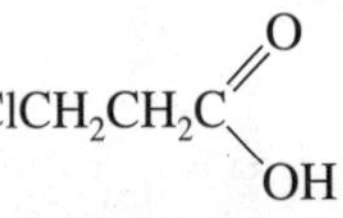

$$ClCH_2CH_2C(=O)OH \quad \text{and} \quad CH_3CH(Cl)C(=O)OH$$

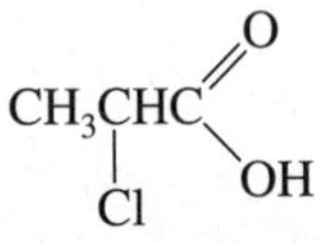

3-Chloropropanoic acid — 2-Chloropropanoic acid

Compound A is determined to be 3-chloropropanoic acid on the basis of its $^1$H NMR spectrum, which shows two triplets at $\delta$ 2.9 and $\delta$ 3.8 ppm.

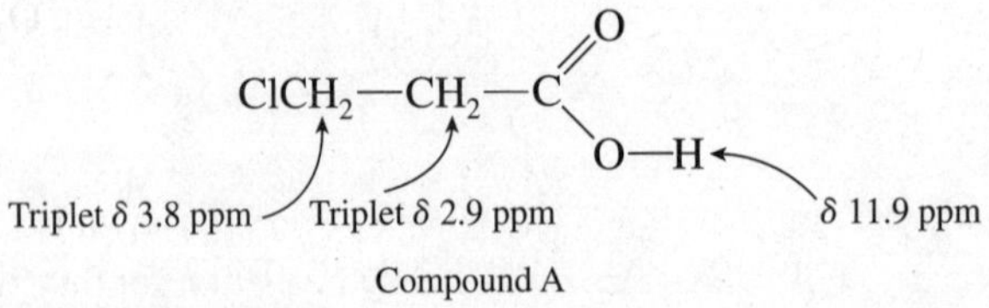

$$ClCH_2—CH_2—C(=O)—O—H$$

Triplet $\delta$ 3.8 ppm ($ClCH_2$); Triplet $\delta$ 2.9 ppm ($CH_2$); $\delta$ 11.9 ppm (O—H)

Compound A

Compound A cannot be 2-chloropropanoic acid, because that compound's $^1$H NMR spectrum would show a three-proton doublet for the methyl group and a one-proton quartet for the methine proton.

(*b*) The formula of compound B ($C_9H_9NO_4$) corresponds to an index of hydrogen deficiency of 6. The presence of an aromatic ring, as evidenced by the $^1$H NMR absorptions at $\delta$ 7.5 and

8.2 ppm, accounts for four of the unsaturations. The appearance of the aromatic protons as a pair of doublets with a total area of 4 suggests a *para*-disubstituted ring.

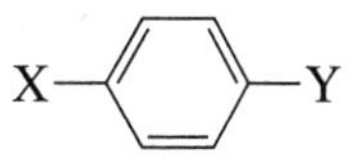

X—$C_6H_4$—Y

That compound B is a carboxylic acid is evidenced by the singlet (area = 1) at $\delta$ 12.1 ppm. The remaining $^1H$ NMR signals—a quartet at $\delta$ 3.9 ppm (1H) and a doublet at $\delta$ 1.6 ppm (3H)—suggest the fragment CH—$CH_3$. All that remains of the molecular formula is —$NO_2$. Combining this information identifies compound B as 2-(4-nitrophenyl)propanoic acid.

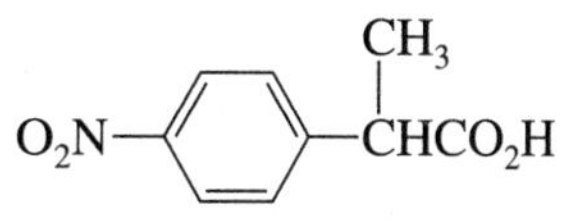

$O_2N$—$C_6H_4$—$CH(CH_3)CO_2H$

2-(4-Nitrophenyl)propanoic acid
(compound B)

# SELF-TEST

## PART A

**A-1.** Provide an acceptable IUPAC name for each of the following:

(*a*) $C_6H_5CH(CH_3)CH(CH_3)CH_2CH_2CO_2H$

(*b*) cyclohexyl—$CO_2H$

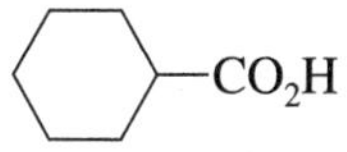

(*c*) $CH_3CH(Br)CH(CH_2CH_3)CO_2H$

**A-2.** Both of the following compounds may be converted into 4-phenylbutanoic acid by one or more reaction steps. Give the reagents and conditions necessary to carry out these conversions.

$C_6H_5CH_2CH_2CH(CO_2H)_2$ $\quad$ $C_6H_5CH_2CH_2CH_2Br$

(Two methods)

**A-3.** The species whose structure is shown is an intermediate in an esterification reaction. Write the complete, balanced equation for this process.

$C_6H_5CH_2C(OH)_2OCH_2CH_3$

**A-4.** Give the correct structures for compounds A through C in the following reactions:

(*a*) $(CH_3)_2CHCH_2CH_2CO_2H \xrightarrow[P]{Br_2} A \xrightarrow[\text{acetone}]{KI} B$

(*b*) $C \xrightarrow{\text{heat}} C_6H_5\overset{\overset{\displaystyle O}{\|}}{C}CH(CH_3)_2 + CO_2$

**A-5.** Give the missing reagent(s) and the missing compound in each of the following:

(*a*) $C_6H_5Br \xrightarrow{?} ? \xrightarrow[2.\ H_3O^+]{1.\ CO_2} C_6H_5CO_2H$

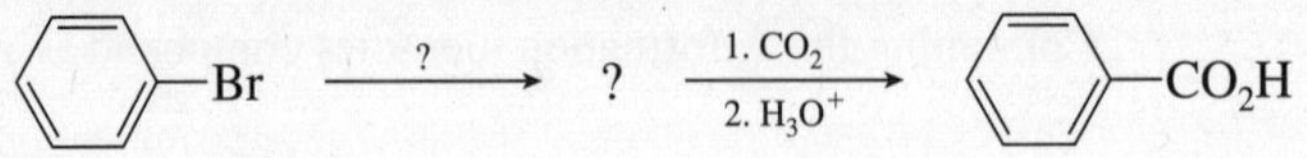

(*b*) $CH_3CH_2CH_2\overset{\overset{\displaystyle O}{\|}}{C}OH \xrightarrow{?} ? \xrightarrow[CH_2Cl_2]{PCC} CH_3CH_2CH_2\overset{\overset{\displaystyle O}{\|}}{C}H$

(*c*) $CH_3CH_2CH_2\overset{\overset{\displaystyle O}{\|}}{C}OH \xrightarrow{?} ? \xrightarrow{NaSCH_3} CH_3CH_2\underset{\underset{\displaystyle SCH_3}{|}}{C}H\overset{\overset{\displaystyle O}{\|}}{C}OH$

(*d*) $CH_3CH_2\overset{\overset{\displaystyle O}{\|}}{C}H \xrightarrow[HCl]{NaCN} ? \xrightarrow{?} CH_3CH_2\underset{\underset{\displaystyle OH}{|}}{C}H\overset{\overset{\displaystyle O}{\|}}{C}OH$

**A-6.** Identify the carboxylic acid ($C_4H_7BrO_2$) having the $^1H$ NMR spectrum consisting of

$\delta$ 1.1 ppm, 3H (triplet)
$\delta$ 2.0 ppm, 2H (pentet)
$\delta$ 4.2 ppm, 1H (triplet)
$\delta$ 12.1 ppm, 1H (singlet)

**A-7.** Draw the structure of the tetrahedral intermediate in the esterification of formic acid with 1-butanol.

**A-8.** Write a mechanism for the esterification reaction shown.

$$CH_3\overset{\overset{\displaystyle O}{\|}}{C}OH + CH_3OH \xrightarrow{H^+} CH_3CO_2CH_3 + H_2O$$

## PART B

**B-1.** Which of the following is a correct IUPAC name for the compound shown?

$$(CH_3CH_2)_2\overset{\overset{\displaystyle CO_2H}{|}}{C}CH_2CH(CH_2CH_3)_2$$

(*a*) 1,1,3-Triethylhexanoic acid
(*b*) 2,2,4-Triethylhexanoic acid
(*c*) 3,5-Diethyl-3-heptylcarboxylic acid
(*d*) 3,5,5-Triethyl-6-hexanoic acid

**B-2.** Rank the following substances in order of decreasing acid strength (strongest → weakest):

$CH_3CH_2CH_2CO_2H$ (1) $CH_3CH{=}CHCO_2H$ (2) $CH_3CH_2CH_2CH_2OH$ (3) $CH_3C{\equiv}CCO_2H$ (4)

(*a*) 4 > 2 > 1 > 3
(*b*) 1 > 2 > 4 > 3
(*c*) 3 > 1 > 2 > 4
(*d*) 2 > 4 > 1 > 3

**B-3.** Which of the following compounds will undergo decarboxylation on heating?

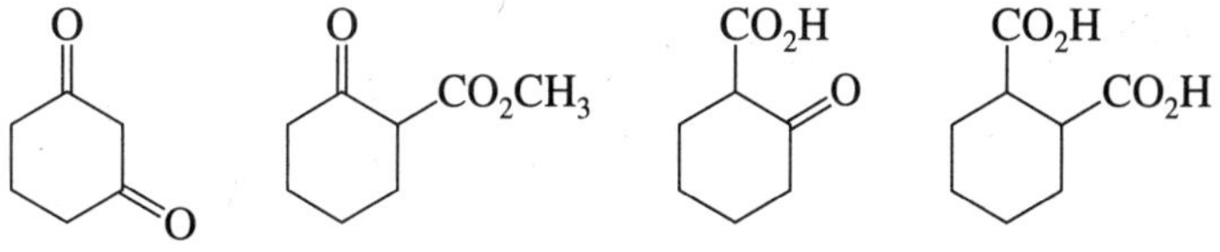

O, O, $CO_2CH_3$, $CO_2H$, O, $CO_2H$, $CO_2H$

1 2 3 4

(*a*) 2 and 3
(*b*) 3 and 4
(*c*) 3 only
(*d*) 1 and 4

**B-4.** Which of the following is *least* likely to form a lactone?

(*a*) $CH_3CH(OH)CH_2CH_2CO_2H$
(*b*) OH, $CH_2CO_2H$
(*c*) OH, $CO_2H$
(*d*) OH, $CO_2H$

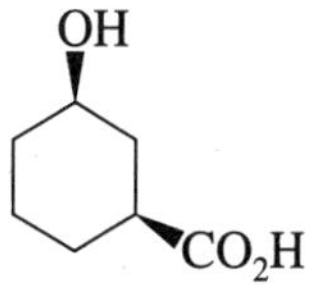

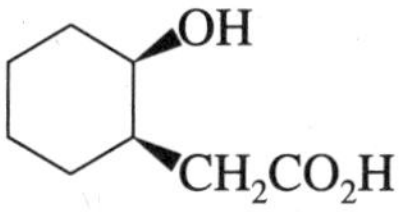

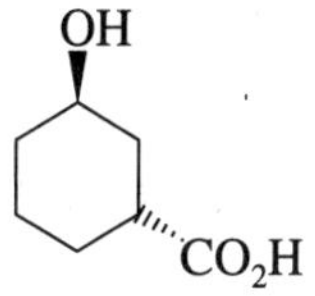

**B-5.** Compare the two methods shown for the preparation of carboxylic acids:

**Method 1:**

$$RBr \xrightarrow[\text{diethyl ether}]{Mg} RMgBr \xrightarrow[2.\ H_3O^+]{1.\ CO_2} RCO_2H$$

**Method 2:**

$$RBr \xrightarrow{NaCN} RCN \xrightarrow[\text{heat}]{H_2O,\ HCl} RCO_2H$$

Which one of the following statements correctly describes this conversion?

Br → $CO_2H$

(*a*) Both method 1 and method 2 are appropriate for carrying out this conversion.
(*b*) Neither method 1 nor method 2 is appropriate for carrying out this conversion.
(*c*) Method 1 will work well, but method 2 is not appropriate.
(*d*) Method 2 will work well, but method 1 is not appropriate.

**B-6.** Which one of the following is *not* an intermediate in the generally accepted mechanism for the reaction shown?

$$CF_3\overset{O}{\overset{\|}{C}}OH + CH_3\underset{OH}{\underset{|}{C}}HCH_3 \xrightarrow{H_2SO_4} CF_3\overset{O}{\overset{\|}{C}}O\underset{CH_3}{\underset{|}{C}}HCH_3 + H_2O$$

(*a*) $CF_3\overset{^+\ddot{O}H}{\overset{\|}{C}}{-}\ddot{O}H$

(*b*) $CF_3\overset{:\ddot{O}H}{\overset{|}{\underset{:\ddot{O}H}{\underset{|}{C}}}}{-}\overset{+}{\ddot{O}}\underset{H}{\underset{|}{}}CH(CH_3)_2$

(*c*) $CF_3\overset{:\ddot{O}H}{\overset{|}{\underset{:\ddot{O}H}{\underset{|}{C}}}}{-}\ddot{O}CH(CH_3)_2$

(*d*) $CF_3\overset{:\ddot{O}:^-}{\overset{|}{\underset{:\ddot{O}H}{\underset{|}{C}}}}{-}\ddot{O}CH(CH_3)_2$

(*e*) $CF_3\overset{:\ddot{O}H}{\overset{|}{\underset{^+\ddot{O}H_2}{\underset{|}{C}}}}{-}\ddot{O}CH(CH_3)_2$

**B-7.** Identify compound C in the following sequence:

$$(CH_3)_2CHCH_2C{\equiv}N \xrightarrow[\text{heat}]{HCl,\ H_2O} \text{compound A} \xrightarrow[2.\ H_2O]{1.\ LiAlH_4} \text{compound B} \xrightarrow[CH_2Cl_2]{PCC} \text{compound C}$$

(*a*) $(CH_3)_2CH\overset{O}{\overset{\|}{C}}CH_3$

(*b*) $(CH_3)_2CHCH{=}C(OH)_2$

(*c*) $(CH_3)_2CHCH_2\overset{O}{\overset{\|}{C}}H$

(*d*) $(CH_3)_2CHCH_2\overset{O}{\overset{\|}{C}}OH$

(*e*) $(CH_3)_2CH\overset{O}{\overset{\|}{C}}CH_2OH$

**B-8.** What is the final product (B) of this sequence?

$$C_6H_5CH_3 \xrightarrow[\text{light}]{Br_2} A \xrightarrow[2.\ H_3O^+,\ \text{heat}]{1.\ KCN} B$$

(*a*) ortho-$CH_3C_6H_4CO_2H$ + para

(*b*) meta-$CH_3C_6H_4CO_2H$

(*c*) $C_6H_5CO_2H$

(*d*) $C_6H_5CH_2CO_2H$

(*e*) None of these

**B-9.** Which one of the following undergoes decarboxylation (loses carbon dioxide) most readily on being heated?

(*a*)

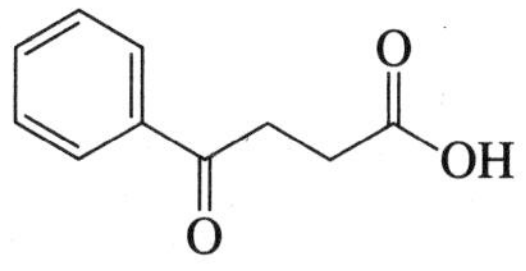

(*b*)

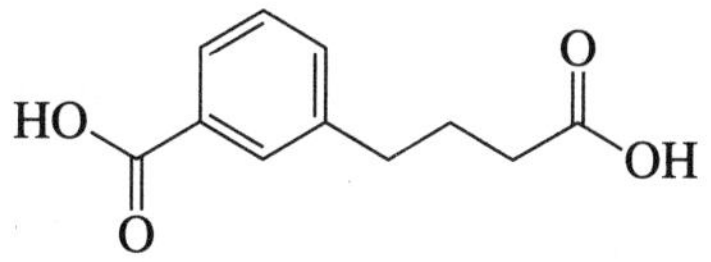

(*c*)

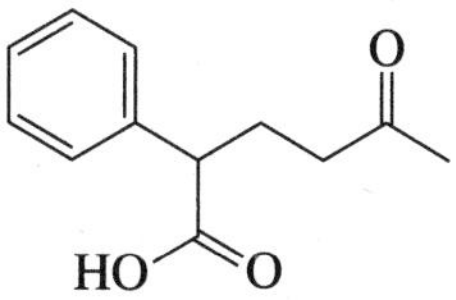

(*d*)

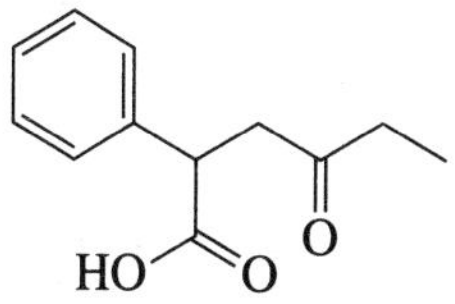

(*e*)

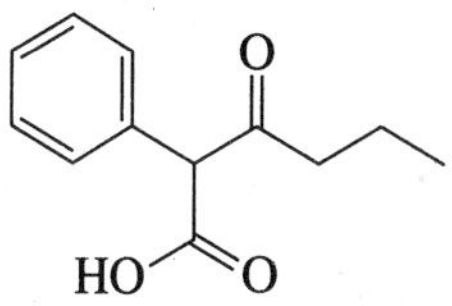

**B-10.** Which of the compounds in the previous problem yields a δ-lactone on being reduced with sodium borohydride?

**B-11.** What is compound Z?

$$CH_3CH_2CH_2Br \xrightarrow{NaCN} X \xrightarrow[heat]{H_3O^+} Y \xrightarrow[H^+]{CH_3CH_2OH} Z$$

(*a*) $CH_3CH{=}CH\overset{O}{\overset{\|}{C}}OH$

(*b*) $CH_3CH_2CH_2CH(OCH_2CH_3)_2$

(*c*) $CH_3CH_2\underset{OCH_2CH_3}{\underset{|}{C}H}C{\equiv}N$

(*d*) $CH_3CH_2CH_2CH{=}NOCH_2CH_3$

(*e*) $CH_3CH_2CH_2\overset{O}{\overset{\|}{C}}OCH_2CH_3$

# CHAPTER 20

## CARBOXYLIC ACID DERIVATIVES: NUCLEOPHILIC ACYL SUBSTITUTION

## SOLUTIONS TO TEXT PROBLEMS

**20.1** (*b*) Carboxylic acid anhydrides bear two acyl groups on oxygen, as in $\overset{O}{\overset{\|}{R C}}O\overset{O}{\overset{\|}{C}}R$. They are named as derivatives of carboxylic acids.

$$\begin{array}{c} O \\ \| \\ CH_3CH_2CHCOH \\ | \\ C_6H_5 \end{array} \qquad \begin{array}{c} O\ \ O \\ \|\ \ \| \\ CH_3CH_2CHCOCCHCH_2CH_3 \\ |\qquad\ | \\ C_6H_5\quad C_6H_5 \end{array}$$

2-Phenylbutanoic acid — 2-Phenylbutanoic anhydride

(*c*) Butyl 2-phenylbutanoate is the butyl ester of 2-phenylbutanoic acid.

$$\begin{array}{c} O \\ \| \\ CH_3CH_2CHCOCH_2CH_2CH_2CH_3 \\ | \\ C_6H_5 \end{array}$$

Butyl 2-phenylbutanoate

(*d*) In 2-phenylbutyl butanoate the 2-phenylbutyl group is an alkyl group bonded to oxygen of the ester. It is not involved in the acyl group of the molecule.

$$\begin{array}{c} O \\ \| \\ CH_3CH_2CH_2COCH_2CHCH_2CH_3 \\ | \\ C_6H_5 \end{array}$$

2-Phenylbutyl butanoate

(*e*) The ending *-amide* reveals this to be a compound of the type $RCNH_2$ (with C=O).

$$CH_3CH_2CH(C_6H_5)C(=O)NH_2$$

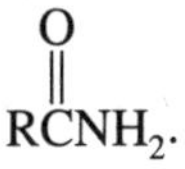

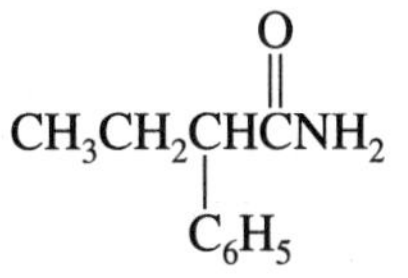

2-Phenylbutanamide

(*f*) This compound differs from 2-phenylbutanamide in part (*e*) only in that it bears an ethyl substituent on nitrogen.

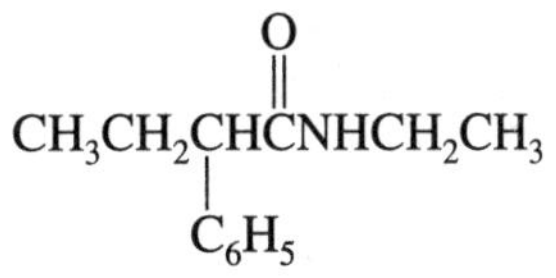

$$CH_3CH_2CH(C_6H_5)C(=O)NHCH_2CH_3$$

*N*-Ethyl-2-phenylbutanamide

(*g*) The *-nitrile* ending signifies a compound of the type RC≡N containing the same number of carbons as the alkane $RCH_3$.

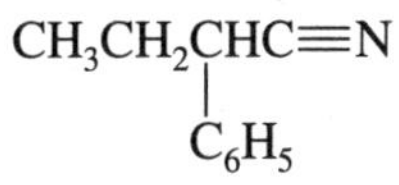

$$CH_3CH_2CH(C_6H_5)C{\equiv}N$$

2-Phenylbutanenitrile

**20.2** The methyl groups in *N*,*N*-dimethylformamide are nonequivalent; one is cis to oxygen, the other is trans. The two methyl groups have different chemical shifts.

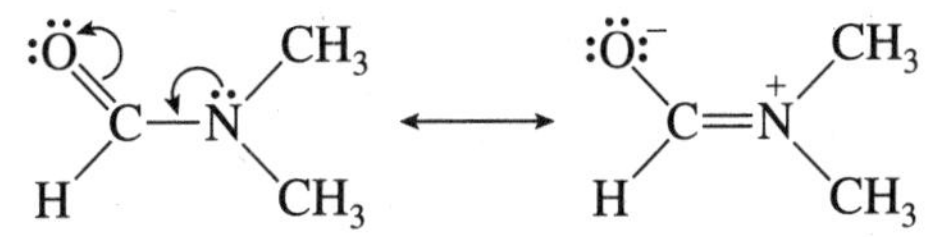

$$H(\ddot{O}{=})C{-}\ddot{N}(CH_3)_2 \longleftrightarrow H(:\ddot{O}:^-)C{=}\overset{+}{N}(CH_3)_2$$

Rotation about the carbon–nitrogen bond is required to average the environments of the two methyl groups, but this rotation is relatively slow in amides as the result of the double-bond character imparted to the carbon–nitrogen bond, as shown by these two resonance structures.

**20.3** (*b*) Benzoyl chloride reacts with benzoic acid to give benzoic anhydride.

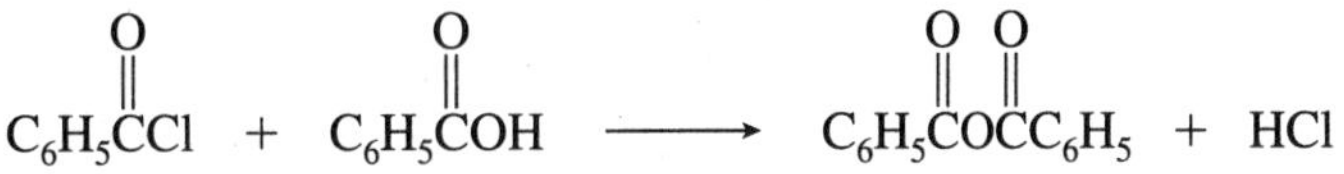

$$C_6H_5C(=O)Cl + C_6H_5C(=O)OH \longrightarrow C_6H_5C(=O)OC(=O)C_6H_5 + HCl$$

Benzoyl chloride — Benzoic acid — Benzoic anhydride — Hydrogen chloride

(*c*) Acyl chlorides react with alcohols to form esters.

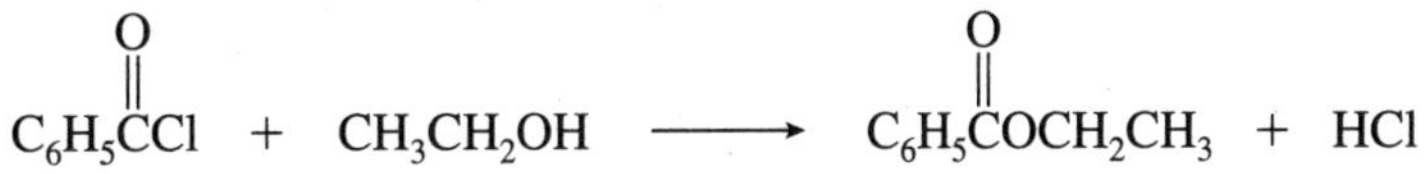

$$C_6H_5C(=O)Cl + CH_3CH_2OH \longrightarrow C_6H_5C(=O)OCH_2CH_3 + HCl$$

Benzoyl chloride — Ethanol — Ethyl benzoate — Hydrogen chloride

The organic product is the ethyl ester of benzoic acid, ethyl benzoate.

(*d*) Acyl transfer from benzoyl chloride to the nitrogen of methylamine yields the amide *N*-methylbenzamide.

$$\underset{\text{Benzoyl chloride}}{C_6H_5\overset{O}{\overset{\|}{C}}Cl} + \underset{\text{Methylamine}}{2CH_3NH_2} \longrightarrow \underset{N\text{-Methylbenzamide}}{C_6H_5\overset{O}{\overset{\|}{C}}NHCH_3} + CH_3\overset{+}{N}H_3\ Cl^-$$

(*e*) In analogy with part (*d*), an amide is formed. In this case the product has two methyl groups on nitrogen.

$$\underset{\text{Benzoyl chloride}}{C_6H_5\overset{O}{\overset{\|}{C}}Cl} + \underset{\text{Dimethylamine}}{2(CH_3)_2NH} \longrightarrow \underset{N,N\text{-Dimethylbenzamide}}{C_6H_5\overset{O}{\overset{\|}{C}}N(CH_3)_2} + (CH_3)_2\overset{+}{N}H_2\ Cl^-$$

(*f*) Acyl chlorides undergo hydrolysis on reaction with water. The product is a carboxylic acid.

$$\underset{\text{Benzoyl chloride}}{C_6H_5\overset{O}{\overset{\|}{C}}Cl} + \underset{\text{Water}}{H_2O} \longrightarrow \underset{\text{Benzoic acid}}{C_6H_5\overset{O}{\overset{\|}{C}}OH} + \underset{\text{Hydrogen chloride}}{HCl}$$

**20.4** (*b*) Nucleophilic addition of benzoic acid to benzoyl chloride gives the tetrahedral intermediate shown.

$$\underset{\text{Benzoyl chloride}}{C_6H_5\overset{O}{\overset{\|}{C}}Cl} + \underset{\text{Benzoic acid}}{C_6H_5\overset{O}{\overset{\|}{C}}OH} \longrightarrow \underset{\text{Tetrahedral intermediate}}{C_6H_5\overset{HO}{\overset{|}{\underset{Cl}{\underset{|}{C}}}}O\overset{O}{\overset{\|}{C}}C_6H_5}$$

Dissociation of the tetrahedral intermediate occurs by loss of chloride and of the proton on the oxygen.

$$\underset{\text{Tetrahedral intermediate}}{C_6H_5\overset{H\text{–}O}{\overset{|}{\underset{Cl}{\underset{|}{C}}}}O\overset{O}{\overset{\|}{C}}C_6H_5} \longrightarrow \underset{\text{Benzoic anhydride}}{C_6H_5\overset{O}{\overset{\|}{C}}O\overset{O}{\overset{\|}{C}}C_6H_5} + \underset{\text{Hydrogen chloride}}{HCl}$$

(*c*) Ethanol is the nucleophile that adds to the carbonyl group of benzoyl chloride to form the tetrahedral intermediate.

$$\underset{\text{Benzoyl chloride}}{C_6H_5\overset{O}{\overset{\|}{C}}Cl} + \underset{\text{Ethanol}}{CH_3CH_2OH} \longrightarrow \underset{\text{Tetrahedral intermediate}}{C_6H_5\overset{OH}{\overset{|}{\underset{Cl}{\underset{|}{C}}}}OCH_2CH_3}$$

In analogy with parts (*a*) and (*b*) of this problem, a proton is lost from the hydroxyl group along with chloride to restore the carbon–oxygen double bond.

$$C_6H_5C(OH)(Cl)OCH_2CH_3 \longrightarrow C_6H_5C(=O)OCH_2CH_3 + HCl$$

Tetrahedral intermediate — Ethyl benzoate — Hydrogen chloride

(*d*) The tetrahedral intermediate formed from benzoyl chloride and methylamine has a carbon–nitrogen bond.

$$C_6H_5C(=O)Cl + CH_3NH_2 \longrightarrow C_6H_5C(OH)(Cl)NHCH_3$$

Benzoyl chloride — Methylamine — Tetrahedral intermediate

The dissociation of the tetrahedral intermediate may be shown as

$$C_6H_5C(OH)(Cl)NHCH_3 \longrightarrow C_6H_5C(=O)NHCH_3 + HCl$$

Tetrahedral intermediate — *N*-Methylbenzamide — Hydrogen chloride

More realistically, it is a second methylamine molecule that abstracts a proton from oxygen.

$$CH_3\ddot{N}H_2 + C_6H_5C(OH)(Cl)NHCH_3 \longrightarrow C_6H_5C(=O)NHCH_3 + CH_3\overset{+}{N}H_3\ Cl^-$$

*N*-Methylbenzamide — Methylammonium chloride

(*e*) The intermediates in the reaction of benzoyl chloride with dimethylamine are similar to those in part (*d*). The methyl substituents on nitrogen are not directly involved in the reaction.

$$C_6H_5C(=O)Cl + (CH_3)_2NH \longrightarrow C_6H_5C(OH)(Cl)N(CH_3)_2$$

Benzoyl chloride — Dimethylamine — Tetrahedral intermediate

Then

$$(CH_3)_2\ddot{N}H + C_6H_5C(OH)(Cl)N(CH_3)_2 \longrightarrow C_6H_5C(=O)N(CH_3)_2 + (CH_3)_2\overset{+}{N}H_2\ Cl^-$$

*N*,*N*-Dimethylbenzamide — Dimethylammonium chloride

(*f*) Water attacks the carbonyl group of benzoyl chloride to form the tetrahedral intermediate.

$$\underset{\text{Benzoyl chloride}}{C_6H_5\overset{O}{\overset{\|}{C}}Cl} + \underset{\text{Water}}{H_2O} \longrightarrow \underset{\text{Tetrahedral intermediate}}{C_6H_5\overset{OH}{\overset{|}{\underset{OH}{\underset{|}{C}}}}Cl}$$

Dissociation of the tetrahedral intermediate occurs by loss of chloride and the proton on oxygen.

$$\underset{\text{Tetrahedral intermediate}}{C_6H_5\overset{H-O}{\overset{|}{\underset{OH}{\underset{|}{C}}}}-Cl} \longrightarrow \underset{\text{Benzoic acid}}{C_6H_5\overset{O}{\overset{\|}{C}}OH} + \underset{\text{Hydrogen chloride}}{HCl}$$

**20.5** One equivalent of benzoyl chloride reacts rapidly with water to yield benzoic acid.

$$\underset{\text{Benzoyl chloride}}{C_6H_5\overset{O}{\overset{\|}{C}}Cl} + \underset{\text{Water}}{H_2O} \longrightarrow \underset{\text{Benzoic acid}}{C_6H_5\overset{O}{\overset{\|}{C}}OH} + \underset{\text{Hydrogen chloride}}{HCl}$$

The benzoic acid produced in this step reacts with the remaining benzoyl chloride to give benzoic anhydride.

$$\underset{\text{Benzoyl chloride}}{C_6H_5\overset{O}{\overset{\|}{C}}Cl} + \underset{\text{Benzoic acid}}{C_6H_5\overset{O}{\overset{\|}{C}}OH} \longrightarrow \underset{\text{Benzoic anhydride}}{C_6H_5\overset{O}{\overset{\|}{C}}O\overset{O}{\overset{\|}{C}}C_6H_5} + \underset{\text{Hydrogen chloride}}{HCl}$$

**20.6** Acetic anhydride serves as a source of acetyl cation.

$$CH_3\overset{O}{\overset{\|}{C}}O-\overset{O}{\overset{\|}{C}}CH_3 \longrightarrow \ddot{O}=\overset{+}{C}CH_3 \longleftrightarrow \underset{\text{Acetyl cation}}{:\overset{+}{O}\equiv CCH_3}$$

**20.7** (*b*) Acyl transfer from an acid anhydride to ammonia yields an amide.

$$\underset{\text{Acetic anhydride}}{CH_3\overset{O}{\overset{\|}{C}}O\overset{O}{\overset{\|}{C}}CH_3} + \underset{\text{Ammonia}}{2NH_3} \longrightarrow \underset{\text{Acetamide}}{CH_3\overset{O}{\overset{\|}{C}}NH_2} + \underset{\text{Ammonium acetate}}{CH_3\overset{O}{\overset{\|}{C}}O^-\ \overset{+}{N}H_4}$$

The organic products are acetamide and ammonium acetate.

(*c*) The reaction of phthalic anhydride with dimethylamine is analogous to that of part (*b*). The organic products are an amide and the carboxylate salt of an amine.

Phthalic anhydride + $2(CH_3)_2NH$ (Dimethylamine) ⟶ $C_6H_4(CN(CH_3)_2)(CO^-)$ $H_2\overset{+}{N}(CH_3)_2$

Product is an amine salt and contains an amide function.

In this case both the amide function and the ammonium carboxylate salt are incorporated into the same molecule.

(*d*) The disodium salt of phthalic acid is the product of hydrolysis of phthalic acid in excess sodium hydroxide.

Phthalic anhydride + 2NaOH (Sodium hydroxide) ⟶ $C_6H_4(CO^-\ Na^+)_2$ (Sodium phthalate) + $H_2O$ (Water)

**20.8** (*b*) The tetrahedral intermediate is formed by nucleophilic addition of ammonia to one of the carbonyl groups of acetic anhydride.

$$CH_3\overset{O}{\overset{\|}{C}}O\overset{O}{\overset{\|}{C}}CH_3 + :NH_3 \longrightarrow CH_3\underset{NH_2}{\overset{HO}{C}}O\overset{O}{\overset{\|}{C}}CH_3$$

Tetrahedral intermediate

Dissociation of the tetrahedral intermediate occurs by loss of acetate as the leaving group.

$$H_3N: + CH_3\underset{NH_2}{\overset{HO}{C}}O\overset{O}{\overset{\|}{C}}CH_3 \longrightarrow CH_3\overset{O}{\overset{\|}{C}}NH_2 + H_4\overset{+}{N}\ {}^-O\overset{O}{\overset{\|}{C}}CH_3$$

Ammonia + tetrahedral intermediate ⟶ Acetamide + Ammonium acetate

(*c*) Dimethylamine is the nucleophile; it adds to one of the two equivalent carbonyl groups of phthalic anhydride.

Phthalic anhydride + $H\ddot{N}(CH_3)_2$ (Dimethylamine) ⟶ Tetrahedral intermediate (HO, $N(CH_3)_2$)

A second molecule of dimethylamine abstracts a proton from the tetrahedral intermediate.

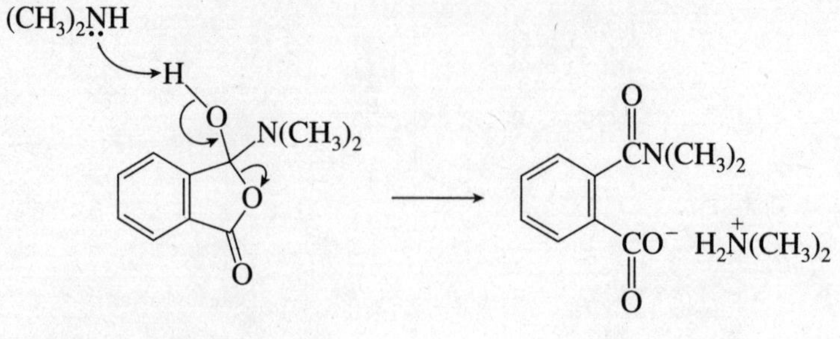

$(CH_3)_2NH$ · H · O · $N(CH_3)_2$ · O · O · O · $CN(CH_3)_2$ · $CO^-$ $H_2\overset{+}{N}(CH_3)_2$ · O

Tetrahedral intermediate + second molecule of dimethylamine

Product of reaction

(*d*) Hydroxide acts as a nucleophile to form the tetrahedral intermediate and as a base to facilitate its dissociation.

**Formation of tetrahedral intermediate:**

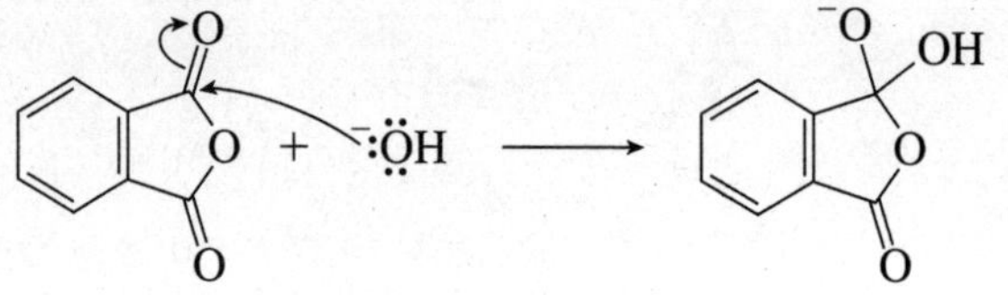

O · O + $^-$:ÖH → $^-$O OH · O · O

Phthalic anhydride

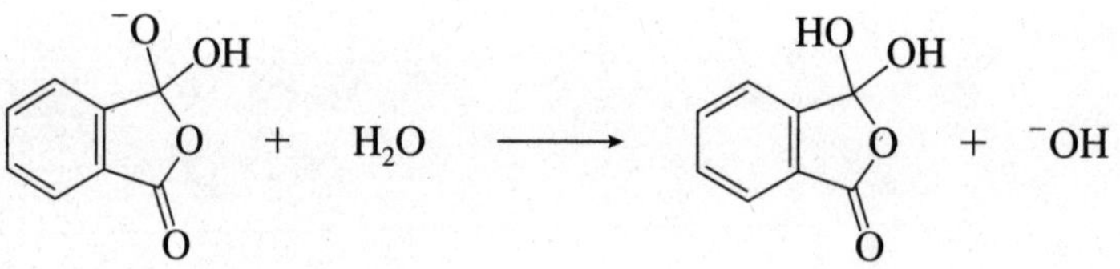

$^-$O OH · O · O + $H_2O$ → HO OH · O · O + $^-$OH

Tetrahedral intermediate

**Dissociation of tetrahedral intermediate:**

HÖ: H · O OH · O · O → O · COH · $CO^-$ · O + $H_2O$

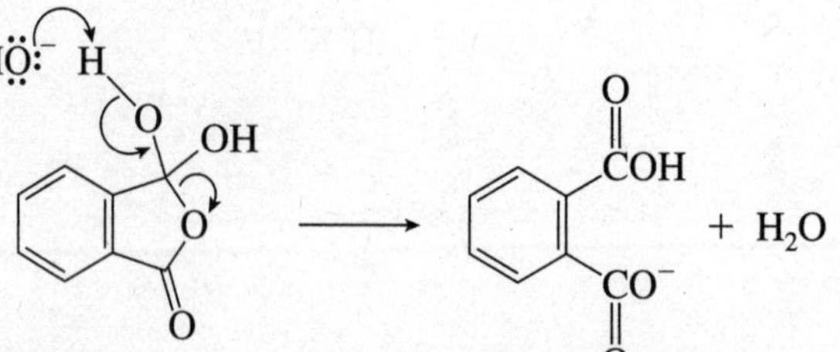

In base, the remaining carboxylic acid group is deprotonated.

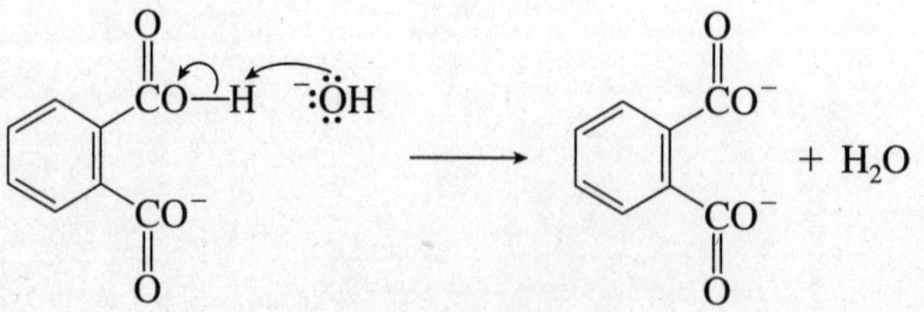

O · CO—H $^-$:ÖH · $CO^-$ · O → O · $CO^-$ · $CO^-$ · O + $H_2O$

**20.9** The starting material contains three acetate ester functions. All three undergo hydrolysis in aqueous sulfuric acid.

$$CH_3\overset{O}{\overset{\|}{C}}OCH_2\underset{\underset{\underset{O}{\|}}{OCCH_3}}{CH}CH_2CH_2CH_2O\overset{O}{\overset{\|}{C}}CH_3 \xrightarrow[H^+]{H_2O} HOCH_2\underset{OH}{CH}CH_2CH_2CH_2OH + 3CH_3\overset{O}{\overset{\|}{C}}OH$$

1,2,5-Pentanetriol ($C_5H_{12}O_3$) Acetic acid

The product is 1,2,5-pentanetriol. Also formed in the hydrolysis of the starting triacetate are three molecules of acetic acid.

**20.10** **Step 1: Protonation of the carbonyl oxygen**

$$C_6H_5C(=O)OCH_2CH_3 + H{-}\overset{+}{O}H_2 \rightleftharpoons C_6H_5C(=\overset{+}{O}H)OCH_2CH_3 + H_2O$$

Ethyl benzoate; Hydronium ion; Protonated form of ester; Water

**Step 2: Nucleophilic addition of water**

$$H_2O + C_6H_5C(=\overset{+}{O}H)OCH_2CH_3 \rightleftharpoons C_6H_5C(OH)(OCH_2CH_3)(\overset{+}{O}H_2)$$

Water; Protonated form of ester; Oxonium ion

**Step 3: Deprotonation of oxonium ion to give neutral form of tetrahedral intermediate**

$$C_6H_5C(OH)(OCH_2CH_3)(\overset{+}{O}H_2) + H_2O \rightleftharpoons C_6H_5C(OH)(OCH_2CH_3)(OH) + H{-}\overset{+}{O}H_2$$

Tetrahedral intermediate; Hydronium ion

**Step 4: Protonation of ethoxy oxygen**

$$C_6H_5C(OH)(OCH_2CH_3)(OH) + H{-}\overset{+}{O}H_2 \longrightarrow C_6H_5C(OH)(\overset{+}{O}HCH_2CH_3)(OH) + H_2O$$

Tetrahedral intermediate; Hydronium ion; Oxonium ion; Water

**Step 5: Dissociation of protonated form of tetrahedral intermediate**

This step yields ethyl alcohol and the protonated form of benzoic acid.

$$C_6H_5C(OH)_2-\overset{+}{O}(H)CH_2CH_3 \rightleftharpoons C_6H_5C(=\overset{+}{O}H)OH + HOCH_2CH_3$$

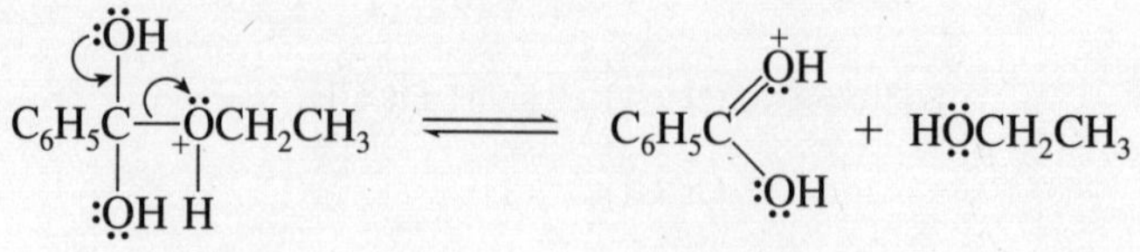

Oxonium ion; Protonated form of benzoic acid; Ethyl alcohol

**Step 6: Deprotonation of protonated form of benzoic acid**

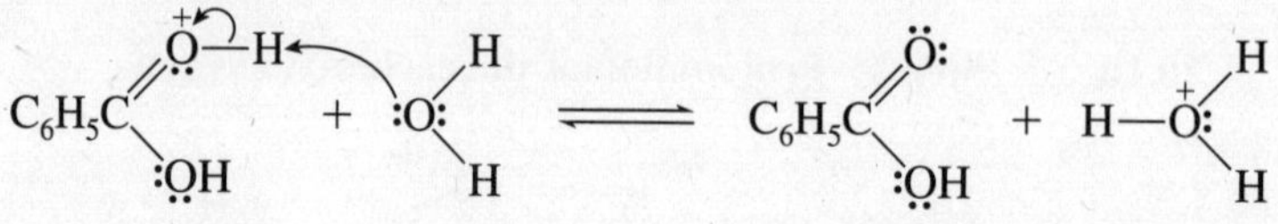

$$C_6H_5C(=\overset{+}{O}H)OH + H_2O \rightleftharpoons C_6H_5C(=O)OH + H_3O^+$$

Protonated form of benzoic acid; Water; Benzoic acid; Hydronium ion

**20.11** To determine which oxygen of 4-butanolide becomes labeled with $^{18}O$, trace the path of $^{18}O$-labeled water (● = $^{18}O$) as it undergoes nucleophilic addition to the carbonyl group to form the tetrahedral intermediate.

4-Butanolide + $H_2$● $\xrightleftharpoons{H^+}$ Tetrahedral intermediate (OH, ●H)

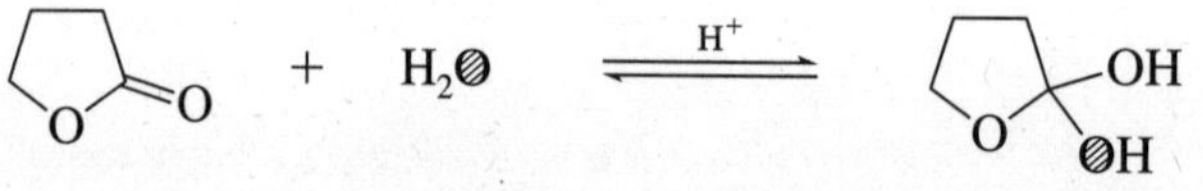

4-Butanolide; $^{18}O$-labeled water; Tetrahedral intermediate

The tetrahedral intermediate can revert to unlabeled 4-butanolide by loss of $^{18}O$-labeled water. Alternatively it can lose ordinary water to give $^{18}O$-labeled lactone.

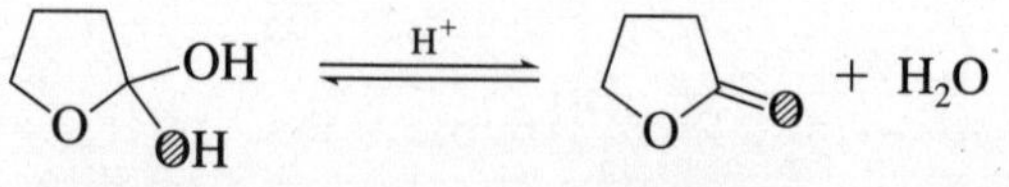

Tetrahedral intermediate $\xrightleftharpoons{H^+}$ $^{18}O$-labeled 4-butanolide + $H_2O$

Tetrahedral intermediate; $^{18}O$-labeled 4-butanolide; Water

The carbonyl oxygen is the one that is isotopically labeled in the $^{18}O$-enriched 4-butanolide.

**20.12** On the basis of trimyristin's molecular formula $C_{45}H_{86}O_6$ and of the fact that its hydrolysis gives only glycerol and tetradecanoic acid $CH_3(CH_2)_{12}CO_2H$, it must have the structure shown.

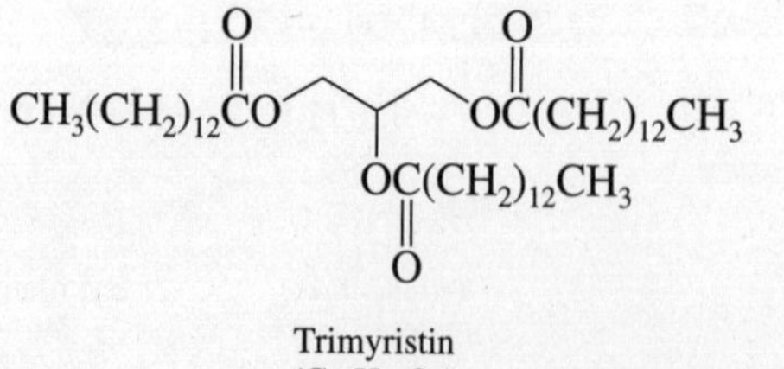

$$CH_3(CH_2)_{12}\overset{O}{\overset{\|}{C}}O-CH_2-CH(O\overset{O}{\overset{\|}{C}}(CH_2)_{12}CH_3)-CH_2-O\overset{O}{\overset{\|}{C}}(CH_2)_{12}CH_3$$

Trimyristin ($C_{45}H_{86}O_6$)

**20.13** Because ester hydrolysis in base proceeds by acyl–oxygen cleavage, the $^{18}O$ label becomes incorporated into acetate ion (⊘ = $^{18}O$).

$$CH_3CH_2CH_2CH_2CH_2O\text{—}\overset{O}{\overset{\|}{C}}CH_3 + {}^{-}\!:\!\ddot{O}H \longrightarrow CH_3CH_2CH_2CH_2CH_2OH + {}^{-}O_2CCH_3$$

Pentyl acetate — Hydroxide ion — 1-Pentanol — Acetate ion

**20.14** **Step 1: Nucleophilic addition of hydroxide ion to the carbonyl group**

$$H\ddot{O}:^{-} + C_6H_5C(=\ddot{O}:)\ddot{O}CH_2CH_3 \rightleftharpoons C_6H_5C(\text{—}\ddot{O}:^{-})(:\ddot{O}H)\text{—}\ddot{O}CH_2CH_3$$

Hydroxide ion — Ethyl benzoate — Anionic form of tetrahedral intermediate

**Step 2: Proton transfer from water to give neutral form of tetrahedral intermediate**

$$C_6H_5C(\text{—}\ddot{O}:^{-})(:\ddot{O}H)\text{—}\ddot{O}CH_2CH_3 + H\text{—}\ddot{O}H \rightleftharpoons C_6H_5C(:\ddot{O}H)(:\ddot{O}H)\text{—}\ddot{O}CH_2CH_3 + {}^{-}\!:\!\ddot{O}H$$

Anionic form of tetrahedral intermediate — Water — Tetrahedral intermediate — Hydroxide ion

**Step 3: Dissociation of tetrahedral intermediate**

$$H\ddot{O}:^{-} + C_6H_5C(\text{H—}\ddot{O}:)(:\ddot{O}H)\text{—}\ddot{O}CH_2CH_3 \rightleftharpoons H\ddot{O}H + C_6H_5C(=\ddot{O}:)\ddot{O}H + {}^{-}\!:\!\ddot{O}CH_2CH_3$$

Hydroxide ion — Tetrahedral intermediate — Water — Benzoic acid — Ethoxide ion

**Step 4: Proton transfer from benzoic acid**

$$C_6H_5C(=\ddot{O}:)\ddot{O}\text{—}H + {}^{-}\!:\!\ddot{O}H \longrightarrow C_6H_5C(=\ddot{O}:)\ddot{O}:^{-} + H\ddot{O}H$$

Benzoic acid — Hydroxide ion — Benzoate ion — Water

**20.15** The starting material is a lactone, a cyclic ester. The ester function is converted to an amide by nucleophilic acyl substitution.

$$CH_3\ddot{N}H_2 + \text{(4-pentanolide)} \longrightarrow CH_3NH\overset{O}{\overset{\|}{C}}CH_2CH_2\underset{OH}{\underset{|}{C}}HCH_3$$

Methylamine — 4-Pentanolide — 4-Hydroxy-*N*-methylpentanamide

**20.16** Methanol is the nucleophile that adds to the carbonyl group of the thioester.

$$CH_3\overset{O}{\overset{\|}{C}}SCH_2CH_2OC_6H_5 + CH_3\ddot{O}H \longrightarrow CH_3\overset{:\ddot{O}H}{\underset{OCH_3}{C}}-\ddot{S}CH_2CH_2OC_6H_5 \longrightarrow CH_3\overset{O}{\overset{\|}{C}}OCH_3 + HSCH_2CH_2OC_6H_5$$

*S*-2-Phenoxyethyl ethanethiolate; Methanol; Tetrahedral intermediate; Methyl acetate; 2-Phenoxyethanethiol

**20.17** (*b*) Acetic anhydride is the anhydride that must be used; it transfers an acetyl group to suitable nucleophiles. The nucleophile in this case is methylamine.

$$CH_3\overset{O}{\overset{\|}{C}}O\overset{O}{\overset{\|}{C}}CH_3 + 2CH_3NH_2 \longrightarrow CH_3\overset{O}{\overset{\|}{C}}NHCH_3 + CH_3\overset{O}{\overset{\|}{C}}O^- \; CH_3\overset{+}{N}H_3$$

Acetic anhydride; Methylamine; *N*-Methylacetamide; Methylammonium acetate

(*c*) The acyl group is $H\overset{O}{\overset{\|}{C}}-$. Because the problem specifies that the acyl transfer agent is a methyl ester, methyl formate is one of the starting materials.

$$H\overset{O}{\overset{\|}{C}}OCH_3 + HN(CH_3)_2 \longrightarrow H\overset{O}{\overset{\|}{C}}N(CH_3)_2 + CH_3OH$$

Methyl formate; Dimethylamine; *N,N*-Dimethylformamide; Methyl alcohol

**20.18** Phthalic anhydride reacts with excess ammonia to give the ammonium salt of a compound known as **phthalamic acid.**

$$\text{Phthalic anhydride} + 2NH_3 \longrightarrow o\text{-}C_6H_4(\overset{O}{\overset{\|}{C}}NH_2)(\overset{O}{\overset{\|}{C}}O^- \; \overset{+}{N}H_4)$$

Phthalic anhydride; Ammonia; Ammonium phthalamate ($C_8H_{10}N_2O_3$)

Phthalimide is formed when ammonium phthalamate is heated.

$$o\text{-}C_6H_4(\overset{O}{\overset{\|}{C}}NH_2)(\overset{O}{\overset{\|}{C}}O^- \; \overset{+}{N}H_4) \xrightarrow{\text{heat}} \text{Phthalimide (NH)} + NH_3 + H_2O$$

Ammonium phthalamate; Phthalimide; Ammonia; Water

**20.19** **Step 1: Protonation of the carbonyl oxygen**

$$CH_3C(=\ddot{O}:)\ddot{N}HC_6H_5 + H-\overset{+}{\ddot{O}}H_2 \rightleftharpoons CH_3C(=\overset{+}{\ddot{O}}H)\ddot{N}HC_6H_5 + :\ddot{O}H_2$$

Acetanilide; Hydronium ion; Protonated form of amide; Water

**Step 2: Nucleophilic addition of water**

$$H_2\ddot{O}: + CH_3C(=\overset{+}{O}H)\ddot{N}HC_6H_5 \rightleftharpoons CH_3C(OH)(\ddot{N}HC_6H_5)\overset{+}{O}H_2$$

Water | Protonated form of amide | Oxonium ion

**Step 3: Deprotonation of oxonium ion to give neutral form of tetrahedral intermediate**

$$CH_3C(OH)(\ddot{N}HC_6H_5)\overset{+}{O}H_2 + H_2\ddot{O}: \rightleftharpoons CH_3C(OH)_2\ddot{N}HC_6H_5 + H_3\overset{+}{O}:$$

Oxonium ion | Water | Tetrahedral intermediate | Hydronium ion

**Step 4: Protonation of amino group of tetrahedral intermediate**

$$CH_3C(OH)_2\ddot{N}HC_6H_5 + H_3\overset{+}{O}: \rightleftharpoons CH_3C(OH)_2\overset{+}{N}H_2C_6H_5 + H_2\ddot{O}:$$

Tetrahedral intermediate | Hydronium ion | *N*-Protonated form of tetrahedral intermediate | Water

**Step 5: Dissociation of *N*-protonated form of tetrahedral intermediate**

$$CH_3C(OH)_2\overset{+}{N}H_2C_6H_5 \rightleftharpoons CH_3C(=\overset{+}{O}H)\ddot{O}H + H_2\ddot{N}C_6H_5$$

*N*-Protonated form of tetrahedral intermediate | Protonated form of acetic acid | Aniline

**Step 6: Proton-transfer processes**

$$H_3\overset{+}{O}: + H_2\ddot{N}C_6H_5 \rightleftharpoons H_2\ddot{O}: + H_3\overset{+}{N}C_6H_5$$

Hydronium ion | Aniline | Water | Anilinium ion

$$CH_3C(=\overset{+}{O}H)\ddot{O}H + H_2\ddot{O}: \rightleftharpoons CH_3C(=\ddot{O}:)\ddot{O}H + H_3\overset{+}{O}:$$

Protonated form of acetic acid | Water | Acetic acid | Hydronium ion

**20.20** **Step 1: Nucleophilic addition of hydroxide ion to the carbonyl group**

$$HO^- + HC(=O)N(CH_3)_2 \longrightarrow HC(O^-)(OH)N(CH_3)_2$$

Hydroxide ion; *N,N*-Dimethylformamide; Anionic form of tetrahedral intermediate

**Step 2: Proton transfer to give neutral form of tetrahedral intermediate**

$$HC(O^-)(OH)N(CH_3)_2 + H{-}OH \longrightarrow HC(OH)(OH)N(CH_3)_2 + {}^-OH$$

Anionic form of tetrahedral intermediate; Water; Tetrahedral intermediate; Hydroxide ion

**Step 3: Proton transfer from water to nitrogen of tetrahedral intermediate**

$$HC(OH)(OH)N(CH_3)_2 + H{-}OH \rightleftharpoons HC(OH)(OH)\overset{+}{N}H(CH_3)_2 + {}^-OH$$

Tetrahedral intermediate; Water; *N*-Protonated form of tetrahedral intermediate; Hydroxide ion

**Step 4: Dissociation of *N*-protonated form of tetrahedral intermediate**

$$HO^- + HC(OH)(OH)\overset{+}{N}H(CH_3)_2 \rightleftharpoons H_2O + HC(=O)OH + HN(CH_3)_2$$

Hydroxide ion; *N*-Protonated form of tetrahedral intermediate; Water; Formic acid; Dimethylamine

**Step 5: Irreversible formation of formate ion**

$$HC(=O)O{-}H + {}^-OH \longrightarrow HC(=O)O^- + HOH$$

Formic acid; Hydroxide ion; Formate ion; Water

**20.21** A synthetic scheme becomes apparent when we recognize that a primary amine may be obtained by Hofmann rearrangement of the primary amide having one more carbon in its acyl group. This amide may, in turn, be prepared from the corresponding carboxylic acid.

$$CH_3CH_2CH_2NH_2 \Longrightarrow CH_3CH_2CH_2\overset{O}{\overset{\|}{C}}NH_2 \Longrightarrow CH_3CH_2CH_2CO_2H$$

The desired reaction scheme is therefore

$$\underset{\text{Butanoic acid}}{CH_3CH_2CH_2CO_2H} \xrightarrow[\text{2. } NH_3]{\text{1. } SOCl_2} \underset{\text{Butanamide}}{CH_3CH_2CH_2\overset{O}{\overset{\|}{C}}NH_2} \xrightarrow[H_2O,\ NaOH]{Br_2} \underset{\text{1-Propanamine}}{CH_3CH_2CH_2NH_2}$$

**20.22** (*a*) Ethanenitrile has the same number of carbon atoms as ethyl alcohol. This suggests a reaction scheme proceeding via an amide.

$$\underset{\text{Ethyl alcohol}}{CH_3CH_2OH} \longrightarrow \underset{\text{Acetamide}}{CH_3\overset{O}{\overset{\|}{C}}NH_2} \xrightarrow{P_4O_{10}} \underset{\text{Ethanenitrile}}{CH_3C{\equiv}N}$$

The necessary amide is prepared from ethanol.

$$\underset{\text{Ethyl alcohol}}{CH_3CH_2OH} \xrightarrow[H_2SO_4,\ \text{heat}]{Na_2Cr_2O_7,\ H_2O} \underset{\text{Acetic acid}}{CH_3\overset{O}{\overset{\|}{C}}OH} \xrightarrow[\text{2. } NH_3]{\text{1. } SOCl_2} \underset{\text{Acetamide}}{CH_3\overset{O}{\overset{\|}{C}}NH_2}$$

(*b*) Propanenitrile may be prepared from ethyl alcohol by way of a nucleophilic substitution reaction of the corresponding bromide.

$$\underset{\text{Ethyl alcohol}}{CH_3CH_2OH} \xrightarrow[\text{or } HBr]{PBr_3} \underset{\text{Ethyl bromide}}{CH_3CH_2Br} \xrightarrow{NaCN} \underset{\text{Propanenitrile}}{CH_3CH_2CN}$$

**20.23** **Step 1: Protonation of the nitrile**

$$\underset{\text{Nitrile}}{RC{\equiv}N{:}} + \underset{\text{Hydronium ion}}{H{-}\overset{+}{\ddot{O}}H_2} \rightleftharpoons \underset{\text{Protonated form of nitrile}}{RC{\equiv}\overset{+}{N}H} + \underset{\text{Water}}{H_2O}$$

**Step 2: Nucleophilic addition of water**

$$\underset{\text{Water}}{H_2\ddot{O}{:}} + \underset{\text{Protonated form of nitrile}}{RC{\equiv}\overset{+}{N}H} \rightleftharpoons \underset{\text{Protonated form of imino acid}}{RC({=}NH)\overset{+}{O}H_2}$$

**Step 3: Deprotonation of imino acid**

$$\underset{\text{Protonated form of imino acid}}{RC({=}NH)H\overset{+}{O}{-}H} + \underset{\text{Water}}{{:}\ddot{O}H_2} \rightleftharpoons \underset{\text{Imino acid}}{RC({=}NH)OH} + \underset{\text{Hydronium ion}}{H_3O^+}$$

**Steps 4 and 5: Proton transfers to give an amide**

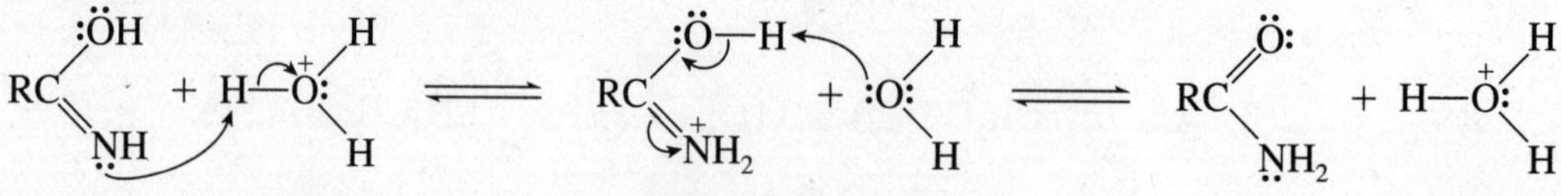

$$\mathrm{RC(=NH)\ddot{O}H} + \mathrm{H{-}\overset{+}{O}H_2} \rightleftharpoons \mathrm{RC(=\overset{+}{N}H_2)\ddot{O}{-}H} + \mathrm{H_2O} \rightleftharpoons \mathrm{RC(=O)NH_2} + \mathrm{H{-}\overset{+}{O}H_2}$$

Imino acid | Hydronium ion | Conjugate acid of amide | Water | Amide | Hydronium ion

**20.24** Ketones may be prepared by the reaction of nitriles with Grignard reagents. Nucleophilic addition of a Grignard reagent to a nitrile produces an imine. The imine is not normally isolated, however, but is hydrolyzed to the corresponding ketone. Ethyl phenyl ketone may be prepared by the reaction of propanenitrile with a phenyl Grignard reagent such as phenylmagnesium bromide, followed by hydrolysis of the imine.

$$CH_3CH_2C{\equiv}N + C_6H_5MgBr \xrightarrow{\text{Diethyl ether}} \left[C_6H_5\overset{NH}{\overset{\|}{C}}CH_2CH_3\right] \xrightarrow[\text{heat}]{H_2O,\ H^+} C_6H_5\overset{O}{\overset{\|}{C}}CH_2CH_3$$

Propanenitrile | Phenylmagnesium bromide | Imine (not isolated) | Ethyl phenyl ketone

**20.25** (*a*) The halogen that is attached to the carbonyl group is identified in the name as a separate word following the name of the acyl group.

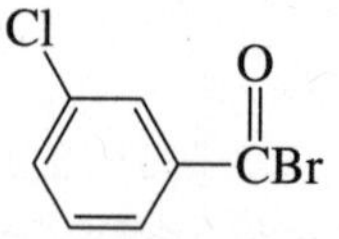

*m*-Chlorobenzoyl bromide ($m\text{-}ClC_6H_4\overset{O}{\overset{\|}{C}}Br$)

(*b*) Trifluoroacetic anhydride is the anhydride of trifluoroacetic acid. Notice that it contains six fluorines.

$$CF_3\overset{O}{\overset{\|}{C}}O\overset{O}{\overset{\|}{C}}CF_3$$

Trifluoroacetic anhydride

(*c*) This compound is the cyclic anhydride of *cis*-1,2-cyclopropanedicarboxylic acid.

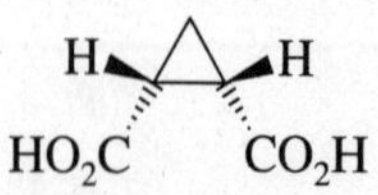

*cis*-1,2-Cyclopropanedicarboxylic acid 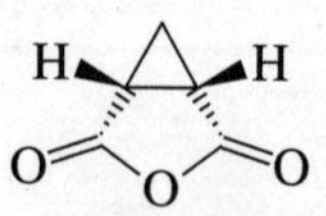 | *cis*-1,2-Cyclopropanedicarboxylic anhydride

(*d*) Ethyl cycloheptanecarboxylate is the ethyl ester of cycloheptanecarboxylic acid.

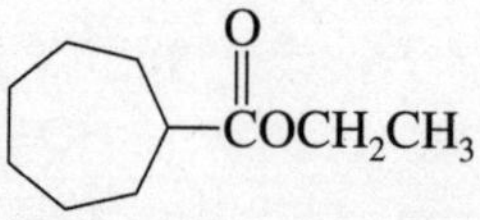

$$C_7H_{13}\overset{O}{\overset{\|}{C}}OCH_2CH_3$$

Ethyl cycloheptanecarboxylate

(*e*) 1-Phenylethyl acetate is the ester of 1-phenylethanol and acetic acid.

$$CH_3\overset{O}{\overset{\|}{C}}O\underset{CH_3}{\underset{|}{C}H}-C_6H_5$$

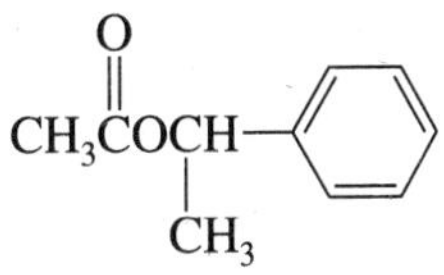

1-Phenylethyl acetate

(*f*) 2-Phenylethyl acetate is the ester of 2-phenylethanol and acetic acid.

$$CH_3\overset{O}{\overset{\|}{C}}OCH_2CH_2-C_6H_5$$

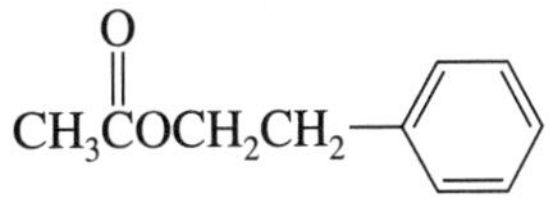

2-Phenylethyl acetate

(*g*) The parent compound in this case is benzamide. *p*-Ethylbenzamide has an ethyl substituent at the ring position para to the carbonyl group.

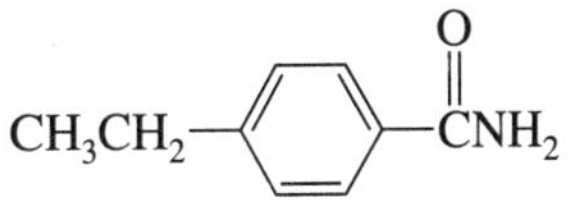

$$CH_3CH_2-C_6H_4-\overset{O}{\overset{\|}{C}}NH_2$$

*p*-Ethylbenzamide

(*h*) The parent compound is benzamide. In *N*-ethylbenzamide the ethyl substituent is bonded to nitrogen.

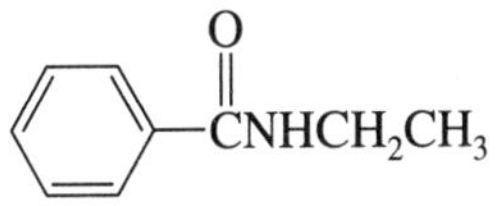

$$C_6H_5-\overset{O}{\overset{\|}{C}}NHCH_2CH_3$$

*N*-Ethylbenzamide

(*i*) Nitriles are named by adding the suffix *-nitrile* to the name of the alkane having the same number of carbons. Numbering begins at the nitrile carbon.

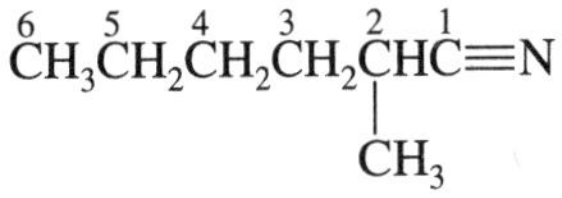

$$\overset{6}{C}H_3\overset{5}{C}H_2\overset{4}{C}H_2\overset{3}{C}H_2\underset{CH_3}{\underset{|}{\overset{2}{C}H}}\overset{1}{C}\equiv N$$

2-Methylhexanenitrile

**20.26** (*a*) This compound, with a bromine substituent attached to its carbonyl group, is named as an acyl bromide. It is 3-chlorobutanoyl bromide.

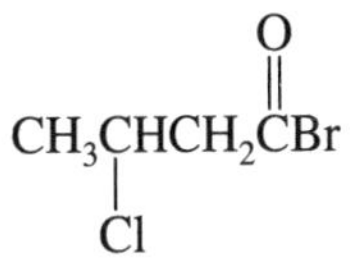

$$CH_3\underset{Cl}{\underset{|}{C}H}CH_2\overset{O}{\overset{\|}{C}}Br$$

3-Chlorobutanoyl bromide

(*b*) The group attached to oxygen, in this case **benzyl,** is identified first in the name of the ester. This compound is the benzyl ester of acetic acid.

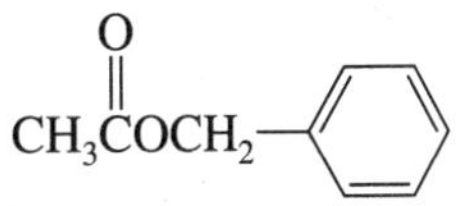

$$CH_3\overset{O}{\overset{\|}{C}}OCH_2-C_6H_5$$

Benzyl acetate

(*c*) The group attached to oxygen is methyl; this compound is the methyl ester of phenylacetic acid.

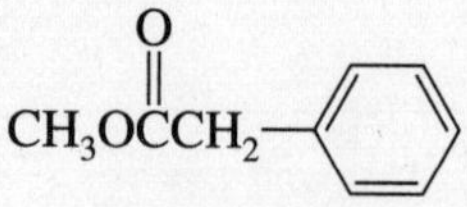

$$CH_3O\overset{O}{\overset{\|}{C}}CH_2-C_6H_5$$

Methyl phenylacetate

(*d*) This compound contains the functional group $-\overset{O}{\overset{\|}{C}}O\overset{O}{\overset{\|}{C}}-$ and thus is an anhydride of a carboxylic acid. We name the acid, in this case 3-chloropropanoic acid, drop the *acid* part of the name, and replace it by *anhydride.*

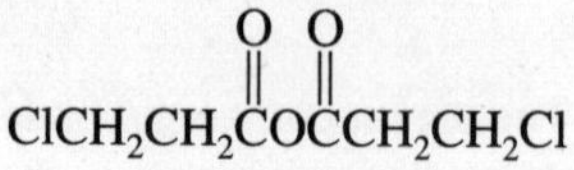

$$ClCH_2CH_2\overset{O}{\overset{\|}{C}}O\overset{O}{\overset{\|}{C}}CH_2CH_2Cl$$

3-Chloropropanoic anhydride

(*e*) This compound is a cyclic anhydride, whose parent acid is 3,3-dimethylpentanedioic acid.

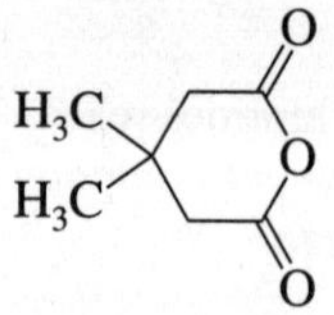

3,3-Dimethylpentanedioic anhydride

(*f*) Nitriles are named by adding *-nitrile* to the name of the alkane having the same number of carbons. Remember to count the carbon of the C≡N group.

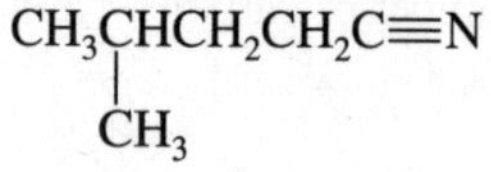

$$CH_3\underset{\underset{CH_3}{|}}{C}HCH_2CH_2C\equiv N$$

4-Methylpentanenitrile

(*g*) This compound is an amide. We name the corresponding acid and then replace the *-oic acid* suffix by *-amide.*

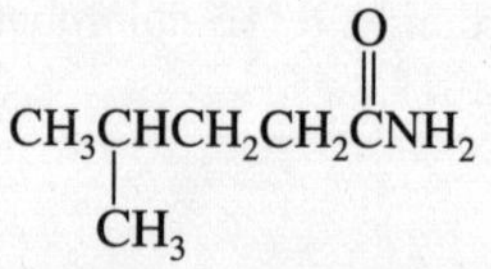

$$CH_3\underset{\underset{CH_3}{|}}{C}HCH_2CH_2\overset{O}{\overset{\|}{C}}NH_2$$

4-Methylpentanamide

(*h*) This compound is the *N*-methyl derivative of 4-methylpentanamide.

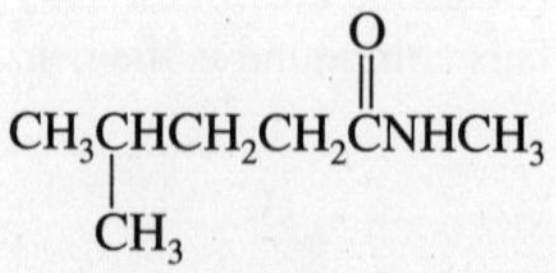

$$CH_3\underset{\underset{CH_3}{|}}{C}HCH_2CH_2\overset{O}{\overset{\|}{C}}NHCH_3$$

*N*-Methyl-4-methylpentanamide

(*i*) The amide nitrogen bears two methyl groups. We designate this as an *N*,*N*-dimethyl amide.

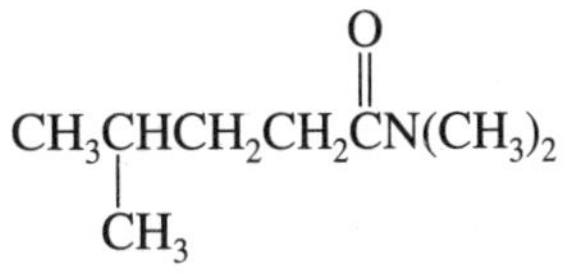

$$CH_3\underset{\substack{|\\CH_3}}{C}HCH_2CH_2\overset{\overset{\large O}{\|}}{C}N(CH_3)_2$$

*N*,*N*-Dimethyl-4-methylpentanamide

**20.27** (*a*) Acetyl chloride acts as an acyl transfer agent to the aromatic ring of bromobenzene. The reaction is a Friedel–Crafts acylation. Bromine is an ortho, para-directing substituent.

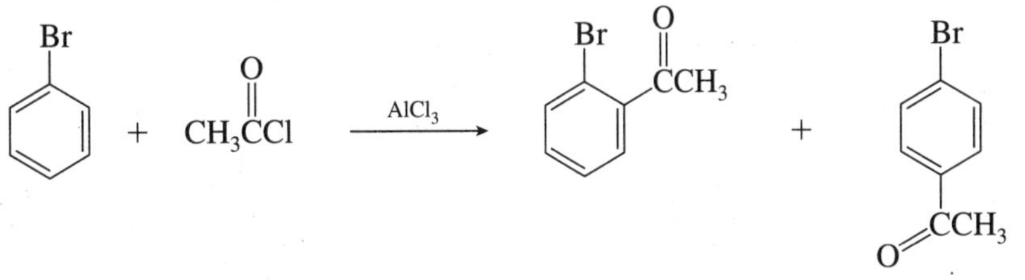

$$C_6H_5Br + CH_3\overset{\overset{\large O}{\|}}{C}Cl \xrightarrow{AlCl_3} o\text{-}BrC_6H_4\overset{\overset{\large O}{\|}}{C}CH_3 + p\text{-}BrC_6H_4\overset{\overset{\large O}{\|}}{C}CH_3$$

Bromobenzene    Acetyl chloride    *o*-Bromoacetophenone    *p*-Bromoacetophenone

(*b*) Acyl chlorides react with thiols to give thioesters.

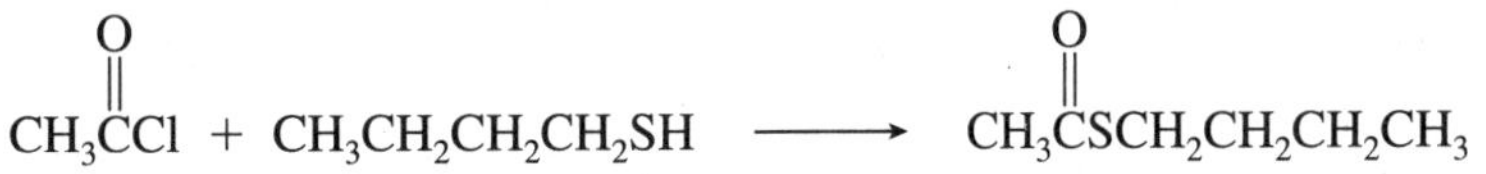

$$CH_3\overset{\overset{\large O}{\|}}{C}Cl + CH_3CH_2CH_2CH_2SH \longrightarrow CH_3\overset{\overset{\large O}{\|}}{C}SCH_2CH_2CH_2CH_3$$

Acetyl chloride    1-Butanethiol    *S*-Butyl ethanethioate

(*c*) Sodium propanoate acts as a nucleophile toward propanoyl chloride. The product is propanoic anhydride.

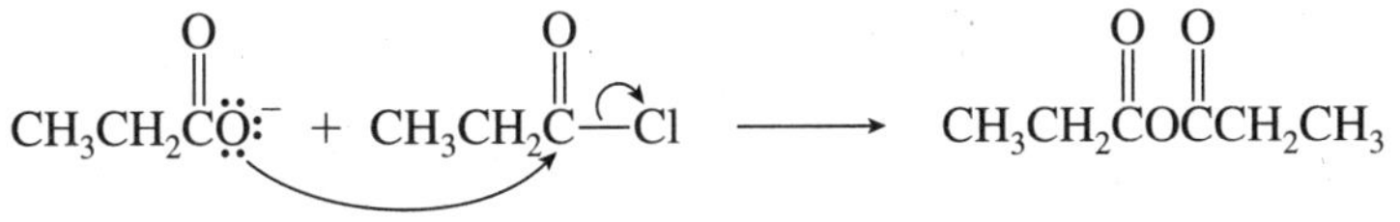

$$CH_3CH_2\overset{\overset{\large O}{\|}}{C}\ddot{\underset{..}{O}}{:}^- + CH_3CH_2\overset{\overset{\large O}{\|}}{C}\text{—}Cl \longrightarrow CH_3CH_2\overset{\overset{\large O}{\|}}{C}O\overset{\overset{\large O}{\|}}{C}CH_2CH_3$$

Propanoate anion    Propanoyl chloride    Propanoic anhydride

(*d*) Acyl chlorides convert alcohols to esters.

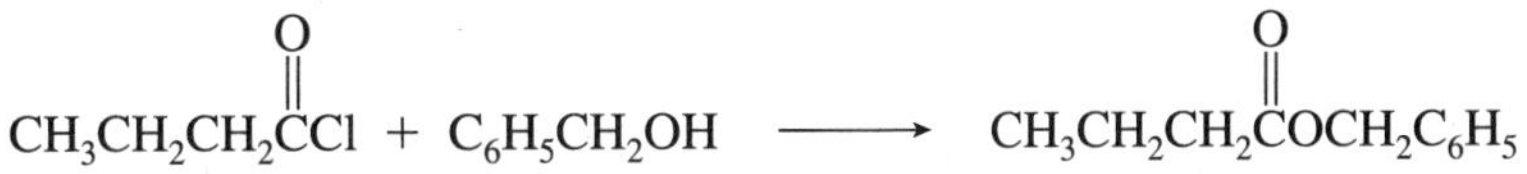

$$CH_3CH_2CH_2\overset{\overset{\large O}{\|}}{C}Cl + C_6H_5CH_2OH \longrightarrow CH_3CH_2CH_2\overset{\overset{\large O}{\|}}{C}OCH_2C_6H_5$$

Butanoyl chloride    Benzyl alcohol    Benzyl butanoate

(*e*) Acyl chlorides react with ammonia to yield amides.

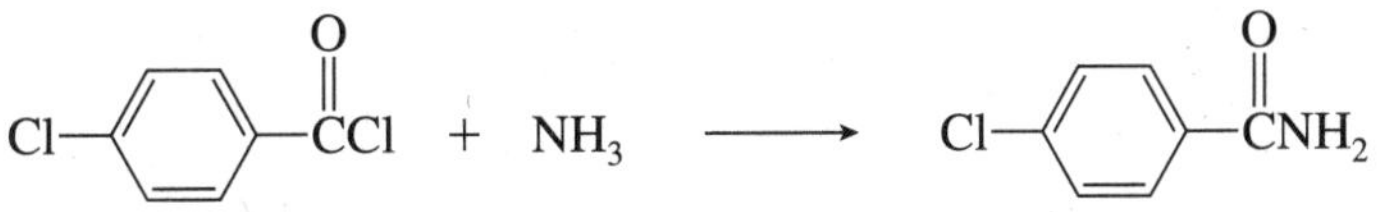

$$Cl\text{—}C_6H_4\text{—}\overset{\overset{\large O}{\|}}{C}Cl + NH_3 \longrightarrow Cl\text{—}C_6H_4\text{—}\overset{\overset{\large O}{\|}}{C}NH_2$$

*p*-Chlorobenzoyl chloride    Ammonia    *p*-Chlorobenzamide

(*f*) The starting material is a cyclic anhydride. Acid anhydrides react with water to yield two carboxylic acid functions; when the anhydride is cyclic, a dicarboxylic acid results.

$$\text{Succinic anhydride} + H_2O \longrightarrow HO\overset{O}{\overset{\|}{C}}CH_2CH_2\overset{O}{\overset{\|}{C}}OH$$

Succinic anhydride    Water    Succinic acid

(*g*) In dilute sodium hydroxide the anhydride is converted to the disodium salt of the diacid.

$$\text{Succinic anhydride} + 2NaOH \xrightarrow{H_2O} Na^+\ {}^-O\overset{O}{\overset{\|}{C}}CH_2CH_2\overset{O}{\overset{\|}{C}}O^-\ Na^+$$

Succinic anhydride    Sodium hydroxide    Sodium succinate

(*h*) One of the carbonyl groups of the cyclic anhydride is converted to an amide function on reaction with ammonia. The other, the one that would become a carboxylic acid group, is converted to an ammonium carboxylate salt.

$$\text{Succinic anhydride} + 2NH_3 \xrightarrow{H_2O} \overset{+}{N}H_4\ {}^-O\overset{O}{\overset{\|}{C}}CH_2CH_2\overset{O}{\overset{\|}{C}}NH_2$$

Succinic anhydride    Ammonia    Ammonium succinamate

(*i*) Acid anhydrides are used as acylating agents in Friedel–Crafts reactions.

$$\text{Succinic anhydride} + \text{Benzene} \xrightarrow{AlCl_3} C_6H_5\overset{O}{\overset{\|}{C}}CH_2CH_2\overset{O}{\overset{\|}{C}}OH$$

Succinic anhydride    Benzene    3-Benzoylpropanoic acid

(*j*) The reactant is maleic anhydride; it is a good dienophile in Diels–Alder reactions.

$$\text{1,3-Pentadiene} + \text{Maleic anhydride} \longrightarrow \text{3-Methylcyclohexene-4,5-dicarboxylic anhydride}$$

1,3-Pentadiene    Maleic anhydride    3-Methylcyclohexene-4,5-dicarboxylic anhydride

(*k*) Acid anhydrides react with alcohols to give an ester and a carboxylic acid.

$$CH_3\overset{O}{\overset{\|}{C}}O\overset{O}{\overset{\|}{C}}CH_3 + CH_3CH_2\underset{OH}{\underset{|}{C}H}CH_2CH_3 \longrightarrow CH_3CH_2\underset{O\overset{}{C}(=O)CH_3}{\underset{|}{C}H}CH_2CH_3 + CH_3CO_2H$$

Acetic anhydride    3-Pentanol    1-Ethylpropyl acetate    Acetic acid (*l*) The starting material is a cyclic ester, a lactone. Esters undergo saponification in aqueous base to give an alcohol and a carboxylate salt.

$$\text{4-Butanolide} + \underset{\text{Sodium hydroxide}}{NaOH} \xrightarrow{H_2O} \underset{\text{Sodium 4-hydroxybutanoate}}{HOCH_2CH_2CH_2\overset{O}{\overset{\|}{C}}O^-\ Na^+}$$

(*m*) Ammonia reacts with esters to give an amide and an alcohol.

$$\text{4-Butanolide} + \underset{\text{Ammonia}}{NH_3} \xrightarrow{H_2O} \underset{\text{4-Hydroxybutanamide}}{HOCH_2CH_2CH_2\overset{O}{\overset{\|}{C}}NH_2}$$

(*n*) Lithium aluminum hydride reduces esters to two alcohols; the one derived from the acyl group is a primary alcohol. Reduction of a cyclic ester gives a diol.

$$\text{4-Butanolide} \xrightarrow[\text{2. } H_2O]{\text{1. } LiAlH_4} \underset{\text{1,4-Butanediol}}{HOCH_2CH_2CH_2CH_2OH}$$

(*o*) Grignard reagents react with esters to give tertiary alcohols.

$$\text{4-Butanolide} \xrightarrow[\text{2. } H_3O^+]{\text{1. } 2CH_3MgBr} \underset{\text{4-Methyl-1,4-pentanediol}}{HOCH_2CH_2CH_2\underset{CH_3}{\underset{|}{\overset{OH}{\overset{|}{C}}}}CH_3}$$

(*p*) In this reaction methylamine acts as a nucleophile toward the carbonyl group of the ester. The product is an amide.

$$\underset{\text{Methylamine}}{CH_3\ddot{N}H_2} + \underset{\text{Ethyl phenylacetate}}{C_6H_5CH_2\overset{O}{\overset{\|}{C}}OCH_2CH_3} \longrightarrow \underset{\textit{N}\text{-Methylphenylacetamide}}{C_6H_5CH_2\overset{O}{\overset{\|}{C}}NHCH_3} + \underset{\text{Ethyl alcohol}}{CH_3CH_2OH}$$

(*q*) The starting material is a lactam, a cyclic amide. Amides are hydrolyzed in base to amines and carboxylate salts.

$$\textit{N}\text{-Methylpyrrolidone} + \underset{\text{Sodium hydroxide}}{NaOH} \xrightarrow{H_2O} \underset{\text{Sodium 4-(methylamino)butanoate}}{CH_3NHCH_2CH_2CH_2\overset{O}{\overset{\|}{C}}O^-\ Na^+}$$

(*r*) In acid solution amides yield carboxylic acids and ammonium salts.

$$\text{N-Methylpyrrolidone} + H_3O^+ \longrightarrow CH_3\overset{+}{N}H_2CH_2CH_2CH_2\overset{O}{\overset{\|}{C}}OH$$

*N*-Methylpyrrolidone; Hydronium ion; 4-(Methylammonio)butanoic acid

(*s*) The starting material is a cyclic imide. Both its amide bonds are cleaved by nucleophilic attack by hydroxide ion.

$$\text{N-Methylsuccinimide} + 2NaOH \xrightarrow{H_2O} Na^+\ {}^-O\overset{O}{\overset{\|}{C}}CH_2CH_2\overset{O}{\overset{\|}{C}}O^-\ Na^+ + CH_3NH_2$$

*N*-Methylsuccinimide; Sodium hydroxide; Disodium succinate; Methylamine

(*t*) In acid the imide undergoes cleavage to give a dicarboxylic acid and the conjugate acid of methylamine.

$$\text{N-Methylsuccinimide} + 2H_2O + HCl \longrightarrow HO\overset{O}{\overset{\|}{C}}CH_2CH_2\overset{O}{\overset{\|}{C}}OH + CH_3\overset{+}{N}H_3\ Cl^-$$

*N*-Methylsuccinimide; Water; Hydrogen chloride; Succinic acid; Methylammonium chloride

(*u*) Acetanilide is hydrolyzed in acid to acetic acid and the conjugate acid of aniline.

$$C_6H_5NH\overset{O}{\overset{\|}{C}}CH_3 + H_2O + HCl \longrightarrow C_6H_5\overset{+}{N}H_3\ Cl^- + CH_3\overset{O}{\overset{\|}{C}}OH$$

Acetanilide; Water; Hydrogen chloride; Anilinium chloride; Acetic acid

(*v*) This is another example of amide hydrolysis.

$$C_6H_5\overset{O}{\overset{\|}{C}}NHCH_3 + H_2O + H_2SO_4 \longrightarrow C_6H_5\overset{O}{\overset{\|}{C}}OH + CH_3\overset{+}{N}H_3\ HSO_4^-$$

*N*-Methylbenzamide; Water; Sulfuric acid; Benzoic acid; Methylammonium hydrogen sulfate

(*w*) One way to prepare nitriles is by dehydration of amides.

$$C_5H_9\overset{O}{\overset{\|}{C}}NH_2 \xrightarrow{P_4O_{10}} C_5H_9C{\equiv}N + H_2O$$

Cyclopentanecarboxamide; Cyclopentyl cyanide

(*x*) Nitriles are hydrolyzed to carboxylic acids in acidic media.

$$(CH_3)_2CHCH_2C{\equiv}N \xrightarrow[\text{heat}]{HCl,\ H_2O} (CH_3)_2CHCH_2\overset{O}{\overset{\|}{C}}OH$$

3-Methylbutanenitrile → 3-Methylbutanoic acid

(*y*) Nitriles are hydrolyzed in aqueous base to salts of carboxylic acids.

$$CH_3O{-}C_6H_4{-}C{\equiv}N \xrightarrow[\text{heat}]{NaOH,\ H_2O} CH_3O{-}C_6H_4{-}\overset{O}{\overset{\|}{C}}O^-\ Na^+ + NH_3$$

*p*-Methoxybenzonitrile → Sodium *p*-methoxybenzoate + Ammonia

(*z*) Grignard reagents react with nitriles to yield ketones after addition of aqueous acid.

$$CH_3CH_2C{\equiv}N \xrightarrow[2.\ H_3O^+]{1.\ CH_3MgBr} CH_3CH_2\overset{O}{\overset{\|}{C}}CH_3$$

Propanenitrile → 2-Butanone

(*aa*) Amides undergo the Hofmann rearrangement on reaction with bromine and base. A methyl carbamate is the product isolated when the reaction is carried out in methanol.

(7,7-dimethylbicyclo[2.2.1]heptane-1-carboxamide, bridgehead $C(=O)NH_2$) $+ Br_2 \xrightarrow[CH_3OH]{NaOCH_3}$ (7,7-dimethylbicyclo[2.2.1]heptane with bridgehead $NH\overset{O}{\overset{\|}{C}}OCH_3$)

(*bb*) Saponification of the carbamate in part (*aa*) gives the corresponding amine.

(bridgehead $NH\overset{O}{\overset{\|}{C}}OCH_3$ compound) $\xrightarrow[H_2O]{KOH}$ (bridgehead $NH_2$ compound)

**20.28** (*a*) Acetyl chloride is prepared by reaction of acetic acid with thionyl chloride. The first task then is to prepare acetic acid by oxidation of ethanol.

$$CH_3CH_2OH \xrightarrow[H_2O]{K_2Cr_2O_7,\ H_2SO_4} CH_3\overset{O}{\overset{\|}{C}}OH \xrightarrow{SOCl_2} CH_3\overset{O}{\overset{\|}{C}}Cl$$

Ethanol → Acetic acid → Acetyl chloride

(*b*) Acetic acid and acetyl chloride, available from part (*a*), can be combined to form acetic anhydride.

$$CH_3\overset{O}{\overset{\|}{C}}OH + CH_3\overset{O}{\overset{\|}{C}}Cl \longrightarrow CH_3\overset{O}{\overset{\|}{C}}O\overset{O}{\overset{\|}{C}}CH_3 + HCl$$

Acetic acid + Acetyl chloride → Acetic anhydride + Hydrogen chloride (*c*) Ethanol can be converted to ethyl acetate by reaction with acetic acid, acetyl chloride, or acetic anhydride from parts (*a*) and (*b*).

$$\underset{\text{Ethanol}}{CH_3CH_2OH} + \underset{\text{Acetic acid}}{CH_3\overset{O}{\overset{\|}{C}}OH} \xrightarrow{H^+} \underset{\text{Ethyl acetate}}{CH_3\overset{O}{\overset{\|}{C}}OCH_2CH_3} + \underset{\text{Water}}{H_2O}$$

or

$$\underset{\text{Ethanol}}{CH_3CH_2OH} + \underset{\text{Acetyl chloride}}{CH_3\overset{O}{\overset{\|}{C}}Cl} \xrightarrow{\text{pyridine}} \underset{\text{Ethyl acetate}}{CH_3\overset{O}{\overset{\|}{C}}OCH_2CH_3}$$

or

$$\underset{\text{Ethanol}}{CH_3CH_2OH} + \underset{\text{Acetic anhydride}}{CH_3\overset{O}{\overset{\|}{C}}O\overset{O}{\overset{\|}{C}}CH_3} \xrightarrow{\text{pyridine}} \underset{\text{Ethyl acetate}}{CH_3\overset{O}{\overset{\|}{C}}OCH_2CH_3}$$

(*d*) Ethyl bromoacetate is the ethyl ester of bromoacetic acid; thus the first task is to prepare the acid. We use the acetic acid prepared in part (*a*), converting it to bromoacetic acid by the Hell–Volhard–Zelinsky reaction.

$$\underset{\text{Acetic acid}}{CH_3CO_2H} \xrightarrow[P]{Br_2} \underset{\text{Bromoacetic acid}}{BrCH_2CO_2H} \xrightarrow[H^+]{CH_3CH_2OH} \underset{\text{Ethyl bromoacetate}}{BrCH_2\overset{O}{\overset{\|}{C}}OCH_2CH_3}$$

Alternatively, bromoacetic acid could be converted to the corresponding acyl chloride, then treated with ethanol. It would be incorrect to try to brominate ethyl acetate; the Hell–Volhard–Zelinsky method requires an acid as starting material, not an ester.

(*e*) The alcohol $BrCH_2CH_2OH$, needed in order to prepare 2-bromoethyl acetate, is prepared from ethanol by way of ethylene.

$$\underset{\text{Ethanol}}{CH_3CH_2OH} \xrightarrow[\text{heat}]{H_2SO_4} \underset{\text{Ethylene}}{CH_2{=}CH_2} \xrightarrow[H_2O]{Br_2} \underset{\text{2-Bromoethanol}}{BrCH_2CH_2OH}$$

Then

$$\underset{\text{2-Bromoethanol}}{BrCH_2CH_2OH} \xrightarrow[\text{or } CH_3\overset{O}{\overset{\|}{C}}Cl]{CH_3\overset{O}{\overset{\|}{C}}O\overset{O}{\overset{\|}{C}}CH_3} \underset{\text{2-Bromoethyl acetate}}{CH_3\overset{O}{\overset{\|}{C}}OCH_2CH_2Br}$$

(*f*) Ethyl cyanoacetate may be prepared from the ethyl bromoacetate obtained in part (*d*). The bromide may be displaced by cyanide in a nucleophilic substitution reaction.

$$\underset{\text{Ethyl bromoacetate}}{BrCH_2\overset{O}{\overset{\|}{C}}OCH_2CH_3} \xrightarrow[S_N2]{NaCN} \underset{\text{Ethyl cyanoacetate}}{N{\equiv}CCH_2\overset{O}{\overset{\|}{C}}OCH_2CH_3}$$

(*g*) Reaction of the acetyl chloride prepared in part (*a*) or the acetic anhydride from part (*b*) with ammonia gives acetamide.

$$\underset{\text{Acetyl chloride}}{CH_3\overset{O}{\overset{\|}{C}}Cl} \quad \text{or} \quad \underset{\text{Acetic anhydride}}{CH_3\overset{O}{\overset{\|}{C}}O\overset{O}{\overset{\|}{C}}CH_3} \xrightarrow{NH_3} \underset{\text{Acetamide}}{CH_3\overset{O}{\overset{\|}{C}}NH_2}$$

(*h*) Methylamine may be prepared from acetamide by a Hofmann rearrangement.

$$\underset{\substack{\text{Acetamide} \\ \text{[prepared as in part (g)]}}}{CH_3\overset{O}{\overset{\|}{C}}NH_2} \xrightarrow{Br_2,\ HO^-,\ H_2O} \underset{\text{Methylamine}}{CH_3NH_2}$$

(*i*) The desired hydroxy acid is available from hydrolysis of the corresponding cyanohydrin, which may be prepared by reaction of the appropriate aldehyde with cyanide ion.

$$CH_3\underset{OH}{\underset{|}{C}}H\overset{O}{\overset{\|}{C}}OH \Longrightarrow CH_3\underset{OH}{\underset{|}{C}}HC{\equiv}N \Longrightarrow CH_3\overset{O}{\overset{\|}{C}}H$$

In this synthesis the cyanohydrin is prepared from ethanol by way of acetaldehyde.

$$\underset{\text{Ethanol}}{CH_3CH_2OH} \xrightarrow[CH_2Cl_2]{PCC} \underset{\text{Acetaldehyde}}{CH_3\overset{O}{\overset{\|}{C}}H} \xrightarrow[H^+]{KCN} \underset{\text{2-Hydroxypropanenitrile}}{CH_3\overset{OH}{\overset{|}{C}}HC{\equiv}N}$$

$$\underset{\text{2-Hydroxypropanenitrile}}{CH_3\underset{OH}{\underset{|}{C}}HC{\equiv}N} \xrightarrow[\substack{\text{or} \\ 1.\ HO^-,\ H_2O,\ \text{heat} \\ 2.\ H^+}]{H_2O,\ H^+,\ \text{heat}} \underset{\text{2-Hydroxypropanoic acid}}{CH_3\underset{OH}{\underset{|}{C}}H\overset{O}{\overset{\|}{C}}OH}$$

**20.29** (*a*) Benzoyl chloride is made from benzoic acid. Oxidize toluene to benzoic acid, and then treat with thionyl chloride.

$$\underset{\text{Toluene}}{C_6H_5CH_3} \xrightarrow[H_2O,\ \text{heat}]{K_2Cr_2O_7,\ H_2SO_4} \underset{\text{Benzoic acid}}{C_6H_5\overset{O}{\overset{\|}{C}}OH} \xrightarrow{SOCl_2} \underset{\text{Benzoyl chloride}}{C_6H_5\overset{O}{\overset{\|}{C}}Cl}$$

(*b*) Benzoyl chloride and benzoic acid, both prepared from toluene in part (*a*), react with each other to give benzoic anhydride.

$$\underset{\text{Benzoic acid}}{C_6H_5\overset{O}{\overset{\|}{C}}OH} + \underset{\text{Benzoyl chloride}}{C_6H_5\overset{O}{\overset{\|}{C}}Cl} \longrightarrow \underset{\text{Benzoic anhydride}}{C_6H_5\overset{O}{\overset{\|}{C}}O\overset{O}{\overset{\|}{C}}C_6H_5}$$

(*c*) Benzoic acid, benzoyl chloride, and benzoic anhydride have been prepared in parts (*a*) and (*b*) of this problem. Any of them could be converted to benzyl benzoate on reaction with benzyl alcohol. Thus the synthesis of benzyl benzoate requires the preparation of benzyl alcohol from toluene. This is effected by a nucleophilic substitution reaction of benzyl bromide, in turn prepared by halogenation of toluene.

$$\underset{\text{Toluene}}{C_6H_5CH_3} \xrightarrow[\text{or } Br_2\text{, light}]{N\text{-bromosuccinimide (NBS)}} \underset{\text{Benzyl bromide}}{C_6H_5CH_2Br} \xrightarrow[HO^-]{H_2O} \underset{\text{Benzyl alcohol}}{C_6H_5CH_2OH}$$

Alternatively, recall that primary alcohols may be obtained by reduction of the corresponding carboxylic acid.

$$\underset{\text{Benzoic acid}}{C_6H_5\overset{O}{\overset{\|}{C}}OH} \xrightarrow[2.\ H_2O]{1.\ LiAlH_4} \underset{\text{Benzyl alcohol}}{C_6H_5CH_2OH}$$

Then

$$\underset{\text{Benzoyl chloride}}{C_6H_5\overset{O}{\overset{\|}{C}}Cl} + \underset{\text{Benzyl alcohol}}{C_6H_5CH_2OH} \xrightarrow{\text{pyridine}} \underset{\text{Benzyl benzoate}}{C_6H_5\overset{O}{\overset{\|}{C}}OCH_2C_6H_5}$$

(*d*) Benzamide is prepared by reaction of ammonia with either benzoyl chloride from part (*a*) or benzoic anhydride from part (*b*).

$$\underset{\text{Benzoyl chloride}}{C_6H_5\overset{O}{\overset{\|}{C}}Cl} \quad \text{or} \quad \underset{\text{Benzoic anhydride}}{C_6H_5\overset{O}{\overset{\|}{C}}O\overset{O}{\overset{\|}{C}}C_6H_5} \xrightarrow{NH_3} \underset{\text{Benzamide}}{C_6H_5\overset{O}{\overset{\|}{C}}NH_2}$$

(*e*) Benzonitrile may be prepared by dehydration of benzamide.

$$\underset{\text{Benzamide}}{C_6H_5\overset{O}{\overset{\|}{C}}NH_2} \xrightarrow[\text{heat}]{P_4O_{10}} \underset{\text{Benzonitrile}}{C_6H_5C{\equiv}N}$$

(*f*) Benzyl cyanide is the product of nucleophilic substitution by cyanide ion on benzyl bromide or benzyl chloride. The benzyl halides are prepared by free-radical halogenation of the toluene side chain.

$$\underset{\text{Toluene}}{C_6H_5CH_3} \xrightarrow[\text{light or heat}]{Cl_2} \underset{\text{Benzyl chloride}}{C_6H_5CH_2Cl} \xrightarrow{NaCN} \underset{\text{Benzyl cyanide}}{C_6H_5CH_2C{\equiv}N}$$

or

$$\underset{\text{Toluene}}{C_6H_5CH_3} \xrightarrow[\text{or } Br_2\text{, light}]{NBS} \underset{\text{Benzyl bromide}}{C_6H_5CH_2Br} \xrightarrow{NaCN} \underset{\text{Benzyl cyanide}}{C_6H_5CH_2C{\equiv}N}$$

(*g*) Hydrolysis of benzyl cyanide yields phenylacetic acid.

$$C_6H_5CH_2C\equiv N \xrightarrow[\text{or 1. NaOH, heat; 2. }H^+]{H_2O,\ H^+,\ \text{heat}} C_6H_5CH_2\overset{O}{\overset{\|}{C}}OH$$

Benzyl cyanide → Phenylacetic acid

Alternatively, the Grignard reagent derived from benzyl bromide may be carboxylated.

$$C_6H_5CH_2Br \xrightarrow[\text{diethyl ether}]{Mg} C_6H_5CH_2MgBr \xrightarrow[2.\ H_3O^+]{1.\ CO_2} C_6H_5CH_2\overset{O}{\overset{\|}{C}}OH$$

Benzyl bromide → Benzylmagnesium bromide → Phenylacetic acid

(*h*) The first goal is to synthesize *p*-nitrobenzoic acid because this may be readily converted to the desired acyl chloride. First convert toluene to *p*-nitrotoluene; then oxidize. Nitration must precede oxidation of the side chain in order to achieve the desired para orientation.

$$C_6H_5CH_3 \xrightarrow[H_2SO_4]{HNO_3} O_2N\text{-}C_6H_4\text{-}CH_3 \xrightarrow[H_2O,\ \text{heat}]{K_2Cr_2O_7,\ H_2SO_4} O_2N\text{-}C_6H_4\text{-}\overset{O}{\overset{\|}{C}}OH$$

Toluene → *p*-Nitrotoluene (separate from ortho isomer) → *p*-Nitrobenzoic acid

Treatment of *p*-nitrobenzoic acid with thionyl chloride yields *p*-nitrobenzoyl chloride.

$$O_2N\text{-}C_6H_4\text{-}\overset{O}{\overset{\|}{C}}OH \xrightarrow{SOCl_2} O_2N\text{-}C_6H_4\text{-}\overset{O}{\overset{\|}{C}}Cl$$

*p*-Nitrobenzoic acid → *p*-Nitrobenzoyl chloride

(*i*) In order to achieve the correct orientation in *m*-nitrobenzoyl chloride, oxidation of the methyl group must precede nitration.

$$C_6H_5CH_3 \xrightarrow[H_2O,\ \text{heat}]{K_2Cr_2O_7,\ H_2SO_4} C_6H_5CO_2H \xrightarrow[H_2SO_4]{HNO_3} m\text{-}O_2N\text{-}C_6H_4\text{-}CO_2H$$

Toluene → Benzoic acid → *m*-Nitrobenzoic acid

Once *m*-nitrobenzoic acid has been prepared, it may be converted to the corresponding acyl chloride.

$$m\text{-}O_2N\text{-}C_6H_4\text{-}\overset{O}{\overset{\|}{C}}OH \xrightarrow{SOCl_2} m\text{-}O_2N\text{-}C_6H_4\text{-}\overset{O}{\overset{\|}{C}}Cl$$

*m*-Nitrobenzoic acid → *m*-Nitrobenzoyl chloride

(*j*) A Hofmann rearrangement of benzamide affords aniline.

$$C_6H_5CNH_2\ (\text{C=O}) + Br_2 \xrightarrow[H_2O]{HO^-} C_6H_5NH_2$$

Benzamide [prepared as in part (*d*)] | Bromine | Aniline

**20.30** The problem specifies that $CH_3CH_2C(=O)\text{Ø}CH_2CH_3$ is to be prepared from $^{18}O$-labeled ethyl alcohol (Ø = $^{18}O$).

$$CH_3CH_2C(=O)Cl + CH_3CH_2\text{Ø}H \longrightarrow CH_3CH_2C(=O)\text{Ø}CH_2CH_3$$

Propanoyl chloride | Ethyl alcohol | Ethyl propanoate

Thus, we need to prepare $^{18}O$-labeled ethyl alcohol from the other designated starting materials, acetaldehyde and $^{18}O$-enriched water. First, replace the oxygen of acetaldehyde with $^{18}O$ by the hydration–dehydration equilibrium in the presence of $^{18}O$-enriched water.

$$CH_3CH(=O) + H_2\text{Ø} \rightleftharpoons CH_3CH(OH)(\text{Ø}H) \rightleftharpoons CH_3CH(=\text{Ø}) + H_2O$$

Acetaldehyde | $^{18}O$-enriched water | Hydrate of acetaldehyde | $^{18}O$-enriched acetaldehyde | Water

Once $^{18}O$-enriched acetaldehyde has been obtained, it can be reduced to $^{18}O$-enriched ethanol.

$$CH_3CH(=\text{Ø}) \xrightarrow[\text{or 1. } LiAlH_4 \text{ 2. } H_2O]{NaBH_4,\ CH_3OH} CH_3CH_2\text{Ø}H$$

**20.31** (*a*) The rate-determining step in basic ester hydrolysis is nucleophilic addition of hydroxide ion to the carbonyl group. The intermediate formed in this step is negatively charged.

$$CH_3C(=O)OCH_2CH_3 + HO^- \longrightarrow CH_3C(O^-)(OH)OCH_2CH_3$$

Ethyl acetate | Hydroxide ion | Rate-determining intermediate

The electron-withdrawing effect of a $CF_3$ group stabilizes the intermediate formed in the rate-determining step of ethyl trifluoroacetate saponification.

$$CF_3C(=O)OCH_2CH_3 + HO^- \longrightarrow CF_3C(O^-)(OH)OCH_2CH_3$$

Ethyl trifluoroacetate | Hydroxide ion | Rate-determining intermediate

Because the intermediate is more stable, it is formed faster than the one from ethyl acetate.

(*b*) Crowding is increased as the transition state for nucleophilic addition to the carbonyl group is approached. The carbonyl carbon undergoes a change in hybridization from $sp^2$ to $sp^3$.

$$(CH_3)_3C{-}C(=O)OCH_2CH_3 \;(sp^2) + {}^{-}OH \longrightarrow (CH_3)_3C{-}C(O^-)(OH)OCH_2CH_3 \;(sp^3)$$

Ethyl 2,2-dimethylpropanoate — Hydroxide ion — Rate-determining intermediate; crowded

The *tert*-butyl group of ethyl 2,2-dimethylpropanoate causes more crowding than the methyl group of ethyl acetate; the rate-determining intermediate is less stable and is formed more slowly.

(*c*) We see here another example of a steric effect of a *tert*-butyl group. The intermediate formed when hydroxide ion adds to the carbonyl group of *tert*-butyl acetate is more crowded and less stable than the corresponding intermediate formed from methyl acetate.

$$CH_3C(=O)OC(CH_3)_3 + HO^- \longrightarrow CH_3C(O^-)(HO)OC(CH_3)_3$$

*tert*-Butyl acetate — Hydroxide ion — Rate-determining intermediate; crowded

$$CH_3C(=O)OCH_3 + HO^- \longrightarrow CH_3C(O^-)(HO)OCH_3$$

Methyl acetate — Hydroxide ion — Rate-determining intermediate; less crowded than intermediate from *tert*-butyl acetate

(*d*) Here, as in part (*a*), we have an electron-withdrawing substituent increasing the rate of ester saponification. It does so by stabilizing the negatively charged intermediate formed in the rate-determining step.

$$m\text{-}O_2N{-}C_6H_4{-}C(O^-)(OH)OCH_3 \quad \text{more stable than} \quad C_6H_5{-}C(O^-)(OH)OCH_3$$

Rate-determining intermediate from methyl *m*-nitrobenzoate — Rate-determining intermediate from methyl benzoate

(*e*) Addition of hydroxide to 4-butanolide introduces torsional strain in the intermediate because of eclipsed bonds. The corresponding intermediate from 5-butanolide is more stable because the bonds are staggered in a six-membered ring.

Eclipsed bonds — Staggered bonds

Less stable; formed more slowly — More stable; formed faster

(*f*) Steric crowding increases more when hydroxide adds to the axial carbonyl group.

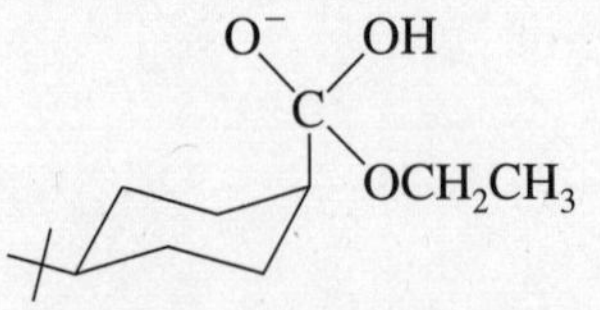

Cis diastereomer: greater increase in crowding when carbon changes from $sp^2$ to $sp^3$; formed more slowly

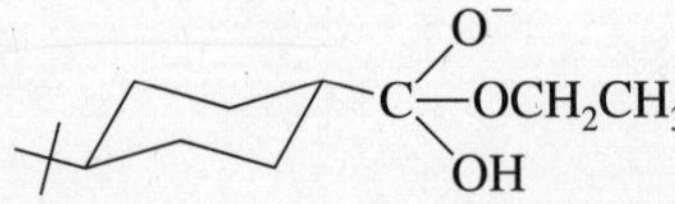

Trans diastereomer: smaller increase in crowding when carbon changes from $sp^2$ to $sp^3$; formed more rapidly

**20.32** Compound A is the *p*-toluenesulfonate ester (tosylate) of *trans*-4-*tert*-butylcyclohexanol. The oxygen atom of the alcohol attacks the sulfur of *p*-toluenesulfonyl chloride, and so the reaction proceeds with retention of configuration.

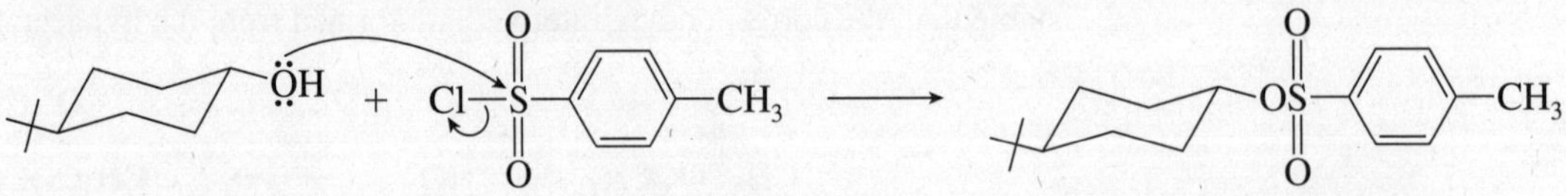

*trans*-4-*tert*-Butylcyclohexanol + *p*-Toluenesulfonyl chloride ⟶ *trans*-4-*tert*-Butylcyclohexyl *p*-toluenesulfonate (compound A)

The second step is a nucleophilic substitution in which benzoate ion displaces *p*-toluenesulfonate with inversion of configuration.

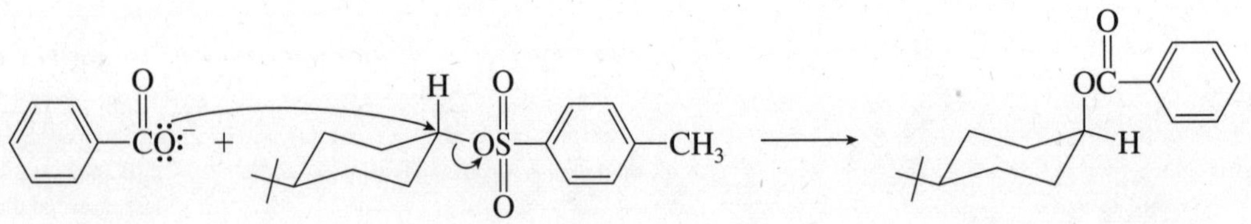

Benzoate ion + *trans*-4-*tert*-Butylcyclohexyl *p*-toluenesulfonate (compound A) ⟶ *cis*-4-*tert*-Butylcyclohexyl benzoate (compound B)

Saponification of *cis*-4-*tert*-butylcyclohexyl benzoate in step 3 proceeds with acyl–oxygen cleavage to give *cis*-4-*tert*-butylcyclohexanol.

**20.33** Reaction of ethyl trifluoroacetate with ammonia yields the corresponding amide, compound A. Compound A undergoes dehydration on heating with $P_4O_{10}$ to give trifluoroacetonitrile, compound B. Grignard reagents react with nitriles to form ketones. *tert*-Butyl trifluoromethyl ketone is formed from trifluoroacetonitrile by treatment with *tert*-butylmagnesium chloride followed by aqueous hydrolysis.

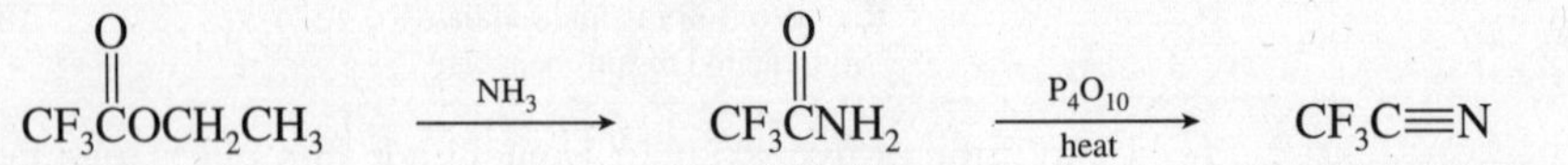

$$CF_3\overset{O}{\overset{\|}{C}}OCH_2CH_3 \xrightarrow{NH_3} CF_3\overset{O}{\overset{\|}{C}}NH_2 \xrightarrow[\text{heat}]{P_4O_{10}} CF_3C{\equiv}N$$

Ethyl trifluoroacetate; Trifluoroacetamide (compound A); Trifluoroacetonitrile (compound B)

$$CF_3C{\equiv}N + (CH_3)_3CMgCl \xrightarrow[\text{2. } H_3O^+]{\text{1. diethyl ether}} CF_3\overset{O}{\overset{\|}{C}}C(CH_3)_3$$

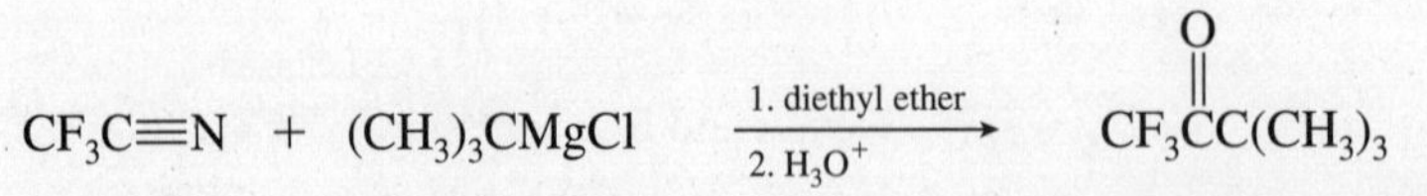

Compound B; *tert*-Butylmagnesium chloride; *tert*-Butyl trifluoromethyl ketone

**20.34** The first step is acid hydrolysis of an acetal protecting group.

**Step 1:**

$$\text{Compound A} \xrightarrow[\text{heat}]{H_2O,\ H^+} \overset{\overset{\displaystyle O}{\|}}{HOC}(CH_2)_5\underset{\underset{\displaystyle HO}{|}}{CH}-\underset{\underset{\displaystyle OH}{|}}{CH}(CH_2)_7CH_2OH$$

Compound B
($C_{16}H_{32}O_5$)

All three alcohol functions are converted to bromide by reaction with hydrogen bromide in step 2.

**Step 2:**

$$\text{Compound B} \xrightarrow{HBr} \overset{\overset{\displaystyle O}{\|}}{HOC}(CH_2)_5\underset{\underset{\displaystyle Br}{|}}{CH}-\underset{\underset{\displaystyle Br}{|}}{CH}(CH_2)_7CH_2Br$$

Compound C
($C_{16}H_{29}Br_3O_2$)

Reaction with ethanol in the presence of an acid catalyst converts the carboxylic acid to its ethyl ester in step 3.

**Step 3:**

$$\text{Compound C} \xrightarrow[H_2SO_4]{\text{ethanol}} CH_3CH_2O\overset{\overset{\displaystyle O}{\|}}{C}(CH_2)_5\underset{\underset{\displaystyle Br}{|}}{CH}-\underset{\underset{\displaystyle Br}{|}}{CH}(CH_2)_7CH_2Br$$

Compound D
($C_{18}H_{33}Br_3O_2$)

The problem hint points out that zinc converts vicinal dibromides to alkenes. Of the three bromine substituents in compound D, two of them are vicinal. Step 4 is a dehalogenation reaction.

**Step 4:**

$$\text{Compound D} \xrightarrow[\text{ethanol}]{Zn} CH_3CH_2O\overset{\overset{\displaystyle O}{\|}}{C}(CH_2)_5CH{=}CH(CH_2)_7CH_2Br$$

Compound E
($C_{18}H_{33}BrO_2$)

Step 5 is a nucleophilic substitution of the $S_N2$ type. Acetate ion is the nucleophile and displaces bromide from the primary carbon.

**Step 5:**

$$\text{Compound E} \xrightarrow[CH_3CO_2H]{NaO\overset{\overset{O}{\|}}{C}CH_3} CH_3CH_2O\overset{\overset{\displaystyle O}{\|}}{C}(CH_2)_5CH{=}CH(CH_2)_7CH_2O\overset{\overset{\displaystyle O}{\|}}{C}CH_3$$

Compound F
($C_{20}H_{36}O_4$)

Step 6 is ester saponification. It yields a 16-carbon chain having a carboxylic acid function at one end and an alcohol at the other.

**Step 6:**

$$\text{Compound F} \xrightarrow[\text{2. H}^+]{\text{1. KOH, ethanol}} \text{HO}\overset{\text{O}}{\overset{\|}{\text{C}}}(\text{CH}_2)_5\text{CH}{=}\text{CH}(\text{CH}_2)_7\text{CH}_2\text{OH}$$

Compound G
($C_{16}H_{30}O_3$)

In step 7, compound G cyclizes to ambrettolide on heating.

**Step 7:**

$$\text{Compound G} \xrightarrow{\text{heat}} \text{Ambrettolide}$$

Compound G — Ambrettolide

20.35 (*a*) This step requires the oxidation of a primary alcohol to an aldehyde. As reported in the literature, pyridinium dichromate in dichloromethane was used to give the desired aldehyde in 84% yield.

$$\text{HOCH}_2\text{CH}{=}\text{CH}(\text{CH}_2)_7\text{CO}_2\text{CH}_3 \xrightarrow[\text{CH}_2\text{Cl}_2]{\text{PDC}} \text{H}\overset{\text{O}}{\overset{\|}{\text{C}}}\text{CH}{=}\text{CH}(\text{CH}_2)_7\text{CO}_2\text{CH}_3$$

Compound A — Compound B

(*b*) Conversion of $-\overset{\text{O}}{\overset{\|}{\text{C}}}\text{H}$ to $-\text{CH}{=}\text{CH}_2$ is a typical case in which a Wittig reaction is appropriate.

$$\text{H}\overset{\text{O}}{\overset{\|}{\text{C}}}\text{CH}{=}\text{CH}(\text{CH}_2)_7\text{CO}_2\text{CH}_3 \xrightarrow{(\text{C}_6\text{H}_5)_3\overset{+}{\text{P}}-\overset{..}{\overset{-}{\text{C}}}\text{H}_2} \text{H}_2\text{C}{=}\text{CHCH}{=}\text{CH}(\text{CH}_2)_7\text{CO}_2\text{CH}_3$$

Compound B — Compound C (observed yield, 53%)

(*c*) Lithium aluminum hydride was used to reduce the ester to a primary alcohol in 81% yield.

$$\text{H}_2\text{C}{=}\text{CHCH}{=}\text{CH}(\text{CH}_2)_7\text{CO}_2\text{CH}_3 \xrightarrow[\text{2. H}_2\text{O}]{\text{1. LiAlH}_4} \text{H}_2\text{C}{=}\text{CHCH}{=}\text{CH}(\text{CH}_2)_7\text{CH}_2\text{OH}$$

Compound C — Compound D

(*d*) The desired sex pheromone is the acetate ester of compound D. Compound D was treated with acetic anhydride to give the acetate ester in 99% yield.

$$\text{H}_2\text{C}{=}\text{CHCH}{=}\text{CH}(\text{CH}_2)_7\text{CH}_2\text{OH} \xrightarrow[\text{pyridine}]{\text{CH}_3\overset{\text{O}}{\overset{\|}{\text{C}}}\text{O}\overset{\text{O}}{\overset{\|}{\text{C}}}\text{CH}_3} \text{H}_2\text{C}{=}\text{CHCH}{=}\text{CH}(\text{CH}_2)_7\text{CH}_2\text{O}\overset{\text{O}}{\overset{\|}{\text{C}}}\text{CH}_3$$

Compound D — (*E*)-9,11-Dodecadien-1-yl acetate

Acetyl chloride could have been used in this step instead of acetic anhydride.

20.36 (*a*) The reaction given in the problem is between a lactone (cyclic ester) and a difunctional Grignard reagent. Esters usually react with 2 moles of a Grignard reagent; in this instance both Grignard functions of the reagent attack the lactone. The second attack is intramolecular, giving rise to the cyclopentanol ring of the product.

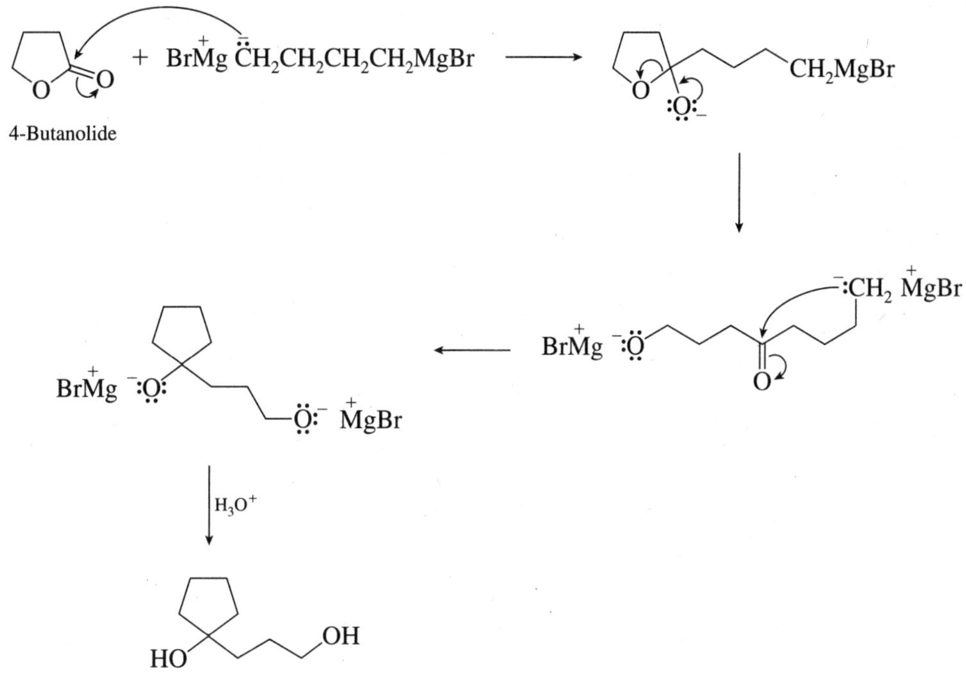

$+\ Br\overset{+}{Mg}\ \ddot{C}H_2CH_2CH_2CH_2MgBr \longrightarrow$ $CH_2MgBr$

4-Butanolide

$Br\overset{+}{Mg}\ {}^-{:}\ddot{O}$ $^-{:}CH_2\ \overset{+}{Mg}Br$

$Br\overset{+}{Mg}\ {}^-{:}\ddot{O}{:}$ $\ddot{O}{:}^-\ \overset{+}{Mg}Br$

$H_3O^+$

HO OH

1-(3-Hydroxypropyl)-cyclopentanol (88%)

(*b*) An intramolecular acyl transfer process takes place in this reaction. The amine group in the thiolactone starting material replaces sulfur on the acyl group to form a lactam (cyclic amide).

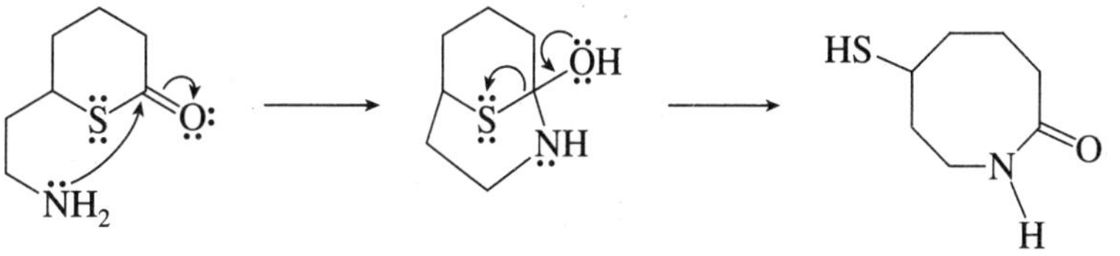

Thiolactone | Tetrahedral intermediate | Lactam

**20.37** (*a*) Acyl chlorides react with alcohols to form esters.

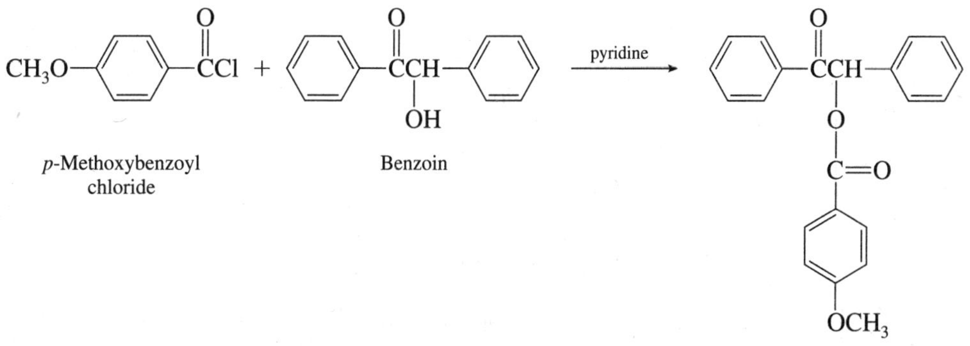

$CH_3O-C_6H_4-\overset{O}{\overset{\|}{C}}Cl + C_6H_5-\overset{O}{\overset{\|}{C}}\underset{OH}{\underset{|}{C}H}-C_6H_5 \xrightarrow{\text{pyridine}} C_6H_5-\overset{O}{\overset{\|}{C}}\underset{O-\overset{}{C}(=O)-C_6H_4-OCH_3}{\underset{|}{C}H}-C_6H_5$

*p*-Methoxybenzoyl chloride | Benzoin | Benzoin *p*-methoxybenzoate (compound A; 95%)

(*b*) Of the two carbonyl groups in the starting material, the ketone carbonyl is more reactive than the ester. (The ester carbonyl is stabilized by electron release from oxygen.)

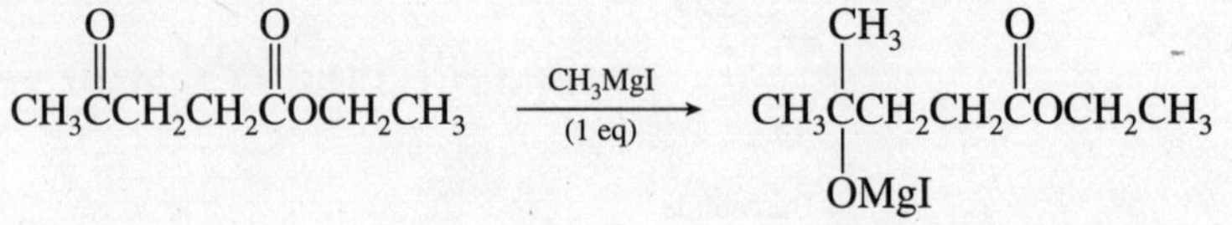

$$CH_3\overset{O}{\overset{\|}{C}}CH_2CH_2\overset{O}{\overset{\|}{C}}OCH_2CH_3 \xrightarrow[\text{(1 eq)}]{CH_3MgI} CH_3\underset{OMgI}{\underset{|}{\overset{CH_3}{\overset{|}{C}}}}CH_2CH_2\overset{O}{\overset{\|}{C}}OCH_2CH_3$$

Compound B has the molecular formula $C_6H_{10}O_2$. The initial product forms a cyclic ester (lactone), with elimination of ethoxide ion.

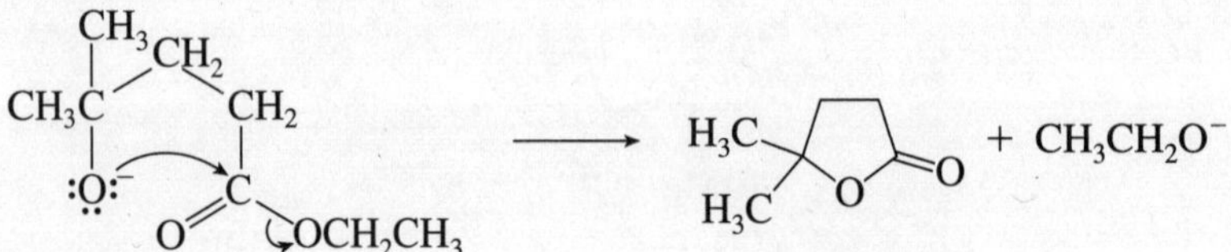

$$\text{CH}_3\text{C(CH}_3\text{)(O}^-\text{)CH}_2\text{CH}_2\text{C(=O)OCH}_2\text{CH}_3 \longrightarrow \text{(H}_3\text{C)}_2\text{-lactone} + CH_3CH_2O^-$$

Compound B

(*c*) Only carboxyl groups that are ortho to each other on a benzene ring are capable of forming a cyclic anhydride.

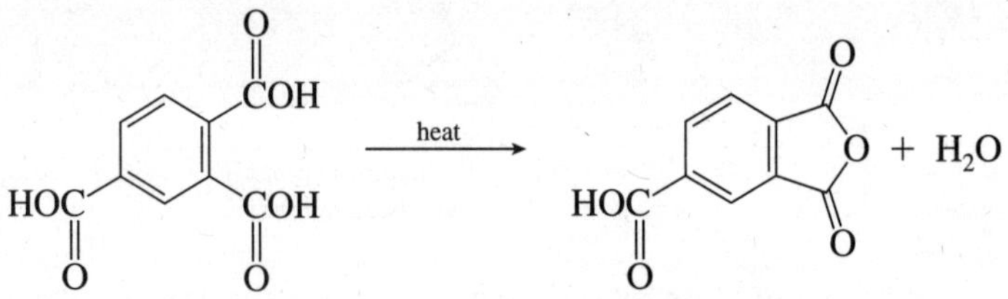

$$\text{HOC(=O)–C}_6\text{H}_3\text{(COH=O)}_2 \xrightarrow{\text{heat}} \text{Compound C} + H_2O$$

Compound C

(*d*) The primary amine can react with both acyl chloride groups of the starting material to give compound D.

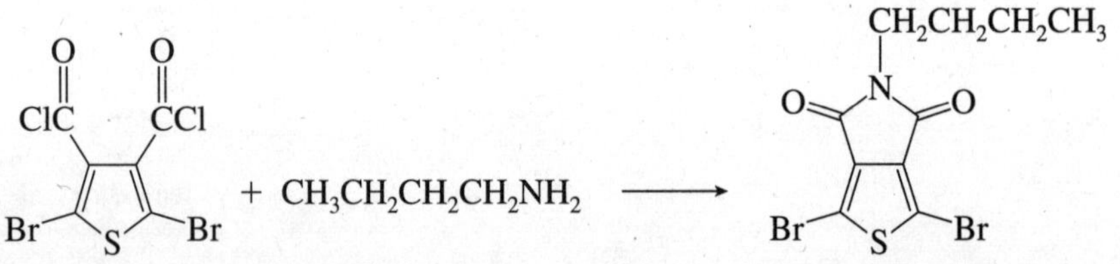

$$\text{2,5-dibromothiophene-3,4-(ClC=O)}_2 + CH_3CH_2CH_2CH_2NH_2 \longrightarrow \text{Compound D}$$

Compound D

**20.38** Compound A is an ester but has within it an amine function. Acyl transfer from oxygen to nitrogen converts the ester to a more stable amide.

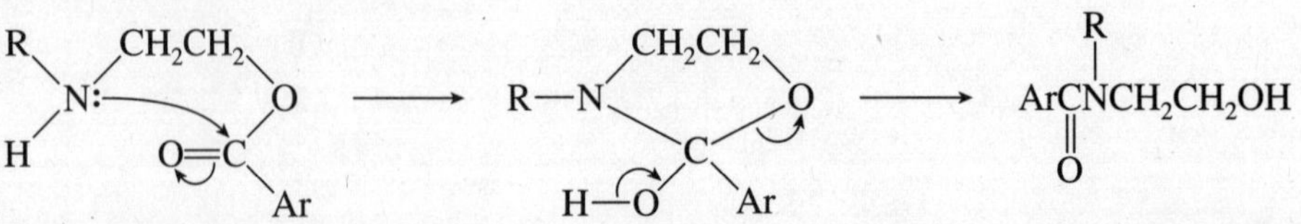

$$\text{Compound A} \longrightarrow \text{Tetrahedral intermediate} \longrightarrow Ar\overset{O}{\underset{}{\overset{\|}{C}}}\overset{R}{\overset{|}{N}}CH_2CH_2OH$$

Compound A (Ar = *p*-nitrophenyl) | Tetrahedral intermediate | Compound B (Ar = *p*-nitrophenyl)

The tetrahedral intermediate is the key intermediate in the reaction.

**20.39** (*a*) The rearrangement in this problem is an acyl transfer from nitrogen to oxygen.

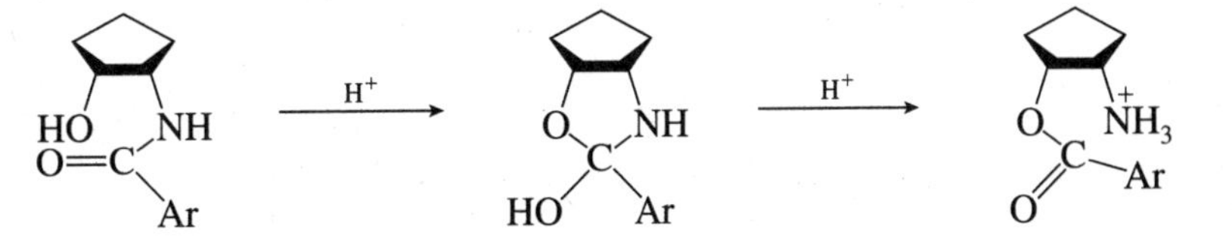

$$\text{Compound A (Ar = } p\text{-nitrophenyl)} \xrightarrow{H^+} \text{Tetrahedral intermediate} \xrightarrow{H^+} \text{Compound B (Ar = } p\text{-nitrophenyl)}$$

This rearrangement takes place in the indicated direction because it is carried out in acid solution. The amino group is protonated in acid and is no longer nucleophilic.

(*b*) The trans stereoisomer of compound A does not undergo rearrangement because when the oxygen and nitrogen atoms on the five-membered ring are trans, the necessary tetrahedral intermediate cannot form.

**20.40** The ester functions of a polymer such as poly(vinyl acetate) are just like ester functions of simple molecules; they can be cleaved by hydrolysis under either acidic or basic conditions. To prepare poly(vinyl alcohol), therefore, polymerize vinyl acetate to poly(vinyl acetate), and then cleave the ester groups by hydrolysis.

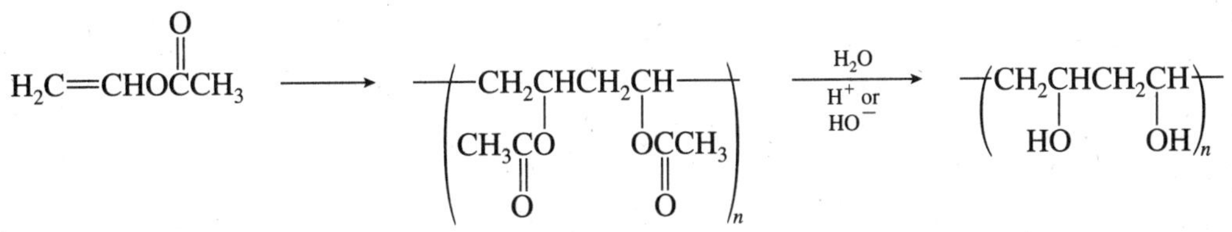

$$H_2C{=}CHO\overset{O}{\overset{\|}{C}}CH_3 \longrightarrow \left(CH_2\underset{CH_3\overset{\|}{\underset{O}{C}}O}{CH}CH_2\underset{O\overset{\|}{\underset{O}{C}}CH_3}{CH}\right)_n \xrightarrow[H^+ \text{ or } HO^-]{H_2O} \left(CH_2\underset{HO}{CH}CH_2\underset{OH}{CH}\right)_n$$

Vinyl acetate — Poly(vinyl acetate) — Poly(vinyl alcohol)

**20.41** (*a*) Each propagation step involves addition of the free-radical species to the $\beta$-carbon of a molecule of methyl methacrylate.

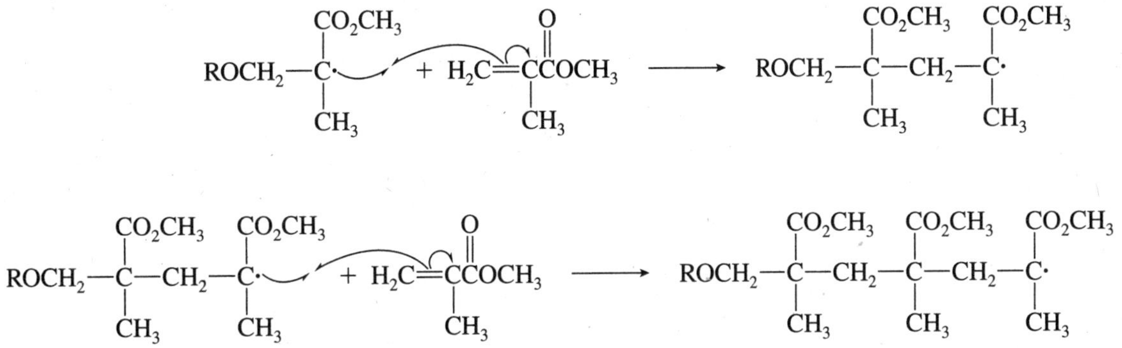

$$ROCH_2{-}\overset{CO_2CH_3}{\underset{CH_3}{C}}\cdot + H_2C{=}\underset{CH_3}{C}\overset{O}{\overset{\|}{C}}OCH_3 \longrightarrow ROCH_2{-}\overset{CO_2CH_3}{\underset{CH_3}{C}}{-}CH_2{-}\overset{CO_2CH_3}{\underset{CH_3}{C}}\cdot$$

$$ROCH_2{-}\overset{CO_2CH_3}{\underset{CH_3}{C}}{-}CH_2{-}\overset{CO_2CH_3}{\underset{CH_3}{C}}\cdot + H_2C{=}\underset{CH_3}{C}\overset{O}{\overset{\|}{C}}OCH_3 \longrightarrow ROCH_2{-}\overset{CO_2CH_3}{\underset{CH_3}{C}}{-}CH_2{-}\overset{CO_2CH_3}{\underset{CH_3}{C}}{-}CH_2{-}\overset{CO_2CH_3}{\underset{CH_3}{C}}\cdot$$

(*b*) The correct carbon skeleton can be constructed by treating acetone with sodium cyanide in the presence of $H_2SO_4$ to give acetone cyanohydrin.

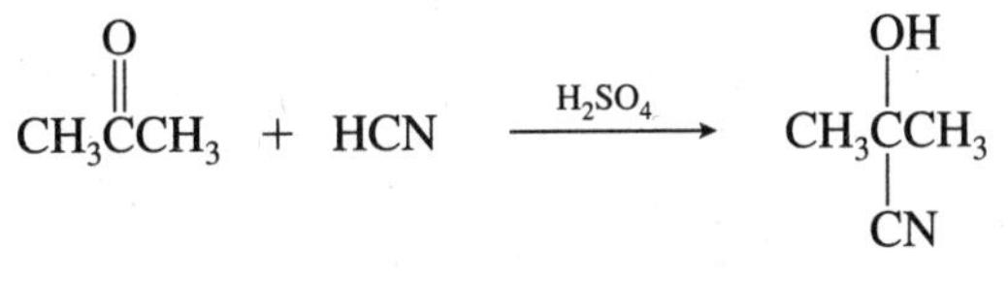

$$CH_3\overset{O}{\overset{\|}{C}}CH_3 + HCN \xrightarrow{H_2SO_4} CH_3\overset{OH}{\underset{CN}{C}}CH_3$$

Acetone — Hydrogen cyanide — Acetone cyanohydrin

Dehydration of the cyanohydrin followed by hydrolysis of the nitrile group and esterification of the resulting carboxylic acid yields methyl methacrylate.

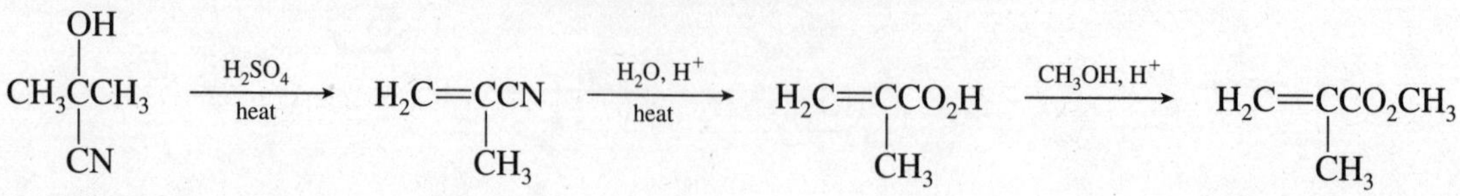

$$\underset{\text{Acetone cyanohydrin}}{CH_3\overset{OH}{\underset{CN}{C}}CH_3} \xrightarrow[\text{heat}]{H_2SO_4} H_2C{=}\underset{CH_3}{C}CN \xrightarrow[\text{heat}]{H_2O,\ H^+} H_2C{=}\underset{CH_3}{C}CO_2H \xrightarrow{CH_3OH,\ H^+} \underset{\text{Methyl methacrylate}}{H_2C{=}\underset{CH_3}{C}CO_2CH_3}$$

**20.42** The compound contains nitrogen and exhibits a prominent peak in the infrared spectrum at 2270 $cm^{-1}$; it is likely to be a nitrile. Its molecular weight of 83 is consistent with the molecular formula $C_5H_9N$. The presence of four signals in the $\delta$ 10 to 30-ppm region of the $^{13}C$ NMR spectrum suggests an unbranched carbon skeleton. This is confirmed by the presence of two triplets in the $^1H$ NMR spectrum at $\delta$ 1.0 ppm ($CH_3$ coupled with adjacent $CH_2$) and at $\delta$ 2.3 ppm ($CH_2CN$ coupled with adjacent $CH_2$). The compound is pentanenitrile.

$$CH_3CH_2CH_2CH_2C{\equiv}N$$

Pentanenitrile

**20.43** The compound has the characteristic triplet–quartet pattern of an ethyl group in its $^1H$ NMR spectrum. Because these signals correspond to 10 protons, there must be two equivalent ethyl groups in the molecule. The methylene quartet appears at relatively low field ($\delta$ 4.1 ppm), which is consistent with ethyl groups bonded to oxygen, as in —$OCH_2CH_3$. There is a peak at 1730 $cm^{-1}$ in the infrared spectrum, suggesting that these ethoxy groups reside in ester functions. The molecular formula $C_8H_{14}O_4$ reveals that if two ester groups are present, there can be no rings or double bonds. The remaining four hydrogens are equivalent in the $^1H$ NMR spectrum, and so two equivalent $CH_2$ groups are present. The compound is the diethyl ester of succinic acid.

$$CH_3CH_2O\overset{O}{\overset{\|}{C}}CH_2CH_2\overset{O}{\overset{\|}{C}}OCH_2CH_3$$

Diethyl succinate

**20.44** Compound A ($C_4H_6O_2$) has an index of hydrogen deficiency of 2. With two oxygen atoms and a peak in the infrared at 1760 $cm^{-1}$, it is likely that one of the elements of unsaturation is the carbon–oxygen double bond of an ester. The $^1H$ NMR spectrum contains a three-proton singlet at $\delta$ 2.1 ppm, which is consistent with a $CH_3\underset{O}{\underset{\|}{C}}$ unit. It is likely that compound A is an acetate ester.

The $^{13}C$ NMR spectrum reveals that the four carbon atoms of the molecule are contained in one each of the fragments $CH_3$, $CH_2$, and CH, along with the carbonyl carbon. In addition to the two carbons of the acetate group, the remaining two carbons are the $CH_2$ and CH carbons of a vinyl group, $CH{=}CH_2$. Compound A is vinyl acetate.

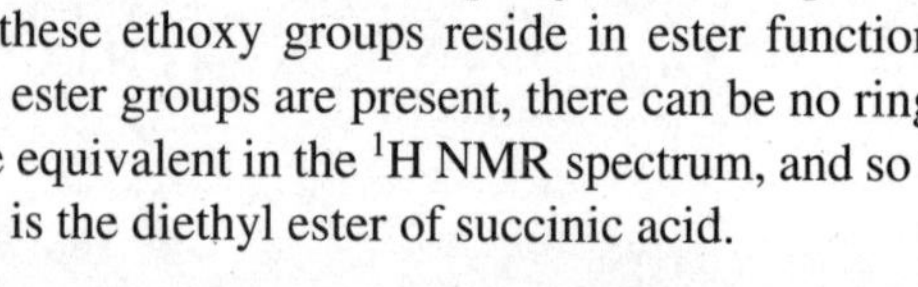

$$CH_3C(=O)OCH{=}CH_2$$

$\delta$ 20.2 ppm ($CH_3$); $\delta$ 167.6 ppm (C=O); $\delta$ 141.8 ppm (OCH); $\delta$ 96.8 ppm ($=CH_2$)

Each vinyl proton is coupled to two other vinyl protons; each appears as a doublet of doublets in the $^1H$ NMR spectrum.

**20.45** Solutions to molecular modeling exercises are not provided in this *Study Guide and Solutions Manual.* You should use *Learning By Modeling* for this exercise.

## SELF-TEST

### PART A

**A-1.** Give a correct IUPAC name for each of the following acid derivatives:

(*a*) $CH_3CH_2CH_2O\overset{O}{\overset{\|}{C}}CH_2CH_2CH_3$

(*b*) $C_6H_5\overset{O}{\overset{\|}{C}}NHCH_3$

(*c*) $(CH_3)_2CHCH_2CH_2\overset{O}{\overset{\|}{C}}Cl$

**A-2.** Provide the correct structure of
(*a*) Benzoic anhydride
(*b*) *N*-(1-Methylpropyl)acetamide
(*c*) Phenyl benzoate

**A-3.** What reagents are needed to carry out each of the following conversions?

(*a*) $C_6H_5CH_2CO_2H \xrightarrow{?} C_6H_5CH_2\overset{O}{\overset{\|}{C}}Cl$

(*b*) $(CH_3)_3C\overset{O}{\overset{\|}{C}}NH_2 \xrightarrow{?} (CH_3)_3CNH_2$

(*c*) $(CH_3)_2CHCH_2NH_2 \xrightarrow{?} C_6H_5\overset{O}{\overset{\|}{C}}NHCH_2CH(CH_3)_2 + CH_3OH$

**A-4.** Write the structure of the product of each of the following reactions:

(*a*) Cyclohexyl acetate $\xrightarrow[2.\ H^+]{1.\ NaOH,\ H_2O}$ ? (two products)

(*b*) Cyclopentanol + benzoyl chloride $\xrightarrow{\text{pyridine}}$ ?

(*c*) (phthalic anhydride) + $CH_3CH_2OH \xrightarrow{H^+(\text{cat})}$ ?

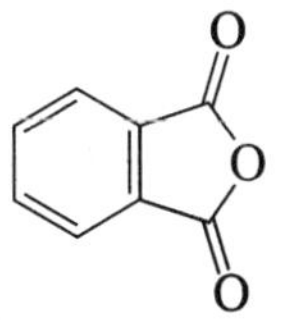

(*d*) Ethyl propanoate + dimethylamine $\longrightarrow$ ? (two products)

(*e*) $H_3C-C_6H_4-\overset{O}{\overset{\|}{C}}NHCH_3 \xrightarrow[\text{heat}]{H_2O,\ H_2SO_4}$ ? (two products)

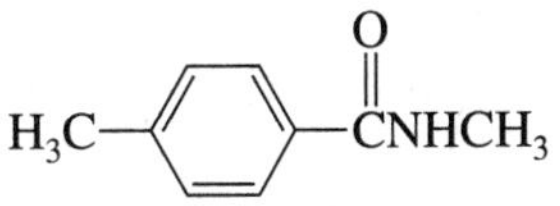

**A-5.** The following reaction occurs when the reactant is allowed to stand in pentane. Write the structure of the key intermediate in this process.

$$C_6H_5\overset{O}{\overset{\|}{C}}OCH_2CH_2NHCH_3 \longrightarrow CH_3\underset{C_6H_5C{=}O}{\underset{|}{N}}CH_2CH_2OH$$

**A-6.** Give the correct structures, clearly showing stereochemistry, of each compound, A through D, in the following sequence of reactions:

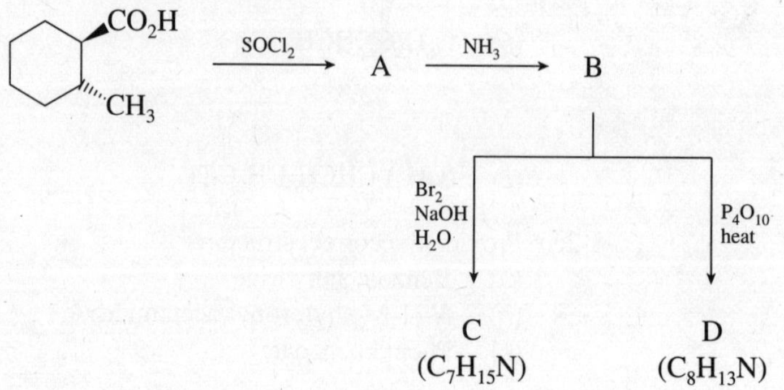

(cyclohexane bearing $CO_2H$ (wedge) and $CH_3$ (hash) on adjacent carbons) $\xrightarrow{SOCl_2}$ A $\xrightarrow{NH_3}$ B

B $\xrightarrow{Br_2,\ NaOH,\ H_2O}$ C ($C_7H_{15}N$)

B $\xrightarrow{P_4O_{10},\ heat}$ D ($C_8H_{13}N$)

**A-7.** Write the structure of the neutral form of the tetrahedral intermediate in the

(*a*) Acid-catalyzed hydrolysis of methyl acetate
(*b*) Reaction of ammonia with acetic anhydride

**A-8.** Write the steps necessary to prepare $H_3C$—$C_6H_4$—$NH_2$ from $H_3C$—$C_6H_4$—Br.

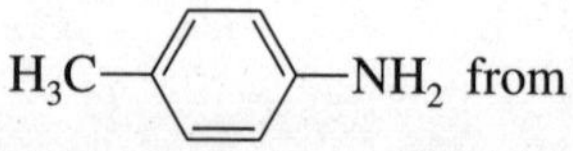

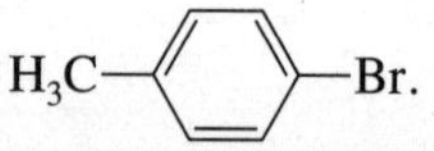

**A-9.** Outline a synthesis of benzyl benzoate using toluene as the source of all the carbon atoms.

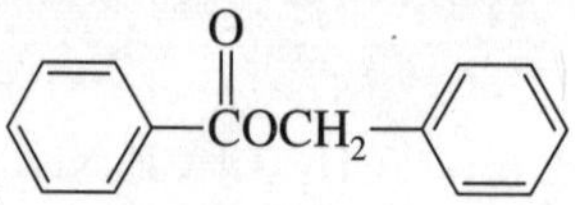

$$C_6H_5\overset{O}{\overset{\|}{C}}OCH_2-C_6H_5$$

Benzyl benzoate

**A-10.** The infrared spectrum of a compound ($C_3H_6ClNO$) has an intense peak at 1680 $cm^{-1}$. Its $^1H$ NMR spectrum consists of a doublet (3H, $\delta$ 1.5 ppm), a quartet (1H, $\delta$ 4.1 ppm), and a broad singlet (2H, $\delta$ 6.5 ppm). What is the structure of the compound? How would you prepare it from propanoic acid?

## PART B

**B-1.** What are the products of the most favorable mode of decomposition of the intermediate species shown?

$$C_6H_5-\overset{OH}{\underset{Cl}{C}}-OH$$

(*a*) Benzoic acid and HCl
(*b*) Benzoyl chloride and $H_2O$
(*c*) Both (*a*) and (*b*) equally likely
(*d*) Neither (*a*) nor (*b*)

**B-2.** What is the correct IUPAC name for the compound shown?

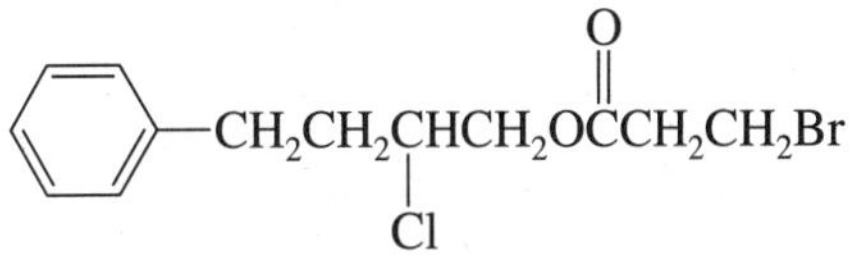

(*a*) 3-Bromopropyl 2-chloro-4-butylbutanoate
(*b*) 2-Chloro-4-phenylbutyl 3-bromopropanoate
(*c*) 3-Chloro-1-phenylbutyl 1-bromopropanoate
(*d*) 3-Chloro-1-phenylbutyl 3-bromopropanoate
(*e*) 7-Bromo-3-chloro-1-phenylbutyl propanoate

**B-3.** Rank the following in order of increasing reactivity (least → most) toward acid hydrolysis:

$CH_3\overset{O}{\overset{\|}{C}}OCH_2CH_3$ (1) $\quad CH_3\overset{O}{\overset{\|}{C}}Cl$ (2) $\quad CH_3\overset{O}{\overset{\|}{C}}NHCH_3$ (3)

(*a*) $1 < 2 < 3$ (*c*) $1 < 3 < 2$
(*b*) $3 < 1 < 2$ (*d*) $2 < 1 < 3$

**B-4.** The structure of *N*-propylacetamide is

(*a*) $CH_3CH_2\overset{O}{\overset{\|}{C}}NHCH_3$ (*c*) $CH_3\overset{O}{\overset{\|}{C}}NHCH_2CH_2CH_3$

(*b*) $CH_3\overset{O}{\overset{\|}{C}}N(CH_2CH_2CH_3)_2$ (*d*) $CH_3CH{=}NCH_2CH_2CH_3$

**B-5.** Choose the response that matches the correct functional group classification with the following group of structural formulas.

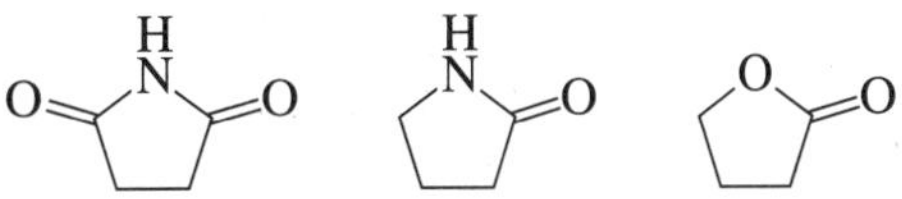

| | | |
|---|---|---|
| (*a*) Anhydride | Lactam | Lactone |
| (*b*) Lactam | Imide | Lactone |
| (*c*) Imide | Lactone | Anhydride |
| (*d*) Imide | Lactam | Lactone |

**B-6.** Choose the best sequence of reactions for the transformation given. Semicolons indicate separate reaction steps to be used in the order shown.

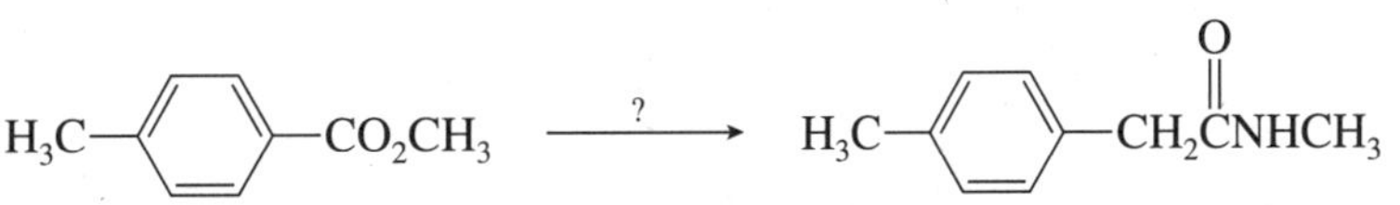

(*a*) $H_3O^+$; $SOCl_2$; $CH_3NH_2$
(*b*) $HO^-/H_2O$; $PBr_3$; Mg; $CO_2$; $H_3O^+$; $SOCl_2$; $CH_3NH_2$
(*c*) $LiAlH_4$; $H_2O$; HBr; Mg; $CO_2$; $H_3O^+$; $SOCl_2$; $CH_3NH_2$
(*d*) None of these would yield the desired product.

**B-7.** A key step in the hydrolysis of acetamide in aqueous acid proceeds by nucleophilic addition of

(*a*) $H_3O^+$ to $CH_3\overset{\overset{O}{\|}}{C}NH_2$

(*b*) $H_3O^+$ to $CH_3\overset{\overset{^+OH}{\|}}{C}NH_2$

(*c*) $HO^-$ to $CH_3\overset{\overset{O}{\|}}{C}NH_2$

(*d*) $H_2O$ to $CH_3\overset{\overset{^+OH}{\|}}{C}NH_2$

(*e*) $HO^-$ to $CH_3\overset{\overset{^+OH}{\|}}{C}NH_2$

**B-8.** Which reaction is *not* possible for acetic anhydride?

(*a*) $(CH_3\overset{\overset{O}{\|}}{C})_2O + 2HN(CH_3)_2 \longrightarrow CH_3\overset{\overset{O}{\|}}{C}N(CH_3)_2 + CH_3CO_2^-\ H_2\overset{+}{N}(CH_3)_2$

(*b*) $(CH_3\overset{\overset{O}{\|}}{C})_2O + CH_3CH_2OH \longrightarrow CH_3CO_2CH_2CH_3 + CH_3CO_2H$

(*c*) $(CH_3\overset{\overset{O}{\|}}{C})_2O + C_6H_6 \xrightarrow{AlCl_3} CH_3\overset{\overset{O}{\|}}{C}C_6H_5 + CH_3CO_2H$

(*d*) $(CH_3\overset{\overset{O}{\|}}{C})_2O + NaCl \longrightarrow CH_3\overset{\overset{O}{\|}}{C}Cl + CH_3CO_2^-\ Na^+$

**B-9.** All but one of the following compounds react with aniline to give acetanilide. Which one does *not?*

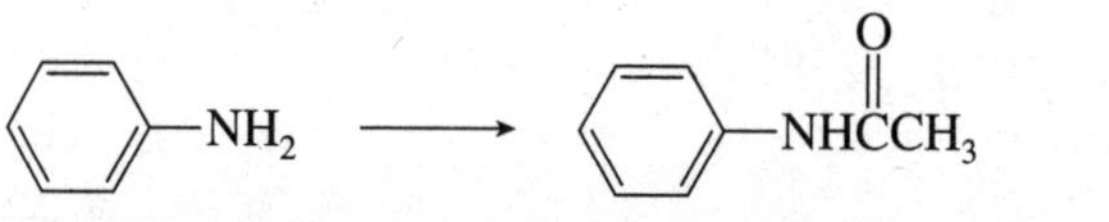

$$C_6H_5{-}NH_2 \longrightarrow C_6H_5{-}NH\overset{\overset{O}{\|}}{C}CH_3$$

Aniline  Acetanilide

(*a*) $CH_3\overset{\overset{O}{\|}}{C}Cl$

(*b*) $CH_3\overset{\overset{O}{\|}}{C}H$

(*c*) $H_3C\overset{\overset{O}{\|}}{C}{-}O{-}\overset{\overset{O}{\|}}{C}CH_3$

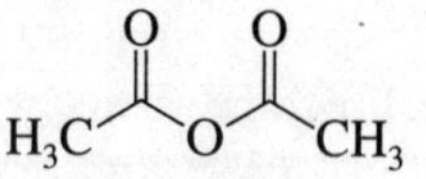

(*d*) $H_3C\overset{\overset{O}{\|}}{C}{-}O{-}CH_2CH_3$

(*e*) $C_6H_5{-}O\overset{\overset{O}{\|}}{C}CH_3$

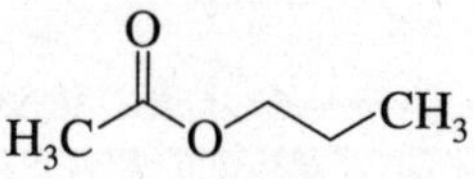

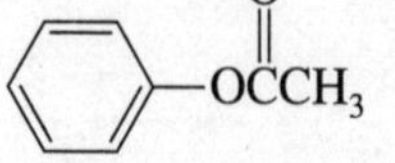

**B-10.** Identify product Z in the following reaction sequence:

$$H_2C{=}CHCH_2Br \xrightarrow{NaCN} Y \xrightarrow[2.\ H_3O^+]{1.\ C_6H_5MgBr,\ \text{diethyl ether}} Z$$

(*a*) $H_2C{=}CHCH_2CC_6H_5$ (C=O)

(*b*) $H_2C{=}CHCH_2CHC_6H_5$ (OH on CH)

(*c*) $H_2C{=}CHCH_2CNHC_6H_5$ (C=O)

(*d*) $H_2C{=}CHCH_2NHCC_6H_5$ (C=O)

(*e*) $H_2C{=}CHCH_2CHC_6H_5$ ($NH_2$ on CH)

**B-11.** Which of the following best describes the nucleophilic addition step in the acid-catalyzed hydrolysis of acetonitrile ($CH_3CN$)?

(*a*) $H_3C{-}C{\equiv}N{:}$ attacked by $HO^-$

(*b*) $H_3C{-}C{\equiv}N{:}$ attacked by $H_2O$

(*c*) $H_3C{-}C{\equiv}\overset{+}{N}H$ attacked by $HO^-$

(*d*) $H_3C{-}C{\equiv}\overset{+}{N}H$ attacked by $H_2O$

(*e*) $H_3C{-}C{\equiv}N{:}$ with $H_2O$

**B-12.** Saponification (basic hydrolysis) of $C_6H_5C\text{Ø}CH_3$ (C=O) will yield: [Ø = mass-18 isotope of oxygen]

(*a*) $C_6H_5CO^-$ (C=Ø) + $H\text{Ø}CH_3$

(*b*) $C_6H_5CO^-$ (C=O) + $H\text{Ø}CH_3$

(*c*) $C_6H_5CO^-$ (C=Ø) + $HOCH_3$

(*d*) $C_6H_5CO^-$ (C=O) + $HOCH_3$

(*e*) $C_6H_5C(OH)(\text{Ø}CH_3)O^-$

**B-13.** An unknown compound, $C_9H_{10}O_2$, did not dissolve in aqueous NaOH. The infrared spectrum exhibited strong absorption at 1730 $cm^{-1}$. The $^1H$ NMR spectrum had signals at $\delta$ 7.2 ppm (multiplet), 4.1 ppm (quartet), and 1.3 ppm (triplet). Which of the following is most likely the unknown?

(*a*) $p$-$CH_3CH_2C_6H_4COH$ (C=O)

(*b*) $C_6H_5COCH_2CH_3$ (C=O)

(*c*) $C_6H_5CH_2OCCH_3$ (C=O)

(*d*) $C_6H_5CH_2COCH_3$ (C=O)

(*e*) $C_6H_5OCH_2CCH_3$ (C=O)

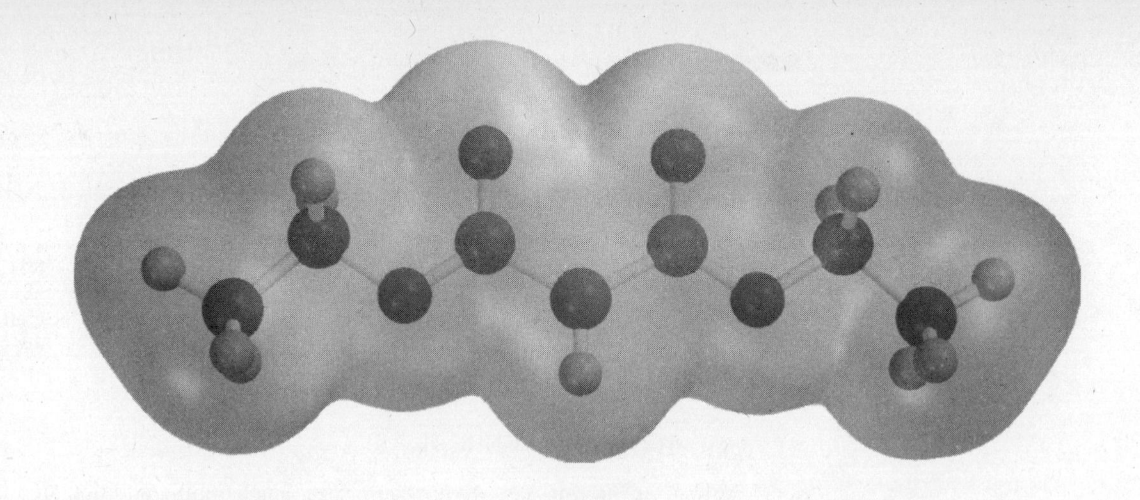

# CHAPTER 21

## ESTER ENOLATES

## SOLUTIONS TO TEXT PROBLEMS

**21.1** Ethyl benzoate cannot undergo the Claisen condensation, because it has no protons on its $\alpha$-carbon atom and so cannot form an enolate. Ethyl pentanoate and ethyl phenylacetate can undergo the Claisen condensation.

$$2CH_3CH_2CH_2CH_2\overset{O}{\overset{\|}{C}}OCH_2CH_3 \xrightarrow[2.\ H_3O^+]{1.\ NaOCH_2CH_3} CH_3CH_2CH_2CH_2\overset{O}{\overset{\|}{C}}\underset{CH_2CH_2CH_3}{\underset{|}{C}H}\overset{O}{\overset{\|}{C}}OCH_2CH_3$$

Ethyl pentanoate → Ethyl 3-oxo-2-propylheptanoate

$$2C_6H_5CH_2\overset{O}{\overset{\|}{C}}OCH_2CH_3 \xrightarrow[2.\ H_3O^+]{1.\ NaOCH_2CH_3} C_6H_5CH_2\overset{O}{\overset{\|}{C}}\underset{C_6H_5}{\underset{|}{C}H}\overset{O}{\overset{\|}{C}}OCH_2CH_3$$

Ethyl phenylacetate → Ethyl 3-oxo-2,4-diphenylbutanoate

**21.2** (*b*) The enolate formed by proton abstraction from the $\alpha$-carbon atom of diethyl 4-methylheptanedioate cyclizes to form a six-membered $\beta$-keto ester.

$$CH_3CH_2O\overset{O}{\overset{\|}{C}}CH_2CH_2\underset{CH_3}{\underset{|}{C}H}CH_2CH_2\overset{O}{\overset{\|}{C}}OCH_2CH_3 \xrightarrow{NaOCH_2CH_3}$$

[Enolate: $CH_3CH_2O{-}\overset{O}{\overset{\|}{C}}$ attacked intramolecularly by $:\overset{-}{C}H$ bearing $C(=O)OCH_2CH_3$; ring chain $H_2C$, $CH_2$, $CHCH_3$, $CH_2$] → [cyclohexanone ring with $C(=O)OCH_2CH_3$ at C-1 and $CH_3$ at C-5]

Diethyl 4-methylheptanedioate → Ethyl (5-methyl-2-oxocyclohexane)-carboxylate (*c*) The two $\alpha$ carbons of this diester are not equivalent. Cyclization by attack of the enolate at C-2 gives

$$CH_3CH_2O\overset{O}{\overset{\|}{C}}\underset{CH_3}{\underset{|}{C}}HCH_2CH_2CH_2\overset{O}{\overset{\|}{C}}OCH_2CH_3 \xrightarrow{NaOCH_2CH_3}$$ Ethyl (1-methyl-2-oxocyclopentane)carboxylate

Site of carbanion — Enolate attacks this carbon.

Ethyl (1-methyl-2-oxocyclopentane)-carboxylate

This $\beta$-keto ester cannot form a stable enolate by deprotonation. It is present in only small amounts at equilibrium. The major product is formed by way of the other enolate.

$$CH_3CH_2O\overset{O}{\overset{\|}{C}}\underset{CH_3}{\underset{|}{C}}HCH_2CH_2CH_2\overset{O}{\overset{\|}{C}}OCH_2CH_3 \xrightarrow{NaOCH_2CH_3}$$

Enolate attacks this carbon. — Site of carbanion

Ethyl (3-methyl-2-oxocyclopentane)-carboxylate

This $\beta$-keto ester is converted to a stable enolate on deprotonation, causing the equilibrium to shift in its favor.

**21.3** (*b*) Both carbonyl groups of diethyl oxalate are equivalent. The enolate of ethyl phenylacetate attacks one of them.

$$C_6H_5\bar{C}H\overset{O}{\overset{\|}{C}}OCH_2CH_3 + CH_3CH_2O-\overset{O}{\overset{\|}{C}}\overset{O}{\overset{\|}{C}}OCH_2CH_3 \longrightarrow C_6H_5\underset{\underset{O}{\|}{COCH_2CH_3}}{\underset{|}{C}H}\overset{O}{\overset{\|}{C}}\overset{O}{\overset{\|}{C}}OCH_2CH_3$$

Diethyl 2-oxo-3-phenylbutanedioate

(*c*) The enolate of ethyl phenylacetate attacks the carbonyl group of ethyl formate.

$$C_6H_5\bar{C}H\overset{O}{\overset{\|}{C}}OCH_2CH_3 + H\overset{O}{\overset{\|}{C}}-OCH_2CH_3 \longrightarrow C_6H_5\underset{\underset{O}{\|}{COCH_2CH_3}}{\underset{|}{C}H}\overset{O}{\overset{\|}{C}}H$$

Ethyl 3-oxo-2-phenylpropanoate

**21.4** In order for a five-membered ring to be formed, C-5 must be the carbanionic site that attacks the ester carbonyl.

$H_3C-\bar{C}H$, $CH_2$, $CH_3CH_2O$, $CH_2$ → $H_3C$, $CH_3CH_2O$, $:\ddot{O}:^-$

Enolate of ethyl 4-oxohexanoate

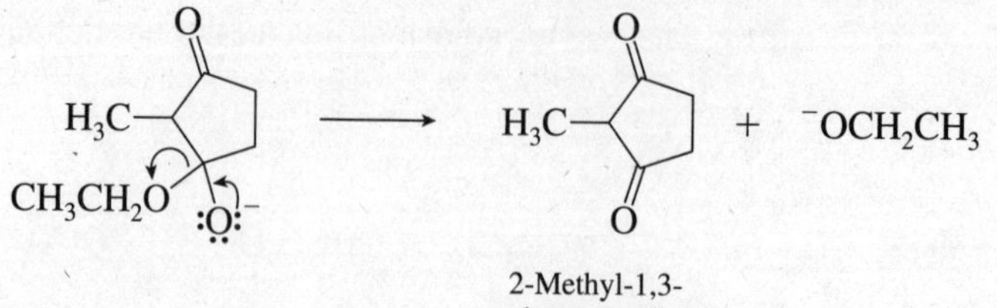

$H_3C$ $CH_3CH_2O$ :Ö:⁻ ⟶ $H_3C$ + $^{-}OCH_2CH_3$

2-Methyl-1,3-cyclopentanedione

**21.5** The desired ketone, cyclopentanone, is derived from the corresponding $\beta$-keto ester. This key intermediate is obtained from a Dieckmann cyclization of the starting material, diethyl hexanedioate.

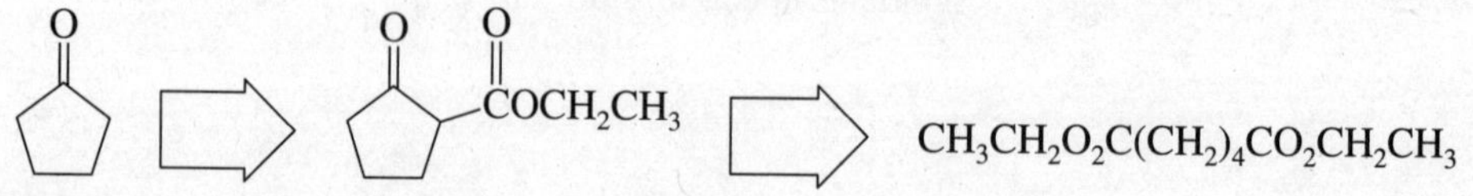

$COCH_2CH_3$ ⟹ $CH_3CH_2O_2C(CH_2)_4CO_2CH_2CH_3$

First treat the diester with sodium ethoxide to effect the Dieckmann cyclization.

$$CH_3CH_2O\overset{O}{\overset{\|}{C}}CH_2CH_2CH_2CH_2\overset{O}{\overset{\|}{C}}OCH_2CH_3 \xrightarrow[2.\ H_3O^+]{1.\ NaOCH_2CH_3}$$

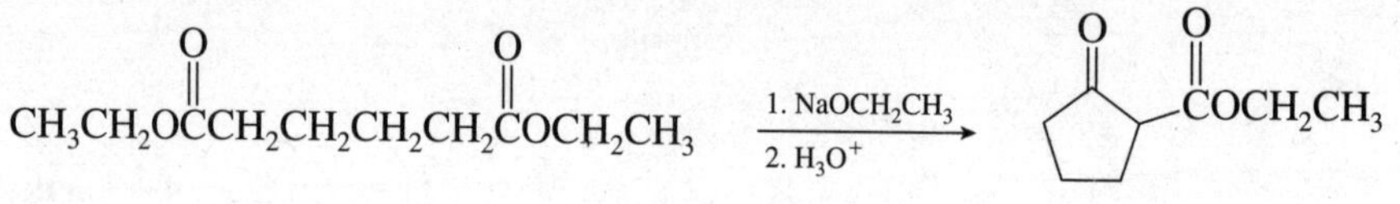

Diethyl hexanedioate

Ethyl (2-oxocyclopentane)-carboxylate

Next convert the $\beta$-keto ester to the desired product by saponification and decarboxylation.

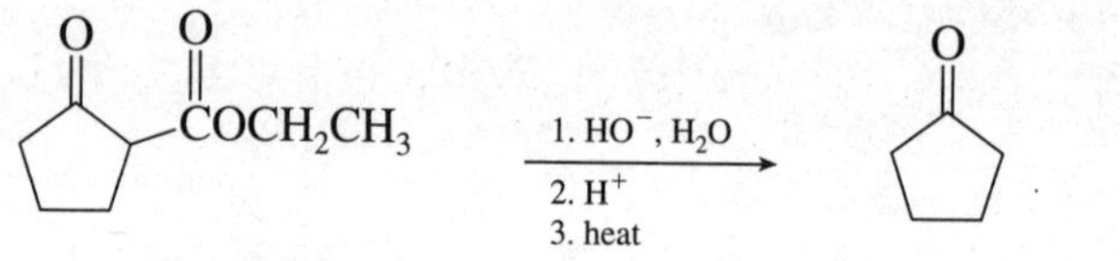

1. $HO^-$, $H_2O$
2. $H^+$
3. heat

Ethyl (2-oxocyclopentane)-carboxylate

Cyclopentanone

**21.6** (*b*) Write a structural formula for the desired product; then disconnect a bond to the $\alpha$-carbon atom.

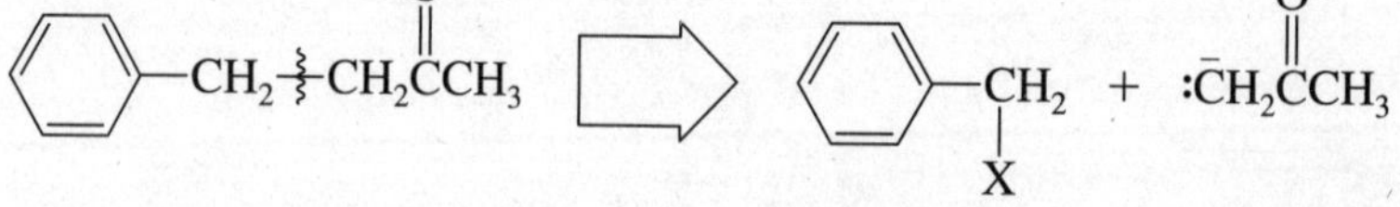

$C_6H_5CH_2$–$CH_2\overset{O}{\overset{\|}{C}}CH_3$ ⟹ $C_6H_5CH_2X$ + $:\bar{C}H_2\overset{O}{\overset{\|}{C}}CH_3$

Required alkyl halide

Derived from ethyl acetoacetate

Therefore

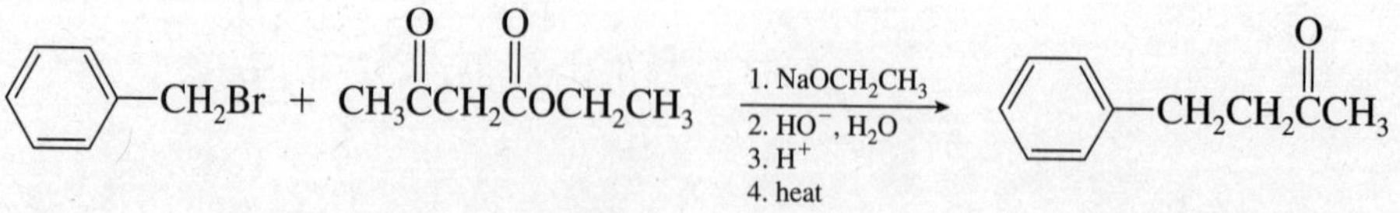

$$C_6H_5CH_2Br + CH_3\overset{O}{\overset{\|}{C}}CH_2\overset{O}{\overset{\|}{C}}OCH_2CH_3 \xrightarrow[\substack{2.\ HO^-,\ H_2O \\ 3.\ H^+ \\ 4.\ heat}]{1.\ NaOCH_2CH_3} C_6H_5CH_2CH_2\overset{O}{\overset{\|}{C}}CH_3$$

Benzyl bromide

Ethyl acetoacetate

4-Phenyl-2-butanone

(*c*) The disconnection approach to retrosynthetic analysis reveals that the preparation of 5-hexen-2-one by the acetoacetic ester synthesis requires an allylic halide.

$$H_2C{=}CHCH_2 \wr CH_2\overset{O}{\overset{\|}{C}}CH_3 \Longrightarrow \underset{\text{Required alkyl halide}}{H_2C{=}CH\underset{X}{\underset{|}{C}}H_2} + \underset{\text{Derived from ethyl acetoacetate}}{:\bar{C}H_2\overset{O}{\overset{\|}{C}}CH_3}$$

$$\underset{\text{Allyl bromide}}{H_2C{=}CHCH_2Br} + \underset{\text{Ethyl acetoacetate}}{CH_3\overset{O}{\overset{\|}{C}}CH_2\overset{O}{\overset{\|}{C}}OCH_2CH_3} \xrightarrow[\text{2. }HO^-,\ H_2O\text{; 3. }H^+\text{; 4. heat}]{\text{1. }NaOCH_2CH_3} \underset{\text{5-Hexen-2-one}}{H_2C{=}CHCH_2CH_2\overset{O}{\overset{\|}{C}}CH_3}$$

**21.7** (*b*) Nonanoic acid has a $CH_3(CH_2)_5CH_2$— unit attached to the $CH_2\overset{O}{\overset{\|}{C}}OH$ synthon.

$$CH_3(CH_2)_5CH_2 \wr CH_2\overset{O}{\overset{\|}{C}}OH \Longrightarrow \underset{\text{Required alkyl halide}}{CH_3(CH_2)_5CH_2X} + \underset{\text{Derived from diethyl malonate}}{:\bar{C}H_2\overset{O}{\overset{\|}{C}}OH}$$

Therefore the anion of diethyl malonate is alkylated with a 1-haloheptane.

$$\underset{\text{1-Bromoheptane}}{CH_3(CH_2)_5CH_2Br} + \underset{\text{Diethyl malonate}}{CH_2(COOCH_2CH_3)_2} \xrightarrow[\text{ethanol}]{NaOCH_2CH_3} \underset{\text{Diethyl 2-heptylmalonate}}{CH_3(CH_2)_5CH_2CH(COOCH_2CH_3)_2}$$

$$\xrightarrow{\text{1. }HO^-,\ H_2O\text{; 2. }H^+\text{; 3. heat}} \underset{\text{Nonanoic acid}}{CH_3(CH_2)_5CH_2CH_2CO_2H}$$

(*c*) Disconnection of the target molecule adjacent to the $\alpha$ carbon reveals the alkyl halide needed to react with the enolate derived from diethyl malonate.

$$CH_3CH_2\underset{CH_3}{\underset{|}{C}}HCH_2 \wr CH_2\overset{O}{\overset{\|}{C}}OH \Longrightarrow \underset{\text{Required alkyl halide}}{CH_3CH_2\underset{CH_3}{\underset{|}{C}}HCH_2X} + \underset{\text{Derived from diethyl malonate}}{:\bar{C}H_2\overset{O}{\overset{\|}{C}}OH}$$

The necessary alkyl halide in this synthesis is 1-bromo-2-methylbutane.

$$CH_3CH_2\underset{\displaystyle CH_3}{\underset{|}{C}}HCH_2Br + CH_2(COOCH_2CH_3)_2 \xrightarrow[\text{ethanol}]{NaOCH_2CH_3} CH_3CH_2\underset{\displaystyle CH_3}{\underset{|}{C}}HCH_2CH(COOCH_2CH_3)_2$$

1-Bromo-2-methylbutane; Diethyl malonate; Diethyl 2-(2-methylbutyl)malonate

$$\xrightarrow{\text{1. } HO^-, H_2O;\ \text{2. } H^+;\ \text{3. heat}} CH_3CH_2\underset{\displaystyle CH_3}{\underset{|}{C}}HCH_2CH_2\overset{\displaystyle O}{\overset{\|}{C}}OH$$

4-Methylhexanoic acid

(*d*) Once again disconnection reveals the necessary halide, which is treated with diethyl malonate.

$$C_6H_5CH_2 \wr CH_2\overset{\displaystyle O}{\overset{\|}{C}}OH \Longrightarrow C_6H_5CH_2X + :\overset{-}{C}H_2\overset{\displaystyle O}{\overset{\|}{C}}OH$$

Required halide; Derived from diethyl malonate

Alkylation of diethyl malonate with benzyl bromide is the first step in the preparation of 3-phenylpropanoic acid.

$$C_6H_5CH_2Br + CH_2(COOCH_2CH_3)_2 \xrightarrow[\text{ethanol}]{NaOCH_2CH_3} C_6H_5CH_2CH(COOCH_2CH_3)_2 \xrightarrow{\text{1. } HO^-, H_2O;\ \text{2. } H^+;\ \text{3. heat}} C_6H_5CH_2CH_2\overset{\displaystyle O}{\overset{\|}{C}}OH$$

Benzyl bromide; Diethyl malonate; Diethyl 2-benzylmalonate; 3-Phenylpropanoic acid

**21.8** Retrosynthetic analysis of the formation of 3-methyl-2-butanone is carried out in the same way as for other ketones.

$$CH_3\overset{\displaystyle O}{\overset{\|}{C}}\underset{\displaystyle CH_3}{\underset{\wr}{C}}H \wr CH_3 \Longrightarrow CH_3\overset{\displaystyle O}{\overset{\|}{C}}\ddot{\underset{..}{C}}H + 2CH_3X$$

3-Methyl-2-butanone (two disconnections as shown); Derived from ethyl acetoacetate

The two alkylation steps are carried out sequentially.

$$CH_3\overset{\displaystyle O}{\overset{\|}{C}}CH_2\overset{\displaystyle O}{\overset{\|}{C}}OCH_2CH_3 \xrightarrow[CH_3Br]{NaOCH_2CH_3} CH_3\overset{\displaystyle O}{\overset{\|}{C}}\underset{\displaystyle CH_3}{\underset{|}{C}}H\overset{\displaystyle O}{\overset{\|}{C}}OCH_2CH_3 \xrightarrow[CH_3Br]{NaOCH_2CH_3} CH_3\overset{\displaystyle O}{\overset{\|}{C}}\underset{H_3C\ \ CH_3}{C}\overset{\displaystyle O}{\overset{\|}{C}}OCH_2CH_3 \xrightarrow{\text{1. } HO^-, H_2O;\ \text{2. } H^+;\ \text{3. heat}} CH_3\overset{\displaystyle O}{\overset{\|}{C}}CH(CH_3)_2$$

Ethyl acetoacetate; Ethyl 2-methyl-3-oxobutanoate; Ethyl 2,2-dimethyl-3-oxobutanoate; 3-Methyl-2-butanone

**21.9** Alkylation of ethyl acetoacetate with 1,4-dibromobutane gives a product that can cyclize to a five-membered ring.

$$CH_3\overset{O}{\overset{\|}{C}}CH_2\overset{O}{\overset{\|}{C}}OCH_2CH_3 + BrCH_2CH_2CH_2CH_2Br \xrightarrow[\text{ethanol}]{NaOCH_2CH_3} BrCH_2CH_2CH_2CH_2\underset{\underset{O}{\overset{|}{C}}OCH_2CH_3}{CH}\overset{O}{\overset{\|}{C}}CH_3$$

Ethyl acetoacetate 1,4-Dibromobutane

$\downarrow NaOCH_2CH_3$

Ethyl 1-acetylcyclopentane-carboxylate

Saponification followed by decarboxylation gives cyclopentyl methyl ketone.

$$\xrightarrow[\text{2. } H^+ \text{ 3. heat}]{\text{1. } HO^-, H_2O}$$

Ethyl 1-acetylcyclopentane-carboxylate Cyclopentyl methyl ketone

**21.10** The last step in the synthesis of pentobarbital is the reaction of the appropriately substituted derivative of diethyl malonate with urea.

$$+ H_2N\overset{O}{\overset{\|}{C}}NH_2 \longrightarrow$$

Diethyl 2-ethyl-2-(1-methylbutyl)malonate Urea Pentobarbital

The dialkyl derivative of diethyl malonate is made in the usual way. It does not matter whether the ethyl group or the 1-methylbutyl group is introduced first.

$$CH_2(COOCH_2CH_3)_2 \xrightarrow[\text{2. } NaOCH_2CH_3,\ CH_3CH_2Br]{\text{1. } NaOCH_2CH_3,\ CH_3CH_2CH_2\underset{}{\overset{Br}{\overset{|}{C}}}HCH_3}$$

Diethyl malonate Diethyl 2-ethyl-2-(1-methylbutyl)malonate

**21.11** The carbonyl oxygen at C-2 of pentobarbital is replaced by sulfur in Pentothal (thiopental).

Pentobarbital; prepared from urea, $(H_2N)_2C{=}O$

Pentothal; prepared from thiourea, $(H_2N)_2C{=}S$

The sodium salt of Pentothal is formed by removal of a proton from one of the N—H groups by sodium hydroxide.

Pentothal sodium

**21.12** The synthesis of phenobarbital requires diethyl 2-phenylmalonate as the starting material.

$$C_6H_5CH(COOCH_2CH_3)_2 \xrightarrow[CH_3CH_2Br]{NaOCH_2CH_3} C_6H_5C(COOCH_2CH_3)_2(CH_2CH_3) \xrightarrow{H_2NC(O)NH_2} \text{Phenobarbital}$$

Diethyl 2-phenylmalonate; Diethyl 2-ethyl-2-phenylmalonate; Phenobarbital

Diethyl 2-phenylmalonate is prepared by a mixed Claisen condensation between ethyl phenylacetate and diethyl carbonate.

$$C_6H_5CH_2\overset{O}{\overset{\|}{C}}OCH_2CH_3 + CH_3CH_2O\overset{O}{\overset{\|}{C}}OCH_2CH_3 \xrightarrow{NaOCH_2CH_3} C_6H_5CH(COOCH_2CH_3)_2$$

Ethyl phenylacetate; Diethyl carbonate; Diethyl 2-phenylmalonate

**21.13** Like diethyl malonate, ethyl acetoacetate undergoes Michael addition to an $\alpha$, $\beta$-unsaturated ketone.

$$\text{cycloheptenone} + CH_3\overset{O}{\overset{\|}{C}}CH_2\overset{O}{\overset{\|}{C}}OCH_2CH_3 \xrightarrow[CH_3CH_2OH]{NaOCH_2CH_3} \text{3-substituted cycloheptanone: } -CH(\overset{O}{\overset{\|}{C}}OCH_2CH_3)(C(=O)CH_3)$$

Basic ester hydrolysis followed by acidification and decarboxylation gives the diketone 3-(2-oxopropyl)cycloheptanone as the major product of the reaction sequence.

$$\text{2-oxocycloheptyl–}\underset{\underset{\text{O=C–CH}_3}{|}}{\text{CH}}\overset{\text{O}}{\overset{\|}{\text{C}}}\text{OCH}_2\text{CH}_3 \xrightarrow[\text{2. H}^+ \text{ 3. heat}]{\text{1. KOH, ethanol–water}} \text{3-oxocycloheptyl–CH}_2\overset{\text{O}}{\overset{\|}{\text{C}}}\text{CH}_3$$

3-(2-Oxopropyl)-cycloheptanone (52%)

**21.14** (*b*) The $\alpha$-carbon atom of the ester bears a phenyl substituent and a methyl group. Only the methyl group can be attached to the $\alpha$ carbon by nucleophilic substitution. Therefore generate the enolate of methyl phenylacetate with lithium diisopropylamide (LDA) in tetrahydrofuran (THF) and then alkylate with methyl iodide.

$$C_6H_5CH_2CO_2CH_3 \xrightarrow[\text{THF}]{\text{LDA}} C_6H_5CH{=}C(OLi)(OCH_3) \xrightarrow{CH_3I} C_6H_5\underset{\underset{CH_3}{|}}{C}HCO_2CH_3$$

Methyl phenylacetate — Enolate of methyl phenylacetate — Methyl 2-phenylpropanoate

(*c*) The desired product corresponds to an aldol addition product.

$$\text{2-[CH(OH)C}_6\text{H}_5\text{]cyclohexanone} \Rightarrow \text{cyclohexanone enolate (carbanion at C-2)} + \overset{\text{O}}{\overset{\|}{\text{H}C}}\text{C}_6\text{H}_5$$

Therefore convert cyclohexanone to its enolate and then treat with benzaldehyde.

$$\text{Cyclohexanone} \xrightarrow[\text{2. C}_6\text{H}_5\text{CHO} \text{ 3. H}_3\text{O}^+]{\text{1. LDA, THF}} \text{2-[CH(OH)C}_6\text{H}_5\text{]cyclohexanone}$$

Cyclohexanone — 1-(2-Oxocyclohexyl)-1-phenylmethanol

(*d*) This product corresponds to the addition of the enolate of *tert*-butyl acetate to cyclohexanone.

$$\text{1-(OH)-1-[CH}_2\overset{\text{O}}{\overset{\|}{\text{C}}}\text{OC(CH}_3)_3\text{]cyclohexane} \Rightarrow \text{cyclohexanone (C=O)} + :\bar{C}H_2\overset{\text{O}}{\overset{\|}{\text{C}}}OC(CH_3)_3$$

Generate the enolate of *tert*-butyl acetate with lithium diisopropylamide; then add cyclohexanone.

$$CH_3CO_2C(CH_3)_3 \xrightarrow[\text{2. cyclohexanone} \text{ 3. H}_3\text{O}^+]{\text{1. LDA, THF}} \text{1-(OH)-1-[CH}_2\text{CO}_2\text{C(CH}_3)_3\text{]cyclohexane}$$

*tert*-Butyl acetate — *tert*-Butyl (1-hydroxycyclohexyl)acetate

**21.15** To undergo a Claisen condensation, an ester must have at least two protons on the $\alpha$ carbon:

$$2RCH_2COCH_2CH_3 + NaOCH_2CH_3 \longrightarrow \left[ RCH_2C\overset{-}{\ddot{C}}COCH_2CH_3 \right] Na^+ + 2CH_3CH_2OH$$

The equilibrium constant for condensation is unfavorable unless the $\beta$-keto ester can be deprotonated to form a stable anion.

(*a*) Among the esters given, ethyl pentanoate and ethyl 3-methylbutanoate undergo the Claisen condensation

$$CH_3CH_2CH_2CH_2COCH_2CH_3 \xrightarrow[2.\ H^+]{1.\ NaOCH_2CH_3} CH_3CH_2CH_2CH_2CCHCOCH_2CH_3 \text{ (with } CH_2CH_2CH_3 \text{ on C-2)}$$

Ethyl pentanoate → Ethyl 3-oxo-2-propylheptanoate

$$(CH_3)_2CHCH_2COCH_2CH_3 \xrightarrow[2.\ H^+]{1.\ NaOCH_2CH_3} (CH_3)_2CHCH_2CCHCOCH_2CH_3 \text{ (with } CH(CH_3)_2 \text{ on C-2)}$$

Ethyl 3-methylbutanoate → Ethyl 2-isopropyl-5-methyl-3-oxohexanoate

(*b*) The Claisen condensation product of ethyl 2-methylbutanoate cannot be deprotonated; the equilibrium constant for its formation is less than 1.

$$CH_3CH_2CHCOCH_2CH_3 \text{ (with } CH_3\text{)} \underset{K<1}{\overset{NaOCH_2CH_3}{\rightleftharpoons}} CH_3CH_2CHCCCOOCH_2CH_3 \text{ (O and } CH_3 \text{ above; } CH_3,\ CH_2CH_3 \text{ below)}$$

Ethyl 2-methylbutanoate

No protons on $\alpha$-carbon atom; cannot form stabilized enolate by deprotonation

(*c*) Ethyl 2,2-dimethylpropanoate has no protons on its $\alpha$ carbon; it cannot form the ester enolate required in the first step of the Claisen condensation.

$$CH_3C(CH_3)_2COCH_2CH_3 + {}^-OCH_2CH_3 \longrightarrow \text{no reaction}$$

Ethyl 2,2-dimethylpropanoate

**21.16** (*a*) The Claisen condensation of ethyl phenylacetate is given by the equation

$$C_6H_5CH_2COCH_2CH_3 \xrightarrow[2.\ H^+]{1.\ NaOCH_2CH_3} C_6H_5CH_2CCHCOCH_2CH_3 \text{ (with } C_6H_5 \text{ on C-2)}$$

Ethyl phenylacetate → Ethyl 3-oxo-2,4-diphenylbutanoate

(*b*) Saponification and decarboxylation of this $\beta$-keto ester gives dibenzyl ketone.

$$C_6H_5CH_2\overset{O}{\overset{\|}{C}}\underset{C_6H_5}{\underset{|}{C}H}\overset{O}{\overset{\|}{C}}OCH_2CH_3 \xrightarrow[\text{2. } H^+ \text{ 3. heat}]{\text{1. } HO^-,\ H_2O} C_6H_5CH_2\overset{O}{\overset{\|}{C}}CH_2C_6H_5$$

Ethyl 3-oxo-2,4-diphenylbutanoate — Dibenzyl ketone

(*c*) This process illustrates the alkylation of a $\beta$-keto ester with subsequent saponification and decarboxylation.

$$C_6H_5CH_2\overset{O}{\overset{\|}{C}}\underset{C_6H_5}{\underset{|}{C}H}\overset{O}{\overset{\|}{C}}OCH_2CH_3 \xrightarrow[H_2C{=}CHCH_2Br]{NaOCH_2CH_3} C_6H_5CH_2\overset{O}{\overset{\|}{C}}\underset{C_6H_5}{\underset{|}{\overset{CH_2CH=CH_2}{\overset{|}{C}}}}COOCH_2CH_3 \xrightarrow[\text{2. } H^+ \text{ 3. heat}]{\text{1. } HO^-,\ H_2O} C_6H_5CH_2\overset{O}{\overset{\|}{C}}\underset{C_6H_5}{\underset{|}{C}H}CH_2CH{=}CH_2$$

Ethyl 3-oxo-2,4-diphenylbutanoate — 1,3-Diphenyl-5-hexen-2-one

(*d*) The enolate ion of ethyl phenylacetate attacks the carbonyl carbon of ethyl benzoate.

$$C_6H_5\overset{O}{\overset{\|}{C}}{-}OCH_2CH_3 \;+\; C_6H_5\ddot{C}H\underset{O}{\underset{\|}{C}}OCH_2CH_3 \longrightarrow C_6H_5\overset{O}{\overset{\|}{C}}\underset{C_6H_5}{\underset{|}{C}H}\overset{O}{\overset{\|}{C}}OCH_2CH_3$$

Ethyl 2,3-diphenyl-3-oxopropanoate

(*e*) Saponification and decarboxylation yield benzyl phenyl ketone.

$$C_6H_5\overset{O}{\overset{\|}{C}}\underset{C_6H_5}{\underset{|}{C}H}\overset{O}{\overset{\|}{C}}OCH_2CH_3 \xrightarrow[\text{2. } H^+ \text{ 3. heat}]{\text{1. } HO^-,\ H_2O} C_6H_5\overset{O}{\overset{\|}{C}}CH_2C_6H_5$$

Ethyl 3-Oxo-2,3-diphenylpropanoate — Benzyl phenyl ketone

(*f*) This sequence is analogous to that of part (*c*).

$$C_6H_5\overset{O}{\overset{\|}{C}}\underset{C_6H_5}{\underset{|}{C}H}\overset{O}{\overset{\|}{C}}OCH_2CH_3 \xrightarrow[H_2C{=}CHCH_2Br]{NaOCH_2CH_3} C_6H_5\overset{O}{\overset{\|}{C}}\underset{C_6H_5}{\underset{|}{\overset{CH_2CH=CH_2}{\overset{|}{C}}}}COOCH_2CH_3 \xrightarrow[\text{2. } H^+ \text{ 3. heat}]{\text{1. } HO^-,\ H_2O} C_6H_5\overset{O}{\overset{\|}{C}}\underset{C_6H_5}{\underset{|}{C}H}CH_2CH{=}CH_2$$

1,2-Diphenyl-4-penten-1-one

**21.17** (*a*) The Dieckmann reaction is the intramolecular version of the Claisen condensation. It employs a diester as starting material.

$$CH_3CH_2O\overset{O}{\overset{\|}{C}}(CH_2)_5\overset{O}{\overset{\|}{C}}OCH_2CH_3 \xrightarrow[\text{2. } H^+]{\text{1. } NaOCH_2CH_3} \text{(2-oxocyclohexyl)}{-}\overset{O}{\overset{\|}{C}}OCH_2CH_3$$

Diethyl heptanedioate — Ethyl (2-oxocyclohexane)-carboxylate (*b*) Acylation of cyclohexanone with diethyl carbonate yields the same $\beta$-keto ester formed in part (*a*).

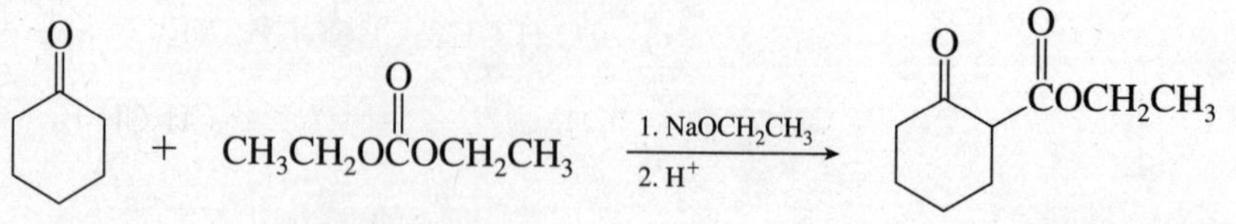

$$\text{Cyclohexanone} + CH_3CH_2OCOCH_2CH_3 \xrightarrow[2.\ H^+]{1.\ NaOCH_2CH_3} \text{Ethyl (2-oxocyclohexane)carboxylate}$$

Cyclohexanone Diethyl carbonate Ethyl (2-oxocyclohexane)-carboxylate

(*c*) The two most stable enol forms are those that involve the proton on the carbon flanked by the two carbonyl groups.

COCH$_2$CH$_3$ ⇌ COCH$_2$CH$_3$ ⇌ COCH$_2$CH$_3$

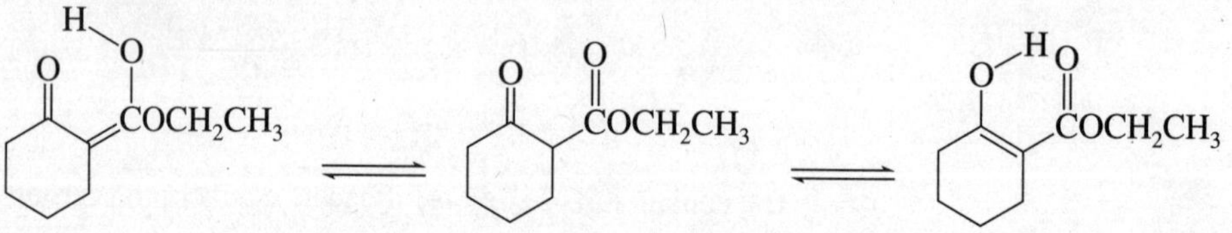

(*d*) Deprotonation of the $\beta$-keto ester involves the acidic proton at the carbon flanked by the two carbonyl groups

COCH$_2$CH$_3$ ⟷ COCH$_2$CH$_3$ ⟷ COCH$_2$CH$_3$

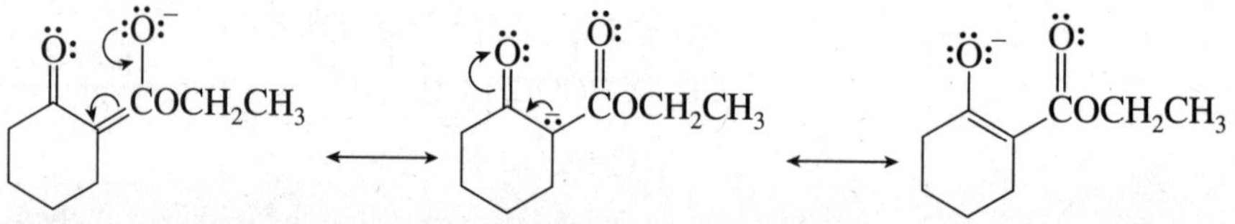

(*e*) The methyl group is introduced by alkylation of the $\beta$-keto ester. Saponification and decarboxylation complete the synthesis.

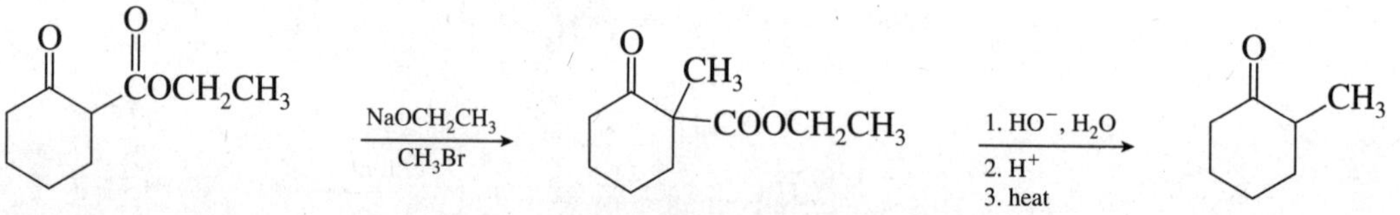

$$\text{Ethyl (2-oxocyclohexane)carboxylate} \xrightarrow[CH_3Br]{NaOCH_2CH_3} \text{Ethyl (1-methyl-2-oxocyclohexane)carboxylate} \xrightarrow[2.\ H^+;\ 3.\ heat]{1.\ HO^-,\ H_2O} \text{2-Methylcyclohexanone}$$

Ethyl (2-oxocyclohexane)-carboxylate Ethyl (1-methyl-2-oxocyclohexane)-carboxylate 2-Methylcyclohexanone

(*f*) The enolate ion of the $\beta$-keto ester [see part (*d*)] undergoes Michael addition to the carbon–carbon double bond of acrolein.

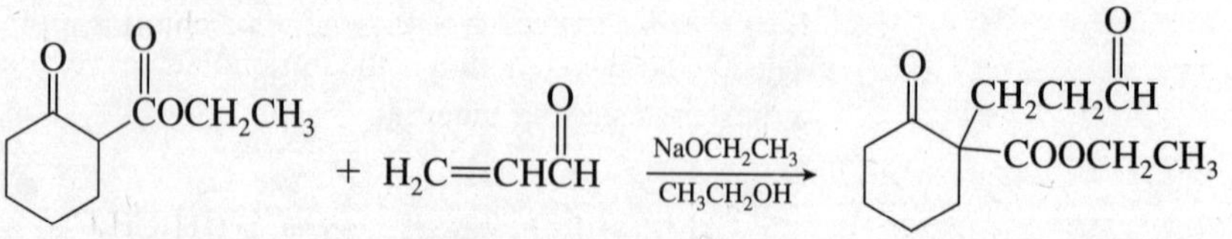

$$\text{Ethyl (2-oxocyclohexane)carboxylate} + H_2C{=}CHCH{=}O \xrightarrow[CH_3CH_2OH]{NaOCH_2CH_3} \text{Michael adduct (}CH_2CH_2CHO,\ COOCH_2CH_3\text{)}$$

Ethyl (2-oxocyclohexane)-carboxylate Acrolein Michael adduct

This reaction has been reported in the chemical literature and proceeds in 65–75% yield.

**21.18** (*a*) Ethyl acetoacetate is converted to its enolate ion with sodium ethoxide; this anion then acts as a nucleophile toward 1-bromopentane.

$$\underset{\text{Ethyl acetoacetate}}{CH_3\overset{O}{\overset{\|}{C}}CH_2\overset{O}{\overset{\|}{C}}OCH_2CH_3} + \underset{\text{1-Bromopentane}}{CH_3CH_2CH_2CH_2CH_2Br} \xrightarrow{NaOCH_2CH_3} \underset{\text{Ethyl 2-acetylheptanoate}}{CH_3\overset{O}{\overset{\|}{C}}\underset{CH_2CH_2CH_2CH_2CH_3}{\underset{|}{C}}H\overset{O}{\overset{\|}{C}}OCH_2CH_3}$$

(*b*) Saponification and decarboxylation of the product in part (*a*) yields 2-octanone.

$$\underset{\text{Ethyl 2-acetylheptanoate}}{CH_3\overset{O}{\overset{\|}{C}}\underset{(CH_2)_4CH_3}{\underset{|}{C}}H\overset{O}{\overset{\|}{C}}OCH_2CH_3} \xrightarrow[\text{2. } H^+ \quad \text{3. heat}]{\text{1. } HO^-, H_2O} \underset{\text{2-Octanone}}{CH_3\overset{O}{\overset{\|}{C}}CH_2CH_2CH_2CH_2CH_2CH_3}$$

(*c*) The product derived from the reaction in part (*a*) can be alkylated again:

$$\underset{\text{Ethyl 2-acetylheptanoate}}{CH_3\overset{O}{\overset{\|}{C}}\underset{CH_2CH_2CH_2CH_2CH_3}{\underset{|}{C}}H\overset{O}{\overset{\|}{C}}OCH_2CH_3} + CH_3I \xrightarrow{NaOCH_2CH_3} \underset{\text{Ethyl 2-acetyl-2-methylheptanoate}}{CH_3\overset{O}{\overset{\|}{C}}\underset{CH_2CH_2CH_2CH_2CH_3}{\underset{|}{\overset{CH_3}{\overset{|}{C}}}}COOCH_2CH_3}$$

(*d*) The dialkylated derivative of acetoacetic ester formed in part (*c*) can be converted to a ketone by saponification and decarboxylation.

$$\underset{\text{Ethyl 2-acetyl-2-methylheptanoate}}{CH_3\overset{O}{\overset{\|}{C}}\underset{CH_2CH_2CH_2CH_2CH_3}{\underset{|}{\overset{CH_3}{\overset{|}{C}}}}COOCH_2CH_3} \xrightarrow[\text{2. } H^+ \quad \text{3. heat}]{\text{1. } HO^-, H_2O} \underset{\text{3-Methyl-2-octanone}}{CH_3\overset{O}{\overset{\|}{C}}\underset{CH_2CH_2CH_2CH_2CH_3}{\underset{|}{C}}HCH_3}$$

(*e*) The anion of ethyl acetoacetate acts as a nucleophile toward 1-bromo-3-chloropropane. Bromide is a better leaving group than chloride and is displaced preferentially.

$$\underset{\text{Ethyl acetoacetate}}{CH_3\overset{O}{\overset{\|}{C}}CH_2\overset{O}{\overset{\|}{C}}OCH_2CH_3} + \underset{\text{1-Bromo-3-chloropropane}}{BrCH_2CH_2CH_2Cl} \xrightarrow{NaOCH_2CH_3} \underset{\text{Ethyl 2-acetyl-5-chloropentanoate}}{CH_3\overset{O}{\overset{\|}{C}}\underset{CH_2CH_2CH_2Cl}{\underset{|}{C}}H\overset{O}{\overset{\|}{C}}OCH_2CH_3}$$

(*f*) Treatment of the product of part (*e*) with sodium ethoxide gives an enolate ion that cyclizes by intramolecular nucleophilic substitution of chloride.

$$\underset{\text{Ethyl 2-acetyl-5-chloropentanoate}}{CH_3\overset{O}{\overset{\|}{C}}\underset{CH_2CH_2CH_2Cl}{\underset{|}{C}H}\overset{O}{\overset{\|}{C}}OCH_2CH_3} \xrightarrow{NaOCH_2CH_3} \underset{\text{Ethyl 1-acetylcyclobutanecarboxylate}}{CH_3\overset{O}{\overset{\|}{C}}\text{-}C(\text{cyclobutane})\text{-}\overset{O}{\overset{\|}{C}}OCH_2CH_3}$$

(*g*) Cyclobutyl methyl ketone is formed by saponification and decarboxylation of the product in part (*f*).

$$\underset{\text{Ethyl 1-acetylcyclobutanecarboxylate}}{CH_3\overset{O}{\overset{\|}{C}}\text{-}C(\text{cyclobutane})\text{-}\overset{O}{\overset{\|}{C}}OCH_2CH_3} \xrightarrow[\text{2. } H^+ \text{ 3. heat}]{\text{1. } HO^-,\ H_2O} \underset{\text{Cyclobutyl methyl ketone}}{CH_3\overset{O}{\overset{\|}{C}}\text{-cyclobutyl}}$$

(*h*) Ethyl acetoacetate undergoes Michael addition to phenyl vinyl ketone in the presence of base.

$$\underset{\text{Ethyl acetoacetate}}{CH_3\overset{O}{\overset{\|}{C}}CH_2\overset{O}{\overset{\|}{C}}OCH_2CH_3} + \underset{\text{Phenyl vinyl ketone}}{H_2C{=}CH\overset{O}{\overset{\|}{C}}C_6H_5} \xrightarrow[\text{ethanol}]{NaOCH_2CH_3} \underset{\text{Ethyl 2-acetyl-5-oxo-5-phenylpentanoate}}{CH_3\overset{O}{\overset{\|}{C}}\underset{CH_2CH_2\underset{O}{\underset{\|}{C}}C_6H_5}{\underset{|}{C}H}\overset{O}{\overset{\|}{C}}OCH_2CH_3}$$

(*i*) A diketone results from saponification and decarboxylation of the Michael adduct.

$$\underset{\text{Ethyl 2-acetyl-5-oxo-5-phenylpentanoate}}{CH_3\overset{O}{\overset{\|}{C}}\underset{CH_2CH_2\underset{O}{\underset{\|}{C}}C_6H_5}{\underset{|}{C}H}\overset{O}{\overset{\|}{C}}OCH_2CH_3} \xrightarrow[\text{2. } H^+ \text{ 3. heat}]{\text{1. } HO^-,\ H_2O} \underset{\text{1-Phenyl-1,5-hexanedione}}{CH_3\overset{O}{\overset{\|}{C}}CH_2CH_2CH_2\overset{O}{\overset{\|}{C}}C_6H_5}$$

**21.19** Diethyl malonate reacts with the reagents given in the preceding problem in a manner analogous to that of ethyl acetoacetate.

(*a*)

$$\underset{\text{Diethyl malonate}}{CH_2(COOCH_2CH_3)_2} + \underset{\text{1-Bromopentane}}{CH_3CH_2CH_2CH_2CH_2Br} \xrightarrow{NaOCH_2CH_3} \underset{\text{Diethyl 1,1-hexanedicarboxylate (diethyl pentylmalonate)}}{CH_3CH_2CH_2CH_2CH_2CH(COOCH_2CH_3)_2}$$

(*b*)

$$\underset{\text{Diethyl 1,1-hexanedicarboxylate}}{CH_3CH_2CH_2CH_2CH_2CH(COOCH_2CH_3)_2} \xrightarrow[\text{2. } H^+ \text{ 3. heat}]{\text{1. } HO^-,\ H_2O} \underset{\text{Heptanoic acid}}{CH_3CH_2CH_2CH_2CH_2CH_2\overset{O}{\overset{\|}{C}}OH}$$

(*c*)

$$CH_3CH_2CH_2CH_2CH_2CH(COOCH_2CH_3)_2 \xrightarrow[NaOCH_2CH_3]{CH_3I} CH_3CH_2CH_2CH_2CH_2\underset{\displaystyle CH_3}{\underset{|}{C}}(COOCH_2CH_3)_2$$

Diethyl 1,1-hexanedicarboxylate — Diethyl 2,2-heptanedicarboxylate

(*d*)

$$CH_3CH_2CH_2CH_2CH_2\underset{\displaystyle CH_3}{\underset{|}{C}}(COOCH_2CH_3)_2 \xrightarrow[\text{2. } H^+ \text{; 3. heat}]{\text{1. } HO^-,\ H_2O} CH_3CH_2CH_2CH_2CH_2\underset{\displaystyle CH_3}{\underset{|}{C}}H\overset{\displaystyle O}{\overset{\|}{C}}OH$$

Diethyl 2,2-heptanedicarboxylate — 2-Methylheptanoic acid

(*e*)

$$CH_2(COOCH_2CH_3)_2 + BrCH_2CH_2CH_2Cl \xrightarrow{NaOCH_2CH_3} ClCH_2CH_2CH_2CH(COOCH_2CH_3)_2$$

Diethyl malonate — 1-Bromo-3-chloropropane — Diethyl 4-chloro-1,1-butanedicarboxylate

(*f*)

$$ClCH_2CH_2CH_2CH(COOCH_2CH_3)_2 \xrightarrow{NaOCH_2CH_3} \text{cyclobutane-1,1-}(\overset{\displaystyle O}{\overset{\|}{C}}OCH_2CH_3)_2$$

Diethyl 4-chloro-1,1-butanedicarboxylate — Diethyl cyclobutane-1,1-dicarboxylate

(*g*)

$$\text{cyclobutane-1,1-}(\overset{\displaystyle O}{\overset{\|}{C}}OCH_2CH_3)_2 \xrightarrow[\text{2. } H^+ \text{; 3. heat}]{\text{1. } HO^-,\ H_2O} \text{cyclobutyl}-\overset{\displaystyle O}{\overset{\|}{C}}OH$$

Diethyl cyclobutane-1,1-dicarboxylate — Cyclobutanecarboxylic acid

(*h*)

$$CH_2(COOCH_2CH_3)_2 + C_6H_5\overset{\displaystyle O}{\overset{\|}{C}}CH{=}CH_2 \xrightarrow[CH_3CH_2OH]{NaOCH_2CH_3} C_6H_5\overset{\displaystyle O}{\overset{\|}{C}}CH_2CH_2CH(COOCH_2CH_3)_2$$

Diethyl malonate — Phenyl vinyl ketone — Diethyl 4-oxo-4-phenylbutane-1,1-dicarboxylate

(*i*)

$$C_6H_5\overset{\displaystyle O}{\overset{\|}{C}}CH_2CH_2CH(COOCH_2CH_3)_2 \xrightarrow[\text{2. } H^+ \text{; 3. heat}]{\text{1. } HO^-,\ H_2O} C_6H_5\overset{\displaystyle O}{\overset{\|}{C}}CH_2CH_2CH_2\overset{\displaystyle O}{\overset{\|}{C}}OH$$

Diethyl 4-oxo-4-phenylbutane-1,1-dicarboxylate — 5-Oxo-5-phenylpentanoic acid

**21.20** (*a*) Both carbonyl groups of diethyl malonate are equivalent, and so enolization can occur in either direction.

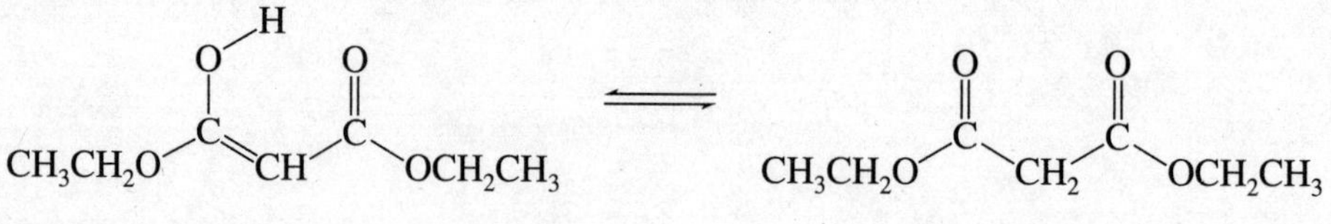

$$CH_3CH_2O\text{–}C(OH){=}CH\text{–}C(=O)OCH_2CH_3 \rightleftharpoons CH_3CH_2OC(=O)CH_2C(=O)OCH_2CH_3 \rightleftharpoons CH_3CH_2OC(=O)CH{=}C(OH)OCH_2CH_3$$

Diethyl malonate

(*b*) Ethyl acetoacetate can give three constitutionally isomeric enols:

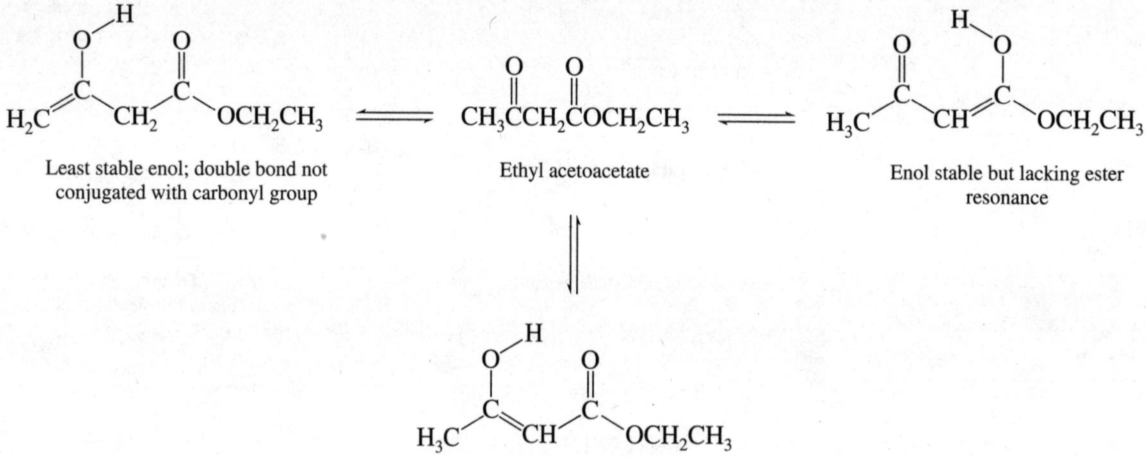

$$H_2C{=}C(OH)CH_2C(=O)OCH_2CH_3 \rightleftharpoons CH_3\overset{O}{\overset{\|}{C}}CH_2\overset{O}{\overset{\|}{C}}OCH_2CH_3 \rightleftharpoons H_3C\text{–}C(=O)CH{=}C(OH)OCH_2CH_3$$

Least stable enol; double bond not conjugated with carbonyl group

Ethyl acetoacetate

Enol stable but lacking ester resonance

$$H_3C\text{–}C(OH){=}CH\text{–}C(=O)OCH_2CH_3$$

Most stable enol; double bond conjugated with carbonyl group; ester carbonyl stabilized by resonance

(*c*) Bromine reacts with diethyl malonate and ethyl acetoacetate by way of the corresponding enols:

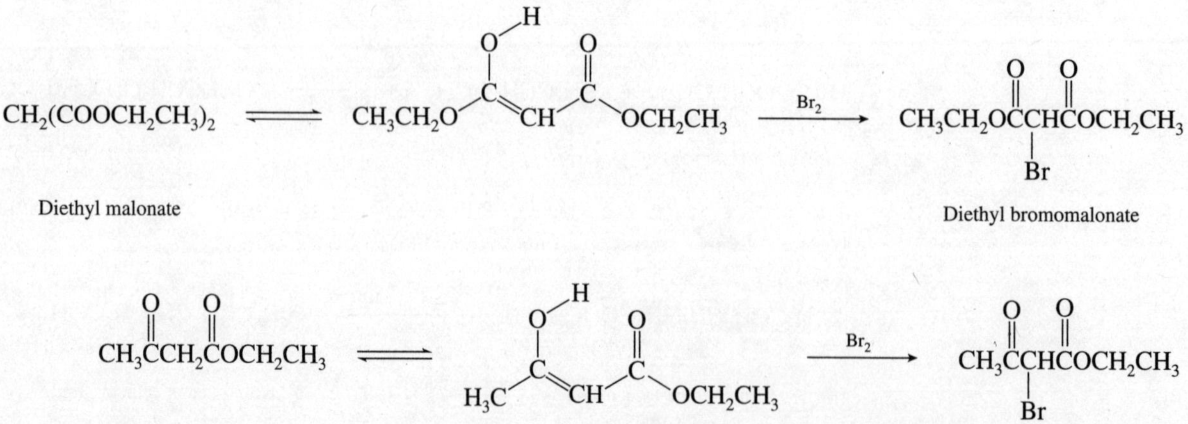

$$CH_2(COOCH_2CH_3)_2 \rightleftharpoons CH_3CH_2O\text{–}C(OH){=}CH\text{–}C(=O)OCH_2CH_3 \xrightarrow{Br_2} CH_3CH_2O\overset{O}{\overset{\|}{C}}\underset{Br}{\underset{|}{C}H}\overset{O}{\overset{\|}{C}}OCH_2CH_3$$

Diethyl malonate

Diethyl bromomalonate

$$CH_3\overset{O}{\overset{\|}{C}}CH_2\overset{O}{\overset{\|}{C}}OCH_2CH_3 \rightleftharpoons H_3C\text{–}C(OH){=}CH\text{–}C(=O)OCH_2CH_3 \xrightarrow{Br_2} CH_3\overset{O}{\overset{\|}{C}}\underset{Br}{\underset{|}{C}H}\overset{O}{\overset{\|}{C}}OCH_2CH_3$$

Ethyl acetoacetate

Ethyl α-bromoacetoacetate

**21.21** (*a*) Recall that Grignard reagents are destroyed by reaction with proton donors. Ethyl acetoacetate is a stronger acid than water; it transfers a proton to a Grignard reagent.

$$\underset{\text{Ethyl acetoacetate}}{CH_3\overset{O}{\overset{\|}{C}}CH_2\overset{O}{\overset{\|}{C}}OCH_2CH_3} + \underset{\text{Methylmagnesium iodide}}{CH_3MgI} \longrightarrow \underset{\text{Methane}}{CH_4} + \underset{\text{Iodomagnesium salt of ethyl acetoacetate}}{CH_3\overset{O}{\overset{\|}{C}}\underset{MgI}{\underset{|}{C}H}\overset{O}{\overset{\|}{C}}OCH_2CH_3}$$

(*b*) Adding $D_2O$ and DCl to the reaction mixture leads to $D^+$ transfer to the $\alpha$-carbon atom of ethyl acetoacetate.

$$\underset{\text{Iodomagnesium salt of ethyl acetoacetate}}{CH_3\overset{O}{\overset{\|}{C}}\underset{MgI}{\underset{|}{C}H}\overset{O}{\overset{\|}{C}}OCH_2CH_3} + \underset{\text{Deuterium oxide}}{D_2O} \xrightarrow{DCl} \underset{\text{Ethyl }\alpha\text{-deuterioacetoacetate}}{CH_3\overset{O}{\overset{\|}{C}}\underset{D}{\underset{|}{C}H}\overset{O}{\overset{\|}{C}}OCH_2CH_3}$$

**21.22** (*a*) Ethyl octanoate undergoes a Claisen condensation to form a $\beta$-keto ester on being treated with sodium ethoxide.

$$\underset{\text{Ethyl octanoate}}{CH_3(CH_2)_5CH_2\overset{O}{\overset{\|}{C}}OCH_2CH_3} \xrightarrow[2.\ H^+]{1.\ NaOCH_2CH_3} \underset{\text{Ethyl 2-hexyl-3-oxodecanoate}}{CH_3(CH_2)_5CH_2\overset{O}{\overset{\|}{C}}\underset{(CH_2)_5CH_3}{\underset{|}{C}H}\overset{O}{\overset{\|}{C}}OCH_2CH_3}$$

(*b*) Saponification and decarboxylation of the $\beta$-keto ester yields a ketone.

$$\underset{\text{Ethyl 2-hexyl-3-oxodecanoate}}{CH_3(CH_2)_5CH_2\overset{O}{\overset{\|}{C}}\underset{(CH_2)_5CH_3}{\underset{|}{C}H}\overset{O}{\overset{\|}{C}}OCH_2CH_3} \xrightarrow[2.\ H^+\ \ 3.\ heat]{1.\ NaOH,\ H_2O} \underset{\text{8-Pentadecanone}}{CH_3(CH_2)_5CH_2\overset{O}{\overset{\|}{C}}CH_2(CH_2)_5CH_3}$$

(*c*) On treatment with base, ethyl acetoacetate is converted to its enolate, which reacts as a nucleophile toward 1-bromobutane.

$$\underset{\text{Ethyl acetoacetate}}{CH_3\overset{O}{\overset{\|}{C}}CH_2\overset{O}{\overset{\|}{C}}OCH_2CH_3} + \underset{\text{1-Bromobutane}}{CH_3CH_2CH_2CH_2Br} \xrightarrow[ethanol]{NaOCH_2CH_3} \underset{\text{Ethyl 2-acetylhexanoate}}{CH_3\overset{O}{\overset{\|}{C}}\underset{CH_2CH_2CH_2CH_3}{\underset{|}{C}H}\overset{O}{\overset{\|}{C}}OCH_2CH_3}$$

(*d*) Alkylation of ethyl acetoacetate, followed by saponification and decarboxylation, gives a ketone. The two steps constitute the acetoacetic ester synthesis.

$$\underset{\text{Ethyl 2-acetylhexanoate}}{CH_3\overset{O}{\overset{\|}{C}}\underset{CH_2CH_2CH_2CH_3}{\underset{|}{C}H}\overset{O}{\overset{\|}{C}}OCH_2CH_3} \xrightarrow[2.\ H^+\ \ 3.\ heat]{1.\ NaOH,\ H_2O} \underset{\text{2-Heptanone}}{CH_3\overset{O}{\overset{\|}{C}}CH_2CH_2CH_2CH_2CH_3}$$

(*e*) An alkylated derivative of ethyl acetoacetate is capable of being alkylated a second time.

$$\underset{\text{Ethyl 2-acetylhexanoate}}{CH_3\overset{O}{\overset{\|}{C}}\underset{CH_2CH_2CH_2CH_3}{\underset{|}{C}H}\overset{O}{\overset{\|}{C}}OCH_2CH_3} + \underset{\text{1-Iodobutane}}{CH_3CH_2CH_2CH_2I} \xrightarrow[\text{ethanol}]{NaOCH_2CH_3} \underset{\text{Ethyl 2-acetyl-2-butylhexanoate}}{CH_3\overset{O}{\overset{\|}{C}}\underset{COOCH_2CH_3}{\underset{|}{C}}(CH_2CH_2CH_2CH_3)_2}$$

(*f*) The dialkylated derivative of acetoacetic ester formed in part (*e*) is converted to a ketone by saponification and decarboxylation.

$$\underset{\text{Ethyl 2-acetyl-2-butylhexanoate}}{CH_3\overset{O}{\overset{\|}{C}}\underset{COOCH_2CH_3}{\underset{|}{C}}(CH_2CH_2CH_2CH_3)_2} \xrightarrow[\text{2. H}^+\text{; 3. heat}]{\text{1. NaOH}} \underset{\text{3-Butyl-2-heptanone}}{CH_3\overset{O}{\overset{\|}{C}}CH(CH_2CH_2CH_2CH_3)_2}$$

(*g*) The enolate of acetophenone attacks the carbonyl group of diethyl carbonate.

$$\underset{\text{Acetophenone}}{C_6H_5\overset{O}{\overset{\|}{C}}CH_3} + \underset{\text{Diethyl carbonate}}{CH_3CH_2O\overset{O}{\overset{\|}{C}}OCH_2CH_3} \xrightarrow[\text{2. H}^+]{\text{1. NaOCH}_2\text{CH}_3} \underset{\text{3-Oxo-3-phenylpropanoate}}{C_6H_5\overset{O}{\overset{\|}{C}}CH_2\overset{O}{\overset{\|}{C}}OCH_2CH_3}$$

(*h*) Diethyl oxalate acts as an acylating agent toward the enolate of acetone.

$$\underset{\text{Acetone}}{CH_3\overset{O}{\overset{\|}{C}}CH_3} + \underset{\text{Diethyl oxalate}}{CH_3CH_2O\overset{O}{\overset{\|}{C}}\overset{O}{\overset{\|}{C}}OCH_2CH_3} \xrightarrow[\text{2. H}^+]{\text{1. NaOCH}_2\text{CH}_3} \underset{\text{Ethyl 2,4-dioxopentanoate}}{CH_3\overset{O}{\overset{\|}{C}}CH_2\overset{O}{\overset{\|}{C}}\overset{O}{\overset{\|}{C}}OCH_2CH_3}$$

(*i*) The first stage of the malonic ester synthesis is the alkylation of diethyl malonate with an alkyl halide.

$$\underset{\text{Diethyl malonate}}{CH_2(COOCH_2CH_3)_2} + \underset{\text{1-Bromo-2-methylbutane}}{BrCH_2\underset{CH_3}{\underset{|}{C}H}CH_2CH_3} \xrightarrow[\text{ethanol}]{NaOCH_2CH_3} \underset{\text{Diethyl 3-methylpentane-1,1-dicarboxylate}}{CH_3CH_2\underset{CH_3}{\underset{|}{C}H}CH_2CH(COOCH_2CH_3)_2}$$

(*j*) Alkylation of diethyl malonate is followed by saponification and decarboxylation to give a carboxylic acid.

$$\underset{\text{Diethyl 3-methylpentane-1,1-dicarboxylate}}{CH_3CH_2\underset{CH_3}{\underset{|}{C}H}CH_2CH(COOCH_2CH_3)_2} \xrightarrow[\text{2. H}^+\text{; 3. heat}]{\text{1. NaOH, H}_2\text{O}} CH_3CH_2\underset{CH_3}{\underset{|}{C}H}CH_2CH_2\overset{O}{\overset{\|}{C}}OH$$

4-Methylhexanoic acid
(57% yield from 1-bromo-2-methylbutane)

(*k*) The anion of diethyl malonate undergoes Michael addition to 6-methyl-2-cyclohexenone.

$$CH_2(COOCH_2CH_3)_2 + \text{6-methyl-2-cyclohexenone} \xrightarrow[\text{ethanol}]{NaOCH_2CH_3} \text{(structure with } H_3C\text{, C=O, and } CH(COOCH_2CH_3)_2 \text{ on cyclohexanone ring)}$$

Diethyl malonate — 6-Methyl-2-cyclohexenone — Diethyl 2-(4-methyl-3-oxocyclohexyl)malonate (isolated yield, 50%)

(*l*) Acid hydrolysis converts the diester in part (*k*) to a malonic acid derivative, which then undergoes decarboxylation.

$\xrightarrow[\text{heat}]{H_2O,\ HCl}$

Diethyl 2-(4-methyl-3-oxocyclohexyl)malonate → (4-Methyl-3-oxocyclohexyl)acetic acid (isolated yield, 80%)

(*m*) Lithium diisopropylamide (LDA) is used to convert esters quantitatively to their enolate ions. In this reaction the enolate of *tert*-butyl acetate adds to benzaldehyde.

$$CH_3\overset{O}{\overset{\|}{C}}OC(CH_3)_3 \xrightarrow{LDA} H_2C{=}C(OLi)OC(CH_3)_3 \xrightarrow[2.\ H^+]{1.\ C_6H_5\overset{O}{\overset{\|}{C}}H} C_6H_5\underset{OH}{\underset{|}{C}}HCH_2\overset{O}{\overset{\|}{C}}OC(CH_3)_3$$

*tert*-Butyl acetate — Lithium enolate of *tert*-butyl acetate — *tert*-Butyl 3-hydroxy-3-phenylpropanoate

**21.23** (*a*) Both ester functions in this molecule are $\beta$ to a ketone carbonyl. Hydrolysis is followed by decarboxylation.

$\xrightarrow[\text{heat}]{H_2O,\ H_2SO_4}$

Diethyl 3-ethylcyclopentanone-2,5-dicarboxylate → 3-Ethylcyclopentanone ($C_7H_{12}O$)

(*b*) Examine each carbon that is $\alpha$ to an ester function to see if it can lead to a five-, six-, or seven-membered cyclic $\beta$-keto ester by a Dieckmann cyclization.

Cyclization to a five-membered ring possible, but $\beta$-keto ester cannot be deprotonated to give a stable anion.

Cyclization not likely; resulting ring is four-membered and highly strained.

Cyclization gives a five-membered ring; $\beta$-keto ester deprotonated under reaction conditions; this is the observed product ($C_{12}H_{18}O_5$).

(*c*) Both ester function undergo hydrolysis in acid, but decarboxylation occurs only at the carboxyl group that is $\beta$ to the ketone carbonyl.

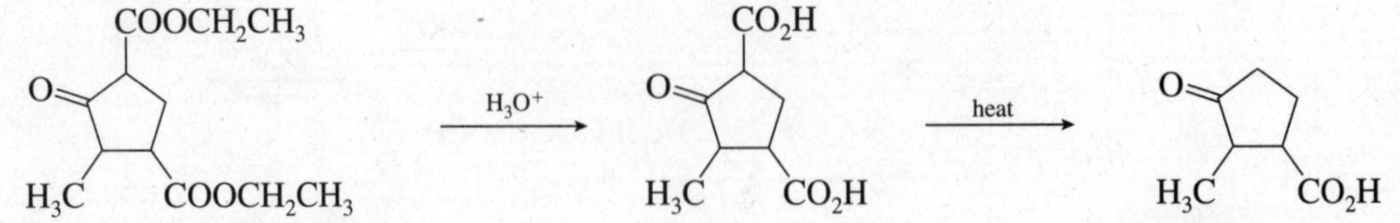

$COOCH_2CH_3$ / O / $H_3C$ / $COOCH_2CH_3$ —$H_3O^+$→ $CO_2H$ / O / $H_3C$ / $CO_2H$ —heat→ O / $H_3C$ / $CO_2H$

Diethyl 2-methylcyclopentanone-3,5-dicarboxylate

2-Methylcyclopentanone-3-carboxylic acid ($C_7H_{10}O_3$)

(*d*) A Dieckmann cyclization occurs, giving a five-membered ring fused to the original three-membered ring.

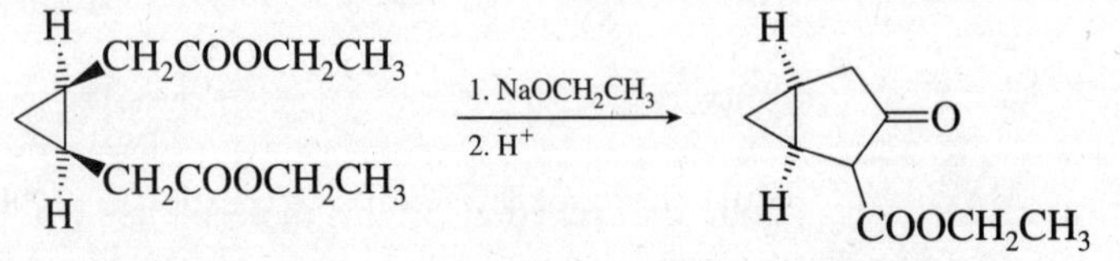

H / $CH_2COOCH_2CH_3$ / $CH_2COOCH_2CH_3$ / H —1. $NaOCH_2CH_3$, 2. $H^+$→ H / O / H / $COOCH_2CH_3$

Diethyl *cis*-1,2-cyclopropanediacetate

Ethyl bicyclo[3.1.0]-hexan-3-one-2-carboxylate ($C_9H_{12}O_3$, 79%)

(*e*) Saponification and decarboxylation convert the $\beta$-keto ester to a ketone.

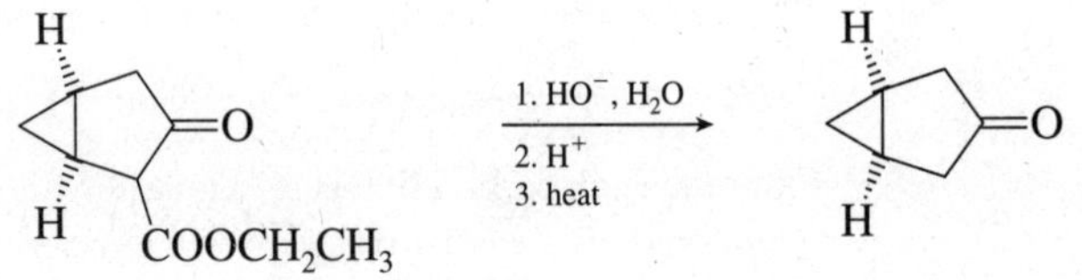

H / O / H / $COOCH_2CH_3$ —1. $HO^-$, $H_2O$, 2. $H^+$, 3. heat→ H / O / H

Ethyl bicyclo[3.1.0]hexan-3-one-2-carboxylate

Bicyclo[3.1.0]hexan-3-one ($C_6H_8O$, 43%)

**21.24** The heart of the preparation of capsaicin is a malonic ester synthesis. The first step is bromination of the primary alcohol by phosphorous tribromide. The resulting primary alkyl bromide is used to alkylate the sodium salt of diethyl malonate. A substituted malonic acid derivative is obtained following basic hydrolysis of the ester groups.

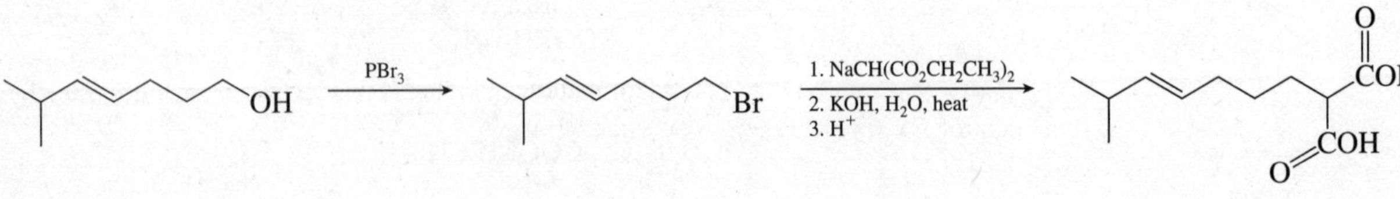

OH —$PBr_3$→ Br —1. $NaCH(CO_2CH_2CH_3)_2$, 2. KOH, $H_2O$, heat, 3. $H^+$→ O ‖ COH / COH ‖ O

$C_8H_{15}Br$

$C_{11}H_{18}O_4$

Malonic acid derivatives undergo decarboxylation on heating.

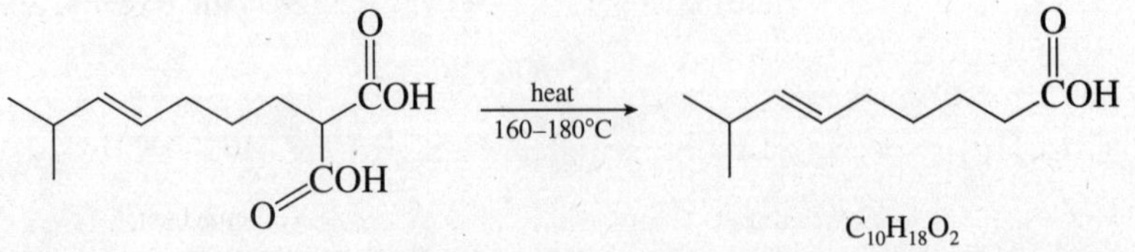

O ‖ COH / COH ‖ O —heat, 160–180°C→ O ‖ COH

$C_{10}H_{18}O_2$

Formation of the amide completes the synthesis of capsaicin.

$$\text{(CH}_3)_2\text{CHCH=CH(CH}_2)_4\overset{\text{O}}{\overset{\|}{\text{C}}}\text{OH} \xrightarrow{SOCl_2} \text{(CH}_3)_2\text{CHCH=CH(CH}_2)_4\overset{\text{O}}{\overset{\|}{\text{C}}}\text{Cl}$$

$$\downarrow \text{4-HO-3-CH}_3\text{O-C}_6\text{H}_3\text{CH}_2\text{NH}_2$$

$$\text{(CH}_3)_2\text{CHCH=CH(CH}_2)_4\overset{\text{O}}{\overset{\|}{\text{C}}}\text{NHCH}_2\text{-C}_6\text{H}_3\text{(OH)(OCH}_3)$$

Capsaicin ($C_{18}H_{27}NO_3$)

**21.25** (*a*) First write out the structure of 4-phenyl-2-butanone and identify the synthon that is derived from ethyl acetoacetate.

$$C_6H_5CH_2 \nmid CH_2\overset{O}{\overset{\|}{C}}CH_3 \Longrightarrow C_6H_5CH_2X + :\overset{-}{C}H_2\overset{O}{\overset{\|}{C}}CH_3$$

Therefore carry out the acetoacetic ester synthesis using a benzyl halide as the alkylating agent.

$$\underset{\text{Benzyl alcohol}}{C_6H_5CH_2OH} \xrightarrow[\text{or } PBr_3]{HBr} \underset{\text{Benzyl bromide}}{C_6H_5CH_2Br}$$

$$\underset{\text{Ethyl acetoacetate}}{CH_3\overset{O}{\overset{\|}{C}}CH_2\overset{O}{\overset{\|}{C}}OCH_2CH_3} + \underset{\text{Benzyl bromide}}{C_6H_5CH_2Br} \xrightarrow[\text{ethanol}]{NaOCH_2CH_3} \underset{\text{Ethyl 2-benzyl-3-oxobutanoate}}{CH_3\overset{O}{\overset{\|}{C}}\underset{CH_2C_6H_5}{\underset{|}{C}H}\overset{O}{\overset{\|}{C}}OCH_2CH_3} \xrightarrow[\text{2. } H^+ \text{ 3. heat}]{\text{1. } HO^-, H_2O} \underset{\text{4-Phenyl-2-butanone}}{CH_3\overset{O}{\overset{\|}{C}}CH_2CH_2C_6H_5}$$

(*b*) Identify the synthon in 3-phenylpropanoic acid that is derived from malonic ester by disconnecting the molecule at its $\alpha$-carbon atom.

$$C_6H_5CH_2 \nmid CH_2\overset{O}{\overset{\|}{C}}OH \Longrightarrow C_6H_5CH_2X + :\overset{-}{C}H_2\overset{O}{\overset{\|}{C}}OH$$

Here, as in part (*a*), a benzyl halide is the required alkylating agent.

$$\underset{\text{Diethyl malonate}}{CH_2(COOCH_2CH_3)_2} + \underset{\text{Benzyl bromide}}{C_6H_5CH_2Br} \xrightarrow[\text{ethanol}]{NaOCH_2CH_3} \underset{\text{Diethyl benzylmalonate}}{C_6H_5CH_2CH(COOCH_2CH_3)_2} \xrightarrow[\text{2. } H^+ \text{ 3. heat}]{\text{1. } HO^-, H_2O} \underset{\text{3-Phenylpropanoic acid}}{C_6H_5CH_2CH_2COOH}$$

(*c*) In this synthesis the desired 1,3-diol function can be derived by reduction of a malonic ester derivative. First propene must be converted to an allyl halide for use as an alkylating agent.

$$H_2C{=}CHCH_2CH(CH_2OH)_2 \Longrightarrow H_2C{=}CHCH_2X + :\bar{C}H(COOCH_2CH_3)_2$$

$$\underset{\text{Propene}}{H_2C{=}CHCH_3} \xrightarrow[\text{heat}]{Cl_2} \underset{\text{Allyl chloride}}{H_2C{=}CHCH_2Cl}$$

$$\underset{\text{Diethyl malonate}}{CH_2(COOCH_2CH_3)_2} + \underset{\text{Allyl chloride}}{H_2C{=}CHCH_2Cl} \xrightarrow[\text{ethanol}]{NaOCH_2CH_3} \underset{\text{Diethyl 2-allylmalonate}}{H_2C{=}CHCH_2CH(COOCH_2CH_3)_2}$$

$$\xrightarrow{1.\ LiAlH_4;\ 2.\ H_2O} \underset{\text{2-Allyl-1,3-propanediol}}{H_2C{=}CHCH_2CH(CH_2OH)_2}$$

(*d*) The desired primary alcohol may be prepared by reduction of the corresponding carboxylic acid, which in turn is available from the malonic ester synthesis using allyl chloride, including saponification and decarboxylation of the diester [prepared in part (*c*)].

$$\underset{\text{4-Penten-1-ol}}{H_2C{=}CHCH_2CH_2CH_2OH} \Longrightarrow H_2C{=}CHCH_2CH_2CO_2H \Longrightarrow H_2C{=}CHCH_2CH(CO_2CH_2CH_3)_2$$

The correct sequence of reactions is

$$\underset{\substack{\text{Diethyl 2-allylmalonate}\\ \text{[prepared as in part (c)]}}}{H_2C{=}CHCH_2CH(COOCH_2CH_3)_2} \xrightarrow[2.\ H^+;\ 3.\ \text{heat}]{1.\ HO^-,\ H_2O} \underset{\text{4-Pentenoic acid}}{H_2C{=}CHCH_2CH_2COOH} \xrightarrow[2.\ H_2O]{1.\ LiAlH_4} \underset{\text{4-Penten-1-ol}}{H_2C{=}CHCH_2CH_2CH_2OH}$$

(*e*) The desired product is an alcohol. It can be prepared by reduction of a ketone, which in turn can be prepared by the acetoacetic ester synthesis.

$$H_2C{=}CHCH_2CH_2\overset{OH}{\overset{|}{C}}HCH_3 \Longrightarrow H_2C{=}CHCH_2CH_2\overset{O}{\overset{\|}{C}}CH_3 \Longrightarrow H_2C{=}CHCH_2X + :\bar{C}H_2\overset{O}{\overset{\|}{C}}CH_3$$

Therefore

$$\underset{\text{Ethyl acetoacetate}}{CH_3\overset{O}{\overset{\|}{C}}CH_2\overset{O}{\overset{\|}{C}}OCH_2CH_3} + \underset{\text{Allyl chloride}}{H_2C{=}CHCH_2Cl} \xrightarrow[\text{ethanol}]{NaOCH_2CH_3} CH_3\overset{O}{\overset{\|}{C}}\underset{CH_2CH=CH_2}{\underset{|}{C}H}\overset{O}{\overset{\|}{C}}OCH_2CH_3 \xrightarrow[2.\ H^+;\ 3.\ \text{heat}]{1.\ HO^-,\ H_2O} CH_3\overset{O}{\overset{\|}{C}}CH_2CH_2CH{=}CH_2$$

$$\xrightarrow[CH_3OH]{NaBH_4} \underset{\text{5-Hexen-2-ol}}{CH_3\underset{OH}{\underset{|}{C}H}CH_2CH_2CH{=}CH_2}$$

(*f*) Cyclopropanecarboxylic acid may be prepared by a malonic ester synthesis, as retrosynthetic analysis shows.

$$\text{(cyclopropyl)}-CO_2H \Rightarrow \text{(cyclopropane-1,1-)}(CO_2CH_2CH_3)_2 \Rightarrow CH_2X-CH_2X + CH_2(CO_2CH_2CH_3)_2$$

The desired reaction sequence is

$$\underset{\text{Diethyl malonate}}{CH_2(COOCH_2CH_3)_2} + \underset{\text{1,2-Dibromoethane}}{BrCH_2CH_2Br} \xrightarrow[\text{ethanol}]{NaOCH_2CH_3} \text{(cyclopropane-1,1-)}(COOCH_2CH_3)_2 \xrightarrow[\text{2. } H^+ \text{ 3. heat}]{\text{1. } HO^-, H_2O} \underset{\text{Cyclopropanecarboxylic acid}}{\text{(cyclopropyl)}-\overset{O}{\overset{\|}{C}}OH}$$

(*g*) Treatment of the diester formed in part (*f*) with ammonia gives a diamide.

$$\underset{\substack{\text{Diethyl cyclopropane-}\\ \text{1,1-dicarboxylate}\\ \text{[prepared as in part (f)]}}}{\text{(cyclopropane-1,1-)}(\overset{O}{\overset{\|}{C}}OCH_2CH_3)_2} \xrightarrow{NH_3} \underset{\substack{\text{Cyclopropane-}\\ \text{1,1-dicarboxamide}}}{\text{(cyclopropane-1,1-)}(\overset{O}{\overset{\|}{C}}NH_2)_2}$$

(*h*) We need to extend the carbon chain of the starting material by *four* carbons. One way to accomplish this is by way of a malonic ester synthesis at each end of the chain.

$$\underset{\text{Octanedioic acid}}{HO\overset{O}{\overset{\|}{C}}(CH_2)_6\overset{O}{\overset{\|}{C}}OH} \xrightarrow[\text{2. } H_2O]{\text{1. } LiAlH_4} HOCH_2(CH_2)_6CH_2OH \xrightarrow[\text{or } PBr_3]{HBr} \underset{\text{1,8-Dibromooctane}}{BrCH_2(CH_2)_6CH_2Br}$$

$$\underset{\text{Diethyl malonate}}{2CH_2(COOCH_2CH_3)_2} + \underset{\text{1,8-Dibromooctane}}{Br(CH_2)_8Br} \xrightarrow{NaOCH_2CH_3} (CH_3CH_2OOC)_2CH(CH_2)_8CH(COOCH_2CH_3)_2$$

$$\xrightarrow{\text{1. } HO^-, H_2O \quad \text{2. } H^+ \quad \text{3. heat}} \underset{\text{Dodecanedioic acid}}{HO\overset{O}{\overset{\|}{C}}CH_2(CH_2)_8CH_2\overset{O}{\overset{\|}{C}}OH}$$

**21.26** The problem states that diphenadione is prepared from 1,1-diphenylacetone and dimethyl 1,2-benzenedicarboxylate. Therefore, disconnect the molecule in a way that reveals the two reactants.

$$\underset{\text{Diphenadione}}{\text{(indane-1,3-dione-2-yl)}-\overset{O}{\overset{\|}{C}}CH(C_6H_5)_2} \Rightarrow C_6H_4(\overset{O}{\overset{\|}{C}}X)_2 \quad \text{and} \quad CH_3\overset{O}{\overset{\|}{C}}CH(C_6H_5)_2$$

Thus all that is required is to treat dimethyl 1,2-benzenedicarboxylate and 1,1-diphenylacetone with base. Two successive acylations of a ketone enolate occur; the first is intermolecular, the second intramolecular.

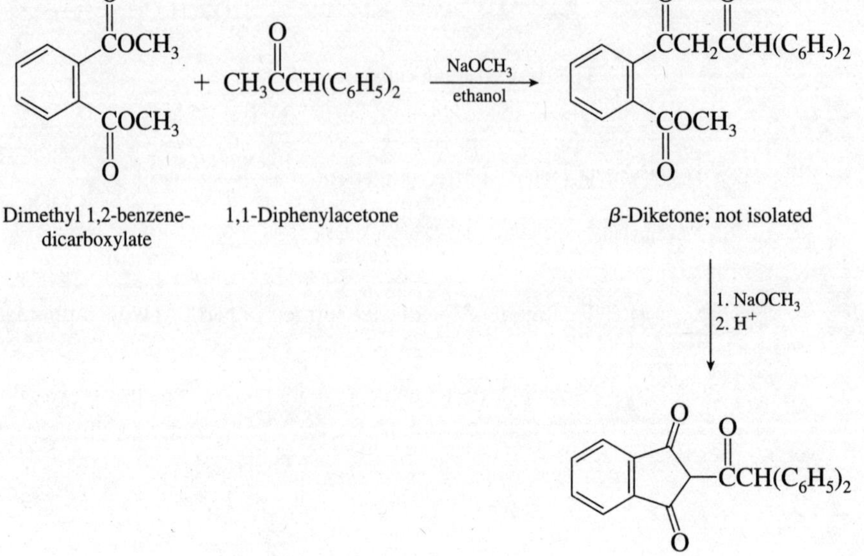

$$C_6H_4(COOCH_3)_2 + CH_3\overset{O}{\overset{\|}{C}}CH(C_6H_5)_2 \xrightarrow[\text{ethanol}]{NaOCH_3} C_6H_4(\overset{O}{\overset{\|}{C}}CH_2\overset{O}{\overset{\|}{C}}CH(C_6H_5)_2)(COOCH_3)$$

Dimethyl 1,2-benzenedicarboxylate; 1,1-Diphenylacetone; β-Diketone; not isolated

1. $NaOCH_3$
2. $H^+$

$$\text{indane-1,3-dione}-\overset{O}{\overset{\|}{C}}CH(C_6H_5)_2$$

Diphenadione

**21.27** Esters react with amines to give amides. Each nitrogen of 1,2-diphenylhydrazine reacts with a separate ester function of diethyl 2-butylmalonate.

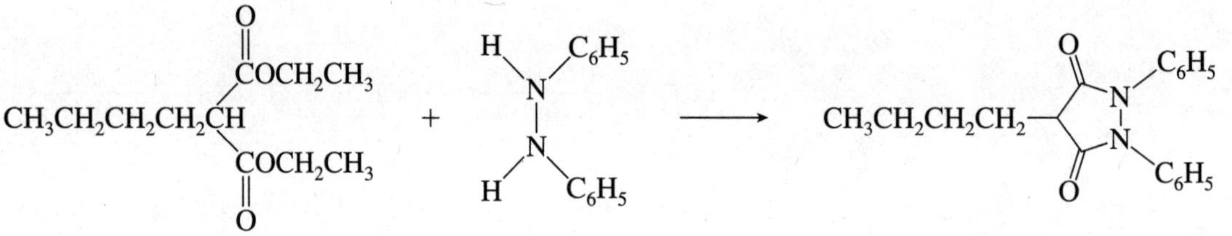

$$CH_3CH_2CH_2CH_2CH(COOCH_2CH_3)_2 + C_6H_5NH{-}NHC_6H_5 \longrightarrow CH_3CH_2CH_2CH_2{-}\text{(3,5-dioxo-1,2-diphenylpyrazolidin-4-yl)}$$

Diethyl 2-butylmalonate; 1,2-Diphenylhydrazine; Phenylbutazone ($C_{19}H_{20}N_2O_2$)

**21.28** Styrene oxide will be attacked by the anion of diethyl malonate at its less hindered ring position.

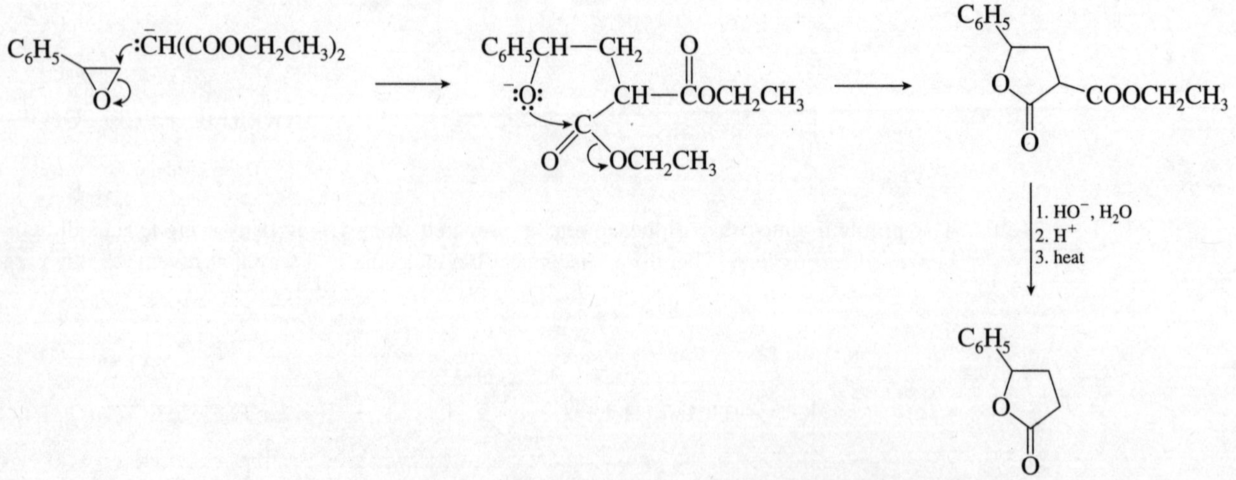

$$C_6H_5\text{-epoxide} + {:}\bar{C}H(COOCH_2CH_3)_2 \longrightarrow C_6H_5CH(\ddot{O}^{-})-CH_2-CH(COOCH_2CH_3)-\overset{O}{\overset{\|}{C}}OCH_2CH_3 \longrightarrow \text{5-phenyl-2-oxo-tetrahydrofuran-3-}COOCH_2CH_3$$

1. $HO^-$, $H_2O$
2. $H^+$
3. heat

$$\text{5-phenyl-tetrahydrofuran-2-one}$$

The product is 4-phenylbutanolide. It has been prepared in 72% yield by this procedure.

**21.29** The first task is to convert acetic acid to ethyl chloroacetate.

$$\underset{\text{Acetic acid}}{CH_3\overset{O}{\overset{\|}{C}}OH} \xrightarrow[P]{Cl_2} \underset{\text{Chloroacetic acid}}{ClCH_2\overset{O}{\overset{\|}{C}}OH} \xrightarrow[H^+]{CH_3CH_2OH} \underset{\text{Ethyl chloroacetate}}{ClCH_2\overset{O}{\overset{\|}{C}}OCH_2CH_3}$$

Chlorination must precede esterification, because the Hell–Volhard–Zelinsky reaction requires a carboxylic acid, not an ester, as the starting material. The remaining step is a nucleophilic substitution reaction.

$$\underset{\text{Ethyl chloroacetate}}{ClCH_2\overset{O}{\overset{\|}{C}}OCH_2CH_3} \xrightarrow{NaCN} \underset{\text{Ethyl cyanoacetate}}{N{\equiv}CCH_2\overset{O}{\overset{\|}{C}}OCH_2CH_3}$$

**21.30** From the hint given in the problem, it can be seen that synthesis of 2-methyl-2-propyl-1,3-propanediol is required. This diol is obtained by a sequence involving dialkylation of diethyl malonate.

$$(CH_3CH_2CH_2)(H_3C)C(CH_2OCNH_2)_2\ [\text{C=O}] \Rightarrow (CH_3CH_2CH_2)(H_3C)C(CH_2OH)_2 \Rightarrow (CH_3CH_2CH_2)(H_3C)C(CO_2CH_2CH_3)_2$$

Begin the synthesis by dialkylation of diethyl malonate.

$$\underset{\text{Diethyl malonate}}{CH_2(COOCH_2CH_3)_2} \xrightarrow[\text{2. } CH_3Br,\ NaOCH_2CH_3]{\text{1. } CH_3CH_2CH_2Br,\ NaOCH_2CH_3} \underset{\text{Diethyl 2-methyl-2-propylmalonate}}{(CH_3CH_2CH_2)(H_3C)C(COOCH_2CH_3)_2}$$

Convert the ester functions to primary alcohols by reduction.

$$\underset{\text{Diethyl 2-methyl-2-propylmalonate}}{(CH_3CH_2CH_2)(H_3C)C(COOCH_2CH_3)_2} \xrightarrow[\text{2. } H_2O]{\text{1. } LiAlH_4} \underset{\text{2-Methyl-2-propyl-1,3-propanediol}}{(CH_3CH_2CH_2)(H_3C)C(CH_2OH)_2}$$

Conversion of the primary alcohol groups to carbamate esters completes the synthesis.

$$\underset{\text{2-Methyl-2-propyl-1,3-propanediol}}{(CH_3CH_2CH_2)(H_3C)C(CH_2OH)_2} \xrightarrow[\text{2. } NH_3,\ H_2O]{\text{1. } COCl_2} \underset{\text{Meprobamate}}{(CH_3CH_2CH_2)(H_3C)C(CH_2O\overset{O}{\overset{\|}{C}}NH_2)_2}$$

**21.31** The compound given in the problem contains three functionalities that can undergo acid-catalyzed hydrolysis: an acetal and two equivalent ester groups. Hydrolysis yields 3-oxo-1,1-cyclobutanedicarboxylic acid and 2 moles each of methanol and 2-propanol. The hydrolysis product is a malonic acid derivative that decarboxylates on heating. The final product of the reaction is 3-oxocyclobutanecarboxylic acid ($C_5H_6O_3$).

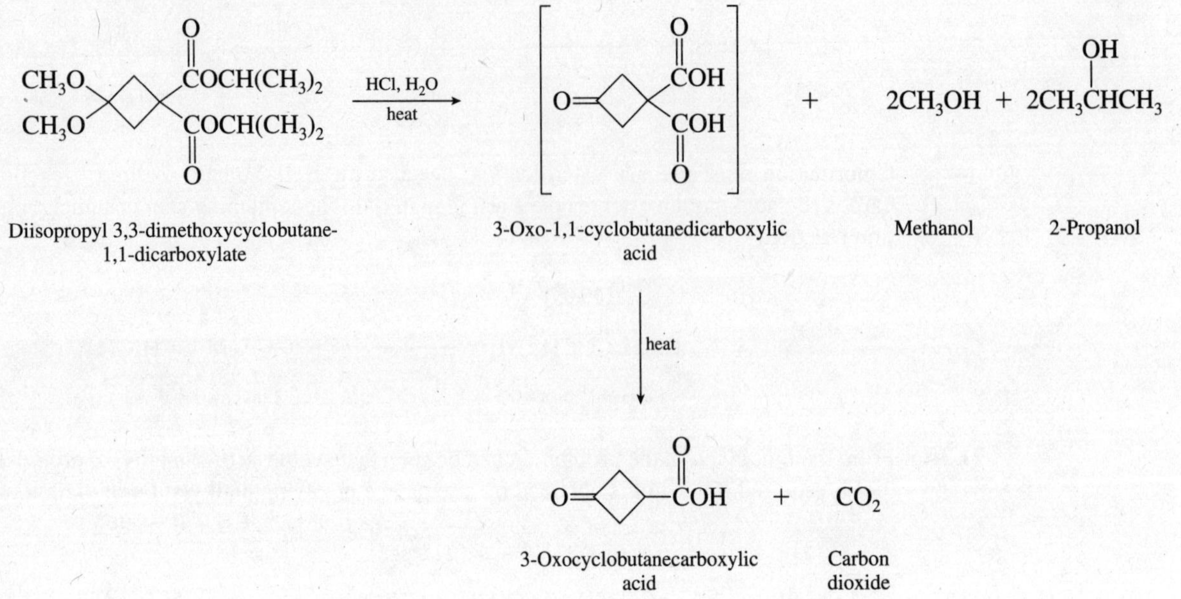

$$\text{Diisopropyl 3,3-dimethoxycyclobutane-1,1-dicarboxylate} \xrightarrow[\text{heat}]{HCl,\ H_2O} \left[\text{3-Oxo-1,1-cyclobutanedicarboxylic acid}\right] + 2CH_3OH\ (\text{Methanol}) + 2CH_3CH(OH)CH_3\ (\text{2-Propanol})$$

$$\text{3-Oxo-1,1-cyclobutanedicarboxylic acid} \xrightarrow{\text{heat}} \text{3-Oxocyclobutanecarboxylic acid} + CO_2\ (\text{Carbon dioxide})$$

## SELF-TEST

### PART A

**A-1.** Give the structure of the reactant, reagent, or product omitted from each of the following:

(*a*) $CH_3CH_2CH_2\overset{O}{\overset{\|}{C}}OCH_2CH_3 \xrightarrow[2.\ H_3O^+]{1.\ NaOCH_2CH_3} ?$

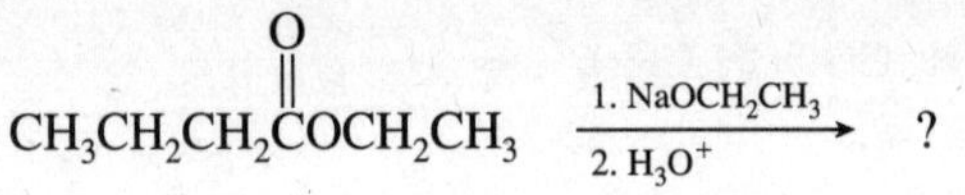

(*b*) $H\overset{O}{\overset{\|}{C}}OCH_2CH_3 + ? \xrightarrow[2.\ H_3O^+]{1.\ NaOCH_2CH_3} C_6H_5CH(CHO)\overset{O}{\overset{\|}{C}}OCH_2CH_3$

(*c*) 3-Chloro-1,1-bis(ethoxycarbonyl)cyclobutane ($COCH_2CH_3$ groups) $\xrightarrow[2.\ H_3O^+;\ 3.\ \text{heat}]{1.\ HO^-,\ H_2O}$ ? (two isomeric products; $C_5H_7ClO_2$)

(*d*) $(CH_3CH_2OOC)_2CH_2 + H_2C{=}CH\overset{O}{\overset{\|}{C}}OCH_2CH_3 \xrightarrow[\text{ethanol}]{NaOCH_2CH_3} ?$

(*e*) $CH_3\overset{O}{\overset{\|}{C}}CH_2\overset{O}{\overset{\|}{C}}OCH_2CH_3 \xrightarrow[2.\ C_6H_5CH_2Br]{1.\ NaOCH_2CH_3} ?$

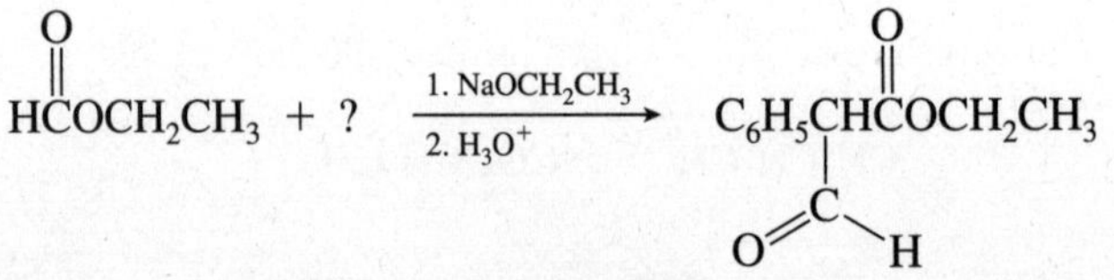

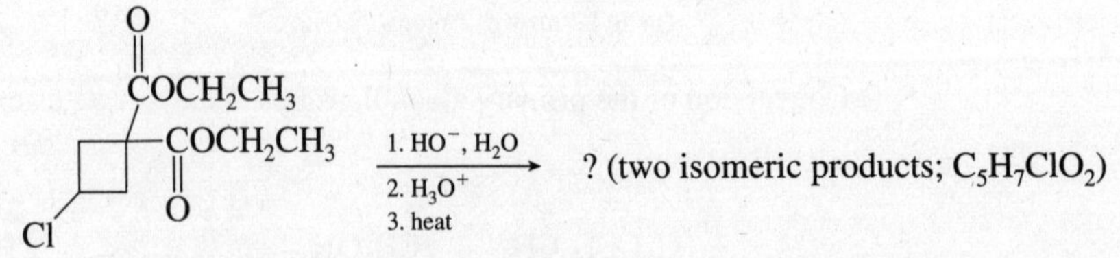

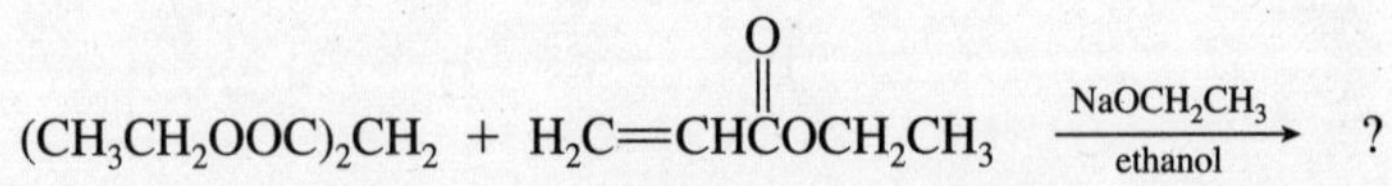

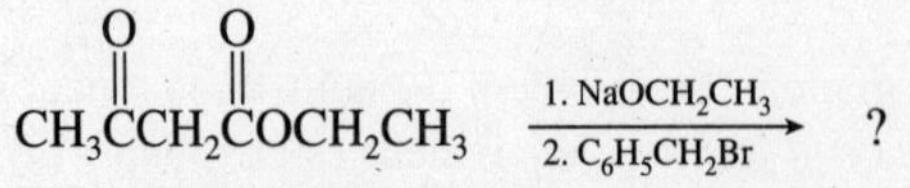

(*f*) Product of part (*e*) $\xrightarrow{?}$ $C_6H_5CH_2CH_2\overset{O}{\overset{\|}{C}}CH_3$

(*g*) $CH_3\overset{O}{\overset{\|}{C}}\underset{CO_2H}{\underset{|}{C}H}CH_2CO_2H \xrightarrow{\text{heat}} CO_2 + ?$

**A-2.** Provide the correct structures of compounds A through E in the following reaction sequences:

(*a*) A $\xrightarrow[\text{2. } H_3O^+]{\text{1. } NaOCH_2CH_3}$ (2-oxocyclopentyl)$\overset{O}{\overset{\|}{C}}OCH_2CH_3$ $\xrightarrow[\text{2. } CH_3CH_2I]{\text{1. } NaOCH_2CH_3}$ B $\xrightarrow[\text{2. } H_3O^+ \text{ 3. heat}]{\text{1. } HO^-, H_2O}$ C

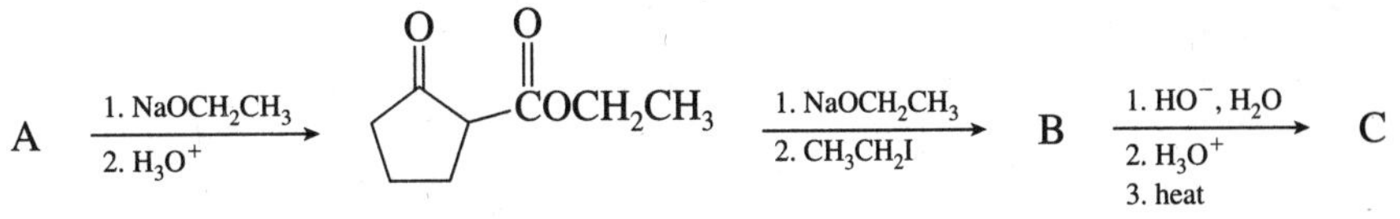

(*b*) $CH_3CH_2CH_2\overset{O}{\overset{\|}{C}}OCH_2CH_3 \xrightarrow[\text{2. } H_3O^+]{\text{1. } NaOCH_2CH_3}$ D $\xrightarrow[\text{2. } H_3O^+ \text{ 3. heat}]{\text{1. } HO^-, H_2O}$ E + $CO_2$

**A-3.** Give a series of steps that will enable preparation of each of the following compounds from the starting material(s) given and any other necessary reagents:

(*a*) $CH_3\overset{O}{\overset{\|}{C}}CH_2CH_2\overset{O}{\overset{\|}{C}}OH$ from ethyl acetoacetate

(*b*) $C_6H_5\overset{O}{\overset{\|}{C}}CH_2CH_2CH_2\overset{O}{\overset{\|}{C}}CH_3$ from $C_6H_5\overset{O}{\overset{\|}{C}}CH_3$ and diethyl carbonate

**A-4.** Write a stepwise mechanism for the reaction of ethyl propanoate with sodium ethoxide in ethanol.

**A-5.** Ethyl 2-methylpropanoate does not undergo a Claisen condensation, whereas ethyl 3-methylbutanoate does. Provide a mechanistic explanation for this observation.

## PART B

**B-1.** Which of the following compounds is the strongest acid?

(*a*) $HCO_2CH_2CH_3$
(*b*) $CH_3CH_2O_2CCH_2CO_2CH_2CH_3$
(*c*) $CH_3CH_2O_2CCH_2CH_2CO_2CH_2CH_3$
(*d*) $CH_3CO_2CH_2CH_3$

**B-2.** Which of the following will yield a ketone and carbon dioxide following saponification, acidification, and heating?

(*a*) $CH_3CH_2\underset{\overset{|}{C(=O)OCH_2CH_3}}{CH}CH_2\overset{O}{\overset{\|}{C}}CH_3$

(*b*) $CH_3CH_2\underset{\overset{|}{C(=O)OCH_2CH_3}}{CH}\overset{O}{\overset{\|}{C}}OCH_2CH_3$

(*c*) $CH_3CH_2\underset{\overset{|}{C(=O)CH_2CH_3}}{CH}\overset{O}{\overset{\|}{C}}CH_3$

(*d*) $CH_3CH_2\underset{\overset{|}{C(=O)OCH_2CH_3}}{CH}\overset{O}{\overset{\|}{C}}CH_3$

**B-3.** Which of the following keto esters is *not* likely to have been prepared by a Claisen condensation?

(*a*) $CH_3CH_2C(=O)CH(CH_3)C(=O)OCH_2CH_3$

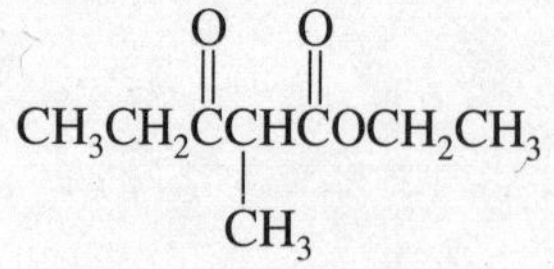

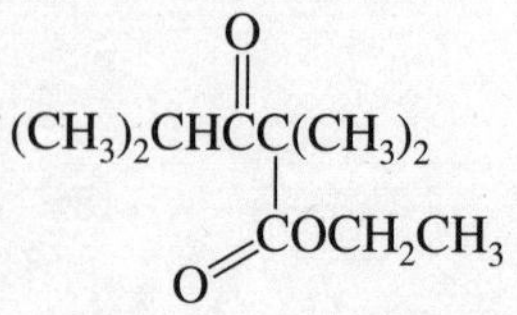

(*b*) $C_6H_5C(=O)CH(CH_3)C(=O)OCH_2CH_3$

(*c*) $(CH_3)_2CHC(=O)C(CH_3)_2C(=O)OCH_2CH_3$

(*d*) $(CH_3)_2CHCH_2C(=O)CH(CH(CH_3)_2)C(=O)OCH_2CH_3$

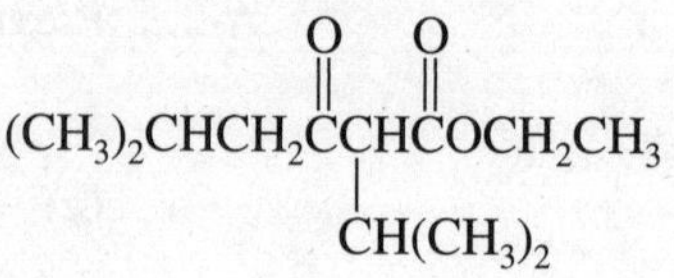

**B-4.** Dieckmann cyclization of $CH_3CH_2OC(=O)(CH_2)_5C(=O)OCH_2CH_3$ will yield

(*a*) cyclohexanone with $C(=O)OCH_2CH_3$ at C-3

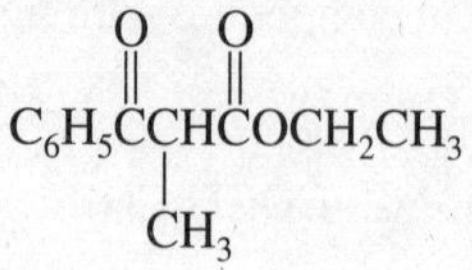

(*b*) cyclohexanone with $C(=O)OCH_2CH_3$ at C-2

(*c*) cyclopentane with two $C(=O)OCH_2CH_3$ groups on adjacent carbons

(*d*) cyclohexane with $C(=O)OCH_2CH_3$

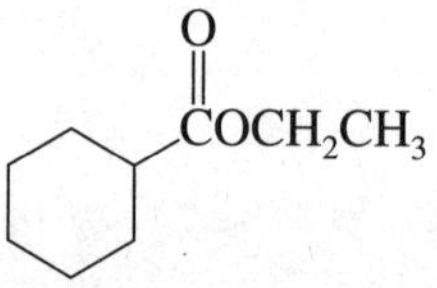

**B-5.** What is the final product of this sequence?

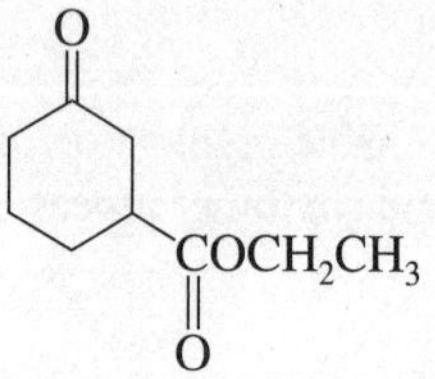

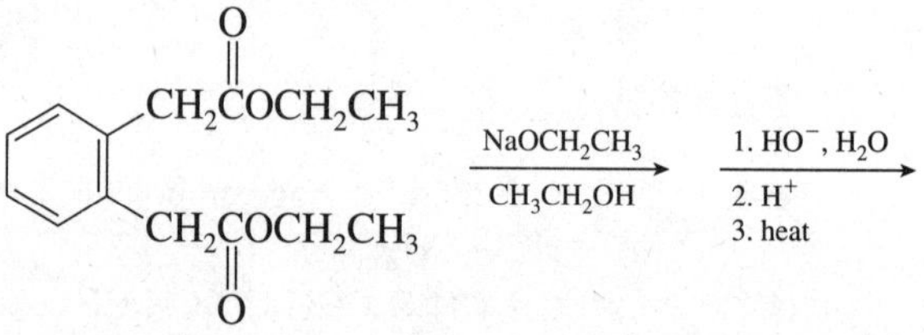

1,2-benzene with two $CH_2C(=O)OCH_2CH_3$ groups $\xrightarrow[CH_3CH_2OH]{NaOCH_2CH_3}$ $\xrightarrow[\text{2. } H^+ \text{ 3. heat}]{\text{1. } HO^-,\ H_2O}$

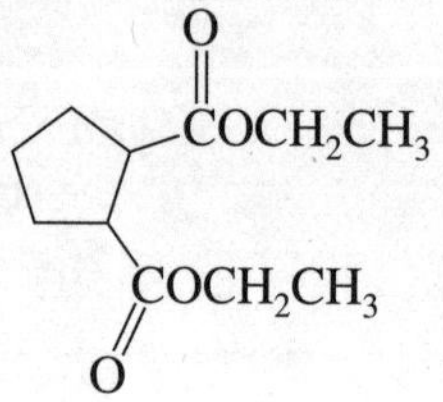

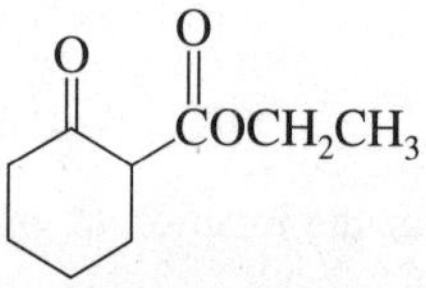

(*a*) 1-indanone

(*b*) 2-tetralone

(*c*) 2-indanone

(*d*) benzene with $C(=O)H$ and $CH{=}CH_2$ ortho

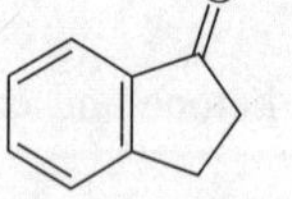

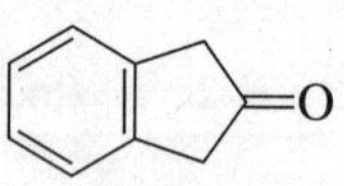

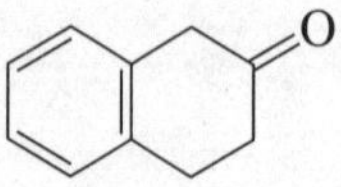

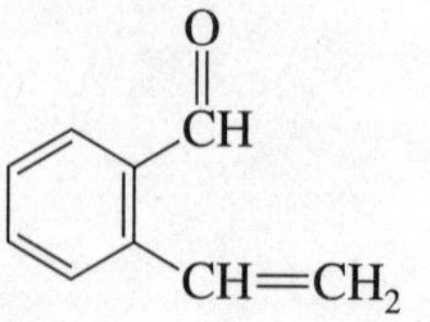

**B-6.** What is the final product of the following sequence of reactions?

$$CH_2(CO_2CH_2CH_3)_2 + (CH_3)_2C{=}CH\overset{O}{\overset{\|}{C}}CH_3 \xrightarrow[\text{ethanol}]{NaOCH_2CH_3} \xrightarrow[\text{2. } H^+ \text{ 3. heat}]{\text{1. KOH}} ?$$

(*a*) O O $OCH_2CH_3$  (*c*) O O OH

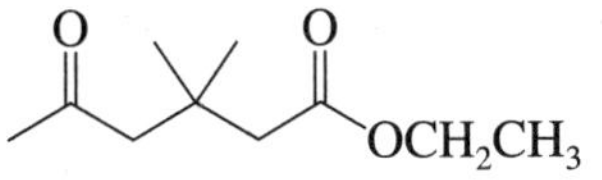

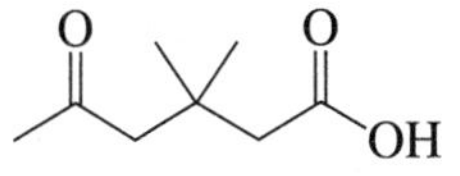

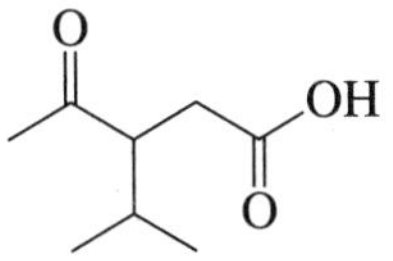

(*b*) O OH O  (*d*) O OH O

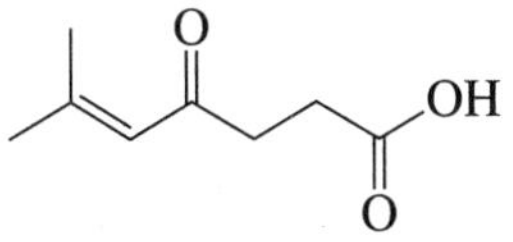

**B-7.** Which of the following would be a suitable candidate for preparation by a mixed Claisen condensation?

(*a*) $CH_3CH_2\overset{O}{\overset{\|}{C}}CH_2\overset{O}{\overset{\|}{C}}OCH_2CH_3$  (*c*) $C_6H_5CH_2\overset{O}{\overset{\|}{C}}CH_2\overset{O}{\overset{\|}{C}}OCH_2CH_3$

(*b*) $C_6H_5\overset{O}{\overset{\|}{C}}CH_2\overset{O}{\overset{\|}{C}}OCH_2CH_3$  (*d*) $C_6H_5\overset{O}{\overset{\|}{C}}{-}\underset{CH_3}{\overset{CH_3}{C}}{-}\overset{O}{\overset{\|}{C}}OCH_2CH_3$

**B-8.** What is the major product of the following reaction?

$$CH_3CH_2\overset{O}{\overset{\|}{C}}OCH_3 + H\overset{O}{\overset{\|}{C}}OCH_3 \xrightarrow[\text{2. } H^+]{\text{1. } NaOCH_3} ?$$

(*a*) $CH_3O\overset{O}{\overset{\|}{C}}\underset{CH_3}{CH}\overset{O}{\overset{\|}{C}}OCH_3$  (*c*) $HOCH_2\underset{CH_3}{CH}\overset{O}{\overset{\|}{C}}OCH_3$  (*e*) $CH_3O\underset{CH_3}{CH}\overset{O}{\overset{\|}{C}}OCH_3$

(*b*) $H\overset{O}{\overset{\|}{C}}O\underset{CH_3}{CH}\overset{O}{\overset{\|}{C}}OCH_3$  (*d*) $H\overset{O}{\overset{\|}{C}}\underset{CH_3}{CH}\overset{O}{\overset{\|}{C}}OCH_3$

# CHAPTER 22

## AMINES

## SOLUTIONS TO TEXT PROBLEMS

**22.1** (*b*) The amino and phenyl groups are both attached to C-1 of an ethyl group.

$$\begin{array}{c} C_6H_5CHCH_3 \\ | \\ NH_2 \end{array}$$

1-Phenylethylamine, or
1-phenylethanamine

(*c*)

$$H_2C{=}CHCH_2NH_2$$

Allylamine, or
2-propen-1-amine

**22.2** *N*,*N*-Dimethylcycloheptylamine may also be named as a dimethyl derivative of cycloheptanamine.

$$C_7H_{13}{-}N(CH_3)_2$$

*N*,*N*-Dimethylcycloheptanamine

**22.3** Three substituents are attached to the nitrogen atom; the amine is tertiary. In alphabetical order, the substituents present on the aniline nucleus are ethyl, isopropyl, and methyl. Their positions are specified as *N*-ethyl, 4-isopropyl, and *N*-methyl.

$$(CH_3)_2CH{-}C_6H_4{-}N(CH_3){-}CH_2CH_3$$

*N*-Ethyl-4-isopropyl-*N*-methylaniline

**22.4** The electron-donating amino group and the electron-withdrawing nitro group are directly conjugated in *p*-nitroaniline. The planar geometry of *p*-nitroaniline suggests that the delocalized resonance form shown is a major contributor to the structure of the compound.

**22.5** The p$K_b$ of an amine is related to the equilibrium constant $K_b$ by

$$pK_b = -\log K_b$$

The p$K_b$ of quinine is therefore

$$pK_b = -\log (1 \times 10^{-6}) = 6$$

the values of $K_b$ and p$K_b$ for an amine and $K_a$ and p$K_a$ of its conjugate acid are given by

$$K_a \times K_b = 1 \times 10^{-14}$$

and

$$pK_a + pK_b = 14$$

The values of $K_a$ and p$K_a$ for the conjugate acid of quinine are therefore

$$K_a = \frac{10^{-14}}{K_b} = \frac{1 \times 10^{-14}}{1 \times 10^{-6}} = 1 \times 10^{-8}$$

and

$$pK_a = 14 - pK_b = 14 - 6 = 8$$

**22.6** The Henderson–Hasselbalch equation described in Section 19.4 can be applied to bases such as amines, as well as carboxylic acids. The ratio $[CH_3NH_3^+]/[CH_3NH_2]$ is given by

$$\frac{[CH_3NH_3^+]}{[CH_3NH_2]} = \frac{[H^+]}{K_a}$$

The ionization constant of methylammonium ion is given in the text as $2 \times 10^{-11}$. At pH = 7 the hydrogen ion concentration is $1 \times 10^{-7}$. Therefore

$$\frac{[CH_3NH_3^+]}{[CH_3NH_2]} = \frac{1 \times 10^{-7}}{2 \times 10^{-11}} = 5 \times 10^3$$

**22.7** Nitrogen is attached directly to the aromatic ring in tetrahydroquinoline, making it an arylamine, and the nitrogen lone pair is delocalized into the $\pi$ system of the aromatic ring. It is less basic than tetrahydroisoquinoline, in which the nitrogen is insulated from the ring by an $sp^3$-hybridized carbon.

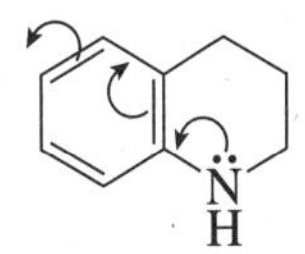

Tetrahydroisoquinoline (an alkylamine): more basic, $K_b$ $2.5 \times 10^{-5}$ (p$K_b$ 4.6)

Tetrahydroquinoline (an arylamine): less basic, $K_b$ $1.0 \times 10^{-9}$ (p$K_b$ 9.0)

See *Learning By Modeling* for the calculated charges on nitrogen.

**22.8** (*b*) An acetyl group attached directly to nitrogen as in acetanilide delocalizes the nitrogen lone pair into the carbonyl group. Amides are weaker bases than amines.

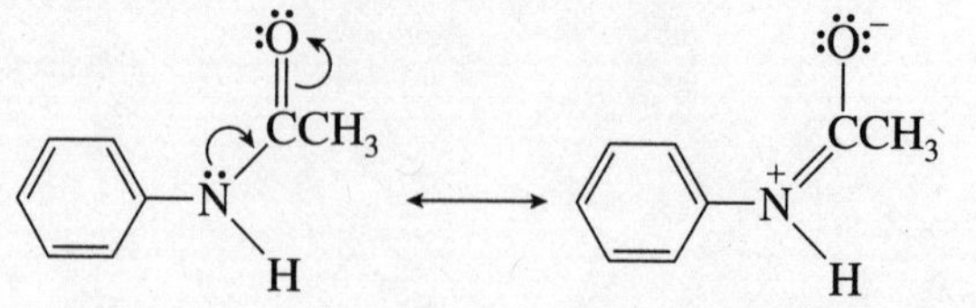

(*c*) An acetyl group in a position para to an amine function is conjugated to it and delocalizes the nitrogen lone pair.

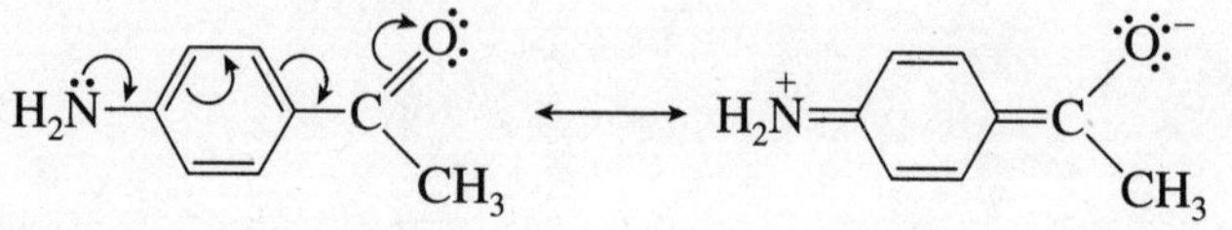

**22.9** The reaction that leads to allylamine is nucleophilic substitution by ammonia on allyl chloride.

$$H_2C{=}CHCH_2Cl + 2NH_3 \longrightarrow H_2C{=}CHCH_2NH_2 + NH_4Cl$$

Allyl chloride; Ammonia; Allylamine; Ammonium chloride

Allyl chloride is prepared by free-radical chlorination of propene (see text page 371).

$$H_2C{=}CHCH_3 + Cl_2 \xrightarrow{400°C} H_2C{=}CHCH_2Cl + HCl$$

Propene; Chlorine; Allyl chloride; Hydrogen chloride

**22.10** (*b*) Isobutylamine is $(CH_3)_2CHCH_2NH_2$. It is a primary amine of the type $RCH_2NH_2$ and can be prepared from a primary alkyl halide by the Gabriel synthesis.

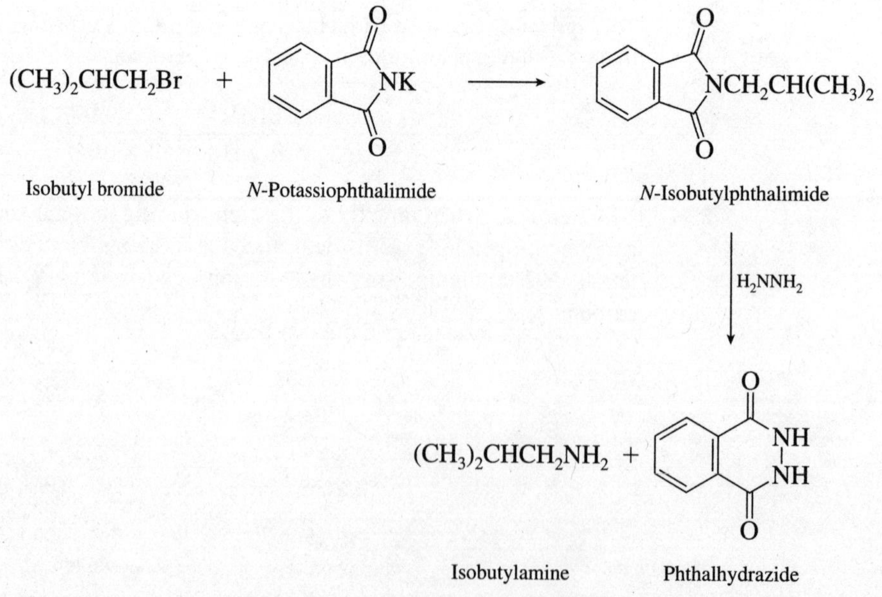

$(CH_3)_2CHCH_2Br$ + N-Potassiophthalimide → $NCH_2CH(CH_3)_2$

Isobutyl bromide; *N*-Potassiophthalimide; *N*-Isobutylphthalimide

$H_2NNH_2$

$(CH_3)_2CHCH_2NH_2$ + Phthalhydrazide

Isobutylamine; Phthalhydrazide

(*c*) Although *tert*-butylamine $(CH_3)_3CNH_2$ is a primary amine, it cannot be prepared by the Gabriel method, because it would require an $S_N2$ reaction on a tertiary alkyl halide in the first step. Elimination occurs instead.

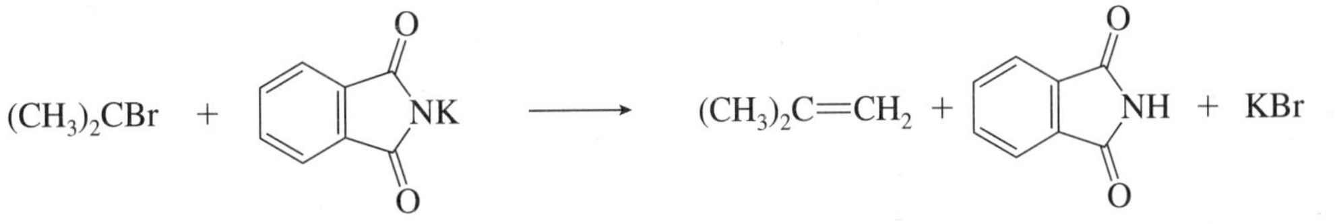

$(CH_3)_3CBr$ + N-Potassiophthalimide ⟶ $(CH_3)_2C{=}CH_2$ + Phthalimide + KBr

*tert*-Butyl bromide | *N*-Potassiophthalimide | 2-Methylpropene | Phthalimide | Potassium bromide

(*d*) The preparation of 2-phenylethylamine by the Gabriel synthesis has been described in the chemical literature.

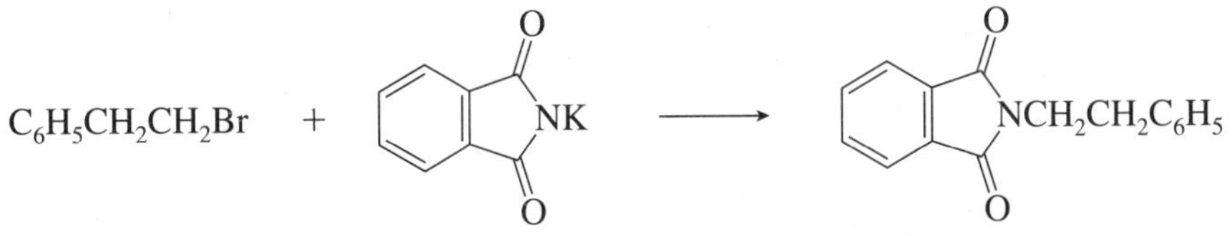

$C_6H_5CH_2CH_2Br$ + NK ⟶ $NCH_2CH_2C_6H_5$

2-Phenylethyl bromide | *N*-Potassiophthalimide | *N*-(2-Phenylethyl)phthalimide

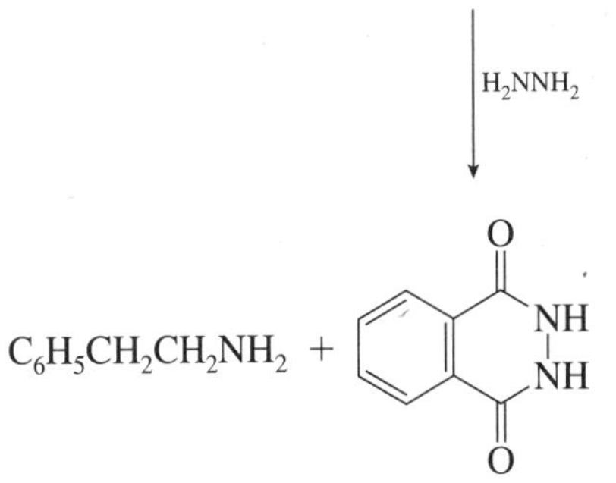

$\downarrow H_2NNH_2$

$C_6H_5CH_2CH_2NH_2$ + Phthalhydrazide

2-Phenylethylamine | Phthalhydrazide

(*e*) The Gabriel synthesis leads to primary amines; *N*-methylbenzylamine is a secondary amine and cannot be prepared by this method.

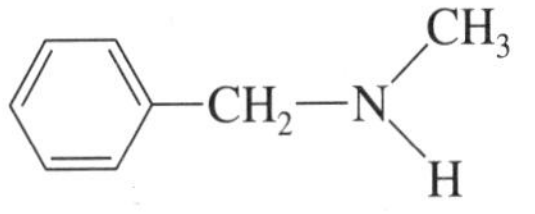

$C_6H_5$—$CH_2$—N($CH_3$)H

*N*-Methylbenzylamine
(two carbon substituents on nitrogen; a secondary amine)

(*f*) Aniline cannot be prepared by the Gabriel method. Aryl halides do not undergo nucleophilic substitution under these conditions.

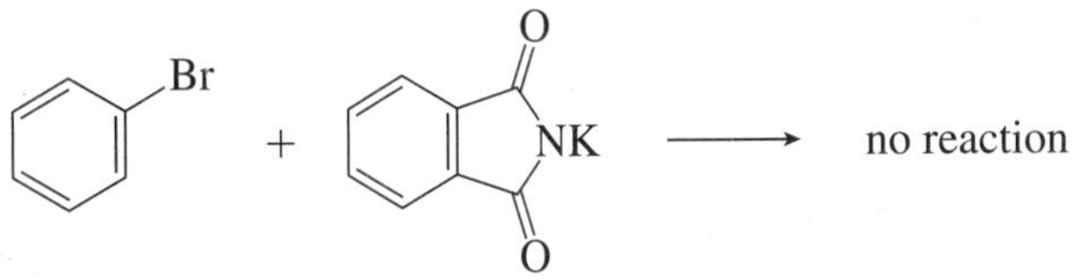

Bromobenzene + *N*-Potassiophthalimide ⟶ no reaction

Bromobenzene | *N*-Potassiophthalimide

**22.11** For each part of this problem, keep in mind that aromatic amines are derived by reduction of the corresponding aromatic nitro compound. Each synthesis should be approached from the standpoint of how best to prepare the necessary nitroaromatic compound.

$$\text{Ar—NH}_2 \Rightarrow \text{Ar—NO}_2 \Rightarrow \text{Ar—H}$$

(Ar = substituted aromatic ring)

(*b*) The para isomer of isopropylaniline may be prepared by a procedure analogous to that used for its ortho isomer in part (*a*).

Benzene $\xrightarrow[AlCl_3]{(CH_3)_2CHCl}$ Isopropybenzene $\xrightarrow[H_2SO_4]{HNO_3}$ *o*-Isopropylnitrobenzene + *p*-Isopropylnitrobenzene

After separating the ortho, para mixture by distillation, the nitro group of *p*-isopropylnitrobenzene is reduced to yield the desired *p*-isopropylaniline.

$p$-$CH(CH_3)_2C_6H_4NO_2$ $\xrightarrow[\text{1. Fe, HCl; 2. HO}^- \text{ or 1. Sn, HCl; 2. HO}^-]{H_2\text{, Ni; or}}$ $p$-$CH(CH_3)_2C_6H_4NH_2$

(*c*) The target compound is the reduction product of 1-isopropyl-2,4-dinitrobenzene.

1-Isopropyl-2,4-dinitrobenzene $\xrightarrow{\text{reduce}}$ 4-Isopropyl-1,3-benzenediamine

This reduction is carried out in the same way as reduction of an arene that contains only a single nitro group. In this case hydrogenation over a nickel catalyst gave the desired product in 90% yield.

The starting dinitro compound is prepared by nitration of isopropylbenzene.

Isopropylbenzene $\xrightarrow[80°C]{HNO_3, H_2SO_4}$ 1-Isopropyl-2,4-dinitrobenzene (43%)

(*d*) The conversion of *p*-chloronitrobenzene to *p*-chloroaniline was cited as an example in the text to illustrate reduction of aromatic nitro compounds to arylamines. *p*-Chloronitrobenzene is prepared by nitration of chlorobenzene.

$$\text{Benzene} \xrightarrow[FeCl_3]{Cl_2} \text{Chlorobenzene} \xrightarrow[H_2SO_4]{HNO_3} o\text{-Chloronitrobenzene} + p\text{-Chloronitrobenzene}$$

The para isomer accounts for 69% of the product in this reaction (30% is ortho, 1% meta). Separation of *p*-chloronitrobenzene and its reduction completes the synthesis.

$$p\text{-Chloronitrobenzene} \xrightarrow[\text{H}_2\text{, catalyst; or 1. Sn, HCl; 2. HO}^-]{\text{1. Fe, HCl; 2. HO}^- \text{ or}} p\text{-Chloroaniline}$$

Chlorination of nitrobenzene would not be a suitable route to the required intermediate, because it would produce mainly *m*-chloronitrobenzene.

(*e*) The synthesis of *m*-aminoacetophenone may be carried out by the scheme shown:

$$\text{Benzene} \xrightarrow[AlCl_3]{CH_3\overset{O}{\overset{\|}{C}}Cl} \text{Acetophenone} \xrightarrow[H_2SO_4]{HNO_3} m\text{-Nitroacetophenone} \xrightarrow{\text{reduce}} m\text{-Aminoacetophenone}$$

The acetyl group is attached to the ring by Friedel–Crafts acylation. It is a meta director, and its nitration gives the proper orientation of substituents. The order of the first two steps cannot be reversed, because Friedel–Crafts acylation of nitrobenzene is not possible (Section 12.16). Once prepared, *m*-nitroacetophenone can be reduced to *m*-nitroaniline by any of a number of reagents. Indeed, all three reducing combinations described in the text have been employed for this transformation.

| | Reducing agent | Yield (%) |
|---|---|---|
| *m*-Nitroacetophenone | $H_2$, Pt | 94 |
| ↓ | Fe, HCl | 84 |
| *m*-Aminoacetophenone | Sn, HCl | 82 |

**22.12** (*b*) Dibenzylamine is a secondary amine and can be prepared by reductive amination of benzaldehyde with benzylamine.

$$\underset{\text{Benzaldehyde}}{C_6H_5\overset{O}{\overset{\|}{C}}H} + \underset{\text{Benzylamine}}{C_6H_5CH_2NH_2} \xrightarrow{H_2,\ Ni} \underset{\text{Dibenzylamine}}{C_6H_5CH_2NHCH_2C_6H_5}$$

(*c*) *N,N*-Dimethylbenzylamine is a tertiary amine. Its preparation from benzaldehyde requires dimethylamine, a secondary amine.

$$\underset{\text{Benzaldehyde}}{C_6H_5\overset{O}{\overset{\|}{C}}H} + \underset{\text{Dimethylamine}}{(CH_3)_2NH} \xrightarrow{H_2,\ Ni} \underset{N,N\text{-Dimethylbenzylamine}}{C_6H_5CH_2N(CH_3)_2}$$

(*d*) The preparation of *N*-butylpiperidine by reductive amination is described in the text in Section 22.11. An analogous procedure is used to prepare *N*-benzylpiperidine.

$$\underset{\text{Benzaldehyde}}{C_6H_5\overset{O}{\overset{\|}{C}}H} + \underset{\text{Piperidine}}{\text{(piperidine, N–H)}} \xrightarrow{H_2,\ Ni} \underset{N\text{-Benzylpiperidine}}{C_6H_5CH_2\text{—N(piperidine ring)}}$$

**22.13** (*b*) First identify the available $\beta$ hydrogens. Elimination must involve a proton from the carbon atom adjacent to the one that bears the nitrogen.

$$(CH_3)_3C\overset{\beta}{C}H_2\text{—}\underset{^+N(CH_3)_3}{\overset{\overset{\beta}{CH_3}}{C}}\text{—}\overset{\beta}{C}H_3$$

A methylene group; Two equivalent methyl groups

It is a proton from one of the methyl groups, rather than one from the more sterically hindered methylene, that is lost on elimination.

$$(CH_3)_3CCH_2\text{—}\underset{\overset{+}{N}(CH_3)_3}{\overset{CH_3}{C}}\text{—}CH_2\text{—}H\ \ {}^-{:}\ddot{O}H \longrightarrow (CH_3)_3CCH_2\overset{CH_3}{C}{=}CH_2 + (CH_3)_3N{:}$$

(1,1,3,3-Tetramethylbutyl)-trimethylammonium hydroxide → 2,4,4-Trimethyl-1-pentene (only alkene formed, 70% isolated yield) + Trimethylamine

(*c*) The base may abstract a proton from either of two $\beta$ carbons. Deprotonation of the $\beta$ methyl carbon yields ethylene.

$$H\ddot{O}{:}^-\ H\text{—}CH_2\text{—}CH_2\text{—}\underset{CH_3}{\overset{CH_3}{\overset{+}{N}}}CH_2CH_2CH_2CH_3 \xrightarrow[(-H_2O)]{\text{heat}} H_2C{=}CH_2 + (CH_3)_2\ddot{N}CH_2CH_2CH_2CH_3$$

*N*-Ethyl-*N,N*-dimethylbutylammonium hydroxide → Ethylene + *N,N*-Dimethylbutylamine

Deprotonation of the $\beta$ methylene carbon yields 1-butene.

$$CH_3CH_2\text{—}\underset{CH_3}{\overset{CH_3}{\overset{+}{N}}}\text{—}CH_2\text{—}\underset{H\ \leftarrow\ {:}\ddot{O}H}{C}HCH_2CH_3 \xrightarrow[(-H_2O)]{\text{heat}} CH_3CH_2\ddot{N}(CH_3)_2 + H_2C{=}CHCH_2CH_3$$

*N*-Ethyl-*N,N*-dimethylbutylammonium hydroxide → *N,N*-Dimethylethylamine + 1-Butene

The preferred order of proton removal in Hofmann elimination reactions is $\beta$ $CH_3$ > $\beta$ $CH_2$ > $\beta$ CH. Ethylene is the major alkene formed, the observed ratio of ethylene to 1-butene being 98 : 2.

**22.14** (*b*) The pattern of substituents in 2,4-dinitroaniline suggests that they can be introduced by dinitration. Since nitration of aniline itself is not practical, the amino group must be protected by conversion to its *N*-acetyl derivative.

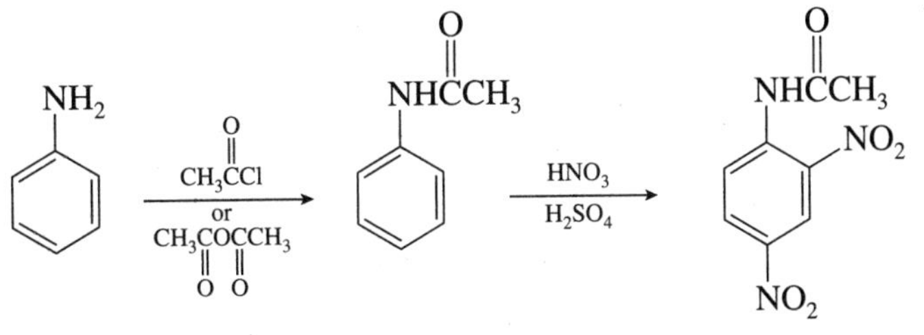

$CH_3CCl$ (O) or $CH_3COCCH_3$ (O O); $HNO_3$, $H_2SO_4$

Aniline → Acetanilide → 2,4-Dinitroacetanilide

Hydrolysis of the amide bond in 2,4-dinitroacetanilide furnishes the desired 2,4-dinitroaniline.

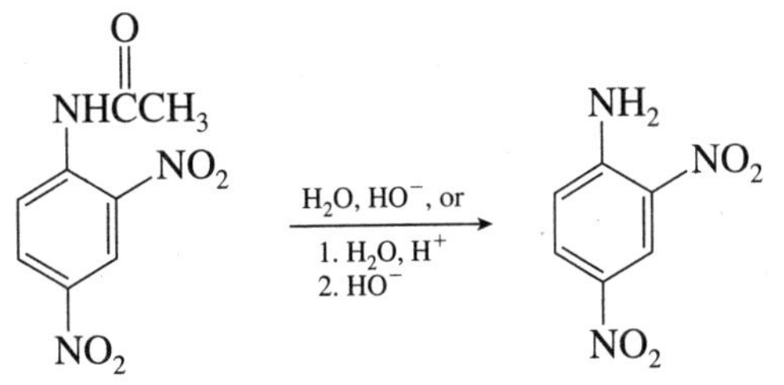

$H_2O$, $HO^-$, or 1. $H_2O$, $H^+$ 2. $HO^-$

2,4-Dinitroacetanilide → 2,4-Dinitroaniline

(*c*) Retrosynthetically, *p*-aminoacetanilide may be derived from *p*-nitroacetanilide.

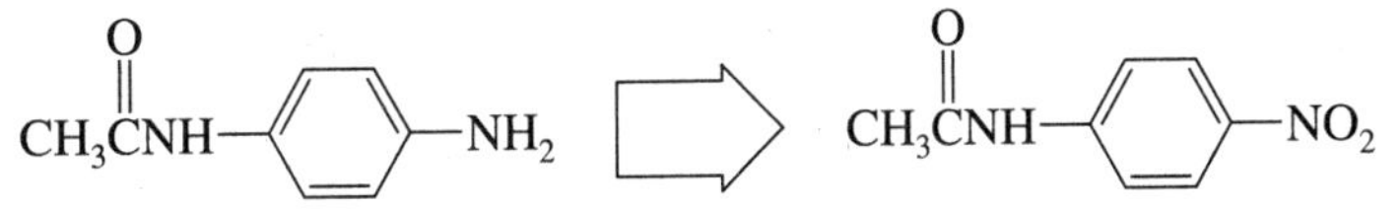

*p*-Aminoacetanilide ⇒ *p*-Nitroacetanilide

This suggests the sequence

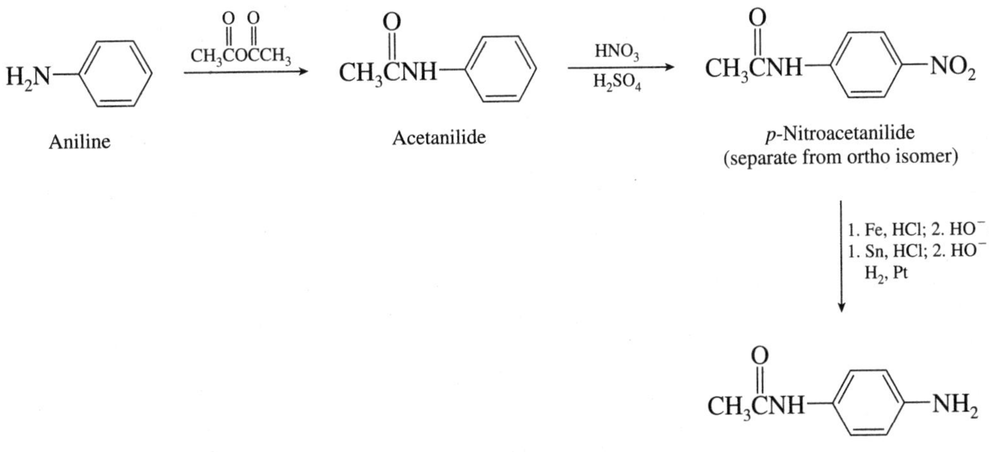

$CH_3COCCH_3$ (O O); $HNO_3$, $H_2SO_4$

Aniline → Acetanilide → *p*-Nitroacetanilide (separate from ortho isomer)

1. Fe, HCl; 2. $HO^-$ or 1. Sn, HCl; 2. $HO^-$ or $H_2$, Pt

→ *p*-Aminoacetanilide

**22.15** The principal resonance forms of *N*-nitrosodimethylamine are

$$(H_3C)_2\ddot{N}-\ddot{N}=\ddot{O}: \longleftrightarrow (H_3C)_2\overset{+}{N}=\ddot{N}-\ddot{O}:^-$$

All atoms (except hydrogen) have octets of electrons in each of these structures. Other resonance forms are less stable because they do not have a full complement of electrons around each atom.

**22.16** Deamination of 1,1-dimethylpropylamine gives products that result from 1,1-dimethylpropyl cation. Because 2,2-dimethylpropylamine gives the same products, it is likely that 1,1-dimethylpropyl cation is formed from 2,2-dimethylpropylamine by way of its diazonium ion. A carbocation rearrangement is indicated.

$$CH_3C(CH_3)_2CH_2NH_2 \xrightarrow{HONO} CH_3C(CH_3)_2-CH_2-\overset{+}{N}\equiv N \xrightarrow{-N_2} CH_3\overset{+}{C}(CH_3)CH_2CH_3$$

2,2-Dimethylpropylamine; 2,2-Dimethylpropyldiazonium ion; 1,1-Dimethylpropyl cation

Once formed, 1,1-dimethylpropyl cation loses a proton to form an alkene or is captured by water to give an alcohol.

$$CH_3\overset{+}{C}(CH_3)CH_2CH_3 \xrightarrow{-H^+} H_2C=C(CH_3)CH_2CH_3 + (CH_3)_2C=CHCH_3$$

1,1-Dimethylpropyl cation → 2-Methyl-1-butene + 2-Methyl-2-butene

$$CH_3\overset{+}{C}(CH_3)CH_2CH_3 \xrightarrow{H_2O} (CH_3)_2C(OH)CH_2CH_3$$

2-Methyl-2-butanol

**22.17** Phenols may be prepared by diazotization of the corresponding aniline derivative. The problem simplifies itself, therefore, to the preparation of *m*-bromoaniline. Recognizing that arylamines are ultimately derived from nitroarenes, we derive the retrosynthetic sequence of intermediates:

*m*-Bromophenol ⇒ *m*-Bromoaniline ⇒ *m*-Bromonitrobenzene ⇒ Nitrobenzene

The desired reaction sequence is straightforward, using reactions that were discussed previously in the text.

Benzene $\xrightarrow[H_2SO_4]{HNO_3}$ nitrobenzene $\xrightarrow[Fe]{Br_2}$ *m*-bromonitrobenzene $\xrightarrow[2.\ NaOH]{1.\ Fe,\ HCl}$ *m*-bromoaniline $\xrightarrow[2.\ H_2O,\ heat]{1.\ NaNO_2,\ H_2SO_4,\ H_2O,\ 0\text{–}5°C}$ *m*-bromophenol

**22.18** The key to this problem is to recognize that the iodine substituent in *m*-bromoiodobenzene is derived from an arylamine by diazotization.

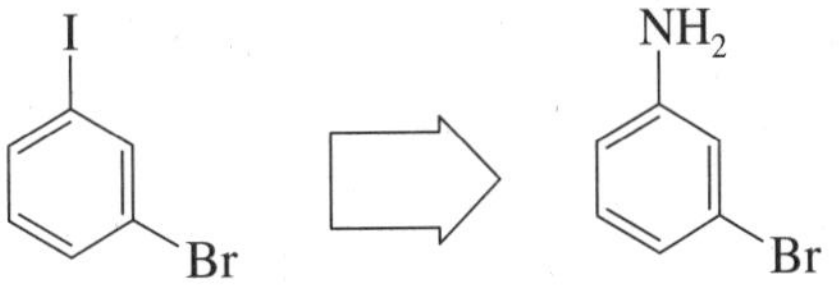

*m*-Bromoiodobenzene ⟹ *m*-Bromoaniline

The preparation of *m*-bromoaniline from benzene has been described in Problem 22.17. All that remains is to write the equation for its conversion to *m*-bromoiodobenzene.

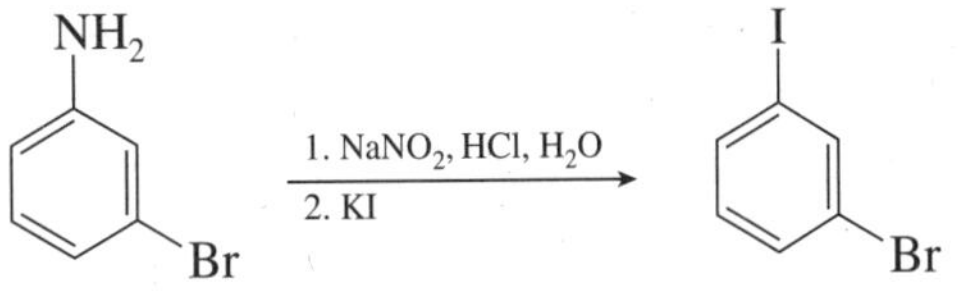

$$m\text{-Bromoaniline} \xrightarrow[\text{2. KI}]{\text{1. NaNO}_2\text{, HCl, H}_2\text{O}} m\text{-Bromoiodobenzene}$$

**22.19** The final step in the preparation of ethyl *m*-fluorophenyl ketone is shown in the text example immediately preceding this problem, therefore all that is necessary is to describe the preparation of *m*-aminophenyl ethyl ketone.

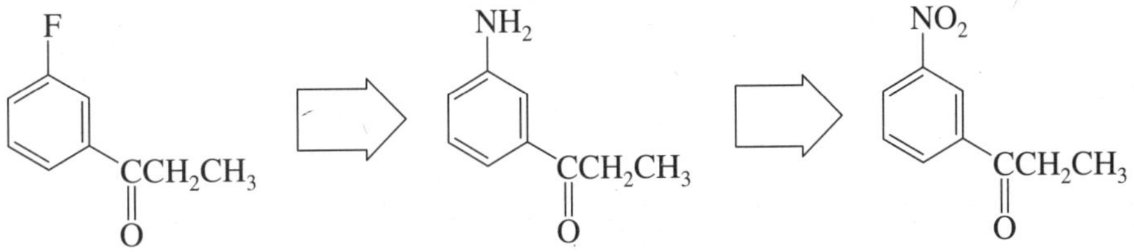

Ethyl *m*-fluorophenyl ketone ⟹ *m*-Aminophenyl ethyl ketone ⟹ Ethyl *m*-nitrophenyl ketone

Recalling that arylamines are normally prepared by reduction of nitroarenes, we see that ethyl *m*-nitrophenyl ketone is a pivotal synthetic intermediate. It is prepared by nitration of ethyl phenyl ketone, which is analogous to nitration of acetophenone, shown in Section 12.16. The preparation of ethyl phenyl ketone by Friedel–Crafts acylation of benzene is shown in Section 12.7.

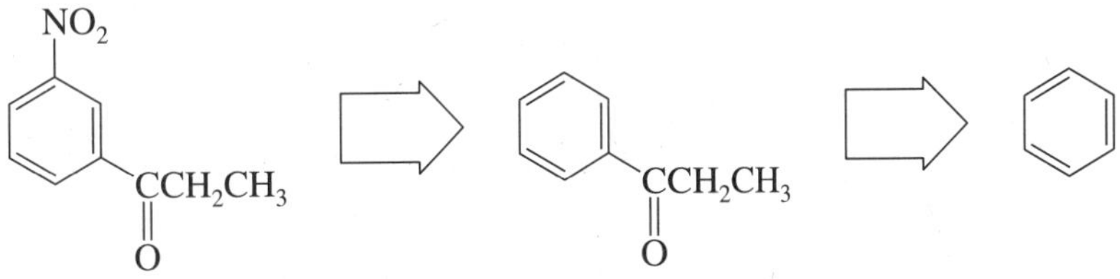

Ethyl *m*-nitrophenyl ketone ⟹ Ethyl phenyl ketone ⟹ Benzene

Reversing the order of introduction of the nitro and acyl groups is incorrect. It is possible to nitrate ethyl phenyl ketone but not possible to carry out a Friedel–Crafts acylation on nitrobenzene, owing to the strong deactivating influence of the nitro group.

**22.20** Direct nitration of the prescribed starting material cumene (isopropylbenzene) is not suitable, because isopropyl is an ortho, para-directing substituent and will give the target molecule *m*-nitrocumene as only a minor component of the nitration product. However, the conversion of 4-isopropyl-2-nitroaniline to *m*-isopropylnitrobenzene, which was used to illustrate reductive deamination of arylamines in the text, establishes the last step in the synthesis.

$CH(CH_3)_2$ / $NO_2$ — *m*-Nitrocumene ⇒ $CH(CH_3)_2$ / $NO_2$ / $NH_2$ — 4-Isopropyl-2-nitroaniline ⇒ $CH(CH_3)_2$ — Cumene

Our task simplifies itself to the preparation of 4-isopropyl-2-nitroaniline from cumene. The following procedure is a straightforward extension of the reactions and principles developed in this chapter.

Cumene ($CH(CH_3)_2$) $\xrightarrow{HNO_3,\ H_2SO_4}$ *p*-Nitrocumene ($CH(CH_3)_2$, $NO_2$) $\xrightarrow{1.\ Fe,\ HCl;\ 2.\ HO^-}$ *p*-Isopropylaniline ($CH(CH_3)_2$, $NH_2$) $\xrightarrow{CH_3CCl\ (C{=}O)}$ *p*-Isopropylacetanilide ($CH(CH_3)_2$, $NHCCH_3$ with C=O)

*p*-Isopropylacetanilide ($CH(CH_3)_2$, $NHCCH_3$) $\xrightarrow{HNO_3,\ H_2SO_4}$ 4-Isopropyl-2-nitroacetanilide ($CH(CH_3)_2$, $NO_2$, $NHCCH_3$) $\xrightarrow{HO^-,\ H_2O,\ \text{or}\ 1.\ H_2O,\ H^+;\ 2.\ HO^-}$ 4-Isopropyl-2-nitroaniline ($CH(CH_3)_2$, $NO_2$, $NH_2$)

Reductive deamination of 4-isopropyl-2-nitroaniline by diazotization in the presence of ethanol or hypophosphorous acid yields *m*-nitrocumene and completes the synthesis.

**22.21** Amines may be primary, secondary, or tertiary. The $C_4H_{11}N$ primary amines, compounds of the type $C_4H_9NH_2$, and their systematic names are

$CH_3CH_2CH_2CH_2NH_2$
Butylamine (1-butanamine)

$(CH_3)_2CHCH_2NH_2$
Isobutylamine (2-methyl-1-propanamine)

$CH_3CHCH_2CH_3$ (with $NH_2$ on C-2)
*sec*-Butylamine (2-butanamine)

$(CH_3)_3CNH_2$
*tert*-Butylamine (2-methyl-2-propanamine)

Secondary amines have the general formula $R_2NH$. Those of molecular formula $C_4H_{11}N$ are

$$(CH_3CH_2)_2NH \qquad CH_3\underset{\displaystyle H}{\underset{|}{N}}CH_2CH_2CH_3 \qquad CH_3\underset{\displaystyle H}{\underset{|}{N}}CH(CH_3)_2$$

Diethylamine (*N*-ethylethanamine) — *N*-Methylpropylamine (*N*-methyl-1-propanamine) — *N*-Methylisopropylamine (*N*-methyl-2-propanamine)

There is only one tertiary amine ($R_3N$) of molecular formula $C_4H_{11}N$:

$$(CH_3)_2NCH_2CH_3$$

*N,N*-Dimethylethylamine
(*N,N*-dimethylethanamine)

**22.22** (*a*) The name 2-ethyl-1-butanamine designates a four-carbon chain terminating in an amino group and bearing an ethyl group at C-2.

$$CH_3CH_2\underset{\displaystyle CH_2CH_3}{\underset{|}{C}}HCH_2NH_2$$

2-Ethyl-1-butanamine

(*b*) The prefix *N*- in *N*-ethyl-1-butanamine identifies the ethyl group as a substituent on nitrogen in a secondary amine.

$$CH_3CH_2CH_2CH_2\underset{\displaystyle H}{\underset{|}{N}}CH_2CH_3$$

*N*-Ethyl-1-butanamine

(*c*) Dibenzylamine is a secondary amine. It bears two benzyl groups on nitrogen.

$$C_6H_5CH_2\underset{\displaystyle H}{\underset{|}{N}}CH_2C_6H_5$$

Dibenzylamine

(*d*) Tribenzylamine is a tertiary amine.

$$(C_6H_5CH_2)_3N$$

Tribenzylamine

(*e*) Tetraethylammonium hydroxide contains a quaternary ammonium ion.

$$(CH_3CH_2)_4\overset{+}{N}\ HO^-$$

Tetraethylammonium
hydroxide

(*f*) This compound is a secondary amine; it bears an allyl substituent on the nitrogen of cyclohexylamine.

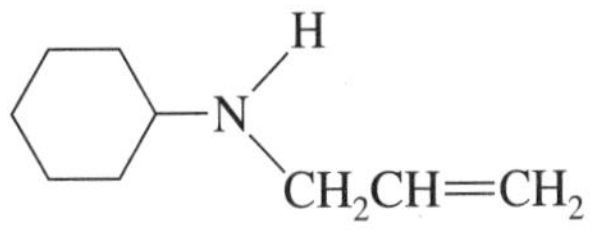

$$C_6H_{11}\text{—}N(H)CH_2CH\text{=}CH_2$$

*N*-Allylcyclohexylamine

(*g*) Piperidine is a cyclic secondary amine that contains nitrogen in a six-membered ring. *N*-Allylpiperidine is a tertiary amine.

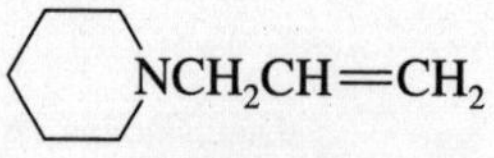

$$\text{(piperidine ring)}NCH_2CH{=}CH_2$$

*N*-Allylpiperidine

(*h*) The compound is the benzyl ester of 2-aminopropanoic acid.

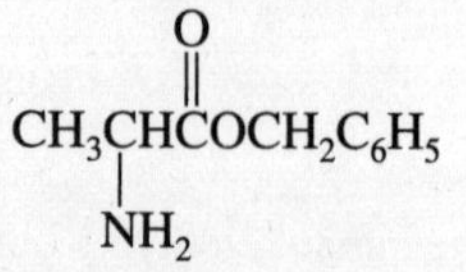

$$CH_3\underset{\displaystyle NH_2}{\underset{|}{C}}H\overset{\displaystyle O}{\overset{\|}{C}}OCH_2C_6H_5$$

Benzyl 2-aminopropanoate

(*i*) The parent compound is cyclohexanone. The substituent $(CH_3)_2N$— group is attached to C-4.

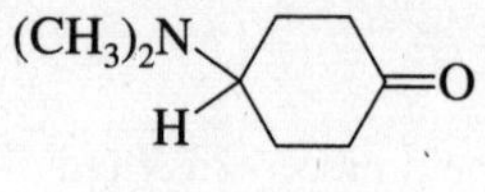

$$(CH_3)_2N(H)\text{-cyclohexane}{=}O$$

4-(*N*,*N*-Dimethylamino)-cyclohexanone

(*j*) The suffix *-diamine* reveals the presence of two amino groups, one at either end of a three-carbon chain that bears two methyl groups at C-2.

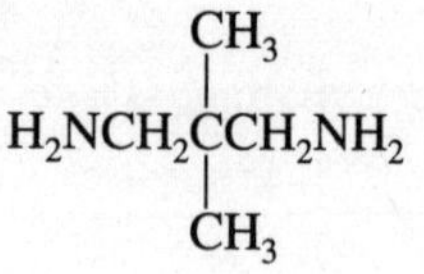

$$H_2NCH_2\underset{\displaystyle CH_3}{\underset{|}{\overset{\displaystyle CH_3}{\overset{|}{C}}}}CH_2NH_2$$

2,2-Dimethyl-1,3-propanediamine

**22.23** (*a*) A phenyl group and an amino group are trans to each other on a three-membered ring in this compound.

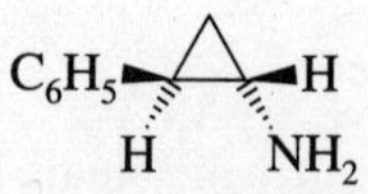

$$C_6H_5,\ H,\ H,\ NH_2 \text{ on cyclopropane ring}$$

*trans*-2-Phenylcyclopropylamine (tranylcypromine)

(*b*) This compound is a tertiary amine. It bears a benzyl group, a methyl group, and a 2-propynyl group on nitrogen.

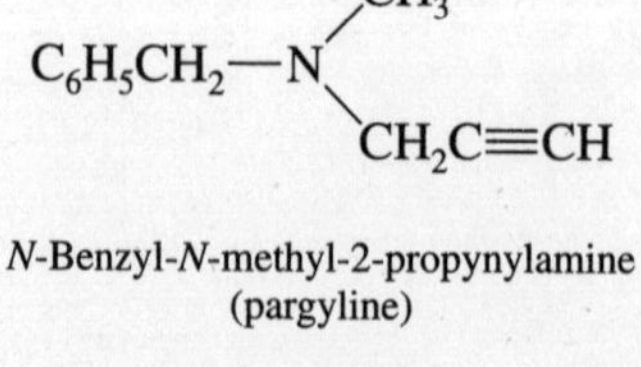

$$C_6H_5CH_2{-}N(CH_3)CH_2C{\equiv}CH$$

*N*-Benzyl-*N*-methyl-2-propynylamine (pargyline)

(*c*) The amino group is at C-2 of a three-carbon chain that bears a phenyl substituent at its terminus.

$$C_6H_5CH_2CHCH_3 \text{ (with } NH_2 \text{ on the CH)}$$

1-Phenyl-2-propanamine
(amphetamine)

(*d*) Phenylephrine is named systematically as an ethanol derivative.

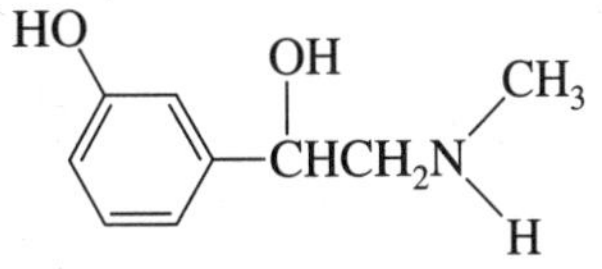

HO (meta) – $C_6H_4$–$CH(OH)CH_2N(CH_3)H$

1-(*m*-Hydroxyphenyl)-
2-(methylamino)ethanol

**22.24** (*a*) There are five isomers of $C_7H_9N$ that contain a benzene ring.

$C_6H_5CH_2NH_2$ — Benzylamine

$C_6H_5NHCH_3$ — *N*-Methylaniline

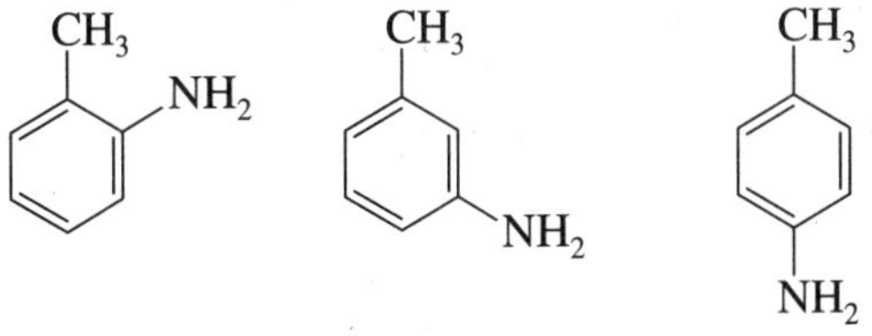

*o*-Methylaniline, *m*-Methylaniline, *p*-Methylaniline (benzene rings bearing $CH_3$ and $NH_2$)

(*b*) Benzylamine is the strongest base because its amine group is bonded to an $sp^3$-hybridized carbon. Benzylamine is a typical alkylamine, with a $K_b$ of $2 \times 10^{-5}$. All the other isomers are arylamines, with $K_b$ values in the $10^{-10}$ range.

(*c*) The formation of *N*-nitrosoamines on reaction with sodium nitrite and hydrochloric acid is a characteristic reaction of secondary amines. The only $C_7H_9N$ isomer in this problem that is a secondary amine is *N*-methylaniline.

$$C_6H_5NHCH_3 \xrightarrow[NaNO_2]{HCl,\ H_2O} C_6H_5N(N{=}O)CH_3$$

*N*-Methylaniline → *N*-Methyl-*N*-nitrosoaniline

(*d*) Ring nitrosation is a characteristic reaction of tertiary arylamines.

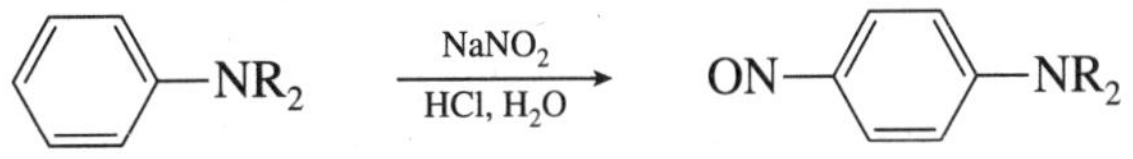

$$C_6H_5NR_2 \xrightarrow[HCl,\ H_2O]{NaNO_2} p\text{-}ON{-}C_6H_4{-}NR_2$$

Tertiary arylamine → *p*-Nitroso-*N*,*N*-dialkylaniline

None of the $C_7H_9N$ isomers in this problem is a tertiary amine; hence none will undergo ring nitrosation.

**22.25** (*a*) Basicity decreases in proceeding across a row in the periodic table. The increased nuclear charge as one progresses from carbon to nitrogen to oxygen to fluorine causes the electrons to be bound more strongly to the atom and thus less readily shared.

$$H_3\bar{C}: > H_2\ddot{\bar{N}}: > H\ddot{\underset{..}{O}}:^- > :\ddot{\underset{..}{F}}:^-$$

| | Strongest base | | | Weakest base |
|---|---|---|---|---|
| $K_a$ of conjugate acid | $10^{-60}$ | $10^{-36}$ | $10^{-16}$ | $3.5 \times 10^{-4}$ |

(*b*) The strongest base in this group is amide ion, $H_2N^-$, and the weakest base is water, $H_2O$. Ammonia is a weaker base than hydroxide ion; the equilibrium lies to the left.

$$:NH_3 + H_2O \rightleftharpoons \overset{+}{N}H_4 + OH^-$$

Weaker base; Weaker acid; Stronger acid; Stronger base

The correct order is

$$H_2\ddot{\bar{N}}: > H\ddot{\underset{..}{O}}:^- > :NH_3 > H_2\ddot{O}:$$

Strongest base — Weakest base

(*c*) These anions can be ranked according to their basicity by considering the respective acidities of their conjugate acids.

| Base | Conjugate acid | $K_a$ of conjugate acid |
|---|---|---|
| $H_2N^-$ | $H_3N$ | $10^{-36}$ |
| $HO^-$ | $H_2O$ | $10^{-16}$ |
| $^-:C{\equiv}N:$ | $HC{\equiv}N:$ | $7.2 \times 10^{-10}$ |
| $^-O{-}\overset{+}{N}(=O){-}O^-$ | $HO\overset{+}{N}(=O){-}O^-$ | $2.5 \times 10^1$ |

The order of basicities is the opposite of the order of acidities of their conjugate acids.

$$H_2N^- > HO^- > :\bar{C}{\equiv}N: > NO_3^-$$

Strongest base — Weakest base

(*d*) A carbonyl group attached to nitrogen stabilizes its negative charge. The strongest base is the anion that has no carbonyl groups on nitrogen; the weakest base is phthalimide anion, which has two carbonyl groups.

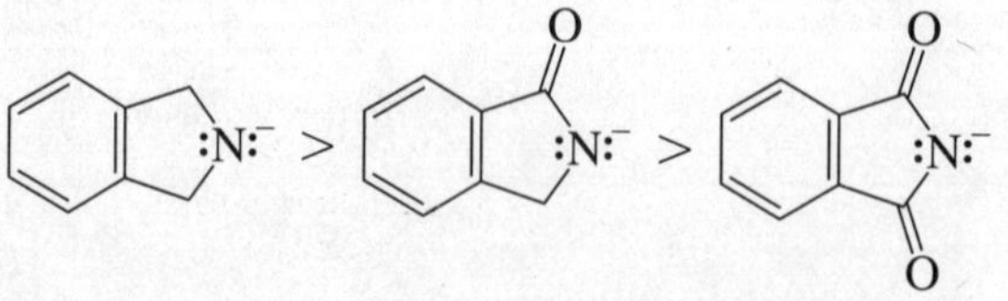

[isoindoline anion :N:⁻] > [isoindolinone anion :N:⁻] > [phthalimide anion :N:⁻]

Strongest base — Weakest base

**22.26** (*a*) An alkyl substituent on nitrogen is electron-releasing and base-strengthening; thus methylamine is a stronger base than ammonia. An aryl substituent is electron-withdrawing and base-weakening, and so aniline is a weaker base than ammonia.

$$CH_3NH_2 > NH_3 > C_6H_5NH_2$$

| Methylamine, strongest base: $K_b$ 4.4 × $10^{-4}$ p$K_b$ 3.4 | Ammonia: $K_b$ 1.8 × $10^{-5}$ p$K_b$ 4.7 | Aniline, weakest base: $K_b$ 3.8 × $10^{-10}$ p$K_b$ 9.4 |
|---|---|---|

(*b*) An acetyl group is an electron-withdrawing and base-weakening substituent, especially when bonded directly to nitrogen. Amides are weaker bases than amines, and thus acetanilide is a weaker base than aniline. Alkyl groups are electron-releasing; *N*-methylaniline is a slightly stronger base than aniline.

$$C_6H_5NHCH_3 > C_6H_5NH_2 > C_6H_5NH\overset{O}{\overset{\|}{C}}CH_3$$

| *N*-methylaniline, strongest base: $K_b$ 8 × $10^{-10}$ p$K_b$ 9.1 | Aniline: $K_b$ 3.8 × $10^{-10}$ p$K_b$ 9.4 | Acetanilide, weakest base: $K_b$ 1 × $10^{-15}$ p$K_b$ 15.0 |
|---|---|---|

(*c*) Chlorine substituents are slightly electron-withdrawing, and methyl groups are slightly electron-releasing. 2,4-Dimethylaniline is therefore a stronger base than 2,4-dichloroaniline. Nitro groups are strongly electron-withdrawing, their base-weakening effect being especially pronounced when a nitro group is ortho or para to an amino group because the two groups are then directly conjugated.

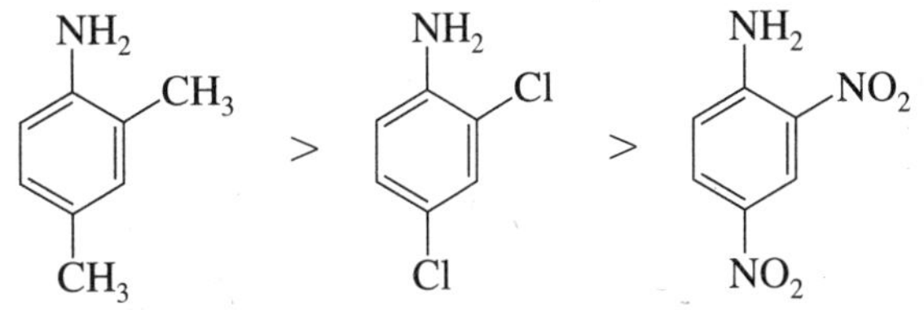

[Structures: 2,4-dimethylaniline (NH₂, CH₃, CH₃) > 2,4-dichloroaniline (NH₂, Cl, Cl) > 2,4-dinitroaniline (NH₂, NO₂, NO₂)]

| 2,4-Dimethylaniline, strongest base: $K_b$ 8 × $10^{-10}$ p$K_b$ 9.1 | 2,4-Dichloroaniline: $K_b$ 1 × $10^{-12}$ p$K_b$ 12.0 | 2,4-Dinitroaniline, weakest base: $K_b$ 3 × $10^{-19}$ p$K_b$ 18.5 |
|---|---|---|

(*d*) Nitro groups are more electron-withdrawing than chlorine, and the base-weakening effect of a nitro substituent is greater when it is ortho or para to an amino group than when it is meta to it.

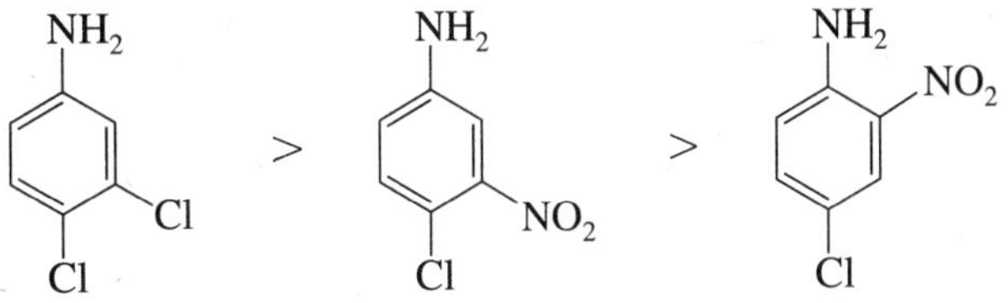

[Structures: 3,4-dichloroaniline (NH₂, Cl, Cl) > 4-chloro-3-nitroaniline (NH₂, NO₂, Cl) > 4-chloro-2-nitroaniline (NH₂, NO₂, Cl)]

| 3,4-Dichloroaniline, strongest base: $K_b \approx 10^{-11}$ p$K_b \approx 11$ | 4-Chloro-3-nitroaniline: $K_b$ 8 × $10^{-13}$ p$K_b$ 12.1 | 4-Chloro-2-nitroaniline, weakest base: $K_b$ 1 × $10^{-15}$ p$K_b$ 15.0 |
|---|---|---|

(*e*) According to the principle applied in part (*a*) (alkyl groups increase basicity, aryl groups decrease it), the order of decreasing basicity is as shown:

$$(CH_3)_2NH > C_6H_5NHCH_3 > (C_6H_5)_2NH$$

| Dimethylamine, strongest base: | *N*-Methylaniline: | Diphenylamine, weakest base: |
|---|---|---|
| $K_b$ 5.1 × $10^{-4}$ | $K_b$ 8 × $10^{-10}$ | $K_b$ 6 × $10^{-14}$ |
| p$K_b$ 3.3 | p$K_b$ 9.1 | p$K_b$ 13.2 |

**22.27** Nitrogen ⓐ is the most basic and the most nucleophilic of the three nitrogen atoms of physostigmine and is the one that reacts with methyl iodide.

Physostigmine + $CH_3I$ ⟶ "Physostigmine methiodide" $I^-$

Physostigmine — Methyl iodide — "Physostigmine methiodide"

The nitrogen that reacts is the one that is a tertiary alkylamine. Of the other two nitrogens, ⓑ is attached to an aromatic ring and is much less basic and less nucleophilic. The third nitrogen, ⓒ, is an amide nitrogen; amides are less nucleophilic than amines.

**22.28** (*a*) Looking at the problem retrosynthetically, it can be seen that a variety of procedures are available for preparing ethylamine from ethanol. The methods by which a primary amine may be prepared include

$CH_3CH_2NH_2$ ⟹

- N-ethylphthalimide ($NCH_2CH_3$) — Gabriel synthesis
- $CH_3CH_2N_3$ — Reduction of an azide
- $CH_3CH$=O — Reductive amination
- $CH_3C(=O)NH_2$ — Reduction of an amide

Two of these methods, the Gabriel synthesis and the preparation and reduction of the corresponding azide, begin with ethyl bromide.

$$\underset{\text{Ethanol}}{CH_3CH_2OH} \xrightarrow[\text{or HBr}]{PBr_3} \underset{\text{Ethyl bromide}}{CH_3CH_2Br}$$

$$\underset{\text{Ethyl bromide}}{CH_3CH_2Br} + \underset{N\text{-Potassiophthalimide}}{C_6H_4(CO)_2N^- K^+} \longrightarrow \underset{N\text{-Ethylphthalimide}}{C_6H_4(CO)_2NCH_2CH_3} \xrightarrow{H_2NNH_2} \underset{\text{Ethylamine}}{CH_3CH_2NH_2}$$

$$\underset{\text{Ethyl bromide}}{CH_3CH_2Br} \xrightarrow{NaN_3} \underset{\text{Ethyl azide}}{CH_3CH_2N_3} \xrightarrow[2.\ H_2O]{1.\ LiAlH_4} \underset{\text{Ethylamine}}{CH_3CH_2NH_2}$$

To use reductive amination, we must begin with oxidation of ethanol to acetaldehyde.

$$\underset{\text{Ethanol}}{CH_3CH_2OH} \xrightarrow[CH_2Cl_2]{\text{PCC or PDC}} \underset{\text{Acetaldehyde}}{CH_3\overset{O}{\overset{\|}{C}}H}$$

$$\underset{\text{Acetaldehyde}}{CH_3\overset{O}{\overset{\|}{C}}H} \xrightarrow{NH_3,\ H_2,\ Ni} \underset{\text{Ethylamine}}{CH_3CH_2NH_2}$$

Another possibility is reduction of acetamide. This requires an initial oxidation of ethanol to acetic acid.

$$\underset{\text{Ethanol}}{CH_3CH_2OH} \xrightarrow[H_2O,\ \text{heat}]{K_2Cr_2O_7,\ H_2SO_4} \underset{\text{Acetic acid}}{CH_3CO_2H} \xrightarrow[2.\ NH_3]{1.\ SOCl_2} \underset{\text{Acetamide}}{CH_3\overset{O}{\overset{\|}{C}}NH_2} \xrightarrow[2.\ H_2O]{1.\ LiAlH_4} \underset{\text{Ethylamine}}{CH_3CH_2NH_2}$$

(*b*) Acylation of ethylamine with acetyl chloride, prepared in part (*a*), gives the desired amide.

$$\underset{\text{Acetyl chloride}}{CH_3\overset{O}{\overset{\|}{C}}Cl} + \underset{\text{Ethylamine}}{2CH_3CH_2NH_2} \longrightarrow \underset{N\text{-Ethylacetamide}}{CH_3\overset{O}{\overset{\|}{C}}NHCH_2CH_3} + \underset{\text{Ethylammonium chloride}}{CH_3CH_2\overset{+}{N}H_3\ Cl^-}$$

Excess ethylamine can be allowed to react with the hydrogen chloride formed in the acylation reaction. Alternatively, equimolar amounts of acyl chloride and amine can be used in the presence of aqueous hydroxide as the base.

(*c*) Reduction of the *N*-ethylacetamide prepared in part (*b*) yields diethylamine.

$$\underset{N\text{-Ethylacetamide}}{CH_3\overset{O}{\overset{\|}{C}}NHCH_2CH_3} \xrightarrow[2.\ H_2O]{1.\ LiAlH_4} \underset{\text{Diethylamine}}{CH_3CH_2NHCH_2CH_3}$$

Diethylamine can also be prepared by reductive amination of acetaldehyde [from part (*a*)] with ethylamine.

$$\underset{\text{Acetaldehyde}}{CH_3\overset{O}{\overset{\|}{C}}H} + \underset{\text{Ethylamine}}{CH_3CH_2NH_2} \xrightarrow[\text{or } NaBH_3CN]{H_2,\ Ni} \underset{\text{Diethylamine}}{CH_3CH_2NHCH_2CH_3}$$

(*d*) The preparation of *N,N*-diethylacetamide is a standard acylation reaction. The reactants, acetyl chloride and diethylamine, have been prepared in previous parts of this problem.

$$\underset{\text{Acetyl chloride}}{CH_3\overset{O}{\overset{\|}{C}}Cl} + \underset{\text{Diethylamine}}{(CH_3CH_2)_2NH} \xrightarrow{HO^-} \underset{N,N\text{-Diethylacetamide}}{CH_3\overset{O}{\overset{\|}{C}}N(CH_2CH_3)_2}$$

(*e*) Triethylamine arises by reduction of *N,N*-diethylacetamide or by reductive amination.

$$\underset{N,N\text{-Diethylacetamide}}{CH_3\overset{O}{\overset{\|}{C}}N(CH_2CH_3)_2} \xrightarrow[2.\ H_2O]{1.\ LiAlH_4} \underset{\text{Triethylamine}}{(CH_3CH_2)_3N}$$

$$\underset{\text{Acetaldehyde}}{CH_3\overset{O}{\overset{\|}{C}}H} + \underset{\text{Diethylamine}}{(CH_3CH_2)_2NH} \xrightarrow[\text{or } NaBH_3CN]{H_2,\ Ni} \underset{\text{Triethylamine}}{(CH_3CH_2)_3N}$$

(*f*) Quaternary ammonium halides are formed by reaction of alkyl halides and tertiary amines.

$$\underset{\text{Ethyl bromide}}{CH_3CH_2Br} + \underset{\text{Triethylamine}}{(CH_3CH_2)_3N} \longrightarrow \underset{\text{Tetraethylammonium bromide}}{(CH_3CH_2)_4\overset{+}{N}\ Br^-}$$

**22.29** (*a*) In this problem a primary alkanamine must be prepared with a carbon chain extended by one carbon. This can be accomplished by way of a nitrile.

$$RCH_2NH_2 \Rightarrow RCN \Rightarrow RBr \Rightarrow ROH$$

$$(R— = CH_3CH_2CH_2CH_2—)$$

The desired reaction sequence is therefore

$$\underset{\text{1-Butanol}}{CH_3CH_2CH_2CH_2OH} \xrightarrow[\text{or } HBr]{PBr_3} \underset{\text{Butyl bromide}}{CH_3CH_2CH_2CH_2Br} \xrightarrow{NaCN} \underset{\text{Pentanenitrile}}{CH_3CH_2CH_2CH_2CN}$$

$$\xrightarrow[2.\ H_2O]{1.\ LiAlH_4} \underset{\text{1-Pentanamine}}{CH_3CH_2CH_2CH_2CH_2NH_2}$$

(*b*) The carbon chain of *tert*-butyl chloride cannot be extended by a nucleophilic substitution reaction; the $S_N2$ reaction that would be required on the tertiary halide would not work. The sequence employed in part (*a*) is therefore not effective in this case. The best route is carboxylation of the Grignard reagent and subsequent conversion of the corresponding amide to the desired primary amine product.

$$(CH_3)_3CCH_2NH_2 \Rightarrow (CH_3)_3C\overset{O}{\overset{\|}{C}}NH_2 \Rightarrow (CH_3)_3CCO_2H \Rightarrow (CH_3)_3CCl$$

The reaction sequence to be used is

$$\underset{\text{tert-Butyl chloride}}{(CH_3)_3CCl} \xrightarrow[\text{2. } CO_2 \text{ 3. } H_3O^+]{\text{1. Mg, diethyl ether}} \underset{\text{2,2-Dimethylpropanoic acid}}{(CH_3)_3CCO_2H}$$

Once the carboxylic acid has been obtained, it is converted to the desired amine by reduction of the corresponding amide.

$$\underset{\text{2,2-Dimethylpropanoic acid}}{(CH_3)_3CCO_2H} \xrightarrow[\text{2. } NH_3]{\text{1. } SOCl_2} \underset{\text{2,2-Dimethylpropanamide}}{(CH_3)_3C\overset{O}{\overset{\|}{C}}NH_2} \xrightarrow[\text{2. } H_2O]{\text{1. } LiAlH_4} \underset{\text{2,2-Dimethyl-1-propanamine}}{(CH_3)_3CCH_2NH_2}$$

(*c*) Oxidation of cyclohexanol to cyclohexanone gives a compound suitable for reductive amination.

$$\underset{\text{Cyclohexanol}}{C_6H_{11}\text{—OH}} \xrightarrow[H_2SO_4,\ H_2O]{K_2Cr_2O_7} \underset{\text{Cyclohexanone}}{C_6H_{10}\text{=O}} \xrightarrow[\text{or } CH_3NH_2,\ NaBH_3CN]{CH_3NH_2,\ H_2,\ Ni} \underset{\text{N-Methylcyclohexylamine}}{C_6H_{11}\text{—NHCH}_3}$$

(*d*) The desired product is the reduction product of the cyanohydrin of acetone.

$$\underset{\text{Acetone cyanohydrin}}{CH_3C(OH)(CN)CH_3} \xrightarrow[\text{2. } H_2O]{\text{1. } LiAlH_4} \underset{\text{1-Amino-2-methyl-2-propanol}}{CH_3C(OH)(CH_2NH_2)CH_3}$$

The cyanohydrin is made from acetone in the usual way. Acetone is available by oxidation of isopropyl alcohol.

$$\underset{\text{Isopropyl alcohol}}{CH_3CH(OH)CH_3} \xrightarrow[H_2O]{K_2Cr_2O_7,\ H_2SO_4} \underset{\text{Acetone}}{CH_3\overset{O}{\overset{\|}{C}}CH_3} \xrightarrow[H_2SO_4]{KCN} \underset{\text{Acetone cyanohydrin}}{CH_3C(OH)(CN)CH_3}$$

(*e*) The target amino alcohol is the product of nucleophilic ring opening of 1,2-epoxypropane by ammonia. Ammonia attacks the less hindered carbon of the epoxide function.

$$CH_3CH\underset{O}{-}CH_2 \xrightarrow{NH_3} CH_3\underset{OH}{CH}CH_2NH_2$$

1,2-Epoxypropane → 1-Amino-2-propanol

The necessary epoxide is formed by epoxidation of propene.

$$CH_3\underset{OH}{CH}CH_3 \xrightarrow[heat]{H_2SO_4} CH_3CH{=}CH_2 \xrightarrow{CH_3\overset{O}{\overset{\|}{C}}OOH} CH_3CH\underset{O}{-}CH_2$$

Isopropyl alcohol → Propene → 1,2-Epoxypropane

(*f*) The reaction sequence is the same as in part (*e*) except that dimethylamine is used as the nucleophile instead of ammonia.

$$CH_3CH\underset{O}{-}CH_2 + (CH_3)_2NH \longrightarrow CH_3\underset{OH}{CH}CH_2N(CH_3)_2$$

1,2-Epoxypropane [prepared as in part (*e*)] + Dimethylamine → 1-(*N,N*-Dimethylamino)-2-propanol

(*g*) The key to performing this synthesis is recognition of the starting material as an acetal of acetophenone. Acetals may be hydrolyzed to carbonyl compounds.

$$\text{2-Methyl-2-phenyl-1,3-dioxolane} \xrightarrow{H_3O^+} C_6H_5\overset{O}{\overset{\|}{C}}CH_3 + HOCH_2CH_2OH$$

2-Methyl-2-phenyl-1,3-dioxolane → Acetophenone + 1,2-Ethanediol

Once acetophenone has been obtained, it may be converted to the required product by reductive amination.

$$C_6H_5\overset{O}{\overset{\|}{C}}CH_3 + \text{piperidine} \xrightarrow[\text{or } H_2,\ Ni]{NaBH_3CN} \text{N-}C_6H_5CHCH_3\text{-piperidine}$$

Acetophenone + Piperidine → *N*-(1-Phenylethyl)-piperidine

**22.30** (*a*) The reaction of alkyl halides with *N*-potassiophthalimide (the first step in the Gabriel synthesis of amines) is a nucleophilic substitution reaction. Alkyl bromides are more reactive than alkyl fluorides; that is, bromide is a better leaving group than fluoride.

$$\text{Phthalimide-NK} + FCH_2CH_2Br \longrightarrow \text{Phthalimide-}NCH_2CH_2F$$

*N*-Potassiophthalimide + 1-Bromo-2-fluoroethane → 2-Phthalimidoethyl fluoride (*b*) In this example one bromine is attached to a primary and the other to a secondary carbon. Phthalimide anion is a good nucleophile and reacts with alkyl halides by the $S_N2$ mechanism. It attacks the less hindered primary carbon.

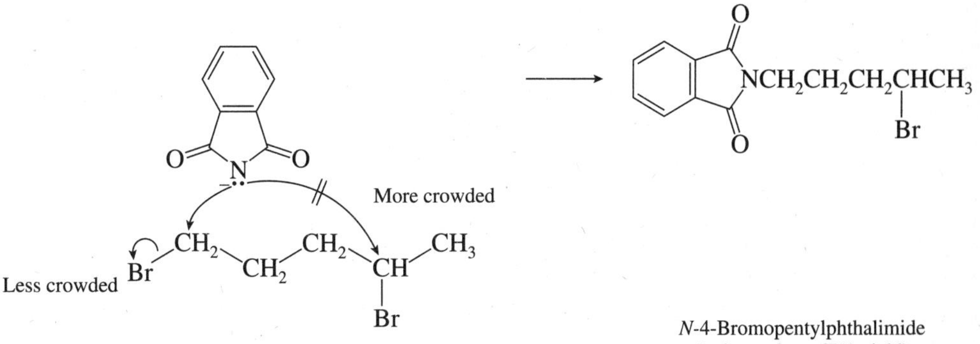

More crowded

Less crowded

$$BrCH_2CH_2CH_2CH(Br)CH_3 \longrightarrow \text{Phthalimide}-NCH_2CH_2CH_2CH(Br)CH_3$$

1,4-Dibromopentane

*N*-4-Bromopentylphthalimide (only product, 67% yield)

(*c*) Both bromines are bonded to primary carbons, but branching at the adjacent carbon hinders nucleophilic attack at one of them.

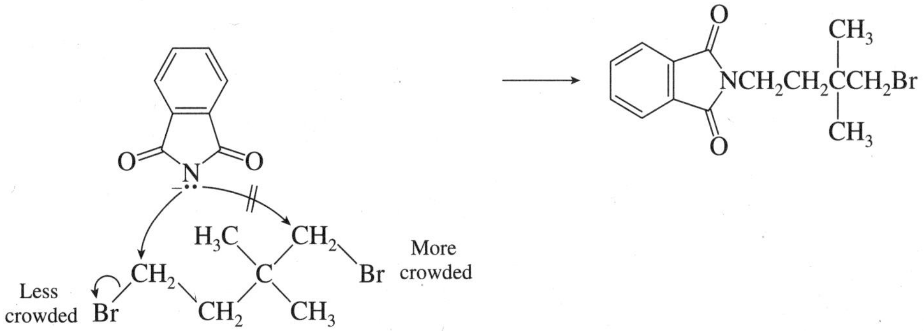

Less crowded

More crowded

$$BrCH_2CH_2C(CH_3)_2CH_2Br \longrightarrow \text{Phthalimide}-NCH_2CH_2C(CH_3)_2CH_2Br$$

1,4-Dibromo-2,2-dimethylbutane

*N*-4-Bromo-3,3-dimethylphthalimide (only product, 53% yield)

**22.31** (*a*) Amines are basic and are protonated by hydrogen halides.

$$C_6H_5CH_2NH_2 + HBr \longrightarrow C_6H_5CH_2\overset{+}{N}H_3\ Br^-$$

Benzylamine — Benzylammonium bromide

(*b*) Equimolar amounts of benzylamine and sulfuric acid yield benzylammonium hydrogen sulfate as the product.

$$C_6H_5CH_2NH_2 + HOSO_2OH \longrightarrow C_6H_5CH_2\overset{+}{N}H_3\ ^-OSO_2OH$$

Benzylamine — Sulfuric acid — Benzylammonium hydrogen sulfate

(*c*) Acetic acid transfers a proton to benzylamine.

$$C_6H_5CH_2NH_2 + CH_3\overset{O}{\overset{\|}{C}}OH \longrightarrow C_6H_5CH_2\overset{+}{N}H_3\ ^-O\overset{O}{\overset{\|}{C}}CH_3$$

Benzylamine — Acetic acid — Benzylammonium acetate

(*d*) Acetyl chloride reacts with benzylamine to form an amide.

$$2C_6H_5CH_2NH_2 + CH_3\overset{O}{\overset{\|}{C}}Cl \longrightarrow CH_3\overset{O}{\overset{\|}{C}}NHCH_2C_6H_5 + C_6H_5CH_2\overset{+}{N}H_3\ Cl^-$$

Benzylamine | Acetyl chloride | *N*-Benzylacetamide | Benzylammonium chloride

(*e*) Acetic anhydride also gives an amide with benzylamine.

$$2C_6H_5CH_2NH_2 + CH_3\overset{O}{\overset{\|}{C}}O\overset{O}{\overset{\|}{C}}CH_3 \longrightarrow CH_3\overset{O}{\overset{\|}{C}}NHCH_2C_6H_5 + C_6H_5CH_2\overset{+}{N}H_3\ {}^-O\overset{O}{\overset{\|}{C}}CH_3$$

Benzylamine | Acetic anhydride | *N*-Benzylacetamide | Benzylammonium acetate

(*f*) Primary amines react with ketones to give imines.

$$C_6H_5CH_2NH_2 + CH_3\overset{O}{\overset{\|}{C}}CH_3 \longrightarrow (CH_3)_2C{=}NCH_2C_6H_5$$

Benzylamine | Acetone | *N*-Isopropylidenebenzylamine

(*g*) These reaction conditions lead to reduction of the imine formed in part (*f*). The overall reaction is reductive amination.

$$C_6H_5CH_2NH_2 + CH_3\overset{O}{\overset{\|}{C}}CH_3 \xrightarrow{H_2,\ Ni} (CH_3)_2CHNHCH_2C_6H_5$$

Benzylamine | Acetone | *N*-Isopropylbenzylamine

(*h*) Amines are nucleophilic and bring about the opening of epoxide rings.

$$C_6H_5CH_2NH_2 + H_2C\overset{O}{—}CH_2 \longrightarrow C_6H_5CH_2NHCH_2CH_2OH$$

Benzylamine | Ethylene oxide | 2-(*N*-Benzylamino)ethanol

(*i*) In these nucleophilic ring-opening reactions the amine attacks the less sterically hindered carbon of the ring.

$$C_6H_5CH_2NH_2 + H_2C\overset{O}{—}CHCH_3 \longrightarrow C_6H_5CH_2NHCH_2\underset{OH}{\underset{|}{C}}HCH_3$$

Benzylamine | 1,2-Epoxypropane | 1-(*N*-Benzylamino)-2-propanol

(*j*) With excess methyl iodide, amines are converted to quaternary ammonium iodides.

$$C_6H_5CH_2NH_2 + 3CH_3I \longrightarrow C_6H_5CH_2\overset{+}{N}(CH_3)_3\ I^-$$

Benzylamine | Methyl iodide | Benzyltrimethylammonium iodide (*k*) Nitrous acid forms from sodium nitrite in dilute hydrochloric acid. Nitrosation of benzylamine in water gives benzyl alcohol via a diazonium ion intermediate.

$$\underset{\text{Benzylamine}}{C_6H_5CH_2NH_2} \xrightarrow[H_2O]{NaNO_2,\ HCl} \underset{\text{Benzyldiazonium ion}}{C_6H_5CH_2\overset{+}{N}{\equiv}N} \xrightarrow[H_2O]{-N_2} \underset{\text{Benzyl alcohol}}{C_6H_5CH_2OH}$$

Benzyl chloride will also be formed by attack of chloride on the diazonium ion.

**22.32** (*a*) Aniline is a weak base and yields a salt on reaction with hydrogen bromide.

$$\underset{\text{Aniline}}{C_6H_5NH_2} + \underset{\text{Hydrogen bromide}}{HBr} \longrightarrow \underset{\text{Anilinium bromide}}{C_6H_5\overset{+}{N}H_3\ Br^-}$$

(*b*) Aniline acts as a nucleophile toward methyl iodide. With excess methyl iodide, a quaternary ammonium salt is formed.

$$\underset{\text{Aniline}}{C_6H_5NH_2} + \underset{\text{Methyl iodide}}{3CH_3I} \longrightarrow \underset{N,N,N\text{-Trimethylanilinium iodide}}{C_6H_5\overset{+}{N}(CH_3)_3\ I^-}$$

(*c*) Aniline is a primary amine and undergoes nucleophilic addition to aldehydes and ketones to form imines.

$$\underset{\text{Aniline}}{C_6H_5NH_2} + \underset{\text{Acetaldehyde}}{CH_3\overset{O}{\overset{\|}{C}}H} \longrightarrow \underset{N\text{-Phenylacetaldimine}}{C_6H_5N{=}CHCH_3} + \underset{\text{Water}}{H_2O}$$

(*d*) When an imine is formed in the presence of hydrogen and a suitable catalyst, reductive amination occurs to give an amine.

$$\underset{\text{Aniline}}{C_6H_5NH_2} + \underset{\text{Acetaldehyde}}{CH_3\overset{O}{\overset{\|}{C}}H} \xrightarrow{H_2,\ Ni} \underset{N\text{-Ethylaniline}}{C_6H_5NHCH_2CH_3}$$

(*e*) Aniline undergoes *N*-acylation on treatment with carboxylic acid anhydrides.

$$\underset{\text{Aniline}}{2C_6H_5NH_2} + \underset{\text{Acetic anhydride}}{CH_3\overset{O}{\overset{\|}{C}}O\overset{O}{\overset{\|}{C}}CH_3} \longrightarrow \underset{\text{Acetanilide}}{C_6H_5NH\overset{O}{\overset{\|}{C}}CH_3} + \underset{\text{Anilinium acetate}}{C_6H_5\overset{+}{N}H_3\ {}^-O\overset{O}{\overset{\|}{C}}CH_3}$$

(*f*) Acyl chlorides bring about *N*-acylation of arylamines.

$$\underset{\text{Aniline}}{2C_6H_5NH_2} + \underset{\text{Benzoyl chloride}}{C_6H_5\overset{O}{\overset{\|}{C}}Cl} \longrightarrow \underset{\text{Benzanilide}}{C_6H_5NH\overset{O}{\overset{\|}{C}}C_6H_5} + \underset{\text{Anilinium chloride}}{C_6H_5\overset{+}{N}H_3\ Cl^-}$$

(*g*) Nitrosation of primary arylamines yields aryl diazonium salts.

$$\underset{\text{Aniline}}{C_6H_5NH_2} \xrightarrow[H_2O,\ 0\text{–}5°C]{NaNO_2,\ H_2SO_4} \underset{\text{Benzenediazonium hydrogen sulfate}}{C_6H_5\overset{+}{N}{\equiv}N\colon\ HSO_4^-}$$

The replacement reactions that can be achieved by using diazonium salts are illustrated in parts (*h*) through (*n*). In all cases molecular nitrogen is lost from the ring carbon to which it was attached and is replaced by another substituent.

$C_6H_5\overset{+}{N}{\equiv}N{:}\ HSO_4^-$

Benzenediazonium hydrogen sulfate

(*h*) $\xrightarrow[\text{heat}]{H^+,\ H_2O}$ $C_6H_5OH$ — Phenol

(*i*) $\xrightarrow{CuCl}$ $C_6H_5Cl$ — Chlorobenzene

(*j*) $\xrightarrow{CuBr}$ $C_6H_5Br$ — Bromobenzene

(*k*) $\xrightarrow{CuCN}$ $C_6H_5CN$ — Benzonitrile

(*l*) $\xrightarrow{H_3PO_2}$ $C_6H_6$ — Benzene

(*m*) $\xrightarrow{KI}$ $C_6H_5I$ — Iodobenzene

(*n*) $\xrightarrow[\text{2. heat}]{\text{1. } HBF_4}$ $C_6H_5F$ — Fluorobenzene

(*o*) The nitrogens of an aryl diazonium salt are retained on reaction with the electron-rich ring of a phenol. Azo coupling occurs.

$$C_6H_5\overset{+}{N}{\equiv}N{:}\ HSO_4^- + C_6H_5OH \longrightarrow C_6H_5N{=}N{-}C_6H_4{-}OH$$

Benzenediazonium hydrogen sulfate — Phenol — *p*-(Azophenyl)phenol

(*p*) Azo coupling occurs when aryl diazonium salts react with *N*,*N*-dialkylarylamines.

$$C_6H_5\overset{+}{N}{\equiv}N{:}\ HSO_4^- + C_6H_5N(CH_3)_2 \longrightarrow C_6H_5N{=}N{-}C_6H_4{-}N(CH_3)_2$$

Benzenediazonium hydrogen sulfate — *N*,*N*-Dimethylaniline — *p*-(Azophenyl)-*N*,*N*-dimethylaniline

**22.33** (*a*) Amides are reduced to amines by lithium aluminum hydride.

$$C_6H_5NH\overset{O}{\overset{\|}{C}}CH_3 \xrightarrow[\text{2. } H_2O]{\text{1. } LiAlH_4\text{, diethyl ether}} C_6H_5NHCH_2CH_3$$

Acetanilide — *N*-Ethylaniline

(*b*) Acetanilide is a reactive substrate toward electrophilic aromatic substitution. An acetamido group is ortho, para-directing.

$$C_6H_5NHCCH_3 \xrightarrow[H_2SO_4]{HNO_3} o\text{-}O_2NC_6H_4NHCCH_3 + p\text{-}O_2NC_6H_4NHCCH_3$$

Acetanilide — *o*-Nitroacetanilide — *p*-Nitroacetanilide

(*c*) Sulfonation of the ring occurs.

$$C_6H_5NHCCH_3 \xrightarrow[H_2SO_4]{SO_3} p\text{-}HO_3SC_6H_4NHCCH_3 + \text{ortho isomer}$$

Acetanilide — *p*-Acetamidobenzenesulfonic acid

(*d*) Bromination of the ring takes place.

$$C_6H_5NHCCH_3 \xrightarrow[\text{acetic acid}]{Br_2} p\text{-}BrC_6H_4NHCCH_3 + \text{ortho isomer}$$

Acetanilide — *p*-Bromoacetanilide

(*e*) Acetanilide undergoes Friedel–Crafts alkylation readily.

$$C_6H_5NHCCH_3 + (CH_3)_3CCl \xrightarrow{AlCl_3} p\text{-}(CH_3)_3CC_6H_4NHCCH_3 + \text{ortho isomer}$$

Acetanilide — *tert*-Butyl chloride — *p*-*tert*-Butylacetanilide

(*f*) Friedel–Crafts acylation also is easily carried out.

$$C_6H_5NHCCH_3 + CH_3CCl \xrightarrow{AlCl_3} p\text{-}CH_3CC_6H_4NHCCH_3 + \text{ortho isomer}$$

(each C bonded to O in the acetyl groups is C=O)

Acetanilide — Acetyl chloride — *p*-Acetamidoacetophenone

(*g*) Acetanilide is an amide and can be hydrolyzed when heated with aqueous acid. Under acidic conditions the aniline that is formed exists in its protonated form as the anilinium cation.

$$C_6H_5NHCCH_3 \text{ (C=O)} + H_2O + HCl \longrightarrow C_6H_5\overset{+}{N}H_3\ Cl^- + HOCCH_3 \text{ (C=O)}$$

Acetanilide, Water, Hydrogen chloride → Anilinium chloride, Acetic acid

(*h*) Amides are hydrolyzed in base.

$$C_6H_5NHCCH_3 \text{ (C=O)} + NaOH \xrightarrow{H_2O} C_6H_5NH_2 + Na^+\ ^-OCCH_3 \text{ (C=O)}$$

Acetanilide, Sodium hydroxide → Aniline, Sodium acetate

**22.34** (*a*) The reaction illustrates the preparation of a secondary amine by reductive amination.

Cyclohexanone (=O) + $H_2N$–cyclohexyl $\xrightarrow{H_2,\ Ni}$ cyclohexyl–NH–cyclohexyl

Cyclohexanone, Cyclohexylamine → Dicyclohexylamine (70%)

(*b*) Amides are reduced to amines by lithium aluminum hydride.

$\xrightarrow{\text{1. } LiAlH_4\text{; 2. } H_2O,\ HO^-}$

O, $NCH_2CH_3$ → $NCH_2CH_3$

6-Ethyl-6-azabicyclo[3.2.1]octan-7-one → 6-Ethyl-6-azabicyclo[3.2.1]octane

(*c*) Treatment of alcohols with *p*-toluenesulfonyl chloride converts them to *p*-toluenesulfonate esters.

$$C_6H_5CH_2CH_2CH_2OH + H_3C\text{–}C_6H_4\text{–}SO_2Cl \xrightarrow{\text{pyridine}} C_6H_5CH_2CH_2CH_2OS(=O)_2\text{–}C_6H_4\text{–}CH_3$$

3-Phenyl-1-propanol, *p*-Toluenesulfonyl chloride → 3-Phenylpropyl *p*-toluenesulfonate

*p*-Toluenesulfonate is an excellent leaving group in nucleophilic substitution reactions. Dimethylamine is the nucleophile.

$$C_6H_5CH_2CH_2CH_2OSO_2\text{–}C_6H_4\text{–}CH_3 + (CH_3)_2NH \longrightarrow C_6H_5CH_2CH_2CH_2N(CH_3)_2$$

3-Phenylpropyl *p*-toluenesulfonate, Dimethylamine → *N*,*N*-Dimethyl-3-phenyl-1-propanamine (86%)

(*d*) Amines are sufficiently nucleophilic to react with epoxides. Attack occurs at the less substituted carbon of the epoxide.

$CH_3O$, $OCH_3$ (aryl)–CH–$CH_2$ (O) + $H_2\ddot{N}CH(CH_3)_2$ ⟶ $CH_3O$, $OCH_3$ (aryl)–CH(OH)$CH_2NHCH(CH_3)_2$

2-(2,5-Dimethoxyphenyl)oxirane, Isopropylamine → 1-(2,5-Dimethoxyphenyl)-2-(isopropylamino)ethanol (67%)

(*e*) α-Halo ketones are reactive substrates in nucleophilic substitution reactions. Dibenzylamine is the nucleophile.

$$(C_6H_5CH_2)_2\ddot{N}H + CH_3\overset{O}{\overset{\|}{C}}CH_2{-}Cl \longrightarrow CH_3\overset{O}{\overset{\|}{C}}CH_2N(CH_2C_6H_5)_2$$

Dibenzylamine — 1-Chloro-2-propanone — 1-(Dibenzylamino)-2-propanone (87%)

Because the reaction liberates hydrogen chloride, it is carried out in the presence of added base—in this case triethylamine—so as to avoid converting the dibenzylamine to its hydrochloride salt.

(*f*) Quaternary ammonium hydroxides undergo Hofmann elimination when they are heated. A point to be considered here concerns the regioselectivity of Hofmann eliminations: it is the less hindered β proton that is removed by the base giving the less substituted alkene.

$H_3C$, H, $^-$:ÖH, $CH_2$, $\overset{+}{N}(CH_3)_3$ $\xrightarrow{-H_2O}$ $H_3C$, $CH_3$, $CH_2$ + $(CH_3)_3N$:

*trans*-1-Isopropenyl-4-methylcyclohexane (98%) — Trimethylamine

Elimination to give $H_3C$, $CH_3$, $CH_3$ does not occur.

(*g*) The combination of sodium nitrite and aqueous acid is a nitrosating agent. Secondary alkylamines react with nitrosating agents to give *N*-nitroso amines as the isolated products.

$$(CH_3)_2CHNHCH(CH_3)_2 \xrightarrow[HCl,\ H_2O]{NaNO_2} (CH_3)_2CHN(N{=}O)CH(CH_3)_2$$

Diisopropylamine — *N*-Nitrosodiisopropylamine (91%)

**22.35** (*a*) Catalytic hydrogenation reduces nitro groups to amino groups.

$CH_3CH_2$, $CH_3CH_2$, $NO_2$ $\xrightarrow[ethanol]{H_2,\ Pt}$ $CH_3CH_2$, $CH_3CH_2$, $NH_2$

1,2-Diethyl-4-nitrobenzene — 3,4-Diethylaniline (93–99%)

(*b*) Nitro groups are readily reduced by tin(II) chloride.

$CH_3$, $NO_2$, $CH_3$ $\xrightarrow[2.\ HO^-]{1.\ SnCl_2,\ HCl}$ $CH_3$, $NH_2$, $CH_3$

1,3-Dimethyl-2-nitrobenzene — 2,6-Dimethylaniline

This reaction is the first step in a synthesis of the drug **lidocaine.**

(*c*) The amino group of arylamines is nucleophilic and undergoes acylation on reaction with chloroacetyl chloride.

$$\text{2,6-Dimethylaniline} + \underset{\text{Chloroacetyl chloride}}{ClCH_2\overset{O}{\overset{\|}{C}}Cl} \longrightarrow \text{N-(Chloroacetyl)-2,6-dimethylaniline } (2,6\text{-}(CH_3)_2C_6H_3NH\overset{O}{\overset{\|}{C}}CH_2Cl)$$

2,6-Dimethylaniline — Chloroacetyl chloride — *N*-(Chloroacetyl)-2,6-dimethylaniline

Chloroacetyl chloride is a difunctional compound—it is both an acyl chloride and an alkyl chloride. Acyl chlorides react with nucleophiles faster than do alkyl chlorides, so that acylation of the amine nitrogen occurs rather than alkylation.

(*d*) The final step in the synthesis of lidocaine is displacement of the chloride by diethylamine from the $\alpha$-halo amide formed in part (*c*) in a nucleophilic substitution reaction.

$$2,6\text{-}(CH_3)_2C_6H_3NH\overset{O}{\overset{\|}{C}}CH_2Cl + (CH_3CH_2)_2NH \longrightarrow 2,6\text{-}(CH_3)_2C_6H_3NH\overset{O}{\overset{\|}{C}}CH_2N(CH_2CH_3)_2$$

*N*-(Chloroacetyl)-2,6-dimethylaniline — Diethylamine — Lidocaine

The reaction is carried out with excess diethylamine, which acts as a base to neutralize the hydrogen chloride formed.

(*e*) For use as an anesthetic, lidocaine is made available as its hydrochloride salt. Of the two nitrogens in lidocaine, the amine nitrogen is more basic than the amide.

$$2,6\text{-}(CH_3)_2C_6H_3NH\overset{O}{\overset{\|}{C}}CH_2N(CH_2CH_3)_2 \xrightarrow{HCl} 2,6\text{-}(CH_3)_2C_6H_3NH\overset{O}{\overset{\|}{C}}CH_2\overset{H}{\overset{|}{\underset{+}{N}}}(CH_2CH_3)_2\ Cl^-$$

Lidocaine — Lidocaine hydrochloride

(*f*) Lithium aluminum hydride reduction of amides is one of the best methods for the preparation of amines, including arylamines.

$$C_6H_5NH\overset{O}{\overset{\|}{C}}CH_2CH_2CH_3 \xrightarrow[2.\ H_2O]{1.\ LiAlH_4} C_6H_5NHCH_2CH_2CH_2CH_3$$

*N*-Phenylbutanamide — *N*-Butylaniline (92%)

(*g*) Arylamines react with aldehydes and ketones in the presence of hydrogen and nickel to give the product of reductive amination.

$$C_6H_5NH_2 + CH_3(CH_2)_5\overset{O}{\overset{\|}{C}}H \xrightarrow{H_2,\ Ni} C_6H_5NHCH_2(CH_2)_5CH_3$$

Aniline — Heptanal — *N*-Heptylaniline (65%)

(*h*) Acetanilide is a reactive substrate toward electrophilic aromatic substitution. On reaction with chloroacetyl chloride, it undergoes Friedel–Crafts acylation, primarily at its para position.

$$CH_3\overset{O}{\overset{\|}{C}}NH-C_6H_5 + ClCH_2\overset{O}{\overset{\|}{C}}Cl \xrightarrow{AlCl_3} CH_3\overset{O}{\overset{\|}{C}}NH-C_6H_4-\overset{O}{\overset{\|}{C}}CH_2Cl$$

Acetanilide — Chloroacetyl chloride — *p*-Acetamidophenacyl chloride (79–83%)

Acylation, rather than alkylation, occurs. Acyl chlorides are more reactive than alkyl chlorides toward electrophilic aromatic substitution reactions as a result of the more stable intermediate (acylium ion) formed.

(*i*) Reduction with iron in hydrochloric acid is one of the most common methods for converting nitroarenes to arylamines.

$$Br-C_6H_4-C_6H_4-NO_2 \xrightarrow[2.\ HO^-]{1.\ Fe,\ HCl} Br-C_6H_4-C_6H_4-NH_2$$

4-Bromo-4′-nitrobiphenyl — 4-Amino-4′-bromobiphenyl (94%)

(*j*) Primary arylamines are converted to aryl diazonium salts on treatment with sodium nitrite in aqueous acid. When the aqueous acidic solution containing the diazonium salt is heated, a phenol is formed.

$$Br-C_6H_4-C_6H_4-NH_2 \xrightarrow[H_2O]{NaNO_2,\ H_2SO_4} Br-C_6H_4-C_6H_4-\overset{+}{N}\equiv N: \xrightarrow[(-N_2)]{H_2O,\ heat} Br-C_6H_4-C_6H_4-OH$$

4-Amino-4′-bromobiphenyl — 4-Bromo-4′-hydroxybiphenyl (85%)

(*k*) This problem illustrates the conversion of an arylamine to an aryl chloride by the Sandmeyer reaction.

$$2,6\text{-}(O_2N)_2C_6H_3NH_2 \xrightarrow[H_2O]{NaNO_2,\ H_2SO_4} 2,6\text{-}(O_2N)_2C_6H_3\overset{+}{N}\equiv N: \xrightarrow[(-N_2)]{CuCl} 2,6\text{-}(O_2N)_2C_6H_3Cl$$

2,6-Dinitroaniline — 2-Chloro-1,3-dinitrobenzene (71–74%)

(*l*) Diazotization of primary arylamines followed by treatment with copper(I) bromide converts them to aryl bromides.

$$m\text{-}BrC_6H_4NH_2 \xrightarrow[H_2O]{NaNO_2,\ HBr} m\text{-}BrC_6H_4\overset{+}{N}\equiv N: \xrightarrow[(-N_2)]{CuBr} m\text{-}BrC_6H_4Br$$

*m*-Bromoaniline — *m*-Dibromobenzene (80–87%)

(*m*) Nitriles are formed when aryl diazonium salts react with copper(I) cyanide.

$$o\text{-}O_2NC_6H_4NH_2 \xrightarrow[H_2O]{NaNO_2,\ HCl} o\text{-}O_2NC_6H_4\overset{+}{N}\equiv N: \xrightarrow[(-N_2)]{CuCN} o\text{-}O_2NC_6H_4CN$$

*o*-Nitroaniline — *o*-Nitrobenzonitrile (87%)

(*n*) An aryl diazonium salt is converted to an aryl iodide on reaction with potassium iodide.

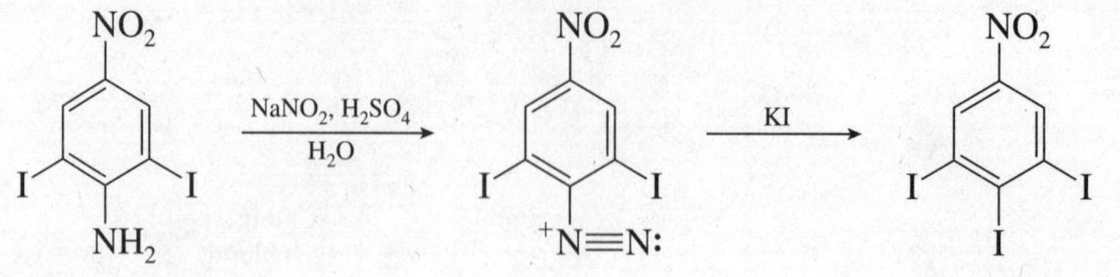

2,6-Diiodo-4-nitroaniline $\xrightarrow[H_2O]{NaNO_2,\ H_2SO_4}$ (diazonium, $^+N{\equiv}N{:}$) $\xrightarrow{KI}$ 1,2,3-Triiodo-5-nitrobenzene (94–95%)

(*o*) Aryl diazonium fluoroborates are converted to aryl fluorides when heated. Both diazonium salt functions in the starting material undergo this reaction.

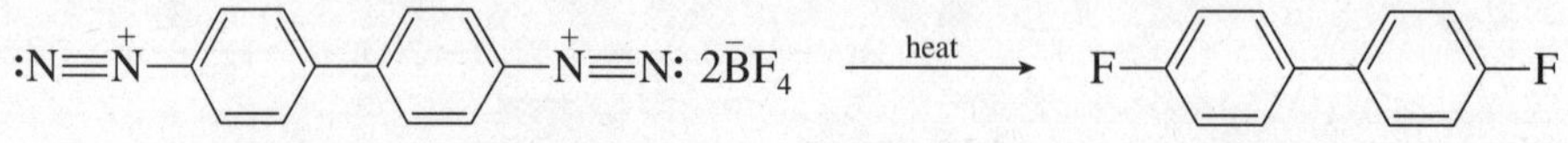

$:N{\equiv}\overset{+}{N}{-}C_6H_4{-}C_6H_4{-}\overset{+}{N}{\equiv}N{:}\ 2\overset{-}{B}F_4 \xrightarrow{heat} F{-}C_6H_4{-}C_6H_4{-}F$

4,4′-Bis(diazonio)biphenyl fluoroborate → 4,4′-Difluorobiphenyl (82%)

(*p*) Hypophosphorous acid ($H_3PO_2$) reduces aryl diazonium salts to arenes.

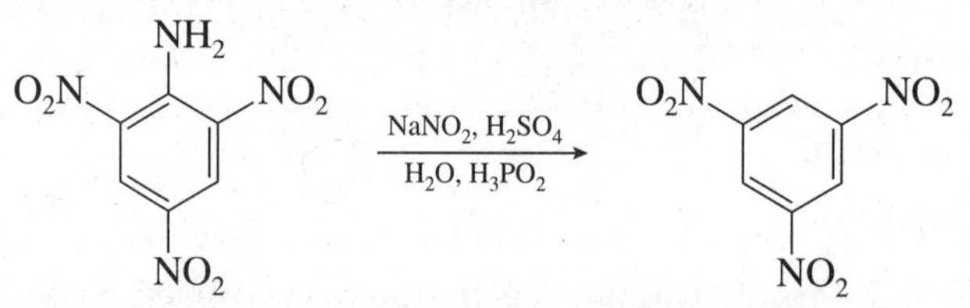

2,4,6-Trinitroaniline $\xrightarrow[H_2O,\ H_3PO_2]{NaNO_2,\ H_2SO_4}$ 1,3,5-Trinitrobenzene (60–65%)

(*q*) Ethanol, like hypophosphorous acid, is an effective reagent for the reduction of aryl diazonium salts.

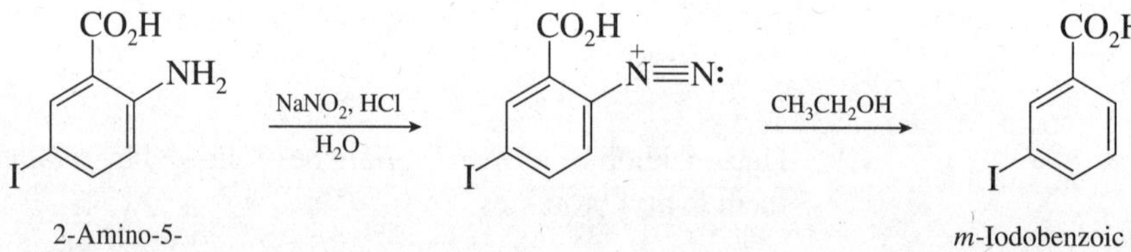

2-Amino-5-iodobenzoic acid $\xrightarrow[H_2O]{NaNO_2,\ HCl}$ (diazonium, $\overset{+}{N}{\equiv}N{:}$) $\xrightarrow{CH_3CH_2OH}$ *m*-Iodobenzoic acid (86–93%)

(*r*) Diazotization of aniline followed by addition of a phenol yields a bright-red diazo-substituted phenol. The diazonium ion acts as an electrophile toward the activated aromatic ring of the phenol.

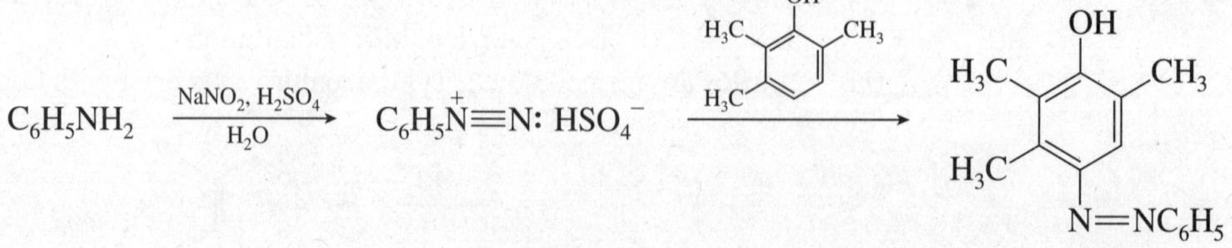

$C_6H_5NH_2 \xrightarrow[H_2O]{NaNO_2,\ H_2SO_4} C_6H_5\overset{+}{N}{\equiv}N{:}\ HSO_4^-$ $\xrightarrow{\text{2,3,6-trimethylphenol}}$ 2,3,6-Trimethyl-4-(phenylazo)phenol (98%) ($N{=}NC_6H_5$)

Aniline → Benzenediazonium hydrogen sulfate → 2,3,6-Trimethyl-4-(phenylazo)phenol (98%)

(*s*) Nitrosation of *N,N*-dialkylarylamines takes place on the ring at the position para to the dialkylamino group.

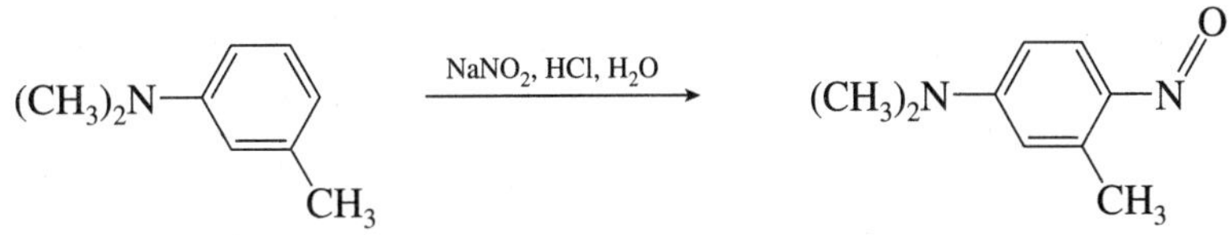

$(CH_3)_2N$–(m-tolyl) $\xrightarrow{NaNO_2,\ HCl,\ H_2O}$ $(CH_3)_2N$–(3-methyl-4-nitrosophenyl)

*N,N*-Dimethyl-*m*-toluidine

3-Methyl-4-nitroso-*N,N*-dimethylaniline (83%)

**22.36** (*a*) 4-Methylpiperidine can participate in intermolecular hydrogen bonding in the liquid phase.

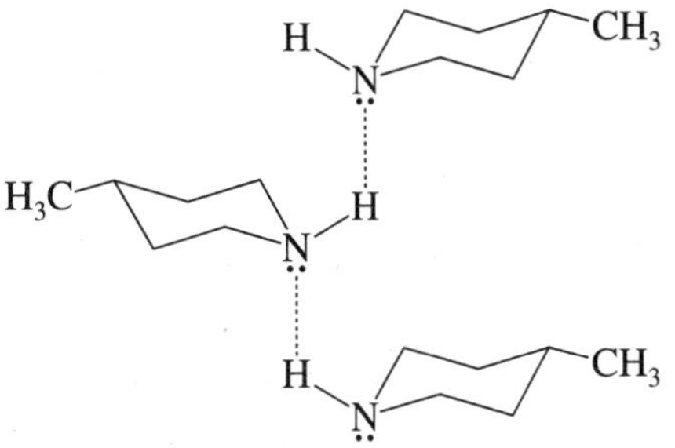

These hydrogen bonds must be broken in order for individual 4-methylpiperidine molecules to escape into the gas phase. *N*-Methylpiperidine lacks a proton bonded to nitrogen and so cannot engage in intermolecular hydrogen bonding. Less energy is required to transfer a molecule of *N*-methylpiperidine to the gaseous state, and therefore it has a lower boiling point than 4-methylpiperidine.

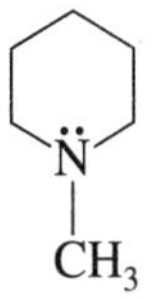

*N*-Methylpiperidine;
no hydrogen bonding possible
to other *N*-methylpiperidine molecules

(*b*) The two products are diastereomeric quaternary ammonium chlorides that differ in the configuration at the nitrogen atom.

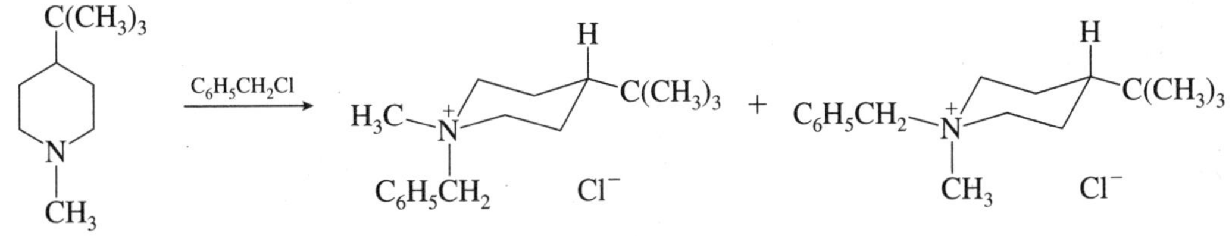

4-*tert*-Butyl-*N*-methylpiperidine $\xrightarrow{C_6H_5CH_2Cl}$ (N-benzyl axial, $H_3C$ equatorial) $Cl^-$ + ($C_6H_5CH_2$ equatorial, $CH_3$ axial) $Cl^-$

(*c*) Tetramethylammonium hydroxide cannot undergo Hofmann elimination. The only reaction that can take place is nucleophilic substitution.

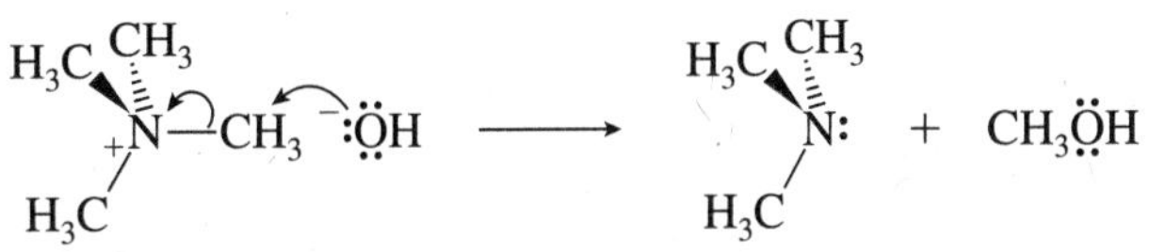

$(CH_3)_3\overset{+}{N}$–$CH_3$ $\ {}^-\!:\ddot{O}H$ $\longrightarrow$ $(CH_3)_3N:$ + $CH_3\ddot{O}H$

Tetramethylammonium hydroxide — Trimethylamine — Methanol

(*d*) The key intermediate in the reaction of an amine with nitrous acid is the corresponding diazonium ion.

$$CH_3CH_2CH_2NH_2 \xrightarrow[H_2O]{NaNO_2,\ HCl} CH_3CH_2CH_2—\overset{+}{N}{\equiv}N:$$

1-Propanamine Propyldiazonium ion

Loss of nitrogen from this diazonium ion is accompanied by a hydride shift to form a secondary carbocation.

$$CH_3CH(H)CH_2—\overset{+}{N}{\equiv}N: \longrightarrow CH_3\overset{+}{C}HCH_3 + :N{\equiv}N:$$

Propyldiazonium ion Isopropyl cation Nitrogen

Capture of isopropyl cation by water yields the major product of the reaction, 2-propanol.

$$CH_3\overset{+}{C}HCH_3 + H_2O \longrightarrow CH_3CH(\overset{+}{O}H_2)CH_3 \longrightarrow CH_3CH(OH)CH_3 + H^+$$

Isopropyl cation Water 2-Propanol

**22.37** Alcohols are converted to *p*-toluenesulfonate esters by reaction with *p*-toluenesulfonyl chloride. None of the bonds to the stereogenic center is affected in this reaction.

$$CH_3(CH_2)_5(H)(H_3C)C—OH + H_3C—C_6H_4—SO_2Cl \xrightarrow{pyridine} CH_3(CH_2)_5(H)(H_3C)C—OSO_2—C_6H_4—CH_3$$

(*S*)-2-Octanol *p*-Toluenesulfonyl chloride (*S*)-1-Methylheptyl *p*-toluenesulfonate (compound A)

Displacement of the *p*-toluenesulfonate leaving group by sodium azide in an $S_N2$ process and proceeds with inversion of configuration.

$$:\overset{-}{N}{=}\overset{+}{N}{=}\overset{-}{N}: + CH_3(CH_2)_5(H)(H_3C)C—OSO_2—C_6H_4—CH_3 \longrightarrow :\overset{-}{N}{=}\overset{+}{N}{=}N—C(H)(CH_2)_5CH_3(CH_3) + {}^{-}OSO_2—C_6H_4—CH_3$$

(*S*)-1-Methylheptyl *p*-toluenesulfonate (compound A) (*R*)-1-Methylheptyl azide (compound B)

Reduction of the azide yields a primary amine. A nitrogen–nitrogen bond is cleaved; all the bonds to the stereogenic center remain intact.

$$:\overset{-}{N}{=}\overset{+}{N}{=}\ddot{N}—C(H)(CH_2)_5CH_3(CH_3) \xrightarrow[2.\ H_2O,\ HO^-]{1.\ LiAlH_4} H_2\ddot{N}—C(H)(CH_2)_5CH_3(CH_3)$$

(*R*)-1-Methylheptyl azide (compound B) (*R*)-2-Octanamine (compound C)

**22.38** (*a*) The overall transformation can be expressed as $RBr \rightarrow RCH_2NH_2$. In many cases this can be carried out via a nitrile, as $RBr \rightarrow RCN \rightarrow RCH_2NH_2$. In this case, however, the substrate is 1-bromo-2,2-dimethylpropane, an alkyl halide that reacts very slowly in nucleophilic substitution processes. Carbon–carbon bond formation with 1-bromo-2,2-dimethylpropane can be achieved more effectively by carboxylation of the corresponding Grignard reagent.

$$(CH_3)_3CCH_2Br \xrightarrow[\text{2. } CO_2 \text{ 3. } H_3O^+]{\text{1. Mg}} (CH_3)_3CCH_2CO_2H$$

1-Bromo-2,2-dimethylpropane → 3,3-Dimethylbutanoic acid (63%)

The carboxylic acid can then be converted to the desired amine by reduction of the derived amide.

$$(CH_3)_3CCH_2CO_2H \xrightarrow[\text{2. } NH_3]{\text{1. } SOCl_2} (CH_3)_3CCH_2\overset{O}{\overset{\|}{C}}NH_2 \xrightarrow[\text{2. } H_2O]{\text{1. } LiAlH_4} (CH_3)_3CCH_2CH_2NH_2$$

3,3-Dimethylbutanoic acid → 3,3-Dimethylbutanamide (51%) → 3,3-Dimethyl-1-butanamine (57%)

The yields listed in parentheses are those reported in the chemical literature for this synthesis.

(*b*) Consider the starting materials in relation to the desired product.

$$H_2C{=}CH(CH_2)_8CH_2 \text{—} N(\text{pyrrolidine}) \Longrightarrow H_2C{=}CH(CH_2)_8\overset{O}{\overset{\|}{C}}OH + \text{pyrrolidine (N–H)}$$

*N*-(10-Undecenyl)pyrrolidine ⇒ 10-Undecenoic acid + Pyrrolidine

The synthetic tasks are to form the necessary carbon–nitrogen bond and to reduce the carbonyl group to a methylene group. This has been accomplished by way of the amide as a key intermediate.

$$H_2C{=}CH(CH_2)_8\overset{O}{\overset{\|}{C}}OH \xrightarrow[\text{2. pyrrolidine}]{\text{1. } SOCl_2} H_2C{=}CH(CH_2)_8\overset{O}{\overset{\|}{C}}\text{—N(pyrrolidine)} \xrightarrow[\text{2. } H_2O]{\text{1. } LiAlH_4} H_2C{=}CH(CH_2)_8CH_2\text{—N(pyrrolidine)}$$

10-Undecenoic acid → *N*-(10-Undecenoyl)pyrrolidine (75%) → *N*-(10-Undecenyl)pyrrolidine (66%)

A second approach utilizes reductive amination following conversion of the starting carboxylic acid to an aldehyde.

$$H_2C{=}CH(CH_2)_8\overset{O}{\overset{\|}{C}}OH \xrightarrow[\text{2. } H_2O]{\text{1. } LiAlH_4} H_2C{=}CH(CH_2)_8CH_2OH \xrightarrow[CH_2Cl_2]{\text{PCC or PDC}} H_2C{=}CH(CH_2)_8\overset{O}{\overset{\|}{C}}H$$

10-Undecenoic acid → 10-Undecen-1-ol → 10-Undecenal

The reducing agent in the reductive amination process cannot be hydrogen, because that would result in hydrogenation of the double bond. Sodium cyanoborohydride is required.

$$H_2C{=}CH(CH_2)_8\overset{O}{\overset{\|}{C}}H + \text{pyrrolidine (N–H)} \xrightarrow{NaBH_3CN} H_2C{=}CH(CH_2)_8CH_2\text{—N(pyrrolidine)}$$

10-Undecenal + Pyrrolidine → *N*-(10-Undecenyl)pyrrolidine

(*c*) It is stereochemistry that determines the choice of which synthetic method to employ in introducing the amine group. The carbon–nitrogen bond must be formed with inversion of configuration at the alcohol carbon. Conversion of the alcohol to its *p*-toluenesulfonate ester ensures that the leaving group is introduced with exactly the same stereochemistry as the alcohol.

$$\text{cis-2-Phenoxycyclopentanol} + H_3C{-}C_6H_4{-}SO_2Cl \xrightarrow{\text{pyridine}} \text{cis-2-Phenoxycyclopentyl } p\text{-toluenesulfonate}$$

*cis*-2-Phenoxycyclopentanol; *p*-Toluenesulfonyl chloride; *cis*-2-Phenoxycyclopentyl *p*-toluenesulfonate

Once the leaving group has been introduced with the proper stereochemistry, it can be displaced by a nitrogen nucleophile suitable for subsequent conversion to an amine.

$$\text{cis-}C_6H_5O{-}C_5H_8{-}OSO_2C_6H_4CH_3 \xrightarrow{NaN_3} \text{trans-}C_6H_5O{-}C_5H_8{-}N_3 \xrightarrow[\text{2. } H_2O]{\text{1. } LiAlH_4} \text{trans-}C_6H_5O{-}C_5H_8{-}NH_2$$

*cis*-2-Phenoxycyclopentyl *p*-toluenesulfonate; *trans*-2-Phenoxycyclopentyl azide (90%); *trans*-2-Phenoxycyclopentylamine

(As actually reported, the azide was reduced by hydrogenation over a palladium catalyst, and the amine was isolated as its hydrochloride salt in 66% yield.)

(*d*) Recognition that the primary amine is derivable from the corresponding nitrile by reduction,

$$C_6H_5CH_2N(CH_3)CH_2CH_2CH_2CH_2NH_2 \Rightarrow C_6H_5CH_2N(CH_3)CH_2CH_2CH_2C{\equiv}N$$

and that the necessary tertiary amine function can be introduced by a nucleophilic substitution reaction between the two given starting materials suggests the following synthesis.

$$C_6H_5CH_2NHCH_3 + BrCH_2CH_2CH_2CN \longrightarrow C_6H_5CH_2N(CH_3)CH_2CH_2CH_2CN \xrightarrow[\text{2. } H_2O]{\text{1. } LiAlH_4} C_6H_5CH_2N(CH_3)CH_2CH_2CH_2CH_2NH_2$$

*N*-Methylbenzylamine; 4-Bromobutanenitrile; *N*-Benzyl-*N*-methyl-1,4-butanediamine

Alkylation of *N*-methylbenzylamine with 4-bromobutanenitrile has been achieved in 92% yield in the presence of potassium carbonate as a weak base to neutralize the hydrogen bromide produced. The nitrile may be reduced with lithium aluminum hydride, as shown in the equation, or by catalytic hydrogenation. Catalytic hydrogenation over platinum gave the desired diamine, isolated as its hydrochloride salt, in 90% yield.

(*e*) The overall transformation may be viewed retrosynthetically as follows:

$$ArCH_2N(CH_3)_2 \Rightarrow ArCH_2Br \Rightarrow ArCH_3$$

$$Ar = NC{-}C_6H_4{-}$$

The sequence that presents itself begins with benzylic bromination with *N*-bromosuccinimide.

$$NC{-}C_6H_4{-}CH_3 \xrightarrow[\text{benzoyl peroxide, } CCl_4\text{, heat}]{N\text{-bromosuccinimide}} NC{-}C_6H_4{-}CH_2Br$$

*p*-Cyanotoluene; *p*-Cyanobenzyl bromide

The reaction shown in the equation has been reported in the chemical literature and gave the benzylic bromide in 60% yield.

Treatment of this bromide with dimethylamine gives the desired product. (The isolated yield was 83% by this method.)

NC–C₆H₄–$CH_2Br$ + $(CH_3)_2NH$ ⟶ NC–C₆H₄–$CH_2N(CH_3)_2$

*p*-Cyanobenzyl bromide · Dimethylamine · *p*-Cyano-*N*,*N*-dimethylbenzylamine

**22.39** (*a*) This problem illustrates the application of the Sandmeyer reaction to the preparation of aryl cyanides. Diazotization of *p*-nitroaniline followed by treatment with copper(I) cyanide converts it to *p*-nitrobenzonitrile.

*p*-Nitroaniline ($NH_2$, $NO_2$) —1. $NaNO_2$, HCl, $H_2O$; 2. CuCN→ *p*-Nitrobenzonitrile (CN, $NO_2$)

(*b*) An acceptable pathway becomes apparent when it is realized that the amino group in the product is derived from the nitro group of the starting material. Two chlorines are introduced by electrophilic aromatic substitution, the third by a Sandmeyer reaction.

3,4,5-trichloroaniline (Cl, Cl, Cl, $NH_2$) ⇒ 1,2,3-trichloro-5-nitrobenzene (Cl, Cl, Cl, $NO_2$) ⇒ 2,6-dichloro-4-nitroaniline ($NH_2$, Cl, Cl, $NO_2$) ⇒ *p*-nitroaniline ($NH_2$, $NO_2$)

Two of the required chlorine atoms can be introduced by chlorination of the starting material, *p*-nitroaniline.

*p*-Nitroaniline ($NH_2$, $NO_2$) —$Cl_2$→ 2,6-Dichloro-4-nitroaniline ($NH_2$, Cl, Cl, $NO_2$)

The third chlorine can be introduced via the Sandmeyer reaction. Reduction of the nitro group completes the synthesis of 3,4,5-trichloroaniline.

2,6-Dichloro-4-nitroaniline ($NH_2$, Cl, Cl, $NO_2$) —1. $NaNO_2$, HCl, $H_2O$; 2. CuCl→ 1,2,3-Trichloro-5-nitrobenzene (Cl, Cl, Cl, $NO_2$) —reduce→ 3,4,5-Trichloroaniline (Cl, Cl, Cl, $NH_2$)

The reduction step has been carried out by hydrogenation with a nickel catalyst in 70% yield.

(*c*) The amino group that is present in the starting material facilitates the introduction of the bromine substituents, and is then removed by reductive deamination.

p-Nitroaniline $\xrightarrow[\text{acetic acid}]{Br_2}$ 2,6-Dibromo-4-nitroaniline (95%) $\xrightarrow[\text{2. ethanol}]{\text{1. } NaNO_2, H_2O, H^+}$ 1,3-Dibromo-5-nitrobenzene (70%)

Hypophosphorous acid has also been used successfully in the reductive deamination step.

(*d*) Reduction of the nitro group of the 1,3-dibromo-5-nitrobenzene prepared in the preceding part of this problem gives the desired product. The customary reducing agents used for the reduction of nitroarenes would all be suitable.

1,3-Dibromo-5-nitrobenzene [prepared from *p*-nitroaniline as in part (*c*)] $\xrightarrow{H_2, Ni}$ 3,5-Dibromoaniline (80%)

(*e*) The synthetic objective is

$HO-C_6H_4-NHCCH_3$ (C=O)

*p*-Acetamidophenol

This compound, known as **acetaminophen** and used as an analgesic to reduce fever and relieve minor pain, may be prepared from *p*-nitroaniline by way of *p*-nitrophenol.

*p*-Nitroaniline $\xrightarrow[\text{2. heat}]{\text{1. } NaNO_2, H_2O, H_2SO_4}$ *p*-Nitrophenol $\xrightarrow[\text{2. acetylate}]{\text{1. reduce}}$ *p*-Acetamidophenol

Any of the customary reducing agents suitable for converting aryl nitro groups to arylamines (Fe, HCl; Sn, HCl; $H_2$, Ni) may be used. Acetylation of *p*-aminophenol may be carried out with acetyl chloride or acetic anhydride. The amino group of *p*-aminophenol is more nucleophilic than the hydroxyl group and is acetylated preferentially.

**22.40** (*a*) Replacement of an amino substituent by a bromine is readily achieved by the Sandmeyer reaction.

*o*-Anisidine $\xrightarrow[\text{2. CuBr}]{\text{1. } NaNO_2, HBr, H_2O}$ *o*-Bromoanisole (88–93%)

(*b*) This conversion demonstrates the replacement of an amino substituent by fluorine via the Schiemann reaction.

$OCH_3$, $NH_2$ — 1. $NaNO_2$, HCl, $H_2O$; 2. $HBF_4$ → $OCH_3$, $\overset{+}{N}\equiv N:$ $\bar{B}F_4$ — heat → $OCH_3$, F

*o*-Anisidine — *o*-Methoxybenzenediazonium fluoroborate (57%) — *o*-Fluoroanisole (53%)

(*c*) We can use the *o*-fluoroanisole prepared in part (*b*) to prepare 3-fluoro-4-methoxyacetophenone by Friedel–Crafts acylation.

$OCH_3$, $NH_2$ — as in part (*b*) → $OCH_3$, F — $CH_3\overset{O}{\overset{\|}{C}}O\overset{O}{\overset{\|}{C}}CH_3$, $AlCl_3$ → $OCH_3$, F, $\overset{O}{\overset{\|}{C}}CH_3$

*o*-Anisidine — *o*-Fluoroanisole — 3-Fluoro-4-methoxyacetophenone (70–80%)

Remember from Section 12.16 that it is the more activating substituent that determines the regioselectivity of electrophilic aromatic substitution when an arene bears two different substituents. Methoxy is a strongly activating substituent; fluorine is slightly deactivating. Friedel–Crafts acylation takes place at the position para to the methoxy group.

(*d*) The *o*-fluoroanisole prepared in part (*b*) serves nicely as a precursor to 3-fluoro-4-methoxybenzonitrile via diazonium salt chemistry.

$OCH_3$, F, CN ⇒ $OCH_3$, F, $NH_2$ ⇒ $OCH_3$, F, $NO_2$ ⇒ $OCH_3$, F [from part (*b*)]

The desired sequence of reactions to carry out the synthesis is

$OCH_3$, $NH_2$ — as in part (*b*) → $OCH_3$, F — $HNO_3$ → $OCH_3$, F, $NO_2$ — $H_2$, Pt → $OCH_3$, F, $NH_2$

*o*-Anisidine — *o*-Fluoroanisole — 2-Fluoro-4-nitroanisole (53%) — 4-Amino-2-fluoroanisole (85%)

↓ 1. $NaNO_2$, HCl, $H_2O$; 2. CuCN

$OCH_3$, F, CN

3-Fluoro-4-methoxybenzonitrile (46%)

Conversion of *o*-fluoroanisole to 4-amino-2-fluoroanisole proceeds in the conventional way by preparation and reduction of a nitro derivative. Once the necessary arylamine is at hand, it is converted to the nitrile by a Sandmeyer reaction.

(*e*) Diazotization followed by hydrolysis of the 4-amino-2-fluoroanisole prepared as an intermediate in part (*d*) yields the desired phenol.

*o*-Anisidine → (as in part (*d*)) → 4-Amino-2-fluoroanisole → (1. $NaNO_2$, $H_2SO_4$, $H_2O$; 2. heat) → 3-Fluoro-4-methoxyphenol (70%)

**22.41** (*a*) The carboxyl group of *p*-aminobenzoic acid can be derived from the methyl group of *p*-methylaniline by oxidation. First, however, the nitrogen must be acylated so as to protect the ring from oxidation.

*p*-Aminobenzoic acid ⇒ $p$-$CH_3CONHC_6H_4CO_2H$ ⇒ *p*-Methylaniline

The sequence of reactions to be used is

*p*-Methylaniline → ($CH_3COCCH_3$, acetic anhydride) → *p*-Methylacetanilide → ($K_2Cr_2O_7$, $H_2SO_4$, $H_2O$, heat) → *p*-Acetamidobenzoic acid → (1. HCl, $H_2O$; 2. neutralize) → *p*-Aminobenzoic acid

(*b*) Attachment of fluoro and propanoyl groups to a benzene ring is required. The fluorine substituent can be introduced by way of the diazonium tetrafluoroborate, the propanoyl group by way of a Friedel–Crafts acylation. Because the fluorine substituent is ortho, para-directing, introducing it first gives the proper orientation of substituents.

Ethyl *p*-fluorophenyl ketone ⇒ Fluorobenzene ⇒ Aniline

Fluorobenzene is prepared from aniline by the Schiemann reaction, shown in Section 22.18. Aniline is, of course, prepared from benzene via nitrobenzene. Friedel–Crafts acylation of fluorobenzene has been carried out with the results shown and gives the required ethyl *p*-fluorophenyl ketone as the major product.

$$\text{Fluorobenzene} + CH_3CH_2\overset{O}{\overset{\|}{C}}Cl \xrightarrow{AlCl_3} \text{Ethyl } p\text{-fluorophenyl ketone } (86\%)$$

Fluorobenzene    Propanoyl chloride    Ethyl *p*-fluorophenyl ketone (86%)

(*c*) Our synthetic plan is based on the essential step of forming the fluorine derivative from an amine by way of a diazonium salt.

1-Bromo-2-fluoro-3,5-dimethylbenzene ⇒ 2-Bromo-4,6-dimethylaniline ⇒ 2,4-Dimethylaniline

The required substituted aniline is derived from *m*-xylene by a standard synthetic sequence.

$$m\text{-Xylene} \xrightarrow[H_2SO_4]{HNO_3} \text{1,3-Dimethyl-4-nitrobenzene (98\%)} \xrightarrow[2.\ HO^-]{1.\ Fe,\ HCl} \text{2,4-Dimethylaniline}$$

$$\text{2,4-Dimethylaniline} \xrightarrow{Br_2} \text{2-Bromo-4,6-dimethylaniline}$$

$$\text{2-Bromo-4,6-dimethylaniline} \xrightarrow[\text{2. } HBF_4;\ \text{3. heat}]{\text{1. } NaNO_2,\ HCl,\ H_2O,\ 0°C} \text{1-Bromo-2-fluoro-3,5-dimethylbenzene (60\%)}$$

(*d*) In this problem two nitrogen-containing groups of the starting material are each to be replaced by a halogen substituent. The task is sufficiently straightforward that it may be confronted directly.

**Replace amino group by bromine:**

$$\text{2-Methyl-4-nitro-1-naphthylamine} \xrightarrow[\text{2. } CuBr]{\text{1. } NaNO_2,\ HBr,\ H_2O} \text{1-Bromo-2-methyl-4-nitronaphthalene (82\%)}$$

2-Methyl-4-nitro-1-naphthylamine    1-Bromo-2-methyl-4-nitronaphthalene (82%)

**Reduce nitro group to amine:**

1-Bromo-2-methyl-4-nitronaphthalene → (1. Fe, HCl; 2. $HO^-$) → 4-Bromo-3-methyl-1-naphthylamine

**Replace amino group by fluorine:**

4-Bromo-3-methyl-1-naphthylamine → (1. $NaNO_2$, HCl, $H_2O$, 0–5°C; 2. $HBF_4$; 3. heat) → 1-Bromo-4-fluoro-2-methylnaphthalene (64%)

(*e*) Bromination of the starting material will introduce the bromine substituent at the correct position, that is, ortho to the *tert*-butyl group.

*p*-*tert*-Butylnitrobenzene → ($Br_2$, Fe) → 2-Bromo-1-*tert*-butyl-4-nitrobenzene

The desired product will be obtained if the nitro group can be removed. This is achieved by its conversion to the corresponding amine, followed by reductive deamination.

2-Bromo-1-*tert*-butyl-4-nitrobenzene → ($H_2$, Ni (or other appropriate reducing agent)) → 3-Bromo-4-*tert*-butylaniline → (1. $NaNO_2$, $H^+$; 2. $H_3PO_2$) → *o*-Bromo-*tert*-butylbenzene

(*f*) The proper orientation of the chlorine substituent can be achieved only if it is introduced after the nitro group is reduced.

m-chloro-tert-butylbenzene ($C(CH_3)_3$, Cl) ⇒ 3-chloro-4-tert-butylaniline ($C(CH_3)_3$, Cl, $NH_2$) ⇒ p-tert-butylaniline ($C(CH_3)_3$, $NH_2$) ⇒ p-tert-butylnitrobenzene ($C(CH_3)_3$, $NO_2$)

The correct sequence of reactions to carry out this synthesis is shown.

$p$-*tert*-Butylnitrobenzene $\xrightarrow{H_2,\ Ni}$ $p$-*tert*-Butylaniline $\xrightarrow{\text{acetic anhydride}}$ $p$-*tert*-Butylacetanilide $\xrightarrow{Cl_2}$ 4-*tert*-Butyl-2-chloroacetanilide

4-*tert*-Butyl-2-chloroacetanilide $\xrightarrow{\text{hydrolysis to remove acetyl group}}$ 4-*tert*-Butyl-2-chloroaniline $\xrightarrow{1.\ NaNO_2,\ H^+;\ 2.\ H_3PO_2}$ $m$-*tert*-Butylchlorobenzene

(*g*) The orientation of substituents in the target molecule can be achieved by using an amino group to control the regiochemistry of bromination, then removing it by reductive deamination.

1-Bromo-3,5-diethylbenzene ⇒ 2-Bromo-4,6-diethylaniline ⇒ 2,4-Diethylaniline

The amino group is introduced in the standard fashion by nitration of an arene followed by reduction.

This analysis leads to the synthesis shown.

$m$-Diethylbenzene $\xrightarrow{HNO_3}$ 2,4-Diethyl-1-nitrobenzene (75–80%) $\xrightarrow{H_2,\ Ni,\ \text{ethanol}}$ 2,4-Diethylaniline (80–90%)

2,4-Diethylaniline $\xrightarrow{Br_2}$ 2-Bromo-4,6-diethylaniline (40%) $\xrightarrow{1.\ NaNO_2,\ H^+,\ H_2O;\ 2.\ H_3PO_2}$ 1-Bromo-3,5-diethylbenzene (70%)

(*h*) In this exercise the two nitrogen substituents are differentiated; one is an amino nitrogen, the other an amide nitrogen. By keeping them differentiated they can be manipulated independently. Remove one amino group completely before deprotecting the other.

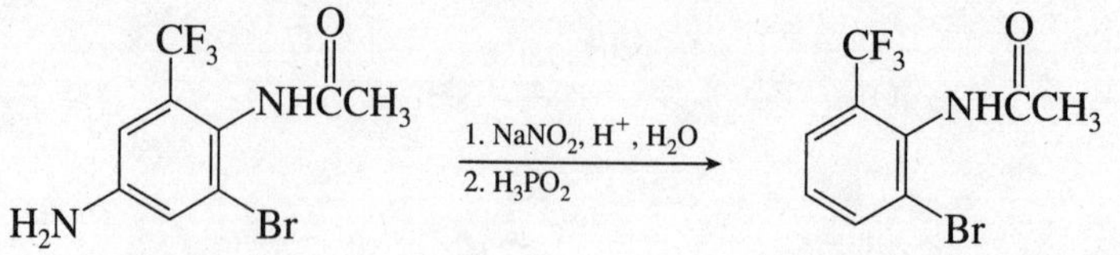

4-Amino-2-bromo-6-(trifluoromethyl)acetanilide $\xrightarrow[\text{2. } H_3PO_2]{\text{1. } NaNO_2,\ H^+,\ H_2O}$ 2-Bromo-6-(trifluoromethyl)-acetanilide (92%)

Once the acetyl group has been removed by hydrolysis, the molecule is ready for introduction of the iodo substituent by way of a diazonium salt.

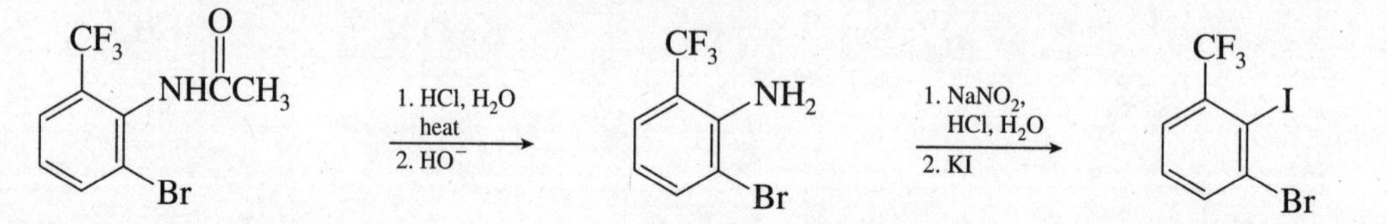

2-Bromo-6-(trifluoromethyl)-acetanilide $\xrightarrow[\text{2. } HO^-]{\text{1. } HCl,\ H_2O,\ \text{heat}}$ 2-Bromo-6-(trifluoromethyl)-aniline (69%) $\xrightarrow[\text{2. KI}]{\text{1. } NaNO_2,\ HCl,\ H_2O}$ 1-Bromo-2-iodo-3-(trifluoromethyl)benzene (87%)

(*i*) To convert the designated starting material to the indicated product, both the nitro group and the ester function must be reduced and a carbon–nitrogen bond must be formed. Converting the starting material to an amide gives the necessary carbon–nitrogen bond and has the advantage that amides can be reduced to amines by lithium aluminum hydride. The amide can be formed intramolecularly by reducing the nitro group to an amine, then heating to cause cyclization.

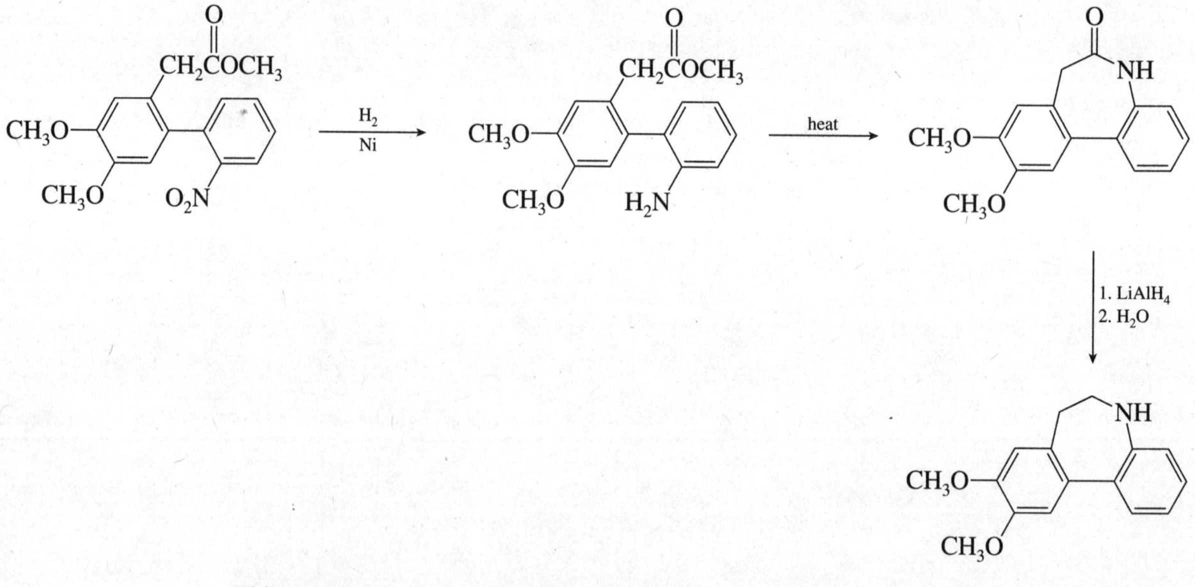

Nitro ester ($CH_2COCH_3$, $CH_3O$, $CH_3O$, $O_2N$) $\xrightarrow[\text{Ni}]{H_2}$ amino ester ($CH_2COCH_3$, $CH_3O$, $CH_3O$, $H_2N$) $\xrightarrow{\text{heat}}$ lactam (NH, $CH_3O$, $CH_3O$) $\xrightarrow[\text{2. } H_2O]{\text{1. } LiAlH_4}$ amine (NH, $CH_3O$, $CH_3O$)

This synthesis is the one described in the chemical literature. Other routes are also possible, but the one shown is short and efficient.

**22.42** Weakly basic nucleophiles react with $\alpha,\beta$-unsaturated carbonyl compounds by conjugate addition.

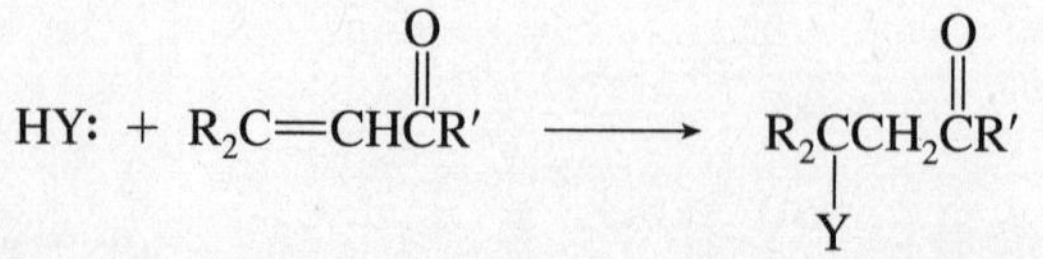

$$HY\!: + R_2C{=}CH\overset{O}{\overset{\|}{C}}R' \longrightarrow R_2\underset{Y}{\underset{|}{C}}CH_2\overset{O}{\overset{\|}{C}}R'$$

Ammonia and its derivatives are very prone to react in this way; thus conjugate addition provides a method for the preparation of $\beta$-amino carbonyl compounds.

(*a*)

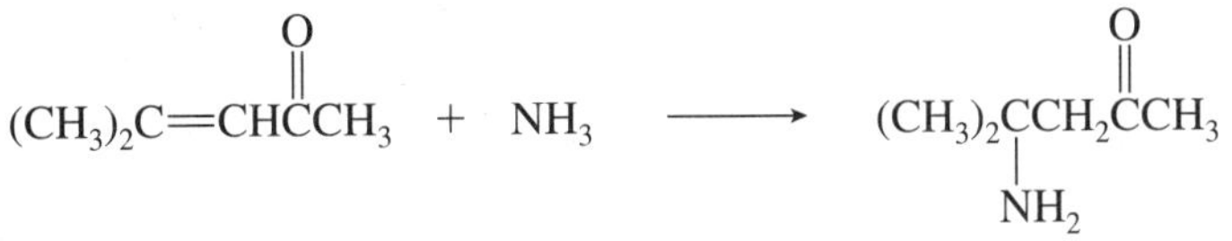

$$(CH_3)_2C{=}CHCCH_3 \ (C{=}O) + NH_3 \longrightarrow (CH_3)_2C(NH_2)CH_2CCH_3 \ (C{=}O)$$

4-Methyl-3-penten-2-one — Ammonia — 4-Amino-4-methyl-2-pentanone (63–70%)

(*b*)

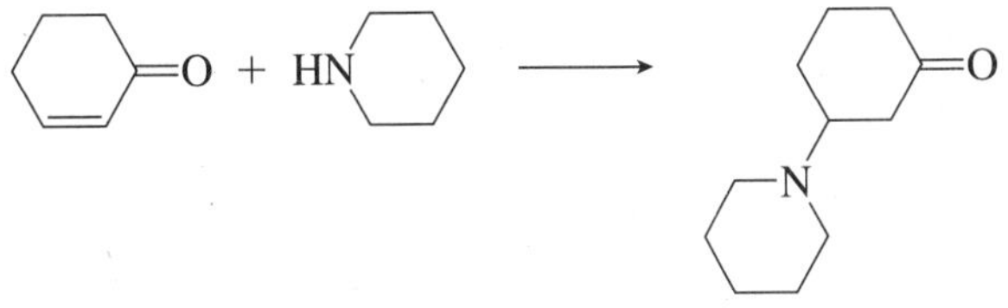

2-Cyclohexenone + Piperidine ⟶ 3-Piperidinocyclohexanone (45%)

(*c*)

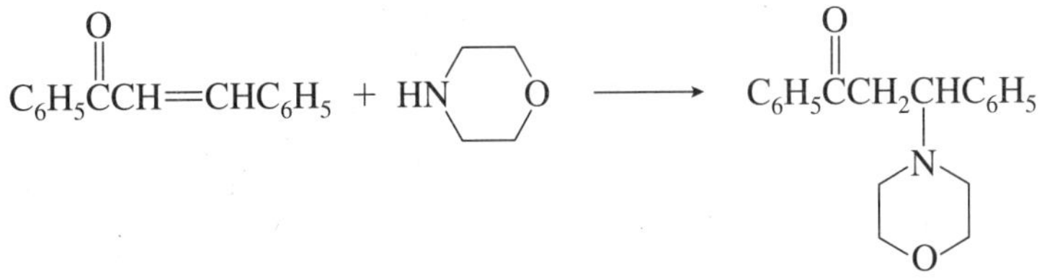

$$C_6H_5CCH{=}CHC_6H_5 \ (C{=}O) + HN\text{(morpholine)} \longrightarrow C_6H_5CCH_2CHC_6H_5 \ (C{=}O), \text{ N-morpholino on CH}$$

1,3-Diphenyl-2-propen-1-one — Morpholine — 3-Morpholino-1,3-diphenyl-1-propanone (91%)

(*d*) The conjugate addition reaction that takes place in this case is an intramolecular one and occurs in virtually 100% yield.

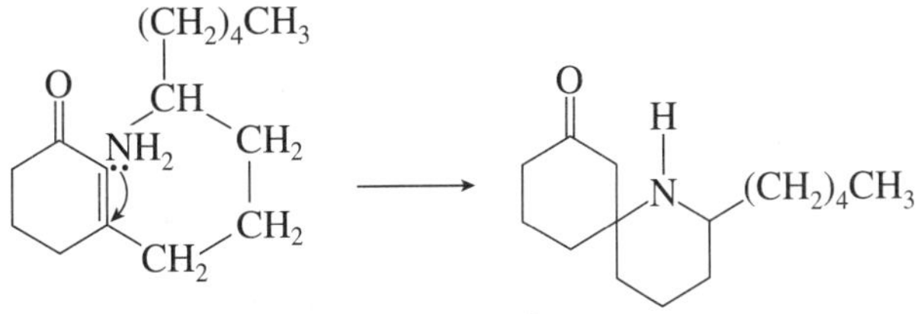

2-[$CH_2CH_2CH_2CH(NH_2)(CH_2)_4CH_3$]-2-cyclohexenone ⟶ spiro product with N–H and $(CH_2)_4CH_3$

**22.43** The first step in the synthesis is the conjugate addition of methylamine to ethyl acrylate. Two sequential Michael addition reactions take place.

$$CH_3NH_2 + H_2C{=}CHCOCH_2CH_3 \ (C{=}O) \longrightarrow CH_3NHCH_2CH_2COCH_2CH_3 \ (C{=}O)$$

Methylamine — Ethyl acrylate

$$\xrightarrow{H_2C{=}CHCO_2CH_2CH_3} CH_3N(CH_2CH_2CO_2CH_2CH_3)_2$$

Conversion of this intermediate to the desired *N*-methyl-4-piperidone requires a Dieckmann cyclization followed by decarboxylation of the resulting $\beta$-keto ester.

$CH_3CH_2OC(=O)CH_2CH_2N(CH_3)CH_2CH_2C(=O)OCH_2CH_3$ —(1. $NaOCH_2CH_3$; 2. $H^+$)→ 3-ethoxycarbonyl-*N*-methyl-4-piperidone —(1. $HO^-$, $H_2O$; 2. $H^+$; 3. heat)→ *N*-Methyl-4-piperidone

Treatment of *N*-methyl-4-piperidone with the Grignard reagent derived from bromobenzene gives a tertiary alcohol that can be dehydrated to an alkene. Hydrogenation of the alkene completes the synthesis.

*N*-Methyl-4-piperidone + $C_6H_5MgBr$ (Phenylmagnesium bromide) —(1. diethyl ether; 2. $H_3O^+$)→ 4-$C_6H_5$-4-OH-*N*-$CH_3$-piperidine —($H^+$, heat)→ 4-$C_6H_5$-*N*-$CH_3$-tetrahydropyridine —($H_2$, Pt)→ *N*-Methyl-4-phenylpiperidine (compound A)

**22.44** Sodium cyanide reacts with alkyl bromides by the $S_N2$ mechanism. Reduction of the cyano group with lithium aluminum hydride yields a primary amine. This reveals the structure of mescaline to be 2-(3,4,5-trimethoxyphenyl)ethylamine.

3,4,5-Trimethoxybenzyl bromide ($(CH_3O)_3C_6H_2CH_2Br$) —NaCN→ 2-(3,4,5-Trimethoxyphenyl)ethanenitrile ($(CH_3O)_3C_6H_2CH_2CN$) —(1. $LiAlH_4$; 2. $H_2O$)→ 2-(3,4,5-Trimethoxyphenyl)ethylamine (mescaline) ($(CH_3O)_3C_6H_2CH_2CH_2NH_2$)

**22.45** Reductive amination of a ketone with methylamine yields a secondary amine. Methamphetamine is *N*-methyl-1-phenyl-2-propanamine.

$C_6H_5CH_2C(=O)CH_3$ (Benzyl methyl ketone) + $CH_3NH_2$ (Methylamine) —($H_2$, Ni)→ $C_6H_5CH_2CH(NHCH_3)CH_3$ (*N*-Methyl-1-phenyl-2-propanamine (methamphetamine))

**22.46** There is no obvious reason why the dimethylamino group in 4-(*N*,*N*-dimethylamino)pyridine should be appreciably more basic than it is in *N*,*N*-dimethylaniline; it is the ring nitrogen of

4-(*N*,*N*-dimethylamino)pyridine that is more basic. Note that protonation of the ring nitrogen permits delocalization of the dimethylamino lone pair and dispersal of the positive charge.

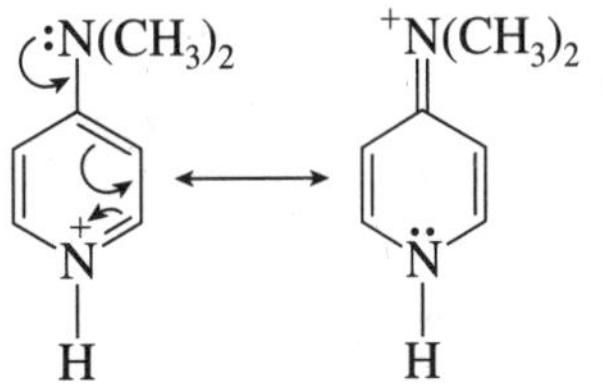

Most stable protonated form of 4-(*N*,*N*-dimethylamino)pyridine

**22.47** The $^1$H NMR spectrum of each isomer shows peaks corresponding to five aromatic protons, so compounds A and B each contain a monosubstituted benzene ring. Only four compounds of molecular formula $C_8H_{11}N$ meet this requirement.

| $C_6H_5CH_2NHCH_3$ | $C_6H_5NHCH_2CH_3$ | $C_6H_5CH(NH_2)CH_3$ | $C_6H_5CH_2CH_2NH_2$ |
|---|---|---|---|
| *N*-Methylbenzylamine | *N*-Ethylaniline | 1-Phenylethylamine | 2-Phenylethylamine |

Neither $^1$H NMR spectrum is consistent with *N*-methylbenzylamine, which would have two singlets due to the methyl and methylene groups. Likewise, the spectra are not consistent with *N*-ethylaniline, which would exhibit the characteristic triplet–quartet pattern of an ethyl group. Although a quartet occurs in the spectrum of compound A, it corresponds to only one proton, not the two that an ethyl group requires. The one-proton quartet in compound A arises from an H—C—$CH_3$ unit. Compound A is 1-phenylethylamine.

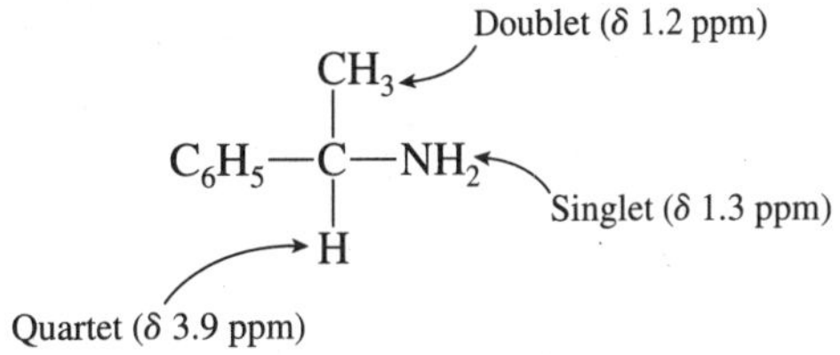

$C_6H_5$—C($CH_3$)(H)—$NH_2$

Doublet ($\delta$ 1.2 ppm) ($CH_3$)

Singlet ($\delta$ 1.3 ppm) ($NH_2$)

Quartet ($\delta$ 3.9 ppm) (H)

Compound B has an $^1$H NMR spectrum that fits 2-phenylethylamine.

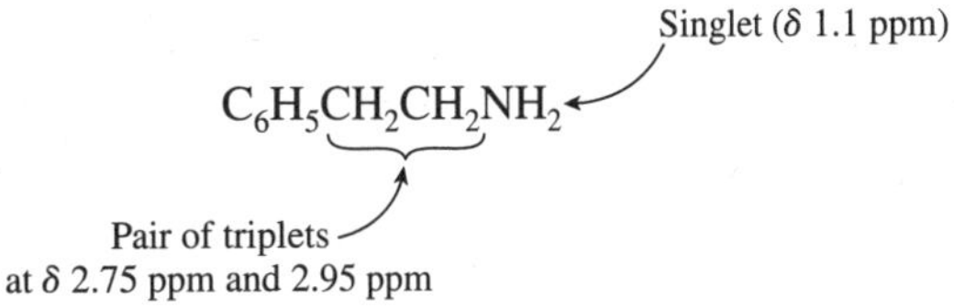

$C_6H_5CH_2CH_2NH_2$

Singlet ($\delta$ 1.1 ppm) ($NH_2$)

Pair of triplets at $\delta$ 2.75 ppm and 2.95 ppm ($CH_2CH_2$)

**22.48** Only the unshared electron pair on nitrogen that is not part of the $\pi$ electron cloud of the aromatic system will be available for protonation. Treatment of 5-methyl-$\gamma$-carboline with acid will give the salt shown.

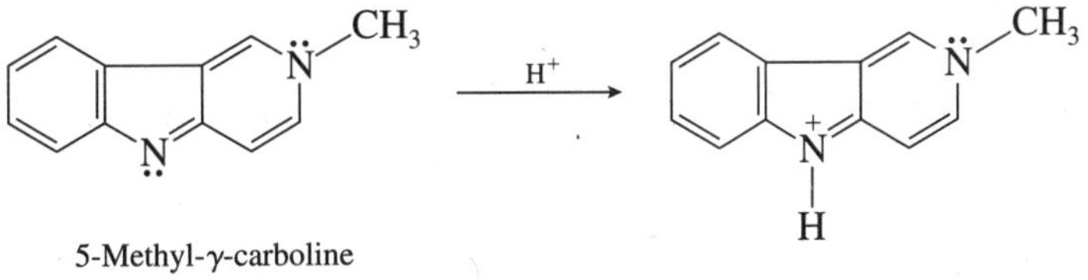

$\xrightarrow{H^+}$

5-Methyl-$\gamma$-carboline

**22.49** Write the structural formulas for the two possible compounds given in the problem and consider how their $^{13}$C NMR spectra will differ from each other. Both will exhibit their $CH_3$ carbons at high field signal, but they differ in the positions of their $CH_2$ and quaternary carbons. A carbon bonded to nitrogen is more shielded than one bonded to oxygen, because nitrogen is less electronegative than oxygen.

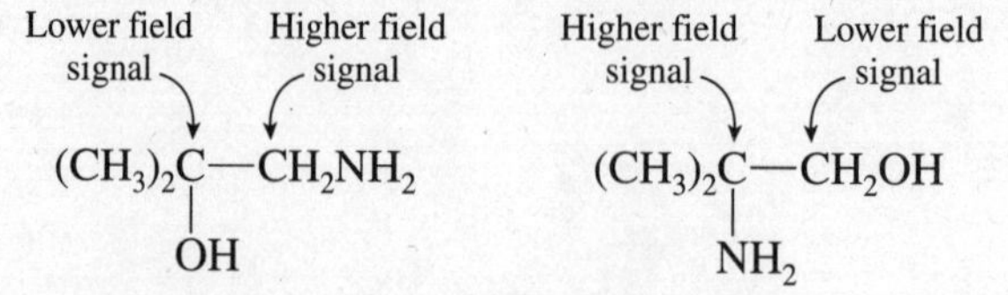

Lower field signal → $(CH_3)_2\underset{\displaystyle OH}{\underset{|}{C}}{-}CH_2NH_2$ ← Higher field signal

1-Amino-2-methyl-2-propanol

Higher field signal → $(CH_3)_2\underset{\displaystyle NH_2}{\underset{|}{C}}{-}CH_2OH$ ← Lower field signal

2-Amino-2-methyl-1-propanol

In one isomer the lowest field signal is a quaternary carbon; in the other it is a $CH_2$ group. The spectrum shown in Figure 22.10 shows the lowest field signal as a $CH_2$ group. The compound is therefore 2-amino-2-methyl-1-propanol, $(CH_3)_2\underset{\displaystyle NH_2}{\underset{|}{C}}CH_2OH$.

This compound *cannot* be prepared by reaction of ammonia with an epoxide, because in basic solution nucleophiles attack epoxides at the less hindered carbon, and therefore epoxide ring opening will give 1-amino-2-methyl-2-propanol rather than 2-amino-2-methyl-1-propanol.

$$(CH_3)_2C\!-\!CH_2 \text{ (epoxide, O bridge)} + \ddot{N}H_3 \longrightarrow (CH_3)_2\underset{\displaystyle OH}{\underset{|}{C}}CH_2NH_2$$

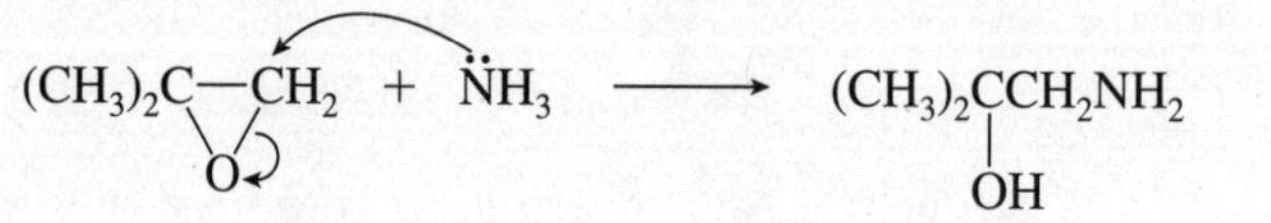

2,2-Dimethyloxirane — Ammonia — 1-Amino-2-methyl-2-propanol

# SELF-TEST

## PART A

**A-1.** Give an acceptable name for each of the following. Identify each compound as a primary, secondary, or tertiary amine.

(*a*) $CH_3CH_2\overset{\displaystyle CH_3}{\overset{|}{\underset{\displaystyle NH_2}{\underset{|}{C}}}}CH_3$

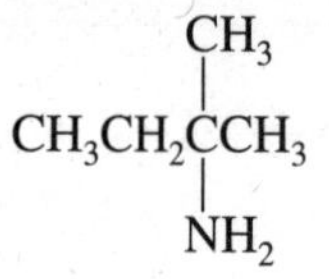

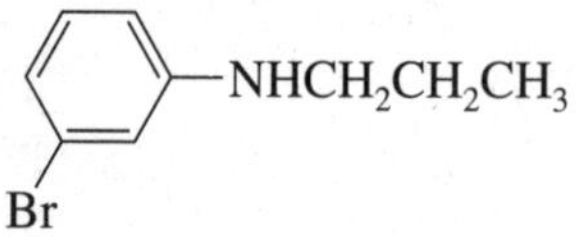

(*b*) cyclopentyl—$NHCH_3$

(*c*) 3-bromophenyl—$NHCH_2CH_2CH_3$ (Br on ring)

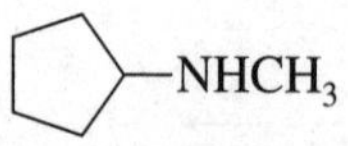

**A-2.** Provide the correct structure of the reagent omitted from each of the following reactions:

(*a*) $C_6H_5CH_2Br \xrightarrow[\text{2. LiAlH}_4 \text{ 3. } H_2O]{\text{1. ?}} C_6H_5CH_2NH_2$

(*b*) $C_6H_5CH_2Br \xrightarrow[\text{2. LiAlH}_4 \text{ 3. } H_2O]{\text{1. ?}} C_6H_5CH_2CH_2NH_2$

(*c*) $C_6H_5CH_2Br \xrightarrow[\text{2. } H_2NNH_2]{\text{1. ?}} C_6H_5CH_2NH_2 +$ phthalhydrazide (benzene ring fused to ring with two C=O and NH—NH)

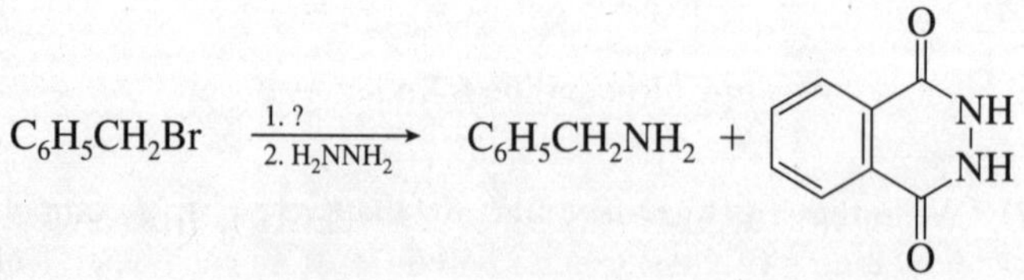

**A-3.** Provide the missing component (reactant, reagent, or product) for each of the following:

(*a*) $H_3C-C_6H_4-NH_2 \xrightarrow[H_2O]{NaNO_2,\ HCl}$ ?

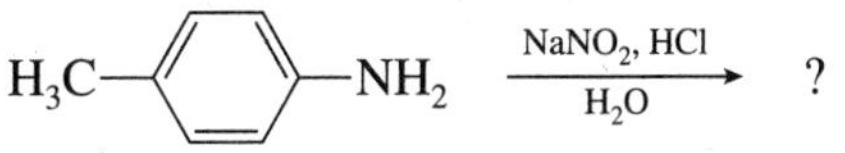

(*b*) Product of part (*a*) $\xrightarrow{CuBr}$ ?

(*c*) Product of part (*a*) $\xrightarrow{?}$ toluene

(*d*) $H_3C-C_6H_4-NH_2 \xrightarrow{?} H_3C-C_6H_4-NHCCH_3$ (with C=O)

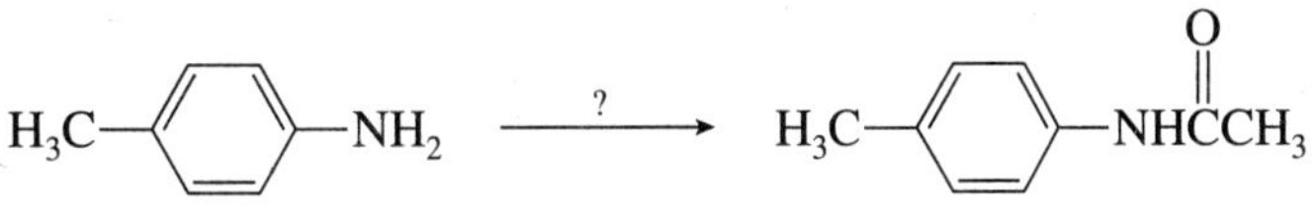

(*e*) $m$-$CH_3CH_2C_6H_4NHCCH_3$ (with C=O) $\xrightarrow[H_2SO_4]{HNO_3}$ ?

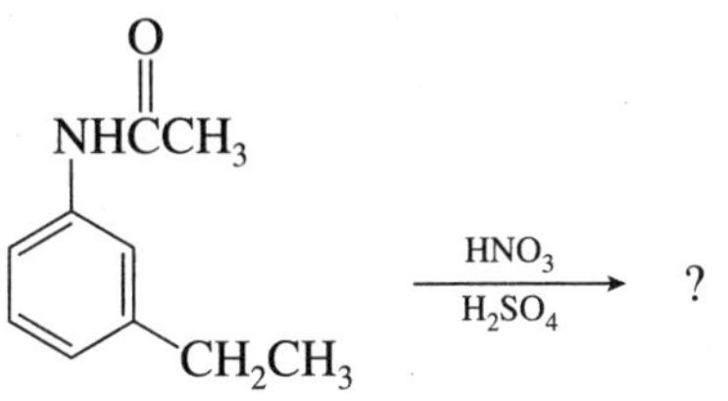

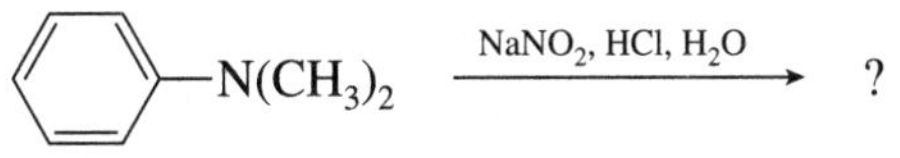

(*f*) $C_6H_5-N(CH_3)_2 \xrightarrow{NaNO_2,\ HCl,\ H_2O}$ ?

(*g*) $C_6H_5-NHCH_2CH_3 \xrightarrow{NaNO_2,\ HCl,\ H_2O}$ ?

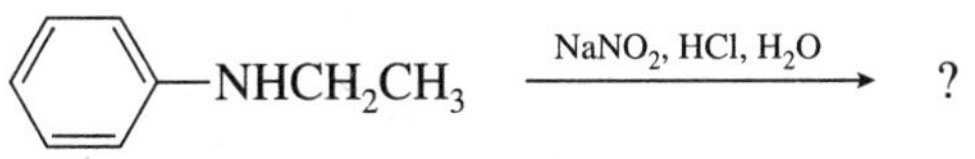

**A-4.** Provide structures for compounds A through E in the following reaction sequences:

(*a*) A $\xrightarrow{CH_3I}$ B $\xrightarrow[H_2O]{Ag_2O}$ C $\xrightarrow{heat}$ $H_2C{=}CHCH_2CH_2N(CH_3)CH_2CH_3$

(*b*) Cyclohexanone + $CH_3CH_2NH_2 \xrightarrow[CH_3OH]{NaBH_3CN}$ D $\xrightarrow[H_2O]{NaNO_2,\ HCl}$ E

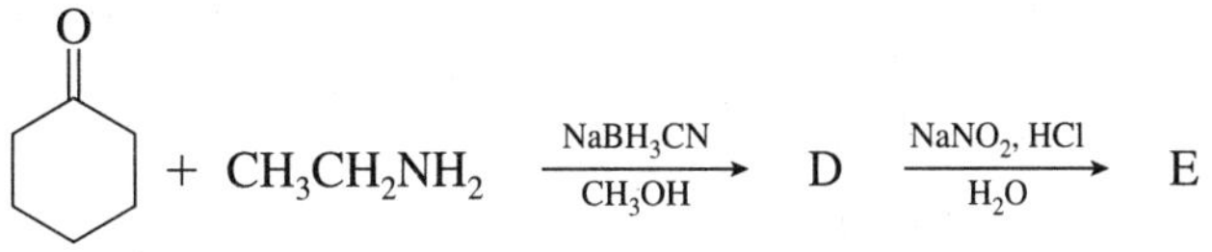

**A-5.** Give the series of reaction steps involved in the following synthetic conversions:

(*a*) $p$-$(CH_3)_3C-C_6H_4-I$ from benzene

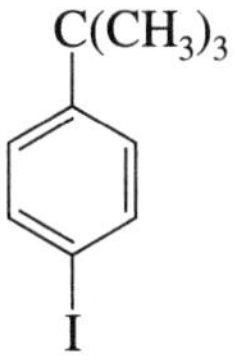

(*b*) *m*-Chloroaniline from benzene

(*c*) $C_6H_5N{=}N-C_6H_4-N(CH_3)_2$ from aniline

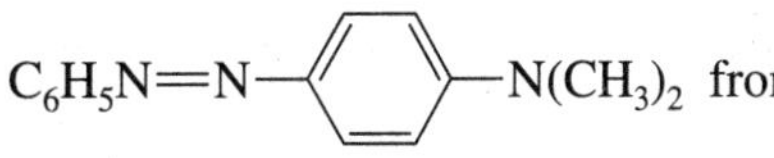

**A-6.** *p*-Nitroaniline (A) is less basic than *m*-nitroaniline (B). Using resonance structures, explain the reason for this difference.

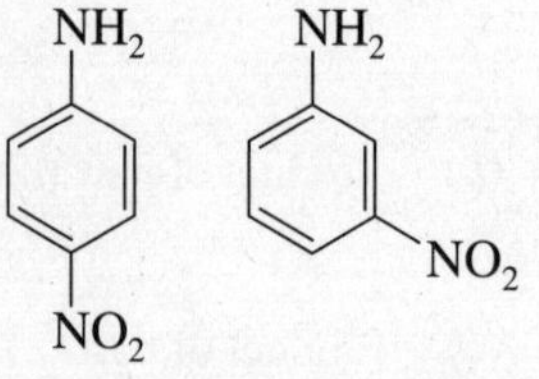

$NH_2$ $NH_2$

$NO_2$

$NO_2$

A B

**A-7.** Identify the strongest and weakest bases among the following:

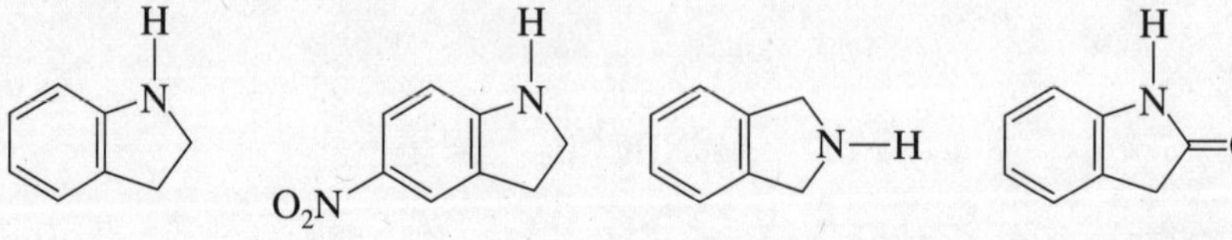

H H H

N N N—H N =O

$O_2N$

A B C D

**A-8.** Write the structures of the compounds A–D formed in the following reaction sequence:

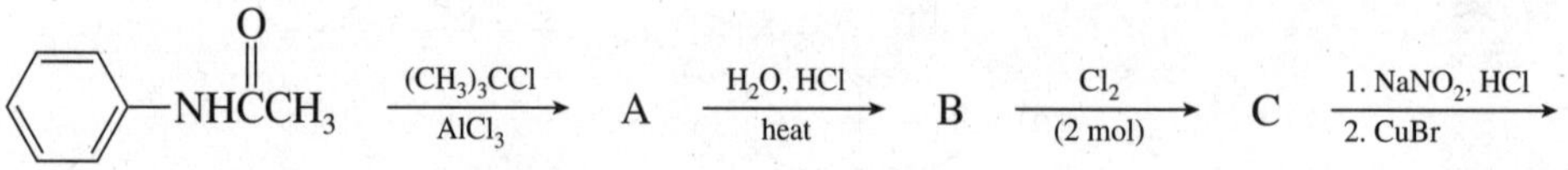

$$C_6H_5NHCCH_3 \text{ (C=O)} \xrightarrow[AlCl_3]{(CH_3)_3CCl} A \xrightarrow[heat]{H_2O,\ HCl} B \xrightarrow[(2\ mol)]{Cl_2} C \xrightarrow[2.\ CuBr]{1.\ NaNO_2,\ HCl} D$$

## PART B

**B-1.** Which of the following is a secondary amine?
(*a*) 2-Butanamine
(*b*) *N*-Ethyl-2-pentanamine
(*c*) *N*-Methylpiperidine
(*d*) *N,N*-Dimethylcyclohexylamine

**B-2.** Which of the following $C_8H_9NO$ isomers is the weakest base?
(*a*) *o*-Aminoacetophenone
(*b*) *m*-Aminoacetophenone
(*c*) *p*-Aminoacetophenone
(*d*) Acetanilide

**B-3.** Rank the following compounds in order of increasing basicity (weakest → strongest):

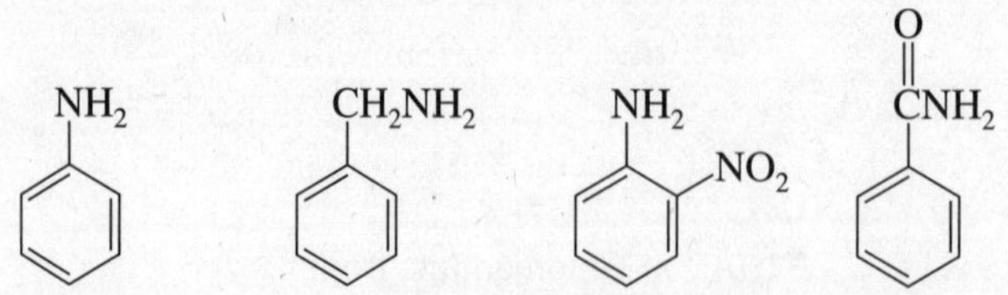

$NH_2$ $CH_2NH_2$ $NH_2$ O

$NO_2$ $\overset{\|}{C}NH_2$

1 2 3 4

(*a*) $4 < 2 < 1 < 3$
(*b*) $4 < 1 < 3 < 2$
(*c*) $4 < 3 < 1 < 2$
(*d*) $2 < 1 < 3 < 4$

**B-4.** Which of the following arylamines will *not* form a diazonium salt on reaction with sodium nitrite in hydrochloric acid?
(*a*) *m*-Ethylaniline
(*b*) 4-Chloro-2-nitroaniline
(*c*) *p*-Aminoacetophenone
(*d*) *N*-Ethyl-2-methylaniline

**B-5.** The amines shown are isomers. Choose the one with the lowest boiling point.

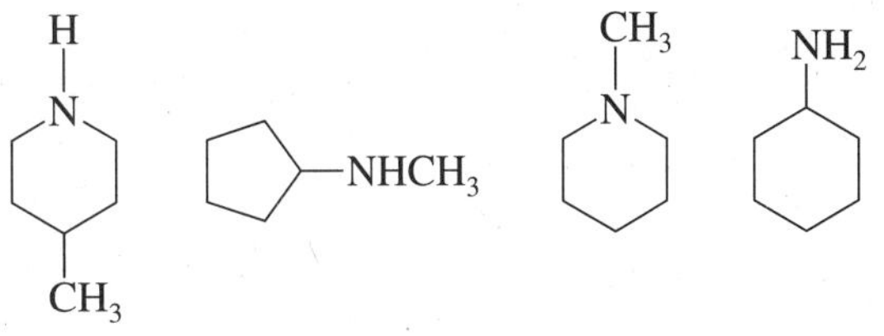

H, N, $CH_3$   —$NHCH_3$   $CH_3$, N   $NH_2$

(*a*) (*b*) (*c*) (*d*)

**B-6.** Which of the following is the strongest acid?

(*a*) 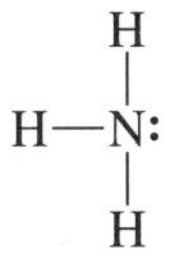 H—N: (with H above and below)
(*d*) H—$N^+$—H (with H above and below)
(*b*) 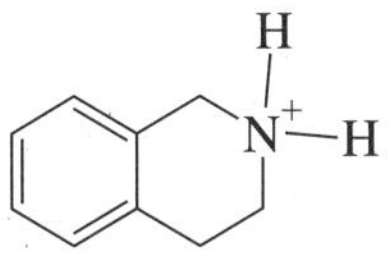 H, $N^+$—H
(*e*) 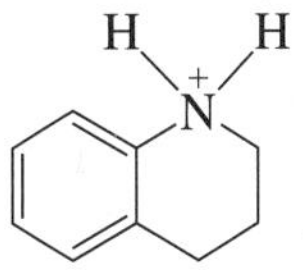 H H, $N^+$
(*c*) H H, $N^+$

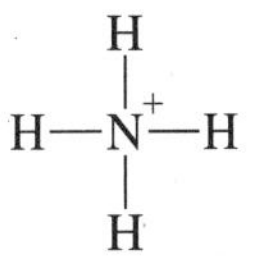

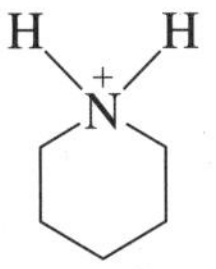

**B-7.** The reaction

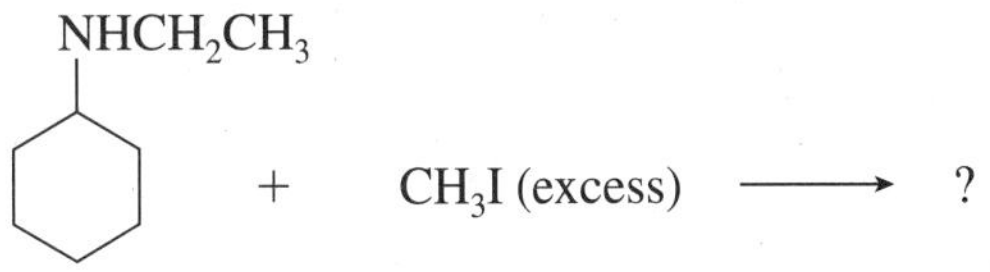

$NHCH_2CH_3$ + $CH_3I$ (excess) ⟶ ?

gives as final product
(*a*) A primary amine
(*b*) A secondary amine
(*c*) A tertiary amine
(*d*) A quaternary ammonium salt

**B-8.** A substance is soluble in dilute aqueous HCl and has a single peak in the region 3200–3500 $cm^{-1}$ in its infrared spectrum. Which of the following best fits the data?

(*a*) —$N(CH_3)_2$
(*c*) —$NHCH_3$
(*b*) —$NH_2$
(*d*) —$CO_2H$

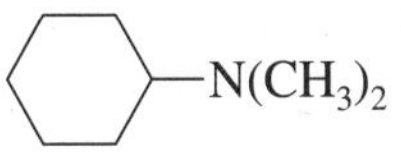

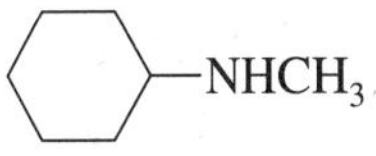

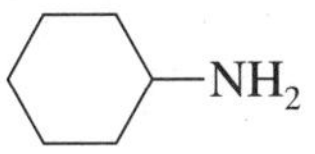

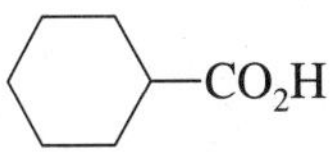

**B-9.** Identify product D in the following reaction sequence:

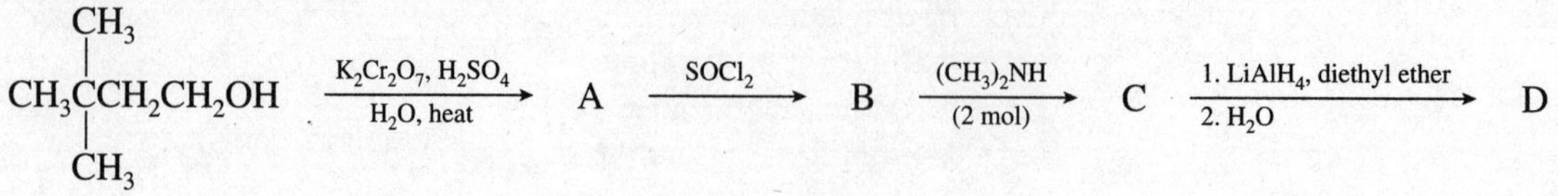

$$CH_3C(CH_3)_2CH_2CH_2OH \xrightarrow[H_2O,\ heat]{K_2Cr_2O_7,\ H_2SO_4} A \xrightarrow{SOCl_2} B \xrightarrow[(2\ mol)]{(CH_3)_2NH} C \xrightarrow[2.\ H_2O]{1.\ LiAlH_4,\ diethyl\ ether} D$$

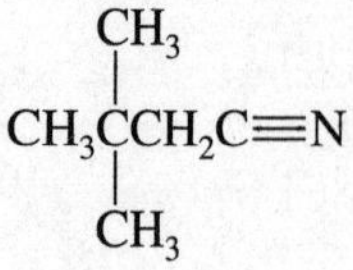

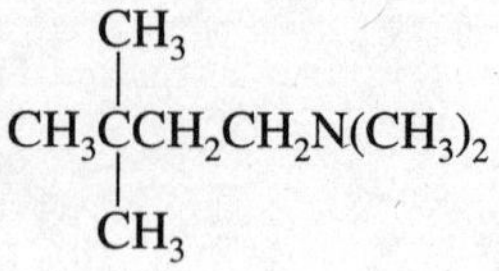

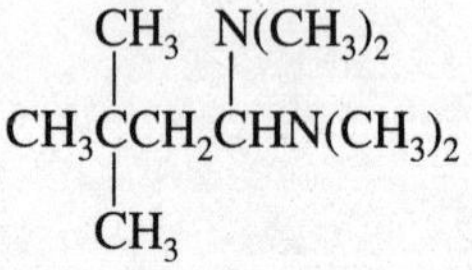

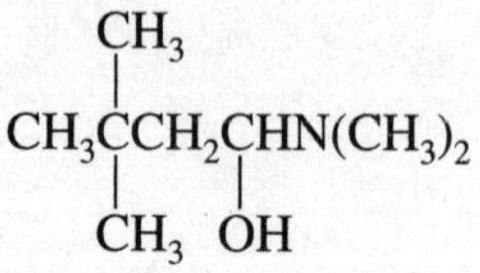

(*a*) $CH_3C(CH_3)_2CH_2C{\equiv}N$

(*b*) $CH_3C(CH_3)_2CH_2CH(N(CH_3)_2)N(CH_3)_2$

(*c*) $CH_3C(CH_3)_2CH_2C({=}O)N(CH_3)_2$

(*d*) $CH_3C(CH_3)_2CH_2CH_2N(CH_3)_2$

(*e*) $CH_3C(CH_3)_2CH_2CH(OH)N(CH_3)_2$

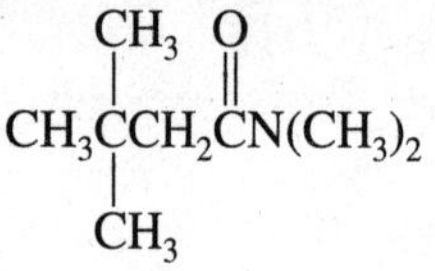

**B-10.** Which one of the following is the best catalyst for the reaction shown?

$$CH_3(CH_2)_8CH_2Br \xrightarrow[benzene]{KCN} CH_3(CH_2)_8CH_2CN$$

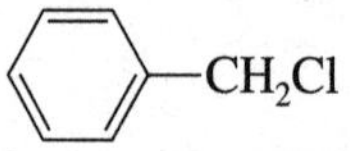

(*a*) 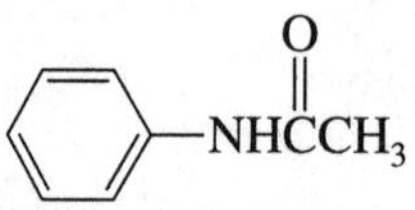 $C_6H_5{-}CH_2Cl$

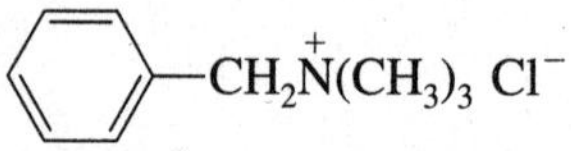

(*b*) $C_6H_5{-}NH_2$

(*c*) $C_6H_5{-}NHC({=}O)CH_3$

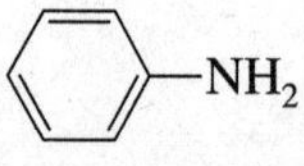

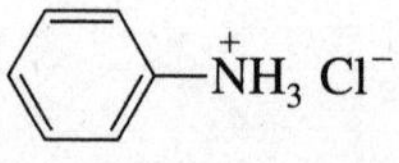

(*d*) $C_6H_5{-}\overset{+}{N}H_3\ Cl^-$

(*e*) $C_6H_5{-}CH_2\overset{+}{N}(CH_3)_3\ Cl^-$

**B-11.** What will be the *major* product of each of the two reactions shown?

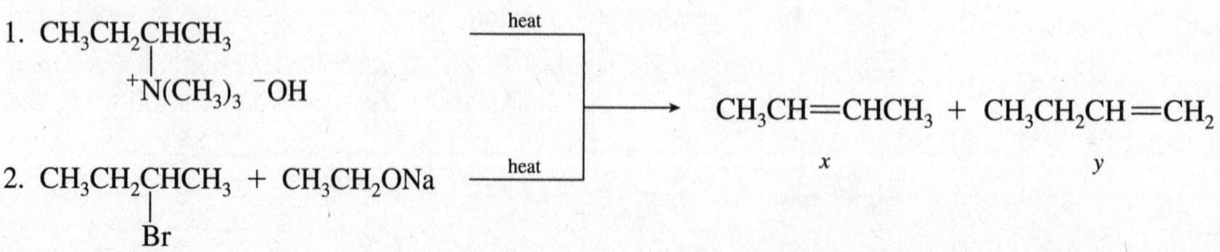

1. $CH_3CH_2CH(\overset{+}{N}(CH_3)_3)CH_3\ ^{-}OH \xrightarrow{heat}$

2. $CH_3CH_2CH(Br)CH_3 + CH_3CH_2ONa \xrightarrow{heat}$

$$\longrightarrow \underset{x}{CH_3CH{=}CHCH_3} + \underset{y}{CH_3CH_2CH{=}CH_2}$$

(*a*) 1*x*, 2*x* (*b*) 1*x*, 2*y* (*c*) 1*y*, 2*x* (*d*) 1*y*, 2*y*

**B-12.** Which sequence represents the best synthesis of 4-isopropylbenzonitrile?

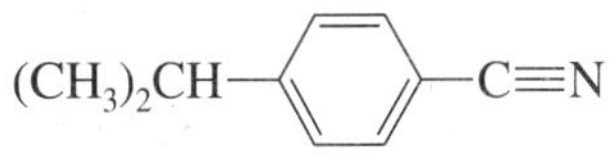

$(CH_3)_2CH-C_6H_4-C\equiv N$

4-Isopropylbenzonitrile

(*a*) 1. Benzene + $(CH_3)_2CHCl$, $AlCl_3$; 2. $Br_2$, $FeBr_3$; 3. KCN
(*b*) 1. Benzene + $(CH_3)_2CHCl$, $AlCl_3$; 2. $HNO_3$, $H_2SO_4$; 3. Fe, HCl; 4. NaOH; 5. $NaNO_2$, HCl, $H_2O$; 6. CuCN
(*c*) 1. Benzene + $(CH_3)_2CHCl$, $AlCl_3$; 2. $HNO_3$, $H_2SO_4$; 3. Fe, HCl; 4. NaOH; 5. KCN
(*d*) 1. Benzene + $HNO_3$, $H_2SO_4$; 2. $(CH_3)_2CHCl$, $AlCl_3$; 3. Fe, HCl; 4. NaOH; 5. $NaNO_2$, HCl, $H_2O$; 6. CuCN
(*e*) 1. Benzene + $HNO_3$, $H_2SO_4$; 2. Fe, HCl; 3. NaOH; 4. $NaNO_2$, HCl, $H_2O$; 5. CuCN; 6. $(CH_3)_2CHCl$, $AlCl_3$

**B-13.** The major products from the following sequence of reactions are

$$(CH_3)_2CHCH_2N(CH_2CH_3)_2 \xrightarrow{CH_3I} \xrightarrow[H_2O]{Ag_2O} \xrightarrow{heat} ?$$

(*a*) $(CH_3)_2CHCH_2NH_2 + H_2C{=}CH_2$
(*b*) $(CH_3)_2NCH_2CH_3 + H_2C{=}C(CH_3)_2$
(*c*) $(CH_3)_2CHCH_2N(CH_3)CH_2CH_3 + H_2C{=}CH_2$
(*d*) $(CH_3)_3\overset{+}{N}CH_2CH_3\ I^- + H_2C{=}CH_2$
(*e*) None of these combinations of products is correct.

**B-14.** Which compound yields an *N*-nitrosoamine after treatment with nitrous acid ($NaNO_2$, HCl)?

(*a*) $C_6H_5-CH_2NH_2$
(*b*) $C_6H_5-N(CH_2)_5$ (N-phenylpiperidine)
(*c*) $C_6H_5-NHCH_3$
(*d*) $H_3C-C_6H_4-NH_2$
(*e*) $C_6H_5-\overset{O}{\overset{\|}{C}}NH_2$

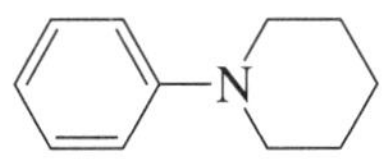

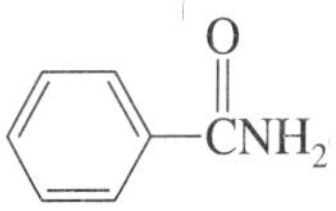

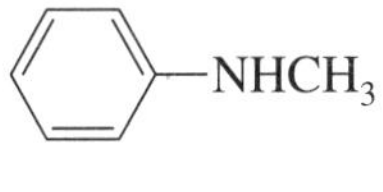

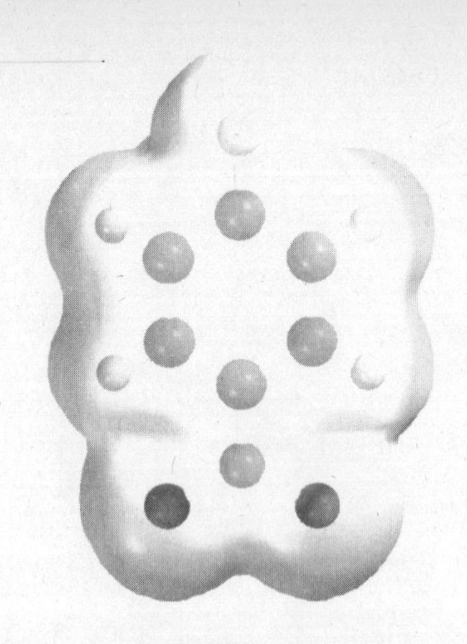

# CHAPTER 23
## ARYL HALIDES

## SOLUTIONS TO TEXT PROBLEMS

**23.1** There are four isomers of $C_7H_7Cl$ that contain a benzene ring, namely, *o*, *m*, and *p*-chlorotoluene and benzyl chloride.

$CH_3$ / Cl — *o*-Chlorotoluene; $CH_3$ / Cl — *m*-Chlorotoluene; $CH_3$ / Cl — *p*-Chlorotoluene; $CH_2Cl$ — Benzyl chloride

Of this group only benzyl chloride is not an aryl halide; its halogen is not attached to the aromatic ring but to an $sp^3$-hybridized carbon. Benzyl chloride has the weakest carbon–halogen bond, its measured carbon–chlorine bond dissociation energy being only 293 kJ/mol (70 kcal/mol). Homolytic cleavage of this bond produces a resonance-stabilized benzyl radical.

$$C_6H_5-CH_2-\ddot{C}l: \longrightarrow C_6H_5-\dot{C}H_2 + \cdot\ddot{C}l:$$

Benzyl chloride → Benzyl radical + Chlorine atom

**23.2** (*b*) The negatively charged sulfur in $C_6H_5CH_2\ddot{S}:^- Na^+$ is a good nucleophile, which displaces chloride from 1-chloro-2,4-dinitrobenzene.

$$\text{1-Chloro-2,4-dinitrobenzene} \xrightarrow{C_6H_5CH_2\ddot{S}:^-\ Na^+} \text{Benzyl 2,4-dinitrophenyl sulfide } (SCH_2C_6H_5,\ NO_2,\ NO_2) + Cl^-$$

1-Chloro-2,4-dinitrobenzene (Cl, $NO_2$, $NO_2$) → Benzyl 2,4-dinitrophenyl sulfide

(*c*) The nitrogen in ammonia has an unshared electron pair and is nucleophilic; it displaces chloride from 1-chloro-2,4-dinitrobenzene.

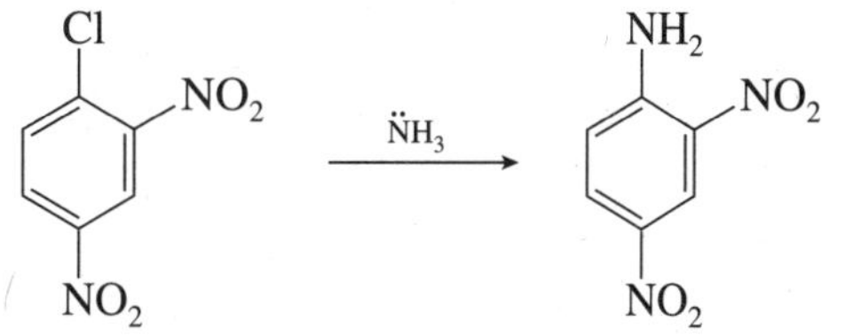

$\xrightarrow{\ddot{N}H_3}$

1-Chloro-2,4-dinitrobenzene — 2,4-Dinitroaniline

(*d*) As with ammonia, methylamine is nucleophilic and displaces chloride.

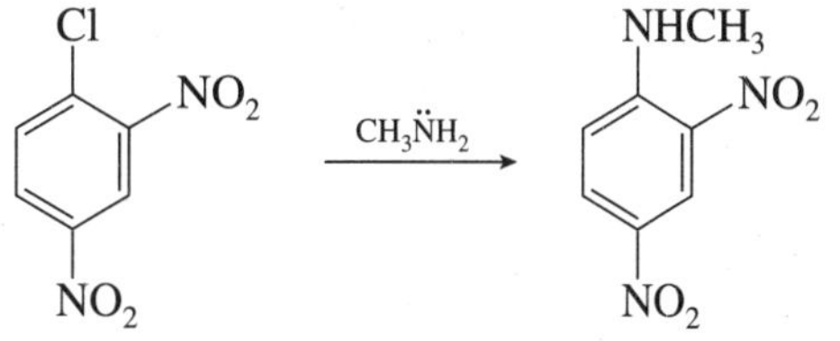

$\xrightarrow{CH_3\ddot{N}H_2}$

1-Chloro-2,4-dinitrobenzene — *N*-Methyl-2,4-dinitroaniline

**23.3** The most stable resonance structure for the cyclohexadienyl anion formed by reaction of methoxide ion with *o*-fluoronitrobenzene involves the nitro group and has the negative charge on oxygen.

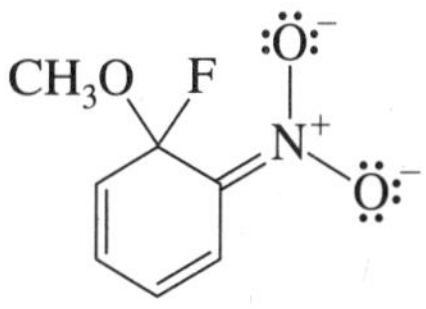

**23.4** The positions that are activated toward nucleophilic attack are those that are ortho and para to the nitro group. Among the carbons that bear a bromine leaving group in 1,2,3-tribromo-5-nitrobenzene, only C-2 satisfies this requirement.

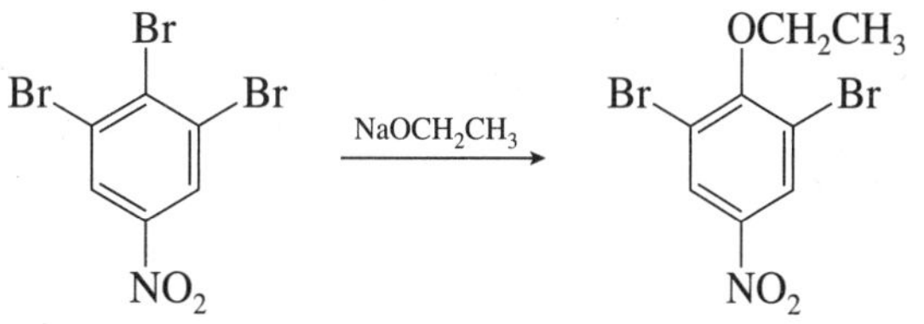

$\xrightarrow{NaOCH_2CH_3}$

1,2,3-Tribromo-5-nitrobenzene — 1,3-Dibromo-2-ethoxy-5-nitrobenzene

**23.5** Nucleophilic addition occurs in the rate-determining step at one of the six equivalent carbons of hexafluorobenzene to give the cyclohexadienyl anion intermediate.

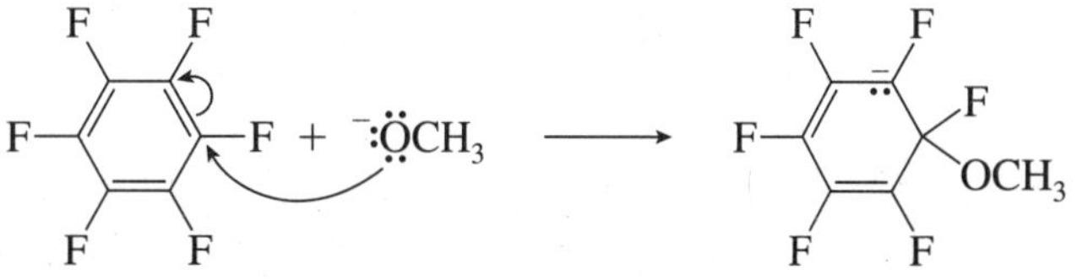

Hexafluorobenzene + Methoxide ion ⟶ Cyclohexadienyl anion intermediate

Elimination of fluoride ion from the cyclohexadienyl anion intermediate restores the aromaticity of the ring and completes the reaction.

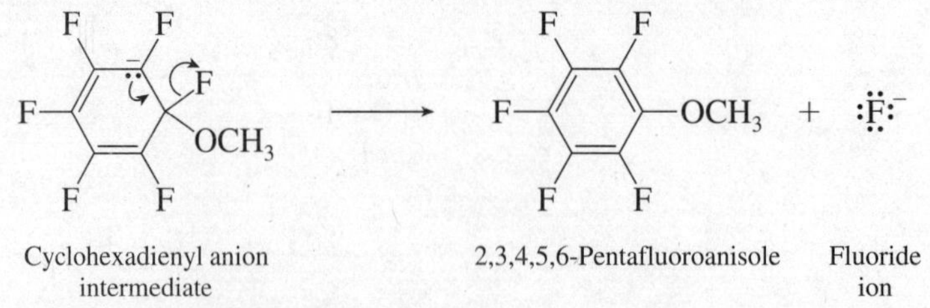

Cyclohexadienyl anion intermediate → 2,3,4,5,6-Pentafluoroanisole + Fluoride ion

**23.6** 4-Chloropyridine is more reactive toward nucleophiles than 3-chloropyridine because the anionic intermediate formed by reaction of 4-chloropyridine has its charge on nitrogen. Because nitrogen is more electronegative than carbon, the intermediate is more stable.

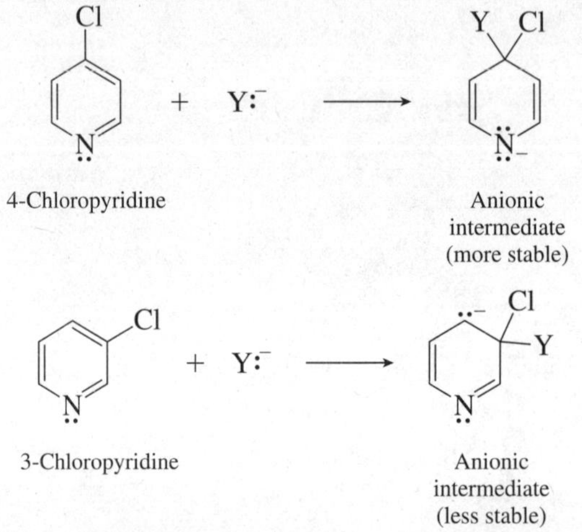

4-Chloropyridine + $Y:^-$ → Anionic intermediate (more stable)

3-Chloropyridine + $Y:^-$ → Anionic intermediate (less stable)

**23.7** The aryl halide is incapable of elimination and so cannot form the benzyne intermediate necessary for substitution by the elimination–addition pathway.

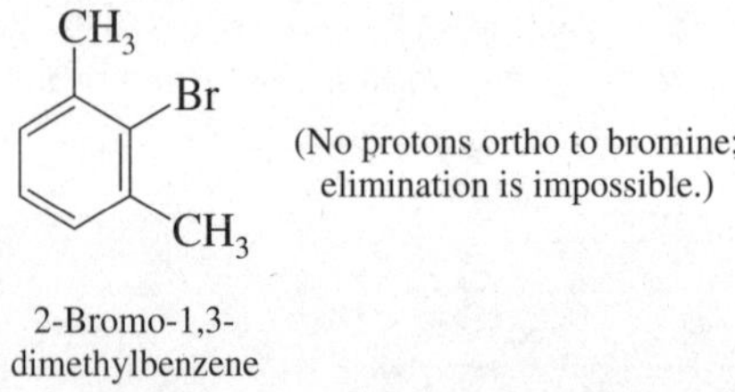

2-Bromo-1,3-dimethylbenzene

(No protons ortho to bromine; elimination is impossible.)

**23.8** The aryne intermediate from *p*-iodotoluene can undergo addition of hydroxide ion at the position meta to the methyl group or para to it. The two isomeric phenols are *m*- and *p*-methylphenol.

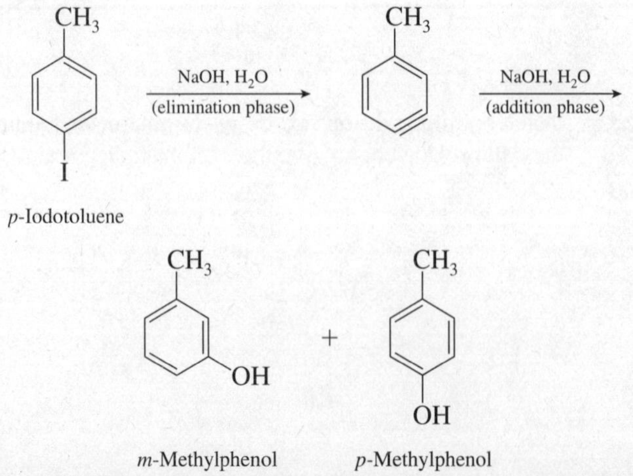

*p*-Iodotoluene $\xrightarrow[\text{(elimination phase)}]{\text{NaOH, } H_2O}$ aryne $\xrightarrow[\text{(addition phase)}]{\text{NaOH, } H_2O}$ *m*-Methylphenol + *p*-Methylphenol

**23.9** The "triple bond" of benzyne adds to the diene system of furan.

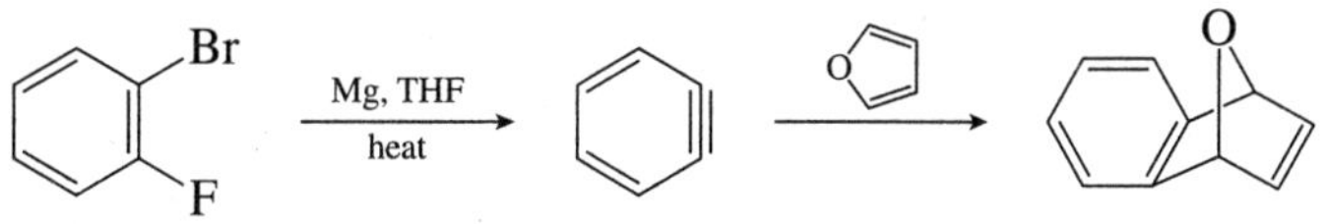

Br, F; Mg, THF, heat; O

**23.10** (*a*) $CH_3$, Cl

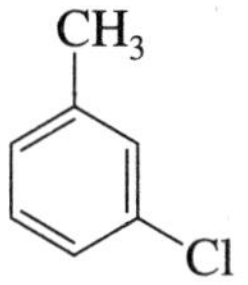

*m*-Chlorotoluene

(*b*) $OCH_3$, Br, Br

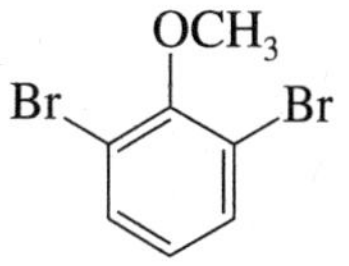

2,6-Dibromoanisole

(*c*) $CH{=}CH_2$, F

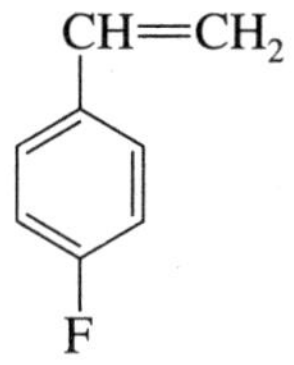

*p*-Fluorostyrene

(*d*) I, I; 3, 2, 2′, 3′, 4, 1, 1′, 4′, 5, 6, 6′, 5′

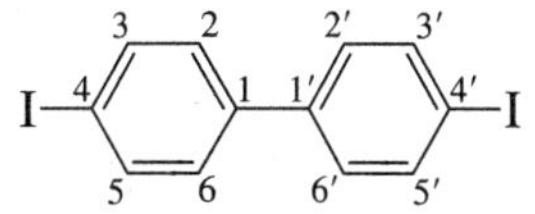

4,4′-Diiodobiphenyl

(*e*) Br, Cl, $O_2N$

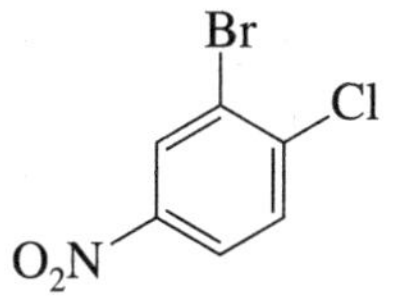

2-Bromo-1-chloro-4-nitrobenzene

(*f*) $ClCHCH_3$

1-Chloro-1-phenylethane (*Note*: This compound is not an aryl halide.)

(*g*) $CH_2Cl$, Br

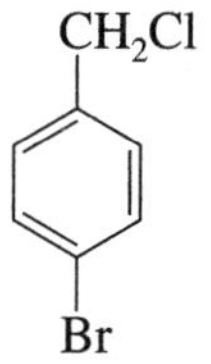

*p*-Bromobenzyl chloride

(*h*) 8, 1, 2, Cl, 7, 6, 3, 5, 4

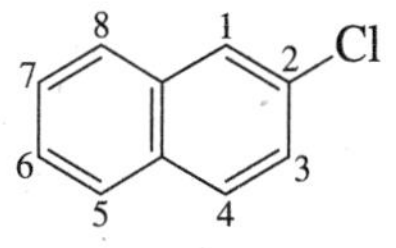

2-Chloronaphthalene

(*i*) Cl, Cl, 8, 1, 7, 2, 6, 3, 5, 4

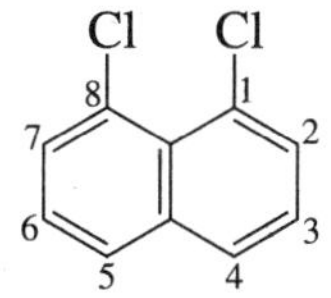

1,8-Dichloronaphthalene

(*j*) 6, 5, 7, 4, 8, 3, 2, 9, 1, 10, F

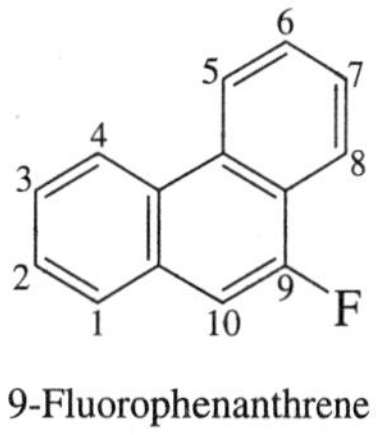

9-Fluorophenanthrene

**23.11** (*a*) Chlorine is a weakly deactivating, ortho, para-directing substituent.

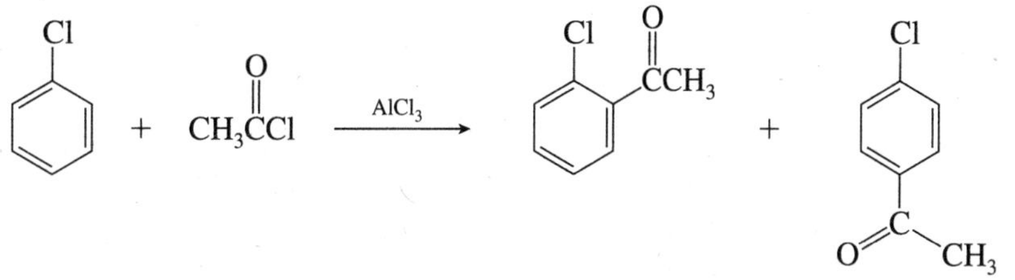

Chlorobenzene + Acetyl chloride ($CH_3\overset{O}{\overset{\|}{C}}Cl$) $\xrightarrow{AlCl_3}$ *o*-Chloroacetophenone + *p*-Chloroacetophenone

Cl; Cl, $\overset{O}{\overset{\|}{C}}CH_3$; Cl, O, C, $CH_3$

(*b*) Bromobenzene reacts with magnesium to give a Grignard reagent.

$$C_6H_5Br + Mg \xrightarrow{\text{diethyl ether}} C_6H_5MgBr$$

Bromobenzene — Phenylmagnesium bromide

(*c*) Protonation of the Grignard reagent in part (*b*) converts it to benzene.

$$C_6H_5MgBr \xrightarrow[HCl]{H_2O} C_6H_6$$

Phenylmagnesium bromide — Benzene

(*d*) Aryl halides react with lithium in much the same way that alkyl halides do, to form organolithium reagents.

$$C_6H_5I + 2Li \xrightarrow{\text{diethyl ether}} C_6H_5Li + LiI$$

Iodobenzene — Lithium — Phenyllithium — Lithium iodide

(*e*) With a base as strong as sodium amide, nucleophilic aromatic substitution by the elimination–addition mechanism takes place.

Bromobenzene $\xrightarrow[NH_3]{NaNH_2}$ Benzyne $\xrightarrow[NH_3]{NaNH_2}$ Aniline

(*f*) The benzyne intermediate from *p*-bromotoluene gives a mixture of *m*- and *p*-methylaniline.

*p*-Bromotoluene $\xrightarrow[NH_3]{NaNH_2}$ 4-Methylbenzyne $\xrightarrow[NH_3]{NaNH_2}$ *m*-Methylaniline + *p*-Methylaniline

(*g*) Nucleophilic aromatic substitution of bromide by ammonia occurs by the addition–elimination mechanism.

1-Bromo-4-nitrobenzene $\xrightarrow{NH_3}$ *p*-Nitroaniline

(*h*) The bromine attached to the benzylic carbon is far more reactive than the one on the ring and is the one replaced by the nucleophile.

$$Br-C_6H_4-CH_2Br \xrightarrow{NaCN} Br-C_6H_4-CH_2CN$$

*p*-Bromobenzyl bromide — *p*-Bromobenzyl cyanide

(*i*) The aromatic ring of *N*, *N*-dimethylaniline is very reactive and is attacked by *p*-chlorobenzenediazonium ion.

$(CH_3)_2N-C_6H_5$ + $:N{\equiv}\overset{+}{N}-C_6H_4-Cl$ ⟶ $(CH_3)_2N-C_6H_4-N{=}N-C_6H_4-Cl$

*N*,*N*-Dimethylaniline — *p*-Chlorobenzenediazonium ion — 4-(4′-Chlorophenylazo)-*N*,*N*-dimethylaniline

(*j*) Hexafluorobenzene undergoes substitution of one of its fluorines on reaction with nucleophiles such as sodium hydrogen sulfide.

$C_6F_6$ + NaSH ⟶ $C_6F_5SH$

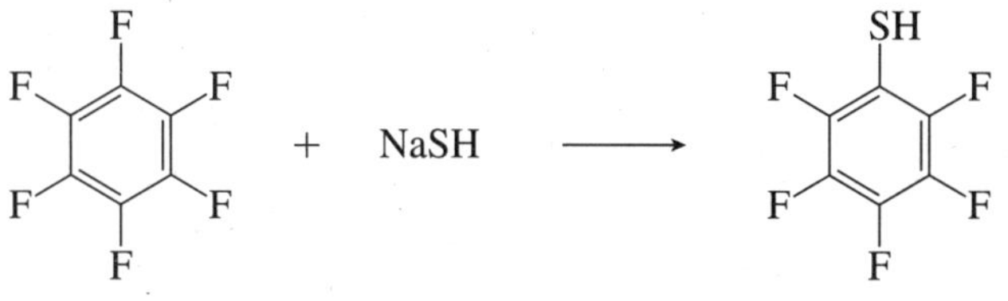

Hexafluorobenzene — Sodium hydrogen sulfide — 2,3,4,5,6-Pentafluorobenzenethiol

**23.12** (*a*) Since the *tert*-butoxy group replaces fluoride at the position occupied by the leaving group, substitution likely occurs by the addition–elimination mechanism.

$CH_3C_6H_4F$ + $(CH_3)_3CO^-$ ⟶ [anionic intermediate bearing F and $OC(CH_3)_3$] $\xrightarrow{-F^-}$ $CH_3C_6H_4OC(CH_3)_3$

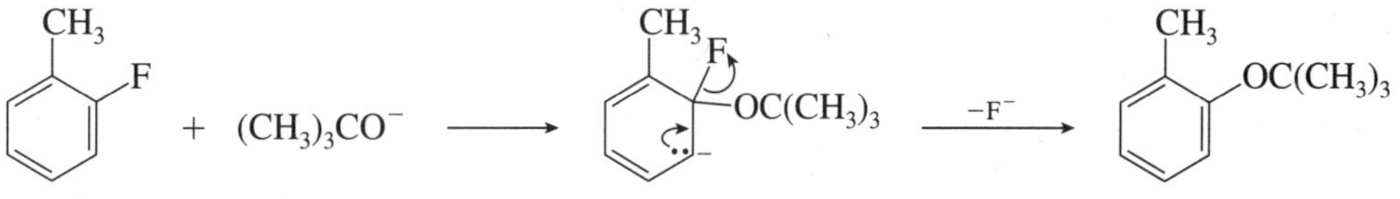

*o*-Fluorotoluene — *tert*-Butoxide ion — *tert*-Butyl *o*-methylphenyl ether

(*b*) In nucleophilic aromatic substitution reactions that proceed by the addition–elimination mechanism, aryl fluorides react faster than aryl bromides. Because the aryl bromide is more reactive in this case, it must be reacting by a different mechanism, which is most likely elimination–addition.

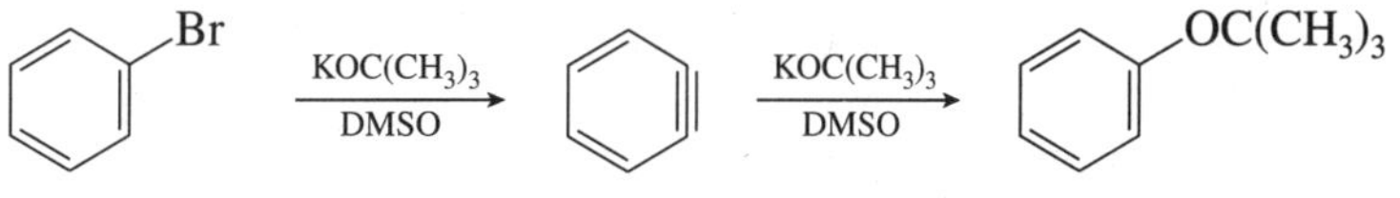

$C_6H_5Br$ $\xrightarrow[DMSO]{KOC(CH_3)_3}$ $C_6H_4$ $\xrightarrow[DMSO]{KOC(CH_3)_3}$ $C_6H_5OC(CH_3)_3$

Bromobenzene — Benzyne — *tert*-Butyl phenyl ether

**23.13** (*a*) Two benzyne intermediates are equally likely to be formed. Reaction with amide ion can occur in two different directions with each benzyne, giving three possible products. They are formed in a 1 : 2 : 1 ratio.

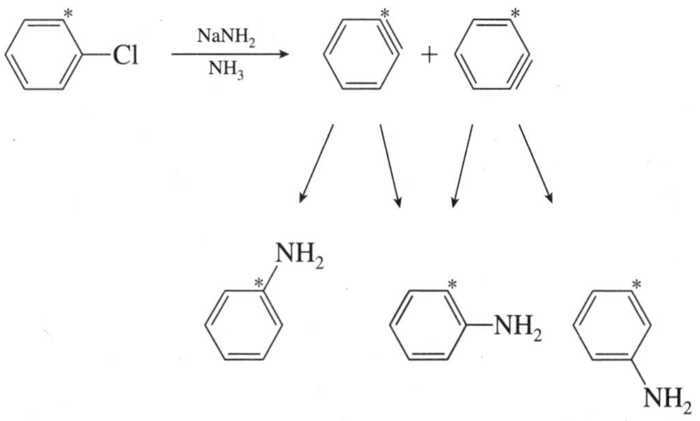

$C_6H_5Cl$ (*) $\xrightarrow[NH_3]{NaNH_2}$ benzyne (*) + benzyne (*) ⟶ aniline products ($NH_2$ on C*; $NH_2$ ortho to C*; $NH_2$ meta to C*)

**Ratio:** 1 : 2 : 1

Asterisk (*) refers to $^{14}C$.

(*b*) Only one benzyne intermediate is possible, leading to two products in a 1:1 ratio.

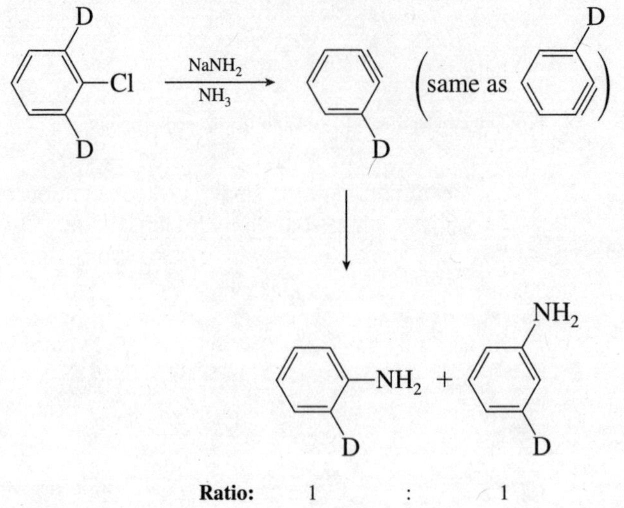

$NaNH_2$, $NH_3$ (same as)

$NH_2$ + $NH_2$

**Ratio:** 1 : 1

D refers to $^2H$ (deuterium).

**23.14** (*a*) *o*-Chloronitrobenzene is more reactive than chlorobenzene, because the cyclohexadienyl anion intermediate is stabilized by the nitro group.

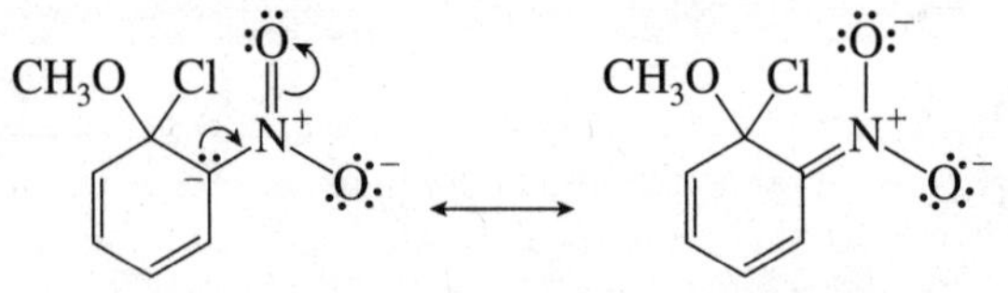

Comparing the rate constants for the two aryl halides in this reaction reveals that *o*-chloronitrobenzene is more than 20 billion times more reactive at 50°C.

(*b*) The cyclohexadienyl anion intermediate is more stable, and is formed faster, when the electron-withdrawing nitro group is ortho to chlorine. *o*-Chloronitrobenzene reacts faster than *m*-chloronitrobenzene. The measured difference is a factor of approximately 40,000 at 50°C.

(*c*) 4-Chloro-3-nitroacetophenone is more reactive, because the ring bears two powerful electron-withdrawing groups in positions where they can stabilize the cyclohexadienyl anion intermediate.

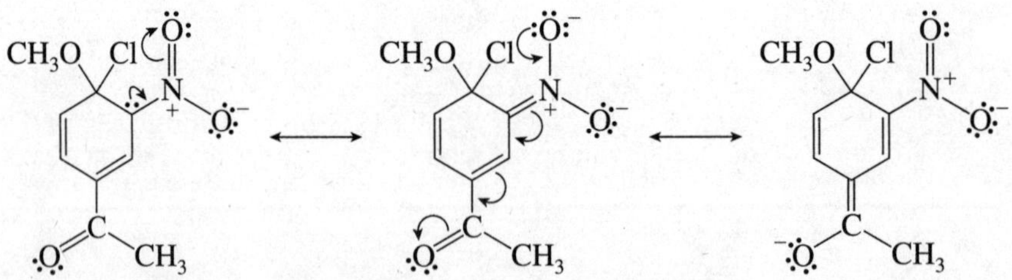

(*d*) Nitro groups activate aryl halides toward nucleophilic aromatic substitution best when they are ortho or para to the leaving group.

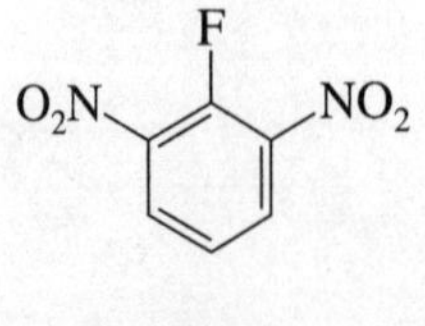

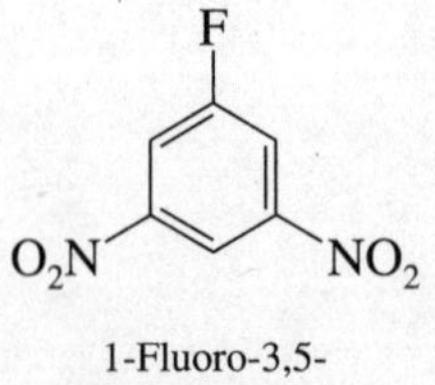

2-Fluoro-1,3-dinitrobenzene is more reactive than 1-Fluoro-3,5-dinitrobenzene

(*e*) The aryl halide with nitro groups ortho and para to the bromide leaving group is more reactive than the aryl halide with only one nitro group.

1-Bromo-2,4-dinitrobenzene is more reactive than 1,4-Dibromo-2-nitrobenzene

**23.15** (*a*) The nucleophile is the lithium salt of pyrrolidine, which reacts with bromobenzene by an elimination–addition mechanism.

$$\text{Bromobenzene} + \text{Lithium pyrrolidide} \longrightarrow N\text{-Phenylpyrrolidine (observed yield, 84\%)} + LiBr$$

(*b*) The nucleophile in this case is piperidine. The substrate, 1-bromo-2,4-dinitrobenzene, is very reactive in nucleophilic aromatic substitution by the addition–elimination mechanism.

$$\text{1-Bromo-2,4-dinitrobenzene} + \text{Piperidine} \longrightarrow N\text{-(2,4-Dinitrophenyl)piperidine}$$

(*c*) Of the two bromine atoms, one is ortho and the other meta to the nitro group. Nitro groups activate positions ortho and para to themselves toward nucleophilic aromatic substitution, and so it will be the bromine ortho to the nitro group that is displaced.

$$\text{1,4-Dibromo-2-nitrobenzene} + \text{Piperidine} \longrightarrow N\text{-(4-Bromo-2-nitrophenyl)piperidine}$$

**23.16** Because isomeric products are formed by reaction of 1- and 2-bromonaphthalene with piperidine at elevated temperatures, it is reasonable to conclude that these reactions do not involve a common intermediate and hence follow an addition–elimination pathway. Piperidine acts as a nucleophile and substitutes for bromine on the same carbon atom from which bromine is lost.

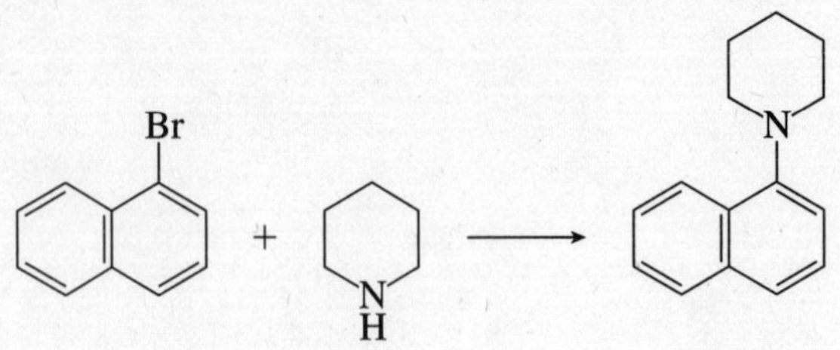

When the strong base sodium piperidide is used, reaction occurs by the elimination–addition pathway via a "naphthalyne" intermediate. Only one mode of elimination is possible from 1-bromonaphthalene.

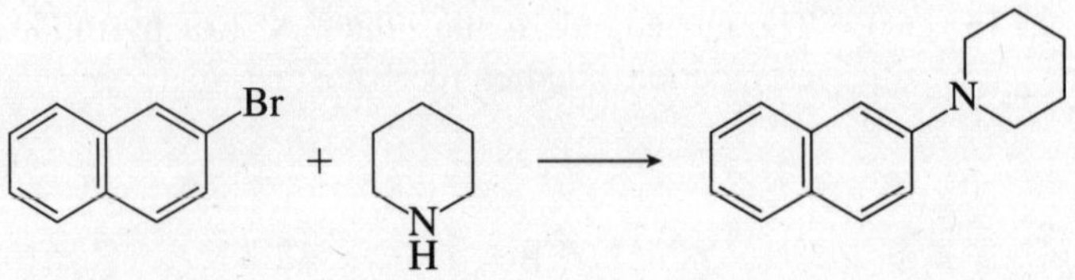

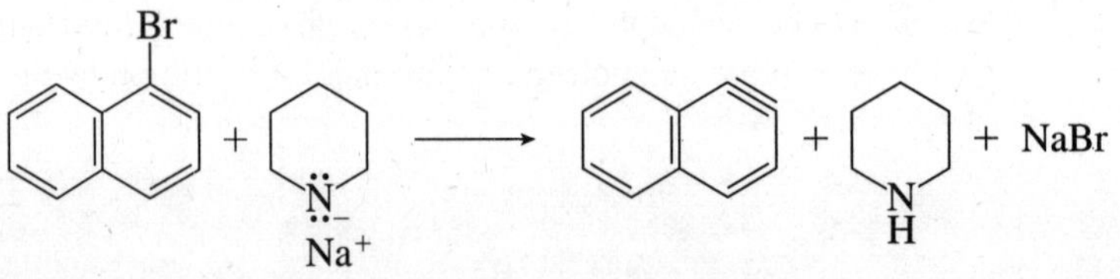

This intermediate can yield both A and B in the addition stage.

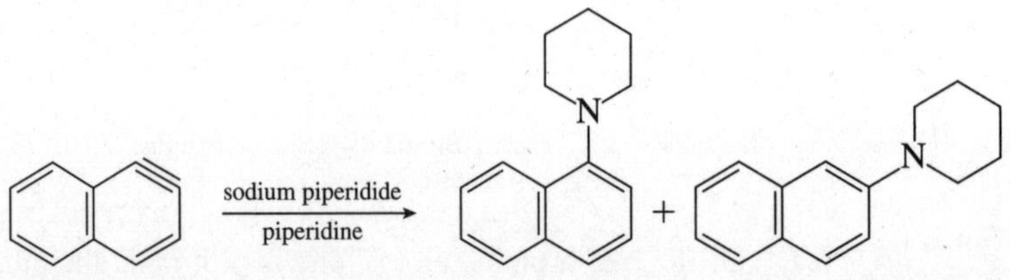

Two modes of elimination are possible from 2-bromonaphthalene:

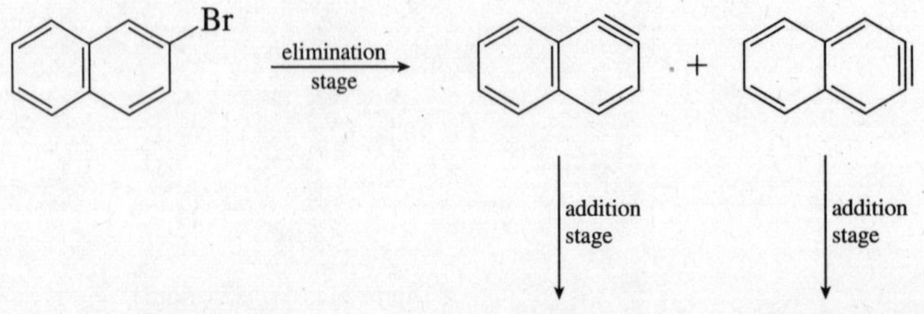

Both naphthalyne intermediates are probably formed from 2-bromonaphthalene because there is no reason to expect elimination to occur only in one direction.

**23.17** Reaction of a nitro-substituted aryl halide with a good nucleophile leads to nucleophilic aromatic substitution. Methoxide will displace fluoride from the ring, preferentially at the positions ortho and para to the nitro group.

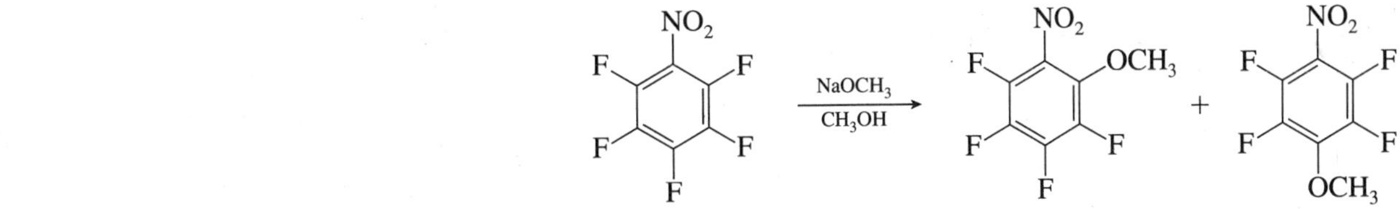

$\xrightarrow[CH_3OH]{NaOCH_3}$ +

1,2,3,4,5-Pentafluoro-6-nitrobenzene | 2,3,4,5-Tetrafluoro-6-nitroanisole | 2,3,5,6-Tetrafluoro-4-nitroanisole

**23.18** (*a*) This reaction is nucleophilic aromatic substitution by the addition–elimination mechanism.

+ $C_6H_5CH_2SK$ ⟶

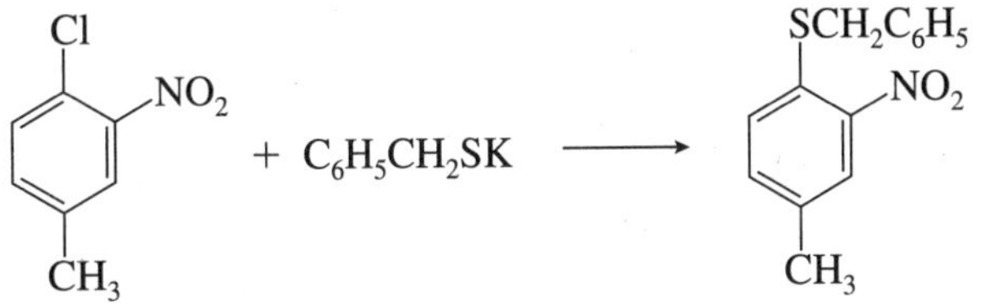

4-Chloro-3-nitrotoluene | 4-(Benzylthio)-3-nitrotoluene

The nucleophile, $C_6H_5CH_2\ddot{S}:^-$, displaces chloride directly from the aromatic ring. The product in this case was isolated in 57% yield.

(*b*) The nucleophile, hydrazine, will react with 1-chloro-2,4-dinitrobenzene by an addition–elimination mechanism as shown.

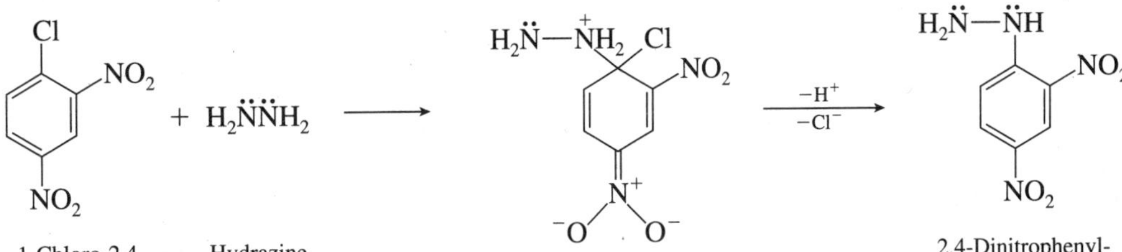

+ $H_2\ddot{N}\ddot{N}H_2$ ⟶ $\xrightarrow[-Cl^-]{-H^+}$

1-Chloro-2,4-dinitrobenzene | Hydrazine | 2,4-Dinitrophenylhydrazine

The nitrogen atoms of hydrazine each has an unshared electron pair and hydrazine is fairly nucleophilic. The product, 2,4-dinitrophenylhydrazine, is formed in quantitative yield.

(*c*) The problem requires you to track the starting material through two transformations. The first of these is nitration of *m*-dichlorobenzene, an electrophilic aromatic substitution reaction.

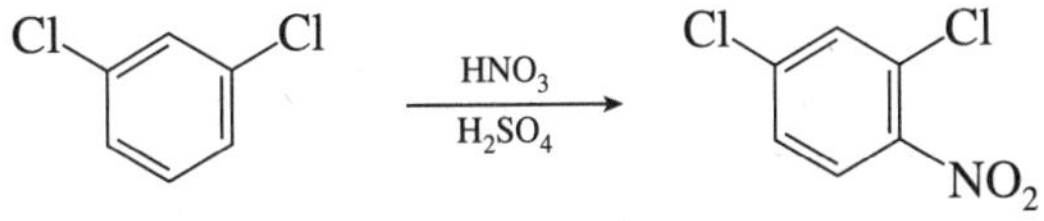

$\xrightarrow[H_2SO_4]{HNO_3}$

*m*-Dichlorobenzene | 2,4-Dichloro-1-nitrobenzene

Because the final product of the sequence has four nitrogen atoms ($C_6H_6N_4O_4$), 2,4-dichloro-1-nitrobenzene is an unlikely starting material for the second transformation. Stepwise nucleophilic aromatic substitution of both chlorines is possible but leads to a compound with the wrong molecular formula ($C_6H_7N_3O_2$).

2,4-Dichloro-1-nitrobenzene → ($NH_3$) → 2,4-Diamino-1-nitrobenzene

To obtain a final product with the correct molecular formula, the original nitration reaction must lead not to a mononitro but to a dinitro derivative. This is reasonable in view of the fact that this reaction is carried out at elevated temperature (120°C).

*m*-Dichlorobenzene → ($HNO_3$, $H_2SO_4$, 120°C) → → ($NH_3$, ethylene glycol, 140°C) → 1,5-Diamino-2,4-dinitrobenzene ($C_6H_6N_4O_4$)

This two-step sequence has been carried out with product yields of 70–71% in the first step and 88–95% in the second step.

(*d*) This problem also involves two transformations, nitration and nucleophilic aromatic substitution. Nitration will take place ortho to chlorine (meta to trifluoromethyl).

1-Chloro-4-(trifluoromethyl)benzene → ($HNO_3$, $H_2SO_4$) → 1-Chloro-2-nitro-4-(trifluoromethyl)benzene → ($NaOCH_3$, $CH_3OH$) → 2-Nitro-4-(trifluoromethyl)anisole

(*e*) The primary alkyl halide is more reactive toward nucleophilic substitution than the aryl halide. A phosphonium salt forms by an $S_N2$ process.

$$\text{I}-\text{C}_6\text{H}_4-\text{CH}_2\text{Br} + (\text{C}_6\text{H}_5)_3\text{P} \longrightarrow \text{I}-\text{C}_6\text{H}_4-\text{CH}_2\overset{+}{\text{P}}(\text{C}_6\text{H}_5)_3\ \text{Br}^-$$

*p*-Iodobenzyl bromide; Triphenyl phosphine; (*p*-Iodobenzyl)triphenylphosphonium bromide (86%)

(*f*) *N*-Bromosuccinimide (NBS) is a reagent used to substitute benzylic and allylic hydrogens with bromine. The benzylic bromide undergoes $S_N2$ substitution with the nucleophile, methanethiolate. As in part (*e*), the alkyl halide is more reactive toward substitution than the aryl halide.

2-Bromo-5-methoxytoluene → (NBS, benzoyl peroxide, $CCl_4$, heat) → 2-Bromo-5-methoxybenzyl bromide ($CH_2Br$) → ($NaSCH_3$) → 2-Bromo-5-methoxybenzyl methyl sulfide ($CH_3SCH_2$)

**23.19** The reaction of *p*-bromotoluene with aqueous sodium hydroxide at elevated temperature proceeds by way of a benzyne intermediate.

$CH_3$ / Br — NaOH, $H_2O$, 300°C → [$CH_3$ benzyne] → $CH_3$ / OH + $CH_3$ / OH

*m*-Methylphenol *p*-Methylphenol

The same benzyne intermediate is formed when *p*-chlorotoluene is the reactant, and so the product ratio must be identical regardless of whether the leaving group is bromide or chloride.

**23.20** Dinitration of *p*-chloro(trifluoromethyl)benzene will take place at the ring positions ortho to the chlorine. Compound A is 2-chloro-5-(trifluoromethyl)-1,3-dinitrobenzene. Trifluralin is formed by nucleophilic aromatic substitution of chlorine by dipropylamine. Trifluralin is *N, N*-dipropyl-4-(trifluoromethyl)-2,6-dinitroaniline.

$CF_3$ / Cl — $HNO_3$, $H_2SO_4$, heat → $CF_3$ / $O_2N$, $NO_2$ / Cl — $(CH_3CH_2CH_2)_2NH$ → $CF_3$ / $O_2N$, $NO_2$ / $N(CH_2CH_2CH_3)_2$

*p*-Chloro-(trifluoromethyl)-benzene

2-Chloro-5-(trifluoromethyl)-1,3-dinitrobenzene (compound A)

*N,N*-dipropyl-4-(trifluoromethyl)-2,6-dinitroaniline (trifluralin)

**23.21** *p*-Chlorobenzenethiolate reacts with *p*-nitrobenzyl chloride by an $S_N2$ process to give compound A. Reduction of the nitro group yields the aniline derivative, compound B. Chlorbenside is then formed by a Sandmeyer reaction in which the diazonium ion is replaced by chlorine.

$O_2N$—C₆H₄—$CH_2Cl$ + NaS—C₆H₄—Cl → $O_2N$—C₆H₄—$CH_2S$—C₆H₄—Cl

*p*-Nitrobenzyl chloride

Sodium *p*-chlorobenzenethiolate

*p*-Chlorophenyl *p*-nitrobenzyl sulfide (compound A)

$O_2N$—C₆H₄—$CH_2S$—C₆H₄—Cl — 1. Fe, HCl 2. NaOH → $H_2N$—C₆H₄—$CH_2S$—C₆H₄—Cl

Compound A

Compound B

↓ 1. $NaNO_2$, HCl 2. CuCl

Cl—C₆H₄—$CH_2S$—C₆H₄—Cl

Chlorbenside

**23.22** *p*-Chloro(trifluoromethyl)benzene undergoes nucleophilic substitution by the alkoxide anion to give compound A.

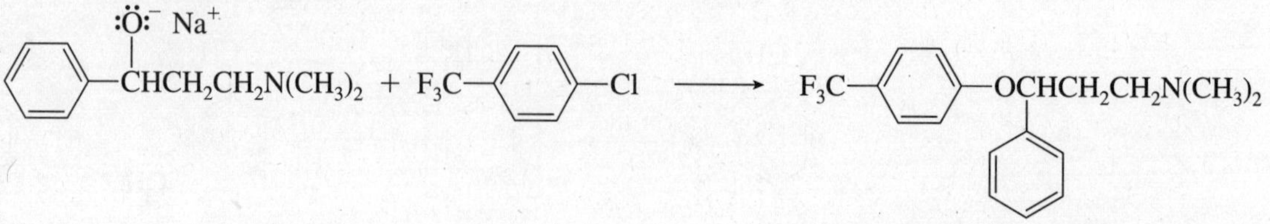

$$C_6H_5CH(\ddot{O}:^- Na^+)CH_2CH_2N(CH_3)_2 + F_3C\text{—}C_6H_4\text{—}Cl \longrightarrow F_3C\text{—}C_6H_4\text{—}OCH(C_6H_5)CH_2CH_2N(CH_3)_2$$

3-(*p*-(Trifluoromethyl)phenoxy)-*N*,*N*-dimethyl-3-phenyl-1-propanamine (Compound A)

Prozac (Fluoxetine hydrochloride) differs from compound A in having an $\text{—NHCH}_3$ group in place of $\text{—N(CH}_3)_2$.

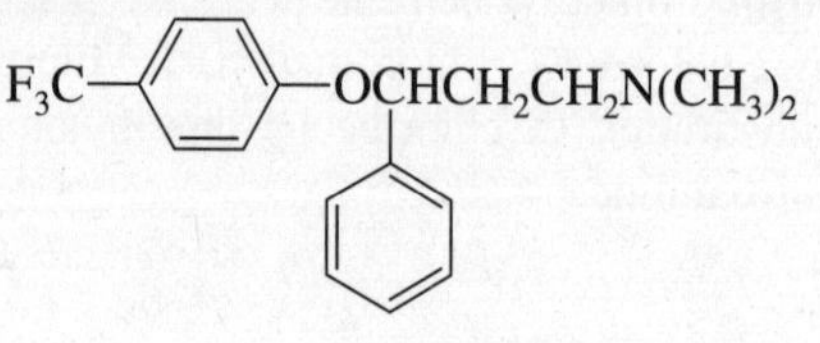

$$F_3C\text{—}C_6H_4\text{—}OCH(C_6H_5)CH_2CH_2N(CH_3)_2 \qquad F_3C\text{—}C_6H_4\text{—}OCH(C_6H_5)CH_2CH_2NHCH_3$$

Compound A 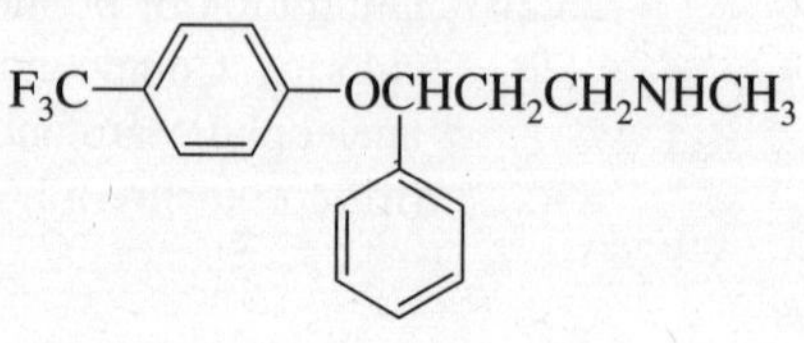 Prozac

**23.23** Benzyne is formed by loss of nitrogen and carbon dioxide.

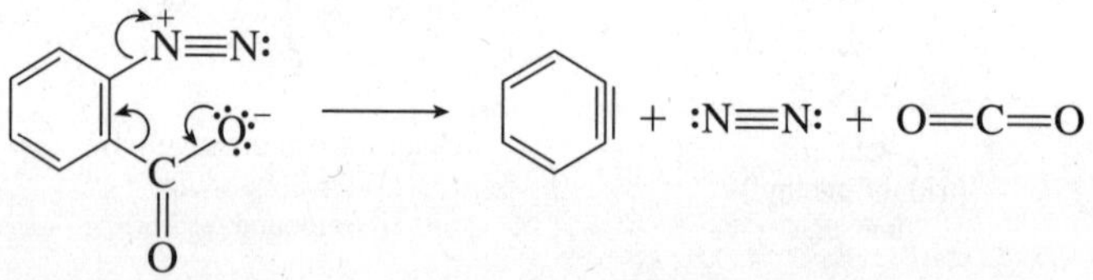

$$o\text{-}(\overset{+}{N}{\equiv}N:)C_6H_4CO_2^- \longrightarrow C_6H_4 + :N{\equiv}N: + O{=}C{=}O$$

Benzenediazonium-2-carboxylate Benzyne Nitrogen Carbon dioxide

**23.24** *o*-Bromofluorobenzene yields benzyne on reaction with magnesium (see text Section 23.9). Triptycene is the Diels–Alder cycloaddition product from the reaction of benzyne with anthracene (compound A). Although anthracene is aromatic, it is able to undergo cycloaddition at the center ring with a dienophile because the adduct retains the stabilization energy of two benzene rings.

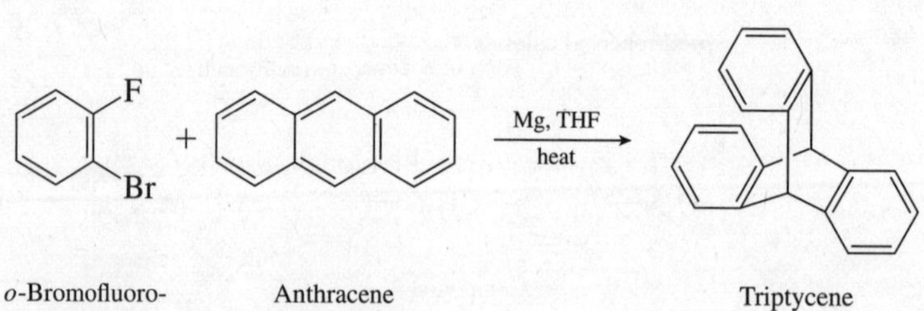

$$o\text{-}BrC_6H_4F + \text{Anthracene} \xrightarrow[\text{heat}]{\text{Mg, THF}} \text{Triptycene}$$

*o*-Bromofluorobenzene Anthracene (compound A) Triptycene

**23.25** (*a*) Ethoxide ion adds to the aromatic ring to give a cyclohexadienyl anion.

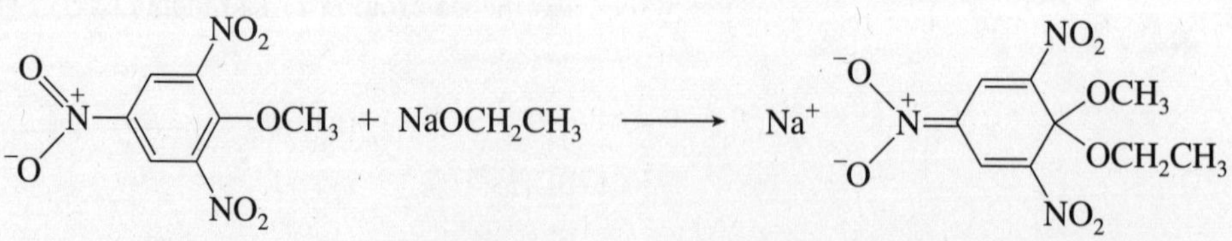

$$2,4,6\text{-}(NO_2)_3C_6H_2OCH_3 + NaOCH_2CH_3 \longrightarrow Na^+ \text{ [}(^-O)_2\overset{+}{N}{=}C_6H_2(NO_2)_2C(OCH_3)(OCH_2CH_3)\text{]}$$

2,4,6-Trinitroanisole Sodium ethoxide Meisenheimer complex

(*b*) The same Meisenheimer complex results when ethyl 2,4,6-trinitrophenyl ether reacts with sodium methoxide.

$$\text{Ethyl 2,4,6-trinitrophenyl ether} + NaOCH_3 \text{ (Sodium methoxide)} \longrightarrow \text{Meisenheimer complex}$$

Ethyl 2,4,6-trinitrophenyl ether   Sodium methoxide   Meisenheimer complex

**23.26** Methoxide ion may add to 2,4,6-trinitroanisole either at the ring carbon that bears the methoxyl group or at an unsubstituted ring carbon.

$$\text{2,4,6-Trinitroanisole} \xrightarrow[CH_3OH]{NaOCH_3} \text{A} + \text{B}$$

2,4,6-Trinitroanisole   A   B

The two Meisenheimer complexes are the sodium salts of the anions shown. It was observed that compound A was the more stable of the two. Compound B was present immediately after adding sodium methoxide to 2,4,6-trinitroanisole but underwent relatively rapid isomerization to compound A.

**23.27** (*a*) The first reaction that occurs is an acid–base reaction between diethyl malonate and sodium amide.

$$CH_2(COOCH_2CH_3)_2 + NaNH_2 \longrightarrow Na^+ \ :\bar{C}H(COOCH_2CH_3)_2 + NH_3$$

Diethyl malonate   Sodium amide   Diethyl sodiomalonate   Ammonia

A second equivalent of sodium amide converts bromobenzene to benzyne.

$$C_6H_5Br + NaNH_2 \longrightarrow C_6H_4 + NH_3 + NaBr$$

Bromobenzene   Sodium amide   Benzyne   Ammonia   Sodium bromide

The anion of diethyl malonate adds to benzyne.

$$C_6H_4 + {}^-:CH(COOCH_2CH_3)_2 \longrightarrow {}^-C_6H_4CH(COOCH_2CH_3)_2$$

Benzyne   Anion of diethyl malonate

This anion then abstracts a proton from ammonia to give the observed product.

$${}^-C_6H_4CH(COOCH_2CH_3)_3 + H{-}NH_2 \longrightarrow C_6H_5CH(COOCH_2CH_3)_2 + :\ddot{N}H_2^-$$

Ammonia   Diethyl 2-phenylmalonate   Amide anion

(*b*) The ester is deprotonated by the strong base sodium amide, after which the ester enolate undergoes an elimination reaction to form a benzyne intermediate. Cyclization to the final product occurs by intramolecular attack of the ester enolate on the reactive triple bond of the aryne.

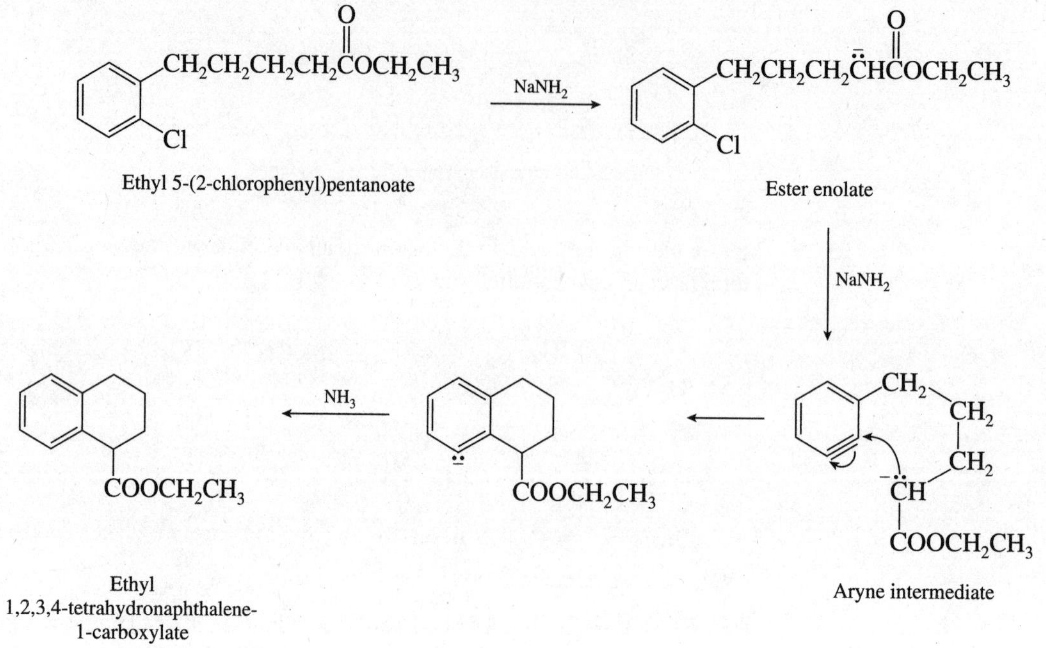

Ethyl 5-(2-chlorophenyl)pentanoate —$NaNH_2$→ Ester enolate —$NaNH_2$→ Aryne intermediate → (cyclized anion) —$NH_3$→ Ethyl 1,2,3,4-tetrahydronaphthalene-1-carboxylate

(*c*) In the presence of very strong bases, aryl halides undergo nucleophilic aromatic substitution by an elimination–addition mechanism. The structure of the product indicates that a nitrogen of the side chain acts as a nucleophile in the addition step.

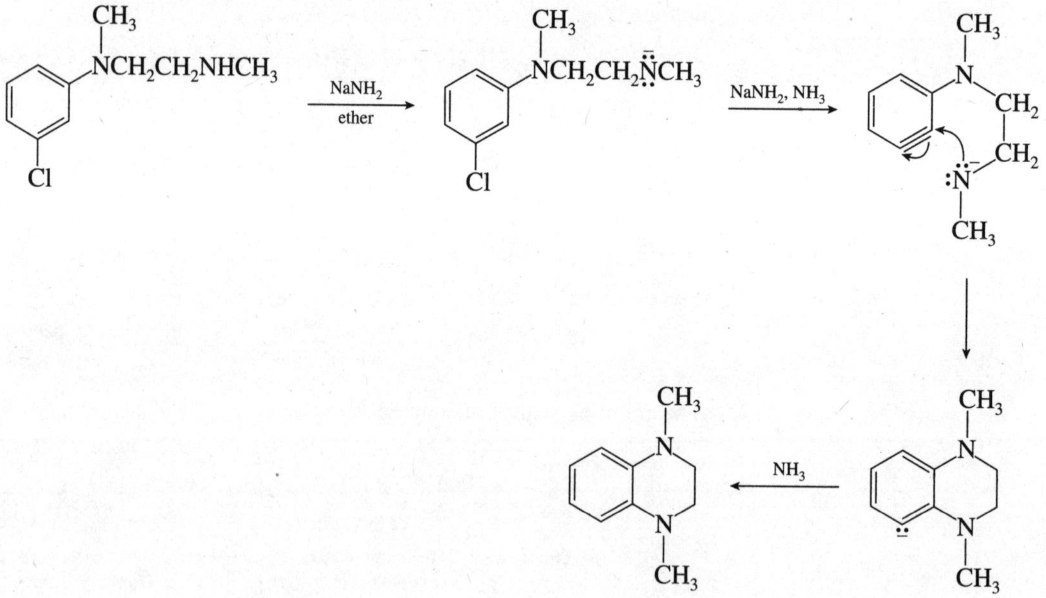

(*d*) On treatment with base, intramolecular nucleophilic aromatic substitution leads to the observed product.

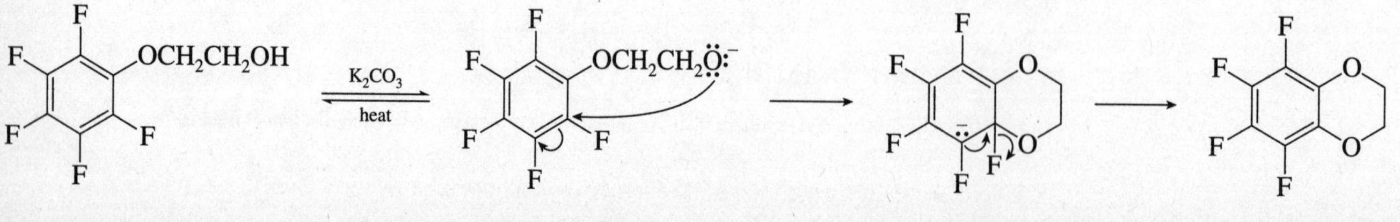

**23.28** Polychlorinated biphenyls (PCBs) are derived from biphenyl as the base structure. It is numbered as shown.

3′ 2′ 2 3
4′ 1′ 1 4
5′ 6′ 6 5

(*a*) There are three monochloro derivatives of biphenyl:

2-Chlorobiphenyl (*o*-chlorobiphenyl) | 3-Chlorobiphenyl (*m*-chlorobiphenyl) | 4-Chlorobiphenyl (*p*-chlorobiphenyl)

(*b*) The two chlorine substituents may be in the same ring (six isomers):

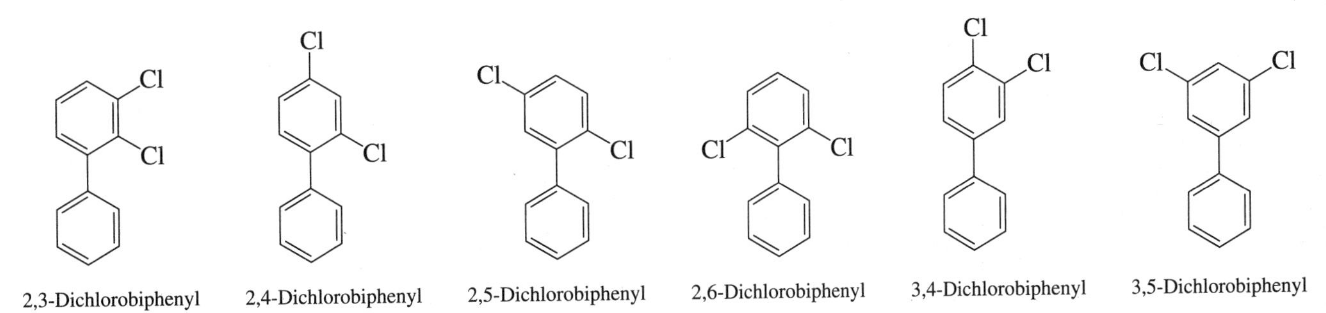

2,3-Dichlorobiphenyl 2,4-Dichlorobiphenyl 2,5-Dichlorobiphenyl 2,6-Dichlorobiphenyl 3,4-Dichlorobiphenyl 3,5-Dichlorobiphenyl

The two chlorine substituents may be in different rings (six isomers):

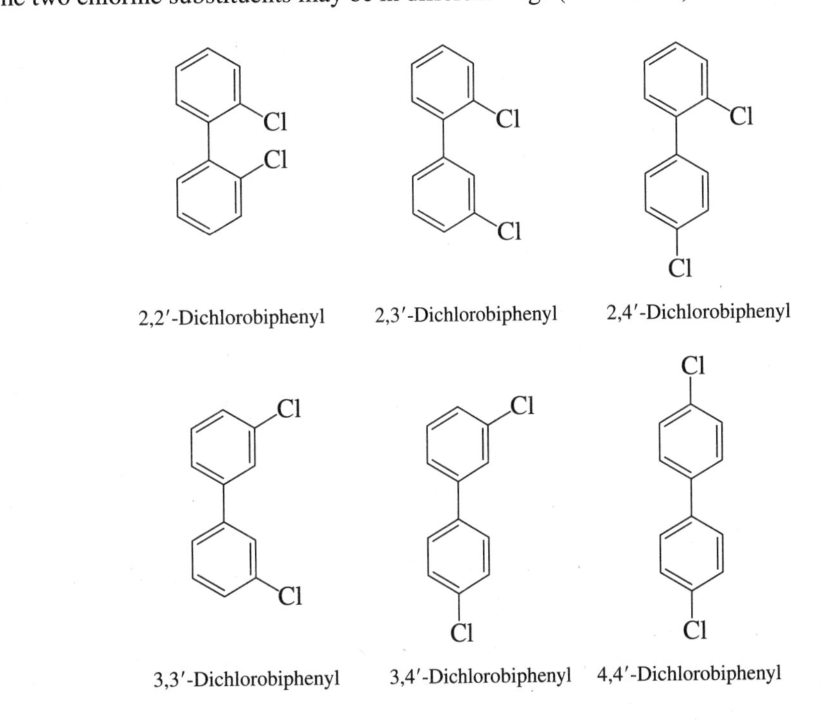

2,2′-Dichlorobiphenyl 2,3′-Dichlorobiphenyl 2,4′-Dichlorobiphenyl

3,3′-Dichlorobiphenyl 3,4′-Dichlorobiphenyl 4,4′-Dichlorobiphenyl

There are therefore a total of 12 isomeric dichlorobiphenyls.

(*c*) The number of octachlorobiphenyls will be equal to the number of dichlorobiphenyls (12). In both cases we are dealing with a situation in which eight of the ten substituents of the biphenyl system are the same and considering how the remaining two may be arranged. In the dichlorobiphenyls described in part (*b*), eight substituents are hydrogen and two are chlorine; in the octachlorobiphenyls, eight substituents are chlorine and two are hydrogen.

(*d*) The number of nonachloro isomers (nine chlorines, one hydrogen) must equal the number of monochloro isomers (one chlorine, nine hydrogens). There are therefore three nonachloro derivatives of biphenyl.

**23.29** The principal isotopes of chlorine are $^{35}Cl$ and $^{37}Cl$. A cluster of five peaks indicates that dichlorodiphenyldichloroethane (DDE) contains *four* chlorines.

| *m/z* for $C_{14}H_8Cl_4$ | | | | |
|---|---|---|---|---|
| 316 | $^{35}Cl$ | $^{35}Cl$ | $^{35}Cl$ | $^{35}Cl$ |
| 318 | $^{35}Cl$ | $^{35}Cl$ | $^{35}Cl$ | $^{37}Cl$ |
| 320 | $^{35}Cl$ | $^{35}Cl$ | $^{37}Cl$ | $^{37}Cl$ |
| 322 | $^{35}Cl$ | $^{37}Cl$ | $^{37}Cl$ | $^{37}Cl$ |
| 324 | $^{37}Cl$ | $^{37}Cl$ | $^{37}Cl$ | $^{37}Cl$ |

The peak at *m/z* 316 therefore corresponds to a compound $C_{14}H_8Cl_4$ in which all four chlorines are $^{35}Cl$. The respective molecular formulas indicate that DDE is the dehydrochlorination product of dichlorodiphenyltrichloroethane (DDT).

$$\underset{\text{DDT}}{C_{14}H_9Cl_5} \xrightarrow{-HCl} \underset{\text{DDE}}{C_{14}H_8Cl_4}$$

The structure of DDT was given in the statement of the problem. This permits the structure of DDE to be assigned.

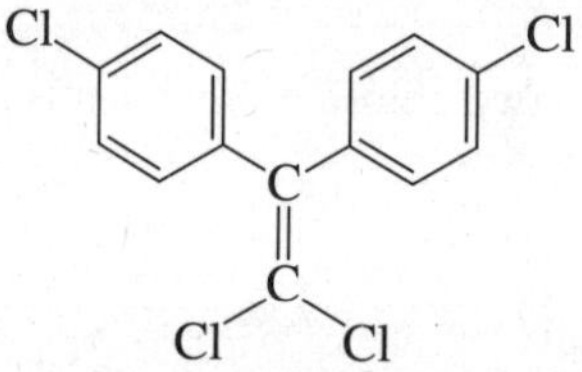

$$(p\text{-}ClC_6H_4)_2C{=}CCl_2$$

DDE (only reasonable dehydrochlorination product of DDT)

## SELF-TEST

### PART A

**A-1.** Give the product(s) obtained from each of the following reactions:

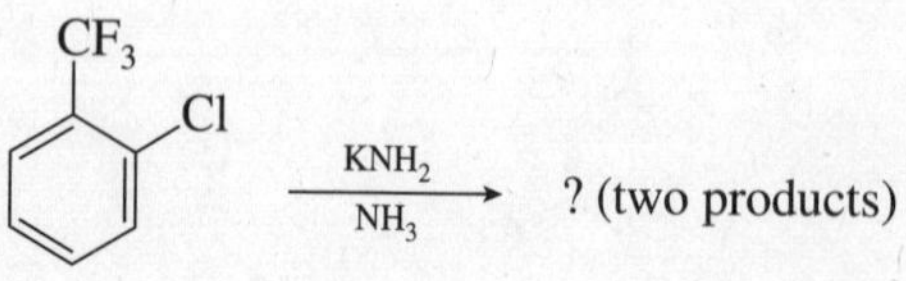

(*a*) $o\text{-}ClC_6H_4CF_3 \xrightarrow[NH_3]{KNH_2}$ ? (two products) (*b*) 3,4-Dichloronitrobenzene $\xrightarrow[CH_3OH]{CH_3O^-}$ ? (monosubstitution)

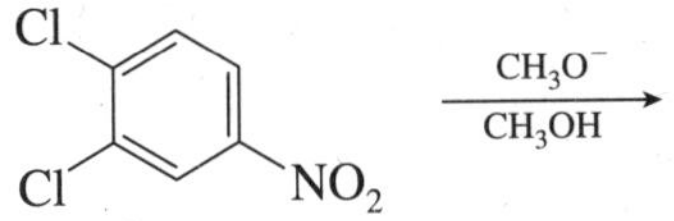

(*c*) 1-tert-Butyl-4-iodobenzene ($C(CH_3)_3$, I) $\xrightarrow[NH_3]{NaNH_2}$ ? (two products)

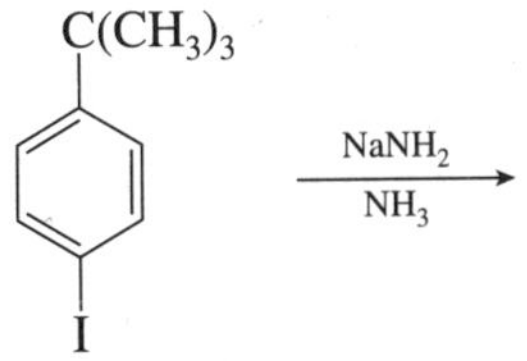

**A-2.** Draw the structure of the intermediate formed in each reaction of problem A-1.

**A-3.** Suggest synthetic schemes by which chlorobenzene may be converted into
(*a*) 2,4-Dinitroanisole (1-methoxy-2,4-dinitrobenzene)
(*b*) *p*-Isopropylaniline

**A-4.** Write a mechanism using resonance structures to show how a nitro group directs ortho, para in nucleophilic aromatic substitution.

**A-5.** What is the cycloaddition product of the following reaction? What is the structure of the short-lived intermediate formed in this reaction?

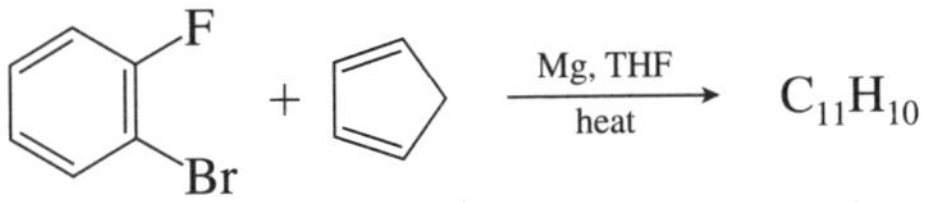

1-Bromo-2-fluorobenzene (F, Br) + cyclopentadiene $\xrightarrow[heat]{Mg, THF}$ $C_{11}H_{10}$

## PART B

**B-1.** The reaction

2-Chloro-1,3-dinitrobenzene ($O_2N$, Cl, $NO_2$) $\xrightarrow{CH_3O^-}$ 2-methoxy-1,3-dinitrobenzene ($O_2N$, $OCH_3$, $NO_2$)

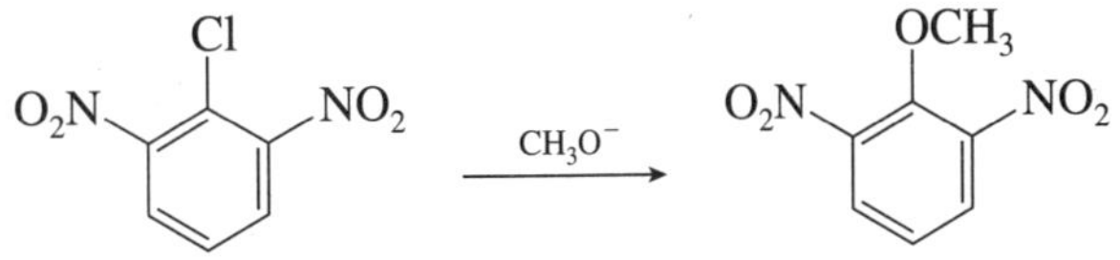

most likely occurs by which of the following mechanisms?
(*a*) Addition–elimination
(*b*) Elimination–addition
(*c*) Both (*a*) and (*b*)
(*d*) Neither of these

**B-2.** Rank the following in order of decreasing rate of reaction with ethoxide ion ($CH_3CH_2O^-$) in a nucleophilic aromatic substitution reaction:

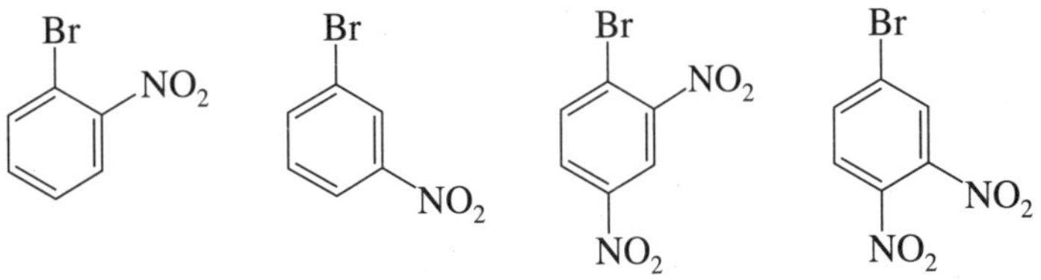

1: 1-bromo-2-nitrobenzene (Br, $NO_2$)
2: 1-bromo-3-nitrobenzene (Br, $NO_2$)
3: 1-bromo-2,4-dinitrobenzene (Br, $NO_2$, $NO_2$)
4: 1-bromo-3,4-dinitrobenzene (Br, $NO_2$, $NO_2$)

(*a*) 3 > 4 > 1 > 2
(*b*) 2 > 1 > 4 > 3
(*c*) 3 > 4 > 2 > 1
(*d*) 4 > 3 > 2 > 1

**B-3.** The reaction

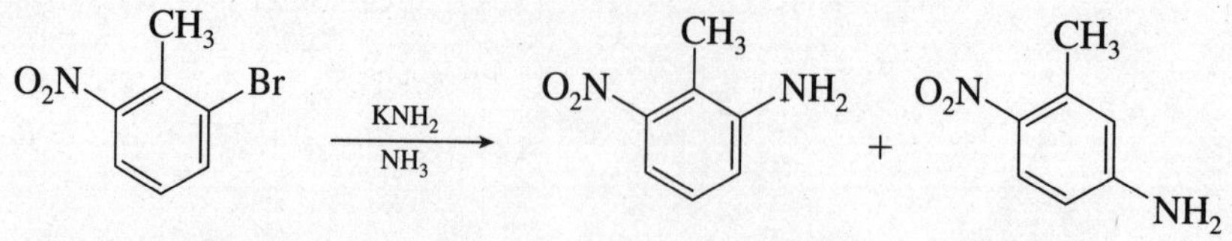

$CH_3$, $O_2N$, Br (ring) $\xrightarrow[NH_3]{KNH_2}$ $CH_3$, $O_2N$, $NH_2$ (ring) + $CH_3$, $O_2N$, $NH_2$ (ring)

most likely involves which of the following aromatic substitution mechanisms?

(*a*) Addition–elimination
(*b*) Electrophilic substitution
(*c*) Elimination–addition
(*d*) Both (*a*) and (*c*)

**B-4.** Identify the principal organic product of the following reaction:

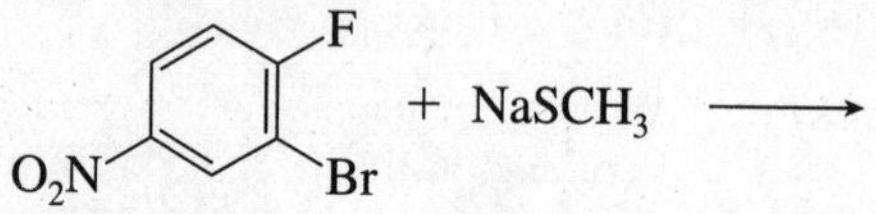

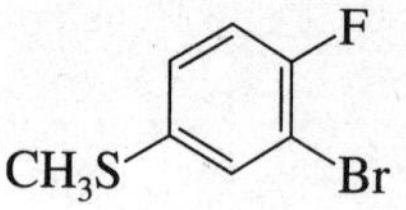

$O_2N$, F, Br (ring) + $NaSCH_3$ ⟶

(*a*) $CH_3S$, F, Br

(*d*) $O_2N$, $SCH_3$, Br

(*b*) $CH_3S$, F, $O_2N$, Br

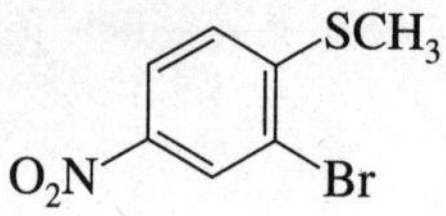

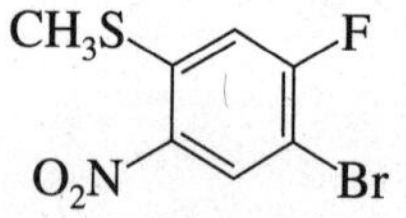

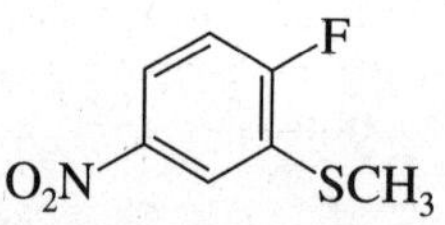

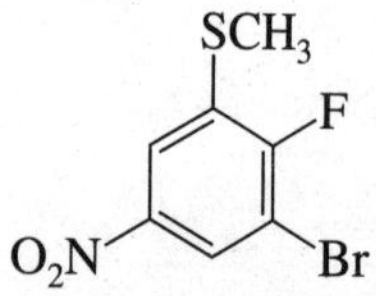

(*e*) $O_2N$, F, $SCH_3$

(*c*) $SCH_3$, F, $O_2N$, Br

**B-5.** Which of the following compounds gives a single benzyne intermediate on reaction with sodium amide?

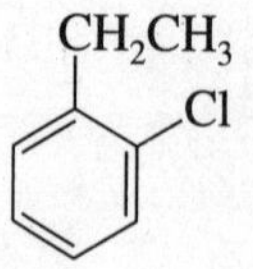

1: $CH_2CH_3$, Cl
2: $CH_2CH_3$, Cl
3: $CH_2CH_3$, Cl

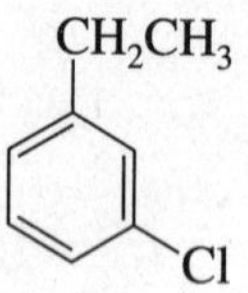

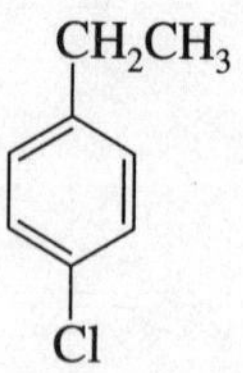

(*a*) 1 only
(*b*) 1 and 3
(*c*) 3 only
(*d*) 1 and 2

**B-6.** Which one of the following compounds can be efficiently prepared by a procedure in which nucleophilic aromatic substitution is the last step?

(*a*) $OCH_3$ / $NO_2$ (*b*) $SO_3H$ / $CH_3$ (*c*) $NH_2$, $H_3C$, Br / $CH_3$ (*d*) $OCH_2CH{=}CH_2$ / $C(CH_3)_3$

**B-7.** Which one of the following undergoes nucleophilic aromatic substitution at the fastest rate?

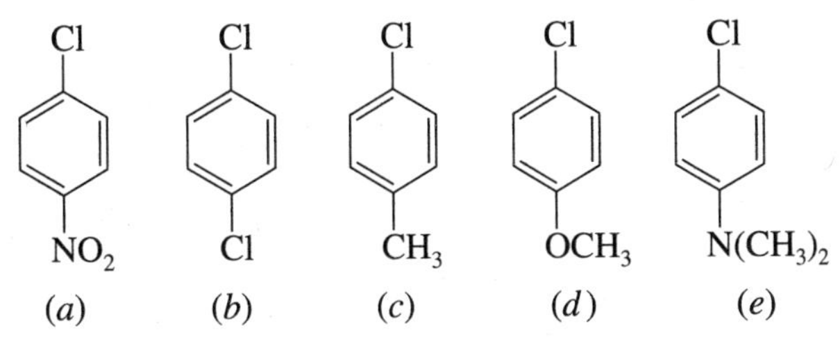


**B-8.** What combination of reactants will give the species shown as a reactive intermediate?

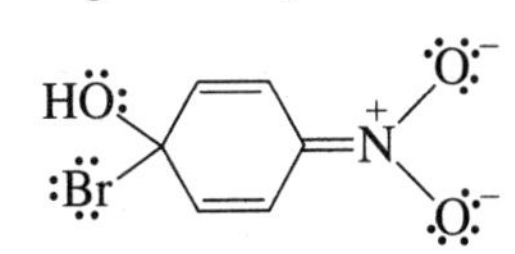


(*a*) 1-Bromo-4-nitrobenzene and NaOH
(*b*) 4-Nitrophenol and HBr
(*c*) 4-Nitrophenol, $Br_2$, and $FeBr_3$
(*d*) Bromobenzene and $HONO_2$
(*e*) Nitrobenzene, $Br_2$, and water

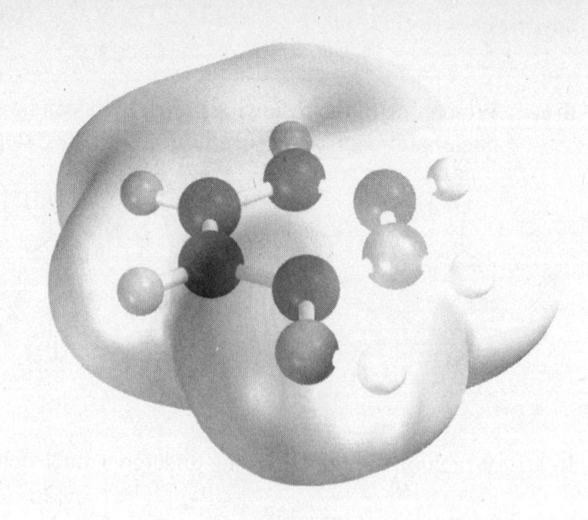

# CHAPTER 24
## PHENOLS

## SOLUTIONS TO TEXT PROBLEMS

**24.1** (*b*) A benzyl group ($C_6H_5CH_2$—) is ortho to the phenolic hydroxyl group in *o*-benzylphenol.

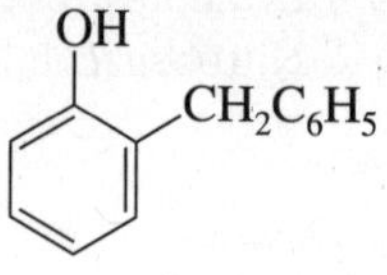

OH
$CH_2C_6H_5$

(*c*) Naphthalene is numbered as shown. 3-Nitro-1-naphthol has a hydroxyl group at C-1 and a nitro group at C-3.

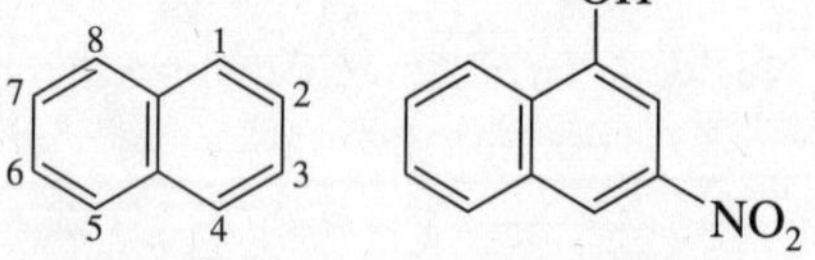

8 1 7 2 6 3 5 4 OH $NO_2$

Naphthalene 3-Nitro-1-naphthol

(*d*) Resorcinol is 1,3-benzenediol. 4-Chlororesorcinol is therefore

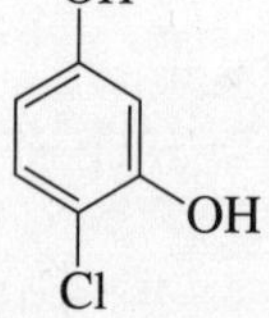

OH
OH
Cl

**24.2** Intramolecular hydrogen bonding between the hydroxyl group and the ester carbonyl can occur when these groups are ortho to each other.

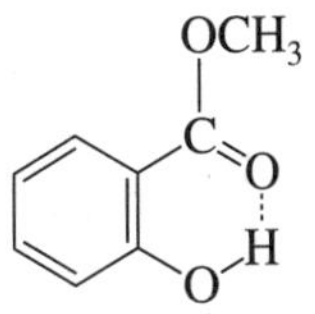

Methyl salicylate

Intramolecular hydrogen bonds form at the expense of intermolecular ones, and intramolecularly hydrogen-bonded phenols have lower boiling points than isomers in which only intermolecular hydrogen-bonding is possible.

**24.3** (*b*) A cyano group withdraws electrons from the ring by resonance. A *p*-cyano substituent is conjugated directly with the negatively charged oxygen and stabilizes the anion more than does an *m*-cyano substituent.

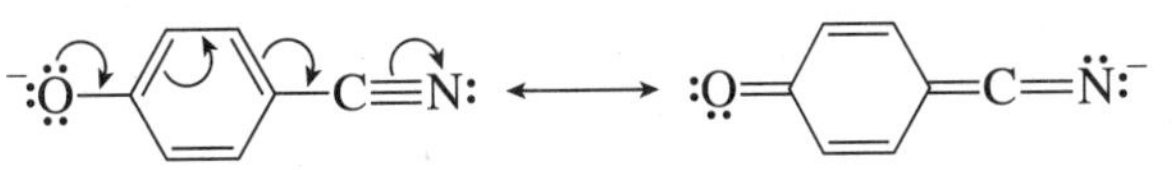

*p*-Cyanophenol is slightly more acidic than *m*-cyanophenol, the $K_a$ values being $1.0 \times 10^{-8}$ and $2.8 \times 10^{-9}$, respectively.

(*c*) The electron-withdrawing inductive effect of the fluorine substituent will be more pronounced at the ortho position than at the para. *o*-Fluorophenol ($K_a = 1.9 \times 10^{-9}$) is a stronger acid than *p*-fluorophenol ($K_a = 1.3 \times 10^{-10}$).

**24.4** The text points out that the reaction proceeds by the addition–elimination mechanism of nucleophilic aromatic substitution.

Under the strongly basic conditions of the reaction, *p*-toluenesulfonic acid is first converted to its anion.

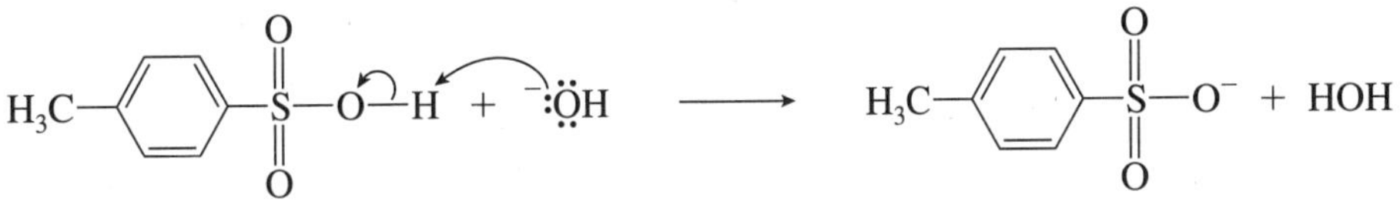

*p*-Toluenesulfonic acid + Hydroxide ion ⟶ *p*-Toluenesulfonate ion + Water

Nucleophilic addition of hydroxide ion gives a cyclohexadienyl anion intermediate.

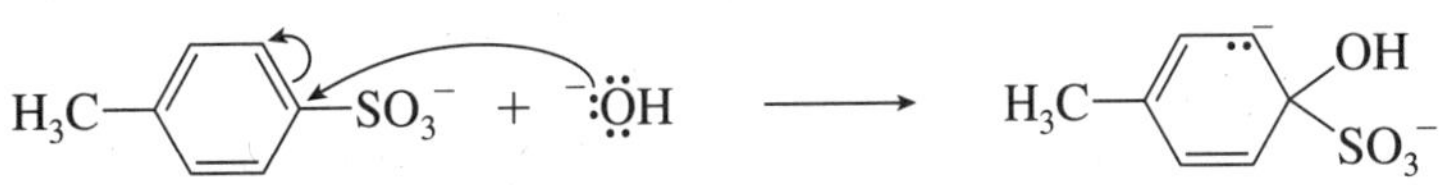

*p*-Toluenesulfonate ion + Hydroxide ⟶ Cyclohexadienyl anion

Loss of sulfite ion ($SO_3^{2-}$) gives *p*-cresol.

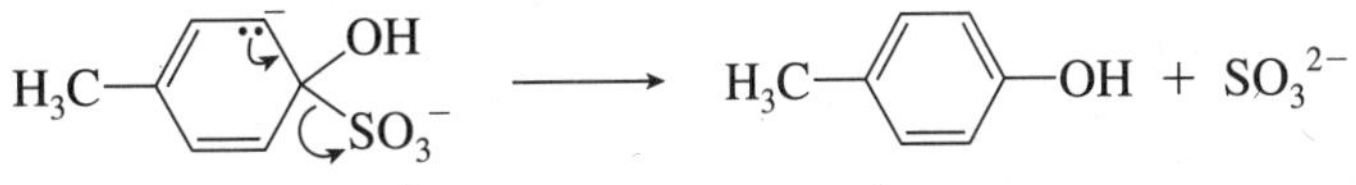

Cyclohexadienyl anion ⟶ *p*-Cresol + $SO_3^{2-}$

It is also possible that the elimination stage of the reaction proceeds as follows:

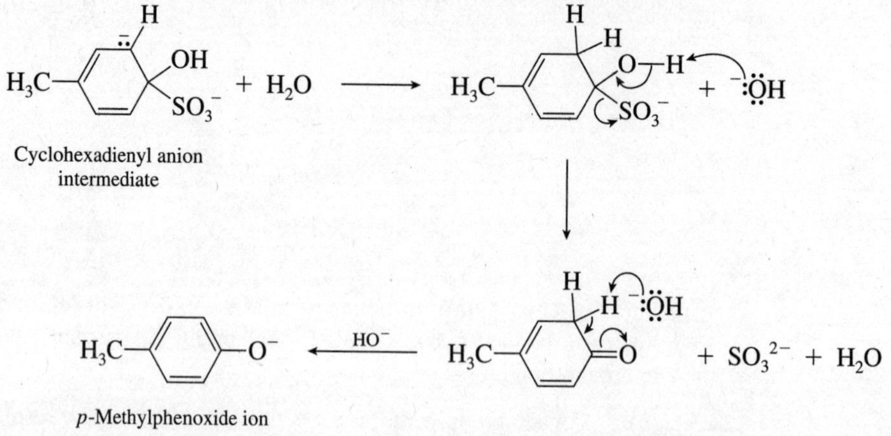

Cyclohexadienyl anion intermediate

p-Methylphenoxide ion

**24.5** The text states that the hydrolysis of chlorobenzene in base follows an elimination–addition mechanism.

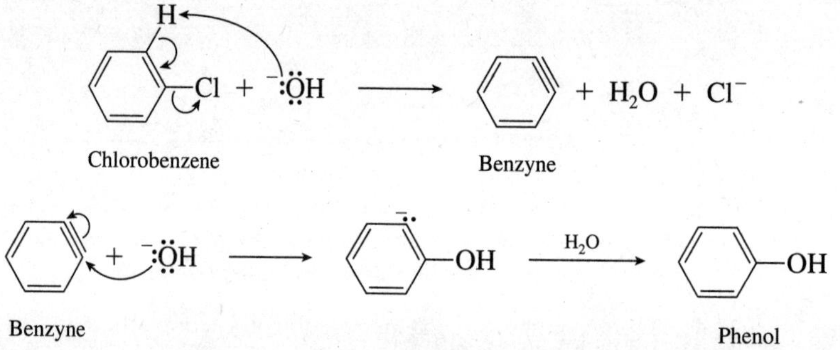

Chlorobenzene

Benzyne

Benzyne

Phenol

**24.6** (*b*) The reaction is Friedel–Crafts alkylation. Proton transfer from sulfuric acid to 2-methylpropene gives *tert*-butyl cation. Because the position para to the hydroxyl substituent already bears a bromine, the *tert*-butyl cation attacks the ring at the position ortho to the hydroxyl.

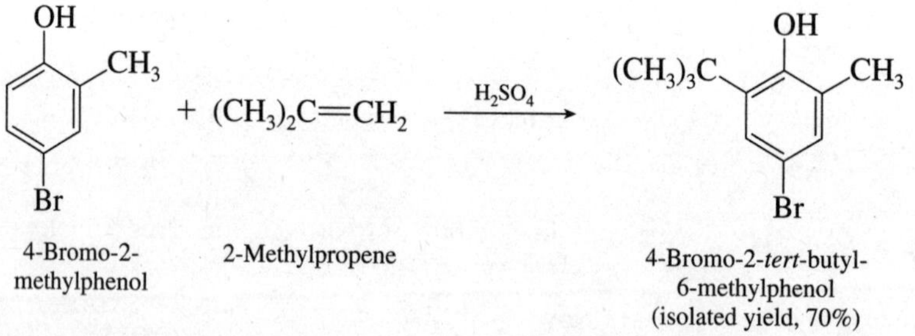

4-Bromo-2-methylphenol + $(CH_3)_2C{=}CH_2$ (2-Methylpropene) $\xrightarrow{H_2SO_4}$ 4-Bromo-2-*tert*-butyl-6-methylphenol (isolated yield, 70%)

(*c*) Acidification of sodium nitrite produces nitrous acid, which nitrosates the strongly activated aromatic ring of phenols.

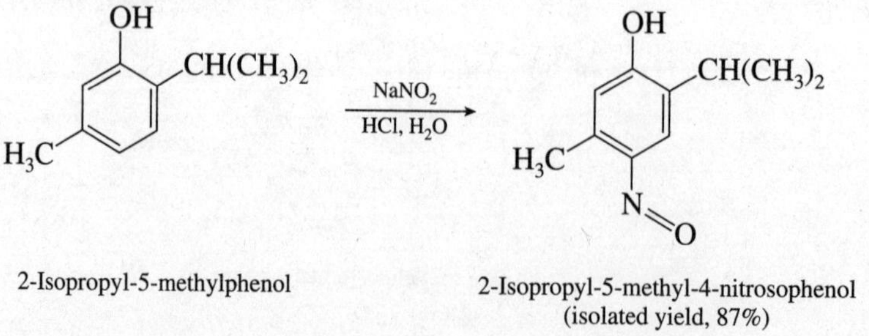

2-Isopropyl-5-methylphenol $\xrightarrow[HCl,\ H_2O]{NaNO_2}$ 2-Isopropyl-5-methyl-4-nitrosophenol (isolated yield, 87%)

(*d*) Friedel–Crafts acylation occurs ortho to the hydroxyl group.

$$p\text{-Cresol} + CH_3CH_2\overset{O}{\overset{\|}{C}}Cl \xrightarrow{AlCl_3} \text{1-(2-Hydroxy-5-methylphenyl)-1-propanone}$$

p-Cresol — Propanoyl chloride — 1-(2-Hydroxy-5-methylphenyl)-1-propanone (isolated yield, 87%)

**24.7** (*b*) The hydroxyl group of 2-naphthol is converted to the corresponding acetate ester.

$$\text{2-Naphthol} + CH_3\overset{O}{\overset{\|}{C}}O\overset{O}{\overset{\|}{C}}CH_3 \xrightarrow{NaOH} \text{2-Naphthyl acetate } (O\overset{O}{\overset{\|}{C}}CH_3) + CH_3\overset{O}{\overset{\|}{C}}ONa$$

2-Naphthol — Acetic anhydride — 2-Naphthyl acetate — Sodium acetate

(*c*) Benzoyl chloride acylates the hydroxyl group of phenol.

$$\text{Phenol} + C_6H_5\overset{O}{\overset{\|}{C}}Cl \longrightarrow C_6H_5O\overset{O}{\overset{\|}{C}}C_6H_5 + HCl$$

Phenol — Benzoyl chloride — Phenyl benzoate — Hydrogen chloride

**24.8** Epoxides are sensitive to nucleophilic ring-opening reactions. Phenoxide ion attacks the less hindered carbon to yield 1-phenoxy-2-propanol.

$$C_6H_5\ddot{O}:^- + H_2C\underset{O}{—}CHCH_3 \xrightarrow{HO^-,\ H_2O} C_6H_5OCH_2\underset{OH}{\underset{|}{C}}HCH_3$$

Phenoxide ion — 1,2-Epoxypropane — 1-Phenoxy-2-propanol

**24.9** The aryl halide must be one that is reactive toward nucleophilic aromatic substitution by the addition–elimination mechanism. *p*-Fluoronitrobenzene is far more reactive than fluorobenzene. The reaction shown yields *p*-nitrophenyl phenyl ether in 92% yield.

$$C_6H_5OK + F-C_6H_4-NO_2 \xrightarrow{150°C} C_6H_5-O-C_6H_4-NO_2$$

Potassium phenoxide — *p*-Fluoronitrobenzene — *p*-Nitrophenyl phenyl ether

**24.10** Substituted allyl aryl ethers undergo a Claisen rearrangement similar to the reaction described in text Section 24.13 for allyl phenyl ether. 2-Butenyl phenyl ether rearranges on heating to give *o*-(1-methyl-2-propenyl)phenol.

2-Butenyl phenyl ether $\xrightarrow{\text{rearrangement}}$ (dienone intermediate, H) $\xrightarrow{\text{enolization}}$ *o*-(1-Methyl-2-propenyl)phenol

**24.11** (*a*) The parent compound is benzaldehyde. Vanillin bears a methoxy group ($CH_3O$) at C-3 and a hydroxyl group (HO) at C-4.

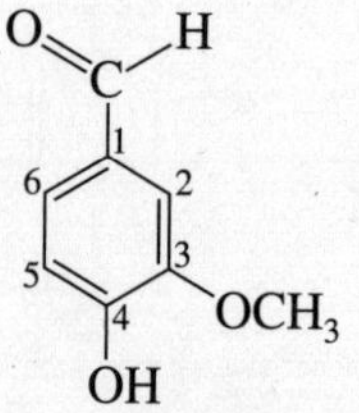

Vanillin
(4-hydroxy-3-methoxybenzaldehyde)

(*b, c*) Thymol and carvacrol differ with respect to the position of the hydroxyl group.

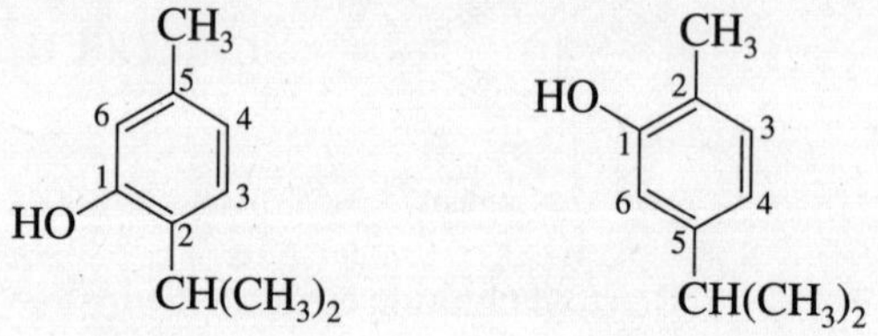

Thymol
(2-isopropyl-5-methylphenol)

Carvacrol
(5-isopropyl-2-methylphenol)

(*d*) An allyl substituent is $-CH_2CH{=}CH_2$.

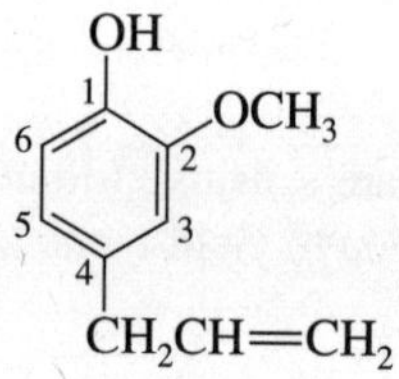

Eugenol
(4-allyl-2-methoxyphenol)

(*e*) Benzoic acid is $C_6H_5CO_2H$. Gallic acid bears three hydroxyl groups, located at C-3, C-4, and C-5.

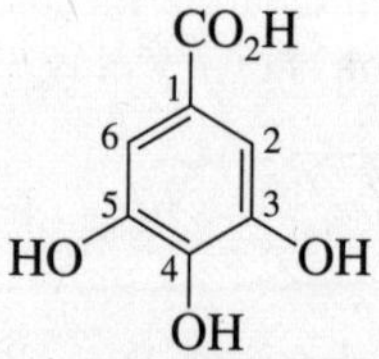

Gallic acid
(3,4,5-trihydroxybenzoic acid)

(*f*) Benzyl alcohol is $C_6H_5CH_2OH$. Salicyl alcohol bears a hydroxyl group at the ortho position.

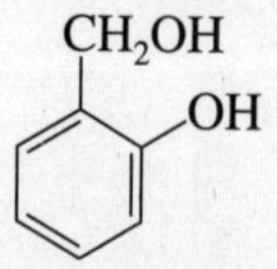

Salicyl alcohol
(*o*-hydroxybenzyl alcohol)

**24.12** (*a*) The compound is named as a derivative of phenol. The substituents (ethyl and nitro) are cited in alphabetical order with numbers assigned in the direction that gives the lowest number at the first point of difference.

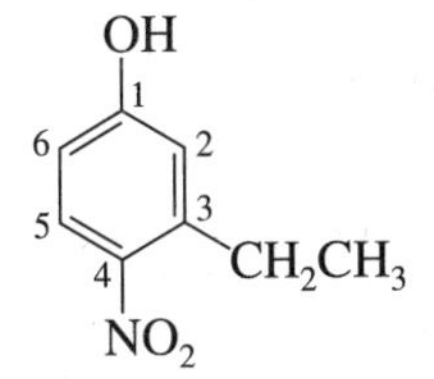

3-Ethyl-4-nitrophenol

(*b*) An isomer of the compound in part (*a*) is 4-ethyl-3-nitrophenol.

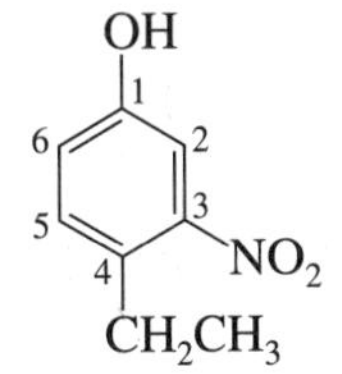

4-Ethyl-3-nitrophenol

(*c*) The parent compound is phenol. It bears, in alphabetical order, a benzyl group at C-4 and a chlorine at C-2.

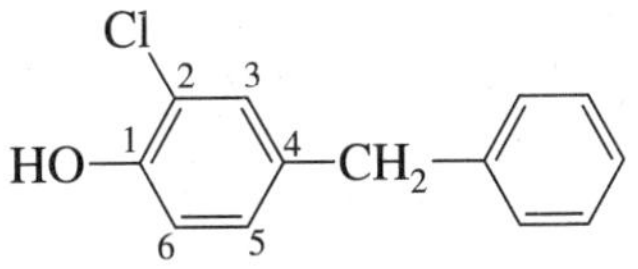

4-Benzyl-2-chlorophenol

(*d*) This compound is named as a derivative of anisole, $C_6H_5OCH_3$. Because multiplicative prefixes (di, tri-, etc.) are not considered when alphabetizing substituents, isopropyl precedes dimethyl.

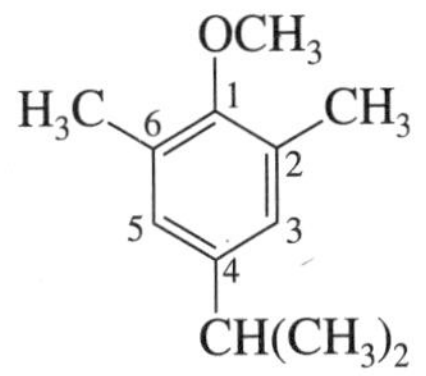

4-Isopropyl-2,6-dimethylanisole

(*e*) The compound is an aryl ester of trichloroacetic acid. The aryl group is 2,5-dichlorophenyl.

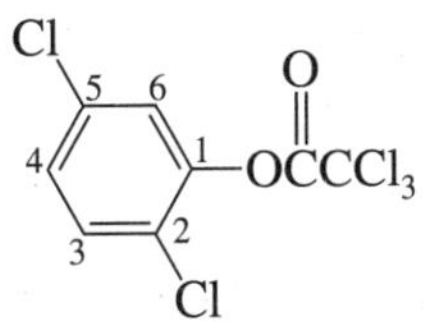

2,5-Dichlorophenyl trichloroacetate

**24.13** (*a*) The reaction is an acid–base reaction. Phenol is the acid; sodium hydroxide is the base.

$$C_6H_5OH + NaOH \longrightarrow C_6H_5ONa + H_2O$$

Phenol (stronger acid) | Sodium hydroxide (stronger base) | Sodium phenoxide (weaker base) | Water (weaker acid)

(*b*) Sodium phenoxide reacts with ethyl bromide to yield ethyl phenyl ether in a Williamson reaction. Phenoxide ion acts as a nucleophile.

$$C_6H_5ONa + CH_3CH_2Br \longrightarrow C_6H_5OCH_2CH_3 + NaBr$$

Sodium phenoxide | Ethyl bromide | Ethyl phenyl ether | Sodium bromide

(*c*) *p*-Toluenesulfonate esters behave much like alkyl halides in nucleophilic substitution reactions. Phenoxide ion displaces *p*-toluenesulfonate from the primary carbon.

$$C_6H_5ONa + CH_3CH_2CH_2CH_2OSO_2C_6H_4CH_3 \longrightarrow C_6H_5OCH_2CH_2CH_2CH_3 + NaOSO_2C_6H_4CH_3$$

Sodium phenoxide | Butyl *p*-toluenesulfonate | Butyl phenyl ether | Sodium *p*-toluenesulfonate

(*d*) Carboxylic acid anhydrides react with phenoxide anions to yield aryl esters.

$$C_6H_5ONa + CH_3COCCH_3 \longrightarrow C_6H_5OCCH_3 + CH_3CONa$$

(with C=O on each carbonyl carbon)

Sodium phenoxide | Acetic anhydride | Phenyl acetate | Sodium acetate

(*e*) Acyl chlorides convert phenols to aryl esters.

$$o\text{-}CH_3C_6H_4OH + C_6H_5CCl \longrightarrow o\text{-}CH_3C_6H_4OCC_6H_5 + HCl$$

*o*-Cresol | Benzoyl chloride | 2-Methylphenyl benzoate | Hydrogen chloride

(*f*) Phenols react as nucleophiles toward epoxides.

$$m\text{-}CH_3C_6H_4OH + H_2C\text{—}CH_2 (\text{O}) \longrightarrow m\text{-}H_3CC_6H_4OCH_2CH_2OH$$

*m*-Cresol | Ethylene oxide | 2-(3-Methylphenoxy)ethanol

The reaction as written conforms to the requirements of the problem that a balanced equation be written. Of course, the reaction will be much faster if catalyzed by acid or base, but the catalysts do not enter into the equation representing the overall process.

(*g*) Bromination of the aromatic ring of 2,6-dichlorophenol occurs para to the hydroxy group. The more activating group (—OH) determines the orientation of the product.

2,6-Dichlorophenol + $Br_2$ ⟶ 4-Bromo-2,6-dichlorophenol + HBr

2,6-Dichlorophenol — Bromine — 4-Bromo-2,6-dichlorophenol — Hydrogen bromide

(*h*) In aqueous solution bromination occurs at all the open positions that are ortho and para to the hydroxyl group.

*p*-Cresol + $2Br_2$ $\xrightarrow{H_2O}$ 2,6-Dibromo-4-methylphenol + 2HBr

*p*-Cresol — Bromine — 2,6-Dibromo-4-methylphenol — Hydrogen bromide

(*i*) Hydrogen bromide cleaves ethers to give an alkyl halide and a phenol.

$C_6H_5OCH(CH_3)_2$ + HBr $\xrightarrow{heat}$ $C_6H_5OH$ + $(CH_3)_2CHBr$

Isopropyl phenyl ether — Hydrogen bromide — Phenol — Isopropyl bromide

**24.14** (*a*) Strongly electron-withdrawing groups, particularly those such as $—NO_2$, increase the acidity of phenols by resonance stabilization of the resulting phenoxide anion. Electron-releasing substituents such as $—CH_3$ exert a very small acid-weakening effect.

2,4,6-Trinitrophenol, more acidic ($K_a = 3.8 \times 10^{-1}$, $pK_a = 0.4$)

2,4,6-Trimethylphenol, less acidic ($K_a = 1.3 \times 10^{-11}$, $pK_a = 10.9$)

Picric acid (2,4,6-trinitrophenol) is a stronger acid by far than 2,4,6-trimethylphenol. All three nitro groups participate in resonance stabilization of the picrate anion.

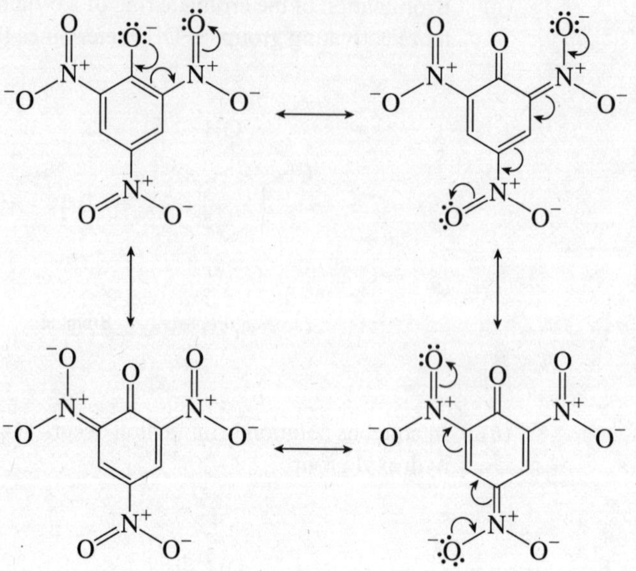

(*b*) Stabilization of a phenoxide anion is most effective when electron-withdrawing groups are present at the ortho and para positions, because it is these carbons that bear most of the negative charge in phenoxide anion.

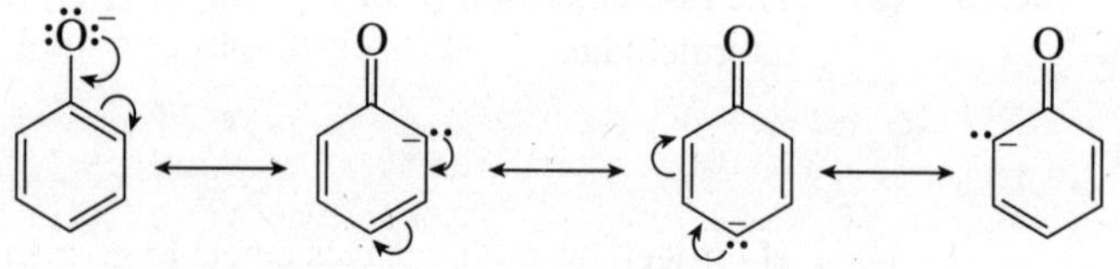

2,6-Dichlorophenol is therefore expected to be (and is) a stronger acid than 3,5-dichlorophenol.

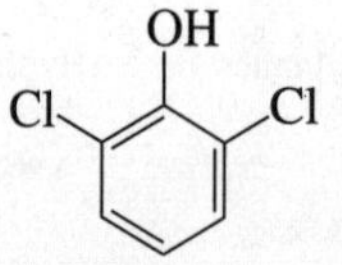

2,6-Dichlorophenol, more acidic ($K_a = 1.6 \times 10^{-7}$, p$K_a$ = 6.8)

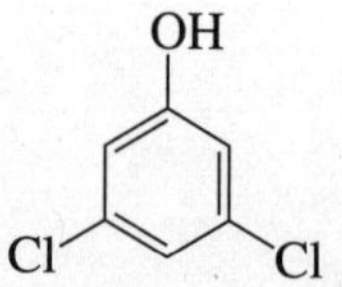

3,5-Dichlorophenol, less acidic ($K_a = 6.5 \times 10^{-9}$, p$K_a$ = 8.2)

(*c*) The same principle is at work here as in part (*b*). A nitro group para to the phenol oxygen is directly conjugated to it and stabilizes the anion better than one at the meta position.

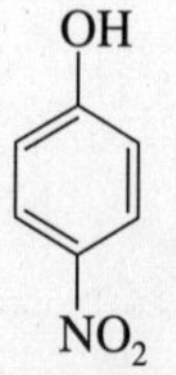

4-Nitrophenol, stronger acid ($K_a = 1.0 \times 10^{-8}$, p$K_a$ = 7.2)

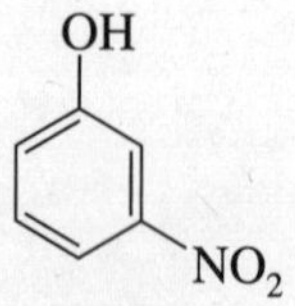

3-Nitrophenol, weaker acid ($K_a = 4.1 \times 10^{-9}$, p$K_a$ = 8.4)

(*d*) A cyano group is strongly electron-withdrawing, and so 4-cyanophenol is a stronger acid than phenol.

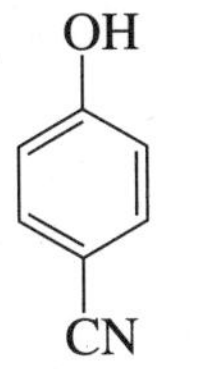

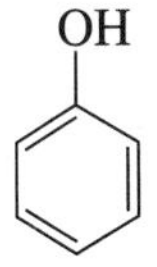

4-Cyanophenol, more acidic ($K_a = 1.1 \times 10^{-8}$, $pK_a = 8.0$)

Phenol, less acidic ($K_a = 1 \times 10^{-10}$, $pK_a = 10$)

There is resonance stabilization of the 4-cyanophenoxide anion.

(*e*) The 5-nitro group in 2,5-dinitrophenol is meta to the hydroxyl group and so does not stabilize the resulting anion as much as does an ortho or a para nitro group.

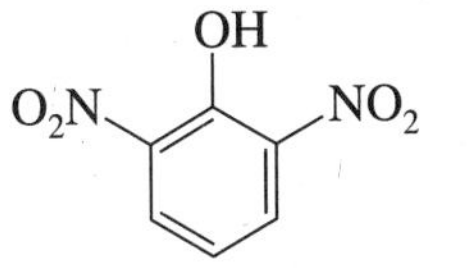

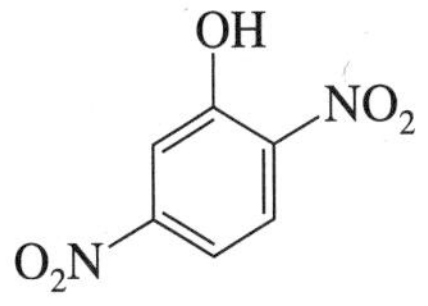

2,6-Dinitrophenol, more acidic ($K_a = 2.0 \times 10^{-4}$, $pK_a = 3.7$)

2,5-Dinitrophenol, less acidic ($K_a = 6.0 \times 10^{-6}$, $pK_a = 5.2$)

**24.15** (*a*) The rate-determining step of ester hydrolysis in basic solution is formation of the tetrahedral intermediate.

$$\text{ArOCCH}_3 \text{ (C=O)} + \text{HO}^- \xrightleftharpoons{\text{slow}} \text{ArOC(O}^-\text{)(OH)CH}_3$$

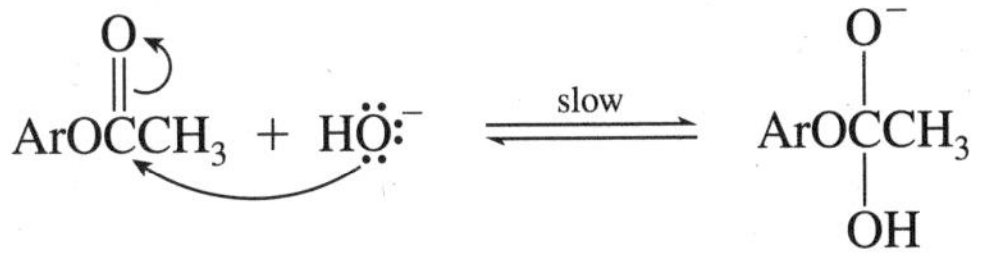

Because this intermediate is negatively charged, there will be a small effect favoring its formation when the aryl group bears an electron-withdrawing substituent. Furthermore, this intermediate can either return to starting materials or proceed to products.

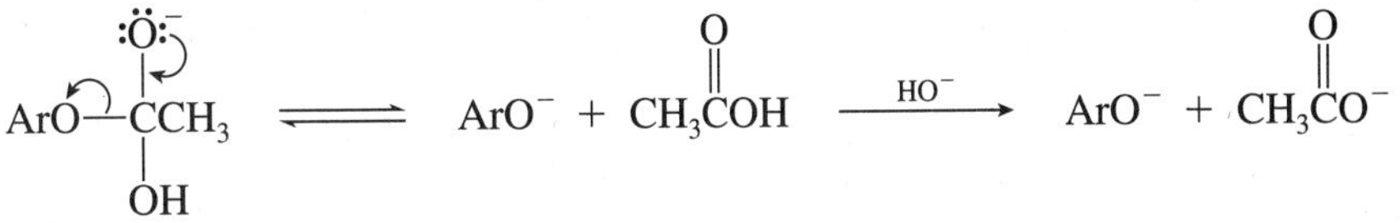

$$\text{ArO–C(O}^-\text{)(OH)CH}_3 \rightleftharpoons \text{ArO}^- + \text{CH}_3\text{COH} \xrightarrow{\text{HO}^-} \text{ArO}^- + \text{CH}_3\text{CO}^-$$

The proportion of the tetrahedral intermediate that goes on to products increases as the leaving group $ArO^-$ becomes less basic. This is strongly affected by substituents; electron-withdrawing groups stabilize $ArO^-$. The prediction is that *m*-nitrophenyl acetate undergoes hydrolysis in basic solution faster than phenol. Indeed, this is observed to be the case; *m*-nitrophenyl acetate reacts some ten times faster than does phenyl acetate at 25°C.

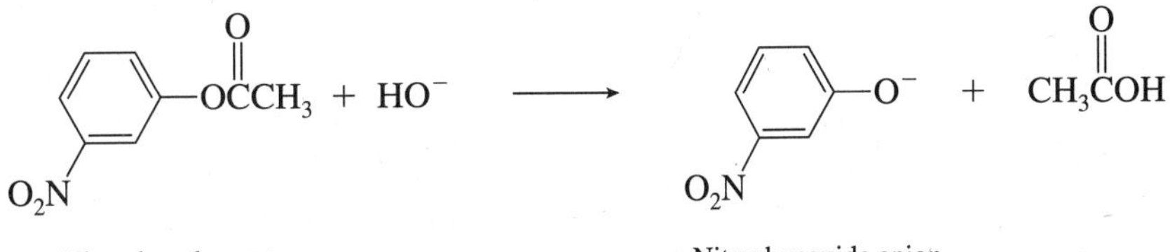

*m*-Nitrophenyl acetate (more reactive) $+ \text{HO}^- \longrightarrow$ *m*-Nitrophenoxide anion (a better leaving group than phenoxide because it is less basic) $+ \text{CH}_3\text{COH}$

(*b*) The same principle applies here as in part (*a*). *p*-Nitrophenyl acetate reacts faster than *m*-nitrophenyl acetate (by about 45%) largely because *p*-nitrophenoxide is less basic and thus a better leaving group than *m*-nitrophenoxide.

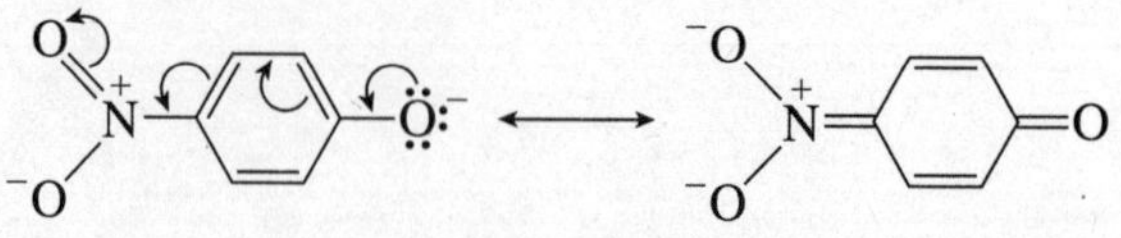

Resonance in *p*-nitrophenoxide is particularly effective because the *p*-nitro group is directly conjugated to the oxyanion; direct conjugation of these groups is absent in *m*-nitrophenoxide.

(*c*) The reaction of ethyl bromide with a phenol is an $S_N2$ reaction in which the oxygen of the phenol is the nucleophile. The reaction is much faster with sodium phenoxide than with phenol, because an anion is more nucleophilic than a corresponding neutral molecule.

**Faster reaction:**

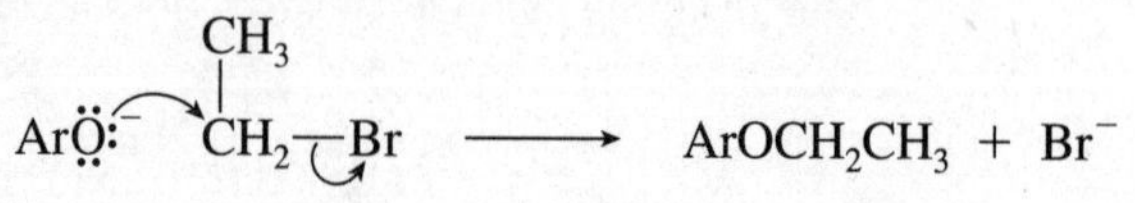

$$ArO^- + CH_3CH_2Br \longrightarrow ArOCH_2CH_3 + Br^-$$

**Slower reaction:**

$$ArOH + CH_3CH_2Br \longrightarrow Ar\overset{+}{O}(H)CH_2CH_3 + Br^-$$

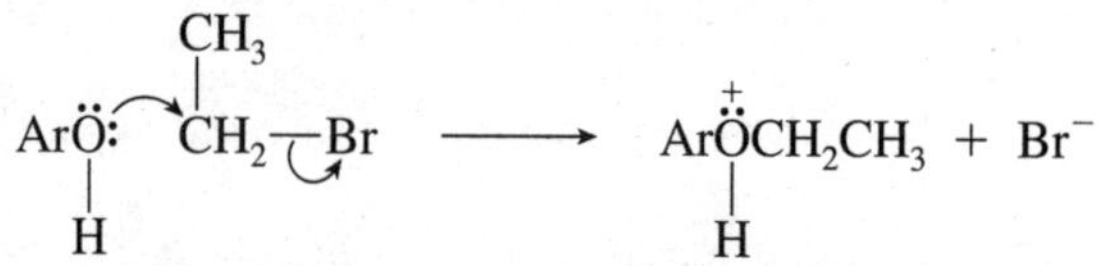

(*d*) The answer here also depends on the nucleophilicity of the attacking species, which is a phenoxide anion in both reactions.

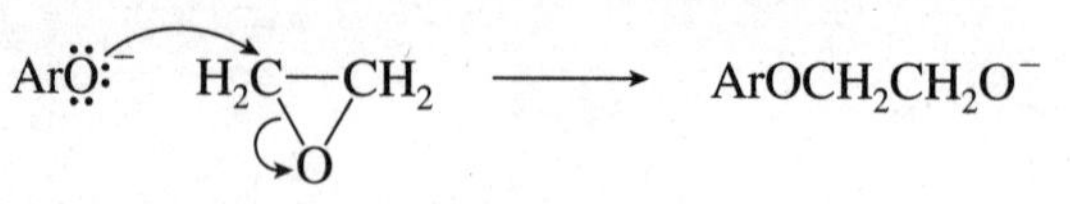

$$ArO^- + H_2C{-}CH_2 \text{ (ethylene oxide)} \longrightarrow ArOCH_2CH_2O^-$$

The more nucleophilic anion is phenoxide ion, because it is more basic than *p*-nitrophenoxide.

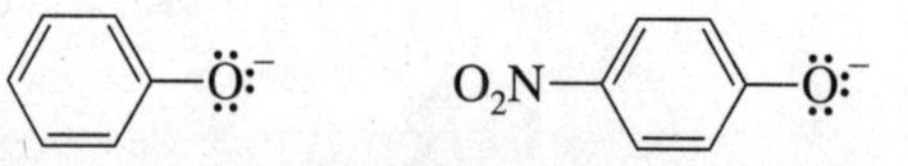

More basic; better nucleophile

Better delocalization of negative charge makes this less basic and less nucleophilic.

Rate measurements reveal that sodium phenoxide reacts 17 times faster with ethylene oxide (in ethanol at 70°C) than does its *p*-nitro derivative.

(*e*) This reaction is electrophilic aromatic substitution. Because a hydroxy substituent is more activating than an acetate group, phenol undergoes bromination faster than does phenyl acetate.

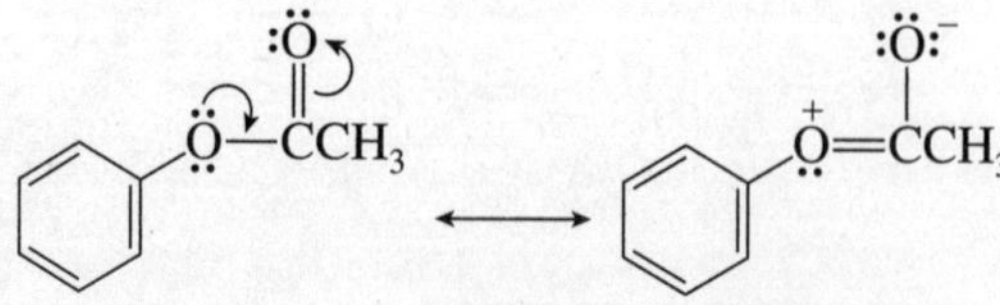

Resonance involving ester group reduces tendency of oxygen to donate electrons to ring.

**24.16** Nucleophilic aromatic substitution by the elimination–addition mechanism is impossible, owing to the absence of any protons that might be abstracted from the substrate. The addition–elimination pathway is available, however.

$C_6F_6 + HO^- \xrightarrow{\text{slow}}$ [cyclohexadienyl anion intermediate] $\xrightarrow{\text{fast}} C_6F_5OH + F^-$

Hexafluorobenzene — Pentafluorophenol

This pathway is favorable because the cyclohexadienyl anion intermediate formed in the rate-determining step is stabilized by the electron-withdrawing inductive effect of its fluorine substituents.

**24.17** (*a*) Allyl bromide is a reactive alkylating agent and converts the free hydroxyl group of the aryl compound (a natural product known as *guaiacol*) to its corresponding allyl ether.

Guaiacol + $H_2C{=}CHCH_2Br$ $\xrightarrow[\text{acetone}]{K_2CO_3}$ $o$-($OCH_2CH{=}CH_2$)($OCH_3$)benzene

Guaiacol — Allyl bromide — 2-Allyloxyanisole (80–90%)

(*b*) Sodium phenoxide acts as a nucleophile in this reaction and is converted to an ether.

$C_6H_5ONa + ClCH_2CH(OH)CH_2OH \longrightarrow C_6H_5OCH_2CH(OH)CH_2OH$

Sodium phenoxide — 3-Chloro-1,2-propanediol — 3-Phenoxy-1,2-propanediol (61–63%)

(*c*) Orientation in nitration is governed by the most activating substituent, in this case the hydroxyl group.

Vanillin $\xrightarrow[\text{acetic acid, heat}]{HNO_3}$ 4-Hydroxy-3-methoxy-5-nitrobenzaldehyde (83%)

Vanillin — 4-Hydroxy-3-methoxy-5-nitrobenzaldehyde (83%)

(*d*) Allyl aryl ethers undergo a Claisen rearrangement on heating. Heating *p*-acetamidophenyl allyl ether gave an 83% yield of 4-acetamido-2-allylphenol.

$CH_3CONH{-}C_6H_4{-}OCH_2CH{=}CH_2 \xrightarrow{\text{heat}} CH_3CONH{-}C_6H_3(CH_2CH{=}CH_2){-}OH$

*p*-Acetamidophenyl allyl ether — 4-Acetamido-2-allylphenol

(*e*) The hydroxyl group, as the most activating substituent, controls the orientation of electrophilic aromatic substitution. Bromination takes place ortho to the hydroxyl group.

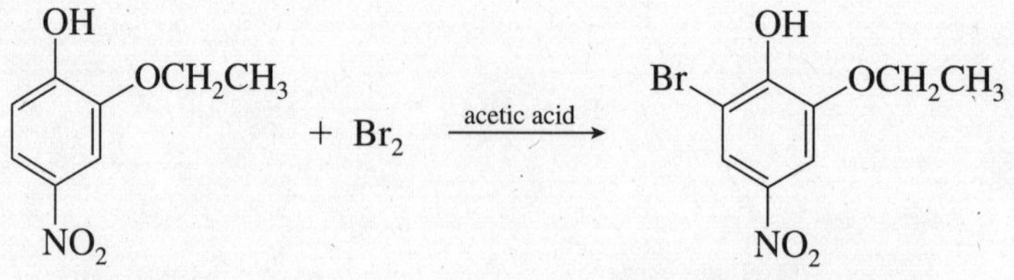

2-Ethoxy-4-nitrophenol + $Br_2$ $\xrightarrow{\text{acetic acid}}$ 2-Bromo-6-ethoxy-4-nitrophenol (65%)

(*f*) Oxidation of hydroquinone derivatives (*p*-dihydroxybenzenes) with Cr(VI) reagents is a method for preparing quinones.

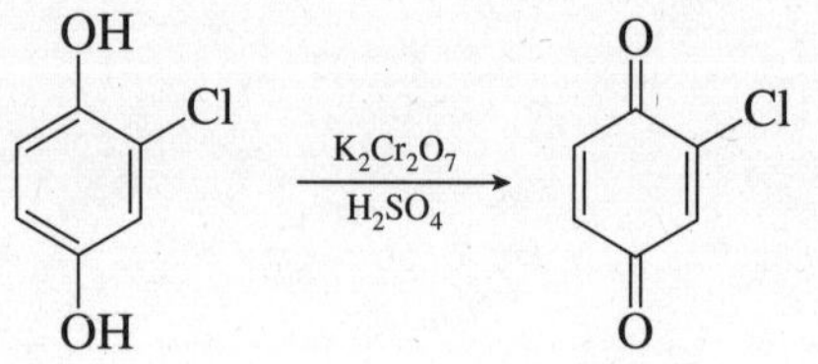

2-Chloro-1,4-benzenediol $\xrightarrow[H_2SO_4]{K_2Cr_2O_7}$ 2-Chloro-1,4-benzoquinone (88%)

(*g*) Aryl esters undergo a reaction known as the **Fries rearrangement** on being treated with aluminum chloride, which converts them to acyl phenols. Acylation takes place para to the hydroxyl in this case.

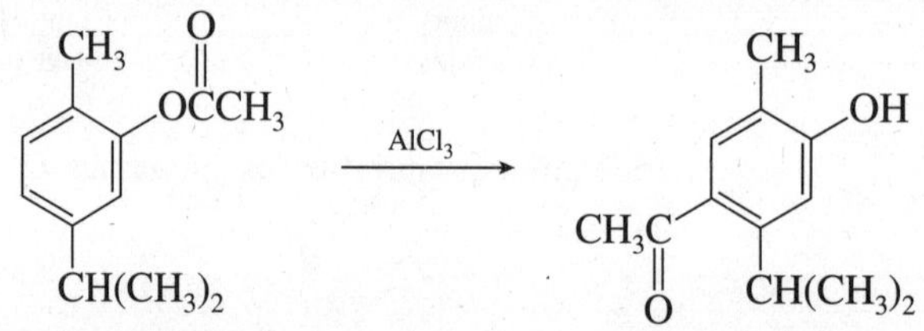

5-Isopropyl-2-methylphenyl acetate $\xrightarrow{AlCl_3}$ 4-Hydroxy-2-isopropyl-5-methylacetophenone (90%)

(*h*) Nucleophilic aromatic substitution takes place to yield a diaryl ether. The nucleophile is the phenoxide ion derived from 2,6-dimethylphenol.

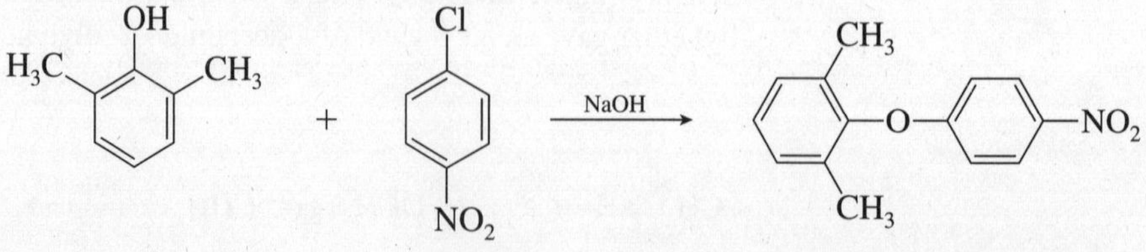

2,6-Dimethylphenol + *p*-Chloronitrobenzene $\xrightarrow{NaOH}$ 2,6-Dimethylphenyl *p*-nitrophenyl ether (82%)

(*i*) Chlorination with excess chlorine occurs at all available positions that are ortho and para to the hydroxyl group.

2,5-Dichlorophenol + $2Cl_2$ (Chlorine) $\xrightarrow{\text{acetic acid}}$ 2,3,4,6-Tetrachlorophenol (isolated yield, 100%) + 2HCl (Hydrogen chloride)

(*j*) Amines react with esters to give amides. In the case of a phenyl ester, phenol is the leaving group.

*o*-Methylaniline + Phenyl salicylate $\xrightarrow{\text{heat}}$ *N*-(*o*-Methylphenyl)salicylamide (isolated yield, 73–77%) + Phenol

(*k*) Aryl diazonium salts attack electron-rich aromatic rings, such as those of phenols, to give the products of electrophilic aromatic substitution.

2,4,5-Trichlorophenol + $C_6H_5\overset{+}{N}{\equiv}N{:}\ Cl^-$ (Benzenediazonium chloride) $\longrightarrow$ 2-Benzeneazo-3,4,6-trichlorophenol (80%)

**24.18** In the first step *p*-nitrophenol is alkylated on its phenolic oxygen with ethyl bromide.

$$O_2N\text{—}C_6H_4\text{—}OH + CH_3CH_2Br \xrightarrow{HO^-} O_2N\text{—}C_6H_4\text{—}OCH_2CH_3$$

*p*-Nitrophenol, Ethyl bromide, Ethyl *p*-nitrophenyl ether

Reduction of the nitro group gives the corresponding arylamine.

$$O_2N\text{—}C_6H_4\text{—}OCH_2CH_3 \xrightarrow[2.\ HO^-]{1.\ Fe,\ HCl} H_2N\text{—}C_6H_4\text{—}OCH_2CH_3$$

Ethyl *p*-nitrophenyl ether, *p*-Ethoxyaniline

Treatment of *p*-ethoxyaniline with acetic anhydride gives phenacetin.

$$H_2N\text{—}C_6H_4\text{—}OCH_2CH_3 + CH_3\overset{O}{\overset{\|}{C}}O\overset{O}{\overset{\|}{C}}CH_3 \longrightarrow CH_3\overset{O}{\overset{\|}{C}}NH\text{—}C_6H_4\text{—}OCH_2CH_3$$

*p*-Ethoxyaniline, Acetic anhydride, *p*-Ethoxyacetanilide (phenacetin)

**24.19** The three parts of this problem make up the series of steps by which *o*-bromophenol is prepared.

(*a*) Because direct bromination of phenol yields both *o*-bromophenol and *p*-bromophenol, it is essential that the para position be blocked prior to the bromination step. In practice, what is done is to disulfonate phenol, which blocks the para and one of the ortho positions.

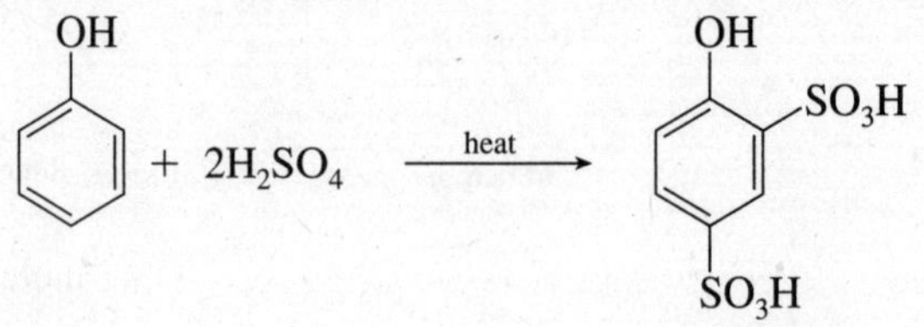

OH ... + $2H_2SO_4$ —heat→ OH, $SO_3H$, $SO_3H$

Phenol

4-Hydroxy-1,3-benzenedisulfonic acid (compound A)

(*b*) Bromination then can be accomplished cleanly at the open position ortho to the hydroxyl group.

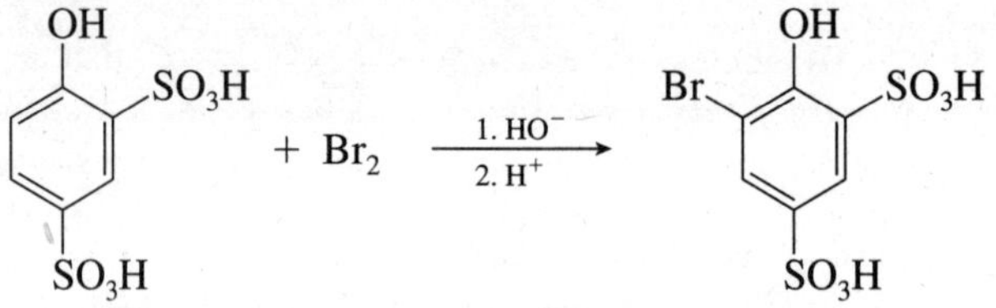

OH, $SO_3H$, $SO_3H$ + $Br_2$ —1. $HO^-$ 2. $H^+$→ OH, Br, $SO_3H$, $SO_3H$

Compound A

5-Bromo-4-hydroxy-1,3-benzenedisulfonic acid (compound B)

(*c*) After bromination the sulfonic acid groups are removed by acid-catalyzed hydrolysis.

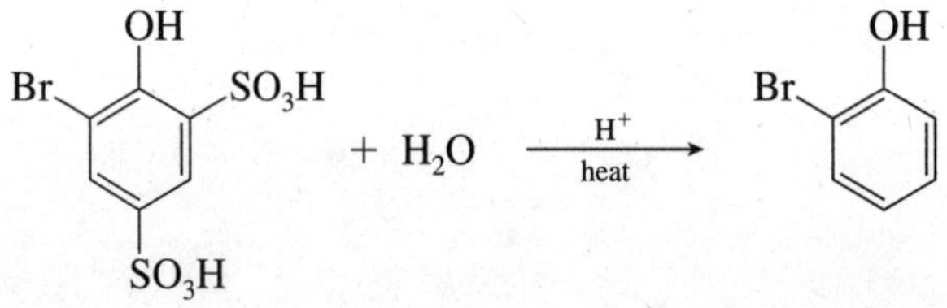

OH, Br, $SO_3H$, $SO_3H$ + $H_2O$ —$H^+$, heat→ OH, Br

Compound B

*o*-Bromophenol (compound C)

**24.20** Nitration of 3,5-dimethylphenol gives a mixture of the 2-nitro and 4-nitro derivatives.

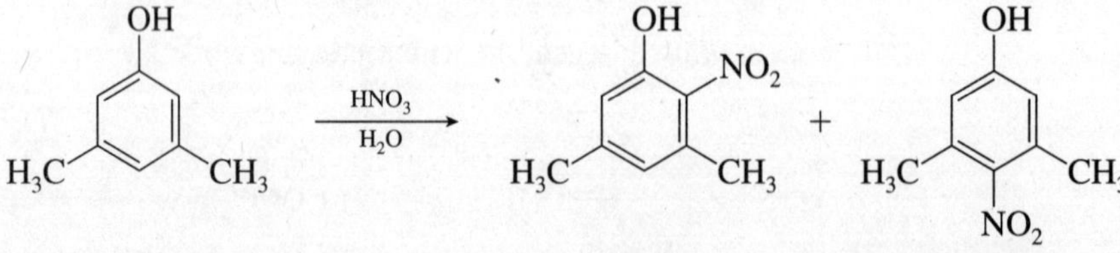

OH, $H_3C$, $CH_3$ —$HNO_3$, $H_2O$→ OH, $NO_2$, $H_3C$, $CH_3$ + OH, $H_3C$, $CH_3$, $NO_2$

3,5-Dimethylphenol

3,5-Dimethyl-2-nitrophenol

3,5-Dimethyl-4-nitrophenol

The more volatile compound (compound A), isolated by steam distillation, is the 2-nitro derivative. Intramolecular hydrogen bonding is possible between the nitro group and the hydroxyl group.

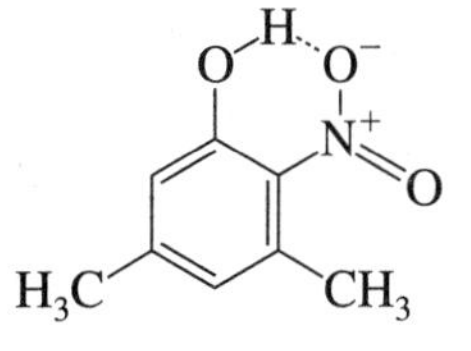

Intramolecular hydrogen bonding in 3,5-dimethyl-2-nitrophenol

The 4-nitro derivative participates in intermolecular hydrogen bonds and has a much higher boiling point; it is compound B.

**24.21** The relationship between the target molecule and the starting materials tells us that two processes are required, formation of a diaryl ether linkage and nitration of an aromatic ring. The proper order of carrying out these two separate processes is what needs to be considered.

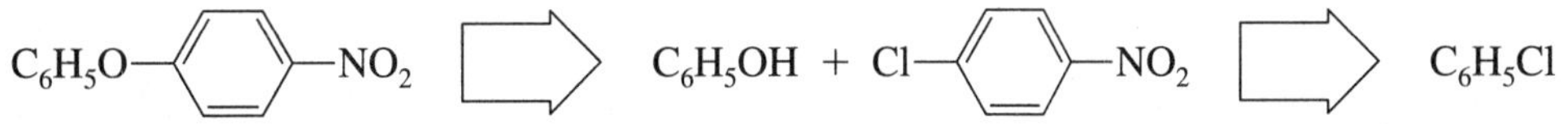

$C_6H_5O$–$C_6H_4$–$NO_2$ ⇒ $C_6H_5OH$ + Cl–$C_6H_4$–$NO_2$ ⇒ $C_6H_5Cl$

The critical step is ether formation, a step that is feasible for the reactants shown:

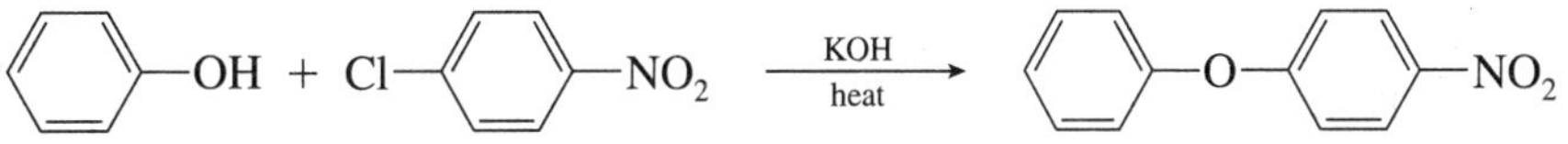

Phenol + p-Chloronitrobenzene $\xrightarrow[\text{heat}]{\text{KOH}}$ 4-Nitrophenyl phenyl ether

The reason this reaction is suitable is that it involves nucleophilic aromatic substitution by the addition–elimination mechanism on a *p*-nitro-substituted aryl halide. Indeed, this reaction has been carried out and gives an 80–82% yield. A reasonable synthesis would therefore begin with the preparation of *p*-chloronitrobenzene.

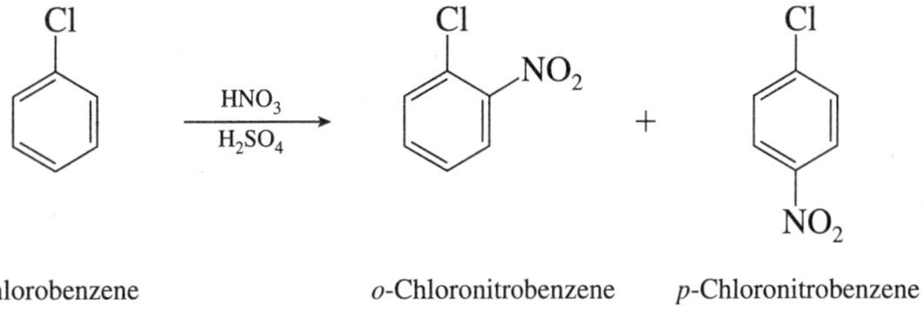

Chlorobenzene $\xrightarrow[H_2SO_4]{HNO_3}$ *o*-Chloronitrobenzene + *p*-Chloronitrobenzene

Separation of the *p*-nitro-substituted aryl halide and reaction with phenoxide ion complete the synthesis.

The following alternative route is less satisfactory:

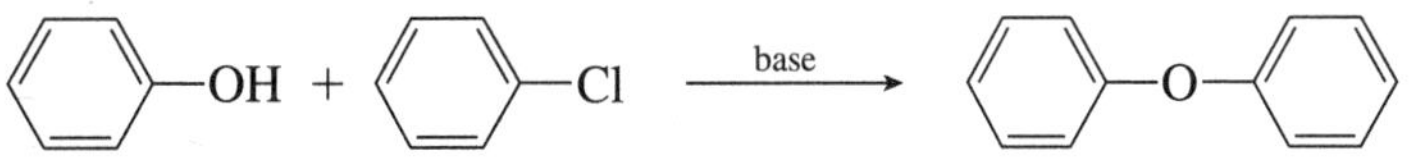

Phenol + Chlorobenzene $\xrightarrow{\text{base}}$ Diphenyl ether

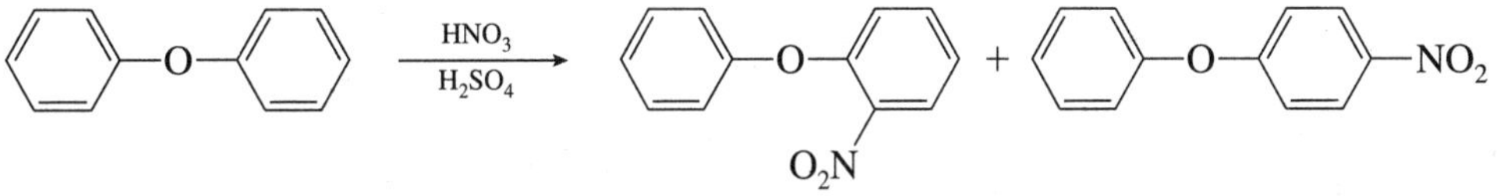

Diphenyl ether $\xrightarrow[H_2SO_4]{HNO_3}$ 2-Nitrophenyl phenyl ether + 4-Nitrophenyl phenyl ether

The difficulty with this route concerns the preparation of diphenyl ether. Direct reaction of phenoxide ion with chlorobenzene is very slow and requires high temperatures because chlorobenzene is a poor substrate for nucleophilic substitution.

A third route is also unsatisfactory because it, too, requires nucleophilic substitution on chlorobenzene.

Phenol $\xrightarrow{HNO_3}$ o-Nitrophenol + p-Nitrophenol

p-Nitrophenol + Chlorobenzene $\xrightarrow{\text{base}}$ 4-Nitrophenyl phenyl ether

**24.22** The overall transformation that needs to be effected is

2,3-Dimethoxybenzaldehyde $\longrightarrow$ 3-Pentadecylcatechol ($(CH_2)_{14}CH_3$)

A reasonable place to begin is with the attachment of the side chain. The aldehyde function allows for chain extension by a Wittig reaction.

$$Ar{-}CH_2CH_2R \Longrightarrow Ar{-}CH{=}CHR \Longrightarrow Ar{-}\overset{O}{\overset{\|}{C}}H + (C_6H_5)_3\overset{+}{P}{-}\ddot{\overset{-}{C}}HR$$

2,3-Dimethoxybenzaldehyde + $CH_3(CH_2)_{12}\ddot{\overset{-}{C}}H{-}\overset{+}{P}(C_6H_5)_3 \longrightarrow$ 2,3-dimethoxy-1-($CH{=}CH(CH_2)_{12}CH_3$)benzene

Hydrogenation of the double bond and hydrogen halide cleavage of the ether functions complete the synthesis.

$CH{=}CH(CH_2)_{12}CH_3$ (dimethoxy) $\xrightarrow[Pt]{H_2}$ $CH_2CH_2(CH_2)_{12}CH_3$ (dimethoxy) $\xrightarrow[\text{heat}]{HBr}$ $CH_2CH_2(CH_2)_{12}CH_3$ (dihydroxy)

3-Pentadecylcatechol

Other synthetic routes are of course possible. One of the earliest approaches used a Grignard reaction to attach the side chain.

CH$_3$O, OCH$_3$, CH=O + $CH_3(CH_2)_{12}CH_2MgBr$ —(1. diethyl ether; 2. $H_3O^+$)→ CH$_3$O, OCH$_3$, $CHCH_2(CH_2)_{12}CH_3$ (OH)

2,3-Dimethoxy-
benzaldehyde

The resulting secondary alcohol can then be dehydrated to the same alkene intermediate prepared in the preceding synthetic scheme.

CH$_3$O, OCH$_3$, $CHCH_2(CH_2)_{12}CH_3$ (OH) —($H^+$)→ CH$_3$O, OCH$_3$, $CH{=}CH(CH_2)_{12}CH_3$

Again, hydrogenation of the double bond and ether cleavage leads to the desired 3-pentadecylcatechol.

**24.23** Recall that the Claisen rearrangement converts an aryl allyl ether to an ortho-substituted allyl phenol. The presence of an allyl substituent in the product ortho to an aryl ether thus suggests the following retrosynthesis:

OCH$_3$, CH$_3$O, $CH_2CH{=}CH_2$, H$_3$C, OH ⟹ $OCH_2CH{=}CH_2$, CH$_3$O, H$_3$C, $OCCH_3$ (=O)

As reported in the literature synthesis, the starting phenol may be converted to the corresponding allyl ether by reaction with allyl bromide in the presence of base. This step was accomplished in 80% yield. Heating the allyl ether yields the *o*-allyl phenol.

OH, CH$_3$O, H$_3$C, $OCCH_3$ (=O) —($H_2C{=}CHCH_2Br$, $K_2CO_3$)→ $OCH_2CH{=}CH_2$, CH$_3$O, H$_3$C, $OCCH_3$ (=O) —(200°C, 3 h)→ OH, CH$_3$O, $CH_2CH{=}CH_2$, H$_3$C, $OCCH_3$ (=O)

The synthesis is completed by methylation of the phenolic oxygen and saponification of the acetate ester. The final three steps of the synthesis proceeded in an 82% overall yield.

OH, CH$_3$O, $CH_2CH{=}CH_2$, H$_3$C, $OCCH_3$ (=O) —($CH_3I$, $K_2CO_3$)→ OCH$_3$, CH$_3$O, $CH_2CH{=}CH_2$, H$_3$C, $OCCH_3$ (=O) —(1. KOH, $CH_3OH$; 2. $H^+$)→ OCH$_3$, CH$_3$O, $CH_2CH{=}CH_2$, H$_3$C, OH

**24.24** The driving force for this reaction is the stabilization that results from formation of the aromatic ring. A reasonable series of steps begins with protonation of the carbonyl oxygen.

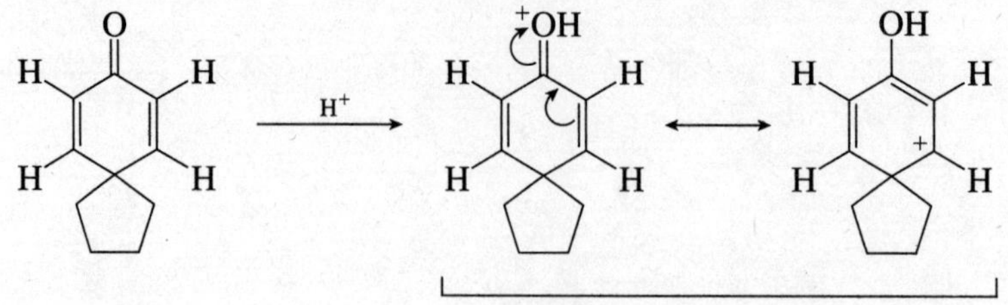

$H^+$

Resonance forms of protonated ketone

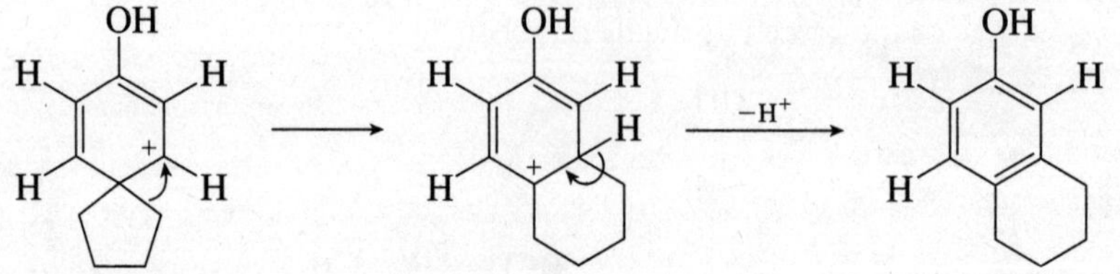

$-H^+$

Protonated ketone can rearrange by alkyl migration.

Aromatization of this intermediate occurs by loss of a proton.

**24.25** Bromination of *p*-hydroxybenzoic acid takes place in the normal fashion at both positions ortho to the hydroxy group.

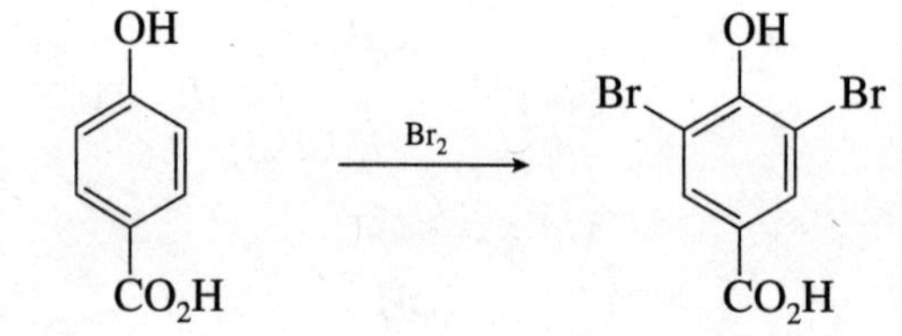

$Br_2$

*p*-Hydroxybenzoic acid

3,5-Dibromo-4-hydroxybenzoic acid

A third bromination step, this time at the para position, leads to the intermediate shown.

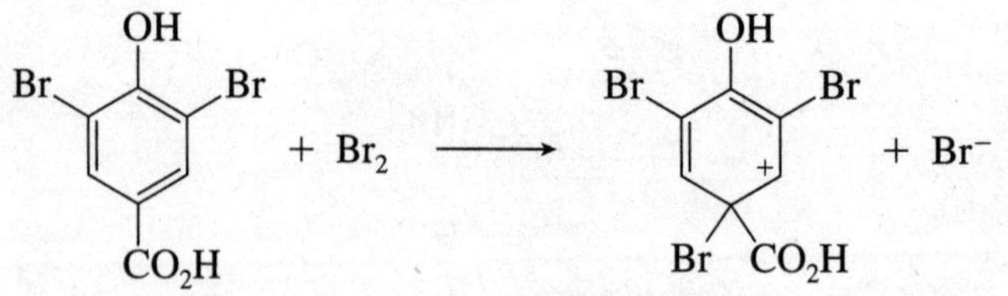

$+ Br_2 \longrightarrow \quad + Br^-$

Aromatization of this intermediate occurs by decarboxylation.

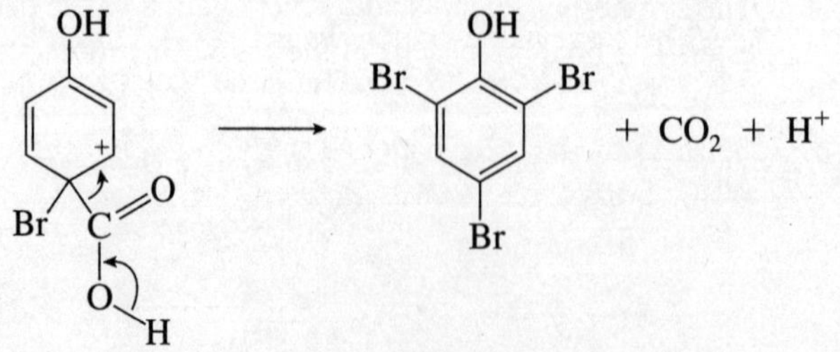

$\longrightarrow \quad + CO_2 + H^+$

2,4,6-Tribromophenol

**24.26** Electrophilic attack of bromine on 2,4,6-tribromophenol leads to a cationic intermediate.

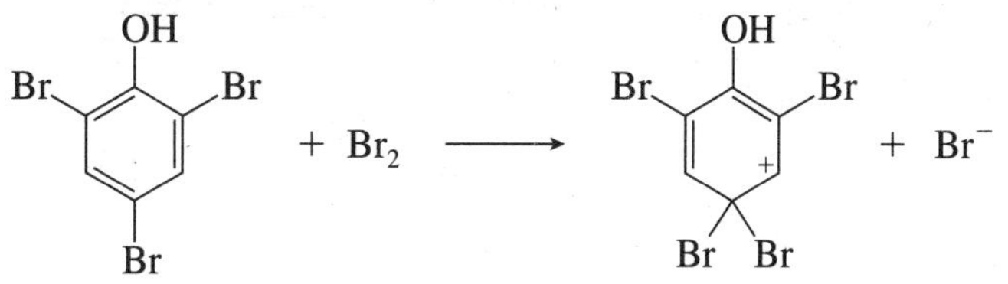

2,4,6-Tribromophenol

Loss of the hydroxyl proton from this intermediate generates the observed product.

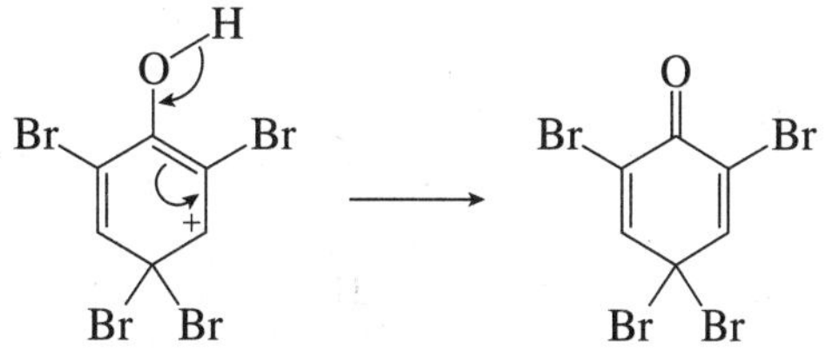

2,4,4,6-Tetrabromo-cyclohexadienone

**24.27** A good way to approach this problem is to assume that bromine attacks the aromatic ring of the phenol in the usual way, that is, para to the hydroxyl group.

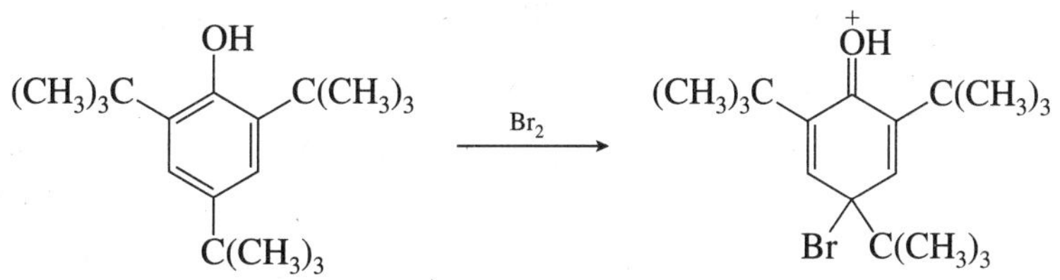

2,4,6-Tri-*tert*-butylphenol

This cation cannot yield the product of electrophilic aromatic substitution by loss of a proton from the ring but can lose a proton from oxygen to give a cyclohexadienone derivative.

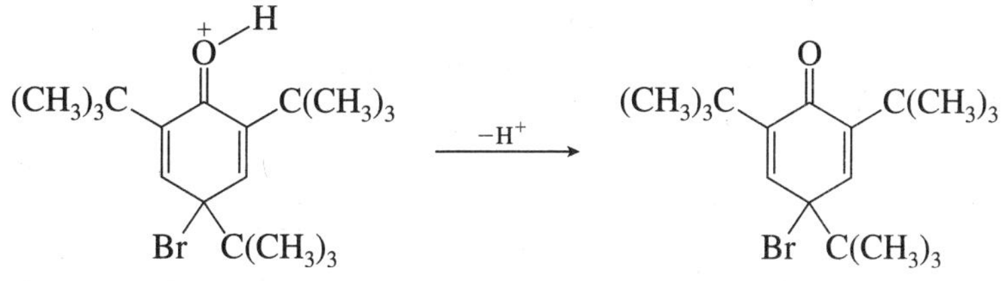

4-Bromo-2,4,6-tri-*tert*-butyl-2,5-cyclohexadienone

This cyclohexadienone is the compound $C_{18}H_{29}BrO$, and the peaks at 1655 and 1630 $cm^{-1}$ in the infrared are consistent with C═O and C═C stretching vibrations. The compound's symmetry is consistent with the observed $^1H$ NMR spectrum; two equivalent *tert*-butyl groups at C-2 and C-6 appear as an 18-proton singlet at $\delta$ 1.3 ppm, the other *tert*-butyl group is a 9-proton singlet at $\delta$ 1.2 ppm, and the 2 equivalent vinyl protons of the ring appear as a singlet at $\delta$ 6.9 ppm.

**24.28** Because the starting material is an acetal and the reaction conditions lead to hydrolysis with the production of 1,2-ethanediol, a reasonable reaction course is

$$\text{Compound A} \xrightarrow{H_2O,\ H^+} HOCH_2CH_2OH + \text{Compound B}$$

Compound A 1,2-Ethanediol Compound B

Indeed, dione B satisfies the spectroscopic criteria. Carbonyl bands are seen in the infrared spectrum, and compound B has two sets of protons to be seen in its $^{1}H$ NMR spectrum. The two vinyl protons are equivalent and appear at low field, $\delta$ 6.7 ppm; the 4 methylene protons are equivalent to each other and are seen at $\delta$ 2.9 ppm.

Compound B is the doubly ketonic tautomeric form of hydroquinone, compound C, to which it isomerizes on standing in water.

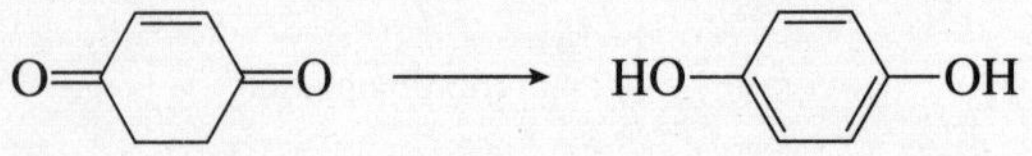

Compound B → Compound C (hydroquinone)

**24.29** A reasonable first step is protonation of the hydroxyl oxygen.

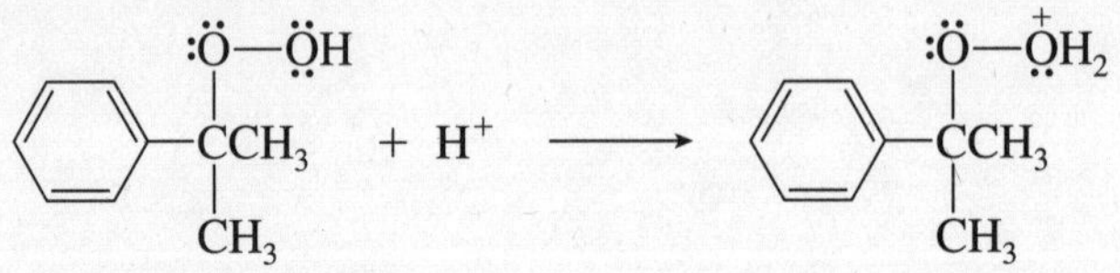

Cumene hydroperoxide $+ H^+$ →

The weak oxygen–oxygen bond can now be cleaved, with loss of water as the leaving group.

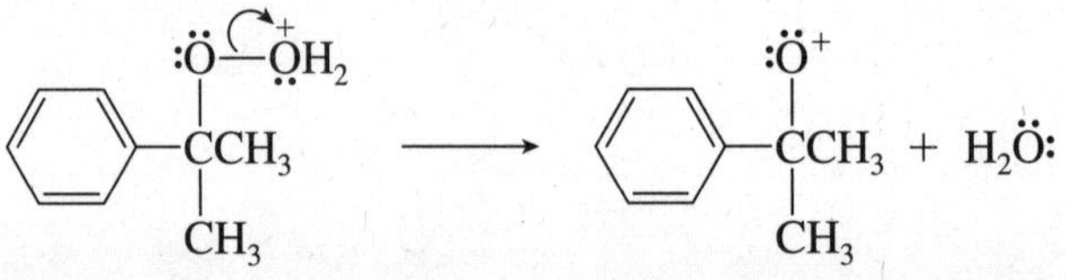

$+ H_2\ddot{O}:$

This intermediate bears a positively charged oxygen with only six electrons in its valence shell. Like a carbocation, such a species is highly electrophilic. The electrophilic oxygen attacks the $\pi$ system of the neighboring aromatic ring to give an unstable intermediate.

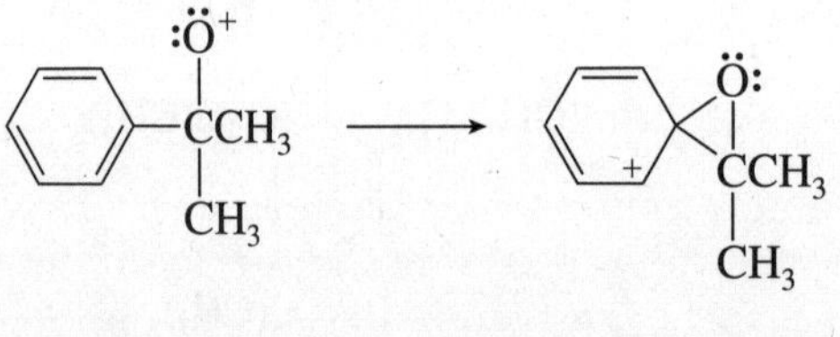

Ring opening of this intermediate is assisted by one of the lone pairs of oxygen and restores the aromaticity of the ring.

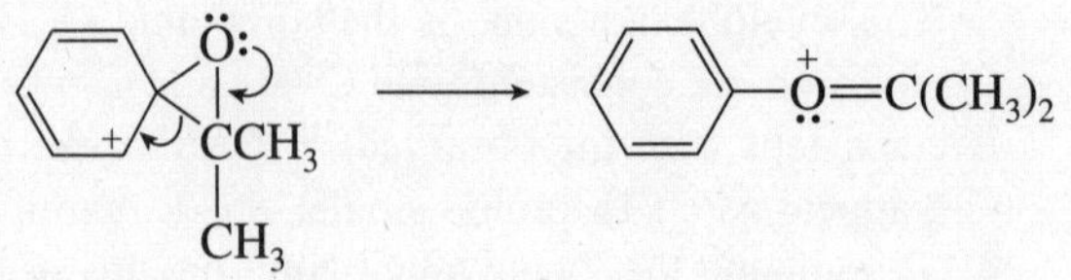

$\overset{+}{O}=C(CH_3)_2$

The cation formed by ring opening is captured by a water molecule to yield the hemiacetal product.

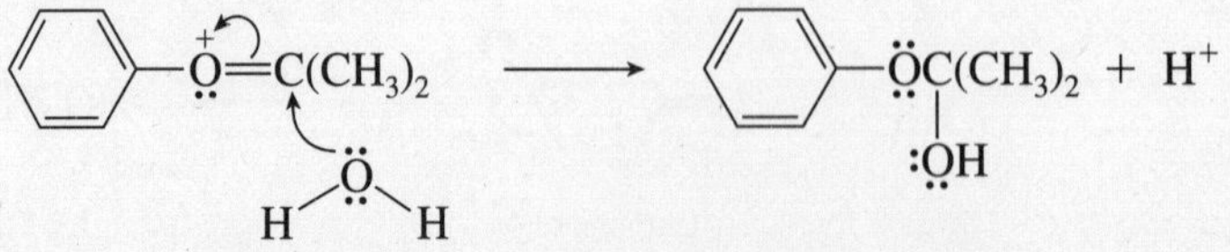

$\ddot{O}C(CH_3)_2 + H^+$ with :ÖH

**24.30** (*a*) The molecular formula of the compound ($C_9H_{12}O$) tells us that it has a total of four double bonds and rings (index of hydrogen deficiency = 4). The prominent peak in the infrared spectrum is the hydroxyl absorption of an alcohol or a phenol at 3300 $cm^{-1}$.

Peaks in the $\delta$ 110–160 ppm region of the $^{13}C$ NMR spectrum suggest an aromatic ring, which accounts for six of the nine carbon atoms and all its double bonds and rings. The presence of four peaks in this region, two of which are C and two CH, indicates a para-disubstituted aromatic derivative. That the remaining three carbons are $sp^3$-hybridized is indicated by the upfield absorptions at $\delta$ 15, 26, and 38 ppm. None of these carbons has a chemical shift below $\delta$ 40 ppm, and so none of them can be bonded to the hydroxyl group. Thus the hydroxyl group must be bonded to the aromatic ring. The compound is 4-propylphenol.

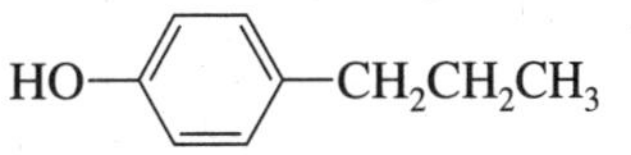

HO—$C_6H_4$—$CH_2CH_2CH_3$

4-Propylphenol

(*b*) Once again the molecular formula ($C_9H_{11}BrO$) indicates a total of four double bonds and rings. The four peaks in the $\delta$ 110–160 ppm region of the spectrum, three of which represent CH, suggest a monosubstituted aromatic ring.

The remaining atoms to be accounted for are O and Br. Because all the unsaturations are accounted for by the benzene ring and the infrared spectrum lacks any hydroxyl absorption, the oxygen atom must be part of an ether function. The three $CH_2$ groups indicated by the absorptions at $\delta$ 32, 35, and 66 ppm in the $^{13}C$ NMR spectrum allow the compound to be identified as 3-bromopropyl phenyl ether.

$C_6H_5$—$OCH_2CH_2CH_2Br$

3-Bromopropyl phenyl ether

# SELF-TEST

## PART A

**A-1.** Which is the stronger acid, *m*-hydroxybenzaldehyde or *p*-hydroxybenzaldehyde? Explain your answer, using resonance structures.

**A-2.** The cresols are methyl-substituted phenols. Predict the major products to be obtained from the reactions of *o*-, *m*-, and *p*-cresol with dilute nitric acid.

**A-3.** Give the structure of the product from the reaction of *p*-cresol with propanoyl chloride, $CH_3CH_2\overset{O}{\overset{\|}{C}}Cl$, in the presence of $AlCl_3$. What product is obtained in the absence of $AlCl_3$?

**A-4.** Provide the structure of the reactant, reagent, or product omitted from each of the following:

(*a*) $C_6H_5$—$OCH(CH_3)_2$ $\xrightarrow{HBr}$ ? (two products)

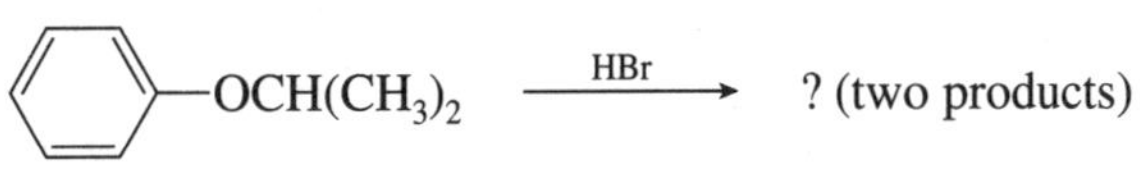

(*b*) ? (two compounds) $\longrightarrow$ (benzene ring bearing $OCH_2CH(CH_3)_2$ and, meta to it, $CH_3$)

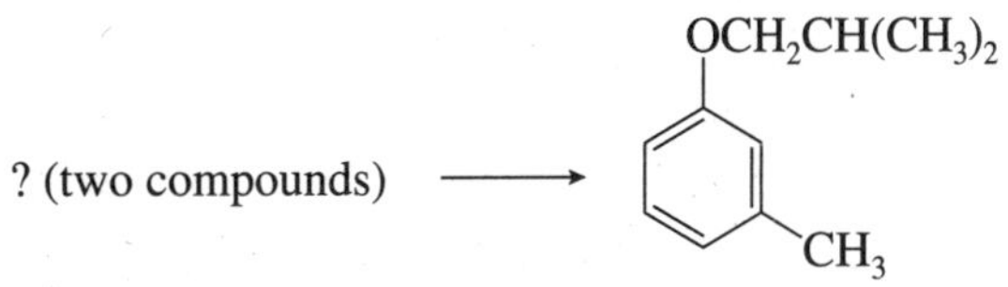

(*c*) $O^-\ Na^+$ $\xrightarrow[2.\ H^+]{1.\ ?}$ OH, $CO_2H$

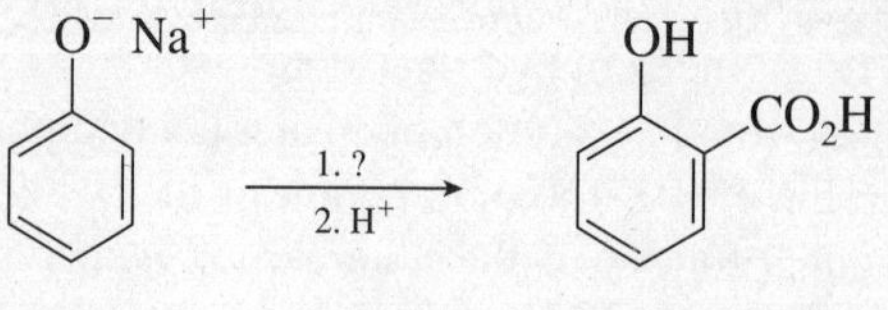

(*d*) O $\xrightarrow[\text{heat}]{\text{HBr}}$ ? ($C_8H_9BrO$)

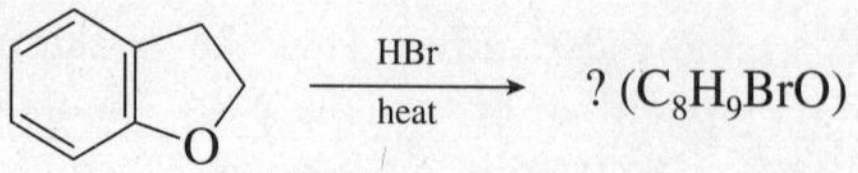

**A-5.** Provide the structures of compounds A and B in the following sequence of reactions:

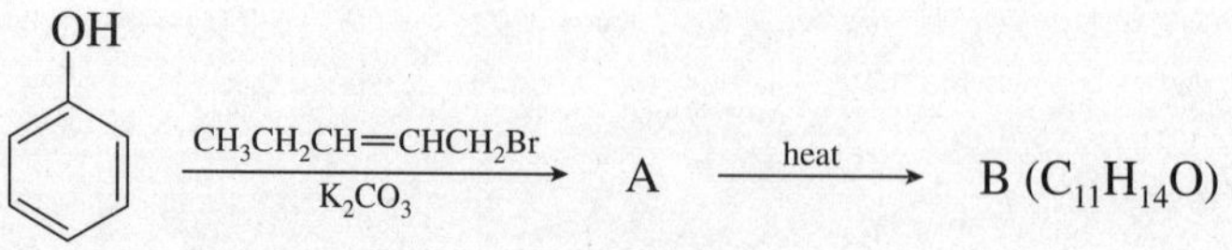

$$\text{OH} \xrightarrow[K_2CO_3]{CH_3CH_2CH{=}CHCH_2Br} A \xrightarrow{\text{heat}} B\ (C_{11}H_{14}O)$$

**A-6.** Prepare *p-tert*-butylphenol from *tert*-butylbenzene using any necessary organic or inorganic reagents.

## PART B

**B-1.** Rank the following in order of decreasing acid strength (most acidic → least acidic):

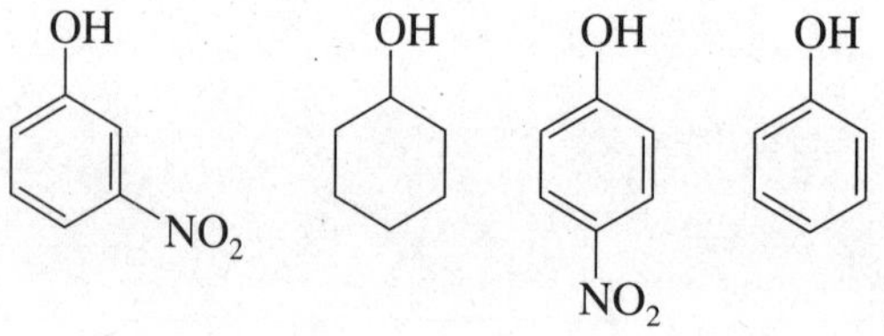

OH, $NO_2$ (1); OH (2); OH, $NO_2$ (3); OH (4)

(*a*) 2 > 4 > 1 > 3
(*b*) 3 > 1 > 2 > 4
(*c*) 1 > 3 > 4 > 2
(*d*) 3 > 1 > 4 > 2

**B-2.** Rank the following compounds in order of increasing acidity (weakest acid first).

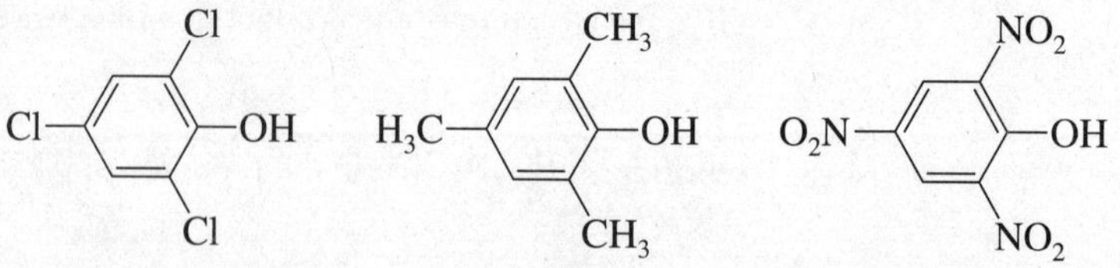

Cl, Cl—, —OH, Cl (1); $CH_3$, $H_3C$—, —OH, $CH_3$ (2); $NO_2$, $O_2N$—, —OH, $NO_2$ (3)

(*a*) 2 < 3 < 1
(*b*) 3 < 2 < 1
(*c*) 3 < 1 < 2
(*d*) 2 < 1 < 3
(*e*) 1 < 2 < 3

**B-3.** Which of the following phenols has the largest p$K_a$ value (i.e., is least acidic)?

(*a*) 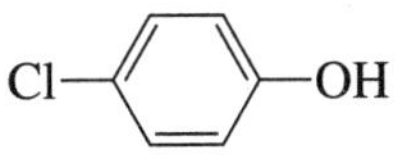 Cl–$C_6H_4$–OH  (*c*) $O_2N$–$C_6H_4$–OH

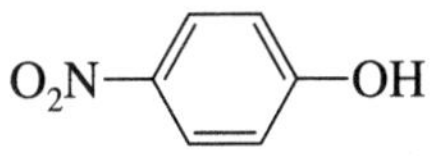

(*b*) $H_3C$–$C_6H_4$–OH  (*d*) N≡C–$C_6H_4$–OH

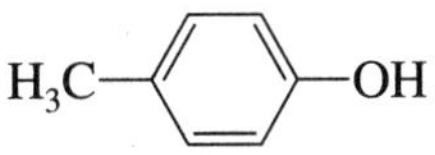

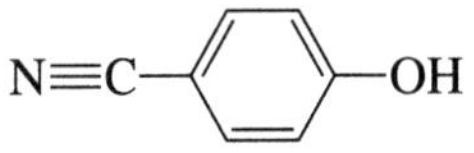

**B-4.** Which of the following reactions is a more effective method for preparing phenyl propyl ether?

I: $C_6H_5ONa + CH_3CH_2CH_2Br \longrightarrow$

II: $CH_3CH_2CH_2ONa + C_6H_5Br \longrightarrow$

(*a*) Reaction I is more effective.
(*b*) Reaction II is more effective.
(*c*) Both reactions I and II are effective.
(*d*) Neither reaction I nor reaction II is effective.

**B-5.** What reactant gives the product shown on heating with aluminum chloride?

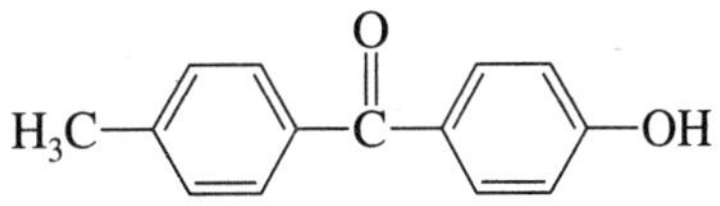

$H_3C$–$C_6H_4$–C(=O)–$C_6H_4$–OH

(*a*) $H_3C$–$C_6H_4$–OC(=O)–$C_6H_5$  (*c*) $H_3C$–$C_6H_4$–C(=O)O–$C_6H_5$

(*b*) $H_3C$–$C_6H_3$(–O–$C_6H_5$)–C(=O)OH  (*d*) $C_6H_4$(–OH)–C(=O)–$C_6H_4$–$CH_3$

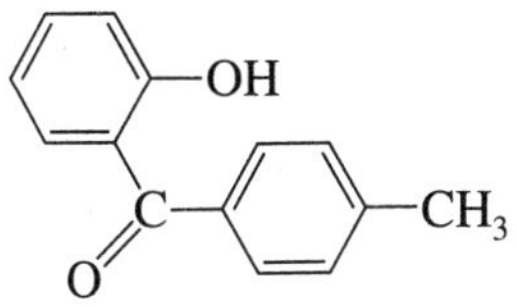

**B-6.** What are the products of the following reaction?

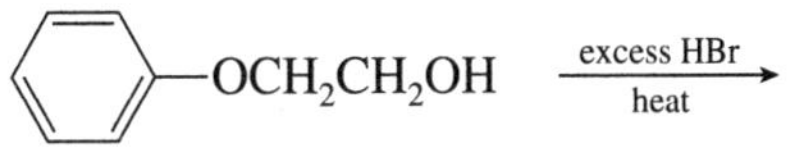

$$C_6H_5\text{–}OCH_2CH_2OH \xrightarrow[\text{heat}]{\text{excess HBr}}$$

(*a*) Br–$C_6H_4$–$OCH_2CH_2Br$  (*d*) $C_6H_5$–Br + $BrCH_2CH_2OH$

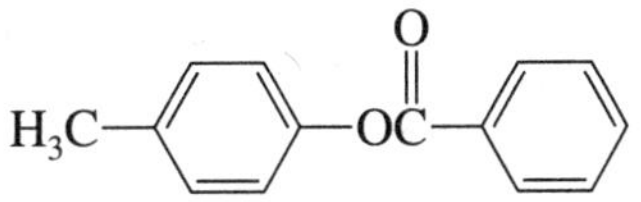

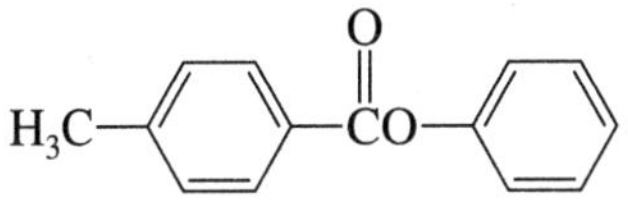

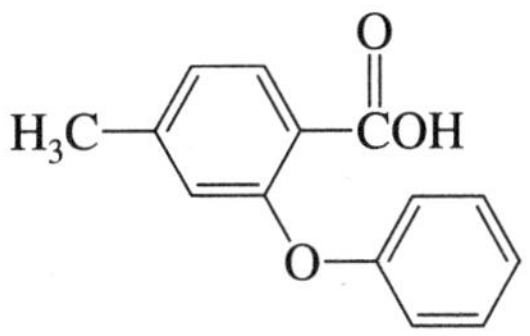

(*b*) $C_6H_5$–OH + $BrCH_2CH_2Br$  (*e*) $C_6H_5$–Br + $BrCH_2CH_2Br$

(*c*) Br–$C_6H_4$–OH + $BrCH_2CH_2Br$

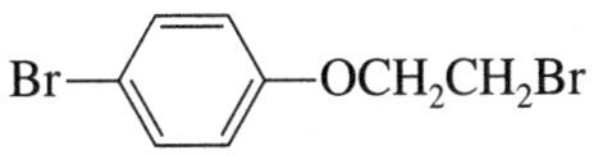

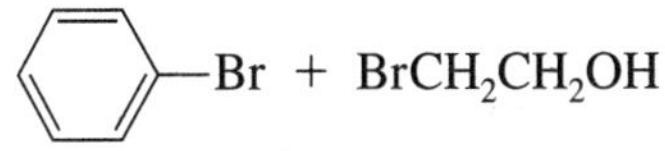

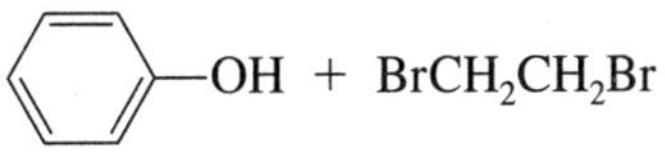

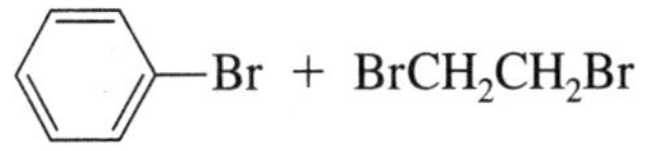

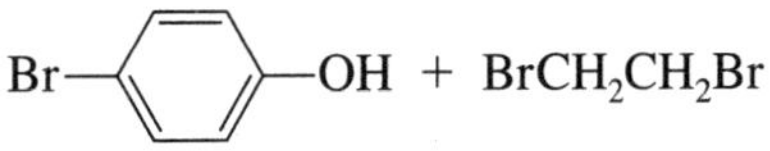

**B-7.** Which of the following sets of reagents, used in the order shown, would enable preparation of *p*-chlorophenol from *p*-chloronitrobenzene?

(*a*) 1. Fe, HCl; 2. NaOH; 3. $NaNO_2$, $H_2SO_4$; 4. $H_3PO_2$
(*b*) 1. Fe, HCl; 2. NaOH; 3. $NaNO_2$, $H_2SO_4$; 4. $H_2O$, heat
(*c*) 1. Fe, HCl; 2. NaOH; 3. $NaNO_2$, $H_2SO_4$; 4. ethanol
(*d*) 1. NaOH, heat; 2. HCl

**B-8.** What is the product obtained by heating the following allylic ether of phenol?

$C_6H_5OCH_2CH{=}CHC_6H_5 \xrightarrow{200°C} ?$

(*a*) *o*-$HOC_6H_4CH_2CH{=}CHC_6H_5$ (OH and $CH_2CH{=}CHC_6H_5$ on adjacent ring carbons)

(*b*) *o*-$C_6H_5OC_6H_4CH_2CH{=}CH_2$ ($OC_6H_5$ and $CH_2CH{=}CH_2$ on adjacent ring carbons)

(*c*) *o*-$HOC_6H_4CH(C_6H_5)CH{=}CH_2$ (OH and $CH(C_6H_5)CH{=}CH_2$ on adjacent ring carbons)

(*d*) *p*-$HO{-}C_6H_4{-}CH(C_6H_5)CH{=}CH_2$

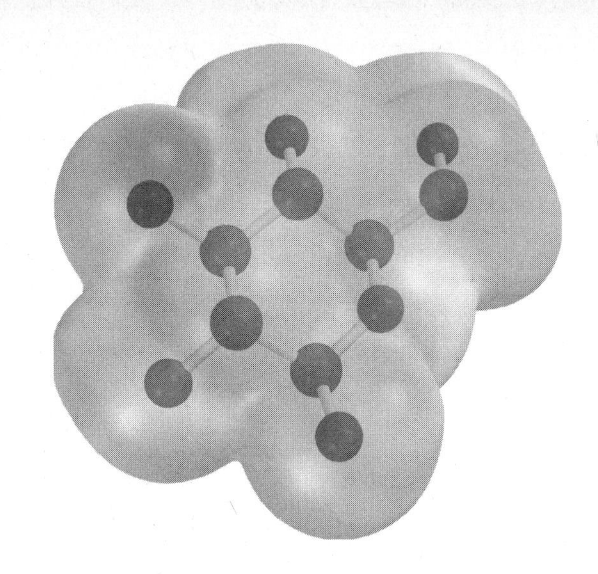

# CHAPTER 25

## CARBOHYDRATES

## SOLUTIONS TO TEXT PROBLEMS

**25.1** (*b*) Redraw the Fischer projection so as to show the orientation of the groups in three dimensions.

H
$HOCH_2$—┼—CHO (Fischer projection, OH at bottom) is equivalent to $HOCH_2$►C◄CHO (H and OH on hashed bonds, above and below)

Reorient the three-dimensional representation, putting the aldehyde group at the top and the primary alcohol at the bottom.

$HOCH_2$►C◄CHO (H up, OH down, hashed) —turn 90°→ CHO (top, wedge) / H····C····OH (hashed) / $CH_2OH$ (bottom, wedge)

What results is not equivalent to a proper Fischer projection, because the horizontal bonds are directed "back" when they should be "forward." The opposite is true for the vertical bonds. To make the drawing correspond to a proper Fischer projection, we need to rotate it 180° around a vertical axis.

CHO / H····C····OH / $CH_2OH$ (rotate 180°) ⟶ CHO / HO►C◄H / $CH_2OH$ (vertical bonds hashed) is equivalent to CHO / HO—┼—H / $CH_2OH$

Now, having the molecule arranged properly, we see that it is L-glyceraldehyde.

(*c*) Again proceed by converting the Fischer projection into a three-dimensional representation.

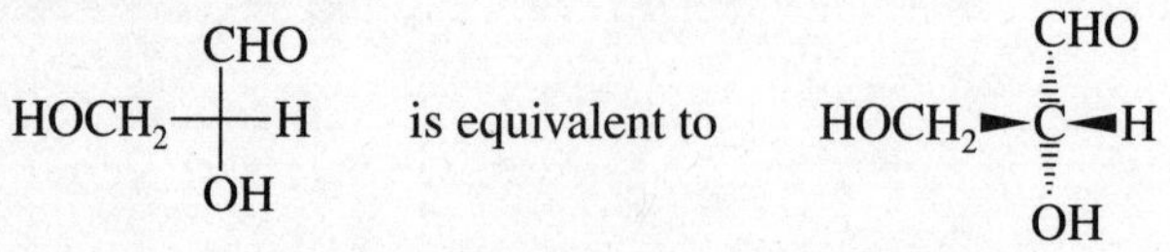

HOCH$_2$—C(CHO)(OH)—H (Fischer projection) is equivalent to HOCH$_2$►C◄H with CHO (top, back) and OH (bottom, back)

Look at the drawing from a perspective that permits you to see the carbon chain oriented vertically with the aldehyde at the top and the $CH_2OH$ at the bottom. Both groups should point away from you. When examined from this perspective, the hydrogen is to the left and the hydroxyl to the right with both pointing toward you.

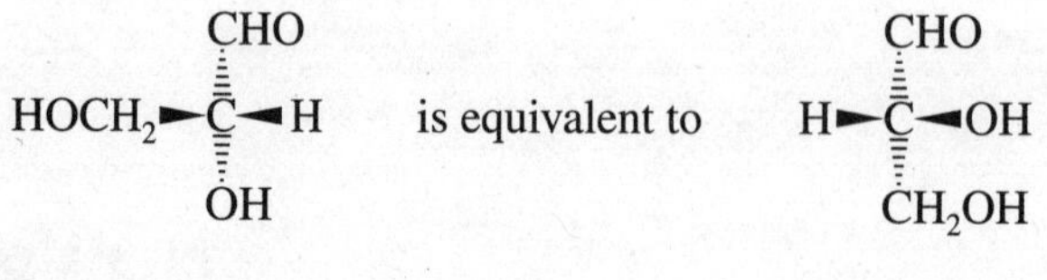

HOCH$_2$►C◄H (CHO top, OH bottom) is equivalent to H►C◄OH (CHO top, $CH_2OH$ bottom)

The molecule is D-glyceraldehyde.

**25.2** Begin by drawing a perspective view of the molecular model shown in the problem. To view the compound as a Fischer projection, redraw it in an eclipsed conformation.

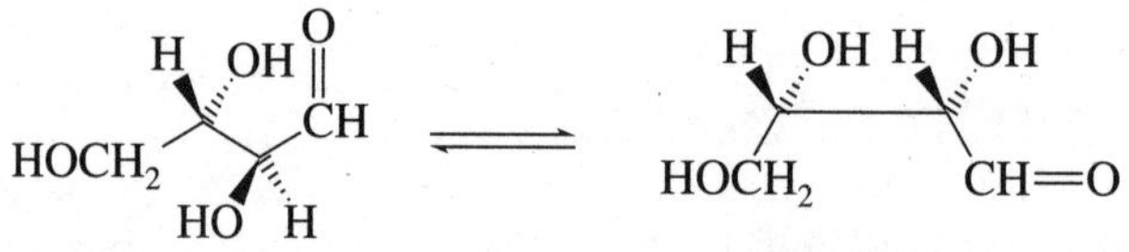

Staggered conformation ⇌ Same molecule in eclipsed conformation

The eclipsed conformation shown, when oriented so that the aldehyde carbon is at the top, vertical bonds back, and horizontal bonds pointing outward from their stereogenic centers, is readily transformed into the Fischer projection of L-erythrose.

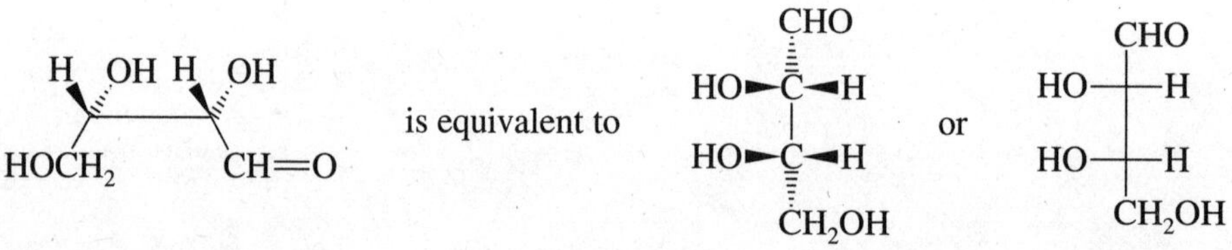

Eclipsed conformation (H, OH, H, OH; HOCH$_2$, CH=O) is equivalent to

CHO
HO►C◄H
HO►C◄H
$CH_2OH$

or

CHO
HO—H
HO—H
$CH_2OH$

L-Erythrose

**25.3** L-Arabinose is the mirror image of D-arabinose, the structure of which is given in text Figure 25.2. The configuration at *each* stereogenic center of D-arabinose must be reversed to transform it into L-arabinose.

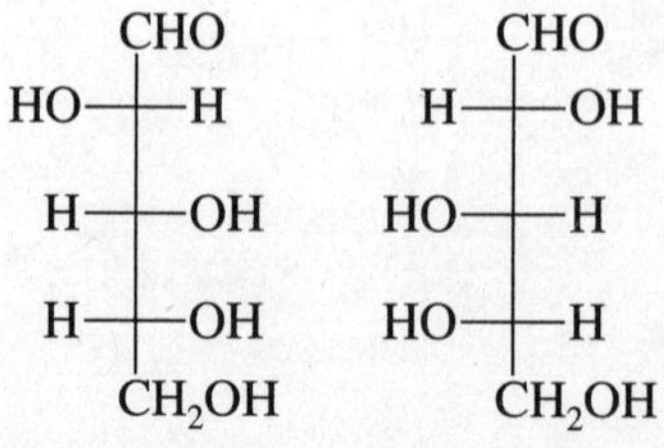

| D-(−)-Arabinose | L-(+)-Arabinose |
|---|---|
| CHO | CHO |
| HO—H | H—OH |
| H—OH | HO—H |
| H—OH | HO—H |
| $CH_2OH$ | $CH_2OH$ |

**25.4** The configuration at C-5 is opposite to that of D-(+)-glyceraldehyde. This particular carbohydrate therefore belongs to the L series. Comparing it with the Fischer projection formulas of the eight D-aldohexoses reveals it to be in the mirror image of D-(+)-talose; it is L-(−)-talose

**25.5** (*b*) The Fischer projection formula of D-arabinose may be found in text Figure 25.2. The Fischer projection and the eclipsed conformation corresponding to it are

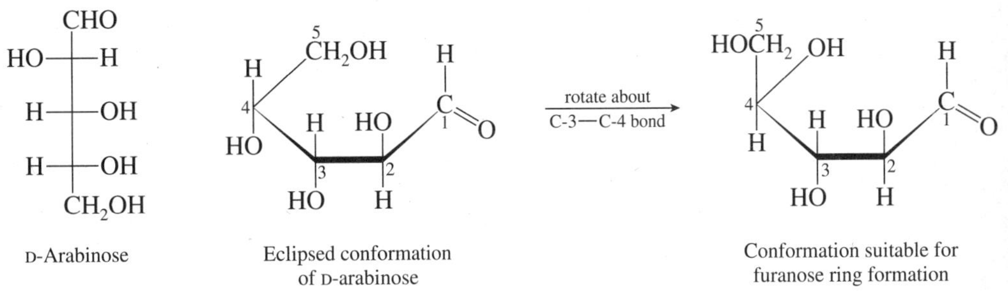

D-Arabinose

Eclipsed conformation of D-arabinose

rotate about C-3—C-4 bond

Conformation suitable for furanose ring formation

Cyclic hemiacetal formation between the carbonyl group and the C-4 hydroxyl yields the α- and β-furanose forms of D-arabinose.

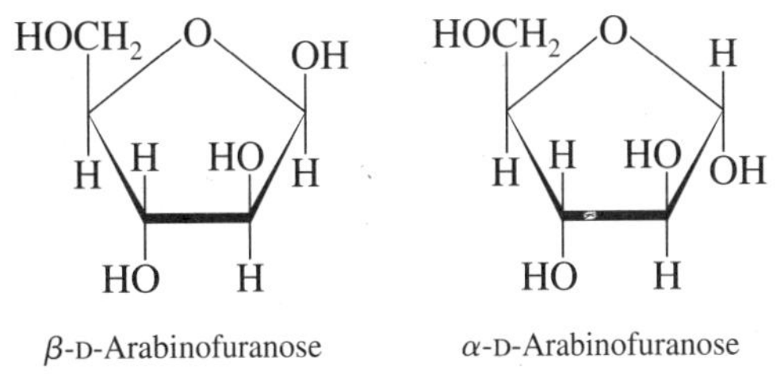

β-D-Arabinofuranose

α-D-Arabinofuranose

(*c*) The mirror image of D-arabinose [from part (*b*)] is L-arabinose.

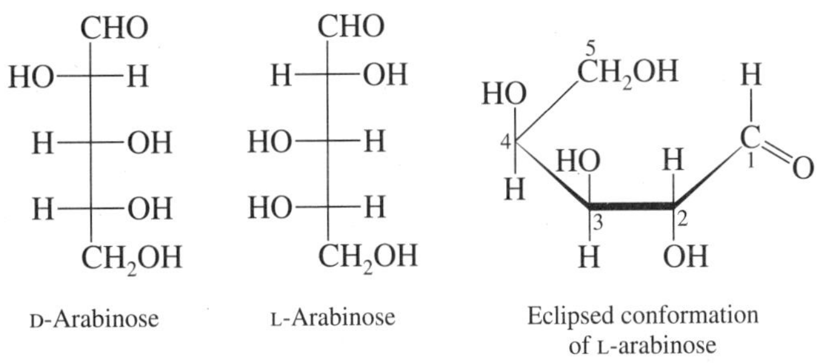

D-Arabinose

L-Arabinose

Eclipsed conformation of L-arabinose

The C-4 atom of the eclipsed conformation of L-arabinose must be rotated 120° in a clockwise sense so as to bring its hydroxyl group into the proper orientation for furanose ring formation.

rotate about C-3—C-4 bond

Original eclipsed conformation of L-arabinose

Conformation suitable for furanose ring formation

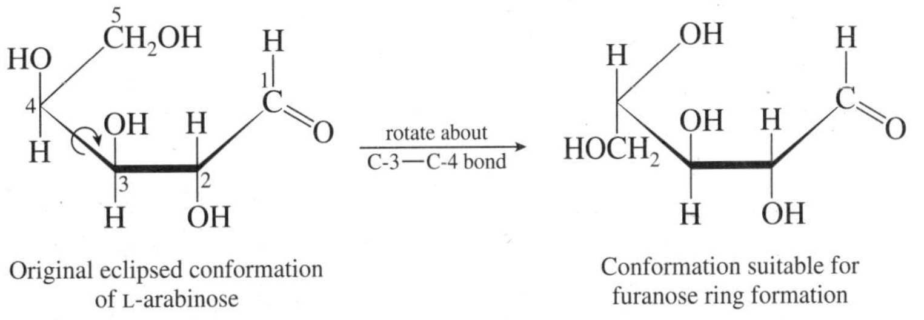

Cyclization gives the $\alpha$- and $\beta$-furanose forms of L-arabinose.

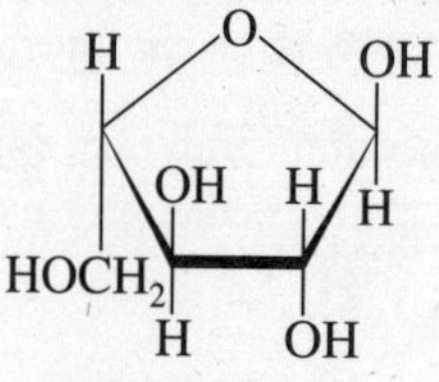

α-L-Arabinofuranose 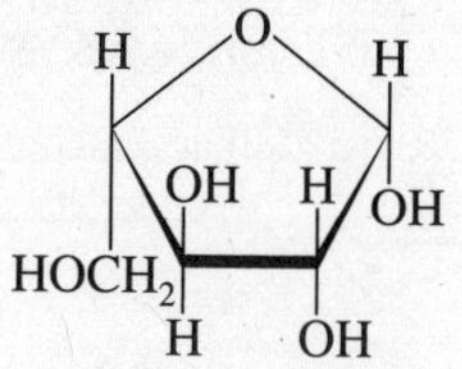 β-L-Arabinofuranose

In the L series the anomeric hydroxyl is up in the $\alpha$ isomer and down in the $\beta$ isomer.

(*d*) The Fischer projection formula for D-threose is given in the text Figure 25.2. Reorientation of that projection into a form that illustrates its potential for cyclization is shown.

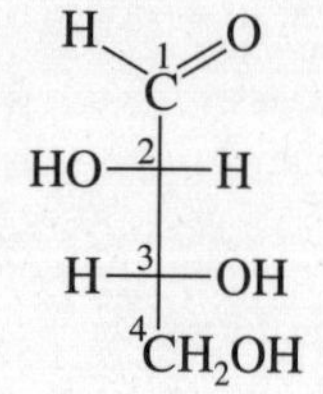

D-Threose

is equivalent to

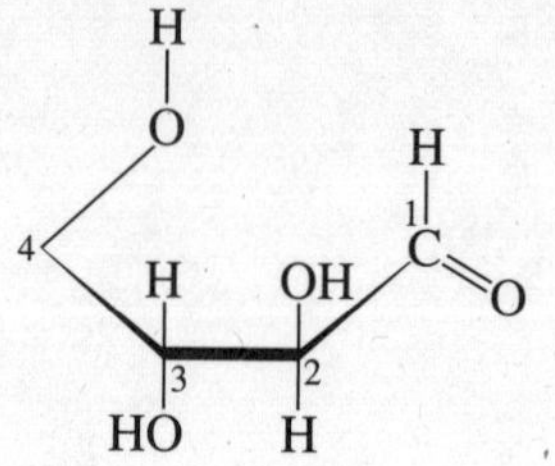

Cyclization yields the two stereoisomeric furanose forms.

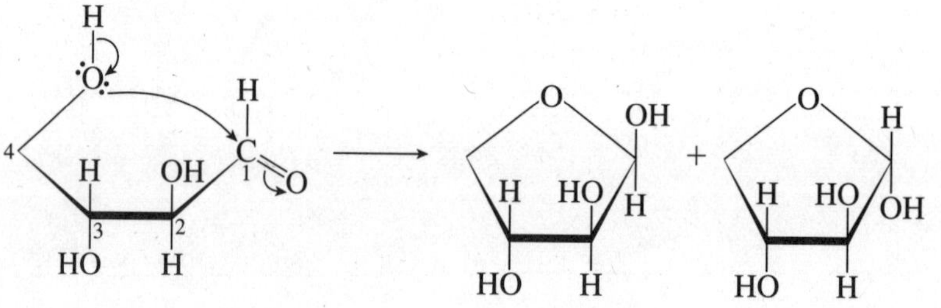

β-D-Threofuranose α-D-Threofuranose

**25.6** (*b*) The Fischer projection and Haworth formula for D-mannose are

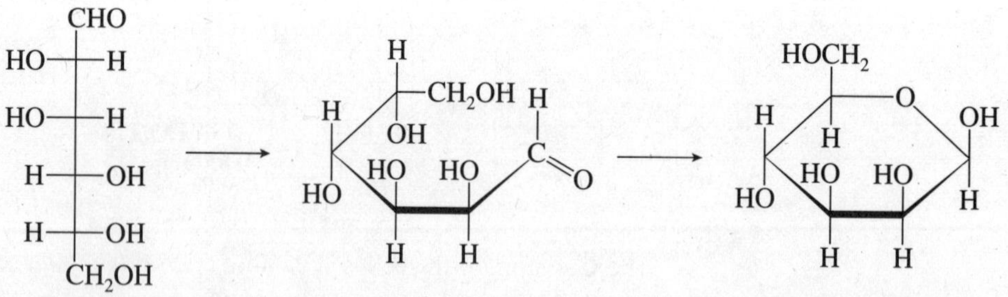

D-Mannose

β-D-Mannopyranose (Haworth formula)

The Haworth formula is more realistically drawn as the following chair conformation:

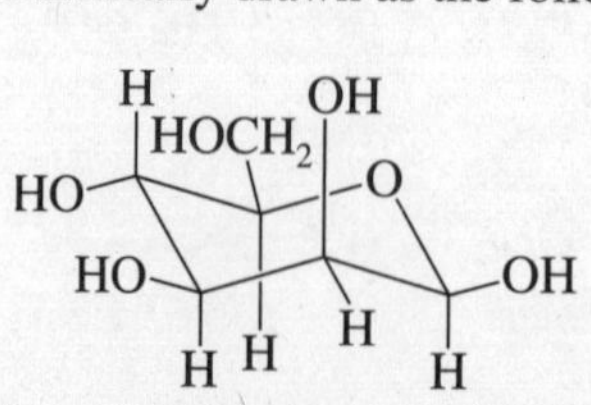

β-D-Mannopyranose

Mannose differs from glucose in configuration at C-2. All hydroxyl groups are equatorial in β-D-glucopyranose; the hydroxyl at C-2 is axial in β-D-mannopyranose.

(*c*) The conformational depiction of β-L-mannopyranose begins in the same way as that of β-D-mannopyranose. L-Mannose is the mirror image of D-mannose.

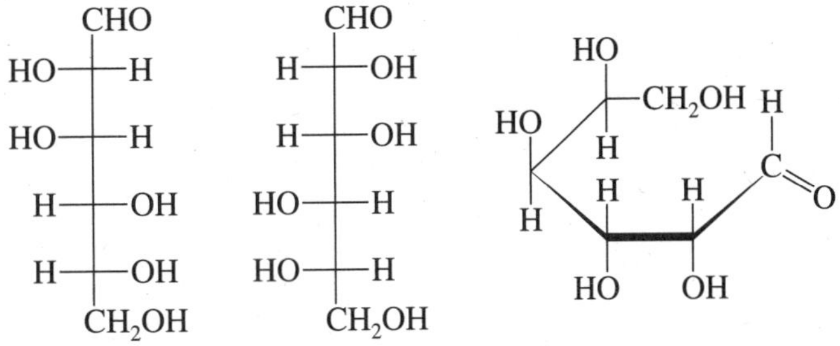

D-Mannose L-Mannose Eclipsed conformation of L-mannose

To rewrite the eclipsed conformation of L-mannose in a way that permits hemiacetal formation between the carbonyl group and the C-5 hydroxyl, C-5 is rotated 120° in the clockwise sense.

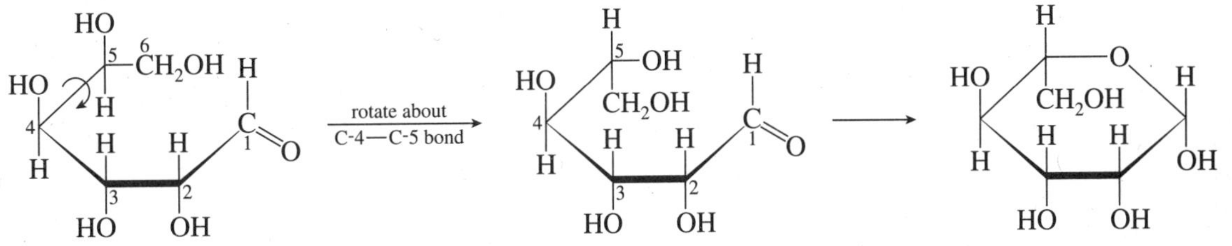

rotate about C-4—C-5 bond

β-L-Mannopyranose (remember, the anomeric hydroxyl is down in the L series)

Translating the Haworth formula into a proper conformational depiction requires that a choice be made between the two chair conformations shown.

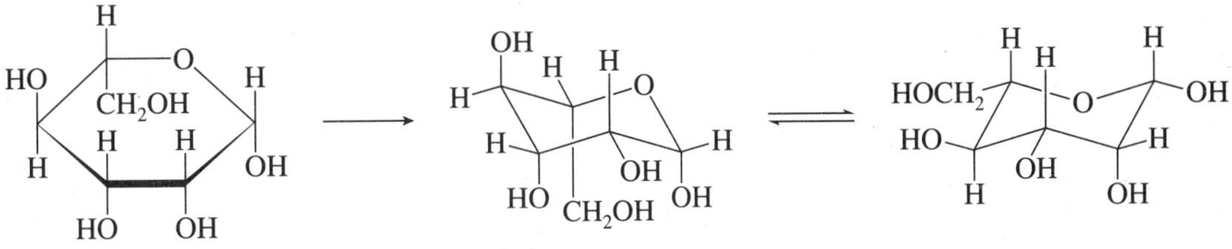

Haworth formula of β-L-mannopyranose

Less stable chair conformation; $CH_2OH$ is axial

More stable chair conformation; $CH_2OH$ is equatorial

(*d*) The Fischer projection formula for L-ribose is the mirror image of that for D-ribose.

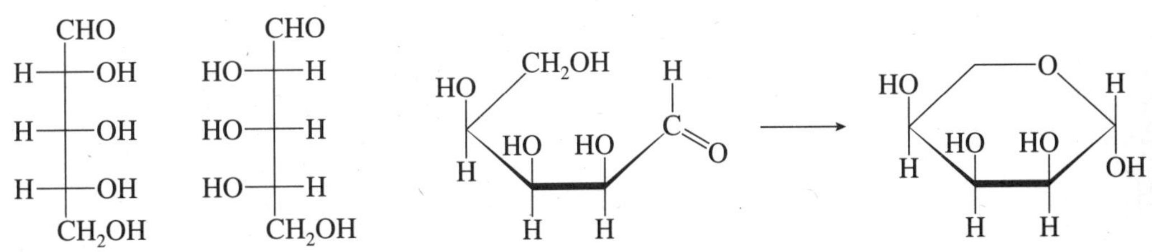

D-Ribose L-Ribose

Eclipsed conformation of L-ribose is oriented properly for ring closure.

Haworth formula of β-L-ribopyranose

Of the two chair conformations of β-L-ribose, the one with the greater number of equatorial substituents is more stable.

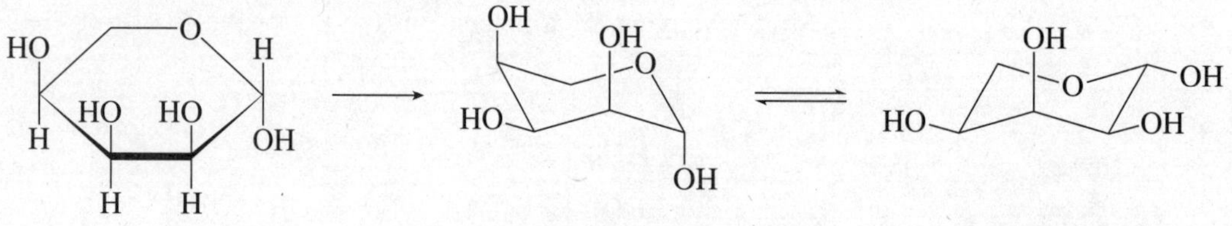

Less stable chair conformation of β-L-ribopyranose

More stable chair conformation of β-L-ribopyranose

**25.7** The equation describing the equilibrium is

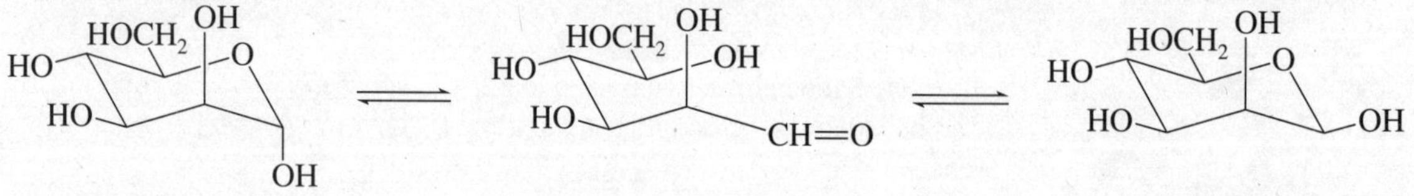

α-D-Mannopyranose $[\alpha]_D^{20}$ + 29.3°

Open-chain form of D-mannose

β-D-Mannopyranose $[\alpha]_D^{20}$ − 17.0°

Let $A$ = percent α isomer; $100 - A$ = percent β isomer. Then

$$A(+29.3°) + (100 - A)(-17.0°) = 100(+14.2°)$$
$$46.3A = 3120$$
$$\text{Percent } \alpha \text{ isomer} = 67\%$$
$$\text{Percent } \beta \text{ isomer} = (100 - A) = 33\%$$

**25.8** Review carbohydrate terminology by referring to text Table 25.1. A **ketotetrose** is a four-carbon ketose. Writing a Fischer projection for a four-carbon ketose reveals that only one stereogenic center is present, and thus there are only two ketotetroses. They are enantiomers of each other and are known as D- and L-erythrulose.

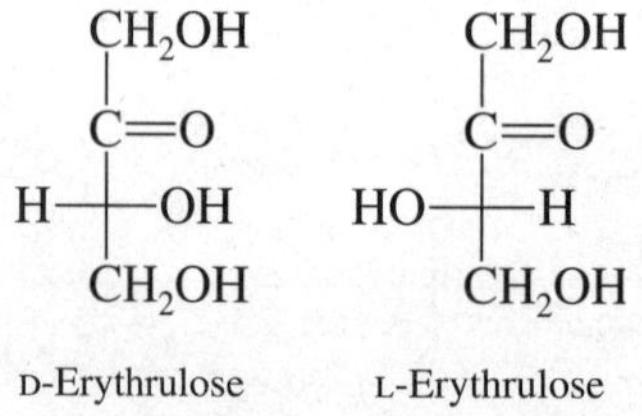

D-Erythrulose L-Erythrulose

**25.9** (*b*) Because L-fucose is 6-deoxy-L-galactose, first write the Fischer projection formula of D-galactose, and then transform it to its mirror image, L-galactose. Transform the C-6 $CH_2OH$ group to $CH_3$ to produce 6-deoxy-L-galactose.

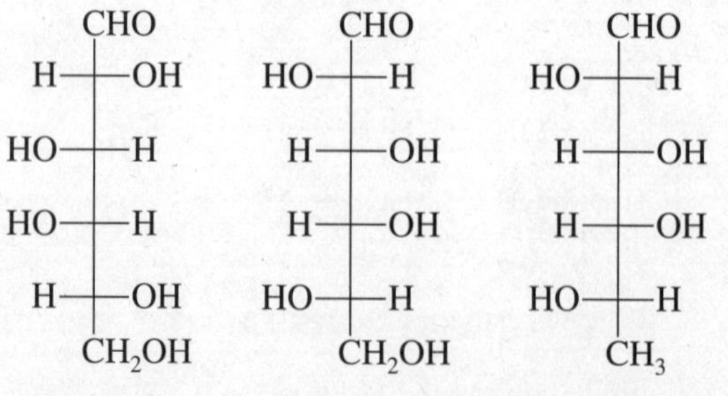

D-Galactose (from Figure 25.2)

L-Galactose

6-Deoxy-L-galactose (L-fucose)

**25.10** Reaction of a carbohydrate with an alcohol in the presence of an acid catalyst gives mixed acetals at the anomeric position.

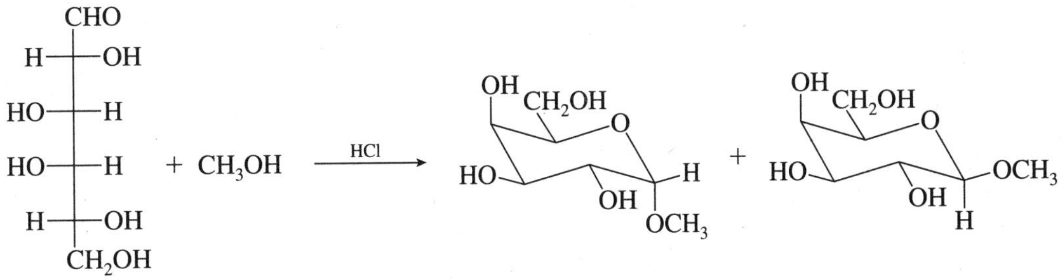

D-Galactose + Methanol $\xrightarrow{HCl}$ Methyl α-D-galactopyranoside + Methyl β-D-galactopyranoside

**25.11** Acid-catalyzed addition of methanol to the glycal proceeds by regioselective protonation of the double bond in the direction that leads to the more stable carbocation. Here again, the more stable carbocation is the one stabilized by the ring oxygen.

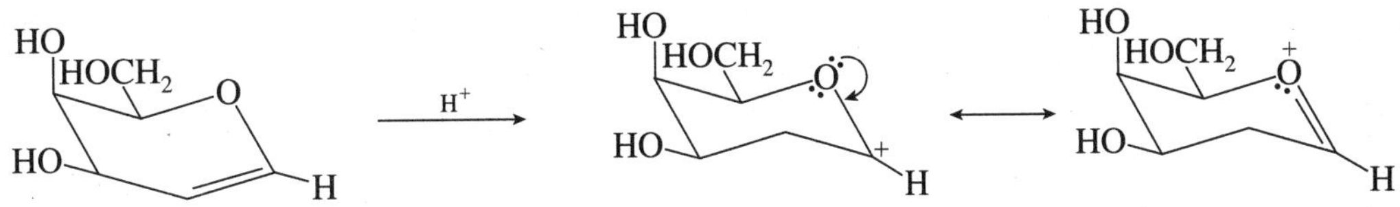

Capture on either face of the carbocation by methanol yields the α and β methyl glycosides.

**25.12** The hemiacetal opens to give an intermediate containing a free aldehyde function. Cyclization of this intermediate can produce either the α or the β configuration at this center. The axial and equatorial orientations of the anomeric hydroxyl can best be seen by drawing maltose with the pyranose rings in chair conformations.

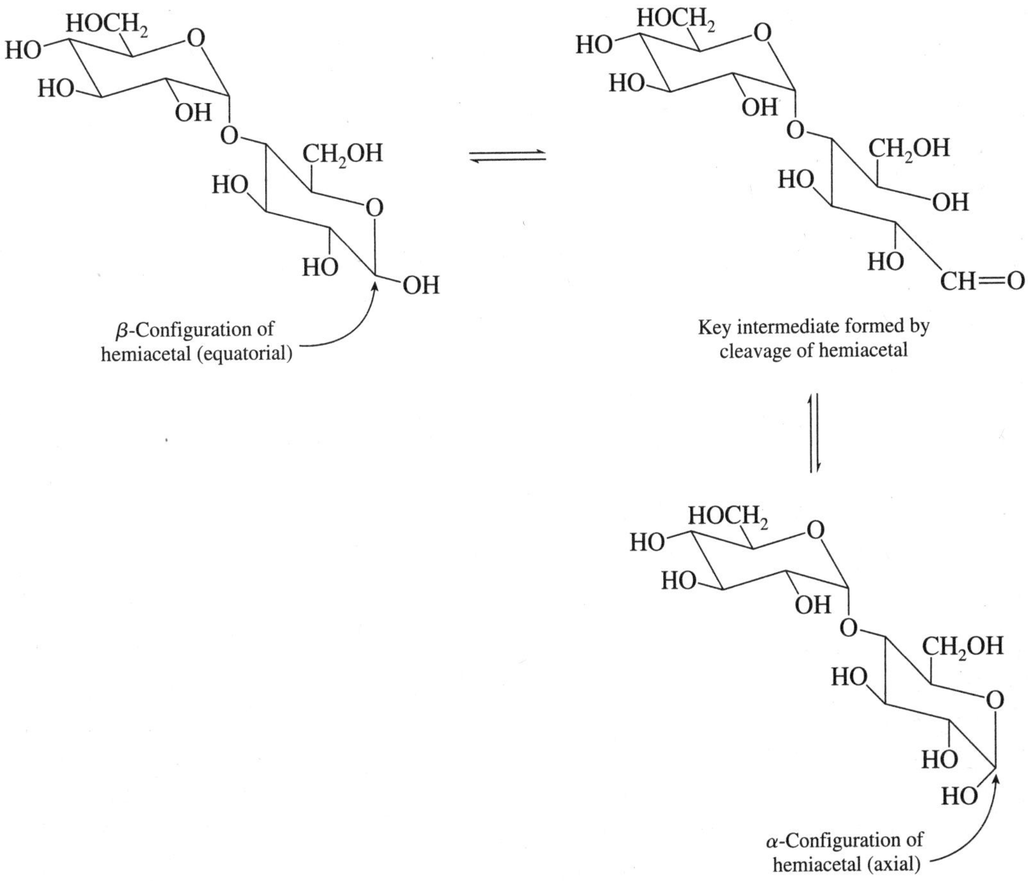

β-Configuration of hemiacetal (equatorial)

Key intermediate formed by cleavage of hemiacetal

α-Configuration of hemiacetal (axial)

Only the configuration of the hemiacetal function is affected in this process. The $\alpha$ configuration of the glycosidic linkage remains unchanged.

**25.13** Write the chemical equation so that you can clearly relate the product to the starting material.

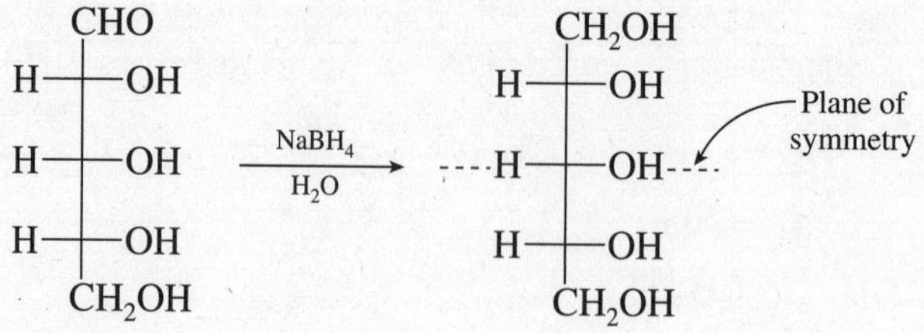

D-Ribose $\xrightarrow[H_2O]{NaBH_4}$ Ribitol (Plane of symmetry)

Ribitol is a meso form; it is achiral and thus not optically active. A plane of symmetry passing through C-3 bisects the molecule.

**25.14** (*b*) Arabinose is a reducing sugar; it will give a positive test with Benedict's reagent, because its open-chain form has a free aldehyde group capable of being oxidized by copper(II) ion.

(*c*) Benedict's reagent reacts with $\alpha$-hydroxy ketones by way of an isomerization process involving an enediol intermediate.

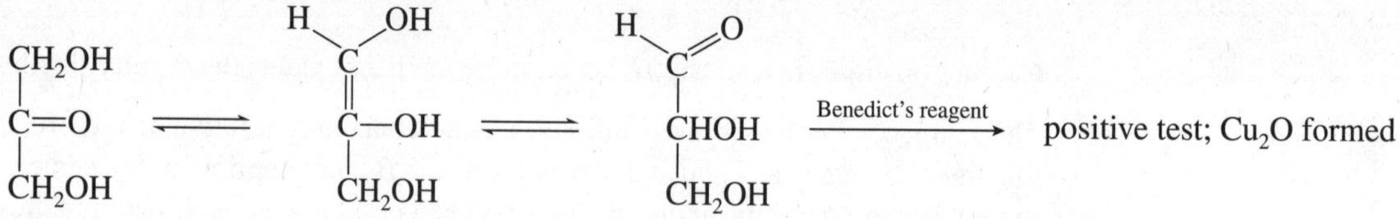

1,3-Dihydroxyacetone $\rightleftharpoons$ Enediol $\rightleftharpoons$ Glyceraldehyde $\xrightarrow{\text{Benedict's reagent}}$ positive test; $Cu_2O$ formed

1,3-Dihydroxyacetone gives a positive test with Benedict's reagent.

(*d*) D-Fructose is an $\alpha$-hydroxy ketone and will give a positive test with Benedict's reagent.

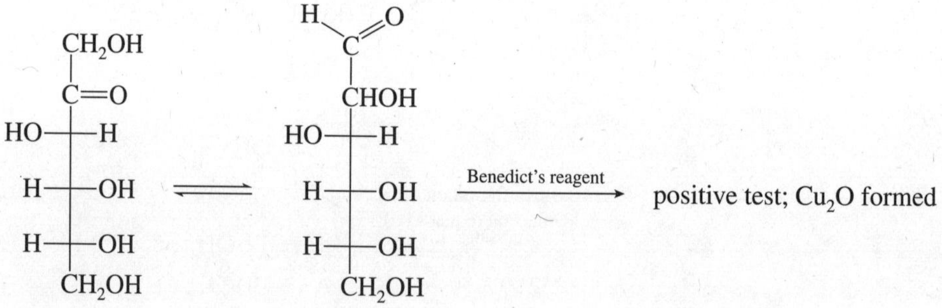

D-Fructose $\rightleftharpoons$ (aldehyde form) $\xrightarrow{\text{Benedict's reagent}}$ positive test; $Cu_2O$ formed

(*e*) Lactose is a disaccharide and will give a positive test with Benedict's reagent by way of an open-chain isomer of one of the rings. Lactose is a reducing sugar.

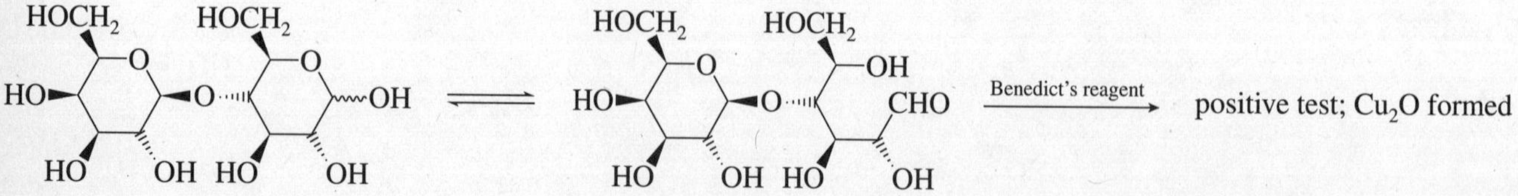

Lactose (structure presented in Section 25.14) $\rightleftharpoons$ Open-chain form $\xrightarrow{\text{Benedict's reagent}}$ positive test; $Cu_2O$ formed

(*f*) Amylose is a polysaccharide. Its glycoside linkages are inert to Benedict's reagent, but the terminal glucose residues at the ends of the chain and its branches are hemiacetals in equilibrium with open-chain structures. A positive test is expected.

**25.15** Because the groups at both ends of the carbohydrate chain are oxidized to carboxylic acid functions, two combinations of one $CH_2OH$ with one CHO group are possible.

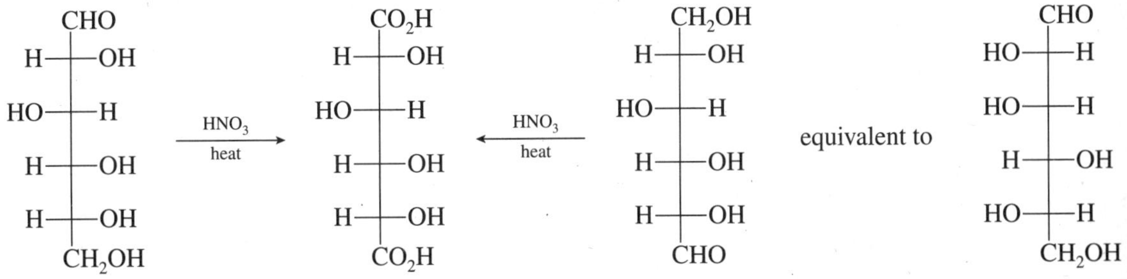

D-Glucose $\xrightarrow[\text{heat}]{HNO_3}$ D-Glucaric acid $\xleftarrow[\text{heat}]{HNO_3}$ (CH$_2$OH/CHO form) equivalent to L-Gulose

D-Glucose D-Glucaric acid L-Gulose

L-Gulose yields the same aldaric acid on oxidation as does D-glucose.

**25.16** In analogy with the D-fructose ⇌ D-glucose interconversion, dihydroxyacetone phosphate and D-glyceraldehyde 3-phosphate can equilibrate by way of an enediol intermediate.

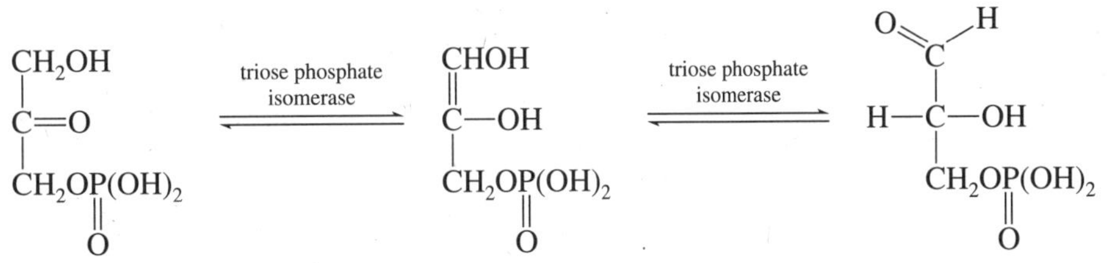

Dihydroxyacetone phosphate $\underset{\text{isomerase}}{\overset{\text{triose phosphate}}{\rightleftharpoons}}$ Enediol $\underset{\text{isomerase}}{\overset{\text{triose phosphate}}{\rightleftharpoons}}$ D-Glyceraldehyde 3-phosphate

Dihydroxyacetone phosphate Enediol D-Glyceraldehyde 3-phosphate

**25.17** (*b*) The points of cleavage of D-ribose on treatment with periodic acid are as indicated.

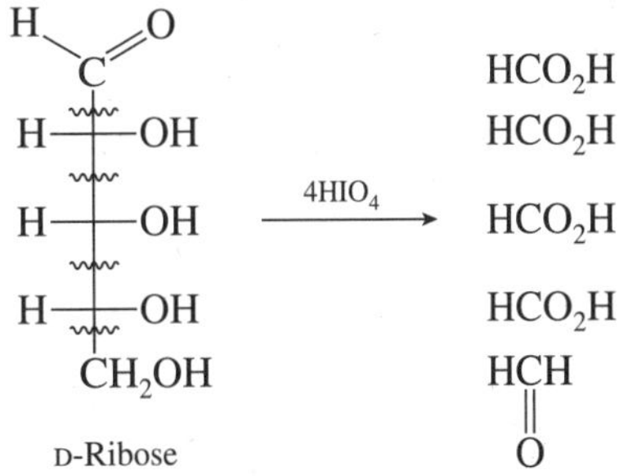

D-Ribose $\xrightarrow{4HIO_4}$ $HCO_2H$ + $HCO_2H$ + $HCO_2H$ + $HCO_2H$ + $HCH{=}O$

D-Ribose

Four moles of periodic acid per mole of D-ribose are required. Four moles of formic acid and one mole of formaldehyde are produced.

(*c*) Write the structure of methyl $\beta$-D-glucopyranoside so as to identify the adjacent alcohol functions.

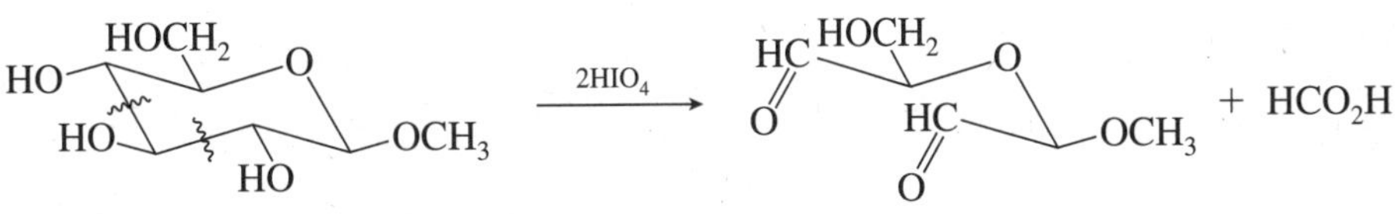

Methyl $\beta$-D-glucopyranoside $\xrightarrow{2HIO_4}$ (dialdehyde) + $HCO_2H$

Methyl $\beta$-D-glucopyranoside

Two moles of periodic acid per mole of glycoside are required. One mole of formic acid is produced.

(*d*) There are two independent vicinal diol functions in this glycoside. Two moles of periodic acid are required per mole of substrate.

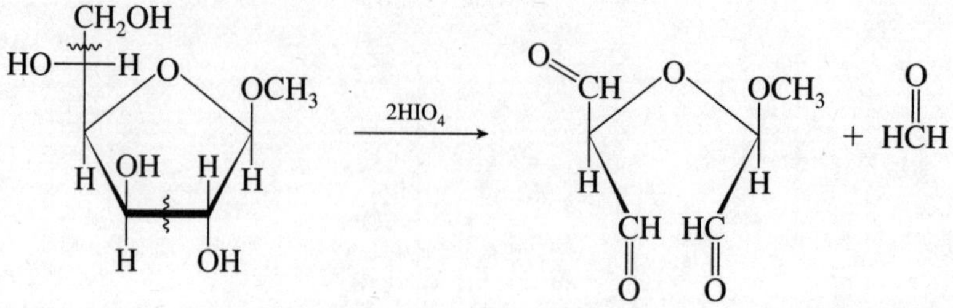

$\xrightarrow{2HIO_4}$ + HCH(=O)

**25.18** (*a*) The structure shown in Figure 25.2 is D-(+)-xylose; therefore (−)-xylose must be its mirror image and has the L-configuration at C-4.

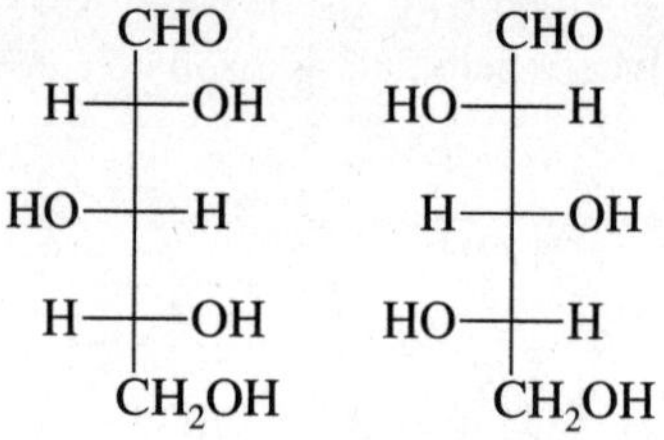

D-(+)-Xylose L-(−)-Xylose

(*b*) Alditols are the reduction products of carbohydrates; D-xylitol is derived from D-xylose by conversion of the terminal —CHO to —$CH_2$OH.

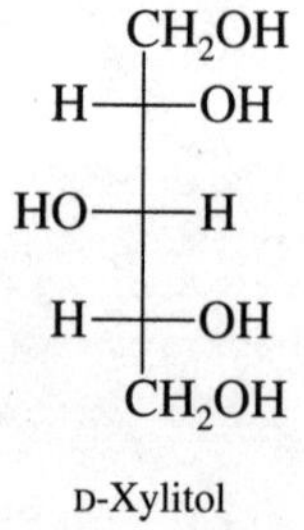

D-Xylitol

(*c*) Redraw the Fischer projection of D-xylose in its eclipsed conformation.

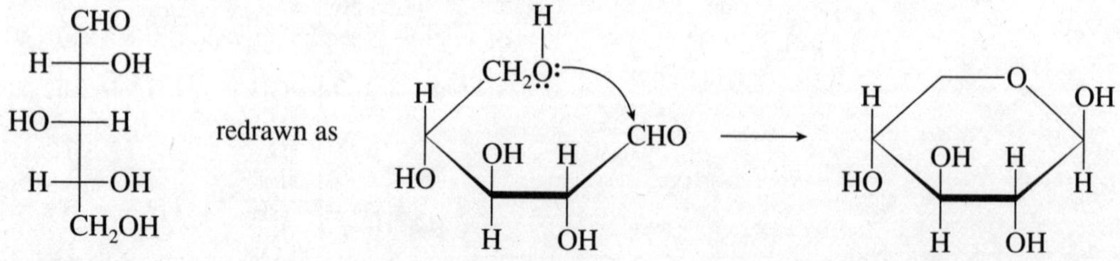

D-Xylose redrawn as

Eclipsed conformation of D-xylose

Haworth formula of $\beta$-D-xylopyranose

The pyranose form arises by closure to a six-membered cyclic hemiacetal, with the C-5 hydroxyl group undergoing nucleophilic addition to the carbonyl. In the β-pyranose form of D-xylose the anomeric hydroxyl group is up.

The preferred conformation of β-D-xylopyranose is a chair with all the hydroxyl groups equatorial.

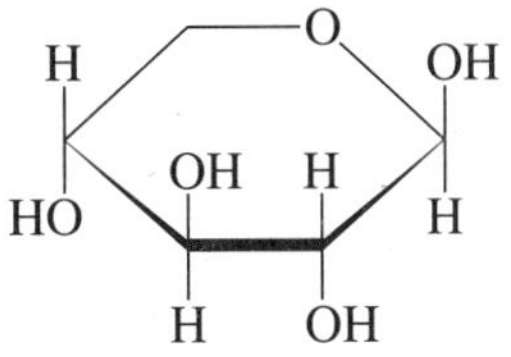

is better represented as

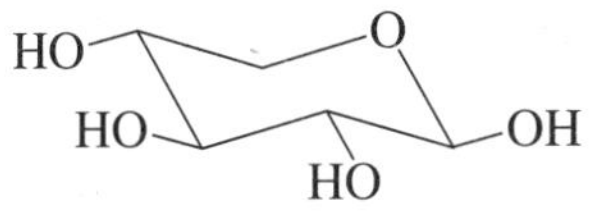

Haworth formula of β-D-xylopyranose

Chair conformation of β-D-xylopyranose

(*d*) L-Xylose is the mirror image of D-xylose.

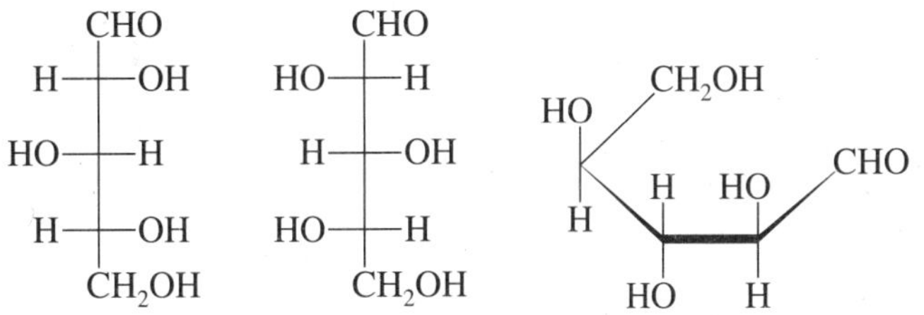

D-Xylose

L-Xylose

Eclipsed conformation of L-xylose

To construct the furanose form of L-xylose, the hydroxyl at C-4 needs to be brought into the proper orientation to form a five-membered ring.

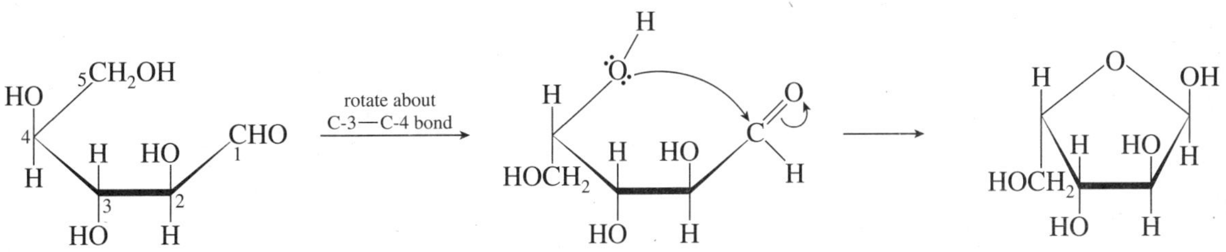

rotate about C-3—C-4 bond

The α anomeric hydroxyl group is up in the L series.

(*e*) Methyl α-L-xylofuranoside is the methyl glycoside corresponding to the structure just drawn.

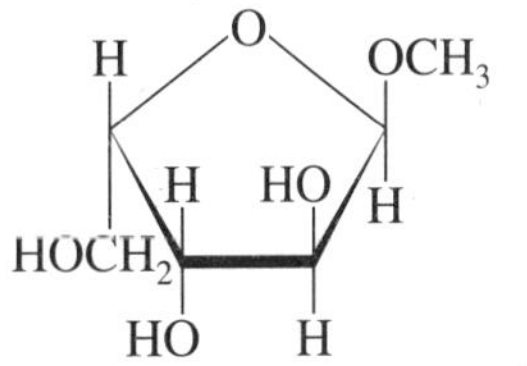

(*f*) Aldonic acids are derived from aldoses by oxidation of the terminal aldehyde to a carboxylic acid.

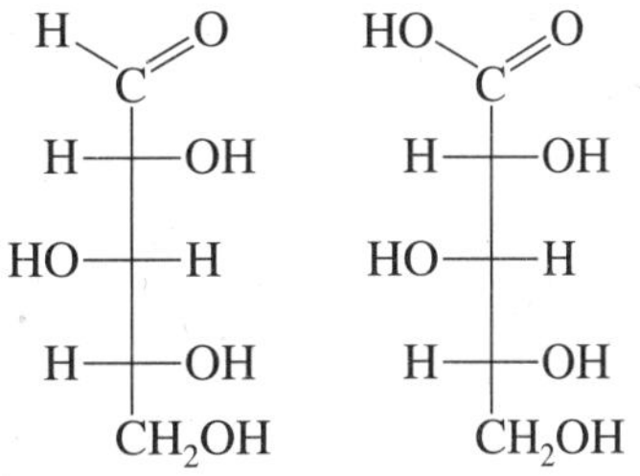

D-Xylose

D-Xylonic acid

(*g*) Aldonic acids tend to exist as lactones. A δ-lactone has a six-membered ring.

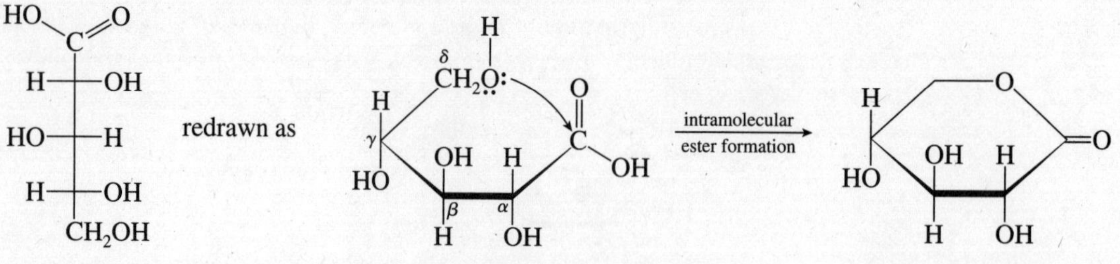

D-Xylonic acid — redrawn as — Eclipsed conformation of D-xylonic acid — intramolecular ester formation → δ-Lactone of D-xylonic acid

(*h*) A γ-lactone has a five-membered ring.

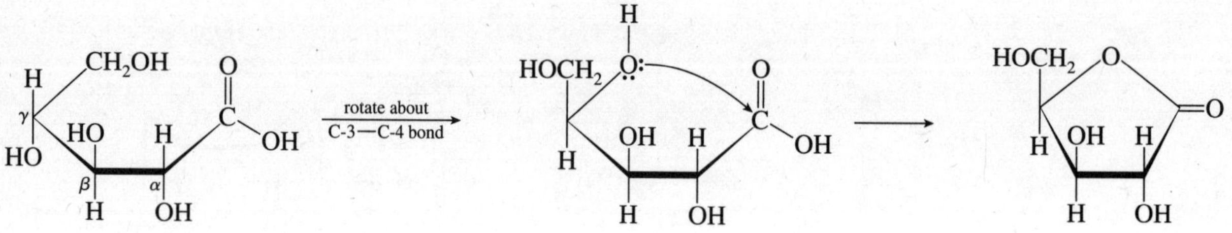

D-Xylonic acid — rotate about C-3—C-4 bond → → γ-Lactone of D-xylonic acid

(*i*) Aldaric acids have carboxylic acid groups at both ends of the chain.

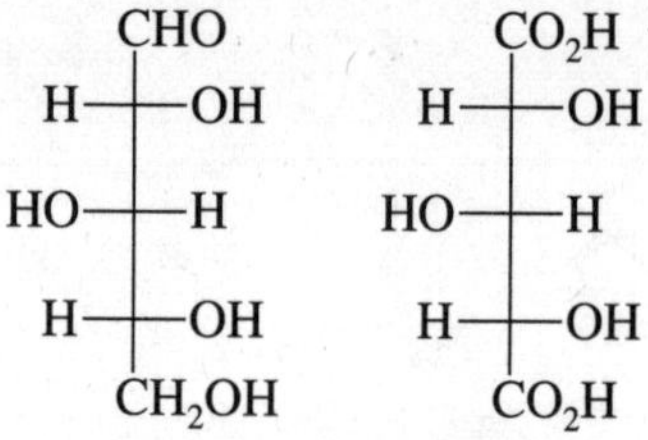

D-Xylose D-Xylaric acid

**25.19** (*a*) Reduction of aldoses with sodium borohydride yields polyhydroxylic alcohols called **alditols.** Optically inactive alditols are those that have a plane of symmetry, that is, those that are meso forms. The D-aldohexoses that yield optically inactive alditols are D-allose and D-galactose.

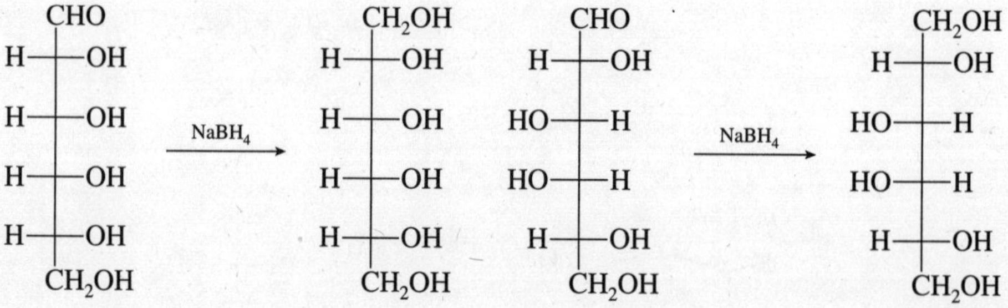

D-Allose $\xrightarrow{NaBH_4}$ Allitol (meso compound)

D-Galactose $\xrightarrow{NaBH_4}$ Galactitol (meso compound)

(*b*) All the aldonic acids and their lactones obtained on oxidation of the aldohexoses are optically active. The presence of a carboxyl group at one end of the carbon chain and a $CH_2OH$ at the other precludes the existence of meso forms.

(*c*) Nitric acid oxidation of aldoses converts them to aldaric acids. The same D-aldoses found to yield optically inactive alditols in part (*a*) yield optically inactive aldaric acids.

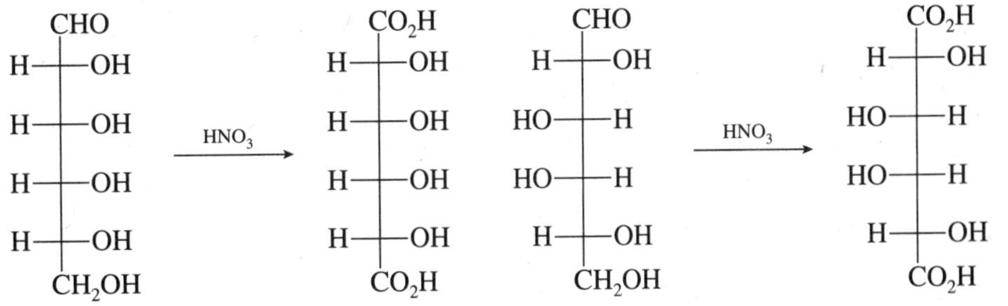

D-Allose $\xrightarrow{HNO_3}$ Allaric acid (meso compound)

D-Galactose $\xrightarrow{HNO_3}$ Galactaric acid (meso compound)

(*d*) Aldoses that differ in configuration only at C-2 enolize to the same enediol.

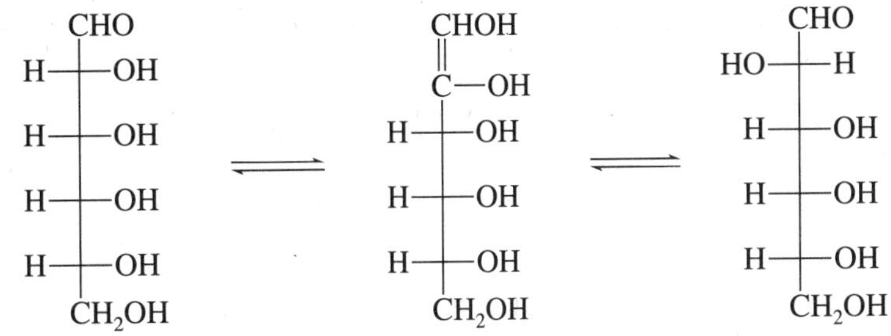

D-Allose ⇌ Enediol ⇌ D-Altrose

The stereogenic center at C-2 in the D-aldose becomes $sp^2$-hybridized in the enediol.

The other pairs of D-aldohexoses that form the same enediols are

D-Glucose and D-mannose
D-Gulose and D-idose
D-Galactose and D-talose

**25.20** (*a*) To unravel a pyranose form, locate the anomeric carbon and mentally convert the hemiacetal linkage to a carbonyl compound and a hydroxyl function.

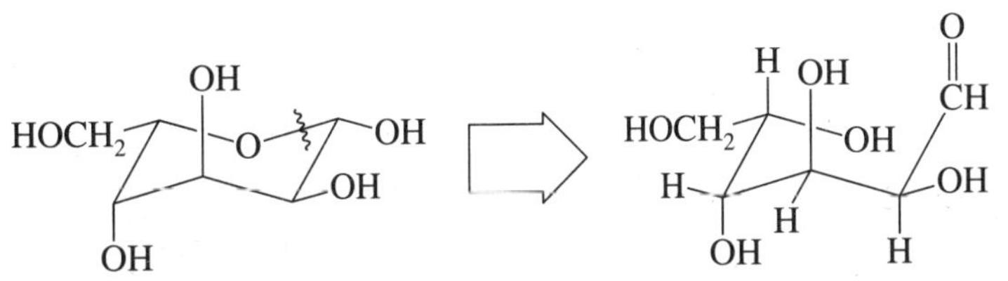

Convert the open-chain form to a Fischer projection.

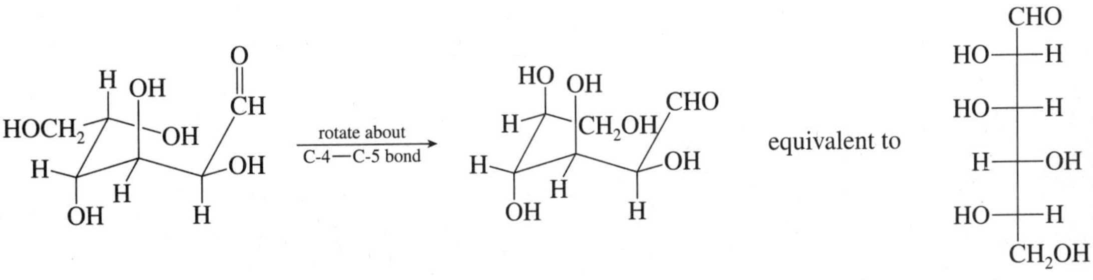

rotate about C-4—C-5 bond

equivalent to

(*b*) Proceed in the same manner as in part (*a*) and unravel the furanose sugar by disconnecting the hemiacetal function.

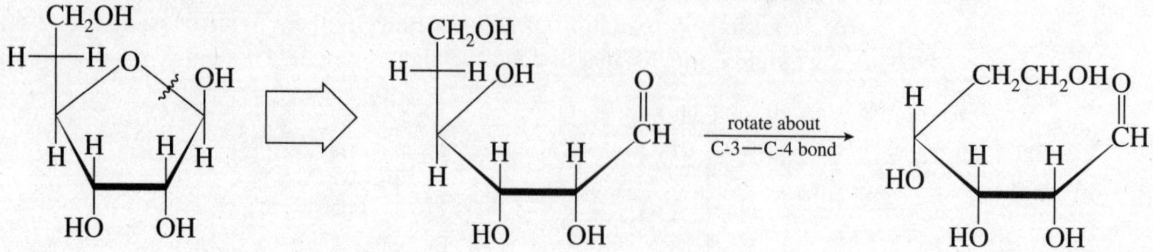

rotate about C-3—C-4 bond

The Fischer projection is

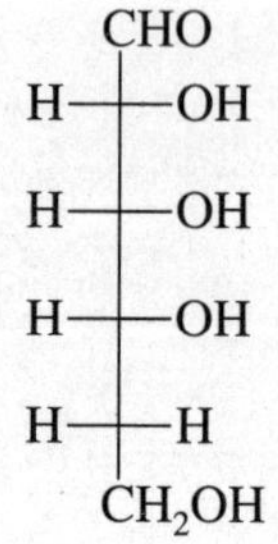

(*c*) By disconnecting and unraveling as before, the Fischer projection is revealed.

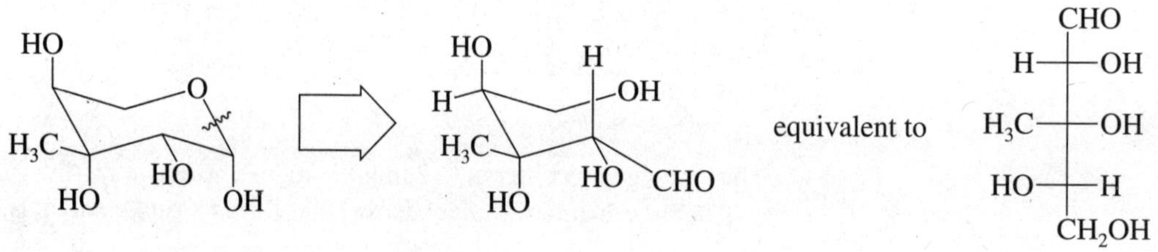

equivalent to

(*d*) Remember in disconnecting cyclic hemiacetals that the anomeric carbon is the one that bears two oxygen substituents.

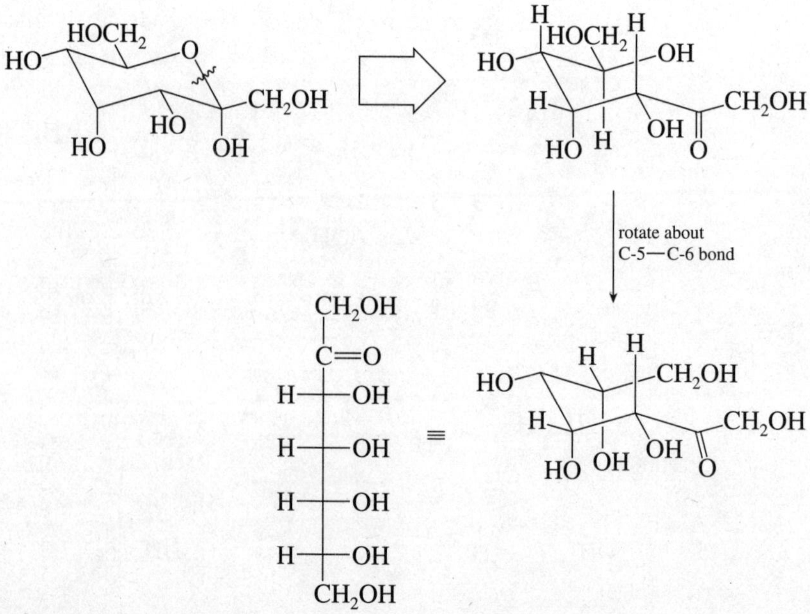

rotate about C-5—C-6 bond

**25.21** Begin the problem by converting the Fischer projection of D-ribose to a perspective view. Remember that the horizontal lines of a Fischer projection represent bonds coming toward you, and the vertical lines are going away from you.

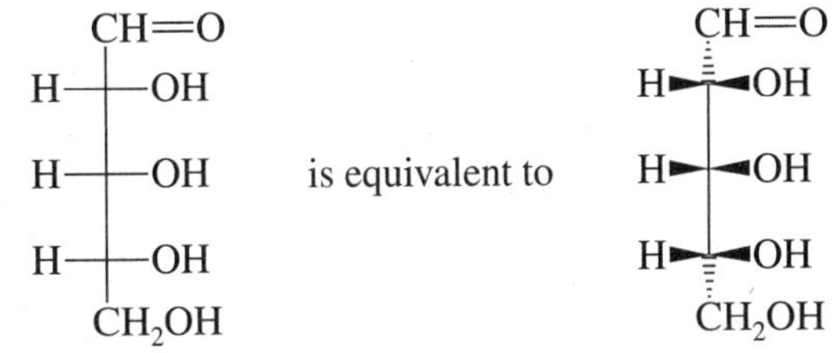
CH=O
H—OH
H—OH
H—OH
CH2OH
is equivalent to
CH=O
H—OH
H—OH
H—OH
CH2OH

Rank the groups attached to each stereogenic center. Identify each stereogenic center as either *R* or *S* according to the methods described in Chapter 7. Remember that the proper orientation of the lowest ranked group (usually H) is away from you. Molecular models will be helpful here. Each of the stereogenic centers in D-ribose has the *R* configuration. The IUPAC name of D-ribose is (2*R*,3*R*,4*R*)-2,3,4,5-tetrahydroxypentanal.

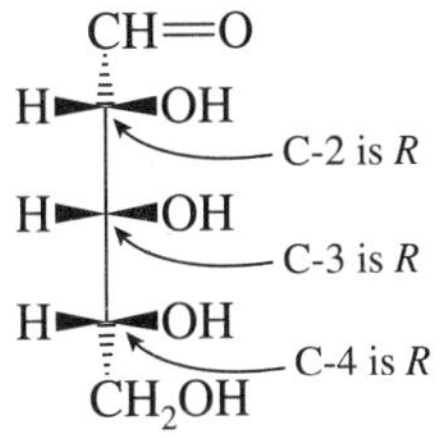
CH=O
H—OH
C-2 is R
H—OH
C-3 is R
H—OH
C-4 is R
CH2OH

**25.22** (*a*) The L sugars have the hydroxyl group to the left at the highest numbered stereogenic center in their Fischer projection. The L sugars are the ones in Problem 25.20*a* and *c*.

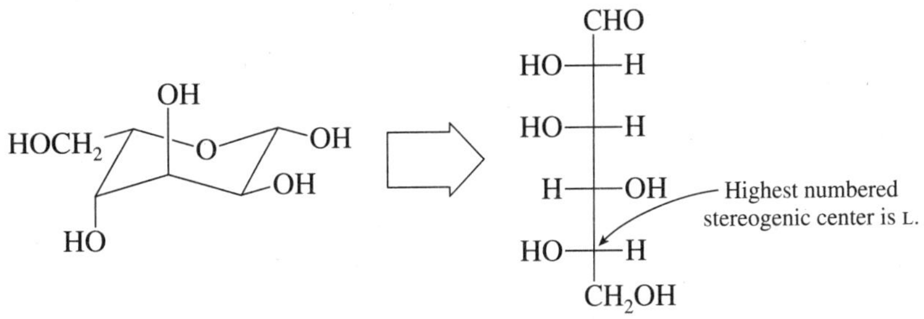
OH
HOCH2
O
OH
OH
HO
CHO
HO—H
HO—H
H—OH
HO—H
CH2OH
Highest numbered stereogenic center is L.

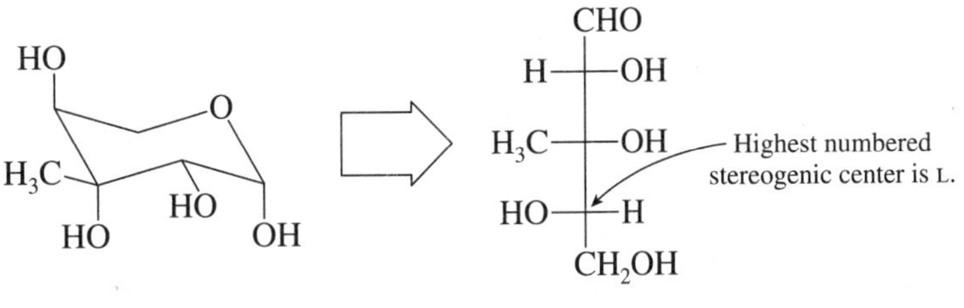
HO
O
H3C
HO
HO
OH
CHO
H—OH
H3C—OH
HO—H
CH2OH
Highest numbered stereogenic center is L.

(*b*) Deoxy sugars are those that lack an oxygen substituent on one of the carbons in the main chain. The carbohydrate in Problem 25.20*b* is a deoxy sugar.

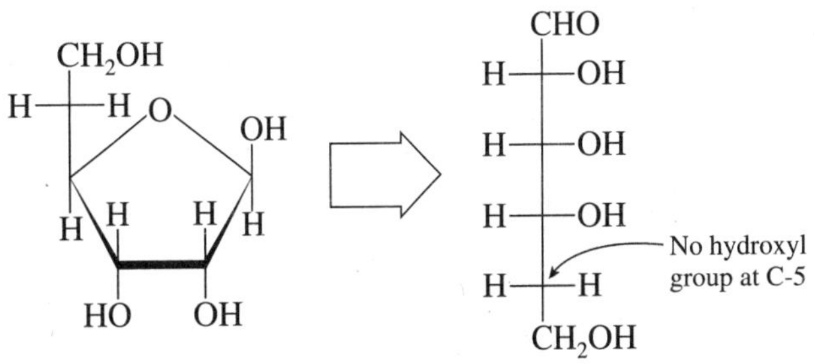
CH2OH
H—H
O
OH
H
H
H
H
HO
OH
CHO
H—OH
H—OH
H—OH
H—H
CH2OH
No hydroxyl group at C-5

(*c*) Branched-chain sugars have a carbon substituent attached to the main chain; the carbohydrate in Problem 25.20*c* fits this description.

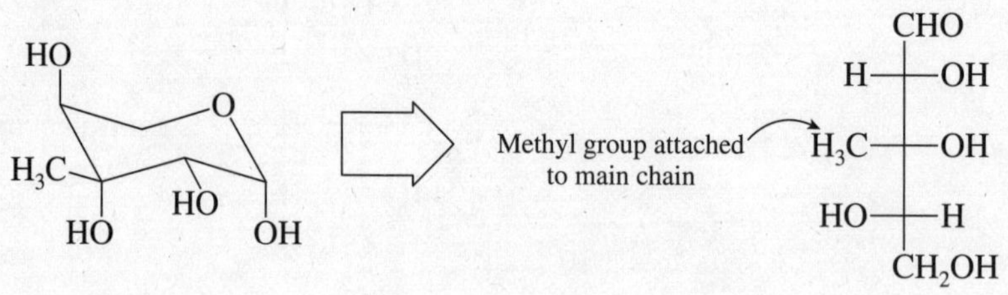

HO, O, $H_3C$, HO, HO, OH → Methyl group attached to main chain

CHO
H—OH
$H_3C$—OH
HO—H
$CH_2OH$

(*d*) Only the sugar in Problem 25.20*d* is a ketose.

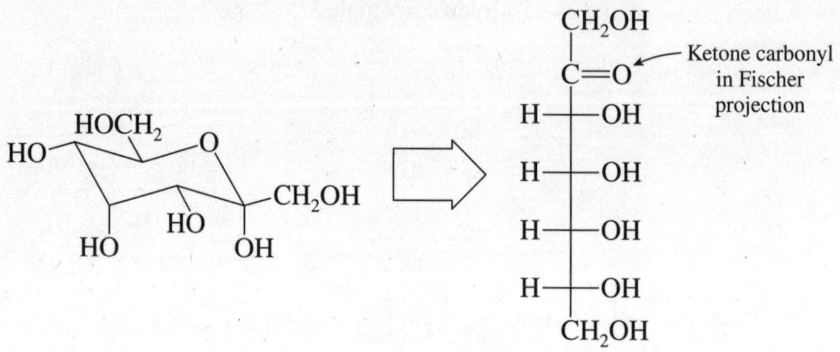

$HOCH_2$, O, HO, $CH_2OH$, HO, HO, OH →

$CH_2OH$
C=O ← Ketone carbonyl in Fischer projection
H—OH
H—OH
H—OH
H—OH
$CH_2OH$

(*e*) A furanose ring is a five-membered cyclic hemiacetal. Only the compound in Problem 25.20*b* is a furanose form.

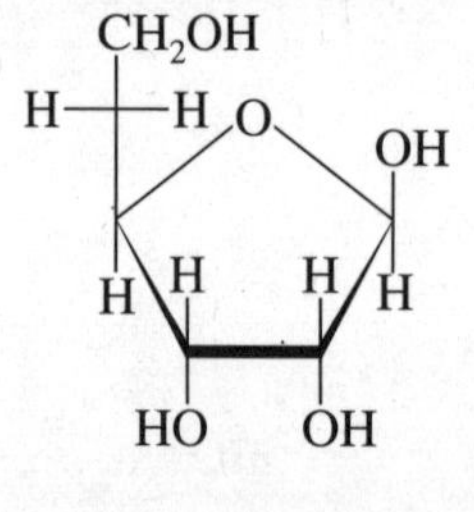

$CH_2OH$, H—H, O, OH, H, H, H, H, HO, OH

(*f*) In D sugars, the $\alpha$ configuration corresponds to the condition in which the hydroxyl group at the anomeric carbon is down. The $\alpha$-D sugar is that in Problem 25.20*d*.

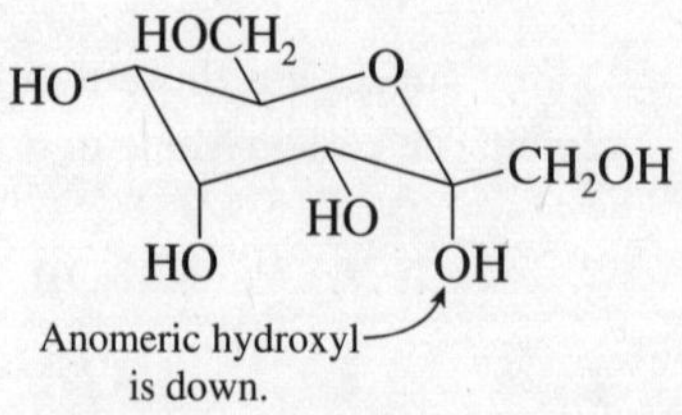

$HOCH_2$, O, HO, $CH_2OH$, HO, HO, OH

Anomeric hydroxyl is down.

$\alpha$-Pyranose form of a D-ketose

In the $\alpha$-L series the anomeric hydroxyl is up. Neither of the L sugars—namely, those of Problem 25.20*a* and *c*—is $\alpha$; both are $\beta$.

**25.23** There are seven possible pentuloses, that is, five-carbon ketoses. The ketone carbonyl can be located at either C-2 or C-3. When the carbonyl group is at C-2, there are two stereogenic centers, giving rise to four stereoisomers (two pairs of enantiomers).

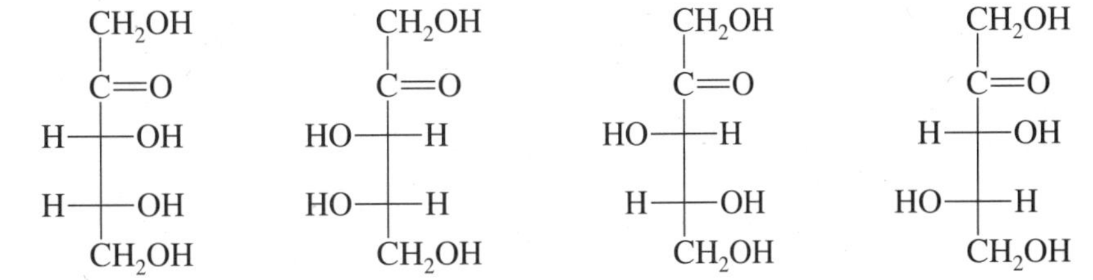

D-Erythropentulose (D-ribulose) | L-Erythropentulose (L-ribulose) | D-Threopentulose (D-xylulose) | L-Threopentulose (L-xylulose)

When the carbonyl group is located at C-3, there are only three stereoisomers, because one of them is a meso form and is superposable on its mirror image.

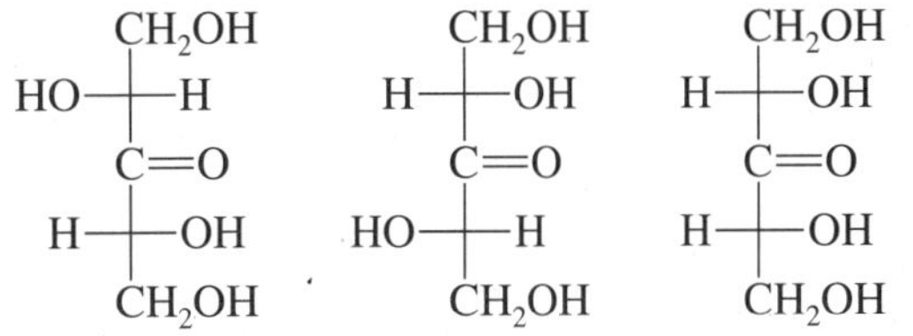

**25.24** (*a*) Carbon-2 is the only stereogenic center in D-apiose.

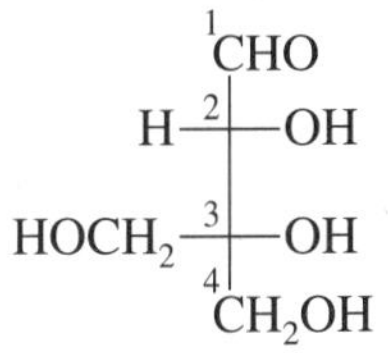

D-Apiose

Carbon-3 is not a stereogenic center; it bears two identical $CH_2OH$ substituents.

(*b*) The alditol obtained on reduction of D-apiose retains the stereogenic center. It is chiral and optically active.

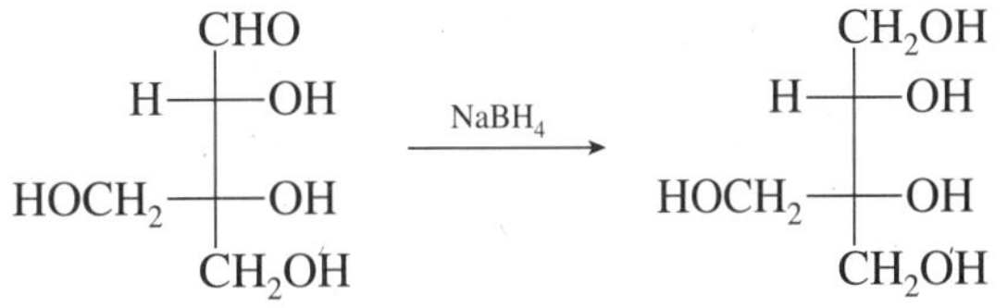

$NaBH_4$

D-Apiose (optically active) → D-Apiitol (optically active)

(*c, d*) Cyclic hemiacetal formation in D-apiose involves addition of a $CH_2OH$ hydroxyl group to the aldehyde carbonyl.

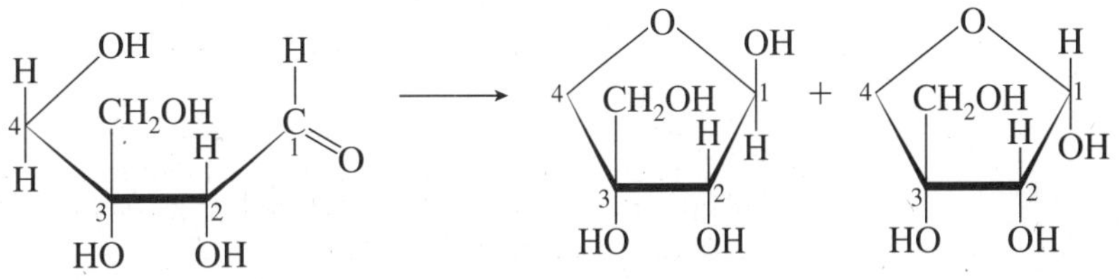

Three stereogenic centers occur in the furanose form, namely, the anomeric carbon C-1 and the original stereogenic center C-2, as well as a new stereogenic center at C-3.

In addition to the two furanose forms just shown, two more are possible. Instead of the reaction of the $CH_2OH$ group that was shown to form the cyclic hemiacetal, the other $CH_2OH$ group may add to the aldehyde carbonyl.

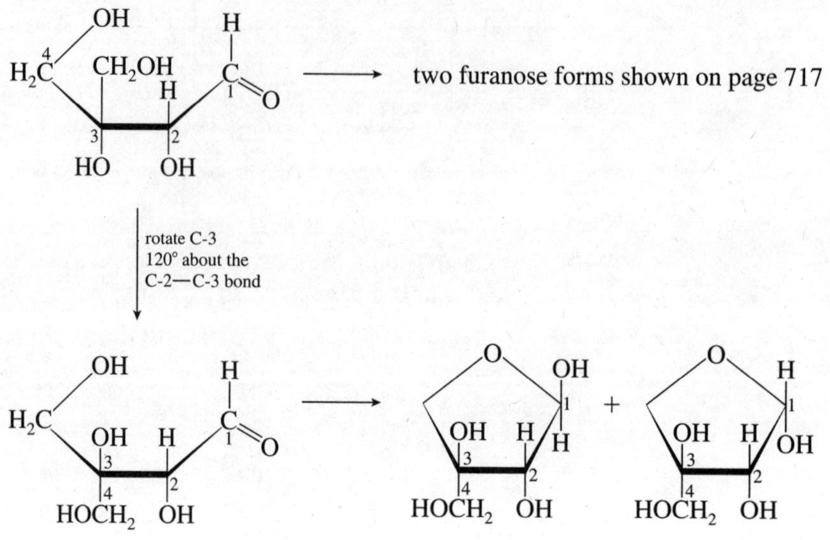

two furanose forms shown on page 717

rotate C-3 120° about the C-2—C-3 bond

**25.25** The most reasonable conclusion is that all four are methyl glycosides. Two are the methyl glycosides of the $\alpha$- and $\beta$-pyranose forms of mannose and two are the methyl glycosides of the $\alpha$- and $\beta$-furanose forms.

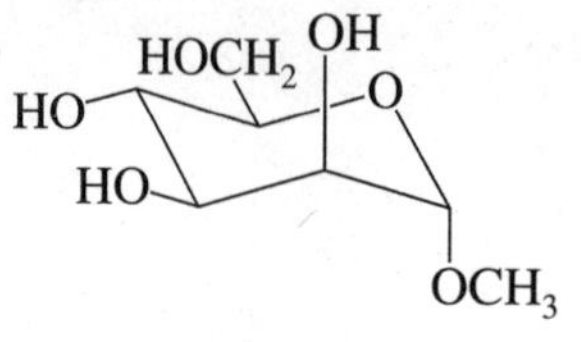

Methyl $\alpha$-D-mannopyranoside

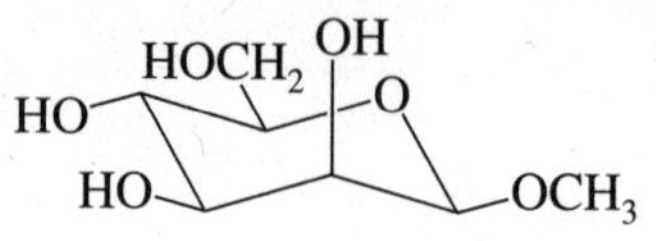

Methyl $\beta$-D-mannopyranoside

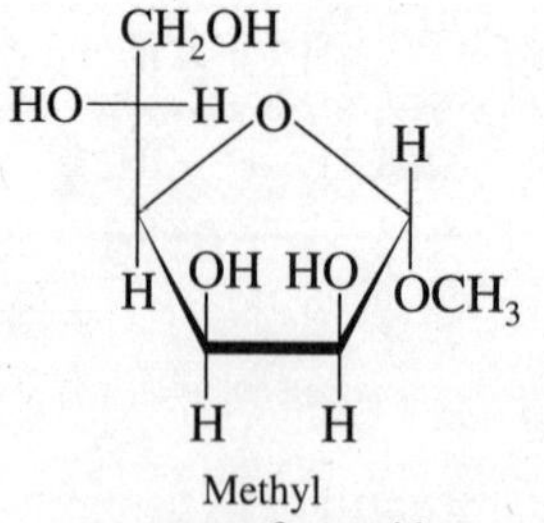

Methyl $\alpha$-D-mannofuranoside

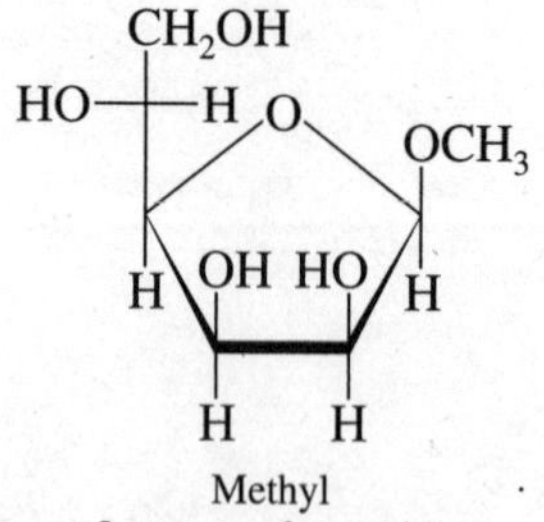

Methyl $\beta$-D-mannofuranoside

In the case of the methyl glycosides of mannose, comparable amounts of pyranosides and furanosides are formed. The major products are the $\alpha$ isomers.

**25.26** (*a*) Disaccharides, by definition, involve an acetal linkage at the anomeric position; thus all the disaccharides must involve C-1. The bond to C-1 can be $\alpha$ or $\beta$. The available oxygen atoms in the second D-glucopyranosyl unit are located at C-1, C-2, C-3, C-4, and C-6. Thus, there are 11 possible disaccharides, including maltose and cellobiose, composed of D-glucopyranosyl units.

| | | |
|---|---|---|
| $\alpha,\alpha(1,1)$ | $\alpha,\beta(1,1)$ | $\beta,\beta(1,1)$ |
| $\alpha(1,2)$ | | $\beta(1,2)$ |
| $\alpha(1,3)$ | | $\beta(1,3)$ |
| $\alpha(1,4)$ (maltose) | | $\beta(1,4)$ (cellobiose) |
| $\alpha(1,6)$ | | $\beta(1,6)$ |

(*b*) To be a reducing sugar, one of the anomeric positions must be a free hemiacetal. All except $\alpha,\alpha(1,1)$, $\alpha,\beta(1,1)$, and $\beta,\beta(1,1)$ are reducing sugars.

**25.27** Because gentiobiose undergoes mutarotation, it must have a free hemiacetal group. Formation of two molecules of D-glucose indicates that it is a disaccharide and because that hydrolysis is catalyzed by emulsin, the glycosidic linkage is $\beta$. The methylation data, summarized in the following equation, require that the glucose units be present in pyranose forms and be joined by a $\beta(1,6)$-glycoside bond.

Site subject to mutarotation when R = H

R = H: gentiobiose

↓

R = $CH_3$: gentiobiose octamethyl ether

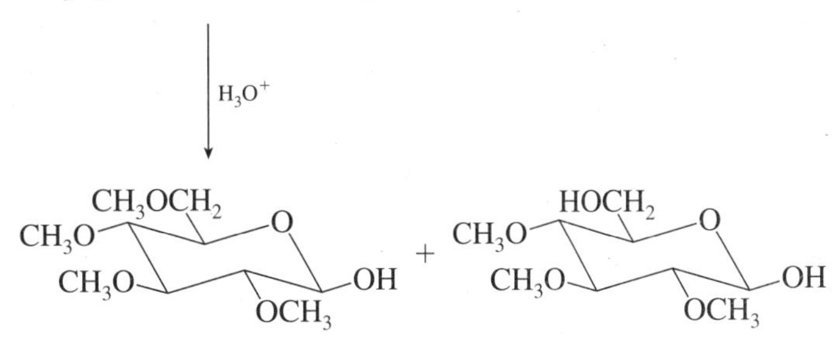


2,3,4,6-Tetra-*O*-methyl-D-glucose 2,3,4-Tri-*O*-methyl-D-glucose

**25.28** Like other glycosides, cyanogenic glycosides are cleaved to a carbohydrate and an alcohol on hydrolysis.

(*a*) In the case of linamarin the alcohol is recognizable as the cyanohydrin of acetone. Once formed, this cyanohydrin dissociates to hydrogen cyanide and acetone.

Linamarin + $H_2O$ $\xrightarrow{H^+ \text{ or enzyme}}$ D-Glucose + Acetone cyanohydrin ($HOC(CH_3)_2CN$)

Acetone cyanohydrin → $CH_3CCH_3$ (O) + HCN

Acetone Hydrogen cyanide

(*b*) Laetrile undergoes an analogous hydrolytic cleavage to yield the cyanohydrin of benzaldehyde.

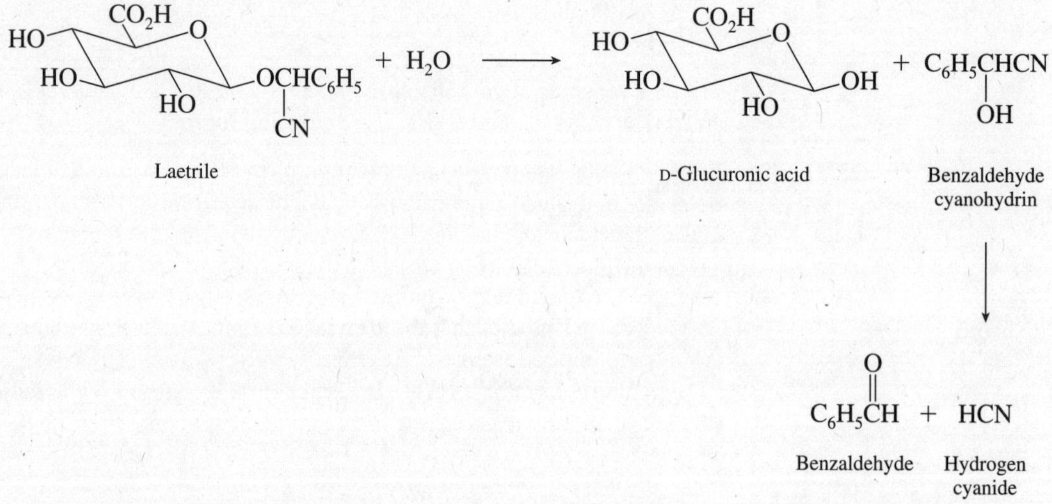

Laetrile + $H_2O$ ⟶ D-Glucuronic acid + $C_6H_5CHCN$ (OH) Benzaldehyde cyanohydrin ⟶ $C_6H_5CH$(=O) + HCN

Benzaldehyde Hydrogen cyanide

**25.29** Comparing D-glucose, D-mannose, and D-galactose, it can be said that the configuration of C-2 has a substantial effect on the relative energies of the $\alpha$- and $\beta$-pyranose forms, but that the configuration of C-4 has virtually no effect. With this observation in mind, write the structures of the pyranose forms of the carbohydrates given in each part.

(*a*) The $\beta$-pyranose form of D-gulose is the same as that of D-galactose except for the configuration at C-3.

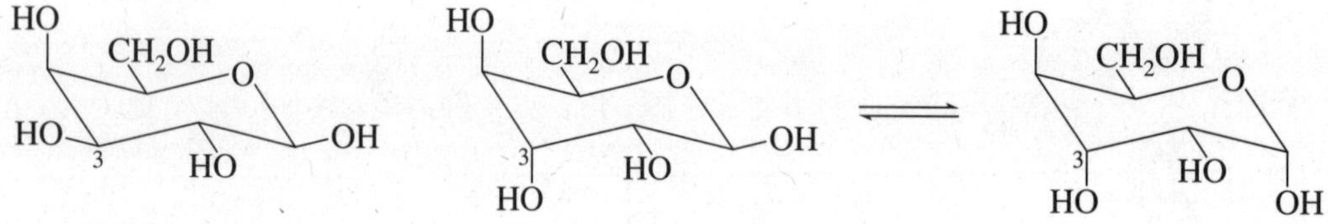

$\beta$-D-Galactopyranose (64% at equilibrium)

$\beta$-D-Gulopyranose ⇌ $\alpha$-D-Gulopyranose (1,3 diaxial repulsion between hydroxyl groups)

The axial hydroxyl group at C-3 destabilizes the $\alpha$-pyranose form more than the $\beta$ form because of its repulsive interaction with the axially disposed anomeric hydroxyl group. There should be an even higher $\beta/\alpha$ ratio in D-gulopyranose than in D-galactopyranose. This is so; the observed $\beta/\alpha$ ratio is 88 : 12.

(*b*) The $\beta$-pyranose form of D-talose is the same as that of D-mannose except for the configuration at C-4.

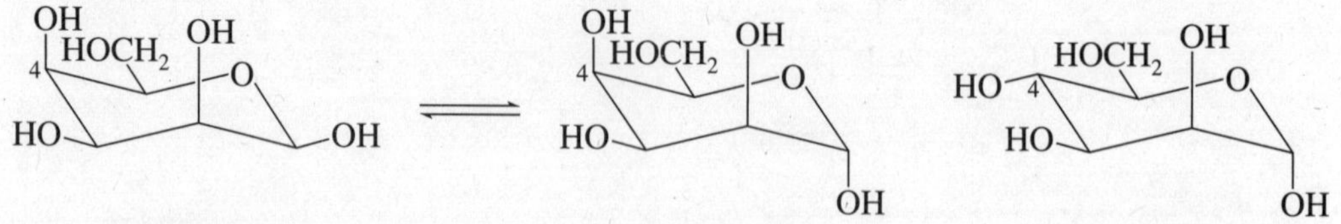

$\beta$-D-Talopyranose ⇌ $\alpha$-D-Talopyranose

$\alpha$-D-Mannopyranose (68% at equilibrium)

Because the configuration at C-4 has little effect on the $\alpha$- to $\beta$-pyranose ratio (compare D-glucose and D-galactose), we would expect that talose would behave very much like mannose and that the $\alpha$-pyranose form would be preferred at equilibrium. This is indeed the case; the $\alpha$-pyranose form predominates at equilibrium, the observed $\alpha/\beta$ ratio being 78 : 22.

(*c*) The pyranose form of D-xylose is just like that of D-glucose except that it lacks a $CH_2OH$ group.

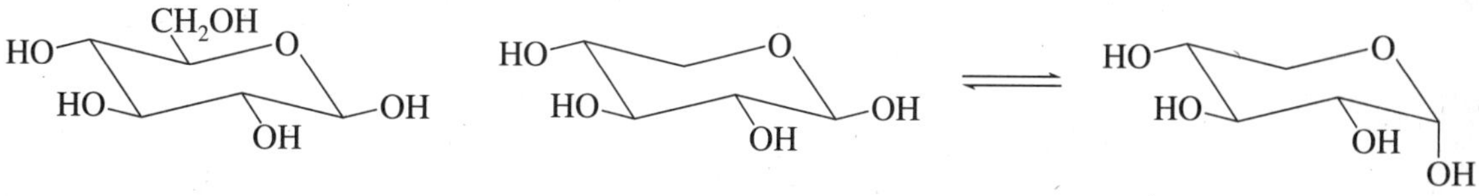

β-D-Glucopyranose (64% at equilibrium)  β-D-Xylopyranose  α-D-Xylopyranose

We would expect the equilibrium between pyranose forms in D-xylose to be much like that in D-glucose and predict that the β-pyranose form would predominate. It is observed that the $\beta/\alpha$ ratio in D-xylose is 64 : 36, exactly the same as in D-glucose.

(*d*) The pyranose form of D-lyxose is like that of D-mannose except that it lacks a $CH_2OH$ group. As in D-mannopyranose, the α form should predominate over the β.

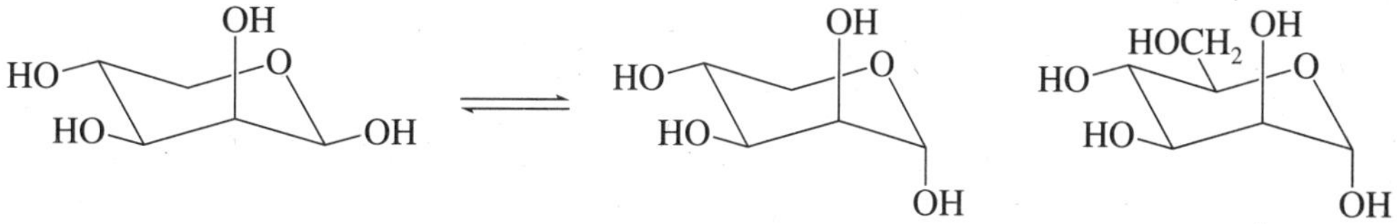

β-D-Lyxopyranose  α-D-Lyxopyranose  α-D-Mannopyranose (68% at equilibrium)

The observed $\alpha/\beta$ distribution ratio in D-lyxopyranose is 73 : 27.

**25.30** (*a*) The rate-determining step in glycoside hydrolysis is carbocation formation at the anomeric position. The carbocation formed from methyl α-D-fructofuranoside (compound A) is tertiary and therefore more stable than the one from methyl α-D-glucofuranoside (compound B), which is secondary. The more stable a carbocation is, the more rapidly it will be formed.

**Faster:**

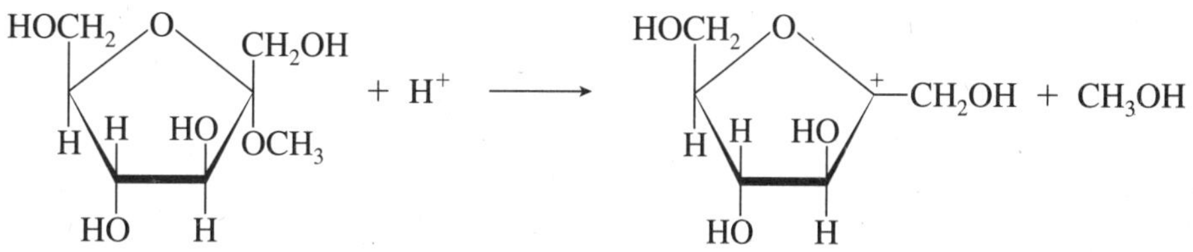

$$A + H^+ \longrightarrow \text{Tertiary carbocation} + CH_3OH$$

A  Tertiary carbocation

**Slower:**

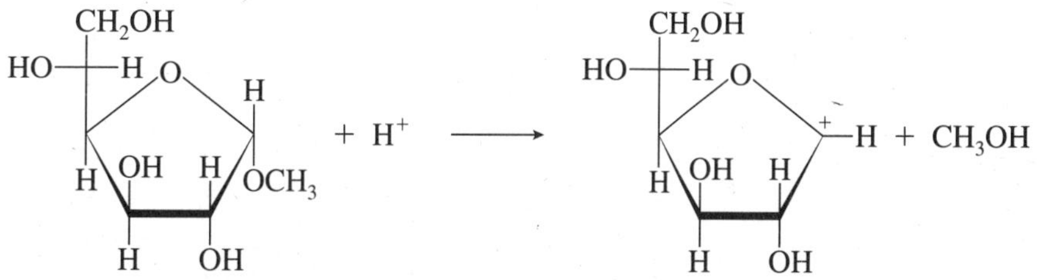

$$B + H^+ \longrightarrow \text{Secondary carbocation} + CH_3OH$$

B  Secondary carbocation

(*b*) The carbocation formed from methyl $\beta$-D-glucopyranoside (compound D) is less stable than the one from its 2-deoxy analog (compound C) and is formed more slowly. It is destabilized by the electron-withdrawing inductive effect of the hydroxyl group at C-2.

**Faster:**

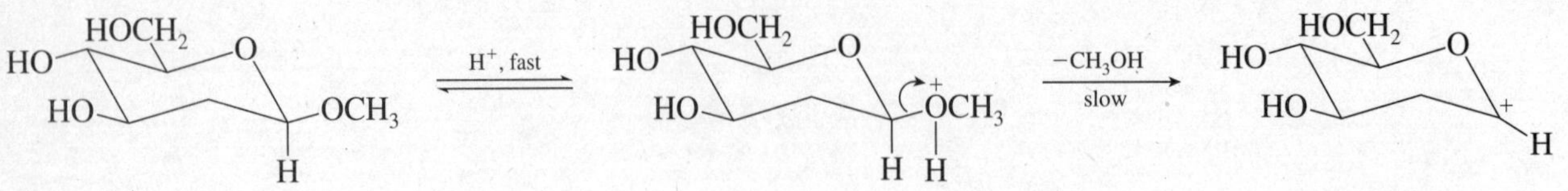

HOCH$_2$ ... OCH$_3$ (C) $\xrightleftharpoons{H^+, \text{fast}}$ protonated acetal $\xrightarrow[\text{slow}]{-CH_3OH}$ carbocation

C

More stable

**Slower:**

HOCH$_2$ ... OH ... OCH$_3$ (D) $\xrightleftharpoons{H^+, \text{fast}}$ protonated acetal $\xrightarrow[\text{slow}]{-CH_3OH}$ carbocation

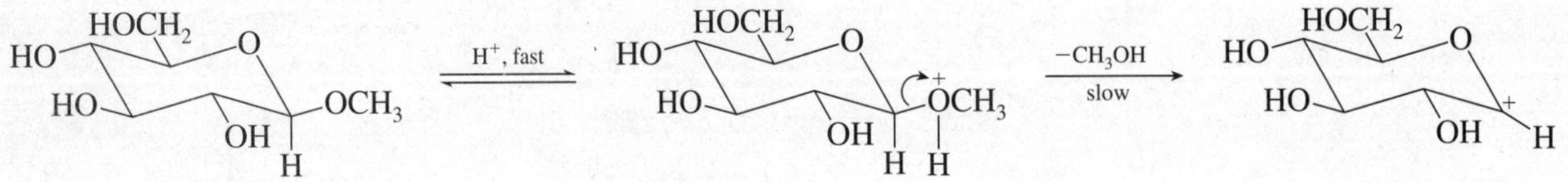

D

Less stable

**25.31** D-Altrosan is a glycoside. The anomeric carbon—the one with two oxygen substituents—has an alkoxy group attached to it. Hydrolysis of D-altrosan follows the general mechanism for acetal hydrolysis.

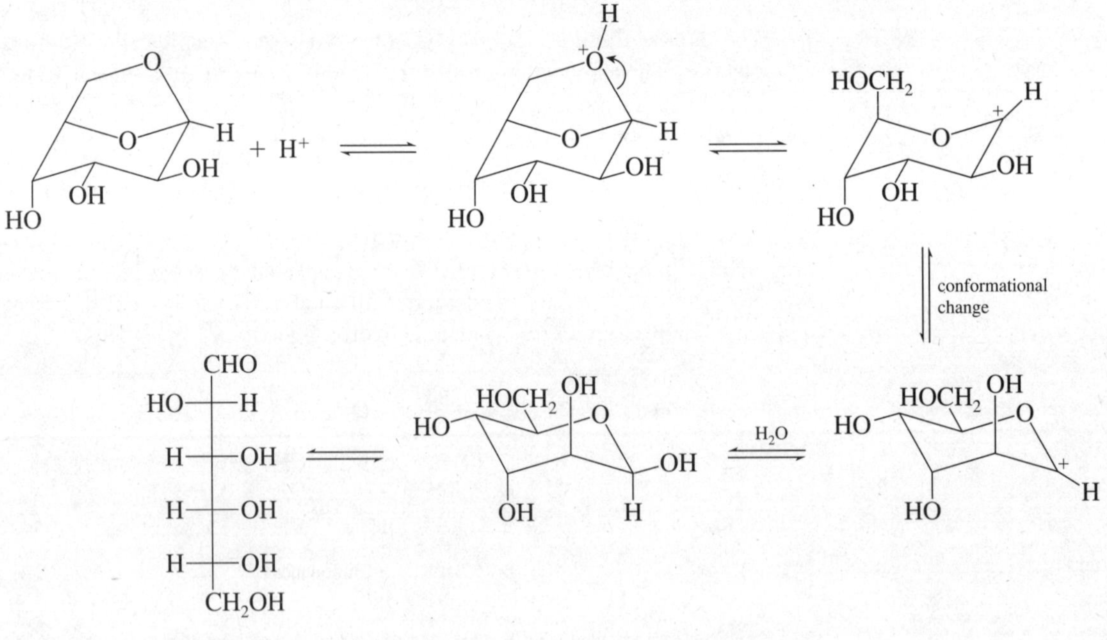

D-Altrosan + H$^+$ $\rightleftharpoons$ protonated altrosan $\rightleftharpoons$ carbocation

conformational change

$\xrightleftharpoons{H_2O}$

CHO

HO—H

H—OH

H—OH

H—OH

$CH_2OH$

D-Altrose

**25.32** Galactose has hydroxyl groups at carbons 2, 3, 4, 5, 6. Ten trimethyl ethers are therefore possible.

| | | | |
|---|---|---|---|
| 2,3,4 | 2,4,5 | 3,4,5 | 4,5,6 |
| 2,3,5 | 2,4,6 | 3,4,6 | |
| 2,3,6 | 2,5,6 | 3,5,6 | |

To find out which one of these is identical with the degradation product of compound A, carry compound A through the required transformations.

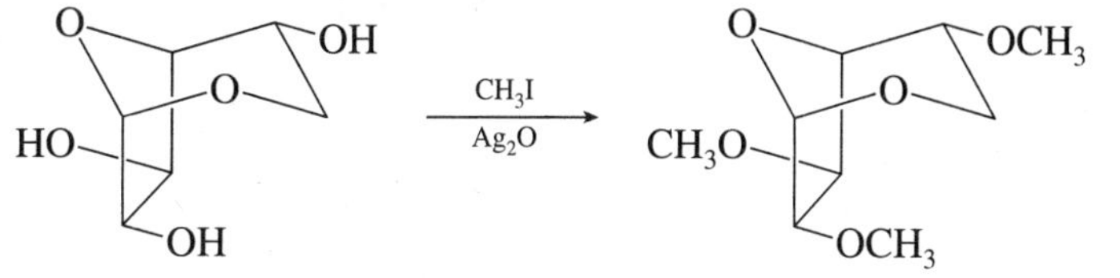

$\xrightarrow[Ag_2O]{CH_3I}$

Compound A

Tri-*O*-methyl ether of compound A

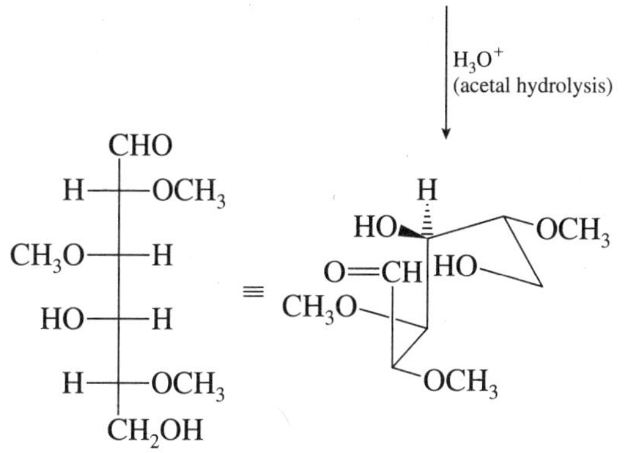

$\downarrow$ $H_3O^+$ (acetal hydrolysis)

2,3,5-Tri-*O*-methyl-D-galactose

**25.33** The fact that phlorizin is hydrolyzed to D-glucose and compound A by emulsin indicates that it is a $\beta$-glucoside in which D-glucose is attached to one of the phenolic hydroxyls of compound B.

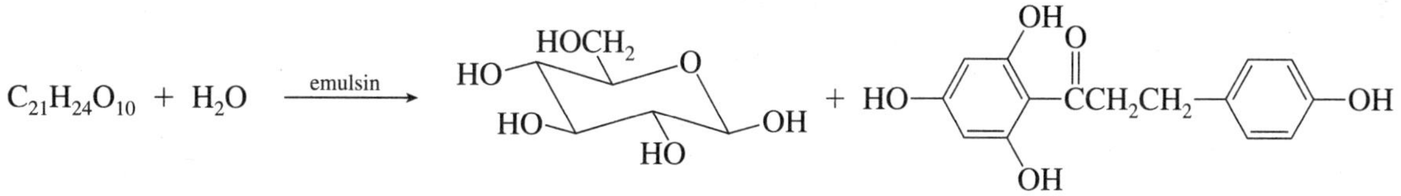

$C_{21}H_{24}O_{10} + H_2O \xrightarrow{\text{emulsin}}$ D-Glucose ($C_6H_{12}O_6$) + Compound A ($C_{15}H_{14}O_5$)

The methylation experiment reveals to which hydroxyl glucose is attached. Excess methyl iodide reacts with all the available phenolic hydroxyl groups, but the glycosidic oxygen is not affected. Thus when the methylated phlorizin undergoes acid-catalyzed hydrolysis of its glycosidic bond, the oxygen in that bond is exposed as a phenolic hydroxyl group.

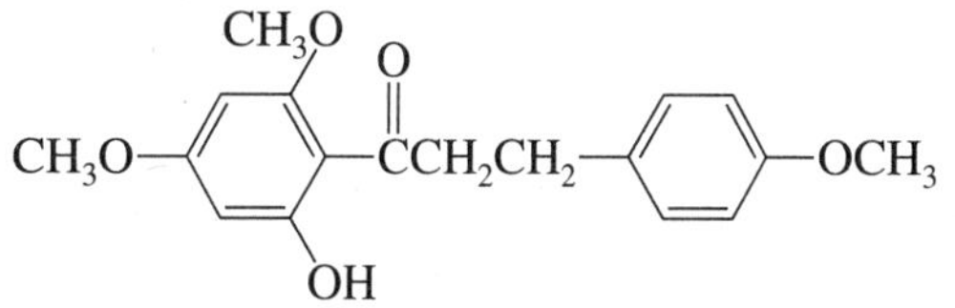

Compound B

This compound must arise by hydrolysis of

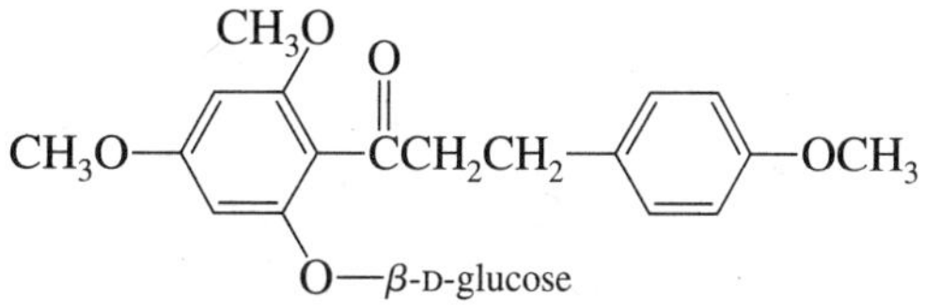

O—$\beta$-D-glucose

The structure of phlorizin is therefore

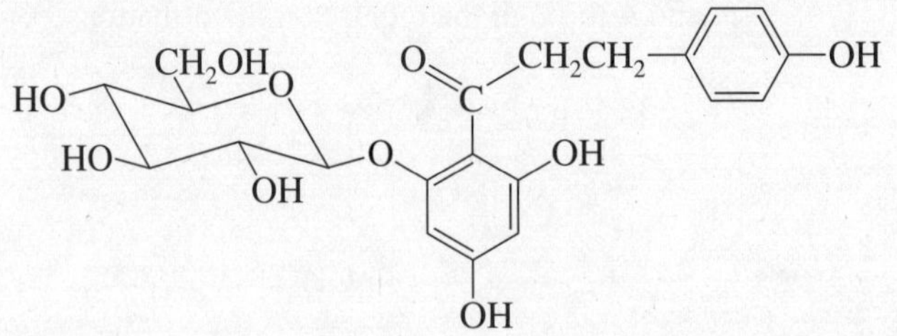

[Structure of phlorizin: a glucopyranose ring (HO, HO, OH, $CH_2OH$) linked through O to a benzene ring bearing two OH groups and a $C(=O)CH_2CH_2$–C$_6$H$_4$–OH side chain]

**25.34** Consider all the individual pieces of information in the order in which they are presented.

1. *Chain extension of the aldopentose (−)-arabinose by way of the derived cyanohydrin gave a mixture of (+)-glucose and (+)-mannose.*

   Chain extension of aldoses takes place at the aldehyde end of the chain. The aldehyde function of an aldopentose becomes C-2 of an aldohexose, which normally results in two carbohydrates diastereomeric at C-2. Thus, (+)-glucose and (+)-mannose have the same configuration at C-3, C-4, and C-5; they have opposite configurations at C-2. The configuration at C-2, C-3, and C-4 of (−)-arabinose is the same as that at C-3, C-4, and C-5 of (+)-glucose and (+)-mannose.

2. *Oxidation of (−)-arabinose with warm nitric acid gave an optically active aldaric acid.*

   Because the hydroxyl group at C-4 of (−)-arabinose is at the right in a Fischer projection formula (evidence of step 1), the hydroxyl at C-2 must be to the left in order for the aldaric acid to be optically active.

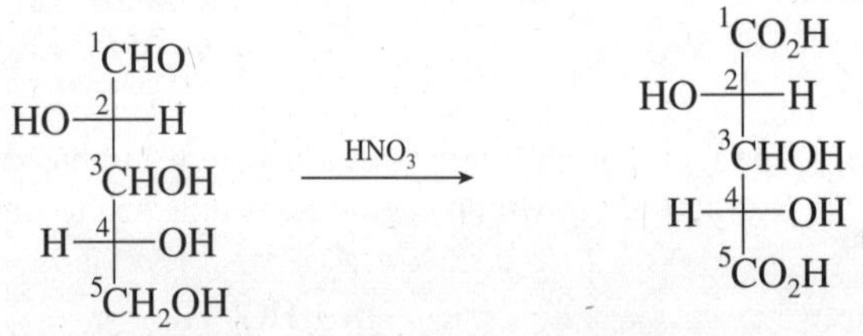

$$\begin{array}{c} {}^{1}CHO \\ HO\overset{2}{-}\!\!\!+\!\!\!-H \\ {}^{3}CHOH \\ H\overset{4}{-}\!\!\!+\!\!\!-OH \\ {}^{5}CH_2OH \end{array} \xrightarrow{HNO_3} \begin{array}{c} {}^{1}CO_2H \\ HO\overset{2}{-}\!\!\!+\!\!\!-H \\ {}^{3}CHOH \\ H\overset{4}{-}\!\!\!+\!\!\!-OH \\ {}^{5}CO_2H \end{array}$$

Partial stereostructure of (−)-arabinose

Aldaric acid from (−)-arabinose; optically active irrespective of configuration at C-3

If the C-2 hydroxyl group had been to the right, an optically inactive meso aldaric acid would have been produced.

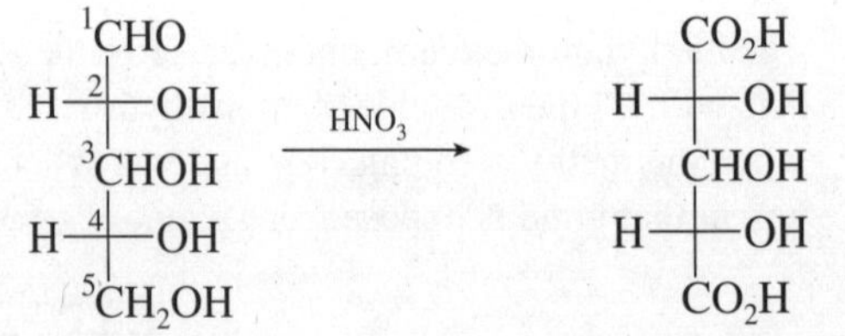

$$\begin{array}{c} {}^{1}CHO \\ H\overset{2}{-}\!\!\!+\!\!\!-OH \\ {}^{3}CHOH \\ H\overset{4}{-}\!\!\!+\!\!\!-OH \\ {}^{5}CH_2OH \end{array} \xrightarrow{HNO_3} \begin{array}{c} CO_2H \\ H-\!\!\!+\!\!\!-OH \\ CHOH \\ H-\!\!\!+\!\!\!-OH \\ CO_2H \end{array}$$

Achiral meso form; cannot be optically active

Therefore we now know the configurations of C-3 and C-5 of (+)-glucose and (+)-mannose and that these two aldohexoses have opposite configurations at C-2, but the same (yet to be determined) configuration at C-4.

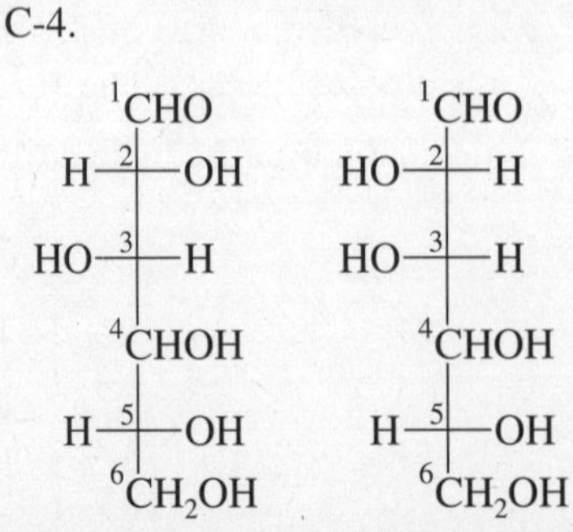

$$\begin{array}{c} {}^{1}CHO \\ H\overset{2}{-}\!\!\!+\!\!\!-OH \\ HO\overset{3}{-}\!\!\!+\!\!\!-H \\ {}^{4}CHOH \\ H\overset{5}{-}\!\!\!+\!\!\!-OH \\ {}^{6}CH_2OH \end{array} \qquad \begin{array}{c} {}^{1}CHO \\ HO\overset{2}{-}\!\!\!+\!\!\!-H \\ HO\overset{3}{-}\!\!\!+\!\!\!-H \\ {}^{4}CHOH \\ H\overset{5}{-}\!\!\!+\!\!\!-OH \\ {}^{6}CH_2OH \end{array}$$

[One of these is (+)-glucose, the other is (+)-mannose.]

**3.** *Both (+)-glucose and (+)-mannose are oxidized to optically active aldaric acids with nitric acid.*

Because both (+)-glucose and (+)-mannose yield optically active aldaric acids and both have the same configuration at C-4, the hydroxyl group must lie at the right in the Fischer projection at this carbon.

```
      1CHO              1CHO
  H ─2─ OH          HO ─2─ H
 HO ─3─ H           HO ─3─ H
  H ─4─ OH           H ─4─ OH
  H ─5─ OH           H ─5─ OH
      6CH2OH            6CH2OH
```

[One of these is (+)-glucose, the other is (+)-mannose.]

The structures of the corresponding aldaric acids are

```
      CO2H              CO2H
  H ──── OH         HO ──── H
 HO ──── H          HO ──── H
  H ──── OH          H ──── OH
  H ──── OH          H ──── OH
      CO2H              CO2H
```

Both are optically active. Had the C-4 hydroxyl group been to the left, one of the aldaric acids would have been a meso form.

```
      CO2H              CO2H
  H ──── OH         HO ──── H
 HO ──── H          HO ──── H
 HO ──── H          HO ──── H
  H ──── OH          H ──── OH
      CO2H              CO2H
```

(This aldaric acid is optically inactive.)

**4.** *There is another sugar, (+)-gulose, that gives the same aldaric acid on oxidation as does (+)-glucose.*

This is the last piece in the puzzle, the one that permits one of the Fischer projections shown in the first part of step 3 to be assigned to (+)-glucose and the other to (+)-mannose. Consider first the structure

```
      CHO                              CH2OH
 HO ──── H                         H ──── OH
 HO ──── H                        HO ──── H
  H ──── OH      equivalent to     H ──── OH
 HO ──── H                         H ──── OH
      CH2OH                            CHO
```

Oxidation gives the aldaric acid

$$\begin{array}{rcl} & CO_2H & \\ H & - & OH \\ HO & - & H \\ H & - & OH \\ H & - & OH \\ & CO_2H & \end{array}$$

This is the same aldaric acid as that provided by one of the structures given as either (+)-glucose or (+)-mannose. That Fischer projection therefore corresponds to (+)-glucose.

$$\begin{array}{rcl} & CHO & \\ H & - & OH \\ HO & - & H \\ H & - & OH \\ H & - & OH \\ & CH_2OH & \end{array}$$

This must be (+)-glucose.

The structure of (+)-mannose is therefore

$$\begin{array}{rcl} & CHO & \\ HO & - & H \\ HO & - & H \\ H & - & OH \\ H & - & OH \\ & CH_2OH & \end{array}$$

A sugar that yields the same aldaric acid is

$$\begin{array}{rcl} & CH_2OH & \\ HO & - & H \\ HO & - & H \\ H & - & OH \\ H & - & OH \\ & CHO & \end{array}$$

This is, in fact, not a different sugar but simply (+)-mannose rotated through an angle of 180°.

# SELF-TEST

## PART A

**A-1.** Draw the structures indicated for each of the following:

(*a*) The enantiomer of D-erythrose

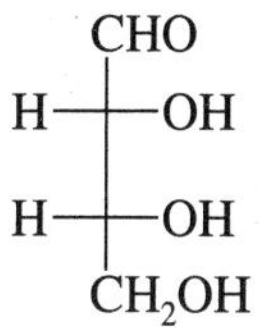

CHO
H—OH
H—OH
$CH_2OH$

(*b*) A diastereomer of D-erythrose
(*c*) The $\alpha$-furanose form of D-erythrose (use a Haworth formula)
(*d*) The anomer of the structure in part (*c*)
(*e*) Assign the configuration of each stereogenic center of D-erythrose as either *R* or *S*.

**A-2.** The structure of D-mannose is

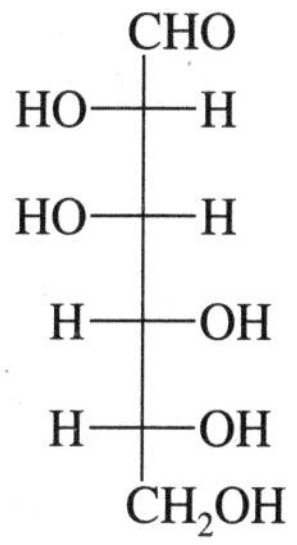

CHO
HO—H
HO—H
H—OH
H—OH
$CH_2OH$

D-Mannose

Using Fischer projections, draw the product of the reaction of D-mannose with

(*a*) $NaBH_4$ in $H_2O$
(*b*) Benedict's reagent
(*c*) Excess periodic acid

**A-3.** Referring to the structure of D-arabinose shown, draw the following:

(*a*) The $\alpha$-pyranose form of D-arabinose
(*b*) The $\beta$-furanose form of D-arabinose
(*c*) The $\beta$-pyranose form of L-arabinose

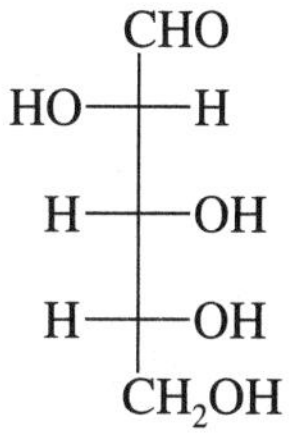

CHO
HO—H
H—OH
H—OH
$CH_2OH$

D-Arabinose

**A-4.** Using text Figure 25.2, identify the following carbohydrate:

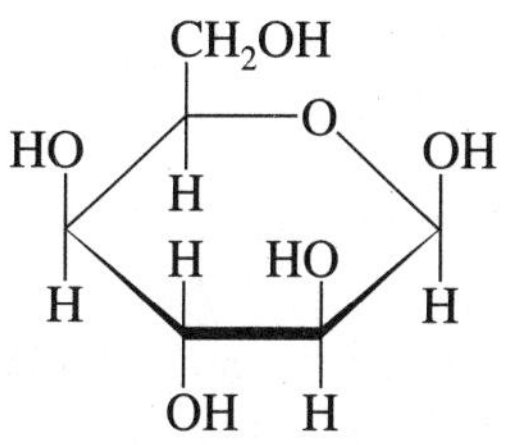

$CH_2OH$
O
HO H OH
H HO
H H
OH H

**A-5.** Write structural formulas for the $\alpha$- and $\beta$-methyl pyranosides formed from the reaction of D-mannose (see Problem A-2 for its structure) with methanol in the presence of hydrogen chloride. How are the two products related—are they enantiomers? Diastereomers?

## PART B

**B-1.** Choose the response that provides the best match between the terms given and the structures shown.

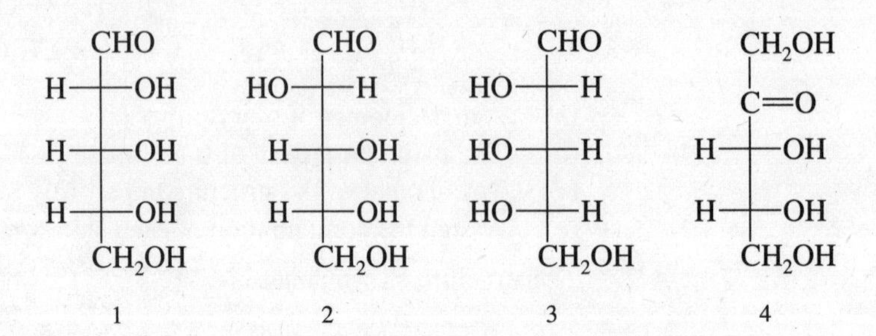


| | Diastereomers | Enantiomers |
|---|---|---|
| (*a*) | 1, 3, and 4 | 1 and 3 |
| (*b*) | 1 and 2 | 1 and 3 |
| (*c*) | 1, 2, and 3 | 1 and 3 |
| (*d*) | 1 and 4 | 1 and 2 |

**B-2.** A D carbohydrate is
(*a*) Always dextrorotatory
(*b*) Always levorotatory
(*c*) Always the anomer of the corresponding L carbohydrate
(*d*) None of the above

**B-3.** Two of the three compounds shown yield the same product on reaction with warm $HNO_3$. The *exception* is

(*a*) CHO; HO—H; H—OH; H—OH; $CH_2OH$

(*b*) CHO; HO—H; HO—H; H—OH; $CH_3$

(*c*) CHO; HO—H; HO—H; H—OH; $CH_2OH$

(*d*) None of these (all yield the same product)

**B-4.** The optical rotation of the $\alpha$ form of a pyranose is $+150.7°$; that of the $\beta$ form is $+52.8°$. In solution an equilibrium mixture of the anomers has an optical rotation of $+80.2°$. The percentage of the $\alpha$ form at equilibrium is
(*a*) 28% (*b*) 32% (*c*) 68% (*d*) 72%

**B-5.** Which of the following represents the anomer of the compound shown?

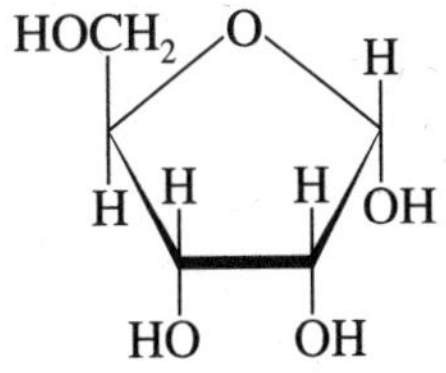

(*a*)

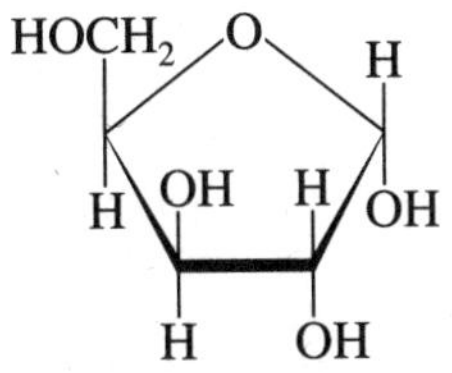

(*c*)

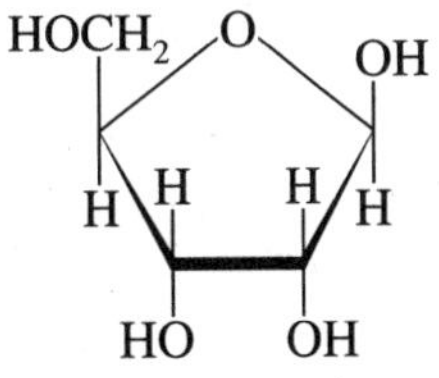

(*b*)

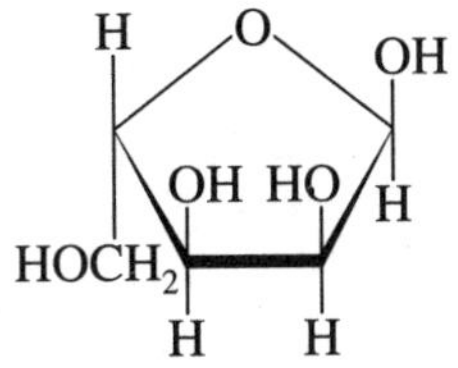

(*d*) None of these

**B-6.** Which of the following aldoses yields an optically inactive substance on reaction with sodium borohydride?

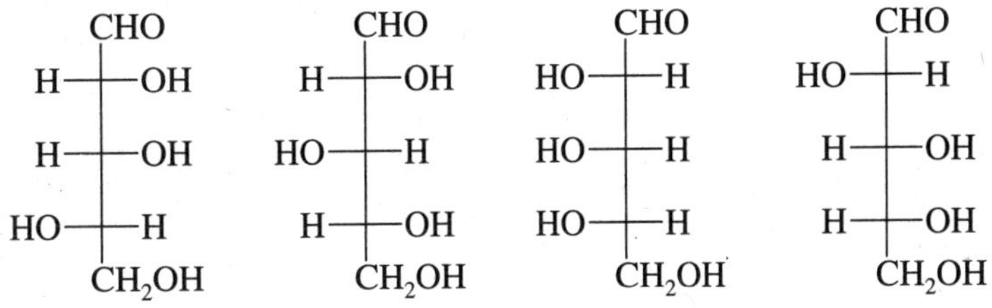

| 1 | | 2 | | 3 | | 4 | |
|---|---|---|---|---|---|---|---|
| CHO | | CHO | | CHO | | CHO | |
| H | OH | H | OH | HO | H | HO | H |
| H | OH | HO | H | HO | H | H | OH |
| HO | H | H | OH | HO | H | H | OH |
| $CH_2OH$ | | $CH_2OH$ | | $CH_2OH$ | | $CH_2OH$ | |

(*a*) 3 only
(*b*) 1 and 4
(*c*) 2 and 3
(*d*) All (1, 2, 3, and 4)

**B-7.** Which set of terms correctly identifies the carbohydrate shown?

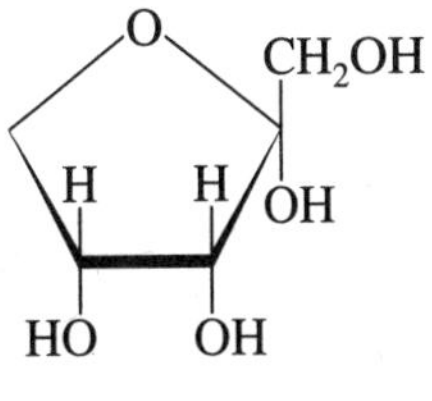

1. Pentose
2. Pentulose
3. Hexulose
4. Hexose
5. Aldose
6. Ketose
7. Pyranose
8. Furanose

(*a*) 2, 6, 8
(*b*) 2, 6, 7
(*c*) 1, 5, 8
(*d*) A set of terms other than these

**B-8.** The structure of D-arabinose is shown in Problem A-3. Which of the following is L-arabinose?

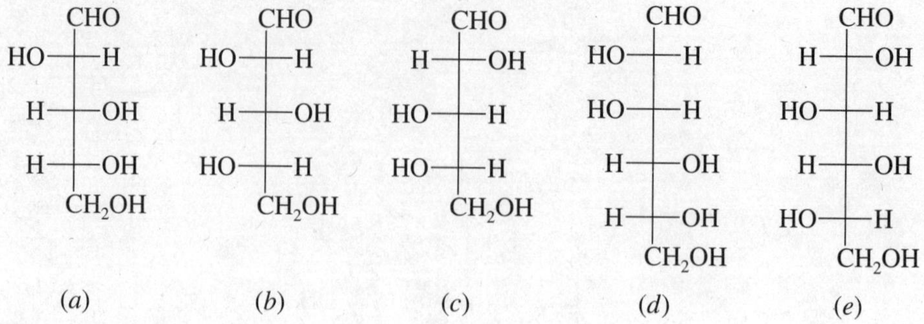

**B-9.** Which one of the statements concerning the equilibrium shown is true?

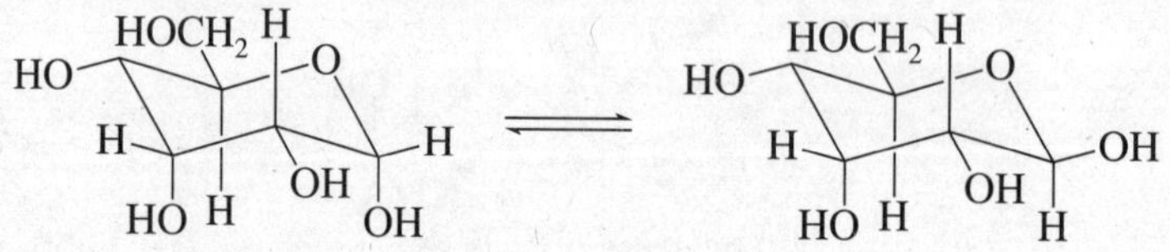

(*a*) The two structures are enantiomers of each other. They have equal but opposite optical rotations and racemize slowly at room temperature.
(*b*) The two structures are enantiomers of each other. They racemize too rapidly at room temperature for their optical rotations to be measured.
(*c*) The two structures are diastereomers of each other. Their interconversion is called mutarotation.
(*d*) The two structures are diastereomers of each other. Their interconversion does not require breaking and making bonds, only a change in conformation.
(*e*) The two structures are diastereomers of each other. One is a furanose form, the other a pyranose form.

**B-10.** The configurations of the stereogenic centers in D-threose (shown) are

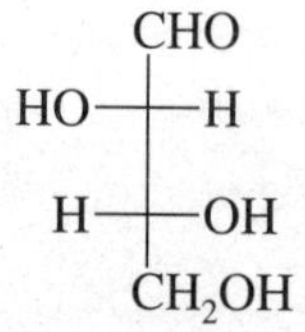

(*a*) 2*R*,3*R* (*b*) 2*R*,3*S* (*c*) 2*S*,3*R* (*d*) 2*S*,3*S*

# CHAPTER 26

## LIPIDS

### SOLUTIONS TO TEXT PROBLEMS

**26.1** The triacylglycerol shown in text Figure 26.2*a*, with an oleyl group at C-2 of the glycerol unit and two stearyl groups at C-1 and C-3, yields stearic and oleic acids in a 2 : 1 molar ratio on hydrolysis. A constitutionally isomeric structure in which the oleyl group is attached to C-1 of glycerol would yield the same hydrolysis products.

$$CH_3(CH_2)_7CH{=}CH(CH_2)_7\overset{O}{\overset{\|}{C}}OCH(CH_2O\overset{O}{\overset{\|}{C}}(CH_2)_{16}CH_3)_2 \quad \text{or} \quad CH_3(CH_2)_{16}\overset{O}{\overset{\|}{C}}OCH(CH_2O\overset{O}{\overset{\|}{C}}(CH_2)_7CH{=}CH(CH_2)_7CH_3)(CH_2O\overset{O}{\overset{\|}{C}}(CH_2)_{16}CH_3)$$

$$\xrightarrow{3H_2O}$$

$$\underset{\text{Oleic acid}}{CH_3(CH_2)_7CH{=}CH(CH_2)_7\overset{O}{\overset{\|}{C}}OH} + \underset{\text{Glycerol}}{HOCH(CH_2OH)_2} + \underset{\text{Stearic acid}}{2HO\overset{O}{\overset{\|}{C}}(CH_2)_{16}CH_3}$$

**26.2** The sulfur of acyl carrier protein acts as a nucleophile and attacks the acetyl group of acetyl coenzyme A.

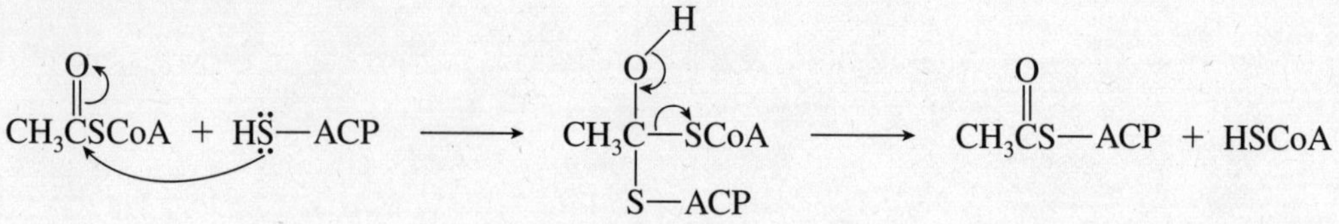

$$\underset{\text{Acetyl coenzyme A}}{CH_3\overset{O}{\overset{\|}{C}}SCoA} + \underset{\text{Acyl carrier protein}}{H\ddot{S}{-}ACP} \longrightarrow \underset{\text{Tetrahedral intermediate}}{CH_3\underset{S-ACP}{\underset{|}{\overset{OH}{\overset{|}{C}}}}{-}SCoA} \longrightarrow \underset{S\text{-Acetyl acyl carrier protein}}{CH_3\overset{O}{\overset{\|}{C}}S{-}ACP} + \underset{\text{Coenzyme A}}{HSCoA}$$

**26.3** Conversion of acyl carrier protein–bound tetradecanoate to hexadecanoate proceeds through the series of intermediates shown.

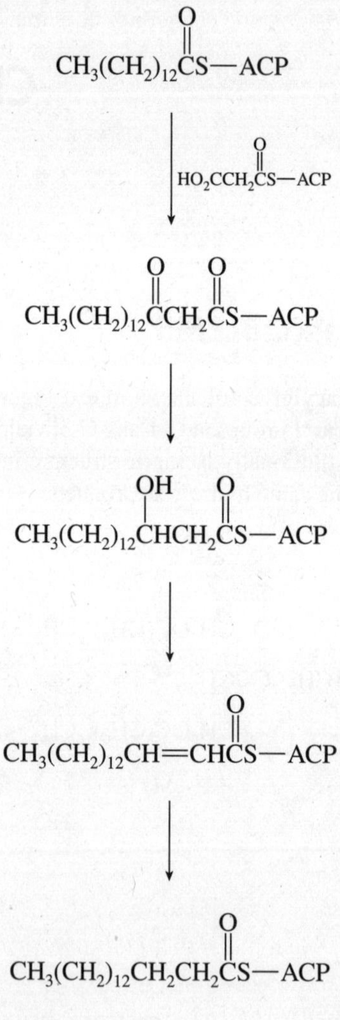

$$CH_3(CH_2)_{12}\overset{O}{\overset{\|}{C}}S{-}ACP$$

$$\Big\downarrow\ HO_2CCH_2\overset{O}{\overset{\|}{C}}S{-}ACP$$

$$CH_3(CH_2)_{12}\overset{O}{\overset{\|}{C}}CH_2\overset{O}{\overset{\|}{C}}S{-}ACP$$

$$\Big\downarrow$$

$$CH_3(CH_2)_{12}\overset{OH}{\overset{|}{C}}HCH_2\overset{O}{\overset{\|}{C}}S{-}ACP$$

$$\Big\downarrow$$

$$CH_3(CH_2)_{12}CH{=}CH\overset{O}{\overset{\|}{C}}S{-}ACP$$

$$\Big\downarrow$$

$$CH_3(CH_2)_{12}CH_2CH_2\overset{O}{\overset{\|}{C}}S{-}ACP$$

**26.4** The structure of L-glycerol 3-phosphate is shown in a Fischer projection. Translate the Fischer projection to a three-dimensional representation.

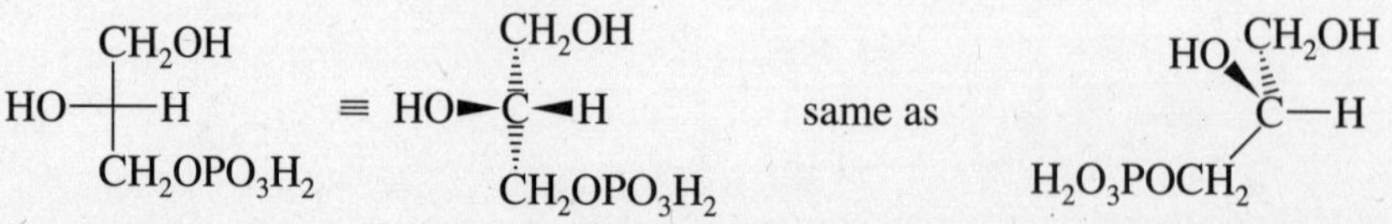

$$HO\overset{CH_2OH}{\underset{CH_2OPO_3H_2}{+}}H \equiv HO\blacktriangleright\overset{CH_2OH}{\underset{CH_2OPO_3H_2}{C}}\blacktriangleleft H \quad \text{same as} \quad \overset{HO\ \ CH_2OH}{C{-}H}\ (H_2O_3POCH_2)$$

The order of decreasing sequence rule precedence is

$$HO— > H_2O_3POCH_2— > HOCH_2— > H—$$

When the three-dimensional formula is viewed from a perspective in which the lowest ranked substituent is away from us, we see

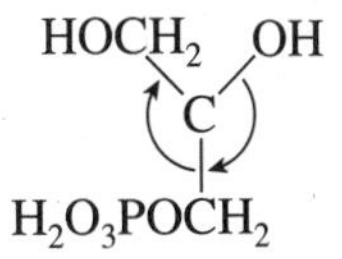

Order of decreasing rank is clockwise, therefore *R*.

The absolute configuration is *R*.

The conversion of L-glycerol 3-phosphate to a phosphatidic acid does not affect any of the bonds to the stereogenic center, nor does it alter the sequence rule ranking of the substituents.

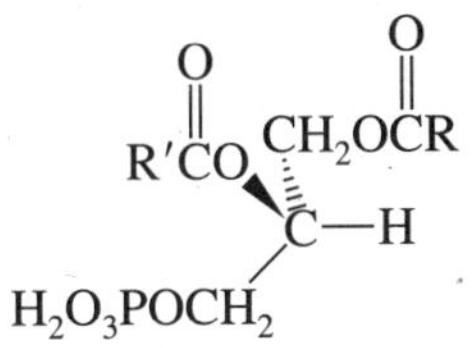

$$R'\overset{O}{\overset{\|}{C}}O— > H_2O_3POCH_2— > R\overset{O}{\overset{\|}{C}}OCH_2— > H—$$

The absolute configuration is *R*.

**26.5** Cetyl palmitate (hexadecyl hexadecanoate) is an ester in which both the acyl group and the alkyl group contain 16 carbon atoms.

$$CH_3(CH_2)_{14}\overset{O}{\overset{\|}{C}}O(CH_2)_{15}CH_3$$

Hexadecyl hexadecanoate

**26.6** The structure of $PGE_1$ is found in text Figure 26.5.

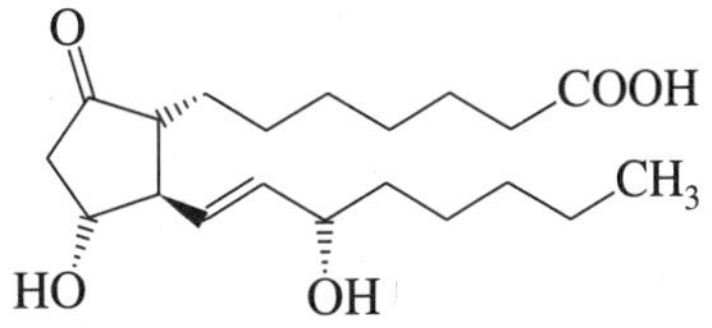

$PGE_1$

The problem states that $PGE_2$ has one more double bond than $PGE_1$ and that it is biosynthesized from arachidonic acid. Arachidonic acid (text Table 26.1) has a double bond at C-5, and thus $PGE_2$ has the structure shown.

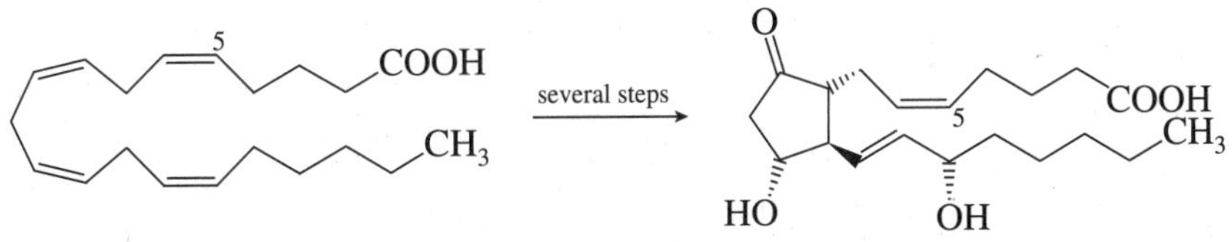

Arachidonic acid —several steps→ $PGE_2$

**26.7** Isoprene units are [isoprene skeleton] fragments in the carbon skeleton. Functional groups and multiple bonds are ignored when structures are examined for the presence of isoprene units.

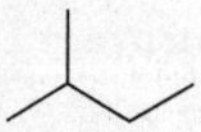

**α-Phellandrene** (two equally correct answers):

or

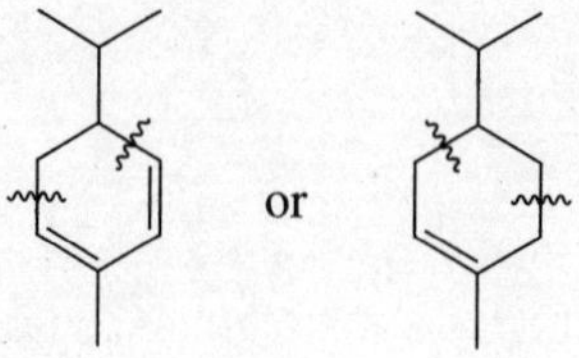

**Menthol** (same carbon skeleton as α-phellandrene but different functionality):

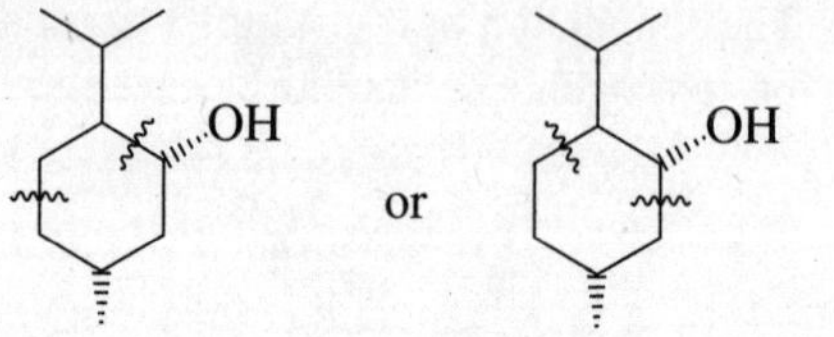

OH or OH

**Citral:**

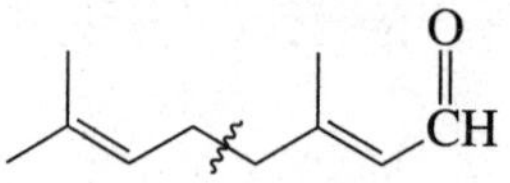

O ‖ CH

**α-Selinene** is shown in text Section 26.7.

**Farnesol:**

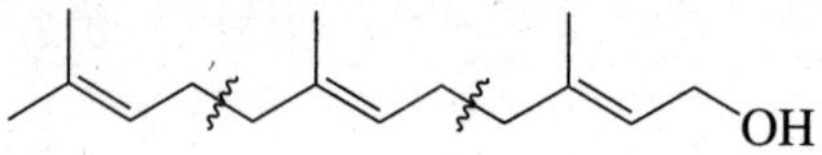

OH

**Abscisic acid:**

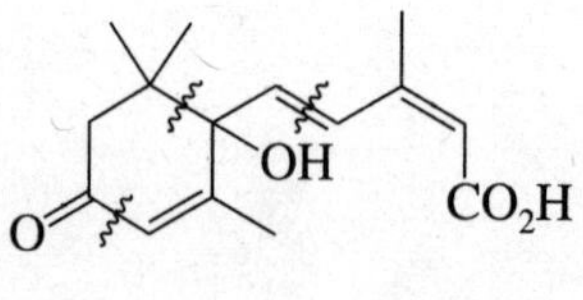

O OH $CO_2H$

**Cembrene** (two equally correct answers):

or

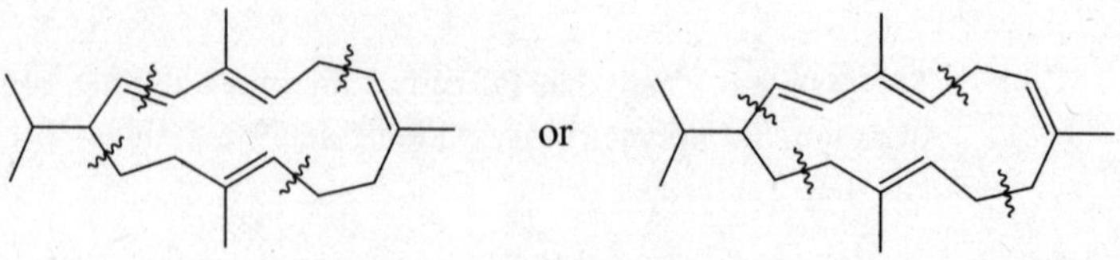

**Vitamin A:**

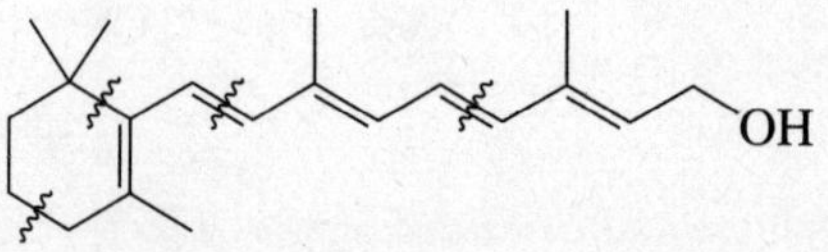

OH

**26.8** $\beta$-Carotene is a tetraterpene because it has 40 carbon atoms. The tail-to-tail linkage is at the midpoint of the molecule and connects two 20-carbon fragments.

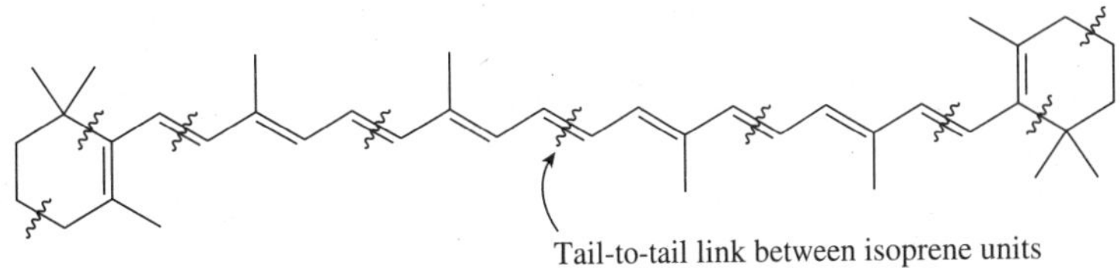

Tail-to-tail link between isoprene units

**26.9** Isopentenyl pyrophosphate acts as an alkylating agent toward farnesyl pyrophosphate. Alkylation is followed by loss of a proton from the carbocation intermediate, giving geranylgeranyl pyrophosphate. Hydrolysis of the pyrophosphate yields geranylgeraniol.

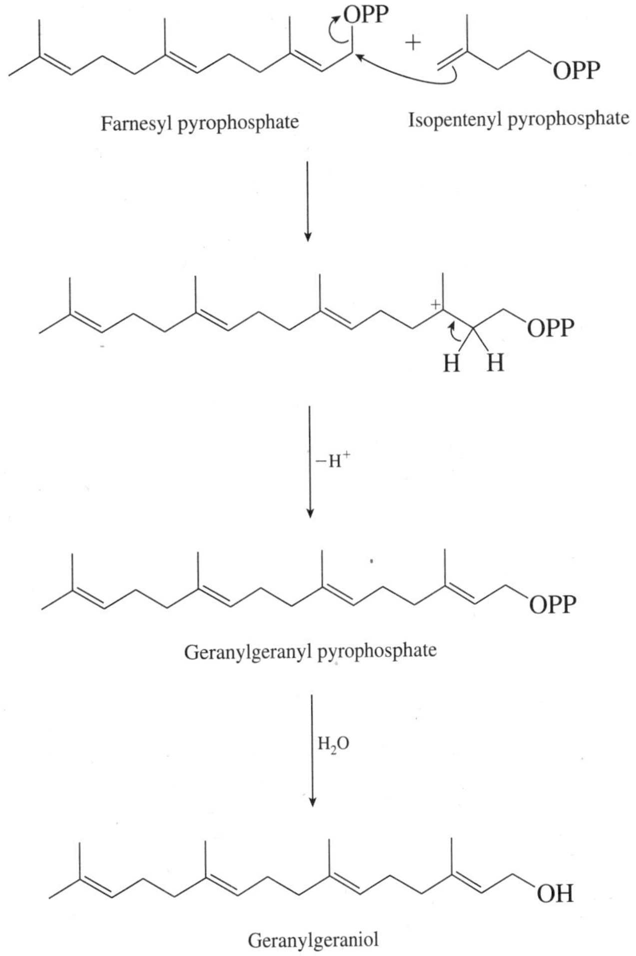

OPP + OPP

Farnesyl pyrophosphate    Isopentenyl pyrophosphate

OPP

H H

$-H^+$

OPP

Geranylgeranyl pyrophosphate

$H_2O$

OH

Geranylgeraniol

**26.10** Borneol, the structure of which is given in text Figure 26.7, is a secondary alcohol. Oxidation of borneol converts it to the ketone camphor.

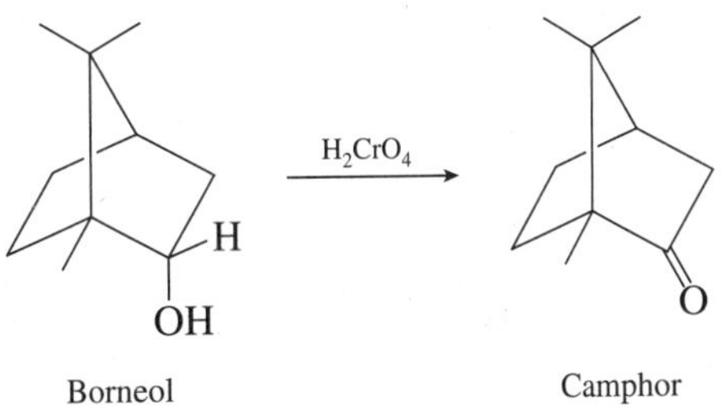

H OH  $\xrightarrow{H_2CrO_4}$  O

Borneol    Camphor

Reduction of camphor with sodium borohydride gives a mixture of stereoisomeric alcohols, of which one is borneol and the other isoborneol.

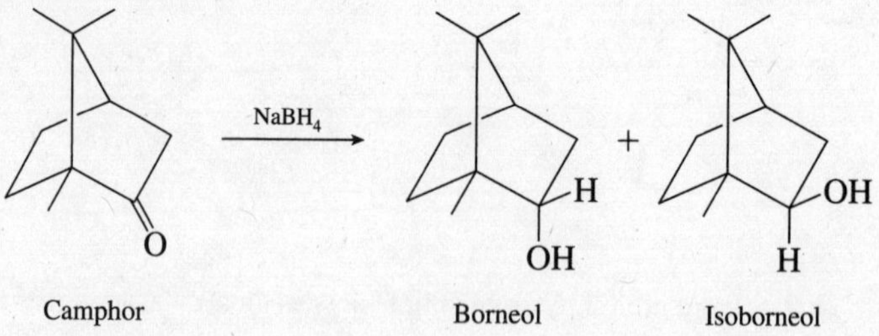

$NaBH_4$

Camphor Borneol Isoborneol

**26.11** Figure 26.8 in the text describes the distribution of $^{14}C$ (denoted by *) in citronellal biosynthesized from acetate enriched with $^{14}C$ in its methyl group.

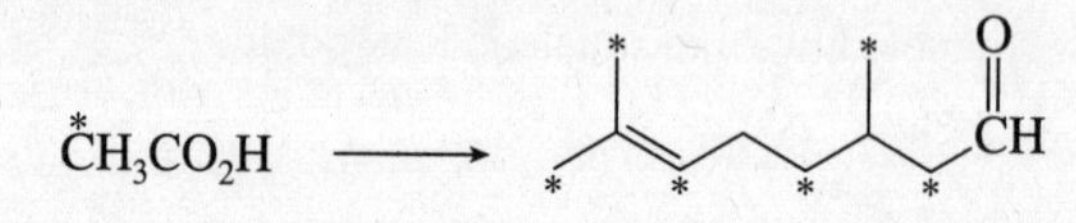

$\overset{*}{C}H_3CO_2H \longrightarrow$

If, instead, acetate enriched with $^{14}C$ at its carbonyl carbon were used, exactly the opposite distribution of the $^{14}C$ label would be observed.

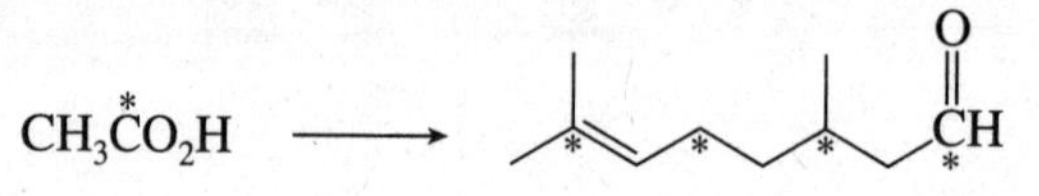

$CH_3\overset{*}{C}O_2H \longrightarrow$

When $^{14}CH_3CO_2H$ is used, C-2, C-4, C-6, C-8, and both methyl groups of citronellal are labeled. When $CH_3{}^{14}CO_2H$ is used, C-1, C-3, C-5, and C-7 are labeled.

**26.12** (*b*) The hydrogens that migrate in step 3 are those at C-13 and C-17 (steroid numbering).

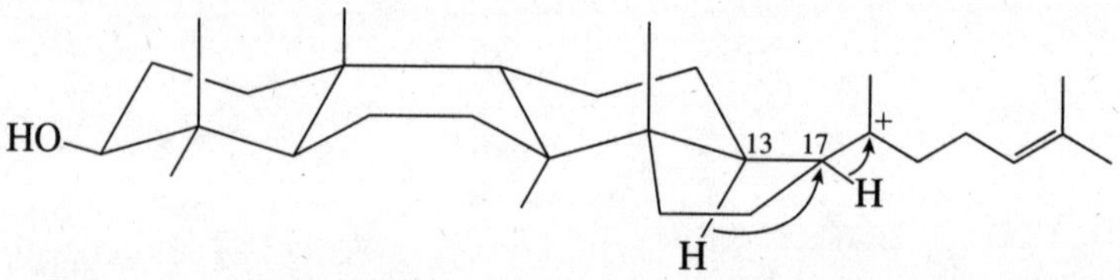

As shown in the coiled form of squalene 2,3-epoxide, these correspond to hydrogens at C-14 and C-18 (systematic IUPAC numbering).

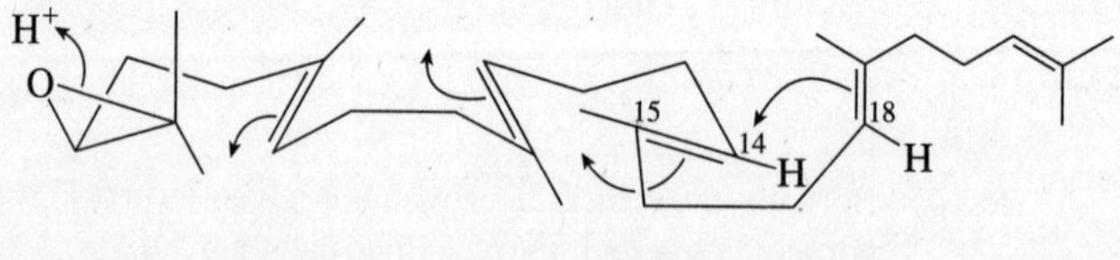

(*c*) The carbon atoms that form the C, D ring junction in cholesterol are C-14 and C-15 of squalene 2,3-epoxide. It is the methyl group at C-15 of squalene 2,3-epoxide that becomes the methyl group at this junction in cholesterol.

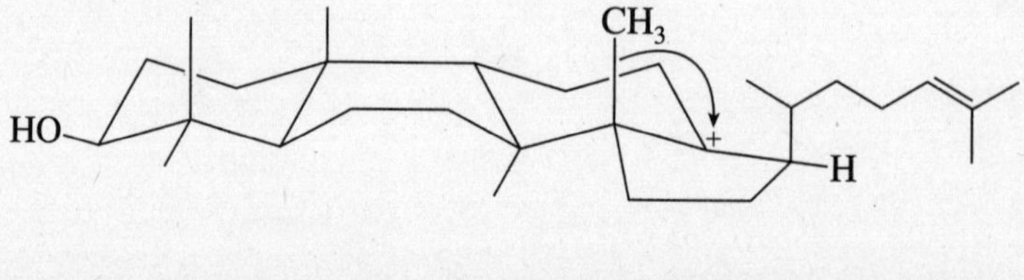

(*d*) The methyl groups that are lost are the methyl substituents at C-2 and C-10 plus the methyl group that is C-1 of squalene 2,3-epoxide.

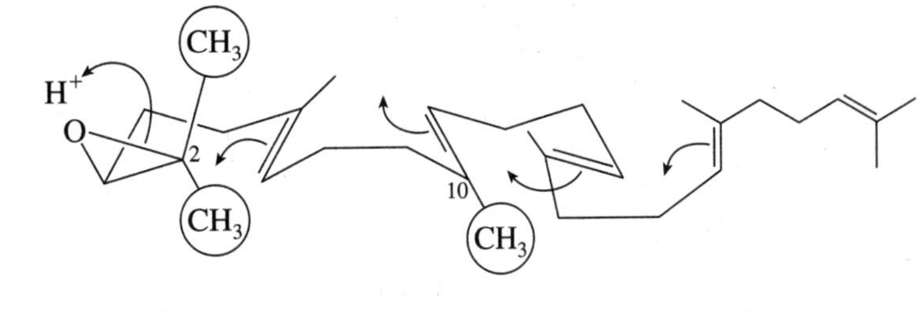

**26.13** Tracking the $^{14}C$ label of $^{14}CH_3CO_2H$ through the complete biosynthesis of cholesterol requires a systematic approach. First, by analogy with Problem 26.11, we can determine the distribution of $^{14}C$ (denoted by *) in squalene 2,3-epoxide.

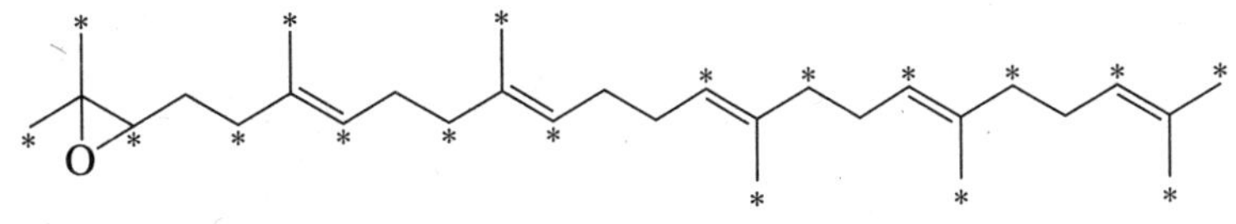

Next, follow the path of the $^{14}C$-enriched carbons in the cyclization of squalene 2,3-epoxide to lanosterol.

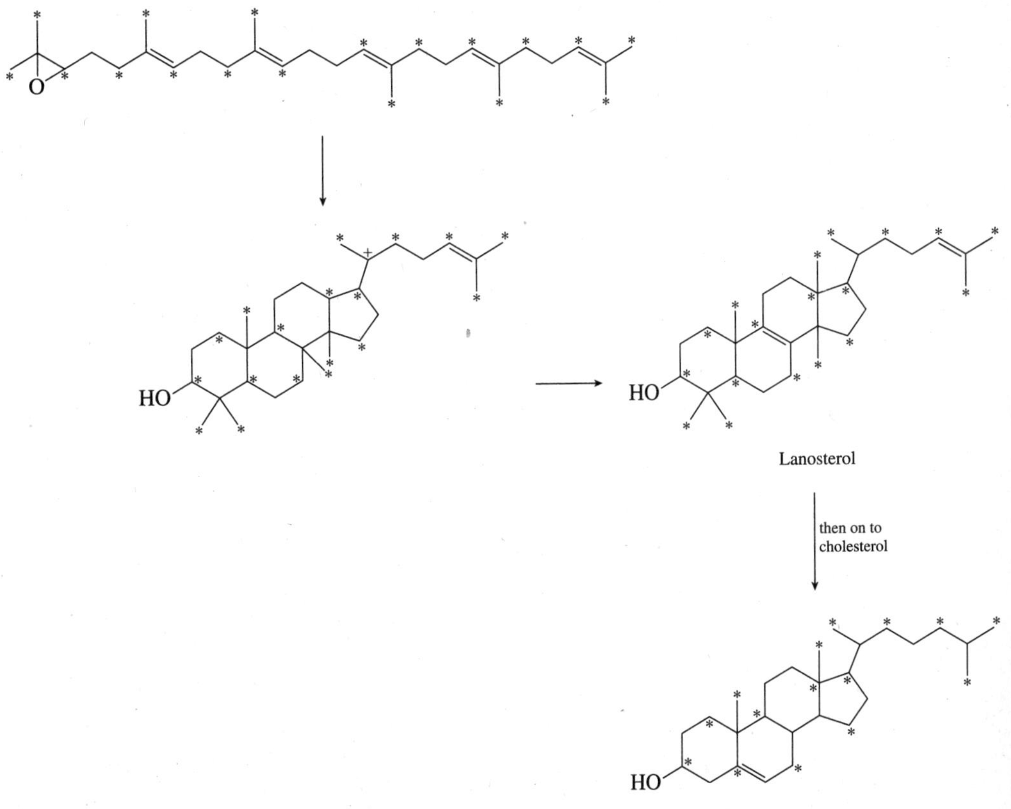

**26.14** By analogy to the reaction in which 7-dehydrocholesterol is converted to vitamin $D_3$, the structure of vitamin $D_2$ can be deduced from that of ergosterol.

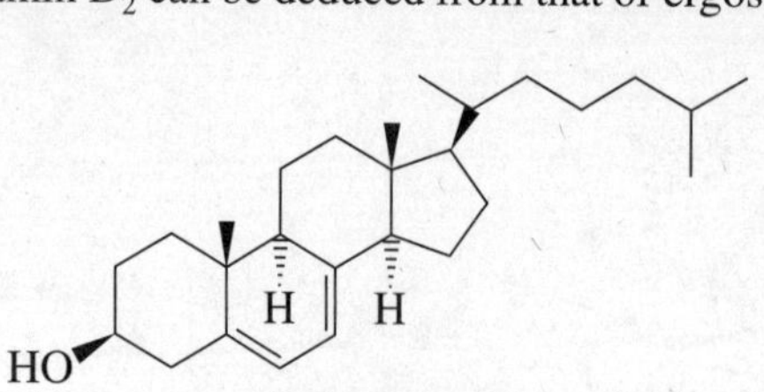

7-Dehydrocholesterol

Ergosterol

light

light

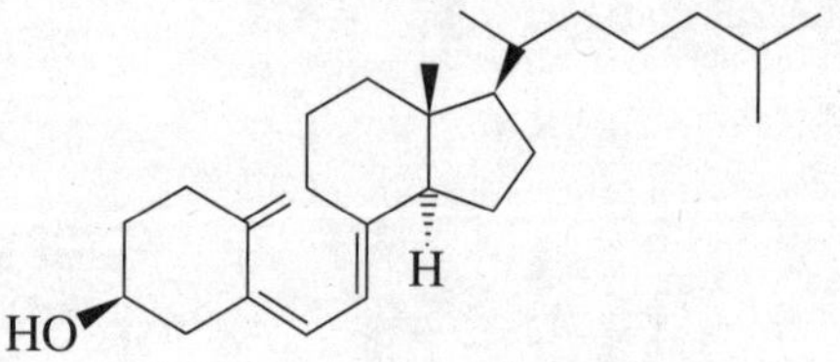

Vitamin $D_3$

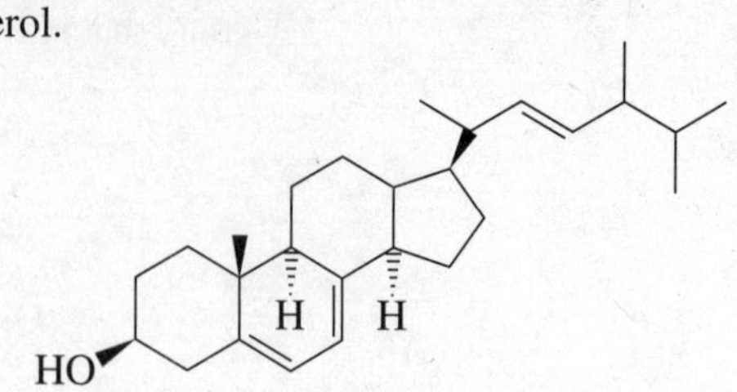

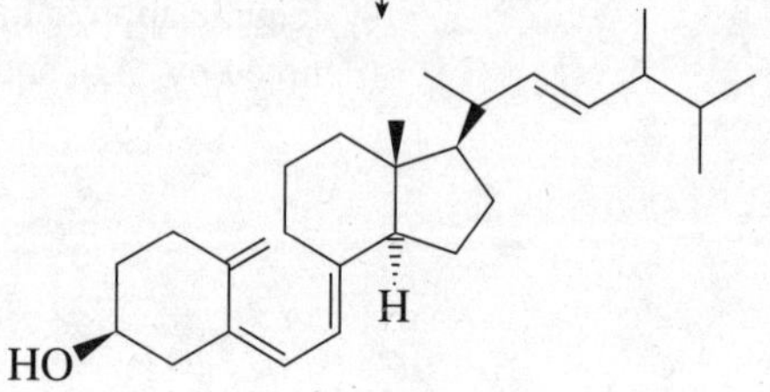

Vitamin $D_2$

**26.15** (*a*) Fatty acid biosynthesis proceeds by the joining of acetate units.

$CH_3\overset{O}{\overset{\|}{C}}SCoA$ — Acetyl coenzyme A

$CH_3\overset{O}{\overset{\|}{C}}CH_2\overset{O}{\overset{\|}{C}}SCoA$ — Acetoacetyl coenzyme A

$CH_3(CH_2)_{14}\overset{O}{\overset{\|}{C}}SCoA$ — Palmitoyl coenzyme A

Thus, the even-numbered carbons will be labeled with $^{14}C$ when palmitic acid is biosynthesized from $^{14}CH_3CO_2H$.

$$\overset{*}{C}H_3CH_2\overset{*}{C}H_2CH_2\overset{*}{C}H_2CH_2\overset{*}{C}H_2CH_2\overset{*}{C}H_2CH_2\overset{*}{C}H_2CH_2\overset{*}{C}H_2CH_2\overset{*}{C}H_2CO_2H$$

(*b*) As noted in Problem 26.6, arachidonic acid (Table 26.1) is the biosynthetic precursor of $PGE_2$. The distribution of the $^{14}C$ label in $PGE_2$ biosynthesized from $^{14}CH_3CO_2H$ reflects the fatty acid origin of the prostaglandins.

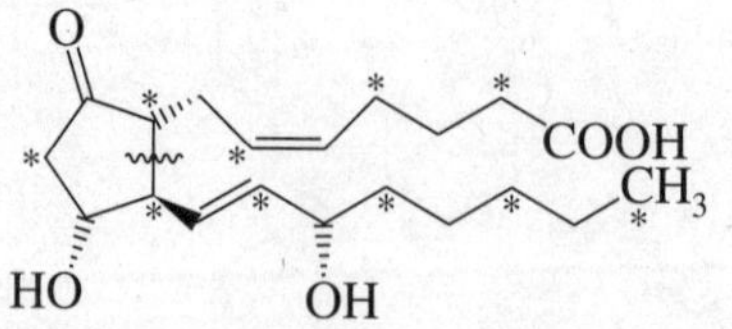

O, COOH, $CH_3$, HO, OH

$PGE_2$

(*c*) Limonene is a monoterpene, biosynthesized from acetate by way of mevalonate and isopentenyl pyrophosphate.

$^{14}CH_3CO_2H$ ⟶ OPP ⟶

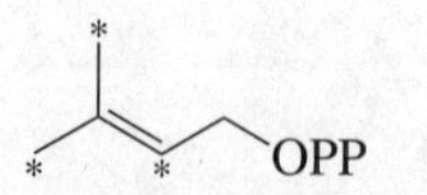

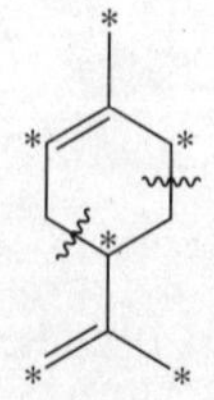

Acetic acid

Isopentenyl pyrophosphate

Limonene

(*d*) The distribution of the $^{14}C$ label in $\beta$-carotene becomes evident once its isoprene units are identified.

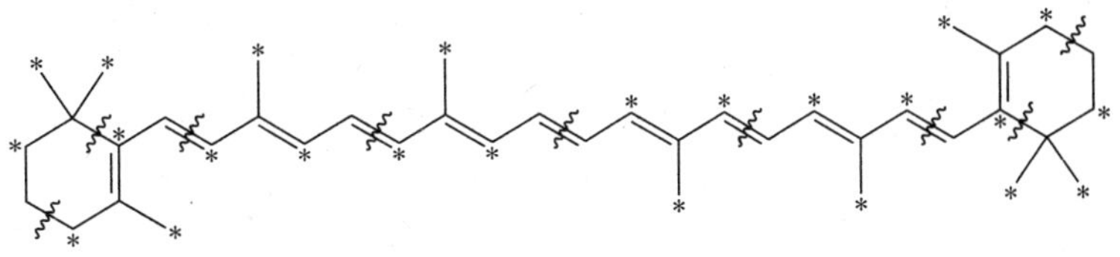

$\beta$-Carotene

**26.16** The carbon chain of prostacyclin is derived from acetate by way of a $C_{20}$ fatty acid. Trace a continuous chain of 20 carbons beginning with the carboxyl group. Even-numbered carbons are labeled with $^{14}C$ when prostacyclin is biosynthesized from $^{14}CH_3CO_2H$.

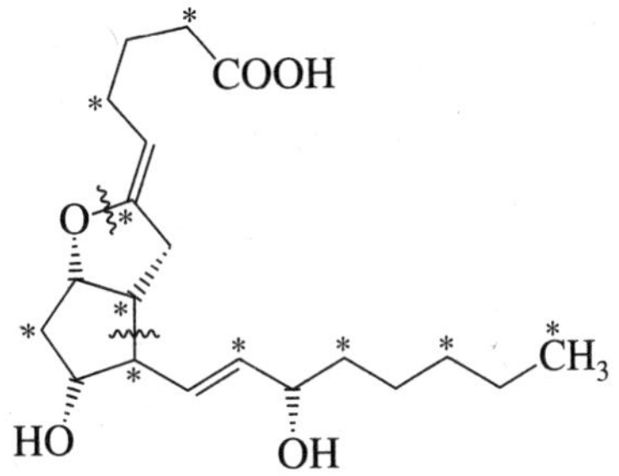

Prostacyclin

**26.17** The isoprene units in the designated compounds are shown by disconnections in the structural formulas.

(*a*) Ascaridole:

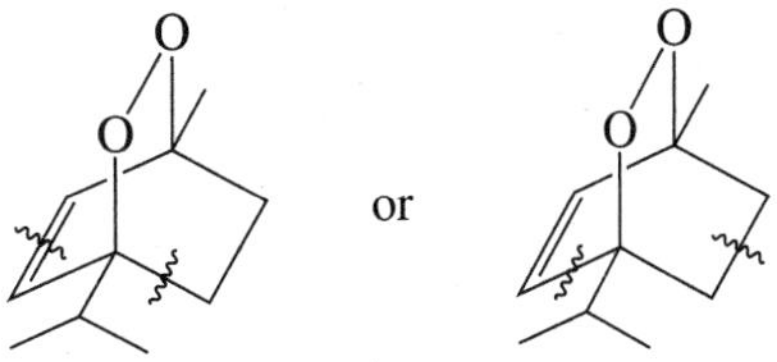

or

(*b*) Dendrolasin:

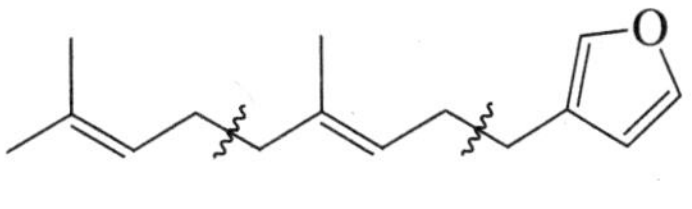

(*c*) $\gamma$-Bisabolene

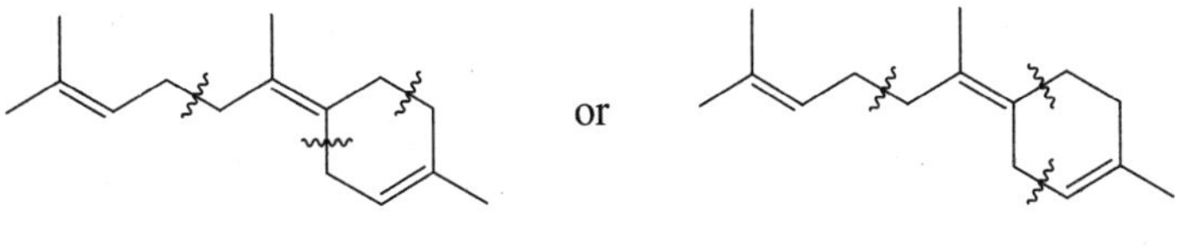

or

(*d*) $\alpha$-Santonin

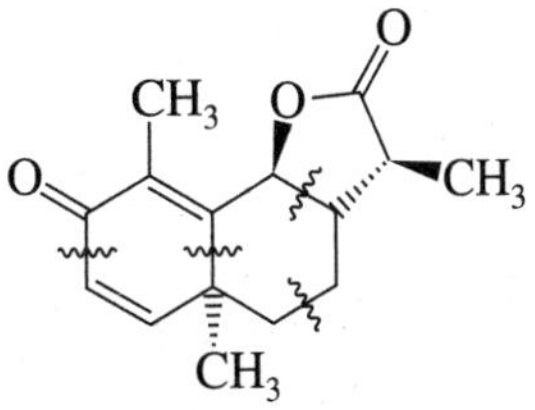

(*e*) Tetrahymanol

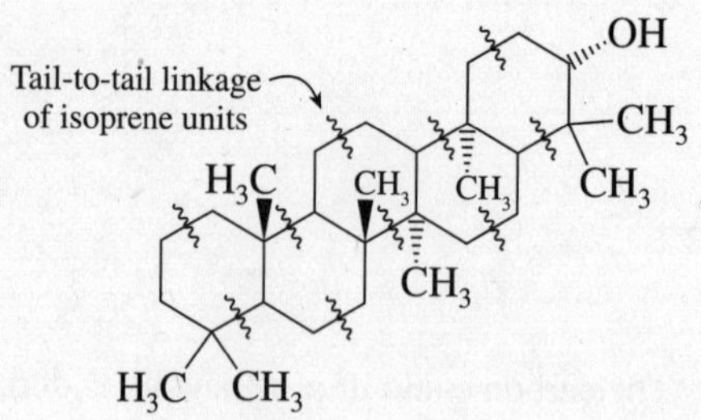

Tail-to-tail linkage of isoprene units

**26.18** Of the four isoprene units of cubitene, three of them are joined in the usual head-to-tail fashion, but the fourth one is joined in an irregular way.

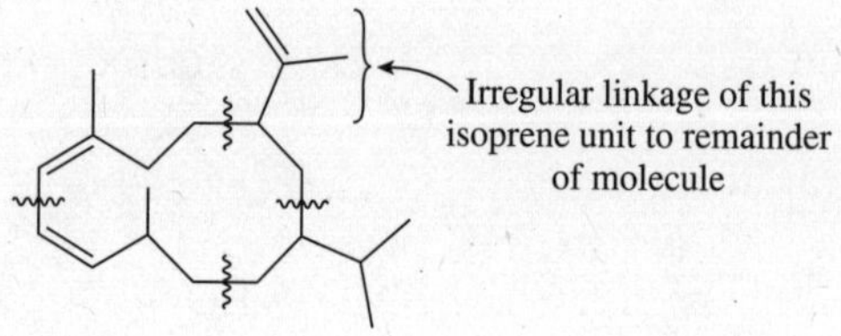

Irregular linkage of this isoprene unit to remainder of molecule

**26.19** (*a*) Cinerin I is an ester, the acyl portion of which is composed of two isoprene units, as follows:

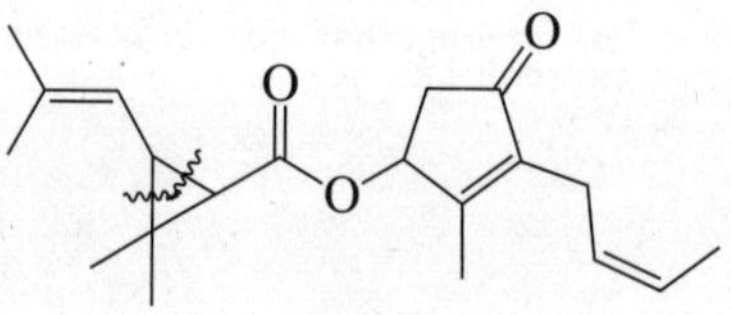

Cinerin I

(*b*) Hydrolysis of cinerin I involves cleavage of the ester unit.

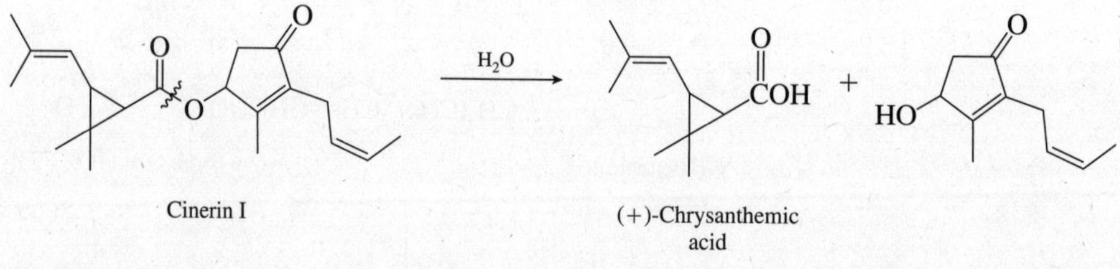

Cinerin I $\xrightarrow{H_2O}$ (+)-Chrysanthemic acid + HO–(cyclopentenone)

Chrysanthemic acid has the constitution shown in the equation. Its stereochemistry is revealed by subsequent experiments.

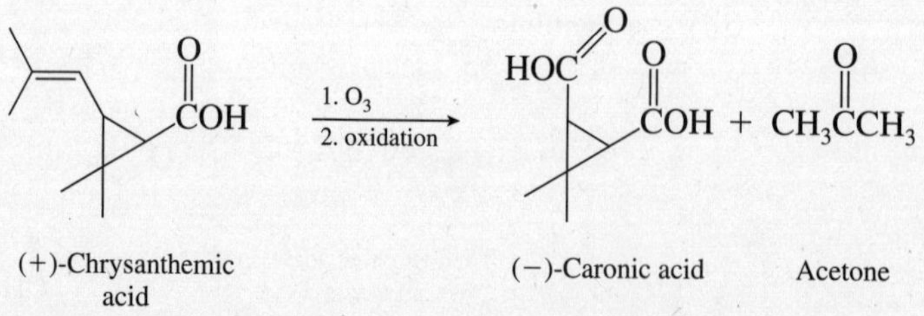

(+)-Chrysanthemic acid $\xrightarrow[\text{2. oxidation}]{\text{1. } O_3}$ (−)-Caronic acid + $CH_3CCH_3$ (Acetone)

Because caronic acid is optically active, its carboxyl groups must be trans to each other. (The cis stereoisomer is an optically inactive meso form.) The structure of (+)-chrysanthemic acid must therefore be either the following or its mirror image.

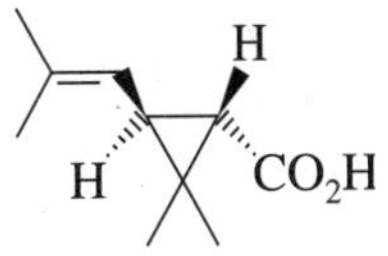

H H $CO_2H$

The carboxyl group and the 2-methyl-1-propenyl side chain must be trans to each other.

**26.20** (*a*) Hydrolysis of phrenosine cleaves the glycosidic bond. The carbohydrate liberated by this hydrolysis is D-galactose.

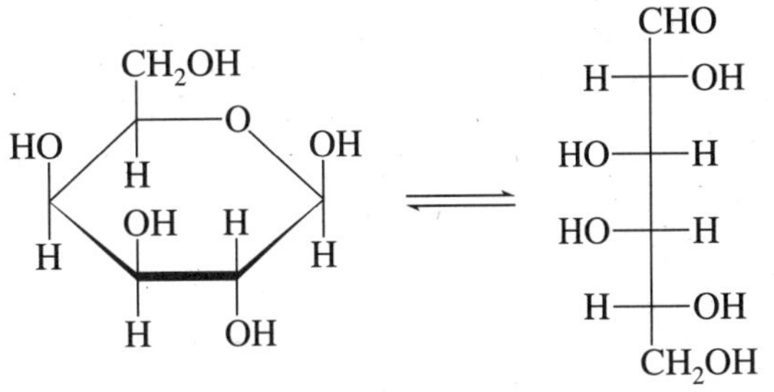

$CH_2OH$ O HO OH H H OH H H H H OH ⇌ CHO, H—OH, HO—H, HO—H, H—OH, $CH_2OH$

Phrenosine is a $\beta$-glycoside of D-galactose.

(*b*) The species that remains on cleavage of the galactose unit has the structure

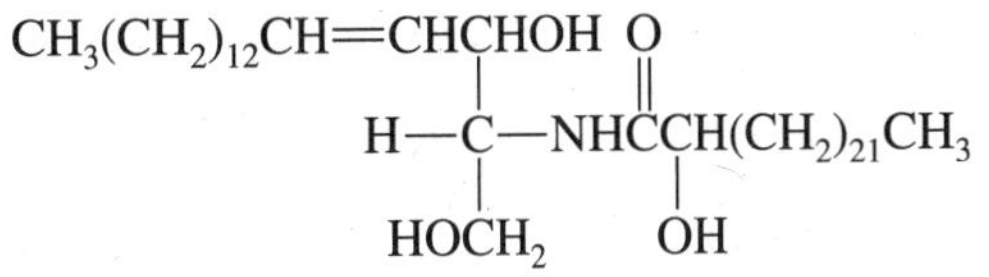

$$CH_3(CH_2)_{12}CH{=}CHCHOH\text{—}\underset{HOCH_2}{\overset{}{C}}H\text{—}NH\overset{O}{\overset{\|}{C}}\underset{OH}{CH}(CH_2)_{21}CH_3$$

The two substances, sphingosine and cerebronic acid, that are formed along with D-galactose arise by hydrolysis of the amide bond.

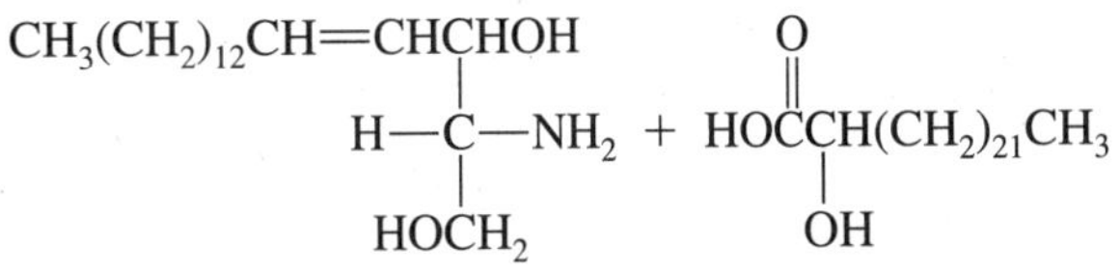

$$CH_3(CH_2)_{12}CH{=}CHCHOH\text{—}\underset{HOCH_2}{C}H\text{—}NH_2 + HO\overset{O}{\overset{\|}{C}}\underset{OH}{CH}(CH_2)_{21}CH_3$$

Sphingosine    Cerebronic acid

**26.21** (*a*) Catalytic hydrogenation over Lindlar palladium converts alkynes to cis alkenes.

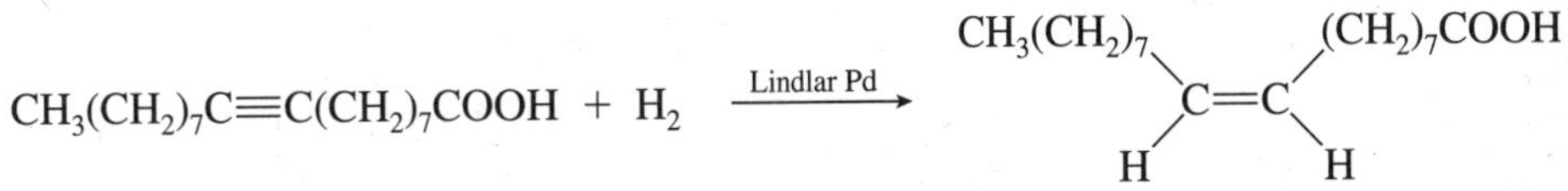

$$CH_3(CH_2)_7C{\equiv}C(CH_2)_7COOH + H_2 \xrightarrow{\text{Lindlar Pd}} \underset{H}{\overset{CH_3(CH_2)_7}{}}C{=}C\underset{H}{\overset{(CH_2)_7COOH}{}}$$

9-Octadecynoic acid (stearolic acid)    (Z)-9-Octadecenoic acid (74%) (oleic acid)

(*b*) Carbon–carbon triple bonds are converted to trans alkenes by reduction with lithium and ammonia.

$$CH_3(CH_2)_7C{\equiv}C(CH_2)_7COOH \xrightarrow[2.\ H^+]{1.\ Li,\ NH_3} \underset{H}{\overset{CH_3(CH_2)_7}{}}\!\!\!>C{=}C<\!\!\!\underset{(CH_2)_7COOH}{\overset{H}{}}$$

9-Octadecynoic acid (stearolic acid) → (*E*)-9-Octadecenoic acid (97%) (elaidic acid)

(*c*) The carbon–carbon double bond is hydrogenated readily over a platinum catalyst. Reduction of the ester function does not occur.

$$(Z)\text{-}CH_3(CH_2)_7CH{=}CH(CH_2)_7\overset{O}{\overset{\|}{C}}OCH_2CH_3 \xrightarrow{H_2,\ Pt} CH_3(CH_2)_{16}\overset{O}{\overset{\|}{C}}OCH_2CH_3$$

Ethyl (*Z*)-9-octadecenoate (ethyl oleate) → Ethyl octadecanoate (91%) (ethyl stearate)

(*d*) Lithium aluminum hydride reduces the ester function but leaves the carbon–carbon double bond intact.

$$(Z)\text{-}CH_3(CH_2)_5\underset{OH}{\underset{|}{C}}HCH_2CH{=}CH(CH_2)_7\overset{O}{\overset{\|}{C}}OCH_3 \xrightarrow[2.\ H_2O]{1.\ LiAlH_4} (Z)\text{-}CH_3(CH_2)_5\underset{OH}{\underset{|}{C}}HCH_2CH{=}CH(CH_2)_7CH_2OH + CH_3OH$$

Methyl (*Z*)-12-hydroxy-9-octadecenoate (methyl ricinoleate) → (*Z*)-9-Octadecen-1,12-diol (52%) + Methanol

(*e*) Epoxidation of the double bond occurs when an alkene is treated with a peroxy acid. The reaction is stereospecific; substituents that are cis to each other in the alkene remain cis in the epoxide.

$$(Z)\text{-}CH_3(CH_2)_7CH{=}CH(CH_2)_7COOH + C_6H_5CO_2OH \longrightarrow \underset{\text{H}\quad \text{O}\quad \text{H}}{\overset{CH_3(CH_2)_7\qquad (CH_2)_7COOH}{C{-}C}} + C_6H_5\overset{O}{\overset{\|}{C}}OH$$

Oleic acid + Peroxybenzoic acid → *cis*-9,10-Epoxyoctadecanoic acid (62–67%) + Benzoic acid

(*f*) Acid-catalyzed hydrolysis of the epoxide yields a diol; its stereochemistry corresponds to net anti hydroxylation of the double bond of the original alkene.

$$\underset{\text{H}\quad \text{O}\quad \text{H}}{\overset{CH_3(CH_2)_7\qquad (CH_2)_7COOH}{C{-}C}} \xrightarrow{H_3O^+} \underset{\text{HO}\qquad (CH_2)_7COOH}{\overset{CH_3(CH_2)_7\quad\ \ OH}{H\cdots C{-}C\cdots H}}$$

*cis*-9,10-Epoxyoctadecanoic acid → 9,10-Dihydroxyoctadecanoic acid

The product is chiral but is formed as a racemic mixture containing equal amounts of the 9*R*,10*R* and 9*S*,10*S* stereoisomers when the starting epoxide is racemic.

(*g*) Hydroxylation of carbon–carbon double bonds with osmium tetraoxide proceeds with syn addition of hydroxyl groups.

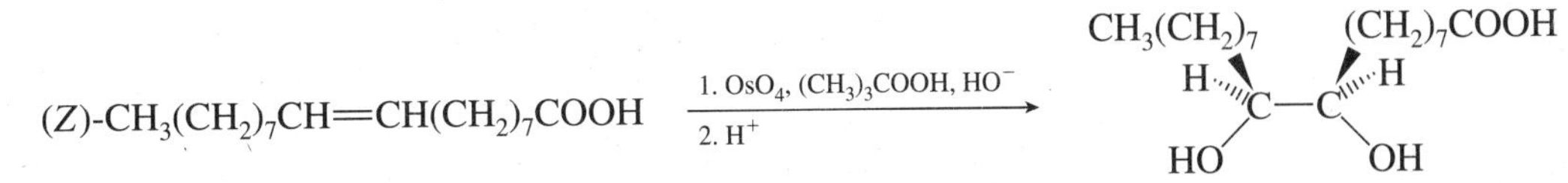

$$\text{(Z)-}CH_3(CH_2)_7CH{=}CH(CH_2)_7COOH \xrightarrow[2.\ H^+]{1.\ OsO_4,\ (CH_3)_3COOH,\ HO^-}$$

$CH_3(CH_2)_7$ $(CH_2)_7COOH$

H C—C H

HO OH

9,10-Dihydroxyoctadecanoic acid (70%)

The product is chiral but is formed as a racemic mixture containing equal amounts of the 9*R*,10*S* and 9*S*,10*R* stereoisomers.

(*h*) Hydroboration–oxidation gives syn hydration of carbon–carbon double bonds with a regioselectivity contrary to Markovnikov's rule. The reagent attacks the less hindered face of the double bond of α-pinene.

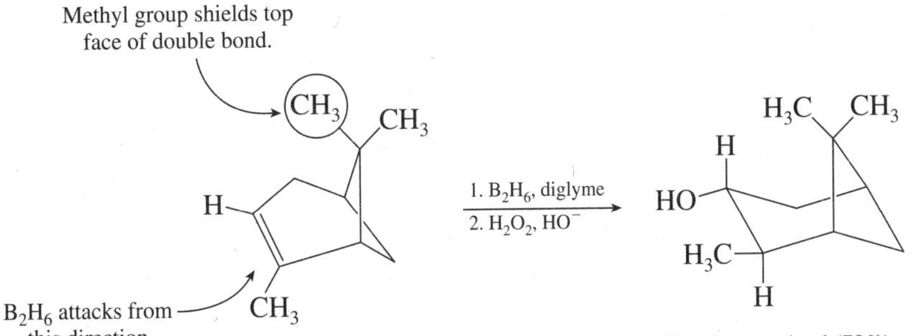

Methyl group shields top face of double bond.

$CH_3$ $CH_3$

H

$B_2H_6$ attacks from this direction.

$CH_3$

$$\xrightarrow[2.\ H_2O_2,\ HO^-]{1.\ B_2H_6,\ \text{diglyme}}$$

$H_3C$ $CH_3$

H

HO

$H_3C$

H

Isopinocampheol (79%)

(*i*) The starting alkene in this case is β-pinene. As in the preceding exercise with α-pinene, diborane adds to the bottom face of the double bond.

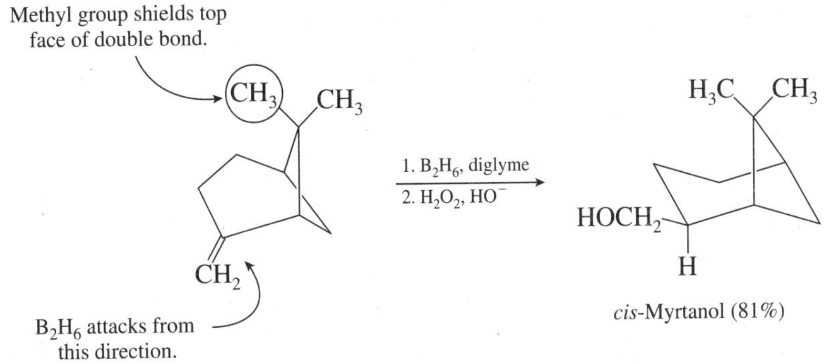

Methyl group shields top face of double bond.

$CH_3$ $CH_3$

$CH_2$

$B_2H_6$ attacks from this direction.

$$\xrightarrow[2.\ H_2O_2,\ HO^-]{1.\ B_2H_6,\ \text{diglyme}}$$

$H_3C$ $CH_3$

$HOCH_2$

H

*cis*-Myrtanol (81%)

(*j*) The starting material is an acetal. It undergoes hydrolysis in dilute aqueous acid to give a ketone.

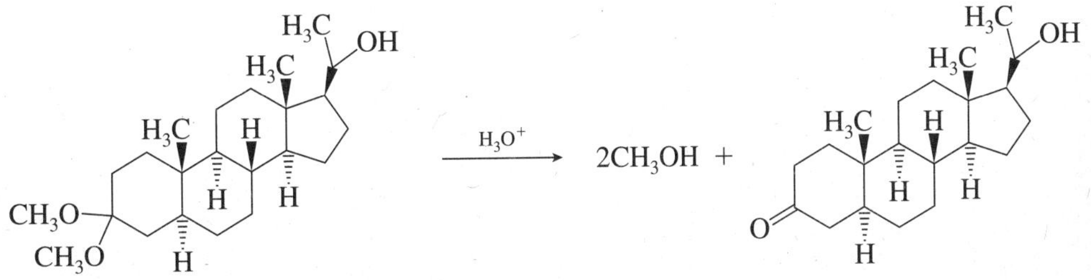

$H_3C$ OH

$H_3C$

$H_3C$ H

H H

$CH_3O$

$CH_3O$ H

$$\xrightarrow{H_3O^+} 2CH_3OH +$$

$H_3C$ OH

$H_3C$

$H_3C$ H

H H

O

H

(95% yield)

**26.22** (*a*) There are no direct methods for the reduction of a carboxylic acid to an alkane. A number of indirect methods that may be used, however, involve first converting the carboxylic acid to an alkyl bromide via the corresponding alcohol.

$$CH_3(CH_2)_{16}CO_2H \xrightarrow[2.\ H_2O]{1.\ LiAlH_4} CH_3(CH_2)_{16}CH_2OH \xrightarrow[\text{or } PBr_3]{\text{HBr, heat}} CH_3(CH_2)_{16}CH_2Br$$

Octadecanoic acid — 1-Octadecanol — 1-Bromooctadecane

Once the alkyl bromide is in hand, it may be converted to an alkane by conversion to a Grignard reagent followed by addition of water.

$$CH_3(CH_2)_{16}CH_2Br \xrightarrow[\text{diethyl ether}]{Mg} CH_3(CH_2)_{16}CH_2MgBr \xrightarrow{H_2O} CH_3(CH_2)_{16}CH_3$$

1-Bromooctadecane — Octadecane

Other routes are also possible. For example, E2 elimination from 1-bromooctadecane followed by hydrogenation of the resulting alkene will also yield octadecane.

(*b*) Retrosynthetic analysis reveals that the 18-carbon chain of the starting material must be attached to a benzene ring.

$$C_6H_5{-}(CH_2)_{17}CH_3 \Longrightarrow C_6H_5{-} + {-}(CH_2)_{17}CH_3$$

1-Phenyloctadecane

The desired sequence may be carried out by a Friedel–Crafts acylation, followed by Clemmensen or Wolff–Kishner reduction of the ketone.

$$CH_3(CH_2)_{16}CO_2H \xrightarrow{SOCl_2} CH_3(CH_2)_{16}\overset{O}{\overset{\|}{C}}Cl \xrightarrow[AlCl_3]{\text{benzene}} C_6H_5{-}\overset{O}{\overset{\|}{C}}(CH_2)_{16}CH_3$$

Octadecanoic acid — Octadecanoyl chloride — 1-Phenyl-1-octadecanone

$$\xrightarrow{Zn(Hg),\ HCl} C_6H_5{-}CH_2(CH_2)_{16}CH_3$$

1-Phenyloctadecane

(*c*) First examine the structure of the target molecule 3-ethylicosane.

$$CH_3(CH_2)_{16}\underset{\displaystyle CH_2CH_3}{\underset{|}{C}H}CH_2CH_3$$

Retrosynthetic analysis reveals that two ethyl groups have been attached to a $C_{18}$ unit.

$$CH_3(CH_2)_{16}\underset{\displaystyle CH_2CH_3}{\underset{|}{C}H}{-}CH_2CH_3 \Longrightarrow CH_3(CH_2)_{16}\underset{|}{C}H{-} + 2CH_3CH_2{-}$$

The necessary carbon–carbon bonds can be assembled by the reaction of an ester with two moles of a Grignard reagent.

$$\underset{\text{Ethyl octadecanoate (from octadecanoic acid and ethanol)}}{CH_3(CH_2)_{16}\overset{O}{\overset{\|}{C}}OCH_2CH_3} + \underset{\text{Ethylmagnesium bromide}}{2CH_3CH_2MgBr} \xrightarrow[\text{2. } H_3O^+]{\text{1. diethyl ether}} \underset{\text{3-Ethyl-3-icosanol}}{CH_3(CH_2)_{16}\overset{OH}{\overset{|}{\underset{CH_2CH_3}{\underset{|}{C}}}}CH_2CH_3}$$

With the correct carbon skeleton in place, all that is needed is to convert the alcohol to the alkene. This can be accomplished by dehydration and reduction.

$$\underset{\text{3-Ethyl-3-icosanol}}{CH_3(CH_2)_{16}\overset{OH}{\overset{|}{\underset{CH_2CH_3}{\underset{|}{C}}}}CH_2CH_3} \xrightarrow[\text{heat}]{H_2SO_4} \underset{\text{3-Ethyl-2-icosene}}{CH_3(CH_2)_{16}\underset{CH_2CH_3}{\underset{|}{C}}{=}CHCH_3} + \underset{\text{3-Ethyl-3-icosene}}{CH_3(CH_2)_{15}CH{=}\underset{CH_2CH_3}{\underset{|}{C}}CH_2CH_3}$$

$$\xrightarrow{H_2,\ Pt} \underset{\text{3-Ethylicosane}}{CH_3(CH_2)_{16}\underset{CH_2CH_3}{\underset{|}{C}}HCH_2CH_3}$$

(*d*) Icosanoic acid contains two more carbon atoms than octadecanoic acid.

$$\underset{\text{Icosanoic acid}}{CH_3(CH_2)_{18}CO_2H} \Longrightarrow CH_3(CH_2)_{16}CH_2Br + \ddot{\bar{C}}H_2CO_2H$$

A reasonable approach utilizes a malonic ester synthesis.

$$\underset{\text{1-Bromooctadecane [prepared as in part (a)]}}{CH_3(CH_2)_{16}CH_2Br} + \underset{\text{Diethyl malonate}}{CH_2(CO_2CH_2CH_3)_2} \xrightarrow{NaOCH_2CH_3} \underset{\text{Diethyl 2-octadecylmalonate}}{CH_3(CH_2)_{16}CH_2CH(CO_2CH_2CH_3)_2}$$

$$\xrightarrow{\text{1. } HO^-,\ H_2O;\ \text{2. } H^+;\ \text{3. heat}} \underset{\text{Icosanoic acid}}{CH_3(CH_2)_{16}CH_2CH_2CO_2H}$$

(*e*) The carbon chain must be shortened by one carbon atom in this problem. A Hofmann rearrangement (text Section 20.17) is indicated.

$$CH_3(CH_2)_{16}CO_2H \xrightarrow[2.\ NH_3]{1.\ SOCl_2} CH_3(CH_2)_{16}\overset{O}{\overset{\|}{C}}NH_2 \xrightarrow{Br_2,\ HO^-} CH_3(CH_2)_{16}NH_2$$

Octadecanoic acid — Octadecanamide — 1-Heptadecanamine

(*f*) Lithium aluminum hydride reduction of octadecanamide gives the corresponding amine.

$$CH_3(CH_2)_{16}\overset{O}{\overset{\|}{C}}NH_2 \xrightarrow[2.\ H_2O]{1.\ LiAlH_4} CH_3(CH_2)_{16}CH_2NH_2$$

Octadecanamide [from part (*e*)] — 1-Octadecanamine

(*g*) Chain extension can be achieved via cyanide displacement of bromine from 1-bromooctadecane. Reduction of the cyano group completes the synthesis.

$$CH_3(CH_2)_{16}CH_2Br \xrightarrow{KCN} CH_3(CH_2)_{16}CH_2C{\equiv}N \xrightarrow[2.\ H_2O]{1.\ LiAlH_4} CH_3(CH_2)_{16}CH_2CH_2NH_2$$

1-Bromooctadecane [from part (*a*)] — Nonadecanenitrile — 1-Nonadecanamine

**26.23** First acylate the free hydroxyl group with an acyl chloride.

$$\text{(acetonide, } CH_2OH\text{; } H_3C\ CH_3\text{)} + R\overset{O}{\overset{\|}{C}}Cl \longrightarrow \text{(acetonide, } CH_2O\overset{O}{\overset{\|}{C}}R\text{; } H_3C\ CH_3\text{)}$$

Treatment with aqueous acid brings about hydrolysis of the acetal function.

$$\text{(acetonide, } CH_2O\overset{O}{\overset{\|}{C}}R\text{; } H_3C\ CH_3\text{)} \xrightarrow{H_3O^+} HOCH_2\underset{OH}{\underset{|}{C}}HCH_2O\overset{O}{\overset{\|}{C}}R + CH_3\overset{O}{\overset{\|}{C}}CH_3$$

The two hydroxyl groups of the resulting diol are then esterified with 2 moles of the second acyl chloride.

$$HOCH_2\underset{OH}{\underset{|}{C}}HCH_2O\overset{O}{\overset{\|}{C}}R + 2R'\overset{O}{\overset{\|}{C}}Cl \longrightarrow R'\overset{O}{\overset{\|}{C}}OCH_2\underset{R'\underset{\|}{\underset{O}{C}}O}{\underset{|}{C}}HCH_2O\overset{O}{\overset{\|}{C}}R$$

**26.24** The overall transformation

$$\text{(3-methyl-3-butenyl)}\ OH \quad \text{to} \quad \text{(3-methyl-3-butenyl)}\ SCH_3$$

requires converting the alcohol function to some suitable leaving group, followed by substitution by an appropriate nucleophile.

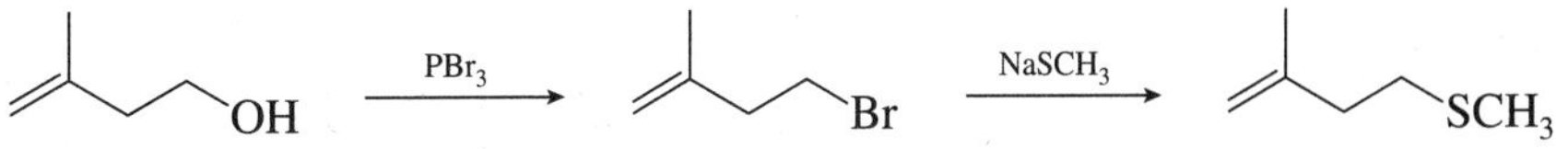

$\xrightarrow{PBr_3}$ $\xrightarrow{NaSCH_3}$

3-Methyl-3-buten-1-ol — 4-Bromo-2-methyl-1-butene — 3-Methyl-3-butenyl methyl sulfide

As reported in the literature, the alcohol was converted to its corresponding *p*-toluenesulfonate ester and this substance was then used as the substrate in the nucleophilic substitution step to produce the desired sulfide in 76% yield.

**26.25** The first transformation is an intramolecular aldol condensation. This reaction was carried out under conditions of base catalysis.

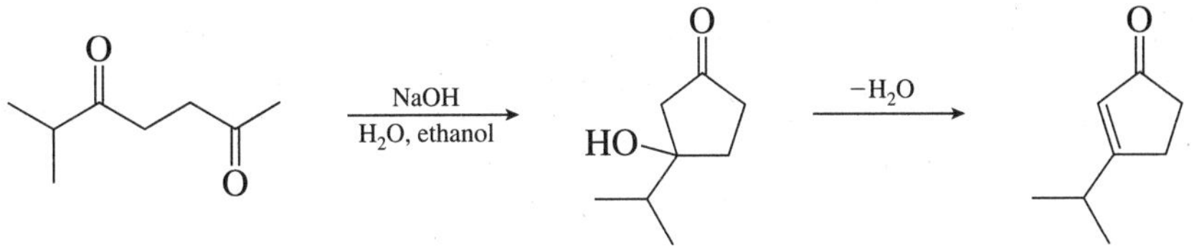

$\xrightarrow[H_2O, \text{ ethanol}]{NaOH}$ $\xrightarrow{-H_2O}$

6-Methyl-2,5-heptanedione — (Not isolated) — 3-Isopropyl-2-cyclopentenone (71%)

The next step is reduction of a ketone to a secondary alcohol. Lithium aluminum hydride is suitable; it reduces carbonyl groups but leaves the double bond intact.

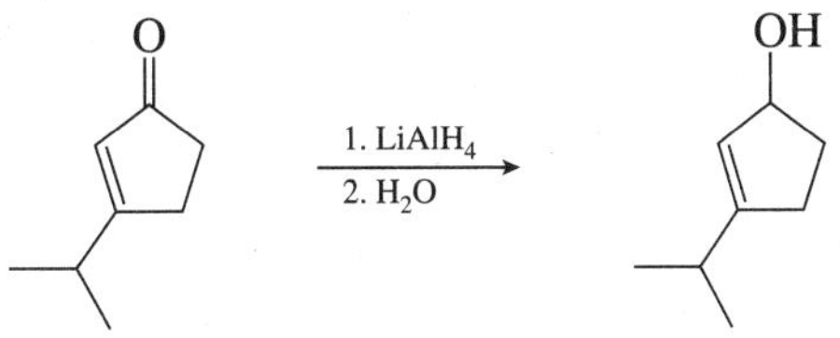

$\xrightarrow[2.\ H_2O]{1.\ LiAlH_4}$

3-Isopropyl-2-cyclopentenone — 3-Isopropyl-2-cyclopenten-1-ol (97%)

Conversion of an alkene to a cyclopropane can be accomplished to using the Simmons–Smith reagent (iodomethylzinc iodide).

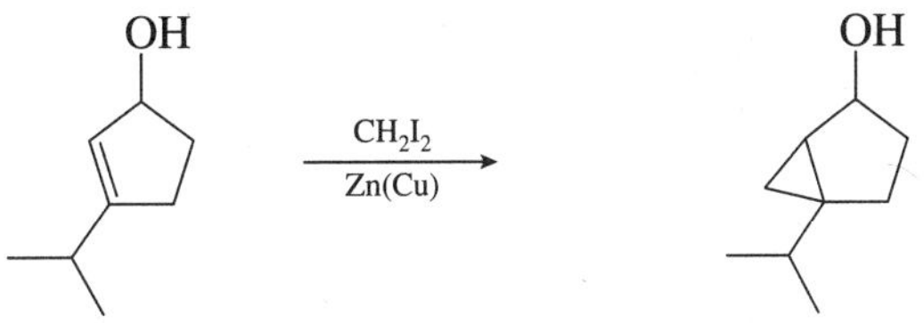

$\xrightarrow[Zn(Cu)]{CH_2I_2}$

3-Isopropyl-2-cyclopenten-1-ol — 5-Isopropylbicyclo[3.1.0]hexan-2-ol (66%)

Oxidation of the secondary alcohol to the ketone can be accomplished with any of a number of oxidizing agents. The chemists who reported this synthesis used chromic acid.

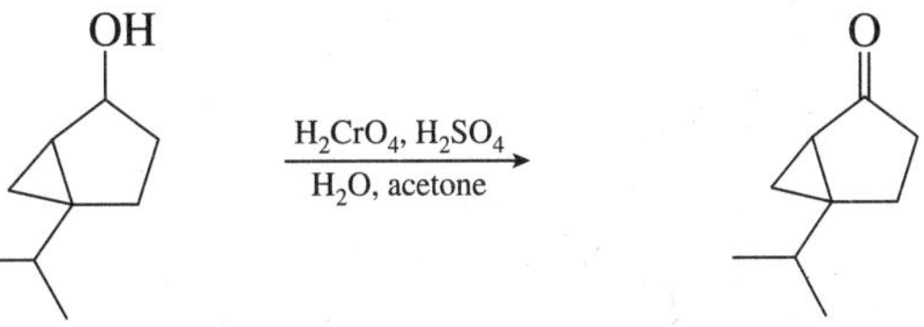

$\xrightarrow[H_2O, \text{ acetone}]{H_2CrO_4,\ H_2SO_4}$

5-Isopropylbicyclo-[3.1.0]hexan-2-ol — 5-Isopropylbicyclo[3.1.0]hexan-2-one (89%)

A Wittig reaction converts the ketone to sabinene.

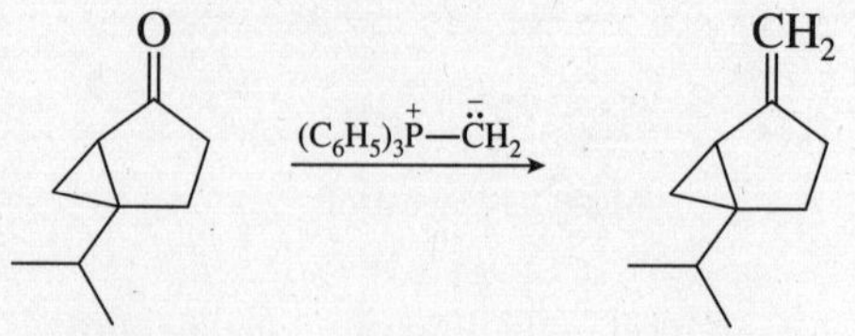

$$\xrightarrow{(C_6H_5)_3\overset{+}{P}-\overset{-}{\ddot{C}}H_2}$$

5-Isopropylbicyclo-[3.1.0]hexan-2-one

$CH_2$

Sabinene (70%)

**26.26** The first step is a 1,4 addition of hydrogen bromide to the conjugated diene system of isoprene.

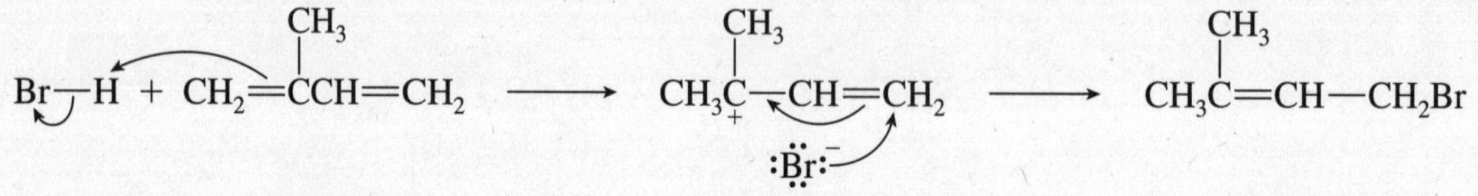

$$Br-H + CH_2=\overset{CH_3}{\overset{|}{C}}CH=CH_2 \longrightarrow CH_3\overset{CH_3}{\overset{|}{\underset{+}{C}}}-CH=CH_2 \;\; :\ddot{Br}:^- \longrightarrow CH_3\overset{CH_3}{\overset{|}{C}}=CH-CH_2Br$$

Hydrogen bromide — 2-Methyl-1,3-butadiene — 1-Bromo-3-methyl-2-butene

This is followed by Markovnikov addition of hydrogen bromide to the remaining double bond.

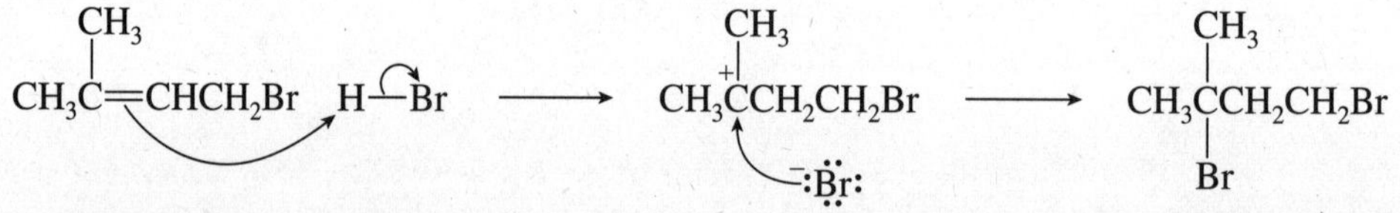

$$CH_3\overset{CH_3}{\overset{|}{C}}=CHCH_2Br \;\; H-Br \longrightarrow CH_3\overset{CH_3}{\overset{|}{\overset{+}{C}}}CH_2CH_2Br \;\; ^-:\ddot{Br}: \longrightarrow CH_3\overset{CH_3}{\overset{|}{\underset{Br}{\underset{|}{C}}}}CH_2CH_2Br$$

1-Bromo-3-methyl-2-butene — Hydrogen bromide — 1,3-Dibromo-3-methylbutane

**26.27** A reasonable mechanism is protonation of the isolated carbon–carbon double bond, followed by cyclization.

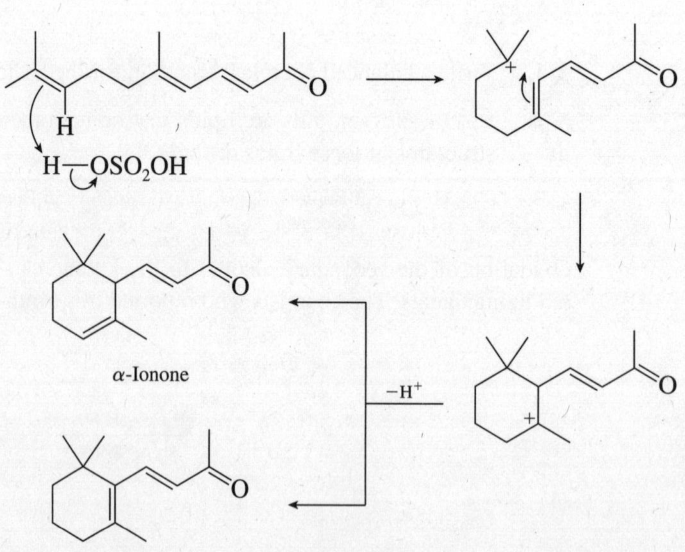

H
$H-OSO_2OH$

O

$-H^+$

α-Ionone

β-Ionone

**26.28** The double bond has a tendency to become conjugated with the carbonyl group. Two mechanisms are more likely than any others under conditions of acid catalysis. One of these involves protonation of the double bond followed by loss of a proton from C-4.

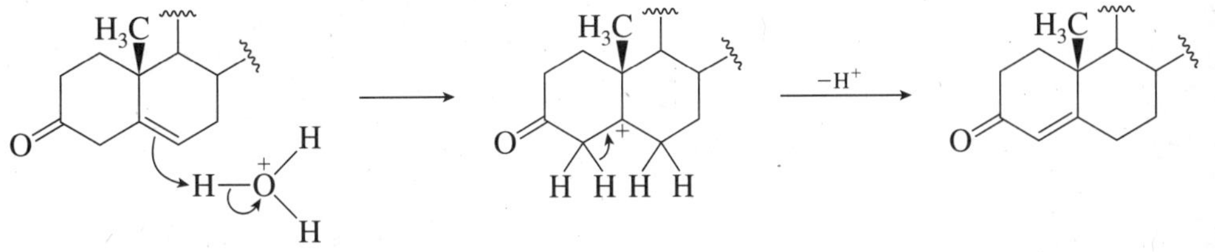

$H_3C$ — $H_3C$ — $-H^+$ — $H_3C$

H H H H

The other mechanism proceeds by enolization followed by proton-induced double-bond migration.

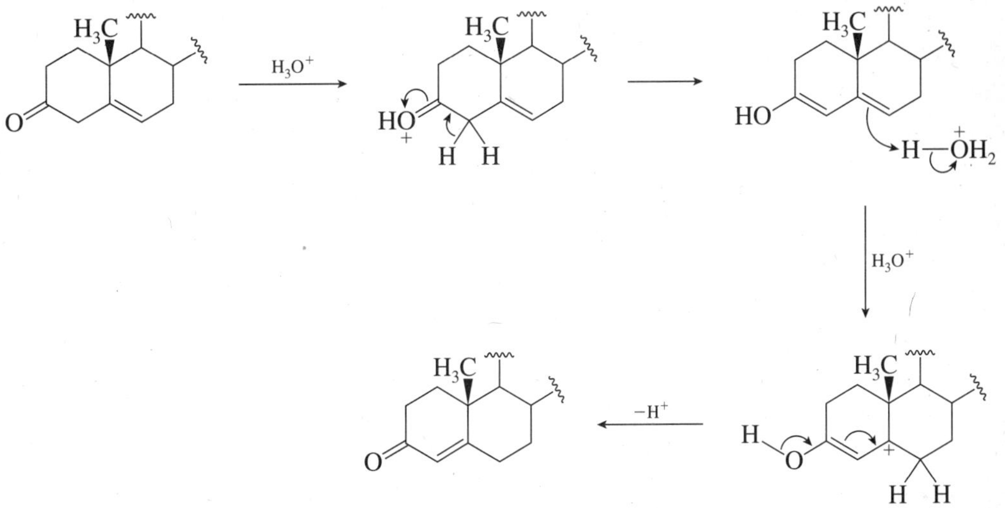

$H_3O^+$ — HO — H H — HO — $H\text{—}\overset{+}{O}H_2$ — $H_3O^+$ — $-H^+$ — H H

**26.29** See the June, 1995, issue of the *Journal of Chemical Education,* pages 541–542, for the solution to this problem.

**26.30** Solutions to molecular modeling exercises are not provided in this *Study Guide and Solutions Manual*. You should use *Learning By Modeling* for this exercise.

# SELF-TEST

## PART A

**A-1.** Write a balanced chemical equation for the basic hydrolysis of tristearin.

**A-2.** Both waxes and fats are lipids that contain the ester functional group. In what way do the structures of these lipids differ?

**A-3.** Classify each of the following isoprenoid compounds as a monoterpene, a diterpene, and so on. Indicate with dashed lines the isoprene units that make up each structure.

(*a*) α-Pinene:

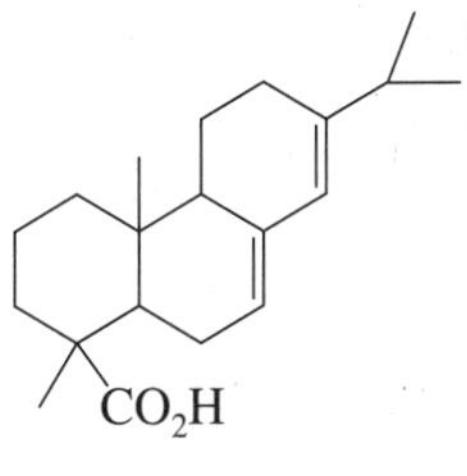

(*b*) Caryophyllene:

(*c*) Abietic acid: $CO_2H$

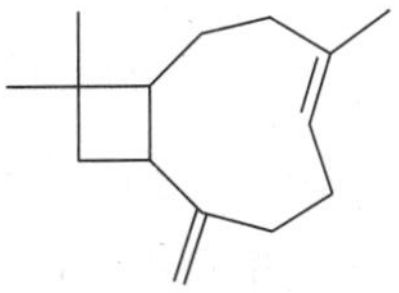

**A-4.** Propose a series of synthetic steps to carry out the preparation of oleic acid [(*Z*)-9-octadecenoic acid] from compound A. You may use any necessary organic or inorganic reagents.

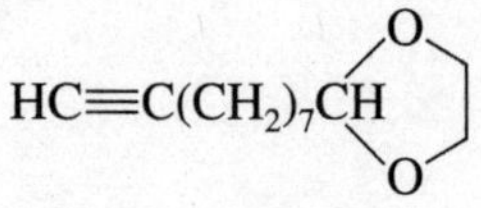

$HC{\equiv}C(CH_2)_7CH$ (1,3-dioxolane ring: O, O)

A

**A-5.** Write a mechanism for the biosynthetic pathway by which limonene is formed from geranyl pyrophosphate.

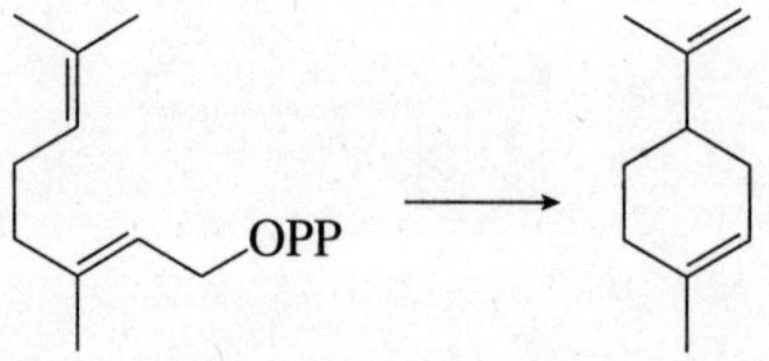

OPP →

Geranyl pyrophosphate    Limonene

## PART B

**B-1.** A major component of a lipid bilayer is
(*a*) A triacylglycerol such as tristearin
(*b*) Phosphatidylcholine, also known as lecithin
(*c*) A sterol such as cholesterol
(*d*) A prostaglandin such as $PGE_1$

**B-2.** Compare the following two triacylglycerols:

| A | B |
|---|---|
| $CH_2O_2CC_{17}H_{35}$ | $CH_2O_2CC_{17}H_{35}$ |
| $CHO_2CC_{17}H_{35}$ | $CHO_2CC_{17}H_{31}$ |
| $CH_2O_2CC_{17}H_{35}$ | $CH_2O_2CC_{17}H_{31}$ |

(*a*) The melting point of A will be higher.
(*b*) The melting point of B will be higher.
(*c*) The melting points of A and B will be the same.
(*d*) No comparison of melting points can be made.

**B-3.** Lanosterol, a biosynthetic precursor of cholesterol, exists naturally as a single enantiomer. How many *possible* stereoisomers having the lanosterol skeleton are there?

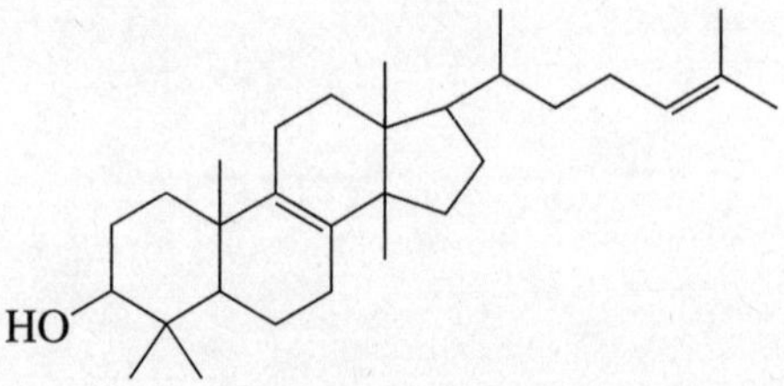

HO

Lanosterol

(*a*) 7    (*b*) 64    (*c*) 128    (*d*) 256

**B-4.** The compound whose carbon skeleton is shown, known as selinene, is found in celery.

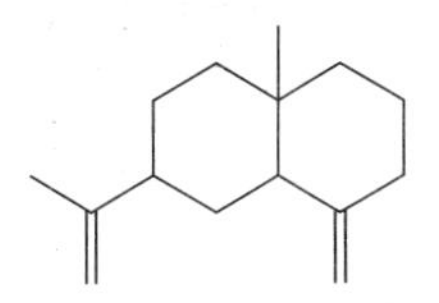

This substance is an example of a

(*a*) Monoterpene
(*b*) Diterpene
(*c*) Sesquiterpene
(*d*) Triterpene

**B-5.** Which of the following correctly represents the isoprenoid units of selinene?

(*a*)

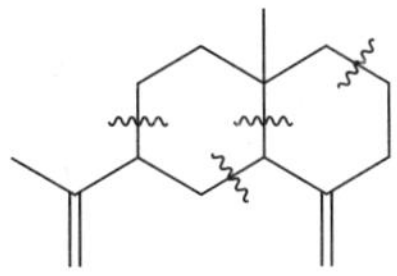

(*b*)

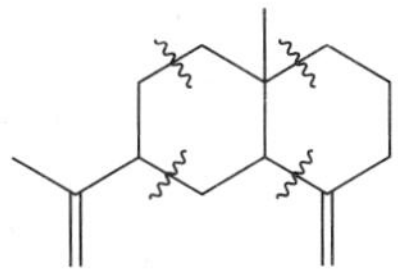

(*c*) Both of these are acceptable

(*d*) Neither of these is acceptable

**B-6.** What is the distribution of radioactive carbon ($^{14}C$) in isopentenyl pyrophosphate biosynthesized from acetic acid labelled with $^{14}C$ at its carboxyl carbon ($CH_3\overset{*}{C}O_2H$)? $^{14}C$ is indicated by an asterisk (*) in the structures.

(*a*) OPP

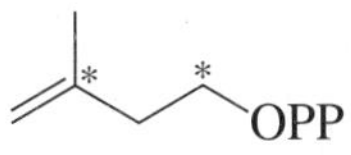

(*b*) OPP

(*c*) OPP

(*d*) OPP

(*e*) OPP

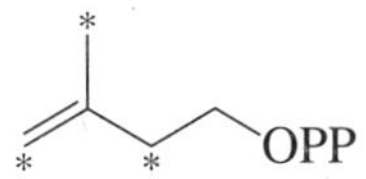

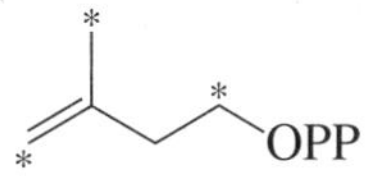

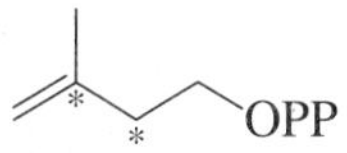

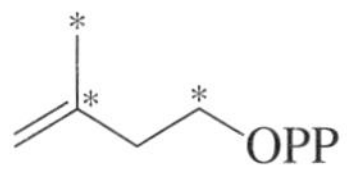

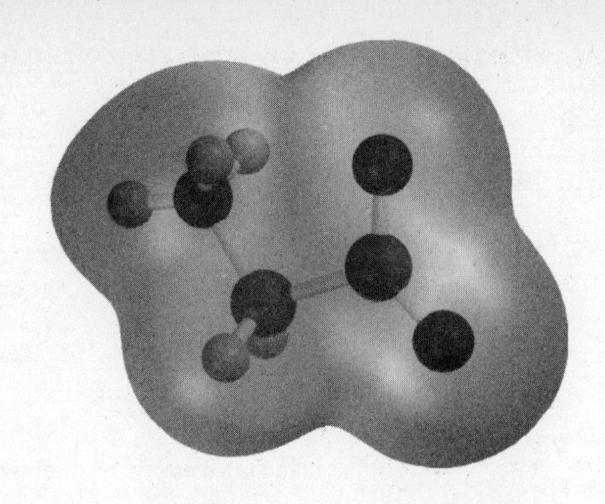

# CHAPTER 27
## AMINO ACIDS, PEPTIDES, AND PROTEINS. NUCLEIC ACIDS

## SOLUTIONS TO TEXT PROBLEMS

**27.1** (*b*) L-Cysteine is the only amino acid in Table 27.1 that has the *R* configuration at its stereogenic center.

$$\text{Fischer: } CO_2^- \text{ (top)},\ H_3\overset{+}{N}\text{—(left)},\ \text{—H (right)},\ CH_2SH \text{ (bottom)} \;\equiv\; H,\ \overset{+}{N}H_3,\ C\text{—}CO_2^-,\ HSCH_2 \;\equiv\; HSCH_2,\ H,\ C\text{—}CO_2^-,\ H_3\overset{+}{N}$$

L-Cysteine

The order of decreasing sequence rule precedence is

$$H_3\overset{+}{N}\text{—} > HSCH_2\text{—} > \text{—}CO_2^- > H\text{—}$$

When the molecule is oriented so that the lowest ranked substituent (H) is held away from us, the order of decreasing precedence traces a clockwise path.

$$H_3\overset{+}{N} \rightarrow CH_2SH \rightarrow CO_2^-$$

Clockwise; therefore *R*

The reason why L-cysteine has the *R* configuration while all the other L-amino acids have the *S* configuration lies in the fact that the —$CH_2SH$ substituent is the only side chain that outranks —$CO_2^-$ according to the sequence rule. Remember, rank order is determined by atomic number at the first point of difference, and —C—S outranks —C—O. In all the other amino acids —$CO_2^-$ outranks the substituent at the stereogenic center. The reversal in the Cahn–Ingold–Prelog descriptor comes not from any change in the spatial arrangement of substituents at the stereogenic center but rather from a reversal in the relative ranks of the carboxylate group and the side chain.

(*c*) The order of decreasing sequence rule precedence in L-methionine is

$$H_3\overset{+}{N}— > —CO_2^- > —CH_2CH_2SCH_3 > H—$$

Sulfur is one atom further removed from the stereogenic center, and so C—O outranks C—C—S.

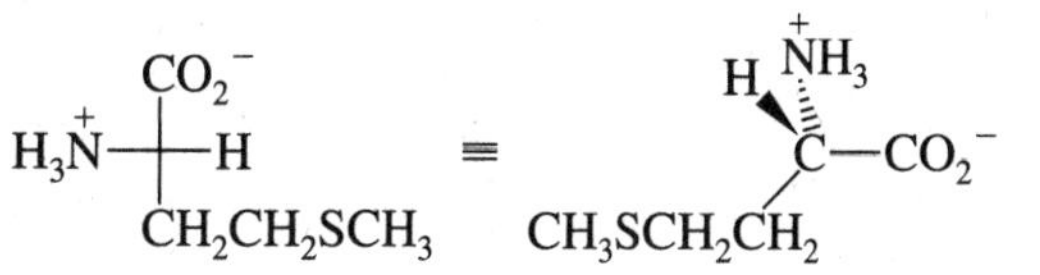

$$H_3\overset{+}{N}—C(CO_2^-)(H)—CH_2CH_2SCH_3 \equiv (H)(\overset{+}{N}H_3)C(CO_2^-)—CH_2CH_2SCH_3$$

The absolute configuration is *S*.

**27.2** The amino acids in Table 27.1 that have more than one stereogenic center are isoleucine and threonine. The stereogenic centers are marked with an asterisk in the structural formulas shown.

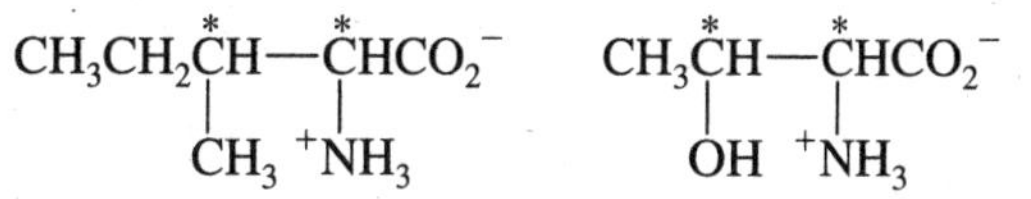

$$CH_3CH_2\overset{*}{C}H(CH_3)—\overset{*}{C}H(\overset{+}{N}H_3)CO_2^- \qquad CH_3\overset{*}{C}H(OH)—\overset{*}{C}H(\overset{+}{N}H_3)CO_2^-$$

Isoleucine — Threonine

**27.3** (*b*) The zwitterionic form of tyrosine is the one shown in Table 27.1.

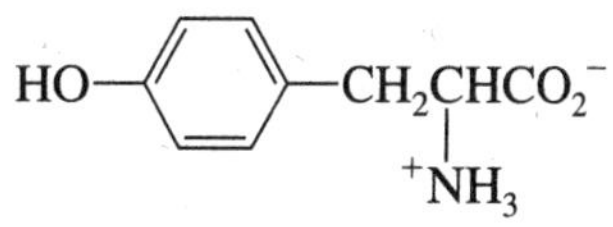

$$HO—C_6H_4—CH_2CH(\overset{+}{N}H_3)CO_2^-$$

(*c*) As base is added to the zwitterion, a proton is removed from either of two positions, the ammonium group or the phenolic hydroxyl. The acidities of the two sites are so close that it is not possible to predict with certainty which one is deprotonated preferentially. Thus two structures are plausible for the monoanion:

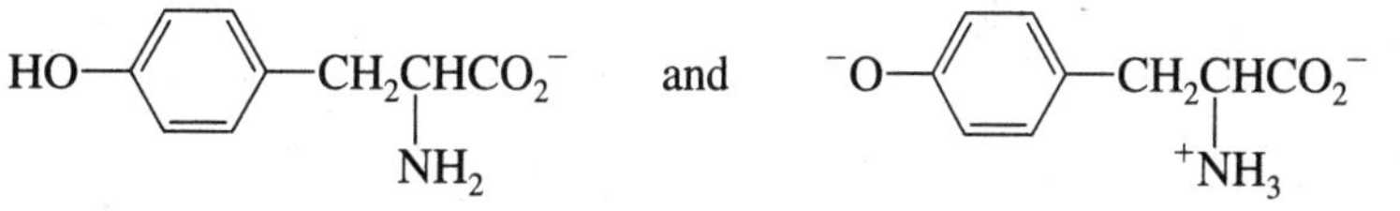

$$HO—C_6H_4—CH_2CH(NH_2)CO_2^- \quad \text{and} \quad {}^-O—C_6H_4—CH_2CH(\overset{+}{N}H_3)CO_2^-$$

In fact, the proton on nitrogen is slightly more acidic than the phenolic hydroxyl, as measured by the p$K_a$ values of the following model compounds:

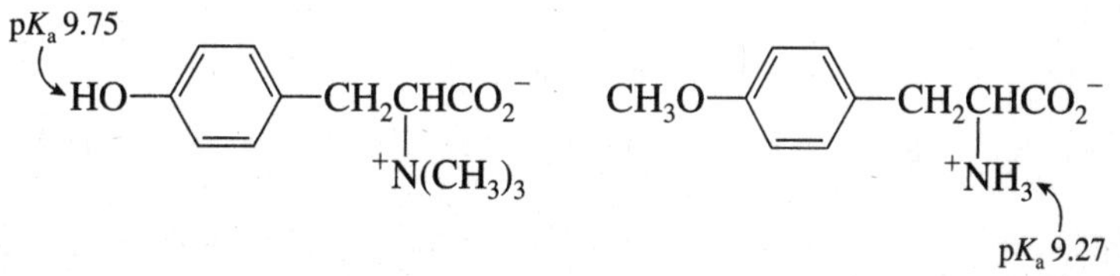

$$\underset{\text{p}K_a\ 9.75}{HO}—C_6H_4—CH_2CH(\overset{+}{N}(CH_3)_3)CO_2^- \qquad CH_3O—C_6H_4—CH_2CH(\underset{\text{p}K_a\ 9.27}{\overset{+}{N}H_3})CO_2^-$$

(*d*) On further treatment with base, both the monoanions in part (*c*) yield the same dianion.

$$^{-}O-C_6H_4-CH_2CH(NH_2)CO_2^-$$

**27.4** At pH 1 the carboxylate oxygen and both nitrogens of lysine are protonated.

$$H_3\overset{+}{N}CH_2CH_2CH_2CH_2CH(\overset{+}{N}H_3)CO_2H$$

(Principal form at pH 1)

As the pH is raised, the carboxyl proton is removed first.

$$H_3\overset{+}{N}CH_2CH_2CH_2CH_2CH(\overset{+}{N}H_3)CO_2H + HO^- \longrightarrow H_3\overset{+}{N}CH_2CH_2CH_2CH_2CH(\overset{+}{N}H_3)CO_2^- + H_2O$$

The $pK_a$ value for the first ionization of lysine is 2.18 (from Table 27.3), and so this process is virtually complete when the pH is greater than this value.

The second $pK_a$ value for lysine is 8.95. This is a fairly typical value for the second $pK_a$ of amino acids and likely corresponds to proton removal from the nitrogen on the $\alpha$ carbon. The species that results is the predominant one at pH 9.

$$H_3\overset{+}{N}CH_2CH_2CH_2CH_2CH(\overset{+}{N}H_3)CO_2^- + HO^- \longrightarrow H_3\overset{+}{N}CH_2CH_2CH_2CH_2CH(NH_2)CO_2^- + H_2O$$

(Principal form at pH 9)

The $pK_a$ value for the third ionization of lysine is 10.53. This value is fairly high compared with those of most of the amino acids in Tables 27.1 to 27.3 and suggests that this proton is removed from the nitrogen of the side chain. The species that results is the major species present at pH values greater than 10.53.

$$H_3\overset{+}{N}CH_2CH_2CH_2CH_2CH(NH_2)CO_2^- + HO^- \longrightarrow H_2NCH_2CH_2CH_2CH_2CH(NH_2)CO_2^-$$

(Principal form at pH 13)

**27.5** To convert 3-methylbutanoic acid to valine, a leaving group must be introduced at the $\alpha$ carbon prior to displacement by ammonia. This is best accomplished by bromination under the conditions of the Hell–Volhard–Zelinsky reaction.

$$(CH_3)_2CHCH_2CO_2H \xrightarrow[\text{or } Br_2,\ PCl_3]{Br_2,\ P} (CH_3)_2CHCH(Br)CO_2H \xrightarrow{NH_3} (CH_3)_2CHCH(\overset{+}{N}H_3)CO_2^-$$

3-Methylbutanoic acid — 2-Bromo-3-methylbutanoic acid — Valine

Valine has been prepared by this method. The Hell–Volhard–Zelinsky reaction was carried out in 88% yield, but reaction of the $\alpha$-bromo acid with ammonia was not very efficient, valine being isolated in only 48% yield in this step.

**27.6** In the Strecker synthesis an aldehyde is treated with ammonia and a source of cyanide ion. The resulting amino nitrile is hydrolyzed to an amino acid.

$$(CH_3)_2CHCH{=}O \xrightarrow[HCN]{NH_3} (CH_3)_2CHCH(NH_2)C{\equiv}N \xrightarrow[2.\ HO^-]{1.\ H_3O^+,\ heat} (CH_3)_2CHCH(\overset{+}{N}H_3)CO_2^-$$

2-Methylpropanal; 2-Amino-3-methylbutanenitrile; Valine

As actually carried out, the aldehyde was converted to the amino nitrile by treatment with an aqueous solution containing ammonium chloride and potassium cyanide. Hydrolysis was achieved in aqueous hydrochloric acid and gave valine as its hydrochloride salt in 65% overall yield.

**27.7** The alkyl halide with which the anion of diethyl acetamidomalonate is treated is 2-bromopropane.

$$CH_3\overset{O}{\overset{\|}{C}}NHCH(CO_2CH_2CH_3)_2 + (CH_3)_2CHBr \xrightarrow[CH_3CH_2OH]{NaOCH_2CH_3} CH_3\overset{O}{\overset{\|}{C}}NHC(CO_2CH_2CH_3)_2CH(CH_3)_2$$

Diethyl acetamidomalonate; 2-Bromopropane; Diethyl acetamidoisopropylmalonate

This is the difficult step in the synthesis; it requires a nucleophilic substitution of the $S_N2$ type involving a secondary alkyl halide. Competition of elimination with substitution results in only a 37% observed yield of alkylated diethyl acetamidomalonate.

Hydrolysis and decarboxylation of the alkylated derivative are straightforward and proceed in 85% yield to give valine.

$$CH_3\overset{O}{\overset{\|}{C}}NHC(CO_2CH_2CH_3)_2CH(CH_3)_2 \xrightarrow[heat]{HBr,\ H_2O} H_3\overset{+}{N}C(CO_2H)_2CH(CH_3)_2 \xrightarrow[-CO_2]{heat} H_3\overset{+}{N}CHCO_2^-CH(CH_3)_2$$

Diethyl acetamidoisopropylmalonate; 2-Aminoisopropylmalonic acid; Valine

The overall yield of valine (31%) is the product of 37% $\times$ 85%.

**27.8** Ninhydrin is the hydrate of a triketone and is in equilibrium with it.

Hydrated form of ninhydrin $\rightleftharpoons$ Triketo form of ninhydrin + $H_2O$

An amino acid reacts with this triketone to form an imine.

Triketo form of ninhydrin + $RCH(\overset{+}{N}H_3)CO_2^-$ ($\alpha$-Amino acid) $\xrightarrow{HO^-}$ Imine (=NCH(R)$CO_2^-$)

This imine then undergoes decarboxylation.

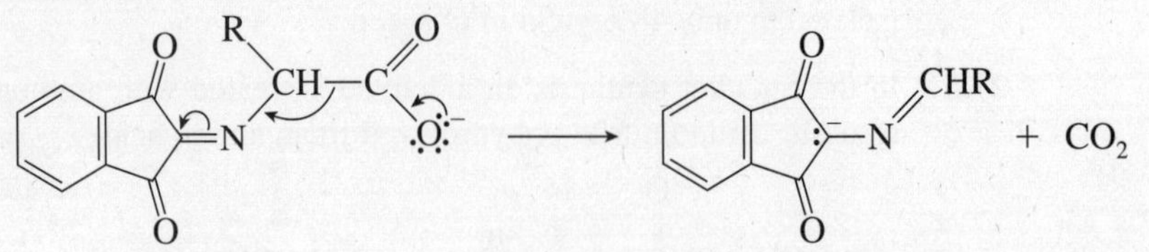

O R O CH—C O N :Ö:⁻ O ⟶ O CHR :⁻N O + $CO_2$

The anion that results from the decarboxylation step is then protonated. The product is shown as its diketo form but probably exists as an enol.

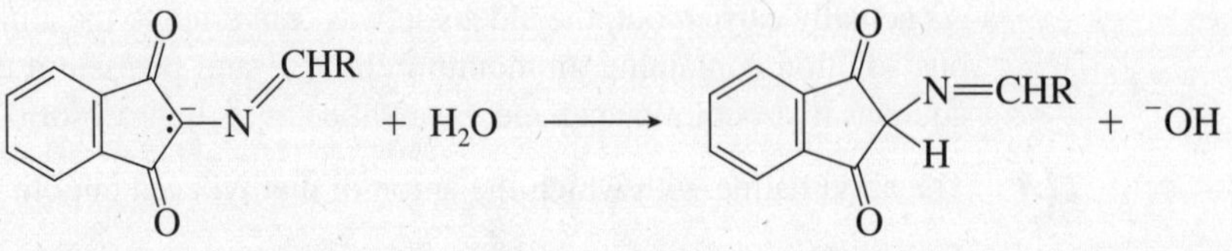

O CHR :⁻N O + $H_2O$ ⟶ O N=CHR H O + ⁻OH

Hydrolysis of the imine function gives an aldehyde and a compound having a free amino group.

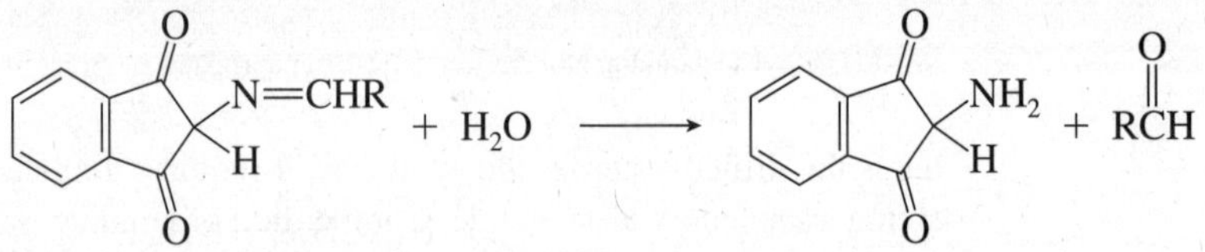

O N=CHR H O + $H_2O$ ⟶ O $NH_2$ H O + RCH (=O)

This amine then reacts with a second molecule of the triketo form of ninhydrin to give an imine.

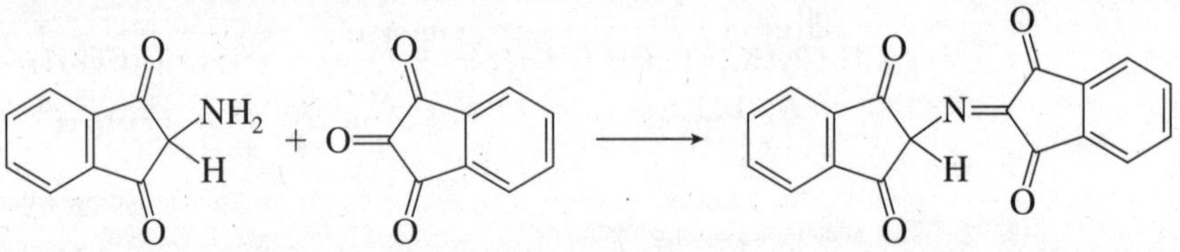

O $NH_2$ H O + O= O O ⟶ O N= O H O O

Proton abstraction from the neutral imine gives its conjugate base, which is a violet dye.

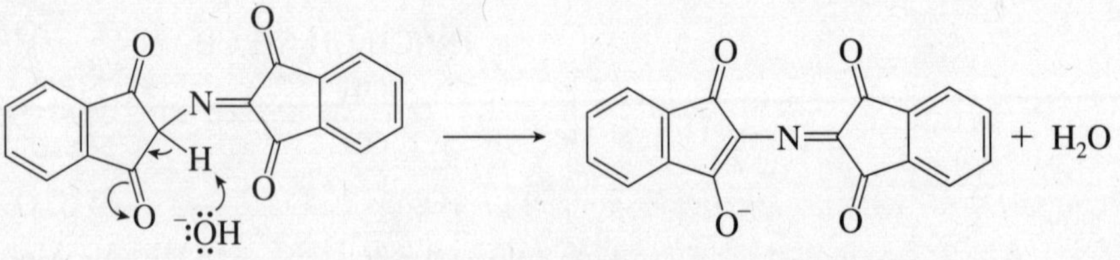

O O N= H O O ⁻:ÖH ⟶ O O N= O⁻ O + $H_2O$

Violet dye

**27.9** The carbon that bears the amino group of 4-aminobutanoic acid corresponds to the $\alpha$ carbon of an $\alpha$-amino acid.

$$\underset{\text{4-Aminobutanoic acid}}{\underset{^{+}NH_3}{CH_2}CH_2CH_2CO_2^-} \quad \text{arises by decarboxylation of} \quad \underset{\text{Glutamic acid}}{^-O_2C\underset{^{+}NH_3}{CH}CH_2CH_2CO_2^-}$$

**27.10** (*b*) Alanine is the N-terminal amino acid in Ala-Phe. Its carboxyl group is joined to the nitrogen of phenylalanine by a peptide bond.

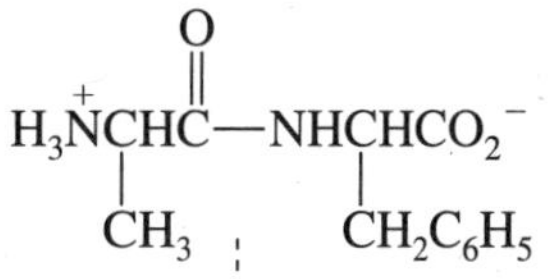

$$\overset{+}{H_3N}CH(CH_3)C(=O)—NHCH(CH_2C_6H_5)CO_2^- \qquad \text{AF}$$

Alanine | Phenylalanine

(*c*) The positions of the amino acids are reversed in Phe-Ala. Phenylalanine is the N terminus and alanine is the C terminus.

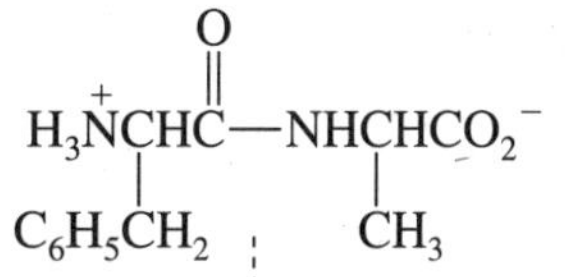

$$\overset{+}{H_3N}CH(CH_2C_6H_5)C(=O)—NHCH(CH_3)CO_2^- \qquad \text{FA}$$

Phenylalanine | Alanine

(*d*) The carboxyl group of glycine is joined by a peptide bond to the amino group of glutamic acid.

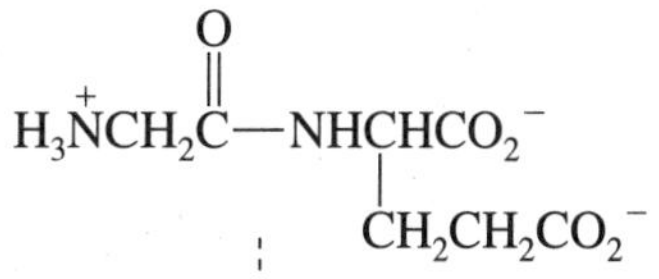

$$\overset{+}{H_3N}CH_2C(=O)—NHCH(CH_2CH_2CO_2^-)CO_2^- \qquad \text{GE}$$

Glycine | Glutamic acid

The dipeptide is written in its anionic form because the carboxyl group of the side chain is ionized at pH 7. Alternatively, it could have been written as a neutral zwitterion with a $CH_2CH_2CO_2H$ side chain.

(*e*) The peptide bond in Lys-Gly is between the carboxyl group of lysine and the amino group of glycine.

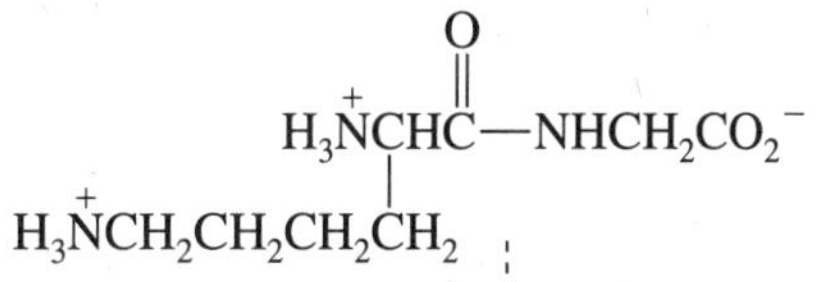

$$\overset{+}{H_3N}CH(CH_2CH_2CH_2CH_2\overset{+}{N}H_3)C(=O)—NHCH_2CO_2^- \qquad \text{KG}$$

Lysine | Glycine

The amino group of the lysine side chain is protonated at pH 7, and so the dipeptide is written here in its cationic form. It could have also been written as a neutral zwitterion with the side chain $H_2NCH_2CH_2CH_2CH_2$.

(*f*) Both amino acids are alanine in D-Ala-D-Ala. The fact that they have the D configuration has no effect on the constitution of the dipeptide.

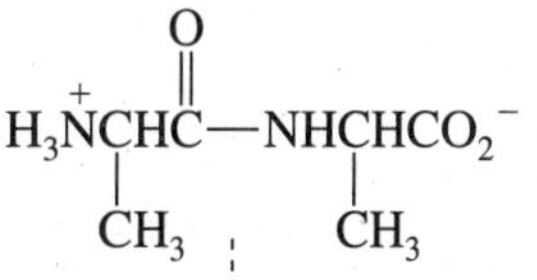

$$\overset{+}{H_3N}CH(CH_3)C(=O)—NHCH(CH_3)CO_2^- \qquad \text{D-A-D-A}$$

Alanine | Alanine

**27.11** (*b*) When amino acid residues in a dipeptide are indicated without a prefix, it is assumed that the configuration at the $\alpha$ carbon atom is L. For all amino acids except cysteine, the L configuration corresponds to *S*. The stereochemistry of Ala-Phe may therefore be indicated for the zigzag conformation as shown.

$H_3\overset{+}{N}$ O $CH_2C_6H_5$ H N H $CO_2^-$ $H_3C$ H

The L configuration corresponds to *S* for each of the stereogenic centers in Ala-Phe.

(*c*) Similarly, Phe-Ala has its substituent at the N-terminal amino acid directed away from us, whereas the C-terminal side chain is pointing toward us, and the L configuration corresponds to *S* for each stereogenic center.

$H_3\overset{+}{N}$ O $H_3C$ H N H $CO_2^-$ $C_6H_5CH_2$ H

(*d*) There is only one stereogenic center in Gly-Glu. It has the L (or *S*) configuration.

$H_3\overset{+}{N}$ O $CH_2CH_2CO_2^-$ H N H $CO_2^-$

(*e*) In order for the N-terminal amino acid in Lys-Gly to have the L (or *S*) configuration, its side chain must be directed away from us in the conformation indicated.

$H_3\overset{+}{N}$ O N H $CO_2^-$ $H_3\overset{+}{N}CH_2CH_2CH_2CH_2$ H

(*f*) The configuration at both $\alpha$-carbon atoms in D-Ala-D-Ala is exactly the reverse of the configuration of the stereogenic centers in parts (*a*) through (*e*). Both stereogenic centers have the D (or *R*) configuration.

$H_3\overset{+}{N}$ O H $CH_3$ N H $CO_2^-$ H $CH_3$

**27.12** Figure 27.7 in the text gives the structure of leucine enkephalin. Methionine enkephalin differs from it only with respect to the C-terminal amino acid. The amino acid sequences of the two pentapeptides are

| Tyr-Gly-Gly-Phe-Leu | Tyr-Gly-Gly-Phe-Met |
|---|---|
| Leucine enkephalin | Methionine enkephalin |

The peptide sequence of a polypeptide can also be expressed using the one-letter abbreviations listed in text Table 27.1. Methionine enkephalin becomes YGGFM.

**27.13** Twenty-four tetrapeptide combinations are possible for the four amino acids alanine (A), glycine (G), phenylalanine (F), and valine (V). Remember that the order is important; AG is not the same peptide as GA. Using the one-letter abbreviations for each amino acid the possibilities are

| | | | | | |
|---|---|---|---|---|---|
| AGFV | AGVF | AFGV | AFVG | AVGF | AVFG |
| GAFV | GAVF | GFAV | GFVA | GVFA | GVAF |
| FAGV | FAVG | FVAG | FVGA | FGAF | FGFA |
| VAGF | VAFG | VGAF | VGFA | VFAG | VFGA |

**27.14** Chymotrypsin cleaves a peptide selectively at the carboxyl group of amino acids that have aromatic side chains. The side chain of phenylalanine is a benzyl group, $C_6H_5CH_2$—. If the dipeptide isolated after treatment with chymotrypsin contains valine (V) and phenylalanine (F), its sequence must be VF.

$$H_3\overset{+}{N}CH(CH(CH_3)_2)C(=O)\text{—}NHCH(CH_2C_6H_5)C(=O)\text{—}[\text{Rest of peptide}] \xrightarrow{\text{chymotrypsin}} H_3\overset{+}{N}CH(CH(CH_3)_2)C(=O)\text{—}NHCH(CH_2C_6H_5)CO^- + [\text{Rest of peptide}]$$

Valine Phenylalanine Valinylphenylalanine (VF)

The possible sequences for the unknown tetrapeptide are VFAG and VFGA.

**27.15** The Edman degradation removes the N-terminal amino acid, which is identified as a phenylthiohydantoin derivative. The first Edman degradation of Val-Phe-Gly-Ala gives the phenylthiohydantoin derived from valine; the second gives the phenylthiohydantoin derived from phenylalanine.

$$\text{Val-Phe-Gly-Ala} \xrightarrow{\text{first Edman degradation}} \text{Phe-Gly-Ala} + \text{phenylthiohydantoin}(C_6H_5, CH(CH_3)_2) \xrightarrow{\text{second Edman degradation}} \text{Gly-Ala} + \text{phenylthiohydantoin}(C_6H_5, CH_2C_6H_5)$$

**27.16** Lysine has two amino groups. Both amino functions are converted to amides on reaction with benzyloxycarbonyl chloride.

$$H_2NCH_2CH_2CH_2CH_2CH(NH_2)CO_2^- + 2C_6H_5CH_2OC(=O)Cl \longrightarrow C_6H_5CH_2OC(=O)NHCH_2CH_2CH_2CH_2CH(NHC(=O)OCH_2C_6H_5)CO_2H$$

**27.17** The peptide bond of Ala-Leu connects the carboxyl group of alanine and the amino group of leucine. We therefore need to protect the amino group of alanine and the carboxyl group of leucine.

Protect the amino group of alanine as its benzyloxycarbonyl derivative.

$$H_3\overset{+}{N}CH(CH_3)CO_2^- + C_6H_5CH_2OC(=O)Cl \longrightarrow C_6H_5CH_2OC(=O)NHCH(CH_3)CO_2H$$

Alanine Benzyloxycarbonyl chloride Z-Protected alanine

Protect the carboxyl group of leucine as its benzyl ester.

$$\underset{\text{Leucine}}{H_3\overset{+}{N}CHCO_2^- \atop (CH_3)_2CHCH_2} + \underset{\text{Benzyl alcohol}}{C_6H_5CH_2OH} \xrightarrow[2.\ HO^-]{1.\ H^+,\ heat} \underset{\text{Leucine benzyl ester}}{H_2NCHCO_2CH_2C_6H_5 \atop (CH_3)_2CHCH_2}$$

Coupling of the two amino acids is achieved by *N,N′*-dicyclohexylcarbodiimide (DCCI)-promoted amide bond formation between the free amino group of leucine benzyl ester and the free carboxyl group of Z-protected alanine.

$$\underset{\text{Z-Protected alanine}}{C_6H_5CH_2O\overset{O}{\overset{\|}{C}}NHCHCO_2H \atop CH_3} + \underset{\text{Leucine benzyl ester}}{H_2NCH\overset{O}{\overset{\|}{C}}OCH_2C_6H_5 \atop (CH_3)_2CHCH_2} \xrightarrow{DCCI} \underset{\text{Protected dipeptide}}{C_6H_5CH_2O\overset{O}{\overset{\|}{C}}NHCH\overset{O}{\overset{\|}{C}}NHCH\overset{O}{\overset{\|}{C}}OCH_2C_6H_5 \atop CH_3 \qquad CH_2CH(CH_3)_2}$$

Both the benzyloxycarbonyl protecting group and the benzyl ester protecting group may be removed by hydrogenolysis over palladium. This step completes the synthesis of Ala-Leu.

$$\underset{\text{Protected dipeptide}}{C_6H_5CH_2O\overset{O}{\overset{\|}{C}}NHCH\overset{O}{\overset{\|}{C}}NHCH\overset{O}{\overset{\|}{C}}OCH_2C_6H_5 \atop CH_3 \qquad CH_2CH(CH_3)_2} \xrightarrow{H_2,\ Pd} \underset{\text{Ala-Leu}}{H_3\overset{+}{N}CH\overset{O}{\overset{\|}{C}}NHCHCO_2^- \atop CH_3 \qquad CH_2CH(CH_3)_2}$$

**27.18** As in the DCCI-promoted coupling of amino acids, the first step is the addition of the Z-protected amino acid to DCCI to give an *O*-acylisourea.

$$\underset{\text{Z-Protected amino acid}}{ZNHCH\overset{O}{\overset{\|}{C}}OH \atop R} + \underset{\text{DCCI}}{C_6H_{11}N{=}C{=}NC_6H_{11}} \longrightarrow \underset{\textit{O}\text{-Acylisourea}}{ZNHCH\overset{O}{\overset{\|}{C}}O{-}C\begin{matrix} {\nearrow\!\!=} NC_6H_{11} \\ \searrow NHC_6H_{11} \end{matrix} \atop R}$$

This *O*-acylisourea is attacked by *p*-nitrophenol to give the *p*-nitrophenyl ester of the Z-protected amino acid.

$$O_2N{-}C_6H_4{-}\ddot{O}H + R'\overset{O}{\overset{\|}{C}}{-}O{-}C(=NC_6H_{11}\cdot H^+)NHC_6H_{11} \longrightarrow O_2N{-}C_6H_4{-}O\overset{O}{\overset{\|}{C}}R' + C_6H_{11}NH\overset{O}{\overset{\|}{C}}NHC_6H_{11}$$

**27.19** To add a leucine residue to the N terminus of the ethyl ester of Z-Phe-Gly, the benzyloxycarbonyl protecting group must first be removed. This can be accomplished by hydrogenolysis.

$$\underset{\text{Z-Protected ethyl ester of Phe-Gly}}{C_6H_5CH_2O\overset{O}{\overset{\|}{C}}NHCH\overset{O}{\overset{\|}{C}}NHCH_2\overset{O}{\overset{\|}{C}}OCH_2CH_3 \atop C_6H_5CH_2} \xrightarrow{H_2,\ Pd} \underset{\text{Phe-Gly ethyl ester}}{H_2NCH\overset{O}{\overset{\|}{C}}NHCH_2\overset{O}{\overset{\|}{C}}OCH_2CH_3 \atop C_6H_5CH_2}$$

The reaction shown has been carried out in 100% yield. Alternatively, the benzyloxycarbonyl protecting group may be removed by treatment with hydrogen bromide in acetic acid. This latter route has also been reported in the chemical literature and gives the hydrobromide salt of Phe-Gly ethyl ester in 82% yield.

Once the protecting group has been removed, the ethyl ester of Phe-Gly is allowed to react with the *p*-nitrophenyl ester of Z-protected leucine to form the protected tripeptide. Hydrogenolysis of the Z-protected tripeptide gives Leu-Phe-Gly as its ethyl ester.

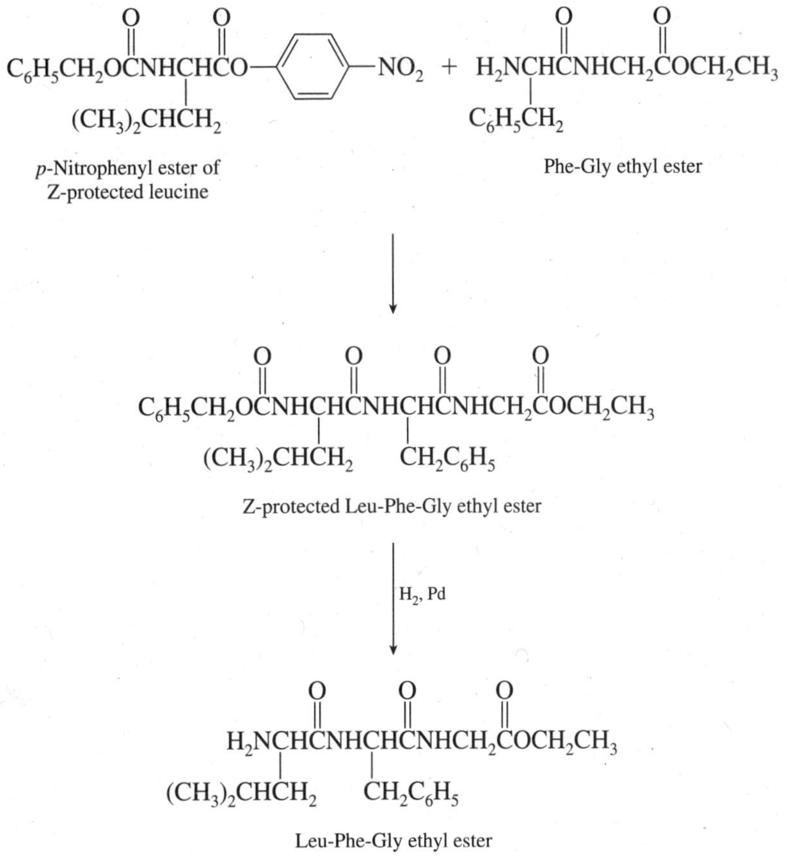

$$C_6H_5CH_2O\overset{O}{\overset{\|}{C}}NH\underset{(CH_3)_2CHCH_2}{\underset{|}{C}H}\overset{O}{\overset{\|}{C}}O\text{—}C_6H_4\text{—}NO_2 + H_2N\underset{C_6H_5CH_2}{\underset{|}{C}H}\overset{O}{\overset{\|}{C}}NHCH_2\overset{O}{\overset{\|}{C}}OCH_2CH_3$$

*p*-Nitrophenyl ester of Z-protected leucine; Phe-Gly ethyl ester

$$\downarrow$$

$$C_6H_5CH_2O\overset{O}{\overset{\|}{C}}NH\underset{(CH_3)_2CHCH_2}{\underset{|}{C}H}\overset{O}{\overset{\|}{C}}NH\underset{CH_2C_6H_5}{\underset{|}{C}H}\overset{O}{\overset{\|}{C}}NHCH_2\overset{O}{\overset{\|}{C}}OCH_2CH_3$$

Z-protected Leu-Phe-Gly ethyl ester

$$\downarrow H_2, Pd$$

$$H_2N\underset{(CH_3)_2CHCH_2}{\underset{|}{C}H}\overset{O}{\overset{\|}{C}}NH\underset{CH_2C_6H_5}{\underset{|}{C}H}\overset{O}{\overset{\|}{C}}NHCH_2\overset{O}{\overset{\|}{C}}OCH_2CH_3$$

Leu-Phe-Gly ethyl ester

**27.20** Amino acid residues are added by beginning at the C terminus in the Merrifield solid-phase approach to peptide synthesis. Thus the synthesis of Phe-Gly requires glycine to be anchored to the solid support. Begin by protecting glycine as its *tert*-butoxycarbonyl (Boc) derivative.

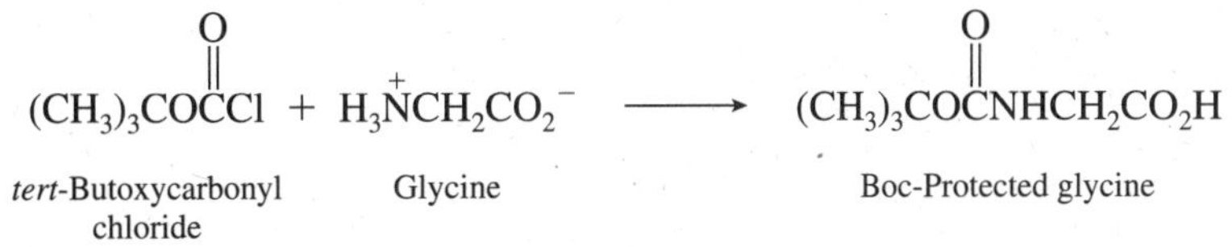

$$(CH_3)_3CO\overset{O}{\overset{\|}{C}}Cl + H_3\overset{+}{N}CH_2CO_2^- \longrightarrow (CH_3)_3CO\overset{O}{\overset{\|}{C}}NHCH_2CO_2H$$

*tert*-Butoxycarbonyl chloride; Glycine; Boc-Protected glycine

The protected glycine is attached via its carboxylate anion to the solid support.

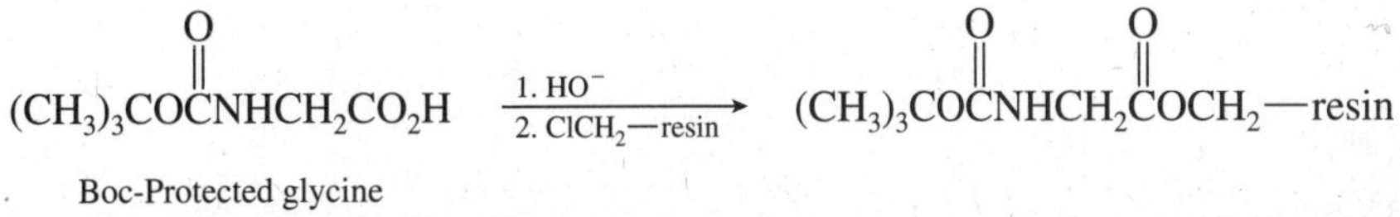

$$(CH_3)_3CO\overset{O}{\overset{\|}{C}}NHCH_2CO_2H \xrightarrow[\text{2. ClCH}_2\text{—resin}]{\text{1. HO}^-} (CH_3)_3CO\overset{O}{\overset{\|}{C}}NHCH_2\overset{O}{\overset{\|}{C}}OCH_2\text{—resin}$$

Boc-Protected glycine

The amino group of glycine is then exposed by removal of the protecting group. Typical conditions for this step involve treatment with hydrogen chloride in acetic acid.

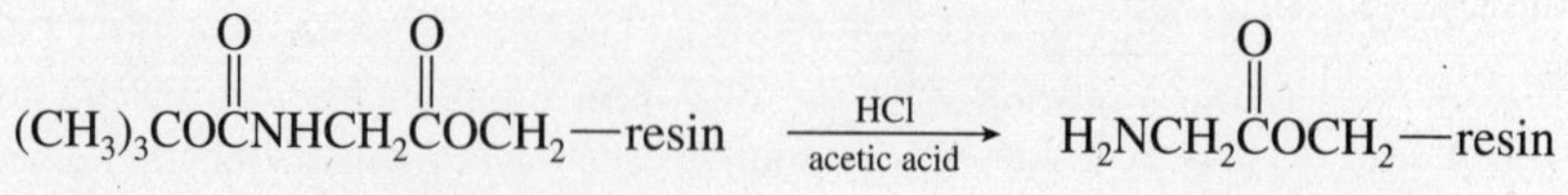

$$(CH_3)_3CO\overset{O}{\overset{\|}{C}}NHCH_2\overset{O}{\overset{\|}{C}}OCH_2\text{—resin} \xrightarrow[\text{acetic acid}]{\text{HCl}} H_2NCH_2\overset{O}{\overset{\|}{C}}OCH_2\text{—resin}$$

Boc-Protected, resin-bound glycine — Resin-bound glycine

To attach phenylalanine to resin-bound glycine, we must first protect the amino group of phenylalanine. A Boc protecting group is appropriate.

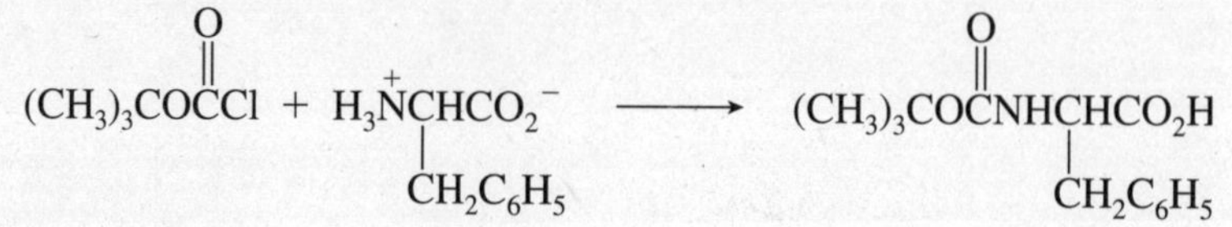

$$(CH_3)_3CO\overset{O}{\overset{\|}{C}}Cl + H_3\overset{+}{N}\underset{CH_2C_6H_5}{\underset{|}{C}}HCO_2^- \longrightarrow (CH_3)_3CO\overset{O}{\overset{\|}{C}}NH\underset{CH_2C_6H_5}{\underset{|}{C}}HCO_2H$$

*tert*-Butoxycarbonyl chloride — Phenylalanine — Boc-Protected phenylalanine

Peptide bond formation occurs when the resin-bound glycine and Boc-protected phenylalanine are combined in the presence of DCCI.

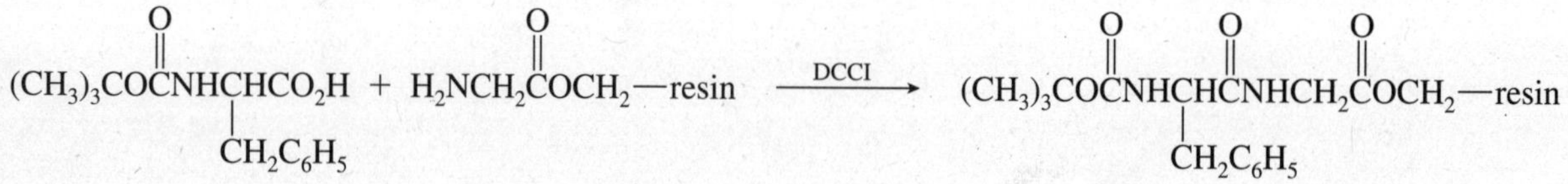

$$(CH_3)_3CO\overset{O}{\overset{\|}{C}}NH\underset{CH_2C_6H_5}{\underset{|}{C}}HCO_2H + H_2NCH_2\overset{O}{\overset{\|}{C}}OCH_2\text{—resin} \xrightarrow{\text{DCCI}} (CH_3)_3CO\overset{O}{\overset{\|}{C}}NH\underset{CH_2C_6H_5}{\underset{|}{C}}H\overset{O}{\overset{\|}{C}}NHCH_2\overset{O}{\overset{\|}{C}}OCH_2\text{—resin}$$

Boc-Protected phenylalanine — Resin-bound glycine — Boc-Protected, resin-bound Phe-Gly

Remove the Boc group with HCl and then treat with HBr in trifluoroacetic acid to cleave Phe-Gly from the solid support.

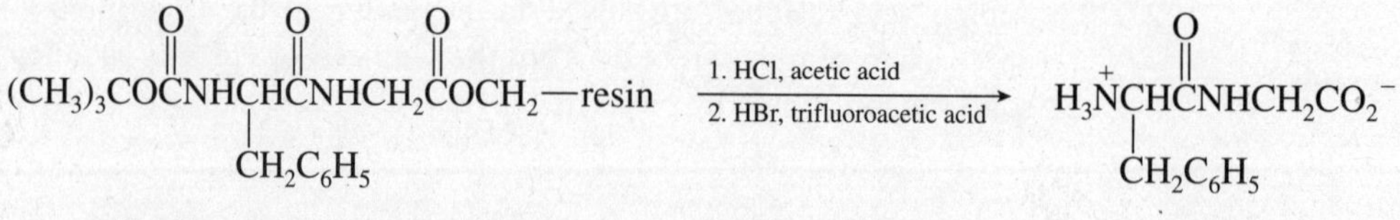

$$(CH_3)_3CO\overset{O}{\overset{\|}{C}}NH\underset{CH_2C_6H_5}{\underset{|}{C}}H\overset{O}{\overset{\|}{C}}NHCH_2\overset{O}{\overset{\|}{C}}OCH_2\text{—resin} \xrightarrow[\text{2. HBr, trifluoroacetic acid}]{\text{1. HCl, acetic acid}} H_3\overset{+}{N}\underset{CH_2C_6H_5}{\underset{|}{C}}H\overset{O}{\overset{\|}{C}}NHCH_2CO_2^-$$

Boc-Protected, resin-bound Phe-Gly — Phe-Gly

**27.21** The numbering of the ring in uracil and its derivatives parallels that in pyrimidine.

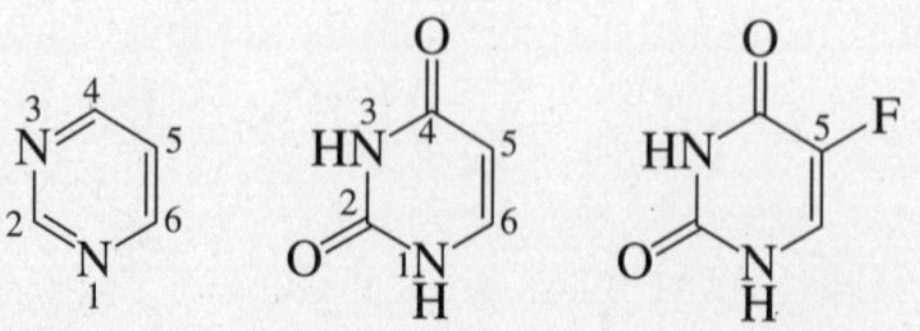

Pyrimidine — Uracil — 5-Fluorouracil

**27.22** (*b*) Cytidine is present in RNA and so is a nucleoside of D-ribose. The base is cytosine.

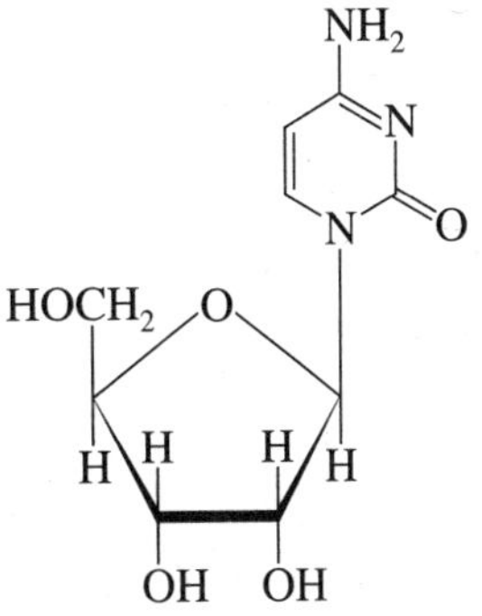

$NH_2$ N N O $HOCH_2$ O H H H H OH OH

(*c*) Guanosine is present in RNA and so is a guanine nucleoside of D-ribose.

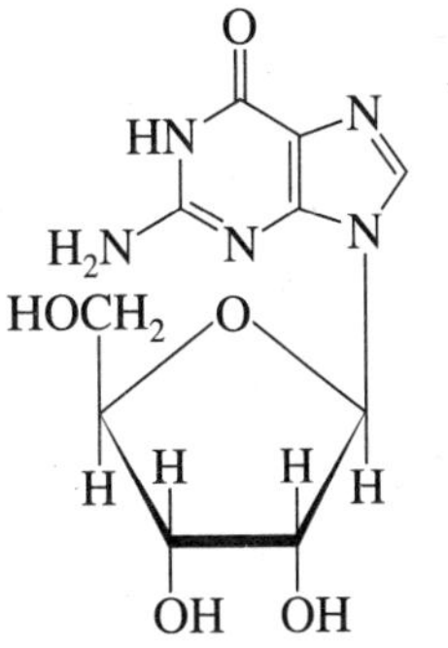

O HN N $H_2N$ N N $HOCH_2$ O H H H H OH OH

**27.23** Table 27.4 in the text lists the messenger RNA codons for the various amino acids. The codons for valine and for glutamic acid are:

| | | | | |
|---|---|---|---|---|
| Valine: | GUU | GUA | GUC | GUG |
| Glutamic acid: | | GAA | | GAG |

As can be seen, the codons for glutamic acid (GAA and GAG) are very similar to two of the codons (GUA and GUG) for valine. Replacement of adenine in the glutamic acid codons by uracil causes valine to be incorporated into hemoglobin instead of glutamic acid and is responsible for the sickle cell trait.

**27.24** The protonated form of imidazole represented by structure A is stabilized by delocalization of the lone pair of one of the nitrogens. The positive charge is shared by both nitrogens.

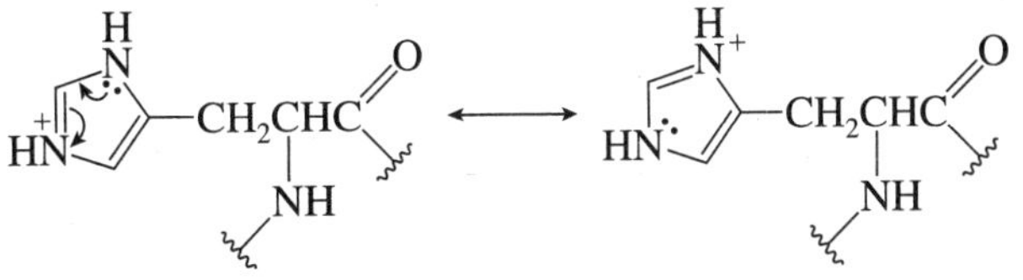

H N $HN^+$ $-CH_2CHC$ O NH ⟷ H $N^+$ HN $-CH_2CHC$ O NH

A

The positive charge in structure B is localized on a single nitrogen. Resonance stabilization of the type shown in structure A is not possible.

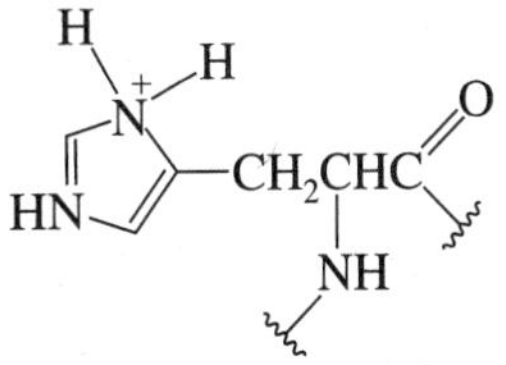

H H $N^+$ HN $-CH_2CHC$ O NH

B

Structure A is the more stable protonated form.

**27.25** The following outlines a synthesis of $\beta$-alanine in which conjugate addition to acrylonitrile plays a key role.

$$\underset{\text{Acrylonitrile}}{H_2C{=}CHC{\equiv}N} \xrightarrow{NH_3} \underset{\text{3-Aminopropanenitrile}}{H_2NCH_2CH_2C{\equiv}N} \xrightarrow[\text{heat}]{H_2O,\ HO^-} \underset{\beta\text{-Alanine}}{H_3\overset{+}{N}CH_2CH_2CO_2^-}$$

Addition of ammonia to acrylonitrile has been carried out in modest yield (31–33%). Hydrolysis of the nitrile group can be accomplished in the presence of either acids or bases. Hydrolysis in the presence of $Ba(OH)_2$ has been reported in the literature to give $\beta$-alanine in 85–90% yield.

**27.26** (*a*) The first step involves alkylation of diethyl malonate by 2-bromobutane.

$$\underset{\text{2-Bromobutane}}{CH_3CH_2\underset{Br}{\underset{|}{C}}HCH_3} + \underset{\text{Anion of diethyl malonate}}{:\bar{C}H(COOCH_2CH_3)_2} \longrightarrow \underset{\text{Compound A}}{CH_3CH_2\underset{CH_3}{\underset{|}{C}}HCH(COOCH_2CH_3)_2}$$

In the second step of the synthesis, compound A is subjected to ester saponification. Following acidification, the corresponding diacid (compound B) is isolated.

$$\underset{\text{Compound A}}{CH_3CH_2\underset{CH_3}{\underset{|}{C}}HCH(COOCH_2CH_3)_2} \xrightarrow[\text{2. HCl}]{\text{1. KOH}} \underset{\text{Compound B }(C_7H_{12}O_4)}{CH_3CH_2\underset{CH_3}{\underset{|}{C}}HCH(COOH)_2}$$

Compound B is readily brominated at its $\alpha$-carbon atom by way of the corresponding enol form.

$$\underset{\text{Compound B}}{(HOOC)_2CH{-}CH(CH_3)CH_2CH_3} \rightleftharpoons \underset{\text{Enol form}}{HOOC{-}C(CH(CH_3)CH_2CH_3){=}C(OH)OH} \xrightarrow{Br_2} \underset{\text{Compound C }(C_7H_{11}BrO_4)}{(HOOC)_2CBr{-}CH(CH_3)CH_2CH_3}$$

When compound C is heated, it undergoes decarboxylation to give an $\alpha$-bromo carboxylic acid.

$$\underset{\text{Compound C}}{CH_3CH_2\underset{CH_3}{\underset{|}{C}}H{-}\underset{Br}{\underset{|}{C}}(COOH)_2} \xrightarrow{\text{heat}} \underset{\text{Compound D}}{CH_3CH_2\underset{CH_3}{\underset{|}{C}}H{-}\underset{Br}{\underset{|}{C}}HCOOH} + \underset{\text{Carbon dioxide}}{CO_2}$$

Treatment of compound D with ammonia converts it to isoleucine by nucleophilic substitution.

$$\underset{\text{Compound D}}{CH_3CH_2\underset{CH_3}{\underset{|}{C}}H{-}\underset{Br}{\underset{|}{C}}HCO_2H} + NH_3 \longrightarrow \underset{\text{Isoleucine (racemic)}}{CH_3CH_2\underset{CH_3}{\underset{|}{C}}H{-}\underset{^+NH_3}{\underset{|}{C}}HCO_2^-}$$

(*b*) The procedure just described can be adapted to the synthesis of other amino acids. The group attached to the $\alpha$-carbon atom is derived from the alkyl halide used to alkylate diethyl malonate. Benzyl bromide (or chloride or iodide) would be appropriate for the preparation of phenylalanine.

$$\underset{\text{Benzyl bromide}}{C_6H_5CH_2Br} \longrightarrow \underset{\text{Phenylalanine (racemic)}}{C_6H_5CH_2\underset{\overset{|}{^+NH_3}}{C}HCO_2^-}$$

**27.27** Acid hydrolysis of the triester converts all its ester functions to free carboxyl groups and cleaves both amide bonds.

$$\text{(phthalimido)}N{-}\overset{CH_2COOCH_2CH_3}{\overset{|}{C}}(COOCH_2CH_3)_2 + 5H_2O \xrightarrow{H^+} H_3\overset{+}{N}{-}\overset{CH_2COOH}{\overset{|}{C}}(COOH)_2 + C_6H_4(COOH)_2$$

The hydrolysis product is a substituted derivative of malonic acid and undergoes decarboxylation on being heated. The product of this decarboxylation is aspartic acid (in its protonated form under conditions of acid hydrolysis).

$$H_3\overset{+}{N}{-}\overset{CH_2COOH}{\overset{|}{C}}(COOH)_2 \xrightarrow{\text{heat}} \underset{\text{Aspartic acid}}{H_3\overset{+}{N}{-}\overset{CH_2COOH}{\overset{|}{C}}HCOOH} + \underset{\text{Carbon dioxide}}{CO_2}$$

Aspartic acid is chiral, but is formed as a racemic mixture, so the product of this reaction is not optically active. The starting triester is achiral and cannot give an optically active product when it reacts with optically inactive reagents.

**27.28** The amino acids leucine, phenylalanine, and serine each have one stereogenic center.

$$R\underset{\overset{|+}{NH_3}}{C}HCO_2^-$$

Leucine: $R{-} = (CH_3)_2CHCH_2{-}$
Phenylalanine: $R{-} = C_6H_5CH_2{-}$
Serine: $R{-} = HOCH_2{-}$

When prepared by the Strecker synthesis, each of these amino acids is obtained as a racemic mixture containing 50% of the D enantiomer and 50% of the L enantiomer.

$$R\overset{O}{\overset{\|}{C}}H + NH_3 + HCN \longrightarrow R\underset{\overset{|}{NH_2}}{C}HC{\equiv}N \xrightarrow[\text{heat}]{H_2O,\ H^+} \underset{\text{Chiral, but racemic}}{R\underset{\overset{|}{^+NH_3}}{C}HCO_2^-}$$

Thus, preparation of the tripeptide Leu-Phe-Ser will yield a mixture of $2^3$ (eight) stereoisomers.

| | |
|---|---|
| D-Leu-D-Phe-D-Ser | L-Leu-L-Phe-L-Ser |
| D-Leu-D-Phe-L-Ser | L-Leu-L-Phe-D-Ser |
| D-Leu-L-Phe-D-Ser | L-Leu-D-Phe-L-Ser |
| D-Leu-L-Phe-L-Ser | L-Leu-D-Phe-D-Ser |

**27.29** Bradykinin is a nonapeptide but contains only five different amino acids. Three of the amino acid residues are proline, two are arginine, and two are phenylalanine. Five peaks will appear on the strip chart after amino acid analysis of bradykinin.

$$\text{Arg-Pro-Pro-Gly-Phe-Ser-Pro-Phe-Arg} \longrightarrow 2\text{Arg} + 3\text{Pro} + \text{Gly} + 2\text{Phe} + \text{Ser}$$

**27.30** Asparagine and glutamine each contain an amide function in their side chain. Under the conditions of peptide bond hydrolysis that characterize amino acid analysis, the side-chain amide is also hydrolyzed, giving ammonia.

$$\underset{\text{Asparagine}}{H_2N\overset{O}{\overset{\|}{C}}CH_2\underset{{}^+NH_3}{\underset{|}{C}}HCO_2^-} + \underset{\text{Water}}{H_2O} \longrightarrow \underset{\text{Ammonia}}{NH_3} + \underset{\text{Aspartic acid}}{HO\overset{O}{\overset{\|}{C}}CH_2\underset{{}^+NH_3}{\underset{|}{C}}HCO_2^-}$$

$$\underset{\text{Glutamine}}{H_2N\overset{O}{\overset{\|}{C}}CH_2CH_2\underset{{}^+NH_3}{\underset{|}{C}}HCO_2^-} + \underset{\text{Water}}{H_2O} \longrightarrow \underset{\text{Ammonia}}{NH_3} + \underset{\text{Glutamic acid}}{HO\overset{O}{\overset{\|}{C}}CH_2CH_2\underset{{}^+NH_3}{\underset{|}{C}}HCO_2^-}$$

**27.31** (*a*) 1-Fluoro-2,4-dinitrobenzene reacts with the amino group of the N-terminal amino acid in a nucleophilic aromatic substitution reaction of the addition–elimination type.

$$\underset{\text{1-Fluoro-2,4-dinitrobenzene}}{2,4\text{-}(O_2N)_2C_6H_3F} + \underset{\text{Leu-Gly-Ser}}{H_3\overset{+}{N}\underset{(CH_3)_2CHCH_2}{\underset{|}{C}}H\overset{O}{\overset{\|}{C}}NHCH_2\overset{O}{\overset{\|}{C}}NH\underset{CH_2OH}{\underset{|}{C}}HCO_2^-} \longrightarrow \underset{\text{DNP-Leu-Gly-Ser}}{2,4\text{-}(O_2N)_2C_6H_3\text{-}NH\underset{(CH_3)_2CHCH_2}{\underset{|}{C}}H\overset{O}{\overset{\|}{C}}NHCH_2\overset{O}{\overset{\|}{C}}NH\underset{CH_2OH}{\underset{|}{C}}HCO_2H}$$

(*b*) Hydrolysis of the product in part (*a*) cleaves the peptide bonds. Leucine is isolated as its 2,4-dinitrophenyl (DNP) derivative, but glycine and serine are isolated as the free amino acids.

$$\underset{\text{DNP-Leu-Gly-Ser}}{2,4\text{-}(O_2N)_2C_6H_3\text{-}NH\underset{(CH_3)_2CHCH_2}{\underset{|}{C}}H\overset{O}{\overset{\|}{C}}NHCH_2\overset{O}{\overset{\|}{C}}NH\underset{CH_2OH}{\underset{|}{C}}HCO_2H} \xrightarrow{\text{hydrolysis}} \underset{\text{DNP-Leu}}{2,4\text{-}(O_2N)_2C_6H_3\text{-}NH\underset{(CH_3)_2CHCH_2}{\underset{|}{C}}H\overset{O}{\overset{\|}{C}}OH} + \underset{\text{Gly}}{H_3\overset{+}{N}CH_2CO_2^-} + \underset{\text{Ser}}{H_3\overset{+}{N}\underset{CH_2OH}{\underset{|}{C}}HCO_2^-}$$

(*c*) Phenyl isothiocyanate is a reagent used to identify the N-terminal amino acid of a peptide by the Edman degradation. The N-terminal amino acid is cleaved as a phenylthiohydantoin (PTH) derivative, the remainder of the peptide remaining intact.

$$H_3\overset{+}{N}CH(CH(CH_3)CH_2CH_3)\overset{O}{\overset{\|}{C}}NHCH(CH_2CH_2CO_2H)\overset{O}{\overset{\|}{C}}NHCH(CH_2C_6H_5)CO_2^- \xrightarrow[\text{2. HBr, nitromethane}]{\text{1. } C_6H_5N{=}C{=}S} \text{PTH derivative} + H_3\overset{+}{N}CH(CH_2CH_2CO_2H)\overset{O}{\overset{\|}{C}}NHCH(CH_2C_6H_5)CO_2^-$$

Ile-Glu-Phe

PTH derivative of isoleucine (S=C ring with N–$C_6H_5$, C=O, HN, and $CHCH_2CH_3$ / $CH_3$ substituent)

Glu-Phe

(*d*) Benzyloxycarbonyl chloride reacts with amino groups to convert them to amides. The only free amino group in Asn-Ser-Ala is the N terminus. The amide function of asparagine does not react with benzyloxycarbonyl chloride.

$$H_3\overset{+}{N}CH(CH_2CONH_2)\overset{O}{\overset{\|}{C}}NHCH(CH_2OH)\overset{O}{\overset{\|}{C}}NHCH(CH_3)CO_2^- + C_6H_5CH_2O\overset{O}{\overset{\|}{C}}Cl \longrightarrow C_6H_5CH_2O\overset{O}{\overset{\|}{C}}NHCH(CH_2CONH_2)\overset{O}{\overset{\|}{C}}NHCH(CH_2OH)\overset{O}{\overset{\|}{C}}NHCH(CH_3)CO_2H$$

Asn-Ser-Ala

Benzyloxycarbonyl chloride

Z-Asn-Ser-Ala

(*e*) The Z-protected tripeptide formed in part (*d*) is converted to its C-terminal *p*-nitrophenyl ester on reaction with *p*-nitrophenol and *N,N′*-dicyclohexylcarbodiimide (DCCI).

$$C_6H_5CH_2O\overset{O}{\overset{\|}{C}}NHCH(CH_2CONH_2)\overset{O}{\overset{\|}{C}}NHCH(CH_2OH)\overset{O}{\overset{\|}{C}}NHCH(CH_3)CO_2H + O_2N{-}C_6H_4{-}OH$$

Z-Asn-Ser-Ala

*p*-Nitrophenol

$$\Big\downarrow \text{DCCI}$$

$$C_6H_5CH_2O\overset{O}{\overset{\|}{C}}NHCH(CH_2CONH_2)\overset{O}{\overset{\|}{C}}NHCH(CH_2OH)\overset{O}{\overset{\|}{C}}NHCH(CH_3)\overset{O}{\overset{\|}{C}}O{-}C_6H_4{-}NO_2$$

Z-Asn-Ser-Ala *p*-nitrophenyl ester

(*f*) The *p*-nitrophenyl ester prepared in part (*e*) is an "active" ester. The *p*-nitrophenyl group is a good leaving group and can be displaced by the amino nitrogen of valine ethyl ester to form a new peptide bond.

$$C_6H_5CH_2OC(=O)NHCH(CH_2C(=O)NH_2)C(=O)NHCH(CH_2OH)C(=O)NHCH(CH_3)C(=O)O\text{—}C_6H_4\text{—}NO_2 + H_2NCH(CH(CH_3)_2)C(=O)OCH_2CH_3$$

Z-Asn-Ser-Ala *p*-nitrophenyl ester — Valine ethyl ester

$$\downarrow$$

$$C_6H_5CH_2OC(=O)NHCH(CH_2C(=O)NH_2)C(=O)NHCH(CH_2OH)C(=O)NHCH(CH_3)C(=O)NHCH(CH(CH_3)_2)C(=O)OCH_2CH_3$$

Z-Asn-Ser-Ala-Val ethyl ester

(*g*) Hydrogenolysis of the Z-protected tetrapeptide ester formed in part (*f*) removes the Z protecting group.

$$C_6H_5CH_2OC(=O)NHCH(CH_2C(=O)NH_2)C(=O)NHCH(CH_2OH)C(=O)NHCH(CH_3)C(=O)NHCH(CH(CH_3)_2)C(=O)OCH_2CH_3 \xrightarrow{H_2,\ Pd} H_2NCH(CH_2C(=O)NH_2)C(=O)NHCH(CH_2OH)C(=O)NHCH(CH_3)C(=O)NHCH(CH(CH_3)_2)C(=O)OCH_2CH_3$$

Z-Asn-Ser-Ala-Val ethyl ester — Asn-Ser-Ala-Val ethyl ester

**27.32** Consider, for example, the reaction of hydrazine with a very simple dipeptide such as Gly-Ala. Hydrazine cleaves the peptide by nucleophilic attack on the carbonyl group of glycine.

$$H_3\overset{+}{N}CH_2C(=O)\text{—}NHCH(CH_3)CO_2^- + H_2\ddot{N}NH_2 \longrightarrow H_2NCH_2C(=O)NHNH_2 + H_3\overset{+}{N}CH(CH_3)CO_2^-$$

Gly-Ala — Hydrazine — Hydrazide of glycine — Alanine

It is the C-terminal residue that is cleaved as the free amino acid and identified in the hydrazinolysis of peptides.

**27.33** Somatostatin is a tetradecapeptide and so is composed of 14 amino acids. The fact that Edman degradation gave the PTH derivative of alanine identifies this as the N-terminal amino acid. A major piece of information is the amino acid sequence of a hexapeptide obtained by partial hydrolysis:

Ala-Gly-Cys-Lys-Asn-Phe

Using this as a starting point and searching for overlaps with the other hydrolysis products gives the entire sequence.

```
Ala-Gly-Cys-Lys-Asn-Phe
                Asn-Phe-Phe-Trp-Lys
                        Phe-Trp
                                Lys-Thr-Phe
                                    Thr-Phe-Thr-Ser-Cys
                                            Thr-Ser-Cys
Ala-Gly-Cys-Lys-Asn-Phe-Phe-Trp-Lys-Thr-Phe-Thr-Ser-Cys
 1   2   3   4   5   6   7   8   9  10  11  12  13  14
```

The disulfide bridge in somatostatin is between cysteine 3 and cysteine 14. Thus, the primary structure is

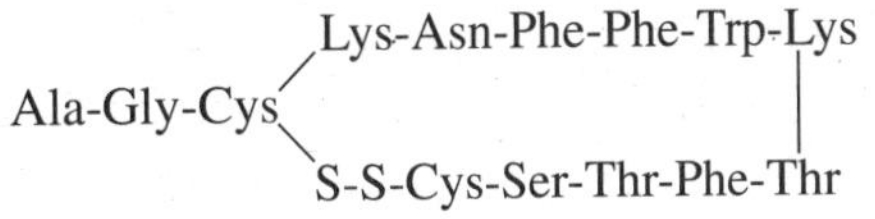

```
             Lys-Asn-Phe-Phe-Trp-Lys
            /                    |
Ala-Gly-Cys                      |
            \                    |
             S-S-Cys-Ser-Thr-Phe-Thr
```

**27.34** It is the C-terminal amino acid that is anchored to the solid support in the preparation of peptides by the Merrifield method. Refer to the structure of oxytocin in Figure 27.8 of the text and note that oxytocin, in fact, has no free carboxyl groups; all the acyl groups of oxytocin appear as amide functions. Thus, the carboxyl terminus of oxytocin has been modified by conversion to an amide.

There are three amide functions of the type $\mathrm{\overset{O}{\overset{\|}{C}}NH_2}$, two of which belong to side chains of asparagine and glutamine, respectively. The third amide belongs to the C-terminal amino acid, glycine, $\mathrm{—NHCH_2\overset{O}{\overset{\|}{C}}OH}$, which in oxytocin has been modified so that it appears as $\mathrm{—NHCH_2\overset{O}{\overset{\|}{C}}NH_2}$. Therefore, attach glycine to the solid support in the first step of the Merrifield synthesis. The carboxyl group can be modified to the required amide after all the amino acid residues have been added and the completed peptide is removed from the solid support.

**27.35** Purine and its numbering system are as shown:

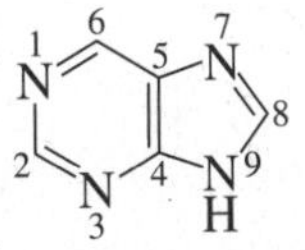

In nebularine, D-ribose in its furanose form is attached to position 9 of purine. The stereochemistry at the anomeric position is $\beta$.

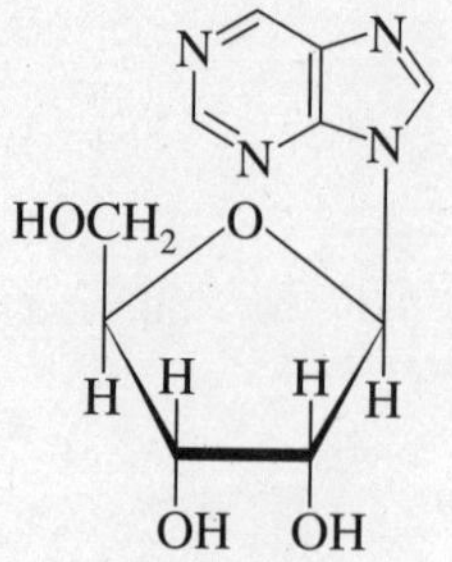

9-$\beta$-D-Ribofuranosylpurine
(nebularine)

**27.36** The problem states that vidarabine is the arabinose analog of adenosine. Arabinose and ribose differ only in their configuration at C-2.

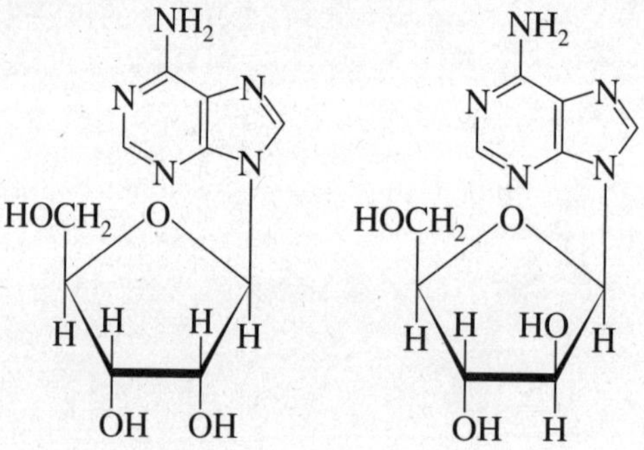

Adenosine

Vidarabine

**27.37** Nucleophilic aromatic substitution occurs when 6-chloropurine reacts with hydroxide ion by an addition–elimination pathway.

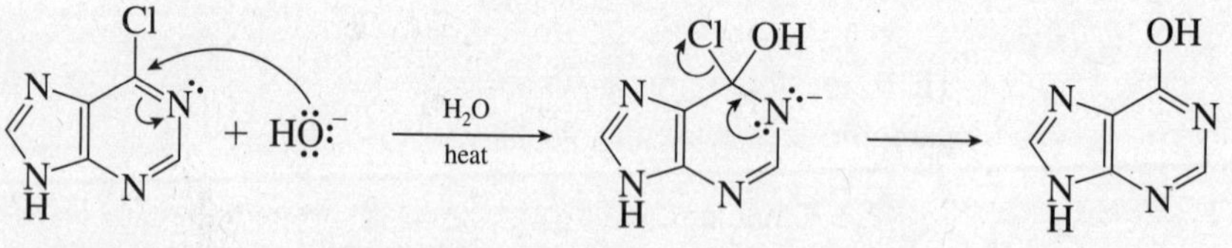

6-Chloropurine

The enol tautomerizes to give hypoxanthine.

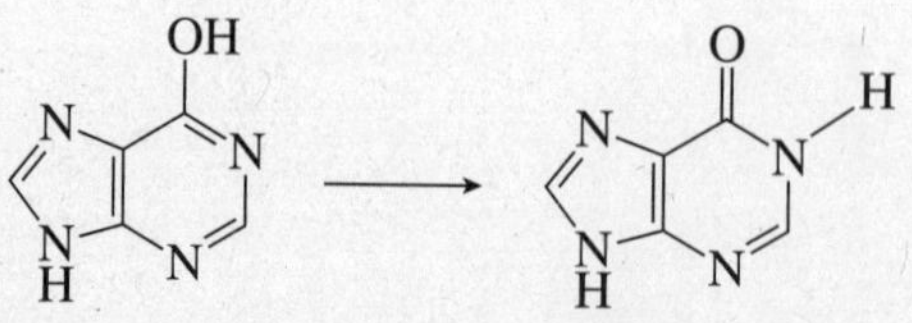

Hypoxanthine

**27.38** Nitrous acid reacts with aromatic primary amines to yield diazonium ions.

Adenosine $\xrightarrow{HONO, H^+}$ (diazonium ion, $N_2^+$ in place of $NH_2$)

Treatment of the diazonium ion with water yields a phenol. Tautomerization gives inosine.

(diazonium ion) $\xrightarrow{H_2O}$ (phenol, OH in place of $N_2^+$) $\longrightarrow$ Inosine

**27.39** The carbon atoms of the ribose portion of a nucleoside are numbered as follows:

(ribose ring: $HOCH_2$ carbon 5′, ring carbons 4′, 3′, 2′, 1′; OH on 3′ and 2′)

(*a*) A 5′-nucleotide has a phosphate group attached to the C-5′ hydroxyl.

(structure: $HO{-}P(=O)(OH){-}OCH_2$ attached to the ribose of inosine)

Inosinic acid

(*b*) Deoxy nucleosides have hydrogens in place of hydroxyl groups at the positions indicated with **boldface.**

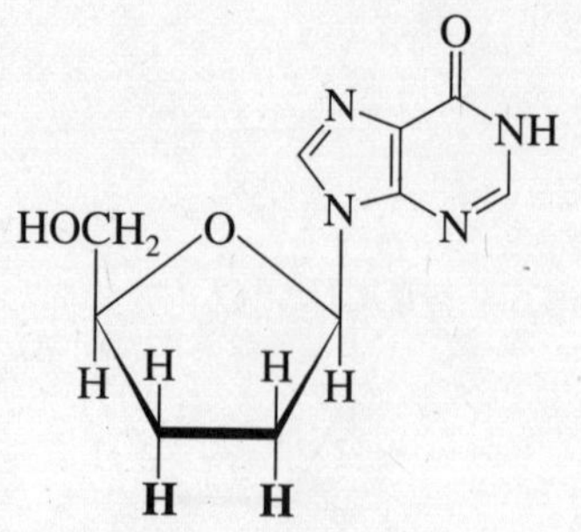

2′,3′-Dideoxyinosine

**27.40** All the bases in the synthetic messenger RNA prepared by Nirenberg were U; therefore, the codon is UUU. By referring to the codons in Table 27.4, we see that the UUU codes for phenylalanine. A polypeptide in which all the amino acid residues were phenylalanine was isolated in Nirenberg's experiment.

# SELF-TEST

## PART A

**A-1.** Give the structure of the reactant, reagent, or product omitted from each of the following:

(*a*) $? \xrightarrow[\text{2. } H_3O^+\text{, heat; 3. neutralize}]{\text{1. } NH_4Cl\text{, NaCN}} C_6H_5CH_2\overset{\overset{+}{N}H_3}{|}{C}HCO_2^-$

(*b*) $C_6H_5CH_2OCCl$ (C=O) + valine $\xrightarrow[\text{2. } H^+]{\text{1. } HO^-\text{, } H_2O}$ ?

(*c*) Boc-Phe + $H_2NCH_2CO_2CH_2CH_3$ $\xrightarrow{?}$ Boc—$NHCH(CH_2C_6H_5)C(=O)NHCH_2CO_2CH_2CH_3$

**A-2.** Give the structure of the derivative that would be obtained by treatment of Phe-Ala with Sanger's reagent followed by hydrolysis.

**A-3.** Outline a sequence of steps that would allow the following synthetic conversions to be carried out:

(*a*) $(CH_3)_2CHCH_2CH(\overset{+}{N}H_3)CO_2^-$ (leucine) from $CH_3C(=O)NHCH(CO_2CH_2CH_3)_2$

(*b*) Leu-Val from leucine and $(CH_3)_2CHCH(\overset{+}{N}H_3)CO_2^-$ (valine)

**A-4.** The carboxypeptidase-catalyzed hydrolysis of a pentapeptide yielded phenylalanine (Phe). One cycle of an Edman degradation gave a derivative of leucine (Leu). Partial hydrolysis yielded the fragments Leu-Val-Gly and Gly-Ala among others. Deduce the structure of the peptide.

**A-5.** Consider the following compound:

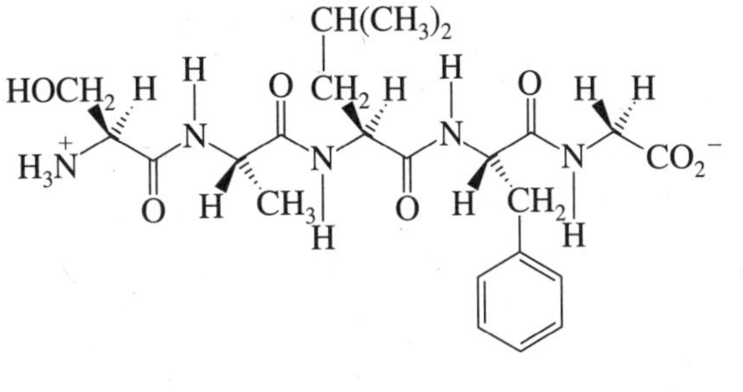

$CH(CH_3)_2$

$HOCH_2$ H H O $CH_2$ H H O H H

$H_3\overset{+}{N}$ N N N N $CO_2^-$

O H $CH_3$ H O H $CH_2$ H

(*a*) What kind of peptide does this structure represent? (For example, dipeptide)
(*b*) How many peptide bonds are present?
(*c*) Give the name for the N-terminal amino acid.
(*d*) Give the name for the C-terminal amino acid.
(*e*) Using three-letter abbreviations, write the sequence.

**A-6.** Consider the tetrapeptide Ala-Gly-Phe-Leu. What are the products obtained from each of the following? Be sure to account for all the amino acids of the peptide.
(*a*) Treatment with 1-fluoro-2,4-dinitrobenzene followed by hydrolysis in concentrated HCl at 100°C.
(*b*) Treatment with chymotrypsin.
(*c*) Treatment with carboxypeptidase
(*d*) Reaction with benzyloxycarbonyl chloride

## PART B

**B-1.** Which phrase correctly completes the statement?
Except for glycine, which is achiral, all the amino acids present in proteins ...
(*a*) are chiral, but racemic
(*b*) are meso forms
(*c*) have the L configuration at their $\alpha$ carbon
(*d*) have the *R* configuration at their $\alpha$ carbon
(*e*) have the *S* configuration at their $\alpha$ carbon

**B-2.** Which statement correctly describes the difference in the otherwise similar chemical constituents of DNA and RNA?
(*a*) DNA contains uracil; RNA contains thymine.
(*b*) DNA contains guanine but not adenine; RNA contains both.
(*c*) DNA contains thymine; RNA contains uracil.
(*d*) None of these applies—the chemical constitution is the same.

**B-3.** Assume that a particular amino acid has an isoelectric point of 6.0. In a solution of pH 1.0, which of the following species will predominate?

(*a*) $H_3\overset{+}{N}\overset{R}{\overset{|}{C}}HCO_2H$ (*c*) $H_3\overset{+}{N}\overset{R}{\overset{|}{C}}HCO_2^-$

(*b*) $H_2N\overset{R}{\overset{|}{C}}HCO_2H$ (*d*) $H_2N\overset{R}{\overset{|}{C}}HCO_2^-$

**B-4.** Choose the response which provides the best match of terms.

| | Purine | Pyrimidine |
|---|---|---|
| (*a*) | Adenine | Guanine |
| (*b*) | Thymine | Cytosine |
| (*c*) | Cytosine | Adenine |
| (*d*) | Guanine | Cytosine |

**B-5.** Which of the following reagents would be combined in the synthesis of Phe-Ala?

[In phenylalanine (Phe), R in the generalized amino acid formula $H_2N\underset{|}{C}HCO_2H$ (R on the α-carbon) is $CH_2C_6H_5$, and in alanine (Ala) it is $CH_3$.]

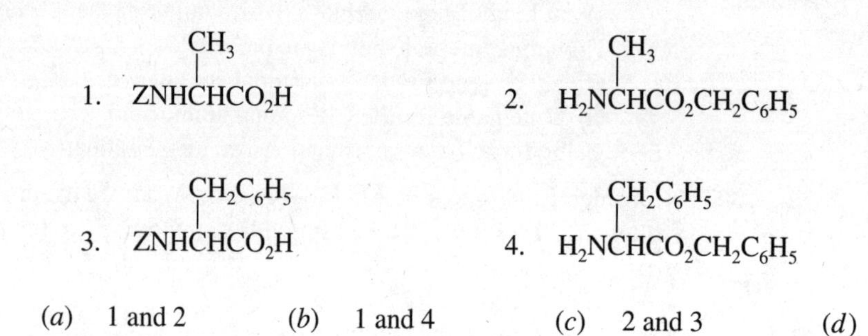

1. $ZNHCH(CH_3)CO_2H$
2. $H_2NCH(CH_3)CO_2CH_2C_6H_5$
3. $ZNHCH(CH_2C_6H_5)CO_2H$
4. $H_2NCH(CH_2C_6H_5)CO_2CH_2C_6H_5$

(*a*) 1 and 2 (*b*) 1 and 4 (*c*) 2 and 3 (*d*) 3 and 4

**B-6.** A nucleoside is a
(*a*) Phosphate ester of a nucleotide
(*b*) Unit having a sugar bonded to a purine or pyrimidine base
(*c*) Chain whose backbone consists of sugar units connected by phosphate groups
(*d*) Phosphate salt of a purine or pyrimidine base

**B-7.** What are the products obtained following treatment of Ser-Tyr-Val-Ala with chymotrypsin?
(*a*) Serine + Tyr-Val-Ala
(*b*) Ser-Tyr + Valine + Alanine
(*c*) Ser-Tyr + Val-Ala
(*d*) Ser-Tyr-Val + Alanine
(*e*) Serine + Tyrosine + Val-Ala

**B-8.** The first cycle of the Edman degradation of the tetrapeptide Gly-Ala-Ile-Leu would give a PTH derivative of
(*a*) Glycine
(*b*) Alanine
(*c*) Isoleucine
(*d*) Leucine

# APPENDIX A

## ANSWERS TO THE SELF-TESTS

## CHAPTER 1

**A-1.** (*a*) P; $1s^22s^22p^63s^23p^3$ (*b*) $S^{2-}$; $1s^22s^22p^63s^23p^6$

**A-2.** (*a*) $:\ddot{N}{=}C{=}\ddot{S}:$

Formal charge: −1 0 0 Net charge: −1

(*b*) $:O{\equiv}N{-}\ddot{\underset{..}{O}}:$

Formal charge: +1 +1 −1 Net charge: +1

(*c*) $HC(-\ddot{\underset{..}{O}}:){=}NH_2$

Formal charge: −1 (O), 0 (C), +1 (N) Net charge: 0

**A-3.** (*a*) $:N{\equiv}C{-}\ddot{\underset{..}{S}}:$

Formal charge: 0 0 −1 Net charge: −1

(*b*) $:\ddot{O}{=}N{=}\ddot{O}:$

Formal charge: 0 +1 0 Net charge: +1

(*c*) $HC(=\ddot{O}:){-}\ddot{N}H_2$

Formal charge: 0 (O), 0 (C), 0 (N) Net charge: 0

The more stable Lewis structures are

(*a*) $^{-}:\ddot{N}{=}C{=}\ddot{S}:$ (*b*) $:\ddot{O}{=}\overset{+}{N}{=}\ddot{O}:$ (*c*) $HC(=\ddot{O}:){-}\ddot{N}H_2$

**A-4.** (*a*) $H{-}C(H)(H){-}\ddot{N}(H){-}H$ (*b*) $H{-}C(H)(H){-}C(H){=}\ddot{O}:$

**A-5.** (*a*) $C_{12}H_{20}O$ :OH (*c*) $C_{14}H_{24}O$ Ö:

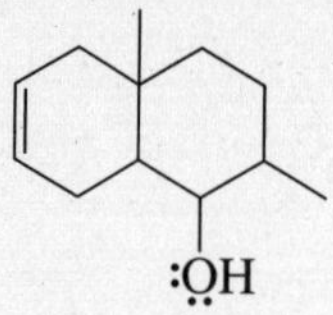

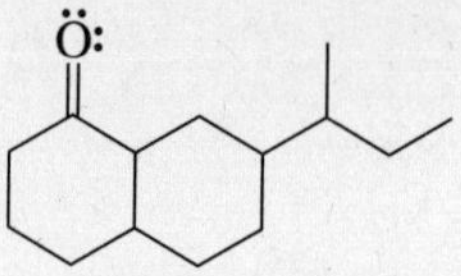

(*b*) $C_{10}H_{22}$ (*d*) $C_9H_6BrN$ N Br:

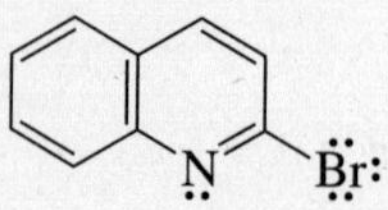

**A-6.** (*a*) 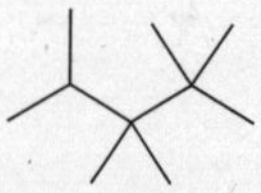 has only $sp^3$-hybridized carbon atoms

(*b*) N Br: has only $sp^2$-hybridized carbon atoms

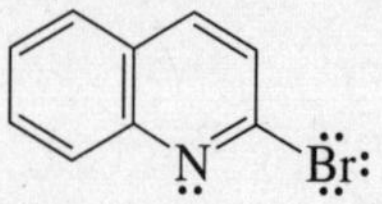

(*c*) Ö: has only one $sp^2$-hybridized carbon atom

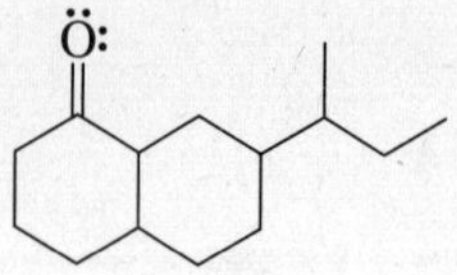

**A-7.** :O: $S^{2+}$ $^-$:Ö: :Ö:$^-$ ⟷ :Ö:$^-$ $S^{2+}$ $^-$:Ö: Ö: ⟷ :Ö:$^-$ $S^{2+}$ Ö: :Ö:$^-$

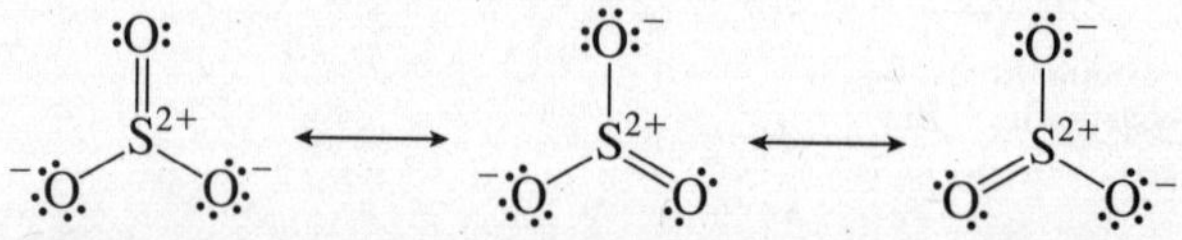

**A-8.** :Ö—C≡N:

Formal charge: −1 0 0 Net charge: −1

**A-9.** (*a*) H—C(H)(H)—Ö—C(H)=C(H)—C(H)(H)—H

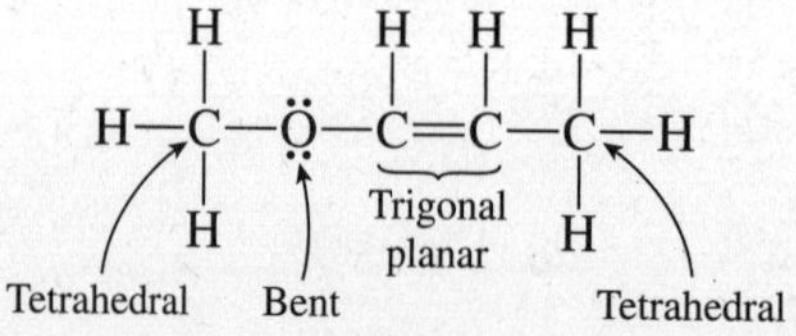

Tetrahedral Bent Trigonal planar Tetrahedral

(*b*) Pyramidal; :N(Cl)(Cl)(Cl) yes, it is polar.

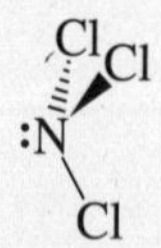

**A-10.** (*a*) Linear (*b*) Linear (*c*) Bent

**A-11.** (*a*) D (*c*) None (*e*) None (*g*) A
(*b*) A, B (*d*) B (*f*) A, D (*h*) C

**A-12.** H—N̈(H)—C(=Ö:)—N̈(H)—H

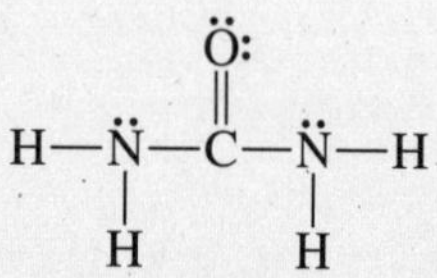

**A-13.** (*a*) 11 $\sigma$; 1 $\pi$ (*b*) 9 $\sigma$; 2 $\pi$ (*c*) 12 $\sigma$; 4 $\pi$ (*d*) 13 $\sigma$; 4 $\pi$

**A-14.** (*a*) $H_3C-CH=CH-CH_3$ (terminal carbons $sp^3$; CH=CH carbons $sp^2$)

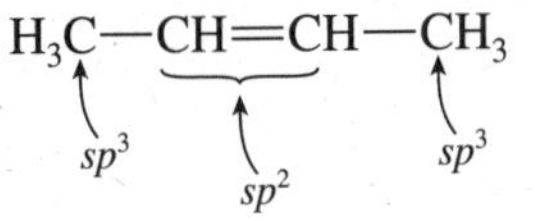

(*b*) $H-C\equiv C-CH_2-CH_3$ ($C\equiv C$ carbons $sp$; $CH_2-CH_3$ carbons $sp^3$)

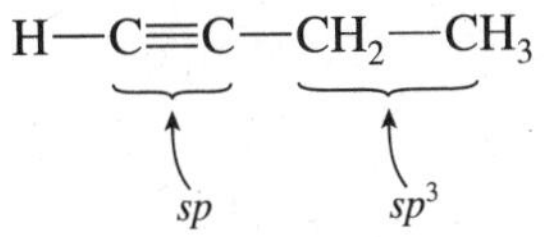

(*c*) All carbons are $sp^2$.

(*d*) Cyclopentenone bearing C≡N: carbonyl carbon $sp^2$; C≡N carbon $sp$; ring C=C carbons $sp^2$; ring carbon bearing CN $sp^3$

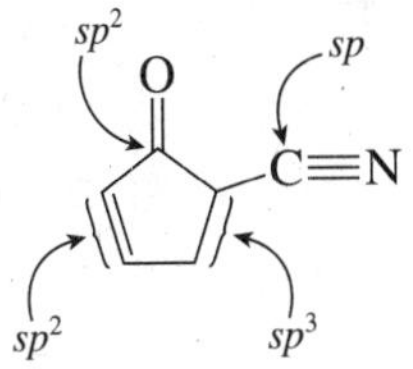

**B-1.** (*b*) **B-2.** (*b*) **B-3.** (*c*) **B-4.** (*d*)
**B-5.** (*a*) **B-6.** (*b*) **B-7.** (*a*) **B-8.** (*d*)
**B-9.** (*b*) **B-10.** (*d*) **B-11.** (*b*) **B-12.** (*e*)
**B-13.** (*d*) **B-14.** (*b*) **B-15.** (*d*)

# CHAPTER 2

**A-1.**

| | $CH_3CH_2CH_2CH_2-$ | $CH_3CH_2CHCH_3$ (attached at CH) |
|---|---|---|
| **Common:** | *n*-Butyl | *sec*-Butyl |
| **Systematic:** | Butyl | 1-Methylpropyl |

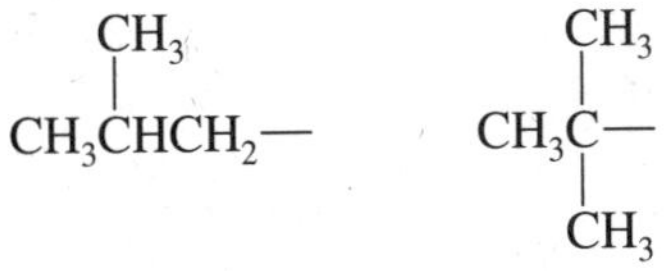

| | $(CH_3)_2CHCH_2-$ | $(CH_3)_3C-$ |
|---|---|---|
| **Common:** | Isobutyl | *tert*-Butyl |
| **Systematic:** | 2-Methylpropyl | 1,1-Dimethylethyl |

**A-2.** (*a*) 28 (8 C—C; 20 C—H) (*b*) 27(9 C—C; 18 C—H)

**A-3.** (*a*) Oxidized (*b*) Neither (*c*) Neither (*d*) Reduced

**A-4.** (*a*) $CH_3CHCHCHCH_3$ with $CH_3CHCH_3$ on C-3 and $CH_3$ on C-2 and C-4 (*b*) Six methyl groups, three isopropyl groups

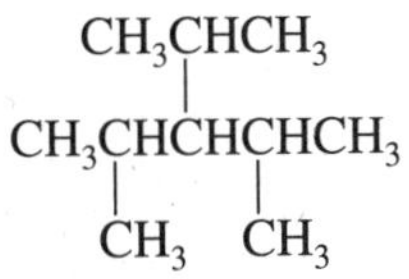

**A-5.** (*a*) 3,4-Dimethylheptane (*b*) (1,2-Dimethylpropyl)cyclohexane

**A-6.**

| | Primary | Secondary | Tertiary |
|---|---|---|---|
| (*a*) | 4 | 3 | 2 |
| (*b*) | 3 | 5 | 3 |

**A-7.** (*a*) 1,3-Dimethylbutyl; secondary
(*b*) 1,1-Diethylpropyl; tertiary
(*c*) 2,2-Diethylbutyl; primary

**A-8.** $\underset{\displaystyle CH_3}{CH_3CH}\overset{\displaystyle CH_3}{CH}CH_2CH_3 \equiv C_7H_{16}$ $\qquad C_7H_{16} + 11O_2 \longrightarrow 7CO_2 + 8H_2O$

**A-9.**

Cyclopentane

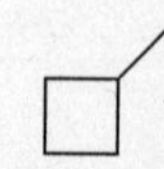

Methylcyclobutane

Ethylcyclopropane

1,1-Dimethylcyclopropane

1,2-Dimethylcyclopropane

**A-10.** (*a*)

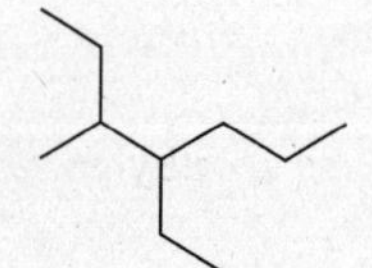

4-Ethyl-3-methylheptane

(*c*)

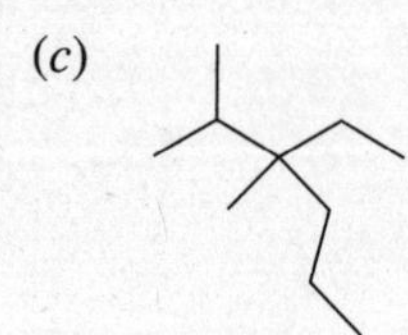

3-Ethyl-2,3-dimethylhexane

(*b*)

(2-Methylbutyl)cyclohexane

**A-11.** (*a*) $CH_3CH_2CH_2CH_2CH_2CH_2CH_2CH_3$

(*b*) $(CH_3)_3CC(CH_3)_3$

(*c*) $(CH_3)_2CH\underset{\displaystyle CH_3}{C}HCH(CH_3)_2$

(*d*) $(CH_3)_3CC(CH_3)_3$

**A-12.**

2,2-Dimethylpentane

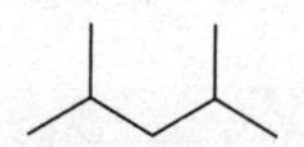

2,4-Dimethylpentane

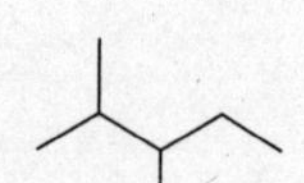

2,3-Dimethylpentane

3-Ethylpentane

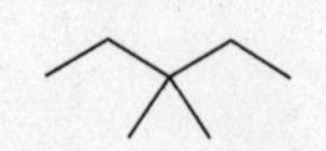

3,3-Dimethylpentane

**A-13.** Alcohol, alkene, ester, ketone

**A-14.** 10,049 kJ/mol

| | | | |
|---|---|---|---|
| **B-1.** (*a*) | **B-2.** (*d*) | **B-3.** (*d*) | **B-4.** (*c*) |
| **B-5.** (*b*) | **B-6.** (*a*) | **B-7.** (*c*) | **B-8.** (*c*) |
| **B-9.** (*a*) | **B-10.** (*a*) | **B-11.** (*b*) | **B-12.** (*e*) |
| **B-13.** (*d*) | **B-14.** (*d*) | | |

# CHAPTER 3

**A-1.** Cl; H, $CH_3$; H, H; H — Gauche. Cl; H, H; H, H; $CH_3$ — Anti

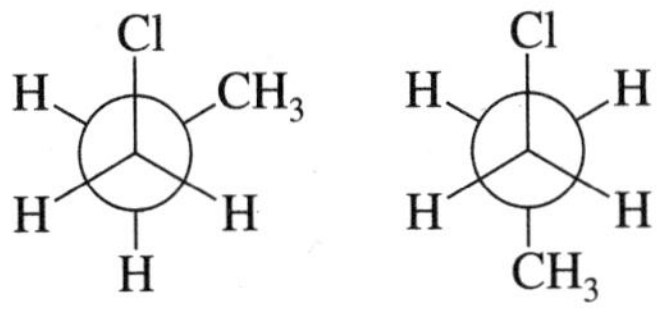

**A-2.** (*a*) $H_3C$ $CH_3$; H, H; H, H (Eclipsed) (*b*) Cl; Cl, Cl; Cl, H; H — Cl; H, Cl; Cl, H; Cl

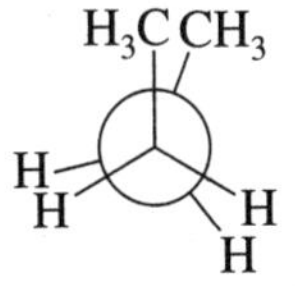

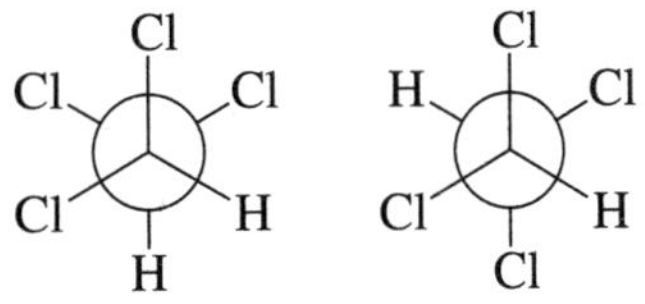

**A-3.** $(CH_3)_3CCH_2C(CH_3)_3$ = 2,2,4,4-tetramethylpentane

**A-4.** $(CH_3)_3C$; H; H; $CH_3$

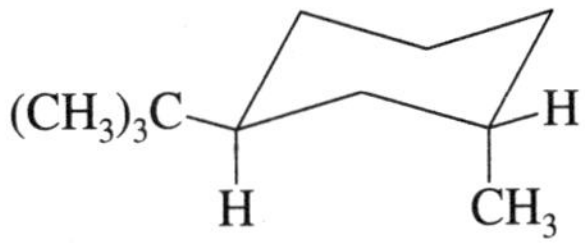

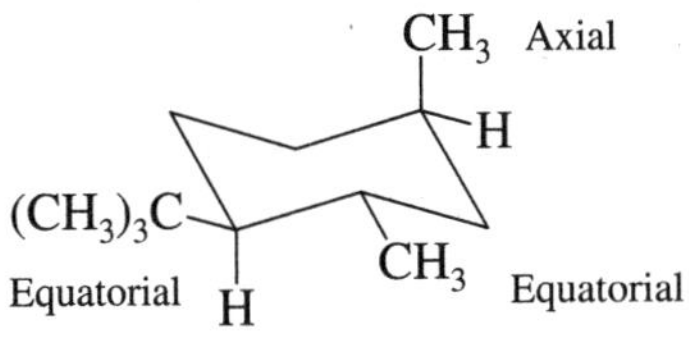

**A-5.** $CH_3$ Axial; H; $(CH_3)_3C$ Equatorial; H; $CH_3$ Equatorial

**A-6.** H; H; O; HO; H; OH; OH; HO; H

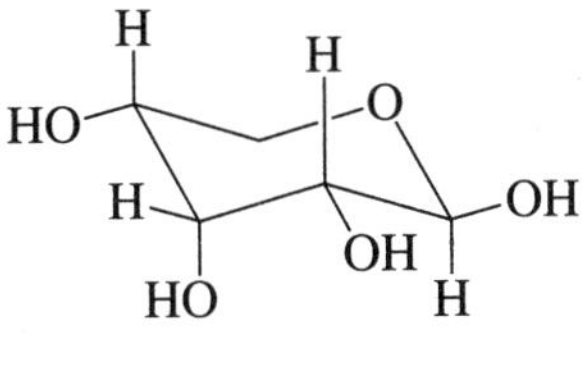

**A-7.** (*a*) C (*b*) A and B (*c*) D (*d*) A

**A-8.** $CH(CH_3)_2$; H; $H_3C$; H — $CH_3$; H; $CH(CH_3)_2$; H — More stable

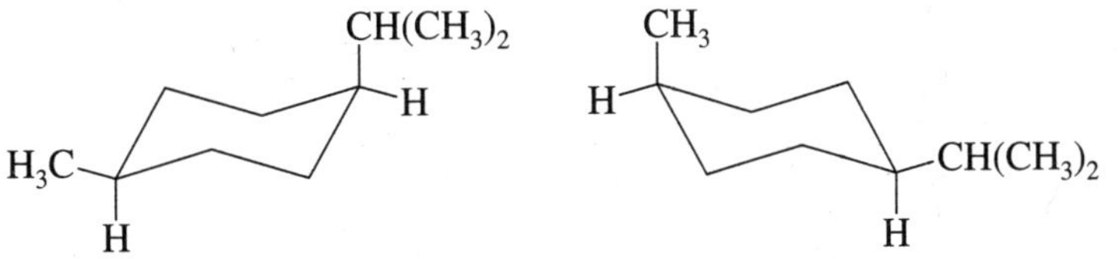

**A-9.** *cis*-1-Ethyl-3-methylcyclohexane has the lower heat of combustion.

**A-10.** Tricyclic; $C_{10}H_{16}$

**A-11.** The form of the curve more closely resembles ethane than butane.

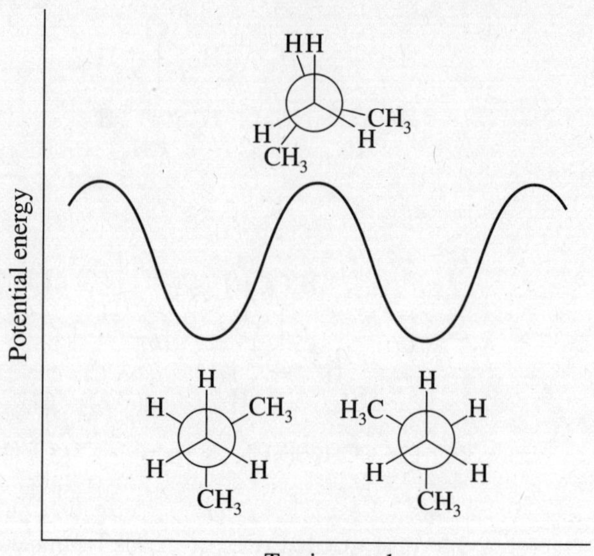


**A-12.** (structure: five-membered ring containing S)

| | | | | | | | |
|---|---|---|---|---|---|---|---|
| **B-1.** | (*d*) | **B-2.** | (*b*) | **B-3.** | (*c*) | **B-4.** | (*a*) |
| **B-5.** | (*c*) | **B-6.** | (*a*) | **B-7.** | (*d*) | **B-8.** | (*e*) |
| **B-9.** | (*c*) | **B-10.** | (*e*) | **B-11.** | (*b*) | **B-12.** | (*a*) |
| **B-13.** | (*d*) | **B-14.** | (*b*) | | | | |

# CHAPTER 4

**A-1.** (*a*) *trans*-1-Bromo-3-methylcyclopentane
(*b*) 2-Ethyl-4-methyl-1-hexanol

**A-2.** (*a*) $ICH_2C(CH_3)(Cl)CH_2CH_2CH_2CH_2CH_3$ (b) (cyclohexane ring bearing OH and $CH(CH_3)_2$)

**A-3.** (*a*) **Functional class:** 1-ethyl-3-methylbutyl alcohol
**Substitutive:** 5-methyl-3-hexanol
(*b*) **Functional class:** 1,1,2-trimethylbutyl chloride
**Substitutive:** 2-chloro-2,3-dimethylpentane

**A-4.** Conjugate acid $CH_3\ddot{O}H_2^+$; conjugate base $CH_3\ddot{O}:^-$

**A-5.** (*a*) $CH_3CH_2CH_2Cl$ (*b*) $CH_3CH_2C(OH)(CH_3)_2$

**A-6.** (*a*) $CH_3CH_2O^- + NH_3 \rightleftharpoons CH_3CH_2OH + NH_2^-$ ($K < 1$)

Conjugate base, Conjugate acid; Stronger acid, Stronger base

(*b*) $CH_3CH_2\overset{\delta-}{\ddot{O}}\cdots H \cdots \overset{\delta-}{\ddot{N}}H_2$

**A-7.** (*a*) Three

(*b, c*) $CH_3\dot{C}(CH_3)CH_2CH(CH_3)CH_3$ (Most stable); $CH_3CH(CH_3)\dot{C}HCH(CH_3)CH_3$; $\dot{C}H_2CH(CH_3)CH_2CH(CH_3)CH_3$ (Least stable)

**A-8.** $CH_3C(CH_3)_2CH_3 + Cl_2 \longrightarrow CH_3C(CH_3)_2CH_2Cl + HCl$

**A-9.** Br· + methylcyclohexane ⟶ HBr + 1-methylcyclohexyl radical

1-methylcyclohexyl radical + $Br_2$ ⟶ 1-bromo-1-methylcyclohexane + Br·

**A-10.** $\Delta H^\circ = -57$ kJ ($-13.5$ kcal)

**A-11.** (*a*) $(CH_3)_3C{-}\ddot{O}H + H{-}Br \rightleftharpoons (CH_3)_3C{-}\overset{+}{O}H_2 + Br^-$

$(CH_3)_3C{-}\overset{+}{O}H_2 \rightleftharpoons H_2O + (CH_3)_3C^+$

$(CH_3)_3C^+ + :\ddot{Br}:^- \longrightarrow (CH_3)_3C{-}Br$

(*b*) $(H_3C)_3\overset{\delta+}{C}\cdots\overset{\delta+}{O}H_2$

(*c*) Water is displaced directly from the oxonium ion of 1-butanol by bromide ion. A primary carbocation is not involved.

$:\ddot{Br}:^- + CH_3CH_2CH_2CH_2{-}\overset{+}{O}H_2 \rightleftharpoons H_2O + CH_3CH_2CH_2CH_2Br$

**A-12.** (*a*) 3-Methyl-3-pentanol (*b*) $KOC(CH_3)_3$ (*c*) Fluorine ($F_2$) (*d*) Ethyl radical, $CH_3\dot{C}H_2$ (*e*) $Cl_2$

**B-1.** (*e*) **B-2.** (*c*) **B-3.** (*b*) **B-4.** (*c*)
**B-5.** (*e*) **B-6.** (*c*) **B-7.** (*d*) **B-8.** (*a*)
**B-9.** (*c*) **B-10.** (*d*) **B-11.** (*c*) **B-12.** (*e*)
**B-13.** (*a*) **B-14.** (*c*) **B-15.** (*c*) **B-16.** (*c*)

# CHAPTER 5

**A-1.** (*a*) 2,4,4-Trimethyl-2-pentene

(*b*) (*E*)-3,5-Dimethyl-4-octene

(*c*) (*E*)-2,7-Dibromo-3-(2-methylpropyl)-2-heptene

(*d*) 5-Methyl-4-hexen-3-ol

**A-2.** (*a*)

2,3-Dimethyl-2-pentene

(*b*)

5-Chloro-2-methyl-1-hexene

(*c*)

1,6-Dimethylcyclohexene

(*d*)

4-Methyl-4-penten-2-ol

**A-3.** (*a*)

1 2 3 4 5 6

(*b*) Isomer 5 (*c*) Isomers 1 and 4 (*d*) Isomers 2 and 3

**A-4.** Two $sp^2$ C atoms; four $sp^3$ C atoms; three $sp^2$—$sp^3$ $\sigma$ bonds

**A-5.** (*a*) (*b*) (*c*)

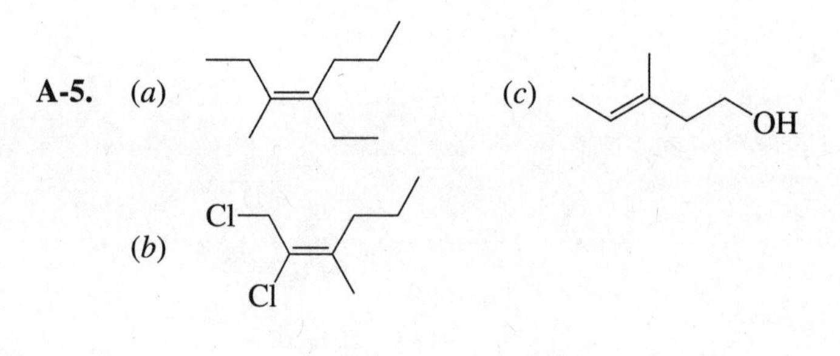


**A-6.**

(*Z*)-3-Methyl-3-hexene (*E*)-3-Methyl-3-hexene

**A-7.** (*a*) $H_2C{=}\underset{\displaystyle CH_3}{C}CH_2CH_2CH_3$ + $(CH_3)_2C{=}CHCH_2CH_3$ (major)

(*b*) $CH_2$ $CH_3$ + (major)

(*c*) $(CH_3)_2CH\overset{\displaystyle X}{C}HCH(CH_3)_2$

(X = Cl, Br, I)

(*d*) $H_2C{=}\overset{\displaystyle CH_3}{C}C(CH_3)_3$

**A-8.** $CH_3CH_2C(CH_3)_2CH_2Br$

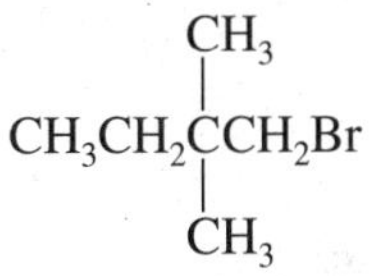

**A-9.** **Step 1:** Protonation

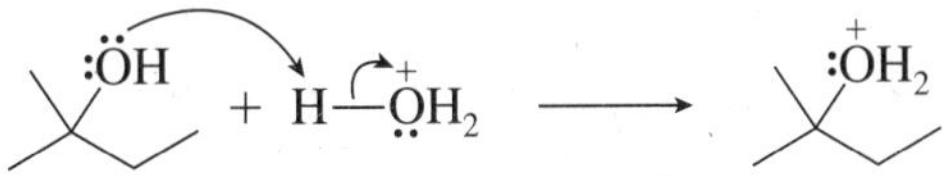

+ $H-\ddot{O}H_2^+$ → $:\overset{+}{O}H_2$

**Step 2:** Dissociation

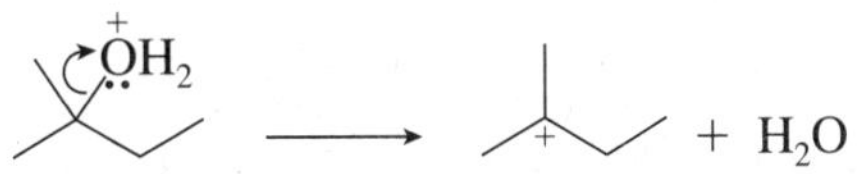

→ + $H_2O$

**Step 3:** Deprotonation

H H + $:\ddot{O}H_2$ → + $H_3O^+$

**A-10.** Br $\xrightarrow[CH_3CH_2OH]{NaOCH_2CH_3}$

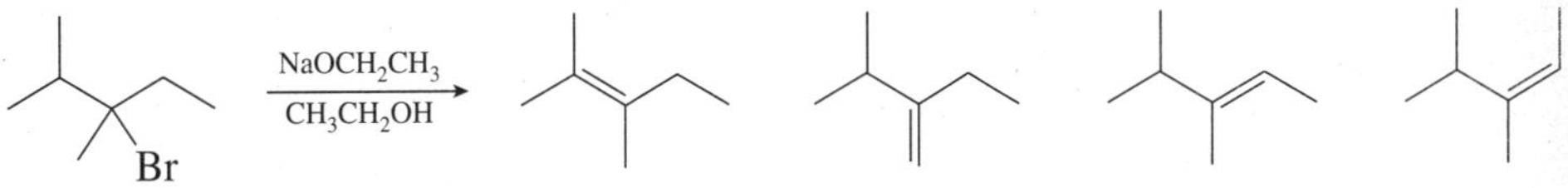

(major)

**A-11.** $CH_3\ddot{O}:^-$ H $CH_3$ Br → $CH_3$ + $Br^-$

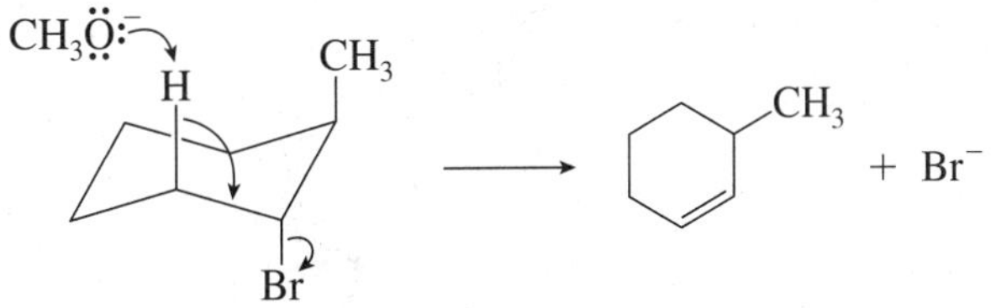

**A-12.** Cis isomer:

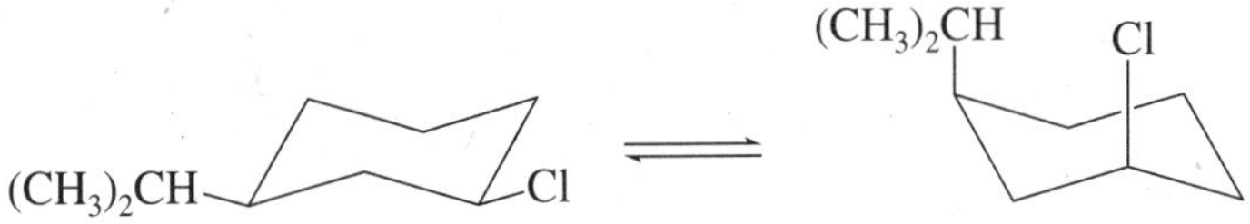

$(CH_3)_2CH$ Cl ⇌ $(CH_3)_2CH$ Cl

Trans isomer:

$(CH_3)_2CH$ Cl ⇌ $CH(CH_3)_2$ Cl

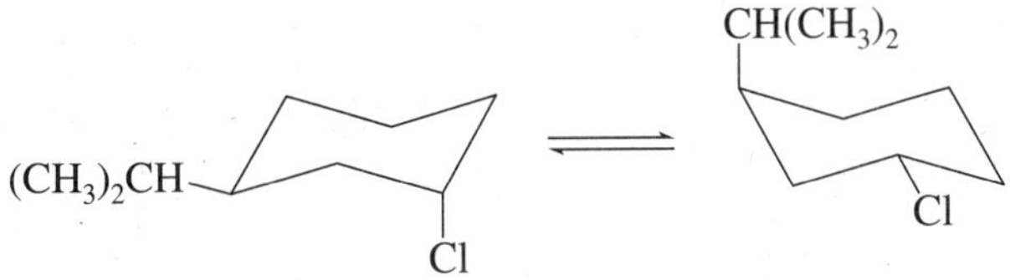

The trans isomer will react faster because its most stable conformation (with the isopropyl group equatorial) has an axial Cl able to undergo E2 elimination.

**A-13.** Rearrangement (hydride migration) occurs to form a more stable carbocation.

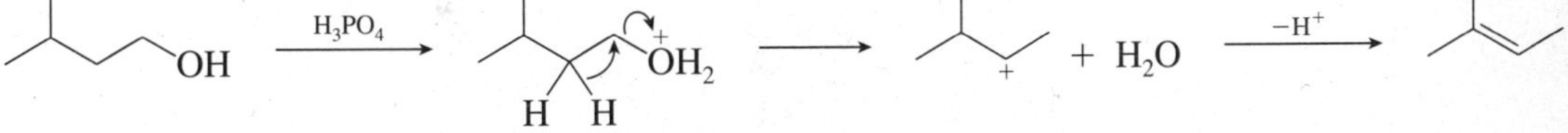

OH $\xrightarrow{H_3PO_4}$ H H $\overset{+}{O}H_2$ → + $H_2O$ $\xrightarrow{-H^+}$

**A-14.**

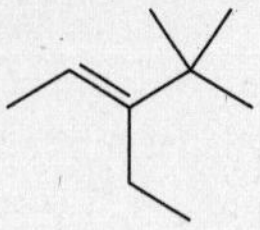

3-Ethyl-4,4-dimethyl-2-pentene

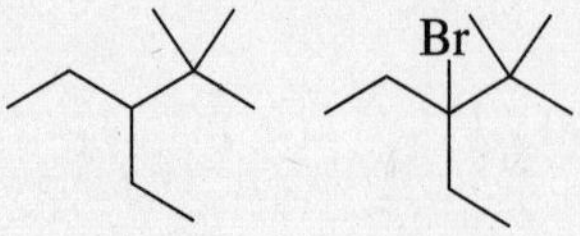

Br

A B

| | | | |
|---|---|---|---|
| **B-1.** (*c*) | **B-2.** (*c*) | **B-3.** (*d*) | **B-4.** (*c*) |
| **B-5.** (*a*) | **B-6.** (*b*) | **B-7.** (*a*) | **B-8.** (*a*) |
| **B-9.** (*a*) | **B-10.** (*d*) | **B-11.** (*b*) | **B-12.** (*c*) |
| **B-13.** (*a*) | **B-14.** (*c*) | **B-15.** (*a*) | |

# CHAPTER 6

**A-1.** Five;

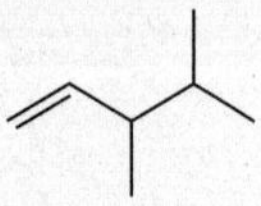

3,4-Dimethyl-1-pentene

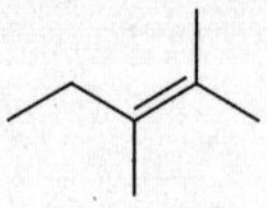

2,3-Dimethyl-2-pentene

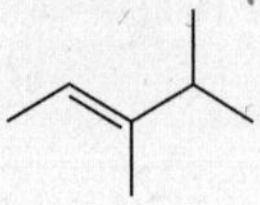

(*E*)-3,4-Dimethyl-2-pentene

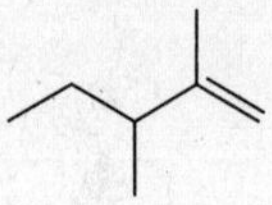

2,3-Dimethyl-1-pentene

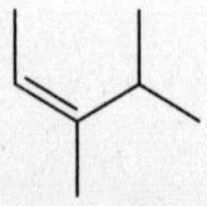

(*Z*)-3,4-Dimethyl-2-pentene

**A-2.** (*a*) $(CH_3)_2C(OH)CH_2CH_3$

(*c*)

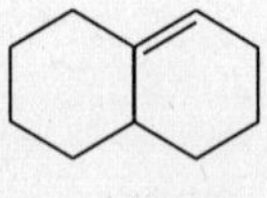

(*b*) HBr, peroxides

(*d*) $CH_3$ OH Br

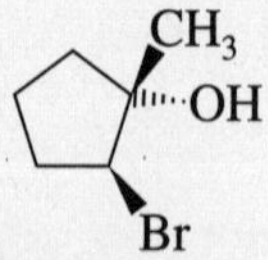

**A-3.** (*a*) $CH_3$ OH $\xrightarrow[\text{heat}]{H_2SO_4\text{ (conc)}}$ $CH_3$ $\xrightarrow[2.\ H_2O_2,\ HO^-]{1.\ B_2H_6}$ $CH_3$ OH

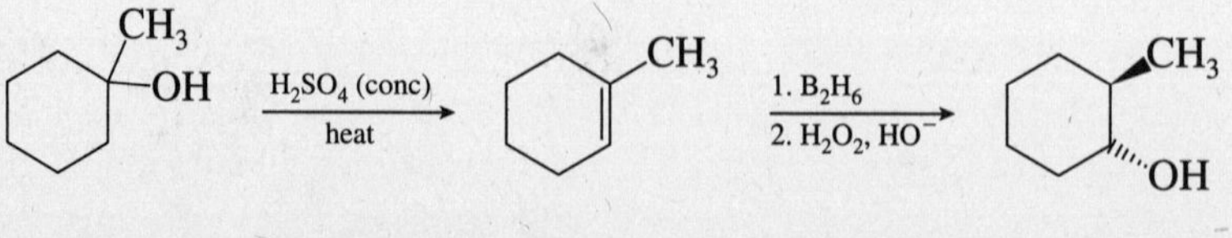

(*b*)

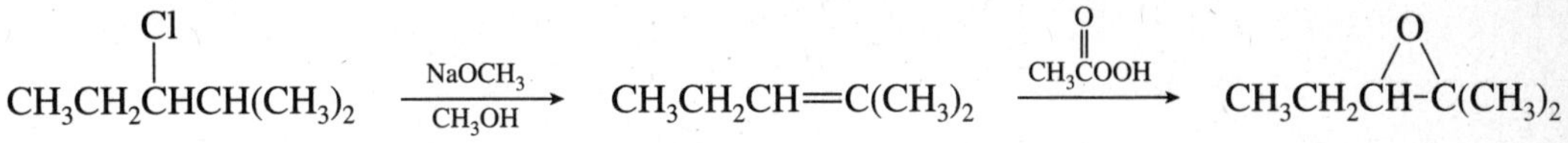

$$CH_3CH_2CHClCH(CH_3)_2 \xrightarrow[CH_3OH]{NaOCH_3} CH_3CH_2CH{=}C(CH_3)_2 \xrightarrow{CH_3COOOH} CH_3CH_2CH\text{-}C(CH_3)_2 \text{ (epoxide, O bridging)}$$

(*c*)

$$(CH_3)_3CCHBrCH_3 \xrightarrow[CH_3OH]{CH_3O^-Na^+} (CH_3)_3CCH{=}CH_2 \xrightarrow[Peroxides]{HBr} (CH_3)_3CCH_2CH_2Br$$

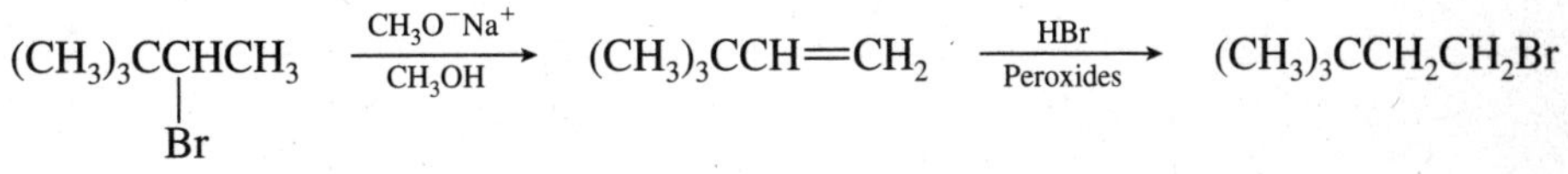

**A-4.** **Initiation:**

$$ROOR \xrightarrow[\text{or heat}]{\text{light}} 2RO\cdot$$

$$RO\cdot + HBr \longrightarrow ROH + Br\cdot$$

**Propagation:**

$$Br\cdot + CH_3CH_2CH{=}CH_2 \longrightarrow CH_3CH_2\dot{C}HCH_2Br$$

$$CH_3CH_2\dot{C}HCH_2Br + HBr \longrightarrow CH_3CH_2CH_2CH_2Br + Br\cdot$$

**A-5.**

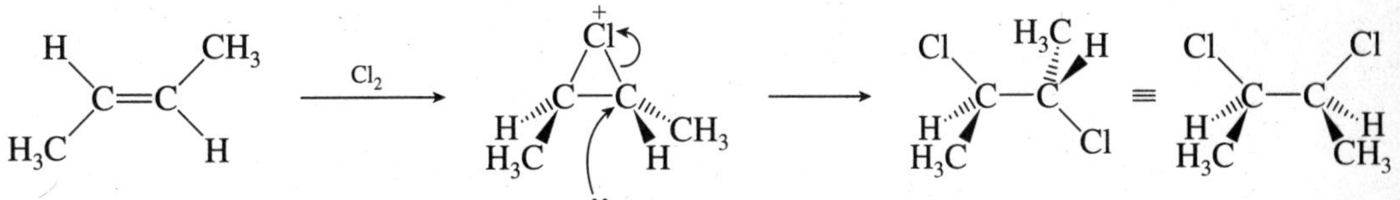

(*E*)-2-Butene $\xrightarrow{Cl_2}$ cyclic chloronium ion (attacked by $:\ddot{Cl}:^-$) $\longrightarrow$ 2,3-dichlorobutane ≡ (rotated conformation)

**A-6.**

$$CH_3CH_2CH{=}CHCH_2CH_3 \text{ (cis)} \xrightarrow{CH_3COOOH} \text{cis-epoxide } CH_3CH_2\text{–}C(H)\text{–O–}C(H)\text{–}CH_2CH_3$$

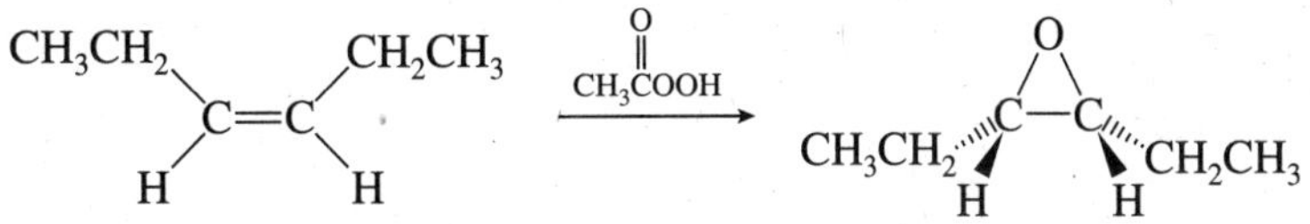

**A-7.** **Step 1:** Protonation to form a carbocation

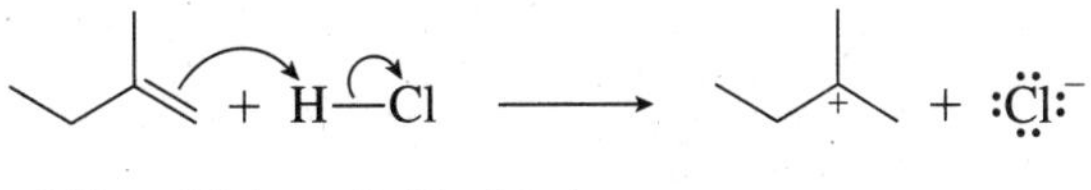

2-methyl-1-butene + H–Cl ⟶ tertiary carbocation + $:\ddot{Cl}:^-$

**Step 2:** Nucleophilic addition of chloride ion

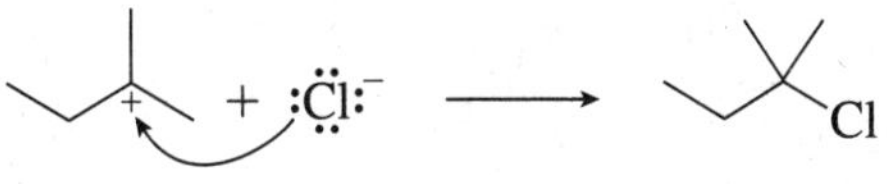

carbocation + $:\ddot{Cl}:^-$ ⟶ 2-chloro-2-methylbutane

**A-8.**

$$H_2C{=}C(CH_3)CH_2CH_3 \text{ or } (CH_3)_2C{=}CHCH_3 \xrightarrow{HCl} CH_3CCl(CH_3)CH_2CH_3$$

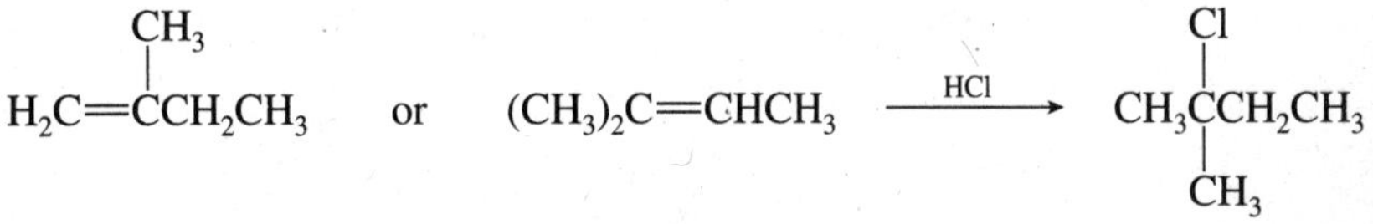

2-Methyl-1-butene — 2-Methyl-2-butene — 2-Chloro-2-methylbutane

**A-9.**

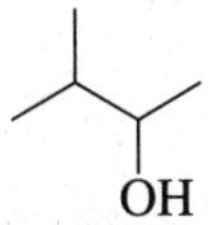

(3-methyl-2-butanol, skeletal structure with OH)

**A-10.** + HCl ⟶ (carbocation) $\xrightarrow{\text{hydride shift}}$ (carbocation)

$:\ddot{Cl}:^-$ ↓      $:\ddot{Cl}:^-$ ↓

Cl      Cl

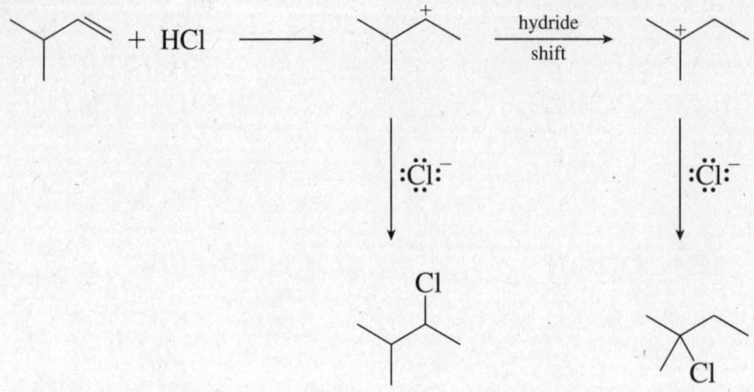

**A-11.** $(CH_3)_2CHC(CH_3){=}CH_2$ $\xrightarrow{HBr}$ $(CH_3)_2CHC(CH_3)(Br)CH_3$ $\xrightarrow[CH_3CH_2OH]{NaOCH_2CH_3}$ $(CH_3)_2C{=}C(CH_3)_2$

A      B      C

C $\xrightarrow[\text{2. } H_2O,\ Zn]{\text{1. } O_3}$ $(CH_3)_2C{=}O$ (2 mol)

| | | | | | | | |
|---|---|---|---|---|---|---|---|
| **B-1.** | (*c*) | **B-2.** | (*a*) | **B-3.** | (*c*) | **B-4.** | (*d*) |
| **B-5.** | (*d*) | **B-6.** | (*e*) | **B-7.** | (*b*) | **B-8.** | (*b*) |
| **B-9.** | (*b*) | **B-10.** | (*b*) | **B-11.** | (*a*) | **B-12.** | (*e*) |
| **B-13.** | (*e*) | | | | | | |

# CHAPTER 7

**A-1.** (*a*) 1 and 2, both achiral; identical
(*b*) 3 and 4, both chiral; enantiomers
(*c*) 5 chiral, 6 achiral (meso); diastereomers
(*d*) 7 and 8, both chiral; diastereomers
(*e*) 9 and 10, both chiral; diastereomers

**A-2.** 3: (*R*)-2-Chlorobutane;      4: (*S*)-2-Chlorobutane

5: 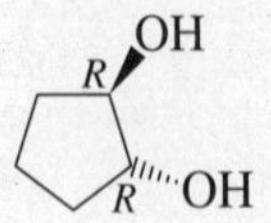 OH *R*, *R* OH      6: HO *S*, HO *R*

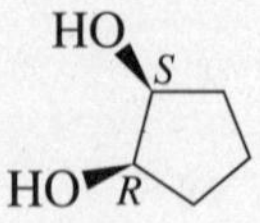

7: (2*S*,3*R*)-2,3-Dibromopentane;      8: (2*R*,3*R*)-2,3-Dibromopentane

9: (2*E*,5*R*)-5-Chloro-2-hexene;      10: (2*Z*,5*S*)-5-Chloro-2-hexene

**A-3.** (*a*) Three; meso form is possible.
(*b*) Eight; no meso form possible.
(*c*) Four; no meso form possible.

**A-4.** (*a*) 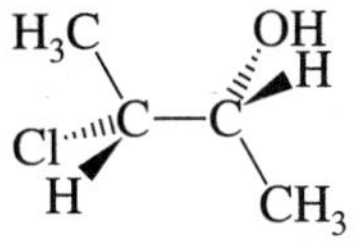 $H_3C$, OH, H, Cl, C—C, H, $CH_3$ (*b*) $CH_3$, H, H, H, Br, $CH_3$

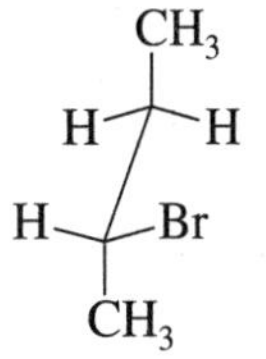

(*c*) $CH_3$; HO—H; H—Cl; $CH_3$ $CH_3$; H—H; H—Br; $CH_3$

**A-5. Chiral stereoisomers:**

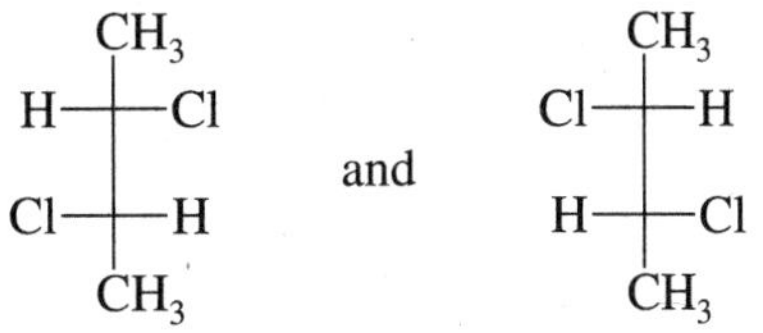

$CH_3$; H—Cl; Cl—H; $CH_3$ and $CH_3$; Cl—H; H—Cl; $CH_3$

(2*S*,3*S*)-2,3-Dichlorobutane (2*R*,3*R*)-2,3-Dichlorobutane

Meso stereoisomer (achiral); plane of symmetry indicated with dashed line

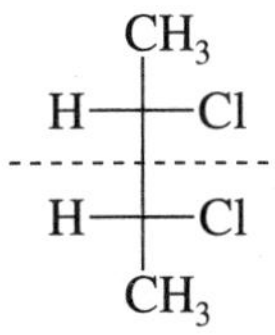

$CH_3$; H—Cl; H—Cl; $CH_3$

*meso*-2,3-Dichlorobutane

**A-6.** (*a*) $[\alpha] = -31.2°$ (*b*) 30% *S*

**A-7.** (*a*) 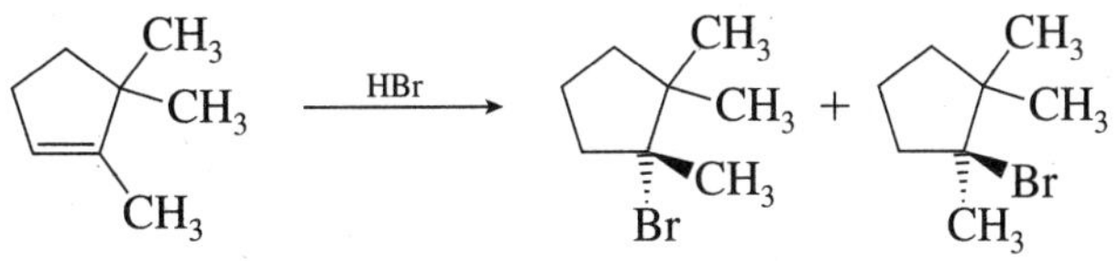 $CH_3$, $CH_3$, $CH_3$ —HBr→ $CH_3$, $CH_3$, $CH_3$, Br + $CH_3$, $CH_3$, Br, $CH_3$

(*b*) $H_3C$, H, C=C, H, $CH_3$ + $Cl_2$ ⟶ Cl, H, $CH_3$, $H_3C$, C—C, H, Cl

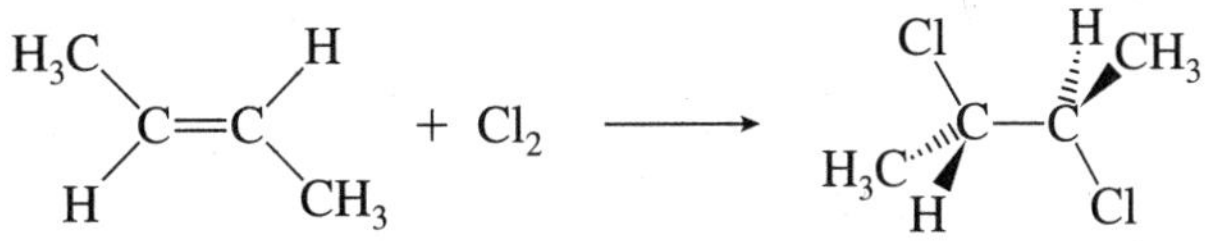

Meso form (only stereoisomer)

(*c*) 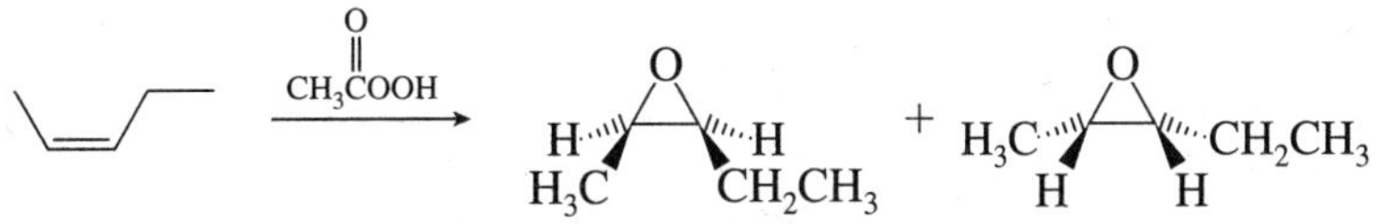 —$CH_3\overset{O}{\overset{\|}{C}}OOH$→ O, H, H, $H_3C$, $CH_2CH_3$ + O, $H_3C$, $CH_2CH_3$, H, H

**A-8.** (*a*) (2*S*,3*S*)-1,3-Dibromo-2-chlorobutane
(*b*) (*R*)-1-Ethylcyclohex-2-en-1-ol

**A-9.** Two: (2*R*,3*S*)-2-bromo-3-chlorobutane and (2*S*,3*S*)-2-bromo-3-chlorobutane; they are diastereomers.

**A-10.**

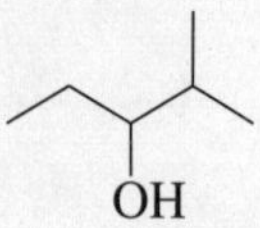

OH

Racemic mixture

| | | | |
|---|---|---|---|
| **B-1.** (*c*) | **B-2.** (*c*) | **B-3.** (*b*) | **B-4.** (*d*) |
| **B-5.** (*b*) | **B-6.** (*c*) | **B-7.** (*d*) | **B-8.** (*d*) |
| **B-9.** (*b*) | **B-10.** (*c*) | **B-11.** (*d*) | **B-12.** (*d*) |
| **B-13.** (*e*) | **B-14.** (*b*) | | |

# CHAPTER 8

**A-1.** (*a*) $CH_3CH_2CH_2CH_2OCH_2CH_3$

(*b*) X, $CH_3$ (X = OTs, Br, I)

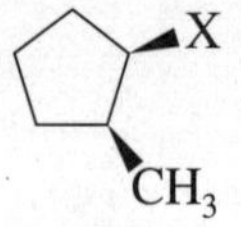

(*c*) $CH_3CH(CH_3)CH_2CH_2I$

(*d*)

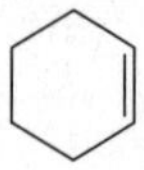

(*e*) 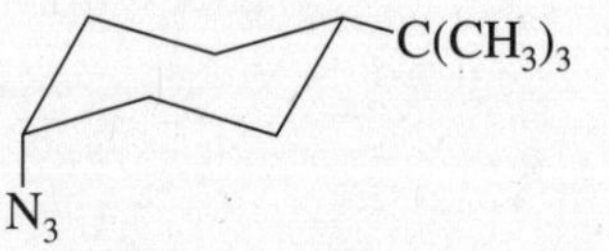 $C(CH_3)_3$, $N_3$

(*f*) $CH_3S$ H

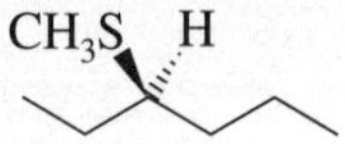

(*g*) $CH_3$; HS—H; H—F; $CH_2CH_3$

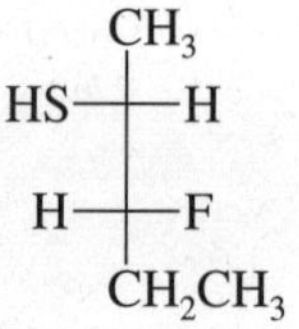

**A-2.** $(CH_3)_2CHO^-Na^+ + CH_3CH_2CH_2Br$

**A-3.** (*a*) $CH_3$; H—OTs; $CH_2CH_3$ $\xrightarrow[\text{DMSO}]{\text{NaBr}}$ $CH_3$; Br—H; $CH_2CH_3$ $\xrightarrow{\text{NaCN}}$ $CH_3$; H—CN; $CH_2CH_3$

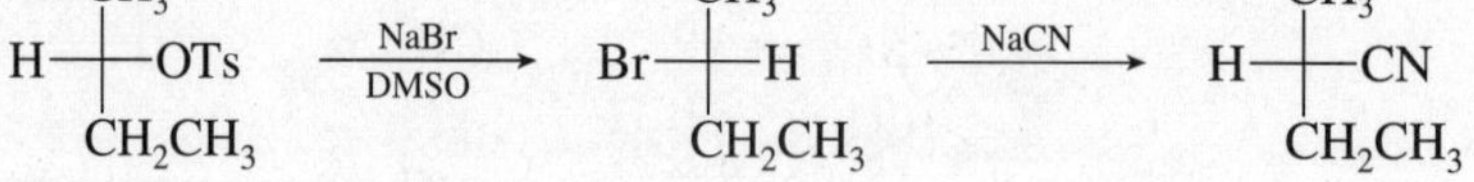

(*b*)

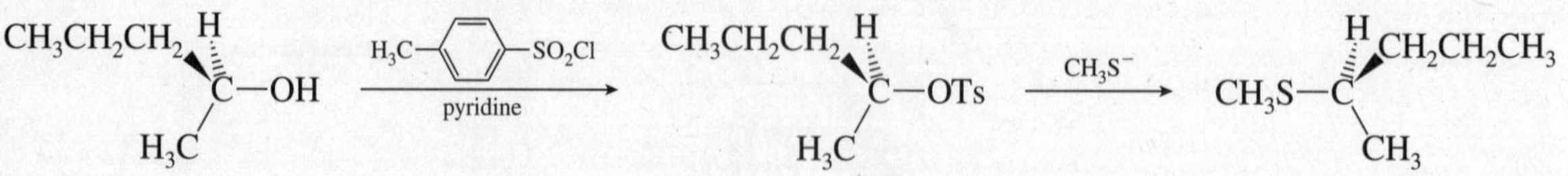

$CH_3CH_2CH_2$, H, C—OH, $H_3C$ $\xrightarrow[\text{pyridine}]{H_3C\text{-}C_6H_4\text{-}SO_2Cl}$ $CH_3CH_2CH_2$, H, C—OTs, $H_3C$ $\xrightarrow{CH_3S^-}$ $CH_3S$—C, H, $CH_2CH_2CH_3$, $CH_3$

**A-4.** **Step 1:** Ionization to form a secondary carbocation

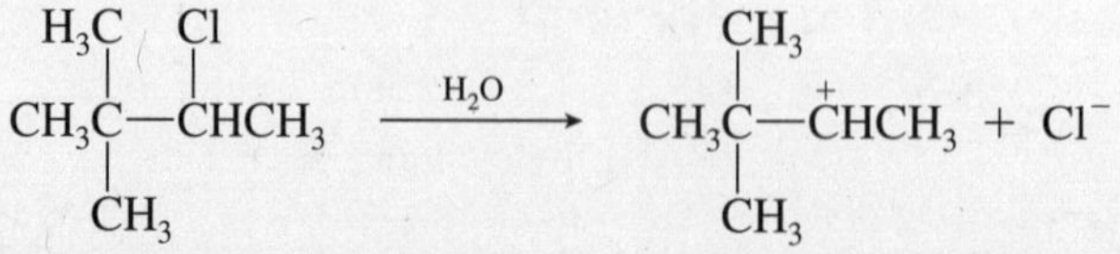

$$CH_3C(CH_3)_2\text{—}CHClCH_3 \xrightarrow{H_2O} CH_3C(CH_3)_2\text{—}\overset{+}{C}HCH_3 + Cl^-$$

**Step 2:** Rearrangement by methyl migration to form a more stable tertiary carbocation

$$CH_3\overset{CH_3}{\underset{CH_3}{C}}-\overset{+}{C}HCH_3 \longrightarrow CH_3\overset{+}{\underset{CH_3}{C}}-\overset{CH_3}{C}HCH_3$$

**Step 3:** Capture of the carbocation by water, followed by deprotonation

$$CH_3\overset{+}{\underset{CH_3}{C}}-\overset{CH_3}{C}HCH_3 \xrightarrow{H_2O} CH_3\overset{H_2\overset{+}{O}}{\underset{CH_3}{C}}-\overset{CH_3}{C}HCH_3 \xrightarrow{-H^+} (CH_3)_2\overset{OH}{C}-CH(CH_3)_2$$

**A-5.** (*a*) $(CH_3)_3CBr \xrightarrow{CH_3OH} (CH_3)_3COCH_3$

$S_N1$, unimolecular substitution; rate $= k[(CH_3)_3CBr]$

(*b*) $C_6H_{11}-Cl \xrightarrow{NaN_3} C_6H_{11}-N_3$ (cyclohexyl chloride → cyclohexyl azide)

$S_N2$, bimolecular substitution; rate $= k[C_6H_{11}Cl][NaN_3]$

**A-6.** (*a*) Sodium iodide is soluble in acetone, whereas the byproduct of the reaction, sodium bromide, is not. According to Le Chatelier's principle, the reaction will shift in the direction that will replace the component removed from solution, in this case toward product.

(*b*) Protic solvents such as water form hydrogen bonds to anionic nucleophiles, thus stabilizing them and decreasing their nucleophilic strength. Aprotic solvents such as DMSO do not solvate anions very strongly, leaving them more able to express their nucleophilic character.

**A-7.** A: cyclohexanol (OH) B: sodium cyclohexoxide ($O^- Na^+$) C: $CH_3CH_2OH$ D: $CH_3CH_2Br$

**A-8.**

1-bromo-1-methylcyclohexane ($H_3C$, Br) $\xrightarrow{\text{slow}}$ 1-methylcyclohexyl cation ($CH_3$, +)

1-methylcyclohexyl cation ($CH_3$, +) $+ H\ddot{O}CH_2CH_3 \xrightarrow{\text{fast}}$ $H_3C$, $\overset{H}{\overset{+}{O}}CH_2CH_3$ on cyclohexane

$H_3C$, $\overset{H}{\overset{+}{O}}CH_2CH_3$ on cyclohexane $\xrightarrow[-H^+]{\text{fast}}$ $H_3C$, $OCH_2CH_3$ on cyclohexane

**A-9.** Dissociation to give a secondary carbocation

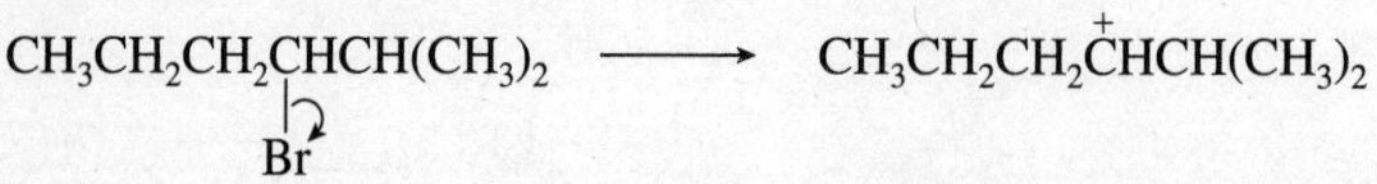

$$CH_3CH_2CH_2CHCH(CH_3)_2 \ (\text{Br on C}) \longrightarrow CH_3CH_2CH_2\overset{+}{C}HCH(CH_3)_2$$

Rearrangement by hydride migration to give a tertiary carbocation

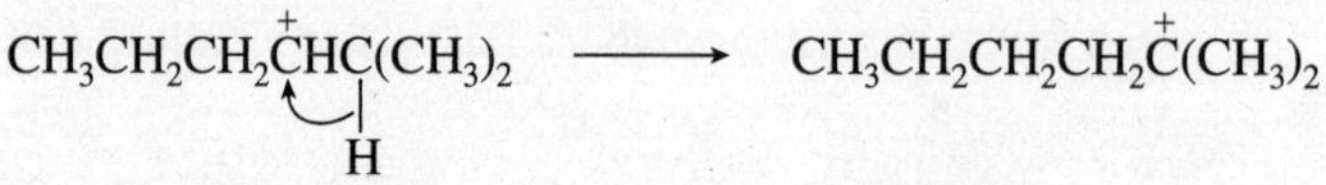

$$CH_3CH_2CH_2\overset{+}{C}HC(CH_3)_2 \ (\text{H on C}) \longrightarrow CH_3CH_2CH_2CH_2\overset{+}{C}(CH_3)_2$$

Capture of the carbocation by water to give product

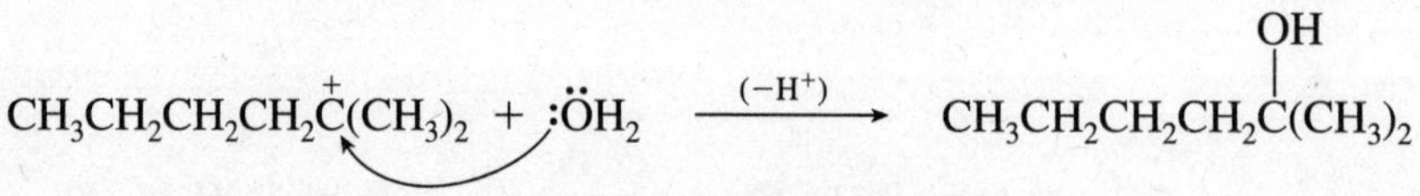

$$CH_3CH_2CH_2CH_2\overset{+}{C}(CH_3)_2 + :\ddot{O}H_2 \xrightarrow{(-H^+)} CH_3CH_2CH_2CH_2\underset{}{\overset{OH}{C}}(CH_3)_2$$

**B-1.** (*b*) **B-2.** (*c*) **B-3.** (*d*) **B-4.** (*c*)
**B-5.** (*d*) **B-6.** (*a*) **B-7.** (*c*) **B-8.** (*d*)
**B-9.** (*c*) **B-10.** (*a*) **B-11.** (*a*) **B-12.** (*c*)
**B-13.** (*c*) **B-14.** (*c*)

# CHAPTER 9

**A-1.** (*a*) 4,5-Dimethyl-2-hexyne
(*b*) 4-Ethyl-3-propyl-1-heptyne
(*c*) 6,6-Dimethylcyclodecyne

**A-2.** (*a*) $CH_3CH_2CH_2\overset{Cl}{C}{=}CH_2$

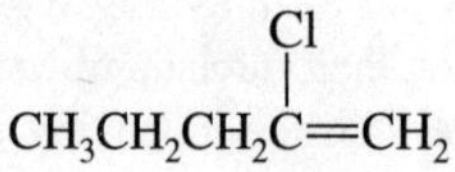

(*b*) $CH_3CH_2CH_2C(Cl)_2CH_3$

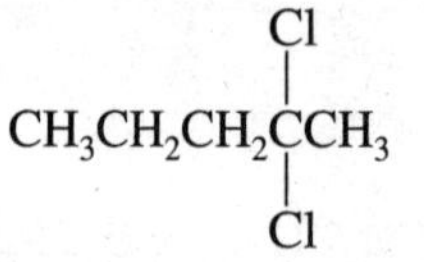

(*c*) $H_2O$, $H_2SO_4$, $HgSO_4$

(*d*) $(H_3C)(H)C{=}C(CH_3)(H)$ — $H_3C$ and $CH_3$ on the same side, H and H on the same side

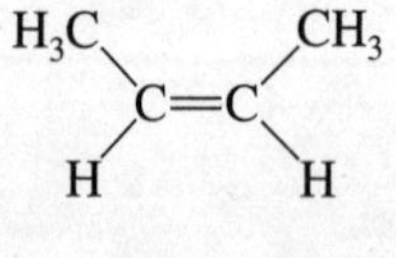

(*e*) $(CH_3)_2CHC{\equiv}CH$

(*f*) Na, $NH_3(l)$

(*g*) $(H_3C)(Cl)C{=}C(Cl)(CH_2CH_3)$ — $H_3C$ and Cl on top, Cl and $CH_2CH_3$ on bottom

(*h*) $CH_3CH_2CH_2CH_2\overset{CH_3}{C}HCO_2H + CH_3CH_2CO_2H$

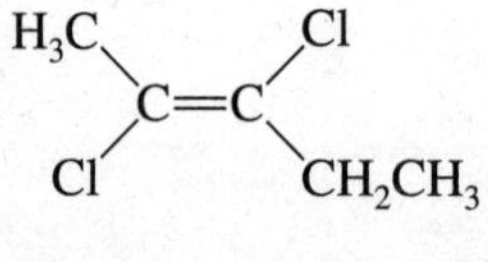

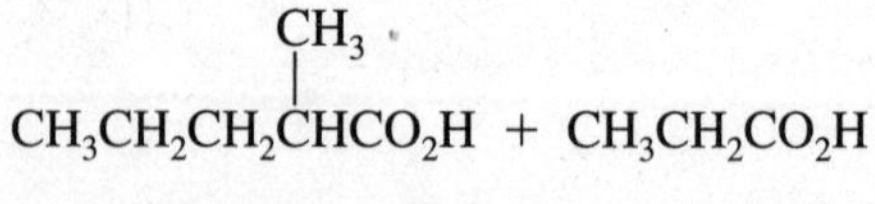

**A-3.** Reaction (2) is effective; the desired product is formed by an $S_N2$ reaction.

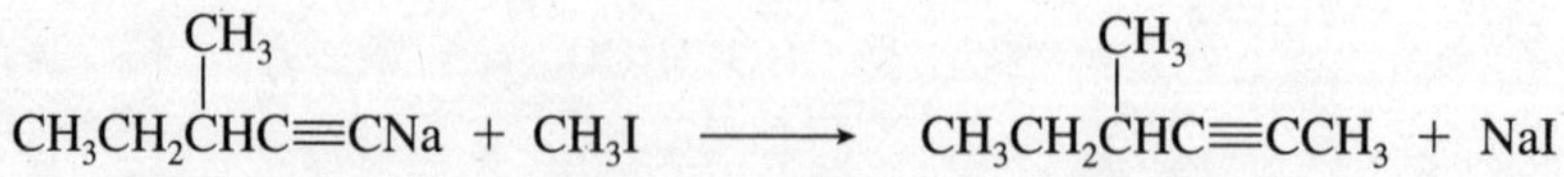

$$CH_3CH_2\overset{CH_3}{C}HC{\equiv}CNa + CH_3I \longrightarrow CH_3CH_2\overset{CH_3}{C}HC{\equiv}CCH_3 + NaI$$

Reaction (1) is not effective, owing to E2 elimination from the secondary bromide.

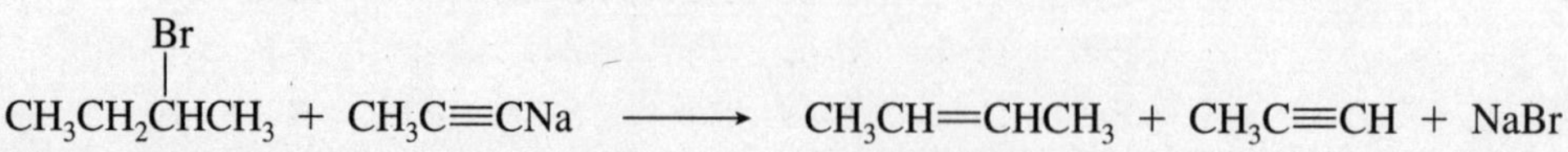

$$CH_3CH_2\overset{Br}{C}HCH_3 + CH_3C{\equiv}CNa \longrightarrow CH_3CH{=}CHCH_3 + CH_3C{\equiv}CH + NaBr$$

**A-4.** (*a*)

$$CH_3CH_2Br \xrightarrow[\text{DMSO}]{KOC(CH_3)_3} H_2C{=}CH_2 \xrightarrow{Br_2} BrCH_2CH_2Br$$

$$BrCH_2CH_2Br \xrightarrow[2.\ H_2O]{1.\ NaNH_2,\ NH_3} HC{\equiv}CH \xrightarrow[2.\ CH_3CH_2Br]{1.\ NaNH_2} HC{\equiv}CCH_2CH_3$$

(*b*)

$$HC{\equiv}CCH_2CH_3 \xrightarrow{NaNH_2} NaC{\equiv}CCH_2CH_3 \xrightarrow{CH_3CH_2Br} CH_3CH_2C{\equiv}CCH_2CH_3$$

(*c*)

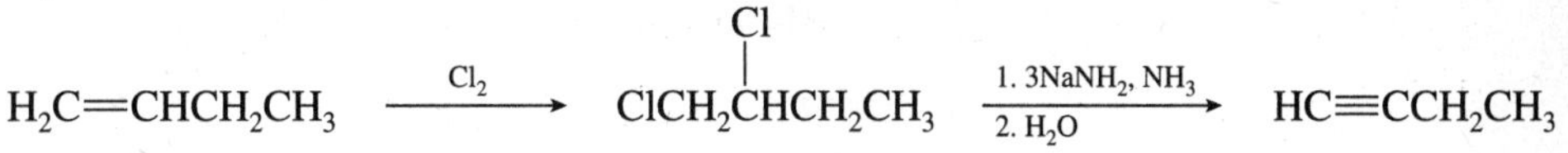

$$H_2C{=}CHCH_2CH_3 \xrightarrow{Cl_2} ClCH_2CH(Cl)CH_2CH_3 \xrightarrow[2.\ H_2O]{1.\ 3NaNH_2,\ NH_3} HC{\equiv}CCH_2CH_3$$

$$HC{\equiv}CCH_2CH_3 \xrightarrow{\text{As in part }(b)} CH_3CH_2C{\equiv}CCH_2CH_3$$

(*d*)

$$HC{\equiv}CH \xrightarrow{NaNH_2} HC{\equiv}CNa \xrightarrow{(CH_3)_2CHCH_2Br} HC{\equiv}CCH_2CH(CH_3)_2$$

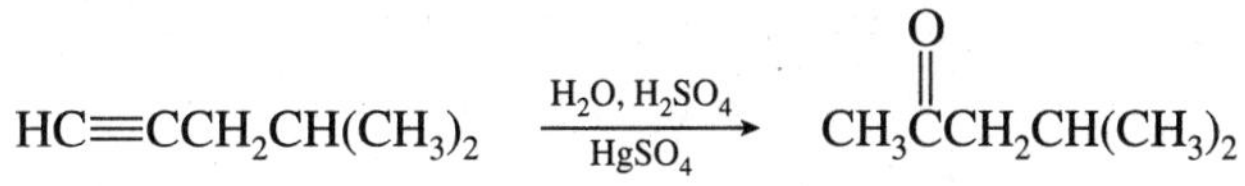

$$HC{\equiv}CCH_2CH(CH_3)_2 \xrightarrow[HgSO_4]{H_2O,\ H_2SO_4} CH_3\overset{O}{\overset{\|}{C}}CH_2CH(CH_3)_2$$

**A-5.** $H_3C(H)C{=}C(H)CH_2CH_2CH_2CH_3$

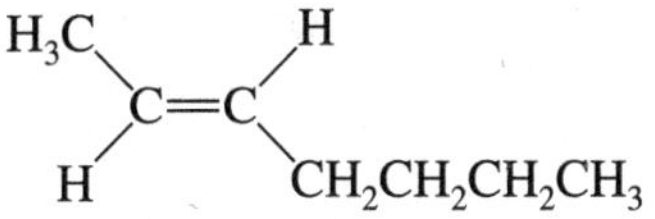

(*E*)-2-Heptene

**A-6.**

A: $(CH_3)_3CC{\equiv}CH$

B: $(CH_3)_3CC{\equiv}C{:}^-\ Na^+$

C: $CH_3CH_2CH_2OH$

D: $CH_3CH_2CH_2Br$

**A-7.** **E:** $HC{\equiv}CCH_2CH_3$ **F:** $CH_3CH_2C{\equiv}CCH_2CH_3$

**A-8.**

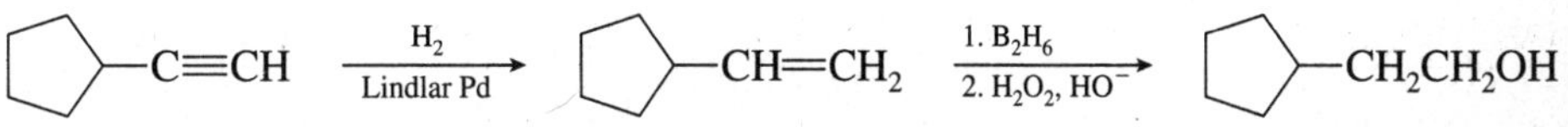

$$\text{cyclopentyl}{-}C{\equiv}CH \xrightarrow[\text{Lindlar Pd}]{H_2} \text{cyclopentyl}{-}CH{=}CH_2 \xrightarrow[2.\ H_2O_2,\ HO^-]{1.\ B_2H_6} \text{cyclopentyl}{-}CH_2CH_2OH$$

**B-1.** (*a*) **B-2.** (*c*) **B-3.** (*a*) **B-4.** (*d*)
**B-5.** (*b*) **B-6.** (*b*) **B-7.** (*e*) **B-8.** (*c*)
**B-9.** (*b*) **B-10.** (*a*) **B-11.** (*d*) **B-12.** (*b*)

## CHAPTER 10

**A-1.** $H_2C{=}CHCH_2CH{=}CH_2$ $H_2C{=}CHCH{=}CHCH_3$ (Conjugated)

$H_2C{=}CHC(CH_3){=}CH_2$ (Conjugated)

$H_2C{=}C{=}CHCH_2CH_3$
$H_2C{=}C{=}C(CH_3)_2$
$CH_3CH{=}C{=}CHCH_3$
Allenes

**A-2.** (3*Z*)-1,3-Pentadiene; (3*E*)-1,3-Pentadiene; $H_2C{=}CHC(CH_3){=}CH_2$ 2-Methyl-1,3-butadiene

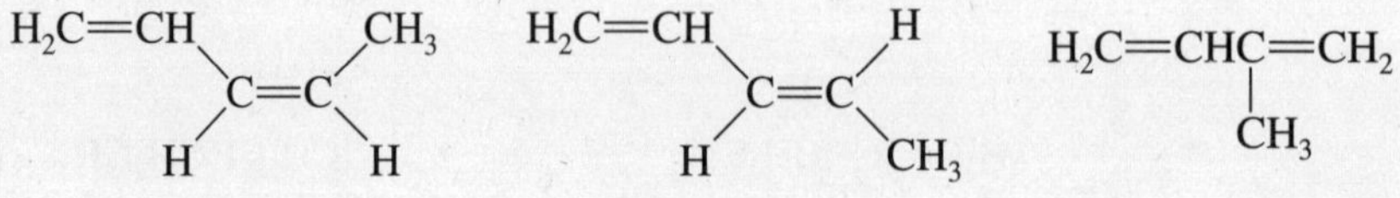

**A-3.** $\xrightarrow{H_2O}$ [ allylic cation $CH_3$ ⟷ allylic cation $CH_3$ ] $\xrightarrow{H_2O}$ OH, $CH_3$ + OH, $CH_3$

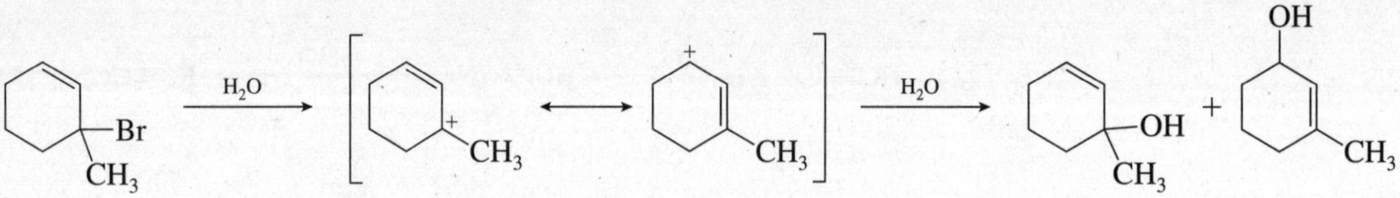

**A-4.** (*a*) $CH_3CH{=}CHCHBrCHBrCH_3$ (Direct addition) + $CH_3CHBrCH{=}CHCHBrCH_3$ (Conjugate addition)

(*b*) $H_2C{=}CHCHClCH_3$ + $ClCH_2CH{=}CHCH_3$

(*c*) + (*d*) N—Br (NBS), heat

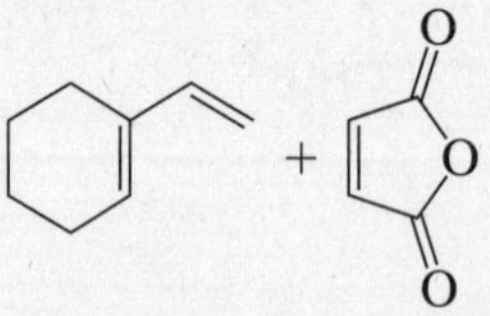

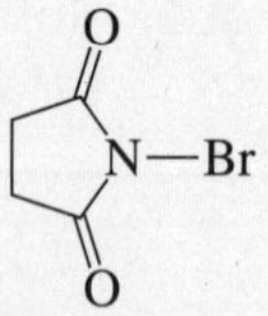

(*e*) $CO_2CH_3$, $CO_2CH_3$

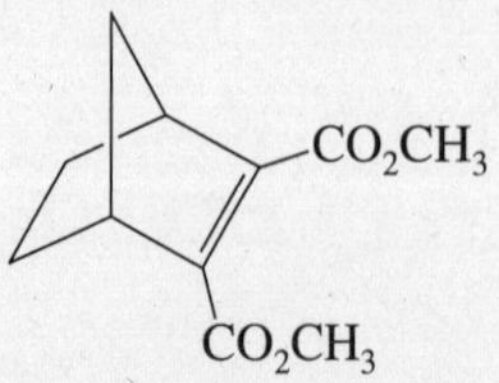

**A-5.** 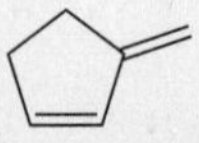 (cannot adopt the required *s*-cis conformation)

A-6. [structures: 1-(1-bromo-1-methylethyl)cyclopentene $\xrightarrow{-Br^-}$ allylic cation $\leftrightarrow$ resonance form; 2-bromo-1-isopropylidenecyclopentane $\xrightarrow{-Br^-}$ same allylic cation]

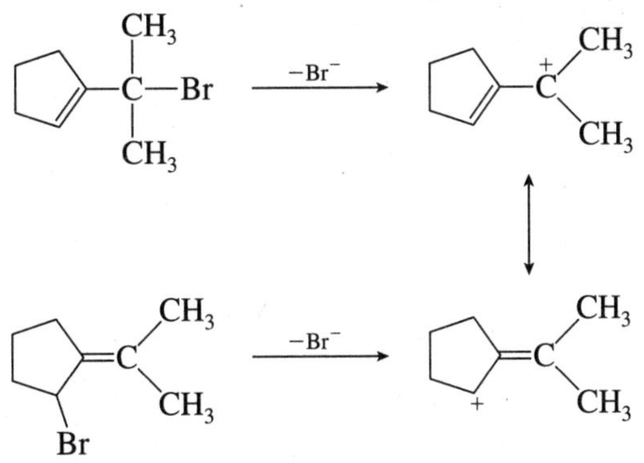

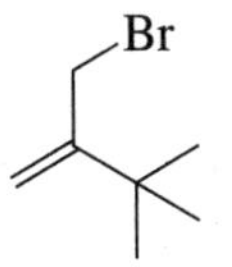

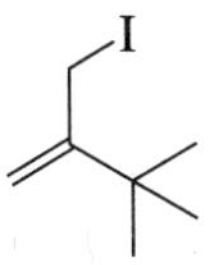

A-7. A: [structure, Br] B: [structure, I]

A-8. [structures] $\xrightarrow[\text{light, heat}]{Br_2}$ [structure, Br] $\xrightarrow[CH_3CH_2OH]{CH_3CH_2ONa}$ [structure] $\xrightarrow[\text{heat}]{NBS}$ [structure, Br]

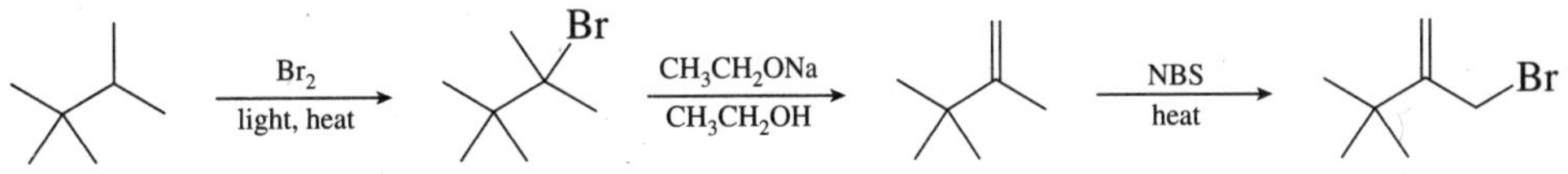

| | | | | | | | |
|---|---|---|---|---|---|---|---|
| **B-1.** | (*b*) | **B-2.** | (*c*) | **B-3.** | (*a*) | **B-4.** | (*c*) |
| **B-5.** | (*a*) | **B-6.** | (*d*) | **B-7.** | (*a*) | **B-8.** | (*a*) |
| **B-9.** | (*a*) | **B-10.** | (*d*) | | | | |

# CHAPTER 11

**A-1.** (*a*) *m*-Bromotoluene
(*b*) 2-Chloro-3-phenylbutane
(*c*) *o*-Chloroacetophenone
(*d*) 2,4-Dinitrophenol

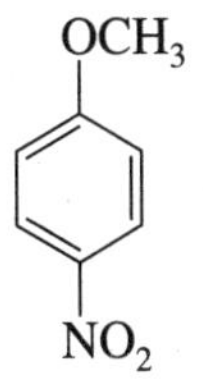

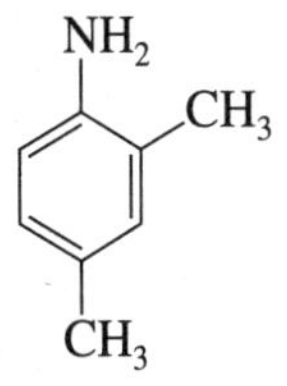

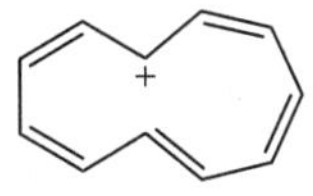

**A-2.** (*a*) [structure: $CO_2H$, Cl, Cl] (*b*) [structure: $OCH_3$, $NO_2$] (*c*) [structure: $NH_2$, $CH_3$, $CH_3$] (*d*) [structure: $CH_2Cl$, Br]

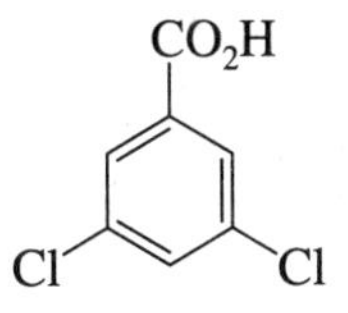

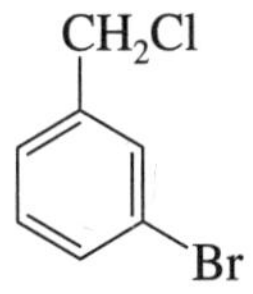

**A-3.** (*a*) [structure] (10 $\pi$ electrons) (*b*) [structure] (14 $\pi$ electrons)

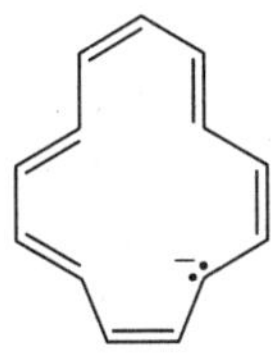

**A-4.** (*a*) Eight $\pi$ electrons. No, the substance is not aromatic.
(*b*) 6 $\pi$ electrons. Yes, it is aromatic.
(*c*) 14 $\pi$ electrons. Yes, it is aromatic.

**A-5.** [structures: azulene ⟷ charge-separated resonance form ≡ cyclopentadienide–tropylium representation]

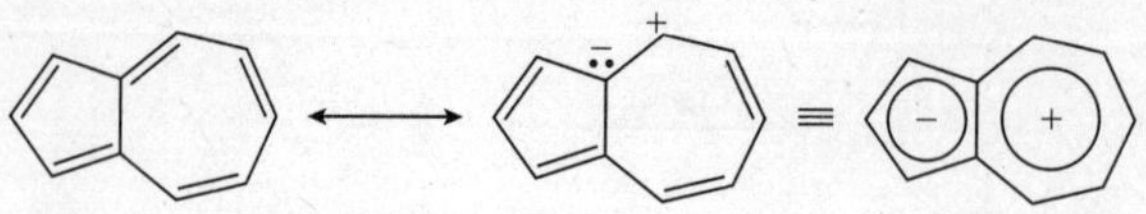

**A-6.** (*a*) [structure: 1-bromotetralin, Br]

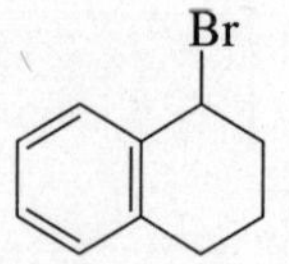

(*b*) $C_6H_5CH_2X$ (X = Cl, Br, I, OTs)

(*c*) [structure: 1-indanol, OH]

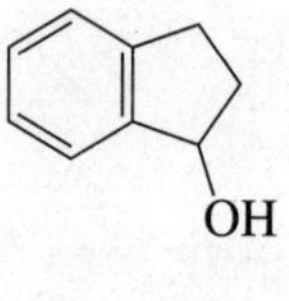

(*d*) $Na_2Cr_2O_7$, $H_2SO_4$, $H_2O$, heat

(*e*) [structure: epoxide fused to tetrahydronaphthalene, O]

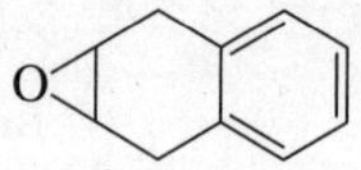

(*f*) $C_6H_5CH(OH)CH(Br)CH_3$ (OH, Br)

**A-7.** (I) $C_6H_5CH{=}CHCH_3 \xrightarrow{HBr} C_6H_5CH(Br)CH_2CH_3$

(II) $C_6H_5CH_2CH_2CH_3 \xrightarrow[\text{light (or NBS, heat)}]{Br_2} C_6H_5CH(Br)CH_2CH_3$

**A-8.** [resonance structures: $^+CH{-}CH_3$ on benzene ⟷ CH—CH₃ with ortho + ⟷ CH—CH₃ with para + ⟷ CH—CH₃ with ortho +]

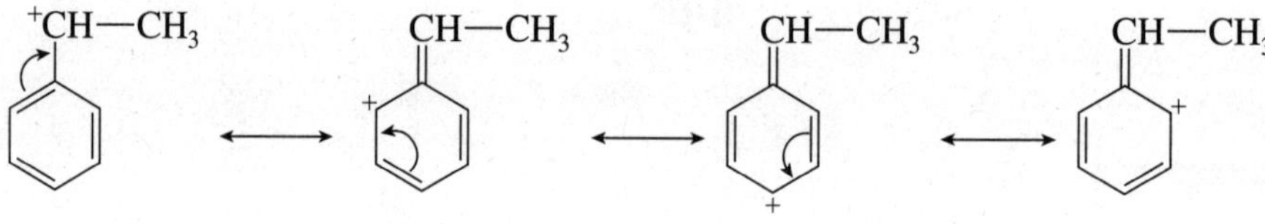

**A-9.** [resonance structures of allylic/benzylic cation: five contributors]

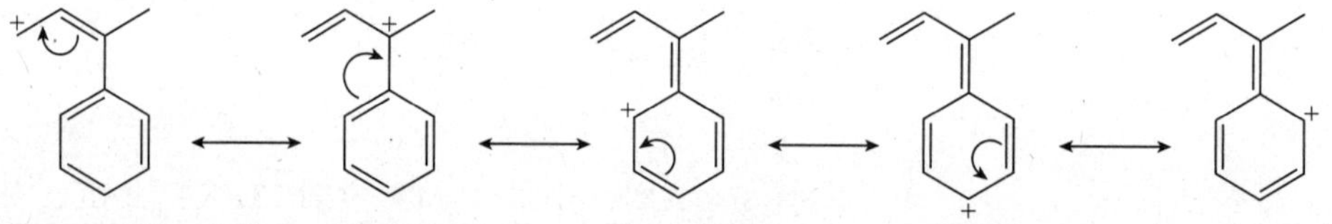

**A-10.** $(CH_3)_3C$—$C_6H_4$—$CH_2CH_3$ (para)

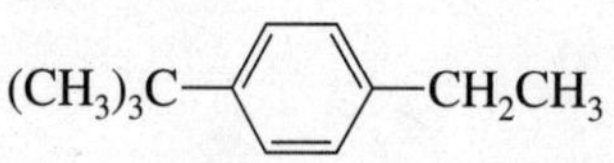

| | | | |
|---|---|---|---|
| **B-1.** (*c*) | **B-2.** (*c*) | **B-3.** (*a*) | **B-4.** (*b*) |
| **B-5.** (*a*) | **B-6.** (*d*) | **B-7.** (*b*) | **B-8.** (*d*) |
| **B-9.** (*b*) | **B-10.** (*d*) | **B-11.** (*a*) | **B-12.** (*b*) |
| **B-13.** (*c*) | **B-14.** (*d*) | **B-15.** (*c*) | |

# CHAPTER 12

**A-1.** [resonance structures: arenium ion with $CH_3$, H, Br; + at three positions ⟷ ⟷]

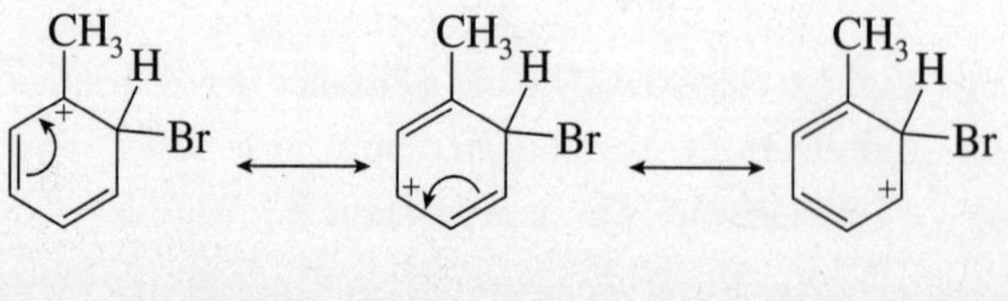

**A-2.** (*a*) $NO_2$ $NO_2$ Slower (*c*) Cl $SO_3H$ + $HO_3S$ Cl Slower

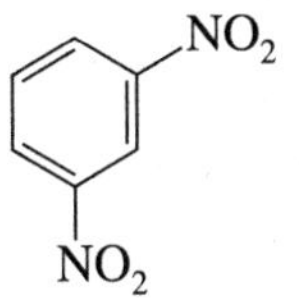

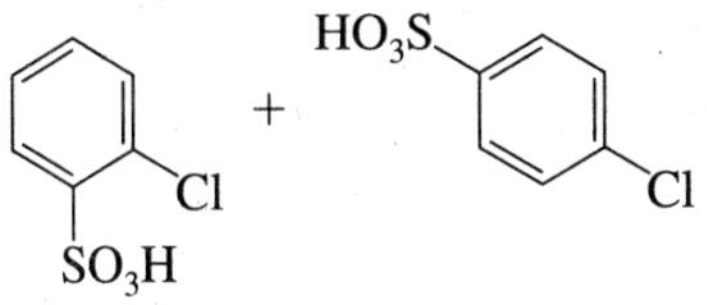

(*b*) $CH_2CH_3$ Br + $CH_2CH_3$ Br Faster

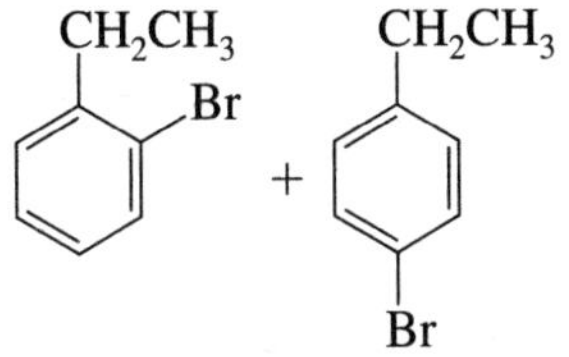

**A-3.** (*a*) $NO_2^+$ (*b*) $Br—\overset{+}{Br}—\overset{-}{Fe}Br_3$ (*c*) $SO_3$

**A-4.** (*a*) $\overset{O}{\overset{\|}{C}}CH(CH_3)_2$ $C(CH_3)_3$ (*d*) O $NO_2$

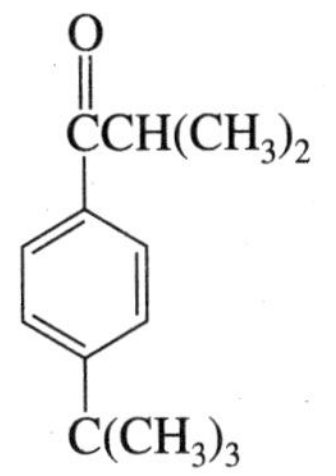

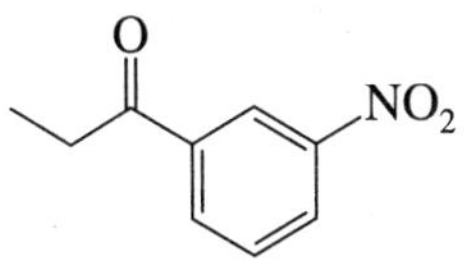

(*b*) $H_3C$ $CH_3$ (*e*) S Cl

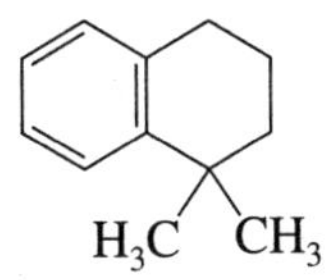

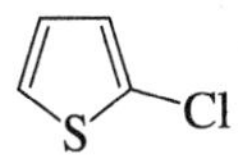

(*c*) $C_6H_5\overset{O}{\overset{\|}{C}}Cl$, $AlCl_3$ or $CH_3\overset{O}{\overset{\|}{C}}O\overset{O}{\overset{\|}{C}}CH_3$, $AlCl_3$

**A-5.** (*a*) $NO_2$ Cl $CH_3$ (*c*) O Cl + O Cl

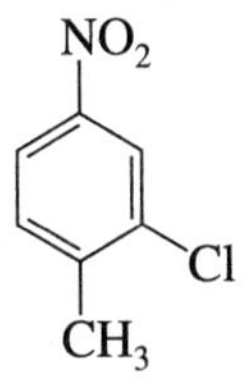

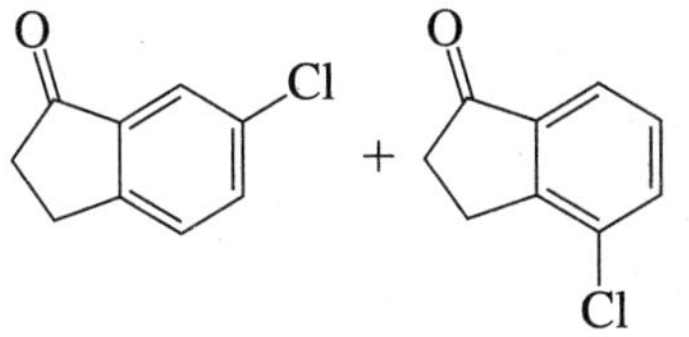

(*b*) CN Cl $OCH_3$ + Cl CN $OCH_3$ (*d*) $CH_3CH_2$— Cl —OH

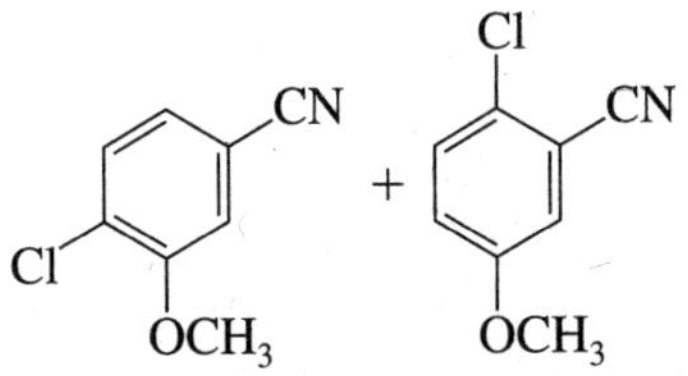

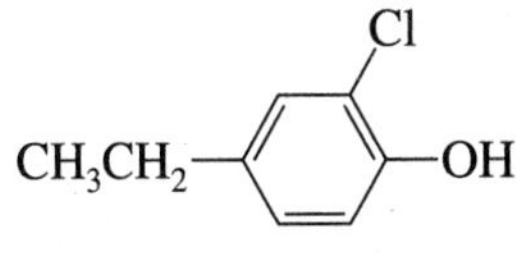

**A-6.** (*a*)

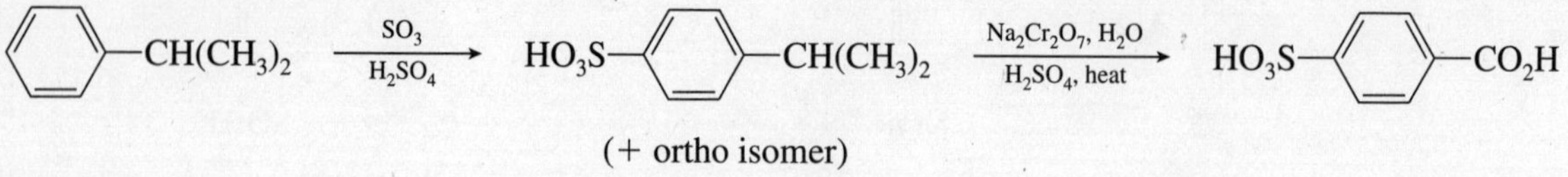

$$C_6H_5CH(CH_3)_2 \xrightarrow[H_2SO_4]{SO_3} HO_3S\text{-}C_6H_4\text{-}CH(CH_3)_2 \xrightarrow[H_2SO_4,\ heat]{Na_2Cr_2O_7,\ H_2O} HO_3S\text{-}C_6H_4\text{-}CO_2H$$

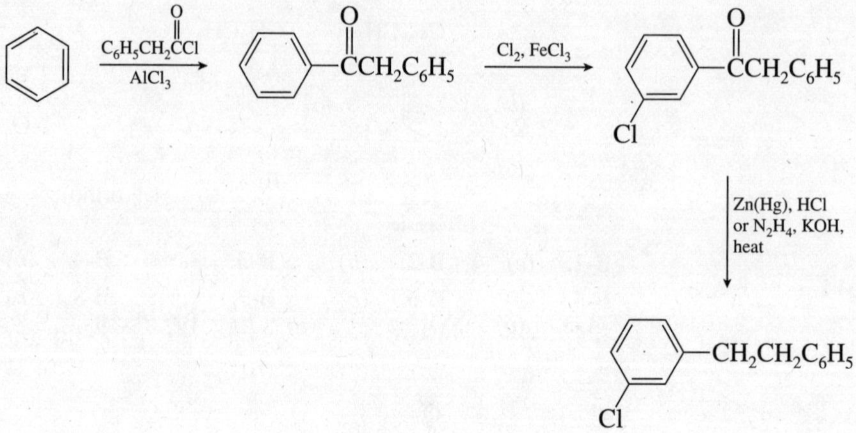

(+ ortho isomer)

(*b*)

$$C_6H_6 \xrightarrow[AlCl_3]{C_6H_5CH_2\overset{O}{\overset{\|}{C}}Cl} C_6H_5\overset{O}{\overset{\|}{C}}CH_2C_6H_5 \xrightarrow{Cl_2,\ FeCl_3} m\text{-}ClC_6H_4\overset{O}{\overset{\|}{C}}CH_2C_6H_5$$

$$\xrightarrow{Zn(Hg),\ HCl \text{ or } N_2H_4,\ KOH,\ heat} m\text{-}ClC_6H_4CH_2CH_2C_6H_5$$

(*c*)

$$C_6H_5\overset{O}{\overset{\|}{C}}H \xrightarrow[\text{or } Zn(Hg),\ HCl]{N_2H_4,\ KOH,\ heat} C_6H_5CH_3 \xrightarrow[AlCl_3]{(CH_3\overset{O}{\overset{\|}{C}})_2O} p\text{-}CH_3\overset{O}{\overset{\|}{C}}C_6H_4CH_3$$

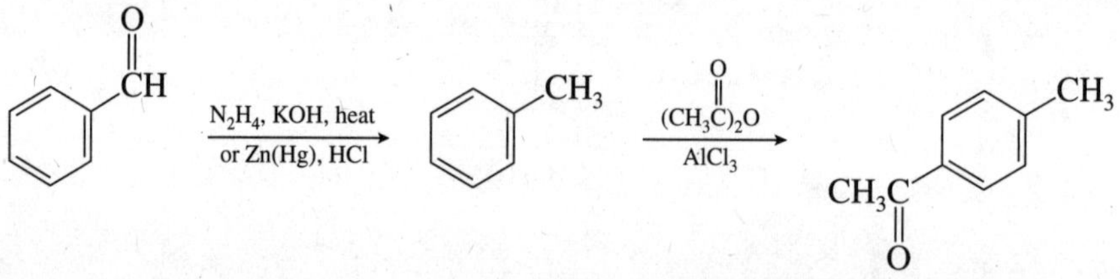

(*d*)

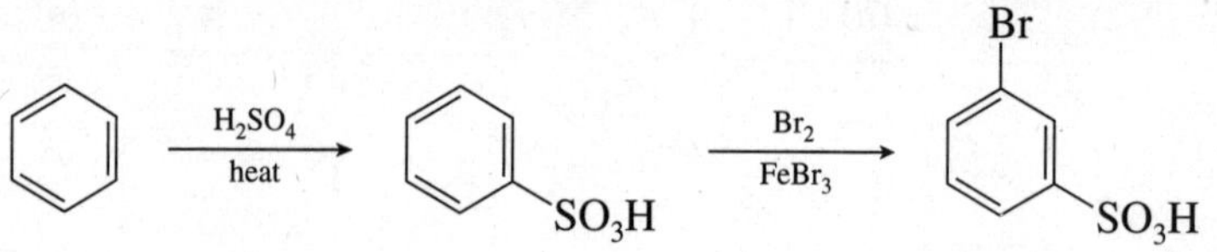

$$C_6H_6 \xrightarrow[heat]{H_2SO_4} C_6H_5SO_3H \xrightarrow[FeBr_3]{Br_2} m\text{-}BrC_6H_4SO_3H$$

(*e*)

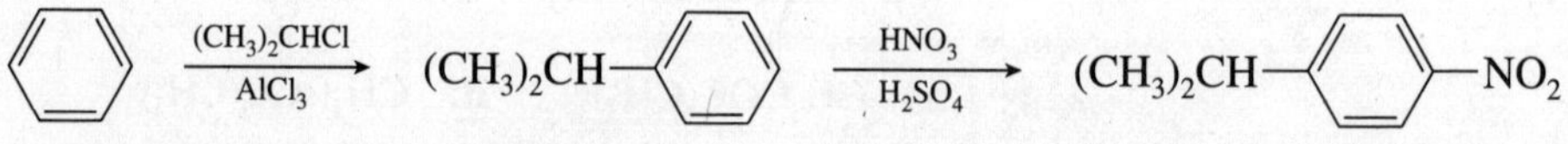

$$C_6H_6 \xrightarrow[AlCl_3]{(CH_3)_2CHCl} (CH_3)_2CH\text{-}C_6H_5 \xrightarrow[H_2SO_4]{HNO_3} (CH_3)_2CH\text{-}C_6H_4\text{-}NO_2$$

**A-7.** (*a*)

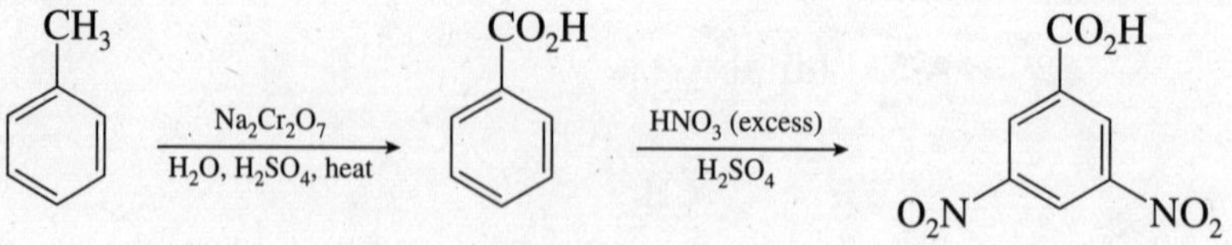

$$C_6H_5CH_3 \xrightarrow[H_2O,\ H_2SO_4,\ heat]{Na_2Cr_2O_7} C_6H_5CO_2H \xrightarrow[H_2SO_4]{HNO_3\ (excess)} 3,5\text{-}(O_2N)_2C_6H_3CO_2H$$

(*b*)

$$(CH_3)_2CH\text{-}C_6H_4\text{-}NO_2 \xrightarrow[light]{Br_2} (CH_3)_2\overset{Br}{\overset{|}{C}}\text{-}C_6H_4\text{-}NO_2 \xrightarrow{NaOCH_2CH_3} H_2C{=}\overset{CH_3}{\overset{|}{C}}\text{-}C_6H_4\text{-}NO_2$$

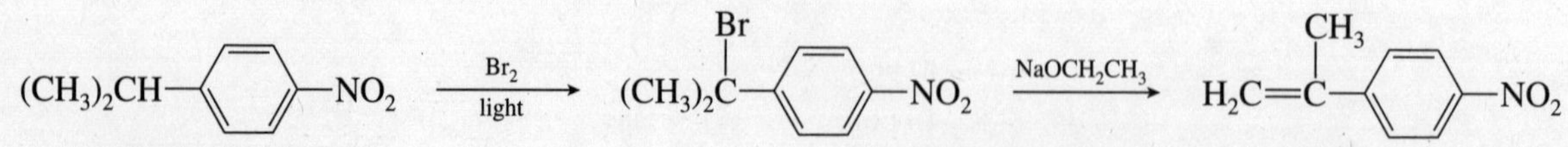

[Prepared from benzene as in Problem A-6(*e*)]

(*c*)

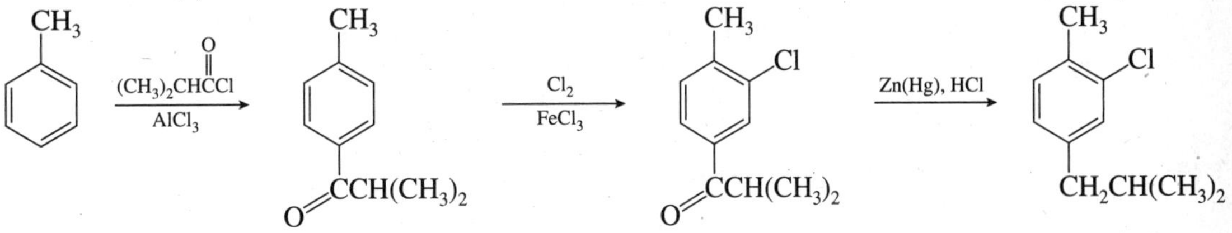

$$\text{toluene} \xrightarrow[\text{AlCl}_3]{(CH_3)_2CHCCl\,(\text{C=O})} p\text{-}CH_3C_6H_4CCH(CH_3)_2\,(\text{C=O}) \xrightarrow[\text{FeCl}_3]{\text{Cl}_2} \text{3-Cl-4-}CH_3C_6H_3CCH(CH_3)_2\,(\text{C=O}) \xrightarrow{\text{Zn(Hg), HCl}} \text{3-Cl-4-}CH_3C_6H_3CH_2CH(CH_3)_2$$

**A-8.**

$$CH_3O\text{–}C_6H_5 \xrightarrow[H_2SO_4]{HNO_3} CH_3O\text{–}C_6H_4\text{–}NO_2 \text{ (+ ortho)} \xrightarrow[FeBr_3]{Br_2} CH_3O\text{–}C_6H_3(Br)\text{–}NO_2$$

(Br ortho to $CH_3O$)

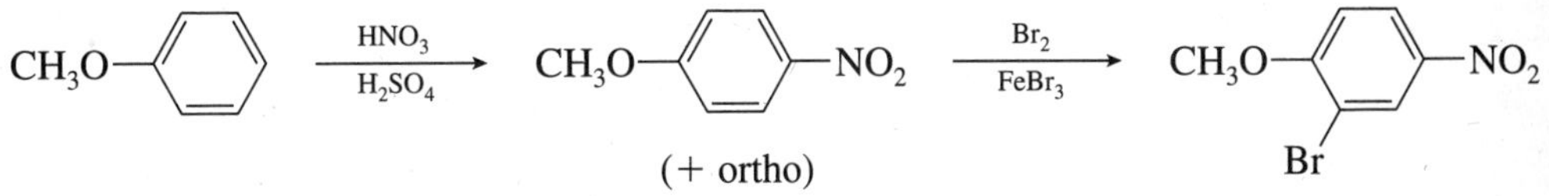

| | | | |
|---|---|---|---|
| **B-1.** (*c*) | **B-2.** (*b*) | **B-3.** (*c*) | **B-4.** (*b*) |
| **B-5.** (*a*) | **B-6.** (*b*) | **B-7.** (*c*) | **B-8.** (*b*) |
| **B-9.** (*c*) | **B-10.** (*a*) | **B-11.** (*e*) | **B-12.** (*c*) |
| **B-13.** (*c*) | **B-14.** (*c*) | **B-15.** (*c*) | |

# CHAPTER 13

**A-1.** 1: 6.10 ppm  3: 200 MHz
2: 1305 Hz  4: 0.00 ppm

**A-2.** (*a*) Two signals $BrCH_2CH_2CH_2Br$ (a b a)

*a*: triplet *b*: pentet

(*b*) Two signals $CH_3CH_2CCl_2CH_2CH_3$ (a b b a)

*a*: triplet *b*: quartet

(*c*) Three signals, all singlets

**A-3.** **A**: $CH_3\overset{O}{\overset{\|}{C}}OC(CH_3)_3$ **B**: $CH_3O\overset{O}{\overset{\|}{C}}C(CH_3)_3$

**A-4.** (*a*) $C_6H_5\text{–}CH_2\overset{O}{\overset{\|}{C}}CH_2CH_3$  (*c*) $HC(\overset{O}{\overset{\|}{C}}OCH_2CH_3)_3$

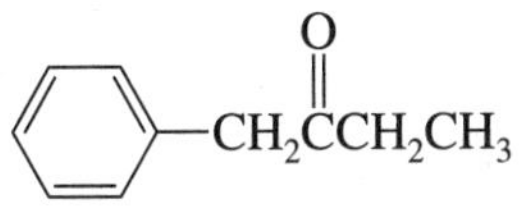

(*b*) $(CH_3)_2C(OH)\text{—}C(OH)(CH_3)_2$  (*d*) $(CH_3)_2C(OH)\text{—}C{\equiv}N$

**A-5.** Seven signals:
a: $\delta$ 10–30 ppm
b: $\delta$ 20–40 ppm
c: $\delta$ 190–220 ppm
d–g: $\delta$ 110–175 ppm

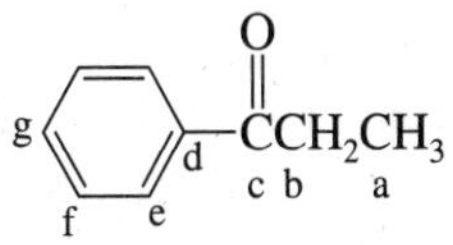

$C_6H_5\overset{O}{\overset{\|}{C}}CH_2CH_3$ (ring carbons d, e, f, g; C=O c; $CH_2$ b; $CH_3$ a)

**A-6.** Pentane: three signals; 2-methylbutane: four signals; 2,2-dimethylpropane: two signals

**A-7.** 2,3-Dimethylbutane: $(CH_3)_2CHCH(CH_3)_2$

**B-1.** (*d*) **B-2.** (*a*) **B-3.** (*b*) **B-4.** (*b*)
**B-5.** (*b*) **B-6.** (*a*) **B-7.** (*b*) **B-8.** (*a*)
**B-9.** (*c*) **B-10.** (*a*) **B-11.** (*c*) **B-12.** (*c*)
**B-13.** (*a*) **B-14.** (*a*) **B-15.** (*d*)

## CHAPTER 14

**A-1.** (*a*) (cyclohexyl–X) + 2Li ⟶ (cyclohexyl–Li) + LiX

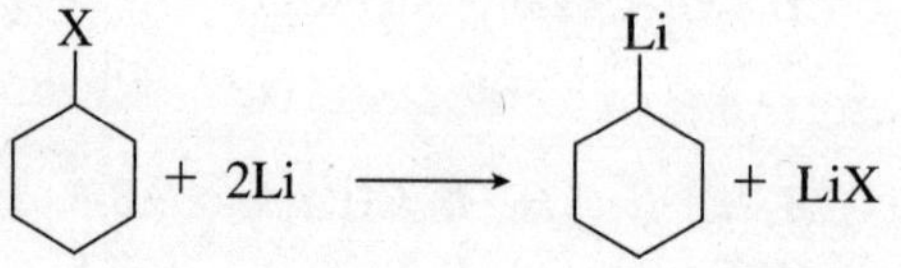

(X = Cl, Br, I)

(*b*) $(CH_3)_3CBr + Mg \longrightarrow (CH_3)_3CMgBr$

(*c*) $2C_6H_5CH_2Li + CuX \longrightarrow (C_6H_5CH_2)_2CuLi + LiX$

(X = Cl, Br, I)

**A-2.** (*a*) $(C_6H_5)_2C(OH)CH_3 + CH_3CH_2OH$

(*b*) $(CH_3)_2CHCH_2D$

(*c*) $H_3C$–$C_6H_4$–$CH_2OH$ (para-substituted benzene)

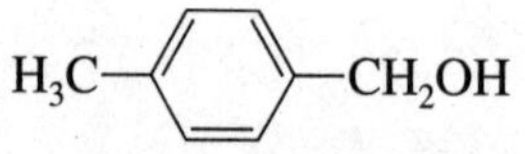

(*d*) 1-ethylcyclohexanol (cyclohexane ring bearing HO and $CH_2CH_3$ on the same carbon)

(*e*) cyclopropane ring bearing Br, Br on one carbon; $H_3C$ and H on the second carbon; H and $CH_3$ on the third carbon

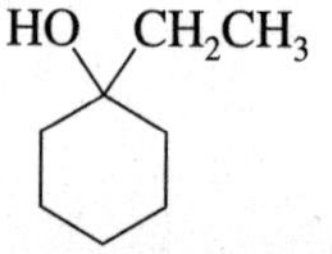

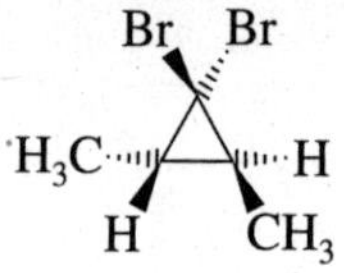

**A-3.** (*a*) $C_6H_5CH{=}O + (CH_3)_3CMgX$ and $C_6H_5MgX + (CH_3)_3CCH{=}O$

(X = Cl, Br, I) (X = Cl, Br, I)

(*b*)

$CH_3CH_2CH_2CH{=}O + CH_3CH_2CH_2CH_2MgX$ and $CH_3CH_2CH_2CH_2CH{=}O + CH_3CH_2CH_2MgX$

(X = Cl, Br, I) (X = Cl, Br, I)

**A-4.** (*a*) $(CH_3CH_2CH_2)_2CuLi$

(*b*) $(CH_3)_2CHMgX$
(X = Cl, Br, I)

(*c*) $CH_2I_2$, Zn(Cu)

**A-5.** Solvents A, B, and E are suitable; they are all ethers. Solvents C and F have acidic hydrogens and will react with a Grignard reagent. Solvent D is an ester which will react with a Grignard reagent.

**A-6.** $CH_3(CH_2)_3OH \xrightarrow[\text{or HBr}]{PBr_3} CH_3(CH_2)_3Br \xrightarrow{2Li} CH_3(CH_2)_3Li + LiBr$

$2CH_3(CH_2)_3Li + CuBr \longrightarrow (C_4H_9)_2CuLi \xrightarrow{CH_3(CH_2)_3Br} CH_3(CH_2)_6CH_3$

**A-7.** (I) $(CH_3)_2CHCCH_3$ (with C=O) $+ CH_3MgBr$

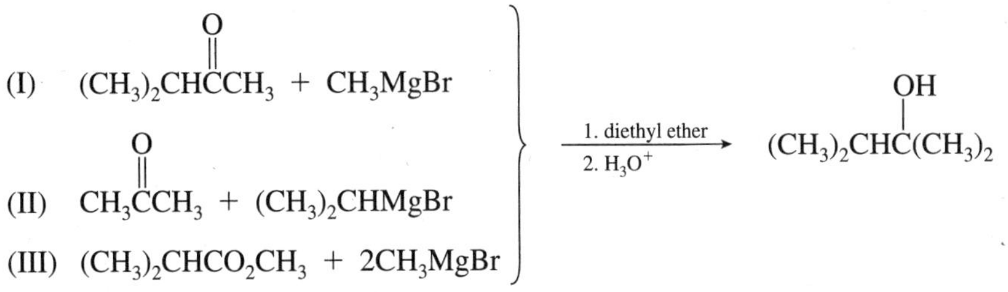

(II) $CH_3CCH_3$ (with C=O) $+ (CH_3)_2CHMgBr$

(III) $(CH_3)_2CHCO_2CH_3 + 2CH_3MgBr$

$\xrightarrow[\text{2. } H_3O^+]{\text{1. diethyl ether}} (CH_3)_2CHC(CH_3)_2$ (with OH on C)

**A-8.** $C_6H_5CH_2CH_3 \xrightarrow[\text{peroxides, heat}]{NBS} C_6H_5CHCH_3$ (Br on CH) $\xrightarrow[\text{diethyl ether}]{Mg} C_6H_5CHMgBr$ ($CH_3$ on CH)

$C_6H_5CHMgBr$ ($CH_3$ on CH) $\xrightarrow[\text{2. } H_3O^+]{\text{1. } CH_3CH(=O)} C_6H_5CHCHCH_3$ ($CH_3$ on first CH, OH on second CH)

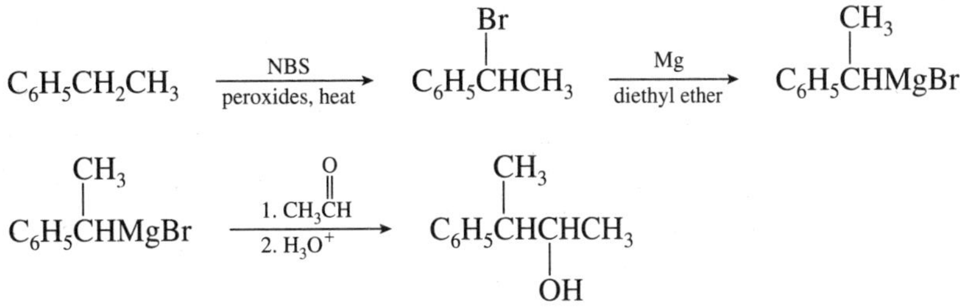

**A-9.** (*a*) 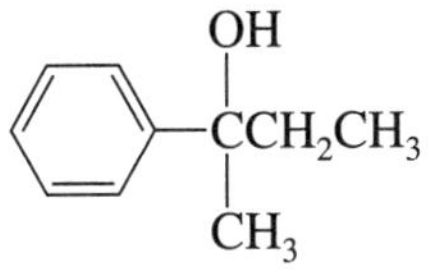 $C_6H_5C(OH)(CH_3)CH_2CH_3$

(*b*) [skeletal structure: 3-methyl-2-pentanol, OH]

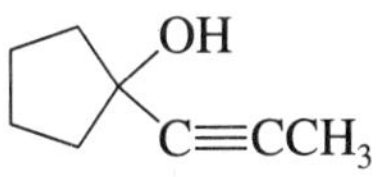

(*c*) 1-(1-propynyl)cyclopentanol: cyclopentane bearing OH and $C{\equiv}CCH_3$

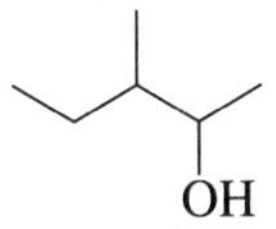

| | | | | | | | |
|---|---|---|---|---|---|---|---|
| **B-1.** | (*c*) | **B-2.** | (*a*) | **B-3.** | (*d*) | **B-4.** | (*a*) |
| **B-5.** | (*e*) | **B-6.** | (*c*) | **B-7.** | (*b*) | **B-8.** | (*a*) |
| **B-9.** | (*e*) | **B-10.** | (*b*) | **B-11.** | (*b*) | **B-12.** | (*b*) |

## CHAPTER 15

**A-1.** (*a*) cyclohexanone (ring with C=O)

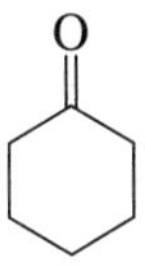

(*b*) $C_6H_5CO_2CH_2CH_3$

(*c*) 1. $B_2H_6$; 2. $H_2O_2$, $HO^-$

(*d*) $OsO_4$, $(CH_3)_3COOH$, $(CH_3)_3COH$, $HO^-$

(*e*) $H_2NCNH_2$ (with C=S)

**A-2.** (*a*) $C_6H_5CH_2\overset{O}{\overset{\|}{C}}H$

(*b*) $CH_3\overset{O}{\overset{\|}{C}}Cl$, pyridine; or

$(CH_3\overset{O}{\overset{\|}{C}})_2O$; or $CH_3CO_2H$, $H^+$

(*c*) $(C_6H_5CH_2CH_2)_2O$

(*d*) $K_2Cr_2O_7$, $H^+$, $H_2O$, heat

**A-3.** (*a*) $(CH_3)_2CHO^-Na^+$

(*b*) $(CH_3)_2C{=}O$

(*c*) $(CH_3)_2C{=}O$

(*d*) $CH_3\overset{O}{\overset{\|}{C}}OCH(CH_3)_2$

(*e*) $(CH_3)_2CHOSO_2-C_6H_4-CH_3$ (para)

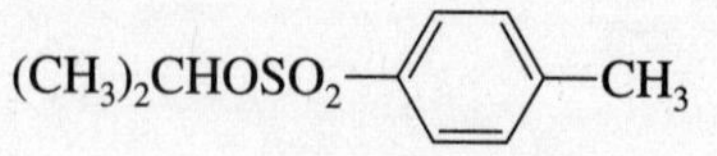

(*f*) $CH_3CH_2-C_6H_4-\overset{O}{\overset{\|}{C}}OCH(CH_3)_2$ (para)

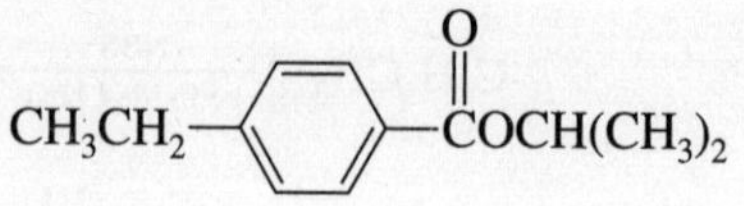

(*g*) $CH_3\overset{O}{\overset{\|}{C}}OCH(CH_3)_2$

**A-4.** (I)

$$(CH_3)_2CHBr + Mg \longrightarrow (CH_3)_2CHMgBr \xrightarrow[2.\ H_3O^+]{1.\ H_2C\overset{O}{-}CH_2} (CH_3)_2CHCH_2CH_2OH$$

(II) $(CH_3)_2CHCH_2Br + Mg \longrightarrow (CH_3)_2CHCH_2MgBr$

$$(CH_3)_2CHCH_2MgBr \xrightarrow[2.\ H_3O^+]{1.\ H_2C{=}O} (CH_3)_2CHCH_2CH_2OH$$

**A-5.** (*a*) $CH_3CO(CH_2)_4CHO$ (O, H)

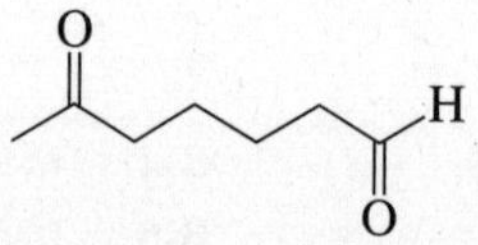

(*b*) $HOCH_2(CH_2)_3CH(OH)CH_3$ (HO, OH)

(*c*) $H_3C$, H, C=C, H, $CH_3$

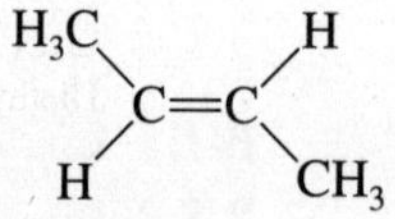

**A-6.** (*a*) PCC or PDC in $CH_2Cl_2$
(*b*) $Na_2Cr_2O_7$, $H^+$, $H_2O$, heat
(*c*) 1. $LiAlH_4$; 2. $H_2O$
(*d*) $OsO_4$, $(CH_3)_3COOH$, $(CH_3)_3COH$, $HO^-$

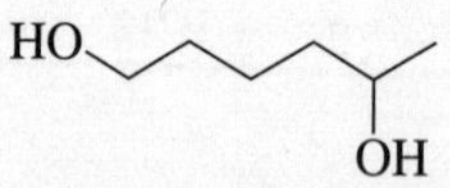

**A-7.**

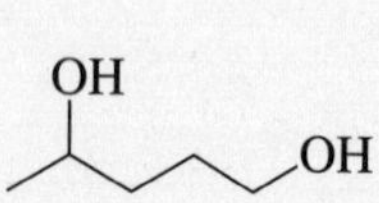

A: $CH_3CH(OH)CH_2CH_2CH_2OH$ (OH, OH)

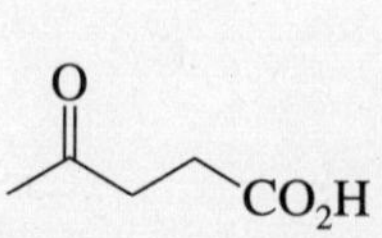

B: $CH_3COCH_2CH_2CO_2H$ (O, $CO_2H$)

C: $CH_3COCH_2CH_2CO_2CH_3$ (O, $CO_2CH_3$)

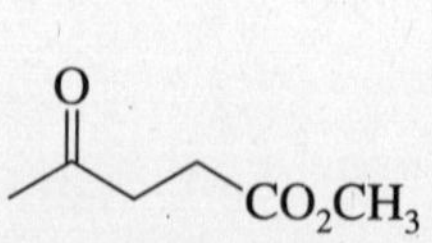

**A-8.** (*a*)

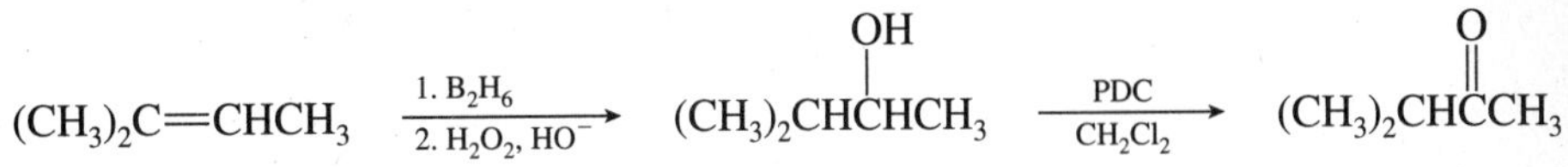

$$(CH_3)_2C{=}CHCH_3 \xrightarrow[2.\ H_2O_2,\ HO^-]{1.\ B_2H_6} (CH_3)_2CH\overset{OH}{\overset{|}{C}}HCH_3 \xrightarrow[CH_2Cl_2]{PDC} (CH_3)_2CH\overset{O}{\overset{\|}{C}}CH_3$$

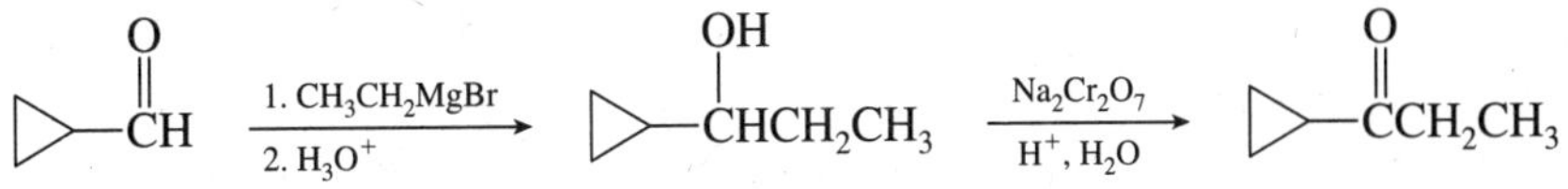

(*b*)

$$\text{cyclopropyl}-\overset{O}{\overset{\|}{C}}H \xrightarrow[2.\ H_3O^+]{1.\ CH_3CH_2MgBr} \text{cyclopropyl}-\overset{OH}{\overset{|}{C}}HCH_2CH_3 \xrightarrow[H^+,\ H_2O]{Na_2Cr_2O_7} \text{cyclopropyl}-\overset{O}{\overset{\|}{C}}CH_2CH_3$$

(*c*)

$$C_6H_5CH_3 \xrightarrow[\text{peroxides, heat}]{NBS} C_6H_5CH_2Br \xrightarrow{Mg} C_6H_5CH_2MgBr \xrightarrow[2.\ H_3O^+]{1.\ H_2C\overset{O}{—}CH_2} C_6H_5CH_2CH_2CH_2OH$$

$$C_6H_5CH_2CH_2CH_2OH \xrightarrow{K_2Cr_2O_7,\ H^+,\ H_2O,\ \text{heat}} C_6H_5CH_2CH_2CO_2H$$

$$C_6H_5CH_2CH_2CO_2H \xrightarrow[H^+]{CH_3CH_2OH} C_6H_5CH_2CH_2CO_2CH_2CH_3$$

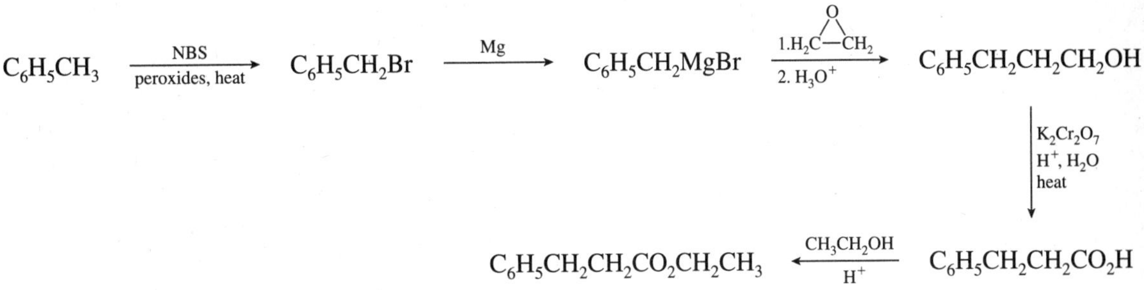

| | | | |
|---|---|---|---|
| **B-1.** (*e*) | **B-2.** (*d*) | **B-3.** (*c*) | **B-4.** (*c*) |
| **B-5.** (*b*) | **B-6.** (*b*) | **B-7.** (*a*) | **B-8.** (*a*) |
| **B-9.** (*d*) | **B-10.** (*a*) | **B-11.** (*b*) | **B-12.** (*d*) |
| **B-13.** (*e*) | **B-14.** (*a*) | **B-15.** (*c*) | |

# CHAPTER 16

**A-1.** $CH_3OCH_2CH_2CH_3$
Methyl propyl ether

$CH_3OCH(CH_3)_2$
Isopropyl methyl ether

$CH_3CH_2OCH_2CH_3$
Diethyl ether

**A-2.** (*a*) H, $CH_3CH_2$, $CH_3$ on C—$OCH_3$ (stereo structure)

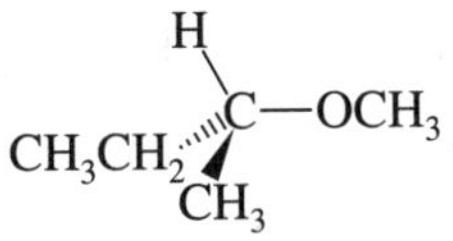

(*d*) Cyclohexane ring bearing $CH_3$, OH, $SCH_2CH_3$, $CH_3$ (stereo structure)

(*b*) HO, $H_3C$, H on C—C bearing $H_3C$, H, OH (stereo structure)

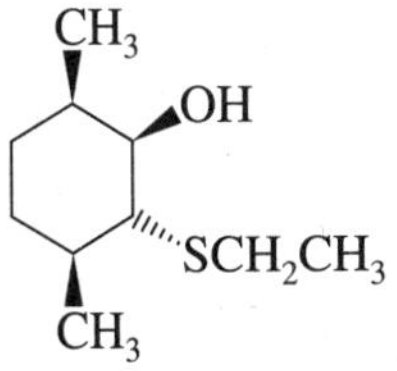

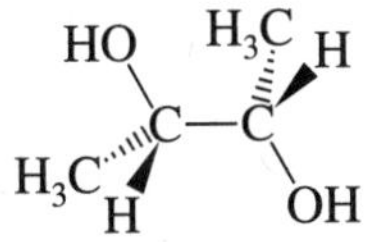

(*e*) $C_6H_5SCH_2CH_3$

(*c*) $C_6H_5\overset{I}{\overset{|}{C}}HCH_2OH$

(*f*) $C_6H_5\overset{O^-}{\overset{|}{S^+}}CH_2CH_3$

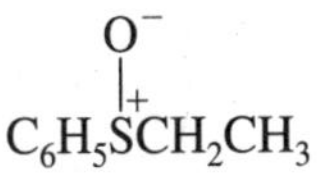

**A-3.** (*a*) Cyclohexanol (OH) $\xrightarrow{Na}$ cyclohexyl $O^- Na^+$ $\xrightarrow{CH_3CH_2I}$ cyclohexyl $OCH_2CH_3$

**A-4.** $CH_3CH_2OH \xrightarrow[heat]{H_2SO_4} H_2C{=}CH_2 \xrightarrow{CH_3\overset{O}{\overset{\|}{C}}OOH} H_2C{-}CH_2$ (epoxide, O bridging)

$CH_3CH_2OH \xrightarrow{Na} CH_3CH_2O^- Na^+ \xrightarrow[CH_3CH_2OH]{H_2C{-}CH_2 \text{ (epoxide)}} CH_3CH_2OCH_2CH_2OH$

**A-5.** (*a*) trans-2-bromocyclopentanol (HO, Br) $\xrightarrow{NaOH}$ cyclopentene oxide (O) $\xrightarrow{NaSCH_3}$ trans-2-(methylthio)cyclopentanol (HO, $SCH_3$)

(*b*) $C_6H_5C(CH_3)_2OH \xrightarrow[heat]{H_2SO_4} C_6H_5C(CH_3){=}CH_2 \xrightarrow{CH_3\overset{O}{\overset{\|}{C}}OOH} C_6H_5\overset{O}{C}(CH_3){-}CH_2$ (epoxide)

**A-6.** **A:** $C_6H_5CH_2OH$ **B:** $C_6H_5CH_2OCH_2CH_3$

**A-7.** Cyclohexanone (O) $\xrightarrow[2.\ H_3O^+]{1.\ CH_3MgBr}$ 1-methylcyclohexanol ($H_3C$, OH) $\xrightarrow[heat]{H_2SO_4}$ 1-methylcyclohexene ($CH_3$)

1-methylcyclohexene ($CH_3$) $\xrightarrow{CH_3\overset{O}{\overset{\|}{C}}OOH}$ 1-methylcyclohexene oxide ($H_3C$, O) $\xrightarrow{H_3O^+}$ trans-1-methylcyclohexane-1,2-diol ($H_3C$, OH, OH)

**A-8.** Cyclopentane bearing H, $H_2N$ on one carbon and $H_3C$, HO on the adjacent carbon

**B-1.** (*a*) **B-2.** (*a*) **B-3.** (*c*) **B-4.** (*d*) **B-5.** (*d*)
**B-6.** (*e*) **B-7.** (*a*) **B-8.** (*c*) **B-9.** (*d*) **B-10.** (*d*)

# CHAPTER 17

**A-1.** (*a*) 3,4-Dimethylhexanal
(*b*) 2,2,5-Trimethylhexan-3-one
(*c*) *trans*-4-Bromo-2-methylcyclohexanone
(*d*) 5-Methyl-4-hexen-3-one

**A-2.** (*a*) $CH_3\overset{O}{\overset{\|}{C}}(H)C{=}C(H)CH_2CH_3$ (cis; $CH_3CO$ and H on one carbon, H and $CH_2CH_3$ on the other)

(*b*) $CH_3\overset{O}{\overset{\|}{C}}\underset{\text{cyclopropyl}}{CH}\overset{O}{\overset{\|}{C}}CH_3$

(*c*) $C_6H_5\overset{CH_3}{CH}\underset{CH_2CH_3}{CH}CH_2\overset{O}{\overset{\|}{C}}H$

**A-3.** (*a*) 1-hydroxycyclohexanecarbonitrile (cyclohexane ring bearing OH and CN on the same carbon)

(*e*) 2-methylcyclohexanone phenylhydrazone: ring $=NNHC_6H_5$, with $CH_3$ on the adjacent carbon

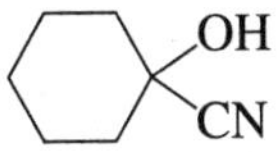

(*b*) $NH_2OH$

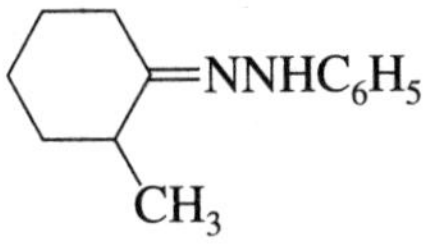

(*f*) $CH_3CH_2CH_2CH(OCH_2CH_3)_2$

(*c*) $(CH_3)_2CHCH{=}O$ + $HOCH_2CH_2CH_2OH$

(*g*) $C_6H_5C({=}O)CH_2CH_3$ + $(CH_3)_2NH$

(*d*) $(C_6H_5)_3\overset{+}{P}{-}\ddot{\overset{-}{C}}HCH_2CH_3$

(*h*) $CH_3CH_2CH_2C({=}O)OH$

**A-4.** (*a*) $(C_6H_5)_3\overset{+}{P}{-}CH_2CH(CH_3)_2\ Br^-$ (A) $(C_6H_5)_3\overset{+}{P}{-}\ddot{\overset{-}{C}}HCH(CH_3)_2$ (B)

$C_6H_5CH{=}CHCH(CH_3)_2$ (C)

(*b*) $C_6H_5CH_2C({=}O)CH_3$ (D) $C_6H_5CH_2OC({=}O)CH_3$ (E)

**A-5.** (*a*) (1) $CH_3MgI$; (2) $H_3O^+$; (3) $H_2SO_4$, heat

(*b*) $(C_6H_5)_3\overset{+}{P}{-}\ddot{\overset{-}{C}}H_2$ [from $(C_6H_5)_3P + CH_3I \longrightarrow \xrightarrow{C_4H_9Li}$ ]

(*c*) $HOCH_2CH_2OH$, $H^+$(cat), heat

(*d*) $CH_3C({=}O)OOH$

**A-6.** (*a*) chair cyclohexanone bearing $H_3C$ and $CH_3$ substituents + $HOCH_2C(CH_3)_2CH_2OH$

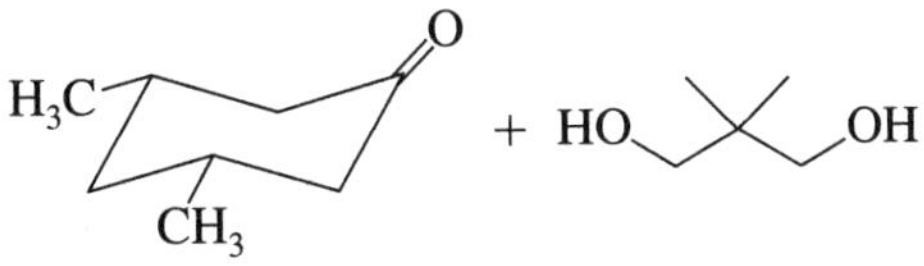

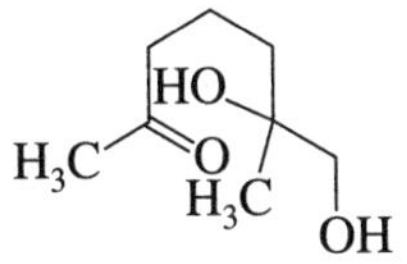

(*b*) (ring-opened structure with $H_3C$, HO, O, $H_3C$, OH) $\equiv CH_3C({=}O)CH_2CH_2CH_2CH_2C(OH)(CH_3)CH_2OH$

**A-7.** (*a*)

$$CH_3CH_2I + (C_6H_5)_3P \longrightarrow (C_6H_5)_3\overset{+}{P}{-}CH_2CH_3\ I^- \xrightarrow{C_4H_9Li} (C_6H_5)_3\overset{+}{P}{-}\ddot{\overset{-}{C}}HCH_3$$

$$(CH_3)_2C{=}O + (C_6H_5)_3\overset{+}{P}{-}\ddot{\overset{-}{C}}HCH_3 \longrightarrow (CH_3)_2C{=}CHCH_3 \xrightarrow{CH_3C(=O)OOH} (CH_3)_2\overset{O}{C{-}}CHCH_3$$ (epoxide)

(*b*) [Reaction scheme: 4-hydroxy-2-methylcyclohexanone $\xrightarrow[H^+]{HOCH_2CH_2OH}$ ethylene acetal with OH $\xrightarrow[CH_2Cl_2]{PCC}$ acetal ketone $\xrightarrow[2.\ H_3O^+]{1.\ CH_3MgI}$ 4-hydroxy-2,4-dimethylcyclohexanone (HO, $CH_3$)]

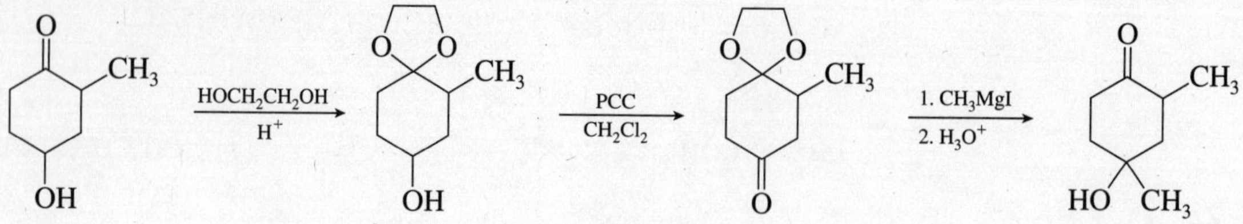

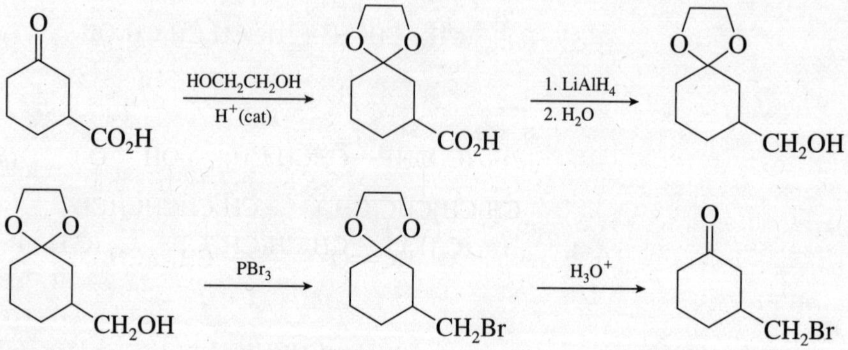

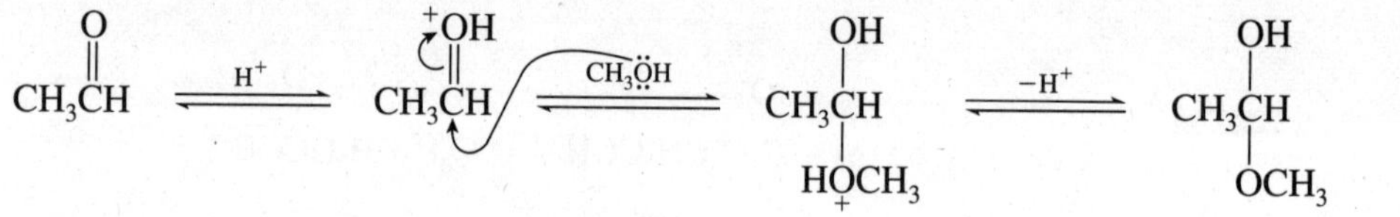

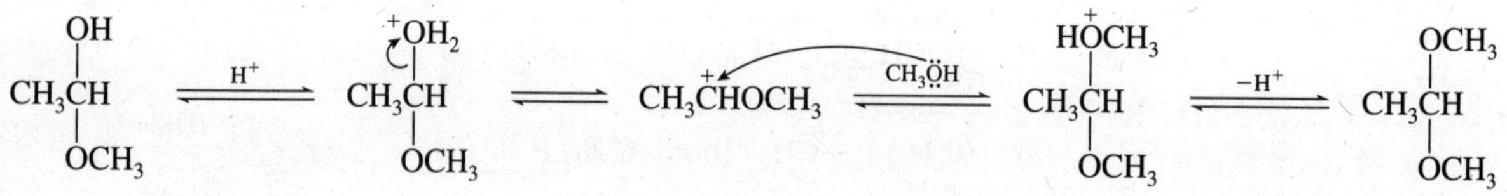

(*c*) [Reaction scheme: 3-oxocyclohexanecarboxylic acid ($CO_2H$) $\xrightarrow[H^+(cat)]{HOCH_2CH_2OH}$ acetal ($CO_2H$) $\xrightarrow[2.\ H_2O]{1.\ LiAlH_4}$ acetal ($CH_2OH$) $\xrightarrow{PBr_3}$ acetal ($CH_2Br$) $\xrightarrow{H_3O^+}$ 3-(bromomethyl)cyclohexanone ($CH_2Br$)]

**A-8.** $CH_3\overset{O}{\overset{\|}{C}}H \underset{}{\overset{H^+}{\rightleftharpoons}} CH_3\overset{+OH}{\overset{\|}{C}}H \overset{CH_3\ddot{O}H}{\rightleftharpoons} CH_3\overset{OH}{\underset{H\overset{+}{O}CH_3}{CH}} \overset{-H^+}{\rightleftharpoons} CH_3\overset{OH}{\underset{OCH_3}{CH}}$

$CH_3\overset{OH}{\underset{OCH_3}{CH}} \overset{H^+}{\rightleftharpoons} CH_3\overset{+OH_2}{\underset{OCH_3}{CH}} \rightleftharpoons CH_3\overset{+}{C}HOCH_3 \overset{CH_3\ddot{O}H}{\rightleftharpoons} CH_3\overset{H\overset{+}{O}CH_3}{\underset{OCH_3}{CH}} \overset{-H^+}{\rightleftharpoons} CH_3\overset{OCH_3}{\underset{OCH_3}{CH}}$

**A-9.** $C_6H_5-CH_2\overset{O}{\overset{\|}{C}}CH_3$

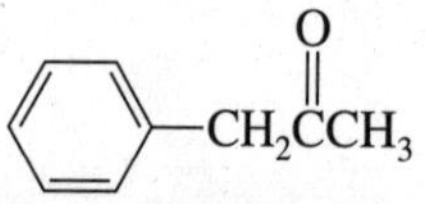

| | | | |
|---|---|---|---|
| **B-1.** (*c*) | **B-2.** (*d*) | **B-3.** (*a*) | **B-4.** (*c*) |
| **B-5.** (*b*) | **B-6.** (*b*) | **B-7.** (*a*) | **B-8.** (*b*) |
| **B-9.** (*e*) | **B-10.** (*c*) | **B-11.** (*c*) | **B-12.** (*c*) |
| **B-13.** (*d*) | **B-14.** (*e*) | **B-15.** (*a*) | **B-16.** (*c*) |

# CHAPTER 18

**A-1.** (*a*) $H_2C{=}\overset{OH}{C}CH_2CH_3$ and $CH_3\overset{OH}{C}{=}CHCH_3$ (*c*) [structure: 2-methylbutanal, H–C(=O)–CH(CH_3)CH_2CH_3]

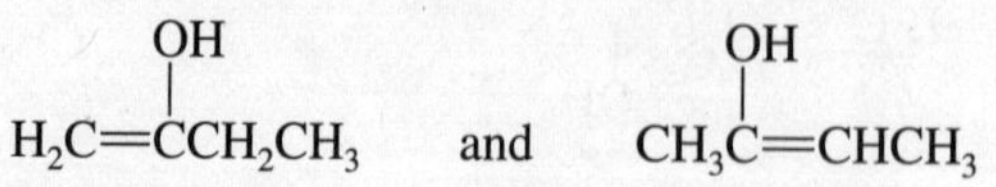

(*b*) [Resonance structures of the enolate of 1,3-cyclohexanedione: :Ö: / carbanion ⟷ :Ö:⁻ enolate, :O:] $Na^+$

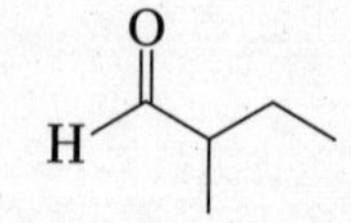

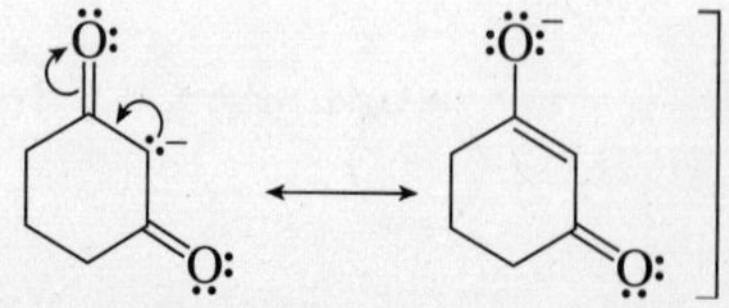

**A-2.** $C_6H_5CH_2CH{=}C(C_6H_5)CH{=}O$ $\quad$ $(CH_3CH_2)_2CHCH_2C(=O)CH_2CH_3$

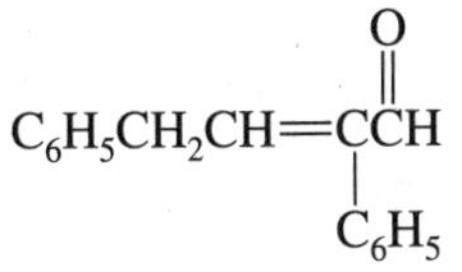

A 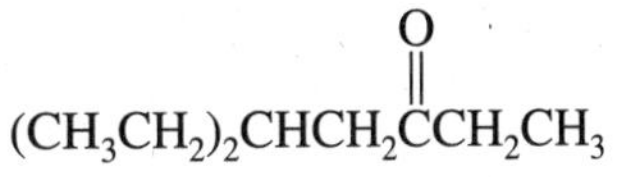 $\qquad$ B

**A-3.** $CH_3CH_2CH(OH)CH(CH_3)CH{=}O$ $\quad$ $CH_3CH(OH)C(CH_3)_2CH(CH_3)$ … printed as: $CH_3CH(CH_3)CH(OH)C(CH_3)_2$ with $HC{=}O$ on C(CH₃)₂

$CH_3CH_2CH(OH)C(CH_3)_2HC{=}O$ $\quad$ $CH_3CH(CH_3)CH(OH)CH(CH_3)CH{=}O$

**A-4.** $CH_3CH_2OH \xrightarrow[CH_2Cl_2]{PCC} CH_3CH{=}O$

$2CH_3CH{=}O \xrightarrow{NaOH} CH_3CH(OH)CH_2CH{=}O \xrightarrow[\text{or 1. } LiAlH_4 \text{; 2. } H_2O]{NaBH_4,\ CH_3OH} CH_3CH(OH)CH_2CH_2OH$

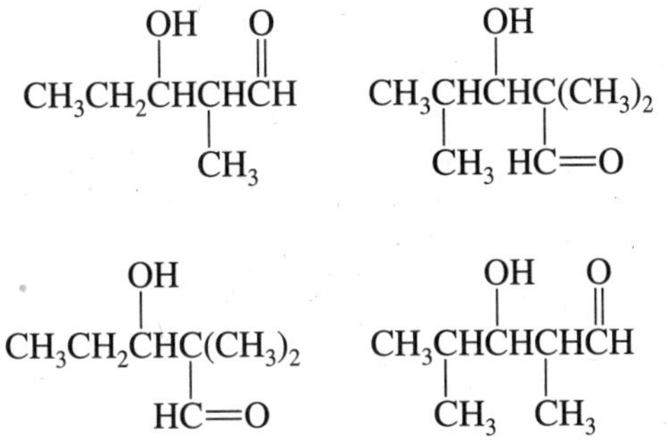

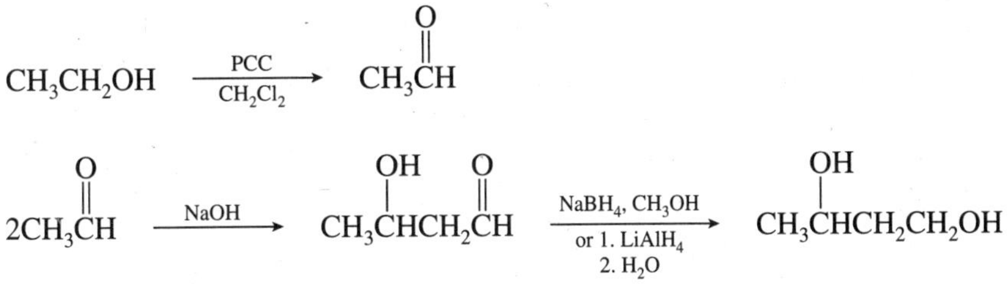

**A-5.**

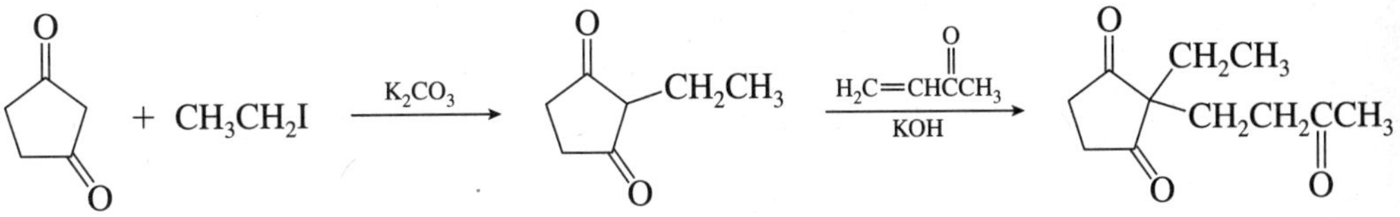

1,3-cyclopentanedione $+ CH_3CH_2I \xrightarrow{K_2CO_3}$ 2-ethyl-1,3-cyclopentanedione ($CH_2CH_3$) $\xrightarrow[KOH]{H_2C{=}CHC(=O)CH_3}$ 2-ethyl-2-($CH_2CH_2C(=O)CH_3$)-1,3-cyclopentanedione

**A-6.** (*a*) $CH_3CH_2CH_2CH(Br)CH{=}O$

(*b*) $CH_3CH_2CH_2CH_2CH(OH)CH(CH_2CH_2CH_3)CH{=}O$

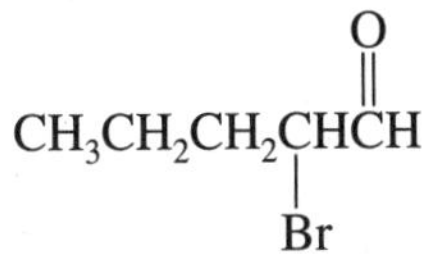

(*c*) $C_6H_5$–CH= (2,2-dimethylcyclopentanone, at C-5) : $C_6H_5CH{=}$ ring with $C{=}O$ and $C(CH_3)_2$

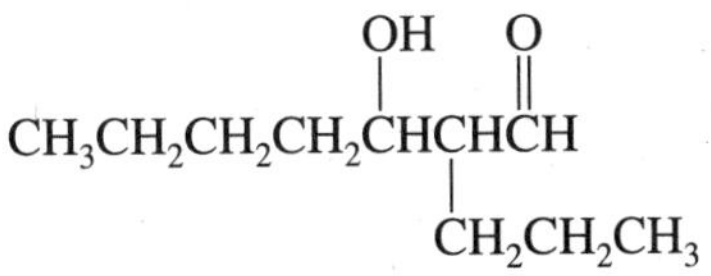

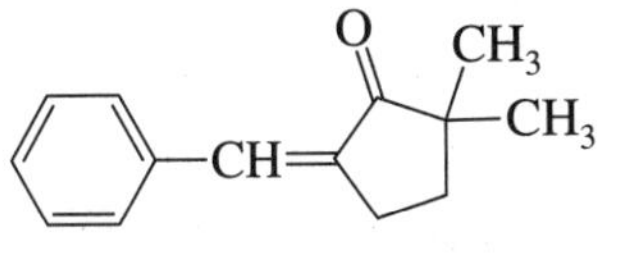

(*d*) $C_6H_5CH(SCH_3)CH_2C(=O)C_6H_5$

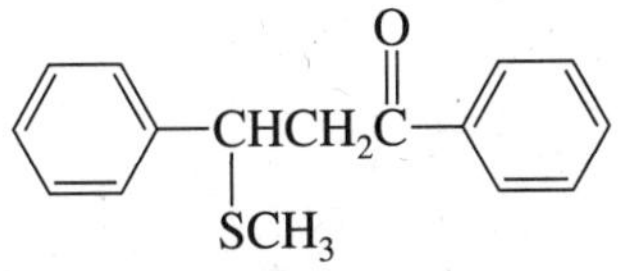

**A-7.**

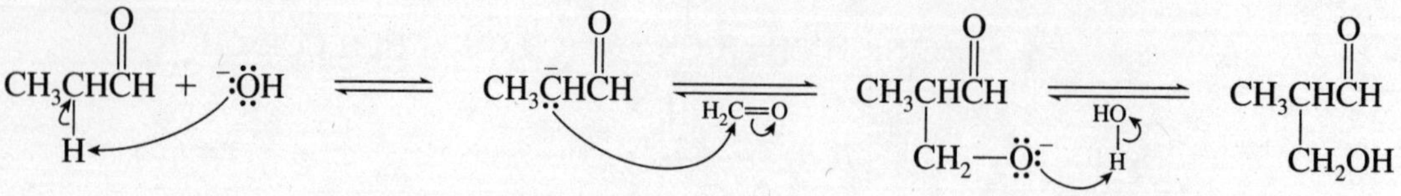

$$CH_3\overset{O}{\overset{\|}{C}}HCH + {}^-\ddot{O}H \rightleftharpoons CH_3\bar{C}HCH(=O) \xrightleftharpoons{H_2C=O} CH_3CH(CH_2-\ddot{O}^-)CH(=O) \xrightleftharpoons{HOH} CH_3CH(CH_2OH)CH(=O)$$

**A-8.** $C_6H_5CH(=O) + CH_3CH_2\overset{O}{\overset{\|}{C}}CH_2CH_3$

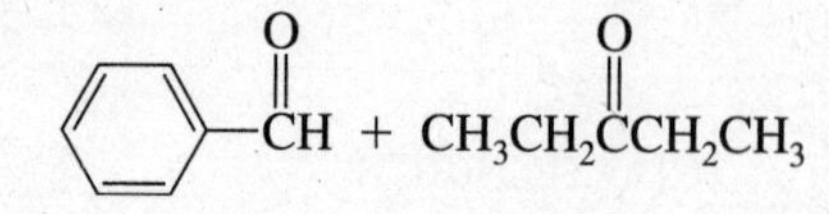

**B-1.** (*a*) **B-2.** (*c*) **B-3.** (*b*) **B-4.** (*b*)
**B-5.** (*a*) **B-6.** (*c*) **B-7.** (*c*) **B-8.** (*e*)
**B-9.** (*c*) **B-10.** (*b*) **B-11.** (*a*) **B-12.** (*a*)

# CHAPTER 19

**A-1.** (*a*) 4-Methyl-5-phenylhexanoic acid
(*b*) Cyclohexanecarboxylic acid
(*c*) 3-Bromo-2-ethylbutanoic acid

**A-2.** 4-Phenylbutanoic acid is $C_6H_5CH_2CH_2CH_2CO_2H$.

$$C_6H_5CH_2CH_2CH(CO_2H)_2 \xrightarrow{\text{heat}}$$

$$C_6H_5CH_2CH_2CH_2Br \xrightarrow[\text{2. } H^+, H_2O, \text{ heat}]{\text{1. } CN^-}$$

$$C_6H_5CH_2CH_2CH_2Br \xrightarrow[\text{2. } CO_2 \text{ 3. } H_3O^+]{\text{1. Mg}}$$

**A-3.** $C_6H_5CH_2CO_2H + CH_3CH_2OH \xrightarrow{H^+(\text{cat})} C_6H_5CH_2\overset{O}{\overset{\|}{C}}OCH_2CH_3 + H_2O$

**A-4.**

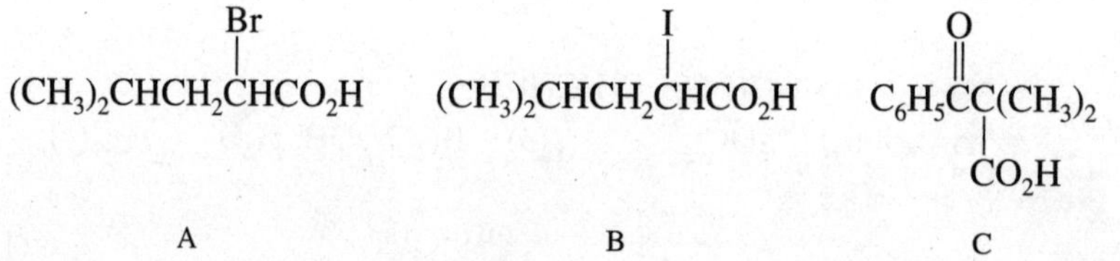

A: $(CH_3)_2CHCH_2CH(Br)CO_2H$

B: $(CH_3)_2CHCH_2CH(I)CO_2H$

C: $C_6H_5\overset{O}{\overset{\|}{C}}C(CH_3)_2(CO_2H)$

**A-5.** (*a*) $\xrightarrow{\text{Mg}} C_6H_5MgBr$

(*b*) $\xrightarrow[\text{2. } H_2O]{\text{1. LiAlH}_4} CH_3CH_2CH_2CH_2OH$

(*c*) $\xrightarrow{Br_2, P} CH_3CH_2CH(Br)\overset{O}{\overset{\|}{C}}OH$

(*d*) $CH_3CH_2CH(CN)(OH) \xrightarrow[\text{heat}]{H^+, H_2O}$

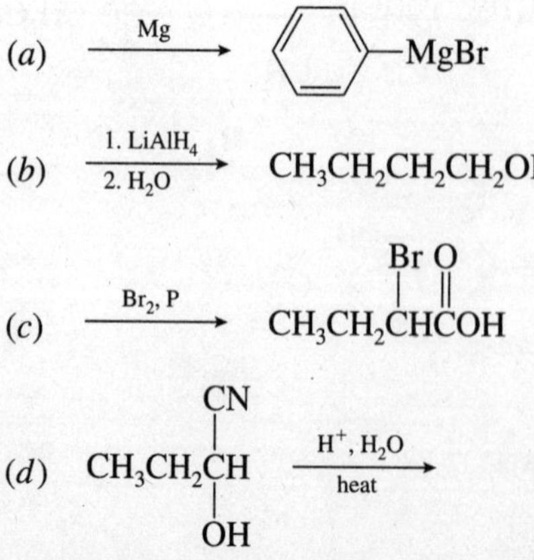

**A-6.** $CH_3CH_2CHCO_2H$ (with Br on the CH)

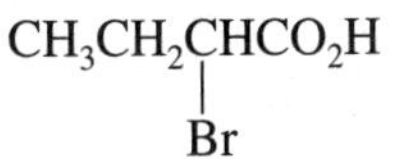

**A-7.** 
$$HC(OH)(OH)OCH_2CH_2CH_2CH_3$$

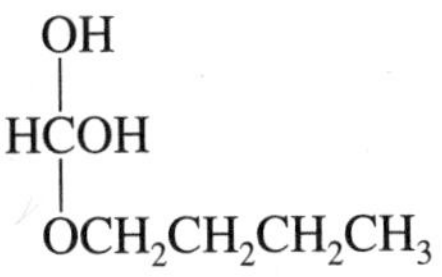

**A-8.**
$$CH_3\overset{O}{\overset{\|}{C}}OH \underset{}{\overset{H^+}{\rightleftharpoons}} CH_3\overset{+OH}{\overset{\|}{C}}OH + H\ddot{O}CH_3 \rightleftharpoons CH_3C(OH)(OH)H\overset{+}{O}CH_3 \underset{}{\overset{-H^+}{\rightleftharpoons}} CH_3C(OH)(OH)OCH_3$$

$$CH_3C(OH)(OH)OCH_3 \overset{H^+}{\rightleftharpoons} CH_3C(\ddot{O}H)(\overset{+}{O}H_2)OCH_3 \rightleftharpoons CH_3\overset{+OH}{\overset{\|}{C}}OCH_3 + H_2O \overset{-H^+}{\rightleftharpoons} CH_3\overset{O}{\overset{\|}{C}}OCH_3$$

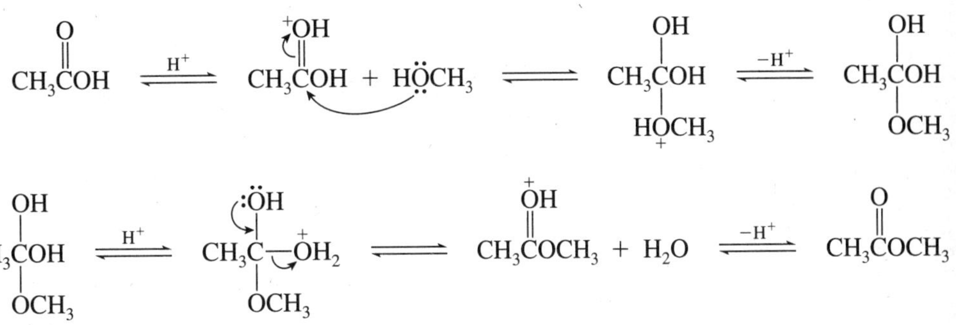

| | | | | | | | |
|---|---|---|---|---|---|---|---|
| **B-1.** | (*b*) | **B-2.** | (*a*) | **B-3.** | (*c*) | **B-4.** | (*d*) |
| **B-5.** | (*c*) | **B-6.** | (*d*) | **B-7.** | (*c*) | **B-8.** | (*d*) |
| **B-9.** | (*e*) | **B-10.** | (*c*) | **B-11.** | (*e*) | | |

# CHAPTER 20

**A-1.** (*a*) Propyl butanoate
(*b*) *N*-Methylbenzamide
(*c*) 4-Methylpentanoyl chloride

**A-2.** (*a*) $C_6H_5\overset{O}{\overset{\|}{C}}O\overset{O}{\overset{\|}{C}}C_6H_5$ (*b*) $CH_3\overset{O}{\overset{\|}{C}}NHCH(CH_3)CH_2CH_3$ (*c*) $C_6H_5\overset{O}{\overset{\|}{C}}OC_6H_5$

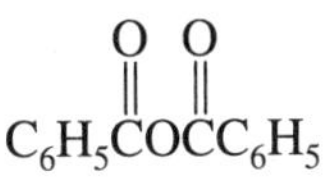

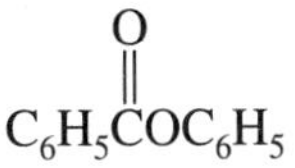

**A-3.** (*a*) $SOCl_2$ (*b*) $Br_2$, NaOH, $H_2O$ (*c*) $C_6H_5\overset{O}{\overset{\|}{C}}OCH_3$

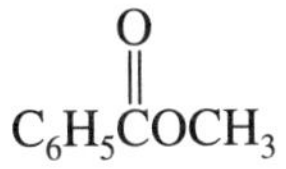

**A-4.** (*a*) $CH_3CO_2H$ + cyclohexanol (cyclohexyl–OH)

(*b*) $C_6H_5\overset{O}{\overset{\|}{C}}O$–cyclopentyl

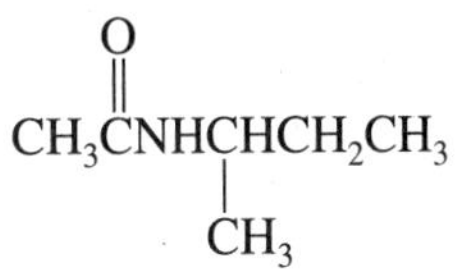

(*c*) benzene ring bearing ortho $\overset{O}{\overset{\|}{C}}OCH_2CH_3$ and $\overset{O}{\overset{\|}{C}}OH$

(*d*) $CH_3CH_2\overset{O}{\overset{\|}{C}}N(CH_3)_2 + CH_3CH_2OH$

(*e*) $H_3C$–(p-phenylene)–$CO_2H + CH_3\overset{+}{N}H_3\ HSO_4^-$

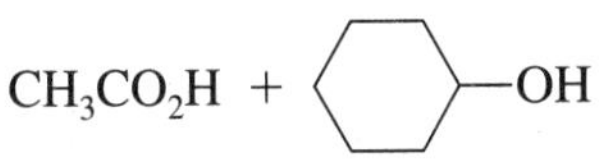

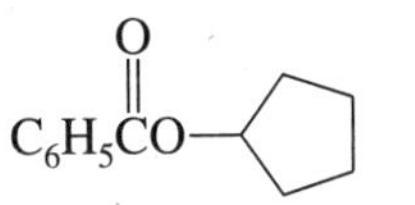

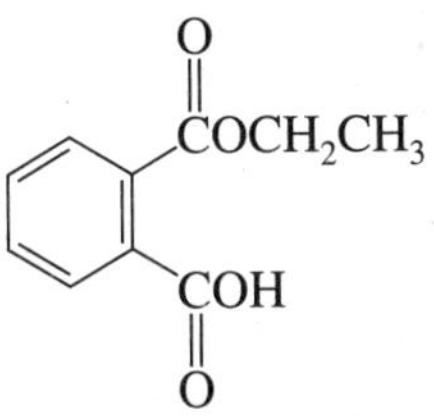

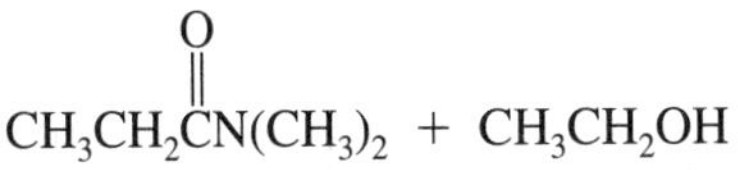

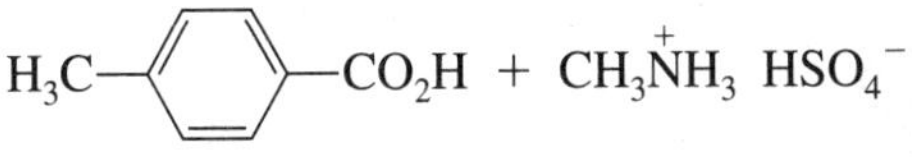

**A-5.** 
$$C_6H_5\text{-}C(OH)\text{-}O\text{-}CH_2\text{-}CH_2\text{-}N(CH_3)\ (\text{ring}) \quad \text{from} \quad C_6H_5\overset{O}{\overset{\|}{C}}\text{-}O\text{-}CH_2\text{-}CH_2\text{-}\ddot{N}H\text{-}CH_3$$

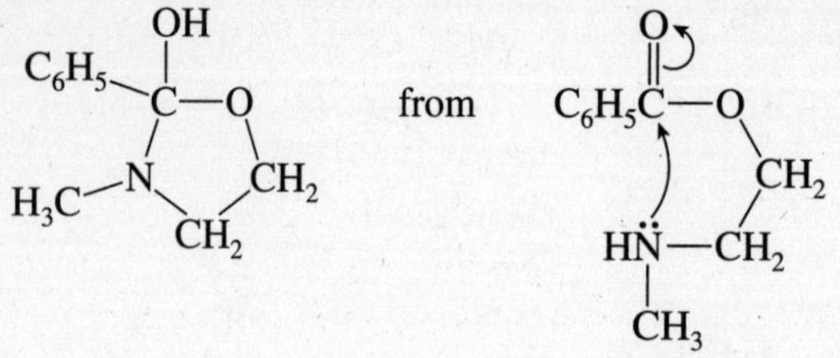

**A-6.**

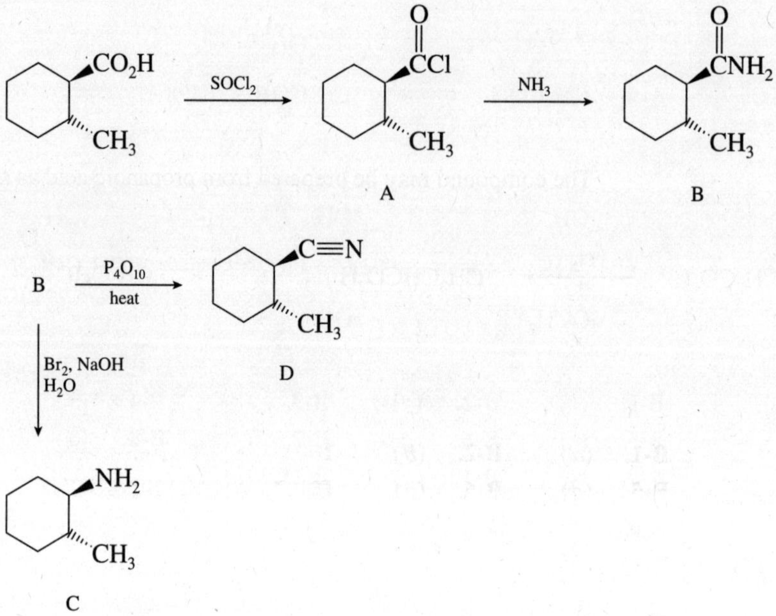

$$\text{2-methylcyclohexanecarboxylic acid } (CO_2H) \xrightarrow{SOCl_2} \underset{\text{A}}{\text{acyl chloride } (\overset{O}{\overset{\|}{C}}Cl)} \xrightarrow{NH_3} \underset{\text{B}}{\text{amide } (\overset{O}{\overset{\|}{C}}NH_2)}$$

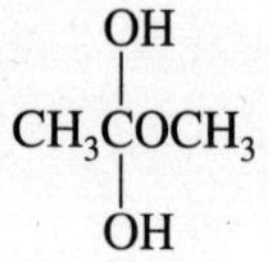

$$\text{B} \xrightarrow[\text{heat}]{P_4O_{10}} \underset{\text{D}}{\text{nitrile } (C\equiv N)}$$

$$\text{B} \xrightarrow{Br_2,\ NaOH,\ H_2O} \underset{\text{C}}{\text{2-methylcyclohexylamine } (NH_2)}$$

**A-7.** (*a*) $CH_3C(OH)_2OCH_3$ (*b*) $CH_3\overset{O}{\overset{\|}{C}}OC(OH)(NH_2)CH_3$

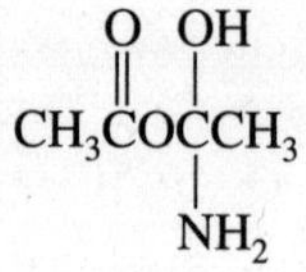

**A-8.**

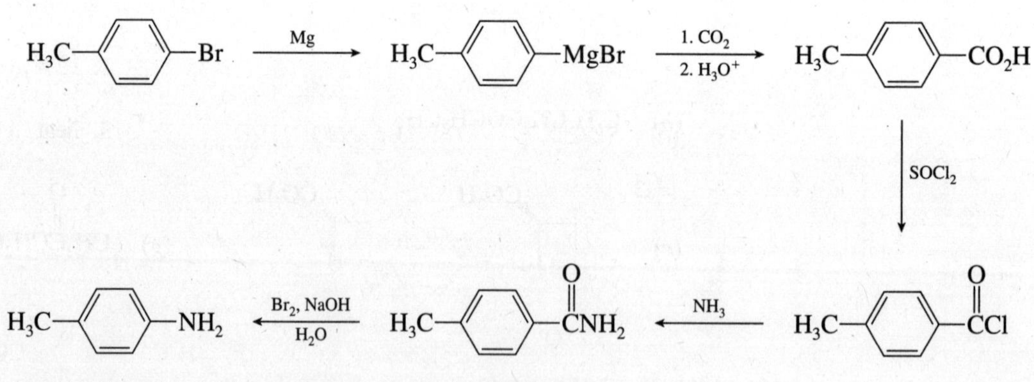

$$H_3C\text{-}C_6H_4\text{-}Br \xrightarrow{Mg} H_3C\text{-}C_6H_4\text{-}MgBr \xrightarrow[2.\ H_3O^+]{1.\ CO_2} H_3C\text{-}C_6H_4\text{-}CO_2H$$

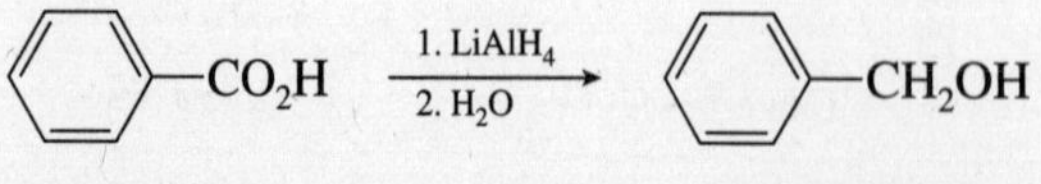

$$H_3C\text{-}C_6H_4\text{-}CO_2H \xrightarrow{SOCl_2} H_3C\text{-}C_6H_4\text{-}\overset{O}{\overset{\|}{C}}Cl \xrightarrow{NH_3} H_3C\text{-}C_6H_4\text{-}\overset{O}{\overset{\|}{C}}NH_2 \xrightarrow[H_2O]{Br_2,\ NaOH} H_3C\text{-}C_6H_4\text{-}NH_2$$

**A-9.**

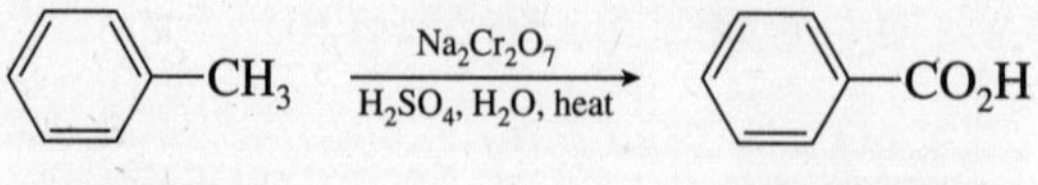

$$\underset{\text{Toluene}}{C_6H_5\text{-}CH_3} \xrightarrow[H_2SO_4,\ H_2O,\ \text{heat}]{Na_2Cr_2O_7} \underset{\text{Benzoic acid}}{C_6H_5\text{-}CO_2H}$$

$$\underset{\text{Benzoic acid}}{C_6H_5\text{-}CO_2H} \xrightarrow[2.\ H_2O]{1.\ LiAlH_4} \underset{\text{Benzyl alcohol}}{C_6H_5\text{-}CH_2OH}$$

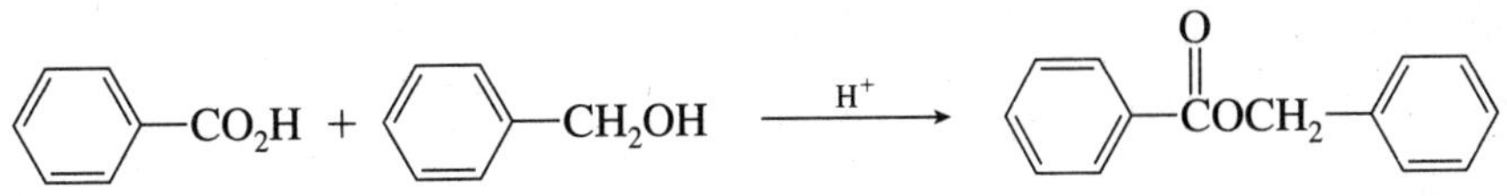

$$C_6H_5CO_2H + C_6H_5CH_2OH \xrightarrow{H^+} C_6H_5\overset{O}{\overset{\|}{C}}OCH_2C_6H_5$$

Benzoic acid · Benzyl alcohol · Benzyl benzoate

**A-10.** The compound is 2-chloropropanamide.

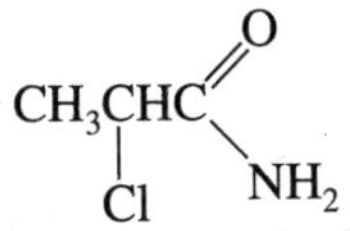

$$CH_3\underset{Cl}{\underset{|}{C}}HC(=O)NH_2$$

2-Chloropropanamide

The compound may be prepared from propanoic acid as shown.

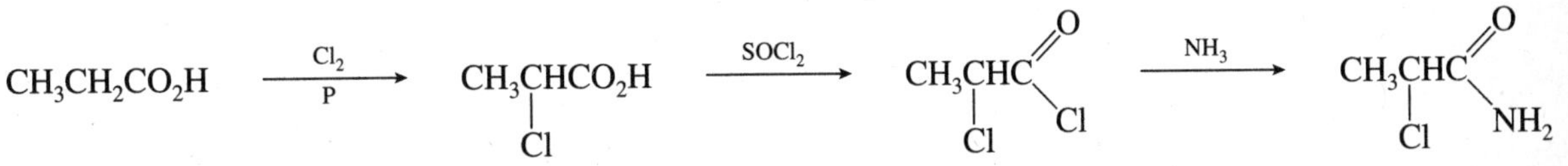

$$CH_3CH_2CO_2H \xrightarrow[P]{Cl_2} CH_3\underset{Cl}{\underset{|}{C}}HCO_2H \xrightarrow{SOCl_2} CH_3\underset{Cl}{\underset{|}{C}}HC(=O)Cl \xrightarrow{NH_3} CH_3\underset{Cl}{\underset{|}{C}}HC(=O)NH_2$$

Propanoic acid · 2-Chloropropanoic acid · 2-Chloropropanoyl chloride · 2-Chloropropanamide

**B-1.** (*a*) **B-2.** (*b*) **B-3.** (*b*) **B-4.** (*c*)
**B-5.** (*d*) **B-6.** (*c*) **B-7.** (*d*) **B-8.** (*d*)
**B-9.** (*b*) **B-10.** (*a*) **B-11.** (*d*) **B-12.** (*b*)
**B-13.** (*b*)

# CHAPTER 21

**A-1.** (*a*) $CH_3CH_2CH_2\overset{O}{\overset{\|}{C}}\underset{CH_2CH_3}{\underset{|}{C}H}\overset{O}{\overset{\|}{C}}OCH_2CH_3$

(*b*) $C_6H_5CH_2\overset{O}{\overset{\|}{C}}OCH_2CH_3$

(*c*) *cis*- and *trans*-3-chlorocyclobutanecarboxylic acid: Cl and $CO_2H$ on cyclobutane ring (wedge/wedge) + Cl and $CO_2H$ on cyclobutane ring (wedge/hash)

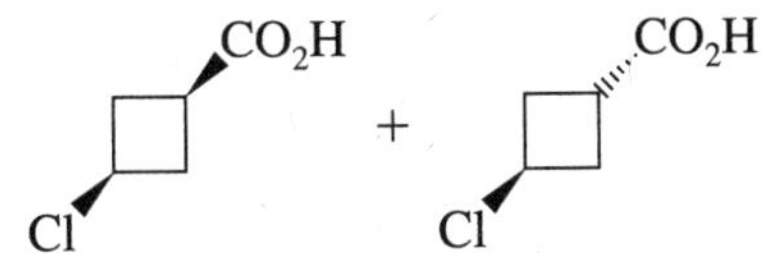

(*d*) $(CH_3CH_2O_2C)_2CHCH_2CH_2\overset{O}{\overset{\|}{C}}OCH_2CH_3$

(*e*) $CH_3\overset{O}{\overset{\|}{C}}\underset{CH_2C_6H_5}{\underset{|}{C}H}\overset{O}{\overset{\|}{C}}OCH_2CH_3$

(*f*) 1. $HO^-$, $H_2O$
2. $H_3O^+$
3. heat

(*g*) $CH_3\overset{O}{\overset{\|}{C}}CH_2CH_2CO_2H$

**A-2.** $CH_3CH_2O\overset{O}{\overset{\|}{C}}(CH_2)_4\overset{O}{\overset{\|}{C}}OCH_2CH_3$ (A)

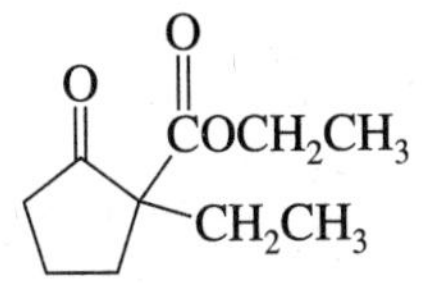

B: 2-oxocyclopentane bearing $\overset{O}{\overset{\|}{C}}OCH_2CH_3$ and $CH_2CH_3$ on C-2 (B)

C: 2-ethylcyclopentanone (cyclopentanone ring with $CH_2CH_3$ substituent)

D: $CH_3CH_2CH_2\overset{O}{\overset{\|}{C}}\underset{CH_2CH_3}{\underset{|}{C}H}\overset{O}{\overset{\|}{C}}OCH_2CH_3$

E: $CH_3CH_2CH_2\overset{O}{\overset{\|}{C}}CH_2CH_2CH_3$

**A-3.** (*a*)

$$CH_3\overset{O}{\overset{\|}{C}}CH_2\overset{O}{\overset{\|}{C}}OCH_2CH_3 \xrightarrow[\text{2. } BrCH_2\overset{\|}{\underset{O}{C}}OCH_2CH_3]{\text{1. } NaOCH_2CH_3} CH_3\overset{O}{\overset{\|}{C}}\underset{CH_2\underset{\underset{O}{\|}}{C}OCH_2CH_3}{\underset{|}{C}H}\overset{O}{\overset{\|}{C}}OCH_2CH_3 \xrightarrow[\text{2. } H_3O^+ \text{ 3. heat}]{\text{1. } HO^-,\ H_2O} CH_3\overset{O}{\overset{\|}{C}}CH_2CH_2\overset{O}{\overset{\|}{C}}OH$$

(*b*)

$$C_6H_5\overset{O}{\overset{\|}{C}}CH_3 + CH_3CH_2O\overset{O}{\overset{\|}{C}}OCH_2CH_3 \xrightarrow{NaOCH_2CH_3} C_6H_5\overset{O}{\overset{\|}{C}}CH_2\overset{O}{\overset{\|}{C}}OCH_2CH_3 \xrightarrow[\text{2. } H_2C{=}CH\overset{O}{\overset{\|}{C}}CH_3]{\text{1. } NaOCH_2CH_3} C_6H_5\overset{O}{\overset{\|}{C}}\underset{CH_2CH_2\underset{\underset{O}{\|}}{C}CH_3}{\underset{|}{C}H}\overset{O}{\overset{\|}{C}}OCH_2CH_3$$

$$\xrightarrow[\text{2. } H_3O^+ \text{ 3. heat}]{\text{1. } HO^-,\ H_2O} C_6H_5\overset{O}{\overset{\|}{C}}CH_2CH_2CH_2\overset{O}{\overset{\|}{C}}CH_3$$

**A-4.**

$$2CH_3CH_2\overset{O}{\overset{\|}{C}}OCH_2CH_3 \underset{}{\overset{NaOCH_2CH_3}{\rightleftharpoons}} CH_3\overset{..}{\underset{-}{C}}H\overset{O}{\overset{\|}{C}}OCH_2CH_3 + CH_3CH_2\underset{\underset{:O:}{\|}}{C}OCH_2CH_3 \longrightarrow CH_3\underset{CH_3CH_2\underset{\underset{:\ddot{O}:^-}{|}}{C}{-}OCH_2CH_3}{\underset{|}{C}H}\overset{O}{\overset{\|}{C}}OCH_2CH_3$$

$$\xrightarrow{-CH_3CH_2O^-} CH_3\underset{CH_3CH_2C{=}O}{\underset{|}{C}H}\overset{O}{\overset{\|}{C}}OCH_2CH_3 \xrightarrow{-H^+} CH_3\underset{CH_3CH_2C{=}O}{\underset{|}{\overset{..}{\underset{-}{C}}}}\overset{O}{\overset{\|}{C}}OCH_2CH_3$$

**A-5.** Enolization of the Claisen condensation product is necessary for completion of the reaction. The condensation product of ethyl 3-methylbutanoate can enolize; the product from condensation of ethyl 2-methylpropanoate cannot.

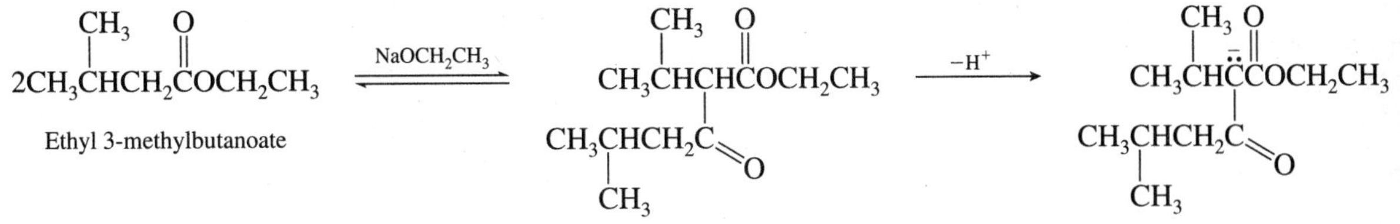

$$2CH_3\underset{}{\overset{CH_3}{|}}CHCH_2\overset{O}{\|}COCH_2CH_3 \xrightleftharpoons{NaOCH_2CH_3} CH_3\overset{CH_3}{C}HCH\overset{O}{C}OCH_2CH_3 \ (CH_3\underset{CH_3}{C}HCH_2C{=}O) \xrightarrow{-H^+} CH_3\overset{CH_3}{C}H\ddot{C}\overset{O}{C}OCH_2CH_3 \ (CH_3\underset{CH_3}{C}HCH_2C{=}O)$$

Ethyl 3-methylbutanoate

$$2CH_3CH_2\underset{CH_3}{C}H\overset{O}{C}OCH_2CH_3 \xrightleftharpoons{NaOCH_2CH_3} CH_3CH_2\overset{H_3C\ O}{C}COCH_2CH_3 \ (CH_3CH_2\underset{CH_3}{C}HC{=}O)$$

Ethyl 2-methylbutanoate — Claisen product cannot enolize

**B-1.** (*b*) **B-2.** (*d*) **B-3.** (*c*) **B-4.** (*b*)
**B-5.** (*c*) **B-6.** (*c*) **B-7.** (*b*) **B-8.** (*d*)

# CHAPTER 22

**A-1.** (*a*) 1,1-Dimethylpropylamine or 2-methyl-2-butanamine; primary
(*b*) *N*-Methylcyclopentylamine or *N*-methylcyclopentanamine; secondary
(*c*) *m*-Bromo-*N*-propylaniline; secondary

**A-2.** (*a*) $NaN_3$ (*b*) KCN (*c*) Potassium phthalimide ($N^- K^+$)

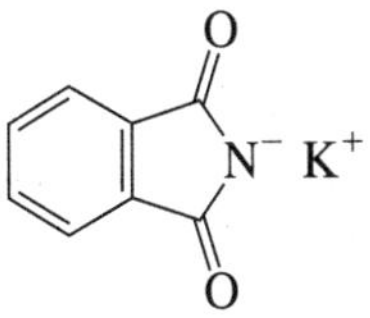

**A-3.** (*a*) $H_3C-C_6H_4-N_2^+ Cl^-$ (para)

(*b*) $H_3C-C_6H_4-Br$ (para)

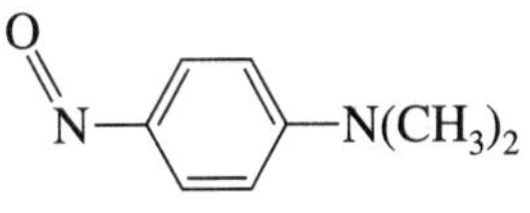

(*c*) $H_3PO_2$

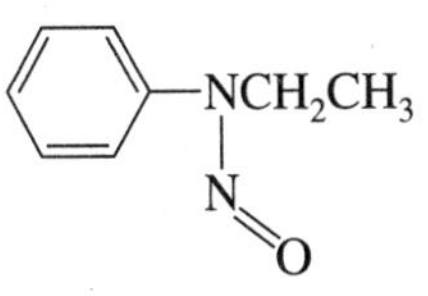

(*d*) $(CH_3\overset{O}{\|}C)_2O$ or $CH_3\overset{O}{\|}CCl$

(*e*) 2-nitro-4-ethylacetanilide ($NH\overset{O}{\|}CCH_3$, $O_2N$, $CH_2CH_3$) + 3-nitro-4-acetamido... ($NH\overset{O}{\|}CCH_3$, $CH_2CH_3$, $NO_2$)

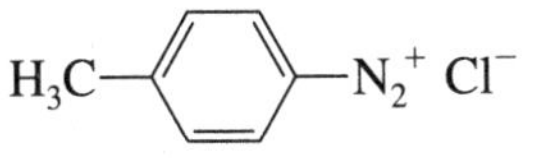

(*f*) $O{=}N-C_6H_4-N(CH_3)_2$ (para)

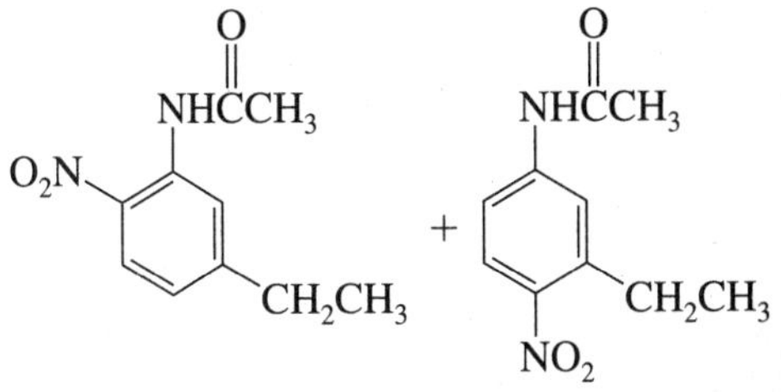

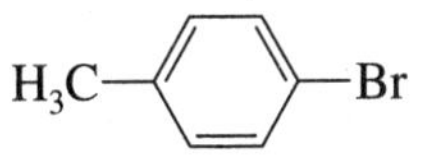

(*g*) $C_6H_5-NCH_2CH_3$ with $N{=}O$ on nitrogen

**A-4.** (*a*) A: N-ethylpyrrolidine ($N$–$CH_2CH_3$); B: pyrrolidinium ion with $H_3C$ and $CH_2CH_3$ on $\overset{+}{N}$, $I^-$; C: pyrrolidinium ion with $H_3C$ and $CH_2CH_3$ on $\overset{+}{N}$, $OH^-$

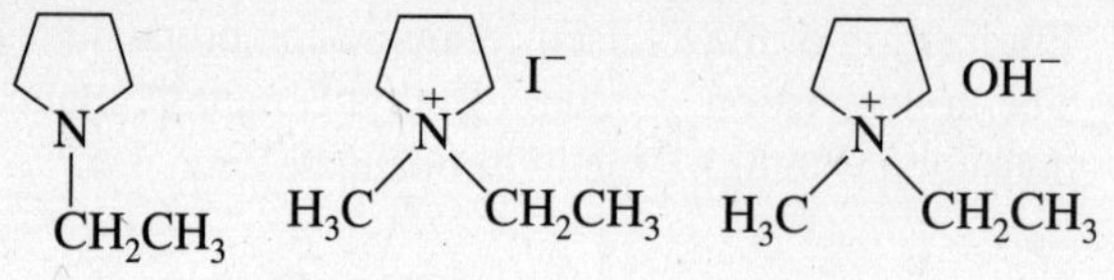

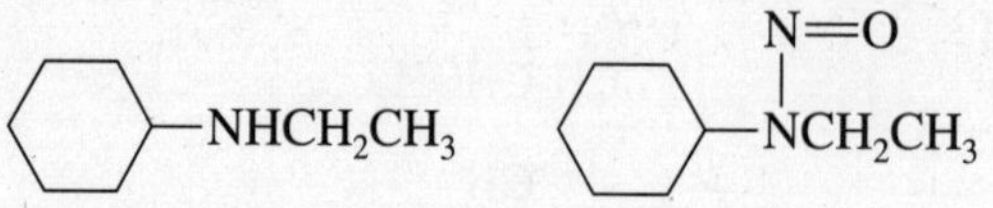

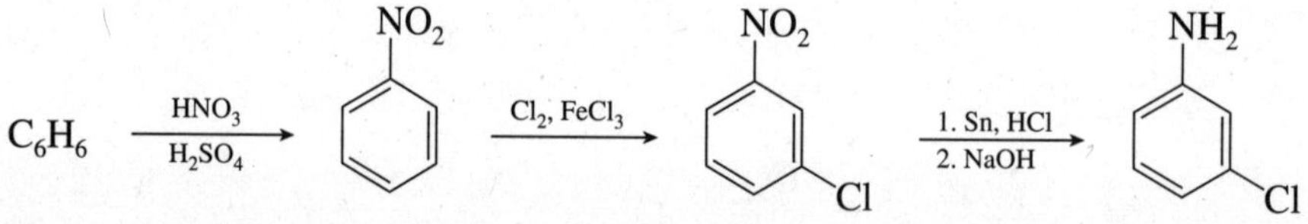

(*b*) D: cyclohexyl–$NHCH_2CH_3$; E: cyclohexyl–$N(N{=}O)CH_2CH_3$

**A-5.** (*a*)

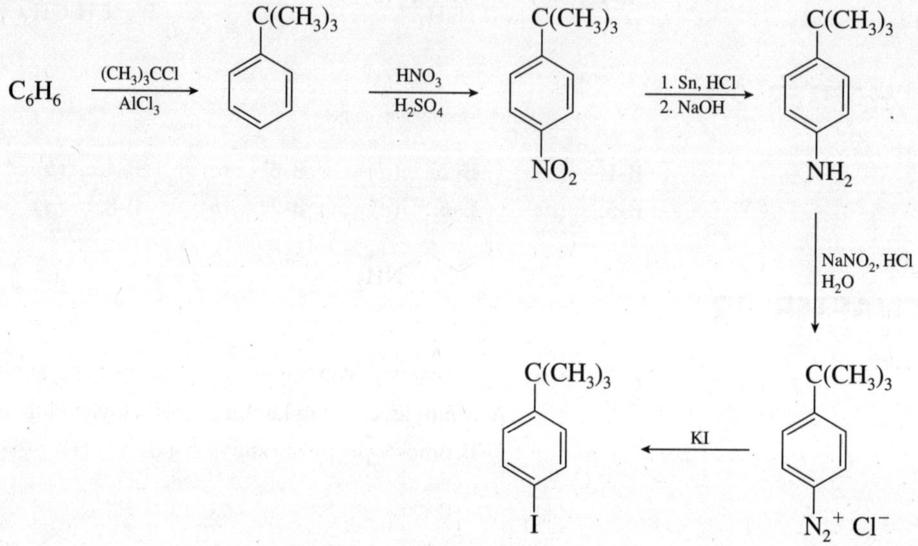

$$C_6H_6 \xrightarrow[AlCl_3]{(CH_3)_3CCl} C_6H_5C(CH_3)_3 \xrightarrow[H_2SO_4]{HNO_3} p\text{-}(CH_3)_3CC_6H_4NO_2 \xrightarrow[2.\ NaOH]{1.\ Sn,\ HCl} p\text{-}(CH_3)_3CC_6H_4NH_2$$

$$p\text{-}(CH_3)_3CC_6H_4NH_2 \xrightarrow[H_2O]{NaNO_2,\ HCl} p\text{-}(CH_3)_3CC_6H_4N_2^+\ Cl^- \xrightarrow{KI} p\text{-}(CH_3)_3CC_6H_4I$$

(*b*)

$$C_6H_6 \xrightarrow[H_2SO_4]{HNO_3} C_6H_5NO_2 \xrightarrow{Cl_2,\ FeCl_3} m\text{-}ClC_6H_4NO_2 \xrightarrow[2.\ NaOH]{1.\ Sn,\ HCl} m\text{-}ClC_6H_4NH_2$$

(*c*)

$$C_6H_5NH_2 \xrightarrow[H_2O]{NaNO_2,\ HCl} C_6H_5N_2^+\ Cl^- \xrightarrow{C_6H_5N(CH_3)_2} C_6H_5N{=}N{-}C_6H_4{-}N(CH_3)_2$$

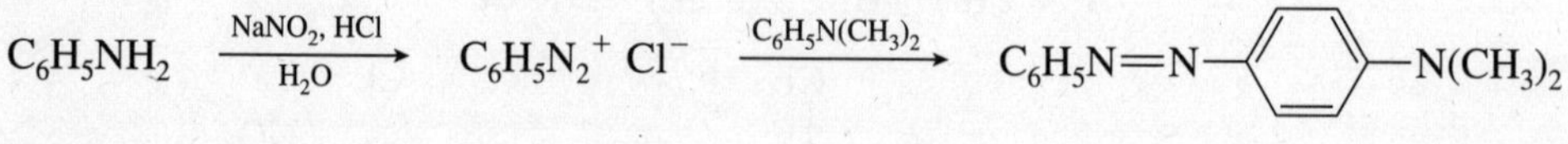

**A-6.** In the para isomer, resonance delocalization of the electron pair of the amine nitrogen involves the nitro group.

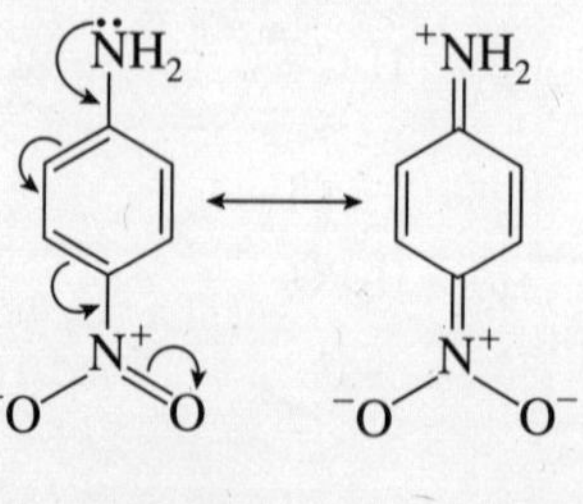

$$\ddot{N}H_2\text{–}C_6H_4\text{–}\overset{+}{N}(O^-){=}O \longleftrightarrow {}^+NH_2{=}C_6H_4{=}\overset{+}{N}(O^-)O^-$$

**A-7.** Strongest base: C, an alkylamine

Weakest base: D, a lactam (cyclic amide)

**A-8.** A: $p$-$(CH_3)_3C$-$C_6H_4$-$NHCCH_3$ (O) B: $p$-$(CH_3)_3C$-$C_6H_4$-$NH_2$ C: Cl, Cl, $NH_2$, $C(CH_3)_3$ D: Cl, Cl, Br, $C(CH_3)_3$

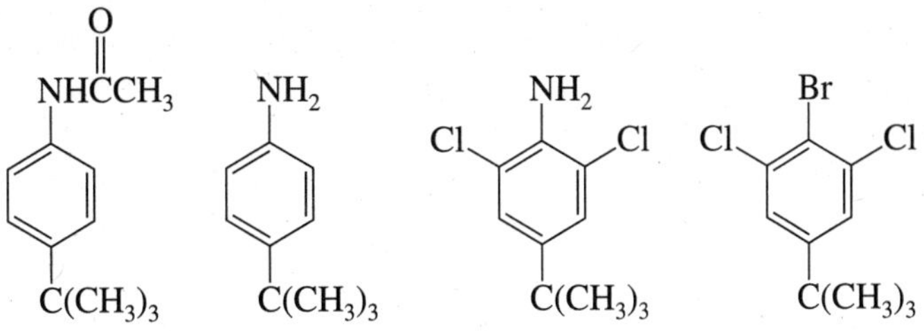

| | | | | | | | |
|---|---|---|---|---|---|---|---|
| **B-1.** | (*b*) | **B-2.** | (*d*) | **B-3.** | (*c*) | **B-4.** | (*d*) |
| **B-5.** | (*c*) | **B-6.** | (*e*) | **B-7.** | (*d*) | **B-8.** | (*c*) |
| **B-9.** | (*d*) | **B-10.** | (*e*) | **B-11.** | (*c*) | **B-12.** | (*b*) |
| **B-13.** | (*c*) | **B-14.** | (*c*) | | | | |

# CHAPTER 23

**A-1.** (*a*) $CF_3$, $NH_2$ + $CF_3$, $NH_2$ (*c*) $C(CH_3)_3$, $NH_2$ + $C(CH_3)_3$, $NH_2$

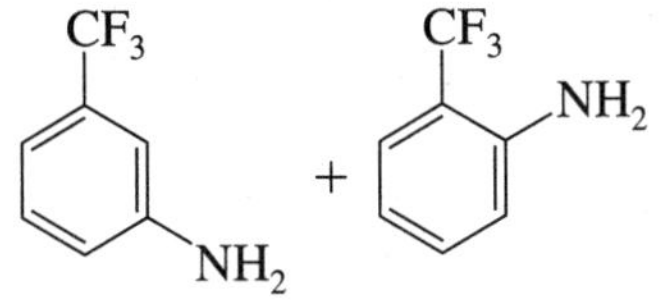

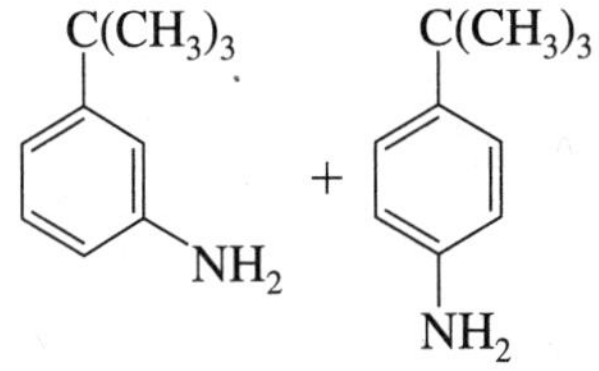

(*b*) $CH_3O$, Cl, $NO_2$

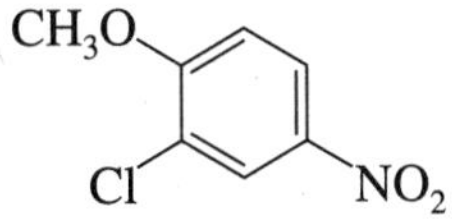

**A-2.** (*a*) $CF_3$ (*c*) $C(CH_3)_3$

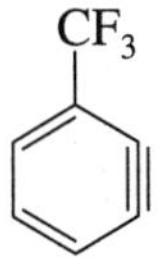

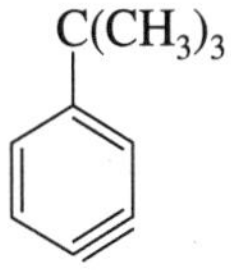

(*b*) $CH_3O$, Cl, Cl, $\overset{+}{N}$, $O^-$, $^-O$

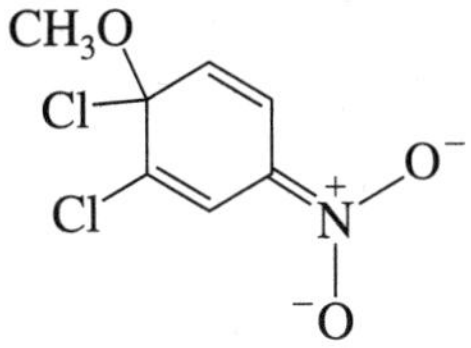

**A-3.** (*a*) Cl $\xrightarrow[\text{heat}]{HNO_3,\ H_2SO_4}$ Cl, $NO_2$, $NO_2$ $\xrightarrow{NaOCH_3}$ $OCH_3$, $NO_2$, $NO_2$

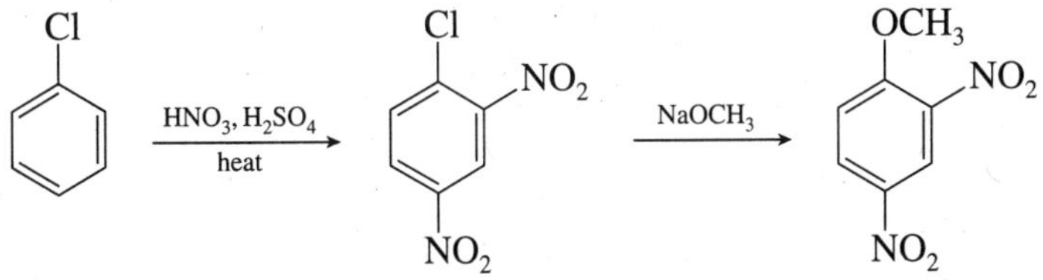

(*b*) Cl $\xrightarrow[AlCl_3]{(CH_3)_2CHCl}$ Cl, $CH(CH_3)_2$ (+ ortho isomer) $\xrightarrow{NaNH_2,\ NH_3}$ $NH_2$, $CH(CH_3)_2$ (+ meta isomer)

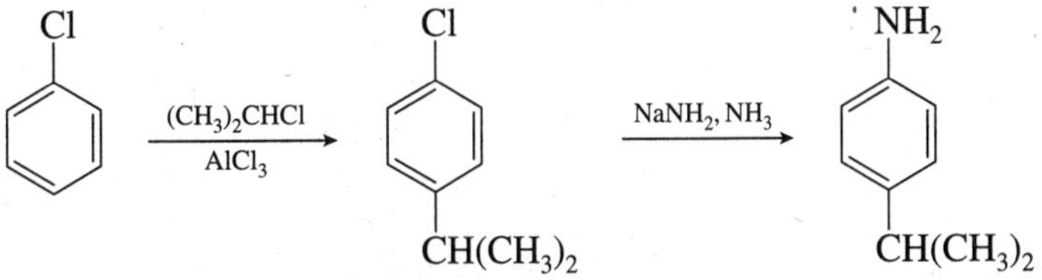

**A-4.** $NO_2$, X; Y:⁻ → ... ⟷ ...

$NO_2$ ... → $NO_2$, Y + $X^-$

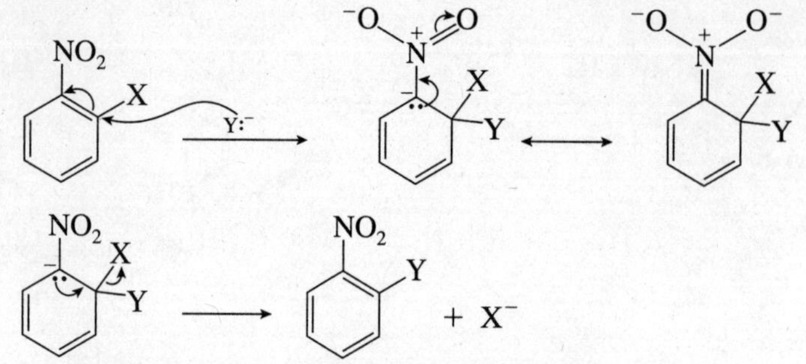

The mechanism for para substitution is similar.

**A-5.** **Product:** 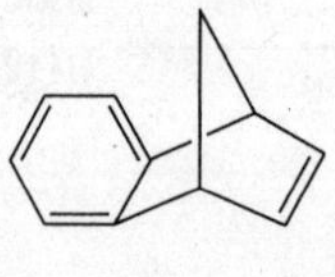 **Intermediate:**

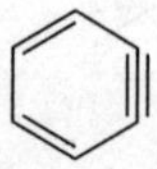

| | | | | | | | |
|---|---|---|---|---|---|---|---|
| **B-1.** | (*a*) | **B-2.** | (*a*) | **B-3.** | (*c*) | **B-4.** | (*d*) |
| **B-5.** | (*b*) | **B-6.** | (*a*) | **B-7.** | (*a*) | **B-8.** | (*a*) |

# CHAPTER 24

**A-1.** *p*-Hydroxybenzaldehyde is the stronger acid. The phenoxide anion is stabilized by conjugation with the aldehyde carbonyl.

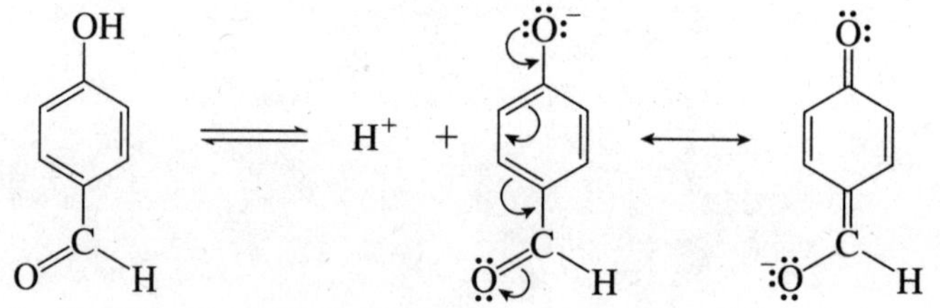

OH ... ⇌ $H^+$ + :Ö:⁻ ... ⟷ Ö: ... H

**A-2.** 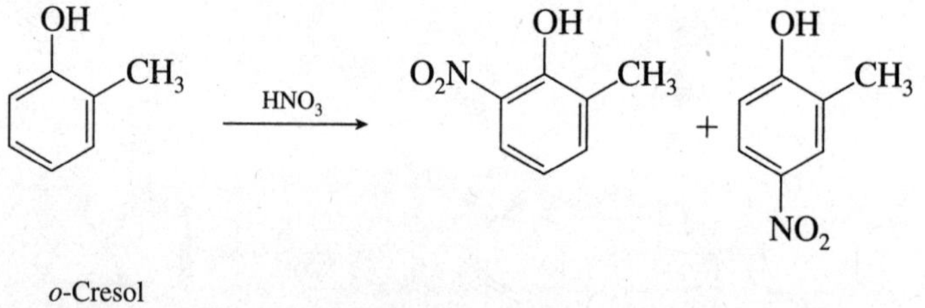 *o*-Cresol $\xrightarrow{HNO_3}$ (OH, $O_2N$, $CH_3$) + (OH, $CH_3$, $NO_2$)

*m*-Cresol $\xrightarrow{HNO_3}$ (OH, $O_2N$, $CH_3$) + (OH, $CH_3$, $NO_2$)

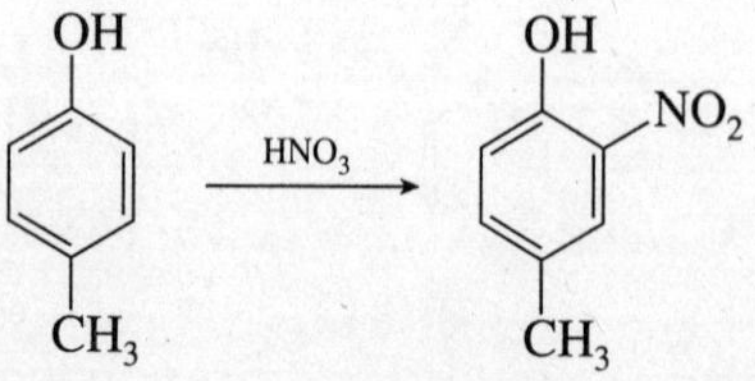

*p*-Cresol $\xrightarrow{HNO_3}$ (OH, $NO_2$, $CH_3$)

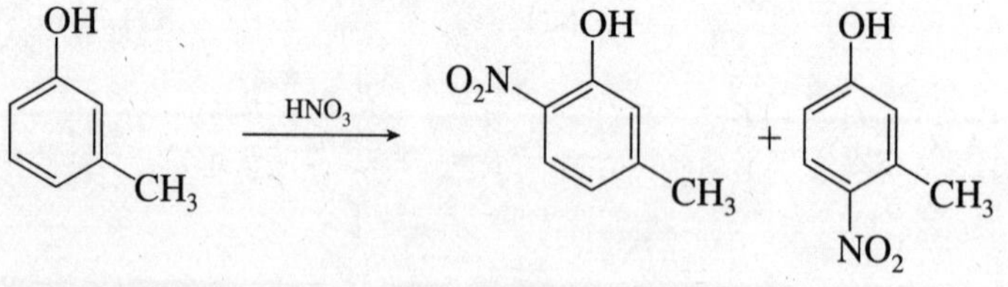

**A-3.** (4-methylphenol) + $CH_3CH_2\overset{O}{\overset{\|}{C}}Cl$ $\xrightarrow{AlCl_3}$ (2-hydroxy-5-methylphenyl) $\overset{O}{\overset{\|}{C}}CH_2CH_3$

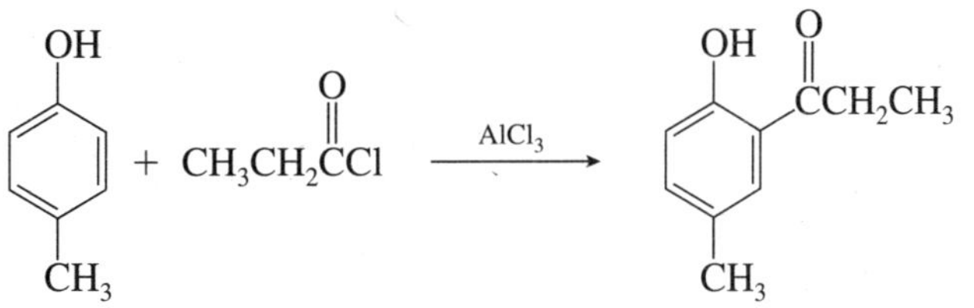

(Friedel–Crafts acylation)

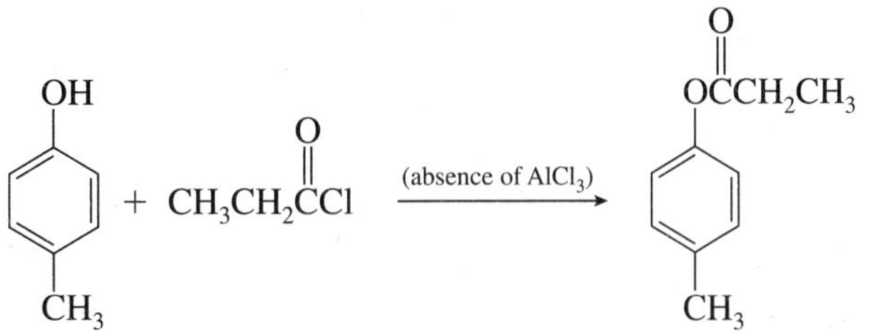

(4-methylphenol) + $CH_3CH_2\overset{O}{\overset{\|}{C}}Cl$ $\xrightarrow{\text{(absence of } AlCl_3)}$ $O\overset{O}{\overset{\|}{C}}CH_2CH_3$ (4-methylphenyl ester)

(Esterification)

**A-4.** (*a*) $C_6H_5$—OH + $BrCH(CH_3)_2$

(*b*) (3-methylphenoxide, $O^- Na^+$, $CH_3$) + $BrCH_2CH(CH_3)_2$

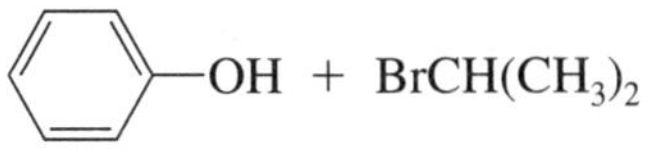

(*c*) $CO_2$, 125°C, 100 atm

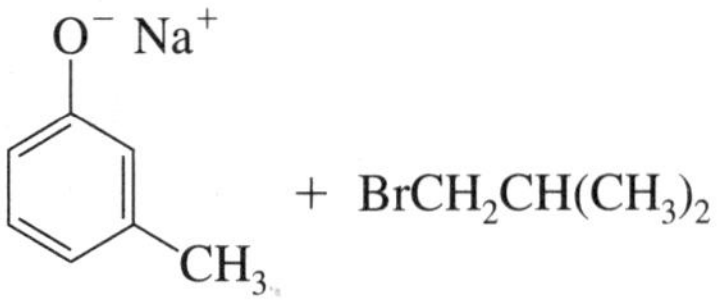

(*d*) (2-(2-bromoethyl)phenol: Br, OH)

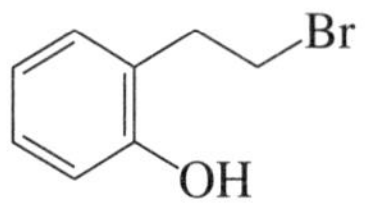

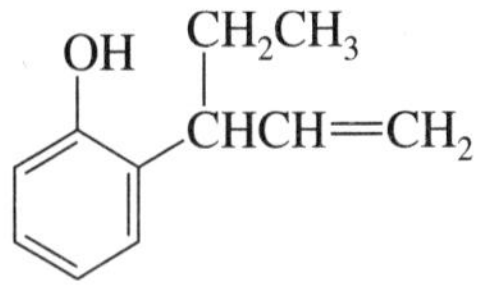

**A-5.** 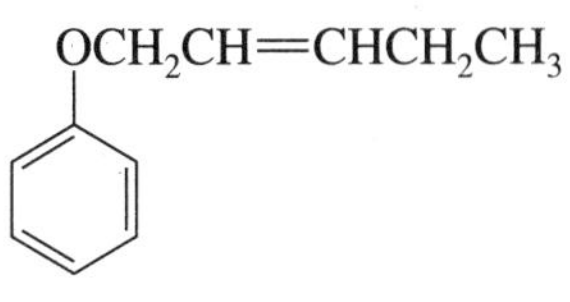 A: $C_6H_5OCH_2CH{=}CHCH_2CH_3$ B: (2-substituted phenol: OH, $\underset{|}{\overset{CH_2CH_3}{}}$ $CHCH{=}CH_2$)

**A-6.**

$C(CH_3)_3$-benzene $\xrightarrow[H_2SO_4]{HNO_3}$ $C(CH_3)_3$-benzene-$NO_2$ $\xrightarrow[\text{2. NaOH}]{\text{1. Sn, HCl}}$ $C(CH_3)_3$-benzene-$NH_2$ $\xrightarrow[\text{2. } H_2O\text{, heat}]{\text{1. } NaNO_2\text{, } H_2SO_4\text{, } H_2O}$ $C(CH_3)_3$-benzene-OH

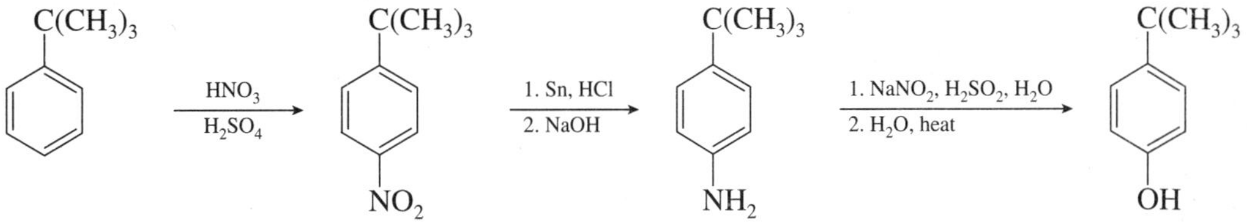

**B-1.** (*d*) **B-2.** (*d*) **B-3.** (*b*) **B-4.** (*a*)
**B-5.** (*c*) **B-6.** (*b*) **B-7.** (*b*) **B-8.** (*c*)

# CHAPTER 25

**A-1.** (*a*)

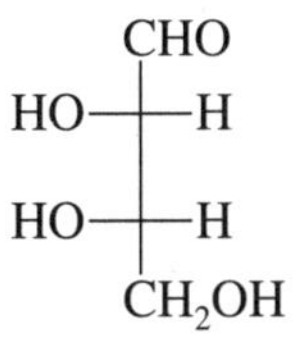

CHO
HO—H
HO—H
$CH_2OH$

L-Erythrose

(*b*)

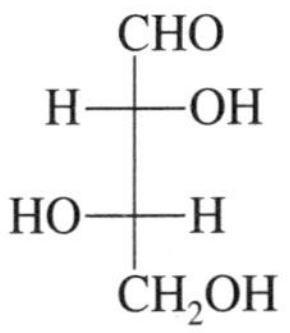

CHO
H—OH
HO—H
$CH_2OH$

L-Threose

or

CHO
HO—H
H—OH
$CH_2OH$

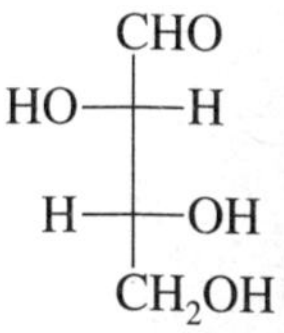

D-Threose

(*c*)

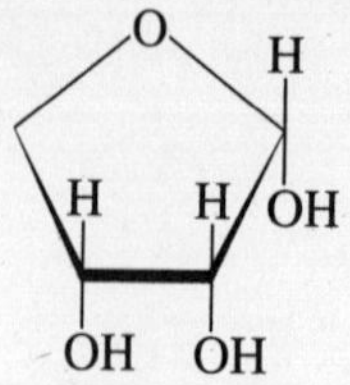

O, H, H, H, OH, OH, OH

α-D-Erythrofuranose

(*e*)

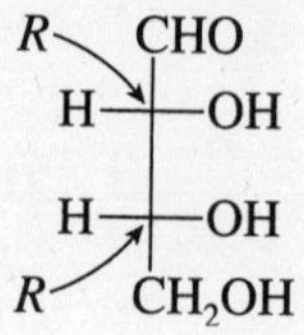

R, CHO
H—OH
H—OH
R, $CH_2OH$

(*d*)

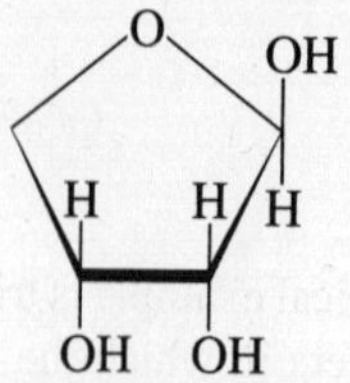

O, OH, H, H, H, OH, OH

β-D-Erythrofuranose

**A-2.** (*a*)

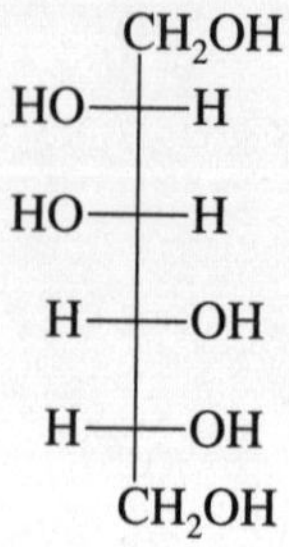

$CH_2OH$
HO—H
HO—H
H—OH
H—OH
$CH_2OH$

(*b*)

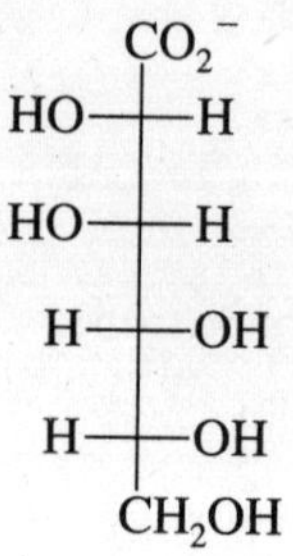

$CO_2^-$
HO—H
HO—H
H—OH
H—OH
$CH_2OH$

(*c*) $5HCO_2H + H_2C{=}O$

**A-3.** (*a*)

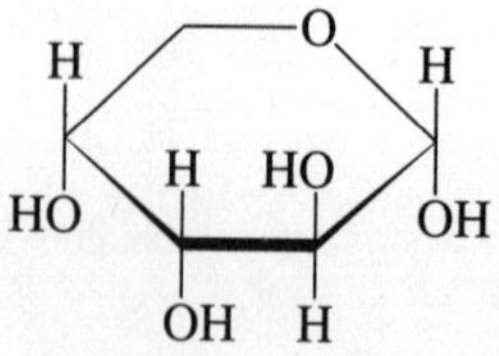

O, H, H, H, HO, HO, OH, OH, H

(*c*)

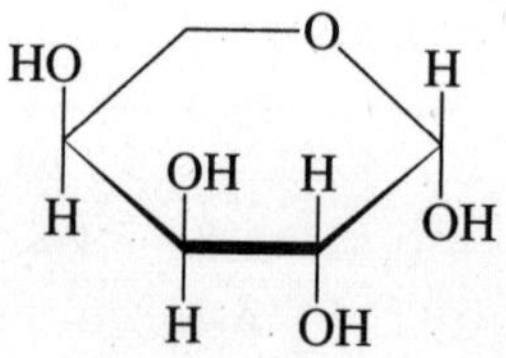

O, HO, H, OH, H, H, OH, H, OH

(*b*)

$HOCH_2$, O, OH, H, H, HO, H, OH, H

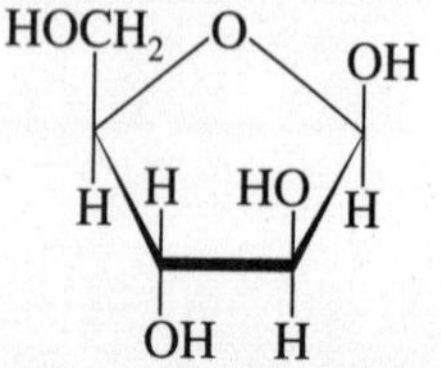

**A-4.** β-D-Idopyranose (β-pyranose form of D-idose)

**A-5.** The products are diastereomers.

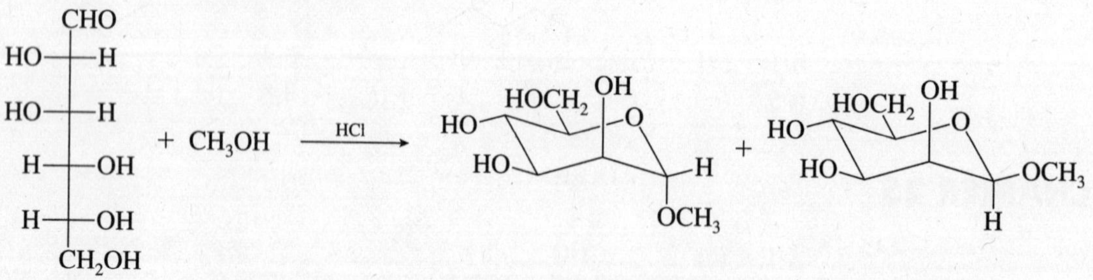

CHO
HO—H
HO—H
H—OH
H—OH
$CH_2OH$

D-Mannose

+ $CH_3OH$ —HCl→

Methanol

$HOCH_2$, OH, O, HO, HO, H, $OCH_3$

Methyl
α-D-mannopyranoside

+

$HOCH_2$, OH, O, HO, HO, $OCH_3$, H

Methyl
β-D-mannopyranoside

**B-1.** (*b*) **B-2.** (*d*) **B-3.** (*b*) **B-4.** (*a*) **B-5.** (*c*)

**B-6.** (*c*) **B-7.** (*a*) **B-8.** (*c*) **B-9.** (*c*) **B-10.** (*c*)

# CHAPTER 26

**A-1.**

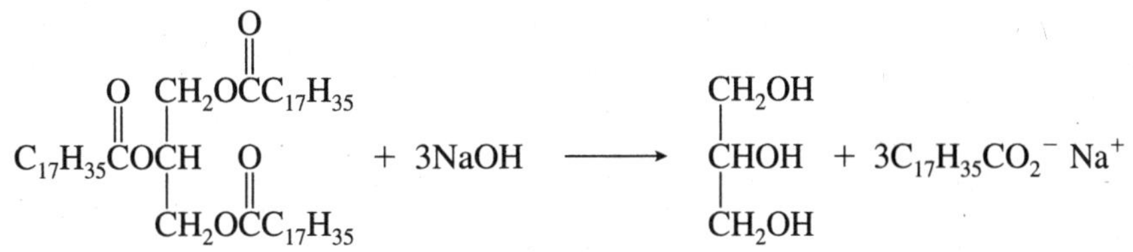

$$\begin{array}{l} \text{O}\ \ \text{CH}_2\text{O}\overset{\text{O}}{\overset{\|}{\text{C}}}\text{C}_{17}\text{H}_{35} \\ \| \ \ \ \ | \\ \text{C}_{17}\text{H}_{35}\text{COCH}\ \ \text{O} \\ \ \ \ \ \ \ \ \ |\ \ \ \ \| \\ \ \ \ \ \ \ \ \ \text{CH}_2\text{OCC}_{17}\text{H}_{35} \end{array} + 3\text{NaOH} \longrightarrow \begin{array}{l}\text{CH}_2\text{OH}\\ | \\ \text{CHOH} \\ | \\ \text{CH}_2\text{OH}\end{array} + 3\text{C}_{17}\text{H}_{35}\text{CO}_2^-\ \text{Na}^+$$

Tristearin

**A-2.** Fats are triesters of glycerol. A typical example is tristearin, shown in the preceding problem. A wax is usually a mixture of esters in which the alkyl and acyl group each contain 12 or more carbons. An example is hexadecyl hexadecanoate (cetyl palmitate).

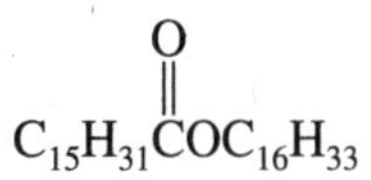

$$C_{15}H_{31}\overset{O}{\overset{\|}{C}}OC_{16}H_{33}$$

**A-3.** (*a*) Monoterpene;

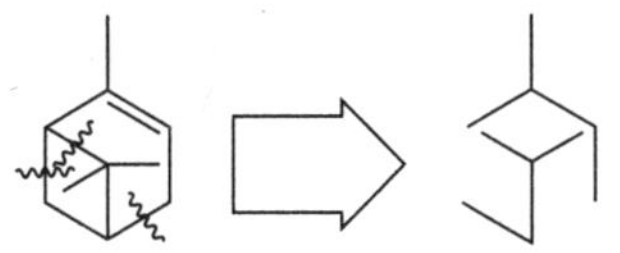

(*b*) Sesquiterpene;

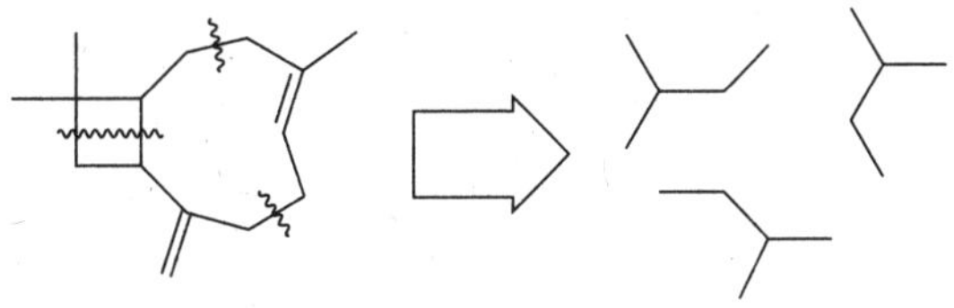

(*c*) Diterpene;

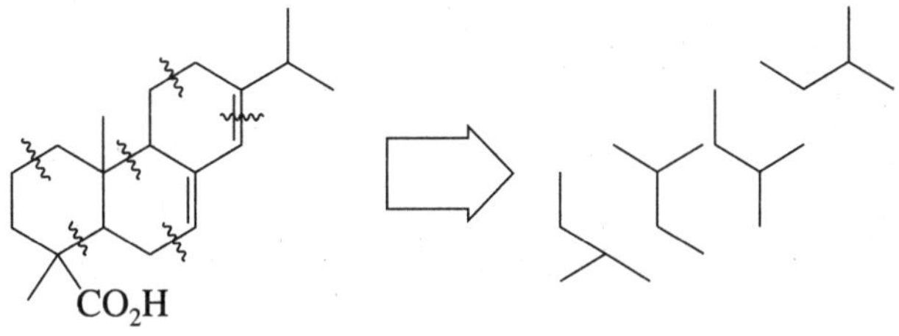

$CO_2H$

**A-4.**

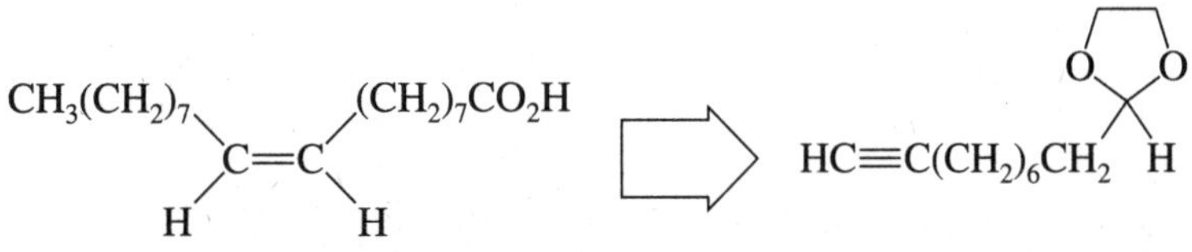

$$\underset{\text{H}}{\overset{\text{CH}_3(\text{CH}_2)_7}{}}\text{C}=\text{C}\underset{\text{H}}{\overset{(\text{CH}_2)_7\text{CO}_2\text{H}}{}} \Longrightarrow \text{HC}\equiv\text{C}(\text{CH}_2)_6\text{CH}_2\text{–}\overset{\text{O}\ \ \text{O}}{\text{C}}\text{–H}$$

Oleic acid

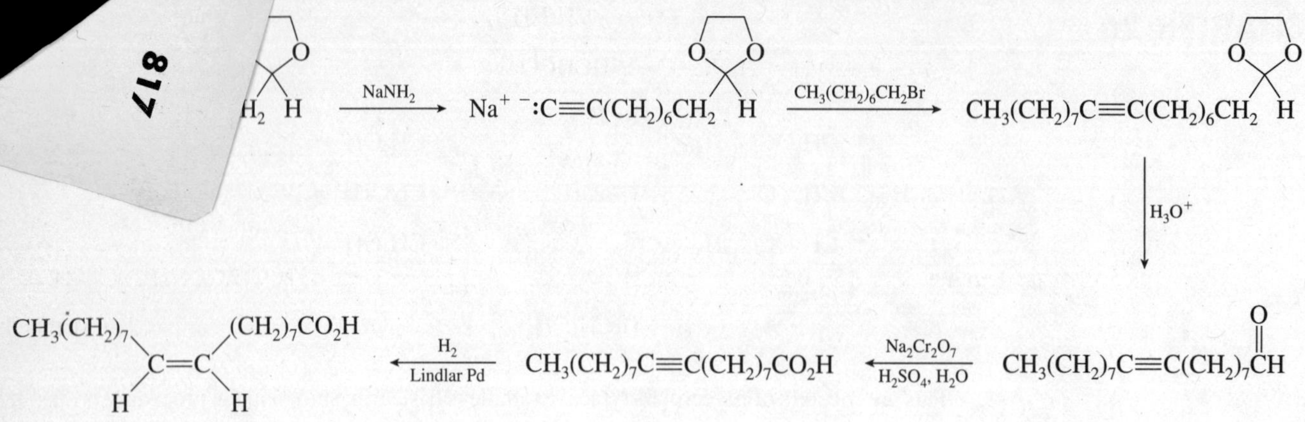

$$\text{(dioxolane)}\ \ldots H_2\ H \xrightarrow{NaNH_2} Na^{+}\ {}^{-}{:}C{\equiv}C(CH_2)_6CH_2\ H\ \text{(dioxolane)} \xrightarrow{CH_3(CH_2)_6CH_2Br} CH_3(CH_2)_7C{\equiv}C(CH_2)_6CH_2\ H\ \text{(dioxolane)}$$

$$\xrightarrow{H_3O^+} CH_3(CH_2)_7C{\equiv}C(CH_2)_7\overset{O}{\overset{\|}{C}}H \xrightarrow{Na_2Cr_2O_7,\ H_2SO_4,\ H_2O} CH_3(CH_2)_7C{\equiv}C(CH_2)_7CO_2H \xrightarrow{H_2,\ \text{Lindlar Pd}} \underset{H}{\overset{CH_3(CH_2)_7}{}}C{=}C\underset{H}{\overset{(CH_2)_7CO_2H}{}}$$

**A-5.** Geranyl pyrophosphate (OPP) → (OPP) → (cation, +) $\xrightarrow{-H^+}$ Limonene

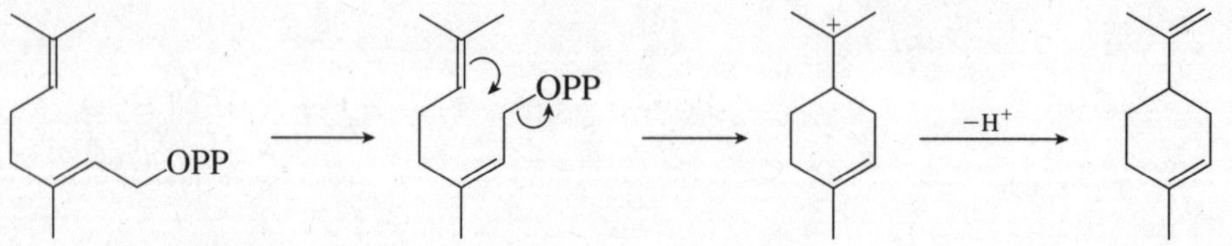

Geranyl pyrophosphate

Limonene

**B-1.** (*b*) **B-2.** (*a*) **B-3.** (*c*) **B-4.** (*c*)
**B-5.** (*a*) **B-6.** (*a*)

# CHAPTER 27

**A-1.** (*a*) $C_6H_5CH_2\overset{O}{\overset{\|}{C}}H$ (*b*) $C_6H_5CH_2O\overset{O}{\overset{\|}{C}}NH\underset{CH(CH_3)_2}{CH}CO_2H$ (*c*) DCCI

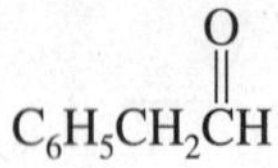

**A-2.** $O_2N{-}C_6H_3(NO_2){-}NH\underset{CH_2C_6H_5}{CH}CO_2H$

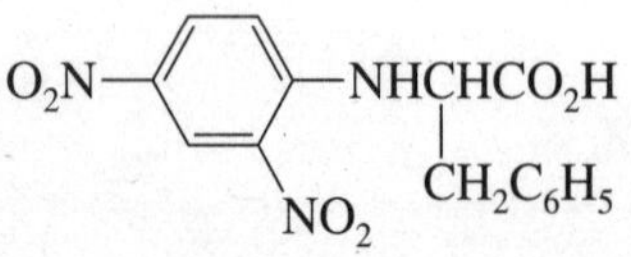

**A-3.** (*a*)

$$CH_3\overset{O}{\overset{\|}{C}}NHCH(CO_2CH_2CH_3)_2 \xrightarrow[\text{ethanol}]{NaOCH_2CH_3} CH_3\overset{O}{\overset{\|}{C}}NH\ddot{C}^{-}(CO_2CH_2CH_3)_2 \xrightarrow{(CH_3)_2CHCH_2Br} CH_3\overset{O}{\overset{\|}{C}}NH\underset{CH_2CH(CH_3)_2}{C}(CO_2CH_2CH_3)_2$$

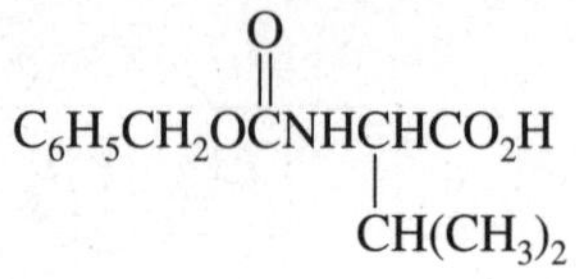

$$\xrightarrow{1.\ H_3O^+\quad 2.\ \text{heat}} H_3\overset{+}{N}\underset{CH_2CH(CH_3)_2}{CH}CO_2^{-}$$

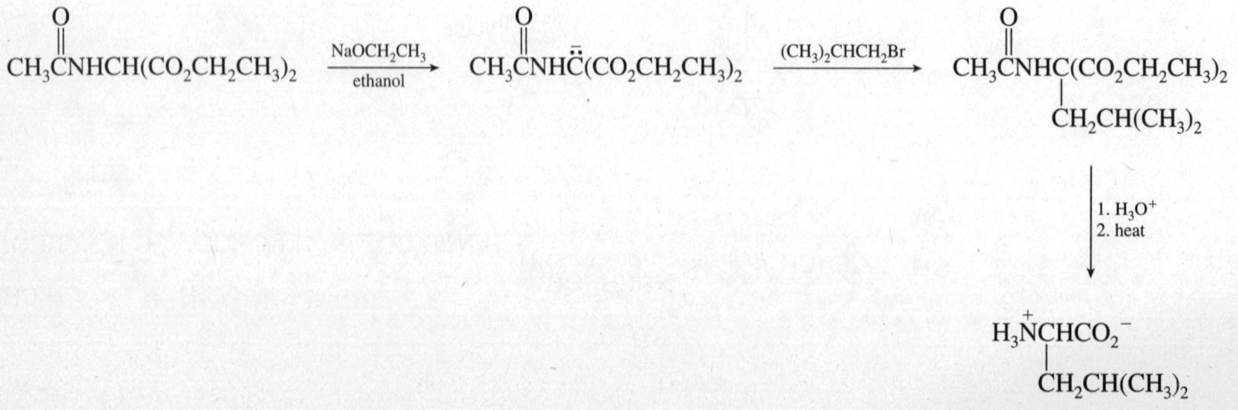

(*b*) Leu-Val = $H_3\overset{+}{N}CH(CH_2CH(CH_3)_2)C(=O)—NHCH(CH(CH_3)_2)CO_2^-$

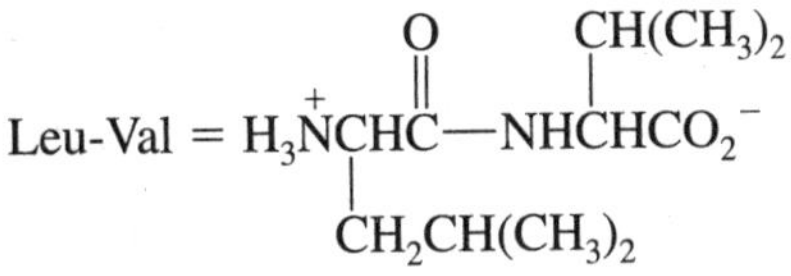

*N*-Protect leucine: $C_6H_5CH_2OC(=O)Cl + H_3\overset{+}{N}CH(CH_2CH(CH_3)_2)CO_2^- \xrightarrow[2.\ H^+]{1.\ NaOH,\ H_2O} C_6H_5CH_2OC(=O)NHCH(CH_2CH(CH_3)_2)CO_2H$

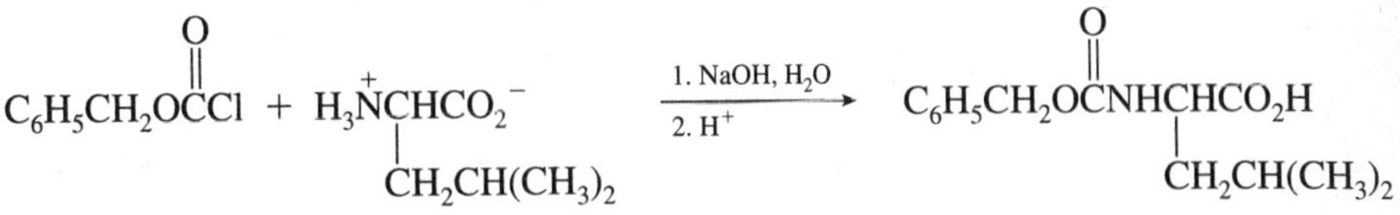

(Z-Leu)

*C*-Protect valine: $C_6H_5CH_2OH + H_3\overset{+}{N}CH(CH(CH_3)_2)CO_2^- \xrightarrow{H^+} H_3\overset{+}{N}CH(CH(CH_3)_2)C(=O)OCH_2C_6H_5$

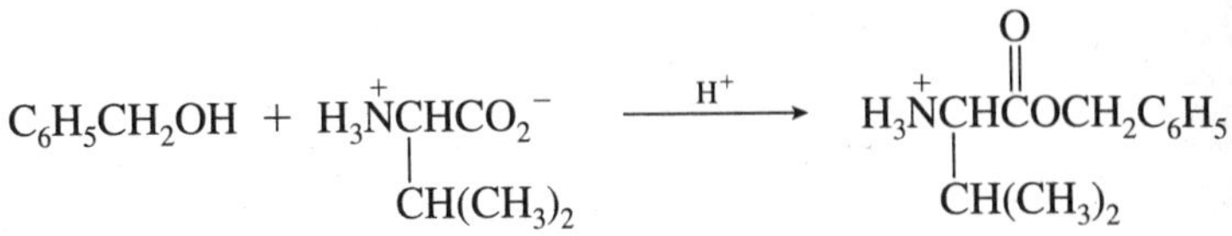

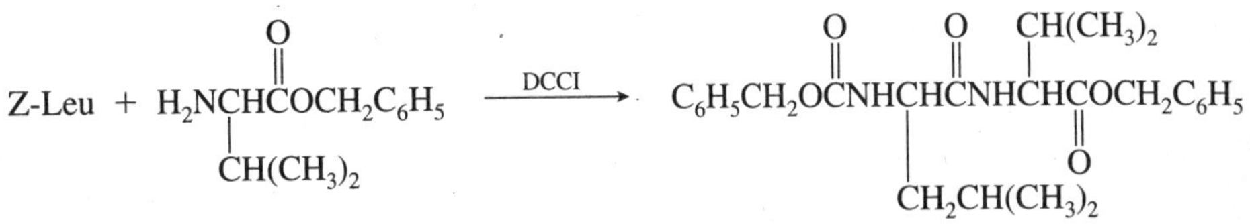

**Couple:** $\text{Z-Leu} + H_2NCH(CH(CH_3)_2)C(=O)OCH_2C_6H_5 \xrightarrow{DCCI} C_6H_5CH_2OC(=O)NHCH(CH_2CH(CH_3)_2)C(=O)NHCH(CH(CH_3)_2)C(=O)OCH_2C_6H_5$

**Deprotect:** $C_6H_5CH_2OC(=O)NHCH(CH_2CH(CH_3)_2)C(=O)NHCH(CH(CH_3)_2)C(=O)OCH_2C_6H_5 \xrightarrow{H_2,\ Pd} H_3\overset{+}{N}CH(CH_2CH(CH_3)_2)C(=O)NHCH(CH(CH_3)_2)CO_2^-$

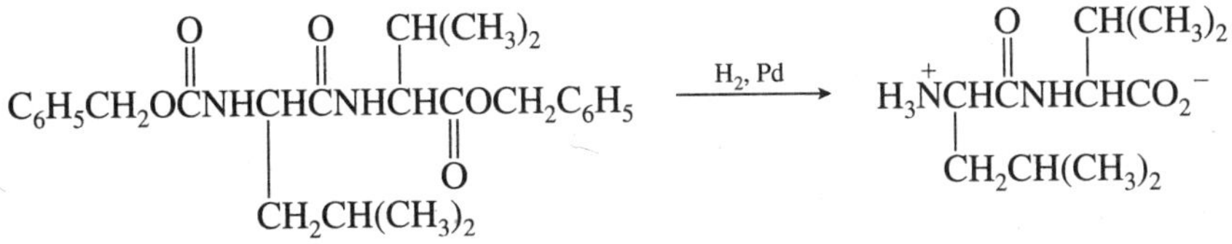

**A-4.** Leu-Val-Gly-Ala-Phe

**A-5.** (*a*) Pentapeptide (*b*) Four (*c*) Serine (*d*) Glycine (*e*) Ser-Ala-Leu-Phe-Gly

**A-6.** (*a*)

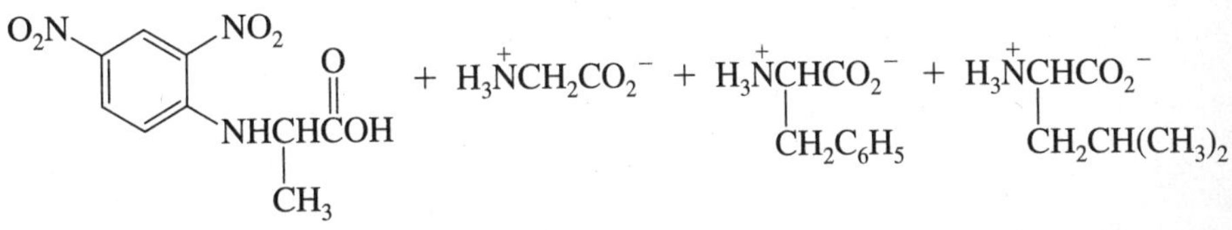

$$\underset{\text{DNP-Ala}}{2,4\text{-}(O_2N)_2C_6H_3NHCH(CH_3)C(=O)OH} + \underset{\text{Gly}}{H_3\overset{+}{N}CH_2CO_2^-} + \underset{\text{Phe}}{H_3\overset{+}{N}CH(CH_2C_6H_5)CO_2^-} + \underset{\text{Leu}}{H_3\overset{+}{N}CH(CH_2CH(CH_3)_2)CO_2^-}$$

(*b*) $H_3\overset{+}{N}CH(CH_3)C(=O)NHCH_2C(=O)NHCH(CH_2C_6H_5)CO_2^- + H_3\overset{+}{N}CH(CH_2CH(CH_3)_2)CO_2^-$

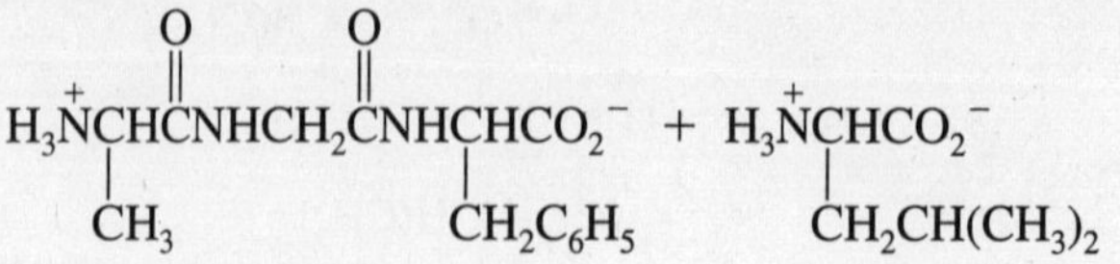

Ala-Gly-Phe  Leu

(*c*) Same as part *b*; Ala-Gly-Phe + Leu

(*d*) $C_6H_5CH_2OC(=O)NHCH(CH_3)C(=O)NHCH_2C(=O)NHCH(CH_2C_6H_5)C(=O)NHCH(CH_2CH(CH_3)_2)CO_2H$

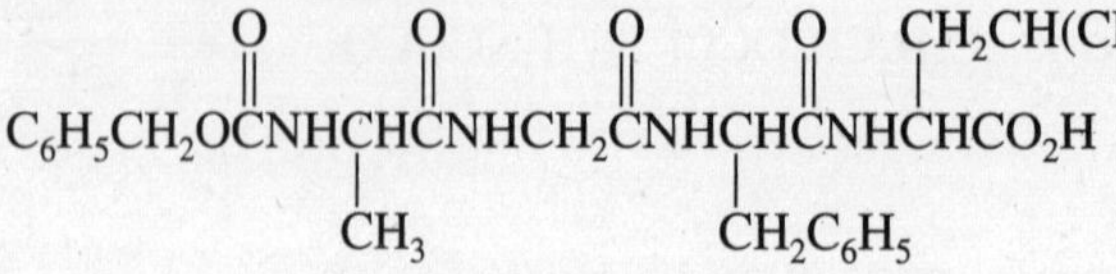

Z-Ala-Gly-Phe-Leu

**B-1.** (*c*) **B-2.** (*c*) **B-3.** (*a*) **B-4.** (*d*)

**B-5.** (*c*) **B-6.** (*b*) **B-7.** (*c*) **B-8.** (*a*)

# APPENDIX B

## TABLES

**Table B-1 Bond Dissociation Energies of Some Representative Compounds***

| Bond | Bond dissociation energy, kJ/mol (kcal/mol) | Bond | Bond dissociation energy, kJ/mol (kcal/mol) |
|---|---|---|---|
| **Diatomic molecules** | | | |
| H—H | 435 (104) | H—F | 568 (136) |
| F—F | 159 (38) | H—Cl | 431 (103) |
| Cl—Cl | 242 (58) | H—Br | 366 (87.5) |
| Br—Br | 192 (46) | H—I | 297 (71) |
| I—I | 150 (36) | | |
| **Alkanes** | | | |
| $CH_3$—H | 435 (104) | $CH_3$—$CH_3$ | 368 (88) |
| $CH_3CH_2$—H | 410 (98) | $CH_3CH_2$—$CH_3$ | 355 (85) |
| $CH_3CH_2CH_2$—H | 410 (98) | $(CH_3)_2CH$—$CH_3$ | 351 (84) |
| $(CH_3)_2CH$—H | 397 (95) | $(CH_3)_3C$—$CH_3$ | 334 (80) |
| $(CH_3)_3C$—H | 380 (91) | | |
| **Alkyl halides** | | | |
| $CH_3$—F | 451 (108) | $(CH_3)_2CH$—F | 439 (105) |
| $CH_3$—Cl | 349 (83.5) | $(CH_3)_2CH$—Cl | 339 (81) |
| $CH_3$—Br | 293 (70) | $(CH_3)_2CH$—Br | 284 (68) |
| $CH_3$—I | 234 (56) | $(CH_3)_3C$—Cl | 330 (79) |
| $CH_3CH_2$—Cl | 338 (81) | $(CH_3)_3C$—Br | 263 (63) |
| $CH_3CH_2CH_2$—Cl | 343 (82) | | |
| **Water and alcohols** | | | |
| HO—H | 497 (119) | $CH_3CH_2$—OH | 380 (91) |
| $CH_3O$—H | 426 (102) | $(CH_3)_2CH$—OH | 385 (92) |
| $CH_3$—OH | 380 (91) | $(CH_3)_3C$—OH | 380 (91) |

**Note:* Bond dissociation energies refer to bonds indicated in structural formula for each substance.

### e B-2 Acid Dissociation Constants*

| cid | Formula | Conjugate base | Dissociation constant | $pK_a$ |
|---|---|---|---|---|
| Hydrogen fluoride | **H**—F | $F^-$ | $3.5 \times 10^{-4}$ | 3.5 |
| Acetic acid | $CH_3CO_2$—**H** | $CH_3CO_2^-$ | $1.8 \times 10^{-5}$ | 4.7 |
| Hydrogen cyanide | **H**—CN | $CN^-$ | $7.2 \times 10^{-10}$ | 9.1 |
| Phenol | $C_6H_5O$—**H** | $C_6H_5O^-$ | $1.3 \times 10^{-10}$ | 9.8 |
| Water | HO—**H** | $HO^-$ | $1.8 \times 10^{-16}$ | 15.7 |
| Ethanol | $CH_3CH_2O$—**H** | $CH_3CH_2O^-$ | $10^{-16}$ | 16 |
| Alkyne (terminal; R = alkyl) | RC≡C—**H** | $RC{\equiv}C^-$ | $10^{-26}$ | 26 |
| Ammonia | $NH_2$—**H** | $NH_2^-$ | $10^{-36}$ | 36 |
| Alkene C—H | RCH=CH—**H** | $RCH{=}CH^-$ | $10^{-45}$ | 45 |
| Alkane C—H | $RCH_2CH_2$—**H** | $RCH_2CH_2^-$ | $10^{-62}$ | 62 |

**Note:* Acid strength decreases from top to bottom of the table; conjugate base strength increases from top to bottom.

### Table B-3 Chemical Shifts of Representative Types of Protons

| Type of proton | Chemical shift (δ), ppm* | Type of proton | Chemical shift (δ), ppm* |
|---|---|---|---|
| H—C—R | 0.9–1.8 | H—C—NR | 2.2–2.9 |
| H—C—C=C | 1.6–2.6 | H—C—Cl | 3.1–4.1 |
| H—C—C(=O)— | 2.1–2.5 | H—C—Br | 2.7–4.1 |
| H—C≡C— | 2.5 | H—C—O | 3.3–3.7 |
| H—C—Ar | 2.3–2.8 | H—NR | 1–3† |
| H—C=C< | 4.5–6.5 | H—OR | 0.5–5† |
| H—Ar | 6.5–8.5 | H—OAr | 6–8† |
| H—C(=O)— | 9–10 | H—OC(=O)— | 10–13† |

*These are approximate values relative to tetramethylsilane; other groups within the molecule can cause a proton signal to appear outside of the range cited.

† The chemical shifts of protons bonded to nitrogen and oxygen are temperature- and concentration-dependent.

**Table B-4 Chemical Shifts of Representative Carbons**

| Type of carbon | Chemical shift ($\delta$), ppm* | Type of carbon | Chemical shift ($\delta$), ppm* |
|---|---|---|---|
| $RCH_3$ | 0–35 | $>C=C<$ | 100–150 |
| $R_2CH_2$ | 15–40 | | |
| $R_3CH$ | 25–50 | (benzene ring) | 110–175 |
| $RCH_2NH_2$ | 35–50 | | |
| $RCH_2OH$ | 50–65 | $>C=O$ | 190–220 |
| —C≡C— | 65–90 | | |

* Approximate values relative to tetramethylsilane.

**Table B-5 Infrared Absorption Frequencies of Some Common Structural Units**

| Structural unit | Frequency, $cm^{-1}$ | Structural unit | Frequency, $cm^{-1}$ |
|---|---|---|---|
| **Stretching vibrations** | | | |
| *Single bonds* | | *Double bonds* | |
| —O—H (alcohols) | 3200–3600 | $>C=C<$ | 1620–1680 |
| —O—H (carboxylic acids) | 2500–3600 | $>C=O$ | |
| $>N—H$ | 3350–3500 | Aldehydes and ketones | 1710–1750 |
| *sp* C—H | 3310–3320 | Carboxylic acids | 1700–1725 |
| $sp^2$ C—H | 3000–3100 | Acid anhydrides | 1800–1850 and 1740–1790 |
| $sp^3$ C—H | 2850–2950 | Acyl halides | 1770–1815 |
| | | Esters | 1730–1750 |
| $sp^2$ C—O | 1200 | Amides | 1680–1700 |
| $sp^3$ C—O | 1025–1200 | | |
| | | *Triple bonds* | |
| | | —C≡C— | 2100–2200 |
| | | —C≡N | 2240–2280 |
| **Bending vibrations of diagnostic value** | | | |
| *Alkenes* | | *Substituted derivatives of benzene* | |
| Cis-disubstituted | 665–730 | Monosubstituted | 730–770 and 690–710 |
| Trans-disubstituted | 960–980 | Ortho-disubstituted | 735–770 |
| Trisubstituted | 790–840 | Meta-disubstituted | 750–810 and 680–730 |
| | | Para-disubstituted | 790–840 |